Human Physiology

FIFTH EDITION

Gillian Pocock
Senior Lecturer in Clinical Science, Canterbury Christ Church University, Canterbury, UK

Christopher D. Richards
Emeritus Professor of Experimental Physiology, Division of Physiology, Pharmacology and Neuroscience, Faculty of Life Sciences, University College London, London, UK

and

David A. Richards
Associate Professor, Department of Basic Pharmaceutical Sciences, Husson University School of Pharmacy, Bangor, Maine, USA

OXFORD
UNIVERSITY PRESS

OXFORD
UNIVERSITY PRESS

Great Clarendon Street, Oxford, OX2 6DP,
United Kingdom

Oxford University Press is a department of the University of Oxford.
It furthers the University's objective of excellence in research, scholarship,
and education by publishing worldwide. Oxford is a registered trade mark of
Oxford University Press in the UK and in certain other countries

Second edition 2004
Third edition 2006
Fourth edition 2013

Impression: 1

Published in the United States of America by Oxford University Press
198 Madison Avenue, New York, NY 10016, United States of America

British Library Cataloguing in Publication Data
Data available

Library of Congress Control Number: 2017951003

ISBN 978-0-19-873722-3

Printed in Great Britain by
Bell & Bain Ltd., Glasgow

Preface to the fifth edition

Our aim in the fifth edition of this book remains that of the first edition: to provide clear explanations of the basic principles that govern the physiological processes of the human body and to show how these principles can be applied to the practice of medicine. Clinical material illustrating physiological points is integrated into the main text, which has been revised and updated throughout. The online resource we have prepared to accompany the text has links to useful websites and a wide selection of self-assessment materials (true/false, multiple choice, and matching questions) as well as clinical and numerical problems.

The layout of the book has been substantially altered. Many of the larger chapters have been subdivided to facilitate navigation around each topic, and this has inevitably led to a substantial increase in the size of the book as new material has been added. This is particularly true of the part devoted to the nervous system, which now includes separate chapters on the general principles of sensory physiology, the somatosensory system, the eye and visual system, the ear and hearing, the vestibular system, and the chemical senses. There is also a new chapter on the limbic system and its role in the physiology of emotion. Other new material includes a discussion of the physiology of the skin and its role in thermoregulation, more detail on pulmonary defence mechanisms, and an expanded chapter on the physiology of the liver.

We have included more than 70 new figures and redrawn many others. As in previous editions, each chapter concludes with a bulleted checklist of key learning points. The 'Recommended reading' lists have been extended to include review articles of interest. A glossary of physiological terms has been added to help readers become familiar with the technical language of the field. As a result of these changes, the book is now around 20 per cent longer than before.

We remain grateful for the comments we have received from our readers, and for the helpful and detailed advice we have received from our colleagues on a number of topics: we give thanks to Professor Tim Arnett (UCL), who provided us with access to a large bank of histological material and research images; to Drs Greg Fitzharris (UCL) and Tony Michael (St George's Hospital Medical School) for their detailed comments on the chapters devoted to the physiology of reproduction; to Dr Geraint Thomas (UCL) for help with protein structures; to Dr Roger Phipps (Husson University) for his detailed comments on the the physiology of bone; and to Dr Barrie Higgs and Professor W. John Russell, who advised us on clinical matters. We also wish to thank our editors, Jonathan Crowe, Jessica White, and Lucy Wells, who have given us much useful advice and support, and the staff of Oxford University Press for their help and encouragement.

G.P., C.D.R., and D.A.R.
London and Bangor (Maine), January 2017

Preface to the first edition

The idea for this book grew out of regular discussions between the senior authors (GP and CDR) when we were both on the staff of the Department of Physiology, Royal Free Hospital School of Medicine in London. We felt that there was a need for a modern, concise textbook of physiology which covered all aspects of the preclinical course in physiology. The text is written primarily for students of medicine and related subjects, so that the clinical implications of the subject are deliberately emphasized. Nevertheless, we hope that the book will also prove useful as core material for first- and second-year science students. We have assumed a knowledge of chemistry and biology similar to that expected from British students with 'AS' levels in these subjects. Our intention has been to provide clear explanations of the basic principles that govern the physiological processes of the human body and to show how these principles can be applied to the understanding of disease processes.

The book begins with cell physiology (including some elementary biochemistry), and proceeds to consider how cells interact both by direct contact and by longer-distance signaling. The nervous system and endocrine system are dealt with at this point. The physiology of the main body systems is then discussed. These extensive chapters are followed by a series of shorter chapters describing integrated physiological responses including the control of growth, the regulation of body temperature, the physiology of exercise, and the regulation of body fluid volume. The final chapters are mainly concerned with the clinical applications of physiology, including acid–base balance, heart failure, hypertension, liver failure, and renal failure. This structure is not a reflection of the organization of a particular course but is intended to show how, by understanding the way in which cells work and how their activity is integrated, one can arrive at a satisfying explanation of body function.

In providing straightforward accounts of specific topics, it has occasionally been necessary to omit some details or alternative explanations. Although this approach occasionally presents a picture that is more clear cut than the evidence warrants, we believe that this is justified in the interests of clarity. Key points are illustrated by simple line drawings as we have found that they are a useful aid to students in understanding and remembering important concepts. We have not included extensive accounts of the experimental techniques of physiology but have tried to make clear the importance of experimental evidence in elucidating underlying mechanisms. Normal values have been given throughout the text in SI units but important physiological variables have also been given in traditional units (e.g. mmHg for pressure measurements).

Each chapter is organized in the same way. In answer to the frequently heard plea 'what do I need to know?' we have set out the key learning objectives for each chapter. This is followed, where appropriate, by a brief account of the physical and chemical principles required to understand the physiological processes under discussion. The essential anatomy and histology are then discussed, as a proper appreciation of any physiological process must be grounded on a knowledge of the main anatomical features of the organs involved. Detailed discussion of the main physiological topics then follows.

To aid student learning, short numbered summaries are given after each major section. From time to time we have set out important biological questions or major statements as section headings. We hope that this will help students to identify more clearly why a particular topic is being discussed. The reading material given at the end of each chapter is intended both to provide links with other subjects commonly studied as part of the medical curriculum and to provide sources from which more detailed information can be obtained. Self-testing is encouraged by the provision of multiple-choice questions or quantitative problems (or both) at the end of each chapter. Annotated answers to the questions are given. Some numerical problems have also been given which are intended to familiarize students with the key formulae and to encourage them to think in quantitative terms. *[Note that these have now been moved to the accompanying website.]*

We are deeply indebted to Professor Michael de Burgh Daly and Dr Ted Debnam, who not only advised us on their specialist topics but also read through and constructively criticized the entire manuscript. Any remaining obscurities or errors are entirely our responsibility. Finally, we wish to thank the staff of Oxford University Press for their belief in the project, their forbearance when writing was slow, and their help in the realization of the final product.

G.P. and C.D.R.
London, February 1999

Acknowledgements

We wish to acknowledge the many colleagues who have helped us to clarify our thinking on a wide variety of topics. Those to whom especial thanks are due for detailed criticisms on particular chapters are listed here.

Professor J.F. Ashmore, FRS, UCL Ear Institute, Gray's Inn Road, London, UK

Professor S. Bevan, Wolfson Centre for Age-Related Diseases, King's College London, UK

Dr T.V.P. Bliss, FRS, Division of Neurophysiology, National Institute for Medical Research, London, UK

Professor M. de Burgh Daly, Department of Physiology, University College London, UK

Dr E.S. Debnam, Department of Physiology, University College London, UK

Professor D.A. Eisner, Unit of Cardiac Physiology, University of Manchester, UK

Dr D.G. Fitzharris, Department of Reproductive Medicine, University College London, UK

Dr B.D. Higgs, Division of Anaesthesia, Royal Free Hospital, London, UK

Professor R. Levick, Department of Physiology, St George's Hospital Medical School, London, UK

Professor A.A. Mathie, Medway School of Pharmacy, Chatham, Kent, UK

Dr A.E. Michael, Department of Biomedical Sciences, St George's Hospital Medical School, London, UK

Dr R. Phipps, Department of Pharmacology, School of Pharmacy, Husson University, Bangor, Maine, USA

Professor I.C.A.F. Robinson, Laboratory of Endocrine Physiology, National Institute for Medical Research, London, UK

Professor W.J. Russell, Director, Research and Development (retired), Department of Anaesthesia and Intensive Care, Royal Adelaide Hospital, North Terrace, Adelaide, South Australia

Dr A.H. Short, Department of Physiology and Pharmacology, Medical School, Queen's Medical Centre, Nottingham, UK

The sources from which the figures are derived are acknowledged in the accompanying captions.

GP: For Chris, Rebecca, David, and James.

CDR: For Joan *in memoriam* and for Sue, Andrew, and Joanne.

DAR: For Jeannette, Quinn, and Iliana.

Summary of contents

Contents

About the authors

Gillian Pocock read Physiology at the University of Oxford before moving to King's College London to study for her PhD under the supervision of Professor P.F. Baker, FRS. She held a postdoctoral position at King's College before taking up an appointment in the Department of Physiology, Royal Free Hospital School of Medicine. She is now Senior Lecturer in Clinical Science in the Department of Health, Wellbeing and Family, Canterbury Christ Church University, UK. Her research interests have focused on the role of calcium in secretion and pH regulation in neurons.

Christopher Richards read Biological Chemistry at the University of Bristol. He completed his PhD in the Department of Zoology, University of Bristol, under the supervision of Professor P.C. Caldwell, FRS before taking up a position at the Institute of Psychiatry in London. He subsequently moved to the National Institute for Medical Research where he was a member of the scientific staff. He later held posts in the Departments of Physiology at the Royal Free Hospital School of Medicine and University College London, where he is currently Emeritus Professor of Experimental Physiology. He has published over 100 scientific articles on a variety of topics including synaptic transmission, the cellular and molecular mechanisms of anaesthesia, pH regulation in neurons, and two photon imaging.

David Richards studied Biochemistry at the University of Bristol before beginning his PhD at the National Institute for Medical Research and University College London under the supervision of Professor T.V.P. Bliss FRS. After postdoctoral positions at the University of Colorado School of Medicine, the University of Zurich Brain Research Institute, and the University of Wisconsin School of Medicine, he was appointed Assistant Professor at the University of Cincinnati College of Medicine, before moving to Cincinnati Children's Hospital Medical Center. He is now Associate Professor of Neuropharmacology at Husson University School of Pharmacy in Maine. His research focuses on the cell biology that supports synaptic transmission.

A note to the reader

The chapters in this book cover the physiological material normally taught in the first and second years of the medical curriculum and degree courses in physiology. While each chapter can be read on its own, the book has been divided into 11 parts. The clinical applications of physiology are discussed throughout the text where appropriate. Part 1 is a broad introduction to the subject and covers some essential anatomy (Chapter 1) and chemistry (Chapters 2 and 3). Part 2 (Chapters 4–6) presents information on the basic properties of cells and how they communicate. Part 3 (Chapters 7 and 8) discusses the basic physiology of excitable cells (nerve and muscle). Part 4 (Chapters 9–19) is an extensive discussion of the role of the nervous system in the coordination and regulation of the activities of the body. Parts 5–10 (Chapters 20–47) include much of the core material of traditional courses in physiology and discusses the functioning of the principal organ systems. These chapters include discussion of various aspects of integrated physiology: the regulation of the internal environment, the responses to exercise, life at high altitude, control of metabolic rate, and body temperature regulation. Part 11 (Chapters 48–51) is concerned with the physiology of reproduction and that of the neonate as well as growth more generally.

Each chapter begins with a list of learning objectives which sets out the principal points that we think you, the reader, should try to assimilate. We have assumed a basic knowledge of chemistry, physics, and biology, but, where necessary, important topics are covered in feature boxes within the appropriate chapters. Important vocabulary is given in bold, and key terms are also defined in the Glossary at the end of the book (Appendix 1). The contents of each chapter are arranged in numbered sections in the same order as the learning objectives, and each major section ends with a summary of the main points. A checklist of key points is also given at the end of each chapter.

We have tried to avoid repetition as far as possible by cross-referencing. Many chapters have feature boxes which contain material that is more advanced or deal with numerical examples. It is not necessary to read these boxes to understand the core material. At the end of each chapter there is a reading list which is intended to link physiology with other key subjects in the medical curriculum (particularly with anatomy, biochemistry, and pharmacology). These writings have been chosen for their clarity of exposition, but many other good sources are available. For those who wish to study a particular physiological topic in greater depth, we have also included in our reading lists some monographs and review articles which we have found helpful in preparing this book. We have also suggested specific chapters in more advanced texts. In addition to the sources listed, the *American Handbook of Physiology* has more detailed articles on specific topics of physiological interest. These sources will provide you with a guide to the primary source literature which, like other areas of biomedical science, is still advancing rapidly. Articles in mainstream review journals such as the *Annual Review of Physiology* and *Physiological Reviews* will provide an introduction in the most recent developments in particular fields. Many specialist journals now also regularly carry review articles relating to their areas of interest.

The online resource has multiple-choice, true/false, numerical, and clinical questions for you to test your knowledge. Answers are provided, along with brief explanations and links to the relevant sections in the main text.

If you find that a particular topic is difficult to understand, break it down into its components to identify where your difficulties lie. This is the first step towards resolving them. If, after further study, you still have difficulty, seek help from your tutor or lecturer.

What a piece of work is a man! How noble in reason! How infinite in faculty! In form, in moving, how express and admirable! ... The paragon of animals!

William Shakespeare, *Hamlet*, Act 2

PART ONE

Basic concepts in physiology

CHAPTER 1

What is physiology?

Chapter contents

This chapter should help you to understand:

* The subject matter of physiology
* The hierarchical organization of the body
* The terms used in anatomical descriptions
* The concept of homeostasis
* Control mechanisms in physiology

1.1 Introduction

For anyone concerned with medicine or the health sciences more generally, a sound knowledge of the normal structure and function of the body is essential as it provides a foundation on which to build strategies for diagnosis, treatment, and prevention of disease.

Physiology is the study of the functions of living matter. It is concerned with *how* an organism performs its varied activities: how it feeds, how it moves, how it adapts to changing circumstances, how it spawns new generations. The subject is vast and embraces the whole of life. The success of physiology in explaining how organisms perform their daily tasks is based on the notion that they are intricate and exquisite machines whose operation is governed by the laws of physics and chemistry. Although some processes are similar across the whole spectrum of biology—the replication of the genetic code, for example—many are specific to particular groups of organisms. For this reason it is necessary to divide the

subject into various parts such as bacterial physiology, plant physiology, and animal physiology. The focus of this book is the physiology of mammals, particularly that of humans.

To study how any animal works it is first necessary to know how it is built. This is the study of **anatomy**, which embraces not only gross structural anatomy but also the microanatomy of structures too small to be visible to the naked eye—the study of cells and tissues (**histology**). A full knowledge of human anatomy requires examination of the human body, both in life through observation of those structures that are easily visible on the surface, and by close examination of dissected bodies after death. These studies are complemented by modern imaging techniques such as magnetic resonance imaging (MRI), which can visualize the internal structures of the body in life. Similar studies have been carried out on other animals. Once the basic anatomy is established, physiological experiments can be carried out to discover how the various parts perform their functions.

Although there have been many important physiological investigations on human volunteers, the need for precise control over the experimental conditions has meant that much of our present knowledge of physiology has been derived from studies on other animals such as frogs, rabbits, cats, and dogs. When it is clear that a specific physiological process has a common basis in a wide variety of animal species, it is reasonable to assume that the same principles will apply to man. The knowledge gained from this approach has given us great insight into human physiology and endowed us with a solid foundation for the effective treatment of many diseases.

1.2 The organization of the body

The building blocks of the body are the **cells**, which are grouped together to form **tissues**. Cells differ widely in form and function, but they all have certain common characteristics.

* First, they are bounded by a limiting membrane, the plasma membrane. This encloses the **cytoplasm**, which contains a number of structures called **organelles** that perform specific functions within the cell. (The mature red cells of the blood are an important exception, as their cytoplasm contains no organelles.)

- Second, cells have the ability to break down large molecules to smaller ones to liberate **energy** for their activities.
- Third, at some point in their life history, all cells possess a nucleus that contains genetic information in the form of deoxyribonucleic acid (DNA).

Living cells continually transform materials. They break down **glucose** and fats to provide energy for other activities such as motility and the synthesis of **proteins** for growth and repair. These chemical changes are collectively called **metabolism**, the study of which forms much of the subject matter of **biochemistry**. The breakdown of large molecules to smaller ones is called **catabolism** and the synthesis of large molecules from smaller ones is called **anabolism**.

In the course of evolution, cells progressively became more specialized in order to serve different functions. Some developed the ability to contract (muscle cells), others to conduct electrical signals (nerve cells). A further group developed the ability to secrete different substances such as enzymes for the digestion of food (e.g. the acinar cells of the salivary glands), while endocrine cells developed the ability to secrete chemical signals (hormones) that regulate the activity of the body. During embryological development, this process of **differentiation** is re-enacted as many different types of cell are formed from the fertilized egg.

The principal types of tissue are:

- blood and lymph
- connective tissue
- nervous tissue
- muscle
- epithelia and glandular tissue.

Blood and lymph are sometimes classified as connective tissue.

Each tissue contains a mixture of cell types that determines its characteristic functions (see Chapter 4). For example, blood consists of red cells, white cells, and platelets suspended in a fluid medium (the plasma). The red cells transport oxygen around the body, the white cells play an important role in the defence against infection, while the platelets are vital components in the process of blood clotting. There are a number of different types of connective tissue, but all are characterized by having cells distributed within an extensive non-cellular matrix. In contrast, muscle consists of densely packed layers of muscle cells with a scanty extracellular matrix, while epithelia consist of continuous sheets of cells joined together.

The fine structure of tissues requires investigation with a microscope. In conventional histology, tissues are prepared for microscopic examination in a series of steps. First they are treated with formaldehyde or another chemical to preserve their constituents. This is known as 'fixation'. The fixed tissue is then cut into thin (10–20 μm) slices with an instrument called a microtome before being mounted on a microscope slide. The mounted thin slices (**histological sections**) are then treated with dyes to permit the fine details of a tissue to be seen with a light microscope. Different parts of the tissues are preferentially stained by different dyes, and this permits the identification of discrete structures within the mounted specimen. Detail that cannot be resolved with the light microscope can be revealed by **electron microscopy**, in which the tissue is first treated with a heavy metal before being imaged with a beam of electrons. By looking at a sequence of such sections it is possible to build up a three-dimensional image of a structure. This can also be achieved by **scanning electron microscopy** and **confocal microscopy**, examples of which will be found throughout this book.

In the past few decades there has been increasing use of **cryofixation**, in which a block of tissue is rapidly frozen before being cut into thin sections with an instrument called a **cryotome**. This method is used to preserve the natural structure of cellular constituents such as proteins, which can then be precisely located in the tissue with antibodies labelled with a fluorescent marker. In the last decade of the twentieth century it became possible to follow changes in the internal chemistry of living cells with fast imaging techniques and fluorescent dyes and markers.

Organs such as the brain, the heart, the lungs, the intestines, and the liver are formed by the aggregation of different kinds of tissue. The organs are themselves parts of distinct physiological **systems**. The heart and blood vessels form the **cardiovascular system**; the lungs, trachea and bronchi together with the chest wall and diaphragm form the **respiratory system**; the skeleton and skeletal muscles form the **musculo-skeletal system**; the brain, **spinal cord**, autonomic nerves and ganglia, and peripheral somatic nerves form the **nervous system**; and so on. This organization is viewed as hierarchical—complexity of function increases from cells through to systems, which function in an integrated fashion within the body as a whole. Each system is adapted to perform a specific set of functions that enable the body to perform all the vital activities of living such as breathing, feeding, and reproduction.

Summary

The body has a hierarchical organization of cells, tissues, organs, and systems. The cells are the basic building blocks of the body. Different kinds of cells aggregate to form the tissues and organs of the body.

1.3 Terms used in anatomical descriptions

For ease of communication and to accurately describe the location and relationships of body parts, it is conventional to use a prescribed set of terms that relate to a stereotyped posture known as the **anatomical position** (see Figure 1.1). For present purposes, these terms may be divided into three groups:

- anatomical planes
- anatomical direction
- the body cavities.

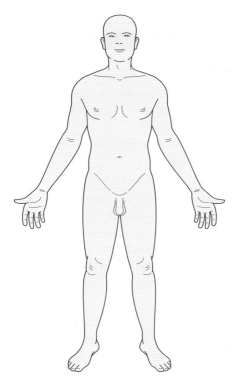

Figure 1.1 An outline of the body in the anatomical position. Note that the palms of the hands face forwards.

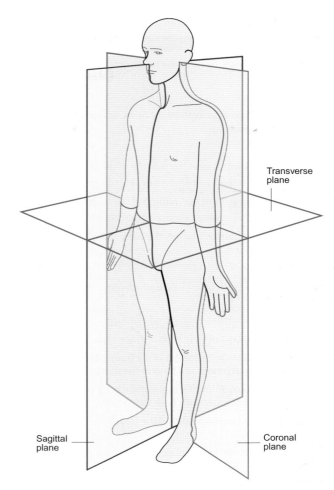

Transverse plane

Sagittal plane

Coronal plane

Figure 1.2 The three principal anatomical planes (sections along which the body may be divided to show its internal structures).

The anatomical position

Unless it is stated otherwise, for the purposes of description the body is always considered to be erect (upright) and facing ahead, with the feet slightly apart. The arms are by the sides with the palms facing forwards and the fingers extended. This is the so-called anatomical position and is shown in Figure 1.1. Of course, in reality, a person may be in any position including lying face up (**supine**), or lying face down (**prone**).

Anatomical planes

Textbooks (including this one) often include diagrams of structures that have been cut (sectioned) in various ways to reveal their internal organization. The three main types of section that are regularly used in this way are illustrated in Figure 1.2. They are also called anatomical planes (imaginary lines along which the section has been made) and lie at right angles to one another.

- **Sagittal** section (or sagittal plane)—a vertical (longitudinal) plane that divides the body or internal structure into a right and left portion. A vertical section through the midline is sometimes called a mid-sagittal or median section. If the section does not fall on the midline it is often called a **parasagittal** section.
- **Coronal** section (or coronal plane)—a vertical plane at right angles to the sagittal. This is also called a frontal section and divides the

body or internal structure into a front (anterior) and back (posterior) portion.
- **Transverse** section—also called a **horizontal section** or **horizontal plane**. This section is at right angles to both coronal and sagittal planes. This divides the body or internal structure into an upper (superior) and lower (inferior) portion. It is also known as a **cross-section**.

Any section or plane that is not parallel to one of the above is called an **oblique** section.

Anatomical directional terms

When describing parts of the body it is often necessary to make reference to their relative positions within the body as a whole or in relation to other structures close by. A number of directional terms have been established to make this easier. These are summarized in Table 1.1. Note that the terms 'inferior' and 'superior' relate only to anatomical position and not to the relative importance of the structures described.

Table 1.1 Commonly used directional terms and their definitions

Term	Definition	Example of use
Superior	Above, towards the head, or towards the top of a structure	Superior vena cava Superior surface of ... The adrenal glands lie superior to the kidneys
Inferior	Below, away from the head, or towards the lower part of a structure	Inferior vena cava Inferior rectus muscle The parotid glands are inferior to the ear
Medial	Towards or at the midline of the body On the inner surface or side of a given structure	The medial surface of each lung The medial border of the kidney
Lateral	Away from the midline of the body On the outer surface of a structure	The kidneys are lateral to the vertebral column The lateral surface of the lung
Dorsal (posterior)	Behind or towards the back of the body or of a structure	The dorsal (or posterior) columns of the spinal cord The posterior abdominal wall As seen from the posterior aspect
Ventral (anterior)	In front of or towards the front of the body or of a structure	The ventral roots of the spinal nerves
Superficial	Towards or at the body surface	The superficial muscles of the leg
Deep	Away from the body surface	The gall bladder lies deep in the abdominal cavity
Proximal	Close to the origin of the body part or the point of attachment of a limb to the trunk	The proximal tubule of the nephron (the section nearest the renal glomerulus)
Distal	Away from the origin of the body part or the point of attachment of a limb to the trunk	The distal tubule of the nephron (the part furthest from the glomerulus)
Ipsilateral	On the same side of the body	The fibres of the reticulospinal tracts synapse with interneurons in the ipsilateral ventral columns
Contralateral	On the opposite side of the body	The right cerebral hemisphere controls the muscles on the contralateral side (those of the left arm and leg)
Afferent	Towards	Afferent fibres of the nervous system carry impulses towards the CNS Afferent arteriole carries blood towards the glomerulus (of the nephron)
Efferent	Away from	Efferent fibres of the nervous system carry impulses away from the CNS Efferent vessels carry blood from the hypothalamus

Body cavities

Many of the internal structures of the body are contained within spaces known as cavities that offer them some protection and support. There are two major body cavities, the dorsal and ventral cavities. The **dorsal cavity** is made up of the **cranial cavity**, which houses the brain, and the **spinal cavity**, which houses the spinal cord. The **ventral cavity** consists of the **thoracic cavity** and the **abdomino-pelvic cavity**. These are shown in Figure 1.3.

The thoracic (chest) cavity houses a number of important structures including the heart and lungs. The abdomino-pelvic cavity is very large and houses the kidneys and urinary tract, most of the structures of the gastrointestinal tract including the liver and pancreas, and most of the reproductive system. Because of its size and complexity, the abdomino-pelvic cavity is often further subdivided in one of two ways:

- into four quadrants (the right and left upper and right and left lower quadrants), as illustrated in Figure 1.4(a);
- into nine regions, as illustrated in Figure 1.4(b).

1.4 The principal organ systems

The cardiovascular and lymphatic systems. The oxygen and nutrients needed by the cells of large animals cannot be obtained directly from the external environment. Instead they must be transported to the cells via the blood, which circulates around the body by virtue of the pumping action of the heart. The heart, blood vessels, and associated tissues form the cardiovascular system, while the lymphatic vessels return fluid from the tissues to the general circulation.

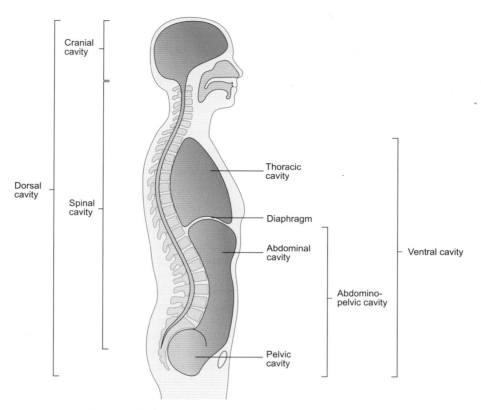

Figure 1.3 The major body cavities and their subdivisions.

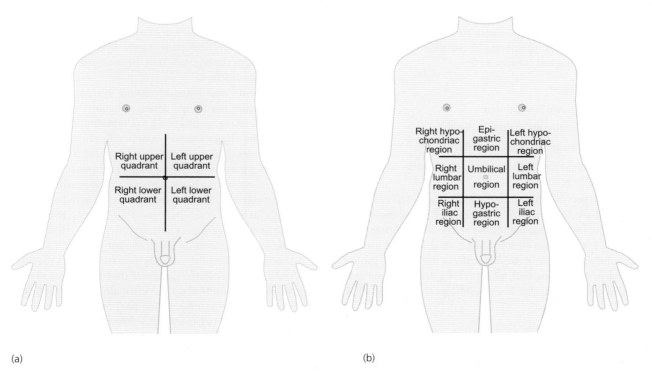

(a)

(b)

Figure 1.4 (a) The four abdominal quadrants. (b) The nine regions of the abdomen.

The heart consists of four chambers, two atria and two ventricles, which form a pair of pumps arranged side by side. The right ventricle pumps deoxygenated blood to the lungs, where it absorbs oxygen from the air, while the left ventricle pumps oxygenated blood returning from the lungs to the rest of body to supply the tissues. Physiology is concerned with establishing the factors responsible for the beating of the heart, its pumping action, the flow of blood around the circulation, and the distribution of blood to the tissues according to their needs.

The respiratory system. The energy required for performing the various activities of the body is ultimately derived from respiration. This process involves the oxidation of foodstuffs (principally sugars and fats) to release the energy they contain. The oxygen needed for this process is absorbed from the air in the lungs and carried to the tissues by the blood. The carbon dioxide produced by the respiratory activity of the tissues is carried to the lungs by the blood in the pulmonary artery prior to its excretion in the expired air. The important physiological questions to be answered include the following:

- How is the air moved in and out of the lungs?
- How is the volume of air breathed adjusted to meet the requirements of the body?
- What limits the rate of oxygen uptake in the lungs?
- How are the lungs protected from airborne particles that might damage them?

The digestive system. The nutrients needed by the body are derived from the diet. Food is taken in by the mouth and broken down into its component parts by enzymes in the **gastrointestinal tract** (also called the **alimentary canal** or **gut**). The products of digestion are then absorbed into the blood across the wall of the intestine and pass to the liver via the portal vein. The liver makes nutrients available to the tissues both for their growth and repair and for the production of energy. In the case of the **digestive system**, key physiological questions are:

- How is food ingested?
- How is it broken down and digested?
- How are the individual nutrients absorbed?
- How is the food moved through the gut?
- How are the indigestible remains eliminated from the body?

The renal system (the kidneys and urinary tract). The chief function of the kidneys is the control of the composition of the extracellular fluid (the fluid which bathes the cells). In the course of this process, the kidneys also eliminate non-volatile waste products from the blood. To perform these functions, the kidneys produce urine of variable composition, which is temporarily stored in the bladder before voiding. The key physiological questions in this case are:

- How do the kidneys regulate the composition of the blood?
- How do they eliminate toxic waste?
- How do they respond to stresses such as dehydration?
- What mechanisms allow the storage and elimination of the urine?

The reproductive system. Reproduction is one of the fundamental characteristics of living organisms. The **gonads** (the testes in the male and ovaries in the female) produce specialized sex cells known as gametes. At the core of sexual reproduction is the creation and fusion of the male and female gametes, the sperm and ova (eggs), with the result that the genetic characteristics of two separate individuals are mixed to produce offspring that differ genetically from their parents. Key questions are:

- How are the sperm and eggs produced?
- What is the mechanism of fertilization?
- How does the embryo grow and develop?
- How is the fully developed fetus delivered, and how is a newborn baby nourished until it is weaned?

The musculo-skeletal system. This consists of the bones of the skeleton, the skeletal muscles, joints, and their associated tissues. Its primary function is to provide a means of movement, which is required for locomotion, for the maintenance of posture, and for breathing. The musculo-skeletal system also provides physical support for the internal organs. Here the mechanism of muscle contraction and its regulation are central issues.

The endocrine and nervous systems. The activities of the different organ systems need to be coordinated and regulated so that they act together to meet the needs of the body. Two coordinating systems have evolved: the nervous system and the endocrine system. The nervous system uses electrical signals to transmit information very rapidly to specific cells or groups of cells. For example, the nerves pass electrical signals to the skeletal muscles to control and coordinate their contractions. The endocrine system secretes chemical agents, **hormones**, which travel in the bloodstream to the cells upon which they exert a regulatory effect. Hormones play a major role in the regulation of many different organ systems. They are particularly important in growth, in the regulation of metabolism, and in the regulation of the menstrual cycle and other aspects of reproduction.

The immune and integumentary systems. The **immune system** provides the body's defences against infection both by killing invading organisms and by eliminating diseased or damaged cells. The **integumentary system** refers to the skin and its related structures (hair, nails etc). In addition to its obvious role in covering and protecting the internal structures of the body from damage caused by mechanical forces, the integumentary system prevents water loss and provides a barrier to invading organisms. The skin also has an important role in regulating body **temperature**, while its sense organs and nerve endings are an important source of information about the local environment.

Although it is helpful to study how each organ performs its functions, it is essential to recognize that the activity of the body as a whole is dependent on the intricate interactions between the various organ systems. If one part fails, the consequences are found in other organ systems throughout the whole body. For example, if the kidneys begin to fail, the regulation of the internal environment is impaired, which in turn leads to disorders of function in other body systems (e.g. erratic beating of the heart).

Summary

Each physiological system consists of a number of organs that work together to carry out particular functions. The functions of the body are dependent on complex interactions between the various organ systems.

1.5 Homeostasis

It is common knowledge that the internal temperature of the body of a healthy person is maintained at around 37°C and that it varies very little despite wide variations in the external environmental temperature. Many other physiological parameters show a similar degree of constancy: e.g. blood glucose, blood volume, the concentrations of sodium and potassium inside and outside the cells, and the concentrations of oxygen and carbon dioxide in the arterial blood. This maintenance of a stable internal environment is essential for the normal healthy function of the body's cells, tissues, and organs. It is called **homeostasis** (literally 'staying the same' or 'standing still'). A loss of homeostasis is reflected in ill health, and much of medicine is concerned with helping the body to maintain or restore homeostasis. The term was coined in 1932 by W. Cannon, who wrote that '*The word does not imply something set and immutable, a stagnation. It means a condition which may vary, but which is relatively constant*'.

The beating of the heart provides a good example of the importance of homeostasis. It depends on the rhythmical and coordinated contractions of the cardiac muscle cells. This activity is governed by electrical signals that, in turn, depend on the concentration of sodium and potassium ions inside and outside the cardiac muscle cells (i.e. it depends on the composition of the intracellular and extracellular fluids). If there is an excess or deficit of potassium in the extracellular fluid, the excitability of the cardiac muscle cells will be affected, which will cause the heart muscle to contract erratically rather than in a coordinated manner. Accordingly, the homeostatic mechanisms of the body act to maintain the concentration of potassium in the extracellular fluid within a narrow range. Note that the heart rate itself can, and does, vary considerably during the day; it is its ionic environment that is maintained within close limits.

Virtually every organ system has a role in homeostasis but the endocrine and nervous systems are particularly important as they allow communication within the body, so integrating the functions of the various body systems. This integration enables the body both to maintain a stable internal environment and to adapt to changing circumstances.

How does the body maintain homeostasis?

The concept of **balance** is central to the understanding of homeostasis. In the course of a day, an adult consumes approximately 1 kg of food. In a month, this amounts to around 30 kg. Yet, in general, body weight remains remarkably constant. Such individuals are said to be *in balance*; the intake of food and drink matches the amounts required for normal bodily activities plus the losses in the urine and faeces. In some circumstances, such as starvation, intake does not match the needs of the body and muscle tissue is broken down to provide glucose for the generation of energy. Here, the intake of protein is less than the rate of breakdown and the individual is said to have a **negative nitrogen balance** (nitrogen being a characteristic component of the **amino acids** that make up the protein—see Chapter 3). Equally, if the body tissues are being built up—as is the case for growing children, pregnant women, and athletes in the early stages of training—the daily intake of protein must be greater than the normal body turnover so that the individual is in **positive nitrogen balance**.

Another well-known example is the balance achieved between fluid intake and urine output. The thirst mechanism and the production of urine by the kidneys are carefully controlled to ensure that the volume of body fluid is kept largely constant. If a person becomes dehydrated they will drink to restore body fluid volume as soon as water becomes available. Conversely, the consumption of a large quantity of water or beer is followed very soon by a large increase in the production of very dilute urine, which serves to eliminate the excess water. In both cases, the volume of the body fluids is maintained within narrow limits.

The concept of balance can be applied to any of the body constituents and is important in considering how the body regulates its own composition. For balance to be maintained, intake must match or exceed requirements and any excess must be excreted. Additionally, for each chemical constituent of the body there is a desirable concentration range, which the control mechanisms are adapted to maintain. For example, the concentration of glucose in the plasma (the fluid part of the blood) is about 4–5 mmol l^{-1} between meals. Shortly after a meal, plasma glucose rises above this level and this rise stimulates the secretion of the hormone insulin by the pancreas, which acts to bring the concentration down. As the concentration of glucose falls, so does the secretion of insulin. Both in the normal fasting state and after a meal, the changes in the circulating level of insulin act together with other mechanisms to maintain the plasma glucose at an appropriate level. This type of regulation is known as **negative feedback**. During the period of insulin secretion, the glucose is being stored either as glycogen (mainly in the liver and muscles) or as fat (in specialized fat cells of the **adipose** tissue).

A **negative feedback loop** is a control system that acts to maintain the level of some variable within a given range following a disturbance. Although the example given above refers to plasma glucose, the basic principle can be applied to other physiological variables such as body temperature, blood pressure, and the osmolality of the plasma. A negative feedback loop requires a **sensor** or **receptor** of some kind that responds to the variable in question but not to other physiological variables. Thus, an osmoreceptor should respond to changes in osmolality of the body fluids, but not to changes in body temperature or blood pressure, for example. The information from the sensor must be compared in some way to the desired level (known as the 'set point' of the system) by some form of **comparator** or **integrator**. If the two do not match, an error signal is transmitted to an **effector**, a system that can act to restore the variable to its desired level. The basic features of a negative feedback loop are summarized in Figure 1.5. These features of negative feedback can be appreciated by examining a simple home heating system: the controlled variable is room temperature, which is sensed by a thermostat. The effector is a heater of some kind. When the room temperature falls below the set point, the temperature difference is detected by the thermostat, which switches on the heater. This heats the room until the temperature reaches the pre-set level, whereupon the heater is switched off.

A physiological example is shown in Figure 1.6. This is a simplified scheme for the restoration of normal water balance after a large intake of water. Here the osmoreceptors of the hypothalamus monitor the osmolarity of the blood. When they detect a decrease in osmolarity, they initiate a response that results in a decrease in water reabsorption by the kidneys and an increase in urine production.

Although negative feedback is an important mechanism for maintaining a more or less constant internal environment, it does have certain disadvantages.

- First, negative feedback control can be exerted only after the controlled variable has been disturbed.

- Second, the correction to be applied can be assessed only by the magnitude of the error signal (the difference between the desired value and the displaced value of the variable in question). In practice, this means that simple negative feedback systems will provide incomplete correction.

- Third, there is inevitably a delay between the change in the controlled variable and the response of the effector that applies the correction.

- Fourth, over-correction has the potential for causing oscillations in the controlled variable.

These disadvantages are largely overcome in physiological systems by means of multiple regulatory processes. In the case of glucose regulation, blood glucose is maintained within a narrow range by two mechanisms that act in opposition (push–pull). Insulin acts to lower plasma glucose while another pancreatic hormone, **glucagon**, acts to increase plasma glucose by mobilizing glucose from the body's stores (see Figure 1.7). Other hormones are also involved in maintaining plasma glucose within a narrow range under a variety of conditions (see Chapter 24). A similar control process is used by the body to regulate body temperature: heat generation

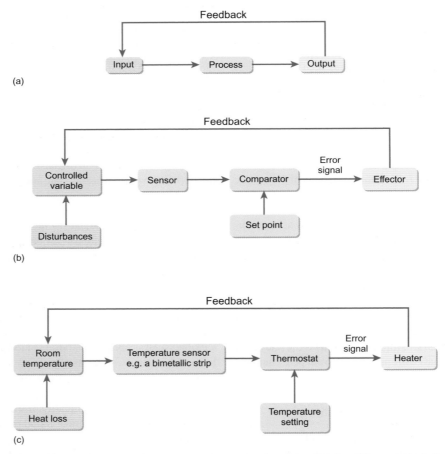

Figure 1.5 Schematic drawing of feedback control loops. (a) The main components of a feedback loop in which the output influences the input either to increase its magnitude (positive feedback) or to diminish it (negative feedback). (b) The elements of a negative feedback loop in more detail. (c) Negative feedback control of a simple heating system in which heat is lost to the surroundings. This results in a fall in room temperature below the set point triggering activation of the heat source, which restores the temperature to its desired value.

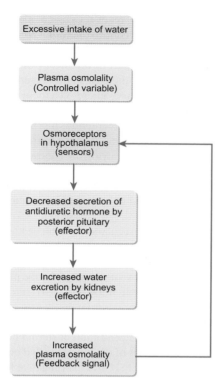

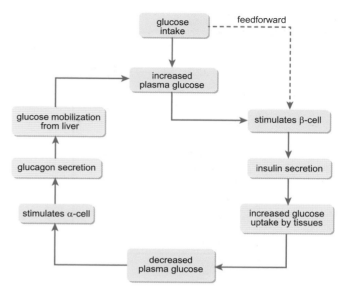

Figure 1.6 A simplified scheme for the restoration of water balance after a water load. When water intake is excessive, the osmolality (concentration) of the plasma falls. This change is sensed by receptors in the brain known as osmoreceptors which regulate the secretion of antidiuretic hormone (ADH) by the posterior pituitary gland. In the case of a water load, ADH secretion decreases and this results in an increase in the excretion of water by the kidneys, so counteracting the fall in plasma osmolarity.

Figure 1.7 An outline of the mechanisms that regulate the concentration of glucose in the plasma. Here two hormones (insulin and glucagon) act in opposition. A rise in plasma glucose leads to secretion of insulin by the β cells of the pancreas. Insulin then stimulates the uptake of glucose by the liver, muscle, and adipose tissue. As a result, the plasma glucose falls. This fall triggers the secretion of glucagon from the α cells of the pancreas. The glucagon then mobilizes glucose from the glycogen stores in the liver. This push–pull regulation of plasma glucose maintains plasma glucose within a relatively narrow range.

increases in response to a fall in body temperature, and heat loss mechanisms are activated if body temperature rises (e.g. in response to **exercise**). More detail of the control of body temperature is given in Chapter 42.

To minimize the lag between a change in a controlled variable and the action of the effector, it is advantageous for the body to anticipate any likely change. This is achieved in many instances by **feedforward control** in which the change is anticipated and action taken to minimize its magnitude. In the case of glucose regulation, insulin secretion begins before the ingested food causes a substantial rise in plasma glucose. This secretion is initiated by nerve impulses arising in the brain as part of the body's response in anticipation of a meal (the cephalic phase of secretion—see Chapter 24).

While it is difficult to over-emphasize the importance of negative feedback control loops in homeostatic mechanisms, they are frequently reset or overridden in stresses of various kinds. For example, arterial blood pressure is monitored by receptors (known as baroreceptors) in the walls of the aortic arch and **carotid sinus**. These receptors are the sensors for a negative feedback loop that maintains the arterial blood pressure within close limits. If the blood pressure rises, compensatory changes occur that tend to restore it to normal. In exercise, however, this mechanism is reset

(see Chapter 38). Indeed, if it were not, the amount of exercise we could undertake would be very limited.

The combination of negative feedback and feedforward regulation provides effective control of many physiological variables over the short term, but the body also needs to regulate many functions over the longer term in response to changing circumstances. This type of regulation is known as **adaptive control**. Examples are the increased secretion of thyroid hormone in response to prolonged periods of cold stress, to increase metabolic rate and heat generation, and the increase in muscle mass that occurs during physical training. Although adaptive mechanisms may have beneficial effects, as in the examples given above, they may also exacerbate some pathologies—for example, the walls of the arterioles become thickened in response to persistent high blood pressure (chronic hypertension). This increases the resistance to blood flow and raises the blood pressure further, adding to the stress on the vessel walls. To make matters worse, the hypertension increases the work done by the heart and so the thickness of the wall of the left ventricle increases, impairing the diffusion of oxygen and glucose to the contracting fibres. In time this can lead to ventricular failure.

Negative feedback loops operate to maintain a particular variable within a specific range. They are a stabilizing force in the economy of the body. In some circumstances, however, **positive feedback** occurs. In this case, the feedback loop is inherently unstable as the error signal acts to increase the initial deviation. An example from everyday life is the howling that occurs when a microphone is placed near one of the loudspeakers of a public

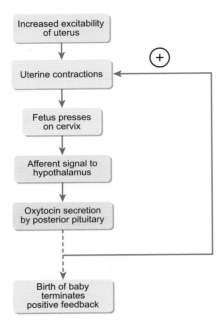

Figure 1.8 An example of a positive feedback loop—the secretion of the hormone oxytocin during childbirth. Small contractions occur at the start of labour which cause the fetus to press on the cervix. This stimulates stretch receptors in the uterus which send nerve impulses to the hypothalamus, the part of the brain that houses the neurons that secrete oxytocin. The oxytocin-containing nerve endings in the posterior pituitary secrete oxytocin into the blood, which transports it to the uterus reinforcing the strength of the uterine contractions. The increased pressure on the cervix results in a higher rate of nerve impulse discharge that results in a further increase in oxytocin secretion. This cycle progressively intensifies the uterine contractions until they are so powerful that the baby is born.

address system. The microphone picks up the initial sound and this is amplified by the electronic circuitry. This drives the loudspeaker to emit a louder sound, which is again picked up by the microphone and amplified so that the loudspeaker makes an even louder sound—and so on until the amplifying circuitry reaches the limit of its power (and the hearers run for cover!). Positive feedback control mechanisms are limited by a definite 'end point' which terminates the control loop.

An example of how positive feedback plays a role in physiology is provided by the hormonal control of uterine contractions (see Figure 1.8). Small contractions occur at the start of labour that put pressure on the cervix (the part of the uterus that leads to the birth canal). This leads to an increase in the secretion of the hormone oxytocin by the pituitary gland. Oxytocin increases the excitability of the uterus so that the strength of its contractions increases. As a result, oxytocin secretion is increased further and the uterine contractions become stronger, putting further pressure on the cervix and so intensifying the contractions until they are so powerful that the baby is born. The birth eliminates the pressure on the cervix. The stimulus to secrete large quantities of oxytocin is thereby lost, and its secretion returns to the normal resting level.

In certain circumstances regulation is achieved by an interaction between negative and positive feedback mechanisms. A striking example is provided by the hormonal control of the menstrual cycle, which is regulated by two hormones from the pituitary gland known as **follicle stimulating hormone** (FSH) and **luteinizing hormone** (LH). Steroid hormones from the ovaries can exert both negative and positive feedback control on the output of FSH and LH, depending upon the concentration of steroid hormone present. Low or moderate levels of a hormone called oestradiol-17β tend to inhibit secretion of FSH and LH (negative feedback). If, however, oestradiol-17β is present in high concentrations for several days, it stimulates the secretion of FSH and LH instead of inhibiting their secretion (positive feedback). As a result, there is a sharp increase in the output of both FSH and LH just before mid-cycle, when **oestrogen** levels are very high. This rise is responsible for the shedding of an egg by the ovary (ovulation). Once ovulation has taken place, oestrogen levels fall sharply and the output of FSH and LH drops as negative feedback reasserts control. These changes are described in greater detail in Chapter 48.

Summary

Many physiological parameters are kept within narrow limits by negative feedback and feedforward mechanisms. This is known as homeostasis. Certain physiological functions involve a self-limiting positive feedback so that a change is enhanced rather than counteracted.

✳ Checklist of key terms and concepts

Areas of study

- Anatomy is the study of the physical structure of the body.
- Biochemistry is the study of the chemical compounds of living matter and their transformations.
- Histology is the study of the minute structure of cells and tissues by microscopy.

- Molecular biology is the study of the structure and functions of biological molecules (especially of proteins and nucleic acids). It overlaps extensively with biochemistry.
- Physiology is the study of the functions of living matter.

Organization of the body

- The cells are the basic units of the body. They are bounded by a limiting membrane and contain small structures (organelles) that perform specific functions within the cell.
- During development, cells differentiate to perform a wide variety of specialized functions.
- Different kinds of cell aggregate to form tissues.
- Organs are formed by the aggregation of different kinds of tissue to perform a specific function or functions.
- Physiological systems are groups of organs adapted to perform particular functions within the body.

Anatomical terminology

- The anatomical position: a posture in which the body is erect (upright) and facing ahead, with the feet slightly apart, the arms by the sides with palms facing forwards, and the fingers extended.
- Supine: lying face up.
- Prone: lying face down.
- Sagittal plane or section: a vertical division of the body or internal structure into a right and left portion.
- Coronal (or frontal) plane or section: a vertical division of the body or structure that divides the body or internal structure into a front (anterior) and back (posterior) portion.

- Transverse (or horizontal) section: a horizontal division that divides the body or structure into an upper (superior) and lower (inferior) portion, also known as a cross-section.

Physiological processes and their regulation

- Metabolism: the sum total of the chemical changes that occur within the body.
- Catabolism: the breakdown of large molecules to smaller ones, usually for the release of energy.
- Anabolism: the synthesis of large molecules from smaller ones.
- Homeostasis: the maintenance of a stable internal environment for the preservation of the normal functioning of the cells.
- Negative feedback loop: a control system that acts to oppose a change in the level of a variable and restore it to its original value.
- Positive feedback: a control system that acts to reinforce a physiological change.
- Feedforward control: the use of a physiological receptor to detect a potential change in a physiological variable and institute a response that will minimize the magnitude of any change.
- Adaptive control: a physiological process that changes to meet new circumstances.

 To check that you have mastered the key concepts presented in this chapter, complete the accompanying online self-assessment questions. Go to www.oup.com/uk/pocock5e/

CHAPTER 2

Key concepts in chemistry

Chapter contents

This chapter should help you to understand:

- The basic structure of matter: atoms and molecules
- How molecules are bound together: covalent and ionic bonds
- How chemical reactions can transform one kind of molecule into another
- The properties of water as a biological solvent and the nature of polar and non-polar substances
- The ionization of molecules: strong and weak electrolytes
- Weak and strong acids
- The pH scale
- The idea of molarity to express concentrations
- The osmotic pressure of aqueous solutions

2.1 Introduction

This chapter presents a brief account of the chemistry necessary to understand the metabolic reactions of the body. It should provide enough information to enable you to understand what is meant by an element and a compound; to differentiate between atoms and molecules; to understand molecular structures; and to calculate the concentration of a substance in solution. The chapter also briefly discusses key properties of solutions such as ionization, pH, and osmosis. A few numerical exercises are given in the online resource website to help consolidate these important concepts.

All matter is composed of **chemical elements**, which are substances that cannot be broken down into simpler materials by chemical means. Each element has a specific abbreviation or **chemical symbol** of one or two letters. Thus carbon is written as C, calcium as Ca, hydrogen as H, oxygen as O, iron as Fe, and so on. However, most of the material encountered in everyday life is made from combinations of elements called **chemical compounds**. Examples are chalk (calcium carbonate), which is a combination of calcium, carbon, and oxygen; common table salt (sodium chloride), which is a combination of sodium and chlorine; and table sugar (sucrose), which is a combination of carbon, hydrogen, and oxygen. The main elements that make up the human body, with their chemical symbols, are given in Table 2.1.

Table 2.1 The principal chemical elements of the body

Element	Chemical symbol	Atomic number	Relative atomic mass	Percentage of body weight
Oxygen	O	8	16	65
Carbon	C	6	12	18
Hydrogen	H	1	1	10
Nitrogen	N	7	14	3.4
Calcium	Ca	20	40.1	1.5
Phosphorus	P	15	31	1.2
Potassium	K	19	39.1	0.28
Sulphur	S	16	32.1	0.25
Sodium	Na	11	23	0.17
Chlorine	Cl	17	35.4	0.16
Magnesium	Mg	12	24.3	0.05
Iron	Fe	26	55.8	0.007
Zinc	Zn	30	65.4	0.002
Iodine	I	53	126.9	4×10^{-5}

The body contains trace amounts of other elements in addition to those listed above.

Each element consists of minute particles of the same type known as **atoms**. These are composed of a dense nucleus made up of **protons** and **neutrons**, surrounded by a cloud of **electrons**. The protons are positively charged while the electrons are negatively charged. The neutrons have the same mass as the protons but carry no charge. In an atom of a given element, the number of electrons is equal to the number of protons.

There are 91 naturally occurring elements, each of which has a particular number of protons (the **atomic number**) that determines its chemical characteristics. The mass of each atom (its **atomic mass**) is determined by the number of protons and neutrons in its nucleus; the electrons have negligible mass. Atoms with the same number of protons (and therefore the same atomic number) may have different numbers of neutrons. As a result they have different atomic masses. Such atoms are called **isotopes**. A good example is carbon, which has an atomic number of 6 but has isotopes with mass numbers of 12, 13, and 14. These are known as carbon 12, carbon 13, and carbon 14, written as ^{12}C, ^{13}C, and ^{14}C. Carbon 12 has 6 protons and 6 neutrons ($6 + 6 = 12$), carbon 13 has 6 protons and 7 neutrons ($6 + 7 = 13$), and carbon 14 has 6 protons and 8 neutrons ($6 + 8 = 14$). The relative mass of an atom is measured on a scale in which a single atom of carbon 12 has a mass of 12 exactly. (As mentioned above, carbon 12 is an atom which has 6 protons and 6 neutrons.) An **atomic mass unit** (a.m.u. or a.u.) is one-twelfth of the mass of an atom of carbon 12.

Many of the elements have isotopes that have unstable nuclei that undergo a process called radioactive decay. For example, carbon 14 (^{14}C) is radioactive and breaks down at a constant rate, so that half of the ^{14}C in a sample is lost every 5,568 years (its half-life). Many radioactive isotopes (including ^{14}C) are used in biomedical research to follow the transformations of chemical compounds that occur in living organisms, while others such as technetium 99 (^{99}Tc) are used in diagnostic tests or in the treatment of particular diseases (such as the use of strontium 89, ^{89}Sr, in the palliative treatment of bone cancer). The radioactive isotopes of elements used in medicine have half-lives that vary from hours to years. Examples are ^{99}Tc, which has a half-life of 6 hours; ^{89}Sr, with a

half-life of 50.5 days; ^{22}Na (sodium 22), with a half-life of 2.58 years; and ^{3}H (tritium), with a half-life of 12.36 years.

Although the nucleus is responsible for almost all the mass of an atom, it occupies a very small volume. Most of the space of an individual atom is occupied by the electrons, as shown in Figure 2.1. Nevertheless, the simple, widely held view that an atom is rather like a miniature solar system is very misleading. The electrons are arranged in a series of energy levels known as **shells**. Within each shell the electrons occupy discrete energy levels known as **orbitals**, each of which can hold no more than two electrons. There are four different types of orbital, known as s, p, d, and f. The electron shells each contain a particular combination of orbitals. The number of shells an atom has depends on its atomic number—the higher the atomic number, the greater the number of shells and orbitals. Thus hydrogen and helium (atomic numbers 1 and 2 respectively) have a single shell (shell 1) that has a single orbital known as an s orbital. Carbon, nitrogen, and oxygen have atomic numbers of 6, 7, and 8 and each atom has two shells: shell 1 and shell 2. Shell 2 has one s orbital and three p orbitals and can accommodate 8 electrons when all its orbitals are filled. Potassium and calcium have atomic numbers of 19 and 20 and each of their atoms has three electron shells, and so on. The space occupied by the electron shells defines the **atomic volume** of an atom. When an electron shell has all its orbitals fully occupied by electrons, it is in a stable configuration and does not readily form chemical bonds. This is the situation for the inert (or noble) gases such as helium, argon, and neon. For other elements, electrons can be shared between atoms to create stable configurations. This is the basis of the formation of molecules by chemical bonding, which is discussed in Section 2.2.

Summary

All matter is formed of atoms, which consist of smaller particles known as protons, neutrons, and electrons. The mass of an atom is equal to the number of protons and neutrons in its nucleus.

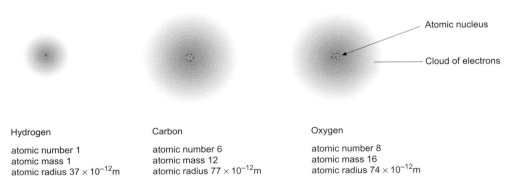

Atomic nucleus

Cloud of electrons

Hydrogen

atomic number 1
atomic mass 1
atomic radius 37×10^{-12}m

Carbon

atomic number 6
atomic mass 12
atomic radius 77×10^{-12}m

Oxygen

atomic number 8
atomic mass 16
atomic radius 74×10^{-12}m

Figure 2.1 A simple diagram to show the atomic structures of hydrogen, carbon, and oxygen. The relative sizes are in proportion to the different atomic radii, but the atomic nuclei are enlarged for clarity. The filled red circles represent protons and the filled green circles represent neutrons. The hydrogen nucleus has a single proton (atomic mass = 1), that of carbon has six protons and six neutrons (atomic mass = 12, ^{12}C), and that of oxygen has eight protons and eight neutrons (atomic mass = 16). Each atomic nucleus is surrounded by a negatively charged cloud of electrons shown here in blue. For each element the number of electrons equals the number of protons.

2.2 Molecules are specific combinations of atoms

As indicated above, atoms of different elements are combined in many ways to form the material world around us. A specific combination of two or more atoms is known as a **molecule**. Sometimes two identical atoms combine to form a molecule—for example, molecular hydrogen consists of two hydrogen atoms (H_2: see Figure 2.2). The oxygen in the air we breathe is a combination of two oxygen atoms (molecular oxygen) with the chemical formula O_2. However, if a molecule consists of more than one kind of atom, it is called a **chemical compound**. Thus a water molecule is a compound of that consists of two atoms of hydrogen and one of oxygen, as shown in Figure 2.2. This is commonly written as H_2O. A molecule of sucrose (table sugar) consists of 12 atoms of carbon, 22 of hydrogen, and 10 of oxygen, and is written as $C_{12}H_{22}O_{10}$. Such formulae give information concerning the number of each type of atom in a molecule. For this reason they are known as **molecular formulae**. Although useful for many purposes, a molecular formula gives no indication of the way in which the atoms of a particular molecule are arranged. This information is provided by **structural formulae**. Some simple examples are shown in Figure 2.3.

Relative molecular mass

The relative molecular mass of a molecule (M_r) is equal to the sum of the relative atomic masses of the atoms of which it is composed. Here are some examples of how the relative molecular mass of a molecule is calculated.

The molecular structure of water is H_2O, so:

There are two hydrogen atoms each with a mass of 1, so their combined mass is 2 atomic mass units (a.u.).

There is one oxygen atom with a mass of 16 a.u.

The molecular mass of water is therefore 18 a.u. ($= [2 \times 1] + [1 \times 16]$).

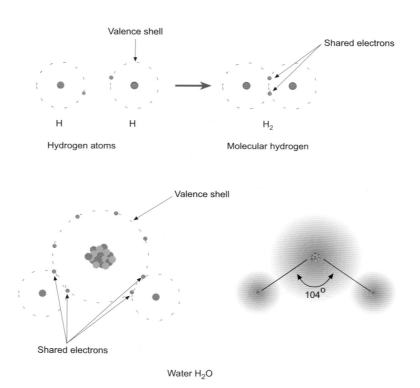

Water H_2O

Figure 2.2 A simplified diagram to show how valence shells are involved in chemical bonding. The top left panel shows two hydrogen atoms; each has a single shell and one s orbital. By sharing the two available electrons (top right), each atom can fill its s orbital to form molecular hydrogen. This configuration is favoured because it is more stable (i.e. it has a lower energy) than two non-bonded hydrogen atoms. The bottom panels show how hydrogen atoms are bonded to oxygen to form water. Each oxygen atom has one inner shell with two electrons in an s orbital (not shown) and six electrons in shell 2, the valence shell for oxygen. (This shell has six electrons distributed between one s orbital and three p orbitals.) As the s orbitals have a lower energy, they fill first. The configuration of electrons in shell 2 would be more stable if all its p orbitals had two electrons each so that the shell would have 8 electrons in total. This configuration can be achieved if two hydrogen atoms share their electrons with an oxygen atom, which is the arrangement shown in the bottom left panel. The two hydrogen–oxygen bonds are at an angle of approximately 104° to each other, as shown in the bottom right of the figure. Thus a water molecule has a distinct molecular shape. The figure also shows that the bond length is determined by the sum of the atomic radii of the atoms making the bond. Similar considerations apply to the formation of bonds between other atoms.

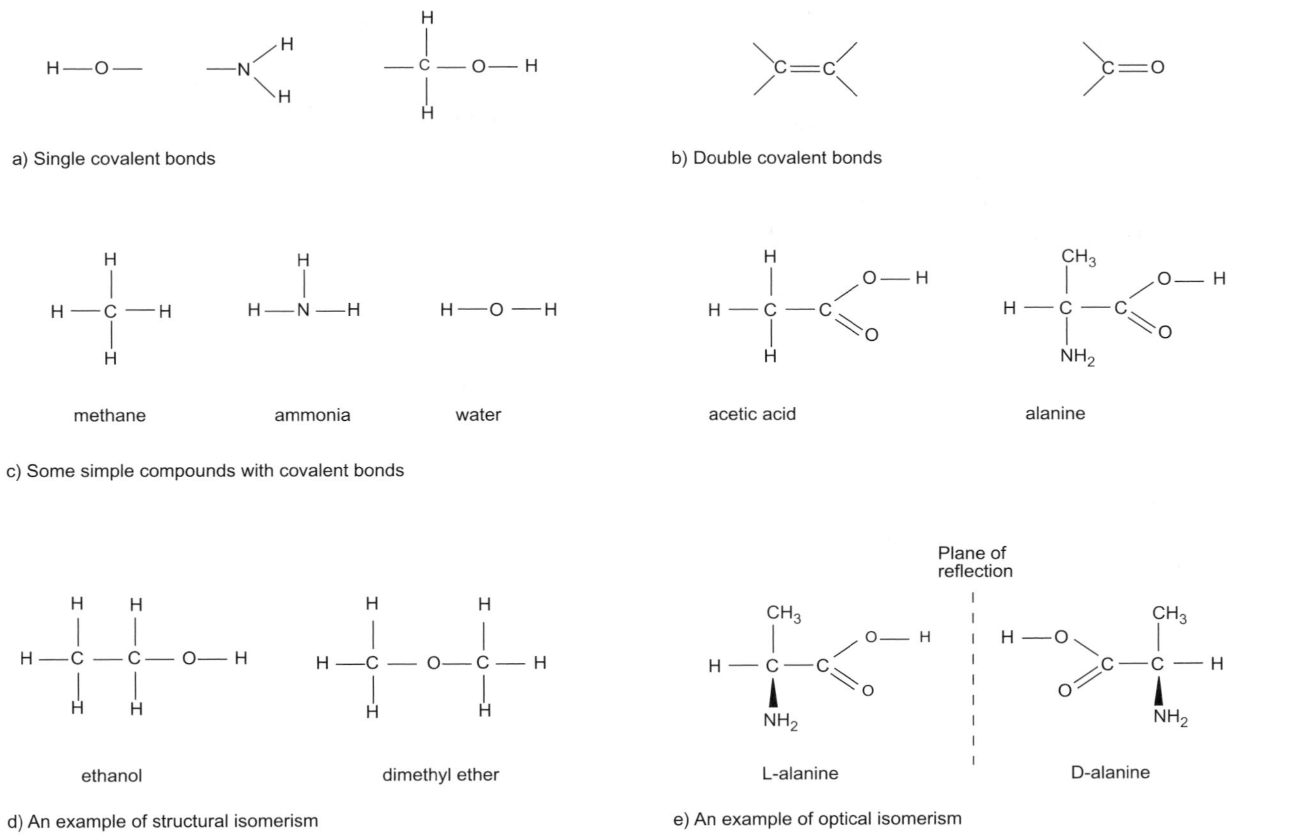

a) Single covalent bonds

b) Double covalent bonds

methane   ammonia   water   acetic acid   alanine

c) Some simple compounds with covalent bonds

ethanol   dimethyl ether   L-alanine   D-alanine

Plane of reflection

d) An example of structural isomerism

e) An example of optical isomerism

Figure 2.3 Examples of structural formulae showing how single and double covalent bonds are represented: panels (a) and (b). Panel (c) shows the complete structural formulae of some simple chemical compounds. Panel (d) shows an example of structural isomerism. Both ethanol and dimethyl ether (also known as methoxymethane) have the molecular formula C_2H_6O, but the arrangement of their constituent atoms is different with the result that their physical and chemical properties also differ. Panel (e) is an example of optical isomerism. In this case, the two enantiomers of the amino acid alanine have their constituent atoms linked in the same way but the two forms are mirror images and the three-dimensional structures cannot be superimposed. (The amino group (—NH_2) is represented as sticking out of the page.) The D (or +) form rotates polarized light to the right, while the L (or −) form rotates it to the left.

The relative molecular mass of carbon dioxide (CO_2) may be calculated in the same way.

There is one carbon atom with a mass of 12 a.u.

There are two oxygen atoms each with a mass of 16 a.u.; the oxygen atoms thus contribute 32 a.u.

The relative molecular mass of CO_2 is therefore 44 a.u. ($= [1 \times 12] + [2 \times 16]$).

The relative molecular mass of glucose ($C_6H_{12}O_6$) is 180 a.u. and is calculated as before.

There are six carbon atoms each with a mass of 12 a.u., total = 72.

There are twelve hydrogen atoms each with a mass of 1 a.u., total = 12.

There are six oxygen atoms each with a mass of 16 a.u., total = 96.

The relative molecular mass of glucose is therefore 180 a.u. ($= [6 \times 12] + [12 \times 1] + [6 \times 16]$).

As all the calculations of relative molecular mass are always based on the atomic mass unit, this is normally understood and the relative molecular mass of water is simply given as 18, that of carbon dioxide as 32, and so on. Nevertheless, the relative molecular mass (in a.u.) is sometimes expressed in units called Daltons (abbreviated D or Da). One Dalton is equal to one atomic mass unit. This usage is particularly common in biochemistry where the molecular masses are large—often many thousands of Da (kDa).

A mole of any substance has the same number of molecules

In chemistry and biology it is often useful to have some idea of the number of molecules in a particular volume of a sample. The amount of substance expressed in this way is given by a quantity called the **mole**. *One mole of atoms has a mass in grams that is equal to the atomic mass of the atom in atomic mass units.* The same applies to compounds—a mole of a compound is equal to its relative molecular mass in grams. A mole of any element or compound has exactly the same number of atoms or molecules—6.022×10^{23}. This is known as the **Avogadro constant** or **Avogadro's number**.

For most medical and physiological purposes, the mole is inconveniently large and quantities are given in thousandths of a mole, a quantity known as a **millimole** (1/1000 or 10^{-3} mole, often written mmol). Some substances, such as hormones, are present in very small amounts and their concentration is often expressed in **micromoles** (1/1,000,000 or 10^{-6} mole, often written μmol) or **nanomoles** (1/1,000,000,000 or 10^{-9} mole, often written nmol).

The nature of the chemical bond

The formation of chemical bonds is governed by the configuration of the electrons of the elements concerned. For any given atom, the electron shell that is farthest from the nucleus is called the **valence shell**. It is this shell that determines the number of chemical bonds that can be formed by an element. Chemical bonds are formed when valence electrons are transferred from one atom to another (**ionic bonds**) or shared between different atoms (**covalent bonds**—see Figure 2.2). The number of bonds an element can form is called its **valence** or **valency**. Many elements have only one valency state, for example carbon always has a valency of four, but some elements—especially metals—have more than one valency state. For example, iron has two valency states: ferrous (Fe^{2+}—which can form two bonds) and ferric (Fe^{3+}—which can form three bonds).

A chemical compound has properties quite distinct from its constituent elements. In this fundamental way, a chemical compound differs from a mixture. A mixture is matter containing two or more elements or chemical compounds that have not undergone any chemical bonding. Unlike the constituents of a chemical compound, the constituents of a mixture can, in principle, be separated by purely physical means. For example, a mixture of iron filings and sulphur can be separated into its constituents with the aid of a magnet, which will attract the iron filings but not the sulphur. It is not possible to separate iron and sulphur by this means when they have combined together to form the compound ferrous sulphide (FeS).

Ions are formed when an atom or molecule gains or loses electrons

When an electron is passed from one atom to another, an ionic bond is formed. The atom donating the electron becomes positively charged while that receiving the electron becomes negatively charged. It is the attraction between the opposite electrical charges of the constituent atoms that holds the material together. A charged atom or molecule is called an **ion** (hence the term 'ionic bond'). A positively charged ion is known as a **cation**, while a negatively charged ion is called an **anion**.

The number of electrons an atom may gain or lose in forming an ionic bond is characteristic of that atom. The resulting ions are represented by their chemical symbols with the charge indicated. An atom of sodium loses a single electron to become a sodium ion (written as Na^+). Potassium atoms also lose a single electron to form potassium ions (K^+), while calcium atoms lose two electrons

to form calcium ions (written as Ca^{2+} because calcium ions have two positive charges). Sodium, potassium, and calcium ions are all cations as they have a positive charge. When chlorine atoms gain an electron they form **chloride ions**, which are written as Cl^-. Since chloride ions carry a negative charge they are anions. So, when metallic sodium reacts with chlorine gas to form sodium chloride (common table salt) there is a transfer of an electron from the sodium atom (which becomes a cation) to the chlorine atom (which becomes an anion).

Covalent bonds and molecular shape

Covalent bonds are formed when two atoms share electrons. When two atoms share one pair of electrons, a single covalent bond is formed. Such bonds are represented on paper by a single line linking the two atoms, as shown in Figure 2.3 (a). When two atoms share two pairs of electrons a double bond is formed, which is represented by two lines linking the neighbouring atoms as in Figure 2.3(b). If two atoms share three pairs of electrons a triple bond is formed (as is the case for molecular nitrogen, N_2). Note that the atoms of different elements are able to make different numbers of bonds.

Individual molecules have shapes that are determined by the way their bonds are arranged in space. Oxygen atoms make two single bonds that are at an angle of approximately 104°. Nitrogen atoms make three single bonds distributed in the shape of a three-sided pyramid. Carbon atoms make four single bonds that are distributed symmetrically in space in the form of a tetrahedron, and so on. Examples of some simple compounds formed by these elements are shown in Figure 2.3(c). Single covalent bonds linking two atoms permit the individual atoms to rotate about the axis of the bond, so molecules can change their shape to some degree without breaking their chemical bonds. A double bond, however, prevents rotation around that axis and gives rigidity to a molecule.

The atoms that make up a molecule may be arranged in more than one way. Chemical compounds that have exactly the same number and kind of atoms (i.e. they have the same molecular formula) but different structures are called **isomers**. Figure 2.3d shows the structural formulae for ethanol and dimethyl ether (or methoxymethane), two chemical compounds that have exactly the same number of carbon, hydrogen, and oxygen atoms but quite different properties.

Molecular shape and optical isomers

When a carbon atom is bonded to four different chemical groups the resulting molecule is asymmetric, so that it can occur in one of two different shapes as shown in Figure 2.3e. Nevertheless, the two forms have exactly the same arrangement of their atoms. The two forms are mirror images of each other—just as our left and right hands are mirror images of each other (you cannot exactly superimpose your right and left hands, gloves, or shoes). Molecules of this kind are able to rotate a beam of polarized light. One form, designated the (+) or D- form, rotates light to the right while the other, the (−) or L- form, rotates it to the left. For this reason such molecules are known as **optical isomers** or **enantiomers**. Many biological molecules can exist in either the D- or L- configuration but, although they have the same molecular components and chemical properties, the two isomers do not have the same biochemical properties. For example, the glucose utilized to provide the energy for bodily activities is strictly called D-glucose (sometimes it is called dextrose because it rotates polarized light to the right). Chemists can make another version called L-glucose which is chemically identical but is the mirror image of D-glucose. (It causes polarized light to rotate to the left). However, L-glucose cannot participate in cellular energy production or other biochemical reactions as it has the wrong molecular shape.

Hydrogen bonds

Some covalent bonds do not have the shared electrons equally distributed between the two atoms, so that the electrons tend to be more associated with one of the two atoms. Such chemical bonds are called **polar bonds**. Examples of polar bonds are the bond between oxygen and hydrogen (—O—H) and the bond between nitrogen and hydrogen (—N—H). In these cases the oxygen and nitrogen atoms tend to attract the electrons of the bond more than the hydrogen atoms.

When a hydrogen atom of a polar bond is attracted to a neighbouring oxygen or nitrogen atom, a type of bond known as a **hydrogen bond** is formed. Hydrogen bonds are much weaker than covalent or ionic bonds but are very important in holding large biological molecules such as proteins in their correct shape. Common types of hydrogen bond found in biological systems are shown in Figure 2.4.

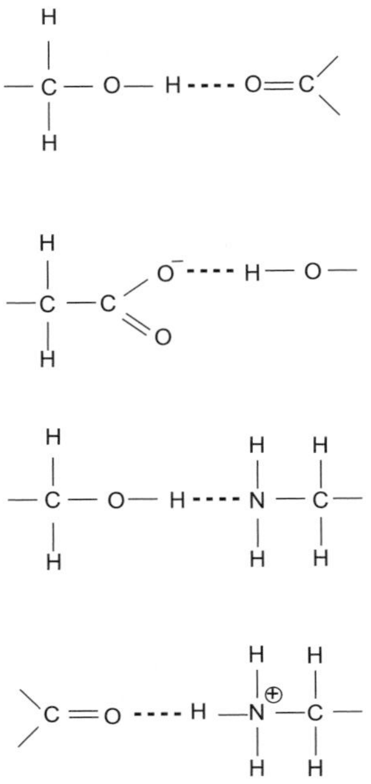

Figure 2.4 Typical examples of hydrogen bond formation. Note that the bonding hydrogen atom is linked to either an oxygen atom or a nitrogen atom. The dotted lines indicate the hydrogen bonds.

Chemical reactions

In the body, molecules are frequently converted from one type into another by way of transformations known as chemical reactions. These reactions constitute the **metabolism** of the body and involve the input or release of energy. In a chemical reaction the total number of atoms remains the same, but their arrangement changes. This is known as the **law of conservation of matter**.

Chemical reactions may be represented by chemical equations such as the following:

$$C_6H_{12}O_6 + 6O_2 \rightarrow 6CO_2 + 6H_2O + \text{heat (energy)}$$

The forward-facing arrow indicates the direction of the chemical transformation. Note that on each side of the equation there are the same number of carbon atoms (6), hydrogen atoms (12), and oxygen atoms (18). The equation is said to balance. This particular equation shows the transformation of glucose to carbon dioxide and water by reaction with oxygen, with the liberation of heat. To make the reaction go in the opposite direction (i.e. to form glucose from carbon dioxide and water) requires energy. Plants use sunlight to provide the energy necessary for this reaction via the process of photosynthesis. This graphically illustrates the point that **molecules can act as a store of energy**. Different molecules are able to store different amounts of energy that can be used by cells to perform their functions.

Summary

Molecules are combinations of atoms that are held together by chemical bonds. The relative molecular mass of a molecule (M_r) is equal to the sum of the atomic masses of the atoms of which it is composed. A mole of a compound is equal to its molecular mass in grams. Molecules can undergo chemical reactions in which the atoms are rearranged.

2.3 Water and solutions

Water is the principal constituent of the human body and is essential for life. It is the chief solvent in living cells. A **solvent** is a liquid that can dissolve a substance (known as the **solute**) to form a **solution**. The amount of a substance in a given volume of solution is known as its **concentration**. It may be simply stated in grams per litre or grams per decilitre of solution (g l^{-1} or g dl^{-1}), but in physiology and medicine it is often much more informative to express the concentration in terms of the number of moles or millimoles per litre of solution. This is known as the **molarity** of the solution, which is denoted by the letter M. A 0.1M solution of glucose will contain 0.1 moles of glucose per litre of solution. This could also be expressed as 100 millimoles per litre of solution (100 mmol/litre or 100 mM). Solutions of the same molarity have the same number of molecules of solute per unit volume of solution. Box 2.1 explains how to calculate the molarity of a solution.

$1_2 3$ Box 2.1 How to calculate the molarity of a solution

Very often the concentration of a solution is expressed as so many grams per unit volume. Although this is a common way to express the concentration of a solution, it gives little information about the number of molecules present in a given volume of the solution. This is important for many purposes—including for calculation of the osmolarity of a solution and for calculation of electrolyte balance (i.e. the number of cations compared to the number of anions). As different molecules may differ greatly in mass, this should be taken into account when we state the concentration of a solution. A mole of a chemical compound (or element) is equal to its molecular mass in grams. So to calculate the molarity of a solution it is necessary to calculate the number of moles in each litre of solution as follows:

$$\text{Molar concentration} = \frac{\text{weight of substance in one litre } (\textit{in grams})}{\text{relative molecular mass}}$$

Here are some examples of molar calculations:

A solution of 0.9 per cent sodium chloride in water (also known as normal saline) has 0.9 grams of sodium chloride per 100 ml (per decilitre or dl) of solution. This is the same as 9 grams per litre. The relative molecular mass (M_r) of sodium chloride is 58.4 so the molarity of this solution is $9 \div 58.4 = 0.154$ moles per litre. This is said to be a 0.154 molar solution of sodium chloride. Rather than express the concentration as a fraction, it is more usual to express

concentrations of this magnitude as so many millimoles per litre—in this case 154 millimoles per litre or 154 mmol l^{-1}. (To convert from moles to millimoles, multiply by 1000; to convert millimoles to moles, divide by 1000.)

In normal blood, the glucose concentration is around 85 mg of glucose per 100 ml of blood. The molecular mass (M_r) of glucose ($C_6H_{12}O_6$) is 180 (= $[6 \times 12] + [1 \times 12] + [6 \times 16]$). To calculate the molarity of glucose in blood, first calculate how many milligrams (mg) there would be per litre of solution. In this case it is $85 \times 10 = 850$ mg per litre or 0.85 g per litre, and the molar concentration is $0.85 \div 180 = 0.00472$ moles per litre or 4.72 millimoles per litre (abbreviated as 4.72 mmol l^{-1}).

To calculate how much potassium chloride (KCl; $M_r = 74.6$) is required to make 100 ml of a 50 mmol l^{-1} solution, first work out how much potassium chloride is required to make a litre of a 50 millimolar solution:

$$(74.6 \times 50) \div 1000 = 3.730 \text{ g}$$

Then multiply by the fraction of a litre that is required (remember that 1 litre = 1000 millilitres):

$$100 \div 1000 = 0.1$$

The final quantity needed is:

$$3.730 \times 0.1 = 0.373 \text{ grams (or 373 milligrams)}$$

Polar and non-polar molecules

Molecules of biological interest can be divided into those that dissolve readily in water and those that do not. Substances that dissolve readily in water are called **polar** or **hydrophilic**, while those that are insoluble in water are called **non-polar** or **hydrophobic**. Examples of polar substances are sodium chloride, sucrose, ethanol, and acetic acid (the pungent ingredient of vinegar). Examples of non-polar materials are fats (e.g. butter), olive oil, and waxes. Many molecules of biological interest have mixed properties so that one part is polar while another part is non-polar. These are known as **amphiphilic** or **amphipathic** substances.

Summary

Water is the chief solvent of the body. Substances that dissolve readily in water are polar (or hydrophilic); those that are insoluble in water are non-polar (or hydrophobic).

Ionization

Pure water can dissociate so that one molecule is transformed into one hydrogen ion (H^+) and one hydroxyl ion (OH^-), as shown in the following chemical equation:

$$H_2O \rightleftharpoons H^+ + OH^- \qquad [1]$$

However, as the two ions tend to attract one another to reform water, an equilibrium is established (indicated here by the double arrow) in which the tendency of water molecules to dissociate is balanced by the tendency of the hydrogen and hydroxyl ions to bind together to form a neutral water molecule. In this case, there are vastly more water molecules than hydrogen and hydroxyl ions, so the equilibrium is heavily biased to the left side.

Compounds held together by ionic bonds such as sodium chloride, potassium chloride, and calcium chloride generally dissolve readily in water and dissociate completely into their constituent ions. So, when sodium chloride dissolves in water there are no molecules of NaCl in the solution, but a mixture of equal numbers of sodium ions (Na^+) and chloride ions (Cl^-).

$$NaCl \rightarrow Na^+ + Cl^- \qquad [2]$$

A solution of potassium chloride contains potassium ions (K^+) and chloride ions (Cl^-), not molecules of KCl.

$$KCl \rightarrow K^+ + Cl^- \qquad [3]$$

However, when calcium chloride ($CaCl_2$) dissolves in water, the resulting solution has twice as many chloride ions as calcium ions (Ca^{2+}), because each calcium ion has two positive charges and so can bind two chloride ions, each of which has a single negative charge.

$$CaCl_2 \rightarrow Ca^{2+} + 2Cl^- \qquad [4]$$

Solutions containing ions conduct electricity. Because of this property, ions in solution are collectively called **electrolytes**. Those substances that dissociate completely into their constituent ions (e.g. NaCl) form solutions that conduct electricity easily, and the electrolytes themselves are known as **strong electrolytes**.

Compounds possessing polar bonds are often able to interact with water and dissociate, giving rise to ions. An example of covalent molecules undergoing ionization is provided by the reaction of carbon dioxide with water, in which carbonic acid is formed followed by its dissociation into bicarbonate ions (HCO_3^-) and hydrogen ions:

$$CO_2 + H_2O \rightleftharpoons H_2CO_3 \rightleftharpoons H^+ + HCO_3^- \qquad [5]$$

A further example is the reaction of ammonia with water to form ammonium hydroxide, which then dissociates into ammonium ions (NH_4^+) and hydroxyl ions:

$$NH_3 + H_2O \rightleftharpoons NH_4OH \rightleftharpoons NH_4^+ + OH^- \qquad [6]$$

This process of **ionization** occurs very frequently in biochemical reactions. The solutions formed are not very good conductors of electricity, and the electrolytes are called **weak electrolytes**. In physiology, the principal strong electrolytes are sodium (Na^+), potassium (K^+), calcium (Ca^{2+}), magnesium (Mg^{2+}), and chloride (Cl^-). The main weak electrolytes are bicarbonate (HCO_3^-) and phosphate (PO_4^{3-}) ions.

Sodium and potassium are the main cations of the body fluids, while chloride and bicarbonate are the main anions. In clinical practice these ions are routinely measured together with urea (urea and electrolytes, often abbreviated U&E).

Acids and bases

In physiological terms an **acid** is a substance that generates hydrogen ions in solution, while a **base** is a substance that absorbs hydrogen ions. Acids and bases are further classified as **weak** or **strong** according to how completely they dissociate in solution. Strong acids such as hydrochloric acid (HCl) are completely dissociated in solution and there are no neutral hydrogen chloride molecules in the solution—only hydrogen ions (H^+) and chloride ions (Cl^-). Similarly, when sodium hydroxide (NaOH) is added to water, only sodium (Na^+) and hydroxyl (OH^-) ions are present. In contrast, when weak acids such as carbonic acid (H_2CO_3) or weak bases such as ammonium hydroxide (NH_4OH) are in solution they are only partly dissociated—as shown by chemical equations [5] and [6]. A neutral solution is one in which the concentrations of hydrogen and hydroxyl ions are equal. When pure water dissociates, both hydrogen ions and hydroxyl ions are produced in equal quantities, as shown in chemical equation [1].

As weak acids and weak bases can react with hydrogen ions, they can 'mop up' any hydrogen ions that may arise during the chemical reactions that occur in the body. They thus can limit the resulting changes in hydrogen ion concentration. For this reason, solutions of weak acids and bases are called **buffers**. The action of the principal physiological buffers is considered in greater detail in Chapter 41.

Hydrogen ion concentration and the pH scale

In the chemical equation [5], a neutral molecule (carbon dioxide) gives rise to a positively charged hydrogen ion (H^+) when it reacts with water. Hydrogen ions are hydrogen atoms that have lost their electron. Another name for a hydrogen ion is a proton, because the hydrogen atom consists of one proton and one electron. Hydrogen ions combine readily with molecules that have polar groups and alter the chemical properties of such molecules. For this reason the concentration of hydrogen ions in a solution is very important in biology. It may be expressed in moles of hydrogen ions per litre, but it is more often expressed on a scale known as the **pH scale**.

The pH of a solution is the logarithm to the base 10 of the reciprocal of the hydrogen ion concentration (the H^+ in square brackets indicates hydrogen ion concentration in moles l^{-1}).

$$pH = \log_{10}\left[\frac{1}{[H^+]}\right] \qquad [7]$$

Since, however,

$$\log_{10}\left[\frac{1}{[H^+]}\right] = -\log_{10}[H^+]$$

then:

$$pH = -\log_{10}[H^+] \qquad [8]$$

A common alternative definition is, therefore, that the pH of a solution is the negative logarithm of the hydrogen ion concentration. A change of one pH unit corresponds to a tenfold change in hydrogen ion concentration (because $\log_{10}10 = 1$). An explanation of how to calculate the pH of a solution of known hydrogen ion concentration is given in Box 2.2.

While the pH notation is convenient for expressing a wide concentration range, it is potentially confusing as *a decrease in pH reflects an increase in hydrogen ion concentration and vice versa*. Nevertheless, the pH scale is widely used to express how acid or alkaline a solution is. Pure water has a pH of 7.0 at 25°C. At this temperature, acidic solutions have a pH value of less than 7, while alkaline solutions have a pH value greater than 7: the lower the pH value, the more acid the solution; the higher the pH, the more alkaline it would be. Blood pH is normally between 7.35 and 7.4 (i.e. very mildly alkaline), while urine pH can range from 4.7 to 8.2 although it is usually around pH 6 and thus is mildly acid. The chemical factors that determine the pH of a solution are set out in Box 41.1.

Summary

The concentration of hydrogen ions in a solution determines its pH value. The lower the pH, the more acid the solution. At 25°C a neutral solution has a pH of 7.0 so that at this temperature, acid solutions have a pH value less than 7 while alkaline solutions have a pH value greater than 7.

$1_2 3$ Box 2.2 How to calculate the pH of a solution

In the age of calculators and computers, this is now a relatively straightforward procedure.

$$pH = -\log_{10}[H^+]$$

First, enter the hydrogen ion concentration $[H^+]$ *expressed as moles per litre* into the calculator. Then press the $\log_{10}$ key. Finally change the sign (press the +/− key). The resulting number is the pH of the solution.

To convert from pH to free $[H^+]$ use the following relationship:

$$[H^+] = 10^{-pH}$$

Enter the pH value and change sign, then press the inverse function key followed by the log key. The resulting number is the hydrogen ion concentration in moles per litre.

Some worked examples:

If a solution has a hydrogen ion concentration of 5×10^{-6} moles per litre, what is its pH?

$$pH = -\log_{10}[5 \times 10^{-6}] = 5.3$$

If the pH of a blood sample is 7.4 (a normal value), what is the hydrogen ion concentration?

$$[H^+] = 10^{-7.4} = 39.8 \times 10^{-9} \text{ moles } l^{-1}$$

Using the same method of calculation, show that if a urine sample has a pH of 5.5, the $[H^+]$ concentration is 3.16×10^{-6} moles l^{-1}. (Note that the 1.9 unit difference in pH between these two samples corresponds to an 80-fold difference in hydrogen ion concentration.)

Key points about logarithms

The common logarithm of a number is the power of 10 that will give that number. If our number is x we can write its logarithm (y) as $y = \log_{10}(x)$, which is equivalent to saying that $x = 10^y$. Logarithms can be calculated using any positive number as a base, but in biology only logarithms to base 10 and natural logarithms are used. Natural logarithms use the mathematical constant e (2.718 approximately) as their base and are denoted as $\ln(x)$ or $\log_e(x)$. Both natural logarithms and logarithms to the base 10 are given on many calculators.

Diffusion

In a solution, the molecules of both solvent and solute are in continuous random motion with frequent collisions between them. The individual solute molecules are free to move in a random way and become evenly dispersed within the solvent. This is the process of **diffusion**. If a drop of ink is added to a volume of pure water, the ink particles slowly disperse throughout the whole volume. Moreover, if the ink drop had been added to a dilute solution of ink, the same process of dispersion of the ink particles would occur until the whole solution was of a uniform concentration. Equally, when a drop of a concentrated solution (e.g. 5 per cent w/v glucose) is added to a volume of pure water, the random motion of the glucose molecules results in their slow dispersion throughout the whole volume. If the drop of 5 per cent solution had been added to a 1 per cent solution of glucose, the same process of dispersion of the glucose molecules would occur until the whole solution was of uniform concentration. There is always a tendency for the glucose (or any other solute) to diffuse from a region of high concentration to one of a lower concentration (i.e. down its concentration gradient).

The rate of diffusion in a solvent depends on the temperature (it is faster at higher temperatures), the magnitude of the concentration gradient, and the area over which diffusion can occur. The molecular characteristics of the solute and solvent also affect the rate of diffusion. These characteristics are reflected in a physical constant known as the **diffusion coefficient**. The role of these different factors is expressed in Fick's law of diffusion, which is briefly discussed in Box 31.1. In general, large molecules diffuse more slowly than small ones. Note that diffusion is not confined to the fluids of the body but also occurs through cell membranes, which are largely made of **lipids** (as discussed in Chapter 4).

Filtration

When a fluid passes through a permeable membrane, it leaves behind those particles that are larger in diameter than the pores of the membrane. This process is **filtration**, which is driven by the pressure gradient between the two sides of the membrane. The pumping action of the heart causes a pressure gradient across the walls of the capillaries, which tends to force fluid from the capillaries into the interstitial space. As the walls of the capillaries are not very permeable to the plasma proteins (e.g. albumin) but are permeable to small solutes such as glucose and inorganic ions (Na^+, K^+, Cl^-, etc.), the fluid of the interstitial space has only a small amount of plasma protein although the concentration of the small solutes is the same as that of plasma. This process is called **ultrafiltration** and occurs in all vascular beds, but it is particularly important in the glomerular capillaries of the kidney, which filter large volumes of plasma each day (as discussed in Chapter 39).

The osmotic pressure of the body fluids

When an aqueous solution is separated from pure water by a membrane that is permeable to water but not to the solute, water moves across the membrane into the solution by a process known as **osmosis**. This is shown in Figure 2.5. The movement of water can be opposed by applying a hydrostatic pressure to the solution. The pressure that is just sufficient to prevent the uptake of water is

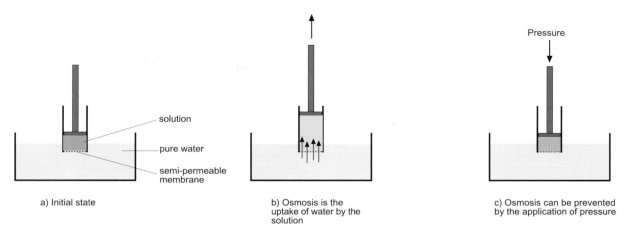

Figure 2.5 Osmosis and osmotic pressure. The passage of water across a semi-permeable membrane into a solution can be opposed by applying pressure. The pressure that just prevents the uptake of water is known as the osmotic pressure of the solution (see text for further details).

known as the **osmotic pressure** of the solution. Rather than measuring osmotic pressure directly, it is more convenient to state the **osmolarity** (moles per litre of solution) or **osmolality** (moles per kg of water). In clinical medicine, the osmotic pressure of a body fluid is generally expressed as its osmolality. The osmotic pressure (π) of a solution of known molar composition (M) can be calculated from the following simple equation:

$$\pi = MRT$$

where R is the universal gas constant (8.31 J K^{-1} mol^{-1}) and T is the absolute temperature (310 K at normal body temperature). The osmotic pressure is thus directly related to the number of particles present in a solution and is independent of their chemical nature.

Since the osmotic pressure depends on the number of particles present in a given volume of water, solutions that have the same **molality** will have the same osmolality. This is a direct consequence of the two solutions having the same number of molecules per unit volume. Despite the large difference in their relative molecular mass, the osmotic pressure exerted by a millimole of glucose (M_r 180) is the same as that exerted by a millimole of albumin (M_r 69,000). However, *aqueous salt solutions are an important exception to this rule*: the salts separate into their constituent ions so that a solution of sodium chloride will exert an osmotic pressure double that of its molal concentration. Hence a 100 mmol kg^{-1} solution of sodium chloride in water will have an osmotic pressure of 200 mOsmol kg^{-1} of which half is due to the sodium ions and half to the chloride ions.

The total osmolality of a solution is the sum of the osmolality due to each of the constituents. The blood plasma has an osmolality of around 0.3 Osm kg^{-1} (300 mOsmol kg^{-1}). The principal ions (Na$^+$, K$^+$, Cl$^-$ etc.) contribute about 290 mOsmol kg^{-1} (about 96 per cent) while glucose, amino acids and other small non-ionic substances contribute approximately 10 mOsmol kg^{-1}. Proteins contribute only around 0.5 per cent to the total osmolality of plasma. This is made clear by the following calculations:

- Blood plasma has about 6.76 g of sodium chloride and 47.4 g of albumin per kg of plasma water. The osmotic pressure of a solution of 6.76 g of sodium chloride (M_r 58.4) is:

$$(6.76 \div 58.4) \times 2 = 0.231 \text{ Osmol kg}^{-1} \text{ or } 231 \text{ mOsmol kg}^{-1}$$

- The osmotic pressure exerted by 47.4 g of albumin is:

$$47.4 \div 69,000 = 6.87 \times 10^{-4} \text{ Osmol kg}^{-1} \text{ or approximately}$$
$$0.69 \text{ mOsmol kg}^{-1}$$

Thus the osmotic pressure exerted by 47 g of albumin is only about 0.3 per cent of the pressure exerted by 6.76 g of sodium chloride. This makes clear that *the osmotic pressure exerted by proteins is far less than that exerted by the principal ions of the biological fluids.* Nevertheless, the small osmotic pressure that the proteins do exert (known as the **colloid osmotic pressure** or **oncotic pressure**) plays an important role in the exchange of fluids between body compartments (see Chapter 31).

The tonicity of solutions

The **tonicity** of a solution refers to the influence of its osmolality on the volume of cells. For example, red blood cells placed in a solution of 0.9 per cent sodium chloride in water (i.e. 0.9 g sodium chloride in 100 ml of water) will neither swell nor shrink. This concentration has an osmolality $\approx$ 310 mOsmol kg^{-1} and is said to be **isotonic** with the cells. (This solution is sometimes referred to as 'normal saline' but would be better called isotonic saline.) If the same cells were to be added to a solution of sodium chloride with an osmolality of 260 mOsmol kg^{-1}, they would swell as they took up water to equalize the osmotic pressure across their cell membranes. This concentration of sodium chloride is said to be **hypotonic** with respect to the cells. Conversely, red blood cells placed in a solution of sodium chloride that has an osmolality of 360 mOsmol kg^{-1} would shrink as water was drawn from the cells. In this case the

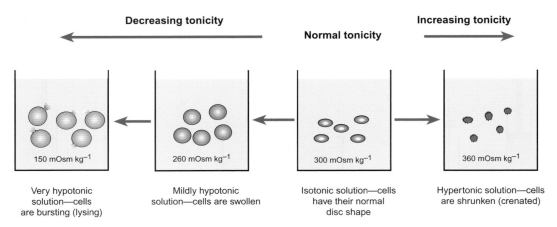

Decreasing tonicity ← **Normal tonicity** → **Increasing tonicity**

150 mOsm kg⁻¹

260 mOsm kg⁻¹

300 mOsm kg⁻¹

360 mOsm kg⁻¹

Very hypotonic solution—cells are bursting (lysing)

Mildly hypotonic solution—cells are swollen

Isotonic solution—cells have their normal disc shape

Hypertonic solution—cells are shrunken (crenated)

Figure 2.6 The changes in cell volume that occur when red blood cells are placed in solutions of different osmolality. In hypotonic solutions, the cells swell and may burst (lyse) if the solution is too dilute. In hypertonic solutions, water is drawn from the cells and they shrink and become irregular in shape (they become crenated). Isotonic solutions maintain the normal volume of the red cells. Other kinds of cell placed in hypotonic or hypertonic solutions would behave in the same way.

fluid is said to be **hypertonic**. These distinctions are illustrated in Figure 2.6. As cells normally maintain a constant volume, it is clear that the osmolality of the fluid inside the cells (the **intracellular fluid**) is the same as that outside the cells (the **extracellular fluid**). The two fluids are said to be **iso-osmotic** or (more accurately) iso-tonic with each other. If the osmotic pressure in one compartment is higher than the other, water will move from the region of low osmotic pressure to that of the higher osmotic pressure until the two become equalized.

Not all solutions that are iso-osmotic with respect to the intracellular fluid are isotonic with cells. A solution containing 310 mOsmol. kg⁻¹ of urea is iso-osmotic with both normal saline and the intracellular fluid but it is not isotonic, as cells placed in such a solution would swell and burst (or **lyse**). This behaviour is explained by the fact that urea can penetrate the cell membrane relatively freely. When it does so, it diffuses down its concentration gradient and

water will follow (otherwise the osmolality of the intracellular fluid would increase and become *hyper-osmotic*). Since there is an excess of urea outside the cells, it will continue to diffuse into the cells attracting water via osmosis and the cells will progressively swell until they burst.

Summary

When a substance dissolves in water, it exerts an osmotic pressure that is related to the number of particles present per kg of water, independent of their chemical nature. The total osmolality of a solution is the sum of the osmolality due to each of the constituents. The tonicity of a solution refers to the influence of its osmolality on the volume of cells. Hypotonic solutions have an osmolality that is less than that of the cells and hypertonic solutions have an osmolality that is greater than that of the cells.

✱ Checklist of key terms and concepts

The chemical elements and the structure of matter

- All matter is composed of atoms. An atom consists of a dense nucleus formed from protons and neutrons, surrounded by a cloud of electrons.

- The mass of an atom is determined by the number of protons and neutrons in its nucleus.

- The chemical elements are substances that cannot be broken down into simpler substances by chemical means. All the atoms in a pure sample of an element have the same number of protons and electrons. This is their atomic number.

- Each element has been assigned a one- or two-letter symbol that is used when describing chemical structures and reactions.

- Many elements have atoms with different numbers of neutrons in their nuclei and therefore different atomic masses. The atoms of differing mass of an element are called isotopes, some of which are unstable and undergo radioactive decay.

The formation of chemical bonds

- Most atoms can donate or receive electrons to form chemical bonds, which hold the component atoms of a chemical compound together.

- When two or more atoms have become linked via chemical bonds they form a molecule.
- There are two main types of chemical bond: ionic and covalent.
- An ionic bond is formed when one or more electrons are transferred from one atom to another.
- When an atom donates or receives an electron it becomes an ion.
- A covalent bond is formed when two atoms share one or more pairs of electrons.

Molecules, moles, and molecular mass

- A specific combination of two or more atoms is known as a molecule.
- The relative molecular mass of a molecule is equal to the sum of the relative molecular masses of the atoms of which it is composed.
- A mole of an element or compound is the mass in grams that is equal to the atomic or relative molecular mass of that substance.
- Molecular formulae provide information about the number of each kind of atom that has participated in the formation of the molecule.
- Structural formulae provide information about the way in which the constituent atoms of a molecule are arranged in space.
- Isomers are molecules that have the same atomic composition with different arrangements of the constituent atoms within the molecule.
- Optical isomers are molecules that are asymmetric and mirror images of each other. Such molecules have the ability to rotate polarized light.

Chemical reactions

- Molecules can be converted into other molecules by means of chemical reactions.
- Such transformations constitute the metabolism of the body.
- In a chemical reaction the total number of atoms remains the same but the atoms are rearranged.
- Chemical reactions are driven by changes in the chemical energy stored within the molecules or by energy provided by an external source.

Molecules and solutions

- A solvent is a liquid that dissolves another substance (the solute) to form a solution.
- The concentration of solute is often expressed in terms of its molarity, which indicates the number of moles dissolved per litre of solution.

- Molecules that dissolve readily in water are called polar or hydrophilic substances; those that do not are non-polar or hydrophobic.
- Some molecules have mixed properties—part polar and part non-polar; these are called amphiphilic or amphipathic substances.
- In polar covalent bonds, a hydrogen atom can be attracted to the electrons of a neighbouring oxygen or a nitrogen atom. This association is called a hydrogen bond.
- Pure water and many polar organic compounds become ionized in aqueous solutions.

Acids, bases, and the pH scale

- An acid is a substance that generates hydrogen ions in an aqueous solution.
- A base is a substance that can bind (or absorb) free hydrogen ions.
- Acids and bases can be either strong or weak. Strong acids and bases are fully ionized (dissociated) in solution, while weak acids and bases are only partially ionized.
- The concentration of hydrogen ions in a solution is often expressed on the pH scale.
- The pH of a solution is the logarithm of the reciprocal of the hydrogen ion concentration.
- Solutions with a high hydrogen ion concentration have a low pH and vice versa.
- Pure water at 25°C has a pH of 7.00 and is neutral.
- Solutions which limit changes in pH when acids or bases are added are called buffer solutions.

Osmotic pressure and tonicity

- Osmosis is the movement of water from a less to a more concentrated solution via a semi-permeable membrane (one that is permeable to water but not to the solute).
- The osmotic pressure is the hydrostatic pressure that is just sufficient to prevent the osmotic transfer of water.
- It directly depends on the number of particles (molecules or ions) present per unit volume of solution.
- Osmotic balance is very important in regulating the volume of the cells and in the transport of water from one body compartment to another.
- An isotonic solution is one that has the same osmotic pressure as that of the cells.
- Hypertonic solutions have a greater osmotic pressure than the cells while hypotonic solutions have an osmotic pressure less than that of the cells.

Recommended reading

Alberts, B., Johnson, A., Lewis, J., Morgan, D., Raff, M., Roberts, K., and Walter, P. (2014) *Molecular biology of the cell* (6th edn), Chapter 2. Garland, New York.

The first part of this chapter provides an alternative brief survey of the chemical concepts required to understand basic biochemical processes.

Crowe, J., and Bradshaw, T. (2010) *Chemistry for the biosciences: The essential concepts*, Chapters 1–4, 8, 10, and 12. Oxford University Press, Oxford.

These chapters offer more detailed but very lucid explanations of the basic chemical concepts covered in this chapter.

 To check that you have mastered the key concepts presented in this chapter, complete the accompanying online self-assessment questions. Go to www.oup.com/uk/pocock5e/

CHAPTER 3
The chemical constitution of the body

Chapter contents

This chapter should help you to understand:

*   The distribution of body water
*   The structure and functions of the carbohydrates
*   The chemical nature and functions of lipids
*   The structure of the amino acids and proteins
*   The structure of the nucleotides and the nucleic acids
*   Gene transcription and translation

3.1 Introduction

The human body consists largely of four elements: oxygen, carbon, hydrogen, and nitrogen. These are combined in many different ways to make a huge variety of chemical compounds. About 70 per cent of the lean body tissues is water, the remaining 30 per cent being made up of organic (i.e. carbon-containing) molecules and minerals. The principal organic constituents of mammalian cells are the **carbohydrates**, **fats**, **proteins**, and **nucleic acids**, which are built from smaller molecules belonging to four classes of chemical compounds: the sugars, the fatty acids, the amino acids, and the nucleotides respectively. The principal minerals found in the tissues are, in order of abundance, calcium, phosphorus, potassium, and sodium. Figure 3.1 gives an approximate indication of the chemical composition of the body for a young adult male, but note that there is much individual variation and that the proportions of the various constituents vary between tissues and change during development.

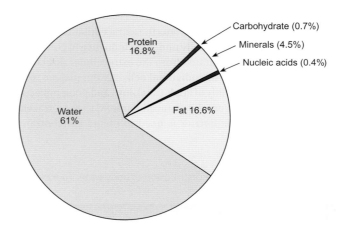

Figure 3.1 The approximate composition of the body of a young adult male. Note that such estimates are subject to some uncertainty and that there is considerable variation between individuals. In healthy females of the same age, there is a higher proportion of body fat.

3.2 Body water

Water is the principal constituent of the human body. It is essential for life and is the chief solvent in all living cells. The proportion of total body weight contributed by water varies with the age and sex of an individual. In both men and women, the water content of the lean body mass (i.e. the non-adipose tissues) is about 70–75 per cent. However, as adipose tissue (i.e. body fat) only contains about 10 per cent water, the proportion of body weight contributed by water varies between the sexes and with age: it is highest in **neonates** (c. 75 per cent body weight), around 60 per cent in adult males and around 50 per cent in adult females—see Table 3.1.

Body water can be divided into that located within the cells, the **intracellular water**, and that which lies outside the cells, the **extracellular water**. As the body water contains many different substances in solution, the liquid portion (i.e. the water plus the dissolved materials) of cells and tissues is known as fluid. The fluid in the space that lies outside the cells is called the **extracellular fluid** while that inside the cells is the **intracellular fluid**. The extracellular fluid is further subdivided into the **plasma** and the **interstitial fluid**. The plasma is the liquid fraction of the blood, while the interstitial fluid lies outside the blood vessels and bathes the cells. The small contribution from the lymph is included in the interstitial fluid. The extracellular fluid in the serosal spaces such as the ventricles of the brain, the abdominal cavity, the joint capsules, and the ocular fluids is called **transcellular fluid** (see Figure 3.2).

The interstitial space (or **interstitium**) consists of connective tissue, chiefly collagen, hyaluronate, and proteoglycan filaments together with an ultrafiltrate of plasma. The water of the interstitial fluid hydrates the proteoglycan filaments to form a gel (much like a thin jelly) and in normal tissues there is very little free liquid. This important adaptation prevents the extracellular fluid flowing to the lower regions of the body under the influence of gravity. The intracellular fluid is separated from the extracellular fluid by the plasma membrane of the individual cells. As the plasma membrane is composed mainly of lipids (fats), ions and polar molecules cannot readily cross from the extracellular fluid to the intracellular fluid. Indeed, this barrier is used to create concentration

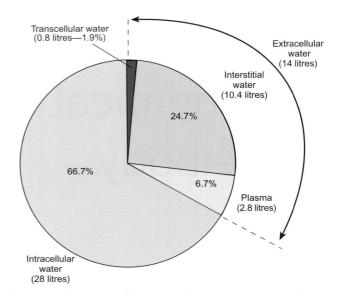

Figure 3.2 The approximate distribution of water between the various body compartments for a 70 kg man. Note that two-thirds of the total is found in the cells and that the blood plasma accounts for only about 7 per cent of the total. This distribution excludes the water of ossified bone (which contributes around 8 per cent to total body water).

gradients that the cells exploit to perform various functions (see Chapters 4–6).

The distribution of body water between compartments

The amount of water in the main fluid compartments can be determined by the dilution of specific markers. For a marker to permit the accurate measurement of the volume of a particular compartment it must be evenly distributed throughout that compartment and it should be physiologically inert (i.e. it should not be metabolized or alter any physiological variable). In practice, it is necessary to correct for the loss of the markers in the urine. Fortunately, it is not difficult to make the appropriate corrections.

The *plasma volume* can be estimated from the dilution of the dye Evans Blue, which does not readily pass across capillary walls into the interstitial space. Radio-labelled albumin (albumin with a radioactive atom such as ^{131}I attached) has also been used to measure plasma volume. Since the amount of marker injected is known, it is a simple matter to calculate the volume in which it has been diluted (the principle is explained in Box 3.1).

To determine the *total body water*, a known amount of radioactive (tritiated) water (^3H_2O) or deuterium oxide (^2H_2O) is injected and sufficient time allowed for the label to distribute throughout the body. A sample of blood is then taken and the concentration of label measured. Measurement of the *extracellular fluid volume* requires a substance that passes freely between the circulation and the interstitial fluid but does not enter the cells. These requirements are met by the plant polysaccharide inulin (note: not the hormone insulin) and by mannitol, although several other markers

Table 3.1 The approximate distribution of body water (per cent total body weight)

	Adult males	Adult females	Neonates
Total body water	60	50	75
Intracellular water	40	30	40
Extracellular water which consists of:	20	20	35
plasma	4	4	5
interstitial fluid*	16	16	30

*The interstitial fluid includes the lymph and transcellular fluid.

Box 3.1 The use of dilution methods to estimate the volume of fluid compartments

Evans Blue does not enter the red cells and is largely retained within the circulation as it binds to plasma albumin. This dye is therefore useful for estimating the plasma volume. As an example, assume that an individual with a body weight of 70 kg was injected with 10 ml of a 1 per cent (w/v) solution of the dye. Further assume that a sample of blood was taken after 10 min, and the plasma was found to contain 0.037 mg ml^{-1} of dye. What is the plasma volume?

$$\text{concentration} = \frac{\text{amount of dye}}{\text{volume}}$$

$$\text{volume} = \frac{\text{amount of dye}}{\text{concentration}}$$

The total amount of dye injected was 0.1 g (or 100 mg) and the concentration in the plasma 10 min after injection was 0.037 mg ml^{-1}. Therefore:

$$\text{plasma volume} = \frac{100}{0.037} = 2702 \text{ ml}$$

Note that this calculation assumes 1) that the dye is evenly distributed and 2) that all of the dye remains in the circulation. In practice, some dye is lost from the circulation and corrections for the lost dye need to be applied to improve the accuracy of the estimate. Similar limitations apply to estimates of the extracellular fluid using inulin (and other markers) and to estimates of total body water using tritiated water (^3H_2O). After allowing sufficient time for equilibration, the volume of a fluid compartment to a first approximation is given by:

$$\text{volume} = \frac{\text{amount of marker infused} - \text{amount excreted}}{\text{concentration in plasma}}$$

Note that the water in bone and dense connective tissue (tendons and cartilage) equilibrates very slowly with the extracellular fluid and will generally not be included in the above estimates.

have been used. The volume of the intracellular fluid is simply the difference between the total body water and the volume of the extracellular fluid. Thus:

$$\text{total body water} = \text{extracellular fluid} + \text{intracellular fluid}$$

and

$$\text{extracellular fluid} = \text{plasma} + \text{interstitial water}$$

Summary

Water is the chief solvent of the body and accounts for 50–60 per cent of body mass. The solutes and water inside the cells constitute the intracellular fluid, while the solutes and water outside the cells constitute the extracellular fluid.

3.3 The carbohydrates

The carbohydrates (also called saccharides or sugars) are the principal source of energy for cellular reactions. They are composed of carbon, hydrogen, and oxygen and have the general formula $(CH_2O)_n$ (the amino and deoxy sugars are considered separately below). Some examples are shown in Figure 3.3. Sugars containing three carbon atoms ($n = 3$) are known as **trioses**, those with five carbons ($n = 5$) are **pentoses**, and those containing six ($n = 6$) are **hexoses**. Examples are glyceraldehyde (a triose), ribose (a pentose), fructose, and glucose (both hexoses). When two sugar molecules are joined together with the elimination of one molecule of water, they form a glycosidic bond. The resulting molecule is known as a **disaccharide**, as shown in Figure 3.4. Fructose and glucose combine to form sucrose, while glucose and galactose (another hexose) form lactose, the principal sugar of milk. When a number of monosaccharides (up to nine or ten) are chemically linked, the resulting compound is called an **oligosaccharide**, and a chain of more than ten monosaccharides is generally called a **polysaccharide**. Examples of polysaccharides are starch, which is an important constituent of the diet, and **glycogen**, which is the main store of carbohydrate within the muscles and liver. Part of the structure of glycogen is shown in the bottom panel of Figure 3.3. By forming glycogen, cells can store large quantities of glucose without making their interior hypertonic. Glycogen is broken down by **hydrolysis** when glucose is required for energy production (see Figure 3.5).

Many of the carbon atoms in a monosaccharide have four different chemical groups attached. This makes these molecules asymmetric so that two different kinds of each monosaccharide exist, which are mirror images of each other—just as our left and right hands are mirror images. Molecules of this kind are known as **optical isomers**. The two isomers are known as D-isomers and L-isomers (see Chapter 2). All naturally occurring carbohydrates are D-isomers. Thus the glucose we find in nature is more correctly known as D-glucose and, for this reason, it is sometimes called dextrose.

Although sugars are the major source of energy for cells, they are also constituents of a number of molecules that play important parts in other cellular activities. The nucleic acids DNA and RNA contain the pentose sugars 2-deoxyribose and ribose. Ribose is also one of the components of the purine nucleotides that play a central role in cellular metabolism. The structure of the nucleotides is given below in section 3.6. Some hexoses have an amino group in place of one of the hydroxyl groups. These are known as the **amino sugars** or **hexosamines**. The amino sugars are found in the **glycoproteins** (= sugar + protein) and the **glycolipids** (= sugar + lipid). In the glycoproteins, a polysaccharide chain is linked to a protein by a covalent bond. The glycoproteins are important constituents of bone and connective tissue. The glycolipids consist of a polysaccharide chain linked to the glycerol residue of a

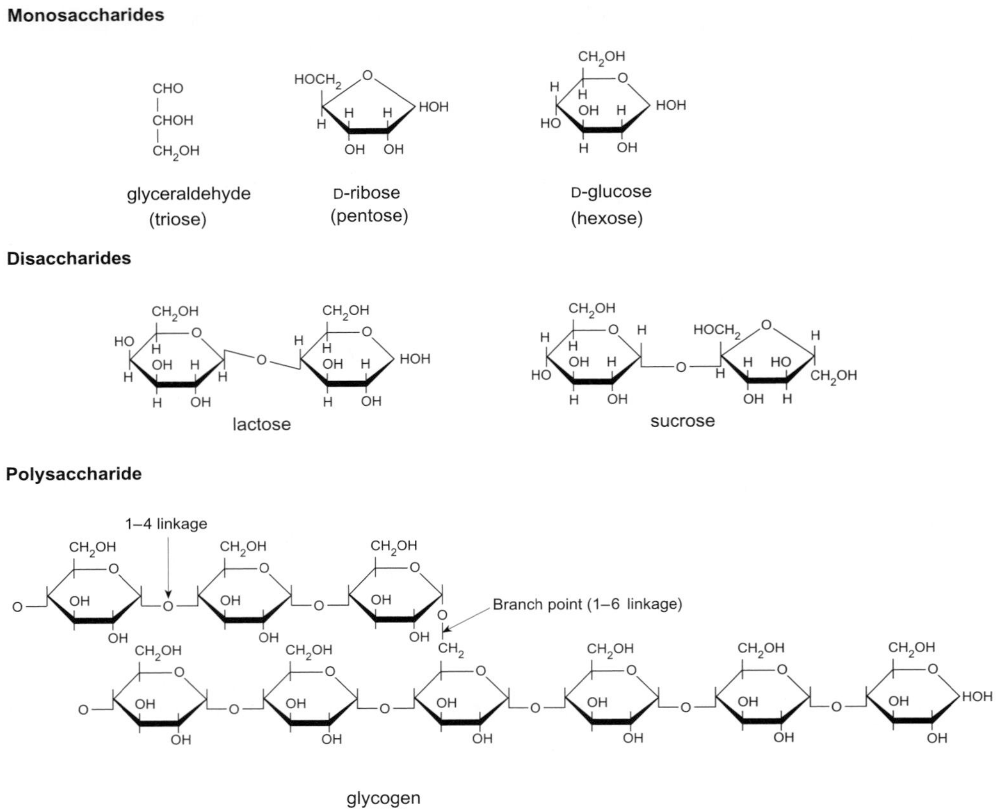

Monosaccharides

glyceraldehyde
(triose)

D-ribose
(pentose)

D-glucose
(hexose)

Disaccharides

lactose

sucrose

Polysaccharide

1–4 linkage

Branch point (1–6 linkage)

glycogen

Figure 3.3 The structures of representative members of the carbohydrates. The polysaccharide glycogen consists of many glucose molecules joined together by 1–4 linkages, known as glycosidic bonds, to form a long chain. A number of glucose chains are joined together by 1–6 linkages to form a single glycogen molecule. Only one such linkage is shown in the figure.

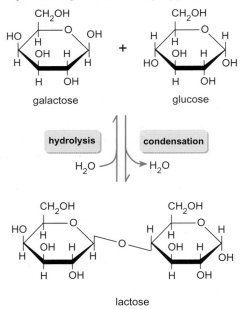

galactose   +   glucose

hydrolysis   condensation

H_2O ← → H_2O

lactose

Figure 3.4 If glucose and galactose undergo a condensation reaction (in which a molecule of water is eliminated) they become linked by a glycosidic bond and form the disaccharide lactose. Similar condensation reactions occur when the polysaccharide glycogen is synthesized from glucose (a process called glycogenesis). By adding a molecule of water, lactose can be broken down to release glucose and galactose in a process called hydrolysis.

sphingosine lipid (see Section 3.4). Glycolipids are found in the cell membranes, particularly those of the white matter of the brain and spinal cord.

Summary

Glucose and other sugars (carbohydrates) are broken down to pro-vide energy for cellular reactions. Sugars are also constituents of many molecules of biological importance (e.g. the purine nucleo-tides and the nucleic acids).

3.4 The lipids

The lipids are a chemically diverse group of substances that share the property of being insoluble in water but soluble in organic sol-vents such as ether and chloroform (see Figure 3.6). They include the fatty acids and the glycerides, the phospholipids, and the ster-oids (e.g. cholesterol). As befits their widely differing structures, the lipids serve a wide variety of functions.

- They are the main structural element of cell membranes (see Chapter 4).

- They are an important reserve of energy.

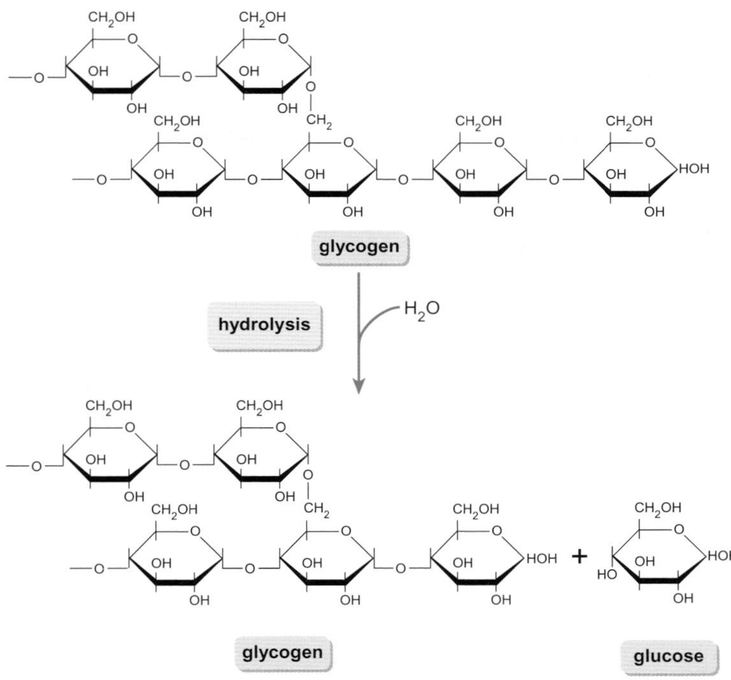

Figure 3.5 Glycogen consists of many glucose residues linked together to form a single large molecule. This is how the liver and muscles store glucose without significantly adding to the osmolality of the cells. When the stored glucose is required for energy production, glycogen is hydrolysed to release individual glucose molecules as shown. This process is called glycogenolysis.

- Some act as chemical signals (e.g. the steroid hormones and prostaglandins).

- Others provide supporting fat pads around many organs and a layer of heat insulation beneath the skin.

- Finally, the lipids of the myelinated nerves provide electrical insulation for the conduction of nerve impulses.

The **fatty acids** have the general formula $CH_3(CH_2)_nCOOH$. Typical fatty acids are acetic acid (with 2 carbon atoms, so in this case n = 0), butyric acid (with 4 carbon atoms, n = 2), palmitic acid (with 16 carbon atoms, n = 14), and stearic acid (with 18 carbon atoms, n = 16). **Triglycerides** or **triacylglycerols** consist of three fatty acids joined by ester bonds to glycerol, as shown in Figures 3.6 and 3.7. **Diglycerides** have two fatty acids linked to glycerol, while **monoglycerides** have only one. In the digestive system, the triglycerides of the diet are first hydrolysed to diglycerides, which have two fatty acids linked to glycerol and then to monoglycerides which have only one, as shown in Figure 3.7. A similar sequence occurs when the body utilizes its reserves of fats for energy production, a process called **lipolysis**.

The triglycerides are the body's main store of energy and can be laid down in adipose tissue in virtually unlimited amounts. They generally contain fatty acids with many carbon atoms, e.g. palmitic and stearic acids, and the middle fatty acid chain is frequently unsaturated. **Oleic acid** (18 carbons with a single double bond), **linoleic acid** (18 carbon atoms with two double bonds) and **arachidonic acid** (20 carbon atoms with four double bonds) commonly occur in triglycerides. Arachidonic acid is a precursor for an important group of lipids known as the prostaglandins, discussed later in this section. Although they play an important role in cellular metabolism, mammals, including man, are unable to synthesize either linolenic acid or linoleic acid. They must be provided by the diet and are therefore called **essential fatty acids**.

The **structural lipids** are the main component of the cell membranes. They fall into three main groups: **phospholipids, glycolipids**, and **cholesterol**. The basic chemical structures of these key constituents can be seen in Figure 3.8. The phospholipids fall into two groups: those based on glycerol and those based on sphingosine. The glycero-phospholipids are the most abundant in mammalian plasma membranes and are classified on the basis of the type of polar group attached to the phosphate. Phosphatidylcholine, phosphatidylserine, phosphatidylethanolamine, and phosphatidylinositol are all examples of glycero-phospholipids. The glycerophosphate head groups are linked to long chain fatty acid residues via ester linkages. However, there is another class of phospholipid, the plasmalogens, in which one hydrocarbon chain is linked to the glycerol of the head group via an ether linkage.

The **glycolipids** are based on sphingosine, which is linked to a fatty acid to form ceramide. There are two classes of glycolipid: the cerebrosides, in which the ceramide is linked to a monosaccharide such as glucose (see Figure 3.8), and the gangliosides, in which it is linked to an oligosaccharide containing amino sugar residues such as N-acetyl galactosamine.

The **steroids** are lipids with a structure based on four carbon rings known as the steroid nucleus (see Figure 3.6). The most abundant steroid is **cholesterol**, which is a major constituent of cell membranes and which acts as the precursor for the synthesis of many steroid hormones such as cortisol, progesterone, and

(a) Fatty acids

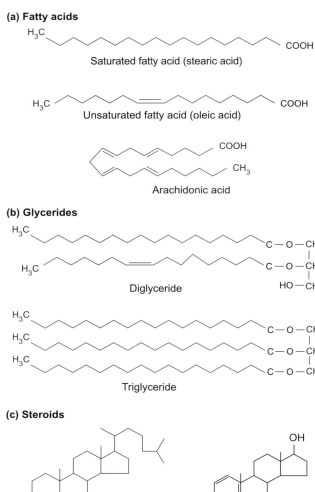

Saturated fatty acid (stearic acid)

Unsaturated fatty acid (oleic acid)

Arachidonic acid

(b) Glycerides

Diglyceride

Triglyceride

(c) Steroids

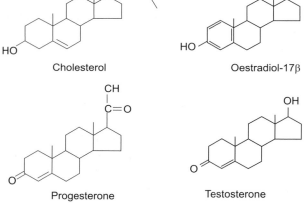

Cholesterol

Oestradiol-17β

Progesterone

Testosterone

Figure 3.6 The chemical structures of some characteristic lipids. Note that the carbon chain of long chain fatty acids (shown in panels (a) and (b)) is represented by a series of lines thus: ⋁⋀⋁⋀. Each angle represents a —CH$_2$— group. Such formulae are known as skeletal structures. Note that individual fatty acids have carbon chains of different lengths. Stearic acid has no double bonds in its long chain and is a saturated fatty acid. Oleic acid and other fatty acids that have double bonds between adjacent carbon atoms are unsaturated fatty acids. (b) shows the structure of glycerides (compounds of fatty acids and glycerol), while the structures of four steroids are shown in (c).

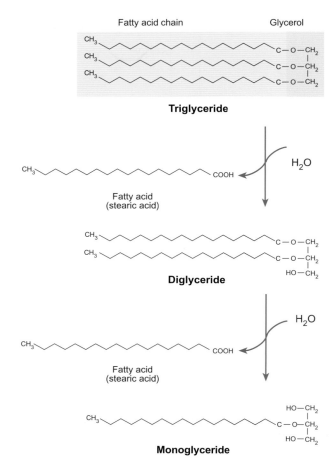

Triglyceride

Fatty acid
(stearic acid)

Diglyceride

Fatty acid
(stearic acid)

Monoglyceride

Figure 3.7 The structure of the mono-, di-, and triglycerides and their interconversion by hydrolysis. Triglycerides can be converted to diglycerides by the addition of a water molecule with the release of a fatty acid (stearic acid in this case). A similar process converts a diglyceride to a monoglyceride.

testosterone. The **prostaglandins** and **leukotrienes** are lipids that are derived from the unsaturated fatty acid arachidonic acid and play an important role in cell signalling. Their biosynthesis and physiological roles are discussed in Chapter 6.

The long chain fatty acids and steroids are insoluble in water but they naturally form **micelles** (aggregates of large insoluble molecules) in which the polar head groups face outwards towards the water (the aqueous phase) and the long hydrophobic chains associate together in the centre. The fatty acids are transported in the blood and body fluids in association with proteins as **lipoprotein** particles. Each particle consists of a lipid micelle protected by a coat of protein. The proteins forming the coat are known as **apoproteins** or **apolipoproteins**.

In cell membranes, the lipids form bilayers, which are arranged so that their polar headgroups are orientated towards the aqueous phase while the hydrophobic fatty acid chains face inwards to form a central hydrophobic region. The outer membrane of the cells (the

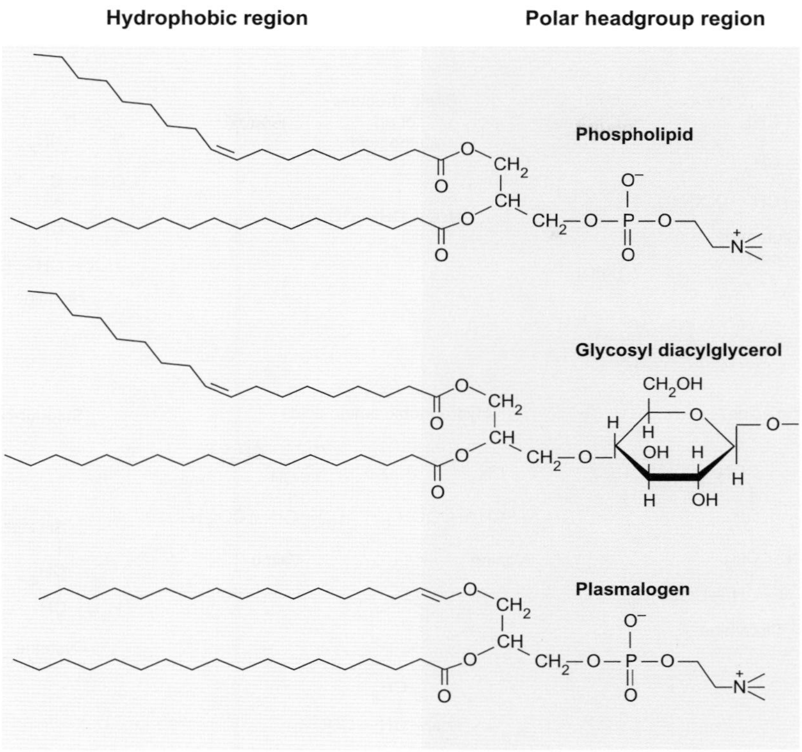

Figure 3.8 The structure of some of the structural lipids (lipids that form the cell membranes). Note that each type has a polar head group region (highlighted in blue) and a long hydrophobic tail (highlighted in light yellow). In reality the hydrophobic region is more extensive than the polar region.

plasma membrane) provides a barrier to the diffusion of polar molecules (e.g. glucose) and ions but not to small non-polar molecules such as carbon dioxide and oxygen. The internal membranes divide the cell into discrete compartments that provide the means of storage of various materials and permit the segregation of different metabolic processes. This compartmentalization of cells by lipid membranes is discussed in greater detail in Chapter 4.

Summary

Lipids are a chemically diverse group of substances that are insoluble in water but soluble in certain organic solvents. The phospholipids form the main structural element of cell membranes, triglycerides are an important reserve of energy, while steroids and prostaglandins act as chemical signals.

3.5 The amino acids and proteins

Proteins serve an extraordinarily wide variety of functions in the body.

- They form the enzymes that catalyse the chemical reactions of living things.

- They are involved in the transport of molecules and ions around the body.

- They bind ions and small molecules for storage inside cells.

- They are responsible for the transport of molecules and ions across cell membranes.

- Proteins such as tubulin form the cytoskeleton that provides the structural strength of cells.

- They form the motile components of muscle and of **cilia**.

- They form the connective tissues that bind cells together and transmit the force of muscle contraction to the skeleton.

- Proteins known as immunoglobulins play an important part in the body's defence against infection.

- As if all this were not enough, some proteins act as signalling molecules—the hormone insulin is one example of this type of protein.

Proteins are assembled from a set of twenty α-amino acids

The basic structural units of proteins are the α-amino acids. An α-amino acid is a carboxylic acid that has an amine group and a side chain attached to the carbon atom next to the carboxyl group (the α carbon atom), as shown in Figure 3.9. With the exception of the

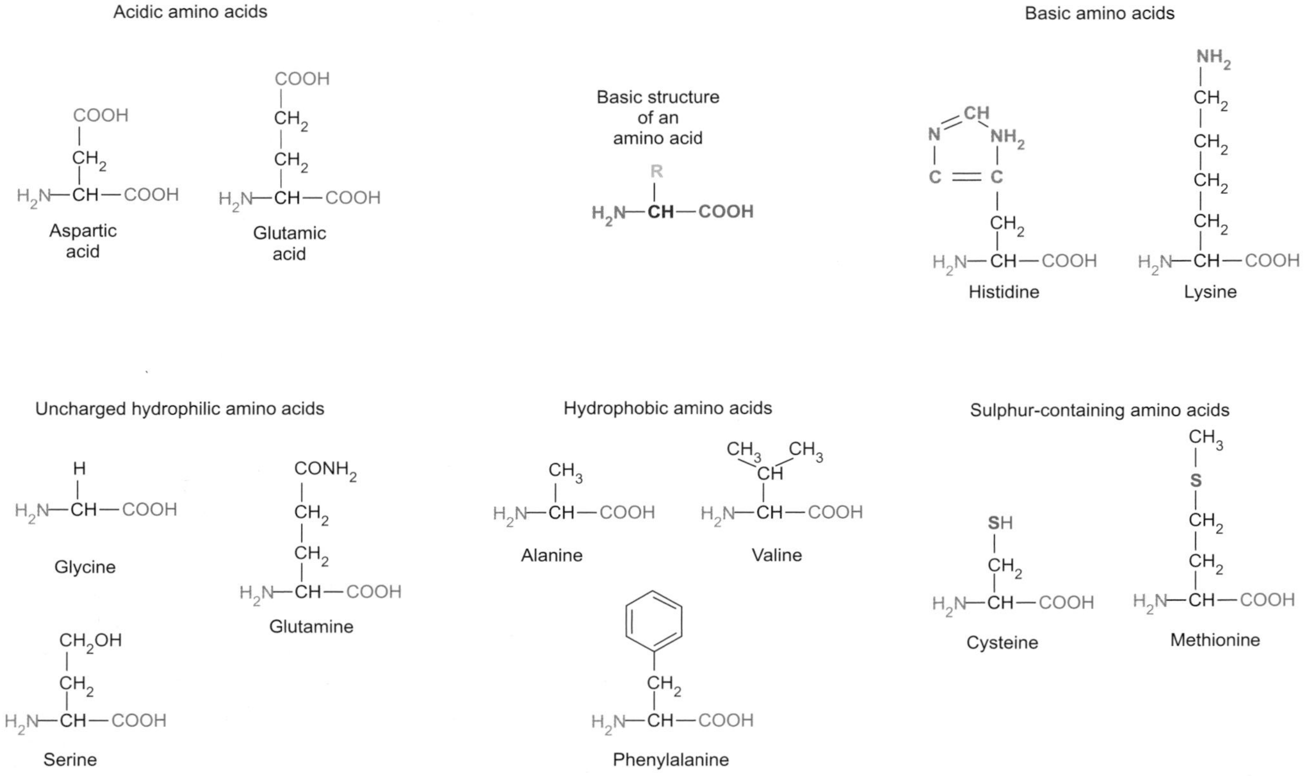

Figure 3.9 The chemical structures of representative α-amino acids. The general structure of the α-amino acids is shown at the top centre of the figure. R represents the side chains of the different amino acids. The α-amino groups are shown in blue and the carboxyl groups in red. The acid amino acids are shown on the top left of the figure with the excess carboxyl group (acid) highlighted in red. Basic amino acids are shown top right with the basic groups highlighted in blue. Bottom right shows the structure of the sulphur-containing amino acids. Bottom centre shows examples of the hydrophobic amino acids—note that the side chains have no oxygen, nitrogen, or sulphur atoms. Bottom left shows uncharged amino acids, two of which have polar groups on their side chains (glutamine and serine).

smallest amino acid, glycine, the α-carbon atom of the amino acids is attached to four different groups. As for the carbohydrates, this makes them asymmetric with L- and D-isomers which are mirror images of each other (i.e. they are optical isomers). The amino acids that occur naturally in the proteins of the body belong to the L-series.

Proteins are built from 20 different L-α-amino acids, which may be grouped into five different classes.

1. Acidic amino acids (aspartic acid and glutamic acid)

2. Basic amino acids (arginine, histidine, and lysine)

3. Uncharged hydrophilic amino acids (asparagine, glycine, glutamine, serine, and threonine)

4. Hydrophobic amino acids (alanine, leucine, isoleucine, phenylalanine, proline, tyrosine, tryptophan, and valine)

5. Sulphur-containing amino acids (cysteine and methionine)

Amino acids can be combined together by linking the amine group of one with the carboxyl group of another and eliminating water to form a **dipeptide**, as shown in Figure 3.10. The linkage between two amino acids joined in this way is known as a **peptide bond**. The addition of a third amino acid would give a tripeptide, a fourth a tetrapeptide, and so on. Peptides with large numbers of amino

acids linked together are known as **polypeptides**. Proteins are large polypeptides. By convention, the naming of a peptide begins at the end with the free amine group (the amino terminus) on the left and ends with the free carboxyl group on the right, and the order in which the amino acids are arranged is known as the **peptide sequence**. Since proteins and most peptides are large structures, the sequence of amino acids would be tedious to write out in full so a single letter or three-letter code is used, as shown in Table 3.2.

Since proteins are made from 20 L-amino acids and there is no specific limit to the number of amino acids that can be linked together, the number of possible protein structures is essentially infinite. Different proteins have different shapes and different physical and biological properties. It is this that makes them so versatile. The fact that some amino acid side chains are hydrophilic while others are hydrophobic results in different proteins having differing degrees of hydrophobicity. As a result, some are soluble in water while others are not.

Although proteins are formed from a continuous sequence of amino acids, they are typically made up from a number of semi-autonomous regions, termed **protein domains**. Protein domains vary in length (i.e. in the number of amino acids in their sequence), and the same sequence may appear in a variety of different proteins. This 'mix and match' aspect of protein structure is derived

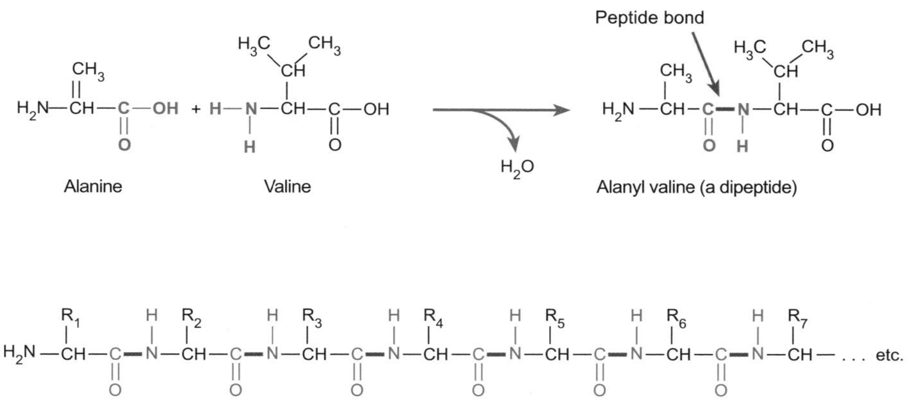

The amino-terminus region of a peptide

Figure 3.10 The formation of a peptide bond and the structure of the amino terminus region of a polypeptide showing the peptide bonds. R_1, R_2, etc. represent different amino acid side chains. The peptide bonds are shown in magenta.

Table 3.2 The α-amino acids of proteins and their customary abbreviations

Name	Three-letter code	Single-letter code
Alanine	Ala	A
Cysteine	Cys	C
Aspartic acid	Asp	D
Glutamic acid	Glu	E
Phenylalanine	Phe	F
Glycine	Gly	G
Histidine	His	H
Isoleucine	Ile	I
Lysine	Lys	K
Leucine	Leu	L
Methionine	Met	M
Asparagine	Asn	N
Proline	Pro	P
Glutamine	Gln	Q
Arginine	Arg	R
Serine	Ser	S
Threonine	Thr	T
Valine	Val	V
Tryptophan	Trp	W
Tyrosine	Tyr	Y

The amino acids are arranged in alphabetical order of their single-letter codes. Asparagine and glutamine are amides of aspartic and glutamic acids.

from the way in which proteins have evolved to carry out tasks of greater and greater complexity. Different protein domains have distinctive properties that determine their biochemical properties. For example, membrane proteins have extensive regions that lack polar amino acids so that they have hydrophobic domains that are associated with the lipid region of cell membranes. Commonly, proteins have several different domains that serve particular functions: one domain of a protein might bind an ion (such as Ca^{2+}) and a separate hydrophobic domain might anchor the protein in a cellular membrane, while a third might be involved in signalling to other proteins the fact that it has bound a calcium ion. This last process is called **signal transduction** and is discussed in more detail in Chapter 6.

Many cellular structures consist of protein assemblies, i.e. units made up of several different kinds of protein. Examples are the myofilaments of the skeletal muscle fibres, which contain the proteins actin, myosin, troponin, and tropomyosin. Actin molecules also assemble together to form microfilaments in other cells. Enzymes are frequently arranged so that the product of one enzyme can be passed directly to another and so on. These multi-enzyme assemblies increase the efficiency of cell metabolism.

Some important amino acids are not found in proteins

Some amino acids of physiological importance are not found in proteins but have other important functions. **Coenzyme A** contains a structural isomer of alanine called β-alanine. The amino acid γ-aminobutyric acid (GABA) plays a major role as a neurotransmitter in the brain and spinal cord. Creatine is phosphorylated in muscle to form **creatine phosphate**, which is an important source of energy in muscle contraction. Ornithine is an intermediate in the urea cycle. Their structures are shown in Figure 3.11.

Summary

Proteins are assembled from a set of 20 α-amino acids, which are linked together by peptide bonds. They serve a wide variety of functions in the body both as structural elements and as biological signals.

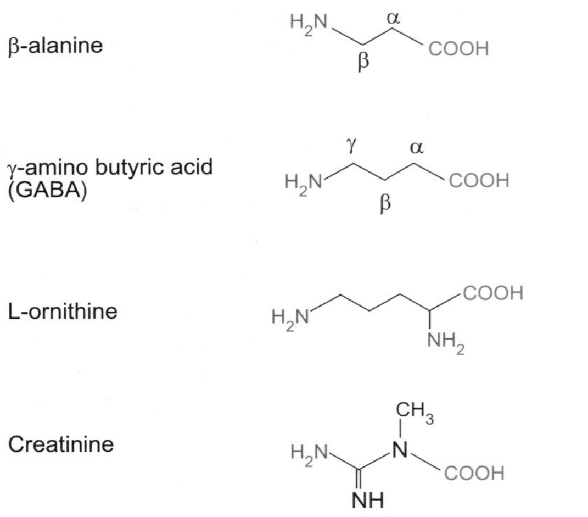

β-alanine

γ-amino butyric acid
(GABA)

L-ornithine

Creatinine

Figure 3.11 The structures of some non-protein amino acids of physiological importance.

3.6 The nucleosides, nucleotides, and nucleic acids

The genetic information of the body resides in its DNA (deoxyribonucleic acid), which is stored in the chromosomes of the nucleus (see Chapter 4). DNA is made by assembling smaller components known as **nucleotides** into a long chain. Ribonucleic acid (RNA) has a similar primary structure. Each nucleotide consists of a base linked to a pentose sugar, which is in turn linked to a phosphate group, as shown in Figure 3.12. The nucleic acids bases are either **pyrimidine bases** (cytosine, thymine, and uracil) or **purine bases** (adenine and guanine).

Nucleosides and nucleotides

When a base combines with a pentose sugar it forms a **nucleoside**. Thus, the combination of adenine and ribose forms adenosine; the combination of thymine with ribose forms thymidine; and so on. When a nucleoside becomes linked to one or more phosphate

Thymine

Cytosine

Uracil

Pyrimidine

Adenine

Guanine

Purine

β-D-Ribose

β-D-2-Deoxyribose

Pentose

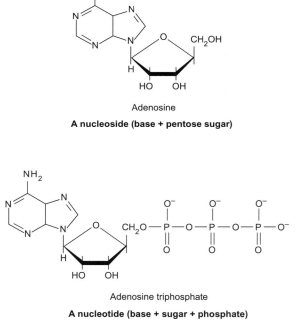

Adenosine

A nucleoside (base + pentose sugar)

Adenosine triphosphate

A nucleotide (base + sugar + phosphate)

Figure 3.12 The structural components of the nucleotides and nucleic acids. The left side of the figure shows the structures of the pyrimidine and purine bases, and the pentose sugars ribose and deoxyribose. The right side shows typical examples of a nucleoside (adenosine) and a nucleotide (adenosine triphosphate).

groups it forms a **nucleotide**. Thus adenosine may become linked to one phosphate to form adenosine monophosphate, uridine will form uridine monophosphate, and so on. The nucleotides are thus the building blocks of the nucleic acids.

The nucleotide coenzymes

Nucleotides can be combined together or with other molecules to form coenzymes. Adenosine monophosphate (AMP) may become linked to a further phosphate group to form adenosine diphosphate (**ADP**) or to two further phosphate groups to form adenosine triphosphate (**ATP**) (see Figure 3.12). Similarly guanosine may form guanosine mono-, di-, and triphosphate, and uridine can form uridine mono-, di-, and triphosphate. The higher phosphates of the nucleotides play a vital role in cellular **energy metabolism** and are important carriers of chemical energy in cells. Indeed, the metabolic breakdown of glucose and fatty acids is directed to the formation of ATP which is used as a source of energy for a host of important cellular processes.

The phosphate group of nucleotides is attached to the 5′ position of the ribose residue and has two negative charges. It can link with the hydroxyl of the 3′ position to form 3′, 5′ cyclic adenosine monophosphate or **cyclic AMP**, which plays an important role as an intracellular messenger. Similarly, guanosine can form 3′, 5′ cyclic guanosine monophosphate or **cyclic GMP**. The structures of cyclic AMP (cAMP) and cyclic GMP (cGMP) are shown in Figure 3.13. Both play important roles as intracellular chemical signals known as second messengers.

The nicotinamide nucleotides are dinucleotides in which adenosine becomes linked to nucleotides based on nicotinamide to form the nicotinamide nucleotide coenzymes (abbreviated as NAD and NADP). These coenzymes are important electron carriers in the oxidation of fuels for cellular energy production (see Chapter 4). Other important nucleotide-based coenzymes are flavine mononucleotide (FMN), flavine adenine dinucleotide (FAD), and coenzyme A. The chemical structures of NAD and FAD are shown in Figure 3.13.

Summary

Nucleotides consist of a base, a pentose sugar, and a phosphate residue. They can be combined with other molecules to form coenzymes (e.g. NAD) and nucleic acids. ATP is the most important carrier of chemical energy in cells. DNA and RNA play a crucial role in protein synthesis.

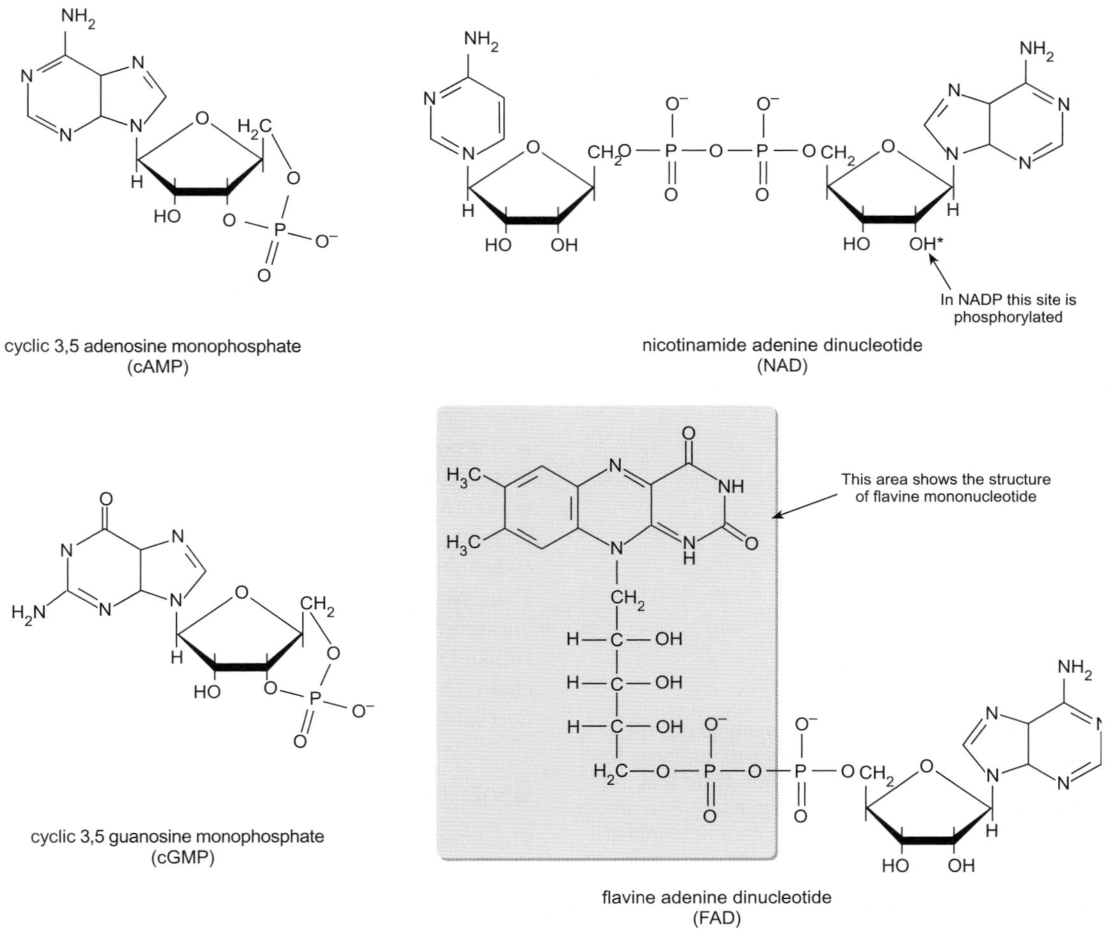

cyclic 3,5 adenosine monophosphate
(cAMP)

nicotinamide adenine dinucleotide
(NAD)

In NADP this site is phosphorylated

This area shows the structure of flavine mononucleotide

cyclic 3,5 guanosine monophosphate
(cGMP)

flavine adenine dinucleotide
(FAD)

Figure 3.13 The molecular structures of the cyclic nucleotides cAMP and cGMP, and the nicotinamide and flavine coenzymes (NAD and FAD).

The nucleic acids

In nature there are two main types of nucleic acid: DNA and RNA. In DNA the sugar of the nucleotides is deoxyribose and the bases are adenine, guanine, cytosine, and thymine (abbreviated A, G, C, and T). In RNA the sugar is ribose and the bases are adenine, guanine, cytosine, and uracil (A, G, C, and U). In both DNA and RNA the nucleotides are joined by phosphate linkages between the 5′ position of one nucleotide and the 3′ position of the next pentose ring, as shown in Figure 3.14(a).

A molecule of DNA consists of a pair of nucleotide chains linked together by hydrogen bonds in such a way that adenine links with thymine and guanine links with cytosine (see Figure 3.15). This is

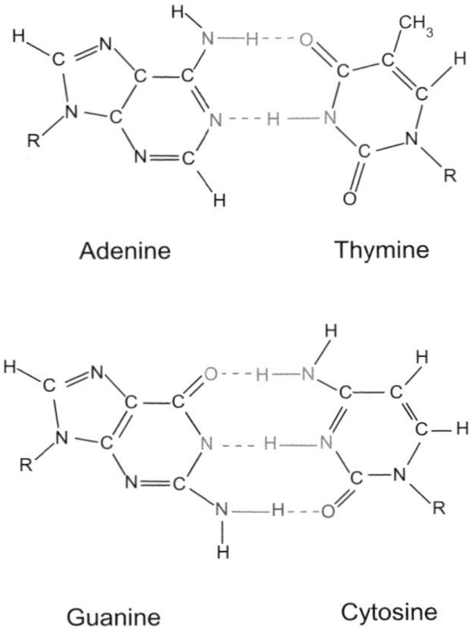

Figure 3.15 The hydrogen bonding between adenosine and thymine, and between guanine and cytosine, that is the basis of base pairing in the complementary strands of DNA. Note that there are three such bonds between guanine and cytosine but only two between adenosine and thymine.

known as base pairing. The hydrogen bonding between the two chains is so precise that the sequence of bases on one chain automatically determines that of the second. The pair of chains is twisted to form a double helix in which the complementary strands run in opposite directions (Figure 3.14(b)). The discovery of this base pairing was crucial to the understanding of the three-dimensional structure of DNA and to the subsequent unravelling of the genetic code. For their work in this area J.D. Watson, F. Crick, and M. Wilkins were awarded the Nobel Prize in 1962.

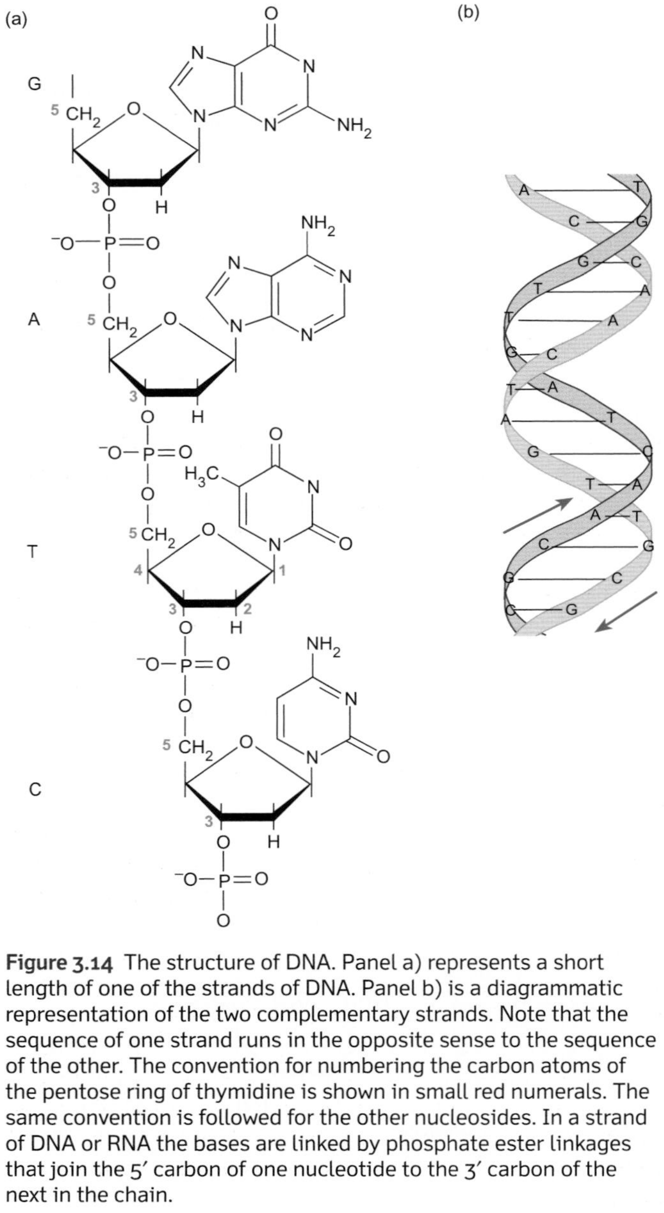

Watson, Crick, and their colleagues proposed that the sequence of bases in a length of DNA or RNA codes for the sequence of amino acids in a specific protein. Subsequent work proved this conjecture to be correct and it is now known that the position of each amino acid is coded by a sequence of three bases called a **codon**. Since there are four different bases in DNA, there are 64 possible codons available ($4 \times 4 \times 4 = 64$) to code for the 20 amino acids found in proteins. This means that several different triplet sequences could code for the same amino acid. This is known as redundancy. In fact a number of amino acids have multiple codons: for example there are six different codons for the amino acid leucine—see Table 3.3. Unlike DNA, each RNA molecule has only one polynucleotide chain. This property is exploited during protein synthesis (see the next subsection).

Figure 3.14 The structure of DNA. Panel a) represents a short length of one of the strands of DNA. Panel b) is a diagrammatic representation of the two complementary strands. Note that the sequence of one strand runs in the opposite sense to the sequence of the other. The convention for numbering the carbon atoms of the pentose ring of thymidine is shown in small red numerals. The same convention is followed for the other nucleosides. In a strand of DNA or RNA the bases are linked by phosphate ester linkages that join the 5′ carbon of one nucleotide to the 3′ carbon of the next in the chain.

Gene transcription and translation

The DNA of an organism contains the sequence information required to synthesize all of the proteins it requires. The DNA sequence is arranged into small regions of DNA, termed **genes**, each of which codes for a particular protein. The totality of all the genes present in an animal is called its **genome**. The genome contains all

Table 3.3 The genetic code of mRNA

First position ↓	← Second position →				Third position ↓
	U	C	A	G	
U	Phenylalanine	Serine	Tyrosine	Cysteine	U
	Phenylalanine	Serine	Tyrosine	Cysteine	C
	Leucine	Serine	**Stop**	**Stop**	A
	Leucine	Serine	**Stop**	Tryptophan	G
C	Leucine	Proline	Histamine	Arginine	U
	Leucine	Proline	Histamine	Arginine	C
	Leucine	Proline	Glutamine	Arginine	A
	Leucine	Proline	Glutamine	Arginine	G
A	Isoleucine	Threonine	Asparagine	Serine	U
	Isoleucine	Threonine	Asparagine	Serine	C
	Isoleucine	Threonine	Lysine	Arginine	A
	Methionine§	Threonine	Lysine	Arginine	G
G	Valine	Alanine	Aspartate	Glycine	U
	Valine	Alanine	Aspartate	Glycine	C
	Valine	Alanine	Glutamate	Glycine	A
	Valine	Alanine	Glutamate	Glycine	G

This table summarizes the genetic code for mRNA. The first base of the triplet that codes for an amino acid (a codon) is given by the column on the left, the second is given by the row across the top of the table and the third is given by the column on the right. To take methionine as an example, the first base is A, the second is U and the third is G, so the mRNA code for methionine is AUG. (§ This codon is also the start signal for synthesis of a peptide chain.) Note that most amino acids are coded by more than one triplet. For example, lysine is coded by AAA and by AAG. Stop codons tell the transcription process when a peptide chain is completed.

the information required for a fertilized egg to make another individual of the same species.

Genes make up a very small portion of all of the DNA of an animal (for example, only about 2 per cent of the human genome actually codes for proteins). The rest of the genetic material is a mixture of regulatory elements, that tell the cell when to turn on and off particular genes in response to complex sets of intracellular messages, and genetic material that has been duplicated within the genome, and which is sometimes called 'junk DNA'. As will be discussed further in the context of antibody production (Chapter 26), even a gene itself does not consist solely of a coding sequence. There is a region that tells the protein synthesis machinery where an amino acid sequence begins. Prior to this 'start' signal are various regulatory sequences, termed promoter elements, and enhancer elements which bind special signalling proteins (accessory proteins). The control exerted by the promoter and enhancer regions provides precise control of when a gene is active or inactive ('on' or 'off'). The coding section itself is generally fragmented into a number of regions, termed **introns** and **exons**. Introns are non-coding, but allow genes to create different proteins as **splice variants**—where particular coding portions of the gene (the exons) may be skipped under certain conditions, in order to synthesize proteins with different properties.

The term **genotype** is also used to describe the genetic makeup of an individual. It refers to the presence of specific forms of individual genes (known as **alleles**). The **phenotype** is the physical expression of the genes in an organism's body, including specific behaviour patterns. The distinction between genotype and phenotype is important, as not all the genes present will be expressed (i.e. active). During the closing years of the last century, a huge international effort was put into determining the sequence of the entire DNA in human cells. This project was known as the **human genome project**, which discovered that there are about 19,000 genes in the human genome (see Box 3.2). The **polymerase chain reaction** (PCR) has allowed researchers and clinicians to use the sequence data obtained from the human genome project to track the association between variations in gene sequence and human disease (see Box 3.3)

For a gene to perform its task, it must instruct a cell to make a specific protein. It does this by generating a copy of the genetic information encoded by the gene in a form that is able to leave the nucleus. The genetic information is copied through a process termed **transcription**, which takes advantage of the single-stranded nature of RNA to make a template from a strand of DNA. This template RNA is termed messenger RNA (**mRNA**). As it contains only the coding portions of the gene and not the regulatory elements or the introns, mRNA is small enough to leave the nucleus. Once in the cytoplasm, it associates with small subcellular particles called **ribosomes** (see Chapter 4), which are the site of protein synthesis. This step is necessary for protein synthesis to take place.

Box 3.2 The Human Genome Project

The **Human Genome Project** began in 1989 and was completed in 2000. It succeeded in its aim of determining the full sequence of human DNA. In the course of this work it was discovered that there are some 3.2 billion (3,200,000,000) base pairs in human DNA and that the DNA sequence of different individuals differs by only 1 base in every 1000 or so. From this we can say that 99.9 per cent of the human DNA sequence is shared between all members of the human population. The 0.1 per cent difference between individuals corresponds to about 3 million variations in sequence. As each of these variations is independent and arises from random changes, each person is genetically unique (except monozygotic (identical) twins who share exactly the same DNA sequence). Recent estimates suggest that there are around 19,000 coding genes in the human genome, far fewer than originally thought.

The regions of DNA involved in coding for proteins (exons) account for only 1.5 to 2 per cent of the total. The remainder includes the regulatory elements and introns as well as large stretches that have no known coding function (accounting for perhaps 97 per cent of the total). As explained in the main text, each amino acid is coded by a sequence of three bases (a sequence triplet or **codon**) and errors occur during the DNA replication that accompanies cell division. The substitution of one base for another in a sequence triplet will result in the substitution of one amino acid for another in the amino acid sequence of the protein encoded by the gene. Depending on the position of the base substitution, the effect of the mutation on the function of the protein may either be trivial or severe. Sickle cell disease and haemophilia B are examples of genetic disorders that arise from the substitution of one base for another (a **single nucleotide polymorphism**—a DNA base is equivalent to nucleotide).

Knowledge of the human genome and its variation between different individuals will, it is hoped, help in the diagnosis and treatment of a range of genetic disorders. For diseases such as sickle cell disease and cystic fibrosis where a single gene is at fault, there is the distant possibility of gene therapy to provide a cure. For many other diseases, however, it is the balance of activity between different proteins that matters. A good example is the regulation of lipid transport by the blood. A protein known as the low-density lipoprotein (LDL) receptor is responsible for the removal of some lipoprotein complexes (including those containing cholesterol) from the blood. A single error in the sequence causes the substitution of the amino acid serine in place of asparagine. This mutation is associated with increased cholesterol levels in the blood and a greater risk of coronary heart disease. In this case, genetic profiling of an individual can be of help in deciding whether a modification to the diet will be beneficial in reducing a major risk factor.

There is also the prospect that knowledge of the genetic profile of an individual will be of benefit in the selection of drugs for the treatment of particular non-genetic disorders and the avoidance of undesirable side effects. For example, certain liver enzymes are responsible for breaking down particular drugs before they are excreted. One of these enzymes (a cytochrome P450 enzyme known as P450dbl) plays an important role in the breakdown of the drug debrisoquine, which is used in the treatment of high blood pressure. A single mutation can result in the production of a defective version of the enzyme. Individuals who have a defective form of the enzyme are unable to break down debrisoquine in the normal way (as well as other drugs that are broken down by the same enzyme). If an individual with the defective form of the enzyme is given debrisoquine for their high blood pressure, the drug will accumulate in their body and cause the same symptoms as a debrisoquine overdose. Prior knowledge of the mutation would allow a better choice of drug for the affected individuals.

Box 3.3 The polymerase chain reaction

The polymerase chain reaction (PCR), developed by Kary Mullis in the 1980s, has become a fundamental technique of modern molecular biology. It provides a simple method for cloning genes and identifying genetic markers of disease. It is even used for personal identification in forensic cases. At its heart, PCR is based upon two facts.

- The two complementary strands of DNA can be separated by warming (known as **denaturing**).

- A heat-resistant DNA polymerase (the enzyme that synthesizes DNA from its component bases to match a complementary strand) can survive this warming to work again and again as DNA strands are synthesized and denatured.

In practice, this means that increasing the amount of DNA with a specific sequence (amplification) is as simple as creating a mixture in a small tube, and then placing it into a PCR machine (also known as a thermal cycler) which will automate the heating and cooling steps.

The first step in setting up a PCR reaction is the choice of short sequences of DNA known as oligonucleotide primers, which are complementary to a sequence at the 5′ (start) and 3′ ends of the DNA sequence that is to be amplified. These are added to the reaction vessel, together with large numbers of individual DNA nucleotides. The mixture is then heated to separate the strands of the initial DNA template. Next, the reaction is cooled, to allow hydrogen bonds to re-form between the DNA bases, allowing complementary DNA strands to form double-stranded DNA. However, as there are far more primer DNA strands than template strands, each half of the original template will anneal to a primer. The DNA polymerase then uses the long template strand as a template for the synthesis of a complete

Box 3.3 (Continued)

double-stranded length of DNA. (A polymerase abbreviated as TAQ is probably the best known of these enzymes. It comes from a bacterium called *Thermus aquaticus* that lives in hot springs.) When this is completed, a single double-stranded section of DNA has generated two double-stranded sections of DNA, where one strand of each comes from the original template. At this point, the cycle is repeated; denaturation separates the double-stranded DNA into single strands, cooling then allows primers to bind (a process called annealing), and the DNA polymerase continues its work. Repeated cycles of heating (denaturation) and cooling (annealing) combined with the large excess of primer DNA permits a large number of copies of the target DNA sequence to be synthesized (amplification).

In DNA fingerprinting, short regions of DNA that vary greatly between individuals are amplified. While two individuals may match about 5 to 20 per cent of the time for a single one of these regions, when large numbers of such regions are analysed simultaneously, the match becomes far more precise. Genetic analysis in disease is similar, but in this case attention is focused on the specific forms of those genes that have been statistically linked to the occurrence of the particular disorder. Genes that exist in more than one form are called alleles (see main text). To identify which of a pair of alleles (specific forms of individual genes) is present in a particular person, the PCR uses primers that contain the region where the sequence of two alleles differ.

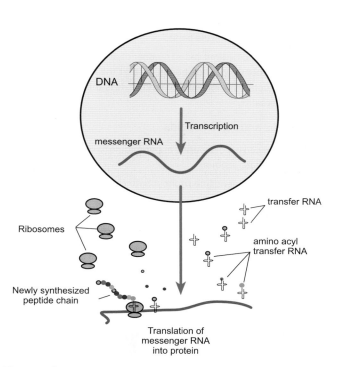

Figure 3.16 The principal steps in the conversion of the genetic information encoded by DNA to the synthesis of specific proteins.

The ribosomes themselves are made of another form of RNA, ribosomal RNA, and certain proteins. Once an mRNA strand has become attached to a ribosome, a new protein is synthesized by progressive elongation of a peptide chain. For this to occur, each amino acid must be arranged in the correct order. The position of each amino acid is coded by an mRNA **codon** that is complementary to that of the original DNA sequence. As in the case of DNA, with four different bases, and a triplet structure, the number of

possible codons is 64. The synthesis of protein from mRNA is called **translation**.

To form a polypeptide chain, the amino acids are assembled in the correct order by a ribosome–mRNA complex as follows: each amino acid must first be bound to a specific kind of RNA molecule known as transfer RNA (**tRNA**). Transfer RNA molecules exist in many forms, but a particular amino acid will bind to only one specific form of tRNA: so, for example, the transfer RNA for alanine will bind only alanine, that for glutamate will bind only glutamate, and so on. Each tRNA–amino acid complex (amino acyl tRNA) has a specific coding triplet of bases (an **anticodon**) which matches the complementary sequence (codon) on the mRNA strand. The ribosome assembles the peptide by first allowing one amino acyl tRNA to bond with the mRNA strand. Then the next amino acyl tRNA is allowed to bind to the mRNA. The ribosome then catalyses the formation of a peptide bond between the two amino acids, moves one codon along the mRNA strand, and binds the next amino acyl tRNA-molecule to extend the peptide chain. This process continues until the ribosome reaches a signal telling it to stop adding amino acids (a stop codon). At this point the protein is complete and is released into the cell. The ribosome is then able to catalyse the synthesis of another peptide chain. The main stages of protein synthesis are summarized in Figure 3.16.

Summary

The DNA of a cell contains the genetic information for making proteins. DNA is made by assembling nucleotides into a long chain that has a specific sequence. Each DNA molecule consists of two complementary helical strands linked together by hydrogen bonds. RNA has a similar primary structure to a single DNA strand and exists in three different forms known as messenger RNA, transfer RNA, and ribosomal RNA. The various forms of RNA play a central role in the synthesis of proteins. In short, DNA makes RNA makes proteins.

✳ Checklist of key terms and concepts

Body water

- Water accounts for 50–60 per cent of body weight.

- As body fat contains little water, the proportion of body weight contributed by water varies both between the sexes and with age.

- Body water is divided into that within the cells, the **intracellular water**, and that which lies outside the cells, the **extracellular water**.

- In the cells and tissues, the water contains dissolved materials and the resulting mixtures are known as fluids. The fluid in the space inside the cells is the **intracellular fluid**, while that outside the cells is called the **extracellular fluid**.

- The extracellular fluid is made up of the **plasma** (the liquid fraction of the blood) and the **interstitial fluid**, which bathes the tissues.

- The volume of the main fluid compartments can be measured by the dilution of specific markers.

Carbohydrates

- The carbohydrates (also called sugars or saccharides) are the principal source of energy for cellular reactions.

- They are classified according to the number of carbon atoms they contain, into trioses, pentoses, and hexoses. These are all monosaccharides.

- When two sugar molecules are joined together with the elimination of one molecule of water, they form a disaccharide.

- If monosaccharides are joined together in chains they form oligosaccharides and polysaccharides.

- Glycogen is the main store of carbohydrate within the muscles and liver. It is formed from chains of glucose molecules and is a large polysaccharide.

- Two pentose sugars (ribose and deoxyribose) are constituents of the nucleic acids DNA and RNA.

- Some hexoses are amino sugars or hexosamines which are key constituents of glycoproteins (= sugar + protein) and glycolipids (= sugar + lipid).

Lipids

- The lipids are a chemically diverse group of substances that are insoluble in water but soluble in organic solvents such as ether and chloroform.

- The triglycerides consist of three fatty acids joined by ester linkages to a glycerol molecule. They are the body's main store of energy.

- Phospholipids are the main structural elements of cell membranes.

- The body uses certain steroids and prostaglandins as chemical signals.

Amino acids and proteins

- All naturally occurring amino acids are L-amino acids.

- The naturally occurring amino acids belong to five different classes: acidic amino acids, basic amino acids, uncharged hydrophilic amino acids, uncharged hydrophobic amino acids, and the sulphur-containing amino acids.

- Proteins are assembled from a set of 20 α-amino acids, which are linked together by peptide bonds.

- There is no specific limit to the number of amino acids that can be linked together via peptide bonds, so that the number of possible protein structures is essentially infinite. It is this that makes proteins so versatile.

Nucleotides and nucleic acids

- Nucleotides consist of a purine or pyrimidine base, a pentose sugar, and a phosphate residue.

- The nucleotide adenosine triphosphate (ATP) is the main form of chemical energy in cells.

- Nucleotides in combination with other molecules form coenzymes, which play a crucial role in cellular metabolism.

- DNA consists of nucleotides arranged into a long chain.

- Each DNA molecule consists of two complementary helical strands linked together by hydrogen bonds.

- Ribonucleic acid (RNA) has a similar primary structure to a single DNA strand and exists in three different forms known as messenger RNA, transfer RNA, and ribosomal RNA.

- The various forms of RNA play a central role in the synthesis of proteins.

- The conveying of the genetic code of DNA to mRNA is called transcription.

- The synthesis of protein from mRNA is called translation.

Recommended reading

Alberts, B., Johnson, A., Lewis, J., Morgan, D., Raff, M., Roberts, K., and Walter, P. (2014) *Molecular biology of the cell* (6th edn), Chapter 2, pp. 45–65, 110–21; Chapter 3, pp. 125–52. Garland, New York.

A well paced introduction to the chemistry of molecules of biological importance.

Berg, J.M., Tymoczko, J.L., and Stryer, L. (2002) *Biochemistry* (5th edn), Chapters 1–5, 27–9. Freeman, New York.

Papachristodoulou, D., Snape, A., Elliott, W.H.. and Elliott, D.C. (2014) *Biochemistry and molecular biology* (5th edn), Chapters 4, 23, and 24. Oxford University Press, Oxford.

A clearly written, well illustrated account of the basic ideas of the roles of DNA and RNA in protein synthesis.

To check that you have mastered the key concepts presented in this chapter, complete the accompanying online self-assessment questions. Go to www.oup.com/uk/pocock5e/

PART TWO

The organization and basic functions of cells

CHAPTER 4
Introducing cells and tissues

Chapter contents

This chapter should help you to understand:

* The structural organization of cells
* The functions of the different cellular organelles
* The structure and role of the cytoskeleton
* The cell cycle and cell division: mitosis and meiosis
* The cellular organization of different types of epithelia
* The principles of cellular energy metabolism

4.1 Introduction

The cells are the building blocks of the body. However, there are many different types of cell, each with its own characteristic size and shape. Some cells are very large. For example, the cells of skeletal muscle (skeletal muscle fibres) may extend for up to 30 cm along the length of a large muscle and be up to 100 μm in diameter. Other cells are very small—for example the red cells of the blood, which are small biconcave discs with diameters in the region of 7 μm. Skeletal muscle fibres and red cells represent some of the more striking variations in cell morphology, but all cells have certain characteristics, some of which may only be evident during differentiation. The structure of a typical mammalian cell is illustrated in Figure 4.1,

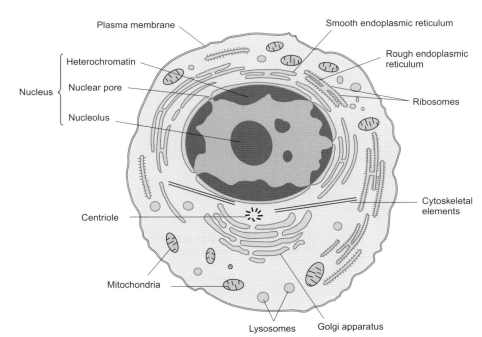

Figure 4.1 A diagram of a typical mammalian cell, showing the major organelles. (After Fig 1.6 in J.M. Austyn and K.J. Wood (1993) *Principles of cellular and molecular immunology.* Oxford University Press, Oxford.)

Labels: Plasma membrane; Smooth endoplasmic reticulum; Rough endoplasmic reticulum; Heterochromatin; Nucleus; Nuclear pore; Nucleolus; Ribosomes; Centriole; Cytoskeletal elements; Mitochondria; Lysosomes; Golgi apparatus

which shows it to be bounded by a **cell membrane**, also called the **plasma membrane** or **plasmalemma**. The cell membrane is a continuous sheet that separates the watery phase inside the cell, the **cytoplasm**, from that outside the cell, the extracellular fluid. The shape of an individual cell is maintained by an internal network of protein filaments known as the **cytoskeleton**.

At some stage of their life cycle, all cells possess a prominent structure called the **nucleus**, which contains the hereditary material, DNA. Most cells have just one nucleus, but skeletal muscle cells have many nuclei, reflecting their embryological origin from the fusion of large numbers of progenitor cells known as myoblasts. In contrast, the red cells of the blood lose their nucleus as they mature. Cells possess other structures that perform specific functions such as energy production, protein synthesis, and the secretion of various materials. The internal structures of a cell are collectively known as **organelles** and include the nucleus, the **mitochondria**, the **Golgi apparatus**, the **endoplasmic reticulum**, and various membrane-bound **vesicles**; these will all be described and discussed in this chapter.

4.2 The structure and functions of the cellular organelles

The plasma membrane is a lipid bilayer containing proteins

The plasma membrane or cell membrane regulates the movement of substances into and out of a cell. It is also responsible for regulating a cell's response to a variety of signals such as hormones and neurotransmitters. An intact plasma membrane is, therefore, essential for the normal functioning of a cell. When viewed at high power in an electron microscope, the plasma membrane appears as a sandwich-like structure 5–10 nm thick (i.e. $5-10 \times 10^{-9}$ metres). A layer of fine filaments, which form the **glycocalyx** or cell coat, covers the outer surface. The membranes of the intracellular organelles (e.g. endoplasmic reticulum, Golgi apparatus, lysosomes, and mitochondria) have a similar structure to that of the plasma membrane.

Chemical analysis shows that the plasma membrane is made of lipid and protein in approximately equal amounts by weight, although there are many more lipid molecules than protein molecules. The lipids are arranged so that their polar head groups are oriented towards the aqueous phase and the hydrophobic fatty acid chains face inwards to form a central hydrophobic region, as shown in Figure 4.2. This arrangement is called a lipid bilayer and is an inherently stable configuration. The membrane proteins are surrounded by large numbers of lipid molecules, and they either span the lipid bilayer or they are anchored to it in various ways.

The principal lipids of the plasma membrane belong to one of three classes: phospholipids, glycolipids, and cholesterol. The composition of the outer and inner layers of the plasma membrane differs significantly. The outer leaflet consists of glycolipids, phosphatidylcholine, and sphingomyelin. The inner leaflet is richer in negatively charged phospholipids such as phosphatidylinositol and phosphatidylserine. Cholesterol is present in both leaflets of the bilayer. The presence of phosphatidylinositol in the inner leaflet of the bilayer is significant as inositol phosphates play an important role in the transmission of certain signals from the cell membrane to the interior of the cell (see Chapter 6). Each

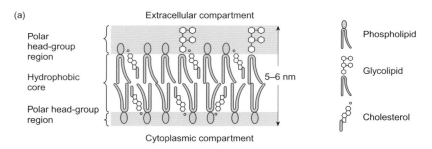

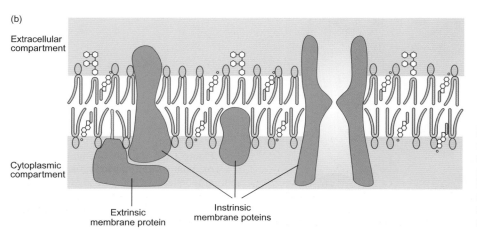

Figure 4.2 The structure of the plasma membrane. Panel a) shows the basic arrangement of the lipid bilayer. Note the presence of glycolipid only in the outer leaflet of the bilayer. Panel b) is a simplified model of the plasma membrane showing the arrangement of some of the membrane proteins.

phospholipid molecule is able to diffuse freely in one plane of the bilayer, but phospholipids rarely flip from one leaf of the bilayer to the other. Loss of membrane asymmetry, especially the appearance of phosphatidylserine on the outer leaflet, appears to be an important indicator of cell ageing.

Artificial lipid membranes are not very permeable to ions or polar molecules

Artificial lipid membranes are relatively permeable to carbon dioxide, oxygen, and lipid-soluble molecules, but they are almost impermeable to polar molecules and ions such as glucose and sodium. Moreover, such membranes are relatively impermeable to water. Natural cell membranes, however, are permeable to a wide range of polar materials, which cross the hydrophobic barrier formed by the lipid bilayer via specific protein molecules—the aquaporins, ion channels, and carrier proteins.

Membrane proteins

The membrane proteins may be divided into two broad groups: *intrinsic*—those which are embedded in the bilayer itself—and *extrinsic*—those which are external to the bilayer but linked to it in some way such as via a hydrophobic chain. The proteins that facilitate the movement of ions and other polar materials across the plasma membrane are all intrinsic membrane proteins. Some extrinsic proteins link the cell to its surroundings (the extracellular matrix) or to neighbouring cells. Others play a role in the transmission of signals from the plasma membrane to the interior of the cell. The physiological roles of the membrane proteins are discussed in greater detail in the following chapters.

To summarize, the plasma membrane consists of roughly equal amounts by weight of protein and lipid. The lipid is arranged as a bilayer whose inner and outer leaflets have a different composition. The lipid bilayer forms a barrier to the passage of polar materials so that these substances must enter or leave a cell via specialized transport proteins. The combination of lipid bilayer and transport proteins allows cells to maintain an internal composition that is very different to that of the extracellular fluid.

The nucleus

The nucleus is separated from the rest of the cytoplasm by the **nuclear membrane** (also known as the **nuclear envelope**), which consists of two lipid bilayers separated by a narrow space. The nuclear membrane is furnished with small holes known as **nuclear pores** that provide a means of communication between the nucleus and the cytoplasm.

The nucleus contains the DNA of a cell, which associates with proteins called histones to form **nucleosomes**. The nucleosomes are arranged along a strand of DNA to form **chromatin fibres**. As shown in Figure 4.3, the chromatin fibres are supported on a protein scaffold to form the individual chromatids that make up a **chromosome**.

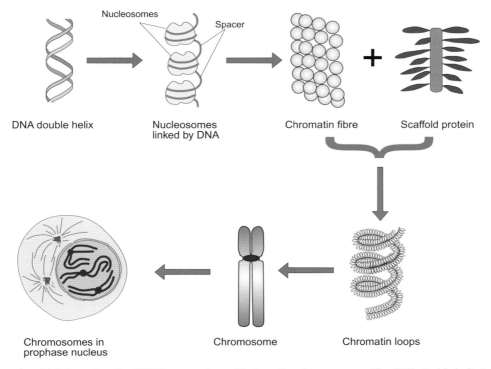

Figure 4.3 The stages by which long strands of DNA are condensed to form the chromosomes. The DNA double helix is first wrapped 1¾ turns around protein molecules known as histones to form nucleosomes that are arranged rather like beads on a string. The nucleosomes then form chromatin fibres which are supported by a protein scaffold to form the loops of chromatin that make up the individual chromatids of a chromosome. The chromosomes are clearly seen in the cell nucleus only during cell division. In this case the nucleus is shown in prophase, just after the nuclear chromatin has become condensed into well-defined chromosomes.

In light microscopy, the appearance of the chromatin of a nucleus depends on the stage of the cell cycle in which the cell happens to be. During the period between cell divisions, the chromatin fibres have a relatively loose arrangement to facilitate **gene expression**. This form of chromatin is known as **euchromatin**. By way of comparison, DNA that is not actively involved in gene expression is more tightly packaged. This form of chromatin is called **heterochromatin**, and it can be readily visualized by basic dyes. During cell division, the chromatin becomes distributed into pairs of chromosomes which attach to a structure known as the mitotic spindle before they separate during cell division (see Section 4.4).

A structure called the **nucleolus** is the most prominent feature visible within the nucleus. It is concerned with the manufacture of organelles called **ribosomes**. Within the nucleolus are one or more weakly staining regions formed from a type of DNA called nucleolar organizer DNA, which codes for ribosomal RNA. Associated with these regions are densely packed fibres, which are formed from the primary transcripts of the ribosomal RNA and protein. A further region known as the pars granulosa consists of maturing ribosomes, which are subsequently released into the cytoplasm where they play an important part in the synthesis of new protein molecules (see Chapter 3). Prominent nucleoli are seen in cells that are synthesizing large amounts of protein, such as embryonic cells, secretory cells, and cells in rapidly growing malignant tumours.

The organelles of the cytoplasm

The cytoplasm contains many different organelles that perform specific tasks within the cell. Some of these organelles are bounded by a membrane that separates them from the rest of the cytoplasm, so that the cytoplasm is divided into various compartments. Examples of membrane-bound organelles are the mitochondria, the endoplasmic reticulum, the Golgi apparatus, and vesicles such as lysosomes.

Mitochondria

The mitochondria are 2–6 μm in length and about 0.2 μm in diameter. They have two distinct membranes, an outer membrane that is smooth and regular in appearance and an inner membrane that is thrown into a large number of folds known as **cristae**, as shown in Figure 4.4. The space between the cristae is the mitochondrial matrix, which contains a rich mixture of enzymes. The mitochondrial matrix also contains a small amount of DNA—**mitochondrial DNA (mtDNA)**—which is maternal in origin (i.e. it comes from the egg rather than the sperm). Mitochondrial DNA takes the form of a small, round chromosome that has 37 genes. These genes contain the genetic instructions for making transfer RNA, ribosomal RNA, and a number of the enzymes involved in ATP generation.

Mitochondria are not assembled from scratch from their molecular constituents but increase in number by the division of existing mitochondria. When cells divide, the mitochondria of the original cell are shared out between the daughter cells. The numbers of mitochondria in a cell are regulated according to the metabolic requirements and may increase substantially if required, for example in skeletal muscle cells subject to prolonged periods of contractile activity. Changes in number of mitochondria are due to fusion and

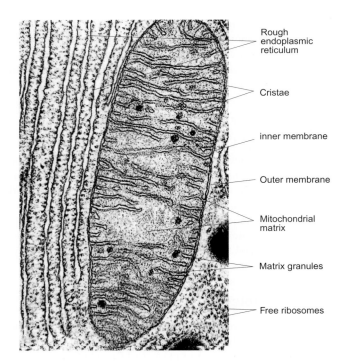

Rough endoplasmic reticulum

Cristae

inner membrane

Outer membrane

Mitochondrial matrix

Matrix granules

Free ribosomes

Figure 4.4 An electron micrograph of a cell from the pancreas of a bat. It shows a longitudinal section through a mitochondrion revealing its internal structure and an extensive region of rough endoplasmic reticulum. Note that the ribosomes are located on the cytoplasmic surface of the endoplasmic reticulum and not on the inside. (From Fig 218 in D.W. Fawcett (1981) *The cell.* Saunders, Philadelphia. Republished with permission of Elsevier.)

fission (division) of existing mitochondria in response to the changing needs of their host cell. Cells that have a high demand for ATP have many mitochondria, which are often located close to the site of ATP utilization. The arrangement of mitochondria in cardiac muscle is a striking example (see Figure 8.5). Here, the mitochondria lie close to the contractile elements of the cells, the myofibrils.

As discussed in Chapter 1, mammalian cells continually transform materials. They oxidize glucose and fats to provide energy as ATP for other activities, such as motility and the synthesis of proteins for growth and repair. The synthesis of ATP takes place in the mitochondrial matrix and on the inner membrane of the mitochondria. In most cells, the bulk of ATP synthesis occurs by a process known as oxidative phosphorylation (see Section 4.6 pp. 61–62). In the course of this process, damaging reactive oxygen species (such as superoxide) are generated but are normally rapidly destroyed by the enzymes superoxide dismutase and catalase. However, cells with defective mitochondria are unable to prevent the release of these intermediates and are vulnerable to oxidative damage. This is now thought to cause a number of human diseases, including Parkinson's disease and vascular disorders.

In addition to their role in energy production, mitochondria can also accumulate significant amounts of calcium. They thus play an important role in regulating the ionized calcium concentration within cells (see Section 5.3 pp.71–72) and in intracellular signalling (see Chapter 6 pp. 90–91).

Endoplasmic reticulum

The endoplasmic reticulum is a system of membranes that extends throughout the cytoplasm of most cells. These membranes are continuous with the nuclear membrane and enclose a significant space within the cell. The endoplasmic reticulum is classified as rough or smooth according to its appearance under the electron microscope. As Figure 4.4 shows, the rough endoplasmic reticulum has many polyribosomes attached to its cytoplasmic surface. It plays an important role in the synthesis of certain proteins and is important for the addition of carbohydrates to membrane proteins (glycosylation), which takes place on its inner surface. The smooth endoplasmic reticulum lacks ribosomes and is involved in the synthesis of lipids and other aspects of metabolism. Like the mitochondria it accumulates calcium to provide an intracellular store for calcium signalling, which is regulated in many cells by the inositol lipids of the inner leaflet of the plasma membrane (see Chapter 6 p. 90). In muscle cells the endoplasmic reticulum is called the **sarcoplasmic reticulum**. It plays an important role in the initiation of muscle contraction (see Chapter 8 p. 126).

The Golgi apparatus

The Golgi apparatus (also called the Golgi complex or the Golgi membranes) is a system of flattened membranous sacs that are involved in modifying and packaging proteins for secretion. In cells adapted for secretion as their main function (e.g. the acinar cells of the pancreas and mucus-secreting goblet cells), the Golgi apparatus lies between the nucleus and the apical surface (where the secretion takes place). While the Golgi apparatus is effectively an extension of the endoplasmic reticulum, it is not in direct continuity with it. Rather, small vesicles known as transport vesicles bud off from the endoplasmic reticulum and migrate to the Golgi apparatus, with which they fuse, as shown in Figure 4.5. The Golgi apparatus itself gives rise to specific secretory vesicles that migrate to the plasma membrane. The *cis* face of the Golgi apparatus is the site that receives transport vesicles from the endoplasmic reticulum, and the *trans* face is the site from which secretory vesicles bud off. During their passage from the *cis* face to the *trans* face, membrane proteins are modified in various ways including the addition of oligosaccharide chains (glycosylation), the addition of phosphate groups (phosphorylation), and/or the addition of sulphate groups (sulphation). The chemical changes to newly synthesized proteins are called **post-translational modification**. Further details of the mechanism of secretion are discussed in Chapter 5 (pp. 76–79).

Membrane-bound vesicles

Cells contain a variety of membrane-bound vesicles that are integral to their function. Secretory vesicles are formed by the Golgi apparatus (see previous subsection). Under the electron microscope, some vesicles appear as simple round profiles while others have an electron-dense core. After they have discharged their contents, the secretory vesicles are retrieved to form endocytotic vesicles (see Chapter 5). Other cytoplasmic vesicles include the **lysosomes** and the **peroxisomes**. Peroxisomes contain enzymes that can synthesize and destroy hydrogen peroxide. The lysosomes contain hydrolytic enzymes, which allow cells to digest materials that they are

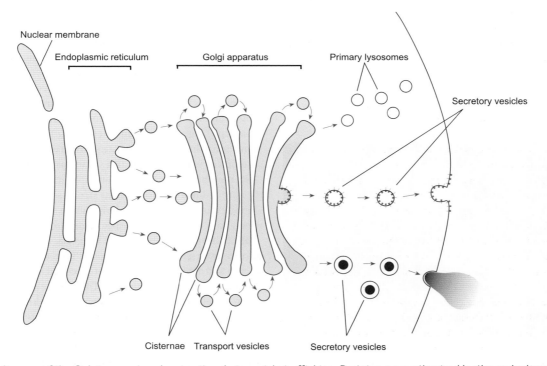

Figure 4.5 A diagram of the Golgi apparatus showing its role in vesicle trafficking. Proteins are synthesized by the endoplasmic reticulum and migrate to the *cis* face of the Golgi apparatus in ER transport vesicles. They then undergo post-translational modification as they pass through the Golgi membranes and leave in another set of transport vesicles via the *trans* face of the Golgi before migrating to their final destination. The specific kinds of vesicle are routed to their various destinations according to the proteins present in their outer coat.

recycling (e.g. plasma membrane components) or that they have taken up during endocytosis.

The importance of the lysosomes to the economy of a cell is strikingly illustrated by Tay-Sachs disease, in which gangliosides taken up into the lysosomes are not degraded in the normal way. The sufferers lack an enzyme known as beta-hexosaminidase A, which is responsible for the breakdown of gangliosides derived from recycled plasma membrane. Consequently, the lysosomes accumulate lipid and become swollen. Although other cells are affected, gangliosides are especially abundant in nerve cells, which show severe pathological changes as their lysosomes become swollen with undegraded lipid. This leads to the premature death of neurons, resulting in muscular weakness and retarded development. The condition is fatal; those with the disease usually die before they are three years old.

Ribosomes

Ribosomes consist of proteins and ribonucleic acid (RNA). They are formed in the nucleolus and migrate to the cytoplasm, where they may occur free, as shown in Figure 4.4, or in groups called polyribosomes. Ribosomes play an important role in the synthesis of new proteins, as outlined in Chapter 3. Some ribosomes become attached to the outer membrane of the endoplasmic reticulum to form the rough endoplasmic reticulum (Figure 4.4), which is the site of synthesis of membrane proteins.

The cytoskeleton

While the plasma membrane forms the boundary of the cell, by itself a lipid bilayer has very little structural strength. The structural properties of cells are generated by a cytoskeleton that serves a similar purpose to our own skeleton. This cytoskeleton is formed from three different types of support: the actin cytoskeleton (also known as microfilaments), intermediate filaments, and microtubules. Each has its own purpose within the cell. A picture of a PC12 cell is shown in Figure 4.6, in which the actin filaments and microtubules are labelled with different colours to reveal their separate networks.

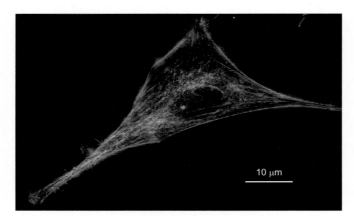

Figure 4.6 A confocal image of a cultured PC12 cell labelled to reveal actin filaments (green) and microtubules (red). Note that the networks formed by the two kinds of filament are quite separate. PC12 cells are a cell line derived from a tumour of rat adrenal medulla called a phaeochromocytoma.

Actin filaments are the narrowest filaments in the cytoskeleton, at about 6 nm across. The actin cytoskeleton is formed from monomers of the protein actin. Actin monomers and filaments possess polarity referred to as the plus and minus ends. Actin filaments grow when actin monomers bind to the plus (+) end and shrink when monomers are lost from the minus (−) end. Actin filaments can grow in both a linear and branched fashion. Many actin filaments are very dynamic, and the rapid remodelling of actin filaments enables cells to undergo rapid changes in shape. Often, actin filaments show a molecular behaviour termed '**treadmilling**' in which a filament loses monomers at one end at the same rate as new monomers bind at the other end. The end result is a dynamic filament that remains at a constant size. In muscle, however, the actin filaments are stabilized by the actin-binding protein α-actinin (see Chapter 8).

The key regulators of the actin cytoskeleton are proteins, which promote branching, promote formation of new filaments (which is termed *nucleation*), or cap one of the filaments' ends. If the growing (plus) end of a dynamic filament is capped, the result is that the filament will shrink, because it will now only show one-half of the treadmilling behaviour (loss from the minus end). Equally, if the minus end is capped, the filament will grow, as it will no longer lose monomers from that end. The proteins which control nucleation, branching, and capping are all actively regulated. In this way processes as diverse as the formation of the acrosomal process of sperm, phagocytic engulfment, and cell migration can all be controlled via the actin cytoskeleton.

Intermediate filaments were originally so called because their diameter, estimated from electron micrographs, lay between that of the thin actin filaments and that of the thick myosin filaments of skeletal muscle. They play an important role in the mechanical stability of cells. Their chief distinguishing feature is their relative stability (unlike actin filaments they do not undergo continuous recycling). Instead they form a stable network within the cell. Intermediate filaments are formed from a wide range of component proteins (there are around 70 proteins in the intermediate protein family)—unlike actin filaments and microtubules (see next paragraph). Recent work indicates that different intermediate filaments form different structural networks within the cell, providing support for particular structures, such as the plasma membrane, Golgi membrane stacks, or the endoplasmic reticulum. Those cells that are subject to mechanical stresses (such as epithelia and smooth muscle) are particularly rich in intermediate filaments, which link cells together via specialized junctions.

Microtubules, as their name suggests, are hollow tubes. They have an external diameter of about 23 nm and a wall thickness of 5–7 nm and are formed from a protein called **tubulin**. Microtubules are the railway system of the cell. They form an organized network, along which specific motor proteins of the **kinesin** and **dynein** families move cell components around the cell in a directional fashion. Microtubules are a major feature of the long processes of neural cells (called axons), where they are responsible for axonal transport (see Chapter 7). They also play a major role in the movement of cilia and flagella. Microtubules originate from a complex structure known as a **centrosome**. Between cell divisions, the centrosome is located at the centre of a cell near the nucleus.

Embedded in the centrosome are two **centrioles**, which are cylindrical structures, arranged at right angles to each other. At the beginning of cell division the centrosome divides into two and the daughter centrosomes move to opposite poles of the nucleus to form the mitotic spindle (see Section 4.4).

Summary

The cells have many organelles that carry out specific cellular functions. The largest is the nucleus, which is bounded by the nuclear membrane. The nucleus contains the cell's DNA and is the site at which the ribosomes are synthesized before they migrate to the cytoplasm, where they play an essential part in protein synthesis. The cytoplasm is divided into a number of separate membrane-bound compartments: the endoplasmic reticulum, the Golgi apparatus, the lysosomes, and the mitochondria. The Golgi apparatus is involved in the post-translational modification of proteins and the formation of secretory vesicles. Mitochondria provide ATP for the energy requirements of the cell by oxidizing carbohydrates and fatty acids. Cell shape is maintained by a cytoskeleton of protein filaments.

4.3 Cell motility

Many cells are motile—the ability of cells to move is not confined to muscle cells. Cell motility is particularly important during embryological development, but even in adults many cells migrate from place to place. White blood cells move from the blood into the tissues during normal immunological surveillance. Fibroblasts invade damaged areas of skin to repair wounds. Both of these cell types move by crawling over their neighbours. The swimming motion of spermatozoa results from the whip-like movement of their flagella. Other cells move material over their surface by the beating of cilia—as in the clearing of particles from the airways. All these forms of movement are due to the activity of specific motor proteins that form the cytoskeleton.

When a cell crawls from one place to another, it does so over a specific substrate. At first it extends a process called a **lamellipodium** in the direction of movement. The lamellipodium then attaches itself to the substrate, which provides the traction for the cell to be pulled along. The process is driven by a cycle of actin polymerization and depolymerization in which single actin subunits bind ATP before linking together to form a web of actin filaments. The filaments form at the leading edge of a lamellipodium and become attached to the substrate by other proteins. Once the actin filaments are formed, they slowly hydrolyse their bound ATP and individual actin subunits are released at the trailing edge to start the cycle again. In this way the cycle of actin polymerization and depolymerization acts as a motor within the cell to pull it along.

In all mammals, including humans, ciliated cells line the upper airways (see Chapter 32) and the Fallopian tubes (oviducts). They beat in an orderly, wave-like motion to propel material over the

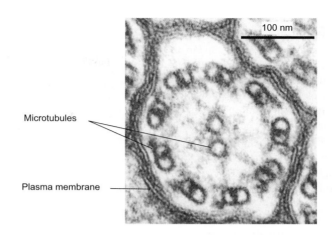

Figure 4.7 An electron micrograph of a cross-section through a cilium of the respiratory epithelium. Note the characteristic 9 + 2 array of microtubules that forms the core of the cilium. (From Fig 3.24 in L. Weiss and R.O. Greep (1977) *Histology*. McGraw Hill, New York.)

surface of an epithelial layer. Flagella are similar in structure to cilia but are much longer. While they are common in single-cell organisms, flagella are found in mammals only as the motile part of the sperm (see Chapter 48). Cilia and flagella have a characteristic internal structure, which is clearly shown in Figure 4.7. The outer surface is an extension of the plasma membrane, which encloses a central core known as the **axoneme**. The axoneme consists of an array of nine doublet microtubules arranged around a central pair of microtubules and is firmly linked to the main cytoskeleton of the cell via a basal body. The doublet microtubules are attached to their neighbours by protein links at regular intervals.

The rapid motion of cilia and flagella is due to a large motor protein called **ciliary dynein**. The ciliary dynein is firmly linked by a polypeptide chain to one side of each doublet microtubule, while the free head end is able to interact with the neighbouring microtubule in an ATP-dependent manner. When the dynein head region binds to the neighbouring microtubule and hydrolyses its bound ATP, it is able to bind to the next protein subunit of the microtubule and so on—rather like the interaction of actin and myosin described in Chapter 8 for the contraction of muscle. Unlike the filaments of muscle cells, the microtubules are linked together, so instead of shortening the cilium, the dynein bends it. The bent region progresses along the cilium and results in a whip-like motion that is coordinated between adjacent cilia to produce a rhythmic wave-like motion across the ciliated surface. Furthermore, this wave is coordinated between adjacent ciliated epithelial cells so that material can be swept along an epithelial surface. The mechanisms responsible for this coordination are still not fully understood.

A defective dynein gene results in Kartagener's syndrome. People with this condition suffer from frequent respiratory infections as they cannot clear the mucus and debris from their airways. Moreover, men with the condition are infertile as their sperm cannot swim through the Fallopian tubes (oviducts).

Summary

Many cells are motile. Sperm progress through the female reproductive tract by means of flagella. Other cells migrate through the tissues by extending lamellipodia and pulling themselves over their neighbours (e.g. white blood cells). Some epithelial cells are specialized to transport material across their surface by the action of cilia.

4.4 Cell division

During life animals grow by two processes: (1) through the addition of new material to pre-existing cells and (2) by increasing the number of cells by division. Cell division occurs by one of two processes:

- **mitosis**, in which each daughter cell has the same number of chromosomes as the parent cell;

- **meiosis**, in which each daughter cell has half the number of chromosomes as its parent.

Most cells that divide do so by mitosis. Meiosis occurs only in the germinal cells of the ovary and testes during the formation of the eggs and sperm. Since mitosis is responsible for cell replication, it is normally tightly regulated; lack of appropriate regulation is responsible for tumour genesis and cancer.

Mitosis and the cell cycle

The processes involved in cell growth, replication of the genetic material, and cell division by mitosis constitute the **cell cycle**. The process of cell division is broadly divided into three phases: the **interphase**, during which DNA replication occurs; the mitotic phase (**M-phase**), when the nucleus divides; and **cytokinesis**, during which the cytoplasm is divided between the two daughter cells. The timing of each phase determines the rate at which the cell reproduces and will influence the rate of tissue growth. Figure 4.8 summarizes the main stages of a normal cell cycle.

The interphase portion of the cell cycle has three main stages: Gap 1 (G_1), synthesis (S), and Gap 2 (G_2).

- G_1 (**the first growth phase**) is the growth phase of the cell, during which the cells begin to synthesize the RNA, enzymes, and proteins that will be needed in subsequent phases. At the end of G_1, just before entry into S-phase, there is a checkpoint that determines whether the cell should divide, delay its division, or enter a resting state called G_0 (which is discussed below).

- S (**the synthesis phase**): **S-phase** is the stage at which the chromosomes of the cell are duplicated, and is generally short, lasting for between 10 and 20 hours.

- G_2 (**the second growth phase**): cells in G_2 are preparing to begin the process of cell division. During G_2 there is a sharp increase in the production of tubulin as microtubule-based forces play a major role in separating the duplicated chromosomes during the M-phase.

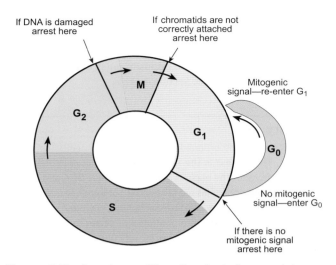

Figure 4.8 The four stages of the cell cycle. As it proceeds to divide, a cell has to pass through a number of checkpoints to prevent errors in cell division (see text for further explanation). (Redrawn after Fig 30.1 in W.H. Elliott and D.C. Elliott, *Biochemistry and molecular biology*, 3rd edition. Oxford University Press, Oxford.)

The cell continues to grow and microtubules are assembled to form the mitotic spindle. This phase lasts for between 1 and 12 hours. At the end of G_2 there is a second checkpoint that ensures that everything is ready for the cell to undergo the final stage of cell division, the M-phase.

- M (**the mitosis phase**): cell division occurs over a period of 1–2 hours. The end result of this phase is the creation of two genetically identical daughter cells.

Cells in G_1 can also enter a non-proliferative stage termed G_0, during which they can be either **stem cells** that can re-enter the cell cycle in response to specific biochemical signals, or post-mitotic cells that have differentiated into a cell type that is no longer capable of mitosis (see Box 4.1). Haematopoietic stem cells and liver cells are examples of cells that remain in G_0 for long periods but can subsequently be induced to divide at times of physiological need. Others, such as the neurons of the central nervous system, are post-mitotic and remain permanently arrested in G_0.

The stages of mitosis

The mitotic phase of the cell cycle, the M-phase, is generally divided into six stages, as shown in Figure 4.9. These are:

- prophase
- prometaphase
- metaphase
- anaphase
- telophase
- cytokinesis

During the early part of **prophase** the nuclear chromatin condenses into well-defined chromosomes, each of which consists of two

Box 4.1 Stem cells: their role in development and tissue maintenance

Following the fusion of sperm and egg, an organism is formed through a succession of cell divisions. Initially the dividing cells grow into a ball, but as the number of cells increases, the ball invaginates to form a hollow sphere. The shell of the sphere is composed of cells called **trophoblasts**, which eventually form the placenta (see Chapter 48), while the **inner cell mass** develops into the embryo. The generation of tissues (histogenesis) depends not just on the ability of cells to divide (which simply increases the number of cells), but also on their ability to **differentiate** into specific cell types.

The cells of the inner cell mass progressively form into three layers. The inner layer is the **endoderm**, the middle layer is the **mesoderm**, and the outer layer is the **ectoderm**. Each gives rise to a different set of tissues.

- Cells from the ectoderm go on to separate into two further layers: the **surface ectoderm** and the **neuroectoderm**. The cells of the surface ectoderm develop into the skin and tooth enamel as well as the lens, cornea, and conjunctiva of the eye. The cells of the neuroectoderm become the neural crest and neural tube. Neural crest cells go on to form the peripheral nervous system while the cells of the neural tube form the central nervous system.

- The mesoderm gives rise to the dermis layer of the skin, the muscles, the connective tissues, bone, the heart and circulatory system, and the lymphatic system.

- The endoderm eventually gives rise to the epithelium of the liver, the gastrointestinal tract, parts of the respiratory tract, the endocrine system, the auditory system, and the urinary bladder.

The development of these lineages restricts the ability of the body to regenerate. As cells progress through rounds of differentiation they become more and more specialized. Although this specialization is crucial for developing the complexity of the mammalian body, it comes at a price. A newt can regrow a limb if it is lost; mammals cannot. Even the ability to repair tissue damage is dependent on specific populations of stem cells and progenitor cells.

A **stem cell** is one that can undergo unlimited rounds of division. A **pluripotent stem cell** is a stem cell that can not only undergo unlimited rounds of division, but can differentiate into any type of tissue. At some point (which is different for each cell lineage) cells become committed to a particular path. At this point they may remain stem cells, but they are no longer pluripotent. Embryonic stem cells from the inner cell mass are pluripotent, but other stem cells can only develop into a limited range of cell types. For instance, the haematopoietic stem cells found in bone marrow can replicate indefinitely, but will only generate blood cells. They are committed to this function, and can no longer generate neurons, or muscle, or skin. One step further are cells termed **progenitors**. Progenitor cells can divide to generate new cells, but only a limited number of times; they are also **unipotent**, which means their daughter cells have a prescribed developmental path.

In the adult, the majority of cells have undergone what is called **terminal differentiation**: they are in their final cell type. Many tissues contain small numbers of stem cells (and progenitors) which are responsible for maintaining turnover of the tissue. This can occur at a very robust rate as part of normal function (for example, in the skin or the gut), or as part of wound healing (e.g. myosatellite cells in muscle). However, they are too committed to their developmental pathway to generate anything but a limited number of cell types.

In both adult stem cells and progenitor cells, cell division occurs asymmetrically. This means that the parent cell gives rise to two daughter cells which are not the same—they have different fates. This is important because it is the mechanism by which populations of progenitors and stem cells are maintained—one daughter cell can be a stem cell (keeping the number consistent) while the other is specified to become the desired cell type.

The discovery that there are stem cells in the brain is one of the more remarkable recent findings in this area. Previously, it had been thought that all neurons were present at birth or shortly after but, over the last decade or so, it has become clear that there are at least two concentrations of neural stem cells in the brain. These are located adjacent to lateral ventricles (in the **sub-ventricular zone**) and in the **hippocampus**. Although they are not capable of replacing the entire brain, they do provide a slow steady route for the creation of new neurons. Although the functions of these neurons are not yet fully understood, it is thought that they play a role in learning and memory.

Recently, there has been a great deal of interest in the therapeutic potential of stem cells. In principle, these cells can grow to become any tissue in the body. Consequently, efforts have been directed to seeking ways to use them to grow artificial organs for transplantation, and to replace diseased or dead cell types. For example, patients with Parkinson's disease have lost most of the dopaminergic neurons in the substantia nigra (see Chapter 10). It is hoped that injection of pluripotent stem cells might provide a means to replace these neurons, and so provide a path to recovery for these patients.

At present there are many obstacles to be overcome. In addition to the considerable scientific and clinical difficulties presented by this approach, there are also significant ethical considerations. Human embryonic stem cells are harvested from early embryos which come from *in vitro* fertilization, where more embryos are produced than are needed for implantation. Since they are still potentially viable, there is an ethical consideration as to whether it is acceptable to use them for research or treatment. This has led to the creation of **induced pluripotent stem cells** which are adult cells that have been reprogrammed to act as though they were embryonic stem cells. This is achieved by the injection of DNA or transcription factors into the cells. The DNA and transcription factors are then able to change the regulatory state of the chromosomes. Such cells have been successfully incorporated into growing mice, but there are many issues that are not yet fully understood. So, while these are promising developments, they do not provide a perfect solution. Another solution is to harvest stem cells from the umbilical cord, which contains embryonic stem cells that are genetically identical to those of the baby. They can be stored frozen in liquid nitrogen for long periods of time, but medical treatments based on this approach would require the systematic removal and archiving of tissue from every baby born—an enormous administrative and logistical problem.

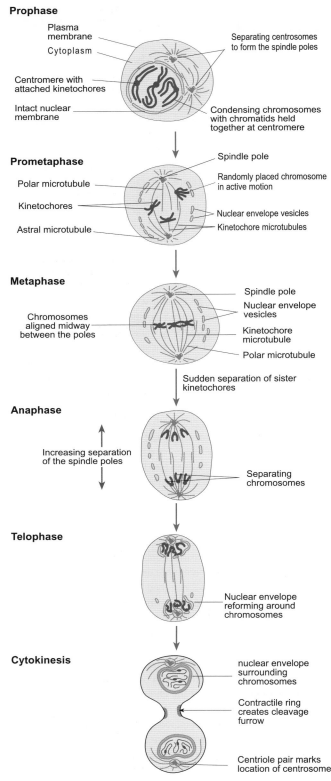

Prophase

- Plasma membrane
- Cytoplasm
- Centromere with attached kinetochores
- Intact nuclear membrane
- Separating centrosomes to form the spindle poles
- Condensing chromosomes with chromatids held together at centromere

Prometaphase

- Polar microtubule
- Kinetochores
- Astral microtubule
- Spindle pole
- Randomly placed chromosome in active motion
- Nuclear envelope vesicles
- Kinetochore microtubules

Metaphase

- Chromosomes aligned midway between the poles
- Spindle pole
- Nuclear envelope vesicles
- Kinetochore microtubule
- Polar microtubule

Sudden separation of sister kinetochores

Anaphase

- Increasing separation of the spindle poles
- Separating chromosomes

Telophase

- Nuclear envelope reforming around chromosomes

Cytokinesis

- nuclear envelope surrounding chromosomes
- Contractile ring creates cleavage furrow
- Centriole pair marks location of centrosome

Figure 4.9 The principal stages of cell division by mitosis. (Adapted from Fig 21-5 in B. Alberts, J. Lewis, M. Raff, K. Roberts, and P. Walter (2008) *Molecular biology of the cell*, 4th edition. Garland, New York. Republished with permission. Permission conveyed through Copyright Clearance Center, Inc.)

identical **chromatids** (sister chromosomes) linked by a specific sequence of DNA known as a **centromere**. As prophase proceeds, the cytoplasmic microtubules disassemble and the mitotic spindle begins to form outside the nucleus between a pair of separating centrioles. **Prometaphase** begins with the dissolution of the nuclear membrane. It is followed by the movement of the microtubules of the mitotic spindle into the nuclear region. The chromosomes then become attached to the mitotic spindle by their centromeres. During **metaphase** the chromosomes become aligned along the central region of the mitotic spindle. **Anaphase** begins with the separation of the two chromatids to form the chromosomes of the daughter cells. The poles of the mitotic spindle move further apart. In **telophase** the separated chromosomes reach the poles of the mitotic spindle, which begins to disappear. A new nuclear membrane is formed around each daughter set of chromosomes, and mitosis proper is complete. **Cytokinesis** is the division of the cytoplasm between the two daughter cells. It begins during anaphase but is only completed after the end of telophase. The cell membrane invaginates in the centre of the cell at right angles to the long axis of the mitotic spindle to form a cleavage furrow. The furrow deepens until only a narrow neck of cytoplasm joins the two daughter cells. This finally breaks, separating the two daughter cells.

The cell cycle is strictly controlled

To preserve tissue homeostasis and to avoid uncontrolled growth, cells must not be allowed to divide unchecked. Indeed, the sequence and timing of the various phases of the cell cycle is very strictly controlled. Cells can only enter mitosis when triggered to do so by specific signalling molecules (the cytokines and growth factors) that bind to plasma membrane receptors to control the activation of growth-promoting genes. Even those cells that have been triggered to enter the active cell cycle only proceed to complete mitosis if the following conditions are met:

1) The DNA of the cell is undamaged.

2) The DNA has been completely replicated (once only) before the start of mitosis.

3) At metaphase, the two sets of chromosomes are correctly aligned along the equator of the mitotic spindle.

Two sets of proteins play a part in this aspect of the control of the cell cycle. They are the cyclins and the cyclin-dependent protein kinases (Cdk). The cyclins bind to Cdk molecules and control their ability to phosphorylate their target proteins. These control proteins trigger cells to move from one phase of the cell cycle to the next and act as 'checkpoints', as illustrated in Figure 4.8. The checkpoints interrupt the cell cycle if an error, such as DNA damage, has occurred. Checkpoints for DNA damage exist before the cell enters the S-phase (the late G_1 checkpoint) and in late G_2 just before the M-phase. Damaged DNA is detected by a tumour suppressor protein called p53 (sometimes referred to as the 'guardian of the genome') which can initiate apoptosis (programmed cell death—see Section 51.7) if the damage cannot be repaired. A further checkpoint, called the 'spindle (or metaphase) checkpoint', detects any failure of spindle formation or chromosomal attachment and stops the cell in metaphase of mitosis.

Meiosis

In humans, somatic cells (i.e. those of the body) are **diploid** with 46 chromosomes (23 pairs). The gametes (eggs and sperm) have 23 (unpaired) chromosomes. They are **haploid** cells. In diploid cells half the chromosomes originated from the father (the paternal chromosomes) and half from the mother (maternal chromosomes). The maternal and paternal forms of a specific chromosome are known as **homologues**. With the exception of the X and Y chromosomes (the chromosomes that determine sex), homologous chromosomes carry identical sets of genes arranged in precisely the same order along their length. This arrangement is crucial for the genetic recombination that takes place during meiosis.

During meiosis the chromosome number is halved to form the gametes (the ova and spermatozoa). This process consists of two successive cell divisions. As for mitosis, meiosis begins with DNA replication. The first stage (**prophase I**) of the first cell division is divided into five substages:

- leptotene
- zygotene
- pachytene
- diplotene
- diakinesis

During **leptotene** the chromosomes condense in a similar manner to that seen in mitosis. The homologous pairs of chromosomes then become aligned along their length so that the order of their genes exactly matches (**zygotene**). Each pair of chromosomes is bound together by proteins to form a **bivalent** (i.e. a complex with four chromosomes—see Figure 4.10). When all 23 pairs of chromosomes have become aligned, new combinations of maternal and paternal genes occur by the crossing over of chromosomal segments (**pachytene**). The pachytene ends with the disassembly of the protein linkage between the homologous pairs of chromosomes, which begin to separate. They do, however, remain joined by two DNA linkages. It is at this stage (**diplotene**) that the eggs (ova) arrest their meiotic division and start to accumulate material. (For this reason the diplotene is also called the synthesis stage). The final stage of prophase I is similar to mitosis with spindle formation and repulsion of the chromosome pairs (**diakinesis**).

The dissolution of the nuclear envelope marks the start of **metaphase I**. The bivalents align themselves so that each pair of homologues faces opposite poles of the spindle. This process is random so that the homologues that face any one pole of the spindle are a mixture of maternal and paternal chromosomes. At **anaphase I** the chromosome pairs separate. As sister chromatids remain joined, they move towards the spindle poles as a unit so that the daughter cells each receive two copies of one of the two homologues.

Formation of the gametes now proceeds by a second cell division (**cell division II**) that resembles a normal mitotic cell division. It differs in that (a) it occurs without further DNA replication and (b) the daughter cells have half the number of chromosomes of the parent. Occasionally, some chromosomes do not separate properly and the daughter cells will either lack one or more chromosomes or

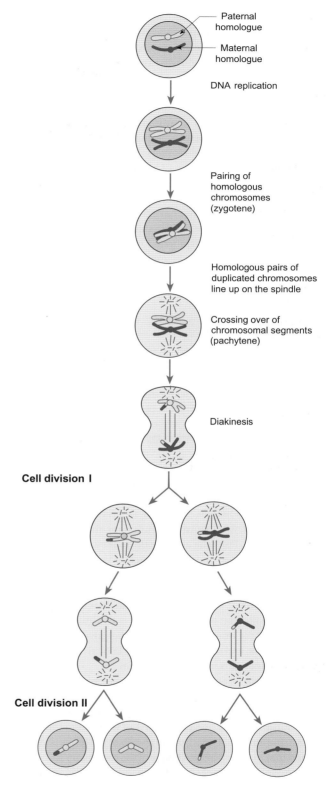

Figure 4.10 An outline of the main stages of gamete formation by meiosis. (Adapted from Fig 21-5 in B. Alberts, J. Lewis, M. Raff, K. Roberts, and P. Walter (2008) *Molecular biology of the cell*, 4th edition. Garland, New York. Republished with permission. Permission conveyed through Copyright Clearance Center, Inc.)

have a greater number than normal. This is known as **nondisjunction**—the normal separation is called **disjunction**. Nondisjunction causes a number of genetic diseases, of which **Down syndrome** is perhaps the best known. In this disease, the sufferer usually has inherited an extra copy of chromosome 21 (trisomy 21).

The X and Y chromosomes determine the sex of an individual. An individual with two X chromosomes is female, while an individual with one X and one Y chromosome is male. As male cells have one X and one Y chromosome, how do chromosomes of such different size form homologous pairs during spermiogenesis? The answer resides in a small region of homology shared by both X and Y chromosomes that enables them to pair and go through meiotic cell division I in a similar manner to other chromosomes. The daughter cells will then either have two X chromosomes or two Y chromosomes and will give rise to gametes with a single X or a single Y chromosome. Nondisjunction of the X or Y chromosomes gives rise to abnormal sexual differentiation (see Box 50.1).

Summary

Most cells divide by mitosis and each daughter cell is genetically identical to its parent. The cell cycle encompasses all the processes involved in cell growth, replication of the genetic material, and cell division. It is divided into interphase, M-phase (the mitotic phase), and cytokinesis, and is strictly controlled. Cell division by meiosis occurs only in the germinal cells during the formation of the gametes (eggs and sperm) but permits genetic recombination to occur so that the offspring possess genes from both parents.

4.5 Epithelia

The interior of the body is physically separated from the outside world by the skin, which forms a continuous sheet of cells known as an **epithelium**. Epithelia also line the hollow organs of the body such as the gut, lungs, and urinary tract, as well as the fluid-filled spaces such as the peritoneal cavity. Any continuous cell layer that separates an internal space from the rest of the body is an epithelium. Note, however, that the cell layer that lines the blood vessels is generally called an **endothelium**, as is the lining of the fluid-filled spaces of the brain, while the epithelial coverings of the pericardium, **pleura**, and peritoneal cavity are known as **mesothelium**, reflecting their origins from the mesoderm.

Epithelia have three main functions: protective, secretory, and absorptive. Their structures reflect their differing functional requirements. For example, the epithelium of the skin is thick and tough to resist abrasion, and to prevent the loss of water from the body. In contrast, the epithelial lining of the alveoli of the lungs is very delicate and thin to permit free exchange of the respiratory gases.

Despite such differences in form and function, all epithelia share certain features.

- They are composed entirely of cells.
- Their cells are tightly joined together via specialized cell–cell junctions to form a continuous sheet.

- Epithelia lie on a matrix of connective tissue fibres called the **basement membrane**, which is 0.1–2.0 μm thick, depending on tissue type. The basement membrane consists of a **basal lamina** 30–70 nm thick overlying a matrix of collagen fibrils. The basement membrane provides physical support and separates the epithelium from the underlying vascular connective tissue, which is known as the **lamina propria**. The ependymal lining of the cerebral ventricles is an exception to this rule.

- To replace damaged and dead cells, all epithelia undergo continuous cell replacement. The natural loss of dead epithelial cells is known as **desquamation**. The rate of replacement depends on the physiological role of the epithelium and is highest in the skin and gut, both of which are continually subject to abrasive forces.

- Unlike cells scattered throughout a tissue, the arrangement of cells into epithelial sheets permits the directional transport of materials either into or out of a compartment. In a word, epithelia show *polarity*. In the gut, kidney, and many glandular tissues this feature of epithelia is of great functional significance. The surface of an epithelial layer that is oriented towards the central space of a **gland** or hollow organ is known as the **apical surface**. The surface that is orientated towards the basement membrane and the interior of the body is called the **basolateral surface**.

Close to the apical surface of an epithelial cell lies a characteristic structure known as the **junctional complex** (see Figure 4.11). This consists of three structural components: the **tight junction** (also known as the **zonula occludens**), the **adherens junction** (or zonula adherens), and the **desmosomes**. Within the junctional complexes specialized regions of contact called **gap junctions** are found. Gap junctions allow small molecules to diffuse between

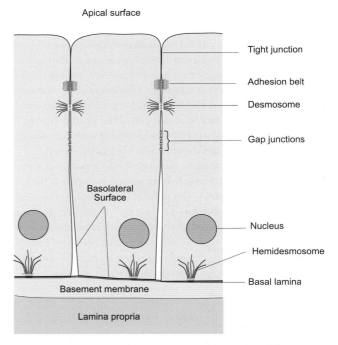

Figure 4.11 A diagrammatic representation of the main features of the junctional complexes of epithelia.

adjacent cells. In this way they play a role in communication between neighbouring cells (see Chapter 6 p. 96).

The tight junction is a continuous region of contact between the membranes of adjacent cells. It links an epithelial cell to each of the surrounding cells to seal off the space above the apical surface from that surrounding the basolateral surface. In a tight junction, the plasma membranes of the adjacent cells are held so closely apposed that the extracellular space is greatly reduced. This prevents ions and molecules from leaking between the cells. The proteins responsible for linking the epithelial cells together in this way are transmembrane proteins known as the **claudins** and **occludins**.

The mechanical strength of an epithelium is provided by the adherens junctions and desmosomes. The adherens junctions form a continuous band around each cell. On the cytoplasmic side, intracellular anchor proteins form a distinct zone of attachment for actin and intermediate filaments of the cytoskeleton. This is known as an attachment plaque and appears as a densely staining band in electron micrographs. The anchor proteins are connected to transmembrane adhesion proteins (**cadherins**), which bind neighbouring cells together.

The desmosomes are points of contact between the plasma membranes of adjacent cells. They are distributed in clusters along the lateral surfaces of the epithelial cells. The attachment plaques consist of various anchoring proteins that link intermediate filaments of the cytoskeleton to the cadherins. The cadherins of one cell bind to those of its neighbours to form focal attachments of great mechanical strength.

Hemidesmosomes, as their name implies, have a similar appearance to half a desmosome as seen in an electron micrograph. However, they are formed by different anchoring proteins that bind cytoskeletal intermediate filaments to transmembrane adhesion proteins known as **integrins**. The integrins fix the epithelial cells to the basal lamina, so linking the cell layer to the underlying connective tissue.

The classification of epithelia

Epithelia vary considerably in their morphology. The main types are known as **simple, stratified**, and **pseudostratified**. Simple epithelia consist of a single cell layer and are classified according to the shape of the constituent cell type. Stratified epithelia have more than one cell layer, while pseudostratified epithelia consist of a single layer of cells in contact with the basement membrane but the varying height and shape of the constituent cells gives the appearance of more than one cell layer.

The morphological characteristics of the main types of epithelium are shown in Figure 4.12 and are summarized below.

- Simple squamous epithelium (squamous = flattened) consists of thin and flattened cells as shown in Figure 4.12(i) and (ii). These epithelia are adapted for the exchange of small molecules between the separated compartments. Squamous epithelium forms the walls of the alveoli of the lungs. It also lines the abdominal cavity and forms the endothelium of the blood vessels.

- Simple cuboidal epithelium, as the name implies, consists of a single layer of cuboidal cells whose width is approximately equal to their height. An example is the cuboidal epithelium that forms the walls of the small collecting ducts of the kidneys.

- Simple columnar epithelium is adapted to perform secretory or absorptive functions. In this form of epithelium, the height of the cells is much greater than their width, as shown in Figure 4.12(iv). It occurs in the large-diameter collecting ducts of the kidneys. It is also found lining the small intestine, where the apical surface is covered with minute projections of the cell surface called microvilli which increase the surface area available for absorption.

- Ciliated epithelium consists of cells which may be cuboidal or columnar in shape. The apical surface is ciliated, although non-ciliated cells are also interspersed between the ciliated cells, as illustrated in Figure 4.12(v). The non-ciliated cells may have a secretory role. Examples of ciliated epithelia are those of the Fallopian tubes and the ependymal lining of the cerebral ventricles.

- Pseudostratified columnar ciliated epithelium consists of cells of differing shapes and heights, as shown in Figure 4.12(vi). This type of epithelium predominates in the upper airways (trachea and bronchi). Epithelia of this type combine a mechanical and secretory function: the goblet cells secrete mucus to trap airborne particles, while the columnar ciliated cells move the mucus film towards the mouth.

- Stratified squamous epithelium is adapted to withstand chemical and physical stresses. The best known stratified epithelium is the epidermis of the skin. In this case, the flattened epithelial cells form many layers, only the lowest layer being in direct contact with the basement membrane (see Figure 4.12(vii)). The more **superficial** cells are filled with a special protein called keratin, which renders the skin almost impervious to water and provides an effective barrier against invading organisms such as bacteria. If the outer layers of the skin are damaged, for example by burns, there is a loss of fluid and a risk of infection. Moreover, if the area involved is very large, the loss of fluid may become life-threatening.

- Transitional epithelium is found in the bladder and ureters. It is similar in structure to stratified squamous epithelium except that the superficial cells are larger and rounded (Figure 4.12(viii)). This adaptation allows stretching of the epithelial layer as the bladder fills.

Serous membranes enclose a fluid-filled space such as the peritoneal cavity. The peritoneal membranes consist of a layer of epithelial cells (called the mesothelium) overlying a thin basement membrane, beneath which is the narrow band of connective tissue. The fluid that separates the two epithelial surfaces is formed from the plasma by the same mechanism as the interstitial fluid (see Chapter 31). The pleural membranes that separate the lungs from the wall of the thoracic cavity are another example of serous epithelia.

Glandular epithelia are specialized for secretion. The epithelia that line the airways and part of the gut are covered with a thick secretion called mucus that is provided by specialized secretory cells known as goblet cells, which discharge their contents directly onto an epithelial surface. Other glands secrete material via a specialized duct onto an epithelial surface, as shown in Figure 4.12(ix). These are known as **exocrine glands**. Examples are the pancreas, salivary glands, and sweat glands. An exocrine gland may be a simple coiled tube (as in the case of the sweat glands) or it may consist

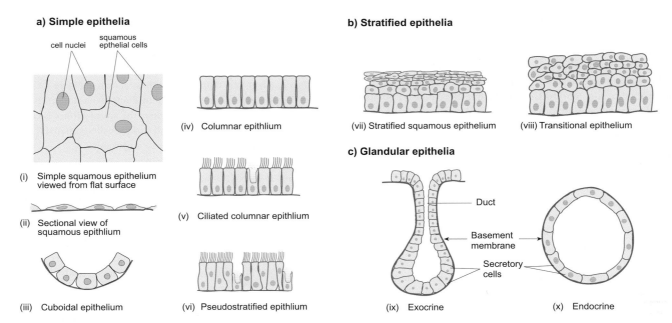

a) Simple epithelia

cell nuclei   squamous epthelial cells

(i) Simple squamous epithelium viewed from flat surface

(ii) Sectional view of squamous epithlium

(iii) Cuboidal epithelium

(iv) Columnar epithlium

(v) Ciliated columnar epithlium

(vi) Pseudostratified epithlium

b) Stratified epithelia

(vii) Stratified squamous epithelium

(viii) Transitional epithelium

c) Glandular epithelia

Duct

Basement membrane

Secretory cells

(ix) Exocrine

(x) Endocrine

Figure 4.12 Diagrammatic representation of the principal types of epithelium. Panel (a) shows the general structure of the common types of simple epithelia. Note that when viewed from the apical surface there are no gaps between the cells (i). Pseudostratified epithelia (iv) have cells of differing shape and size, so giving a false appearance of multiple cell layers. The nuclei are found at many levels within the epithelium. Panels (c) and (d) show the general structure of stratified, transitional, and glandular epithelia. See text for further details.

of a complex set of branching ducts linking groups of cells together to form assemblies called **acini** (singular **acinus**). The acinar cells of an individual acinus are held in a tight ball by a capsule of connective tissue. This type of organization is found in the pancreas and salivary glands. Cells that produce a copious watery secretion when they are stimulated are called serous cells; examples are the acinar cells of the pancreas and salivary glands. Other glandular epithelia lack a duct and secrete material across their basolateral surfaces, from where it passes into the blood. These are the **endocrine glands**. Examples include the thyroid gland and the irregular clusters of epithelial cells that constitute the islets of Langerhans of the pancreas.

Summary

Epithelia act both to separate the interior of the body from the external environment and to separate one compartment of the body from another. Individual epithelia are formed entirely from a sheet of cells and consist of one or more cell layers. Glandular epithelia are specialized for secretion. If their secretion is via a duct, they form part of an exocrine gland. If their secretion passes directly into the blood, they form part of an endocrine gland.

4.6 Energy metabolism in cells

Animals take in food as carbohydrate, fats, and protein. These complex molecules are broken down in the gut into simple molecules, which are then absorbed. The main carbohydrate of the diet is starch, which is broken down into glucose by the action of enzymes. The fats are broken down into fatty acids and glycerol, while dietary protein is broken down into its constituent amino acids. These breakdown products can then be used by the cells of the body to make ATP, which provides a convenient way of harnessing chemical energy.

ATP can be synthesized in two ways: first by the glycolytic breakdown of glucose to pyruvate, and second by the oxidative metabolism of pyruvate and acetate via the tricarboxylic acid cycle or TCA cycle. The utilization of glucose, fatty acids, and amino acids for the synthesis of ATP is summarized in Figure 4.13. In each case the synthesis of ATP is accompanied by the production of carbon dioxide and water.

The generation of ATP by glycolysis

Of the simple sugars, glucose is the most important in the synthesis of ATP. It is transported into cells, where it is phosphorylated to form glucose 6-phosphate. This is the first stage of its breakdown by the process of glycolysis, in which glucose is broken down to form pyruvate. Glycolysis takes place in the cytoplasm of the cell, outside the mitochondria, and does not require the presence of oxygen. For this reason, the glycolytic breakdown of glucose is said to be **anaerobic**. The glycolytic pathway is summarized in Figure 4.14. Further details can be found in any standard textbook on biochemistry.

The breakdown of a molecule of glucose by glycolysis yields two molecules of pyruvate and two molecules of ATP. In addition, two molecules of reduced nicotinamide adenine dinucleotide (NADH) are produced. When oxygen is present, the NADH generated by

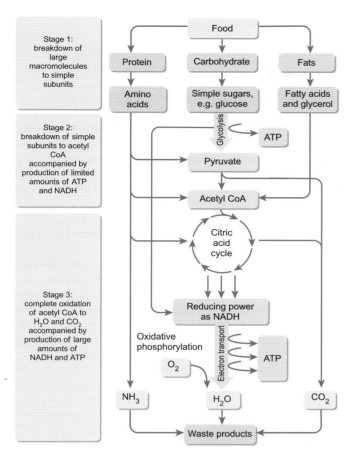

Figure 4.13 A schematic diagram showing the three main stages by which foodstuffs are broken down to yield ATP for cellular metabolism. (Adapted from Fig 2-18 in B. Alberts, D. Bray, J. Lewis, M. Raff, K. Roberts, and J.D. Watson (1989) *Molecular biology of the cell*, 2nd edition. Garland, New York.)

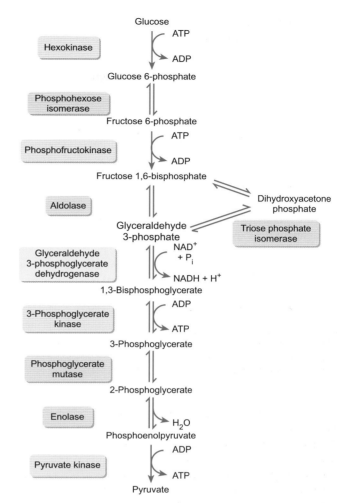

Figure 4.14 A schematic chart showing the principal steps in the glycolytic breakdown of glucose. (From Fig 12.7 in W.H. Elliott and D.C. Elliott (2005) *Biochemistry and cell biology*, 3rd edition. Oxford University Press, Oxford.)

glycolysis is oxidized by the mitochondria via the **electron transport chain**, resulting in the synthesis of about three molecules of ATP and the regeneration of two molecules of nicotinamide adenine dinucleotide (NAD).

In normal circumstances, the pyruvate that is generated during glycolysis combines with coenzyme A to form **acetyl coenzyme A** (acetyl CoA), which is oxidized via the TCA cycle to yield ATP (discussed in the next subsection). In the absence of sufficient oxygen, however, some of the pyruvate is reduced by NADH in the **cytosol** to generate lactate. This step regenerates the NAD used in the early stages of glycolysis so that it can participate in the breakdown of a further molecule of glucose. Thus *glycolysis can generate ATP even in the absence of oxygen* and becomes an important source of ATP for skeletal muscle during heavy exercise (Chapter 38).

ATP generation by oxidative metabolism in the mitochondrion—the tricarboxylic acid cycle

As described earlier in this chapter, mitochondria are small cellular organelles that superficially resemble the bacteria from which they are thought to be descended. They are typically oval or rod-shaped in cross-section, and have a characteristic double membrane structure. The outer membrane is smooth and contains a large number of proteins, called porins, that allow free diffusion of molecules of up to about 10 kDa in size. In contrast, the inner membrane is intensely invaginated to form structures called cristae (see Figure 4.4). The inner membrane is impermeable to most hydrophilic substances—including small ions—and so contains carrier proteins to allow the factors necessary for ATP generation to pass through, as well as to allow the exchange of cytosolic ADP with the ATP generated within the mitochondrion. Enclosed within the inner membrane is the matrix of the mitochondrion, which can be thought of as the cytoplasm of the mitochondrion itself.

The pyruvate formed by the glycolytic breakdown of glucose is transported into the mitochondrial matrix, where its acetyl group is combined with coenzyme A to form acetyl CoA, in a reaction that reduces NAD to NADH. Acetyl CoA is then combined with a molecule of oxaloacetate to form the tricarboxylic acid, citrate, which provides the common name for this metabolic cycle: the **citric acid cycle**, or **tricarboxylic acid cycle** (TCA). A series of reactions then converts the acetate group of acetyl CoA to CO_2 while generating GTP, free protons, and reduced flavine adenine dinucleotide (FADH$_2$) and NAD (NADH). The end products are oxaloacetate and free CoA, which can then react with new molecules of pyruvate to start the cycle once more. These reactions are summarized in Figure 4.15. During each turn of the TCA cycle three molecules of NADH, one of FADH$_2$ and one molecule of GTP are generated.

How does this process help the mitochondrion (and hence the cell) generate ATP? The answer lies in the way in which the inner mitochondrial membrane uses the subsequent reoxidation of NADH and FADH$_2$, via a process called electron transport. This process requires molecular oxygen and for this reason it is called **aerobic metabolism**. Aerobic metabolism is very efficient and yields about 10 molecules of ATP for each molecule of acetyl CoA that enters the TCA cycle. There are four large multi-protein complexes located in the inner mitochondrial membrane which act through a series of redox (oxidation-reduction) reactions to move electrons from NADH and FADH$_2$ to generate water from oxygen and protons, while using the energy released by these oxidative reactions to pump protons out across the inner mitochondrial membrane. Although a full description of this process is beyond the scope of this book, it has been mapped out in great detail, and it worth noting that the second of the four complexes in the electron transport chain is the protein responsible for converting succinate to fumarate in the citric acid cycle—therefore the electron transport chain and the TCA cycle are intimately linked.

The final step in the generation of ATP from oxidative metabolism uses the proton gradient that has been generated by oxidation of NADH and FADH$_2$. This involves a protein complex called the F_0/F_1 complex. Both F_0 and F_1 components are comprised of multiple protein subunits, but in essence, sub-complex F_1 generates ATP from ADP by movement of protons down their electrochemical gradient (see Chapter 5 p. 67). Overall, the complete oxidation of one molecule of glucose to carbon dioxide and water yields about 30 molecules of ATP. This should be compared with the generation of ATP by anaerobic metabolism, where only two molecules of ATP are generated for each molecule of glucose used.

Fatty acids are the body's largest store of food energy. They are stored in fat cells (adipocytes) as triglycerols. Fat cells are widespread but are most abundant in the adipose tissues. Stored fat is broken down to fatty acids and glycerol by **lipases** in a process known as lipolysis, which takes place in the cytoplasm of the cell. The glycolytic pathway metabolizes the glycerol, while the fatty

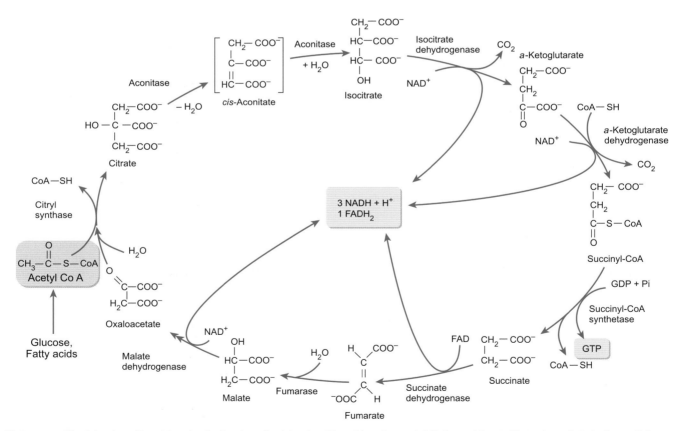

Figure 4.15 The tricarboxylic acid cycle. As it enters the tricarboxilic acid cycle, acetyl CoA combines with oxaloacetate to form citric acid. This reaction is catalysed by citryl synthase. Each complete cycle results in the production of one molecule of GTP, three of NADH, and one of FADH$_2$. In the process two molecules of carbon dioxide are produced. The NADH and FADH$_2$ are then used by the electron transport chain to generate approximately 10 molecules of ATP. (From Fig 12.13 in W.H. Elliott and D.C. Elliott (2005) *Biochemistry and cell biology*, 3rd edition. Oxford University Press, Oxford.)

acids combine with coenzyme A to form acyl-CoA before being broken down by a process known as β-oxidation. The breakdown of fatty acids to provide energy takes place within the mitochondria. The first step is the formation of acyl-CoA, followed by its transport from outside the mitochondria to the mitochondrial matrix, where it undergoes a series of metabolic steps that result in the formation of acetyl CoA, NADH, and $FADH_2$, as shown in Figure 4.16. The acetyl CoA is oxidized via the TCA cycle, and the NADH and $FADH_2$ are oxidized via the electron transport chain to provide ATP, as described above. For each two carbon unit metabolized, 13 molecules of ATP and one molecule of GTP are produced. Overall, the complete oxidation of a molecule of palmitic acid (which has 16 carbon atoms) yields over 100 molecules of ATP.

Although animals can synthesize fats from carbohydrates via acetyl CoA, they cannot synthesize carbohydrates from fatty acids. When glucose reserves are low, many tissues preferentially utilize fatty acids liberated from the fat reserves. Under these circumstances, the liver relies on the oxidation of fatty acids for energy production and may produce more acetyl CoA than it can utilize in the TCA cycle. Under such conditions, it synthesizes acetoacetate and D-3-hydroxybutyrate (also called β-hydroxybutyrate). These compounds are known as **ketone bodies** and are produced in large amounts when the utilization of glucose by the tissues is severely restricted, as in starvation or in poorly controlled diabetes mellitus (see Chapter 24). Acetoacetate and β-hydroxybutyrate

are not waste products but can be utilized for ATP production by the heart and the kidney. In severe uncontrolled diabetes mellitus, significant amounts of acetone form spontaneously from acetoacetate and this gives the breath a characteristic sweet smell, which can be a useful aid in diagnosis.

Proteins are the main structural components of cells, and the principal use for the protein of the diet is the synthesis of new protein. Those amino acids that are not required for protein synthesis are deaminated. This process results in the replacement of the amino group ($-NH_2$) by a keto group ($>C=O$) and the liberation of ammonia, which is subsequently metabolized to urea. The carbon skeleton of most amino acids can be used to synthesize glucose (a process known as **gluconeogenesis**, from the Greek words *glykys*, meaning 'sweet', *neo*, meaning 'new', and *genesis*, meaning 'origin'). These steps occur in the liver (see Figure 4.17). Those amino

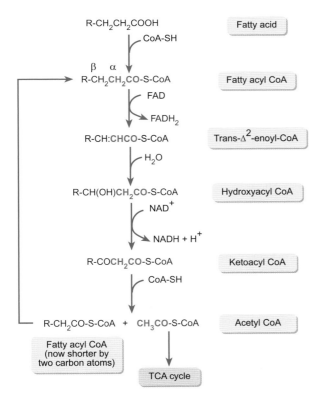

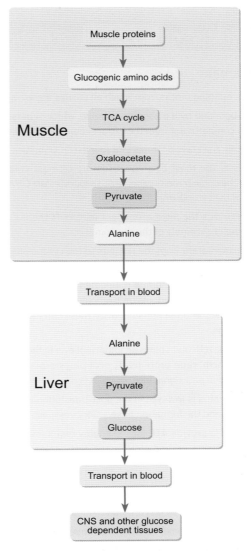

Figure 4.16 The pathway by which fatty acids are broken down to form acetyl CoA which can be metabolized via the tricarboxylic acid cycle to produce ATP. In these reactions, each two-carbon unit generates one molecule of NADH and one molecule of $FADH_2$ in addition to that generated by the tricarboxylic acid cycle.

Figure 4.17 An outline of the pathways by which amino acids are broken down to form glucose that can used for energy (ATP) production (gluconeogenesis). TCA = tricarboxylic acid; CNS = central nervous system.

acids that can be so utilized are known as the glucogenic amino acids. Of the 20 amino acids found in proteins, only leucine and lysine cannot be used for gluconeogenesis and are oxidized via the same pathway as fats. Such amino acids are called ketogenic. The glucogenic amino acids ultimately form pyruvate or oxaloacetate, which can be oxidized via the tricaboxylic acid cycle to produce ATP or to generate glucose. The carbon skeleton of the ketogenic amino acids is oxidized by β-oxidation to form acetyl CoA which, like the fatty acids, is metabolized via the TCA cycle to form ATP. Although the amino acids are generally considered to be either glucogenic or ketogenic, several, including the aromatic amino acids phenylalanine, tryptophan, and tyrosine, can be either glucogenic or ketogenic depending on the metabolic pathway by which they are broken down.

Summary

Animals generate the energy they require for their activities by the progressive breakdown of foodstuffs. Carbohydrates are first broken down by the glycolytic pathway to form pyruvate. Fats are broken down by β-oxidation to form acetyl CoA. Proteins are first broken down to amino acids, which may then generate pyruvate (glucogenic amino acids) or acetyl CoA (ketogenic amino acids). Pyruvate and acetyl CoA are utilized by the TCA cycle and electron transport chain to synthesize ATP. When glucose reserves are low, many tissues preferentially utilize fatty acids as a source of energy. During starvation, the glucose required by the brain is synthesized in the liver by gluconeogenesis.

(✳) Checklist of key terms and concepts

The cell organelles

- Cells are the building blocks from which all tissues are assembled.
- Each cell is bounded by a plasma membrane, which separates the intracellular compartment from the extracellular compartment.
- The cell membrane plays an important role in cell signalling.
- The nucleus is the most prominent feature of most cells. It is bounded by the nuclear membrane, which separates the DNA of the nucleus from the cytoplasm.
- Within the nucleus the most prominent feature is the nucleolus. This is the site of assembly of the ribosomes.
- The ribosomes migrate from the nucleus to the cytoplasm, where they play an essential part in protein synthesis.
- Within the cytoplasm, membrane-delineated structures known as organelles are found: the endoplasmic reticulum, the Golgi apparatus, the lysosomes, and the mitochondria. These carry out specific cellular functions.
- The endoplasmic reticulum plays an important role in the synthesis of lipids and membrane proteins. It also plays a major role in calcium regulation and signalling.
- The Golgi apparatus is involved in the post-translational modification of proteins and the formation of secretory vesicles.
- The mitochondria provide ATP for the energy requirements of the cell by metabolizing carbohydrates and fatty acids to carbon dioxide and water.

The cytoskeleton and cell motility

- Cell shape is maintained by a cytoskeleton of protein filaments.
- Many cells are motile. Some have cilia and flagella that provide the means of movement, while other use amoeboid movement to migrate through the tissues (e.g. white blood cells).
- The internal structure of the cilia and flagella is formed from a characteristic array of microtubules called the axoneme.

Cell division

- Cell division occurs either by mitosis or by meiosis.
- Most cells divide by mitosis, in which each daughter cell is genetically identical to its parent.
- Mature cells that cannot divide are known as post-mitotic cells.
- Cells that have the potential to undergo further division are known as stem cells.
- Meiosis occurs only in the germinal cells during the formation of the gametes (eggs and sperm).
- Meiosis permits genetic recombination to occur so that the offspring possess genes from both parents.

Epithelia

- Epithelia act both to separate the interior of the body from the external environment and to separate one compartment of the body from another.
- All epithelia are formed entirely from sheets of cells and consist of one or more cell layers.
- An epithelium consists of cells joined together by specialized junctions to form a continuous sheet. A simple epithelium consists of a single sheet of cells, while those epithelia with more than one layer are called stratified epithelia.
- Some epithelia have the appearance of being stratified because they contain cells of differing height and shape. These are known as pseudostratified epithelia.
- Glandular epithelia are specialized for secretion. If their secretion is via a duct they form part of an exocrine gland. If their secretion passes directly into the blood they form part of an endocrine gland.

Energy metabolism in cells

- Animals generate the energy they require for movement and growth of body tissues by the progressive breakdown of foodstuffs.

- Carbohydrates are first broken down by the glycolytic pathway to form pyruvate.
- Fats are broken down by β-oxidation to form acetyl CoA.
- Proteins are first broken down to amino acids, which may then generate pyruvate (glucogenic amino acids) or acetyl CoA (ketogenic amino acids).

- Pyruvate and acetyl CoA are utilized by the tricarboxylic acid (TCA) cycle and electron transport chain to synthesize ATP.
- Although animals can synthesize fats from carbohydrates via acetyl CoA, they cannot synthesize carbohydrates from fatty acids.
- When glucose reserves are low, many tissues preferentially utilize fatty acids for energy metabolism.
- During starvation, the liver synthesizes the glucose required by the brain from amino acids. This is called gluconeogenesis.

Recommended reading

Biochemistry and cell biology

Alberts, B., Johnson, A., Lewis, J., Morgan, D., Raff, M., Roberts, K., and Walter, P. (2014) *Molecular biology of the cell* (6th edn), Chapters 2, 10, 16, and 17. Garland, New York.

Berg, J.M., Tymoczko, J.L., and Stryer, L. (2011) *Biochemistry* (7th edn), Chapters 12–14, 16–18, and 22. Freeman, New York.

Mescher, A.L. (2010) *Junquieira's (2003) Basic histology* (12th edn), Chapters 2–4. McGraw-Hill Medical, New York.

Papachristodoulou, D., Snape, A., Elliott, W.H., and Elliott, D.C. (2014) *Biochemistry and molecular biology* (5th edn), Chapters 3, 7, 12–14, and 30. Oxford University Press, Oxford.

To check that you have mastered the key concepts presented in this chapter, complete the accompanying online self-assessment questions. Go to www.oup.com/uk/pocock5e/

CHAPTER 5
The transport functions of the plasma membrane

This chapter should help you to understand:

- The principal types of proteins involved in membrane transport

- The difference between passive and active transport

- How polar molecules and ions cross the plasma membrane

- The role of metabolic pumps in generating and maintaining ionic gradients

- How the sodium pump generates the Na$^+$ and K$^+$ gradients across the plasma membrane

- The importance of the sodium pump in maintaining constant cell volume

- How cells regulate their intracellular free Ca^{2+}

- How cells make use of the Na$^+$ gradient to regulate intracellular [H$^+$]

- How cells exploit the ionic gradients established by the sodium pump for secondary active transport across epithelial layers

- The origin of the membrane potential

- The nature of ion channels: ligand- and voltage-gated channels

- The mechanisms of secretion: constitutive and regulated secretion by exocytosis

- Endocytosis and the retrieval of membrane constituents

- The mechanism by which certain cells ingest cell debris and foreign material—phagocytosis

5.1 Introduction

As explained in Chapter 4, each cell is bounded by a plasma membrane. The region outside the cell (the **extracellular compartment**) is thereby separated from the inside of the cell (the **intracellular compartment**). This physical separation allows each cell to regulate its internal composition independently of other cells. Chemical analysis has shown that the composition of the intracellular fluid is very different to that of the extracellular fluid (Table 5.1). It is rich in potassium ions (K$^+$) but relatively poor in both sodium ions (Na$^+$) and chloride ions (Cl$^-$). It is also rich in proteins (enzymes and structural proteins) and the small organic molecules that are involved in metabolism and signalling (amino

Table 5.1 The approximate ionic composition of the intracellular and extracellular fluid of mammalian skeletal muscle

Ionic species	Extracellular fluid	Intracellular fluid	Nernst equilibrium potential
Na$^+$	145	20	+53 mV
K$^+$	4	150	−97 mV
Ca^{2+}	1.8	$1-2 \times 10^{-4}$	c. +120 mV
Cl$^-$	114	3	−97 mV
HCO$_3^-$	31	10	−30 m V

Values are given in mmol l^{-1} of cell water and the equilibrium potentials were calculated from the Nernst equation (see Box 5.1). Note that the resting membrane potential is about −90 mV—close to the equilibrium potentials for potassium (the principal intracellular cation) and Cl$^-$.

acids, ATP, fatty acids, etc.). The first part of this chapter is concerned with the mechanisms responsible for establishing and maintaining the difference in ionic composition between the intracellular and the extracellular compartments. This is followed by a discussion of the ways in which cells utilize ionic gradients to perform their essential physiological roles, and then by a discussion of the mechanisms by which proteins and other large molecules cross the cell membrane—secretion and endocytosis.

5.2. The permeability of cell membranes to ions and uncharged molecules

The plasma membrane consists of a lipid bilayer in which many different proteins are embedded (see Section 4.2). Both natural membranes and artificial lipid bilayers are permeable to gases and lipid-soluble molecules (**hydrophobic molecules**). However, compared to artificial lipid bilayers, natural membranes have a high permeability to water and to water-soluble molecules (**hydrophilic** or **polar molecules**) such as glucose, and to ions such as sodium, potassium, and chloride. The relatively high permeability of natural membranes to ions and polar molecules can be ascribed to the presence of two classes of integral membrane proteins: the **channel proteins** and the **carrier proteins**.

What determines the direction in which molecules and ions move across the cell membrane?

For uncharged molecules such as carbon dioxide, oxygen, and urea, the direction of movement across the plasma membrane is simply determined by the prevailing concentration gradient. For charged molecules and ions, however, the situation is more complicated. Measurements have shown that mammalian cells have an electrical potential across their plasma membrane called the **membrane**

potential, which is discussed in greater detail in Section 5.4. Although the magnitude of the membrane potential varies from one type of cell to another (from about –35 to –90 mV), the inside of a cell is always negative with respect to the outside. The existence of the membrane potential influences the diffusion of charged molecules and ions. Positively charged chemical species will tend to be attracted into the cell, while negative ones will tend to be repelled. Overall, the direction in which ions and charged molecules move across the cell membrane is determined by three factors:

1. the concentration gradient
2. the charge of the molecule or ion
3. the membrane potential.

These factors combine to give rise to the **electrochemical gradient**, which can be calculated from the difference between the **equilibrium potential** for the ion in question and the membrane potential. If the intracellular and extracellular concentrations for a particular ion are known, the equilibrium potential can be calculated using the Nernst equation (Box 5.1).

Molecules and ions diffuse across the plasma membrane down their electrochemical gradients. This is called **passive transport**. However, studies have shown that polar materials such as ions and small organic molecules such as glucose and amino acids cross the plasma membrane much more readily than they cross artificial lipid bilayers. This is sometimes called **facilitated diffusion**. It is now clear that ions and polar molecules, including water, pass across the plasma membrane by way of specific types of membrane protein known as **ion channels** and **carrier proteins**. Furthermore, cells can transport molecules and ions against their prevailing electrochemical gradients. This uphill transport requires a cell to expend metabolic energy either directly or indirectly and is called **active transport**.

While the direction of passive movement of a particular substance is determined by its electrochemical gradient, the rate at

---

(1₂3) Box 5.1 The Nernst equation and the resting membrane potential

The flow of any ion across the membrane via an ion channel is governed by its electrochemical gradient, which reflects the charge and the concentration gradient for the ion in question together with the prevailing membrane potential. The potential at which the tendency of the ion to move down its concentration gradient is exactly balanced by the membrane potential is known as the equilibrium potential for that ion so that, at the equilibrium potential, the rate at which ions enter the cell is exactly balanced by the rate at which they leave. The equilibrium potential can be calculated from the Nernst equation:

$$E = \frac{RT}{zF} \ln \frac{[C]_o}{[C]_i}$$

where E is the equilibrium potential, ln is the natural logarithm (log_e), $[C]_o$ and $[C]_i$ are the extracellular (outside) and intracellular

(inside) concentrations of the ion in question. R is the gas constant (8.31 J K⁻¹ mol⁻¹), T the absolute temperature, F the Faraday constant (96,487 C mol⁻¹), and z the charge of the ion (+1 for Na⁺, +2 for Ca²⁺, –1 for Cl⁻, etc.). At 37°C the term $RT/F = 26.7$ mV.

The equilibrium potential for K⁺ using the data shown in Table 5.1 may be calculated as:

$$E_K = \frac{RT}{zF} \ln \frac{[K^+]_{out}}{[K^+]_{in}}$$

$$E_K = 26.7 \times \ln \frac{4}{150} = 26.7 \times (-3.62)$$

hence :

$$E_K = -96.8 \text{ mV}$$

which it can cross the plasma membrane is determined by the number and properties of the channels and/or carrier proteins present. If there are few channels or carriers for a particular molecule or ion, the permeability of the membrane to that substance will be low: the more channels or carriers that are available, the greater the permeability of the membrane to that substance.

Channel proteins and carrier proteins permit polar molecules to cross the plasma membrane

Water channels

The relatively high permeability of natural membranes to water can be attributed to the presence of specialized **water channels** known as **aquaporins**. These are proteins that possess pores allowing water to pass from one side of the membrane to the other according to the prevailing osmotic gradient. Currently, at least nine different types of aquaporin are known in mammals, several of which are found in tissues which transport large volumes of water in the course of a day—examples of such tissues are the tubules of the kidney and the secretory glands of the gut. In the collecting ducts of the kidney, aquaporins 2 and 3 play a central role in regulating water reabsorption (see Chapter 39) while aquaporin 5 plays a significant role in the production of salivary secretions (Chapter 44).

Ion channels

Ion channels are specific membrane proteins that have a pore which spans the membrane to provide a route for a particular ion to diffuse down its electrochemical gradient. They are responsible for the relatively high permeability of natural membranes to various ions such as sodium, potassium, and chloride (see Figure 5.1a). Ion channels can exist in one of two main states: they are either open and allow the passage of the appropriate ion from one side of the membrane to the other, or they are closed, preventing such movements. Many ion channels have a very high capacity for transport—for example, some potassium channels permit as many as 10^8 ions (i.e. a hundred million ions) to cross the membrane each second. Despite this, they are able to discriminate one type of ion from another so that they show **selectivity** with respect to the kind of ion they allow through their pore.

Ion channels are named after the principal ion to which they are permeable. For example, sodium channels allow sodium ions to cross the membrane but do not allow the passage of other cations such as potassium and calcium. Potassium channels are very permeable to potassium ions but not to sodium or calcium ions. Chloride ions cross the membrane via chloride channels, and so on. The permeability of a cell membrane to a particular ion will depend on how many channels of the appropriate type are open and the number of ions that can pass through each channel in a given period of time.

Some ion channels allow a variety of positively charged ions to cross the plasma membrane but do not allow the passage of negatively charged ions. Such channels are known as **non-selective cation channels**. An important example of this type of channel is the **acetylcholine receptor** (**AChR**) of the neuromuscular junction. Many channel proteins have now been identified and their peptide sequence determined with the aid of the techniques of

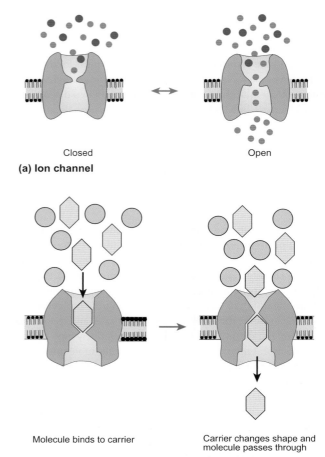

(a) Ion channel

Closed   Open

(b) Carrier protein

Molecule binds to carrier   Carrier changes shape and molecule passes through

Figure 5.1 Schematic drawings illustrating the differing modes of action of an ion channel and a membrane carrier. (a) When an ion channel is activated, a pore in the plasma membrane is opened and ions are able to diffuse from one side of the membrane to the other in a continuous stream. (b) Membrane carrier proteins operate quite differently. A molecule that is transported across the plasma membrane first binds to a specific site on the carrier which then undergoes a conformational change to release the bound molecule on the other side of the membrane. Both ions channels and carriers show selectivity for particular ions or molecules.

molecular biology. This has revealed that they can be grouped into large families according to certain features of their structure. For example, the peptide chains of many types of potassium channel have a similar sequence of amino acids in certain regions and make the same number of loops across the plasma membrane.

Carrier proteins

Carrier proteins (carriers) bind specific substances—usually small organic molecules, such as glucose, or inorganic ions (Na^+, K^+, etc.)—and then undergo a change of shape (known as a **conformational change**) to move the solute from one side of the membrane to the other (see Figure 5.1b). The capacity of a cell to transport a molecule is limited both by the number of carrier molecules and by the number of molecules each carrier is able to translocate in a given period of time (the 'turnover number'). Carriers tend to transport

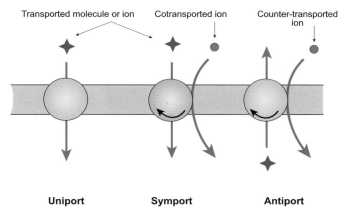

Uniport   Symport   Antiport

Figure 5.2 A diagrammatic representation of the main types of carrier proteins employed by mammalian cells: uniports act to transfer a molecule or ion from one side of the membrane to the other; symports link the transport of two ions or molecules; and antiports link the inward movement of one molecule or ion to the outward movement of another (counter-transport). Unidirectional transport and counter-transport may be linked to hydrolysis of ATP (and thus play a role in active transport). Symports exploit existing ionic gradients for secondary active transport.

fewer ions or molecules than channels. The fastest carrier proteins each transport about 10^4 molecules each second, but more typical values are between 10^2 and 10^3 molecules a second. Carriers are selective for a particular type of molecule and can even discriminate between optical isomers. For example, the natural form of glucose (D-glucose) is readily transported by specific carrier proteins of the GLUT family, but the synthetic L-isomer is not transported. Although both isomers have the same chemical constitution, the D- and L-isomers of glucose are mirror images—like our left and right hands. This proves that the carrier distinguishes between optical isomers by their shape—a property known as **stereoselectivity**.

Carrier proteins (also called **transporters**) are subdivided into three main groups according to the way in which they permit molecules to cross the plasma membrane. **Uniports** bind a specific molecule on one face of the membrane and then transfer it to the other side, as shown in Figures 5.1b and 5.2. Many substances are, however, transported across the membrane only in association with a second molecule or ion. Carriers of this type are called cotransporters, and the transport itself is called **cotransport** or **coupled transport**. When both molecular entities move in the same direction across the membrane, the carrier is called a **symport**; when the movement of an ion or molecule into a cell is coupled to the movement of a second ion or molecule out of the cell, the carrier is called an **antiport**. Figure 5.2 shows schematic representations of the different kinds of transport proteins.

Summary

While lipid-soluble molecules can cross pure lipid membranes relatively easily, water-soluble molecules cross only with difficulty. Two groups of membrane proteins facilitate the movement of water-soluble molecules into and out of cells: the channel proteins

and the carrier proteins. Ion channels permit the passage of ions from one side of the membrane to the other via a pore. Carrier proteins translocate a molecule from one side of the membrane to the other by binding the molecule on one face of the membrane and undergoing a conformational change to release it on the other.

5.3 The active transport of ions and other molecules across cell membranes

Active transport requires a cell to expend metabolic energy either directly or indirectly and involves carrier proteins. In many cases, the activity of a carrier protein is directly dependent on metabolic energy derived from the hydrolysis of ATP. This is known as **primary active transport** (e.g. the sodium pump discussed in the next subsection). In other cases, the transport of a substance (e.g. glucose) can occur against its concentration gradient by coupling its 'uphill' movement to the 'downhill' movement of sodium ions into the cell. This type of active transport is known as **secondary active transport**. It depends on the ability of the sodium pump to keep the intracellular concentration of sodium significantly lower than that of the extracellular fluid.

The sodium pump is present in all cells and exchanges intracellular sodium for extracellular potassium

As shown in Table 5.1, the intracellular concentration of sodium, calcium, and chloride in skeletal muscle cells is much lower than the extracellular concentration. Conversely, the intracellular concentration of potassium is much greater than that of the extracellular fluid. These differences in composition are common to all healthy mammalian cells, although the precise values for the concentrations of intracellular ions vary from one kind of cell to another.

What mechanisms are responsible for these differences in ionic composition? The first clues about the mechanisms by which cells regulate their intracellular sodium came from the problems associated with the storage of blood for transfusion. Like other cells, the red cells of human blood have a high intracellular potassium concentration and low intracellular sodium. When blood is stored at a low temperature in a blood bank, the red cells lose potassium and gain sodium over a period of time—a trend that can be reversed by warming the blood to body temperature (37°C). If red cells that have lost their potassium during storage are incubated at 37°C in an artificial solution similar in ionic composition to that of plasma, they only re-accumulate potassium if glucose is present. This glucose-dependent uptake of potassium and extrusion of sodium by the red cells occurs against the concentration gradients for these ions. It is therefore clear that the movement of these ions is dependent on the activity of a membrane pump driven by the energy liberated by the metabolic breakdown of glucose—in this case the **sodium pump**.

The sodium pump is found in all mammalian cells as well as in those of other animals, and it plays a central role in regulating the intracellular environment. But how does it work? Important clues were provided by experiments on the giant axon of the squid. This tissue was chosen for these experiments because it has a large diameter for a single cell (1–2 mm), making it possible to take direct measurements of ionic movements across the plasma membrane. By injecting radioactive sodium (^{22}Na$^+$) into the axon, the rate at which the axon pumped out sodium ions could be followed by measuring the appearance of radioactive sodium in the bathing solution. When the metabolic inhibitor cyanide was used to block ATP generation, the rate of pumping declined. When ATP was subsequently injected into an axon that had been poisoned by cyanide, the sodium pumping was restored to almost normal levels (Figure 5.3). This experiment showed that ATP was required for sodium to be pumped out of the axon against its electrochemical gradient. In the same series of experiments, it was found that the sodium efflux was also inhibited if potassium was removed from the extracellular fluid. This demonstrates that the uptake (influx) of potassium is closely coupled to the efflux of sodium. The sodium pump can be inhibited by a glycoside called ouabain, which binds to the extracellular face of the protein.

Subsequent studies on red blood cells showed that the hydrolysis of ATP is tightly coupled to the efflux of sodium and to the influx of potassium in such a way that, for each ATP molecule hydrolysed, a cell pumps out three sodium ions in exchange for two potassium ions. For this reason, the sodium pump is also called the Na$^+$, K$^+$-**ATPase**. A schematic diagram of the operation of the sodium pump is given in Figure 5.4.

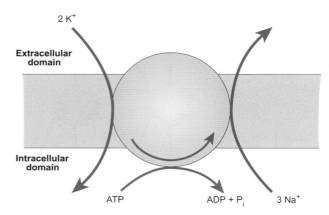

Figure 5.4 A schematic diagram of the sodium pump (Na$^+$, K$^+$-ATPase). Note that for every molecule of ATP hydrolysed, three sodium ions are pumped out of the cell and two potassium ions are pumped into the cell. The pump is therefore electrogenic in its action (it generates a voltage).

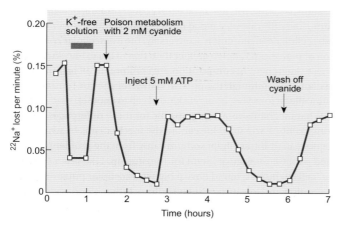

Figure 5.3 An experiment showing that the sodium pump requires intracellular ATP. In this experiment, which was conducted on a large axon isolated from a squid, normal oxidative metabolism was blocked by cyanide 1½ hours after the start of the experiment. As ATP levels fell, the axon rapidly became less and less able to pump sodium ions from its cytoplasm until ATP was injected 3 hours after the beginning of the experiment. Removal of the cyanide 6 hours after the start of the experiment gradually restored the ability of the axon to pump sodium ions. The first part of the experiment, shown on the left, demonstrates that sodium can only be actively pumped out of the axon if it can be exchanged for extracellular potassium. (Based on Fig 13 in P.C. Caldwell et al. (1960) *Journal of Physiology* **152**, 561–90.)

The sodium pump stabilizes cell volume by maintaining a low intracellular sodium concentration

The total number of particles present in a solution determines its osmotic pressure. *Outside the cell* the osmolality is due chiefly to the large number of small inorganic ions present in the extracellular fluid (Na$^+$, K$^+$, Cl$^-$, etc.). *Inside the cell* the osmolality is due to inorganic ions and to a large number of membrane-impermeant molecules such as ATP, amino acids, and proteins. Over time, there is a propensity for the intracellular concentration of ions to become equal to that of the extracellular fluid as they diffuse down their electrochemical gradients. If this tendency were not countered, the total osmolality of the intracellular fluid would tend to increase because the large impermeant molecules cannot pass out of the cell to compensate for the inward movement of small ions. The increase in osmolality would cause the cells to take up water and swell. By keeping the intracellular sodium ion concentration low, the sodium pump maintains the total osmolality of the intracellular compartment equal to that of the extracellular fluid. As a result, cell volume is kept relatively constant.

Cells use a number of transport proteins to regulate the intracellular hydrogen ion concentration

All respiring cells continually produce metabolic acids (e.g. CO_2 and carboxylic acids) which are capable of altering the intracellular concentration of hydrogen ions (H$^+$). Measurements have shown that cells maintain the hydrogen ion concentration of their cytoplasm close to 10^{-7} moles l^{-1}. Although this is higher than that of the extracellular fluid (which is usually about 4×10^{-8} moles l^{-1}), it is about a tenth of that expected if hydrogen ions were simply at electrochemical equilibrium. Therefore cells must actively regulate their intracellular hydrogen ion concentration.

Since hydrogen ions are very reactive and readily bind to a wide variety of proteins, changes in the intracellular concentration of hydrogen ions can have major consequences for cellular activity. For example, many enzymes work best at a particular hydrogen ion concentration (their pH optimum). Changes in intracellular hydrogen ion concentration also influence the function of other proteins, such as ion channels and the contractile proteins actin and myosin.

Some intracellular compartments maintain their internal pH at a different value to that seen in the cytoplasm. These include lysosomes, which have a low (acidic) internal pH value of around 5. The hydrolytic enzymes they contain are active at this low pH, and this enables the lysosomes to break down cellular debris such as damaged organelles and ingested bacteria to recycle their components. Secretory vesicles are also acidic, and in this case, the low pH establishes a concentration gradient of protons which is used to drive the neurotransmitter transporters that fill vesicles with very high concentrations of neurotransmitter.

When the hydrogen ion concentration rises during cellular activity, many hydrogen ions will bind to intracellular molecules, thus limiting the extent of the rise. This phenomenon is known as **buffering** (see Chapter 41 for further details). However, buffering can only limit the change in hydrogen ion concentration—to restore its original hydrogen ion concentration, a cell must pump out the excess hydrogen ions. To enable them to do so, cells have evolved a number of regulatory mechanisms, three of which are discussed here:

1. sodium–hydrogen ion exchange
2. co-transport of Na^+ and HCO_3^- into cells to increase intracellular HCO_3^-
3. chloride–bicarbonate exchange.

Sodium–hydrogen ion exchange occurs via an antiport that couples the outward movement of hydrogen ions against their electrochemical gradient with the inward movement of sodium ions down their electrochemical gradient (see Figure 5.5). It is thus an example of secondary active transport, the driving force for the transport of hydrogen ions out of the cell being provided by the sodium gradient established by the sodium pump. This carrier is found in a wide variety of different cell types and is particularly important in the epithelial cells of the kidney.

Many cells also possess a symporter that transports both sodium and bicarbonate ions into the cell. By linking the influx of bicarbonate to that of sodium ions, the bicarbonate ions are transported into a cell against their electrochemical gradient. Once inside, these ions bind excess hydrogen ions to form carbonic acid, which is in equilibrium with dissolved carbon dioxide that can cross the cell membrane by diffusion. For each molecule of carbon dioxide that leaves the cell as a result of this transport, one hydrogen ion is used up to form one molecule of water (see Figure 5.5).

Under most circumstances the cells are acting to prevent an increase in their hydrogen ion concentration. At high altitude, however, the extra breathing required to keep the tissues supplied with oxygen leads to a fall in the carbon dioxide concentration of the blood and tissues. This makes the cells more alkaline than they should be (i.e. they have a hydrogen ion deficit). To maintain the intracellular hydrogen ion concentration within the normal range, intracellular bicarbonate is exchanged for extracellular chloride.

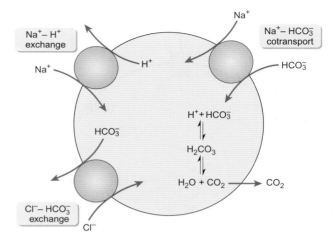

Figure 5.5 Cells regulate their internal H^+ concentration by a number of mechanisms, three of which are illustrated. They utilize Na/H exchange in which internal H^+ is exchanged for external Na^+. They also use a sodium-linked symport to raise the intracellular concentration of HCO_3^- ions. These subsequently combine with H^+ to form carbonic acid and ultimately CO_2, which is able to cross the cell membrane. When the cell is alkaline (deficient in H^+), HCO_3^- can be exchanged for Cl^-.

This mechanism (known as chloride–bicarbonate exchange) provides a means of defence against a fall in the intracellular hydrogen ion concentration. Chloride–bicarbonate exchange is freely reversible and plays an important role in the carriage of carbon dioxide by the red cells (see Chapter 25).

Many processes regulate intracellular calcium

The intracellular fluid of mammalian cells has a very low concentration of free calcium ions—typical values for a resting cell are about 10^{-7} mol l^{-1} while that of the extracellular fluid is 1–2×10^{-3} mol l^{-1}. There is, therefore, a very steep concentration gradient for calcium ions across the plasma membrane and this gradient is exploited by a wide variety of cells to provide a means of transmitting signals to the cell interior. A rise in intracellular Ca^{2+} is the trigger for many processes, including the contraction of muscle and the initiation of secretion. It is therefore essential that resting cells maintain a low level of intracellular Ca^{2+}. As shown in Figure 5.6, this is accomplished in a rather interesting and intricate way.

1. Calcium ions can be pumped from the inside of the cell across the plasma membrane via a Ca^{2+}-ATPase. This is known as the **calcium pump**, which, like the sodium pump, uses energy derived from ATP to pump calcium against its concentration gradient.
2. Intracellular calcium can be exchanged for extracellular sodium (the Na^+–Ca^{2+} exchanger). In this case the inward movement of sodium ions down their electrochemical gradient provides the energy for the uphill movement of calcium ions from the inside to the outside of the cell against their electrochemical gradient (another example of secondary active transport—in this case via an antiport).

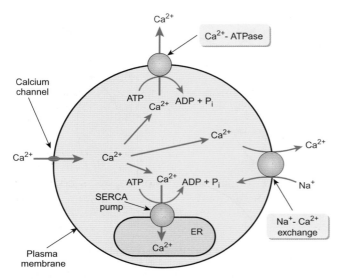

Figure 5.6 The regulation of intracellular calcium. Because the level of intracellular Ca^{2+} is widely employed by cells as a signal to initiate a wide variety of physiological responses, it is essential that intracellular Ca^{2+} is closely regulated. The figure is a schematic drawing illustrating the main ways by which this is achieved. Calcium ions enter cells by way of calcium channels. Calcium is taken up by the SERCA pump of the endoplasmic reticulum (sarcoplasmic reticulum in muscle). It is also taken up by the mitochondria (not shown). It is removed from the cell either by the calcium pump (Ca^{2+} – ATPase) or by Na^+ – Ca^{2+} exchange.

3. Another type of calcium pump (the SERCA pump—**S**arcoplasmic/**E**ndoplasmic **R**eticulum **C**alcium **A**TPase) is used to pump calcium into the space enclosed by the endoplasmic reticulum to provide a store of calcium within the cell itself while keeping the calcium concentration of the cytosol very low. (In muscle the endoplasmic reticulum is called the sarcoplasmic reticulum—see Chapter 8.) This store of calcium can be released in response to signals from the plasma membrane (see Chapter 6 for further details).

4. The mitochondria also take up calcium from the cytosol by two distinct mechanisms: a high-affinity but low-capacity uptake that is postulated to be a Ca^{2+}/H^+ antiporter, and a calcium uniport that has a low-affinity uptake but a large capacity.

The transepithelial transport of glucose and amino acids occurs by secondary active transport

Glucose and amino acids are required to generate energy and for cell growth. Both are obtained by the digestion of food, so that the cells that line the **small intestine** (the enterocytes) must be able to transport glucose and amino acids from the central cavity of the gut (the **lumen**) to the blood. To take full advantage of all the available food, the enterocytes must be able to continue this transport even when the concentration of these substances in the lumen has fallen below that of the blood. This requires active transport across the intestinal epithelium. How is this transepithelial transport achieved?

Like all epithelia, those of the gut are polarized. Their apical surfaces face the lumen and their basolateral surfaces are oriented towards the blood stream. The two regions are separated by the zonula occludens, which is made up of tight junctions that are relatively impermeable to glucose and other small solutes (see Chapter 4 and Figure 5.7). To be absorbed, these substances must therefore pass through the cells. This is called **transcellular absorption** to distinguish it from absorption across the tight junctions, which is called **paracellular absorption**.

The plasma membranes of the apical and basolateral regions of the enterocytes possess different carriers. The **apical membrane** contains a symport called SGLT1 (sodium-linked glucose transporter 1) that binds both sodium and glucose. As the concentration of sodium in the enterocytes is about 15 mmol l^{-1}, the movement of sodium from the lumen (where the sodium concentration is about 140 mmol l^{-1}) into these cells is favoured by the concentration gradient. The carrier links the inward movement of sodium down its electrochemical gradient to the uptake of glucose against its concentration gradient, and this enables the enterocytes to accumulate glucose. The **basolateral membrane** has a different type of glucose carrier—a uniport called GLUT2 (glucose transporter type 2)—that permits the movement of glucose from the cell interior down its concentration gradient into the space surrounding the basolateral surface. This is an example of facilitated diffusion. As GLUT2 has a low affinity for glucose and does not link glucose uptake to that of sodium, the reverse transport of glucose from the blood into the lumen is effectively prevented. The sodium absorbed with the glucose is removed from the enterocyte by the sodium pump of the basolateral membrane. Thus the energy for the uptake of glucose is provided by the sodium gradient established by the sodium pump. This, together with the asymmetric arrangement of the

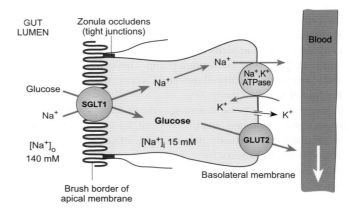

Figure 5.7 The secondary active transport of glucose across the epithelium of the small intestine occurs in two stages. First, glucose entry into the enterocyte is coupled to the movement of sodium ions down their concentration gradient via a carrier called SGLT1. This coupled transport (a symport) allows the cell to accumulate glucose until its concentration inside the cell exceeds that bathing the basolateral surface of the cell. Glucose then crosses the basolateral membrane down its concentration gradient via another carrier protein (GLUT2) that is not dependent on extracellular sodium. The sodium pump removes the sodium ions accumulated during glucose uptake. (Courtesy of Dr E.S. Debnam.)

glucose carriers on the cell, permits the secondary active transport of glucose across the wall of the intestine from where it can be absorbed into the blood (see Figure 5.7). Similar mechanisms exist for the transport of amino acids in the renal tubules and small intestine (see Chapters 27 and 30).

Summary

Molecules and ions that cross the plasma membrane by diffusion down their electrochemical gradients are said to undergo passive transport which may be mediated by ion channels or by carrier molecules. Active transport occurs when a carrier molecule transports an ion or molecule against its prevailing electrochemical gradient and requires the cell to expend metabolic energy, either directly or indirectly.

The activity of some carrier molecules is directly linked to the hydrolysis of ATP to permit the active transport of a substance uphill against its electrochemical gradient (active transport). The sodium pump is the prime example. Certain carriers use the ionic gradient established by the sodium pump to provide the energy to move another molecule (e.g. glucose) against its concentration gradient. This process is known as secondary active transport. Cells use carriers in combination to regulate the intracellular concentration of certain ions.

5.4 The potassium gradient determines the resting membrane potential of cells

The activity of the sodium pump leads to an accumulation of potassium ions inside the cell. The plasma membrane, however, is not totally impermeable to potassium ions, so that some are able to diffuse out of the cell down their concentration gradient via potassium channels. The membrane is much less permeable to sodium ions (the membrane at rest is between 10 and 100 times more permeable to potassium ions than it is to sodium ions), so the lost potassium ions cannot be readily replaced by sodium ions. The leakage of potassium ions from the cell leads to the build-up of negative charge on the inside of the membrane. This negative charge gives rise to a potential difference across the membrane known as the **membrane potential**.

The membrane potential is a physiological variable that is used by many cells to control various aspects of their activity. The **resting membrane potential** is the membrane potential of cells that are not engaged in a major physiological response that involves the plasma membrane, such as contraction or secretion. In many cases physiological responses are triggered by a fall in the membrane potential (i.e. the membrane potential becomes less negative). This fall is known as a **depolarization**. If the membrane potential becomes more negative, the change is called a **hyperpolarization**.

The negative value of the membrane potential tends to attract (positively charged) potassium ions into the cell. Thus, on one hand, potassium ions tend to diffuse out of the cell down their concentration gradient and, on the other, the negative charge on the inside of the membrane tends to attract potassium ions from the external medium into the cell. The potential at which these two opposing tendencies are exactly balanced is known as the **potassium equilibrium potential**, which is very close to the resting membrane potential of many cells. Indeed, if the intracellular and extracellular concentrations of potassium ions are known, the approximate value of the resting membrane potential can be calculated using the Nernst equation (see Box 5.1). As the resting membrane potential is negative, the inward movement of positively charged ions such as sodium and calcium is favoured and they are able to diffuse down their respective concentration gradients into the cell. In contrast, the negative value of the membrane potential opposes the inward movement of negatively charged ions such as chloride, even though their concentration gradient favours net inward movement (Table 5.1).

The membrane potential of cells can be measured with fine glass electrodes (microelectrodes) that have a very small tip diameter (c. 0.2 μm), enabling them to puncture the cell membrane without destroying the cell. The magnitude of the resting membrane potential varies from one type of cell to another but is a few tens of millivolts (1 mV = 1/1000 volt). It is greatest in nerve and muscle cells (excitable cells), where it is generally −70 to −90 mV (the minus sign indicates that the inside of the cell is negative with respect to the outside). In non-excitable cells the membrane potential may be significantly lower. For example, the membrane potential of the hepatocytes of the liver is about −35 mV.

The importance of the sodium pump in establishing the membrane potential is illustrated during embryonic development. Early in development, the sodium pump is not very active and embryonic cells have low membrane potentials. As they develop, sodium pump activity increases, the potassium gradient becomes established, and membrane potentials reach the levels seen in mature tissues.

The distribution of chloride across the plasma membrane is determined largely by the potassium ion gradient

Intracellular chloride is much lower than extracellular chloride (see Table 5.1), yet few cells are able to pump chloride ions across their plasma membrane against their electrochemical gradient. What factors regulate intracellular chloride? Intracellular fluid contains molecules such as proteins, amino acids, and ATP that are negatively charged. As these molecules cannot readily cross the cell membrane, the inside of the cell contains a fixed quantity of negative charge. Moreover, to keep the cell volume constant, the osmolality of the cell must be kept close to that of the extracellular environment. As the sum of the charges of all the ionized groups inside the cell must be zero (i.e. the cell interior must be electroneutral), the negative charges of the fixed anions balance the positive charge due to the cations—principally potassium. The difference between the total negative charge due to the intracellular fixed anions (ATP, proteins, etc.) and the positive charge due to potassium ions is made up of small diffusible anions (chiefly chloride and bicarbonate). If the potassium gradient is altered, chloride (the principal extracellular anion) is passively redistributed according to its equilibrium potential to maintain electroneutrality.

Ion channel activity can be regulated by membrane voltage or by chemical signals

Some ion channels open when they bind a specific chemical agent known as an **agonist** or **ligand** (Figure 5.8). This kind of channel is known as a **ligand-gated ion channel**. (A ligand is a molecule that binds to another; an agonist is a molecule that *both binds to and activates* a physiological system.) Other ion channels open when the membrane potential changes—usually when it becomes depolarized. These are known as **voltage-gated ion channels**. Both kinds of channel are widely distributed throughout the cells of the body, but a particular type of cell will possess a specific set of channels that are appropriate to its function. Some ion channels are regulated both by a ligand and by changes in membrane potential.

Ligand-gated ion channels are widely employed by cells to send signals from one to another. The plasma membrane of a cell has many different kinds of ion channel. As each type of channel exhibits different properties, one cell can respond to different agonists in a variety of ways. Often, however, a cell is specialized to perform a specific function (e.g. contraction or secretion) and is adapted to respond to a specific agonist. To take one important example, the activity of skeletal muscles is controlled by acetylcholine released from motor nerve endings. The acetylcholine activates non-selective cation channels, which leads to depolarization of the muscle membrane and muscle contraction (see Chapter 8).

Voltage-gated ion channels generally open in response to depolarization of the membrane (i.e. they open when the membrane potential becomes less negative). When they are open they allow certain ions to cross the membrane. Quite commonly, the channels spontaneously close after a brief period of time even though the membrane remains depolarized—a property known as **inactivation**. Thus this type of channel can exist in three distinct states: it

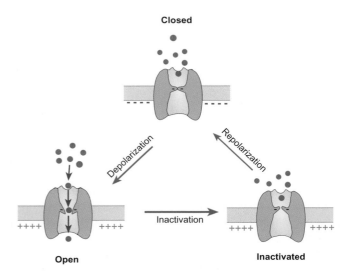

Figure 5.9 Voltage-gated ion channels are opened by changes in membrane potential (usually by a depolarization) and ions are then able to pass through the pore down their concentration gradient. The open state of most voltage-gated channels is unstable and the channels close spontaneously even though the membrane remains depolarized. This is known as channel inactivation. When the membrane is repolarized the channel returns to its normal closed state from which it can again be opened by membrane depolarization.

can be primed ready to open (the closed or resting state); it can be open and allow ions to cross the membrane; or it can be in an inactivated state from which it must first return to the closed state before it can reopen (Figure 5.9). Examples are voltage-gated sodium channels, which are employed by nerve and muscle cells to generate action potentials, and voltage-gated calcium channels. Both of these are utilized by cells to control a variety of cell functions including secretion.

As the passage of ions represents the movement of charge from one side of the membrane to the other, ions moving through a channel generate a small electrical current, which, for a single ion channel, is of the order of a picoampere (10^{-12} A). These minute currents can be detected with modern patch-clamp recording techniques (see Box 5.2). It is now clear that ion channels are either in a state that permits specific ions to cross the membrane (the open or conducting state) or the path for ion movement is blocked (the closed or inactivated state—see Figure 5.10). The switching from one state to the other is extremely rapid.

How do changes in ionic permeability alter the membrane potential?

As mentioned earlier, the membrane potential of a cell is an important physiological variable that is used to control many aspects of cell behaviour. Since the equilibrium potentials of the various ions present on either side of the plasma membrane differ (see Table 5.1), the electrochemical gradients for ion movements also differ. At the resting membrane potential, the potassium ion

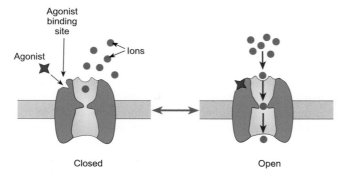

Figure 5.8 A diagrammatic representation of a ligand-gated ion channel. Agonists bind to specific sites on the channels they activate. Once they are bound they cause their associated channel to open and permit ions to pass through the central pore. When the agonist dissociates from its binding site, the channel is able to revert to the closed state.

Box 5.2 Patch-clamp recording techniques

Recent developments in electrophysiological methods have enabled physiologists to study how ions move through channels. A small glass pipette filled with a solution containing electrolytes (e.g. NaCl) is pressed against the surface of a cell. By applying a little suction, the electrode becomes sealed against the cell membrane and a small patch of membrane is isolated from the remainder of the cell by the electrode. The movement of ions through channels in the isolated patch of membrane can then be detected as weak electrical signals using a suitable amplifier. This is called patch-clamp recording.

The recordings made with this method unequivocally showed that channels open in discrete steps. Thus they either open fully (and can pass ions) or they are closed. The switch between the two states is extremely rapid. If the patch of membrane under the electrode is ruptured it is then possible to record all the current that passes across the cell membrane. This is called *whole-cell recording* and has proved to be of great value in understanding how many different types of cell perform their specific functions. Other variants of the method exist in which the patches of membrane are pulled from the cell.

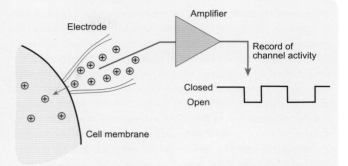

Figure 1 A schematic drawing of a typical patch-clamp recording circuit. The cell (bottom left) has an electrode pressed against its surface to isolate a small patch of membrane. The currents passing through the electrode are measured with a sensitive amplifier. When single-ion channels open and close, small step-like currents can be detected.

distribution is close to the equilibrium potential, and the tendency of potassium ions to diffuse out of the cell down their concentration gradient is balanced by their inward movement due to the membrane potential. For sodium ions the situation is very different. At the resting membrane potential, the electrochemical gradient strongly favours sodium influx but few sodium channels are open, so the permeability of the membrane to sodium is low. When sodium channels open, the permeability of the membrane to sodium increases dramatically and the membrane potential adopts a value nearer to the sodium equilibrium potential (i.e. the membrane becomes more depolarized). This shows how *the membrane potential depends both on the ionic gradients across the cell membrane and on*

the permeability of the membrane to the different kinds of ion present. The precise relationship between the membrane potential, the ionic gradients, and the permeability of the membrane to specific ions is given by the **Goldman equation** (see Box 5.3).

As an example of the way in which ionic permeabilities control cell activity, consider how a secretory cell responds to stimulation by an agonist. When the agonist activates the cell, the membrane depolarizes and this depolarization triggers secretion (see Section 6.3). At rest, the membrane potential is close to the potassium equilibrium potential (about −70 mV) because there are more open potassium channels than there are open sodium channels. Consequently, the permeability of the membrane to potassium is

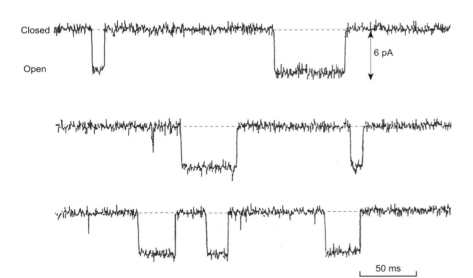

Figure 5.10 Examples of single-channel currents recorded by the patch-clamp technique. In this example, the ion channel opens (downward deflection) when it has bound acetylcholine. The record is of the activity of a nicotinic receptor channel in a chromaffin cell. (From an original figure of Dr P. Charlesworth.)

(1₂3) Box 5.3 The Goldman equation and changes in membrane potential during cell activity

If the membrane potential was determined solely by the distribution of potassium across the membrane it should be exactly equal to the potassium equilibrium potential. In practice it is found that while the resting membrane potential is close to the potassium equilibrium potential it is seldom equal to it. Instead, it is usually somewhat less negative. Moreover, during periods of activity the membrane potential may be very different to the potassium equilibrium potential. For example, during the secretory response of an exocrine cell the membrane potential may be very low (i.e. it will be *depolarized*) and during the peak of a nerve action potential it is actually positive.

A modified form of the Nernst equation known as the Goldman constant field equation (or, more simply, the Goldman equation) can explain these dynamic aspects of membrane potential behaviour. The equation takes into account not only the ionic gradients that exist across the membrane but also the permeability of the membrane to the different ions. The equation is:

$$E = \frac{RT}{F} \ln \frac{P_{Na}[Na^+]_o + P_K[K^+]_o + P_{Cl}[Cl^-]_i}{P_{Na}[Na^+]_i + P_K[K^+]_i + P_{Cl}[Cl^-]_o}$$

R, T, and F are physical constants (the gas constant, the absolute temperature, and the Faraday constant), while P_{Na}, P_K, and P_{Cl} are the permeability coefficients of the membrane to Na^+, K^+, and Cl^- respectively. $[Na^+]_o$, $[Na^+]_i$, etc. are the extracellular and intracellular concentrations of Na^+, K^+, and Cl^-. Note that as Cl^- has a negative charge, $[Cl^-]_i$ and $[Cl^-]_o$ are reversed compared to the positively charged ions: $[Na^+]_i$, $[Na^+]_o$ and $[K^+]_i$, $[K^+]_o$.

As the resting membrane is much more permeable to K^+ than to Na^+ ($P_{Na}/P_K \approx 0.01$), the resting membrane potential lies close to the potassium equilibrium potential. (Remember that the Cl^- gradient is determined by the K^+ gradient, as discussed earlier in Section 5.4). During an action potential, the membrane becomes much more permeable to Na^+ than to K^+ (at the peak of the action potential $P_{Na}/P_K \approx 20$) and the membrane potential is closer to the sodium equilibrium potential, which is positive (+53 mV). Thus the Goldman equation shows how the membrane potential of a cell can be altered by changes in the relative permeability of its membrane to Na^+, K^+, Cl^-, and other ions without any significant change in the ionic gradients themselves.

---

much greater than its permeability to sodium. When the cell is stimulated, more sodium channels open and the permeability of the membrane to sodium increases relative to that of potassium. As the sodium equilibrium potential is positive (about +50 mV), the membrane potential becomes depolarized and this triggers the opening of voltage-gated calcium channels. The calcium concentration within the cell rises and this initiates the secretory response.

Summary

The potassium gradient generated by the activity of the sodium pump gives rise to a membrane potential: the cell interior is negative with respect to the outside. The membrane potential of a resting cell generally lies between –40 and –90 mV. The exact value of the membrane potential is determined by the ion gradients across the plasma membrane and by the number and type of ion channels that are open. The activity of many cells is governed by changes in their membrane potential. Ion channels may be opened by a chemical signal (ligand-gated channels) or by a change in membrane potential (voltage-gated channels).

5.5 Secretion, exocytosis, and endocytosis

Many cells release molecules that they have synthesized into the extracellular environment. This is known as **secretion**. It may be more or less continuous (constitutive secretion) or triggered by a specific physiological event (calcium-dependent secretion, sometimes called regulated or stimulated secretion). Constitutive

secretion is common to all cells and is the process by which they are able to insert newly synthesized lipids and proteins, such as carriers and ion channels, into their plasma membranes (see Figure 5.11). Although the term 'constitutive secretion' implies that it is continuous, constitutive secretion is nevertheless tightly regulated. The key difference between constitutive secretion and calcium-dependent secretion is the time frame over which their regulation occurs: many minutes for constitutive secretion, compared to seconds or fractions of a second for regulated secretion.

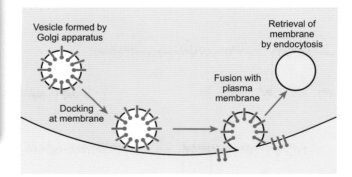

Figure 5.11 The principal phases of constitutive secretion. The Golgi apparatus forms vesicles. The fully formed vesicles move towards the plasma membrane, become closely apposed to the plasma membrane ('docking'), and finally fuse with it. After fusion of the two membranes, the proteins embedded in the vesicular membrane and their associated lipids are able to diffuse into the plasma membrane itself. Note that the orientation of the proteins is preserved so that the protein on the inside of the vesicle becomes exposed to the extracellular space.

Secretion can occur by simple diffusion through the plasma membrane—the secretion of steroid hormones by the cells of the adrenal cortex occurs in this way. This method of secretion is, however, limited to those molecules that can penetrate the hydrophobic lipid barrier of the cell membrane, such as the steroid and thyroid hormones. Polar molecules (e.g. digestive enzymes) are packaged in membrane-bound vesicles, which can fuse with the plasma membrane to release their contents (often called their cargo) into the extracellular space in a process called **exocytosis.** Exocytosis leads to the insertion of new material into the plasma membrane, increasing the surface area of the cell. This increase in surface area is balanced by the subsequent retrieval of membrane components via a process called **endocytosis** (discussed later in this section). An overview of how exocytosis and endocytosis relate to each other in the vesicle cycle is shown in Box 5.4.

The mechanism of exocytosis

Exocytosis of secretory vesicles is a complex process that is currently the subject of much ongoing research. It can be broken down into the following stages:

1. the formation of vesicles

2. the filling of vesicles with their cargo (i.e. the kind of molecule they transport)

3. the movement of filled vesicles to the plasma membrane ('docking')

 Box 5.4 The secretory vesicle cycle

Secretory vesicles undergo a series of changes as they form, empty their contents, and are retrieved. The main stages, illustrated in Figure 1, are as follows.

1. **Docking.** Filled vesicles are actively delivered to the site of their eventual exocytosis. This involves both cytoskeletal-based transport, and tethering of the vesicle at the eventual site of exocytosis through interactions between vesicle and plasma membrane proteins.

2. **Triggering.** Calcium-dependent exocytosis is triggered by an elevation of calcium in the cytosol surrounding the release site. This calcium activates the release machinery.

3. **Formation of fusion pore.** The SNARE complex forms, pulling vesicle and plasma membranes into close proximity, and permitting the formation of a pore linking the inside of the vesicle to the extracellular space. Release of the vesicle contents can then begin.

4. **Full collapse of vesicle.** In many cases, the vesicle will collapse into the plasma membrane. If this happens immediately following a fusion pore, the bulk of the vesicle contents will be released into the extracellular space.

5. **Endocytic trigger.** Endocytosis following secretion and in response to signalling events follows a similar mechanism. A plasma membrane protein undergoes a conformational change, often involving phosphorylation, which provides the trigger for endocytosis.

6. **Adaptor recruitment.** This triggering protein recruits adaptor proteins that bind to the cytoplasmic domain of the protein in response to the conformational change.

7. **Coat formation.** In turn, the adapter proteins recruit clathrin heavy and light chains which assemble to form clathrin-coated pits budding inwards on the plasma membrane.

8. **Endocytosis.** The clathrin-coated pit completes its budding and separates from the plasma membrane to form a clathrin-coated vesicle. A protein known as dynamin is responsible for the final separation of the two membranes.

9. **Uncoating.** The clathrin coat dissolves to leave an uncoated endocytic vesicle which can be refilled for another round of secretion (in the case of synaptic vesicles) or directed to an endosome for sorting of its plasma membrane components (in the case of other endocytic vesicles.)

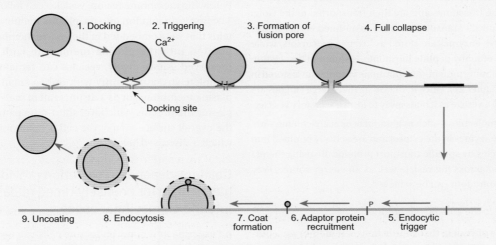

Figure 1 Changes in secretory vesicles as they form, empty their contents, and are retrieved.

4. the release of the vesicles' contents by fusion with the plasma membrane

5. the retrieval of the vesicle components after membrane fusion.

Vesicle formation

Cells are full of different types of vesicles, shuttling between different membrane compartments such as the cisternae of the Golgi apparatus. (See Chapter 4 for more details of the Golgi apparatus and other aspects of cell structure.) Vesicles that are destined for the plasma membrane are considered to be part of the secretory pathway.

The formation of secretory vesicles begins when part of the endoplasmic reticulum buds off to form specialized transport vesicles that carry newly synthesized proteins to the Golgi apparatus, where they may undergo further modification. The composition of vesicles destined to become part of the plasma membrane is determined during these steps and defines the different kinds of vesicle. The secretory vesicles subsequently bud off from the *trans*-face of the Golgi apparatus and migrate towards the plasma membrane, where they are stored prior to their secretion in response to an appropriate stimulus. The delivery of plasma membrane proteins such as glucose transporters or growth factor receptors occurs by constitutive secretion, while the secretion of digestive enzymes, neurotransmitters, and many hormones occurs in response to specific signals, i.e. by stimulated secretion.

There are many different types of secretory vesicle, which differ in size from around 40 nm (e.g. the small, clear, synaptic vesicles that deliver neurotransmitters like acetylcholine and glutamate) to a micron or more in diameter, such as the histamine-containing granules of mast cells (see Section 26.5).

Vesicle contents

Many different cargoes are delivered to the plasma membrane by vesicles. Membrane proteins, as mentioned above, are inserted in the endoplasmic reticulum. The cargoes contained within the sack-like boundary of the vesicle membrane (the vesicle lumen) enter by two routes.

- Large proteins and peptide hormones are synthesized as precursor molecules by the ribosomes and are then transported to the Golgi apparatus via specific transport vesicles, as outlined earlier in this section. Digestive enzymes are stored as inactive precursors, while the active form of many peptide hormones is stored in association with a specific binding protein—the hormone vasopressin is stored in this way (see Chapter 21). Once in the Golgi, these large cargoes are encapsulated by a vesicle as it buds away to form a secretory vesicle.

- Small secreted molecules—such as glutamate or acetylcholine—are synthesized by enzymes in the cytosol and are actively pumped into preformed vesicles by specific transport proteins. By using the pH gradient present across the vesicle wall as an energy source, very high final concentrations can be achieved.

Vesicle transport to the plasma membrane

The various kinds of vesicle that shuttle between membrane compartments are identified by protein labels present on their outer surface and routed to the appropriate region of the cell. The secretory vesicles are transported towards the plasma membrane by an active process in which they interact with the actin cytoskeleton and are eventually positioned close to the plasma membrane in preparation for eventual secretion, a process often called 'docking'. In a typical small cell, secretory vesicle transport involves a short movement of a few microns. In contrast, nerve cells make contact with their target tissues via a long, thin process called an axon. This poses a particular difficulty for nerve cells, as the secretory vesicles (called synaptic vesicles) must be transported considerable distances to their point of secretion; this distance may be up to a metre in the case of the motor nerves serving the muscles of the lower regions of the leg. To do this, nerve cells make use of a process called **axonal transport**, in which a molecular motor called kinesin carries precursor synaptic vesicles to the nerve ending along the microtubules that run the length of the axon (see Chapters 4 and 7).

Membrane fusion

The final stage of vesicle secretion is the process of exocytosis, in which the vesicles fuse with the plasma membrane to release their contents, a process that is tightly regulated. In response to fast calcium signals, or to slower kinase-dependent signalling cascades, membrane fusion occurs both *when* and *where* it should (i.e. at the right time, in the right place). This is achieved through the interactions of proteins that catalyse the membrane fusion reaction itself, which form the **SNARE** complex (**S**oluble **N**SF **A**ttachment **P**rotein **RE**ceptor); this is formed from proteins contributed by both the vesicle and the plasma membranes. The action of this 'fusion machine' enables the two lipid bilayers to overcome the considerable repulsive electrostatic forces that push membranes apart (the headgroups of the lipids of both membranes carry negative charges), and allows the vesicle and plasma membrane to fuse together. Certain cells use additional proteins to provide this complex with a responsiveness to calcium, and a greater speed of response. After membrane fusion, the contents of the vesicles are released into the extracellular space, where they are able to influence their target cells.

The fate of vesicles after secretion

Following membrane fusion, vesicles can follow one of two paths. The first is dissolution into the plasma membrane, where any proteins that were embedded in the vesicle membrane become incorporated into the plasma membrane with their orientation preserved. Alternatively, vesicles can remain intact, either as a complete structural entity with only a small connection to the plasma membrane, or as a distinct raft of material 'floating' in the plasma membrane. Whichever short-term pathway a vesicle takes, the overall surface area of the cell is maintained by endocytosis, which is discussed below.

Calcium-dependent secretion provides cells with a mechanism for precisely timed release of molecules into the extracellular space

All exocrine and many endocrine cells use regulated secretion to control the timing and rate of release of their vesicles into the extracellular space. The signal that triggers the secretion of a particular substance (e.g. a hormone or digestive enzyme) may be precisely timed or it may vary continuously. One way of achieving precise timing of secretion is to use the activity of a specific nerve. For example, the enzyme amylase is secreted by the acinar cells of the salivary glands in response to nerve impulses in the salivary nerves. In some other cases, a hormone may act to stimulate secretion. An example is the regulation of the secretion of the digestive enzymes of the pancreas by the hormone cholecystokinin (CCK)—see Chapter 44. Many hormonal secretions are themselves regulated by the concentration of a substance that is continuously circulating in the plasma, which may be another hormone or some other chemical constituent of the blood. For example, the glucose concentration in the plasma regulates the secretion of insulin by the β-cells of the pancreas.

To control the secretory event, the extracellular signal must be translated into an intracellular signal that can regulate the rate at which the preformed secretory vesicles will fuse with the plasma membrane. In many types of cell, regulated exocytosis is triggered by an increase in the concentration of ionized calcium within the cytosol. This occurs via two processes:

1. entry of calcium through calcium channels in the plasma membrane

2. release of calcium from intracellular stores (mainly the endoplasmic reticulum).

The entry of calcium ions into the cytosol usually occurs through voltage-gated calcium channels that open following depolarization of the plasma membrane in response to a chemical signal. Calcium is then able to diffuse into the cell down its electrochemical gradient. However, it may also be mobilized from internal stores following activation of **G protein** coupled receptor systems (see Chapter 6). In both cases the intracellular free calcium is increased and this triggers the fusion of docked secretory vesicles with the plasma membrane. As fusion proceeds, a pore is formed that connects the extracellular space with the interior of the vesicle. This pore provides a pathway for the contents of the vesicle to diffuse into the extracellular space (see Figure 5.12).

Many secretory cells are polarized so that one part of their membrane is specialized to receive a signal (e.g. a hormone) while another region is adapted to permit the fusion of secretory vesicles. The acinar cells of exocrine glands (e.g. those of the pancreas) provide a clear example of this zonation. The basolateral surface receives chemical signals from circulating hormones or from nerve endings. These signals activate receptors that control the secretory response (see Chapter 6). Exocytosis occurs at the apical surface of the cells, which is in direct communication with the secretory duct.

Endocytosis is used by cells to retrieve components of the plasma membrane and to take up macromolecules from the extracellular space

When cells undergo exocytosis, their cell membrane increases in area as the vesicular membrane fuses with the plasma membrane. This increase in area is offset by membrane retrieval known as

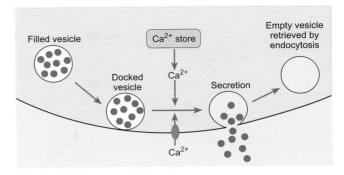

Figure 5.12 A rise in intracellular Ca²⁺ can trigger regulated secretion. In exocrine cells and many endocrine cells, secretion is dependent on extracellular Ca²⁺. Following stimulation, the cell depolarizes and this opens Ca²⁺ channels. The Ca²⁺ ions enter and cause docked vesicles to fuse with the plasma membrane and release their contents. Calcium ions can also be released into the cytoplasm from intracellular stores to initiate secretion—this calcium is derived mainly from the endoplasmic reticulum.

endocytosis, in which small areas of the plasma membrane are pinched off to form endocytotic vesicles (also called endocytic vesicles). This compensatory endocytosis occurs in response to high levels of exocytosis (e.g. at a synapse) and is required to keep the surface area more or less constant. Endocytotic vesicles are generally small (less than 150 nm in diameter) and may contain macromolecules derived from the extracellular space.

All the cells of the body continuously undergo membrane retrieval via endocytosis, as this is an important process for the regulation of cell surface receptors. Since the formation of the vesicles traps some of the extracellular fluid, this process is also known as **pinocytosis** ('cell drinking'). Pinocytosis and endocytosis are often employed as interchangeable terms. Some proteins and other macromolecules are absorbed with the extracellular fluid (**fluid phase endocytosis**). In other cases, proteins bind to specific surface receptors. For example, the cholesterol required for the formation of new membranes is absorbed by cells via the binding of specific carrier protein–cholesterol complexes (low density lipoproteins) to a surface receptor (**receptor-mediated endocytosis**). In receptor-mediated endocytosis, activation of a cell surface receptor causes structural changes in the protein on the cytosolic side, which induce internalization.

Pathways of endocytosis

Endocytosis proceeds by multiple mechanisms, but the best-characterized route is called **clathrin-mediated endocytosis**. This form of endocytosis has several distinct steps which are summarized in Figure 5.13, and, as the name suggests, it uses a scaffolding protein called clathrin to provide a physical framework to help the membrane bud into the cell.

The endocytotic vesicles fuse with larger vesicles known as **endosomes**, which are located either just beneath the plasma membrane (the peripheral endosomes) or near the cell nucleus (the perinuclear endosomes). The perinuclear endosomes fuse with vesicles containing lysosomal enzymes derived from the Golgi

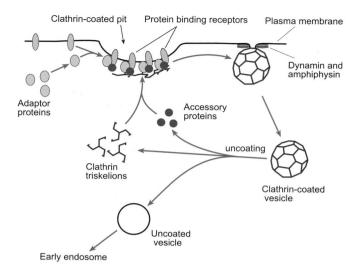

Figure 5.13 The stages of clathrin-mediated endocytosis. The structure of the cytoplasmic domain of a cell-surface receptor protein changes when it binds its specific ligand. This change is followed by recruitment of adaptor proteins which connect the membrane receptor to the clathrin scaffold. The adaptor proteins then recruit clathrin, which forms a molecular scaffold for inward budding of the cell membrane. As the clathrin assembles into a meshwork, the membrane buds into the cell to form a clathrin-coated pit, after which the protein dynamin acts to cut the membrane off from the plasma membrane. The endocytosed membrane is now a clathrin-coated vesicle. Once separated from the plasma membrane, chemical changes in the membrane cause the dissolution of the clathrin complex, leaving an endocytic vesicle, which is then trafficked by the appropriate pathway.

apparatus to form endolysosomes. Through a process of maturation which involves the removal (through budding and separation) of membrane components that are to be recycled back to the plasma membrane, combined with acidification of the vesicle contents, the endolysosomes become transformed into lysosomes. The interior of the lysosomes is acidified by the activity of an ATP-driven proton pump. This acidic environment allows the lysosomal enzymes to break down the macromolecules trapped during endocytosis into their constituent amino acids and sugars. These small molecules are then transported out of the lysosomes to the cytosol, where they may be re-incorporated into newly synthesized proteins.

Not all endocytosed molecules are broken down by the lysosomes. Some elements of the plasma membrane that are internalized during endocytosis are eventually returned to the plasma membrane by transport vesicles. This is often the case with membrane lipids, carrier proteins, and receptors. Indeed in many cases, cells regulate the number of active carriers or receptors by adjustment of the rate at which these proteins are removed and reinserted into the plasma membrane. For example, when the body needs to conserve water, the cells of the collecting duct of the kidney increase their permeability to water by increasing the rate at which water channels (aquaporins) are inserted into the apical membrane by transport vesicles (see Chapter 39 p. 608).

Endocytosis allows the proteins expressed on the cell surface to be sorted according to the cellular need. For example, during growth (when there is a need for cells to divide in response to external cues), growth factor receptors are expressed on the cell surface, but at other times the number of receptors is reduced to eliminate over-growth. This explains why both growth factor receptors and endocytic proteins are implicated in various cancers.

Phagocytosis is important both for defence against infection and for the retrieval of material from dead and dying cells

Phagocytosis ('cell eating') is a specialized form of endocytosis in which large particles (e.g. bacteria or cell debris) are ingested by cells. Within the mammalian body, this activity is confined to specialized cells called **phagocytes**. The principal cells that perform this function in humans and other mammals are the **neutrophils** of the blood and the **macrophages**, which are widely distributed throughout the body (see Chapters 25 and 26). Together these cells form the **reticulo-endothelial system**.

Unlike endocytosis, phagocytosis is triggered only when receptors on the surface of the cell bind to the particle to be engulfed. Both macrophages and neutrophils have a rich repertoire of receptors, but any particle coated with an antibody is avidly ingested. Phagocytosis occurs by the extension of processes known as pseudopodia around the particle. The extension of the pseudopodia is guided by contact between the particle and the cell surface. When the pseudopodia have engulfed the particle, a membrane-bound **phagosome** is formed around the particle; its size is determined by that of the ingested particle. The phagosomes fuse with lysosomes to form **phagolysosomes**, which are able to digest the ingested material (Figure 5.14) and transport its metabolically useful components to the cytosol. The material that cannot be digested remains in the phagocyte as a membrane-bound particle called a residual body.

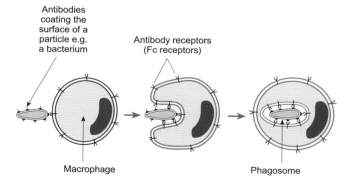

Figure 5.14 The stages of phagocytosis. A bacterium or other particle must first adhere to the surface of a phagocyte before it can be engulfed. This is achieved by linking surface receptors on the phagocyte with complementary binding sites on the microbe (e.g. the Fc region of antibodies manufactured by the host—see Chapter 26). The microbe is then engulfed by a zipper-like action until it becomes fully enclosed in a vacuole before being digested in a phagolysosome (not shown).

Not all the surface protein of the foreign particle is digested. Some is combined with cell surface glycoproteins and inserted into the plasma membrane of the phagocyte. This exposes part of the foreign protein to scrutiny by the T-lymphocytes, which are then able to signal the immune system to increase the production of the appropriate antibody (see Chapter 26).

Although phagocytes play an important role in protecting the body from infection by ingesting bacteria, this activity is dwarfed by their role in the scavenging of dying cells and cellular debris. For example, about 1 per cent of red cells die and are replaced each day. This requires the phagocytes of the reticulo-endothelial system to ingest about 10^{11} red cells each day. The signal that triggers this response appears to be a change in the composition of the outer leaflet of the lipid bilayer. As red cells age, phosphatidyl serine appears in the outer leaflet (it is normally found in the inner leaflet) and induces the macrophages to remove these senescent cells from the circulation.

Summary

Cells release materials that they have synthesized into the extracellular space by means of secretion. Lipid-soluble materials such as steroid hormones are secreted by diffusion across the plasma membrane, but water-soluble molecules are secreted by exocytosis—a process by which vesicles containing the secretory material fuse with the plasma membrane to release their contents. Two pathways exist: constitutive secretion, which operates virtually continuously, and regulated secretion, which occurs in response to specific signals. The principal signal for regulated secretion is a rise in intracellular Ca^{2+}. The vesicular membrane that became incorporated into the plasma membrane during fusion is later retrieved by endocytosis. Specialized cells (phagocytes) engulf foreign particles and cell debris by phagocytosis. The ingested material is digested in vacuoles called phagolysosomes.

✳ Checklist of key terms and concepts

Transport of molecules across the plasma membrane

- Lipid-soluble molecules are able to cross lipid membranes.
- Lipid membranes are relatively impermeable to water-soluble molecules.
- Two groups of membrane proteins facilitate the movement of water and water-soluble molecules from one side of the plasma membrane to the other: the membrane channels and the carrier proteins.
- Ion channels permit the passage of ions from one side of the membrane to the other via a pore.
- Ion channels show selectivity for specific ions. They are named after the principal ions to which they are permeable.
- Carrier molecules permit relatively large molecules to pass from one side of the membrane to the other by binding a molecule on one face of the membrane and undergoing a conformational change to release it on the other.
- Carrier molecules exhibit stereoselectivity for the molecules they are able to transport.

Active and passive transport across cell membranes

- Passive transport occurs when molecules and ions cross the plasma membrane by diffusion down their electrochemical gradients.
- Passive transport may be mediated by ion channels or by carrier molecules.
- The net movement of ions across the membrane via ion channels is always down an electrochemical gradient via a pore. This is an example of passive transport.

- Active transport occurs when a cell transports an ion or molecule against the prevailing electrochemical gradient.
- Active transport requires the expenditure of metabolic energy, either directly or indirectly.
- The sodium pump is a carrier protein that hydrolyses ATP to provide the energy for moving Na^+ against the prevailing electrochemical gradient. This is an example of primary active transport.
- The use of the Na^+ gradient to provide the energy to move another molecule (e.g. glucose) against its concentration gradient is known as secondary active transport.
- Cells use carriers in combination to regulate the intracellular concentration of certain ions. Thus, the intracellular concentration of calcium ions is regulated by Na^+/Ca^{2+} exchange, by a Ca^{2+} pump, and by uptake into intracellular stores.
- Intracellular pH is regulated by many mechanisms, including Na^+/H^+ exchange, $Na^+ - HCO_3^-$ cotransport, and $Cl^- - HCO_3^-$ exchange.

The membrane potential

- The activity of the sodium pump allows cells to accumulate potassium ions.
- The resulting potassium gradient gives rise to a potential difference across the plasma membrane where the cell interior is negative with respect to the outside.
- At any instant, the membrane potential is determined by the ion gradients across the plasma membrane and by the number and kinds of ion channels that are open.
- The membrane potential of a resting cell varies with the cell type but generally lies between −40 and −90 mV.

- The direction in which ions and charged molecules diffuse across the cell membrane is determined by their electrochemical gradients.

- The electrochemical gradient of a molecule or ion is determined by the concentration gradient of the molecule or ion in question, its electrical charge, and the membrane potential.

- The electrochemical gradient can be calculated from the difference between the equilibrium potential for the ion in question and the membrane potential.

- The activity of many cells is governed by changes in their membrane potential, which are determined by the opening or closing of various ion channels.

- Ion channels may be opened as a result of binding a chemical (ligand-gated channels) or as a result of depolarization of the cell membrane (voltage-gated channels).

Secretion, endocytosis, and phagocytosis

- Secretion is the process by which cells release newly synthesized materials into the extracellular space.

- Secretion of lipid-soluble materials such as steroid hormones occurs by diffusion across the cell membrane.

- Small water-soluble molecules and macromolecules are secreted by exocytosis—a process by which vesicles containing the secreted material fuse with the cell membrane to release their contents.

- Secretion may be constitutive, i.e. it operates virtually continuously, or it may be triggered by a specific signal.

- When secretion is initiated by an agonist or a change in membrane potential, it leads to a rise in intracellular Ca^{2+} which then triggers the process of exocytosis.

- The vesicular membrane that became incorporated into the plasma membrane during fusion is later retrieved by endocytosis.

- Specialized cells called phagocytes engulf foreign particles and cell debris.

- The process of phagocytosis is triggered when receptors on the cell surface recognize specific proteins on the surface of a foreign particle.

- After internalization, the ingested material is digested in vacuoles called phagolysosomes.

Recommended reading

Biochemistry and cell biology

Alberts, B., Johnson, A., Lewis, J., Morgan, D., Raff, M., Roberts, K., and Walter, P. (2014) *Molecular Biology of the Cell* (6th edn), Chapters 11–13. Garland, New York.

Berg, J.M., Tymoczko, J.L., and Stryer, L. (2011) *Biochemistry* (7th edn), Chapter 13. Freeman, New York.

Biophysics

Kew, J., and Davies, C. (eds) (2010) *Ion channels: From structure to function*. Oxford University Press, Oxford.
An encyclopaedic tour of ion channels and their properties.

Zheng, J., and Trudeau, M.C. (2015) *Handbook of ion channels*. CRC Press, Boca Raton, FL.
A comprehensive review of ion channels including basic principles, methods of study, and channel regulation.

To check that you have mastered the key concepts presented in this chapter, complete the accompanying online self-assessment questions. Go to www.oup.com/uk/pocock5e/

CHAPTER 6
Principles of cell signalling

Chapter contents

This chapter should help you to understand:

- The need for cell signalling
- The differing roles of paracrine, endocrine, and synaptic signalling
- How receptors in the plasma membrane regulate the activity of target cells
- The roles of cyclic AMP and inositol trisphosphate (IP_3) as second messengers
- The roles of inositol lipids and monomeric G proteins in cell signalling
- How steroid and thyroid hormones control gene expression via intracellular receptors
- The role of cell surface proteins in cell–cell adhesion and cell recognition
- The functions of gap junctions between cells

6.1 Introduction

Individual cells are specialized to carry out specific physiological roles such as secretion or contraction. In order to coordinate their activities they need to receive and transmit signals of various kinds. As the plasma membrane forms a barrier between the signalling components within the cell and the extracellular environment, various ways around this have evolved. These adaptations fall into five groups:

1) the generation of diffusible chemical signals;
2) the expression of receptors in the plasma membrane that are able to convey the external signals across the plasma membrane;
3) direct contact between the plasma membrane proteins of adjacent cells;
4) direct cytoplasmic contact via gap junctions;
5) the use of signals generated by a cell itself at particular stages of development.

Diffusible chemical signals allow cells to communicate at a distance, while direct contact between cells is particularly important in cell–cell recognition during development and during the passage of lymphocytes through the tissues (diapedesis—see Chapter 25). Direct cytoplasmic contact between neighbouring cells via gap junctions permits the electrical coupling of cells and plays an important role in the spread of excitation between adjacent cardiac muscle cells. It also allows the direct exchange of chemical signals between adjacent cells.

6.2 Cells use diffusible chemical signals for paracrine, endocrine, and synaptic signalling

Cells release a variety of chemical signals. Some are local mediators that act on neighbouring cells, reaching their targets by diffusion over relatively short distances (up to a few mm). This is known as **paracrine signalling**. When the secreted chemical also acts on the cells that secreted it, the signal is said to be an **autocrine signal**. Frequently, substances are secreted into the blood by specialized glands—the **endocrine glands**—to act on various tissues around the body. The secreted chemicals themselves are called **hormones**. Finally, nerve cells release chemicals at their endings to affect the cells they contact. This is known as **synaptic signalling**.

Local chemical signals—paracrine and autocrine secretions

Paracrine secretions are derived from individual cells rather than from a collection of similar cells or a specific endocrine gland. They act on cells close to the point of secretion and have a local effect (see Figure 6.1a). The signalling molecules are rapidly destroyed by extracellular enzymes or by uptake into the target cells. Consequently very little of the secreted material enters the blood. For this reason they are sometimes called local hormones.

One example of paracrine signalling is provided by mast cells which are found in connective tissues all over the body. They have large secretory granules that contain histamine, which is secreted in response to injury or infection. The secreted histamine dilates the local arterioles and this results in an increase in the local blood flow. In addition, the histamine increases the permeability of the nearby capillaries to proteins such as **immunoglobulins**. However, this increase in capillary permeability must be local and not widespread, otherwise there would be a significant loss of protein from the plasma to the interstitial fluid, which could result in circulatory failure (see Chapter 40 p. 625). The mast cells also secrete small peptides that are released with the histamine to stimulate the invasion of the affected tissue by white blood cells—phagocytes and eosinophils. These actions form part of the inflammatory response and play an important role in halting the spread of infection (see Chapter 26 p. 395).

The inflammatory response is also associated with an increase in the synthesis and secretion of a group of local chemical mediators called the prostaglandins (see Section 6.5). The prostaglandins secreted by a cell act on neighbouring cells to stimulate them to produce more prostaglandins. This is another example of paracrine signalling. In addition, the secreted prostaglandin stimulates further prostaglandin production by the cell that initiated the response. Here the prostaglandin is acting as an **autocrine signal**. This autocrine action amplifies the initial signal and may help it to spread throughout a population of cells so ensuring a rapid mobilization of the body's defences in response to injury or infection.

Hormonal secretions provide a means of diffuse, long-distance signalling to regulate the activity of distant tissues

Hormones play an extensive and vital role in regulating many physiological processes and their physiology will be discussed at length in subsequent chapters. Here their role as cell signals will be discussed only in general terms.

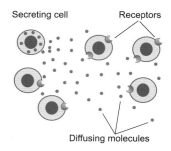

(a)

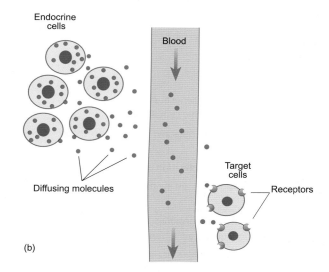

(b)

Figure 6.1 A comparison between paracrine and endocrine signalling. (a) Schematic diagram illustrating paracrine secretion. The secreting cell releases a signalling molecule (e.g. a prostaglandin) into the extracellular fluid from where it reaches the receptors of neighbouring cells by diffusion. (b) The chief features of endocrine signalling. Endocrine cells secrete a hormone into the extracellular space in sufficient quantities for it to enter the bloodstream from where it can be distributed to other tissues. The target tissues may be at a considerable distance from the secreting cells.

Endocrine cells synthesize hormones and secrete them into the extracellular space, from where they are able to diffuse into the blood. Once in the general circulation, a hormone will be distributed throughout the body and thus is able to influence the activity of tissues remote from the gland that secreted it. In this important respect hormones differ from local chemical mediators. The secretion and distribution of hormones is shown schematically in Figure 6.1b.

Endocrine glands secrete hormones in response to a variety of signals.

1) They may respond to the level of some constituent of the blood. For example, insulin secretion from the β-cells of the islets of Langerhans is regulated by the blood glucose concentration.

splanchnic nerve binds to nicotinic receptors on the plasma membrane of the chromaffin cells. This increases the permeability of the membrane to sodium ions (Na⁺), causing the membrane to become depolarized. The depolarization results in the opening of voltage-gated calcium channels, allowing calcium ions to flow down their concentration gradient into the cell via these channels. The intracellular Ca^{2+} concentration rises and triggers the secretion of adrenaline. This complex sequence of events is summarized in Figure 6.4.

Activation of catalytic receptors

Catalytic receptors are membrane-bound protein kinases that become activated when they bind their specific ligand. (A kinase is an enzyme that adds a phosphate group to its substrate—which can be another enzyme.) A typical example of a catalytic receptor is the insulin receptor that is found in liver, muscle, and fat cells. This receptor is activated when insulin binds it and in turn it activates other enzymes by adding a phosphate group to tyrosine residues. Consequently, these receptors are called receptor tyrosine kinases. In the case of the insulin receptor, activation results in an increase in the activity of the affected enzymes which culminates in an increase in the rate of glucose uptake. Many peptide hormone and growth factor receptors are tyrosine specific kinases.

G protein linked receptors

G proteins link receptor activation directly to the control of a second messenger. Second messengers are synthesized in response to the activation of specific receptors to transmit the signal from the plasma membrane to particular enzymes (e.g. cyclic AMP) or intracellular receptors (e.g. inositol trisphosphate, which is abbreviated as IP_3). The series of events linking the change in the level of the second messenger to the final response is called a signalling cascade.

GTP binding regulatory proteins or G proteins are a specific class of membrane-bound regulatory proteins that are activated when a receptor binds its specific ligand. The receptor-linked G proteins

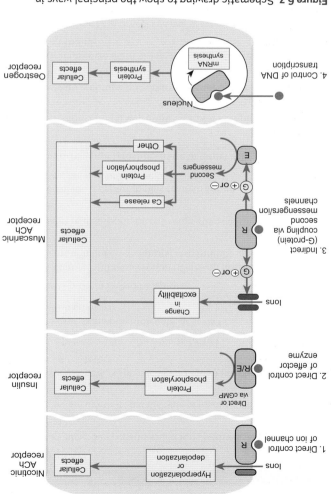

Figure 6.3 Schematic drawing to show the principal ways in which chemical signals affect their target cells. Examples of each type of coupling are shown at the foot of the figure. R = receptor, E = enzyme, G = G protein; + indicates increased activity, − = decreased activity. (Based on Fig 2.3 in H.P. Rang, M.M. Dale, and J.M. Ritter (1995) *Pharmacology*, 3rd edition. Churchill-Livingstone, Edinburgh.)

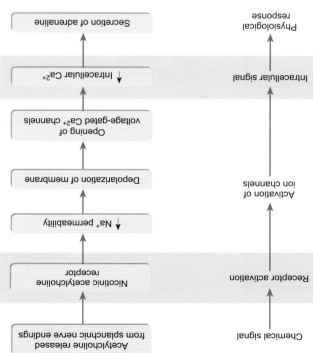

Figure 6.4 The sequence of events that links activation of a nicotinic receptor to the secretion of the hormone adrenaline (epinephrine) by adrenal chromaffin cells. Acetylcholine binds to a nicotinic receptor on the plasma membrane of the chromaffin cell and opens an ion channel. This increases the permeability of the membrane to Na⁺ and the membrane depolarizes. The depolarization activates voltage-gated Ca^{2+} channels and Ca^{2+} ions enter the cell to trigger the secretion of adrenaline by exocytosis.

Table 6.1 **Examples of signalling molecules used in cell–cell communication**

Class of molecule	Specific example	Physiological role and mode of action
A—Signalling molecules secreted by fusion of membrane-bound vesicles (exocytosis)		
Ester	Acetylcholine	Synaptic signalling molecule; opens ligand-gated ion channels; also activates G protein linked receptors
Amino acid	Glycine	Synaptic signalling molecule; opens a specific type of ligand-gated ion channel
	Glutamate	Synaptic signalling molecule; affects specific ligand-gated ion channels; also activates G protein linked receptors
Amine (biogenic)	Adrenaline (epinephrine)	Hormone; wide variety of effects; acts via G protein linked receptors
	5-Hydroxytryptamine (5-HT, serotonin)	Local mediator and synaptic signalling molecule; acts via G protein linked receptors and activates ion channels
	Histamine	Local mediator; acts via G protein linked receptors
Peptide	Somatostatin	Hormone and local mediator; inhibits secretion of growth hormone by the anterior pituitary via G protein linked receptors; also inhibits gastrin secretion by the stomach
	Vasopressin (antidiuretic hormone: ADH)	Hormone; increases water reabsorption by collecting ducts of the kidney via G protein linked receptor
Protein	Insulin	Hormone; activates a catalytic receptor in plasma membrane; increases uptake of glucose by liver, fat, and muscle cells
	Growth hormone	Hormone; activates a non-receptor tyrosine kinase in target cell
B—Signalling molecules that can diffuse through the plasma membrane		
Steroid	Oestradiol-17β	Hormone; binds to nuclear receptors; hormone–receptor complex regulates gene expression
Thyroid hormone	Tri-iodothyronine (T_3)	Hormone; binds to nuclear receptors; hormone–receptor complex regulates gene expression
Eicosanoid	Prostaglandin E_2 (PGE_2)	Local mediator; diverse actions on many tissues; activates G protein linked receptors in plasma membrane
Inorganic gas	Nitric oxide	Local mediator; acts by binding to guanylyl cyclase in target cell

How do receptors control the activity of the target cells?

Cells respond to chemical signals by initiating an appropriate physiological response. This process is called *transduction*. There are four basic ways in which activation of a receptor can alter the activity of a cell (see Figure 6.3).

- First, it may open a linked ion channel and transduce the signal through a change in membrane potential, or through the entry of calcium ions (which have specific signalling functions independent of their effect on membrane potential).
- Second, it may activate a membrane-bound enzyme (a catalytic receptor).
- Third, it may activate a G protein linked receptor that may modulate an ion channel or change the intracellular concentration of a specific chemical called a **second messenger**—the original signalling molecule is the first messenger.
- Finally, the signal may act on an intracellular receptor to modulate the transcription of specific genes.

Ion channel modulation

Many receptors are directly coupled to ion channels. These receptor-channel complexes are called **ligand-gated ion channels** (see Chapter 5) and they are employed by cells to regulate a variety of functions. In general, ligand-gated channels open for a short period of time following the binding of their specific agonist, and this transiently alters the membrane potential of the target cell and thereby modulates its physiological activity. In some instances, a change in the membrane potential of the target cell is the final response. This is the situation when one nerve cell inhibits the activity of another (see Chapter 7, p. 110). More often the change in membrane potential often triggers some further event. Thus, in many cases, activation of ligand-gated channels causes the target cell to depolarize. This depolarization then activates voltage-gated ion channels that trigger the appropriate cellular response. In this way the activation of a ligand-gated ion channel can be used to control the activity of even the largest of cells.

This pattern of events is illustrated by the stimulatory effect of acetylcholine on adrenaline secretion by the chromaffin cells of the adrenal gland. Acetylcholine released by the terminals of the

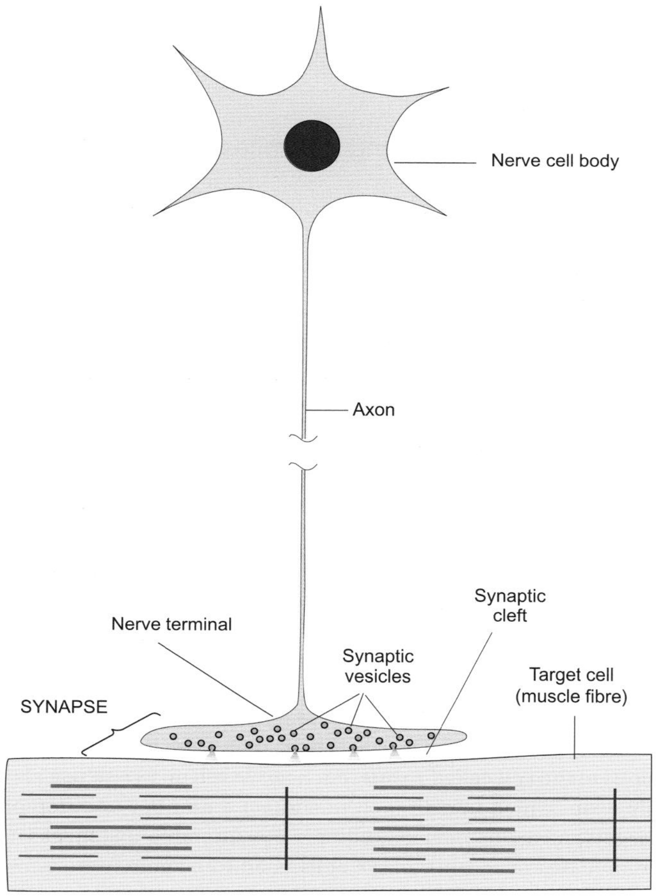

Water-soluble signalling molecules are derived from amino acids: examples are peptides, proteins, and the biological amines. Hydrophobic signalling molecules include the prostaglandins (see Section 6.5), the steroid hormones such as testosterone, and thyroid hormone.

Many chemical signalling molecules are stored in membrane-bound vesicles prior to secretion by exocytosis (see Chapter 5). This is the case with many hormones (e.g. adrenaline (epinephrine) and vasopressin) and neurotransmitters (e.g. acetylcholine). Other signalling molecules are so lipid-soluble that they cannot be stored in vesicles on their own but must be bound to a specific storage protein. This is the case with thyroid hormone. Finally, some chemical mediators are secreted as they are formed. This happens with the steroid hormones and the prostaglandins. As lipid-soluble hormones (thyroid hormone and the steroids) are not very soluble in water, they are carried to their target tissues bound to plasma proteins. Table 6.1 lists the principal families of signalling molecule and their chief modes of action.

Figure 6.2 Synaptic signalling is performed by nerve cells. Electrical signals (action potentials) originating in the cell body pass along the axon and trigger the secretion of a signalling molecule by the nerve terminal. As the nerve terminal is closely apposed to the target cell (in this case a skeletal muscle fibre), the signal is highly localized. The junction between the nerve cell and its target is called a synapse.

Figure labels: Nerve cell body — Axon — Nerve terminal — Synaptic cleft — Synaptic vesicles — Target cell (muscle fibre) — SYNAPSE

Summary

In order to coordinate their activities, cells need to send and receive signals of various kinds. They do this in three main ways: by the secretion of specific chemical signals, by direct cell–cell contact, and by gap junctions. Diffusible chemical signals reach their targets by local diffusion (paracrine signalling) or via the blood-stream (endocrine signalling). In each case they influence the activity of certain populations of cells. However, diffusible chemical signals can also be highly localized and discrete in their effects (as in synaptic signalling). Chemical signals range in size from small, highly diffusible molecules to large proteins. Water-soluble signalling molecules are secreted by exocytosis (e.g. acetylcholine), while lipid-soluble molecules are secreted by diffusion across the plasma membrane (e.g. prostaglandins).

6.3 Chemical signals are detected by specific receptor molecules

Cells are exposed to a wide variety of chemical signals and need to have a means of detecting those signals that are intended for them. They do so by means of molecules known as **receptors**, which are specific for particular chemical signals. For example, an acetylcholine receptor binds acetylcholine but does not bind adrenaline, histamine, or nitric oxide. When a receptor has detected a chemical signal, it initiates an appropriate cellular response. The link between the detection of the signal and the response is called **transduction**. Ligands that bind to and activate a particular receptor are called **agonists**, while those drugs that block the effect of an agonist are called **antagonists**.

All receptors are proteins, and many are located in the plasma membrane where they are able to bind the water-soluble signalling molecules that are present in the extracellular fluid. Hydrophobic signalling molecules such as the steroid hormones can cross the plasma membrane, and these bind to cytoplasmic and nuclear receptors. Finally, some intracellular organelles possess receptors for molecules that are generated within the cell (second messengers, discussed later in this section).

In recent years it has become clear that an individual cell may possess many different types of receptor so that it is able to respond to a variety of extracellular signals. The response of a cell to a specific signal depends on which receptors are activated. Consequently, a particular chemical mediator can produce different responses in different cell types. For example, acetylcholine released from the nerve terminals of motor nerve fibres onto skeletal muscle causes the muscle to contract, but when acetylcholine is released from the endings of the vagus nerve it slows the rate at which the heart beats (the heart rate). The acetylcholine has different effects in these two tissues because it acts on different receptors. Acetylcholine receptors of skeletal muscle are known as **nicotinic receptors** because the alkaloid nicotine can also activate them, while those of the heart have a different structure and are called **muscarinic receptors** as they can be activated by another chemical—muscarine.

2) The circulating levels of other hormones may closely regulate their activity. This is the case for the secretion of the sex hormones from the ovaries (oestrogens) and testes (testosterone), which respond to hormonal signals from the **anterior pituitary gland**.

3) They may be directly regulated by the activity of nerves. This is how oxytocin secretion from the **posterior pituitary gland** is controlled during lactation.

Since hormones are distributed throughout the body via the circulation, they are capable of affecting widely dispersed populations of cells. The hormones of the hypothalamus (a small region in the base of the brain) are an important exception to this rule. These hormones are secreted into the hypophyseal portal blood vessels in minute quantities and travel a few millimetres to the anterior pituitary, where they control the secretion of the anterior pituitary hormones (see Chapter 21). While cells are exposed to almost all the hormones secreted into the bloodstream, a particular cell will only respond to a hormone if it possesses receptors of the appropriate type (see Section 6.3). Thus the ability of a cell to respond to a particular hormone depends on whether it has the right kind of receptor. This diffuse signalling is beautifully adapted to regulate a wide variety of cellular activities in different tissues.

As the endocrine glands secrete their products into the general circulation, the concentrations of the various hormones in the blood are generally very low indeed—typically between 10^{-6} to 10^{-9} mol l^{-1}. This means that the receptors must bind hormones very effectively. For this reason, the individual receptors must have a high affinity for their particular hormone.

Compared to the signals mediated by nerve cells (synaptic signalling), the effects of hormones are usually relatively slow in onset (ranging from seconds to hours). Nevertheless, the effects of hormones can be very long-lasting, as in the control of growth (Chapter 51). Perhaps the most striking example of a permanent change triggered by a hormone is the effect of testosterone on the development of the male reproductive system. In its absence a genetically male fetus will fail to develop male genitalia (see Chapter 50 p. 832).

The difference between endocrine and paracrine signalling fundamentally depends on the quantity of the chemical signal secreted. If sufficient quantities of a signalling molecule are secreted for it to enter the blood, it is employed as a hormone. If, however, the amounts secreted are sufficient only to affect neighbouring cells, the chemical signal acts as a paracrine signal. Consequently, a substance can be a hormone in one situation and a paracrine signal in another. For example, the peptide somatostatin is found in the hypothalamus. It is secreted into the hypophyseal portal blood vessels; these carry it to the anterior pituitary gland, where it acts to inhibit the release of growth hormone (see Chapter 21). As somatostatin has entered the blood to be carried to its target tissue, it is acting as a hormone. Somatostatin is also found in the D-cells of the gastric mucosa. It is secreted when the hydrogen ion concentration in the stomach rises to inhibit the secretion of gastrin by the adjacent G-cells. (Gastrin is the hormone that stimulates acid secretion by the parietal cells of the gastric mucosa—see Chapter 44). In this situation somatostatin acts on neighbouring cells as a paracrine signal rather than as a hormone.

Fast signalling over long distances is accomplished by nerve cells

Chemical signals that are released into the extracellular space and need to diffuse some distance to reach their target cells have two disadvantages:

1. their effect cannot be restricted to an individual cell
2. their speed of signalling is relatively slow—particularly if there is any great distance between the secreting cell and its target.

For many purposes these factors are not important, but there are circumstances where the signalling needs to be both rapid (occurring within a few milliseconds) and discrete. For example, during locomotion different sets of muscles are called into play at different times to provide coordinated movement of the limbs. This type of rapid signalling is performed by nerve cells. The contact between the nerve ending and the target cell is called a **synapse**, and the overall process is known as synaptic signalling.

To perform their role, nerve cells need to make direct contact with their target cells. They do this via long, thin, hair-like extensions of the cell called **axons**. Each nerve cell gives rise to a single axon, which may branch to contact a number of different targets. As axons may extend over considerable distances (in some cases over 1 m), nerve cells need to be able to transmit their signals at relatively high speeds. This is achieved by means of an electrical signal (an **action potential**) that passes along the length of the axon from the cell body to its terminal (the **nerve terminal**). When an action potential reaches a nerve terminal it triggers the release of a small quantity of a chemical (a **neurotransmitter**), which then acts on the target cell. The nerve terminal is usually very closely apposed to its target, and so the neurotransmitter released by the nerve terminal has to diffuse a very short distance (about 20 nm) to reach its point of action. Since the receptors for the neurotransmitter are located directly under the nerve terminal, only very small quantities of neurotransmitter are required to activate the target cell and neighbouring cells will not be affected. The combination of electrical signalling and a very short diffusion time therefore permits both rapid and discrete activation of the target. The essential features of synaptic signalling are summarized in Figure 6.2. The organization and properties of nerve cells and synapses will be more fully discussed in Chapter 7.

Signalling molecules are very diverse in structure

The chemical signals employed by cells are very diverse. Cells use both water-soluble molecules and hydrophobic molecules for signalling. Some, such as nitric oxide and γ-aminobutyric acid (GABA), have a small molecular mass ($M_r \sim 100$), while others are much larger molecules such as growth hormone (which has 191 amino acids linked together: $M_r \sim 21,500$). Many of the

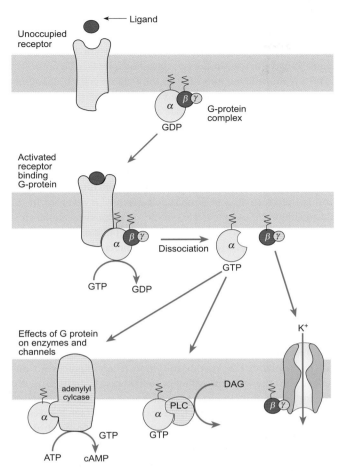

Figure 6.5 Receptor activation of heterotrimeric G proteins leads to activation of enzymes and ion channels. The top two panels show a schematic representation of a receptor–G protein interaction. The ligand binds to its receptor, which is then able to associate with a G protein. When this occurs, the alpha (α) subunit exchanges bound GDP for GTP and dissociates from the beta (β) and gamma (γ) subunits (centre). The dissociated α subunit can then interact either with adenylyl cyclase (bottom left) or with phospholipase C (bottom centre). The $\beta\gamma$ subunits can directly activate certain channels (bottom right).

have three subunits (α, β, and γ), each with a different amino acid composition. They are therefore called **heterotrimeric G proteins**. When the largest subunit (the α subunit) binds GDP, the three subunits associate together. Activation of a G protein linked receptor results in the G protein exchanging bound GDP for GTP, and this causes the G protein to dissociate into two parts: the α subunit and the $\beta\gamma$ subunit complex. The α and $\beta\gamma$ subunits can then migrate laterally in the plasma membrane to modulate the activity of ion channels or membrane-bound enzymes (Figure 6.5). There are many different kinds of heterotrimeric G proteins, but they all act as biochemical switches: they may modulate the activity of an ion channel or alter the rate of production of a second messenger—for example, cyclic AMP or IP_3. In their turn, the second messengers regulate a variety of intracellular events. Changes in the level of cyclic AMP alter the activity of a variety of enzymes via protein kinase A, while

IP_3 triggers the release of Ca^{2+} from intracellular stores. There are other GTP binding proteins which consist of a single polypeptide chain, and are not restricted to the cell membrane. These are called monomeric G proteins, and they play an important role in the control of cell signalling cascades. They will be discussed below.

Summary

Chemical signals are detected by specific receptor molecules, which are able to affect the function of the target cell. Activated receptors may open an ion channel or activate a membrane-bound enzyme.

6.4 Second messenger activation of signalling cascades

Adenylyl cyclase and phosphodiesterase regulate cyclic AMP concentrations inside cells

Cyclic AMP is generated when adenylyl cyclase (often incorrectly called adenylate cyclase) is activated by binding the α-subunit of a G protein called G_s. The cyclic AMP formed as a result of receptor activation then binds to other proteins (enzymes and ion channels) within the cell and thereby alters their activity. The exact response elicited by cyclic AMP in a particular type of cell will depend on which enzymes are expressed by that cell. Only one molecule of hormone or other chemical mediator is required to activate the membrane receptor, and once adenylyl cyclase is activated it can produce many molecules of cyclic AMP, with the result that activation of adenylyl cyclase allows a cell to amplify the initial signal many times. The signal is terminated by conversion of cyclic AMP to AMP by enzymes known as phosphodiesterases. Consequently, the balance of activity between adenylyl cyclase and phosphodiesterase controls the intracellular concentration of cyclic AMP. The main steps of G protein activation of adenylyl cyclase are summarized in Figure 6.5.

The action of adrenaline (epinephrine) on skeletal muscle highlights the role of G proteins in the regulation of cyclic AMP. Skeletal muscle stores glucose as glycogen (a large polysaccharide—see Chapter 3). ATP is required during exercise to power muscle contraction, and its synthesis necessitates the breakdown of glycogen to glucose. This change in metabolism is triggered by the hormone adrenaline that is secreted into the blood from the adrenal medulla. Increased levels of circulating adrenaline activate a particular kind of adrenergic receptor on the muscle membrane called a β-**adrenergic receptor** or β-adrenoceptor. These receptors are linked to G_s, and when the α subunit of G_s dissociates, it activates adenylyl cyclase (see Figure 6.5). The activation of adenylyl cyclase leads to an increase in the intracellular concentration of cyclic AMP. In turn, cyclic AMP activates another enzyme called protein kinase A that, in its turn, activates another enzyme called glycogen phosphorylase that breaks glycogen down to glucose. The individual steps progressively amplify the initial signal and lead to the rapid mobilization of glucose. The main features of this cascade are summarized in Figure 6.6.

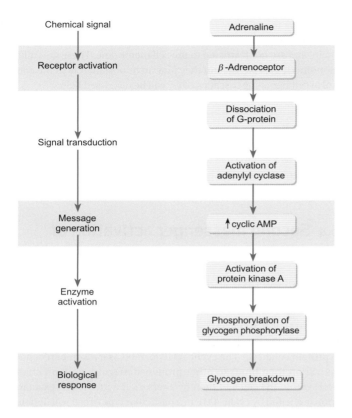

Chemical signal

Receptor activation

Signal transduction

Message
generation

Enzyme
activation

Biological
response

Adrenaline

↓

β-Adrenoceptor

↓

Dissociation
of G-protein

↓

Activation of
adenylyl cyclase

↓

↑cyclic AMP

↓

Activation of
protein kinase A

↓

Phosphorylation of
glycogen phosphorylase

↓

Glycogen breakdown

Figure 6.6 A simplified diagram to show the transduction pathway for the action of adrenaline (epinephrine) on the glycogen stores of skeletal muscle. On the left is a key to the main stages in the signalling pathway. The details of the individual steps are shown on the right. Adrenaline binds to β-adrenoceptors which are linked to a specific type of G protein (G_s) that can activate adenylyl cyclase. This enzyme increases the intracellular concentration of the second messenger cyclic AMP, which in turn leads to the activation of enzymes that break glycogen down to glucose.

While G_s activates adenylyl cyclase and so stimulates the production of cyclic AMP, another G protein (G_i) inhibits adenylyl cyclase. Activation of receptors coupled to G_i causes the intracellular level of cyclic AMP to fall. This, for example, is how somatostatin inhibits the release of gastrin by the G-cells of the gastric mucosa.

Certain membrane lipids can be the precursor for second messenger generation, or even a message in their own right

The plasma membrane contains many different phospholipids, some of which are starting points for the production of signalling molecules including arachidonic acid, diacylglycerol (DAG), and IP_3. The inner leaflet of the plasma membrane contains a small quantity of a phospholipid called phosphatidyl inositol which is important for several different signalling pathways. The inositol group on the lipid can be phosphorylated on up to three hydroxyl groups. While phosphatidyl inositol 4 phosphate (PIP, which has a

phosphate group attached to the 4 position on the inositol ring) does not appear to have a signalling function, phosphatidyl inositol 4,5 bisphosphate (PIP_2) and phosphatidyl inositol 3,4,5 trisphosphate (PIP_3) both have important signalling roles (see Figure 6.7).

PIP_2 is the starting point for an important second messenger cascade. Certain G protein linked receptors (e.g. the muscarinic receptor for acetylcholine) activate an enzyme known as phospholipase C. This enzyme hydrolyses PIP_2 to produce diacylglycerol and IP_3, both of which act as intracellular mediators.

IP_3 is a water-soluble molecule that can diffuse through the cytoplasm and bind to a specific receptor (the IP_3 receptor) to mobilize Ca^{2+} from the store within the endoplasmic reticulum. IP_3 generation is therefore able to couple activation of a receptor in the plasma membrane to the release of Ca^{2+} from an intracellular store (see Figure 6.8). Many cellular responses depend on this pathway. Examples are enzyme secretion by the pancreatic acinar cells and smooth muscle contraction. Furthermore, calcium ions themselves are capable of activating or inhibiting signalling cascades within the cell so that activation of the IP_3 signalling pathway can also regulate the pattern of enzymatic activity within the cell.

The diacylglycerol (DAG) that is generated by hydrolysis of PIP_2 is a hydrophobic molecule and is retained in the membrane when IP_3 is formed. However, like other membrane lipids, it is able to diffuse in the plane of the membrane where it can interact with, and activate, another enzyme called protein kinase C. In its turn, this enzyme activates other enzymes and thereby regulates a variety of cellular responses including DNA transcription (see Figures 6.7 and 6.8). DAG can also be metabolized to form **2-arachidonyl glycerol**, an endogenous agonist for the cannabinoid receptors of the nervous system, and **arachidonic acid**, which is discussed below.

Inositol phospholipids play an important role in the regulation of cell activity

Although phosphatidyl inositol is reasonably abundant in the plasma membrane, PIP_2 and especially PIP_3 are far less abundant, and their presence is isolated both spatially (they are found in small patches of membrane) and temporally (they are rapidly turned over). PIP_2 is localized to small 'rafts' in the plasma membrane where it binds to and modulates the activity of membrane proteins such as ion channels. It also acts as a docking site for proteins that move between the cytosol and the plasma membrane—for example, synaptotagmin, the calcium sensor for exocytosis, has a calcium-dependent PIP_2 binding activity that promotes membrane fusion.

Receptor tyrosine kinases (catalytic receptors) activate a membrane-bound enzyme called phosphatidyl inositol-3-kinase, to phosphorylate PIP_2 and form another inositol phospholipid, PIP_3 (Figure 6.7). The best-understood pathway regulated by PIP_3 levels is signalling through protein kinase B, also known as Akt. Activation of this enzyme occurs following phosphorylation by specific kinases. Activated Akt acts on its intracellular targets through its own kinase activity. It is necessary for the insulin-dependent uptake of glucose in the tissues and promotes cell survival, cell cycle progression, cell migration, cell proliferation, as well as changes in cellular

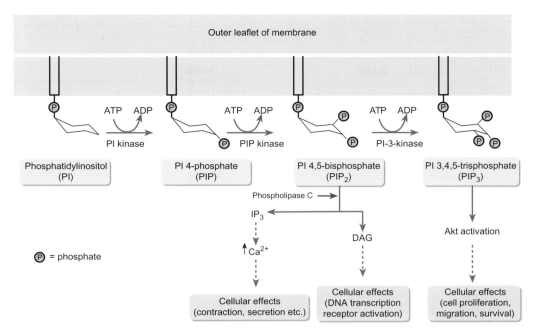

Figure 6.7 Phosphatidyl inositol and its derivatives. Phosphatidyl inositol (PI) is converted to phosphatidyl inositol 4-phosphate (PIP) by the enzyme PI kinase. In turn this can be converted to phosphatidyl inositol 4,5-bisphosphate (PIP_2) by another kinase called PIP kinase. Following receptor activation, PIP_2 is hydrolysed by membrane-bound phospholipase C to inositol 1,4,5-phosphate (IP_3) and diacylglycerol (DAG), both of which have a signalling role. PIP_2 can also be converted to phosphatidyl inositol 3,4,5-trisphosphate (PIP_3) by another kinase known as PI-3-kinase. This conversion is stimulated by the activation of tyrosine kinase catalytic receptors, which respond to growth factors and hormones such as insulin. PIP_3 activates another enzyme, protein kinase B, usually known as Akt, that regulates the insulin-dependent uptake of glucose by the tissues and other important aspects of cell function.

metabolism. In healthy cells, Akt activity is regulated by specific phosphorylases (enzymes that remove phosphate groups), but inappropriate activation of Akt is one of the common changes found in human cancer cells.

Calcium ions can act as a second messenger

Calcium ions are unusual second messengers, as their concentration can be directly modulated both by the activity of plasma membrane ion channels and by second messengers such as IP_3. In general, the effects of calcium ions are exerted by calcium binding proteins within the cell. When these proteins bind calcium ions they either undergo a change of shape (called a conformational change), or the calcium enables the protein to bind to other partners. One of the most important intermediary proteins in calcium signalling is the protein **calmodulin**. On binding four calcium ions, calmodulin becomes an activator for a large number of signalling cascades, including protein phosphorylation and dephosphorylation, regulation of the cytoskeleton, and the metabolism of second messengers.

Monomeric G proteins

As for the heterotrimeric G proteins, there are many different monomeric G proteins (there are five subfamilies of monomeric G proteins with a combined total of around 150 members). Each consists

of a single subunit with a relative molecular mass of around 20 kDa. They are not membrane-bound and can diffuse locally to transmit a signal from a surface receptor to the cell interior. In many cases they act as the second step in signalling by receptor tyrosine kinases and can be thought of as enzymatic second messengers which regulate the activity of many different signalling cascades. Like heterotrimeric G proteins, they require GTP for their activity; but unlike the heterotrimeric G proteins, they are entirely dependent on additional proteins to switch them between their active and inactive states. These additional regulatory proteins are known as GTPase activating proteins (GAPs) and GDP exchange factors (GEFs).

Monomeric G proteins are only active when GTP is bound, and so they can be turned off by GAPs (because this triggers the catalytic activity which converts GTP to GDP). Conversely, GEFs activate small G proteins, because they allow the exchange of GTP and GDP. In their role as switches of signalling networks, monomeric G proteins control many different signalling pathways, especially those that are linked to growth and development. Amongst their many functions is the regulation of the actin cytoskeleton, especially during cell migration. The activation of protein kinase cascades by monomeric G proteins has been implicated in the regulation of **gene transcription** and the control of the cell cycle. A failure of small G protein regulation has also been firmly linked to the development of various cancers.

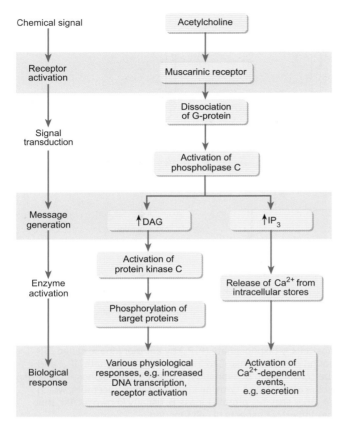

Figure 6.8 A simplified diagram to show the transduction pathway for the formation of inositol trisphosphate (IP$_3$) and diacylglycerol (DAG). Both IP$_3$ and DAG act as second messengers. The main stages in the signal transduction pathway are shown on the left of the figure, while the detailed steps are shown on the right. In this example, acetylcholine acts on muscarinic receptors that are linked to a G protein that can activate phospholipase C. This enzyme breaks down membrane phosphoinositides to form DAG and IP$_3$.

Summary

Many transmembrane receptors are linked to a heterotrimeric G protein, which acts to modulate the concentration of a second messenger. Second messengers diffuse through the cytosol of the target cell to reach their targets. Second messengers include cyclic AMP, IP$_3$, diacylglycerol (DAG), calcium ions, and monomeric G proteins.

6.5 Some local mediators are synthesized as they are needed

Certain chemical signals are very lipid-soluble and, unlike peptides and amino acids, cannot be stored in vesicles. Instead, the cells synthesize them as they are required. Important examples are the eicosanoids and nitric oxide, which regulate a wide variety of physiological processes.

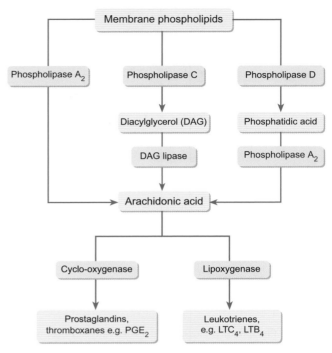

Figure 6.9 A simple schematic diagram illustrating the pathways that lead to the formation of arachidonic acid from membrane phospholipids and the subsequent synthesis of prostaglandins, leukotrienes, and lipoxins. LTB$_4$ = leukotriene B$_4$; LTC$_4$ = leukotriene C$_4$; PGE$_2$ = Prostaglandin E$_2$.

Arachidonic acid is formed directly from membrane phospholipids by the action of phospholipase A2 and indirectly via diacylglycerol and phosphatidic acid by the action of phospholipase C and D (Figure 6.9). It then acts as a precursor for the synthesis of important signalling molecules by one of two pathways:

- via cyclo-oxygenase to generate prostaglandins, thromboxanes, and prostacylin;

- via lipoxygenases to give rise to leukotrienes and lipoxins.

These metabolites form a group of 20-carbon compounds known as the **eicosanoids**; the chemical structures of some of the members of this family of molecules are shown in Figure 6.10. The secretion of the eicosanoids is continuously regulated by increasing or decreasing their rate of synthesis from membrane phospholipids. Once formed, the eicosanoids are rapidly degraded by enzyme activity.

Eicosanoid synthesis is initiated in response to stimuli that are specific for a particular type of cell, and different cell types produce different kinds of eicosanoid—over 16 different kinds are known. As local chemical mediators, the eicosanoids have many effects throughout the body (Table 6.2) and the specific effect exerted by a particular eicosanoid depends on the individual tissue. For example, prostaglandins PGE$_1$ and PGE$_2$ relax vascular smooth muscle and are powerful vasodilators. In the gut and uterus, however, they cause contraction of the smooth muscle. The diversity of effects in response to a particular prostaglandin is explained by the presence of different prostaglandin receptors in different tissues. These

Table 6.2 Some actions of eicosanoids

Eicosanoid	Effect on blood vessels	Effect on platelets	Effects on the lung
Prostaglandin E$_1$ (PGE$_1$)	Vasodilatation	Inhibition of aggregation	Bronchodilatation
Prostaglandin E$_2$ (PGE$_2$)	Vasodilatation	Variable effects	Bronchodilatation
Prostacyclin (PGI$_2$)	Vasodilatation	Inhibition of both aggregation and adhesion	Bronchodilatation
Thromboxane A$_2$ (TXA$_2$)	Vasoconstriction	Aggregation	Bronchoconstriction
Leukotriene C$_4$ (LTC$_4$)	Vasoconstriction	—	Bronchoconstriction, increased secretion of mucus

receptors are located in the plasma membrane of the target cells and are linked to second messenger cascades via G proteins.

Thromboxane A$_2$ (TXA$_2$) plays an important role in haemostasis (blood clotting—see Chapter 25) by causing platelets to aggregate (i.e. to stick together). It is produced by platelets in response to a blood clotting factor called thrombin (which is formed in response to tissue damage). Thrombin acts on a receptor in the cell membrane which activates phospholipase C. In turn phospholipase C liberates diacylglycerol, from which TXA$_2$ is synthesized. The TXA$_2$ then acts on cell surface receptors via an autocoid action to give rise to more TXA$_2$ (a positive feedback effect). It also diffuses to neighbouring platelets, inducing them to generate more TXA$_2$. The TXA$_2$ also

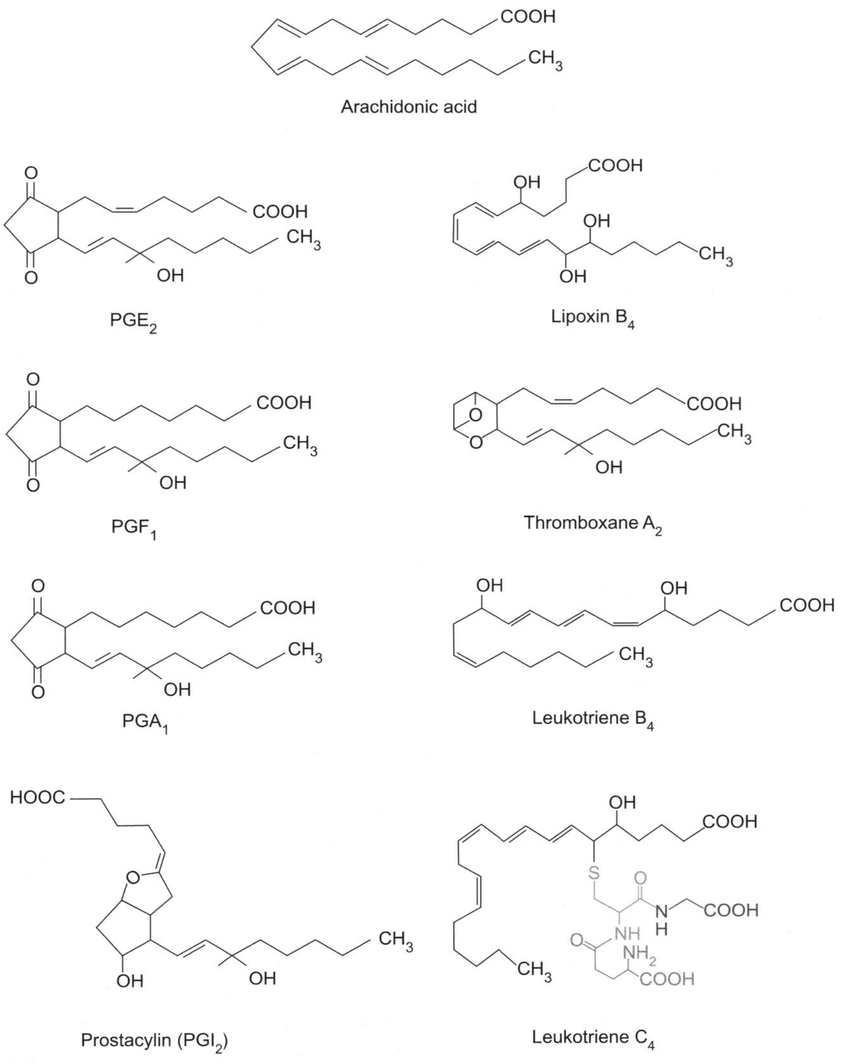

Figure 6.10 The chemical structures of various prostaglandins, leukotrienes, and lipoxins. Leukotriene C$_4$ is the product of leukotriene A$_4$ and the tripeptide glutathione (whose constituent amino acid residues are shown in colour).

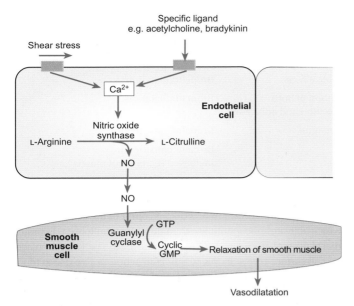

Figure 6.11 Diagram to show the synthesis of nitric oxide (NO) by endothelial cells and its action on vascular smooth muscle. The trigger for increased synthesis of NO is a rise in Ca²⁺ in the endothelial cell. This can occur as a result of stimulation by chemical signals (acetylcholine, bradykinin, ADP, etc.) acting on receptors in the plasma membrane or as a result of the opening of ion channels by shear stress (stretching of the plasma membrane caused by the flow of blood). The NO diffuses across the plasma membrane of the endothelial cell into the neighbouring smooth muscle cells and converts guanylyl cyclase into its active form. The increased production of cyclic GMP leads to relaxation of the smooth muscle and dilatation of the blood vessels.

activates a protein called β₃ integrin, which enables the platelets to stick to each other and to the blood clotting protein fibrin. In this way, tissue damage leads to the formation of a blood clot. This process is normally held in check by another eicosanoid, prostacyclin (PGI₂), which is secreted by the endothelial cells that line the blood vessels.

Both prostaglandins and leukotrienes play a complex role in regulating the inflammatory response to injury and infection. When the tissues become inflamed, the affected region reddens, becomes swollen, and feels hot and painful. These effects are, in part, the result of the actions of prostaglandins and leukotrienes, which cause vasodilatation in the affected region. These substances also increase the permeability of the capillary walls to immunoglobulins, and this leads to local accumulation of tissue fluid and swelling. Leukotriene B₄ (LTB₄) also attracts phagocytes. Non-steroidal anti-inflammatory drugs such as aspirin are used when the inflammatory response becomes excessively painful or persistent (as in arthritis). They act as inhibitors of cyclo-oxygenase to prevent the synthesis of prostaglandins. The inflammatory response is described in greater detail in Chapter 26.

Nitric oxide dilates blood vessels by increasing the production of cyclic GMP in smooth muscle

Acetylcholine is able to relax the smooth muscle of the walls of certain blood vessels, and this leads to vasodilatation. If the vascular

endothelium (the layer of cells lining the blood vessels) is first removed, acetylcholine causes contraction of the smooth muscle rather than relaxation. This experiment indicates that acetylcholine must release another substance in intact blood vessels—endothelium derived relaxing factor or EDRF—which is now known to be the highly reactive gas nitric oxide (NO). Many other vasoactive materials including adenine nucleotides, bradykinin, and histamine also act by releasing nitric oxide. It is now thought that the vasodilatation that occurs when the walls of blood vessels are subjected to stretch is also attributable to the release of nitric oxide by the endothelial cells and that this may play an important role in the local regulation of blood flow (see Chapter 30 p. 469).

How do the endothelial cells form nitric oxide, and how does nitric oxide cause the smooth muscle to relax? Nitric oxide is derived from the amino acid arginine through the action of an enzyme called nitric oxide synthase. This enzyme is activated when the intracellular free Ca²⁺ concentration in the endothelial cells is increased by various ligands (acetylcholine, bradykinin, etc.) or by the opening of stress-activated ion channels (i.e. ion channels that are activated by stretching of the plasma membrane). As it is a gas, the newly synthesized nitric oxide readily diffuses across the plasma membrane of the endothelial cell and into the neighbouring smooth muscle cells. In the smooth muscle cells, nitric oxide binds to, and activates, an enzyme called guanylyl cyclase (sometimes wrongly called guanylate cyclase). This enzyme converts GTP into cyclic GMP. Thus stimulation of the endothelial cells leads to an increase in cyclic GMP within the smooth muscle and this in turn activates other enzymes to bring about muscle relaxation. This sequence of events is summarized in Figure 6.11.

Nitric oxide synthase is not normally present in macrophages, but when these cells are exposed to bacterial toxins, the gene controlling the synthesis of this enzyme is switched on (a process known as induction) and the cells begin to make nitric oxide. In this case the nitric oxide is not used as a signalling molecule but as a lethal agent to kill invading organisms.

Organic nitrites and nitrates such as amyl nitrite and nitroglycerine have been used for over a hundred years to treat the pain that occurs when the blood flow to the heart muscle is insufficient. (This pain is called **angina pectoris** or angina.) These compounds promote the relaxation of the smooth muscle in the walls of blood vessels. Detailed investigation has revealed that this effect can be attributed to the formation of nitric oxide by enzymatic conversion of nitrite ions derived from the organic nitrates. This exogenous nitric oxide then acts in a similar way to that derived from normal metabolism.

Summary

Eicosanoids such as the prostaglandins are synthesized from arachidonic acid. Although they act as autocrine and paracrine signalling molecules, they are not stored in vesicles, but are synthesized from membrane phospholipids as required. Nitric oxide is a short-lived paracrine signalling molecule that is a powerful vasodilator. It acts by increasing the synthesis of cyclic GMP in the smooth muscle of blood vessels and this, in turn, leads to vasodilatation.

6.6 Steroid and thyroid hormones bind to intracellular receptors to regulate gene transcription

The steroid hormones are themselves lipids, so they are able to pass through the plasma membrane freely—unlike more polar, water-soluble signalling molecules such as peptide hormones. Thus, steroid hormones are not only able to bind to specific receptors in the plasma membrane of the target cell but they can also bind to receptors within its cytoplasm and nucleus.

The existence of cytoplasmic receptors for steroid hormones was first shown for oestradiol. This hormone is accumulated by its specific target tissues (the uterus and vagina) but not by other tissues. It was found that the target tissues possess a cytoplasmic receptor protein for oestradiol which, when it has bound the hormone, increases the synthesis of specific proteins.

The full sequence of events for the action of oestradiol can be summarized as follows. The hormone first crosses the plasma membrane by diffusing through the lipid bilayer and binds to its cytoplasmic receptor. The receptor–hormone complex then migrates to the nucleus, where it increases the transcription of DNA into the appropriate mRNA. The new mRNA is then used as a template for protein synthesis (see Chapter 3). Other steroid hormones such as the glucocorticoids and aldosterone are now known to act in a similar way. This scheme is outlined in Figure 6.12. Thyroid hormones regulate gene expression in a similar manner to steroid hormones but gain entry to their target tissues via specific transport proteins.

The receptors for the steroid and thyroid hormones are members of a large group of proteins involved in regulating gene expression, the **nuclear receptor super family**. However, not all nuclear receptors are located in the cytoplasm prior to binding their hormone. Some are bound to DNA in the nucleus even in the absence of their normal ligand. For example, the thyroid hormone receptor is bound to DNA in the nucleus even in the absence of thyroid hormone.

In the absence of their specific ligand, all members of the nuclear receptor superfamily are bound to other regulatory proteins to form inactive complexes. When they bind their ligand, the receptors undergo a conformational change that leads to the dissociation of the regulatory proteins from the receptor. The activated receptor protein then regulates the transcription of a specific gene or set of genes. This occurs in successive steps. Initially, the hormone–receptor complex activates a small set of genes relatively quickly. This is called the primary response and occurs in less than an hour. The proteins synthesized in this early phase then activate other genes—this is called the secondary response. Since many proteins are involved in gene activation, the initial hormonal signal is able to initiate very complex changes in the pattern of protein synthesis within its target cells.

The physiological response to thyroid hormone, the steroid hormones, and other ligands that bind to members of the nuclear receptor super family is determined by the nature of the target

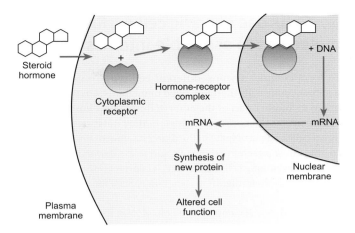

Figure 6.12 A simplified diagram to show how steroid hormones regulate gene transcription in their target cells. Steroid hormones are lipophilic and are able to pass through the plasma membrane to bind to specific receptor proteins in the cytoplasm of the target cells. The hormone–receptor complex diffuses to the cell nucleus, where it binds to a specific region of DNA to regulate gene transcription. Some steroid and thyroid hormones bind directly to receptors in the nucleus.

cells themselves. It depends on the presence of the appropriate intracellular receptor and the particular set of regulatory proteins present, both of which are specific to certain cell populations. The final physiological effects are slow in onset and long-lasting. For example, sodium retention by the kidney occurs after a delay of two to four hours following the administration of aldosterone (see Chapter 39) and may last for many hours. The effect of testosterone on male development lasts a lifetime (see Chapter 50).

In recent years it has become clear that there are G protein linked cell surface receptors for oestrogens (G protein coupled estrogen receptor 1 or GPER) which stimulate the production of cAMP and activation of protein kinase A as well as other signalling pathways. In addition to the GPER, there is evidence for plasma membrane receptors for androgens, progesterone, glucocorticoids, and mineralocorticoids. While the physiological role and signalling pathways of these newly discovered receptors is still being elucidated, the presence of plasma membrane receptors offers an explanation for the ability of many steroid hormones to influence cell activity within a few minutes.

Summary

Steroid and thyroid hormones are very hydrophobic and are carried in the blood bound to specific carrier proteins. They enter cells by diffusing across the plasma membrane and bind to cytoplasmic and nuclear receptors to modulate the transcription of specific genes. Plasma membrane G protein linked receptors for oestrogens have been described that stimulate the production of cAMP.

6.7 Cells use specific cell surface molecules to assemble into tissues

To form complex tissues, different cell types must aggregate together. Some cells therefore need to migrate from their point of origin to another part of the embryo during development. When they arrive at the appropriate region, they must recognize their target cells and participate in the differentiation of the tissue. To do so, they must attach to other cells and to the extracellular matrix. How do developing cells establish their correct positions, and why do they cease their migrations when they have found their correct target?

Unlike adult cells, embryonic cells do not form strong attachments to each other. Instead, when they interact, their cell membranes become closely apposed to each other leaving a very small gap of only 10–20 nm. Exactly how cells are able to recognize their correct associations is not known, but it is likely that each type of cell has a specific marker on its surface. When cell membranes touch each other the marker proteins on the surfaces of the cells can interact. If the cells have complementary proteins they are then able to cross-link and the cells will adhere. This must be an early step in tissue formation. It has been shown that cells will associate only if they recognize the correct surface markers. Thus, if differentiated embryonic liver cells are dispersed by treatment with enzymes and grown in culture with cells from the retina, the two cell types aggregate with others of the same kind. Thus liver cells aggregate together and exclude the retinal cells, and vice versa.

The various kinds of cell–cell and cell–matrix junctions have already been discussed in Chapter 4 and many of the proteins involved have been characterized. These may be grouped into several families including the cadherins that form desmosomes, the connexins that form gap junctions, the immunoglobulin-like (Ig-like) cell adhesion molecules (e.g. N-CAM), and the integrins that form hemidesmosomes that attach cells to the extracellular matrix. The integrins also play an important role in development and wound repair. The cadherins, integrins, and Ig-like cell adhesion molecules are also involved in non-junctional cell–cell adhesion, which must play an important role in the formation of integrated tissues.

While the integrins are important in maintaining attachments between the majority of cells, those of platelets are not normally adhesive. If they were, blood clots would form spontaneously with disastrous consequences. During haemostasis (blood clotting), however, platelets adhere to fibrin and to the damaged wall of the blood vessel (see Chapter 25). This adhesion results from a change in the properties of the platelets when a non-adhering integrin precursor present in the platelet membrane is transformed into an adhesive protein. This transformation is triggered by factors released from the walls of damaged blood vessels that activate second messenger cascades within the platelets which, in turn, trigger modification of the structure of preformed integrins so that they act as receptors for extracellular molecules including fibrin. The final result is an increase in platelet adhesion and clotting of the blood.

Summary

For cells to assemble into tissues, they must adhere to other cells of the correct type. This recognition requires tissue specific cell surface marker molecules. Cell–cell adhesion and cell–matrix interactions play an important role in tissue maintenance and development. Several different families of proteins are involved in these processes including the cadherins, the Ig-like cell adhesion molecules, and the integrins.

6.8 Gap junctions permit the exchange of small molecules and ions between neighbouring cells

Some cells are joined together by a specific type of junction known as a **gap junction**. Specific membrane proteins associate to form doughnut-shaped structures known as **connexons**. When the connexons of two adjacent cells are aligned, the cells become joined by a water-filled pore. Because the connexons jut out above the surface of the plasma membrane, the cell membranes of the two cells forming the junction are separated by a small gap—hence the name gap junction.

Unlike ion channels, the pores of gap junctions are kept open most of the time so that small molecules (less than 1500 Da) and inorganic ions can readily pass from one cell to another (see Figure 6.13). Consequently, gap junctions form a low-resistance pathway between the cells and electrical current can spread from one cell to another. The cells are thus *electrically coupled*. This property is exploited by the cardiac muscle cells (cardiac myocytes), which are connected by gap junctions. Since they are electrically coupled, depolarization of one myocyte causes current to pass between it and its immediate neighbours, which become depolarized. In their turn, these cells cause the depolarization of their neighbours, and so on. Consequently, current from a single point

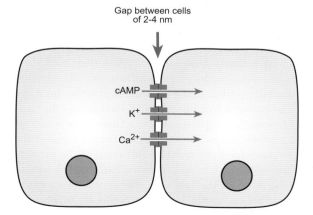

Figure 6.13 A simple diagram showing two cells linked by gap junctions. Ions can pass between the neighbouring cells so the cells are electrically coupled. In addition, small organic molecules ($M_r < 1500$) can also pass through gap junctions, allowing for the spread of second messenger molecules (e.g. cyclic AMP) between adjacent cells.

of excitation spreads across the whole of the heart via the gap junctions and the heart muscle behaves as a **syncytium** (a collection of cells fused together). By this means the electrical and contractile activity of individual myocytes is coordinated. This allows the muscle of the heart to provide the wave of contraction (the heart beat) that propels the blood around the body (see Chapter 28).

In the liver, the gap junctions between adjacent liver cells (hepatocytes) allow the exchange of intracellular signals (second messengers) between cells. For example, the hormone glucagon stimulates the breakdown of glycogen to glucose by increasing the level of cyclic AMP. The cyclic AMP diffuses through the water filled pores of the gap junctions from one cell to its neighbours so that cells not directly activated by glucagon can be stimulated to initiate

glycogen breakdown. The gap junctions provide a means of spreading the initial stimulus from one cell to another.

Summary

Gap junctions are formed by membrane proteins called connexins. When a gap junction is formed, the extracellular space between the plasma membranes of the adjacent cells is eliminated. Gap junctions permit the diffusion of small molecules and ions directly from the cytoplasm of one cell to its neighbour. They also enable electrical currents to flow from one cell to another and thereby allow the electrical coupling of cells.

✳ Checklist of key terms and concepts

Principles of cell signalling

- In order to coordinate their activities, cells need to send and receive signals of various kinds.
  - They interact with their neighbours by direct cell–cell contact.
  - They exchange small solutes with their neighbours via gap junctions.
  - They secrete specific chemical signals (diffusible signals) which they use in three ways:
    - as local signals (paracrine signalling),
    - as diffuse signals that reach their target tissues via the bloodstream (endocrine signalling), and
    - for rapid, discrete signalling (synaptic signalling).
- A wide variety of molecules are employed as chemical signals ranging in size from small, highly diffusible molecules such as nitric oxide to large proteins such as growth hormone.
- Signalling molecules may be secreted by exocytosis (e.g. acetylcholine and peptide hormones) or they may be secreted by diffusion across the plasma membrane (e.g. prostaglandins and steroid hormones such as testosterone).

Cell receptors, second messengers, and intracellular signalling cascades

- Diffusible chemical signals are detected by specific receptor molecules.
- When a receptor has bound a signalling molecule it must have a means of altering the behaviour of the target cell. It does this in one of two ways: by opening an ion channel or by activating a membrane-bound enzyme.
- Some receptors are membrane-bound protein kinases which become activated when they bind a ligand. These are known as catalytic receptors.
- Some receptors activate heterotrimeric G proteins, which alter the level of a second messenger (e.g. cyclic AMP) within the cell.

- Second messengers diffuse through the cytosol of the target cell and exert their effects via a sequence of enzymes called a signalling cascade.
- Cyclic AMP acts as second messenger and its intracellular concentration is determined by the balance of activity of adenylyl cyclase and phosphodiesterase.
- Certain membrane lipids can be the precursor for second messenger generation, or even a message in their own right.
- The inositol phospholipids are hydrolysed by phospholipase C to liberate inositol triphosphate (IP_3) and diacylglycerol (DAG), both of which act as second messengers.
- Calcium ions are released from intracellular stores by IP_3 and exert a wide variety of cellular effects.
- Calcium ions can also directly enter cell via ion channels.

Some local mediators are synthesized as they are needed

- Certain chemical signals are very lipid-soluble and cannot be stored in vesicles but must be synthesized as required.
- The eicosanoids are lipid signalling molecules synthesized from arachidonic acid that include the prostaglandins, thromboxanes, and leukotrienes.
- Different cell types produce different eicosanoids, and their effects are specific to a particular tissue.
- Nitric oxide (NO), an inorganic gas, is a paracrine signalling agent.
- NO is a powerful vasodilator that is synthesized by the endothelial cells of blood vessels in response to a variety of stimuli.
- NO acts by increasing the synthesis of cyclic GMP in the smooth muscle of the blood vessels and this, in turn, leads to muscle relaxation and dilatation.
- NO has many other actions including employment by macrophages as a lethal agent to kill invading organisms.

Nuclear receptors and the regulation of gene transcription

- Steroid and thyroid hormones are very hydrophobic and are carried in the blood bound to specific carrier proteins.
- They enter cells by diffusing across the plasma membrane and bind to cytoplasmic and nuclear receptors to modulate the transcription of specific genes.
- This leads to the increased synthesis of specific proteins and alterations in cell activity.

Cell–cell adhesion and gap junctions

- For cells to assemble into tissues they need to adhere to other cells of the correct type. This recognition requires tissue specific cell surface marker molecules.
- Cell–cell adhesion and cell–matrix interaction play an important role in tissue maintenance and development.
- Several different families of proteins are involved in these processes.
- Gap junctions between adjacent cells permit the diffusion of small molecules and ions directly from the cytoplasm of one cell to its neighbour.
- Gap junctions allow the electrical coupling of cells.

 Recommended reading

Biochemistry and cell biology

Alberts, B., Johnson, A., Lewis, J., Morgan, D., Raff, M., Roberts, K., and Walter, P. (2014) *Molecular biology of the cell* (6th edn), Chapters 15 and 19. Garland, New York.
  A well-written and well-illustrated introduction to cell signalling.
Berg, J.M., Tymoczko, J.L., and Stryer, L. (2011) *Biochemistry* (7th edn), Chapter 14. Freeman, New York.
Gomperts, B.D., Kramer, I.M., and Tatham, P.E.R. (2009) *Signal transduction* (2nd edn). Academic Press, San Diego.
  A very comprehensive account of mechanisms of intracellular signalling.

Pharmacology

Rang, H.P., Dale, M.M., and Ritter, J.M. (2011) *Pharmacology* (7th edn), Chapters 2–4. Churchill-Livingstone, Edinburgh.

Review articles

Alabi, A.A., and Tsien, R.W. (2013) Perspectives on kiss-and-run: Role in exocytosis, endocytosis, and neurotransmission. *Annual Review of Physiology* **75**, 393–422, doi:10.1146/annurev-physiol-020911-153305.
Berridge, M.J. (2014) *Cell signalling biology*. www.cellsignalling.org./, doi:10.1042/csb0001001.
  An excellent free resource written by a major figure in the field.
Sigismund, S., Confalonieri, S., Ciliberto, A., Polo, S., Scita, G., and Di Fiore, P.P. (2012) Endocytosis and signaling: Cell logistics shape the eukaryotic cell plan. *Physiological Reviews* **92**, 273–366, doi:10.1152/physrev.00005.2011.
Wu, L.-G., Hamid, E., Shin, W., and Chiang, H.-C. (2014) Exocytosis and endocytosis: Modes, functions, and coupling mechanisms. *Annual Review of Physiology* **76**, 301–31, doi:10.1146/annurev-physiol-021113-170305.

 To check that you have mastered the key concepts presented in this chapter, complete the accompanying online self-assessment questions. Go to www.oup.com/uk/pocock5e/

PART THREE

The excitable tissues—nerve and muscle

CHAPTER 7

Nerve cells and their connections

Chapter contents

This chapter should help you to understand:

- The structure of nerve cells and axons
- The ionic basis of the action potential in neurons and axons
- Action potential conduction in myelinated and unmyelinated axons
- The principal features of synaptic transmission between nerve cells—the synaptic basis of excitation and inhibition
- Neuromuscular transmission: the neuromuscular junction as a model chemical synapse
- The effects of denervation of skeletal muscle
- Axonal transport

7.1 Introduction

The nervous system is adapted to provide rapid and discrete (point-to-point) signalling over long distances (from millimetres to a metre or more). It is divided into the **central nervous system** (CNS) comprising the brain and spinal cord, and the **peripheral nervous system** comprising the **peripheral nerves**, the **enteric nervous system**, and the **autonomic nervous system** (see Chapter 9). The nervous system is made up of two main types of cell: the **nerve cells** or **neurons** and the satellite cells. In the brain and spinal cord the satellite cells are called **glial cells** or **neuroglia**, while elsewhere they have different names (Schwann cells, Müller cells, pituicytes, etc.).

The neurons are the fast signalling elements of the nervous system and this chapter discusses the basis of the signalling processes they employ to control their target cells. They are responsible for the ability of the CNS to control the muscles and secretory glands of the body and for receiving information from the senses.

Two key observations demonstrate that the neurons are the principal functional units of the nervous system. Crushing or cutting a peripheral nerve prevents the CNS from controlling the target organ supplied by that nerve. In the case of nerves supplying a skeletal muscle, such injuries result in paralysis of the muscle even though the nerve sheath may remain intact. Control of movement is regained only if the appropriate axons regenerate. Similarly, the death of neurons within the brain following a **stroke** can cause paralysis even though the peripheral nerves remain intact. (A stroke is a loss of cerebral function due to a cerebral thrombosis—i.e. a blockage of a cerebral blood vessel caused by a blood clot—or, more rarely, due to a cerebral **haemorrhage** caused by rupture of a blood vessel).

7.2 The structure of the neurons and their axons

Nerve cell bodies are located within the brain, the spinal cord, and the autonomic and enteric ganglia. They are very varied in both size and shape, but all stain strongly with basic dyes. They all possess certain morphological characteristics (see Figure 7.1): each neuron has a set of fine branches called **dendrites** that receive information from other neurons and a threadlike extension of the cell body called an **axon** (or nerve fibre) that transmits information

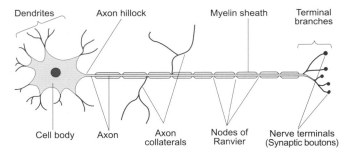

Figure 7.1 A diagrammatic representation of a CNS neuron. The appearance of individual neurones is quite variable—the one illustrated shows the general features of a spinal motoneuron. Others may have fewer dendrites or show a different polarity such as distinct populations of apical and basal dendrites, as seen in Figure 7.2.

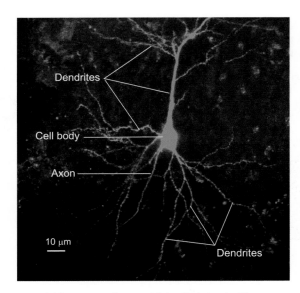

Figure 7.2 A hippocampal pyramidal cell stained with a green fluorescent dye (Oregon green BAPTA-1). Note the extensive branching of the dendrites and the single axon.

to its target cells (which may be other neurons). Axons and dendrites are collectively called neuronal processes.

In the brain, the dendrites are frequently covered with small projections known as **dendritic spines** that give the dendrites a roughened or spiny appearance. Each spine is a point of contact between an axon terminal and the dendrite. The dendrites of neurons within the CNS generally branch extensively: a typical example (a hippocampal pyramidal cell) is shown in Figure 7.2. The dendrites are therefore able to receive information from many different sources. Each nerve cell gives rise to a single axon, which subsequently gives off side branches (axon collaterals) to contact a number of different target cells. The parent axon and its collateral branches end in a small swelling where it meets its target cell—an **axon terminal** (also called a **nerve terminal** or **synaptic bouton**). The contact between an axon terminal and its target is called

a **synapse**. This contact may be made between nerve cells or between an axon and a non-neuronal cell such as a muscle fibre.

The structure of peripheral nerve trunks

Axons are delicate structures that may traverse considerable distances to reach their target organs. Outside the CNS they run in peripheral nerve trunks alongside the major blood vessels, where they are protected from damage by three layers of connective tissue (Figure 7.3). The outermost layer of a peripheral nerve trunk

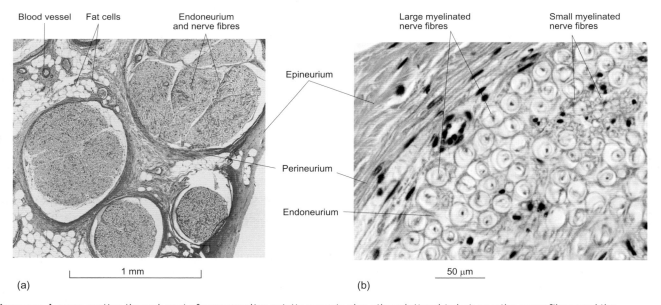

Figure 7.3 A cross-section through part of a mammalian sciatic nerve to show the relationship between the nerve fibres and the surrounding layers of connective tissue (the epineurium, perineurium, and endoneurium). The low-power view (a) shows a number of nerve fascicles, while the high-power view (b) shows a part of one fascicle in which individual large nerve fibres are clearly resolved. The outer myelinated regions appear pale, while the axon itself is densely stained. Smaller-diameter fibres are not so clearly resolved.

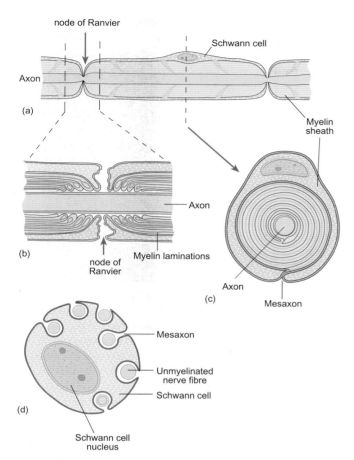

Figure 7.4 Drawing to show the relationship between myelinated and unmyelinated nerve fibres and the surrounding Schwann cells. Panel (a) shows a short length of myelinated nerve. Panel (b) illustrates the detailed structure of a single node of Ranvier. Panel (c) is a diagrammatic view of a cross-section of a myelinated nerve fibre to show how the Schwann cell forms the myelin layers. Panel (d) illustrates the arrangement of a number of unmyelinated nerve fibres within a single Schwann cell. (Redrawn from Fig 3-5 in M.L. Barr (1974) *The human nervous system.* Harper International, Hagerstown.)

Axons may be either myelinated or unmyelinated. **Myelinated axons** are covered by a thick layer of fatty material called **myelin** that is formed by layers of plasma membrane derived from specific satellite cells. In peripheral nerves, the myelin is derived from Schwann cells (see Figure 7.4), while in the CNS the myelin is formed by processes of a specific kind of glial cells called oligodendrocytes. Although the myelin sheath extends along the length of an axon, it is interrupted at regular intervals by gaps known as the **nodes of Ranvier**. At the nodes of Ranvier the axon membrane is not covered by myelin but is in direct contact with the extracellular fluid. The distance between adjacent nodes varies with the axon diameter, larger fibres having a greater distance between the nodes (the internodal distance). In peripheral nerves, **unmyelinated axons** are also covered by Schwann cells but, in this case, there is no layer of myelin and a number of nerve fibres are covered by a single Schwann cell (see Figure 7.4). Unmyelinated axons are in direct communication with the extracellular fluid via a longitudinal cleft in the Schwann cell called the **mesaxon**.

Summary

The principal functional units of the nervous system are the nerve cells (neurons), which possess two types of process: dendrites and axons. Dendrites are highly branched and receive information from other nerve cells. Each nerve cell gives rise to a single axon, which subsequently branches to make contact with other cells. Axons transmit information to other neurons or to non-neuronal cells such as muscles.

7.3 Axons transmit information via a sequence of action potentials

In the late eighteenth century, Galvani showed that electrical stimulation of the nerve in a frog's leg caused the muscles of the leg to twitch. This key observation led to the discovery that the excitation of nerves was accompanied by an electrical wave that passed along the nerve. This wave of excitation is now called the **nerve impulse** or **action potential**. To generate an action potential, an axon requires a stimulus of a certain minimum strength known as the threshold stimulus or **threshold**. An electrical stimulus that is below the threshold (a subthreshold stimulus) will not elicit an action potential, while a stimulus that is above threshold (a suprathreshold stimulus) will do so. With stimuli above threshold, each action potential has approximately the same magnitude and duration irrespective of the strength of the stimulus. This is known as the 'all or none' law of action potential transmission.

In a mammalian axon, each action potential lasts about 0.5–1.0 milliseconds (ms). If a stimulus is given immediately after an action potential has been elicited, a second action potential is not generated until a certain minimum time has elapsed. The interval during which it is impossible to elicit a second action potential is known as the **absolute refractory period**. This determines the upper limit to the number of action potentials a particular axon

(or nerve) is a loose aggregate of connective tissue called the **epineurium**, which serves to anchor the nerve trunk to the adjacent tissue. Within the epineurium, axons run in bundles called fascicles and each bundle is surrounded by a tough layer of connective tissue called the **perineurium**. Within the perineural sheath, individual nerve fibres are supported by a further layer of connective tissue called the **endoneurium**. Individual axons are covered by specialized cells called **Schwann cells**.

Some nerve fibres transmit information from specific sensory end organs to the CNS (**sensory nerves**) while others transmit signals from the CNS to specific effectors (**motor nerves**). Nerve trunks that contain both sensory and motor fibres (such as the spinal nerves) are called **mixed nerves** (see Chapter 9). The spinal nerves also contain sympathetic postganglionic fibres, which innervate the blood vessels and sweat glands (see Chapter 11).

can transmit in a given period. Following the absolute refractory period (which in mammals is usually 0.5–1.0 ms in duration) a stronger stimulus than normal is required to elicit an action potential. This phase of reduced excitability lasts for about 5 ms and is called the **relative refractory period**.

What mechanisms are responsible for the generation of the action potential? The answer to this important question was provided by a series of experiments carried out on the giant axon of the squid between 1939 and 1950 by K.C. Cole in the USA, and by A.L. Hodgkin and A.F. Huxley in England. Like mammalian nerve cells, the resting membrane potential of the squid axon is negative (about –70 mV), close to the equilibrium potential for potassium ions (see Chapter 5). Hodgkin and Huxley found that, during the action potential, the membrane potential briefly reversed in polarity, reaching a peak value of +40 to +50 mV before falling back to its resting level of about –70 mV (see Figure 7.5). This discovery provided an important clue about the underlying mechanism. At the peak of the action potential, the membrane potential is positive and close to the equilibrium potential for sodium ions, unlike the resting membrane potential, which is negative and lies close the equilibrium potential for potassium ions. This suggested that the action potential resulted from a large increase in the permeability of the axon membrane to sodium ions. This was confirmed when it was shown that removal of sodium ions from the extracellular solution prevented the axon from generating an action potential. In addition, if the sodium concentration in the bathing medium was reduced by two-thirds, the action potential was slower and smaller than normal (see Figure 7.5).

What underlies the change in the permeability of the axon membrane to sodium ions during an action potential? The axonal membrane contains ion channels of a specific type called **voltage-gated sodium channels**. Following a stimulus, the membrane depolarizes; this causes some of the sodium channels to open, permitting

sodium ions to move into the axon down their electrochemical gradient. This inward movement of sodium depolarizes the membrane further, so leading to the opening of more sodium channels. This results in a greater influx of sodium, which causes a greater depolarization. This process continues until, at the peak of the action potential, the highest proportion of available sodium channels is open and the membrane is, for a brief time, highly permeable to sodium ions. Indeed, at the peak of the action potential, the membrane is more permeable to sodium ions than to potassium ions and this explains the fact that, at the peak of the action potential, the membrane potential is positive (see Chapter 4).

Why does the membrane repolarize after the action potential? The open state of the sodium channels is unstable and the channels show a time-dependent inactivation. Immediately after the action potential has peaked, the sodium channels begin to inactivate (i.e. they cease to allow the passage of sodium ions—see Chapter 4) so that the permeability of the axon membrane to sodium begins to fall. At the same time, voltage-activated potassium channels begin to open in response to the depolarization and potassium ions leave the axon down their electrochemical gradient. This increases the permeability of the membrane to potassium at the same time as the permeability of the membrane to sodium begins to fall. The decrease in sodium permeability and the increase in potassium permeability combine to drive the membrane potential from its positive value at the peak of the action potential towards the equilibrium potential for potassium ions. As the membrane potential approaches its resting level, the voltage-activated potassium channels close and the membrane potential assumes its resting level. At the resting membrane potential, the inactivated sodium channels revert to their closed state and the axon is primed to generate a fresh action potential. The time course of the changes in the ionic permeability of the axon membrane during the action potential is shown in Figure 7.6.

This sequence of events explains both the threshold and the refractory period. When a weak stimulus is given, the axon does not depolarize sufficiently to allow enough voltage-gated sodium channels to open to depolarize the membrane further. Such a stimulus is subthreshold. At the threshold, sufficient sodium channels open to permit further depolarization of the membrane as a result of the increased permeability to sodium. The resulting depolarization causes the opening of more sodium channels and the process continues until all available sodium channels are open (this is an example of a self-limiting positive feedback). Immediately after the action potential, most sodium channels are in their inactivated state and cannot reopen until they have returned to their closed state. To do so, they must spend a brief period at the resting membrane potential. This accounts for the absolute refractory period. One plausible explanation of the relative refractory period of a single nerve fibre is that an action potential can be supported when enough sodium channels have returned to the closed state provided that they are all activated together. This will require a stronger stimulus than normal to ensure that all the available sodium channels open at the same time.

The effect of the passage of an action potential along an axon is to leave it with a little more sodium and a little less potassium than it had before. Nevertheless, *the quantities of ions exchanged during*

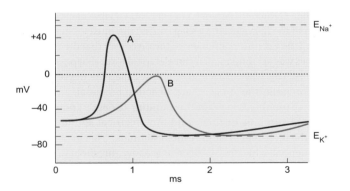

Figure 7.5 The membrane potential changes that occur during the action potential of the giant axon of the squid in normal sea water (A) and when external sodium is reduced by two-thirds (B). Note that in normal sea water, the membrane potential is positive at the peak of the action potential and is close to the equilibrium potential for sodium (E_{Na}). Reduction of extracellular sodium both reduces the maximum amplitude of the action potential and slows its time course. The effect is reversible (not shown). (Based on Fig 17 in A.L. Hodgkin and B. Katz (1949) *Journal of Physiology* **108**, 37–77.)

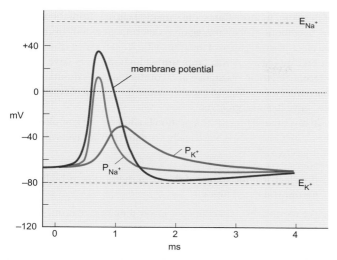

Figure 7.6 The permeability changes that are responsible for the action potential of the squid giant axon. During the upstroke of the action potential there is a very abrupt but short-lived rise in the membrane's permeability to sodium (P_{Na}). This is followed by a rise in the membrane permeability to potassium, which returns to normal as the membrane potential returns to its resting level (P_K). (Based on Fig 17 in A.L. Hodgkin and A. F. Huxley (1952) *Journal of Physiology* **117**, 500–40.)

a single action potential are very small and are not sufficient to alter the ionic gradients across the membrane. In a healthy axon, the ionic gradients are maintained by the continuous activity of the sodium pump. If the sodium pump is blocked by a metabolic poison, a nerve fibre is able to conduct impulses only until its ionic gradients are dissipated. Thus, the sodium gradient established by the sodium pump indirectly powers the action potential.

How does an action potential start?

While it is convenient to use electrical stimuli to provide a precise and controlled way of exciting an isolated nerve, action potentials in living animals do not arise in this way. So how do they start? In many cases, they arise as a result of specific stimuli arriving at the sense organs. The sense organs convert the external stimulus into a sequence of action potentials that is transmitted to the CNS. The process that activates a particular receptor is known as **sensory transduction**. The details differ from one kind of receptor to another (see Chapter 12). Within the CNS, the action potentials in the sensory nerves activate the neurons involved in processing the appropriate sensory information. This kind of direct activation of neurons occurs continuously as the sense organs report the current state of the environment. As the brain is constantly processing this information, it is perhaps not surprising to discover that the membrane potential of most neurons in the CNS is not stable but fluctuates according to the level of activity in the excitatory and inhibitory nerves impinging on the cell. When all the influences on a particular neuron cause its membrane potential to reach threshold, an action potential results. There is also some evidence that certain nerve cells have an intrinsic pacemaker activity so that they generate action potentials spontaneously.

How is the action potential propagated along an axon?

When an action potential has been initiated, it is transmitted rapidly along the entire length of an axon. How does this happen? At rest the membrane potential is about –70 mV, while during the peak of the action potential, the membrane potential is positive (about +50 mV). Thus, when an action potential is being propagated along an axon, the active zone and the resting membrane will be at different potentials and a small electrical current will flow between the two regions. This forms a local circuit that links the active zone to the neighbouring resting membrane (see Figure 7.7), which then depolarizes. This causes sodium channels to open; when sufficient channels have opened, the action potential invades this part of the membrane. This in turn spreads the excitation farther along the axon, where the process is repeated until the action potential has traversed the length of the axon.

In the intact nervous system, an action potential is propagated from its point of origin in only one direction along an axon (orthodromic propagation). Why is this the case? What prevents the retrograde (or antidromic) propagation of an action potential? When an action potential has invaded one part of the membrane and has moved on, it leaves the sodium channels in an inactivated state from which they cannot be reactivated until they have passed through their closed (or resting) state. The time taken for this transition is the absolute refractory period of the axon during which the membrane cannot support another action potential (as discussed earlier in this section). By the time the sodium channels have reverted to their closed state, the peak depolarization has passed further along the axon. It is then no longer capable of depolarizing the regions through which it has passed.

The precise way in which an action potential is propagated depends on whether the axon is myelinated or unmyelinated. In

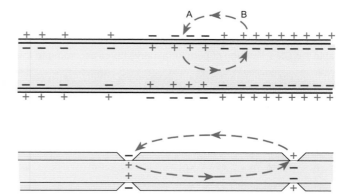

Figure 7.7 A simplified diagram illustrating the local circuit theory of action potential propagation in unmyelinated and myelinated axons. In unmyelinated axons (top) the local circuit current passes from the active region (point A) to the neighbouring resting membrane (point B). The current flow progresses smoothly along the axon membrane without discontinuous jumps. In myelinated fibres (bottom), the current can only cross the axon membrane at the nodes of Ranvier where there are breaks in the insulating layer of myelin. As a result, the action potential is propagated in a series of jumps (this is called saltatory conduction).

unmyelinated axons, the axonal membrane is in direct electrical contact with the extracellular fluid for the whole of its length via the mesaxon (as discussed in Section 7.2 above). Each region depolarized by the action potential results in the spread of current to the immediately adjacent membrane, which then becomes depolarized. This ensures that the action potential propagates smoothly along the axon in a continuous wave. In myelinated axons, the axon membrane is insulated from the extracellular fluid by the layers of myelin except at the nodes of Ranvier, where the axon membrane is in contact with the extracellular fluid. In this case, an action potential at one node of Ranvier completes its local circuit via the next node. This enables the action potential to jump from one node to the next—a process called **saltatory conduction**. This adaptation serves to increase the rate at which the action potential is propagated so that, size for size, myelinated nerve fibres have a much higher conduction velocity than unmyelinated fibres.

What factors determine the conduction velocity of axons?

The speed with which mammalian axons conduct action potentials (i.e. their **conduction velocity**) varies from less than 0.5 m s^{-1} for small unmyelinated fibres to over 100 m s^{-1} for large myelinated fibres (see Table 7.1). Why are there such wide variations in conduction velocity, and what factors determine the speed of conduction?

The spread of current along an axon from an active region to an inactive region mainly depends on three factors:

1. the resistance of the axon to the flow of electrical current along its length (its internal electrical resistance)
2. the electrical resistance of the axon membrane
3. the electrical capacitance of the axon membrane.

In unmyelinated fibres, the electrical resistance and capacitance are the same for each unit area of membrane regardless of the diameter

of the axon. In these fibres, the spread of current from the active to the neighbouring inactive region is determined mainly by the internal electrical resistance of the axon, which decreases as the axon diameter increases. Thus, large unmyelinated fibres have a lower internal electrical resistance and conduct action potentials faster than small unmyelinated axons because the spread of current from the active region is greater.

In myelinated fibres, the situation is a little more complicated as the axon membrane is electrically insulated by the layers of myelin. This has two effects:

1. the electrical resistance of the axon membrane is higher than for unmyelinated fibres
2. the membrane capacitance is lower than that of unmyelinated fibres.

These two factors combine to allow the depolarizing influence of an action potential to spread much farther from the active region. To take advantage of the greater current spread, the axon membrane is exposed only at the nodes of Ranvier. Thus, the action potential in a myelinated axon must jump from node to node as described above. The thickness of the myelin and the internodal distance are directly related to fibre size. Large axons have the thickest myelin and the greatest internodal distance. As a result, large myelinated axons have the highest conduction velocity.

If myelination of axons confers such advantages in terms of conduction velocity, why are some axons unmyelinated? Myelin is largely made of fats and represents a considerable investment in terms of metabolic energy, so a balance needs to be struck between the requirements for rapid signalling and the cost of maintaining such a sophisticated structure. Not all information needs to be transmitted rapidly, and some axons are very short (e.g. those that remain wholly within a particular region of the brain or spinal cord). In these cases, the speed advantages conferred by myelination are not great. Where the speed of conduction is not of paramount

Table 7.1 Classification of peripheral nerve fibres according to their function and conduction velocity

Diameter (µm)	Conduction velocity (m s^{-1})	Fibre classification	Cutaneous nerve classification	Function	
15–20	70–120	Aα	–	motor	control of skeletal muscles
15–20	70–120	Aα	Ia	sensory	innervation of primary muscle spindles
			Ib		innervation of Golgi tendon organs
5–10	30–70	Aβ	II	sensory	cutaneous senses; secondary innervation of muscle spindles
3–6	15–30	Aγ	–	motor	control of intrafusal muscle fibres
2–5	12–30	Aδ	III	sensory	cutaneous senses, especially pain and temperature
≈3	3–15	B	–	motor	autonomic preganglionic fibres
0.5–1.0	0.5–2.0	C	–	motor	autonomic postganglionic fibres
0.5–1.0	0.5–2.0	C	IV	sensory	cutaneous senses—pain and temperature

Note: As the table indicates, two separate classifications have been used for axons. The classification shown in the third column was developed by Erlanger and Gasser between 1922 and 1935 to label the peaks seen in the compound action potential of peripheral nerves and classifies axons solely on the basis of their conduction velocity. The classification shown in the fourth column was introduced by Lloyd and Chang in 1948. It applies to afferent (sensory) fibres only and is intended to indicate the position and innervation of specific types of receptor. Both systems of classification are in current use.

importance, axons have either a thin layer of myelin (Aδ fibres) or they are completely unmyelinated (C-fibres) (see Table 7.1). Some invertebrates, such as the squid, have very large unmyelinated nerve fibres that can conduct as rapidly as Aα (Group Ia) mammalian nerve fibres. These fibres are 0.5 to 1 mm in diameter, compared to about 18 μm for a mammalian Aα fibre with a similar conduction velocity. A mammalian nerve trunk of the same cross-sectional area as one squid giant axon can therefore contain more than 400 nerve fibres. Thus, the evolution of myelinated nerve fibres has permitted mammals to pack many more rapidly conducting fibres into the amount of space that would be occupied by one giant axon.

Action potentials can be recorded from intact nerves

While it is possible to record action potentials from individual axons, for diagnostic purposes it is more useful to stimulate a nerve trunk through the skin and record the summed action potentials of all the different fibres present (Figure 7.8a). This summed signal is called a **compound action potential**. Unlike individual nerve fibres, which show an all-or-none response, the action potential of a nerve trunk is graded with stimulus strength (Figure 7.8b). For weak stimuli below the threshold, no action potentials are elicited. Above threshold, those axons with the lowest threshold are excited first. As the stimulus intensity is increased, more axons are excited and the action potential becomes progressively larger. In this way, the compound action potential grows with increasing stimulus strength until all the available nerve fibres have been excited.

Compound action potentials exhibit both absolute and relative refractory periods. Just as with a single axon, a second stimulus applied within 1–2 ms of the first will fail to elicit an action potential (absolute refractory period). As the interval between successive stimuli is increased, a second action potential is generated which progressively grows in amplitude until it reaches the amplitude of the first (Figure 7.8c). During the relative refractory period, more

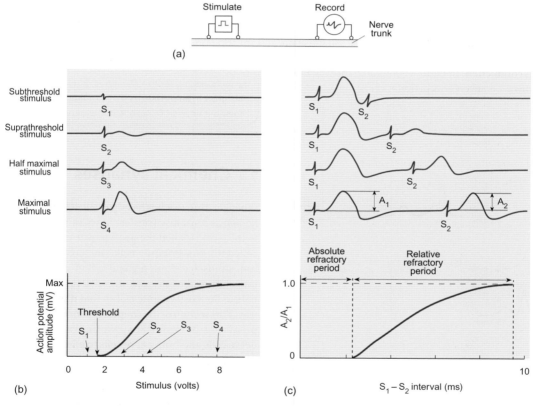

Figure 7.8 Some characteristic properties of compound action potentials recorded from nerve trunks. Panel (a) shows a simple diagram of the arrangement of the stimulating and recording electrodes. Panel (b) illustrates the progressive increase in the amplitude of a compound action potential with increasing stimulus strength. As the strength of the electrical stimulus given to the nerve increases, the resulting action potential becomes progressively larger in amplitude until it reaches a maximum value. S_1 is a very weak shock that does not elicit an action potential (a *subthreshold* stimulus). S_2–S_4 elicit action potentials of progressively larger amplitude until a maximum is reached (S_4 is called a *maximal* stimulus). Panel (c) shows the change in excitability that follows the passage of an action potential when two shocks (S_1 and S_2) of equal intensity are given at various intervals. If S_2 follows S_1 within 2 ms, no fibres are excited (top record—this is the absolute refractory period). As the interval between S_1 and S_2 is increased, more and more fibres are excited by S_2 (middle two records) until all are excited (bottom record). The change in the ratio of the amplitudes of the first and second compound action potentials with the interval between the stimuli is shown at the bottom of the panel. Note the relative durations of the absolute and relative refractory periods.

and more of the fibres in the nerve trunk recover their excitability as the interval between the two stimuli increases.

When a long length of nerve is stimulated and the compound action potential is recorded well away from the stimulating electrode, a number of peaks become visible in the record (see Figure 7.9). These different peaks reflect differences in the conduction velocity of the different axons present within the nerve trunk. Axons can be classified on the basis of their conduction velocity and physiological role (see Table 7.1). The axons with the highest conduction velocity are motor fibres and the sensory fibres involved in motor control (see Chapter 10). Unmyelinated fibres (sometimes called C-fibres) have the slowest conduction velocity of all and serve to transmit sensory information (e.g. pain and temperature) to the CNS.

Some diseases, such as Guillain–Barré syndrome, are characterized by a loss of myelin from axons. These diseases, which are known as **demyelinating diseases,** may be diagnosed by measuring the conduction velocity of peripheral nerves as the affected nerves have an abnormally slow conduction velocity. The loss of myelin affects conduction velocity because the normal pattern of current spread is disrupted. Action potentials are propagated through the affected region by continuous conduction similar to that seen in small unmyelinated fibres rather than by saltatory conduction as seen in healthy fibres. In severe cases, there may even be a total failure of conduction. In either eventuality, the function of the affected pathways is severely impaired.

Summary

Nerve cells transmit information along their axons by means of action potentials. This enables them to transmit signals rapidly over considerable distances. Large-diameter axons conduct action potentials more rapidly than small-diameter axons, and myelinated axons conduct impulses faster than unmyelinated axons.

Action potentials are generated when a neuron is activated by a depolarization of its membrane potential that reaches a certain level (c. −55 mV) known as the threshold. Each action potential has approximately the same magnitude and duration. An action potential is caused by a large, short-lived increase in the permeability of the plasma membrane to sodium ions caused by the opening of voltage-gated sodium channels. These channels spontaneously close a short time after they have opened (inactivation). At the same time, voltage-gated potassium channels begin to open so that the duration of the action potential is limited to about 1 ms.

After the passage of an action potential, an axon cannot propagate another until sufficient sodium channels have returned to their resting state. This period of inexcitability is known as the absolute refractory period. The axon is less excitable than normal for a few milliseconds after the absolute refractory period has ended. This is known as the relative refractory period.

An electrical stimulus applied to a nerve trunk elicits a compound action potential, which is the summed activity of all the nerve fibres present in the nerve trunk. Compound action potentials grow in amplitude as the stimulus strength is increased above threshold until all the axons in the nerve are recruited.

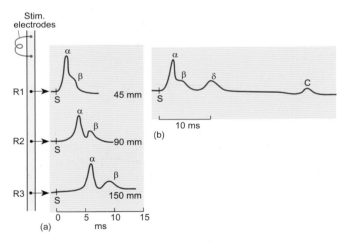

Figure 7.9 (a) shows a compound action potential recorded at different points along an intact nerve (shown as R1 to R3 in the diagram on the left of the figure). Note that as the action potential travels along the nerve, it breaks up into two distinct components that propagate at different velocities (marked α and β—the slower components are omitted for clarity). Panel (b) shows all of the components of the compound action potential recorded at position R1. Each wave reflects the activity of a group of fibres with a similar conduction velocity. The fastest are the α fibres, which have a maximum conduction velocity of about 100 m s⁻¹, and the slowest are the C fibres, which have a conduction velocity of about 1 to 1.5 m s⁻¹. (Adapted from J. Erlanger and H.S. Gasser (1937) *Electrical signs of nervous activity.* University of Pennsylvania Press, Philadelphia.)

7.4 Chemical synapses

When an axon reaches its target cell, it forms a specialized junction known as a synapse (see Section 6.2). The nerve cell that gave rise to the axon is called the **presynaptic neuron** and the target cell is called the **postsynaptic cell**. The target cell may be another neuron, a muscle cell, or a gland cell. A synapse has the function of transmitting the information coded in a sequence of action potentials to the postsynaptic cell so that it responds in an appropriate way. This section of the chapter is concerned with the fundamental properties of chemical **synaptic transmission**, both between nerve cells and from nerve cells to effector organs such as skeletal muscle and secretory glands.

Synaptic transmission is a one-way signalling mechanism—information flows from the presynaptic to the postsynaptic cell and not in the other direction. When the activity of the postsynaptic cell is increased, the synapse is called an **excitatory synapse**. Conversely, when activity in the presynaptic neuron leads to a fall in activity of the postsynaptic cell, the synapse is called an **inhibitory synapse**. In mammals, including man, most synapses operate by the secretion of a small quantity of a chemical (a neurotransmitter) from the nerve terminal. This type of synapse is called a **chemical synapse**.

In some instances, a synapse operates by transmitting the electrical current generated by the action potential to the postsynaptic cell via gap junctions. This type of synapse is called an **electrical synapse**. Electrical synapses are found in the retina and in some other areas of the brain.

The structure of chemical synapses

When an axon reaches its target cell, it loses its myelin sheath (if it has one) and ends in a small swelling known as a nerve terminal or synaptic bouton (see Figure 7.10). A nerve terminal together with the underlying membrane on the target cell constitutes a synapse. The nerve terminals contain mitochondria and a large number of small vesicles known as **synaptic vesicles**. The membrane immediately under the nerve terminal is called the **postsynaptic membrane**. It contains electron-dense material that makes it appear thicker than the plasma membrane outside the synaptic region. This is known as the postsynaptic thickening (see the lower panel of Figure 7.10). The postsynaptic membrane contains specific receptor molecules for the neurotransmitter released by the nerve terminal. Between a nerve terminal and the postsynaptic membrane there is a small gap of about 20 nm, which is known as the **synaptic cleft**.

Nerve cells employ a wide variety of signalling molecules as neurotransmitters, including **acetylcholine, dopamine, noradrenaline** (norepinephrine), glutamate, γ-aminobutyric acid (GABA), **serotonin** (5-HT), and many peptides such as substance P and the enkephalins (see Table 7.2). Chemical analysis of nerve terminals isolated from the brain has shown that they contain high concentrations of neurotransmitters. By careful subcellular fractionation

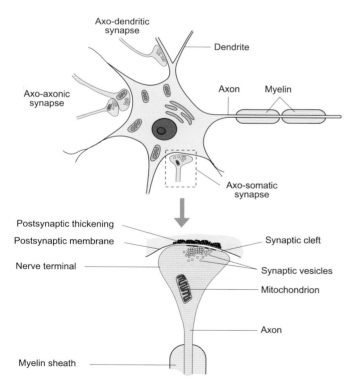

Figure 7.10 A diagrammatic view of the principal types of synaptic contact in the central nervous system (CNS); below, the detailed structure of a typical CNS chemical synapse.

of nerve terminals, the synaptic vesicles can be separated from the other intracellular organelles. Analysis of the contents of the isolated vesicles has shown that they contain almost all of the neurotransmitter present in the terminal.

How does a chemical synapse work?

The transmission of information across a synapse occurs when an action potential reaches the presynaptic nerve terminal. The nerve terminal becomes depolarized and this depolarization causes voltage-gated calcium channels in the presynaptic membrane to open. Calcium ions flow into the nerve terminal down their electrochemical gradient, and the subsequent rise in free calcium triggers the fusion of one or more synaptic vesicles with the presynaptic membrane, resulting in the secretion of neurotransmitter into the synaptic cleft. In nerve terminals, this secretory process is extremely rapid and occurs in less than 0.25 ms after the arrival of the action potential. The secreted transmitter diffuses across the synaptic cleft and binds to receptors on the postsynaptic membrane. Subsequent events depend on the kind of receptor present. If the transmitter activates a ligand-gated ion channel, synaptic transmission is usually both rapid and short-lived. This type of transmission is called **fast synaptic transmission**; an example is the action of acetylcholine at the neuromuscular junction (see Section 7.5 below). If the neurotransmitter activates a G protein linked receptor, the change in the postsynaptic cell is much slower in onset and lasts much longer. This type of synaptic transmission is called **slow synaptic transmission**. A typical example is the excitatory action of noradrenaline (norepinephrine) on α_1-adrenoceptors in the peripheral blood vessels.

Fast excitatory synaptic transmission occurs when a neurotransmitter (e.g. acetylcholine or glutamate) is released from the presynaptic nerve ending and is able to bind to and open non-selective cation channels. The opening of these channels causes a brief **depolarization** of the postsynaptic cell. This shifts the membrane potential closer to the threshold for action potential generation and so renders the postsynaptic cell more excitable. When the postsynaptic cell is a neuron, the depolarization is called an **excitatory postsynaptic potential** or **epsp**. A single epsp occurring at a fast synapse reaches its peak value within 1–5 ms of the arrival of the action potential in the nerve terminal and decays to nothing over the ensuing 20–50 ms (see Figure 7.11).

How does activation of a non-selective cation channel lead to depolarization of the postsynaptic membrane? The membrane potential is determined by the distribution of ions across the plasma membrane and the permeability of the membrane to those ions (see Chapter 5). At rest, the membrane is much more permeable to potassium ions than it is to sodium ions. The membrane potential (about –70 mV) is therefore close to the equilibrium potential for potassium (about –90 mV). If, however, the membrane were to be equally permeable to sodium and potassium ions, the membrane potential would be close to zero (i.e. the membrane would be depolarized). Consequently, when a neurotransmitter such as acetylcholine opens a non-selective cation channel in the postsynaptic membrane, the membrane depolarizes at the point of excitation. The exact value of the depolarization will depend on

Table 7.2 Some neurotransmitters and their receptors

Class of compound	Specific example	Receptor types	Physiological role	Mechanism of action
ester	acetylcholine	nicotinic	fast excitatory synaptic transmission, especially at neuromuscular junction	activates ion channels
		muscarinic	both excitatory and inhibitory slow synaptic transmission, depending on tissue: e.g. slowing of heart; contraction of visceral smooth muscle	acts via G protein
monoamine	noradrenaline	various α and β adrenoceptors	slow synaptic transmission in CNS and smooth muscle	acts via G protein
	serotonin (5-HT)	various 5-HT receptors (e.g. 5-HT_{1A}, 5-HT_{2A}, etc)	slow synaptic transmission in CNS and periphery (smooth muscle and gut)	acts via G protein
		$5HT_3$	fast excitatory synaptic transmission	activates ion channels
	dopamine	D1, D2 receptors	slow synaptic transmission in CNS and periphery (blood vessels and gut)	acts via G protein
amino acid	glutamate	AMPA	fast excitatory synaptic transmission in CNS	activates ion channels
		NMDA	slow excitatory synaptic transmission in CNS	activates ion channels
		metabotropic	neuromodulation	acts via G protein
	GABA	$GABA_A$	fast inhibitory synaptic transmission in CNS	activates ion channels
		$GABA_B$	slow inhibitory synaptic transmission in CNS	acts via G protein
peptide	substance P	NK_1	slow excitation of smooth muscle and neurons in CNS	acts via G protein
	enkephalins	μ/δ-opioid	slow synaptic signalling (reduction in excitability)	act via G protein
			decrease in gut motility	
			reduction in sensitivity to pain (analgesia)	
	β-endorphin	μ-opioid	slow synaptic signalling analgesia	acts via G protein

how many channels have been opened, as this will determine how far the membrane's permeability to sodium ions has increased relative to that of potassium. In neurons, single epsps rarely exceed a few mV, while at the neuromuscular junction the synaptic potential (known as the **endplate potential**—see Section 7.5) has an amplitude of about 40 mV.

Fast inhibitory synaptic transmission occurs when a neurotransmitter such as GABA or glycine is released from a presynaptic nerve ending and is able to activate chloride channels in the postsynaptic membrane. The opening of these channels causes the postsynaptic cell to become **hyperpolarized** (i.e. more negative) for a brief period. This negative shift in membrane potential is called an **inhibitory postsynaptic potential** or **ipsp** as the membrane potential is moved farther away from threshold. A single ipsp occurring at a fast synapse reaches its peak value within 1–5 ms of the arrival of the action potential in the nerve terminal and decays to nothing within a few tens of milliseconds (see Figure 7.12).

Why does the membrane hyperpolarize during an ipsp? The resting membrane potential is less (at about –70 mV) than the equilibrium potential for potassium ions (which is about –90 mV) as the membrane has a small permeability to sodium ions. The distribution of chloride ions across the plasma membrane, however, mirrors that of potassium (see Chapter 5), so that the chloride equilibrium potential lies close to that of potassium. When an inhibitory neurotransmitter such as GABA opens chloride channels, the permeability of the membrane to chloride ions is increased, and chloride ions flow down their electrochemical gradient into the cell. This increase in chloride permeability shifts the membrane potential towards a more negative value that is closer to the equilibrium potential for chloride (and potassium) ions. The extent of the hyperpolarization will depend on how much the chloride permeability has increased relative to that of both potassium and sodium. Ipsps are generally small in amplitude—about 1–5 mV. Nevertheless, as they last for tens of milliseconds they play an important role in determining the excitability of neurons.

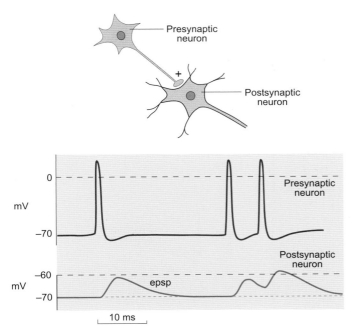

Figure 7.11 Diagram to show the relationship between an action potential in a presynaptic neuron and the generation of an excitatory postsynaptic potential (epsp) in the postsynaptic neuron. Note that if two epsps occur in rapid succession they summate, and the final degree of depolarization is greater than either epsp could achieve on its own (this is an example of temporal summation—see also Figure 7.13). If the depolarization reaches threshold, an action potential will occur in the post-synaptic neuron.

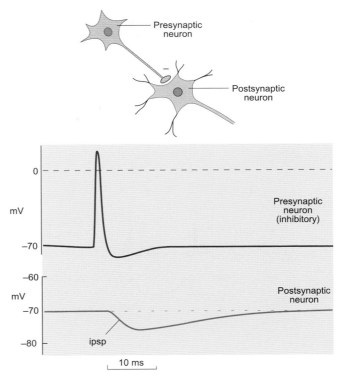

Figure 7.12 Diagram to show the relationship between an action potential in a presynaptic neuron and the generation of an inhibitory postsynaptic potential (ipsp) in the postsynaptic neuron. Note that in this case the membrane potential hyperpolarizes (it becomes more negative) and thus moves further away from the threshold for action potential generation. This leads to a decrease in the excitability of the postsynaptic neuron.

Epsps and ipsps are not all-or-none events but are graded with the intensity of activation. They have a duration of tens of milliseconds so that, if a synapse is activated repeatedly, the resulting synaptic potentials become superimposed on one another, a phenomenon called **temporal summation** (see Figure 7.13 i–iii). Moreover, as individual neurons receive very many synaptic contacts, it is possible for two synapses on different parts of a neuron to be activated at the same time. The resulting synaptic potentials also summate, but, because they were activated at different points on the cell, this type of summation is called **spatial summation** (see Figure 7.13 iv, v).

The amplitude of an epsp at a given synapse is not fixed but depends on the interval between successive action potentials in the presynaptic neuron. At many synapses, if the interval is short, successive epsps tend to increase in amplitude, a phenomenon known as **frequency potentiation**. This is a presynaptic phenomenon and is distinct from temporal summation. It is the result of an increase in the amount of transmitter secreted in response to the action potential in the presynaptic axon. At certain synapses, the amplitude of an epsp after a rapid sequence of action potentials is greater than it was before, and this increase in amplitude may last many hours. This phenomenon is called **long-term potentiation** (see Chapter 18) and is an example of synaptic plasticity (the ability of the strength of a synaptic contact to be modified by previous activity).

The changes in the membrane potential of a neuron that occur as a result of the synaptic activity occurring all over its surface determine its ability to generate an action potential. Within the CNS and autonomic ganglia, the epsps that occur as a result of activation of individual synapses are small—often only a few mV in amplitude. If the cell is already relatively depolarized by the activity of various excitatory synapses, activation of a further set of excitatory synapses may depolarize the cell sufficiently for the membrane potential to reach the threshold for action potential generation. The cell will be excited and will transmit an action potential to its target cells. If the cell is already hyperpolarized by ipsps following the activation of inhibitory synapses, the same stimulus may not be sufficient to trigger an action potential.

Presynaptic inhibition

As Figure 7.10 shows, synapses may occur between a nerve terminal and the cell body of the postsynaptic cell (axo-somatic synapses), between a nerve terminal and a dendrite on the postsynaptic cell (axo-dendritic synapses), and between a nerve terminal and the terminal region of another axon (axo-axonic synapses).

Axo-axonic synapses are generally neuromodulatory, acting to increase or decrease the magnitude of the action potential depolarization invading the presynaptic terminal. This means that an inhibitory input will reduce the likelihood of neurotransmitter release, while an

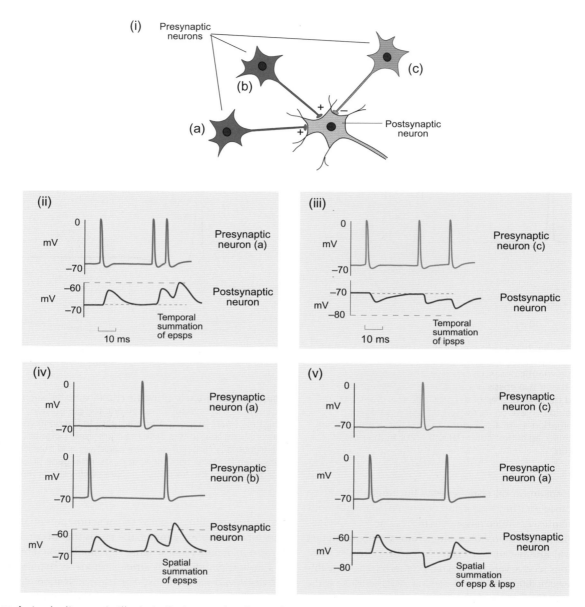

Figure 7.13 A simple diagram to illustrate the temporal and spatial summation of synaptic potentials. Consider a group of four neurons such as those at the top of the figure (i). Action potentials in the presynaptic neurons (a–c) elicit excitatory postsynaptic potentials (epsps) and inhibitory postsynaptic potentials (ipsps) in the postsynaptic neuron. If two epsps occur in quick succession at a particular synaptic contact they summate, as shown in panel (ii) (temporal summation). Ipsps also show temporal summation—panel (iii). If epsps occur at different synaptic contacts within a short time of one another, they exhibit spatial summation —panel (iv). Epsps and ipsps can also undergo spatial summation, as shown in panel (v).

excitatory input will increase the release of neurotransmitter. The best-studied type of such connections are inhibitory inputs onto excitatory nerve terminals, leading to a reduction in synaptic excitation. This impairment of synaptic transmission is called **presynaptic inhibition** to distinguish it from the inhibition that results from an ipsp occurring in a postsynaptic neuron (**postsynaptic inihibition**). Unlike postsynaptic inhibition, which changes the excitability of the postsynaptic cell, presynaptic inhibition permits the selective blockade of a specific synaptic connection without altering the excitability of the postsynaptic neuron.

Neurotransmitters may directly activate ion channels or act via second messenger systems

As with other kinds of chemical signalling, synaptic signalling is mediated by a wide variety of substances. They can be grouped into six main classes:

1. esters—acetylcholine (ACh);

2. monoamines—such as noradrenaline, dopamine, and serotonin;

3. amino acids—such as glutamate and GABA;

4. purines—such as adenosine and ATP

5. peptides—such as the enkephalins, substance P, and vasoactive intestinal polypeptide (VIP)

6. inorganic gases—nitric oxide (NO).

Most of these transmitters are able to activate both ion channels and G protein linked receptors as shown in Table 7.2, which also includes a brief outline of the role of various types of receptor. A particular synapse may utilize more than one neurotransmitter (a process called cotransmission, discussed later in this section).

Slow synaptic transmission plays a major role in regulating the internal environment

Like fast synaptic transmission, slow synaptic excitation and inhibition result from changes in the relative permeability of the membrane to the principal extracellular ions. They differ in their time course because the channels that are responsible for slow synaptic transmission are regulated over a longer timescale, both by second messengers such as cyclic AMP and by G protein subunits (see Chapter 6).

Fast and slow synaptic transmission serve different functions. In those synapses that support fast synaptic excitation or inhibition, the ligand-gated channels lie under the nerve terminal and are rapidly activated by high concentrations of neurotransmitter. The high concentration is achieved by the release of a small quantity of neurotransmitter into the narrow synaptic cleft. As the transmitter diffuses away from the synaptic cleft, its concentration falls rapidly and is too low to affect neighbouring cells. This type of transmission is therefore highly specific for the contact between a presynaptic neuron and its target cell and is well adapted to serve a role in the rapid processing of sensory information and the control of locomotion.

In slow synaptic transmission, the neurotransmitters are usually not secreted onto a specific point on the postsynaptic cell. Instead, the nerve fibres have a number of swellings (varicosities) along their length that secrete neurotransmitters into the extracellular fluid close to a number of cells. The synapses of the autonomic nervous system are of this kind (Figure 7.14). Slow synaptic transmission is of great importance for the control of such varied functions as the cardiac output, the calibre of blood vessels, and the secretion of hormones. Within the CNS, slow synaptic transmission may underlie changes of mood and the control of appetites, e.g. **hunger** and thirst.

Cotransmission

Some nerve terminals are known to contain two different kinds of neurotransmitter. When such a nerve ending is activated, both neurotransmitters may be released. This is called cotransmission. The parasympathetic nerves of the salivary glands are an example of cotransmission, as they release both acetylcholine and vasoactive intestinal polypeptide (VIP) when they are activated. In this case, the acetylcholine acts on the acinar cells to increase secretion and the VIP acts on the smooth muscle of the arterioles to increase the local blood flow (see Chapter 11, Figure 11.7).

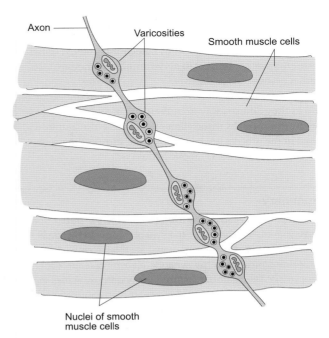

Figure 7.14 A diagram of the axonal varicosities of adrenergic autonomic nerve fibres. Once an axon has reached its target, it courses through the tissue (in this case smooth muscle) and forms a series of varicosities, each of which contains neurotransmitter (noradrenaline) packaged in vesicles. Unlike most CNS synapses or the skeletal neuromuscular junction, the autonomic fibres release their neurotransmitter into the extracellular space in a diffuse manner, where it affects a number of smooth muscle cells.

What limits the duration of action of a neurotransmitter?

Neurotransmitters are highly potent chemical signals that are secreted in response to very specific stimuli. If the effect of a particular neurotransmitter is to be restricted to a particular synapse at a given time, there needs to be some means of terminating its action. This can be achieved in one of three different ways:

1. by rapid enzymatic destruction

2. by uptake either into the secreting nerve terminals or into neighbouring cells

3. by diffusion away from the synapse followed by enzymatic destruction, uptake, or both.

Of the currently known neurotransmitters only acetylcholine is inactivated by rapid enzymatic destruction. It is hydrolysed to acetate and choline by the enzyme acetylcholinesterase. The acetate and choline are not effective in stimulating the cholinergic receptors but they may be taken up by the nerve terminal and resynthesized into new acetylcholine. The role of acetylcholine as a fast neurotransmitter at the neuromuscular junction is discussed below in more detail.

The monoamines (such as noradrenaline) are inactivated by uptake into the nerve terminals where they may be reincorporated into synaptic vesicles for subsequent release. Any monoamine that is not removed by uptake into a nerve terminal is metabolized either by monoamine oxidase or by catechol-O-methyl transferase. These enzymes are present in nerve terminals. They are also found in other tissues such as the liver.

Peptide neurotransmitters become diluted in the extracellular fluid as they diffuse away from their site of action and are subsequently destroyed by extracellular peptidases. The amino acids released in the process are taken up by the surrounding cells, where they enter the normal metabolic pathways.

Summary

At a chemical synapse, the axon terminal is separated from the postsynaptic membrane by the synaptic cleft. In response to an action potential, the nerve terminal secretes a small quantity of neurotransmitter into the synaptic cleft by calcium-dependent exocytosis. The neurotransmitter rapidly diffuses across the synaptic cleft and binds to specific receptors to cause a short-lived change in the membrane potential of the postsynaptic cell.

Activation of an excitatory synapse elicits an excitatory postsynaptic potential (epsp) in the postsynaptic cell (a depolarization). Activation of an inhibitory synapse elicits an inhibitory postsynaptic potential (ipsp) in the postsynaptic cell (a hyperpolarization). Epsps and ipsps are graded in intensity and outlast the action potentials that initiated them. As a result, synaptic potentials can be superimposed on each other leading to temporal and spatial summation.

Many different kinds of chemical can serve as neurotransmitters. Some nerve terminals secrete more than one kind of neurotransmitter and some neurotransmitters activate more than one kind of receptor at the same synapse.

7.5 Neuromuscular transmission is an example of fast synaptic signalling at a chemical synapse

The neurons that are directly responsible for controlling the activity of skeletal muscle fibres are known as **motoneurons**, and the nerves that transmit signals from the CNS to the skeletal muscles are known as **motor nerves**. The process of transmitting a signal from a motor nerve to a skeletal muscle to cause it to contract is called **neuromuscular transmission**. The individual axons of the motor nerves (the motor axons) that supply skeletal muscle are myelinated and branch as they enter a muscle, each branch making contact with a single muscle fibre (see Figure 7.15). As a result, each individual motoneuron controls the activity of a number of muscle fibres. The motoneuron and the muscle fibres it controls form a **motor unit**, and an action potential in the motoneuron will cause the contraction of all the muscle fibres to which it is connected (see Chapter 8).

The region of contact between a motor axon and a muscle fibre is called the **neuromuscular junction**. As a motor axon

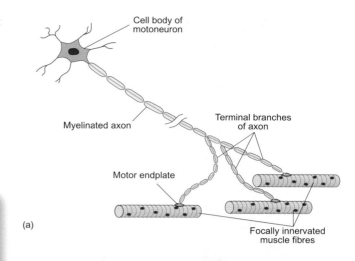

(a)

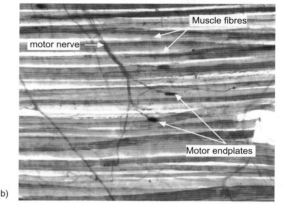

(b)

Figure 7.15 A diagram of the innervation of mammalian skeletal muscle and the structure of the neuromuscular junction. Panel (a) shows the general organization of a motor unit. A motoneuron in the CNS gives rise to a motor axon, which branches to supply a number of muscle fibres. The motoneuron and the muscle fibres it supplies form a single motor unit. In panel (b), a photomicrograph shows the motor innervation of a mammalian muscle. Note the branching of the motor nerve trunk and the enlargement of the individual axon terminals that form the motor endplates.

approaches its terminal, it loses its myelin sheath and runs along the surface of the muscle membrane to form a complex nerve terminal called the **motor endplate** (Figure 7.16a). The terminals of motor axons contain mitochondria and large numbers of synaptic vesicles, which contain the neurotransmitter acetylcholine (ACh). Beneath the axon terminal, the muscle membrane is thrown into elaborate folds known as junctional folds (see Figure 7.16b). This is the postsynaptic region of the muscle fibre membrane and it contains the nicotinic receptors that bind the acetylcholine. Finally, as with other chemical synapses, there is a small gap of about 20 nm, the **synaptic cleft**, separating the axon membrane from that of the muscle fibre. The synaptic cleft contains acetylcholinesterase, which rapidly inactivates acetylcholine by breaking it down to acetate and choline. Figure 7.17 shows a transmission electron micrograph of a section through the neuromuscular junction.

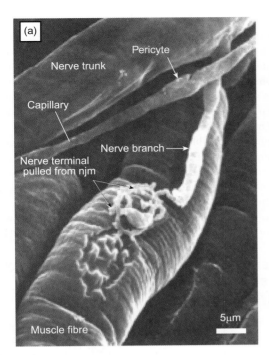
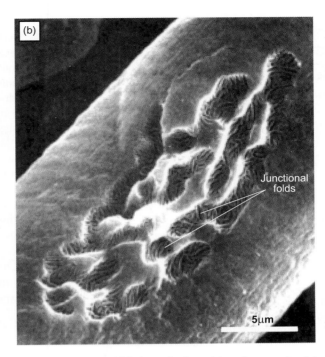

Figure 7.16 Scanning electron micrograph of the endplate region of mouse muscle. Panel (a) shows the branching of a nerve trunk to innervate a muscle fibre. In this example the axon terminal has become almost totally detached from the muscle fibre, revealing the grooves in the muscle membrane where it is normally situated. Panel (b) is a higher-power view of the grooves in the muscle membrane, revealing the junctional folds. (Figs 1 and 2 in Y. Matsuda, et al. (1988) *Muscle & Nerve* **11**, 1266–71. Reproduced with permission of John Wiley and Sons.)

The neuromuscular junction operates in a similar way to other chemical synapses.

1) First, an action potential in the motor axon invades the nerve terminal and depolarizes it.

2) This depolarization opens voltage-gated calcium channels in the terminal membrane, and calcium ions flow into the nerve terminal down their electrochemical gradient.

3) This leads to a local rise in free calcium within the terminal, which triggers the fusion of docked synaptic vesicles with the plasma membrane.

4) The acetylcholine contained within the synaptic vesicles is released into the synaptic cleft.

5) Acetylcholine molecules diffuse across the cleft and bind to the nicotinic receptors on the postsynaptic membrane.

6) When the receptors bind acetylcholine, non-selective cation channels open and this depolarizes the muscle membrane in the endplate region. This depolarization is called the **endplate potential** or **epp**.

7) Finally, when the epp has reached threshold, the muscle membrane generates an action potential that propagates along the length of the muscle fibre and triggers its contraction. The muscle action potential is the first stage of a process called excitation–contraction coupling (see Chapter 8).

The epp is confined to the endplate region of the muscle fibre (Figure 7.18). If the electrical activity of the muscle membrane

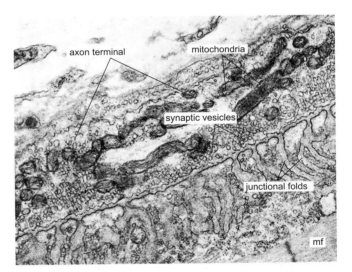

Figure 7.17 A transmission electron micrograph of a neuromuscular junction, showing the principal structures. Note the folds under the axon terminal (the junctional folds) and the large number of synaptic vesicles in the nerve endings (mf = myofibril). (Fig 4 in H.A. Padykula and G.F. Gauthier (1970) *Journal of Cell Biology* **46**, 27–41.)

adjacent to the endplate is examined closely, small spontaneous depolarizations of about 1 mV are observed even in the absence of action potentials in the motor axon. These small depolarizations are similar in time course to the epp and are called **miniature**

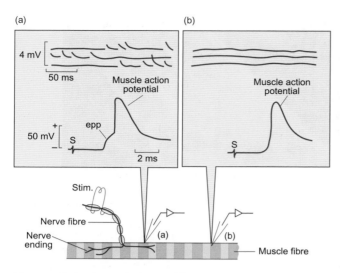

Figure 7.18 A schematic drawing illustrating the electrical activity in the membrane of frog skeletal muscle fibres. Stimulating electrodes are applied to the motor nerve fibre, and intracellular recording electrodes record the changes in the membrane potential of the muscle at the neuromuscular junction (a) and at a point remote from the junctional region (b). Note that small, spontaneous, randomly occurring miniature endplate potentials (mepps) occur in the junctional region but not elsewhere, and that the muscle action potential is preceded by the endplate potential (epp) only in the junctional region (the upstroke of the mepps was too fast to be recorded). (Based on Fig 1 of P. Fatt and B. Katz (1952) *Journal of Physiology* **117**, 109–28.)

endplate potentials or **mepps**. Each mepp is thought to reflect the release of the acetylcholine contained within a single synaptic vesicle. The mepps occur only in the junctional region and are both random and relatively infrequent in the resting muscle membrane. If the nerve terminal is artificially depolarized, however, the frequency of mepps increases. In 1952, this led B. Katz and P. Fatt to suggest that depolarization of the motor nerve terminal triggers the simultaneous release of many synaptic vesicles to give rise to the epp. This is known as the vesicular hypothesis of transmitter release. Subsequent work has strongly supported this idea. Indeed, it has proved possible to capture a picture of the fusion of synaptic vesicles with the plasma membrane of the nerve terminal by rapidly freezing the neuromuscular junction at the moment of transmitter release.

After the acetylcholine has activated the nicotinic receptors, it is rapidly broken down by acetylcholinesterase to acetate and choline. This enzymatic breakdown limits the action of acetylcholine to a few milliseconds. The acetate and choline can be taken up by the nerve terminal and resynthesized into acetylcholine for recycling as a neurotransmitter. If acetylcholinesterase is inhibited by a specific blocker such as **neostigmine**, activation of the motor nerve leads to a prolonged epp and to disruption of cholinergic transmission (depolarization block). Nerve gas (e.g. sarin) and organophosphates are poisons that also act on acetylcholinesterase.

The nicotinic acetylcholine receptors of the neuromuscular junction are among the most well characterized of all receptors. It is now clear that the binding site for acetylcholine is part of the same molecular complex as the ion channel. Thus, the binding of acetylcholine to the receptor directly opens the ion channel with no intervening step—an ideal adaptation for fast synaptic transmission. The receptors can be blocked by a number of specific drugs and poisons (called neuromuscular blockers), the best-known of which is curare—an arrow poison originally used by indigenous South American peoples for hunting.

Neuromuscular blockers such as pancuronium, rocuronium, and succinylcholine (suxamethonium) have long been used in general anaesthesia to prevent muscle contractions during complex surgery. (These drugs are often called muscle relaxants.) Since the respiratory muscles also rely on acetylcholine to trigger their contractions, this way of providing muscle relaxation is only possible while the patient's breathing is being supported by artificial respiration.

Denervation of skeletal muscle leads to atrophy and supersensitivity of the membrane to acetylcholine

If the motor nerve to a muscle is cut or crushed, the muscle becomes paralysed. After a time the nerve terminals degenerate and disappear and the endplate region becomes less sensitive to acetylcholine. The muscle becomes weak and atrophies.

Except at the endplate region, the muscle membrane is not normally sensitive to acetylcholine, but following denervation and degeneration of the endplate, the non-junctional region becomes sensitive to acetylcholine as the receptors diffuse away from the endplate. This is known as **denervation supersensitivity**. If the nerve regenerates and the endplate is reformed, the supersensitivity disappears and the muscle progressively regains its original strength. This type of influence of a nerve on its target cell is known as a **trophic action**. The main factor responsible for this type of interaction appears to be the pattern of the contractile activity of the muscle itself.

Functional denervation occurs in a disease of the neuromuscular junction called **myasthenia gravis**. Patients with this disease make antibodies against their own nicotinic receptors. The disease itself is characterized by progressive muscular weakness—especially of the cranial muscles—and drooping of the eyelids is a characteristic early sign. The circulating antibodies bind to the nicotinic receptors of the neuromuscular junction and reduce the number available for neuromuscular transmission. The decline in the number of active receptors leads to a progressive failure of neuromuscular transmission and progressive paralysis of the affected muscles. The symptoms of the disease can be ameliorated by giving the sufferer a low dose of an anticholinesterase such as neostigmine to prolong the action of the acetylcholine.

Fast and slow axonal transport are used to provide the distal parts of the axon with newly synthesized proteins and organelles

In an adult man or woman there is often a great distance between a motor endplate and the cell body of the motoneuron that gives rise to the axon. This raises an important issue: how do proteins and

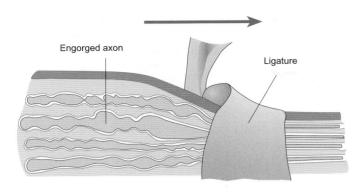

Figure 7.19 The effect of compression of a localized region of the sciatic nerve of a rat caused by a ligature. Note the engorged and distorted appearance of the axons proximal to the ligature. The arrow indicates the direction of axoplasmic flow.

other membrane constituents that are synthesized in the cell body reach the nerve terminal? The first experiments to answer this question were performed by P. Weiss in 1948, when he tied a ligature around the sciatic nerve of a rat so that it occluded the nerve without totally crushing it. After a few weeks, the axons had become swollen on the proximal side of the ligature (the side nearest the cell body), indicating that material had accumulated at the constriction (see Figure 7.19). When the ligature was removed, the accumulated material was found to progress towards the axon terminal at 1–2 mm per day. Weiss concluded that the cell body exported material to the distal parts of the axon. Subsequent studies have confirmed the original discovery and shown that axons transport materials in three ways:

1) by fast axonal transport from the cell body towards the nerve terminal at rates up to 400 mm per day, which is called anterograde axonal transport;

2) by slow anterograde axonal transport (0.2–4 mm per day);

3) by fast transport from the terminal to the cell body at rates of 200–300 mm per day, which is called retrograde axonal transport.

All newly synthesized organelles are exported to the axon and the distal dendrites by fast anterograde transport. This includes the synaptic vesicles and precursor molecules for peptide neurotransmission. Slow anterograde axonal transport is used to export both soluble cytoplasmic and cytoskeletal proteins. Both slow and fast anterograde axonal transport are much too rapid to occur by diffusion along the length of the axon, and it is known that axonal transport occurs along the microtubules. Drugs such as colchicine that disrupt the microtubules also inhibit axonal transport.

Retrograde axonal transport from the nerve terminal to the cell body enables neurons to recycle materials used by the nerve terminal during synaptic transmission. While retrograde axonal transport allows the recycling of structural components, it also allows the transport of growth factors (e.g. nerve growth factor) from the nerve terminal to the cell body. This permits two-way traffic between nerves and their target tissues and may play an important role during the development of the nervous system. Retrograde transport is exploited by some viruses to gain entry to the CNS (e.g. herpes simplex, polio, and rabies). Tetanus toxin also gains entry to the CNS via retrograde transport.

Summary

The neuromuscular junction of skeletal muscle is a classic example of a fast chemical synapse. The motor nerve endings release acetylcholine, which activates nicotinic cholinergic receptors to generate an endplate potential (epp). The epp triggers an action potential in the muscle membrane and causes the affected muscle fibre to contract.

The effect of acetylcholine is terminated by acetylcholinesterase. If this enzyme is inhibited, the muscle membrane becomes depolarized and neuromuscular transmission is blocked. Neuromuscular transmission can also be blocked by drugs such as curare that block the acetylcholine receptor.

Axons transport materials that are required for the normal function of the nerve terminal by anterograde axoplasmic flow and recover expended materials by retrograde transport.

✳ Checklist of key terms and concepts

Structure of nerve cells and axons

- The principal signalling units of the nervous system are the neurons.
- The cell bodies of neurons give rise to two types of process:
  - the dendrites, which receive contacts from other neurons;
  - a single axon, which transmits information via its branches to other neurons or to non-neuronal cells such as muscles.
- Individual axons may be either myelinated or unmyelinated.

- Myelin is formed by oligodendrocytes in the CNS and by Schwann cells in the periphery.
- After leaving the CNS, axons run in peripheral nerve trunks, which provide structural support.

Action potentials

- Nerve cells transmit information along their axons by means of action potentials. This enables them to transmit signals rapidly over considerable distances.

- Action potentials are generated when a neuron is activated by a depolarization of the membrane potential below about −55 mV. The membrane potential at which an action potential is initiated is known as the threshold. Each action potential has approximately the same magnitude and duration. This is known as the 'all or none' law.
- The action potential is caused by a large, short-lived increase in sodium permeability that results from the opening of voltage-gated sodium channels.
- Sodium channels spontaneously inactivate a short time after they have opened. This limits the duration of the action potential to about 1 ms.
- Voltage-gated sodium channels cannot reopen until they have been reprimed by spending a period at the resting membrane potential.
- After the passage of an action potential, an axon has a period of inexcitability known as the absolute refractory period.
- After the absolute refractory period has ended, an axon is less excitable than normal for a short time. This period is known as the relative refractory period.

Compound action potentials

- An electrical stimulus applied to a nerve trunk elicits a compound action potential, which is the summed activity of all the nerve fibres present in the nerve trunk.
- Compound action potentials grow in amplitude as the stimulus strength is increased above threshold until all the axons in the nerve are recruited.

Conduction of the action potential

- Large-diameter axons conduct faster than small-diameter axons, and myelinated axons conduct impulses faster than unmyelinated axons.
- Myelinated axons conduct impulses by saltatory conduction.

Synaptic transmission

- Nerve cells communicate with their targets via specialized contacts called synapses, in which a nerve terminal is separated from its target cell membrane by a small gap—the synaptic cleft.
- Synaptic transmission is a one-way process from an axon terminal to its target cell.
- At a chemical synapse the axon terminal is separated from the postsynaptic membrane by a small gap—the synaptic cleft.
- In response to an action potential the nerve terminal secretes a neurotransmitter into the synaptic cleft. The secretion of neurotransmitter occurs by calcium-dependent exocytosis of synaptic vesicles.

- The neurotransmitter rapidly diffuses across the synaptic cleft and binds to specific receptors to cause a short-lived change in the membrane potential of the postsynaptic cell.

Synaptic excitation and inhibition

- Activation of an excitatory synapse causes depolarization of the membrane potential, making the target more excitable. For this reason these depolarizations are called excitatory postsynaptic potentials (epsps).
- Activation of an inhibitory synapse leads to hyperpolarization of the membrane potential, rendering the target less excitable. These hyperpolarizing potentials are therefore called inhibitory postsynaptic potentials (ipsps).
- Both epsps and ipsps are graded in intensity and greatly outlast the action potentials that initiated them. As a result, synaptic potentials can be superimposed on each other, leading to temporal and spatial summation.
- Many different kinds of chemicals are utilized as neurotransmitters.
- Some nerve terminals secrete more than one kind of neurotransmitter and some neurotransmitters activate more than one kind of receptor at the same synapse.

Neuromuscular transmission

- The neuromuscular junction of skeletal muscle is a classic example of a fast chemical synapse.
- The motor nerve endings release acetylcholine, which activates nicotinic cholinergic receptors to depolarize the muscle membrane. The depolarization is known as an endplate potential or epp.
- The effect of acetylcholine is terminated by the enzyme acetylcholinesterase.
- Neuromuscular transmission can also be blocked by drugs such as curare that compete with acetylcholine for binding sites on the nicotinic receptors. Such drugs are known as neuromuscular blockers.
- The epp triggers an action potential in the muscle membrane that leads to contraction of the muscle. This is a key step in excitation–contraction coupling.

Axonal transport

- Axons transport materials required for the normal function of the nerve terminal by anterograde axonal transport.
- Axonal transport has a fast and a slow component.
- Axons utilize retrograde axonal transport to recycle material from their nerve endings.

Recommended reading

Pharmacology of synaptic transmission

Rang, H.P., Dale, M.M., Ritter, J.M., Flower, R.J., and Henderson, G. (2011) *Pharmacology* (7th edn), Chapters 36–38. Elsevier, London.

Physiology

Katz, B. (1966) *Nerve, muscle, and synapse*. McGraw-Hill Inc., New York.

A classic account of the physiology of nerves and muscle by one of the great figures in the field.

Kandel, E.R., Schwartz, J.H., and Jessell, T.M. (eds) (2000) *Principles of neural science* (4th edn), Chapters 7–16. McGraw-Hill Medical, New York.

A clear account with nice illustrations. A good starting point for advanced studies.

Levitan, I.B., and Kaczmarek, L.K. (2001) *The neuron*, Chapters 3–6 and 9–12. Oxford University Press, New York.

Nicholls, J.G., Martin, A.R., Fuchs, P.A., Brown, D.A.,, Diamond, M.E., and Weisblat, D.A. (2012) *From neuron to brain* (5th edn), Chapters 5–7 and 12–14. Sinauer, Sunderland, MA.

A slightly idiosyncratic approach but provides good clear explanations and nice illustrations. Another good starting point for advanced studies in neurobiology.

Squire, L.R., Bloom, F.E., Spitzer, N.C., Du Lac, S., Ghosh, A., and Berg, D. (2008) *Fundamental neuroscience* (3rd edn), Chapters 5–9 and 11. Academic Press, San Diego.

Review articles

Bender, K.J., and Trussell, L.O. (2012) The physiology of the axon initial segment. *Annual Review of Neuroscience* **35**, 249–65.

Davis, G.W., and Muller, M. (2015) Homeostatic control of presynaptic neurotransmitter release. *Annual Review of Physiology* 77, 251–70.

Kaeser, P.S., and Regehr, W.G. (2014) Molecular mechanisms for synchronous, asynchronous, and spontaneous neurotransmitter release. *Annual Review of Physiology* 76, 333–63.

Maday, S., Twelvetrees, A.E., Moughamian, A.J., and Holzbaur, E.L.F. (2014) Axonal transport: Cargo-specific mechanisms of motility and regulation. *Neuron* **84**, 292–309, http://dx.doi.org/10.1016/j.neuron.2014.10.019

Pereda, A.E. and Purpura D.P. (2014) Electrical synapses and their functional interactions with chemical synapses. *Nature Reviews in Neuroscience* 15, 250–263. doi:10.1038/nrn3708

Wu, L.-G., Hamid, E., Shin, W., and Chiang, H.-C. (2014) Exocytosis and endocytosis: Modes, functions, and coupling mechanisms. *Annual Review of Physiology* **76**, 301–31.

 To check that you have mastered the key concepts presented in this chapter, complete the accompanying online self-assessment questions. Go to www.oup.com/uk/pocock5e/

CHAPTER 8
Muscle

This chapter should help you to understand:

- The different morphological characteristics of the three principal types of muscle

- The detailed structure of skeletal muscle

- How an action potential leads to the contraction of a skeletal muscle (excitation–contraction coupling)

- The mechanism of contraction in skeletal muscle

- The mechanical properties of skeletal muscle

- The detailed structure of cardiac muscle

- The cardiac action potential and pacemaker activity

- Excitation–contraction coupling in cardiac muscle

- The mechanical properties of cardiac muscle

- The role of smooth muscle in different organs

- The detailed structure of smooth muscle cells

- The distinction between single- and multi-unit types of smooth muscle

- Excitation–contraction coupling in smooth muscle

- The mechanical and electrical properties of smooth muscle

8.1 Introduction

One of the distinguishing characteristics of most multicellular animals is their ability to use coordinated movement to explore their environment. This is achieved by the use of muscles, which consist of cells (**myocytes**) that can change their length by a specific contractile process. In vertebrates, including man, three types of muscle can be identified on the basis of their structure and function: skeletal muscle, cardiac muscle, and smooth muscle.

As its name implies, **skeletal muscle** is the muscle directly attached to the bones of the skeleton, and its role is both to maintain posture and to move the limbs by contracting in a coordinated way. **Cardiac muscle** is the muscle of the heart, while **smooth muscle** is the muscle that lines the blood vessels and the hollow organs of the body. When the cells of skeletal and cardiac muscle are viewed at high magnification, they are seen to have characteristic striations—small regular stripes running across the individual muscle cells. For this reason, skeletal and cardiac muscles are sometimes called **striated muscles**. Smooth muscle lacks striations and consists of sheets of spindle-shaped cells linked together. The microscopical appearance of the three different kinds of muscle is shown in Figure 8.1. Note that some muscles, such as those that act as sphincters surrounding the mouth and anus, as well as those of the upper oesophagus, have the same microscopic appearance as skeletal muscle but are not attached to the skeleton.

Together the three kinds of muscle account for nearly half of body weight, the bulk of which is contributed by skeletal muscle (about 40 per cent of total body weight). This chapter describes the physiological properties of the different kinds of muscle and their distinctive roles in the economy of the body. Despite their differences in appearance, the cellular and molecular basis of contraction is very similar for all types of muscle.

8.2 The structure of skeletal and cardiac muscle

Skeletal muscle

The main structural features of skeletal muscle are summarized in Figure 8.2. Each skeletal muscle is made up of a large number of skeletal **muscle fibres**, which are long, thin, cylindrical cells that contain many nuclei. The length of individual muscle fibres varies according to the length of the muscle, ranging from a few

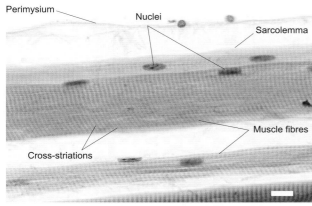

Skeletal muscle

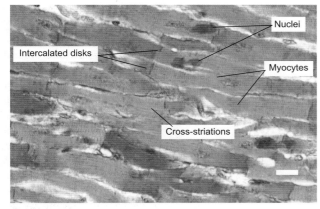

Cardiac muscle

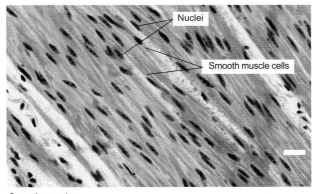

Smooth muscle

Figure 8.1 The microscopic appearance of skeletal, cardiac, and smooth muscle at similar magnification. At this magnification the striations of cardiac muscle are only just visible. Scale bar—10 μm.

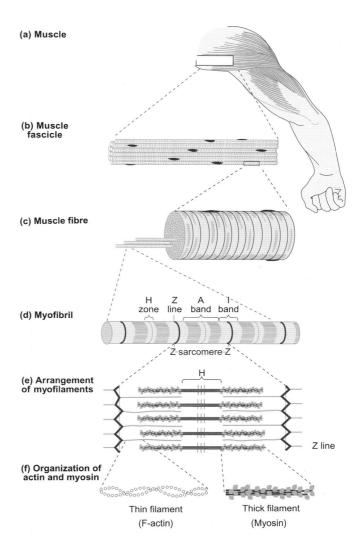

Figure 8.2 The organization of skeletal muscle at various degrees of magnification. Panel (a) is a diagram of the deltoid muscle of the shoulder. The area indicated by the white rectangle is shown below panel (b) to illustrate the appearance of a muscle fascicle (a bundle of muscle fibres). The nuclei of the muscle fibres are shown in blue. Panel (c) shows an individual muscle fibre and its constituent myofibrils, and panel (d) shows the structure of an individual myofibril showing the arrangement of the individual sarcomeres. Panel (e) represents the organization of an individual sarcomere, showing the arrangement of the thin and thick filaments to form the A and I bands. Panel (f) illustrates the arrangement of actin and myosin molecules in the thin and thick filaments. In panels (e) and (f), the head groups of the myosin molecules are shown in green. Note the absence of head groups in the central H zone of the sarcomere, which contains the tail region of the myosin molecules. (Based on Fig 11.19 in W. Bloom and D.W. Fawcett (1975) *Textbook of histology.* W.B. Saunders & Co, Philadelphia.)

millimetres to 10 cm or more. Their diameter also depends on the size of the individual muscle and ranges from about 50 to 100 μm. Despite their great length, few muscle fibres extend for the full length of a large muscle. The individual muscle fibres are embedded in connective tissue called the **endomysium**, and groups of muscle fibres are bound together by connective tissue called the **perimysium** to form bundles called muscle **fascicles**. Surrounding the whole muscle is a coat of connective tissue called the **epimysium** or **fascia** that binds the individual fascicles together. The connective tissue matrix of the muscle is secreted by **fibroblasts** that lie between the individual muscle fibres. It contains collagen and elastic fibres that merge with the connective tissue of the tendons, which transmit the mechanical force generated by the whole muscle to the skeleton. Finally, within the body of a skeletal muscle there are specialized sense organs known as **muscle spindles** that play an important role in the regulation of muscle length (see Chapter 10).

The individual muscle fibres are made up of filamentous protein bundles that run along the length of the fibre. These bundles are called **myofibrils** and have a diameter of about 1 μm. Each myofibril consists of a series of repeating units called **sarcomeres**, which are the fundamental contractile units of both skeletal and cardiac muscle. Each sarcomere is only about 2 μm in length and consists of protein filaments, as shown in Figure 8.2. The alignment of the sarcomeres between adjacent myofibrils gives rise to the characteristic striations of skeletal and cardiac muscle.

When a muscle fibre from a skeletal muscle is viewed by polarized light, the sarcomeres are seen as alternating dark and light zones. The regions that appear dark do so because they refract the polarized light. This property is called **anisotropy** and the corresponding band is known as an **A band**. The regions that appear light do not refract polarized light and are said to be **isotropic**. These regions are called **I bands**. Each I band is divided by a characteristic line known as a **Z line** and the unit between successive Z lines is a sarcomere (see Figure 8.2d). At high magnification in the electron microscope, the A bands are seen to be composed of thick filaments arranged in a regular order surrounded by a hexagonal array of thin filaments which form the I bands. At the normal resting length of a muscle, a paler area can be seen in the centre of the A band. This is known as the H zone, which corresponds to the region where the thick and thin filaments do not overlap (see Figure 8.3b). In the centre of each H zone is the M line, at which links are formed between adjacent thick filaments by proteins of the cytoskeleton.

The principal protein of the thick filaments is **myosin**, while that of the thin filaments is **actin**. The interaction between these proteins is fundamental to the contractile process (see Section 8.3). The actin filaments of the I band are made by joining many G (globular) actin subunits together by polymerization to form F (filamentous) actin, which is stabilized by binding to proteins that form the Z line. The thick filaments are made by assembling myosin molecules together. Each myosin molecule consists of two heavy chains, each of which has two light chains associated with its head region, which is globular in shape. The junction between the head region and the long tail contains the hinge that allows the myosin to generate the force required for muscle contraction. The tail regions of the myosin molecules associate together to form the thick filaments. Each thick filament consists of several hundred myosin molecules. This elaborate cellular architecture is maintained by a number of structural proteins, one of which, titin, is a long molecule that links the myosin filaments to the Z line. The Z line itself contains a number of proteins including α-actinin, which binds the actin filaments, and desmin, which links adjacent Z lines and serves to keep the Z lines in register.

Like all cells, skeletal muscle fibres are bounded by a plasma membrane which, in the case of muscle, is known as the

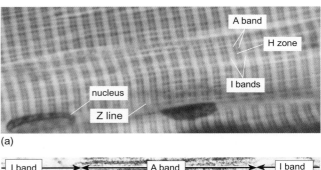

(a)

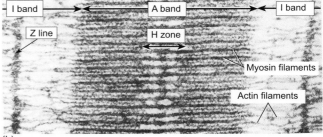

(b)

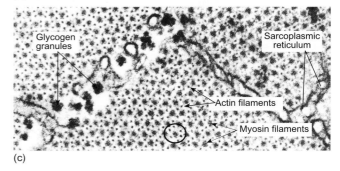

(c)

Figure 8.3 The detailed organization of skeletal muscle. Panel (a) shows part of three muscle fibres at high magnification. The individual sarcomeres are clearly visible, as is the H zone in the centre of the A band. The Z line is very faint. These fibres are somewhat stretched. Panel (b) is an electron micrograph of an individual sarcomere of rabbit muscle in which the individual thick (myosin) and thin (actin) filaments are clearly visible. Panel (c) is a cross-section of frog muscle showing the orderly array of thick and thin filaments. The circled area shows the hexagonal arrangement of thin filaments surrounding a thick filament. Note the presence of the sarcoplasmic reticulum and glycogen granules between the myofibrils. (Panels (b) and (c) from Plate 210 in H.E. Huxley (1957) *Journal of Biophysics and Biochemical Cytology* **3**, 631–48.)

sarcolemma. At its normal resting length, the sarcolemma is folded, forming small indentations known as **caveolae**. These are probably important in permitting the muscle fibre to be stretched without causing damage to the sarcolemma, but they may also serve other functions. Beneath the sarcolemma lie the nuclei and many mitochondria. In mammalian muscle, narrow tubules run from the sarcolemma transversely across the fibre near the junction of the A and I bands. These are known as **T-tubules**. The lumen of each tubule is continuous with the extracellular space. Each myofibril is surrounded by the **sarcoplasmic reticulum**, which is a membranous sac that is homologous with the

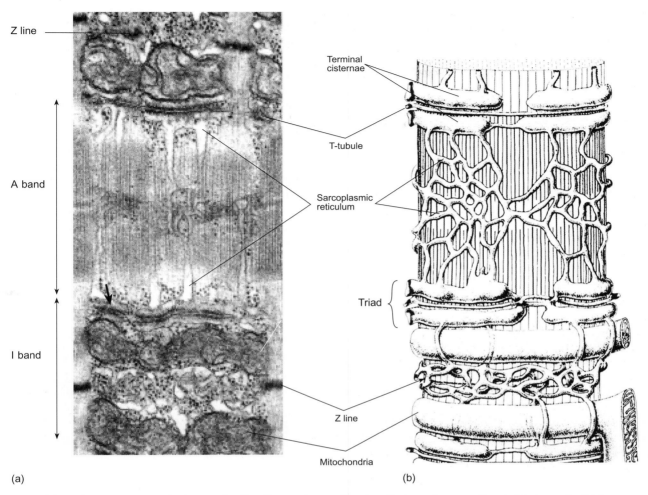

Z line

Terminal
cisternae

T-tubule

A band

Sarcoplasmic
reticulum

Triad

I band

Z line

Mitochondria

(a)   (b)

Figure 8.4 (a) An electron micrograph showing the T-system and sarcoplasmic reticulum of mammalian skeletal muscle. (b) A diagrammatic representation of the arrangement of the T-tubules and sarcoplasmic reticulum of skeletal muscle. The sarcoplasmic reticulum is a hollow network of membranes that form a distinct intracellular compartment, which stores and releases calcium ions to regulate muscle contraction. The T-tubules have their origin at the sarcolemma (plasma membrane) and course across the muscle fibre near the junction of the A and I bands. When a T-tubule comes into close apposition with the terminal cisternae of the sarcoplasmic reticulum, it forms a structure known as a triad. This arrangement allows the muscle action potential to be transmitted from the sarcolemma deep into the fibre to initiate a contraction. (Modified from L. Weiss, ed. *Cell and tissue biology*, 6th edition, p. 267, Urban and Schwartzenberg, Baltimore (1988). Republished with permission of Elsevier.)

endoplasmic reticulum of other cell types. Where the T-tubules and the sarcoplasmic reticulum come into contact, the sarcoplasmic reticulum is enlarged to form the **terminal cisternae**. Each T-tubule is in close contact with the cisternae of two regions of sarcoplasmic reticulum, and the whole complex is called a **triad** (see Figure 8.4). The T-tubules and triads play an important role in excitation–contraction coupling (see Section 8.3).

Cardiac muscle

Cardiac muscle consists of individual cells (**cardiac myocytes**) with a single nucleus linked together by junctions called **intercalated discs**. The intercalated discs are not aligned but cross the fibres in a characteristic irregular pattern (see Figure 8.5a). Individual cardiac muscle cells are aligned so that

they run in arrays that often branch to link adjacent groups of fibres together. The individual cells are about 15 μm in diameter and up to 100 μm in length. Adjacent cells are coupled electrically by gap junctions, which permit electrical activity to spread from one cell to another.

The arrangement of the contractile elements of cardiac myocytes is very similar to that of skeletal muscle, but the individual sarcomeres are not as regularly aligned. Compared to skeletal muscle, cardiac muscle is well endowed with mitochondria, which lie in longitudinal rows alongside the individual myofibrils (see Figure 8.5a). The sarcoplasmic reticulum is not as clearly organized as that of skeletal muscle and does not form clearly defined cisternae (see Figure 8.5b). The ventricular myocytes (but not those of the atria) have T-tubules at the level of the Z lines where they contact the sarcoplasmic reticulum to form a small diad.

Summary

Skeletal muscle is made up of long, cylindrical, multinucleated cells called muscle fibres, which contain myofibrils that are made up of repeating units called sarcomeres. Each sarcomere is separated from its neighbours by Z lines and consists of two half I bands, one at each end, separated by a central A band. The myofibrils are surrounded by a membranous sac known as the sarcoplasmic reticulum.

Cardiac muscle is made up of many individual cells (cardiac myocytes) linked together via intercalated discs and electrically coupled by gap junctions. Unlike skeletal muscle fibres, cardiac myocytes generally have a single nucleus. Cardiac myocytes have a striated appearance similar to that seen in skeletal muscle. The principal contractile proteins of the sarcomeres of both skeletal and cardiac muscle are actin and myosin.

8.3 How does a skeletal muscle contract?

Skeletal muscle, like nerve, is an excitable tissue, and stimulation of a muscle fibre at one point will rapidly lead to excitation of the whole cell. The process by which a muscle action potential triggers a contraction is known as **excitation–contraction coupling**.

In the body, each skeletal muscle consists of a collection of motor units in which a single motoneuron controls the contractile activity of a number of muscle fibres. As it enters a muscle, an individual motor axon branches and each branch supplies a separate muscle fibre. Furthermore, each muscle fibre receives innervation from only one motoneuron. An action potential in the motor neuron leads to the generation of a motor endplate potential (epp) in each of the muscle fibres to which it is connected (see Chapter 7).

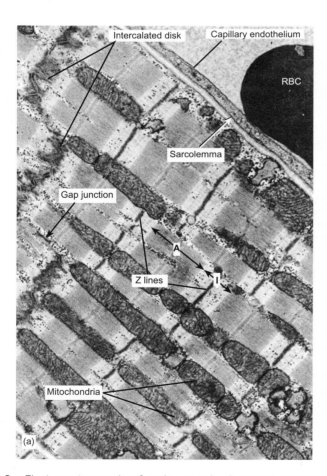

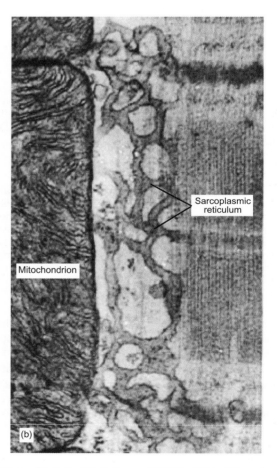

Figure 8.5 Electron micrographs of cardiac muscle, showing the organization myofibrils in detail. In panel (a), the field of view shows two myocytes joined by an intercalated disk which courses irregularly across the upper part of the field. Note the membrane thickening at the gap junction, the large numbers of mitochondria oriented with the myofibril, and the irregular arrangement of the Z lines. This field of view shows a capillary adjacent to the myocyte (top right; RBC—red blood cell). Panel (b) is an oblique section at higher magnification showing the arrangement of the sarcoplasmic reticulum in relation to an individual sarcomere. (From Plates 1 and 39 in D.W. Fawcett and N.S. McNutt (1969) *Journal of Cell Biology* **42**, 1–45.)

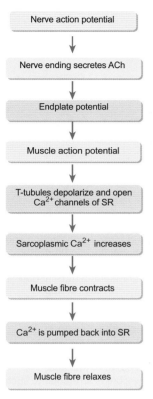

Figure 8.6 The sequence of events leading to the contraction and subsequent relaxation of a skeletal muscle fibre. SR = sarcoplasmic reticulum.

filament theory of muscle contraction, and it is is well supported by both physiological and biochemical observations. The individual filaments do not contract.

The individual myosin molecules of the thick filaments have two globular heads and a long, thin tail region. They are so arranged that the thin tail regions associate together to form the backbone of the thick filaments, while the thicker head regions project outwards to form cross-bridges with the neighbouring thin filaments (see Figure 8.2 (e) and (f)). The actin molecules link together to form a long polymer chain (F actin). Each actin molecule in the chain is able to bind one myosin head region. Actin and myosin molecules dissociate when a molecule of ATP is bound by the myosin. The breakdown of ATP and the subsequent release of inorganic phosphate cause a change in the angle of the head region of the myosin molecule that enables it to move relative to the thin filament (the power stroke—see Figure 8.7). Once again, ATP causes

The epp depolarizes the muscle fibre membrane in the region adjacent to the endplate and this depolarization, in its turn, triggers an action potential that propagates away from the endplate region along the whole length of the muscle fibre. The passage of the muscle action potential is followed by the contraction of the muscle fibre and the development of tension (force).

What steps link the muscle action potential to the contractile response? It has been known for a long time that injection of calcium ions (Ca^{2+}) into a muscle fibre causes it to contract. It was later discovered that significant amounts of Ca^{2+} are stored in the sarcoplasmic reticulum and that much of this Ca^{2+} is released during contraction. It is now thought that the depolarization of the plasma membrane during the muscle action potential spreads along the T-tubules, where it causes voltage-gated Ca^{2+} channels in the sarcoplasmic reticulum to open. As a result, Ca^{2+} stored in the sarcoplasmic reticulum is released and the level of Ca^{2+} in the sarcoplasm rises. This rise in Ca^{2+} triggers the contraction of the muscle fibre by the mechanism described below. Relaxation of the muscle occurs as the Ca^{2+} in the sarcoplasm is pumped back into the sarcoplasmic reticulum by a Ca^{2+} pump of the kind described in Chapter 4. These events are summarized in Figure 8.6.

How is tension generated and how does a rise in Ca^{2+} lead to the contractile response? All muscles contain two proteins—actin and myosin. The thick filaments are chiefly composed of myosin, while the thin filaments contain F actin (the principal protein) and lesser quantities of two other proteins known as troponin and tropomyosin. Muscle contraction is now known to occur through interactions between actin and myosin, which cause the thick and thin filaments to slide past each other. This is known as the **sliding**

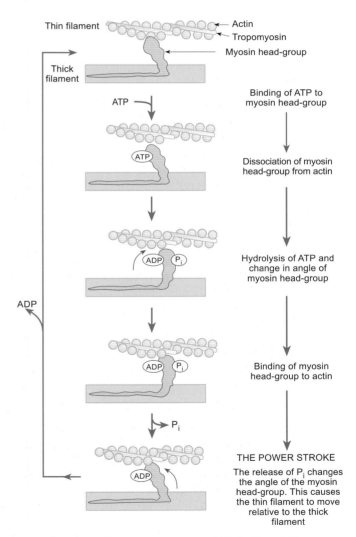

Figure 8.7 The molecular events responsible for the relative movement of the thin and thick filaments of a striated muscle. The change in the angle of the myosin head group allows it to bind to successive actin molecules on the thin filaments and causes the filaments to move relative to each other. P_i = inorganic phosphate.

the dissociation between the actin and myosin, and the cycle is repeated. This process is known as **cross-bridge cycling** and results in the thick and thin filaments sliding past each other, thus shortening the fibre. The formation of cross-bridges between the actin and myosin head groups is not synchronized across the myofibril. While some myosin head groups are dissociating from the actin, others are binding or developing their power stroke. Consequently, tension is developed smoothly.

The role of calcium ions in muscle contraction

If purified actin and myosin are mixed in a test tube, they form a gel (a jelly-like mixture). If ATP is then added to this mixture, the gel shrinks and the ATP is hydrolysed. So what prevents actin and myosin continuously hydrolysing ATP in an intact muscle fibre? The answer to this problem came when it was discovered that muscle contained the two additional proteins **troponin** and **tropomyosin** as mentioned earlier. These proteins form a complex with actin that prevents it binding the myosin head groups. When Ca^{2+} is released from the sarcoplasmic reticulum, the concentration of free Ca^{2+} in the sarcoplasm is transiently raised from the low levels found in resting muscle (0.1–0.2 µM) to a peak value of about 10 µM at the beginning of a contraction. The troponin complex binds the released Ca^{2+} and, in doing so, it changes its position on the actin molecule, so permitting actin to interact with the myosin head groups as described above.

The sliding filament theory of muscle contraction provides a clear explanation for the length–tension relationship of skeletal muscle (see Figures 8.8 and 8.12). When the muscle is at its natural resting length, the thin and thick filaments overlap optimally and form the maximum number of cross-bridges. When the muscle is stretched, the degree of overlap between the thin and the thick filaments is reduced and the number of cross-bridges falls. This leads to a decline in the ability of the muscle to generate tension. When the muscle is shorter than its natural resting length, the thin filaments already fully overlap the thick filaments but those from each end of the sarcomere touch in the centre of the A band and each interferes with the motion of the other. As a result, tension development declines. When the thin and thick filaments fully overlap, the A bands abut the Z lines and further shortening is no longer possible.

The role of ATP and creatine phosphate

At rest, a skeletal muscle can readily be stretched. In this state, although the myosin head groups have bound ATP, there are no cross-bridges being formed between the thick and thin filaments because the troponin complex prevents the interaction between actin and myosin. When ATP levels fall to zero after death, the cross-bridges between actin and myosin do not dissociate so the muscles become stiff—a state known as **rigor mortis**.

The energy for contraction is derived from the hydrolysis of ATP. As with other tissues, the ATP is mainly derived from the oxidative metabolism of glucose and fats (see Chapter 4). During the contractile cycle, however, it is important for the levels of ATP to be maintained. Since the blood flow through a muscle during

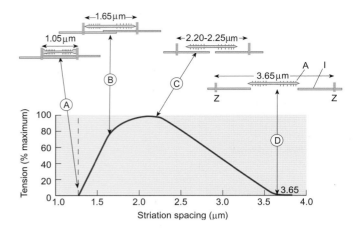

Figure 8.8 The sliding filament theory of muscle contraction and the length–tension relationship of a skeletal muscle fibre. The sequence A–D represents the overlap between the thick and thin filaments as the resting length of a single skeletal muscle fibre is increased. When the thin and thick filaments fully overlap, the A band is compressed against the Z line (point A) so the muscle is unable to develop tension. As the fibre is stretched so that the thin and thick filaments overlap without compressing the A band, active tension is generated when the muscle is stimulated (point B). Further stretching provides optimal overlap between the thin and thick filaments, leading to maximal development of tension (point C). If the fibre is stretched to such a degree that the thin and thick filaments no longer overlap, no tension is developed (point D). (Adapted from Figs 12 and 14 in A.M. Gordon, A.F. Huxley, and F.J. Julian (1966) *Journal of Physiology* **184**, 170–92.)

contraction may be intermittent (as in the leg muscles when walking or running), a store of metabolic energy is needed to maintain contraction, as the available ATP (about 3 mmol l⁻¹) would all be hydrolysed within a few seconds. This need is met by **creatine phosphate** (also known as **phosphocreatine**), which is present in muscle at high concentrations (about 15–20 mmol l⁻¹) and has a phosphate group that can be readily transferred to ATP. This reaction is catalysed by the enzyme **creatine kinase**:

$$\text{Creatine phosphate} + \text{ADP} + \text{H}^+ \rightleftharpoons \text{creatine} + \text{ATP}$$

During heavy exercise there may be insufficient oxygen delivered to the exercising muscles for oxidative metabolism. In this situation, the generation of ATP from glucose and fats via the tricarboxylic acid cycle is compromised and ATP is generated from glucose via the glycolytic pathway instead. This anaerobic phase of muscle contraction is much less efficient in generating ATP (see Chapter 4), and hydrogen ions, lactate, and phosphate ions are produced in increasing quantities. As these ions accumulate, muscle pH falls, the muscular effort becomes progressively weaker, and the muscle relaxes more slowly. This is known as **fatigue**. During muscle fatigue, the muscle fibres remain able to propagate action potentials but their ability to develop tension is impaired.

Summary

The link between the electrical activity of a muscle and its contractile response is called excitation–contraction coupling. The action potential of the sarcolemma depolarizes the T-tubules, causing Ca^{2+} channels in the sarcoplasmic reticulum to open. This allows Ca^{2+} to diffuse from the sarcoplasmic reticulum into the sarcoplasm. The resulting increase in Ca^{2+} around the myofibrils leads to the development of tension by the muscle fibre.

The A band of each sarcomere consists mainly of myosin molecules arranged in thick filaments, and the I band consists mainly of F actin, tropomyosin, and troponin arranged in thin filaments. Tension development occurs when actin and myosin interact. This is a calcium-dependent process in which actin and myosin form a series of cross-bridges that break and reform as ATP is hydrolysed. As a result, the thick and thin filaments slide past each other and force is generated.

The energy for muscle contraction is provided by the hydrolysis of ATP. As it is used up, ATP is rapidly replenished from the reserves of creatine phosphate and by the oxidative metabolism of glucose and fatty acids.

8.4 The activation and mechanical properties of skeletal muscle

Under normal circumstances, a skeletal muscle will contract only when the motor nerve to that muscle is activated. The signal for contraction of a skeletal muscle therefore originates in the CNS—the contraction is said to be **neurogenic** in origin. In contrast, provided it is placed in a suitable nutrient medium, the heart will continue to beat spontaneously for long periods even when it is isolated from the body. Some smooth muscles behave in the same way. Contractions that arise from activity within the muscle itself are said to be **myogenic** in origin.

The innervation of skeletal muscles

The nerve fibres supplying a mammalian skeletal muscle are myelinated. As a motor nerve enters the muscle, it branches to supply different groups of muscle fibres. Within the motor nerve bundle, individual axons also branch, so that one motor fibre makes contact with a number of muscle fibres (see Figure 7.15). When a motoneuron is activated, all the muscle fibres it supplies contract in an all-or-none fashion so they act as a single unit—a **motor unit**.

The size of motor units varies from muscle to muscle according to the degree of control required. Where fine control of a movement is not required, the motor units are large. Thus, in the gastrocnemius muscle of the lower leg, a motor unit may consist of up to 2000 muscle fibres. In contrast, the motor units of the extraocular muscles of the eye are much smaller (as few as 6–10 muscle fibres being supplied by a single motoneuron). This allows a fine degree of control over the direction of the gaze. The detailed mechanism by which a motoneuron activates skeletal muscle (neuromuscular transmission) is considered in Chapter 7.

The mechanical properties of skeletal muscle

Like nerve, skeletal muscle is an excitable tissue that can be activated by direct electrical stimulation. When a muscle is activated, it shortens and, in doing so, exerts a force on the tendons to which it is attached. The amount of force exerted depends on many factors:

1. the number of active muscle fibres
2. the frequency of stimulation
3. the rate at which the muscle shortens
4. the initial resting length of the muscle
5. the cross-sectional area of the muscle.

Imagine a situation where the muscle is being artificially activated by a single electrical shock to its nerve. The response of such a muscle is a brief contraction called a **muscle twitch**. The force developed by the muscle is called the **twitch tension**. If the electrical shock is very weak, no fibres will be excited (a subthreshold stimulus). As the strength is increased above threshold, a small proportion of the nerve fibres will be activated. Consequently, only a small number of the muscle fibres will contract and the amount of force generated will be small. If the intensity of the stimulus is increased, more nerve fibres will be recruited and there will be a corresponding increase both in the number of active muscle fibres and in the total tension developed. When all the muscle fibres are activated, the total tension developed will reach its maximum. A stimulus that is sufficient to activate all the fibres in a muscle is called a **maximal stimulus**.

Muscles do not elongate when they relax unless they are stretched. In the body, this stretching is achieved for skeletal muscles by the arrangement of muscles acting on particular joints. Muscles acting together to support a particular movement are known as **synergists**. Pairs of muscles having opposing actions are known as **antagonists**. For example, in the upper arm the major muscles flexing and extending the arm are, respectively, the biceps and triceps, although other muscles are also involved.

The force developed by a muscle increases during repetitive activation

When a muscle fibre is activated, the process of contraction begins with an action potential passing along its length. This is followed shortly after by the contractile response, which consists of an initial phase of acceleration during which the tension in the muscle increases until it equals that of the load to which it is attached. The fibre then begins to shorten. At first, the fibre shortens at a relatively constant rate but, as the muscle continues to contract, the rate of shortening progressively falls—the phase of deceleration. Finally, the muscle relaxes and the tension it exerts declines to zero. The whole cycle of shortening and relaxation takes several tens of milliseconds, while the muscle action potential is over in a matter of two or three milliseconds. Consequently, the mechanical response lasts much longer than the electrical signal that initiated it (see Figure 8.9a).

If a muscle is activated by two stimuli that are so close together that the second stimulus arrives before the muscle has fully relaxed after the first, the total tension developed following the second

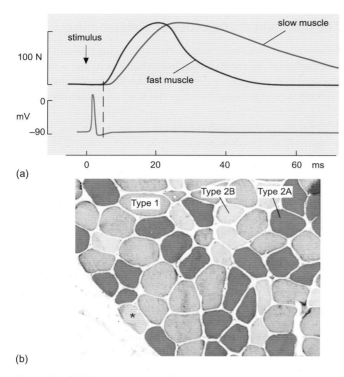

(a)

(b)

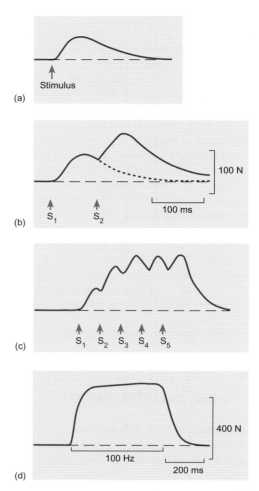

(a)

(b)

(c)

(d)

Figure 8.9 (a) The time course of the contractile response (a twitch) to a single stimulus for two different types of skeletal muscle in which either fast or slow fibres predominate. The timing of a muscle fibre action potential is shown for comparison. The contractile response begins after the action potential and lasts much longer, notably in the case of the slow muscle. Note the fast development of tension and rapid relaxation in the fast muscle (which has a high proportion of type 2 fibres) compared to the time course of the slow muscle (which has a high proportion of type 1 fibres). Force is expressed as newtons (N). (b) A cross-section through a skeletal muscle that has been processed to reveal the myosin heavy chain (MHC) isoforms found in the various types of fibre. The type 1 fibres are stained brown and reflect the presence of MHC 1; the type 2A fast twitch fibres are stained red; (MHC 2A) while the type 2B fibres are stained blue (MHC2X). The fibre marked with an asterisk contains both MHC2A and MHC2X isoforms. (Lower panel from Fig 1a in O. Raheem et al. (2010) *Acta Neuropathologica* **119**, 495–500.)

Figure 8.10 The summation and fusion of the mechanical response of a muscle in response to repetitive stimulation. Panel (a) shows the response to a single electrical stimulus (arrow). Panel (b) illustrates the summation of tension when a second stimulus (S_2) is given before the mechanical response to the first has decayed to zero, and panel (c) shows the summation of tension during a brief train of five stimuli. Panel (d) shows the smooth development of tension during a high-frequency train of stimuli (giving a fused tetanus). Note the different time and tension calibrations for (a) and (b) compared to (c) and (d).

stimulus is greater than that developed in response to a single stimulus. This increase in developed tension is called **summation** and is illustrated in Figure 8.10b. The degree of summation is at its maximum when the intervals between the stimuli are short, and it declines as the interval between the stimuli increases. If a number of stimuli are given in quick succession, the tension developed progressively summates and the response is called a **tetanus**. At low frequencies, the tension oscillates at the frequency of stimulation as shown in Figure 8.10c but, as the frequency of stimulation rises, the development of tension proceeds more and more smoothly. When tension develops smoothly (Figure 8.10d), the contraction is known as a **fused tetanus**.

Why is a muscle able to develop more tension when two or more stimuli are given in rapid succession? Consider first how tension develops during a single twitch. When the muscle is activated,

it must transmit the tension developed to the load. To do this, the tension of the tendons and of the connective tissue of the muscle itself must be raised to that of the load. As the tendons are to some degree elastic, they need to be stretched a little until the tension they exert on the load is equal to that of the muscle. This takes a short, but finite, amount of time during which the response of the tension-generating machinery begins to decline. The transmission of force to the load is accordingly reduced in efficiency. The time taken to match the tension to the load also accounts for the initial acceleration in the rate of contraction described above. During a train of impulses, however, the contractile machinery is activated repeatedly and the tension in the tendons has little time to decay between successive contractions. Transmission of tension to the load is more efficient, and a greater tension is developed.

Furthermore, the concentration of Ca^{2+} in the sarcoplasm that is attained in response to a single action potential is insufficient for maximal troponin binding. Consequently, not all actin molecules can interact with the myosin heads. However, when stimuli are given in rapid succession, the contractile machinery will be primed because not all the Ca^{2+} released during one stimulus will have been pumped back into the sarcoplasmic reticulum before the arrival of the next. As a result, more of the troponin binds Ca^{2+}. This permits the formation of a greater number of cross-bridges and the development of greater tension.

Fast and slow twitch muscle fibres

As Figure 8.9a shows, different muscles have different rates of contraction; they also have different susceptibilities to fatigue. Broadly speaking, skeletal muscle fibres can be classified as slow (type 1) or fast (type 2) according to their rate of contraction. Most muscles contain a mixture of both kinds of fibre, but some muscles have a predominance of one type. The soleus muscle of the lower leg is an example of a muscle composed chiefly of type 1 fibres, while the extensor digitorum longus muscle has a predominance of type 2 fibres. The different fibre types can be distinguished by various histochemical procedures and by the particular kind of myosin they contain. (The different kinds of myosin are called **isoforms** and can be identified by specific antibody stains—see Figure 8.9b.)

The type 1 fibres are thin and rich in both mitochondria and the oxygen-binding protein myoglobin (see Chapter 25). This gives them a reddish appearance. They rely mainly on the oxidative metabolism of fats for their energy supply and, as they have both a copious blood supply and significant oxygen reserves, they are very resistant to fatigue. Muscles that have a predominance of type 1 (slow) muscle fibres contract at about 15 mm s^{-1} and relax relatively slowly. They are activated by their motoneurons at a continuous, steady rate, which enables them to play an important role in the maintenance of posture.

Muscles that have a predominance of type 2 (fast) muscle fibres shorten rapidly (at about 40–45 mm s^{-1}) and relax relatively quickly. Two types of fast fibre (type 2A and type 2B) exist. Type 2A are fast fibres which have a high myosin ATPase activity compared to the slow fibres discussed above. The type 2A fibres are relatively thin, have a good blood supply, and, like the type 1 fibres, are rich in both mitochondria and myoglobin. They are relatively resistant to fatigue. For their energy requirements, they rely on oxidative metabolism and utilize either glucose or fats as their source of energy.

Type 2B fibres are fast fibres with a large diameter. They have a high myosin ATPase activity, contain large quantities of glycogen, and have high concentrations of glycolytic enzymes. This enables type 2B fibres to develop great tension very rapidly. However, as they have a limited blood supply, few mitochondria, and little myoglobin, they are easily fatigued. They are therefore well adapted to provide short periods of high tension development during anaerobic exercise. Unsurprisingly, these fibres are recruited for short periods of intense muscular activity (e.g. during sprinting). During heavy exercise, they rely chiefly on the glycolytic breakdown of glycogen and glucose for their energy source. Their relative lack of mitochondria and myoglobin gives them a pale appearance.

Smooth tension development in intact muscle is due to asynchronous activation of motor units

In the absence of organic disease, the development of tension during normal movement is smooth and progressive. This arises because the CNS recruits motoneurons progressively. Consequently, the motor units comprising a muscle are activated at different times—in contrast to the experimental situations described in the previous sections. Individual motor units may be activated by relatively low frequencies of stimulation, but the maintenance of a steady tension ensures the efficient transmission of force to the load. The ability of a muscle to maintain a smooth contraction is further enhanced by the presence of both fast and slow muscle fibres in most muscles.

The power of a muscle depends on the rate at which it shortens

The rate at which a muscle can shorten depends on the load against which it acts. If there is no external load, a muscle shortens at its maximum rate. With progressively greater loads, the rate of shortening decreases until the load is too great for the muscle to move. The relationship between the load imposed on a muscle and its rate of shortening is known as the **force–velocity curve** (see Figure 8.11). If a muscle contracts against a load which prevents shortening, the muscle is said to undergo an **isometric contraction**, while if it shortens against a constant load it is said to undergo an **isotonic contraction**. Thus, an isometric contraction and an isotonic contraction with no external load represent the extreme positions of the relationship between the force developed by a muscle and the rate at which it shortens.

The work of a muscle is determined by the distance it is able to move a given load, and the power of a muscle is the rate at which it performs work (see Box 8.1). Thus:

$$\text{Power} = \text{Force} \times \text{Velocity}$$

From the curve relating the velocity of shortening to the power developed (Figure 8.11), it is clear that the power developed by a

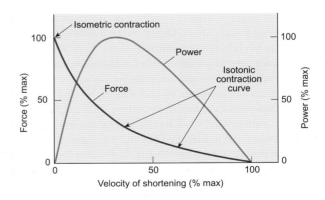

Figure 8.11 The force–velocity relation for a skeletal muscle. Note that maximum force is developed during an isometric contraction, but maximum speed of shortening occurs in an unloaded muscle. Maximum power is developed when the muscle shortens at about one-third of maximum velocity.

(1₂3) Box 8.1 The efficiency and power of muscles

It is a matter of common experience that it is more difficult to move heavy objects than light ones, but how efficient are muscles in converting chemical energy into useful work?

The **force** exerted on a given load is defined as:

$$\text{Force} = \text{Mass} \times \text{Acceleration} \qquad [1]$$

and is given in **newtons** (N). One newton is the force that will give a mass of 1 kg an acceleration of 1 m s⁻¹. The **work** performed on a load is the product of the load and the distance through which the load is moved. Thus:

$$\text{Work} = \text{Force} \times \text{Distance} \qquad [2]$$

The unit for work is therefore newtons per metre (N m), and one N m is a **joule** (J). **Power** is defined as the capacity to do work, or the work per unit time, and is expressed in joules s⁻¹ or **watts** (W).

$$\begin{aligned} \text{Power} &= \text{Work/Time} \\ &= \text{Force} \times \text{Distance/Time} \\ &= \text{Force} \times \text{Velocity} \qquad [3] \end{aligned}$$

The key to understanding the power and efficiency of a muscle is its force–velocity curve. For an isometric contraction, maximum force is exerted but the load is not moved, so that no external work is done and the power is also zero. When the muscle shortens against zero load, no useful work is done. Between these two extremes, the work is given by equation 2 above and the power by equation 3. The power is usually at a maximum when the muscle is shortening at about one-third of the maximum possible rate (see Figure 8.11).

The mechanical efficiency of muscular activity or work is expressed as a percentage of work done relative to the increase in metabolic rate attributable to the activity of the muscles employed in the task.

$$\text{Efficiency} = (\text{Work done/Energy expended in task}) \times 100$$

For our examples above, both isometric contraction and contraction with no external load have zero efficiency. When a muscle does external work, for example walking up stairs or cycling, its efficiency is about 20–25 per cent.

---

muscle passes through a definite maximum. When a muscle shortens isometrically, it does no external work, as the load is not moved through a distance. Consequently, no power is developed. Equally, if the muscle contracts while it is not acting on an external load, no work is done and no power is developed. In between these two extremes, the muscle performs useful work and develops power. In general, the greatest power is developed when the muscle is shortening at about one-third of its maximum rate.

The effect of muscle length on the development of tension

If the force generated by a muscle during isometric contraction is measured for different initial resting lengths, a characteristic relationship is found: in the absence of stimulation, the tension increases progressively as the muscle is stretched beyond its normal resting length. This is known as the **passive tension** and is due to the stretching of the muscle fibres themselves, and of the associated connective tissue and tendons. The extra tension developed as a result of stimulation (called the **active tension**) is at its maximum when the muscle is close to its resting length (i.e. the length it would have had in its resting state in the body). If the muscle is stimulated when it is shorter than normal, it develops less tension; if it is stretched beyond normal resting length, the tension developed during contraction is also less than normal. Overall, the relationship between the initial length of a muscle and the active tension is described by a bell-shaped curve such as that shown in Figure 8.12. This curve is very similar to the length–tension relationship seen for individual skeletal muscle fibres (see Figure 8.8). In the body, the range over which a muscle can shorten is determined by the anatomical arrangement of the joint on which it acts. Muscles attached to the

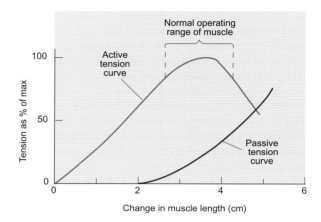

Figure 8.12 Isometric force–tension relationship at different muscle lengths. The data is for a human triceps muscle, which is about 20 cm in length. As the muscle is stretched, passive tension increases. Active tension increases from zero to a maximum and then declines with further stretching. The total tension developed during a contraction is the sum of the active and passive tensions.

skeleton operate at between 70 per cent and 120 per cent of their equilibrium length.

Effect of cross-sectional area on the power of muscles

The force generated by a single skeletal muscle fibre does not depend on its length but on its cross-sectional area (which determines the number of myofibrils that can act in parallel).

Box 8.2 Diseases of skeletal muscle

Like any other complex tissue, skeletal muscle is subject to many disorders. Muscle diseases (or **myopathies**) usually become apparent either because of evident weakness or because there is inappropriate contractile activity. They may be broadly divided into myopathies that have been acquired and those that are due to abnormal genes. In the case of the genetic disorders, the specific gene loci responsible are now known, offering the distant prospect of a cure.

The **acquired myopathies** may be divided into three groups:

a) inflammatory myopathies such as polymyositis (inflammation of muscle tissue), parasitic infection, and malignancy;

b) disorders of the neuromuscular junction such as myasthenia gravis, in which antibodies to the muscle nicotinic receptor are present in the blood resulting in impaired neuromuscular transmission, and Lambert–Eaton syndrome, in which there is defective release of acetylcholine from the motor nerve endings;

c) acquired metabolic and endocrine myopathies arise from a number of causes, such as Cushing's syndrome in which there are high concentrations of corticosteroids circulating in the blood, and thyrotoxicosis in which there are high circulating levels of thyroid hormone; myopathy of the proximal muscles occurs in hypocalcaemia, rickets, and osteomalacia.

The **genetic myopathies** can be divided into four main groups.

a) In muscular dystrophies there is muscle destruction, as in Duchenne muscular dystrophy where the gene for the structural protein dystrophin is faulty or even missing.

b) In myotonias, there is sustained contraction with slow relaxation. Myotonic dystrophy is associated with weakness of the distal muscles and is caused by an abnormal trinucleotide repeat in a protein kinase.

c) In channelopathies, the function of certain ion channels is impaired. These disorders are associated with intermittent loss of muscle tone. Though rare, two principal types are known: hypokalaemic periodic paralysis in which plasma potassium falls below 3 mmol l^{-1}, and hyperkalaemic periodic paralysis in which serum potassium is elevated above its normal level of 5 mmol l^{-1}. In hypokalaemic periodic paralysis, there are a number of possible genetic defects, the most common of which is a defect in a type of calcium channel called the dihydropyridine-sensitive (or L-type) calcium channel. In the case of the hyperkalaemic form of the disease there is a mutation in the voltage-gated sodium channel of the neuromuscular junction (see Chapter 40 p. 632).

d) Specific metabolic disorders of muscle are due to defects of specific enzymes. The glycogen storage disease McArdle disease is perhaps the best known. In this case there is a defect in muscle phosphorylase, which is responsible for mobilization of glucose from muscle glycogen. Another disorder, which is of considerable practical importance, is malignant hyperpyrexia in which there is a generalized muscle rigidity and elevation of body temperature. This is provoked by administration of certain general anaesthetics and it is caused by a defect in the ryanodine receptor of skeletal muscle, which is responsible for regulating the release of calcium from the sarcoplasmic reticulum. Malignant hyperpyrexia is discussed in more detail in Chapter 42.

Consider the force generated by a myofibril consisting of two sarcomeres. The force generated by the two central half sarcomeres cancel out as one pulls to the left and the other to the right. Consequently, it is the two end sarcomeres that generate useful force. This remains true whether the myofibril has a hundred or a thousand sarcomeres arranged in series (i.e. end to end). For both males and females between the ages of 12 and 20 years, the maximal isotonic force in **flexor** muscles is approximately 60 N cm^{-2}. The difference in strength between individuals is due to the difference in cross-sectional area of the individual muscles. Thicker fibres have more myofibrils arranged in parallel (i.e. side by side). They therefore develop more tension. When muscles hypertrophy (enlarge) in response to training, the number of muscle fibres does not increase. Rather, there is an increase in the number of myofibrils in the individual fibres and this leads to an increase in their cross-sectional area. It is the loss of muscle mass in muscle diseases (**myopathies**) and starvation that is the cause of the associated muscle weakness (see Box 8.2).

As long muscle fibres have more sarcomeres than short ones, they can shorten to a greater degree. In addition, they shorten more rapidly. Consider a muscle 1 cm long in which each fibre runs the complete length of the muscle. Each fibre will have approximately 4000 sarcomeres arranged end to end. If each sarcomere were to shorten from 2.5 to 2.0 μm, the muscle would shorten by 2.0 mm (20 per cent of the resting length). If this shortening occurred in 100 ms, the rate of shortening would be 0.2 ÷ 0.10 = 2 cm/s since each sarcomere shortens at approximately the same rate. Following the same line of argument, a muscle 30 cm long can shorten at 60 cm s^{-1}, i.e. 30 times faster.

The power of a muscle is equal to the force generated multiplied by the rate of contraction. As the force of contraction depends on the cross-sectional area of the muscle and the rate of contraction depends on the length of the muscle, the power of a muscle is proportional to its volume. A short, thick muscle will therefore develop the same power as a long, thin one of the same volume. The thick muscle will develop more force but will shorten more slowly than the long, thin one.

Summary

Skeletal muscles are innervated by myelinated nerve fibres that branch within each muscle to make contact with a number of muscle fibres. A motor neuron and its associated muscle fibres form a motor unit. A skeletal muscle contracts in response to action potentials in its motor nerve. Its activity is therefore said to be neurogenic in origin.

A single action potential gives rise to a contractile response called a twitch. During repeated activation of a muscle, the tension summates and the muscle is said to undergo tetanic contraction. There are two principal types of skeletal muscle fibre: type 1, which develop tension slowly but are able to maintain tension for long periods, and type 2, which are fast (or twitch) fibres that develop tension rapidly but fatigue quickly. The force developed by a muscle depends on the number of active motor units, its cross-sectional area, and the frequency of activation.

8.5 Cardiac muscle

In the body, skeletal muscle contracts only in response to activity in the appropriate motor nerve. Denervated skeletal muscle does not contract. In contrast, denervated cardiac muscle continues to contract rhythmically. It is this intrinsic or **myogenic activity** that is responsible for the steady beating of the heart and enables the organ to be transplanted from one individual to another. The myogenic activity of the heart has its origin in cells found at the junction between the great veins and the right atrium. This region is called the sinoatrial (SA) node, and it is the activity of the **pacemaker cells** of the SA node that sets the heart rate. Action potentials from the SA node spread across the atria to the atrioventricular (AV) node and thence to the ventricular fibres themselves.

The action potential of cardiac myocytes is of long duration and is maintained by a prolonged inward movement of calcium ions

As in skeletal muscle, the contractile response of a cardiac muscle fibre is associated with an action potential. However, cardiac action potentials are of far longer duration than those of nerve and skeletal muscle and last between 150 and 300 ms. The characteristics of the cardiac action potential depend on the position of the myocytes within the heart (see Figures 8.13 and 28.4).

- In the SA node, action potentials have a slow rise time and a duration of 150–200 ms, and they are preceded by a slow depolarization, the **pacemaker potential**, which sets the overall heart rate.

- In the atria, there is no pacemaker potential and the action potential has a fast rise time and a duration of about 200 ms.

- In the AV node, the action potential has a rapid rise time and a duration of about 150 ms. It is preceded by a pacemaker potential that has a significantly slower rate of depolarization than that of the SA node.

- In the ventricular fibres, the action potential has a fast rise time and a duration of about 200–250 ms.

- The action potential of the Purkinje fibres, which form the conducting system of the ventricles, is similar in appearance to that of the ventricular fibres but it has a longer duration (c. 300 ms). Purkinje fibres may also have a slow pacemaker potential, though this is much slower than that of the SA node.

The pacemaker potential of the sinoatrial node

The membrane potential of the cells of the SA node fluctuates spontaneously (see Figure 8.13). It is at its most negative (about −60 mV) immediately after the action potential and slowly becomes less negative until it reaches a value of about −50 mV, which is the threshold for action potential generation. The action potential of the SA node cells has a slow rise time (time to peak is about 50 ms), and the whole action potential lasts for 150–200 ms. The rate at which the slow depolarization of the pacemaker potential falls towards threshold (i.e. its slope) is an important factor in setting the heart rate.

The pacemaker potential of the SA node cells arises because a non-selective cation channel is slowly activated when the membrane potential becomes more negative than about −50 mV following **repolarization**. This results in a net inward current (called the pacemaker or funny current, I_f—see Box 8.3). The inward current opposes the outward potassium current responsible for repolarization of the SA node cells, and the membrane potential progressively becomes less negative. The resulting slow depolarization activates a calcium current which sums with the pacemaker current (or funny current) to accelerate the rate of depolarization until an action potential is triggered. The upstroke of the action potential of SA node cells is caused by a large increase in the permeability of the membrane to calcium ions (not sodium ions, as is the case for the myocytes of the atria and ventricles). Since the equilibrium potential for calcium ions is positive (just as it is for sodium ions), the rise in calcium permeability leads to a reversal of the membrane potential. Repolarization occurs as the permeability of the membrane to potassium increases while the permeability to calcium falls, so that the membrane potential becomes more negative; this activates the pacemaker current, and the cycle is repeated. In disease states, myocytes in other parts of the heart can show pacemaker activity, which causes irregular beating of the heart known as **arrhythmia** (or **dysrhythmia**).

Nerves and hormones may alter the heart rate as follows.

- They may change the slope of the pacemaker potential (e.g. during stimulation of the cardiac sympathetic nerves), so decreasing or increasing the time for the pacemaker potential to reach the threshold for action potential generation.

- They may hyperpolarize the membrane of the SA node cells (e.g. inhibition following brief stimulation of the vagus nerve), so that it takes longer for the pacemaker potential to reach the threshold for action potential generation.

- They may both hyperpolarize the membrane and reduce the slope of the pacemaker potential (as seen following strong vagal stimulation).

These effects on heart rate are discussed in more detail in Chapter 28.

(a)

(b)

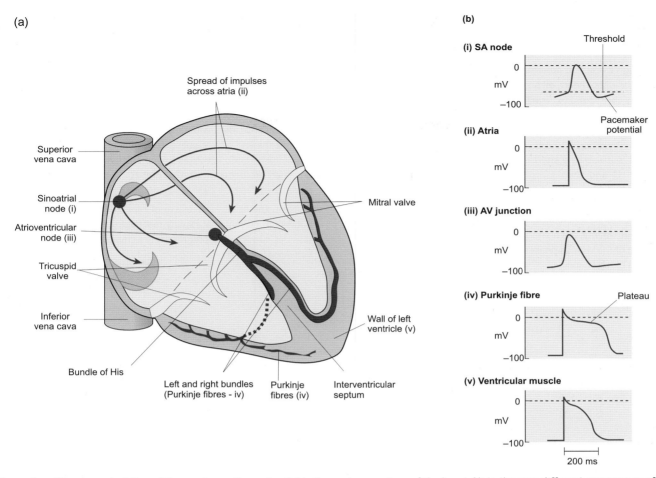

Figure 8.13 The characteristics of the cardiac action potential in the various regions of the heart. Note the very different appearance of the action potential in the different areas and the presence of the pacemaker potential in the cells of the SA and AV nodes.

The action potential of atrial and ventricular myocytes

As for other cells, the resting membrane potential of atrial and ventricular myocytes is determined mainly by the permeability of the membrane to potassium ions. The upstroke of the action potential is due to a rapid increase in the permeability of the membrane to sodium ions, similar to that seen in nerve axons and skeletal muscle fibres. The initial rapid phase of repolarization is due to inactivation of these sodium channels and to the transient opening of potassium channels. The long phase of depolarization (called the plateau phase) is due to slowly activating calcium channels, which increase the permeability of the membrane to calcium ions. As these channels progressively inactivate, the membrane repolarizes and assumes a value close to the potassium equilibrium potential. Box 8.3 discusses the ionic basis of cardiac action potentials in greater detail.

The contractile response of cardiac muscle

The long duration of the cardiac action potential has the important consequence that the mechanical response of the muscle occurs while the muscle membrane remains depolarized (see

Figure 8.14). Since a second action potential cannot occur until the first has ended, and since the mechanical response of the cardiac muscle largely coincides with the action potential, cardiac muscle cannot be tetanized. This is an important adaptation, as the heart needs to relax fully between beats in order to allow time for filling. Fibrillation (rapid and irregular contractions) may occur if the duration of the cardiac action potential (and therefore that of the refractory period) is substantially decreased.

Calcium activates the contractile machinery of cardiac muscle in much the same way as it does in skeletal muscle but there is one important difference: if heart muscle is placed in a physiological solution lacking calcium, it quickly stops contracting. This is in contrast to skeletal muscle, which will continue to contract each time it is stimulated. In cardiac muscle, the rise in Ca^{2+} within the myocyte during the plateau phase of the action potential is mainly derived from the calcium store within the sarcoplasmic reticulum, but the action potential also causes L-type voltage-gated calcium channels of the sarcolemma to open. The resulting calcium influx activates calcium release channels (ryanodine receptors) found on the sarcoplasmic reticulum. Once these channels have opened, much of the calcium stored in the

Box 8.3 The ionic basis of the cardiac action potential

Although the action potential of cardiac myocytes differs in different parts of the heart, all show similar characteristics. It is conventionally considered to have 5 phases numbered 0–4 (see Figure 1a). In ventricular myocytes, each action potential begins with a large depolarization (phase 0) in which the membrane potential becomes positive by around 20 mV (i.e. a membrane potential of +20 mV). This is followed by a short initial phase of repolarization of 10 mV (phase 1) and then a prolonged phase of slow repolarization (phase 2) during which the membrane potential slowly approaches zero mV. Phase 3 begins around 80–180 ms after the start of the action potential and continues until the membrane potential reaches about −85 mV, the resting membrane potential (phase 4).

Each of the various phases reflects that activity of a specific population of ion channels.

- Phase 0 reflects a large inward current caused by the opening of fast sodium channels (I_{Na}), with the result that the membrane potential depolarizes rapidly by a process similar to that described for nerve action potentials. As the equilibrium potential for sodium ions is positive, the membrane potential becomes positive (c.+20 mV). At the end of this phase, the channels inactivate. However, the depolarization activates L-type calcium channels which produces an prolonged inward current (I_{CaL}) that inactivates much more slowly.

- Phase 1, the brief, early phase of repolarization, is caused by a potassium current that rapidly inactivates (I_{Kto}), while other potassium currents are progressively activated such as the slowly activated outward rectifying current I_{Ks}.

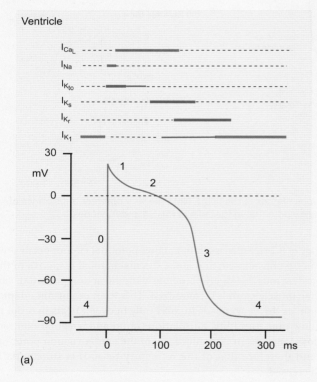

(a)

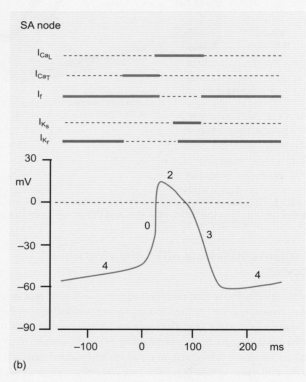

(b)

Figure 1 The various phases of action potentials recorded in ventricular myocytes and in SA node cells. In ventricular myocytes (a), the action potential is rapidly activated from a stable membrane potential of around −85 mV (phase 0). Repolarization proceeds rapidly at first (phase 1) before a period of slow repolarization (phase 2), after which complete repolarization (phase 3) proceeds fairly rapidly. The membrane currents that underlie the various phases of the action potential are shown above the voltage record. Inward (depolarizing) currents(I_{Na} and I_{CaL}) are shown in red, while outward (repolarizing) potassium currents (I_{Kto}, I_{Ks}, I_{Kr}, and I_{K1}) are shown in green. In SA node cells (b) the membrane potential is not stable (phase 4) but constantly varies, reflecting the activity of the pacemaker current (I_f). When the membrane potential declines from its maximum value (about −55 to −60 mV) to approximately −50 mV, T-type calcium channels open, increasing the inward current and depolarizing the membrane further. In turn, this causes L-type calcium channels to open resulting in a pronounced inward calcium current (I_{CaL} and I_{CaT}) and an action potential (phase 0) that is not due to the activity of sodium channels. The T-type calcium channels inactivate relatively quickly, allowing the membrane to repolarize slowly (phase 2). Outward potassium currents (I_{Ks} and I_{Kr}) are activated and the membrane potential reaches a value of about −60 mV. This increase in membrane potential activates the pacemaker current (I_f) and the cycle starts again (phase 4). The various membrane currents underlying the action potential of the SA node are shown above the voltage record. Outward currents are shown in green and inward currents in red.

Box 8.3 (*Continued*)

- Phase 2 is known as the plateau phase, during which the inward current carried by L type calcium channels (I_{CaL}) slowly declines while various outward potassium currents become activated (I_{Ks}, I_{Kr}, and I_{K1}). The net effect is a slow outward current leading to a progressive repolarization.

- During phase 3 the membrane potential repolarizes fully as there is a significant net outward current due to the activity of various potassium channels, while the inward calcium current (I_{CaL}) declines as the L-type calcium channels become inactivated.

- In phase 4 the resting membrane potential is restored and is stabilized by an outward current known as the inward rectifying current (I_{K1}) which is carried by potassium channels called $K_{ir}2.1$.

The different characteristics of the action potential in different parts of the heart largely reflect the levels of expression of the various kinds of ion channel.

As Figure 1b shows, the appearance of the action potential in the pacemaker cells of the SA node is somewhat different. Phase 0 has a significantly slower rise time than that seen elsewhere in the heart, phase 1 is absent, while phase 4 is characterized by the presence of the pacemaker potential.

- The pacemaker potential is caused by the opening of a type of non-selective cation channel that activates slowly as the membrane potential becomes more negative and inactivates slowly as the membrane depolarizes. These unusual properties have led to the naming of the resulting ionic current the funny (= strange) current (I_f). It is also known as the pacemaker current.

- As the membrane depolarizes below −50 mV, T-type calcium channels open, further depolarizing the membrane. This results in the opening of L-type calcium channels and the full depolarization of the membrane (phase 0).

- In phase 2 the T-type channels close, and the membrane begins to repolarize. (Note that phase 1 is absent in pacemaker cells.)

- The depolarization opens two types of potassium channels known as I_{Kr} (rapidly activating delayed outward rectifying current) and I_{Ks} (slowly activating delayed outward rectifying current). The I_{Ks} channels deactivate as the membrane repolarizes (phase 3) so are closed during the later phase of repolarization.

- The I_{Kr} channels remain open and permit the repolarization of the membrane to around −60 mV. This activates the I_f channels and the pacemaker potential starts again.

The heart rate is set by the slope of the pacemaker potential, which is profoundly influenced by the autonomic nerves. The sympathetic nerves increase the heart rate while parasympathetic (vagus) nerves decrease the heart rate (see Chapter 27). It is now clear that these influences are mediated by modulating the rate of activation of I_f which is facilitated by a rise, and inhibited by a fall, in intracellular cAMP. The sympathetic nerves secrete noradrenaline, which acts on β_1 adrenergic receptors to stimulate the formation of cAMP. In turn, the cAMP increases the slope of the pacemaker potential resulting in an increase in heart rate. In contrast, action potentials in the vagus nerves result in the secretion of acetylcholine by the parasympathetic nerve terminals, which acts on M2 muscarinic receptors to inhibit the formation of cAMP (Chapter 11). The fall in cAMP decreases the slope of the pacemaker potential and slows its rate of activation, resulting in a decrease in heart rate. The changes in cAMP that follow activity in the autonomic nerves also affect I_{Ks}, which is markedly increased by activation of β_1 adrenoceptors, facilitating repolarization and so shortening the duration of the action potential in both pacemaker and other cardiac cells.

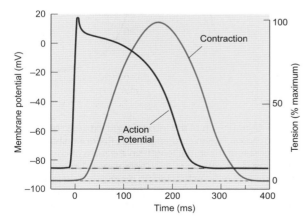

Figure 8.14 The relationship between the time course of a ventricular action potential and the development of isometric tension in ventricular muscle.

sarcoplasmic reticulum is rapidly released. This is known as **Ca²⁺-induced Ca²⁺ release** or CICR. During contraction, the intracellular calcium rises about tenfold from around $0.1\ \mu mol\ l^{-1}$ to a peak of about $1\ \mu mol\ l^{-1}$.

Relaxation of cardiac muscle occurs as calcium ions are pumped from the sarcoplasm either back into the sarcoplasmic reticulum or out of the cells. The majority is taken up into the sarcoplasmic reticulum by a Ca^{2+}-ATPase **metabolic pump** that is regulated by a protein called phospholamban. When this protein is in its dephosphorylated state, it has an inhibitory effect on the pump. When it is phosphorylated, phospholamban has no inhibitory effect and the uptake of calcium by the sarcoplasmic reticulum is accelerated, speeding relaxation. This phosphorylation occurs during beta-adrenergic stimulation of the heart (the **lusitropic action** of adrenaline and noradrenaline).

The force of contraction in cardiac muscle is even more closely linked to the initial length of the sarcomeres than it is in skeletal

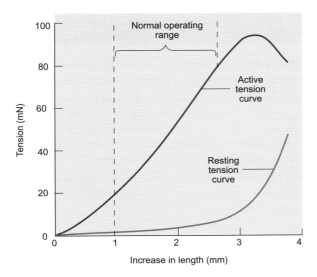

Figure 8.15 The relationship between the resting length of a papillary muscle from the ventricle of the heart and the maximum isometric force generated in response to stimulation. The passive tension increases steeply as the muscle is stretched beyond 3 mm. Note that cardiac muscle normally operates on the ascending phase of the active tension curve, unlike skeletal muscle, which operates around the peak (as shown in Figure 8.12).

muscle. As cardiac muscle is stretched beyond its normal resting length, more force is generated until it is about 40 per cent longer than normal. Further stretching of the muscle then leads to a decline in tension development. These characteristics are summarized in Figure 8.15.

In the normal course of events, the degree to which the muscle fibres of the heart are stretched is determined by the amount of blood returning to the heart (the venous return). If the venous return is increased, the ventricular muscle will be stretched to a greater degree as the ventricle fills with blood and will respond with a more forceful contraction (see Figure 8.16a). Similarly, if the work of the left ventricle is increased by a rise in blood pressure while the venous return remains constant, the muscle will again be stretched to a greater degree as the ventricle fills. The ventricle will respond with a more forceful contraction. This principle is enshrined in **Starling's law** of the heart (see Chapter 28). This property of cardiac muscle ensures that, in normal circumstances, the heart will pump out all the blood it receives.

The force with which the heart contracts varies according to the needs of the circulation. During exercise, the heart beats more strongly and frequently. These changes are mediated by the sympathetic nerves that innervate the SA node and ventricles, and by circulating adrenaline (epinephrine) secreted by the adrenal medulla. The change in rate is called a **chronotropic effect**, while the change in the force of contraction (or contractility) is called an **inotropic effect**. The intrinsic contractility of the heart (i.e. its inotropic state) determines its efficiency as a pump. The increased contractility of the heart is seen following stimulation of the sympathetic nerves, with increased circulating adrenaline (epinephrine), and with certain drugs such as digoxin. This increase in contractility is called a **positive inotropic effect**, while a decrease in contractility is a **negative inotropic effect**.

The positive inotropic effect occurs without any change in the length of the cardiac muscle fibres (see Figure 8.16b). There is no similar effect in skeletal muscle. The increase in contractility is caused by an increase in calcium influx following activation of β-adrenergic receptors and the subsequent generation of cyclic AMP. The cyclic AMP activates protein kinase A, which phosphorylates the calcium channels of the plasma membrane. The phosphorylated channels remain open for longer following depolarization and this, in turn, leads to an increased calcium influx that results in an increase in the force of contraction.

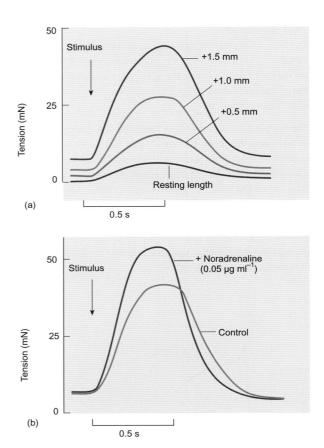

(a)

(b)

Figure 8.16 Comparison of the intrinsic and extrinsic regulation of the force of contraction in cardiac muscle. Panel (a) shows the effect of increasing the initial length on the force of contraction. Note the increase in resting tension with each 0.5 mm increment of the muscle above its resting length. As the muscle is stretched, the force developed in response to stimulation increases. Panel (b) shows the effect of noradrenaline on the force of contraction. Note the increased rate of contraction (the rising phase), the increased tension, and the greater speed of relaxation (the lusitropic effect). In this case the increased force of contraction occurs for the same initial resting length.

Summary

The beating of the heart is due to an intrinsic (myogenic) rhythm. Pacemaker cells in the sinoatrial node set the rate at which the heart beats. In these cells, the action potential is preceded by a pacemaker potential, which sets the frequency of action potentials and thus the intrinsic rhythm of the normal heart.

In the atria and ventricles, the upstroke of the action potential is due to a rapid increase in the permeability of the membrane to sodium ions, while the long plateau phase is maintained by an influx of calcium ions. Repolarization is due to an increase in the permeability of the membrane to potassium ions. The action potential of cardiac muscle varies according to the position of the cells in the heart, but it is always of long duration (150–300 ms). The contractile response of cardiac muscle largely overlaps the action potential. Consequently, for much of the contractile response, cardiac muscle cannot be re-excited and this prevents it from undergoing tetanic contractions.

As in skeletal muscle, the contractile response of cardiac muscle is triggered by a rise in intracellular free Ca^{2+}. Shortening occurs by the relative movement of actin and myosin filaments by the same mechanism as that seen in skeletal muscle. The long duration of each contraction is sustained by a steady influx of Ca^{2+} during the plateau phase of the action potential.

The force of contraction of cardiac muscle is determined by the initial length of the cardiac fibres and their inotropic state. In positive inotropy, cardiac muscle develops a greater force of contraction for the same initial fibre length.

8.6 Smooth muscle

Smooth muscle is the muscle of the internal organs such as the gut, blood vessels, bladder, and uterus. It forms a heterogeneous group with a range of physiological properties. In some cases the muscle must maintain a steady contraction for long periods of time and then rapidly relax (as in the case of the sphincter muscles controlling the emptying of the bladder and rectum). In others, such as those of the stomach and small intestine, the muscles are constantly active. Smooth muscle may express different properties at different times—as in the case of uterine muscle, which must be quiescent during pregnancy but contract forcefully during labour. It is therefore not surprising that the smooth muscle serving a specific function will have distinct properties. For this reason, rather than classifying smooth muscles into particular types, it is more useful to determine how the properties of particular smooth muscles are adapted to serve their function in the body.

Each smooth muscle consists of sheets of many small spindle-shaped cells (Figure 8.1 and 8.17a) linked together by two types of junctional contact as shown in Figure 8.17b. These are mechanical attachments between neighbouring cells and gap junctions, which provide electrical continuity between cells and thus provide a pathway for the passage of electrical signals between cells. Each smooth muscle cell has a single nucleus, is about 2–5 μm in diameter at its widest point and is 50–200 μm in length. In some tissues—such as the alveoli of the mammary gland and some small blood vessels—the smooth muscle cells are arranged in a single layer known as **myoepithelium**. Myoepithelial cells have broadly similar physiological properties to other smooth muscle cells.

In smooth muscle, no cross-striations are visible under the microscope (hence its name) but, like skeletal and cardiac muscle, smooth muscle contains both actin and myosin filaments. However, these are not arranged in a regular manner like those of skeletal and cardiac muscle but are arranged in a loose, three-dimensional lattice with the filaments running obliquely across the smooth muscle cells, as shown in Figure 8.17a. In comparison to

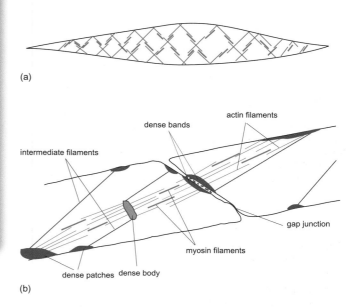

(a)

(b)

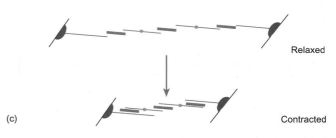

Relaxed

Contracted

(c)

Figure 8.17 The organization of the contractile and cytoskeletal elements of smooth muscle fibres. Panel (a) is a diagrammatic representation of the three-dimensional lattice of actin and myosin filaments in smooth muscle (intermediate filaments are omitted for simplicity). Panel (b) illustrates how thin (actin) filaments greatly outnumber the thick (myosin) filaments in smooth muscle. The thin filaments are also much longer than the thick filaments, allowing a greater degree of shortening. The intermediate filaments and the actin filaments attach to the plasma membrane at dense patches. Note the points of close contact for mechanical coupling (dense bands) and the gap junction for electrical signalling between cells. Individual groups of filaments run obliquely across smooth muscle cells and do not form regular arrays as they do in striated muscles. Panel (c) is a simple model of the contraction of smooth muscle. As the obliquely running contractile elements contract, the muscle shortens.

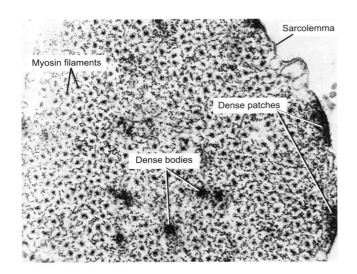

Figure 8.18 An electron micrograph of smooth muscle from the *taenia coli* muscle of the colon. This image shows myosin filaments surrounded by large numbers of thin filaments arranged very haphazardly. (Compare this with the orderly array of thin and thick filaments in skeletal muscle, as shown in Figure 8.3c.) Dense bodies are clearly seen scattered throughout the cytoplasm, with dense patches along the plasma membrane. (From Fig 12 in J.V. Small (1977) *Journal of Cell Science* **24**, 327–49.)

skeletal muscle, smooth muscle has many more actin filaments and fewer myosin filaments: the ratio of thin to thick filaments in smooth muscle is greater than 10:1 (in skeletal muscle the ratio is 2:1). Nevertheless, as for cardiac and skeletal muscle, the relative movement of the actin and myosin filaments is the basis of smooth muscle contraction (Figure 8.17c).

In addition to actin and myosin filaments, the cytoskeleton of smooth muscle cells also has intermediate filaments. During contraction, these assist in transmitting the force generated to the neighbouring smooth muscle cells and connective tissue. While there are no Z lines in smooth muscle, they have a functional counterpart in **dense bodies** which are distributed throughout the cytoplasm (see Figure 8.18) and serve as attachments for both the thin and intermediate filaments. Both actin filaments and intermediate filaments are anchored to the plasma membrane (sarcolemma) at **dense patches** (also called dense plaques) as well as at the junctional complexes between neighbouring cells (see Figures 8.17b and 8.18).

Smooth muscle cells do not have a T-system, and the sarcoplasmic reticulum is not as extensive as that found in other types of muscle. However, they have a large surface area relative to their volume as well as small membrane infoldings known as caveolae, which may serve to increase their surface area for ion fluxes. A comparison of some of the properties of smooth, cardiac, and skeletal muscle is given in Table 8.1.

Smooth muscle is innervated by fibres of the autonomic nervous system which have varicosities along their length that correspond to the nerve endings of the motor axons of the neuromuscular junction (see Chapter 7). In some tissues, each varicosity is closely associated with an individual muscle cell (e.g. the arrector pili muscles of the hairs of the skin), while in others, the axon varicosities remain in small bundles within the bulk of the muscle and are not closely associated with individual fibres (e.g. the smooth muscle of the gut). The varicosities release their neurotransmitter into the space surrounding the muscle fibres rather than onto a clearly defined synaptic region, as is the case at the neuromuscular junction of skeletal muscle. Moreover, the neurotransmitter receptors are distributed over the surface of the cells instead of being concentrated at one region of the membrane as they are at the motor endplate.

In many tissues, particularly those of the viscera, the individual smooth muscle cells are grouped loosely into clusters that extend in three dimensions. Gap junctions connect the cells so that the whole

Table 8.1 A comparison of the properties of skeletal, cardiac, and smooth muscle

Property	Skeletal muscle	Cardiac muscle	Smooth muscle
cell characteristics	very long cylindrical cells with many nuclei	irregular rod-shaped cells, usually with a single nucleus	spindle-shaped cells with a single nucleus
maximum cell size (length × diameter)	30 cm × 100 μm	100 μm × 15 μm	200 μm × 5μm
visible striations?	yes	yes	no
initiation of contraction	neurogenic	myogenic	mostly myogenic, some neurogenic
motor innervation	somatic	autonomic (sympathetic and parasympathetic)	autonomic (sympathetic and parasympathetic)
type of contracture	phasic	rhythmic	mostly tonic, some phasic
basis of muscle tone	neural activity	none	intrinsic and extrinsic factors
cells electrically coupled?	no	yes	yes
T-system?	yes	only in ventricular muscle	no
mechanism of e/c coupling	action potential and T-system	action potential and T-system	action potential, Ca^{2+} channels, and second messengers
force of contraction regulated by hormones?	no	yes	yes

muscle behaves as a functional syncytium. In this type of muscle, activity originating in one part spreads throughout the rest of the muscle. This is known as **single-unit smooth muscle**. The smooth muscle of the gut, uterus, and bladder are good examples of single-unit smooth muscle. In some tissues, such as the gut, there are regular spontaneous contractions (**myogenic contractures**) that originate in specific pacemaker areas (see Figure 8.19). Evidence is accumulating to show that in the gastrointestinal tract this pacemaker activity originates in populations of cells known as **interstitial cells of Cajal**. As interstitial cells are found in other smooth muscles, they may play a similar role in other organs.

The activity of many single-unit muscles is strongly influenced by hormones circulating in the bloodstream as well as by the activity of autonomic nerves. For example, during pregnancy the motor activity of the uterine muscle (the myometrium) is much reduced due to the presence of high circulating levels of the hormone progesterone. In this instance, the progesterone decreases the expression of certain proteins involved in the formation of gap junctions (e.g. connexin 43) and this reduces the excitability of the myometrium. Another steroid hormone, oestriol, antagonizes this effect and increases the expression of gap junction proteins, so increasing the excitability of the muscle. As the plasma concentration of oestriol rises steeply towards the end of pregnancy (see Figure 49.12), it is presumed to be linked to the increase in myometrial excitability that occurs prior to the onset of labour.

Certain smooth muscles do not contract spontaneously and are normally activated by motor nerves. These muscles are known as **multi-unit smooth muscle** and are organized into motor units similar to those of skeletal muscle, although their motor units are more diffuse. The intrinsic muscles of the eye (e.g. the smooth muscle of the iris), the arrector pili muscles of the skin, and the smooth muscle of the larger blood vessels are all examples of the multi-unit type. Nevertheless, the distinction between the two types of muscle is not rigid as, for example, the smooth muscle of certain arteries and veins shows spontaneous activity but also responds to stimulation of the appropriate sympathetic nerves.

Excitation–contraction coupling in smooth muscle

The membrane potential of smooth muscle is often quite low—typically from about −50 to −60 mV, which is some 30 mV more positive than the potassium equilibrium potential. This low value of the resting membrane potential arises because the sodium ion permeability of the cell membrane is about one-fifth that of potassium (compared to a Na$^+$/K$^+$ permeability ratio of about 1:100 for skeletal muscle). When a smooth muscle fibre generates an action potential, the depolarization depends on an influx of both sodium and calcium ions, although the exact contribution of each ion depends on the individual muscle. For example, in the smooth muscle of the vas deferens and the gut, the action potential appears to be mainly dependent on an influx of calcium ions. In contrast, the action potentials of the smooth muscle of the bladder and ureters depends on an influx of sodium ions in just the same way as the action potential of skeletal muscle. The action potential of this

smooth muscle lasts 10–50 ms (which is 5–10 times longer than that of a skeletal muscle fibre). In addition, in some smooth muscles the action potential may develop a prolonged plateau phase that is similar in appearance to that seen in cardiac muscle, but much longer in duration.

In single-unit smooth muscle, certain cells act as pacemaker cells and these show spontaneous fluctuations of the membrane potential known as slow waves. During an excitatory phase, slow wave activity builds up progressively until the membrane potential falls below about −35 mV, when a series of action potentials is generated. These are propagated throughout the muscle via gap junctions, and the muscle slowly contracts. This pattern of electrical activity and force generation is seen in the gut during peristalsis and in uterine muscle during parturition (see Figure 8.19a).

In many smooth muscles, the pacemaker activity is regulated by the activity of the sympathetic and parasympathetic nerves. In the intestine, the release of acetylcholine from the parasympathetic nerve varicosities causes a depolarization that results in the slow waves occurring at a more depolarized membrane potential. Consequently, during the slow wave cycle, the membrane potential exceeds the threshold for action potential generation for a greater period of time, making the muscle more active. Conversely, activity in the sympathetic nerves results in membrane hyperpolarization, so maintaining the membrane potential below threshold for a longer period of time and inhibiting contractile activity. Neither acetylcholine nor noradrenaline (norepinephrine) appear to have a direct action on the pacemaker activity, which is intrinsic to the muscle itself. In other smooth muscles, the role of the sympathetic and parasympathetic innervation is reversed, sympathetic

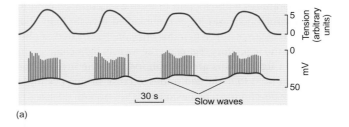

(a)

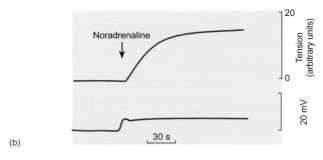

(b)

Figure 8.19 Patterns of electrical activity recorded from different kinds of smooth muscle. Panel (a) illustrates the electrical and mechanical response of uterine smooth muscle isolated from a rat giving birth. Note the slow waves leading to burst of action potentials and the slow and sustained development of tension. Panel (b) shows the development of tension in a sheep carotid artery following application of noradrenaline. Note that in this case the development of tension is not preceded by an action potential.

activity resulting in excitation and parasympathetic activity resulting in inhibition (see Chapter 11).

Like skeletal and cardiac muscle, smooth muscle contracts when intracellular Ca^{2+} rises. As smooth muscle does not possess a T-system, the rise in intracellular Ca^{2+} is triggered by events occurring at the plasma membrane: it can occur as a direct result of calcium influx through calcium channels in the plasma membrane, or by the release of calcium from the sarcoplasmic reticulum following activation of receptors linked to phospholipase C, in which case the increase the formation of IP_3 stimulates the release of calcium from the stores in the sarcoplasmic reticulum (see Chapter 6). Ryanodine receptors are also present in the sarcoplasmic reticulum, and there is also evidence for calcium-induced calcium release similar to that seen in cardiac muscle.

The contractile response of smooth muscle is slower and much longer-lasting than that of skeletal and cardiac muscle (see Figure 8.19). Furthermore, not all smooth muscles require an action potential to occur before they contract. In some large blood vessels, for example the carotid and pulmonary arteries, noradrenaline causes a strong contraction but only a small change in membrane potential (see Figure 8.19b). In this case, the contractile response is initiated by a rise in intracellular Ca^{2+} in response to the generation of IP_3 following the activation of α-adrenoceptors (see Chapter 11).

The excitation–contraction coupling of smooth muscle is regulated in a different manner to that of cardiac and skeletal muscle. The thin filaments of smooth muscle do not possess the regulatory protein troponin. Other proteins regulate the cross-bridge cycling of actin and myosin. In most smooth muscles, the regulation is performed by a calcium binding protein called **calmodulin**. This protein combines with calcium to form a calcium–calmodulin complex which activates an enzyme called myosin light chain kinase (see Figure 8.20). When this enzyme is activated, it phosphorylates the regulatory region on the myosin light chains. Once the myosin light chains are phosphorylated, the myosin head groups can bind to actin and undergo cross-bridge cycling, so causing the development of tension. In other smooth muscles, the phosphorylation of the myosin light chains is regulated in a more complex manner by other proteins that are associated with the thin filaments such as caldesmon.

The slow rate at which smooth muscle is able to hydrolyse ATP explains the slowness of the contractile process in smooth muscle compared to that seen in cardiac and skeletal muscle. Smooth muscle relaxation occurs when intracellular free calcium falls. This leads to dephosphorylation of myosin light chain kinase by a specific phosphatase. The direct dependence of the activity of the myosin light chain kinase on the availability of the calcium–calmodulin complex permits the contractile response to be smoothly graded. Moreover, the slow and steady generation of tension enables smooth muscle to generate and maintain tension with relatively little expenditure of energy.

Smooth muscle is able to maintain a steady level of tension called tone

The smooth muscle of the hollow organs maintains a steady level of contraction that is known as **tone** or tonus. Tone is important in maintaining the capacity of the hollow organs. For example, the flow of blood through a particular tissue depends on the calibre of the arterioles and this, in turn, is determined by the tone in the smooth muscle of the vessel wall (i.e. by the degree of contraction of the smooth muscle). Smooth muscle tone depends on many factors, which may be either extrinsic or intrinsic to the muscle. Extrinsic factors include activity in the autonomic nerves and circulating hormones, while intrinsic factors include the slow rate of cross-bridge cycling, the response to stretch, local metabolites, locally secreted chemical agents (e.g. nitric oxide in the blood vessels), and temperature. Thus, smooth muscle tone does not depend solely on activity in the autonomic nerves or on circulating hormones.

Length–tension relationships in single-unit smooth muscle

If a smooth muscle is stretched, there is a corresponding increase in tension immediately following the stretch. This is followed by a progressive relaxation of the tension towards its initial value. This property is unique to smooth muscle and is called stress relaxation or plasticity. The converse happens if the tension on smooth muscle is decreased (e.g. by voiding the contents of a hollow organ such

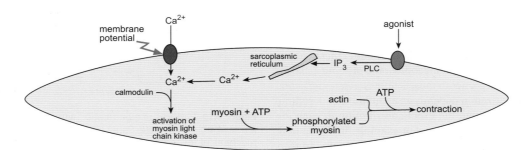

Figure 8.20 Excitation–contraction coupling in smooth muscle. As in cardiac and skeletal muscle, the development of tension is regulated by calcium, which can enter the sarcoplasm either via voltage-gated calcium channels in the plasma membrane or from the intracellular calcium stores of the sarcoplasmic reticulum. Calcium binds to calmodulin, which regulates the interaction between actin and myosin via myosin light chain kinase.

as the colon or bladder). In this case the tension initially falls but returns to its original level after a short period of time. This is called reverse stress relaxation. By adjusting its tension in this way, smooth muscle is able to maintain a level of tone in the wall of a hollow organ, permitting the internal diameter of the organ to be adjusted to suit the volume of material it contains.

Compared to skeletal or cardiac muscle, smooth muscle can shorten to a far greater degree. A stretched striated muscle can shorten by as much as a third of its resting length, while a normal resting muscle would shorten by perhaps a fifth—perfectly adequate for it to perform its normal physiological role. In contrast, a smooth muscle may be able to shorten by more than two-thirds of its initial length. This unusual property is conferred by the loose arrangement of the thick and thin myofilaments in smooth muscle cells. It is a crucial adaptation, as the volume contained by an organ such as the bladder depends on the cube of the length of the individual muscle fibres. Thus the ability of smooth muscle to change its length to such a large degree permits the hollow organs to adjust to much wider variations in the volume of their contents than would be possible for skeletal muscle. A simple calculation shows the advantage conferred by this property of smooth muscle: if a full bladder contains about 400 ml of urine, almost all the urine is expelled when the bladder wall contracts. If the bladder were a simple sphere, its circumference would be about 30 cm when it contained 400 ml of urine but would be only about 6 cm if 4.0 ml of urine were left after it had emptied (i.e. if it contained only 1 per cent of the original volume). This corresponds to a change in muscle length of about 80 per cent. If the bladder were made of skeletal muscle, however, the maximum length change would be only about 30 per cent and it would only be able to void about 70 per cent of its contents, leaving behind about 120 ml of urine.

Summary

Smooth muscle consists of sheets of small spindle-shaped cells linked together at specific junctions. Smooth muscle cells contain actin and myosin, but these proteins are not arranged in regular sarcomeres. Instead, each smooth muscle cell has a loose matrix of contractile proteins that is attached to the plasma membrane at the dense patches.

Smooth muscle is of two types: single-unit (or visceral) and multi-unit. Fibres from the autonomic nervous system innervate both types of smooth muscle. Single-unit smooth muscle shows myogenic activity and behaves as a syncytium. Multi-unit smooth muscle has little spontaneous activity and is activated by impulses in specific motor nerves.

The contractile response of smooth muscle is slow and is regulated by the activity of myosin light chain kinase, which in turn depends on the level of free calcium in the sarcoplasm. Smooth muscle maintains a steady level of tension, known as tone, which may be increased or decreased by circulating hormones, by local factors, or by activity in autonomic nerves.

✳ Checklist of key terms and concepts

The structure of muscle tissue

- There are three distinct types of muscle tissue: skeletal muscle, cardiac muscle, and smooth muscle.

- Muscle cells are also known as myocytes or muscle fibres. The fibres of skeletal and cardiac muscle have regular striations that are visible when viewed under a microscope. Smooth muscle has no striations.

- Skeletal muscle is made up of long, cylindrical, multi-nucleated cells, each of which contains myofibrils that are made up of repeating units called sarcomeres.

- The sarcomeres are the fundamental contractile units.

- Each sarcomere is separated from its neighbours by Z lines and consists of two sets of thin (F actin) filaments, one set at each end, separated by a central set of thick (myosin) filaments. The thin and thick filaments overlap.

- The myofibrils are surrounded by a membranous structure known as the sarcoplasmic reticulum.

- Cardiac muscle is made up of many individual cells (cardiac myocytes) linked together via intercalated discs. Unlike skeletal muscle fibres, cardiac myocytes generally have a single nucleus.

- The striated appearance of cardiac myocytes is due to the presence of an orderly array of sarcomeres similar to that seen in skeletal muscle.

- The principal contractile proteins of the sarcomeres of both skeletal and cardiac muscle are actin and myosin.

- The A band of each sarcomere consists mainly of myosin molecules arranged in thick filaments, and the I band consists mainly of actin arranged in thin filaments. The thin filaments also contain the regulatory proteins tropomyosin and troponin.

The mechanism of contraction

- The link between the electrical activity of a muscle and the contractile response is called excitation–contraction coupling.

- Tension development occurs when actin and myosin interact. This is a calcium-dependent process in which actin and myosin form a series of cross-bridges that break and reform as ATP is hydrolysed. As a result, the thick and thin filaments slide past each other and force is generated.

- In skeletal muscle, the action potential of the sarcolemma depolarizes the T-tubules and this causes Ca^{2+} channels in the sarcoplasmic reticulum to open. This leads to an increase in Ca^{2+} around the myofibrils, resulting in their contraction and the development of tension.

- The energy for muscle contraction is provided by the hydrolysis of ATP.
- As ATP is used up, it is rapidly replenished from the reserves of creatine phosphate.
- In prolonged exercise, the ATP for muscle contraction is derived either from glucose breakdown to lactate (anaerobic activity) or from the oxidative metabolism of glucose and fats (aerobic activity).
- During phases of anaerobic activity, muscles accumulate hydrogen ions, lactate, and phosphate ions. The increased concentration of these metabolites causes a decline in the development of tension known as muscular fatigue.

The mechanical properties of skeletal muscle

- Skeletal muscles are innervated by motoneurons, which give rise to myelinated nerve fibres known as motor nerve fibres. Each motor nerve fibre branches within a muscle to make contact with a number of muscle fibres.
- A motoneuron and its associated muscle fibres form a motor unit.
- A skeletal muscle contracts in response to action potentials in its motor nerve. Its activity is therefore said to be neurogenic in origin.
- A single action potential gives rise to a contractile response called a twitch.
- During repeated activation of a muscle, the tension summates and the muscle is said to undergo tetanic contraction.
- Maximum tension is developed during a fused tetanus.
- The force developed by a muscle depends on the number of active motor units, its cross-sectional area, and the frequency of stimulation.
- The speed with which a muscle contracts depends on the properties of the individual fibres that make up that muscle.
- There are two principal types of skeletal muscle fibre known as type 1 and type 2.
- Type 1 are slow fibres that develop tension slowly but are able to maintain tension for long periods.
- Type 2 are fast twitch fibres that develop tension quickly but rapidly fatigue.

The properties of cardiac muscle

- Cardiac myocytes have an intrinsic or myogenic rhythm.
- In the intact heart, pacemaker cells in the sinoatrial node (SA node) set the rate at which the heart beats.
- SA node cells exhibit a pacemaker potential that precedes the action potential.
- The upstroke of the action potential of the SA node cells is relatively slow and is due to an influx of calcium ions.
- In the atria and ventricles, the upstroke of the action potential is due to a rapid increase in the permeability of the membrane to sodium ions, while the long plateau phase is maintained by an influx of calcium ions.
- Repolarization is due to an increase in the permeability of the membrane to potassium ions.
- The action potential of cardiac muscle varies according to the position of the cells in the heart but it is always of long duration (150–300 ms) compared to that of skeletal muscle.
- The contractile response of cardiac muscle largely overlaps the action potential. Consequently, for much of the contractile response, cardiac muscle cannot be re-excited and this prevents it from undergoing tetanic contractions.
- As in skeletal muscle, the contractile response of cardiac muscle is triggered by a rise in intracellular free Ca^{2+}. Shortening occurs by the relative movement of actin and myosin filaments by the same mechanism as that seen in skeletal muscle.
- The long duration of each contraction is sustained by the influx of Ca^{2+} during the plateau phase of the action potential.
- The force of contraction of cardiac muscle is determined by the initial length of the cardiac fibres and by the inotropic state of the cardiac myocardium.
- The inotropic state of the myocardium determines the force the cardiac fibres can develop. When the inotropic state is enhanced in response to activation of the sympathetic nerves, the cardiac fibres develop an increased force for the same initial fibre length (positive inotropy).

The properties of smooth muscle

- Smooth muscle consists of sheets containing many small spindle-shaped cells linked together.
- Smooth muscle is of two types: single-unit (or visceral) and multi-unit. Fibres from the autonomic nervous system innervate both types of smooth muscle.
- Single-unit smooth muscle shows myogenic activity and behaves as a syncytium.
- Multi-unit smooth muscle has little spontaneous activity and is activated by impulses in specific autonomic nerves.
- Although smooth muscle cells contain actin and myosin, these proteins are not arranged in regular sarcomeres. Instead, each smooth muscle cell has a loose matrix of contractile proteins that lie within the cytoskeleton.
- The actin filaments are attached to dense bodies within the cytoplasm and are ultimately anchored to the plasma membrane at the dense patches.
- The contractile response of smooth muscle is slow and is regulated by the activity of myosin light chain kinase, which in turn depends on the level of free calcium in the sarcoplasm.
- Smooth muscle maintains a steady level of tension known as tone. The tone exhibited by a particular muscle may be increased or decreased by circulating hormones, by local factors, or by activity in autonomic nerves.
- Smooth muscle is much more plastic in its properties than other types of muscle and is able to adjust its length over a much wider range than skeletal or cardiac muscle.

Recommended reading

Histology

Mescher, A.L. (2009) *Junquieira's basic histology* (12th edn), Chapter 10. McGraw-Hill, New York.

A well-illustrated account of the histology of skeletal and cardiac muscle.

Physiology

Åstrand, P.-O., Rodahl, K., Dahl, H., and Stromme, S.B. (2003) *Textbook of work physiology* (4th edn), Chapter 3. Human Kinetics, Champaign, IL.

A well-referenced, authoritative discussion of the physiology of muscle.

Jones, D.A., Round, J.M., and De Haan, A. (2004) *Skeletal muscle from molecules to movement*. Churchill-Livingstone, London.

A straightforward, nicely paced account of muscle physiology, beginning with the molecular organization of skeletal muscle fibres before discussing all aspects of contraction including fatigue, muscle damage, and efficiency. Well worth reading.

Levick, J.R. (2010) *An introduction to cardiovascular physiology* (5th edn), Chapter 3. Hodder Arnold, London.

An eminently readable account of the detailed physiology of cardiac muscle.

Medicine

Ledingham, J.G.G., and Warrell, D.A. (eds) (2000) Concise Oxford textbook of medicine, Chapters 13.32–13.41. Oxford University Press, Oxford.

A comprehensive introduction to the disorders of skeletal muscle and neuromuscular transmission.

Review articles

Amin, A.S., Tan, H.L., and Wilde, A.A.M. (2010) Cardiac ion channels in health and disease. *Heart Rhythm* **7**, 117–26.

Berchtold, M.W., Brinkmeier, H., and Muntener, M. (2000) Calcium ion in skeletal muscle: Its crucial role for muscle function, plasticity, and disease. *Physiological Reviews* **80**, 1215–65.

Gunst, S.J., and Zhang, W. (2008) Actin cytoskeletal dynamics in smooth muscle: A new paradigm for the regulation of smooth muscle contraction. *American Journal of Physiology: Cell Physiology* **295**, C576–87.

Hong, F., Haldeman, B.D., Jackson, D., Carter, M., Baker, J.E., and Cremo, C.R. (2011) Biochemistry of smooth muscle myosin light chain kinase. *Archives of Biochemistry and Biophysics* **510**, 135–46.

Nerbonne, J.M., and Kass, R.S. (2005) Molecular physiology of cardiac repolarization. *Physiological Reviews* **85**, 1205–53.

Orchard, C., and Brette, F. (2008) T-tubules and sarcoplasmic reticulum function in cardiac ventricular myocytes. *Cardiovascular Research* **77**, 237–44.

Parton, R.G., and Simons, K. (2007) The multiple faces of caveolae. *Nature Reviews, Molecular Cell Biology* **8**, 185–94.

Schiaffino, S., and Reggiani, C. (2011) Fiber types in mammalian skeletal muscles. *Physiological Reviews* **91**, 1447–531.

Sweeney, H.L., and Houdusse, A. (2010) Structural and functional insights into the myosin motor mechanism. *Annual Review of Biophysics* **39**, 539–57.

Wray, S., and Burdyga, T. (2010) Sarcoplasmic reticulum function in smooth muscle. *Physiological Reviews* **90**, 113–78.

To check that you have mastered the key concepts presented in this chapter, complete the accompanying online self-assessment questions. Go to www.oup.com/uk/pocock5e/

PART FOUR

The nervous system and special senses

CHAPTER 9

Introduction to the nervous system

This chapter should help you to understand:

* The main divisions of the nervous system

* The basic anatomical organization of the brain and spinal cord

* The meninges

* The cranial nerves and their principal functions

* The organization of the spinal nerves

* The segmental innervation of the body by the spinal nerves

* The various cell types that make up the nervous system

9.1 Introduction

The nervous system is adapted to provide rapid and discrete (i.e. point-to-point) signalling over long distances (from millimetres to a metre or more). The key cellular events involved in communication between nerve cells have been discussed in Chapter 7, while motor control and the basis of sensation are discussed in the following chapters. This chapter is chiefly concerned with the organization of the nervous system and the nature of its constituent cells.

The nervous system may be divided into five main parts:

* the brain
* the spinal cord
* the peripheral nerves
* the autonomic nervous system
* the enteric nervous system.

The brain and spinal cord constitute the **central nervous system** (CNS), while the peripheral nerves, autonomic nervous system, and enteric nervous system make up the **peripheral nervous system**. The **autonomic nervous system** is the part of the nervous system that is concerned with the innervation of blood vessels and the internal organs. It includes the autonomic ganglia that run parallel to the spinal column (the paravertebral ganglia) and their associated nerves. The organization and functions of the autonomic nervous system are discussed in Chapter 11. The **enteric nervous system** controls the activity of the gut. Its organization and function are discussed in Chapter 43.

9.2 The organization of the brain and spinal cord

As Figure 9.1 shows, the brain and spinal cord lie within a bony case formed by the skull and vertebral canal of the spinal column. Both are covered by three membranes called the **meninges**. Immediately beneath the skull is a tough membrane of dense connective tissue called the **dura mater** (or dura) that envelops the whole brain and extends in the form of a tube over the spinal cord. Attached to the inner face of the dura is the **arachnoid membrane**, which is so named because of the fine web-like fibres with which it connects to the underlying **pia mater** (or pia). The pia is a delicate, highly vascular membrane, that follows every contour of the surface of the brain and spinal cord (Figure 9.2). The narrow space between the pia and arachnoid membranes (the **subarachnoid space**) is filled with a clear fluid called the **cerebrospinal fluid (CSF)**. This fluid is actively secreted by the choroid plexuses, which are vascular structures situated in the **cerebral ventricles**

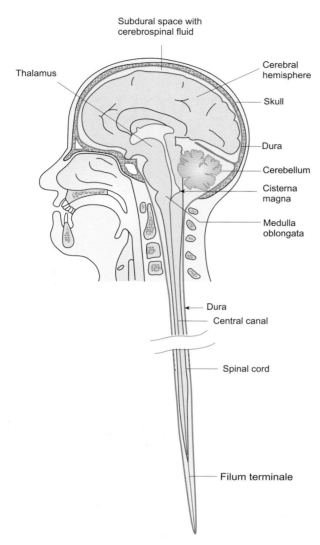

Figure 9.1 A diagrammatic representation of a mid-sagittal view of the central nervous system. The spinal cord is shown somewhat larger than it would be in reality. (Based on Fig 3.1 in P. Brodal (2004) *The central nervous system: Structure and function*, 3rd edition. Oxford University Press, New York.)

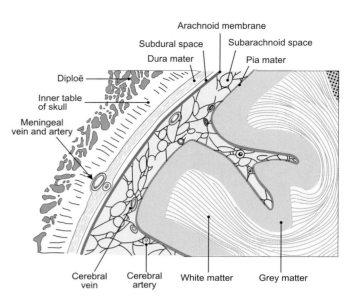

Figure 9.2 A diagrammatic section of the cerebral cortex to show the relationships between the brain, the meninges, and the inner face of the skull. The meningeal arteries are found in the endocranium (i.e. on the inside surface of the cranium), while the cerebral blood vessels lie in the subarachnoid space. The diploë is a layer of spongy bone that separates the inner and outer layers of the compact bone of the skull. (Redrawn from Fig 182 in G.J. Romanes (1986) *Cunningham's manual of practical anatomy*, Vol. 3, *Head and neck*. Oxford University Press, Oxford.)

(the internal fluid-filled spaces of the brain). The CSF provides a hydrostatic support for the brain within the skull and plays an important role in the regulation of the extracellular environment of nerve cells. The composition, formation, and circulation of the CSF are discussed in Chapter 30.

Since the space between the skull and the brain is filled with CSF, the brain floats in a fluid-filled container. Moreover, deep infoldings of the dura divide the fluid-filled space between the skull and the brain into smaller compartments. This arrangement restricts the displacement of the brain within the skull during movements of the head and limits the stresses on the blood vessels and the cranial nerves.

The surface of the human brain has many folds called **sulci** (singular: **sulcus**). The smooth regions of the brain surface that lie between the folds are known as **gyri** (singular: **gyrus**). Viewing the

brain from above reveals a deep cleft known as the longitudinal cerebral fissure that divides the brain into two **cerebral hemispheres**, each of which can be broadly divided into four lobes: the frontal, parietal, occipital, and temporal lobes (see Figures 9.3 and 9.4). Below and beneath the cerebral hemispheres lies a smaller, highly convoluted structure known as the **cerebellum**.

If the brain is cut in half along the midline (a mid-sagittal section) between the cerebral hemispheres, some details of its internal organization can be seen (see Figure 9.5). On the medial surface is a broad white band known as the **corpus callosum** that consists of a vast array of nerve fibres that interconnect the two hemispheres. Immediately below the corpus callosum is a membranous structure called the **septum pellucidum** that separates two internal spaces known as the **lateral cerebral ventricles**, which are filled with CSF.

Beneath the septum pellucidum and lateral ventricles is the **thalamus**, a major site for the processing of information from the sense organs. Lying just in front of and below the thalamus is the **hypothalamus**, which plays a vital role in the regulation of the endocrine system via its control of the **pituitary gland** (see Chapter 21). Posterior and ventral to the thalamus lies the midbrain (Figure 9.5), which merges into the **pons**, a large swelling of the brainstem that contains fibres connecting the two halves of the cerebellum. Below and behind the pons lies the **medulla oblongata** (or medulla), which merges with the spinal cord. The spinal cord passes through the spinal canal of the vertebral column. In adults it is about 45 cm in length and terminates at around the level of the first or second lumbar vertebra. As it

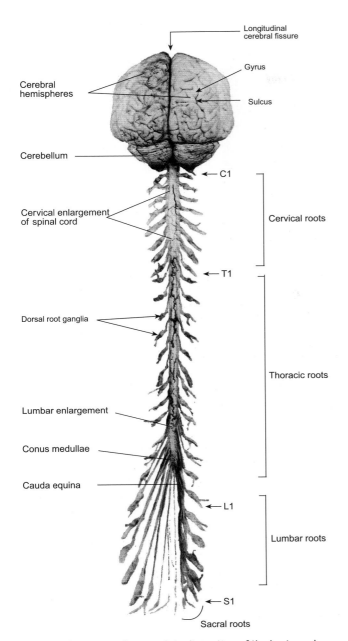

Figure 9.3 An image of a complete dissection of the brain and spinal cord showing their anatomical relationship as seen from the dorsal (posterior) aspect. Note the paired spinal nerves, which extend the whole length of the spinal cord, and the cervical and lumbar enlargements. (Adapted from Fig 8.88 in *Gray's anatomy*, 38th edition (1995), Churchill Livingstone, London, p. 177. Republished with permission of Elsevier.)

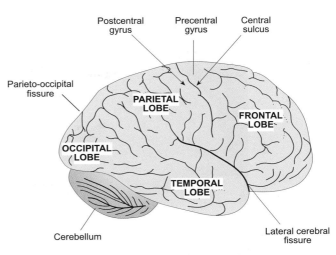

Figure 9.4 Side view of the human brain, showing the lobes of the right cerebral hemisphere and the position of the cerebellum.

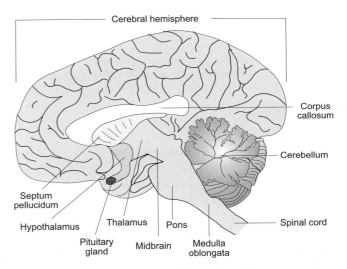

Figure 9.5 Mid-sagittal view of the right side of the human brain, showing the relationships of the main structures. (Based on Fig 2.6 in P. Brodal (1992) *The central nervous system: Structure and function*, 2nd edition. Oxford University Press, New York.)

passes down the vertebral canal, the spinal cord gives rise to a series of paired nerves (the spinal nerves) that connect the CNS to peripheral organs (see Figure 9.3).

If the brain is cut at right angles to the midline, a coronal section is obtained which reveals the internal structure of the brain. In a fresh brain the outer part has a greyish appearance beneath which is a region that has a paler appearance—see Figure 9.6. (In the figure, the brownish coloration is caused by the fixative used to preserve the brain.) The outer region is known as the **grey matter** and the inner region is the **white matter**. The grey matter contains large numbers of nerve cell bodies, while the white matter consists of large numbers of myelinated nerve fibres such as those of the **corpus callosum** and the **internal capsule**. Grey matter is distributed throughout the brain and spinal cord, often in discrete regions called **nuclei** (not to be confused with the nuclei of individual cells). Aggregates of white matter connecting one part of the CNS with another are called **nerve tracts**. Bundles of nerve fibres outside the CNS are simply called nerves.

An oblique section through the brain reveals a number of other important structures, which are shown diagrammatically in Figure 9.7. The **caudate nucleus**, the **putamen**, and the **globus pallidus** together form the **corpus striatum**. Between the caudate

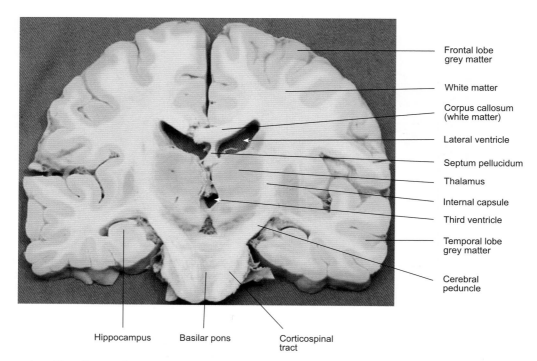

Frontal lobe
grey matter

White matter

Corpus callosum
(white matter)

Lateral ventricle

Septum pellucidum

Thalamus

Internal capsule

Third ventricle

Temporal lobe
grey matter

Cerebral
peduncle

Hippocampus   Basilar pons   Corticospinal
tract

Figure 9.6 A coronal section of human brain showing the difference in the appearance of grey and white matter and the anatomical relationships of various internal structures, including the cerebral ventricles. This brain section is treated with fixative but unstained. (From University of British Columbia: http://www.neuroanatomy.ca/cross_sections/sections_coronal.html)

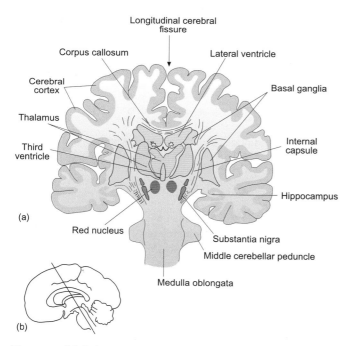

Longitudinal cerebral
fissure

Corpus callosum

Lateral ventricle

Cerebral
cortex

Basal ganglia

Thalamus

Third
ventricle

Internal
capsule

Hippocampus

(a)

Red nucleus

Substantia nigra

Middle cerebellar peduncle

Medulla oblongata

(b)

Figure 9.7 (a) A diagram of a coronal section through the human brain at a similar level to that of Figure 9.6, showing the spatial relationship of the cerebral cortex, basal ganglia, and thalamus. (b) The plane of section. (Redrawn from Fig 3.27 in P. Brodal (2004) *The central nervous system: Structure and function*, 3rd edition. Oxford University Press, New York.)

nucleus and putamen runs the internal capsule, which contains nerve fibres connecting the **cerebral cortex** to the spinal cord. A small region known as the **substantia nigra** lies beneath the thalamus. (The substantia nigra is so named because in adults it contains the black pigment melanin—see Figure 9.6). All of these structures play an important role in the control of movement (see Chapter 10 for further details).

The cranial nerves

On the base of the brain there are twelve pairs of nerves that serve the motor and sensory functions of the head (see Figure 9.8). These are the **cranial nerves**, which are numbered I to XII. Some contribute to the parasympathetic division of the autonomic nervous system (Chapter 11). These are the oculomotor (III), the facial (VII), the glossopharyngeal (IX), and the vagus (X). The names and the main functions of all the cranial nerves are given in Table 9.1.

The organization of the spinal cord and spinal nerves

The spinal cord is contained within a tubular sac of dura mater which is separated from the periosteum that lines the vertebral canal by soft fatty connective tissue (areolar tissue). The space between the periosteum of the vertebral canal and the dura mater is known as the **epidural space**. Like the brain, the inner face of the dura is lined by the arachnoid membrane, which is separated from the pia mater covering the spinal cord itself by the sub-arachnoid

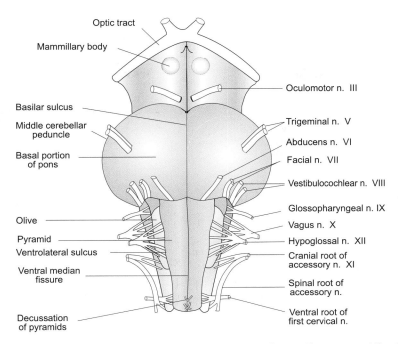

Optic tract
Mammillary body
Oculomotor n. III
Basilar sulcus
Middle cerebellar peduncle
Trigeminal n. V
Abducens n. VI
Basal portion of pons
Facial n. VII
Vestibulocochlear n. VIII
Glossopharyngeal n. IX
Olive
Vagus n. X
Pyramid
Hypoglossal n. XII
Ventrolateral sulcus
Cranial root of accessory n. XI
Ventral median fissure
Spinal root of accessory n.
Ventral root of first cervical n.
Decussation of pyramids

Figure 9.8 A ventral view of the human brain showing the cranial nerves, the optic chiasm, the pons, and the decussation of the pyramids (crossing over of the nerve fibres of the pyramidal tracts). The first cranial nerve does not emerge from the brainstem but from the olfactory bulb and is not shown here. The fourth cranial nerve, the trochlear nerve, emerges on the dorsal side of the brainstem below the inferior colliculi. It is the only cranial nerve to emerge on the dorsal aspect of the brainstem. (Redrawn from Fig 6-1 in M.L. Barr (1974) *The human nervous system*. Harper International, Hagerstown.)

Table 9.1 The functions of the cranial nerves

Number	Name	Chief functions
I	Olfactory	Sensory nerve subserving the sense of smell
II	Optic	Sensory nerve subserving vision (output of the retina)
III	Oculomotor	Chiefly motor control of the extrinsic muscles of the eye and the parasympathetic supply for the intrinsic muscles of the iris and ciliary body
IV	Trochlear	Chiefly motor control of the extrinsic muscles of the eye
V	Trigeminal	Sensory and motor: motor control of the jaw and facial sensation
VI	Abducens	Chiefly motor control of the extrinsic muscles of the eye
VII	Facial	Sensory and motor: motor control of the facial muscles and parasympathetic supply to the salivary glands; subserves the sense of taste via the chorda tympani
VIII	Vestibulo-cochlear	Sensory: subserves the sense of hearing and balance
IX	Glossopharyngeal	Sensory and motor: subserves the sense of taste from the back of the tongue (bitter sensations) and controls the muscles of swallowing and the parasympathetic supply to the salivary glands
X	Vagus	Sensory and motor: it is the major parasympathetic outflow to the chest and abdomen and receives afferent inputs from the viscera
XI	Spinal accessory	Motor control of neck muscles and larynx
XII	Hypoglossal	Motor control of the tongue

space. The movement of the spinal cord within the dural sac is restricted by thin bands of connective tissue known as the **denticulate ligaments**; these are attached to the dura at 21 positions on each side (see Figure 9.9).

The spinal cord runs through the vertebral canal of the spinal column from the first cervical vertebra to the lumbar region. It is a delicate structure around 45 cm in length in men and 42 cm in women. The cervical segments have a maximum transverse diameter of about 12 mm, the thoracic segments are somewhat thinner and have a maximum diameter of 7–8 mm (about the thickness of a pencil), while the maximum diameter of the lumbar segments is 9–10 mm (the lumbar enlargement). The terminal segments of the

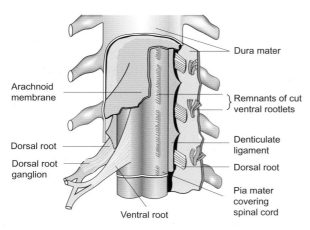

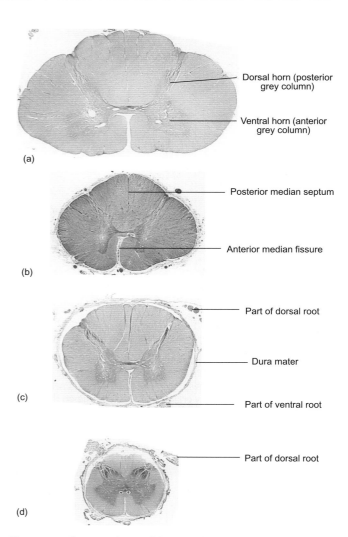

Figure 9.9 A diagrammatic representation of a ventral (anterior) view of the spinal cord, showing the origin of the spinal nerves and the arrangement of the spinal meninges. The denticulate ligaments run along each side of the spinal cord and stabilize the position of the spinal cord within the dural sac with a series of links between the pia mater and the dura mater. (Redrawn from Fig 169 in G.J. Romanes (1986) *Cunningham's manual of practical anatomy*, Vol. 3, *Head and neck*. Oxford University Press, Oxford.)

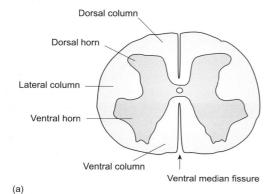

(a)

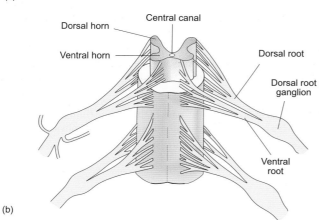

(b)

Figure 9.10 Diagrams illustrating the structure of the spinal cord at the level of the lumbar enlargement (a) and the arrangement of the spinal roots (b). In the lower figure, part of the white matter of the spinal cord is cut away to show the direct entry of the spinal roots into the central gray matter. (Based on Figs 3.4 and 3.5 in P. Brodal (2004) *The central nervous system: Structure and function*, 3rd edition. Oxford University Press, New York.)

Figure 9.11 Cross-sections of the spinal cord at four different levels: (a) cervical, (b) thoracic, (c) lumbar, and (d) sacral. (Approximately 5× magnification.)

sacral region have a diameter of only 4–6 mm. The spinal cord ends at the **conus medullae**, which is generally located at the level of the first lumbar vertebra in adults, although in certain individuals it may be as high as the twelfth thoracic segment (T12) or as low as the third lumbar vertebra (L3). The lower end of the spinal cord is tethered to the coccyx by a thin cord of connective tissue called the filum terminale.

A cross-section of the spinal cord shows that it has a central region of grey matter surrounded by white matter (see Figure 9.10a, b). The white matter of the spinal cord is arranged in columns that contain the nerve fibres connecting the brain and spinal cord, while the central column of grey matter is roughly shaped in the form of a butterfly (or a letter H) around a central canal. The grey matter is broadly divided into two **dorsal horns** and two **ventral horns**, but its exact appearance depends on whether the spinal cord has been cut across at the cervical, thoracic, lumbar, or sacral level (see Figure 9.11). (The dorsal and ventral horns are also known as the posterior and anterior horns respectively.)

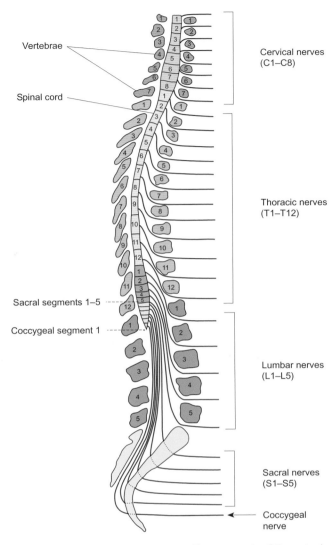

Vertebrae

Spinal cord

Sacral segments 1–5

Coccygeal segment 1

Cervical nerves
(C1–C8)

Thoracic nerves
(T1–T12)

Lumbar nerves
(L1–L5)

Sacral nerves
(S1–S5)

Coccygeal
nerve

Figure 9.12 The relationship between the segments of the spinal cord and the spinal nerves with the vertebral column. Note that while the first four cervical nerves leave the spinal canal above the corresponding vertebrae, all the others leave below the corresponding vertebrae. The spinal cord occupies the upper two-thirds of the spinal canal. The position of the final segment varies from person to person, but in adults it lies between the twelfth thoracic and second or third lumbar vertebrae.

At each segmental level, the spinal cord gives rise to a pair of spinal nerves. Each of these is formed by the fusion of nerve segments known as the **dorsal** and **ventral roots**, as shown in Figure 9.10(b). In human beings, the spinal cord has thirty-one pairs of spinal nerves: eight cervical, twelve thoracic, five lumbar, five sacral, and one coccygeal (Figure 9.3). The first seven pairs of cervical nerves leave the vertebral canal above the corresponding vertebrae, while the eighth cervical nerve and subsequent thoracic, lumbar, and sacral nerves leave below their corresponding vertebrae (see Figure 9.12). The spinal nerves are generally numbered according to the number of vertebrae above each pair of nerves, but since the first cervial nerve leaves above the first vertebra (the atlas), there are

eight cervical nerves even though there are only seven cervical vertebrae. In adults, the vertebral canal below the second lumbar vertebra houses the **cauda equina** (= horse's tail) which consists of the spinal nerves of the lower lumbar (L2–L5), sacral, and coccygeal segments. This anatomical arrangement permits samples of CSF to be obtained safely in adults by inserting a needle into the subarachnoid space at a level below the third or fourth lumbar vertebrae (lumbar puncture—see Chapter 30 p. 485).

Each dorsal root has an enlargement known as a **dorsal root ganglion** that contains the cell bodies of the nerve fibres making up the dorsal root (see Figures 9.3 and 9.10). The fibres of the ventral root originate from nerve cells in the ventral horn of the spinal grey matter. To leave the spinal canal, the spinal nerves pierce the dura mater between the vertebrae. Thereafter, they form the peripheral nerve trunks that innervate the muscles and organs of the body.

Sensory information enters the spinal cord via the dorsal root ganglia. Since the sensory fibres carry information from sense organs to the spinal cord, they are known as **afferent nerve fibres**. The ventral root fibres are known as **efferent nerve fibres**. They carry motor information from the spinal cord to the muscles and secretory glands (the **effectors**). The nerves that leave the spinal cord to supply the body wall and the skeletal muscles are known as **somatic nerves**, while those that supply the blood vessels and viscera are **autonomic efferent fibres** (see Chapter 11 for further details of the organization and function of the autonomic nervous system). Each spinal nerve innervates a particular region of the body known as a **dermatome**, although there is a considerable degree of overlap between the nerves supplying adjacent segments. The cutaneous (sensory) innervation of the individual dermatomes by the spinal nerves is illustrated in Figure 9.13.

In addition to their sensory function, nerves arising from cervical segments C1 to C6 provide motor innervation of the neck muscles, while those arising from C3 to C5 supply the diaphragm. Nerves arising from C5 to T1 (thoracic segment 1) supply the muscles of the shoulder, arm, and hand. Those arising from T1 to T6 supply the intercostal muscles and the muscles of the trunk above the waist. The abdominal muscles are supplied by nerves from segments T7 to L1. The nerves of the lumbar and sacral segments S1 to S3 supply the motor innervation of the leg and foot, while those of the fourth and fifth sacral segments together with the coccygeal nerve form the **pudendal** and **coccygeal plexi**. Short nerves arising from these plexi innervate the muscles of the pelvic region (see Chapters 39, 44, and 49).

9.3 The cellular constituents of the nervous system

The CNS is made up of two main types of cell, the **nerve cells** or **neurons** and the **glial cells** or **neuroglia**. The cell bodies of the neurons are found throughout the grey matter of the brain and spinal cord. They are very varied in both size and shape, but their cell bodies all stain strongly with basic dyes (see the example in Figure 9.14). The stained material is called **Nissl substance** and contains a high proportion of RNA, reflecting the continuous high

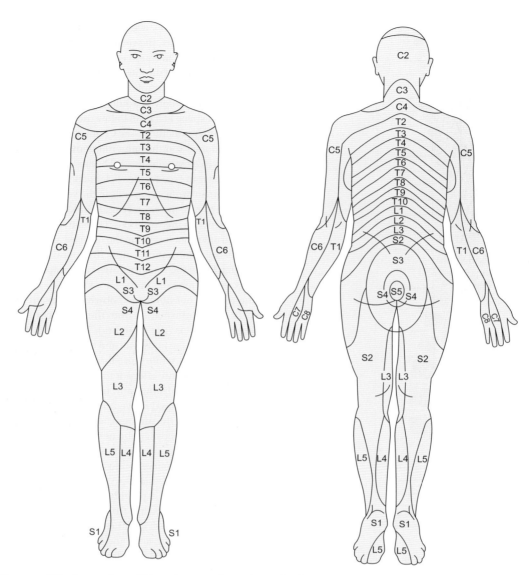

Figure 9.13 A diagram of the innervation of the cutaneous dermatomes by the spinal nerves. Although each dermatome is shown with a clear boundary, there is overlap in the innervation between adjacent segments.

level of protein synthesis in neurons. The space between the neuronal cell bodies appears rather amorphous in light microscopy and is known as the **neuropil**. It consists of the cell bodies of glial cells and the fine processes of both neurons and glia.

The morphology of neurons

As explained in Chapter 7, each neuron receives information via fine processes called dendrites and transmits information to its target cells via its axon. Neurons serve a variety of functions. Those that directly transmit information about the environment are called **sensory** or **afferent neurons**, while those that directly control the activity of the glands and muscles of the body are classed as **efferent neurons** or **motoneurons**. (Sensory and motoneurons

are sometimes also called projection neurons.) The neurons that lie completely within the CNS are called **interneurons**. The axons of small interneurons are short and they course for tens or perhaps a few hundred microns before reaching their targets, but the axons of some of the large sensory neurons and those of spinal motoneurons may exceed a metre in length.

Although most neurons contain all the structural elements shown in Figure 7.1, individual neurons show a remarkable diversity of size and shape. Their cell bodies range in size from a transverse diameter of 5 μm in small interneurons to 100 μm or more for large spinal motoneurons. As Figure 9.15 shows, their appearance is rather variable, but the cell bodies can be roughly classified as unipolar, bipolar, or multipolar. Dorsal root ganglion neurons are pseudo-unipolar as their cell body connects to

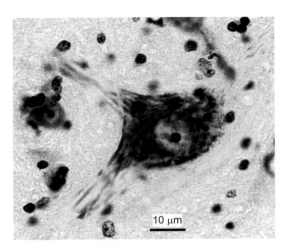

Figure 9.14 A large spinal neuron stained with a basic dye to show the characteristic staining of the Nissl substance. Two smaller neurons are present in this field of view, one to the left of the large neuron and one just coming into view at the very top of the field.

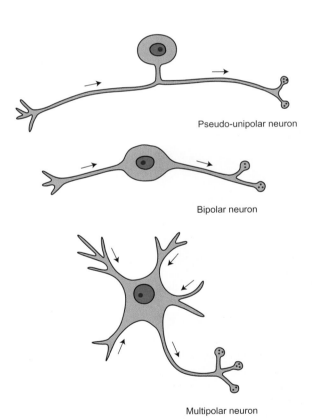

Pseudo-unipolar neuron

Bipolar neuron

Multipolar neuron

Figure 9.15 Diagram to show three different arrangements of neural processes. Dorsal root ganglion cells have a pseudo-unipolar arrangement of their dendrites and axons. Bipolar neurons are found in many places but are prominent in the retina. Neurons in which the processes have a multipolar distribution are common throughout the nervous system. Examples are spinal motoneurons, stellate cells, and cortical pyramidal cells. (Redrawn from Fig 1.5 in P. Brodal (2004) *The central nervous system: Structure and function*, 3rd edition. Oxford University Press, New York.)

a long axon near to its termination in the spinal cord. Certain other neurons are bipolar in shape (e.g. the bipolar cells of the retina—see Chapter 14), while most neurons are multipolar, as they have many processes arising from their cell bodies. Pyramidal cells have a distinctive pattern of apical and basal dendrites (Figure 9.16a), while cerebellar **Purkinje cells** have a profuse set of apical dendrites but no basal dendrites at all (see Figure 9.16b). Hippocampal basket cells have a very restricted dendritic tree but have an extensive axonal arborization that embraces the neighbouring neurons (Figure 9.16c). The dendrites of the spiny neurons of the **basal ganglia** are distributed in a more or less radial pattern (Figure 9.16d). In contrast, the peripheral neurons of the enteric nervous system (which innervates the gastrointestinal tract) have no extensive dendritic branching (Figure 9.16e).

The non-neuronal cells of the CNS

There are four main classes of non-neuronal cell that occur in the brain and spinal cord.

1. **Astroglia** or **astrocytes**. These are cells with long processes that make firm attachments to blood vessels. The ends of the astrocytic processes seal closely together and form an additional barrier between the blood and the extracellular fluid of the brain and spinal cord (see Figure 9.17c). This barrier is known as the **blood–brain barrier** and it serves to prevent changes in the composition of the blood influencing the activity of the nerve cells within the CNS.

2. **Oligodendroglia** or **oligodendrocytes**. Oligodendrocytes account for about 75 per cent of all glial cells in white matter, where they form the myelin sheaths of axons. Unlike the Schwann cells of the peripheral nervous system which form the myelin sheath of a single segment of one axon (see Chapter 7), each oligodendrocyte forms the myelin sheaths of a number of axons.

3. **Microglia**. These are scattered throughout the grey and white matter. They are phagocytes and rapidly converge on a site of injury or infection within the CNS.

4. **Ependymal cells**. These are ciliated cells that line the central fluid-filled spaces of the brain (the cerebral ventricles) and the central canal of the spinal cord. They form a cuboidal-columnar epithelium called the ependyma.

In the human brain the numbers of glial cells and neurons are roughly equal, although the neuron-to-neuroglia ratio varies from region to region. Although glial cells are generally considered to play no direct role in the processing of neural information, they are not simply supporting cells that hold the neurons in place as their name might imply. The most numerous, the astrocytes, have an important homeostatic role in which they regulate the extracellular environment of the neurons by restricting the exchange of solutes between the blood and the neurons (the blood–brain barrier described above). They also take up excess potassium and remove excess neurotransmitters from the extracellular space. During the development of the cerebellum, certain glia (called Bergmann glia) play a key role in guiding developing neurons to their correct positions.

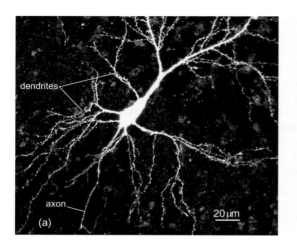
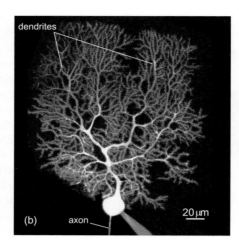

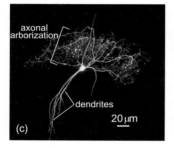
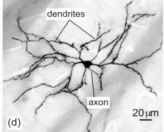
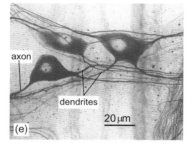

Figure 9.16 Examples of the morphological diversity shown by neurons. Panel (a) shows a hippocampal pyramidal cell with its extensive apical and basal dendritic arborizations. Panel (b) shows a cerebellar Purkinje cell with its extensive apical dendritic tree (note the absence of basal dendrites). The microelectrode used to inject the fluorescent dye is visible at the bottom right of the Purkinje cell body. Panel (c) shows a hippocampal basket cell injected with a fluorescent dye to reveal the extensive axon arborization. Note the extensive branching of the axonal arborization compared to the relative paucity of dendritic branching. Panels (d) and (e) show neurons stained with silver: (d) a spiny neurone from the striatum and (e) a small group of enteric neurons. The horizontal scale bars are approximately 20 μm. (Panel (b) is courtesy of Dr A. Batchelor, University College London.)

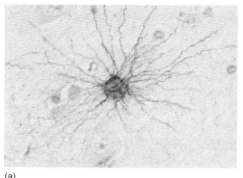
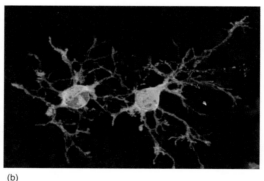

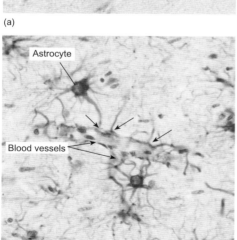
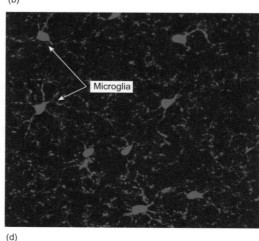

Figure 9.17 Examples of the three classes of glial cell that are found in the central nervous system. Panels (a) and (c) show astrocytes in the cerebral cortex. In panel (a) an astrocyte has been stained to show glial fibrillary acidic protein (GFAP). Note the radial distribution of its processes. In panel (c) the stain has been chosen to reveal the end-feet of the astrocyte processes. The arrows indicate the close association of astrocyte end feet with small blood vessels. Panel (b) shows two oligodendrocyte precursors tagged with two fluorescent marker proteins. Panel (d) shows microglia in the cerebral cortex labelled with green fluorescent protein. (Courtesy of D. Agamanolis (a and c) and F. Guillemot (b); (d) is from Fig 3(a) in F. Zhang et al. (2008) *Molecular Pain* **4**, 15.)

✳ Checklist of key terms and concepts

Overall structure of the CNS

- The nervous system may be broadly divided into five main parts:
  - the brain
  - the spinal cord
  - the autonomic nervous system
  - the enteric nervous system
  - the peripheral nerves.
- The brain and spinal cord are protected within the skull and vertebral canal by the meninges:
  - a tough outer layer, the dura mater;
  - the delicate arachnoid membrane, which lines the dura mater and is separated from the pia mater by the subarachnoid space;
  - the pia mater, which is highly vascular and covers the entire brain and spinal cord, following every contour.
- The subarachnoid space contains cerebrospinal fluid.
- The brain and spinal cord consist of grey and white matter.
- The grey matter contains the cell bodies of the neurons.
- The white matter contains the myelinated axons of the CNS neurons.
- The brain is connected to the head and neck by twelve pairs of cranial nerves.
- The spinal cord is connected to the peripheral muscles and organs by 31 pairs of spinal nerves.

- The cranial and spinal nerves serve both sensory and motor functions.

The cells of the CNS

- Within the CNS two distinct types of tissue can be discerned on the basis of their appearance: grey matter and white matter. Grey matter contains the cell bodies of the neurons. White matter mainly consists of myelinated nerve axons.
- The neuron is the principal functional unit of the nervous system. The cell bodies of neurons give rise to two types of process: dendrites and axons.
- The dendrites are highly branched and receive information from many other nerve cells.
- Each cell body gives rise to a single axon, which subsequently branches to make contact with a number of other cells.
- There are four main classes of non-neuronal cell that occur in the brain and spinal cord: the astroglia or astrocytes; the oligodendroglia or oligodendrocytes; the microglia; and the ependymal cells.
- The astrocytes play an important role in regulating the extracellular environment of the neurons, both by forming the blood–brain barrier and by removing excess potassium and neurotransmitters from the extracellular space.
- Oligodendrocytes form the myelin sheaths of the axons in the brain and spinal cord.
- The microglia are phagocytes.

Recommended reading

Brodal, P. (2004) *The central nervous system: Structure and function* (3rd edn). Oxford University Press, New York.
A very comprehensive introduction to the anatomy and physiology of the CNS, with many helpful drawings.

Kiernan, J.A. (2005) *Barr's The human nervous system: An anatomical viewpoint* (8th edn). Lippincott, Williams & Wilkins, Baltimore.
A good introduction to the anatomy of the nervous system, with many original plates showing cross-sections of the brain and spinal cord.

To check that you have mastered the key concepts presented in this chapter, complete the accompanying online self-assessment questions. Go to www.oup.com/uk/pocock5e/

CHAPTER 10

The physiology of motor systems

This chapter should help you to understand:

- The hierarchical nature of control within the motor systems of the body

- The organization of the spinal cord and its role in reflex activity

- The contribution of spinal reflexes to postural control

- The descending pathways that regulate the motor output of the spinal cord

- The role of the motor cortical regions in the programming and execution of voluntary activity

- The organization of the cerebellum and its role in refined, coordinated movements

- The role of the basal ganglia in the planning and execution of defined motor patterns

- The effects of lesions at various levels of the motor hierarchy

10.1 Introduction

Intrauterine movement is detectable by ultrasound from a very early stage of gestation and is felt by the mother for the first time between 16 and 20 weeks ('quickening' of the **fetus**). By the time of birth, a baby is capable of some coordinated movement. During the first two years of life, as the brain and spinal cord continue to develop and mature, a child learns to defy gravity, by first sitting, then standing. Later, it learns to walk, run, jump, and climb. At the same time, the capacity to perform the precise movements needed for complex manipulations and for speech are acquired. In short, coordinated purposeful movement is a fundamental aspect of human existence.

The simplest form of motor act controlled by the nervous system is called a **reflex**. This is a rapidly executed, automatic, and stereo-typed response to a given stimulus. As reflexes are not under the conscious control of the brain, they are described as involuntary motor acts. Nevertheless, most reflexes involve extensive coordination between groups of muscles and this is achieved by interconnections between various groups of central neurons. The neurons forming the pathway taken by the nerve impulses responsible for a reflex make up a **reflex arc**. Many voluntary motor acts are guided by the intrinsic properties of reflex arcs but are modified by commands from higher centres in the brain and signals from sensory inputs.

Two kinds of voluntary motor function can be distinguished: (i) the maintenance of position (posture) and (ii) goal-directed movements. In practice, they are inextricably linked. A goal-directed movement will only be performed successfully if the moving limb is first correctly positioned. Similarly, a posture may only be maintained if appropriate compensatory movements are made to counteract any force tending to oppose that posture.

Physiological understanding is still insufficient to allow a full explanation of the events occurring within the central nervous system during the execution of a voluntary movement. There is no doubt that voluntary movement is impaired whenever there is an interruption of the afferent pathways to the brain arising from sense organs such as the labyrinth, eyes, proprioceptors, and mechanoreceptors. However, what happens between the arrival of afferent information and the execution of an appropriate movement is unclear. Similarly, the processes of motor learning and the relationship between the intention to carry out a movement and the movement itself remain poorly understood.

Investigation of the ways in which the CNS controls movement is far more difficult than the study of sensory systems. There are a number of reasons for this. Firstly, movements themselves are difficult to describe in a precise, quantitative manner. Secondly, while experimentation on anaesthetized animals can give useful information about the organization of reflexes, it is less useful in investigating voluntary movements. Thirdly, the many complex motor pathways operate in parallel making it hard to assign particular roles to each. Furthermore, every motor action results in sensory feedback that may modify it further.

Despite these difficulties, it is possible to define some key questions which must be addressed if we are to make progress towards a greater understanding of motor systems.

1. What structural components of the central and peripheral nervous systems participate in the maintenance of posture and the movement of the head, trunk, and limbs?

2. How are reflexes organized within the spinal cord?

3. How do 'higher centres' influence the fundamental motor patterns contained within the spinal cord?

4. How is information from the peripheral sense organs used to plan and refine both postural mechanisms and voluntary movements?

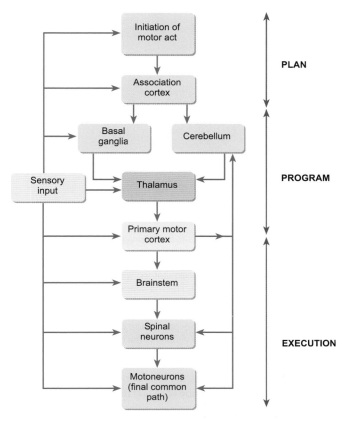

Figure 10.1 A diagrammatic representation of the hierarchy of the motor systems of the body, showing possible interactions between them.

10.2 The hierarchical nature of motor control systems

The term **motor system** refers to the neural pathways that control the sequence and pattern of muscle contractions. The structures responsible for the neural control of posture and movement are distributed throughout the brain and spinal cord, as indicated in Figure 10.1.

As skeletal muscles can only contract in response to excitation of the motoneurons that supply them, all motor acts depend on neural circuits that eventually impinge on the α-motoneurons that form the output of the motor system. As discussed in Chapter 7, each motoneuron supplies a number of skeletal muscle fibres; an α-motoneuron together with the skeletal muscle fibres it innervates constitute a **motor unit**, which is the basic element of motor control. For this reason α-motoneurons are often referred to as the final common pathway of the motor system. The processes that underlie neuromuscular transmission and muscle contraction are discussed in Chapters 7 and 8.

The motoneurons are found in the brainstem and spinal cord. Their excitability is influenced by neural pathways that may form local circuits or that arise in a variety of brain areas. Thus, there is a hierarchical arrangement of so-called 'motor centres' from the spinal cord to the cerebral cortex. Many reflexes are controlled by

neural circuits within the spinal cord that determine the basic motor patterns of posture and movement. In turn, these are influenced by commands from higher centres in the brain: postural control is exerted largely at the level of the brainstem, while goal-directed movements require the participation of the cerebral cortex. The basal ganglia and cerebellum both play an important role in motor control, though neither is directly connected with the spinal motoneurons. Instead, they influence the motor cortex by way of the thalamic nuclei (see Figure 10.1).

10.3 Organization of the spinal cord

The α-motoneurons (somatic motor neurons) are large neurons whose cell bodies lie in columns within the ventral horn of the spinal cord and in the motor nuclei of the brainstem. A cross-section of the spinal cord reveals the positions of the motoneuron columns in the ventral horn (Figure 10.2). Each motoneuron innervates a motor unit that may consist of anything between six and 2000 skeletal muscle fibres. Motoneurons have long dendrites over which they receive many synaptic connections, including afferents from interneurons and proprioceptors as well as descending fibres from higher levels of the CNS. The axons of α-motoneurons collect in bundles that leave the ventral horn and pass through the ventral white matter of

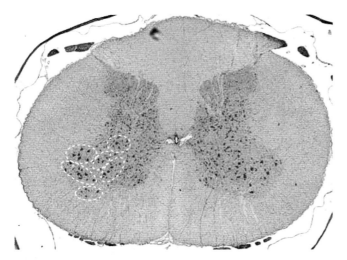

Figure 10.2 A cross-section of the lumbar region of the human spinal cord. The section has been stained with a basic dye to show the neurons (Nissl substance). The motoneurons are located in the anterior (ventral) horn and appear to be clumped into discrete populations. In fact the clumps (which are indicated by the dotted outlines) reflect the columnar organization of motoneurons. A fragment of a dorsal rootlet is visible at the top right of this section.

The spinal cord receives afferent input from proprioceptors in muscles and joints

For movements to be carried out in an appropriate way, it is essential for sensory and motor information to be integrated. All the neural structures involved in the execution of movements are continually informed of the position of the body and of the progress of the movement by sensory receptors within the muscles and joints. These provide information regarding the position of the limbs and their movements relative to each other and to the surroundings. These receptors are called **proprioceptors**, and the information they provide is used to control muscle length and posture (as discussed in this section and in Section 10.8).

The main proprioceptors are the **muscle spindles** and **Golgi tendon organs**. Both provide information with regard to the state of the musculature. Muscle spindles lie in parallel with the skeletal muscle fibres within the individual muscles. They can therefore respond both to muscle length and to its rate of change. In contrast, Golgi tendon organs lie within the tendons and are in series with the contractile elements of the muscle. They are sensitive to the force generated within that muscle during contraction.

Muscle spindles

Although they are found in all human skeletal muscles, muscle spindles are particularly numerous in those muscles that are responsible for fine motor control, especially those of the eyes, neck, and hands. Each spindle consists of a small bundle of 2 to 12 thin muscle fibres enclosed within a connective tissue capsule about 1.5 mm in length with a maximum diameter of about 0.5 mm. They are innervated by both motor (efferent) and sensory (afferent) nerve fibres (see Figure 10.4). The muscle fibres of the muscle spindles are called **intrafusal fibres**, while those of the main body of the muscle are the **extrafusal fibres**.

There are two different types of intrafusal fibres: nuclear bag fibres and nuclear chain fibres, so called because of the arrangement of their nuclei, which are shown diagrammatically in Figure 10.5.

the spinal cord before entering the ventral root. Some axons send off branches that turn back into the cord and make excitatory synaptic contact with small interneurons called **Renshaw cells**. These cells in turn have short axons that synapse with the pool of motoneurons by which they are stimulated. These synapses are inhibitory and bring about recurrent, or feedback, inhibition.

The ventral horn motoneurons have an orderly topographical arrangement: motoneurons supplying the muscles of the trunk are situated in the medial ventral horn, while those supplying more distal muscle groups tend to be situated more laterally, as shown in Figure 10.3. Muscles that flex the limbs (**flexors**) are under the control of neurons that lie dorsal to those that control the muscles that extend the limbs (**extensors**).

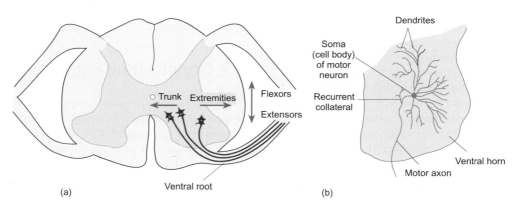

(a)   (b)

Figure 10.3 (a) A transverse section of the spinal cord. The localization of motoneurons corresponding to various groups of muscles are indicated, with flexors represented more dorsally and extensors represented ventrally within the cord. The muscles of the trunk are represented medially, and the extremities are represented laterally. (b) An α-motoneuron in the ventral horn of the spinal cord, illustrating its elaborate dendritic tree. Although not shown in this diagram, numerous synaptic connections are made with these dendrites.

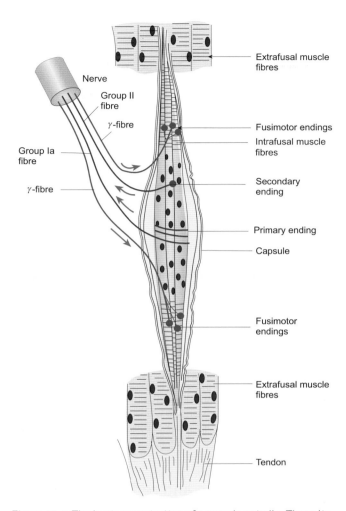

Figure 10.4 The basic organization of a muscle spindle. These lie in parallel with the extrafusal muscle fibres and are therefore adapted to monitor muscle length rather than muscle tension. Note that each muscle spindle is innervated by both motor (γ-efferent) and sensory nerve fibres, which are shown in magenta and blue respectively.

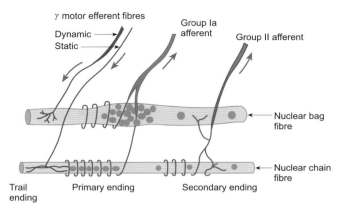

Figure 10.5 A nuclear bag and a nuclear chain fibre of a muscle spindle, with their sensory and motor nerve supplies. Note the different positions of the annulospiral (primary) and flower spray (secondary) sensory endings, which are connected to group I and group II afferents respectively.

Nuclear bag fibres have a cluster of nuclei near their midpoint. Nuclear chain fibres are smaller and have a single row of nuclei near their midpoint. The central regions of both bag and chain fibres contain no myofibrils and are the most elastic parts of the fibres, so that this central region stretches preferentially when the muscle spindle is stretched. This central region is innervated by afferent fibres that are sensitive to stretch.

The afferent fibres of muscle spindles are of two types. Each spindle receives fibres from a primary afferent (group Ia fibres). These wind around the middle section of both bag and chain fibres forming so-called **annulospiral endings** (also known as primary sensory endings). Many muscle spindles are also innervated by one or more afferent fibres of group II (see Chapter 7 for the classification of nerve fibres). They terminate more peripherally than the primary endings, almost exclusively on the nuclear chain fibres. They are known as **flower-spray endings** because of their multiple branched nature (and are also called secondary sensory endings).

The two types of afferent nerve endings respond to muscle stretch in different ways. The rate of firing of both kinds is proportional to the degree of stretch of the muscle spindle at any moment. Although the primary (Ia) fibres respond in proportion to the degree of stretch, they are much more sensitive to rapid changes in muscle length and, for this reason, they are classed as rapidly adapting or dynamic endings. The secondary endings are non-adapting and are said to be static endings. The differing nature of the responses to stretch of the primary and secondary endings is illustrated in Figure 10.6.

The motoneurons innervating the intrafusal fibres are known as **gamma motoneurons** (γ- motoneurons) to distinguish them from the large α-motoneurons, which innervate the extrafusal fibres. The cell bodies of the gamma motoneurons lie in the ventral horn of the spinal cord and their axons, which are also known as **fusimotor fibres** (or as γ-efferents), leave the spinal cord via the ventral roots. The fusimotor fibres are myelinated and have diameters in the range 3–6 μm with conduction velocities of 15–30 m s^{-1}. The α-motoneurons have large-diameter myelinated fibres that range in size between 15 and 20 μm and have conduction velocities of 70–120 m s^{-1} (see Table 7.1).

Within a muscle, the fusimotor fibres branch to supply several muscle spindles, and within these they branch further to supply several intrafusal fibres. The γ-motoneurons innervate the contractile regions of both the nuclear chain and nuclear bag fibres to bring about contraction of the peripheral regions of the muscle spindles. Note, however, that contraction of the intrafusal fibres is far too weak to effect movements of the muscle as a whole; rather, the tension in the intrafusal fibres determines both the operating range and the sensitivity of the spindle. When the intrafusal fibres contract in response to fusimotor stimulation, their sensory endings are stimulated by the stretch. The excitation of the Ia fibres that results is superimposed on the firing that results from the degree of stretch imposed by the extrafusal muscle fibres.

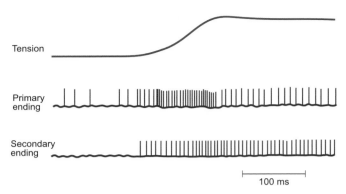

Tension

Primary ending

Secondary ending

100 ms

Figure 10.6 Responses of primary (Ia) and secondary (II) muscle spindle afferent fibres to muscle stretch. Note the very intense period of activity of the primary ending during the stretch (the dynamic component of the response). The response of the secondary ending reflects the maintained increase in muscle tension. (Redrawn from Fig 1 in A. Crowe and P.B.C. Matthews (1964) *Journal of Physiology* **174**, 109–31.)

To sum up, muscle spindles may be stimulated in two ways:

1. by stretching the entire skeletal muscle; or
2. by causing the intrafusal fibres to contract while the extrafusal muscle fibres remain at the same length.

In either case, stretching a muscle spindle will increase the rate of discharge of the group Ia and group II afferent nerve fibres to which it is connected.

Golgi tendon organs

The Golgi tendon organs are mechanoreceptors that lie within the tendons of muscles immediately beyond their attachments to the muscle fibres (see Figure 10.7). Around 10 or 15 muscle fibres from several different motor units are usually connected in series to each Golgi tendon organ, which responds to the tension produced by this bundle of fibres rather than to that produced by the whole muscle. Impulses are carried from the tendon organs to the central nervous system (particularly the spinal cord and the cerebellum) by group Ib afferent fibres.

Summary

The motoneurons are located in the ventral horn of the spinal cord and their axons innervate skeletal muscle fibres. The motoneurons receive numerous synaptic connections from proprioceptors and from higher levels of the CNS. Proprioceptors are mechanoreceptors situated within muscles and joints. They provide the CNS with information regarding muscle length, position, and tension (force). Muscle spindles are innervated by γ-motoneurons and by afferent fibres. Their afferents respond to muscle stretch, while γ-efferent activity regulates the sensitivity of the spindles. Golgi tendon organs respond to the degree of tension within the muscle.

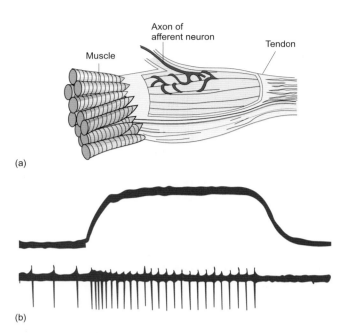

(a)

(b)

Figure 10.7 The Golgi tendon organ. Panel (a) shows the basic organization of a Golgi tendon organ; panel (b) illustrates the response of a Golgi tendon organ to tension in the muscle. The upper trace shows the tension in the muscle to which the tendon organ is attached, and the lower trace shows the firing pattern in the Ib fibre supplying the receptor. Note that the Golgi tendon organ lies in series with the muscle and so is adapted to monitor muscle tension. (Redrawn from data of Houk and Henneman.)

10.4 Reflex action and reflex arcs

Reflexes represent the simplest form of motor activity elicited by the nervous system. The neurons that control a given reflex are said to constitute a **reflex arc**, which must include at least two neurons: an **afferent** or **sensory** neuron and an **efferent** or **motor** neuron. The afferent fibre carries information from a receptor towards the central nervous system, while the efferent fibre transmits nerve impulses from the central nervous system to an effector (commonly a muscle or group of muscles). The latency of a reflex refers to the time between the stimulus and the response. Reflexes may be (and often are) subject to modulation by activity arising in the higher centres of the central nervous system.

In the simplest reflex arc, there are two neurons and just one synapse. Such reflexes are therefore known as **monosynaptic reflexes**. Other reflex arcs have one or more neurons interposed between the afferent and efferent neurons. These neurons are called **interneurons** or internuncial neurons. If there is one interneuron, the reflex arc will have two synaptic relays and the reflex is called a **disynaptic reflex**. If there are two interneurons, there will be three synaptic relays so that the associated reflex would be trisynaptic. If many interneurons are involved, the reflex would be called a **polysynaptic reflex**. Examples are the stretch reflex (monosynaptic), the withdrawal reflex (disynaptic), and the scratch reflex (polysynaptic).

The knee-jerk is an example of a dynamic stretch reflex

A classic example of a stretch reflex (also known as the myotactic reflex) is the **knee-jerk reflex** or **tendon-tap reflex**, which is used routinely in clinical neurophysiology as a tool for the diagnosis of certain neurological conditions. Hinge joints such as the knee and ankle are extended and flexed by extensor and flexor muscles, which act in an antagonistic manner. A sharp tap applied to the patellar tendon stretches the quadriceps muscle. The stretch stimulates the 'dynamic' nuclear bag receptors of the muscle spindles. As a result, there is an increase in the rate of firing of the group Ia afferents of the muscle spindles within the quadriceps. This informs the spinal cord that the quadriceps muscle has been stretched. The afferent fibres branch as they enter the spinal cord. Some of these nerve branches enter the grey matter of the cord and make monosynaptic contact with the α-motoneurons supplying the quadriceps muscle, causing them to discharge in synchrony. The resulting contraction of this muscle abruptly extends the lower leg (hence the name knee-jerk). Collaterals of the Ia fibres make synaptic contact with inhibitory interneurons, which in turn inhibit the antagonistic (flexor) muscles of the knee joint.

The stretch reflex arc is illustrated diagrammatically in Figure 10.8a. It is a very fast (i.e. short latency) reflex that is lost if damage occurs to the lower lumbar dorsal roots of the spinal cord through which the afferents from the quadriceps pass. A similar reflex occurs when the Achilles tendon is struck (the ankle-jerk reflex). In this case, there is a plantar flexion of the foot produced by contraction of the calf muscles.

The tonic stretch reflex

This reflex contributes to muscle tone and helps to maintain posture. Even when the length of a muscle is kept more or less constant, muscle spindles continue to relay information to the spinal cord. When standing upright, for example, the slightest bending of the knee joint will stretch the quadriceps muscle and increase the activity in the primary muscle spindle endings. This will result in stimulation of the α-motoneurons supplying the quadriceps. The tone of the muscle will increase and will counteract the bending so that the posture will be maintained. The converse will occur when there is excessive contraction of the muscle. The tonic stretch reflex therefore helps to stabilize the length of a muscle when it is under a constant load.

The flexion reflex

In this protective reflex, a limb is rapidly withdrawn from a threatening or damaging stimulus. It is more complex than the stretch reflex and usually involves large numbers of interneurons and proprio-spinal connections arising from many segments of the spinal cord. Withdrawal may be elicited by noxious stimuli applied to a large area of skin or deeper tissues (muscles, joints, and viscera) rather than from a single muscle as in the stretch reflex. The sensory receptors responsible are called **nociceptors** (see Chapters 12 and 13) and they give rise to the afferent impulses that are responsible for the flexion reflex.

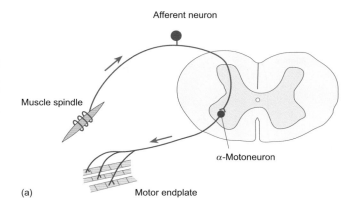

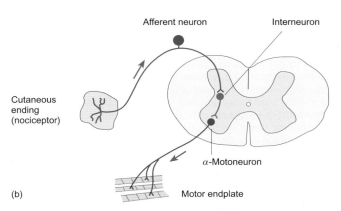

Figure 10.8 (a) A diagrammatic representation of the stretch reflex arc. Note that this reflex arc comprises only two neurons and one synapse. It is therefore a monosynaptic reflex. The knee-jerk is a typical example. (b) A diagrammatic representation of the basic flexor (withdrawal) reflex arc. In this case, the basic arc is disynaptic with three neurons and two synapses. This reflex has a protective role.

To achieve withdrawal of a limb, the flexor muscles of one or more joints in the limb must contract while the extensor muscles relax. Afferent volleys cause excitatory interneurons to activate α-motoneurons that supply flexor muscles in the affected limb. At the same time, the afferent volleys activate interneurons, which inhibit the α-motoneurons supplying the extensor muscles. The excitation of the flexor motoneurons coupled with inhibition of the extensor motoneurons is known as **reciprocal inhibition**.

The synaptic organization of the flexor reflex is illustrated in Figure 10.8b. As the figure shows, the basic reflex is disynaptic but in a powerful withdrawal reaction it is likely that several spinal segments would be involved and that all the major joints of the limb will show movement. The flexion reflex has a longer latency than the stretch reflex because it arises from a disynaptic arc. It is also non-linear, insofar as a weak stimulus will elicit no response while a powerful withdrawal is seen when the stimulus reaches a certain level of intensity.

The crossed extensor reflex

Stimulation of the flexion reflex (see previous subsection) frequently elicits extension of the contralateral limb about 250 ms

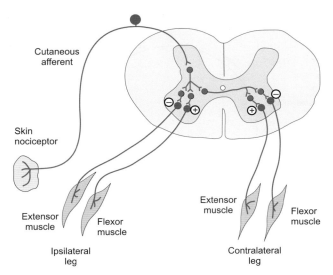

Figure 10.9 The neuronal pathways participating in a crossed extensor reflex arc. Many neurons and synapses are involved in this reflex, which is therefore polysynaptic. When the skin nociceptor is activated, the limb on the same (ipsilateral) side is withdrawn (its flexors are activated). Interneurons send impulses to inhibit the flexor motoneurons of the contralateral limb and, in the case of the legs, the motoneurons of the contralateral extensors are activated to stabilize the posture. (− denotes synaptic inhibition and + synaptic excitation.)

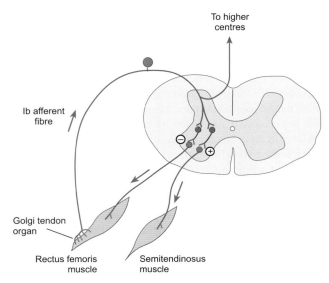

Figure 10.10 A diagrammatic representation of the Golgi tendon organ reflex arc. This shows the synaptic basis of reciprocal inhibition, which is also seen in the crossed extensor reflex illustrated in Figure 10.9. The motoneurons controlling the semitendinosus muscle are activated, while those of the rectus femoris muscle are inhibited. (− denotes synaptic inhibition and + synaptic excitation.)

later. This crossed extension reflex assists in the maintenance of posture and balance. The long latency between flexion and crossed extension represents the time taken to recruit interneurons. The reflex arc for crossed extension is illustrated in Figure 10.9.

The Golgi tendon reflex

This reflex, which is also known as the **inverse mytotactic reflex,** is activated by a load on the Golgi tendon organs. It complements the tonic stretch reflex and contributes to the maintenance of posture. Its organization is illustrated for the knee joint in Figure 10.10. In this example, the tendon organ is located in the tendon of the rectus femoris muscle and its afferent fibre branches as it enters the spinal cord. One set of branches excites interneurons which inhibit the discharge of the α-motoneurons supplying the rectus femoris muscle, while another set activates interneurons that stimulate activity in the α-motoneurons innervating the antagonistic semitendinosus (hamstring) muscles.

How does this reflex operate to maintain posture? During a maintained posture such as standing, the rectus femoris muscle will start to fatigue. As it does so, the force in the patellar tendon, monitored by Golgi tendon organs, will decline. As a result, activity in the afferent Ib fibres will decline and the normal inhibition of the motoneurons supplying the rectus femoris will be removed. Consequently, the muscle will be stimulated to contract more strongly, thereby increasing the force in the patellar tendon once more.

Summary

The neurons participating in a reflex form a reflex arc. This includes a receptor, an afferent neuron that synapses in the CNS, and an efferent neuron that sends a nerve fibre to an effector. Interneurons may be present between the afferent and efferent neurons. The number of synapses in a reflex arc is used to define the reflex as monosynaptic, disynaptic, or polysynaptic.

The simplest reflexes are the monosynaptic stretch reflexes such as the patellar tendon-tap reflex (the knee-jerk reflex). Here, stretching of a muscle stimulates the muscle spindle afferents. These excite the α-motoneurons supplying that muscle and cause it to contract. Stretch reflexes play an important role in the control of posture.

The protective flexion (withdrawal) reflexes are elicited by noxious stimuli. Their reflex arc possesses at least one interneuron, so the most basic flexion reflex is disynaptic. Reciprocal inhibition ensures that the extensor muscles acting on a joint will relax while the flexor muscles contract.

10.5 The role of muscle proprioceptors in voluntary motor activity

Goal-directed movements often need to be adjusted as a specific task proceeds. These adjustments are made in the light of feedback information from the proprioceptors in the muscles and tendons: the muscle spindles and the Golgi tendon organs. This information enables people to accurately sense both the position of their limbs in space (the kinaesthetic sense—see Chapter 13 p. 205) and the forces exerted by their muscles.

The activity of the muscle spindles enables the nervous system to compare the lengths of the extrafusal and intrafusal muscle fibres. Whenever the length of the extrafusal fibres exceeds that of the intrafusal fibres, the afferent discharge of a muscle spindle will increase, and whenever the length of the extrafusal fibres is less than that of the intrafusal fibres, the discharge of the spindle afferents will decline. This decline is possible because the spindle afferents normally show a tonic level of discharge. In this way, the muscle spindles can provide feedback control of muscle length.

Studies of various movements (e.g. chewing and finger movements) have shown that when voluntary changes in muscle length are initiated by motor areas of the cerebral cortex (see Section 10.9), the motor command includes changes to the set point of the muscle spindle system. To achieve this, both α- and γ-motoneurons are activated simultaneously by way of the neuronal pathways descending from higher motor centres. The simultaneous activation of extrafusal fibres (by way of α-motoneurons) and intrafusal fibres (by way of γ-motoneurons) is called **alpha-gamma co-activation**. Its physiological importance lies in the fact that it allows the muscle spindles to remain functional at all times during a muscle contraction.

The benefits of incorporating length sensitivity into voluntary activities may be illustrated by the following example. Suppose a heavy weight needs to be lifted. Before lifting, and from previous experience, the brain will estimate roughly how much force will be required to lift the weight and the motor centres will transmit the command to begin the lift. Both extrafusal and intrafusal fibres will be activated together. If the initial estimate is accurate, the extrafusal fibres will be able to shorten as rapidly as the intrafusal fibres and the activity of the spindle afferents will not change much during the lift. If, however, the weight turns out to be heavier than expected, the estimate of required force will be insufficient and the rate of shortening of the extrafusal fibres will be slower than expected. Nevertheless, the intrafusal fibres will continue to shorten and their central region will become stretched. As a result, the activity in the spindle afferents will increase and will summate with the excitatory drive already arriving at the α-motoneurons via the descending motor pathways. The increased activity in the α-motoneurons will increase the force generated by the muscle until it matches that required to lift the load. Conversely, if the weight is lighter than predicted, the muscle will shorten rapidly and unload the muscle spindles whose discharge will decline. There will be less summation, with the excitatory drive already arriving at the α-motoneurons via the descending motor pathways, and the tension developed by the muscle will fall. In either case, the excitatory drive to the α-motoneurons is matched to the load.

The muscle spindles lie in parallel with the extrafusal fibres and cannot sense the tension exerted by the muscle on the load. This role is performed by the Golgi tendon organs which lie in series with the extrafusal muscle fibres (see Figure 10.7). Each tendon organ responds to the tension developed by its associated muscle fibres. The information from the tendon organs is relayed back to the motor cortex, which can then adjust the excitatory drive to the appropriate muscles via the α-motoneurons. In addition to this role in proprioception, the Golgi tendon reflex can protect the muscle from damage caused by cramp or by an excessive load.

Summary

The role of the muscle spindles as comparators for the maintenance of muscle length is important during goal-directed voluntary movements. When voluntary changes in muscle length are initiated by motor areas of the brain, the motor command includes changes to the set point of the muscle spindle system. To achieve this, both α- and γ-motoneurons are activated simultaneously (α–γ co-activation) by way of the neuronal pathways descending from higher motor centres. Co-activation of α- and γ-efferent motoneurons enables the sensitivity of muscle spindles to be continuously adjusted as the muscle shortens.

10.6 Effects of injury to the spinal cord

Despite the protection afforded to the spinal cord by the vertebral column, spinal injuries are still relatively common. Motor accidents are the most frequent cause of spinal cord injury, followed by falls and sports injuries (particularly diving). The alterations in body function that result from such injuries depend on the level of injury and the extent of the damage to the spinal cord. Following damage, below the level of the lesion there is a loss of sensation due to interruption of ascending spinal pathways and loss of the voluntary control over muscle contraction as a result of damage to the descending motor pathways. The loss of voluntary motor control is known as **muscle paralysis**. Both spinal reflex activity and functional activities such as breathing, micturition, and defecation may be affected. Paralysis may be **spastic** (in which the degree of muscle tone is increased above normal) or **flaccid** (in which the level of tone is reduced and the muscles are 'floppy'). The characteristics of motor dysfunction due to lesions of the corticospinal tract ('**upper motoneuron lesions**') and those due to dysfunction of spinal motoneurons ('**lower motoneuron lesions**') are discussed further in Box 10.1.

Immediately following injury to the spinal cord, there is a loss of spinal reflexes (**areflexia**). This is known as **spinal shock** or **neurogenic shock** and involves the descending motor pathways. The clinical manifestations are flaccid paralysis, a lack of tendon reflexes, and loss of autonomic function below the level of the lesion. Spinal shock is probably the result of the loss of the normal continuous excitatory input from higher centres such as the vestibulospinal tract, parts of the reticulospinal tract, and the corticospinal tract, as will be discussed later in this chapter. In humans, spinal shock may last for several weeks. After this time, there is usually some return of reflex activity as the excitability of the undamaged spinal neurons increases. Occasionally this excitability becomes excessive and then **spasticity** of affected muscle groups is seen.

Box 10.1 Upper and lower motoneuron lesions and sensory deficits following spinal lesions

Voluntary motor acts involve a large number of structures including the cerebral cortex, corticospinal tract, basal ganglia, cerebellum, and spinal cord. When any of these regions are damaged by traumatic injuries or strokes, characteristic changes in the activity of the motor system are apparent.

Damage to the primary motor cortex or corticospinal tract is often referred to as an **upper motoneuron lesion**. Such lesions are characterized by spastic (rigid) paralysis without muscle wasting. There is an increase in muscle tone and the tendon reflexes are exaggerated (**hyperreflexia**). The Babinski sign is positive so that in response to stimulation of the sole of the foot, the big toe extends upwards and the other toes fan out rather than flexing downwards as in a normal person (see Figure 1). The muscles themselves are weak, although the flexors of the arms are stronger than the extensors. In the legs, the reverse is true; the extensors are stronger than the flexors. The muscles do not show fasciculation (muscle twitching, in which groups of muscle fibres contract together).

Damage to the spinal cord that involves the motoneurons is called a **lower motoneuron lesion**. A lower motoneuron lesion results in denervation of the affected muscles, which show a loss of muscle tone. Tendon reflexes may be weak or absent. There is a loss of muscle bulk, and fasciculation (brief spontaneous contractions of individual muscle fascicles) is usually present. Coordination is impaired. The Babinski sign is absent (i.e. the toes flex in response to stimulation of the sole of the foot, as in a normal person).

Lesions to the spinal cord affect sensation as well as motor activity. If the spinal cord is completely transected, there is a loss of both motor and sensory function below the level of the lesion. Partial loss of motor activity and sensation occurs when the spinal cord is compressed, either by the protrusion of an intervertebral disc or by a spinal tumour. The spinal cord may also be partially cut across as a result of certain traumatic injuries such as stabbing. In such partial transections, the pattern of impaired sensation provides important information regarding the site of the lesion.

In clinical examination, the primary senses tested are light touch (which is tested with a wisp of cotton wool), pinprick, temperature, vibration, and position sense. Reflex activity is tested with a patellar hammer. On the side of the lesion, below the area of damage, there is a loss of voluntary motor activity, which results from damage to the corticospinal tract. The immobility is associated with the signs of upper motoneuron damage. Motor activity is unimpaired on the unaffected side.

The spinothalamic tract is crossed at the segmental level and the dorsal column pathway crosses in the brainstem. The loss of sensation following hemisection of the spinal cord is therefore characteristic. On the side of the lesion below the area of damage, there is a loss of those sensations that reach the brain via the dorsal column pathway, namely proprioception, vibration sense, and fine touch discrimination. Pain and thermal sensations are transmitted to the brain via the crossed spinothalamic tract and so are preserved. On the side opposite the lesion, position and vibration sense below the area of damage are normal, as is tactile discrimination. Pain and thermal sensations are lost, as the spinothalamic pathways are transected.

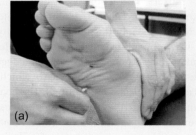
(a)

(b)

(c)

Figure 1 In the normal plantar reflex elicited by stroking the sole of the foot as shown in (a), the toes flex downward. If there is damage to the corticospinal tract (an upper motoneuron lesion) the toes extend upward and fan out (b) just as they do in a normal young baby (c). This response is called a positive Babinski sign.

The term **tetraplegia** (also called **quadriplegia**) refers to impairment or loss of motor and sensory function in the arms, trunk, legs, and pelvic organs. **Paraplegia** refers to impairment of function of the legs and pelvic organs; the arms are spared and the degree of functional impairment will depend on the exact level of the spinal injury. Injuries above the level of around T12 will normally result in spastic paralysis of the affected skeletal muscle groups; control of bowel, bladder, and sexual functions are also affected. Injuries below the level of T12, especially those that involve damage to the peripheral nerves leaving the spinal cord, generally result in flaccid paralysis of the affected muscle groups as well as those controlling bowel, bladder, and sexual function.

Arteriolar tone is maintained by activity in axons that pass from the vasomotor regions in the brainstem to the intermedio-lateral column of the spinal cord via the lateral column of the white matter. Damage to the upper segments of the spinal cord will affect the sympathetic vasoconstrictor tone, causing a marked decrease in blood pressure. Damage to the lumbar region of the spinal cord, however, will have a relatively small effect.

Summary

Damage to the spinal cord causes a loss both of sensation and of voluntary control over muscle contraction (paralysis) below the level of the lesion. Immediately following such an injury, there is a loss of spinal reflexes (areflexia) and flaccidity of the muscles controlled by nerves leaving the spinal cord below the level of the injury. This is known as spinal shock or neurogenic shock. Both reflex somatic motor activity, and involuntary activities involving bladder and bowel functions may be affected. Spinal shock may last for several weeks. As the excitability of the undamaged neurons below the damaged area increases, reflex activity returns and may result in the spasticity of the paralysed muscle groups.

10.7 Descending pathways involved in motor control

Although the spinal cord contains the neural networks required for reflex actions, more complex motor behaviours are initiated by pathways that originate at various sites within the brain. Furthermore, the activity of the neural circuitry within the spinal cord is modified and refined by the descending motor control pathways. There are five important brain areas which give rise to descending tracts, four of which lie within the brainstem and medulla. These are the **reticular formation**, the vestibular nuclei, the red nucleus and the **tectum** (the dorsal part of the midbrain). Their fibres constitute a set of descending pathways that are sometimes referred to as the **extrapyramidal tracts**. The fifth area lies within the cerebral cortex and gives rise to the **pyramidal** or **corticospinal tract**. Figure 10.11 shows the approximate positions of the major descending tracts and their location within the spinal cord.

Extrapyramidal pathways

The reticular system (a poorly defined set of interconnected nuclei in the brainstem) gives rise to two important descending tracts within the cord: the lateral reticulospinal tract and the medial reticulospinal tract. These tracts are largely uncrossed, terminating mainly on spinal interneurons rather than the α-motoneurons themselves. They mainly influence the muscles of the trunk and the proximal parts of the extremities. They are believed to be important in the control of certain postural mechanisms and in the startle reaction ('jumping' in response to a sudden and unexpected stimulus).

The vestibular nuclei are located just below the floor of the fourth ventricle close to the cerebellum. The medial and lateral vestibular nuclei give rise to a descending motor pathway called the vestibulospinal tract. Most of the axons of this tract, like those of the reticulospinal tract, synapse with interneurons on the ipsilateral side of the cord. The vestibulospinal tracts are concerned mostly with the activity of extensor muscles and are important in the control of posture (acting to make adjustments in response to

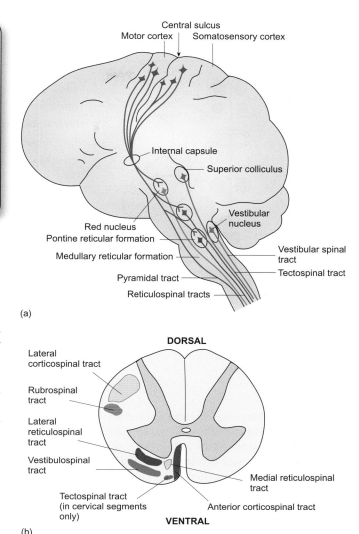

Figure 10.11 The principal motor pathways arising in the brain. Panel (a) shows the arrangement of the major descending motor tracts and the approximate location of the main motor nuclei. Panel (b) shows the position of the major pyramidal and extrapyramidal descending pathways within the spinal cord. (Redrawn from Fig 14.14 in P. Brodal (1997) *The central nervous system: Structure and function*, 2nd edition. Oxford University Press, New York.)

vestibular signals). Lesions of reticulospinal or vestibulospinal tracts affect the ability to maintain a normal erect posture.

The red nucleus is one of the most important 'extrapyramidal' structures. It has a topographic organization in which the upper limbs are represented dorsomedially and the lower limbs ventrolaterally. It receives afferent inputs from both the cortex and the cerebellum as well as from the globus pallidus, which is the major output nucleus of the basal ganglia. It gives rise to the **rubrospinal tract** (see Figure 10.11). The fibres of this tract decussate (cross) and then travel to the spinal cord where they terminate in the lateral part of the grey matter. Some make monosynaptic contact with

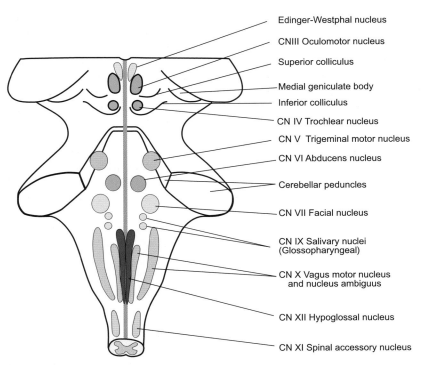

Edinger-Westphal nucleus

CNIII Oculomotor nucleus

Superior colliculus

Medial geniculate body

Inferior colliculus

CN IV Trochlear nucleus

CN V Trigeminal motor nucleus

CN VI Abducens nucleus

Cerebellar peduncles

CN VII Facial nucleus

CN IX Salivary nuclei
(Glossopharyngeal)

CN X Vagus motor nucleus
and nucleus ambiguus

CN XII Hypoglossal nucleus

CN XI Spinal accessory nucleus

Figure 10.12 A dorsal view of the human brainstem showing the location of the motor nuclei of the cranial nerves (CN). The cerebellum is not shown, but the position of the cerebellar peduncles indicates its approximate position. This map does not show the sensory nuclei.

α-motoneurons, but most terminate on interneurons which excite both flexor and extensor motoneurons supplying the contralateral limb muscles. Lesions of the red nucleus or rubrospinal tract impair the ability to make voluntary limb movements while having relatively little effect on the control of posture.

The tectospinal tract has its origin in the tectum, which forms the roof of the fourth ventricle and comprises the superior and inferior colliculi (see Figure 10.12). These areas seem to be concerned with the integration of visual and auditory signals and may have a role in orientation. The tectospinal tract projects to cervical regions of the spinal cord. Its fibres are crossed and end on interneurons that influence the movements of the head and eyes.

In general, the extrapyramidal tracts that influence the axial muscles (the muscles of the neck, back, abdomen, and pelvis) are largely crossed, while those influencing the muscles of the limbs are mostly uncrossed. This arrangement permits the independent control of the limbs and the axial muscles so that manipulation can proceed while posture is maintained.

Cranial nerve motor nuclei

The neuronal circuitry of the spinal cord is responsible only for the motor activity of parts of the body below the upper part of the neck. Movements of the head, eyes, and facial muscles are under the control of cranial nerve (CN) motor nuclei. These include the oculomotor nuclei (CN III), the trigeminal motor outflow (CN V), the facial nuclei (CN VII), and the nucleus ambiguus (CN X), all of which supply muscles that are part of a bilateral motor control system. Figure 10.12 illustrates the position of these nuclei in the

brainstem. Fibres from these nuclei make contact with interneurons in the brainstem that are thought to be organized in much the same way as those of the spinal cord which supply the axial and limb muscles.

The corticospinal (pyramidal) tract

The corticospinal tract has traditionally been described as the predominant pathway for the control of fine, skilled, manipulative movements of the extremities. While it is undoubtedly of considerable significance, it is now evident that many of its actions appear to be duplicated by the rubrospinal tract, which can, if necessary, take over a large part of its motor function.

The corticospinal tract originates in the cerebral cortex and runs down into the spinal cord. Many of its axons are, therefore, very long. Around 80 per cent of the fibres cross as they run through the extreme ventral surface of the medulla (the decussation of the pyramids—see Figure 9.8) and pass down to the spinal cord as the lateral corticospinal tract. The uncrossed fibres descend as the anterior (or ventral) corticospinal tract and eventually cross in the spinal cord. As the fibres of the corticospinal tract pass the thalamus and basal ganglia, they fan out into a sheet called the internal capsule (see Figures 9.6 and 9.7). Here they seem particularly susceptible to damage following a cerebrovascular accident (also called a CVA or stroke), and the resulting ischaemic damage gives rise to a characteristic form of paralysis (see Box 10.1).

The axons that give rise to the pyramidal tract originate from a wide area of the cerebral cortex. About 40 per cent of them come from the motor cortex proper, the rest originating either in the frontal or in the parietal lobes. The parietal lobe forms part of the sensorimotor (or somaesthetic) cortex, and the function of fibres from this region may be to provide feedback control of sensory input. About 3 per cent of the fibres in the pyramidal tract are derived from large neurons called **Betz cells,** which were once thought to give rise to the entire pyramidal tract. In fact, the great majority of the fibres of the corticospinal tract are of small diameter and may be either myelinated or unmyelinated. These fibres conduct nerve impulses relatively slowly.

Most axons of the pyramidal tract synapse with interneurons in the contralateral spinal cord. Many also send collaterals to other areas of the brain including the red nucleus, the basal ganglia, the thalamus, and the reticular formation of the brainstem. A few make monosynaptic contact with contralateral α- and γ-motoneurons. This is most noticeable in those supplying the hand and finger muscles and may reflect the large variety of movements of which these muscles are capable. In monkeys, lesions to the pyramidal tract result, for example, in a loss of the precision grip, although the power grip and gross movements of the limbs and trunk remain unaffected. In humans with corticospinal tract damage, for example following a stroke, there are few overt motor deficits. There is, however, weakness of the hand and finger muscles and a **positive Babinski sign.** This is a pathological reflex in which stroking the sole of the foot causes dorsiflexion of the big toe (i.e. the big toe turns upward) and the other toes fan out. Note that in infants less than a year old, stroking the sole of the foot also results in the smaller toes fanning out and a slow dorsiflexion of the big toe (see Figure 1 of Box 10.1). This reflects the immaturity of the corticospinal tract. This extensor response changes to the adult pattern between the ages of 12 and 24 months, after which the toes show a flexor response (i.e. the toes turn downward—the plantar reflex).

Summary

The activity of the neural circuitry within the spinal cord is modified and refined by descending motor control pathways. The reticulospinal tracts mainly influence the muscles of the trunk and proximal parts of the limbs and are important for the maintenance of certain postures. The vestibulospinal tract terminates on interneurons in the ipsilateral spinal cord that control the activity of extensor muscles and are important in the maintenance of an erect posture. The red nucleus receives afferent information from the cortex, cerebellum, and basal ganglia.

10.8 The control of posture

Maintenance of a stable upright posture is an active process. The skeletal muscles that maintain body posture (the axial muscles) function largely unconsciously. They continually make one tiny adjustment after another to enable the body to maintain a seated or erect posture despite the constant downward pull of gravity. When we stand, our centre of gravity must lie within the area bounded by our feet; if it lies outside, we fall over. Furthermore, every voluntary movement we make must be accompanied by a postural adjustment to compensate for the shift in our centre of gravity. The postural muscles therefore maintain a certain level of tone so that they are in a constant state of partial contraction. It is therefore not surprising to find that these muscles contain a high proportion of slow twitch (type 1) fibres (see Chapter 8).

As with most forms of motor activity, assuming a posture and maintaining it depends upon an internally generated central program for action, which is then modified and regulated by peripheral feedback. The 'command' program for action is assembled within the CNS (particularly the basal ganglia, brainstem, and reticular formation), while information from four peripheral sources provides feedback about the action. These sources are:

1. pressure receptors in the feet
2. the vestibular system
3. the eyes
4. proprioceptors in the neck and spinal column.

The last three provide information concerning the position of the head relative to the environment, while the pressure receptors in the feet relay information about the distribution of weight relative to the centre of gravity.

Pressure receptors in the feet

While it is clearly very difficult to carry out a detailed mechanistic study of motor activity in human subjects, it is possible to do so in experimental animals. By using decerebrate preparations (animals in whom the cerebral cortex has been removed), it has been possible to demonstrate a variety of reflex mechanisms that act both to maintain postural stability and to track the centre of gravity. One of these is the **positive supporting reaction**. Here, if the animal is held in mid-air and the sole of one of its feet is pressed, there is a reflex extension of the corresponding limb. This extension acts to support the weight of the body and is seen in newborn babies. Similarly, if the animal's body is pushed to the left, the changing pressures sensed by the feet result in an extension of the limbs on the left side and a retraction of the right limbs so that an upright posture can be maintained. If the animal were to be pushed to the right, the limbs on the right side would be extended. This is the **postural sway reaction**.

In the **righting reflex**, a decerebrate animal placed on its side will move its limbs and head in an attempt to right itself. In this case, cutaneous sensation of unequal pressures on the two sides of the body initiates the reflex activity. Other reflexes, which contribute to the maintenance of posture, include stepping reactions that can be demonstrated in humans very easily. It is impossible to fall over intentionally by leaning to one side, forwards or backwards. Once the centre of gravity has moved outside a critical point, reflex stepping or hopping will occur to prevent a fall.

Role of the vestibular system in the control of posture

The vestibular apparatus is the sensory organ that detects stimuli concerned with balance. Its anatomy and physiology is described in Chapter 16. Patterns of stimulation of the different hair cells within the maculae of the utricles and saccules inform the nervous system of the position of the head in relation to the pull of gravity. Sudden changes in the orientation of the body in space elicit reflexes that help to maintain balance and posture. These are the vestibular reflexes which fall into three categories: the tonic labyrinthine reflexes, the labyrinthine righting reflexes, and the dynamic vestibular reactions.

- **Tonic reflexes** are elicited by changes in the spatial orientation of the head and result in specific responses (contraction or relaxation) of the extensor muscles of the limbs. In this way, extensor tone is altered in such a way as to maintain an erect posture.

- **Labyrinthine righting reflexes** act to restore the body to the standing posture, for example from the reclining position. The first part of the body to move into position is always the head, in response to information from the vestibular system. As the head then assumes an abnormal position in relation to the rest of the body, further reflexes (notably the neck reflexes, as discussed in the next subsection) operate and the trunk follows the head into position.

- **Dynamic vestibular reactions**—because the semicircular canals of the vestibular apparatus are sensitive to angular velocity (see Chapter 16), they can give advance warning that one is about to fall over and avoiding action can be taken by the movement of the feet, as in the stepping reaction.

Neck reflexes

The neck is very flexible, and for this reason, the position of the head is almost independent of that of the body. The head has its own mechanisms for preserving a constant position in space, as discussed in the previous subsection. It is therefore important that there is a mechanism for informing the head and body about their relative positions. This is the role of the proprioceptors in the neck region. Movements of the head relative to the body result in a set of reflex movements of the limbs. For example, if the head is moved to the left, the left limbs extend and the right limbs flex to restore the axis of the head to the vertical. These reflexes modulate the basic stretch and tension reflexes described earlier to adjust the length and tone of the appropriate sets of muscles.

The role of visual information in control of posture

Information regarding the position of the head is provided by the eyes as well as by the vestibular system and the neck proprioceptors. Visual information is used to make appropriate postural adjustments and responses. The visual system contains certain neurons that respond specifically to an image moving across the retina. When the head moves in relation to its surroundings, the relative position of objects in the visual field will change and this will excite those neurons sensitive to movement. This information is transmitted to the brain and used to supplement information originating in the semicircular canals. When people with damage to their vestibular system close their eyes, they tend to sway and risk falling.

Summary

Tension in the axial muscles is continuously adjusted to maintain and alter posture. The 'command' programme for the posture required is assembled within the CNS (particularly the basal ganglia) and modified according to feedback from the vestibular system, the eyes, proprioceptors in the neck and vertebrae, and pressure receptors in the skin. A variety of reflex reactions help to adjust posture and prevent falling. Examples are the positive supporting reaction, righting reflexes, and stepping reactions. The vestibular apparatus provides information regarding the position of the head in relation to the pull of gravity. Vestibular reflexes, elicited by sudden changes in body position, help to maintain posture. Proprioceptors in the neck and vertebral column provide information about the positions of the head and body in relation to one another.

10.9 Goal-directed movements

This chapter has so far been concerned largely with the kinds of movements that rely heavily on reflexes. Throughout life, however, we perform a host of movements that are voluntary and goal-directed in nature. Regions of the brain which are of particular importance in the initiation and refinement of such movements are the motor regions of the cerebral cortex, the cerebellum, and the basal ganglia.

The motor cortex, its organization, and functions

More than 120 years ago it was found that electrical stimulation of particular areas of the cerebral cortex of dogs could elicit movements of the contralateral limbs. Subsequent work by W. Penfield and his colleagues in the 1930s showed that stimulation of the **precentral gyrus** of the cerebral cortex elicited movements in the contralateral muscles of conscious human subjects. They went on to establish in more detail the exact location of the areas of cortex in which stimulation led to coordinated movements.

Three important motor regions are recognized within the cortex.

1. The primary motor cortex (MI, also known as Brodmann's area 4) is located in the frontal lobe of the brain anterior to the primary somatosensory area of the parietal lobe. It extends into the depths of the central sulcus, over the medial edge of the hemisphere, and a little way anterior to the precentral gyrus.

2. The premotor cortex (PMA or Brodmann's area 6) is situated in front of the primary motor cortex.

3. The secondary motor cortex (also called the supplementary motor cortex or SMA, MII or Brodmann's areas 6 and 8) lies anterior to the premotor area.

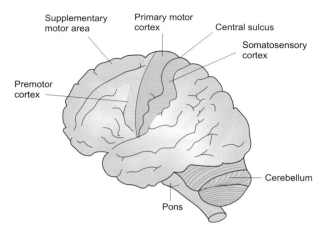

Figure 10.13 The location of the motor areas of the cerebral cortex seen from the left side.

Both the primary and secondary motor areas also possess a sensory projection of the periphery of the body that is sometimes referred to as the sensorimotor cortex. Figure 10.13 shows the positions of the sensory and motor areas of the cerebral cortex.

Somatotopic organization of the motor cortex

Experimental observations using a variety of methods (direct stimulation of the exposed brain, functional magnetic resonance imaging or fMRI, and transcranial stimulation) have made it clear that both the primary and secondary motor cortices are arranged in a **somatotopic** map (i.e. the different regions are represented by specific areas of cortex). The somatosensory cortex has a similar somatotopic organization. The motor representation of the body in the primary motor cortex is illustrated in Figure 10.14, and it is apparent that those areas of the body with especially refined and complex motor abilities, such as the fingers, lips, and tongue, have a disproportionately larger representation in the primary motor cortex than those with less precise motor control such as the trunk.

While stimuli applied to the surface of the primary motor cortex will evoke discrete contralateral movements that involve several muscles, micro-stimulation (i.e. highly localized stimulation) within the cortex itself can elicit movement of single muscles. The motor cortex is made up of an overlapping array of motor points that control the activity of particular muscles or muscle groups. Although the primary motor cortex controls the muscles of the opposite side of the body, the secondary motor cortex controls muscles on both sides, and stimulation here may evoke vocalization or complex postural movements. The orderly representation of parts of the body in the motor cortex is dramatically illustrated in patients suffering from a form of epilepsy known as Jacksonian epilepsy (after the neurologist Hughlings Jackson). Jacksonian convulsions are characterized by twitching movements that begin at an extremity such as the tip of a finger and show a progressive and systematic 'march': after the initial twitching is seen, there is clonic movement of the affected finger, followed by movement of the hand, then the arm. The process culminates in a generalized

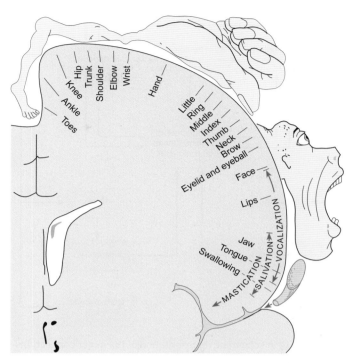

Figure 10.14 The somatotopic organization of the motor cortex (motor homunculus) showing the relative size of the regions representing various parts of the body. Note the large representation of the face, tongue, and hand relative to the trunk and legs. (Redrawn from P. Brodal (1997) *The central nervous system: Structure and function*, 2nd edition. Oxford University Press, New York.)

convulsion. This progression of abnormal movements reflects the spread of excitation over the cortex from the point at which the overactivity began (the epileptic focus).

Connections of the motor cortex

Outflow from the motor cortex

The corticospinal (or pyramidal) tract discussed earlier is one of the major pathways by which motor signals are transmitted from the motor cortex to the anterior motoneurons of the spinal cord. Nevertheless, the extrapyramidal tracts also carry a significant proportion of the outflow from the motor cortex to the spinal cord. The pyramidal tract itself also contributes to these extrapyramidal pathways via numerous collaterals, which leave the tract within the brain (see Figure 10.11). Indeed, each time a signal is transmitted to the spinal cord to elicit a movement, these other brain areas receive strong signals from the pyramidal tract.

Inputs to the motor cortex

The motor areas of the cerebral cortex are interconnected and receive inputs from a number of other regions. The most prominent sensory input is from the somatosensory (or somaesthetic) system. Information reaches the motor areas of the cerebral cortex, either directly from the thalamus or indirectly by way of the somatosensory cortex of the postcentral gyrus. The posterior parietal cortex relays both visual and somatosensory information to the motor

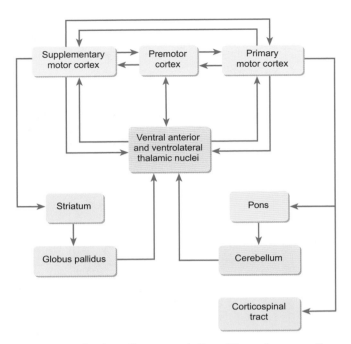

Figure 10.15 A schematic representation of the major connections of the motor areas of the cerebral cortex with the principal subcortical structures.

areas. The cerebellum and basal ganglia also send afferents to the motor cortex by way of the ventrolateral and ventral anterior thalamic nuclei. The schematic chart shown in Figure 10.15 illustrates connections between these pathways.

The pyramidal cells of the motor cortex receive information from a wide range of sensory modalities including joint receptors, tendon organs, cutaneous receptors, and spindle afferents. It is thought that integration of this information from the skin and muscles is carried out in the motor cortex and is used to enhance grasping, touching, or manipulative movements. The motor cortex is also believed to be involved in the generation of 'force commands' to muscles being used to counter particular loads. These commands adjust the force generated to match that of the load on the basis of information from pressure receptors in the skin and the load sensed by the Golgi tendon organs. Certainly, the rates of firing of pyramidal cells during voluntary movement correlate well with the force being produced to generate the movement.

Role of the motor areas in movement programming

For a voluntary movement to take place, an appropriate neuronal impulse pattern must first be generated. This pattern links the initial drive to perform the movement to the execution of the movement itself. Although the mechanisms underlying the generation of such patterns are not understood, some information regarding the kinds of neuronal activity that precede a movement has been obtained using conscious human subjects.

If a subject is told to make a voluntary movement such as bending a finger, an electrode placed over the cerebral cortex can record a slowly rising surface negative potential beginning some 800 ms before the start of the movement. This is known as the 'readiness potential'. This potential is particularly associated with the premotor area. As the movement gets under way, the electrode records a series of more rapid potentials which are particularly pronounced over the contralateral motor areas and may reflect their involvement in the execution of the movement.

Blood flow to the supplementary motor cortex increases both before and during complex voluntary movements (Figure 10.16). This further suggests a 'central command' role for this region in the planning or programming of actions as well as their execution. Motor activity is partly mediated by direct corticospinal connections. Furthermore, individual corticospinal neurons have also been shown to fire both before and during a movement (see Figure 10.17).

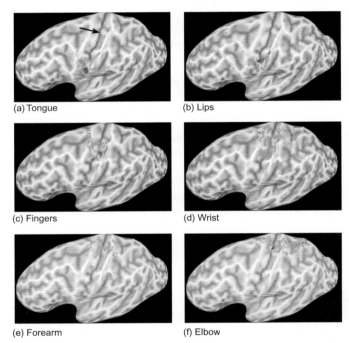

Figure 10.16 The areas of the left cerebral cortex that are activated during simple movements. The grey area represents the left cortex and the coloured areas show the local increases in blood flow that occur in response to the parts of the body indicated. The colour changes from blue to red indicate increasing levels of local blood flow. The central sulcus is clearly shown by the arrow in panel (a). The motor tasks were as follows. (a) The tongue was moved back and forth in the mouth without moving the lips or jaw. (b) The lips were alternately pursed forward and drawn back over the teeth. (c) The fingers of the right hand were flexed towards each other and then extended. (d) The wrist was flexed and then extended. (e) The forearm was repeatedly rotated, first showing the palm uppermost and then rotated downward. (f) The elbow was repeatedly flexed and extended. Note the similarity between the areas activated and the somatotopic map shown in Figure 10.14 with the additional involvement of areas outside the primary motor cortex. (Data from Fig 2 in J.D. Meier et al. (2008) *Journal of Neurophysiology* **100**, 1800–12.)

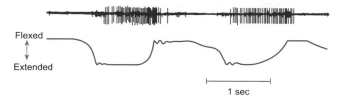

Figure 10.17 The response pattern of a pyramidal tract neuron (PTN) before and during the voluntary flexion of the wrist. The experiment was performed on a trained monkey and shows that the PTN fibres discharge *before* the onset of the wrist flexion. (Based on data of Fig 6 in E.V. Evarts (1968) *Journal of Neurophysiology* **31**, 14–27.)

Effects of lesions of the motor cortex

The motor areas of the cortex are often damaged as a result of strokes (which result in a loss of blood supply to the cortex) and the muscles controlled by the damaged region show a corresponding loss of function (see Box 10.1). Rather than a total loss of movement, however, there is a loss of voluntary control of fine movements, with clumsiness and slowness of movement and an unwillingness to use the affected muscles.

Lesions of parts of the motor cortex (particularly the supplementary and premotor areas) impair the ability to prepare for voluntary movements, giving rise to a condition similar to apraxia in which affected patients are unable to perform complex motor acts despite retaining both sensation and the ability to perform simple movements.

Many strokes cause widespread damage not only to the primary motor cortex but also to sensorimotor areas both anterior and posterior to it. These areas relay inhibitory signals to the spinal motoneurons via the extrapyramidal pathways. When the sensorimotor areas are damaged, there is a release of inhibition that can result in the affected contralateral muscles going into spasm. Spasm is particularly intense if the basal ganglia are also damaged, since strong inhibitory signals are transmitted via this region, and will be discussed later.

Summary

The cerebral cortex contains motor areas in which the stimulation of cells will elicit contralateral movements. The primary and secondary areas are arranged somatotopically in such a way that those areas of the body that are capable of especially refined and complex movements have a disproportionately large area of representation.

Outflow from the motor cortex to the spinal cord is carried by the corticospinal (pyramidal) and extrapyramidal tracts. The motor cortex also sends numerous collaterals to the basal ganglia, cerebellum, and brainstem. The motor areas receive inputs from many sources. The predominant sensory input is from the somatosensory system, which receives its input from the thalamus. Afferent information is also received from the visual system, cerebellum, and basal ganglia. This afferent information is used to refine

movements, particularly to match the force generated in specific muscle groups to given load. The secondary motor cortex is thought to play a role in the programming of movements.

When motor areas of the cerebral cortex are damaged by strokes, the muscles controlled by the damaged areas show a corresponding loss of function. However, recovery is often good, resulting in little more than clumsiness and a loss of fine muscle control. Lesions to the secondary and premotor areas may give rise to apraxia, a loss of the ability to prepare for voluntary movement although the ability to execute simple movements is retained.

10.10 The role of the cerebellum in motor control

Electrical stimulation of the cerebellum causes neither sensation nor significant movement. Loss of this area of the brain, however, is associated with severe abnormalities of motor function. The cerebellum appears to play a particularly vital role in the coordination of postural mechanisms and in the control of rapid muscular activities such as running, playing a musical instrument, and typing.

Anatomical structures of the cerebellum

The cerebellum (literally the 'little brain') is located dorsal to the pons and medulla and lies under the occipital lobes of the cerebral hemispheres (see Figures 9.3 and 10.13). The cerebellum is usually subdivided into two major lobes, the anterior and posterior lobes, separated by the primary fissure, as shown in Figure 10.18. The lobes are further subdivided into nine transversely orientated lobules, each of which is folded extensively, the folds being known as folia. The flocculonodular lobe is a small, propeller-shaped structure, caudally situated with respect to the major lobes. As it is tucked under the main body of the cerebellum, it cannot be seen in a surface view. Along the midline is the vermis, from which the cerebellar hemispheres extend laterally.

Like the cerebral cortex, the cerebellum is composed of a thin outer layer of grey matter, the cerebellar cortex, overlying internal white matter. Embedded within the white matter are the paired cerebellar nuclei. These are the dentate (which are the most lateral), globose, emboliform, and fastigial (the most medial) nuclei. They receive afferents from the cerebellar cortex as well as sensory information via the spinocerebellar tracts. The cerebellar nuclei give rise to all the efferent tracts from the anterior and posterior lobes of the cerebellum. The flocculus and nodulus send efferent fibres to the lateral vestibular nucleus (Deiter's nucleus).

The cerebellum is attached to the brainstem by nerve fibres that run in three pairs of **cerebellar peduncles**: the inferior, the middle, and the superior peduncles. The inferior peduncles connect the cerebellum to the medulla. They carry afferent fibres conveying sensory information from muscle proprioceptors and the vestibular nuclei. These inform the cerebellum of the disposition of the limbs and the state of whole body equilibrium and balance. The cerebellum receives information from the pons concerning

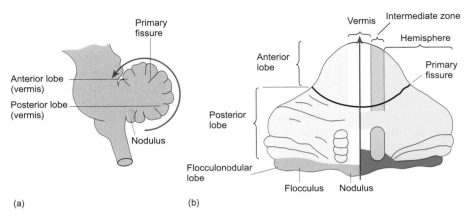

Figure 10.18 The major lobes of the cerebellum. Panel (a) shows a sectional view of the brainstem and cerebellum. Panel (b) shows the surface of the cerebellum 'unfolded' along the axis of the arrow indicated in (a). The right hand half indicates the areas that receive projections from the motor cortex (the cerebro-cerebellum: orange), the vestibular system (the vestibulo-cerebellum: green), and the spinal cord (the spino-cerebellum: white). In the intermediate zone (turquoise), afferents from the motor cortex and spinal cord overlap.

voluntary motor activities initiated by the motor cortex via the middle cerebellar peduncles. The superior cerebellar peduncles connect the cerebellum and the midbrain. Fibres in these peduncles originate from neurons in the deep cerebellar nuclei and communicate with the motor cortex via the thalamus. (The cerebellum has no direct connections with the cerebral cortex.) In contrast to the cerebral cortex, the two halves of the cerebellum each control and receive input from muscles on the ipsilateral side of the body.

Cellular organization of the cerebellar cortex

The cerebellar cortex is largely uniform in structure consisting of three layers of cells: the granular, Purkinje cell, and molecular layers. The functional organization of the cerebellar cortex is relatively simple and is shown in Figure 10.19. The most prominent cells of the cerebellar cortex are the Purkinje cells of the middle layer. A human brain contains about 15 million of these large neurons, each of which possesses an extensively branched dendritic tree which can be seen in Figure 9.16b. The axons of the Purkinje cells

form the only output from the cerebellar cortex. They pass to the deep cerebellar nuclei and, to a lesser extent, to the vestibular nuclei. The Purkinje cell output is entirely inhibitory, the neurotransmitter being GABA.

Inputs to the cerebellar cortex are of two types, climbing fibres and mossy fibres. **Climbing fibres** originate in the inferior olive of the medulla and form part of the olivo-cerebellar tract, which crosses in the midline and enters the cerebellum via the contralateral inferior cerebellar peduncle. **Mossy fibres** form the greater proportion of the afferent input to the cerebellar cortex, originating in all the cerebellar afferent tracts apart from the inferior olive. Mossy fibres send collaterals to deep nuclear cells before ascending to the granular layer of the cortex where they branch to form large terminal structures making synaptic contact with the dendrites of many granule cells. These in turn give rise to the **parallel fibres**, which synapse with the dendrites of Purkinje cells.

The climbing fibres excite the Purkinje cells while the mossy fibres excite the granule cells which, in turn, make excitatory contact with the Purkinje cells via the parallel fibres. The actions of all

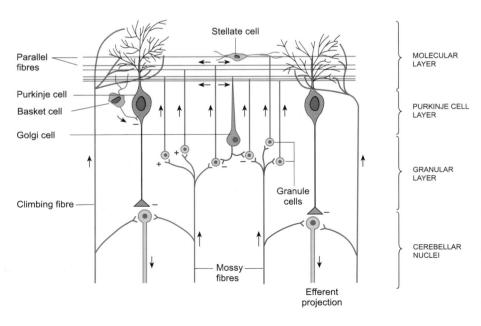

Figure 10.19 The neural organization of the cerebellar cortex. The basket cells, Golgi cells, and stellate cells are inhibitory interneurons. The basket, Golgi, and stellate cells are inhibitory interneurons. The minus symbol represents the inhibitory effect of the Purkinje cells on the neurons of the deep cerebellar nuclei. The arrows indicate the direction of information flow.

the other cell types within the cerebellar cortex are inhibitory. Basket cells and stellate cells make inhibitory contact with the Purkinje cells, while the Golgi cells inhibit the granule cells and so reduce the excitation of the Purkinje cells. The Purkinje cells exert a tonic inhibition on the activity of the neurons of the cerebellar nuclei. Since the axons of the Purkinje cells are the output of the cerebellar cortex, all excitatory inputs to the cerebellum will be converted to inhibition after two synapses at most. This has the effect of removing the excitatory influence of the cerebellar input and ensures that the neurons of the deep cerebellar nuclei are ready to process a new input. It is believed that this 'erasing' of inputs is important to the participation of the cerebellum in rapid movements.

Afferent connections of the cerebellum

The cerebellum receives information from the motor cortex, vestibular system, and proprioceptors via the cortico-cerebellar, vestibulo-cerebellar, reticulo-cerebellar, and spino-cerebellar pathways. The cortico-cerebellar tract originates mainly in the motor cortex (but also from other cortical areas) and passes via the pontine nuclei to the cerebellum in the ponto-cerebellar tract. It is thought that this pathway informs the cerebellum of intended movements.

Sensory signals from the muscle spindles, tendon organs, and touch receptors of the skin and joints are received by the cerebellum directly by way of the ventral and dorsal spino-cerebellar tracts. These signals inform the cerebellum of the moment-by-moment status of the muscles and joints. All of this information reaches the cerebellar cortex by way of the mossy fibres, which make excitatory contacts with the granule cells which excite Purkinje cells by way of their parallel fibres (see Figure 10.19).

The climbing fibre input to the cerebellar Purkinje cells originates in the inferior olive, a large nucleus in the ventral medulla. Neurons in this region receive their input from a number of sources, including the motor cortex and the spino-olivary tracts (which relay information from cutaneous receptors as well as muscle and joint proprioceptors). The climbing fibres make direct excitatory contact with Purkinje cell dendrites.

Input reaching the cerebellum via the spino-cerebellar pathways is somatotopically organized. There are two maps of the body in the spino-cerebellum, one in the anterior and one in the posterior lobe, which are shown in Figure 10.20. In each map, the head is facing the primary fissure so that the two maps are inverted with respect to each other. Signals from the motor cortex, which is also arranged somatotopically, project to corresponding points in the sensory maps of the cerebellum.

Outputs from the cerebellum

The output neurons of the cerebellum lie in the deep cerebellar nuclei. The cells of the cerebellar nuclei receive inhibitory synapses from the Purkinje cells of the cerebellar cortex and excitatory synapses from collaterals of the mossy and climbing fibres. The final output from the cerebellum is, therefore, the net result of the interaction between these excitatory and inhibitory inputs to the nuclei.

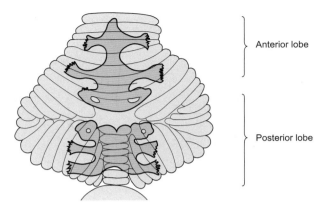

Figure 10.20 Somatotopic maps in the anterior and posterior lobes of the cerebellum.

Crude topographical maps of the peripheral musculature can be demonstrated within the cerebellar nuclei. Neurons of the dentate nucleus project contralaterally through the superior cerebellar peduncle to neurons in the contralateral thalamus. From the thalamus, neurons project to the premotor and primary motor cortex, where they influence the planning and initiation of voluntary movements. A diagram of the major output pathways of the cerebellar nuclei is shown in Figure 10.21. The red nucleus gives rise to the rubrospinal tract, which is particularly important in the control of proximal limb muscles. The neurons of the fastigial nucleus project to the vestibular nuclei and to the pontine and medullary reticular formation which, in turn, give rise to the vestibulospinal and reticulospinal tracts respectively.

How does the cerebellum contribute to the control of voluntary movement?

Although the neuronal architecture of the cerebellum is relatively well understood, the functions of the cerebellum are less clear-cut. Much of our information concerning its role in the control of motor activity has been gathered from the effects of lesions and other damage to the cerebellum and from experimental stimulation of, and recording from, cerebellar neurons. On the basis of such work, it is now believed that the primary role of the cerebellum is to supplement and correlate the activities of the other motor areas. More specifically, it seems to play a role in the control of posture and the correction of rapid movements initiated by the cerebral cortex. The cerebellum may also contribute to certain forms of motor learning, as the frequency of nerve impulses in the climbing fibres has been shown to double when a monkey is learning a new task.

Damage to the posterior cerebellum results in an impairment of postural coordination which is similar to that seen when the vestibular apparatus itself is damaged. For example, a patient with a lesion in the area will feel dizzy and have difficulty standing upright, and may develop a staggering gait (**cerebellar ataxia**). Damage to other regions of the cerebellum produces more generalized impairment of muscle control. Ataxia is again a problem, coupled with a loss of muscle tone (**hypotonia**) and a lack of coordination (**asynergia**). Figure 10.22 shows an example of impaired

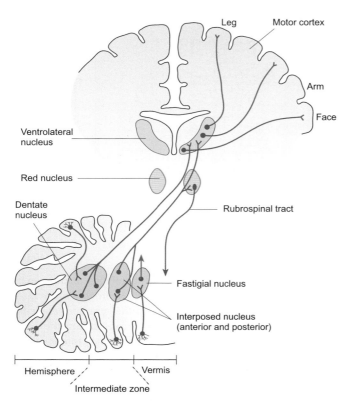

Figure 10.21 The connections that ascend from the cerebellum. The Purkinje neurons of the cerebellar cortex project to the deep cerebellar nuclei. The interposed nucleus (nucleus interpositus) receives its Purkinje cell input from the intermediate zone of the cerebellum. The Purkinje cells that innervate the fastigial nucleus lie in the medial zone of the cerebellum. The neurons of the dentate, interposed, and fastigial nuclei project to the cerebral cortex via the ventrobasal nucleus of the thalamus and give off collaterals to the red nucleus. (Redrawn from Fig 14.14 in P. Brodal (2004) *The central nervous system: Structure and function*, 3rd edition. Oxford University Press, New York.)

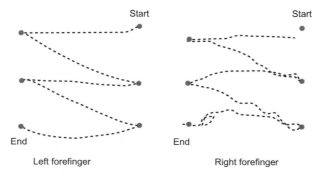

Figure 10.22 Records illustrating cerebellar ataxia. In this task, a flashing light was fixed to each index finger in turn and the subject asked to trace a path as accurately as possible between two sets of three lights spaced 75 cm apart (shown here as red circles). The movement was captured on photographic film. The patient, who had a right-sided cerebellar lesion, had great difficulty tracing the path with his right hand but no difficulty with his left.

motor control from a patient with a right-sided cerebellar ataxia. The patient was asked to trace a path between a set of lights with the forefinger of each hand. The left forefinger, which was unaffected, completes the task with ease, but the right forefinger traces the path in a very uncertain manner, with looping and obvious tremor.

The cerebellum exerts its control over motor activity on a moment-by-moment basis, using sensory information (especially from proprioceptors) to monitor body position, muscle tension, and muscle length. Inputs from the motor cortex to the cerebellum inform the cerebellum of an intended movement before it is initiated. Sensory information regarding the progress of the movement will then be received continually by the cerebellum via the spinocerebellar tract. This information is processed by the cerebellum to generate an error signal, which is fed back to the cortex so that the movement can be adjusted to meet the precise circumstances of the situation. The cerebellum itself has no direct connection with the spinal motoneurons. It exerts its effect on motor performance indirectly.

Much of the clumsiness of movement seen in patients with cerebellar lesions seems to be the result of a slowness to respond to sensory information about the progress of the movement. There is often an intention tremor preceding goal-directed movements, in which the movement appears to oscillate around the desired position, or the movement may overshoot when the patient reaches for an object, as in the example shown in Figure 10.22. Speech may also become staccato ('scanning speech'), more laboured, and less 'automatic'. In general, patients with cerebellar damage seem to require a great degree of conscious control over movements, which normally require little thought. This may be because the cerebellum plays a role in the learning of complex motor skills.

Summary

The cerebellum is located dorsal to the pons and medulla and protrudes from under the occipital lobes. It is attached to the brainstem by nerves running in three pairs of cerebellar peduncles. The cerebellum receives information about intended movements by way of the corticocerebellar tract.

The cerebellum is divided into anterior, posterior, and flocculonodular lobes and consists of an outer cortical layer of grey matter and an inner layer of white matter in which the deep cerebellar nuclei are embedded. The internal neuronal circuitry of the cerebellar cortex is largely uniform, comprising three layers of cells: the Purkinje cell layer, the granular layer, and the molecular layer.

Inputs to the cerebellum are via climbing and mossy fibres. The climbing fibres originate in the inferior olive and make excitatory contact with the Purkinje cell dendrites. Mossy fibres form the bulk of the afferent input to the cerebellum and excite the Purkinje cells indirectly by way of the parallel fibres of the granule cells.

Efferent tracts from the cerebellum originate in the deep nuclei and pass to the thalamic nuclei (and thence to the motor cortex), the red nucleus, the vestibular nuclei, the pons, and the medulla. The cerebellum plays a vital role in the coordination of postural mechanisms and the control of rapid muscular activities, supplementing and correlating the activities of other motor areas. Lesions to the cerebellum may result in dizziness and postural difficulties or in a more generalized loss of muscle control involving intention tremor, clumsiness, and difficulty with speech.

10.11 The basal ganglia

The basal ganglia are the deep nuclei of the cerebral hemispheres. They consist of the **caudate nucleus** and **putamen** (known collectively as the **corpus striatum**), the **globus pallidus**, and the **claustrum**. The putamen and the globus pallidus are separated from the caudate nucleus by the anterior limb of the internal capsule. Functionally, the basal ganglia are associated with the **subthalamic nuclei**, the **substantia nigra** of the midbrain (so called because many of its cells are pigmented with melanin), and the **red nucleus**. Figures 10.23 and 10.24 shows the location of these structures within the brain.

The basal ganglia form an important subcortical link between the frontal lobes and the motor cortex. Their importance in the control of motor function is clear from the severe disturbances of movement seen in patients with lesions of the basal ganglia. These deep structures are, however, relatively inaccessible to experimental study, and a detailed understanding of their actions is yet to emerge.

Afferent and efferent connections of the basal ganglia

The circuitry of the basal ganglia is very complex and not fully understood. There are numerous connections between the various structures, but their significance is unknown. However, a large feedback loop seems to exist between the basal ganglia and the cerebral cortex. All parts of the cerebral cortex project to the caudate nucleus and putamen of the basal ganglia (see Figure 10.25).

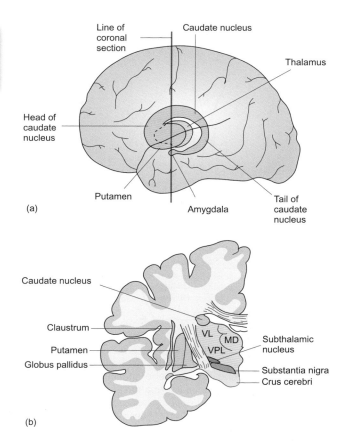

(a)

(b)

Figure 10.24 The location of the basal ganglia and their associated nuclei. Panel (a) shows the position of the basal ganglia within the cerebral hemispheres. Panel (b) illustrates a coronal section showing the detailed location of the various nuclei of the basal ganglia. MD, VL and VPL are the mediodorsal, ventrolateral, and posterolateral thalamic nuclei. (Redrawn from Fig 13.3 in P. Brodal (2004) *The central nervous system: Structure and function*, 3rd edition. Oxford University Press, New York.)

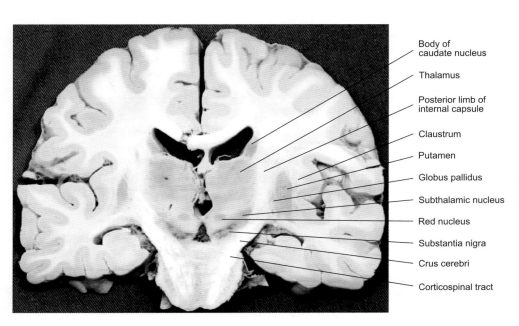

Figure 10.23 A coronal section through the brain to show the location of the basal ganglia and their associated nuclei. The reddish tint of the red nucleus and black deposits in the substantia nigra are clearly visible. (From University of British Columbia: http://www.neuroanatomy.ca/cross_sections/sections_coronal.html)

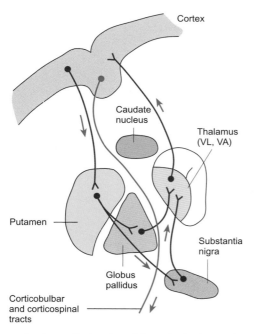

Figure 10.25 A schematic drawing of the principal neural connections of the basal ganglia with other brain areas. VL and VA are the ventrolateral and ventro-anterior thalamic nuclei. The arrows indicate the direction of information flow. (Redrawn from Fig 13.4 in P. Brodal (2004) *The central nervous system: Structure and function*, 3rd edition. Oxford University Press, New York.)

This input is topographically organized and excitatory, and probably uses glutamate as its transmitter. From the putamen and caudate nucleus, neurons project to the globus pallidus (a predominantly inhibitory projection), then to the thalamus, and from there back to the cerebral cortex (mainly to the motor area) to complete the loop. Such an arrangement suggests that the basal ganglia, like the cerebellum, are involved in the processing of sensory information, which is then used to regulate motor function.

The substantia nigra receives an input from the putamen and projects to both the putamen and the caudate nucleus (the nigrostriatal pathway). The pigmented neurons in the substantia nigra synthesize and store the neurotransmitter dopamine, which is subsequently transported via the nigrostriatal fibres to receptors in the corpus striatum. It is believed that this pathway is inhibitory to the striatal neurons that in turn project to the motor cortex via the thalamus, as shown in Figure 10.25.

There are many other neuronal connections between the basal ganglia and other regions of the brain, such as the limbic system (which is concerned with motivation and emotion), but their role is unknown.

The role of the basal ganglia in the control of movement

Most of the information that has accumulated concerning the actions of the basal ganglia in humans has been derived from studies of the effects of damage to those brain areas. A range of movement disorders (**dyskinesias**) result from damage to the basal ganglia or their connections. The best-known disease of the basal ganglia is **Parkinson's disease** (shaking palsy), a degenerative disorder that results in variable combinations of slowness of movement (**bradykinesia**), increased muscle tone (rigidity), resting (or 'pill-rolling') tremor, and impaired postural responses. Patients suffering from Parkinson's disease have difficulty both starting and finishing a movement. They also tend to have a mask-like facial expression. Parkinsonism appears to result primarily from a defect in the nigrostriatal pathway following degeneration of the dopaminergic neurons of the substantia nigra. Considerable success in relieving the symptoms of these patients has been achieved by the administration of carefully controlled doses of L-DOPA, a precursor of the neurotransmitter dopamine.

In contrast to Parkinson's disease, in which movements are restricted, other disorders of the basal ganglia result in the spontaneous production of unwanted movements. These spontaneous movements include **hemiballismus** (a sudden flinging out of limbs on one side of the body as if to prevent a fall or to grab something), **athetosis** (snakelike writhing movements of parts of the body, particularly the hands, which are often seen in cerebral palsy), and **chorea** (a series of rapid, uncontrolled movements of muscles all over the body). In all cases, the disease process results in inappropriate or repetitive execution of normal patterns of movement. The hereditary disorder Huntington's chorea is perhaps the best known of this group of diseases.

While observations of the various movement disorders associated with damage to the basal ganglia cannot give direct information about their role in the control of normal movements, some tentative assumptions have been made. The basal ganglia seem to be involved in very basic, innate patterns of movement, possibly elaborating relatively crude movement 'programs' in response to cues from the cortical association areas. These programs would then provide the basic structure of a movement which could subsequently be refined and updated in the light of sensory input from the periphery, this latter function being carried out by the cerebellum. Furthermore, the basal ganglia may contain the blueprints for the particular sets of muscle contractions needed to effect or adjust certain postures, for example, the throwing out of an arm to prevent a fall.

Summary

The basal ganglia include the corpus striatum (caudate nucleus, globus pallidus, and putamen), the substantia nigra of the midbrain, and the red nucleus. All parts of the cerebral cortex project to the corpus striatum. The basal ganglia in turn project to the thalamus which, in turn, projects to the motor cortex, thus completing a loop. The activity of the neurons in this loop is influenced by inhibitory connections between the substantia nigra and the corpus striatum.

A range of movement disorders (dyskinesias) result from damage to the basal ganglia or its connections. Parkinson's disease is the most familiar and is due to degeneration of the dopaminergic neurons of the substantia nigra. Other disorders of the basal ganglia (e.g. Huntington's chorea) are characterized by the inappropriate or repetitive execution of movement patterns that are themselves often normal. The basal ganglia are believed to generate basic patterns of movement, possibly representing motor 'programs' formed in response to cues from cortical association areas.

10.12 Concluding remarks

Figure 10.26 summarizes current ideas concerning the interactions between the various motor areas of the brain. While certain aspects of the mechanisms by which the nervous system controls movement are reasonably well understood, our understanding of the ways in which motor areas plan and initiate movements is far from complete. Nevertheless, a possible sequence of events might be as follows. Voluntary movements are planned in areas of the brain outside the primary motor cortex. This motor planning is likely to involve the processing of sensory information and its integration with memory. An appropriate motor program will then be selected and preliminary commands relayed to the basal ganglia and cerebellum. These areas of the brain are thought to organize and refine the crude motor program and relay it back to the primary motor cortex which, in turn, activates both α- and γ-motoneurons to bring about the desired sequence of muscle contractions. During the execution of the movement, information from both cortical and peripheral regions constantly informs the cerebellum and other subcortical motor areas of the progress of the movement so that it can be modified appropriately to take account of changing circumstances.

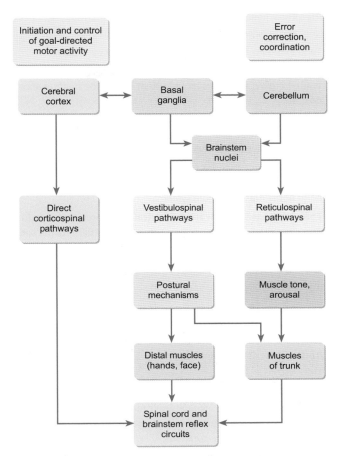

Figure 10.26 A summary of possible interactions between the various motor areas of the brain.

(✱) Checklist of key terms and concepts

The components of the motor system

- The basic element of motor control is the motor unit, which consists of an α-motoneuron and the skeletal muscle fibres it innervates.

- The α-motoneurons receive numerous synaptic connections from proprioceptors and from higher levels of the CNS including the brainstem, cerebellum, and motor cortex.

- Proprioceptors are mechanoreceptors situated within muscles and joints. They provide the CNS with information regarding muscle length, position, and tension.

- Muscle spindles lie in parallel with extrafusal muscle fibres. They are innervated by γ-motoneurons and by afferent fibres. The afferents respond to muscle stretch, while γ-efferent activity regulates the sensitivity of the spindles.

- Golgi tendon organs respond to the degree of tension within the muscle.

Reflexes

- Reflexes are the simplest form of irritability that involve the nervous system.

- The neurons participating in a reflex form a reflex arc, which includes a receptor, an afferent neuron that synapses in the CNS, and an efferent neuron that sends a nerve fibre to an effector.

- Interneurons may be present between the afferent and efferent neurons.

- The number of synapses in a reflex arc is used to define the reflex as monosynaptic, disynaptic, or polysynaptic.

- The simplest reflexes are the monosynaptic stretch reflexes such as the knee-jerk reflex.

- Stretch reflexes play an important role in the control of posture.

- The protective flexion (withdrawal) reflexes are elicited by noxious stimuli. Their reflex arc possesses one or more interneurons.

- Reciprocal inhibition ensures that the extensor muscles acting on a joint will relax when its flexor muscles contract.
- Tendon organs monitor force in the tendons they supply.
- The inverse myotactic reflex is activated by discharge in the Golgi tendon organs and plays an important role in the maintenance of posture.

Muscle spindles

- Muscle spindles sense muscle length.
- When voluntary changes in muscle length are initiated by motor areas of the brain, the motor command includes changes to the set point of the muscle spindle system.
- The simultaneous activation of extrafusal fibres and intrafusal fibres is called alpha–gamma co-activation and permits the sensitivity of muscle spindles to be continuously adjusted as a muscle shortens.

Central control of motor function

- The activity of the neural circuitry within the spinal cord is modified and refined by descending motor control pathways that include the reticulo-spinal tracts, the vestibulospinal tract, the rubrospinal tract, and the corticospinal tract.
- The corticospinal tract originates over a wide area of cerebral cortex, including both motor and somatosensory areas.
- The fibres of the corticospinal tract form the internal capsule as they pass the thalamus and basal ganglia. They cross to the contralateral side at the pyramids.
- Lesions to the corticospinal tract result in loss of precise hand movements.

Control of posture

- Tension in the axial muscles is continuously adjusted to maintain and alter posture.
- The 'command' program for the posture required is assembled by the basal ganglia and modified according to feedback from the vestibular system, the eyes, proprioceptors in the neck, and vertebrae, as well as pressure receptors in the skin.
- A variety of reflex reactions help to adjust posture and prevent falling. They occur in response to information regarding changes in the balance of pressures experienced by the feet.
- The vestibular apparatus provides information regarding the position of the head in relation to the pull of gravity.
- Vestibular reflexes, elicited by sudden changes in body position, help to maintain posture.
- Proprioceptors in the neck and vertebral column provide information about the relative positions of the head and body.
- Information from the eyes concerning head position supplements information from the semicircular canals of the vestibular system.

Goal-directed movements

- The cerebral cortex contains three motor areas (primary, secondary, and premotor), in which the stimulation of cells will elicit contralateral movements.
- The primary and secondary areas are somatotopically arranged, and those areas of the body that perform complex movements have a disproportionately large area of representation.
- Outflow from the motor cortex to the spinal cord is carried by the corticospinal and extrapyramidal tracts.
- The motor cortex receives a large input from the somatosensory system via the thalamus.
- This afferent information is used to refine movements, particularly to match the force generated in specific muscle groups to an imposed load.
- The secondary motor cortex is thought to initiate voluntary movements by generating the appropriate neural patterns.

The role of the cerebellum

- The cerebellum protrudes from under the occipital lobes and is dorsal to the pons and medulla.
- It is divided into three lobes and consists of an outer cortical layer of grey matter and an inner layer of white matter in which are embedded the deep cerebellar nuclei.
- The cerebellar cortex comprises three layers of cells: the Purkinje cell layer, the granular layer, and the molecular layer.
- Climbing and mossy fibres form the inputs to the cerebellum.
- The cerebellum is attached to the brainstem by nerves running in three pairs of cerebellar peduncles.
- The cerebellum has no direct connections with the motor cortex but receives information about intended movements by way of the cortico-cerebellar tract.
- Efferent tracts from the cerebellum originate in the deep nuclei and pass to the thalamic nuclei and thence to the motor cortex and other motor areas.
- The cerebellum plays a vital role in the coordination of postural mechanisms and the control of rapid muscular activities, supplementing and correlating the activities of other motor areas. It may also contribute to motor learning.
- The two halves of the cerebellum control and receive inputs from ipsilateral muscles.

The basal ganglia

- The basal ganglia are deep cerebral nuclei involved in motor control. They include the caudate nucleus, globus pallidus, and putamen (collectively called the corpus striatum).
- All parts of the cerebral cortex project to the corpus striatum. In turn, the basal ganglia project to the thalamus, which sends fibres to the motor cortex, thus completing a loop. The activity of neurons

within this loop is influenced by inhibitory connections between the substantia nigra and the corpus striatum.

- The basal ganglia are believed to generate basic patterns of movement, possibly representing motor 'programs' formed in response to cues from cortical association areas.

Disorders of the motor system

- Section of the spinal cord causes a loss both of sensation and of voluntary control over muscle contraction (paralysis) below the level of the lesion.

- Immediately following such an injury, there is a loss of spinal reflexes (areflexia) and flaccidity of the muscles controlled by nerves leaving the spinal cord below the level of the injury.

- This is known as spinal shock or neurogenic shock and involves both reflex somatic motor activity and the autonomic reflexes concerned with bladder and bowel functions.

- Spinal shock may last for several weeks.

- As the excitability of the undamaged neurons below the damaged area increases, reflex activity returns and may result in the spasticity of the paralysed muscle groups.

- Lesions to the cerebellum may result in dizziness and postural difficulties or in a more generalized loss of muscle control resulting in intention tremor, clumsiness, and speech problems.

- Damage to the basal ganglia or their connections are responsible for a range of movement disorders known as dyskinesias.

- Parkinson's disease, in which movement is impoverished, is the most familiar dyskinesia. It is caused by degeneration of the dopaminergic neurons of the substantia nigra.

- Other disorders of the basal ganglia (for example the inherited disease Huntington's chorea) are characterized by the inappropriate or repetitive execution of movement patterns that are themselves often normal.

- Motor areas of the cerebral cortex are often damaged by strokes, and muscles controlled by the damaged areas show a corresponding loss of function. However, recovery is often good, resulting in little more than clumsiness and a loss of fine muscle control.

- Lesions to the secondary and premotor areas may give rise to apraxia, a loss of the ability to prepare for voluntary movement. The ability to execute simple movements is retained.

Recommended reading

Neuroanatomy of the motor system

Brodal, P. (2004) *The central nervous system: Structure and function* (3rd edn), Chapters 11–14. Oxford University Press, New York.
A clear and detailed account of the anatomy of the motor pathways of the CNS.

Pharmacology of neurodegenerative disorders

Rang, H.P., Ritter, J.M., Flower, R.J., and Henderson, G. (2016) *Pharmacology* (8th edn), Chapter 40, pp. 491–6. Churchill-Livingstone, Edinburgh.

Physiology

Carpenter, R., and Reddi, B. (2012) *Neurophysiology, a conceptual approach* (5th edn), Chapters 9–12. Hodder Arnold, London.

Squire, L.R., Berg, D., Bloom, F.E., Du Lac, S., Ghosh, A., and Spitzer, N.C. (2012) *Fundamental neuroscience* (4th edn), Chapters 27–31. Academic Press, San Diego.
These chapters provide a useful introduction to the earlier scientific literature concerned with motor control.

Medicine

Donaghy, M. (2005) *Neurology* (2nd edn), Chapters 5–9, 26, and 27. Oxford University Press, Oxford.
These chapters provide concise accounts of the principal motor disorders.

Review articles

Eisenberg, M., Shmuelof, L., Vaadia, E., and Zohary, E. (2010) Functional organization of human motor cortex: Directional selectivity for movement. *Journal of Neuroscience* **30**, 8897–905.

Lotze, M., Erb, M., Flor, H., Huelsmann, E., Godde, B., and Grodd, W. (2000) fMRI evaluation of somatotopic representation in human primary motor cortex. *NeuroImage* **11**, 473–81.

Meier, J.D., Aflalo, T.N., Kastner, S., and Graziano, M.S.A. (2008) Complex organization of human primary motor cortex: A high-resolution fMRI study. *Journal of Neurophysiology* **100**, 1800–12.

Pramstaller, P.P., and Marsden, C.D. (1996) The basal ganglia and apraxia. *Brain* **119**, 319–40.

Proske, U., and Gandevia, S.C. (2012) The proprioceptive senses: Their roles in signaling body shape, body position and movement, and muscle force. *Physiological Reviews* **92**, 1651–97.

Toma, K., and Nakai, T. (2002) Functional MRI in human motor control studies and clinical applications. *Magnetic Imaging in Medical Sciences* **1**, 109–20.

Ullsperger, M., Danielmeier, C., and Jocham, G. (2014) Neurophysiology of performance monitoring and adaptive behavior. *Physiological Reviews* **94**, 35–79.

To check that you have mastered the key concepts presented in this chapter, complete the accompanying online self-assessment questions. Go to www.oup.com/uk/pocock5e/

CHAPTER 11
The autonomic nervous system

This chapter should help you to understand:

- The anatomical organization of the autonomic nervous system and its separation into the sympathetic and parasympathetic divisions

- How the sympathetic division regulates the activity of the cardiovascular system, visceral organs, and secretory glands

- How the parasympathetic nerves regulate the activity of the gut, heart, and secretory glands

- The role of autonomic nerves in regulating the activity of the enteric nervous system

- Chemical transmission in the autonomic nervous system—the distribution of cholinergic and adrenergic synapses

- The roles of nicotinic and muscarinic receptors in both divisions of the autonomic nervous system

- The role of α- and β-adrenoceptors in the sympathetic nervous system

- The regulation of the autonomic nervous system by the CNS

- Some disorders of the autonomic nervous system

11.1 Introduction

The autonomic nervous system (or ANS) regulates the operation of the internal organs, and in this way it supports the activity of the body as a whole. The autonomic nervous system is not a separate nervous system but is the efferent (motor) pathway that links those areas within the brain concerned with the regulation of the internal environment to specific effectors such as blood vessels, the heart, the gut, and so on. Unlike the efferent fibres of the skeletal muscles, those of the autonomic nervous system do not pass directly to the effector organs; rather they synapse in **autonomic ganglia**, which are located outside the CNS. Furthermore, the autonomic nervous system is not under voluntary control; for this reason it is sometimes called the involuntary nervous system. The fibres that project from the CNS to the autonomic ganglia are called **preganglionic fibres**, and those that connect the ganglia to their target organs are called **postganglionic fibres**. The sensory nerves of the internal organs are known as **visceral afferents**. Their cell bodies lie in the dorsal root ganglia and they are morphologically indistinguishable from other sensory neurons. Their peripheral processes follow the same sympathetic or parasympathetic nerves as the efferent autonomic nerve fibres serving a particular organ.

11.2 Organization of the autonomic nervous system

The autonomic nervous system is divided into two parts.

- The **sympathetic division** broadly acts to prepare the body to meet a challenge (which may be physical or psychological)—the 'fight or flight' response. Increased sympathetic activity is associated with an increased heart rate, vasoconstriction in the visceral organs, and vasodilatation in skeletal muscle.

- The **parasympathetic division**, which is more discrete in its actions and tends to promote restorative functions such as digestion and a slowing of the heart rate.

The sympathetic nervous system

The **sympathetic nervous system** originates in the cells of the intermedio-lateral column of the thoracic and lumbar regions of the spinal cord between segments T1 and L2 or L3. These neurons are called **sympathetic preganglionic neurons**. The axons of

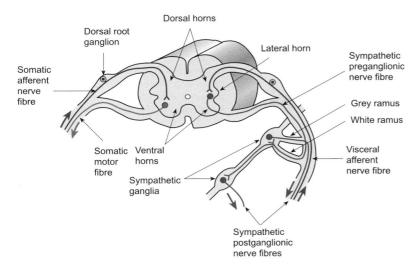

Figure 11.1 A simple diagram comparing the arrangement of the sympathetic preganglionic and postganglionic neurons with the organization of the somatic motor nerves. Note that sympathetic preganglionic fibres may terminate in the sympathetic ganglion of the same segment or pass to another ganglion in the sympathetic chain. Some also terminate in prevertebral ganglia such as the coeliac ganglion.

these neurons (sympathetic preganglionic fibres) pass from the spinal cord via the ventral root together with the somatic motor fibres. Shortly after the dorsal and ventral roots fuse, the sympathetic preganglionic fibres leave the spinal nerve trunk to travel to sympathetic ganglia via the white rami communicantes, as shown in Figure 11.1. (A **ramus communicans** is a nerve bundle connecting two nerve trunks.) The preganglionic fibres synapse on sympathetic neurons within the ganglia, and the postganglionic sympathetic fibres project to their target organs mainly via the grey rami communicantes and the segmental spinal nerves.

The majority of sympathetic ganglia are found on each side of the vertebral column (the paravertebral ganglia) and are linked together by longitudinal bundles of nerve fibres to form the two sympathetic trunks. With the exception of the cervical region, the sympathetic ganglia are distributed segmentally down as far as the coccyx. The sympathetic preganglionic fibres to the abdominal organs join to form the splanchnic nerves; these pass to the coeliac, superior, and inferior mesenteric ganglia, where they synapse. Postganglionic sympathetic fibres then pass to the various abdominal organs, as shown in Figure 11.2.

The sympathetic innervation to the adrenal medulla is an exception to the general rule. The preganglionic fibres from the thoracic spinal cord pass to the adrenal glands via the splanchnic nerves. There they synapse directly with the chromaffin cells of the adrenal medulla. The chromaffin cells of the adrenal medulla are homologous with sympathetic postganglionic neurons and share many of their physiological properties, including the generation of action potentials and the secretion of catecholamines (see Section 11.3).

Although the sympathetic preganglionic fibres are myelinated, the sympathetic postganglionic fibres are unmyelinated. This explains the difference in appearance of grey and white rami. As shown in Figure 11.2, sympathetic postganglionic fibres innervate many organs including the eye, the salivary glands, the gut, the heart, and the lungs. They also innervate the smooth muscle of the blood vessels, the sweat glands, and the piloerector muscles of the skin hairs. As the sympathetic ganglia are located close to the

spinal cord, most sympathetic preganglionic fibres are relatively short, while the postganglionic fibres are much longer, as indicated in Figure 11.3.

The parasympathetic nervous system

The preganglionic neurons of the **parasympathetic nervous system** have their cell bodies in two regions of the CNS: the brainstem and the sacral segments S3–S4 of the spinal cord. Thus the parasympathetic preganglionic fibres emerge either as part of the **cranial outflow** in cranial nerves III (oculomotor), VII (facial), IX (glossopharyngeal), and X (vagus) or from the **sacral outflow**.

The parasympathetic ganglia are usually located close to the target organ or even embedded within it. Thus the parasympathetic innervation is characterized by long preganglionic fibres and short postganglionic fibres, in contrast to the organization of the sympathetic nervous system (see Figure 11.3). Parasympathetic postganglionic fibres innervate the eye, the salivary glands, the genitalia, the gut, the heart, the lungs, and other visceral organs, as shown on the right-hand side of Figure 11.2. Parasympathetic innervation of blood vessels is confined to vasodilator fibres supplying the salivary glands, the exocrine pancreas, the gastrointestinal mucosa, the genital erectile tissue, and the cerebral and coronary arteries. Other blood vessels are innervated exclusively by sympathetic fibres.

Many organs receive a dual innervation from sympathetic and parasympathetic fibres

Most visceral organs (but not all) are innervated by both divisions of the autonomic nervous system. The specific actions of the autonomic nerves on the various organ systems of the body are discussed at length in the relevant chapters of this book. In many cases, the actions of the sympathetic and parasympathetic divisions are antagonistic, so that the actions of the two divisions

Sympathetic innervation

Parasympathetic innervation

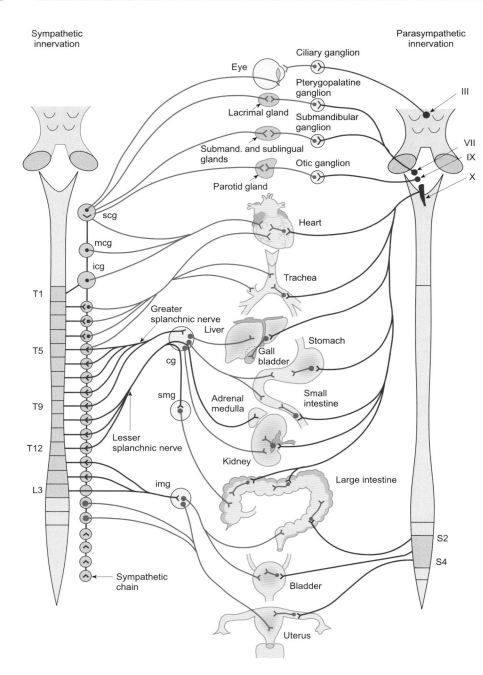

Figure 11.2 A schematic drawing illustrating the organization of the autonomic nervous system seen from the dorsal surface. In addition to the innervation of the principal organ systems, segmental sympathetic fibres also innervate blood vessels, piloerector muscles, and sweat glands. Parasympathetic preganglionic fibres are found in cranial nerves III (oculomotor), VII (facial), IX (glossopharyngeal), and X (vagus). Abbreviations: scg, mcg, and icg = superior, middle, and inferior cervical ganglion; cg = coeliac ganglion; smg and img = superior and inferior mesenteric ganglia. Postganglionic fibres are shown in green.

provide a delicate control over the functions of the viscera. Thus activation of the sympathetic nerves to the heart increases heart rate and the force of contraction of the heart muscle, while activation of the vagus nerve (parasympathetic) slows the heart. Activation of the parasympathetic supply to the gut enhances its motility and secretory functions, while activation of the sympathetic supply inhibits the digestive functions of the gut and constricts its sphincters.

Some organs have only a sympathetic supply. Examples are the adrenal medulla, the **pilomotor** muscles (the arrector pili) of the skin hairs, the sweat glands, and the spleen. In humans, it is

probable that most blood vessels are also exclusively innervated by the sympathetic nerves. The parasympathetic supply has exclusive control over the focusing of the eyes by the ciliary muscles and of pupillary constriction by the constrictor pupillae muscle of the iris. However, sympathetic stimulation dilates the pupil of the eyes by its action on the dilator pupillae muscle of the iris. Thus the iris provides an example of functional antagonism rather than dual antagonistic innervation of a specific smooth muscle. The effects of activation of the sympathetic and parasympathetic divisions of the autonomic nervous system on various organs are summarized in Table 11.1.

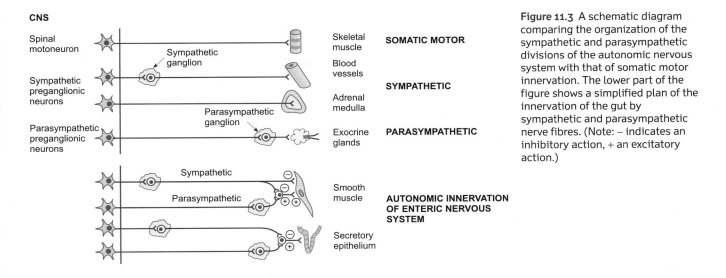

Figure 11.3 A schematic diagram comparing the organization of the sympathetic and parasympathetic divisions of the autonomic nervous system with that of somatic motor innervation. The lower part of the figure shows a simplified plan of the innervation of the gut by sympathetic and parasympathetic nerve fibres. (Note: − indicates an inhibitory action, + an excitatory action.)

Table 11.1 The main actions of sympathetic and parasympathetic stimulation on various organ systems

Organ	Effect of sympathetic activation	Receptor subtype	Effect of parasympathetic activation	Receptor subtype
Eye	Pupillary dilatation	α	Pupillary constriction & accommodation	M_3
Lacrimal gland	No effect (vasoconstriction?)	—	Secretion of tears	M_3
Salivary glands	Vasoconstriction, secretion of viscous fluid	α, β	Vasodilatation and copious secretion of saliva	M_3
Heart	Increased heart rate and force of contraction	β_1	Decreased heart rate, no effect on force of contraction	M_2
Blood vessels	Mainly vasoconstriction, vasodilatation in skeletal muscle	α β_2	No effect except for vasodilatation of certain exocrine glands and the external genitalia	—
Lungs	Bronchial dilatation via circulating adrenaline	β_2	Bronchial constriction, secretion of mucus	M_3
Liver	Glycogenolysis, gluconeogenesis, & release of glucose into blood	α, β	No effect on liver, secretion of bile by gall bladder	— M_3
Adrenal medulla	Secretion of adrenaline and noradrenaline (mediated via nicotinic ACh receptors)	N	No innervation	—
Gastrointestinal tract	Decreased motility and secretion; constriction of sphincters; vasoconstriction	α_1, α_2, β_2 α_2, β_2	Increased motility and secretion; relaxation of sphincters; increased gastric acid secretion	M_3 M_1
Kidneys	Renin secretion	β_1	No effect	—
Urinary bladder	Inhibition of micturition Relaxation of wall	α_1 β_2	Initiation of micturition	M_3
Genitalia	Ejaculation	α	Erection*	M_2 and M_3
Sweat glands	Secretion of sweat by eccrine glands (mediated by ACh acting on muscarinic receptors)	M_3	No innervation	—
Hair follicles	Piloerection	α	No innervation	—
Metabolism	Increase	α, β_2	No effect	—

Notes: This table summarizes the main effects of activation of the sympathetic or parasympathetic nerves to various organs and the receptors that mediate these effects. The details of specific autonomic reflexes can be found in the relevant chapters of this book. Key: α_1, $\beta1$, etc. $= \alpha_1$- and β_1-adrenoceptors; M = muscarinic receptors (M_1, M_2, M_3); N = nicotinic receptor; ACh = acetylcholine.

*See also Figure 49.1.

Autonomic nerves maintain a basal level of tonic activity

The autonomic innervation generally provides a basal level of activity, called **tone**, in the tissues it innervates (see also the discussion of smooth muscle in Chapter 8). The autonomic tone can be either increased or decreased to modulate the activity of specific tissues. For example, the blood vessels are generally in a partially constricted state as a result of **sympathetic tone**. This partial constriction restricts the flow of blood. If sympathetic tone is increased, the affected vessels become more constricted and this results in a decrease in blood flow. Conversely, if sympathetic activity is inhibited, tone decreases and the affected vessels dilate, so increasing their blood flow (see Chapter 30).

The heart in a resting person is normally under the predominant influence of **vagal tone**. If the vagus nerves are cut, the heart rate rises. During the onset of exercise, the tonic parasympathetic inhibition of the heart declines and sympathetic activation increases, with a resulting elevation in heart rate.

The enteric nervous system

Sympathetic and parasympathetic nerve fibres act on neurons that are present in the walls of the gastrointestinal tract. These neurons are considered to form a separate division of the nervous system—the **enteric nervous system** or ENS. The enteric neurons are organized as two interconnected plexuses (see Figures 43.6 and 43.7):

- the **submucosal plexus** (also known as Meissner's plexus), which lies in the submucosal layer beneath the muscularis mucosae;
- the **myenteric plexus** (or Auerbach's plexus), which lies between the outer longitudinal and the inner circular smooth muscle layers of the muscularis.

The enteric nervous system can function independently of its autonomic supply, and its neurons play an important part in the regulation of the motility and secretory activity of the digestive system. Its organization and functions are discussed in greater detail in Chapter 43. Other tissues such as the airways, bladder, and heart also possess intrinsic neurons, although their physiological role is not well understood.

Summary

The autonomic nervous system is a system of efferent nerves that act to control the activity of the internal organs. It is divided into the sympathetic and parasympathetic divisions, which have different functions and different anatomical origins. The sensory nerves that accompany the autonomic nerves are known as visceral afferents.

The sympathetic preganglionic neurons pass from the spinal cord via the ventral root and synapse on sympathetic neurons within the ganglia. Postganglionic sympathetic fibres project to their target organs mainly via the segmental spinal nerves. The parasympathetic preganglionic fibres emerge either as part of the cranial nerves or from the sacral outflow. Both divisions of the ANS innervate the gut via the enteric nervous system.

The sympathetic division broadly acts to prepare the body for activity ('fight or flight'), while the parasympathetic division tends to promote restorative functions ('rest and digest'). Many organs receive innervation from both sympathetic and parasympathetic nerve fibres, which regulate the activity of the internal organs according to the needs of the body at the time.

11.3 Chemical transmission in the autonomic nervous system

Within the autonomic ganglia, the synaptic contacts are highly organized and similar in structure to the other neuronal synapses described in Chapter 7. In the target tissues, however, the synaptic contacts are more diffuse than those of the CNS or those of the neuromuscular junction of skeletal muscle. There is not a simple one-to-one correspondence between a nerve ending and a point of contact on the target tissue. Instead, as they course through their target tissue, the postganglionic fibres have a number of varicosities (beadlike swellings) along their length which secrete neurotransmitters into the space adjacent to the target cells rather than onto a clearly defined synaptic region (see Figure 11.4). There are no specialized postsynaptic thickenings or junctional folds as there are at the neuromuscular junction (Figure 7.15). In multiunit smooth muscle, however, each varicosity is closely associated with an individual smooth muscle cell.

Activation of autonomic nerves can elicit both synaptic excitation and inhibition in smooth muscle. Figure 11.5(a) shows both excitatory and inhibitory junction potentials in smooth muscle fibres following electrical stimulation of the postganglionic sympathetic nerves supplying the smooth muscle of the vas deferens (a) and that of the taenia coli muscle of the large intestine (b). In common with other nerve terminals, the varicosities of the autonomic postganglionic fibres are subject to presynaptic modulation by a variety of chemical mediators.

The main neurotransmitters secreted by the neurons of the autonomic nervous system are **acetylcholine (ACh)** and **noradrenaline** (known as **norepinephrine** in North America). Within the ganglia of both the sympathetic and parasympathetic divisions, the principal transmitter secreted by the preganglionic fibres is acetylcholine. The postganglionic fibres of the parasympathetic nervous system also secrete ACh onto their target tissues. The postganglionic sympathetic fibres secrete noradrenaline, except for those fibres that innervate the sweat glands. These fibres secrete ACh. The principal transmitters utilized by the different neurons of the autonomic nervous system are summarized in Figure 11.6.

Since the original description of synaptic transmission in the ANS it has become clear that autonomic nerves utilize a wide variety of chemical transmitters in addition to ACh and noradrenaline. A number of these are listed in Table 11.2. The first non-adrenergic non-cholinergic (NANC) neurotransmitter to be identified was ATP, which mediates purinergic transmission via P_{2X} ionotropic and P_{2Y} metabotropic receptors. Purinergic transmission is employed by the autonomic nerves of the gastrointestinal tract and in

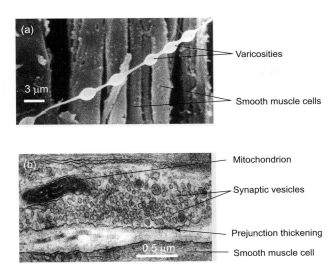

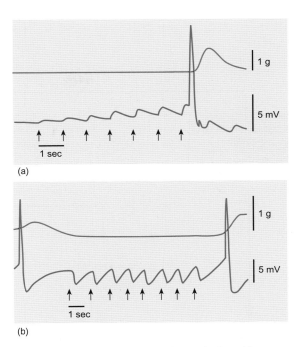

Figure 11.4 The structure of the terminal region of autonomic nerve fibres. (a) A scanning electron micrograph of a single autonomic fibre passing over smooth muscle cells of the small intestine. Note the lack of precise correspondence between the varicosities and the individual muscle cells. The intestine was treated with enzymes to remove the connective tissue. (b) A transmission electron micrograph of an autonomic varicosity. As in other synaptic terminals, the varicosity is filled with synaptic vesicles. There is a mitochondrion and two areas showing thickening of the presynaptic membrane, which may represent active zones for secretion. Note the wide separation of the pre- and postsynaptic cells (~0.1–0.2 μm) and the absence of postsynaptic specializations. (Compare this with Figure 7.15c, which shows a comparable image of the neuromuscular junction of skeletal muscle.) Bars: 3 μm (a), 0.5 μm (b). (From panels (a) and (c) of Fig 1 in G. Burnstock (2009) *Annual Review of Pharmacology* **49**, 1–30.)

Figure 11.5 Excitatory and inhibitory transmission at two autonomic neuromuscular junctions. (a) The excitatory junction potentials (EJPs) in the guinea-pig vas deferens recorded by the sucrose-gap method. The bottom trace (green) shows the EJPs evoked by stimulation of postganglionic sympathetic nerves, and the top trace (red) is the associated contraction. Note the summation of the EJPs and their increasing amplitude with successive stimuli (arrows) culminating in an action potential. (b) Inhibitory junctional potentials recorded in the intramural nerves of the taenia coli muscle of the large intestine of the guinea-pig. Note that repetitive simulation (arrows) of the muscle fibre bundle caused inhibition of the spontaneous action potential activity (green trace) and relaxation of the muscle (red trace). In this case, the preparation has been deprived of its adrenergic nerves. (Redrawn from Fig 2 in G. Burnstock (2009) *Annual Review of Pharmacology* **49**, 1–31).

the CNS. Since this discovery, a large number of neuropeptides have been shown to be employed as neurotransmitters by autonomic nerves. Other neurotransmitters secreted by autonomic nerves are 5-hydroxytryptamine, dopamine, γ-aminobutyric acid, glutamate, and nitric oxide (NO).

It is now known that autonomic nerves (as well as some in the CNS) can secrete more than one neurotransmitter at a given site. This is called **cotransmission** (see Chapter 7). Combinations of neurotransmitters secreted by individual nerve terminals have been identified for a number of synapses, particularly those of the enteric nerves. An example of cotransmission is provided by the parasympathetic innervation of the salivary glands. Here the secretion is elicited by ACh acting on muscarinic receptors, while the accompanying vasodilatation is mediated by vasoactive intestinal polypeptide (VIP) secreted from the same nerve terminals. ACh and VIP are stored in separate vesicles within the nerve terminals: ACh is found within small clear vesicles similar to those seen at the neuromuscular junction (see Figure 7.15), while VIP is found in dense-cored vesicles. The two vesicle populations are differentially released by different frequencies of stimulation; ACh is preferentially released at low frequencies of stimulation, while VIP is released in response to high frequencies (see Figure 11.7). The

presence of two distinct transmitters in the same nerve terminal thus allows for additional chemical coding within the target tissue.

Acetylcholine activates nicotinic receptors in the autonomic ganglia but muscarinic receptors in the target tissues

The ACh receptors of the postganglionic neurons in autonomic ganglia are called **nicotinic receptors** because they can also be activated by the alkaloid **nicotine**. They are similar in structure to the nicotinic receptors of the neuromuscular junction but have a different response to various drugs and toxins. For example, they can be blocked by mecamylamine, which has no action at the neuromuscular junction, but not by the snake toxin α-bungarotoxin—which is a potent blocker of the nicotinic receptors of the neuromuscular junction. Activation of nicotinic receptors leads to the opening of an ion channel and rapid excitation, as described in Chapter 7.

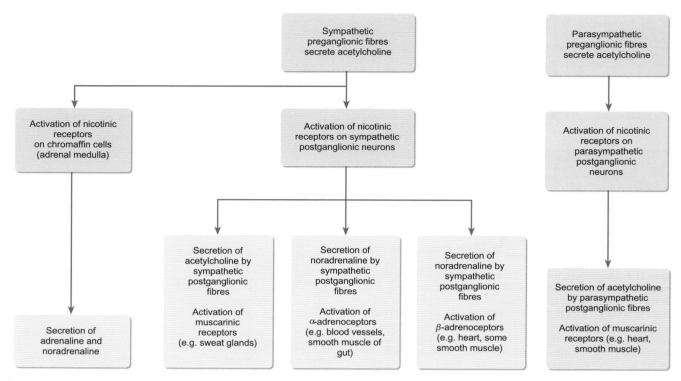

Figure 11.6 A flow chart illustrating the role of the cholinergic and adrenergic innervation in the autonomic nervous system.

The ACh receptors of the target tissues of both parasympathetic and sympathetic postganglionic fibres are known as **muscarinic receptors**, as they can be activated by the alkaloid **muscarine** (the poisonous constituent of certain fungi such as fly agaric).

Table 11.2 Some neurotransmitters of the autonomic nervous system

Small molecules	Acetylcholine (ACh)
	Noradrenaline (NA)
	ATP (adenosine triphosphate) and other purines
	5-HT (5-hydroxytryptamine, serotonin)
	Dopamine (2(3,4–dihydroxyphenyl) ethylamine, DA)
	GABA (γ-aminobutyric acid)
	Glutamate
	Nitric oxide
Neuropeptides	Enkephalin, endorphin, and dynorphin
	Vasoactive intestinal peptide (VIP)
	Neuropeptide Y
	Substance P
	Neurotensin
	Somatostatin
	Endothelin

Muscarinic receptors are also present at sympathetic nerve endings in sweat glands, pilomotor muscles of the skin, and, in some animal species, at the nerve endings of vasodilator fibres in skeletal muscle. They can be inhibited by low concentrations of **atropine**.

Muscarinic receptors can be divided into five subtypes based on their molecular structures and responses to various antagonists. All are linked to G proteins: the receptors designated M_1, M_3, and M_5 are coupled to the inositol trisphosphate (IP_3) pathway via G_q/G_{11}, while M_2 and M_4 receptors act by inhibiting adenylyl cyclase via G_i/G_o and reducing intracellular cAMP (see Chapter 6).

Of the five subtypes that have been characterized, the functions of the M_1, M_2, and M_3 subtypes are well understood.

1. **M_1 receptors** are mainly located on neurons in the CNS and peripheral ganglia. They are also found on gastric parietal cells. Activation of these receptors generally has an excitatory effect. For example, the secretion of gastric acid that follows stimulation of the vagus nerve is mediated by M_1 receptors (see Chapter 43).

2. **M_2 receptors** occur in the heart and on the nerve terminals of both CNS and peripheral neurons. The slowing of the heart following stimulation of the vagus nerves is mediated by this class of muscarinic receptor.

3. **M_3 receptors** are located in secretory glands and on smooth muscle. Activation of these receptors generally has an excitatory effect. In visceral smooth muscle, for example, activation of M_3 receptors by ACh leads to contraction.

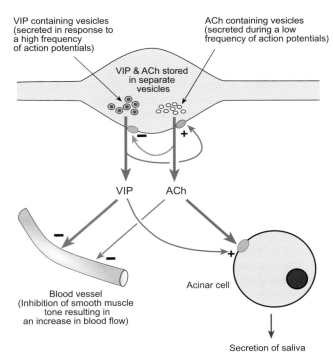

VIP containing vesicles
(secreted in response to
a high frequency
of action potentials)

ACh containing vesicles
(secreted during a low
frequency of action potentials)

VIP & ACh stored
in separate
vesicles

VIP   ACh

Blood vessel
(Inhibition of smooth muscle
tone resulting in
an increase in blood flow)

Acinar cell

Secretion of saliva

Figure 11.7 A diagram showing synaptic transmission in the salivary glands during parasympathetic nerve stimulation. Both acetylcholine (ACh) and vasoactive intestinal polypeptide (VIP) are present in the synaptic terminals (they are cotransmitters), but they are stored in separate populations of vesicles that are secreted in response to different action potential frequencies. Both transmitters inhibit the tone of the local blood vessels, thereby increasing blood flow; VIP is the more effective of the two. ACh also stimulates the acinar cells to secrete saliva, an effect that is enhanced by the binding of VIP to their M_3 receptors (a neuromodulatory effect). VIP also acts on the nerve varicosities to enhance the secretion of ACh. However, ACh exerts an inhibitory effect on the secretion of VIP. + indicates stimulation; − indicates inhibition (also indicated by red arrows).

Both M_4 and M_5 receptors have been shown to be present in the CNS, but apart from some evidence for a role for M_4 receptors in locomotion, their physiological roles remain to be established.

Adrenergic receptors belong to two main classes

The receptors for noradrenaline are called **adrenoceptors**. Two main classes are known: α-**adrenoceptors** and β-**adrenoceptors**. These were originally distinguished by the relative potency of adrenaline, noradrenaline, and isoprenaline (isoprotorenol—a synthetic catecholamine) on specific tissues. Those receptors for whom the order was noradrenaline ≥ adrenaline > isoprenaline were designated α-receptors, while those that had the potency order isoprenaline > adrenaline ≥ noradrenaline were designated β-receptors.

Alpha-adrenoceptors are further divided into two main classes: α_1 and α_2; each of these is subdivided into three subtypes: α_{1A}, α_{1B}, α_{1D}, and α_{2A}, α_{2B}, α_{2C}. Beta-adrenoceptors are subdivided into

three subtypes: β_1, β_2, β_3. As for muscarinic receptors, the adrenoceptors are coupled to second-messenger systems via G proteins: α_1 receptors activate the phospholipase C cascade and increase intracellular Ca^{2+}; α_2 receptors are coupled to G_i and their activation leads to a reduction in intracellular cAMP; all three β-receptor subtypes stimulate adenylyl cyclase and increase intracellular cAMP.

The distribution and properties of the main adrenoceptor subtypes are as follows.

1. α_1-adrenoceptors are found in the smooth muscle of the blood vessels, bronchi, gastrointestinal tract, uterus, and bladder. Activation of these receptors is mainly excitatory and results in the contraction of smooth muscle. However, the smooth muscle of the gut wall (but not that of the sphincters) becomes relaxed after activation of these receptors.

2. α_2-adrenoceptors are found in the smooth muscle of the blood vessels, where their activation causes vasoconstriction.

3. β_1-adrenoceptors are found in the heart, where their activation results in an increased heart rate and an increase in the force of contraction (a positive inotropic effect—see Chapter 28). They are also present in the sphincter muscle of the gut, where their activation leads to relaxation.

4. β_2-adrenoceptors are found in the smooth muscle of certain blood vessels, where their activation leads to vasodilatation. They are also present in the bronchial smooth muscle, where they mediate bronchodilatation.

5. β_3-adrenoceptors are present in adipose tissue, where they initiate lipolysis to release free fatty acids and glycerol into the circulation.

Although the diversity in both adrenoceptors and cholinergic receptors is somewhat bewildering, the development of agonists and antagonists that act on specific subtypes of these receptors has been of considerable clinical benefit in the treatment of diseases such as asthma and hypertension. Selective blockers of α_1 adrenoceptors such as **prazosin** are used to treat hypertension and **heart failure**. Blockers of β-adrenoceptors such as **propranolol** and **atenolol** are used to treat hypertension, some arrhythmias, and heart failure; β-adrenoceptor agonists such as **salbutamol** are used to relieve the bronchoconstriction of asthma.

The adrenal medulla secretes adrenaline and noradrenaline into the circulation

Although the activation of the autonomic nerves provides a mechanism for the discrete regulation of specific organs, activation of the splanchnic nerve results in the secretion of adrenaline and noradrenaline (also known as epinephrine and norepinephrine) from the adrenal medulla into the circulation. About 80 per cent of the secretion is adrenaline and 20 per cent is noradrenaline. These catecholamines exert a hormonal action on a variety of tissues (see Chapters 22 and 23), which forms part of the overall sympathetic response. Their release is always associated with an increase in the secretion of noradrenaline from sympathetic nerve terminals.

Broadly, adrenaline activates β-adrenoceptors more strongly than α-adrenoceptors, while noradrenaline is more effective at activating α-adrenoceptors than β-adrenoceptors. The potency of adrenaline and noradrenaline at the different receptor subtypes is:

α_1-adrenoceptors   noradrenaline > adrenaline

α_2-adrenoceptors   noradrenaline ≥ adrenaline

β_1-adrenoceptors   adrenaline = noradrenaline

β_2-adrenoceptors   adrenaline >> noradrenaline

β_3-adrenoceptors   noradrenaline > adrenaline

Consequently, the response of a tissue to circulating adrenaline or noradrenaline will depend on the relative proportions of the different types of adrenoceptor it possesses. For example, during exercise, increased activity in the sympathetic nerves will result in an increased heart rate (mediated via β_1 receptors) and vasoconstriction in the renal and splanchnic circulations (via α-receptors). Circulating noradrenaline will have a similar effect, but the actions of circulating adrenaline will lead to relaxation of the smooth muscle that possesses a high proportion of β_2-adrenoceptors, such as that of the blood vessels of skeletal muscle. The increased levels of circulating adrenaline also cause bronchodilatation, thus favouring increased gas flow to the alveoli. The combination of bronchodilatation and vasodilatation in the skeletal muscle enhances the delivery of oxygen to the exercising muscles.

Circulating catecholamines affect virtually every tissue, with the result that the metabolic rate of the body is increased. Indeed, maximal sympathetic stimulation may double the metabolic rate. The major metabolic effect of adrenaline and noradrenaline is to increase the rate of **glycogenolysis** within cells by activating adenylyl cyclase via β-adrenoceptors (see Chapter 6). The result is a rapid mobilization of glucose from glycogen and an increased availability of fatty acids for oxidation as a result of lipolysis occurring in adipose tissue. The increased availability of substrates for oxidative metabolism is important both in exercise and during cold stress, where an increase in metabolic rate is important for generating the heat required to maintain body temperature (see Chapter 42).

Summary

ACh is the principal transmitter secreted by the preganglionic fibres within the autonomic ganglia of both divisions. The parasympathetic postganglionic fibres also secrete ACh onto their target tissues, while noradrenaline is the principal neurotransmitter secreted by the postganglionic sympathetic fibres.

ACh acts on nicotinic receptors in the autonomic ganglia and on muscarinic receptors in the target tissues. Circulating adrenaline and noradrenaline affect virtually every tissue and exert their effects via α- and β-adrenoceptors. Activation of α-adrenceptors generally leads to contraction of smooth muscle, while activation of β-adrenoceptors leads to its relaxation. In the heart, activation of β-adrenoceptors results in an increased heart rate and force of contraction.

11.4 Central nervous control of autonomic activity

The activity of the autonomic nervous system varies according to the information it receives from both visceral and somatic afferent fibres. It is also subject to regulation by the higher centres of the brain, notably the hypothalamus.

Autonomic reflexes

The internal organs are innervated by afferent fibres that respond to mechanical and chemical stimuli. Some visceral afferents reach the spinal cord by way of the dorsal roots and enter the dorsal horn together with the somatic afferents. These fibres synapse at the segmental level, and the second-order fibres ascend the spinal cord in the spinothalamic tract. They project to the thalamus and hypothalamus, the nucleus of the solitary tract, and various motor nuclei in the brainstem. Other visceral afferents, such as those from the arterial baroreceptors, reach the brainstem by way of the vagus nerves.

Information from the visceral afferents elicits specific visceral reflexes which, like the reflexes of the somatic motor system, may either be segmental or involve the participation of neurons in the brain. **Autonomic reflexes** play an essential role in maintaining the homeostasis of the body and are discussed in detail in the relevant chapters of this book. Examples are the baroreceptor reflex (Chapter 30), the lung inflation reflex (Chapter 32), the salivary, vomiting, and defecation reflexes (Chapter 44), and the micturition reflex (Chapter 39).

In response to a perceived danger, there is behavioural alerting that may result in aggressive or defensive behaviour. This is known as the **defence reaction** (the 'fight or flight' response), which has its origin in the hypothalamus. During the defence reaction there are marked changes in the activity of the autonomic nerves in which normal reflex control is overridden. There is an increase in heart rate, cardiac output, and blood pressure. In addition, there is a redistribution of blood flow with vasoconstriction in the viscera and skin and an increased blood flow to the muscles. These cardiovascular responses involve resetting of the baroreceptor reflex to a higher level of pressure (see Chapter 28).

The hypothalamus regulates the homeostatic activity of the autonomic nervous system

Both the activity of the autonomic nervous system and the functions of the endocrine system are under the control of the hypothalamus, which is the part of the brain mainly concerned with maintaining the homeostasis of the body. If the hypothalamus is damaged or destroyed, the homeostatic mechanisms fail. The hypothalamus receives afferents from the retina, the chemical sense organs, somatic senses, and visceral afferents. It also receives many inputs from other parts of the brain including the limbic system and cerebral cortex. Hypothalamic neurons play important roles in thermoregulation, in the regulation of tissue osmolality, in the control of feeding and drinking, and in reproductive activity.

11.5 Disorders of autonomic function

Autonomic failure is a relatively rare condition that generally affects older people. It is more common in men than in women and its clinical signs are dizziness, fatigue, and blackouts during exercise or after meals. The key feature is a persistent postural **hypotension** (low blood pressure) which results from the inability of the sympathetic nerves to constrict the peripheral blood vessels. The disease can occur on its own or in combination with neurodegenerative diseases, specifically Parkinson's disease and multiple system atrophy.

Interruption of the sympathetic supply to the head is associated with a characteristic set of symptoms known as **Horner's syndrome**, in which the facial skin on the affected side is flushed and dry due to loss of sympathetic tone in the blood vessels and lack of sweat production. The pupil is constricted (miotic), reflecting paralysis of the pupillary sphincter muscle, and there is slight drooping of the eyelid (ptosis), which is the result of paralysis of the superior tarsal muscle. Horner's syndrome may be congenital in origin or it may reflect damage to the sympathetic trunk arising from surgical intervention, trauma, or a tumour of the apex of the lung. It may also be caused by a lesion in the brainstem that interrupts the descending fibres to the intermediolateral column of the spinal cord.

Summary

The hypothalamus is the part of the brain concerned with maintaining homeostasis. It regulates the activity of the autonomic nervous system and coordinates its activity with the endocrine system. Autonomic reflexes play an essential role in maintaining homeostasis.

Information from the visceral afferents elicits specific visceral reflexes, which may be segmental or may involve the participation of neurons in the brain. Disorders of autonomic function generally affect older people and may result in persistent postural hypotension.

✳ Checklist of key terms and concepts

The organization of the autonomic nervous system

- The autonomic nervous system is a system of efferent nerves that act to control the activity of the internal organs.
- It is divided into the sympathetic and parasympathetic divisions, which have different functions and different anatomical origins.
- The sympathetic division broadly acts to prepare the body for activity ('fight or flight').
- Increased sympathetic activity is associated with an increased heart rate, vasodilatation in skeletal muscle, but vasoconstriction in the visceral organs.
- Increased sympathetic activity is also accompanied by bronchodilatation and by glycogenolysis and gluconeogenesis in the liver.
- The parasympathetic division tends to promote restorative functions ('rest and digest').
- Increased parasympathetic activity is associated with increased motility and secretion by the gastrointestinal tract and slowing of the heart rate.
- Many organs receive innervation from both sympathetic and parasympathetic nerve fibres, which act in opposing ways. Nevertheless, the two divisions act together to regulate the activity of the internal organs according to the needs of the body at the time.
- The activity of the autonomic nerves is under the control of neurons in the hypothalamus and brainstem.
- The autonomic functions of the gastrointestinal tract are largely served by the enteric nervous system.

Chemical transmission in the ANS

- The main neurotransmitters secreted by the neurons of the autonomic nervous system are acetylcholine (ACh) and noradrenaline.

- ACh is the principal transmitter secreted by the preganglionic fibres within the autonomic ganglia of both sympathetic and parasympathetic divisions.
- The parasympathetic postganglionic fibres also secrete ACh onto their target tissues.
- Noradrenaline is the principal neurotransmitter secreted by the postganglionic sympathetic fibres.
- ACh acts on nicotinic receptors in the autonomic ganglia and on muscarinic receptors in the target tissues.
- Circulating adrenaline and noradrenaline affect virtually every tissue and exert their effects via α- and β-adrenoceptors.
- Activation of α-adrenoceptors generally leads to contraction of smooth muscle, while activation of β-adrenoceptors leads to its relaxation.
- In the heart, activation of β-adrenoceptors results in an increased heart rate and force of contraction.

Central control of autonomic function

- The hypothalamus is the part of the brain concerned with maintaining the homeostasis of the body.
- The hypothalamus regulates the activity of the autonomic nervous system and coordinates its activity with that of the endocrine system.
- Information from the visceral afferents elicits specific autonomic reflexes.
- These reflexes may be segmental or may involve the participation of neurons in the brain.
- Disorders of autonomic function generally affect older people.
- A key feature is a persistent postural hypotension.

Recommended reading

Anatomy

Brodal, P. (2004) *The central nervous system: Structure and function* (3rd edn), Chapters 18 and 19. Oxford University Press, New York.

These chapters provide a detailed account of the anatomy of the autonomic nervous system as well as a brief summary of its physiology.

Pharmacology

Rang, H.P., Ritter, J.M., Flower, R.J., and Henderson, G. (2015) *Pharmacology* (8th edn), Chapters 12–16. Churchill-Livingstone, Edinburgh.

A very clear and detailed account of the pharmacology of cholinergic and adrenergic transmission.

Physiology

Brading, A.S. (1999) *Autonomic nervous system and its effects.* Blackwell Science, Oxford.

A detailed monograph describing the mechanisms by which autonomic control is achieved.

Jänig, W. (2006) *Integrative action of the autonomic nervous system: Neurobiology of homeostasis.* Cambridge University Press, Cambridge.

Robertson, D., Biaggioni, I., Burnstock, G., Low, P.A., and Paton, J.F.R. (eds) (2011) *Primer on the autonomic nervous system* (3rd edn). Academic Press, London.

Not really a primer, rather it is a comprehensive source on all aspects of the biology of the autonomic nervous system and associated pathological states.

Squires, L., Berg, D., Bloom, F.E., du Lac, S., Ghosh, A., and Spitzer, C.D. (eds) (2012) *Fundamental neuroscience* (4th edn), Chapters 34 and 35. Academic Press, London.

An account of the organization of the autonomic nervous system and its hypothalamic control, followed by an account of autonomic regulation of the cardiovascular system.

Medicine

Bannister, R. (2000) The autonomic nervous system, Chapter 13.7, in: *Concise Oxford textbook of medicine* (ed J.G.G. Ledingham and D.A. Warrell). Oxford University Press, Oxford.

A short chapter discussing the origin and treatment of autonomic failure.

Review article

Burnstock, G. (2009) Autonomic neurotransmission: 60 years since Sir Henry Dale. *Annual Review of Pharmacological Toxicology* **49**,1–30; doi:10.1146/annurev.pharmtox.052808.102215.

To check that you have mastered the key concepts presented in this chapter, complete the accompanying online self-assessment questions. Go to www.oup.com/uk/pocock5e/

CHAPTER 12

General principles of sensory physiology

This chapter should help you to understand:

- The need for sensory information
- The classification of sensory receptors
- The principles of sensory transduction
- Sensory adaptation
- The organization of sensory pathways within the CNS

12.1 Introduction

We smell the air, taste our food, feel the earth under our feet, hear and see what is around us. To do all this, and more, we must have some means of converting the physical and chemical properties of the environment into nerve impulses, which are the means of signalling between the neurons of the nervous system. Once a change in the environment is identified, the central nervous system (CNS) can then determine the appropriate response and initiate the required course of action. The process by which specific properties of the environment become encoded as nerve impulses is called **sensory transduction**, which is carried out by specialized structures called **sensory receptors**, often simply called receptors. This chapter discusses the general principles that are involved

in forming the sensations that are a normal part of life, while the subsequent four chapters discuss the application of these principles to the somatosensory, visual, auditory, and vestibular systems. The final chapter in this section discusses the **chemical senses**: taste and smell (**olfaction**).

Different sensory receptors are specialized to respond to particular environmental factors. This was established early in the nineteenth century and expressed in Muller's **law of specific nerve energies**, which states that the nature of a sensory stimulus is determined by the neural pathway over which the sensory information is carried. In other words the differences in sensory quality—the difference between seeing and hearing, between hearing and touch, and so on—are not caused by differences in the stimuli themselves but by the different nervous structures that these stimuli excite. For example, a vibrating tuning fork placed on the skull will elicit a sensation of sound, but if it is placed on the elbow the sensation is one of vibration.

Although most receptors are adapted to respond to a specific sort of stimulus, others respond to more than one. Those that do so are known as **polymodal receptors**. One example is the vanilloid receptor TRPV1, which is activated both by an increase in temperature and by chemical stimuli that cause pain such as capsaicin, the active component of chilli peppers. Nevertheless, most receptors are usually especially sensitive to one kind of stimulus.

The type of stimulus to which a receptor is especially sensitive is known as its **adequate stimulus**. The airborne vibrations of sound provide the adequate stimulus for the hair cells of the inner ear; skin **thermoreceptors** respond to small changes in temperature; the photoreceptors of the eye respond to light; and so on. Each of these stimuli gives rise to sensations that may vary in intensity but that nevertheless have the same overall quality. A group of similar sensations is called a **sensory modality**. The best known sensory modalities are the classical 'five senses': taste, smell, touch, sight, and hearing. However, other modalities are also recognized: the senses of balance, of limb position, of temperature, itch, pain, and vibration. Moreover, there are the vague sensations that arise

Table 12.1 Classification of sensory receptors

Receptor type		Other classification	Examples
Mechanoreceptors	Special senses (ear)	Teleceptor	Cochlear hair cells
		Interoceptor	The hair cells of the vestibular system
	Muscle and joints	Proprioceptor	Muscle spindles
			Golgi tendon organs
	Skin and viscera	Exteroceptors	Pacinian corpuscle
			Ruffini ending
			Meissner's corpuscle
			Bare nerve endings
	Cardiovascular	Interoceptor	Arterial baroreceptors (sense high pressures)
			Atrial volume receptors (low pressure receptors)
Chemoreceptors	Special senses	Teleceptor	Olfactory receptors
		Exteroceptors	Taste receptors
	Skin and viscera	Exteroceptors	Nociceptors
		Interoceptors	Nociceptors
			Glomus cells (carotid body, sense arterial Po_2)
			Hypothalamic osmoreceptors and glucose receptors
Photoreceptors	Special senses	Teleceptor	Retinal rods and cones
Thermoreceptors	Skin	Exteroceptor	Warm and cold receptors
	CNS	Interoceptor	Temperature-sensing hypothalamic neurons

from our internal organs such as the sense of fullness following a large meal, the feeling of breathlessness during heavy exercise, or the desire to empty the bladder or the bowel. All these sensory impressions eventually reach consciousness as specific perceptions.

Sensory receptors may be classified in two principal ways. They may be classified on the basis of the specific environmental qualities to which they are sensitive—chemoreceptors, mechanoreceptors, nociceptors, photoreceptors, and thermoreceptors. Alternatively, they may be classified according to the source of the quality that they sense (see Table 12.1). Thus, there are receptors that sense events that originate at some distance from the body—the eye, ear, and nose, which are sometimes called **teleceptors**. Others sense changes occurring in the immediate external environment—touch, pressure and temperature. These are called **exteroceptors**. Then there are receptors that signal changes in the internal environment—the **interoceptors**. These sense blood pressure (baroreceptors), the oxygen and carbon dioxide levels of the blood (chemoreceptors), as well as substances released following tissue damage (nociceptors). Other receptors provide information about our position in space and the disposition of our limbs—the gravitational receptors and proprioceptors.

12.2 Principles of sensory transduction

Sensory receptors respond to four principal triggers: mechanical stimuli, chemical stimuli, temperature, and electromagnetic radiation (light). The detailed structure of individual receptors

optimizes their response to particular characteristics of stimuli. Many cutaneous nerve endings lie within a capsule of non-neuronal cells. These specialized structures enable particular nerve endings to respond to very specific kinds of mechanical stimulation. Examples are the nerve endings associated with **Pacinian corpuscles**, which respond best to vibrations with frequencies between 50 and 600 Hz. Other skin receptors known as Meissner's corpuscles respond best to vibrations of 5–80 Hz. The rods and cones of the retina have stacks of membranes packed with the photo-pigment rhodopsin to optimize their chance of absorbing **photons** (see Chapter 14). The apical membranes of the olfactory cells in the nose are covered by cilia, an arrangement that increases the area available for absorption of odoriferous molecules (see Chapter 17). The elaborate structure of the **cochlea**, discussed in Chapter 15, allows hair cells to respond to sound waves of a particular frequency. Some simple bare nerve endings act as receptors for skin temperature, itch, and pain, while those associated with Merkel cells respond to sustained mechanical stimuli (see Chapter 13).

The basic steps in sensory transduction are illustrated in Figure 12.1. Although sensory receptors respond to different environmental stimuli, in all cases their adequate stimulus ultimately leads to a change in membrane potential called a **receptor potential** (also known as a generator potential). Unlike action potentials, receptor potentials increase with the intensity of the factor responsible for their activation, and the magnitude and duration of a receptor potential governs the number and frequency of action potentials transmitted by the afferent nerve fibres to the CNS

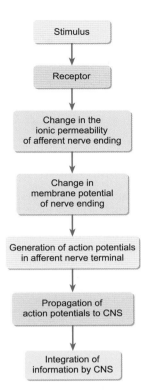

Figure 12.1 Flow chart illustrating the main steps in sensory transduction for cutaneous receptors.

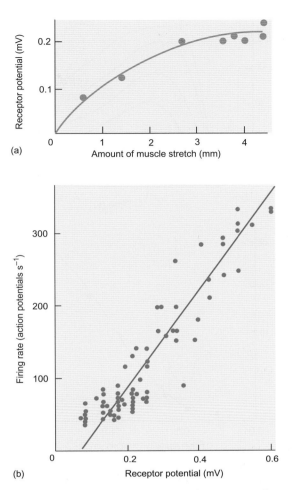

Figure 12.2 The response of a muscle spindle to varying degrees of stretch. Panel (a) shows the increase in the receptor potential, recorded with an extracellular electrode, as a muscle (the sartorius muscle of the frog) was subjected to increasing degrees of stretch. Note the progressive increase in amplitude. Panel (b) shows the increase in the frequency of action potentials in the afferent fibre as the receptor potential increases in amplitude. (Adapted from Figs 9 and 10 in B. Katz (1950) *Journal of Physiology* 111, 261–82.)

(see Figure 12.2). The action potentials that result from stimulation of sensory receptors are relayed to the CNS (the brain and spinal cord) via afferent nerve fibres (see Chapter 9).

The coding of stimulus intensity and duration

As Figure 12.2 shows, increasing the strength of a stimulus results in a larger receptor potential and an increase in the number of action potentials in the associated afferent nerve fibre. More generally, the intensity of a stimulus is also coded by the number of active receptors and by the number of action potentials it elicits from each receptor. However, there is an upper limit to the number of action potentials that can be carried by an axon each second: around 200–500 s^{-1} for the fastest fibres. This poses the problem of how to code for changes in the intensity of stimuli that may vary over many orders of magnitude. In the example shown in Figure 12.3, a single photoreceptor (a rod cell) is able to signal changes in the intensity of a brief flash of light over nearly three orders of magnitude, and the intact ear is able to signal sound intensities over an even greater range of intensities (see Figures 14.1 and 15.1).

How is such a feat achieved? It appears that subjective judgements of the intensity are approximately proportional to the logarithm of the physical stimulus. So as the stimulus increases in intensity, the difference in the absolute magnitude of a stimulus that can just be detected increases in proportion to the magnitude of that stimulus. This has been confirmed in experiments in which it is been possible to measure both the rate of action potential

discharge and the intensity of the sensation elicited in conscious subjects. As the example in Figure 12.4 shows for taste receptors, it has been found that as the strength of the stimulus increases (in this case the concentrations of citric acid and sucrose in pure solutions), both the rate of action potential discharge and the subjective judgement of concentration are highly correlated and follow a power law. This particular set of data was obtained in subjects who were undergoing ear surgery in which the nerve that subserves the sense of taste was exposed (the chorda tympani nerve, which is a branch of the facial nerve, CN VII; see Chapter 9).

The timing of the sequence of action potentials signals its onset and duration. Many receptors generate action potentials when they are first stimulated, but the action potential frequency then falls with time even though the intensity of the stimulus is unchanged. This property is known as **adaptation**. A good example

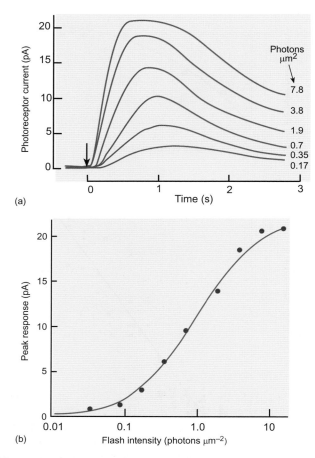

(a)

(b)

Figure 12.3 The responses of a single photoreceptor (a rod cell) to brief flashes of light of varying intensity. Panel (a) shows the current (in pA) elicited in a single rod cell by 200 ms flashes of light. The intensity of each flash in photons per μm² is indicated on the right-hand side of each trace. In panel (b), the peak response of the same rod cell is plotted as a function of flash intensity. Note the progressive increase in response with increased intensity. (Adapted from Fig 3 in D.A. Baylor, T. Lamb, and K.-W. Yau (1979) *Journal of Physiology* **288**, 589–611.)

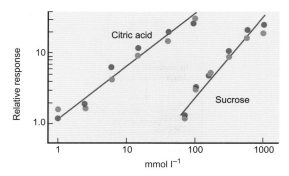

Figure 12.4 The correlation between nerve activity and subjective assessment of taste sensation in two subjects. The average values of nerve responses (green circles) and subjective taste perception (red circles) to varying concentrations of citric acid and sucrose are shown. Increasing concentration of either substance increases the activity in the chorda tympani nerve (the nerve subserving taste) and the intensity of the taste sensation judged by the subjects themselves. In each case, the stimuli were presented in pairs of a standard with a comparison stimulus. The same pairs of stimuli were presented in a random order for both taste sensation and nerve recordings. (Adapted from Fig 7 in G. Borg, H. Diamant, L. Ström, and Y. Zotterman (1967) *Journal of Physiology* **192**, 13–20.)

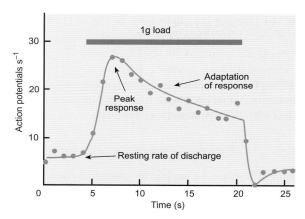

Figure 12.5 The adaptation of a muscle stretch receptor to a steady load of 1 gram. Before loading, the receptor nerve has a firing rate of around 5 action potentials a second. It rises rapidly to around 28 per second immediately following loading and then slowly declines over the next 15 seconds until the load is removed. Thereafter the firing rate falls briefly to zero before resuming a rate of around 2 per second. (Adapted from Fig 10B in E.D. Adrian and Y. Zotterman (1926) *Journal of Physiology* **61**, 151–71.)

is shown in Figure 12.5. This is the response of a muscle spindle following the loading of the muscle in which it is embedded (see Chapter 10).

Adaptation is a matter of everyday experience: when we first step into warm or cold water we notice the temperature change immediately but subsequently we become used to it. We are initially aware of the contact between our skin and our clothes when we first put them on. In a very short time we cease to notice them unless they become snagged on an obstacle or they become a source of irritation. This also emphasizes the point that the nervous system responds most strongly to changes in the environment rather than to steady conditions, especially to the onset and termination of a given stimulus (which elicit 'on' and 'off' responses respectively).

Some receptors respond to the onset of a stimulus with a few action potentials and then become quiescent. This type of receptor is called a **rapidly adapting** receptor. Other receptors maintain a

steady flow of action potentials for as long as the stimulus is maintained. These are known as **slowly adapting** or **non-adapting** receptors. Examples of the response to various kinds of mechanical stimulus by rapidly adapting and slowly adapting mechanoreceptors are illustrated in Figure 12.6. Pacinian corpuscles are rapidly adapting receptors that respond best to high-frequency vibrations (50–600 Hz). Other rapidly adapting mechanoreceptors such as Meissner's corpuscles (see Chapter 13) signal the onset and offset

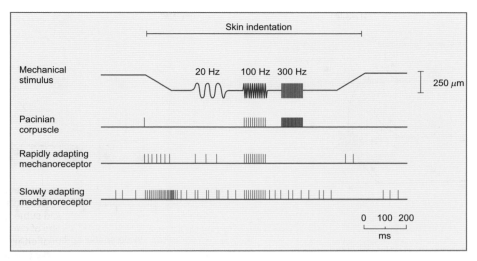

Figure 12.6 The response of three different types of mechanoreceptor to pressure stimuli applied to the skin. The top record shows the characteristics of the mechanical stimulus. The lower three records show the response of a Pacinian corpuscle, a rapidly adapting mechanoreceptor, and a slowly adapting mechanoreceptor. Each vertical spike represents an action potential. Note that the Pacinian corpuscle responds to maintained skin indentation with a single action potential, while the slowly adapting mechanoreceptor continues to generate action potentials. Only the Pacinian corpuscle is able to respond to the 300 Hz vibration of the skin. (Based on Fig 17.2 in A. Iggo (1982) Chapter 17 in *The Senses*, (ed. H.B. Barlow and J.D. Mollon). Cambridge University Press, Cambridge.)

of a stimulus as well as responding to lower frequencies of vibration. Slowly adapting receptors, such as those associated with Merkel cells, show 'on' and 'off' responses with an increased rate of action potential discharge when a mechanical stimulus is first applied. They fall silent immediately after the stimulus is withdrawn. Furthermore, slowly adapting receptors continue to respond as long as the stimulus is present, as shown in the bottom record of Figure 12.6.

Summary

The external and internal environment is continuously monitored by sensory receptors, each of which is excited most effectively by a specific type of stimulus known as its adequate stimulus. Sensory receptors respond to four principal kinds of stimulus: mechanical, chemical, temperature, and light. Some receptors signal the onset and offset of a stimulus (rapidly adapting receptors), while others (slowly adapting receptors) continuously signal the intensity of the stimulus. The process by which a stimulus becomes encoded as a sequence of nerve impulses in an afferent nerve fibre is called sensory transduction, and the nerve fibres that convey this information to the CNS are called afferent or sensory nerve fibres.

12.3 The organization of sensory pathways in the CNS

The nervous system needs to establish the location, physical nature, and intensity of all kinds of stimuli. Since most sensory receptors respond to very specific stimuli, the afferent fibres to

which they are connected can be considered as 'labelled lines'. For example, the skin has receptors that respond selectively to touch and others that respond to a small fall or a small rise in temperature. The activation of a specific population of receptors will therefore inform the CNS of the nature and location of the stimulus. The afferent nerve fibres of the somatosensory system (i.e. those providing information about the body and its immediate environment) are often called **primary afferents**.

As a given afferent nerve fibre often serves a number of receptors of the same kind, it will respond to a stimulus over a certain area of space and intensity that is called its **receptive field**. The concept of a receptive field can be applied to any modality of sensation. In the somatosensory system, the receptive fields of neighbouring afferents innervating the skin often overlap, as shown in Figure 12.7. The extent of the overlap determines how accurately the position of a stimulus can be determined (i.e. the spatial resolution of the particular group of receptors). In the ear, the receptive fields of the hair cells of the cochlea correspond to the specific frequencies to which they are sensitive. In this case, louder sounds (i.e. stronger stimuli) elicit responses in an overlapping population of hair cells (see Chapter 15). In the eye, a single bipolar cell may be connected to a small number of photoreceptors in the central part of the visual field, while in the peripheral field, a single bipolar cell may be connected to many photoreceptors and its receptive field will be correspondingly wide. (Bipolar cells are the retinal cells that receive a direct synaptic contact from their associated rods or cones.)

Individual nerve cells in the dorsal horn of the spinal cord may receive inputs from many primary afferent fibres, so that the receptive field of a particular sensory neuron is usually larger than that of the individual afferent fibres to which it is connected. This

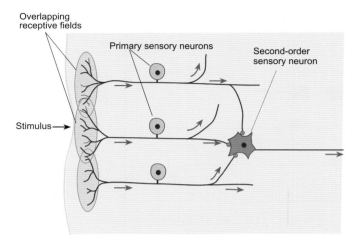

Figure 12.7 A simple diagram illustrating overlap between receptive fields of neighbouring primary afferents and their convergence onto a second-order sensory neuron. The primary afferents also send axon branches to other spinal neurons, as indicated by the arrows.

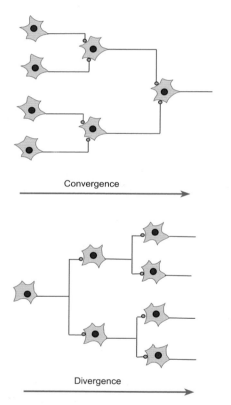

Convergence

Divergence

Figure 12.8 A diagram illustrating the principles of neural convergence (top) and divergence (bottom). See text for further details.

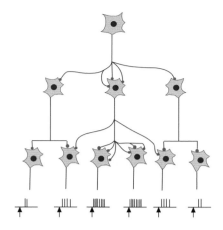

(a)   Wide spatial spread of activity

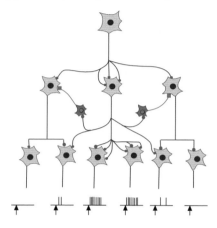

(b)   Narrow spread of activity due to lateral inhibition by interneurons

Figure 12.9 Lateral inhibition limits the spread of excitation within neural networks. (a) The spread of excitation across a population as a consequence of divergent neural connections. (b) This spread can be limited by the activity of inhibitory interneurons (shown in red). The black arrows indicate the timing of the stimulus.

neurons, and so on. The second-order and third-order neurons then make contact with many other nerve cells as information is processed by the CNS. Thus a specific piece of sensory information tends to be spread among more and more nerve cells as it is relayed towards the higher centres. This is called **divergence**—see Figure 12.8. Convergence and divergence are an essential part of the processing of sensory information. Ultimately, this information is incorporated into an internal representation of the world that the brain can use to determine appropriate patterns of behaviour.

To avoid the excessive spread of excitation in response to a single stimulus, nerve cells in the CNS receive inhibitory connections from neighbouring cells via small neurons (known as interneurons), as shown in Figure 12.9. In this way, a strongly

is called **convergence** and is illustrated in Figures 12.7 and 12.8. As the dorsal horn neurons receive information from the primary afferent fibres, they are celled second-order neurons (Figure 12.7); the neurons to which they relay information are third-order

excited cell will exert a powerful inhibitory effect on those of its neighbours that were less intensely excited. This is known as **lateral inhibition** or **surround inhibition**, and it features prominently in the processing of sensory information at all levels of the CNS. At the highest levels, in the cerebral cortex for example, this type of neural interaction allows the brain to extract information about the specific features of a stimulus. For example, the visual cortex is able to discriminate the position of an object in space, its illumination, and its relation to other nearby objects. All of this complex processing is achieved by the interplay of excitatory and inhibitory synaptic connections of the kind illustrated in Figure 12.9.

The need for transformation of stimulus intensity makes clear that the nervous system does not have a direct impression of the environment. Similarly, there is no one-to-one correspondence between a sensory surface such as the skin and its representation in the brain. The fingers have more of the cerebral cortex devoted to processing information from them than the whole of the forearm (see Figure 13.14). Similarly, the representation of the central visual field in the visual cortex is much greater than that of the peripheral regions. The areas of representation reflect their importance in resolving detail. It is known that certain nerve cells in the brain respond to specific details of an image (e.g. a moving edge),

and there is evidence that similar feature extraction occurs for sounds in the auditory cortex. Despite this, it is not known how these various transformations give rise to our conscious perceptions. However, it is clear from various illusions that an individual's perceptions are personal interpretations made in the light of previous experiences.

Summary

An individual sensory receptor responds to an area of sensory space known as its receptive field. As each afferent fibre often serves more than one receptor, the receptive fields of sensory fibres are usually greater than those of the individual receptors they serve. Moreover, the receptive fields of neighbouring fibres generally overlap. The degree of overlap determines the resolution of the system.

As sensory information passes through the nervous system it influences a progressively greater number of neurons via the process of divergence. The process of lateral inhibition restricts the spread of such divergence. Sensory information is relayed via ascending nerve tracts to the cerebral cortex, where it is used to form an internal representation of the world.

✻ Checklist of key terms and concepts

The common features of sensory receptors

- The external and internal environment is continuously monitored by the sensory receptors.

- Activation of a particular sensory receptor gives rise to a sensation that may vary in intensity but has the same overall quality.

- A group of similar sensations is called a sensory modality.

- Sensory receptors respond to four principal triggers: mechanical stimuli, chemical stimuli, temperature, and light.

- Each kind of receptor is excited most effectively by a specific type of stimulus known as its adequate stimulus.

Sensory transduction and adaptation

- Different kinds of receptor are activated in different ways, but the first stage in sensory transduction is the generation of a receptor potential (or generator potential).

- Receptor potentials progressively increase in intensity as the strength of the initiating stimulus increases. This is then encoded by the frequency of action potentials in the associated afferent nerve fibre.

- The process by which the intensity of an environmental stimulus becomes encoded as a sequence of nerve impulses in an afferent nerve fibre is called sensory transduction.

- Different types of receptor show differing degrees of adaptation. Some (the rapidly adapting receptors) signal the onset and offset of a stimulus, while others (the slowly adapting receptors) continuously signal the intensity of the stimulus.

Receptive fields and sensory resolution

- Receptors send their information to the CNS via afferent nerve fibres.

- Any given afferent nerve fibre responds to a stimulus over a certain area called its receptive field.

- To avoid the excessive spread of excitation in response to a single stimulus, nerve cells in the CNS receive inhibitory connections from neighbouring cells in a process called lateral inhibition, in which strongly excited cells exert a powerful inhibitory effect on those that were less intensely excited.

- The representation of a region of the external world in the brain reflects its importance in resolving detail.

Recommended reading

Barlow, H.B. (1982) General principles: The senses considered as physical instruments, in: *The senses*, Chapter 1. Cambridge University Press, Cambridge.

Carpenter, R., and Reddi, B. (2012) *Neurophysiology, a conceptual approach* (5th edn), Chapters 4–8. Hodder Arnold, London.

Review articles

Katta, S., Krieg, M., and Goodman, M.B. (2015) Feeling force: Physical and physiological principles enabling sensory mechanotransduction. *Annual Review of Cell and Developmental Biology* **31**, 347–37.

Proske, U., and Gandevia, S.C. (2012) The proprioceptive senses: Their roles in signaling body shape, body position and movement, and muscle force. *Physiological Reviews* **92**, 1651–97, doi:10.1152/p.

Wicher, D. (2010) Design principles of sensory receptors. *Frontiers in Cellular Neuroscience* **4**, 25, http://doi.org/10.3389/fncel.2010.00025.

To check that you have mastered the key concepts presented in this chapter, complete the accompanying online self-assessment questions. Go to www.oup.com/uk/pocock5e/

CHAPTER 13

The somatosensory system

This chapter should help you to understand:

- The physiological basis of skin sensation
  - Mechanoreceptors and their role in touch sensation
  - Active exploration of objects (haptic touch)
  - Kinaesthesia—position sense
  - Thermoreceptors and temperature sensation
- The role of visceral receptors in body sensation
- The nerve pathways subserving somatic sensations
- The role of the cerebral cortex in sensation
- The pathophysiology of pain
  - Classification of pain sensations
  - The properties of pain receptors (nociceptors)
  - The triple response to injury
  - The central pathways activated by pain sensations
  - Causes of visceral pain and aberrations of pain sensations
- The pathophysiology of itch

13.1 Introduction

The skin is the interface between the body and the outside world. It is not uniformly sensitive over its surface. Instead, the distribution of the various sensations is punctate: specific points of the skin are sensitive to touch, others are sensitive to cooling, warming, or noxious stimuli. There is little overlap between the different modalities of cutaneous sensation. In addition to the skin, the muscles and joints also possess sensory receptors that provide information concerning the disposition and movement of the limbs. All of this information is relayed to the spinal cord and brain by the afferent nerves of the somatosensory system.

Some of the receptors present in smooth (or glabrous) and in hairy skin are illustrated in Figure 13.1. Both bare nerve endings and encapsulated receptors are present. This prompts three key questions: Why are there so many different kinds of receptor? What are their functions? How do the various receptors convert the different kinds of stimuli into nerve impulses?

To answer these questions, it is necessary to examine the responses of particular receptors to specific stimuli. Ideally, it should also be possible to correlate the nerve activity with the subjective experience elicited by those stimuli. This latter requirement is clearly impossible in animal experiments, but in 1968 Hagbarth and Valbö introduced the technique of microneurography. This involved recording the activity of afferent fibres in human volunteers (often the experimenters themselves) by inserting fine metal electrodes into the peripheral nerves serving a particular area (Figure 13.2). The activity of a specific sensory nerve fibre could then be recorded, its receptive field mapped, and its sensitivity to particular stimuli established. By stimulating the afferent fibre and asking the subject to identify the receptive field and the nature of the sensation they experience, it is possible to determine which receptors respond to particular kinds of stimuli. For example, by identifying a point sensitive to touch, then cutting it out and subjecting it to histological examination, it has been possible to associate particular types of receptor with specific modalities of sensation and their physiological properties (i.e. the particular stimulus required to excite a specific type of receptor and whether it responds to constant or changing (dynamic) stimuli).

These studies have shown that each kind of sensory receptor subserves a specific sub-modality of skin sensation. Furthermore, the structure of these receptors permits the nervous system to discriminate the specific features of a stimulus. Each kind of receptor is innervated by a particular type of nerve fibre. For example, Pacinian corpuscles are innervated by relatively large Aβ myelinated nerve fibres, while the bare nerve endings that subserve temperature and pain sensations are derived either

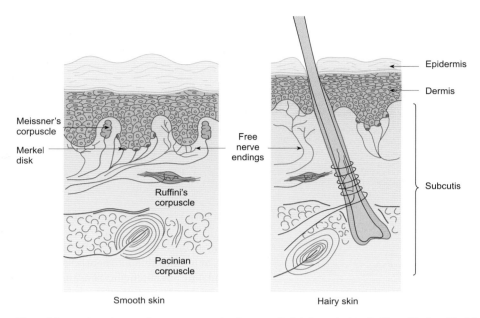

Figure 13.1 The disposition of the various types of sensory receptor in smooth (glabrous) skin (left) and hairy skin (right). Note the thickness of the epidermis in glabrous skin and the location of the Meissner's corpuscles between the dermal ridges. (Based on Fig 6.1 in P. Brodal (2004) *The central nervous system: Structure and function*, 3rd edition. Oxford University Press, New York.)

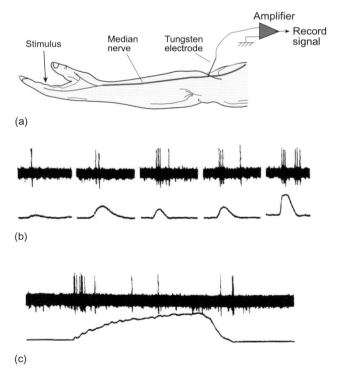

Figure 13.2 The principle of microneurography. (a) A tungsten wire electrode is introduced through the skin into a peripheral nerve (here the median nerve of the forearm) to record action potentials from an individual nerve fibre. The hand and arm are then explored with various stimuli until the receptor served by the nerve fibre has been identified. The characteristics of the receptor and the extent of its receptive field can then be examined. (b) The response of a rapidly adapting receptor to a series of brief touch stimuli. Note how the fibre responds with a short burst of action potentials at the beginning of each stimulus. A strong stimulus also elicits a response as the stimulus ends. (c) The response of the same receptor to a maintained stimulus. Once again, note the burst of action potentials at the onset and termination of the stimulus. (Panels (b) and (c) adapted from Fig 4 in A.B. Vallbo and K.E. Hagbarth (1968) *Experimental Neurology* **21**, 271–89.)

Table 13.1 The receptor types and modalities of the somatosensory system

Modality	Receptor type	Afferent nerve fibre type and conduction velocity
Touch	Rapidly adapting mechanoreceptors: hair follicle receptors, bare nerve endings, Meissner's corpuscles, Pacinian corpuscles	Aβ 6–12 μm diameter 33–75 m sec⁻¹
Touch and pressure	Slowly adapting mechanoreceptors, e.g. Merkel's cells, Ruffini end organs	Aβ 6–12 μm diameter 33–75 m sec⁻¹
	Bare nerve endings	Aδ 1–5 μm diameter 5–30 m sec⁻¹
Vibration	Meissner's corpuscles Pacinian corpuscles	Aβ 6–12 μm diameter 33–75 m sec⁻¹
Temperature	Cold receptors (bare nerve endings)	Aδ 1–5 μm diameter 5–30 m sec⁻¹
	Warm receptors (bare nerve endings)	C fibres 0.2–1.5 μm diameter 0.5–2.0 m sec⁻¹
Pain	Bare nerve endings— fast 'pricking' pain	Aδ 1–5 μm diameter 5–30 m sec⁻¹
	Bare nerve endings— slow burning pain	C fibres 0.2–1.5 μm diameter 0.5–2.0 m sec⁻¹
Itch	Bare nerve endings	C fibres 0.2–1.5 μm diameter 0.5–2.0 m sec⁻¹

from small Aδ myelinated fibres or from slowly conducting unmyelinated C-fibres (Table 13.1). (For the classification of nerve fibres, see Chapter 7.)

13.2 Skin mechanoreceptors

Receptors sensitive to touch respond to the physical deformation of the skin or a change in the position of hairs. They are called **mechanoreceptors**. They are of two main types: low-threshold mechanoreceptors (which mediate the sense of touch) and high-threshold mechanoreceptors (which mediate the sensations of pain). The low threshold receptors may be further classified as rapidly adapting receptors and slowly adapting receptors. Rapidly adapting receptors are those associated with hair follicles, the end

bulbs of Krause, Meissner's corpuscles, and Pacinian corpuscles. (The end bulbs of Krause were once, wrongly, thought to be cold receptors, but modern studies have proved that they are, in fact, mechanoreceptors.) The slowly adapting receptors include Ruffini endings, the nerve endings associated with Merkel cells, and mechanosensitive free nerve endings (C-mechanoreceptors).

Sensitivity to touch

The sensitivity of the skin to touch can be assessed by measuring the smallest indentation that can be detected, either subjectively or by recording the action potentials in a single afferent fibre. For the fingertips, an indentation of as little as 6–7 μm can be detected, which corresponds to the diameter of a single red cell. Elsewhere on the hand, the skin is less sensitive to deformation so that, for example, an indentation of about 20 μm is required to evoke an action potential in an afferent fibre serving the sense of touch on the palm. The skin of the back or that of the soles of the feet is even less sensitive to touch.

Encapsulated receptors

The skin possesses a variety of elaborate sensory receptors including the end bulbs of Krause, Meissner's corpuscles, Pacinian corpuscles, and Ruffini corpuscles. In each case, the sensory nerve ending is encapsulated by an organized structure composed of fibroblasts or other non-neuronal cells.

What is the role of these associated structures? This important question was first tackled by examining the role of the layers of fibroblasts that surround a Pacinian corpuscle (see Figure 13.3a). An intact Pacinian corpuscle responds to a maintained deformation of the skin with a brief depolarization when the stimulus is first applied and when it is subsequently removed ('on' and 'off' responses). If the capsule is treated with enzymes to remove the associated fibroblasts, the naked nerve ending is left intact and responds to a mechanical stimulus with a depolarization that is sustained for as long as the mechanical stimulus is applied (see Figure 13.3b). Thus, the fibroblast layers are a structural feature that allows the intact corpuscle to respond to rapid tissue movement (such as vibration) rather than maintained pressure.

The Meissner corpuscles are rapidly adapting touch receptors found in the dermal ridges of smooth (glabrous) human skin (see Figure 13.1). The capsule is formed of layers of non-neuronal cells that surround the nerve ending which is arranged in a helical sheet that is at right angles to the long axis of the corpuscle. Collagen fibres connect the most superficial part of the corpuscle to the epidermal cells surrounding the dermal ridge, forming a mechanical arrangement that efficiently transmits rapid movements of the epidermis to the sensory nerve endings. Similar structural specializations are seen in other encapsulated receptors.

The role of accessory cells

Some touch-sensitive nerve endings in the skin are closely associated with specialized epithelial cells known as Merkel cells (see Figure 13.4), which are attached to neighbouring epidermal cells in

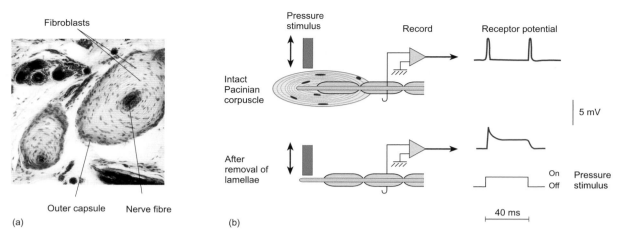

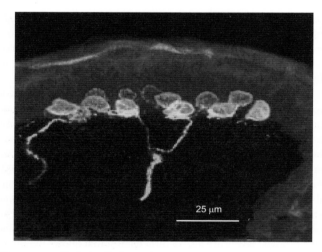

Figure 13.3 The role of the fibroblast lamellae in shaping the response of the Pacinian corpuscle to a pressure stimulus. (a) A cross-section of two Pacinian corpuscles showing the outer capsule and the layers of flattened fibroblasts surrounding the afferent nerve fibres. (b) The response of a Pacinian corpuscle before and after the removal of the layers of fibroblasts. The experimental arrangement is shown on the left and the receptor potentials are shown on the right. Note that the intact corpuscle only signals the onset and offset of the stimulus, while the desheathed corpuscle remains depolarized for the duration of the stimulus. (Panel (b) is based on data in W.R. Lowenstein and M. Mendelson (1965) *Journal of Physiology* **177**, 377–97.)

Figure 13.4 Merkel cells (green) and their associated nerve afferents (orange) in a touch dome of mouse skin. The overlying epidermis is visible as a pale green area above the Merkel cells. (From Fig 1B in S. Maksimovic, Y. Baba, and E.A. Lumpkin (2013) *Annals of the New York Academy of Sciences* **1279**, 13–21. Reproduced with permission of John Wiley and Sons.)

such a way that any movement of the skin will subject them to a certain degree of mechanical stress. Functionally, the Merkel cells are accessory cells associated with the fine afferent fibres that act as slowly adapting receptors and respond to gentle touch. When the skin is stretched, the Merkel cells become depolarized and trigger action potentials in the fine nerve branches with which they are associated. It appears that the nature of the contact between the nerve ending and a Merkel cell is similar to that of a synaptic contact, with the Merkel cell acting as the presynaptic element. When they are subjected to mechanical stress, mechanosensitive ion channels known as PIEZO2 channels open, causing the Merkel

cells to depolarize. (The PIEZO2 channels are non-selective cation channels.) The depolarization of the Merkel cell leads to the release of a neurotransmitter onto its associated nerve ending, which initiates a series of action potentials. It is likely that the nerve endings themselves also contain mechanosensitive ion channels.

Receptive fields and two-point discrimination

In general, the receptive fields of touch receptors overlap considerably, as illustrated in Figure 13.5. The finer the discrimination required, the higher the density of receptors, the smaller their receptive fields, and the greater the degree of overlap. The receptive fields for touch are particularly small at the tips of the fingers and tongue (about 1 mm^2) where fine tactile discrimination is required (Figure 13.6). In other areas such as the small of the back, the buttocks, and the calf, the receptive fields are much larger.

The distance between two points on the skin that can just be detected as separate stimuli is closely allied to the density of touch receptors and the size of their receptive fields. This is known as the two-point discrimination threshold. Not surprisingly, the tips of the fingers, the tip of the tongue, and the lips are able to discriminate between points very close together. The two-point discrimination shown by the skin of the abdomen, neck, and chest is much less precise (see Figure 13.7). A loss of precision in two-point discrimination can be used to help localize specific neurological lesions.

Haptic touch

People move through their environment; they lift objects, rearrange them, and feel their texture. The constant stimulation of different receptors prevents them adapting. As a result, the brain is provided with more information about an object than would be possible with a single contact. The active exploration of an object

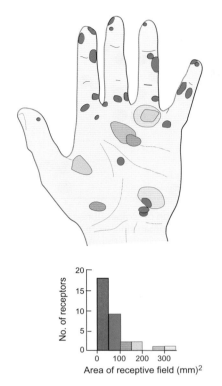

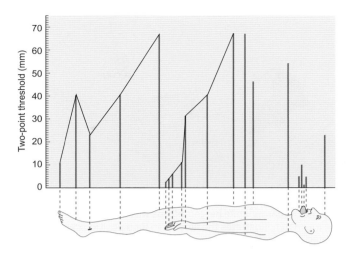

Figure 13.7 The variation of two-point discrimination across the body surface. Each vertical line represents the minimum distance that two points can be distinguished as being separate when they are simultaneously stimulated. Note the fine discrimination achieved by the fingertips, lips, and tongue, and compare this to the poor discrimination on the thigh, chest, and neck.

Figure 13.5 Receptive field sizes and distribution for slowly adapting touch receptors in the left hand of a single subject. Note that the small receptive fields are found mainly on the fingers and thumb. Larger receptive fields are located on the palm with considerable overlap between different receptive fields. (Adapted from Fig 1B in M. Knibestol and A.B. Vallbo (1970) *Acta Physiologica Scandinavica* 80, 178–95.)

to determine its shape and texture is known as **haptic touch**, which is used to great effect by the blind.

Kinaesthesia

We know the position of our limbs even when we are blindfold. This sense is called **kinaesthesia**. Two sources of information provide the brain with information about the disposition of the limbs. These are the corollary discharge of the motor efferents to the sensory cortex (see Chapter 10), which provides information about the intended movement, and the sensory feedback that directly informs the sensory cortex of the actual progress of the movement. As the load on the muscles moving the limbs cannot be known by the brain in advance, both of these sources of information are required (see Chapter 10).

The importance of the muscle spindles in kinaesthesia is revealed by the vibration illusion. A normal subject can accurately replicate the position of one arm by moving the other to the same angle. The application of vibration to a muscle such as the biceps can give the illusion that the arm has been moved. If both arms are initially placed at 90° to the horizontal and the left biceps is then vibrated, a blindfolded subject will have the illusion that the arm has moved and will alter the position of the right arm to report the perceived (but incorrect) position of the left arm (see Figure 13.8). This occurs because the vibration stimulates the stretch receptors. The CNS interprets the increased afferent discharge as indicating that the muscle is longer than it actually is.

Visceral receptors

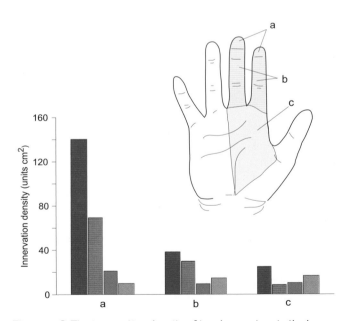

Figure 13.6 The innervation density of touch receptors in the human hand. The greatest density is found in the tips of the fingers (a), while the lowest is in the palm (c). This correlated closely with the two-point discrimination in the same subject. (Based on data of R.S. Johansson and A.B. Vallbo (1983) *Trends in Neuroscience* 6, 27–32.)

The internal organs are much less well innervated than the skin. Nevertheless, all of the internal organs have an afferent innervation, although the activity of these afferents rarely reaches consciousness except as a vague sense of 'fullness' or as pain (see Section 13.5).

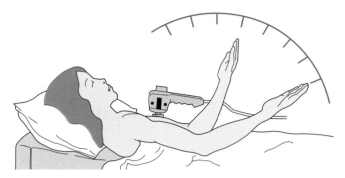

Figure 13.8 An illustration of limb position mismatch induced by vibratory stimulation. The blindfolded subject was asked to align her forearms while her left biceps was being vibrated. The mismatch illustrates the illusion in limb position sense elicited by the vibration.

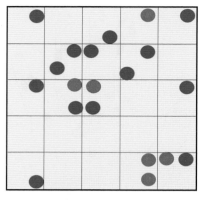

Figure 13.9 Distribution of temperature-sensitive spots in a square centimetre of human skin. Note the different distribution of those spots sensitive to cooling (blue) and those sensitive to warming (red). There is no overlap. (Based on data in K.M. Dallenbach (1927) *American Journal of Psychology* **39**, 402–17.)

The afferent fibres reach the spinal cord by way of the visceral nerves which carry the sympathetic and parasympathetic fibres that provide motor innervation to the viscera (see Chapter 11).

The visceral receptors include both rapidly adapting and slowly adapting mechanoreceptors and chemoreceptors. Many of these afferents are an essential component of the visceral reflexes that control vital body functions. Examples are the baroreceptors of the aortic arch and carotid sinus, which monitor arterial blood pressure, and the chemoreceptors of the carotid bodies which detect the PO_2, PCO_2, and pH of the arterial blood.

Summary

The somatosensory system provides the CNS with information concerning touch, peripheral temperature, limb position (kinaesthesia), and tissue damage (nociception). Individual skin receptors respond to stimuli applied to specific areas of the body surface, known as their receptive field. The position, size, and degree of overlap of the receptive fields permit the localization of a stimulus.

13.3 Skin thermoreceptors

Exploration of the surface of the skin with small metal probes maintained at various temperatures reveals that that the skin is not uniformly sensitive to temperature. Temperature-sensitive spots are scattered over the surface in an apparently random distribution (Figure 13.9). Such experiments also show that the skin has at least two kinds of thermoreceptor: one type specifically responds to cooling of the skin, another responds to warming. The receptive fields of these temperature-sensitive spots are small—less than 1 mm²—and do not appear to overlap. Histological examination of a cold- or warm-sensitive point shows no encapsulated receptors, only bare nerve endings, which are therefore considered to be the cutaneous thermoreceptors.

Cutaneous thermoreceptors are generally insensitive to mild mechanical and chemical stimuli, and they maintain a constant

rate of discharge for a particular skin temperature. The cold receptors are innervated by Aδ myelinated afferents while the warm receptors are innervated by C-fibres. It is now known that specific ion channels of the TRP family are activated by small changes in temperature (see Box 13.1).

The cold receptors may respond to a change in temperature either with an increase or with a decrease in firing rate, and have a maximal rate of discharge around 25–30°C. The warm receptors have a maximal rate of discharge around 40°C (see Figure 13.10).

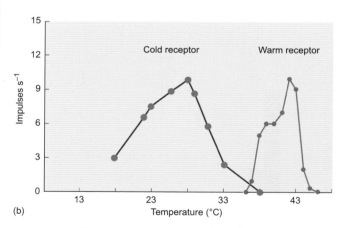

Figure 13.10 Typical response patterns of a warm and a cold skin thermoreceptor. (a) The action potential discharge of a cold receptor is seen to increase as the skin is cooled from 28° to 23°C. (b) The rate of action potential discharge as a function of temperature for a single cold and a single warm thermoreceptor. (Based on data in Fig 17.6 of A. Iggo (1982) *The senses* (eds H.B. Barlow and J.D. Mollon). Cambridge University Press, Cambridge, Chapter 17.)

Box 13.1 TRP channels and thermosensation

It has been known for a long time that people can sense temperature at two levels—simple awareness, which is termed **innocuous** hot or cold, and a painful sensation, which is termed **noxious** hot or cold. Most people have an acute ability to detect small changes in temperature that are close to body temperature—in some studies, they have been found to reliably detect cooling by as little as 1°C. It is now known that most of this exquisite temperature sensitivity is due to ion channels of the transient receptor potential (TRP) family.

TRP channels are a diverse superfamily of calcium-permeable ion channels. Those responsible for temperature sensation fall into the TRPV, TRPM, and TRPA families (see Figure 1). The TRPV1 channel was the first thermo-sensing channel to be identified and was shown to detect noxious heat (in the range of 42–48°C). An interesting side note to this research is that we feel spicy food to be 'hot', because the TRPV1 channel also responds to capsaicin, the chemical in chilli peppers that makes them feel hot—a curry feels hot because it activates the same receptors as heat does. Different members of the TRPV family detect different heat ranges. TRPV2

detects very high (i.e. painful) skin temperatures (up to 52°C), while TRPV3 and TRP4 detect warmth (temperatures close to normal skin temperatures, 27 to 42°C).

Detection of cool and cold temperatures relies on other TRP channels: TRPM8 is activated by modest cooling (> 25°C—see Figure 1a), and TRPA1 is activated by noxious (painful) cold (< 15°C.) These two have chemical counterparts that we encounter every day. The TRPA1 channel (also called ANKTM) is sensitive to allyl isothiocyanate, the component of mustard and horseradish that gives them their bite. TRPM8 is responsive to menthol, explaining the cooling sensation caused by applying menthol to the skin.

All of these channels act by controlling the entry of calcium ions into the cell, which seems contradictory until you realize that their expression is tightly controlled. Specific sensory nerve endings are responsible for particular elements of temperature sensation; different nerve endings detect different temperature ranges. The activations of these neurons is then integrated by the nervous system to provide our overall sensation of cold to warm to hot.

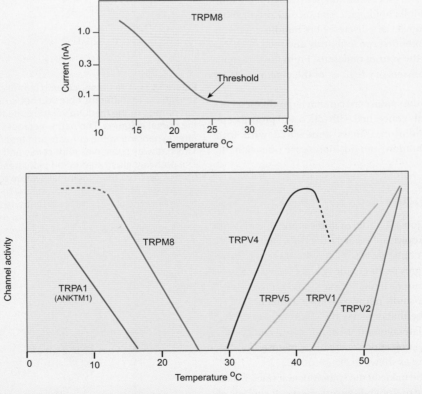

Figure 1 The top panel shows the response of a cultured Chinese hamster ovary cell (a CHO cell) that has been transfected with the DNA for the TRPM8 gene. Cooling the cell below 25°C progressively activates the TRPM8 channels, as shown by the steadily rising membrane current. The lower panel shows how temperature changes activate different members of the TRP channel family. Noxious heat activates TRPV1 and TRPV2 over somewhat different temperature ranges. TRPV4 is activated by warm temperatures and appears to be inhibited by excessive heat (> 40°C), while TRPM8 and ANKTM1 are activated by cooling. Additional channels may also be involved in thermosensation. (Based on Fig 3b in A. Patapoutian et al. (2003) *Nature Reviews Neuroscience* 4, 529–39.)

Thus, a given frequency of discharge from the cold receptors may reflect a temperature that is either above or below the maximum firing rate for a particular receptor. This ambiguity helps to explain the well-known paradoxical sense of cooling when cold hands are being rapidly warmed by immersion in hot water.

Summary

The skin has receptors that respond to changes in temperature. Two kinds of thermoreceptor have been identified: those that respond to cooling of the skin, which are nerve endings of small myelinated (Aδ) nerve fibres, and those that respond to warming of the skin, which are nerve endings of unmyelinated C-fibres.

13.4 Somatosensory pathways

The dorsal column pathway

The cutaneous and visceral afferent fibres enter the spinal cord via the dorsal (posterior) roots. The large-diameter afferents branch after they have entered the spinal cord and travel in the dorsal columns to synapse in the dorsal column nuclei (the cuneate and gracile nuclei) of the medulla oblongata. The second-order fibres leave the dorsal column nuclei as a discrete fibre bundle called the medial lemniscus. The fibres first run anteriorly and then cross the midline before reaching the ventral thalamus. From the thalamus they project to the somatosensory regions of the cerebral cortex (see Figure 13.11).

The dorsal columns of the spinal cord contain large-diameter afferent fibres that are mainly concerned with touch and proprioception. They relay information that is concerned with fine discriminatory touch, vibration, and position sense (kinaesthetic information).

The spinothalamic pathway

The small afferent fibres join a bundle of fibres at the dorsolateral margin of the spinal cord known as Lissauer's tract (see Figure 13.12). These thin afferent fibres only travel for a few spinal segments at most before entering the grey matter of the spinal cord, where they synapse on spinal interneurons in the substantia gelatinosa. These interneurons then synapse on neurons whose axons cross the midline and project to the thalamus via the spinothalamic tract, which runs in the anterolateral quadrant of the spinal cord. From the thalamus, the sensory projections reach the somatosensory regions of the cerebral cortex, as shown in Figure 13.13.

Finally, mention must be made of the spinoreticular tract, which receives afferents from the peripheral sensory nerves. It consists of a chain of short fibres that synapse many times as they ascend in the anterolateral spinal cord, and it terminates in various nuclei on the same side of the reticular formation of the brainstem. The spinothalamic and spinoreticular tracts receive information from the smaller afferent fibres and transmit information concerning crude touch, temperature, and pain to the brain (see Table 12.1).

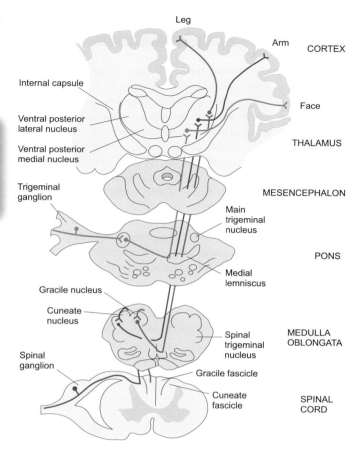

Figure 13.11 The dorsal-column lemniscal pathway for cutaneous sensation. The afferent fibres do not cross at the segmental level but project to the medulla via the cuneate and gracile fascicles of the dorsal columns. The pathway crosses to the contralateral side after synapsing in the cuneate and gracile nuclei of the medulla. This pathway subserves joint sense and fine discriminatory touch. The pathway illustrated in red indicates the route that sensory fibres take from the trigeminal system to reach the cerebral cortex. The mauve and green pathways show the routes taken by sensory fibres from the upper and lower limbs, respectively. (Based on Fig 6.17 in P. Brodal (2004) *The central nervous system: Structure and function*, 3rd edition. Oxford University Press, New York.)

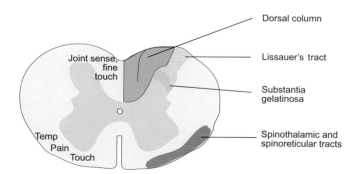

Figure 13.12 Diagrammatic representation of the spinal cord, showing the positions of the main sensory tracts and the substantia gelatinosa of the dorsal horn.

and project to the intralaminar nuclei and the posterior thalamus. The sensory cells of the thalamus project to two specific regions of the cerebral cortex, the primary and secondary sensory cortex, and the limbic system.

Somatosensory information is represented on the postcentral gyrus of the cerebral cortex

The exploration of the human somatosensory cortex by Wilder Penfield and his colleagues in the 1930s is one of the most remarkable investigations in neurology. While treating patients for epilepsy, Penfield attempted to localize the site of the lesions responsible for the condition by electrically stimulating the cerebral cortex of his patients. During this procedure, the cut edges of the scalp, skull, and dura were infiltrated with local anaesthetic but the patients remained conscious. Since the brain has no nociceptive fibres, this procedure did not elicit pain. However, it did elicit specific sensations which the patients were able to report. Systematic exploration of the **postcentral gyrus** revealed that it was organized somatotopically, with the different regions of the opposite side of the body represented as shown in Figure 13.14. The genitalia and feet are mapped on to an area adjacent to the central fissure, while the face, tongue, and lips are mapped on the lateral aspect of the postcentral gyrus; other areas of the body are mapped in between.

The extent of each area of representation reflects the degree of importance of that area in sensation rather than the area of the

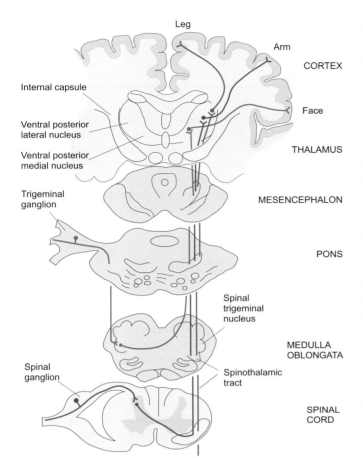

Figure 13.13 The spinothalamic tract and its projection to the cerebral cortex. The afferents for this pathway synapse at the segmental level and cross to the contralateral side before projecting to the thalamus. The spinothalamic tract sends information concerning coarse touch, pain, and temperature to the brain. The colour coding of the sensory pathways is as in Figure 13.11. (Based on Fig 6.18 in P. Brodal (2004) *The central nervous system: Structure and function*, 3rd edition. Oxford University Press, New York.)

The afferent pathways have a somatotopic organization

All of the afferents of a particular type that enter one dorsal root tend to run together in the lower regions of the spinal cord. Initially this segmental organization is preserved but, as they ascend, the fibres from the different segments become rearranged so that those from the leg run together as do those of the trunk, hand, and so on. Thus, for example, the afferents from the hand project to cells in a particular part of the dorsal column nuclei and thalamus, while those from the forearm project to an adjacent group of cells. This orderly arrangement provides topographical maps of the body in those parts of the brain that are responsible for integrating information from the different sensory receptors.

Although the cells of the dorsal column nuclei project chiefly to the specific sensory nuclei in the ventrobasal region of the thalamus, those from the spinothalamic tract have a wider distribution

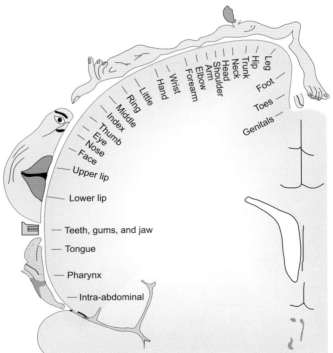

Figure 13.14 The representation of the body surface on the postcentral gyrus revealed by electrical stimulation of the cerebral cortex of conscious subjects. (Based on Fig 6.24 in P. Brodal (2004) *The central nervous system: Structure and function*, 3rd edition. Oxford University Press, New York.)

body surface from which a sensation has come. Thus the hands, lips, and tongue all have a relatively large area of cortex devoted to them when compared to the areas devoted to the legs, upper arm, and back. The body is also mapped on to the superior wall of the Sylvian gyrus. This area is called the SII region. Unlike the postcentral gyrus (the SI region), the SII region has a representation of both sides of the body surface.

The trigeminal system

The sensory inputs from the face are relayed to the brain via the fifth cranial nerve, the trigeminal nerve. The trigeminal nerve is mixed, having both somatic afferent and efferent fibres, although the afferent innervation dominates. It arises in the pontine region of the brainstem (see Figure 9.8) and shortly after its origin, it expands to form the semilunar ganglion. This contains the primary sensory neurons, which are analogous to the dorsal root ganglion neurons of the spinal cord. Three large nerves leave the ganglion to innervate the face. These are the opthalmic, maxillary, and mandibular nerves, which relay information to the brain concerning touch, temperature, and pain from the face (see Figure 13.15). They also relay information from the mucous membranes and the teeth. The large afferent fibres transmit information from the mechanoreceptors to the thalamus via the medial lemniscus (see Figure 13.11), while the Aδ and C-fibre afferents (which are mainly concerned with temperature and nociception) join the spinothalamic tract, as

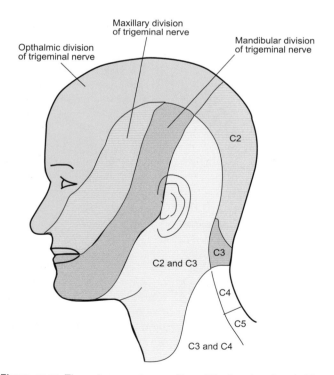

Figure 13.15 The cutaneous innervation of the head and neck. The three principal divisions of the trigeminal nerve innervate the face, while the upper cervical spinal nerves innervate the back of the head and the neck.

shown in Figure 13.13. From the thalamus, information from both large and small afferents is transmitted to the face region of the primary sensory cortex.

Summary

Information from the somatosensory receptors reaches the cerebral cortex by way of the dorsal columns and spinothalamic tract. The dorsal column pathway is primarily concerned with fine discriminatory touch and position sense, while the spinothalamic and spinoreticular tracts are concerned with crude touch, temperature, and pain. The postcentral gyrus of the cerebral cortex possesses a topographical map of the contralateral surface of the body.

13.5 The sense of pain

Pain is an unpleasant experience associated with acute tissue damage. It is the main sensation experienced by most people following injury. It also accompanies certain organic diseases such as advanced cancer. The weight of evidence now suggests that pain is conveyed by specific sets of afferent nerve fibres and is not simply the result of a massive stimulation of afferent fibres. Nevertheless, pain may arise spontaneously without an apparent organic cause or in response to an earlier injury, long since healed. This kind of pain (called neuropathic pain) often has its origin within the CNS itself. Although it is not obviously associated with tissue damage, pain of central origin is no less real to the sufferer. Unlike most other sensory modalities, pain is almost invariably accompanied by an emotional reaction of some kind such as fear or anxiety (see Chapter 18). If it is intense, pain elicits autonomic responses such as sweating, pallor, and increases in blood pressure and heart rate.

Pain that is short-lasting and directly related to the injury that elicited it is known as **acute pain**. Pain that persists for many days or even months is known as **chronic pain**.

Pain can be classified based on its underlying physiology (nociceptive, neuropathic, or inflammatory), its intensity, its duration (whether it is short- or long-lasting), or its site of origin (superficial, deep somatic, or visceral). Clearly there is some significant overlap between these classifications. Here the discussion will be based on the classification of pain according to its specific characteristics, which provides some insight into the underlying processes:

- Pricking pain is sometimes called 'first' or 'fast' pain and is transmitted to the CNS via small myelinated Aδ fibres. It is rapidly appreciated and accurately localized, but it elicits little by way of autonomic responses. This kind of pain is usually transient and, as its name implies, has a sharp pricking quality. In normal physiology, it serves an important protective function as activation of these fibres triggers the reflex withdrawal of the affected region of the body from the source of injury.

- Burning pain is of slower onset and greater persistence than fast pain. It reaches the CNS via non-myelinated C-fibres and is sometimes called 'second' or 'slow' pain. It is more intense and less easy to endure than pricking pain. This kind of pain has a diffuse quality, is difficult to localize, and readily evokes autonomic responses causing an increase in heart rate and blood pressure, dilation of the pupils, and sweating (see Chapter 11). During severe episodes of burning pain, there are changes to the pattern of breathing, with rapid shallow breaths that may be interrupted with periods of apnoea (breath holding). The difference in quality between fast pricking pain and slow burning pain will be known to anyone who has inadvertently stepped into an excessively hot bath: the initial response is to withdraw the affected foot quickly, but an intense burning pain slowly develops which is also slow to subside.

- Deep pain. This arises when deep structures such as muscles or visceral organs are diseased or injured. Deep pain has an unpleasant, aching quality, sometimes with the additional feeling of burning. Like burning pain it evokes autonomic responses, particularly an increased heart rate, elevated blood pressure, dilation of the pupils, and sweating. As with burning pain, severe episodes alter the pattern of breathing, which becomes irregular with rapid shallow breaths and periods of apnoea. Deep pain is usually difficult to localize and, when it arises from visceral organs, it may be felt at a site other than that at which it originates. This is known as **referred pain** (which is discussed later in this section).

Nociceptors are activated by specific substances released from damaged tissue

As with the sense of touch and temperature, the distribution of the pain receptors in the skin is punctate. Histological examination of a pain spot reveals a dense innervation with bare nerve endings, which are believed to be the nociceptors. The adequate stimulus for the nociceptors is not known with certainty, but the application of pain-provoking stimuli such as radiant heat elicits reddening of the skin and other inflammatory changes. It is probable that a number of chemical agents called **pain-producing substances** (**algogens**) are released following injury to the skin and cause the pain endings to discharge. These include ATP, bradykinin, histamine, serotonin (5-HT), hydrogen ions, and a number of inflammatory mediators such as prostaglandins.

The triple response

If a small area of skin is injured, for example by a burn, there is an inflammatory response with local vasodilatation causing a reddening of the skin. This is followed by a swelling (a weal or wheal) that is localized to the site of the injury and its immediate surroundings. The original site of injury is then surrounded by a much wider area of less intense vasodilatation known as the 'flare'. The local reddening ('red reaction'), flare, and weal formation comprise the **triple response** that was first described by Thomas Lewis. The weal is a local oedema caused by the accumulation of fluid in the damaged area. The red reaction is due to arteriolar dilatation in response to vasodilator substances released from the damaged skin, such as histamine, while the flare is due to dilatation of arterioles in the area surrounding the site of injury.

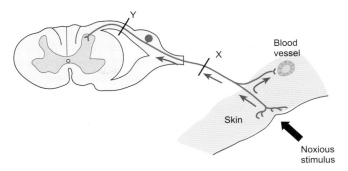

Figure 13.16 A schematic diagram of the axon reflex that gives rise to the flare of the triple response. Nociceptive fibres from the skin branch and send collateral fibres to nearby blood vessels. If the skin is injured, action potentials pass to the spinal cord via the dorsal root and to the axon collaterals, where they cause a vasodilatation that gives rise to the flare. If the nerve trunk is cut at point 'X' and sufficient time is allowed for the nerve to degenerate, the flare reaction is abolished. If the cut is made beyond the dorsal root ganglion at point 'Y', the segmental spinal reflexes are abolished but the flare reaction is not.

Within the injured area and across the surrounding weal, the sensitivity to mildly painful stimuli such as a pinprick is much greater than before the injury. This is known as **primary hyperalgesia**, which may persist for many days. In the region covered by the flare, outside the area of tissue damage, there is also an increased sensitivity to pain, which may last for some hours. This is known as **secondary hyperalgesia**.

Unlike the weal and local red reaction, the flare is abolished by infiltration of the skin with local anaesthetic. However, it is not blocked if the nerve trunk supplying the affected region is anaesthetized. Moreover, if the nerve trunk is cut (e.g. at point X in Figure 13.16) and allowed to degenerate before eliciting the triple response, only the local reddening and weal formation occur; the flare is absent. The triple response can be elicited in animals that have had complete removal of their sympathetic innervation. Furthermore, cutting the nerve supply to the dorsal root of the appropriate segment (point Y of Figure 13.16) does not block the flare reaction. These experiments show that the flare is a local **axon reflex** rather than a reflex vasodilatation involving the spinal cord or brainstem.

CNS pathways in pain perception

Nociceptive fibres are a specific set of small-diameter dorsal root afferents that subserve pain sensation in a particular region. These fibres enter Lissauer's tract and synapse in the substantia gelatinosa close to their site of entry into the spinal cord (see Figure 13.12). The second-order fibres cross the midline and ascend to the brainstem reticular formation and thalamus via the spinoreticular and spinothalamic tracts. The fibres of the spinoreticular tract may be concerned with cortical arousal mechanisms and with eliciting the defence reaction (see Chapters 11 and 30).

Although the neurons of the ventrobasal thalamus project to the primary sensory cortex, there is no compelling evidence that noxious stimuli evoke neural responses in this region. However, electrical stimulation of the posterior thalamic nuclei in conscious human subjects does elicit pain. Moreover, degenerative lesions of the posterior thalamic nuclei can give rise to a severe and intractable pain of central origin known as **thalamic pain**. It is known that the neurons of the posterior thalamus project to the secondary sensory cortex on the upper wall of the lateral fissure. Electrical stimulation of the white matter immediately beneath the secondary somatosensory cortex (SII) also elicits the sensation of pain. This suggests that pain sensations are relayed to the SII region. Other areas that appear to be involved in the emotional response to pain experience include the reticular formation, the structures of the limbic system such as the amygdala, and the frontal cortex (see Chapter 18).

The perception of pain may be greatly modified by circumstances. It is widely known that the pain from a bruise can be relieved by vigorous rubbing of the skin in the affected area. Pain can also be relieved by the electrical stimulation of the peripheral nerves. In both cases, the large-diameter afferent fibres are activated. This activation appears to inhibit the transmission of pain signals by the small unmyelinated nociceptive afferents. This inhibition occurs at the segmental level in the local networks of the spinal cord and prevents the onward transmission of pain signals to the brain. Transcutaneous electrical nerve stimulation (TENS) is now frequently used to control pain during childbirth and for some other conditions where prolonged use of powerful pain-suppressing drugs is undesirable.

Stressful situations can enhance the emotional component of pain (its effect—see Chapter 18) but can also produce a profound loss of sensation to pain (analgesia). It has been known for many years that soldiers with very severe battlefield injuries are often surprisingly free of pain. This suggests that the brain is able to control the level of pain in some way. Evidence for this view has come from experiments in which specific brain regions of conscious animals were stimulated electrically. When the central mass of grey matter surrounding the aqueduct was stimulated, the animals were almost immune to pain. Fibres from this region (known as the periaqueductal grey matter) project to the spinal cord via a series of small nuclei in the brainstem called the raphe nuclei. Axons from the largest of these, the raphe magnus nucleus, terminate on neurons in the dorsal horn of the spinal cord, as shown in Figure 13.17, and inhibit the transmission of pain to the brain. The **analgesic** ('pain-suppressing') effect of stimulation of these brain regions has subsequently been demonstrated in humans.

The brain and spinal cord synthesize a number of peptides that have a powerful analgesic effect: the **enkephalins**, the **dynorphins**, and the **endorphins**. In animal studies, injection of these peptides into certain areas of the brain such as the periaqueductal grey matter or the spinal cord has a powerful analgesic effect. β-endorphin is also secreted by the anterior hypothalamus and by the adrenal medulla, along with adrenaline and noradrenaline, as part of the body's overall response to stress. Morphine and other opioid drugs such as fentanyl, pethidine, and tramadol act on the endogenous opioid receptors and are able to provide significant pain relief during childbirth and following surgery.

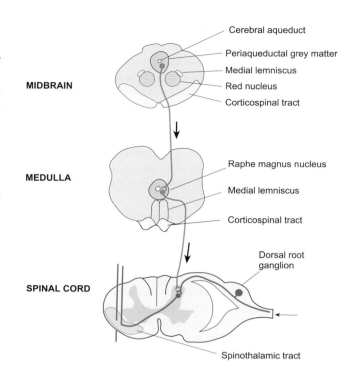

Figure 13.17 A schematic diagram of the descending pathways that inhibit the activity of nociceptive neurons in the dorsal horn. The different sections are not drawn to scale. Descending fibres are shown in red, ascending fibres in green.

Visceral pain

Although the viscera are not as densely innervated with nociceptive afferents as the skin, they are nevertheless capable of transmitting pain signals. The majority of the visceral afferent fibres that signal pain have their cell bodies in the dorsal root ganglia of the thoracic and upper two or three lumbar segments. Their axons may run in the sympathetic trunk for several segments before leaving to innervate the viscera via the cardiac, pulmonary, and splanchnic nerves. The nociceptors of the fundus of the uterus and the bladder run in the sympathetic nerves of the hypogastric plexus. As a general rule, the fibres that convey information from the nociceptors of the viscera follow the course of the sympathetic nerves and enter the spinal cord via the dorsal roots. They synapse in the substantia gelatinosa of the dorsal horn and the second-order axons project to the brain, mainly via the spinothalamic tract. The nociceptive afferents of the neck of the bladder, the prostate, the cervix, and the rectum are an exception to this general rule. They reach the spinal cord via the pelvic parasympathetic nerves prior to ascending to the brain in the spinothalamic tract.

The nociceptive nerve endings are stimulated by a variety of stimuli. When the walls of hollow structures such as the bile duct, intestine, and ureter are stretched, the nociceptive afferents are activated. Thus, when a gallstone is being passed along the bile duct from the gall bladder it causes an obstruction that the smooth muscle of the bile duct tries to overcome by a forceful contraction. The nerve endings are stretched and the nociceptive afferents are activated. Each time the smooth muscle contracts, the sufferer will experience a bout of severe pain known as **colic**, until the stone is passed into the intestine. Similar bouts of spasmodic pain accompany the passage of a kidney stone along a ureter. Ischaemia also causes intense pain. The classic example is the severe pain felt following narrowing or occlusion of one of the coronary arteries. Chemical irritants are also a potent source of pain originating in the viscera. This is commonly caused by gastric acid coming into direct contact with the gastric or oesophageal mucosa, giving rise to the unpleasant sensation known as heartburn.

Referred pain

The pain arising from the viscera is often scarcely felt at the site of origin but may be felt as a diffuse pain at the body surface in the region innervated by the same spinal segments. This separation between the site of origin of the pain and the site at which it is apparently felt is known as referral, and the pain is called **referred pain**. The site from which the pain is reported, together with its characteristics (sharp, dull, throbbing etc.) provide valuable information for the diagnosis of organic disease. For example, the pain arising from acute cardiac ischaemia is frequently referred to the upper left quadrant of the chest and the base of the neck, and it may radiate down the inner aspect of the left arm (see Figure 13.18a). It may also be felt directly under the sternum. The passage of gallstones along the bile duct causes an intermittent but intense pain that appears to travel along the base of the right shoulder (see Figure 13.18b). The pains arising from the small intestine are localized to the area around the umbilicus. The pain of appendicitis is also initially localized to the area around the umbilicus but, as the inflammation of the appendix becomes more severe, the pain becomes localized to the lower right quadrant of the abdomen. Pain that has its origin in other internal organs is usually referred to the somatic segments innervated by the spinal roots

that provide their sympathetic innervation. That from the lower colon, rectum, and bladder neck is relayed via the parasympathetic nerves, as summarized in Table 13.2.

The mechanism of pain referral remains uncertain. One plausible hypothesis proposes that the pain afferents from the viscera converge on the same set of spinal neurons as the somatic afferents arising from the same segment. Activity in these neurons is normally associated with somatic (i.e. cutaneous) sensation and, as the brain does not normally receive much information from the visceral nociceptive afferents, it interprets the afferent discharge arising from the visceral nociceptors as arising from that region of the body surface innervated by the same spinal segment.

When a nerve trunk is stimulated somewhere along its length rather than at its termination, there is an unpleasant confused sensation known as **paraesthesia**. This presumably arises because the CNS attempts to interpret the sequence of action potentials in terms of its normal physiology. The somatic nerve fibres are specific labelled lines. Each afferent is connected to a particular set of sensory receptors. The abnormal pattern of activation is then projected to the area of the body served by the nerve in question. Sometimes the sensation is one of explicit pain and is projected to the body surface (**projected pain**). A well-known example is the pain felt when the ulnar nerve is knocked, giving rise to a tingling sensation in the third and fourth fingers of the affected hand.

Causalgia, neuralgia, tic douloureux, and phantom limb pain

Gunshot wounds and other traumatic injuries that damage, but do not sever, major peripheral nerves such as the sciatic nerve may subsequently give rise to a severe burning pain in the area served by the nerve. This is known as **causalgia**, which fortunately occurs in less than 5 per cent of patients receiving such an injury. During

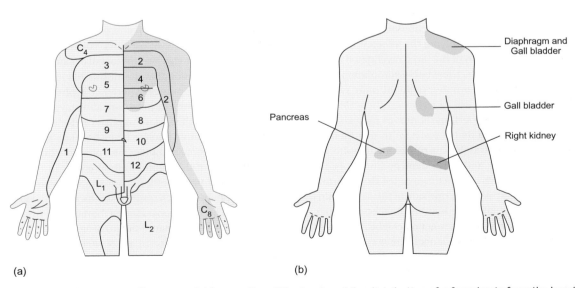

(a)　　　　　　　　　　　　　　　(b)

Figure 13.18 (a) The overlap between the segmental innervation of the trunk and the distribution of referred pain from the heart. The left-hand side shows those regions innervated by nerves from C4, T1 (inner arm), T3, T5, T7, T9, T11, and L1, while the right-hand side illustrates the innervation for T2, T4, T6, T8, T10, and T12. L2 innervates the thigh. The blue shading shows the distribution of referred cardiac pain. (b) The sites of referral for pain arising from the gall bladder, pancreas, and right kidney.

Table 13.2 The segmental sensory innervation of the viscera and sites of pain referral

Organ	Site of pain referral	Site of entry of visceral afferents	Afferent nerve from viscus
Heart	Left upper chest and inner part of left arm	T1–T5	Cervical and thoracic cardiac nerves
Lungs	No referred pain	T2–T7	Inferior cervical and thoracic nerves
Liver and gall bladder	Right upper quadrant and right scapula	T5–T9	Greater splanchnic nerve
Stomach	Epigastrum	T6–T8	Greater splanchnic nerve
Small intestine	Umbilicus	T9–T10	Greater splanchnic nerve
Appendix	Umbilicus, spreading to lower right quadrant	T9–T10	Greater splanchnic nerve
Ascending and transverse colon	Front of abdomen	T11–L1	Lumbar sympathetic chain
Splenic flexure of colon	Upper right quadrant of abdomen	T11–L1	Lumbar sympathetic chain
Sigmoid colon and rectum	Deep pelvis and anus	S2–S4 (parasympathetic)	Pelvic nerves
Kidneys	Mid-back and groin	T1–L1	Renal plexus
Ureter	Lower mid-back and groin	T1–L2	Renal plexus
Fundus of bladder	Above the pubis	T11–T12	Hypogastric plexus
Bladder neck	Perineum and penis	S2–S4 (parasympathetic)	Pelvic nerves
Uterus and cervix	Below the pubis, lower back	T10–T11	Hypogastric plexus
Testes, prostate, and associated structures	Pelvis and perineum	T10–L1	Hypogastric plexus

Note: As the sensations arising from the viscera are vague, the sites of pain referral are often diffuse and not easily localized. Therefore over-reliance on the classic sites of pain referral could be misleading. Persistent severe pain requires thorough investigation to avoid misdiagnosis.

an acute episode, the affected limb is initially warm and dry but later it becomes cool and the affected area sweats profusely. The pain is so great that the patient cannot bear the mildest of stimuli to the affected area. Even a puff of air elicits unbearable pain. Treatment is often ineffective, but injection of local anaesthetic into the sympathetic ganglia supplying the affected region may provide relief, as may activation of the large myelinated afferents by means of transcutaneous electrical nerve stimulation (TENS). Causalgia is also known as complex regional pain syndrome (CRPS).

Other kinds of damage to peripheral nerves may also cause severe and unremitting pain that is resistant to treatment. This kind of pain is known as **neuralgia**. The causes include viral infections, neural degeneration, and nerve damage from poisons. In **trigeminal neuralgia**, which is sometimes called **tic douloureux**, pain of a lacerating quality is felt on one side of the face within the distribution of the trigeminal nerve. Between attacks, the patient is free of pain and no abnormalities of function can be detected. The attacks are not provoked by thermal or nociceptive stimuli but may be triggered by light mechanical stimulation of the face or lip. An attack may also be triggered during eating by food coming into contact with a trigger point on the mucous membranes of mouth or the throat. In this case, the pain may be relieved by section of the appropriate branch of the trigeminal nerve, although this will leave the affected area without normal sensation.

After limb amputations, the nerve fibres in the stump begin to sprout and eventually form a tangle of fibres, fibroblasts, and Schwann cells called a **neuroma**. The nerve fibres in a neuroma often become very sensitive to mechanical stimulation that may give rise to severe shooting pains, which may be blocked by injection of local anaesthetic into the mass of the neuroma.

People who have survived the loss of a major limb frequently develop a strong illusion that the amputated limb is still present. In addition, most amputees experience a phenomenon known as **phantom limb pain**. The phantom limb may appear to be held in a very uncomfortable or painful position, for example one patient described a phantom hand as being held as a tightly clenched fist. Intriguingly, if some amputees with phantom limb pain are shown the reflection of their good limb in a mirror, they interpret the image as their phantom limb. For some, the ability to move the supposed phantom limb results a significant reduction in phantom limb pain. This visual observation therapy can, in some individuals, be more effective than standard drug therapy, revealing the extraordinary complexity of the processes that underpin the interpretation of pain stimuli. Although the pain evoked by noxious stimuli serves an obvious protective function by warning that a specific action can lead to tissue damage, much pain is of pathological origin. It is this pathological pain that drives many patients to seek medical help. Although the characteristics of the pain may be helpful in reaching a diagnosis, it is essential to understand that severe unremitting pain will colour every aspect of the sufferer's life. Prompt and effective treatment is therefore required. This terrifying aspect of chronic pain was very well put in 1872 by S.W. Mitchell, who wrote: *'Perhaps few persons who are not physicians can realize the influence which long-continued and unendurable pain may have upon the mind and body. Under such torments the temper changes, the most amiable grow irritable, the soldier becomes a coward and the strongest man is scarcely less nervous than the most hysterical girl.'*

Summary

Pain is an unpleasant experience that is generally associated with actual or impending tissue damage. The pain receptors (nociceptors) relay their information to the CNS via Aδ and C-fibre afferents that synapse on many neurons within the same segment of the spinal cord. Their second-order fibres project to the CNS by way of the spinothalamic and spinoreticular tracts.

Pain can be classified as pricking pain, which is accurately localized but elicits little by way of autonomic or emotional responses; as burning pain, which is difficult to endure and readily elicits both autonomic and emotional reactions; or as deep pain, which elicits both autonomic and emotional reactions but is difficult to localize accurately. Deep pain is often felt at a site distant from its site of origin (referred pain). Pain may also reflect hyperexcitability in the nociceptive pathways, particularly after a severe injury (e.g. phantom limb pain). Other pains may be elicited by stimulation of specific trigger points (e.g. trigeminal neuralgia).

13.6 Itch

Itch or **pruritus** is a well-known sensation that is associated with the desire to scratch. There are many causes of itch, including skin parasites such as scabies and ringworm, insect bites, chemical irritants derived from plants e.g. 'itching powder', and contact with rough cloth or with inorganic materials such as fibreglass. A variety of common skin diseases such as eczema also gives rise to the sensation of itch, as do obstructive jaundice and **renal failure**. The intensity of the sensation is highly variable; it may be so mild as to be scarcely noticeable, or so severe as to be as unendurable as chronic pain.

What is the physiological basis of this sensation? Exploration of the skin with itch-provoking stimuli has shown that there are specific itch points with a punctate distribution. If one of these points is cut out and examined histologically, no specific sensory structures are found but, as for pain points, there is an increased density of bare nerve endings. These are thought to be the receptors that give rise to the sensation of itch.

Several pieces of evidence suggest that itch is transmitted to the CNS via a subset of peripheral C-fibres. The first is the fact that the sensation disappears, as does burning pain, following the blocking of cutaneous C-fibres by local anaesthetics. Second, the sensation persists when a nerve trunk is blocked by pressure at a time when only C-fibres remain active. Finally, the reaction time to pruritic stimuli is characteristically slow, being around 1–3 seconds after the stimulus. In these respects, there is a similarity to cutaneous pain. Itch, however, differs from superficial pain in a number of important respects: it can only be elicited from the most superficial layers of the skin, the mucous membranes, and the cornea. Increasing the frequency of electrical stimulation of a cutaneous itch spot increases the intensity of the itch without eliciting the sensation of pain. Finally, the reflexes evoked by pruritic (itch-provoking) stimuli and nociceptive stimuli are different. Pruritic stimuli elicit the well-known scratch reflex, while nociceptive stimuli elicit withdrawal reflexes. On these grounds, it is reasonable to conclude that itch is a distinct modality of cutaneous sensation.

Exactly how the afferent fibres subserving the sensation of itch are excited remains unknown. However, a number of naturally occurring substances give rise to itch, including certain proteolytic enzymes and histamine. Histamine is of interest because it is released by mast cells when they become activated by antigens. Moreover, when histamine is applied to the superficial regions of the skin, it produces a distinct sensation of itch that is graded with its local concentration. In a recent study on human volunteers, a small subset of unmyelinated fibres were found to respond with sustained discharge to localized injections of histamine into the superficial regions of the skin and the time course of their discharge corresponded with the time course of the itch sensations. Nevertheless, as antihistamine drugs are not always effective in controlling itch, other chemical mediators are likely to be involved.

Pruritic afferents travel to the spinal cord in nerve trunks along with other sensory fibres. They enter the cord via the dorsal roots in a similar manner to those fibres that subserve pain sensation in the same region, and they synapse close to their site of entry into the spinal cord. The second-order fibres cross the midline and ascend to the brainstem reticular formation and thalamus via the spinothalamic tract. In addition to projecting centrally, activity in pruritic afferents elicit the scratch reflex, which is a polysynaptic spinal reflex that is clearly directed towards the removal of the itch-provoking stimulus. This reflex has been thoroughly examined in dogs, but the principle also applies to humans. In the first stage of the reflex, a limb (in humans usually the hand) is positioned above the irritated area. This is followed by the rhythmic alternation of extension and flexion of the limb such that the extremity repeatedly passes over the affected area in a manner calculated to dislodge the offending item.

Summary

Itch (pruritus) is a specific cutaneous sensation chiefly arising in the superficial regions of the skin. Specific bare nerve endings appear to respond to stimuli, such as histamine, that cause itch. Itch is transmitted to the CNS via specific afferents, where it may elicit a scratch reflex. Fibres conveying information about pruritic stimuli travel to the brain via the spinothalamic tract.

✳ Checklist of key terms and concepts

The somatosensory system

- The somatosensory system is concerned with stimuli arising from the skin, joints, muscles, and viscera.
- It provides the CNS with information concerning touch, peripheral temperature, limb position (kinaesthesia), and tissue damage (nociception).
- Individual skin receptors respond to particular stimuli—their adequate stimulus.
  - Mechanoreceptors respond to distortion of the skin and internal organs. They play a major role in the sense of touch.
  - Thermoreceptors respond either to cooling or warming of the skin.
  - Nociceptors are stimulated by local tissue damage and give rise to the sensation of pain.
- The area within which a stimulus will excite a specific afferent fibre is known as the receptive field of that fibre. The receptive fields of adjacent fibres often overlap.
- The size of the receptive fields and the degree of overlap between adjacent receptive fields both play an important role in the spatial discrimination of a stimulus.
- The visceral receptors include both rapidly adapting and slowly adapting mechanoreceptors and chemoreceptors, which are an essential component of the visceral reflexes that control vital body functions.

Sensory pathways

- Information from the somatosensory receptors reaches the cerebral cortex by way of the dorsal column–medial lemniscal pathway and by the spinothalamic tract.
- The dorsal column pathway is primarily concerned with fine discriminatory touch and position sense, while the spinothalamic tract is concerned with crude touch, temperature, and nociception.
- The postcentral gyrus of the cerebral cortex receives somatosensory information and possesses a topographical map of the contralateral surface of the body.

Pain

- Pain is an unpleasant experience that is generally experienced following tissue damage. It also accompanies severe organic disease such as cancer.
- The pain receptors (nociceptors) appear to be the bare nerve endings of a specific set of Aδ and C-fibre afferents that are triggered by strong thermal, mechanical, or chemical stimuli.
- Pain fibres synapse extensively within the spinal cord at a segmental level and project to the CNS by way of the spinothalamic and spinoreticular tracts.
- Pain arising from the body may be classified as:
  - pricking pain, which is accurately localized but elicits little by way of autonomic or emotional responses;
  - burning pain, which arises following activation of C-fibre nociceptors—this type of pain is difficult to endure and readily elicits both autonomic and emotional reactions;
  - deep pain which, like burning pain, elicits both autonomic and emotional reactions—it is difficult to localize accurately and is often referred to a site distant to its site of origin (referred pain).
- Many pains are of central origin (neuropathic pain) and reflect hyperexcitability in the nociceptive pathways, which can follow a severe injury. This is the case with phantom limb pain. Other pains may be elicited by stimulation of specific trigger points, as in trigeminal neuralgia.

Itch

- Itch is a specific cutaneous sensation chiefly arising in the superficial regions of the skin.
- The end-organs responsible for the sensation of itch appear to be bare nerve endings that are excited by histamine and other chemical mediators.
- Itch is transmitted to the CNS via specific afferents where it may elicit a scratch reflex.
- Fibres conveying information about pruritic stimuli travel to the brain via the spinothalamic tract.

📄 Recommended reading

Anatomy of sensory systems

Brodal, P. (2004) *The central nervous system: Structure and function* (3rd edn), Chapter 6. Oxford University Press, New York.

Physiology

Carpenter, R., and Reddi, B. (2012) *Neurophysiology: A conceptual approach* (5th edn), Chapters 4 and 5. Hodder Arnold, London.

Squire, L.R., Berg, D., Bloom, F.E., Du Lac, S., Ghosh, A., and Spitzer, N.C. (2012) *Fundamental Neuroscience* (3rd edn), Chapter 24. Academic Press, San Diego.

Medicine

Donaghy, M. (2005) *Neurology* (2nd edn), Chapters 6, 14–17. Oxford University Press, Oxford.

Review articles

Abraira, V.E., and Ginty, D.D. (2013) The sensory neurons of touch. *Neuron* **79**, 618–39.

Clapham, D.E. (2003) TRP channels as cellular sensors. *Nature* **426**, 517–24.

Costigan, M., Scholz, J., and Woolf, C.J. (2009) Neuropathic pain: A maladaptive response of the nervous system to damage. *Annual Review of Neuroscience* **32**, 1–32.

Gu, Y., and Gu, C. (2014) Physiological and pathological functions of mechanosensitive ion channels. *Molecular Neurobiology* **50**(2), 339–47.

Patapoutian, A., Peier, A.M., Story, G.M., and Viswanath, V. (2003) ThermoTRP channels and beyond: Mechanisms of temperature sensation. *Nature Reviews Neuroscience* **4**, 529–39.

Pethö, G., and Reeh, P.W. (2012) Sensory and signaling mechanisms of bradykinin, eicosanoids, platelet-activating factor, and nitric oxide in peripheral nociceptors. *Physiological Reviews* **92**, 1699–775.

Proske, U., and Gandevia, S.C. (2012) The proprioceptive senses: their roles in signaling body shape, body position and movement, and muscle force. *Physiological Reviews* **92**, 1651–97.

Roudaut, Y., Lonigro, A., Coste, B., Hao, J., Delmas, P., and Crest, M. (2012) Touch sense. *Channels* **6**, 234–45.

Vallbo, Å.B., Hagbarth, K.-E., and Wallin, B.G. (2004) Microneurography: How the technique developed and its role in the investigation of the sympathetic nervous system. *Journal of Applied Physiology* **96**: 1262–9.

To check that you have mastered the key concepts presented in this chapter, complete the accompanying online self-assessment questions. Go to www.oup.com/uk/pocock5e/

CHAPTER 14

The eye and visual pathways

This chapter should help you to understand:

- The anatomy of the eye and organization of the retina
- The general physiology of the eye
  - Tear formation
  - Intraocular pressure
  - The reflexes of the optic pathway
  - The process of image formation and common refractive errors
  - The extent of the visual field and diagnostic value of various defects
  - Light and dark adaptation: photopic and scotopic vision
- The organization of the visual pathways
- The process of phototransduction
- The processing of visual information in the visual cortex
- The basic features of colour vision and causes of colour confusion
- Eye movements and their role in vision

14.1 Introduction

Man is pre-eminently a visual animal. Those blessed with normal vision largely react to the world as they see it, rather than to its feel, sounds, or smells. Light itself is a component of the electromagnetic spectrum, and the human eye responds to radiation with wavelengths between about 380 nm (deep violet) to 750 nm (deep red). This is a relatively narrow part of the total spectrum, and other animals have a slightly different range of sensitivities. The particles that carry the radiation we call light are known as photons, which have no mass or electric charge but carry a very small amount of energy (a quantum) that depends on the wavelength of light. The units of light intensity are explained in Box 14.1.

Although the eye is able to respond to light of intensities over 15 orders of magnitude, the range of light intensities that are normally encountered in daily life is somewhat smaller (see Figure 14.1). Nevertheless, this range extends from bright daylight to near total darkness. The different phases of vision in bright, low-level light, and dark conditions are classified as **photopic**, **mesopic**, and **scotopic** respectively. During the day, photopic vision is dominant, but at night, especially in the absence of artificial light, the ability to see depends entirely on scotopic vision. At dusk and during the night in towns and cities, we rely largely on a transitional zone called mesopic vision.

14.2 The anatomy of the eye

The eyes are protected by their location in the bony cavities of the orbits. Only about a third of the eyeball is unprotected by bone. The eyeball itself its roughly spherical and its wall consists of three layers (see Figure 14.2): a tough outer coat, the **sclera**, which is white in appearance; a pigmented layer called the **choroid**, which is highly vascular; and the **retina**, which contains the photoreceptors (the rods and cones) together with an extensive network of nerve cells. The retinal ganglion cells are the output cells of the retina, and they send their axons to the brain via the optic nerves.

At the front of the eye, the sclera gives way to the transparent **cornea**, which consists of a transparent layer of connective tissue that lacks blood vessels. The health and transparency of the cornea is maintained by the tear fluid secreted by the lacrimal glands, and by the **aqueous humour** that is secreted by the ciliary body within the eye itself. The pigmented **iris** covers much of the transparent opening of the eye. At the centre of the iris is an opening called the **pupil**, which admits light to the photoreceptors of the retina.

(1₂3) Box 14.1 Units of light measurement

Expressing the amount of light is complicated by the fact that it may be emitted from a source or reflected by a surface. Emitted light is measured in *candelas*, while reflected light is measured in *lux*. The luminous intensity of a surface, its luminance, is a measure of the intensity of the light emitted or reflected in all directions per unit of area of the surface.

A **candela** is the base SI unit of light and is defined as the intensity of a light source emitting light at a wavelength of 555 nm with a power of 18.3988 milliwatts over a complete sphere centred at the light source.

A **lumen** is the SI unit for measuring the flux of light being produced by a light source or received by a surface. One lumen represents the total flux of light emitted, and is equal to the intensity in candelas multiplied by the solid angle into which the light is emitted. Thus the total flux of a one-candela light emitting uniformly in all directions is 4π lumens (= 12.56 lumens). The SI unit for measuring the illumination of a surface is called a **lux**, which is defined as an illumination of one lumen per square metre.

One **lambert** is the luminance of a surface that emits or reflects one lumen per square centimetre but, as this is a large unit, luminance is generally expressed in millilamberts (3.183 candelas m⁻²).

The fraction of incident light that is reflected by a surface is called its **albedo** or reflection coefficient. Objects with a high albedo reflect a large amount of the incident light: for example, fresh snow has an albedo of approximately 0.9, while charcoal has a very low albedo of around 0.04.

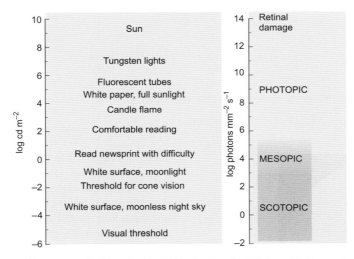

Figure 14.1 Illustration of the range of luminance (left) and retinal illumination (right) found in the environment. Note that the logarithmic scales represent a range of 15 orders of magnitude. Photopic, mesopic, and scotopic refer to vision in bright, low-level light, and dark conditions. (Based on Fig 7.3 in R.H.S. Carpenter (1996) *Neurophysiology*, 3rd edition. Edward Arnold, London.)

The appearance of the retina as seen through the cornea with an ophthalmoscope is shown in Figure 14.3.

The pupil diameter is controlled by two muscles in the iris, the circular **sphincter pupillae** and the radial **dilator pupillae.** Both are innervated by the autonomic nervous system: the sphincter pupillae receives parasympathetic innervation via the ciliary ganglion, while the dilator pupillae receives sympathetic innervation via the superior cervical ganglion.

Behind the iris lies the ciliary body, which contains smooth muscle fibres. The **lens** of the eye is attached to the ciliary body by a circular array of fibres called the **zonule of Zinn** or the **suspensory ligament.** The lens is formed as a series of cell layers, which arise from the cuboidal epithelial cells that cover its anterior surface. The cells of the lens synthesize proteins known as crystallins that are important for maintaining its transparency. Like the cornea, the lens has no blood vessels and depends on the diffusion of nutrients from the aqueous humour for its nourishment. The lens itself is elastic and can change its shape according to the tension placed on it by the zonular fibres. This ability of the lens to change its shape is an essential part of the mechanism by which the eye can bring images into focus on the retina. This process is controlled by the ciliary muscles and is called **accommodation.**

The organization of the retina

The retina is the sensory region of the eye. It consists of eight layers. Starting from the vascular choroid layer, the first, most outward, component of the retina is the **pigmented epithelium**. The next

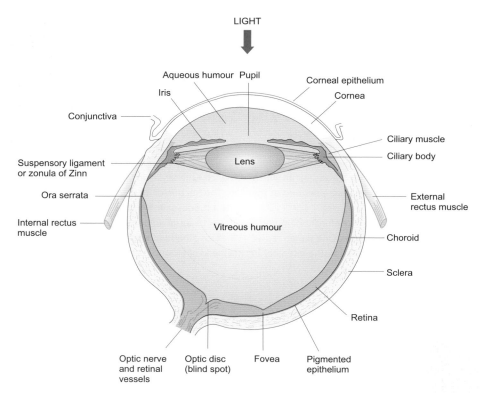

Figure 14.2 A cross-sectional view of the eye to show its principal structures. (Based on Fig 2.1 of H.B. Barlow and J.D. Mollon (eds) (1982) *The senses*. Cambridge University Press, Cambridge.)

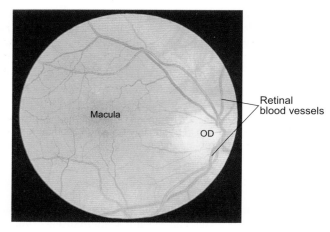

Figure 14.3 The appearance of a normal retina as seen through an ophthalmoscope. Note the origin of the main blood vessels from the optic disc (OD) or papilla, and their clear outline. The region of the retina that has the greatest resolution is the macula, which is devoid of large blood vessels. (From Plate 1 in J.G.G. Ledingham and D.A. Warrell (2000), *Concise Oxford textbook of medicine*. Oxford University Press, Oxford.)

three layers contain the photoreceptors—the **rods** and **cones** and the terminal regions of the photoreceptors where they make synaptic contact with other retinal cells. Above this are two layers that consist of the cell bodies of the bipolar cells, horizontal cells, and

amacrine cells and their processes. The final two layers contain the output cells of the retina, the **ganglion cells** and their axons. Individual photoreceptors consist of an outer segment, which contains the photosensitive pigment, an inner segment where the cell nucleus is located, and a photoreceptor terminal where the photoreceptors make synaptic contact with the bipolar and horizontal cells of the retina.

A highly schematic diagram of the organization of the retina is shown in Figure 14.4. Note that light passes through the cell layers to reach the photoreceptors, which are located next to the pigmented epithelium. Rods and cones are distributed throughout the retina, but in the central region, known as the **fovea centralis**, the retina is very thin and consists of a densely packed layer of cones, the other retinal cells being displaced to the area surrounding the fovea itself. In the surrounding region, the parafoveal region, both rods and cones are present in abundance together with the bipolar, amacrine, and horizontal cells, which are connected to the cones of the fovea. In Figure 14.4, one cone is shown connected to one bipolar cell and one ganglion cell. This is only the case for the central region of the retina; elsewhere the signals from a number of photoreceptors converge on a single bipolar cell, as in the examples seen in Figure 14.5. In the extreme periphery, many rods are connected to a single ganglion cell. The region where the ganglion cell axons pass out of the eye to form the optic nerve along with the retinal blood vessels is called the **papilla** or **optic disc**—see Figure 14.3. This region is devoid of photoreceptors and, for this reason, it is known as the **blind spot**.

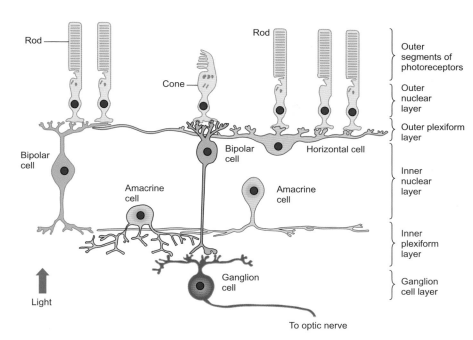

Figure 14.4 A simple diagram to show the cellular organization of the retina. Note that light has to pass through the innermost layers of the retina to reach the photoreceptors. Note also that the rods are significantly larger and have a much higher sensitivity to light than the cones, which are adapted for colour vision in relatively bright light. (Based on Fig 20.6 of E.R. Kandel, J.H. Schwartz, and T.M. Jessell (eds) (1991) *Principles of neuroscience*, 3rd edition. Elsevier Science, New York.)

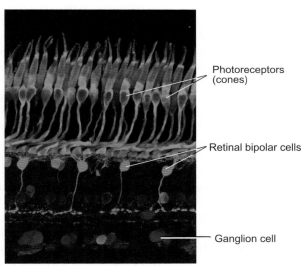

Figure 14.5 A section of monkey retina labelled to show the photoreceptors (green) and bipolar cells (red). The pale outlines of amacrine cells and retinal ganglion cells can also be distinguished. (Courtesy of Dr N. Cuenca, University of Alicante.)

Summary

The eyes are protected by their location in the bony cavities of the orbits. The eyeball itself its roughly spherical and its wall consists of three layers: the sclera, the choroid, and the retina. Light is admitted to the eye via the pupil. Behind the cornea is the pigmented iris which possesses muscles that control the size of the pupil and thereby the amount of light reaching the photosensitive layer, the retina. The retina contains the photoreceptors (the rods and cones) and nerve cells that participate in the analysis of the retinal image. The region where the optic nerve leaves the eye lacks photoreceptors and is known as the blind spot.

14.3 The general physiology of the eye

The lacrimal glands and tear fluid

Each orbit is endowed with lacrimal glands, which provide a constant secretion of tear fluid that serves to lubricate the movement of the eyelids and to keep the outer surface of the cornea moist, so providing a good optical surface. The lacrimal glands are innervated by the parasympathetic outflow of the facial nerve (CN VII). Tear fluid has a pH similar to that of plasma (pH 7.4) and is isotonic with blood. It possesses a mucolytic enzyme called **lysozyme** that has a bactericidal action. Under normal circumstances about 1 ml of tear fluid is produced each day, most of which is lost by evaporation; the remainder is drained into the nasal cavity via the tear duct. Irritation of the corneal surface (the conjunctiva) increases the production of tear fluid and this helps to flush away noxious agents.

The intraocular pressure is maintained by the balance between the rate of production and absorption of the aqueous humour

The space behind the cornea and surrounding the lens is filled with a clear fluid called the aqueous humour, which supplies the lens and cornea with nutrients. The aqueous humour is secreted by the processes of the ciliary body that lie just behind the iris. It flows into the anterior chamber, from where it drains though a fibrous mesh (the trabecular meshwork) situated at the junction between the cornea and the sclera into the **canal of Schlemm** (see Figure 14.6). From the canal of Schlemm the fluid is returned to the venous blood.

The aqueous humour is not a simple ultrafiltrate of plasma, but its production depends on the active transport of Na^+, Cl^-, and bicarbonate ions across the ciliary epithelium, accompanied by the passive movement of water through aquaporins (water

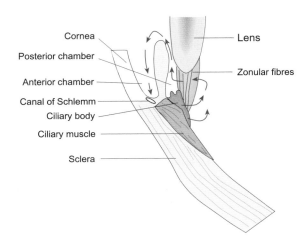

Cornea
Posterior chamber
Anterior chamber
Canal of Schlemm
Ciliary body
Ciliary muscle
Sclera
Lens
Zonular fibres

Figure 14.6 Diagrammatic representation of the circulation of the aqueous humour in the anterior chamber of the eye.

channels). It has a similar tonicity to that of plasma but it contains much less protein.

The constant production of aqueous humour generates a pressure within the eye known as the **intraocular pressure**. In normal individuals the intraocular pressure is about 2 kPa (15 mmHg) and serves to maintain the rigidity of the eye, which is essential for clear image formation. This pressure is sustained by keeping the balance between production and drainage of aqueous humour constant. If the drainage is obstructed, the intraocular pressure rises and a condition known as **glaucoma** results. If glaucoma is not treated promptly, it can result in progressive and permanent loss of vision. In some cases of glaucoma (defined by progressive loss of vision) the intraocular pressure is found to be normal. This is known as low pressure or low tension glaucoma. Its cause remains unknown.

Obstructions to drainage of aqueous humour may arise in a variety of ways.

- In **open-angle glaucoma**, the obstruction lies in the canal of Schlemm or the trabecular meshwork itself.

- In **closed-angle glaucoma**, there is forward displacement of the iris towards the cornea such that the angle between the iris and the cornea (the irido-corneal angle) is narrowed and the trabecular meshwork becomes covered by the root of the iris. Consequently, the outflow of fluid from the anterior chamber is impeded. This form of obstruction is occasionally seen following a blunt injury to the eye. It may develop slowly and progressively over many years or it may be sudden in onset, with acute closure of the irido-corneal angle. In such cases there is an abrupt rise in intraocular pressure, pain, and visual disturbances which, if left untreated, may lead to blindness within days or even hours.

- Occasionally the characteristic thickening of the lens seen in old age is sufficient to put pressure on the canal of Schlemm and impede drainage of aqueous humour from the anterior chamber.

- Glaucoma may be the result of abnormal blood vessel formations within the eye (which are often genetic in origin) or it may result from degenerative changes to the drainage area or swelling of the iris (uveitis).

In all cases of glaucoma, peripheral vision is normally lost first, followed by central vision impairment. Total loss of vision may occur if the condition remains undiagnosed and untreated. Loss of visual acuity is the result of pressure on the optic nerve. Over time, this pressure results in a reduced blood supply, disrupting the delivery of nutrients to the retinal neurons and the optic nerve fibres where they leave the back of the eye. Eventually there is retinal degeneration and widespread cell death. Available treatments for glaucoma focus on trying to reduce the rate of production of aqueous humour or to increase its rate of drainage. Drug treatments, surgical procedures, or a combination of both are used. While these treatments may slow, or even halt, the degeneration of vision they cannot restore vision that has already been lost.

Drugs that block β-adrenoceptors (β-antagonists) reduce the rate of fluid production by the ciliary body, but these are not suitable for people with heart conditions as they can alter cardiac and lung function. Other drugs with similar actions are prostaglandin analogues and α_2-agonists. Occasionally carbonic anhydrase inhibitors (e.g. acetazolamide) are used to reduce the production of aqueous humour. Acetylcholine agonists such as pilocarpine are also used to treat patients with glaucoma, as these drugs cause pupillary constriction through contraction of the sphincter pupillae muscle of the iris. This increases the ability of aqueous humour to drain from the anterior chamber, thereby lowering the intraocular pressure.

Surgical treatment options for glaucoma include trabeculoplasty and trabeculectomy. In the former procedure a highly focused laser beam is used to create a series of tiny holes (usually between 40 and 50) in the trabecular meshwork to enhance drainage. Trabeculectomy may be performed in cases of advanced glaucoma. In this procedure an artificial opening is made in the sclera (sclerostomy) through which excess fluid is allowed to escape. A tiny piece of the iris may also be removed so that fluid can flow backwards into the eye.

If the glaucoma is unresponsive to other forms of treatment, an artificial device may be implanted in the eye to facilitate the drainage of aqueous humour from the anterior chamber. A thin silicon tube is inserted into the anterior chamber and fluid drains onto a tiny plate sewn to the side of the eye. The fluid is eventually absorbed from the plate by the tissues of the eye. In the case of an acute closed-angle glaucoma that threatens to destroy vision within hours, emergency treatment is required and, in such cases, surgical removal of the iris (**iridectomy**) may be performed. Very occasionally, a procedure known as **cyclophotocoagulation** may be used to reduce the rate at which fluid is secreted by the ciliary body. Essentially the ciliary body is destroyed using a laser beam, so this procedure is normally reserved for end-stage glaucoma or for patients whose disease has failed to respond satisfactorily to any of the treatments described above.

The blink reflex and the dazzle reflex

The eyelids are closed by relaxation of the levator palpabrae muscles, which are supplied by the oculomotor nerve (CN III) coupled with contraction of the orbicularis oculi muscles, which are supplied by the facial nerve (CN VII). Reflex closure of the eyelids can result from corneal irritation due to specks of dust and other

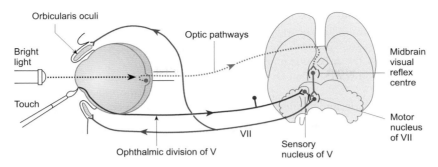

Figure 14.7 The neural pathways that subserve the blink and dazzle reflexes. A bright light shone directly in the eye is sensed by the retina and transmitted to the midbrain via the optic nerve. There it excites the dazzle reflex, in which motoneurons in the motor nucleus of the facial nerve cause the obicularis oculi muscles of the eyelid to contract. This shields the retina from the excessive light. The blink reflex is a response to irritation of the corneal surface which also leads to closure of the eyelid by the same efferent pathway. (Based on Fig 7.4 in vol. 3 of P.C.B McKinnon and J.F. Morris (2005) *Oxford textbook of functional anatomy.* Oxford University Press, Oxford.)

debris. This is known as the **blink reflex** or **corneal reflex**. In this case, the afferent fibres run in the trigeminal nerve (CN V). Very bright light shone directly into the eyes also elicits closure of the eyelids. Lid closure will cut off more than 99 per cent of incident light. In this case, the afferent arm of the reflex originates in the retina and collateral fibres pass to the oculomotor nuclei. This is known as the **dazzle reflex**, and a similar response can be elicited by an object that rapidly approaches the eye—the 'menace' or 'threat' reflex. The neural pathways involved in these reflexes are shown in Figure 14.7.

Pupillary reflexes

The pupils are the central dark regions of the eyes. They define the area over which light can pass to reach the retina. If a light is shone directly into one eye, its pupil constricts. This response is known as the **direct pupillary response**. The pupil of the other eye also constricts and this is known as the **consensual response**. These reflexes have a reaction time of about 0.2 s. As Figure 14.8 shows, the afferent arm of the reflex is a collateral projection from the optic nerve to the oculomotor nucleus in the brainstem. Preganglionic parasympathetic fibres originating in the Edinger–Westphal part of the oculomotor nucleus then pass to the ciliary ganglion. The short postganglionic fibres then travel from the ciliary ganglion to the sphincter pupillae muscle of the iris, which constricts the pupil when it is activated. The pupils also constrict when the eye is focused on a close object. This is known as the **accommodation reflex**, which has the effect of improving the depth of focus.

Dilation of the pupils occurs when the activity in the sympathetic nerves increases. This increased activity causes contraction of the radial dilator muscles. Thus the iris has a dual antagonistic innervation by the two divisions of the autonomic nervous system (see Chapter 11). Nevertheless, the muscles of the iris can exert only a slight degree of control over the amount of light admitted to the eye. Under steady-state conditions, the pupil is about 4–5 mm in diameter. In extremely bright light, it may constrict to 2 mm and in very dim illumination, it may assume a diameter of about 8 mm. Thus, the maximal change in light transmission that can be

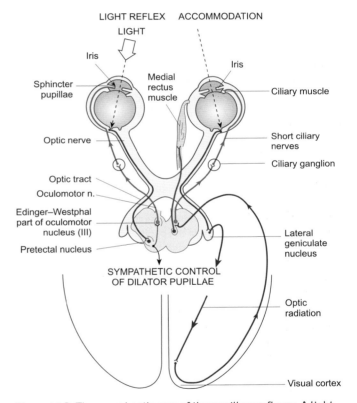

Figure 14.8 The neural pathways of the pupillary reflexes. A light shone directly in the eye is sensed by the retina and collateral fibres of the optic nerve carry this information to the Edinger–Westphal region of the oculomotor nucleus. Here the afferent fibres synapse and activate a parasympathetic reflex in which the sphincter pupillae muscle of the iris contracts and the pupil constricts. This is the direct pupillary response. The contralateral pupil also constricts (the consensual response). The pupil also constricts when a near object is focused on the retina (right-hand side of illustration). Here the afferent pathway involves the visual cortex (to permit pattern recognition) prior to activation of the parasympathetic reflex. (Based on Fig 7.2 in vol. 3 of P.C.B. McKinnon and J.F. Morris (2005) *Oxford textbook of functional anatomy.* Oxford University Press, Oxford.)

controlled by the pupil is only about 16-fold—far less than the total range of sensitivity of the eye. This indicates that the wide range of light intensity to which the eye can respond is due to the adaptation of the photoreceptors themselves. The changes in pupil diameter do, however, help the eye to adjust to sudden changes in illumination.

While the benefit of the corneal, dazzle, and pupillary reflexes in protecting the eye from damage is obvious, these reflexes are also of value in the assessment of the physiological state of the brainstem nuclei and in diagnosing brain death following traumatic head injury. In a condition known as **Argyll–Robertson pupil**, the pupils constrict during accommodation but the direct and consensual pupillary responses to light are absent. It is an important clinical sign indicating CNS disease such as tertiary syphilis.

Summary

The cornea is kept moist and clear of dust and infection by the secretion of tear fluid from the lacrimal glands. The shape of the eye is maintained by hydrostatic pressure arising from the secretion of aqueous fluid by the ciliary body. Excessive secretion or restricted drainage of the aqueous fluid leads to a rise in the intraocular pressure (glaucoma).

Light enters the eye via the cornea and pupil and is focused on the retina by the lens. The cornea is protected from damage by the blink reflex, which is elicited by irritants on the corneal surface. The eyelids also close in response to excessively bright light (the dazzle reflex). The amount of light falling on the retina is controlled by the pupil. If light is shone into one eye, the pupil constricts (the direct pupillary response). The pupil of the other eye also constricts (the consensual pupillary response). The diameter of the pupils is controlled by the autonomic nervous system. Sympathetic stimulation causes dilation of the pupils, while parasympathetic stimulation elicits pupillary constriction.

14.4 The optics of the eye—image formation

The eye behaves much as a camera. The image is inverted so that light falling on the retina nearest the nose (the **nasal retina**) comes from the lateral part of the visual field (the **temporal field**) while the more lateral part of the retina (the **temporal retina**) receives light from the central region of the visual field (the **nasal field**).

Since objects may lie at different distances from the eye, the optical elements of the eye must be able to vary its focus in order to maintain a clear image on the retina. The eyes of a young adult with normal vision are able to focus on objects as distant as the stars and as close as 25 cm (the **near point**). The optical power of the eye is at its minimum when distant objects are brought into focus and at its maximum when it is focused on the near point. This variation in optical power is achieved by the lens. (Optical power is also called refractive power or focusing power.)

As for any lens system, the optical power of the eye is measured in **dioptres** (see Box 14.2 for further details). For a normal relaxed eye the total optical power is about 60 dioptres, most of which is provided by the cornea (about 43 dioptres) while the lens contributes a further 17 dioptres or so when the eye is focused on a distant object, and about 30 dioptres when it is focused on the near point. When the lens loses its elasticity, which commonly occurs with advancing age, the power of accommodation is reduced and the nearest point of focus recedes. This condition is known as **presbyopia** (age-related long-sightedness).

The ciliary muscles control the focusing of the eye

The lens is suspended from the ciliary muscle by the zonular fibres (or suspensory ligament), as shown in Figure 14.9. When the ciliary muscles are relaxed, there is a constant tension on the zonular fibres exerted by the effect of the intraocular pressure on the sclera.

(1₂3) Box 14.2 Calculation of the power of lenses

The refractive power of a lens is measured in *dioptres*. The **dioptric power** (D) of a lens is the reciprocal of the focal length measured in metres:

$$\text{dioptric power} = \frac{1}{\text{focal length}}$$

A converging lens with a focal length of 1 metre has a power of 1 D. One with a focal length of 10 cm will have a power of 10 D. A lens with a focal length of 17 mm (the approximate focal length of the lens system of the human eye) has a dioptric power of:

$$\frac{1}{17 \times 10^{-3}} = 58.8\,D$$

Expressing the power of lenses in this way has the advantage that the focal length of lenses in combination is given by adding the dioptric power of each lens together. For diverging lenses, the dioptric power is negative, so that a diverging lens with a focal length of 10 cm has a dioptric power of −10 D.

For example, a person has an eye with a focal length of 59 D but the retina is only 16 mm behind the lens instead of the usual 17 mm. (The individual is affected by hyperopia, also called hypermetropia.) What power of spectacle lens is required for correction? To bring parallel light into focus in 16 mm requires a power of approximately 62.5 D. Thus, in this case, a converging lens of 62.5 − 58.8 = 3.7 D is needed.

For another person, whose retina is 18 mm behind the lens, the lens system would need to be 55.5 D to bring parallel light into focus on the retina. Thus a lens of 55.5 − 58.8 = −3.3 D would be needed for correction (i.e. a diverging lens of 3.3 D).

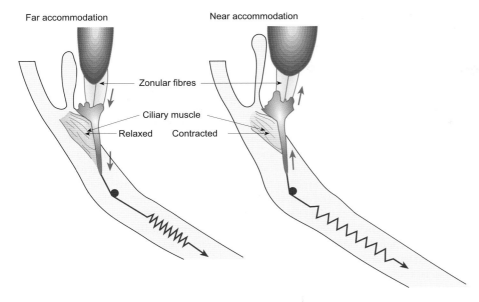

Far accommodation   Near accommodation

Zonular fibres

Ciliary muscle

Relaxed   Contracted

Figure 14.9 A diagram to illustrate the mechanism of accommodation. When the ciliary muscles are relaxed, the tension in the wall of the eye is transmitted to the lens via the zonular fibres. The lens is stretched and adapted for distant accommodation (left). When the ciliary muscles contract, they relieve the tension on the zonular fibres and the lens assumes a more rounded shape suited to near accommodation (right). (Based on Fig 5–5 in R.F. Schmidt (ed.) (1986) *Fundamentals of sensory physiology*, 3rd edition. Springer-Verlag, Berlin.)

This tension stretches the lens and minimizes its curvature. When the eye switches its focus from a distant to a near object, the ciliary muscle contracts and this opposes the tension in the sclera. As a result, the tension on the zonular fibres decreases and the lens is able to assume a more spheroidal shape due to its inherent elasticity: the more rounded the lens, the greater its optical power. The ciliary muscle is innervated by parasympathetic fibres from the ciliary ganglion.

Refractive errors—myopia, hyperopia, and astigmatism

When the ciliary muscle of a normal eye is relaxed, the eye itself is focused on infinity so that the parallel rays of light from a distant object will be brought into sharp focus on the retina. This is called **emmetropia**.

If the eyeball is too long, the parallel rays of light from a distant object will be brought into focus in front of the retina and vision will be blurred. This situation can also arise if the lens system is too powerful. In both cases, the eye is able to focus objects much nearer than normal. This optical condition is known as **myopia** (short-sightedness). It may be corrected with a diverging (or concave) lens of appropriate power, as shown in Figure 14.10.

If the eye is too short, then parallel light from a distant object is brought into focus behind the retina. This can also arise if the lens system is insufficiently powerful. People with this optical condition have difficulty in bringing near objects into focus and are said to suffer from **hyperopia** or **hypermetropia** (long-sightedness). It may be corrected with a converging (or convex) lens, as shown in Figure 14.10.

In some cases, the curvature of the cornea or lens is not uniform in all directions. As a result, the power of the optical system of the eye is different in different planes. This is known as **astigmatism** and it may be corrected by a cylindrical lens placed so that the refraction of light is the same in all planes.

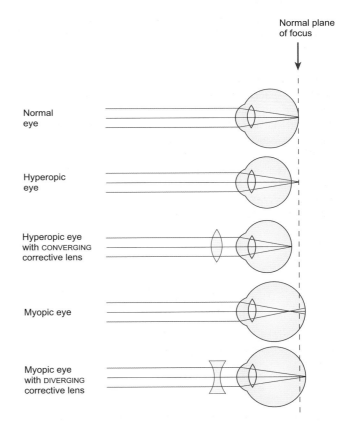

Normal plane of focus

Normal eye

Hyperopic eye

Hyperopic eye with CONVERGING corrective lens

Myopic eye

Myopic eye with DIVERGING corrective lens

Figure 14.10 A simple diagram illustrating the principal refractive errors of the eye and their correction by external lenses.

The visual field

The visual field of an eye is that area of space that can be seen at any instant of time. It is measured in a routine clinical procedure called **perimetry**. The eye to be tested is kept focused ('fixated') on a central point, and a test light appears at various points across the visual field in a random sequence generated by computer; the subject is asked to signal when they see it. In normal healthy subjects, the field of view of each eye is restricted by the nose and by the roof of the orbit so that the visual field is at its maximum laterally and inferiorly. The visual fields of the two eyes overlap extensively in the nasal region. This is the basis of binocular vision, which confers the ability to judge distances precisely. Defects in the visual field can be used to diagnose damage to different parts of the optical pathways, as described in Box 14.3.

Photopic and scotopic vision

The cones are the main photoreceptors used during the day when the ambient light levels are high. This is called **photopic vision**. In photopic conditions, visual acuity (discussed in next subsection) is high and there is full colour vision. At night, when light levels are low, the rods are the main photoreceptors. This is known as **scotopic vision** and is characterized by high sensitivity but poor acuity and a lack of colour vision. As dusk falls, colours gradually become less distinct and during twilight, both rods and cones are used. This phase is called **mesopic vision**.

Visual acuity varies across the retina

It is a commonplace observation that objects fixated at the centre of the visual field are seen in great detail while those in the peripheral regions of the field are seen less distinctly. The capacity of the eye to resolve the detail of an object is its **visual acuity**, which is measured in terms of the angle subtended by two points that can just be distinguished as being separate entities. Visual acuity is best in bright light (i.e. in photopic conditions). If visual acuity is plotted against the distance from the main optical axis of the eye, it is found to be greatest at the central region and to fall progressively towards the periphery. In dim light (i.e. under scotopic conditions), the cones are not stimulated and visual acuity is zero at the centre of

 Box 14.3 The effect of lesions in different parts of the visual pathways on the visual field

The visual field of an eye covers that part of the external world that can be seen without changing the point of fixation. It is measured by **perimetry**—which is a useful clinical tool for determining the health of the eye and visual pathways. The extent of the visual field is measured with an instrument called a perimeter. The visual field of each eye is measured in turn by moving a small spot of light along a meridian until the subject signals that they have seen it. The point is marked and another meridian tested until the full 360° have been

tested. More recent methods use a computer-generated random sequence of lights which are flashed at various points across the visual field (see main text).

If the eye were not protected within the orbit, the visual field would be a perfect circle. In reality, the features of the face—the nose, eyebrows and cheekbones—limit the peripheral extent of the visual field, as shown in Figure 1. The visual fields of the two eyes overlap extensively in the central region.

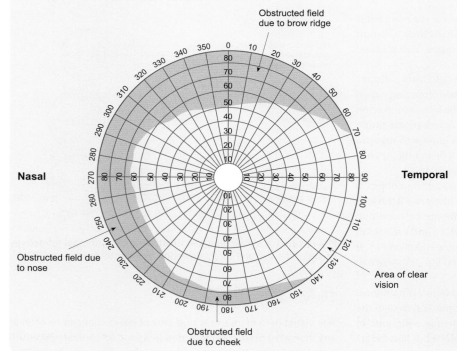

Figure 1 A perimeter chart to show the field of vision for the right eye. The nasal field (left) is obstructed by the nose, while the temporal field extends a full 90°. Note the obstruction produced by the brow ridge.

Box 14.3 (Continued)

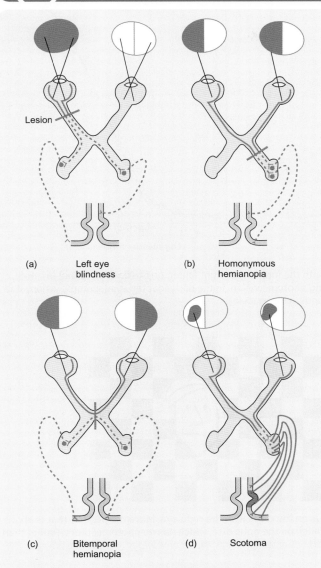

(a) Left eye blindness

(b) Homonymous hemianopia

(c) Bitemporal hemianopia

(d) Scotoma

Figure 2 The effect on the visual field of lesions in (a) the left optic nerve; (b) the right optic tract; (c) the optic chiasm; (d) the visual cortex.

Since the visual system is organized in a highly characteristic way, defects in specific parts of the visual pathways give rise to characteristic regions of blindness in the visual field (Figure 2). If one optic nerve is cut or compressed as in panel (a), the eye served by that nerve is blind. In contrast, if the optic nerve fibres are damaged after the optic chiasm, then the patient will be blind in the visual field on the opposite side to the lesion (**homonymous hemianopia**—panel b). If the optic chiasm is damaged, there will be a loss of the crossing fibres resulting in **bitemporal hemianopia** or **tunnel vision**, as shown in panel (c). This commonly occurs as a result of the growth of a pituitary tumour, which initially compresses the middle part of the optic chiasm. Damage to the optic radiation produces an area of blindness (a **scotoma**) in the corresponding region of the contralateral visual field, as shown in panel (d).

It is important to note that patients suffering from partial loss of vision do not experience 'darkness'; they have a loss of vision in the affected part of the visual field. Often the first signs of a visual deficit are an apparent clumsiness on the part of the patient. They fail to notice things in a particular region of the visual field. Proper testing of the visual field should therefore be carried out on people who have suffered a traumatic head injury or if there are signs of abnormal pituitary function.

the visual field and at its (modest) maximum in the region surrounding the fovea—the parafoveal region (Figure 14.11).

Light from objects in the centre of the visual field falls onto the fovea centralis, which has the highest density of cones and the smallest degree of convergence between adjacent photoreceptors. Outside the fovea, the degree of convergence increases and visual acuity falls. The rods are specialized for the detection of low levels of light and have a high degree of convergence. They thus provide the eye with a high sensitivity at the expense of visual acuity. Figure 14.11(b) shows that visual acuity declines as light intensity falls.

Light falling on the area of the optic disc (the point where the optic nerve leaves the eye) will not be detected, as there are no rods or cones in this region. As previously mentioned, this is known as the blind spot, which can be readily demonstrated as described in the caption to Figure 14.12.

Dark adaptation

When one first moves from a well-lit room into a garden lit only by the stars, it is difficult to see anything. After a short time, the surrounding shrubs and trees become increasingly visible.

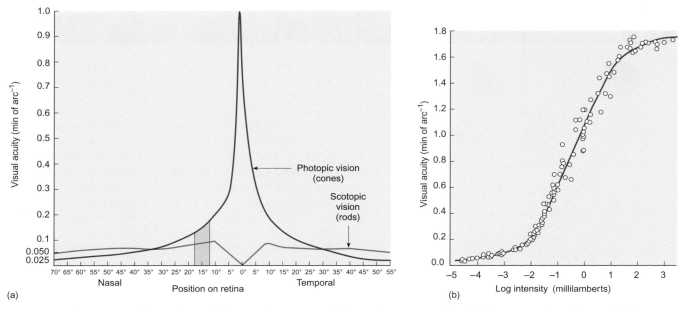

(a)

(b)

Figure 14.11 The variation of visual acuity (a) across the retina and (b) with the intensity of light falling on the fovea. In panel (a), note that while visual acuity is greatest for an image falling on the fovea during photopic vision, in the dark adapted eye (scotopic vision) it is least at the fovea and highest in the parafoveal region (the region surrounding the macula).

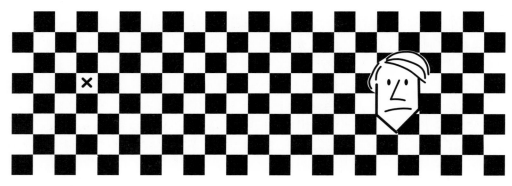

Figure 14.12 Demonstration of the blind spot. Close the left eye and focus on the cross with the right eye. Move the page so that is about 25 cm from you and the face will disappear. At this point the image has fallen on the optic disc, which has no photoreceptors. Rather than leaving a gap in the image, the brain fills in the background. (Fig 7.3 in R.H.S. Carpenter (1996) *Neurophysiology*, 3rd edition. Edward Arnold, London.)

The process that occurs during this change in the sensitivity of the eye is called **dark adaptation** and its time course can be followed by measuring the threshold intensity of light that can be detected at different times following the switch from photopic conditions to scotopic conditions. The dark adaptation curve obtained in this way shows that the greatest sensitivity is attained after about 20–30 minutes in the dark (Figure 14.13). The adaptation curve shows two distinct phases. The first shows a relatively rapid change in threshold, which appears to reach its limit after about 5–10 minutes. This is attributed to adaptation of the cones. The second phase is much slower, taking a further 10–20 minutes to reach its limit. This phase is attributed to adaptation of the rods. The change from scotopic vision to photopic vision is called **light adaptation**. This process is faster than dark adaptation and is readily appreciated on moving from a darkened area into bright light. The initial sense of being blinded by the light rapidly wears off.

Summary

The optics of the eye are very like those of a camera. The image of the world is brought into focus on the retina by the action of the lens system of the eye. The total optical power of the eye when it is fixated on a distant object is about 60 dioptres (D). Of this, the refractive power of the cornea accounts for some 43 D and that of the lens accounts for about 17 D. The lens is able to alter its refractive power by a further 14 or 15 D to bring near objects into focus.

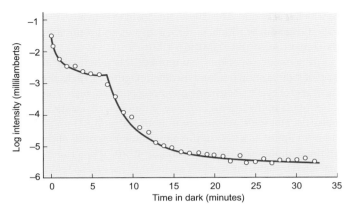

Figure 14.13 The dark adaptation curve obtained by plotting the visual threshold as a function of time spent in the dark following a period spent in bright light. The first segment of the curve is due to the adaptation of the cones, which is complete in a little over 5 min. The adaptation of the rods takes a further 15–20 min before maximum sensitivity is achieved.

The focusing of the lens is controlled by the ciliary muscle, which modulates the tension in the suspensory ligament. Near accommodation occurs by contraction of the ciliary muscle, which relieves the tension in the suspensory ligament, allowing the lens to become more rounded and so increasing its optical power. The most common defects of image formation are myopia (short-sightedness), hypermetropia (or hyperopia—long-sightedness), and astigmatism. Each can be corrected with an external lens of appropriate power.

The capacity of the eye to resolve the detail of an object is its visual acuity. In good light, visual acuity is best in the central region of the visual field, but in poor light, visual acuity is best in the area surrounding the central region. This difference reflects the distribution of cones and rods in the retina and their differing roles in photopic and scotopic vision.

14.5 The neurophysiology of vision

The organization of the visual pathways

From the retina, the axons of the ganglion cells pass out of the eye in the optic nerve. At the optic chiasm, the medial bundle of fibres (which carries information from the nasal side of the retina) crosses to the other side of the brain. This partial decussation of the optic nerve allows all the information arriving in one field of vision to project to the opposite side of the brain. Thus, the left visual field projects to the right visual cortex and the right visual field projects the left visual cortex (Figure 14.14). From the optic chiasm, the fibres pass to the lateral geniculate bodies, giving off collateral fibres that pass to the superior colliculi and oculomotor nuclei in the brainstem. From the lateral geniculate bodies, the main optic radiation passes to the primary visual cortex located in the occipital lobes. Nerve fibres connecting the two hemispheres pass through the corpus callosum.

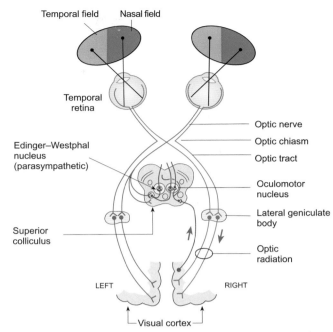

Figure 14.14 A schematic drawing of the visual pathways. Note that the nasal field projects to the temporal retina and that fibres from this region do not cross. Thus each half of the visual field projects to the opposite hemisphere. The connections between the two hemispheres are not shown. The projections to the brainstem mediate the corneal (blink) and pupillary reflexes.

The mechanism of phototransduction

The receptors of the retina are the rods and cones. Both contain photosensitive pigments. In the rods the pigment, known as rhodopsin, is made up of the aldehyde of vitamin A (**11-*cis*-retinal**) and a protein called **opsin**. The photosensitive pigments of the cones also contain 11-*cis*-retinal but it is conjugated to one of three different opsins. The combination of 11-*cis*-retinal with a particular opsin determines the colour sensitivity of individual cones. Colour vision depends on populations of cones that are sensitive to light of different wavelengths.

Both rods and cones are specialized neurons that can be divided into two portions, termed inner and outer segments (see Figure 14.4). The inner segment consists of the cell body and contains the nucleus and other organelles, including the photoreceptor terminal. The outer segment of both rods and cones is specialized for the capture of light. In the rods, the visual pigment is located in the densely packed membranous discs of the outer segment, which contains all the necessary biochemical apparatus for phototransduction. The densely packed discs are able to capture the photons of the dimmest light. In the cones, although the process of transduction is essentially the same as in the rods, the outer segment is much smaller than that of the rods; the discs are fewer in number and are formed by infoldings of the plasma membrane. The cones are therefore less able to capture photons at low levels of light.

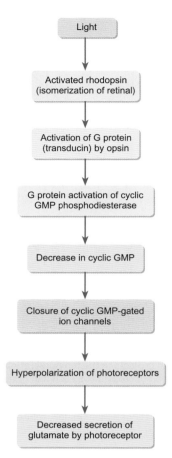

Figure 14.15 Flow chart of the principal steps in phototransduction.

When a photon of light is captured by one of the visual pigment molecules, the retinal changes shape (isomerizes) from 11-*cis*-retinal to all-*trans*-retinal (see Figure 14.16) and this causes a change in the three-dimensional shape of the photoreceptor protein (opsin) to which it is bound. This allows the opsin to activate a G protein called **transducin** which, in turn, activates a phosphodiesterase that breaks down cyclic GMP to 5′ GMP.

The sequence of events that immediately follows the absorption of light is shown schematically in Figure 14.15. In the dark, the levels of cyclic GMP in the photoreceptors are high. The cyclic GMP binds to the internal surface of non-selective cation channels (i.e. channels permeable to both Na$^+$ and K$^+$) and causes them to open. As a result, the membrane potential of the dark adapted photoreceptors is low, about −40 mV. When rhodopsin is activated by light, the levels of cyclic GMP fall and so fewer of the non-selective ion channels are open. The resulting fall in Na$^+$ permeability causes a hyperpolarization of the photoreceptor and a decrease in the secretion of neurotransmitter (glutamate) by the photoreceptor terminals. The bipolar cells to which the photoreceptors are connected respond with a change in their membrane potential.

The role of the retina in visual processing

Within the retina there is a considerable degree of neural processing. There are two types of bipolar cell. One type depolarizes in response to light and the other type hyperpolarizes. Moreover, the horizontal cells and amacrine cells (see Figure 14.4) permit the spread of excitation and inhibition across the retina. As a result, the responses of an individual retinal ganglion cell to light will depend on the properties of the cells to which it is connected. In general, retinal ganglion cells have receptive fields with a centre-surround organization of the kind shown in Figure 14.17. Two types are generally recognized: the 'on-centre off-surround' and the 'off-centre on-surround'. In the first type, the ganglion cell discharge is increased when a spot of light is shone in the centre of its receptive field and decreased when a light is shone in the surround. The second type has the reverse arrangement: the action potential frequency falls when light is shone in the central region of the field but rises when light falls on the surround. The

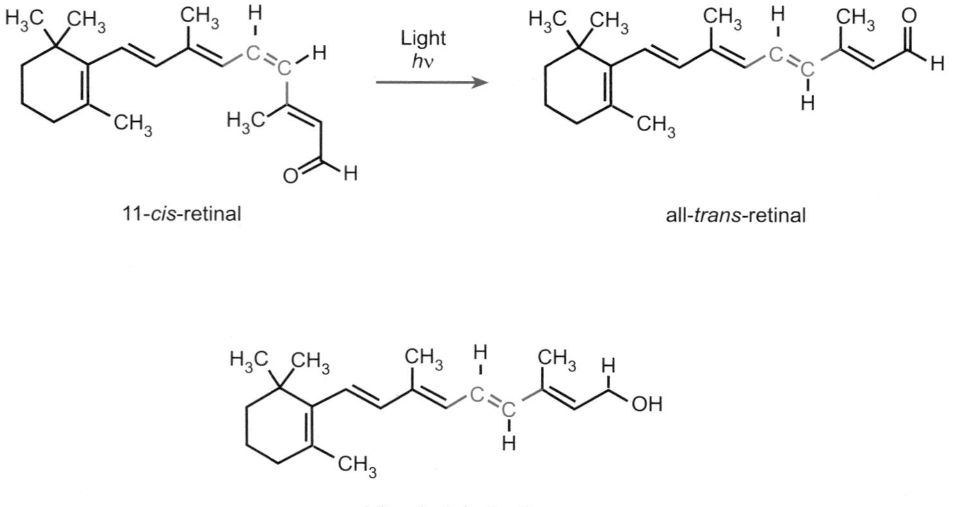

11-*cis*-retinal

all-*trans*-retinal

Vitamin A (retinol)

Figure 14.16 The structure of retinal and vitamin A. The isomerization from 11-*cis*-retinal to 11-*trans*-retinal is the initial step in phototransduction following the absorption of a photon of light. When the retinal is bound to an opsin, this change in structure causes a change in the conformation of the associated opsin, which then activates the G protein transducin.

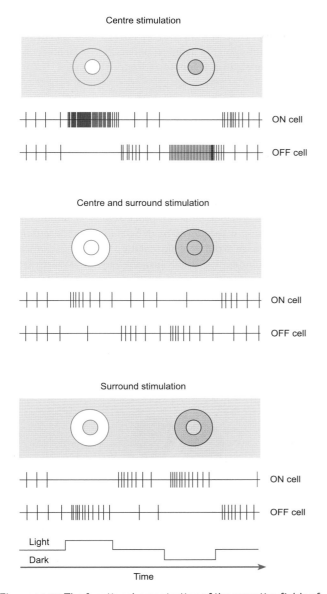

Figure 14.17 The functional organization of the receptive fields of two types of retinal ganglion cells. For 'on-centre' cells, a spot of light shone into the centre of the receptive field excites the ganglion cell. A similar spot of light shone into the surround region inhibits ganglion cell discharge. For 'off-centre' ganglion cells, the organization is reversed. For both kinds, diffuse illumination of the whole of the receptive field leads to weak excitation. (Based on P.H. Schiller (1992) *Trends in Neurosciences* 15, 87, with permission of Elsevier Science.)

receptive fields of the neurons of the lateral geniculate body to which the ganglion cells project have a similar organization.

The neurons of the primary visual cortex respond to specific features of an image

Unlike the receptive fields of cells in the early part of the visual pathway, neurons in the visual cortex often respond to particular features of objects in the visual field. Some respond optimally to a

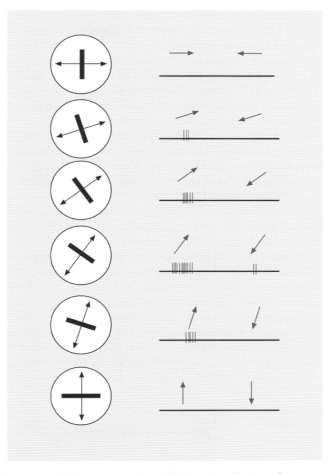

Figure 14.18 The response of a cell in the visual cortex of a macaque to a black bar passing across an illuminated region of the visual field (shown by the white circles). The response of the neuron is shown in the right-hand panel by a series of action potential records. The cortical neuron shows no response to a vertically or horizontally oriented bar but gives a strong response to a bar placed at 36 degrees to the horizontal. In this example, the cell shows a distinct directional preference, firing strongly to a rightward movement but weakly or not at all to a leftward movement. (Redrawn from Fig 2 in D.H. Hubel and T.N. Wiesel (1968) *Journal of Physiology* 195, 215–43.)

bright or dark bar passing across the retina at a specific angle (see Figure 14.18). Others respond to a light–dark border, and so on. In many cases, the direction of movement is also critical. In this way, the brain is able to identify the specific features of an object, which it then uses to form its internal representation of that object in the visual field. Such representations are crucial to our ability to interpret the visual aspects of the world around us.

Many neurons in the visual cortex respond to stimuli presented to both eyes. Moreover, as the visual cortices of the two hemispheres are interconnected via the corpus callosum, some cortical neurons receive information from both halves of the visual field and can detect small differences between the images in each retina. This provides cues about the distance of objects and is the basis of stereoscopic vision.

Vision is an active process. We learn to interpret the images we see. We associate particular properties with specific visual objects and have a remarkable capacity to identify objects, particularly other people, from quite subtle visual cues. We are able to recognize a person we know at a distance. Yet we do not notice that they appear larger as they approach us, rather we know that they have a particular size and use that fact as a means of judging how far away they are. This phenomenon is known as **size constancy**.

The importance of the way the brain uses visual cues to form its internal representation of an object is illustrated by many **visual illusions**. Perhaps the best known is the Müller–Lyer illusion shown in Figure 14.19(a). Here, the central line in each half of the figure is the same length but the line with the arrowheads appears shorter. Knowledge of this fact does not diminish the subjective illusion. The fact that the brain judges the size of an object by its context is revealed by the illusion shown in Figure 14.19(b). In this illusion, although the central circle is the same size in both parts of the figure, it appears larger when surrounded by small circles. Another illusion that illustrates how the brain interprets what the eye sees is **form vision**. In the case of the example given in Figure 14.19(c) a white square is 'seen' when, in fact, no such image is present. It appears that the brain uses the information from the missing segments of the surrounding black circles to complete the square. In Figure 14.19(d) the displacement of the second row of tiles gives rise to a gross distortion that gives the impression of a series of long wedges. In fact, the black and white tiles are all exactly the same shape and size. However, illusions are not just related to the shape of an object; the apparent lightness or darkness of an object is also judged by its context, as Figure 14.20 shows.

The brain processes colour, motion, and shape by parallel pathways

As the illusions mentioned above show, the brain processes visual information in a very different manner to a camera system. Detailed neurophysiological studies have shown that different aspects of a visual image are processed by different parts of the brain. The shape of an object is processed by one pathway while another processes its colour. Movement is processed by yet another pathway. This notion is supported by deficits in visual function that arise from damage to different parts of the brain, some of which are discussed in Chapter 19. Exactly how the various visual aspects of an object are subsequently synthesized into a unified image is not known.

The failure to name and recognize a common object is known as **object agnosia** (agnosia means not knowing). This occurs when there is damage to certain visual association areas in the left hemisphere. Damage to nearby areas can lead to difficulty in distinguishing colours (a condition known as **cerebral achromatopsia**). Failure to discern the movement of an object is associated with

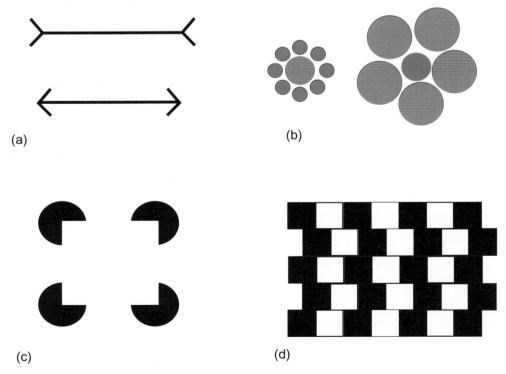

(a)   (b)

(c)   (d)

Figure 14.19 Some visual illusions. (a) The Müller-Leyer illusion. Here the central line appears longer in the figure with outwardly directed fins, although it is the same length as that with inwardly directed fins. The length of the lines can be confirmed by measurement with a ruler, yet the illusion does not disappear. (b) Errors of size perception—the Tichener illusion. Here the central circle is the same diameter in both parts of the figure, though it looks much larger when surrounded by small circles. (c) Form perception in a Kanizsa figure. Here the brain completes the outline of a square, although none is present. (d) The café-wall illusion. The displacement of the second row of tiles creates a false impression of a wedge. In fact, the black and white tiles are all of exactly the same size.

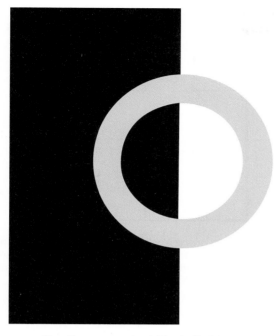

Figure 14.20 Simultaneous contrast. The half of the grey circle abutting the black rectangle appears lighter than that over the white background. The effect is even more marked if a black thread is placed across the circle at the border with the black area.

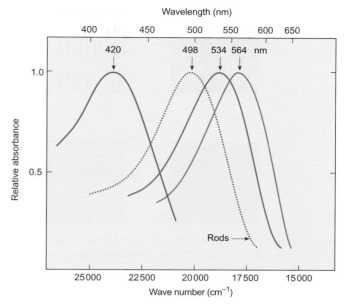

Figure 14.21 The absorbance spectra of the photo-pigments in the three types of human cones and the absorbance spectrum of rhodopsin (which is found in the rods). The absorption spectra measure the likelihood that a pigment will absorb a photon of light at a given wavelength. Thus light with a wavelength of 500 nm is much more effectively absorbed by those cones most sensitive to green light than it is by those cones more sensitive to blue or red light. Similarly, light of 450 nm is more effectively absorbed by those cones most sensitive to blue light than by those cones sensitive to green or red light.

bilateral lesions at the border of the occipital and temporal lobes. An inability to recognize faces (**prosopagnosia**) results from damage to the fusiform gyrus on the base of the brain (see Chapter 19).

Summary

The photoreceptors contain photosensitive pigments and a photoreceptor protein called an opsin. When light is absorbed, the photosensitive pigment alters the **conformation** of opsin, which then activates a cascade of reactions that culminates in the generation of action potentials by the retinal ganglion cells.

The visual pathways are arranged so that each half of the visual field is represented in the visual cortex of the contralateral hemisphere. To achieve this, the fibres arising from the nasal retina cross to the other side while those of the temporal retina do not. The right visual field is therefore represented in the left hemisphere and vice versa. Retinal ganglion cells project to the lateral geniculate bodies via the optic tracts. The axons of the lateral geniculate neurons project to the visual cortex, where the specific features of objects in the visual field are resolved.

14.6 Colour vision

In full daylight, the cones are the principal photoreceptors. Most people experience colour vision—individual objects have their own intrinsic colour. An unripe tomato appears to us as green, while a ripe tomato appears red. Similar colour changes occur when other fruits ripen, so there is an obvious advantage in distinguishing the colour of different objects in the environment. How is this remarkable feat achieved? People with normal colour vision can match the colour of an object by mixing varying amounts of just three coloured lights—blue, green, and red. This suggests that there should be three different cone pigments, one sensitive to blue light, one to green light, and one to red light. This supposition has now been validated by direct measurements of the absorption of pigments found in individual human cones. Each photoreceptor utilizes the same photo-pigment (11-*cis*-retinal) which has an absorption maximum of approximately 370 nm. When it is bound to an opsin to form a photo-pigment, the absorption maximum of retinal is shifted to longer wavelengths (see Figure 14.21):

- the blue-sensitive cones show an absorption maximum at 420 nm;
- the green-sensitive cones have a maximum absorption at about 534 nm;
- the red-sensitive cones have an absorption maximum at 564 nm; and
- the rhodopsin of the rods has an absorption maximum at 498 nm.

The different absorption maxima reflect the different protein sequences of the four opsins.

The absorption spectra measure the likelihood that a pigment will absorb a photon of light at a given wavelength. Thus at least

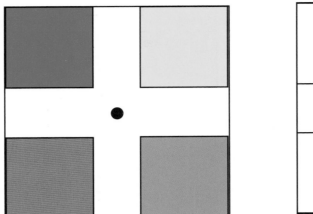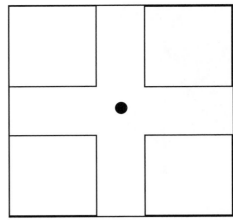

Figure 14.22 Colour after-images show complementary colours. Close one eye and focus on the black spot in the centre of the colour panel on the left for about a minute. Then switch your eye to the black spot of the right-hand figure. A yellow after-image will be seen in the top left square, a blue after-image in the top right, a green after-image in the bottom left and a red after-image in the bottom right.

two different pigments are required for any colour vision, as the brain must be able to compare the intensity of the signals emanating from different sets of cones. For normal human colour vision, a green light is seen when the green-sensitive cones are more strongly stimulated than the red- and blue-sensitive cones, and so on. White light reflects equal intensity of stimulation of all three types of cone. This is the basis of the trichromatic theory of colour vision proposed by Thomas Young at the beginning of the nineteenth century and confirmed by James Clerk Maxwell some 50 years later.

Useful as this theory is, it fails to explain some well-known observations. First, certain colour combinations, such as a reddish green or a bluish yellow, do not occur. Yet it is possible to see a reddish yellow (orange) or a bluish green (cyan). Second, if one stares at a blue spot for a short time and then looks at a white page, a yellow after-image is seen. Similarly, a green after-image will be seen after looking at a red spot (see Figure 14.22). To address these difficulties, Ewald Hering proposed the existence of neural processes in which blue and yellow were considered opponent colours, as were green and red. This colour opponent theory (with some later modifications) together with the trichromatic principle enunciated by Young and Maxwell provide the basis for understanding colour vision. Moreover, experimental evidence in favour of the colour-opponent theory is found in the retina. For example, some retinal ganglion cells are excited by a red light in the centre of their receptive field but inhibited by a green light in the surround.

One of the more remarkable aspects of colour vision is the phenomenon of **colour constancy**. An unripe tomato will appear green, and a ripe tomato red, whatever the time of day. This distinction is observed even when they are both illuminated with artificial light. As the mixture of wavelengths making up the light (its spectral content) varies considerably under these differing circumstances, it follows that the amount of green, blue, and red light reflected by the unripe and ripe tomatoes must be different

under these different circumstances. How does the brain compensate for the changing illumination? If they are illuminated with (say) red light, the green tomato will reflect less light that the red one so it will appear darker. If the illuminating light is green, the situation will be reversed: the green tomato will appear bright and the red tomato dark. Thus the comparative intensity of the light detected by the red and green cones provides information about the colour of the two fruits regardless of the spectral content of the illumination. The brain is thus able to compensate for the effect of changes in the quality of the illuminating light.

Colour blindness (colour confusion)

More than 90 per cent of the human race can match a given colour by mixing the appropriate proportions of red, green, and blue light. However, some people (mostly males) can match any colour with only two colours. These people are **dichromats** (rather than the **trichromats** of the general mass of the population). Although they are generally considered 'colour blind', they do distinguish different colours. It is just that their colour matching is abnormal: they are colour confused rather than colour blind. About 6 per cent of trichromats match colours with abnormal proportions of one or other of the primary colours. These people are called **anomalous trichromats**.

Dichromats are classified as **protanopes** (a relative insensitivity to red), **deuteranopes** (a relative insensitivity to green), and **tritanopes** (a relative insensitivity to blue). Protanopes and deuteranopes each constitute about 1 per cent of the male population. Tritanopes are exceedingly rare, as are those who have no cones at all (rod monochromats). Apart from the rod monochromats, those suffering from **colour blindness** are not debilitated but, in certain industries where colour codes are important, it is necessary to be aware of any potential colour confusions. This diagnosis is generally done with the Ishihara card test, one example of which is shown in Figure 14.23.

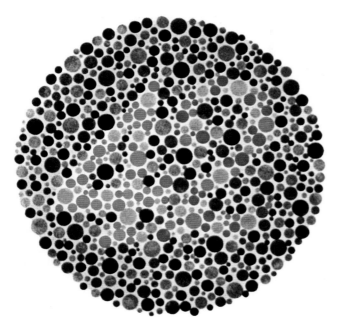

Figure 14.23 An Ishihara card. People with normal colour vision see '42', a red-blind person (a protanope) will see only the 2, while a green-blind person (a deuteranope) will see only the 4. A complete test uses many such cards; different cards use different combinations of colours.

Summary

There are three types of cone, each of which has a photo-pigment sensitive to light of particular wavelengths. This confers the ability to discriminate between different colours. Some people have defective cone pigments and are colour blind. The commonest are the anomalous trichromats, who match colours with unusual proportions of red and green. True colour blind people lack one or other of the cone pigments. The commonest are those who lack either the red pigment (protanopes) or the green pigment (deuteranopes).

14.7 Eye movements

Our eyes constantly scan the world around us. An image appearing in the peripheral field of vision is rapidly centred onto the fovea by a jerky movement of the eyes. These rapid eye movements are called **saccades**. During a saccade, the eyes move at angular velocities of between 200 and 600° s^{-1}. In contrast, when watching a race or when playing a ball game, the eye follows the object of interest, keeping its position on the retina fairly constant. This smooth tracking of an object is called a **pursuit movement**, which can have a maximum angular velocity of about 50° s^{-1}. These two types of eye movement can be combined, for example when looking out of a moving vehicle. One object is first fixated and followed until the eyes reach the limit of their travel. The eyes then flick to fixate and follow another object and so on. The continuous switching of the point of fixation is seen as a pursuit movement followed by a

saccade and then another pursuit movement. This pattern is known as **optokinetic nystagmus**. A **nystagmus** is a rapid involuntary movement of the eyes.

If an object such as a pencil is fixated and then moved around the visual field, both eyes track the object. These are known as **conjugate eye movements**. If the pencil is moved first away from the face and then towards it, the two eyes move in mirror-image fashion to keep the image in focus on each retina. When the object approaches the eyes, the visual axes converge and as the object moves away they progressively diverge until they are parallel with each other. These eye movements are called **vergence movements**. If the movements of the two eyes are not properly coordinated, double vision (**diplopia**) results.

The position of each eye within the orbit is controlled by six **extraocular muscles**, which are shown in Figure 14.24(a). They are innervated by the third, fourth, and sixth cranial nerves (the oculomotor, trochlear, and abducent nerves). The lateral and medial rectus muscles control the sideways movements, the superior and inferior oblique muscles control diagonal movements, and the superior and inferior rectus muscles control the up and down movements, as shown in Figure 14.24(b). The trochlear nerve

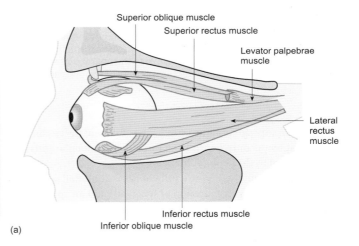

(a)

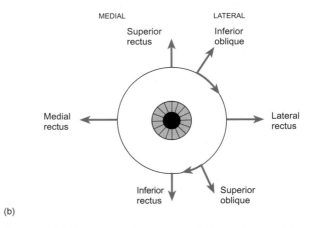

(b)

Figure 14.24 The extraocular muscles of the human eye (a) and the direction of eye movement controlled by each (b).

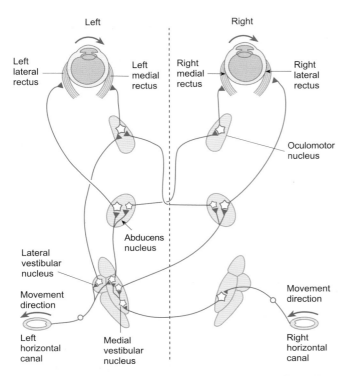

Figure 14.25 The neural basis of the vestibulo-ocular reflexes. The semicircular canals send afferents to the oculomotor nuclei to keep the visual field stable during motion by regulating the activity of the extraocular muscles. In this diagram, only the neural pathways that control the medial and lateral rectus muscles of the eyes are shown.

controls the superior oblique muscle, the abducent nerve controls the lateral rectus muscle, while all the other extraocular muscles are controlled by the oculomotor nerve.

The eye movements require full coordination of all of the extraocular muscles with activation of synergists and an appropriate degree of inhibition of antagonists. If we look to one side, the right for example, the right lateral rectus and the left medial rectus are both activated while the right medial rectus and left lateral rectus are inhibited, as shown in Figure 14.25. Diagonal movements involve more muscles. They consequently require an even greater

complexity of control. It is not surprising, therefore, that about 10 per cent of all motoneurons are employed in serving the extraocular eye muscles. These muscles have the smallest motor units in the body, with each motoneuron controlling 5–10 muscle fibres. This permits a high degree of precision in the control of eye position.

Although the eye movements allow the eye to centre the image of any item of interest on the fovea, they also have the important function of maintaining a stable visual image during locomotion. The necessary eye movements are reflexes mainly driven by the semicircular canals, saccule, and utricle of the vestibular system. Further details of the vestibulo-ocular reflexes are discussed in Chapter 16.

Defective control of the eye movements is often revealed by a squint, in which the eyes fail to look in the same direction and the sufferer experiences double vision. As the majority of the extraocular muscles are innervated by the oculomotor nerve (CN III), damage to this nerve is the most common source of such problems. There are a number of causes, including meningitis and compression of the nerve by a tumour. Less common is loss of ocular abduction, in which the affected eye cannot traverse in the temporal direction. This results from a lesion of the abducent nerve (CN VI). Very rarely, head injury can result in damage to the trochlear nerve (CN IV) and the resulting paralysis affects the downward and inward movement of the affected eye.

Summary

The movements of the eyes are controlled by six paired extraocular eye muscles that are innervated by the third, fourth, and sixth cranial nerves. Eye movements are driven by visual information and by the vestibular system. Their role is to keep objects of importance in the centre of the visual field.

The eye movements are classified as saccades, smooth pursuit movements, and vergence movements. In saccades and pursuit movements, the eyes move together. These are known as conjugate eye movements. Nystagmus occurs when saccadic movements are repeatedly followed by a smooth pursuit movement. In vergence movements, the two eyes move in mirror-image fashion so that their optical axes converge.

✱   Checklist of key terms and concepts

The physiology of the eye and visual pathways

- The eye consists of an outer coat of connective tissue, the sclera; a highly vascular layer, the choroid; and a photoreceptive layer, the retina.

- Light enters the eye via the transparent cornea and is focused on the retina.

- The cornea is protected from damage by the blink reflex, which is elicited by irritants on the corneal surface.

- The eyelids also close in response to excessively bright light. This is known as the dazzle reflex.

- The amount of light falling on the retina is controlled by the pupil.

- If light is shone into one eye, the pupil constricts (the direct pupillary response). The pupil of the other eye also constricts (the consensual pupillary response).

- The diameter of the pupils is controlled by the autonomic nervous system. Sympathetic stimulation leads to dilation of the pupils as a result of contraction of the radial smooth muscle of the iris. Parasympathetic stimulation elicits pupillary constriction as a result of contraction of the circular smooth muscle of the iris.

The optics of the eye

- The optics of the eye are very like those of a camera. The image of the world is brought into focus on the retina by the action of the lens system of the eye.
- The total optical power of the eye when it is fixated on a distant object is about 58 dioptres (D). Of this, the refractive power of the cornea accounts for some 43 D and that of the lens accounts for about 15 D.
- The lens is able to alter its refractive power by a further 14 or 15 D to bring near objects into focus.
- The focusing of the lens is controlled by the ciliary muscle, which modulates the tension in the zonular fibres (the suspensory ligament of the lens).
- Near accommodation occurs by contraction of the ciliary muscle, which relieves the tension in the zonular fibres. The reduction in tension allows the lens to become more rounded, so increasing its optical power.
- The major problems in image formation are due to malformation of the eyeball. The most common defects are myopia, hypermetropia (hyperopia), and astigmatism. They may be corrected with an external lens of appropriate power.
- The capacity of the eye to resolve the detail of an object is its visual acuity.
- In good light, visual acuity is best in the central region of the visual field, but in poor light, visual acuity is best in the area surrounding the central region.
- This difference reflects the distribution of cones and rods in the retina and their differing roles in photopic and scotopic vision.

The neurophysiology of vision

- The photoreceptors contain photosensitive pigments containing vitamin A aldehyde (11-*cis*-retinal) bound to a photoreceptor protein (opsin).
- When light is absorbed by a rod or cone cell, the 11-*cis*-retinal isomerizes to all-*trans*-retinal and dissociates from the opsin.
- This activates a cascade of reactions that leads to the hyperpolarization of the photoreceptor, which is the first step in a sequence of events that leads to the generation of action potentials by the retinal ganglion cells.

- The visual pathways are arranged so that each half of the visual field is represented in the visual cortex of the contralateral hemisphere. To achieve this, the fibres arising from the ganglion cells of the nasal retina cross to the other side while those of the temporal retina do not.
- The right visual field is therefore represented in the left hemisphere and vice versa.
- The axons of the ganglion cells reach the lateral geniculate bodies via the optic tracts. The axons of the lateral geniculate neurons project in their turn to the visual cortex, where the specific features of objects in the visual field are resolved.

Colour vision

- There are three kinds of cone, each of which is sensitive to light of different wavelengths. This confers the ability to discriminate between different colours.
- Some people (most of whom are male) have defective cone pigments and are said to be colour blind.
  - Anomalous trichromats match colours with unusual proportions of red and green.
  - True colour blind people lack one or other of the cone pigments.
  - Those who lack the red pigment are protanopes, while those lacking the green pigment are deuteranopes. People lacking the blue pigment are tritanopes, although this is rare.

Eye movements

- The eye movements are controlled by six extraocular eye muscles that are innervated by the third, fourth, and sixth cranial nerves.
- The movements of the eyes are driven both by visual information and by information arising from the vestibular system. Their role is to keep objects of importance centred on the central region of the retina.
- The eye movements are classified as saccades (in which the eyes move at very high angular velocities), smooth pursuit movements, and vergence movements.
- In saccades and pursuit movements the eyes move together; these are known as conjugate eye movements.
- In vergence movements, the two eyes move in mirror-image fashion so that their optical axes converge or diverge.
- Nystagmus occurs when saccadic movements are repeatedly followed by a smooth pursuit movement.

 Recommended reading

Anatomy

Brodal, P. (2004) *The central nervous system: Structure and function* (3rd edn), Chapter 7. Oxford University Press, New York.

MacKinnon, P., and Morris, J. (2005) *Oxford textbook of functional anatomy* (2nd edn), Volume 3, Chapter 6. Oxford University Press, Oxford.
A clear account of the anatomy of the eye and orbit, with many good drawings.

Physiology

Berg, J.M., Tymoczko, J.L., and Stryer, L. (2011) *Biochemistry* (7th edn), Chapter 33. Freeman, New York.

Gomperts, B.D., Tatham, P.E.R., and Kramer, I.J. (2009) *Signal transduction* (2nd edn), Chapter 6. Academic Press, San Diego.

Gregory, R.L. (1998) *Eye and brain: The psychology of seeing* (5th edn). Oxford University Press, Oxford.

A straightforward non-technical introduction to the process of visual perception.

Squire, L.R., Berg, D., Bloom, F.E., Du Lac, S., Ghosh, A., and Spitzer, N.C. (2012) *Fundamental neuroscience* (4th edn), Chapter 26. Academic Press, San Diego.

Medicine

Donaghy, M. (2005) *Neurology* (2nd edn), Chapters 6, 14–17. Oxford University Press, Oxford.

Review articles

Burns, M.E., and Pugh, E.N. (2010) Lessons from photoreceptors: Turning off G-protein signaling in living cells. *Physiology (Bethesda)* **25**, 72–84.

Gegenfurtner, K.R., and Kiper, D.C. (2003) Color vision. *Annual Review of Neuroscience* **26**, 181–206.

Hamer, R.D., Nicholas, S.C., Tranchina, D., Lamb, T.D., and Jarvinen, J.L.P. (2005) To a unified model of vertebrate rod phototransduction. *Visual Neuroscience* **22**, 417–36.

Kourtzi, Z., and Connor, C.E. (2011) Neural representations for object perception: Structure, category, and adaptive coding. *Annual Review of Neuroscience* **34**, 45–67.

Palczewski, K.(2014) Chemistry and biology of the initial steps in vision: The Friedenwald Lecture. *Investigative Ophthalmology and Visual Science* **55**, 6651–72.

Solomon, S.G., and Lennie, P. (2007) The machinery of colour vision. *Nature Reviews Neuroscience* **8,** 276–86.

To check that you have mastered the key concepts presented in this chapter, complete the accompanying online self-assessment questions. Go to www.oup.com/uk/pocock5e/

CHAPTER 15

The ear and auditory pathways

This chapter should help you to understand:

- The physical properties of sound
- The structure of the ear and auditory system
- The mechanism of sound transduction
- The auditory pathways and the processing of sound by the brain
- The cues that are used by the brain to locate a sound
- Hearing deficits and their clinical evaluation

15.1 Introduction

Most higher animals, including primates, use **sound** to detect danger, raise the alarm, claim territory, attract a mate, and so on. Human beings have developed the ability to communicate via sounds to a high degree by using speech and language. Interpreting these sounds is therefore vital to our existence, so it is hardly surprising to find that large areas of the brain are devoted to decoding what we hear. This chapter is concerned with the processes involved in the detection of sounds and with the nervous pathways subserving the process of hearing. It concludes with a brief discussion of the causes of deafness.

15.2 The physical nature of sound

Sounds arise from small fluctuations in the pressure of air (or of water in the case of sounds we hear when submerged). These pressure variations originate at a point in space and radiate outwards as a series of waves. The sensitivity of the human ear to the small pressure changes of sounds is truly astounding: at the threshold of hearing, the ear can detect pressure changes of as little as 20 µPa, while those sounds that cause a feeling of severe discomfort have pressure changes of the order of 5 Pa (recall that atmospheric pressure is about 101 kPa). Still greater sound intensities cause pain and damage to the ear itself. This huge intensity range is normally expressed on the decibel (dB) scale, which is explained in Box 15.1. The loudest sounds to which the ear can be exposed without irreversible damage are about 10 Pa (around 110–115 dB above threshold).

A **pure tone** is a sound consisting of just one frequency. Playing a series of pure tones of varying intensity to a subject using an instrument called an **audiometer**, it is possible to determine the threshold of hearing for different frequencies. The results can be plotted as intensity in dB SPL (SPL = sound pressure level) against the logarithm of the frequency to produce a graphical representation of the auditory threshold known as an **audiogram** (Figure 15.1). For healthy young people, the frequency of airborne vibrations that can be heard ranges from about 20 Hz to 20 kHz, but the threshold of hearing varies considerably with the frequency.

The ear is very sensitive to sounds of between 200 Hz and 5 kHz. These frequencies correspond to those encountered during normal speech. The ear is less sensitive to sounds of lower and higher frequencies. In the most sensitive range, between 1 and 3 kHz, the threshold of hearing is close to 0 dB SPL (equivalent to a pressure change of 20 µPa). At 100 Hz and 15 kHz the ear is approximately 30 dB less sensitive than it is at 3 kHz. As people get older, they tend to become less and less sensitive to the higher frequencies (a condition called **presbyacusis** or **presbycusis**—see Section 15.6).

The ability to interpret a sound obviously depends upon its characteristics. A sound may be sharp and brief (impulsive or percussive) when it begins abruptly and dies away quickly. The pressure

(1₂3) Box 15.1 Sound waves and the decibel scale of intensity

When the vibrations of a solid, liquid, or gas can be detected by the human ear, they are considered to be sound waves. Essentially sound waves are the result of a mechanical force displacing molecules from their normal resting position. The molecules collide with their neighbours, so transmitting the energy through the medium without themselves moving significantly from their starting position. The sound waves are therefore a series of compressions and rarefactions of the conducting medium. In the case of human hearing the medium is usually air. When a sound is propagated, the air pressure alternately increases and decreases above the mean atmospheric pressure. There is no net flow of air when a sound wave is transmitted. Figure 1 shows the pressure variation at a given point during the propagation of a sound which has a distinct periodicity T. The frequency of a sound (f) is determined by the periodicity of the vibration (i.e. the number of cycles of occurring in a second). The wavelength of a sound wave is the distance between successive pressure maxima or two pressure minima. In a given medium, frequency and wavelength are related by the formula:

$$\lambda = c \times T = \frac{c}{f}$$

where λ is the wavelength in metres (m), c is the sound speed in metres per second (m/s), T is the interval in seconds (s) and f is frequency (= $1/T$) of successive phases of the cycle measured in cycles per sec or Hertz (Hz). For a pure tone with a frequency of 1000 Hz and a propagation speed in air of 343 m/s, the wavelength is equal to 0.343 metres (34 cm). Note that the wavelength of a sound is inversely proportional to its frequency: the higher the frequency, the shorter the wavelength. The larger the pressure variation, the higher the energy of the wave and the louder the sound will appear to be.

While it is perfectly possible to express loudness in terms of the peak pressure change, it is much more convenient to express the intensity of sounds in relation to an arbitrary standard using the **decibel scale** (abbreviated as dB):

$$dB = 10 \cdot \log_{10} \frac{\text{(sound intensity)}}{\text{(reference intensity)}} \qquad [1]$$

Hence the decibel unit represents *the ratio* of the power of two sounds, not the absolute sound intensity. The decibel scale has the advantage that the subjective assessment of sound intensity is roughly proportional to the logarithm of the stimulus power (see Section 12.2). Since the energy of a sound wave depends on the square of the pressure change, the decibel scale is more often written:

$$dB = 20 \cdot \log_{10} \frac{\text{(sound pressure)}}{\text{(reference pressure)}} \qquad [2]$$

The reference pressure used in auditory physiology is 20 μPa, which is close to the average threshold for human hearing at 1 kHz. Because the dB scale is expressed relative to an arbitrary standard, it is sometimes expressed as dB SPL (sound pressure level). A doubling of the sound pressure equals a 6 dB increase. If a sound pressure is ten times that of the reference pressure, this will correspond to 20 dB SPL. A sound pressure one thousand times that of the reference pressure will be 60 dB SPL, and so on.

When a sound is initiated, it begins at a specific point in space and then the energy is radiated evenly in all directions. As the wave front spreads it covers an increasingly large area, with the effect that the intensity of the sound (i.e. its energy) decreases with distance from the source. This decrease in intensity is governed by the inverse square law:

$$I \propto \frac{1}{r^2}$$

If the distance from a sound source is doubled, the sound intensity is reduced to one-quarter.

$$\frac{I_r}{I_s} = \frac{1}{2^2} = \frac{1}{4}$$

From equation 1] the calculated intensity loss in dB SPL is:

$$\text{Intensity loss} = 10 \log_{10} \frac{1}{4} = 10 \times (-0.60) = -6 \; dB$$

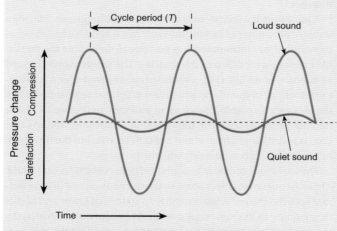

Wavelength = speed of sound x T
Frequency = $1/T$

Cycle period (T)

Loud sound

Pressure change
Compression
Rarefaction

Quiet sound

Time

Figure 1 Diagram of the pressure changes occurring during a sound wave. The pressure oscillates around atmospheric pressure in a repetitive manner with a period of T seconds. Upward deflections correspond to slight increases in pressure (compression) and downward deflections to slight decreases in pressure (rarefaction). Loud sounds cause larger pressure changes than soft ones without changing the period of the cycle.

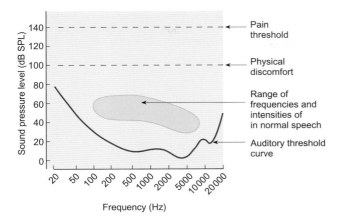

Figure 15.1 A plot of the auditory threshold for human hearing. Note that greatest sensitivity is found in the 2–3 kHz region. The range of frequencies and intensities of sounds encountered in speech are indicated by the area in green.

changes of percussive sounds are irregular. The snapping of a twig and the clicking of the fingers are good examples. Other impulsive sounds may occur randomly as acoustic 'noise' of the kind often encountered in older sound recordings as a background hiss. Finally, there are sounds which are relatively continuous and have a regular waveform, such as the sounds of speech and music (periodic sounds).

Individual frequencies (pure tones) give rise to distinct sensations of **pitch**. Pitch is a feature of all continuous sounds and is especially clear in music and speech, despite the fact that these sounds contain a mixture of frequencies. A high-frequency pure tone will subjectively be heard as high-pitched, while a low-frequency tone will be heard to have a low pitch. The apparent pitch of a continuous sound such as a musical note is related to its fundamental frequency—the lowest number of vibrations or pressure waves that occur each second. If the frequency increases, the pitch of the sound will be higher, and vice versa. This is easily demonstrated with a child's Swanee slide whistle. In the most sensitive part of the auditory spectrum, the ear is able to distinguish between frequencies that differ by only 3 cycles in 1 kHz. It is also able to distinguish between sounds of the same frequency that differ in intensity by only 1–2 dB.

If a particular note is sung by different people, or the same note is played by two different musical instruments, a flute and a violin for example, it is easy to distinguish between them because they sound different despite having the same pitch. The specific quality of a sound, known as its **timbre**, arises from the mixture of frequencies present and their relative intensities. In addition, their sound envelope plays an important role. This term encompasses how a sound begins and ends. For example, it may begin and end relatively slowly (as in a note on a church organ) or begin abruptly but end relatively slowly (as in the case of the same note played on the piano). These characteristics together with the mixture of frequencies determine the distinctive quality of a sound. The **loudness** or intensity of a sound depends on the amplitude of the vibrations affecting the ear. The greater their amplitude, the louder the sound.

Summary

Sounds consist of pressure variations in the air. The intensity of a sound is expressed in decibels (dB) relative to a standard intensity. Sounds are characterized by their duration, pitch, timbre (their specific qualities), and loudness. The pitch of a sound is correlated with frequency, and a pure tone is a sound consisting of just one frequency. The human ear is most sensitive to frequencies between 1 and 3 kHz, although it can detect sounds ranging from 20 Hz to 20 kHz. Most sounds are made up of a mixture of frequencies, and their timbre is correlated with the relative amplitudes of the different frequencies that make up the sound. The amplitude of a sound wave correlates with its loudness.

15.3 Structure of the auditory system

The auditory system may be arbitrarily divided into the **peripheral auditory system**, which consists of the ears, the auditory nerves, and the neurons of the two spiral ganglia; and the **central auditory system**, which consists of the neural pathways concerned with the analysis of sound from the cochlear nuclei to the auditory cortex. The ear itself is conventionally divided into the outer ear (the pinna and external auditory meatus), the middle ear (the eardrum, the ossicles, and their associated muscles), and the inner ear (the cochlea and auditory nerve). These structures are illustrated in Figure 15.2.

The visible part of the **outer ear** is the **pinna** or **auricle**, which is a highly convoluted structure formed of cartilage and covered by a closely fitting layer of skin. The lower part of the outward surface of the pinna leads to the **external auditory meatus** or **auditory canal**, which, in adults, is about 25 mm in length, about 7 mm in diameter, and lined with a layer of stratified squamous epithelium that is continuous with the skin. Like the skin, this epithelium contains hair follicles and sebaceous glands. It also contains modified sweat glands known as **ceruminous glands** that produce **cerumen**, or earwax, which traps debris, has antibacterial properties, and helps to maintain the flexibility of the eardrum.

The shape of the head, the convolutions of the pinna, and the auditory canal act together to form an acoustic resonator that increases the intensity of sound waves with frequencies in the range 2–5 kHz by 10–20 dB. In this way they effectively increase sound pressure at the tympanic membrane for those frequencies that are important in speech. In addition, the convolutions of the pinna play an important role in helping to determine the direction of a sound (see Section 15.5).

The **middle ear** or tympanic cavity is an air-filled space bounded laterally by the **tympanic membrane**, or **eardrum**, and medially by the **oval window** of the cochlea. These two membranes are coupled via three small bones, the **ossicles**. The middle ear is connected to the pharynx via the **Eustachian tube** (Figure 15.2). As long as the Eustachian tube is not blocked, the middle ear will be maintained at atmospheric pressure and the pressures on either side of the tympanic membrane will be equalized—a process aided by swallowing. If the pressures are not equalized, then some degree of hearing loss and discomfort will occur. This is a common experience during the ascent or descent of an aircraft.

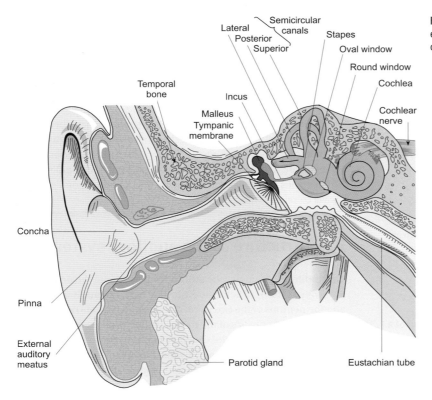

Figure 15.2 The principal structures of the human ear. The tensor tympani and stapedius muscles are omitted for clarity.

The tympanic membrane (also called the tympanum) is slightly ovoid in shape, 9–10 mm in its longest and 8–9 mm in its shortest diameter. It forms the boundary between the outer and middle ear. The main region of the tympanic membrane (the pars tensa) is translucent and consists of three layers: the cuticular striatum, which is continuous with the skin of the auditory meatus; the fibrous stratum, which consists of fibres radiating from the centre of the tympanic membrane and a less extensive array of circular fibres; and an inner layer called the mucous striatum, which is an extension of the mucosa of the middle ear. The pars tensa is bounded by a thick fibro-cartilaginous ring—the annulus tympanicus. The fibrous layer is richly supplied with blood vessels and sensory nerve fibres.

When viewed during normal otoscopic examination, the healthy tympanic membrane is translucent and greyish-pink in appearance (see Figure 15.3). It is attached to an elongated extension (the manubrium) of the first of the ossicles, the malleus, which can also be seen in the figure. The tympanic membrane takes the form of a shallow cone, with the apex protruding towards the middle ear. (The point of attachment to the end of the manubrium is called the **umbo**). The head of the malleus (or hammer) is connected to the second of the ossicles, the **incus** (or anvil), which is itself connected to the **stapes** (or stirrup), the smallest bone in the body (Figure 15.4). The footplate of the stapes is connected to the margin of the oval window by a ring of fibres, the annular ligament.

The middle ear has two small muscles: the **tensor tympani** and the **stapedius**. The tensor tympani emerges from a bony canal above the opening of the Eustachian tube. Its tendon is attached to the upper part (the root) of the malleus so that when it contracts, it pulls the malleus inward and increases the tension of the tympanic membrane. The stapedius is shorter and broader than the tensor tympani and arises from the back of the middle ear cavity, extending forwards to the neck of the stapes, where it is attached. Contraction

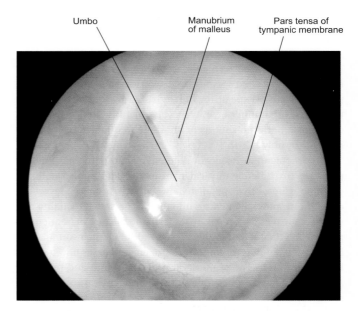

Figure 15.3 The appearance of the auditory canal of a healthy individual viewed through an otoscope. The manubrium of the malleus can be readily seen through the translucent tympanic membrane. (Image courtesy of health-advisors.org.)

of the stapedius acts to retract the footplate of the stapes from the oval window. The two muscles acting together reduce the efficiency by which sounds are transmitted through the middle ear (see Section 15.4). The tensor tympani is innervated by a branch of the mandibular division of the trigeminal nerve (CN V), while the stapedius is innervated by a branch of the facial nerve (CN VII). Indeed, one of the symptoms of paralysis of the facial nerve (**Bells' palsy**) is an increased sensitivity to sound.

The **inner ear** is located in the temporal bone of the skull and consists of a series of passages called the **bony labyrinth** containing a further series of fluid-filled sacs and tubes, the **membranous labyrinth**. The fluid of the bony labyrinth is called **perilymph** and is similar in composition to plasma in having a high sodium concentration, a low potassium concentration, and little protein. The fluid within the membranous labyrinth is called **endolymph** and has a composition similar to that of intracellular fluid, with a high potassium concentration and a low sodium concentration.

As Figure 15.2 shows, the cochlea is a coiled structure which, in humans, is about 1 cm in diameter at the base and 5 mm in height. In cross-section (Figure 15.4, p. 244) it is seen to consist of three tubes which spiral together for about 2¾ turns. These are the **scala vestibuli** (filled with perilymph), the **scala media** (filled with endolymph), and the **scala tympani** (also filled with perilymph). The scala vestibuli and scala tympani communicate at the apex of the cochlear spiral via the **helicotrema**. The endolymph is secreted by a highly vascular region in the wall of the scala media known as the **stria vascularis**. The secretion of endolymph is dependent on the active transport of ions which generates a large positive potential (c. +80 mV)—the **endocochlear potential**. The scala media is separated from the scala vestibuli by **Reissner's membrane** (also called the **vestibular membrane**) and from the scala tympani by the **basilar membrane**, on which the **organ of Corti**, the site of sound transduction, rests.

The movement of the footplate of the stapes is transmitted to the fluid of the scala vestibuli via the **oval window**. The flexible membrane of the **round window** separates the fluid of the scala tympani from the cavity of the middle ear. The central axis around which the cochlea is coiled is known as the modiolus and contains the auditory nerve and spiral ganglion.

Summary

The auditory system consists of the ear and the auditory pathways. The ear is divided into the outer ear, the middle ear, and the inner ear. The inner ear consists of a series of passages called the bony labyrinth that contains the fluid-filled sacs and tubes of the membranous labyrinth. The fluid of the bony labyrinth is perilymph, which is similar in ionic composition to plasma ($Na^+ > K^+$), while that of the membranous labyrinth is endolymph, which resembles intracellular fluid in ionic composition ($K^+ > Na^+$). The part of the labyrinth known as the cochlea is the true organ of hearing. The outer and middle ear ensure the efficient transfer of sound energy to the cochlea.

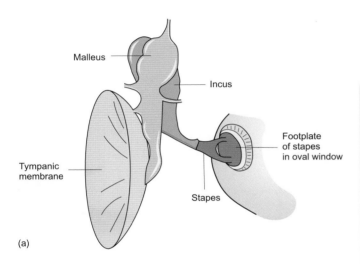

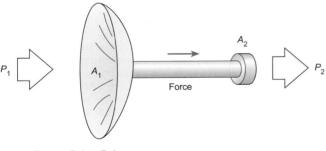

Force = $P_1 A_1 = P_2 A_2$
Therefore $P_2/P_1 = A_1/A_2$

(b)

Figure 15.4 The transmission of sound energy from the tympanic membrane to the oval window of the inner ear. As shown in panel (a), the pressure waves are collected over the whole of the area of the tympanic membrane and become concentrated over the much smaller area of the footplate of the stapes. This generates a higher pressure at the oval window than would otherwise be the case, as shown in panel (b). (Based on Fig 2.5 in J.O. Pickles (1982) *An introduction to the physiology of hearing.* Academic Press, London).

15.4 Mechanism of sound transduction

Impedance matching by the middle ear

Acoustic impedance can be thought of as resistance to the vibration of a medium such as air or water. The higher the impedance, the greater the energy required to set up a sound wave. The acoustic impedance of a given medium depends on its density, so a dense medium such as water has a much higher impedance than air. Indeed, the impedance of water is about 3,600 times that of air. As a result, airborne sounds striking the surface of water are almost totally reflected. Only about 0.1 per cent of the energy is transmitted as waterborne sound. Similar considerations apply to the hearing process, as the hair cells that are responsible for sound transduction are bathed in the aqueous fluids of the inner ear.

The role of the middle ear is to act as an **impedance transformer** to maximize the transfer of sound energy from the air to the fluids of the inner ear. The main factor in this **impedance matching** is the difference in area between the tympanic membrane and the footplate of the stapes. Since force is equal to pressure times area, the pressure collected over the tympanic membrane is applied to the much smaller area of the stapes footplate and a 21-fold net pressure gain is achieved (see Figure 15.5). This is further assisted by the lever action of the ossicles, which provides a mechanical advantage of 1:1.3, and by the broad acoustic resonance of the auditory canal mentioned earlier. These factors result in a theoretical 27-fold (c. 29 dB) increase in pressure at the oval window, which largely compensates for the loss of acoustic energy at the air/liquid interface. Overall, the efficient operation of the middle ear permits most of the energy of an airborne sound to be transmitted to the cochlea.

If the function of the middle ear is impaired, for example by fluid accumulation or as a result of fixation of the footplate of the stapes to the bone surrounding the oval window, the threshold of hearing will be elevated. This type of deafness is known as **conductive hearing loss** and will be discussed further in Section 15.6. The effectiveness of the operation of the middle ear is clinically assessed by means of **tympanometry**, in which a probe (a tympanometer) is placed in the ear canal. The tympanometer generates a pure tone at different intensities and measures the eardrum responses to the different pressures to indicate the compliance of the ear canal (the reciprocal of the impedance). A value for compliance above the normal range indicates abnormal mobility of the eardrum, which is most likely to be due to disruption of the ossicular chain. A decrease in compliance indicates an increase in the stiffness of the ossicular chain, which can be caused by otosclerosis or by an accumulation of fluid in the middle ear.

As mentioned above, the transmission of energy through the middle ear can be affected by the **tensor tympani** and **stapedius** muscles of the middle ear. Contraction of these muscles reduces the efficiency with which the ossicles transmit energy to the cochlea and attenuates the perceived sound. The middle ear muscles contract in response to loud sounds (the acoustic middle ear reflex), when chewing, and when speaking. While the latency of the middle ear reflex is too great (c. 150 ms) for protection against sudden loud sounds such as gunfire, the reflex is able to protect the ear from damage caused by the continuous loud sounds of noisy environments such as factories and rock concerts. The normal threshold for triggering the acoustic middle ear reflex is around 80dB SPL. The reflex affects both ears, so a loud sound in one ear will also reduce the transmission of sound in the other.

The travelling wave of the basilar membrane

The movement of the stapes in response to sound results in pressure changes within the cochlea. If the basilar membrane were rigid, the pressure waves would be transmitted from the oval window along the scala vestibuli to the helicotrema and down the scala tympani to be dissipated at the round window (Figure 15.6a). However, the basilar membrane is not rigid but flexible, so that pressure waves arriving at the oval window set up a series of travelling waves in the basilar membrane itself. In humans, the basilar membrane is about 34 mm long and varies in width from about

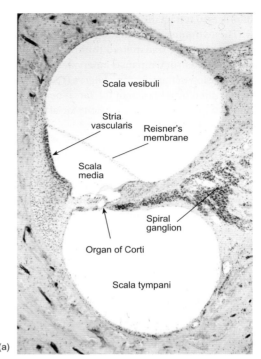

(a)

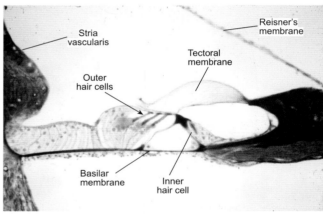

(b)

Figure 15.5 The structure of the cochlea. Panel (a) shows the three scalae and the position of the organ of Corti on the basilar membrane, which separates the scala tympani from the scala media. Part of the spiral ganglion is visible on the right. Panel (b) shows the structure of the organ of Corti in greater detail. Three outer hair cells are clearly visible in this section as is a single inner hair cell. The images in panels (a) and (b) are from different specimens.

100 μm at the base to 500 μm at the helicotrema. The stiffness of the basilar membrane also changes along its length. It is stiffest near the base (at the oval window) and becomes progressively less stiff as it spirals towards the helicotrema. These mechanical properties are important in the propagation of the travelling wave, which begins at the basal end by the oval window. As it travels along the basilar membrane, the amplitude of the wave increases until it

reaches a peak, whereupon it is rapidly attenuated (Figure 15.6b). The position of the peak amplitude of the wave depends on the frequency of the sound acting on the ear. High frequencies reach their peak amplitude near to the base of the cochlea, while low-frequency sounds elicit their peak nearer to the helicotrema (Figure 15.6c). The frequency of a sound wave is thus mapped onto the basilar membrane by the point of maximum displacement. Furthermore, the constituent frequencies of complex sounds are separated as they are mapped onto different regions of the basilar membrane. The peak displacement of the basilar membrane as it propagates the travelling wave is very small—less than a tenth of a micron (< 0.1 μm) even at high sound pressures (c. 100 dB SPL) and proportionally less at lower sound pressures. However, this displacement is about 30 times greater than the movement of the stapes.

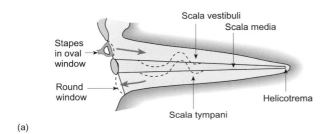

(a)

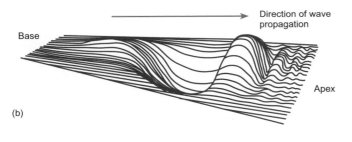

(b)

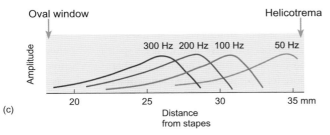

(c)

Figure 15.6 Panel (a) is a diagrammatic representation of the way in which sound energy elicits the travelling wave in the basilar membrane. Movement of the stapes footplate produces pressure waves in the scala vestibuli, which are transmitted to the scala media and scala tympani, setting the basilar membrane in motion. Panel (b) shows how the travelling wave progressively reaches a maximum as it passes along the basilar membrane before abruptly dying out. The peak amplitude of the travelling wave is reached at different places along the basilar membrane for different frequencies of sound, shown in panel (c).

The hair cells and the initiation of nerve impulses

How is the displacement of the basilar membrane by the travelling wave converted into nerve impulses? The cells responsible for this process are the **hair cells** of the organ of Corti. If these are absent (as in some genetic defects) or destroyed by ototoxic antibiotics such as kanamycin, the affected individual will be deaf (their vestibular function will also be disturbed—see Chapter 16).

Cochlear hair cells are arranged in two groups: outer hair cells, which are arranged in three rows, and inner hair cells, which are arranged in a single row (see Figure 15.7). A bundle of **stereocilia** projects from the upper surface of each hair cell. The stereocilia of the inner hair cells are arranged in a roughly linear fashion, while those of the outer hair cells are arranged in a characteristic V- or W-formation. In immature cochlear hair cells, a **kinocilium** is present in the front of each bundle of stereocilia. The stereocilia are modified microvilli with a core of fibrils, while the kinocilium is similar in structure to the motile cilia with the usual 9+2 arrangement of microtubules found elsewhere in the body (see Chapter 4). The kinocilium has no known motile function in the cochlea and is lost during development.

Within each bundle, the stereocilia are arranged in rows of descending size, the smaller ones being linked to those of the row in front by fine threads of protein known as tip links (see Figure 15.8). It is now believed that these links control the opening of specific mechano-sensitive channels in the tips of the stereocilia.

In the absence of stimulation, the inner hair cells have a membrane potential of about −60 mV. Displacement of the stereocilia elicits a **receptor potential** in the hair cells. This occurs by the following process: bending the stereocilia towards the tall edge of the bundle generates tension in the links, and this increases the chances of a transduction channel being open.

Conversely, when the stereocilia are bent away from the tall edge of the bundle, the tension in the links is reduced and more of the transduction channels become closed. In this way, the movement of the stereocilia controls the permeability of the hair cells to potassium ions.

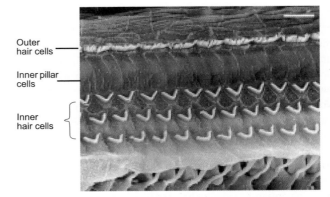

Figure 15.7 A scanning electron micrograph showing the arrangement of the three rows of the outer hair cells and the single row of inner hair cells of the organ of Corti. Note the different shapes formed by the stereocilia of the outer and inner hair cells. The pillar cells form the roof of the tunnel of Corti. (Courtesy of 'Journey into the World of Hearing', www.cochlea.eu, by R. Pujol et al.

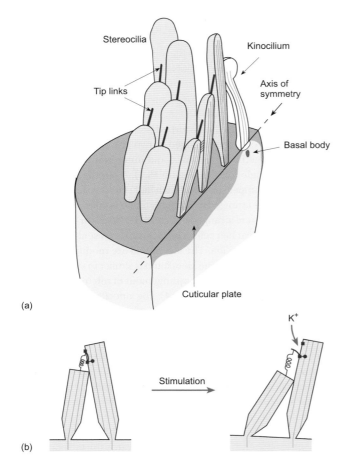

(a)

(b)

Figure 15.8 (a) The organization of the stereocilia on hair cells. (b) The likely mechanism by which the fine tip links activate ion channels to depolarize the hair cells in response to sound waves. Although the kinocilium (shown here in white) is present in the apical membrane of vestibular hair cells, it is not present in the rows of stereocilia found on mature cochlear hair cells. (Based on J.O. Pickles and D.P. Corey (1992) *Trends in Neurosciences* **15**, 255.)

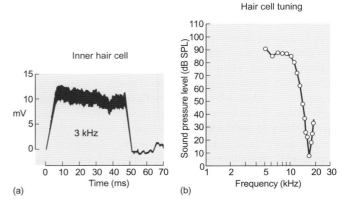

(a)

(b)

Figure 15.9 (a) A receptor potential observed in a guinea pig inner hair cell for a 3 kHz tone of 80 dB SPL. (b) The variation in sound intensity needed to elicit a receptor potential of a fixed amplitude for different frequencies. In this case the cell was most sensitive to a pure tone at 18 kHz. (Based on Fig 5 in I.J. Russell and P.M. Sellick (1978) *Journal of Physiology* **284**, 261–90.)

The transduction channels are non-selective cation channels, so their opening depolarizes the hair cells (cf. the action of acetylcholine at the neuromuscular junction—see Section 7.5). As the stereocilia are bent farther in the direction of the tall edge of the bundle, more and more channels open, resulting in a greater depolarization of the hair cell (i.e. a larger receptor potential). For any particular hair cell, the specific frequency that most easily elicits a receptor potential, its **characteristic frequency**, will depend on its position on the basilar membrane (see Figure 15.9). More intense sounds of frequencies higher and lower than the characteristic frequency will also elicit a receptor potential. The range of frequencies and intensities that excite a particular hair cell constitute its receptive field.

Why do the hair cells depolarize when the transduction channels are open? The apical surface and stereocilia are bathed in endolymph, while the basolateral surface of the hair cells is exposed to perilymph, which has a low potassium ion concentration, just like the extracellular fluid elsewhere in the body. Consequently there is a substantial potassium gradient across the basolateral region of the hair cells. At rest, the stereocilia and the apical surface

of the hair cells have a very low permeability to cations so that the resting membrane potential of the hair cells (c. −60 mV) is determined mainly by the potassium gradient across the basolateral surface. When the hair cells are activated, the transduction channels open and the permeability of the stereocilia to cations increases substantially. This results in a depolarization of the hair cells because the stereocilia are bathed in endolymph, which is rich in potassium. The depolarization of the apical region spreads to the rest of the hair cell and causes the voltage-gated calcium channels of the basolateral membrane to open. The subsequent rise in intracellular Ca^{2+} then triggers the secretion of neurotransmitter (glutamate); this diffuses across the synaptic cleft onto the afferent nerve endings of the cochlear nerve, which depolarize. The action potentials that result are then transmitted to the CNS. These events are summarized in Figure 15.10.

The cell bodies of the afferent nerve fibres are located in the spiral ganglion, and a specific set of fibres innervates a particular region of the basilar membrane. Since the basilar membrane is tuned so that a sound of a given frequency will elicit maximum motion in a particular region, specific sounds will excite a specific set of afferent fibres. For a given cochlear nerve fibre, there is a specific frequency to which it is most sensitive. This is known as the characteristic frequency of that fibre (cf. characteristic frequency of a hair cell), as shown in Figure 15.11. Since afferent fibres from different parts of the basilar membrane are connected to different parts of the spiral ganglion, the frequencies of the impinging sound waves are mapped along the ganglion and then on to the higher centres of the brain concerned with hearing. This is known as **cochleotopic** or **tonotopic mapping**.

The nerve fibres of the cochlear nerve show spontaneous activity (i.e. action potentials in the absence of specific stimulation). When they are stimulated by a specific sound, the frequency of action potential discharge is increased and the increased discharge frequency is related to the intensity of the sound. When a sound ends, the action potential frequency falls below its natural spontaneous rate for a short period. This may be important in signalling the timing of specific sounds.

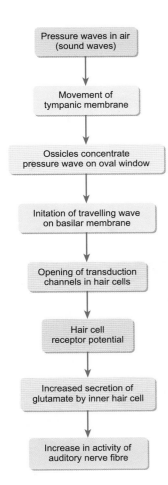

Figure 15.10 A flow chart illustrating the main stages in auditory transduction.

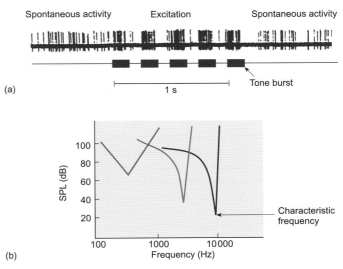

Figure 15.11 (a) The characteristic pattern of action potential discharge in the auditory nerve when the ear is exposed to brief periods of sound consisting of a pure tone (a tone burst). Note the spontaneous firing of the auditory nerve fibre and the increase in action potential discharge during each tone burst. (b) Tuning curves for various auditory nerve fibres. The lines represent the threshold intensity required to elicit an increase in action potential discharge. Note the overlap between the receptive fields of the different fibres at higher intensities of sound. (By kind permission of E.F. Evans.)

The role of the outer hair cells

The human cochlea has around three times as many outer hair cells as inner hair cells. However, they receive only about 5 per cent of the cochlear nerve fibres and their afferents do not send auditory information to the brain. Instead, the outer hair cells act as active filters that enhance the frequency selectivity and the sensitivity of the cochlea. The outer hair cells are excited by K^+ entry into the tips of stereocilia in the same way as the inner hair cells but, once activated, they undergo a fast contraction that relies on a very rapid conformational change in an abundant membrane protein called prestin. In this way they cause a localized amplification of the movement of the basilar membrane, which both sharpens the tuning of the basilar membrane and increases the sensitivity of the inner hair cells by about 60 dB. This action of the outer hair cells is known as the **cochlear amplifier**. The motility of the outer hair cells is lost in mice whose prestin gene has been deleted, and their auditory threshold is greatly increased.

Neural determinants of loudness

The assessment of the loudness of a sound depends at least two basic cues: the number of action potentials in specific nerve fibres and the recruitment of neighbouring afferents with similar, but not

identical, characteristic frequencies. This second cue reflects the fact that the auditory receptive field of a single afferent fibre becomes wider as a sound becomes louder (see Figure 15.11b). The duration of a sound also appears to play a role: to maintain a given subjective loudness, the intensity of a brief tone burst has to be increased by around 3 dB for every halving of its duration.

Summary

The middle ear acts as an impedance transformer to maximize the transfer of sound energy from the air to the fluids of the inner ear. Movements of the ossicles of the middle ear cause pressure waves to be set up in the fluids of the cochlea, which evoke a travelling wave in the basilar membrane. The movement of the basilar membrane activates the inner hair cells, which are responsible for sound transduction. The outer hair cells are motile and sharpen the tuning of the basilar membrane. This enhances the sensitivity of the inner hair cells.

The basilar membrane is tuned so that the constituent frequencies of a sound are mapped along it. High frequencies are represented near the oval window, while low frequencies are represented near to the apex of the cochlea. Since individual auditory fibres terminate in specific regions of the basilar membrane, activation of a specific set of fibres codes for the frequency components of a sound wave. The number of action potentials in a specific population of fibres codes for the intensity of the sound.

15.5 Central auditory processing

The organization of the auditory pathway is shown in Figure 15.12. The first steps in the processing of auditory information take place in the brainstem. The cochlear nerve divides as it enters the cochlear nucleus and sends branches to the three main subdivisions: the antero-ventral division, the postero-ventral division, and the dorsal division. Since the arrangement of the cochlear nerve fibres reflects their origin in the cochlea, this organization is preserved in the cochlear nucleus: the neurons of each subdivision are arranged so that they respond to different frequencies in a tonotopic order. In essence they provide an auditory map of frequencies of the sounds impinging on the ear. Fibres from the ventral cochlear nucleus on each side project to the superior olivary nuclei, whose neurons are the first neurons to receive an input from both ears.

From each olivary nucleus, fibres project to the inferior colliculi via the lateral lemniscus and thence to the medial geniculate bodies of the thalamus. From the medial geniculate body, auditory fibres project to the primary auditory cortex on each side. This is located on the upper aspect of the lateral (or Sylvian) fissure (see Figure 15.13). In most people, the area of cortex devoted to hearing

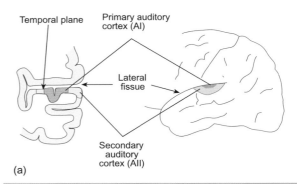

(a)

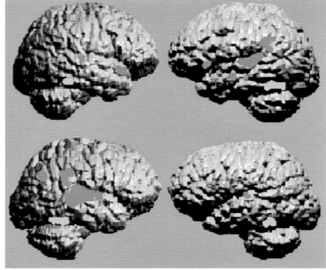

(b)

Figure 15.13 (a) The location of the auditory cortex. The primary auditory cortex is located mainly on the temporal plane (the planum temporale) of the lateral fissure (the Sylvian fissure), where it is represented by the darker blue zone. It is surrounded by the secondary auditory area (light blue). (b) Activation of the auditory areas of the brain, revealed by magneto-electroencephalography. The upper pair of images show the response of a normal brain to speech. The lower images show the equivalent images for a dyslexic subject. (The right side of the brain is shown on the left side of the figure.) The auditory response in the left hemisphere is greater in extent than that in the right. In the dyslexic subject most auditory activity appears to be in the right hemisphere. The area shown in light yellow indicate the short-latency responses of the brainstem. Those areas in red are longer-latency responses, which are all of cortical origin. Courtesy of www.cochlea.eu 'Journey into the World of Hearing', by R. Pujol et al., NeurOreille, Montpellier.)

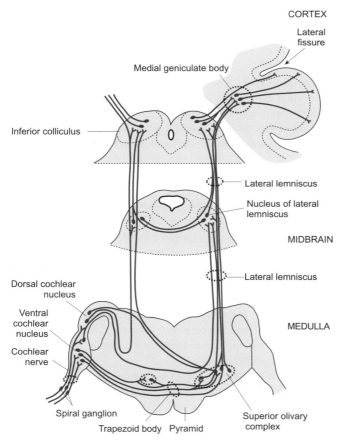

Figure 15.12 The organization of the ascending pathways of the auditory system. Note the crossing of the pathways in the brainstem and midbrain.

in the left hemisphere is greater than that in the right and is closely associated with **Wernicke's area**—a region of the brain specifically concerned with the comprehension of speech (see Chapter 19).

At each stage, auditory information is processed as it travels towards the auditory cortex, often by combining the input from both ears. For example, neurons in the superior olive are influenced by sounds in both ears. The sounds arriving at the ear on the same side (the ipsilateral ear) excite the olivary neurons, while those from the opposite ear (the contralateral ear) inhibit them. Neurons in other parts of the olivary complex show firing patterns that are related to the difference in the times at which a sound arrives at the two ears. It is highly likely that the brain utilizes this kind of information to aid the localization of sounds (as discussed in the next subsection).

The primary auditory cortex (A I) receives its innervation from the auditory nuclei of the medial geniculate body and tonotopic mapping is preserved. Nevertheless, animal studies suggest that the responses of the neurons of the auditory cortex are more complex than those of the earlier stages of the auditory system. Indeed, they show a wide variety of response patterns. In some, a tone burst is simply excitatory, in others it is inhibitory. A further group signal the beginning or end of a tone burst, or both its beginning and end. Yet other groups of auditory neurons are influenced by the modulation of specific frequencies. In these respects the primary auditory cortex appears to identify key features of a sound—rather as neurons in the primary visual cortex respond to particular features of a visual image, such as an edge or a movement in a specific direction (see Section 14.5). The secondary auditory cortex (A II) forms part of the auditory 'belt' that surrounds the primary auditory cortex. Its tonotopic organization is less clear, reflecting inputs from non-auditory thalamic nuclei in addition to that of the medial geniculate nucleus. It appears that this area (A II) is involved in the processing of sensory information arising from several modalities, but exactly how the brain interprets the auditory information it receives remains something of a mystery.

The localization of sound

One important function of the auditory system is to localize the source of a sound. It has been shown that humans, in common with many other animals, are able to do so with considerable accuracy. The head casts a sound shadow which gives rise to an intensity difference for the sounds arriving at the two ears, at least for those sounds whose wavelength is similar to, or smaller than, the dimensions of the head. Furthermore, the sound shadow cast by the pinnae provides a means of distinguishing sounds that originate in front from those that originate behind the head. Finally, the convolutions of the pinnae of the ears provide information regarding the localization of sounds in the vertical plane, although sounds emanating from sources directly above or below the head are the most difficult to locate.

The fact that the two ears are located at different points in space is of considerable significance in localizing a sound. As sound travels through the air at a finite speed (c. 340 m s^{-1}), it reaches one ear before the other, unless the source is directly in front of, or behind, the head. The brain is therefore able to use this time delay as a cue to localizing a sound—as mentioned earlier. A sound originating on the left side has to travel an extra 15 cm or so to reach the right ear. Thus, the sound will reach the right ear approximately 0.4 ms after it has reached the left. In fact, the brain can detect timing delays much shorter than this and, under optimum conditions, can detect time delays as brief as 30 μsec for sounds arriving at the two ears.

Summary

The first steps in the processing of auditory information take place in the brainstem. The neurons of the olivary complex receive information from both ears and may respond to differences in the intensity and timing of sounds arriving in the two ears. Nevertheless, an underlying tonotopic order is preserved. It is likely that many auditory reflexes depend on processing within the brainstem and thalamus.

The auditory cortex is located on the temporal plane of the lateral fissure. The primary auditory cortex (A I) receives its input from the auditory relay nucleus of the medial geniculate body. It is surrounded by the auditory 'belt' region, which includes the secondary auditory cortex (A II). This receives information from other thalamic nuclei in addition to that from the medial geniculate. Neurons in the auditory cortex have a variety of responses to sounds and appear to identify specific features of sounds.

15.6 Hearing deficits and their clinical evaluation

Hearing is a complex physiological process that can be disrupted in a variety of different ways. Broadly, hearing deficits are classified as conductive deafness, sensorineural deafness, and central deafness according to the site of the primary lesion.

Hearing loss

When a person is partially deaf, they have a raised threshold for hearing so that a higher pressure change is required for the ear to detect sound. This represents a loss of sensitivity compared to normal subjects and is called hearing loss. It is expressed as the difference in dB SPL between the threshold of hearing for the person concerned and that of the average sensitivity of normal healthy young subjects. If the threshold intensity of a particular tone were to be 40 dB higher than that of normal subjects, this would be indicated as a 40 dB hearing loss (sometimes expressed as 40 dB HL).

Hearing loss may affect one or both ears, and one may be more affected than the other. Its severity is classified as follows:

- Slight: 16–20 dB HL
- Mild: 20–40 dB HL
- Moderate: 40–70 dB HL
- Severe: 70–90 dB HL
- Profound: > 90 dB HL

Those with mild to moderate hearing loss generally benefit from using a hearing aid, while those with severe or total loss of hearing (**anacusis**) often have recourse to lip reading or sign language. Some people with profound deafness can be treated with a cochlear implant.

Conductive deafness

This type of hearing loss results from a defect in the middle or outer ear. Essentially the cause is a failure of the outer and middle ear to transmit sound efficiently to the inner ear. Conductive deafness is readily distinguished from sensorineural and central deafness by **Rinne's test**, which compares the ability of a patient to hear airborne sounds with their ability to detect those reaching the cochlea via vibrations of the skull ('bone conduction'). In this test, a vibrating tuning fork is moved close to the ear and, when its tone is no longer heard, the base of the fork is applied to the mastoid process behind the ear. If the patient can hear the tone once more, then the middle ear is affected; if not, then the deafness is likely to be due to sensorineural hearing loss. A more precise diagnosis can be obtained by **pure tone audiometry**. Pure tone audiometry is used to determine the auditory threshold of a subject by asking them to respond to a series of tones of progressively lower intensity. Two tests are performed on each ear: one for air conduction and one for bone conduction. The auditory threshold curve is plotted and compared with the average threshold for healthy subjects. A typical audiogram for conductive hearing loss is shown in Figure 15.14a.

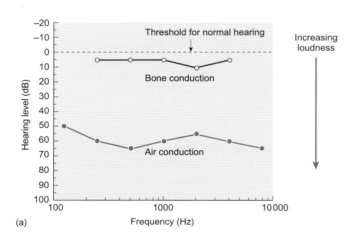

(a)

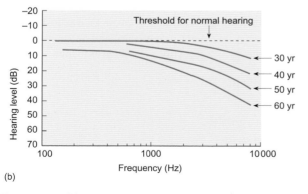

(b)

Figure 15.14 (a) An audiogram for a patient with conduction deafness in one ear. Note that while the auditory threshold is elevated for airborne sounds, bone conduction is nearly normal. (b) Audiogram showing the progressive loss of high frequency hearing with age.

Conductive hearing loss can result from various causes including repeated middle ear infections (which, in children, can lead to otitis media with effusion, commonly known as 'glue ear') and otosclerosis, in which the movement of the footplate of the stapes is impeded by the growth of bone around the oval window.

Sensorineural hearing loss

Sensorineural hearing loss results when some part of the cochlea or auditory nerve is damaged. Loss of hair cells has many different causes. It may be acquired by chronic exposure to certain chemicals, including industrial solvents such as toluene. It may also be the result of treatments by ototoxic drugs such as the aminoglycoside antibiotics (e.g. streptomycin and neomycin) and certain diuretics (e.g. furosemide). Finally, it may be caused by infections, particularly meningitis. Indeed, hearing loss is one of the most common after-effects of meningitis. Normal fetal development of hair cells can be affected if a pregnant woman is infected with rubella in the first half of pregnancy. Fortunately, the widespread adoption of the measles, mumps, and rubella (MMR) vaccine in the United Kingdom and elsewhere has significantly reduced the incidence of deafness caused by rubella.

Sensorineural hearing loss is quite commonly the result of traumatic damage to the cochlea by high-intensity sounds caused by industrial processes such as the riveting of steel plates ('boilermakers disease') or by sounds resulting from over-amplified music. Age-related hearing loss (known as **presbyacusis** or presbycusis) is a type of sensorineural deafness that specifically affects the high frequencies (see Figure 15.13b). The excessive growth of Schwann cells in the cochlear nerve (a Schwannoma or acoustic neuroma) can result in compression of the nerve, causing deafness.

A further very distressing but common condition associated with sensorineural hearing loss is **tinnitus**—the unremitting sensation of sound generated within the auditory system itself. The characteristics of tinnitus vary from subject to subject and include high-pitched continuous notes as well as buzzing or pulsing sounds. The causes of tinnitus are uncertain and may differ from person to person. It is possible that the electromechanical properties of the outer hair cells generate oscillations in the cochlea which are then perceived as sounds but, as cutting the auditory nerve does not always succeed in relieving severe tinnitus, it may simply reflect abnormal spontaneous activity in the auditory pathway. At present, available treatments for tinnitus offer little relief. However, the addition of restful sounds to quiet environments to distract the sufferer may offer some relief. This is known as 'sound therapy' or 'sound enrichment'. Nevertheless, most sufferers simply have to adjust to its continual presence.

Central deafness

Because the auditory pathways are extensively crossed at all levels above the cochlear nuclei, the hearing deficits resulting from damage to the pathways subserving the sense of hearing are usually very subtle. Generally speaking, unilateral damage to the auditory cortex does not result in any obvious hearing deficit although the affected person may have difficulty in localizing sounds. Extensive damage to the auditory cortex of the dominant hemisphere (usually the left hemisphere) will lead to difficulties in the recognition of speech, while extensive damage to the auditory cortex of the minor (right) hemisphere affects recognition of timbre and the interpretation of temporal sequences of sound, both of which are important in music and speech. These aspects of higher sensory processing are discussed in more detail in Chapter 18.

Summary

Hearing deficits are classified as conductive deafness, sensorineural deafness, or central deafness. Conductive and sensorineural deafness may be distinguished from one another by pure tone audiometry: in conductive deafness, the auditory threshold for bone conduction is normal, while that for air conduction is elevated. In sensorineural hearing less, the thresholds of both air and bone conduction are elevated.

✳ Checklist of key terms and concepts

Properties of sound

- Sounds consist of pressure variations in the air.
- The intensity of a sound is measured in decibels (dB).
- The human ear is most sensitive to frequencies between 1 and 3 kHz, although it can detect sounds ranging from 20 Hz to 20 kHz.
- Subjectively, sounds are characterized by their pitch, timbre (specific qualities), and loudness.

The auditory system

- The auditory system consists of the ear and the auditory pathways.
- The ear is divided into the outer ear, the middle ear, and the inner ear.
- The inner ear consists of the cochlea, which is the organ of hearing, and the vestibular system, which is concerned with balance.
- The inner ear is connected to the brain via the auditory branch of the eighth cranial nerve, which sends information to the dorsal and ventral cochlear nuclei.
- Fibres from the cochlear nuclei cross to the contralateral side and ascend via the lateral lemniscus to the inferior colliculi and medial geniculate body before reaching the auditory cortex. A smaller population ascends the brainstem on the ipsilateral side.

The mechanism of sound transduction

- The outer and middle ear serve to collect sound energy and focus it on to the oval window. This ensures the efficient transfer of sound energy to the cochlea.
- Incident sound waves cause pressure waves to be set up in the fluids of the cochlea, which evoke a travelling wave in the basilar membrane.
- The basilar membrane is tuned so that the constituent frequencies of a sound are represented on it as a map, with high frequencies mapped near the oval window and low frequencies towards the apex of the cochlea.
- The travelling wave activates the hair cells of the organ of Corti, which are responsible for sound transduction.

- The inner hair cells directly excite the nerve terminals of the auditory fibres.
- The outer hair cells act as an active filter to enhance the sensitivity of the basilar membrane for particular frequencies.

The coding of sounds

- Since individual auditory fibres terminate in specific regions of the basilar membrane, activation of a specific set of fibres codes for the frequency components of a sound wave.
- The loudness of the sound is indicated by the number of action potentials in specific afferent nerve fibres and by the number of auditory afferents recuited.
- The first steps in the processing of auditory information take place in the cochlear and olivary nuclei of the brainstem.
- The primary auditory cortex receives its input from the auditory relay nucleus of the medial geniculate body.
- Neurons in the auditory cortex have a variety of responses to sounds and appear to identify specific features of sounds.
- The auditory system is able to localize the source of a sound with considerable accuracy. Important cues are the differences in intensity and timing of sounds arriving at each ear.

Hearing deficits

- Hearing deficits are classified as conductive deafness, sensorineural deafness, or central deafness.
- Conductive and sensorineural deafness may be distinguished from one another by pure tone audiometry.
  - In conductive deafness, the auditory threshold for bone conduction is normal while that for air conduction is elevated.
  - In sensorineural hearing less, the thresholds of both air and bone conduction are elevated.
- Individuals suffering from central deafness may have difficulty in localizing sounds or difficulty in understanding speech.

Recommended reading

Anatomy of sensory systems

Brodal, P. (2004) *The central nervous system: Structure and function* (3rd edn), Chapter 8. Oxford University Press, New York.

Physiology

Pickles, J.O. (2013) *An introduction to the physiology of hearing* (4th edn). Brill Publishing, Leiden.

Squire, L.R., Berg, D., Bloom, F.E., Du Lac, S., Ghosh, A., and Spitzer, N.C. (2012) *Fundamental neuroscience* (4th edn) Chapter 25. Academic Press, San Diego.

Medicine

Donaghy, M. (2005) *Neurology* (2nd edn), Chapter 15, Deafness. Oxford University Press, Oxford.

Review articles

Grothe, B., Pecka, M., and McAlpine, D. (2010) Mechanisms of sound localization in mammals. *Physiological Reviews* **90**, 983–1012.

Hudspeth, A.J. (2014) Integrating the active process of hair cells with cochlear function. *Nature Reviews Neuroscience* **15**, 600–14.

Moerel, M., DeMartino, F., and Formisano, E. (2014) An anatomical and functional topography of human auditory cortical areas. *Frontiers in Neuroscience*, doi:10.3389/fnins.2014.00225.

Raphael, Y., and Altschuler, R.A. (2003) Structure and innervation of the cochlea. *Brain Research Bulletin* **60**, 397–422.

Saenz, M., and Langers, D.R.M. (2014) Tonotopic mapping of human auditory cortex. *Hearing Research* 307, 42–52.

Zatorre, R.J., Belin, P., and Penhune, V.B. (2002) Structure and function of auditory cortex: Music and speech. *Trends in Cognitive Sciences* **6**, 37–46.

To check that you have mastered the key concepts presented in this chapter, complete the accompanying online self-assessment questions. Go to www.oup.com/uk/pocock5e/

CHAPTER 16

The vestibular system and the sense of balance

Chapter contents

This chapter should help you to understand:

* The structure of the vestibular system

* How the vestibular system responds to angular and linear acceleration

* The role of the vestibular system in balance

* The vestibular control of eye movements

* Disorders of the vestibular system

16.1 Introduction

The sense of balance plays an important role in the maintenance of normal posture and in stabilizing the retinal image, particularly during running and walking. Indeed, people with impaired vestibular function often have difficulty in walking over irregular or compliant surfaces, particularly when they are deprived of visual cues by a blindfold. Such people also have difficulty with their vision during walking, as the visual world appears to move up and down (a condition called **oscillopsia**) rather than remaining stable as it does for normal individuals. This chapter explores the basic structure and function of the vestibular system and briefly discusses some common vestibular disorders.

16.2 Structure of the vestibular system

The organ of balance is the vestibular portion of the inner ear. It consists of two chambers, the **utricle** and **saccule**, and three **semicircular canals** (Figure 16.1). The utricle and saccule are arranged horizontally and vertically respectively, while the three semicircular canals are arranged at right angles to each other, with

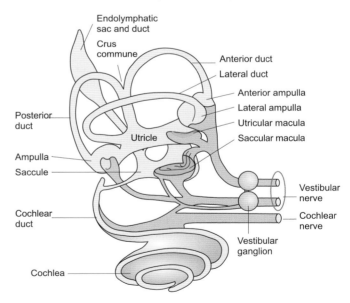

Figure 16.1 A schematic drawing showing the principal structures of the membranous labyrinth (shown in light orange) and their neural connections (shown in blue). Each of the semicircular canals ends with a swelling, the ampulla, which houses their sensory epithelia. (Based on Fig 16.1 of A.J. Benson in H.B. Barlow and J.D. Mollon (eds) (1982) *The senses*. Cambridge University Press, Cambridge, with permission.)

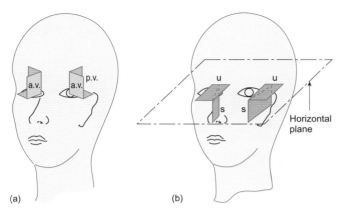

(a)   (b)

Figure 16.2 (a) The principal planes of the three semicircular canals, which are so arranged that they are able to detect any angular movements of the head. (b) The sensory epithelia of the utricle (u) and saccule (s), which are arranged in the horizontal and vertical planes to detect linear accelerations such as gravity. (Based on Fig 16.3 in A.J. Benson in H.B. Barlow and J.D. Mollon (eds) (1982) *The senses*. Cambridge University Press, Cambridge, with permission.)

the lateral canal inclined at an angle of about 30° from the horizontal. The other two canals lie in vertical planes (see Figure 16.2). The fluid within the semicircular canals, utricle, and saccule is endolymph, and the whole structure, known as the **membranous labyrinth**, floats in the perilymph that is contained within the bony labyrinth of the inner ear. (The different properties of endolymph and perilymph are described in Chapter 15.) The sensory portion of the semicircular canals is located near the utricle in a swelling known as the **ampulla**, while those of the utricle and saccule lie in a region of their internal surface called the **macula**. The nerve supply is via the vestibular branch of the eighth cranial nerve (CN VIII) and its associated ganglion (**Scarpa's ganglion**).

The sensory cells of the vestibular system are hair cells, which are similar to those of the cochlea. However, they differ in that their cilia consist of a single large hair known as the **kinocilium** and smaller **stereocilia** which are arranged in rows of descending height, as in the cochlea. (The mature hair cells of the cochlea have no kinocilium.) The orientation of the hair cells is rather specific for different parts of the vestibular apparatus. In the crista ampullaris of the semicircular canals, the hair cells are oriented so that they all have their kinocilium pointing in the same direction. In the utricle and saccule, the orientation of the hair cells is rather more complex and will be described later in the chapter. This morphological polarization coupled with the specific orientation of the semicircular canals, utricle, and saccule enables the vestibular system to interpret any movement of the head, as well as its position in space.

Summary

The vestibular system of each ear consists of three semicircular canals arranged at right angles to each other and two chambers arranged horizontally and vertically: the utricle and the saccule. The vestibular receptors are hair cells. The nerve supply is via the vestibular branch of the eighth cranial nerve.

16.3 The semicircular canals

A cross-section of the ampulla reveals that the wall of the canal projects inward to form a ridge of tissue called the **crista ampullaris** (Figure 16.3). Hair cells are located on the epithelial layer that covers the crista, and the hairs that project from their upper surface are embedded into a large gelatinous mass, the **cupula**, which is in loose contact with the wall of the ampulla at its free end. As a result, the cupula forms a compliant seal that closes the lumen of the canal and prevents the free circulation of the endolymph.

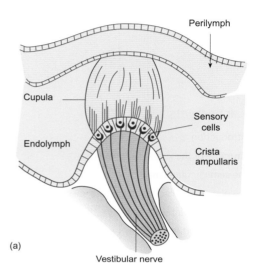

(a)

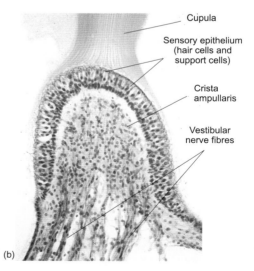

(b)

Figure 16.3 (a) Diagram of the structure of the ampulla of the semicircular canals. (b) A cross-section of the ampulla showing the sensory epithelium in greater detail. Note that the sensory hairs of the crista ampullaris project into the gelatinous mass of the cupula to form a flexible barrier that prevents the endolymph from circulating freely. (Panel (a) is adapted from Fig 13.24 in P. Brodal (1992) *The central nervous system: Structure and function*. Oxford University Press, New York.)

The mechanism of hair cell excitation in the vestibular system is essentially the same as that seen in the cochlea (see Chapter 15). Bending of the hairs towards the kinocilium opens transduction channels, and because the upper surface of the hair cells is bathed in a high potassium medium (endolymph) this results in depolarization of the hair cells and excitation of the vestibular afferent fibres. Bending of the hairs away from the kinocilium leads to closure of the transduction channels and hyperpolarization of the hair cells, resulting in a reduction of the vestibular afferent fibre discharge.

Operation of the crista ampullaris

Although the cupula of the crista effectively seals the semicircular canal at the ampulla, it hinges about the axis of the crista. When the head is turned, the walls of the labyrinth must move as they are attached to the skull, but as the endolymph in the semicircular canals is fluid it is less constrained and tends to lag behind by virtue of its inertia. The effect is to deflect the stereocilia of the cupula away from the direction of the head movement, as shown in Figure 16.4. For a simple turning movement to the right in a horizontal plane, the hair cells in the right horizontal canal excite their afferent fibres while those in the left canal inhibit theirs. As a result, the outputs of the two cristae act in a push–pull manner to signal the angular movement of the head. Similar coupled reactions will occur in the other pairs of canals in response to movements in other planes. The disposition of the semicircular canals in space is such that whatever the angular movement, at least one pair of the semicircular canals will be stimulated.

Signals from the semicircular canals control eye movements

The information derived from the semicircular canals is used to control eye movements (see also Chapter 14). Direct stimulation of the ampullary nerves elicits specific movements of the eyes. Stimulation of the horizontal canals by turning the head to the left results in the eyes turning to the right, as shown in Figure 16.5. This is one of a group of **vestibulo-ocular reflexes**. The eye movements elicited by activation of the ampullary receptors are specifically adapted to permit the gaze to remain steady during movement of the head. Such compensatory movements can also be seen in human subjects seated in a rotating chair. Movement of the chair to the left causes the eyes to move to the right to preserve the direction of the gaze (Figure 16.6). If the subject's eyes are first defocused using a pair of strong lenses, the eyes move in the direction opposite to the induced rotation until they reach their limit of travel, when they flick back to their centre position and the process of drift recurs. This is known as **vestibular nystagmus**. If the chair is rotated at a steady speed, the nystagmus decreases and is finally lost after about 20 s as the position of the cupula slowly reverts to its position at rest. Sudden stopping of the chair then leads to nystagmus in the direction opposite to that of the imposed rotation (**post-rotatory nystagmus**) as the inertia of the

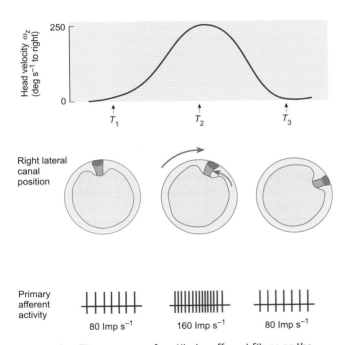

Figure 16.4 The response of vestibular afferent fibres as the head is turned to the *right*. Immediately following the movement, the stereocilia of the cupula hair cells are displaced towards the *left* by virtue of the inertia of the endolymph, and the afferent fibres respond with an increased firing rate. Once the movement ceases, the cupula and stereocilia resume their normal position with respect to the wall of the ampulla and the action potential discharge returns to its resting level. (Based on Fig 16.6 in A.J. Benson in H.B. Barlow and J.D. Mollon (eds) (1982) *The senses*. Cambridge University Press, Cambridge, with permission.)

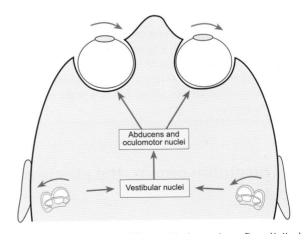

Figure 16.5 The operation of the vestibulo-ocular reflex elicited by rotating the head to the left. The eyes remain fixated on an object in the visual field and rotate to the right. A similar response is elicited by vertical movements of the head, such as those occurring during walking and running.

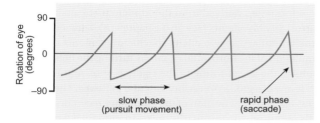

Figure 16.6 A vestibular nystagmus elicited by the steady slow rotation of a subject to the right. In the slow phase, the eye rotates to the left as it remains fixated on an object in the visual field until the eye reaches the limit of its travel, when it rapidly flicks to another object and tracks that before flicking to a third object and so on. The slow phase is called a pursuit movement and the fast phase is a saccade (see Chapter 14).

fluid then deflects the cupula in the opposite direction. Such procedures are sometimes used clinically to assess the function of the vestibular system.

Summary

The semicircular canals signal the angular movement of the head. Turning causes the endolymph of the semicircular canals to lag by virtue of its inertia. This has the effect of deflecting the cupula away from the direction of movement, and this excites the hair cells on the side to which the head has turned and inhibits the activity of those on the opposite side. The information derived from the semicircular canals is used to control eye movements via the vestibulo-ocular reflexes, whose role is to stabilize the visual field on the retina.

16.4 The otolith organs: the utricle and saccule

The sensory epithelium of the utricle is in the horizontal plane, while that of the saccule lies in the vertical plane, as shown in Figure 16.2. The sensory epithelium of both the utricle and saccule consists of a layer of hair cells covered by a gelatinous membrane

(the otolith membrane) over which lie small crystals of calcium carbonate called **otoconia** or **otoliths** (see Figure 16.7a–c). (The otoconia are also known as statoconia or statoliths.) The presence of the otoconia increases the density of the otolith membrane approximately threefold, and it is therefore more responsive to the position of the head than it would be without them. The stereocilia of the hair cells are embedded in the otolith membrane. As the orientation of the head is changed, the stereocilia of the hair cells are deflected from their normal resting position; this results in excitation or inhibition of the vestibular afferents. The principle is illustrated in Figure 16.8.

The receptors of the utricle and saccule signal linear accelerations such as gravity

In the utricle, the hair cells of the outer zone are polarized, and so they are excited when their stereocilia are deflected towards the centre of the macula. Those in the central region are polarized in the opposite sense (see Figure 16.7d). In the saccule, the hair cells are arranged so that they are stimulated when the stereocilia are deflected to the margins of the macula. Since the maculae of the utricle and saccule are arranged perpendicular to each other, they combine to signal linear acceleration in any direction. Unlike the semicircular canals, which signal angular acceleration, the utricle and saccule continuously relay information to the brain about the position of the head with respect to gravity. This is important in the control of posture and can be illustrated with a tilting platform. A subject is blindfolded and positioned over the pivot. As the platform is tilted, the arms extend and flex in an attempt to keep the head level and maintain a stable centre of gravity.

Summary

The utricle and saccule signal linear accelerations such as gravity. The two organs are disposed in such a way that they can signal any position of the head. The information derived from the utricle and saccule plays an important role in maintaining the position of the head and an upright posture.

16.5 The nervous pathways of the vestibular system

The nervous connections from the vestibular organs that permit the postural and oculomotor adjustments are complex and are shown in outline only in Figure 16.9. The afferents from the hair cells send their axons to the vestibular nuclear complex in the brainstem. These afferents give off collaterals to the cerebellar nuclei and play a crucial role in the maintenance of balance. The vestibular nuclei send fibres to the spinal cord via the lateral vestibulospinal tracts, and information from the proprioceptors reaches the vestibular nuclear complex via the spinovestibular tract. Eye movements are controlled via an anterior projection to

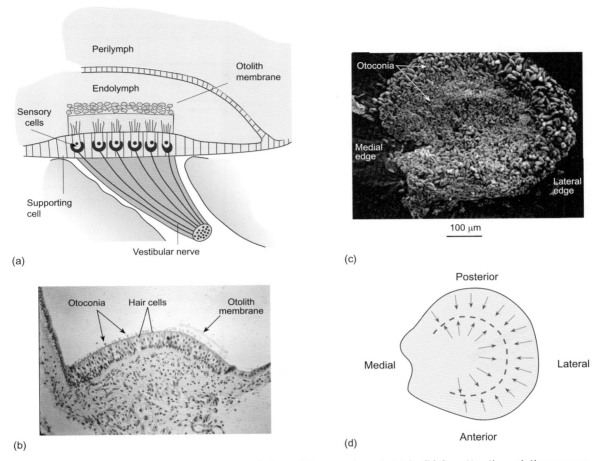

Figure 16.7 (a) Diagram of the structure of the sensory epithelium of the saccule and utricle. (b) A section through the sensory epithelium of the utricle. Much of the gelatinous otolith membrane has been lost in the preparation of the specimen, but small crystals of calcium carbonate (otoliths) can be seen embedded in its surface. (c) Scanning electron micrograph of the macula of the utricle of a mouse. Note that the larger crystals are located on the outer zones of the sensory epithelium. (d) A diagram showing the directional sensitivity of the hair cells of the macula (i.e. their morphological polarization). (Adapted from various sources. Part (c) is from H.P.N.M. Sondag et al. (1995) *Acta Otolaryngologica* (Stockholm) **115**, 227–30. © The Institute of Materials, Minerals and Mining, reprinted by permission of Taylor & Francis Ltd., www.tandfonline.com, on behalf of the Institute of Materials, Minerals and Mining.)

the pons, which ultimately relays information to the oculomotor nuclei which then act to maintain a stable image on the retina via the oculomotor reflexes described in Chapter 14.

16.6 Disorders of the vestibular system

From the previous discussion, it should be clear that the vestibular system is concerned with maintaining our posture and stabilizing the visual field on the retina. People with unilateral damage to the vestibular system have a sense of turning and of **vertigo**; they may also have abnormal eye movements. Vertigo is a disorienting illusion in which the sufferer either feels that they or their surroundings are moving even though they are, in fact, both stationary. For example, the sufferer may feel as though they are being pulled to

one side by an invisible force, or the floor may appear to tilt or sink. Initially, they have difficulty in maintaining their balance, although they eventually learn to compensate. Remarkably, if the vestibular system is damaged on both sides (for example as a result of the ototoxic effects of certain antibiotics), affected subjects are essentially unaware that they have a sensory deficit. They are able to stand, walk, and run in an apparently normal fashion. If, however, they are blindfolded, they become unsure of their posture and will fall over if they are asked to walk over a compliant surface such as a mattress. A normal subject would have no such difficulty.

The vestibular nerve has a basal level of activity signals that the hair cells of the semicircular canals are in their resting position. Head rotation elicits signals that are dependent on the angle and direction of movement, and the eyes move to stabilize the

Head tilt
(acceleration (1 g) constant)

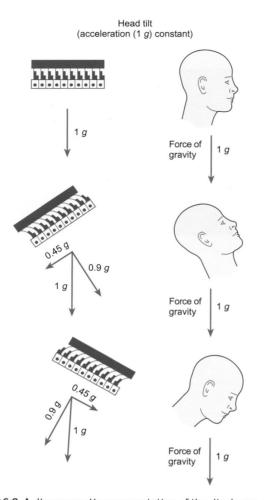

Figure 16.8 A diagrammatic representation of the displacement of the otolith membrane of the utricle to show how the hair cell stereocilia are displaced during tilting movements of the head. (Based on Fig 16.10 in A.J. Benson in H.B. Barlow and J.D. Mollon (eds) (1982) *The senses.* Cambridge University Press, Cambridge, with permission.)

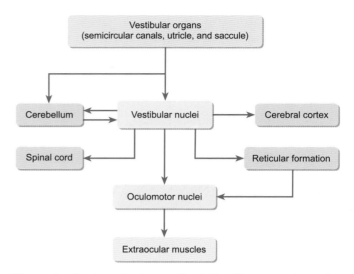

Figure 16.9 A schematic drawing illustrating the principal neural connections of the vestibular system.

direction of gaze (Figure 16.5). If, however, there is some abnormality of the semicircular canals, the signals from the various receptors will be out of balance and this will result in abnormal eye movements. Indeed, vestibular damage is characterized by a persistent nystagmus (see Chapter 14) that is evident even when the head is still.

The vestibulo-ocular reflexes have proved valuable in the diagnosis of vestibular disease. Caloric stimulation of the horizontal canal is generally used. The head is tilted backwards by 60° so that the horizontal canal is in the vertical plane. Then the auditory canal is rinsed with warm water. The heat from the auditory canal is conducted to the horizontal semicircular canal, and this elicits convective currents that cause deflection of the cupula resulting in

a nystagmus. The test can easily be repeated on the other side to establish whether one canal is defective. This test is also one of several that are used to establish whether a comatose patient is brain-dead.

Overproduction of endolymph may result in a condition known as **Menière's disease**, which affects both hearing and balance. There is a loss of sensitivity to low-frequency sounds, accompanied by attacks of dizziness or vertigo that may be so severe that the sufferer is unable to stand. These attacks are frequently accompanied by nausea and vomiting. Since the labyrinth is affected, there is also a pathological vestibular nystagmus.

Labyrinthitis (also known as otitis interna, vestibular neuronitis, and vestibular neuritis) is an inflammation of the inner ear affecting the vestibular organs. It is associated with vertigo, nausea, and vomiting. There may also be hearing loss, tinnitus (see Chapter 15), and a nystagmus. The cause is often unknown but it is thought that around half of all cases are caused by a viral infection of the chest, nose, mouth, or airways.

Motion sickness does not reflect damage to the vestibular system; rather it is caused by a conflict between the information arising from the vestibular system and that provided from other sensory systems such as vision and proprioception. Indeed, subjects with bilateral damage to the vestibular system do not appear to suffer from motion sickness.

Summary

Damage to the vestibular system results in the sufferer having a sense of turning and of vertigo. They have difficulty in maintaining their balance and find walking over uneven surfaces challenging. Common causes are Menière's disease and labyrinthitis.

Checklist of key terms and concepts

The vestibular system and the control of balance

- The vestibular system on each side of the head consists of three semicircular canals and two chambers, the utricle and the saccule.

- The three semicircular canals are arranged at right angles to each other, while the utricle and saccule are arranged horizontally and vertically respectively.

- The vestibular receptors are hair cells, which have specific orientations in each of the vestibular organs. The nerve supply is via the vestibular branch of the vestibulo-cochlear cranial nerve (CN VIII).

- The semicircular canals are so arranged that they are able to detect the angular movement of the head.

- The information derived from the semicircular canals is used to control eye movements via the vestibulo-ocular reflexes, whose role is to stabilize the visual field on the retina.

- The utricle and saccule signal linear accelerations such as gravity.

- Information from the utricle and saccule plays an important role in the maintenance of posture.

 ## Recommended reading

Anatomy of sensory systems

Brodal, P. (2004) *The central nervous system: Structure and function* (3rd edn), Chapter 9. Oxford University Press, New York.

Physiology

Squire, L.R., Berg, D., Bloom, F.E., Du Lac, S., Ghosh, A., and Spitzer, C.N. (2012) *Fundamental neuroscience* (4th edn), Chapter 32. Academic Press, San Diego.

Medicine

Donaghy, M. (2005) *Neurology* (2nd edn), Chapter 16. Oxford University Press, Oxford.

Review articles

Angelaki, D.E., and Cullen, K.E. (2008) Vestibular system: The many facets of a multimodal sense. *Annual Review of Neuroscience* **31**, 125–50.

Proske, U., and Gandevia, S.C. (2012) The proprioceptive senses: Their roles in signaling body shape, body position and movement, and muscle force. *Physiological Reviews* **92**, 1651–97.

To check that you have mastered the key concepts presented in this chapter, complete the accompanying online self-assessment questions. Go to www.oup.com/uk/pocock5e/

CHAPTER 17

The chemical senses—smell and taste

This chapter should help you to understand:

- The anatomy of the olfactory system
- The cellular basis of olfactory transduction
- The main olfactory nerve pathways
- The distribution of taste receptors in the oral cavity and upper oesophagus
- The structure of the taste buds
- The five modalities of taste and their mechanisms of transduction
- The afferent taste pathways

17.1 Introduction

The chemical senses, smell (**olfaction**) and taste (**gustation**), are amongst the most basic responses of higher animals to their environment. Each plays a distinct role in survival: the sense of smell allows for the assessment of foods—whether a fruit is ripe and suitable for consumption, or whether a foodstuff is rotten and unfit to eat—while the sense of taste plays a vital role once food is in the mouth. This is especially important both in the selection of nutritious foods and in the avoidance of poisons. The same considerations apply to humans, where both the sense of smell and that of taste play a significant role in the enjoyment of food. Indeed, the taste of food is profoundly disturbed when the sense of smell is temporarily impaired during a severe head cold.

This chapter first discusses the classification of different odours before describing the cellular basis of the sense of smell. It then proceeds to discuss the physiology of taste receptors and the central pathways involved in taste sensations.

17.2 The sense of smell (olfaction)

The sense of smell plays a crucial role in the behaviour of many mammals, including attracting a mate and claiming and defending territory. Although most primate behaviour, including that of humans, is driven more by visual and auditory information than by smell, the primate sense of smell is not a vestigial sense but is well developed. The difference in olfactory sensitivity between primates and animals such as dogs reflects the smaller number of olfactory receptors and the smaller proportion of the brain devoted to olfaction. Nevertheless, a large part of the primate brain is devoted to the sense of smell, and human beings are able to discriminate between the odours of many thousands of different substances. As odours are detected by receptors high in the nasal cavity, it is thought that only a few molecules are required to excite an individual olfactory receptor.

The smells of foodstuffs and everyday materials such as leather are a complex mixture of many different individual odours, and the ability to distinguish between them is a matter of everyday experience. The reaction of an individual to any odour is inevitably subjective. It often elicits an emotional response, reflecting pleasure or disgust which can depend on the context. Thus, while the smell of rotting vegetation is unacceptable in, say, a fridge or a kitchen, it may be perfectly acceptable (but not necessarily pleasant) near a compost heap or by a marsh.

Classification of odours

While there is no generally agreed classification, some authorities group odours into seven or more categories such as floral, fruity, earthy (e.g. mushroom compost), medicinal (e.g. disinfectant), sweaty, rotten, and pungent. Some regularities can be noted. Substances containing thiol (–SH) groups (known as thiols or mercaptans) all have disagreeable smells and can be detected at very low concentrations. Examples are hydrogen sulphide and methyl mercaptan (also called methane thiol), which can be detected at astonishingly low concentrations, around 0.001 parts per million. Most other odoriferous substances can only be detected at higher concentrations. There is, however, no simple relationship between chemical structure and odour: thus acetic acid (CH_3COOH—the principal ingredient of vinegar) is pungent, propionic acid (CH_3CH_2COOH) has an unpleasant rather pungent smell, while the odour of butyric acid ($CH_3CH_2CH_2COOH$) is not pungent but resembles body odour.

It is common experience that in the continuing presence of some odours they become less and less noticeable. This reflects an adaptation of the olfactory system. A total absence of the sense of smell is called **anosmia**. Some people may be anosmic for a particular odour—a condition called **specific anosmia**. For example, certain individuals are insensitive to the objectionable sweaty smell of isobutyric acid. Others, about one in a thousand, cannot smell butyl mercaptan—a component of the pungent odour of skunk spray. Many other examples of specific anosmia exist and reflect the different genetic makeup of each individual.

The threshold concentration for the detection of a given odour varies considerably between individual people. Moreover, the concentration at which an odour can be *identified* is always higher, sometimes many times higher than the threshold at which it can be *detected* (see Table 17.1). Odour recognition is complicated by the fact that the characteristics of some odours are influenced by the presence of others. Moreover, the characteristics of some odours such as perfumes change with their concentration in the air.

The peripheral olfactory system

The sensory organ for the sense of smell is the **olfactory epithelium**, which lies high in the nasal cavity above the turbinate bones. It is about 2–3 cm^2 in area on each side and is covered by a layer of mucus secreted by Bowman's glands, which lie beneath the epithelial layer (see Figure 17.1). The olfactory epithelium is pseudostratified (see Chapter 4) and consists of sensory receptor cells surrounded by supporting cells. Humans have approximately ten million olfactory sensory cells, which are thought to be constantly renewed. The sensory cells are bipolar neurons. They have a dendrite which has an expansion above the epithelial surface that is densely covered with sensory cilia. Their axons are unmyelinated and aggregate in bundles to form the first cranial nerve, the olfactory nerve (CN I), before passing through the cribriform plate of the ethmoid bone to the olfactory bulbs. Odours can reach the olfactory sensory cells either via inhaled air (the orthonasal route)

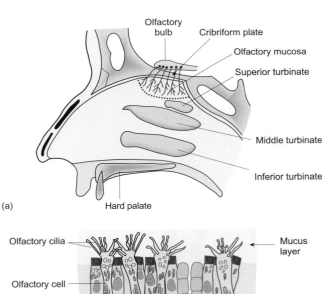

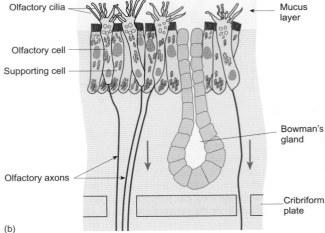

(a)

(b)

Figure 17.1 A sectional view of the nose to show the location (a) and detailed arrangement of the olfactory epithelium (b).

Table 17.1 Threshold concentrations for the detection and recognition of some odours

Compound	Odour detection threshold (ppm v/v)	Recognition threshold (ppm v/v)	Odour characteristics
Acetaldehyde	0.067	0.21	Fruity, pungent
Ammonia	17	37	Pungent, irritating
Chlorine	0.080	0.31	Pungent
Di-isopropyl amine	0.13	0.38	Fishy
Ethyl mercaptan	0.0003	0.001	Rotten cabbage
Hydrogen sulphide	0.0005	0.0047	Rotten eggs
Methyl mercaptan	< 0.0005	0.0010	Rotten cabbage
Skatole	0.001	0.050	Faecal and nauseating
Sulphur dioxide	2.7	4.4	Pungent, irritating

or from the back of the oral cavity (the retronasal route). The latter is the main route by which the complex process of appreciation of food occurs once it is in the mouth.

Olfactory transduction depends on activation of specific G protein linked receptors

To excite an olfactory sensory neuron, a substance must be both volatile and able to dissolve in the layer of mucus that covers the olfactory epithelium. Water-soluble odorants can readily diffuse to the sensory receptors, while hydrophobic molecules are bound by specific odorant-binding proteins which may aid their diffusion in the mucus layer. Odorants that have dissolved in the mucus can then activate their appropriate olfactory receptors on the cilia of the sensory cells. It is very likely that individual sensory cells express only one kind of olfactory receptor, although electrophysiological experiments have shown that individual sensory neurons respond to more than one odorant. Nevertheless, a sensory neuron is usually excited best by one odour.

It is now evident that mammals have more than a thousand genes for olfactory receptor proteins and about 300–400 of these are expressed on the olfactory cilia of human olfactory sensory neurons. Each olfactory receptor protein is coupled with a G protein that activates adenylyl cyclase. Thus, when an odorant molecule is bound by an appropriate receptor molecule, the intracellular concentration of cyclic AMP rises in the receptor cell. This increase in cyclic AMP opens a cyclic nucleotide gated channel, which is cation selective and allows both sodium and calcium ions to enter the cell (see Figure 17.2). As calcium ions enter the cell, intracellular calcium rises and activates a calcium-dependent chloride channel. Unusually, olfactory sensory neurons maintain a high intracellular chloride level and their chloride equilibrium potential is about +12 mV, so an increase in chloride conductance will augment depolarization of an olfactory sensory neuron (unlike most neurons where it inhibits depolarization—see Chapter 7). If the depolarization reaches the threshold for action potential generation, an action potential will be propagated to the olfactory bulb.

Central olfactory pathways

In most vertebrates, the olfactory bulbs are a prominent outgrowth of the forebrain. In humans, the olfactory bulbs and associated regions of the brain are less prominent and are located towards the front of the brain on its inferior aspect on either side of the midline (see Figure 17.3).

The olfactory axons terminate in the superficial regions of the olfactory bulb in spherical structures 100–200 μm in diameter called **olfactory glomeruli**. In the human brain, each olfactory bulb houses around 600 such glomeruli which are formed by the branching terminals of the olfactory nerve fibres and the apical axons of mitral cells with which they make synaptic contact. The mitral cells then send axons to other parts of the brain. Olfactory information therefore reaches the brain directly without passing through a peripheral ganglion.

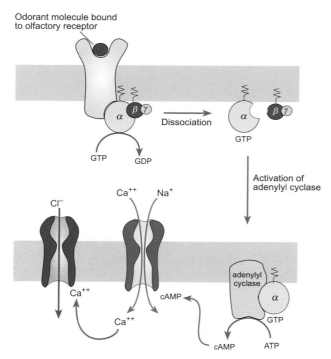

Figure 17.2 The process of transduction in an olfactory cilium. When a molecule of an odorant binds to an appropriate receptor, it activates a G protein which then dissociates, liberating its α-subunit. In turn, the α-subunit activates membrane-bound adenylyl cyclase, which then catalyses the formation of cyclic AMP. The cyclic AMP opens a cation-selective ion channel, which leads to the depolarization of the sensory neuron. As calcium ions enter the cell via the ion channel, intracellular calcium rises and this activates a calcium-dependent chloride channel which further augments the depolarization and the generation of action potentials by the olfactory sensory neuron.

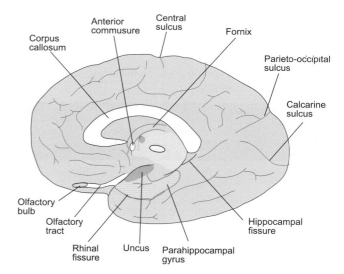

Figure 17.3 A mid-sagittal section of the brain with the brainstem removed to show the olfactory region of the right hemisphere on the *medial aspect* of the temporal lobe. The primary olfactory cortex is located in the uncus, here shown in red. The surrounding area, coloured light red, is the entorhinal area which also receives olfactory information.

In addition to the mitral cells, which are the second-order neurons of the olfactory system, the olfactory bulbs have various interneurons that are involved in processing olfactory information. These include tufted cells, periglomerular cells, and granule cells. The mitral cells are the main output of the olfactory bulb and send axons to other parts of the brain via the ipsilateral olfactory tract and to the contralateral olfactory cortex via the anterior commissure (Figure 17.4). Both electrophysiological and imaging experiments show that olfactory receptors with sensitivity to similar odours send their axons to a distinct subset of glomeruli, where they make synaptic contacts with the interneurons of the olfactory bulb. This suggests that individual olfactory glomeruli may respond selectively to particular kinds of odour.

Most of the olfactory tract fibres terminate in the cortex on the medial side of the temporal lobe in an area known as the **uncus** and in a subcortical structure called the **amygdala**, which forms part of the **limbic system**. Here the olfactory signals play a role in the establishment of emotions, motivational drives, and memory (see Chapter 18). Damage to the uncus may result in seizures that are often preceded by hallucinations of disagreeable odours.

Those areas of cortex that receive fibres from the olfactory tract are called the primary olfactory cortex, and it seems likely that this and its neighbouring areas are where the conscious awareness of olfactory stimuli occurs. In addition, olfactory information reaches the frontal lobes via the thalamus. As well as projecting to the amygdala and other structures of the limbic system, nerve fibres carrying olfactory information also reach the hypothalamus, where they may be important in regulating food intake.

Summary

The receptors for the sense of smell are located in the sensory epithelium above the third turbinate in the nose. The olfactory receptors are bipolar neurons whose apical dendrite terminates in sensory cilia where the olfactory receptors are located. Olfactory transduction involves the binding of an odorant molecule to a receptor protein. Once the odorant is bound, the receptor activates a G protein cascade that results in an increase in cyclic AMP and the depolarization of the primary sensory neurons.

The sensory neurons send their axons to the olfactory bulbs in bundles that form the olfactory nerve (CN I) and make synaptic contacts with the second-order olfactory neurons, the mitral cells, in the glomeruli of the olfactory bulb. The mitral cells project to the uncus of the temporal lobe and to the amygdala, which is part of the limbic system.

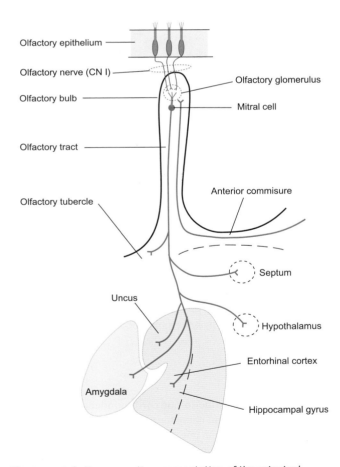

Figure 17.4 A diagrammatic representation of the principal olfactory pathways. Note the wide distribution of the olfactory afferents.

Olfactory epithelium
Olfactory nerve (CN I)
Olfactory bulb
Olfactory glomerulus
Mitral cell
Olfactory tract
Anterior commisure
Olfactory tubercle
Septum
Uncus
Hypothalamus
Entorhinal cortex
Amygdala
Hippocampal gyrus

17.3 The sense of taste (gustation)

Taste plays an important role in survival. It plays a crucial role in the regulation of our appetite and helps in the avoidance of poisons. Unlike the sense of smell, the sense of taste requires that food first be ingested before it can be tasted; only then can its qualities be determined. The sensory receptors for taste, the taste buds, are mainly found on the tongue and scattered around the oral cavity. A smaller number are also present in the epithelia of the soft palate, epiglottis, and pharynx. Although the tongue is the main organ of taste, it does have other important functions such as helping to form the sounds of speech, aiding mastication (chewing) by positioning the food within the mouth, and bolus formation prior to swallowing. All over the surface of the tongue there are small projections called **papillae** that give it its roughness. Four different types can be identified: filiform, folate, fungiform, and vallate. The filiform papillae have no taste buds and play no role in the sensation of taste. They are thought to be important in positioning food on the tongue during chewing and swallowing. The **taste buds** are found on the folate, fungiform, and vallate papillae (see Figure 17.5).

There are five basic modalities of taste

There are five distinct modalities of taste: salty, sour, sweet, bitter, and umami (the meaty, mouth filling taste of L-amino acids such as glutamate). Other qualities of taste have been described, such as fatty and metallic tastes, but there is no clear evidence that they arise from specific receptors. These different modalities serve somewhat different functions. The sweet and umami sensations

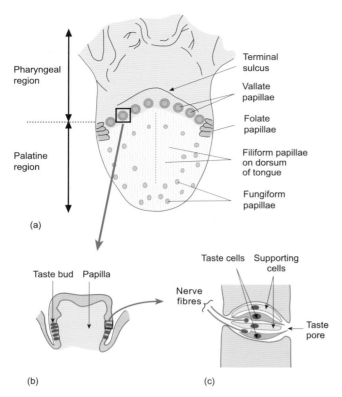

(a)

(b)   (c)

Figure 17.5 (a) The regions of the tongue and the distribution of the various kinds of papillae. (b) A cross-sectional view of a vallate papilla, showing the location of the taste buds. (c) The structure of a single taste bud; the taste cells are coloured green.

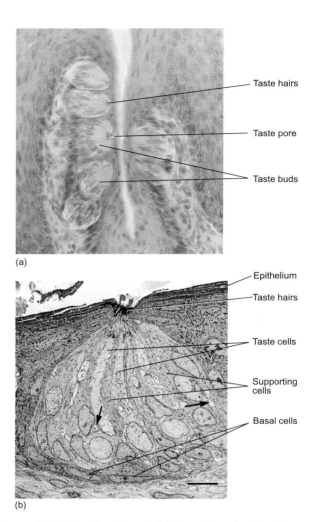

(a)

(b)

Figure 17.6 (a) A section through folate papillae showing the location of the taste buds. The microvilli (taste hairs) are strongly stained and project through a small opening in the surface epithelium (a taste pore) into the gutter between adjacent papillae. The gustatory nerve fibres are not visible in this preparation. (b) An electron micrograph of a taste bud in a folate papilla. The taste cells are lightly stained and are surrounded by darker supporting cells. Basal cells that proliferate to renew taste cells can be seen at the base of the taste bud. The arrows indicate afferent nerve fibres (intragemmal fibres). Scale bar 10 μm. (Panel (b) is from Fig 2a of N. Chaudhari and S.D. Roper (2010) The cell biology of taste. *The Journal of Cell Biology* **190**, 285–96 by Rockefeller University Press. Reproduced with permission of Rockefeller University Press in the format Book via Copyright Clearance Center.)

allow the selection of nutritious foods. Salt is essential for normal body function and the desire for salt is part of the regulation of sodium balance. The bitter and sour sensations are often aversive—most bitter-tasting substances are poisons, while contaminated foods often taste sour. The complexity of the taste of foodstuffs arises partly from the mixed sensations arising from stimulation of the different modalities of taste but chiefly from the additional stimulation of the olfactory receptors.

In humans (though not in some other mammals), all regions of the tongue are sensitive to the five modalities of taste, although there are small regional differences in threshold sensitivity between them. Only the upper surface (dorsum) of the tongue is insensitive to specific taste sensations (see Figure 17.5).

The structure of the taste buds

Each taste bud is located just below the surface epithelium and consists of a group of taste cells and supporting cells (see Figure 17.6). The taste cells have microvilli on their apical surface, visible in the light microscope as 'taste hairs', that communicate with the mucosal surface via a small opening called a **taste pore**. The basal regions of the taste cells of the tongue are innervated by afferent fibres of the facial (CN VII) and glossopharyngeal (CN IX) nerves. Those of the soft palate, pharynx, and upper side of the epiglottis are innervated by the vagus nerve (CN X). Current evidence suggests that there is no one-to-one correspondence between the taste cells and the afferent nerve terminals. Moreover, a taste cell may be served by more than one afferent fibre. Unlike the olfactory receptor cells (which are bipolar neurons), taste cells are epithelial cells that excite the afferent nerve terminals via synaptic contacts. The taste cells have a half-life of around 10 days and are constantly renewed from basal cells at the base of each taste bud.

Mechanisms of transduction

The transduction mechanisms for each modality of taste are summarized in Figure 17.7. Solutions that are salty activate taste cells by opening a specific epithelial ion channel (ENaC) that has a high permeability to sodium ions. This channel is activated by an increase in the concentration of extracellular sodium and can be inhibited by the potassium-sparing diuretic amiloride in most mammals but not, it appears, in humans. The opening of this ion channel depolarizes the taste cell, which leads to exocytosis of neurotransmitter by the taste cells onto the appropriate taste afferent nerve fibres, so exciting them.

Sour solutions are always of low pH. The hydrogen ions enter the taste cell either directly, via a proton channel, or by dissociation from a weak organic acid. This leads to acidification of the inside of the taste cell, closure of potassium channels, and depolarization of the cell membrane. The depolarization opens voltage-gated Ca^{2+} channels which trigger the exocytosis of neurotransmitter by the taste cells, causing excitation of the associated afferent taste fibres. The neurotransmitters secreted by those cells sensing salt and sour substances are not known. (A number of different transmitters are associated with taste buds including acetylcholine, ATP, serotonin (5-HT), and several peptides. The precise role of each is not yet clear.)

Bitter, sweet, and umami tastes are sensed by specific G protein coupled receptors belonging to two families designated T1R and T2R. Members of the T1R family act as receptors for sweet and umami tastes, while those of the T2R family serve as receptors for bitter tastes. They all activate phospholipase C, and the resultant increase in intracellular calcium opens an ion channel (TRPM5) that depolarizes the taste cell to cause the release of an excitatory transmitter (probably ATP) onto the taste afferents to cause excitation.

While individual taste cells respond to one modality of taste, recordings from the chorda tympani nerve show that afferent fibres respond to more than one modality. Nevertheless, the response of

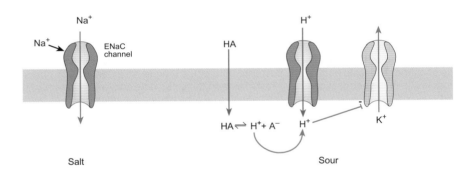

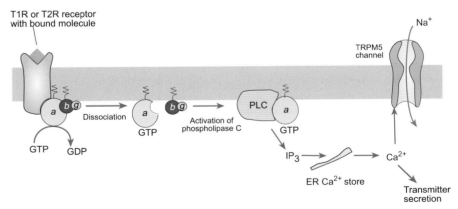

Figure 17.7 A summary of the cellular mechanisms permitting the detection of specific taste sensations. Taste cells responsive to salt open in response to an increase in sodium chloride concentration on their apical surface. This will directly depolarize the cell and trigger the secretion of neurotransmitter onto the afferent nerve terminals. Sour sensations are activated by a fall in pH (increased acidity). Hydrogen ions of strong acids enter the taste cell via a proton channel in the apical membrane, while weak acids cross the cell membrane in their neutral state and dissociate to release hydrogen ions once inside the cell. These then inhibit a potassium channel that is active at the resting membrane potential, depolarizing the membrane and opening a calcium channel (not shown) to initiate transmitter secretion. The bitter-, sweet-, and umami-sensing cells have different G protein coupled receptors but employ the same inositol lipid signalling pathway to release calcium ions from the intracellular stores. The calcium can either open a TRPM5 channel to depolarize the cell or participate directly in initiating transmitter secretion.

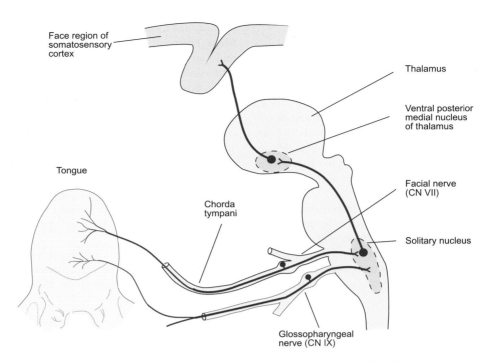

Figure 17.8 Diagrammatic representation of the neural pathways conveying taste information from the tongue. Taste buds on the anterior two thirds of the tongue are innervated by the lingual nerve (a branch of the facial nerve, CN VII) and reach the brainstem via the chorda tympani and the geniculate ganglion. The taste buds of the posterior third of the tongue are innervated by fibres of the glossopharyngeal nerve (CN IX), whose cell bodies lie in the petrosal ganglion. Taste buds on the soft palate and upper surface of the epiglottis are innervated by the vagus nerve (CN X, not shown) and reach the brainstem via the nodose ganglion.

these fibres generally shows a preference for one particular taste modality. So, to determine the taste of a substance that is present in the mouth, the brain presumably interprets the pattern of activity arising from the population of taste cells.

Afferent taste pathways

The anterior two-thirds of the tongue send afferent fibres to the CNS via the chorda tympani branch of the facial nerve (CN VII), while the posterior third of the tongue is served by the glossopharyngeal nerve (CN IX). The taste buds on the walls of the mouth and soft palate are innervated by the vagus nerves (CN X). All the taste fibres converge on the rostral part of the ipsilateral nucleus of the solitary tract—see Figure 17.8. Taste information is relayed to the face region of the cortex via the thalamus, from where it projects to the insular cortex, which is situated deep inside the lateral fissure adjacent to the somatosensory area for the tongue. There is no topographical mapping of taste sensations.

In addition to the main cortical projections, taste sensations are also relayed to the pons, the lateral hypothalamus, and the amygdala. These subcortical projections are thought to play a role in the regulation of the appetite for salt and for sweet-tasting foods. They also probably play a role in the aversion to bitter-tasting substances, many of which are toxic.

The sense of taste plays an important role in regulating motor activity relating to food intake, including mouth and tongue movements, salivation, and swallowing (see Chapter 43). In the case of aversive reactions to ingested material, taste sensations elicit other motor reactions including copious salivation, presumably to dilute any poisons that have been ingested, as well as coughing and breath-holding (apnoea). All of these responses are coordinated by neurons in the brainstem, particularly the nucleus ambiguus and nucleus of the solitary tract. Taste sensations also play a role in the regulation of gastric secretion and of the secretion of enzymes and insulin by the pancreas.

Summary

There are five main modalities of taste: salty, sour, sweet, bitter, and umami. The taste receptors are mainly found on the epithelium of the tongue but others are scattered around the oral cavity, pharynx, and epiglottis. Their afferent nerve fibres reach the brainstem via the facial, glossopharyngeal, and vagus nerves.

Individual taste cells respond preferentially to one modality of taste, but the afferent nerve fibres by which they are innervated tend to respond to more than one taste modality. It therefore appears that, as in the olfactory system, a particular taste sensation is coded in the pattern of information reaching the brain. The sense of taste is important in discriminating between different foodstuffs and in regulating some gastrointestinal secretions.

Checklist of key terms and concepts

The sense of smell

- The chemical senses are olfaction (sense of smell) and gustation (sense of taste).

- The olfactory receptors are located in the epithelium above the third turbinate in the nose.

- The olfactory receptor proteins are located on the cilia of the olfactory neurons.

- Individual olfactory receptor cells respond to more than one odour, so specific odours are encoded in the pattern of information that reaches the brain.

- In many animals, olfactory stimuli play an important role in sexual behaviour, although this is not strikingly evident in humans.

The sense of taste

- There are five modalities of taste: salty, sour, sweet, bitter, and umami.

- The taste receptors are found on the tongue and scattered around the oral cavity.

- Individual taste cells respond preferentially to one modality of taste, although other modalities will excite them so that, as for the olfactory system, the specific sense of taste is coded in the pattern of information reaching the brain.

- The taste receptors are innervated by branches of the facial, glossopharyngeal, and vagus nerves (cranial nerves VII, IX, and X).

- The sense of taste is important in discriminating between different foodstuffs and in regulating some gastrointestinal secretions.

 ## Recommended reading

Anatomy

Brodal, P. (2004) *The central nervous system: Structure and function* (3rd edn), Chapter 10. Oxford University Press, New York.

Physiology

Carpenter, R.H.S., and Reddi, B. (2012) *Neurophysiology: A conceptual approach* (5th edn), Chapter 8. Hodder Arnold, London.

Squire, L.R., Berg, D., Bloom, F.E., Du Lac, S., Ghosh, A., and Spitzer N.C. (2012) *Fundamental neuroscience* (4th edn), Chapter 23. Academic Press, San Diego.

Review articles

Dalton, R.P., and Lomvardas, S. (2015) Chemosensory receptor specificity and regulation. *Annual Review of Neurosciernce* **38**, 331–49.

Liman, E.R., Zhang, Y.V., and Montell, C. (2014) Peripheral coding of taste. *Neuron* **81**, 984–1000.

Spors, H., Albeanu, D.F., Murthy, V.N., Rinberg, D., Uchida, N., Wachowiak, M., and Friedrich, R.W. (2012) Illuminating vertebrate olfactory processing. *Journal of Neuroscience* **32**, 14102–8.

Zou, D.-J., Chesler, A., and Firestein, S. (2009) How the olfactory bulb got its glomeruli: A just so story? *Nature Reviews Neuroscience* **10**, 611–8.

 To check that you have mastered the key concepts presented in this chapter, complete the accompanying online self-assessment questions. Go to www.oup.com/uk/pocock5e/

CHAPTER 18
Emotion, learning, and memory

Chapter contents

This chapter should help you to understand:

- Emotional states and their role in behaviour
- The structures of the limbic system
- The role of the amygdala in emotion
- The role of the hypothalamus in emotion
- The three classes of learning:
  - simple learning
  - associative learning
  - complex learning
- Conditioned reflexes: passive and operant conditioning
- Cellular mechanisms of learning
- The role of the temporal lobe in memory
- Different types of memory: declarative memory and procedural memory

18.1 Introduction

The brain assesses aspects of the world around us so that we may adjust our behaviour to assist our survival. Moreover, human beings are social creatures who enjoy the company of their fellows, with whom they communicate using the complicated processes of speech. To achieve all this requires a certain level of self-awareness that we call consciousness. The way a person responds to a situation will depend on many factors including their previous experiences. This implies that there is a memory of past events which can only be acquired through experience, i.e. learning. Consideration of the underlying physiological processes raises many questions: What is consciousness? How do we form a representation of the external world? How are memories formed and retrieved? How do they trigger emotions? Are specific regions of the brain dedicated to specific functions, or are these complex activities distributed throughout the whole organ? This and the following chapter are concerned with neurophysiological aspects of these very complex issues. This chapter discusses the neural processes that underlie the phenomena of emotion, learning, and memory. The following chapter will discuss the specific functions of the cerebral hemispheres, the basis of speech and language, and concludes with a brief account of the physiological basis of biological rhythms.

18.2 The physiological basis of emotion

Experiences are assimilated into memories, which can be used to determine some future course of action. Indeed, specific events trigger particular emotions that often determine a particular pattern of behaviour.

All mammals show evidence of emotions such as aggression, fear, and pleasure. These basic emotions drive behaviours that serve to preserve the individual's life or to propagate the species. While mere observation cannot give any insight into the qualities of the emotions experienced by other species, we know from our own subjective experiences that life is coloured by many subtle feelings that affect our judgements and tend to drive our behaviour. A moment's reflection will identify major emotions such as anger, contempt, fear, joy, grief, hunger, panic, pleasure, surprise, thirst, and even disgust. From this it appears that emotions do not fall into neatly defined categories. Rather, they encompass a wide variety of important physiological and, in human beings at least, psychological phenomena. Some emotions are very particular responses to a given situation: the love (or contempt) for a particular person, or the disgust elicited by seeing (and no doubt smelling) a particular object such as a rotting corpse. Other emotions, such as happiness or sadness, are very general moods which colour many of the activities of life. Some emotions may persist for long periods, such as the love for a spouse and other family members, while others are transient, such as a moment of embarrassment or a burst of anger.

In everyday speech, an 'emotion' is any strong feeling elicited by a particular circumstance. More formally, an emotion is a complex physiological state with three distinct components: a subjective experience of some kind (often a specific sensation or group of sensations); a physiological response, such as an increase in heart rate; and a behavioural response—for example a change of facial expression: a smile of pleasure at seeing a friend or loved one, a frown in response to an adverse comment, the wrinkling of the nose at an unpleasant smell. In essence, emotions are driven by two opposing tendencies: positive emotions are driven by a reward of some kind, while negative emotions are a response to unpleasant stimuli such as an explicit punishment. Stimuli that elicit pleasurable emotions are termed positive stimuli, while those that are unpleasant or aversive are negative stimuli.

The mental state elicited by an emotional stimulus is known in psychology and psychiatric medicine as an **affective state**. It has three main components: how positive or negative the individual feels; the intensity of the feeling experienced (i.e. the degree of arousal); and the effect the feeling has on motivation—the drive to do something in response to the situation, such as the desire to escape from a threat. Many emotions are evident from the autonomic responses they elicit: an increase in heart rate in anger, or the drying of the mouth, pale skin, and sweating that accompanies fear or dread. Others elicit overt motor acts such as escaping from a major threat. However, all emotions have their origins in the activity of the brain. The various lines of evidence discussed in this chapter indicate that the seat of emotional activity lies in a group of forebrain structures that form what is known as the **limbic system**, which is the subject of the next section.

Summary

An emotion is a strong involuntary feeling that occurs in response to a particular situation or stimulus. It has three distinct components: a subjective experience of some kind, a physiological response and a behavioural response. In simple terms, emotions are driven by two opposing factors—reward and punishment. Emotions elicit autonomic responses appropriate to a given situation.

18.3 The limbic system

The components of the limbic system

The term 'limbic' comes from the Latin *limbus*, meaning 'border' or 'edge', which in this case refers to the cortical structures of the border between the cerebral hemispheres and the brainstem.

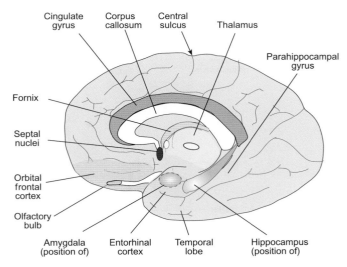

Figure 18.1 Diagram of the right hemisphere viewed from its medial aspect, showing the approximate positions of the principal structures of the limbic system in the human brain. Note that the amygdala and hippocampus lie beneath the temporal cortex.

Functionally, the limbic system as a whole is implicated in emotion, motivation, learning, and memory.

While there is some disagreement about the structures that constitute the limbic system, or even whether they act as a single system, the term generally includes the **olfactory bulbs, amygdala, hypothalamus, anterior thalamic nuclei, hippocampus, parahippocampal gyrus, fornix**, and **cingulate gyrus** (see Figure 18.1). Other structures that are often considered part of the limbic system are the **mammillary bodies**, the **septal nuclei**, and certain areas of the cerebral cortex including the **insular cortex** or **insula** (which lies deep within the lateral fissure) and the **prefrontal cortex** (specifically the orbital frontal cortex, which is that part of the frontal lobes lying above the orbits).

The amygdala

The amygdala of each hemisphere is a mass of grey matter roughly shaped like an almond that is located in the frontal part of the temporal lobe beneath the uncus (see Figure 18.2). (The name is derived from the Greek word *amygdale*, which means almond.) Each amygdala has connections with many different regions of the brain (see Figure 18.3), including the cingulate gyrus, the prefrontal cortex, the thalamus, and the hypothalamus.

The amygdala is subdivided into a number of nuclei that can broadly be considered as a cortico-medial group and a basolateral group. The two groups have somewhat different functions. The cortico-medial group receives information from the olfactory system and hypothalamus. It sends fibres to the hypothalamus and thence to the brainstem and the autonomic nervous system. Its primary function appears to be to mediate the autonomic responses that accompany the various emotions. The basolateral group is connected to the prefrontal cortex and thalamus, where it participates in the conscious perceptions of the temporal lobes.

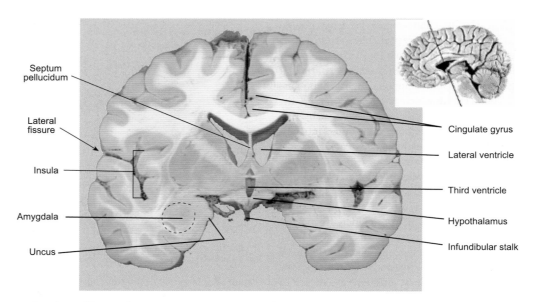

Figure 18.2 A coronal section of human brain showing the positions of key structures of the limbic system including the amygdala, hypothalamus, cingulate gyrus, and insular cortex (insula). The small inset figure shows the level of the section. (Courtesy of Claudia Krebs, University of British Columbia.)

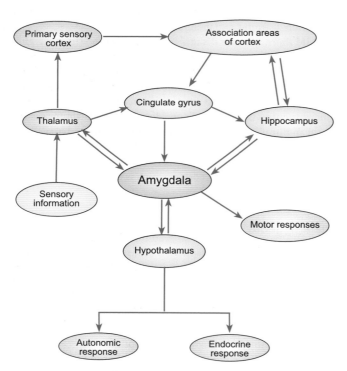

Figure 18.3 A diagrammatic representation of the key connections of the amygdala. The direct efferent projection from the amygdala to the prefrontal cortex is omitted for clarity.

In essence, the amygdala can be considered an important link between sensations and the emotional responses they evoke. As such, it plays an important part in determining social behaviour. In animal studies, bilateral destruction of the amygdala leads to hyper-sexuality, which is perhaps a reflection of the high density of sex hormone receptors in this area.

The amygdala stores memories strongly linked to the emotions. Animals with damaged amygdalae fail to respond to fear conditioning (a process in which a neutral stimulus such as a brief sound or flash of light is associated with a punishment of some kind such as an electric shock). In addition, damage to the amygdala leads to a loss of aggression, which is also seen following destruction of the septal nuclei. In monkeys this loss of aggression is accompanied by a more general reduction in the expression of emotions and in normal social interactions. This pattern is reflected in patients who have suffered bilateral damage to the amygdala—often as a result of viral infections. These people have difficulty in recognizing facial expressions, they suffer memory loss, and they do not appear to experience fear in situations that would elicit such a feeling in a healthy individual. Stimulation of the amygdala in conscious subjects causes them acute anxiety and may elicit hallucinations, with the subject reporting being in a 'dreamy' state.

If the cortico-medial nuclei are stimulated electrically, animals show behaviour appropriate to feeding such as salivation, lip licking, and chewing movements. Electrical stimulation may also initiate defecation and urination—functions mediated by the autonomic nervous system. If the basolateral nuclei are stimulated, animals show increased attention and arousal. Strong electrical stimulation may elicit behaviour appropriate to rage and anger.

The importance of the amygdala in unconsciously processing information has been revealed by functional imaging studies which show that the amygdala is strongly activated when a subject is presented with an image of a face expressing a strong emotion.

Both structural and functional changes in the amygdala have been reported in patients with a variety of psychiatric conditions, including anxiety disorders, depression, schizophrenia, and autism. Investigators have pointed out that while this means that people with such disorders have alterations to the amygdala, it does not mean that these changes *cause* the disorders; they may simply reflect a deficit in the role of the amygdala in the normal processing of emotion.

The role of the hypothalamus in emotion

In relation to its size, it is scarcely possible to exaggerate the importance of the hypothalamus in regulating the internal environment. Its role in controlling the basic emotions of thirst and hunger are discussed in Chapters 40 and 46 respectively while its role in autonomic function is discussed in Chapter 11. The hypothalamus plays a crucial part in the endocrine response to stress, particularly the secretion of cortisol. The underlying processes are outlined briefly in Chapters 20–22. Here the discussion concerns the role of the hypothalamus in mediating other aspects of behaviour relating to the emotions.

Electrical stimulation of the hypothalamus results in aggression, particularly towards other members of the same species, while lesions in these areas reduce aggression. One interpretation suggests that the hypothalamus initiates aggressive or defensive behaviour which is then executed by the motor pathways of the brainstem and spinal cord. Under normal circumstances this kind of behaviour is subject to regulation by higher centres, such as those parts of the limbic system that act to assess any potential threat.

A pair of small round bodies on the lower surface of the brain mark the posterior boundary of the hypothalamus. These are the **mammillary bodies** and are located at the anterior end of the arches of the **fornix** (see Figure 18.4). They form part of the limbic system and receive synaptic inputs from the amygdalae and hippocampi. Their output goes to the thalamus via the mammillo-thalamic tract. These connections appear to be important for recollection of memories, and the degeneration of the mammillary bodies seen in **Korsakoff psychosis** is associated with severely impaired memory.

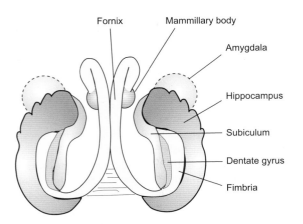

Figure 18.4 Illustration of the relative positions of the hippocampal formation, fornix, amygdala, and mammillary bodies as viewed from above. (Adapted from Fig 20.2 in P. Brodal (2004) *The central nervous system: Structure and function*, 3rd edition. Oxford University Press, New York.)

This syndrome is due to thiamine deficiency (i.e. lack of vitamin B_1), commonly as a result of alcoholism. Atrophy of the mammillary bodies is also seen in a number of other conditions including **Alzheimer's disease** (see Box 18.1).

The role of the cerebral cortex in emotion

Various structures of the cortex have also been shown to be involved in aspects of behaviour related to the emotions, especially the cingulate cortex, the orbital frontal cortex, and the insular cortex. Stimulation of these regions elicits combined autonomic, behavioural, and emotional responses.

The cingulate gyrus makes afferent and efferent connections with the prefrontal cortex, septum, subiculum (a part of the cortex between the hippocampus and the entorhinal cortex), and thalamus. It also sends fibres to the striatum and cerebellum (both of which are concerned with motor control) and to the amygdala. Evidence suggests that the anterior and posterior regions of the cingulate gyrus have somewhat different roles in the generation of emotional states. The anterior region has been implicated in the formation of a conscious awareness of an emotion and in monitoring the success of goal-directed behaviour. Evidence from fMRI studies indicates that it is also involved with the evaluation of the emotional quality of pain. The posterior region of the cingulate gyrus is concerned with learning and memory; in particular it is concerned with autobiographical memory (also called episodic memory; discussed in Section 18.4). When this region is stimulated in a conscious subject, both positive and negative emotions are evoked, although negative emotions such as fear and anxiety predominate. This is also the case when the amygdala is electrically stimulated. Removal of the cingulate gyrus in monkeys has similar effects to damage of the amygdala, with a loss of aggression and withdrawal from normal social interaction.

The hippocampus

The hippocampus and its associated structures of the temporal lobe (the dentate gyrus, entorhinal cortex, and subiculum) form an elongated structure called the **hippocampal formation**. The two hippocampi have extensive reciprocal connections with cortical association areas, and connecting fibres link them with other structures of the limbic system. In addition, the two hippocampi are linked via extensive commissural fibres. While many aspects of their function remain unclear, they play a major role in spatial awareness—knowing one's place in the environment—as well as in learning and memory, which are discussed in Section 18.3.

The limbic system is connected with another forebrain structure, the **nucleus accumbens**, which is part of the basal ganglia. Evidence from experiments on rats with stimulating electrodes implanted into the nucleus accumbens show that they repeatedly press a lever to stimulate this region and will persist in this behaviour rather than eat or drink, even to the point where they can die of exhaustion. For this reason the nucleus accumbens has been considered to be part of the reward system and has been linked more generally to addictive behaviour.

Box 18.1 Alzheimer's disease

Spontaneous and progressive degeneration of neurons in specific areas of the brain or spinal cord is responsible for a number of disorders of the CNS. These include Parkinson's disease, Huntington's chorea, Creutzfeldt-Jakob disease (CJD), and Alzheimer's disease.

Alzheimer's disease is the most common cause of dementia in the elderly, with more than 30 per cent of those over 85 years of age showing signs of the disorder, but it may also affect people in their thirties. Interestingly, degenerative changes characteristic of Alzheimer's disease are also seen in almost all patients with Down syndrome (trisomy 21—see Chapter 4) over the age of 40. Alzheimer's disease is characterized by progressive and inexorable disorientation and impairment of memory (generally over a period of 5–15 years) as well as by other defects such as disturbances of language and of visuospatial and locomotor function.

The precise causes of Alzheimer's disease are not yet established, but post-mortem examination of the brains of patients has revealed some characteristic anatomical and microscopic histological changes. Gross inspection of an affected brain shows significant atrophy with generalized loss of neural tissue, narrowing of the cerebral cortical gyri, and widening of the sulci. There is also dilation of the cerebral ventricles. These changes are most obvious in the frontal, temporal, and parietal lobes of the brain. The neocortex, basal forebrain (which contains the major cholinergic projections for the brain), and hippocampal areas are particularly affected. The latter area is known to be associated with the formation of memories (see main text).

Microscopic examination of post-mortem brain reveals the presence of abnormal protein aggregates, senile (or amyloid) plaques in the extracellular space, and neurofibrillary tangles within the neurons themselves. The plaques are largely made up of fragments of amyloid precursor protein (APP), termed β-amyloid protein (Aβ). The tangles are coarse filamentous aggregates of a hyperphosphorylated form of a protein known as **Tau protein**. Tau is a soluble, microtubule-associated protein, required for the function and stability of microtubules. Plaques and tangles are seen in the brains of normal elderly people, but their increased prevalence and specific location within areas such as the hippocampus is diagnostic of Alzheimer's disease. Similar amyloid deposits are often seen in the blood vessels of the brains (amyloid angiopathy) of patients who have had Alzheimer's disease.

Although the histological changes characteristic of Alzheimer's disease are well-documented, the underlying causes are not established. However, a number of factors have been identified which may have a role in the development of the disease. Around 10 per cent of Alzheimer's disease cases are familial in origin. Furthermore, as previously mentioned, there is a clear association between Down syndrome and neuronal degeneration similar to that seen in Alzheimer's disease. Thus it is likely that there is a genetic component to this disease. Indeed it has been shown that mutations at four chromosomal loci are associated with Alzheimer's disease. Unsurprisingly, one of these is located on chromosome 21. The affected gene encodes the amyloid precursor protein whose abnormal breakdown generates the β-amyloid plaques.

Further gene mutations also occur at loci on chromosomes 14 and 1, the so-called presenilin genes that are apparently associated with increased production of amyloid in the CNS and with an enhanced rate of apoptosis (programmed cell death). A fourth mutation on chromosome 19 occurs at the locus that encodes apolipoprotein E. The mutant form of apoE, expressed in AD, seems to be involved in the transport and processing of APP. It also seems to bind Aβ more effectively than the normal form and may therefore enhance the rate at which amyloid fibrils form.

There is no effective treatment for Alzheimer's disease at present, although treatment with cholinesterase inhibitors (e.g. galantamine) appears to slow its course. Considerable research effort is being directed towards the development of drugs that may be able to inhibit the activity of enzymes involved in the formation of β-amyloid.

Drinking excessive amounts of alcohol over a long period increases the risk of developing some form of dementia, including Alzheimer's disease. The resulting brain damage is also associated with other types of neurological disorder including Korsakoff's psychosis, which is characterized by degeneration of the mammillary bodies resulting from thiamine deficiency. (Thiamine is also known as vitamin B$_1$, and its importance is discussed further in Chapter 46.)

Summary

The limbic system consists of a number of forebrain structures including the amygdala, hypothalamus, and cingulate gyrus. The amygdala is a mass of grey matter located in the frontal part of the temporal lobe. It mediates the autonomic responses that accompany the various emotions and is an important link between sensations and the emotional responses they evoke.

Patients with bilateral damage to the amygdala have difficulty in recognizing facial expressions, suffer memory loss, and apparently do not experience fear in threatening situations. Electrical stimulation of the amygdala in conscious subjects causes acute anxiety and elicits hallucinations.

Stimulation of parts of the hypothalamus initiates aggressive behaviour, which is normally subject to regulation by higher centres, including the cingulate cortex, the orbital frontal cortex, and the insular cortex. Stimulation of these regions elicits combined autonomic, behavioural, and emotional responses. The limbic system is connected with the nucleus accumbens, which is part of the reward system and has been linked more generally to addictive behaviour.

18.4 Learning and memory

To survive in the world, all complex animals, including man, need to learn about their environment so that they can find food and water and avoid danger. The process of learning is concerned with establishing a store of information that can be used to guide future behaviour. The store of information gained through learning is known as memory. The importance of memory to normal human activity is evident in patients suffering from various neurodegenerative diseases, notably Alzheimer's disease (see Box 18.1).

Although learning can, and often does, occur without any obvious immediate change in behaviour, we can only be sure that a task has been learnt (i.e. has been stored in the memory and can be retrieved) if there is some modification of behaviour. For this reason, work on learning and memory requires an experimenter to devise some external measure of performance. The study of the process of learning is therefore bound up with the study of memory as a matter of practical necessity.

The various kinds of learning can be divided into three groups: simple learning, associative learning, and complex learning. **Simple learning** is concerned with the modification of a behavioural response to a repeated stimulus. The response may become weaker as the stimulus is perceived to have no particular importance. This process is called **habituation**. If an unpleasant or otherwise strong stimulus is given, the original reflex response is enhanced. This is called **sensitization**. In **associative learning** an animal makes a connection between a neutral stimulus and a second stimulus that is either rewarding or noxious in some way. **Complex learning** is diverse in its nature. It includes **imprinting**, as when young birds learn to recognize their parents by some specific characteristic; **latent learning**, in which experience of a particular environment can hasten the learning of a specific task, such as finding food in a maze; and **observational learning** (copying).

In all types of learning, there needs to be a change in the strength of specific neural connections. In this way, a behavioural response to a given stimulus can be modified by experience to suit changing circumstances. Not all experiences are remembered. Many are forgotten or only partially remembered, so the important questions to answer therefore include:

a) What are the conditions that lead to learning?

b) Where is memory located? Are specific memories located in particular structures or are they stored in parallel pathways with overlap and significant redundancy?

c) How is memory laid down? What mechanisms are responsible for the changes in neural connectivity?

d) How is information recalled?

Associative learning

It is a matter of everyday experience that people will associate one thing with another. The chiming of a clock can be a reminder that lunch is due, or that it is time to go home. In this type of association, a neutral stimulus is associated with some more important matter. This kind of behaviour is not confined to human beings. It can be found in many animals and was first systematically studied by I.P. Pavlov when he was working on salivary secretion in dogs.

In his experiments Pavlov paired a powerful stimulus, such as food, with a neutral stimulus—classically the ringing of a bell. He discovered that after a number of trials in which a dog was fed immediately after hearing the sound of a bell, the animal would begin to salivate on hearing the bell in anticipation of being fed. This is known as a **conditioned reflex**—normal salivation in response to food is called the **unconditioned response**. After establishment of the conditioned response, the ringing of the bell (the conditional stimulus) is sufficient for the salivary response to occur. The process of establishing a conditioned reflex is known as **passive conditioning**.

In **operant conditioning**, an animal learns to perform a specific task to gain a reward or avoid punishment. In this respect, the protocol is different to the establishment of a classical conditioned response where an animal responds passively to pairs of stimuli provided by the experimenter. During operant conditioning, an animal may learn to press a lever to gain a reward of a sweet drink or to avoid an electric shock. Initially, a naive animal is placed into the experimental chamber which is often an apparatus known as a Skinner box (after the behaviourist B.F. Skinner, who investigated operant conditioning in detail). The animal then explores its new surroundings, and by chance it may find that pressing a lever will lead to the delivery of a food pellet or a drop of a sweet liquid. After a short period, the animal learns to associate the pressing of the lever with the delivery of food or some other powerful stimulus. In this way animals can be taught to perform quite complex tasks.

In both passive conditioning and operant conditioning, the animal learns to **associate** one stimulus with another. This is also true of **aversive learning**, where an animal learns to avoid harmful experiences such as the eating of poisonous foods by associating them with a particular colour, smell, or adverse reaction (e.g. vomiting).

Cellular mechanisms of learning

All animal behaviour is determined by the synaptic activity of the CNS, especially that of the brain. In view of this, it would appear to be a logical step to look for changes in synaptic connectivity following some learnt task. This approach has proved feasible in certain lower animals that have simple nervous systems and stereotyped behavioural patterns. Changes in the strength of certain synaptic connections have been shown, for example during the habituation and sensitization of the gill withdrawal reflex of the sea-slug *Aplysia*.

In mammals, the CNS is so complex that this direct association is less easy to achieve. Nevertheless, in some brain regions, long-lasting changes in the strength of certain synaptic connections have been reliably found following specific patterns of stimulation. In some circumstances synaptic connections are strengthened, while in others they are weakened. These are precisely the kinds of synaptic modulation that would be expected for the remodelling of a neural pathway during learning.

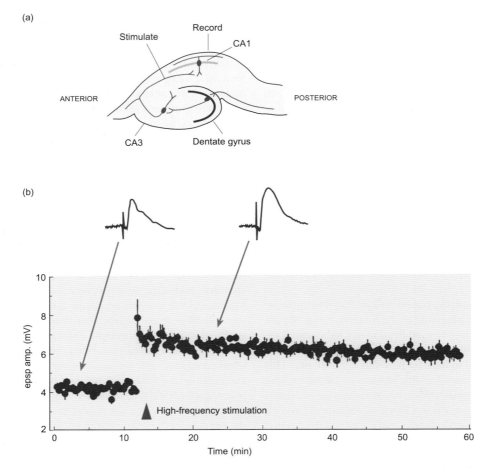

Figure 18.5 Long-term potentiation of synaptic transmission in the hippocampus. The diagram at the top of the figure shows the organization of the hippocampus and the position of the electrodes for the records shown below. Before the brief period of high-frequency stimulation, the epsps evoked by stimulation were about 4 mV in amplitude. After the high-frequency stimulation, they were about 6 mV (see inset examples). The increase in amplitude lasted for the duration of the experiment. (From original data of D.A. Richards.)

One region that has been the subject of intensive study is the **hippocampus**, which has been implicated in human memory (discussed later in this section). The efficacy of synaptic connections between neurons of the hippocampus is readily modified by previous synaptic activity. This is precisely what is required of synaptic connections that are involved in learning. The basic neural circuit of the hippocampus consists of a trisynaptic pathway, as shown in Figure 18.5a. Brief patterned stimulation leads to a very long-lasting increase in the efficacy of synaptic transmission in several of these pathways that lasts for many minutes or even hours (see Figure 18.5b). This phenomenon has become known as **long-term potentiation** or LTP.

Since its discovery by T. Bliss and T. Lømo in 1971, LTP has become the focus of a great deal of experimental work to determine exactly how the changes in synaptic efficacy are brought about. One important factor is the increase in intracellular calcium in the postsynaptic neuron that follows repeated synaptic activation. Sadly, further discussion of the mechanisms involved is outside the scope of this book. Nevertheless, hippocampal LTP provides an example of synaptic transmission behaving as predicted for establishing associative memory.

Memory

Human memory does not act like a tape recorder or a computer hard disc in which experiences are recorded in an orderly sequence that is then available for total recall. Only certain aspects of experience are remembered for long periods of time, other matters may be remembered for a short time and then forgotten. Still other trivial incidents are soon beyond recall—the precise location of your pen or your reading glasses for example. Even our long-term memories are not exact—no two people will give identical versions of an event they both witnessed. More usually, the salient points are recalled. This shows that memory is a representation of our past experience, not a record of it.

Memory must first begin with our sensory impressions—and at this early stage it will consist of the information passing through the sensory pathways. This will include that part of sensory experience that reaches the association cortex, where information from the various senses is integrated to form an image of the world. This type of memory, sometimes called **immediate memory**, is very short-lived as it is constantly being updated. For this reason, it is assumed that it is encoded in the electrical activity of neural networks.

More enduring information storage is classified as **short-term memory** and **long-term memory**. Information stored in short-term memory may either become incorporated into a permanent long-term memory store or discarded. Long-term memory can be disrupted without loss of short-term memory capabilities, but the necessity of a long-term memory store becomes evident when our capacity to store information in short-term memory is exceeded, or when we are distracted in some way.

Examples of short-term memory include the remembering of an appointment, the rehearsal of a telephone number before dialling, or the need to buy our groceries—once the task has been executed then the incident is rapidly forgotten unless it is given some special significance. Our long-term memories include our name, the names and appearances of our family and friends, important events in our lives, and so on. In the absence of brain damage or disease, it lasts for a lifetime. This stage of memory requires the remodelling of specific neural connections. That this is the case is shown by the inability of experimental animals to learn specific tasks if inhibitors of protein synthesis are injected into their cerebral ventricles prior to training. The various stages of establishing a memory are summarized in Figure 18.6. Long-term memory is conveniently divided into procedural memory (also called reflexive memory), declarative memory, and emotional memory.

Procedural memory has an automatic quality. It is acquired over time by the repetition of certain tasks and is evidenced by an improvement in the performance of those tasks. It is knowing *how* to do something. Examples are speech, in which the vocabulary and grammar of a language are acquired through experience; the playing of a game; or the mastery of a musical instrument. Once learnt, the performance of these tasks does not require a conscious effort of recall.

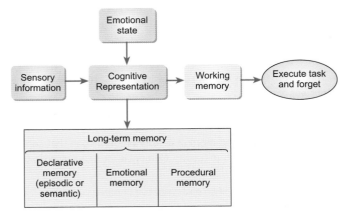

Figure 18.6 A simple schematic drawing showing one view of the organization of memory.

Declarative memory is further subdivided into **semantic memory** (factual knowledge) and memory for events or **episodic memory** (our personal autobiography). Procedural memories are very resistant to disruption, while episodic memories (which relate to specific events) are relatively easily lost.

The neural basis of memory

From the foregoing account, it is clear that memory is a very complex phenomenon; but work carried out over the last half-century has begun to reveal a few of its mysteries. The protocols for studying associative learning and careful clinical investigation of various memory deficits have provided some tools with which to approach the problem of the location of memory in the human brain.

Early animal studies in which the cerebral cortex was partially or totally removed showed no evidence of a specific memory being located at a specific site. In these experiments, animals were taught to find their way around a maze; their performance after part of their cortex was removed was then compared to their performance before the operation. The results indicated that the loss of memory was related to the total area of cortex removed rather than to removal of a specific area. This suggested that it is stored in many parallel pathways rather than at a specific site. This is, perhaps, not surprising as many parts of the brain will be involved in any task that requires extensive sensorimotor coordination.

The role of the temporal lobe

When the great Canadian neurosurgeon Wilder Penfield stimulated the temporal lobes of patients with an electrode prior to removal of brain tissue for the relief of epilepsy, he found that stimulation of points in the temporal cortex, hippocampus, and amygdala apparently evoked specific memories of a vivid nature. These studies have been interpreted as indicating that specific episodic memories are located in the temporal lobes. Careful re-evaluation of this evidence suggests that this is not the case. The reported experiences tended to have a dream-like quality and may have represented confabulations. Moreover, stimulation of the same site did not always elicit the same memory and excision of the area under the electrode did not result in the loss of the memory item that had been elicited by stimulation.

How is memory retrieved?

At present, this question cannot be answered as we do not know where specific memories are laid down in the brain. For procedural memory, it seems plausible that recall begins with the triggering of specific sequences of nerve activity by certain cues. The more a specific act has been rehearsed and refined, the more accurate the performance. This is obviously true of motor acts such as walking and running but is also true of the use of language, reading music, and other similar tasks. The execution of the learnt pattern is therefore the act of recall in this case.

The recall of declarative memory has been studied by observing how subjects recount stories they have been told. Rather than simply retelling the original with all its detail, they tend to shorten the story to its essentials—in essence, they reconstruct the story. This suggests that the recall of such memories is a specific decoding process.

A deficit in recall is called **amnesia** and is often observed in patients who have received a head injury. If the amnesia is of events before the injury it is known as **retrograde amnesia**; if it is of events after the injury it is **anterograde amnesia**. Total amnesia is rare and generally very short-lived, so that the amnesia that follows a head injury gradually passes and the period of amnesia is localized to the period immediately surrounding the incident.

The importance of the hippocampus and amygdala in memory was dramatically revealed by the case of a patient known until his death as H.M. (Henry Molaison), who underwent surgery for temporal lobe epilepsy that required bilateral removal of his hippocampus and amygdala. Following the operation he had a profound anterograde amnesia (an inability to remember events that occurred after the operation). He was only able to remember items by rehearsal. If his repetition was interrupted then he was no longer able to recall any item. He did not know where he lived or what he last ate. In short, he had no short-term memory and effectively always lived in the present. In his own words 'At this moment everything looks clear to me, but what happened just before? That's what worries me. It's like waking from a dream. I just don't remember.' Patients suffering degeneration of other related structures such as the mammillary bodies (which, with the hippocampus and amygdala, form part of the limbic system) also develop a profound anterograde amnesia.

A remarkable feature of H.M.'s condition is that he could learn new tasks, such as how to read mirror writing. As with normal subjects, his performance improved with each training session even though he could not recall ever being trained in the task. The ability of other amnesic patients to learn new skills has been repeatedly demonstrated. This highlights the essential difference between episodic and procedural memory.

Summary

Learning is the laying down of a store of knowledge that can be used as a guide to future activity. Memory is the name given to that store. Learning is a complex process that can be classified as associative or non-associative. All learning must involve changes to the connections between the neurons involved in specific tasks.

Memory is not localized to very specific sites but is distributed through a number of neural networks, each of which is involved in the task at hand. Loss of memory is called amnesia. Inability to recall events that occurred before an incident is retrograde amnesia, while inability to recall events that occurred after an incident is anterograde amnesia.

(✱) Checklist of key terms and concepts

Physiology of emotion

- An emotion is an involuntary strong feeling elicited by a particular situation.
- Emotions have survival value and serve to preserve the life of an individual or to enhance its prospects of successful reproduction.
- Emotions elicit autonomic responses appropriate to a given situation and are driven by two opposing factors—reward and punishment.
- The mental state elicited by an emotional stimulus is known in psychology and psychiatric medicine as an affective state.
- The limbic system plays an important role in establishing an emotion.

The limbic system

- The limbic system encompasses a number of structures in the basal forebrain.
- Areas of the limbic system that are involved in emotions are the amygdala, the hypothalamus, and the cingulate gyrus.
- The amygdala provides a link between sensations and emotional responses.
- It also acts as a store of memories linked to the emotions.
- Electrical stimulation of the amygdala in human subjects causes acute anxiety.

- Patients with a variety of psychiatric conditions show evidence of structural and functional changes in the amygdala.
- The involvement of the hypothalamus in emotions is reflected in its central role in the autonomic nervous system and in its role in mediating stress, which is accompanied by an increase in cortisol secretion.
- Stimulation of the cingulate cortex, the orbital frontal cortex, and the insular cortex elicits complex emotional responses.
- The cingulate gyrus is involved with the evaluation of the emotional quality of pain and with memory of that pain.
- The hippocampus plays a major role in spatial awareness and in learning and memory.
- The limbic system is connected with the reward systems of the nucleus accumbens.

Learning and memory

- Learning is the laying down of a store of knowledge that can be used as a guide to future activity. Memory is the name given to that store.
- Learning can be classified as associative or non-associative.
- All learning must involve changes to the connections between the neurons involved in specific tasks.

- Memory is not, in general, localized to very specific sites but is usually distributed through the neural networks that are involved in the performance of a given task.

- Memory can be divided into short-term memory and long-term memory, both of which require changes in the strength of specific synaptic connections.

- Long-term memory involves structural changes to specific neural pathways.

- Loss of memory is called amnesia. Inability to recall events that occurred before an incident is retrograde amnesia, while inability to recall events that occurred after an incident is anterograde amnesia.

Recommended reading

Andersen, P., Morris, R., Amaral, D., Bliss, T.V.P., and O'Keefe, J. (eds) (2006) *The hippocampus book*. Oxford University Press, Oxford.

Carpenter, R., and Reddi, B. (2012) *Neurophysiology: A conceptual approach* (5th edn), Chapter 13. Hodder Arnold, London.

Kandel, E.R., Schwartz, J.H., Jessell, T.M., Siegelbaum, S.A., and Hudspeth, A.J. (2012) *Principles of neural science* (5th edn), Chapters 48, 62–65, and 67. McGraw-Hill, New York.

Levitan, I.B., and Kaczmarek, L.K. (2015) *The neuron* (4th edn), Chapters 17 and 19. Oxford University Press, Oxford.

Rolls, E.T. (2007) *Emotion explained*. Oxford University Press, Oxford.

Squire, L.P. and Kandel, E.R. (2002) *Memory: From mind to molecules.* Freeman, New York.

Non-technical articles from *Scientific American* that provide a good introduction to the field.

Review articles

Dalgleish, T. (2004) The emotional brain. *Nature Reviews Neuroscience* **5**, 582–9.

Goto, Y,. and Grace, A.A. (2008) Limbic and cortical information processing in the nucleus accumbens. *Trends in Neuroscience* **31**, 552–8.

Lynch, M.A. (2004) Long-term potentiation and memory. *Physiological Reviews* **84**, 87–136.

Phelps, E.A., Lempert, K.M., and Sokol-Hessner, P. (2014) Emotion and decision making: Multiple modulatory neural circuits. *Annual Review of Neuroscience* **37**, 263–87.

Rolls, E.T. (2013) A biased activation theory of the cognitive and attentional modulation of emotion. *Frontiers in Human Neuroscience* **7**, doi:10.3389/fnhum.2013.00074, available from www.frontiersin.org.

To check that you have mastered the key concepts presented in this chapter, complete the accompanying online self-assessment questions. Go to www.oup.com/uk/pocock5e/

CHAPTER 19

The cerebral cortex, sleep, and circadian rhythms

This chapter should help you to understand:

- The functions of the association cortex in the cerebral hemispheres

- The separate roles of the two cerebral hemispheres

- The organization of the pathways that control speech

- How the EEG can be used to monitor cerebral function

- The physiology of sleep

- The basis of circadian rhythms

19.1 Introduction

The cerebral hemispheres are the largest and most obvious structures of the human brain. They receive information from the primary senses (somatic sensations, taste, smell, hearing, and vision) and control coordinated motor activity. It is important to remember that the main somatosensory and motor pathways are crossed, so that the left hemisphere receives information from, and controls the motor activity of, the right side of the body while the right hemisphere is concerned with the left side. However, the two hemispheres are not symmetrical morphologically or in their functions. It is commonly recognized that people prefer to use one hand rather than another. Indeed most people (about 90 per cent of the population) prefer to use their right hand for writing, holding a knife or tennis racquet, and so on. What is not always appreciated is that most people also prefer using their right foot (for example when kicking a ball) and they generally pay more attention to information derived from one eye and one ear. Therefore, the two halves of the brain are not equal with respect to the information to which they pay attention or with respect to the activities they control.

19.2 The specific functions of the left and right hemispheres

In the nineteenth century Paul Broca and Marc Dax described an association between the loss of speech and paralysis of the right side of the body in patients who had suffered a stroke. This showed that the left hemisphere played an important role in the production of speech as well as the control of motor activity of the right side of the body. Later studies showed that specific deficits in the understanding of language, reading, and writing were also associated with damage to the cerebral hemispheres. Thus, as the clinical evidence accumulated, it became apparent that the two hemispheres must have different capacities and the idea developed that one hemisphere, the left in right-handed people, was the leading or **dominant hemisphere**, while the other was subordinate and lacked a specific role in higher nervous functions. This idea has had to be greatly modified in the light of current evidence, and it is now clear that both hemispheres have very specific functions in addition to the role they play in sensation and motor activity.

Association areas of the cerebral cortex

Early exploration of the cerebral cortex of humans showed that while stimulation of some areas elicited specific sensations or motor responses, stimulation of others had no detectable effect. These areas were called 'silent' areas and were loosely considered as **association cortex** (Figure 19.1). The association areas are now considered to be those areas that do not receive primary sensory fibres or connect directly with subcortical motor nuclei. Rather, their connections reflect their role in integrating information from the sensory and limbic areas of the cortex to guide the appropriate behaviour.

The frontal cortex

Compared to the brains of other animals, including those of the higher apes, the frontal lobes of the human brain are very large relative to the size of the brain as a whole. Indeed, as Figure 19.1 shows, the frontal lobes are the largest division of the human cerebral cortex. What is their role? One aspect of their function became clear in the mid-nineteenth century when an American mining engineer called Phineas Gage received a devastating injury to his frontal lobes (see Figure 19.2). He was packing down an explosive charge when it detonated, driving the tamping iron through his upper jaw, into his skull and through his frontal lobes. Remarkably, he survived the accident. Although he was subsequently able to live an apparently normal life, his personality had irrevocably changed. He became irascible, unpredictable, and much less inhibited in his social behaviour. So changed was his behaviour that his friends remarked that he was 'no longer Gage'.

Later work on monkeys showed that lesions to the frontal lobes appeared to reduce anxiety. This discovery was subsequently exploited clinically in an attempt to help patients suffering severe and debilitating depression. By cutting the fibres connecting the frontal lobes to the thalamus and other areas of the cortex (a **frontal leucotomy**) it was hoped that this procedure would reduce the feelings of desperate anxiety and so permit patients to resume a normal life. Early results were encouraging in that the affected

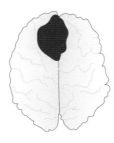

Figure 19.2 The course of the iron bar (left) that penetrated the skull of Phineas Gage and the likely extent of the damage to his frontal cortex (right). (Based on Fig 13.4 in R.H.S. Carpenter (1996) *Neurophysiology* 3rd edition. Edward Arnold, London.)

patients seemed less anxious than before, but it later became clear that the procedure could result in severe personality changes that perhaps might have been predicted from the case of Phineas Gage. Epilepsy (abnormal electrical activity of the brain—see Section 19.4) was another undesirable complication that often developed. In recent years this procedure has become replaced by less drastic and more reversible drug therapies.

Detailed psychological testing of patients who have had a frontal leucotomy showed that, although their general intelligence was little affected, they were less good at problem solving than people with intact frontal lobes. They tended to persevere with failed strategies. They also appeared to be less spontaneous in their behaviour. Thus it appears that the association areas of the frontal lobes are implicated in the determination of the personality of an individual in all its aspects.

The parietal lobes

The parietal lobes extend from the central sulcus to the parieto-occipital fissure and from an imaginary line drawn from the angular gyrus to the parieto-occipital fissure (see Figure 19.1). As with the frontal lobes, much of our knowledge of the functions of the association areas of the parietal lobes in humans is derived from careful clinical observation. Damage to the parietal lobes leads to loss of higher-level motor and sensory performance that is associated with specific deficits known as agnosias and apraxias.

Agnosia is a failure to recognize an object even though there is no specific sensory deficit. It reflects an inability of the brain to integrate the information in a normal way. For this reason, the agnosias are regarded as a failure of 'higher-level' sensory performance. A striking example of agnosia is **astereognosis** in which there is a failure to recognize an object through touch. This disorder is associated with damage to regions in the parietal lobe adjacent to the primary cortical receiving areas of the postcentral gyrus. Another example is **visual object agnosia**, in which the visual pathways appear to be essentially normal but recognition of objects does not occur. If an affected person is allowed to explore the object with another sense, by touch for example, they can often name it. Nevertheless, they may not be able to appreciate its qualities as a physical object. Thus, they may see a chair but not avoid it as they cross a room.

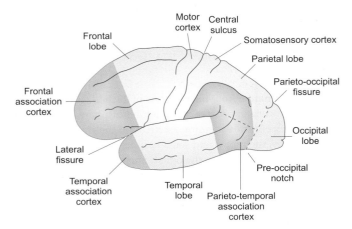

Figure 19.1 A view of the left side of the human brain showing the association areas of the frontal, parietal, and temporal lobes.

Apraxia is the loss of the ability to perform specific purposeful movements even though there is no paralysis or loss of sensation. An affected person may be unable to perform a complex motor task on command e.g. waving someone goodbye, but may be perfectly capable of carrying out the same act spontaneously. Other apraxias may result in failure to use an everyday object appropriately, or the sufferer may be unable to construct or draw a simple object (**constructional apraxia**). A deficit in the control of fine movements of one hand can be caused by damage to the premotor area of the frontal lobe on the opposite side. This is known as **kinetic apraxia**. (The premotor area lies anterior to the area of the primary motor cortex that controls the muscles of the head and face—see Figure 10.13).

The most bizarre effect of lesions to the parietal lobes is seen when the lesion affects the posterior part of the right hemisphere around the border of the parietal and occipital lobes. These lesions lead to neglect of the left side of the body. Affected individuals ignore the left side of their own bodies, leaving them unwashed and uncared for. They ignore the food on the left side of their plates and will only copy the right side of a simple drawing (see Figure 19.3). Many are blind in the left visual field although they are themselves unaware of the fact.

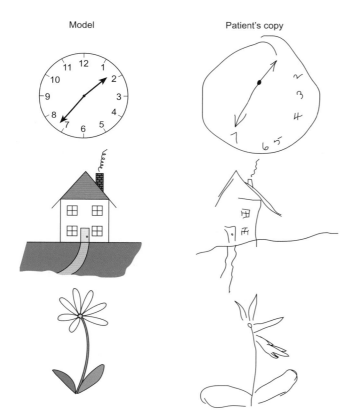

Figure 19.3 A drawing by a patient suffering from the neglect syndrome caused by a stroke in the posterior region of the right hemisphere. The model is on the left of the figure and the patient's copy is shown on the right. Note the failure to complete the left-hand parts of the originals. (Modified from Fig 6.3 in S.P. Springer and G. Deutsch (1989) *Left brain, right brain*, 3rd edition. W.H. Freeman & Co, New York.)

The temporal lobes

The primary auditory cortex is located on the superior temporal gyrus, and the surrounding association areas are concerned with interpreting sounds and with speech (see Chapter 15). The inferior temporal cortex is a visual association area and receives its information from the visual areas adjacent to the primary visual cortex (the striate cortex—see Chapter 14). The association areas of the temporal lobes are also connected to limbic system structures including the amygdala and hippocampal formation. Electrical stimulation of the temporal lobes elicits memories of past events, while bilateral damage to the temporal lobes causes profound amnesia (see Section 18.3).

Experiments in monkeys show that the inferior temporal cortex is involved in the interpretation of complex images. Certain neurons respond specifically to a hand, while others show a clear response to a face. In humans, a selective loss of face recognition may occur after a lesion in one of the temporal lobes. This is known as **prosopagnosia** (from Greek *prosōpon* 'face' + *agnōsia* 'ignorance') and reflects an inability to integrate visual information with the appropriate memories. Individuals with prosopagnosia will fail to recognize the faces of close friends and even those of their family but will instantly recognize the same people by their voices. It appears from functional magnetic resonance imaging (fMRI) that face recognition involves neural networks in a region known as the **fusiform gyrus**, which is located on the underside of the temporal lobe.

The occipital lobes

The properties of the visual association areas were outlined in Chapter 14. In brief, the visual cortex processes different aspects of the visual image in parallel with different areas of cortex processing visual motion, colour, and shape. Lesions to the primary visual cortex (V1) are associated with total blindness. Lesions to an area of the visual cortex designated V4 result in cortical colour blindness (**achromatopsia**), while lesions to the adjacent area (V5) result in a condition called cerebral akinetopsia—losing the perception of motion. The visual association areas extend into the inferior temporal lobe, where they are involved in the interpretation of complex images (as described in the previous subsection).

The corpus callosum plays an essential role in integrating the activity of the two cerebral hemispheres

Sensory information from the right half of the body is represented in the somatosensory cortex of the left hemisphere, and the left motor cortex controls the motor activity of the right side of the body. Equally, the motor and sensory functions of the left half of the body are represented in the right hemisphere. Despite this apparent segregation, the brain acts as a whole, integrating all aspects of neural function. This is possible because, although the primary motor and sensory pathways are crossed, there are many cross-connections between the two halves of the brain; these connections are known as **commissures**. As a result, each side of the brain is constantly informed of the activities of the other. The largest of these commissures is the large band of white matter connecting the two cerebral hemispheres known as the **corpus callosum** (see Chapter 9, Figures 9.4 and 9.5). Damage to the corpus callosum was first reported for shrapnel injuries during the First World War. Amazingly, soldiers with these injuries showed remarkably little by way of a neurological deficit that could be attributed to the severance of such a large nerve tract. However, later work has shown that the deficits are subtle.

Experimental work has shown that most of the nerve fibres that traverse the corpus callosum project to comparable functional areas on the contralateral side. It was known that epileptic discharges could spread from one hemisphere to the other via the corpus callosum, causing a major epileptic attack that involved both sides of the brain. In the search for a cure for the severe bilateral epilepsy experienced by some patients, their corpus callosum was cut—the **split-brain operation** (properly called a **corpus callosotomy**—pronounced callos/otomy). The procedure had the desired end result—a reduction in the frequency and severity of the epileptic attacks. However, it also offered an extraordinary opportunity for a careful and detailed study of the functions of the two hemispheres of the human brain (see next subsection).

In more recent times only the anterior portion of the corpus callosum is cut, as the most common sources of epilepsy are located in the frontal and temporal lobes. This less drastic procedure has much milder effects.

The human split brain operation showed that each hemisphere is specialized to perform a specific set of higher nervous functions

The key to the first experiments on split-brain patients was to exploit the fact that the visual pathway is only partially crossed at the optic chiasm, while speech is located in the left hemisphere. Fibres from the temporal retina remain uncrossed, while those from the nasal region of the retina project to the contralateral visual cortex, as shown in Figure 19.4. The effect of this arrangement is that the right visual field is represented in the left visual cortex while the left visual field is represented in the right visual cortex. Words and images were projected onto a screen in such a way that they would appear in either the right or the left visual field, and the subject was then asked what they had seen. Using this approach, R.W. Sperry and his colleagues were able to investigate the specific capabilities of each hemisphere.

If, say, the word 'BAND' was briefly projected to the right visual field, the patient was able to report that to the investigator (see Figure 19.5), as this word was represented in the visual cortex of the left hemisphere, which also controls speech. If the left visual field had a qualifying word such as 'HAT', the subject was unaware of the fact. If asked what type of band had been mentioned, they were reduced to guessing. If a word such as 'BOOK' was projected to the left field, they could only guess at the answer. If, however, the patient was then allowed to choose something that matched the word that had been projected onto the left visual field, they then made a suitable choice with their left hand (i.e. the hand that is controlled from the right hemisphere). This series of experiments showed that

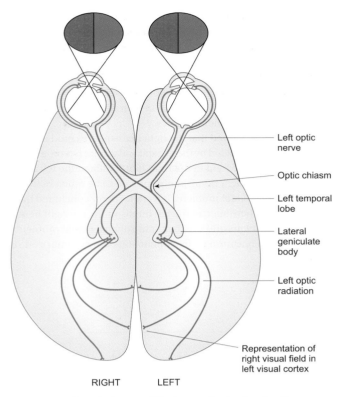

Figure 19.4 A ventral view of the human brain showing the organization of the visual pathways. Note that the left cortex 'sees' the right visual field and vice versa.

while the control of speech was totally lateralized to the left hemisphere, the right hemisphere was aware of the environment, was capable of logical choice, and possessed simple language comprehension.

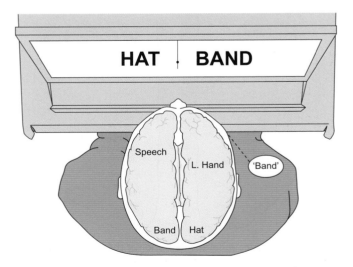

Figure 19.5 The response of a 'split-brain' patient to words projected onto the left and right visual fields. The right image ('HAT') is ignored in the spoken response. (After R.W. Sperry.)

In one patient, when a picture of a nude woman was projected onto the left visual field, there was a strong emotional response—the patient blushed and giggled. When asked what she had seen, she replied 'nothing, just a flash of light'. When asked why she was laughing, she could not explain but replied—'Oh Doctor, you have some machine!' From this observation, it would appear that emotional reactions begin at a lower level of neural integration than the cortex and involve both hemispheres.

Further studies showed that the right hemisphere could identify objects held in the left hand by their shape and texture (**stereognosis**) as efficiently as those held in the right hand. Indeed the left hand (and therefore the right hemisphere) proved to be much better at solving complex spatial problems. Overall, Sperry concluded that, far from being subordinate to the left hemisphere, the right hemisphere was conscious and was better at solving spatial problems and non-verbal reasoning. His conclusions are summarized in Figure 19.6.

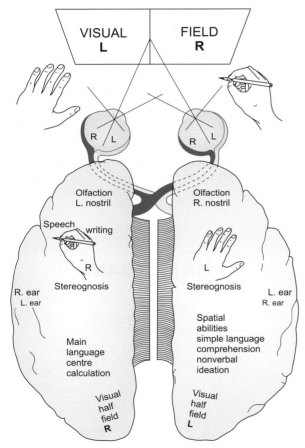

Figure 19.6 The functional specialization of the two cerebral hemispheres revealed by studies of patients after their corpus callosum has been divided. (Note that the areas of the cortex associated with the primary senses and those involved in motor control are not represented in this diagram). (After R.W. Sperry.)

Summary

The frontal lobes play an important role in planning movements, including those of speech. Lesions in the parietal lobe result in defects of sensory integration known as agnosias and in an inability to perform certain purposeful acts (apraxia). Severance of the corpus callosum, which connects the two hemispheres, has shown that the two hemispheres have very specific capabilities in addition to their roles in sensation and motor activity. Speech and language abilities are mainly located in the left hemisphere, together with logical reasoning. The right hemisphere is better at solving spatial problems and non-verbal tasks.

19.3 Speech

Many animals have some means of communicating with their fellows. The production of sound that has no specific meaning is called **vocalization**. What distinguishes human vocal communication from that of other animals is its range and subtlety of expression. Humans use **language**. A language consists of a specific vocabulary and a set of rules of expression (syntax). While most language communication is by means of speech, expression in a given language is independent of the mode of communication. This text is written in English and obeys its grammatical rules, but the written word is only the same as the spoken word in its meaning. The representation of a word on the page is arbitrary, and its meaning may vary with the context.

Speech is a very complex skill. It involves knowledge of the vocabulary and grammatical rules of at least one language. It also requires very precise motor acts to permit the production of specific sounds in their correct order, as well as precise regulation of the flow of air through the larynx and mouth. For these reasons it is perhaps not surprising to find that large areas of the brain are devoted to both the production of speech and its comprehension. Because it is specifically a human characteristic, our knowledge of the systems that govern the production and comprehension of speech has largely come from careful neurological observations and various whole-brain imaging techniques, including functional magnetic resonance imaging (fMRI) and positron emission tomography (PET) scanning.

The principal areas of the brain controlling speech production are located in the frontal and temporal lobes

Earlier in this chapter, the work of Broca and Dax was cited as evidence for the lateralization of speech to the left hemisphere. This was based on the study of patients who had difficulty in producing speech following a stroke that had damaged the left hemisphere. Such disabilities are known as **aphasias**. Broca studied a number of patients and noted that they had difficulty in producing speech but appeared to be quite capable of understanding spoken or written language. Characteristically the patients with this condition have speech that is slow, halting, and telegraphic in quality. They are able to name objects and describe their attributes but have difficulty with the small parts of speech that play such an important role in grammar (e.g. 'if', 'is', 'the', and so on). They also have difficulty in writing. Since these patients apparently have a good understanding of language, this type of aphasia is sometimes called **expressive aphasia**. It is also known as **Broca's aphasia**.

Broca was able to examine the brains of several of his aphasic patients at post-mortem. He discovered that there was extensive damage to the frontal lobe of the left hemisphere, particularly in the region that lies just anterior to the motor area responsible for the control of the lips and tongue (Figure 19.7). This is now known as Broca's area. Patients with Broca's aphasia do not have a paralysis of the lips and tongue, as they can sing wordlessly. What they have lost is the ability to use the apparatus of speech to form words, phrases, and sentences.

Another type of aphasia, which was discovered by Carl Wernicke, is characterized by free-flowing speech that has little or no informational content (technically known as **jargon**). Patients suffering from this kind of aphasia tend to make up words (**neologisms**), e.g. 'lork', 'flieber', and often have difficulty in choosing the appropriate words to describe what they mean. They also have difficulty in comprehending speech and written language. This type of aphasia is called **receptive aphasia** or **Wernicke's aphasia** after its discoverer. Wernicke's aphasia is associated with lesions to the posterior region of the temporal lobe adjacent to the primary auditory cortex and the angular gyrus (Figure 19.7).

The two principal brain regions responsible for speech are interconnected by a set of nerve fibres called the **arcuate fasciculus**. If these fibres are damaged, another type of aphasia occurs called **conduction aphasia**. This is typified by the speech characteristics of Wernicke's aphasia, but comprehension of written and spoken language remains largely intact.

Electrical stimulation of the speech areas of the cerebral cortex of conscious human subjects does not lead to vocalization. However, if a subject is already speaking when one of the areas of cortex concerned with language is stimulated, their speech may be interrupted or there may be an inappropriate choice of words. Stimulation of the supplementary motor area on the medial surface of the left hemisphere may lead to speech which is limited to a few words or syllables that may be repeated over and over, e.g.

'ba-ba-ba-'. Unlike lesions to Broca's area and Wernicke's area, lesions to the supplementary motor area do not result in permanent aphasia.

Although the main areas that control the production of speech are located in the left hemisphere, the right hemisphere does have a very basic language capability. More significantly, the posterior part of the right cerebral hemisphere seems to play an important role in the interpretation of speech which, unlike the written word, has an emotional content that reveals itself in its intonation. Consequently, many patients who have damage to their *left* temporal lobe are still able to understand the intention of something said to them, even though their comprehension of individual words and phrases is poor. Conversely, patients with damage to their right hemispheres will often speak with a flat monotone.

These observations suggest an organization of the pathways that control speech similar to that shown in Figure 19.7. According to this model, speech is initiated in Wernicke's area and is passed to Broca's area via the arcuate fasciculus for execution. The neural pathway involved in naming an object that has been seen is also shown in Figure 19.7. Disconnection of the visual association areas from the angular gyrus will lead to **word blindness** or **alexia**. **Word deafness** (auditory agnosia) occurs when lesions disconnect Wernicke's area from the auditory cortex. Comprehension of spoken, but not written, language is impaired.

Recent studies with imaging techniques have revealed that other areas of the brain are also involved in the production and comprehension of language (see Figure 19.8). As described above, normal speech involves Wernicke's area, Broca's area, and the premotor area. Reading aloud also involves the visual cortex and a region close to the end of the lateral fissure known as the angular gyrus. The angular gyrus interprets visual information that is then converted to speech patterns by Wernicke's area before being transmitted to Broca's area. Silent reading involves the visual cortex, the premotor area, and Broca's area, but not the auditory cortex or Wernicke's area, while silent counting involves the frontal lobe. Unsurprisingly, as Figure 19.8 shows, other structures in the motor pathways are also activated during the production of speech, including the thalamus and basal ganglia, the red nucleus, and the cerebellum.

Careful anatomical examination has found that the upper aspect of the left temporal lobe (the **planum temporale**) is larger than the right in about 70 per cent of people. However, in over 90 per cent of the population language functions are largely dependent on the left hemisphere, so the link between anatomical asymmetry and language functions is not absolute. Despite these insights, it is important to remember that knowing which parts of the brain are involved in the production of speech and the comprehension of language is not the same as knowing how the brain encodes and executes the patterns of neural activity that are responsible for the production of speech.

In the majority of people, speech is controlled from the left hemisphere

It is of some importance that a neurosurgeon is aware of which hemisphere controls speech before embarking on an operation. This can be established by the **Wada test**, which temporarily

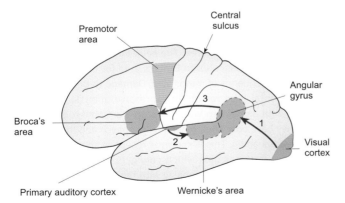

Figure 19.7 A diagram of the left hemisphere to show the location of the principal regions involved in the control of speech.

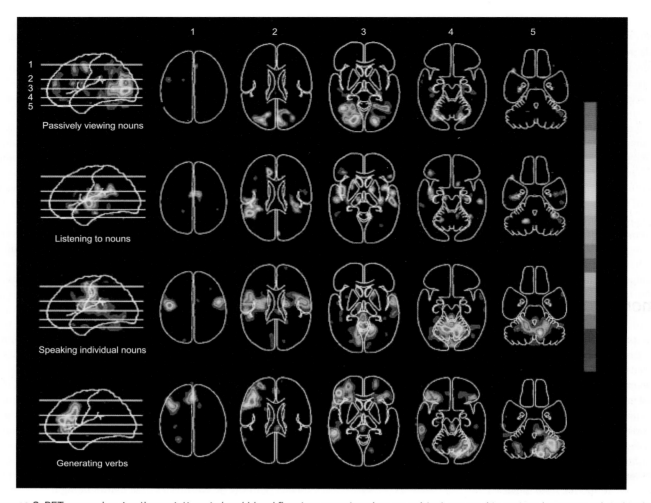

Figure 19.8 PET scans showing the variations in local blood flow in a conscious human subject engaged in various language-related tasks. The images in columns 1–5 show the increase in blood flow in horizontal sections through the brain indicated in the top image of the leftmost column. From this and similar studies, it is clear that extensive areas of the cerebral cortex are involved in language production and comprehension in addition to the classical speech areas. The colour coding indicates the increase in blood flow, with red being the most and deep blue the least intense. (Adapted from M. Raichle (1994) *Scientific American* **270**, 60. By permission of Dr Marcus E. Raichle.)

anaesthetizes one hemisphere. The patient lies on their back while a cannula is inserted into the carotid artery on one side. The patient is then asked to count backwards from 100 and to keep both arms raised. A small quantity of a short-lasting anaesthetic is then injected. As the anaesthetic reaches the brain, the arm on the side opposite that of the injection falls and the patient may stop counting for a few seconds or for several minutes, depending on whether speech is localized to the side receiving the injection.

These tests reveal which hemisphere controls speech in a person who has no aphasia. In 95 per cent of right-handers, speech is controlled from the left hemisphere. This is also true of about 70 per cent of left-handers. Speech is localized to the right hemisphere in about 15 per cent of left-handers, and the remainder show evidence of speech being controlled from both hemispheres (see Table 19.1). This distribution of the speech areas has recently been confirmed using a non-invasive imaging technique called Doppler ultrasonography.

Summary

Speech is localized to the left hemisphere in the majority of the population. The neural patterns of speech originate in the temporal lobe in Wernicke's area. The neural codes for speech pass via the arcuate fasciculus to Broca's area for execution of the appropriate sequence of motor acts. Damage to Broca's area in the left frontal lobe results in expressive aphasia. Although patients with damage to Wernicke's area are able to speak fluently, they have poor comprehension of speech and their speech lacks clear meaning. This is known as receptive aphasia. Damage to the arcuate fasciculus which connects Wernicke's area to Broca's area results in an aphasia with the fluent, largely meaningless speech characteristic of Wernicke's aphasia, but language comprehension is largely intact. This is known as conduction aphasia.

Table 19.1 The distribution of handedness and the location of speech in the cerebral hemispheres

Location of speech	Right-handed individuals	Left-handed individuals	Total
Left hemisphere	85.5%	7%	92.5%
Right hemisphere	4.5%	1.5%	6.0%
Both hemispheres	0%	1.5%	1.5%
Total	90%	10%	100%

Note: The table shows the location of speech as a percentage of the total adult population. Thus, 85.5% of the population are right-handed and have their speech localized to their left hemispheres, and so on.

19.4 How the EEG can be used to monitor the activity of the brain

The brain is constantly involved in the control of a huge range of activities both when awake and during sleep. Its activity can be monitored indirectly by placing electrodes on the scalp. If this is done in such a way as to minimize electrical interference from the muscles of the head and neck, small oscillations are seen that reflect the overall activity of the brain (see Figure 19.9). These electrical oscillations are known as the **electroencephalogram** or **EEG**. For normal subjects the amplitude of the EEG waves ranges from 10 to 150 μV.

The activity of the EEG is continuous throughout life but is not obviously related to any specific sensory stimulus or motor task. It reflects the spontaneous electrical activity of the brain itself. The appearance of the EEG varies according to the position of the electrodes, the behavioural state of the subject (i.e. whether awake or asleep), the subject's age, and whether there is any organic disease.

The specific state of the EEG is classified by the frequency of the electrical waves that are present. When a subject is awake and alert, the EEG consists of high-frequency waves (20–50 Hz), which have a low amplitude (about 10–20 μV). These are known as **beta waves**, and they appear to originate in the cerebral cortex. As a subject closes their eyes, this low-amplitude–high-frequency pattern gives way to a higher-amplitude–lower-frequency pattern known as the alpha rhythm. The **alpha waves** have an amplitude of 20–40 μV and contain one predominant frequency in the range of 8–12 Hz. As the subject becomes more drowsy and falls asleep, the alpha rhythm disappears and is replaced by slower waves of greater amplitude known as **theta waves** (40–80 μV and 4–7 Hz). These are interspersed with brief periods of high-frequency activity known as 'sleep spindles'. In very deep sleep the EEG waves are slower still (**delta waves**, which have a frequency of less than 3 Hz) and they have a relatively high amplitude (100–120 μV). The characteristics of the principal EEG waves are summarized in Table 19.2.

Alpha waves are best seen over the occipital region and they have a characteristic appearance, slowly growing in amplitude to a maximum and then slowly declining ('waxing and waning'). The

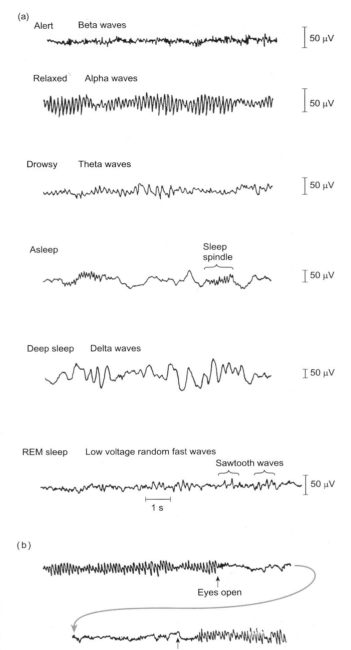

Figure 19.9 Typical stretches of the EEG for various stages of awareness. Panel (a) shows the changes in the EEG as a subject falls asleep. Note that during deep sleep the EEG changes to a pattern similar to that seen in the normal awake state but which has characteristic 'sawtooth' waves. This is the REM phase of sleep. Panel (b) shows alpha block following opening of the eyes. Note the sudden loss of the alpha waves when the eyes open and their resumption when the eyes close again. The two lines in panel (b) are a continuous stretch of record from the occipital region of the left hemisphere.

Table 19.2 The characteristics of the principal waves of the EEG

Wave type	Frequency (Hz)	Amplitude (μV)	Notes
Alpha (α)	8–12	20–40	Best seen over the occipital pole when the eyes are closed.
Beta (β)	20–50	10–20	Normal awake pattern.
Theta (θ)	4–7	40–80	Normal in children and in early sleep. Evidence of organic disease when seen in awake adults.
Delta (δ)	< 3	100–150	Seen during deep sleep. Evidence of organic disease when seen in awake adults.

Note that the higher the frequency the lower the amplitude i.e. the higher frequencies are less synchronized than the low frequencies.

alpha rhythm is disrupted when the subject concentrates their attention on a problem or when they open their eyes (see Figure 19.9). This is known as 'alpha block'. The alpha waves appear to be driven by feedback between the cerebral cortex and the thalamus. Theta waves are believed to originate in the hippocampus, while the delta waves probably originate from activity in the brainstem.

Epilepsy

Although detailed interpretation of the EEG is fraught with difficulties, it has proved to be of great practical value in the diagnosis and localization of organic brain disease. Moreover, the EEG changes are characteristic of particular disease processes. To record the EEG, pairs of electrodes are placed over the scalp according to a standard arrangement (such as the international 10–20 system, which employs 21 electrodes). The electrical activity is recorded between specific electrode pairs and is scrutinized for the presence of abnormal patterns of activity. The greatest use of the EEG is in the diagnosis of **epilepsy**, which may be defined as a disorder of the electrical activity of the brain. Not only is the EEG useful in diagnosing epilepsy but it can also be used to locate a site that triggers an epileptic attack (the **epileptic focus**). Unlike many other disorders, epilepsy is intermittent and attacks generally occur infrequently, although they may be triggered by specific stimuli.

Epileptic seizures may be divided into two broad categories: focal seizures (also called local or partial) and generalized seizures. Focal seizures reflect abnormal electrical activity that is localized to a specific part of the brain, while generalized seizures are caused by abnormal electrical activity that spreads throughout the entire brain.

Focal seizures begin when a group of neurons are activated and paroxysmal activity spreads to other neurons in the same region of one hemisphere. Often these episodes cause no loss of consciousness. The symptoms depend on the location of the initiating site and may be motor (as in Jacksonian epilepsy—see Section 10.9), sensory, aphasic (i.e. affecting speech—see Section 19.3), or

psychological in character. The most common focal seizures arise from abnormal activity in one or other of the temporal lobes and are often accompanied by a stereotypical behaviour pattern involving the lower part of the face, including grimaces and sucking.

Generalized seizures affect both sides of the body with no evidence of a focal onset. Clinically, they may be manifest in various ways. The most striking are tonic-clonic seizures, also known as **grand mal seizures** or **generalized convulsions**, in which the body stiffens and cannot maintain its posture so that the sufferer falls. After a brief time, the rigid phase is followed by clonic muscular spasms (convulsions). The seizure ends in a stupor that gradually clears. Absence seizures are much less dramatic: there is a brief loss of consciousness of which the sufferer is unaware. During myoclonic seizures both sides of the body show brief, jerky movements, while clonic seizures exhibit a pattern with rhythmic, jerky movements affecting both sides of the body. Tonic seizures are characterized by a stiffening of the muscles, while during atonic seizures there is a sudden and general loss of muscle tone, particularly in the arms and legs.

Absence seizures (**petit mal seizures**) are more common in children than in adults and may occur several times a day, with serious implications for their education. During an attack the child stops whatever they were doing and may stare blankly as though daydreaming. These seizures are usually brief and last for no more than 1 to 20 seconds. They begin and end abruptly, so that the suffer is usually unaware that they have had a seizure. Fortunately, petit mal epilepsy that begins in childhood rarely continues in into adult life.

It is known that epilepsy may be caused by traumatic injury, infection, ischaemic damage to the brain during birth, vascular disease, or, more rarely, tumours. However, in most cases (c. 60 per cent) there is no obvious precipitating cause, in which case the condition is called idiopathic epilepsy. Some cases of epilepsy appear to be of familial origin. Some examples of the appearance of the EEG during an epileptic attack are shown in Figure 19.10. Most

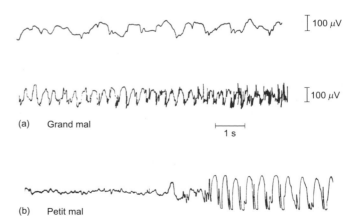

Figure 19.10 The change in the EEG that can be seen during an epileptic attack. Panel (a) shows an attack of grand mal epilepsy in which the EEG has a preponderance of delta wave activity, which is replaced by large spikes during a convulsive seizure. Panel (b) shows the change in the EEG of another patient during a petit mal seizure. Note the characteristic spike and wave complex.

epilepsy can be successfully treated with anti-epileptic drugs such as carbamazepine, sodium valproate, and phenytoin.

Summary

The EEG can be used to monitor the spontaneous electrical activity of the brain. The frequency and amplitude of the EEG waves depend on the state of arousal of a person. The EEG is helpful in the diagnosis and location of abnormalities of the electrical activity of the brain, such as epilepsy.

19.5 Sleep

It is common experience that, under normal circumstances, people are active during the day and sleep at night. In some other animals, the pattern is reversed and the main periods of activity occur during the night. This cyclical variation in activity is called the sleep–wakefulness cycle. Why do we sleep? What controls the sleep–wakefulness cycle? At present, these questions cannot be fully answered, but the work they have engendered has given some important information about the processes that occur during sleep.

The appearance of the EEG can be used to follow the main stages of sleep. As previously discussed, the EEG waves become slower and more synchronized as a subject passes from the awake state to deep sleep (see Figure 19.9). As sleep sets in, the EEG is dominated by theta waves and there may be periods of fast EEG activity with short periods of fast rhythmic waves called sleep spindles. As sleep becomes deeper, the sleep spindles are associated with occasional large waves to form patterns known as K complexes, and in deep sleep the EEG is dominated by large amplitude, low-frequency delta waves.

The depth of sleep, as judged by how easy it is to arouse someone with a standard stimulus (e.g. a sound), correlates well with the state of the EEG. The deep phase of sleep associated with delta wave activity is called **slow-wave sleep** or **ortho-sleep**. It is associated with a slowing of the heart rate, a fall in blood pressure, slowing of respiration, low muscle tone, a fall in body temperature, and an absence of rapid eye movements. However, after one or two hours of slow-wave sleep, the EEG assumes a pattern similar to the awake state, with fast beta wave activity (Figure 19.9). This change in EEG pattern is associated with jerky movements of the eyes and increases in both the heart rate and respiratory rate. In males, penile erection may occur. There may also be clonic movements of the limbs, although most muscles are fully relaxed in this phase of sleep. During these changes, the subject is as difficult to arouse as they are during slow wave sleep. Since the EEG has similarities to the appearance of the EEG in an awake subject, this phase has been called **paradoxical sleep**, although it is more often called rapid eye movement or **REM sleep**. During a single night, slow-wave sleep is interrupted by four to six episodes of REM sleep, each of which lasts about 20 minutes.

The amount of time spent in sleep varies with age, as does the proportion of sleep spent as REM sleep. As Figure 19.11 shows, very young children spend a large part of the day asleep and REM sleep

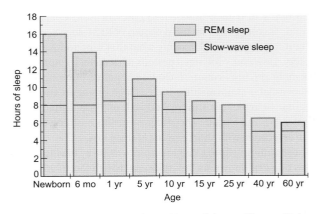

Figure 19.11 The changes in the pattern of sleep with age. Note the decline in the total hours of sleep and in the proportion of REM sleep with increasing age. (Based on data in H.P. Roffwarg, J.N. Muzio and W.C. Dement (1966) *Science* **152**, 604–19.)

accounts for nearly half of all their sleep. By the age of 20, only about eight hours are spent in sleep, of which less than a quarter is spent as REM sleep. The requirement for both kinds of sleep declines further with age and, by the age of 60, many people require as little as six hours.

If a subject is woken during an REM episode, they are much more likely to report having dreams than they are if they are awoken during periods of slow-wave sleep. Thus REM sleep is associated with dreaming. The content of dreams has received much attention over the millennia, from forecasting the future to revealing intimate details of a personality, but little can be said about the role of dreams with certainty. The vast majority are forgotten soon after they occur.

Effects of sleep deprivation

Why do we sleep? The answer to this question still eludes us. The notion that sleep is required to restore the body is an old one, but exactly what needs to be restored is not clear. It is known that the pattern of secretion of many hormones varies with the sleep–wakefulness cycle (see Chapter 20) and that sleep reflects a change in the pattern of brain activity.

Chronic sleep deprivation leads to severe mental changes that can border on psychosis, including delusions and hallucinations. However, full sleep deprivation is uncommon. It is more usual to find that periods of sleep are restricted by particular circumstances—for example, the conflict between the need to meet the normal obligations of family life and those of shift work. If sleep is restricted, over time the affected individuals become irritable, their intellectual performance declines, and their reaction times increase. They also suffer from memory lapses and feel mentally and physically fatigued. Periods of sleep deprivation or restriction lasting a few days are followed by longer periods of sleep to compensate for the lost sleep time.

If subjects are specifically deprived of REM sleep by the simple expedient of waking them every time a REM episode begins, they also become anxious and irritable. Following such a period, their sleep tends to have a higher frequency of REM episodes than normal. This suggests that the episodes of REM sleep serve an important function, though exactly what that function is remains a mystery. However, prolonged periods of sleep restriction have significant health implications including impaired immune function, cardiovascular abnormalities such as high blood pressure and an irregular heartbeat, an increased risk of Type 2 diabetes (see Chapter 24), and obesity. These effects are less easily reversed than those associated with short periods of sleep loss.

Those suffering from sleep loss compensate by having brief episodes of sleep known as 'microsleeps' which last 1–20 seconds. These episodes are associated with the presence of theta waves in the EEG during which the eyelids slowly close followed by relaxation of the neck muscles and nodding of the head—which often serves to terminate the episode. Although microsleeps themselves are harmless, they become dangerous when driving or undertaking some other task that requires constant alertness.

The sleep–wakefulness cycle is actively controlled from the thalamus, hypothalamus, and brain stem

In the 1930s R. Hess found that electrical stimulation of the thalamus initiates sleep behaviour in cats. During stimulation, the animals will circle their chosen resting place, yawn, and stretch their limbs before curling up and going to sleep. Subsequently, G. Morruzi and H. Magoun showed that sleeping animals can be aroused by electrical stimulation of a region of the brain stem known as the ascending reticular formation. This observation suggested that sleep is controlled, at least in part, by the activity of the brain stem.

Later work has suggested that some specific groups of nerve cells containing serotonin and others that contain noradrenaline (norepinephrine) play a role in the regulation of sleep. Serotonergic neurons are found in a series of nuclei called the raphe nuclei that are scattered along the ventral region of the brain stem. If these neurons are depleted of their serotonin, the affected animals are unable to sleep at all. Sleep can, however, be restored if 5-hydroxytryptophan (a precursor of serotonin) is administered. Noradrenergic neurons are located in the locus coeruleus, which lies beneath the cerebellum. If these neurons are depleted of their noradrenaline by administration of a metabolic inhibitor, the affected animals are able to enter slow-wave sleep but do not have normal REM sleep episodes.

An attractive, if somewhat over-simplified, explanation of these results is that activity in the serotonergic neurons initiates sleep by lowering the activity of the brain stem reticular formation. This leads to a lessening of cortical activity and results in sleep. Subsequent activity in the noradrenergic neurons gives rise to the REM episodes. However, continued administration of inhibitors of serotonin synthesis does not cause permanent insomnia even though serotonin levels remain low. This suggests that the system controlling the sleep–wakefulness cycle is much more complex than this simple model implies. Recent work has suggested that many other neurotransmitters are involved in regulating the sleep cycle such as acetylcholine, adenosine, γ-amino butyric acid (GABA), and a class of peptides known as hypocretins or **orexins**.

Sleep-promoting peptide is found in the CSF of sleep-deprived animals

Before the roles of the diencephalon and brain stem in the control of the sleep-wakefulness cycle were recognized, it had been suggested that, during wakefulness, the body might release some substance that accumulated during activity and would initiate sleep when its concentration had reached a certain threshold. In recent years, a nonapeptide called delta sleep-inducing peptide (DSIP) has been isolated from the CSF of sleep-deprived animals. When DSIP is injected into the cerebral ventricles of normal animals, it induces sleep. In addition, it also modulates body temperature and influences the immune response. For this reason, it has been suggested that part of the function of sleep may be to help counter infections. Orexins seem to have the opposite effect and promote wakefulness. Dogs with defective orexin receptors exhibit a marked tendency to sleep (narcolepsy), as do mice in which the gene has been deleted.

Sleep disorders

There are a very large number of sleep disorders which may be broadly classified into dyssomnias, parasomnias, circadian rhythm sleep disorders, and disorders caused by specific medical conditions. By far the most common sleep disorder is **insomnia**, in which the sufferer has persistent difficulties in falling or staying asleep. This inevitably leads to a degree of sleep deprivation with the effects discussed earlier.

Other common sleep disorders include:

- Sleep apnoea, in which breathing temporarily stops during deep sleep (discussed in Chapter 36).
- Sleep-walking.
- Narcolepsy, in which the sufferer is excessively sleepy during the day and may fall asleep at an inappropriate time e.g. during a meal.
- Restless limb syndrome (nocturnal myoclonus), in which there are jerky involuntary movements of the arms or legs during sleep.
- Night terrors and nightmares. These may occur after an individual has experienced a trauma of some kind. They are often suffered by children around 6 to 8 years of age. Fortunately, most children quickly grow out of them.
- Circadian rhythm sleep disorders, in which affected individuals cannot fall asleep and wake at conventional clock times. This is often associated with disruptive shift patterns at work.

Summary

Sleep is a widespread biological phenomenon associated with periods of rest. Its exact role remains unclear, but sleep loss is associated with an impaired ability to carry out normal tasks. Long periods of limited sleep are associated with a number of significant health problems.

The stages of sleep can be followed by monitoring the EEG. During deep sleep (also called slow-wave sleep), the EEG is dominated by slow-wave activity. Several times a night, bouts of REM (rapid eye movement) occur in which the EEG is desynchronized. During slow-wave sleep, there is a slowing of the heart rate and respiratory rate, blood pressure falls, and there is extensive relaxation of the somatic muscles. During REM sleep, there are rapid eye movements, the heart rate and respiratory rate increase, and there may be jerky movements of the limbs. REM episodes are associated with dreaming.

19.6 Circadian rhythms

Many aspects of normal physiology are related to time. Indeed more than 200 measurable physiological parameters have been shown to exhibit rhythmicity. Cycles of periodicity shorter than 24 hours are called **ultradian** rhythms (e.g. the heartbeat, the respiratory rhythm), while those longer than 24 hours are called **infradian** rhythms (e.g. the menstrual cycle, gestation). Most biological variables, however, show a periodicity that roughly approximates to the earth's rotational period—the 24-hour day. These are called **circadian rhythms** (from the Latin *circa*, meaning 'around', and *dies*, meaning 'a day'). The most familiar example is the sleep–wakefulness cycle, but other circadian rhythms are seen with core temperature, pulse rate, systemic arterial blood pressure, renal activity (as measured by the excretion of potassium), and the secretion of a number of hormones (e.g. the secretion of cortisol by the adrenal cortex—see Figure 20.2).

Originally it was believed that the regular cycle of light and darkness provided the fundamental stimulus for synchronization, but tests carried out on animals and humans kept in wholly artificial environments have shown that the situation is rather more complex. If a normal subject is kept isolated from the outside world without any clues regarding the time of day, their sleep–wakefulness cycle tends to adopt a periodicity slightly longer than the 24-hour day. For most subjects the natural cycle appears to be nearer to 25 hours, so that a person isolated for four weeks would effectively lose a whole active day compared with someone living in a normal environment. Curiously, the daily variations in body temperature and cortisol secretion that normally closely follow the sleep–wakefulness cycle deviate from the 24-hour cycle far less, indicating that the mechanisms controlling these physiological processes are not the same as those of the sleep–wakefulness cycle. There is now clear evidence that there is more than one 'biological clock'. Indeed, such clocks exist in tissues other than the brain and regulate many physiological processes with a periodicity of roughly 24 hours. In fact, recent work suggests that almost all the cells of the body possesses intrinsic rhythm generators that are entrained to a 24-hour cycle by external environmental cues which may be either physical or social factors. These external cues are sometimes called **zeitgebers** (the German for 'time-givers').

Cells isolated from various tissues, including the liver, white blood cells, salivary glands, and certain endocrine organs, show circadian periodicity of activities such as mitosis, respiration, and secretion. It is now clear that these 'biological clocks' have a genetic basis and that mutation or loss of these genes can lead to disruption of circadian rhythms. The molecular mechanism responsible for the circadian rhythm of mammals depends on the operation of feedback loops that regulate the expression of specific proteins known as transcription factors. One key gene is known as CLOCK (for **C**ircadian **L**ocomotor **O**utput **C**ycles **K**aput) which binds to another gene BMAL1 to form a complex that encodes a transcription factor—a protein also called CLOCK. This enhances the production of further proteins such as cryptochrome 1, which form complexes that then inhibit the production of the CLOCK transcription factor (a negative feedback loop). This feedback loop is further regulated by other cell signalling cascades to establish the appropriate biological rhythm.

These peripheral 'clocks' are synchronized by neurons in a region of the hypothalamus called the **suprachiasmatic nucleus (SCN)** which is sometimes referred to as the 'master clock'. Brain slices or cultured neurons from the SCN retain synchronized, rhythmical firing patterns even though they are isolated from the rest of the brain. Furthermore, destruction of the SCN in laboratory rats causes a loss of circadian function which may be restored by the implantation in the SCN of cells from fetal rat brain. The intrinsic rhythm of the SCN neurons is linked to the day–night cycle by specific photosensitive retinal ganglion cells which send information directly to the SCN via the retino-hypothalamic pathway. The SCN then sends axons to various nuclei in the hypothalamus, notably to the paraventricular nucleus and the area immediately below it, the subparaventricular zone. Efferent fibres from these nuclei play a significant role in the regulation of pituitary function, including the secretion of adrenocorticotrophic hormone (ACTH). Axons from the SCN also project to several other nuclei in the thalamus and amygdala (Figure 19.12). Since synchronization of the circadian rhythm can be achieved by transplantation of fetal SCN tissue, it is thought that the SCN neurons also secrete a paracrine or endocrine factor. However, the nature of this factor or factors is not yet known.

The pineal gland has also been implicated in the regulation of circadian rhythmicity. This gland, which is attached to the dorsal aspect of the diencephalon by the pineal stalk (see Figure 19.13), secretes a hormone called **melatonin** that is synthesized from serotonin. The synthesis and secretion of melatonin is increased during darkness and inhibited by daylight. This pattern is controlled by sympathetic nerve activity that is itself regulated by signals from the SCN by a polysynaptic pathway. Melatonin appears to induce drowsiness and loss of alertness. Although the evidence still remains equivocal, melatonin is used by some people to treat jet lag. It is also used to help restore the sleep patterns of those who work night shifts. Alterations in the secretion of melatonin, or other related neuromodulators, may be responsible for the condition known as **seasonal affective disorder** (or **SAD**), in which the patient feels depressed and lethargic during the dark winter months and for which light therapy is often a successful treatment.

There is no doubt that biological rhythms are of considerable significance both for normal daily activity and for clinical medicine. Shift work (particularly when it involves rotation of day and night work) and travel through time zones, with its consequent jet lag, both disrupt the normal circadian rhythms. This makes it difficult for an individual to maintain an active state at times when

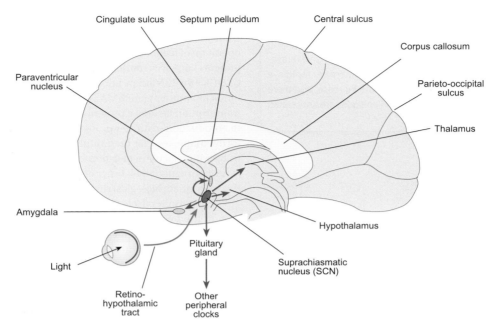

Figure 19.12 A simplified diagram of the midline structures of the brain, showing the principal connections of the suprachiasmatic nucleus (SCN) and their role in synchronizing the various biological clocks via the pituitary gland. Neural connections are not shown in detail.

their natural biological rhythms are preparing the body for inactivity, rest, or sleep. Unsurprisingly, more errors of judgment are made at work in the early hours of the morning than at any other time. Conversely, when a worker is going home to rest or sleep following a night shift, the natural rhythms are stimulating a number of physiological processes in preparation for the activities of the day. Cortisol levels rise, along with core temperature, heart rate, and blood pressure. These changes interfere with normal sleep patterns.

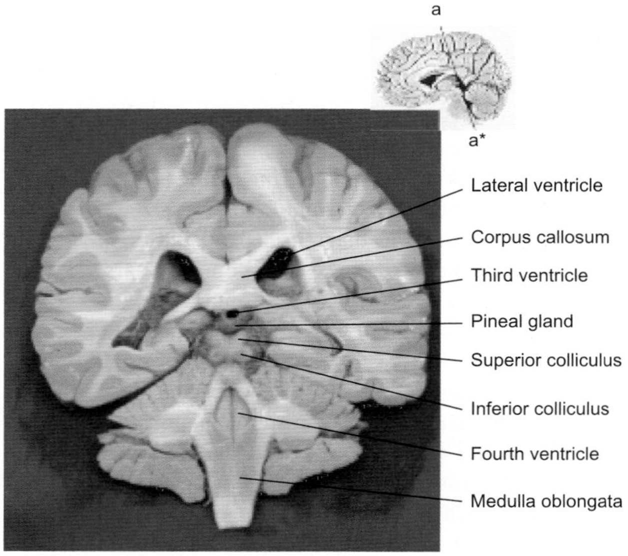

Many of the physiological parameters that are routinely monitored by clinicians are influenced by the time of day. Heart rate, blood pressure, and temperature are obvious examples but there are many others including blood cell numbers, respiratory values, enzyme activities, and **blood gas** levels. In addition, it is now clearly established that the effectiveness of many drugs differs according to the time of administration as a result of differences in gastric emptying, gut motility, renal clearance rates, and the metabolic activity of cells. Indeed, the subject of chronopharmacology is becoming an increasingly important aspect of many drug therapies. Chronopharmacology is concerned with the relationship between the efficacy of a drug at various times during a biological cycle and with the effects of drugs on those cycles.

Figure 19.13 A coronal section of human brain, showing the position of the pineal gland and other key structures. The inset figure shows the line of the section (a–a*). (From University of British Columbia Neuroanatomy website: http://www.neuroanatomy. ca/cross_sections/sections_coronal.html.)

Summary

Internal biological clocks exist in the brain and other tissues and regulate the activity of many physiological processes including heart rate, blood pressure, core body temperature, and the blood levels of many hormones. These biological clocks are entrained to a strict 24-hour cycle, particularly by light signals from the retina. The sleep–wakefulness cycle is an example of a 24-hour or circadian rhythm.

✱ Checklist of key terms and concepts

The functions of the cerebral hemispheres

- In addition to their role in motor control, the frontal lobes appear to play a significant part in shaping the personality of an individual.
- Lesions in the parietal lobe result in defects of sensory integration known as agnosias and in an inability to perform certain purposeful acts (apraxia).
- Severance of the corpus callosum, which interconnects the two hemispheres, has shown that the two hemispheres have very specific capabilities in addition to their roles in sensation and motor activity.
- Speech and language abilities are mainly located in the left hemisphere, together with logical reasoning.
- The right hemisphere is better at solving spatial problems and non-verbal tasks.

The neurophysiology of speech

- Speech is localized to the left hemisphere in the majority of the population, irrespective of whether they are right- or left-handed.
- The neural patterns of speech originate in the temporal lobe in Wernicke's area, which is adjacent to the auditory cortex.
- The neural codes for speech pass via the arcuate fasciculus to Broca's area for execution of the appropriate sequence of motor acts.
- Damage to the speech areas results in aphasia.
- Damage to Broca's area in the left frontal lobe results in expressive aphasia, in which production of speech is difficult and the small parts of speech are lacking.
- Although patients with damage to Wernicke's area are able to speak fluently, they have poor comprehension of speech and their speech lacks clear meaning. This is known as receptive aphasia.
- Lesions to the parietal lobe affect the understanding of the nuances implied in spoken language.

The electroencephalogram or EEG

- The EEG can be used to monitor the spontaneous electrical activity of the brain.
- The waves of the EEG are of low amplitude (10–150 μV), and their frequency and amplitude depend on the state of arousal of a person.
- Slow waves generally are of relatively large amplitude, while fast waves have a low amplitude.
- The slower the wave, the greater the synchrony of the activity of the cortical neurons.
- The EEG is helpful in the diagnosis and location of abnormalities of the electrical activity of the brain, such as epilepsy.

The physiological characteristics of sleep

- The stages of sleep can be followed by monitoring the EEG.
- During deep sleep the EEG is dominated by slow wave activity (slow-wave sleep), but this pattern is interrupted several times a night by bouts of high-frequency wave activity associated with rapid eye movements.
- The rapid eye movement phase is known as REM sleep or paradoxical sleep.
- During slow-wave sleep, there is a slowing of the heart rate and respiratory rate, a decrease in blood pressure, and extensive relaxation of the somatic muscles.
- During REM sleep, the heart rate and respiratory rate increase and penile erection may occur. There are rapid eye movements and there may be clonic movements of the limbs, although most muscles are fully relaxed in this phase of sleep.
- REM episodes are associated with dreaming.
- The sleep–wakefulness cycle is linked to the activity of neurons in the brain stem.

Circadian rhythms

- The sleep–wakefulness cycle is an example of a 24-hour or circadian rhythm.
- Internal biological clocks exist in the brain and other tissues.
- Many physiological processes show a circadian rhythm, including heart rate, blood pressure, body core temperature, and the blood levels of many hormones.
- This intrinsic rhythm of the various biological clocks is synchronized by the neural activity of the suprachiasmatic nucleus (SCN).
- The circadian rhythm of the SCN neurons is linked to the day–night cycle by specific photoreceptive ganglion cells. These relay information about light intensity via the retino-hypothalamic tract.
- The pineal gland has been implicated in the regulation of circadian rhythms.
- This gland secretes a hormone called melatonin that is synthesized from serotonin.
- The synthesis and secretion of melatonin is increased during darkness and inhibited by daylight. This pattern is controlled by sympathetic nerve activity, which is regulated by light signals from the retina.

Recommended reading

Carpenter, R., and Reddi, B. (2012) *Neurophysiology: A conceptual approach* (5th edn), Chapter 13. Hodder Arnold, London.

Donaghy, M. (2005) *Neurology* (2nd edn), Chapters 21–24. Oxford University Press, Oxford.

A series of short chapters that provide a lucid introduction to the disorders mentioned in this chapter.

Kandel, E.R., Schwartz, J.H., and Jessell, T.M. (2000) *Principles of neural science* (4th edn), Chapters 46–48 and 59. McGraw-Hill, New York.

Sacks, O. (1986) *The man who mistook his wife for a hat*. Pan Books, London.

Not really a scientific study but a collection of fascinating neurological case studies; well written for a lay readership. Worth the effort of finding a copy.

Springer, S.P., and Deutsch, G. (1997) *Left brain, right brain* (5th edn). W.H. Freeman & Co., New York.

Squire, L.R., Bloom, F.E., Spitzer, N.C., Du Lac, S., Ghosh, A., and Berg, D. (2013) *Fundamental neuroscience* (4th edn), Chapters 39, 40, 49, and 50. Academic Press, San Diego.

These chapters provide detailed reviews of the principal topics discussed in this chapter. A useful introduction to the specialist literature.

Review articles

Brown, R.E., Basheer, R., McKenna, J.T., Strecker, R.E., McCarley, R.W. (2012) Control of sleep and wakefulness. *Physiological Reviews* **92**, 1087–187.

Demonet, J.F., Thierry, G., and Cardebat, D. (2005) Renewal of the neurophysiology of language: Functional neuroimaging. *Physiological Reviews* **85**, 49–95.

Dibner, C., Schibler, U., and Albrecht, U. (2010) The mammalian circadian timing system: Organization and coordination of central and peripheral clocks. *Annual Review of Physiology* **72**, 517–49.

Friederici, A.D (2011) The brain basis of language processing: From structure to function. *Physiological Reviews* **91**, 1357–92.

Mohawk, J.A., Green, C.B., and Takahashi, J.S. (2012) Central and peripheral circadian clocks in mammals. *Annual Review of Neuroscience* **35**, 445–62.

Richter, C., Woods, I.G., and Schier, A.F. (2014) Neuropeptidergic control of sleep and wakefulness. *Annual Review Neuroscience* **37**, 503–31.

To check that you have mastered the key concepts presented in this chapter, complete the accompanying online self-assessment questions. Go to www.oup.com/uk/pocock5e/

CHAPTER 20

Introduction to the endocrine system

Chapter contents

This chapter should help you to understand:

- The concept of glands, hormones, and target tissues
- The chemical nature of hormones and their mechanisms of action
- Carriage of hormones in the blood
- Patterns of hormone secretion—circadian rhythms and feedback control
- Measurement of hormones in body fluids

20.1 Introduction

Complex multicellular organisms require coordinating systems that can regulate and integrate the functions of different cell types. Two coordinating systems have evolved: the nervous system and the endocrine system. Broadly, the former uses electrical signals to transmit information very rapidly to discrete target cells (see Chapter 9), while the endocrine system uses chemical signalling to regulate the activity of particular cell populations. Chemical agents, **hormones**, are produced by cells within endocrine tissues and travel in the bloodstream to other cells (their targets) upon which they exert a regulatory effect. Therefore, a hormone is usually defined as a blood-borne chemical messenger, and endocrinology is defined as the study of the endocrine glands, their hormones, and their target tissues. The anatomical locations of the principal endocrine glands are shown in Figure 20.1, which also indicates the principal hormones secreted by each gland.

Hormones play a crucial role in the activity of all the major physiological systems of the body and are of particular importance in the regulation of processes requiring the integrated actions of a number of systems. Such processes include growth, development, metabolism, the maintenance of a stable internal environment (homeostasis), and the reproductive processes of both men and women. The various roles played by the major hormones in the regulation of somatic function are discussed in Chapters 21–24, while the endocrine regulation of the gut is discussed in Chapter 44 and that of reproduction is discussed in Chapters 48 and 49.

Originally, hormones were simply considered to be chemical signals secreted into the bloodstream by specific endocrine glands and then transported to their target cells. Although many hormones and glands do fit this classical definition, it is now clear that such a description needs to be widened to include a variety of tissues which, while having other crucial roles within the body, also synthesize and secrete substances that exert effects on other cells. Examples of such organs include the heart, which secretes atrial natriuretic peptide (ANP); the liver, which secretes hepcidin and a number of growth factors; and the brain, which secretes hormones from the hypothalamus and melatonin from the pineal gland. Recently it has been shown that adipose tissue secretes a hormone (**leptin**) that travels to the brain, where it plays a role in the regulation of food intake (see Chapter 47). Some tumours are also known to be capable of secreting polypeptide hormones, which may result in specific disease processes. One example is small cell carcinoma of the lung, which can secrete a variety of hormones such as ACTH (adrenocorticotrophic hormone), vasopressin (antidiuretic hormone or ADH),

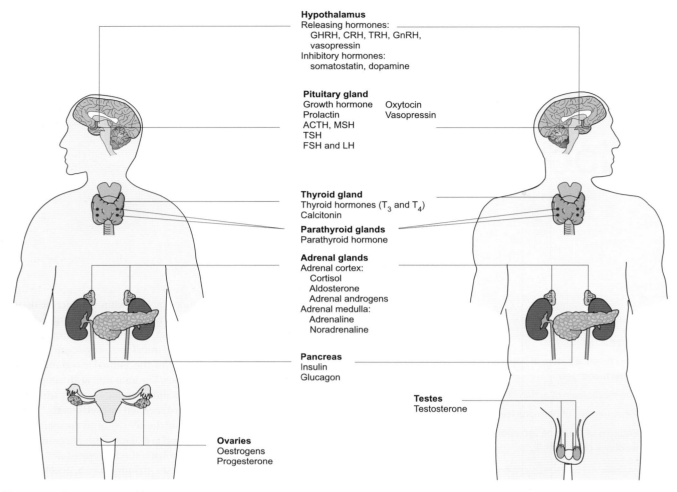

Hypothalamus
Releasing hormones:
   GHRH, CRH, TRH, GnRH,
   vasopressin
Inhibitory hormones:
   somatostatin, dopamine

Pituitary gland
Growth hormone   Oxytocin
Prolactin   Vasopressin
ACTH, MSH
TSH
FSH and LH

Thyroid gland
Thyroid hormones (T_3 and T_4)
Calcitonin

Parathyroid glands
Parathyroid hormone

Adrenal glands
Adrenal cortex:
   Cortisol
   Aldosterone
   Adrenal androgens
Adrenal medulla:
   Adrenaline
   Noradrenaline

Pancreas
Insulin
Glucagon

Testes
Testosterone

Ovaries
Oestrogens
Progesterone

Figure 20.1 The anatomical localization of the main endocrine glands and the principal hormones they secrete.

and a parathyroid-hormone-like agent. A list of the major 'non-classical' endocrine organs is shown in Table 20.1. In this book, the hormones secreted by these organs are discussed alongside the physiology of the appropriate organ systems.

While clearly differing in several respects, the nervous and endocrine systems are closely linked, and many neurons are capable of secreting hormones. This is known as **neurosecretion**. It occurs, for example, in the hypothalamus, where releasing and inhibitory hormones are secreted into specialized portal blood vessels which carry them to the anterior pituitary gland, where they alter the rate of secretion of other hormones. The hormones of the posterior pituitary gland are synthesized and secreted by the neurons of the paraventricular and supraoptic nuclei. Furthermore, certain substances that were originally believed to act only as blood-borne hormones have since been shown to act also as neurotransmitters within the central nervous system. Examples are the gastrointestinal hormones gastrin and cholecystokinin.

Not all hormones enter the general circulation in appreciable concentrations. For example, the **hypothalamic releasing hormones** mentioned above do not reach the systemic circulation in significant amounts. Other hormones exert their effects at a still

more local level, in some cases acting only on contiguous cells (a so-called paracrine action) or even performing an autocrine function in modifying the secretory action of the cells that produce them. One such example is oestradiol 17β, which, in addition to its normal endocrine role, also modifies the activity of the follicular granulosa cells that secrete it (Chapter 48).

20.2 The chemical nature of hormones and their carriage in the blood

Hormones fall into three broad categories according to their chemical properties.

1. The **steroids** are derivatives of cholesterol. They include the hormones of the ovaries and testes, the hormones of the adrenal cortex, and the active metabolites of vitamin D.

2. The **peptides** form the largest group of hormones in the body. They vary considerably in size, from the hypothalamic thyrotrophin releasing hormone, which consists of only three amino acid

Table 20.1 The secretion of hormones by non-classical endocrine tissues

Organ	Hormones secreted
Brain (particularly the hypothalamus)	Corticotrophin releasing hormone (CRF)
	Thyrotrophin releasing hormone (TRH)
	Luteinizing hormone releasing hormone (LHRH)
	Growth hormone releasing hormone (GHRH)
	Somatostatin
	Fibroblast growth factor
Heart	Atrial natriuretic peptide (ANP)
Kidney	Erythropoietin
	1,25 dihydroxycholecalciferol
Liver	Insulin-like growth factors (IGF-1, IGF-2)
	hepcidin
Gastrointestinal tract	Cholecystokinin (CCK)
	Gastrin
	Secretin
	Pancreatic polypeptide
	Gastric inhibitory peptide
	Motilin
	Enteroglucagon
Platelets	Platelet-derived growth factor (PDGF)
	Transforming growth factor-β (TGF-β)
Lymphocytes	Interleukins
Adipose tissue	Leptin
Various sites	Epidermal growth factor
	Transforming growth factor-α (TGF-α)

residues, to growth hormone (GH) and follicle stimulating hormone (FSH), which consist of almost 200.

3. Derivatives of specific **amino acids** include the catecholamines (such as adrenaline), which are derived from tyrosine, and the thyroid hormones, which are formed by the combination of two iodinated tyrosine residues.

The chemical nature of hormones influences the manner in which they are transported in the bloodstream. While the catecholamines and peptide hormones generally travel in free solution in the plasma, the steroids and thyroid hormones are very hydrophobic and are carried in the blood bound to a variety of plasma proteins, including albumin. Many binding proteins have a high affinity for specific hormones. Examples of these include sex hormone binding globulin, cortisol binding globulin, and thyroid hormone binding globulin.

Hormones that are bound to carrier proteins in the plasma are cleared from the circulation much more slowly than those travelling in free solution. Consequently their half-lives in the blood tend to be prolonged. This ensures a relatively constant rate of tissue delivery and a more sustained endocrine response.

20.3 The mechanism of action of hormones

All hormones act by binding to specific receptors situated on the plasma membrane itself, or within the cytoplasm or the nucleus of their target cells The details of the cellular mechanisms of hormone action are discussed in Chapter 6, so only a brief summary will be provided here.

Steroid and thyroid hormones exert their main effects by diffusing into cells and binding to intracellular receptors within the cytoplasm and nucleus. The receptor–hormone complex then alters gene expression. In some cases, however, steroid hormones may bind to receptors on the plasma membranes of their target cells without the need to enter the cell. Examples of this non-genomic action are sperm maturation in response to testosterone, the activation of Na^+/H^+ exchange in kidney cells by aldosterone, and the effects of corticosterone on neural excitability.

Water-soluble protein hormones and catecholamines bind to receptors on the plasma membrane. They subsequently exert their effects either through second messengers such as calcium, cyclic AMP, and inositol trisphosphate (IP_3) or by activating membrane-bound kinases which act to modify cellular behaviour. Although the characteristics of a hormone response will depend on the precise nature of the interaction between hormone and receptor, the actions of many steroid hormones have a relatively long latency compared with protein hormones, whose actions are generally more rapid in onset.

20.4 Measurement of hormone levels in body fluids

In clinical practice, it is often helpful to measure the concentration of hormones in the blood or other body fluids in order to assess endocrine function or make a diagnosis. The earliest hormone assays were mostly based on the biological response of a tissue or whole animal to a hormone—a **bioassay**. An example of a bioassay was the estimation of glucocorticoid concentrations by the measurement of glycogen levels in rat and mouse liver. However, many hormones are present in the plasma and other body fluids in very low concentrations and most bioassays lacked precision, specificity, and sensitivity. Although chemical analysis of body fluids has been used to measure the plasma levels of the catecholamines, it has been of limited value in the measurement of plasma levels of other hormones. The whole field of endocrinology was revolutionized by the development of **competitive binding assays** and **radioimmunoassays**. While these assays measure immunological rather than biological activity, they are relatively quick, sensitive, and specific. It is now possible to detect and quantify extremely low concentrations of hormones and thus to obtain information regarding their rates of secretion, half-lives, and rates of **clearance** from the blood. Radioimmunoassays are now available for all the polypeptide, thyroid, and steroid hormones while, more recently, sensitive chemiluminescence assays have been developed for measuring peptides.

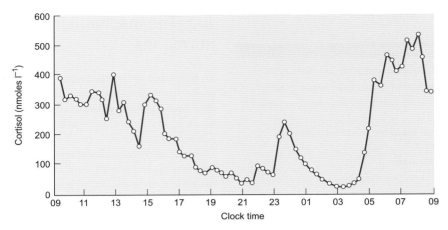

Figure 20.2 The circadian (24-hour) variation in plasma cortisol. Plasma levels rise during the early part of the day and are at their lowest around 3 a.m. This reflects the pattern of secretion of ACTH (adrenocorticotrophic hormone) from the anterior pituitary gland. (Adapted from Fig 4.12 in C. Brook and N. Marshall (1996) *Essential endocrinology.* Blackwell Science, Oxford.)

Gas chromatography allied to mass spectrometry is an extremely sensitive technique for measuring steroid hormones and is used extensively to monitor drug misuse in sports. Some hormone assays have been developed for home use, examples of which include the home pregnancy test, which detects human chorionic gonadotrophin (hCG) in the urine, and fertility and menopause predictors, which measure fluctuating levels of pituitary and ovarian hormones.

20.5 Patterns of hormone secretion— circadian rhythms and feedback control

Many physiological parameters show periodic or rhythmic changes that are controlled by the brain. Some of these patterns follow environmental cues such as the light–dark cycle or the sleep–wakefulness cycle, while others appear to be driven by an internal biological clock that is independent of the environment. Such 24-hour rhythms are called **circadian** or **diurnal rhythms** (see Chapter 19). The secretion of a number of hormones follows a circadian pattern. These include pituitary ACTH (which regulates the secretion of cortisol), growth hormone, and prolactin. Furthermore, many hormones show significant alterations in their rate of secretion in response to stress of all kinds. For these reasons, it is important to take these sources of variation into consideration when interpreting hormone measurements in clinical practice. Figure 20.2 illustrates the 24-hour (circadian) pattern of secretion of cortisol, plasma levels of which are lowest around 3 a.m. but rise sharply during the early part of the waking day to reach a peak at 7–9 a.m.

Many biological systems, including the secretion of hormones, are regulated by negative feedback (see Chapter 1 p 10). In the case of endocrine systems, such feedback regulation may be seen when the target tissue itself secretes a hormone. Many of the anterior pituitary hormones show negative feedback control of this kind. For example, thyroid stimulating hormone (TSH), secreted by the anterior pituitary, stimulates the output of thyroid hormones from the thyroid gland. Once released into the bloodstream, thyroid hormones (T_3 and T_4) exert negative feedback on the anterior pituitary gland to inhibit the release of TSH. In this way, the secretion of TSH is kept within narrow limits.

Although negative feedback is the most widespread form of endocrine regulation, positive feedback is also seen under some conditions. During the ovarian cycle, for example, the preovulatory gonadotrophin surge is brought about by the positive feedback actions of oestrogens on the anterior pituitary (see Chapter 48). Positive feedback is also seen during parturition, when oxytocin is secreted in response to stretching of the walls of the uterus and vagina (see Chapter 49).

Summary

The endocrine and nervous systems coordinate the complex functions of the body. Most hormones are released by endocrine glands and travel in the bloodstream to act on target tissues at a distance. Certain hormones, however, exert more localized (paracrine or autocrine) actions. Some hormones are secreted by 'non-classical' endocrine glands and by tumour cells. Hormones fall into three broad categories:

- steroids (derivatives of cholesterol)
- peptides (the largest group)
- those derived from single amino acids (the thyroid hormones and catecholamines).

The peptides and catecholamines travel in free solution in the plasma, while the steroids and thyroid hormones are largely bound to plasma proteins. Hormones act by binding to specific receptor proteins in their target cells. Peptides and catecholamines bind to receptors on the plasma membrane, while steroids and thyroid hormones exert most of their effects by binding to intracellular receptors. The secretion of most hormones is under negative feedback control, but the secretion of some hormones is subject to positive feedback regulation under certain physiological conditions.

Checklist of key terms and concepts

The endocrine glands, hormones, and target tissues

- Endocrinology is the study of endocrine glands, hormones, and their target cells.

- Hormones are chemical messengers produced by cells within endocrine tissues. Most travel in the bloodstream to other cells (their targets) whose behaviour they regulate.

- Some hormones act more locally, exerting a paracrine or autocrine activity.

- Many hormones are concerned with the control of physiological processes that involve the integrated activity of several body systems (e.g. growth and reproductive function).

- Certain groups of neurons are capable of secreting hormones.

The chemical nature of hormones and their mechanisms of action

- Hormones fall into three chemical categories:
  - the steroids, such as testosterone
  - the peptides, such as insulin
  - derivatives of specific amino acids, including the thyroid hormones and catecholamines.

- Catecholamines and peptide hormones generally travel in the blood in free solution, while more hydrophobic thyroid and steroid hormones are carried in the blood bound to plasma proteins.

- Peptide hormones and catecholamines bind to receptors on the plasma membranes of their target cells, whereas steroid and thyroid hormones exert their main effects via intracellular receptors.

Measurement of hormones

- The earliest hormone assays were bioassays and were relatively inaccurate.

- Modern techniques of hormone measurement such as radioimmunoassay allow the accurate measurement of tiny concentrations of hormone.

- They also yield information regarding rates of secretion, half-lives, and rates of clearance.

Regulation of hormone secretion

- Many hormones exhibit 24-hour (circadian) patterns of secretion. Examples include cortisol and growth hormone.

- Many hormones show alterations to their secretory patterns in response to stress. Such variations are important to consider when interpreting hormone levels in clinical practice.

- The secretion of many hormones is regulated by negative feedback control.

- More rarely, positive feedback regulation of hormone secretion occurs.

Recommended reading

Hall, R., and Evered, D.C. (2009) *A colour atlas of endocrinology* (2nd edn). Wolfe Medical, London.

An informative collection of colour plates illustrating the physical changes associated with various endocrine disorders.

Kovacs, W.J., and Ojeda, S.R. (eds) (2011) *Textbook of endocrine physiology* (6th edn), Chapters 1 and 3. Oxford University Press, Oxford.

Chapter 1 provides a more detailed description of the organization of the endocrine system, and Chapter 3 gives an excellent account of mechanisms of hormone action.

To check that you have mastered the key concepts presented in this chapter, complete the accompanying online self-assessment questions. Go to www.oup.com/uk/pocock5e/

CHAPTER 21

The pituitary gland and hypothalamus

This chapter should help you to understand:

- The anatomy of the pituitary gland and the hypothalamus

- The role of the CNS in the regulation of the endocrine system via the hypothalamo-pituitary axis

- The actions of the hormones of the anterior pituitary gland and the regulation of their secretion

- The actions of the hormones of the posterior pituitary gland, oxytocin and vasopressin (antidiuretic hormone), and the regulation of their secretion

21.1 Introduction

The nervous and endocrine systems are both concerned with communication within the body. Nerves communicate through rapid, specific electrical signals (action potentials—see Chapter 7), whereas the endocrine system uses hormones as chemical signals to regulate physiological processes. Nevertheless, the nervous and endocrine systems are closely linked via the control that the hypothalamus exerts over the pituitary gland. Two different modes of control are employed:

- direct neural connection between the hypothalamus and posterior pituitary via the hypothalamo-hypophyseal tract;

- hormonal regulation via a dedicated portal vascular system that links the hypothalamus to the anterior pituitary—the hypothalamic-hypophyseal portal system.

This chapter explores the hypothalamo-pituitary system and provides a brief description of the principal hormones of the anterior and posterior lobes of the pituitary gland. The actions of the individual hormones are discussed in further detail in the chapters dealing with their target glands: hormonal control of the thyroid and parathyroid glands is discussed in Chapter 22, the adrenal glands in Chapter 23, hormonal control of the reproductive system is discussed in Chapters 48 and 49, and the control of growth is discussed in Chapter 51.

21.2 The hypothalamo-pituitary axis

Figure 21.1 illustrates the anatomical relationship between the pituitary gland and the hypothalamus, the part of the brain to which it is attached by the pituitary stalk. The pituitary gland is situated in a saddle-shaped depression of the sphenoid bone at the base of the skull called the sella turcica. It consists of two

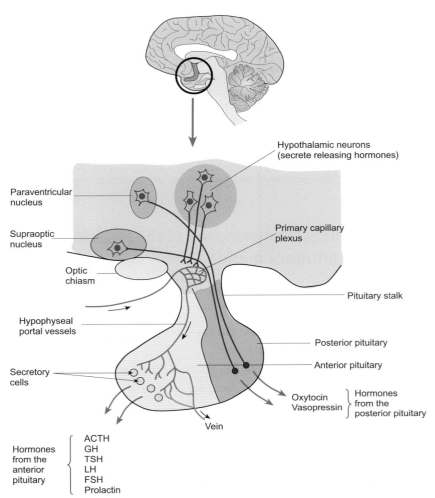

Figure 21.1 The relationship between the hypothalamus and the pituitary gland. Note the prominent portal system that links the hypothalamus to the anterior pituitary gland. The anterior pituitary has no direct neural connection with the hypothalamus. In contrast, nerve fibres from the paraventricular and supraoptic nuclei pass directly to the posterior pituitary, where they secrete the hormones they contain into the bloodstream.

anatomically and functionally distinct regions, the **anterior lobe** or **adenohypophysis** and the **posterior lobe** or **neurohypophysis**. Between these lobes lies a small sliver of tissue called the **intermediate lobe** or pars intermedia. This region is poorly developed in humans but does appear to secrete small amounts of MSH (melanocyte-stimulating hormone) and β-endorphin. MSH may have a role in stimulating the growth and activity of fetal melanocytes.

Figure 21.2 illustrates the way in which the pituitary gland is formed during the first trimester of gestation. The various parts of the pituitary gland have different embryonic origins: the anterior and intermediate lobes are derived from embryonic ectoderm as an upgrowth from the pharynx, while the posterior lobe is neural in origin. In the early embryo, the roof of the mouth lies adjacent to the third ventricle of the brain and both sheets of tissue bulge towards each other: the buccal cavity bulges upwards to form a structure called Rathke's pouch, and the neural ectoderm bulges downwards to form the infundibulum of the hypothalamus. Over a period of several weeks, the base of Rathke's pouch gradually constricts until it finally pinches off from the rest of the pharyngeal ectoderm and folds around the infundibulum to form the pituitary

stalk. Embryogenesis is complete at around 11 or 12 weeks of gestation in humans. The neural tissue, which remains as part of the brain, forms the posterior lobe of the pituitary and the non-neural tissue forms the anterior lobe.

The **adenohypophysis** consists of two portions: the anterior pituitary itself (also called the pars distalis) and the much smaller pars tuberalis, which is wrapped around the infundibular stem to form the pituitary stalk. The **neurohypophysis** strictly consists of three parts: the median eminence, which is the neural tissue of the hypothalamus from which the pituitary protrudes, the posterior pituitary itself, and the infundibular stem, which connects the two. The fully developed adult pituitary gland weighs about 0.5 g.

The anterior pituitary gland plays a central role in the regulation of the endocrine system. It accounts for around 80 per cent of the pituitary by weight and secretes at least seven different hormones, most of which regulate the secretions of other endocrine organs. Although the anterior pituitary receives no direct neural input from the median eminence, it is now known that the hypothalamus plays a key role in regulating pituitary function. Between the hypothalamus and the anterior pituitary, there is a system of blood vessels known as the **hypothalamic-hypophyseal portal**

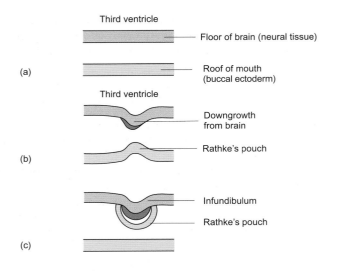

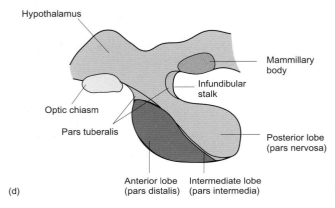

Figure 21.2 The embryonic development of the pituitary gland. Panels (a)–(c) show outgrowths of neural and ectodermal tissue, with separation of Rathke's pouch. Panel (d) shows the structure of an adult gland in which the colouring corresponds to that shown in panels (a)–(c).

system which originates in the primary capillary plexus—a series of capillary loops in the median eminence (the part of the hypothalamus that lies along the midline of its lower surface). Most of the arteries that supply the anterior pituitary do not form a capillary network amongst the epithelial cells, but rather they course upwards into the pituitary stalk, where they empty into a network of capillary sinusoids. The blood leaving this primary capillary plexus then flows in parallel veins, the long portal vessels, down the pituitary stalk to the anterior lobe. Here the portal vessels break up into sinusoids, which form the main blood supply of the anterior pituitary. A diagram of this arrangement is shown in Figure 21.1. The role of the hypophyseal portal system is to transport specific hormones secreted by the neurons of the median eminence to the anterior pituitary. These hypothalamic hormones regulate the output of the hormones of the anterior pituitary (see Section 21.3). The hypothalamus together with the pituitary and the products of its target tissues therefore form a complex functional unit.

Summary

The pituitary gland consists of two principal lobes, the anterior lobe (or adenohypophysis) and the posterior lobe (or neurohypophysis). The intermediate lobe is sandwiched between the two main divisions. The hypothalamus plays a crucial role in regulating the secretion of the pituitary hormones. In the case of the posterior pituitary, this is achieved by a direct neural connection via the hypothalamo-hypophyseal tract. The regulation of the anterior pituitary occurs by the secretion of specific releasing hormones into the hypothalamic-hypophyseal portal system.

21.3 The hormones of the anterior pituitary gland

The cells that line the blood sinusoids of the anterior pituitary gland secrete a variety of hormones. The seven most important are:

1. growth hormone (GH or somatotrophin)
2. prolactin (PRL)
3. adrenocorticotrophic hormone (ACTH or corticotrophin)
4. melanocyte stimulating hormone (MSH)
5. thyroid stimulating hormone (TSH or thyrotrophin)
6. follicle stimulating hormone (FSH)
7. luteinizing hormone (LH).

Follicle stimulating hormone and luteinizing hormone are called **gonadotrophins**, reflecting their essential role in reproduction.

On the basis of microscopic examination of granule size and number, and of immunological staining reactions, the cells secreting certain pituitary hormones can be identified as shown in Figure 21.3. At least five different endocrine cell types may be distinguished:

- somatotrophs
- lactotrophs
- gonadotrophs
- corticotrophs
- thyrotrophs.

The **somatotrophs** secrete **growth hormone** and account for around 50 per cent of the glandular cells. The **lactotrophs** secrete **prolactin** and account for a further 15–20 per cent of glandular cells. Both somatotrophs and lactotrophs are mainly located in the lateral wings of the anterior pituitary. The **gonadotrophs** account for about 10 per cent of the glandular cell population. They secrete **FSH (follicle stimulating hormone)** and **LH (luteinizing hormone)** and are scattered throughout the anterior pituitary, especially the pars tuberalis. The **corticotrophs** account for 15–20 per cent of glandular cells and are mainly located in the central wedge of the anterior pituitary. They secrete **ACTH (adrenocorticotrophic hormone)** as

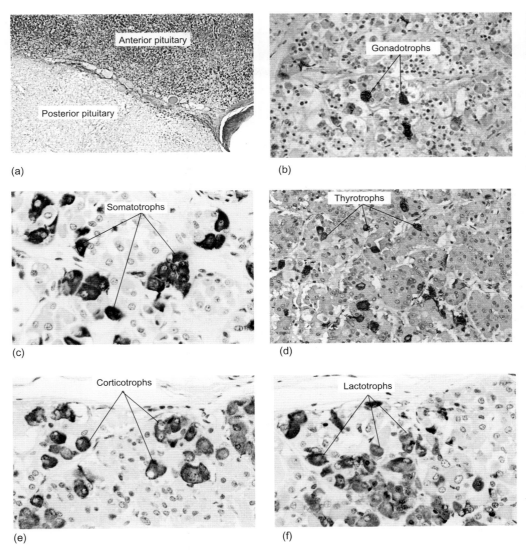

Figure 21.3 Panel (a) shows a section of the anterior pituitary gland stained with Masson's triple stain. The remaining panels show sections of pituitary treated with antibodies to label specific cell populations: (b) gonadotrophs (which secrete FSH and LH), (c) somatotrophs (which secrete growth hormone), (d) thyrotrophs (which secrete thyroid stimulating hormone, TSH), (e) corticotrophs (which secrete ACTH), and (f) lactotrophs (which secrete prolactin). Note the predominance of somatotrophs in panel (c) and the relative scarcity of gonadotrophs in panel (b) and thyrotrophs in panel (d). (Images courtesy of the following: (b) Endotext.org; (c), (e), (f) Laura P. Hale, MD PhD, Duke University Medical Center; (d) Reproduced with permission of Biocare Medical.)

well as some related peptides. Finally, the **thyrotrophs** secrete **TSH (thyroid stimulating hormone)** and contribute around 5 per cent of the glandular cell population. They are mainly found in the anterior part of the central portion of the lobe.

Although many of the cells secrete only one type of hormone (e.g. lactotrophs secrete prolactin and somatotrophs secrete growth hormone), it is now known that some of the pituitary cells are able to produce more than one hormone. The best example of this is provided by the gonadotrophs, many of which secrete both FSH and LH. Moreover, LH and prolactin are also found in the thyrotrophs, while the corticotrophs secrete β-**lipotrophin** and α- **and** β-**MSH (melanocyte stimulating hormones)** in addition to ACTH. Table 21.1 lists the principal hormones of the anterior pituitary along with their major target tissues and the cells that synthesize and secrete them. All the anterior pituitary hormones are proteins or polypeptides.

The secretion of the anterior pituitary hormones is controlled by hormones released by the hypothalamus into the hypophyseal portal blood

The hypothalamus controls the secretory activity of the anterior pituitary gland. The rates of secretion of thyroid stimulating hormone (TSH), follicle stimulating hormone (FSH), luteinizing hormone (LH), melanocyte stimulating hormone (α-MSH), and adrenocorticotrophic hormone (ACTH) and related peptides are all stimulated by hypothalamic hormones (known as **releasing hormones**), while the secretion of prolactin is mainly regulated by the inhibitory effect of dopamine (also known as prolactin inhibitory hormone), which is also secreted by neurons in the hypothalamus. Prolactin secretion may also be stimulated by thyrotrophin releasing hormone (TRH). The release of growth hormone (GH) is under dual control by the hypothalamus: its secretion is stimulated

Table 21.1 The anterior pituitary hormones

Class of hormone	Specific hormones	Synthesized and secreted by	Target tissues	Number of amino acid residues
Somatotrophic hormones				
These hormones have a single peptide chain	Growth hormone (GH) also known as somatotrophin	Somatotrophs	Most tissues except CNS	191
	Prolactin (PRL)	Lactotrophs	Mammary glands	198
Corticotrophin-related peptide hormones				
These are all derived from a single common precursor	Corticotrophin (ACTH)	Corticotrophs	Adrenal cortex	39
	β-lipotrophin (β-LPH)	Corticotrophs	?Adipose tissue	91
	β-endorphin (β-LPH 61-91)	Corticotrophs	Adrenal medulla, gut	31
	α-melanocyte stimulating hormone (α-MSH)	Corticotrophs	Melanocytes	13
Glycoprotein hormones				
These are composed of a common α-peptide chain associated with a variable β-peptide chain	Thyrotrophin (TSH)	Thyrotrophs	Thyroid gland	α-chain 89 β-chain 112
	Follicle stimulating hormone (FSH)	Gonadotrophs	Ovaries (granulosa cells) Testes (Sertoli cells)	α-chain 89 β-chain 115
	Luteinizing hormone (LH)	Gonadotrophs	Ovaries (thecal and granulosa cells) Testes (Leydig cells)	α-chain 89 β-chain 115

by **growth hormone releasing hormone** (GHRH) but suppressed by another peptide, **somatostatin** (also known as growth hormone inhibiting hormone or GHIH).

The hypothalamic releasing and inhibiting hormones are synthesized in the cell bodies of neurons lying within discrete areas (nuclei) of the hypothalamus. They are then transported along the fine axons that constitute the tuberoinfundibular tract to their terminals in the upper part of the hypophyseal stalk (the median eminence), where their endings lie next to the capillaries of the primary plexus. In response to neural activity, the hypothalamic hormones are released from the nerve endings into the hypophyseal portal blood and are then carried down the pituitary stalk by the long portal veins to the anterior lobe. Here they act on specific pituitary cells to modify the rate of secretion of one, or sometimes several, of the anterior pituitary hormones. The major hypothalamic releasing and inhibiting hormones, along with their target hormones and alternative names, are listed in Table 21.2.

Specific hypothalamic nuclei synthesize and secrete the releasing hormones

The neurons that synthesize and store the various hypothalamic releasing hormones have been identified by immunocytochemistry (a specific staining technique that identifies substances by their immunological reactivity). Most of the releasing hormones seem to be produced by relatively discrete groups of neurons in the hypothalamus. These groups of neurons are called **hypothalamic**

nuclei. A simple diagram showing their positions is shown in Figure 21.4.

Neurons in the arcuate nucleus secrete GHRH, while somatostatin is secreted by cells of the periventricular nucleus, which is not shown in Figure 21.4. (The *peri*ventricular nucleus is a thin layer of nerve cells adjacent to the third ventricle and should not be confused with the *para*ventricular nucleus.) **Corticotrophin releasing hormone** (CRH) is synthesized mainly in neurons of the paraventricular nucleus. Luteinizing hormone releasing hormone (LHRH) is found mainly in neurons of the medial preoptic area and in the arcuate nucleus, though gonadotrophs are also found in the pars tuberalis. LHRH is alternatively known as gonadotrophin releasing hormone (GnRH) because it stimulates the secretion of both FSH and LH. Dopamine, which inhibits the secretion of prolactin, is found in neurons of the arcuate region, while thyrotrophin releasing hormone (TRH) is located in both the preoptic and paraventricular nuclei— see Table 21.2. As mentioned earlier, the axons of neurosecretory neurons terminate in the median eminence at the top of the pituitary stalk, from where they secrete their hormones into the bloodstream. Figure 21.5 shows the accumulation of GHRH in the median eminence. Other hypothalamic hormones show similar accumulations.

A further point to note is that some of the hypothalamic hormones are also found in parts of the body other than the hypothalamus, where they act in different ways. Somatostatin, for example, acts in other parts of the brain as a neurotransmitter, throughout the gut as a hormone, and in the pancreas, where it inhibits the release of insulin and glucagon.

Table 21.2 The hypothalamic releasing and inhibitory hormones with their alternative names

Hypothalamic hormone	Alternative names	Source	Stimulates secretion of ...	Inhibits secretion of ...	Structure
Vasopressin	Antidiuretic hormone (ADH)	Paraventricular and supraoptic nuclei	ACTH	—	9 amino acid peptide
Corticotrophin releasing hormone	CRH	Paraventricular nucleus	ACTH	—	41 amino acid peptide
Gonadotrophin releasing hormone (GnRH)	Luteinizing hormone releasing hormone (LHRH) FSH releasing factor	Medial preoptic area and arcuate nucleus	LH, FSH	—	10 amino acid peptide
Thyrotrophin releasing hormone	TRH	Preoptic nucleus and paraventricular nucleus	TSH, prolactin	—	3 amino acid peptide
Growth hormone releasing hormone	GHRH	Arcuate nucleus	GH	—	44 amino acid peptide
Growth hormone inhibiting hormone	Somatostatin GHIH	Periventricular nucleus	—	GH, Prolactin, TSH	14 amino acid peptide
Dopamine	Prolactin inhibiting factor (PIF)	Arcuate nucleus	—	Prolactin	Catecholamine

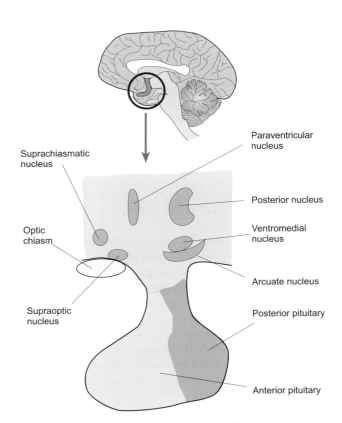

Figure 21.4 A diagrammatic representation of the location of the principal nuclei of the hypothalamus that are associated with the production and secretion of many of the releasing hormones.

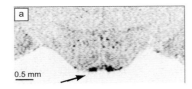
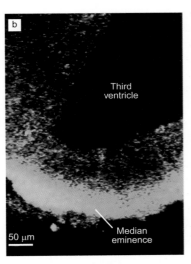

Figure 21.5 A section of mouse brain in which nerve cells and nerve terminals containing GHRH have been labelled. Panel (a) is a low-power view of the hypothalamus in which the mRNA for GHRH has been labelled to show where the hormone is synthesized. The acuate nucleus (arrow) shows a strong signal. Panel (b) shows GHRH tagged with green fluorescent protein. The region of intense green colour in the median eminence reveals where the axons containing GHRH terminate. Bars, 0.5 mm (a), 50 μm (b). (From Fig 4 in N. Balthasar et al. (2003) *Endocrinology* **144**, 2728–40.)

Feedback mechanisms operate within the hypothalamo-pituitary–target tissue axes to ensure fine control of endocrine function

A general discussion of feedback mechanisms can be found in Chapter 1, and Chapter 20 includes a brief discussion of the regulatory role played by both negative and positive feedback in the endocrine system. Such processes play an important role in regulating the secretion of the hypothalamic releasing hormones and contribute to the overall control of the secretion of the anterior pituitary hormones, and of its target glands.

In many cases the output of a pituitary hormone is increased by the removal of its target gland. For example, removal of the thyroid gland and the subsequent loss of the thyroid hormones stimulates an increase in the output of thyroid stimulating hormone (TSH)

from the anterior pituitary. Indeed, patients with hypothyroidism caused by reduced thyroid gland activity characteristically exhibit raised levels of TSH in parallel with their depressed thyroid hormone levels. Conversely, administration of exogenous thyroxine depresses the output of TSH by reducing pituitary sensitivity to TRH. There may also be a direct effect of thyroid hormones on the output of TRH itself. These hormonal interactions are summarized in Figure 21.6.

Other feedback loops are thought to modulate pituitary function in a similar fashion. ACTH secretion, for example, is depressed by an increase in the circulating level of the adrenal steroids. This results from both a direct inhibition of CRH release and a reduction in the responsiveness of the ACTH-secreting cells of the anterior pituitary. Like the other hormones discussed above, prolactin secretion is subject to negative feedback control. In this case, however, prolactin inhibits its own release by binding to dopaminergic neurones and thus stimulating further output of dopamine. Its secretion is stimulated by TSH, so that prolactin secretion is subject to dual regulation.

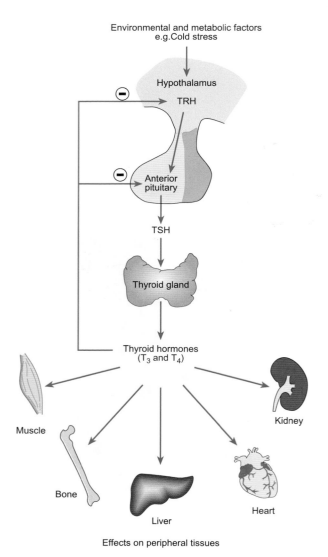

Figure 21.6 Summary of the negative feedback control mechanisms operating within the hypothalamo–pituitary–thyroid axis. Negative feedback is indicated by the red arrows.

Summary

The anterior pituitary secretes seven major hormones, all of which are polypeptides: growth hormone (GH), prolactin (PRL), thyroid stimulating hormone (TSH), adrenocorticotrophic hormone (ACTH), melanocyte stimulating hormone (MSH), follicle stimulating hormone (FSH), and luteinizing hormone (LH).

The secretion of the anterior lobe hormones is regulated by specific releasing and inhibiting hormones that are synthesized and secreted by hypothalamic neurons. The releasing and inhibiting hormones are secreted into the primary capillary network of the hypothalamo-pituitary portal system before being transported to the anterior lobe, where they regulate the secretion of the anterior lobe hormones. The secretion of the anterior lobe hormones is subject to feedback control, which operates both on the hypothalamus and directly on the secretory cells of the anterior pituitary gland.

21.4 Growth hormone (GH)

GH is synthesized and stored in somatotrophs, which are the most abundant pituitary cell type (see Figure 21.3c). The human form of growth hormone (hGH) is a large single-chain peptide consisting of 191 amino acid residues. The anterior pituitary contains around 10 mg of growth hormone which, in adults, is secreted at a rate of around 1.4 mg day^{-1}, which is greater than that of any of the other pituitary hormones. It is important to note, however, that over a 24-hour period the rate of GH secretion fluctuates considerably. This fluctuation is discussed later in this section. In the plasma, about 70 per cent of GH is bound to various proteins including a specific GH binding protein (GHBP), which is derived by cleavage of the extracellular region of the GH receptor present on target cells. Increased amounts of GH binding protein (and thus a reduction in free, biologically active GH) have been linked with the condition known as Laron dwarfism (see Chapter 51).

Actions of growth hormone

Growth hormone exerts a wide range of metabolic actions that involves virtually every type of cell except neurons. Its principal targets, however, are the bones and skeletal muscles. Although it is the principal regulator of growth in children and adolescents, it continues to have important metabolic effects throughout adult life.

For the purposes of this discussion, it is convenient to divide the actions of growth hormone into two categories. These are its direct effects on the metabolism of fats, proteins, and carbohydrates, and its indirect actions that result in skeletal growth. The principal actions of GH are illustrated in Figure 21.7.

Direct metabolic effects of growth hormone

Essentially, growth hormone is anabolic: i.e. it promotes protein synthesis. It is also a glucose-sparing agent with an anti-insulin-like, diabetogenic action (i.e. causing an increase in blood glucose).

Growth hormone stimulates the uptake of amino acids by cells (particularly those of the liver, muscle, and adipose tissue) and the incorporation of amino acids into proteins in many organs of the body. The net effects of GH on protein metabolism are an increase in the rate of protein synthesis, a decrease in plasma amino acid content, and a positive nitrogen balance (defined as the difference between daily nitrogen intake in food and excretion in urine and faeces as nitrogenous wastes). Furthermore, GH also

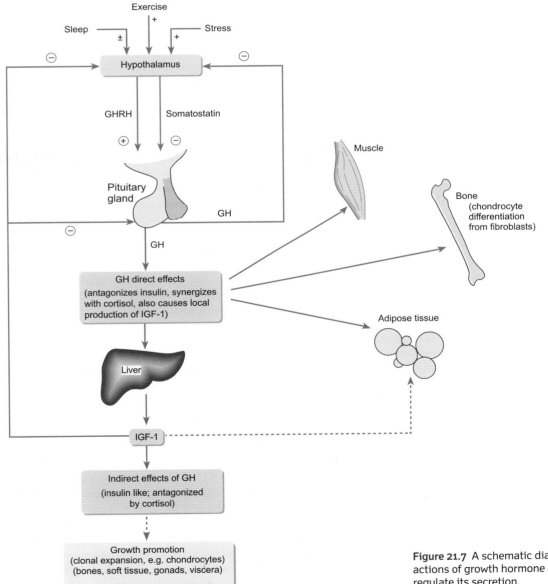

Figure 21.7 A schematic diagram showing the principal actions of growth hormone and the factors that regulate its secretion.

increases the rates of synthesis of RNA and DNA and ultimately of cell division. This effect is particularly important during the growing years, when GH contributes to the increase in bone length and soft tissue mass. GH also stimulates an increase in the rate of chondrocyte differentiation from fibroblasts in cartilage (see Chapter 51).

The actions of GH on lipid and carbohydrate metabolism are essentially diabetogenic, i.e. they cause an increase in blood glucose. GH increases plasma glucose levels in two ways. It decreases the rate of glucose uptake by cells (largely those of muscle and adipose tissue) and increases the rate of glycogenolysis by the liver. GH promotes the breakdown of stored fat in adipose tissue and the release of free fatty acids into the plasma. This action is important for providing a non-carbohydrate source of metabolic substrate for ATP generation by tissues such as muscle, thus conserving glucose for use by the CNS. This action is enhanced during fasting. As a result of the increased oxidation of fat, there is a reduction in the **respiratory quotient** (see Chapter 47).

Indirect actions of GH on skeletal growth

The main indirect physiological actions of GH are the maintenance of tissues and the promotion of linear growth during childhood and adolescence. These actions, particularly the latter, are considered further in Chapter 51. It is important to remember that growth is a highly complex process that is under the control of numerous agents, including a variety of growth factors and hormones, in addition to GH itself.

The growth-promoting effects of GH are mediated partly by its actions on amino acid transport and partly by its stimulation of protein synthesis. Skeletal growth results from the stimulation of mitosis in the epiphyseal discs of cartilage (also called growth plates) present in the long bones of growing children. GH exerts direct actions on both cartilage and bone to stimulate growth and differentiation, aided by polypeptides called **insulin-like growth factors (IGFs)** that are synthesized chiefly by the liver but also by bone and by the growing tissues themselves. These agents encourage the cartilage cells to divide and to secrete more cartilage matrix. As growing cartilage is eventually converted to bone, the growth of cartilage enables the bone to increase in length. Growth of the long bones in response to GH ceases once the epiphyseal discs themselves are converted to bone at the end of adolescence (a process known as fusion of the epiphyses). No further increase in stature can then occur despite the continued, though declining, secretion of GH throughout adulthood.

What are the IGFs and how do they promote growth?

The IGFs show significant amino acid sequence homology with the pancreatic hormone insulin, and they share some of its effects. Two such factors have been identified, IGF-1 and IGF-2, the former having the most potent growth-promoting effect and the latter the strongest insulin-like action. An important action of GH is to enhance IGF-1 production by the liver. IGF-1, in turn, stimulates a variety of the cellular processes that are responsible for tissue growth. In many cells (e.g. fibroblasts, muscle, and liver),

IGF-1 stimulates the production of DNA and increases the rate of cell division. It also encourages the incorporation of sulphate into chondroitin in the **chondrocytes** (the cells which produce and maintain the extracellular matrix of cartilage), and the synthesis of glycosaminoglycan for cartilage and collagen formation. (Chondroitin is a long-chain polysaccharide with a repeating disaccharide unit consisting of N-acetylgalactosamine and glucuronic acid.)

The growth-promoting effects of GH are crucial for growth during childhood, particularly from the age of about 3 to the end of adolescence. IGF-1 levels reflect the rate of growth during this time, and there is a marked increase at **puberty**. GH and IGF-1 seem to be less important to growth during the fetal and neonatal periods. During infancy, thyroid hormones are of great importance in growth and development (both skeletal and neurological—see Chapter 51) whereas IGF-2 is believed to be important for normal fetal growth. Indeed, IGF-2 has been shown to promote the growth and proliferation of cells in many different fetal tissues. Furthermore, although the IGF-2 gene is highly active during fetal development, it is much less active after birth.

Growth hormone secretion is governed by the hypothalamic secretion of GHRH and somatostatin

As described earlier, growth hormone is under dual control by hormones released from hypothalamic neurons. Its secretion is stimulated by GHRH and inhibited by somatostatin. The rate at which GH is released from the somatotrophs of the anterior pituitary is therefore determined by the balance between these two hormones. Since GH secretion declines when all hypothalamic influences are removed (either experimentally or in disease), it may be assumed that the positive effects of GHRH on GH release are normally dominant.

GH, like all the anterior pituitary hormones, is released in discrete pulses, these pulses being most frequent in adolescence. The detailed mechanism of pulsatile release is unclear, but peaks of GH secretion appear to coincide with peaks of GHRH output, while troughs coincide with increased rates of somatostatin release. GH secretion also shows a circadian rhythm, with marked elevations in output associated with periods of deep sleep, during which bursts of secretion occur every 1–2 hours (see Figure 21.8). The detailed control mechanisms that govern this periodicity are not fully understood, but more discussion of biological clocks and their genetic regulation may be found in Chapter 19. It is important to be aware of this pattern when carrying out measurements of plasma GH levels in clinical practice. A single measurement will be insufficient for diagnostic purposes, and it will be necessary to perform frequent serial assays.

A number of physical and psychological stresses promote the secretion of GH. Examples include anxiety, pain, surgery, cold, haemorrhage, fever, and strenuous exercise. Adrenergic and cholinergic pathways in the brain are believed to mediate these effects. The significance of the raised GH output is not fully understood, but it seems likely that the glucose-sparing effect of the hormone would be of value in circumstances of this kind.

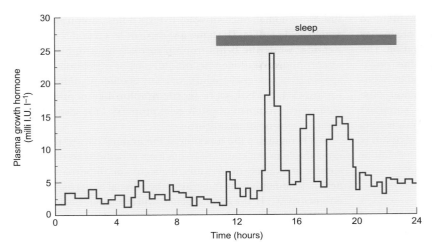

Figure 21.8 The diurnal variation in growth hormone secretion recorded for a normal nine-year-old child. Note the pronounced pulsatile release of GH during sleep. (Adapted from Fig 3.5 in C. Brook and N. Marshall (1996) *Essential endocrinology*, 3rd edition. Blackwell Science, Oxford.)

Metabolic factors influencing GH secretion

The most potent metabolic stimulus for GH release is hypoglycaemia. This is an appropriate homeostatic response, since GH acts as a glucose-sparing hormone, promoting the breakdown of fats to make fatty acids available for oxidation. At the same time, GH inhibits the uptake of glucose by the peripheral tissues and conserves glucose for use by the brain. By contrast with the effect of hypoglycaemia, an oral glucose load rapidly suppresses the release of GH.

Growth hormone is secreted during prolonged fasting. Again, its glucose-sparing effects and its effects on lipid metabolism ensure that those tissues (the CNS and the germinal epithelium of the ovaries and testes) that rely entirely on glucose as their source of energy are adequately supplied. Other metabolic factors known to increase the rate of GH secretion include a rise in plasma amino acid levels and a reduction in the plasma concentration of free fatty acids. All of these metabolic actions are mediated by changes in the output of GHRH and somatostatin.

Feedback actions of GH and the IGFs

The secretion of growth hormone appears to be influenced by the plasma concentration of GH itself. High plasma GH levels inhibit the release of further GH. This is an example of a short negative feedback loop in which GH depresses its own release, either by altering the rates of secretion of GHRH and somatostatin by the hypothalamus, or by altering the sensitivity of the anterior pituitary to these hypothalamic factors. In addition to the direct feedback effects of GH itself, IGF-1 can also inhibit GH release via a feedback action on GH synthesis by the pituitary gland. These interactions are shown in Figure 21.7.

Both hypersecretion and hyposecretion of GH may result in disorders of growth. Hypersecretion in children, which is usually caused by a pituitary tumour, results in **gigantism**, a condition in which growth is exceptionally rapid, while GH-deficiency in children results in **pituitary dwarfism**, in which the growth of the long bones is slowed. These conditions are discussed in more detail in Chapter 51.

GH secretion across the lifespan

The synthesis and secretion of GH decreases with age by approximately 14 per cent per decade after the age of 40 years, and this decrease is exacerbated in obese individuals. Many age-related changes may be linked to this reduced GH secretion, including reduced muscle mass and exercise capacity, increased body fat, reduction in bone mineral density, increased plasma lipid concentrations, and an increased risk of vascular disease. Treatment of healthy older adults with GH results in increased lean body mass and decreased fat mass, but it appears to have little effect on exercise performance.

Summary

Growth hormone (GH) is a peptide hormone secreted by the somatotrophs of the anterior pituitary gland. Its secretion is stimulated by GHRH and suppressed by somatostatin. GH exerts a wide range of metabolic actions. GH secretion is stimulated by hypoglycaemia, increased plasma levels of amino acids, and reduced plasma levels of free fatty acids. GH inhibits glucose uptake by most tissues. This anti-insulin action conserves glucose for use by the brain. GH promotes lipolysis, thus providing a non-carbohydrate source of substrate for ATP generation for other tissues. GH promotes protein synthesis and is crucial for normal skeletal growth between the ages of about 3 years and puberty. Skeletal growth occurs in response to IGFs, whose synthesis and secretion is stimulated by GH. The IGFs encourage cartilage cells to divide and enhance the deposition of cartilage at the epiphyses (growth plates). GH secretion falls in later life.

21.5 Prolactin

Prolactin has considerable structural similarities with growth hormone. It is a large single-chain peptide consisting of 198 amino acid residues and is synthesized by, and stored in, the anterior pituitary lactotrophs. It has a weakly somatotrophic action (reflecting its structural closeness to GH) but its predominant role is to promote growth and maturation of the mammary gland during pregnancy to prepare it for the secretion of milk throughout lactation. The secretion of prolactin is normally inhibited by dopamine (previously known as prolactin inhibitory hormone), secreted by hypothalamic neurons. In lactating women, however, stimulation of the nipple by the baby during breast-feeding (the suckling stimulus) inhibits the secretion of dopamine. Prolactin secretion is thus allowed to rise, and milk synthesis by the mammary tissue is stimulated (galactopoiesis). This is discussed in more detail in Chapter 49.

Summary

Prolactin promotes growth and maturation of the mammary gland to prepare it for lactation following the delivery of an infant. Prolactin secretion is normally inhibited by dopamine, but suckling during breast-feeding inhibits dopamine secretion, allowing prolactin secretion to increase and stimulate galactopoiesis.

21.6 Adrenocorticotrophic hormone and melanocyte stimulating hormone

Both adrenocorticotrophic hormone (ACTH) and melanocyte stimulating hormone (MSH) are synthesized and secreted by the anterior pituitary corticotrophs. ACTH is a polypeptide hormone consisting of 39 amino acid residues. It is derived originally from a much larger molecule, pre-pro-opiomelanocortin, which is also a precursor for several other physiologically active peptides including MSH. The relationships between the peptides derived from the common precursor are illustrated in Figure 21.9.

Pre-pro-opiomelanocortin splits to form β-lipotrophin and a 146-amino acid peptide. The latter gives rise to ACTH and N-terminal peptide, while the β-lipotrophin forms γ-lipotrophin and β-endorphin (an endogenous opioid, part of which may further split to form met-enkephalin). A variety of other peptides including α-MSH, β-MSH, CLIP (corticotrophin-like peptide), and some with unknown physiological properties are also derived from pre-pro-opiomelanocortin.

ACTH regulates the function of the adrenal cortex, playing a crucial role in the stimulation of glucocorticoid secretion in response to a variety of stressors. It also has an important trophic action, maintaining the integrity of the adrenal tissue itself. In the absence of ACTH, the adrenal glands will eventually begin to atrophy. The pattern of ACTH secretion varies during the day, showing a typical circadian rhythm (see Figure 20.2). ACTH secretion is under the control of hypothalamic CRH and is subject to negative feedback regulation: glucocorticoids (steroid hormones secreted from the adrenal cortex—see Chapter 23) inhibit ACTH secretion by a direct action on the anterior pituitary gland and through inhibition of hypothalamic CRH secretion. This is a classic example of negative feedback regulation of hormone secretion, as discussed in Chapter 20.

Melanocyte stimulating hormone (α-MSH) is derived by cleavage of ACTH, but although ACTH has some MSH-like activity, MSH does not appear to share any of the actions of ACTH. In certain species, MSH plays a role in skin pigmentation through the stimulation of melanocytes in the epidermis, as well as in the control of sodium excretion. However, the physiological significance of these effects in humans is unclear. It is known that α-MSH binds to a receptor (MC-1) on the human melanocyte membrane and that this binding activates tyrosinase, an enzyme required for the synthesis of the pigment **melanin**. Furthermore, melanocytes taken from individuals who tan poorly (usually those with red hair) appear to

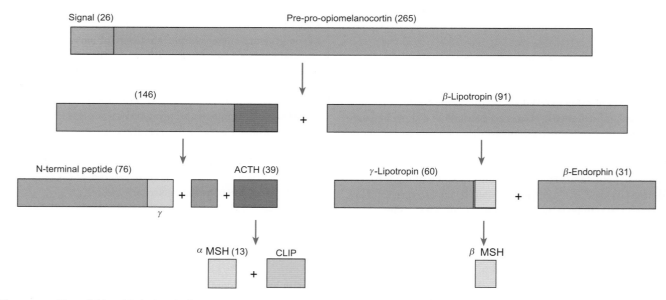

Figure 21.9 The relationship between the amino acid sequences of the various hormones derived from pre-pro-opiomelanocortin. ACTH, β-lipoprotein, β-endorphin, and a 76 amino acid peptide are the end products that are secreted by the anterior lobe of the pituitary gland. The numbers in brackets after the peptide name give the number of amino acid residues in each peptide.

have mutations in the MC-1 receptor. Four additional receptor sub-types for MSH have recently been identified. While their functions remain unclear, they appear to be responsible for some additional effects of MSH, including the mediation of the actions of ACTH on the adrenal gland, effects on food intake and energy expenditure, and control of exocrine secretions.

Summary

Adrenocorticotrophic hormone (ACTH) is secreted by anterior pituitary corticotrophs. It is derived from a much larger precursor molecule called pre-pro-opiomelanocortin. ACTH maintains the structural integrity of the adrenal cortex and regulates the secretion of glucocorticoid steroid hormones in response to stress. ACTH secretion is under the control of hypothalamic corticotrophin releasing hormone (CRH) and is subject to negative feedback regulation. MSH is derived from ACTH and serves to activate tyrosinase, an enzyme required for the synthesis of the pigment melanin.

21.7 Pituitary glycoprotein hormones: thyroid stimulating hormone (TSH), follicle stimulating hormone (FSH), and luteinizing hormone (LH)

The pituitary glycoprotein hormones consist of two interconnected amino acid chains (α- and β-subunits) containing sialic acids (a family of 8- and 9-carbon monosaccharides) and the carbohydrates hexose and hexosamine. The α-subunits of all three hormones are identical, while the β-subunits confer biological specificity (another example of a 'family' of related hormones). The α-subunit is also identical to that of human chorionic gonadotrophin (hCG) (see Chapter 49) and is considered to be the effector region responsible for the activation of adenylyl cyclase and the generation of cAMP in the target cells. All three pituitary glycoproteins are **trophic hormones**, which means that they not only regulate the secretions of their target glands but are also responsible for the maintenance and integrity of the target tissue itself.

TSH is secreted by pituitary thyrotrophs (see panel (d) of Figure 21.3), which contain numerous small secretory granules. It controls the function of the thyroid gland and the output of the thyroid hormones thyroxine and tri-iodothyronine (see Chapter 22). The secretion of TSH is stimulated by hypothalamic thyrotrophin releasing hormone (TRH) and is under strong negative feedback control by thyroxine and tri-iodothyronine (see Figure 21.6).

The gonadotrophins FSH and LH are secreted by the anterior pituitary gonadotrophs, as shown in Figure 21.3(b). Although many gonadotrophs secrete both FSH and LH, some secrete only FSH while others secrete only LH. As their name suggests, the gonadotrophins control the functions of the ovaries and testes (the gonads). The secretion of both hormones is controlled by a single hypothalamic releasing hormone called gonadotrophin releasing hormone (GnRH, also known as LHRH or luliberin), and both

negative and positive feedback control mechanisms may operate to control their release.

In females, FSH stimulates the growth and development of follicles during the first half of each menstrual cycle in preparation for ovulation. It is also needed for the secretion of oestrogens by the developing follicle. LH stimulates ovulation itself and is required for the development and secretory activity of the corpus luteum. In males, FSH is required for normal spermatogenesis and LH is responsible for stimulating the secretion of testosterone by the Leydig cells of the testes. (In males, LH is sometimes called interstitial cell stimulating hormone or ICSH.) The physiological actions of the gonadotrophins are discussed in more detail in Chapters 48 and 49.

Summary

Thyroid stimulating hormone (TSH) maintains the structural integrity of the thyroid gland and regulates the secretion of thyroxine and tri-iodothyronine. Its secretion is regulated by hypothalamic thyrotrophin releasing hormone (TRH).

FSH and LH regulate the functions of the ovaries and testes. Their secretion is under the control of hypothalamic gonadotrophin releasing hormone (GnRH). In females, FSH stimulates growth and development of follicles in preparation for ovulation and the secretion of oestrogens by the maturing follicle, while LH triggers ovulation. In males, FSH is required for spermatogenesis while LH stimulates testosterone secretion by Leydig cells.

21.8 The role of the posterior pituitary gland (neurohypophysis)

The posterior lobe of the pituitary gland develops as a downgrowth from the hypothalamus, as described in Section 21.2, so that, unlike the anterior pituitary, it is directly connected to the hypothalamus via a nerve tract (the hypothalamo-hypophyseal tract). For this reason, the posterior pituitary is also known variously as the neurohypophysis, pars nervosa, or neural lobe.

The posterior pituitary is not glandular and does not synthesize hormones itself. However, it is the storage site of two hormones, **oxytocin** and **vasopressin** (or **ADH**), which are synthesized within the cell bodies of large (magnocellular) neurons lying in the supraoptic and paraventricular nuclei of the hypothalamus. Figure 21.10 is a photomicrograph of the hypothalamus stained to reveal the oxytocin- and vasopressin-secreting cells and the fibres of the hypothalamo-hypophyseal tract.

The posterior pituitary hormones are transported from the hypothalamus to the posterior pituitary in association with specific proteins, the **neurophysins**. This occurs by axonal transport along the axons of the magnocellular neurons to their nerve terminals in the posterior lobe. Prior to secretion, these hormones are stored in secretory granules either in the terminals themselves or in varicosities (called Herring bodies) that are distributed along the length of

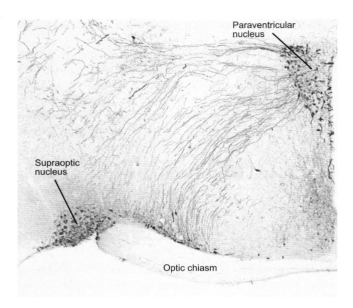

Figure 21.10 A section of hypothalamus treated with antibodies to neurophysin to label neurons and axons that synthesize and secrete oxytocin and vasopressin (ADH). The reaction product is brown to black in colour. Staining is mainly confined to the neurons of the paraventricular and supraoptic nuclei and the fibres of the hypothalamo-hypophyseal tract. (From Fig 8 in J. Preil et al. (2001) *Endocrinology* **142**, 4946–55.)

the axons (see Figure 21.11). In response to nerve impulses originating in the supraoptic and paraventricular nuclei, the hormones are secreted into the blood of the capillaries that perfuse the neural lobe. Both oxytocin and vasopressin are secreted by calcium-dependent exocytosis, which is similar to the secretion of neurotransmitters at other nerve terminals (see Chapter 7).

In addition to neurons, the posterior pituitary also contains specialized glial cells (pituicytes) that may participate in the modulation of hormone release.

Oxytocin and vasopressin (ADH) are closely related structurally but have different functions

Oxytocin and vasopressin are both nonapeptides and differ in only two of their amino acid residues, as shown in Figure 21.12. Although they are secreted along with the neurophysin molecules to which they are bound in the neurons, once released they circulate in the blood largely as free hormones. The kidneys, liver, and brain are the main sites of clearance of both hormones, which have a half-life in the bloodstream of around a minute.

Both oxytocin and vasopressin act on their target cells via G protein linked cell surface receptors (see Chapter 6). Interaction of oxytocin with its receptors stimulates phosphoinositide turnover and thereby raises the level of intracellular calcium in the myoepithelial cells of the mammary gland. In turn, the increased intracellular calcium activates the contractile machinery of these cells to bring about milk ejection or 'let-down' (see Chapter 49). It also acts as a uterine **spasmogen** (a substance that induces a powerful contraction of smooth muscle) and plays a role in parturition (see Chapter 49).

Vasopressin receptors are divided into three subtypes: V_{1A}, V_{1B}, and V_2. Activation of V_{1A} and V_{1B} receptors increases phosphoinositide turnover and elevates intracellular calcium. V_{1A} receptors mediate the effects of vasopressin on vascular smooth muscle. V_{1B} receptors are found throughout the brain and on the corticotrophs of the anterior pituitary, where vasopressin plays a role in the control of ACTH secretion. The renal actions of the hormone are mediated by V_2 receptors, with cyclic AMP as the second messenger (see Chapter 39).

Actions of vasopressin (ADH)

The principal physiological action of vasopressin is as an antidiuretic hormone. For this reason, it is also known as ADH. This role is discussed in more detail in Chapter 39 (p. 607). Briefly, when V_2 receptors are activated they facilitate the reabsorption of water from the final third of the distal tubule and the collecting ducts of the kidney by

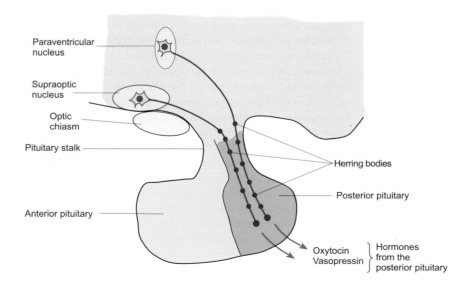

Figure 21.11 A diagrammatic representation of the relationship between the supraoptic and paraventricular nuclei and the posterior pituitary gland (the neurohypophysis). The neurosecretory fibres originate in these nuclei and terminate in the posterior pituitary gland itself. The axons of the hypothalamo-hypophyseal tract exhibit local swellings, known as Herring bodies, that contain oxytocin or vasopressin bound to neurophysin.

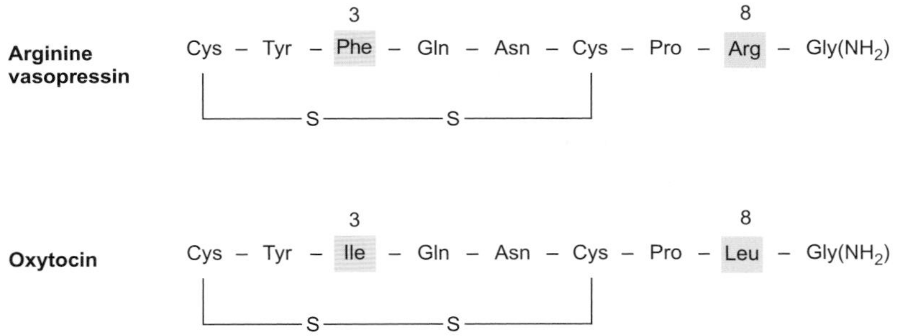

Figure 21.12 The amino acid sequences of arginine vasopressin and oxytocin. The small differences in their amino acid sequence result in molecules that have very different physiological effects.

increasing the permeability of these cells to water. The net result of its actions is an increase in urine osmolality and a decrease in urine flow. Additional renal effects of vasopressin include stimulation of sodium reabsorption and urea transport from lumen to interstitial fluid in the medullary collecting duct. By this action, vasopressin helps to maintain the osmotic gradient from cortex to papilla which is crucial for the elaboration of a concentrated urine.

Vasopressin, as its name suggests, is also a potent vasoconstrictor that acts particularly on the arteriolar smooth muscle of the skin and splanchnic circulation. In spite of this, the increase in blood pressure brought about by vasopressin is small under normal circumstances because the hormone also causes bradycardia and a decrease in cardiac output, both of which tend to offset the increase in total peripheral resistance. However, the vasoconstrictor effect of vasopressin is important during severe haemorrhage or dehydration (see Chapter 40). Vasopressin also exerts a CRH-like activity whereby it stimulates the release of ACTH from the anterior pituitary. It may also play a role in the control of thirst.

The physiological circumstances under which vasopressin (ADH) is secreted are discussed in Chapters 39 and 40. Only a brief resume will be given here. Figure 21.13 illustrates the changes in plasma osmolality and volume that control vasopressin release. The principal physiological stimulus for its release is an increase in the osmolality of the circulating blood. Osmoreceptors located in the hypothalamus detect this increase and activate neurons in the supraoptic and paraventricular nuclei. As a result of the increased rate of action potential discharge of these neurons, vasopressin secretion into the circulation is increased.

Vasopressin is also secreted in response to a fall in the effective circulating volume (ECV)—for example, during haemorrhage—and in response to other factors including pain, stress, and other traumas. The amount of vasopressin secreted when there is a fall in the ECV increases proportionally as the central venous pressure and arterial pressure fall (see Figure 40.3). Central venous pressure is sensed by the low-pressure receptors (volume receptors) of the atria and great veins, while the arterial blood pressure is sensed by the arterial baroreceptors, which are located in the carotid sinuses and aortic arch (see Chapters 30 and 40).

Disorders of vasopressin secretion

The consequences of under- or over-production of vasopressin may easily be predicted from the descriptions of its actions given above. Abnormally high circulating levels of vasopressin may result from

certain drug treatments, brain traumas, or vasopressin-secreting tumours. Such patients will have highly concentrated urine, with water retention, lowered plasma osmolality, and sodium depletion.

A lack of vasopressin leads to a condition known as **diabetes insipidus**, in which the individual is unable to produce a concentrated urine or to limit the production of urine even when the plasma osmolality is raised. This condition may result from developmental abnormalities of the pituitary gland, head injuries, or tumours that damage the posterior pituitary (central diabetes insipidus). Without appropriate treatment, a person with this condition would have to drink a large volume of water (20–25 litres per day) to counteract the loss of water via the kidneys. Fortunately, the deficiency can be treated by the administration of a synthetic analogue of vasopressin (desmopressin). Diabetes insipidus may also arise as a consequence of a loss of vasopressin receptors in the distal nephron (nephrogenic diabetes insipidus). In this case, the administration of exogenous vasopressin will be of no value.

Actions of oxytocin

The main actions of oxytocin are described in some detail in Chapter 49. Briefly, this hormone stimulates the ejection of milk from the mammary glands in response to suckling (the milk ejection or 'let-down' reflex). It causes the myoepithelial cells surrounding the ducts and alveoli of the gland to contract, thus squeezing milk into the lactiferous sinuses and towards the nipple. Oxytocin is known to promote contractions of the uterus and to increase the sensitivity of the myometrium to other spasmogenic agents, and it plays a role in expelling the fetus and placenta during labour. Synthetic analogues of oxytocin (e.g. syntocinon and pitocin) are often administered to women in whom labour has begun but is failing to progress due to inadequate uterine contraction. In males, oxytocin appears to play a role in erection, sperm progression and ejaculation. It may also play a role in sperm progression in the female reproductive tract. A number of behavioural effects have also been suggested for oxytocin in both sexes, including the establishment of maternal behaviour and some aspects of paternal bonding.

Control of oxytocin secretion

Like vasopressin, oxytocin is released in response to afferent neural input to the hypothalamic neurons that synthesize the hormone. Although oxytocin is released in response to vaginal stimulation

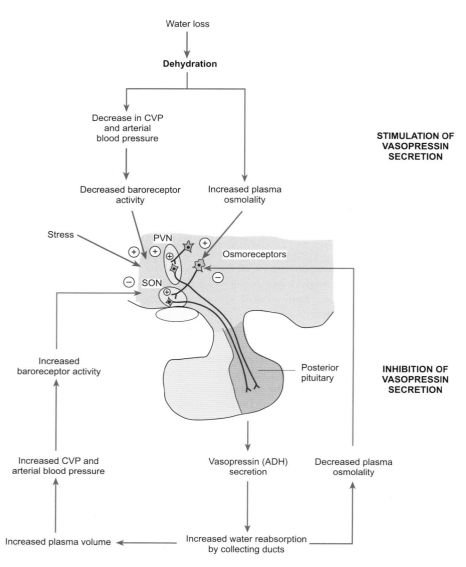

Water loss

Dehydration

Decrease in CVP and arterial blood pressure

Decreased baroreceptor activity

Increased plasma osmolality

Stress

PVN

Osmoreceptors

SON

STIMULATION OF VASOPRESSIN SECRETION

Increased baroreceptor activity

Posterior pituitary

INHIBITION OF VASOPRESSIN SECRETION

Increased CVP and arterial blood pressure

Vasopressin (ADH) secretion

Decreased plasma osmolality

Increased plasma volume

Increased water reabsorption by collecting ducts

Figure 21.13 A schematic diagram showing the factors that regulate vasopressin release in response to changes in plasma osmolality and blood volume. Green arrows directed towards the hypothalamus indicate stimulation; red arrows indicate inhibition. CVP = central venous pressure; PVN = paraventricular nucleus; SON = supraoptic nucleus.

(particularly during labour), the most potent stimulus for release is the mechanical stimulation of the nipple by a suckling baby. Impulses from the breast travel to the hypothalamus via the spinothalamic tract and the brainstem (see Figure 49.22). This explains the so-called 'after pains' experienced by many women when they first breast-feed their babies following delivery. The uterus, which is still highly sensitive to spasmogenic agents, begins to contract once more in response to the oxytocin released during suckling. Certain psychogenic stimuli can also influence the secretion of oxytocin. The milk let-down reflex is known to be inhibited by certain forms of stress and to be stimulated by the cry of the hungry infant or during play prior to feeding.

Disorders of oxytocin secretion

Excessive oxytocin secretion has never been demonstrated, but oxytocin deficiency results in failure to breast-feed an infant because of inefficient milk ejection.

Summary

The posterior pituitary gland (the neurohypophysis) secretes vasopressin (also called antidiuretic hormone or ADH) and oxytocin. These hormones are synthesized in neurons of the paraventricular and supraoptic nuclei and are transported to the posterior pituitary by axonal transport.

Oxytocin and vasopressin are related peptides but have very different actions. Vasopressin is released in response to an increase in the osmotic pressure of the plasma or a fall in blood volume and stimulates the reabsorption of water in the renal collecting ducts. Vasopressin also exerts a pressor (constrictor) effect on vascular smooth muscle. Oxytocin stimulates the ejection of milk from the lactating breast. It also increases the contractile activity of the uterine myometrium and may play a role in expulsion of the fetus during parturition. In males, oxytocin appears to play a role in erection, ejaculation, and sperm progression.

✱ Checklist of key terms and concepts

The hypothalamo-pituitary axis

- The pituitary gland is situated within a depression in the base of the skull.
- It is connected to the brain via the pituitary stalk and consists of two principal lobes: the anterior lobe (or adenohypophysis) and the posterior lobe (or neurohypophysis).
- The anterior lobe is derived from non-neural embryonic tissue, while the posterior lobe is a down-growth from the hypothalamus.
- The anterior pituitary secretes growth hormone (GH), thyroid stimulating hormone (TSH), adrenocorticotrophic hormone (ACTH), follicle stimulating hormone (FSH), luteinizing hormone (LH), prolactin, and melanocyte stimulating hormone (MSH) as well as a number of related peptides.
- A system of blood vessels (the hypophyseal portal vessels) carries regulatory hormones from the hypothalamus to the anterior pituitary. These regulatory hormones control the release of the anterior pituitary hormones.
- The secretion of the hypothalamic hormones, the hormones of the anterior pituitary, and their target organs form a complex feedback system—the hypothalamo-hypophyseal axis.
- The posterior pituitary secretes vasopressin (ADH) and oxytocin, whose secretion is directly controlled by nerve activity in the hypothalamo-hypophyseal tract.

Growth hormone

- Growth hormone is a peptide hormone consisting of 191 amino acid residues.
- GH secretion is stimulated by hypothalamic GHRH and suppressed by hypothalamic somatostatin.
- The secretion of GH shows a circadian rhythm, with the highest rates occurring during deep sleep.
- GH secretion is stimulated by hypoglycaemia, increased plasma levels of amino acids, and reduced plasma levels of free fatty acids.
- GH exerts a wide range of direct metabolic actions in addition to indirect effects that are mediated by insulin-like growth factors (IGFs) produced by the liver.
- GH promotes protein synthesis.
- GH promotes lipolysis, thus providing a non-carbohydrate source of energy. This, together with its anti-insulin action on muscle, spares glucose for use by the CNS and the germinal epithelium.
- GH is crucial to normal skeletal growth between the age of about 3 and puberty. IGFs encourage cartilage cells to divide, and they enhance the deposition of cartilage at the epiphyses (growth plates) of long bones.

Prolactin

- Prolactin is structurally related to GH and is a peptide consisting of 198 amino acid residues.
- Prolactin secretion is normally inhibited by hypothalamic dopamine (prolactin inhibitory hormone) but is stimulated in response to suckling, which inhibits the release of dopamine.
- Prolactin stimulates the synthesis of milk during lactation (galactopoiesis).

ACTH and MSH

- Adrenocorticotrophic hormone (ACTH) and melanocyte stimulating hormone (MSH) are secreted by anterior pituitary corticotrophs. Both are derived from a much larger precursor molecule, pre-pro-opiomelanocortin, by proteolytic cleavage.
- ACTH maintains the structural integrity of the adrenal cortex and regulates the secretion of glucocorticoid hormones in response to stress.
- ACTH secretion is under the control of hypothalamic corticotrophin releasing hormone (CRH) and is subject to negative feedback regulation.
- The role of MSH in humans is unclear, but it may be concerned with skin pigmentation and the regulation of food intake and energy expenditure.

TSH, FSH, and LH

- Thyroid stimulating hormone (TSH), follicle stimulating hormone (FSH), and luteinizing hormone (LH) are all glycoproteins secreted by the anterior pituitary.
- TSH is a trophic hormone that maintains the structural integrity of the thyroid gland and regulates the secretion of thyroxine and tri-iodothyronine.
- The secretion of TSH is regulated by thyrotrophin releasing hormone (TRH) secreted by neurons in the preoptic and paraventricular hypothalamic nuclei.
- FSH and LH (the gonadotrophins) regulate the functions of the ovaries and testes and are themselves under the control of hypothalamic luteinizing hormone releasing hormone (LHRH, also known as GnRH).
- In females, FSH stimulates growth and development of follicles in preparation for ovulation and the secretion of oestrogens by the maturing follicle.
- In males, FSH is required for spermatogenesis.
- In females, LH triggers ovulation and stimulates the secretion of progesterone by the corpus luteum.
- In males, LH stimulates the secretion of testosterone by Leydig cells.

The posterior pituitary hormones

- The posterior pituitary gland (the neurohypophysis) secretes two peptide hormones: vasopressin (also known as antidiuretic hormone or ADH) and oxytocin. They are synthesized in the cell bodies of neurons within the paraventricular and supraoptic nuclei, and they reach the posterior pituitary by axonal transport.

- They are secreted in response to nerve activity in the hypothalamo-hypophyseal tract.

- Oxytocin and vasopressin are structurally similar but have very different actions.

- Vasopressin is secreted in response to an increase in the osmotic pressure of the plasma or a fall in blood volume.

- The antidiuretic action of vasopressin stimulates the reabsorption of water from the collecting ducts of the renal nephrons. As a result, urinary volume is reduced and urinary osmolality is increased.

- Oxytocin stimulates the ejection ('let-down') of milk from the lactating breast.

- It also increases the contractile activity of the uterine myometrium and plays a role in expulsion of the fetus during parturition.

- Oxytocin may play a role in sperm progression through the male (and possibly also the female) reproductive tract.

Recommended reading

Biochemistry

Berg, J.M., Tymoczko, J.L., and Stryer, L. (2011) *Biochemistry* (7th edn), Chapters 14 and 27. Freeman, New York.

A classic text with helpful accounts of signal transduction and the integration of metabolism.

Histology

Mescher, A.(2015) *Junquieira's Basic Histology* (14th edn), Chapter 20. McGraw-Hill, New York.

A recent revision of Junquieira and Carneiro's classic text. Good text, fewer pictures than some alternatives.

Pharmacology

Rang, H.P., Dale, M.M., Ritter, J.M., Flower, R., and Henderson, G. (2015) *Pharmacology* (8th edn), Chapter 33. Churchill-Livingstone, Edinburgh.

The first part of this chapter provides an easily digested introduction to pituitary disorders and their treatment.

Endocrine physiology

Kovacs, W.J., and Ojeda, S.R. (2012) *Textbook of endocrine physiology* (6th edn), Chapters 5 and 6. Oxford University Press, Oxford.

An authoritative account of the pituitary gland. Well referenced.

Medicine

Hall, R., and Evered, D.C. (1990) *A colour atlas of endocrinology* (2nd edn). Woolfe Medical, London.

An informative collection of colour plates illustrating the physical changes associated with various endocrine disorders.

Ledingham, J.G.G., and Warrell, D.A. (2000) *Concise Oxford textbook of medicine*, Chapters 7.1 and 7.2. Oxford University Press, Oxford.

These chapters provide a useful introduction to disorders of the pituitary gland and their treatment.

Maitra, A., and Kumar, V. (2007) Chapter 20, in: *Robins basic pathology* (8th edn) (ed. V. Kumar, A.K. Abbas, N. Fausto, and R. Mitchell). Saunders, New York.

Review articles

Bartke, A., Sun, L.Y., and Longo, V. (2013) Somatotropic signaling: Trade-offs between growth, reproductive development, and longevity. *Physiological Reviews* **93**, 571–98, doi:10.1152/physrev.00006.2012.

Bisset, G.W., and Chowdrey, H.S. (1988) Control of release of vasopressin by neuroendocrine reflexes. *Quarterly Journal of Experimental Physiology* **73**, 811–72.

Perez-Castro, C., Renner, U., Haedo, M.R., Stalla, G.K., and Arzt, E. (2012) Cellular and molecular specificity of pituitary gland physiology. *Physiological Reviews* **92**: 1–38, doi:10.1152/physrev.00003.2011.

Tsutsumi, R., and Webster, N.J.G. (2009) GnRH pulsatility, the pituitary response and reproductive dysfunction. *Endocrine Journal* **56**, 729–37.

To check that you have mastered the key concepts presented in this chapter, complete the accompanying online self-assessment questions. Go to www.oup.com/uk/pocock5e/

CHAPTER 22

The thyroid and parathyroid glands

This chapter should help you to understand:

- The anatomical organization of the thyroid gland
- The synthesis and storage of the thyroid hormones (T_3 and T_4)
- The secretion of the thyroid hormones
- The principal actions of T_3 and T_4
  - Heat production (calorigenesis)
  - Effects on metabolism
  - Effects on the cardiovascular system
  - Whole body effects of T_3 and T_4
- The major disorders of thyroid hormone secretion
- The role of the endocrine system in the regulation of plasma calcium and phosphate
  - Hypocalcaemia and tetany
  - Factors involved in maintaining calcium and phosphate balance
  - The role of vitamin D and its metabolites
- The parathyroid glands, parathyroid hormone and calcium homeostasis
  - Control of parathyroid hormone secretion
  - Disorders of parathyroid hormone secretion
- Calcitonin and its role in regulating plasma calcium

22.1 Introduction

This chapter is concerned with the principal hormones secreted by the thyroid and the parathyroid glands. The thyroid gland secretes two iodine-containing hormones, **thyroxine** (T_4) and **tri-iodothyronine** (T_3), that exert a wide variety of actions throughout the body. They are chiefly concerned with the regulation of metabolism and the promotion of normal growth and development. A third hormone, **calcitonin**, is secreted by cells scattered throughout the gland (**parafollicular cells** or C-cells). Calcitonin is a peptide hormone that plays a role in the regulation of plasma calcium levels.

Lying embedded within the thyroid gland are several (usually four) small glands known as the parathyroid glands. These secrete a polypeptide hormone called **parathyroid hormone** (**PTH**). This exerts its effects principally on bone and on the kidneys, which contribute to the regulation of plasma calcium and phosphate. In addition to calcitonin and PTH, metabolites of vitamin D (cholecalciferol) also play a part in the minute-to-minute regulation of plasma mineral levels.

22.2 The thyroid gland

The thyroid gland lies just below the larynx and adheres to the trachea. It consists of two flat lobes connected by a narrow region called the isthmus (see Figure 22.1) and, in adults, weighs between 10 and 20 g. The thyroid has a rich blood supply (around 5 ml per g of tissue per minute) which is regulated by the autonomic nervous system via the vagus nerve (parasympathetic) and cervical sympathetic trunks.

The thyroid gland develops from embryonic endoderm associated with the pharyngeal gut before migrating to the front of the neck. By week 11 or 12 of gestation, it is capable of synthesizing and secreting thyroid hormones under the influence of fetal thyroid stimulating hormone (TSH). Indeed, fetal thyroid hormones are crucial for the subsequent growth and development of the fetal skeleton and central nervous system.

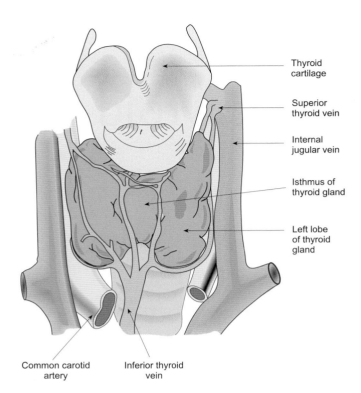

Figure 22.1 A ventral view showing the location of the thyroid gland and its venous drainage.

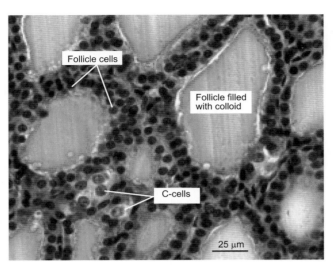

Figure 22.2 The cellular organization of the thyroid gland: a section of human thyroid gland showing thyroid follicles containing colloid. Each follicle is surrounded by a layer of follicular cells that synthesize thyroglobulin and secrete thyroid hormones (i.e. T_3 and T_4). The C-cells are distinguished from the follicular cells by their pale staining and relatively large, pale nuclei. They are scattered between the follicles and secrete the hormone calcitonin.

Histology of the thyroid gland

The chief histological features of the thyroid gland are illustrated in Figure 22.2. The functional unit is the follicle, many thousands of which are present in the gland, each bounded by a basement membrane. Every follicle consists of an epithelial layer of follicular cells surrounding a central colloid-filled cavity. The follicles range from 20 to 900 μm in diameter and vary in appearance according to the prevailing level of thyroid gland activity. When the thyroid is highly active, the follicles contain rather little colloid and the follicular epithelial cells have a tall columnar appearance. During periods of relative inactivity, however, the follicles fill with colloid and the epithelial cells then assume a flattened, cuboidal appearance. Parafollicular cells (also called C-cells) lie scattered between the follicles. These cells secrete calcitonin (see Section 22.3) and do not come into contact with the follicular colloid.

Synthesis of the thyroid hormones

Tri-iodothyronine (T_3) and thyroxine (T_4) are the principal hormones secreted by the thyroid gland. They are iodinated amino acids and their chemical structures, as well as those of their iodinated precursors and of the biologically inactive form of T_3, **reverse T_3**, are shown in Figure 22.3. Thyroid hormones are the only substances in the body that contain iodine, and an adequate iodide intake is essential for normal thyroid hormone synthesis. The minimum **dietary requirement** is about 75 μg a day but, in most parts of the world, intake considerably exceeds this level. The synthesis of

thyroid hormones involves the uptake of iodine from the blood and the incorporation of iodine atoms into tyrosine residues of **thyroglobulin**, the glycoprotein that is synthesized by the follicular cells and forms the bulk of the colloid within the follicles. The principal steps involved in the synthesis and secretion of the thyroid hormones are shown in Figure 22.4 and are discussed below.

Iodide trapping—the iodide pump

Iodine from food and drinking water is absorbed into the blood by the small intestine, as inorganic iodide, before being taken up by the thyroid gland. The basal plasma membrane of the follicular cells contains a sodium/iodide cotransport system (a symporter known as NIS, which is member SLC5A5 of the solute carrier superfamily) that actively transports iodide from the blood into the cells against a steep electrochemical gradient. Two sodium ions are transported for each iodide ion—the sodium gradient providing the energy for the secondary active transport of iodide. In this way, iodide is concentrated 20- to 100-fold by the follicular cells. Iodide uptake is stimulated by TSH and may be inhibited competitively by anions such as perchlorate, thiocyanate, bromide, and nitrite. Iodide ions cross the apical membrane of the follicular cells via an anion channel called pendrin (see Figure 22.4).

Iodide oxidation

Before dietary iodide can enter into organic combination, it must be oxidized to free (reactive) iodine. This conversion to reactive iodine is carried out in the presence of the enzyme peroxidase, which is chiefly located near to the apical membrane of the

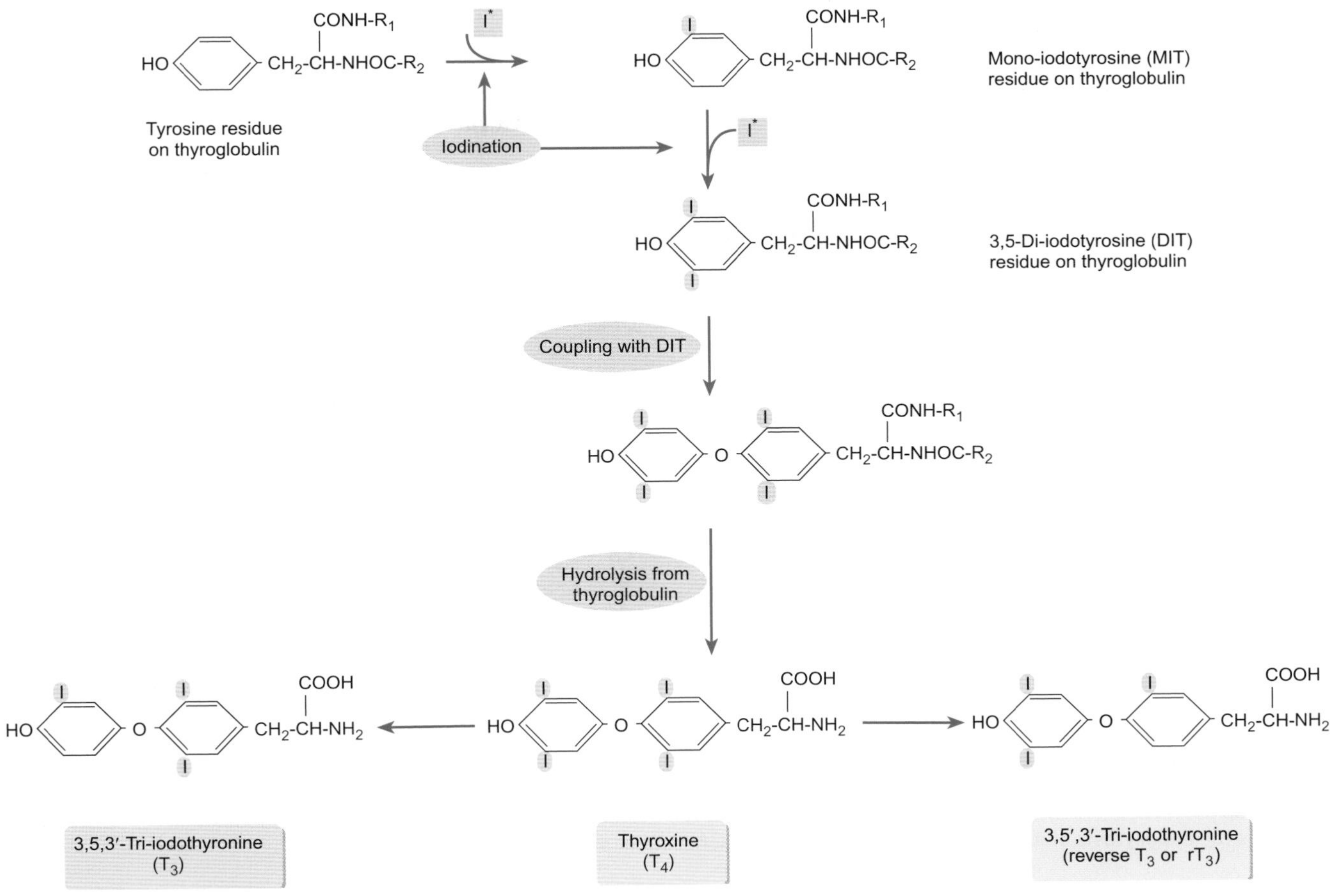

Figure 22.3 A simplified scheme showing the principal steps in the synthesis of the thyroid hormones. Atoms of iodine are highlighted in red. The iodide ion is first oxidized to form an iodide radical (designated as I* in this figure), which reacts with the tyrosine residues on thyroglobulin to form mono- and di-iodothyronine. Note that the thyroid gland secretes about 80 μg of thyroxine (T_4) a day compared to only about 4 μg of T_3 and about 2 μg of reverse T_3 (rT_3). Most T_3 and rT_3 is formed by de-iodination of T4 in the tissues.

follicular cells. Hydrogen peroxide, generated by peroxisomes, acts as the electron acceptor for the oxidation reaction. Free iodine is released at the interface between the follicular cell and the colloid contained within the lumen of the follicles. Subsequent iodination of thyroglobulin takes place within the lumen.

Thyroglobulin synthesis

Thyroglobulin is a glycoprotein with a molecular weight of 670 kD. It is synthesized on the rough endoplasmic reticulum of the follicular cells as peptide units of molecular weight 330 kD. These combine and the carbohydrates are added. The completed protein is packaged into small vesicles which move to the apical plasma membrane, where they are released into the follicular lumen by exocytosis (see Chapter 4) to be stored as colloid.

Iodination of thyroglobulin

Contained within the colloid are tyrosine residues held to the thyroglobulin molecules by peptide linkages. Free iodine becomes attached to the 3 position of a tyrosine residue to form mono-

iodotyrosine (MIT). A second iodination at position 5 gives rise to di-iodotyrosine (DIT). Although each thyroglobulin molecule contains around 125 tyrosine residues, only about a third of these are available for iodination because they are situated at or near the surface of the glycoprotein. Following the iodination, coupling reactions occur between the mono- and di-iodotyrosines to form the hormonally active forms, tri-iodothyronine (T_3) and thyroxine (T_4), as indicated in Figure 22.3.

The iodinated protein is stored within the lumen of the thyroid follicle. In contrast to most endocrine glands, which do not store appreciable amounts of their particular hormones, the thyroid gland contains several weeks' supply of thyroid hormones. This can be important for maintaining normal circulating levels of thyroid hormones should there be a temporary fall in the supply of dietary iodine.

Secretion of thyroid hormones

Thyroglobulin must be hydrolysed before the thyroid hormones T_3 and T_4 can be released into the circulation; thyroglobulin itself is not normally released into the circulation unless inflammation or

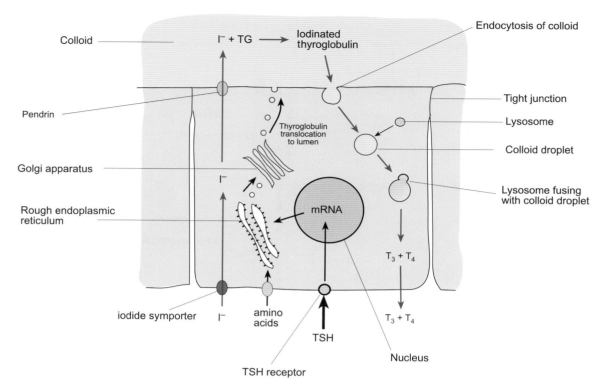

Figure 22.4 The cellular processes involved in the synthesis and subsequent release of the thyroid hormones. In response to TSH, the follicular cells take up iodide via the sodium-iodide transporter and transfer it to the follicle lumen via the anion channel pendrin. They also take up amino acids to synthesize thyroglobulin (TG), which is secreted into the lumen where it becomes iodinated. TSH also stimulates the endocytosis of iodinated thyroglobulin and the subsequent secretion of T_3 and T_4.

damage of the gland occurs. The processes involved in the secretion of thyroid hormone are shown in Figure 22.4. When the thyroid gland is stimulated by thyroid stimulating hormone (TSH) from the anterior pituitary, droplets of colloid are taken up into the follicular cells by endocytosis. The endocytotic vesicles containing colloid then fuse with lysosomes to form endosomes in which the colloid containing thyroglobulin is hydrolysed to liberate iodotyrosines, amino acids, and sugars. The amino acids and sugars are recycled within the gland, while MIT and DIT are de-iodinated and their iodine recycled. T_3 and T_4 are released into the fenestrated capillaries that surround the follicle, probably by diffusion.

Regulation of thyroid hormone secretion

The activity of the thyroid gland is controlled by thyroid stimulating hormone (TSH) secreted by the thyrotrophs of the anterior pituitary gland. The secretion of TSH is, in turn, controlled by the hypothalamic thyrotrophin releasing hormone, TRH. The circulating thyroid hormones influence the rate of TSH secretion by means of their negative feedback effects both on the hypothalamus and on the anterior pituitary (see Chapter 21 p. 306). Other hormones also appear to be able to alter the output of TSH. Oestrogens, for example, increase the responsiveness of the TSH-secreting cells of the anterior pituitary to TRH, while high levels of glucocorticoids inhibit TSH release.

In addition to the hormonal control mechanisms, a variety of nervous inputs to the hypothalamus play a role in regulating the output of TRH and thus of TSH and the thyroid hormones. Cold stress is known to stimulate thyroid hormone secretion. Within 24 hours of entering a cold environment there is a rise in circulating T_4 levels, which reach a peak a few days later.

TSH regulates most aspects of thyroid hormone synthesis and secretion. It acts by binding to specific G protein linked receptors on the follicular cells that activate membrane-bound adenylyl cyclase and increase the level of cyclic AMP in the follicular cells. TSH also activates phospholipase C, thus raising intracellular calcium. The elevated cyclic AMP activates protein kinase A, resulting in an increase in gene transcription and increased synthesis of thyroglobulin, thyroid peroxidase, and the sodium iodide symporter. Activation of phospholipase C regulates iodide efflux, the production of free iodide radicals, and the iodination of thyroglobulin. TSH also stimulates the coupling of mono- and di-iodotyrosine molecules and all the events leading to thyroid hormone secretion.

The net result of the various actions of TSH is an increase in the synthesis of fresh thyroid hormone for storage within the follicles and an increased secretion of thyroid hormones into the circulation. TSH is a trophic hormone, which means that it exerts a tonic maintenance effect on the thyroid gland and its blood supply. In the absence of TSH, the thyroid gland rapidly atrophies.

Transport, tissue delivery, and metabolism of thyroid hormones

T_3 and T_4 travel in the bloodstream largely bound to plasma proteins. Only about 0.5 per cent of circulating T_3 is unbound (free), and thus biologically active, while less than 0.05 per cent of T_4 is in the unbound form. About 75 per cent of the circulating T_4 is bound to thyronine binding globulin, 15–20 per cent is bound to prealbumin, and the remainder is bound to albumin. Virtually all the T_3 is bound to thyronine binding globulin. Because T_4 binds to plasma proteins with an affinity ten times that of T_3, its metabolic clearance rate is slower. Consequently, the half-life of T_4 in the plasma is significantly longer than that of T_3 (seven days for T_4 and less than a day for T_3).

Although T_3 and T_4 are hydrophobic molecules, they enter their target cells by several carrier-mediated processes including organic ion transporters, amino acid transporters, and the monocarboxylate transporter MCT8, which appears to be particularly important for T_3 uptake by neural tissues. In the liver, separate processes have been implicated in the uptake of T_3 and T_4. The uptake of T_3 and T_4 seems to be dependent on ATP and extracellular sodium. Once inside cells, much of the T_4 undergoes de-iodination to T_3, most of this conversion taking place in the liver and kidneys. For this reason, T_4 is sometimes considered to be a pro-hormone for T_3. Nevertheless, it should be remembered that although biologically less active than T_3, T_4 does exert hormonal effects in its own right and probably has its own cellular receptors.

In addition to being de-iodinated to the active form of T_3, T_4 may also undergo conversion to inactive reverse T_3 (rT_3) in the tissues. In this case, the iodide is removed from the inner tyrosine residue of the molecule (see Figure 22.3). The production of rT_3 from T_4 seems to increase when **calorie** intake is restricted. From the following discussion of the effects of the thyroid hormones, it will become clear that this acts as an energy-conserving mechanism.

Actions of the thyroid hormones

Information about the effects of the thyroid hormones has largely been obtained from studies of human and animal subjects with over- or under-active thyroid glands (hyper- and hypothyroidism), and from studies of the actions of T_3 and T_4 on isolated tissues. Thyroid receptors are located in the cellular nuclei, where they regulate gene expression. Three isoforms of the thyroid receptor have been identified:

- TR-α1, which is widely distributed throughout the body;
- TR-β1, which is particularly enriched in the liver but is also found in the brain and kidney;
- TR-β2, which is highly expressed in the hypothalamus and pituitary.

With such a wide distribution of receptors, it is not surprising that thyroid hormones have effects on virtually every tissue. In general terms, the thyroid hormones may be thought of as tissue growth factors: even if growth hormone levels are normal, overall whole body growth and brain development are severely impaired in the absence of thyroid hormones. Thyroid hormones have

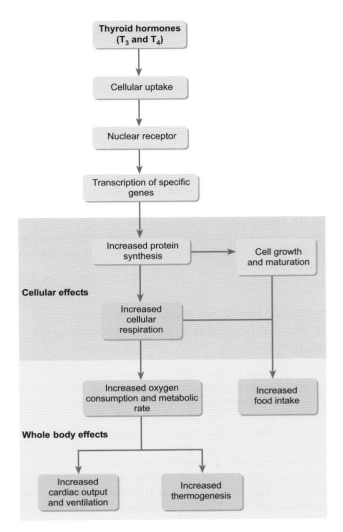

Figure 22.5 A flow chart that summarizes the principal actions of the thyroid hormones. Note the link between the cellular effects and their consequences for the body as a whole.

specific tissue effects as well as increasing the basal rate of oxygen consumption and heat production by the body as a whole (see Chapter 42). The principal effects of the thyroid hormones are summarized in Figure 22.5.

Heat production (calorigenesis)

Most tissues increase their oxygen consumption and heat production (**calorigenesis**) in response to thyroid hormone. This calorigenic action is important in temperature regulation and particularly in adaptation to cold environments, although basal metabolic rate (BMR) does not begin to rise until a day or more after the increase in the secretion of thyroid hormone. BMR is a useful indicator of the thyroid status of an individual. A high BMR may indicate an overactive thyroid gland, while thyroidectomy or hypothyroidism causes BMR to fall. The measurement of BMR and a discussion of its hormonal control can be found in Chapter 47.

What causes the increase in oxygen consumption in response to thyroid hormone?

The increased heat production that occurs during cold exposure depends on the synthesis of mitochondrial uncoupling protein that is induced by T_3 in brown adipose tissue and skeletal muscle. The uncoupling of oxidative phosphorylation stimulates oxygen consumption by the mitochondria and results in the dissipation of metabolic energy as heat, which is reflected in an increase in BMR. In addition, both T_3 and T_4 stimulate the activity of the Na$^+$, K$^+$-ATPase in many tissues, increasing the rate of Na$^+$ and K$^+$ transport across the cell membrane and thus further increasing heat production. The increase in BMR is reflected in an increase in body temperature, though this is somewhat offset by a compensatory increase in heat loss through increased blood flow to the surface vessels, sweating, and respiration (changes that are also mediated by thyroid hormone).

The thyroid hormones modify the metabolism of fats and carbohydrates to provide substrates for oxidation

Thyroid hormones exert a number of effects on the metabolism of carbohydrates. They act directly on the enzyme systems involved, but also work indirectly by potentiating the actions of other hormones such as insulin and catecholamines. Thyroid hormones enhance the rate of intestinal absorption of glucose and increase the rate of glucose uptake by peripheral cells such as muscle and adipose tissue. In animal experiments, low doses of thyroid hormones appear to potentiate the effects of insulin, stimulating both glycogen synthesis and glucose utilization. However, higher thyroid hormone levels seem to enhance the calorigenic effects of adrenaline (epinephrine), increasing the rates of both glycogenolysis and gluconeogenesis.

Thyroid hormones affect most aspects of lipid metabolism. They exert a powerful lipolytic action on fat stores, thereby increasing plasma levels of free fatty acids. Part of this action is probably due to potentiation of catecholamine activity that increases lipolysis via the adenylyl cyclase-cyclic AMP system. There is also increased oxidation of the free fatty acids released, and this contributes to the calorigenic effect discussed above. The overall effect of these changes in fat metabolism is a depletion of the body's fat reserves, a fall in weight, and a fall in plasma levels of cholesterol and other lipids.

Thyroid hormones stimulate both protein synthesis and protein degradation

The actions of the thyroid hormones on protein metabolism are somewhat complex. At low concentrations, T_3 and T_4 stimulate the uptake of amino acids into cells and the incorporation of these into specific structural and functional proteins, many of which are involved in calorigenesis. Protein synthesis is depressed in hypothyroid individuals. Conversely, high levels of thyroid hormones are associated with protein catabolism. This effect is particularly marked in muscle and can lead to severe weight loss and muscle weakness (**thyrotoxic myopathy**). The breakdown of protein results in an increased level of amino acids in the plasma and increased excretion of creatine.

Effects of thyroid hormones on the cardiovascular system

Thyroid hormones promote vasodilatation, thereby lowering systemic vascular resistance. In response to this fall, there is an increase in sodium chloride and water reabsorption secondary to activation of the renin–angiotensin–aldosterone cascade (see Chapter 39). This, together with an increase in the rate of erythropoiesis (see Chapter 25), brings about an increase in blood volume. Thyroid hormones enhance cardiac output both by increasing heart rate and through a positive inotropic action (increase in myocardial contractility). Hypothyroid individuals have poor cardiac function, with a low cardiac output coupled with an increased peripheral resistance. They also often experience **anaemia**. Hyperthyroid individuals, in contrast, have a high cardiac output. Thyroid hormones enhance protein synthesis in the heart, including myosin heavy chain and the Ca^{2+}-ATPase of the sarcoplasmic reticulum (which determines the rate at which the heart is able to relax, the lusitropic effect—see glossary (p. 887)). It seems likely that these changes account for the increase in cardiac output caused by the thyroid hormones. However, if thyroid hormone levels are chronically elevated there may be a loss of muscle mass and a reduction in the strength of the heartbeat. Cardiac tissue lacks the ability to de-iodinate T_4 to T_3, so these effects are attributable to the directly circulating T_3.

Whole body actions of thyroid hormones

Thyroid hormones have major effects on growth and maturation. While linear growth itself appears to be independent of thyroid hormone levels in the fetus, T_3 and T_4 are essential for the normal differentiation and maturation of fetal tissues, particularly those of the skeleton and the nervous system. After birth, the thyroid hormones stimulate the linear growth of bone until puberty. They promote **ossification** of the bones and maturation of the epiphyseal growth regions. More details of skeletal growth and the effects of hormones upon growth are given in Chapter 51.

Adequate thyroid hormone levels during the late fetal and early prenatal periods are vital to normal development of the central nervous system. Inadequate hormone concentrations at this time result in severe mental retardation (**cretinism**), which is irreversible if it is not recognized and treated quickly. The exact role of thyroid hormones in maturation of the nervous system is unclear, but it is known that in their absence there is a reduction in both size and number of cerebral cortical neurons, a reduction in the degree of branching of dendrites, deficient myelination of nerve fibres, and a reduction in the blood supply to the brain.

A number of the effects of the thyroid hormones, including the increase in BMR, calorigenesis, increased heart rate, and central nervous system excitation are shared by the sympathetic system. Indeed, synergism between the catecholamines and thyroid hormones may be essential for maximum **thermogenesis**, lipolysis, glycogenolysis, and gluconeogenesis to occur. This physiological link is emphasized by the fact that the administration of drugs which block adrenergic β-receptors has the effect of lessening many of the cardiovascular and central nervous manifestations of hyperthyroidism.

Mode of action of the thyroid hormones

The mechanism of action of the thyroid hormones is not completely understood. Although there is some evidence of direct hormonal effects on some tissues, the major effects of the thyroid hormones are exerted via their nuclear receptors. After thyroid hormone enters the nucleus and binds to its receptor, the resulting hormone–receptor complex modulates the transcription of specific genes, as outlined in Chapter 6 (p. 95). The genes may either be activated or repressed (as in the negative feedback inhibition of TSH production by the anterior pituitary thyrotrophs). Because thyroid hormones act on the nucleus of their target cells, their effects are generally rather slow in onset and normally take between a few hours and several days to appear following the stimulation of hormone secretion. The first intracellular change to occur is an increase in mRNA, followed by an increase in protein synthesis that, in turn, leads to raised intracellular levels of specific proteins.

Principal disorders of thyroid hormone secretion

Disorders of the thyroid may represent a congenital defect of thyroid development or they may develop later in life, with a gradual or a sudden onset. The thyroid gland may be overactive, leading to **hyperthyroidism**, or underactive, leading to **hypothyroidism**. Defects of thyroid function are among the most common endocrine disorders, affecting 1–2 per cent of the adult population. The chief manifestations of both these extremes may largely be predicted from the preceding account of normal thyroid function.

Since T_4 is more tightly bound to serum proteins than T_3, the fraction of T_4 that is not bound (i.e. free T_4 or FT_4) is much less (0.02 per cent) than the fraction of free T_3 (0.2 per cent). In normal thyroid function, as the concentration of the carrier proteins changes, the total level of thyroid hormone also changes, so that free hormone levels remain more or less constant. In an abnormally functioning thyroid, this is not necessarily so. For this reason, to confirm a diagnosis of hyper- or hypothyroidism the plasma levels of free (i.e. unbound) T_3 and T_4 (FT_3 and FT_4) are measured to take into account any change in the concentration of the carrier proteins. TSH, and possibly also TRH levels will also be monitored.

The clinical features of thyroid hormone excess

Hyperthyroidism or **thyrotoxicosis** results from excessive delivery of thyroid hormone to the peripheral tissues. The most common cause is **Graves' disease** (also called Basedow's disease), which is believed to be an autoimmune disease because the patient's serum often contains abnormal antibodies that mimic

TSH (so-called thyroid-stimulating antibodies, TSAb). Most of the clinical manifestations of hyperthyroidism are related to the increase in oxygen consumption and increased utilization of metabolic fuels associated with a raised metabolic rate, and to a parallel increase in sympathetic nervous activity. The principal symptoms of hyperthyroidism are:

- loss of weight
- excessive sweating
- palpitations and an irregular heartbeat
- anxiety and nervousness
- protrusion of the eyeballs (**exophthalmos**) caused by inflammation and oedema of the extraocular tissues coupled with lid retraction.

On clinical examination, there is a raised BMR and oxygen consumption, increased heart rate, hypertension, and possibly atrial fibrillation. There may also be goitre (swelling of the thyroid gland). The physical appearance of a hyperthyroid man with goitre and exophthalmos is shown in Figure 22.6.

Treatments for hyperthyroidism include surgical removal of all or part of the thyroid gland or ingestion of radioactive iodine (^{131}I),

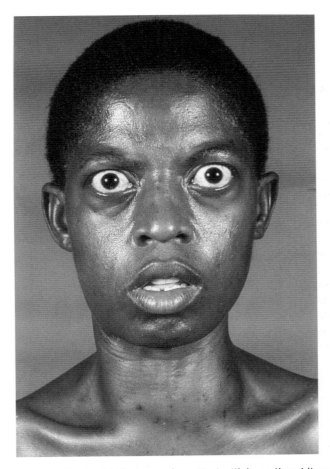

Figure 22.6 The typical features of a patient with hyperthyroidism. Note the prominent exposure of the whites of the eyes (exophthalmos) and retraction of the eyelids. (Courtesy of the Wellcome Library.)

which selectively destroys the most active thyroid cells. The output of thyroid hormones can also be decreased by a number of drugs that are structurally related to thiourea (the **thioureylenes**), such as carbimazole and propylthiouracil. These act by inhibiting the activity of peroxidase (the enzyme that oxidizes dietary iodide to free reactive iodine) and also by reducing the peripheral de-iodination of T_4 to T_3.

Clinical assessment of thyroid status requires the measurement of serum free T_3 and T_4 levels, TSH concentration, and anti-thyroid antibodies.

The clinical features of hypothyroidism

In the absence of sufficient dietary iodine, the thyroid gland cannot produce sufficient amounts of T_3 and T_4. As a result, there is a reduction in the negative feedback inhibition of TSH secretion by the anterior pituitary gland. TSH secretion is therefore abnormally high, and this results in abnormal growth of the thyroid gland due to the trophic effect of TSH. This is known as an **iodine-deficiency** or **endemic goitre**, an example of which is shown in Figure 22.7. It is common in many parts of the world, especially in high mountain regions (e.g. the Andes and Himalayas). However, in most developed countries it has been largely eliminated by the use of iodized table salt and iodized flour.

Hypothyroidism may also result from a primary defect of the thyroid gland or as a secondary consequence of reduced secretion of either TSH or TRH. The most common cause is an autoimmune disorder called **Hashimoto's thyroiditis**, which causes destruction of the thyroid by an autoimmune process.

In adults, the full-blown hypothyroid syndrome is called **myxoedema** (literally 'mucous swelling'). The symptoms include:

- puffiness of the face and swelling around the eyes
- dry, cold skin
- loss of hair (known as alopecia)
- sensitivity to cold
- weight gain, despite a loss of appetite
- constipation
- impaired memory
- mental dullness
- lethargy

On clinical examination, there is a lowered BMR and oxygen consumption, **hypothermia**, a slowed pulse rate, reduced cardiac output, and elevated plasma cholesterol. The diagnosis is confirmed by low plasma levels of free T_3 and T_4 (FT_3 and FT_4). Unless the primary defect is a lack of TSH or TRH, plasma levels of TSH will be high as a result of the removal of negative feedback inhibition by thyroid hormone. Depending upon the cause, hypothyroidism may be reversed by iodine supplements or by hormone replacement therapy. Synthetic T_4 is the sodium salt of **levothyroxine** (L-thyroxine). It is administered orally to hypothyroid patients, and the correct dose regime is determined by monitoring TSH levels, which return to normal when the correct plasma concentration of drug is achieved. The more rapidly acting drug **liothyronine**, which is the sodium salt of T_3, may be used to treat severe acute cases of hypothyroidism when a quicker response is needed. Figure 22.8 shows the typical physical appearance of a woman with hypothyroidism and the dramatic effects of hormone replacement.

Thyroid hormones are essential for normal brain development and growth. Severe hypothyroidism in infants is called **cretinism**. A seventeen year-old girl with untreated hypothyroidism can be seen compared to a normal six-year-old child in Figure 22.9(a). In addition to being mentally retarded, affected children often have a short, disproportionate body, a thick tongue and neck, and obesity. Congenital hypothyroidism is one of the commonest causes of preventable mental retardation, affecting approximately one in every 4000 infants. In Western countries, all neonates are screened for congenital hypothyroidism between the first and fifth days of life. This is done using a small sample of cord blood or blood from a heel-prick. An example of an affected infant who was subsequently treated successfully by hormone replacement therapy is shown in Figure 22.9(b) (note the thick tongue).

Figure 22.7 A Tikar woman (from northern Cameroon) with a prominent goitre (the swelling in the neck cause by an enlarged thyroid gland). This appearance is a characteristic feature of those suffering from endemic goitre due to iodine deficiency. (Courtesy of Dave Price.)

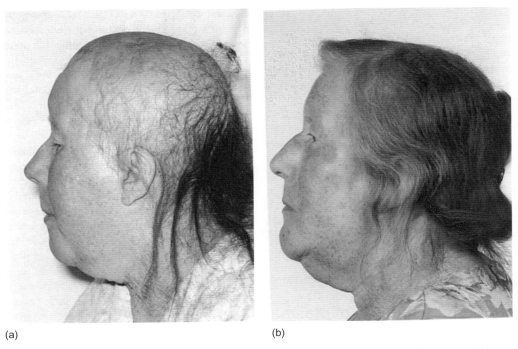

(a) (b)

Figure 22.8 The left panel shows a patient suffering from severe hypothyroidism (myxoedema) manifesting typical facial puffiness, thinning hair, and dull appearance. The right panel shows the same patient after treatment with thyroid hormone, which has reversed the physical symptoms evident in the left panel. (From plates 9.1 and 9.2 in J. Laycock and P. Wise (1996) *Essential Endocrinology,* 3rd edition. Oxford Medical Publications, Oxford.)

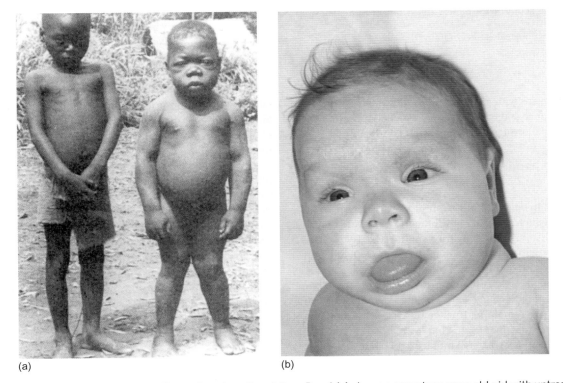

(a) (b)

Figure 22.9 Typical examples of children suffering from hypothyroidism. Panel (a) shows a seventeen-year-old girl with untreated hypothyroidism (cretinism), on the right, standing next to a normal six year old child. Panel (b) shows the appearance of a baby suffering from congenital hypothyroidism which was later successfully treated by hormone replacement. (Panel (a) from F.M. Delange (1996) Endemic cretinism in L.E. Braverman and R.D. Utiger (eds) *Werner and Ingbar's the thyroid,* 7th edition. Lippincott-Raven, Philadelphia. Panel (b) from Plate 9.3 in J. Laycock and P. Wise (1996) *Essential endocrinology,* 3rd edition. Oxford Medical Publications, Oxford.)

Summary

The follicular cells of the thyroid gland secrete tri-iodothyronine (T_3) and thyroxine (T_4). A further hormone, calcitonin, is secreted by parafollicular cells. T_3 and T_4 play an important role in the control of metabolic rate, maturation of the skeleton, and development of the central nervous system.

TSH, secreted by the anterior pituitary, controls all aspects of the activity of the thyroid gland. Its rate of secretion is inhibited by the negative feedback effects of thyroid hormone. Most tissues increase their oxygen consumption in response to thyroid hormones. This increase in metabolic rate is important in the maintenance of body temperature.

Excessive thyroid hormone secretion (hyperthyroidism) results in an increased BMR, sweating, increased heart rate, nervousness, and weight loss. Insufficient secretion of thyroid hormone (hypothyroidism) results in a reduced BMR, bradycardia, lethargy, cold sensitivity, and a range of other metabolic abnormalities.

22.3 The endocrine regulation of plasma calcium: vitamin D, parathyroid hormone, and calcitonin

The minute-to-minute regulation of plasma mineral levels, particularly that of calcium, is achieved by the combined effects of three different hormones acting on the bone, kidney, and intestine. These are:

- metabolites of cholecalciferol (vitamin D)
- parathyroid hormone
- calcitonin.

This section begins with an overview of whole body calcium and phosphate balance and is followed by a detailed discussion of the hormonal regulation of plasma mineral levels.

Whole body handling of calcium and calcium balance

The adult human body contains around 1 kg of calcium, roughly 99 per cent of which is found in the bones in the form of calcium phosphate and as crystals of hydroxyapatite ($Ca_{10}(PO_4)_6(OH)_2$) which give strength and rigidity to the skeleton. Calcium is also a major component of the teeth and connective tissue. About 99 per cent of the skeletal calcium forms so-called stable bone, which is not readily exchangeable with the calcium in the extracellular fluid. The remaining 1 per cent (~10 g) is in the form of simple calcium phosphate salts that form a readily releasable pool of calcium and can act as a buffer system in response to alterations in plasma calcium. Even during adult life, all the bones are in dynamic equilibrium where deposition (accretion) and resorption of bone are balanced while allowing remodelling of the skeleton in response to changing mechanical requirements.

The cells responsible for the accretion of new bone are called **osteoblasts**, while those responsible for the resorption of bone are large multi-nucleated cells called **osteoclasts** ('bone-eating' cells). Osteoblasts secrete the organic constituents of bone (osteoid) that later becomes mineralized, providing that there is sufficient calcium and phosphate in the extracellular fluid. Bone resorption results in the release of calcium and phosphate into the plasma. The physiology of bone is discussed in more detail in Chapter 51.

Although only a relatively small amount of calcium (10 g or so) is present outside structural tissues, it plays a key role in the regulation of many cellular processes. Calcium ions also play a central role in blood clotting. They are important factors in many cellular functions such as stimulus–secretion coupling, muscle contraction, cell–cell adhesion, and the control of neural excitability. They also act as an intracellular second messenger to regulate the activity of many enzymes. Because of the importance of calcium as a regulatory ion, it is essential that the level of free calcium in both the intra- and extracellular fluids is maintained within narrow limits.

Total intracellular calcium (i.e. bound plus free calcium) is around $1–2$ mmol kg^{-1} of tissue, while the intracellular free calcium concentration is much lower (about 0.1 μmol l^{-1}). This low level is maintained by a variety of mechanisms. These include a plasma membrane calcium pump, which extrudes calcium from the intracellular fluid, and the sequestration of calcium by intracellular storage sites such as mitochondria and the endoplasmic (or, in muscle, the sarcoplasmic) reticulum (see Chapter 5).

Total plasma calcium is $2.3–2.4$ mmol l^{-1}, of which around half is ionized and half is bound, either to plasma proteins (albumin and globulins) or as inorganic complexes with anions, particularly phosphate. The concentration of free calcium ions in the extracellular fluid is around 1.4 mmol l^{-1}, more than ten thousand times higher than the intracellular free calcium concentration.

Daily oral intake of calcium is very variable but falls between 200 and 1500 mg per day. In Western diets, the major sources of calcium are dairy products and flour to which calcium salts are often added. Of a daily intake of 1000 mg calcium, roughly 350 mg will be absorbed into the extracellular fluid from the small intestine. Net absorption is, however, reduced to around 150 mg a day because the intestinal secretions themselves contain calcium and around 850 mg of calcium are lost from the body each day in the faeces. Although the absolute amounts of calcium excreted in the urine vary according to the prevailing calcium balance of the body, almost 99 per cent of filtered calcium is normally reabsorbed along the length of the renal tubules. Nevertheless, close to 150 mg of calcium is excreted per day in the form of inorganic salts. Thus, the intestine and the kidneys are the most important organs in the regulation of the entry and exit of calcium from the plasma, although small amounts of calcium are also lost in the saliva and sweat.

When whole body calcium balance is normal:

$$[\text{dietary } Ca^{2+}] + [Ca^{2+} \text{ resorbed from bones}] = [Ca^{2+} \text{ loss in faeces, urine, saliva, and sweat}] + [Ca^{2+} \text{ added to new bone}]$$

Principal effects of hypocalcaemia

Low plasma calcium (**hypocalcaemia**) will be reflected in a low calcium concentration in the extracellular fluid. This may result in **tetany**, an abnormal excitability of the nerves and skeletal muscles which is manifested as muscular spasms, particularly in the feet and hands (which is known as **carpopedal spasm**). The increased neural excitability may even result in convulsions. Tetany occurs when plasma calcium falls from its normal level of 2.3–2.4 mmol l^{-1} to around 1.5 mmol l^{-1}. The hyperexcitability caused by a fall in plasma calcium can be detected before the onset of tetany by tapping the facial nerve as it crosses the angle of the jaw. In normal people this has no effect, but in hypocalcemia, the muscles on that side of the face will twitch or even go into spasm. This is Chvostek's sign of **latent tetany**.

Tetany should be carefully distinguished from tetanus, the maintained contraction of skeletal muscle, and tetanus, the disease caused by infection with the toxin (tetanus toxin) of the bacillus *Clostridium tetani*.

Principal effects of hypercalcaemia

Elevated plasma calcium is characterized by nausea, vomiting, and dehydration. If the plasma calcium level remains chronically elevated, the soft tissue within the kidneys becomes calcified and renal function may be impaired.

Whole body handling of phosphate

The plasma level of inorganic phosphate is around 2.3 mmol l^{-1}, although its concentration is not as closely regulated as that of calcium. The total body phosphate content of a 70 kg man is about 770 g, of which between 75 and 90 per cent is contained within the skeleton in combination with calcium, both as hydroxyapatite crystals and as calcium phosphate, which forms part of the readily exchangeable pool. Daily intake of phosphate is roughly 1200 mg, of which about a third is excreted in the faeces. The principal route of phosphate loss from the plasma is the urine, where it is excreted along with calcium.

Like calcium, phosphate is required for a wide variety of cellular functions. Phosphate metabolites play a central role in energy metabolism, and the activity of numerous enzymes is regulated by phosphorylation. Phosphate is an important component of the phospholipids of the cell membranes and of both DNA and RNA.

A family of hormones called **phosphatonins** have been identified as key regulators of plasma phosphate. The best known of these is fibroblast growth factor 23 (FGF23), which is secreted principally by **osteocytes** and has a phosphaturic action, i.e. it promotes phosphate loss from the body via the urine. In animal studies, the production of FGF23 has been shown to increase when plasma phosphate levels rise.

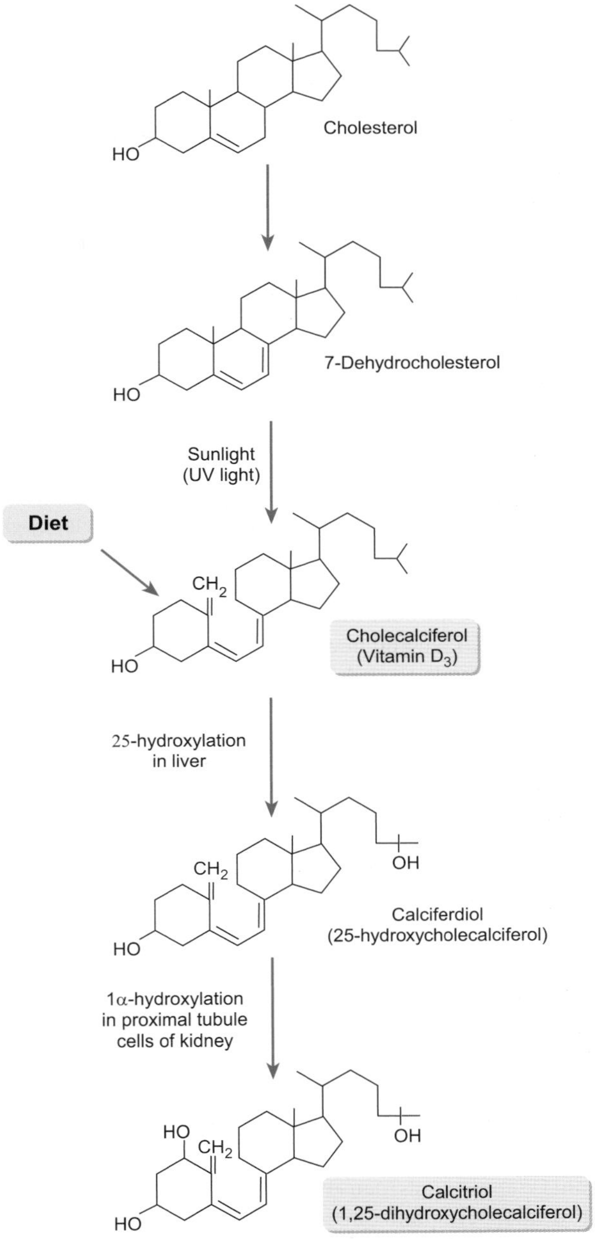

Vitamin D and its metabolites

Vitamin D or **cholecalciferol** acts as the precursor for a group of steroid compounds that behave as hormones and play a crucial role in the regulation of plasma calcium levels. The daily vitamin D requirement is 400 IU for children and 100 IU for adults. Although much of this is usually obtained from the diet, particularly from oily fish and eggs, cholecalciferol can be synthesized in the skin from 7-dehydrocholesterol in the presence of sunlight. Figure 22.10 shows the hydroxylation steps involved in the synthesis of the hormone.

Vitamin D is not itself biologically active but undergoes hydroxylation reactions to form active hormones. The first hydroxylation takes

Figure 22.10 The synthesis of 1,25-dihydroxycholecalciferol from cholesterol and dietary vitamin D.

place in the liver, giving rise to 25-hydroxycholecalciferol, which is the major form of vitamin D in the circulation. A further hydroxyl group is added in the proximal tubule cells of the kidney to give either 1,25-dihydroxycholecalciferol (also called **calcitriol**) or 24,25 dihydroxycholecalciferol, which is inactive. The synthesis of the active 1,25 derivative predominates when plasma calcium levels are low.

Actions of 1,25-dihydroxycholecalciferol

The main action of this hormone is to stimulate the absorption of ingested calcium. This occurs by a direct effect on the intestinal mucosa. The hormone binds to specific nuclear receptors, and this interaction leads to an increase in the expression of the luminal calcium channels, calbindins (the calcium binding proteins that transport calcium across the cell), and the calcium extrusion proteins of the basolateral membrane (the Na^+/Ca^{2+} exchanger and the Ca^{2+}-ATPase). Phosphate absorption is also enhanced by 1,25-dihydroxycholecalciferol.

The actions of vitamin D on bone are rather poorly understood, and much of our current knowledge has been derived from observations of vitamin D deficiency. Nevertheless, it is clear that 1,25-dihydroxycholecalciferol stimulates the calcification of the bone matrix. While part of this effect is probably an indirect result of increased plasma levels of calcium and phosphate, 1,25-dihydroxycholecalciferol also appears to stimulate both osteoblast and osteoclast activity directly. The combined effect of these actions will be to facilitate the remodelling of bone.

Effects of vitamin D deficiency

It has been known for many years that vitamin D deficiency gives rise to the conditions known as **rickets** in children and **osteomalacia** in adults. In both of these disorders there is a failure of calcification of the bone matrix, and bone remodelling, whereby old bone is resorbed and new bone is laid down, is impaired. In children, this leads to skeletal deformities, particularly of the weight-bearing bones, giving rise to the characteristic bowing of the tibia. Figure 51.20 shows a picture of a child with rickets. In adults, the chief features of vitamin D deficiency include a reduction in bone density with large amounts of non-mineralized osteoid, increased susceptibility to fracture, bone pain, and proximal muscle weakness.

Patients with vitamin D deficiency may have low plasma calcium levels because of the reduction in intestinal calcium absorption. This may lead to increased excitability of nervous tissue, causing paraesthesia ('pins and needles') or attacks of tetany. Hypocalcaemia in such patients is, however, unlikely to become severe since there will be a compensatory rise in parathyroid hormone secretion.

Effects of vitamin D excess

Large doses of vitamin D (usually resulting from over-ingestion of vitamin D supplements) can give rise to a condition called vitamin D intoxication. Symptoms are related to hypercalcaemia as a result of increased intestinal absorption and include nausea, vomiting, and dehydration. If the plasma calcium level remains chronically elevated, soft tissues within the kidneys, heart, lungs, and blood vessel walls become calcified so that renal function is impaired and there is an increase in blood pressure.

The parathyroid glands

The parathyroid glands are embedded within the posterior surfaces of the lateral lobes of the thyroid gland, as shown in Figures 22.11 and 22.12(a). Although most people possess two pairs of parathyroid glands, supernumerary glands are not uncommon. Each gland is covered by a fibrous capsule and measures 3–8 mm in length, 2–5 mm in width, and about 1.5 mm in depth. Together, the four glands weigh about 1.6 g. They receive a rich blood supply

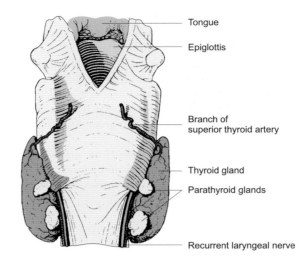

Figure 22.11 The anatomical location of the thyroid and parathyroid glands with their arterial blood supply. This view is from the **dorsal** (posterior) aspect of the pharynx.

which is derived mainly from the inferior thyroid arteries. The adult gland contains two major cell types: chief cells, which secrete parathyroid hormone (PTH), and the somewhat larger oxyphilic cells, whose cytoplasm stains strongly with eosin. The function of the oxyphilic cells is uncertain, but studies suggest that they can also secrete PTH as well as other factors including calcitriol and PTHrP (parathyroid hormone related peptide), a local hormone with effects similar to those of PTH. Oxyphil numbers appear to increase with age. The histological appearance of the parathyroid gland, illustrating these two types of cell, is shown in Figure 22.12(b).

Parathyroid hormone is a polypeptide hormone consisting of 84 amino acid residues: M_r 9500 Da. It is derived initially from a 115 amino acid polypeptide precursor called pre-pro-PTH. Two cleavages of the peptide chain give rise to the active parathyroid hormone, which is packed into secretory granules for storage and eventual secretion. PTH exerts effects on bone, gut, and kidneys which result in an elevation of plasma calcium levels and a reduction in plasma phosphate levels. Together with calcitonin from the parafollicular cells of the thyroid gland, and metabolites of cholecalciferol (vitamin D), PTH makes an important contribution to the regulation of whole body mineral levels. Its actions are summarized diagrammatically in Figure 22.13.

Parathyroid hormone acts to mobilize calcium from bone and increase calcium reabsorption by the kidneys

Normal levels of PTH are necessary for the maintenance of the skeleton. It promotes the production of osteoblasts (osteoblastogenesis) and osteoblast survival, and it enhances the calcification of the bone matrix. When plasma calcium levels fall, however, PTH secretion rises. High levels of PTH have a biphasic action on bone metabolism. There is an initial rapid loss of calcium from the

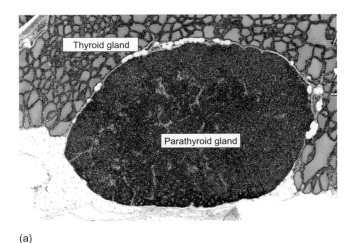

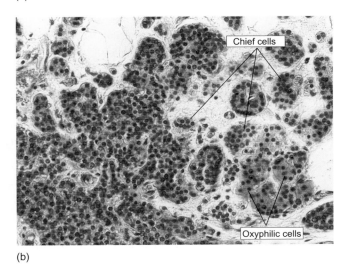

Figure 22.12 The detailed structure of the parathyroid gland. Panel (a) shows a low-power view of the posterior aspect of the thyroid gland, with the densely staining parathyroid gland embedded within it. Each gland has a distinctive histological appearance. Panel (b) shows a section of parathyroid gland at a higher power in which the chief cells with densely staining nuclei can be clearly distinguished from the oxyphil cells, which are strongly eosinophilic with red cytoplasm. The chief cells secrete parathyroid hormone.

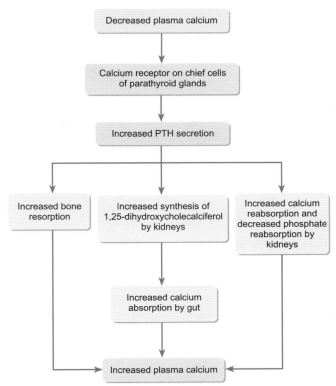

Figure 22.13 A flow diagram summarizing the principal actions of parathyroid hormone.

readily releasable pool of calcium on the bone surface. There is some evidence that PTH acts in combination with vitamin D to bring about this effect. In the longer term (hours to days), high circulating PTH levels stimulate resorption of stable bone by the osteoclasts, thereby adding large amounts of mineral to the extracellular fluid.

In the kidney, PTH stimulates the reabsorption of calcium by the cortical ascending thick limb of the loop of Henle and the distal tubule and decreases the reabsorption of phosphate by the proximal tubule. The net effect of these actions will be an increase in plasma calcium and a fall in plasma phosphate. Although the latter effect may appear to be undesirable, the fall in plasma phosphate will further raise free plasma calcium levels by reducing the amount of phosphate ions available to bind with calcium. It is the rise in calcium which seems to be of primary importance to the body.

PTH does not appear to exert a direct effect on the intestine. It does, however, stimulate the synthesis of 1,25-dihydroxycholecalciferol in the proximal tubules of the renal nephrons. This metabolite of vitamin D enhances the absorption of ingested calcium and phosphate from the proximal small intestine and, to a small extent, the colon.

Control of PTH secretion

Parathyroid hormone is cleared rapidly from the plasma and has a half-life in the blood of around five minutes. In order to maintain its basal concentration in the circulation, PTH is secreted continuously at a low rate. There appears to be a direct negative feedback relationship between plasma calcium ions and PTH secretion, which is mediated via a specific calcium receptor, the calcium sensing receptor or CaSR, found on the plasma membrane of the PTH-secreting chief cells. As a result, the most potent stimulus for increased PTH secretion is hypocalcaemia (lowered plasma calcium concentration), which also encourages the biosynthesis of the hormone and proliferation of parathyroid cells. Conversely, a rise in plasma calcium inhibits the release of PTH. The inverse relationship between plasma calcium levels and PTH secretion is illustrated in Figure 22.14. PTH secretion is also stimulated by catecholamines and dopamine but is suppressed by 1,25-dihydroxycholecalciferol. The secretion of PTH in hypocalcaemia appears to require the presence of normal circulating levels of

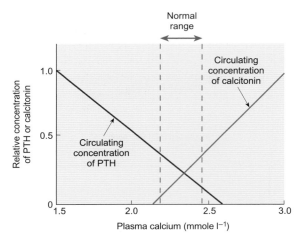

Figure 22.14 The relationship between the plasma calcium concentration and the secretion of both parathyroid hormone and calcitonin. As calcium rises, the secretion of parathyroid hormone falls while that of calcitonin rises.

magnesium. Premature babies, who often have low plasma magnesium levels, may also become hypocalcaemic because of low PTH output.

Disorders of parathyroid hormone secretion

Effects of excess PTH secretion (primary hyperparathyroidism)

Hypersecretion of PTH occurs when a tumour of the parathyroid glands develops. Certain malignant tumours originating in other cell types such as the lungs also secrete PTH and can give rise to the symptoms of PTH excess. The excess secretion of PTH results in hypercalcaemia, which is associated with a reduction in plasma phosphate concentration. Although PTH increases the rate of reabsorption of calcium in the renal tubules, there is a very high filtered calcium load, resulting in an increase in the amount of calcium excreted in the urine. The increased urinary excretion of calcium predisposes the kidney to form stones (renal calculi), which are common in this disorder. When untreated, kidney stones can lead to erosion of the renal tissue, causing severely impaired renal function. This is the most common cause of death in untreated hyperparathyroidism.

The skeletal effects of high PTH levels vary considerably between patients, but demineralization of the skeleton is often found. In such cases, there is bone pain, fractures of the long bones, and compression fractures of the spine. Cysts composed of osteoclasts ('brown tumours') may also be present. The hypercalcaemia that results from over-production of PTH has other consequences including fatigue, weakness, mental aberrations, CNS depression, constipation, and anorexia.

Effects of insufficient secretion of PTH (hypoparathyroidism)

Hypoparathyroidism, due to damage, failure, or removal of the parathyroid glands (which may occasionally occur accidentally during thyroid surgery), results in a gradual decline in plasma calcium, which, if untreated, will eventually prove fatal. The major consequences of hypocalcaemia are described above. Treatment of the disorder includes an initial infusion of calcium, to restore normal levels, followed by the administration of vitamin D metabolites, which stimulate the intestinal absorption of ingested calcium.

The role of calcitonin in the regulation of plasma calcium

Calcitonin (also known as thyrocalcitonin) is a peptide hormone secreted by the parafollicular cells, or C-cells, of the thyroid gland which lie between the follicles (see Figure 22.2). Although it is known that calcitonin is able to lower the level of free calcium in the plasma, its significance in the overall regulation of mineral levels in humans is unclear. Indeed, plasma calcium remains unaffected in patients with tumours of the C-cells in whom calcitonin levels are elevated.

Actions of calcitonin

Figure 22.15 summarizes the actions of calcitonin and the regulation of its secretion. The primary action of this hormone appears to be the inhibition of the activity of the bone osteoclasts. As a result, bone resorption is reduced and calcium and phosphate are not released into the plasma. Calcitonin (administered as a nasal spray) has been used with some success to reduce the incidence of vertebral compression fractures in post-menopausal women with osteoporosis (see below). It is also sometimes administered

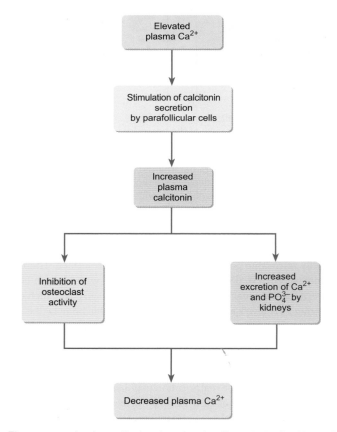

Figure 22.15 A schematic drawing showing the principal actions of calcitonin and the factors thought to regulate its secretion.

to patients with hypercalcaemia associated with malignancy and in Paget's disease, in which there is excessive turnover of bone. Receptors for calcitonin are also present on kidney cells, and calcitonin may produce a transient increase in the rates of excretion of calcium, phosphate, sodium, potassium, and magnesium.

Control of calcitonin secretion

Calcitonin secretion in response to an elevated plasma calcium concentration is mediated by the CaSR (calcium sensing receptor, discussed earlier in this section), possibly via activation of a non-selective cation channel that brings about a depolarization that opens voltage-gated calcium channels. The subsequent rise in intracellular calcium then triggers the secretion of calcitonin by exocytosis—an example of classical excitation–secretion coupling.

Other hormones involved in the regulation of plasma calcium

Although parathyroid hormone, metabolites of vitamin D, and possibly calcitonin are the major hormonal regulators of plasma calcium and phosphate concentrations, a number of other hormones are known to exert effects on the way in which these minerals are regulated by the body. These include growth hormone, the adrenal glucocorticoids, and the thyroid hormones, but of considerably more significance are the oestrogens and androgens, particularly the former.

In adult females, oestrogens appear to inhibit the PTH-mediated resorption of bone and to stimulate the activity of the osteoblasts. They may also reduce the activity of osteoclasts by down-regulating osteoclast precursors and increasing osteoclast

apoptosis. Following removal of the ovaries, or after the menopause, the rapid fall in oestrogens results in an increased rate of bone resorption that can lead to a condition called **osteoporosis** (see Chapter 51). This is characterized by increased bone fragility, susceptibility to vertebral compression fractures, and fractures of the wrist and hip. Oestrogen replacement therapy has been shown to reduce the rate of progress of postmenopausal osteoporosis.

Summary

Calcium plays a vital role in many aspects of cellular function. It is also a major structural component of the bones of the skeleton. Plasma levels of calcium and phosphate are regulated by parathyroid hormone (PTH), dihydroxycholecalciferol (calcitriol, a metabolite of vitamin D), and calcitonin. These hormones act on bone, gut, and kidney to regulate the entry and exit of calcium into and out of the extracellular pool.

Parathyroid hormone (PTH) is a peptide hormone secreted by the parathyroid glands. It acts to maintain normal plasma calcium levels (normocalcaemia) and is released in response to a fall in plasma calcium. PTH also stimulates the reabsorption of calcium in the distal tubules of the kidney. Normal PTH levels are needed for the maintenance of the skeleton. PTH also stimulates the reabsorption of calcium in the distal tubules of the kidney. Calcitriol enhances the absorption of dietary calcium by the intestine and stimulates the turnover of bone mineral. Calcitonin is secreted by the parafollicular cells of the thyroid gland in response to elevated plasma calcium.

✱ Checklist of key terms and concepts

The thyroid gland, T_3, and T_4

- The follicular cells of the thyroid gland secrete two iodine-containing hormones, tri-iodothyronine (T_3) and thyroxine (T_4).
- Iodide is concentrated in the follicular cells, where it is oxidized to iodine for incorporation into thyroglobulin.
- Thyroid hormones are released from the gland following enzymatic hydrolysis of the iodinated thyroglobulin.
- T_3 and T_4 travel in the plasma bound to carrier proteins, including a specific thyronine binding globulin (TBG).
- Thyroid stimulating hormone (TSH), secreted by the anterior pituitary, controls all aspects of the activity of the thyroid gland. Its rate of secretion is inhibited by the negative feedback effects of thyroid hormone.

The actions of T_3 and T_4

- T_3 and T_4 play an important role in the control of metabolic rate, maturation of the skeleton, and development of the central nervous system.

- Most tissues increase their oxygen consumption in response to thyroid hormones. This increase in metabolic rate is important in the maintenance of body temperature in cold environments.
- Thyroid hormones increase cardiac output, the rate of ventilation, and the rate of red blood cell production.
- The metabolic actions of thyroid hormones are dose-dependent. Low concentrations tend to be hypoglycaemic, while higher doses stimulate glycogenolysis and gluconeogenesis, leading to a rise in plasma glucose.
- Thyroid hormones are powerfully lipolytic and stimulate the oxidation of free fatty acids.
- Low levels of thyroid hormone stimulate protein synthesis, while higher concentrations are catabolic.
- Excessive thyroid hormone secretion is known as hyperthyroidism. It results in an increased basal metabolic rate (BMR), sweating, increased heart rate, nervousness, and weight loss.
- Insufficient secretion of thyroid hormone (hypothyroidism) results in a reduced BMR, bradycardia, lethargy, cold sensitivity, and a range of other metabolic abnormalities.

Hormonal regulation of plasma calcium levels

- Calcium plays a vital role in many aspects of cellular function. It is also a major structural component of bone.

- Plasma levels of calcium and phosphate are regulated by the actions of three hormones: calcitonin, parathyroid hormone (PTH), and the active metabolites of vitamin D.

- These hormones act on bone, gut, and kidney to regulate the entry and exit of calcium into and out of the extracellular pool.

- Calcitonin is secreted by the parafollicular cells of the thyroid gland. It is a hypocalcaemic agent, secreted in response to elevated plasma calcium.

- PTH is a peptide hormone secreted by the parathyroid glands. It is released in response to a fall in plasma calcium.

- Normal PTH levels are needed for the maintenance of the skeleton. PTH also stimulates the reabsorption of calcium in the distal tubules of the kidney.

- Dihydroxycholecalciferol is a metabolite of vitamin D, which is active in the regulation of plasma calcium. Its main effect is to enhance the absorption of dietary calcium by the intestine. It also seems to stimulate the turnover of bone.

- Vitamin D deficiency causes rickets in children and osteomalacia in adults.

- Calcitonin from parafollicular cells of the thyroid gland acts to reduce plasma calcium levels through actions on bone and kidney. Its overall importance in humans is unclear.

- A number of other hormones, including the sex steroids, growth hormone, thyroid hormones, and adrenal glucocorticoids, contribute to whole body calcium homeostasis.

 ## Recommended reading

Biochemistry

Berg, J.M., Tymoczko, J.L., and Stryer, L. (2011) *Biochemistry* (7th edn), Chapters 14, 27, and 32. Freeman, New York.

A recent revision of a classic text with helpful accounts of signal transduction, the integration of metabolism, and the regulation of gene expression in eukaryotes.

Histology

Mescher, A. (2015) *Junquieira's basic histology* (14th edn), Chapter 20. McGraw-Hill, New York.

A recent revision of Junquieira and Carneiro's classic text. Good text, fewer pictures than some alternatives.

Pharmacology

Rang, H.P., Ritter, J.M., Flower, R.J., and Henderson, G. (2015) *Pharmacology* (8th edn), Chapter 33. Churchill-Livingstone, Edinburgh.

This provides an easily digested introduction to the endocrine disorders discussed in this chapter and provides a helpful introduction to their treatment.

Endocrine physiology

Kovacs, W.J., and Ojeda, S.R. (2012) *Textbook of endocrine physiology* (6th edn), Chapters 12 and 14. Oxford University Press, Oxford.

Authoritative accounts of the thyroid, parathyroid, and adrenal glands.

Medicine

Hall, R., and Evered, D.C. (1990) *A colour atlas of endocrinology* (2nd edn). Wolfe Medical, London.

An informative collection of colour plates illustrating the physical changes associated with the endocrine disorders mentioned in this chapter.

Jameson, J.L. (ed) (2006) *Harrison's endocrinology*, Chapters 4, 5, and 23–25. McGraw-Hill Medical, New York.

Kumar, V., Abbas, A.K., and Aster, J.C. (eds) (2013) Chapter 19, in: *Robbins basic pathology* (9th edn). Saunders, New York.

Ledingham, J.G.G., and Warrell, D.A. (2000) *Concise Oxford textbook of medicine*, Chapters 7.3, 7.5, and 7.6. Oxford University Press, Oxford.

An authoritative introduction to thyroid, parathyroid, and adrenal disorders and their treatment.

Review articles

Brown, E.M. (2013) Role of the calcium sensing receptor in extracellular calcium homeostasis: Best practice and research. *Clinical Endocrinology and Metabolism* **27**, 333–43.

A useful review of the role of the CaSR that also includes an overview of some important general aspects of calcium homeostasis.

Friesema, E.C.H., Jansen, J., and Visser, T.J. (2005) Membrane transporters for thyroid hormone. *Current Opinion in Endocrinology and Diabetes* **12**, 371–80.

Martin, A., David, V., and Quarles, L.D. (2012) Regulation and function of the fgf23/klotho endocrine pathways. *Physiological Reviews* **92(1)**, 131–55, http://doi.org/10.1152/physrev.00002.2011.

Mullur, R., Liu, Y.-Y., and Brent, G.A. (2014). Thyroid hormone regulation of metabolism. *Physiological Reviews* **94**, 355–82, http://doi.org/10.1152/physrev.00030.2013.

Riccardi, D., Finney, B.A., Wilkinson, W.J., and Kemp, P.J. (2009) Novel regulatory aspects of the extracellular Ca^{2+}-sensing receptor, CaR. *Pflügers Archiv—European Journal of Physiology* **458**, 1007–22.

 To check that you have mastered the key concepts presented in this chapter, complete the accompanying online self-assessment questions. Go to www.oup.com/uk/pocock5e/

CHAPTER 23

The adrenal glands

Chapter contents

This chapter should help you to understand:

- The anatomy of the adrenal glands
- The histology of the adrenal cortex and its functional significance
- The main pathways for the synthesis of the adrenal steroid hormones
- The principal actions of the adrenal steroid hormones
  - The effects of aldosterone (the main mineralocorticoid)
  - The effects of cortisol (the main glucocorticoid)
- The effects of the adrenocortical sex steroids
- The principal disorders of adrenal cortical hormone secretion
- The role of the adrenal medullary hormones in metabolism and their effects on the cardiovascular system
- The regulation of adrenal medullary secretion
- Disorders of adrenal medullary secretion

23.1 Introduction

The adrenal glands have two distinct endocrine functions. The **adrenal medulla** secretes the catecholamines **adrenaline** (epinephrine) and **noradrenaline** (norepinephrine) in response to activation of its sympathetic nerve supply. It thereby acts as part of the sympathetic nervous system and its hormones are concerned with the body's very rapid responses to acute stress (often referred to as the 'fight, fright, or flight' responses). The **adrenal cortex** secretes three classes of steroid hormones; glucocorticoid hormones, mineralocorticoid hormones, and sex steroids. **Cortisol**, the principal glucocorticoid, has a wide range of metabolic effects,

while **aldosterone**, the principal mineralocorticoid, is crucial to the maintenance of normal body fluid volume and composition. This chapter is concerned with the important actions of the catecholamines and the adrenal corticosteroid hormones. The role of cortisol is discussed in detail here, while the actions of aldosterone are discussed briefly in this chapter and in more detail in Chapter 39. Some important conditions arising from imbalances in adrenal cortical hormone secretion are also considered briefly.

23.2 The adrenal glands—location and gross anatomy

The paired adrenal (or suprarenal) glands are roughly pyramid-shaped organs lying above the kidneys (see Figure 23.1). Each gland is enclosed in a fibrous capsule surrounded by fat (which is not shown in the figure) and each is structurally and functionally equivalent to two separate endocrine glands. The inner **adrenal medulla** derives from ectodermal cells of the embryonic neural crest and secretes the catecholamines **adrenaline** (epinephrine) and **noradrenaline** (norepinephrine) in response to activation of its sympathetic nerve supply. The outer **adrenal cortex**, which encapsulates the medulla and forms the bulk (80–90 per cent) of the gland, is derived from embryonic mesoderm and secretes a number of steroid hormones. Unlike those of the adrenal medulla, the secretions of the adrenal cortex are controlled hormonally. Moreover, loss of the adrenal cortical hormones will result in death within a few days.

The adult adrenal glands each weigh between 5 and 10 g. They have an extremely rich blood supply, receiving arterial blood from branches of the aorta, the renal arteries, and the phrenic arteries. The arterial blood enters from the outer cortex, flows through fenestrated capillaries between the cords of cortical cells, and drains inwardly into venules in the medulla. This arrangement is important, as it allows interactions between the different regions of the gland. The right adrenal vein drains directly into the inferior vena cava, while the left drains into the left renal vein. The adrenal glands have a rich nerve supply derived from the coeliac plexus and the thoracic splanchnic nerves that chiefly innervate the adrenal medulla. In the fetus, the adrenal glands are much larger,

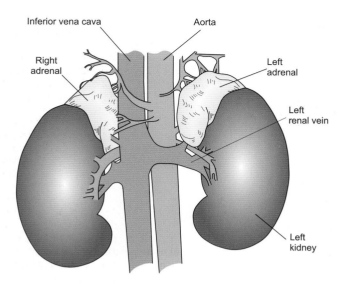

Figure 23.1 The anatomical location of the adrenal glands and the organization of their blood supply. Note that the arterial supply is via many small arteries that originate from the aorta. The venous drainage is via a large central vein. In life, the adrenals are embedded in a large fat pad that is not shown in this diagram.

relative to body size, than in the adult. Their functions during fetal life are described in Chapter 50.

23.3 The adrenal cortex

The adrenal cortex consists of three morphologically distinct zones of cells (see Figures 23.2 and 23.3). These are the outer **zona glomerulosa** (occupying around 10 per cent of the adrenal cortex), the **zona fasciculata** (around 75 per cent), and the **zona reticularis** (around 15 per cent), which lies closest to the adrenal medulla. The zona reticularis does not differentiate fully until between 6 and 8 years of age. In the adult gland, the cells of the glomerulosa continually migrate down through the zona fasciculata to the zona reticularis, changing their secretory pattern as they go. The functional significance of this migration is not clear.

Synthesis and secretion of adrenal cortical hormones

The cells of the three zones secrete different steroid hormones: the cells of the zona glomerulosa secrete the **mineralocorticoids**; those of the zona fasciculata secrete **glucocorticoids**; and the cells of the zona reticularis secrete **sex steroids**.

Cholesterol is the starting point for all steroid biosynthesis. Figure 23.4 summarizes the main pathways of steroid biosynthesis in the adrenal cortex and shows that the cortical hormones are obtained by the modification of cholesterol through a series of hydroxylation reactions. Most of these reactions involve **cytochrome P-450 enzymes**, which are situated in the mitochondrial cristae and the endoplasmic reticulum. Small amounts of cholesterol are synthesized within the adrenal gland from acetyl CoA, but most

is obtained by the receptor-mediated endocytosis of low-density lipoproteins circulating in the bloodstream. The cholesterol required for hormone synthesis is mainly stored within cytoplasmic lipid droplets in the adrenal cortical cells. ACTH activates cholesterol esterase, which in turn liberates cholesterol from the droplets when it is needed for corticosteroid hormone synthesis.

Aldosterone is the principal mineralocorticoid secreted by the adrenal cortex. It is synthesized exclusively by the cells of the zona glomerulosa. It is not stored within the adrenal cortical cells to any significant extent but diffuses rapidly out of the cells after synthesis. Consequently, the rate of aldosterone production must increase whenever there is a need for increased circulating levels of the hormone. Normally between 0.1 and 0.4 μmoles of aldosterone are secreted each day.

In humans, **cortisol** is the dominant glucocorticoid but corticosterone and cortisone are other glucocorticoids that are produced in small amounts. (Cortisone itself is inactive, but it can be converted to cortisol in the liver and other tissues.) The synthesis of the glucocorticoids takes place largely in the zona fasciculata and, like aldosterone, these hormones are released rapidly from the cells by diffusion across the plasma membrane following synthesis. Between 30 and 80 μmoles of cortisol are secreted each day.

The principal sex steroids produced by the cells of the zona reticularis are **dehydroepiandrosterone** and **androstenedione**. Although they are weak androgens themselves, they are converted to the more powerful androgen **testosterone** in the peripheral tissues. Small amounts of the oestrogenic hormones oestrone and oestradiol-17β and of progesterone are also secreted by the adrenal glands of both sexes. In females, the adrenal glands secrete about half the total androgenic hormone requirements, but in males, the amounts produced are insignificant in comparison with the production of testosterone by the testes. Similarly, the oestrogenic hormones secreted by the cortex are quantitatively insignificant in women (at least until after the menopause).

Plasma contains a specific corticosteroid binding globulin (CBG) called **transcortin**, which is a glycoprotein synthesized in the liver. Between 70 and 80 per cent of the circulating cortisol is reversibly bound to transcortin, while a further 15 per cent or so binds to serum albumin. This means that only 5–10 per cent of plasma cortisol is in a free or active form. It follows, then, that changes in circulating plasma levels of transcortin will have significant effects on the free cortisol available for biological activity. Transcortin synthesis is enhanced by oestrogens (transcortin levels tend to be elevated during pregnancy) but is lowered in liver pathologies such as cirrhosis. Transcortin also binds progesterone with a fairly high affinity but it has a much lower affinity for aldosterone, which is carried mainly in combination with albumin.

Principal actions of the adrenocorticosteroids

Effects of aldosterone

The mineralocorticoid secretions of the adrenal cortex, of which aldosterone is the most important, are essential to life. In their absence, death occurs within a few days unless prevented by the therapeutic administration of salt or by the replacement of hormones. The

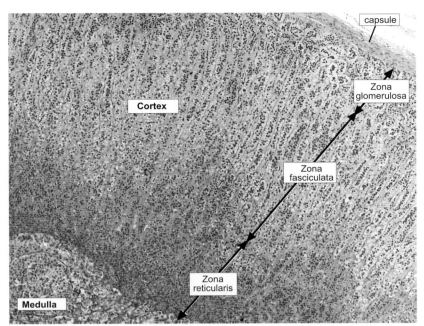

Figure 23.2 A section through an adult human adrenal gland, showing the different zones of the cortex and the distinctive appearance of the medulla. The cords of cells that make up the zona fasciculata are clearly visible, as are the rounded clusters of cells that make up the zona glomerulosa. The main cells of the adrenal medulla are chromaffin cells.

secretion and effects of aldosterone are discussed fully in Chapter 39. Briefly, aldosterone secretion is regulated by the plasma levels of sodium and potassium via the renin–angiotensin system (see Chapter 39 p. 602). Aldosterone acts to conserve body sodium by stimulating the reabsorption of sodium in the distal nephron in exchange for potassium. Failure of aldosterone secretion results in a rise in plasma potassium levels and a fall in sodium and chloride levels.

The extracellular fluid volume and the blood volume fall, with a subsequent drop in cardiac output which, if uncorrected, may prove fatal.

Aldosterone also stimulates the reabsorption of sodium from saliva, sweat, and faeces. A person arriving in a hot climate will, at first, lose large amounts of salt through sweating, particularly during exercise. However, over a period of days, as the renin–angiotensin–aldosterone system becomes increasingly effective,

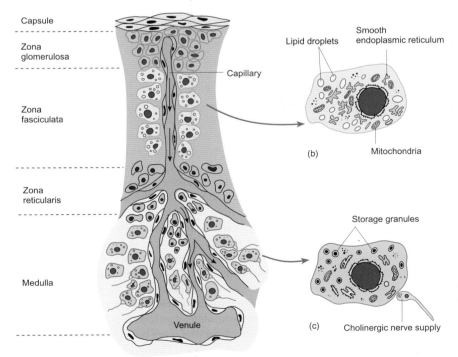

Figure 23.3 A diagram of the cortex and medulla of the adrenal gland showing the relationships between them. Note the three zones of the adrenal cortex, the cells of which secrete steroid hormones. Panel (b) shows the appearance of steroid-secreting cells, while (c) illustrates a single catecholamine-secreting chromaffin cell. (From Fig 4.1 of C. Brook and N. Marshall (1996) *Essential endocrinology.* Blackwell Science, Oxford.)

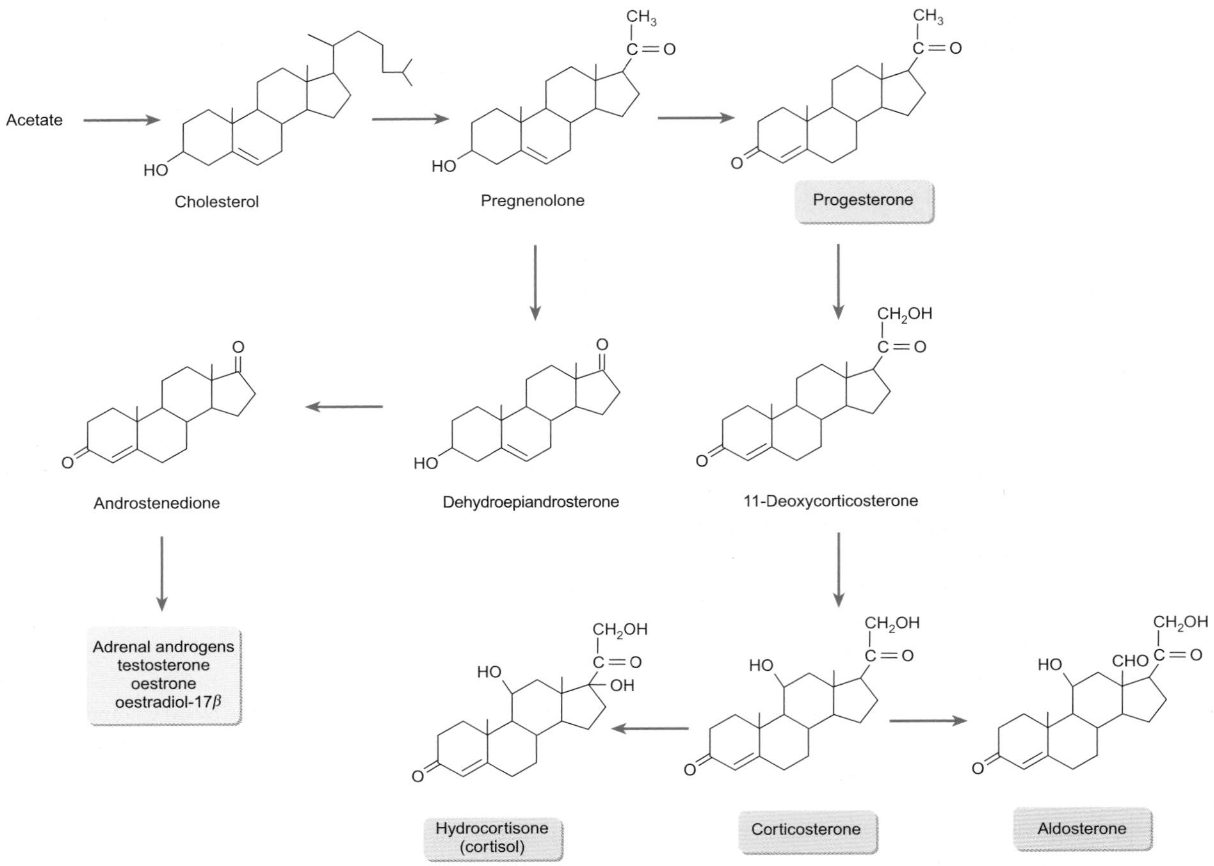

Figure 23.4 A schematic diagram showing the principal steps in the synthesis of the adrenal steroid hormones from cholesterol. Cortisol is the principal glucocorticoid and aldosterone is the principal mineralocorticoid.

the sodium content of sweat falls to practically zero—this is part of the process of acclimatization and is discussed further in Chapter 40.

Effects of cortisol

Cortisol is essential to life. Together with the other glucocorticoids, it has a wide range of different actions throughout the body. It exerts its most important physiological effects on the metabolism of carbohydrates, proteins, and, to a lesser extent, fats. Glucocorticoids also play a crucial role in the responses of the body to a variety of stressful stimuli. They are immunosuppressive, anti-inflammatory, and anti-allergic, and they possess weak mineralocorticoid activity. This mineralocorticoid activity is important, as plasma levels of cortisol are much higher than those of aldosterone. Recent evidence suggests that enzymes in different target tissues can either inactivate specific steroids or convert them to active hormones. For example, the enzyme 11β-hydroxysteroid dehydrogenase 2 (11β-HSD2) catalyses the conversion of cortisol to its inactive metabolite cortisone in the cells of the distal tubule. In this way the mineralocorticoid receptor in renal tubules is protected from excess stimulation by cortisol. Another form of the enzyme, 11β-HSD1, expressed mainly in the liver and in fat, acts predominantly to convert inactive

cortisone to active cortisol, and thereby facilitates the actions of glucocorticoids in these tissues.

Metabolic actions of cortisol

In general terms, the metabolic effects of cortisol may be said to oppose those of insulin. Its effects vary with the target tissue but result in a rise in plasma glucose concentration. The most important overall action of cortisol is to facilitate the conversion of protein to glycogen. It stimulates the mobilization of protein from muscle tissue in particular, thereby increasing the rate at which amino acids are presented to the liver for **gluconeogenesis** (the synthesis of glucose from non-carbohydrate precursors). As a result of this, glycogen stores are initially built up while any excess glucose is released into the plasma. At the same time, cortisol seems to inhibit the uptake and utilization of glucose by those tissues in which glucose uptake is insulin-dependent.

Cortisol also stimulates the appetite and influences the metabolism of fats, particularly when it is released in greater than normal amounts. It directly stimulates lipolysis in adipose tissue and enhances the lipolytic actions of other hormones such as growth hormone and the catecholamines. The net result of its rather complex actions on lipid metabolism is a redistribution of body fat, with a reduction in fat levels in the limbs but increased deposition of fat around the face, back, and abdomen. The major metabolic actions of cortisol are summarized in Figure 23.5.

Other actions of cortisol

Cortisol acts to counteract many of the effects of stress throughout the body. Although the concept of stress is difficult to define, it includes physical trauma, intense heat or cold, infection, and mental or emotional trauma. The exact nature of the glucocorticoid response is unclear but may include cardiovascular, neurological, and anti-inflammatory effects as well as effects on the immune system. Cortisol increases vascular tone, possibly by promoting the actions of catecholamines, and blocks the processes which lead to inflammation in damaged tissues. Although this action is only apparent at high concentrations of the hormone, it has proved to be of value in the treatment of inflammatory conditions such as rheumatoid arthritis and in the treatment of severe asthma.

High levels of cortisol also suppress the normal immune response to infection. There is a gradual destruction of lymphoid tissue, leading to a fall in antibody production and in the number of circulating lymphocytes. It is thought that this action may be important physiologically in preventing the immune system from causing damage to the body. (In autoimmune diseases the immune system responds to molecules that are naturally present in the body and attacks normal healthy cells expressing those molecules.) This action of cortisol explains the efficacy of steroid medications in the treatment of chronic inflammatory conditions such as rheumatoid arthritis and Crohn's disease (p. 715), and in the prevention of transplant rejection.

Glucocorticoid hormones are known to influence the central nervous system. Although the underlying mechanisms are unclear, cortisol acts on the CNS to produce euphoria, an effect that may also have a role in helping to mitigate the effects of stress. This also accounts for the severe depression that can follow the sudden withdrawal of prescribed steroidal drug treatments. Fetal cortisol plays a vital role in the maturation of many organs and is particularly important in the stimulation of pulmonary surfactant production (see Chapter 50).

Effects of adrenocortical sex steroids

The production of sex steroids (androgens in men, oestrogens and progesterone in women) occurs mainly in the gonads (see Chapter 48). Secretion of these hormones by the adrenal cortex is minor by comparison. Indeed, the role of the adrenal sex steroids is not fully understood. They may be important in the development of certain female secondary sexual characteristics, particularly in the growth of pubic and axillary hair. They may also play a role in the growth spurt seen in middle childhood, which occurs around the age of 6 or 7 years. The functions of adrenal oestrogens in males are unclear, but some studies suggest they may play a part in the maintenance of skeletal stability and the normal male libido. The actions of the adrenal sex steroids become more significant in disease, particularly when there is hypersecretion by adrenal tumours or enzyme defects in the pathways that normally synthesize cortisol. Furthermore, effects of adrenal androgens such as male pattern hair loss, skin changes, and an increase in facial hair may become more pronounced in post-menopausal females, in whom the physiological effects of oestrogens are markedly reduced.

The regulation of steroid hormone secretion from the adrenal cortex

The regulation of aldosterone secretion by the renin–angiotensin system is discussed fully in Chapter 40. It is known that the synthesis and secretion of the glucocorticoids and, to a lesser extent, the adrenal sex

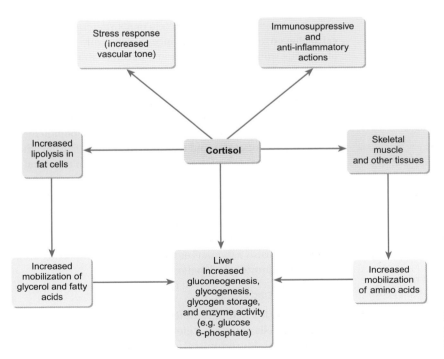

Figure 23.5 A diagram summarizing the main physiological actions of cortisol (the principal glucocorticoid hormone).

steroids are under the control of adrenocorticotrophic hormone (ACTH) secreted by the anterior pituitary in response to hypothalamic corticotrophin releasing hormone (CRH—see Chapter 21). ACTH interacts with the type 2 melanocortin (MC_2) receptor, which is a G protein linked cell-surface receptor found on the cells of the zona fasciculata. Once activated, the MC_2 receptor stimulates the uptake of cholesterol and increases the rate of cortisol synthesis. Glucocorticoid secretion is regulated by a typical negative feedback system (see Figure 23.6). Rising cortisol levels act on the anterior pituitary and probably also the hypothalamus to inhibit the release of CRH and ACTH, thereby reducing the rate of secretion of cortisol. Loss of ACTH or removal of the pituitary (hypophysectomy) results in gradual atrophy of the zona fasciculata and zona reticularis regions of the adrenal cortex, while chronically high levels of ACTH cause the adrenal cortex to hypertrophy. These effects underline the trophic nature of the actions of ACTH on the structure and function of both the zona fasciculata and the zona reticularis.

ACTH secretion shows a distinct circadian rhythm related to the sleep–wakefulness cycle, and this is reflected in the pattern of cortisol secretion (as shown in Figure 20.2 for more discussion of

biological clocks, see Chapter 19). The concentration of cortisol in the plasma is minimal at around 3 a.m. and rises to a maximum between 6 and 8 a.m. before falling slowly during the rest of the day (see Chapter 19). About half of the total daily output of cortisol (i.e. 15–40 µmol) is released during the pre-dawn surge. Superimposed on this cycle is an episodic pattern of release characterized by short-lived fluctuations in output (pulsatile secretion). Around 7–8 larger secretory pulses are normally seen throughout each 24-hour period. The normal rhythm of cortisol release is interrupted by acute stress of any kind as a result of the direct stimulation of CRH secretion. Night shift work appears to blunt, though not abolish, the normal circadian pattern. This means that people having to work and remain alert during the night are doing so while their cortisol levels are at their lowest. Similarly, when they are trying to fall asleep after their shift their cortisol levels are high in preparation for the day ahead. This can result in an increased risk of errors of judgement during the night and disruption to sleep during the day.

Principal disorders of adrenal cortical hormone secretion

Overproduction of cortisol

Excessive cortisol in the plasma (hypercortisolism), also known as **Cushing's syndrome**, may result from a tumour of the adrenal gland itself, from increased output of ACTH resulting from a pituitary tumour (in which case it is known as Cushing's disease), or as a result of hypersecretion of CRH. It may also occur following the over-administration of steroid medications. In some cases, there is ectopic production of ACTH, for example from certain types of lung tumours. Patients with Cushing's syndrome have a characteristic appearance. There is often a redistribution of body fat, which includes increased deposition of fat in the face and trunk with a loss of fat elsewhere. This, combined with the wasting of muscle tissue resulting from increased protein mobilization, gives the typical 'melon on toothpicks' appearance seen in Figure 23.7. The skin of patients with Cushing's syndrome becomes thin and bruises easily, and there may be abnormal pigmentation. The disorder is also characterized by changes in carbohydrate and protein metabolism, hyperglycaemia, and hypertension. Long-term hypercortisolism is associated with an increased risk of diabetes mellitus, dyslipidaemia (i.e. abnormal amounts of lipids and lipoproteins in the blood), osteoporosis, and recurrent infections. The circadian rhythm of cortisol secretion is disturbed, along with the elevated basal level, and this can lead to an increased incidence of sleep disorders. Similar effects are sometimes seen in patients with chronic inflammatory disorders who are receiving prolonged treatment with corticosteroids. Treatment options for patients with Cushing's syndrome will depend on the cause of the problem but include surgery (if the cause is a tumour) or the use of cortisol-lowering drugs such as **ketoconazole** and **metyrapone**.

Mineralocorticoid excess

Over-production of aldosterone can arise either from hyperactivity of the cells of the zona glomerulosa (**Conn's syndrome** or primary hyperaldosteronism), or as a result of excessive renin

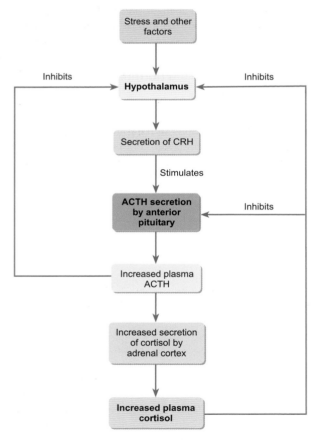

Figure 23.6 A flow chart summarizing the factors that regulate the secretion of glucocorticoids. Note that the red arrows represent negative feedback inhibition. Circulating glucocorticoids inhibit the secretion of corticotrophin releasing hormone (CRH) by the hypothalamus and the secretion of ACTH by the anterior pituitary. Circulating ACTH also probably inhibits the secretion of CRH.

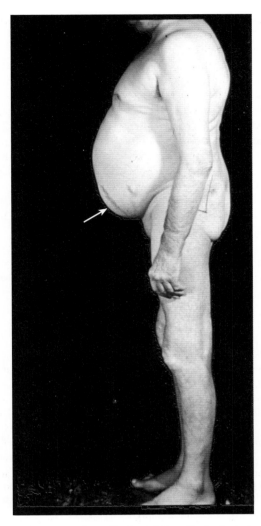

Figure 23.7 A picture of a patient exhibiting the typical physical characteristics associated with Cushing's syndrome. Note the swollen abdomen with characteristic linear bruising (striae—indicated by the white arrow) and thin, wasted legs. (Courtesy of the Wellcome Library.)

secretion (secondary hyperaldosteronism). The consequences of aldosterone excess include retention of sodium, loss of potassium (hypokalaemia), and alkalosis. Hypertension results from the expansion of the plasma volume, which follows the increased reabsorption of sodium. Spironolactone, an antagonist of aldosterone, can be helpful in treating mineralocorticoid excess.

Excessive production of adrenal androgens

The effects of excess androgen secretion can be very distressing to both men and women. It occurs most often because of overproduction of ACTH either by the anterior pituitary itself, or by an ACTH-secreting tumour. Occasionally an adrenal tumour or a fault in steroid biosynthesis as in congenital adrenal hyperplasia will lead to androgen excess. The principal signs of androgen excess are acne,

frontal baldness, and hirsutism (excessive growth of abdominal and facial hair). Males may experience a reduction in testicular volume due to negative feedback inhibition of pituitary gonadotrophins by the androgens, while females may show enlargement of the clitoris and virilization of the secondary sex characteristics. Excessive adrenal sex steroid production in children may bring about precocious puberty.

Deficiency of adrenal cortical hormones

The major deficiency disorder of the adrenal cortex is called **Addison's disease**. It is a comparatively rare condition (around 10–40 cases per million population) that may arise because of damage to the adrenal glands, autoimmune disease, or pituitary damage. It usually involves deficits in both glucocorticoids and mineralocorticoids, and the symptoms of the disease may be predicted from the preceding account of the actions of these steroids. People with Addison's disease tend to show progressive weakness, lassitude, and loss of weight. Their plasma glucose and sodium levels drop, while potassium levels rise. Severe dehydration and hypotension are common, and an increased white blood cell count is often seen. Unless the disease is caused by damage to the pituitary gland, patients with Addison's disease characteristically exhibit hyperpigmentation of the skin and mucosal membranes of the mouth. This is because, freed from the negative feedback effects of the adrenal corticosteroids, the anterior pituitary secretes more and more ACTH. ACTH is derived from a large precursor molecule called pro-opiomelanocortin (see Chapter 21) which also gives rise to melanocyte stimulating hormone (MSH). Consequently, an increase in ACTH secretion is accompanied by a parallel increase in MSH production.

Corticosteroid replacement therapy at physiological doses is the usual treatment. Dose regimes that mirror as far as possible the natural circadian rhythm of cortisol secretion, or sustained release preparations, are the most effective. Figure 23.8 shows the characteristic appearance of a patient suffering from Addison's disease.

Summary

The adrenal gland is a composite gland consisting of the outer cortex and inner medulla. The cortex secretes glucocorticoids, mineralocorticoids, and small quantities of sex steroids. The medulla secretes the catecholamines adrenaline and noradrenaline.

In humans, aldosterone is the chief mineralocorticoid and cortisol is the dominant glucocorticoid. Glucocorticoid secretion is regulated by pituitary ACTH. Aldosterone acts to conserve body sodium by stimulating its reabsorption in exchange for potassium in the distal nephron. Cortisol is essential to life and forms a vital part of the body's response to stress of all kinds. It stimulates gluconeogenesis and glycogen production as well as lipolysis. High levels of cortisol suppress the immune response to infection by reducing the mass of lymphoid tissue. Glucocorticoids are also used clinically as anti-inflammatory drugs.

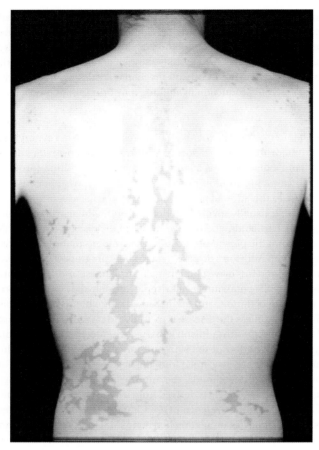

Figure 23.8 The characteristic pigmentation of the skin of a patient suffering from Addison's disease. (Courtesy of the Wellcome Library.)

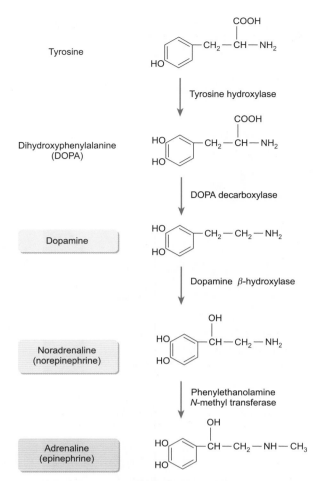

Figure 23.9 A schematic diagram showing the main steps in the synthesis of adrenaline and noradrenaline from tyrosine in the adrenal gland.

23.4 The adrenal medulla

In essence, the adrenal medulla represents an enlarged and specialized sympathetic ganglion that secretes the catecholamine hormones **adrenaline** (epinephrine) and **noradrenaline** (norepinephrine) into the circulation. It is composed chiefly of **chromaffin cells**, which may be considered as specialized sympathetic postganglionic neurons. As for other preganglionic nerves, the transmitter secreted by the splanchnic nerve fibres onto the chromaffin cells is acetylcholine (see Chapter 11).

The chromaffin cells are filled with storage granules that contain either adrenaline or noradrenaline, dopamine β-hydroxylase (one of the enzymes important in the synthesis of catecholamines—see Figure 23.9), ATP, and a variety of opioid peptides such as met-enkephalin and leu-enkephalin. These are secreted by exocytosis into the bloodstream in response to acetylcholine released from the splanchnic nerve terminals. About 80 per cent of the secreted catecholamine is adrenaline (epinephrine), while the remaining 20 per cent is noradrenaline (norepinephrine). Catecholamine release occurs as part of a general sympathetic stimulation and is of particular importance in preparing the body for coping with acute stress (the fight, flight, or fright response).

Adrenaline and noradrenaline secreted from the adrenal medulla are inactivated extremely rapidly and have half-lives in the plasma in the range of only 1–3 minutes. They are either taken up into sympathetic nerve terminals or inactivated by the enzymes catechol-O-methyltransferase and monoamine oxidase in tissues such as the liver, kidneys, and brain.

The actions of adrenal medullary catecholamines

The physiological effects of adrenal medullary catecholamines must be considered as part of an overall sympathetic response, since their release is always associated with an increase in the secretion of noradrenaline from sympathetic nerve terminals. While the adrenal medulla is not vital for survival, it does contribute to the response of the body to stress. Adrenaline and noradrenaline exert slightly different effects, which are summarized in Table 23.1.

Adrenaline and noradrenaline act preferentially on different types of adrenergic receptors (or **adrenoceptors**). These are α- and

Table 23.1 The efficacy of the catecholamine hormones in modulating various physiological processes

Adrenaline > noradrenaline	Noradrenaline > adrenaline
↑ glycogenolysis (β_2)	↑ gluconeogenesis (α_1)
↑ lipolysis (β_3)	
↑ calorigenesis (β_1)	
↑ insulin secretion (β_2)	↓ insulin secretion (α_2)
↑ glucagon secretion (β_2)	
↑ K$^+$ uptake by muscle (β_2)	
↑ heart rate (β_1)	
↑ contractility of cardiac muscle (β_1)	
↓ arteriolar tone in skeletal muscle (β_2)	↑ arteriolar tone in non-muscle vascular beds (α_1) leading to vasoconstriction and ↑ BP
	↑ tone in gastrointestinal sphincters (α_1)
↓ tone in non-sphincter GI smooth muscle (β_1)	↓ tone in non-sphincter smooth muscle (α_1)
↓ tone in bronchial smooth muscle (bronchodilatation) (β_2)	↑ tone in bronchial smooth muscle (bronchoconstriction) (α_1)

Note: The adrenoceptors mediating the various effects are shown in brackets. An **up arrow** indicates an increase, and a **down arrow** a decrease, in the specified physiological process. For a discussion of adrenergic receptor subtypes, see Chapter 11.

β-receptors, which are further subdivided into α_1, α_2, β_1, β_2, and β_3 receptors (see Chapter 11). Adrenaline interacts primarily with β-receptors, while noradrenaline binds preferentially to α- and β_1-receptors. The receptors that mediate the various actions of the adrenal medullary hormones, where known, are indicated in Table 23.1. The various classes of adrenergic receptors provide a mechanism by which the same adrenergic hormone or neurotransmitter can exert differing effects on different effector cells.

Both agonists and antagonists of the α- and β-adrenoceptors are widely used in clinical medicine. For example, β-antagonists ('β-blockers'), such as atenolol (which binds mainly to β_1 receptors), are often prescribed to reduce cardiac output in the treatment of high blood pressure. β_2-agonists, such as salbutamol, are administered to asthmatic patients to bring about bronchodilatation.

Both adrenaline and noradrenaline raise the systolic blood pressure by stimulating heart rate and contractility, thereby increasing cardiac output. Adrenaline, however, reduces diastolic pressure as a result of causing vasodilatation of certain vessels, particularly those of skeletal muscle, while noradrenaline raises diastolic pressure by causing a more generalized vasoconstriction. Both catecholamines cause piloerection and dilatation of the pupils. Adrenaline also acts as a bronchodilator and reduces the motility of the gut.

Adrenaline exerts important metabolic effects: it promotes the breakdown of glycogen in the liver (glycogenolysis), lipolysis, oxygen consumption, and calorigenesis. In this respect, its actions are similar to those of the thyroid hormones with which it synergizes. Noradrenaline is also a potent stimulator of lipolysis but has little effect on glycogenolysis.

Control of adrenal medullary secretion

While basal secretion of adrenal medullary catecholamines is probably very small, the gland may be stimulated to release its hormones in response to a number of stressful situations. These include exercise, hypoglycaemia, cold, haemorrhage, and hypotension. Secretion may also accompany emotional reactions such as fear, anger, pain, and sexual arousal, while the fetal adrenal medulla seems to respond directly to **hypoxia**. With the exception of this direct response, catecholamine secretion is mediated by the activity of the splanchnic nerves. The adrenal medulla becomes non-functional if these nerves are cut.

Disorders of adrenal medullary secretion

While under-production of the adrenal medullary hormones is not a clinical problem, excessive catecholamine output can have serious consequences. It can arise as the result of a tumour of the chromaffin tissue, a **phaeochromocytoma**. The principal symptoms of such a tumour are severe hypertension, which may be episodic or sustained, hyperglycaemia, and a raised metabolic rate. In addition, there is often anxiety, tremor, arrhythmias, and sweating. Surgical removal of the tumour, though a difficult procedure, is the usual treatment, although most of the symptoms can be alleviated by the administration of drugs such as phenoxybenzamine, an adrenergic α-antagonist, and propranolol, a non-selective β-blocker.

Summary

The adrenal medulla is composed of chromaffin cells that secrete adrenaline and noradrenaline as part of a general sympathetic response to stress. They cause an increase in heart rate, contractility, and cardiac output. Adrenaline is also a bronchodilator and reduces gut motility. It promotes glycogenolysis, lipolysis, and oxygen consumption.

✻ Checklist of key terms and concepts

The adrenal cortex

- The adrenal gland is a composite gland consisting of the outer cortex and inner medulla.

- The cortex secretes glucocorticoids, mineralocorticoids, and small quantities of sex steroids.
- Aldosterone is the chief mineralocorticoid and cortisol is the dominant glucocorticoid.

Aldosterone

- Aldosterone is secreted as part of the renin–angiotensin–aldosterone cascade.

- Aldosterone acts to conserve body sodium by stimulating its reabsorption in exchange for potassium in the distal nephron.

- Aldosterone also stimulates sodium reabsorption in salivary glands, sweat glands, and the gastrointestinal tract.

Cortisol

- Cortisol has crucial metabolic actions and forms a vital part of the body's response to stress of all kinds.

- Cortisol stimulates gluconeogenesis and glycogen production as well as lipolysis.

- High levels of cortisol suppress the immune response to infection by reducing the mass of lymphoid tissue.

- Glucocorticoids are used clinically as anti-inflammatory drugs.

- The secretion of the adrenal glucocorticoids is regulated by negative feedback from pituitary ACTH.

The adrenal medulla

- The adrenal medulla is composed of chromaffin cells that secrete the catecholamines adrenaline and noradrenaline as part of a general sympathetic response to stress.

- Adrenaline and noradrenaline cause increases in heart rate, contractility, and cardiac output.

- Adrenaline is also a bronchodilator and reduces gut motility. It promotes glycogenolysis, lipolysis, and oxygen consumption.

 Recommended reading

Histology

Mescher, A. (2015) *Junquieira's basic histology* (14th edn), Chapter 20. McGraw-Hill, New York.
   A recent revision of Junquieira and Carneiro's classic text. Good text, fewer pictures than some alternatives.

Pharmacology

Rang, H.P., Ritter, J.M., Flower, R.J., and Henderson, G. (2015) *Pharmacology* (8th edn), Chapter 32. Churchill-Livingstone, Edinburgh.
   Chapter 32 provides an easily digested introduction to the endocrine disorders discussed above and a helpful introduction to their treatment.

Endocrine physiology

Jameson, J.L. (ed) (2006) *Harrison's endocrinology*, Chapters 4, 5, and 23–25. McGraw-Hill Medical, New York.

Kovacs, W.J., and Ojeda, S.R. (2012) *Textbook of endocrine physiology* (6th edn), Chapter 13. Oxford University Press, Oxford.
   An authoritative account of the anatomy and physiology of the adrenal gland.

Medicine

Hall, R., and Evered, D.C. (1990) *A colour atlas of endocrinology* (2nd edn). Wolfe Medical, London.
   An informative collection of colour plates illustrating the physical changes associated with the endocrine disorders mentioned in this chapter.

Kumar, V., Abbas, A.K., and Aster, J.C. (eds) (2013) Chapter 19, in: *Robbins basic pathology* (9th edn). Saunders, New York.

Ledingham, J.G.G., and Warrell, D.A. (2000) *Concise Oxford textbook of medicine*, Chapters 7.3, 7.5, and 7.6. Oxford University Press, Oxford.
   An excellent introduction to thyroid, parathyroid, and adrenal disorders and their treatment.

Review articles

Chung, S., Son, G.H., and Kim, K. (2011) Circadian rhythm of adrenal glucocorticoid: Its regulation and clinical implications. *BBA—Molecular Basis of Disease* **1812**, 581–91.
   An interesting overview with references to many additional relevant sources.

Groeneweg, F.L., Karst, H., de Kloet, E.R., and Joels, M. (2011) Rapid non-genomic effects of corticosteroids and their role in the central stress response. *Journal of Endocrinology* **209**, 153–67.

Hartmann, K., Koenen, M., Schauer, S., Wittig-Blaich, S., Ahmad, M., Baschant, U., and Tuckermann, J.P. (2016) Molecular actions of glucocorticoids in cartilage and bone during health, disease, and steroid therapy. *Physiological Reviews* **96**, 409–47, doi:10.1152/physrev.00011.2015.

Sacta, M.A., Chinenov, Y., and Rogatsky, I. (2016) Glucocorticoid signaling: An update from a genomic perspective. *Annual Review of Physiology* **78**, 155–80, doi:10.1146/annurev-physiol-021115-105323.

 To check that you have mastered the key concepts presented in this chapter, complete the accompanying online self-assessment questions. Go to www.oup.com/uk/pocock5e/

CHAPTER 24

The endocrine pancreas and the regulation of plasma glucose

Chapter contents

This chapter should help you to understand:

- The mechanisms involved in the maintenance of plasma glucose levels including glycogen storage, glycogenolysis, and gluconeogenesis

- The reciprocal actions of the pancreatic hormones insulin and glucagon that bring about plasma glucose regulation

- The actions of other hormones concerned with the regulation of plasma glucose

- The mechanisms involved in the regulation of plasma glucose during the absorptive and post-absorptive states

- The consequences of hyperglycaemia, with particular reference to type 1 and type 2 diabetes mellitus

- The consequences of hypoglycaemia

24.1 Introduction

To carry out their normal metabolism, the cells of the body must have continuous access to glucose, the major fuel used to produce cellular energy (ATP). Indeed, certain tissues, notably the central nervous system, the retina, and the germinal epithelium, rely almost entirely on glucose metabolism for the generation of ATP. The nervous system alone requires around 110 g of glucose each day to meet its metabolic needs. It is therefore vital that the concentration gradient for glucose between the blood and the extracellular environment of the brain cells is maintained. Blood glucose levels are normally maintained between 70 and 125 mg glucose dl^{-1} blood (3.5–7.5 mmol l^{-1}) despite wide fluctuations in dietary intake (this is referred to as **normoglycaemia**) and regulation takes place continuously over a timescale of minutes. A variety of hormones contributes to this short-term regulation of plasma glucose, the most important of which are insulin and glucagon, synthesized and secreted by cells within the endocrine tissue of the pancreas. Longer-term regulation involves adrenaline, glucocorticoids, growth hormone, and the thyroid hormones.

24.2 Whole body handling of glucose

Following their absorption from the gastrointestinal tract, the nutrient sugars, including glucose, are transported directly to the liver through the portal vein (see Chapter 45). Since it is able both to store glucose as glycogen and to synthesize it from non-carbohydrate precursors, the liver acts as a buffer system to regulate blood sugar levels. When plasma glucose is raised, the liver removes glucose from the blood and stores it for future use, and when plasma glucose falls, the liver releases glucose into the circulation.

Glucose may be stored as glycogen or as fat

Glucose and other nutrient sugars (e.g. fructose) arriving at the liver may be broken down by **glycolysis** to intermediates, which may be used as energy sources, or converted to glycogen (in the process called **glycogenesis**), which forms the liver's store of carbohydrate. The relative amounts utilized and stored will depend on the current plasma level of glucose and the glucose requirements of the tissues at the time. Smaller amounts of glucose are also stored as glycogen in other tissues, particularly kidney, skeletal muscle, skin, and certain glands. The total body glycogen store normally amounts to 1500–2000 kcal (around 6300–8400 kJ). When the body's glycogen stores are fully saturated, any excess plasma glucose is converted in the liver to fatty acids, which are transported by lipoproteins to the adipose tissue to be stored in adipocytes (fat cells) in the form of triglycerides. In cells that do not manufacture glycogen, glucose is metabolized by enzymes involved in the glycolytic pathway to form pyruvate, which enters the tricarboxylic acid cycle and is metabolized to produce ATP.

Liberation of stored glucose by glycogenolysis

All cells that store glycogen are capable of utilizing it for their own metabolism, but the cells of the liver and kidney also release glucose into the general circulation so that it can be used by other cells. When plasma glucose falls during a period of fasting, the glycogen in these cells is broken down by phosphorylase enzymes to glucose, which is then released into the blood to provide a source of energy for the glucose-dependent tissues.

Glucose can be synthesized from non-carbohydrate precursors

Cells of the liver and kidney are able to synthesize glucose from non-carbohydrate precursors such as glycerol, lactate, and certain amino acids (a process known as **gluconeogenesis**). The kidneys, however, only become a significant source of plasma glucose in times of starvation. Under most conditions, the liver is the chief source of glucose for the general circulating pool and plays a crucial role in supplying the brain with the glucose on which it depends. The principal pathways involved in glucose metabolism are discussed in Chapter 4 and summarized in Figure 24.1.

Figure 24.1 also illustrates two pathways of glucose metabolism that may become significant when plasma glucose is elevated. These are the **protein glycosylation pathway**, in which certain proteins, especially haemoglobin, become glycosylated, and the **polyol pathway**. Here, glucose is converted to sorbitol, which is then oxidized slowly to fructose. This occurs chiefly in the retina, lens, Schwann cells, kidney, and aorta.

Many endocrine systems can influence the metabolic pathways in order to ensure that plasma glucose levels are maintained within the normal range. In view of its importance in cerebral metabolism, it is vital that glucose levels should not be allowed to fall too low, and all but one of the hormones concerned with

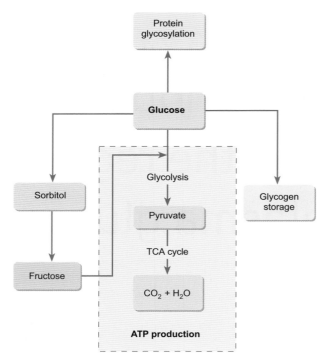

Figure 24.1 A schematic diagram outlining the principal pathways involved in glucose metabolism.

glucose metabolism act to raise plasma glucose. These so-called **glucogenic** hormones include glucagon, the catecholamines (adrenaline and noradrenaline), the glucocorticoids (chiefly cortisol), the thyroid hormones, and certain anterior pituitary hormones, notably growth hormone. By contrast, there is only one hormone whose action elicits a decrease in plasma glucose—pancreatic insulin. Nevertheless, this hormone plays a key role both in the overall homeostasis of plasma glucose and in regulating glucose uptake by non-neural tissues.

Summary

Glucose is an essential metabolic substrate, particularly for the neurons of the CNS. Plasma glucose is closely regulated between 3.5 and 7.5 mmol l^{-1}, by a number of hormones. These include insulin, glucagon, glucocorticoids, catecholamines, GH, and thyroid hormones. All except insulin are hyperglycaemic (or glucogenic) in their actions, i.e. they act to raise the plasma glucose concentration.

Plasma glucose is determined by the balance between intestinal absorption, storage, and synthesis by the liver and uptake by other tissues. Glucose can be stored in the form of glycogen—the only form of carbohydrate storage in the body.

Under appropriate conditions, glucose can be synthesized from non-carbohydrate precursors such as amino acids and glycerol. This is the process of gluconeogenesis and takes place mainly in the liver.

24.3 Insulin and glucagon provide short-term regulation of plasma glucose levels

The pancreas is both an exocrine and an endocrine organ. The exocrine (acinar) tissue, which forms the bulk of the gland, secretes a range of digestive enzymes into the duodenum. These are discussed in Chapter 44, and the location and anatomical relationships of the pancreas are shown in Figure 44.18. Dispersed throughout the pancreas and occupying about 2 per cent of its total mass are small clumps of endocrine tissue known as the **islets of Langerhans**. These are between 100 and 200 μm in diameter and are surrounded by the exocrine (acinar) tissue (see Figure 24.2). The islets secrete insulin and glucagon, which are peptide hormones vital to the regulation of plasma glucose levels. The islets are highly vascular, and are innervated by the sympathetic and parasympathetic branches of the autonomic nervous system. Three main cell types have been identified within the islets. These are the α-, β-, and δ-cells. Glucagon and insulin are synthesized, stored, and secreted by the α- and β-cells respectively, while the δ-cells secrete somatostatin (which acts to inhibit the secretion of both hormones, possibly by a paracrine action). A fourth type of cell within the islets, known as F cells or PP cells, synthesize the hormone **pancreatic polypeptide** which plays a role in regulating the digestive functions of the pancreas (see Chapter 44). Figure 24.2 (b) illustrates pancreatic islet tissue stained for glucagon and insulin respectively. From this it is clear that the most common cells in the islets are the β-cells (about 75 per cent of the total) followed by the α-cells (up to 20 per cent of the total). The δ-cells account for about 5 per cent while the F- or PP-cells are relatively rare, accounting for only 1 per cent or so of the islet cells.

Insulin

Insulin is a small protein hormone (51 amino acids, $M_r \sim 5{,}800$) derived from a larger single-chain precursor called proinsulin, which is synthesized in the rough endoplasmic reticulum of the pancreatic β-cells. Most of this precursor molecule is converted to insulin in the Golgi apparatus as a result of proteolytic enzyme cleavage, to form a structure in which two amino acid chains are joined by two disulphide (S–S) bridges. The hormone is then stored within granules in the β-cells until it is secreted into the bloodstream by exocytosis in response to an appropriate stimulus.

Much of the circulating insulin is loosely bound to a β-globulin, but the half-life of insulin within the plasma is short (around 5 minutes) because it is avidly taken up by tissues, particularly those of the liver, kidneys, muscle, and fat. Very little insulin appears in the urine, despite its low molecular mass.

Insulin secretion in response to glucose

Although a large number of nutrients and hormones can alter the output of insulin from the pancreatic β-cells, glucose is of greatest importance in humans. Glucose is believed to act directly on the islet cells to stimulate the secretion of insulin. At low plasma glucose levels (0–3 mmol l^{-1}) the β-cells have a resting potential of about −60 mV. As plasma glucose is raised above 3 mmol l^{-1}, there is a progressive depolarization of the β-cells as glucose is transported into the cells via the glucose transporters of the GLUT family (GLUT1 and, in particular, GLUT2). The glucose is then rapidly phosphorylated by glucokinase, and intracellular ATP levels rise. The increase in intracellular ATP results in the closure of K^+-ATP channels and depolarization of the β-cells. When this depolarization reaches threshold (around −55 mV), the β-cells generate action potentials superimposed on slow waves. The action potentials

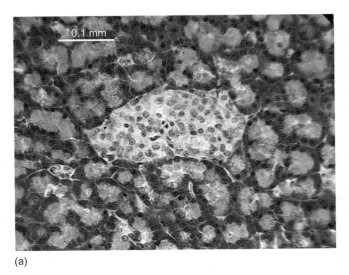

(a)

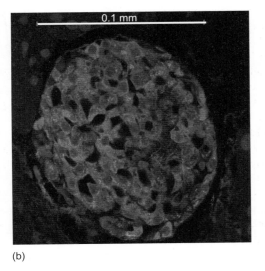

(b)

Figure 24.2 (a) shows a pale-staining pancreatic islet surrounded by typical acinar tissue. The cytoplasm and nuclei of the acinar cells stain strongly, while the central secretory region stains more lightly. (b) shows a pancreatic islet stained for glucagon (red) and insulin (green). The cell nuclei are stained blue. The glucagon containing cells are less numerous than those containing insulin. (Panel (b) is from Wikipedia commons.)

cause voltage-gated calcium channels to open, allowing calcium ions to enter the cells and trigger the release of insulin by exocytosis. A separate K-independent mechanism of stimulus–secretion coupling may also be important but is less well understood.

During fasting, when plasma glucose is relatively low (around 3–4 mmol l^{-1}), insulin is secreted at a very low rate and is barely detectable in the blood; but following a meal, insulin secretion increases as plasma glucose rises (see panels (a) and (b) of Figure 24.3). After a typical meal, plasma insulin rises 3–10-fold and usually peaks 30–60 minutes after eating begins. The secretion of insulin in response to a glucose load is normally biphasic in nature. The early rise in the rate of secretion (phase 1) reflects the release of available insulin, while the later rise (phase 2) is believed to depend on the synthesis of new insulin in response to the glucose load. Plasma glucose reaches a maximum about an hour after ingestion of a meal; it then declines until it falls slightly below the normal fasting level, before returning to normal (see panel (a) of Figure 24.3). The close link between insulin secretion and the plasma glucose concentration prevents the latter from reaching excessively high values.

The regulation of insulin secretion in response to glucose

When glucose is given orally, a greater insulin response is elicited than when it is infused intravenously. This is believed to be the result of the actions of certain other gastrointestinal hormones, sometimes called **incretins**, which seem to enhance the β-cell response to glucose. They include glucagon-like peptides GLP-1, GLP-2, and gastric inhibitory polypeptide (or GIP), which are secreted by cells of the duodenum along with **secretin** and cholecystokinin (CCK) as well as gastrin, secreted by the G cells of the stomach. Glucagon itself also stimulates the secretion of insulin when plasma glucose is high. These hormones are probably important in preventing a large rise in plasma glucose immediately following the absorption of a carbohydrate-rich meal. Somatostatin is a potent inhibitor of insulin secretion, but its exact role is uncertain.

The autonomic nervous inputs to the pancreatic β-cells also seem to play a role in the regulation of insulin release. The major effect of sympathetic stimulation (and circulating catecholamines) is a reduction in insulin release (seen for instance during stress) which is mediated via α_2- adrenoceptors on the pancreatic β-cells. Parasympathetic (vagal) stimulation enhances the rate of insulin secretion.

Certain amino acids (particularly leucine) are able to stimulate the secretion of insulin. Amino acid uptake into cells is stimulated as a result of the increased insulin release and there is a net stimulation of protein synthesis. The major factors that control the secretion of insulin are summarized diagrammatically in Figure 24.4.

The hypoglycaemic actions of insulin

The major targets for insulin action are the liver, the adipose tissue, and the muscle mass. The insulin receptor consists of two extracellular α-subunits and two intracellular β-subunits, which have an intrinsic tyrosine kinase activity. When insulin binds to its receptors on the surfaces of cells, the tyrosine residues of the β-subunits

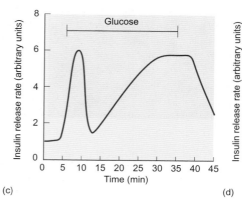

(a)

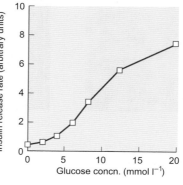

(b)

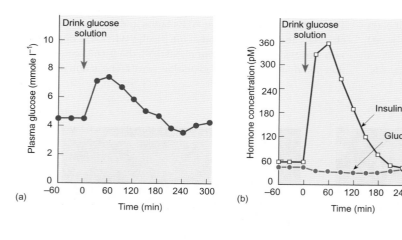

(c)

(d)

Figure 24.3 The relationship between plasma glucose and insulin secretion. Panels (a) and (b) show the changes in plasma glucose, insulin, and glucagon following a carbohydrate-rich meal. Note the initial rise and rapid fall in both plasma glucose and plasma insulin. During the peak insulin secretion, glucagon levels decline. Panel (c) shows the two phases of insulin secretion (expressed in arbitrary units) following a glucose load. The initial phase reflects secretion of preformed insulin, while the secondary rise is due to secretion of newly synthesized insulin. Panel (d) plots the rate of insulin secretion as a function of the plasma glucose concentration.

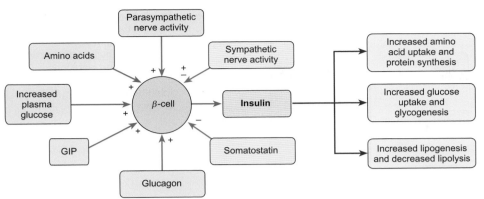

Figure 24.4. A summary of the major factors involved in the regulation of insulin secretion by the β-cells of the pancreatic islets and the principal actions of insulin; (+) = enhancement of secretion; (−) = inhibition of secretion; GIP = gastric inhibitory polypeptide.

undergo auto-phosphorylation (i.e. they are phosphorylated by the β-subunits themselves) together with other cellular proteins. The resulting signalling cascade leads to the insertion of preformed glucose carriers (GLUT4) into the plasma membrane from intracellular vesicles. The overall result is an increase in the rate of uptake of glucose by the insulin responsive cells, notably the adipocytes and muscle cells. Some of the glucose is used for metabolism, while the remainder enters anabolic pathways leading to the conversion of glucose to glycogen and fat. As glucose enters the cells, plasma levels fall once more and, after a short lag, insulin secretion is inhibited and returns to basal levels. The facilitated glucose uptake is terminated by the rapid endocytosis of GLUT4 carriers to return them to their intracellular storage compartments.

Actions of insulin on the nervous system

The consumption of glucose by the central nervous system is independent of insulin. Certain areas of the brain, however, possess dense populations of insulin receptors, and it is possible that insulin plays a role in regulating aspects of normal brain function. Neurons within the so-called 'satiety centre' of the hypothalamus, for example, increase their rates of discharge in response to insulin, before showing a decline in firing as the plasma glucose level falls in response to insulin. This may indicate a role for insulin in the regulation of appetite. Of particular interest is the recent finding that insulin may modulate glucose utilization and contribute to **synaptogenesis** in brain areas such as the hippocampus and frontal cortex. It has been suggested that reduced levels of insulin or of insulin activity may contribute to a number of the pathological processes that bring about the memory impairment characteristic of Alzheimer's disease and other forms of dementia. Intranasal administration of insulin has been shown to benefit some patients with early-stage Alzheimer's disease or mild cognitive impairment with memory loss.

Summary

Insulin is synthesized by the β-cells of the pancreatic islets and is secreted chiefly in response to a rise in plasma glucose and amino acids. Insulin stimulates the uptake and utilization of glucose by cells (particularly liver, adipose tissue, and skeletal muscle), thereby causing a fall in plasma glucose. It stimulates both glycogen and protein synthesis.

Glucagon

Although there are a number of hyperglycaemic hormones, the most potent is glucagon, many of whose actions directly oppose those of insulin. Glucagon is a 29 amino acid, single-chain polypeptide hormone (M_r 3485) synthesized by the α-cells of the pancreatic islets and released from them by exocytosis. Like insulin, it has a short half-life in the plasma (around 6 minutes).

The regulation of glucagon secretion

Glucagon is secreted by the pancreatic α-cells in response to lowered plasma glucose and acts to raise circulating plasma glucose levels. The cellular mechanisms by which glucagon secretion occurs in response to hypoglycaemia are poorly understood. However, insulin appears to modulate the efficacy of glucose as a stimulus for glucagon release—glucagon secretion is stimulated much more by hypoglycaemia if insulin is absent. Indeed, insulin appears to inhibit glucagon secretion directly (see Figure 24.3b). The factors influencing the secretion of glucagon are summarized diagrammatically in Figure 24.5. Glucagon secretion is powerfully stimulated by certain amino acids (especially arginine and alanine), by both parasympathetic and sympathetic stimulation, and by the gastrointestinal hormone CCK. Somatostatin inhibits glucagon release.

The hyperglycaemic actions of glucagon

Glucagon exerts effects on the metabolism of carbohydrate, fat, and protein. Its actions are hyperglycaemic, i.e. they raise plasma levels of glucose. It also acts to maintain adequate circulating levels of other energy substrates during periods of fasting, thus sparing glucose for use by dependent tissues such as the brain. Glucagon receptors are found chiefly in the liver and kidneys but are also present in the heart, adipose tissue, gastrointestinal tract, and brain. However, the major target of glucagon is the liver, where it stimulates glycogenolysis and inhibits glycogen synthesis.

Glucagon binds to a G protein coupled receptor on the plasma membrane of the hepatocytes and activates adenylyl cyclase, which initiates a signal cascade that begins with an increase in intracellular cyclic AMP followed by activation of protein kinase A. In turn, protein kinase A activates another enzyme, phosphorylase kinase. This phosphorylates glycogen phosphorylase, converting it

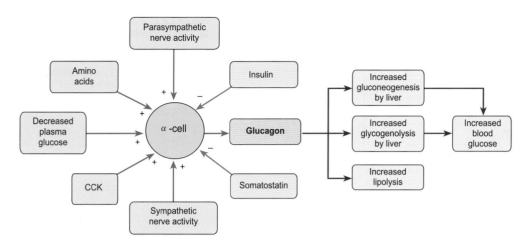

Figure 24.5 A summary of the main factors controlling the secretion of glucagon from the α-cells and the principal actions of glucagon on metabolism; (+) = enhancement of secretion; (−) = inhibition of secretion; CCK = cholecystokinin.

to its active form that is able to break down glycogen and liberate glucose 1-phosphate, which is converted to glucose 6-phosphate by the enzyme phosphoglucomutase. Glucose 6-phosphate can directly enter the glycolytic pathway for ATP production, but its principal fate is dephosphorylation by glucose 6-phosphatase to liberate free glucose that can enter the blood and thus be utilized by other tissues. Note that adrenaline acts on the liver to liberate glucose in a similar manner.

Hepatic uptake of certain amino acids and gluconeogenesis (the synthesis of new glucose from amino acids) are also enhanced, contributing further to the overall increase in plasma glucose. Glucagon has also been shown to exert a significant lipolytic effect, mobilizing fatty acids and glycerol from adipose tissue. This provides a ready supply of metabolic substrates, thus enabling glucose to be spared for use by the brain, as well as

providing glycerol, which can act as a precursor for glucose in the hepatic gluconeogenic pathway. Figure 24.6 presents an overview of the regulation of blood glucose by both insulin and glucagon.

Summary

Glucagon is the most potent hyperglycaemic hormone and acts in opposition to insulin to provide short-term regulation of plasma glucose. It is secreted by the α-cells of the pancreatic islets and acts to promote the release of glucose into the blood. Hypoglycaemia is the principal stimulus for the secretion of glucagon, which then stimulates glycogenolysis, lipolysis, and gluconeogenesis.

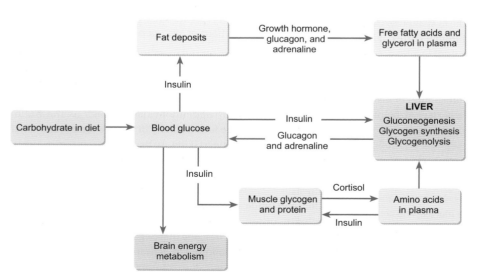

Figure 24.6 An overview of the hormonal regulation of plasma glucose concentration. During the absorptive state, insulin promotes the uptake of glucose by the liver, muscle, and adipose tissue. In the post-absorptive state, glucose levels are maintained by glycogenolysis in the liver (which is stimulated by glucagon and adrenaline) and by gluconeogenesis, which is regulated by cortisol and glucagon. Lipolysis makes fatty acids available for oxidation, and this process is promoted by growth hormone, glucagon, and adrenaline.

24.4 Other hormones involved in the regulation of plasma glucose

Insulin and glucagon play a pivotal role in the fine regulation of plasma glucose levels—indeed, insulin is the only hormone capable of lowering plasma glucose, and glucagon is the most important hyperglycaemic hormone. Nevertheless, a number of other agents also contribute to the maintenance of a stable blood glucose, as well as mobilizing glucose when necessary. These hormones include adrenal glucocorticoids, growth hormone, the catecholamines, and the thyroid hormones. The actions of all these regulators are described in detail elsewhere (particularly in Chapters 22 and 23) and so their effects on plasma glucose levels will be discussed only briefly here.

The role of the glucocorticoid hormones in plasma glucose regulation

The adrenal cortical hormones are steroids, so they are able to cross the plasma membrane of their target cells and bind to their intracellular receptors. The hormone–receptor complex then binds reversibly with DNA and functions as a gene-activator, stimulating the production of appropriate mRNA, as described in Chapter 3. The mRNA then leaves the nucleus to promote the synthesis of specific proteins by the ribosomes. Hormones that exert their effects in this way tend to have a rather prolonged time-course of action, with a lag between stimulus and effect. The glucocorticoids, of which cortisol is the most important, do not, therefore, play a role in minute-to-minute regulation of plasma glucose levels in the same way as insulin and glucagon, but exert a longer-term influence—particularly in pathological conditions of steroid excess.

The primary role of the glucocorticoids in carbohydrate metabolism seems to be to maintain the reserves of glycogen in the liver (and to a lesser extent the heart and skeletal muscle). In so doing they have an overall hyperglycaemic effect and are secreted in response to a fall in plasma glucose. In addition to this, they also exert an anti-insulin action in peripheral tissues, particularly adipose and lymphoid tissue, where they inhibit glucose uptake. At the same time, they promote the release of free fatty acids by lipolysis, thereby increasing the supply of these substrates and reducing glucose utilization by many tissues, so sparing glucose for use by the brain.

The glucocorticoids promote glycogen production by stimulating gluconeogenesis. They do this both by enhancing amino acid release from skeletal muscle protein and by inducing the synthesis of specific glycogenic proteins by the liver (Figure 24.6).

The role of growth hormone in plasma glucose regulation

The actions of growth hormone on glucose metabolism become significant in times of fasting or starvation. Overall it has an anti-insulin, glucose-sparing effect that depresses glucose uptake by muscle tissue in particular. It also stimulates hepatic glycogenolysis, and increases lipolysis in adipose tissue. These actions provide fatty acids for metabolism while increasing the availability of glucose for use by the brain.

The role of thyroid hormones in plasma glucose regulation

As explained in Chapter 22, the effects of the thyroid hormones on carbohydrate metabolism are somewhat complex, and depend upon the levels of circulating hormone. All aspects of glucose metabolism seem to be enhanced by thyroid hormones, including glycogenolysis, gluconeogenesis, and absorption from the gastrointestinal tract. These actions will result in an increase in plasma glucose. At the same time, however, thyroid hormones increase the rate of glucose uptake by cells (particularly muscle) and enhance the rate of insulin-dependent glycogenesis, effects which will tend to reduce plasma glucose. In general, it appears that low concentrations of thyroid hormones are anabolic and tend to reduce plasma glucose, while higher doses are catabolic and hyperglycaemic.

Thyroid hormones are released in response to stresses of all kinds, including hypoglycaemia and starvation. Under such conditions, they are of importance, together with other hormones (particularly the catecholamines), in ensuring that glucose reserves are mobilized and that glucose-dependent tissues receive an adequate supply.

The role of the catecholamines in the regulation of plasma glucose

When plasma glucose falls below 4 mmol l^{-1}, there is an increase in the secretion of catecholamines from the adrenal medulla and an increased output of noradrenaline from sympathetic nerve terminals. Adrenaline and noradrenaline work alongside the glucocorticoids, GH, and glucagon to maintain plasma glucose levels during the period of hypoglycaemia and spare the available glucose for use by the brain. Indeed, catecholamines become very important in this regard if plasma glucose falls to very low levels or if glucagon secretion is impaired.

Adrenaline increases plasma glucose in two ways. It does so directly by stimulating the mobilization of glucose from hepatic glycogen (see Chapter 45) and indirectly by enhancing the rate of glucagon secretion and inhibiting the secretion of insulin. High concentrations of noradrenaline also inhibit insulin secretion (although low concentrations stimulate it), so the net effect of generalized sympathetic stimulation is a stimulation of the secretion of hyperglycaemic hormones and a depression of insulin secretion. The catecholamines exert a potent lipolytic action in adipose tissue, stimulating lipases to break down fats to liberate free fatty acids and glycerol. These metabolites may then be utilized in preference to glucose by many tissues. These changes result in an elevation of plasma glucose (Figure 24.6). Such an action of the catecholamines probably underpins the hyperglycaemia that is characteristic of severe injury.

Summary

During prolonged periods of fasting, hyperglycaemic hormones such as cortisol, catecholamines, GH, and thyroid hormones play a significant role. Their effects are geared to the maintenance of glycogen stores, the stimulation of gluconeogenesis, and the mobilization of fatty acids and proteins, providing other metabolic substrates for those tissues able to use them and sparing glucose for use by the CNS.

24.5 Plasma glucose regulation following a meal

Having discussed the various hormones which play a role in the regulation and cellular handling of glucose within the body, it may be helpful to examine the changes in plasma glucose that follow the ingestion of a glucose-rich meal and during fasting. Immediately prior to a meal, plasma glucose is likely to be relatively low, around 3–4 mmol l^{-1}, and the plasma concentration of insulin will be low. Levels of glucagon and the other hyperglycaemic hormones, however, will be elevated as the body acts to maintain the supply of glucose to the brain. If a glucose-rich meal is then eaten, and glucose is absorbed from the gastrointestinal tract, a glucose load is presented to the body as plasma levels rise (see Figure 24.3).

During the 90 minutes or so following a meal, the body is said to be in the **absorptive state of metabolism** and plasma glucose may rise to 7 or 8 mmol l^{-1}. The exact timing and extent of the rise will, however, depend upon the form of the ingested carbohydrate, the relative proportions of the other constituents of the meal (fats, proteins etc.), the time of day, and so on. Two hours or so after the meal, plasma glucose will start to fall once more, as glucose is utilized by cells or converted to its stored forms. As a rule, roughly 35 per cent of the glucose ingested will be oxidized, chiefly by the brain and muscle tissue, while the remaining 65 per cent will be stored. Glucose is stored principally in the form of glycogen, mostly in the liver but also in renal, muscle, and other tissues. If glycogen stores become saturated, the additional glucose is converted into fatty acids and stored in fat cells as triglycerides.

Insulin is the dominant regulator of the events underlying the changes in plasma glucose concentration in the period following a meal. Insulin is secreted in response to the initial rise in plasma glucose (see Figure 24.3). It brings about a fall in plasma glucose by the actions described earlier, notably an increase in cellular uptake of glucose, stimulation of glycogen production, and a reduction in gluconeogenesis. There is also decreased glycogenolysis, lipolysis, and ketogenesis.

Within 3–4 hours of the completion of a meal, we see the onset of processes whereby the body defends itself against hypoglycaemia. These are known as the **counter-regulatory effects** and, at this time, the body is said to have entered the **post-absorptive state of metabolism**. By now, plasma glucose will have fallen back to around 4 mmol l^{-1}, at or a little below the normal fasting level. Insulin secretion will be low and the secretion of the hyperglycaemic hormones will increase. Glucagon is the most important of these. It acts largely on the liver to stimulate the breakdown of glycogen and the release of glucose into the bloodstream. The glucose stored as glycogen in other tissues, such as muscle, cannot be released into the plasma but is metabolized directly by those tissues, thereby helping to conserve plasma glucose. At the same time, in response to adrenal glucocorticoids, amino acids are mobilized from muscle and used by the liver to create new glucose (gluconeogenesis).

The catecholamines, GH, and glucagon all stimulate lipolysis in adipose cells. The free fatty acids released can be used by the liver and muscle as a metabolic substrate in preference to glucose. Ketone bodies, produced in the liver by the metabolism of fatty acids, provide an energy source for muscle and, in longer periods of fasting, even for the brain. As a result of the counter-regulatory mechanisms described above, and summarized in Figure 24.6, plasma glucose rarely falls below 3 mmol l^{-1} even during quite prolonged periods of starvation.

Summary

In the absorptive phase of metabolism (just after a meal), plasma glucose is high, insulin levels are high, and the secretion of hyperglycaemic hormones is inhibited. During this phase, glycogen stores are replenished and glucose uptake and utilization by cells is high. In the post-absorptive state (3 hours or more after a meal), counter-regulatory mechanisms are initiated. Secretion of the hyperglycaemic hormones is enhanced, resulting in lipolysis (and, in prolonged fasting, gluconeogenesis) to spare glucose for use by the CNS.

24.6 Lack of pancreatic insulin results in diabetes mellitus

Diabetes mellitus (literally 'sweet siphon') is a condition that occurs when inadequate uptake of glucose by the cells of the body causes high levels of glucose in the blood (**hyperglycaemia**). It should be distinguished from diabetes insipidus, which is caused by a lack of vasopressin (ADH, see Chapter 21). In this chapter, any reference to diabetes should be taken to indicate diabetes mellitus, which is a relatively common disorder, affecting around 3.5 per cent of the UK population and 300 million people world-wide. Members of African, Asian, and Afro-Caribbean ethnic groups have an incidence of diabetes of more than 6 per cent. Its prevalence increases with age, and more than 10 per cent of people over 65 years of age have some form of diabetes mellitus.

A diagnosis of diabetes is usually made in patients with a fasting plasma glucose level in excess of 7 mmol l^{-1}. This diagnosis may be confirmed by the **oral glucose tolerance test**, in which a patient drinks 300 ml of a 25 per cent glucose solution (corresponding to an intake of 75 g of glucose). Blood samples are taken before ingestion of the glucose and at 30-minute intervals thereafter, and the plasma glucose concentration is measured. In a normal, healthy individual, blood glucose returns to normal within 2 hours of ingestion of a normal glucose load, while in diabetic individuals plasma glucose exceeds 11 mmol l^{-1} 2 hours after a meal—see Figure 24.7.

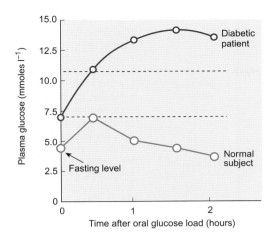

Figure 24.7 Changes in plasma glucose following a glucose load in a diabetic patient compared with those seen in a normal subject. The glucose load was administered at time zero. The dotted lines show the range of plasma glucose levels which prompt a diagnosis of diabetes.

Diabetes mellitus is a metabolic disorder that may result from defects in insulin secretion, insulin action, or both. Different forms of the disease are recognized, and patients are classified according to their disease characteristics. **Type 1 diabetes mellitus** is an organ-specific autoimmune disorder characterized by the destruction of pancreatic β-cells by cytotoxic T-lymphocytes (CTLs: see Chapter 26). CTL activation triggers the release of further cytokines that stimulate the proliferation of macrophages and auto-antibodies that are attracted to the site of inflammation. This exacerbates the destructive processes and results in the eventual death of the islet tissue. As a consequence, there is a complete loss of insulin secretion. The exact triggers for this autoimmune reaction remain unclear, but there is evidence that a viral infection or exposure to a specific chemical agent may be responsible in some cases for triggering the initial CTL activation. Genetic susceptibility may also be a factor. This type of diabetes normally develops before the age of 40 years, and often much earlier. Because type 1 diabetics produce little or no insulin of their own, they require daily treatment with exogenous insulin in order to survive. For this reason, type 1 diabetes was previously known as insulin-dependent diabetes mellitus (IDDM). This term is rarely used nowadays because many people with type 2 disease eventually require exogenous insulin to manage their condition.

A second form of diabetes mellitus known as **type 2 diabetes mellitus** is seen more often in older people and is strongly linked with obesity, particularly when fat is deposited preferentially around the abdomen (~80 per cent of sufferers are overweight). In this form of the disease there is a reduction in pancreatic insulin secretion and/or an increasing failure of the tissues to respond to insulin (i.e. insulin resistance). The link between obesity and type 2 diabetes is uncertain, but some studies have suggested that pro-inflammatory chemicals released from abdominal adipocytes may disrupt the function of insulin receptors and thereby reduce their sensitivity to insulin. Others point to hyperlipidaemia as a primary cause. There is no evidence of a viral trigger for type 2 diabetes, but at least 40 genetic mutations are linked with type 2 diabetes, and

concordance between genetically identical twin pairs is almost 100 per cent. Although this condition often worsens over time so that patients need to begin insulin treatment, type 2 diabetics can often be managed by careful monitoring, regulation of their diet, appropriate exercise, and the administration of oral hypoglycaemic drugs. This form of the disease was therefore known previously as non insulin-dependent diabetes mellitus (NIDDM).

Other, rarer forms of diabetes mellitus are also seen. These include gestational diabetes (which often resolves after delivery of the baby—see Chapter 49), drug-induced diabetes, genetic defects of the β-cells, and diabetes occurring as a consequence of diseases of the exocrine pancreas.

Acute symptoms of diabetes mellitus

Whatever the cause or type of the disease, the end result is the same—insulin deficiency or insulin resistance leads to chronic hyperglycaemia. In addition, since the hyperglycaemic suppression of glucagon secretion requires insulin, there is an increased secretion of glucagon. This tends to exacerbate the effects of the insulin deficiency.

Uncontrolled diabetes is accompanied by the following metabolic changes:

- an increase in glycogen breakdown;
- an increase in the rate of gluconeogenesis, which leads to weight loss as skeletal muscle proteins are catabolized to form glucose (this, together with the increased glycogen breakdown, contributes to the hyperglycaemia);
- an increase in the breakdown of fats (lipolysis) with increased production of ketone bodies (acetoacetate and β-hydroxybutyric acid).

The classical symptoms of type 1 diabetes are the results of these changes. They include the passage of an increased volume of urine (polyuria), intense thirst (polydipsia), and, in many cases, weight loss. The excessive plasma glucose levels lead to saturation of the renal glucose carriers so that glucose appears in the urine (**glycosuria**). Glycosuria induces an osmotic diuresis by reducing the amount of water reabsorbed in the distal part of the nephron (hence the polyuria—see Chapter 39). Not surprisingly, the excessive loss of water induces thirst. Significant dehydration is not normally a feature of type 2 diabetes, as water ingestion usually matches the loss of water via the urine. A severe and potentially life-threatening consequence of elevated plasma glucose levels in uncontrolled diabetes is diabetic ketoacidosis caused by the increased production of ketone bodies. These changes and the dire consequences of failing to treat the disease are summarized in Figure 24.8.

Consequences of chronically elevated plasma glucose

A number of chronic problems are associated with diabetes and are the result of prolonged hyperglycaemia. Tissues such as muscle and fat, which depend on insulin for glucose uptake and utilization, remain relatively unaffected by high levels of plasma glucose. In non-insulin-dependent tissues, however, glucose uptake

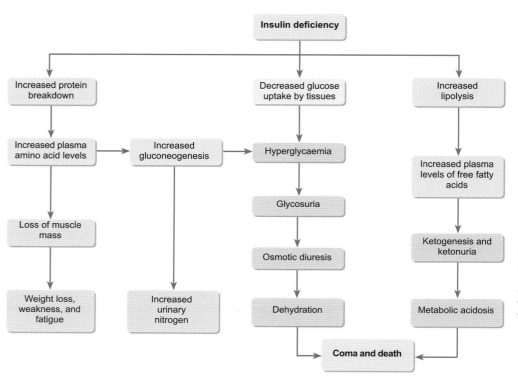

Insulin deficiency

Increased protein breakdown → Increased plasma amino acid levels → Loss of muscle mass → Weight loss, weakness, and fatigue

Increased plasma amino acid levels → Increased gluconeogenesis → Hyperglycaemia

Increased gluconeogenesis → Increased urinary nitrogen

Decreased glucose uptake by tissues → Hyperglycaemia → Glycosuria → Osmotic diuresis → Dehydration

Increased lipolysis → Increased plasma levels of free fatty acids → Ketogenesis and ketonuria → Metabolic acidosis

→ Coma and death

Figure 24.8 A flow chart showing the metabolic consequences of insulin deficiency. Lack of insulin causes hyperglycaemia, glycosuria, and osmotic diuresis leading to dehydration. It also increases the breakdown of triglycerides (lipolysis) to provide an alternative metabolic fuel. An increased production of ketone bodies is associated with the increased lipolysis that can result in severe ketoacidosis (a form of metabolic acidosis—see Chapter 41). As glucose reserves are lost, there is increased breakdown of muscle protein for gluconeogenesis. This leads to weight loss and an increased loss of nitrogen in the urine. Unless promptly treated, the dehydration and ketoacidosis will lead to coma and, eventually, to death.

depends only on the concentration gradient for glucose between the extracellular and intracellular fluid. In hyperglycaemia, this will tend to drive large amounts of glucose into the cells, resulting in over-supply of glucose. Over long periods of time, this over-supply can cause pathological changes in the affected tissues. These may be the result of increased protein glycosylation or they may result from the accumulation of sorbitol in cells, as described in section 24.2 and shown in Figure 24.1.

The most common chronic problems associated with both types of diabetes are changes to the lens of the eye, degenerative changes in the retina (**retinopathy**) and peripheral nerves (**peripheral neuropathy**), thickening of the filtration membrane of the nephron

(**nephropathy**), vascular lesions (micro- and macro-angiopathy) that increase the risks of peripheral vascular disease, cardiovascular disease, and chronic skin infections. The incidence of complications related to diabetes and chronic hyperglycaemia is extremely high. For example, 30 per cent of patients diagnosed before the age of 30 will eventually develop retinopathy, 40 per cent will develop nephropathy, 40 per cent will develop neuropathies, and 50 per cent of men will suffer erectile dysfunction or impotence. It is estimated that diabetes causes the premature death of around 20,000 people in the UK each year. An example of retinopathy is shown in Figure 24.9. This shows characteristic small haemorrhages dotted over the retina that are completely absent from the

Normal retina

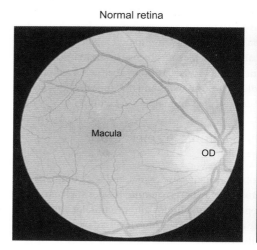

Background diabetic retinopathy

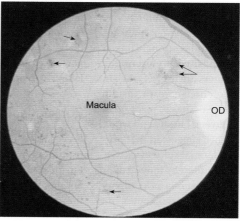

Figure 24.9 A comparison between a normal retina (left) and that of a diabetic. The diabetic retina has a large number of small red lesions resulting from minor haemorrhages. Larger aggregates have been indicated with arrows. 'Background' is a term signifying retinopathy that is not a direct threat to sight. OD = optic disc. (From Plates 1 and 2 of J.G.G. Ledingham and D.A. Warrell (eds) (2000) *Concise Oxford textbook of medicine.* Oxford University Press, Oxford.)

healthy retina. Other examples of conditions caused by chronic hyperglycaemia are illustrated in Figure 24.10 which shows a foot ulcer resulting from diabetic neuropathy with poor wound healing. Panel (b) provides a comparison between a healthy artery and that of an animal with chronic hyperglycaemia.

Management of diabetes mellitus

Strict glycaemic control can delay the onset, and reduce the severity, of many of the complications of diabetes, while parallel control of blood pressure, blood lipids, and weight can improve the outlook still more. Type 1 diabetics produce little or no insulin of their own and so must be treated with exogenous hormone. The treatment strategy is designed to mimic as closely as possible the secretion of endogenous insulin. A number of types of insulin (genetically engineered human, porcine, or bovine) and daily regimes are used to achieve this. Short, intermediate, and long-acting insulin analogues, as well as mixtures of these, may be used to achieve optimum glycaemic control for an individual patient. Bolus injections are widely used but insulin pumps are also used successfully by some patients.

Many people with type 2 diabetes are able to maintain adequate glycaemic control without exogenous insulin (although deterioration of their condition may eventually mean that insulin therapy is needed). These patients can manage their condition through careful diet and the use of oral hypoglycaemic drugs. A number of such drugs are available including sulphonylureas, biguanides, thiazolidinediones (TZDs) or glitazones, post-prandial glucose regulators (PPGRs) such as glucosidase inhibitors, incretin mimetics, and dipeptidyl peptidase-4 (DPP-4) inhibitors (gliptins). The **sulphonylureas** have been used in the treatment of diabetes for many years. They act by stimulating any remaining islet cells to secrete insulin and also reduce both gluconeogenesis and glycogenolysis by the liver. Examples include tolbutamide, chlopropamide, glibenclamide, gliclazide, glipizide, and newer drugs such as nateglinide. All insulin secretagogues rely on the existence of some healthy islet cells. **Biguanides** such as metformin and the **thiazolidinediones** (or **glitazones**) such as rosiglitazone and pioglitazone work by increasing the sensitivity of the body's own insulin receptors and thus increase glucose transport into muscle and liver cells.

Post-prandial glucose regulating drugs (PPGRs) such as the glucosidase inhibitor acarbose (glucobay) may help to maintain glycaemic control in type 2 diabetes, as they inhibit the enzymes that break down complex carbohydrates in the gut and thus delay the absorption of glucose after a meal.

Drugs that mimic the action of incretins may also help to enhance the activity of residual endogenous insulin in type 2 diabetes. Examples of these agents include exenatide (Byetta) and liraglutide (Victoza), which interact with the GLP-1 receptor but are more metabolically stable. These drugs reduce the appetite and help to achieve weight loss. Byetta is often prescribed for obese patients (**body mass index** $> 35 \, \mathrm{kg \, m^2}$).

Finally, **dipeptidyl peptidase-4 (DPP-4) inhibitors** may be of value in the management of type 2 diabetes. DPP-4 is an enzyme that degrades GLP-1 (a naturally-occurring incretin) rapidly in the body. Drugs that inhibit this enzyme therefore prolong the half-life of GLP-1 and thus maximize the effects of insulin. Examples include sitagliptin, vildagliptin, and the newer saxagliptin and alogliptin.

Monitoring glycaemic control in diabetic patients

In order to monitor short-term fluctuations in plasma glucose and to determine the most appropriate insulin regime, diabetics are accustomed to measuring their blood glucose level (the so-called BM measurement) up to several times a day. However, in view of the problems associated with chronically elevated blood glucose, it is also important to monitor long-term glycaemic control in diabetic individuals. This has been made possible by measurements of the blood concentration of **glycosylated haemoglobin** (HbA1). Haemoglobin (Hb) is glycosylated by a non-enzymatic process at a rate proportional to the prevailing level of blood glucose so that, as long as the lifespan of a person's red blood cells is normal, the level of glycosylated Hb will provide an indication of the mean blood glucose concentration over the preceding 60 days or so (half the average lifespan of a red blood cell—see Chapter 25). Although total HbA1 is sometimes measured clinically, the measurement used most widely in the monitoring of diabetics is that of HbA1c, which is produced by glycosylation of the N-terminal valine of the haemoglobin β-chain. Normoglycaemic individuals with normal insulin responses will show a HbA1c value in the range of 3.5–5.5 per cent (or 15–37 mmol/mol). Diabetics generally aim for values below 6.5 per cent (48 mmol/mol), as this is evidence of good long-term control of blood glucose. The figures presented in Table 24.1 show how the mean blood glucose concentration influences the measured level of HbA1c. Values for HbA1c are given as per cent of total Hb and in mmol/mol.

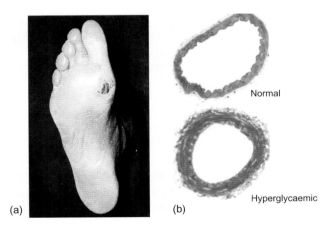

(a)   (b)

Figure 24.10 (a) A foot ulcer resulting from diabetic neuropathy. The position over the head of the first metatarsal is typical. (b) A section though a normal rat mesenteric artery (top) and a similar section taken from an animal suffering chronic hyperglycaemia (bottom). Note the thickening of the vessel wall. ((a) From Plate 11.10 of J. Laycock and P. Wise (1996) *Essential Endocrinology* 3rd edition, Oxford Medical Publications, Oxford. (b) from Figure 1 of K. Sachidanandam et al. (2009) *American Journal of Physiology Regulation Integration and Comparative Physiology* **296**, R952–959.)

Normal

Hyperglycaemic

Table 24.1 The relationship between plasma glucose concentration and glycosylation of haemoglobin

Blood glucose (mmol l^{-1})	HbA1c (as % of total Hb)	HbA1c (mmol/mol)
18	13	119
13	10	86
10	8	64
8	7	53
5	5	31

In some patients, for example children, pregnant women, and those with abnormal erythropoiesis or red blood cell pathologies, it is not appropriate to use HbA1c measurements as an indicator for plasma glucose regulation. In such cases, glycated albumin (in which the fructosamine fraction of the protein is glycated) may be used as an alternative measure. It is less accurate than HbA1c and generally only reflects the mean plasma glucose level over the preceding 2 weeks or so.

Recent years have seen the development of continuous measurement devices that provide 24-hour monitoring of blood glucose levels. In combination with an insulin pump, these sensors allow variable amounts of insulin to be administered throughout the day to match the individual's requirements. Commercial devices for such closed loop systems of glucose monitoring are now available.

Consequences of hypoglycaemia

Hypoglycaemia is defined as a blood glucose level below 2.5 mmol l^{-1}. It may arise from a number of causes, including overdosage of insulin in diabetics, an insulin-secreting tumour of the islet cells, lack of GH or cortisol, severe liver disease, very severe exercise, or inherited defects of gluconeogenic enzymes. The consequences of hypoglycaemia fall into two categories: those associated with activation of the autonomic nervous system and those caused by altered cerebral function.

A hypoglycaemic episode is often heralded by a feeling of hunger mediated by the parasympathetic nervous system. Later, symptoms of sympathetic activation are seen. These include tachycardia, sweating, pallor, and anxiety. Because the brain relies on glucose as its prime energy source, hypoglycaemia also elicits symptoms related to altered cerebral function. Headache, difficulty with problem solving, confusion, irritability, convulsions, and eventually coma occur. Hypoglycaemia represents an acute emergency, and as soon as these symptoms appear, the treatment is to ingest carbohydrate (e.g. as glucose tablets or glucogel). If this is not possible, a glucose solution may be given intravenously. Alternatively, glucagon can be administered via intramuscular or subcutaneous injection. If plasma glucose falls below 1 mmol l^{-1} there may be irreversible neuronal damage and the patient may die.

Summary

A lack of insulin or insensitivity of tissues to insulin results in diabetes mellitus. Type 1 diabetes is an autoimmune disorder that results in the destruction of pancreatic β-cells, while type 2 diabetes is characterized by reduced insulin secretion and a loss of insulin sensitivity. Diabetes results in loss of glycaemic control. Acute hyperglycaemia causes symptoms including thirst, polyuria, glycosuria, and in extreme cases ketoacidosis. Chronically elevated blood glucose leads to microvascular damage, peripheral neuropathy, nephropathy, and retinopathy. Type 1 diabetes must be treated with exogenous insulin, while type 2 diabetes may be controlled by careful management of the diet and the use of oral hypoglycaemic drugs.

A blood glucose level below 2.5 mmol l^{-1} (hypoglycaemia) can be caused by various factors, including over-dosage of insulin in diabetics. It is an acute emergency characterized by tachycardia, sweating, pallor, anxiety, and altered cerebral function. As soon as these symptoms appear, the treatment is to ingest carbohydrate (e.g. as glucose tablets or glucogel).

(✳) Checklist of key terms and concepts

Glucose handling by the body

- Glucose is an essential metabolic substrate, particularly for the CNS.
- Plasma glucose is closely regulated between 3.5 and 7.5 mmol l^{-1}.
- Insulin, glucagon, glucocorticoids, catecholamines, GH, and thyroid hormones all contribute to the regulation of plasma glucose concentration. All except insulin, are hyperglycaemic in their actions.
- Plasma glucose is determined by the balance between intestinal absorption, storage and synthesis by the liver, and uptake by other tissues.

- Glucose is stored in the form of glycogen, particularly in the liver.
- Glucose can be synthesized from non-carbohydrate precursors, chiefly in the liver. This is called gluconeogenesis.

The role of insulin in the regulation of plasma glucose

- Insulin is synthesized by the β-cells of the pancreatic islets. It is the only hormone with a hypoglycaemic action.
- It is secreted chiefly in response to a rise in plasma glucose and amino acids.
- Insulin stimulates the uptake and utilization of glucose by cells (particularly liver, fat, and skeletal muscle), thereby causing a fall in plasma glucose.

- It also stimulates the synthesis of glycogen and protein.

The role of glucagon in the regulation of plasma glucose

- Glucagon is the most potent hyperglycaemic hormone, acting in antagonism to insulin.
- It is secreted by the α-cells of the pancreatic islets in response to hypoglycaemia and acts to promote the release of glucose into the blood.
- Glucagon stimulates glycogenolysis, lipolysis, and gluconeogenesis.

The role of other hormones in the regulation of plasma glucose

- Other hyperglycaemic hormones include cortisol, catecholamines, GH, and thyroid hormones.
- These hormones become important during prolonged periods of fasting.
- Their effects are geared to the maintenance of glycogen stores, the stimulation of gluconeogenesis, and the mobilization of fatty acids and proteins.
- In this way, other metabolic substrates are provided for those tissues able to use them and glucose is spared for use by the tissues reliant upon it, notably the brain.

Plasma glucose regulation after a meal

- In the absorptive phase of metabolism (just after a meal), plasma glucose is high, insulin levels are high, and the hyperglycaemic hormones are inhibited.

- During this phase, glycogen stores are replenished and glucose uptake and utilization by cells is high.
- In the post-absorptive state (from 3 hours or so after a meal until the next intake of food), levels of the hyperglycaemic hormones are enhanced, and there is an increase in the mobilization of fats (and, in prolonged fasting, proteins) so that glucose is spared for use by the CNS.

Diabetes mellitus

- A lack of insulin or insensitivity of tissues to insulin results in diabetes mellitus.
- Type 1 diabetes is an autoimmune disorder that results in the destruction of pancreatic β-cells. It requires treatment with exogenous insulin.
- Type 2 diabetes is characterized by reduced insulin secretion and a loss of insulin-sensitivity of tissues. It may be controlled by careful management of the diet and the use of oral hypoglycaemic drugs.
- Untreated diabetes results in loss of glycaemic control. Acute hyperglycaemia causes symptoms including thirst, polyuria, glycosuria and, in extreme cases, ketoacidosis.
- Chronically elevated blood glucose leads to microvascular damage, peripheral neuropathy, nephropathy, and retinopathy.
- Hypoglycaemia (plasma glucose below 2.5 mmol l^{-1}) results in tachycardia, sweating, pallor, mental confusion and, eventually, coma.

Recommended reading

Physiology

Kovacs, W.J., and Ojeda, S.R. (2012) *Textbook of endocrine physiology* (6th edn), Chapter 15. Oxford University Press, Oxford.
  A detailed account of the hormonal control of glucose, lipid, and protein metabolism.

Medicine

Bell, J.I., and Hockaday, T.D.R. (2000) Diabetes mellitus (Chapter 6.13), in: *Concise Oxford Textbook of Medicine* (ed J.G.G. Ledingham and D.A. Warrell). Oxford University Press, Oxford.
  A detailed introduction to the causes, pathophysiology, and clinical management of diabetes mellitus.

Carton, J., Daly, R., and Ramani, P. (2007) *Clinical pathology*, Chapter 14. Oxford University Press, Oxford.
  A readable discussion of principal features of diabetes mellitus.

Holt, R.I.G., and Hanley, N.A. (2012) *Essential endocrinology and diabetes*. Wiley Blackwell, Oxford.

Ramrakha, P., Moore, K., and Sam, A. (2010) Chapter 9, in: *Oxford handbook of acute medicine* (3rd edn). Oxford University Press, Oxford.
  Provides useful clinical details of diabetic ketoacidosis as a medical emergency.

Wass, V., and Owen, K. (eds) (2014) *Oxford handbook of endocrinology and diabetes* (3rd edn). Oxford University Press, Oxford.

Review articles

Borge, P.D., Moibi, J., Greene, S.R., Trucco, M., Young, R.A., Gao, Z., and Wolf, B.A.. (2002) Insulin receptor signaling and sarco/endoplasmic reticulum calcium ATPase in β-cells. *Diabetes* **51** (Suppl. 3): S427–3.

Cryer, P.E. (2005) Mechanisms of hypoglycemia-associated autonomic failure and its component syndromes in diabetes. *Diabetes* **54**, 3592–601.

Rorsman, P., and Braun, M. (2013) Regulation of insulin secretion in human pancreatic islets. *Annual Review of Physiology* **75**, 155–79, doi 10.1146/annurev-physiol-030212-183754.

To check that you have mastered the key concepts presented in this chapter, complete the accompanying online self-assessment questions. Go to www.oup.com/uk/pocock5e/

PART SIX

Blood and the immune system

CHAPTER 25

The properties of blood

Chapter contents

This chapter should help you to understand:

- The principal roles and chief constituents of the blood
- The physical and chemical characteristics of the plasma
- The haematocrit
- The roles of the red cells, white cells, and platelets
- The origin of blood cells—haematopoiesis
- The main features of the metabolism of iron and its role in the formation of haemoglobin
- The carriage of oxygen and carbon dioxide by the red cells
- Some major disorders of the blood—anaemia, leukaemia, and thrombocytopenia
- Blood clotting (haemostasis), clot retraction, and dissolution
- Blood groups and their importance in blood transfusions

25.1 Introduction

The blood is a vital vehicle of communication between the tissues of multicellular organisms. Its numerous functions include the following:

- The carriage of oxygen from the lungs to the other body tissues, and carbon dioxide from the tissues to the lungs
- The maintenance of ionic balance within the body
- The delivery of nutrients from the gut to the tissues
- The transport of the waste products of metabolism from their sites of production to their sites of disposal
- The carriage of hormones from the endocrine glands to their target tissues
- The protection of the body against invading organisms.

Blood consists of a fluid called **plasma** in which are suspended the so-called formed elements—the **red cells** (or erythrocytes), **white cells** (leukocytes), and **platelets** (thrombocytes). It is possible to demonstrate the nature of the suspension by centrifuging a sample of whole blood in a test-tube at $\approx 2000 \times g$ for 5–10 minutes. After such treatment, the heavier red cells are packed at the bottom of the tube while the plasma is seen above them as a clear pale yellow fluid, as illustrated in Figure 25.1. A thin layer of white cells and platelets (the 'buffy coat') separates the packed red cells from the plasma.

The circulating blood volume is about 7–8 per cent of body weight, so that for a 70 kg man, blood volume will be around 5 litres, but for a newborn baby weighing 3.2 kg (7 lbs), blood volume will only be around 250 ml—an important point to remember when considering blood loss and transfusion for a small baby. At any one time, assuming a blood volume of 5 litres, about 0.6 litre will be in the lungs, about 3 litres in the systemic venous circulation, and the remaining 1.4 litres in the heart, systemic arteries, arterioles, and capillaries (see Figure 30.6).

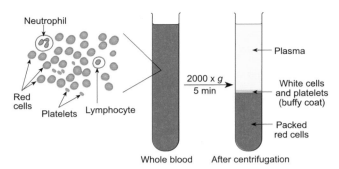

Figure 25.1 The separation of blood into cells and plasma by centrifugation. On the left is a diagrammatic view of the appearance of whole blood showing red cells, leukocytes, and platelets. The right-hand side shows the separation of the blood into a pale fluid (plasma) and a deep red layer (packed red cells). Between the two is the buffy coat (a thin layer consisting of white cells and platelets).

Summary

Blood consists of plasma, red cells, white cells, and platelets. It is the vehicle of communication between the tissues and serves to transport the respiratory gases, nutrients, hormones, and waste materials around the body.

25.2 The physical and chemical characteristics of plasma

The total blood volume and the plasma volume may be measured using the dilution techniques described in Chapter 3. Normal adults have 35–45 ml of plasma per kg body weight, so the plasma volume is 2.8–3.0 litres in men and around 2.4 litres in women. The plasma accounts for about 4 per cent of body weight in both sexes. It consists of 95 per cent water, the remaining 5 per cent being made up by a variety of substances in solution and suspension. These include mineral ions (e.g. sodium, potassium, calcium, and chloride), small organic molecules (e.g. amino acids, fatty acids, and glucose), and plasma proteins (e.g. albumin). Typical values for a number of important constituents of the plasma are given in Table 25.1.

The major constituents of the plasma are normally present at roughly constant levels. These include the inorganic ions and the plasma proteins. However, the plasma also contains a number of substances that are in transit between different cells of the body and may be present in varying concentrations according to their rates of removal or supply from various organs. Such substances include enzymes, hormones, vitamins, products of digestion (e.g. glucose), and dissolved excretory products.

The composition of the plasma is normally kept within biologically safe limits by a variety of homeostatic mechanisms. However, this balance may be disturbed by a variety of disorders, particularly those involving the kidneys, liver, lungs, cardiovascular system, or endocrine organs. For this reason, accurate analysis of the plasma levels of a host of variables can provide important diagnostic information on the function of many organs. In acute renal failure, for example, the plasma levels of urea, creatine, and potassium are all elevated, while the calcium and bicarbonate levels are lower than normal, as is plasma pH (indicating an **acidosis**—see Chapter 39).

The ionic constituents of plasma

The chief inorganic cation of plasma is sodium (see Table 25.1), which has a concentration of 140–145 mmol l^{-1}. There are much smaller amounts of potassium, calcium, magnesium, and hydrogen ions. Chloride is the principal anion of plasma (around 100 mmol l^{-1}). Electroneutrality is achieved by the presence of other anions including bicarbonate, phosphate, sulphate, protein, and various organic anions.

Regulation of the ionic components of the plasma helps to maintain both its osmolality (280–300 mOsm per kg water) and its pH (7.35–7.45) within physiological limits. Further information concerning the homeostatic mechanisms responsible for the regulation of plasma pH, volume, and osmolality may be found in Chapters 39–41.

Plasma proteins

There are a great many different proteins in plasma but the principal proteins can be divided into three main categories: the albumins, the globulins, and the clotting factors (notably fibrinogen and prothrombin). In normal healthy individuals, plasma proteins account for 7–9 per cent of the plasma by weight.

The **albumins** are the smallest and the most abundant, accounting for about 60 per cent of the total plasma protein. They are transport proteins for lipids and steroid hormones that are synthesized by the liver. They are also important in body fluid balance, since they provide most of the colloid osmotic pressure (the **oncotic pressure**) that regulates the passage of water and solutes through the walls of the capillaries (see Chapter 31). Many drugs (such as **warfarin** and sulphonamides) become bound to albumin, and this affects their free concentration in the plasma.

Globulins account for about 40 per cent of total plasma protein and may be further subdivided into **alpha** (α), **beta** (β), and **gamma** (γ) **globulins**. The α- and β-globulins are made in the liver and they transport lipids and fat-soluble vitamins in the blood. The γ-globulins are antibodies produced by lymphocytes in response to exposure to antigens (foreign agents in the body that evoke the formation of specific antibodies). They are crucial in defending the body against infection (see Chapter 26).

Fibrinogen is an important clotting factor produced by the liver (see Section 25.8). It accounts for about 2–4 per cent of the total plasma protein and is generally grouped with the globulins.

Summary

Plasma is an aqueous solution consisting of proteins, mineral ions, small organic molecules, and a number of substances in transit between tissues (e.g. products of digestion). The main plasma proteins are the albumins, α-, β-, and γ-globulins, and clotting factors.

Table 25.1 Reference values for the principal constituents of the plasma (mean value given with the normal range in backets)

Constituent	Quantity	Units	Remarks
Water	945 (930–950)	$g\,l^{-1}$	
Bicarbonate	25 (18–32)	$mmol\,l^{-1}$	Important for the carriage of CO_2 and for H^+ buffering
Chloride	103 (95–105)	$mmol\,l^{-1}$	The principal extracellular anion
Inorganic phosphate	1.0 (0.7–1.25)	$mmol\,l^{-1}$	
Calcium	2.5 (2.1–2.6)	$mmol\,l^{-1}$	This is total calcium; ionized calcium is about $1.25\ mmol\,l^{-1}$
Magnesium	0.8 (0.7–1.0)	$mmol\,l^{-1}$	
Potassium	4.0 (3.8–5.0)	$mmol\,l^{-1}$	
Sodium	144 (136–148)	$mmol\,l^{-1}$	The principal extracellular cation
Hydrogen ions (pH)	40 (38–44) (pH 7.35–7.42)	$nmol\,l^{-1}$	$1\ nmol\,l^{-1} = 10^{-9}\ mol\,l^{-1}$
Glucose	4.5 (2.9–4.8)	$mmol\,l^{-1}$	This value is for a fasting subject. Glucose is the major source of metabolic energy—particularly for the CNS
Cholesterol	5.2 (3–6.5)	$mmol\,l^{-1}$	
Urea	5.0 (3.0–6.5)	$mmol\,l^{-1}$	Waste product of nitrogen metabolism
Osmolality	290 (285–295)	$mOsmol\,kg^{-1}$	
Fatty acids (total)	3.0 (1.0–5.0)	$g\,l^{-1}$	Fatty acids are mainly present as esters (glycerides, phosphatides etc); only about 10% are present in their free form
Total protein	75–85	$g\,l^{-1}$	
Albumin	45 (35–50)	$g\,l^{-1}$	Principal protein of the plasma; binds hormones and fatty acids
α-globulins	7	$g\,l^{-1}$	Transports lipids and fat soluble vitamins
β-globulins	8.5	$g\,l^{-1}$	Transports lipids and fat soluble vitamins
γ-globulins	10.6 (5–15)	$g\,l^{-1}$	Immunoglobulins (antibodies)
Fibrinogen	3	$g\,l^{-1}$	Blood clotting (factor I)
Prothrombin	1	$g\,l^{-1}$	Blood clotting (factor II)
Transferrin	2.4	$g\,l^{-1}$	Iron transport

Note: Values are means with the normal range in parentheses. Note that even in health there is considerable individual variation.

25.3 The formed elements of the blood

The formed elements of blood include red cells, five separate classes of white cells (recognized according to their morphology and staining reactions), and platelets. Of these, the red cells are by far the most numerous (see Figure 25.2). Table 25.2 lists the cellular components and their concentrations in whole blood.

The haematocrit

The **haematocrit ratio** or haematocrit describes the proportion of the total blood volume occupied by the erythrocytes (red cells). For any blood sample, the haematocrit may be obtained by centrifuging a small volume of blood in a capillary tube until the cellular components become packed at the bottom of the tube (see Figure 25.1). For this reason the haematocrit is also known as the **packed cell volume**. By measuring the height of the column of red cells relative to the total height of the column of blood and correcting for the plasma which remains trapped between the packed red cells, it is possible to determine the volume occupied by the packed red cells as a percentage of the total blood volume. In adult males, the average haematocrit determined in this way from a sample of venous blood is around 0.47 litre of red cells per litre of whole blood (range 0.4–0.54) while in females it is closer to 0.42 (range 0.37–0.47). However, the ratio of cells to plasma is not uniform throughout the body. In the capillaries, arterioles, and other small vessels, it is lower than in the larger arteries and veins as a result of **axial streaming** of blood cells in vessels (see Chapter 30). This is the tendency for red cells to remain near the central axis as blood flows through a vessel rather than flowing near to the vessel wall. In large vessels, the ratio of wall surface area to volume is smaller than in the tiny vessels and so the large vessels contain more cells per unit volume of blood. The **viscosity** of the blood (i.e. its resistance to flow) rises progressively as the haematocrit increases.

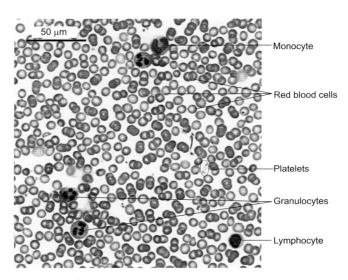

Figure 25.2 A stained blood smear. The field of view shows large numbers of red cells (erythrocytes) here stained purple with a pale centre. Also present are a monocyte (identified by its large size and indented nucleus), several granulocytes, a lymphocyte, and some platelets. The size of the various cells can be judged from that of the red cells, which are approximately 7–8 μm in diameter. (Leishman's stain.)

Red blood cells

The red cells (also called **erythrocytes**) are the most numerous cell type in the blood—each litre of normal blood containing 4.5–6.5 × 10^{12} red cells. Their chief function is to transport the respiratory

gases oxygen and carbon dioxide around the body. The red cells are small, circular, biconcave discs of 7–8 μm diameter and do not possess a nucleus (see Figures 25.2–25.4). They are very thin and flexible and can squeeze through the narrow bore of the capillaries, which have internal diameters of only 5–8 μm. Their shape gives red cells a large surface area to volume ratio which promotes efficient gas exchange (see Section 25.6). In a mature erythrocyte, the principal protein constituent of the cytoplasm is **haemoglobin**, an oxygen binding protein, which is synthesized by the red cell precursors in the bone marrow.

White blood cells—leukocytes

Leukocytes are generally larger than the red blood cells, they possess a nucleus, and they are present in smaller numbers—normal blood contains around 7×10^9 white cells l^{-1} (Table 25.2). These cells have a vital role in the protection of the body against disease—they are the mobile units of the body's immune system, being transported rapidly to specific areas of infection and inflammation to give powerful defence against invading organisms. They possess several characteristics that enhance their efficacy as part of the body's defence system. They are able to pass through the walls of capillaries and enter the tissue spaces in accordance with the local needs. This process is known as **diapedesis**. Once within the tissue spaces, the leukocytes (particularly the granulocytes) have the ability to move through the tissues by an amoeboid motion at speeds of up to 40 μm a minute. They are attracted to sites of infection or inflammation by specific chemical signals (a process called **chemotaxis**). For more information concerning the role of the white cells in the immune system, see Chapter 26.

Table 25.2 The cellular elements of whole blood

Cell type	Site of production	Typical cell count (number per litre)	Comments and function
Erythrocytes (red cells)	Bone marrow	5×10^{12} (men) 4.5×10^{12} (women)	Transport of O_2 and CO_2
Leukocytes (differential count)		7×10^9	
Granulocytes			
neutrophils	Bone marrow	5.0×10^9 (40–75%)	Phagocytes—engulf bacteria and other foreign particles
eosinophils	Bone marrow	100×10^6 (1–6%)	Congregate around sites of inflammation—have antihistamine properties. Very short-lived in blood
basophils	Bone marrow	40×10^6 (<1%)	Implicated in inflammation; produce histamine and heparin
Agranulocytes			
monocytes	Bone marrow	0.4×10^9 (2–10%)	Phagocytes, become macrophages when they migrate to the tissues
lymphocytes	Bone marrow Lymphoid tissue, Thymus, Spleen	1.5×10^9 (20–45%)	Production of antibodies (B-lymphocytes); cell-mediated immunity (T-lymphocytes)
Platelets	Bone marrow	250×10^9	Aggregate at sites of injury and initiate haemostasis

Note: While mean values are given, these are subject to considerable individual variation. The approximate percentages of individual types of leukocyte are given after the number per litre—this is called the **differential white cell count**.

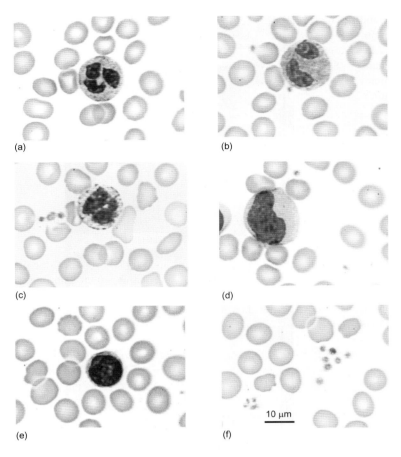

Figure 25.3 The appearance of the major kinds of blood cell after staining; (a) shows a neutrophil in a field of red cells—note its characteristic lobed nucleus; (b) an eosinophil which has a bi-lobed nucleus and red staining granules in its cytoplasm; (c) a basophil—note the strongly staining blue granules in the cytoplasm which make it difficult to see the nucleus; (d) a monocyte—note its prominent indented nucleus and the large size of the cell body (~20–25 μm); (e) a lymphocyte which is identified by its prominent nucleus and thin cytoplasm; (f) a field of red cells and a cluster of platelets—note the densely staining granulomere that can just be discerned in several of the group of platelets on the right.

As Figure 25.3 shows, there are three major categories of white blood cells:

1. the **granulocytes** (also called **polymorphonuclear leukocytes** because their nuclei are divided into lobes or segments)

2. the **monocytes** (macrophages)

3. the **lymphocytes**.

The monocytes and the lymphocytes are sometimes referred to collectively as **agranulocytes** or **mononuclear leukocytes**.

Although all white blood cells are concerned with defending the tissues against disease-producing agents, each class of cell has a different role to play. Consider first the granulocytes, which account for around 70 per cent of the total number of white cells in the blood. These are classified as neutrophils, eosinophils, or basophils, according to their staining reactions.

Neutrophils are by far the most numerous of the granulocytes. They are so named because their cytoplasm does not stain with eosin or basophilic dyes such as methylene blue. They are 12–15 μm in diameter with a nucleus of 2–5 lobes linked together by thin segments of chromatin (see panels (a) and (d) of Figure 25.3). They are phagocytes which are formed in the bone marrow. When they mature they are released into the circulation, where they remain for 6–12 hours before entering the intercellular

spaces by diapedesis; they remain there for a few days, engulfing and destroying any invading bacteria. Enzymes within the cytoplasmic granules are able to digest the phagocytosed particles. This action of the neutrophils forms the first line of defence against infection. They eventually undergo apoptosis (a pre-programmed cell death—see Chapter 51), and their remains are taken up by the tissue phagocytes and digested.

Eosinophils are so called because their granules stain red in the presence of the dye eosin (see Figure 25.3(b)). Like the neutrophils and basophils, they are formed in the bone marrow before being released into the blood, where they survive for only 12–20 hours. They are 12–15 μm in diameter with a characteristic bi-lobed nucleus. Normally they represent only about 1–5 per cent of the total number of white blood cells in adults, but in people with allergic conditions such as asthma or hay fever, their population greatly increases. These cells have antihistamine properties and congregate around sites of inflammation; they also combat worm infestations.

Basophils are 12–15 μm in diameter and their cytoplasm possesses granules that are strongly stained by basic dyes such as methylene blue, as shown in Figure 25.3(c). They have a segmented nucleus that is often difficult to see amongst the strongly staining granules. They are the least common of the granulocytes and account for only about 0.5 per cent of the white cell population. Like the other granulocytes, they are formed in the bone marrow before being released into the blood. They secrete histamine and are responsible for some of the phenomena associated with immunological reactions, such as the limited vasodilatation and increased permeability of blood vessels that results in local oedema. Basophils are stimulated to secrete the contents of their granules by many different factors, including antigens bound to immunoglobulin E (IgE) and various cytokines.

Monocytes are larger than the other classes of white cells, having a diameter of 15–20 µm. Their prominent nuclei are indented as shown in Figure 25.3(d). They are formed in the bone marrow where they mature before being released into the circulation. Within two days, they have migrated to tissues such as the spleen, liver, lungs, and **lymph nodes**, where they mature to become **macrophages**. These cells are long-lived and reside in the tissues for months. They act in much the same way as the neutrophils, ingesting bacteria and other large particles. They also participate in immune responses by presenting antigens so that they will be recognized by the T-lymphocytes.

Lymphocytes represent around 25 per cent of the total white cell population in adults (although in children they are much more numerous). Small lymphocytes 6–12 µm are characterized by a spherical nucleus that is strongly stained and is surrounded by a thin layer of cytoplasm (see panel (e) of Figure 25.3). The lymphocytes are classified as **B-lymphocytes**, which mature in the lymphoid tissues (the lymph nodes, tonsils, and spleen, and to a lesser extent in the bone marrow), and **T-lymphocytes**, which mature in the thymus. The two kinds cannot be distinguished in a normal blood smear but can be identified by the presence of specific cell markers. Large lymphocytes have diameters up to 20 µm with an eccentric ('off centre') nucleus and extensive cytoplasm. They are activated **B-cells** (also called **plasma cells**). B-cells have a very short life in the circulation (a few hours) but T-cells may survive for 200 days or more. Each has an important part to play in the protection of the body against infection, either by producing **antibodies** (B-cells) or by participating in cell-mediated immune responses (T-cells). The functions of the lymphocytes are discussed in greater detail in Chapter 26.

Platelets (thrombocytes)

Platelets have an important role in the control of bleeding (haemostasis—see Section 25.8) and in the maintenance of integrity of the vascular endothelium. They are not strictly speaking cells at all but are cell fragments that are formed in the bone marrow by budding off from the cytoplasm of large (35–150 µm) cells called **megakaryocytes**. They appear in normal blood smears as densely staining small particles (see panel (f) of Figure 25.3). Normal blood contains $150–400 \times 10^9$ platelets l^{-1}. They have a lifespan in the blood of around 10 days.

Platelets are 2–4 µm in diameter and do not normally possess a nucleus. Unstimulated platelets have a discoid shape (see panel (a) of Figure 25.4). The outer region is weakly stained by dyes and is called the **hyalomere**. The hyalomere contains actin filaments that provide the contractile apparatus for the formation of the **filopodia** (the small cellular projections that can be seen in panel (b) of Figure 25.4). The filopodia are implicated in platelet aggregation

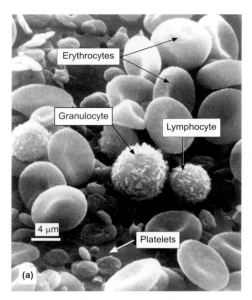

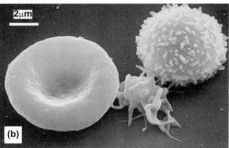

Figure 25.4 Scanning electron micrographs of blood cells and platelets. (a) A view of a field of red cells including a granulocyte, which is distinguished from the adjacent lymphocyte by its greater size. Both have a roughened appearance, unlike the smooth surface of the red cells (erythrocytes) and platelets. (b) A red cell (left), an activated platelet (centre), and a granulocyte (right). Note the pronounced biconcave shape of the red cell, which maximizes the surface area for gas exchange. Bars: 4 µm (a), 2 µm (b). (Courtesy of the National Cancer Institute, USA.)

and clot retraction. The inner region of platelets, known as the **granulomere**, stains strongly with basophilic dyes. As its name implies, this region contains granules which are similar in structure to the large dense cored vesicles of secretory cells, whose secretion is regulated in a similar manner.

Platelets have three different kinds of granule:

1. alpha granules that contain a number of proteins including platelet derived growth factors and proteins involved in the formation and dispersal of blood clots, among them factors V and XII, fibrinogen, and von Willebrand factor (see Section 25.8)

2. delta granules, which contain calcium, ADP, ATP, and serotonin (5-hydroxytryptamine)

3. lambda granules, which appear to be homologous with lysosomes and contain hydrolytic enzymes.

Summary

The formed elements of blood include the erythrocytes, five types of leukocytes, and the platelets. They can be separated from the plasma by centrifugation. The red cells become packed at the bottom of the tube, with the white cells and platelets forming a thin line above them. The packed cell volume (the haematocrit) may be measured in this way.

Red blood cells are small, non-nucleated biconcave discs whose function is to transport oxygen and carbon dioxide between the lungs and tissues. They contain the oxygen binding protein haemoglobin. The white cells (leukocytes) are fewer in number than red cells but play a crucial role in mediating the body's immune responses. The platelets (thrombocytes) play an essential role in blood clotting (haemostasis) and are cell fragments derived from the megakaryocytes of the bone marrow.

25.4 Haematopoiesis—the formation of blood cells

Mature blood cells have a relatively short lifespan and must therefore be renewed continuously. The process by which new blood cells are formed is known as **haematopoiesis** (or **haemopoiesis**) and the tissue in which this occurs is called the **haematopoietic tissue**. The term **erythropoiesis** refers to the formation of erythrocytes (red blood cells), **leucopoiesis** refers to the formation of leukocytes (white blood cells), and **thrombopoiesis** is the name given to the formation of the platelets (thrombocytes). The number of cells required is quite astonishing: around 250 billion (25×10^{10}) newly formed red cells are required each day to replace those that are senescent or have died. At least 20 billion (20×10^9) white cells and around 25 billion (25×10^9) platelets are also needed each day. To keep these large numbers in perspective, recall that in just 20 divisions a single cell can give rise to over a million daughter cells (1,048,576 to be precise).

The blood cells originate in the bone marrow

In early embryos, blood cells arise from the mesoderm of the yolk sac, but as development progresses, the liver and spleen act as haematopoietic tissue. From the second month of gestation the bone marrow is increasingly important, eventually becoming the main site of haematopoiesis after birth. There are two kinds of bone marrow: **red marrow**, which is the site of haematopoiesis, and **yellow marrow**, which contains a large number of adipocytes (fat cells). The cellular organization of the red bone marrow is shown in Figure 25.5.

In newborn babies and young children all of the bone marrow is red, but as they grow, the proportion of red marrow becomes progressively smaller as adipocytes infiltrate the marrow cavity of the long bones to form yellow marrow. In adults, the remaining sites of haematopoiesis are the bone marrow of the ends of the long bones (particularly the humerus and femur) and the flat bones: the pelvis, ribs, sternum, vertebrae, clavicles, scapulae, and skull.

Nevertheless, the total mass of red marrow in adults is substantial at around 1.5 kg, or 3.5 to 3.6 per cent of total body weight. Moreover, the yellow marrow can be reconverted to red marrow if circumstances require it (e.g. after severe haemorrhage).

The bone marrow is entirely surrounded by bone, and small arteries must penetrate the bony casing to provide its blood supply. Small arterial branches communicate directly with venous sinuses, which are thin-walled vessels 50–75 μm in diameter (i.e. there is no intervening capillary network). As Figure 25.5(b) shows, the sinuses branch and join together (anastomose) liberally to form an irregular network. Their walls consist of a simple layer of flattened endothelial cells covered with fibroblasts called **adventitial cells**, whose processes form part of the **stroma** that supports the haematopoietic tissue (see panels (c) and (d) of Figure 25.5). The veins draining the marrow course along with the arterial supply.

The haematopoietic tissue itself lies between the sinuses in irregular cords, as shown in Figure 25.5(a). Nevertheless, the cells are not distributed completely at random. The megakaryocytes lie close to the adventitial surface of the sinuses (indicated by black arrow in panel (c) of Figure 25.5) and cover apertures between the endothelial cells through which they release platelets into the circulation. The erythrocytes develop in association with macrophages near the wall of the sinuses before being delivered to the circulation. Granulocytes are produced in colonies farther from the sinuses, while the lymphocytes tend to develop in clusters surrounding small arteries. The production of all blood cells takes place outside the bloodstream itself and, as they mature, the newly formed cells enter the circulation by a process called **diapedesis** (see Figures 25.5(d) and 25.6). Only rarely do precursor cells appear in the blood.

Stem cells give rise to all blood cells

Despite the fact that the blood contains many different cells with a variety of functions, they are all ultimately generated from a small population of pluripotent **haematopoietic stem cells** in the bone marrow (see Box 4.1 for more detail on stem cells). During growth or regeneration most haematopoietic stem cells are self-renewing and divide symmetrically to produce two identical copies of themselves. This self-renewing property is exploited therapeutically in bone marrow transplantation. Once growth is complete, the self-renewal of stem cells is achieved by each parent cell undergoing an *asymmetric* cell division in which the two daughter cells have different physiological fates: one is a complete copy of the original stem cell, while the other is destined to differentiate into a haemopoietic progenitor cell which is multipotent. Progenitor cells then form the lineages that give rise to the different types of blood cell through a series of cell divisions, as outlined in Figure 25.6.

The formation of the various kinds of blood cell is controlled by **haematopoietic growth factors** which originate from a variety of sources including the bone marrow connective tissue, the liver, the kidneys, and white blood cells. These growth factors are local signalling molecules called **cytokines** which are supplemented by two hormones—**erythropoietin** (EPO) and **thrombopoietin** (TPO). Almost all cytokine haematopoietins are either interleukins (e.g. IL-2, IL-3 etc.) or colony stimulating factors (CSFs—factors that have

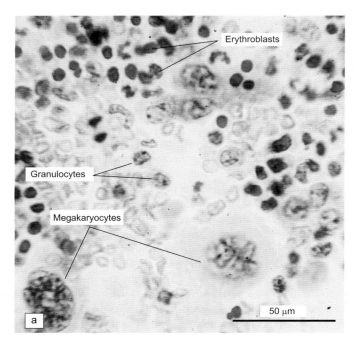

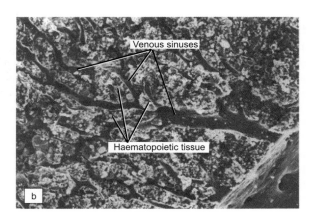

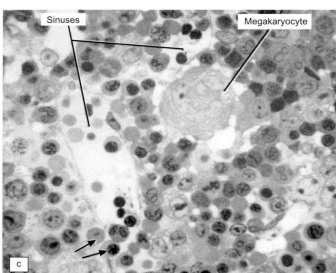

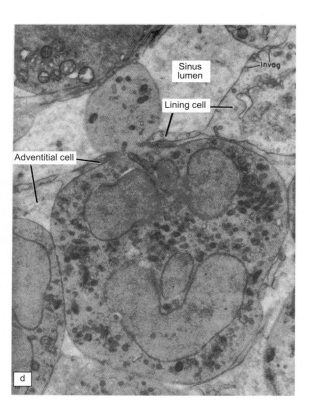

Figure 25.5 The fine structure of the bone marrow. Panel (a) shows a typical section of red marrow—note the large, weakly staining megakaryocytes and the large fat cells (adipocytes). The erythoid cells (which are dark red in appearance) are much more common than other cell type and tend to cluster together. A granulocyte colony can be seen at the top right. Panel (b) is a scanning electron micrograph of bone marrow, showing the cords of haematopoietic tissue and the branching of the venous sinuses. A vein is visible at the bottom right, showing the drainage apertures of smaller venous sinuses. Panel (c) is a transmission electron micrograph of bone marrow showing the thin-walled venous sinuses, a megakaryocyte which is adjacent to a venous sinus, and, at the bottom left, a reticulocyte and erythroblast entering the sinus (arrows). Panel (d) is a transmission electron micrograph showing the passage of a granulocyte from a haematopoietic cord into the lumen of a venous sinus. (Compiled from an original image (a) and (b–d) L. Weiss (1976) *Anatomical Record* **186**, 161–84, Plates 8 and 18.)

been shown to promote the growth of specific types of blood cell in culture). As Figure 25.7 shows, the haematopoietic stem cells give rise to two main kinds of progenitor cells: the **common myeloid progenitor cells** and the **common lymphoid progenitor cells**. These progenitor cells give rise to various **precursor cells** that

become 'committed' to the formation of one or a few types of blood cell. The myeloid stem cells remain within the bone marrow and generate precursor cells that differentiate to form the erythrocytes, granulocytes, monocytes, and megakaryocytes. Undifferentiated lymphoid cells migrate to the lymph nodes, spleen, and thymus,

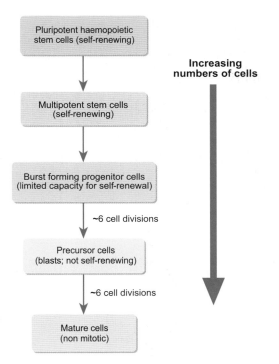

Figure 25.6 An outline of the principal stages of cellular differentiation during haematopoiesis. As cells progress through the stages of development they become committed to forming a specific kind of blood cell.

where they mature to become B- or T-lymphocytes. The origins of each kind of blood cell are outlined in Figure 25.7.

Erythropoiesis is regulated by the hormone erythropoietin

Each litre of blood contains around 5 trillion (5×10^{12}) erythrocytes, although this figure varies according to the age, sex, and state of health of the individual. As most red cells enter the circulation as reticulocytes, the rate of red cell production is indicated by the relative numbers of reticulocytes in the circulation (normally 1–1.5 per cent). To maintain a steady mass of red cells, the rate of erythropoiesis must be closely matched to the requirement for new red cells within the circulation. This regulation is achieved by erythropoietin (EPO), a glycoprotein hormone secreted mainly by the kidneys (probably by the endothelial cells of the peritubular capillaries).

A variety of stimuli may cause an increase in the rate of production of new erythrocytes, including loss of red cells through haemorrhage, donation of blood, and the chronic hypoxia experienced while living at high altitude. In all cases, it appears that the secretion of erythropoietin is stimulated by a fall in tissue oxygenation, and its concentration in plasma is inversely related to the partial pressure of oxygen in the arterial blood (P_aO_2). The normal plasma level of erythropoietin is quite modest at around 10–25 mU ml^{-1} but can increase substantially (over a thousandfold) if tissue oxygen levels are low (1 unit is approximately equivalent to 10 ng of

EPO). Erythropoietin acts by accelerating the differentiation of progenitor cells in the marrow into erythroblasts. In addition to erythropoietin, other factors essential for normal red blood cell production are iron, folic acid, and vitamin B$_{12}$.

Vitamin B$_{12}$ is absorbed from the small intestine in combination with a glycoprotein called **intrinsic factor**, which is secreted by the parietal cells in the gastric mucosa (vitamin B$_{12}$ was formerly known as extrinsic factor). This is discussed in greater detail in Chapter 44. If the diet is deficient in B$_{12}$ or there is a lack of intrinsic factor, red cell development is impaired, resulting in pernicious anaemia.

The formation of erythrocytes

In the first step towards formation of the erythrocytes, pluripotent stem cells divide to produce myeloid multi-potential cells which, under the influence of the growth factors interleukin-3 (IL-3), stem cell factor (SCF), and thrombopoietin (TPO), give rise to progenitor cells called **erythrocyte colony-forming cells** (CFC-E), also called colony forming units (see Figure 25.7). Erythropoietin together with other growth factors then promotes the development of precursor cells called **proerythroblasts** (or pronormoblasts), which subsequently undergo a further series of cell divisions, each of which results in smaller daughter cells until the mature erythrocytes are formed. The sequence of divisions gives rise first to **basophilic erythroblasts** and then to **polychromatic erythroblasts** and **orthochromatic erythroblasts** (these cells are also called early normoblasts, intermediate normoblasts, and late normoblasts). During these divisions (which are not shown in the figure) cell numbers increase substantially and the cells synthesize haemoglobin. Finally, they lose their nuclei to become **reticulocytes** which still have a small number of ribosomes. Development from proerythroblast to reticulocyte normally takes around 7 days.

The reticulocytes are released into the circulation and, within a day or so, they mature into erythrocytes. During this transition, they lose their remaining mitochondria and ribosomes. Consequently, they cannot synthesize haemoglobin or generate ATP by oxidative metabolism. Mature red blood cells therefore rely on glucose and the glycolytic pathway for their metabolic needs. Moreover, their glycolytic pathway has a specific adaptation for the production of 2,3-bisphosphoglycerate (2,3-BPG), which reduces the affinity of the haemoglobin for oxygen, thereby facilitating the release of oxygen in the tissues (see Section 25.6). The percentage of circulating reticulocytes to mature erythrocytes can be estimated from the proportion of cells that stain with the dye brilliant cresyl blue (the reticulocytes). It is normally around 1 per cent (range 0.5–2.5 per cent).

Red blood cells have a lifespan of about 120 days

Once it has entered the general circulation, a red blood cell survives for around 120 days, after which time it is destroyed in the spleen, liver, or lymph nodes by large phagocytic cells known as **macrophages**. The protein portion of the erythrocyte is broken down into its constituent amino acids. The iron of the haem group is stored in the liver or bone marrow as ferritin and may be re-used later (see Section 25.5), while the remainder of the haem group is broken down

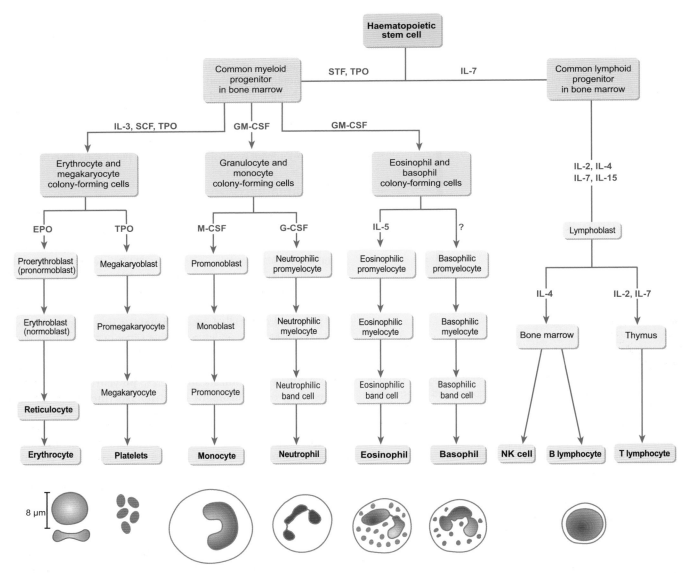

Figure 25.7 The principal steps in the development of the various cellular elements of the blood. The appearance of the different types of mature blood cell after staining is given at the foot of the figure. Pluripotent stem cells give rise to two main lineages, from which all of the main cell types are derived. The development of each cell type depends on specific growth factors, some of which are indicated in the figure. Those cells that are normally encountered in the circulation are indicated by the boxes with a pale red background. Note that not all the stages of development are shown. (Abbreviations: EPO = erythropoietin; TPO = thrombopoietin; SCF = stem cell factor; GM-CSF = granulocyte-macrophage cell stimulating factor; G-CSF = granulocyte stimulating factor; M-CSF = macrophage stimulating factor; IL-2, IL-3 etc. = interleukins 2, 3 etc.; ? = not known.)

into **biliverdin** which is subsequently metabolized to **bilirubin** and combined with glucuronic acid before being excreted into the gut by way of the bile (see Chapter 45). If red cell destruction, and therefore bilirubin production, is excessive, unconjugated bilirubin may build up in the blood and give a yellow colour to the skin—**haemolytic jaundice**. This condition can arise following a haemolytic blood transfusion reaction (see Section 25.9), in haemolytic disease of the newborn, or in genetic disorders such as **hereditary spherocytosis** in which the membrane of the red blood cells is defective.

Maturation of white cells

All of the different kinds of white cell are ultimately derived from haematopoietic stem cells in the bone marrow. The myeloid progenitor cells develop into distinct lineages that give rise to the erythrocytes and to megakaryocytes, granulocytes, and monocytes. The lymphoid progenitor cells give rise to B- and T-lymphocytes as well as the natural killer (NK) cells, which are discussed in Chapter 26.

Maturation of monocytes and granulocytes

In response to stem cell factor (SCF) and thrombopoietin, the haematopoietic stem cells develop into common myeloid progenitor cells. In turn, under the influence of granulocyte-macrophage cell stimulating factor (GM-CSF), these progenitor cells develop into monocyte and granulocyte colony-forming cells which are thought to become committed to forming the granulocytes and monocytes. During an infection, the rate of granulocyte production (especially of neutrophils) increases considerably as stored cells are released from the bone marrow and the mitotic activity of the colony-forming cells is stimulated.

The specific precursor cells of the neutrophils and monocytes are the **myeloblasts** and **monoblasts**, which develop from the granulocyte and monocyte colony-forming cells. The myeloblasts progress through a series of cell divisions to form **neutrophilic promyelocytes**, **myelocytes**, and **band cells** (cells with curved nuclei) before maturing into neutrophils with their characteristic lobular nuclei. This sequence of events is regulated by many growth factors including IL-3 and G-CSF (granulocyte cell stimulating factor). Similarly, the **monoblasts** give rise to **promonocytes** which are large (c. 18 μm) diameter cells containing a large nucleus with prominent nucleoli. Promonocytes subsequently develop into mature monocytes (see Figure 25.7). The development of the monoblasts into monocytes is regulated by at least seven different growth factors including IL-3, GM-CSF, and M-CSF (macrophage cell stimulating factor). After entering the blood, mature monocytes circulate for about 8 hours before entering the connective tissues, where they mature into **macrophages** or myeloid **dendritic cells** whose function is discussed in Chapter 26.

The progenitor cells of the eosinophils and basophils are the eosinophil colony-forming cells and basophil colony-forming cells, which give rise to the eosinophilic and basophilic promyelocytes, myelocytes, and band cells as indicated in Figure 25.7. Before the mature cells are released into the circulation, there is an increase in specific granule content which takes place over a period of 10 days or so. It is known that the maturation of the eosinophils requires IL-3, IL-5, and GM-CSF as growth factors, but the factors regulating the production of the basophils are not well understood.

Maturation of lymphocytes

Until recently the lymphocytes were thought to form a distinct and separate cell lineage, but there is now evidence to suggest that the progenitor cells that give rise to the lymphocytes retain the ability to form myeloid cells, the development of one kind rather than another depending on the specific growth factors to which the cells are exposed. Like the other white cells, the lymphocytes originate from stem cells in the bone marrow and under the influence of IL-7 give rise to lymphoid colony forming cells. The first identifiable progenitor is the **lymphoblast**. These cells divide several times to become smaller **prolymphocytes** and then lymphocytes.

Some immature lymphocytes migrate to the thymus where, under the influence of a number of interleukins, they mature into T-lymphocytes. Others remain in the bone marrow where, under the influence of IL-4, they differentiate into B-lymphocytes before migrating to become resident in the peripheral lymphoid tissues. Circulating lymphocytes originate mainly in the thymus and peripheral lymphoid organs (spleen, lymph nodes, tonsils etc.).

Production of platelets

The precursor cells that ultimately give rise to the platelets are known as **megakaryoblasts**. These cells are 15–20 μm in diameter and possess a large ovoid or kidney-shaped nucleus. Differentiation of the megakaryoblasts gives rise first to the promegakaryocytes and then the **megakaryocytes**, which are giant cells (35–150 μm in diameter) whose large irregular nucleus contains a large excess of DNA (i.e. they are polyploid cells). Their cytoplasm contains numerous mitochondria, rough endoplasmic reticulum, and Golgi complex. The megakaryocytes mature within the bone marrow and, as they mature, invaginations of the plasma membrane become evident, eventually branching throughout the entire cytoplasm. These form the so-called **demarcation membranes** which define the areas that will be shed as platelets.

The maturation of megakaryocytes and the production of platelets are regulated by a hormone called **thrombopoietin (TPO)** that is constitutively secreted by the liver, kidneys, and skeletal muscle. The circulating thrombopoietin is removed from the circulation, when it binds to its receptors on the platelets and megakaryocytes. Once the receptors have bound thrombopoietin, the hormone–receptor complex is internalized and degraded. If the platelet count rises, more thrombopoietin is removed from the circulation, its plasma level falls, and platelet production declines. If the platelet count falls, less TPO is bound, its plasma level rises, and platelet production increases. The platelet–thrombopoietin system thus forms a simple negative feedback control loop, although in practice other factors such as IL-6 secreted by macrophages and fibroblasts can also enhance platelet production.

Summary

Mature blood cells are renewed continuously by haematopoiesis. All cell types are generated ultimately from stem cells in the bone marrow; these divide to form progenitor cells that give rise to committed precursor cells, which mature to form the various types of blood cell.

The red cells are formed in the red bone marrow from myeloid precursor cells called erythroblasts. Through successive divisions, these cells start to synthesize haemoglobin, eventually developing into reticulocytes that are released into the circulation where they become mature erythrocytes. Erythropoiesis is controlled by erythropoietin, a hormone secreted by the kidneys. Red cells have a lifespan of about 120 days, after which they are destroyed by macrophages in the spleen, liver, or lymph nodes.

The white blood cells are derived from the bone marrow and the lymphoid tissue. The maturation of the various kinds of white blood cell is determined by growth factors which include many different interleukins and colony-stimulating factors.

Platelets bud off from giant cells (megakaryocytes), which are themselves derived from the stem cells in the bone marrow. Platelet production is regulated by thrombopoietin secreted by the liver.

25.5 Iron metabolism

Iron is an essential component of haemoglobin and myoglobin, as well as of certain enzymes and respiratory pigments. The body of an adult man contains a total of about 4.5 grams of iron, of which about 65 per cent is within the haemoglobin of red blood cells. A further 5 per cent or so is contained within myoglobin and enzymes, while the remainder is stored in the form of a protein complex called **ferritin**, largely by the liver. Smaller quantities of ferritin are also stored in the spleen and intestine.

When red blood cells become senescent, they are removed by phagocytes in the liver and spleen. Much of the iron derived from haemoglobin is recycled by the body, as illustrated in Figure 25.8. The iron from the digested haemoglobin is either returned to the plasma where it binds to **transferrin** (an iron-carrying protein), or is stored in the liver as ferritin. The iron–transferrin complex is carried to the erythropoietic tissue in the bone, where it is either used immediately in the production of haemoglobin for developing red cells or stored within the marrow itself. If blood loss occurs, the iron stored within the marrow is utilized and there is an increase in the rate of iron uptake from the circulation. The balance is restored by an increased rate of iron absorption in the gut some 3–4 days later.

Because the recycling of iron is so efficient, the need for dietary iron in adults arises mainly from loss caused by bleeding and by the death of intestinal cells. It therefore follows that the dietary requirement for iron is greater in menstruating women than in men, being about 1 mg per day in men and 2 mg per day in women of child-bearing age. Furthermore, children and pregnant women need relatively more iron because of their expanding circulatory volume. Dietary sources of iron include meat (specifically the myoglobin of the muscle), vegetables, and fruits. The normal Western diet contains adequate quantities of iron (around 15 mg day^{-1}) but strict vegetarians and vegans risk iron deficiency, as much of their dietary iron is bound by phytic acid (also called inositol hexaphosphate or IP$_6$) which is found in the husks of cereals. Iron bound to phytic acid is unavailable for absorption.

How is iron absorbed in the intestine?

Ionized iron can exist in two oxidation states, ferrous (Fe^{2+}) and ferric (Fe^{3+}). The low pH of the stomach lumen caused by the secretion of hydrochloric acid by the gastric mucosa solubilizes both ferrous and ferric iron salts. In addition, ascorbate (vitamin C) reduces Fe^{3+} to Fe^{2+}, which is less likely to form insoluble complexes with other constituents of the diet (particularly the fibre of cereal grains).

A simple scheme of the mechanism by which the epithelial cells of the upper intestine absorb iron is shown in Figure 25.9. Any remaining ferric iron (Fe^{3+}) that enters the duodenum is reduced to ferrous iron by an enzyme on the brush border (ferric oxidoreductase). The Fe^{2+} is taken up with hydrogen ions into the enterocytes by a transporter known as DMT1. Iron complexed with haem derived from dietary meat is absorbed directly by a separate pathway. Within the cell, haem oxygenase liberates Fe^{2+} from the haem molecule. Once inside the cell, iron can follow one of two pathways: it either leaves the enterocytes by the basolateral membrane transporter, ferroportin, or becomes bound to a cytoplasmic protein called apoferritin to form

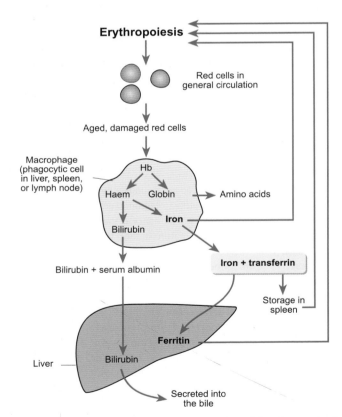

Figure 25.8 The principal stages in the recycling of Fe^{++} from red cell haemoglobin. Senescent or dead red cells are broken down in the liver and spleen and their components recycled. The haemoglobin is broken down by the liver to form bilirubin, which is excreted, and the iron is transported to the bone marrow where it is re-used in the synthesis of haemoglobin. (Courtesy of Dr E.S. Debnam.)

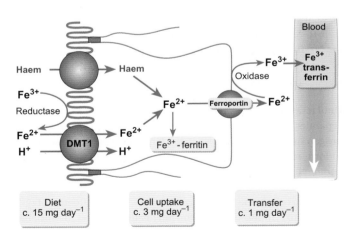

Figure 25.9 An outline of the mechanisms by which iron is absorbed by the intestine. Free iron is absorbed in association with hydrogen ions and can either be stored in the cell complexed with ferritin or exported to the blood, where it binds to transferrin for transport to the haematopoietic tissues. Iron bound to haem is absorbed by a separate pathway.

ferritin. The path taken is determined by the body's demand for iron. Normally, around two-thirds of absorbed iron is bound to ferritin within the enterocytes. Ferrous iron leaving the enterocytes via the basolateral membrane is converted to ferric iron before being absorbed into the blood and transported to the liver bound to its specific transport protein transferrin, which is primarily synthesized in the liver. Apotransferrin (transferrin that is not complexed with iron) binds two ferric ions (Fe^{3+}) and, importantly, renders the iron both soluble and non-toxic. Transferrin is taken up into cells by receptor mediated endocytosis and the iron is separated from transferrin by the acidification of the endosome. The apotransferrin and the transferrin receptor are subsequently recycled back to the cell surface, where they separate. The apotransferrin then re-enters the circulation to participate in a further round of iron transport.

Iron absorption is regulated in accordance with the body's needs

In iron deficiency, or following haemorrhage, the capacity of the small intestine to absorb iron is increased. After severe blood loss, there is a time lag of 3–4 days before absorption is enhanced. This is the time needed for the enterocytes to migrate from their sites of origin in the mucosal glands to the tips of the villi, where they are best able to participate in iron absorption. The absorption of iron from the intestinal lumen is controlled by a hormone synthesized by the liver called **hepcidin**. Hepcidin regulates the number of iron transporters (both DMT1 and ferroportin) expressed in the membrane of the enterocytes. When demand for iron is high, for example following haemorrhage, circulating levels of hepcidin fall and iron uptake from the small intestine is increased. When demand is low, hepcidin levels are high and iron absorption decreases.

Excess iron absorption is as undesirable as iron deficiency, since high levels of iron can be toxic. This situation is normally prevented by the binding of iron to ferritin within the cytoplasm of the enterocyte. This binding is almost irreversible, so any iron bound in this way is unavailable for absorption into the plasma. Instead, it is lost in the faeces when the intestinal cell is shed into the intestinal lumen. The level of iron in the plasma is sensed by specific iron regulatory proteins which control the rate of ferritin synthesis and, to maintain homeostasis, the amount of iron held in the so-called storage pool increases when dietary intake rises. Nevertheless, iron toxicity can become a problem if the diet is excessively rich in iron. In the genetic disease **hereditary haemochromatosis**, excessive amounts are absorbed even from a diet containing normal quantities of iron. This saturates plasma transferrin and leads to high levels of iron in the tissues and organ damage resulting from cell death caused by impaired mitochondrial function.

Summary

About two-thirds of the total body iron is within the haemoglobin of red blood cells, 5 per cent is within myoglobin and enzymes, while the rest is stored as ferritin, mainly in the liver. When red blood cells are destroyed, most of their iron is recycled immediately, or stored as ferritin.

Iron in the diet is absorbed into the enterocytes as ferrous iron (Fe^{2+}) by a carrier-mediated process. Absorbed iron is released into the blood across the basolateral membrane, where it combines with transferrin in the plasma to be transported to the tissues. Excess iron is stored within the enterocytes bound to ferritin. Iron absorption is regulated in accordance with the body's requirements by the hormone hepcidin. Following a haemorrhage, plasma hepcidin levels fall and the capacity of the small intestine to absorb iron is enhanced.

25.6 The carriage of oxygen and carbon dioxide by the blood

The blood transports the respiratory gases around the body. Oxygen is carried from the lungs to all the tissues, while the carbon dioxide produced by metabolizing cells is transported back to the lungs for removal from the body. The principles governing the exchange of gases in the lungs and the tissues are discussed fully in Chapter 32. Briefly, oxygen passes from the alveoli to the pulmonary capillary blood by diffusion because the partial pressure of oxygen in the alveolar air (P_AO_2) is greater than that of the blood in the pulmonary capillaries. In the peripheral tissues, the PO_2 is lower in the cells than in the arterial blood entering the capillaries and so oxygen diffuses out of the blood, through the interstitial spaces and into the cells. Conversely, the partial pressure of carbon dioxide (PCO_2) in metabolizing cells is much higher than that of the capillary blood. Carbon dioxide diffuses down its concentration gradient into the blood, which transports it to the lungs. Here, the PCO_2 of the blood as it reaches the pulmonary capillaries is greater than that of the alveoli and carbon dioxide diffuses from the capillaries to the alveoli. It is removed from the body during expiration. Reference values for the partial pressures of the blood gases are given in Table 25.3.

Haemoglobin increases the capacity of the blood to transport oxygen: each gram of haemoglobin can bind 1.34 ml of oxygen

At rest, oxygen is consumed by the body at a rate of around 250 ml min^{-1} and this must be supplied by the blood. The solubility of oxygen in the plasma water is very low—only 0.225 ml of oxygen are dissolved for each kPa of oxygen per litre of plasma (equivalent to 0.03 ml mm Hg^{-1}). At the normal arterial PO_2 of 13.33 kPa (100 mm Hg), each litre of plasma will therefore contain just 3 ml of dissolved oxygen ($0.225 \times 13.33 = 3.0$ ml). If this were the only means of transporting oxygen to the tissues, the heart would need to pump more than 80 litres of blood each minute to supply the required 250 ml min^{-1}. In fact, the blood is able to carry far more oxygen than this. At a P_AO_2 of 13.33 kPa (100 mm Hg), the oxygen content of whole blood is about 20 ml dl^{-1} blood (200 ml l^{-1}). As a result, the normal resting cardiac output of 5 l min^{-1} is more than sufficient to meet the oxygen requirements of the body at rest.

Table 25.3 Reference values for the partial pressures of blood gases

	Arterial blood	**Mixed venous blood**
Oxygen	13.3 (9.3–13.3) kPa	5.33 (4.0–5.33) kPa
	100 (80–100) mm Hg	40 (30–40) mm Hg
Carbon dioxide	5.33 (4.7–6.0) kPa	6.12 (5.33–6.8) kPa
	40 (35–45) mm Hg	46 (40–51) mm Hg

Note: This table gives the standard value and the range (in brackets) that covers 95% of the healthy population. Note that the capacity of the blood to carry oxygen will depend on its haemoglobin content. In males there is normally about 130–180 g l^{-1} of haemoglobin, while in females the value is usually lower at about 115–160 g l^{-1}. See text for further details.

The vast majority of the oxygen in the blood is carried in chemical combination with **haemoglobin**, an oxygen binding protein contained within the red cells. Each haemoglobin molecule comprises a protein part (globin) consisting of 4 polypeptide chains, and 4 nitrogen-containing pigment molecules called haem groups. Each of the four polypeptide groups is combined with one haem group (see Figure 25.10). In the centre of each haem group is one atom of ferrous (Fe^{2+}) iron that can combine loosely with one molecule of oxygen. Each molecule of haemoglobin (Hb) can therefore combine with four molecules of oxygen, to form **oxyhaemoglobin** (often written as HbO_2). The reaction for the binding of oxygen can be expressed as:

$$Hb + O_2 \rightleftharpoons HbO_2$$

When oxyhaemoglobin dissociates to release oxygen to the tissues, the haemoglobin is converted to **deoxyhaemoglobin**—also called reduced haemoglobin. Combination of oxygen with haemoglobin to form oxyhaemoglobin occurs in the alveolar capillaries of the lungs where the PO_2 is high (13.33 kPa or 100 mm Hg). Where the PO_2 is low (as in the capillaries supplying metabolically active cells), oxygen is released from oxyhaemoglobin and is then able to diffuse down its concentration gradient to the cells via the interstitial space.

Haemoglobin that is fully saturated with oxygen is bright red, while haemoglobin that has lost one or more oxygen molecules (deoxyhaemoglobin) is darker in appearance. When it has lost most of its oxygen, haemoglobin becomes deep purple in colour. As the blood passes through the tissues, it gives up its oxygen and the proportion of deoxyhaemoglobin increases. For this reason systemic venous blood is much darker in colour than systemic arterial blood. When the quantity of deoxyhaemoglobin in the circulating blood exceeds 5 g dl^{-1}, the skin and mucous membranes appear blue—a condition known as **cyanosis**.

The ease with which haemoglobin accepts an additional molecule of oxygen depends on how many of the binding sites are already occupied by oxygen molecules. There is cooperativity between the binding sites such that occupancy of one of the four sites makes it easier for a second oxygen molecule to bind and so on. As a result, the amount of oxygen bound to haemoglobin increases in an S-shaped (sigmoid) fashion as PO_2 increases—see Figure 25.11. This is known as the **oxyhaemoglobin dissociation curve** (or **the oxygen dissociation curve**). The sigmoid nature of the dissociation curve is physiologically significant because as PO_2 falls from 13.33 kPa (100 mm Hg)—the value in arterial blood—to about 8 kPa (60 mm Hg), the saturation of the haemoglobin with oxygen decreases by only about 10 per cent. As the PO_2 falls below

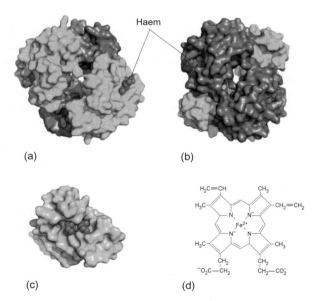

Haem

(a)   (b)

(c)   (d)

Figure 25.10 The structures of haemoglobin and myoglobin. As panels (a) and (b) show, the haemoglobin molecule consists of two α and two β peptide chains (coloured pale blue and dark blue respectively). Each peptide chain in haemoglobin has a single haem group (shown in red), making a total of four. Each haem group binds a single molecule of oxygen. Myoglobin (c) consists of a single peptide chain (coloured green) with a single haem group. Thus, one molecule of haemoglobin can carry four molecules of oxygen and one molecule of myoglobin can bind a single molecule of oxygen. The molecular structure of the haem group is shown in (d).

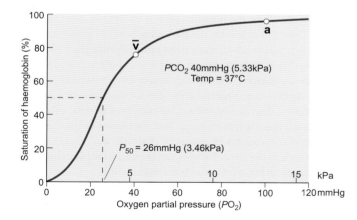

Figure 25.11 The oxyhaemoglobin dissociation curve for a PCO_2 of 5.3 kPa (40 mm Hg) at 37°C. Under these conditions, the P_{50} value is 3.46 kPa (26 mm Hg); point **a** gives the PO_2 in arterial blood (97 per cent saturated) and point $\bar{v}$ is the PO_2 for mixed venous blood (5.33 kPa or 40 mm Hg) at which value the haemoglobin is still about 75 per cent saturated. Note that as the PO_2 falls below 8 kPa (60 mm Hg) the curve becomes progressively steeper.

8 kPa, however, the curve becomes relatively steep so that small changes in PO_2 result in large changes in the degree of haemoglobin saturation.

Summary

The blood must supply oxygen to all tissues of the body and transport the carbon dioxide produced by metabolism to the lungs for removal from the body. Only a small amount of oxygen is carried by the plasma in physical solution. Most is carried loosely bound to haemoglobin within the red blood cells.

Oxygen content, oxygen capacity, and oxygen saturation

The quantity of oxygen in a given volume of blood must be carefully distinguished from the percentage saturation, which only indicates what proportion of the available haemoglobin is saturated. This distinction should be clear from the following definitions.

The **oxygen content** is the quantity of oxygen in a given sample of blood, whether obtained from an artery or a vein. It represents the quantity of oxygen combined with haemoglobin plus that physically dissolved in the plasma.

The **oxygen capacity** is the maximum quantity of oxygen that can combine with the haemoglobin of a given sample of blood. It can be determined by measuring the haemoglobin concentration in a blood sample. This is normally about 15 g dl^{-1} in males and 13.5 g dl^{-1} in females. When fully saturated, each g of haemoglobin will bind 1.34 ml of O_2 at STP; the oxygen capacity in ml O_2 per dl blood is then given by the haemoglobin concentration multiplied by 1.34. The oxygen capacity of a sample of blood depends, therefore, on the haemoglobin content and is independent of the partial pressure of oxygen.

The **oxygen saturation** is the ratio of the quantity of oxygen combined with haemoglobin in a given sample of blood to the oxygen capacity of that sample. It is expressed as a percentage thus:

$$\% \text{ saturation} = \frac{O_2 \text{ content} - \text{dissolved } O_2}{O_2 \text{ capacity}} \times 100$$

For a normal adult male, when the arterial PO_2 is close to 13.3 kPa (100 mm Hg) and the haemoglobin is 97 per cent saturated, the oxygen content of the blood will be $15 \times 1.34 \times 0.97 = 19.5$ ml O_2. dl^{-1} bound to haemoglobin *plus* 13.3×0.0225 ml $= 0.3$ ml O_2 in physical solution, giving a total O_2 content of 19.8 ml dl^{-1} (or 198 ml l^{-1}).

The dependence of the oxygen content of blood on its haemoglobin concentration is strikingly evident for those suffering from anaemia. For example, assume that a person has a haemoglobin concentration of only 7.5 g dl^{-1}. Further assume that the arterial PO_2 and percentage saturation (SpO$_2$) remain at 13.3 kPa and 97 per cent respectively. The amount of oxygen dissolved will be the same as for the previous normal example, 0.3 ml dl^{-1}, but the amount combined with haemoglobin will be much less: $7.5 \times 1.34 \times 0.97 = 9.7$ ml dl^{-1} plus 0.3 ml O_2 in solution, giving a total of only 10 ml dl^{-1} blood—just over half the total oxygen content of arterial blood with a normal haemoglobin concentration.

The important point to realize here is that, despite a normal arterial PO_2 and a normal SpO$_2$, the low haemoglobin content of the blood results in a low oxygen content. Consequently, oxygen delivery to the tissues may be insufficient to meet the requirements of even the mildest exercise. (Oxygen delivery is equal to the oxygen content of arterial blood multiplied by the cardiac output.) Indeed, someone suffering from severe anaemia may even experience breathlessness at rest.

Measurements of SpO$_2$ are made with a non-invasive technique called **pulse oximetry**, which exploits the difference in colour between reduced haemoglobin and oxyhaemoglobin. Although measurement of the SpO$_2$ is a useful indicator of the efficiency of gas exchange occurring in the lungs, particularly in patients with pulmonary diseases and those receiving oxygen therapy, it should be used together with careful observation of the state of the patient. A full picture of the state of tissue oxygenation requires measurement of the blood gases and the haemoglobin content of their blood.

The affinity of haemoglobin for oxygen is influenced by pH, PCO_2, 2,3-BPG, and temperature

So far, the oxyhaemoglobin dissociation curve has been considered as though the percentage saturation of haemoglobin remained constant for a given PO_2. In reality, the position of the curve varies with temperature, pH, PCO_2, and the concentration of certain metabolites such as 2,3-bisphosphoglycerate (2,3-BPG) (formerly known as 2,3-diphosphoglycerate or 2,3-DPG). In view of this, the dissociation curve is usually given for a pH of 7.4, a PCO_2 of 5.3 kPa (40 mm Hg), and a temperature of 37°C. It is worth noting that, under these conditions, haemoglobin in normal red cells is 50 per cent saturated with oxygen at a PO_2 of 3.4 kPa (26 mm Hg). This is known as the P_{50} value (the partial pressure at which the haemoglobin is half saturated with oxygen).

An increase in the PCO_2 (above 5.3 kPa or 40 mm Hg) and a reduction in pH (i.e. an increase in H$^+$ ion concentration) both shift the haemoglobin dissociation curve to the right (Figure 25.12). This effect is known as the **Bohr shift** or the **Bohr effect**. Physiologically this effect is important, as the affinity of haemoglobin for oxygen becomes less as the PCO_2 rises. Thus, in the tissues where the PCO_2 is relatively high, the affinity of haemoglobin for oxygen is lower than in the lungs (i.e. less oxygen is bound for a given PO_2). Consequently, oxygen delivery to actively metabolizing tissues is facilitated. In the lungs, as the PCO_2 of the pulmonary capillary blood falls, the Bohr shift acts to increase the affinity of haemoglobin for oxygen. In this way, the uptake of oxygen is facilitated during the passage of blood through the lungs.

As the temperature increases, the affinity of the haemoglobin for oxygen is also reduced and the dissociation curve for haemoglobin shifts to the right. Consequently, for a given level of PO_2, the percentage saturation of haemoglobin will be less than at 37°C. This may be of benefit during heavy muscular exercise, for example, since oxygen will be unloaded more readily from the blood to the active tissues as body temperature rises.

The affinity of purified haemoglobin for oxygen is much greater than that seen in whole blood—indeed, purified haemoglobin has an affinity for oxygen similar to that of myoglobin, which has a P$_{50}$ of

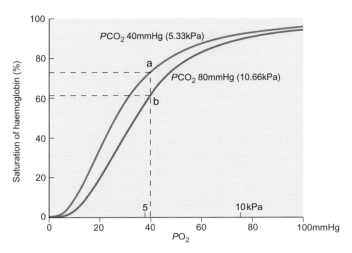

Figure 25.12 The effect of increasing PCO_2 on the oxyhaemoglobin dissociation curve. As the PCO_2 increases the P_{50} value for the dissociation curve is shifted to the right (blue curve). This is known as the Bohr shift. The dissociation curve is affected in a similar manner by a fall in pH or an increase in 2,3-BPG or temperature. The effect of the rightward shift is to decrease the affinity of haemoglobin for oxygen. This is shown by the difference in haemoglobin saturation when PO_2 is 5.33 kPa (40 mm Hg) as PCO_2 increases from 5.33 kPa (40 mm Hg), point **a** on the red curve, to 10.66 kPa (80 mm Hg), point **b** on the blue curve.

only 0.13 kPa (1 mm Hg). In normal red cells, however, haemoglobin has a P_{50} of 3.4 kPa at a PCO_2 of 5.3 kPa (i.e. P_{50} is 26 mm Hg at a PCO_2 of 40 mm Hg). This difference in the affinity of haemoglobin for oxygen is attributable to 2,3-BPG which is synthesized by the red cells during glycolysis. The 2,3-BPG binds strongly to haemoglobin and decreases its affinity for oxygen (i.e. it causes the oxyhaemoglobin dissociation curve to be shifted to the right). The concentration of 2,3-BPG is about 4 mmol l^{-1} in normal red cells but may be increased in anaemia or when living at high altitude, where PO_2 of the inspired air is significantly reduced. Fetal haemoglobin (HbF) has a somewhat higher affinity for oxygen than adult haemoglobin.

Summary

The amount of oxygen carried in the blood depends on the partial pressure of oxygen and on the concentration of haemoglobin. Oxygen binding is described by the oxyhaemoglobin dissociation curve, which has a sigmoid shape. The oxygen dissociation curve is shifted to the right by an increase in PCO_2, an increase in the level of 2,3-BPG, an increase in temperature, or a fall in pH. This change in the affinity of haemoglobin for oxygen is known as the Bohr effect.

Myoglobin

Myoglobin is an oxygen binding protein, present in cardiac and skeletal muscle, that has a much higher affinity for oxygen than the haemoglobin of the red cells. It is half saturated at a PO_2 of only 0.13 kPa (1 mm Hg). The shape of the oxygen dissociation curve for

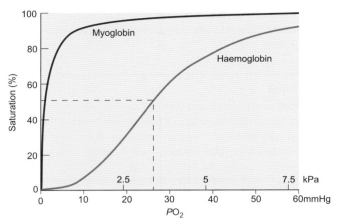

Figure 25.13 A comparison between the oxygen dissociation curves for myoglobin (blue curve) and haemoglobin (red curve). Note that myoglobin has a P_{50} value of only about 0.13 kPa (1 mm Hg) while the haemoglobin of arterial blood has a P_{50} of 3.46 kPa (26 mm Hg).

myoglobin is shown in Figure 25.13. Oxygen is not liberated in significant quantities until the PO_2 falls below 0.65 kPa (5 mm Hg). Nevertheless, myoglobin acts as a store of oxygen for situations where the oxygen supply from the capillaries is insufficient to meet the demands of aerobic metabolism in an exercising muscle. This situation may arise both in skeletal muscles during heavy exercise and during the contraction of the heart when the capillary circulation is temporarily interrupted. During periods of severe tissue hypoxia, the oxygen bound by myoglobin is used to maintain the production of ATP by the mitochondria until the local circulation is restored. It is therefore not surprising that the muscles of diving mammals such as seals contain very large quantities of myoglobin.

Carbon monoxide binds strongly to haemoglobin

Carbon monoxide is able to bind to haemoglobin and other oxygen binding proteins such as myoglobin and cytochrome oxidase. Indeed, the affinity of carbon monoxide for haemoglobin is more than 200 times that of oxygen. This would mean that breathing air containing a PCO of only 0.13 kPa (1 mm Hg) would quickly result in virtually all of the haemoglobin in the blood being bound to carbon monoxide (as carboxyhaemoglobin). Moreover, carbon monoxide tends to shift the oxygen-haemoglobin dissociation curve to the left and this impairs the unloading of oxygen from the blood. For these reasons, carbon monoxide is a highly toxic gas. Carboxyhaemoglobin is cherry red in colour, and patients suffering from carbon monoxide poisoning remain pink in colour despite the poor oxygen delivery to the tissues.

The standard treatment for carbon monoxide poisoning is to administer 100 per cent O_2 via a tight-fitting mask, but in certain severe cases, for example following extensive CO exposure with suspected nerve damage, the patient may be given 100 per cent oxygen at a pressure greater than one atmosphere (hyperbaric oxygen treatment). When oxygen is administered at a pressure of three atmospheres the half-life of carboxyhaemoglobin in the blood is reduced from around four hours at ambient pressure to 30 min. In both treatment regimes, the high partial pressure of oxygen helps to displace the carbon monoxide from haemoglobin. However, the standard oxygen therapy is usually the recommended treatment option as the use of hyperbaric oxygen remains controversial. Treatment is continued until the carboxyhaemoglobin level is less than 10 per cent of total haemoglobin.

Carbon dioxide is carried in the blood in three different forms: as dissolved gas, as bicarbonate, and as carbamino compounds

Chemical determination shows that arterial blood contains much more CO_2 than O_2 (49 ml dl^{-1} compared to 19.8 ml dl^{-1} for O_2). The CO_2 is carried in the blood in several forms. These are:

1. in physical solution as dissolved CO_2

2. as bicarbonate ions

3. as carbamino compounds—a combination between CO_2 and free amino groups on proteins.

At first sight, this gives the impression that the carriage of carbon dioxide by the blood is far more complex than that of oxygen. In reality, the principles involved are quite straightforward. In what follows, each form of transport will be considered briefly.

As with all gases, the concentration of dissolved CO_2 in the blood is determined by its solubility and its partial pressure. For plasma at normal body temperature, the solubility of CO_2 is 0.526 ml dl^{-1} kPa^{-1} (0.07 ml dl^{-1} mm Hg^{-1}). Therefore, at a normal arterial PCO_2 of 5.3 kPa (40 mm Hg), the amount of CO_2 transported in solution is $5.3 \times 0.526 = 2.8$ ml dl^{-1} (this is equivalent to 1.2 mmol CO_2 per litre of blood). Mixed venous blood has a PCO_2 of around 6.12 kPa (46 mm Hg) and will therefore contain $6.12 \times 0.526 = 3.2$ ml CO_2 dl^{-1}. Because of its high solubility, between 5 and 7 per cent of total blood carbon dioxide is in physical solution (in normal arterial blood only 1.5 per cent of oxygen is in solution).

The carbon dioxide which is produced as a result of tissue metabolism combines with water to form **carbonic acid** in the reaction:

$$CO_2 + H_2O \rightleftharpoons H_2CO_3 \qquad [1]$$

This readily dissociates to form hydrogen ions (H^+) and bicarbonate ions (HCO_3^-), as follows:

$$H_2CO_3 \rightleftharpoons H^+ + HCO_3^- \qquad [2]$$

Reaction 1 takes place only very slowly in the plasma, but in the red cells it is catalysed by an enzyme called **carbonic anhydrase**. Consequently, as carbon dioxide diffuses into the red blood cells, carbonic acid is formed which immediately dissociates to yield bicarbonate and hydrogen ions. The latter are mainly buffered by haemoglobin, while much of the bicarbonate moves back out of the cell in exchange for chloride ions (see panel (a) of Figure 25.14). This is known as the **chloride shift** or **Hamburger effect** and accounts for the fact that the concentration of chloride in the plasma of venous blood is lower than that of arterial blood. About 90 per cent of the total blood CO_2 is transported in the form of bicarbonate ions.

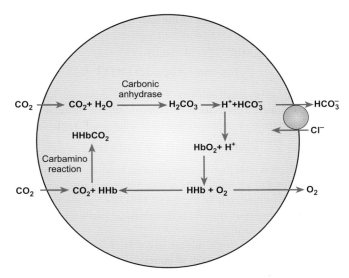

(a) CO_2 uptake by red cells as the blood perfuses active tissues

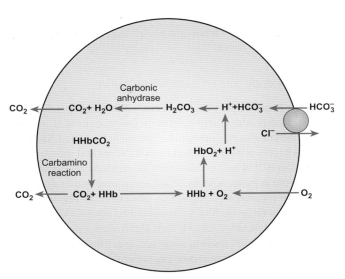

(b) O_2 uptake by red cells as the blood passes through the lungs

Figure 25.14 A schematic representation of CO_2 and O_2 transport in the blood; panel (a) shows the exchange of CO_2 and O_2 that occurs between the blood and the tissues. Note that as CO_2 is taken up, bicarbonate ions exit the red cells in exchange for Cl^{--}. In panel (b) the process is reversed in the lungs as the blood takes up O_2 from the alveolar air, CO_2 leaves the cells, and bicarbonate ions are taken up in exchange for Cl^-.

The buffering of the hydrogen ions formed by dissociation of carbonic acid by haemoglobin is extremely important, since it allows large amounts of carbon dioxide to be carried in the blood (as HCO_3^-) without the pH of the blood altering by more than about 0.05 pH unit (see Chapter 41 p. 638).

Although the majority of the carbon dioxide that enters the red blood cells from the tissues is hydrated to form carbonic acid which subsequently dissociates into H^+ and HCO_3^-, about a third combines with amino groups on the haemoglobin molecules in the reaction:

$$HbNH_2 + CO_2 \rightleftharpoons HbNHCOO^- + H^+$$

(carbaminohaemoglobin)

In addition, a very small amount of carbon dioxide is carried in the blood combined with α-amino groups on plasma proteins in the form of carbamino compounds formed by the general reaction:

$$R\text{-}NH_2 + CO_2 \rightleftharpoons R\text{-}NHCOO^- + H^+$$

The reactions involved in the carriage of carbon dioxide in the form of both bicarbonate ions and carbamino compounds are illustrated diagrammatically in Figure 25.14a. The reverse processes occur in the lungs as the red cells exchange carbon dioxide for oxygen (Figure 25.14b).

To summarize: arterial blood has a PCO_2 of approximately 5.3 kPa (40 mm Hg) and each dl contains about 2.8 ml of CO_2 in solution, 43.9 ml as HCO_3^-, and 2.3 ml as carbamino compounds, making a total of 49 ml CO_2 per dl^{-1}. Mixed venous blood has a PCO_2 of 6.1 kPa (46 mm Hg) and each dl contains approximately 3.2 ml of CO_2 in solution, 47 ml as HCO_3^-, and 3.8 ml as carbamino compounds (mainly carbamino haemoglobin), equivalent to a total of 54 ml CO_2 per dl.

The carbon dioxide dissociation curve

The amount of carbon dioxide present in solution depends on the PCO_2, and this in turn will determine the amount of HCO_3^- and carbamino compounds that will be formed in the blood. The relationship between the PCO_2 (in kPa or mm Hg) and the total CO_2 (ml CO_2 per dl^{-1} blood) is called the CO_2 dissociation curve. It differs from the oxyhaemoglobin dissociation curve in that it does not become saturated even at high PCO_2 (see Figure 25.15). Across the physiological range of PCO_2 for whole blood from 5.3 kPa (40 mm Hg) in arterial blood to 6.13 (46 mm Hg) in mixed venous blood, the CO_2 dissociation curve is roughly linear. The quantity of CO_2 carried in the blood is, however, dependent on the degree of oxygenation of haemoglobin. This is called the **Haldane effect**, which is also illustrated in Figure 25.15.

Two main factors are responsible for the changes in carbon dioxide affinity of the blood seen when HbO_2 levels vary:

1. oxyhaemoglobin is less able to form carbamino compounds than reduced haemoglobin

2. oxyhaemoglobin is a less efficient buffer of hydrogen ions than reduced haemoglobin.

Consequently, hydrogen ions are more readily buffered by haemoglobin in the tissues (where less of the haemoglobin is in the form of HbO_2). This favours the formation of bicarbonate ions in the blood by driving the reaction:

$$H_2O + CO_2 \rightleftharpoons HCO_3^- + H^+$$

to the right and permits more CO_2 to be carried as bicarbonate ion.

In the lungs, where about 97 per cent of the haemoglobin is in the form of oxyhaemoglobin, the carbon dioxide content of the blood is relatively lower than it is in the tissues, where

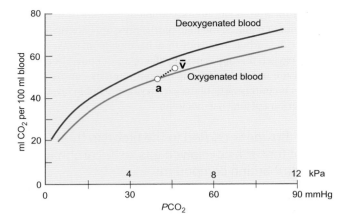

Figure 25.15 The carbon dioxide dissociation curve for whole blood and the Haldane effect. Note that the curve for deoxygenated blood (blue) lies above that for oxygenated blood (red) indicating that deoxygenated blood contains more carbon dioxide for a given PCO_2. The dotted line linking points **a** and **v̄** show the change in carbon dioxide content of oxygenated arterial blood (**a**) as it leaves the lungs compared to that of mixed venous blood **v̄** returning to the lungs.

oxyhaemoglobin makes up around 75 per cent of the total Hb. In other words, more carbon dioxide may be carried when HbO_2 is low. This adaptation is physiologically highly significant, as a major purpose of blood gas transport is to load the blood with CO_2 in the tissues and unload it for expiration in the lungs.

Summary

Carbon dioxide (CO_2) is carried in the blood in three forms: in physical solution, as bicarbonate, and bound to haemoglobin as carbamino compounds. The carbon dioxide dissociation curve is virtually linear in the physiological range of blood PCO_2. The affinity of the blood for CO_2 is determined by the degree of oxygenation of the haemoglobin (the Haldane effect). In the red blood cells, CO_2 combines with water to form carbonic acid. This reaction is catalysed by carbonic anhydrase. The carbonic acid dissociates to H^+ and HCO_3^-. The H^+ is buffered by haemoglobin and other blood buffers, while the HCO_3^- diffuses out of the red cells in exchange for Cl^- (the chloride shift).

25.7 Major disorders of the red and white blood cells

This section focuses on the consequences arising from changes in the rate of production or destruction of the cellular elements of the blood. Broadly, blood cell disorders fall into two categories, proliferative disorders (where there is an excess of cells, often with abnormal function) and deficiency disorders (where there are too few). Red and white cells will be considered here, while abnormalities of platelet function will be considered in the section on haemostasis (Section 25.8).

Red cell abnormalities

Anaemia

This term covers a variety of blood disorders characterized by a reduced number of red cells, a reduced haemoglobin concentration, or both. All types of anaemia result in a reduction in the oxygen-carrying capacity of the blood. Anaemia may arise for a number of reasons.

1. A reduction in red cell number—this can arise as a result of acute haemorrhage, after which plasma volume is restored in a short time but red cell production takes much longer.

2. A reduction in the haemoglobin content of the red cells—for example as a result of iron deficiency due to chronic blood loss or to pregnancy.

3. Occasionally the bone marrow fails to function normally. This results in so-called **aplastic anaemia** and can arise spontaneously or as a consequence of damage to the marrow by irradiation or exposure to cytotoxic drugs.

4. The normal volume of the red cells (also called the mean corpuscular volume or MCV) lies between 80 and 95×10^{-15} l (80–95 fl). A reduction in red cell size is known as **microcytic anaemia** (i.e. a MCV of less than 80 fl). This condition is seen in cases of iron deficiency and in thalassaemia (see point 7 below).

5. Anaemia with an MCV greater than 96 fl is called **macrocytic anaemia**. It has many causes but in Western countries is most commonly seen in patients lacking vitamin B_{12} (cyanocobalamin) or folate, both of which are essential for the normal maturation of erythrocytes in the bone marrow (see Section 25.4). This type of macrocytic anaemia is known as **pernicious anaemia** and is characterized by abnormally large red cells called megaloblasts (MCV > 110 fl) that are present in greatly reduced numbers. A deficiency of vitamin B_{12} may arise when its absorption is inadequate due to a lack of intrinsic factor in the gastric mucosa (see Chapter 44) or, more rarely, when the final part of the ileum (small intestine) is diseased—as in Crohn's disease. It can also arise as a result of diet lacking vitamin B_{12}. A poor diet may also lead to a deficiency of folate, a cause of macrocytic anaemia in alcoholism.

6. Abnormalities in haemoglobin structure can lead to acceleration of red cell destruction (**haemolysis**). One such abnormality is **sickle cell anaemia**, in which there is a defect in one of the chains of the haemoglobin molecule. Sickle haemoglobin (HbS) is transmitted by recessive inheritance, and the disease is prevalent in people descended from black African populations. In homozygous individuals, the HbS becomes sickled when deoxygenated, causing deformation of the red cells (see Figure 25.16). The deformed cells clump together and obstruct the blood flow in the capillaries, causing tissue hypoxia with subsequent damage and intense pain. Virtually every organ is affected, but the liver, spleen, heart, and kidneys are especially vulnerable to damage because of the increased risk of blood clot formation caused by the sluggish blood flow. The HbS in the red cells of those suffering from sickle cell anaemia has a $P_{50} \approx 50$ mm Hg so that the oxygen dissociation curve is shifted to the right, facilitating the unloading of oxygen in the tissues but potentially reducing the haemoglobin saturation in the lungs. People who carry the HbS gene have a higher resistance to malaria because the parasite that causes malaria cannot live in blood cells containing HbS. This is true for both homozygotes (where the affected individual inherits two copies of the defective gene from their parents) and heterozygotes (where only a single copy is inherited, the second copy being normal). Possession of the affected gene is a significant biological advantage where malaria is endemic and explains why such a disadvantageous mutation has survived in a high proportion of the population.

7. **Thalassaemia** is the name given to a group of anaemias in which there is a reduced rate of synthesis of one or both of the α or β globin chains. This is a consequence of mutations in the genes that synthesize haemoglobin (many different mutations have been recorded). The defects are inherited in a simple Mendelian fashion and are recessive, so that heterozygotes do not normally exhibit the disease. The two most common forms are the α-thalassaemias and the β-thalassaemias, the latter group being clinically the most important. Thalassaemias occur predominantly amongst Mediterranean, African, and black American populations. In healthy people the synthesis of both haemoglobin chains are balanced 1:1 so there no excess of one over the other. In β-thalassaemia, however, there is an excess production of α chains which form insoluble precipitates that impair the normal maturation of red cells. Those cells that are formed and released into the circulation are fragile and short-lived, which compounds the anaemia caused by the reduction in red cell production. In α-thalassaemia no insoluble complexes are formed but red cell survival is shorter than normal. This then leads to anaemia and a reduced oxygen carrying capacity.

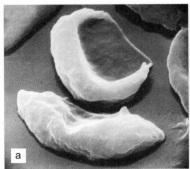 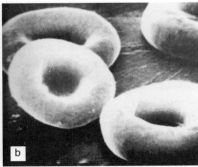

Figure 25.16 Scanning electron micrographs that show the appearance of sickled and normal red cells (panels a and b respectively). The images are similar in scale and the normal red cell (right) has a diameter of about 8 μm. (Courtesy of NASA.)

Polycythaemia

This condition is the result of over-stimulation of red blood cell production. It brings about an increase in the haematocrit value (to as much as 60–80 per cent) and a rise in blood viscosity that is not associated with dehydration. It is often seen in people living at high altitude, who experience chronic hypoxia because of the low prevailing atmospheric oxygen tension (see Chapter 37 for further details), though it can also arise under other circumstances (e.g. **polycythaemia rubra vera**—increased red cell production caused by neoplasia of the bone marrow). The increase in red blood cell numbers increases the oxygen-carrying capacity of the blood but the resulting increase in the viscosity of the blood places an extra load upon the heart. Over time, the heart hypertrophies (enlarges) to adapt to the increased work load.

White cell abnormalities

As with the red cells, disorders of the leukocytes fall into two broad categories: deficiency disorders and proliferative disorders.

Deficiency disorders

Leukopenia is the term that describes an absolute reduction in the numbers of white blood cells. It may affect any of the different types of leukocyte but most often involves the neutrophils, which are the predominant type of granulocyte. In this case, the disorder is known as **neutropenia**. It can result from defective neutrophil production or from an increase in the rate of removal of neutrophils from the circulation. The former may arise as part of a genetic impairment of the regulation of neutrophil production, aplastic anaemia, in which all the myeloid stem cells are affected, or following certain types of chemotherapy. It may also be a consequence of the overgrowth of neoplastic cells characteristic of some forms of leukaemia, which suppresses the function of the neutrophil precursor cells.

Occasionally, leukopenia arises because of an accelerated rate of neutrophil removal from the circulation rather than a reduction in the rate of production. This is most usually a consequence of chemotherapy but may also be seen in certain infections or auto-immune disorders in which neutrophils are destroyed. The neutrophils are essential in the inflammatory response. Infections are, therefore, common in people with neutropenia and these may be severe or even life threatening.

Proliferative disorders of the white blood cells

An increased white cell count ($> 10 \times 10^9$ cells l^{-1}) is known as **leukocytosis**. It is often a sign of an inflammatory response to infection, which is normally self-limiting—as in infectious mononucleosis (glandular fever)—but it may be evidence of a malignant proliferative disease such as a leukaemia.

Leukaemia is characterized by greatly increased numbers of abnormal white cells circulating in the blood. There are several different types of leukaemia, classified according to their cells of origin (lymphocytic or myelocytic) and whether the disease is acute or chronic. Lymphocytic leukaemias, which are most commonly seen in children, involve the lymphoid precursors that originate in the bone marrow. Cancerous production of lymphoid cells then spreads to other tissues such as the spleen, lymph nodes, and CNS. Myelocytic disease, which is more common in adults, involves the myeloid stem cells in the bone marrow. The maturation of all the blood cell types, including granulocytes, erythrocytes, and thrombocytes, is affected.

Leukaemic cells are usually non-functional and therefore cannot provide the normal protection associated with white blood cells. Common consequences of the disease include the development of infections, severe anaemia, and an increased tendency to bleed because of a lack of platelets (thrombocytopenia). Furthermore, the leukaemic cells of the bone marrow may proliferate so rapidly that they invade the surrounding bone itself. This causes pain and an increased risk of fractures.

Almost all forms of leukaemia spread to other tissues, particularly those that are highly vascular, such as the spleen, liver, and lymph nodes. As they invade these regions, the growing cancerous cells cause extensive tissue damage and place heavy demands on the metabolic substrates of the body, especially amino acids and vitamins. The energy reserves of the sufferer are thus depleted and the body protein is broken down. Hence, weight loss and excessive fatigue are characteristic symptoms of leukaemia.

Lymphomas are malignant diseases of the lymphoid tissue that are characterized by abnormal numbers of lymphocytes. There are a number of different lymphomas including Hodgkin's disease, Burkitt's lymphoma, and follicular lymphoma. All are characterized by enlargement of lymph nodes, often in more than one region of the body. **Myelomas** are malignant diseases of the bone marrow most commonly seen in the elderly. They are characterized by the presence of an abnormal immunoglobulin (called paraprotein) which is produced by a single clone of abnormal plasma cells. These cells infiltrate the bone marrow and may cause a reduction in the number of neutrophils, platelets, and red cells. There is also destruction of bone tissue that may cause pain and result in fractures.

Summary

Anaemia is a general term used to describe disorders of the red blood cells characterized by a reduced haematocrit. It may arise from a reduction in the number of red blood cells or in their haemoglobin content, or from abnormalities of haemoglobin structure. An important consequence of all types of anaemia is a reduction in the oxygen-carrying capacity of the blood. An excess of red cells is known as polycythaemia and causes an increase in blood viscosity. It is commonly seen in people who reside at high altitudes, but may reflect abnormal red cell production by the bone marrow.

Leukopenia is defined as an absolute reduction in the number of white blood cells. It may be due either to defective production or to accelerated removal of white cells from the circulation. Infections are common in patients suffering from leukopenia. Proliferation disorders of the white blood cells include leukaemias, lymphomas, and myelomas. In leukaemia, there are high numbers of abnormal white blood cells which are usually non-functional. Patients suffer from severe anaemia, infections, weight loss, and excessive fatigue.

25.8 Mechanisms of haemostasis

When a blood vessel is damaged by mechanical injury of some kind, excessive blood loss from the wound is prevented by a process called **haemostasis**. This involves a series of events—vasoconstriction, platelet aggregation, and blood coagulation (clot formation). Later, blood vessel repair, and clot retraction and dissolution, complete the healing process.

Vasoconstriction

When the vascular endothelium is damaged, there is a localized contractile response by the vascular smooth muscle causing the vessel to narrow. This may be mediated by humoral factors or directly by mechanical stimulation. In arterioles and small arteries, closure may be virtually complete. However, this response lasts for only a short time and, to prevent serious loss of blood, further haemostatic mechanisms are initiated that lead to the formation of a blood clot.

The role of platelets

Within seconds of a vascular injury, platelets start to accumulate and adhere to the site of damage. This occurs in a number of stages: first the platelets become attached to the extracellular matrix surrounding the injured vessel via integrin receptors that are activated by collagen and other extracellular proteins (see Chapter 6). This activates the platelets and, once activated, platelets change from their normal small disc-like shape and become more elongated with fine surface projections (filopodia—see Figure 25.4b). Activated platelets aggregate in a positive feedback process mediated by ADP, 5-hydroxytryptamine (5-HT), prostaglandins, and thromboxane A_2 (TXA_2)—all of which are secreted by the platelets themselves. In addition to its role in platelet aggregation, the 5-HT also causes vasoconstriction which results in a reduction of blood flow. The activated platelets adhere to each other and to the walls of damaged vessels via their integrin receptors. This adhesion requires von Willebrand factor and other adhesive glycoproteins (e.g. fibronectin and fibrinogen) and results in the formation of a **platelet plug**, which may be sufficient to stem the flow of blood from minor wounds. Furthermore, the activated platelets provide a surface that facilitates the reactions that lead to the formation of a fibrin network which strengthens the platelet plug (see next subsection).

In addition to sealing damaged vessels, the platelets play a continuous role in maintaining normal vascular integrity. This is illustrated by the increased capillary permeability seen in people suffering from platelet deficiency (**thrombocytopenia**). Such individuals often develop spontaneous tiny haemorrhages in the skin (**petechiae**) and mucous membranes, giving the skin a curious blotchy appearance, particularly in the lower extremities. There may also be bleeding into subcutaneous tissue, giving the appearance of severe bruising (**ecchymoses**).

Blood coagulation

This is the process by which fibrin strands create a mesh that binds blood components together to form a blood clot. It is a complex process that involves the sequential activation of a number of factors (**clotting factors**) that are normally present in the plasma in an inactive form. Most of the clotting factors are serine proteases related to trypsin that are synthesized in the liver before undergoing post-translational modification by a process requiring vitamin K. Throughout the medical and scientific literature, these enzymes are known by a variety of names and/or roman numerals (see Table 25.4). In this account, the factors are assigned the nomenclature by which they are most commonly known. A cascade of reactions occurs by which one activated factor activates another according to the following scheme:

$$\text{Activated clotting factor}$$
$$\downarrow$$
$$\text{Inactive clotting factor} \rightarrow \text{Active clotting factor}$$

Activated factors are designated by the letter a after the numeral, e.g. activated factor X is called Xa. In the scheme shown above, an enzyme is released from an inactive precursor; it then complexes with its substrate (usually another inactive clotting factor) and an appropriate cofactor on an organizing surface, which is usually a cell surface (e.g. the platelet membrane). The formation of these complexes speeds up the activation of the individual factors several thousandfold.

The major reactions in the clotting process are shown in Figure 25.17, from which it is evident that there are two pathways implicated in the formation of a fibrin clot. These are known as the **extrinsic and intrinsic pathways**; these are not totally separate

Table 25.4 The nomenclature of the clotting factors of blood

Factor	Names and synonyms
I	Fibrinogen
II	Prothrombin
IIa	Thrombin
III	Tissue factor, Tissue thromboplastin
IV	Calcium (Ca^{2+})
V	Proaccelerin, labile factor, accelerator globulin
VI	Not assigned
VII	Proconvertin, SPCA, stable factor, autoprothrombin I
VIII	Antihaemophilic factor, antihaemophilic globulin, antihaemophillic factor A, platelet cofactor I
IX	Plasma thromboplastic component, Christmas factor, antihaemophillic factor B, platelet cofactor II
X	Stuart-Prower factor, autoprothrombin III
XI	Plasma thromboplastin antecedent (PTA), antihaemophillic factor C
XII	Hageman factor
XIII	Fibrin stabilizing factor, Laki-Lorand factor, fibrin ligase

Note: Factors I–IV are generally known by their names, while factors V–XIII are generally referred to by their Roman numeral. Activated factors are designated by the letter a after the numeral, e.g. activated factor X is called Xa.

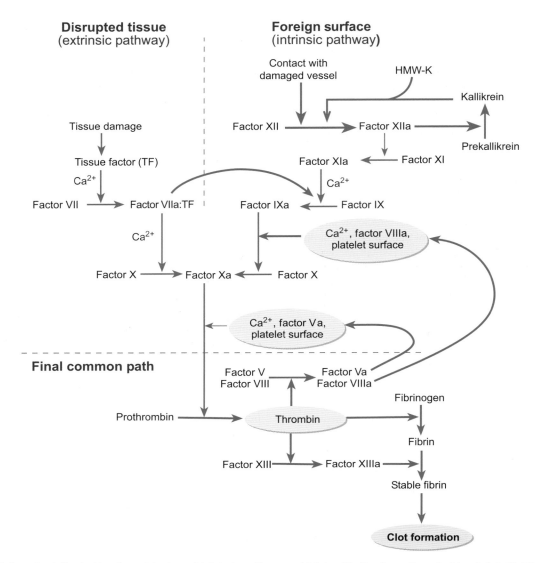

Figure 25.17 A flow chart illustrating the extrinsic and intrinsic pathways which lead to the formation of a blood clot. Clotting is initiated by contact of blood with tissue factor (the extrinsic pathway) or by contact with collagen exposed following damage to a blood vessel (the intrinsic pathway). Both lead to the conversion of factor X to factor Xa and to activation of thrombin and the formation of a blood clot. Moreover, the complex formed by tissue factor and factor VIIa participates in the activation of factor IX.

and both are needed for normal haemostasis. They are activated when a blood vessel is damaged and blood leaks out of the vascular system. The intrinsic system (which is the slower of the two) is activated as blood comes into contact with the injured vessel wall, while the extrinsic system is activated when blood comes into contact with a glycoprotein called **tissue factor** (formerly called tissue thromboplastin). The intrinsic pathway is so called because all of the elements required to activate it are present in normal blood, while the extrinsic pathway is activated by a factor from outside the blood, i.e. tissue factor. Both pathways contribute to the formation of activated factor X (factor Xa) at the end of the first stage of coagulation. Further steps in the clotting reaction are common to both pathways and involve the enzymatic conversion of inactive prothrombin to **thrombin**. This then initiates the polymerization of **fibrinogen** to form **fibrin** strands within which

plasma, blood cells and platelets become trapped to form a clot—see Figure 25.18.

The extrinsic pathway

The extrinsic pathway is initiated when blood comes into contact with tissue surrounding a damaged blood vessel. It requires Ca^{2+} and occurs in three basic steps (Figure 25.17 top left).

1. Damage to a tissue exposes the blood to tissue factor, which is expressed on the surface of smooth muscle cells and the fibroblasts of the tunica adventitia (the outer layer of the vessel wall) which are normally separated from the blood by the vascular endothelium. The exposed tissue factor usually remains at the site of injury and this localizes the formation of the blood clot.

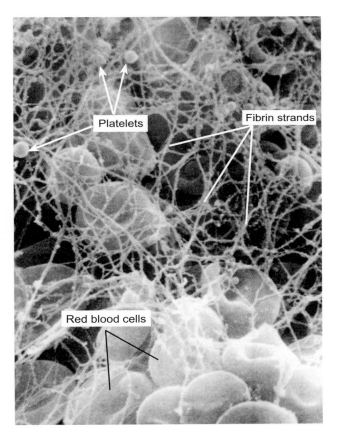

Figure 25.18 A scanning electron micrograph of a blood clot showing the entrapment of the red blood cells by a mesh of cross-linked fibrin strands. (From W. Shelley (1985) *Journal of the American Medical Association* **249**, 3089.)

2. The contact of blood with tissue factor sets the clotting process in motion. Tissue factor combines with factor VII. Proteases present in small amounts in the plasma activate the bound factor VII by the cleavage of a single peptide bond. The tissue factor/factor VIIa complex then activates factors IX and X.

3. On the surface of the platelets, factor Xa combines with factor Va and converts prothrombin to thrombin, as described below for the final common pathway. It also speeds up the activation of factor VII that is complexed with tissue factor (a positive feedback loop).

The intrinsic pathway

The initial step in this series of reactions is dependent upon the activation of factor XII (Hageman factor). When there is vascular damage and blood is exposed to collagen in the surrounding tissues, factor XII is converted to activated factor XII (XIIa) which can act on a plasma protein called prekallikrein to release **kallikrein**. This is a protease that forms a complex with another plasma protein called high molecular weight kininogen (HMW-K). This complex then activates more factor XIIa in a positive feedback loop.

Factor XIIa initiates a cascade in which factor XI becomes activated (factor XIa). Factor XIa then converts factor IX (Christmas factor) to factor IXa by a calcium-dependent process. Factor IXa then associates with factor VIIIa on the surface of the activated platelets to generate factor Xa. Factor VIII (anti-haemophilic factor) is the factor missing in people (chiefly males) who suffer from haemophilia A (discussed later in this section). Factor Xa combines with factor Va on the surface of the activated platelets to form a complex which catalyses the formation of thrombin from prothrombin (an inactive enzyme precursor).

The final common pathway

In the next step, thrombin cleaves fibrinogen to release fibrin monomers, which undergo end-to-end polymerization to form long strands of an insoluble protein, **fibrin**. Thrombin also activates factor XIII which is synthesized in the liver and is also secreted by the α-granules of the platelets. Factor XIIIa stabilizes the fibrin by forming cross links between the long polymerized strands of fibrin to make a mesh-like structure that traps the blood constituents (plasma and formed elements), forming the clot and binding the edges of the damaged vessel together (see Figure 25.18). Note that thrombin activates factors V and VIII, which leads to the formation of more thrombin (a positive feedback loop). The importance of this feedback can be judged from the fact that the blood of people lacking factor VIII fails to clot normally.

Clot retraction

Following the coagulation of blood, the clot gradually shrinks as **serum** is extruded from it. (Serum is plasma lacking the clotting factors.) The exact mechanism of this process is not understood, but it appears to be initiated by the action of thrombin on platelets. One idea is that thrombin causes the release of intracellularly stored calcium into the platelet cytoplasm. This calcium then triggers the contraction of contractile proteins within the platelets by a process resembling the contraction of muscle. The contractile process may then cause the extrusion of filopodia from the platelets. These stick to the fibrin strands within the clot and as they contract, the fibrin strands are pulled together, at the same time squeezing out the entrapped fluid as serum.

The role of calcium in haemostasis

As Figure 25.17 shows, calcium ions are required for each step in the clotting process except for the first two reactions of the intrinsic pathway. Adequate levels of calcium ions are therefore necessary for normal clotting. In reality, plasma calcium levels never fall low enough to impair the clotting processes since death would have resulted from other causes (most notably tetany of the respiratory muscles) long before. It is, however, possible to prevent the coagulation of blood removed from the body and stored *in vitro* by reducing the calcium ion concentration of the plasma. This may be achieved by the addition of substances which bind calcium, such as EDTA or citrate.

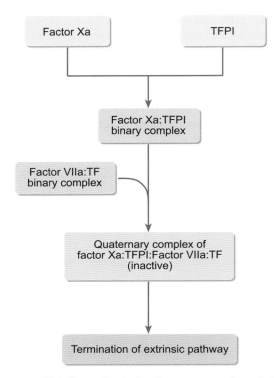

Figure 25.19 This figure illustrates the sequence of events by which tissue factor pathway inhibitor (TFPI) terminates the extrinsic pathway of blood coagulation by forming a complex, first with factor Xa and then with the factor VIIa–tissue factor complex.

Termination of the clotting cascade

While preventing excessive blood loss is clearly desirable, it is essential that the clotting process remains localized and does not continue longer than necessary. Since thrombin plays a central role in coagulation, it is not surprising that its activity is kept in check by a number of inhibitors, most notably by **antithrombin**, a plasma protein secreted by the liver. (It was formerly called antithrombin III but is now generally referred to as antithrombin.) Antithrombin also inhibits factor Xa and kallikrein. When antithrombin binds to one of the activated clotting factors, it forms an inactive complex that is eventually removed from the circulation and broken down.

The extrinsic pathway is inhibited when factor Xa forms a complex with a protein called tissue factor pathway inhibitor (**TFPI**) that is synthesized by the endothelial cells. This complex binds and inhibits factor VIIa and tissue factor, as shown in Figure 25.19. The resulting quaternary (four-way) complex is inactive.

A third mechanism for limiting blood clotting is the **protein C system**, which is activated when thrombin binds to **thrombomodulin** on the surface of the vascular endothelial cells (see Figure 25.20). The thrombin–thrombomodulin complex activates protein C which binds to the membranes of platelets and endothelial cells, where it breaks down factors Va and VIIIa to smaller inactive peptides. The binding of protein C to the membranes is

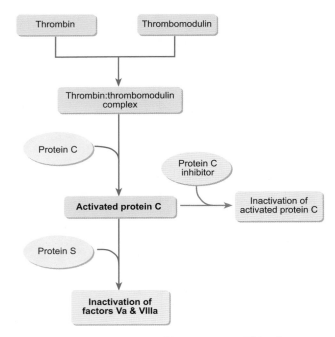

Figure 25.20 The inactivation of factors V and VIII by the protein C pathway. Thrombin forms a complex with thrombomodulin on the surface of the endothelial cells and this activates protein C circulating in the plasma. Protein C with protein S as cofactor then inactivates factors Va and VIIIa, so inhibiting the clotting cascade.

enhanced by another plasma protein called protein S which is secreted by the platelet α-granules. (Protein S is also synthesized by the liver.)

Dissolution of the clot

Once the wall of the damaged blood vessel is repaired, the blood clot is dissolved. Activated factor XII stimulates the production of kallikrein. In turn, kallikrein promotes the conversion of inactive **plasminogen** into active **plasmin**, an enzyme that digests fibrin and thus brings about dissolution of the clot.

A variety of other plasminogen activators may be used clinically to promote the dissolution of clots. These include **streptokinase**, a substance produced naturally by certain bacteria, and an endogenous substance called **tissue plasminogen activator** (**TPA**), which can now be produced commercially by genetic engineering (a product called alteplase). These substances can be injected either into the general circulation or into a specific blood vessel to promote the dispersion of a clot. The use of streptokinase is limited by its propensity to generate antibodies.

It is sometimes necessary to prevent premature dissolution of blood clots in cases of heavy blood loss, such as bleeding caused by traumatic injury or heavy menstrual bleeding. This can be achieved by inhibiting the action of plasmin with **tranexamic acid**. Tranexamic acid is also used to prevent and treat blood loss in other situations, including dental procedures for haemophiliacs, and surgical procedures that carry a high risk of blood loss.

Inappropriate clotting of blood is prevented by endogenous anticoagulants

Normally, blood is prevented from clotting inappropriately by a variety of mechanisms. The preceding account has shown that clotting is initiated when blood encounters an abnormal or damaged tissue surface. By contrast, undamaged vascular endothelial cells generally prevent clotting by releasing substances that inhibit coagulation—**anticoagulants**. These are:

1. **Prostacyclin**, a potent inhibitor of platelet aggregation which acts as an antagonist of thromboxane A_2 (which causes platelet aggregation—see Chapter 6).

2. **Heparin**, a negatively charged proteoglycan that is present in the plasma and on the surface of the endothelial cells of the blood vessels. It inhibits platelet aggregation and interacts with antithrombin to inhibit the action of thrombin.

3. Normal endothelial cells express a protein called **thrombomodulin**, which binds thrombin. The thrombomodulin–thrombin complex activates protein C, which inhibits the actions of factors Va and VIIa as described earlier. In addition, protein C stimulates the production of the proteolytic enzyme **plasmin** from plasminogen. Plasmin rapidly dissolves fibrin to disperse any clots that have begun to form.

Clot formation at inappropriate sites within the circulation is potentially lethal

It is clearly highly undesirable for blood clots to form in the circulation at inappropriate sites. If such a clot does occur, it is known as a **thrombus** and may block the vessel in which it forms. If a thrombus forms or lodges within a coronary artery, the result is a heart attack, where a part of the heart muscle becomes ischaemic (i.e. receives insufficient blood to meet its oxygen requirements) and dies. A clot forming in one of the blood vessels supplying the brain deprives the affected area of oxygen. The ischaemia that results leads to the death of the neurons in the affected area, giving rise to a stroke. Sometimes small clots (emboli) break off from larger thrombi and travel to other parts of the circulation, where they may block small vessels. Such a block is called an **embolism**. Clots that form in the systemic vessels often travel to the lungs where they lodge in pulmonary vessels to produce pulmonary emboli, as discussed in Box 25.1.

Certain conditions favour the formation of clots within blood vessels. These include damage to the vascular wall, sluggish blood flow (stasis), and alterations in one or more of the components of blood that renders it more easily coagulated. One of the most common causes of thrombosis is a condition known as **atherosclerosis**. This is characterized by the formation of fibro-fatty lesions called

 Box 25.1 Deep vein thrombosis

Blood normally flows freely around the circulation. The integrity of the walls of the blood vessels, the anticoagulation proteins of the plasma, and normal flow dynamics all ensure that unwanted thromboses do not occur. However, if the vascular wall is damaged or defective, or if blood flow is low, or even static, a local accumulation of platelets and thrombin may occur and lead to the formation of a blood clot.

In arteries, blood flow is normally streamline except in the vicinity of an atheromatous plaque, where it becomes turbulent, facilitating the formation of a clot (see Figure 30.2). If pieces of the clot break off, they will be carried along by the blood and may block its passage through smaller vessels. If a coronary artery is affected, the myocardium supplied by that artery becomes ischaemic and the heart's ability to pump blood is impaired. This is commonly known as a heart attack, which is usually serious and may be fatal. The blocking of a cerebral artery results in local brain ischaemia, neuronal death, and impairment of brain function (a stroke or cerebrovascular accident).

During long periods of immobility such as car journeys and intercontinental flights, the leg muscles do not pump blood back to the heart by the normal rhythmic contraction that accompanies walking. (This is known as the muscle pump—see Figure 30.25.) As a result, blood flow is sluggish with pooling of the blood in the deep veins of the legs, and there is an increased likelihood of clot formation. The commonest sites of thrombosis are the deep veins of the calf.

Imaging studies indicate that the clots first form in the valve pockets, where blood flow is essentially static.

Other circumstances that favour the development of a deep vein thrombosis include the oral contraceptive pill, polycythaemia, malignant cancer, and surgery. If pieces of the clot break off, they can lodge in the lungs, causing a pulmonary embolism which may be life-threatening. The best treatment for deep vein thrombosis is prevention, which is easily achieved by gentle exercise involving the legs. The use of elastic stockings is also beneficial—particularly for the periods just before and after surgery, and during long flights. These reduce swelling of the legs and help to prevent venous pooling.

Dispersion of venous thromboses relies on anticoagulants such as heparin, and on fibrinolytic agents such as recombinant tissue plasminogen activator (alteplase, reteplase). Oral anticoagulants are used for the long-term management of patients at risk of developing deep vein thromboses. Anticoagulants such as warfarin act as vitamin K inhibitors and reduce the synthesis of clotting factors by the liver (particularly prothrombin, and factors VII, IX, and X). The effectiveness of the therapy is monitored by measuring the prothrombin clotting time, which is expressed as the International Normalized Ratio (or INR). Other newer oral anticoagulants such as dabigatran act as thrombin inhibitors and do not require continuous monitoring of the prothrombin clotting time.

plaques on the intima of large or medium-sized arteries such as the aorta, the coronaries, and the large vessels that supply the brain (the intima is the endothelial lining of the larger blood vessels—see Chapter 27). Major risk factors in atherosclerosis (apart from genetic factors, being male, and getting older) include cigarette smoking, high blood pressure, and a high blood cholesterol level. As the plaques increase in size, they gradually occlude the vessel and cause a reduction in blood flow. Moreover, disruption of the plaque initiates platelet adhesion and aggregation on the exposed vascular surface which leads to activation of the clotting cascade. (This is called the atherothrombotic process.) Plaques are not static features once they have been formed. They can be stabilized by the administration of statins (e.g. simvastatin) which reduce proliferation of macrophages and inflammation around a plaque. The statins also lower plasma cholesterol.

Defective clotting is the result of a deficiency of platelets or one of the clotting factors

Bleeding disorders or failure of coagulation may result from defects in any of the factors that are involved in the normal process of haemostasis, i.e. platelets, clotting factors, or vascular integrity. These will be considered briefly in turn.

Platelet deficiency

A decrease in the number of circulating platelets is known as **thrombocytopenia**. This may arise as a consequence of autoimmune reactions which may be initiated by a viral infection or by certain commonly used drugs including carbamazepine (an anticonvulsant), heparin, paracetamol (acetaminophen), ranitidine, and antibiotics including rifampin and vancomycin. It seems that the binding of these drugs to the platelets forms a new antigen that the immune system regards as foreign. This 'foreign' antigen then initiates an immune reaction that causes the formation of antibodies to the host's platelets, resulting in their destruction. Thrombocytopenia may also be caused by aplastic anaemia (where bone marrow function is impaired) or by invasion of the bone marrow by malignant cells—as in leukaemia—both of which result in decreased platelet production.

The depletion of platelets must be relatively severe before problems with clotting are seen. Haemorrhagic tendencies become evident when the platelet count falls to around $25 \times 10^9 \, l^{-1}$ (normal platelet counts are in the range $150–400 \times 10^9 \, l^{-1}$). Characteristics of the condition include the appearance of bruised areas, tiny reddish spots on the arms and legs (petechiae), and bleeding from the mucous membranes of the nose, mouth, and gastrointestinal tract.

In addition to a reduction in platelet numbers, impairment of the clotting process may result from a deficiency of platelet function—**thrombocytopathia**. Such a defect may be inherited, as with the disorder of platelet adhesion known as von Willebrand's disease (see next subsection), or acquired following disease or drug treatment.

Hereditary disorders of blood clotting—the haemophilias

As may be deduced from the cascade mechanism of clotting illustrated in Figure 25.17, impairment of blood coagulation can result from deficiencies in one or more of the clotting factors involved. Such disorders may be inherited, or they may arise from a reduction in synthesis of one or more of the clotting factors. Although there are known to be hereditary defects associated with each of the protein clotting factors, most are extremely rare. By far the most common forms of haemophilia are factor VIII deficiencies (affecting approximately 1 in 5,000 males) and von Willebrand's disease.

Haemophilia A is a sex-linked recessive trait primarily affecting males. The affected gene is located on the X chromosome, which explains why the disease is mainly seen in males. (Males have a single X chromosome and a single Y chromosome (i.e. they are XY) while females have two X chromosomes (they are XX) and so have a second copy of the gene.) The disease may be mild or severe in form. In severe cases, spontaneous bleeding into soft tissues, joints, and the gastrointestinal tract occurs and can lead to serious disability. In such cases, factor VIII replacement therapy is essential. Recent advances in recombinant DNA technology have enabled pure factor VIII to be produced, thereby eliminating the risk of disease transmission from factor VIII extracted from donated blood. **Haemophilia B** is also sex-linked but is much rarer than haemophilia A (its incidence is about 1 in 30,000). It is clinically indistinguishable from haemophilia A but is caused by a lack of factor IX.

Like haemophilia A and haemophilia B, **von Willebrand's disease** is associated with a prolonged bleeding time. A mild form of the disease is the commonest of the bleeding disorders, with an estimated 125 million sufferers worldwide. Fortunately, severe cases are rare. It is caused by a deficiency of von Willebrand factor (vWF) which is a glycoprotein synthesized by megakaryocytes and stored in the α granules of the platelets. It is also synthesized by the endothelial cells, which secrete it directly into the subendothelial tissue and the plasma. In the plasma it binds circulating factor VIII and protects it from degradation by proteolytic enzymes. After factor VIII has been activated by thrombin, factor VIIIa dissociates from vWF and forms a complex with factor IXa on the platelet surface. A second function of vWF at the site of injury is to bind platelets together or anchor them to exposed collagen.

Impaired synthesis of clotting factors

Prothrombin, fibrinogen, and factors V, VII, IX, X, XI, and XII are all synthesized in the liver. Furthermore, the activity of factors VII, IX, and X and prothrombin requires the presence of vitamin K. Liver disease or vitamin K deficiency may therefore result in a loss of clotting factors or a reduction in their availability. Either of these will produce impairment of the clotting mechanism, with abnormal bleeding.

Vascular disorders

Abnormal bleeding may occur from vessels that are structurally weak or that have been damaged by inflammation or immune responses. Examples include vitamin C deficiency (scurvy), in which the vessels become fragile due to a lack of adhesion between the endothelial cells, and Cushing's disease, in which the excess corticosteroid hormones cause a loss of protein and reduction in support for the vascular tissue.

Summary

Following damage to the vascular endothelium, a cascade of events is initiated leading ultimately to the formation of a blood clot (haemostasis). Platelet aggregation at the site of damage occurs within seconds of an injury, forming a platelet plug. A blood clot then forms as soluble fibrinogen is converted into insoluble fibrin, which traps blood cells and plasma. The conversion of fibrinogen to fibrin is catalysed by thrombin derived from an inactive precursor (prothrombin), by the action of factor Xa. The clotting cascade requires calcium ions.

Following coagulation, the blood clot retracts by shrinkage and is then dissolved by an enzyme called plasmin. Undamaged vascular endothelial cells prevent inappropriate clotting by synthesizing anticoagulants such as heparin, prostacyclin, and thrombomodulin. Should a clot form in an undamaged blood vessel, it will become obstructed, a potentially lethal event if it occurs in vessels such as the coronary arteries.

Failure of the normal clotting reactions may occur for a variety of reasons. These include thrombocytopenia (a reduction in platelet numbers), structural disorders of the vasculature, and hereditary deficiency of clotting factors such as lack of factor VIII in haemophilia A.

25.9 Blood transfusions and the ABO system of blood groups

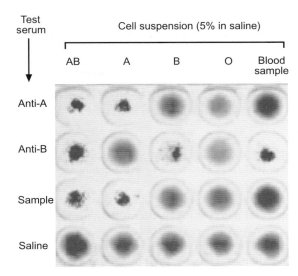

Figure 25.21 The determination of the blood group of an individual. Drops of serum containing anti-A and anti-B antibodies are placed in shallow wells as shown, together with plasma from the blood sample and a saline control. Drops of blood with known blood groups and the test sample are added to the wells, as shown along the top row, and mixed with the serum. If the blood samples are compatible, the mixed blood sample appears uniform but, if the blood is incompatible with the serum, the red cells agglutinate and precipitate as in the top left wells. In this case, the blood sample is agglutinated by anti-B serum and its serum agglutinates the red cells of groups A and AB. So this blood sample is group B, as both group A and group AB have group A antigen on their red cells.

Early attempts to restore heavy loss of blood by transfusion of blood from another person were frequently disastrous. The transfused cells aggregated together in large clumps that were sufficiently large to block minor blood vessels. This clumping is known as **agglutination**. Following an agglutination reaction, the red cell membranes break down and haemoglobin is liberated into the plasma (this is known as **haemolysis**). This reaction is associated with an acute fever and a fall in blood pressure. The liberated haemoglobin is converted to bilirubin by the liver and results in jaundice (see Chapter 45). In addition, the high plasma levels of bilirubin adversely affect the glomerular filtration rate and urine production by the kidney. When such clinical signs follow the transfusion of blood, the transfused blood is said to be **incompatible** with that of the recipient and it becomes imperative to stop the transfusion. Death frequently follows the transfusion of incompatible blood.

What is the basis of this incompatibility and why is some blood compatible while other blood is not? It is now known that agglutination results from an antibody–antigen interaction. Normal human plasma (and the corresponding serum) may contain antibodies that cause red cells to stick together in large clumps (i.e. agglutinate—see Figure 25.21). Clearly, if red cells agglutinate in response to a particular kind of plasma or serum, they must possess the corresponding antigen (or agglutinogen) on their red cells. The antibodies that cause the red cells to agglutinate occur naturally but are not usually present in neonates. They develop in the first years of life and are thought to result from

exposure to antigens in food, bacteria, or viruses that have similar structural motifs (epitopes) to the antigens found on the surface of the red cells.

To account for the known cross-reactivity of blood from different people, K. Landsteiner proposed that two kinds of antigen are present on human red cells. These antigens are called A and B and they may be present separately or together, or be completely absent. The gene that determines which blood group an individual will possess codes for an enzyme called glycosyltransferase, which modifies an intrinsic antigen (the H antigen) that is found on the surface of the red cell membrane. The gene has three alleles (different forms). The A and B forms modify the H antigen in different ways, but the O form is inactive. Since a child can inherit only two forms of the gene (one from each parent) and since A and B forms can both be expressed at the same time (they are co-dominant) while O is recessive, there are four blood groups in the ABO system. Table 25.5 gives the relationships between the different groups and their approximate frequency of occurrence in the general population of the United Kingdom and the United Sates.

Human plasma may contain antibodies to one or both antigens that are known as anti-A and anti-B (or α and β). Where the blood contains red cells with a particular antigen, the corresponding antibody is absent from the plasma. Thus, people with antigen A on their red cells do not have anti-A in their plasma, as they do not agglutinate their own blood. Nevertheless, this group of people

Table 25.5 ABO blood group characteristics

Blood group	% of population	Antigen on red cells	Antibody in plasma
A	42	A	Anti-B (β)
B	10	B	Anti-A (α)
AB	4	A and B	none
O	44	none	Anti-A and Anti-B (α and β)

Note: The frequencies are approximate values for the United Kingdom and the United States. Other populations have different frequencies.

does have anti-B in their plasma. Conversely, people with group B have antigen B on their red cells but anti-A in their plasma. People with group AB have both antigens A and B on their red cells but no corresponding antibodies in their plasma, and those of group O have neither antigen but both anti-A and anti-B antibodies.

The rhesus blood group system

In 1940 Landsteiner and Wiener found that the serum of rabbits that had been immunized against the blood of rhesus monkeys could agglutinate human blood. Using this antibody, they identified two groups in the general population: those whose blood could be agglutinated by this serum—rhesus (or Rh) positive—and those whose blood could not be agglutinated—Rh negative. Rh positive persons have a specific antigen on their red cells known as the **D-antigen** (also known as the **rhesus factor**).

About 85 per cent of the European population and over 90 per cent of people of African and Asian descent are Rh positive. If Rh-positive fetal red blood cells leak into the circulation of a Rh-negative mother, she may develop anti-Rh antibodies. This immunization of the mother by the baby's red blood cells may happen at any time during pregnancy but is most likely to occur when the placenta separates from the wall of the uterus while the mother is giving birth. This means that, although her first baby is unlikely to be affected, any subsequent Rh-positive fetus may be exposed to any anti-Rh antibodies that developed in the mother's blood during an earlier pregnancy. The anti-Rh antibodies are IgG antibodies of about 150 kDa and are sufficiently small to pass across the placenta into the fetal circulation. If this happens, they may cause a severe agglutination reaction in the baby—a condition called **haemolytic disease of the newborn**. In the absence of suitable preventative measures, it occurs about 1 in every 160 births. Around half of the babies affected will require a partial replacement of their blood by transfusion. However, in many countries, this problem is now largely avoided by preventing the formation of anti-Rh antibodies by Rh-negative mothers. This is achieved by injecting them with anti-D immunoglobulin (IgG) which coats any fetal Rh-positive cells present, so preventing the formation of anti-Rh antibodies. This may be done after delivery but is commonly done as a prophylactic measure at weeks 28 and 34 of gestation.

Although haemolytic disease can also occur as a result of an anti-A antibody in the blood of group O mothers, ABO blood group incompatibility generally causes no problems during pregnancy. This reflects the fact that the plasma agglutinins are IgM antibodies of high relative molecular mass (about 900 kDa) and proteins of this size do not readily cross the placenta.

Other blood group types

Other blood group antigens are found on the surface of the red cell membrane, and over 30 major blood group systems are now recognized by the International Society of Blood Transfusion in addition to the fundamental ABO system. For example, soon after the original description of the ABO system of blood groups it was discovered that group A could be further subdivided into two groups, A_1 and A_2. Nevertheless, the A_1 and A_2 subdivisions and other blood groups are not generally of significance in blood transfusion. However, blood types that may cause transfusion reactions include the Kell, Duffy, and Kidd groups.

Blood must be cross-matched for safe transfusions

To prevent the problems of blood group incompatibility, blood for transfusion is **cross-matched** to that of the recipient. In this process, serum from the recipient is tested against the donor's cells. If there is no reaction to the cross-match test, the transfusion will be safe. Note that *this test screens for all serum antibodies and not just those of the ABO system*. Although correct matching of blood groups of both donor and recipient is always preferable, in emergencies group O, Rh negative blood can be transfused into people of other groups because group O red cells have neither A nor B antigens. For this reason a group O person who is rhesus negative is sometimes called a **universal donor**. As the plasma of group AB has neither anti-A nor anti-B antibodies, other blood groups can be transfused into a group AB patient. Such a patient is known as a **universal recipient**. The plasma antibodies present in the blood of a donor do not generally cause adverse reactions because they become diluted in the recipient's circulation.

Summary

Two kinds of antigen are present on human red cells called A and B which may be present separately or together, or be completely absent, so giving rise to four groups: A, B, AB, and O. In addition, human plasma may contain antibodies (anti-A and anti-B) to one or both antigens. In about 85 per cent of the population a further antigen called the **D-antigen** (rhesus factor) is present on the red cells, so each of the main blood groups may be either rhesus positive or rhesus negative. For successful blood transfusion, the blood of the donor must be compatible with that of the recipient. If it is not compatible, the red cells will agglutinate following transfusion. To avoid transfusion of incompatible blood the donated blood should be cross-matched to that of the recipient.

✳ Checklist of key terms and concepts

General properties of blood

- Blood consists of plasma and the red cells, white cells, and platelets.
- Plasma is a solution containing mineral ions, small organic molecules (e.g. glucose), and a variety of proteins including the albumins, globulins, and fibrinogen.
- Some plasma proteins serve to transport lipids and fat-soluble materials (including steroid hormones) around the body.
- The γ-globulins are antibodies and play an essential role in defence against infection.

The blood cells

- Red blood cells (erythrocytes) are small, non-nucleated biconcave discs whose function is to transport oxygen and carbon dioxide between the lungs and tissues. They have a lifespan of about 120 days.
- Leukocytes (white cells) can be grouped as granulocytes, monocytes and lymphocytes. They are present in fewer numbers than red cells but play a crucial role in mediating the body's immune responses.
- Platelets (or thrombocytes) play an essential role in haemostasis. They are cell fragments derived from the megakaryocytes of the bone marrow.

Haematopoiesis

- Mature blood cells are renewed continuously by haematopoiesis.
- All cell types are generated ultimately from a common population of multipotent stem cells in the bone marrow.
- The stem cells divide to form committed precursor cells, which differentiate into one of the mature cell types via a series of cell divisions.
- The formation of the various types of blood cell is controlled by specific growth factors including erythropoietin, stem cell factor, thrombopoietin, and a variety of interleukins.
- Platelets bud off from giant cells (megakaryocytes) and their rate of production is regulated by a hormone called thrombopoietin, which is secreted by the liver.

Iron metabolism

- Most of the iron of the body is within the haemoglobin of the red blood cells.
- When red blood cells are destroyed, most of their iron is recycled immediately, or stored as ferritin.
- The iron in the diet is absorbed as ferrous iron (Fe^{++}) by a carrier-mediated process.
- Absorbed iron is transported in the plasma to the tissues bound to transferrin.
- Iron absorption is regulated in accordance with the body's requirement by the hormone hepcidin.

Gas transport

- The amount of oxygen carried in the blood depends on the partial pressure of oxygen and the concentration of haemoglobin.
- The oxyhaemoglobin dissociation curve has a sigmoid shape.
- The curve is shifted to the right by an increase in the concentration of 2,3-BPG in the red cells, a rise in temperature and PCO_2, and a decrease in pH. This shift is known as the Bohr effect.
- Carbon dioxide is carried in the blood: in physical solution, as bicarbonate ions, and as carbamino compounds.
- The affinity of the blood for CO_2 is determined by the degree of oxygenation of the haemoglobin. This is known as the Haldane effect.
- At sea level normal arterial blood has a PO_2 of 13.3 kPa (100 mmHg) and an oxygen content of around 20 ml O_2 per dl^{-1}. It has a PCO_2 of 5.33 kPa (40 mm Hg) and a CO_2 content of 49 ml dl^{-1}.

Blood cell disorders

- Anaemia is a general term to describe disorders of the red blood cells characterized by a reduced number of red cells.
- An important consequence of all types of anaemia is a reduction in the oxygen-carrying capacity of the blood.
- Over-stimulation of red blood cell production results in polycythaemia, which increases the work of the heart by virtue of a rise in the viscosity of the blood.
- Leukopenia is defined as an absolute reduction in white blood cell numbers and may be due either to defective production or to accelerated removal of white cells from the circulation.
- Proliferation disorders of the white blood cells include leukaemias, lymphomas, and myelomas. In leukaemia there are high numbers of abnormal white blood cells which are usually non-functional.
- Thrombocytopenia is a decrease in the number of circulating platelets which is accompanied by spontaneous bleeding into the tissues.

Haemostasis

- Following damage to the vascular endothelium, a cascade of events is initiated leading ultimately to the formation of a blood clot (haemostasis).
- The initial response is the formation of a platelet plug which is followed by the formation of a blood clot.
- In blood clotting, fibrinogen is converted into insoluble threads of fibrin by thrombin. The fibrin threads then trap blood components to form a clot.
- The clotting mechanism requires calcium ions and the phospholipids present in the membranes of the platelets. It is terminated by activation of protein C and by the inhibition of the action of thrombin by antithrombin.
- Clot retraction and dissolution complete the healing process.

- Inappropriate clotting is prevented by endogenous anticoagulants including prostacylin and heparin.

- A thrombus is a blood clot that occurs at an inappropriate site.

Blood groups

- There are many different blood groups, some of which are very rare. They are characterized by the presence of specific antigens on the surface of the red cells.

- In the ABO system, two kinds of antigen called A and B may be present on human red cells. They may be present separately or together, or be completely absent, so giving rise to four groups: A, B, AB, and O.

- In addition, human plasma may contain antibodies (anti-A and anti-B) to one or both antigens. Transfusion of incompatible blood will lead to agglutination of the recipient's red cells and may result in their death.

- To avoid blood group incompatibility, blood for transfusion is cross-matched to that of the recipient. In this process, serum from the recipient is tested against the donor's cells. If there is no reaction to the cross-match test, the transfusion will be safe.

Recommended reading

Biochemistry of haemoglobin and myoglobin

Berg, J.M., Tymoczko, J.L., Gatto, G.J., and Stryer, L. (2015) *Biochemistry* (8th edn), Chapter 7. W.H. Freeman, New York.

A useful introductory account of the biochemical properties of haemoglobin.

Haematopoesis and histology

Alberts, B., Johnson, A., Lewis, J., Morgan, D., Raff, M., Roberts, K., and Walter, P. (2014) *Molecular biology of the cell* (6th edn), Chapter 22, pp 1239–47. Garland, New York.

Provides a simple of account of haematopoiesis from the perspective of cell biology.

Mescher, A. (2009) *Junquieira's basic histology* (12th edn), Chapter 13. McGraw-Hill, New York.

A recent revision of Junquieira and Carneiro's classic text. A good basic account of the characteristic properties of blood cells and their formation in the bone marrow.

Pharmacology of the blood

Rang, H.P., Ritter, J.M., Flower, R.J., and Henderson, G. (2015) *Rang & Dale's pharmacology* (8th edn), Chapter 24. Churchill-Livingstone, Edinburgh.

A helpful account of the drugs used to treat clotting disorders.

Haematology and immunology

Hoffbrand, A.V., and Moss, P.A.H. (2016) *Hoffbrand's essential hematology* (7th edn). Blackwell Science, Oxford.

A well-established introductory text covering all aspects of haematology.

Ledingham, J.G.G., and Warrell, D.A. (eds) (2000) *Concise Oxford textbook of medicine*, Chapters 3.1–3.41. Oxford University Press, Oxford.

A very comprehensive introduction to the physiology of the blood and the pathophysiology of blood disorders.

Review articles

Garrett, R.W., and Emerson, S.W. (2009) Bone and blood vessels: The hard and the soft of hematopoietic stem cell niches. *Cell Stem Cell* **4**, 503–6.

Kaushansky, K. (2005) The molecular mechanisms that control thrombopoiesis. *Journal of Clinical Investigation* **115**, 3339–47.

Kawamoto, H., Wada, H., and Katsura, Y. (2010) A revised scheme for developmental pathways of hematopoietic cells: The myeloid-based model. *International Immunology* **22**, 65–70.

Manly, D.A., Boles, J., and Mackman, N. (2011) Role of tissue factor in venous thrombosis. *Annual Review of Physiology* **73**, 515–25.

Monroe, D.M., and Hoffman, M. (2006) What does it take to make the perfect clot? *Arteriosclerosis, Thrombosis, and Vascular Biology* **26**, 41–8.

To check that you have mastered the key concepts presented in this chapter, complete the accompanying online self-assessment questions. Go to www.oup.com/uk/pocock5e/

CHAPTER 26

Defence against infection: the immune system

Chapter contents

This chapter should help you to understand:

- The passive mechanisms by which the body resists infection
- How the body recognizes invading organisms—distinguishing self from non-self
- The natural immune system
- The inflammatory response
- The adaptive immune system and the role of the lymphocytes
- Disorders of the immune system
- The need for tissue matching in transplantation

26.1 Introduction

As animals move through their environment to feed and reproduce, they inevitably come into contact with other organisms. Some of these will be food, while others may attempt to invade the body. Of those that invade the body, some will live in harmony with the host. When this is of benefit to both the host and the other organism, it is known as **mutualism**. If the association is neither beneficial nor harmful, it is called **commensalism**. When the presence of the invading organism compromises the health of the host, the relationship is known as **parasitism**. All infectious diseases are due to parasites of one kind or another. In the developed countries, most infections are caused by bacteria, fungi, and viruses but infections by protozoa and worms of various kinds are also very common especially in the poorer regions of the world. This chapter discusses the various ways in which the body defends itself against infection.

26.2 The principal features of the immune system

To defend themselves against infections, animals have two basic strategies: they use passive barriers to protect the body from parasites, and they actively attack any organisms that have evaded the defences and become lodged in the tissues. However, to eliminate an invading organism, the host must first be able to distinguish it from its own cells before neutralizing or killing it. Finally, the body must dispose of the remains in such a way that they do no further harm. These functions are performed by the immune system which, for the purposes of discussion, may be conveniently divided into the **natural immune system** (also known as the **innate immune system**) and the **adaptive immune system**.

The principal organs of the immune system are the bone marrow, the thymus, the spleen, the lymph nodes, and the lymphoid tissues associated with the epithelia that line the gut and airways (see Figure 26.1). Lymphoid tissues associated with the mucosa of the airways are known as MALT (mucosa associated lymphoid tissue), while those of the gut mucosa are usually called gut associated lymphoid tissue (GALT). The cells of the immune system include the **leukocytes** of the blood (see Chapter 25), **mast cells,**

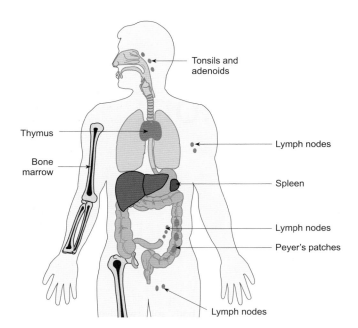

Figure 26.1 The location of the major lymphoid organs. The thymus and bone marrow constitute the primary lymphoid tissues; the remainder are secondary lymphoid tissues. (Based on Fig 9.1 in J.H. Playfair (1995) *Infection and immunity*. Oxford University Press, Oxford.)

and various accessory cells that are scattered throughout the body. The accessory cells include phagocytic cells that are found in many organs including the lungs, liver, spleen, and kidneys. **Dendritic cells** are specialized phagocytic cells that engulf bacteria and other material in the tissues before migrating to the lymphoid tissues, where they play an important role in the initiation of the immune response. The main proteins of the immune system are **antibodies** and **complement**.

The cells of the immune system recognize foreign materials by their surface molecules. Those molecules that generate an immune response are called **antigens**. The immune response may be relatively non-specific (as with the natural immune system) or it may be highly specific, in which a small part of a particular antigen is precisely recognized. This second type of interaction is characteristic of the antibodies secreted by the cells of the adaptive immune system in response to an infection.

Summary

The immune system has two major divisions: the natural immune system (or innate immune system) and the adaptive immune system. Its principal organs are the bone marrow, the thymus, the spleen, the lymph nodes, and the lymphoid tissues associated with the epithelia. The cells of the immune system include the leukocytes, mast cells, phagocytic cells, and dendritic cells, while its major proteins are antibodies and complement. Antigens are molecules that stimulate an immune response.

26.3 Passive barriers to infection

The first barrier encountered by most invading organisms is the skin. Its pseudostratified and keratinized epidermis forms an effective physical barrier to infiltration by microorganisms. In addition, the sweat glands and sebaceous glands secrete fatty acids that inhibit the growth of bacteria on the skin surface. However, when the skin is broken either by abrasion or by burns, infection may become a significant problem.

The skin is continuous with the mucous membranes that line the airways, gut, and urogenital tract. Although the epithelia of the mucous membranes are less rugged than that of the skin, they are protected by secretions that help to provide an effective barrier to invasion by microorganisms. For example, the epithelia that line the airways are protected by a thick layer of mucus which traps many bacteria and viruses and prevents them adhering to the underlying cells. The mucus is then eliminated via the mucociliary escalator and coughed up (see Chapter 36).

Other regions that are vulnerable to infection are regularly flushed by sterile fluid (e.g. the urinary tract) or by fluids that contain bactericidal agents. For example, the external surface of the eyes is washed by fluid from the tear glands, which both flushes the surface to remove foreign materials (dust particles etc.) and contains the bactericidal enzyme lysozyme. Other body secretions such as semen and milk also contain bactericidal substances as well as antibodies (IgA) that coat invading particles to inactivate them.

The food we eat is inevitably contaminated by bacteria and other microorganisms. Indeed, some organisms are deliberately introduced into certain foods such as cheese and yogurt to help preserve and flavour them. The gut has several stratagems to combat infection arising from such sources. The mucous membranes of the mouth and upper gastrointestinal tract are protected by lysozyme and antibodies of the IgA class (discussed in Section 26.7) which are secreted by the salivary glands. Many bacteria are killed by the low pH of the gastric juice. The mucosal surface of the gut also possesses mucous glands that secrete a layer of mucus that both lubricates the passage of food and protects the surface epithelium from infection. Despite these barriers, the lumen of the intestine contains a healthy bacterial population that provides the body with a further line of defence. The normal bacterial flora both competes with potential pathogens for essential nutrients and secretes inhibitory factors (**bactericidins**) that kill invading pathogens.

Clearly, these passive mechanisms do not always prevent the ingress of pathogens. The skin can be penetrated by ectoparasites such as ticks and mosquitoes, which may themselves be infected with smaller organisms such as *Plasmodium vivax* (one of the parasites that causes malaria). Small pathogens such as bacteria and viruses may penetrate the body's defences via the internal epithelia such as those of the airways. Those that enter the gut may overwhelm the defences afforded by the natural commensal bacteria, as happens in typhoid fever, for example. When an infection occurs, the active processes of immunity come into play.

Summary

The body protects itself from infection by means of passive barriers and the immune system. The barriers to infection include bactericidal secretions and the mucus layer covering the mucous membranes.

26.4 How does the immune system identify an invading organism?

Before the body can mount a defence against infection, it first needs to know the difference between the normal cells of the body and those of invading parasites (i.e. bacteria, viruses, worms, etc.). So how does the immune system distinguish 'self' from 'non-self'? It is now known that mammalian cells possess markers on their surface that identify them as host cells. Just as red cells possess surface proteins that determine particular blood groups (see Chapter 25), other cells possess integral membrane proteins that identify them as being host cells. By using these markers, the immune system can distinguish host cells from those of invading organisms. The components of the immune system that detect a general 'non-self' characteristic are called **non-specific** while those that can detect a particular invading organism amongst the many thousands of possible candidates are called **specific** recognition molecules. As we shall see, non-specific antigen recognition is characteristic of the natural immune system, while the adaptive immune system can identify and destroy a specific type of invading organism.

The proteins that identify host cells are known as the major histocompatibility complex or MHC. Their rather unfortunate name arises from the history of their discovery. They were first detected as the proteins responsible for the rejection of tissue grafts between a donor and a recipient animal. In human immunology the MHC complex is known as the HLA complex (for human leukocyte antigen). It is now known that the MHC (or HLA) complex consists of more than 200 genes on the short arm of chromosome 6. These genes encode three classes of proteins:

- MHC class I
- MHC class II
- MHC class III.

MHC class I proteins are subdivided into three groups known as HLA A, HLA B, and HLA C. They are integral membrane proteins found on the plasma membrane of all nucleated cells and on platelets. However, they are not found on red cells.

MHC class II proteins are similarly subdivided into three groups known as HLA DP, HLA DQ, and HLA DR. MHC class II proteins are expressed by dendritic cells, macrophages, and a class of lymphocytes known as B-cells (discussed in Section 26.7). Both MHC class I and class II proteins function to expose parts of foreign proteins to a class of lymphocytes known as T-cells to stimulate an immune response to infection. This is known as **antigen presentation** and it permits the T-cell population to identify infected cells prior to initiating an appropriate immune response.

As there are many genetic variants of the MHC class I and II proteins, and as each MHC protein is a combination of two different polypeptide chains, there are around 10^{13} possible variations in the MHC. With such a large degree of variation, every individual has a unique signature in their MHC except identical twins (who share the same genetic makeup). This natural variation in MHC proteins was a major problem that had to be overcome to permit successful organ transplantation (see Section 26.9). Defects in certain MHC genes are implicated in some autoimmune disorders such as multiple sclerosis, ankylosing spondylitis (an inflammatory joint disease of the spine), and diabetes.

MHC class III includes proteins of the complement system (e.g. C4), certain cytokines (e.g. tumour necrosis factors—TNF-α and TNF-β), and two proteins known as **heat shock proteins** which are expressed by cells under conditions of stress, for example when combating viral infections.

Summary

Host cells have cell-surface markers known as the major histocompatibility complex (MHC). The MHC allows the cells of the immune system to distinguish between host cells and those of invading pathogens.

26.5 The natural immune system

The natural immune system provides the innate defence mechanisms which do not change very much either with age or following infections. It consists of five kinds of cells and three different classes of proteins.

The cells of the natural immune system are:

- phagocytes (neutrophils and macrophages)
- natural killer cells
- eosinophils
- basophils
- mast cells.

The proteins of the natural immune system are:

- complement
- interferons
- acute phase proteins.

The principal phagocytes of the natural immune system are the neutrophils and the macrophages

The **neutrophils** are the most common white cell in the blood (see Chapter 25 p. 363). They contain two main types of granule: the primary azurophil granules and the secondary specific granules. As their name implies, the primary azurophil granules can be stained with azure or similar dyes. They contain an enzyme called myeloperoxidase, a range of bactericidal proteins, and a protease

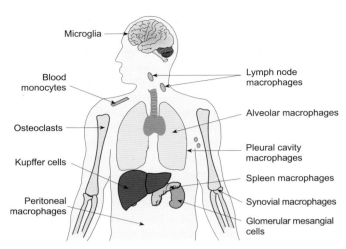

Figure 26.2 The mononuclear phagocyte system. Monocytes are formed in the bone marrow where they mature before being released into the circulation. They migrate to tissues such as the spleen, liver, lungs, and lymph nodes where they take up residence as mature macrophages. The neutrophils (the other main class of phagocyte) remain in the circulation until they participate in an inflammatory reaction. (Based on Fig 9.2 in J.H. Playfair (1995) *Infection and immunity.* Oxford University Press, Oxford.)

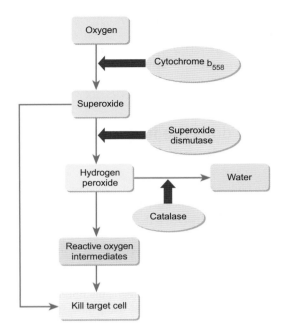

Figure 26.3 The processes leading to the generation of reactive oxygen intermediates by the phagocytes. These reactive free radicals fatally damage invading organisms, while the host cell is protected from damage by various mechanisms (e.g. vitamins C and E, glutathione, and the enzyme catalase).

called cathepsin G. The secondary specific granules contain lysozyme, alkaline phosphatase, and a form of cytochrome (cytochrome b_{558}) that can be inserted in the plasma membrane.

Neutrophils are able to pass from the blood into the intercellular spaces by diapedesis (see Figure 26.6 p. 395) and actively phagocytose and engulf disease-producing bacteria. The enzymes within the cytoplasmic granules then kill the invading organisms and digest them. As a result of this action, the neutrophils form the first line of defence against infection.

The **macrophages** are formed in the bone marrow and released into the blood as monocytes. Within two days, they migrate to tissues such as the spleen, lungs, and lymph nodes where they mature. As Figure 26.2 indicates, macrophages are found in all tissues, even in the brain, where they are known as **microglia**. In the liver they are known as **Kupffer cells**, and in the kidneys they are known as **mesangial cells**. Macrophages are situated around the basement membrane of small blood vessels. They also line both the spleen sinusoids and the medullary sinuses of the lymph nodes, where they are able to remove particulate matter from the circulation. Macrophages contain a large number of lysosomes and phagocytic vesicles containing the remains of ingested materials.

The phagocytosis and killing of bacteria

Phagocytes are non-specific immune cells that will attack a wide variety of invading organisms and cell debris. However, before a phagocyte can kill an invading bacterium, it must first recognize it as a foreign body and engulf it. This is the process of **phagocytosis**, which is described in detail in Chapter 5. After a bacterium has been engulfed, the phagocyte will kill it. This is achieved by a variety of methods. Following phagocytosis, the cytochrome b_{558} present in macrophages and neutrophils liberates a number of

reactive oxygen intermediates as follows: molecular oxygen is first converted to the superoxide anion (O_2^-) which, under the influence of an enzyme called superoxide dismutase, gives rise to hydrogen peroxide. The hydrogen peroxide then gives rise to a number of highly reactive intermediates that kill the bacteria trapped within the phagosome (see Figure 26.3). These events cause a marked increase in oxygen uptake by the activated cell called the **respiratory burst**. Both macrophages and neutrophils also produce nitric oxide and other reactive nitrogen-containing intermediates to kill bacteria. In addition, bactericidal proteins (called **defensins**) are inserted into the bacterial cell membrane and cause it to rupture. Protease enzymes then digest the remains.

Natural killer cells

Viruses lack the ability to replicate by themselves. Instead, they subvert the genetic machinery of host cells to make copies of themselves. For this reason, it is important that those cells that become infected with a virus are destroyed before the virus has had time to replicate and infect neighbouring cells. The cells that perform this vital function are known as **natural killer cells (NK cells)** which are large granular lymphocytes that have their origin in the bone marrow and are thought to mature in the secondary lymphoid tissues. Their granules contain membrane-penetrating proteins called **perforins** and proteases known as **granzymes**. NK cells can be distinguished from cytotoxic T-cells (see Section 26.7) by their surface markers. Unlike T-cells, NK cells do not require prior sensitization to become activated.

NK cells have a number of different cell surface receptors that regulate their activity: some inhibit the cytotoxic effects of NK cells while others lead to their activation. It is the balance between the activity engendered by the two types of receptor that determine whether an individual NK cell is activated when it comes into contact with another cell as it circulates round the body. Host cells bearing intact MHC class I molecules inhibit the activity of NK cells, and this is an important mechanism for preventing the destruction of healthy cells. However, a NK cell is activated if it encounters a host cell lacking MHC class I molecules on its surface (a tumour cell) or if it encounters a host cell with MHC class I molecules incorporating part of a foreign peptide (such as a cell infected with a virus). Once activated, a NK cell positions its cytotoxic granules between its nucleus and the target cell. It then secretes their contents onto the target cell which responds by undergoing a pre-programmed cell death (**apoptosis**) which prevents further viral replication in the infected cell.

Although red blood cells lack MHC class I cell surface molecules, they possess other cell surface markers that identify them as host cells. For this reason, they are not attacked by NK cells.

Immune system cells secrete interferons

The interferons are proteins secreted in response to viral infection. They are classified as type I (interferon α variants and interferon β) and type II (interferon γ). Type I interferons are cytokines (small signalling proteins) secreted into the extracellular fluid by many cell types, including lymphocytes and macrophages, when a cell infected by a virus is detected. The interferons bind to receptors on neighbouring cells, which respond by reducing their rate of mRNA translation. This results in an infected cell being surrounded by a layer of cells that cannot replicate the virus, forming a barrier that prevents the spread of the infection. The infected cells are subsequently destroyed by natural killer cells.

Interferon γ (type II interferon) is a cytokine secreted by activated natural killer cells and by certain T-cells. It activates the macrophages of the immune system to respond to an intracellular infection or to the growth of a tumour. During the process of activation, macrophages acquire the ability to kill microorganisms and secrete proinflammatory cytokines. This results in inflammation, the recruitment of immune cells, and the subsequent elimination of the infecting microbe by phagocytosis or by the secretion of toxic metabolites.

Eosinophils

Eosinophils are amongst the least numerous of the white cells of the blood (typical blood contains around 100×10^6 eosinophils per litre—see Table 25.2). They are also found in the connective tissues underlying the epithelia of the skin, bronchi, gut, and other hollow organs. The granules that take up the dye eosin (eosinophil specific granules) contain a protein rich in arginine residues called **major basic protein**. Eosinophils also secrete **perforins** and a battery of enzymes including peroxidase and phospholipase D. Furthermore, they also secrete a protein known as eosinophil derived neurotoxin which kills nerve cells within the invading parasite.

Eosinophils appear to play an important role in combating helminth (worm) infections. These organisms are too large to be phagocytosed by a single cell, so they must be attacked extracellularly. Eosinophils are particularly attracted to parasites whose outer membranes have been coated with antibody of the IgE class (see Section 26.7). Major basic protein, perforins, peroxidase, and phospholipase D attack the outer membrane of the parasites to inactivate or kill them. Eosinophils are attracted to sites of infection or inflammation by chemical signals (interleukins) released from basophils and mast cells.

Basophils and mast cells

Basophils are the least common of the granulocytes and represent only about 0.5 per cent of the white cell population (typically there are only around 40×10^6 basophils per litre of blood—see Table 25.2). Although they share many of the properties of mast cells, they have a different origin and mature in the bone marrow before being released into the circulation. Like the eosinophils, the basophils are believed to play an important role in combating parasites too large to be ingested by the phagocytes. Basophils are stimulated to secrete the contents of their granules in many different ways: by cytokines, by IgE, by proteases, and by antigens associated with parasites. Once activated, they secrete histamine and leukotriene C_4, which acts as a powerful attractant for eosinophils and neutrophils. It also promotes the development of naive T-cells into T helper cells (see Section 26.7).

Mast cells are formed in the bone marrow and migrate to the tissues where they mature. They are found associated with nerves, blood vessels, skin, and the epithelia of the airways and gut. Mast cells are oval in shape with a length of around 20 μm (see Figure 26.4). The nucleus is located in the centre of the cell and is surrounded by cytoplasm rich in granules that stain strongly with basic dyes—a feature they share with basophils. There are two populations of mast cell which are differentiated on the basis of the contents of their granules. One type is mostly found in the mucosa of the gut and respiratory tract, while the other predominates in connective tissue such as the dermis and the submucosa of the gut. The granules of both types contain histamine and various protease enzymes.

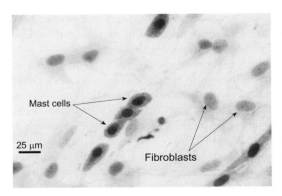

Figure 26.4 The typical appearance of mast cells in connective tissue. Note the round nucleus and dark staining cytoplasm. Scale bar: 25 μm.

Mast cells can be activated by direct injury, by IgE, or by the complement proteins C3a and C5a (see next subsection). When stimulated, they rapidly release the contents of their granules (a process called degranulation) into the surrounding space. Mast cell activation also leads to the formation of various cytokines and eicosanoids (e.g. prostaglandin D_2 and leukotriene C_4) that help recruit eosinophils, neutrophils, and monocytes to the site of injury or infection.

The role of mast cells is gradually being elucidated and there is evidence to suggest that they play an important role in the normal physiology of the gastrointestinal tract, where they have been shown to regulate gastric acid secretion, local blood flow, and smooth muscle function. They also help mount a defence against those parasites that are too large to be ingested by the phagocytes by rendering the local environment hostile to invading organisms (compare this with the action of the basophils and eosinophils discussed earlier). By releasing histamine and other vasoactive substances to increase local blood flow they play an important role in the acute inflammatory response (see Section 26.6). As several of these vasoactive substances—notably histamine and leukotriene C_4—increase the permeability of the capillary walls to plasma proteins, this action facilitates the diffusion of complement and antibodies into the tissues, where they play a crucial role in combating the infection.

Complement

Discovered at the beginning of the last century, complement was initially identified as a heat-sensitive factor that enhanced the effects of specific antisera on bacteria and red cells. This complementary action gave rise to the name now in use. The complement system is now known to comprise a group of about 30 proteins that play an important role in innate immunity and in the regulation of the adaptive immune system. These proteins are particularly important in the defence against bacterial and fungal infections.

Activation of the complement system occurs by one of three pathways known as the classical, alternative, and lectin pathways, discussed in the paragraphs that follow. In each case, a component of the complement system known as C3 is cleaved to form two fragments known as C3a and C3b. Events that follow the formation of C3a and C3b include local inflammation, induction of phagocytosis, and destruction of the invading organism by cell lysis. To prevent unnecessary activation of the complement system, the plasma also contains a number of regulatory proteins that limit the action of key complement components including C3b.

The **classical pathway** was the first to be discovered. It acts to complement the actions of the antibody molecules secreted by the B-lymphocytes and is activated by IgG or IgM antibodies bound to the surface of the invading organism. These antibodies bind to a complement component known as C1q which combines with two other components C1r and C1s to form a complex that activates two further components, C2 and C4. In turn, these activate C3 convertase which cleaves C3 to form C3a and C3b, which have the actions that are described below and summarized in Figure 26.5.

The **alternative pathway** depends on the spontaneous formation of low levels of C3b and C3a from C3. This process is markedly accelerated at the surfaces of invading organisms such as bacteria

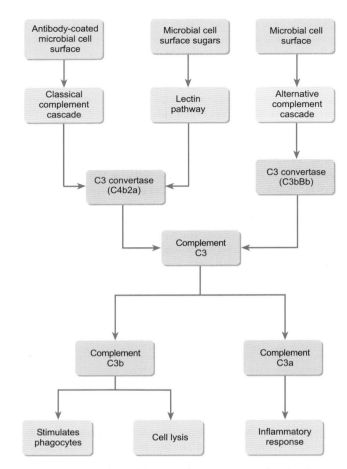

Figure 26.5 An outline of the complement system showing the three pathways that lead to activation of complement C3 and the cellular responses to deal with infection.

and fungi. Normal host cells produce specific proteins that inhibit the activation of the complement system on their surfaces.

The **lectin pathway** is activated when a protein known as mannose binding lectin (MBL) becomes attached to the mannose or fucose residues present on bacterial cell walls. (Mannose and fucose are carbohydrates not normally present on mammalian cell surfaces.) The MBL binds to repeated sequences of these sugars and activates two proteases (MASP-1 and MASP-2) that cleave C2 and C4, which then form a complex known as C4b2a that acts as a C3 convertase.

The binding of C3b to antigens provides the body with a mechanism to mark them for uptake by follicular dendritic cells, permitting recognition of the antigen by the appropriate B-cells (see Section 26.7). Tagging of the antigens by C3b also serves as coating that facilitates their uptake by phagocytes which are activated by C3a. When C3b binds C5, C3 convertase is able to cleave C5 into C5a and C5b, which then recruit other components of the complement system to form a membrane attack complex. This attack complex is able to destroy invading bacteria by cell lysis. Furthermore, C3a and C5a are pro-inflammatory peptides that induce the release of inflammatory mediators from mast cells (discussed in Section 26.6).

Acute phase proteins

The acute phase proteins are a group of plasma proteins synthesized by the liver that show a marked change in concentration during infection or inflammation (the **acute phase response**). They include C-reactive protein, serum amyloid P component, fibrinogen, and a number of complement components—including C3. As described above, C3b binds to the surface of invading organisms, a process known as **opsonization**. Both C-reactive protein and serum amyloid P component can also opsonize foreign organisms. As the phagocytes have receptors for the coating proteins (including the Fc region of the immunoglobulins—see Section 26.7), they are able to recognize opsonized particles and engulf them as described in Chapter 5.

The plasma level of C reactive protein may increase by as much as a thousandfold during inflammation. Other conditions that lead to marked changes in C reactive protein include infection, trauma, burns, and cancer. Less dramatic changes occur after strenuous exercise, heatstroke, and childbirth. Although testing for C-reactive protein is of clinical value, an increase in its plasma concentration is not diagnostic of a specific disease.

Summary

The natural immune system provides the innate defence mechanisms. It consists of five kinds of cells: the phagocytes, natural killer cells, mast cells, basophils, and eosinophils; plus three different classes of proteins: complement, interferons, and acute phase proteins. The macrophages and neutrophils engulf small invading organisms (e.g. bacteria) and kill them with highly reactive oxygen and nitrogen intermediates. Cells that become infected with a virus are destroyed by natural killer cells before the virus can replicate.

26.6 The acute inflammatory response

A number of physiological changes occur when the body is subjected to traumatic injury, infection, or cellular necrosis. There is local vasodilatation, increased permeability of the capillaries to plasma proteins, and infiltration of the damaged tissues by white cells. These changes constitute the **inflammatory response**. For an injury to the skin, the stages of the inflammatory response are as follows. There is a reddening of the skin at the site of injury, which results from vasodilatation (the **acute vascular response**). This is rapidly followed by local tissue swelling due to the accumulation of fluid by the affected tissues (local oedema). Then the skin of the surrounding area becomes flushed (the flare). These three components of the inflammatory response constitute **the triple response** (see also Chapter 13).

If the infection or trauma is sufficiently extensive, the acute vascular phase is followed by the **acute cellular phase** in which the injured tissues become infiltrated by leukocytes, particularly neutrophils, by means of diapedesis. The vascular endothelium of the injured area becomes modified and the neutrophils attach themselves to the capillary wall in a process called **margination**. They then squeeze between the endothelial cells and pass into the tissues (see Figure 26.6).

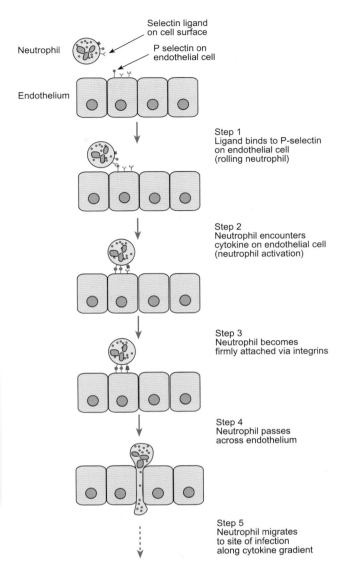

Step 1
Ligand binds to P-selectin on endothelial cell (rolling neutrophil)

Step 2
Neutrophil encounters cytokine on endothelial cell (neutrophil activation)

Step 3
Neutrophil becomes firmly attached via integrins

Step 4
Neutrophil passes across endothelium

Step 5
Neutrophil migrates to site of infection along cytokine gradient

Figure 26.6 An outline of the principal stages of diapedesis. When an infection triggers a mast cell or macrophage to secrete cytokines such as IL-1, the endothelial cells in the adjacent venules translocate P-selectin from intracellular vesicles to the luminal surface. When a circulating neutrophil (or other white cell) encounters the P-selectins on the endothelial wall, it slows down and rolls along the epithelial surface where it will encounter the secreted cytokines. The cytokines activate adhesion proteins called integrins which then tether the cell to the endothelium. The white cell then passes between the adjacent endothelial cells by amoeboid movement and migrates to the site of infection (dotted arrow). The green symbols in the figure refer to P selectin and its ligand, the red to the cytokines, and the blue to the integrins.

This brings them into direct contact with any invading organism or cell debris, where they can undertake their normal phagocytic role.

A **chronic cellular response** then follows in which macrophages and lymphocytes invade the damaged area. Like the neutrophils, the macrophages dispose of the cellular debris and play an important role in the healing process.

Finally, the inflammatory response declines as the damaged tissue becomes healed. This phase is known as **resolution**. If it has not been possible to eliminate the invading organism or any particles that triggered the inflammatory response, the offending material is sealed off by a layer of macrophages, lymphocytes, and other cells to form a **granuloma**. Injury to internal organs is accompanied by a similar sequence of events.

What triggers the inflammatory response?

The processes involved in the inflammatory response are summarized in Figure 26.7. In the first stages of a response to infection, the invading organism becomes coated with complement C3b which is present in small amounts in normal plasma. The immobilized C3b then generates a form of complement called C3 convertase which splits C3 into C3a and C3b as described earlier. When C3b binds to another complement component called C5, C3 convertase is able to cleave C5 to form two peptides: C5a and C5b. Together C3a and C5a stimulate the mast cells in the vicinity to degranulate.

Activated mast cells secrete histamine and synthesize prostaglandins, leukotrienes, and interleukins, as well as other cytokines,

all of which are released into the interstitial space. The histamine and interleukins induce the endothelial cells of the affected area to move cell adhesion proteins (known as **selectins**) to their apical surface (the cell surface facing the blood). The selectins and other cell adhesion molecules allow passing neutrophils to adhere to the capillary wall before squeezing between the endothelial cells and migrating to the site of infection, as outlined in Figure 26.6. Once in position, these neutrophils begin engulfing the invaders and killing them by means of the mechanisms described earlier. In addition, the histamine and prostaglandins elicit a local vasodilatation and cause the capillary endothelial cells to retract. This allows plasma proteins, including complement and antibodies, to pass into the interstitial space. (The interstitial fluid normally has very little plasma protein.) The presence of these immune system proteins further aids the counter-measures against infection.

The inflammatory mediators released by mast cells have a powerful influence on the surrounding tissues. If they are secreted inappropriately, they may cause severe pathological changes such as inflammatory bowel disease and rheumatoid arthritis (which is the commonest autoimmune disease, affecting around 1 per cent of the population in developed countries). The fluid that accumulates in the affected joints in rheumatoid arthritis contains a high concentration of cytokines which causes the synovial membrane of the joint surface to become infiltrated by lymphocytes and macrophages. The resulting chronic inflammation results in erosion of the joint cartilage, the adjacent bone, and, eventually, the joint ligaments. The disease is usually treated with anti-inflammatory drugs such as aspirin and ibuprofen, but recently some success has been achieved with drugs that block the action of cytokines such as tumour necrosis factor α (TNF-α) and interleukin 1 (IL-1).

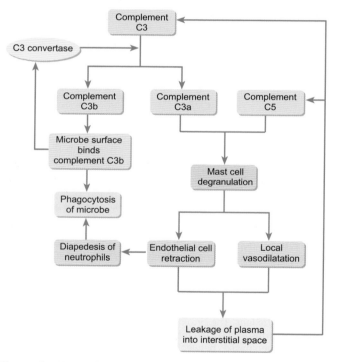

Figure 26.7 The mechanism by which complement C3 triggers an inflammatory response after it has been activated by an invading organism. The initiation of this series of events begins with the activation of C3 convertase. This leads to a positive feedback loop in which complement C3b binds to the surface of the invading organism and activates more C3 convertase to act on the C3 that enters the interstitial space from the plasma. The surface of the microbe binds complement C3b and this facilitates its phagocytosis. The other fragment of C3, C3a, acts with complement C5 to initiate mast cell degranulation; this facilitates the retraction of the endothelial cells and the diapedesis of neutrophils, which also participate in the phagocytosis of the invading microbe.

Summary

When the body becomes injured or infected, there is a local vasodilatation, local oedema, and infiltration of the damaged tissues by white cells. These changes constitute the inflammatory response, which brings the proteins and cells of the immune system to the point of injury. The trigger for the inflammatory response is mast cell degranulation, which leads to the secretion of histamine and prostaglandins.

26.7 The adaptive immune system

It is well known that while exposure to infective agents (e.g. the chicken pox virus) will cause disease on the first occasion, a subsequent exposure will not generally result in a recurrence of the disease. Nevertheless, the resistance to that infection does not extend to other diseases—for example, experience of chicken pox does not prevent infection by the measles virus. These facts highlight two important features of the immune system. First, resistance can be acquired by one exposure to an invading organism and then lasts for many years—even for a whole lifetime. Second, the resistance is specific for that organism. In immunological terminology, the response has **memory** and is **specific**. These characteristics distinguish the response of the adaptive immune system from that of the natural immune system.

Lymphocytes

The cells of the adaptive immune system are the lymphocytes. The lymphoid system consists of the total mass of tissue associated with the lymphocytes and their functions. It is disseminated throughout the body (see Figure 26.1). The tissues in which the lymphocytes mature (the bone marrow and thymus) are known as **primary lymphoid tissue** while the lymph nodes, spleen, and other lymphoid tissues are **secondary lymphoid tissue**. The two principal classes of lymphocyte of the adaptive immune system are known as B-cells and T-cells. The B-cells largely mature in the bone marrow and secrete antibody, while the T-cells mature in the thymus gland and secrete cytokines, cytotoxic substances, or both.

Lymphocytes are stimulated by antigens that bind to their surface receptors. Individual lymphocytes respond only to one antigen, and when they are stimulated they proliferate by mitosis to form a population of cells with an identical specificity called a **clone**. As shown in Figure 26.8 for B-cells (although T-cells behave in a similar manner), some cells continue to proliferate and carry out their specific immunological function (discussed in the following subsections) while others remain in the lymphoid tissue as **memory cells**, able to respond quickly to the same antigen in the future.

Lymphocytes continuously circulate through the tissues

To monitor the tissues of the body for invading organisms, the lymphocytes continuously circulate throughout the tissues—a process known as immunological surveillance. They migrate from the blood, through the tissues, to the lymph nodes, which they enter either by way of the afferent lymphatic vessels or directly from the post-capillary venules (see Chapter 30). After they have entered the lymph nodes, they pass into the efferent lymphatics and return to the blood via the thoracic duct.

The first stage of a lymphocyte's migration from the blood is its adhesion to the wall of the blood vessel. Normally, like the red cells, the lymphocytes remain in the centre of a blood vessel, but when they reach a target tissue some of them become attached to the vessel wall. This process is guided by homing receptors that are specific for particular tissues. The cells then flatten and squeeze between the endothelial cells and move into the surrounding tissues (diapedesis), as described in Figure 26.6 for neutrophils. It is the particular combination of cytokine, selectin, and integrin that determines the homing of the various immune cells to specific tissues and sites of infection.

Summary

The adaptive immune system provides a mechanism for defending the body against an extraordinarily wide range of organisms. Unlike the response of the natural immune system, that of the adaptive immune system is specific and has memory. Indeed, once resistance has been acquired it usually lasts for many years. The cells of the adaptive immune system are the lymphocytes. There are two principal classes of lymphocyte: B-cells and T-cells, which mature in the bone marrow and thymus respectively.

B-lymphocytes and antibody production

A resting B-lymphocyte has little cytoplasm, a darkly staining nucleus, and few mitochondria or ribosomes. After it has been stimulated by antigen, it becomes transformed into a **plasma cell** in which the cytoplasm is greatly expanded (see Figure 26.9) and full of ribosomes. It also has a well-developed Golgi apparatus. The ribosomes are the site of synthesis of the antibodies, which are then secreted into the plasma. As the antibody has the same specificity as the B-cell

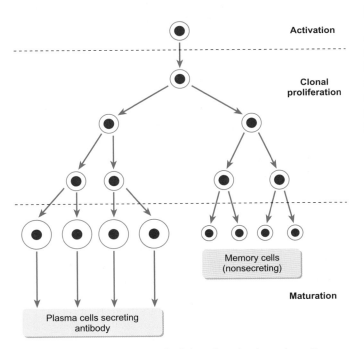

Figure 26.8 The generation of a B-lymphocyte clone. An antigen activates a naïve lymphocyte, which then proliferates. Some of the clonal cells mature and secrete antibody of the same specificity as that of the receptor (i.e. it is capable of binding to the initiating antigen) while other cells remain in the lymphoid tissues as memory cells.

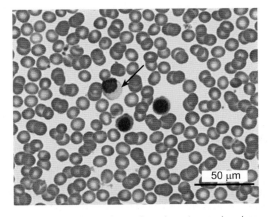

Figure 26.9 A blood film with two lymphocytes and a plasma cell (indicated by the arrow). Note the large eccentric nucleus and extensive cytoplasm compared to the appearance of the lymphocytes.

antigen receptor, a particular antigen will stimulate only those B-cells that will respond by secreting an appropriate antibody. In this way, the secretion of antibody is tailored to the nature of the infection.

There are five classes of antibody

The antibodies are globulins (immunoglobulins) which have the same basic structure, consisting of two identical light chains and two identical heavy chains linked to give a Y-shaped molecule, as shown in Figure 26.10. There are five main types of heavy chain (α, δ, ϵ, γ, and μ) which determine the particular class of antibody, and there are two main types of light chain (κ and λ). The antigen binding domains (known as Fab from 'fragment antibody binding') are formed by one heavy and one light chain and are located at the two tips of the Y so that each antibody molecule has two identical antigen binding sites. This part of the molecule (the N-terminal end) is highly variable in structure. It is this variability that permits an antibody to bind to one particular antigen rather than another. An antibody will recognize and bind to a particular specific structural feature known as an **epitope**. The stem of an immunoglobulin molecule is known as the Fc region (which stands for 'fragment crystallizable'). Its sequence is constant for an antibody of a specific class. The Fc region is used by the cells of the immune system to recognize particles that have been coated with antibody, e.g. viruses or bacteria. The Fab and Fc regions are linked by a hinge region, which gives flexibility to the antibody and which allows the binding regions to become attached to separate antigens on the surface of various microbes.

The five classes of antibody (isotypes) are IgA, IgD, IgE, IgG, and IgM.

1. IgA is formed from two light chains and two α heavy chains. It the most common immunoglobulin in secretions such as saliva and bile. It is also found in colostrum (the milk secreted during the first week of lactation—see Chapter 49). IgA is predominantly found as a monomer in the blood, but when it is secreted it occurs

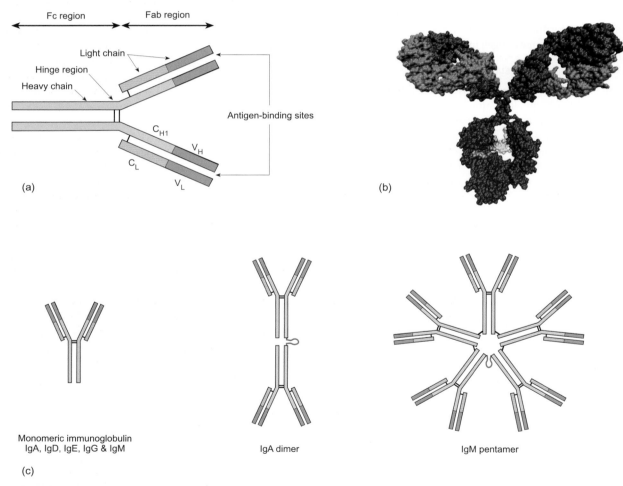

(a)

(b)

(c)

Monomeric immunoglobulin
IgA, IgD, IgE, IgG & IgM

IgA dimer

IgM pentamer

Figure 26.10 The structure of immunoglobulins. (a) An immunoglobulin molecule, which consists of four peptide chains: two heavy chains and two light chains linked together by disulphide bonds, in the form a letter Y. The antigen binding sites located on the tips of the Y show high specificity for a particular antigen. C_L and C_H = constant regions of the light and heavy chains, respectively; V_L and V_H = variable regions. (b) An image of the true molecular shape of an immunoglobulin; blue = heavy chains; red = light chains; yellow = the position of a glycan (carbohydrate chain) which is linked to the protein. (c) The structures of monomeric immunoglobulins (IgA, IgD, IgE, IgG, and IgM), of dimeric IgA, and of pentameric IgM; red = the J chain joining immunoglobulin molecules together.

as a dimer in which two IgA heavy chains are linked together by a peptide known as the J chain (Figure 26.10c). When IgA is secreted it is first bound to an epithelial receptor called the polymeric Ig receptor. This complex becomes internalized by the epithelial cells and is transported across the epithelial layer. On the lumen side, the complex is cleaved by a protease to release the IgA still bound to the part of the receptor called the secretory component.

2. IgD is formed from two light chains and two δ heavy chains. It is a cell surface receptor that is expressed when B-cells leave the bone marrow and migrate to the secondary lymphoid tissues. It is co-expressed with IgM, but the role of the two isotypes is not understood. IgD is absent from mature antibody-producing cells.

3. IgE is formed from two light chains and two ε heavy chains. It binds to Fc receptors on basophils and mast cells to facilitate the inflammatory response to antigen. It appears to be especially important in combating worm infections.

4. IgG consists of two light chains and two γ heavy chains. It is the most abundant immunoglobulin in plasma. In addition, this protein is small enough to cross the placental membrane and so provide the fetus with ready-made antibodies to protect it *in utero* and for some months after its birth. In humans, four different isotypes of IgG are present which are known as IgG1, IgG2, IgG3, and IgG4. Although the order indicates their abundance in blood, the different isotypes serve different functions, IgG1 and IgG4 being particularly effective in activating complement.

5. IgM consists of two light chains and two μ heavy chains. It is the first antibody to be produced during development and during the primary immune response. In germ line cells, the gene segment encoding the μ constant region of the heavy chain is positioned first among other constant region gene segments. This appears to account for IgM being the first immunoglobulin to be secreted by mature B-cells. In the blood IgM is present as a pentamer in which the individual IgM molecules are linked by disulphide bonds and a J chain. It is very effective in the activation of complement. Together with IgD it also forms a component of the B-cell antigen receptor.

Antibodies have two main functions: to bind an antigen, and to elicit a response that results in the removal of that antigen from the body. The antibody acts together with complement to stimulate phagocytosis by neutrophils and macrophages, with the result that the organism carrying the antigen is killed and digested. The combination of antibody, complement, and phagocyte is very effective. A lack of any one component seriously compromises the ability of the body to mount an adequate immune response. In some cases, the binding of the antibody to its antigen is sufficient to inactivate the invading organism. Thus, when viruses are coated with antibody, they cannot infect the host cells and so are prevented from proliferating.

Antibody diversity

Although antibodies can be grouped into five main classes, the number of different antibody molecules is immense. Estimates suggest that there may be more than a thousand million million (10^{15}) different kinds of antibody molecule, each with its own specificity. It would seem that this number would be sufficient to ensure that the body could mount an antibody response to virtually any potential antigen.

How is such a large number of different proteins generated when human DNA only codes for about 19,000 (1.9×10^4) genes? The answer to this problem lies in the way in which antibody molecules are encoded by DNA. As described above, each antibody consists of four polypeptide chains: two light chains and two heavy chains. The genes for these peptides are located on three separate chromosomes. In humans, the genes that code for the heavy chains are found on chromosome 14 while those for the two kinds of light chain are found on chromosomes 2 and 22. On each chromosome, there are many gene segments known as **exons** that code for proteins, separated by non-coding regions called **introns**. These non-coding regions are, nevertheless, important in gene regulation and in the recombination of gene segments as outlined below.

The DNA sequence that codes for each polypeptide is determined by the combination of different gene segments, which are shown diagrammatically in Figure 26.11. In the case of the heavy chains there are three sets of exons that are known as V, D, and J segments, which code for particular peptide domains (regions of the protein). Each type of gene segment exists in many different forms, and an individual B-cell will assemble a DNA sequence that codes for a heavy chain by combining one V (variable), one D (diversity), and one J (joining) segment with the segment that codes for the constant (Fc) region (Figure 26.11). The same cell will assemble the sequence for a light chain in a similar manner except that no D segment is incorporated. For both light and heavy chains the exact sites of gene combination can vary, and additional bases may be inserted or deleted. This increases antibody diversity still further. These remarkable adaptations allow a relatively small number of gene segments to code for a vast number of different immunoglobulins.

Once a B-cell has assembled a given sequence for antibody synthesis, it is committed to the production of that particular immunoglobulin. However, as point mutations occur in the DNA during the normal clonal proliferation of B-cells, antibody diversity can increase still further. This fact explains why the affinity of antibody for a particular antigen increases with time after the initial exposure. The antibody produced by a single clone is called a monoclonal antibody and in recent years these have been used for both diagnosis and treatment of a number of diseases (see Box 26.1).

Summary

After it has been stimulated by antigen, a B-cell becomes transformed into a plasma cell, which secretes antibody into the circulation. The antibody acts together with complement to stimulate the phagocytes, with the result that the organism carrying the antigen is eliminated.

The genes for the production of antibodies are located on three separate chromosomes. On each chromosome, the gene segments that code for antibodies (exons) are separated by non-coding regions called introns. The DNA sequence that codes for each antibody is determined by the combination of different gene segments. In this way a vast number of different antibodies can be produced from a relatively small number of genes.

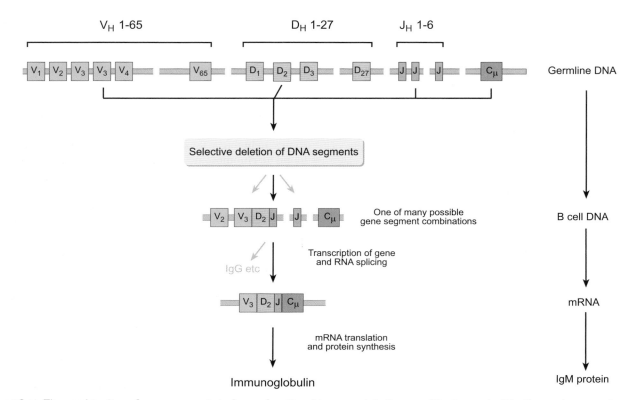

Figure 26.11 The combination of gene segments to form a functional immunoglobulin gene. The top part of the figure shows a schematic representation of the germ line chromosomal DNA that encodes a heavy chain for IgM. During maturation of a B-cell one D_H segment is linked to a J_H segment by deletion of the intervening DNA. This is followed by the linking of a V_H gene segment to the newly formed $D_H J_H$ pairing by deletion of a further DNA segment. The mRNA that is transcribed undergoes further splicing to join the $V_H D_H J_H$ segment to the segment coding for the constant region (C_μ for IgM). The resulting mRNA is then used by the B-cell to synthesize the corresponding protein. A similar process occurs during the synthesis of a light chain except that there is no D segment in the light chains.

T-lymphocytes and cell-mediated immunity

Unlike B-cells, T-cells do not enlarge greatly or secrete antibody when they encounter their specific antigen. Instead, they secrete cytokines or cytotoxic molecules (or both) onto neighbouring cells. Since these substances act in a paracrine fashion, the effects of T-cell activation are local and usually only one cell responds. Those T-cells that secrete cytokines to stimulate B-cells to proliferate and secrete antibody are called **helper T-cells**. Those that secrete cytotoxic substances to kill a target cell are called **cytotoxic T-cells**. Each type has a different role to play.

Antigen processing and presentation

Unlike B-cells, which respond to circulating antigens, T-cells respond to cells whose chemistry has altered—such as a tumour cell, a cell infected with a virus, or a dendritic cell bearing an antigen. T-cells respond to MHC molecules that have bound a foreign peptide. An individual T-cell therefore recognizes only a particular MHC molecule incorporating a particular foreign peptide. In effect the MHC molecules are exposing the foreign peptide to the T-cell, i.e. they are *presenting* it to the T-cell for inspection.

The sequence of events leading to **antigen presentation** is as follows. When cells become infected by viruses or intracellular bacteria, foreign proteins are expressed within their cytoplasm.

These foreign proteins are broken down into peptides by the cell's lysosomes. The resulting peptides are transported to the endoplasmic reticulum, where they become complexed with MHC class I proteins before being inserted in the plasma membrane where they can be recognized by cytotoxic T-cells.

Dendritic cells play a central role in presenting antigens to the lymphocyte population. They are derived from the monocyte colony-forming cell line (see Chapter 25) and take their name from their characteristic appearance. They are covered with numerous thin processes similar to those of astrocytes (see Figure 9.17). This gives them a large surface area with which to interact with foreign antigens and with other cells of the immune system. Dendritic cells that have captured a foreign antigen migrate to the various lymphoid organs, where they present the antigen to helper T-cells.

Although all dendritic cells perform an essentially similar role, they are known by different names according to their location. They are termed **Langerhans cells** in the skin, **veiled cells** in the lymph, **interdigitating dendritic cells** in the medulla of the thymus and secondary lymphoid tissue, and **interstitial dendritic cells** in other tissues.

All dendritic cells express high levels of class II MHC molecules, which complex with foreign antigen as follows. Extracellular

Box 26.1 Monoclonal antibodies

Antibodies are proteins specialized for selective high affinity binding to specific chemical structures (known as chemical moieties). As discussed in the main text, antibodies are formed from the splicing together of different gene elements to give rise to a huge variety of potential structures. Antibodies can be generated to recognize almost any chemical structure, and so they are a unique biological resource for identifying particular extracellular markers. In principle, this means that they can be generated to bind to any extracellular signalling moiety, whether identifying cancerous cells for attack by natural killer cells, or as a means to modulate specific signalling pathways (for example in the treatment of autoimmune diseases). The difficulty in using antibodies lies in the fact that *individual* antibodies are present at low levels in serum. One of the best known examples for the therapeutic use of mixed populations of antibodies (*poly*clonal antibodies) is in antivenin, where host animals are hyperimmunized to a particular snake venom, and the resultant serum is used to treat cases of snake bite. However, for use as medication, polyclonal antibodies are far from ideal as they are time-consuming and expensive to prepare. Moreover, each preparation will differ.

These difficulties can be largely resolved through the use of *mono*clonal antibodies, where only a single antibody species is used. Our ability to produce monoclonal antibodies results from a number of important experimental findings. The first of these was the observation that patients with a rare cancer, multiple myeloma, had very high levels of a single antibody in their blood. This is due to clonal expansion of a single line of B-cell precursors. Furthermore, myeloma-based cell lines are essentially immortal and can be used for producing large amounts of a specific antibody.

The next step was to generate an immortal cell line which produced an antibody with targeted specificity. This was eventually achieved through fusion of a myeloma cell and a specific B-cell clone taken from a mouse immunized against a particular antigen (sheep red blood cells). Kohler and Milstein, who carried out this work (published in 1975, and recognized with the Nobel Prize in Physiology or Medicine in 1984), termed these fusion cells **hybridomas**, and the singular antibodies they produced in abundance, **monoclonal antibodies**.

Monoclonal antibodies have achieved great clinical significance in the years since their initial development. The first patents for the use of monoclonal antibodies were claimed by groups using the Kohler and Milstein technique to establish antibodies against viruses and tumours. Despite this, the first approved clinical use for monoclonal antibodies was muromonab-CD3, which was used to prevent kidney transplant rejection by blocking the activity of CD3, a cell surface receptor expressed on T-lymphocytes that is required for their activation.

Early clinical trials revealed the problem with this type of monoclonal antibody—severe side effects due to the mouse origin of the antibodies. For this reason, transgenic mouse lines were developed with the mouse immunoglobulin genes replaced with human ones, providing either humanized or chimeric (i.e. hybrid) antibodies which are more easily tolerated. By mid-2016, more than 60 monoclonal antibodies had been approved for therapeutic use by the FDA. These have applications ranging from cancer treatment, to targeted immunosuppression, to treatment of diseases as varied as cardiovascular disorders and osteoporosis. Monoclonal antibody binding can be used to kill targeted cells either through activation of the endogenous cytotoxic T-cells, or by bringing radioisotopes or specific toxins to cells expressing the targeted cell surface marker. They can also be used to modulate receptor signalling by blocking ligand binding, or by inducing the internalization and degradation of receptors. In Alzheimer's disease, there is even hope that they may be able to stimulate the phagocytosis and breakdown of amyloid plaques, potentially greatly slowing disease progression.

In addition to their direct therapeutic value, monoclonal antibodies are also widely used in diagnostic testing. Because the antibodies react to a specific chemical structure, when they are tagged with a specific marker such as a fluorescent molecule they can be used to identify particular proteins. This approach is widely used within clinical diagnostics for the identification of pathogens and tumour cell types, and also for staining pathological sections to observe which normal cell types are exhibiting aberrant behaviour, such as overgrowth or selective loss. These same features are also invaluable in basic scientific research.

---

antigens are taken up by endocytosis and degraded to form small peptides. The endosomes then fuse with vesicles containing MHC class II proteins, which complex with the foreign peptides before becoming inserted in the plasma membrane where they can be scrutinized by helper T-cells.

A further class of dendritic cells, the **follicular dendritic cells**, behave in a rather different fashion. They do not express class II MHC and do not present antigen to helper T-cells. Instead they remain located within lymphoid follicles, the areas of a lymph node populated by B-lymphocytes. Follicular dendritic cells express membrane receptors for antibody and complement (known as FcRs and CRs respectively). Binding of immune complexes to these

cells appears to promote the activation of B-lymphocytes and secretion of the appropriate antibody.

T-cell receptors

The T-cell receptors consist of eight polypeptide chains, two of which have large and variable extracellular regions that are responsible for recognizing antigens. The remaining six peptide chains are concerned with intracellular signalling. The variability of the extracellular T-cell receptors (often abbreviated as TCR) is generated by rearrangements of gene segments in the DNA similar to those described for the generation of antibody diversity, as discussed earlier in this section.

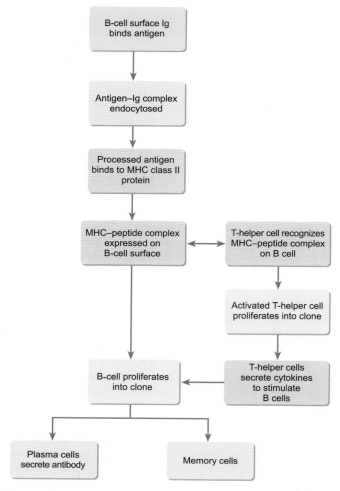

Figure 26.12 The cellular basis of the antibody response. Helper T-cell activation by an activated B-cell is shown, although activation of antigen presenting cells (APC) would be important in the primary response. Memory cells are long-lived T-cells and B-cells that permit a very rapid response to a subsequent encounter with the priming antigen.

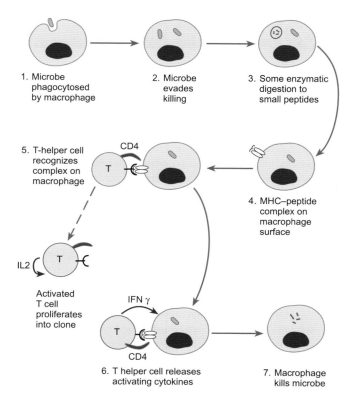

Figure 26.13 A simple outline of the process by which a helper T-cell (T) enhances the ability of a macrophage to counter infection by a microbe that has evaded killing. CD4 is a cell surface marker that acts as a co-receptor for the invariant part of the MHC class II protein and plays an essential role in activation of the helper T-cell. IFNγ is interferon γ. (Based on Fig 19.2 in J.H. Playfair and G.J. Bancroft (2004) *Infection and immunity*, 2nd edition. Oxford University Press, Oxford.)

Antigen recognition by T-cells requires binding of the T-cell receptor to the antigen and to specific cell markers known as CD4 and CD8 which act as co-receptors. The identifying label CD stands for cluster of differentiation, and individual markers are characteristic of particular cells and stages of development. Those cells bearing CD4 are T helper cells and those that bear CD8 are cytotoxic T-cells.

T helper cells identify cells bearing MHC class II molecules—B-cells and macrophages. When they detect one of these cells bearing a foreign peptide, they secrete cytokines to stimulate them. The B-cells proliferate, so increasing the size of the particular clone and the amount of antibody secreted (see Figure 26.12) with the result that the antibody response is enhanced. Those T helper cells that respond to the MHC complex on macrophages do so by stimulating the affected macrophages to activate their normal killing mechanisms, as shown in Figure 26.13. This process is important

for those cells that have become infected with intracellular bacteria (e.g. the bacillus that causes tuberculosis).

Cytotoxic T-cells respond to MHC class I molecules bearing an antigenic peptide. The binding of the T-cell receptor to its specific antigen on the MHC I molecule initiates changes that result in the reorganization of the T-cell cytoskeleton so that its cytotoxic granules face towards the target cell. The activated T-cell may then either activate surface receptors that lead to apoptosis of the target cell or directly kill the target by exocytosis of the contents of its cytotoxic granules which contain perforin, granulysin, and granzymes (a potent mixture of proteases). The perforin and granulysin form pores in the plasma membrane of the target cell which allows the entry of calcium ions and collapses the potassium gradient. This results in the death of the target cell (see Figure 26.14). Activated cytotoxic T-cells kill their targets within a few minutes and a single T-cell can kill many infected cells within a few hours. However, as their cytotoxic effect depends on direct cell–cell contact, they are able to do so while sparing the neighbouring healthy cells. The T-cells themselves are protected from their own cytotoxic proteins by a protease called cathepsin B, which is inserted into the plasma membrane when the cytotoxic granule is secreted. As for the natural killer cells, the T-cells have inhibitory receptors on their surface to prevent inappropriate activation in healthy tissues.

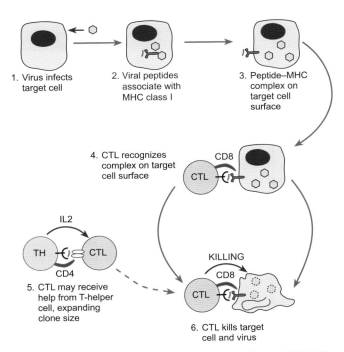

1. Virus infects target cell

2. Viral peptides associate with MHC class I

3. Peptide–MHC complex on target cell surface

4. CTL recognizes complex on target cell surface

5. CTL may receive help from T-helper cell, expanding clone size

6. CTL kills target cell and virus

IL2   CD4   CD8   KILLING

Figure 26.14 The main steps by which cytotoxic T-cells (CTL) kill cells infected with a virus. CD8 is a co-receptor for the invariant part of the MHC class I protein on the host cell. It plays an essential role in the activation of the cytotoxic T-cells. IL2 is the cytokine interleukin 2 released from a helper T-cell (TH). (Based on Fig 19.4 in J.H. Playfair and G.J. Bancroft (2004) *Infection and immunity*, 2nd edition. Oxford University Press, Oxford.)

Summary

T-cells respond to those cells that possess MHC molecules that have bound a foreign peptide, such as a cell infected with a virus. Activated T-cells secrete cytokines (helper cells) or cytotoxic molecules. The effects of T-cell activation are local and usually only one target cell responds.

The antibody response to infection

When people are first exposed to an infectious agent, they rely on the natural immune system for their defence. If the infection persists, B-cells begin to make antibody to the specific bacterium or virus but this only occurs after a lag phase during which the appropriate B- and T-cells are activated. Initially the B-cells secrete mostly IgM, which reaches a peak after about a week (see Figure 26.15). This antibody is able to activate complement and so enhances the ability of the natural immune system to combat the infection. If the disease continues, some B-cells switch to the production of IgG and the level of this antibody in plasma continues to rise after the IgM level begins to decline. This is known as class switching. IgG levels peak after about two weeks and then slowly decline as the infection wanes. This pattern of antibody secretion is known as the **primary response**.

A subsequent infection by the same organism is met by a much more prompt response—the **secondary response**. IgG levels rapidly rise above those seen during the primary response and remain elevated for a longer period (Figure 26.15). In contrast, the IgM levels follow approximately the same time course as in the primary response. The rapid and augmented secondary response is due to the recruitment of specific memory cells—cells that arose during the development of the initial clone but which did not mature into antibody-secreting plasma cells. The B-lymphocyte memory cells, in particular, respond to very small levels of antigen once they have been primed. This memory function of the adaptive immune system makes possible artificial immunization against specific diseases such as poliomyelitis (see Box 26.2) and smallpox ('vaccination').

Summary

On initial exposure to an antigen, the B-cells secrete IgM and IgG. Plasma levels of these antibodies peak after 1–2 weeks and then slowly decline. A subsequent infection by the same organism is met by a much more prompt and long-lasting increase in the plasma levels of the appropriate IgG antibody.

26.8 Disorders of the immune system

Like other organ systems, the function of the immune system may become disturbed. Broadly speaking, the disorders are due to inappropriate immune activity (**autoimmunity** and **hypersensitivity**) or to a failure of the immune system (**immunodeficiency**).

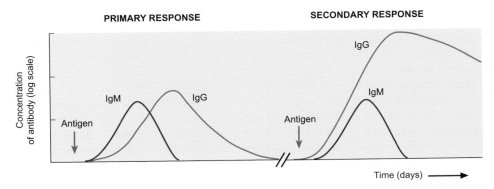

Figure 26.15 Primary and secondary antibody responses to an infection. Note the relatively slow IgG response to the first exposure and the early secretion of IgM. Compare this with the very rapid IgG response to the second challenge. (From Fig 8.8 in J.M. Austyn and K.J. Wood (1993) *Principles of cellular and molecular immunology*. Oxford University Press, Oxford.)

Box 26.2 Peyer's patches and the oral polio vaccine

One component of the mucosal-associated lymphoid tissue (MALT) is the organized gut-associated lymphoid tissue (GALT) that includes the Peyer's patches. These are lymphoid aggregates in the gut mucosa that are found predominantly in the ileum. They house large numbers of naive T- and B-lymphocytes and are covered by an epithelial layer containing cuboidal cells known as M-cells, which are specialized antigen-transporting cells that do not express MHC II. They lie over the phagocytic dendritic cells, which take up the transported antigens and then migrate to the local lymphoid tissue where they activate the lymphocytes. Once exposed to the antigen, lymphocytes within the Peyer's patch become sensitized and begin to secrete IgA on to the epithelial surface that binds to, and inactivates, the antigen. In this way the Peyer's patches play an important part in the immunological homeostasis of the gastrointestinal tract.

A further example of the importance of the gut-associated lymphoid tissue is provided by the effectiveness of the oral polio vaccine in developing countries. The polio virus is normally transmitted via the faecal–oral route (usually by contaminated water) and may cause paralysis through the death of neurons in the anterior horn of the spinal cord. The polio vaccine is now administered routinely to children in three doses during the first year of life (with boosters in childhood and adolescence). A non-virulent (or attenuated) form of the polio virus is given to the child orally. Once administered, the attenuated virus colonizes the intestine where it triggers an immune response mediated by cells of the Peyer's patches and other GALT cells. As a result, the plasma will contain antibodies to the polio virus and the lymph glands will have acquired memory cells that can be rapidly activated in the event of a subsequent infection.

In developed countries inactivated poliomyelitis vaccine is usually administered, as there is a slight risk of the attenuated virus reverting to a virulent strain and the risk is considered to outweigh the benefits. In developing counties, however, the benefits of low cost, ease of administration, and high risk of infection outweigh the slight risk of using an attenuated live vaccine.

Hypersensitivity

The adaptive immune system is very powerful and capable of responding to a wide variety of antigens. In normal people, the response is appropriate and correctly targeted against the invading organism. Sometimes, however, the activity of the immune system leads to pathological changes in the host tissues. Such reactions are called **hypersensitivity** or **immunopathology**. Hypersensitive reactions may be grouped under one of four types.

Type I: allergic reactions

These are mediated by IgE antibodies and mast cells. Examples are hay fever and asthma. Four steps lead to an allergic response.

1. Initial sensitization in which an IgE is generated that is specific for a particular allergen (e.g. pollen).
2. Subsequent exposure resulting in the secretion of IgE which binds to mast cells and basophils.
3. The binding of further molecules of allergen to the *bound* IgE.
4. The binding of IgE to its receptor, resulting in the degranulation of the mast cells and the secretion of histamine as well as various cytokines and eicosanoids.

This sequence of events leads to the generation of the inflammatory response by the mechanisms discussed earlier in the chapter. The subsequent events depend very much on the site at which the allergen initiates the reaction. In the airways, the release of histamine and the formation of leukotrienes causes the broncho-constriction characteristic of asthma. If the allergen reaches the circulation, it may provoke a generalized inflammatory reaction that results in a circulatory collapse known as **anaphylactic shock**. This is a potentially fatal condition (see Chapter 40).

Type II: antibody-dependent cytotoxic hypersensitivity

In this case, IgG reacts with host cells and initiates the processes that lead to their destruction. This unwanted reaction may follow a transfusion of incompatible blood (see Chapter 25). It may also follow skin or organ transplants, leading to the rejection of the graft or transplant (discussed in Section 26.9). Occasionally the lymphocytes attack the host cells themselves (i.e. they respond to the normal host antigens). This is known as **autoimmunity**. If antibodies react with normal hormone receptors, they may either stimulate the particular receptor as in hyperthyroidism (Grave's disease) or they may block the action of the normal hormone as in hypothyroidism (myxoedema).

Type III: immune complex–mediated hypersensitivity

When an antibody reacts with an antigen, it forms an immune complex that may precipitate. If these complexes are not phagocytosed rapidly (their usual fate), they may accumulate in the small blood vessels. When this happens, they are attacked by complement and neutrophils. The ensuing reactions may cause damage to the endothelium at a critical point in the circulation and compromise the function of the affected organ. One example is **glomerulonephritis** (inflammation of the renal glomeruli), which results from the deposition of immune complexes in the glomerular capillaries and is a common cause of renal failure.

Type IV: cell-mediated delayed hypersensitivity

This occurs when helper T-cells or cytotoxic T-cells become activated by certain antigens. It can manifest itself as an allergic reaction to certain parasites, as a contact dermatitis, or as a reaction to specific components of the diet; or it may result from sensitization to a chemical agent. The antigen is taken up by dendritic cells and transported to the adjacent lymph nodes where it activates memory cells. These proliferate and either release inflammatory cytokines or directly cause tissue damage. It is the involvement of the memory cells that gives rise to the delay, which may make it difficult to identify the antigen responsible. Examples are the dermatitis elicited by contact with certain plants (e.g. poison ivy) and coeliac disease, in which there is a reaction to the glutens present in wheat.

Autoimmune diseases

An autoimmune disease can be caused by antibodies directed at normal host proteins, by inappropriate activity of T-cells, or both. When a self-reacting antibody is formed, it may directly affect a specific function as in myasthenia gravis and Grave's disease, in which the antibodies bind directly to cell surface receptors (the nicotinic receptor in myasthenia gravis and the TSH receptor in the case of Grave's disease). Alternatively, the antibody may activate tissue phagocytes directly or via the formation of immune complexes, as in the type II and type III hypersensitivity reactions described above. Examples of T-cell-mediated autoimmune disease are type I diabetes, in which T-cells attack the pancreatic β cells, and multiple sclerosis, in which cells of the immune system attack the myelin sheath of axons in the nervous system.

Self-tolerance and autoimmunity

In autoimmunity, the lymphocytes of the immune system mount an attack on normal host cells. Although these reactions are thankfully rare, they are frequently triggered by an infection. Their very existence raises an important question: why are the cells of the immune system normally tolerant of the host cells? The explanation is as follows: only those T-cells that react to self-MHC molecules incorporating a foreign peptide are released into the circulation. If a T-cell reacts with unaltered MHC molecules, it will not be released from the thymus but will be destroyed. It is thought that many B-cells do express antibody receptors that react with host cells. Those that have a high affinity for the host cell surface markers are deleted in the bone marrow and never reach the circulation (this is comparable with the fate of T-cells that react with self-MHC molecules). Any remaining B-cells that react with host antigens will do so only weakly, and even these are ineffective because they do not get help from the T-cells. Such B-cells are said to be **anergic**.

Defects of the complement system

These appear to be mainly due to deficiency of particular components of the complement system. Low levels of C1, C2, or C4 lead to difficulty in clearing antigen–antibody complexes from the plasma. They are a common feature of systemic lupus erythematosus, in which deposition of immune complexes in the renal glomeruli leads to acute renal failure (see Chapter 39). A deficiency of C3 increases susceptibility to bacterial and viral infections. If complement components C5–C9 are low or even absent, there is a predisposition to serious infections by bacteria of the genus *Neisseria*. These organisms cause disseminated gonorrhoea and menigococcal meningitis. Reduced levels of C1s esterase inhibitor can result in overactivity of the classical complement cascade. In this situation, there are recurrent inflammatory attacks of mucosal tissues that may cause obstruction of the intestine or larynx.

Immunodeficiency and human immunodeficiency virus (HIV)

Immunodeficiency of genetic origin is called **primary immunodeficiency**. If it is due to some other cause, it is **secondary immunodeficiency**. Primary immunodeficiences are rare and are usually due to a defect in a single gene. They include defects in phagocyte function, defects in complement, and defects of lymphocyte function. In all cases, there is an increased susceptibility to infection. Secondary immunodeficiency is far more common, particularly in adults. The commonest cause is malnutrition, but damage to the immune system by certain infections (notably HIV), tumours, traumas, and some medical interventions may also impair the function of the immune system (e.g. treatment with immunosuppressive drugs or exposure of the bone marrow to high levels of X-irradiation).

The human immunodeficiency virus (HIV) is perhaps the best-known infectious agent that can compromise the function of the immune system. Unlike other agents, the HIV attacks the cells of the immune system, particularly the T-cells. The explanation for this alarming state of affairs is that the T-cell receptor is also the receptor to which HIV binds. HIV is a retrovirus that can insert its genetic material into the host cell DNA. The stimulation of an infected T-cell therefore results in replication of the virus. This leads to a slow depletion of the T-cell population and an increased susceptibility to infection. The resulting disease is now known as acquired immunodeficiency syndrome or AIDS.

Summary

Host cells are not normally attacked by the cells of the immune system, which are able to differentiate self from non-self. The immune system may react powerfully to an antigen (hypersensitivity) or it may fail to mount an adequate immune response (immunodeficiency). Hypersensitive reactions may be grouped under one of four types: type I (allergic reactions, e.g. hay fever and asthma); type II (antibody-dependent cytotoxicity hypersensitivity); type III (immune complex hypersensitivity); and type IV (cell-mediated delayed hypersensitivity). Immunodeficiency may be of genetic origin (primary immunodeficiency) or due to some other cause (secondary immunodeficiency).

26.9 Tissue transplantation and the immune system

Tissue transplantation from one person to another to treat organ failure has long been a major goal of medicine. The first such procedure to be wholly successful was the transfusion of blood. This depends on the correct identification of the antibodies (agglutinins) and antigens (agglutinogens) present in both donor and recipient blood, as described in Chapter 25. In the same way, successful skin grafts or organ transplants require a close match between the specific cell markers of both donor and recipient.

The problems associated with transplantation are very well illustrated by the difficulties experienced in successfully grafting skin. If the skin becomes severely damaged, for example as a result of extensive burns, the healing process cannot make good the lost germinal tissue and the wound contracts as it becomes infiltrated by connective tissue. This results in disfigurement and distortion of the neighbouring tissue. If the affected area is large, there may also be a continued loss of fluid from the damaged area, which will become increasingly susceptible to infection. For these reasons it is sometimes desirable to graft some healthy skin from another part of the body on to the site of injury. Such grafts (which are known as **autografts**) are usually successful. The transplanted tissue is quickly infiltrated by blood vessels and heals into place. The donor area also heals rapidly. If the damaged area is very extensive, however, it may be impossible to find sufficient undamaged skin to act as a source for the grafts. In this case, it is necessary to consider grafting skin from someone else--to use an **allograft**.

A skin graft from another individual will initially take quite well, but after about a week it will become rejected (type IV hypersensitivity). Moreover, a second graft from the same donor will be rejected immediately. If the donor is an identical twin, however, the initial graft will not be rejected, as both of the twins have the same genetic makeup and their tissue antigens (MHC or, in humans, HLA) will be identical. These discoveries provided crucial evidence that tissue rejection is an immunological phenomenon and paved the way for successful skin grafts and organ transplants between individuals of different genetic background.

To avoid graft rejection, the HLA proteins of the donor are matched as closely as possible to those of the recipient (tissue typing). In addition, immunosuppressant drugs are administered to inhibit the activity of the immune system. This approach has proved very successful with skin grafts as well as with kidney and heart transplants. More than 80 per cent of kidney transplants now survive for more than five years provided the HLA antigens are well matched. For heart transplants the figure is more than 70 per cent. The success rate for the transplantation of other organs is also showing considerable improvement. Nearly half of all liver transplant patients will survive for more than five years, although lung transplants are, at present, less successful. Close tissue matching is not required for corneal grafts because the cornea does not become vascularized and so is not subject to attack by the lymphocytes.

✳ Checklist of key terms and concepts

General concepts

- The immune system is divided into the natural immune system and the adaptive immune system.

- The cells of the immune system distinguish host cells from those of invading organisms by cell surface markers known as the major histocompatibility complex (MHC), also called the human leukocyte antigen (HLA) complex.

The natural immune system

- The natural immune system provides innate immunity by recognizing molecular features characteristic of common infecting organisms. It is able to mount an immediate defence.

- The natural immune system comprises five kinds of cells and three different classes of proteins.

- Macrophages and neutrophils phagocytose invading organisms and kill them with reactive oxygen intermediates.

- Cells that become infected with a virus are destroyed by natural killer cells before the virus can replicate.

- When the body becomes injured or infected, the site of injury undergoes changes known as the inflammatory response.

- The inflammatory response brings the proteins and cells of the immune system to the point of injury.

- The trigger for the inflammatory response is mast cell degranulation.

The adaptive immune system

- The adaptive immune system is specific and has memory.

- It provides a mechanism for responding to infections by organisms not previously encountered.

- The cells of the adaptive immune system are the B-cells and T-cells.

- The B-cells provide humoral immunity by secreting antibodies, while the T-cells provide cellular immunity.

- Antibodies act together with complement to stimulate the phagocytes and attack invading microorganisms.

- On initial exposure to an antigen, antibody levels in the plasma peak after one to two weeks and then slowly decline.

- A subsequent infection by the same organism is met by a prompt and long-lasting increase in the appropriate antibody. The is the basis of immunization.

- T-cells respond to MHC molecules that have bound a foreign peptide.

- Activated T-cells secrete cytokines (helper T-cells) or cytotoxic molecules (killer T-cells).

- The activated T-cells target individual cells via close contact.

Disorders of the immune system

- Disorders of the immune system may be due to inappropriate immune activity (autoimmunity and hypersensitivity) or to a failure of the immune system to respond to an infection (immunodeficiency).

- In autoimmunity host cells are attacked by the immune system.

- The immune system may react powerfully to an antigen (hypersensitivity) or it may fail to mount an adequate immune response (immunodeficiency).

- Hypersensitive reactions may be grouped under one of four types: type I (allergic reactions, e.g. hay fever and asthma); type II (antibody-dependent cytotoxicity hypersensitivity); type III (immune complex hypersensitivity); and type IV (cell-mediated delayed hypersensitivity).

- Immunodeficiency may be of genetic origin (primary immunodeficiency) or due to some other cause such as malnutrition (secondary immunodeficiency).

 Recommended reading

Biochemistry

Berg, J.M., Tymoczko, J.L., Gatto, G.J., and Stryer, L. (2015) *Biochemistry* (8th edn), Chapter 33. Freeman, New York.

Immunology

Alberts, B., Johnson, A., Lewis, J., Morgan, D., Raff, M., Roberts, K., and Walter, P. (2014) *Molecular biology of the cell* (6th edn), Chapters 23 and 24. Garland, New York.

A straightforward introduction to the immune system with simple and clear illustrations.

DeFranco, A.L., Locksley, R.M., and Robertson, M. (2007) *Immunity: The immune response in infectious and inflammatory disease.* Oxford University Press, Oxford.

A very comprehensive introduction to the field with many good illustrations.

Nossal, G.J.V. (1993) Life, death and the immune system. *Scientific American* **249**, 53–62.

This and the other articles in this issue (September 1993) give an excellent, non-technical introduction to immunology and diseases of the immune system.

Playfair, J.H.L., and Bancroft, G.J. (2004) *Infection and immunity* (2nd edn). Oxford University Press, Oxford.

A gentle and highly readable introduction to a very complex subject.

Tonegawa, S. (1987) Somatic generation of immune diversity (Nobel lecture), http://nobelprize.org/nobel_prizes/medicine/laureates/1987/tonegawa.html

A very readable account of the discovery of how antibody diversity is generated.

Pharmacology

Rang, H.P., Ritter, J.M., Flower, R.J., and Henderson, G. (2015) *Pharmacology* (8th edn), Chapters 6 and 26. Churchill-Livingstone, Edinburgh.

Review articles

Caligiuri, M.A. (2008) Human natural killer cells. *Blood* **112**, 461–9.

Carroll, M.C. (2004) The complement system in the regulation of adaptive immunity. *Nature Immunology* **5**, 981–6.

Min, B. (2008) Basophils: What they 'can do' versus what they 'actually do'. *Nature Immunology* **9**, 1333–9.

Stone, K.D., Prussin, C., and Metcalfe, D.D. (2010) IgE, mast cells, basophils, and eosinophils. *Journal of Allergy and Clinical Immunology* **125**, S73–80.

 To check that you have mastered the key concepts presented in this chapter, complete the accompanying online self-assessment questions. Go to www.oup.com/uk/pocock5e/

CHAPTER 27

Introduction to the cardiovascular system

Chapter contents

This chapter should help you to understand:

* The basic organization of the circulation
* The arrangement of the chambers of the heart and the heart valves
* The structure of the blood vessels
* The nerve supply to the heart and blood vessels

the cells and removes the products of metabolism. Second, it regulates the volume and composition of the extracellular fluid, which is essential for the normal function of the cells. In higher animals this regulation is achieved via the renal circulation (see Chapter 39). Third, because the blood is distributed to all parts of the body, the circulation plays an important role in a wide variety of physiological functions including the following.

* It acts as the vehicle for distributing hormones and thereby contributes to hormonal control.
* The circulating white cells and immunoglobulins of the blood provide the principal means of defence against infection (see Chapter 26).
* By adjusting blood flow to the skin, the circulation also plays a significant role in the regulation of body temperature (see Chapter 42).

This chapter describes the organization of the circulation, its gross anatomy, the structure of the blood vessels, and their innervation. The detailed functioning of the heart and circulation as well as the common disorders of the cardiovascular system are discussed in Chapters 28–31.

27.1 Introduction

In unicellular organisms and simple animals such as sponges, the exchange of nutrients and waste products between the cells and the environment can be accomplished by simple diffusion across the cell membranes. However, since diffusion is a random process in three dimensions, the time required for equilibration increases rapidly with increasing distance. Consequently, in more complex animals, diffusion of nutrients by itself would not suffice to permit adequate exchange of nutrients and waste products, as most cells are separated from the external environment by a considerable distance.

To overcome this problem, higher animals have evolved a circulatory system which serves three primary functions. First, it promotes the carriage of oxygen and nutrients such as glucose to

27.2 The gross anatomy and organization of the circulation

The circulation consists of a pump (the heart) and a series of interconnected pipes (the blood vessels). As the blood is pumped from the right side of the heart through the lungs (the **pulmonary circulation**) and then from the left side of the heart to the rest of the body (the **systemic circulation**), the overall arrangement is of two circulations in series (see Figure 27.1). The general anatomical arrangement can be seen from the beautiful image of the circulation of a healthy young adult male shown in Figure 27.2, which was obtained by functional magnetic imaging (fMRI). The heart, the aorta, and the major arteries are all clearly visible. Figure 27.3 shows the heart and major vessels in greater detail.

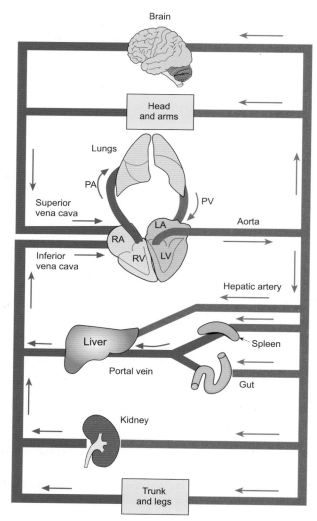

Figure 27.1 A schematic drawing of the circulation. The arrows indicate the direction of blood flow. Note that the blood returning to the heart enters the right atrium. It then enters the right ventricle, which pumps the blood through the lungs. After leaving the lungs, the blood enters the left atrium and then passes to the left ventricle, which pumps it through the rest of the body via the systemic circulation. Thus, the pulmonary circulation is in series with the systemic circulation. PA, PV = pulmonary artery and pulmonary vein; RA, LA = right and left atria; RV, LV = right and left ventricles.

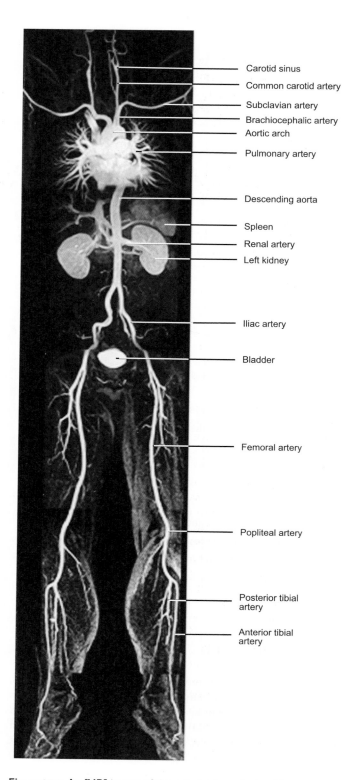

Figure 27.2 An fMRI image of the arterial tree of a healthy young adult male. A magnetic contrast medium was injected prior to acquiring the magnetic resonance image, in which the heart and the main arterial vessels are clearly seen. The contrast medium has accumulated in the bladder during the imaging process. The veins are much less distinct. (Courtesy of Dr S. Ruehm.)

The heart lies in the thoracic cavity. In an adult, it is about the size of a clenched fist and consists of four muscular chambers: two atria and two ventricles (see Figure 27.4) which are arranged in pairs so that each half of the heart forms a functionally separate pump. In the adult heart, the two halves are completely separated by a sheet of tissue known as the **septum** which prevents mixing of oxygenated and deoxygenated blood. The pumping activity of the heart raises the pressure in the aorta above that of the large veins, where the pressure is close to that of the atmosphere. It is this pressure difference that causes the flow of blood around the systemic circulation. Equally, blood flows through the lungs because the

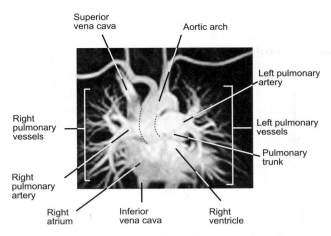

Figure 27.3 An enlarged image of the chest region from Figure 27.2 to show the position of the heart and the origin of the aorta and pulmonary arteries more clearly. Note the extensive branching of the pulmonary vessels. (Courtesy of Dr S. Ruehm.)

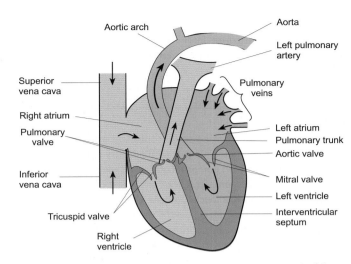

Figure 27.4 A simple diagram illustrating the arrangement of the chambers and the direction of blood flow through the heart (arrows). The blood passing through the right side of the heart contains deoxygenated blood (shown in blue) while that passing through the left side of the heart contains oxygenated blood (shown in red).

pressure in the pulmonary arteries is greater than that in the pulmonary veins. The pressure in the major arteries is known as the **systemic arterial blood pressure** or, more commonly, the **blood pressure**.

The atria are thin-walled chambers that receive blood from the large veins and deliver it to the ventricles, which have much thicker walls—the wall of the left ventricle being thickest (shown diagrammatically in Figure 27.4). The atria are separated from the ventricles by the atrioventricular valves, which ensure that the blood flows from the atria to the ventricles but not in the other direction.

On the right side of the heart, the atrioventricular valve (the **tricuspid valve**) consists of three roughly triangular flaps of fibrous connective tissue. On the left side, the atrioventricular valve consists of two such flaps and is known as the **mitral valve** (it is also called the **bicuspid valve**). Back flow of blood from the pulmonary artery into the right ventricle is prevented by the **pulmonary valve**, while back flow of blood from the aorta into the left ventricle is prevented by the **aortic valve**. The pulmonary and aortic valves are also known as the **semilunar valves**. The valves and chambers of the heart are lined by a cellular layer known as the **endocardium** which is continuous with the lining of the aorta, pulmonary artery, and great veins—the **vascular endothelium**. The muscle of the heart is called the **myocardium**.

The heart lies in a tough fibrous sac known as the **pericardium** that prevents it from expanding excessively due to overfilling with blood. The pericardium is attached to the diaphragm so that the apex of the heart is relatively fixed. When the ventricles contract, the atria move towards the apex. This has the effect of expanding the atria as the ventricles contract, and this aids atrial filling.

The systemic circulation

The aorta emerges from beneath the right side of the pulmonary trunk and ascends, giving rise to the brachiocephalic artery which subsequently divides to form the common carotid and subclavian arteries (see Figure 27.2). The ascending aorta curves backwards to the left over the right pulmonary artery to form the **aortic arch** (Figures 27.3 and 27.4). When the aorta reaches the left side of the fourth thoracic vertebra, it descends to supply the lower regions of the body. This is known as the descending aorta and supplies all the major organs of the abdomen, as well as the skeletal muscles of the trunk and lower limbs.

Blood returning to the heart from the head, neck, and arms reaches the right side of the heart via the **superior vena cava**, while blood returning from the lower part of the body (the lower limbs and abdomen) reaches the right side of the heart via the **inferior vena cava** (Figures 27.3 and 27.4). More blood is contained within the veins than in the arteries, partly because the walls of the veins are thinner and more easily distended than those of the arteries and partly because there are more veins than there are arteries. (There are approximately two veins for every artery of moderate size.) The veins thus act as a store of blood that can be mobilized when required (see Chapter 30 p. 466).

The pulmonary circulation

The right ventricle occupies most of the anterior surface of the heart. It gives rise to the pulmonary trunk, which divides after a short distance to give rise to the left and right pulmonary arteries (Figures 27.3 and 27.4). The pulmonary circulation is discussed in detail in Chapter 34, but a brief outline follows. The output of the right ventricle passes into the **pulmonary trunk**, which divides into the right and left **pulmonary arteries**. The two pulmonary arteries divide to supply the individual lobes of the lung (two on the left and three on the right). The arteries then branch along with the bronchial tree until they reach the respiratory bronchioles, where they form a dense capillary network that drains into pulmonary venules. The venules merge to form small veins that progressively merge to form larger veins until two large **pulmonary veins** emerge from each lung to empty into the left atrium, as shown in Figure 27.4.

Summary

The circulation is organized so that the right side of the heart pumps blood through the lungs (the pulmonary circulation) and the left side of the heart pumps blood around the rest of the body (the systemic circulation). Thus, the two circulations are arranged in series. The heart has four chambers: two atria and two ventricles. The atria are separated from the ventricles by the mitral and tricuspid valves, which prevent the reflux of blood into the atria when the ventricles contract. Reflux of blood from the pulmonary artery and aorta into the ventricles is prevented by the pulmonary and aortic valves.

27.3 The structure of the systemic blood vessels

As the blood leaves the heart and proceeds around the body, it first flows through a series of arteries of progressive smaller diameter until it reaches the tissues, where it passes though the arterioles to the capillaries, which are the exchange vessels. From the capillaries, the blood first enters postcapillary venules and then the true venules, before finally entering the veins themselves. The veins merge to form progressively larger vessels that culminate in the superior and inferior vena cavae, which return the blood to the right atrium.

The walls of the blood vessels consist of more or less distinct layers whose composition and thickness vary according to the type of vessel (see Figure 27.5 and Table 27.1). Starting from the inside, the main layers of the larger vessels are the **tunica intima**, the **tunica media**, and the **tunica adventitia**.

The tunica intima consists of a layer of flat endothelial cells overlying a thin layer of connective tissue. The endothelial cells of the tunica intima are in direct contact with the blood. In the arteries and larger arterioles, the tunica intima is separated from the tunica media by the **internal elastic lamina**, which is not present in the capillaries, venules, or small veins. The tunica media provides much of the mechanical strength of the larger blood vessels and consists of concentric layers of smooth muscle cells arranged in a spiral manner, with fibres of elastin and collagen interspersed between the muscle cells. These connective tissue fibres are secreted by the smooth muscle cells themselves. The tunica media is completely absent in capillaries and postcapillary venules. In the arteries, the tunica media is separated from the tunica adventitia by a prominent external elastic lamina which can be clearly seen in the cross-section of femoral artery shown in Figure 27.5. The tunica

adventitia consists of a loosely formed layer of elastic and collagenous fibres oriented along the length of the vessel. It serves to anchor the blood vessels in place and provides much of the mechanical strength of the capillaries and venules.

The circular layer of smooth muscle of the tunica media is innervated by an extensive network of sympathetic nerve fibres. In the larger blood vessels there is an appropriate blood supply provided by the vasa vasorum (meaning the 'vessels of the vessel'). Some of these vessels originate directly from the lumen of an artery and pass into the vessel wall to supply the tunica intima and tunica media. Others arise from smaller arterial branches that penetrate the wall of the vessel they supply.

The arteries are the primary distribution vessels and may be subdivided into two groups:

a) the **elastic arteries** which, in humans, are large vessels of 1–2.5 cm in diameter;

b) the **muscular arteries**, which range in size from about 1 mm to 1 cm in diameter.

The elastic arteries include the aorta and pulmonary arteries together with their major branches. Their walls are very distensible because their tunica media contains a high proportion of elastin (up to 40 per cent compared to about 10 per cent for a muscular artery—see Table 27.1). The elastin is arranged in a series of thin sheets known as lamellae (see the image of the femoral artery wall in Figure 27.5).

The muscular arteries are the main distribution vessels and include the cerebral, popliteal (in the legs), and brachial (upper arm) arteries. Their tunica media contains a higher proportion of smooth muscle and a smaller proportion of elastic tissue than that of elastic arteries. Moreover, it is thicker relative to the diameter of the lumen (Figure 27.5 and Table 27.1). This makes muscular arteries very resistant to collapse at the joints during movements of the limbs. The muscular arteries ensure the efficient distribution of blood to the various vascular beds.

The muscular arteries give rise to the **arterioles**, which are the resistance vessels responsible for regulating blood flow through particular vascular beds. The distinction between a small muscular artery and an arteriole is arbitrary, but any artery smaller than about 0.3 mm is generally considered to be an arteriole (Figure 27.5c). The arterioles branch repeatedly and the final branches (the terminal arterioles) give rise to capillaries, which are short, thin-walled vessels of about 5–8 µm diameter and 200 µm to 1 mm in length. The **capillaries** are the principal exchange vessels. Their walls consist of a single layer of flattened endothelial cells (see Figure 27.6) surrounded by a tunica adventitia consisting of a thin layer of connective tissue (the basal lamina). Satellite cells called **pericytes** (also known as Rouget or adventitial cells) partly surround the capillaries. These are contractile cells which are important in capillary development and may play a role in the regulation of blood flow. The capillary wall is so thin (about 0.5 µm) that the diffusion path between plasma and tissue fluid is extremely short.

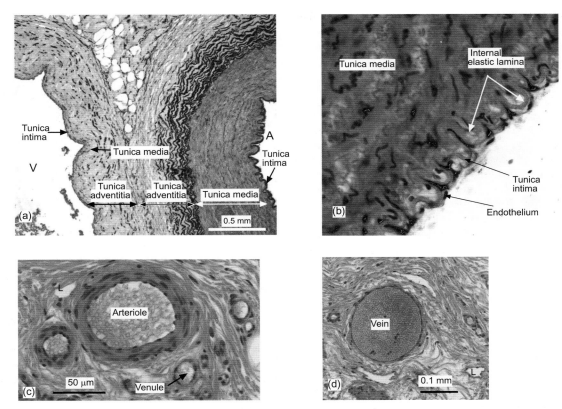

Figure 27.5 Blood vessels seen in cross-section after staining. (a) Part of the walls of the femoral artery (A) on the right and femoral vein (V) on the left. The tunica media is stained purple and is very thick in the artery but thin in the vein. The intense black staining surrounding the tunica media of the artery reveals the presence of the elastic fibres that form external elastic lamina. Some scattered elastic tissue is visible in the vein but there is no distinct layer. Note the marked difference in wall structure between the artery and the vein. (b) The wall of a muscular artery enlarged to show the tunica intima and the endothelium with its flattened nuclei. The wavy red line is the internal elastic lamina, which is also visible as a dense black line in the tunica intima of the femoral artery shown in panel (a). (c) A cross-section of a small and a large arteriole, which should be compared to that of the venule. Note the prominent smooth muscle (tunica media) of the arterioles compared to the thin layer found in the venule. (d) A cross-section of a small vein. Note the thin wall. A small lymphatic vessel (L) can be seen on the lower right of the field. Bars, 0.5 mm (a), 50 μm (c), 0.1 mm (d).

Table 27.1 Structural characteristics of blood vessels

	Elastic artery	Muscular artery	Arteriole	Capillary	Venule	Vein
Diameter	1–2.5 cm	0.3–1.0 cm	30–300 μm	5–8 μm	50–200 μm	0.5–3 cm
Wall thickness	1–2 mm	1 mm	~20 μm	0.5–1.0 μm	2 μm	0.5–1.5 mm
Composition of wall (% wall thickness)						
Endothelium	5%	10%	10%	95%	20%	5%
Smooth muscle	25%	40%	60%	0%	20%	30%
Elastic tissue	40%	10%	10%	0%	0%	0%*
Collagenous tissue	30%	25%	20%	5%	60%	65%

Note: The various kinds of blood vessel merge into each other as the vascular system first divides and then coalesces. Consequently, there is often no clear distinction between one type of vessel and another (e.g. a small artery and a large arteriole). The classification shown here is based on the generally accepted characteristics of the main kinds of vessel. It follows that the dimensions given above are approximate and are intended only to illustrate the characteristics of each kind of vessel.

* Some inconspicuous elastic tissue is present, particularly in the larger veins.

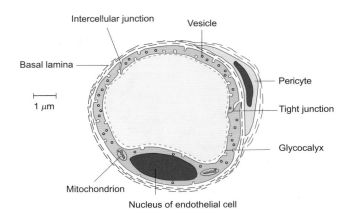

(a) Continuous capillary

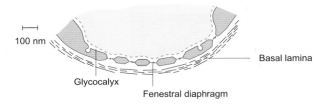

(b) Section of the wall of a fenestrated capillary

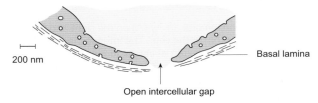

(c) Section of the wall of a discontinuous capillary

Figure 27.6 A diagrammatic representation of the structure of the three principal kinds of capillary. (a) Cross-section of a continuous capillary. (b) Part of the wall of a fenestrated capillary. (c) Part of the wall of a discontinuous capillary. (Based on Fig 9.2 in J.R. Levick (1995) *An introduction to cardiovascular physiology*, 2nd edition. Butterworth-Heinemann, Oxford.)

There are three types of capillary.

1. **Continuous capillaries** are by far the most common type of capillary. Their walls consists of a continuous endothelial layer perforated only by narrow clefts between the cells.

2. **Fenestrated capillaries** are mainly found in tissues that are specialized for bulk fluid exchange, such as exocrine glands and kidneys. Their endothelial cells are perforated by small circular pores, the **fenestrae**, which permit a relatively free passage of salts and water from the plasma to the tissues. Except in the capillaries of the renal glomerulus, each fenestra is closed by a fenestral diaphragm which has a central knob linked to the edge by filaments, rather like the hub and spokes of a wheel.

3. **Discontinuous capillaries** are found in the liver, spleen, and bone marrow. They are also known as **sinusoids** and are thin-walled

vessels with an irregular outline that is determined by the neighbouring cells. There are gaps between the endothelial cells that are sufficiently large to permit the passage of blood cells and plasma proteins.

The capillaries coalesce to form postcapillary **venules** about 20 μm in diameter, which also lack smooth muscle. In turn, these give rise to the true venules and **veins** which merge to form the great veins that return the blood to the heart. The walls of the veins and venules are similar in structure to those of the arteries, but they are much thinner in relation to the overall diameter of the vessel (see Figure 27.5d and Table 27.1). The tunica media is much thinner and the tunica adventitia correspondingly thicker than in the arteries (see Figure 27.5a). Consequently, the veins are much more distensible than the arteries. They are also more liable to collapse. In the mesenteric and other large abdominal veins the adventitia contains *longitudinal* bundles of smooth muscle which serve to strengthen the vessel wall. The walls of the pulmonary veins and venae cavae near to the atria contain cardiac muscle rather than the smooth muscle characteristic of most vessels, while those of the meninges, the retina, and the dural sinuses are devoid of any smooth muscle.

Unlike other blood vessels, the small and medium-sized veins of the limbs possess valves at intervals along their length. These are formed by folds of the tunica intima strengthened by elastic connective tissue. They are arranged so that blood can pass freely towards the heart while back flow is prevented. The large central veins and those of the head and neck do not possess functional valves.

In a few tissues, notably the skin, there are some direct connections between the arterioles and venules. These specialized vessels are known as arterio-venous shunt vessels (or **anastomoses**) and they have relatively thick muscular walls that are richly supplied by sympathetic nerve fibres. When these vessels are open, some blood can pass directly from the arterioles to the venules without passing through the capillaries.

Summary

The circulatory system consists of the arteries, the arterioles, the capillaries, the venules, and the veins. Except for the capillaries and the smallest venules, the walls of the blood vessels have three layers: the tunica intima, the tunica media, and the tunica adventitia.

27.4 The nerve supply of the cardiovascular system

The heart and blood vessels are innervated by postganglionic fibres of the autonomic nervous system. The heart receives its parasympathetic supply via the cardiac branches of the vagus nerves and its sympathetic supply from the superior, middle, and inferior cervical ganglia. The preganglionic sympathetic fibres of these ganglia originate in spinal segments T1 to T5 (see Chapter 11). The capillaries and the pericytic venules are not innervated, while the majority of blood vessels are innervated by sympathetic

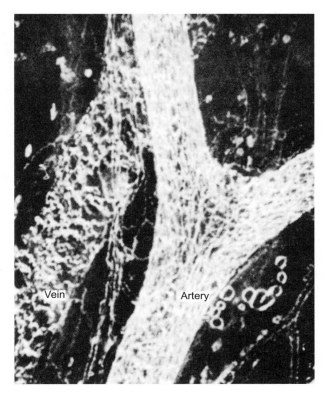

Figure 27.7 An image of a small artery and vein from the mesentery of the rat treated by the Falck-Hillarp technique to show the adrenergic nerves. The nerve plexus is more dense in the artery than in the vein, though both are richly innervated. A much less dense innervation pattern can be seen surrounding a small vessel at the bottom right of the figure. (From B. Falck (1962) *Acta Physiologica Scandinavica* **56**, suppl 197.)

postganglionic fibres originating in the paravertebral ganglia. These fibres provide a rich adrenergic innervation to the smooth muscle of the tunica media (Figure 27.7).

When the nerves supplying the blood vessels are activated they cause contraction of the smooth muscle, which results in a narrowing of the vessel and an increase in its resistance to flow. This is called **vasoconstriction** and the corresponding nerves are vasoconstrictor nerves. Certain blood vessels (e.g. those of the sweat glands) receive cholinergic sympathetic nerve fibres which relax their smooth muscle and cause the vessels to dilate. These are therefore called vasodilator nerves, and the effect of activating these nerves is called **vasodilatation** or **vasodilation**. Although the majority of blood vessels do not receive a parasympathetic innervation, some are innervated by parasympathetic vasodilator fibres. Examples are found in the salivary glands, intestinal mucosa, and genital tissue (see Chapter 30). In addition to this motor supply, the blood vessels also possess afferent fibres which respond to stretching of the vessel wall (baroreceptors) or to the chemical composition of the blood (chemoreceptors). The role of these receptors is discussed in Chapters 30 and 35.

Summary

The heart and blood vessels are innervated by postganglionic fibres of the autonomic nervous system. The heart receives its parasympathetic supply via the cardiac branches of the vagus nerves and its sympathetic supply from the cervical ganglia. The majority of blood vessels are innervated by sympathetic postganglionic fibres. The capillaries and the pericytic venules have no smooth muscle and are not innervated.

✱ Checklist of key terms and concepts

General properties of the circulation

- The blood circulates around the body in a closed circuit formed by the blood vessels.
- The circulation through the lungs is known as the pulmonary circulation, and the circulation to the other organs of the body is the systemic circulation.
- The right side of the heart pumps blood through the lungs and the left side of the heart pumps blood around the rest of the body.
- The two circulations are arranged in series.

The main anatomical features of the heart

- The heart consists of four chambers: two atria and two ventricles.
- The left atrium and ventricle are separated from those of the right by the septum.
- The atria are separated from the ventricles by the mitral and tricuspid valves.

- The heart valves prevent the reflux of blood into the atria when the ventricles contract.
- Reflux of blood from the pulmonary artery and aorta into the ventricles is prevented by the pulmonary and aortic valves.

The blood vessels

- The principal types of blood vessel are the arteries, the arterioles, the capillaries, the venules, and the veins.
- Except for the capillaries and the smallest venules, the walls of the blood vessels have three layers: the tunica intima, the tunica media, and the tunica adventitia.

Innervation of the cardiovascular system

- The heart and blood vessels are innervated by the autonomic nervous system.
- The heart receives both a sympathetic and a parasympathetic nerve supply.

- The majority of blood vessels are innervated only by sympathetic nerve fibres.

- The sympathetic nerves are predominantly vasoconstrictor nerves.

- Certain blood vessels have a vasodilator sympathetic innervation. These are sympathetic cholinergic nerves.

- A few tissues are innervated by parasympathetic vasodilator fibres.

Recommended reading

Mescher, A. (2015) *Junqueira's basic histology* (14th edn), Chapter 11. McGraw-Hill, New York.

  A recent revision of Junquieira and Carneiro's classic text. A good basic account of the structure of the blood vessels and the heart.

Levick, J.R (2010) *An introduction to cardiovascular physiology* (5th edn), Chapter 1. Arnold, London.

  A lucid introduction to the cardiovascular system.

To check that you have mastered the key concepts presented in this chapter, complete the accompanying online self-assessment questions. Go to www.oup.com/uk/pocock5e/

CHAPTER 28

The heart

Chapter contents

This chapter should help you to understand:

- The detailed anatomy of the heart
- How the heartbeat is initiated and regulated
- The cardiac cycle and the heart sounds
- The causes of abnormal heart sounds
- The measurement of cardiac output
- The factors that regulate the cardiac output (cardiodynamics)
- The pathophysiology of heart failure

28.1 Introduction

The heart lies in the **mediastinum** of the thoracic cavity beneath a tough membrane called the **fibrous pericardium**. (The mediastinum is the region of the chest located between the lungs that contains all the principal thoracic organs except the lungs). The fibrous pericardium is attached to the central tendon of the diaphragm and to the outer coats of the aorta and venae cavae. The heart itself is covered by a pair of thin membranes, the **serous pericardium**, that enclose a narrow, fluid-filled space called the **pericardial sac**. The pericardium restricts the movement of the diaphragm and prevents overfilling of the heart. The overall shape of the heart is approximately that of a blunt cone which has its base around the midline and its apex facing obliquely downward towards the left side of the chest (see Figure 28.1). As previously discussed in Chapter 27, the heart consists of four hollow chambers: two atria and two ventricles. Their walls follow the same three-layered pattern as that of the blood vessels, with an inner **endocardium**, a thicker layer of **myocardium**, and an outer covering of **epicardium**. The endocardium consists of a single layer of squamous endothelial cells covering a layer of connective tissue. The myocardium consists chiefly of cardiac muscle cells (cardiac myocytes), while the epicardium is formed by the mesothelial cells that make up the visceral pericardium. They cover a thin region containing elastic and collagen fibres.

The atria are separated from the ventricles by fibrous tissue that forms the skeleton of the heart and the four heart valves. It also provides a point of insertion for cardiac muscle fibres. The atria and ventricles are arranged in pairs so that each half of the heart forms a functionally separate pump. In the adult heart, the two halves are separated by a sheet of tissue known as the septum (see Figure 28.1). The portion separating the two atria is the **atrial septum**, which is relatively thin and largely made up of fibrous tissue. The two ventricles are separated by the **interventricular septum**, which consists of muscle tissue except for a small region of fibrous tissue near the base of the ventricle. Both atria have thin walls of cardiac muscle and receive blood from the venae cavae and pulmonary veins prior to delivering it to the ventricles which have much thicker walls, the wall of the left ventricle being thickest. The right ventricle pumps blood through the lungs (the pulmonary circulation) at a relatively low pressure, while the left ventricle pumps the blood around the rest of the body (the systemic circulation) at a much higher pressure.

The detailed structure and contractile properties of the cardiac muscle cells are discussed in Chapter 8 (Sections 8.2 and 8.5). The cardiac muscle fibres consist of discrete cells that are arranged end to end and linked by their intercalated discs. The individual fibres branch and form bundles (fasciculi) whose arrangement and course depends on their position in the heart. Superficial fibres run across both atria, although many of the deeper fibres insert into the atrial septum. In the ventricles, the muscle fasciculi are arranged in loops that originate at the fibrous skeleton and course around the ventricles in a helical pattern before finally ending at another part

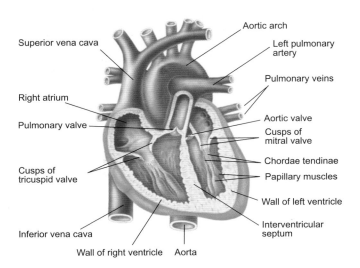

Figure 28.1 A cut-away drawing of the heart in diastole showing the papillary muscles, the chordae tendinae and their attachments to the mitral and tricuspid valves. Note that the pulmonary and aortic valves are closed while the tricuspid and mitral valves are open. (From Plate 2 in C. Blakemore and S. Jennett (eds) (2001) *Oxford companion to the body*. Oxford University Press, Oxford.)

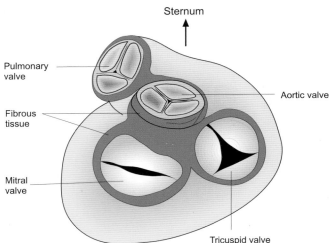

Figure 28.2 The valves of the heart viewed from above after removal of the atria, aorta and pulmonary trunk. The fibrous tissue (mauve) that separates the atria from the ventricles is continuous with that of the heart valves. (Based on Fig 5.3.7 in P.C.B. McKinnon and J.F. Morris (2005) *Oxford textbook of functional anatomy.* Oxford University Press, Oxford.)

of the fibrous skeleton. As in the atria, the orientation of these fibres depends on their position in the wall of the heart. For example, in the wall of the left ventricle the fibre bundles closest to the endocardium and epicardium tend to course along the length of the ventricle (i.e. longitudinally), those in the centre of the muscle mass run around the ventricle (circumferentially), while those occupying intermediate positions in the wall tend to run obliquely. This complex arrangement serves to ensure the even development of tension in the ventricular wall during systole (ventricular contraction).

The ability of the ventricles to fill under low pressure and to eject blood against high arterial pressures is critically dependent on the precise operation of the heart valves—the atrioventricular (or AV) and semilunar valves (see Figure 28.2). The atrioventricular valves (the bicuspid or mitral valve on the left and the tricuspid on the right) are composed of membranous leaflets or cusps that are flexible flaps of fibrous connective tissue which protrude into the ventricle. They are covered by a layer of endothelium and their free borders are attached to a set of fibrous tendons (the **chordae tendinae**) that connect them to a set of conical muscles on the ventricular wall known as the **papillary muscles** (see Figure 28.1). This arrangement prevents the valves from being pushed into the atria during systole. The aortic and pulmonary valves (the semilunar valves) consist of three crescent-shaped pockets of connective tissue that are located at the origins of the aorta and pulmonary artery. When the valves are closed, the three cusps are firmly pressed against each other as shown in Figure 28.2.

The heart is supplied with oxygenated blood by the coronary arteries, which arise at the root of the aorta just above the aortic valve as shown in Figure 28.3. The right coronary artery principally supplies the right atrium and right ventricle, while the left supplies the left atrium and left ventricle. Most of the venous blood from the coronary circulation drains into the right atrium via the coronary sinus, but a small amount drains directly into the left ventricle via the Thebesian veins. The myocardium itself is richly endowed with capillaries. The detailed physiology of the coronary circulation is discussed in Chapter 30.

28.2 The initiation of the heartbeat

At rest, the heart beats at a steady rate that varies between individuals and with age. The resting rate is at its highest in newborns and small infants (80–150 beats per minute or b.p.m.), in children it is between 70 and 130 b.p.m., while in adults it is 60–100 b.p.m. Note that in athletes who are in full training the resting heart rate is lower still, at 40–60 b.p.m. In all cases heart rate increases during exercise (see Chapter 38).

The rhythmic pulsation of the heart is maintained by excitatory signals generated within the heart itself. Indeed, under appropriate conditions, the heart will continue to beat rhythmically for a considerable time following its removal from the body. This **autorhythmicity** is due to the presence of pacemaker cells in the sinoatrial node of the right atrium. It is this intrinsic pacemaker activity that permits successful heart transplantation. Nevertheless, for the heart to be an effective pump, the contraction of the atria and ventricles must be coordinated. This is achieved by means of specialized conducting tissue found at the junction of the right atrium and the right ventricle (the atrioventricular node) and extending throughout the two ventricles (the bundle of His and Purkinje fibre system).

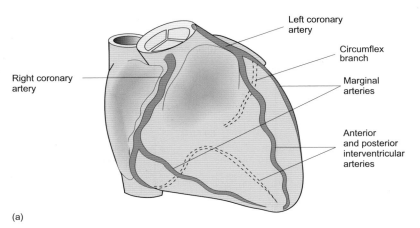

(a)

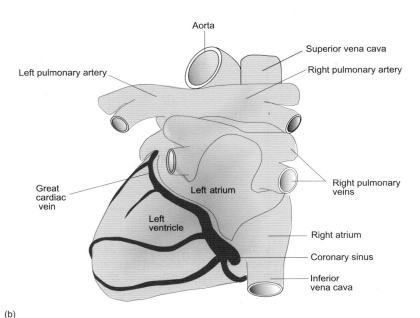

(b)

Figure 28.3 The coronary blood supply. (a) The arterial supply viewed from the front; note the origin of the coronary arteries at the base of the aorta just above the aortic valve. (b) The venous drainage of the coronary circulation viewed from the posterior aspect. (Based on Fig 5.4.10 in P.C.B. McKinnon and J.F. Morris (2005) *Oxford textbook of functional anatomy*. Oxford University Press, Oxford).

Pacemaker potentials and myocardial excitation

All the cells of the myocardium can show spontaneous electrical activity given suitable conditions, i.e. they are all potential pacemaker cells. Normally, however, only the cells of the **sinoatrial (SA) node** (or pacemaker region) show such activity. These cells are located within the wall of the right atrium near the junction with the superior vena cava (see Figure 28.4a). In the normal heart it is the activity of these cells that initiates each beat of the heart. The SA node cells spontaneously generate action potentials that are subsequently conducted throughout the whole myocardium.

In the absence of any extrinsic nervous input, the SA node cells will drive the heart at a rate of around 100 b.p.m. In other words, the cells of the SA node spontaneously generate an action potential every 600 ms or so. The action potentials of the sinoatrial node cells are relatively small, slowly rising depolarizations brought about by the inward movement of calcium ions. The action potentials of

atrial and ventricular myocytes are quite different, having a fast initial rise followed by a prolonged period of depolarization known as the plateau phase, which is particularly prominent and long lasting in Purkinje fibres (see Figure 28.4b). The fast initial rise is due to a rapid influx of sodium ions, while the plateau phase is mainly due to the inward movement of calcium ions. The ionic basis of the pacemaker activity and the action potential of the various cardiac cells is discussed in detail in Chapter 8 (pp. 133–135).

The conduction of the impulse throughout the myocardium

The action potential initiated in the SA node is propagated throughout the atria, the conducting system, and the ventricles via gap junctions, in a manner similar to that employed by unmyelinated nerve fibres (Chapter 7). This process continues until the entire myocardium has been excited. From the SA node, excitation spreads first across the whole of both atria via bundles of atrial

(a)

(b)

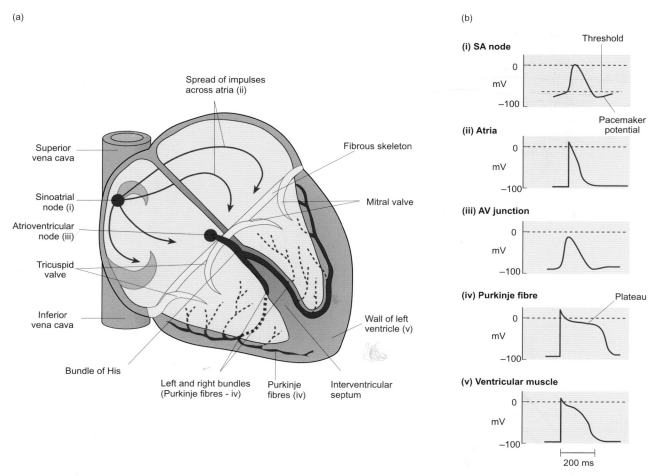

Figure 28.4 (a) The arrangement of the specialized conducting fibres of the heart. (b) The characteristic appearance of action potentials recorded from various types of cardiac myocyte. Note the pacemaker potential in the cells of the SA node and the prominent plateau of the action potentials recorded from the Purkinje and ventricular fibres. (Based on Fig 5.4.6 in P.C.B. McKinnon and J.F. Morris (2005) *Oxford textbook of functional anatomy*. Oxford University Press, Oxford).

muscle fibres. It then passes to the ventricles via the **atrioventricular (AV) node**, which forms the only bridge of conducting tissue between the atria and ventricles. The AV node consists of a narrow bundle of small-diameter cardiac myocytes which have relatively few gap junctions linking them to their neighbours. As a result, conduction through the AV node is relatively slow and the spread of excitation to the ventricles is effectively delayed for around 0.1 s at this point. This adaptation ensures that the atria have time to contract fully before the ventricular muscle is excited.

Conduction through the remainder of the system is rapid and occurs via the **bundle of His**, which divides into the left and right bundle branches to supply the left and right ventricles respectively, as shown in Figure 28.4a. The bundle fibres are specialized large-diameter cardiac myocytes that are arranged end to end to permit the rapid conduction of the wave of excitation from the AV node to an extensive network of large fibres (**Purkinje fibres**) that lies just beneath the endocardium, as shown in Figure 28.5. These fibres then spread the excitation to the ventricular myocytes via gap junctions. Conduction through the Purkinje fibre network is much

faster ($3-5$ m s^{-1}) than that through the myocardium itself, with the result that all parts of the ventricles are excited at much the same time. Figure 28.4b shows the action potentials of myocytes in various parts of the heart and illustrates some important differences between them. The myocyte membrane, like that of a nerve axon, is refractory during and immediately following an action potential (see Chapter 7). Consequently, it cannot be re-excited during the relaxation phase of the heart and this ensures that the conduction of the cardiac impulse is unidirectional.

The spread of excitation through the myocardium generates small electrical currents in the extracellular fluid, which create small differences in electrical potential that may be detected by appropriately positioned electrodes on the surface of the body. The small voltage changes that are recorded are known as an **electrocardiogram** or ECG which can be used to follow the phases of the cardiac cycle.

The origin of the ECG and its use in the diagnosis and treatment of heart disease is extensively discussed in Chapter 29. In brief, the normal ECG shows three main deflections in each cardiac cycle.

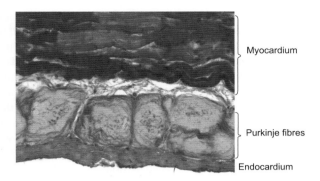

Figure 28.5 A section through the wall of a ventricle that shows the innermost layer, the endocardium, some Purkinje fibres in cross-section and a mass of densely staining muscle tissue. Note the large diameter of the Purkinje fibres and their pale staining.

These are the **P wave** which corresponds to atrial depolarization, the **QRS complex** which corresponds to ventricular depolarization, and the **T wave** which corresponds to ventricular repolarization. Atrial repolarization occurs during ventricular depolarization and is masked by the QRS complex. Figure 28.6 shows the timing of the ECG waves in comparison to the mechanical events of the heart. Note that the electrical events of the ECG occur *before* the corresponding mechanical events, so that the P wave precedes atrial contraction, the QRS complex precedes ventricular contraction, and the T wave precedes ventricular relaxation.

Summary

The heart shows an inherent rhythmicity that is independent of any extrinsic nerve supply. Excitation is initiated by a group of specialized pacemaker cells in the sinoatrial node, which generate spontaneous action potentials that are conducted throughout the myocardium as a wave of depolarization. The action potentials of the cardiac myocytes have a fast initial upstroke followed by a plateau phase of depolarization prior to repolarization. The plateau phase ensures that the action potential lasts almost as long as the contraction of the cell and ensures the unidirectional excitation of the myocardium.

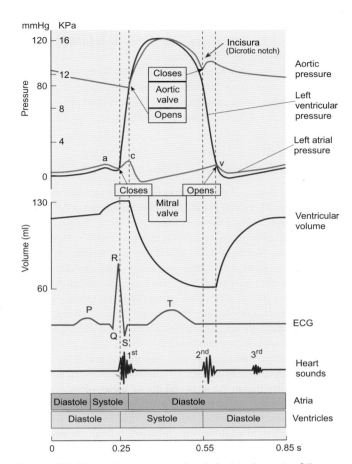

Figure 28.6 The major mechanical and electrical events of the cardiac cycle. The pressure changes shown are for the left side of the heart and reflect the underlying mechanical events. A phonogram of the heart sounds is shown at the bottom of the figure together with an ECG trace. Note the relative timing of the various events. Thus, for example, the QRS complex (which reflects ventricular depolarization) largely precedes ventricular contraction and the timing of the first and second heart sounds which are caused by closure of the atrioventricular valves and the aortic and pulmonary valves respectively. The interval between the first pair of vertical dotted lines is the period of isometric contraction and the interval between the second pair is the period of isometric relaxation. See text for further explanation.

28.3 The heart as a pump—the cardiac cycle

The electrical events described in the preceding section govern the beating of the heart. Depolarization during the action potential leads to an influx of calcium ions that causes the myocardial cells to contract, while the subsequent phase of repolarization permits their relaxation. This alternating cycle of contraction and relaxation of the myocardium allows the heart to act as a pump that propels blood around the pulmonary and systemic vessels. This repeating pattern of contraction and relaxation is known as the **cardiac cycle**. It consists of two major phases: **diastole**, during which the chambers relax and fill with blood, and **systole**, during which the heart contracts, ejecting blood into the pulmonary and systemic circulations. The heart has a two-step pumping action. The right and left atria contract virtually simultaneously (**atrial systole**), followed 0.1–0.2 s later by contraction of the right and left ventricles (**ventricular systole**).

During diastole, both the atria and ventricles are relaxed. The right atrium and ventricle fill with the partially deoxygenated blood of the venous return while the left atrium and ventricle fill with oxygenated blood from the lungs. The volume of blood that fills a ventricle just before it contracts is known as the **end-diastolic volume**. When a ventricle contracts, it eject nearly two-thirds of the blood it contains. This is called the **stroke volume**. The volume of blood remaining in a ventricle after systole is the **end-systolic volume**. The proportion of blood ejected during systole is called the **ejection fraction**. When an adult human is at rest, the stroke volume is about 70 ml and the end-systolic volume is about 50 ml, giving an ejection fraction of about 0.6 or, to put it another way, each ventricle ejects about 60 per cent of the blood that entered it during diastole.

The action of the heart valves is driven by pressure differences

Opening and closing of the AV valves occurs as a result of the pressure differences between the atria and ventricles that occur during the cardiac cycle. When the ventricles are relaxed, the pressure in the atria is slightly greater than that in the ventricles and the AV valves are open (see Figure 28.6). Blood in the systemic circulation drains into the right atrium from the superior and inferior venae cavae, while that returning from the lungs drains into the left atrium via the pulmonary veins. From the atria, the blood flows into the ventricles, increasing the volume of blood they contain. When the atria contract, atrial pressure rises and some 10–20 per cent more blood is forced into the ventricles (seen as a bump in the ventricular filling curve that occurs after the P wave of the ECG in Figure 28.6). As the ventricles start to contract, the ventricular pressure rises above that in the atria and the AV valves close. Since the total surface area of the cusps of the AV valves is much greater than that of the opening they cover, the upper (atrial) surfaces are firmly pressed together. In addition, the chordae tendinae are tensioned by contraction of the papillary muscles during systole, which prevents the valves being inverted into the atria as the pressure within the ventricles increases. These adaptations ensure reliable closure of the AV valves during ventricular systole. At this point in the cardiac cycle the ventricular pressure exceeds that of the atria.

During ventricular systole, the pressure in the ventricles rises until it exceeds that in the aorta and pulmonary arteries and, once this occurs, the aortic and pulmonary valves open. At the commencement of diastole, the ventricles begin to relax and the pressures in the right and left ventricles fall below those in the aorta and pulmonary artery, forcing the corresponding semilunar valves to close. Closure of the valves prevents reflux of blood from the aorta and pulmonary arteries into the ventricles during diastole.

Note that during diastole, the AV valves are open and the aortic and pulmonary valves are closed. In the ejection phase of systole the situation is reversed: the AV valves close and the aortic and pulmonary valves open. Thus the heart valves ensure the one-way flow of blood that is essential to the efficient operation of the circulatory system.

The pressure changes of the cardiac cycle

The cardiac cycle is the period from the end of one contraction of the heart to the end of the next. In a resting adult, the heart rate is around 70 b.p.m. so that each cardiac cycle lasts for about 0.8 s, where systole lasts around 0.3 s and diastole 0.5 s. The upper part of Figure 28.6 shows the major events occurring during a single cardiac cycle and the associated pressure changes for the *left side* of the heart. A similar pattern of pressure changes is seen in the right side of the heart, although the pressures are lower. The bottom half of the figure shows the associated changes in ventricular volume, the ECG, and the **phonocardiogram** (a recording of the heart sounds).

Blood normally flows continually from the pulmonary veins into the left atrium and most of this blood flows directly into the left ventricle. Only around 10–20 per cent is pumped actively into the ventricle during atrial contraction. Three major increases in pressure can be seen in the left atrium, the a-, c-, and v-waves (see the top panel of Figure 28.6). The a-wave occurs during atrial contraction, which itself is preceded by atrial depolarization (indicated by the P-wave of the ECG). When ventricular pressure exceeds atrial pressure, the AV valves close and there is a second 'bump' in the atrial pressure curve, the c-wave. This is mainly due to the bulging of the mitral valve into the left atrium and occurs just after the start of ventricular contraction. While the ventricle is contracting, the atrium relaxes and atrial pressure falls. As blood flows into the left atrium from the pulmonary veins, the atrial pressure slowly rises. This rise is reflected in the v-wave of the atrial pressure curve. Pressure falls once more as soon as the AV valves open at the end of ventricular contraction.

Ventricular contraction starts at the peak of the R wave of the ECG (note that the QRS complex represents ventricular *depolarization* which precedes ventricular contraction). The AV valves close at the start of contraction. This produces the first heart sound. For a short time (0.02–0.03 s) the aortic valve at the entrance to the aorta remains closed so this is a period of **isovolumetric contraction**—the intraventricular pressure rises but the volume does not change. This may be seen in the ventricular volume curve of Figure 28.6, which remains steady.

The **ejection phase** of ventricular systole begins when the aortic valve opens (the point at which left ventricular pressure exceeds aortic pressure). There is a period of rapid ejection followed by a period of rather slower emptying (see the ventricular volume curve in Figure 28.6). During the rapid ejection phase, both ventricular and aortic pressures rise steeply. For about the last quarter of ventricular systole, very little blood flows from the left ventricle to the aorta even though the ventricle remains contracted.

At the end of ventricular systole, the ventricle repolarizes (this is reflected in the T-wave of the ECG) and begins to relax so that intraventricular pressure falls rapidly. The higher pressure in the aorta causes aortic valves to close, preventing back flow of blood into the ventricle. There then follows a brief period of **isovolumetric relaxation** during which the ventricle continues to relax (with a concomitant fall in ventricular pressure) while the ventricular volume remains constant because the mitral valve is not yet open. This phase lasts for 0.03–0.06 s before the ventricular pressure falls below that of the left atrium and the mitral valve opens. Ventricular relaxation continues, but now blood enters the ventricle from the atrium and the volume of blood in the ventricle increases rapidly. About two-thirds of the way through ventricular diastole, the atria depolarize (the P-wave of the ECG) and subsequently contract to thrust additional blood into the ventricles shortly before the onset of ventricular systole. This may be seen clearly in Figure 28.6, which shows a second component to the rising phase of the ventricular volume curve corresponding to atrial systole.

The aortic pressure curve

Figure 28.6 shows that when the left ventricle contracts, the intraventricular pressure rises quickly. When this pressure exceeds that in the aorta, the aortic valve opens and blood begins to flow rapidly into the aorta. As it does so, the pressure within the aorta increases to around 16 kPa (120 mm Hg) (**systolic pressure**) and the wall of the aorta becomes stretched. When the left ventricle starts to relax, blood no longer flows into the aorta, and aortic pressure falls a little. The closure of the aortic valve causes a brief surge in aortic pressure which, like the primary pressure wave, is transmitted down the arterial tree. This secondary wave is known as the dicrotic wave and is preceded by the **incisura** or **dicrotic notch**. After this, aortic pressure falls slowly during diastole, as the stretched elastic tissue of the artery wall forces the blood into the systemic circulation (the maintenance of arterial pressure after closure of the aortic valves is sometimes called the **Windkessel effect**). By the time the ventricles contract again, the aortic pressure has fallen to around 10.6 kPa (80 mm Hg) (**diastolic pressure**).

A similar pattern of pressure changes is seen in the right side of the heart, but note that the pressure changes in the right ventricle are smaller than those in the left, the peak systolic pressure being about 3.3 kPa (25 mm Hg) rather than the 16 kPa (120 mm Hg) seen in the left ventricle and aorta. The pressure curve for the pulmonary artery has similar characteristics to that of the aorta except that the pressures are much lower: systolic pressure in the pulmonary artery is around 3.3 kPa (25 mm Hg) and diastolic pressure is around 1 kPa (8 mm Hg). The pressure changes in the right atrium are reflected in the jugular pulse.

Jugular pulse

As there are no valves between the right atrium and the internal jugular vein, the pressure changes of the right atrium can be inferred from the jugular pulse. Four components can be detected in the pressure waveform (see Figure 28.7) recorded from the internal jugular vein. The first is called the **a-wave**, which reflects contraction of the right atrium while the tricuspid valve is open. The **c-wave** appears as a hump at the base of the a-wave and reflects the bulging of the tricuspid valve into the right atrium during ventricular systole and the pressure of the carotid pulse on the jugular vein. This is followed by a rapid fall in pressure—the **x descent**—which is associated with the relaxation and initial rapid filling of the right atrium while the tricuspid valve remains closed. Following the x descent there is a rise in pressure called the **v-wave**, which occurs as the right atrium fills. As atrial systole begins, the atrium empties into the ventricle following the opening of the tricuspid valve. The jugular pressure falls—the **y descent**—and then rises as blood flows into the right atrium and the next cycle begins.

As the pressures in the jugular veins are normally very low (they average around 0.4 kPa (4 cm H$_2$O)), no pulse can be felt (palpated). However, during heart failure (see Section 28.7) and in certain other conditions in which the venous pressure is elevated, pulsation of the jugular vein can be clearly observed if the patient's upper body is supported at 45° to the horizontal with their head

slightly inclined to the left. The shape of the pulse wave is changed in certain disease states. For example, if the tricuspid valve does not close properly (**tricuspid incompetence**), the regurgitation of blood into the right atrium causes the c- and v-waves to merge, creating one prominent wave, while the amplitude of the x descent is reduced.

Stroke work and the pressure–volume loop

The pumping action of the heart is achieved by the mechanical work of the myocardium. In Chapter 8 we saw that to perform mechanical work, a load must be moved through a distance. For a three-dimensional system such as the heart, the work done is equal to the change in pressure multiplied by the change in volume so that the work performed by the heart each time it beats is given by the area of the pressure–volume curve for ventricular contraction. This is known as the **stroke work**. Ventricular pressure varies during the ejection phase of the cardiac cycle. So, in order to determine stroke work accurately, it is necessary to construct a graph of ventricular pressure against volume. The total external work carried out by the left ventricle during a single cardiac cycle is given by the area of the pressure–volume loop ABCDA shown in Figure 28.8.

Between points A and B the ventricle is filling. Pressure falls at first due to the suction effect of the relaxing ventricular muscle. This is analogous to the way a rubber bulb fills with liquid after it has been emptied of air. Subsequently, the pressure starts to rise passively as the volume of blood in the ventricle increases and the tension in the ventricular wall rises. Between points B and C the ventricle contracts but, since the aortic valve is closed, there is no change in volume and the pressure rises steeply. This is the

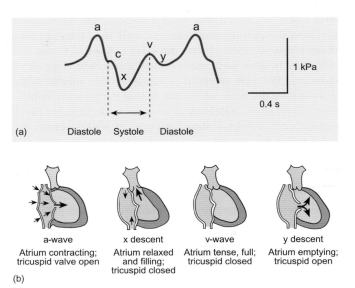

Figure 28.7 The origin of pressure waves recorded from the internal jugular vein. (a) A normal jugular pulse showing the characteristic waveform. (b) The mechanical events that give rise to the component waves of the jugular pulse shown in (a).

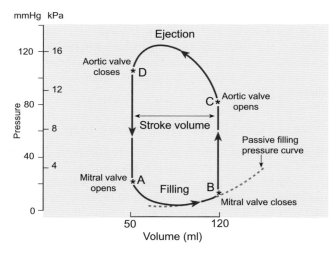

Figure 28.8 The pressure-volume cycle of the human left ventricle. The area bounded by the pressure-volume curve gives the stroke work of the heart. The lower confine of the pressure-volume loop shown by the dotted red line is defined by the passive stretch of the relaxed ventricular muscle by the returning blood. Note the points at which the aortic and mitral valves open and close.

phase of isovolumetric contraction and the heart is performing no external work (as there is no change in volume). The aortic valve opens at point C. Between points C and D, blood is being ejected, the volume of the ventricle decreases, and the heart performs external work, driving the blood around the circulation. The valve closes again at point D, and the phase between points D and A represents isovolumetric relaxation.

The shape of the pressure–volume curve is similar for the right and left ventricles, though the left has slightly higher filling pressures because it has thicker, less distensible walls. As it has to eject blood against a higher arterial pressure, the left ventricle performs significantly more external work than the right even though both ventricles pump the same volume of blood. The difference in work performed by the left and right ventricles is illustrated in Figure 28.9(a). If either the stroke volume or the systolic pressure is increased, the area of the pressure–volume loop will increase, indicating an increase in stroke work. This is true for both ventricles, although Figure 28.9(b) illustrates this only for the left ventricle.

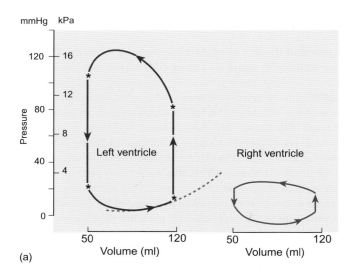

(a)

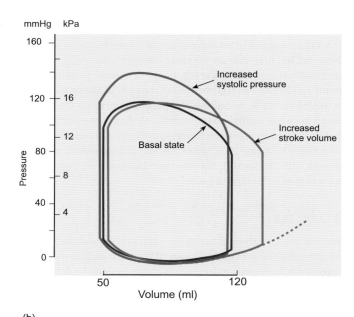

(b)

Figure 28.9 A comparison of the pressure-volume relationship for the human left and right ventricles (a). Note that the area of the pressure-volume curve for the right ventricle is much smaller than that for the left ventricle as both the filling and systolic pressures are significantly lower. (b) This panel shows an idealized representation of the increase in stroke work that results from an increase in end-diastolic volume and stroke volume (increased preload—red curve) and an increase in aortic pressure (increased afterload–green curve). The abscissa is expanded and the individual curves are slightly displaced for clarity. The dotted line represents the passive filling pressure curve for the left ventricle.

Summary

The alternating contraction and relaxation of the heart muscle is known as the cardiac cycle. The chambers of the heart relax and fill with blood during diastole, and contract to eject blood during systole. During early diastole, both the atria and ventricles fill with blood. Towards the end of diastole, the atria contract, forcing blood into the ventricles. The total amount of blood filling the ventricles at the end of diastole is called the end-diastolic volume (EDV). Nearly two-thirds of the blood in the ventricles is expelled during systole (the stroke volume or SV). The work performed by the heart during each beat (the stroke work) is the product of the rise in ventricular pressure occurring during systole and the stroke volume.

28.4 The heart sounds

Normal heart sounds

During routine clinical examination with a stethoscope, two heart sounds are generally audible in normal adults. These sounds are caused by vibrations in the heart wall and by turbulence in the blood largely generated by closure of the heart valves. As the sounds made by the heart valves radiate outwards in the direction of the blood flow, each sound is most clearly audible in a particular region of the chest (see Figure 28.10 and accompanying caption). The heart sounds can be recorded with a microphone applied to the chest wall (a **phonocardiogram**) as shown in Figures 28.6 and 28.11, which show their timing in relation to the electrical and mechanical events of the cardiac cycle.

The first heart sound begins immediately after the R-wave of the ECG, while the second heart sound coincides with the end of the T-wave. The second sound is quieter and has a more snapping quality than the first (hence the commonly used terms 'lub' and 'dup' for the first and second heart sounds). Two further heart sounds may be present during a normal cardiac cycle, although they are much quieter and more difficult to hear than the first two sounds.

The first two heart sounds are caused by the closure of the heart valves as the sudden tension in the valve cusps sets up vibrations in the heart wall and generates eddy currents in the blood. The first heart sound (S1) is caused by closure of the tricuspid and mitral valves at the start of ventricular systole, while the second sound (S2) occurs as the aortic and pulmonary valves close at the end of ventricular systole. During inspiration, the second heart sound may be audibly split into aortic (A_2) and pulmonary (P_2) components which are heard at slightly different times, the aortic component fractionally preceding the pulmonary component. This is caused by the changes in venous return during the respiratory cycle and is quite normal in healthy young people.

The third heart sound (S3 in Figure 28.11) occurs as a result of opening of the AV valves at the end of the period of isovolumetric relaxation. It is thought to be due to rapid turbulent entry of blood into the ventricles at the onset of filling and occurs 140–160 ms after the second heart sound. It is heard best at the apex and is often heard in children in whom the chest wall is thin and in young adults. It is also heard during pregnancy due to the increase in blood flow. A fourth heart sound (S4) is occasionally audible. It occurs just before the first heart sound and is a very low-frequency sound thought to be caused by oscillations in the blood flow that follow atrial contraction. It is a common finding in healthy elderly individuals, but in younger people it usually indicates an abnormal increase in the stiffness of the ventricular wall. When either S3 or S4 occur they give rise to a triple rhythm that can sound like the

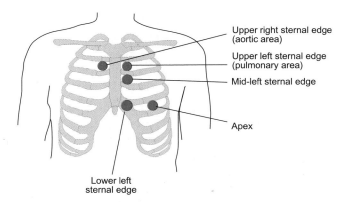

Figure 28.10 The key auscultation points for listening to the heart sounds. The aortic sounds are best heard in the second intercostal space on the right side; the pulmonary sounds are best heard in the second intercostal space on the left side. Aortic and pulmonary regurgitation are best heard in the third intercostal space on the left side (mid-left sternal edge). Tricuspid regurgitation is best heard at the lower left sternal edge while mitral regurgitation, mitral valve prolapse, and mitral valve stenosis are best heard in the mid clavicular line of the fifth intercostal space (apex).

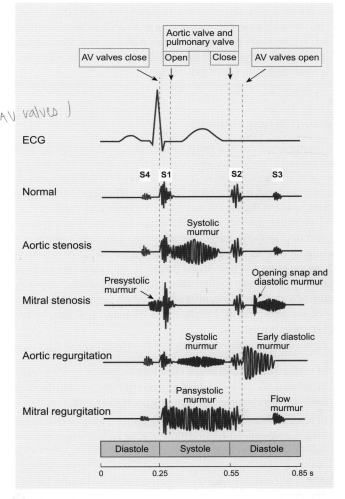

Figure 28.11 The top two traces provide a comparison of the ECG with a phonocardiogram of the heart sounds for the normal heart. The lower four traces show phonograms for some common abnormalities of the heart valves. The phases of the cardiac cycle are shown at the bottom of the figure.

hooves of a galloping horse. For that reason they are called 'gallop' sounds—S3 is referred to as a ventricular gallop and S4 as an atrial gallop. When S3 is audible the rhythm set up is said to sound like KEN-TU-CKY (the sequence is S1–S2–S3) while S4 sounds like TEN-NES-SEE (the sequence is S4–S1–S2).

Heart sounds occurring at other times are known as **murmurs**. Some of these are normal and are referred to as 'innocent' or 'functional' murmurs, as they arise from physiological rather than pathological causes. Such additional heart sounds may be audible during pregnancy, during heavy exercise, and in young children. In such cases, the sounds are the result of an increased rate of blood flow through the normal heart and, as for S3 discussed earlier, they are not associated with any underlying cardiac pathology. Abnormal heart sounds of pathological origin are discussed in the following section.

Summary

During the cardiac cycle, various sounds can be heard using a stethoscope applied to the chest. The first heart sound (S1) arises as the AV valves close and the second heart sound (S2) corresponds to closure of the pulmonary and aortic valves. The third and fourth heart sounds can occasionally be heard in normal subjects. They arise from the rapid flow of blood into the ventricles in diastole (S3) and atrial systole (S4).

Abnormal heart sounds

Murmurs are abnormal heart sounds that are often the result of deformities of the valves, alterations in the diameter of the aortic or pulmonary arterial roots, or holes in the septum that separates the right and left sides of the heart (these are called septal defects). An ability to assess the significance of such sounds and to differentiate between benign and clinically significant murmurs is an important aspect of diagnostic cardiology. Murmurs sometimes give rise to strong vibrations that can be felt during palpation of the chest. These are known as **precordial thrills**. Careful palpation and auscultation of the chest is often sufficient to permit a diagnosis of the cause of a murmur, but confirmation and assessment of the severity of the abnormality usually relies on echocardiography and other imaging methods.

Abnormalities of the valves fall into two categories.

- In **valvular incompetence**, a valve fails to close completely and is therefore leaky. This allows regurgitation of blood through the valve, resulting in eddy currents and turbulence in the blood.
- In **valvular stenosis**, a valve is narrowed so that a high-pressure gradient is needed to push blood through it. As a result, a high-pressure stream of turbulent blood is created.

These abnormalities may be congenital or arise in later life as a result of disease. Either kind of pathology can affect any of the heart valves. However, the most common problems are acquired lesions of the valves of the left side of the heart, which include aortic stenosis, mitral stenosis, aortic regurgitation, and mitral regurgitation. Tricuspid and pulmonary valve disease is less common.

Murmurs are described in a variety of ways according either to the type of sound they create or to their timing within the cardiac cycle. For example, the murmur may be of a blowing or swishing quality or it may be low-pitched and rumbling. It is generally true that the volume of the sound reflects the severity of the underlying disease, so murmurs are often graded in terms of how loud they sound. They may be heard during either the systolic or diastolic phase of the cardiac cycle.

Systolic murmurs are subdivided into three categories:

- **ejection murmurs**, which are heard at the start of ventricular systole
- **late systolic murmurs**, which are audible towards the end of systole
- **pansystolic murmurs**, which are present throughout the whole of the systolic phase.

Diastolic murmurs are usually quieter than systolic murmurs but always indicate pathology. They are most often audible during either the early or the middle part of diastole but they may be heard throughout the diastolic phase of the cardiac cycle. Some examples of the most common pathologies that give rise to murmurs are described below, and diagrammatic representations of the phonocardiograms showing the magnitude and timing of these murmurs is illustrated in Figure 28.11.

Aortic stenosis

This condition is most commonly caused by inflammatory damage to the valve leaflets (following rheumatic heart disease for example), congenital malformation, or degenerative thickening and calcification of the valve leaflets. In each case there is narrowing of the orifice of the valve and increased pressure in the left ventricle that creates a jet of turbulent blood as it enters the root of the aorta. The murmur associated with this condition is often very loud and, although it may persist throughout systole, it begins at the start of this phase (an ejection murmur), first becoming louder, then quieter (crescendo–decrescendo). As the narrowing of the aortic valve requires a higher pressure to drive the blood into the aorta, there is an increase in cardiac work that will eventually result in hypertrophy of the left ventricle. This reduces the efficiency of oxygen diffusion and may result in ischaemia and heart failure.

Mitral stenosis

Narrowing of the mitral valve is most often the result of thickening and scarring following inflammation. This impedes the flow of blood from the left atrium to the left ventricle, thus creating turbulent blood flow that can produce low-frequency rumbling sounds in mid- to late-diastole as the valve cusps flutter in the turbulent stream. Murmurs of this kind can be mistaken for a third heart sound and can be missed. As the cusps of the mitral valve are more rigid than normal, they open rather suddenly. This gives rise to an opening snap. There is also a presystolic murmur produced by the **turbulent flow** of blood into the left ventricle following atrial contraction.

Aortic valve incompetence

Aortic valve incompetence ('leakiness') causes the ejected blood to flow back into the left ventricle (**aortic regurgitation**). It may be caused by many conditions including hypertension, inflammation of the valve, atherosclerosis, or connective tissue disorders such as Marfan syndrome. Aortic regurgitation causes a 'blowing' murmur during diastole as blood flows backwards from the aorta into the left ventricle, creating turbulence. The murmur is usually loudest in early diastole and then fades towards the end of the diastolic phase. As the stroke volume is increased there may also be a softer systolic murmur. (The stroke volume increases because blood enters the left ventricle both from the pulmonary circulation, via the left atrium, and from the aorta, by regurgitation through the aortic valve.)

Mitral valve incompetence

Incomplete closure of the mitral valve causes a typical pansystolic murmur as regurgitation occurs into the left atrium throughout ventricular systole. This murmur is heard most loudly over the mitral area and obscures both S1 and S2. Incompetence of the tricuspid valve produces a similar murmur. The regurgitated volume plus the normal venous return from the pulmonary circulation passing through the mitral valve orifice can increase blood flow sufficiently to cause turbulence during the rapid filling phase. This can give rise to a flow murmur that occurs around the same time as S3. Prolapse of the mitral valve (MVP or floppy valve syndrome) may give rise to a murmur that is heard towards the end of systole (a late systolic murmur). MVP is a very common cardiac condition which is often asymptomatic and harmless.

A variety of other pathologies can give rise to abnormal heart sounds, but a detailed account of these is beyond the scope of this book. Further information can be found in specialist texts on cardiology.

Summary

In various heart diseases, the flow of blood may become very turbulent and the resulting vibrations give rise to abnormal heart sounds known as murmurs. Murmurs commonly arise as a result of defects in the heart valves and usually affect the left side of the heart. Cardiac incompetence occurs when a valve fails to close completely, allowing the regurgitation of blood. A valve may be narrowed (stenosis) so that an abnormally high pressure is needed to push blood through it. This increases the work of the heart and eventually results in hypertrophy of the myocardium on the affected side.

28.5 The measurement of cardiac output

The volume of blood pumped from one ventricle each minute is known as the **cardiac output**. It is the product of the heart rate (b.p.m.) and the stroke volume. Thus:

$$\text{cardiac output} = \text{heart rate} \times \text{stroke volume}$$

During normal daily life the cardiac output (C.O.) varies continually according to the oxygen requirements of the body tissues. In an adult human at rest, the cardiac output is 4–7 litres min^{-1}, it is reduced in sleep, but is raised following a heavy meal, in fear, or during periods of excitement. A much larger rise in cardiac output (by as much as 6 times) is seen during periods of strenuous exercise.

In clinical medicine, cardiac output is often described in terms of the **cardiac index** (CI) which relates the cardiac output to the size of the individual as indicated by their body surface area, hence CI = C.O. ÷ body surface area. The normal range is 2.6–4.2 l min^{-1} m^{-2}.

The **venous return** is the volume of blood returning to the heart from the central veins every minute, and it is inextricably linked with cardiac output. For the closed circulatory system to function efficiently it is essential that, except for very small transient alterations, the heart must be able to pump a volume equivalent to that which it receives, i.e. cardiac output must be equal to venous return and vice versa over any significant period of time (Starling's law—see Section 28.6).

Since cardiac output is determined by heart rate and stroke volume, changes in either (or more often both) of these variables will bring about alterations in cardiac output. Regulation of cardiac output is achieved both by autonomic nerves and hormones and by mechanisms that are intrinsic to the heart itself. These will be discussed in Section 28.6.

Techniques for measuring cardiac output

A variety of methods have been devised to measure the cardiac output. They either measure the cardiac output as a whole (the Fick principle and dilution methods), or they measure stroke volume and heart rate separately (pulsed ultrasound and radionuclide methods).

The **Fick principle** states that the total uptake or release of a substance by an organ is equal to the blood flow through that organ multiplied by the difference between the arterial and venous concentrations of the substance. To measure cardiac output, this principle is applied to the uptake of oxygen by blood flowing through the lungs (remember that the output of the right and left sides of the heart are the same and that the entire cardiac output flows through the lungs). The cardiac output (C.O.) is then given by the following equation:

$$\text{C.O.} = \frac{\text{oxygen uptake} \, (\text{ml min}^{-1})}{C\bar{v}O_2 - CaO_2}$$

—where CaO_2 is the *oxygen content* of the blood in the arterial blood (which is the same as that of the pulmonary veins) and $C\bar{v}O_2$ is the oxygen content of the blood in the pulmonary artery, which is a measure of the oxygen content of mixed venous blood (including that of the cardiac circulation). An example of this way of calculating the cardiac output is given in Box 28.1. It is worth noting that the Fick principle is quite general and applies to any organ through which blood flows and in which a substance is exchanged at a steady rate. It is widely used to calculate the rate at which an organ consumes substances such as fats or glucose

1₂3 **Box 28.1** Application of the Fick principle to the determination of cardiac output

The Fick equation for the uptake of oxygen by the blood as it passes through the lungs is as follows:

$$C.O. = \frac{\text{Oxygen uptake (ml min}^{-1})}{C\bar{v}O_2 - CaO_2}$$

where $C\bar{v}O_2$ is the oxygen content of the blood in the pulmonary veins (which is the same as that of normal arterial blood) and CaO_2 is the oxygen content of the blood in the pulmonary arteries (which is the same as fully mixed venous blood). Oxygen consumption is determined using spirometry (see Chapter 33) and the oxygen content of pulmonary venous blood can be obtained from a sample of blood obtained from the radial, brachial, or femoral artery. The oxygen content of pulmonary arterial blood is more difficult to determine since it requires a sample of fully mixed venous blood, which can only be obtained from the right ventricle or pulmonary artery

itself. Normally a catheter is introduced through the antecubital vein and passed into the outflow of the right ventricle or the pulmonary artery. The blood samples are then analysed and the cardiac output calculated using the equation given above.

A worked example

Assume that a person consumes 250 ml of oxygen each minute from the inspired air, that the oxygen content of the pulmonary arterial blood is 15 ml per 100 ml of blood, and that the oxygen content of the pulmonary venous blood is 20 ml per 100 ml blood. Each 100 ml of blood passing through the lungs must therefore have taken up 5 ml oxygen; each litre would have taken up 50 ml. To supply an oxygen demand of 250 ml min^{-1} will therefore require a cardiac output of 250 ÷ 50 ml of blood per minute (= 5 litres min^{-1}).

This is done by measuring both the rate of blood flow through the organ in question and the arterio-venous difference in concentration of the substance under investigation.

Dilution methods for estimating cardiac output

Dilution methods are used to measure the volume of the body fluid compartments (see Chapter 3) and have been adapted for measuring cardiac output in humans. In general terms, a known quantity of an indicator (I) is rapidly injected intravenously. The concentration of indicator in the arterial blood is then continuously monitored until it effectively disappears from the circulation as it becomes diluted in the circulating blood volume. This follows an exponential time course from which the average concentration of indicator such as a dye (C) can be determined (see Figures 1 and 2 of Box 28.2). If the time between the first appearance of the indicator in the arterial blood and its disappearance (passage time, t) is known, then cardiac output (C.O.) is given by:

$$C.O. = \frac{I \times 60}{C \times t}$$

If I is measured in milligrams, the mean concentration of indicator (C) in mg l^{-1}, and the passage time (t) in seconds, the cardiac output will be given as litres per minute. Further details of the method are given in Box 28.2.

Thermodilution is a widely used variant of the dilution method. In this case, a known volume of isotonic saline at room temperature is rapidly injected into the right heart and the temperature of the blood is continuously measured in the distal pulmonary artery via a suitable temperature probe. A curve of temperature against time is plotted from which the cardiac output can be calculated. As the cold saline is innocuous and rapidly warms to body temperature, repeated measurements may be made. Lithium ion dilution is another widely used method.

Other methods used for measuring cardiac output

While the Fick principle and the dilution methods described above measure cardiac output directly, there are a variety of other methods available which estimate cardiac output by measuring heart rate and stroke volume separately. These include **echocardiography**, which compares the end-diastolic and end-systolic diameters of the ventricle, and **pulsed Doppler ultrasonography**, which measures the velocity of the blood flowing in the aorta. While these methods are non-invasive and are sufficiently fast to permit the cardiac output to be measured beat by beat, they are not as accurate as the dilution methods. Finally, it is possible to implant an electromagnetic flow probe around the aorta of an experimental animal. This is commonly done under anaesthesia and the animal allowed to recover. The flow probe can then be used to provide an accurate measure of aortic blood flow. Since the heart rate can be measured simultaneously, the cardiac output can be readily calculated. This is the preferred method for experimental studies.

Summary

The cardiac output is the volume of blood pumped each minute by either ventricle. It is the product of heart rate and stroke volume. It varies with the size and sex of the individual and with the metabolic demands of the body. Over any significant period of time, the amount of blood pumped by the ventricles is equal to the amount received (the venous return). Cardiac output is measured clinically using a variety of methods, including indicator dilution techniques, ultrasonography, and methods based on the Fick principle.

(1₂3) Box 28.2 Measurement of cardiac output by dilution methods

One way of measuring the volume of an irregularly shaped vessel is to fill it with liquid and add a known quantity of dye. The concentration of the dye can then be used to calculate the volume in which it has been diluted from the simple formula:

$$Volume = \frac{Mass\ of\ indicator}{Concentration}$$

This principle can be extended to measure the flow of liquid through a tube, as follows. A quantity of dye is injected into the stream of liquid. At a remote point, samples are taken at regular intervals and the concentration of dye measured. At first there is no dye, but after a delay the dye concentration rises to a maximum and then decays exponentially with time, as shown in Figure 1.

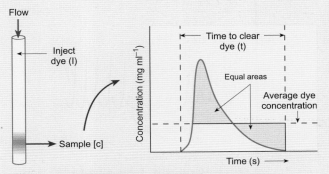

Figure 1 Measuring the flow of liquid through a tube.

The flow can be calculated from the time taken for the dye to be cleared by the flowing liquid from the following formula (remember that flow = volume per unit time):

$$Flow = \frac{I}{C \times t}$$

where I is the mass of dye injected, C is its average concentration while it passes the point of measurement, and t is the time taken for the dye to pass the point of measurement. As indicated above, the method is essentially equivalent to measuring the dilution of a given quantity of dye and using it to calculate the volume in which it has been diluted, with the difference that the time for that volume to accumulate is taken into account.

This principle can be adapted to measure cardiac output with several different kinds of indicator. In one well-established method, a known amount of a dye such as indocyanine green is injected rapidly into a vein and its appearance in the arterial blood is then measured by a micro-colorimeter clipped to the ear lobe. This yields a plot of dye concentration against time that has a characteristic appearance, as shown in Figure 2(a).

Once the injected dye reaches the point of sampling, its concentration rises rapidly to a peak. The concentration then declines exponentially but rises again before reaching zero. This secondary rise is due to the recirculation of dye that took the shortest path through the circulation.

The time taken to clear the dye requires knowledge of the time it would have taken for the concentration to fall to zero had recirculation of the dye not taken place. This requires extrapolation of the falling phase of the dye concentration curve to a value close to zero,

which is greatly aided by plotting the dye concentration on a logarithmic scale as shown in Figure 2(b). This type of plot exploits the fact that an exponential curve plotted on logarithmic coordinates is a straight line which can readily be extrapolated to the time axis, as shown, to give the clearance time (t).

The average concentration of dye is given by the area under the curve, as shown in the figure (the area under the curve has units of mg l⁻¹). From this information the cardiac output can be calculated from the formula:

$$Cardiac\ output = \frac{I \times 60}{C \times t}$$

where the factor 60 converts the cardiac output to litres per minute.

The ideal indicator for measuring cardiac output should possess a variety of characteristics: it should be non-toxic, it should mix completely and virtually instantaneously with the blood, it should remain within the circulation for the period of the determination, and it should not itself alter cardiac output. Several different indicators meet these requirements, including cold saline (thermodilution), low concentrations of lithium chloride, and dyes such as indocyanine green.

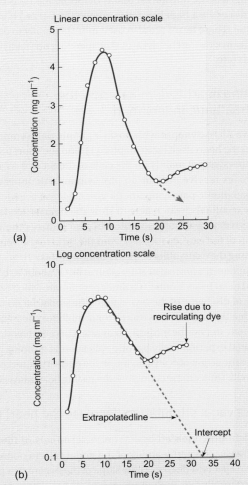

Figure 2 Measuring cardiac output: plot of dye concentration against time.

28.6 Cardiodynamics

At first sight, it seems logical to suppose that an increase in heart rate will necessarily bring about an increase in cardiac output. Up to a point this is true, but as the heart rate increases, the time for filling of the ventricles falls. As a result, the stroke volume does not increase in proportion to the increase in heart rate. In fact, it tends to level off when cardiac output approaches 50 per cent of its maximum value (see Chapter 38). The following section is concerned with **cardiodynamics**, the regulation of the activity of the heart by intrinsic and extrinsic factors.

Nervous and hormonal control of heart rate

Although the SA node has an inherent rhythmicity, it is supplied with parasympathetic and sympathetic autonomic nerves, which are both able to influence the heart rate. Physiological changes in heart rate are known as **chronotropic effects**. The parasympathetic supply to the heart is via the vagus nerves which, when activated, slow the heart (**negative chronotropy**), while stimulation of the sympathetic nerves increases heart rate (**positive chronotropy**).

Clinically, a heart rate at rest that is below 60 b.p.m. is called a **bradycardia** while a high heart rate (above 100 b.p.m.) is called a **tachycardia**. These terms are also used more generally to describe any significant slowing (bradycardia) or speeding up (tachycardia) of the heart rate.

The resting heart is dominated by the parasympathetic innervation

Vagal nerve fibres synapse with postganglionic parasympathetic neurons in the heart itself (see Chapter 11), and the short postganglionic fibres synapse mainly on the cells of the SA and AV nodes. The ventricles receive little or no parasympathetic innervation. The postganglionic parasympathetic fibres release acetylcholine from their terminals, which acts to slow the heart rate via muscarinic M_2 receptors. Vagal stimulation also reduces the rate of conduction of the cardiac impulse from the atria to the ventricles by decreasing the excitability of the AV bundle fibres.

The heart rate of a normal healthy adult at rest is about 70 b.p.m. A denervated heart beats around 100 times per minute, which is the intrinsic rate of discharge of the myocytes of the sinoatrial node. This indicates that the vagus nerves exert a tonic inhibitory action on the sinoatrial node to slow the intrinsic heart rate. This effect may also be demonstrated experimentally by the application of atropine, which antagonizes the action of acetylcholine on the muscarinic receptors of the nodal cells. Under these conditions the resting heart rate will rise.

How does vagal stimulation decrease the heart rate?

The acetylcholine released by the nerve terminals of the vagal nerves increases the permeability of the nodal cells to potassium. This has two effects: it decreases the slope of the pacemaker potential and hyperpolarizes the membrane potential. These changes increase the time taken for the pacemaker potential to reach

threshold, so that the interval between successive action potentials is longer and the heart rate falls. The effect of vagal stimulation on the heart rate is illustrated in Figure 28.12a.

Stimulation of the sympathetic nerves increases the heart rate

The sympathetic preganglionic nerves that supply the heart originate in spinal segments T1–T6 (mainly T1–T3). They synapse in the ganglia of the thoracic sympathetic chain, from where they project to the heart via long postganglionic fibres. All parts of the heart receive a sympathetic innervation. The varicosities of the postganglionic fibres secrete noradrenaline (norepinephrine), and stimulation of these nerves increases the rate at which the heart beats. When physiological circumstances require the heart to beat more rapidly, as in exercise, the activity of the parasympathetic nerves is inhibited,

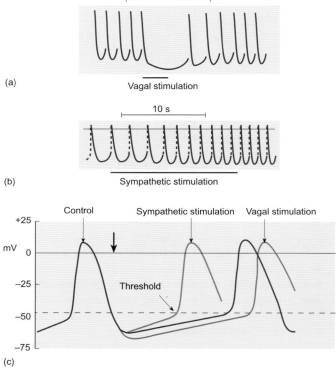

Figure 28.12 The effect of stimulation of the parasympathetic (vagal) nerves (a) and the sympathetic nerves (b) on the pacemaker activity of the frog heart. The vagal stimulation elicits a hyperpolarization of the pacemaker cell that temporarily stops the heart. After the period of stimulation, the slope of the pacemaker potential is reduced and the heart rate remains low compared to the period before stimulation. Stimulating the sympathetic nerves increases the slope of the pacemaker potential and results in an increased heart rate. Panel (c) shows a diagrammatic representation of the effects of sympathetic and parasympathetic stimulation on the pacemaker potential in relation to the threshold for action potential generation. (Based on Figs 4 and 10 in O.F. Hutter and W. Trautwein (1956) *Journal of General Physiology* **39**, 715–33 by permission of the Rockefeller University Press).

while that of the sympathetic nerves is enhanced. Maximal sympathetic stimulation can almost triple the resting heart rate. The heart rate is also increased by circulating catecholamines, i.e. the adrenaline and noradrenaline secreted by the adrenal medulla.

The effects of noradrenaline and adrenaline on the heart are mediated by β_1-adrenoceptors which activate membrane-bound adenylyl cyclase. This in turn increases the level of cAMP within the mycocytes, which has four major effects. First, it acts directly on the pacemaker channels to increase their probability of opening, with the result that the rate at which the pacemaker cells depolarize is accelerated. The increased rate of decay of the pacemaker potential causes the sinoatrial node cells to reach threshold more quickly, reducing the interval between successive action potentials, and so increasing the heart rate as shown in Figure 28.12b. Second, it activates protein kinase A which phosphorylates the delayed rectifier potassium channels. This results in an increase in the rate at which the ventricular myocytes repolarize, shortening their action potentials. Third, it increases the strength of contraction (a positive inotropic action—see the next subsections), and finally it speeds up the rate at which the myocytes relax (this is called a **lusitropic effect**). Conduction time through the AV node is also reduced by sympathetic stimulation (this is known as a positive **dromotropic effect**).

Summary

Heart rate is governed by the influence of the autonomic nerves on the rate of discharge of the pacemaker cells of the sinoatrial node. Parasympathetic stimulation slows the heart rate (negative chronotropy), while sympathetic stimulation increases it (positive chronotropy). Circulating catecholamines from the adrenal medulla also increase the heart rate.

The regulation of stroke volume

The stroke volume is regulated by two different mechanisms:

- *intrinsic regulation* of the force of contraction, which is determined by the degree of stretch of the myocardial fibres at the end of diastole
- *extrinsic regulation*, which is determined by the activity of the autonomic nerves and the circulating levels of various hormones.

Intrinsic regulation of the stroke volume: the Frank–Starling relationship

The fundamental relationship between the initial length of skeletal muscle and the force of contraction, discussed in Chapter 8, was shown by Otto Frank to be valid for cardiac muscle, which also responds to increased stretch with a more forceful contraction. As the blood returns to the heart in diastole, it begins to fill the ventricle so that the intraventricular pressure progressively rises (see the lower curve of Figure 28.13). As it does so, the myocardial fibres in the ventricular wall are stretched and placed under a degree of tension known as **preload**. The strength of the subsequent contraction depends on the extent of the preload: in the normal heart, the greater the preload, the more forceful the contraction and the greater the stroke volume (Figures 28.13 and 28.14).

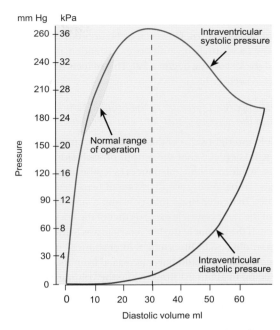

Figure 28.13 The relationship between end-diastolic volume and intraventricular pressure for an isolated dog heart (lower curve). The upper (red) curve shows the maximum pressure generated during the subsequent systole. Note that the systolic pressure progressively increases until the diastolic volume exceeds 30 ml and then declines. (Redrawn after Fig 3 in S.W. Patterson, H. Piper, and E.H. Starling (1914) *Journal of Physiology* **48**, 465–513.)

On the basis of this kind of evidence, Starling formulated his Law of the Heart which states that 'the energy of contraction of the ventricle is a function of the initial length of the muscle fibres comprising its walls'. This is now often referred to as the Frank–Starling relationship. Stated simply, this means that during systole, the ventricle will eject the volume of blood that entered it during diastole.

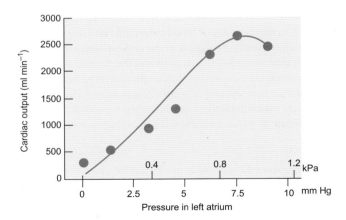

Figure 28.14 The relationship between filling pressure and stroke volume in an isolated dog heart-lung preparation. There is a progressive increase in stroke volume until the filling pressure exceeds 6 mm Hg (0.8 kPa), whereupon the stroke volume begins to fall despite the increase in pressure (compare with the upper curve in Figure 28.13). (Drawn from data in S.W. Patterson and E.H. Starling (1914) *Journal of Physiology* **48**, 357–79.)

Consequently, *the heart automatically adjusts its cardiac output to match its venous return*. However, as Figures 28.13 and 28.14 also show, when a ventricle becomes over-distended, the systolic pressure and stroke volume begin to fall. This reflects the onset of heart failure (see Section 28.7).

The most important function of the Frank–Starling mechanism is to balance the outputs of the right and left ventricles

As the left and right sides of the heart are arranged as two pumps in series, it is crucial that the output of each side of the heart is the same. While transient imbalances (a few beats) may occur, more prolonged imbalances, even very small ones, will quickly prove disastrous. To demonstrate this, consider a situation in which the output of the right ventricle exceeds that of the left ventricle by just 1 per cent. If the output of the left ventricle was 5 litres min^{-1}, that of the right would be 5.05 litres min^{-1}. After about 30 minutes, pulmonary blood volume would have increased from its normal level of about 600 ml to about 2.1 litres, while the volume in the systemic circulation would have fallen from 3.9 litres to 2.4 litres. From this it is clear that unless the output of the two ventricles is matched, the distribution of the blood between the pulmonary and systemic circulations will be altered with potentially serious consequences, including heart failure and pulmonary oedema (see Section 28.7).

The Frank–Starling mechanism normally ensures that the output from the two ventricles is closely matched. If the output of the right ventricle exceeds that of the left, after a few beats the volume of blood in the pulmonary circulation will be increased slightly and pressure in the pulmonary veins will rise. As a result, the venous return to the left side of the heart will be increased, resulting in a greater end-diastolic volume in the left ventricle. This, in turn, will lead to an increase in its stroke volume via the Frank–Starling mechanism, so restoring the balance between the output of the two ventricles.

The pressure in the aorta and arterial system opposes the pressure driving blood from the ventricles and represents the load against which the heart must pump the blood. For this reason, it is known as the **afterload**. A sudden increase in afterload causes only a transient fall in stroke volume. Why does the stroke volume recover? As the quantity of blood returning from the pulmonary circulation remains unchanged, at the end of diastole the left ventricle is more distended than before and the preload is increased. (It will contain the normal end-diastolic volume plus the excess amount that was left after the previous systole.) The increased preload increases the force with which the ventricular muscle contracts (Figure 28.13) and the normal stroke volume is restored.

What factors determine the end-diastolic volume of the ventricles?

From the preceding account it should be clear that the end-diastolic volume (EDV), and therefore the degree of stretch of the ventricles, is an important determinant of both the cardiac output and the work performed by the heart. It is therefore important to understand the factors that affect this volume. Essentially these fall into two categories: factors affecting pressure outside the heart and those affecting pressure inside the heart.

The pressure outside the heart is the **intrathoracic pressure**. This is affected by the depth and rhythm of breathing (see Chapter 33). During inspiration, contraction of the diaphragm increases the volume of the chest and decreases the volume of the abdominal cavity. This results in a fall in the pressure within the thorax and a rise in the pressure within the abdominal cavity. This pressure difference favours the flow of blood from the abdominal to the thoracic veins and so enhances the filling of the right ventricle.

The pressure inside the right ventricle at the end of atrial diastole is equal to the pressure in the superior and inferior venae cavae as they enter the heart. This is known as the **central venous pressure** (commonly abbreviated as CVP). The central venous pressure is the chief factor that determines the filling of the right ventricle. It follows from the Frank-Starling relationship that the central venous pressure is also an important determinant of cardiac output. The relationship between central venous pressure and cardiac output is shown in Figure 28.15.

The central venous pressure and venous return are influenced by a variety of factors. These include gravity, respiration (as described above), the muscle pump (compression of the deep veins during exercise displaces blood into the central veins), peripheral venous tone, and the blood volume (see Table 28.1). Since about two-thirds of the total blood volume is located within the veins, which are highly distensible (they are known as the capacitance vessels—see Chapter 30), any change in venous tone or blood volume will also bring about alterations in venous return and the central venous pressure. Haemorrhage, for example, will lower the blood volume, the venous return, and the central venous pressure. Consequently, cardiac output will fall. In contrast, strong sympathetic activation causes venoconstriction, an increased venous return, and increased cardiac output.

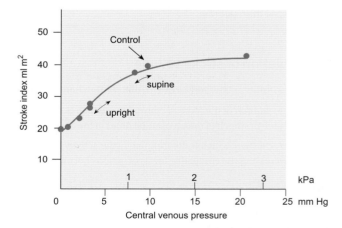

Figure 28.15 The relationship between stroke volume and central venous pressure in a human subject. The stroke volume is expressed as stroke index (the stroke volume divided by body surface area) and central venous pressure was taken as left ventricular end-diastolic pressure. The changes in central venous pressure were achieved by various means including reducing blood volume by taking blood and subsequently re-infusing it, and by expanding blood volume with a dextran infusion. Once more, there is a distinct maximum when CVP is around 10 mm Hg (1.2 kPa). The stroke volume is significantly more sensitive to central venous pressure when upright compared to lying down (i.e. supine). (Redrawn from Fig 4 in J.O. Parker and R.B. Case (1979) *Circulation* **60**, 4–12.)

Table 28.1 Factors that control the venous return

- Pressure at the venous end of the capillaries
- Right atrial pressure (the central venous pressure)
- Total blood volume
- Venous tone
- The skeletal muscle pump
- The respiratory cycle
- Intra-abdominal pressure
- Inotropic state of the heart
- Descent of the A-V fibrous ring, drawing of blood into the atria during ventricular systole
- Suction of the ventricles as they relax during diastole, drawing blood into the heart

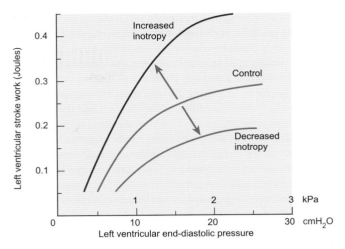

Figure 28.16 A diagrammatic representation of positive and negative inotropic effects on the relationship between stroke work and left end-diastolic pressure. The mauve line indicates an increase in the inotropic state of the myocardium which is reflected in an increase in stroke work for a given end-diastolic pressure. The red line shows a decrease in the inotropic state and a corresponding fall in stroke work.

Summary

The stroke volume is regulated by both intrinsic and extrinsic mechanisms. Intrinsic regulation is expressed by the Frank–Starling Law of the Heart, which states that 'the energy of contraction of the ventricle is a function of the initial length of the muscle fibres comprising its walls' so that the degree of stretch is determined by the venous return. Over any significant period, the output of the heart matches the venous return.

Extrinsic regulation of the stroke volume—the modulation of myocardial contractility by the sympathetic nervous system

The contractile energy (i.e. the force of contraction) of the heart muscle may be altered by factors other than the initial resting length of the fibres. A change in contractile energy that is mediated by such extrinsic factors is referred to as a change in the **inotropic state** of the myocardium. This is independent of the initial length of the cardiac muscle fibres (see Chapter 8). It is often, perhaps somewhat misleadingly, called a change of contractility. Activation of the sympathetic nerves that supply the heart or a rise in the circulating level of adrenaline and noradrenaline result in a **positive inotropic effect**: an increase in the force of contraction of the ventricles during systole. This effect is illustrated in Figure 28.16. A positive inotropic effect results in a more complete emptying of the ventricles and an increase in systolic pressure, with the result that stroke work is increased (see Figure 28.17). The greater the degree of sympathetic stimulation, the greater the increase in stroke work. At the same time as increasing contractility, sympathetic activity shortens the duration of systole—the lusitropic effect mentioned earlier in this section. While this does not affect the stroke volume (because the force of contraction is enhanced), it does help to offset the reduction in the time available for filling during diastole. Without this speeding up of systole, the filling of the heart might otherwise be compromised by the increase in heart rate.

Other circulating positive inotropic agents

Although the catecholamines, particularly adrenaline, are by far the most potent positive inotropic agents acting upon the heart, other

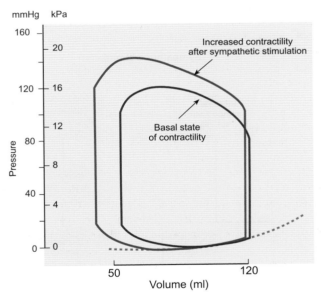

Figure 28.17 A diagrammatic illustration of the increase in stroke work that follows an increase in the contractility (positive inotropy) of the left ventricle. Note that both the stroke volume and systolic pressure are increased. In this instance the enhanced stroke volume reduces the end-systolic volume from just over 50 ml to nearer to 45 ml.

substances circulating in the bloodstream can also exert positive inotropic effects on the cardiac muscle. These include angiotensin II, calcium ions, glucagon, insulin, thyroxine, and certain drugs such as caffeine, theophylline, and the cardiac glycoside digoxin.

Many of these positive inotropic agents work through their actions on the cAMP-protein kinase pathway. The catecholamines activate β_1-adrenoceptors to elevate intracellular cAMP, which leads to the phosphorylation of L-type calcium channels by protein kinase A. This phosphorylation prolongs the open time of the channels, resulting in an increased influx of calcium during the action potential and a stronger contraction. Angiotensin II exerts its effects largely by increasing the secretion of noradrenaline from the sympathetic nerve endings, while caffeine and theophylline inhibit phosphodiesterase and thus slow the enzymic inactivation of cAMP. Digoxin has a quite different mode of action. It inhibits the Na^+/K^+ pump in the plasma membrane and this causes an increase in intracellular sodium. Since calcium extrusion from cardiac myocytes largely depends on Na^+/Ca^{2+} exchange, there is a decrease in calcium extrusion and an increase in calcium seques-tration by the sarcoplasmic reticulum. As there is a larger intracel-lular store available, more calcium is released by each action potential, resulting in a more powerful contraction.

Negative inotropic agents

Substances that reduce the force with which the heart contracts are known as negative inotropic agents. As there is virtually no para-sympathetic innervation to the ventricular myocardium in humans, parasympathetic activation has no direct effect on the inotropic state of the ventricles. However, β-receptor antagonists such as propranolol, calcium-channel blockers such as verapamil, and many general anaesthetics including the barbiturates and some volatile agents are negative inotropic agents. A low blood pH (acidaemia) also has a negative inotropic effect.

Variation of cardiac output in the normal and denervated heart

The cardiac output is the product of the heart rate and the stroke volume. The heart rate can increase from around 70 b.p.m. at rest to as much as 200 b.p.m. during heavy exercise. The stroke volume is normally about 70 ml, but it too can increase during exercise. As a consequence of these changes, cardiac output can vary between $5\,l\,min^{-1}$ at rest up to about $25\,l\,min^{-1}$ in severe exercise. In trained athletes, the range is much wider. The resting heart rate may be as low as 40–50 b.p.m., while stroke volume is about 120 ml and the cardiac output of such individuals can exceed $35\,l\,min^{-1}$ during maximal effort. The effects of the sympathetic and parasympa-thetic nervous divisions of the autonomic nervous system on cardiac output are summarized in Figure 28.18.

Heart transplant patients are able to increase their cardiac out-put to meet the demands of heavy exercise almost as efficiently as healthy people, despite the lack of sympathetic innervation to the heart itself. Part of this increase in heart rate is due to the sympa-thetic effects of adrenaline and noradrenaline released from the

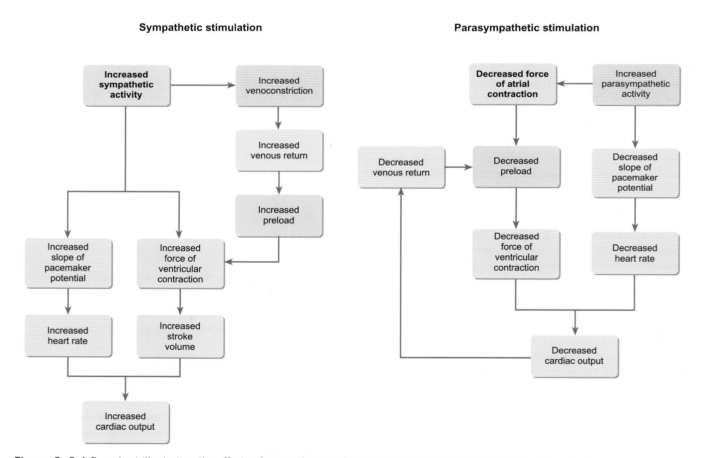

Figure 28.18 A flow chart illustrating the effects of sympathetic and parasympathetic activation on the cardiac output.

adrenal medulla, but to a large extent the increased cardiac output reflects the ability of the heart muscle to contract more forcefully as it is stretched (i.e. to the intrinsic regulation of stroke volume).

Summary

Extrinsic regulation of the stroke volume is mainly due to activity of the cardiac sympathetic nerves and the circulating catecholamines. Increased sympathetic activity increases the force of contraction during systole for any given end-diastolic volume (a positive inotropic effect). There is no significant parasympathetic innervation to the ventricles in humans.

28.7 Heart failure

Heart failure or **cardiac failure** occurs when the heart is unable to pump sufficient blood at normal filling pressures to meet the metabolic demands of the body. Failure may involve either of the two ventricles individually, or both left and right ventricles simultaneously. People with heart failure are unable to exercise normally and complain of excessive fatigue because their cardiac output does not increase in proportion to the work performed, as it would in a healthy person.

Heart failure occurs for a number of reasons. There may be an impairment of myocardial contractility, the heart may not fill effectively, or it may not be able to pump efficiently. These impairments are said to be cardiac in origin. In addition, preload or afterload may become excessive with the result that the ventricles become over-distended. In this case the impairment is said to be extracardiac.

Heart failure due to cardiac impairment

Impairment of myocardial contractility

Loss of healthy muscle mass can occur if the myocardium experiences hypoxia or ischaemia because of the loss of the normal blood supply (e.g. following a coronary thrombosis—a 'heart attack'). The loss of healthy muscle mass is known as **myocardial infarction**. Loss of myocardium from other forms of cardiac myopathy also results in a decreased contractility. Inflammation of the heart muscle is known as **myocarditis** and this, too, impairs the contractile performance of the heart. In all such situations, cardiac performance falls in proportion to the extent of the pathology.

Impairment of the filling or emptying of the heart

Impairment of the filling or emptying of the heart is seen in various conditions. In atrial fibrillation, the ventricles will not be completely filled and cardiac output will fall. The narrow valve opening in mitral stenosis will prevent normal filling of the left ventricle. If the membrane surrounding the heart becomes inflamed (a condition called **pericarditis**) it becomes stiffer, and this prevents normal expansion of the ventricles during filling, reducing the end-diastolic volume. This condition is made worse if the inflammation is followed by an accumulation of fluid in the space surrounding the heart (the pericardial sac), as the pressure outside the heart will increase. This will prevent proper filling of the ventricles during diastole, resulting in a fall in cardiac output and a raised jugular pressure. This condition is known as **cardiac tamponade**. Emptying of the ventricles is compromised in aortic stenosis (narrowing of the aortic valve) and in mitral regurgitation (where some of the blood in the left ventricle re-enters the left atrium instead of the aorta).

Heart failure due to extracardiac factors

Increased preload

The preload is the tension that exists in the walls of the heart as a result of diastolic filling. It is therefore determined by the end-diastolic pressure. The preload may become excessively elevated because of renal failure (in which there is an increase in blood volume due to retention of sodium and water). It may also increase due to valvular regurgitation or myocardial infarction, as discussed above. The heart performs more work (i.e. it is less efficient) when it pumps a given volume of blood from distended ventricles. As a result, a situation can develop where the coronary arteries are unable to supply sufficient oxygen to meet the requirements of the heart (myocardial ischaemia) and the initial heart failure is exacerbated.

Increased afterload

The afterload is the pressure in the aorta during the period that the aortic valve is open. It is the load that the heart must overcome in order to pump the blood from the left ventricle into the aorta and through the vascular system. The afterload is determined by vascular tone and is increased when arterial blood pressure rises. Whenever the afterload is increased, the work of the heart is increased for any given stroke volume (see Figure 28.9b). As with the effects of an increased preload, a situation can arise in which the coronary arteries are unable to supply sufficient oxygen to meet the requirements of the heart. The afterload is persistently elevated in patients with hypertension and in those with aortic stenosis.

Acute heart failure

Consider the situation when a blood clot occludes one of the coronary arteries supplying the left heart (a coronary thrombosis). The loss of the blood supply deprives the affected tissue of oxygen and so prevents the affected region of the myocardium from contracting normally. Consequently, the left ventricle cannot pump the blood it contains as efficiently as it should (left-sided heart failure). The first effects are a fall in cardiac output and arterial blood pressure. These changes are followed by a variety of compensatory mechanisms that act to restore the cardiac output as far as possible (see Figure 28.19).

The fall in arterial blood pressure will result in unloading of the arterial baroreceptors (the sensors that monitor blood pressure), causing a reflex increase in sympathetic stimulation of the heart and the blood vessels—see Section 28.6. The secretion of adrenaline and noradrenaline by the adrenal medulla is also increased. At the same time, vagal tone will be inhibited. Consequently, the heart rate will increase. This effect is supported by an increase in the contractility of that part of the myocardium that is still being normally perfused with blood. As a result of these two mechanisms, cardiac output increases, although it may still be well below normal. The increased vascular sympathetic tone leads to an increased peripheral resistance that acts to support the blood pressure and maintain cerebral perfusion.

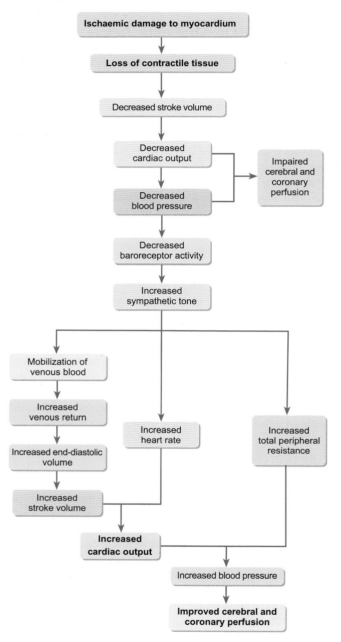

Figure 28.19 A flow chart of the compensatory changes in the circulation that follow occlusion of one of the branches of the left coronary artery. Note that the increase in cardiac output after compensation has a metabolic cost and that oxygen supply will be limited by the coronary artery disease. As a result any further increase in output is likely to be severely limited (see text for further details).

These changes occur within about 30 s of the occurrence of the coronary thrombosis and give rise to the characteristic symptoms of a heart attack: tachycardia, pallor (resulting from vasoconstriction of the skin vessels), and sweating ('cold sweat'). These changes are accompanied by severe pain in the chest which may radiate to the left arm (**angina pectoris**).

The venoconstriction that occurs in response to the increased sympathetic tone leads to mobilization of the blood in the veins and an increase in the venous return to the right heart which, being unaffected, will increase its stroke volume and its output to the pulmonary circulation. In the normal course of events, this would lead to a rapid increase in the output of the left ventricle by the Frank–Starling mechanism. In the present situation, the function of the left ventricle has been impaired by loss of part of the active muscle mass so that the left ventricle is unable to respond normally to the increased venous return. Consequently, both the end-diastolic volume and the end-diastolic pressure of the left ventricle increase (increased preload). Unless the damage to the myocardium is so extensive that the ventricle cannot respond (in which case death will rapidly follow), the increased preload will eventually result in an increase in left ventricular output so that it matches that of the right ventricle.

Chronic compensated heart failure

The previous paragraphs have discussed the basic mechanisms that operate to restore cardiac output following a coronary thrombosis. In chronic heart failure, these same mechanisms act to compensate for the poor cardiac output that results from ischaemic damage to the myocardium, valvular incompetence, or chronic hypertension. This can be a relatively stable condition in which the patient may scarcely be aware of the situation until he or she undertakes exercise, at which point dyspnoea and fatigue set in rapidly. Nevertheless, in all cases of heart failure there is an increase in the end-diastolic pressure required to achieve a given stroke volume, as shown in the ventricular function curves of Figure 28.20.

In heart failure, the cardiac output is diverted away from the skin and visceral organs in favour of the heart, brain, and skeletal muscle. The reduction in renal blood flow has profound consequences: following the vasoconstriction of the afferent arterioles, there is an increase in the secretion of renin leading to elevated levels of angiotensin II (see Chapter 39). This hormone has two important effects: it is a powerful vasoconstrictor, and it stimulates the secretion of aldosterone from the adrenal cortex. The low glomerular filtration rate caused by the constriction of the afferent arterioles, coupled with the increased sodium retention caused by the elevated levels of aldosterone, leads to increased fluid retention and expansion of the plasma volume. Following severe haemorrhage, similar compensatory mechanisms operate to restore cardiac output (see Chapter 40 p. 624) but in this case end-diastolic pressure is not elevated.

Oedema in cardiac failure

In chronic heart failure, the plasma volume is expanded because of fluid retention. This lowers the oncotic pressure and increases the capillary pressure so that the **Starling forces** increasingly favour

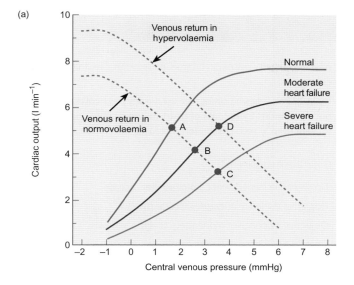

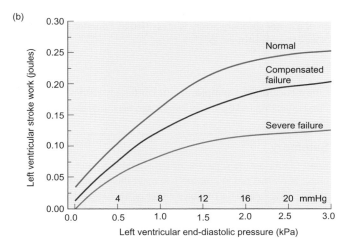

In right-sided heart failure, the increased capillary pressure leads to oedema in the periphery. This is most noticeable in the ankles. In acute left-sided heart failure (the most common form of ischaemic heart disease), there is a raised pressure in the left atrium and pulmonary vessels. The raised pressure in the pulmonary veins causes them to become engorged and there is increased backpressure on the pulmonary capillaries. The increased pressure in the pulmonary capillaries leads to greater transfer of fluid into the pulmonary interstitium and ultimately to pulmonary oedema. If this situation arises, the diffusing capacity of the lungs will be impaired, further compounding the effects of low cardiac output. A vicious cycle can develop, leading to death (Figure 28.21).

Principles of treatment

Since the main problem in cardiac failure is the inability of the heart to pump sufficient blood to meet the needs of the circulation (see Box 28.3), the first requirement is to minimize the demands on

Figure 28.20 (a) The relationship between central venous pressure and cardiac output for the normal heart, during moderate heart failure and during severe heart failure. In normovolaemia (normal blood volume) a cardiac output of 5 litres min⁻¹ is attained when the central venous pressure is less than 0.25 kPa (2 mm Hg) (point A). In heart failure, a higher central venous pressure is required to maintain a smaller cardiac output as shown by points B and C. In chronic heart failure, there is hypervolaemia (expansion of the blood volume) and the normal resting cardiac output is attained only with a central venous pressure of 0.5 kPa (4 mm Hg) as shown by point D. (b) The relationship between left ventricular stroke work and left ventricular end-diastolic pressure for a normal heart, during severe acute failure and in compensated failure. (The end-diastolic pressure is directly related to the end-diastolic volume – see Figure 28.13.) Note that in heart failure the function curve is flattened so that a higher end-diastolic pressure is required for a given level of stroke work. Moreover, the overall capacity of the heart to perform work becomes severely limited. (Based on A.C. Guyton, C.E. Jones, and T.G. Coleman (1973) *Circulatory physiology: output and regulation*. Saunders, Philadelphia.)

filtration. As a result, fluid accumulates in the tissues and oedema results. The oedema may be more evident in the limbs or in the lungs, depending upon which side of the heart is most affected.

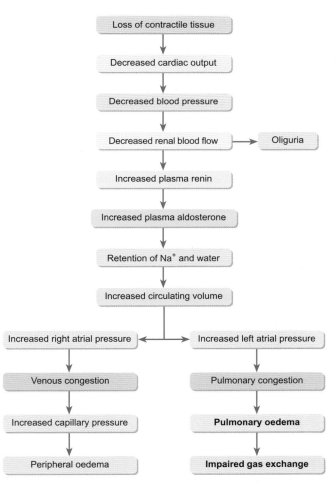

Figure 28.21 A flow chart showing the sequence of events that cause oedema during chronic heart failure. When there is pulmonary oedema, there is a need to increase pulmonary blood flow which increases the demands on the heart. This will lead to a vicious circle that will result in death unless the underlying cause is treated.

Box 28.3 Classification of heart failure

To guide the treatment of patients with a history of cardiac disease, heart failure can be classified according to the remaining functional capacity. This idea was first proposed by the New York Heart Association in 1928 and has since been revised on several occasions. The main features of the most recent classification are summarized below.

Functional Capacity

- Class I. Patients who have had a history of heart disease but have recovered sufficiently to undertake normal physical activity that does not cause undue fatigue, palpitations, breathlessness (dyspnoea), or chest pain.

- Class II. Patients who are comfortable at rest but in whom the ability to undertake physical activity is slightly limited. Ordinary physical activity results in fatigue, palpitation, breathlessness, or chest pain.

- Class III. Patients who are comfortable at rest but are very limited in the amount of physical activity they can undertake. Minor physical activity results in fatigue, palpitation, breathlessness, or chest pain.

- Class IV Patients who cannot undertake any physical activity without discomfort. Symptoms of heart failure or chest pain may be present even at rest. Discomfort increases with physical activity.

Objective Assessment

A. No objective evidence of cardiovascular disease.

B. Objective evidence of cardiovascular disease.

C. Objective evidence of moderately severe cardiovascular disease.

D. Objective evidence of severe cardiovascular disease.

These two classifications are combined to provide a concise clinical description. For example, a patient who is comfortable at rest but is quickly fatigued when exercising normally and has a severe obstruction of a major coronary artery is classified as Functional Capacity II, Objective Assessment D.

---

the circulation. This can be achieved by rest. Following a mild coronary thrombosis, the physiological compensatory mechanisms help to maintain an adequate cardiac output and blood pressure. Over the ensuing months the remaining heart muscle hypertrophies and there is an increase in vascularization in those areas of the myocardium that became hypoxic as a result of the occlusion of part of the blood supply. These adaptations improve the contractility of the heart and eventually a normal pattern of life can be resumed.

In severe heart failure the situation is more complicated, and treatment has three aims.

1. To reduce the work of the heart. This can be achieved by rest, as already mentioned.

2. To reduce the circulating volume and the resulting cardiac dilatation. This can be achieved by administration of diuretics such as frusemide.

3. To improve myocardial contractility—this is often attempted by the administration of drugs that have a positive inotropic effect on the myocardium. Examples are the cardiac glycosides (e.g. digoxin and ouabain) and β_1-adrenoceptor agonists such as dobutamine.

Summary

Heart failure occurs when the heart is unable to pump sufficient blood at normal filling pressures to meet the metabolic demands of the body. Failure may involve either of the two ventricles alone or both the left and right ventricles simultaneously. It may result from ischaemic damage to the myocardium leading to loss of contractility, from incompetence or narrowing of the heart valves resulting in increased preload, or from excessive peripheral resistance resulting in increased afterload. When heart failure occurs, there is increased sympathetic activity—increased heart rate, increased contractility of the unaffected myocardium, and generalized vasoconstriction. The increased venous return leads to a greater end-diastolic volume and restoration of the stroke volume by the Frank–Starling mechanism. These changes help to compensate for the reduced cardiac output.

Checklist of key terms and concepts

The origin of the heartbeat

- The heart beats spontaneously and has an inherent rhythmicity independent of any extrinsic nerve supply.

- Excitation is initiated by the pacemaker potential generated by a group of specialized pacemaker cells in the sinoatrial (SA) node.

- Action potentials initiated by the SA node are then conducted throughout the whole heart.

- In the ventricles, the action potentials are conducted rapidly from the atrioventricular (AV) node to the bundle of His and its branches. In turn, the bundle branch fibres excite Purkinje fibres, which then excite the cardiac myocytes.

- The action potentials differ in different parts of the heart. Those recorded from the atria, ventricles, and conducting system have a fast initial upstroke followed by a prolonged plateau phase.

- The plateau phase ensures that the action potential lasts almost as long as the contraction of the cell and that the excitation of the myocardium is unidirectional.

The cardiac cycle

- The alternating contraction and relaxation of the heart muscle is known as the cardiac cycle.

- The cardiac cycle consists of two phases: diastole and systole.

- In diastole the chambers of the heart relax and fill with blood, which is then pumped around the circulation during systole.

- The total amount of blood filling the ventricles at the end of diastole is called the end-diastolic volume (EDV).

- Nearly two-thirds of the blood in the ventricles at the end of diastole is expelled during systole.

- The volume ejected by each ventricle is known as the stroke volume, and the volume remaining in the ventricles at the end of systole is the end-systolic volume.

- The work performed by the heart during each beat is called the stroke work.

The heart sounds

- During the cardiac cycle, various sounds can be heard using a stethoscope applied to the chest. This is known as auscultation.

- The first heart sound (S1) arises as the atrioventricular (AV) valves close, and the second heart sound (S2) is caused by the closure of the pulmonary and aortic valves.

- Two other heart sounds can occasionally be heard in normal subjects, the third and fourth heart sounds.

- In various heart diseases, the flow of blood may become very turbulent and the resulting vibrations give rise to abnormal heart sounds known as murmurs.

- Murmurs are often due to defects in the heart valves of the left side of the heart.

- If a valve does not close completely it allows regurgitation of blood. This defect is known as incompetence or insufficiency.

- Alternatively a valve may be narrowed so that an abnormally high pressure is needed to push blood through it. This defect is called a stenosis.

Measurement of cardiac output

- The cardiac output is the volume of blood pumped each minute by either ventricle.

- The cardiac output is the product of heart rate and stroke volume and is around 5 l min^{-1} in a resting adult man.

- The resting cardiac output depends on the size and sex of each individual.

- Cardiac output varies according to the metabolic demands of the body.

- A variety of methods are used for measuring cardiac output, including indicator dilution techniques, ultrasonography, and methods based on the Fick principle.

Control of heart rate

- The heart rate is governed by the influence of the autonomic nerves on the rate of discharge of the pacemaker cells of the sinoatrial node.

- Factors that influence heart rate are known as chronotropic influences.

- Parasympathetic stimulation slows the heart rate (negative chronotropy), while sympathetic stimulation increases it (positive chronotropy).

- Circulating catecholamines from the adrenal medulla also increase the heart rate.

- The effects of sympathetic stimulation are mediated by β_1-adrenoceptors.

Regulation of stroke volume

- The stroke volume is regulated by mechanisms intrinsic to the heart and by nerves and hormones (extrinsic factors).

- The intrinsic regulation is expressed by Starling's Law of the Heart which states that 'the energy of contraction of the ventricle is a function of the initial length of the muscle fibres comprising its walls'.

- The degree of stretch is determined by the venous return. This is called the preload.

- The pressure that the heart must overcome in systole is called the afterload. For the left heart it is the aortic pressure.

- Over any significant period, the output of the normal heart matches the venous return.

- The extrinsic regulation of the stroke volume is due to changes in the force of contraction. This is referred to as a change in the inotropic state of the heart.

- Increased sympathetic activity increases the force of contraction during systole for any given end-diastolic volume (a positive inotropic effect).

- Although the parasympathetic innervation has no direct effect on the inotropic state of the ventricles, negative inotropic agents include β-receptor antagonists, calcium-channel blockers, and many general anaesthetics.

Heart failure

- Heart failure occurs when the heart is unable to pump sufficient blood at normal filling pressures to meet the metabolic demands of the body.

- Heart failure has many causes, including ischaemic damage to the heart muscle, incompetence or narrowing of the heart valves resulting in increased preload, or excessive peripheral resistance resulting in increased afterload.

- Moderate heart failure can be compensated by an increased sympathetic activity, which leads to an increase in heart rate and increased contractility of the unaffected myocardium. This is accompanied by a generalized vasoconstriction and an increased venous return.

Recommended reading

Anatomy

MacKinnon, P.C.B., and Morris, J.F. (2005) *Oxford textbook of functional anatomy* (2nd edn), Volume 2: *Thorax and abdomen*, pp. 67–79. Oxford University Press, Oxford.

A lucid account of the anatomy and development of the heart and great vessels.

Histology of the heart and blood vessels

Mesher, A. (2015) *Junqueira's basic histology: Text and atlas* (14th edn), Chapters 10 and 11. McGraw-Hill, New York.

A good account of the microscopical structure of the blood vessels and heart.

Physiology of the circulatory system

Levick, J.R. (2010) *An introduction to cardiovascular physiology* (5th edn). Hodder Arnold, London.

An excellent introduction to the advanced study of the cardiovascular system as a whole.

Pharmacology of the heart and circulation

Grahame-Smith, D.G., and Aronson, J.K. (2002) *Oxford textbook of clinical pharmacology* (3rd edn), Chapter 23. Oxford University Press, Oxford.

A useful account of the practical considerations confronting the drug treatments available for treatment of cardiovascular disorders.

Rang, H.P., Ritter, J.M., Flower, R.J., and Henderson, G. (2015) *Pharmacology* (8th edn), Chapter 21. Churchill-Livingstone, Edinburgh.

A helpful account of the physiology of cardiac function and the drugs that affect it. Well worth reading.

Medicine

Brown, E.M., Leung, T., Collis, W., and Salmon, A.P. (2008) *Heart sounds made easy* (2nd edn). Churchill-Livingstone, Edinburgh.

A short book providing a helpful account of the heart sounds. It also includes a CD of heart sounds which can be played over a computer.

Ledingham, J.G.G., and Warrell, D.A. (2000) *Concise Oxford textbook of medicine*, Chapters 2.2, 2.8, 2.11, 2.13. 2.19, and 2.23. Oxford University Press, Oxford.

An excellent starting point for the detailed study of cardiac disease.

Kumar, P., and Clark, M. (eds) (2012) *Clinical medicine* (8th edn), Chapter 14. Saunders, Edinburgh.

Review articles

Bers, D.M. (2008) Calcium cycling and signaling in cardiac myocytes. *Annual Review of Physiology* **70**, 23–49.

Hinton, R.B., and Yutze, K.E. (2011) Heart valve structure and function in development and disease. *Annual Review of Physiology* **73**, 29–46.

Qu, Z., and Weiss, J.N. (2015) Mechanisms of ventricular arrhythmias: From molecular fluctuations to electrical turbulence. *Annual Review of Physiology* **77**, 29–55.

Reuter, D.A., Huang, C., Edrich, T., Shernan, S.K., and Eltzschig, H.K. (2010) Cardiac output monitoring using indicator-dilution techniques: Basics, limits, and perspectives. *Anesthesia and Analgesia* **110**, 799–811.

Schrier, R.W., and Abraham, W.T. (1999) Hormones and hemodynamics in heart failure. *New England Journal of Medicine* **341**, 577–85.

Schmitt, N., Grunnet, M., and Olesen, S.-P. (2014) Cardiac potassium channel subtypes: New roles in repolarization and arrhythmia. *Physiological Reviews* **94**, 609–53, doi:10.1152/physrev.00022.2013.

To check that you have mastered the key concepts presented in this chapter, complete the accompanying online self-assessment questions. Go to www.oup.com/uk/pocock5e/

CHAPTER 29

The electrocardiogram

This chapter should help you to understand:

- The origin of the electrocardiogram (ECG)
- How the ECG is recorded
- The appearance of the ECG in the standard leads
- How the ECG is used to assess the electrical activity of the heart
- The determination of the main cardiac axis
- The interpretation of pathological changes in the ECG
- The principles underlying treatment of cardiac arrhythmias

29.1 Introduction

The initiation of the heartbeat begins with the pacemaker activity of the sinoatrial (SA) node. It then spreads across the heart from right atrium to left atrium before passing to the ventricles by way of the atrioventricular (AV) node and the bundle of His and its left and right branches (see Chapters 8 and 28). Since the cardiac myocytes form a syncytium, an electrical disturbance in one region can spread across the whole heart in much the same way as the conduction of the action potential in unmyelinated nerve fibres (see Chapter 7). As the wave of excitation proceeds, some parts of the heart remain depolarized while other regions are polarized and at rest. Local electrical circuits are established between these regions which permit the flow of small currents through the extracellular fluid. In turn, these currents generate small potential differences that can be detected by appropriately positioned electrodes and recorded at the surface of the body as an **electrocardiogram** or **ECG**. In acknowledgement of the pioneering work of the Dutch physiologist Einthoven, the ECG is sometimes called the EKG (from the German *elektrokardiogramm*). As it can be recorded from the body surface, the ECG provides a non-invasive means of recording the electrical activity of the heart *in situ* and is of significant clinical value in the investigation of cardiac function.

Irregularities in the cardiac rhythm are known as arrhythmias, the most common of which are atrial fibrillation, extra contractions of the ventricles (called extrasystoles or **ectopic beats**), and progressive stages of heart block in which excitation from the atria to the ventricles is impaired. The ECG is used to determine the cause of an arrhythmia and to determine the principal electrical axis of the heart, which provides information about certain pathological cardiac conditions such as ventricular hypertrophy. The ECG is also used to investigate the condition of the heart after an episode of cardiac ischaemia (a heart attack).

Summary

The electrocardiogram (ECG) is a recording of the electrical activity of the heart, giving insight into both normal and abnormal cardiac function. It is used to gain information about the heart rate and any disorders of conduction, to determine the site of any abnormal pacemaker activity, to investigate cardiac hypertrophy, and to monitor the health of the myocardium.

29.2 Recording the ECG

Although the most direct way of recording the ECG is by placing electrodes directly on the cardiac muscle itself, this is rarely done except during certain surgical procedures in which the thorax is open. Instead, the ECG is recorded by placing electrodes at different points on the body surface and measuring voltage differences between them with the aid of an electronic amplifier. Any pair of electrodes positioned on the body surface will detect a small portion of the current flow during depolarization and repolarization.

For this reason, information from a number of standard electrode placements (called **leads**) is required to provide a complete picture of the spread of excitation throughout the heart.

There are two types of ECG leads, **bipolar** and **unipolar**. Bipolar leads record the voltage between electrodes placed on the wrists and left ankle (with the right ankle acting as the earth). Unipolar leads record the voltage between a single electrode placed on the body surface and an electrode that is maintained at zero potential (earth).

The three standard bipolar limb leads are, by convention, known as limb leads I, II, and III. These are illustrated in Figure 29.1.

- In lead I, the positive terminal of the amplifier is connected to the left arm and the negative terminal to the right arm. With this placement of electrodes, the amplifier records the component of the wave of excitation that is moving along an axis between the right and left sides of the heart.

- In lead II, the right arm is the negative terminal and the left leg the positive, so that the component of the wave of excitation moving from the right upper portion of the heart to the tip of the ventricles is recorded.

- In lead III, the left leg is the positive terminal and the left arm the negative. This lead records the component of the wave of excitation spreading along an axis between the left atrium and the tip of the ventricles.

There are two types of unipolar leads used in electrocardiography, the **chest** (or **precordial**) **leads** and the **augmented limb leads**.

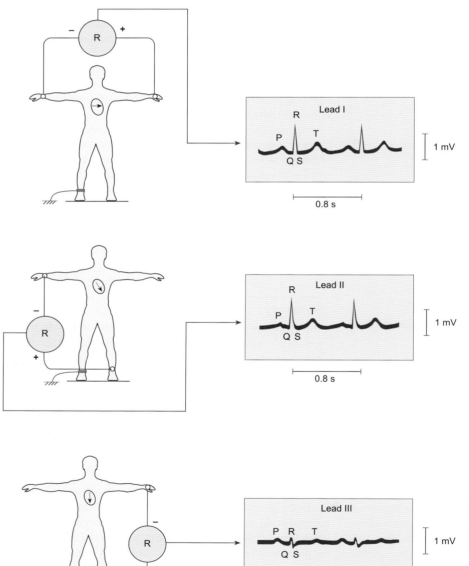

Figure 29.1 The arrangement of the standard limb leads used to record the ECG. The appearance of the principal waves for the various leads is also shown. Lead II is orientated along the main atrioventricular axis of the heart and gives rise to the most prominent R wave. These leads provide information about current flow in the frontal plane.

When the ECG is recorded using the unipolar chest leads, a reference electrode is produced by joining the three limb leads while an exploring electrode records from specific points on the chest at the level of the heart, as shown in Figure 29.2. The exploring electrode is then placed in one of six positions on the chest. These positions are known as leads V1 to V6, and their locations are described in the caption to Figure 29.2.

The voltage that can be recorded from the two arms and the left ankle with a unipolar exploring electrode can be increased using the augmented limb leads. In this configuration, two of the limb leads are used to produce a reference electrode. The signal is then recorded as the difference between these electrodes and the remaining limb electrode, as shown in Figure 29.3. Lead aV_R is recorded from the right arm with the reference electrode formed by adding the signals from left arm and left ankle. Lead aV_L is formed by recording from the left arm with the reference electrode formed by adding the signals from right arm and left ankle. Lead aV_F is formed by recording the signal from the left ankle with the reference electrode formed by adding the signals from the two arms.

The characteristics of the normal ECG recorded with the limb leads

The component waves of the ECG were arbitrarily named P, Q, R, S, and T in the early days of electrocardiography. It is an unfortunate fact that there are many minor variations in the appearance of the ECG recorded by the standard leads from normal individuals. The P and T waves are distinct entities, but the naming of the QRS complex follows these rules: if the first deflection of the QRS complex is downward, it is called a Q wave. An upward deflection is always called an R wave irrespective of whether it is preceded by a Q wave. A negative deflection immediately following the R wave is called an S wave. Figure 29.4 shows a typical ECG trace together with the standard reporting intervals (P, PR, QRS, ST, QT, and T).

The **P wave** reflects the electrical currents generated as the atria depolarize prior to contraction, the **QRS complex** corresponds to ventricular depolarization, and the **T wave** corresponds to ventricular repolarization. Atrial repolarization occurs during ventricular depolarization and is masked by the QRS complex. Figure 29.5

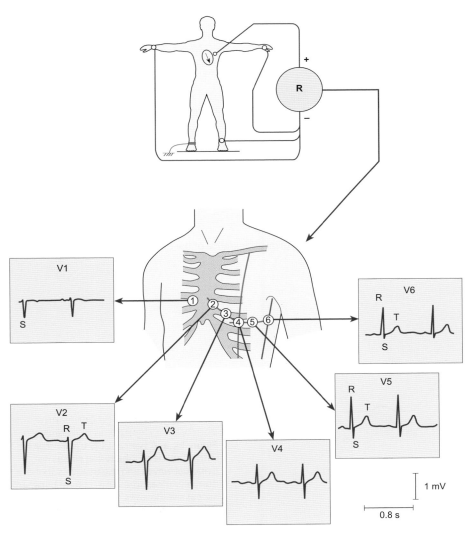

Figure 29.2 Unipolar recording of the ECG with the standard chest leads. The limb leads are connected together to provide a virtual earth, as shown in the inset figure at the top. The recording electrodes are then placed in six positions on the chest (leads V1 to V6), as shown. The chest leads provide information about current flow in the transverse plane. Leads V1 and V2 usually show a pronounced S wave while leads V5 and V6 show a large R wave. Lead V1 is placed at the right margin of the sternum in the fourth intercostal space. Lead V2 is placed at the left margin of the sternum in the fourth intercostal space. Lead V3 is placed midway been leads V2 and V4. Lead V4 is placed on the mid-clavicular line (shown here in red) in the fifth intercostal space. Lead V5 is placed at the same level as lead V4 in the anterior axillary line. Lead V6 is also placed at the same level as lead V4 but in the mid-axillary line (shown here in blue).

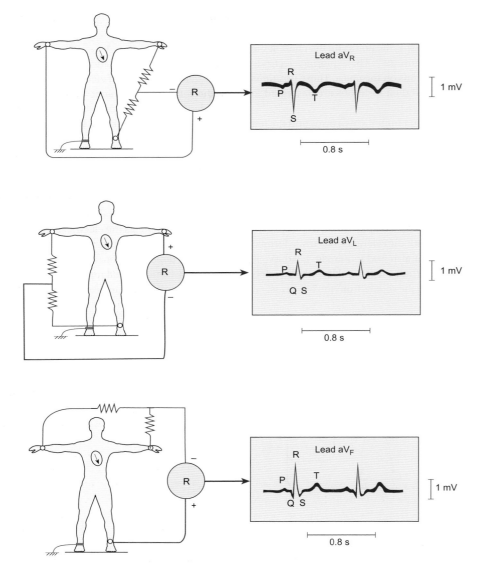

Figure 29.3 The arrangement for recording the augmented limb leads and the typical appearance of the ECG recorded from these leads.

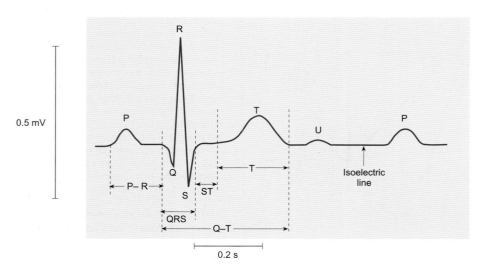

Figure 29.4 A typical ECG trace recorded with a limb lead labelled to show the principal waves and the intervals that are normally measured. The interval between successive R waves can be used to calculate the heart rate. The U wave is usually absent.

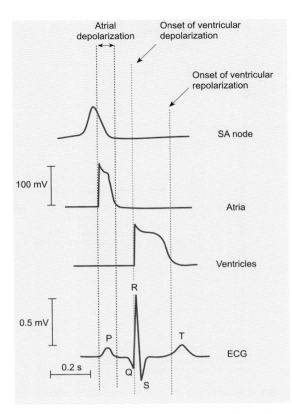

Figure 29.5 The relationship between the onset and duration of the action potentials of cells in the SA node, the atria, and the ventricles during a single cardiac cycle and the corresponding ECG trace.

shows the timing of the ECG waves in comparison with the underlying cardiac action potentials, while the timing of the ECG trace in relation to the mechanical events of the heart is shown in Figure 28.6 (Chapter 28).

The electrical signals that give rise to the ECG are greatly attenuated, as they are conducted through the body to the skin surface and are very small compared with the amplitude of the action potentials recorded from different parts of the heart (which have amplitudes of 70–120 mV). The largest QRS complexes are only of the order of 3 or 4 mV even when one of the recording leads is placed directly over the heart. When the electrodes are placed on the body surface the voltages are smaller still, and the amplitude of a typical R wave recorded with a standard limb lead is 0.5–1 mV.

Although each cardiac cycle is initiated by depolarization of the sinoatrial node, this electrical event is not seen in the ECG trace because the mass of tissue involved is very small. The first discernible electrical event, the P wave, lasts about 0.08 s and coincides with the depolarization of the atria. The isoelectric line (i.e. the part of the record in which there are no measurable deflections) between the P wave and the start of the QRS complex coincides with the depolarization of the AV node, the bundle branches, and the Purkinje system. Atrial contraction occurs during the PR interval, which lasts 0.12–0.2 s.

The next electrical event of the ECG is the QRS complex, which lasts about 0.08–0.1 s. This is normally seen as a large deflection from the isoelectric line because it is produced by the excitation

(depolarization) of the ventricles, which have the largest mass of muscle tissue in the heart. According to convention, the Q and S waves are downward deflections and the R wave is an upward deflection, although the exact shape and size of the components of the complex depend upon the position of the electrodes being used to record the ECG (see Figures 29.1–29.3). The atria repolarize during the QRS complex, but as the muscle mass of the ventricles is so much larger than that of the atria, this event is not seen as a separate wave in the ECG.

During the interval between the S and the T waves the entire ventricular myocardium is depolarized while the ventricles contract. Because all the myocardial cells are at about the same potential, the S–T segment lies on the isoelectric line. This corresponds to the long plateau phase of the cardiac action potential.

The final major event of the ECG trace is the T wave, which normally appears as a broad upward deflection and represents the repolarization of the ventricular myocardium which precedes ventricular relaxation. The action potential of the ventricular myocytes lasts for 0.2–0.3 s, so that the interval between the start of the QRS complex and the end of the T wave is around 0.3 s. The T wave is relatively broad because some ventricular fibres begin to repolarize earlier than others, and thus the whole process of repolarization is rather prolonged.

Occasionally the T wave is followed by a low-amplitude wave known as the U wave of the same polarity. It is usually less than 10 per cent of the preceding T wave and is most obvious in leads V2 or V3. It is more prominent at slower heart rates and when plasma potassium is low (hypokalaemia). A negative U wave in leads V2 and V3 indicates the presence of heart disease. The origin of the U wave remains unclear, but there is evidence to suggest that the U wave results from stretching of ventricular muscle during the initial period of rapid blood inflow into the ventricles. This stretching of the myocardium is known to elicit changes in the membrane potential of cardiac myocytes. Other possibilities are that U waves reflect delayed repolarization of mid-myocardial myocytes ('M cells') or the papillary muscles.

Why does the T wave have the same polarity as the R wave? The polarity of an ECG wave reflects the predominant direction of current flow sensed by a particular ECG electrode. During ventricular depolarization, the cardiac cells closest to the Purkinje cells depolarize first and the wave of depolarization passes from the inside of the ventricles to the outside, as illustrated in Figure 29.6. The cells on the outer surface of the ventricles have shorter action potentials than those on the inside and begin to repolarize first. The wave of repolarization thus passes from the outside to the inside of the ventricles. Consequently, the T wave has the same polarity as the R wave (see next section).

Summary

The ECG is recorded by placing electrodes at different points on the body surface and measuring voltage differences between them. The P wave of the ECG is due to atrial depolarization, the QRS complex to ventricular depolarization, and the T wave to ventricular repolarization. Atrial repolarization is hidden within the QRS complex. The PR interval reflects the time it takes the action potential to propagate through the AV node.

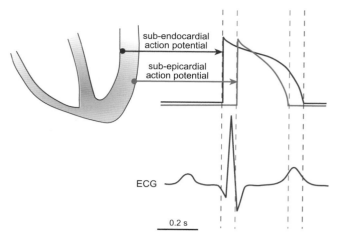

Figure 29.6 Action potentials recorded from cardiac myocytes close to the endocardium (sub-endocardial myocytes) compared to those recorded from myocytes adjacent to the epicardium (sub-epicardial myocytes). The shorter duration of the action potential of the sub-epicardial cells allows them to repolarize first. This difference in the onset of repolarization between the myocytes in the different parts of the ventricular wall accounts for the upright polarity of the T wave.

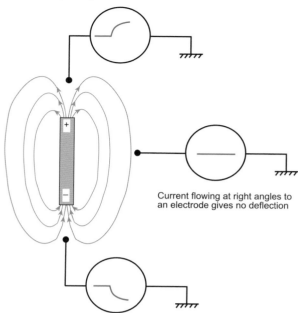

Figure 29.7 A simple diagram to show the pattern of current flow between the positive and negative poles of an electrical dipole in a conducting medium. The positive voltage deflections are recorded when current flows towards an electrode, and negative deflections are recorded when current flows away from an electrode. The changing voltage deflections of the ECG reflect progressive changes in the direction of current flow during the cardiac cycle.

29.3 How the electrical activity of the heart gives rise to the ECG

The ECG arises because, during the cardiac cycle, the myocytes of one part of the heart may be at their normal resting potential while others are in a depolarized state. Since the heart muscle is a syncytium and the myocytes are electrically connected, current will flow between these regions. The extracellular currents that arise because of this state of affairs can be detected by appropriately placed electrodes. The situation is analogous to a battery placed in a conducting medium where current will flow from one pole to the other. Current flowing towards an electrode will give rise to a positive potential, while that flowing away from an electrode will give rise to a negative potential. Current flowing at right angles to an electrode will not be detected (see Figure 29.7).

During the cardiac cycle, the situation is more complex, as the pattern of current flow resulting from depolarization of the heart muscle continually changes. Each electrode placement of the ECG detects the average current flowing towards or away from it at any instant of time. Thus, a particular wave may be more prominent when recorded by one electrode compared to the amplitude recorded by another. For example, the S wave is very prominent in V2 but much less so in V6 (see Figure 29.2).

The heart lies mainly in the lower part of the left side of the chest and is normally oriented so that the ventricles point diagonally downwards away from the atria, which lie close to the midline. In what follows it will be helpful to refer to the various panels of Figure 29.8. For simplicity, only the changes in potential that can be detected by the chest leads (V1–V6) will be considered, but the

same principles apply to the ECG recorded with the limb leads and the augmented limb leads.

- As the atria depolarize, extracellular current flows between the depolarized regions of the atria and the rest of the heart. This current flow can be detected by an electrode placed on the chest wall. The chest leads show a small positive deflection (the P wave), which reaches a peak and then declines as the atrial muscle becomes fully depolarized: Figure 29.8(a).

- After a brief delay at the AV node, the septum depolarizes and current flows from left to right. This is at right angles to V4 and is not sensed by this lead. However, the current flows away from V5 and V6 so the depolarization of the septum is detected as a small negative deflection (the Q wave) in these leads: Figure 29.8(b).

- Depolarization of the ventricular wall follows. It starts from the endocardium and moves outwards. The current flows towards V3–V6 and a large positive wave is recorded in these leads (the R wave): Figure 29.8(c).

- As the depolarization passes up the ventricular wall towards the atria, the current flows away from V2 and V3 and a prominent negative wave is recorded in these leads (the S wave): Figure 29.8(d). The S wave is less prominent in V5 and may be absent in V6. By the end of the S wave, the ventricles are fully depolarized and there is no current flow (the S–T segment—see Figure 29.4).

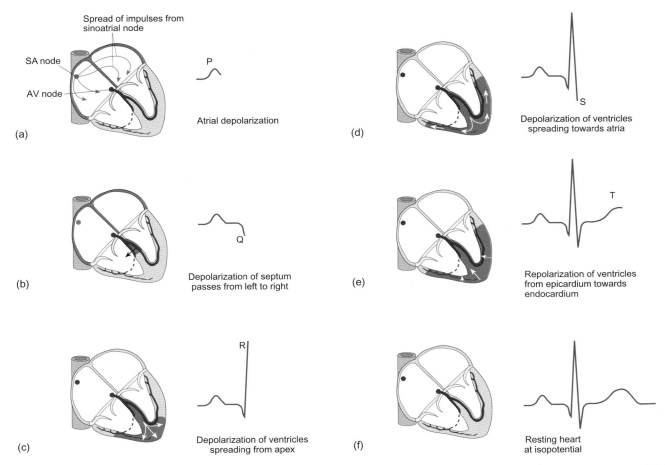

Figure 29.8 The relationship between the electrical activity of the myocardium and the component waves of the ECG. The arrows indicate the direction of current flow at different points of the cardiac cycle. Note that the ECG shown is generic in form and does not represent the trace recorded by a particular lead.

• The duration of the action potential of the myocytes near the endocardium is longer than that of the outer myocytes. As a result, the outermost cells repolarize first and the extracellular current flow is towards V3–V6, which thus records a positive deflection (the T wave): Figure 29.8(e). Note that, depending on the orientation of the heart, the current may flow away from V1 so that the T wave may be inverted in this lead. When the cardiac cycle is complete, no myocytes are depolarized and the ECG returns to baseline (the isoelectric line): Figure 29.8(f).

Summary

The polarity and amplitude of the components of the ECG are determined by the pattern of excitation and size of muscle mass, respectively. Three standard limb leads, three augmented limb leads, and six chest leads are used for this purpose. Each ECG recording electrode detects the average current flowing towards or away from it at any instant of time, so a particular wave may be more prominent when recorded by one electrode compared to its amplitude recorded by another.

29.4 Clinical aspects of electrocardiography

The ECG provides a non-invasive technique for monitoring the spread of excitation throughout the heart. In this section, some common abnormalities of the ECG and their underlying causes are discussed. In a full clinical investigation, a 12-lead ECG record is used, with three limb leads (leads I, II, and III), three augmented limb leads (aV_L, aV_R, and aV_F) and six chest leads (V1–V6). The three limb leads and three augmented limb leads each provide information about the activity of the heart viewed from a specific point on the frontal (vertical) plane, while each of the six chest leads provide information about the activity of the heart from a specific point in the transverse plane, as shown in Figure 29.9. The standard intervals for reporting the ECG and their normal values are given in Table 29.1. In interpreting the ECG it is essential to keep the following two points in mind:

• the polarity of the signal in any lead at a given time reflects the average direction of current flow with respect to that lead

• the amplitude of the signal reflects the mass of cardiac muscle involved.

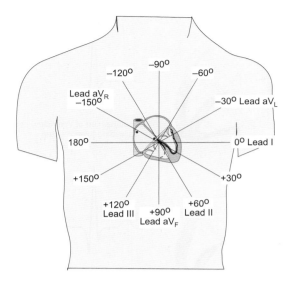

a) The limb leads record the ECG from the frontal plane

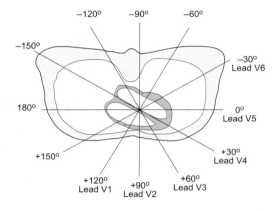

b) The chest leads record the ECG in the transverse plane

Figure 29.9 Each ECG lead records the activity of the heart from a different perspective and the complete array provides detailed information about the origin and spread of excitation within the heart. (a) The angles formed by the limb leads and augmented limb leads give rise to a hexaxial reference system for the frontal plane. (b) The chest leads record the electrical activity of the heart with respect to the transverse plane.

The ECG is used to gain information about the following aspects of cardiac function:

- the electrical rhythm of the heart
- the size of the muscle mass of the individual chambers of the heart
- disorders in the origin or in the conduction of the wave of excitation throughout the heart
- the site of any abnormal pacemaker activity

- abnormal cardiac excitability caused by altered plasma electrolyte levels
- the metabolic state and viability of the myocardium.

Each action potential originating in the SA node initiates one beat of the heart. This is known as **sinus rhythm** and is the normal state of affairs. Any deviation from the normal sinus rhythm is known as an **arrhythmia** or **dysrhythmia**. While some arrhythmias are of no clinical significance, others may reflect serious and life-threatening disorders of the myocardium. Indeed, many arrhythmias arise as a consequence of ischaemic heart disease.

The heart rate commonly varies with the respiratory cycle in healthy young people (sinus arrhythmia—see Section 30.4) and in highly trained athletes the heart rate is usually slower than normal (though stroke volume is higher). This is known as **sinus bradycardia** (less than 60 b.p.m.). Sinus bradycardia is also seen in fainting attacks, hypothermia, and hypothyroidism. A rapid heart rate (> 100 b.p.m.) is known as **sinus tachycardia** and is associated with exercise, stress, haemorrhage, and hyperthyroidism. The

Table 29.1 Standard ECG intervals

Measurement	Duration	Comments
P wave	< 0.12 s	Normally smooth and rounded. Entirely positive or entirely negative in all leads except V1.
PR interval	0.12–0.21 s	Interval is shorter in children and in adolescents. Varies with heart rate.
QRS width	0.07–0.11 s	Tends to be slightly shorter in females. Amplitude and polarity of R wave in different leads used to determine main cardiac axis.
QT interval	0.3–0.4 s (QT$_c$ 0.35–0.43 s)	Varies with heart rate and often expressed as the QT$_c$ interval which is QT divided by the square root of the R–R interval.
R–R interval	c. 0.75–0.85 s	The heart rate can be quickly calculated as follows: b.p.m. = 60 ÷ R–R interval in seconds
ST segment	80–120 ns	Also called the ST interval. It should lie on the isoelectric line, but in cardiac ischaemia it is elevated and in myocardial infarction it is depressed with respect to the baseline.
T wave duration	c.160 ms	Normally the same polarity as the QRS complex. The opposite polarity indicates a past or current infarction.

standard definitions of the terms bradycardia and tachycardia are given in Chapter 28 (p. 432).

If the rhythmical activity of the SA node activity is much slower than normal, the fundamental rhythm may be taken over by another part of the heart. These rhythms are known as **escape rhythms** and are named after their site of origin. Those that originate in the atria are called atrial rhythms, those that originate in or close to the AV node are junctional rhythms, and those that originate in the ventricles are ventricular rhythms.

Sick sinus syndrome

In some people, particularly the elderly, the SA node fails to excite the atria in a regular manner, resulting in a slow resting heart rate that does not increase appropriately with exercise. This is called **sick sinus syndrome** and has many causes. It may be drug induced, or it may reflect abnormally intense vagal activity. However, a non-specific, scar-like degeneration of the pacemaker region following ischaemia of the muscle supplied by the sinus nodal artery accounts for around one-third of all cases. Although the most common manifestation of sick sinus syndrome is bradycardia, it may also appear as alternating fast and slow rhythms or as a tachycardia that originates in the atria. Arrhythmias caused by failure of the SA node to excite the heart can often be treated by the implantation of an artificial pacemaker.

Ventricular hypertrophy and the electrical axis of the heart

As previously mentioned, the heart lies mainly in the lower part of the left side of the chest, with the right ventricle in front of the left ventricle. The ventricles normally point obliquely downwards away from the atria, which lie close to the midline. During the cardiac cycle the pattern of current flow resulting from depolarization and repolarization of the heart muscle continually changes, and, as already mentioned, each ECG lead detects the average current flowing towards or away from it at any instant of time (see Box 29.1).

The electrical axis of the heart (the principal cardiac axis) refers to the direction of the largest electrical dipole recorded from the frontal plane and can be assessed by comparing the amplitude of the R wave in the three limb leads, or by calculating the resultant vector of the activity recorded by lead I and lead aV_F. These two leads record the electrical activity in the horizontal and vertical planes respectively—see Figure 2b of Box 29.1. The cardiac axis depends on body size and shape but normally lies between −30° and +100° to the horizontal. In right axis deviation the cardiac axis lies between +100° and +150°, while in left axis deviation the cardiac axis lies between −30° and −90° (see Figure 29.10). The determination of the cardiac axis is helpful in the diagnosis of a number of conditions including right ventricular hypertrophy, conduction defects, and pulmonary embolus.

Atrial enlargement

Atrial enlargement is associated with changes to the P wave. In right atrial enlargement, which may be the result of pulmonary hypertension or of stenosis (narrowing) of the tricuspid valve, the P wave is larger than normal, because the right atrium has an increased muscle mass (atrial hypertrophy). This alters the appearance of the P wave, which has a peaked appearance. This change occurs because the right atrium depolarizes before the left atrium. In left atrial enlargement, which usually caused by mitral (bicuspid) stenosis, the P wave is broadened and may have a double peak (which is called a bifid P wave). The broadening of the P wave and the appearance of the second peak can be attributed to the later depolarization of the left atrium with its increased muscle mass.

Defects of conduction

Heart block refers to a defect in the conduction of the electrical activity at any point in the pathway between the SA node and the ventricular muscle. Problems with conduction from the atria to the ventricles are known as **atrioventricular block** and are classified according to their severity as first, second, or third degree heart block. Impairment of conduction through the bundle of His is known as **bundle block** and may affect the main bundle or any of its branches (branch bundle block).

In **first degree heart block**, each wave of depolarization originating in the SA node is conducted to the ventricles but there is an abnormally long PR interval (> 0.2s), indicating that somewhere in the pathway a delay has occurred (see panel (a) of Figure 29.11). This is usually at the AV node. First degree heart block is not usually a problem in itself, but it may be a sign of some other disease process (e.g. coronary artery disease or electrolyte disturbance).

In **second degree heart block** there is an intermittent failure of excitation to pass to the ventricles. There are three main types of second degree heart block:

- Mobitz type I, in which the PR interval becomes progressively longer until one P wave fails to elicit a QRS complex (and thus a ventricular contraction), as shown in Figure 29.11(b). The pattern is then repeated. This type of block is also known as the Wenckebach phenomenon or Wenckebach block.

- Mobitz type II, in which most P waves elicit a ventricular contraction but occasionally a P wave fails to be conducted to the ventricles so it is not followed by a QRS complex: Figure 29.11(c). In this type of block there is no progressive prolongation of the PR interval.

- 2:1 or 3:1 block occurs when every second or third P wave elicits a QRS complex: Figure 29.11(d).

In **third degree heart block** (also called complete heart block), the wave of excitation originating in the SA node fails to excite the ventricles (i.e. there is an atrioventricular block). The ventricles then show a slow intrinsic rhythm of their own (sometimes called an escape rhythm) with abnormally shaped QRS complexes that are not associated with P waves—see Figure 29.11(e). While the shape of the abnormal QRS complexes will depend on the site at which the pacemaker activity originates, the QRS complexes will be broader than normal unless a nodal rhythm is established. The broadening of the QRS complex arises because the conduction of the wave of

(1₂3) Box 29.1 Determination of the cardiac axis

The direction of the largest electrical dipole recorded by the ECG during the cardiac cycle is called the mean cardiac axis. Together with the cardiac rhythm (i.e. heart rate), the conduction intervals, and the description of the QRS complex, ST segment, and T waves, it is one of the reported ECG variables. As mentioned in the main text, the normal range for the cardiac axis is wide ($-30°$ to $+100°$). Right axis deviation is normal in children and tall thin adults, but it also occurs in various pathologies: it occurs in right ventricular hypertrophy (which may be caused by pulmonary hypertension or pulmonary stenosis); it also occurs in pulmonary embolism and in left posterior hemiblock. Left axis deviation occurs in left anterior fascicular block (or left anterior hemiblock) but *not* in left ventricular hypertrophy. (Left ventricular hypertrophy is indicated when at least one of the R waves seen in the left chest leads (V4–V6) is abnormally large (> 2.7 mV) and one of the S waves seen in the right chest leads exceeds 3 mV.) A quick assessment of the cardiac axis can be made by examining the polarity of the R wave in the limb leads, as shown in Figure 1.

A more accurate assessment is made by vector analysis of the R wave. Traditionally this has been based on **Einthoven's triangle**. To compute the cardiac axis, the three limb leads are represented by the three sides of an equilateral triangle, as shown in Figure 2(a). Each side is then bisected by a perpendicular line, and the amplitude of the R wave is represented as an arrow on the sides of the triangle for each lead. Positive potentials are plotted from the zero point (where the perpendicular line intersects the side of the triangle) with the head of the arrow towards the positive terminal of the lead, and negative values are plotted in the other direction. (NB the same scale must be used for each lead.) Lines parallel to the original perpendiculars are then drawn from the head of each arrow. An arrow drawn from the centre of the triangle to the point where these lines intersect gives the angle of the cardiac axis. A slightly simpler alternative is to plot the amplitude of the R wave in lead I as a horizontal vector and the amplitude of the R wave in lead aV_F as a vertical vector, as shown in Figure 2(b). The diagonal of the completed rectangle gives the main cardiac axis. This method exploits the fact that lead I records the ECG in the horizontal plane while lead aV_F records the ECG in the vertical plane (see Figure 29.9 of the main text).

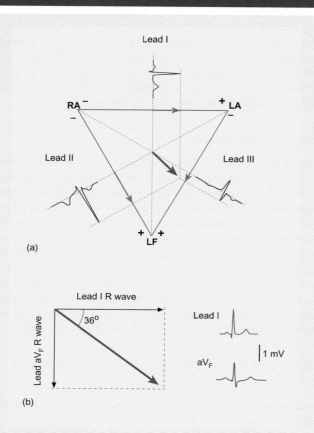

(a)

(b)

Figure 2 The electrical axis of the heart may be precisely determined from vector diagrams of the amplitude of the R wave recorded in the standard limb leads (a) and from its amplitude recorded in lead I and lead aV_F (b). A vector is a variable that has both magnitude and direction. Vectors are represented graphically by lines in which the magnitude of the variable is indicated by the length of the line and its direction is indicated by the angle plotted on suitable coordinates. They can be added by means of vector diagrams, as represented here by the Einthoven triangle (a) or rectangle (b).

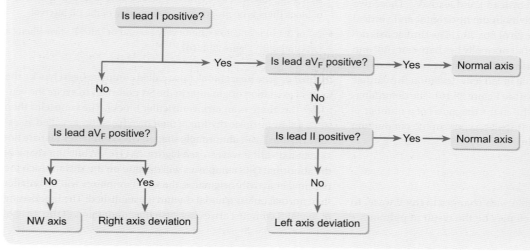

Figure 1 A simple flow diagram to show how the cardiac axis can be quickly estimated by inspection of the polarity of the R wave recorded in leads I, aV_F, and II.

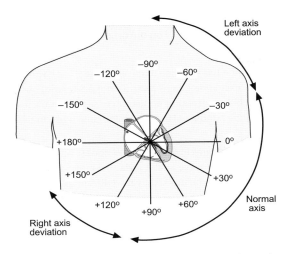

Figure 29.10 The mean QRS axis of the heart. Note the wide range for normal hearts. The region between −90° and +150° is familiarly known as North West territory or no man's land, but the cardiac axis may lie in this region in emphysema, in hyperkalaemia, and during ventricular tachycardia.

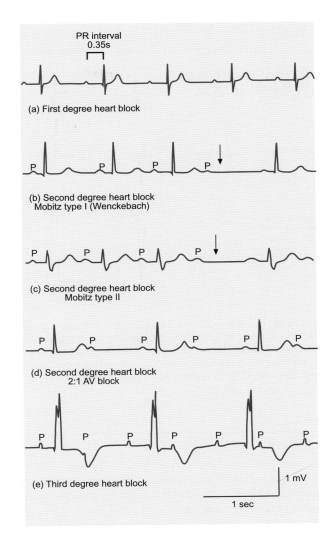

Figure 29.11 The ECG seen in various types of heart block. (a) First degree heart block; note the long PR interval (normal range 0.12–0.2 s). (b) Second degree heart block, showing the Mobitz type I or Wenckebach pattern in which the PR interval progressively lengthens until one beat is missed (arrow). The cycle then repeats itself. (c) Second degree heart block of the Mobitz type II pattern. Here the PR interval is constant but occasionally a beat is missed (arrow). This pattern reflects a failure of the conducting system. (d) Second degree heart block in which every second P wave elicits an R wave—this is 2:1 block. Other regular patterns may also be seen, such as 3:1 or 4:1 block. (e) Third degree heart block—here there is complete dissociation of the QRS complex from the P wave. Moreover, the QRS complex is broader than normal and has an abnormal appearance.

excitation through the cardiac muscle is significantly slower than that through the Purkinje fibres.

In some situations, the excitation reaches the AV node and passes through the bundle of His only to be delayed or fail to pass from one or other branch to the ventricular muscle. This is known as **bundle branch block** and, in this situation, the precise appearance of the ECG is very variable between individuals. So what follows is a general guide rather than a precise description of the ECG changes. However, as the wave of excitation will spread more slowly across the ventricles than it normally would, the duration of QRS is longer than normal (see Figure 29.12): in incomplete block it is between 0.1 and 0.12 s, and in complete block the duration of QRS is > 0.12 s.

The ECG in right bundle branch block shows considerable variation. However, as depolarization of the right ventricle is delayed, there is little effect on the initial part of the QRS complex, which is dominated by events in the left side of the heart. Nevertheless, QRS is broadened because of the late depolarization of the right ventricle, and the QRS complex in lead V1 is distorted with a reduced R wave (written as the r wave) accompanied by a secondary rise in the QRS complex after the S wave. The secondary R wave is called the R′ wave because it follows the earlier r wave. (This is often referred to as an rSR′ complex.) The S wave is also broadened in leads I and V6.

On the left side, the bundle of His branches to form an anterior fascicle and a posterior fascicle. Conduction block can thus occur in the main bundle (left bundle branch block) or in either of the two fascicles. In left bundle branch block there is a significant distortion of the QRS complex because activation proceeds from the right side of the heart (the reverse of the normal sequence). The QRS complex is monophasic in leads I and V6 (i.e. both Q and S are absent) and lead V1 shows a QS sequence with no intervening R

wave. Block of the conduction through the left anterior fascicle will cause the main cardiac axis to deviate towards the left, as the wave of excitation will spread from the septum and base of the left ventricle. This sometimes called left anterior hemiblock, although it is better called **left anterior fascicular block** (LAF). In left posterior fascicle block (LPF) there is a rightward shift in the main cardiac axis (see Figure 29.10).

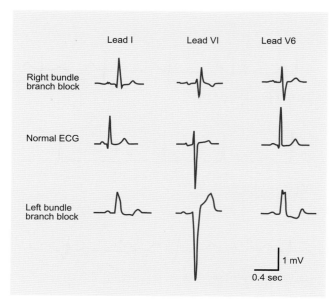

Figure 29.12 The appearance of the ECG in left and right bundle branch block. Note the wide QRS complex and the notched peak in lead V6 in left bundle branch block and the rSR' sequence in lead V1 in right bundle branch block. In both left and right bundle branch block the QRS complex is wider than normal.

beat). For both escape rhythms and extrasystoles, the appearance of the ECG trace will depend on the origin of the abnormal excitation (see Figure 29.13).

- If the excitation has its origin in the atria, the P wave will be abnormal (as the excitation begins away from the SA node) but the QRS complex will be normal in appearance, as the ventricles will be excited via the bundle of His and the Purkinje fibres in the normal way.

- If the excitation begins in or near the AV node (a junctional rhythm), there may be no P wave as the atria may not be excited or their depolarization may be masked by the QRS complex, which will be normal in appearance. Alternatively, a P wave may be seen in the ECG trace but occuring later than normal, either just before or just after the QRS complex.

- If the excitation begins somewhere in the ventricles, the QRS complex will be abnormally wide and distorted in appearance because the wave of excitation will be conducted from the point of excitation via the cardiac muscle fibres, rather than by the conducting system of the heart. There will be no consistent relation between the P waves and the QRS complexes. The appearance of the ECG trace in this situation is similar to that seen for a ventricular escape rhythm, as seen in third degree heart block.

Summary

The mean QRS axis of the heart normally lies between −30° and +100° to the horizontal. In right axis deviation the cardiac axis lies between +100° and +150°, while in left axis deviation the cardiac axis lies between −30° and −90°.

The PR interval is prolonged in first degree heart block, but in second degree heart block the PR interval depends on the type of block. In Mobitz type I (or Wenkebach) block, the PR interval progressively lengthens until a beat is missed. In Mobitz type II block, the PR interval is constant but a beat is occasionally missed. Second degree heart block may also be manifested as 2:1 or 3:1 block. In third degree heart block, there is complete dissociation of the QRS complex from the P wave. In bundle branch block, the wave of excitation fails to pass from one or other branch of the conducting system to the ventricular muscle.

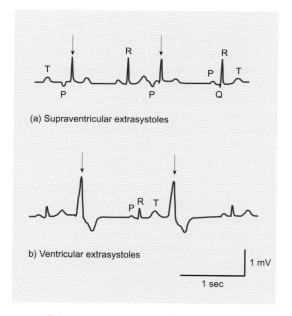

Figure 29.13 Examples of atrial and ventricular extrasystoles recorded via lead II. (a) Two supraventricular extrasystoles (arrows) with preceding inverted P waves. The normal appearance of the QRS complex should be compared with the normal pattern of excitation seen on the right of this trace. The inverted P wave and short PR interval preceding the first and third R waves indicates that the ectopic focus is located near the base of one of the atria or in the nodal tissue. (b) Two ventricular extrasystoles (arrows). Note the broad QRS complexes and the inverted T wave, which have a highly abnormal appearance compared to the normal ECG pattern seen in the middle of this trace (marked P, R, T).

Arrhythmias of atrial or ventricular origin

As mentioned earlier, abnormal rhythms may begin in the atria—away from the SA node, they may begin at the AV node (nodal or junctional rhythms) or they may begin in the ventricles. Arrhythmias originating in the atria or nodal region are often called **supraventricular rhythms**. When the cycle of excitation begins at a site remote from the SA node, the resulting rhythm is slower than normal (bradycardia). Sometimes the atria or ventricles contract earlier than expected. This is known as an **extrasystole** (an ectopic

The normal pattern of excitation in the heart relies on the refractory period of the cardiac cells to ensure that the conduction proceeds in an orderly way: from the SA node to the atria, then to the AV node, the bundle of His, and its branches. Finally, the excitation reaches the Purkinje cells which excite the ventricular myocytes. Retrograde excitation is normally prevented by the long refractory period of the cardiac cells. In some circumstances this inherent regulation breaks down and premature excitation can occur via **re-entry circuits**. These circuits give rise to premature beats and tachycardias, which may be so rapid that they compromise the effective filling of the heart. The site of the abnormal excitability can be deduced using the analysis already discussed.

- In atrial tachycardia the heart rate is higher than normal (100–200 b.p.m.) at rest. Tachycardias may arise from a rapid sinus rhythm or from another site in the atria (an ectopic atrial tachycardia). When the tachycardia originates in the SA node, the rate is usually less than 200 b.p.m. and each P wave is followed by a QRS complex. If the rate is high the P waves tend to be lost in the T wave, as in the example shown in Figure 29.14(a). Atrial tachycardias that arise elsewhere in the atria have normal QRS complexes, but the P waves are often abnormal in appearance.

- When the atria depolarize more frequently than about 290 times a minute, **atrial flutter** is present. The ECG baseline is not flat but often shows a continuous regular sawtooth pattern, as in Figure 29.14(b). As the AV node cannot be activated more than about 200 times a minute, the ventricles are activated at a lower rate. In some people the ventricles are activated in a completely regular manner with, for example, a 2:1 or 4:1 ratio of P waves to QRS complexes. In others, the ventricles may be activated in a very irregular manner.

- If the atria are activated more than about 350 times a minute, they do not contract in a coordinated way. The individual muscle fibre bundles contract asynchronously. The ECG trace shows no P waves, only an irregular baseline (panel (c) of Figure 29.14) and the ventricles show an irregular rhythm. This is known as **atrial fibrillation**.

- If the site of abnormal activity is in the wall of one of the ventricles, the rhythm is known as **ventricular tachycardia** and, for the reasons already discussed, the QRS complex will be wide and abnormal in appearance, as seen in Figure 29.14(d).

- When the ventricular muscle fibres fail to contract in a concerted way, the heart is in a state of **ventricular fibrillation**, and is unable to pump the blood around the body: Figure 29.14(e).

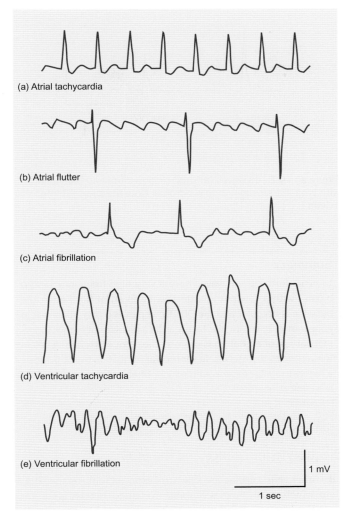

Figure 29.14 Abnormal cardiac rhythms. (a) Atrial tachycardia. (b) Atrial flutter—note the sawtooth pattern of the P waves and the broad QRS complexes. The ventricular rate is 45 b.p.m. (c) Atrial fibrillation. Note the irregular baseline and the highly abnormal QRST pattern. (d) Ventricular tachycardia—a regular but highly abnormal pattern of excitation is seen as the point of excitation lies in the ventricles themselves. The ventricular rate is 136 b.p.m. (e) Ventricular fibrillation—note the highly irregular pattern with no identifiable waves present.

As the quantity of blood forced from the atria into the ventricles contributes only about 20 per cent of the end diastolic volume, neither atrial flutter nor atrial fibrillation is immediately life-threatening. In contrast, ventricular fibrillation requires urgent action to restore normal function by means of a defibrillator, which uses a strong electrical shock to terminate the fibrillation and reset a normal heart beat.

Wolf–Parkinson–White syndrome

An example of a re-entry circuit is provided by the Wolf–Parkinson–White syndrome. Affected individuals have an accessory pathway connecting atrium and ventricle, usually on the left side of the heart. This is known as the bundle of Kent. Unlike the excitation via the AV node and the bundle of His, the excitation via the bundle of Kent is not delayed. Consequently, the left ventricle is excited earlier than in normal individuals. As a result, the PR interval is reduced and there is a small upstroke on the rising phase of the QRS complex called a **delta wave** reflecting this early excitation (see Figure 29.15). The vast majority of people with this anatomical feature have no symptoms of heart disease, but in a few cases an abnormal circuit of excitation is set up in which impulses travel from the ventricle to the atrium via the bundle of Kent. This may result in the early excitation of the AV node and a sustained ventricular tachycardia.

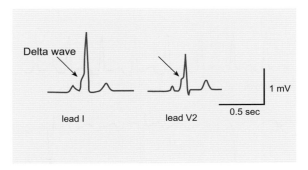

Figure 29.15 Delta waves (arrows) recorded from an individual with Wolf–Parkinson–White syndrome. Note the short PR interval in lead I.

Summary

In atrial flutter the P waves have a very high frequency (> 290 b.p.m.) with a sawtooth pattern, while in atrial fibrillation the baseline is irregular and the QRST complex is highly abnormal in appearance. In ventricular tachycardia a regular but highly abnormal pattern of excitation is seen, as the point of excitation lies in the ventricles themselves. In ventricular fibrillation the ECG shows a highly irregular pattern with no identifiable waves present. This is a clinical emergency.

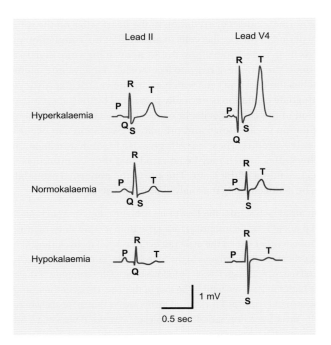

Figure 29.16 The appearance of the ECG in hyperkalaemia (top) and hypokalaemia (bottom). Note the peaked T wave in hyperkalaemia (especially in lead V4) and the low amplitude T wave in hypokalaemia.

Electrolyte imbalance and the ECG

Like other cells, cardiac myocytes are exposed to the extracellular fluid, which makes them vulnerable to changes in their ionic environment. Their excitability is profoundly affected by the distribution of potassium ions across their plasma membrane, and the plateau phase of the cardiac action potential is maintained by calcium influx. Moreover, the threshold for action potential generation depends on the concentrations of calcium and magnesium ions in the extracellular fluid. It is therefore not surprising to find that electrolyte imbalance is reflected in the appearance of the ECG.

When plasma potassium levels are low (**hypokalaemia**), the most characteristic change is a flattening of the T wave (compare the middle and lower ECG records shown in Figure 28.16). A U wave may appear. Other ECG changes to note are an increase in the amplitude and width of the P wave, and a prolongation of the PR interval. These changes have been attributed to a prolongation of the recovery phase of the cardiac action potential.

If plasma potassium is elevated (**hyperkalaemia**), the most prominent change is a T wave that has a peaked appearance with a greater amplitude and width than normal (see Figure 29.16). There is also a prolongation of the PR and QRS intervals with a flattening of the P wave. The elevated potassium accelerates repolarization of the cardiac action potential and is responsible for the peaked appearance of the T wave, while the high potassium partially inactivates the sodium channels and this leads to a slowing of

action potential conduction through the myocardium. The partial inactivation of the sodium channels reduces the amplitude of the action potential and leads to smaller P waves and a widening of the QRS complex.

The plateau phase of the cardiac action potential is maintained by an influx of calcium through voltage-gated calcium channels (see Chapter 5). This calcium entry both triggers contraction of the cardiac myocytes and determines the rate at which they repolarize. If plasma calcium levels are lower than normal (**hypocalcaemia**), the cardiac action potential is prolonged and this is reflected in the ECG trace as a long QT interval. If plasma calcium levels are higher than normal (**hypercalcaemia**), the QT interval is shorter than normal (see Figure 29.17) and the heart is prone to arrhythmias.

When plasma magnesium is low (< 0.7 mmol l^{-1}—**hypomagnesaemia**), the QT interval may be longer than normal and the T wave may be broad and flattened, changes that reflect the prolonged repolarization of the myocytes. **Hypermagnesaemia** (> 1.1 mmol l^{-1}) occurs in renal failure or following the ingestion of large amounts of magnesium-based antacids. When plasma magnesium exceeds 2.5 mmol l^{-1}, there is a prolongation of the PR interval, a widening of the QRS complex, and an increase in the amplitude of the T wave.

ECG changes in ischaemia and infarction

Cardiac ischaemia refers to a situation in which the blood supply is inadequate to meet the metabolic requirements of the myocardium. If this situation is prolonged, the affected tissue becomes

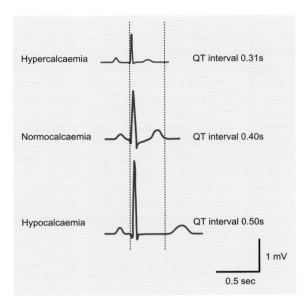

Figure 29.17 The changes in the ECG seen in hypercalcaemia and hypocalcaemia compared to the ECG in normal plasma calcium. Note the shortening of the QT interval in hypercalcaemia and its lengthening in hypocalcaemia.

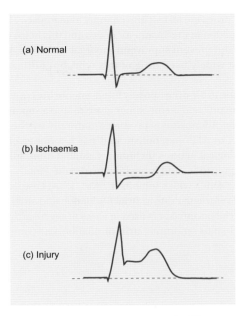

Figure 29.18 Characteristic changes in the QRST waves seen in ischaemia (depressed ST segment) and infarct (injury—elevated ST segment).

damaged and may die. This will result in necrosis of the affected area (a **cardiac infarct**). Since the spread of excitation throughout the myocardium will be affected in both these conditions, the ECG will show characteristic changes. When ischaemia affects the part of the wall of the left ventricle adjacent to the endocardium, the affected myocytes may be unable to maintain the prolonged action potential characteristic of normal cells. They begin to repolarize early, and the T wave is inverted in many of the leads. More generally, during ischaemia the ST segment recorded in long axis leads such as lead II is negative (i.e. it lies below the isoelectric line—see Figure 29.18b). If the ischaemia progresses to cause a cardiac infarct, the ECG shows characteristic changes immediately after the damage has been sustained. The most obvious change is that the ST segment does not return to baseline as it normally does, but is elevated as shown in Figure 29.18(c). A detailed interpretation of the ECG records during cardiac ischaemia and following myocardial infarction is beyond the scope of this book but can be found in specialist texts on electrocardiography.

29.5 Treatment of arrhythmias

Most arrhythmias are treated by anti-arrhythmic drugs (sometimes called anti-dysrhythmic drugs). The choice of which anti-arrhythmic to use will depend on the particular type of arrhythmia to be treated, and those which are most widely used have been divided into four main classes based on their electrophysiological mode of action. However, not all drugs used to treat arrhythmias fit neatly into this scheme. The exceptions are briefly discussed below.

Class I anti-arrhythmic drugs are sodium channel blockers, and they reduce the rate at which the cardiac muscle cells depolarize. The class is divided into three sub-classes. Class IA includes **quinidine**, **procainamide**, and **disopyramide**; Class IB includes the local anaesthetic **lignocaine** (lidocaine in North America). Lignocaine is given intravenously when it is used as an anti-arrhythmic, as it is metabolized rapidly by the liver. Class IC drugs cause a general decrease in cardiac excitability. Examples are **flecainide** and **encainide**.

Class II drugs are β-adrenoceptor antagonists. This group includes **propranolol** (the best-known drug in this class), **atenolol**, and **timolol**. They counter the effects of excessive adrenaline secretion by slowing the heart and decreasing cardiac contractility. As they reduce cardiac work, they are helpful in patients recovering from heart attacks.

Class III drugs are agents that slow the heart rate by prolonging the refractory period of the cardiac action potential. These drugs are used to slow tachycardias. The classic example is **amiodarone** (although it has a number of serious side-effects). **Sotalol** is a β-adrenoceptor antagonist that has a class III action.

Summary

The ECG shows characteristic changes in electrolyte disorders: in hyperkalaemia the T wave is accentuated, while in hypokalaemia it is flattened. In hypercalcaemia the QT interval is shortened, while in hypocalcaemia it is prolonged. In hypomagnesaemia the QT interval is increased and in hypermagnesaemia the PR interval is prolonged with a wider QRS complex. The ECG also shows characteristic changes in ischaemia and after a cardiac infarct.

Class IV drugs are all calcium channel antagonists. These drugs slow conduction in the SA and AV nodes and this helps to slow the heart rate and prevent supraventricular tachycardias. **Verapamil** is the most widely used drug of this type, but **diltiazem** is also used.

In addition to the drugs already listed, **adenosine** is a safe alternative to verapamil and is sometimes given to stop a supraventricular tachycardia. Adenosine is produced naturally by the body as a by-product of ATP metabolism and has a number of powerful pharmacological actions including vasodilatation, inhibition of platelet aggregation, and bronchoconstriction. Finally, digoxin (a cardiac glycoside), which inhibits the Na^+, K^+-ATPase, can be used to slow the ventricular rate in atrial fibrillation and atrial flutter.

An artificial pacemaker is used in cases of severe arrhythmia including sick sinus syndrome, atrial fibrillation, and complete (third degree) heart block. A pacemaker consists of a small electronic device that is implanted just under the skin on the upper chest which is connected to the heart via wires passed through the veins to the heart. Pulses are transmitted via the wires to stimulate the heart electrically and maintain a normal regular rhythm. The pacemaker may be set to deliver pulses at a set rate that is independent of cardiac activity, or a more sophisticated sensing circuit can be used to determine the intrinsic rhythm and then deliver pulses accordingly. Pacing may be via the right atrium or right ventricle, or both. The emergency treatment of serious arrhythmias often requires the use of a defibrillator (e.g. ventricular fibrillation) or direct current cardioversion, which can be used to restore sinus rhythm in atrial fibrillation and ventricular tachycardia, for example. An implantable cardiac defibrillator is used in cases of severe and life-threatening arrhythmias.

(✳) Checklist of key terms and concepts

General properties of the ECG

- The ECG records small potential differences (around 1 mV) arising from the sequential electrical depolarization and repolarization of the heart muscle.
- The ECG is recorded by placing electrodes (leads) at different points on the body surface and measuring voltage differences between them.
- Clinical ECG recording employs 12 leads: three limb leads, three augmented limb leads, and six chest leads.
- The precise appearance of an ECG trace depends on the position of the individual lead with respect to the electrical activity of the heart.
- The P wave of the ECG is due to atrial depolarization, the QRS complex to ventricular depolarization, and the T wave to ventricular repolarization. Atrial repolarization is hidden within the QRS complex.
- Using Einthoven's triangle, the orientation of the heart can be determined from the relative heights of the R wave recorded with the limb leads.
- The mean QRS axis of the heart normally lies between $-30°$ and $+100°$ to the horizontal.
- In right axis deviation the cardiac axis lies between $+100°$ and $+150°$, while in left axis deviation the cardiac axis lies between $-30°$ and $-90°$.

Interpretation of abnormal ECG patterns

- In first degree heart block, the PR interval is prolonged.
- In Mobitz type I block (second degree heart block), the PR interval progressively lengthens until a beat is missed. The cycle then repeats itself.
- In Mobitz type II block, the PR interval is constant but a beat is occasionally missed.
- Second degree heart block may also be manifested as 2:1 or 3:1 block, in which every second or third P wave elicits an R wave.
- In third degree (or complete) heart block there is complete dissociation of the QRS complex from the P wave.
- In atrial flutter the P waves have a very high frequency (> 290 b.p.m.) with a sawtooth pattern.
- In atrial fibrillation the baseline is irregular and the QRST pattern is highly abnormal.
- In ventricular tachycardia there is a regular but highly abnormal ECG pattern, as the point of excitation lies within the ventricles.
- In ventricular fibrillation the ECG shows a highly irregular pattern with no identifiable waves present.
- The ECG shows characteristic changes in electrolyte imbalance.
- In ischaemia, the ST segment lies below baseline. Immediately following an infarct, the ST segment is elevated above baseline.

Recommended reading

Physiology

Levick, J.R. (2010) *An introduction to cardiovascular physiology* (5th edn), Chapter 5. Hodder Arnold, London.

An eminently readable account of the principles of electrocardiography.

Pharmacology and clinical pharmacology of anti-arrhythmic drugs

Grahame-Smith, D.G., and Aronson, J.K. (2002) *Oxford textbook of clinical pharmacology* (3rd edn), Chapter 23, pp. 241–48. Oxford University Press, Oxford.

A straightforward account of the treatment of cardiac arrhythmias with a helpful Key Points table.

Rang, H.P., Dale, M.M., Ritter, J.M., and Flower, R. (2015) *Pharmacology* (8th edn), Chapter 21. Churchill-Livingstone, Edinburgh.

This chapter contains a detailed discussion of the pharmacology of anti-arrhythmics.

Clinical medicine

Hampton, J.R. (2013) *The ECG made easy* (8th edn). Churchill-Livingstone, Edinburgh.

A short book providing a helpful introductory account of the ECG and its changes in disease.

Morris, F., Brady, W.J., and Camm, J. (eds) (2008) *ABC of clinical electrocardiography* (2nd edn). Wiley-Blackwell, Oxford.

A basic introduction to the subject based on review articles previously published in the British Medical Journal.

Wagner, G.S., and Strauss, D.G., (2013) *Marriott's practical electrocardiography* (12th edn). Lippicott, Williams & Wilkins, Philadelphia.

A comprehensive account of the ECG and its changes in disease, well illustrated with clinical case studies.

Review articles

Dobrzynski, H., Boyett, M.R., and Anderson, R.H. (2007) New insights into pacemaker activity: Promoting understanding of sick sinus syndrome. *Circulation* **115**, 1921–32.

El-Sherif, N., and Turitto, G. (2011) Electrolyte disorders and arrhythmogenesis. *Cardiology Journal* **18**, 233–45.

Riera, A.R.P., Ferreira, C., Filho, C.F., Ferreira, M., Meneghini, A., Uchida, A.H., Schapachnik, E., Dubner, S., and Zhang, L. (2008) The enigmatic sixth wave of the electrocardiogram: The U wave. *Cardiology Journal* **15**, 408–21.

To check that you have mastered the key concepts presented in this chapter, complete the accompanying online self-assessment questions. Go to www.oup.com/uk/pocock5e/

CHAPTER 30

The circulation

This chapter should help you to understand:

- Pressure, resistance and flow in the circulation (haemodynamics)

- The significance of the systemic arterial blood pressure and its clinical measurement

- The intrinsic and extrinsic factors controlling the calibre of blood vessels

- The principle of autoregulation

- The role of the CNS in the control of the circulation

- Hypertension

- The specialized features of several regional circulations

- The formation and circulation of the cerebrospinal fluid

30.1 Introduction

The arrangement of the circulation and the physiology of the heart are discussed in Chapters 27 and 28, together with the relevant histology. This chapter is concerned with the detailed physiology of the systemic circulation and deals with the relationship between pressure and flow in the circulation (haemodynamics) and the factors that regulate blood flow through particular tissue beds. This is followed by a brief discussion of hypertension (high blood pressure). The specific features of several regional circulations are also discussed, namely that of the heart (the coronary circulation), the skin (the cutaneous circulation), the skeletal muscles, and the brain (the cerebral circulation). The specific adaptations of the blood supply to the respiratory, renal, and gastrointestinal systems are discussed in subsequent chapters along with their physiology.

30.2 Pressure and flow in the circulation

The flow of blood through any part of the circulation is driven by the difference in pressure between the arteries that supply the region in question and the veins that drain it. This pressure difference is known as the **perfusion pressure**. The resistance offered by the blood vessels to the flow of blood is known as the **vascular resistance**. Mathematically the relationship between perfusion pressure, blood flow, and vascular resistance is described by the following simple equations:

$$\text{perfusion pressure} = \text{arterial pressure} - \text{venous pressure}$$

and

$$\text{blood flow} = \frac{\text{perfusion pressure}}{\text{vascular resistance}}$$

Thus, blood flow will increase if the perfusion pressure is increased or if the vascular resistance is decreased. Conversely, if perfusion pressure falls, or vascular resistance rises, the blood flow will fall. This relationship can be applied to an individual vascular bed as well as to the entire systemic or pulmonary circulations (see also Box 30.1).

For the systemic circulation as a whole, the driving force for blood flow is the difference between the arterial pressure and the pressure in the right atrium, which is called the **central venous pressure**. Since the blood is pumped into the arteries intermittently, the pressure in the arterial system varies with the cardiac cycle. The pressure at the peak of ejection is called the **systolic pressure**, while at its lowest point, during ventricular relaxation, it is known as the **diastolic pressure**.

The cardiac output represents the volume of blood flowing around the systemic circulation each minute (ml min^{-1}), while the

1₂3 Box 30.1 Darcy's Law, Poiseuille's Law, and blood flow

Darcy's Law quantitatively relates the flow of a liquid through a channel to the driving pressure. Darcy found that the flow of liquid is directly proportional to the pressure difference between the inlet and outflow. In symbols:

$$\dot{Q} = K \times (P_i - P_o) \tag{1}$$

where $\dot{Q}$ is the flow rate; P_i and P_o are the pressures at the inlet and outflow respectively (the difference between the two pressures is the perfusion pressure); K is a constant called the hydraulic conductance. The reciprocal of the hydraulic conductance ($1/K$) is the hydraulic resistance R, so equation 1 can also be written as:

$$\dot{Q} = \frac{(P_i - P_o)}{R} \tag{1a}$$

Applying this relationship to the circulation:

$$CO = \frac{(\bar{P}_a - CVP)}{TPR} \tag{2}$$

where CO is the cardiac output, $\bar{P}_a$ is the mean arterial pressure, CVP is the central venous pressure (usually close to zero), and TPR is the total peripheral resistance.

Poiseuille discovered that the resistance to flow of liquid through a straight rigid tube is directly proportional to the viscosity (the resistance to flow) of the liquid (η) and the length of the tube (l), but varies inversely with the fourth power of the tube radius (r):

$$R = \frac{8\eta l}{\pi r^4} \tag{3}$$

This is known as Poiseuille's Law. Combining equations 1 and 3 gives the following mathematical relationship:

$$\dot{Q} = (P_i - P_o)\frac{\pi r^4}{8\eta l} \tag{4}$$

Thus, for a given fluid (e.g. blood), the larger the diameter of a tube the higher the flow rate for a given pressure difference. As the flow depends on the fourth power of the radius, *doubling the diameter will increase the flow rate sixteenfold*. Conversely, halving the diameter will lead to a sixteen-fold reduction in flow rate. In addition, for a given pressure gradient, the longer the length of the tube, or the higher the viscosity of the blood, the smaller the flow rate.

Poiseuille's Law relates strictly to laminar flow (sometimes called streamlined flow), but if the pressure gradient along the tube is

increased, the flow will eventually become irregular and **turbulence** occurs. Once turbulence has occurred, proportionately more pressure is required to achieve a given increase in flow. The critical pressure at which flow ceases to be laminar is determined by a coefficient known as the **Reynolds number**. Studies have shown that turbulence is more likely to occur at high flow rates in wide tubes that have an irregular cross-section (as in a large branching blood vessel such as the aorta). The Reynolds number is not normally reached in healthy blood vessels except in the aorta during peak flow.

When blood flows through the circulation it passes from a few large vessels (the arteries) to a very large number of small vessels (the capillaries) via a series of vessels of progressively smaller diameter. When the hydraulic resistances are in series (i.e. when blood flows directly from one vessel to another of different diameter) the total resistance to flow is equal to the sum of the resistances of the individual vessels. So:

$$R_{total} = R_1 + R_2 + R_3 \ldots \text{etc.} \tag{5}$$

When the blood is distributed among many vessels arranged in parallel, the total hydraulic conductance is the sum (K_T) of the hydraulic conductances of all the vessels ($K_1, K_2, K_3 \ldots K_t$) through which it flows:

$$K_T = \sum_{n=1}^{t} K \tag{6}$$

This follows directly from Darcy's Law (see equation 1). However, as the conductance of any vessel is the reciprocal of the hydraulic resistance (i.e. to $1/R$), equation 6 may be rewritten as:

$$\frac{1}{R} = \frac{1}{R_1} + \frac{1}{R_2} + \frac{1}{R_3} \ldots \frac{1}{R_n} \tag{7}$$

If, for a network of 20 capillaries arranged in parallel, each has a resistance 10 times that of the arteriole that supplies them, their total resistance relative to the arteriole is:

$$1/R = 20 \times 0.1 = 2$$
$$\text{and } R = 0.5$$

i.e. one-half that of the arteriole supplying the capillary network. The increased area available for blood flow more than compensates for the increased resistance of the individual vessels.

velocity of blood flow is the rate at which blood flows through a vessel (cm s⁻¹). For a given flow rate, the velocity varies inversely with the cross-sectional area of the vessel through which it is flowing: the velocity of blood flow in the aorta and major arteries is much greater than that in either the capillaries or veins, both of which have a larger total cross-sectional area (see Figure 30.1 and Table 30.1).

The vascular resistance depends on the radius of the vessels and the viscosity of the blood

The resistance offered by a blood vessel to the flow of blood is described by **Poiseuille's Law** (see Box 30.1), which states that the flow of blood is proportional to the fourth power of the radius of the vessel and is inversely proportional to the viscosity. This means

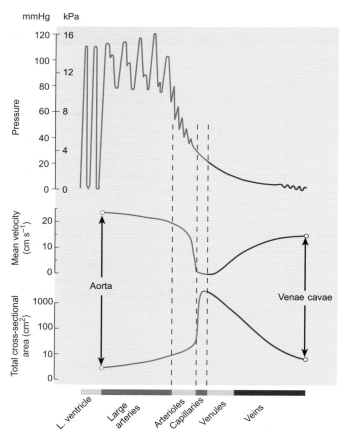

Figure 30.1 The changes in pressure and the velocity of blood flow in the various parts of the systemic circulation. Note that the greatest fall in pressure occurs as the blood traverses the arterioles, which are the main site of vascular resistance. (Based on Fig 1.4 in J.R. Levick (1995) *An introduction to cardiovascular physiology*, 2nd edition. Butterworth-Heinemann, Oxford.)

Table 30.1 The total cross-sectional area of the different types of blood vessel in the systemic and pulmonary circulations relative to that of the aorta

Vessels	Relative area
aorta	1
systemic arteries	14
arterioles	20
capillaries	485
venules	280
veins	225
venae cavae	1.1
pulmonary arteries	50
pulmonary capillaries	485
pulmonary venules and veins	75

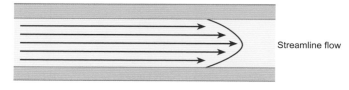

Streamline flow

Turbulent flow

Figure 30.2 A diagram illustrating the difference between streamline (or laminar) flow and turbulent flow. In this case, the turbulence occurs as the blood passes a distortion of the vessel wall such as an atheromatous plaque. The formation of eddies permits accumulation of material just downstream of the obstruction, exacerbating the obstruction.

that if, for example, the radius of a vessel is reduced by half, it will carry one-sixteenth (= $1/2^4$) the quantity of blood for the same pressure difference. In other words, after the radius of a blood vessel has been reduced by half, the resistance to blood flow increases sixteen-fold. If the blood viscosity doubled, the resistance to blood flow would also double.

The flow of blood is not uniform across the diameter of a blood vessel. The layer of fluid next to the vessel wall tends to adhere to it and the neighbouring layer tends to adhere to this static layer, and so on. Thus, the velocity of flow is fastest in the middle of the vessel and slowest next to the wall. When the different layers slip smoothly past each other, flow is said to be **laminar** (or streamline) and the profile of flow across the vessel is described by a smooth curve (see Figure 30.2). This is the situation that normally exists in the blood vessels, and Poiseuille's Law holds. If this smooth pattern of flow is broken up, for example by an irregularity in the vessel wall (such as an atheromatous plaque), eddies form and the flow is said to be turbulent. In most blood vessels, turbulence is undesirable as blood clots are more likely to form, but turbulence occurs naturally in the ventricles and in the aorta during peak flow. In both of these

situations, turbulence promotes the mixing of blood before its distribution to the systemic vessels. When the flow is turbulent, vibrations are set up which can be heard as sounds with a stethoscope. These are known as 'bruits' or murmurs and can be useful in the diagnosis of cardiovascular disease, as discussed in Chapter 28. Laminar flow is silent.

The apparent viscosity of blood falls in vessels of small diameter

In addition to the calibre of the smaller blood vessels, the resistance to flow is affected by the viscosity of the blood. When measured in a conventional viscometer, the apparent viscosity of blood is about 2.5 times that of water. In living tissues, however, the apparent viscosity of the blood is about half this value. This

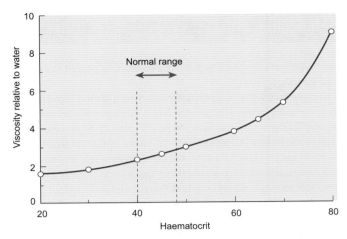

Figure 30.3 The effect of the haematocrit on the viscosity of blood relative to that of water. Note that the viscosity rises progressively with an increase in haematocrit. The increase in viscosity becomes progressively steeper as the haematocrit rises above about 60 per cent.

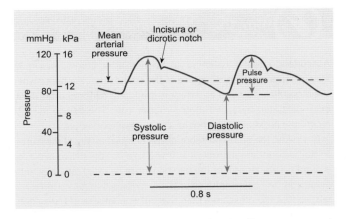

Figure 30.4 The arterial pressure wave showing the measurement of the systolic, diastolic, and pulse pressures. Note that the mean arterial pressure is nearer to diastolic pressure than to systolic pressure, reflecting the relative durations of systole and diastole.

anomalous behaviour is due to the tendency of red blood cells to flow along the central axis of the smaller blood vessels—a phenomenon known as **axial streaming**. While the mechanisms responsible for this behaviour are not fully understood, it appears that flexibility of the red cells is an important factor. At the low velocities found in the capillaries, rigid particles tend to remain uniformly distributed across the vessel while flexible particles migrate towards the central axis.

Since blood is largely made up of plasma and red cells, it is perhaps not surprising to find that its viscosity varies with the haematocrit. The higher the haematocrit, the greater the viscosity (Figure 30.3). The haematocrit can increase both in disease (e.g. in polycythaemia vera) and as a result of physiological adaptation to life at high altitude. The associated increase in viscosity has significant effects on the work needed to pump a given quantity of blood around the circulation and can lead to a persistent elevation of the blood pressure (hypertension) as well as an increase in the thickness of the left ventricle (left ventricular hypertrophy). Conversely, a fall in haematocrit (as a result of anaemia or haemorrhage) will lower the viscosity of the blood.

Blood flow and pressure in the systemic arteries

As Figure 30.1 shows, the systemic arterial pressure fluctuates during the cardiac cycle. It is at its maximum during systole and at its minimum during diastole. Its maximum value depends on the rate of ejection from the ventricles, the distension of the arterial walls and the rate at which it becomes distributed throughout the circulation. During systole, the pressure rises rapidly as the rate at which blood is being pumped into the arterial tree is greater than the rate at which it can be distributed. As a result, the pressure rises and the arterial walls become distended. As the left ventricle begins to relax, the flow of blood into the aorta declines and the pressure falls. When the pressure in the aorta exceeds that in the ventricles,

the aortic valve closes and this generates a small pressure wave known as the dicrotic wave, which is preceded by the **dicrotic notch** or **incisura** (see Figure 30.4). Following this, the pressure declines to its diastolic value before the next systole causes another rise.

The pressure wave in the arteries can be felt as a pulse at various parts of the body. This pressure wave distends the arterial walls, and as the pressure begins to decline at the end of systole, the stretched elastic tissue in arterial walls provides a further source of energy for the propulsion of the blood.

The shape and magnitude of the pulse wave changes as it passes from the aorta to the peripheral arteries. As Figure 30.1 shows, the peak pressure in the large arteries is somewhat higher than in the aorta. This increase in systolic pressure is believed to be due to the reflection of pressure waves from the distal arterial tree reinforcing the aortic pressure wave. Further along the arterial tree, the peak pressure progressively declines as the vessels become narrower.

How is blood pressure measured?

Although it is possible to measure the blood pressure in various parts of the arterial system by direct insertion of a cannula connected to a pressure transducer, this is normally performed in humans only during cardiac catheterization or major surgery. Pressure is much more often measured indirectly by **auscultation**. This method relies on the fact that turbulent blood flow creates sounds within the blood vessels that may be heard by means of a stethoscope, whereas streamline flow is silent. Further details of the auscultatory method for measuring blood pressure are given in Box 30.2. Another method for the indirect measurement of blood pressure is the Finapres. This device permits the continuous non-invasive measurement of systolic and diastolic pressure by monitoring the pressure that needs to be applied to a small artery to keep its diameter constant. It was originally a research instrument employed where fast and accurate measurements of blood pressure are required—for example while studying the effects of

Box 30.2 Measurement of blood pressure by auscultation

Auscultation means 'listening to'. So measuring blood pressure by auscultation means making use of the sounds that are heard when the blood flow through an artery is gradually restored after it has been occluded by an inflatable rubber cuff. The device used to record the pressures is known as a **sphygmomanometer**.

Normally, the blood pressure of the brachial artery is measured. Initially, an inflatable rubber cuff within a cotton sleeve is placed around the upper arm of the person whose blood pressure is to be measured. The cuff is inflated until the radial pulse can no longer be felt, so that the pressure within the cuff is in excess of the systolic pressure. A stethoscope is then positioned over the brachial artery at the antecubital fossa (the inside of the elbow). The precise position can be established by feeling the brachial pulse before inflating the cuff. The pressure in the cuff is then gradually lowered. Initially no sounds will be heard through the stethoscope as the pressure of the cuff occludes blood flow through the artery. When the pressure within the cuff is lowered, there comes a point where the pressure within the artery is just sufficient to overcome the pressure exerted by the cuff and at the peak of systole, there is a brief spurt of blood into the artery below the point at which it is occluded. This causes vibration of the vessel wall, which can be detected as a tapping sound through the stethoscope (**phase 1**). This sound is known as

the first **Korotkoff sound** and is conventionally accepted to represent systolic pressure.

Cuff pressure is then lowered further. As more and more blood passes through the artery, the sounds heard through the stethoscope become louder. As the diastolic pressure is approached, however, the artery remains open for almost all of the cardiac cycle and blood flow starts to become less turbulent and more streamlined. Streamlined flow creates less vibration and therefore less noise in the artery, so the Korotkoff sounds diminish in volume fairly abruptly as diastolic pressure is reached (**phase 4**). The pressure is allowed to fall still further until the sounds disappear (**phase 5**). By convention, the point at which complete silence occurs (phase 5) is taken as diastolic pressure. Between the systolic and diastolic pressures, the Korotkoff sounds may disappear (**phase 2**) and reappear (**phase 3**)—this is known as the auscultatory gap. It is therefore important not to mistake phase 2 for the diastolic pressure or phase 3 for the systolic pressure.

Normal systolic pressure measured in this way is usually less than 150 mm Hg in a healthy adult, and diastolic pressure should be less than 90 mm Hg. In young adults and children, the pressures tend to be lower. In elderly people, there tends to be an increase in systolic pressure without a proportionate increase in diastolic pressure (see Figure 30.18).

G-forces on military pilots—but commercial machines based on the same principle are now available for routine clinical use.

What is normal arterial blood pressure?

In a healthy young adult at rest, systolic pressure during the day is around 16 kPa (120 mm Hg) while diastolic pressure is around 10.7 kPa (80 mm Hg). This is normally written as 16/10.7 kPa and is more usually given in traditional units as 120/80 mm Hg. The difference between the systolic and diastolic pressures (normally about 5.3 kPa or 40 mm Hg) is called the **pulse pressure**—see Figure 30.4.

Although the figure of 16/10.7 kPa (120/80 mm Hg) is a useful one to remember, it is also important to realize that blood pressure constantly fluctuates according to the normal activities of the day (see Figure 30.5a). Moreover, there is a distinct circadian rhythm—blood pressure is highest between about 8 a.m. and 7 p.m. and falls during the night, with the lowest point occurring around 3 a.m. (Figure 30.5b). In addition, systolic blood pressure tends to increase with age so that by the age of 70, blood pressure averages 24/12 kPa (180/90 mm Hg). This increase in arterial pressure is due to a reduction in the elasticity of the arteries (**arteriosclerosis** or hardening of the arteries). Consistently high blood pressure (diastolic pressure >13 kPa or 100 mm Hg) is known as **hypertension** and is very common. The vascular complications associated with hypertension include stroke, heart disease, and chronic renal failure. For this reason, regular screening is essential to avoid serious organ damage. For a more detailed discussion of hypertension and its causes, see Section 30.5.

The **mean arterial pressure** (MAP) is a time-weighted average of the arterial pressure over the whole cardiac cycle. It is not a simple arithmetic average of the diastolic and systolic pressures because the arterial blood spends relatively longer near the diastolic pressure than the systolic (see Figure 30.4). For most working purposes, however, an approximation to mean arterial pressure can be obtained by applying the following equation:

mean arterial pressure = diastolic pressure + 1/3 (pulse pressure)

Using the pressure in the brachial artery as an example, since this is normally the pressure measured in clinical practice, if the systolic pressure is 14.7 kPa (110 mm Hg) and the diastolic pressure is 10.7 kPa (80 mm Hg):

$$\text{mean arterial pressure} = 10.7 + 1/3\,(14.7 - 10.7)\,\text{kPa}$$
$$= 12\,\text{kPa}\,(90\,\text{mm Hg})$$

The central venous pressure is close to zero and does not alter significantly during the cardiac cycle, so it does not need to be averaged. The perfusion pressure of the systemic circulation is thus very nearly equal to the mean arterial pressure.

The flow through the circulation is the cardiac output so that the relationship between the mean blood pressure, the cardiac output, and the **total peripheral resistance** (or TPR) is given by the following equation (see Box 30.1):

$$\text{mean blood pressure} = \text{cardiac output} \times \text{TPR}$$

The total peripheral resistance is the sum of all the vascular resistances within the systemic circulation. It is determined by the

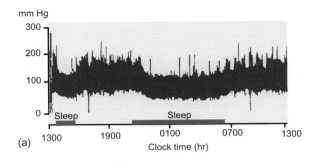

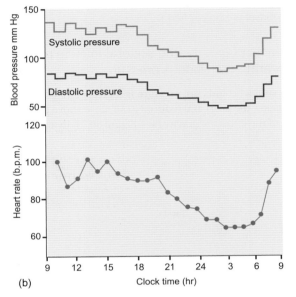

Figure 30.5 The changes in blood pressure seen during a 24- hour period. (a) A continuous measurement of blood pressure in a single individual who had an arterial cannula inserted in their left brachial artery. The initial values on the extreme left of the record were calibration signals. The two sharp downward deflections at 4 hours and 20 hours after the start of the recording were caused by flushing the cannula line and checking the zero position on the measuring transducer. The blood pressure fell during the two periods of sleep and showed considerable minute-to-minute variation during the waking day, reflecting the activities of the subject. (b) The hourly average arterial blood pressure of five subjects measured with implanted arterial catheters over a 48-hour period. Note the fall in blood pressure at night and the early morning rise, which was consistent across all subjects. The changes in heart rate parallel the changes in blood pressure. The two sets of data were from separate studies. (Based on data in M.W. Millar-Craig et al. (1978) *Lancet* **1**, 795–7).

viscosity of the blood and by the total cross-sectional area of those vessels that are being perfused.

Short-term increases in arterial blood pressure can be brought about by pressor stimuli such as pain, fear, anger, and sexual arousal. Conversely, pressure falls significantly during sleep, sometimes to as little as 9.3/5.3 kPa (70/40 mm Hg—see Figure 30.5), and, to a much lesser and more gradual extent, during normal pregnancy. Gravity also affects blood pressure. On rising from a lying to a standing position, there is a transient fall in blood pressure followed by a small reflex rise.

Summary

Blood flows through the systemic circulation from the aorta to the veins because the pressure in the aorta and other arteries (the arterial blood pressure) is higher than the venous pressure. The arterial blood pressure is determined by both the cardiac output and the total peripheral resistance.

Blood flow in the arteries is pulsatile. At the peak of ejection, the pressure (the systolic pressure) is at its maximum while the lowest pressure occurs at the end of diastole (the diastolic pressure). The difference between the systolic and diastolic pressures is called the pulse pressure. In a healthy young adult, the arterial blood pressure at rest will be around 120/80 mmHg (16/11 kPa).

The arterioles are the main source of vascular resistance

Since the flow is the same throughout a given vascular bed, the greatest fall in pressure must occur in the region of greatest resistance. Measurement of the pressures in the different kinds of blood vessel show that the largest fall in pressure in the systemic circulation occurs as the blood passes through the arterioles (see Figure 30.1), showing that they are the principal sites of vascular resistance.

The majority of arterioles are in a state of tonic constriction due to the activity of the sympathetic nerves that supply them. As a result, their effective cross-sectional area is much less than the total cross-sectional area they would offer if they were fully dilated. The resistance of the systemic circulation (i.e. the total peripheral resistance) in humans is around 2.6 Pa ml^{-1} min^{-1} (0.02 mm Hg ml^{-1} min^{-1}). The total resistance of the pulmonary circulation is much lower, around 0.4 Pa ml^{-1} min^{-1} (0.003 mm Hg ml^{-1} min^{-1}), which is why a lower pressure is needed to drive the cardiac output through the lungs (see Chapter 25).

Since the resistance of a vessel is dependent on the fourth power of its radius (see Poiseuille's Law, Box 30.1), major changes in blood flow to a particular region can be achieved by modest adjustment of the calibre of the arterioles. This adaptation is important in regulating the redistribution of the cardiac output between the various vascular beds during exercise (see Chapter 38). The mechanisms by which this regulation is achieved are discussed in Section 30.3.

The capillary pressure

On the basis of Poiseuille's Law, one might expect that, since the capillaries have the smallest diameter, they would be the principal site of vascular resistance. However, the overall resistance to blood flow depends both on the diameter of the vessels and on the total cross-sectional area available for the passage of the blood. The cross-sectional area offered by the capillaries is about 25 times that

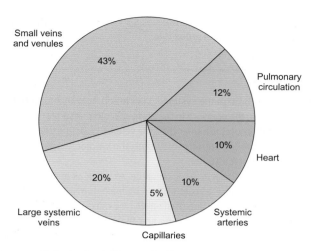

Figure 30.6 A pie chart diagram to illustrate the approximate distribution of the blood between the different parts of the circulation of a man at rest. Note the very high proportion of the blood in the systemic veins (c. 60 per cent).

of the arterioles (see Table 30.1). Since the capillaries have no smooth muscle in their walls, they cannot be constricted. Consequently, they offer relatively little resistance to blood flow despite their small diameter. The blood flow in the capillaries is normally steady, not pulsatile (but fluctuations in capillary blood flow do occur as a result of the vasomotion of the arterioles—see Section 30.4). The pressure at the arteriolar end of the capillaries is about 4.3 kPa (32 mm Hg). This falls to 1.5–2.7 kPa (12–20 mm Hg) by the time the blood has reached the venous end of the capillary bed. The small pressure at the venous end of the capillaries is sufficient to drive the blood back to the heart, because the veins offer comparatively little resistance to blood flow unless they are collapsed.

The veins provide a reservoir of blood

The blood volume of a normal adult is about 5 litres, but the distribution of this blood is not even throughout the circulation (see Figure 30.6). The heart and lungs each contain about 500 ml of blood and the systemic arteries account for a further 600 ml, while the capillaries have still less (about 250 ml). The bulk of the blood (about 3–3.5 litres) is found in the veins. The veins, particularly the large veins, act as a reservoir for blood. For this reason they are sometimes called **capacitance vessels**. As the walls of the veins are relatively thin and possess little elastic tissue (see Figure 20.6 and Table 27.1), blood returning to the heart can pool in the veins simply by distending them.

The degree of venous pooling is regulated by the tone of the smooth muscle in the walls of the veins. This is known as **venomotor tone**. It is determined by the activity of the sympathetic nerves supplying the veins and by local humoral factors. During periods of activity when the cardiac output is high, venomotor tone is increased and the diameter of the veins is correspondingly reduced. Consequently, blood stored in the large veins is mobilized for distribution to the exercising muscles and the flow of blood returning to the heart (i.e. the venous return) is increased.

Venous pressure

Although veins hold much of the blood, the average venous pressure measured at the level of the heart is only around 0.27 kPa (2 mm Hg) compared to the average arterial pressure of about 10.3 kPa (100 mm Hg). Venous pressure is highest in the venules: blood enters them at a pressure of about 1.6–2.7 kPa (12–20 mm Hg) and falls to about 1 kPa (8 mm Hg) by the time the blood reaches larger veins such as the femoral vein. This pressure head is sufficient to drive the blood into the central veins and thence into the right side of the heart, where pressure is essentially zero (i.e. it is equal to that of the atmosphere).

The effects of gravity, the skeletal muscle pump, and breathing on venous pressure

When a subject stands up, pressure is increased in all the blood vessels below the heart and reduced in all those above the heart as a result of the effects of gravity (see Figure 30.7). In an adult, the

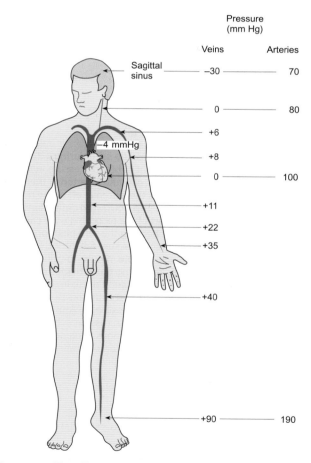

Figure 30.7 The effect of hydrostatic pressure on the venous and arterial pressures in an adult male who is standing quietly. The numerical values are approximate and will depend on the height of the individual. (Based on Fig 19.14 in A.C. Guyton (1986) *Textbook of medical physiology*, 7th edition. W.B. Saunders & Co, Philadelphia.)

pressure in the veins of the foot increases by about 12 kPa (90 mm Hg) on standing. Consequently, the veins in the lower limbs become distended and they accumulate blood (an effect sometimes referred to as **venous pooling**). The additional blood comes mostly from the intrathoracic compartment, so the central venous pressure falls. By the Frank–Starling mechanism (see Chapter 28), stroke volume falls and there is a transient arterial hypotension known as **postural hypotension**. This is rapidly corrected by the baroreceptor reflex (see Section 30.5). Since gravity also affects the arterial pressure in an exactly similar manner, the difference in pressure between the arteries and veins (the perfusion pressure) does not change significantly.

When a skeletal muscle contracts, it compresses the veins within it. Since the limb veins contain valves that prevent the backward flow of blood, the compression of the veins forces the blood towards the heart. This is known as the skeletal muscle pump. During exercise, the central venous pressure may rise slightly due to this effect. As Figure 30.8 shows, the squeezing action of the muscles on the veins leads to a progressive decline in venous pressure measured at the level of the foot. Once exercise ends, venous pressure begins to rise once more. When the muscle pumps are less active, as in a bedridden subject, blood tends to accumulate in the veins and there is a risk of a deep vein thrombosis (see Box 25.1). A similar situation occurs during prolonged periods of standing or sitting. In all three situations there is an increase in the peripheral venous pressure and reduced venous return to the heart. Cardiac output falls as a result.

Venous return is also influenced by respiration. During inspiration, the fall in intrathoracic pressure expands the intrathoracic veins and lowers central venous pressure. In addition, there is an increase in the pressure in the abdominal veins caused by the compression of the abdominal contents. These two factors tend to favour the movement of blood from the abdomen to the thorax. The situation is reversed during expiration. As a result, right ventricular stroke volume increases during inspiration and falls during expiration. Left ventricular stroke volume, on the other hand, falls during inspiration and rises during expiration. Over the course of one complete respiratory cycle, the outputs of the two sides of the heart are equalized. The various factors influencing the venous return are summarized in Table 28.1.

Summary

The total peripheral resistance is determined by the total cross-sectional area offered by the arterioles to blood flow. Hence, the chief determinant of blood flow to a particular capillary bed is the calibre of the arterioles that supply it, as the capillaries themselves offer little resistance to blood flow.

Average venous pressure at the level of the heart is only around 0.3 kPa (2 mmHg), falling to around zero in the right atrium (the central venous pressure). It is the difference in pressure between the arteries and the major veins that drives the blood around the circulation. The veins contain around two-thirds of the total blood volume. Since this blood can be mobilized when required, the veins are called capacitance vessels.

Respiration, gravity, and the pumping action of the skeletal muscles can all influence venous return and the central venous pressure.

30.3 The mechanisms that control the calibre of blood vessels

The smooth muscle of all blood vessels exhibits a degree of resting tension known as 'tone'. Changes in vascular tone alter the calibre of the blood vessels and so alter vascular resistance. If the tone is increased (i.e. if the smooth muscle contracts further), **vasoconstriction** occurs and vascular resistance increases. If tone decreases, there is **vasodilatation** and a fall in vascular resistance. The level of resting or basal tone varies between vascular beds. In areas where it is important to be able to increase blood flow substantially, such as skeletal muscle, basal tone is high, while in the large veins basal tone is much lower.

The tone of a blood vessel is controlled by a variety of factors. These fall into two broad categories.

1. **Intrinsic (or local) control** of blood vessels is brought about by the response of the smooth muscle to stretch, temperature, and locally released chemical factors.

2. **Extrinsic control** is exerted by the autonomic nervous system and by circulating hormones.

The major arteries (except the aorta) and the larger veins are mainly under extrinsic control, while the arterioles and small veins are subject to both local and extrinsic control. As the capillaries and post-capillary venules have no smooth muscle, their diameter is not regulated.

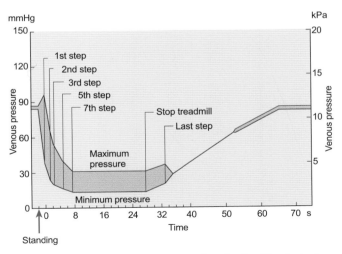

Figure 30.8 The pressure changes in a dorsal vein of the foot when a subject is first standing and then begins to walk on a treadmill. During walking, the active muscles help to 'pump' the blood towards the heart. As a result, venous pressure falls and stabilizes at a lower level where it remains until exercise ceases. The pressure then progressively rises towards its original level. (Based on Fig 7 in A.E. Pollack and E.H. Wood (1949) *Journal of Applied Physiology* 1, 649–62.)

Local control of blood vessels

Autoregulation stabilizes blood flow when perfusion pressure changes

A maintained change in blood pressure has remarkably little effect on the flow of blood through a skeletal muscle. However, when the pressure is first changed, the blood flow changes in the same direction. Thus, if the pressure is raised, blood flow increases at first but then returns close to its original level. Equally, if pressure falls, blood flow initially declines before returning to its previous level (Figure 30.9). A similar stability of blood flow in the face of changes in perfusion pressure is seen in many other tissues. This relative stability of blood flow is known as **autoregulation**. It occurs independently of the nervous system and is the result of direct changes in vascular tone.

The mechanism responsible for this effect is as follows: a rise in pressure within a vessel causes it to distend slightly so that the smooth muscle of the vessel wall is stretched. The muscle responds by contracting (the myogenic response). This narrows the vessels, increases their resistance, and restores blood flow to its previous level. If the pressure falls, the smooth muscle relaxes and the vessels dilate, so restoring blood flow. The myogenic response to stretching depends on the activation of specific mechanoreceptor channels that increase the membrane permeability to sodium and potassium ions. This leads to a depolarization and an increase in action potential frequency of the smooth muscle fibres, which is followed by their contraction. It is likely that chemical factors also play a part in autoregulation.

Although autoregulation is seen in most vascular beds (but not the lungs—see Chapter 34), blood flow in specific organs will vary with the physiological requirements. Indeed, changes in sympathetic drive and metabolic rate frequently act to reset the autoregulatory mechanism so that it will operate at a new level.

Vasodilatation occurs in response to a variety of metabolic by-products

Metabolism within cells gives rise to a number of chemical by-products. Many of these cause relaxation of the vascular smooth muscle and, therefore, vasodilatation. This has the effect of increasing the flow of blood through the vascular bed. The phenomenon is known as **functional hyperaemia** (also called **metabolic** or **active hyperaemia**). It has the important effect of facilitating the removal of waste products from the vicinity of the actively metabolizing cells. It is particularly significant in tissues such as exercising muscle, in the myocardium, and in the brain. Metabolites that are known to induce vasodilatation include carbon dioxide, lactic acid, potassium ions, and the breakdown products of ATP (adenosine and inorganic phosphate). Local tissue hypoxia can also bring about relaxation of the vascular smooth muscle, although this is not the case in the pulmonary circulation (see Chapter 34 p. 529).

If the artery supplying blood to a tissue is compressed, blood flow is interrupted and the tissue becomes **ischaemic**. When the

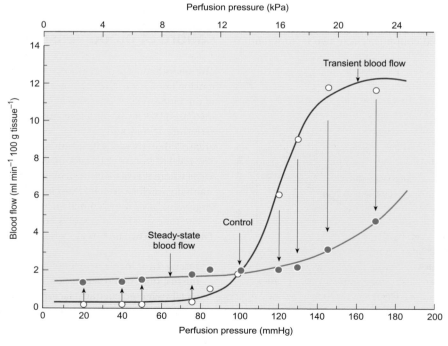

Figure 30.9 The autoregulation of blood flow in an isolated, perfused skeletal muscle of the dog. The open circles represent the blood flow measured immediately after the perfusion pressure had been quickly raised or lowered from the control level (100 mm Hg or 13.3 kPa). Immediately after the perfusion pressure was altered, there was a transient rise or fall in blood flow depending on whether the pressure had been increased or decreased (the blue line). The autoregulatory mechanisms then restored blood flow to levels close to control values within a few seconds (here shown by the red symbols). (Adapted from Fig 2 in R.D. Jones and R.M. Berne (1964) *Circulation Research* **14**, 126–38.)

of the secretion is adrenaline. Both adrenaline and noradrenaline act on **adrenoceptors**, which are of two main types known as α- and β-adrenoceptors (see Chapters 6 and 11). Interaction of the catecholamines with the α-adrenoceptors leads to vasoconstriction, while their interaction with β-adrenoceptors brings about vasodilatation.

Noradrenaline has a much greater affinity for α- than for β-adrenoceptors and will therefore normally cause vasoconstriction. Adrenaline interacts with the β-adrenoceptors present in the vascular smooth muscle of skeletal muscle, the heart, and the liver to produce vasodilatation. For this reason, sympathetic stimulation of the kind seen during the alerting reaction allows the blood vessels of the heart and skeletal muscle to dilate, thereby increasing the blood flow to these tissues, while vessels elsewhere constrict. Blood is thus diverted preferentially to those tissues with an important role to play in the alerting response and any subsequent exercise (i.e. it prepares the body for 'fight or flight').

Vasopressin (anti-diuretic hormone or ADH)

Although vasopressin (ADH) is chiefly concerned with the regulation of fluid excretion by the kidneys, it also exerts powerful effects on the vasculature. It is secreted into the circulation from the posterior pituitary gland in response to the fall in blood pressure that follows a substantial haemorrhage (see Chapter 40). At such times, vasopressin causes a powerful vasoconstriction in many tissues, which helps to maintain arterial blood pressure (hence its name). In the cerebral and coronary vessels, however, vasopressin seems to elicit vasodilatation. The net effect is a redistribution of the blood to the essential organs—the heart and brain. Further details of the secretion and actions of this hormone can be found in Chapters 21 and 39.

The renin–angiotensin–aldosterone system

Renin is a proteolytic enzyme that is secreted by the kidneys in response to a fall in the sodium concentration within the distal tubule. It acts on an inactive peptide in the blood called angiotensinogen, to form angiotensin I, which is then converted in the lungs to its active form, angiotensin II. This hormone has two important actions: it stimulates the secretion of aldosterone from the adrenal cortex, and it causes vasoconstriction. The former will result in the reabsorption of an increased amount of salt and water by the distal tubule, which may be particularly important when the blood volume is low—following a haemorrhage for example. The latter brings about a rise in arterial blood pressure. Further details of this important hormonal system may be found in Chapters 39 and 40.

Atrial natriuretic peptide (ANP)

This hormone (also called atrial natriuretic factor or ANF) is secreted by the atrial myocytes in response to high cardiac filling pressures. Its action is the opposite to that of aldosterone in that it stimulates the excretion of salt and water by the renal tubules. It also has a weak vasodilator action on the resistance vessels. The role of this hormone in the regulation of body fluid volume is discussed in Chapter 40.

Summary

Nerves and hormones exert extrinsic control over the heart and circulation. In most tissues, the calibre of blood vessels is regulated by sympathetic vasoconstrictor fibres. Postganglionic sympathetic vasoconstrictor fibres secrete noradrenaline, which interacts with α-adrenoceptors of vascular smooth muscle to cause vasoconstriction. Widespread activation of α-adrenoceptors leads to an increase in arterial blood pressure. Some arterioles, such as those of muscle, liver, and heart, also possess β-adrenoceptors. Interaction of noradrenaline with these receptors causes vasodilatation. A number of hormones, notably adrenaline and ADH, also influence vascular tone.

30.4 The role of the central nervous system in the control of the circulation

The activity of the heart and the tone of the blood vessels are both regulated by the autonomic nervous system, which is, in turn, subject to control by the higher centres of the brain. The principal regions of the brain that are concerned with the control of the cardiovascular system are the hypothalamus and the medulla oblongata. The areas of the medulla concerned with control of the cardiovascular system are sometimes referred to as cardiac control 'centres', although this term is somewhat misleading since the medullary neurons are themselves regulated by higher brain areas and by afferent input from pressure receptors. Figure 30.10 shows the basic organization of those areas that are concerned with the control of the cardiovascular system.

Afferent information concerning the pressures within the circulation is supplied by two groups of receptors: high pressure receptors (**baroreceptors**), which are located in the walls of the aorta, and carotid arteries and low pressure receptors (or **volume receptors**), which are located in the walls of the atria and ventricles. The afferent fibres of these receptors relay information to the brainstem regarding arterial pressure and cardiac filling pressure. This information, coupled with other inputs from higher brain areas, from the chemoreceptors, and from other sensors, is used to initiate cardiovascular responses via the appropriate parasympathetic and sympathetic nervous pathways.

Stimulation of the arterial baroreceptors elicits a reflex vasodilatation and a fall in heart rate

The arterial blood pressure is very closely regulated by the body. In young adults, systolic pressure is maintained close to 16 kPa (120 mm Hg) during the day, although it falls somewhat during sleep (p. 464). There are two types of regulatory mechanisms: these are the rapid regulation by nerves and hormones, and the longer-term control of blood volume, which is largely mediated by the kidneys (see Chapters 39 and 40).

The pressure in the arterial circulation is monitored by baroceptors, which are most abundant in the walls of the carotid sinuses and the aortic arch. The baroreceptors are mechanoreceptors that sense the degree of stretch of the walls of the vessels in which they are located and are thus able to monitor blood pressure. Impulses from the baroreceptors are transmitted to the brainstem, where they terminate in the nucleus tractus solitarius (NTS—see Figure 30.10). Afferents from the aortic arch baroreceptors travel in the vagus nerves, while those from the carotid sinus receptors are carried by the carotid sinus nerves, which merge with the glossopharyngeal nerves. The afferent nerve supply of the baroreceptor system is illustrated diagrammatically in Figure 30.11.

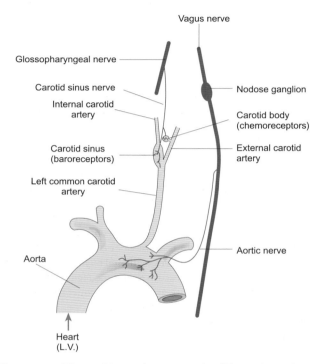

Figure 30.11 The position and nerve supply of the aortic and carotid baroreceptors.

The response of the baroreceptors to a changing arterial pressure is illustrated in Figure 30.12. As the arterial pressure wave rises, so does the discharge frequency of the baroreceptors, and it falls as the pressure declines. Consequently, the frequency of discharge in baroreceptor afferents varies in phase with the arterial pulse wave. Like other mechanoreceptors, the baroreceptors are dynamic receptors that respond best to changes in pressure. At normal levels of blood pressure (MAP = 90–100 mm Hg) the baroreceptor afferents show a modest degree of activity that is in phase with the arterial pulse wave, but below about 8 kPa (60 mm Hg) they exhibit little activity. As pressure rises over the range between 8 and 24 kPa (60 and 180 mm Hg), there is a steep increase in the rate of discharge (see Figure 30.13).

A sharp rise in arterial blood pressure will increase the discharge of the baroreceptor afferents. This afferent information is relayed to the brainstem, where it elicits an inhibition of the sympathetic activity to the heart and vasculature and an increase in vagal activity which slows the heart (bradycardia). The reduced activity of the sympathetic vasoconstrictor fibres results in vasodilatation, which in turn leads to a fall in peripheral resistance. The net result of these effects is a fall in arterial blood pressure that rapidly 'buffers' the initial rise (see Figure 30.14). Conversely, a fall in arterial blood pressure (acute hypotension) brings about a reduction in baroreceptor discharge which increases the activity of the sympathetic nerves and elicits a fall in the activity of the parasympathetic nerves. The heart rate increases and there is an increase in the force of contraction of the myocardium. The vessels supplying skeletal muscle, skin, kidneys, and splanchnic circulation constrict. As a result, blood pressure rises to counteract the initial fall. These changes in peripheral vascular tone and in the force and rate of cardiac contraction constitute the **baroreceptor reflex** or **baroreflex**, which is illustrated in Figure 30.14. The reflex vasomotor responses occurring in response to a change in blood pressure are due entirely to alterations in the frequency of discharge in sympathetic vasoconstrictor fibres. The sympathetic vasodilator fibres are not involved.

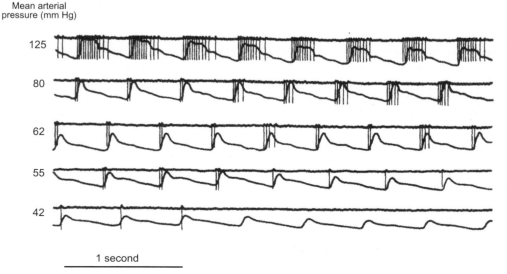

Figure 30.12 The discharge of carotid baroreceptors recorded from the carotid sinus nerve at different arterial pressures. The upper trace of each pair shows the discharge of the baroreceptors and the lower trace shows the pulse wave. Note that the discharge is greatest as the pressure rises towards it maximum value. (This is the dynamic response of the baroreceptors.) As the mean arterial pressure falls, the baroreceptors become less active, and when mean arterial pressure is reduced to 42 mm Hg their activity almost ceases. (From a baroreceptor recording by Professor E. Neil.)

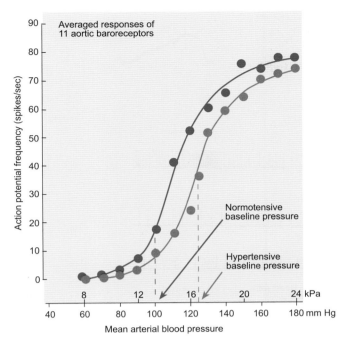

Figure 30.13 The activity of aortic baroreceptors recorded in the aortic nerve of the dog as a function of the mean aortic pressure. The blue (normotensive) curve and associated data points shows the relationship when pressure was abruptly raised or lowered from a value of 100 mm Hg (13.3 kPa). The red (hypertensive) curve and associated data points were recorded after the mean arterial pressure had been kept at 125 mm Hg (16.7 kPa) for 20 minutes. Note the rightward shift in the relationship indicating a reduced baroreceptor sensitivity. (Adapted from Fig 6 in H.M. Coleridge et al. (1984) *Journal of Physiology* **350**, 309–26).

Resetting of the baroreceptor reflex

As Figure 30.13 shows, if arterial blood pressure is raised for a period of 20 minutes or so, the threshold for baroreceptor activity rises to a higher value. This property makes the baroreceptors ineffective monitors of the absolute pressure of the blood passing to the brain—they are short-term regulators of blood pressure rather than long-term controllers.

The resetting of the baroreceptors occurs during exercise, for example. This ensures that the heart rate does not fall in response to the increase in systolic blood pressure and permits a high cardiac output to be maintained throughout a period of exercise. Similarly, during the alerting response in which the body is prepared for intense activity, the baroreceptor mechanism appears to be inhibited and there is a sharp rise in blood pressure.

The baroreceptor reflex stabilizes blood pressure following a change in posture

When a person moves quickly from a lying to a standing position, there is a significant fall in venous return to the heart (as a result of a shift of blood from the chest to veins in the legs). In turn, this leads to a fall in stroke volume (by Starling's Law), cardiac output,

and therefore blood pressure (see Figure 30.15). This is known as **postural hypotension**. The baroreceptors play an important role in the rapid restoration of normal pressure. When pressure falls, baroreceptor discharge is proportionately decreased and this results in an increase in sympathetic discharge to the heart and vasculature. Heart rate is increased and there is a rise in total peripheral resistance. Both these effects restore the blood pressure to its normal level. Because the baroreceptor reflex normally takes a few seconds to become fully effective, the blood supply to the brain is briefly reduced when standing up quickly and, as a result, it is quite common to experience momentary dizziness. This sequence of events is summarized in Figure 30.16.

The Valsalva manoeuvre

The **Valsalva manoeuvre** is essentially an attempt to expire against a closed glottis. It may be carried out voluntarily and is associated with forceful defecation, the lifting of heavy weights, and childbirth. The manoeuvre raises intrathoracic pressure which, in turn, elicits the complex cardiovascular response illustrated in Figure 30.17.

The initial rise in blood pressure is due to the normal contraction of the left ventricle augmented by the additional force of the raised intrathoracic pressure acting on the myocardium. This is followed by a transient fall in heart rate. At this stage, the increased intrathoracic pressure impedes the venous return with the result that cardiac output and mean arterial pressure fall. As the arterial pressure falls, the heart rate increases and this, together with an increase in total peripheral resistance, stabilizes the blood pressure. When the intrathoracic pressure falls following opening of the glottis, the blood pressure initially falls but as soon as the venous return is restored, end-diastolic volume and cardiac output increase and blood pressure rises. This rise is sensed by the baroreceptors, which cause a reflex bradycardia. This, together with a fall in peripheral resistance, restores the blood pressure to normal.

Low-pressure receptors in the atria sense central venous pressure

Stretch receptors are present in the walls of the right and left atria of the heart. The atrial receptors respond to the central venous pressure and to cardiac distension. They are stimulated when venous return is increased. This elicits a reflex rise in heart rate and contractility that is mediated via the sympathetic nervous system. Activation of this reflex rapidly reduces the initial cardiac distension. By contrast, activation of the mechanoreceptors in the left ventricle induces a reflex bradycardia and vasodilatation in response to a rise in end-diastolic pressure. The exact function of this response is unclear, but it may serve to assist the baroreceptor reflex in the regulation of arterial blood pressure.

Long-term regulation of blood pressure depends on hormonal control of blood volume

Although the baroreceptors are concerned with minute-to-minute regulation of the arterial blood pressure, the dynamic nature of their responses makes them unsuitable for long-term regulation of

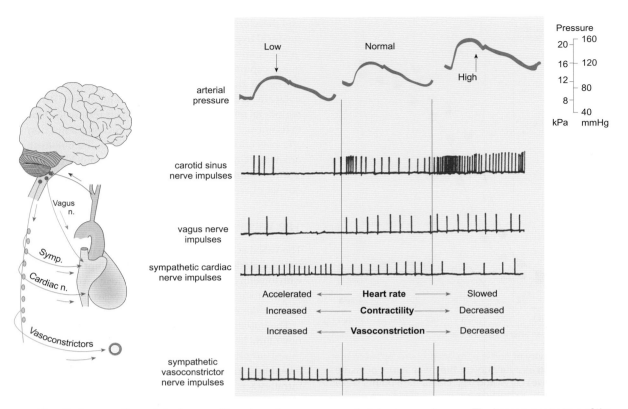

Figure 30.14 The discharge pattern of single nerve fibres in various nerves as arterial pressure changes. The basic organization of the nervous pathways is shown on the left. The top panel on the right shows the aortic pressure wave for low, normal, and elevated (high) arterial pressure. Below is shown the discharge pattern for a baroreceptor afferent in the carotid sinus nerve, the discharge pattern for a cardio-vagal nerve fibre, the discharge pattern for a sympathetic cardiac nerve fibre, and the pattern for a sympathetic vasoconstrictor nerve fibre (bottom trace). At low arterial pressure, vagal activity is inhibited while sympathetic activity is increased, resulting in an increased heart rate and vasoconstriction. At high pressures, vagal activity is increased while sympathetic activity is decreased, resulting in a slowing of the heart rate and vasodilatation. (Based on Fig 5.9 in R.F. Rushmer (1976) *Cardiovascular dynamics*. W.B. Saunders & Co, Philadelphia.)

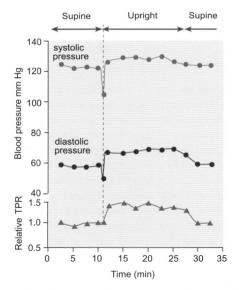

Figure 30.15 The changes in blood pressure and total peripheral resistance (TPR) that occur on rapidly moving from a supine to an upright position. Note the transient fall in pressure that is rapidly corrected by an increase in TPR. (Based on data in J.J. Smith et al. (1970) *J Appl Physiol* **29**, 133–7.)

absolute blood pressure. This depends on the maintenance of a normal extracellular volume by the kidneys. The chief mechanisms involved are hormonal and include the reflex control of pituitary ADH secretion by osmoreceptors in the hypothalamus, the operation of the renin–angiotensin–aldosterone system, and the secretion of atrial natriuretic peptide by the atrial myocytes. All these processes act to maintain a constant circulating volume. The underlying mechanisms are discussed more fully in Chapter 40.

Activation of the peripheral arterial chemoreceptors causes an increase in blood pressure

While the peripheral arterial chemoreceptors are chiefly concerned with the control of respiration (see Chapter 35), they also play a part in the reflex elevation of blood pressure seen during hypoxia. At normal blood gas tensions, they have little influence on the circulation, but during both hypoxia and hypercapnia they elicit a reflex vasoconstriction in the resistance vessels and in the large veins of the splanchnic circulation. The constriction of the arterioles and the mobilization of blood from the splanchnic circulation elevate the arterial blood pressure and increase the amount of oxygen carried to the tissues—particularly the brain. This

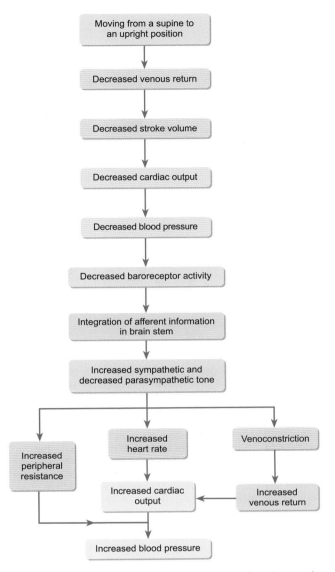

Figure 30.16 A flow chart showing the sequence of cardiovascular changes initiated by the baroreceptor reflex following a fall in arterial blood pressure and resulting in the restoration of normal arterial pressure.

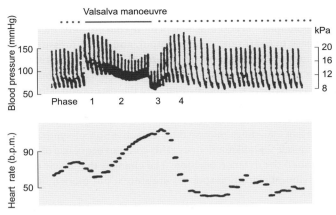

Figure 30.17 Changes in blood pressure and heart rate occurring during and after the Valsalva manoeuvre (expiration against a closed glottis). When the intrathoracic pressure first rises there is an increase in mean arterial pressure as venous return is initially favoured by compression of the great veins (phase 1). Cardiac output and pulse pressure rise. This is followed by a decline in mean arterial pressure and pulse pressure as venous return is impeded and cardiac output falls (phase 2). At the end of the manoeuvre the intrathoracic pressure falls, which briefly results in a drop in blood pressure (phase 3). This is followed by a sharp increase in venous return, leading to an increase in stroke volume and pulse pressure (phase 4). The changes in heart rate are a reflex response to the changes in pressure. When pressure rises, the heart rate falls, and vice versa.

chemoreceptor reflex is particularly powerful at low arterial blood pressures, when the baroreceptor reflex is relatively inactive.

Respiratory sinus arrhythmia

During inspiration there is a small increase in heart rate that is followed by a decrease in heart rate during expiration. This variation in heart rate with the respiratory cycle is called **respiratory sinus arrhythmia** and is almost entirely due to changes in the activity of cardio-vagal motoneurons, whose normal tonic activity is partly dependent on the arterial baroreceptor input (see Figure 30.10). During inspiration, the cardio-vagal motoneurons are inhibited by central inspiratory neurons and by an increased activity of the slowly adapting stretch receptors that are present within the larger airways of the lungs. Consequently, they become less sensitive to their baroreceptor inputs. This results in a decrease in vagal tone and an increase in heart rate. During expiration, this effect is reversed; the excitability of the cardio-vagal motoneurons returns to normal, vagal tone increases, and the heart slows.

Activation of 'work receptors' elicits an increase in cardiac output

The final group of receptors which can activate reflexes to influence the cardiovascular system are the 'work receptors' found in skeletal muscle. They are of two types: chemoreceptors that respond to potassium and hydrogen ions produced during exercise, and mechanoreceptors sensitive to the active tension generated within a muscle. Activation of these receptors during exercise causes an increase in heart rate, vasoconstriction of vessels other than those supplying the working muscles (notably those of the renal and splanchnic circulations), and increased myocardial contractility. All of these serve to raise blood pressure and increase perfusion of the active muscles.

The defence reaction

In response to a perceived danger, all mammals, including humans, show a behavioural alerting reaction that may lead to aggressive or defensive behaviour if the stimulus is strong enough. This is accompanied by marked changes in sympathetic activity in which normal reflex control is overridden. The pupils dilate, the skin hair becomes erect (piloerection), the teeth become bared and a defensive posture is adopted. In addition, and of more concern here, there is an increase in heart rate, cardiac output, and blood pressure. The cardiovascular responses involve resetting of the baroreceptor reflex to a higher level of pressure and there is a redistribution of blood flow: there is vasoconstriction in the viscera and skin, while the blood flow to the muscles is increased. In humans, this increase is due to an inhibition of sympathetic vasoconstrictor tone, while in other species there may also be an activation of sympathetic cholinergic vasodilator activity. This complex sequence of events is termed the defence reaction. The changes that occur during behavioural alerting responses are similar but more modest in scale.

Summary

Arterial blood pressure is monitored by baroreceptors present in the walls of the aortic arch and the carotid sinuses. A short-term rise in blood pressure elicits the baroreceptor reflex in which there is a reflex slowing of the heart, peripheral vasodilatation, and a fall in blood pressure. Arterial chemoreceptors and muscle work receptors also play a role in the short-term control of blood pressure. Long-term regulation of the blood pressure is achieved through the maintenance of normal extracellular volume, largely by means of the renin–angiotensin–aldosterone system and atrial natriuretic peptide (ANP).

30.5 Hypertension

Hypertension, or high blood pressure, is probably the most common of all cardiovascular disorders and accounts for a large number of deaths each year, mainly through vascular complications that lead to stroke, coronary heart disease, and chronic renal failure. The disease is characterized by an arterial blood pressure that is persistently higher than normal (generally taken to be in excess of 18.7 kPa (140 mm Hg) for systolic pressure and 12 kPa (90 mm Hg) for diastolic pressure in a young adult). It should be remembered, however, that the diastolic pressure is the chief determinant of mean blood pressure. Accordingly, the diastolic pressure is a particularly important indicator of hypertension. As Figure 30.18 shows, systolic blood pressure normally rises with age and an approximate value (in mm Hg) of 100 + age in years may be considered to be a 'normal' systolic blood pressure for adults. Nevertheless, this inexorable rise in arterial blood pressure is undesirable.

Although hypertension could be caused either by an increase in cardiac output or by an increase in total peripheral vascular resistance, in practice, cardiac output is generally relatively normal so that the raised blood pressure is due almost entirely to an increase in vascular resistance. The arterioles are particularly affected, with veins, capillaries, and pulmonary vessels usually remaining normal. Hypertension is commonly divided into the categories of primary (or essential) hypertension and secondary hypertension.

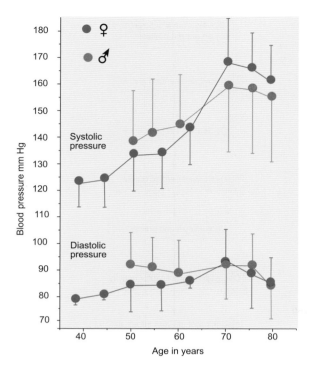

Figure 30.18 The changes in systolic and diastolic pressure with age. The data is from a large number of individuals who had not received any medication for the treatment of high blood pressure. The green symbols are data for women and the red symbols are data for men. Although there is a progressive rise in the mean value of both systolic and diastolic pressures with age, there is considerable variation between subjects of similar age—the vertical lines drawn from each data point indicate one standard deviation. (68 per cent of values will lie within plus or minus one standard deviation and 95 per cent of values will lie within plus or minus two standard deviations.) (Based on data in S. Landhal et al. (1986) *Hypertension* **8**, 1044–9.)

Primary hypertension

About 80 per cent of all patients with high blood pressure fall into this group. Primary hypertension is further subdivided into a number of categories that are classified according to the severity of the condition, as shown in Table 30.2. The most significant types according to the UK classification are:

- High normal or borderline hypertension, with diastolic blood pressures in the range 85–90 mm Hg (12–12.6 kPa) and a systolic pressure less than 139 mm Hg (18.5 kPa).

- Grade 1 hypertension, where diastolic pressures range from 90–99 mm Hg (12.0–13.3 kPa) with a systolic pressure between 140 and 159 mm Hg (18.6–21.2 kPa).

- Grade 2 hypertension, where diastolic pressures range from 100–109 mm Hg (13.3–15.2 kPa) with a systolic pressure between 160 and 179 mm Hg (21.3–23.8 kPa).

- Severe hypertension, where the diastolic pressure exceeds 110 mm Hg (15.2 kPa).

Table 30.2 Classification of hypertension

Blood pressure mm Hg		
Systolic	Diastolic	UK classification
<130	<85	Normal
<139	85–90	High normal
140–159	90–99	Grade 1 (mild hypertension)
160–179	100–109	Grade 2 (moderate hypertension)
>/ = 180	>110	Severe hypertension
140–159	<90	Isolated systolic grade 1 hypertension
>/ = 160	<90	Isolated systolic grade 2 hypertension

Blood pressure mm Hg		
Systolic	Diastolic	US Classification
<120	<80	Normal
120–139	Or 80–90	Pre-hypertensive
140–159	Or 90–99	Stage 1 hypertension
>160	Or >100	Stage 2 hypertension

Systolic pressures less than 130 mm Hg together with a normal diastolic pressure would be considered normal for a healthy young adult. A systolic pressure in excess of 160 mm Hg is considered severely hypertensive. The classification in the United States is broadly similar (see Table 30.2). Nevertheless, it is important to remember that one elevated blood pressure reading should not constitute the diagnosis of hypertension. Such a diagnosis should be based on repeated observations that give consistent readings. Indeed, the method of carrying out continuous 24-hour monitoring of blood pressure on an ambulatory patient is now considered to be of more value in determining the true extent of hypertension. Some people show an automatic elevation in blood pressure when they are confronted by a doctor ('white coat hypertension') while others do so in response to the blood pressure cuff itself ('cuff responders'). These patients are often found to have normal blood pressure when it is monitored over a 24-hour period.

In all cases of primary hypertension there is a chronically elevated arterial blood pressure without evidence of other disease. The causes of this type of hypertension remain unclear. The patients show no evidence of an increase in renin levels and possess active baroreceptor reflexes, although the sensitivity of these reflexes is often reduced. While the cause or causes of primary hypertension are largely unknown, several risk factors have been implicated as contributing to its development. These include advancing age, family history, and obesity. Other factors may also contribute to the development of high blood pressure. These include high salt intake, excessive alcohol consumption, stress, and the use of certain types of oral contraceptives.

Secondary hypertension

About 20 per cent of patients suffering from high blood pressure are classified as having secondary hypertension, i.e. hypertension resulting from another condition. Such conditions include pregnancy, endocrine disorders, renal disease, vascular disorders, and certain types of brain lesions.

Pregnancy-induced hypertension

About 10 per cent of all pregnancies are accompanied by hypertension. It normally occurs after the twentieth week of gestation and is often associated with the appearance of protein in the urine and with oedema. Women showing all of these symptoms are said to be suffering from **pre-eclampsia**. **Eclampsia** is an exaggerated form of pre-eclampsia which has progressed to include convulsions and sometimes coma (see Box 49.3). It has been difficult to define the causes of hypertension in pregnancy, but the elevated levels of oestrogens, progesterone, prolactin, and ADH may result in an increased vascular responsiveness to angiotensin II.

Endocrine causes of hypertension

A variety of different hormonal disturbances can result in a raised blood pressure. The most common (although still comparatively rare) disorders are phaeochromocytoma, Conn's syndrome, and Cushing's syndrome.

A phaeochromocytoma is a tumour of the adrenal medulla which, like normal chromaffin tissue, produces the catecholamines adrenaline and noradrenaline. Hypertension results from the large circulating quantities of these hormones. In many cases, release of the catecholamines occurs in spurts rather than continuously, giving rise to episodes of extreme hypertension, tachycardia, sweating, and anxiety. Diagnosis requires monitoring the excretion of vanilyl mandelic acid (VMA)—a catecholamine metabolite—in the urine over 24 hours, or the measurement of plasma catecholamine levels. Treatment is usually by surgical removal of the adrenal tumour.

Elevated levels of adrenocorticosteroid hormones can also produce secondary hypertension. Excess production of aldosterone by a tumour of the adrenal cortex (Conn's syndrome) leads to the excessive reabsorption of sodium in exchange for potassium in the distal tubules of the kidney. Water and chloride ions accompany the reabsorbed sodium, and the extracellular fluid volume is expanded. In turn, this leads to an expanded plasma volume and an increase in arterial blood pressure.

Excessive secretion of glucocorticoid hormones from the adrenal cortex (Cushing's syndrome) can also lead to hypertension since high concentrations of these hormones have aldosterone-like effects and can promote the retention of salt and water.

Renal causes of hypertension

The kidneys play an important role in the long-term regulation of arterial blood pressure by regulating the volume of the extracellular fluid (see Chapter 40). It is not surprising, therefore, that renal disease accounts for more cases of secondary hypertension than any other disorder. Any condition that restricts the blood supply to the kidney can produce a rise in blood pressure. A reduction in renal blood flow activates the renin–angiotensin system which, in turn, leads to the retention of sodium and water. As a result, the extracellular volume is expanded and hypertension results.

Vascular causes of hypertension

Vascular pathology tends to produce or perpetuate high blood pressure. Arteriosclerosis can bring about narrowing of the vessels, which leads to an increase in total peripheral resistance and a rise in arterial pressure. Coarctation of the aorta (narrowing of the aorta as it leaves the heart) can also give rise to an increase in systolic blood pressure and blood flow, particularly to the upper part of the body. In polycythaemia (a condition in which the red blood cell count is elevated), blood pressure may rise because of the increase in blood viscosity.

Neural causes of hypertension

While hypertension can be caused by some abnormality in the parts of the brain responsible for the control of blood pressure, most cases of neurogenic hypertension are relatively short-lived responses to a reduction in cerebral blood flow (Cushing's reflex). A raised intracranial pressure will tend to compress the vessels supplying the brain and will result in a reduction in cerebral perfusion because the brain is located within the rigid confines of the skull with no room for expansion. In response, there is a rise in arterial pressure brought about by a reflex vasoconstriction in the peripheral vessels, which acts to restore cerebral blood flow. If, however, the blood supply to the brainstem becomes inadequate, then vasoconstrictor tone will be lost and blood pressure will fall.

The effects of high blood pressure

The complications and mortality associated with both essential and secondary hypertension can be explained in terms of increased stresses ('wear and tear') on the heart, blood vessels, and organs—see Figure 30.19.

Cardiac changes in hypertension

When the arterial blood pressure increases, the heart must work harder to eject the same stroke volume (see Chapter 28). As a result of the Starling mechanism, the increase in left ventricular stroke work brings about a rise in left ventricular end-diastolic pressure. At the same time, the heart muscle hypertrophies to compensate for its increased work load. The effect of this is to shift the ventricular function curve upwards so that the ventricle is able to achieve a greater stroke work at the normal filling pressure. As a result of these changes, the increase in ventricular end-diastolic filling pressure which would arise from the operation of the Starling mechanism is minimized. Nevertheless, the hypertrophy of the cardiac muscle fibres that occurs in response to the increased demands placed upon the heart in hypertension can only partially compensate for the increased cardiac work. It is offset by the reduced efficiency of oxygen diffusion from the capillaries to the enlarged cardiac fibres. Gradually, the left ventricular function curve becomes depressed and, with prolonged and severe hypertension, left ventricular failure will eventually occur.

Vascular effects of hypertension

Hypertension is caused by an increase in total peripheral resistance. This increase in vascular resistance affects the vasculature of virtually every organ. Narrowing of small arteries and arterioles is seen in hypertensive patients. This is initially caused by the myogenic response of the vascular smooth muscle in response to the increased degree of stretching and will be responsive to to vasodilator drugs, but when hypertension has been present for some time, the smooth muscle of the tunica media of the vessel starts to hypertrophy. The thickening of the walls of the arterioles results in a permanent narrowing of their lumen which cannot be

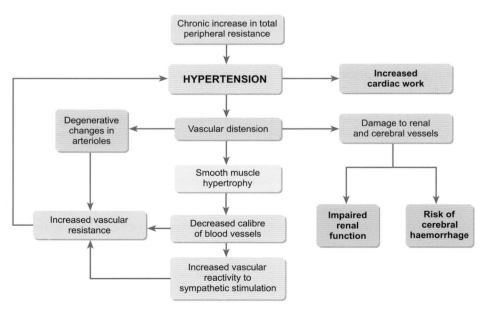

Figure 30.19 A flow chart illustrating the effect of hypertension on the circulation, brain, and kidneys.

fully reversed even during maximal vasodilatation. Moreover, as the resistance of a vessel varies inversely as the fourth power of its radius (Box 30.1), the vascular response of chronically hypertensive patients to sympathetic stimulation is more pronounced than that of normal subjects. Consequently, a given level of sympathetic activity will have a relatively larger effect on the blood vessels of hypertensive people compared to those with a normal blood pressure. A second feature of the vasculature in severe, chronic hypertension is **rarefaction**. This is a reduction in the number of vessels present per unit volume of tissue and has been observed in both the retina and the intestine of individuals suffering from hypertension.

It might be expected that the baroreceptor reflex would act to restore blood pressure to normal in cases of hypertension. However, the baroreceptor reflex is dynamic and rapidly adapting, so although it continues to operate in patients with raised blood pressure, it is reset to operate at higher pressures (see Figure 30.13). Moreover, in severe cases, the sensitivity of the baroreceptor reflex may be reduced by the stiffening of the arterial wall.

The treatment of hypertension

If left untreated for any length of time, there is a high risk that a hypertensive individual will suffer heart failure, renal failure, or cerebral complications such as strokes. The aim of treatment in hypertension is to reduce systolic arterial blood pressure to less than 140 mm Hg (<18.6 kPa) and diastolic blood pressure to less than 90 mm Hg (<12 kPa). A variety of drug treatments are employed. Diuretics are given to reduce the extracellular fluid volume, usually in combination with an inhibitor of angiotensin converting enzyme (an ACE inhibitor e.g. captopril) or an angiotensin-II receptor blocker (e.g. losartan) to lower peripheral resistance. If the ACE inhibitor is not well tolerated, a beta-blocker, e.g. propranolol, may be administered. When older hypertensive patients do not respond adequately to ACE inhibitors, a calcium channel blocker (e.g. amilodipine) is often employed in combination with a thiazide diuretic.

Summary

Hypertension exists when the arterial blood pressure is persistently higher than normal. In primary hypertension, there is a chronic elevation of blood pressure without evidence of other disease. In secondary hypertension, the hypertension is the result of another disease process.

Hypertension increases the work of the heart and stresses the vasculature. It leads to narrowing of the vessels, which exacerbates the initial hypertension and leads to further hypertrophy of the smooth muscle in the walls of the arterioles. Breaking this vicious circle is an important consideration in the treatment of hypertension.

30.6 Regional circulations

This section deals briefly with the circulation of blood to specific organs. It will illustrate the ways in which some of the regulatory mechanisms described in earlier sections are used in localized regions of the cardiovascular system to match blood flow to tissue requirements. However, the specific features of the pulmonary circulation are discussed in Chapter 34, those of the renal circulation are considered in Chapter 39, and those of the gut are described in Chapter 43.

The coronary circulation

The blood supply to the heart is provided by the right and left coronary arteries which arise at the root of the aorta. The venous drainage is principally to the right atrium via the coronary sinus (see Figure 27.3). The right coronary artery sends descending braches to all parts of the right ventricle and to part of the septum. The left coronary artery branches extensively to provide a rich blood supply to the left ventricle as well as sending branches to the septum. The relative blood supply to the different parts of the heart can be readily appreciated from the angiogram shown in Figure 30.20, which shows that the ventricles have a much richer blood supply than the atria, the left ventricle having the richest blood supply.

Blood flow through the coronary arteries varies greatly during the cardiac cycle. Flow to the myocardium is at its peak during early diastole, when the mechanical compression of coronary vessels is minimal and aortic pressure is still high (see Figure 30.21). During the isovolumetric contraction phase of the cycle, the coronary blood vessels become compressed as the ventricular pressure rises and coronary blood flow declines to its minimum value. During the ejection phase of systolic contraction, coronary flow varies according to the rise and fall in aortic pressure. After its initial decline it rises briefly during the peak of the arterial pressure wave before declining once more prior to the relaxation of the ventricles. The bulk of coronary blood flow occurs during diastole. Changes in diastolic or systolic pressure affect the coronary blood

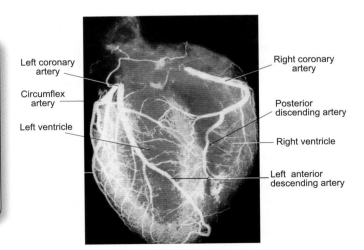

Figure 30.20 A coronary angiogram showing the arterial blood supply to the various parts of the heart. Note the relatively small circulation of the atria compared to the rich blood supply of the ventricles (especially that of the left ventricle). (From Fig 5.4.7 in P.C.B. McKinnon and J.F. Morris (2005) *Oxford textbook of functional anatomy.* Oxford University Press, Oxford.)

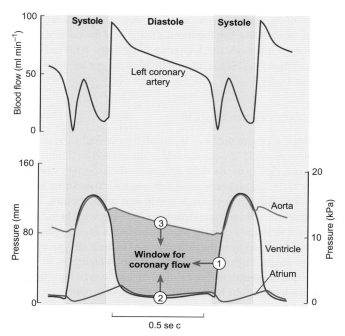

Figure 30.21 The changes in coronary blood flow during the cardiac cycle. The bulk of the coronary blood flow occurs during diastole when the heart is relaxed. It declines sharply at the onset of systole, with a brief surge corresponding to the aortic pressure wave. The lower panel shows that the time for coronary blood flow will be reduced when the diastolic interval becomes shorter as heart rate rises (1). Coronary blood flow will also be adversely affected by a rise in ventricular end-diastolic pressure (2) or by a fall in aortic pressure (3). (Adapted from Fig 13.6 in H.P. Rang, M.M. Dale, and J.M. Ritter (1995) *Pharmacology*, 3rd edition. Churchill-Livingstone, Edinburgh.)

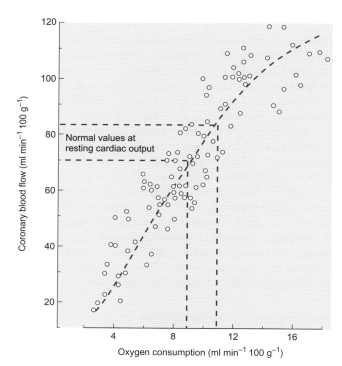

Figure 30.22 The relationship between cardiac metabolism (as measured by oxygen consumption) and coronary blood flow in the dog. In these experiments, the cardiac work was increased by the administration of adrenaline and decreased by producing a controlled haemorrhage. (Based on data of R.M. Berne and R. Rubio (1979) Coronary Circulation In: *American handbook of physiology*, Section 2, The cardiovascular system, Oxford University Press, New York.)

flow and, as the lower panel of Figure 30.21 indicates, an increase in heart rate decreases the time available for perfusing the heart with blood.

It is obvious that the work carried out by the heart must vary considerably with the demands of the circulation and that a greater workload requires a greater supply of oxygen and nutrients to be delivered to the myocardium. Indeed, blood flow through the coronary circulation can increase around five-fold from its resting level of around 75 ml min^{-1} 100 g^{-1} tissue to as much as 400 ml min^{-1} 100 g^{-1} tissue during maximal cardiac work. During periods of high cardiac work, the time available for blood to flow through the coronary circulation will be much reduced because the heart is contracting more powerfully and more often. Under these conditions, it is crucial that blood flow to the myocardium during diastole is adequate to provide sufficient oxygen for the whole of the cardiac cycle.

The coronary sinus blood has a PO_2 of about 2.7 kPa (20 mm Hg). Oxygen extraction is, therefore, very high. While the myoglobin present in the cardiac muscle fibres can provide a further source of oxygen during systole, this needs to be replenished during diastole. So how is the blood flow matched to the metabolic demands? Direct measurement of coronary blood flow and cardiac

oxygen consumption shows that blood flow increases as the metabolic activity of the heart is increased (Figure 30.22). There is therefore a vasodilatation of the coronary vessels which is thought to be mediated principally by adenosine, whose production increases during increased workload. Increased concentrations of other metabolites such as carbon dioxide, hydrogen ions, and potassium ions may also play a role, as may interstitial hypoxia. The sympathetic nerve supply to the coronary arteries is of lesser importance in the control of coronary blood flow. Nevertheless, stimulation of these fibres causes some vasodilatation, as does the interaction of circulating adrenaline on the β_2-adrenoceptors of the vascular smooth muscle.

Summary

The coronary circulation is adapted to provide the myocardium with sufficient blood to meet its metabolic requirements, despite the compression of coronary vessels that occurs during each beat of the heart. Coronary blood flow is regulated largely by metabolic hyperaemia and there is a close parallel between the work of the heart and coronary blood flow.

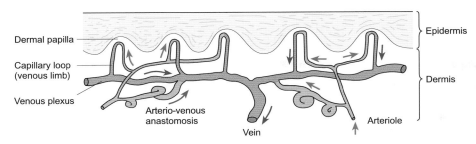

Figure 30.23 A diagrammatic representation of the cutaneous circulation from a region possessing arterio-venous anastomoses. (Based on Fig 13.6 in J.R. Levick (1995) *An introduction to cardiovascular physiology*, 2nd edition. Butterworth-Heinemann, Oxford.)

The cutaneous circulation

The skin possesses two types of resistance vessels: the arterioles, which are similar to those found elsewhere, and **arterio-venous (AV) anastomoses**, which provide a route for blood to pass from the arterioles directly to venules, bypassing the capillaries. The AV anastomoses are found predominantly in the skin of the hands, feet, ears, nose, and lips. A simplified diagram of the cutaneous circulation in such an extremity is shown in Figure 30.23. While the AV anastomoses are richly innervated with sympathetic fibres, which can cause substantial vasoconstriction, they do not appear to be under local metabolic control. By contrast, the arterioles of the skin are regulated both by sympathetic nerves and by local chemical factors.

The most important regulator of blood flow to the skin is not the supply of oxygen and nutrients as it is in most other tissues: indeed, its nutritional requirements are small. Rather, the cutaneous circulation plays a crucial role in the regulation of body temperature (see Chapter 42). In response to information from the temperature receptors in the CNS and skin, the level of sympathetic activity is adjusted according to the need to retain or dissipate heat. As Figure 30.24 shows, local changes in skin temperature have a powerful effect on local blood flow. The cutaneous vessels do not receive any parasympathetic vasodilator fibres. If the ambient temperature falls and heat production is insufficient to maintain normal body temperature, sympathetic activity is increased, the AV anastomoses close, and the arterioles constrict. This vasoconstriction shifts blood away from the body surface. When body temperature rises, the reduction in sympathetic activity opens up the AV anastomoses and the arterioles dilate. Blood flow to the skin is increased and there is greater heat loss. In circumstances that require even greater heat loss, sweating is stimulated. The sweat glands are innervated by cholinergic sympathetic vasodilator fibres. Sweat contains an enzyme whose action produces the potent vasodilator bradykinin. This acts locally to bring about vasodilatation of the arterioles. The cutaneous arterioles also dilate in response to increased local levels of metabolites, a situation likely to occur when body temperature is increased.

The blood flow to the skin varies from as little as 10 ml min^{-1} kg^{-1} under conditions of extreme cold to 2.0 l min^{-1} kg^{-1} in hot environments. Blood flow to the skin under normal (thermoneutral) conditions is around 100 ml min^{-1} kg^{-1}.

Summary

The cutaneous circulation plays an important role in thermoregulation by varying the blood flow to the skin. Specialized vessels, the arteriovenous anastomoses, constrict in response to sympathetic nerve activity to shunt blood from the arterioles directly to the venules. When this occurs, blood bypasses the capillaries and is diverted away from the skin surface. This acts to conserve body heat.

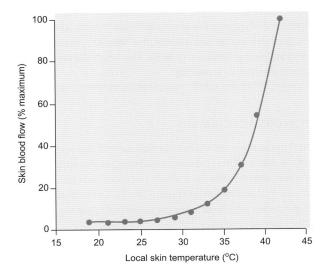

Figure 30.24 Changes in skin blood flow as a function of the local skin temperature. Above 30°C the flow increases substantially as skin temperature increases. (From data in D.P. Stephens et al. (2001) *American Journal of Physiology. Regulatory, Integrative and Comparative Physiology* **281**, R894–R901.)

The circulation of skeletal muscle

During walking or running, the leg muscles must alternately contract and relax. When they contract their blood vessels are compressed, and this drastically reduces their blood flow. As the muscles relax, the compression of the blood vessels is relieved and blood flow is restored. Consequently, in rhythmic exercise, blood flow to individual muscles will show periodic surges as shown in Figure 30.25. In contrast, during static exercise such as weight-lifting the contracted muscles have little blood flow as their blood vessels are compressed and they must rely on oxygen stored in their myoglobin and on anaerobic metabolism to maintain their contraction (see Chapter 38). When blood flow is restored, it must rapidly lead to the replenishment of the energy and tissue oxygen

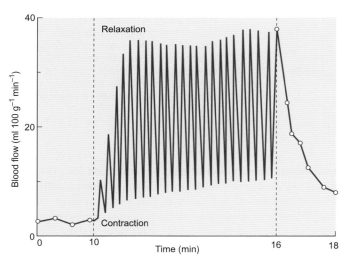

Figure 30.25 The changes in blood flow through the calf muscle during strong rhythmic contractions. During the periods of contraction the inflow of blood to the muscles is greatly reduced, but there is a large increase in blood flow when the muscle is relaxed (functional hyperaemia). After the period of exercise, blood flow rapidly declines towards its resting level. (Redrawn from H. Barcroft (1953) *Sympathetic control of human blood vessels.* Arnold, London.)

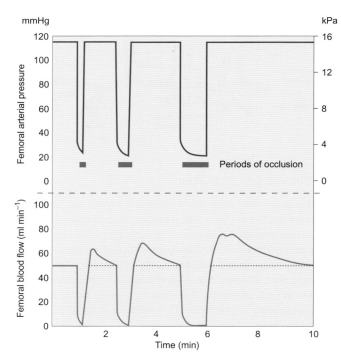

Figure 30.26 Reactive hyperaemia in the hind limb of the dog following periods of occlusion of the femoral artery. Note that the magnitude and duration of the increased blood flow (hyperaemia) is related to the duration of the occlusion. (From Fig 8.6 in R.M. Berne and M.N. Levy (1992) *Cardiovascular physiology.* Mosby, Baltimore.)

stores, and removal of the accumulated tissue metabolites. In view of these considerations, it is not surprising to find that skeletal muscle blood flow is subject to regulation by local, hormonal, and neural factors.

The arterioles of skeletal muscle have a high resting level of tone. This is partly myogenic in origin and partly the result of tonic sympathetic vasoconstrictor activity, which is controlled by the baroreceptor reflex. Resting blood flow is low (30–50 ml min^{-1} kg^{-1}). During rhythmic exercise, there is a marked vasodilatation of the arterioles of the active muscle which greatly increases perfusion of the capillary beds during the periods of relaxation. Overall blood flow can show a forty-fold increase above its resting level. This pronounced dilatation is caused by the increased production of metabolites and is a striking example of **functional hyperaemia**. Superimposed on this is a vasodilatation in response to the activation of β_2 adrenoceptors by circulating adrenaline secreted by the adrenal medulla. Therefore, in continuous rhythmical exercise, such as walking, there is time for the dilated vessels to provide an adequate supply of nutrients to the muscle between contractions.

In contrast, during a sustained strong contraction blood flow will practically cease and metabolites, particularly lactic acid, will quickly build up (see Chapter 38). When the contraction ceases, the accumulation of these metabolites elicits a **reactive hyperaemia**. This is illustrated in Figure 30.26 by the change in blood flow in the hind limb of a dog following a period of vascular compression. After each period of compression, the blood flow is significantly increased above the normal resting level and the extent of this increase is proportional to the duration of the occlusion.

Summary

Blood flow to skeletal muscle varies directly with contractile activity. During exercise, vasodilatation occurs in response to the local build-up of metabolites (functional hyperaemia). A sustained contraction may compress the vessels so much that blood flow to the muscle practically ceases. Metabolites then accumulate quickly and cause a rapid increase in blood flow when the circulation is restored (reactive hyperaemia).

The cerebral circulation

Blood flows to the brain via the internal carotid and vertebral arteries. These join together (anastomose) around the optic chiasm to form the **circle of Willis** from which the anterior, middle, and posterior cerebral arteries arise (Figure 30.27). These branch to form the pial arteries, which run over the surface of the brain. From here, smaller arteries penetrate into the brain tissue itself to give rise to short arterioles and an extensive capillary network. Brain tissue, particularly the grey matter, has a very high capillary density.

In contrast to most other organs, the blood flow to the brain is kept within very closely defined limits (around 0.55–0.60 l min^{-1} kg^{-1} or about 15 per cent of the resting cardiac output). The brain is very sensitive to ischaemia, and loss of consciousness occurs if blood flow is interrupted for only a few seconds. Irreversible brain

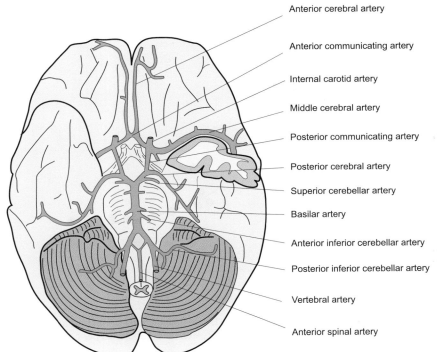

Figure 30.27 A diagrammatic representation of the principal arterial blood supply to the brain. The internal carotid arteries and basilar arteries are linked by the anterior communicating artery and the posterior communicating arteries to form the circle of Willis from which the left and right anterior, middle, and posterior cerebral arteries arise. The left temporal lobe has been shown cut away to reveal the course of the left middle cerebral artery. The basilar artery links the two vertebral arteries to the circle of Willis. (Adapted from Fig 1.15 in P. Brodal (1992) *The central nervous system: Structure and function.* Oxford University Press, New York.)

Labels in figure:
- Anterior cerebral artery
- Anterior communicating artery
- Internal carotid artery
- Middle cerebral artery
- Posterior communicating artery
- Posterior cerebral artery
- Superior cerebellar artery
- Basilar artery
- Anterior inferior cerebellar artery
- Posterior inferior cerebellar artery
- Vertebral artery
- Anterior spinal artery

damage may occur if blood flow is stopped for several minutes. If cardiac output falls, for example in haemorrhage, the cerebral blood flow will be maintained as far as possible even though perfusion of many peripheral organs may be compromised.

Cerebral blood flow is closely controlled by a combination of autoregulation, local metabolic regulatory mechanisms, and reflexes initiated by the brain itself which either act locally or by altering systemic blood pressure (e.g. Cushing's reflex, discussed later in this subsection). The cerebral arteries receive some sympathetic vasoconstrictor innervation, although its effect on cerebral perfusion is fairly weak. There are also some vasodilator fibres (which are probably parasympathetic in origin), but their role is unclear. As the brain is enclosed within the rigid structure of the skull, changes in the volume of blood or extracellular fluid are minimal. Any change in the volume of blood flowing to the brain must be matched by a change in the venous outflow.

Regional functional hyperaemia is well developed in the brain; an increase in neuronal activity evokes a local increase in blood flow. Indeed, modern techniques of computerized mapping of regional blood flow such as functional magnetic resonance imaging (fMRI) and positron emission tomography (PET) have exploited this property to localize sites of mental function or epileptic foci in the brains of conscious subjects. The cause of the metabolite-induced increase in blood flow is thought to be a rise in extracellular potassium ions resulting from the outward potassium currents of active neurons. Local increases in hydrogen ion concentration may also play a role. Cerebral blood flow is also highly sensitive to arterial carbon dioxide. An increase in PCO_2 (hypercapnia) brings about vasodilatation and a rise in blood flow (see Figure 30.28a). Vasoconstriction occurs during hypocapnia (which is a consequence of hyperventilation). The fall in blood flow leads to a

feeling of dizziness and may cause disturbed vision. There is also a degree of vasodilatation in response to hypoxia (Figure 30.28b). It is now thought that the arterial responses to hypoxia and hypercapnia in the brain are mediated by nitric oxide released from the vascular endothelium.

As the brain is encased in the skull, which provides a rigid enclosure except in very young children, anything that increases the volume of the brain tissue or obstructs the drainage of the CSF, such as a cerebral tumour or a cerebral haemorrhage, will increase the pressure within the skull (the **intracranial pressure**). This will tend to reduce the blood flow to the brain and force the brainstem into the opening of the skull through which the spinal cord passes (the foramen magnum). The resulting compression of the brainstem elicits a marked rise in arterial blood pressure, which acts to offset the fall in cerebral blood flow. This is known as **Cushing's reflex** (also called Cushing's response). In addition to the rise in blood pressure, there is a fall in heart rate mediated by the baroreceptor reflex. This combination of a marked rise in blood pressure and a slow heart rate is an important clinical indicator of a space-occupying lesion within the skull such as a glioma (a tumour derived from glial cells). If the increase in intracranial pressure is very high, cerebral blood flow will decline and the sufferer will experience mental confusion leading to coma and death unless prompt action is taken.

The blood–brain barrier

Although lipid-soluble molecules are able to pass freely from the blood to the interstitium of the brain, ionic solutes are unable to do so. Because of this, the neuronal environment can be tightly controlled and the cells protected from fluctuating levels of ions, hormones, and tissue metabolites circulating in the plasma. This

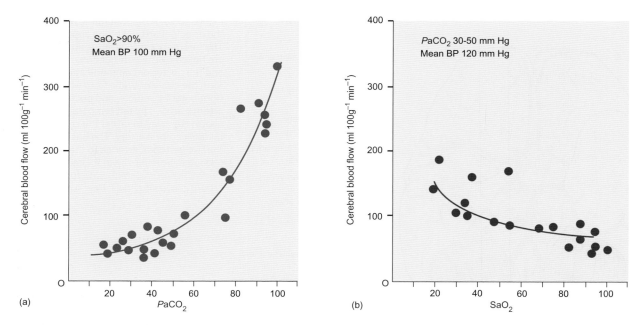

Figure 30.28 The relationship between (a) arterial PCO_2 and cerebral blood flow and (b) SaO_2. Blood flow increases substantially as $PaCO_2$ rises beyond 50 mm Hg. It is less sensitive to SaO_2 but begins to rise significantly as SaO_2 falls below about 60 per cent. (Based on data in E. Haggendal and B. Johansson (1965) *Acta Physiol. Scand.* **66**, Suppl **258**, 27–53.)

restricted movement has been attributed to a specific 'blood–brain' barrier. The restriction appears to be at least partly due to the nature of the tight junctions between the capillary endothelial cells of the cerebral circulation. In addition, astrocytes extend processes that contact the endothelial cells (astrocytic end-feet),

helping to seal the interstitial space of the brain from the circulating blood (Figure 30.29).

The blood–brain barrier is breached at a few sites along the midline of the brain. These sites are located along the third and fourth cerebral ventricles and are known as the **circumventricular**

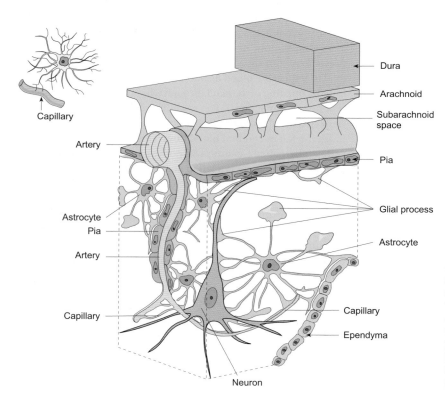

Figure 30.29 Diagrammatic representation of the arrangement of the pial arteries and the grey matter of the cerebral cortex. The inset figure illustrates the relationship between the capillary endothelium and the end-feet of the astocytes, which effectively isolate the neurons from the plasma. This is the cellular basis of the blood–brain barrier. (Adapted from Fig 2.38 in P. Brodal (1992) *The central nervous system: Structure and function.* Oxford University Press, New York.)

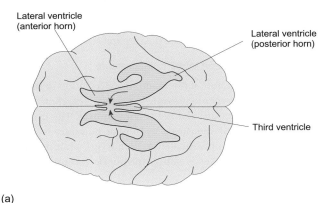

(a)

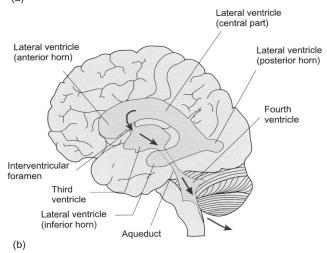

(b)

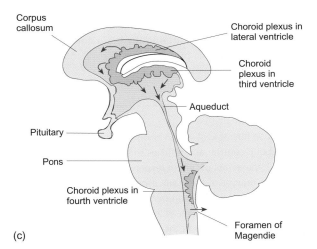

(c)

Figure 30.30 The formation and circulation of the cerebrospinal fluid (CSF). The CSF is formed by the choroid plexuses of the lateral, third, and fourth ventricles (a and c). From the lateral ventricles the CSF flows through the third ventricle and aqueduct into the fourth ventricle (b,) from where it leaves the ventricles via the foramen of Magendie which lies on the midline (c) and the two lateral foramina of Luschka (not shown in the figure). It then flows around the outside of the brain and spinal cord beneath the dura mater before being reabsorbed into the blood via the arachnoid granulations and cranial lymphatic drainage.

organs. Here the capillaries are fenestrated, allowing relatively free exchange between the plasma and the interstitial fluid of the neural space. Their function is still being elucidated, but they appear to play a significant role in the regulation of fluid balance.

The cerebral ventricles and the cerebrospinal fluid

The fluid-filled spaces in the brain (the cerebral ventricles) and the central canal of the spinal cord contain a clear fluid known as cerebrospinal fluid (or CSF). The CSF also fills the space between the dura mater (the tough membranous outer covering of the brain and spinal cord) and the surface of the brain and spinal cord (see Figure 9.1). In all, the total volume of CSF is about 130 ml. About 500 ml of CSF is formed each day by the choroid plexus in the cerebral ventricles, which consists of networks of capillaries surrounded by a layer of epithelial cells.

The CSF flows from the lateral ventricles into the third ventricle and thence through the cerebral aqueduct to the fourth ventricle, as shown in Figure 30.30. It leaves the cerebral ventricles via apertures in the roof of the fourth ventricle (the foramina of Magendie and Luschka) and flows into the subarachnoid space which lies between the arachnoid membrane and the pia mater. From the subarachnoid space, the CSF flows around the brain and spinal cord before draining into the venous blood. The CSF is returned to the venous system via at least two routes:

1. via the **arachnoid granulations** (also called the **arachnoid villi**) which project into the venous sinuses of the CNS, as shown in Figure 30.31

2. via extracranial lymphatics. In this case, the CSF flows out of the cavity of the skull via the cribriform plate (a perforated region of bone through which the olfactory nerves pass to the nasal mucosa). Experiments suggest that perhaps half of the CSF production is returned to the venous system in this way.

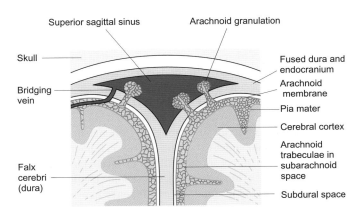

Figure 30.31 Diagrammatic representation of the relationship between the meninges, the arachnoid granulations (villi), and the superior sagittal sinus. The CSF drains into the venous blood via the arachnoid granulations that penetrate the large venous sinuses formed by folds in the dura. (Based on Fig 184 in G.J. Romanes (ed.) (1978) *Cunningham's practical anatomy*, Vol. 3. 14th edition, Oxford University Press, Oxford).

Table 30.3 The composition of the plasma and cerebrospinal fluid

	CSF	Plasma	ratio CSF/plasma
Na^+ (mmol l^{-1})	141	141	1.0
K^+ (mmol l^{-1})	3.0	4.5	0.67
Ca^{2+} (mmol l^{-1})	1.15	1.5	0.77
Mg^{2+} (mmol l^{-1})	1.12	0.8	1.4
Cl^- (mmol l^{-1})	120	105	1.14
HCO_3^- (mmol l^{-1})	23.6	24.9	0.95
glucose (mmol l^{-1})	3.7	4.5	0.82
protein (g l^{-1})	0.18	75	0.002
pH	7.35	7.42	—

The veins connecting the subarachnoid space and the venous sinuses are known as **bridging veins**. They form part of the attachments between the brain and the skull and may become ruptured during head injuries. This results in venous bleeding and the formation of a chronic subdural haematoma. Over time this raises intracranial pressure with consequent alterations to behaviour such as confusion, headache, and impaired motor function.

The composition of the CSF

The CSF is not an ultrafiltrate of plasma as it differs in its ionic composition, having relatively less potassium and calcium, and more chloride than plasma (see Table 30.3). It also has a very low protein content, no red cells, and very few leukocytes. Samples of CSF can be obtained by inserting a needle into the sub-arachnoid space at a level below the third or fourth lumbar vertebrae. This is usually done with the subject is lying on their side. Although the tip of the needle lies in the sub-arachnoid space, there is no risk of damage to the spinal cord, which usually extends as far as the first lumbar vertebra. This procedure is known as lumbar puncture.

Function of the CSF

The CSF serves several functions: it provides buoyancy for the brain and spinal cord, reducing the effective weight of the brain to around 50 g, and reduces the traction on the blood vessels and cranial nerves. As it occupies the space between the brain surface and the inside of the skull, the CSF is able to act as a hydraulic buffer to cushion the brain against damage resulting from movements of the head. In addition, as it can flow through the foramen magnum, it is able, within certain limits, to counter any rise in intracranial pressure. The CSF is in direct contact with the extracellular fluid of the brain and there is evidence to suggest that some chemical agents are actively secreted into the CSF to influence neurons located elsewhere close to the ventricular surface. For example, the hormone melatonin is secreted by the pineal gland into the CSF to influence neurons in the hypothalamus concerned with maintaining circadian rhythms. There is also evidence that sleep-inducing

peptides are secreted into the CSF. Finally, the CSF may also help to provide a stable ionic environment for neuronal function, at least for neurons close to the walls of the cerebral ventricles.

Consequences of raised intracranial pressure

As with the circulation of the blood, the circulation of the CSF is driven by the difference in pressure between the cerebral ventricles (the site of production) and the venous sinuses. The hydrostatic pressure of the CSF measured by lumbar puncture is normally about 0.5–1.5 kPa (approximately 4–12 mm Hg), while the pressure in the cerebral sinuses is sub-atmospheric (around −2 kPa or −30 mm Hg).

Obstruction of the CSF drainage or overproduction of CSF leads to an accumulation of fluid and an increase in intracranial pressure. This is called **hydrocephalus** which, if untreated, can cause brain damage. Hydrocephalus may be either congenital (i.e. present at birth) or it may be acquired in later life following a cerebral haemorrhage, a head injury, inflammation of the aqueduct, or the growth of a tumour. In babies and young children (where the skull sutures are not fused), the increased pressure leads to a disproportionate increase in the size of the head. In an adult, the raised intracranial pressure elicits Cushing's reflex. Hydrocephalus is usually treated by providing a shunt that consists of an appropriately positioned tube and valve to allow the excess CSF to drain from the ventricles into the right atrium or abdominal cavity.

A further cause of raised intracranial pressure is **meningitis**, in which the membranes covering the brain (the meninges) are invaded by viruses or bacteria. Under these circumstances, the meninges become inflamed leading to a potentially life-threatening condition. In meningitis, the intracranial pressure is increased due to restricted drainage of the CSF, giving rise to characteristic clinical signs including headache, vomiting, photophobia, and rigidity of the neck. In severe cases, convulsions and coma may occur. Because the CSF flows freely around the brain and spinal cord, a sample obtained by lumbar puncture can be used to ascertain the nature of the invading organism.

Summary

Cerebral blood flow is kept within narrow limits by autoregulation, local metabolic control, and reflexes initiated by the brain itself. Functional hyperaemia occurs in response to an increase in activity in specific regions of the brain, probably in response to a local increase in potassium and hydrogen ions. Hypercapnia also causes vasodilatation and an increase in blood flow.

The cerebral ventricles and the central canal of the spinal cord are filled with cerebrospinal fluid (CSF), which is formed continuously by the choroid plexus. The CSF provides physical support for the brain and protects it from damage that could result from abrupt movement of the head. Obstruction of the CSF circulation raises intracranial pressure and has serious consequences for brain function.

✳ Checklist of key terms and concepts

Blood pressure

- Blood flow in the arteries is pulsatile, and the maximum and minimum pressures of each pulse wave are the systolic and diastolic pressures respectively.

- The difference between the systolic and diastolic pressures is called the pulse pressure.

- The mean arterial pressure is a time-weighted average, which is calculated as the sum of the diastolic pressure plus one-third of the pulse pressure.

- The arterial blood pressure is determined by both the cardiac output and the total peripheral resistance.

Blood flow and vascular resistance

- Total peripheral resistance is the resistance of the systemic circulation to blood flow.

- The arterioles are the main source of vascular resistance.

- The calibre of the arterioles supplying a particular capillary bed determines capillary blood flow. The capillaries themselves offer little resistance to blood flow.

- Hyperaemia is an increase of blood flow to one part of the body, such as the skin or a specific set of muscles. It is caused by the vasodilatation of a particular vascular bed.

- Ischaemia occurs when local blood flow is not sufficient to supply the required amount of oxygen to the tissues.

- The veins contain around two-thirds of the total blood volume and are the capacitance vessels.

- The flow of blood from the circulation into the heart is called the venous return.

- The venous return is influenced by respiration, gravity, and the skeletal muscle pump.

- The venous pressure at the level of the heart is called the central venous pressure.

- The difference between the arterial blood pressure and the central venous pressure is the source of energy that drives the blood around the body.

Control of the blood vessels

- The tone of the smooth muscle in the wall of a blood vessel controls the vessel's diameter.

- The resting tone of a vessel may be altered by either intrinsic or extrinsic mechanisms to cause vasoconstriction or vasodilatation.

- The relative stability of blood flow when perfusion pressure changes is called autoregulation.

- The factors responsible for autoregulation of blood flow are intrinsic to the local circulation and include the myogenic contraction seen in response to stretch of a vessel, the vasodilator actions of tissue metabolites, and the effects of local vasoactive substances.

- Functional hyperaemia is an increase in local blood flow in response to increased metabolic activity. Functional hyperaemia is also known as active hyperaemia or metabolic hyperaemia.

- Reactive hyperaemia is the marked increase in blood flow that follows a period of ischaemia. The blood flow may be many times the normal resting level. It may arise from occlusion of the blood supply or as a consequence of static exercise.

- Nerves and hormones exert extrinsic control over the heart and circulation, which can override the intrinsic autoregulation.

- This allows for adjustments to the pattern of blood flow according to the physiological needs of the body as a whole.

- Most blood vessels are regulated by sympathetic vasoconstrictor fibres.

- Hormones that play a role in the extrinsic regulation of the circulation include adrenaline and vasopressin (ADH).

The control of the arterial blood pressure

- Arterial blood pressure is regulated by autonomic nerves, by hormones, and by changes in blood volume.

- It is monitored by baroreceptors present in the walls of the aortic arch and the carotid sinuses.

- The baroreceptors are the sensors for a reflex, the baroreceptor reflex, that acts as a short-term regulator of blood pressure.

- A rapid rise in blood pressure activates the baroreceptor reflex, which slows the heart and causes peripheral vasodilatation. As a result, blood pressure falls.

- A fall in blood pressure increases both heart rate and total peripheral resistance, which both contribute to the stabilization of arterial blood pressure.

- The baroreceptor reflex adapts to prolonged changes in blood pressure and can be overridden in situations of stress, such as exercise.

- Long-term regulation of the blood pressure is achieved through the maintenance of normal extracellular volume and composition.

- Arterial chemoreceptors and muscle work receptors also play a role in the control of blood pressure.

Hypertension

- Hypertension exists when the arterial blood pressure is persistently higher than normal.

- In primary hypertension there is a chronic elevation of blood pressure without evidence of other disease.

- In secondary hypertension the hypertension is caused by another condition or disease process.

- Hypertension is highly undesirable as it increases the work of the heart and stresses the vasculature.

- Hypertension leads to hypertrophy of the smooth muscle in the walls of the resistance vessels, which exacerbates the rise in blood pressure.

- Hypertension increases the risk of heart failure, renal failure, and stroke.

The coronary circulation

- The coronary circulation is adapted to provide the heart with sufficient blood to meet its metabolic requirements despite the compression of coronary vessels that occurs during each systole.
- The most important determinant of coronary blood flow is metabolic hyperaemia.
- There is a close parallel between the work of the heart and coronary blood flow.

Blood flow in skeletal muscle

- Blood flow to skeletal muscle varies directly with contractile activity.
- During exercise, vasodilatation occurs in response to the local build-up of metabolites (functional hyperaemia).
- A sustained contraction may compress the vessels so much that blood flow to the muscle practically ceases. Metabolites then accumulate quickly and cause a rapid increase in blood flow when the circulation is restored (reactive hyperaemia).

The circulation of the skin

- The cutaneous circulation plays an important role in thermoregulation by varying the blood flow to the skin.
- The cutaneous circulation is regulated by extrinsic factors, particularly the sympathetic nervous system.
- The skin of the extremities has specialized vessels, the arteriovenous anastomoses, that constrict in response to sympathetic nerve activity.

- In the cold, the AV anastomoses shunt blood from the arterioles directly to the venules to divert blood away from the skin surface. This acts to conserve body heat.
- When the body needs to lose heat, the AV anastomoses close and blood is diverted to the superficial veins. Excess heat can then be radiated to the environment from the skin's surface.

The cerebral circulation

- Cerebral blood flow is kept within narrow limits by autoregulation, local metabolic control, and reflexes initiated by the brain itself.
- Functional hyperaemia occurs in response to an increase in activity in specific regions of the brain.
- Hypercapnia also causes vasodilatation and an increase in blood flow.
- The nerve cells are protected from changes in the composition of the plasma by the blood–brain barrier.
- The cerebral ventricles and the central canal of the spinal cord are filled with cerebrospinal fluid (CSF), which is continuously formed by the choroid plexus.
- The CSF is formed by specific transport processes and has a different composition to plasma.
- The CSF provides hydrostatic support for the brain and, together with the meninges, acts to restrict the displacement of the brain within the skull, protecting it from damage that could result from abrupt movement of the head.
- Obstruction of the circulation of the cerebrospinal fluid raises intracranial pressure and has serious consequences for brain function.

Recommended reading

Anatomy

MacKinnon, P.C.B., and Morris, J.F. (2005) *Oxford textbook of functional anatomy* (2nd edn), Volume 2:. *Thorax and abdomen*, pp. 74–87, 123–6. Oxford University Press, Oxford.

A well-illustrated, straightforward account of the anatomy of principal vessels of the chest and abdomen.

Histology of the heart and blood vessels

Mescher, A.L. (2015) *Junquieira's basic histology* (14th edn), Chapter 11. McGraw-Hill, New York.

A well-illustrated account of the histology of the heart and vascular system.

Physiology of the circulatory system

Brodal, P. (2004) *The central nervous system: Structure and function* (3rd edn), pp. 90–102. Oxford University Press, New York.

A very clear, well-illustrated discussion of the cerebral circulation and the formation and circulation of the CSF.

Levick, J.R. (2010) *An introduction to cardiovascular physiology* (5th edn), Chapters 8 and 13–15. Hodder Arnold, London.

This text, now in its fifth edition, remains eminently readable and the best introduction for advanced studies of the cardiovascular system.

Pharmacology of the heart and circulation

Grahame-Smith, D.G., and Aronson, J.K. (2002) *Oxford textbook of clinical pharmacology* (3rd edn), Chapter 23, pp. 226–33. Oxford University Press, Oxford.

A useful account of the practical considerations confronting the drug treatments available for hypertension.

Rang, H.P., Ritter, J.M., Flower, R.J., and Henderson, G. (2015) *Pharmacology* (8th edn), Chapter 22. Churchill-Livingstone, Edinburgh.

A comprehensive account of the control of vascular smooth muscle and the drugs that affect it. Well worth reading.

Review articles

Damkier, H.H., Brown, P.D., and Praetorius, J. (2013) Cerebrospinal fluid secretion by the choroid plexus. *Physiological Reviews* **93**, 1847–92, doi:10.1152/physrev.00004.2013.

Duncker, D.J., and Bache, R.J. (2008) Regulation of coronary blood flow during exercise. *Physiological Reviews* **88**, 1009–86, doi:10.1152/physrev.00045.2006.

Guyenet, P.G. (2006) The sympathetic control of blood pressure. *Nature Neuroscience Reviews* **7**, 335–46.

Joyner, M.J., and Casey, D.P. (2015) Regulation of increased blood flow (hyperemia) to muscles during exercise: A hierarchy of competing physiological needs. *Physiological Reviews* **95**, 549–601, doi:10.1152/physrev.00045.2006.

Mancia, G., and Mark, A.L. (2011) Arterial baroreflexes in humans. *Comprehensive Physiology* 2011, Supplement 8: *Handbook of physiology, the cardiovascular system, peripheral circulation and organ blood flow*, 755–93, doi:10.1002/cphy.cp020320.

Victor, R.G. (2015) Carotid baroreflex activation therapy for resistant hypertension. *Nature Reviews Cardiology* **12**, 451–63, doi:10.1038/nrcardio.2015.96.

Ward, M., and Langton, J.A. (2007) Blood pressure measurement. *Continuing Education in Anaesthesia, Critical Care & Pain* **7**, 122–6, doi:10.1093/bjaceaccp/mkm022.

Willie, C.K., Tzeng, Y.-C., Fisher, J.A., and Ainslie, P.N. (2014) Integrative regulation of human brain blood flow. *Journal of Physiology* **592**, 841–59, doi:10.1113/jphysiol.2013.268953.

 To check that you have mastered the key concepts presented in this chapter, complete the accompanying online self-assessment questions. Go to www.oup.com/uk/pocock5e/

CHAPTER 31

The microcirculation and lymphatic system

Chapter contents

This chapter should help you to understand:

* The organization of the microcirculation

* The role of arteriolar tone in the regulation of capillary perfusion

* The principles that govern the exchange of solutes between the blood and the tissues

* The difference between diffusion-limited and flow-limited exchange

* The role of Starling forces in tissue fluid balance

* The importance of the lymphatic system in the maintenance of tissue fluid balance

31.1 Introduction

The chief function of the circulation is the exchange of fluid, nutrients, and metabolites between the blood and the tissues. In general terms, blood flows from the arterioles to the venules via the capillaries, which are the main exchange vessels. In the course of tissue exchange a small volume of fluid passes from the capillaries to the interstitial space. While a small fraction of this fluid is returned directly to the circulation, the remainder is returned via the afferent lymphatic vessels. For this reason, the basic organization of the **lymphatic system** will be discussed here as well as its role in the regulation of the volume of the interstitial fluid.

31.2 The organization of the microcirculation

The arterioles that branch directly from the arteries are known as primary arterioles, and they are extensively innervated by sympathetic nerve fibres. The primary arterioles progressively give rise to secondary and tertiary arterioles, which have less smooth muscle and are more sparsely innervated. The final degree of branching gives rise to the **terminal arterioles**, which have a very sparse innervation, their calibre being regulated largely by the local concentration of tissue metabolites (see Chapter 30). The terminal arterioles are 10–40 μm in diameter and each arteriole gives rise directly to a group of capillaries known as a cluster or module, as in the example shown in Figure 31.1, and the flow through a capillary bed is regulated by the calibre of the terminal arterioles that supply it.

The capillaries themselves are 5–8 μm in diameter and about 0.5–1 mm in length. As the flow of plasma is greatest in the centre of the capillary and slowest next to the wall (see Chapter 30), the red cells become deformed into a characteristic 'parachute' shape as they flow through the capillaries and other small vessels, as Figure 31.2 shows. (Recall that the red cells are 7–8 μm in diameter, i.e. comparable in diameter to the capillaries themselves.)

As discussed in Chapter 27, there are three types of capillary: **continuous capillaries**, **fenestrated capillaries**, and **discontinuous capillaries**, each of which have specific structural adaptations in their walls. The walls of all types of capillary consist of a single layer of flattened endothelial cells (see Figure 27.6). Their internal surface is covered by a meshwork of glycoproteins called the

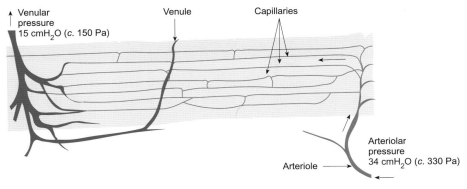

Figure 31.1 A diagram of the capillary bed in a relaxed cremaster muscle of a rat. An arteriole branches to give rise to a terminal arteriole that feeds a module of capillaries. The capillaries empty into post-capillary venules which, in turn, drain into the true venules. Note the difference in pressure between the arteriole and venule. (Based on Fig 5 in L.H. Smaje et al. (1970) *Microvasc. Res.* **2**, 96–110).

glycocalyx which acts as a molecular sieve, permitting the passage of small molecules but impeding that of large molecules such as proteins. The walls of continuous capillaries are partly covered by cells known as **pericytes**, and the whole structure is surrounded by a delicate basement membrane. The capillary wall is so thin (about 0.5 μm) that the diffusion path between plasma and tissue fluid is extremely short.

In most vascular beds the capillaries drain into post-capillary venules whose walls do not contain smooth muscle. As the walls of these vessels are partially enveloped by pericytes, these small venules are also known as **pericytic venules**. Their walls, like those of the capillaries, are sufficiently thin to permit the passage of fluid between the plasma and interstitial space. Thus, like the capillaries, the pericytic venules are able to act as exchange vessels. The post-capillary venules coalesce into larger venules. Those with a diameter greater than about 30 μm have smooth muscle in their walls and are sometimes called **true venules**. These merge progressively and eventually empty into the veins.

Vasomotor activity in the arterioles causes blood flow in the capillaries to vary

Direct observation of the pattern of blood flow in a capillary bed shows that it is constantly varying: the blood flow alternates between near stasis and rapid flow according to the changes in the calibre of the arterioles. This variation in calibre is known as **vasomotion**. When there is increased metabolic activity, the demand

for oxygen and nutrients is increased and the arterioles (especially the terminal arterioles) become dilated by the action of local metabolites. This results in the recruitment of more capillaries and an increase in the surface area for exchange. This local regulation of the terminal arterioles (and of capillary perfusion) can be overridden by the extrinsic factors described in Chapter 30.

Summary

The capillaries have thin walls that consist of a single layer of endothelial cells. They provide a large surface area for the exchange of solutes between the blood and the tissues. Their blood flow is governed continuously by vasomotor activity in the arterioles (vasomotion).

31.3 Solute exchange between the capillaries and the tissues

As the capillaries are the principal exchange vessels, their density determines the total surface area for exchange between tissue and blood. Capillary density varies considerably between tissues, reflecting their differing physiological roles. In the lungs, the alveoli are almost entirely surrounded by capillaries, as shown in Figure 32.6. Capillary density is also very high in both heart muscle and in the cerebral cortex (around 1000 capillaries per mm^2 in a cross-section of tissue). It is much lower in the skin and connective tissue (around 50 capillaries per mm^2). Of more significance is the surface area available for exchange, which is enormous in the lungs (around 3–4 m^2 per gram of tissue), reflecting its crucial role in gas exchange (see Chapter 34). Of the other tissues, the brain and myocardium have a capillary surface area of around 500 cm^2 per gram, while skeletal muscle has a capillary surface area of approximately 100 cm^2 per gram of tissue. Clearly, if the density of capillaries is high there will be a large surface area for exchange and a relatively short distance between capillaries. This ensures the efficient delivery of oxygen, glucose, and other nutrients in these tissues.

In all tissues, Fick's law applies to the rate of diffusion of any substance in the interstitial fluid (see Box 31.1). In free solution, three

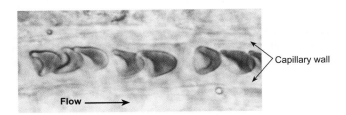

Figure 31.2 The characteristic appearance of red blood cells flowing in single file through a small blood vessel. The shallow depth of field has blurred the outline of the blood vessel. (From Fig 1 in R. Skalak and P.-I. Branemark (1969) *Science* **164** 717–19).

(1₂3) Box 31.1 Fick's Law of Diffusion

Substances move from the plasma to the tissues or from the tissues to the plasma by diffusion down their concentration gradients. Diffusion is a passive process and results from the random movement of molecules. The amount of a substance moving from one region to another can be expressed by Fick's Law of Diffusion, which states that the amount of substance moved depends on the area available for diffusion, the concentration gradient, and a constant known as the diffusion coefficient. Thus:

$$\text{amount moved} = \text{area} \times \text{concentration gradient} \times \text{diffusion coefficient}$$

In symbols:

$$J = -D \cdot A \cdot \frac{dC}{dx}$$

where J is the quantity moved, D the diffusion coefficient, A the area over which diffusion can occur, and dC/dx the concentration gradient.

The negative sign indicates that diffusion occurs from a region of high concentration to one of low concentration. The diffusion coefficient becomes smaller as the molecular size increases, so that large molecules diffuse more slowly than small ones.

In the tissues, the situation is rather more complicated as free diffusion is restricted by the presence of protein filaments, cell processes, etc. These restrict the diffusion path considerably. To correct for both the volume occupied by elements that block diffusion and the convoluted path any solute is obliged to take, the Fick equation is modified by the inclusion of two parameters. This first is called the volume fraction and represents the fraction of the volume through which the solute is diffusing that is occupied by those items that obstruct its path. The second is known as the tortuosity, which takes into account the convoluted path the solute must take as it diffuses through the tissue. These are generally merged with the diffusion coefficient (D) for free solution into a constant called the apparent diffusion coefficient (D^*), which will obviously reflect the characteristics of the individual tissues.

---

factors are sufficient to describe the rate of diffusion of a particular substance: area, concentration gradient, and diffusion coefficient. Substances with a small molecular mass (such as oxygen) diffuse more readily (i.e. have higher diffusion coefficients) than those with a large molecular mass, such as proteins. Furthermore, diffusion between the plasma in the capillaries and the fluid of the surrounding tissues is also limited by the permeability of the capillary wall to individual solutes. Very lipophilic substances, such as carbon dioxide and oxygen, pass through the endothelial cells relatively freely (**transcellular exchange**) and therefore the capillaries are very permeable to them. Water-soluble substances such as electrolytes and glucose are not soluble in the lipids of the cell membranes and do not pass across the endothelial cells themselves. Rather they diffuse through the small spaces between the cells (**paracellular exchange**) and through the fenestrations when these are present (see Figure 31.3).

With the important exception of the cerebral circulation, the exchange of solutes between the blood and the surrounding tissues is entirely passive and occurs by simple physical diffusion across the capillary wall. Measurements with radioactive tracers indicate that the capillaries allow relatively free exchange of both water and small water-soluble molecules between the plasma and the interstitial fluid. Note that water passes into the interstitial space by bulk flow via the paracellular pathway (including the fenestrae when they are present) and not by diffusion, which is a two-way process.

As Table 31.1 shows, the permeability of the capillary wall to solutes falls sharply with increasing molecular mass and the plasma proteins pass from the plasma to the interstitial space very slowly. Except in discontinuous capillaries, the narrowness of the intercellular clefts together with the barrier formed by the glycocalyx restricts their diffusion across the capillary wall. Nevertheless, the passage of proteins from the plasma to the interstitium is functionally important as it permits the delivery of peptide hormones (e.g. growth hormone) and protein-bound substances with low water solubility to their target tissues. It is thought that these proteins are transported across the capillary walls by vesicular transport. The steroid hormones (e.g. oestradiol-17β, testosterone), thyroxine, and the essential fatty acids are examples of substances that are delivered bound to carrier proteins.

The capillaries of the brain and spinal cord are much less permeable to solutes than those of most other tissues. This reflects the tight junctions between endothelial cells and the barrier formed by the end feet of the astrocytes, which are the morphological basis of the blood–brain barrier discussed in Chapter 30. Certain substances, particularly inorganic ions, glucose, and amino acids, are actively transported across the endothelium of brain capillaries. As the brain is dependent on glucose for its principal energy source, efficient glucose transport is crucial. Larger molecules such as growth factors probably reach the neurons by receptor-mediated transport similar to the vesicular transport already discussed. The brain exports organic acids (e.g. lactate) and lipid-soluble molecules to the plasma via specific transporters.

Blood flow and solute transfer

The influence of blood flow on the transfer of solutes between the plasma and the tissues largely depends on the permeability of the capillaries to particular solutes. The fraction of a solute that is

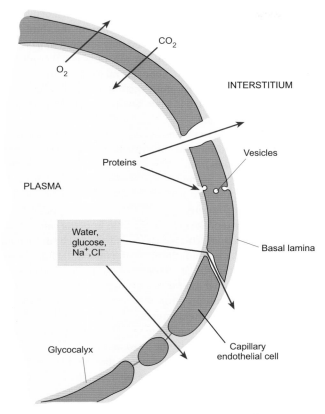

Figure 31.3 A schematic diagram to illustrate how different types of molecule cross the walls of the capillaries. Gases and small lipophilic molecules directly penetrate the endothelial cells of capillary wall to exchange with the extracellular fluid (transcellular exchange). Water and small water-soluble molecules such as glucose and inorganic ions exchange with the extracellular fluid via small pores in the capillary wall (the intercellular junctions and the fenestrae). This is known as paracellular exchange. The permeability of continuous capillaries to large molecules such as proteins is low and may depend on vesicular transport. However, in discontinuous capillaries, such as those of the liver, proteins are able to pass through large intercellular spaces.

Table 31.1 Relative permeability of continuous capillaries to various molecules

Substance	Relative molecular mass	Permeability of solute relative to water
Oxygen	32	~3000
Water	18	1.00
NaCl	58	0.96
Urea	60	0.8
Glucose	180	0.6
Inulin	5000	0.2
Albumin	69,000	0.0001

Note: Water-soluble substances pass through spaces between the endothelial cells, and their permeability is related to their molecular weight. Lipophilic substances such as oxygen and other gases diffuse freely through the walls of the endothelial cells (see Figure 31.3).

removed from the plasma as it passes through the capillaries and venules is known as its **extraction**. Hence:

$$\text{extraction} = (\text{arterial concentration} - \text{venous concentration}) \div \text{arterial concentration}$$

As tissue exchange is passive, the venous concentration cannot be lower than the concentration in the interstitial fluid; but if the capillary wall is relatively impermeable to the solute, its concentration in the venous blood may be significantly higher than that in the interstitial fluid. In this case, the exchange is said to be **diffusion-limited** and an increase in blood flow will have little effect on solute transfer. Many large solutes such as insulin behave in this way. Diffusion-limited exchange is illustrated by the blue line of Figure 31.4. Conversely, when the capillary wall is very permeable to a solute, there is a rapid equilibration between the plasma and the surrounding tissue so that the venous concentration is equal to that in the interstitial fluid. This pattern of exchange is said to be **flow-limited** and is characteristic of small lipophilic molecules such as the respiratory gases (O_2 and CO_2) and certain drugs such as the volatile general anaesthetics. In flow-limited exchange, an increase in blood flow will increase the amount of solute exchanged with the surrounding tissue, as shown Figure 31.5. Flow-limited and diffusion-limited exchange are at the two ends of a continuum, while molecules of intermediate size show transitional behaviour.

Small solutes may exhibit both flow-limited and diffusion-limited exchange depending on the blood flow. At low flow rates there is sufficient time for the solute (e.g. glucose) to equilibrate completely with the surrounding tissue. However, when blood flow is high and blood flow through the capillaries is rapid, there may not be sufficient time for complete equilibration and the amount of solute exchanged becomes limited by the diffusion capacity (Figure 31.5). Note that the quantity of a solute that is exchanged between the plasma and the surrounding tissue is equal to the

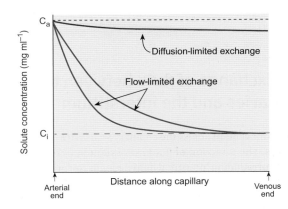

Figure 31.4 The change in plasma concentration of solutes along a capillary for those solutes whose exchange is diffusion-limited (blue curve) and flow-limited (mauve curves). In diffusion-limited exchange, the solute concentration in the plasma does not fully equilibrate with that of the interstitial fluid. In flow-limited exchange, there is rapid equilibration between the plasma and interstitial fluid and an increase in blood flow will increase the delivery of solute to the tissues.

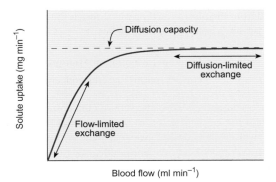

Figure 31.5 The effect of increasing blood flow on solute exchange. At low flow rates, the solute uptake (e.g. glucose) increases in proportion to the increase in blood flow. As the flow rate increases, the transit time through the capillaries falls and the increase in the rate of uptake progressively declines until a point is reached where the diffusion capacity of the capillary bed is reached and uptake becomes diffusion-limited.

difference in its concentration in the arterial and venous blood multiplied by the blood flow (ml min^{-1}). This is the Fick principle that was discussed in Chapter 28 in relation to cardiac output.

Summary

Solute exchange between blood and tissues occurs by diffusion across capillary walls. Lipophilic substances diffuse through the endothelial cells and equilibrate rapidly with the surrounding tissues (transcellular exchange). Water-soluble substances equilibrate with the surrounding tissues by diffusion through the clefts between the cells (paracellular exchange). The capillary wall is highly permeable to small molecules, especially the respiratory gases, but is relatively impermeable to proteins, which reach the interstitial fluid mainly by vesicular transport, thus allowing protein-bound hormones to reach their target tissues.

31.4 Exchange of fluid between the capillaries and the interstitium

Net fluid movement across the capillary wall is driven by pressure gradients. The direction of fluid movement between a capillary and the surrounding interstitial fluid depends on four pressures:

1. pressure within the capillary (the capillary pressure P_c)

2. pressure within the tissue surrounding the capillary (the interstitial pressure P_i)

3. osmotic pressure exerted by the plasma proteins (the oncotic pressure π_p)

4. osmotic pressure exerted by the proteins present within the interstitial fluid (π_i).

The difference between the capillary pressure and that of the interstitial fluid is the hydraulic pressure acting on the fluids. The greater this value, the greater the tendency for fluid to pass from the capillary to the interstitium. For most tissues, this physical movement of fluid amounts to about 0.1 ml each minute for every kg of tissue and must be carefully distinguished from tissue exchange by diffusion, which is a two-way process in which there is no net movement of fluid.

The forces that govern the movement of fluid between the capillaries and the interstitium are often called **Starling forces** (see Box 31.3 for further details). Since the walls of most capillaries have a low permeability to proteins, they behave as semi-permeable membranes and the osmotic pressure exerted by the plasma proteins (the **oncotic pressure**—see Box 31.2) acts to oppose the hydrostatic pressure tending to force the fluid from the capillaries to the surrounding tissues. This is offset to some extent by the osmotic pressure due to the protein present in the interstitial fluid, which acts to draw fluid from the capillaries. The algebraic sum of the various pressures is called the **net filtration pressure**. For most tissues, the net filtration pressure is positive and the hydrostatic forces favour fluid movement from the capillary to the interstitium, a process known as **filtration**. If the net filtration pressure is negative, the osmotic forces favour the movement of fluid from the interstitium to the capillary. Movement of fluid in this direction is **absorption** and usually results from a fall in capillary pressure (e.g. after haemorrhage—see Chapter 40).

The rate of fluid filtration depends both on the net filtration pressure and on the filtration coefficient, which takes into account the permeability of the capillary wall (see Box 31.3). Continuous capillaries have low filtration coefficients and the rate of filtration is correspondingly modest. In those capillary beds where there is a significant transepithelial movement of fluid (e.g. the exocrine secretory glands and the **renal glomerulus**), the capillaries are fenestrated and have large filtration coefficients. This specialization allows the relatively rapid movement of fluid from the plasma to the interstitial space.

Capillary pressure depends on the resistance of the arterioles and the venous pressure

Clearly, while a capillary module is being perfused, the pressure at the arteriolar end of the capillaries will be higher than that at the end terminating in the venules (the venular end). In most capillary beds, such as those of the skin and skeletal muscle, the capillary pressure is greater than the oncotic pressure of the plasma, the net filtration pressure is positive, and Starling forces favour filtration (see Box 31.3 and Figure 31.6). This difference is accentuated by gravity. The pressure in the capillaries of the foot increases dramatically on standing up. When a subject is lying down, the mean capillary pressure in the toes is around 4 kPa (30 mm Hg). On standing, this increases to around 13 kPa (c. 98 mm Hg). Note that the capillary pressures above the heart do not fall by a corresponding amount, as venous pressure remains close to zero (see Figure 30.7).

Since the **microcirculation** of an individual organ has a particular role, it is not surprising to find that the average pressure in the

(1₂3) Box 31.2 The plasma oncotic pressure and Donnan equilibria

The oncotic pressure of the plasma is around 3.5 kPa (25 mm Hg). Simple calculation shows that the osmotic pressure of the plasma proteins can account for only about 60 per cent of the total. Albumin, the smallest and most abundant of the plasma proteins, contributes around 1.6 kPa, while all the other plasma proteins together only contribute around 0.5 kPa at most, partly because of their relatively low abundance, and partly because of their greater molecular mass (for example, the relative molecular mass of the γ-globulins (IgGs) is around 150,000 while that of fibrinogen is 340,000).

How can the missing 40 per cent be accounted for? The plasma proteins, especially albumin, are negatively charged at normal blood pH. To maintain electroneutrality, they must be associated with a positive counter-ion which, in the case of the plasma, is sodium. This is the key to understanding why the oncotic pressure of the plasma is higher than can be accounted for by the plasma proteins alone.

A particular type of equilibrium known as a **Donnan equilibrium** exists when two compartments are separated by a membrane that is permeable to water and to ions (e.g. Na⁺ and Cl⁻), but not to large molecules such as proteins. This is the situation for the equilibrium that exists across the walls of most capillaries. The ions can diffuse freely between the two compartments, so in the absence of proteins, the composition of the two compartments will eventually become the same. If a negatively charged protein is present in one compartment (compartment I) but not the other, the situation changes. As

before, the ions can move freely between the compartments but their distribution is affected by the presence of the protein in compartment I. At equilibrium, the concentration of the diffusible cation (Na⁺ in this example) is greater in compartment I, which contains the protein, while that of chloride is less. In compartment II the situation is reversed so the concentration of sodium is less, while that of chloride is more. It can be shown from thermodynamic arguments that the relationship between the concentrations of the ions in compartments I and II is given by:

$$[Na^+]_I \times [Cl^-]_I = [Na^+]_{II} \times [Cl^-]_{II}$$

In words, the product of the concentrations of any pair of diffusible ions is the same on each side of the membrane. However, the *sum* of the concentrations of the diffusible ions is greater in compartment I, the side containing the protein. Thus the osmotic pressure of compartment I must be greater than that of compartment II, not only because of the presence of the protein but also because of the greater concentration of diffusible ions. In fact, the concentration of diffusible cations is about 0.5 mmol l⁻¹ higher in the plasma than it is in the interstitial fluid. This corresponds to an osmotic pressure of around 1.3 kPa and will bring the total oncotic pressure to around 3.5 kPa. The main text and Box 31.3 explain the importance of the oncotic pressure in the exchange of fluid between the plasma and the tissues.

capillaries varies from one organ to another. In some organs, the capillaries absorb large volumes of fluid. Examples are the capillaries of the intestinal mucosa and the peritubular capillaries of the kidneys. In these tissues, the capillary pressure is low. Moreover, the bulk transport of fluid across the adjacent epithelium keeps the

interstitial oncotic pressure low. As a result, the Starling forces favour fluid uptake by these capillaries (see Box 31.3).

How do the capillaries reabsorb interstitial fluid? For most capillary beds, the net filtration pressure is positive except when the arterioles constrict during vasomotion. During these periods,

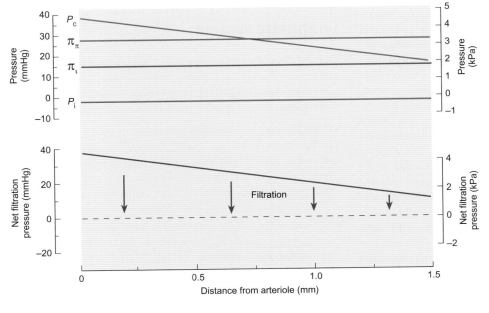

Figure 31.6 A diagram representing the factors that determine the direction of fluid movement across the wall of a capillary in the skin of the hand. At the arteriolar end, the hydrostatic pressure in the capillaries (P_c) is high, but as the blood passes along the capillary the pressure progressively falls. Nevertheless, filtration is favoured along the whole length of the capillary (indicated by the downward arrows). See Box 31.3 for further details.

(1₂3) Box 31.3 Calculation of Starling forces for capillary filtration

The pressures exerted on the capillaries of human skin can be measured and provide a clear example of the forces that determine the net filtration pressure (P_f) in a capillary bed.

At the level of the heart, the hydrostatic pressure at the arteriolar end of the capillaries of the fingers (P_c) is about 5 kPa (c. 37 mm Hg). At the venular end, the pressure is about 2 kPa (c. 15 mm Hg). The oncotic pressure of human plasma (π_p) is about 3.5 kPa (c. 26 mm Hg) and that of the interstitial fluid (π_i) is about 2 kPa (15 mm Hg). The hydrostatic pressure of interstitial fluid (P_i) is about 0.2 kPa (1.5 mm Hg) below that of the atmosphere (i.e. −0.2 kPa or −1.5 mm Hg). The filtration pressure is equal to the difference between the pressure forcing fluid from the capillary and that tending to oppose filtration and is given by the following equation:

$$P_f = (P_c - P_i) - (\pi_p - \pi_i)$$

At the arteriolar end:

$$P_f = (5 - (-0.2)) - (3.5 - 2) = +3.7 \text{ kPa (c.} + 28 \text{ mm Hg)}$$

At the venular end:

$$P_f = (2.0 - (-0.2)) - (3.5 - 2) = +0.7 \text{ kPa (c.} + 5 \text{ mm Hg)}$$

Thus at the level of the heart, the pressure in the capillaries favours filtration along their length (as the net filtration pressure is positive) and so movement of fluid from the capillaries to the interstitial fluid occurs. If the capillary pressure falls to zero (as it does when the arterioles constrict fully during vasomotion or following haemorrhage), the filtration pressure becomes negative (c. −1.5 kPa) and absorption of fluid is favoured. However, fluid reabsorption is quickly limited by the rise in the oncotic pressure of the interstitial fluid. Thus, in this case, fluid absorption is transient.

In those tissues where the continuous absorption of fluid by the capillaries is an important part of their function (e.g. the peritubular capillaries of the nephron or the capillaries of the small intestine), the net filtration pressures are negative, at least at the venular end. For example, when isotonic fluid is being reabsorbed from the gut lumen, the filtration pressure at the venular end of the capillaries is about −0.5 to −1 kPa (c. −7 mm Hg). The interstitial oncotic pressure is kept low because the absorbed fluid is removed by the capillaries and is replaced by fluid transported across the epithelial lining of the intestine.

Calculations such as these reflect the forces tending to move fluid in one direction or another across the capillary wall, assuming a steady state and that the capillary wall is impermeable to the plasma proteins. The quantities of fluid that move into or out of a capillary will depend on the hydraulic permeability of the capillary wall, the area available for fluid movement, and its permeability to plasma proteins. The full relationship can be expressed mathematically as:

$$J_v = L_p A \{(P_c - P_i) - \sigma(\pi_p - \pi_i)\}$$

where J_v is the volume of fluid filtered or absorbed per unit time; L_p is a constant that reflects the hydraulic permeability of the capillary wall; A is the area of the capillary wall; σ is the reflection coefficient, which is a measure of how freely proteins can cross the capillary wall. If the capillary wall were to be completely impermeable to proteins $\sigma = 1$ and if it were fully permeable to proteins $\sigma = 0$. In most capillaries σ is approximately 0.9. P_c, P_i are the capillary and interstitial pressures and π_p, π_i are the oncotic pressures of the plasma and interstitial fluid respectively.

---

the pressure within the capillaries declines towards zero and the net filtration pressure becomes negative. Under these conditions, fluid is reabsorbed into the capillary mainly because the oncotic pressure of the plasma is higher than that of interstitial fluid. However, as fluid is reabsorbed, the oncotic pressure of the interstitial fluid rises and the driving force for absorption declines towards zero. Thus, fluid reabsorption is transient, not sustained. It therefore appears that one function of vasomotion is to provide periods of fluid absorption, so reducing the accumulation of fluid in the interstitial space.

Taking the body as a whole, only a very small fraction of the plasma passing through the continuous capillaries is filtered (around 0.1–0.2 per cent). However, since about 4000 litres of plasma pass through the systemic capillaries every day, this results in the movement of approximately 4–8 litres of fluid from the capillaries to the interstitial space. Estimates suggest that between a half and two-thirds of this fluid is reabsorbed in the lymph nodes and the remainder is returned to the circulation via the efferent lymph (discussed in the next section).

31.5 The organization and role of the lymphatic system

The lymphatic system is intimately linked to both the circulatory system and the immune system. It consists of a series of vessels (lymphatics) that drain fluid (lymph) from the tissues and return it to the blood. The lymphatic system serves three major functions.

- First, it preserves fluid balance by returning fluid from the tissues to the blood. As discussed above, in the majority of capillary beds filtration exceeds fluid absorption by the capillaries and pericytic venules. If the excess fluid was not removed, it would accumulate in the tissues, causing them to swell. Such swelling does occur in some pathological states and is called **oedema** (see Chapter 40).

- Second, the lymphatic circulation plays an important role in the defence against infection, by carrying antigens to the lymph nodes (this is discussed in Chapter 26).

- Third, the lymphatic system plays a crucial role in the absorption of fats via the lacteals of the small intestine (see Chapter 43).

The lymph that is directly derived from the tissues is known as **afferent lymph**, as it flows towards the lymph nodes. The afferent lymph has the same composition as the interstitial fluid—i.e. it has the same ionic composition as the plasma but a lower protein content. It also has few cells. After the lymph has passed through the lymph nodes, it is then known as **efferent lymph** which is ultimately returned to the circulation via the thoracic duct and the right lymphatic duct. These empty into the subclavian veins, as shown in Figures 31.7 and 31.8.

The smallest lymphatic vessels are the **lymphatic capillaries**, which begin either as small tubes closed at their distal end (as in the lacteals of the small intestine) or as a network of small tubes with diameters in the range 10–50 μm. Their walls are very thin and consist only of a single layer of endothelial cells on a basement membrane. Like the discontinuous capillaries of the main circulation, the intercellular junctions of the walls of the lymphatic capillaries have clefts easily large enough to permit the passage of plasma proteins.

The lymphatic capillaries merge to form collecting vessels which, in turn, empty into afferent lymph trunks which are analogous to the veins of the main circulation. The afferent lymphatics differ somewhat in structure from the veins: they have thinner walls and semilunar valves occur more frequently along their length. As the lymph flows towards one of the main lymph trunks, it is supplemented by way of other afferent lymphatics before it passes through the central lymph nodes. The lymphatic system is so arranged that the lymph draining any area of the body will pass through at least one node before it reaches the main circulation. This ensures that there is a very high chance that any antigenic particle present in the tissues (such as a bacterium) will be taken up in a lymph node and trigger the appropriate immune response (see Chapter 26). Eventually the lymphatic vessels merge to form the main lymphatic trunks, each of which drains lymph from a particular region of the body as shown in Figure 31.8.

The main lymph trunks are:

- The right and left lumbar trunks, which drain lymph from the lower body, the kidneys, the reproductive organs, and the viscera of the pelvis.
- The intestinal trunk, which receives the lymph from the stomach and intestine as well as from the major abdominal organs such as the liver, pancreas, and spleen.

- The right and left bronchomediastinal trunks, which drain the bronchi and structures of the mediastinum.
- The right and left subclavian trunks, which receive lymph from the mammary glands, thoracic wall, and arms.
- The right and left jugular trunks, which receive lymph from the head and neck.

The lymph from the lower part of the body and abdomen drains into a dilated region known as the cisterna chyli, which forms the lower part of the thoracic duct. As the thoracic duct ascends it also receives lymph from intercostal lymphatics before it eventually empties into the left subclavian vein at its junction with the left jugular vein. On the right side of the body, lymph from the upper body is returned to the blood via the right jugular trunk, the right subclavian trunk, and the right bronchomediastinal trunk, which sometimes merge to form a single duct called the right lymphatic duct. The lymphatic vessels are generally rather narrow compared to the veins—for example the thoracic duct, which drains lymph from most of the body, is only about 2 mm in diameter. Note that the brain and spinal cord do not have any lymphatic drainage—the circulation of the CSF serves this function instead.

The flow of lymph cannot be a passive process as the interstitial pressure in most tissues is lower than that of the subclavian veins—about −1.5 mm Hg (−200 Pa) compared to +3 mm Hg (400 Pa). Some means of driving the lymph into the central veins is clearly needed. Two processes provide the necessary force.

1. The walls of the large lymphatic vessels contain circular smooth muscle that contracts rhythmically to pump the lymph towards the central veins, while back flow of lymph is prevented by the semilunar valves.

2. Muscular activity of the tissues surrounding the lymphatics compresses them and assists the flow of lymph. The subsequent recoil of the lymphatics as the muscles relax will draw lymph from the more distal regions.

As the lymphatic capillaries are closed at the distal end and can only empty into the larger lymphatic vessels, the process of filling the lymphatic capillaries is somewhat analogous to the filling of a rubber bulb with liquid after first expelling air.

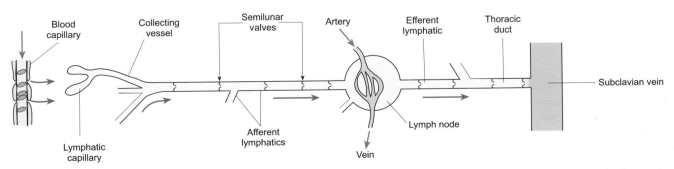

Figure 31.7 A simple diagram illustrating the relationships between the tissues, lymphatic vessels, and lymph nodes. Note the frequent occurrence of semilunar valves along the length of the main lymphatics.

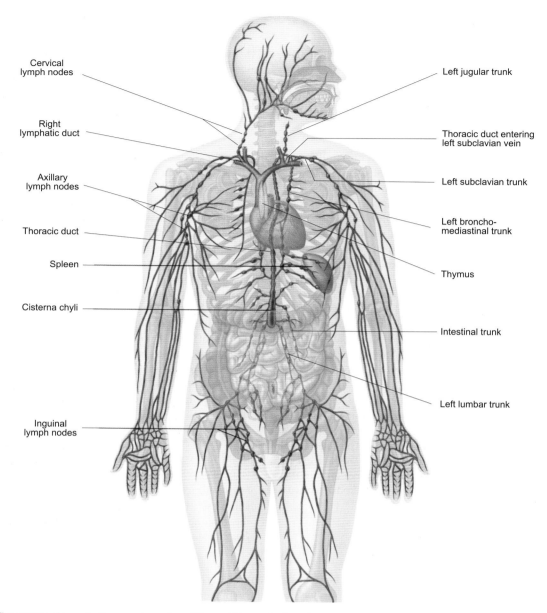

Figure 31.8 The principal lymphatic vessels and positions of the major lymph nodes. Most of the lymph formed in the tissues drains into the blood via the thoracic duct. The lymph from the upper right quadrant is returned to the blood via the right jugular trunk, the right subclavian trunk, and the right bronchomediastinal trunk. In some individuals these merge to form the right lymphatic duct, which is indicated in the figure. Note the wide distribution of the lymph nodes. (From Plate 4 in C. Blakemore and S. Jennett (eds) (2001) *Oxford companion to the body*. Oxford University Press, Oxford.)

The flow of lymph is linked to the capillary filtration rate so that the greater the rate of filtration, the higher the flow of lymph. This matching of filtration rate and lymph flow is necessary for the maintenance of fluid balance in the tissues. However, since the lymphatics have a relatively small diameter, there is a limit to the volume that they can transport in a given time. When this limit is exceeded, fluid will accumulate in the tissues, giving rise to oedema.

The lymph nodes

The lymph nodes are distributed throughout the body with the exception of the brain and spinal cord. As Figure 31.8 shows, the lymph nodes are chiefly located in the neck, axilla, groin, thorax, and abdomen. The nodes themselves are roughly ovoid or bean-shaped and vary considerably in size, the smallest being about 1 mm long and the largest 25 mm. On one side there is a small inden-

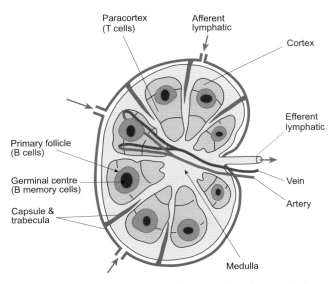

Figure 31.9 A schematic drawing of a typical lymph node. Note that the afferent lymph must pass through the central mass of tissue (the medulla) before leaving via the efferent lymphatic. The B- and T-lymphocytes are located in different regions of the nodal cortex. (Based on Fig 13.1 in J.H. Playfair (1995) *Infection and immunity*. Oxford University Press, Oxford).

centre of each node lies the medulla, which contains antibody-secreting B-cells and macrophages. The mesh-like structure of the lymph nodes allows the afferent lymph to percolate slowly through the nodal tissue until it enters the efferent lymphatic. In this way, any antigens that may be present are exposed to cells that are capable of mounting an appropriate immune response. Individual lymph nodes may enlarge greatly once they encounter an antigen or if they become infiltrated by cancerous cells.

Lymph nodes are known to modify the volume and protein composition of the lymph. The afferent lymph generally has a low oncotic pressure but after passing through the lymph nodes, the protein concentration is increased substantially—probably as a result of absorption of water and electrolytes by the vascular capillaries of the lymph nodes. It has been shown experimentally that as much as half of the afferent lymph is reabsorbed during its passage through the lymph nodes.

tation (the hilus) through which a small artery enters and the corresponding vein leaves together with the efferent lymphatic.

Each lymph node is covered by an outer capsule which sends supporting fingers of connective tissue (trabeculae) into the cortex. Beneath the capsule is a space called the sub-capsular sinus which is fed by the afferent lymphatics (see Figure 31.9). The cortex is organized into primary follicles (which contain B-cells) and secondary follicles, also known as germinal centres, that contain a class of B-cells known as memory cells. The space between the follicles is called the paracortex and is populated by T-cells. At the

Summary

The bulk flow of fluid between the plasma and the interstitial fluid is determined by the net filtration pressure, which varies somewhat from one capillary bed to another. The rate of fluid movement depends on both the net filtration pressure and the permeability of the capillary wall. Filtration occurs when the net filtration pressure is positive, and absorption of interstitial fluid is favoured when the net filtration pressure is negative. In most capillary beds, filtration exceeds absorption and the excess fluid is returned to the circulation via the lymph.

The lymph flowing towards the lymph nodes is known as afferent lymph, while that leaving is efferent lymph. Both afferent and efferent lymph have a similar ionic composition to the interstitial fluid and plasma but have many fewer cells. Efferent lymph has many more cells than afferent lymph. The flow of lymph through the lymph nodes ensures that most, if not all, antigens will stimulate an immune response.

✳ Checklist of key terms and concepts

- Most solute exchange between blood in the capillaries and tissues occurs by diffusion.

- Oxygen and carbon dioxide diffuse through the endothelial cells and equilibrate rapidly by transcellular exchange.

- Water-soluble substances diffuse through the spaces between the capillary endothelial cells. This is called paracellular exchange.

- In most vascular beds the capillary wall is almost impermeable to proteins.

- Capillary blood flow is governed by arteriolar tone.

- It varies continuously as a result of changes in vasomotor activity in the arterioles (vasomotion).

- Bulk flow of fluid occurs between the plasma and the interstitial fluid.

- The rate of fluid movement depends on both the net filtration pressure and the permeability of the capillaries that serve a particular vascular bed.

- Filtration occurs when the net filtration pressure is positive, and absorption of interstitial fluid is favoured when the net filtration pressure is negative.

- In most capillary beds, filtration exceeds absorption and the excess fluid is returned to the circulation via the lymphatic system.

- The afferent lymph has a similar composition to the interstitial fluid, while the efferent lymph (that leaving the lymph nodes) contains more cells and has a higher concentration of proteins.

- Lymph flows from the tissues to the subclavian veins, where it enters the blood.

- The flow of lymph is assisted by rhythmical contraction of the circular muscle present in the larger lymphatic vessels and by the periodic contraction of the somatic muscles.

- Back flow of lymph is prevented by semilunar valves of the afferent and efferent lymphatic trunks.

- Oedema occurs when there is an abnormal accumulation of fluid in the tissues.

 ## Recommended reading

Levick, J.R. (2010) *An introduction to cardiovascular physiology* (5th edn), Chapters 10 and 11. Hodder Arnold, London.
An eminently readable and informative introduction to current research on the microcirculation.

Review articles

Gutterman, D.D., Chabowski, D.S., Kadlec, A.O., Durand, M.J., Freed, J.K., Ait-Aissa, K., and Beyer, A.M. (2016) The human microcirculation: Regulation of flow and beyond. *Circulation Research* **118**, 157–72, doi:10.1161/CIRCRESAHA.115.305364.

Mehta, D., and Malik, A.B. (2006) Signaling mechanisms regulating endothelial permeability. *Physiological Reviews* **86**, 279–67, doi:10.1152/physrev.00012.2005.

Popel, A.S., and Johnson, P.C. (2005) Microcirculation and hemorheology. *Annual Review of Fluid Mechanics* **37**, 43–69, doi:10.1146/annurev.fluid.37.042604.133933.

 To check that you have mastered the key concepts presented in this chapter, complete the accompanying online self-assessment questions. Go to www.oup.com/uk/pocock5e/

PART EIGHT

The respiratory system

CHAPTER 32

Introduction to the respiratory system

Chapter contents

This chapter should help you to understand:

- The distinction between internal and external respiration
- The gas laws and their application to respiratory physiology
- Henry's law and the solubility of the respiratory gases
- The concept of diffusing capacity
- The composition of the alveolar, expired, and inspired air
- The respiratory exchange ratio
- The structure of the respiratory system
  - The detailed structure of the airways
  - The alveolar–capillary unit and its importance in gas exchange
  - The muscles of respiration
  - The pleural membranes
- The somatic and autonomic innervation of the respiratory system

32.1 Introduction

The energy needed by animals for their normal activities is mainly derived from the oxidative breakdown of foodstuffs, particularly that of carbohydrates and fats described in Chapter 4. During this process, which is called **internal** or **cellular respiration**, the mitochondria oxidize carbohydrates and fatty acids to generate ATP. The oxygen needed for this energy metabolism is ultimately derived from the atmosphere by the process of **external respiration**, which also serves to eliminate the carbon dioxide produced by the cells. The key process of external respiration is gas exchange between the air deep in the lungs and the blood that perfuses them. In addition to their role in gas exchange, the lungs have a variety of non-respiratory functions such as their role in trapping blood-borne particles (e.g. small fragments of blood clots) and the metabolism of a variety of vasoactive substances.

A complete understanding of the processes of breathing and gas exchange requires answers to the following questions:

1. What mechanisms are employed to cause air to move in and out of the lungs?
2. How is oxygen taken up in the lungs and carried in the blood?
3. How is carbon dioxide carried in the blood and eliminated from the body via the lungs?
4. How efficient are the lungs in matching their ventilation to their blood flow?
5. How is the respiratory rhythm generated?
6. What factors determine the rate and depth of respiration?
7. What mechanisms prevent the lungs becoming clogged with particles from the air?

This and the following four chapters discuss questions 1 and 4–7, while the transport of oxygen and carbon dioxide by the blood (questions 2 and 3) is discussed in Chapter 31. The role of the lungs in acid–base balance is discussed in Chapter 41.

Respiratory physiology employs a large number of standard abbreviations which are often used to calculate respiratory data. The most common are given in Box 32.1, together with the convention for their use.

Box 32.1 The use of symbols in respiratory physiology

Respiratory physiology makes extensive use of standard symbols to express concepts concisely and allow simple algebraic manipulations. The primary variables are given as capital letters in italics (e.g. pressure P or volume V), while the location to which they apply is given by a suffix (e.g. P_aO_2 is the partial pressure of oxygen in the arterial blood, while P_AO_2 is the partial pressure of oxygen in the alveolar air). In addition, the prefix s is used for specific, e.g. sRaw is the specific airway resistance.

Variable	Symbol used
Pressure, partial pressure or gas tension	P
Volume of gas	V
Flow of gas (l min⁻¹)	$\dot{V}$
Volume of blood	Q
Flow of blood (l min⁻¹)	$\dot{Q}$
Fractional concentration of dry gas	F
Resistance	R
Conductance	G

Variable	Suffix
Arterial	a
Capillary	c
Venous	v
Mixed venous	$\bar{v}$
Alveolar	A
Inspired	I
Expired	E
End-tidal air (≡ alveolar gas)	E′
Tidal	T
Lung	L
Barometric	B
Dead space	D
Pleural space	pl
Airway	aw
Chest wall	w
Elastic	el
Resistive	res
Total	tot

32.2 The application of the gas laws to respiratory physiology

External respiration involves the exchange of oxygen and carbon dioxide between the pulmonary blood and the air in the lungs. To understand the factors that determine the uptake and loss of the respiratory gases from the lungs, it is helpful to have a knowledge of their physical properties.

Avogadro's law states that equal volumes of different gases at the same pressure and temperature contain the same number of molecules. A mole of any substance has a mass that is equal to its molecular mass in grams (see Chapter 2). The number of molecules in a mole (Avogadro's number) is 6.0232×10^{23} which, for an ideal gas, occupies a volume of 22.4 litres at 0°C and 101 kPa (standard temperature and pressure or STP, discussed later in this section).

Boyle's law states that the pressure exerted by a gas is inversely proportional to its volume, so that

$$P \propto 1/V$$

Charles's law states that the volume occupied by a gas is directly related to the absolute temperature (T):

$$V \propto T$$

These two laws are combined with Avogadro's law in the **ideal gas law**, which states that

$$PV = nRT$$

where n is the number of moles of gas and R is the gas constant (8.31 joules K⁻¹ mol⁻¹).

From the ideal gas law, the pressure and volume of a given mass of gas are related to the absolute temperature by the following relationship:

$$\frac{P_1 V_1}{T_1} = \frac{P_2 V_2}{T_2}$$

which makes clear that the volume of gas will depend on both the temperature and the pressure of that gas.

When a liquid or solid is in equilibrium with a volume of gas, it exerts a pressure in the gas phase known as its **vapour pressure** which depends only on the temperature: the warmer the temperature the higher the vapour pressure. In the case of the **respiratory system**, only the vapour pressure of water is of significance. In the lungs, the water vapour pressure is 6.2 kPa (47.1 mmHg) at 37°C (normal body temperature). The vapour pressure of water in the inspired air depends on the prevailing relative humidity, which is usually significantly less than the saturated water vapour pressure.

The air we breathe is a mixture of gases consisting mainly of nitrogen, oxygen, carbon dioxide, and water vapour. **Dalton's law of partial pressures** states that the total pressure is the sum of the pressures that each of the gases would exert if it were present on its own in the same volume. Thus:

$$P_T = P_{N_2} + P_{O_2} + P_{CO_2} + P_{H_2O}$$

where P_T is the total pressure of the gas mixture and $P_{N_2}, P_{O_2}, P_{CO_2}$, and P_{H_2O} are the partial pressures of nitrogen, oxygen, carbon dioxide, and water vapour respectively.

Thus, a volume of gas collected from a subject in a gas sampling bag (known as a Douglas bag) will depend both on the atmospheric pressure, the room temperature, and the water vapour pressure. To be able to compare samples of gas collected at different times, the volume of a gas may be expressed in one of two ways:

- as STPD—standard temperature and pressure dry. This gives the volume of gas after removal of water vapour at standard temperature (273 K or 0°C) and pressure (101 kPa or 760 mm Hg).
- As BTPS—body temperature and pressure saturated with water vapour, i.e. at 37°C (310 K) and a water vapour pressure of 6.2 kPa (47 mm Hg). This would be the volume of gas expired from the lungs.

Because a given mass of gas will occupy significantly different volumes at 0°C and at 37°C, it is essential to state which standard is being employed. Box 32.2 gives the formulae for converting a volume of gas at ambient temperature and pressure (ATP) to BTPS or STPD.

Like liquids, gases flow from regions of high pressure to regions of lower pressure (both are fluids). Moreover, since the partial pressure of a gas is a direct measure of its molar concentration, a gas will also diffuse from a region of high partial pressure to one of a lower partial pressure even though the total pressure in the gas phase is uniform. The rate of diffusion under these circumstances is inversely proportional to the relative molecular mass of the gas, i.e. the greater the molecular mass the slower the rate of diffusion. This is known as **Graham's law**. However, since the relative molecular masses of oxygen, carbon dioxide, and nitrogen are 32, 44, and 28 respectively, all three gases diffuse in the gas phase at very similar rates.

Solubility of gases

The amount of oxygen that is dissolved is proportional to its partial pressure in the gas phase. This is **Henry's law**, which can be written as:

$$V = s.P$$

where s is the solubility coefficient ($ml\,l^{-1}\,kPa^{-1}$ or $ml\,l^{-1}\,mm\,Hg^{-1}$), V is the volume of dissolved gas in a litre of the liquid phase, and P is the partial pressure of the gas under consideration. For oxygen at body temperature (37°C), s is 0.225 $ml\,l^{-1}\,kPa^{-1}$ (0.03 $ml\,l^{-1}\,mm\,Hg^{-1}$). Thus the volume (V) of oxygen dissolved in one litre of water or plasma when PO_2 is 13.33 kPa (100 mm Hg) is:

$$V = 0.225 \times 13.33 = 3\ ml$$

Similar calculations can be made for carbon dioxide, where $s = 5.1\ ml\,l^{-1}\,kPa^{-1}$ (0.68 $ml\,l^{-1}\,mm\,Hg^{-1}$), and for nitrogen, where $s = 0.112\ ml\,l^{-1}\,kPa^{-1}$ (0.015 $ml\,l^{-1}\,mm\,Hg^{-1}$).

Note that this relationship applies only to dissolved gas. Where the gas enters into chemical combination, the total amount in the liquid phase is the *sum* of that chemically bound plus that in physical solution.

(1₂3) Box 32.2 Conversion of gas volumes to BTPS and STPD

Conversion from ambient temperature and pressure to BTPS

As air is breathed in, it is heated and humidified. This leads to an increase in its volume that can be calculated from the universal gas law:

$$\frac{P_{ATP}.V_{ATP}}{T_{ATP}} = \frac{P_{BTPS}.V_{BTPS}}{T_{BTPS}}$$

The following formula can be derived from this relationship:

$$V_{BTPS} = V_{ATP} \times \frac{273+37}{273+T_A} \times \frac{P_B - P_{H_2O}}{P_B - 6}$$

where:

V_{BTPS}	is the volume at body temperature and pressure saturated
V_{ATP}	is the volume of air inhaled at ambient temperature and pressure
T_A	is the ambient temperature
P_B	is the barometric pressure
P_{H_2O}	is the water vapour pressure of the ambient air

The numerical constants 37 and 6 are the body temperature in degrees Celsius and the saturated water vapour pressure in kPa (equivalent to 47 mm Hg). Thus, for a litre of room air inhaled when the temperature is 20°C, the barometric pressure is 100 kPa (c. 750 mm Hg) and the water vapour pressure is 2 kPa (c. 17 mm Hg), the change in the volume of the thorax will be:

$$V_{BTPS} = 1 \times \frac{310}{293} \times \frac{100-2}{100-6} = 1.103\ litres$$

So, the expansion of the chest will be just over 10 per cent greater than the volume of gas inhaled.

Conversion from ambient temperature and pressure to STPD

The volume of oxygen absorbed or carbon dioxide exhaled is expressed as STPD because, in this case, it is necessary to express the number of moles of gas exchanged. The volume of one mole of gas is 22.4 litres at STP (0°C and 101 kPa or 760 mm Hg). In this case, the conversion uses the formula:

$$V_{STPD} = V_{ATP} \times \frac{273}{273+T_A} \times \frac{P_B - P_{H_2O}}{101}$$

Thus 1 litre of humidified oxygen at the same ambient temperature and pressure as the previous example would occupy a volume at STPD of:

$$V_{STPD} = 1 \times \frac{273}{298} \times \frac{100-2}{101} = 0.89\ litres$$

The above calculations are for air saturated with water vapour. This is normally the case for gas samples taken from a spirometer. For ambient air, the percentage saturation of the air (the relative humidity) must be taken into account. To do this, simply multiply the saturated water vapour pressure for the ambient temperature by the relative humidity expressed as a fraction (70 per cent humidity is 0.7, etc.).

In respiratory physiology, the concentration of dissolved gases is usually given as their partial pressures even when they are present in a solution with no gas phase (e.g. in the arterial blood). The partial pressure of a gas can readily be converted to the equivalent molar concentration using Avogadro's law. For example, when carbon dioxide has a partial pressure of 5.33 kPa (40 mm Hg), each litre of plasma will dissolve

$$5.33 \times 5.1 \times 10^{-3} = 2.72 \times 10^{-2} \text{ litres (27.2 ml)}$$

Since at STP, 1 mole of CO_2 occupies 22.4 litres. this corresponds to

$$2.72 \times 10^{-2}/22.4 = 1.2 \times 10^{-3} \text{ moles l}^{-1} \text{ (1.2 mmol l}^{-1}\text{)}$$

Diffusion of dissolved gases in the lungs

When oxygen, for example, is taken up by the blood it must first dissolve in the aqueous phase that lines the lungs and then diffuse across the alveolar membrane into the blood. The rate at which oxygen and carbon dioxide diffuse from the aqueous lining of the alveoli to the blood is governed by **Fick's law of diffusion** (see Chapter 31, Box 31.1).

The extreme thinness of the alveolar membranes and their large area helps to optimize the diffusion of the respiratory gases. In addition to these factors, the rate at which the respiratory gases diffuse will depend on their solubility and their concentration gradient. The importance of solubility in determining the rate of diffusion is evident with carbon dioxide. This gas is about 20 times more soluble in the alveolar membranes than oxygen. Thus, although it has a concentration gradient one-tenth that of oxygen, carbon dioxide diffuses from the blood to the alveolar air about twice as fast as oxygen diffuses from the alveoli to the blood. The ability of a given gas to diffuse between the alveolar air and the blood is measured by its **diffusing capacity**, also known as its **transfer factor**, which is discussed in Chapter 34 (Section 34.6).

The composition of the expired air

The expired air contains less oxygen and more carbon dioxide than the inspired air. Standard values for the partial pressures of the gases present in inspired, expired, and alveolar air are given in Table 32.1. Note that, although nitrogen is not exchanged with the blood, its partial pressure changes as it becomes diluted by the water vapour and carbon dioxide from the lungs.

The ratio of the carbon dioxide produced to the oxygen uptake is called the **respiratory exchange ratio** (R) which, under steady-state conditions, is equal to the metabolic respiratory quotient (RQ)—see Chapter 47 for further details.

$$\text{respiratory exchange ratio} = \frac{[\text{volume of } CO_2 \text{ produced}]}{[\text{volume of oxygen taken up}]}$$

Under normal resting conditions, the respiratory exchange ratio varies according to the type of food being metabolized to produce ATP. It ranges from 0.7 when fats are the principal substrate to 1.0 for

Table 32.1 Standard values for respiratory gases

	N_2	O_2	CO_2	H_2O
Inspired air (kPa)	79.6	21.2	0.04	0.5
(mm Hg)	597	159	0.3	3.7
% total	78.5	20.9	0.04	0.5
Expired air (kPa)	75.5	16	3.6	6.3
(mm Hg)	566	120	27	47
% total	74.5	15.8	3.5	6.2
Alveolar air (kPa)	75.9	13.9	5.3	6.3
(mm Hg)	569	104	40	47
% total	74.9	13.7	5.2	6.2

carbohydrates. Usually R is around 0.75–0.85 as both carbohydrates and fats are metabolized. During starvation, protein becomes an important source of energy and R has a value of about 0.8.

Summary

The volume of a given quantity of gas depends on both its temperature and its pressure. In respiratory physiology, the volume of a gas is given either as standard temperature and pressure dry (STPD) or as body temperature and pressure saturated with water vapour (BTPS). The amount of a gas in solution is proportional to its partial pressure in the gas phase (Henry's law) and its rate of diffusion in body fluids is governed by Fick's law.

The expired air has less oxygen and more carbon dioxide than room air. The ratio of the amount of carbon dioxide expired to the amount of oxygen taken up is known as the respiratory exchange ratio, which depends on the nature of the foodstuffs being metabolized.

32.3 The structure of the respiratory system

The lungs are the principal organs of the respiratory system, and a diagram of their arrangement within the chest is shown in Figure 32.1. They are supplied with blood by the pulmonary circulation (see Chapter 33) and provide the surface over which oxygen is absorbed and carbon dioxide is excreted.

As the lungs are situated in the chest, air from the atmosphere must pass through the nose or mouth and enter the airways before it can be directed to the respiratory surface where gas exchange occurs. During quiet breathing, air is normally taken in via the nose, but during heavy exercise, air is taken in via the mouth, which offers much less resistance to air flow. Although the nasal passages offer a high resistance to airflow, they moisten and warm the air during its passage to the lungs. After entering the nose or mouth, the air passes through the pharynx to the larynx which, like the nose, is a significant source of resistance to the flow of air. This property is exploited in vocalization.

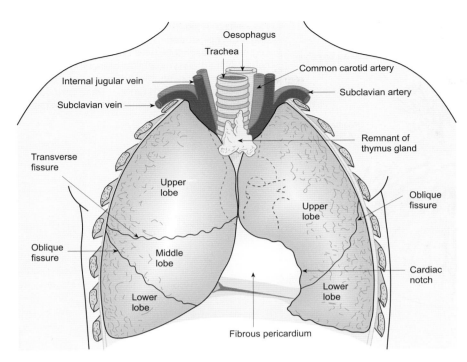

Figure 32.1 Diagrammatic representation of the disposition of the lungs within the thorax. The dotted lines indicate the positions of the aortic arch and pulmonary arteries. Note that the right lung has three lobes while the left lung has two. The heart lies on the left side within the pericardium. (Based on Fig 5.3.2 in vol. 2 of P.C.B. McKinnon and J.F. Morris (2005) *Oxford textbook of functional anatomy*, 2nd edition. Oxford University Press, Oxford.)

The airways begin with the paired nostrils (the **anterior nares**) which lead to the nasal cavities, which are separated along the midline by the nasal septum. Three projections from lateral walls of the nasal cavities form the **nasal conchae** or turbinates, each consisting of a thin layer of bone covered by a layer of ciliated respiratory epithelium. This arrangement divides the airflow from each nostril into four narrow streams and exposes the incoming air to a large area of nasal mucosa. The olfactory epithelium, the peripheral organ serving the sense of smell, is located in the roof of the nasal cavity above the superior conchae. The nasal cavities communicate with the **pharynx** via the posterior nares. The oropharynx forms part of the digestive and respiratory systems and is connected to the **larynx** via an opening known as the glottis, which is protected by a flap of cartilage called the epiglottis that prevents food particles entering the larynx (see Chapter 43). Below the larynx lies the **trachea**, which is the first of the airways of the lower respiratory system; the airways above the trachea form the upper respiratory system.

In an adult man, the trachea is about 1.8 cm in diameter and 12 cm in length. It is the first component of the respiratory tree—the branching set of tubes that link the respiratory surface to the atmosphere. In the upper chest, the trachea branches to form the two main **bronchi**—one for each lung. The right bronchus has a larger diameter than the left. In turn, the bronchi branch to give rise to two smaller branches on the left and three on the right, corresponding to the lobes of the lung. (The right lung has three lobes while the left has two.) Within each lobe, the bronchi divide into two smaller branches and these smaller branches also divide into two, and so on until the final branches reach the respiratory surface. The airways from the trachea to the respiratory bronchioles receive their blood supply via the bronchial circulation, which is discussed in Chapter 34.

In all, there are 23 generations of airways between the atmosphere and the alveoli. The trachea is generation 0. It bifurcates asymmetrically to give rise to the two main bronchi, which form generation 1 (see Figure 32.2). The bronchi serving the lobes of the lungs (the **lobar bronchi**) form generations 2 and 3. Generation 4 serves the segments within the lobes (**segmental bronchi**). Small bronchi form the fifth to the eleventh generations. From the twelfth to the nineteenth generations, the airways are known as **bronchioles**. The sixteenth generation that links the bronchioles to the respiratory surface are known as the **terminal bronchioles**.

The airways as far as the terminal bronchioles are concerned with warming and moistening the air on its way to the respiratory surface. They are known as **conducting airways** and play no significant part in gas exchange. From generation 17 to generation 19, the airways begin to participate in gas exchange and these are known as the **respiratory bronchioles**. These eventually give rise to the **alveolar ducts**, from which the principal gas exchange structures arise. These are the **alveolar sacs**, which consist of two or more **alveoli**. The respiratory bronchioles, alveolar ducts, and alveoli comprise the **transitional** and **respiratory airways**, which provide a total area for gas exchange of about 60–80 m² in an adult.

The airways

The structure of the trachea, bronchi, and bronchioles is illustrated in Figures 32.3 and 32.4. The trachea and the primary bronchi are held open by C-shaped rings of cartilage. In the smaller bronchi, this role is taken by overlapping plates of cartilage. The bronchioles, which are less than 1 mm diameter, have no cartilage. The absence of cartilage from the bronchioles allows them to

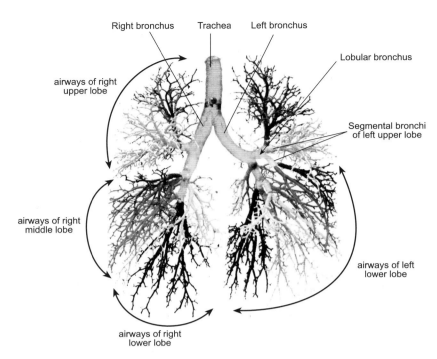

Right bronchus   Trachea   Left bronchus

Lobular bronchus

airways of right
upper lobe

Segmental bronchi
of left upper lobe

airways of right
middle lobe

airways of left
lower lobe

airways of right
lower lobe

Figure 32.2 A plastic cast of the trachea, bronchi, and broncho-pulmonary segments. The different segments of each lobe have been given a different colour. Note the asymmetrical branching of the trachea. (Based on Figure 5.3.7 in vol. 2 of P.C.B. McKinnon and J.F. Morris (2005) *Oxford textbook of functional anatomy*, 2nd edition. Oxford University Press, Oxford.)

be easily collapsed when the pressure outside the lung exceeds the pressure in the airways. This occurs during a forced expiration (see Chapter 33). Smooth muscle is found in the walls of all the lower airways including the alveolar ducts, but not in the walls of the alveoli themselves. In the terminal bronchioles, the smooth

muscle accounts for much of the thickness of the wall. The outermost part of the bronchiolar wall—the adventitial layer—is composed of dense connective tissue including elastic fibres (see Figure 32.4c).

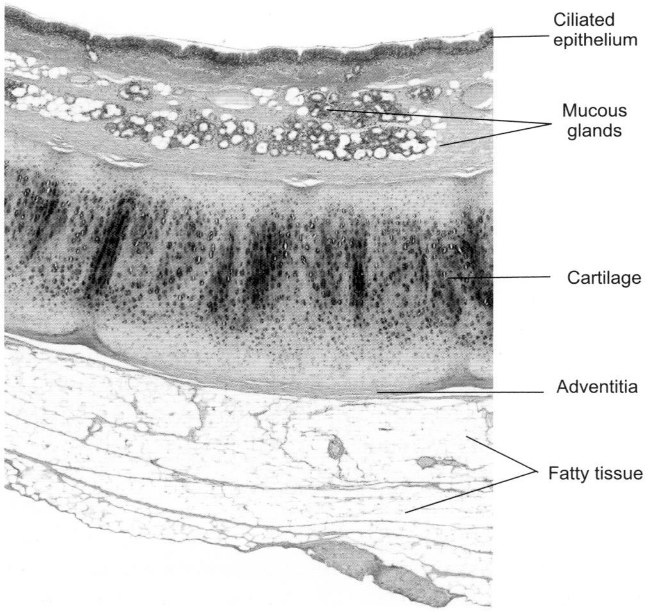

From the nasal passages to the small bronchi, the airways are lined with a pseudo-stratified columnar ciliated epithelium that contains many mucus-secreting goblet cells (Figure 32.5). Beneath the epithelial layer, there are numerous submucosal glands that discharge serous secretions into the bronchial lumen (see Figure 32.3). In the bronchioles, the epithelium progressively changes to become a simple ciliated cuboidal epithelium. The cilia beat continuously and slowly move the mucus secreted by the goblet cells and submucosal glands towards the mouth. This arrangement is known as the **mucociliary escalator**, which plays an important role in the removal of inhaled particles (see Chapter 36). The epithelium of the bronchioles also contains non-ciliated cells that are probably secretory in function. The characteristics of the various airways are summarized in Table 32.2.

The site of gas exchange is the **alveolar–capillary unit**. There are about 300 million alveoli in the adult lung, each of which is almost completely enveloped by pulmonary capillaries (see Figure 32.6) which provides a huge area for gas exchange by diffusion. The walls of the alveoli (the alveolar septa) consist of a thin epithelial layer comprising two types of cell: the alveolar type I and type II cells (Figure 32.7a). The type I cells are squamous epithelial cells, while the type II cells are thicker and produce the fluid layer that lines the alveoli. The type II cells also synthesize and secrete **pulmonary surfactant** (see Section 33.4).

Beneath the alveolar epithelium lie the pulmonary capillaries. The cell membranes of the alveolar epithelial cells and

Ciliated
epithelium

Mucous
glands

Cartilage

Adventitia

Fatty tissue

Figure 32.3 A section through dog trachea that shows the principal tissue layers. Note the thick layer of cartilage (stained green) and the large number of secretory glands just beneath the epithelium.

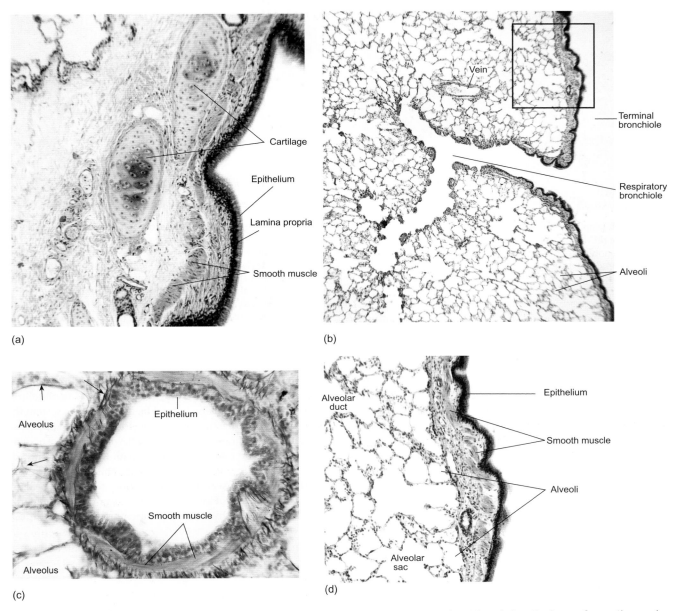

Figure 32.4 The structure of the airways. (a) A cross-section of a bronchus with its plates of cartilage below the layer of smooth muscle. (b) A horizontal section through a terminal bronchiole where it branches to give rise to a respiratory bronchiole; note the absence of cartilage. The outlined area is shown at greater magnification below in panel (d), in which the thin layer of smooth muscle in the terminal bronchiole is clearly visible. The outer layer of connective tissue contains elastic fibres which are more evident in panel (c), where the specimen was treated with Verhoeff's elastic stain. Arrows indicate elastic fibres in the walls of the alveoli and bronchiole.

pulmonary capillary endothelial cells are in close apposition, and the pulmonary blood is separated from the alveolar air by as little as 0.5 μm, as shown in Figure 32.7. Interspersed between the capillaries in the walls of the alveoli are the elastic and collagen fibres that form the connective tissue of the lung. This connective tissue links the alveoli together to form the lung **parenchyma**, which is sponge-like in appearance. Neighbouring alveoli are interconnected by small air passages called the **pores of Kohn**.

The chest wall

The lungs are not capable of inflating themselves; inflation is achieved by changing the dimensions of the chest wall by means of the respiratory muscles (see Section 33.3). The principal respiratory muscles are the **diaphragm** and the **internal and external intercostal muscles**. The external intercostal muscles are arranged in such a way that they lift the ribs upwards and outwards as they contract. The internal intercostal muscles pull the ribs downwards,

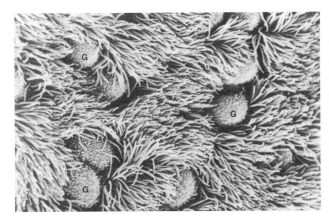

Figure 32.5 A scanning electron micrograph of the ciliated epithelium of rat trachea. Note the abundant cilia and the interspersed goblet cells (labelled G). (From P. Andrews (1974) *American J Anatomy* **139**, 421.)

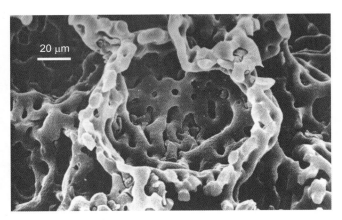

Figure 32.6 Scanning electron micrograph of a resin cast of the capillary network of the alveoli. The cast was prepared by injecting a resin into the pulmonary artery. After the resin had polymerized, the biological tissue was digested to leave a cast of the blood vessels in their original position surrounding the space previously occupied by the alveoli. Note that the capillaries almost completely envelop the alveolar space. (Image from R.G. Kessel and R.H. Kardon, *Tissues and Organs*, WH Freeman (1979). © Drs. Kessel & Kardon/Visuals Unlimited, Inc.)

in opposition to the external intercostal muscles. In addition, further muscles, which are not involved during normal quiet breathing, may be called upon during exercise. These are the **accessory muscles** (the scalenes and sternocleidomastoids), which assist inspiration, and the **abdominal muscles** (the rectus abdominus and external oblique abdominal muscles), which assist in expiration (see Figure 32.8).

The chest wall is lined by a pair of serous membranes known as the pleura, which are continuous at the apex and root of the lungs. They each consist of mesothelial cells overlying a delicate layer of connective tissue containing collagenous and elastic fibres. The **parietal pleura** is the outermost of these membranes (Figure 32.9) and is separated from the inner membrane, the **visceral pleura**, by a thin layer of liquid (the pleural fluid) which serves to lubricate

their movement during respiration. In healthy adults, the total volume of intrapleural fluid is only about 10 ml. It is an ultrafiltrate of plasma and is normally drained by the lymphatic system that lies beneath the visceral pleura. Beneath the visceral pleura lies the limiting membrane of the lung itself, which consists of dense connective tissue that is continuous with the membranes that divide the lungs into the individual lobes. This connective tissue, together with the visceral pleura, provides a gas-tight seal and limits the expansion of the lungs. As the lungs are separated from the chest wall only by the pleural membranes, in health they

Table 32.2   The structure of the various generations of airways

Anatomical name	Airway generation	Cartilage	Epithelium type	Smooth muscle	Gas exchange
Trachea	0	C-shaped rings	Ciliated pseudo-stratified columnar	Spans gap in cartilage rings	none
Bronchi	1–4	C-shaped rings	Ciliated pseudo-stratified columnar	Spiral bundles of fibres	none
Small bronchi	5–11	Cartilage plates	Ciliated pseudo-stratified columnar	Spiral bundles of fibres	none
Bronchioles	12–15	0	Ciliated pseudo-stratified columnar	Spiral bundles of fibres	none
Terminal bronchioles	16	0	Ciliated columnar	Spiral bundles of fibres	none
Respiratory bronchioles	17–19	0	Ciliated cuboidal	Bundles of fibres	+
Alveolar ducts	20–22	0	Squamous	0	++
Alveoli	23	0	Squamous	0	+++

Note: Generations 0–16 play no significant part in gas exchange and form the conducting zone, while generations 17–23 do participate in gas exchange and form the respiratory zone. The transition from one generation of airways to the next is gradual, and the thickness of the smooth muscle layer in the bronchioles becomes progressively less with each successive airway generation. The epithelia of the airways generally contain more than one cell type (i.e. they are heterogeneous in composition). The ciliated epithelium of the upper airways also contains a relatively large population of mucus-producing goblet cells. The walls of the alveoli are lined with squamous alveolar cells (type I cells) but also have cells known as alveolar type II cells, which secrete lung surfactant.

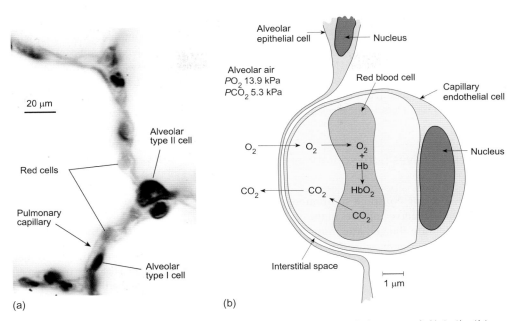

(a)   (b)

Figure 32.7 Panel (a) shows part of the walls of several alveoli at high magnification (scale bar: 20 μm). Note the thinness of the walls and the proximity of the red cells to air spaces. The alveolar type I cells are thin with a flattened nucleus and are difficult to distinguish from the capillary endothelial cells. The type II cells (or great alveolar cells) are roughly cuboidal in shape and have more rounded nuclei. As here, they tend to be located where the alveolar walls join. In thin sections they can be seen to possess a vesicular cytoplasm, consistent with their secretory role. Panel (b) shows a diagrammatic representation of the layers separating the alveolar air space from the blood in the pulmonary capillaries.

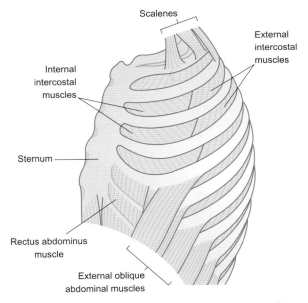

Figure 32.8 The arrangement of the respiratory muscles of the human chest wall. Between each pair of ribs there are two layers of muscle: the external intercostal muscles and the internal intercostal muscles. The figure illustrates the orientation of the muscle fibres in these muscle groups. Note that the angle of the external intercostal muscles allows them to lift the rib cage when they shorten, so expanding the chest. The internal intercostal muscles act to lower the rib cage. The contraction of the accessory muscles also acts to lift the rib cage, while contraction of the abdominal muscles tends to force the diaphragm (not shown) upwards into the chest, assisting expiration.

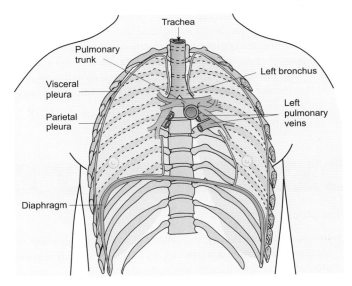

Figure 32.9 The arrangement of the diaphragm and pleural membranes. In life, only a thin layer of fluid normally separates the visceral and parietal pleura. The diaphragm forms a continuous sheet that separates the thoracic and abdominal cavities. It consists of connective tissue in the central region and is muscular in the periphery. Although the diaphragm normally behaves as a single unit, each half (or hemi diaphragm) is separately innervated via the left and right phrenic nerves. (Based on Fig 5.3.3 in Vol. 2 of P.C.B. McKinnon and J.F. Morris (2005) *Oxford textbook of functional anatomy*, 2nd edition. Oxford University Press, Oxford.)

occupy the entire cavity of the chest except the mediastinum (which contains the heart and great vessels). If the chest wall is subjected to a penetrating injury, the pleural cavity on the affected side will expand due to the accumulation of air and the lung on the affected side will partially or totally collapse, a condition known as **pneumothorax**. In certain pathological states, the pleural cavity may expand considerably as fluid accumulates, resulting in a **pleural effusion**.

The respiratory system receives both somatic and autonomic nerves

The respiratory muscles do not contract spontaneously. Rhythmical breathing depends on nerve impulses in the phrenic and intercostal nerves, which are the somatic motor nerves serving the diaphragm and intercostal muscles. The rhythmical discharge of these nerves is determined by the activity of specific groups of nerve cells in the brainstem, whose activity is discussed in Chapter 35.

The smooth muscle of the bronchi and bronchioles is innervated by parasympathetic fibres, which reach the lungs via the vagus nerves. Activation of these nerve fibres causes narrowing of the bronchi (bronchoconstriction), dilatation of the bronchial blood vessels, and mucus secretion. Other substances that constrict the airways include substance P and neurokinin A. In humans, the bronchial smooth muscle is sparsely innervated by sympathetic fibres and the bronchodilatation resulting from stimulation of the sympathetic nerves appears to result from the activity of nonadrenergic noncholinergic (NANC) fibres, which secrete vasoactive intestinal peptide and nitric oxide. However, bronchodilatation does occur in response to circulating adrenaline and noradrenaline (epinephrine and norepinephrine), which act on β-adrenergic receptors to cause relaxation of the smooth muscle. Inhalation of β-adrenergic drugs such as salbutamol is used to overcome the **bronchospasm** that occurs during asthmatic

attacks. Circulating adrenaline and noradrenaline act on α-receptors to cause pulmonary vasoconstriction, but pulmonary arterioles are not subject to significant autonomic control. Rather, they are regulated by the alveolar PO_2 and PCO_2 as discussed in Chapter 34, Section 34.4.

In addition to the somatic motor and autonomic nerves, the lungs also have slowly adapting stretch receptors, irritant receptors, and pulmonary C-fibre endings. These send information to the CNS via visceral afferent fibres in the vagus nerves and play an essential role in the respiratory reflexes, which are discussed in Chapter 34.

Summary

The airways consist of the nasopharynx, larynx, trachea, bronchi, and bronchioles and are lined by a mucociliary epithelium. The trachea and bronchi are kept open by rings or plates of cartilage. The bronchioles have no cartilage and their walls consist mainly of smooth muscle.

From the trachea, the airways branch dichotomously for 23 generations to reach the alveoli. The first 16 generations play no significant part in gas exchange and are known as the conducting airways. The remaining seven generations (the respiratory bronchioles, alveolar ducts, and alveoli) comprise the transitional and respiratory airways. The alveoli are the principal site of gas exchange. Their walls consist of a very thin epithelium beneath which lies a dense network of pulmonary capillaries. The alveolar walls also contain some connective tissue.

The chest wall is formed by the rib cage, the intercostal muscles, and the diaphragm. It is lined by the pleural membranes and forms a large gas-tight compartment that contains the lungs. The chest wall is thus an integral part of the respiratory system. The muscles of respiration receive their motor innervation via the phrenic and intercostal nerves.

✳ Checklist of key terms and concepts

The properties of gases

- In respiratory physiology, the volume of a gas is given either as standard temperature and pressure dry (STPD) or as body temperature and pressure saturated with water vapour (BTPS).
- The amount of a gas in solution is proportional to its partial pressure in the gas phase (Henry's law).
- The expired air has less oxygen and more carbon dioxide than room air.
- The ratio of the amount of carbon dioxide expired to the amount of oxygen taken up is known as the respiratory exchange ratio.
- The respiratory exchange ratio depends on the nature of the foodstuffs being metabolized.

The structure of the respiratory system

- The upper respiratory tract consists of the nose, nasopharynx, and larynx.
- The lower respiratory tract encompasses the trachea, bronchi, bronchioles, alveolar ducts, and alveoli.
- Below the trachea, the airways branch dichotomously for 23 generations to reach the alveoli.
- The first 16 generations play no significant part in gas exchange. They are known as the conducting airways.
- The respiratory bronchioles, alveolar ducts, and alveoli are known as the transitional and respiratory airways.

- The airways are lined by a ciliated epithelium that contains many mucus-secreting cells. This is supported by the lamina propria, which consists of connective tissue. Beneath the lamina propria is a layer of smooth muscle which rests on a submucosa of connective tissue.

- The lamina propria of the trachea and bronchi contains numerous glands which secrete a mucus-rich fluid that serves to trap foreign particles and moisten the air entering the lungs.

- The trachea and bronchi are kept open by rings or plates of cartilage.

- The bronchioles have no cartilage and their walls consist of smooth muscle and connective tissue.

- The alveoli are the principal sites of gas exchange. Their walls consist of a very thin epithelium beneath which lies a dense network of pulmonary capillaries.

- The smooth muscle of the bronchi and bronchioles is innervated by cholinergic parasympathetic fibres.

- The lung parenchyma and bronchial tree have stretch receptors and receptors that respond to irritants.

Recommended reading

Dejours, P. (1966) *Respiration*, Chapters 2 and 3. Oxford University Press, New York.

An old book that gives an excellent quantitative treatment that is suitable for graduate students and those considering research in human respiratory physiology.

Hlastala, M.P., and Berger, A.J. (2001) *Physiology of respiration* (2nd edn), Chapters 1 and 2. Oxford University Press, New York.

A good introduction to the structure of the respiratory system and the physical principles that underlie the process of respiration.

Lumb, A.B. (2017) *Nunn's applied respiratory physiology* (8th edn), Chapter 1. Churchill-Livingstone, London.

Slonim, N.B., and Hamilton, L.H. (1987) *Respiratory physiology*, Chapters 1–3. Mosby, St Louis.

Now rather old but still well worth reading for its excellent coverage and clear explanations.

Review articles

Kummer, W. (2011) Pulmonary vascular innervation and its role in responses to hypoxia— Size matters! *Proceedings of the American Thoracic Society* **8**, 471–6.

Paredi, P., and Barnes, P.J. (2009) The airway vasculature: Recent advances and clinical implications. *Thorax* **64**, 444–50.

 To check that you have mastered the key concepts presented in this chapter, complete the accompanying online self-assessment questions. Go to www.oup.com/uk/pocock5e/

CHAPTER 33

The mechanics of breathing

This chapter should help you to understand:

- The lung volumes: tidal volume, vital capacity, residual volume, and total lung capacity
- The mechanics of inspiration and expiration
- The intrapleural pressure
- Pressure changes during the respiratory cycle
- Airways resistance
- The work of respiration
- Pulmonary surfactant and its role in reducing the work of breathing
- Tests of ventilatory function

33.1 Introduction

It is common knowledge that breathing is associated with changes in the volume of the chest. During inspiration, the chest is expanded and air enters the lungs. During expiration, the volume of the chest decreases and air is expelled from the lungs. This movement of air into and out of the lungs is known as **ventilation**. The volume of air entering and leaving the lungs each minute is called the **minute volume** and is the product of the air taken in each breath (the tidal volume—discussed in the next section) and the number of breaths taken each minute (the frequency of breathing, also called the **respiratory rate**). Like the heart rate, the respiratory rate at rest is highly variable. The resting rate is highest in newborns

(30–60 breaths per minute) and in young children (20–30 breaths per minute). In older children and adults the respiratory rate at rest is 12–20 breaths per minute. In this chapter, the mechanisms responsible for the changes in the dimensions of the chest will be examined followed by a discussion of the factors governing the flow of air in the airways.

33.2 The lung volumes

The volume of air that moves in and out of the chest during breathing can be measured with the aid of a **spirometer**, which consists of an inverted bell that has a water seal to form an airtight chamber. The bell is free to move in the vertical direction, and the movements are recorded onto a chart or logged by a computer (see Figure 33.1).

The relationship between the various lung volumes is shown in Figure 33.2. When the chest is expanded to its fullest extent and the lungs are allowed time to inflate fully, the amount of air they contain is at its maximum. This is the **total lung capacity**. If this is followed by a maximal expiration, the lungs will still contain a volume of air that cannot be expelled. This is called the **residual volume** (RV). The amount of air exhaled during a maximal expiration following a maximal inspiration is called the **vital capacity** (VC).

The air inhaled and exhaled with each breath is known as the **tidal volume** (V_T). At rest, with normal quiet breathing, the tidal volume is much less than the vital capacity (usually 500–600 ml compared to about 5.0 l for the vital capacity in an adult male). The difference in lung volume at the end of a normal inspiration and the vital capacity is known as the **inspiratory reserve volume** (IRV), and the amount of air that can be forced from the lung after a normal expiration is called the **expiratory reserve volume** (ERV). The **functional residual capacity** (FRC) is the volume of air left in the lungs at the end of a normal expiration. The various lung volumes depend on height (they are larger in tall people), age, sex (the volumes tend to be smaller in women than in men of similar body size), and training.

The tidal volume varies according to the requirements of the body for oxygen. Consequently, the inspiratory and expiratory reserve volumes are also variable—the larger the tidal volume, the smaller the inspiratory and expiratory reserve volumes. In

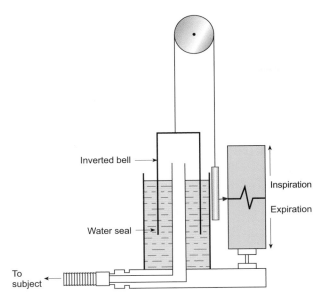

Figure 33.1 A simplified diagram of a recording spirometer. The subject breathes in and out through the mouth via the flexible tube shown bottom left. Inspiration draws air from the bell and the volume of air trapped within the bell decreases. It increases during expiration. These changes in volume are recorded on a calibrated chart as shown, or by a computer. In use, the spirometer would normally be connected to a more elaborate gas circuit with a soda-lime canister to absorb exhaled carbon dioxide.

Summary

Ventilation is the volume of air moved into and out of the lungs. It is driven by changes in the dimensions of the chest arising from the contraction and relaxation of the muscles of respiration. The volume of air entering and leaving the lungs each minute is called the minute volume, and the number of breaths taken each minute is the respiratory rate.

The total lung capacity is the largest volume of air that can be accommodated by the lungs. If the air is expelled from the lungs by a maximal expiration, a residual volume of air remains. The difference between the residual volume and the total lung capacity is the vital capacity. The tidal volume is the volume of air inhaled and exhaled with each breath. The difference in lung volume at the end of a normal inspiration and the vital capacity is known as the inspiratory reserve volume, and the amount of air that can be forced from the lung after a normal expiration is called the expiratory reserve volume.

33.3 The processes of inspiration and expiration

The various muscles of respiration are called on at different times. The diaphragm is the principal respiratory muscle. It forms a continuous sheet that separates the thorax from the abdomen. At rest, it assumes a dome-like shape, but when it contracts during inspiration, the crown of the diaphragm descends, thereby increasing the volume of the chest. During expiration, the diaphragm smoothly relaxes and the elastic recoil of the chest wall and lungs results in passive expiration.

When the demand for oxygen increases, the other muscles of inspiration are called into play. The chest wall is lifted upward and

contrast, the vital capacity and residual volume are relatively fixed as they are limited by the total lung capacity. With the exception of the residual volume and functional residual capacity, all the lung volumes can be directly measured by spirometry. However, the functional residual capacity and residual volume can be measured by the helium dilution method described in Box 33.1.

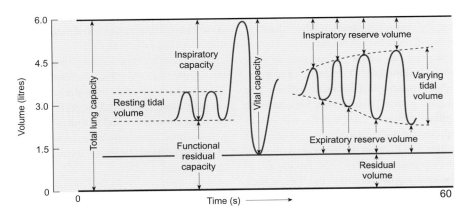

Figure 33.2 The subdivisions of the lung volumes. The record shows an idealized spirometer record of the changes in lung volume during normal breathing at rest (resting tidal volume) followed by a large inspiration to total lung capacity followed by a full expiration to the residual volume. This measures the vital capacity. Note that the residual volume and functional residual capacity cannot be measured by spirometry. They are measured by the helium dilution method (Box 33.1). As the right-hand side of the figure shows, the tidal volume is subject to considerable variation, e.g. during exercise. The inspiratory and expiratory reserve volumes become smaller as tidal volume increases.

Box 33.1 Determination of functional residual capacity and residual volume by the helium dilution method

The various subdivisions of the lung volume are of interest in a number of respiratory diseases. It is therefore desirable that they can be measured with some accuracy. While the *change* in the volume of the lungs during various breathing manoeuvres can be directly measured by a spirometer, the volume of air left in the lungs at the end of a normal expiration (the functional residual capacity or FRC) and the residual volume (RV) cannot be measured in this way. Instead, they are measured by the helium dilution method.

To determine the FRC, a subject is asked to breathe out normally and then is asked to inspire from a spirometer filled with a known volume of air containing a known concentration of the inert gas helium (He). As the gas containing He is breathed in and out, the He is diluted by the volume of air left in the lungs. A subsequent measurement of the He concentration permits this additional volume to be calculated from the following formula:

$$FRC = \left(\frac{initial\ He\ concentration}{final\ He\ concentration} - 1 \right) \times volume\ of\ spirometer \quad [1]$$

To measure the RV, a similar procedure is used, but the subject is asked to perform a maximal expiration before breathing from the spirometer containing the helium.

The difference in volume between the FRC and RV is the expiratory reserve volume (ERV), while the total lung capacity (TLC) is attained after a full inspiration. The volume taken in from the FRC to reach TLC is called the inspiratory capacity (IC). The inspiratory reserve volume (IRV) is the maximum volume of air that can inspired after a normal breath. These quantities are related by the following equations:

$$FRC = ERV + RV \quad [2]$$

$$TLC = FRC + IC \quad [3]$$

$$IC = V_T + IRV \quad [4]$$

where V_T is the tidal volume. The vital capacity (VC) is the volume of air that can be expired from the total lung volume.

$$VC = IC + ERV \quad [5]$$

If either the FRC or the RV is known, all the other subdivisions of the lung volume (including the total lung volume) can be determined with the aid of a spirometer.

---

outward by the activity of the external intercostal muscles and the diaphragm contracts more strongly so that the volume of the chest is increased further. In severe exercise, the accessory muscles (e.g. the scalenes shown in Figure 32.8) are called on, and they act to lift the chest wall still further. The internal intercostal muscles also contract to assist in decreasing the volume of the chest so that, under these conditions, expiration becomes an active process. Powerful expiration may also be assisted by contraction of the abdominal muscles, which force the abdominal contents against the diaphragm, pushing it upwards and reducing the volume of the chest.

The adult human diaphragm has equal proportions of slow (type 1) and fast (type 2) muscle fibres (see Chapter 8), while intercostal muscles have a higher proportion of fast muscle fibres. The fibres of the diaphragm have a small diameter, a high aerobic capacity, and a dense capillary network. This gives them the great resistance to fatigue required by their continuous activity. The higher proportion of type 2 muscle fibres in the intercostal muscles enables them both to develop tension and to relax rapidly as required for increasing and decreasing the volume of the chest at high rates of respiration, but it also makes them less resistant to fatigue.

The intrapleural pressure

In health, the lungs are expanded to fill the thoracic cavity because the pressure between the chest wall and the lungs (the **intrathoracic pressure**) is less than that of the air in the alveoli. The

intrathoracic pressure, also known as the **intrapleural pressure**, is measured clinically as described in Box 33.2. At the end of a quiet expiration, it is found to be about 0.5 kPa (5 cm H$_2$O) *below* that of the atmosphere. By convention, pressures less than that of the atmosphere are called negative pressures and those above atmospheric pressure are called positive pressures. Thus, the intrathoracic pressure at the end of a quiet expiration is −0.5 kPa or −5 cm H$_2$O. During inspiration the chest wall expands and the intrathoracic pressure becomes more negative, reaching a maximum of about −1 kPa (−10 cm H$_2$O).

If the chest wall is punctured by a hollow needle so that the tip of the needle lies in the space between the chest wall and the lungs, there is an inrush of air from the atmosphere into the cavity of the chest (**pneumothorax**). Under these conditions, the intrapleural pressure becomes equal to that of the atmosphere and the lungs will collapse. This shows that the negative value of the intrapleural pressure is due to the elastic recoil of the lungs (i.e. their tendency to collapse).

Pressure changes during the respiratory cycle

As with all liquids and gases, air will flow from a region of high pressure to one of low pressure. It follows, therefore, that air enters the lungs during inspiration because the pressure within the lungs (the **alveolar pressure** or **intrapulmonary pressure**) is less than that of the atmosphere. Conversely, during expiration, the alveolar

Box 33.2 Measurement of intrapleural pressure and the absorption of gas from the intrapleural space

In principle, the intrapleural pressure can be directly measured by inserting a hollow needle into the intrapleural space and recording the pressure. This is both technically difficult and invasive. A simpler and much less invasive method depends on the fact that the pressure in the lumen of the intrathoracic region of the oesophagus is the same as intrapleural pressure. This situation arises because the oesophagus is a collapsible tube and, as it passes though the thorax, it will experience the same pressure as the outside of the lungs—the intrapleural pressure (intrathoracic pressure).

To measure the intrapleural pressure, a small air-filled balloon is usually placed in the lower third of the oesophagus and connected to a manometer. With this technique, the intrapleural pressure at the end of quiet expiration is found to be about -0.5 kPa (-5 cm H_2O) relative to that of the atmosphere (which by convention is taken as zero). During inspiration, the chest wall expands and the intrapleural pressure becomes more negative, reaching a maximum of about -1 kPa (-10 cm H_2O). During forced expiration, the intrapleural pressure may become positive and can reach values of 10 kPa (100 cm H_2O).

It may seem surprising that the intrapleural pressure is always less than that of the alveolar pressure. Why is there no accumulation of gas in the intrapleural space? Why does the gas of a pneumothorax eventually disappear from the intrapleural space? The total pressure of the gases in the venous blood (~93 kPa or 700 mm Hg) is lower than that of the alveoli (101 kPa or 760 mm Hg), as there is a considerable drop in the PO_2 but only a small rise in the PCO_2. The PN_2 is unchanged. In trapped air, the PO_2 and the PCO_2 equilibrate with the surrounding tissue and the PN_2 rises as the oxygen is absorbed. This favours the absorption of the nitrogen down its concentration gradient. The consequence of absorption of the nitrogen is a rise in the partial pressures of oxygen and carbon dioxide, which again equilibrate with the surrounding tissue so raising the PN_2 once more. The cycle repeats itself until all the gas has been reabsorbed. Strictly speaking, once the gas has been absorbed there is no intrapleural pressure, as the pleural membranes are held together by intermolecular forces. However, the concept of intrapleural pressure provides a convenient way of describing the magnitude of the forces acting on the lungs during normal respiration.

The balance of the Starling forces explains why there is normally almost no liquid in the intrapleural space. The osmotic pressure of the plasma proteins is greater than the transmural capillary pressure (the capillary pressure minus the intrapleural pressure). The net filtration pressure therefore favours fluid absorption. In certain disease states, however, the Starling forces favour fluid movement into the intrapleural space, resulting in a pleural effusion.

---

pressure exceeds atmospheric pressure and air is expelled. The **trans-pulmonary pressure** (P_L) is, as its name implies, the pressure difference across the lung—i.e. the difference between the alveolar pressure (P_A) and the intrapleural pressure (P_{pl}):

$$P_L = P_A - P_{pl}$$

This pressure is a measure of the elastic recoil of the lung, so that its magnitude at any level of the chest is determined by the distension and elasticity of the lung at that level. The trans-pulmonary pressure is by convention always positive and is the pressure required to maintain the lungs in an inflated state.

Figure 33.3 shows the variation in the pressure in the lungs during a single respiratory cycle. As the chest expands, the intrapleural pressure falls and this leads to an increase in trans-pulmonary pressure and expansion of the lungs. As the lungs expand, the alveolar pressure falls below that of the atmosphere and air flows into the lungs until the alveolar pressure becomes equal to that of the atmosphere. At this time, the intrapleural pressure remains at its most negative. During expiration, intrapleural pressure rises (i.e. becomes less negative) and the tension in the elastic fibres of the lung parenchyma causes the volume of the lungs to decrease. This results in an increase in alveolar pressure which forces air

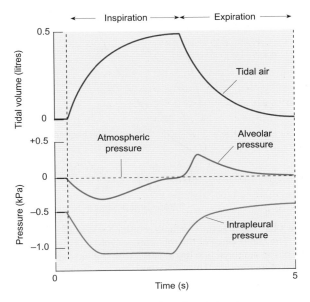

Figure 33.3 The changes in intrapleural and alveolar pressure during a single respiratory cycle. Note that the changes in intrapleural pressure occur before the change in alveolar pressure.

from the lungs. As the air is expelled, the alveolar pressure falls until it reaches atmospheric pressure once more and the cycle begins again with the next breath.

How much does the intrapleural pressure have to change for a given amount of air to enter the lungs?

The change in the volume of the chest that results from a given change in intrapleural pressure is called the **compliance** (C):

$$C = \frac{\Delta V}{\Delta P}$$

Compliance is a measure of the ease with which the chest volume can be changed and is determined when there is no movement of air into or out of the lungs (**static compliance**—see Figure 33.4). If compliance is high, there is little resistance to expansion of the chest; conversely, if it is low, the chest is expanded only with difficulty. The compliance depends on the volume of the chest; it becomes smaller as the chest expands towards its maximum volume (Figure 33.5). This reflects the stretching of the elastic tissue of the chest wall and lung parenchyma. In practice the compliance is measured at normal tidal volumes (i.e. near FRC). For healthy young subjects, the static

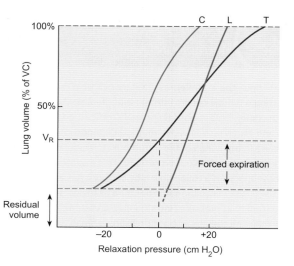

Figure 33.5 The pressure–volume relationships for the intact chest (T), the chest wall (C), and the lungs (L). The ordinate shows their volumes as a percentage of the vital capacity, while the abscissa is the pressure relative to barometric pressure. Note that the inflation of the lungs alone requires positive pressures but the chest wall naturally assumes about two-thirds of its maximum volume. Smaller chest volumes require negative pressures. The pressure required to maintain a particular volume of the intact chest is given by the sum of the pressures required to maintain the same volume of the chest wall and the lungs (i.e. curve T is the sum of curves C and L). At the relaxation volume (V_R), the pressure in the lungs exactly balances that of the chest wall. Smaller chest volumes can only be obtained by a forced expiration. (Based on H. Rahn et al. (1946) *Am J Physiol* **146**, 161.)

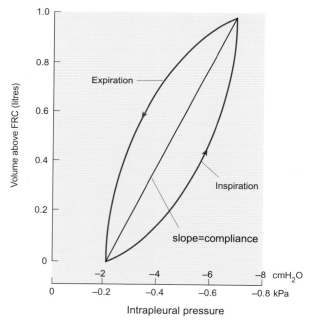

Figure 33.4 The pressure–volume relation for a single respiratory cycle. The compliance of the respiratory system is given by the slope of the diagonal line linking FRC with the lung volume at the end of inspiration. This represents the change in volume if the work of respiration was against purely elastic resistances. In fact, during inspiration, additional pressure is needed to overcome the airflow resistance and other resistive forces. This is shown by the blue curve to the right of the compliance line. The blue curve to the left of the compliance line shows the resistive work done during passive expiration.

compliance has a typical value of $1.0 \; l \; kPa^{-1}$ ($0.1 \; l \; cm \; H_2O^{-1}$), as shown by the straight black line in Figure 33.4.

During normal breathing, a larger change in pressure is required to move a given volume of air into the chest than would be expected from the static compliance of the chest. This is shown by the inspiration curve in Figure 33.4. The additional pressure is required to overcome additional **non-elastic resistances**, which are:

1. the resistance of the airways to the movement of air (the **airways resistance**)

2. the frictional forces arising from the viscosity of the lungs and chest wall (tissue resistance)

3. the inertia of the air and tissues.

Of these, the airways resistance is by far the most significant. The tissue resistance is about a fifth of the total, while the inertia of the respiratory tract (air plus tissues) is only of significance when there are sudden large changes in airflow, such as in coughing or sneezing. During expiration, particular lung volumes are reached at lower pressures. The pressure–volume relationship during a single respiratory cycle is thus a closed loop, as shown in Figure 33.4. The physiological reason for this property (known as **hysteresis**) will be discussed later (Section 33.4).

The compliance of the intact chest is determined by the elasticity of the chest wall and that of the lungs. (An elastic body is one that resumes its original dimensions after the removal of an external force that has deformed it.) If the respiratory muscles are relaxed and the glottis is open, the volume of the intact chest is about 30 per cent of the total vital capacity. This is known as the relaxation volume for the intact chest. Chest volumes greater or smaller than this value can only be attained by muscular effort. If the chest wall is cut open when it is at its relaxation volume, the ribs spring outwards and the lungs collapse inwards. The volume assumed by the chest wall is greater, and that of the lungs is smaller, than that of the intact chest. This shows that the dimensions of the chest at rest reflect the balance of the forces acting on the chest wall and the lungs. The elasticity of the chest wall (and therefore its compliance) is determined mainly by that of its muscles, ligaments, and tendons. That of the lungs is determined by two major factors: the elastic fibres of the lung parenchyma and the **surface tension** of the liquid film that lines the alveoli. For this reason, the pressure–volume relationships of the chest wall and the lungs differ significantly, as shown in Figure 33.5.

What are the sites of airways resistance?

In a highly branched set of tubes such as the human bronchial tree, it is difficult to obtain precise knowledge of the patterns of airflow throughout the whole structure. Like the flow of blood in the circulatory system, the flow of air through the airways may be either laminar or turbulent. Laminar flow occurs at low linear flow rates (flow rate in volume per second divided by the cross-sectional area), but when the linear flow rate increases beyond a critical velocity, the orderly pattern of flow breaks down, eddies form, and the flow becomes turbulent.

Turbulent flow is more likely to occur in large-diameter, irregularly branched tubes when the flow rate is high. Unlike laminar flow, in which resistance to flow is constant for a tube of given dimensions (see Box 30.1), when airflow is turbulent, resistance increases with the flow rate. This is the situation in the upper airways (the nasal cavities, pharynx, and larynx), which account for about a third of the total airways resistance. Upper airway resistance can be significantly reduced by breathing through the mouth—a fact that is widely exploited during heavy exercise. The remaining two-thirds of the airways resistance is located within the tracheobronchial tree. The greatest resistance to airflow in the lower respiratory tract is found in the segmental bronchi (generation 4), where the cross-sectional area is relatively low and the airflow is high (and turbulent). As the airways branch further, their cross-sectional area increases and the linear flow rate falls. By the time air reaches the smallest airways, the flow is laminar and the resistance becomes very small (see Figure 33.6).

Airways resistance falls as the volume of the lungs increases. This change is chiefly due to the forces acting on the bronchioles (which have no cartilage in their walls—see Table 32.2). These airways are attached to the lung parenchyma, and as the lungs expand, the connective tissue of the parenchyma pulls on the bronchioles, increasing their diameter and decreasing their resistance to the flow of air.

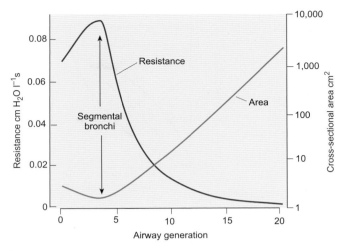

Figure 33.6 The resistance of the airways plotted as a function of airway generation. Note that the resistance is highest in the segmental bronchi, which also have the smallest cross-sectional area. The resistance falls sharply as the cross-sectional area increases.

Summary

Inspiration is an active process that depends on the contraction of the diaphragm and external intercostal muscles. Expiration is largely passive. Air enters the lungs during inspiration because the pressure within the lungs (the alveolar pressure) is less than that of the atmosphere. The transpulmonary pressure is the pressure required to maintain the lungs in an inflated state and is the difference between the alveolar pressure and the intrapleural pressure. It is a measure of the elastic recoil of the lung.

The change in the volume of the chest resulting from a given change in intrapleural pressure is called the compliance and is a measure of the ease with which the chest volume can be changed. It is determined both by the elastic elements in the lung parenchyma and by the surface tension of the air–liquid interface of the alveoli.

During breathing, the total pressure required to inflate the chest is the pressure required to expand the elastic elements of the chest (measured by the compliance) plus the pressure required to overcome the airways resistance.

33.4 How much work is done by the respiratory muscles?

Inflation and deflation of the lungs and chest wall requires that work is performed by the respiratory muscles. Mechanical work is performed when a load is moved through a distance (see Chapter 8) and, for a three-dimensional system such as the respiratory system, the work done is equal to the change in pressure times the change in volume (see Figure 33.7). In quiet breathing, the

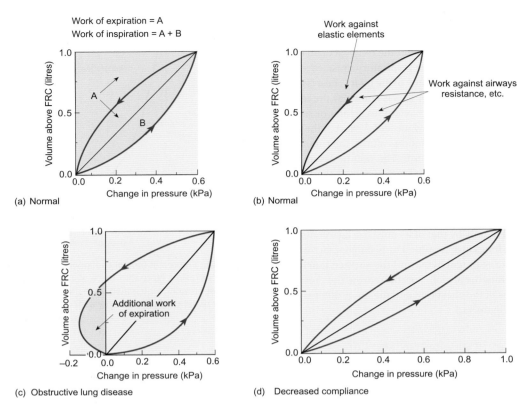

Figure 33.7 The work of respiration. The figure shows the increase in volume above FRC plotted as a function of the change in intrapleural pressure. The work involved in changing the volume of the chest is given by the area of the pressure–volume curve. As shown in panel (a), the work of inspiration is greater than the work of expiration. The area of the loop labelled B represents the energy required to overcome the airways resistance during inspiration. The energy for expiration is largely derived from the stretching of the elastic elements of the lungs and chest during inspiration. Panel (b) shows the two principal components of the work of inspiration. People with obstructive lung disease must breathe against an increased airways resistance. Thus a greater pressure change is required to move a given volume of air into and out of the lungs. This results in the performance of extra work, as shown by the area of the loop in panel (c). If the compliance of the chest is decreased, a larger pressure change is required to move a given volume of air and this requires extra work to be done as shown in panel (d). Compare the area of this pressure–volume curve with that shown in panel (a). (Redrawn after Fig 2.4 in J. Widdicombe and A. Davies (1991) *Respiratory physiology.* Edward Arnold, London.)

volume changes are modest and the work done is small. If the depth of breathing is increased, as in exercise, the pressure–volume loop has a greater area and the energy cost of each breath increases.

During inspiration, the work of breathing consists of two components: the work needed to overcome the elastic forces of the chest wall and lungs, and the work needed to overcome non-elastic resistances. During inspiration, the elastic elements of the lungs and chest wall are stretched; the work done during this phase of respiration is shown by the area marked A in Figure 33.7(a). The additional work required to overcome the airways resistance is shown by area B. In quiet expiration, the respiratory muscles progressively relax, so that much of the energy required is derived from the elastic elements of the chest that were stretched during inspiration. The work of quiet expiration is thus done by the muscles of inspiration. In forced expiration, however, an additional muscular effort will be required such as that shown for obstructive lung disease in

Figure 33.7(c). If the compliance of the chest is decreased, as show in Figure 33.7(d), a greater intrapleural pressure will be required to cause a given change in lung volume, so that the work of breathing is increased.

To move a given volume of air into and out of the lungs, breathing can either be deep and slow or fast and shallow. Deep, slow breathing results in increased work against the elastic forces of the lungs and chest wall (the elastic resistance), while rapid, shallow breathing results in increased airflow resistance. Therefore, the work of breathing is at a minimum where the decrease in the work against the elastic forces gained by shallower breaths is balanced by the increase in work against airways resistance resulting from increased tidal volume, as shown in Figure 33.8. For normal subjects, this occurs when the respiratory rate is about 12–15 breaths a minute. If the elastic resistance increases (as in pulmonary fibrosis) the work of breathing is minimized by increasing the rate and decreasing the depth of respiration. If airflow resistance is

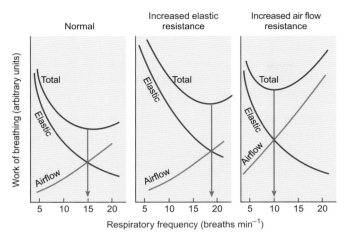

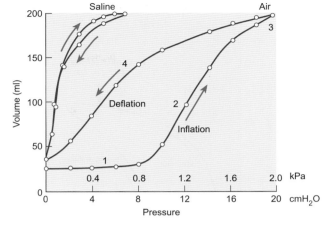

Figure 33.8 The work of breathing at different frequencies in normal subjects, those with increased elastic resistance, and those with increased airflow resistance. In normal subjects the work of breathing is minimized, with a respiratory frequency of about 15 breaths per minute. When the elastic resistance is increased, work is minimized by frequent shallow breaths. In contrast, slow deep breathing minimizes the work required when airflow resistance is increased. (Based on Fig 6.7 in J. F. Nunn (1993), *Nunn's applied respiratory physiology*, 4th edition. Butterworth-Heinemann, Oxford.)

Figure 33.9 Pressure–volume relationships for isolated cat lungs when inflated with air or with isotonic saline. Note the low pressures required to expand the saline-filled lungs and that the curve for inflation is virtually the same as that for deflation. For air-filled lungs, much greater pressures are required for a given volume change. The curve shows pronounced hysteresis, a greater pressure being required to inflate the lungs to a given volume compared to the pressure required to hold that volume during deflation. See text for further explanation.

increased, the work of breathing is minimized by increasing the depth of breathing and decreasing the rate (see Figure 33.8).

Surface tension in the alveoli contributes to the elasticity of the lungs

During the initial stages of inflation with air, collapsed lungs require a considerable pressure before they begin to increase in volume (about 0.8–1.0 kPa: phase 1 in Figure 33.9). The lungs then expand roughly in proportion to the increase in pressure (phase 2) until they approach their maximum capacity (phase 3). Once they have been fully expanded, the volume of the lungs changes slowly during deflation until the pressure holding them open decreases to about 0.8 kPa (phase 4), whereupon their volume declines more steeply as the pressure falls. The unequal pressure required to maintain a given lung volume during inflation with air as opposed to deflation accounts for the hysteresis in the pressure–volume relationship seen during the respiratory cycle (Figure 33.4). If the lungs are inflated with isotonic saline (0.9 per cent NaCl), however, the pressures required to expand the lung to a given volume are much reduced and there is little or no hysteresis.

Why is it more difficult to inflate the lungs with air than with isotonic saline? When the lungs are inflated with saline, the only force opposing expansion is the tension in the elastic elements of the parenchyma that become stretched as the lungs expand. When the lungs are inflated by air, however, the surface tension at the air–liquid interface in the alveoli also opposes their expansion. As with bubbles of gas in a liquid, the magnitude of the surface tension in the alveoli is given by **Laplace's law**, which states that the pressure

(P) inside a hollow sphere is two times the surface tension (T) divided by the radius (r):

$$P = \frac{2T}{r}$$

The alveoli are about 100 μm in diameter after a normal quiet expiration. If they were lined with normal interstitial fluid, which has a surface tension of about 70 mN m^{-1}, Laplace's law predicts that the pressure gradient across the alveolar wall would need to be about 3 kPa just to prevent collapse—expansion would require still higher pressures. As Figure 33.9 shows, however, the lungs can be inflated with much lower pressures than this. Maximum inflation is achieved at less than 2 kPa, and the lungs can be held open during deflation by pressures well below 1 kPa. Thus, the alveoli cannot be lined with interstitial fluid. This anomaly was resolved when it was discovered that the alveoli are lined with a fluid layer containing **lung surfactant** (also called **pulmonary surfactant**), which is secreted by the type 2 cells of the alveoli. The surfactant lowers the surface tension of the liquid lining the air spaces of the lung. Consequently, the pressures needed to hold the alveoli open are much reduced.

Lung surfactant stabilizes the alveoli by reducing the surface tension of the air–liquid interface

By lowering surface tension, lung surfactant minimizes the tendency of the small alveoli to collapse and tends to stabilize the alveolar structure. In addition, the lowering of surface tension

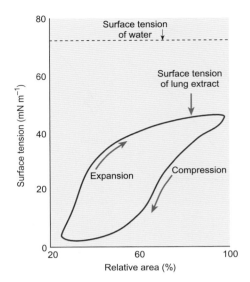

Figure 33.10 The relationship between surface tension and surface area for water and lung surfactant. Note that lung surfactant greatly reduces the surface tension and that the reduction in surface tension is greatest during compression (equivalent to lung deflation). The surface tension of water does not vary with area.

increases the compliance of the lungs, and this reduces the work of breathing. This effect of lung surfactant on surface tension is particularly important at birth, when the lungs first expand (see Figure 50.5). Finally, as surfactant lowers surface tension it helps to prevent fluid accumulating in the alveoli and so plays an important role in keeping the alveolar air space dry.

Lung surfactant consists mainly of phospholipid molecules, which form a separate phase in the air–liquid interface that exists over the alveolar epithelium. The phospholipids are aligned so that their polar headgroups remain in the aqueous phase while their long hydrocarbon chains are oriented to the air space of the alveoli. When alveoli close during expiration, the phospholipid molecules become compressed against each other to form a monolayer, water molecules are excluded, and the surface tension falls dramatically (Figure 33.10). Once the monolayer is formed, a subsequent increase in area causes the surface tension to rise quite rapidly at first. During this stage, the phospholipid molecules tend to remain packed together. As the area increases, they begin to separate and surface tension increases more slowly until it reaches its maximum value. This property of pulmonary surfactant explains the hysteresis of air-inflated lungs seen in Figure 33.9.

Summary

The work of respiration is determined by the change in pressure required for a given change in volume (i.e. the compliance of the intact chest). It is equal to the change in pressure multiplied by the change in volume, and it has two components: the work needed to overcome the elastic forces of the chest wall and lungs, and the work needed to overcome airways resistance. The work of respiration is significantly reduced by the presence of pulmonary surfactant.

33.5 Tests of ventilatory function

In the diagnosis and treatment of respiratory diseases, assessment of pulmonary function is of considerable importance. Key tests of ventilatory function are:

1. The vital capacity
2. The forced vital capacity
3. The maximal (or peak) expiratory flow rate
4. Maximal ventilatory volume.

To measure the **vital capacity** (VC), a subject is asked to make a maximal inspiration and then to breathe out as much air as possible. The volume of air exhaled is measured by spirometry (see Section 33.2). Note that in this test, the time taken to expel the air is not taken into account, and it is usual to estimate the vital capacity during expiration rather than inspiration. The normal values depend on age, sex, and height. For a healthy adult male of average height and 30 years of age, vital capacity is about 5 litres. For women of the same age, the vital capacity is somewhat less at about 3.5 litres.

In the **forced vital capacity** (FVC) test, the subject is asked to make a maximal inspiration and then to breathe out fully as fast as possible. The air forced from the lungs is measured as a function of time (with an instrument called a **pneumotachograph**). After the final quantity of air is forced from the lungs, only the residual volume (RV) is left. In healthy young subjects, around 85 per cent of the vital capacity is forced from the lungs within the first second. This is known as the forced expiratory volume at 1 second or FEV_1. The remaining volume is expired over the next few seconds (see Figure 33.11). The FEV_1 declines with increasing age. Even so, a healthy 60-year-old man should have a value of around 70 per cent. By way of contrast, a person with an obstruction of the airways (e.g. during an asthmatic attack) would have a much lower FEV_1. In severe cases, the FEV_1 can be less than 40 per cent.

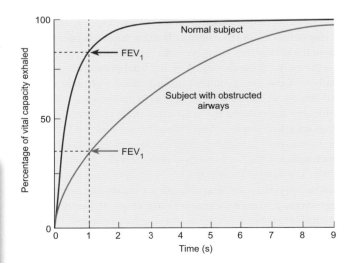

Figure 33.11 Forced vital capacity test for a normal subject and for a subject with obstructed airways. Note the marked difference in the FEV_1 values.

The FVC test is also useful in diagnosis of restrictive lung diseases such as fibrosis of the lung. In restrictive disorders, the ability of the lung to expand normally is compromised. As a result, the FEV_1 may be normal but the vital capacity will be much reduced.

The **maximal expiratory flow rate** (also known as the **peak expiratory flow rate** or **PEFR**) is also used to distinguish between obstructive and restrictive diseases. The maximal airflow is measured with a pneumotachograph during a forced expiration following an inspiration to the total lung volume. The maximum flow rate is normally reached in the first tenth of a second of the forced expiration and is measured in litres per second. Healthy young adults are able to achieve flow rates of 8–10 l s^{-1}. As with the FEV_1, obstructive airway disease results in a reduced peak flow, while this is not the case in restrictive lung disease.

Flow–volume curves

The relationship between airflow and the amount of air remaining in the lungs is described by a curve of the type shown in Figure 33.12. In these curves (called flow–volume curves), the airflow is plotted as a function of the lung volume. To construct such curves, a subject is asked to breathe in to their full lung volume and then asked to breathe out completely with varying degrees of effort. At maximal effort, peak flow rises rapidly to a maximum and then steadily declines. At lesser degrees of effort, the flow increases more slowly to its maximum and then declines to zero. In the first phase, the slope of the flow–volume curve becomes steeper with

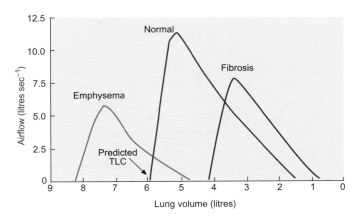

Figure 33.13 Flow–volume curves for maximal effort in a normal individual (blue curve) and in two patients: one suffering from emphysema (red curve) and one from pulmonary fibrosis (purple curve). (Adapted from Fig 7.7 of N.B. Slonim and L.H. Hamilton (1987) *Respiratory physiology*, 5th edition. Mosby, St Louis.)

increasing effort—the airflow is *effort dependent*. As the lung volume declines below about 70 per cent of vital capacity, the airflow falls off at the same rate no matter how much effort is exerted—i.e. it is *effort independent*. This limiting of airflow as lung volume declines can be attributed to a decrease in the diameter of the airways and to collapse of the smaller airways. (Recall that the flow of a fluid—liquid or gas—through a tube is inversely proportional to the fourth power of the radius of the tube: Poiseuille's law, discussed in Box 30.1).

As abnormalities of lung compliance or airways resistance will affect the appearance of the curve, the flow–volume curves provide a sensitive measure of ventilatory function. Some examples are shown in Figure 33.13. When lung compliance decreases (e.g. in pulmonary fibrosis), the expansion of the lungs is compromised, so there is a decrease in vital capacity. This is accompanied by a decrease in residual volume, and the flow–volume curve is displaced to the right (lower lung volumes). Peak flow is also reduced when compared to that of a normal subject of similar stature. In contrast, lung compliance and total lung capacity are increased in emphysema (see Section 36.3). The loss of lung parenchyma results in early closure of the smaller airways, particularly those lacking cartilage, and residual volume is thus substantially increased. Overall there is a shift in the flow–volume curve to the left (i.e. to higher lung volumes). The slope of the rising phase is affected by airway obstruction such as that caused by a tumour, inflammation of the respiratory mucosa, or traumatic damage to the airways themselves.

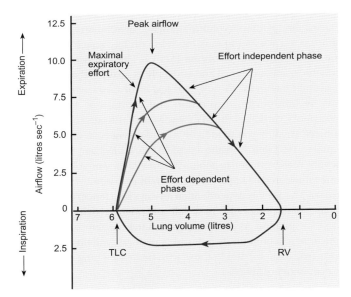

Figure 33.12 Flow–volume curves for a normal individual. The subject is asked to fill their lungs completely and then breathe out to residual volume with maximum effort (blue curve), with medium effort (green curve), and with slight effort (red curve). Note that the initial flow rate depends on the degree of effort but that below 60–70 per cent of total lung volume, the airflow is independent of the effort. TLC = total lung capacity, RV = residual volume. (Based on Fig 4.9 in J.F. Nunn (1993), *Nunn's applied respiratory physiology*, 4th edition. Butterworth-Heinemann, Oxford.)

The closing volume

The airways resistance increases as lung volume decreases. This situation arises because the decrease in lung volume is accompanied by a reduction in the volume of the alveoli and the diameter of the airways. To reach the residual capacity, however, the internal intercostal muscles and the abdominal muscles generate a positive intrapleural pressure, which can be over 10 kPa (100 cm H_2O). This

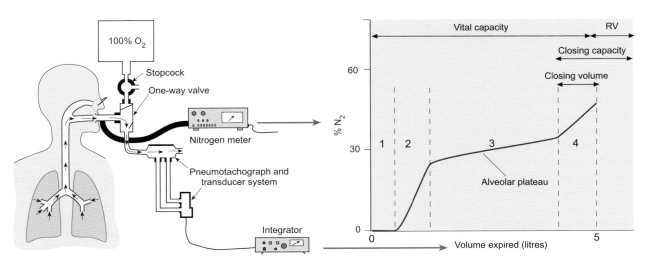

Figure 33.14 The determination of closing volume from the single breath nitrogen washout curve. First, the subject breathes out to the residual volume and is then asked to inspire pure oxygen until they reach their vital capacity. Finally, the subject breathes out to residual volume while the fractional content of nitrogen of the expired air is measured continuously. The closing volume is indicated by the upturn in the nitrogen washout curve (see text for further explanation). (Adapted from Fig 3.8 in M.G. Levitsky (1991) *Pulmonary physiology*, 3rd edition. McGraw-Hill New York.)

pressure sums with the alveolar pressure to drive air from the alveoli. However, when the intrathoracic pressure exceeds the pressure in the airways, the small airways become compressed because they have no supporting cartilage. Consequently, air becomes trapped within the lungs. This is known as **dynamic airways compression**. The lung volume at which the airways begin to collapse during a forced expiration is known as the **closing capacity**. The **closing volume** is equal to the closing capacity less the residual volume.

In normal healthy young people, the closing volume measured by this method is about 10 per cent of the vital capacity (i.e. about 500 ml). It is greater when lying down and increases with age. By the age of 65 years, the closing volume may be as much as 40 per cent of vital capacity. In emphysema, the loss of the parenchymal tissue results in decreased traction on the airways so that the small airways begin to collapse at a higher lung volume than normal. Thus an increased closing volume is an early sign of small airways disease.

The closing volume can be measured as follows: the subject is asked to breathe out to the residual volume and take a breath of pure oxygen to vital capacity. The subject is then asked to breathe out to residual volume once more while the fractional content of nitrogen in his or her expired air is measured, as shown in Figure 33.14. At first, the oxygen is forced from the dead space and the fractional nitrogen content is zero (phase 1). As the oxygen is cleared from the conducting airways, the nitrogen content rises rapidly (phase 2) until it reaches a gently rising plateau (phase 3). The slow rise in the nitrogen content reflects differences in ventilation in different parts of the lungs. As the lung volume approaches the residual volume, the lower airways become compressed and more of the exhaled gas comes from the upper region of the lungs, which have a higher fractional content of nitrogen because they were less well ventilated during the breath of oxygen. (The reasons for the difference in ventilation between the upper and lower parts

of the lung are discussed in Chapter 34—see Section 34.3). The nitrogen content therefore begins to rise more sharply (phase 4). The point at which this begins to occur marks the closing volume.

Maximum breathing capacity

The maximum minute volume attainable by voluntary hyperventilation is known as the maximum breathing capacity (MBC) or **maximal ventilatory volume** (MVV). The subject is asked to breathe in and out as fast and as deeply as possible through a low-resistance circuit for 15–30 s. This test involves the whole respiratory system during inspiration and expiration. As for other respiratory variables, the MVV varies with the age and sex of the subject. Healthy young men of 20 years of age can attain a MVV equal to 150 l min^{-1}. By the age of 60 years, however, the MVV for healthy men has fallen to about 100 l min^{-1}. For women the equivalent values are 100 l min^{-1} at age 20 falling to about 75 l min^{-1} by age 60. The MVV is dependent on airways resistance, on the compliance of the lungs and chest wall, and on the activity of the muscles of respiration. As a result, it is a sensitive measure of ventilatory function. It is profoundly reduced in patients with obstructed airways (e.g. asthma) and in those with decreased lung compliance (e.g. pulmonary fibrosis).

Summary

Clinical assessment of ventilatory function can be made using a variety of tests. These include measurement of the vital capacity, the FEV$_1$, peak expiratory flow rate, flow–volume curves, and maximal ventilatory volume. The extent of small airways disease can be assessed by measuring the closing volume.

Checklist of key terms and concepts

The mechanics of breathing

- The chest cavity is a large gas-tight compartment formed by the rib cage, the intercostal muscles, and the diaphragm. It is an integral part of the respiratory system.

- The movement of air into and out of the lungs occurs in response to changes in the dimensions of the chest that arise from the activity of the muscles of respiration.

- As the diaphragm descends and the external intercostal muscles contract, the chest expands and the pressure in the alveoli falls below that of the atmosphere. This causes air to enter the lungs. The process is reversed in expiration.

- During expiration, alveolar pressure is greater than that of the atmosphere.

- The trans-pulmonary pressure is the difference between the alveolar pressure and the intrapleural pressure. It is the pressure required to maintain the lungs in an inflated state.

- Inspiration is an active process, but expiration is largely passive and is due to the elastic recoil of the lungs and chest wall.

The work of breathing

- The total pressure required to inflate the chest is the pressure required to expand the elastic elements of the chest plus the pressure required to overcome the airways resistance. Most of the resistance to the flow of air is located in the upper airways.

- The work of breathing is equal to the change in pressure times the change in volume.

- The compliance of the respiratory system is a measure of the ease with which the chest can be expanded and the lungs inflated.

- Diseases that reduce respiratory compliance increase the work of breathing.

- Respiratory compliance is determined both by the elasticity of the chest and by the surface tension of the air–liquid interface of the alveoli.

- The surface tension is reduced below that of water by pulmonary surfactant secreted by the type II alveolar cells.

- The presence of pulmonary surfactant in the alveoli significantly reduces the work of breathing.

- Clinical assessment of ventilatory function can be made using a variety of tests including measurement of vital capacity, FEV$_1$, and peak expiratory flow rate (PEFR).

Recommended reading

Dejours, P. (1966) *Respiration*, Chapter 5. Oxford University Press, New York.

An old book that gives an excellent quantitative treatment that is suitable for graduate students and those considering research in human respiratory physiology.

Hlastala, M.P., and Berger, A.J. (2001) *Physiology of respiration* (2nd edn), Chapter 3. Oxford University Press, New York.

Lumb, A.B. (2017) *Nunn's applied respiratory physiology* (8th edn), Chapters 2 and 3. Churchill-Livingstone, London.

Slonim, N.B., and Hamilton, L.H. (1987) *Respiratory physiology* (5th edn), Chapters 4–7. Mosby, St Louis.

Now rather old but still well worth reading for its excellent coverage and clear explanations.

Review articles

Bossé, Y., Riesenfeld, E.P., Paré, P.D., and Irvin, C.G. (2010) It's not all smooth muscle: Non-smooth-muscle elements in control of resistance to airflow. *Annual Review of Physiology* **72**, 437–62.

Perez-Gil, J., and Weaver, T.E. (2010) Pulmonary surfactant pathophysiology: Current models and open questions. *Physiology* **25**, 132–41.

Polla, B., D'Antona, G., Bottinelli, R., and Reggiani, C. (2004) Respiratory muscle fibres: specialisation and plasticity. *Thorax* **59**, 808–17, doi:10.1136/thx.2003.009894.

Widdicombe, J.H., and Wine, J.J. (2015) Airway gland structure and function. *Physiological Reviews* **95**, 1241–319.

To check that you have mastered the key concepts presented in this chapter, complete the accompanying online self-assessment questions. Go to www.oup.com/uk/pocock5e/

CHAPTER 34

Alveolar ventilation and blood gas exchange

Chapter contents

This chapter should help you to understand:

- Alveolar ventilation and dead space
- Non-uniform alveolar ventilation and its causes
- Gas exchange in the alveoli
- The bronchial and pulmonary circulations
- The factors that determine the ratio of ventilation to blood flow in different parts of the lung
- Fluid exchange in the lungs
- The metabolic role of the lungs

34.1 Introduction

The main function of the respiratory system is to provide the cells of the body with oxygen for the generation of metabolic energy and to remove the carbon dioxide produced by oxidative metabolism. To do so, it must have a means of transporting these two gases to and from the tissues and exchanging them with the atmospheric air. This chapter is principally concerned with gas exchange in the alveoli and the matching of blood flow to the availability of oxygen (blood-gas matching). The transport of oxygen and carbon dioxide by the blood is discussed in Chapter 25 (pp. 371–376).

34.2 Alveolar ventilation and dead space

Broadly, the respiratory system can be considered to consist of two parts: the conducting airways and the area of gas exchange. In dividing the respiratory system in this way it becomes obvious that not all the atmospheric air taken in during a breath reaches the alveolar surface, as some of it must occupy the airways that connect the atmosphere to the respiratory surface deep in the lungs. This air does not take part in gas exchange and is known as the **dead space**. The remaining fraction of the tidal volume enters the alveoli. Thus:

tidal volume = dead space + volume of air entering the alveoli

or, in symbols:

$$V_T = V_D + V_A$$

where V_T is tidal volume, V_D is the volume of the dead space, and V_A is the volume of air entering the alveoli.

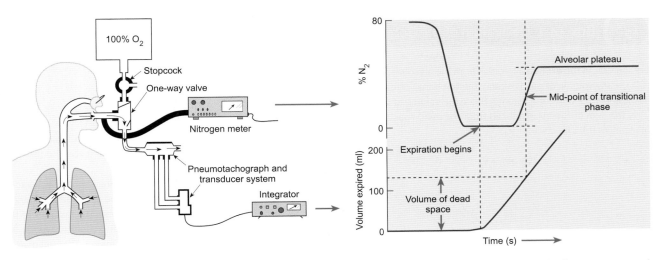

Figure 34.1 Fowler's method for determination of the anatomical dead space. The subject is asked to take a breath of pure oxygen and then breathe out. The concentration of nitrogen in the expired air is continuously measured as the subject breathes out and rises from zero to a plateau value. The expired volume at the midpoint of the transition from pure O_2 to alveolar gas is taken as the volume of the dead space.

Since not all of the air that enters the alveoli takes part in gas exchange, two different types of dead space are distinguished:

1) the **anatomical dead space**, which is the volume of air taken in during a breath that does not mix with the air in the alveoli;

2) the **physiological dead space**, which is the volume of air taken in during a breath that does not take part in gas exchange.

As with other lung volumes, such as vital capacity, the anatomical and physiological dead spaces depend on body size as well as the age and the sex of the individual. In a normal healthy young person the anatomical and physiological dead space are about the same—around 150 ml for a tidal volume of 500 ml. However, in some diseases of the lung such as emphysema, the physiological dead space can greatly exceed the anatomical dead space (see Section 36.3 p. 548).

The anatomical dead space can be measured in a similar manner to the closing volume. The subject is asked to take a breath of pure oxygen and hold it for a second before exhaling. By this simple manoeuvre, the air in the airways has a different composition to that of the alveoli. Air in the conducting system will be pure oxygen, while that in the alveoli will also contain nitrogen. All that is then required to determine the dead space is to ask the subject to breathe out while the nitrogen content and volume of the expired air are continuously monitored (see Figure 34.1). The airways contain pure oxygen, and this volume must first be displaced before the alveolar air is exhaled. Thus, the level of nitrogen will rise from zero to reach a plateau value, which is the same as that in the alveoli. The anatomical dead space is taken as the volume exhaled between the beginning of expiration and the midpoint between the zero level and the plateau value (Figure 34.1).

The physiological dead space is equal to the volume of the non-respiratory airways *plus* the volume of air that enters those alveoli that are not perfused with blood, since these alveoli cannot participate in gas exchange. The physiological dead space can be estimated by measuring the carbon dioxide content of the expired and alveolar air using the **Bohr equation** (see Box 34.1).

The amount of air that is taken in during each breath (the tidal volume V_T) multiplied by the frequency of breathing (f) is known as the **minute volume** ($\dot{V}_E$).

$$\dot{V}_E = f \times V_T$$

The fraction of the minute volume that ventilates the alveoli is known as the **alveolar ventilation** ($\dot{V}_A$). So:

$$f.V_T = f.V_D + f.V_A$$

or

$$\dot{V}_E = \dot{V}_D + \dot{V}_A$$

where $\dot{V}_D$ is the ventilation of the dead space. Note that the dots over the volumes indicate flow in litres per minute.

In the case of a subject breathing a tidal volume of 0.5 litres 12 times a minute,

$$\dot{V}_E = 12 \times 0.5$$
$$= 6 \text{ litres per minute}$$

If the dead space is 150 ml, the alveolar ventilation is:

$$12 \times (0.50 - 0.15) = 12 \times 0.35$$
$$= 4.2 \text{ litres per minute}$$

From this it is clear that only about 70 per cent ($4.2 \div 6$) of the air taken in during normal breathing actually ventilates the alveoli.

(1₂3) Box 34.1 Derivation of the Bohr equation for calculating the physiological dead space

All of the CO_2 in the expired air comes from the alveoli. The amount of CO_2 in any volume of gas is simply the fractional content of CO_2 in the sample (F) times the volume. Since the volume of air expired (V_E) is the sum of the air displaced from the dead space (V_D) plus that expelled from the alveoli (V_A),

$$V_E = V_D + V_A \qquad [1]$$

and

$$F_E.V_E = F_D.V_D + F_A.V_A \qquad [2]$$

Since there is no CO_2 in the dead space (as it is filled with atmospheric air which has a negligible CO_2 content), $F_D = 0$ and equation 2 becomes

$$F_E.V_E = F_A.V_A \qquad [3]$$

This can be rewritten as

$$V_A = \frac{V_E.F_E}{F_A} \qquad [3a]$$

But since $V_A = V_E - V_D$ (from equation 1),

$$V_E - V_D = \frac{V_E.F_E}{F_A} \qquad [4]$$

which can be rewritten as the Bohr equation:

$$V_D = V_E \left(1 - \frac{F_E}{F_A} \right) \qquad [4a]$$

The physiological dead space can therefore be calculated from the volume and fractional CO_2 content of the expired gas and the fractional concentration of CO_2 in the alveolar air.

The volume and CO_2 content of the expired air can be readily measured, and a sample of alveolar air can be obtained by asking a subject to expire fully through a long thin tube. The last part of the expired volume will have the same composition as the alveolar air. This gas can be sampled and its FCO_2 determined. This will give an average for the composition of the alveolar air. A more accurate estimate of the CO_2 content of alveolar air can be obtained by measuring the PCO_2 of arterial blood.

A worked example

If the percentage of CO_2 in the expired and alveolar air were to be 3.6 per cent and 5.2 per cent and the tidal volume was 500 ml, what is the physiological dead space?

Using the Bohr equation:

$$V_D = 500 \times (1 - (3.6 \div 5.2))$$
$$= 500 \times (1 - 0.69)$$
$$= 155\,\text{ml}$$

34.3 Alveolar ventilation is not uniform throughout the lung

The ventilation of the lung is *the change in volume relative to the resting volume during a single respiratory cycle*; the greater the relative change in volume, the greater the ventilation. Measurements with tracer gases show that the inspired air is not distributed evenly to all parts of the lung. The pattern of ventilation depends on posture (i.e. whether the subject is upright or lying down), on the speed of inspiration, and on the amount of air inspired. In an upright subject, during a slow inspiration following a normal expiration the base of each lung is ventilated about 50 per cent more than the apex (Figure 34.2). If the subject inspires after a forced expiration to the residual volume, the base of the lung is ventilated nearly three times as much as the apex. This difference is reduced if the subject lies down. Until recently, the variation in ventilation was attributed to the influence of gravity on the lung parenchyma. However, recent measurements on astronauts during space flight have shown that uneven ventilation persists in zero gravity.

What are the causes of uneven pulmonary ventilation? First, during inspiration, the volume of the lower part of the chest increases significantly more than the upper part. This situation arises

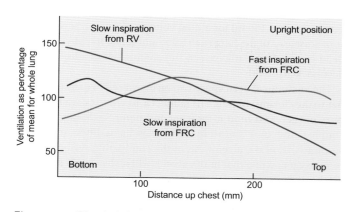

Figure 34.2 The distribution of ventilation in the normal upright human lung. The data show the distribution following a slow inspiration following a normal expiration (FRC—blue curve), a fast inspiration from a normal expiration (red curve), and a slow inspiration from residual volume (mauve curve). In all cases the ventilation is uneven. For slow inspiration it is greater at the base of the lung compared to the apex. For fast inspiration the apex is better ventilated than the base. When the subject is supine a similar pattern is seen, although the difference in ventilation between the apex and the base is smaller.

because the lower ribs are more curved and more mobile than the upper ribs so that this part of chest expands relatively more than the apex. Second, the descent of the diaphragm expands the lower lobes of the lungs more than the upper ones, which are attached to the main bronchi. (The bronchi are much less easily stretched than the lung parenchyma.) Third, the compliance of the lung is not uniform. The peripheral lung tissue is more compliant than the deeper tissue, which is attached to the stiffer airways. Fourth, the effect of gravity on the lungs causes a progressive decrease in intrapleural pressure from the apex to the base of the thorax. Since the alveolar pressure will be equal to that of the atmosphere when the airways are open and there is no airflow, the transpulmonary pressure is smaller in the lower regions of the lungs than in the upper regions. Hence the lower regions are somewhat less inflated at FRC and increase more in volume during inspiration. The combination of these four factors results in differing regions of the lung exhibiting differing amounts of ventilation.

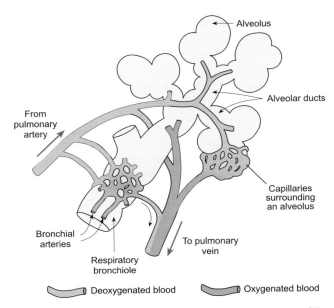

Figure 34.3 A schematic drawing showing the arrangement of the pulmonary and bronchial circulations in relation to the alveoli. Note that the bronchial circulation does not supply the alveoli but drains into the pulmonary veins having bypassed the alveoli. (Courtesy of Professor M. De Burgh Daly.)

Summary

The anatomical dead space is the volume of air taken in during a breath that does not mix with the air in the alveoli, while the physiological dead space is the volume of air taken in during a breath that does not take part in gas exchange. In healthy subjects the physiological and anatomical dead spaces are very similar. In certain diseases, physiological dead space may greatly exceed the anatomical dead space.

The minute ventilation is the tidal volume multiplied by the frequency of breathing. The alveolar ventilation is the volume of air entering the alveoli per minute. In upright subjects, the ventilation of the lung is not uniform, being somewhat greater at the base than at the apex. This difference is reduced when lying down.

34.4 The bronchial and pulmonary circulations

The lungs receive blood from two sources: the bronchial circulation and the pulmonary circulation (see Figure 34.3).

The bronchial circulation

Bronchial arteries arise from the aortic arch, the thoracic aorta, or their branches (mainly the intercostal arteries). These arteries supply oxygenated blood to the smooth muscle of the principal airways (the trachea, bronchi, and bronchioles as far as the respiratory bronchioles), the intrapulmonary nerves, nerve ganglia, and the interstitial lung tissue. The blood draining the airways is deoxygenated. The blood from the upper airways (as far as the second order bronchi) drains into the right atrium on the right side via the bronchial veins. The venous return from the later generations of airways flows into the pulmonary veins, where it mixes with the oxygenated blood from the alveoli. The bronchial circulation normally accounts for only a very small part of the output of the left

ventricle (about 2 per cent), but in certain rare clinical conditions, it can be as much as 20 per cent.

The pulmonary circulation

The output of the right ventricle passes into the pulmonary artery, which subsequently branches to supply the individual lobes of the lung. The pulmonary arteries branch along with the bronchial tree until they reach the respiratory bronchioles. Here they form a dense capillary network to provide a vast area for gas exchange that is similar in extent to that of the alveolar surface, as shown in Figure 34.3 (also see Figure 32.6). The capillaries drain into pulmonary venules, which arise in the septa of the alveoli. Small veins merge with larger veins that are arranged segmentally. Finally, two large pulmonary veins emerge from each lung to empty into the left atrium.

Pulmonary blood flow and its regulation

The output of the right ventricle is virtually equal to that of the left ventricle so that, at rest, about 5 l of blood pass through the pulmonary vessels each minute. Since the pulmonary capillaries contain only about 100 ml of blood, and since the stroke volume is about 70 ml (see Chapter 27), a large part of the blood in the pulmonary capillary bed is replaced with each beat of the heart.

The pulmonary arterioles do not appear to be subject to significant autonomic regulation. Instead, the calibre of the small vessels is regulated by the alveolar PO_2 and PCO_2. In areas of the lung that have a low PO_2 (hypoxia) or a high PCO_2 (hypercapnia), the arterioles constrict and blood is diverted to the well oxygenated areas. This response is the opposite of that seen in other vascular beds (where low PO_2 causes a dilatation of the arterioles) and is not abolished by section of the autonomic nerves. It is therefore a local response.

Blood flow through the upright lung is greatest at the base and least at the apex

As in other vascular beds, the flow of blood through the lungs is determined by the perfusion pressure and the vascular resistance. Compared to the systemic circulation, however, the pressures in the pulmonary arteries are rather low, the systolic and diastolic pressures being about 3.3 and 1.0 kPa respectively (c. 25/8 mm Hg). Since a column of blood 1 cm high will exert a pressure of about 0.1 kPa, the pressure in the pulmonary artery during systole is sufficient to support a column of blood 33 cm high. During diastole, however, the pressure is sufficient only to support a column about 10 cm high. As a result, differences in the hydrostatic pressures of the blood in various parts of the pulmonary circulation will exert a considerable influence on the distribution of pulmonary blood flow when the body is erect. Moreover, because the pressures in the pulmonary circulation are low, the pressure of the air in the alveoli has a marked effect on vascular resistance and hence on blood flow. Thus, the flow of blood to different parts of the lung is determined by three pressures:

1. the hydrostatic pressure in the pulmonary arteries of different parts of the lungs

2. the pressure in the pulmonary veins

3. the pressure of the air in the alveoli.

When the body is erect, the base of the lung lies below the origin of the pulmonary artery and the hydrostatic pressure of the blood

sums with the pressure in the main pulmonary artery. Above the origin of the pulmonary artery, the pressure falls with distance up the lung until, at the apex of the lung, which is about 15 cm above the origin of the pulmonary artery, blood flow occurs only during systole and not during diastole. The influence of these pressures on the distribution of blood flow in the upright lung is shown in Figure 34.4.

For simplicity we will assume that the pressure in the pulmonary veins (P_V) is equal to that of the atmosphere (taken as zero by convention) so that the perfusion pressure will be equal to the pressure in the pulmonary arteries (P_a).

- At the apex of the lung, the alveolar pressure (P_A) is similar to that of the pressure in the branches of the pulmonary arteries (P_a). The pulmonary capillaries will be relatively compressed and vascular resistance will be high. As a result, blood flow is low (zone 1 of Figure 34.4). Indeed, if the arterial pressure falls below the alveolar pressure, the capillaries will be collapsed (as $P_A > P_a$) and there will be no blood flow. This may happen during diastole.

- In the middle zone (zone 2 of Figure 34.4), the pressure in the pulmonary arteries is higher and exceeds the alveolar pressure, so that blood flow progressively increases down the zone.

- At the base of the lung (zone 3 in Figure 34.4), the arterial pressure exceeds the alveolar pressure by a considerable margin and the pulmonary vessels are fully open. Blood flow is relatively high.

It is important to understand that the regional variation in blood flow is mainly due to the effect of gravity. It disappears when the subject lies down, as the anterio-posterior difference in pressure is much smaller. Moreover, when gravity is reduced during simulated space flights in aircraft, apical blood flow is increased.

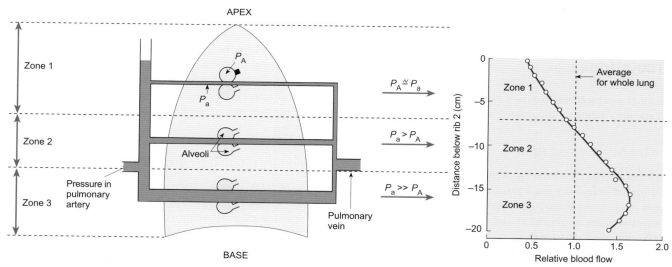

Figure 34.4 The influence of the hydrostatic and alveolar pressures on the distribution of blood flow in the upright lung. Blood flow depends on the perfusion pressure (here assumed to be proportional to the pressure in the pulmonary arteries P_a for simplicity) and on the vascular resistance. When perfusion pressure is low and is similar in value to the alveolar pressure (P_A), the pulmonary vessels will be partially compressed, vascular resistance will be increased (zone 1), and blood flow will be correspondingly low. When the perfusion pressure exceeds the alveolar pressure P_A, vascular resistance is low and blood flow is high (zone 3). Zone 2 shows the transition from zone 1 to zone 3 with distance below rib 2.

When cardiac output increases, the pulmonary vascular resistance falls

As in other vascular beds, the blood flow through the lungs depends on the perfusion pressure and the vascular resistance. During exercise, cardiac output rises considerably but the pressure in the pulmonary arteries shows only a small increase. For example, if cardiac output increases threefold, the mean pressure in the pulmonary artery rises by less than 1 kPa (~7.5 mm Hg). Since pressure = flow × resistance, the small increase in pressure coupled with the large increase in flow indicates that the resistance of the pulmonary vessels must fall with increased cardiac output.

This fall in the resistance of the pulmonary vasculature is not due to an autonomic reflex or to circulating hormones but is thought to be a passive response to the increased perfusion pressure. Two mechanisms are believed to be responsible: the recruitment of additional vessels and the distension of those vessels that were already open.

Passive recruitment of the pulmonary vessels is possible because many are closed at the resting values of pressure in the pulmonary arteries. This occurs because the pressure of the air in the alveoli acts directly on the walls of the pulmonary capillaries, tending to collapse them, as discussed earlier in the chapter. As cardiac output rises during exercise, however, the pressure in the pulmonary vessels exceeds that in the alveoli, with the result that more capillaries are recruited. Moreover, the distribution of blood flow to the various parts of the lungs becomes more even.

Summary

The lungs receive their blood supply via the bronchial and the pulmonary circulations. The bronchial circulation is part of the systemic circulation and supplies oxygenated blood to the trachea, bronchi, and bronchioles as far as the respiratory bronchioles.

The pulmonary circulation is supplied by the right ventricle and participates in gas exchange. The pressures in the pulmonary artery are about 25/8 mmHg (3.3/1.0 kPa)—much lower than those in the aorta. As a result, the effects of gravity on pulmonary blood flow are very significant and there is considerable variation in blood flow in the upright lung, the base of the lung being relatively well perfused compared to the apex.

34.5 The matching of pulmonary blood flow to alveolar ventilation

In the upright lung, both ventilation and perfusion fall with height above the base. Since the local blood flow falls more rapidly than ventilation, the ratio of alveolar ventilation to blood flow ($\dot{V}_A/\dot{Q}$ ratio) will vary. Since the ventilation of the lung serves to promote gas exchange between the blood and the alveolar air, this variation has considerable physiological significance. For the lungs as a whole, the alveolar ventilation is about 4.2 l min^{-1} while the resting cardiac output is about 5.0 l min^{-1}, so that the average value for the $\dot{V}_A/\dot{Q}$ ratio is 4.2 ÷ 5.0 = 0.84. The base of the lung is relatively well

perfused and ventilated, and estimates suggest a $\dot{V}_A/\dot{Q}$ ratio of about 0.6. The ratio rises slowly with distance from the base (measured by rib number in Figure 34.5). About two-thirds of the way up, the ratio is close to one—theoretically perfect matching. Above this, the ratio rises steeply, reaching a value of about three in the apex. These are average figures for the various segments of the lung, but they are not constant. In exercise, for example, the ventilation at the base of the lungs increases significantly more than the blood flow, while at the top of the lungs blood flow increases more than ventilation and the $\dot{V}_A/\dot{Q}$ ratio becomes more uniform (see Chapter 38).

The $\dot{V}_A/\dot{Q}$ ratio can vary considerably, from infinity (ventilated alveoli that are not perfused) to zero (for blood that passes through the lung without coming into contact with the alveolar air). For present purposes it is convenient to distinguish between three situations:

1. Well-ventilated alveoli that are well perfused with blood ($\dot{V}_A/\dot{Q}$ ratio ≈ 1). In this situation, the blood will equilibrate with the alveolar air and will become arterialized (i.e. it will have the PO_2 and PCO_2 of arterial blood). This is the optimum matching of ventilation and perfusion.

2. Poorly ventilated alveoli that are well perfused with blood ($\dot{V}_A/\dot{Q}$ ratio <<1). In this case, the PO_2 and PCO_2 in the alveolar air will tend to equilibrate with the blood. As a result, PO_2 will be lower than normal but the PCO_2 will be close to normal. The extent of this equilibration will obviously depend on the extent of the alveolar ventilation, but when this is zero, the alveolar air will have the same PO_2 and PCO_2 as mixed venous blood.

3. Well-ventilated alveoli that are poorly perfused with blood ($\dot{V}_A/\dot{Q}$ ratio >> 1). In this case, the blood leaving the alveoli will have a low PCO_2 as the pressure gradient favours the loss of CO_2 from the blood. However, while the blood will be fully oxygenated, the *oxygen content* of the blood will not be significantly greater than normal, as the haemoglobin will be fully saturated.

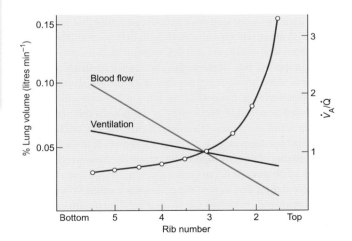

Figure 34.5 The distribution of ventilation, blood flow, and the ventilation–perfusion ratio in the normal upright lung. Straight lines have been drawn through the data for ventilation and blood flow (data shown in Figures 34.2 and 34.4). Note that the ventilation–perfusion ratio rises slowly at first, then rapidly towards the top of the lung (blue line). (Based on Fig 19 in J.B. West (1977) *Ventilation/blood flow and gas exchange*. Blackwell Science, Oxford.)

(1_23) Box 34.2 Calculating venous admixture: the shunt equation

While the lungs are very efficient at oxygenating the blood, even in healthy subjects there is always a small fraction of the blood entering the left atrium that is not fully oxygenated. In various disease states this fraction, which is known as pulmonary shunt, can become very significant. In all situations the addition of this venous blood (i.e. the venous admixture) will lower the oxygen content of arterial blood. The extent of this venous admixture can be calculated from the shunt equation:

$$\frac{\dot{Q}_S}{\dot{Q}_T} = \frac{Cc_{O_2} - Ca_{O_2}}{Cc_{O_2} - C\bar{v}_{O_2}}$$

where Cc_{O_2} is the oxygen content of the blood that has passed though the capillaries of the ventilated alveoli, Ca_{O_2} is the oxygen content of the arterial blood, and $C\bar{v}_{O_2}$ is the oxygen content of the mixed venous blood. Note that this calculation provides a measure of the degree to which the lungs deviate from being totally efficient blood oxygenators.

In healthy individuals, a small amount of venous blood mixes with the pulmonary end-capillary blood as it passes to the left side of the heart. This includes blood from the bronchial circulation, which normally contributes around 1 per cent, and that draining the cardiac muscle via the thebesian veins, which contributes about 0.3 per cent. Nevertheless, the amount of venous admixture can be considerable in congenital heart disease and in disease states such as pulmonary oedema and pneumonia.

A worked example

The total haemoglobin of a blood sample is 14 g dl^{-1}. The mixed venous blood of the same subject is 40 per cent saturated with oxygen, their arterial blood is 75 per cent saturated, and their end capillary blood is 97 per cent saturated. So their oxygen contents are 7.6, 14.2, and 18.3 ml O_2 dl^{-1}. (Recall that each gram of haemoglobin binds 1.34 ml O_2 and at 13.3 kPa each 100 ml of blood has 0.3 ml dissolved oxygen.) The shunt fraction is:

$$\frac{\dot{Q}_S}{\dot{Q}_T} = \frac{18.3 - 14.2}{18.3 - 7.6} = 0.38$$

Another way of looking at venous admixture is to ask what proportion of mixed venous blood (x) would be needed to mix with fully oxygenated blood to account for the observed oxygen content of a sample of arterial blood.

$$Ca_{O_2} = Cc_{O_2} \times (1 - x) + (C\bar{v}_{O_2} \times x)$$

Using the values above:

$$14.2 = 18.3 - 18.3x + 7.6x$$
$$= 18.3 - 10.7x$$
$$10.7x = 18.3 - 14.2 = 4.1$$
$$\therefore x = 0.38$$

This is the same as the shunt fraction calculated earlier.

The lowered PO_2 of blood leaving poorly ventilated parts of the lungs will not be compensated by well-oxygenated blood leaving relatively over-ventilated parts. This situation arises because the oxygen content of blood from the over-ventilated alveoli is not significantly higher than that from well-matched alveoli, while that from poorly ventilated areas will be substantially below normal. Thus, when blood from well-ventilated and poorly ventilated regions becomes mixed in the left side of the heart, the *oxygen contents* of the two streams of blood are averaged but the PO_2 of the mixed arterial blood will be below average because of the shape of the oxygen dissociation curve (see Figure 25.11). Thus the $\dot{V}_A/\dot{Q}$ ratio in the different parts of the lung is the main factor that determines the PO_2 of the arterial blood, and not the average $\dot{V}_A/\dot{Q}$ ratio for the whole of the lung.

The mixing of venous blood with oxygenated blood is known as **venous admixture**. It occurs naturally when blood from the bronchial circulation drains into the pulmonary veins. When venous blood completely bypasses the lungs it is called a right–left shunt; this is commonly seen in congenital heart disease where deoxygenated blood from the right side of the heart mixes with oxygenated blood from the pulmonary veins. As for the case when blood from poorly ventilated alveoli mixes in significant quantities with arterialized blood from well-ventilated alveoli, a right–left shunt will reduce the PO_2 and the oxygen content of the blood reaching the systemic circulation. Box 34.2 shows how to calculate the extent of venous admixture.

The effect of $\dot{V}_A/\dot{Q}$ mismatch and venous admixture is to lower the arterial PO_2 so that it is less than that of the alveolar PO_2. This **alveolar–arterial PO_2 difference** (the A–a difference) normally amounts to no more than 2 kPa (15 mm Hg) in young healthy adults, although it increases with age (discussed in Section 34.9). Any lung disease that causes a mismatch in the $\dot{V}_A/\dot{Q}$ ratio or increases the venous admixture (increased shunting) will increase the alveolar–arterial PO_2 difference.

Summary

The ratio of alveolar ventilation to blood flow ($\dot{V}_A/\dot{Q}$ ratio) is an important measure of respiratory function. In the upright lung both the ventilation of the alveoli and their perfusion decrease from the base to the apex of the lung, with significant effects on the $\dot{V}_A/\dot{Q}$ ratio. The $\dot{V}_A/\dot{Q}$ is approximately 0.6 at the base of the lung, rising to a value of 1.0 around two-thirds of the distance between the base and apex (theoretically perfect matching) before steeply increasing to a value of about 3.0 at the apex. The fully oxygenated blood leaving well-ventilated parts of the lung cannot compensate for the low PO_2 of blood leaving poorly ventilated regions, with the result that the PO_2 of arterial blood is normally slightly lower than that of the alveoli. In certain lung diseases the difference may be considerable. The mixing of venous blood with oxygenated blood is known as venous admixture or shunt.

34.6 Gas exchange across the alveolar membrane occurs by physical diffusion

As the inspired air passes through the passages of the lung, its velocity falls steeply with each airway generation until it arrives at the alveoli where it equilibrates by simple diffusion with the air already in the alveoli. The rate of diffusion across the alveolar membranes is governed by Fick's law of diffusion (see Box 31.1). To be able to oxygenate the blood, a molecule of O_2 must first dissolve in the aqueous layer covering the alveolar epithelium. It then diffuses across the thin membranes that separate the alveolar air spaces from the blood before combining with haemoglobin in the red cells. When a subject is resting, the blood takes about 1 s to pass through the pulmonary capillaries, but during severe exercise it can take as little as 0.3 seconds. Despite the short time available, in healthy subjects the diffusion is so efficient that the blood still becomes fully equilibrated with the alveolar air during its transit through the pulmonary capillaries.

The ability of the lungs to ensure equilibration between the blood of the pulmonary capillaries and the alveolar air is measured by its **diffusing capacity** (sometimes called its **transfer factor**), which is defined as the volume of gas that diffuses through the alveolar membranes per second for a pressure difference of one kPa. The formal definition is given by the following equation:

$$D_{LX} = \frac{V_X}{(P_{AX} - P_{CX})}$$

where D_{LX} is the diffusing capacity of gas X (ml min^{-1} kPa^{-1}), V_X is the volume of gas diffusing between the alveoli and the blood, P_{AX} is the partial pressure of the gas in the alveolar air, and P_{CX} is the average partial pressure of the gas in the pulmonary capillaries.

The numerical value of the diffusing capacity depends on how it is measured. It is normally estimated by asking the subject to take a breath of a gas mixture containing a small amount of carbon monoxide. As this gas binds strongly to haemoglobin, the partial pressure in the alveolar capillaries in non-smokers is effectively zero and the equation for calculating the diffusing capacity becomes:

$$D_{LCO} = \frac{V_{CO}}{P_{ACO}}$$

In normal healthy young people the diffusing capacity for carbon monoxide measured by the single breath method averages 225 ml min^{-1} kPa^{-1} (30 ml min^{-1} mm Hg^{-1}) at rest. The oxygen diffusing capacity is equal to that for carbon monoxide multiplied by 1.23.

The diffusing capacity for CO_2 is about 20 times that for oxygen, as it is much more soluble in the alveolar membranes. Nevertheless, the overall rate of equilibration of CO_2 between the blood and the alveolar air is similar to that for oxygen. This is partly because the pressure gradient for CO_2 is much smaller (0.8 kPa or 6 mm Hg) and partly because most of the CO_2 of the blood is in chemical combination (as carbamino compounds and bicarbonate), from which it is released relatively slowly.

The diffusing capacity increases with body size and with lung volume. (Diffusing capacity depends directly on the area available for gas exchange, while the partial pressures of the respiratory gases do not.) It increases significantly during exercise, when previously closed pulmonary capillaries open. It also increases when a subject lies down. This change probably reflects the more uniform distribution of pulmonary blood flow. Diffusing capacity declines with age. If the alveolar membranes become thickened by disease, as in emphysema or fibrosis, or if they become filled with fluid (pulmonary oedema), the diffusing capacity will be significantly reduced.

Summary

The ability of the lungs to ensure equilibration between the blood of the pulmonary capillaries and the alveolar air is measured by its diffusing capacity. Diffusing capacity depends directly on the area available for gas exchange and increases with body size. While the diffusing capacity for carbon dioxide is about 20 times that for oxygen, the larger partial pressure gradient of oxygen ensures that the overall rate of equilibration is similar for both gases.

34.7 Fluid exchange in the lungs

In health, the alveoli contain no fluid. What prevents fluid leaking from the pulmonary capillaries into the alveoli? Like other vascular beds, the exchange of fluid between the capillaries and the interstitial fluid is governed by Starling forces (see Box 31.3). The pressure in the pulmonary artery is low: the mean pressure is only about 1.5 kPa (12 mm Hg), and the pressure in the pulmonary capillaries is still lower—between 0 and 2 kPa (0–15 mm Hg) depending on the distance above the heart. The interstitial pressure is about 0.5 kPa (3.8 mm Hg) below that of the atmosphere. Although the oncotic pressure of the plasma is the same as that in the systemic circulation (c. 3.6 kPa or 27 mm Hg), that of the pulmonary interstitial fluid is relatively high (c. 2.2 kPa or 16 mm Hg). Consequently, the net filtration pressure is about 1 kPa (~7.5 mm Hg) in the lower part of the lungs. Normally the filtered fluid is returned to the circulation via the pulmonary lymphatics. The pulmonary lymph flow is about 10 ml h^{-1}, keeping the alveoli free of fluid.

As the capillary pressure in the upper part of the lungs is low, there is little fluid formation in this part of the lung. However, as the Starling forces favour filtration in the dependent regions, a small increase in the pressure within the pulmonary capillaries will lead to a greater filtration of fluid. If this exceeds the drainage capacity of the lymphatics, fluid will accumulate in the interstitium, eventually resulting in **pulmonary oedema**. This occurs during left-sided heart failure and is also a consequence of mechanical or chemical damage to the lining of the alveoli. The fluid first accumulates within the pulmonary interstitium and the lymphatic vessels. However, above a critical pressure, fluid will enter the alveoli themselves, flooding them and seriously compromising their ability to participate in gas exchange.

34.8 Metabolic functions of the pulmonary circulation

All of the venous return from the systemic circulation passes through the lungs on its way to the left side of the heart. As a result, the pulmonary circulation is ideally situated to metabolize vasoactive materials released by specific vascular beds. Many substances such as bradykinin, norepinephrine, and prostaglandin E_1 are almost completely removed in a single pass through the lungs. In contrast, the lungs have an enzyme (angiotensin converting enzyme or ACE) that converts angiotensin I to its active form angiotensin II, which stimulates aldosterone secretion by the adrenal cortex (Chapter 39). The site of this metabolic activity is the endothelium of the pulmonary circulation.

34.9 Age-related changes in respiratory function

Although some older people suffer from chronic respiratory illnesses as a result of smoking or industrial disease, alveolar function is relatively unchanged with age. The most noticeable changes to the respiratory system include a gradual reduction in lung compliance associated with a loss of elasticity, and a reduction in the strength of the muscles of the ribcage. Consequently, there is an increase in the work of breathing, a fall in vital capacity, and an increase in residual volume, as shown in Figure 34.6. Closing volume increases with age and may encroach upon the normal tidal volume. Usable lung capacity typically falls to around 82 per cent of its maximum value by the age of 45, to 62 per cent at the age of 65, and to around 50 per cent by the age of 85.

Over time, some alveoli are replaced by fibrous tissue and gas exchange is reduced, so arterial PO_2 declines with age (see Figure 34.7). There may also be a small rise in arterial PCO_2, particularly during exercise. Overall, the respiratory changes seen in the elderly tend to limit the ability to increase ventilation and oxygen delivery to the tissues during periods of increased demand such as exercise.

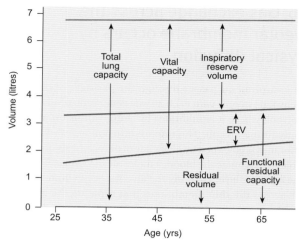

Figure 34.6 The changes in static lung volumes with age in healthy non-smoking adults between the ages of 25 and 70 years. ERV is expiratory reserve volume. Note that total lung volume is relatively unaltered but that the residual volume progressively increases with age. (Adapted from Fig 1 in J.P. Janssens et al. (1999) *Eur Respir J* **13**, 197–205).

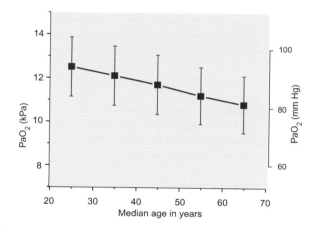

Figure 34.7 The change in arterial PO_2 with age. Note that most healthy young subjects do not achieve the theoretical PO_2 value of 13.3 kPa (100 mm Hg). However, even when PO_2 has fallen to around 10.6 kPa (80 mm Hg), there is adequate oxygenation of the arterial blood. (Based on data in Table 11.3 in J.F. Nunn (1993) *Nunn's applied respiratory physiology*, 4th edition. Butterworth-Heinemann, Oxford.)

✳ Checklist of key terms and concepts

Alveolar ventilation

- The anatomical dead space is the volume of air taken in during a breath that does not mix with the air in the alveoli. It is a measure of the volume of the conducting airways.

- The physiological dead space is the volume of air taken in during a breath that does not take part in gas exchange.

- In healthy subjects, the anatomical and physiological dead spaces are similar.

- The minute ventilation is the tidal volume multiplied by the frequency of breathing.

- The alveolar ventilation is the volume of air entering the alveoli per minute.

- The ventilation of the lung is not uniform, being somewhat greater at the base than at the apex. This situation arises because the base of the lungs expands proportionately more than the apex.

The bronchial and pulmonary circulations

- The lungs receive their blood supply via the bronchial and the pulmonary circulations.

- The bronchial circulation is part of the systemic circulation and supplies oxygenated blood to the trachea, bronchi, and bronchioles as far as the respiratory bronchioles.

- The pulmonary circulation is supplied by the output of the right ventricle, and the blood in this vascular bed participates in gas exchange.

- The pressures in the pulmonary artery are much lower than those in the aorta. Systolic pressure is about 25 mm Hg, while diastolic pressure is only about 8 mm Hg.

- Because the systolic and diastolic pressures in the pulmonary arteries are low, the effects of gravity on regional blood flow in the lungs are very significant and the base of the upright lung has a higher blood flow than the apex.

Ventilation/perfusion ($\dot{V}_A / \dot{Q}$) ratios and diffusing capacity

- In the upright lung, both ventilation ($\dot{V}_A$) and perfusion ($\dot{Q}$) fall with height above the base but the blood flow falls significantly faster than ventilation. This results in an increase in the $\dot{V}_A / \dot{Q}$ ratio with height above the base of the lung.

- In regions that are under-ventilated or over-perfused, the blood will tend to have a lower than normal PO_2.

- In healthy lungs, a low PO_2 tends to cause a local vasoconstriction, diverting the blood to better-ventilated areas.

- The ability of the lungs to ensure equilibration between the blood of the pulmonary capillaries and the alveolar air is measured by its diffusing capacity.

- While the diffusing capacity for carbon dioxide is about 20 times that for oxygen, the overall rate of equilibration is similar for both gases.

- The pulmonary circulation plays a significant role in the metabolism of many vasoactive substances.

 ## Recommended reading

Physiology of the respiratory system

Dejours, P. (1966) *Respiration*, Chapter 9. Oxford University Press, New York.

An old book that gives an excellent quantitative treatment that is suitable for graduate students and those considering research in human respiratory physiology.

Hlastala, M.P., and Berger, A.J. (2001) *Physiology of respiration* (2nd edn). Oxford University Press, New York.

Offers a more detailed account of respiratory physiology than other introductory texts, with good coverage of the underlying neural processes.

Levitzky, M.G. (2003) *Pulmonary physiology* (6th edn). McGraw-Hill, New York.

A well-established introductory text with good coverage of the mechanics of breathing.

Lumb, A.B. (2010) *Nunn's applied respiratory physiology* (7th edn). Churchill-Livingstone, London.

A detailed monograph covering respiratory physiology from the anaesthetist's point of view.

Slonim, N.B., and Hamilton, L.H. (1987) *Respiratory physiology*. Mosby, St Louis.

Now rather old but still well worth reading for its excellent coverage and clear explanations.

West, J.B. (2004) *Respiratory physiology: The essentials* (7th edn). Lippincott, Williams & Wilkins, Philadelphia.

A popular introductory text that offers simple explanations with clear diagrams.

Review articles

Cotes, J.E., Chinn, D.J., Ouanjer, P.H., Roca, J., and Yernault, J.-C. (1993) Standardization of the measurement of transfer factor (diffusing capacity). *European Respiratory Journal* **6**, Suppl. 16, 41–52.

Janssens, J.P., Pache, J.C., and Nicod, L.P. (1999) Physiological changes in respiratory function associated with ageing. *European Respiratory Journal* **13**, 197–205.

Kummer, W. (2011) Pulmonary vascular innervation and its role in responses to hypoxia— Size matters! *Proceedings of the American Thoracic Society* **8**, 471–76.

Paredi, P., and Barnes, P.J. (2009) The airway vasculature: Recent advances and clinical implications. *Thorax* **64**, 444–50.

Prisk, G.K., Fine, J.M., Cooper, T.K., and West, J.B. (2006) Vital capacity, respiratory muscle strength, and pulmonary gas exchange during long-duration exposure to microgravity. *Journal of Applied Physiology* **101**, 439–47.

 To check that you have mastered the key concepts presented in this chapter, complete the accompanying online self-assessment questions. Go to www.oup.com/uk/pocock5e/

CHAPTER 35
The control of respiration

This chapter should help you to understand:

- The origin of the respiratory rhythm
- The respiratory reflexes
- The role of blood gases in the control of ventilation
  - The peripheral arterial chemoreceptors
  - The central chemoreceptors
- The effects of breathing different gas mixtures on ventilation and the importance of arterial PCO_2 in the regulation of respiration

35.1 Introduction

No normal healthy person has to think about when, or how deeply, to breathe. Breathing is an automatic, rhythmical process that is constantly adjusted to meet the everyday requirements of life such as exercise and speech. To account for this remarkable fact, it is necessary to consider three important questions:

1. Where does this rhythmical activity originate?
2. How is it generated?
3. How is the rate and depth of respiration controlled?

35.2 The origin of the respiratory rhythm

The role of the brainstem

The basic respiratory rhythm is maintained even if all the afferent nerves are cut. If the brainstem of an anaesthetized animal is completely cut through above the pons, the basic rhythm of respiration continues. Cutting the spinal cord below the outflow of the phrenic nerve (C3–C5) leads to paralysis of the intercostal muscles but not of the diaphragm (which is innervated by the phrenic nerves). However, cutting through the lower region of the medulla will block all respiratory movements. From these observations, two things are clear.

1. The respiratory muscles themselves have no intrinsic rhythmic activity.
2. The caudal part of the brainstem has all of the neuronal mechanisms required to generate and maintain a basic respiratory rhythm.

After the vagus nerves have been cut, respiration becomes slower and deeper. If the brainstem is subsequently cut across between the medulla and pons, there is little change in the respiratory pattern. Cutting through the pons, however, alters the pattern of respiration so that inspiration becomes relatively prolonged with brief episodes of expiration. Stimulation of specific groups of nerve cells in the pons synchronizes the discharge of the phrenic nerves with the stimulus. From these and other experimental observations, the pons has been shown to have an important role in regulating the respiratory rhythm. It is here that afferent information concerning the state of the lungs is thought to act to modulate the rate and depth of respiration.

The generation of the respiratory rhythm

There are two groups of neurons in the medulla that discharge action potentials with an intrinsic rhythm that corresponds to that of the respiratory cycle. These are known as the **dorsal respiratory**

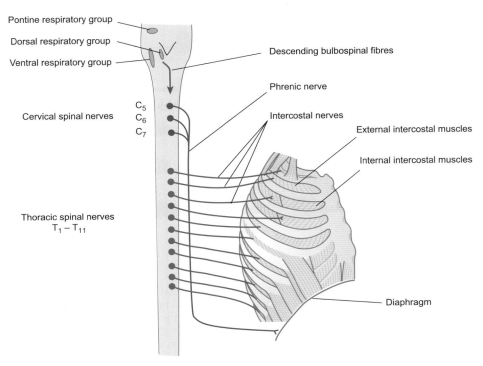

Figure 35.1 A diagrammatic representation of the nerve pathways involved in breathing. Neurons in the pons and medulla oblongata establish the respiratory rhythm, which is then transmitted by descending fibres called bulbospinal fibres to motoneurons in the cervical and thoracic segments of the spinal cord. The phrenic nerves originate from segments C_5, C_6, and C_7 and innervate the diaphragm. They are mainly active during inspiration. The motor fibres that supply the external and internal intercostal muscles originate from thoracic segments T_1 to T_{11}.

group and the **ventral respiratory group**. Those of the dorsal respiratory group are active just prior to and during inspiration and are therefore mainly inspiratory neurons. The ventral respiratory group consists of both inspiratory and expiratory neurons, and they receive inputs from the dorsal respiratory group. The neurons of both groups are upper respiratory motoneurons.

The neurons of the dorsal and ventral respiratory groups receive a variety of inputs from higher centres in the brain, including the cerebral cortex and pons. They also receive inputs from the carotid bodies (which are the peripheral chemoreceptors that sense the PO_2, PCO_2 and pH of the arterial blood—see Section 35.4) and the vagus nerve (which carries afferent nerve fibres from the airways). It appears that inspiration is initiated by neurons of the dorsal respiratory group. This intrinsic activity subsequently sums with afferent activity coming from lung stretch receptors to switch off inspiration and commence expiration.

The neurons of the dorsal respiratory group project to the lower respiratory motoneurons of the phrenic nerve on the opposite side (the contralateral phrenic nerve), while those of the ventral respiratory group are the upper respiratory motoneurons for both the contralateral phrenic and intercostal nerves. Figure 35.1 shows diagrammatic representation of the arrangement of the principal respiratory nerves.

The activity of the expiratory respiratory motoneurons in the spinal cord (the lower respiratory motoneurons) is inhibited during inspiration, while that of the inspiratory motoneurons is inhibited during expiration. This pattern of reciprocal inhibition has its origin in the dorsal and ventral respiratory groups and is not a local spinal reflex. This is illustrated in Figure 35.2, which shows the electrical activity of the thoracic respiratory neurons that supply the intercostal muscles and that of the cervical respiratory motoneurons that supply the diaphragm via the phrenic nerves. Throughout the respiratory cycle, the activity of the inspiratory motoneurons alternates with that of the expiratory motoneurons. During inspiration, the activity of the diaphragm and external intercostal muscles progressively increases as additional motor units are recruited and the muscles shorten progressively, thereby expanding the volume of the chest (see Chapter 33). During expiration, the activity of the phrenic nerves gradually declines and that of the external intercostal nerves ceases. The inspiratory muscles relax, allowing the chest to return smoothly to its resting volume (FRC). The internal intercostal motoneurons show a reciprocal pattern, with activity increasing during expiration but becoming inactive during inspiration. The progressive modulation of the tone of the phrenic nerve provides a smooth transition from expiration to inspiration and forms part of the work of breathing.

Voluntary control of respiration

Normal regular breathing (or **eupnoea**) is an automatic process, although the rate and depth of breathing can be readily adjusted by voluntary means. For example, it is possible to suspend breathing

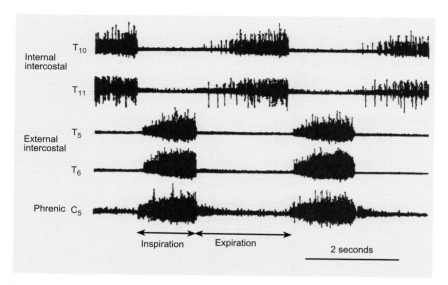

Figure 35.2 An example of reciprocal motor activity in the diaphragm and the internal and external intercostal muscles. The upper two records show the action potential activity of motor nerve fibres from thoracic segments 10 and 11, which supply the internal intercostal muscles between ribs 10 and 12. Note the increased activity during expiration and the sharp cut-off as inspiration commences. The second pair of records shows the electrical activity of motor nerve fibres from segments T5 and T6 that supply the external intercostal muscles between ribs 5 and 7. Note the increased activity during inspiration and the sharp cut-off as expiration commences. The activity of the inspiratory and expiratory nerves alternate. The final record shows the activity of the phrenic nerve, which innervates the diaphragm. Here there is a progressive increase in activity during inspiration which progressively declines during the start of expiration before increasing at the start of the next phase of inspiration. (Adapted from Fig 9.7 in M.P. Hlastala and A.J. Berger (1996) *Physiology of respiration*. Oxford University Press, New York.)

for a short period. This breath-holding is known as **voluntary apnoea**, and its duration is normally limited by the rise in arterial PCO_2. Equally, it is possible to increase the rate and depth of breathing deliberately during **voluntary hyperventilation** (also known as **voluntary hyperpnoea**). This voluntary control affects both lungs—it is not possible voluntarily to rest the left lung while ventilating the right. The pathways involved in voluntary regulation are not known with any certainty but presumably have their origin in the motor cortex. A fine degree of control over the muscles of respiration is possible, e.g. during speech, singing, or the playing of wind instruments.

Summary

The diaphragm and intercostal muscles have no inherent rhythmic activity but contract in response to efferent activity in the phrenic and intercostal nerves. The basic respiratory rhythm originates in the medulla, where the upper respiratory neurons of the dorsal and ventral groups discharge action potentials with an intrinsic rhythm that corresponds to that of the respiratory cycle. The dorsal and ventral respiratory neurons receive a variety of inputs from higher centres in the brain, the peripheral chemoreceptors, and afferent fibres from the lungs. These inputs modulate the rate and depth of respiration.

Normal regular breathing (or eupnoea) is under voluntary control. Breath-holding is known as voluntary apnoea; its duration is normally limited by the progressive rise in arterial PCO_2. The rate and depth of breathing can be deliberately increased during voluntary hyperventilation (voluntary hyperpnoea).

35.3 The reflex regulation of respiration

The smooth muscle of the upper airways (trachea, bronchi, and bronchioles) possesses slowly adapting stretch receptors. When the lung is inflated, these receptors send impulses to the dorsal respiratory group via the vagus nerves. This afferent information tends to inhibit respiratory activity and so acts to limit inspiration. This is known as the **Hering–Breuer lung inflation reflex**. If the lungs are inflated by positive pressure, the frequency of respiratory movements falls and may cease altogether (**apnoea**). The **Hering–Breuer deflation reflex** serves to shorten exhalation when the lung is deflated and is initiated by stimulation of those proprioceptors active during lung deflation. As for the Hering–Breuer lung inflation reflex, action potentials from these receptors reach the brainstem by way of vagus afferent fibres, where they terminate on neurons of the dorsal respiratory group.

In animals such as the cat and rabbit, the Hering–Breuer reflexes appear to play a significant part in the control of the respiratory rhythm. In humans, the inflation reflex is only activated when tidal volumes exceed about 0.8–1 litre. For this reason it is thought that the Hering–Breuer inflation reflex may play a role in regulating inspiration during exercise, rather than in normal quiet breathing. The role of the Hering–Breuer deflation reflex is less clear, although it may be important in maintaining the FRC of newborn infants.

In addition to their stretch receptors, the airways possess receptors that respond to irritants. Activation of these receptors elicits **cough and sneeze** reflexes. When the irritant receptors of the trachea and bronchi are stimulated, they elicit a cough (Figure 35.3).

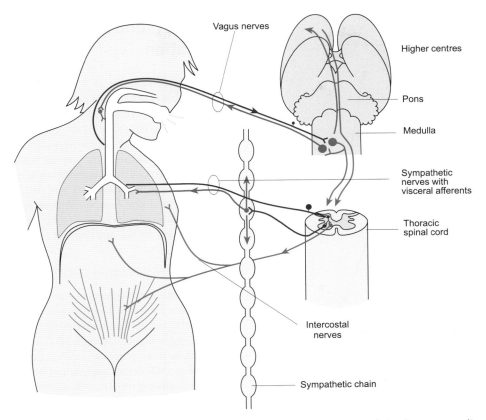

Figure 35.3 The pathways subserving the cough reflex. Irritation of the trachea or bronchi stimulates the nerve endings of the vagus nerve and visceral afferent fibres (which run alongside the sympathetic motor fibres). Impulses pass to the medulla and spinal cord, from where the excitation spreads to the higher regions of the brain. The glottis is closed by the activity in the vagal afferents and then the intercostal and abdominal muscle contract. The glottis is suddenly opened and the airflow resulting from the rapid change in pressure of the airways dislodges the irritant material. (Based on Fig 8.1 of P.C.B. McKinnon and J.F. Morris (2005) *Oxford textbook of functional anatomy*, Vol. 2. Oxford University Press, Oxford.)

Irritants stimulating receptors in the nasal mucosa elicit a sneeze. The initial phase of either response is a deep inspiration followed by a forced expiration against a closed glottis. As the pressure in the airways rises, the glottis suddenly opens and the trapped air is expelled at high speed. This dislodges some of the mucus covering the epithelium of the airways and helps to carry the irritant away with it via the mouth or nose.

Swallowing

During swallowing, respiration is inhibited. This is part of a complex reflex pattern: as food or drink passes into the oropharynx, the nasopharynx is closed by the upward movement of the soft palate and the contraction of the upper pharyngeal muscles. Respiration is inhibited at the same time and the laryngeal muscles contract, closing the glottis. The result is that aspiration of food into the airways is avoided. The act of swallowing is followed by an expiration that serves to dislodge any food particles lying near the glottis. These actions are coordinated by neural networks in the medulla. If particles of food are accidentally inhaled, they stimulate irritant receptors in the upper airways and elicit a cough reflex. If water is aspirated into the larynx, there is a prolonged apnoea, which prevents water entering the lower airways. For further details of the swallowing reflex see Chapter 43.

The pulmonary chemoreflex

Inhalation of smoke and noxious gases such as sulphur dioxide and ammonia stimulates irritant receptors (also known as rapidly adapting receptors) within the tracheo-bronchial tree and elicits a powerful pulmonary reflex, in which there is constriction of the larynx and bronchi and an increase in mucus secretion. If the lungs become congested, breathing becomes shallow and rapid (known as **pulmonary tachypnoea**). The receptors that mediate this response are C-fibre endings located in the interstitial space of the alveolar walls previously known as J-receptors (for juxtapulmonary capillary receptors). The role that these receptors play in normal breathing is not known.

Other reflex modulations of respiration

The normal pattern of breathing is modified by many other factors. For example, passive movement of the limbs results in an increase in ventilation that is believed to occur as a result of stimulation of proprioceptors in the muscles and joints (see Figure 35.4).

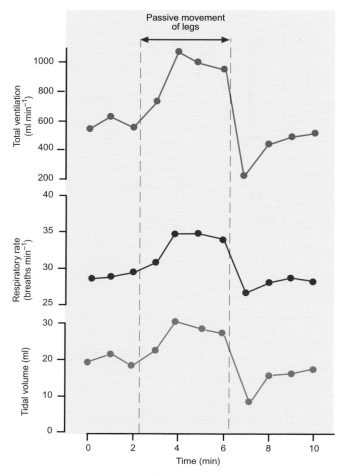

Figure 35.4 Proprioceptive reflexes increase ventilation. In this experiment (on an anesthetized cat), the motor nerves to the legs were cut, leaving the afferent fibres intact. Vigorous movement of the limbs resulted in a marked increase in ventilation, which declined to resting levels after movement ceased. (Based on data in W. Barron and J.H. Coote (1973) *Journal of Physiology* **235**, 423–36.)

This reflex may play an important role in the increase in ventilation during exercise. Pain results in alterations to the normal pattern of respiration. Abdominal pain (e.g. post-operative pain) can be so severe that it causes a reflex inhibition of inspiration and apnoea. Prolonged severe pain is associated with fast shallow breathing along with other signs of stress. Immersion of the face in cold water elicits the diving response, in which there is apnoea, bradycardia, and peripheral vasoconstriction (see Chapter 37).

Summary

A number of reflexes directly influence the pattern of breathing. These include the cough reflex, the Hering–Breuer lung inflation and lung deflation reflexes, and swallowing. These are all protective reflexes. Ventilation is increased during passive movement of the limbs, which stimulates proprioceptors in the muscles and joints. The reflex increase in ventilation during exercise is adaptive and serves to keep blood gases stable.

35.4 The blood gases and the control of ventilation

Respiration regulates the PCO_2 and PO_2 of the arterial blood (i.e. the $PaCO_2$ and PaO_2—often referred to as arterial blood gases or ABG), which are maintained within very close limits throughout life. Indeed, the $PaCO_2$ and PaO_2 vary little between deep sleep and severe exercise where the oxygen consumption and carbon dioxide output of the body may increase more than tenfold. Clearly, to achieve such remarkable stability, the body needs some means of sensing both the $PaCO_2$ and PaO_2. This role is performed by the peripheral and central chemoreceptors. Information from these chemoreceptors is relayed to the respiratory neurons, which then determine the rate and depth of ventilation.

The peripheral arterial chemoreceptors

The peripheral arterial chemoreceptors—the **carotid bodies**—are small organs about 7×5 mm in size that are located just above the carotid bifurcation on each side of the body. The **aortic bodies** are diffuse islets of tissue scattered around the arch of the aorta that have a similar microscopic structure to the carotid bodies. There is, however, no evidence to suggest that they act as chemoreceptors in humans, although they may do so in other species.

The carotid bodies are anatomically and functionally separate from the arterial baroreceptors, which are located in the wall of the carotid sinus (see Figure 35.5). Nevertheless, afferent fibres from the carotid body and the ipsilateral carotid sinus run in the same nerve, the carotid sinus nerve, which is a branch of the glossopharyngeal (IX cranial) nerve. The carotid bodies have a very high blood flow relative to their mass (about 20 litres kg^{-1} min^{-1}), which is derived from the external carotid arteries.

The carotid bodies respond to changes in the PaO_2, $PaCO_2$, and pH of the arterial blood. Afferents from the carotid bodies increase their rate of discharge very significantly as the PaO_2 falls below

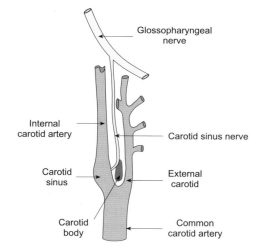

Figure 35.5 A simplified diagram to show the relative positions of the carotid body and carotid sinus.

about 8 kPa (60 mm Hg), and this sensitivity to low PaO_2 is enhanced if it is accompanied by an increase in $PaCO_2$—see Figure 35.6(a). They also increase their rate of discharge in response to an increase in $PaCO_2$ that is independent of the effect of CO2 on pH, as shown in Figure 35.6(b). Moreover, as Figure 35.6(c) shows, they respond to changes in blood hydrogen ion concentration (i.e. blood pH) that are independent of any change in $PaCO_2$.

In human beings, the carotid bodies are the only receptors that are able to elicit a ventilatory response to hypoxia. Thus, if they have been surgically removed for therapeutic reasons, the ventilatory response to hypoxia is lost—even though the aortic bodies remain intact. When breathing normal room air, the influence of the carotid bodies on the rate of ventilation is small. For example, if a subject suddenly switches from breathing room air to breathing 100 per cent oxygen, the minute volume falls by about 10 per cent for a brief period before returning to its previous level. The transient nature of this response can be explained as follows: breathing pure oxygen reduces the respiratory drive from the peripheral chemoreceptors, and this has the effect of reducing the minute volume. During this period, the $PaCO_2$ rises slightly and this acts on the central chemoreceptors to cause the minute volume to return to its original value.

During hypoxia, however, the carotid bodies play an important role in stimulating ventilation. This can be shown in anaesthetized, spontaneously breathing animals. If the arterial PO_2 is lowered by giving an animal a gas mixture of 8 per cent O_2 and 92 per cent N_2 to breathe, the minute volume increases by about 50 per cent. This is a reflex response that can be blocked by cutting the aortic and carotid sinus nerves. The increase in minute volume seen following administration of a gas mixture containing 5 per cent CO_2 (in 21 per cent O_2 and 74 per cent N_2) is largely unaffected by cutting these nerves, showing that the response to hypercapnia is mediated mainly by the central chemoreceptors.

The central chemoreceptors

The central chemoreceptors are located on, or close to, the ventral surface of the medulla near to the origin of the glossopharyngeal and vagus nerves. They respond to changes in the pH of the CSF resulting from alterations in $PaCO_2$ and provide most of the chemical stimulus to breathe under normal resting conditions. The mechanism by which they sense the $PaCO_2$ is illustrated in Figure 35.7: increased $PaCO_2$ results in an increase in the PCO_2 of the CSF and the hydration reaction for carbon dioxide is driven to the right. This leads to the increased liberation of hydrogen ions as follows:

$$CO_2 + H_2O \rightleftharpoons H_2CO_3 \rightleftharpoons H^+ + HCO_3^-$$

Unlike the blood, the CSF has little protein and no red cells, so the hydrogen ions produced by this reaction are not buffered to any great extent. As a result, the pH falls in proportion to the rise in PCO_2 and this fall in pH stimulates the chemoreceptors. Conversely, during hyperventilation, CO_2 is lost from the blood and this causes a reduction in the PCO_2 of the CSF. The hydration reaction is driven to the left, the pH of the CSF rises, and ventilation decreases.

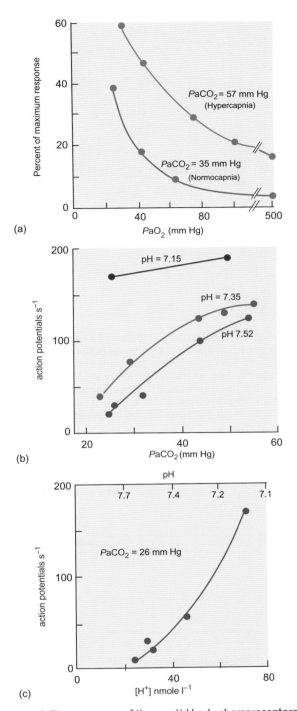

Figure 35.6 The responses of the carotid body chemoreceptors to $PaCO_2$, PaO_2 and pH. (a) The green symbols indicate the response to changes in PaO_2 at normal $PaCO_2$ (normocapnia), while the red symbols show the enhanced response when $PaCO_2$ is elevated (hypercapnia). (b) The response of the carotid chemoreceptors to changes in $PaCO_2$ at normal pH (pH7.35—green symbols), at pH 7.52 (alkalaemia—purple symbols), and at pH 7.15 (acidaemia—blue symbols). The response to $PaCO_2$ is enhanced in acidaemia and reduced in alkalaemia. (c) The response of the carotid chemoreceptors to hydrogen ions at constant $PaCO_2$. (Based on Fig 10.3 in M.P. Hlastala and A.J. Berger (1996) *Physiology of respiration*. Oxford University Press, New York.)

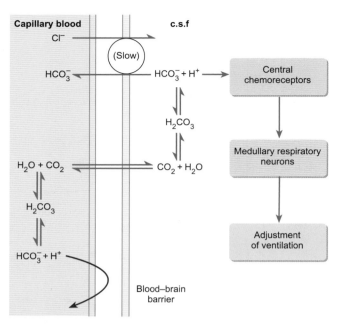

Figure 35.7 A schematic diagram to illustrate how the PCO_2 of the capillary blood in the brain stimulates the central chemoreceptors. A rise in plasma CO_2 leads to increased CO_2 uptake into the brain, where it is converted to bicarbonate and hydrogen ions via carbonic acid. The hydrogen ions stimulate the central chemoreceptors, and this increases the rate and depth of respiration. A fall in plasma CO_2 has the opposite effect.

If the $PaCO_2$ is persistently above or below its normal value of 5.3 kPa (40 mm Hg), the central chemoreceptors become less sensitive to changes in $PaCO_2$ than normal. In these situations, the

bicarbonate concentration of the CSF is regulated by exchange with chloride ions derived from the plasma. This compensation is important during chronic changes in $PaCO_2$ arising from residence at high altitude (where the $PaCO_2$ falls—see Chapter 38) or from chronic respiratory disease (where the $PaCO_2$ rises).

The roles of the stretch receptors of the airways and the peripheral and central chemoreceptors in the regulation of pulmonary ventilation are summarized in Figure 35.8. The pulmonary stretch receptors act to inhibit inspiration, while the chemoreceptors act to maintain blood PaO_2, $PaCO_2$, and pH stable as far as environmental circumstances permit.

Summary

The partial pressures of the blood gases are sensed by the peripheral and central chemoreceptors. The peripheral chemoreceptors in humans are the carotid bodies, which respond to falls in the PaO_2 and pH of the arterial blood or to a rise in $PaCO_2$. The central chemoreceptors respond to the changes in arterial PCO_2 and are responsible for most of the chemical stimulus to breathing.

35.5 The effects of breathing different mixtures of O_2 and CO_2

When air containing a significant amount of CO_2 is inhaled, its partial pressure in the alveoli and arterial blood rises. This is known as **hypercapnia**. If a subject deliberately hyperventilates for a brief period, the partial pressure of carbon dioxide in the alveoli and

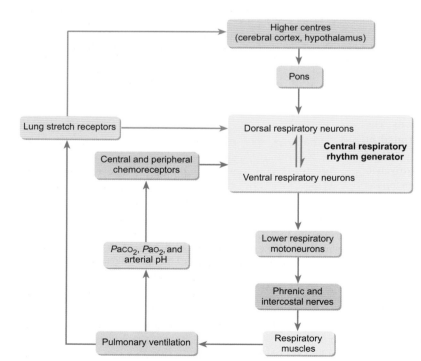

Figure 35.8 A flow diagram showing the interrelations between the main neural elements that regulate the rate and depth of respiration. Activation of the lung stretch receptors inhibits inspiration (the Hering–Breuer lung inflation reflex).

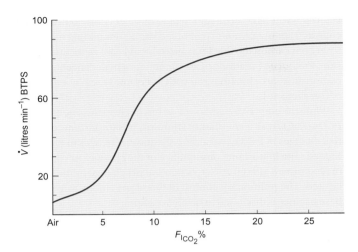

Figure 35.9 The effect of breathing carbon dioxide on ventilation. The figure shows the relationship between the concentration of CO_2 in the inspired air (F_ICO_2) and the total ventilation for a normal subject. Note the steep rise in ventilation as F_ICO_2 increases from 5 per cent to 10 per cent. (Based on Fig 39 in P. Dejours (1966) *Respiration.* Oxford University Press, New York.)

arterial blood falls, as it is lost from the lungs faster than it is being generated in the tissues. This fall in the partial pressure of CO_2 is known as **hypocapnia**.

If subjects breathe a gas mixture that has a PO_2 lower than the normal 21.2 kPa (159 mm Hg), the arterial PO_2 will fall. This is known as **hypoxemia**. If the oxygen content is insufficient for the needs of the body, the subject is said to be **hypoxic**. The total absence of oxygen is **anoxia**.

The relative importance of CO_2 and O_2 in determining the ventilatory volume is readily investigated by asking subjects to breathe different gas mixtures. If a normal healthy subject breathes a gas mixture containing 21 per cent oxygen, 5 per cent carbon dioxide, and 74 per cent nitrogen for a few minutes, their ventilation increases about threefold. A higher fraction of carbon dioxide in the inhaled gas mixture will stimulate breathing even more. Even a single breath of air containing an elevated carbon dioxide is sufficient to increase ventilation for a short time. Conversely, if a subject hyperventilates for a brief period, the subsequent ventilation is temporarily decreased. Thus, any manoeuvre that alters the partial pressure of CO_2 in the alveolar air ($PACO_2$) results in a change in ventilation that tends to restore the $PACO_2$ to its normal value (c. 5.3 kPa or 40 mm Hg). The relationship between the partial pressure of CO_2 in the inspired air and total ventilation is shown in Figure 35.9.

By contrast, if the same subject breathes a mixture of 15 per cent oxygen in 85 per cent nitrogen there is little change to the rate of ventilation at normal barometric pressure. Indeed, hypoxia only tends to stimulate ventilation strongly when the alveolar PO_2 falls below about 8 kPa (60 mm Hg)—see Figure 35.10. As the PAO_2 falls further, ventilation increases steeply.

From these observations, it appears that the principal chemical stimulus for respiration is the PCO_2 of the alveolar air rather than

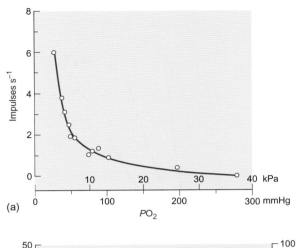

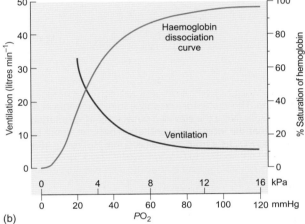

Figure 35.10 (a) The rate of carotid chemoreceptor discharge (impulses s^{-1}) plotted against the inspired PO_2. Note that the rate of discharge progressively increases below 13 kPa (100 mmHg). The discharge shows little adaptation. (b) The effect of acute hypoxia on pulmonary ventilation (blue) compared to the oxyhaemoglobin dissociation curve (red). Ventilation becomes much more sensitive to the inspired PO_2 below about 8 kPa (60 mmHg), which is the point at which the oxyhaemoglobin dissociation curve becomes very steep.

the PO_2. At first, this may appear strange, as the main purpose of gas exchange is to maintain the oxygenation of the tissues. The reason for the relatively small ventilatory effect of mild hypoxia can be understood by looking at the oxyhaemoglobin dissociation curve (see Figure 35.10), which shows that at a PO_2 of 8 kPa (60 mm Hg) the haemoglobin is still about 90 per cent saturated. Below this value, the percentage saturation rapidly falls. Consequently, at normal atmospheric pressure (101 kPa or 760 mm Hg), mild hypoxia would be a relatively weak stimulus to ventilation.

As Figure 35.11 shows, the increase in ventilation with increasing alveolar PCO_2 becomes steeper as the PO_2 falls, so that the sensitivity of the respiratory drive to CO_2 is greater during hypoxia than it is when the PO_2 is normal. The effects of hypoxia and hypercapnia are not simply additive; they have a strong synergistic interaction. This is of some importance during breath-holding and in

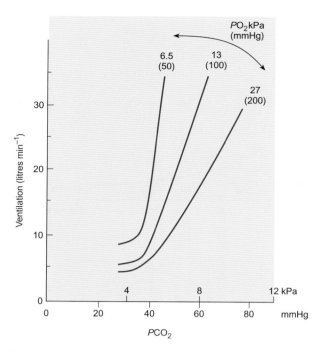

Figure 35.11 The ventilatory response to hypercapnia at different values of P_AO_2. Note the elevated basal level of ventilation as PO_2 falls, and that the increase in ventilation becomes more sensitive to P_ACO_2 as PO_2 falls from 27 kPa (200 mm Hg) to 6.5 kPa (50 mm Hg). (Based on Fig 8.5 in J.G. Widdicombe and A. Davies (1991) *Respiratory physiology*, 2nd edition. Edward Arnold, London.)

asphyxia, where hypoxia and hypercapnia occur together. In this context it should be noted that breathing air containing more than 5 per cent carbon dioxide is unpleasant and causes mental confusion. Prolonged breathing of air containing a high concentration of carbon dioxide or breathing air with a very low PO_2 may lead to loss of consciousness.

While the emphasis in this section has been on the role of arterial PCO_2 ($PaCO_2$) as a ventilatory stimulus, it is important to realize that changes in $PaCO_2$ cause very significant changes to blood pH. Indeed, one important consequence of the changes in ventilation in response to altered $PaCO_2$ is that the pH change in the blood is limited. The pulmonary control of blood pH and its role in acid–base balance is discussed in detail in Chapter 41.

Summary

When a healthy subject breathes a gas mixture containing an elevated concentration of carbon dioxide, their ventilation increases significantly. A modest reduction (< 5 per cent) in the partial pressure of oxygen has little effect on ventilation. Ventilation increases only when the alveolar PO_2 falls below about 8 kPa. This shows that under normal circumstances the principal chemical stimulus for breathing is the PCO_2 of the alveolar air rather than the PO_2.

✱ Checklist of key terms and concepts

The control of respiration

- The diaphragm and intercostal muscles have no inherent rhythmic activity themselves but contract in response to efferent activity in the phrenic and intercostal nerves. The basic respiratory rhythm originates in the medulla oblongata.

- A number of reflexes directly influence the pattern of breathing. These include the cough reflex, the Hering–Breuer lung inflation reflex, and swallowing.

- In broad terms, respiration is stimulated by a lack of oxygen (hypoxia) but more strongly by an increase in carbon dioxide (hypercapnia).

- The partial pressures of the blood gases are sensed by the peripheral and central chemoreceptors.

- The peripheral chemoreceptors in humans are the carotid bodies. They respond to changes in the PaO_2, $PaCO_2$, and pH of the arterial blood and are the only receptors that can respond to hypoxia.

- The central chemoreceptors are located in the brainstem and are responsible for most of the chemical stimulus to breathing. They respond to the changes in the pH of the CSF brought about by alterations in arterial PCO_2.

📄 Recommended reading

Physiology of the respiratory system

Dejours, P. (1966) *Respiration*, Chapters 7 and 8. Oxford University Press, New York.
  An old book that gives an excellent quantitative treatment that is suitable for graduate students and those considering research in human respiratory physiology.

Hlastala, M.P., and Berger, A.J. (2001) *Physiology of respiration* (2nd edn), Chapters 9–11. Oxford University Press, New York.
  Offers a more detailed account of the neural control of respiration than other introductory texts.

Levitzky, M.G. (2013) *Pulmonary physiology* (8th edn), Chapter 9. McGraw-Hill, New York.

A well-established introductory text with good coverage of the mechanics of breathing.

Lumb, A.B. (2016) *Nunn's applied respiratory physiology* (8th edn), Chapter 4. Churchill-Livingstone, London.

Slonim, N.B., and Hamilton, L.H. (1987) *Respiratory Physiology*, Chapters 13–15. Mosby, St Louis.

Now rather old but still well worth reading for its clear explanations.

Review articles

Nishino, T. (2013) The swallowing reflex and its significance as an airway defensive reflex. Mini review. *Frontiers in Physiology* **3**, article 489, doi: 10.3389/fphys.2012.00489.

Teppema, L.J., and Dahan, A. (2010) The ventilatory response to hypoxia in mammals: Mechanisms, measurement, and analysis. *Physiological Reviews* **90**, 675–754, doi:10.1152/physrev.00012.2009.

 To check that you have mastered the key concepts presented in this chapter, complete the accompanying online self-assessment questions. Go to www.oup.com/uk/pocock5e/

CHAPTER 36

Pulmonary defence mechanisms and common disorders of the respiratory system

Chapter contents

This chapter should help you to understand:

- Lung defence systems
  - The airways reflexes
  - Mucus clearance (the mucociliary escalator)
- Some common disorders of respiration
  - Hay fever (allergic rhinitis)
  - Asthma
  - Emphysema
  - Chronic obstructive pulmonary disease
  - Pulmonary fibrosis
  - Cystic fibrosis
  - Sleep apnoea
- Hypoxia, its origins and consequences
- Oxygen therapy
- Respiratory failure

36.1 Introduction

As all city-dwellers know, the air is full of particulate matter, some of which is inhaled with each breath. A respiratory minute volume of 6 l min^{-1} results in the intake of over 8500 litres of air each day. Even if the concentration of particles were to be only 0.001 per cent (10 parts per million), this would include 85 ml of particulate matter. Clearly, unless some mechanism existed for the removal of this material, the lungs would rapidly become clogged with dust and debris. Moreover, not all of the inhaled material is biologically inert. Some will be infectious agents (bacteria, fungal spores, and viral particles) and some will be allergenic (e.g. pollen). The respiratory system therefore needs to remove the inert material and inactivate the infectious and allergenic agents.

36.2 Airway defence mechanisms

Within the nasal cavity are three curved bony plates, the nasal conchae (or tubinates), that are covered with respiratory epithelium (see Section 32.3). These disturb the smooth flow of air and

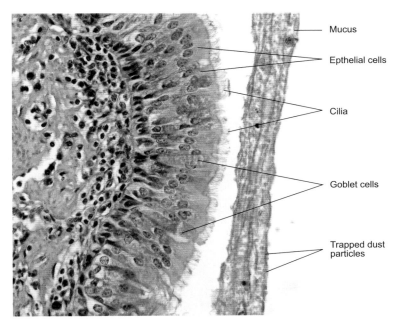

Mucus

Epthelial cells

Cilia

Goblet cells

Trapped dust particles

Figure 36.1 A view of the ciliated epithelium of the trachea showing the large number of mucus-secreting goblet cells and the prominent ciliated epithelium. Trapped dust particles are clearly visible in the overlying mucus layer.

make it turbulent by forcing it into narrow passages where it is warmed and moistened. At the same time, the largest airborne particles (> 10–15 μm) are brought into contact with the mucus covering the nasal epithelium and are entrapped by it, filtering them from the air before it reaches the trachea. As in the nasal cavity, the airflow in the trachea and bronchi is turbulent, and this brings much of the incoming air stream into direct contact with the walls of these airways. As a result, most of the remaining large particles (5–10 μm) become lodged in the mucus lining of the upper respiratory tree, as shown in Figure 36.1. Further down the airways, the airflow becomes slow and laminar. In these regions of the lung, smaller particles (0.2–5 μm) settle on the walls of the airways under the influence of gravity. Only the smallest particles reach the alveoli, where most remain suspended as aerosols which are subsequently exhaled. Nevertheless, about one-fifth of these small particles may become deposited in the alveolar ducts or in the alveoli themselves, where they are phagocytosed by the alveolar macrophages.

The protective role of the airways reflexes

The material that becomes trapped in the mucus of the nasal passages and upper airways may stimulate receptors that elicit a reflex that acts to protect the airways. Irritants affecting the nasal passages elicit a sneeze, while deeper in the airways they generally elicit a cough. Both sneezing and coughing expel the irritant material via the reflex pathways described in Chapter 35 (Section 35.3). The swallowing reflex acts to prevent food particles or liquids entering the respiratory tract. Occasionally it fails to prevent small food particles entering the larynx. This results in reflex coughing to eliminate the offending material.

The mucociliary escalator and airways clearance

As Figure 36.1 shows, the respiratory tract from the upper airways to the terminal bronchioles is lined with a ciliated epithelium that is covered by a layer of mucus. The ciliated cells make up about half of the epithelium of the upper airways (trachea and bronchi), but this proportion declines as the airways branch. By the fifth generation, the ciliated cells account for only about 15 per cent of the total.

The mucus layer is secreted by the goblet cells and lies over a layer of fluid known as the pericellular fluid layer, which is secreted by the epithelial cells (see Figure 36.2). The low viscosity of the pericellular fluid layer allows the cilia to beat freely and propel the overlying layer of mucus towards the mouth. As the mucus reaches the pharynx, it is either swallowed or coughed up and expectorated.

The glycoproteins of the mucus have the ability to bind a very wide range of antigens. This makes the mucociliary escalator very effective at trapping inhaled particles as the air passes from the nose and mouth to the alveoli. By using small particles labelled with a radioactive tracer, it has been shown that almost all of the inhaled particles are removed within 24 hours (see Figure 36.3). The mucus not only acts as a physical trap, it also contains small antimicrobial proteins called **defensins** that protect the airways from bacteria, fungal spores, and viruses (see Section 26.5).

The respiratory bronchioles, alveolar ducts, and alveoli do not possess a ciliated epithelium. Particles reaching these regions of the lung are phagocytosed by alveolar macrophages, which are found in the fluid lining of the respiratory airways. Bacteria and other biologically active materials such as pollen grains are

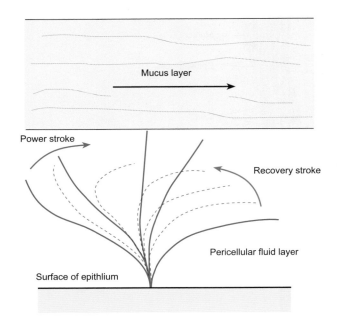

Figure 36.2 Diagrammatic representation of the propulsive action of the cilia of the airways. The pericellular layer is similar in viscosity to the extracellular fluid, allowing the free movement of the cilia. The steady progress of this layer carries the overlying layer of mucus towards the oropharynx. The power stroke of the cilia is shown in green and the recovery stroke in red. (Modified and redrawn from Fig 3d in M.R. Knowles and R.C. Boucher (2002) *J Clin Invest* **109**, 571–77.)

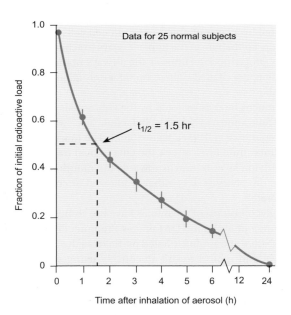

Figure 36.3 The clearance of radio-labelled particles from the airways of 25 normal subjects. The subjects were asked to inhale a finely dispersed aerosol of particles labelled with a radioactive tracer. The radioactivity was then tracked over the next six hours, and also one day later. The initial clearance had a half-time of about 1.5 hours, and all the radioactive material was eliminated 24 hours later. Note the break in the time axis. The error bars represent standard errors of the mean values. (Based on data in D. Pavia et al. (1985) *Thorax* **40**, 175.)

ingested by the macrophages and digested in their lysosomes. Macrophages also ingest small particles of mineral matter (e.g. dust particles of silica), which they cannot digest. In these cases, the material is stored within the cell and is removed via the muco-ciliary escalator when the cell dies. If bacteria or viruses penetrate these defences and enter the interstitial space, they are dealt with by the immune system (see Chapter 26).

Summary

Particulate matter entering the airways becomes lodged in the mucus lining of the respiratory tree. In the upper airways, allergenic material may elicit a sneeze or cough which forces it from the airways. Most of the remaining material is removed by the muco-ciliary escalator. Material that becomes deposited in the alveolar ducts or in the alveoli is ingested by alveolar macrophages.

36.3 Some common disorders of respiration

Normally, respiration continues unnoticed and uneventfully. It is only when things go wrong that we become aware of our breathing. Difficulty in breathing causes distress known as **dyspnoea**. It is a subjective phenomenon during which an individual may report being breathless. It may be quite normal—as in a subject who has rapidly climbed to a high altitude (where the atmospheric PO_2 is low)—or it may reflect some organic disease. In both cases, it is the sense of breathlessness that limits the ability to undertake exercise. The ability of the lungs to provide enough air for the body's needs is known as the **ventilatory capacity**. If this is less than normal, some form of respiratory disorder is present. These disorders can be divided into those in which the airways are obstructed, those in which the expansion of the lungs is restricted, and those in which the respiratory muscles are weakened and unable to expand the chest fully. The following sections discuss the pathophysiological processes underlying some commonly occurring respiratory disorders.

Allergic rhinitis

Allergic rhinitis (hay fever) is commonly the result of an allergic reaction to pollens, fungal spores, or dust mites. It affects a large percentage of the UK population (c.15 per cent) and is characterized by nasal itching with bouts of sneezing, swelling of the nasal mucosa (nasal congestion), and a runny nose (nasal discharge). The symptoms are elicited when an allergen activates mast cells in the nasal mucosa. The activated mast cells degranulate and release a wide range of inflammatory mediators, including histamine, leukotrienes, and prostaglandin D_2. In turn these mediators cause an inflammatory response in the nasal mucosa, resulting in the symptoms described above. For a more detailed account of the process of inflammation see Chapter 26 (Section 26.6).

Asthma

Asthma is a condition in which a person has difficulty in breathing, particularly in expiration. Asthmatic attacks are characterized by the sudden onset of dyspnoea. This is the result of bronchospasm, which usually occurs in response to an allergen that is present in the environment, although it may also occur in response to other factors such as exercise or stress. The initial phase is probably triggered by an interaction between an allergen and the IgE antibodies present on the mast cells of the pulmonary interstitium. Once activated, the mast cells secrete a number of inflammatory mediators including histamine and leukotrienes. These cause spasm of the smooth muscle of the bronchi and increased secretion of mucus into the airway, both of which effectively reduce the diameter of the airways. During an asthmatic attack, the FRC and RV are increased although vital capacity is normal. The FEV_1 and peak flow are markedly reduced, in severe cases by more than half (see Figure 36.4). This limits the ventilatory capacity and results in dyspnoea.

The bronchospasm that occurs in the initial phase of an asthmatic attack is of relatively short duration. For this reason, the airways obstruction in asthma is considered reversible even though the inflammatory condition of the bronchi persists long after the acute attack has subsided. In chronic asthma, there is destruction of the bronchial epithelium, and FRC and RV become permanently elevated. However, if asthma is left untreated, the chronic inflammation of the lungs can result in permanent airway obstruction as a result of irreversible thickening of the bronchial wall.

The drugs used to treat asthma act to reduce one of the following factors that contribute to the increase in airways resistance seen in bronchial asthma:

- excessive constriction of the bronchi (bronchoconstriction)
- secretion of excessive amounts of mucus by the bronchi
- bronchial oedema (swelling of the bronchial mucosa).

β_2-adrenoceptor agonists such as salbutamol and terbutaline act directly on the bronchial smooth muscle to relax it and have little effect on the heart. Salmeterol and formoterol are longer-acting drugs with a similar mode of action. Atropine-derived anticholinergic drugs such as ipratropium and oxitropium inhibit the bronchoconstricting action of vagal nerves and have the additional benefit of reducing mucus secretion. Inflammation of the bronchi can be treated by corticosteroids such as beclomethazone and budesonide. Other drugs that have been used are xanthene derivatives (e.g. theophilline) and cromones (e.g. sodium cromoglicate). Most of these drugs can be given by inhalation, which permits their delivery directly to the airways. This mode of administration helps to confine their action to the lungs and avoids the involvement of other organ systems, particularly the heart. Of those listed, only the xanthenes cannot be administered in this way.

Emphysema

Emphysema is a condition in which the alveoli are increased in size due to destruction of the lung parenchyma. It is frequently (but not invariably) caused by the smoking of tobacco. The pathophysiology

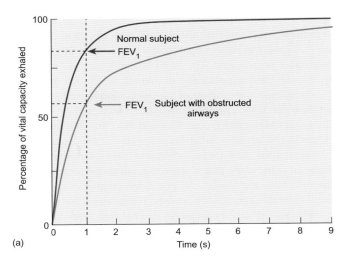

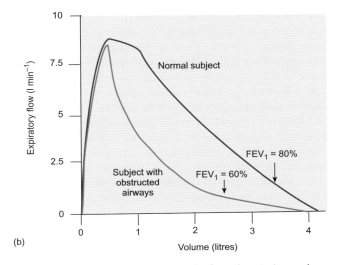

Figure 36.4 The effect of asthma on the forced expiratory volume at 1 second (FEV_1) (a) and the flow–volume relationship (b). Asthma reduces the FEV_1 and alters the shape of the flow–volume relationship.

of the disease is not entirely clear. Bronchial lavage shows that air spaces of the lungs of smokers become invaded by neutrophils. It is now thought that these cells secrete proteolytic enzymes such as elastase that damage the parenchyma of the lungs. In addition, the smoke inhibits the movement of the bronchial cilia, slowing the removal of particulate matter from the airways. The irritant effect of the smoke is the probable cause of the increased secretion of mucus in the larger airways. These effects combine to increase the chances of infection, which results in chronic inflammation of the bronchiolar epithelium. As a result, the diameter of the airways is reduced and, as in asthma, it becomes difficult to exhale, leading to the entrapment of air (from which the disease takes its name). As a result of the loss of parenchymal tissue, the traction on the airways is reduced, and as they become narrower their resistance is increased. By itself, this will limit ventilation, but the problem is compounded by the destruction of the alveoli. The result is a significant reduction in the diffusing capacity of the lungs. Moreover, since the destruction of the alveoli is not uniform, there are abnormal $\dot{V}_A/\dot{Q}$ ratios in different part of the lung and the physiological dead space is increased. The result is inadequate gas exchange, hypoxemia (poor oxygenation of the blood), and chronic dyspnoea.

As in asthma, the progress of the disease is reflected in respiratory function tests. The entrapment of air leads to high lung volumes, but vital capacity, peak flow and FEV_1 are all reduced (see Figure 33.13). Because the underlying cause is the destruction of the parenchyma, the increased resistance is always present—unlike that seen in the acute stages of asthma, where the increase in airflow resistance is reversible. As the disease progresses, the sufferer may expire through pursed lips which provides a back pressure (positive end expiratory pressure) that helps to keep the airways open and prevents their collapse.

Chronic obstructive pulmonary disease

Chronic obstructive pulmonary disease (COPD)—also known as chronic obstructive lung disease (COLD)—is one of the most common respiratory conditions and is a leading cause of death. It is characterized by chronic obstruction of lung airflow that is not fully reversible so that it interferes with normal breathing. COPD includes various progressive lung diseases such as emphysema, chronic bronchitis (inflammation of the bronchi), chronic asthma, and some forms of bronchiectasis. People suffering from COPD have a reduced peak airflow and an increase in mucus production. As for asthma, FEV_1 is significantly below normal (i.e. <70 per cent of vital capacity). The principal aims of treatment are to prevent infections and to relieve both the bronchoconstriction and mucus accumulation in the airways by inhalation of a bronchodilator such as ipratropium.

Pulmonary fibrosis

In pulmonary fibrosis, the alveolar wall becomes thickened and this decreases the diffusing capacity of the lung and impairs gas exchange. The diffusing capacity is also reduced in **pneumonia**. In this case bacterial infection leads to fluid accumulation in the alveoli. In severe cases a pulmonary lobe may become filled with fluid ('consolidated') containing bacterial toxins, with the result that there is a local increase in blood flow and $\dot{V}_A/\dot{Q}$ abnormalities.

Cystic fibrosis

Cystic fibrosis is a recessive inherited disorder in which the mucus of the airways is abnormally thick and difficult to dislodge. It results from the failure of an epithelial chloride channel to open normally in response to cyclic AMP. There is a decreased secretion of both sodium and chloride into the lumen of the airways, and less water passes across the epithelial membranes. Consequently, the pericellular layer is thinner and the viscosity of the mucus layer is greatly increased. This combination results in a reduction in the clearance of particulate matter via the mucociliary escalator shown in Figure 36.5. This causes frequent bronchial infections and obstruction of the small airways.

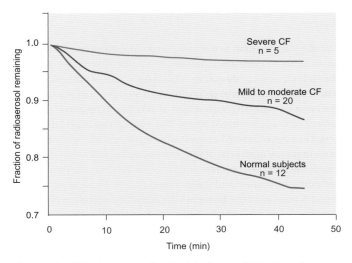

Figure 36.5 The clearance of radio-labelled particles from the airways in 12 normal subjects (green line), 20 patients with mild to moderate cystic fibrosis (mauve line), and 5 patients with severe cystic fibrosis (red line). The subjects with cystic fibrosis were less efficient at clearing the labelled particles. Those with severe disease were much less efficient. The procedure for inhaling the radioactive aerosol was similar to that described for Figure 36.3. (Based on data in J.A. Regnis et al. (1994) *Am J Respir Crit Care Med* **150**, 66–71.)

Sleep apnoea

Sleep apnoea describes a number of conditions in which breathing temporarily stops during deep sleep. Two principal types can be recognized: obstructive sleep apnoea and central sleep apnoea.

Obstructive sleep apnoea

Obstructive sleep apnoea results from physical obstruction of the upper airway. During deep sleep, the muscles of the mouth, larynx, and pharynx relax. As a result, there is a tendency for the pharynx to collapse and obstruct the airway. Periods of increasing obstruction are accompanied by loud snoring followed by a period of silence when obstruction is complete. PaO_2 falls and $PaCO_2$ rises. Both stimulate respiratory effort. Breathing movements then increase in intensity until the intrathoracic pressure overcomes the obstruction and ventilation resumes—often with a loud snore. When the period of apnoea is prolonged (and it may last for up to 2 minutes), the patient wakes and breathing resumes, but the sleep pattern has been interrupted. If this occurs frequently during the night, the characteristic signs of sleep deprivation appear. These include irritability and loss of concentration.

Central sleep apnoea

Central sleep apnoea occurs when the respiratory drive is not able to initiate breathing movements. During the apnoeic phase, the $PaCO_2$ rises and the patient wakes. Once awake the patient resumes breathing, their $PaCO_2$ returns to normal, and they fall asleep once again. As this cycle may be repeated many times during the night, sleep may become severely disturbed.

Periodic breathing

The importance of the chemical control of breathing is revealed by an abnormal pattern known as **Cheyne–Stokes breathing**, where breathing alternates between apnoea and mild hyperventilation. Breathing begins with slow shallow breaths; the frequency and tidal volume gradually increase to a maximum before slowly subsiding into apnoea. This cyclical pattern is seen in patients who are terminally ill or who have suffered brain damage. It is sometimes seen in normal people during sleep, especially at high altitude (see Figure 36.6).

The pattern in terminally ill patients is explained as follows: during the period of low ventilation $PaCO_2$ rises, and this stimulates the central chemoreceptors which progressively increase the respiratory drive so that both frequency and depth increase. The increased ventilation results in a fall in $PaCO_2$ and a reduction of respiratory drive, so both rate and depth of respiration fall. $PaCO_2$ increases and the cycle recommences. In order for this pattern to occur, there must be an abnormal delay in the ability of the central chemoreceptors to respond to the change in $PaCO_2$. This may reflect either a reduced sensitivity to carbon dioxide or a sluggish blood flow to the brain, for example during heart failure. Periodic breathing at high altitude does not indicate a pathological state. It can be induced by a brief period of hyperventilation and is abolished by the inhalation of pure oxygen.

36.4 Insufficient oxygen supply to the tissues—hypoxia and its causes

The respiratory system is principally concerned with gas exchange—with the uptake of oxygen from the air and the elimination of carbon dioxide from the pulmonary blood. If the oxygen content of the blood is reduced, there may be insufficient oxygen to support the aerobic metabolism of the tissues. This condition is known as **hypoxia**. There are four principal types of hypoxia: hypoxic hypoxia, anaemic hypoxia, stagnant hypoxia, and histotoxic hypoxia. Of these, hypoxic hypoxia and stagnant hypoxia are the most commonly seen in clinical medicine.

Hypoxic hypoxia refers to a form of hypoxia that results from a low arterial PO_2. There are a number of causes, but each results in lowering of the oxygen content of the systemic arterial blood. If the alveolar PO_2 is low, the arterial PO_2 will inevitably follow and so will the oxygen content. As a result, more oxygen is extracted from the blood to support the oxidative metabolism of the tissues (see panels (b) and (c) of Figure 36.7). This is quite normal following ascent to high altitude, as the barometric pressure and PO_2 both fall with increasing altitude (see Chapter 37).

As discussed below in the section on **respiratory failure**, reduced ventilation (hypoventilation) will lead to a reduced alveolar PO_2 and an increased PCO_2 (hypercapnia). Hypoventilation may

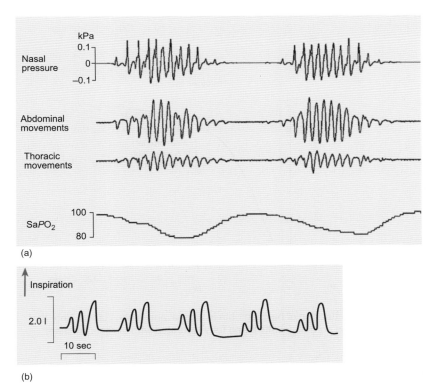

(a)

(b)

Figure 36.6 Records illustrating the pattern of Cheyne–Stokes breathing. (a) A record obtained from a patient suffering from central sleep apnoea. Note the increasing respiratory effort which then declines and is followed by a period of apnoea, during which there is no respiratory effort. The fall in oxygen saturation of the arterial blood ($SaPO_2$) is out of phase with the breathing effort. (b) A record obtained from a healthy male subject who had recently moved to a high altitude. The time calibration applies to both records. (The recording in (a) is taken from part of Fig 2 in T. Bekfani and W.T. Abraham (2016) Current and future developments in the field of central sleep apnoea. *Europace* **18** (8), 1123–34. doi: 10.1093/europoace/evu435. Translated and reprinted with permission of Oxford University Press on behalf of the Euopean Society of Cardiology.)

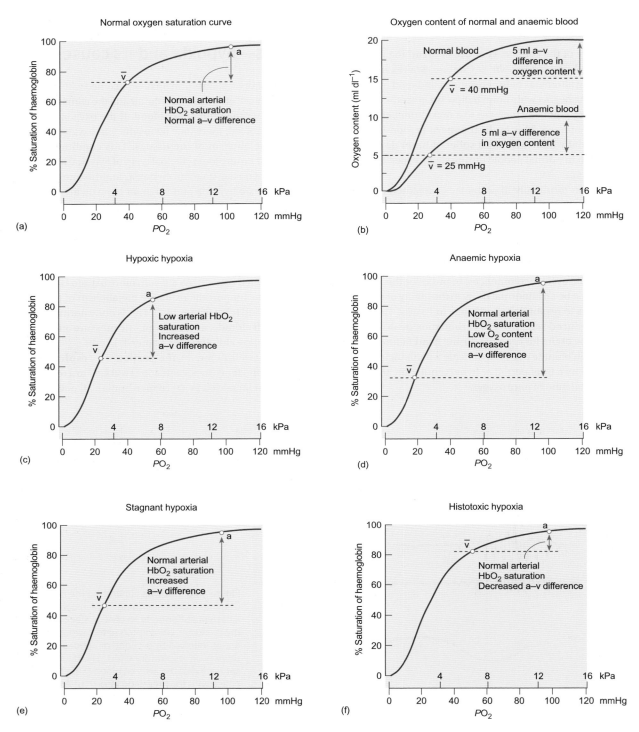

Figure 36.7 The effect of different types of hypoxia on the percentage saturation of haemoglobin in arterial and venous blood. Note that in anaemic hypoxia the total oxygen content of arterial blood is less than normal despite being fully saturated with oxygen: compare panel (c) (oxygen saturation) with panel (b) (oxygen content).

result from respiratory depression caused by a drug overdose (e.g. in barbiturate poisoning). It may also arise because of severe weakness of the muscles that support respiration—for example in poliomyelitis or myasthenia gravis. Airway obstruction will also lead to hypoventilation.

A right–left shunt will allow some venous blood to bypass the lungs completely. (A shunt is the passage of blood through a channel that diverts it away from its normal route.) Although the haemoglobin of the blood that has passed through the alveoli is virtually fully saturated with oxygen, the shunted blood will have the same PO_2 as mixed venous blood. As a result, the PO_2 and oxygen content of the blood in the systemic arteries is reduced. Equally, ventilation–perfusion inequality will lead to hypoxic hypoxia if the $\dot{V}_A/\dot{Q}$ ratio is low in a significant portion of the lung. This occurs in many respiratory diseases and is the most common cause of central cyanosis (i.e. cyanosis caused by inadequate oxygenation of the blood). Another significant cause of hypoxic hypoxia is a reduced diffusing capacity due to fibrosis of the lung parenchyma or to pulmonary oedema.

Anaemic hypoxia is caused by a decrease in the amount of haemoglobin available for binding oxygen, with the result that the oxygen content of the arterial blood is abnormally low (see panels (b) and (d) of Figure 36.7). It may be due to blood loss, to reduced red cell production, or to the synthesis of an abnormal form of haemoglobin because of a genetic defect, as in sickle cell anaemia (Section 25.7). It may also be caused by carbon monoxide poisoning, since the affinity of haemoglobin for this gas is much greater than its affinity for oxygen (see Section 25.6).

In anaemic hypoxia the arterial PO_2 is normal, but since the oxygen content of the blood is low, a higher proportion of the available oxygen will need to be extracted from the haemoglobin to support the metabolism of the tissues. Consequently, the venous PO_2 is much lower than normal (see panels (b) and (d) of Figure 36.7).

Stagnant hypoxia is the result of a low blood flow. It may occur peripherally due to local vasoconstriction (e.g. exposure of the extremities to cold) or it may result from a reduced cardiac output. In this cas, the alveolar and arterial PO_2 may be normal, but since the blood flow through the metabolizing tissues is very slow, excessive extraction of the available oxygen occurs and venous PO_2 is very low (Figure 36.7e). This gives rise to peripheral cyanosis.

Histotoxic hypoxia refers to poisoning of the oxidative enzymes of the cells, e.g. by cyanide. In this situation, the supply of oxygen to the tissues is normal but they are unable to make full use of it. As a result, venous PO_2 will be abnormally high (see panel (f) of Figure 36.7).

Oxygen therapy in hypoxia

The administration of a high partial pressure of oxygen can be beneficial in the treatment of hypoxic hypoxia. By increasing the partial pressure of oxygen in the alveoli, the oxygen content of the blood leaving the lungs is raised. This will lessen any central cyanosis and alleviate any dyspnoea. Oxygen therapy will be of less value in other forms of hypoxia. Note that newborn infants should not be exposed to partial pressures of oxygen greater than about 40 kPa (c. 300 mm Hg) as they are particularly sensitive to its toxic effects (see Section 37.3).

To increase the amount physically dissolved in the plasma, oxygen is occasionally administered for short periods at pressures higher than that of the atmosphere. This is known as **hyperbaric oxygen therapy**, which can be helpful in the treatment of carbon monoxide poisoning. The high PO_2 acts to displace the carbon monoxide bound to the haemoglobin and provides much-needed oxygen for the tissues.

Summary

Hypoxia (reduced PaO_2) can arise from a number of causes: poor ventilation (hypoxic hypoxia), anaemia (anaemic hypoxia), poor blood flow (stagnant hypoxia), and metabolic poisons (histotoxic hypoxia). In hypoxic hypoxia, the low arterial PO_2 may arise from a number of causes but it can be treated by administration of pure oxygen. Oxygen administration is not beneficial in other forms of hypoxia and should be avoided in very young babies.

36.5 Respiratory failure

Respiratory failure occurs when the respiratory system fails to maintain normal values of arterial PO_2 and PCO_2. The PaO_2 depends on the inspired PO_2 (at high altitude the partial pressure of oxygen is reduced). At sea level, normal $PaCO_2$ is 5.1 ± 1.0 kPa (38.5 ± 7.5 mm Hg). This range encompasses 95 per cent of the normal population. The PaO_2 decreases with age: for healthy young subjects (< 30 years of age) inspiring air at sea level, the mean value is 12.5 ± 1.3 kPa (94 ± 10 mm Hg). However, in healthy subjects over 60 years of age, the mean value is noticeably lower at 10.8 ± 1.3 kPa (81 ± 10 mm Hg), although the shape of the oxygen dissociation curve ensures an adequate saturation of haemoglobin (see Figure 36.7a).

A diagnosis of respiratory failure is made if PaO_2 is less than 8 kPa (60 mm Hg) or the $PaCO_2$ is greater than 7 kPa (55 mm Hg).

- In **type I respiratory failure**, PaO_2 is low while $PaCO_2$ is normal or low. This occurs when there is a major right-to-left shunt of deoxygenated blood, or when the $\dot{V}_A/\dot{Q}$ ratio is abnormal (see cases 2 and 3 of the discussion in Section 34.5). Such a situation may arise during pneumonia, pulmonary oedema, and adult respiratory distress syndrome (ARDS), which is discussed in the next subsection.

- In **type II respiratory failure**, PaO_2 is low while $PaCO_2$ is elevated. This situation occurs when alveolar ventilation is not sufficient to excrete the carbon dioxide produced by the normal metabolism of the body (i.e. there is a failure of ventilation). The commonest cause is chronic obstructive pulmonary disease or COPD, discussed in Section 36.3.

Type II respiratory failure and its causes

Ventilatory failure may occur as a result of one or more of the following:

- Failure of neural control of the respiratory muscles
- Neuromuscular disease

- Pneumothorax and pleural effusion
- Decreased compliance of the chest or lungs
- Increased airways resistance
- Increased physiological dead space.

The activity of the respiratory muscles is neurogenic in origin and arises from neurons of the medulla (see Chapter 35). It may be depressed during hypoxia or exposure to respiratory depressants such as general anaesthetics and morphine. In this case, ventilation will inevitably be diminished. Traumatic damage to the spinal column below cervical segment C4 may interrupt the flow of information from the medulla to the lower respiratory motoneurons, resulting in paralysis of the intercostal muscles, while lesions above C4 may damage the phrenic outflow. Loss of lower motoneurons following poliomyelitis or other neurological disorders (e.g. motoneuron disease) may also result in ventilatory failure. A decrease in neuromuscular transmission caused by myasthenia gravis may have the same effect.

Ventilation is reduced following a pneumothorax and during a pleural effusion, as the lungs are unable to expand properly. The compliance of the chest depends on the elasticity of the lungs (see Chapter 33) and on that of the chest wall itself. The compliance of the chest will be reduced if there is a reduction in the elasticity of the lungs, if the pleural cavity becomes infiltrated with fibrous tissue, or if the elasticity of the chest wall itself is reduced. Decreased elasticity of the chest wall can occur in various postural disorders such as **scoliosis** (sideways curvature of the spine) and **kyphosis** (an abnormal backward curvature of the spine). These are examples of restrictive disorders.

Increased airways resistance may arise from the presence of foreign material in the airways (e.g. refluxed gastric contents entering the larynx) or from narrowing of the airways themselves, as in asthma and emphysema.

Respiratory failure following surgery is a subtype of type I failure and occurs in the post-operative period, with atelectasis (partial collapse or incomplete inflation of the lungs) being the most common cause. A further type of respiratory failure occurs in circulatory shock (see Section 40.5). This is initially treated by ventilating the patient to reduce the work of the respiratory muscles until the cause of the shock is identified and corrected.

Adult respiratory distress syndrome (ARDS) is a condition in which the lung parenchyma is severely damaged—so much so that more than half of all cases are fatal. There is no entirely satisfactory definition, but ARDS is characterized by a severe hypoxaemia (hence its alternative names of **acute respiratory distress syndrome** or **acute respiratory failure**), the presence of diffuse shadows in chest radiographs (probably due to patches of fluid accumulation), low pulmonary compliance, and pulmonary oedema that is not due to left-sided heart failure. Precipitating causes include septic shock (see Chapter 40), aspiration of the gastric contents, near drowning, and inhalation of toxic gases or smoke. Despite its name, the condition is not confined to adults and is probably the result of damage to the alveolar–capillary membranes which allows fluid to accumulate in the air spaces. This leads to a redistribution of pulmonary blood flow, partly as a result of the normal response to local hypoxia and partly as a result of compression of the pulmonary vessels by the local oedema. Subsequently, the release of chemical mediators may result in further constriction of the pulmonary vasculature and the development of pulmonary hypertension. Within a week of the onset of the condition, the lungs become infiltrated by fibroblasts, which lay down fibrous tissue in the pulmonary interstitium. There is a loss of elastic tissue and emphysema develops. This is reflected in an increase in the physiological dead space.

Summary

Respiratory failure occurs when the respiratory system fails to maintain normal values of arterial PO_2 and PCO_2. Type I respiratory failure may result from a major right-to-left shunt of deoxygenated blood or because of an abnormal $\dot{V}_A/\dot{Q}$ ratio. Type II respiratory failure (or ventilatory failure) occurs when alveolar ventilation is not sufficient to excrete the metabolically derived carbon dioxide. In adult respiratory distress syndrome, the lung parenchyma is severely damaged. It is characterized by severe hypoxaemia.

(✳) Checklist of key points and concepts

Pulmonary defence mechanisms

- The lungs are protected from particles in the air by the respiratory reflexes and by the secretions of mucus of the submucosal glands and goblet cells.
- Particulate matter entering the airways becomes lodged in the mucus lining of the respiratory tree. In the nasal passages this may be removed by sneezing.
- Most of the inhaled material is removed by the mucociliary escalator and by coughing.

- As well as providing a barrier to infection, mucus has antimicrobial properties.
- The epithelium of the alveoli and that of the respiratory bronchioles have no cilia, and material that becomes deposited in this part of the respiratory tree is ingested by alveolar macrophages.

Common respiratory disorders

- Airborne allergens cause an inflammatory reaction in the nasal mucosa known as allergic rhinitis or hay fever.

- Difficulty in breathing causes distress known as dyspnoea.

- Asthma is a condition in which a person has difficulty in breathing (dyspnoea), particularly expiration. It is the result of bronchospasm and is often triggered by inhaling an allergen.

- During an asthmatic attack, the FEV_1 and peak expiratory flow rate are reduced compared to those of normal individuals.

- Emphysema is a condition in which the alveoli are increased in size due to destruction of the lung parenchyma.

- In pulmonary fibrosis, the alveolar wall becomes thickened and this decreases the diffusing capacity of the lung.

- Chronic obstructive pulmonary disease (COPD) is characterized by chronic obstruction of lung airflow that is not fully reversible. It is the result of various diseases including chronic bronchitis and emphysema.

- Cystic fibrosis is a recessive inherited disorder in which the mucus of the airways is abnormally thick and difficult to dislodge. It results from the failure of an epithelial chloride channel to open normally in response to cyclic AMP.

- Sleep apnoea describes a number of conditions in which breathing temporarily stops during sleep. Two principal types can be recognized: obstructive sleep apnoea and central sleep apnoea.

- Cheyne–Stokes breathing consists of a periodic change in the frequency and tidal volume in which breathing alternates between apnoea and mild hyperventilation. It is also known as periodic breathing.

Hypoxia

- In hypoxic hypoxia there is a low arterial PO_2 which may arise from a number of causes. Hypoxic hypoxia can be treated by administration of pure oxygen.

- In anaemic hypoxia there is a decrease in the amount of haemoglobin available for binding oxygen, so the oxygen content will be low, even with a normal PaO_2. The arterio-venous difference in PO_2 is greater than in normal subjects.

- In stagnant hypoxia there is a low blood flow through the tissues which results in excessive extraction of oxygen from the blood.

- In histotoxic hypoxia the oxidative enzymes of the cells are inhibited and unable to function. Venous PO_2 is abnormally high.

Respiratory failure

- Respiratory failure occurs when the respiratory system fails to maintain normal values of arterial PO_2 and PCO_2.

- Type I respiratory failure may result from a major right-to-left shunt of deoxygenated blood or because of an abnormal $\dot{V}_A/\dot{Q}$ ratio.

- In type II respiratory failure (or ventilatory failure), there is a low PaO_2 and an elevated $PaCO_2$. It occurs when alveolar ventilation is not sufficient to excrete the metabolically derived carbon dioxide.

- Adult respiratory distress syndrome is characterized by a severe hypoxemia that is the result of extensive damage to the lung parenchyma. It is a life-threatening conditon.

Recommended reading

Pharmacology of the respiratory system

Grahame-Smith, D.G., and Aronson, J.K. (2002) *Oxford textbook of clinical pharmacology* (3rd edn), Chapter 24. Oxford University Press, Oxford.

   A useful account of the drug treatments available for treatment of respiratory diseases, including an account of the use of oxygen.

Rang, H.P., Dale, M.M., Ritter, J.M., Flower, R.J., and Henderson, G. (2015) *Pharmacology* (8th edn), Chapter 28. Churchill-Livingstone, Edinburgh.

   A clear account of drugs that are used to treat respiratory disorders.

Physiology of the respiratory system

Levitzky, M.G. (2013) *Pulmonary physiology* (8th edn), Chapter 10. McGraw-Hill, New York.

   A straightforward introduction to the non-respiratory functions of the lungs and pulmonary defence mechanisms.

Lumb, A.B. (2015) *Nunn's applied respiratory physiology* (8th edn), Chapters 11, 22–29. Churchill-Livingstone, London.

Review articles

Dempsey, J.A., Veasey, S.C., Morgan, B.J., and O'Donnell, C.P. (2010) Pathophysiology of sleep apnea. *Physiological Reviews* **90**, 47–112.

Hansell, D.M. (2001) Small airways diseases: Detection and insights with computed tomography. *European Respiratory Journal* **17**, 1294–313.

Knowles, M.R., and Boucher, R.C. (2002) Mucus clearance as a primary innate defense mechanism for mammalian airways. *Journal of Clinical. Investigation* **109**, 571–77.

Pilewski, J.M., and Frizzell, R.A. (1999) Role of CFTR in airway disease. *Physiological Reviews* **79**, Suppl.: S215–55.

To check that you have mastered the key concepts presented in this chapter, complete the accompanying online self-assessment questions. Go to www.oup.com/uk/pocock5e/

CHAPTER 37

The physiology of high altitude and diving

Chapter contents

This chapter should help you to understand:

- The physiological changes that occur following ascent to high altitude

- The effects of acute and chronic hypoxia

- The physiological problems associated with high environmental pressures

- Breath-hold diving and the diving response

37.1 Introduction

The physiology of the cardiovascular and respiratory systems have been discussed in the preceding chapters. This chapter is principally concerned with the physiology of the respiratory system under the stresses imposed by changes to ambient pressure. It will first discuss the physiological adaptations that occur when a person experiences the reduced atmospheric pressure of high altitudes and will conclude with a discussion of the respiratory problems associated with high ambient atmospheric pressures such as those experienced by divers.

37.2 The physiological effects of high altitude

Atmospheric pressure decreases with altitude, and since the fraction of oxygen in the air (20.9 per cent) does not change, the partial pressure of oxygen in the inspired air will also fall progressively with increasing altitude (see Figure 37.1). This follows directly from Dalton's law of partial pressures (see Section 32.2). Consequently, the partial pressure and content of oxygen of the blood will decline and this lack of oxygen will ultimately limit the capacity of the body to perform work.

The hypoxia of altitude can be divided into *acute hypoxia*, which is experienced by subjects who have been exposed to high altitude for a few minutes or hours, and *chronic hypoxia*, which is experienced by people living for long periods at high altitude or by mountaineers who have become **acclimatized** to high altitude. In high-altitude medicine it is often convenient to consider the effects under one of three headings:

- those of high altitude (1500–3000 m or 4900–11,500 ft)

- those of very high altitude (3000–5500 m or 11,500–18,000 ft)

- those of extreme altitude (>5500 m or >18,000 ft).

Acute hypoxia

Although the partial pressure of oxygen in the inspired air falls with altitude, the arterial PO_2 is a relatively weak stimulus to breathing (see Section 35.4). Not until the alveolar partial pressure of oxygen falls below about 8 kPa (60 mm Hg) does the rate and depth of breathing increase substantially. Thus, until altitude exceeds 2500–3000 m (c. 8000–10,000 ft) breathing is largely unaffected by the reduction in atmospheric pressure. However, above this altitude, the respiratory minute volume increases progressively as alveolar

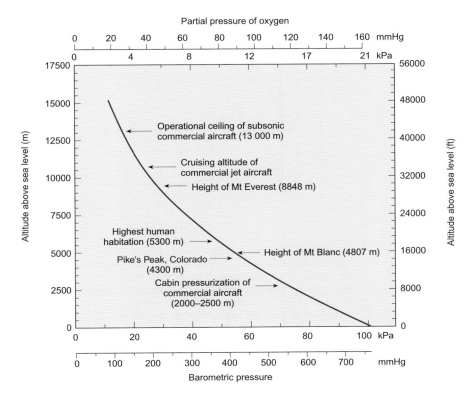

Figure 37.1 The change in barometric pressure and in the partial pressure of oxygen with altitude.

PO_2 declines below 60 mm Hg (8 kPa)—(Figure 37. 2: upper panel, blue line). The stimulus for this increase in ventilation comes from the carotid bodies, which sense the fall in arterial PO_2. As explained in Section 35.4, the carotid bodies are the sole organs able to elicit a ventilatory response to hypoxia, although they also respond to an increase in arterial PCO_2 and a fall in arterial pH.

As ventilation increases, carbon dioxide is lost from the lungs faster than it is produced by the body. Alveolar PCO_2 progressively falls in proportion to the increase in ventilation (Figure 37.2, lower panel, and Table 37.1). This fall in alveolar PCO_2 leads to a diminished respiratory drive from the central chemoreceptors that tends to offset the increased respiratory drive due to the hypoxic stimulation of the carotid bodies. Moreover, the fall in alveolar PCO_2 results in a rise in the pH of the arterial blood (see Figure 37.3, green line). This is known as a **respiratory alkalosis** and is discussed more fully in Chapter 41 (Section 41.5).

While the carotid bodies have little influence on the circulation at normal blood gas tensions, during hypoxia they elicit a reflex vasoconstriction in the resistance vessels and in the large veins of the splanchnic circulation. In addition, there is an increase in heart rate and cardiac output. As a result of these changes, blood is diverted from the skin and splanchnic circulation resulting in an increase in arterial blood pressure. These cardiovascular changes serve to increase the proportion of the available oxygen for use by the brain and exercising muscles.

At very high altitude the normal mental faculties are impaired: there is an initial euphoria but it is accompanied by slow thought processes with elementary mistakes in simple mental tasks. The lack of oxygen results in dyspnoea when undertaking even mild exercise. After a few hours at altitude, the euphoria is followed by mental confusion that is accompanied by headache and, in some individuals, by nausea and vomiting (mountain sickness, discussed later in this section); the more severe the hypoxia, the more marked the symptoms. Even at moderate altitudes, there are psychological changes that often result in poor judgment and elementary mistakes. For this reason, commercial aircraft maintain a cabin pressure equivalent to 2000–2500 m (6500–8200 ft). If cabin pressure falls, oxygen is immediately made available.

Severe acute hypoxia

Rapid ascent in an unpressurized aircraft or balloon is associated with a rapid change in the inspired PO_2, and alveolar PO_2 declines steeply in parallel. Aviators in unpressurized aircraft and balloonists will experience severe acute hypoxia as they ascend unless they are provided with supplementary oxygen. When people are rapidly exposed to low inspired PO_2 and become hypoxic, they first experience physical weakness, which progresses to full paralysis of the limbs when the alveolar PO_2 is less than 2.5 kPa (c.18 mm Hg). This will occur when inspired PO_2 falls below about 8 kPa (60 mm Hg), corresponding to an altitude of 6000 m (about 19,600 ft). As alveolar PO_2 falls further (i.e. at extreme altitude), there is loss of consciousness, and death rapidly ensues unless urgent corrective action is taken. This is a risk first encountered during rapid ascent in a balloon. Acute hypoxia will also occur if the pressurized cabin of an aircraft is ruptured at extreme altitude.

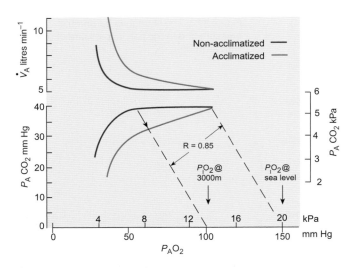

Figure 37.2 Upper panel: alveolar ventilation as a function of the partial pressure of oxygen in alveolar air (P_AO_2) for normal subjects at sea level (blue) and in those acclimatized to an altitude of c. 3000 m (10,000 ft, red). Lower panel: the partial pressure of carbon dioxide in alveolar air as a function of the partial pressure of oxygen in the alveolar air for normal subjects acclimatized to sea level (blue) and those acclimatized to high altitude (3000 m, red). Note that in those individuals not acclimatized to high altitude, the increase in ventilation only occurs when P_AO_2 falls below about 60 mm Hg (8 kPa), which occurs when the inspired partial pressure of oxygen (P_IO_2) falls below about 100 mm Hg (13.3 kPa). These values assume that the respiratory exchange ratio is 0.85. On arrival at high altitude there is an increase in ventilation in response to the low P_IO_2 which results in a change in the composition of the alveolar air. After a few days residence at altitude, people become acclimatized (here shown by the red curves) and ventilation is increased for a given P_AO_2. This increased ventilation increases the partial pressure of oxygen in the alveoli, a crucial adaptation for undertaking exercise at altitude. (Based on Fig 48 in P. Dejours (1966) *Respiration*. Oxford University Press, Oxford.)

Table 37.1 The effect of altitude on the partial pressures of carbon dioxide and oxygen in the alveoli

Place	Altitude (m)	Barometric pressure kPa (mm Hg)	P_ACO_2 kPa (mm Hg)	P_AO_2 kPa (mm Hg)
Sea level	0	101 (760)	5.3 (40)	13.3 (100)
Colorado Springs	1800	82 (620)	4.8 (36)	10.5 (79)
Pike's Peak, Colorado	4300	61 (460)	3.7 (28)	7 (53)
North Col of Everest	6400	45 (355)	2.5 (19)	5.1 (39)
Summit of Everest	8848	32 (240)	2.0 (15)	3.2 (24)

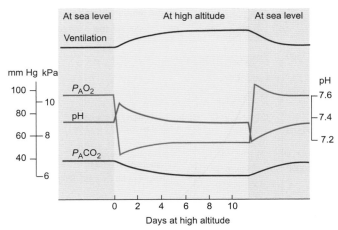

Figure 37.3 The time course of changes in ventilation, arterial pH (green), and alveolar PO_2 (red) and PCO_2 (blue) with time spent at an altitude of approximately 3000 m. Note the rapid initial fall in P_AO_2 and the rise in pH compared to the slower change in P_ACO_2 and ventilation. The initial increase in ventilation occurs in response to the fall in P_AO_2, but this is offset by the fall in P_ACO_2 (which reduces the stimulation of the central chemoreceptors) and the rise in blood pH. As bicarbonate ions are transported from the CSF, the drive to the central chemoreceptors is restored and an elevated ventilation rate is maintained. The decline in P_ACO_2 permits a small rise in P_AO_2 from its initial low level. (Based on Fig 51 in P. Dejours (1966) *Respiration*. Oxford University Press, Oxford.)

The extent of the hypoxia following rapid ascent to any given altitude can be calculated from the alveolar gas equation (see Box 37.1) assuming the alveolar PCO_2 has not changed from its value at sea level—5.3 k Pa (40 mm Hg).

The effects of reduced P_IO_2 on the oxygen content of the blood

Although the fall in P_IO_2 experienced following rapid ascent to very high altitude reduces the oxygen saturation and oxygen content of the blood, it does not alter its oxygen-carrying capacity (Figure 37.4). Nevertheless, the stimulation of ventilation by low arterial PO_2 results in a progressive fall of the alveolar PCO_2 because CO_2 is being lost from the body faster than it is being produced by the tissues. As Dalton's law predicts, this fall in alveolar PCO_2 permits a small increase in alveolar PO_2 (see Figures 37.2 and 37.3 and Box 37.1). This modest increase in alveolar PO_2, together with the fall in PCO_2 and rise in arterial pH, increases the ability of haemoglobin to bind oxygen because of the reversal of the Bohr shift. (As $PaCO_2$ falls and pH rises, the position of the oxyhaemoglobin dissociation curve shifts to the left—see Figure 25.12.) Consequently, the haemoglobin is more saturated at a given PaO_2 than it would be if the $PaCO_2$ had remained at 5.3 kPa (40 mm Hg), the value it would normally be at sea level. Despite this, above about 3000 m (c. 10,000 ft), the oxygen content of the blood is significantly less than it would be at sea level (see Figure 37.4). To meet the demands of the tissues, more oxygen is taken up from the blood and this reduces the PO_2 of the venous blood, increasing the arterio-venous difference in PO_2.

(1₂3) Box 37.1 The alveolar gas equation

The relationship between the alveolar PO_2 and PCO_2 is given by the alveolar gas equation:

$$PA_{O_2} = FI_{O_2}(PB - 47) - PA_{CO_2}\left[FI_{O_2} + \frac{(1 - FI_{O_2})}{R_A}\right]$$

where FI_{O_2} is the fraction of oxygen in the inspired air (normally 0.209), PA_{O_2} is the alveolar partial pressure of oxygen, PA_{CO_2} is the alveolar partial pressure of carbon dioxide, and R_A is the alveolar respiratory exchange ratio; 47 mm Hg (6.26 kPa) is the saturated water vapour pressure at body temperature (which is taken to be 37°C).

A worked example

At 3050 m (10,000 ft) the barometric pressure is 523 mm Hg (69.7 kPa). If we assume that the respiratory exchange ratio is 0.85 (a common value for a normal diet) and that PA_{CO_2} is 40 mm Hg (5.33 kPa), the PA_{O_2} can be calculated as follows:

$$PA_{O_2} = 0.209 \times (523 - 47) - 40 \times (0.209 + 0.791/0.85)$$

$$= 99.484 - 40 \times 1.14$$

$$= 53.9 \text{ mm Hg (7.18 kPa)}$$

If after several days' acclimatization the PA_{CO_2} has fallen to 30 mm Hg:

$$PA_{O_2} = 0.209 \times (523 - 47) - 30 \times (0.209 + 0.791/0.85)$$

$$= 65.3 \text{ mm Hg (8.7 kPa)}.$$

Mountain sickness and other effects of very high altitude

When people initially experience the hypoxia of altitude, they often develop an illness known as **mountain sickness**. The typical symptoms are headache, nausea, giddiness, gastrointestinal disturbances, lassitude, and psychological disturbances that include loss of appetite. Severe dyspnoea occurs in response to mild exercise, as the circulation is unable to supply the tissues with adequate amounts of oxygen. Mountain sickness can rapidly progress to pulmonary oedema (discussed later in this subsection) in susceptible individuals.

Sleep apnoea, which is usually associated with periodic breathing, occurs at altitudes above 4000 m (c. 13,000 ft). In one study at higher altitudes (above c. 6000 m—19,500 ft), periodic breathing during sleep was seen in all subjects. The primary cause of the periodic breathing is the hypoxaemia (i.e. low P_aO_2) due to the low inspired PO_2. At sea level, respiration is principally driven by the response of the central chemoreceptors to arterial PCO_2 (i.e. P_aCO_2). As explained above, the increase in ventilation that occurs following

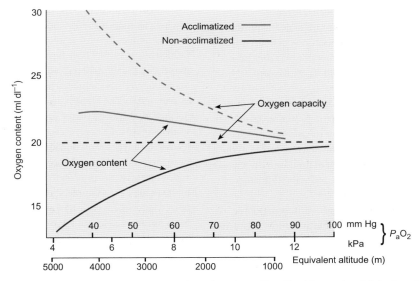

Figure 37.4 Total oxygen capacity of blood and oxygen content for subjects acclimatized to sea level (blue) and those acclimatized to high altitude (>3000 m or 10,000 ft, red). Note that the oxygen capacity for non-acclimatized individuals is not affected by altitude, unlike their blood oxygen content, which falls. In altitude-acclimatized individuals, the blood oxygen capacity and the oxygen content of the blood are both increased for a given altitude. Thus, for a non-acclimatized individual at 4000 m the oxygen content of the blood is around 15 ml dl⁻¹ while a fully acclimatized person has an oxygen content of around 22 ml dl⁻¹—which is higher than that of a normal subject at sea level. This increase in oxygen content is due to two factors: an increase in blood haemoglobin and an increase in ventilation. (The bottom scale shows the equivalent PaO_2 for a non-acclimatized individual at various altitudes.) (Based on Fig 52 in P. Dejours (1966) *Respiration*. Oxford University Press, Oxford.)

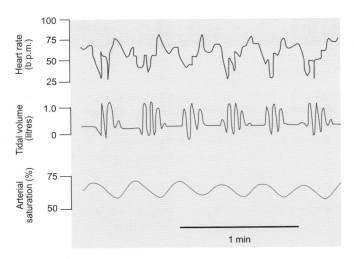

Figure 37.5 An example of periodic breathing during sleep recorded from a subject at an altitude of 6300 m (c. 20,500 ft). Note the periods of apnoea and the variations in heart rate and oxygen saturation. The heart rate was calculated from the R–R interval of the ECG and the oxygen saturation was measured with an ear oximeter. The changes in oxygen saturation are out of phase with the breathing cycle, with the lowest arterial saturation seen just after the point of maximum ventilation. (Based on data in Fig 1 in J. Wes et al (1986) *J Appl Physiol* **61**, 280–7.)

ascent to high altitude leads to a reduction in alveolar PCO_2 (i.e. P_ACO_2) and the ventilatory drive from the central chemoreceptors falls. As a result, breathing is not adequately stimulated by the $PaCO_2$ and ventilation becomes periodic. The periods of apnoea exacerbate the hypoxemia that is already present as a result of the low inspired PO_2 and there are substantial cyclical changes in oxygen saturation, as Figure 37.5 shows. In these subjects, the periodic breathing was also associated with cyclical variations in heart rate.

Pulmonary oedema can occur when unacclimatized climbers reach altitudes in excess of 3000 m (c. 10,000 ft). The cause of this oedema is not clear and it is likely that several factors are involved. Perhaps the most important factor is pulmonary hypertension resulting from the hypoxic vasoconstriction of the pulmonary vessels (see Chapter 34). Moreover, pulmonary hypertension is increased during exercise. As a result, the balance of the Starling forces in the pulmonary capillaries favours the movement of fluid into the interstitial space and, eventually, into the alveoli, reducing diffusing capacity and impairing gas exchange.

There is wide variation in the susceptibility of different people to mountain sickness; some are affected at relatively low altitude (c. 2500 m or 8200 ft) while others show few signs. Nevertheless, very few subjects can venture to extreme altitudes (i.e. those in excess of 5500 m or 18,000 ft) without experiencing severe symptoms. The causes of mountain sickness are not entirely clear but include hypoxia and dehydration. In severe cases the primary steps in treatment require immediate descent to lower altitude and the administration of oxygen.

To reduce the incidence of acute mountain sickness, acetazolamide (an inhibitor of carbonic anhydrase) is sometimes taken by climbers, starting a few days before their ascent to high altitude. This is often recommended for those climbing to 3000 m (~10,000 ft) from sea level within a day. It is also recommended for those starting above 2500 m (~8200 ft) who intend to climb more than 600 m (~2000 ft) in a single day. Acetazolamide is also used as a prophylactic for those climbers who have a history of acute mountain sickness. The use of acetazolamide was originally predicated on the assumption that it would increase excretion of excess bicarbonate by the kidneys and restore plasma pH to normal values, but recent work suggests that a number of factors contribute to its beneficial effects.

Acclimatization to chronic hypoxia

Once at altitude, the body progressively becomes acclimatized to the new circumstances. Indeed, many people live permanently at high altitude and their physiology has become so well adapted that they are able to perform the everyday tasks of life as easily as those who live near sea level. Even for subjects who have newly arrived at high altitude, the process of acclimatization begins almost immediately. The principal changes that occur during acclimatization to high altitude are as follows.

- Increase in respiratory minute volume: this occurs despite the low $PaCO_2$. The respiratory alkalosis associated with the hyperventilation is gradually compensated by the excretion of excess bicarbonate by the kidneys. Similar compensation occurs in the CSF, and this is believed to restore the respiratory drive provided by the central chemoreceptors in response to carbon dioxide. As a result, the minute volume increases rapidly at first and then more slowly to reach its final level (Figure 37.3). The adjustment of CSF pH occurs within about 24 hours, while the pH of the plasma is restored to normal levels within a week.

- Increase in the number of red cells and haemoglobin: following ascent to a high altitude, the red blood cell count increases significantly. This trend continues for several weeks (see Figure 37.6) and red cell counts as high as 8×10^{12} l^{-1} have been recorded after long periods of acclimatization (the normal red cell count is $4–5 \times 10^{12}$ l^{-1}). The increased haematopoiesis is stimulated by the hormone **erythropoietin**, which is secreted by the kidneys in response to the low blood PO_2 (see Section 25.4). The increased haemoglobin content of the blood leads to an increase in its oxygen-carrying capacity. In addition, the low PO_2 leads to an increased production of 2,3 bisphosphoglycerate (2,3-BPG) by the red cells, which enhances the rightward shift in the oxygen dissociation curve (the Bohr shift) and so facilitates the release of oxygen in the tissues. For example, during a simulated expedition in a hypobaric chamber (Everest II) the concentration of 2,3-BPG increased from around 1.7 mmol l^{-1} at a P_IO_2 of 150 mm Hg (sea level) to 3.8 mmol l^{-1} at a simulated altitude of 8540 m (c. 28,000 ft).

- Increase in cardiac output: the rise in heart rate seen in the early stages of acute hypoxia is maintained during acclimatization and gives rise to an increased cardiac output. This results in a greater blood flow through the tissues, so improving their oxygenation. The enhanced oxygen delivery to the tissues is so efficient that the PO_2 of mixed venous blood of people acclimatized to very high altitude is only slightly less than that of those who live at sea level (see Figure 37.7). During a prolonged sojourn at high altitude, the cardiac output slowly declines to near normal values.

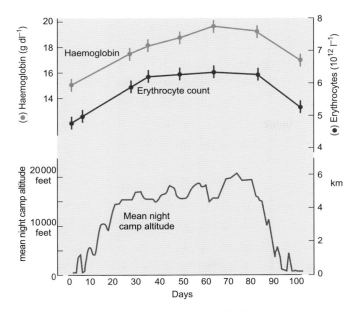

Figure 37.6 The change in red cell count with time spent at altitude. The data were collected during a period of 62 days spent continuously above 4600 m (15,000 ft). Note that the red cell count (blue) and blood haemoglobin (orange) increased over several weeks after reaching high altitude. Recovery on descent to sea level was much more rapid. The symbols and vertical bars indicate mean ± standard error of the mean. (Adapted from Fig 1 in N. Pace, B. Meyer, and B.E. Vaughan (1956) *Journal of Applied Physiology* **9**, 141–4.)

- Increase in the vascularization of the tissues: the number of the capillaries is increased and they become more tortuous in their course through the tissues. Blood volume is also increased. These changes greatly enhance the ability of the blood to supply the tis-

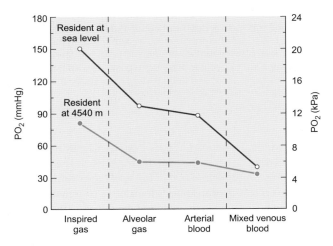

Figure 37.7 The partial pressure of oxygen in the inspired air, alveoli, arterial blood, and mixed venous blood for subjects acclimatized to sea level and 4540 m (14,755 ft). Although the inspired air at high altitude has a much lower PO_2 than at sea level, acclimatized individuals show a smaller arterio-venous difference than those resident at sea level. This reflects the higher haemoglobin content of their blood.

sues with the oxygen they require for metabolism. The high red cell count will increase blood viscosity (see Figure 23.3) and this, coupled with the increase in cardiac output, might be expected to result in an increase in arterial blood pressure. In fact, people who live permanently at high altitudes have no elevation in their blood pressure compared to people living at sea level. The increased cross-sectional area available for blood flow through the tissues leads to a fall in total peripheral resistance, and it is this change that accounts for the normal value of arterial blood pressure.

Despite these adaptations, the ability to perform exercise is increasingly compromised the greater the altitude. In the simulated ascent of Mount Everest (Everest II) mentioned earlier, subjects were able to attain a maximal exercise level of 300–360 watts with an inspired PO_2 of 150 mm Hg (equivalent to that at sea level). When they were exercising with an inspired PO_2 of 80 mm Hg (equivalent to an altitude of c. 4270 m –14,000 ft) they could achieve only a maximum of 240–270 watts. When PO_2 was 43 mm Hg (equivalent to the inspired PO_2 at the summit of Mount Everest—8848 m or 29,028 ft), they were able to attain only a maximum exercise level of 120 watts. This reduction in the capacity for exercise is, of course, compounded both by the low ambient temperature and by the severe weather experienced by mountaineers.

Summary

In acute hypoxia, the fall in inspired PO_2 with altitude leads to an increase in ventilation and an increase in cardiac output. The increased ventilation results in a decrease in alveolar PCO_2 and a respiratory alkalosis. Following acclimatization to chronic hypoxia, there is a maintained increased in ventilation, increased vascularization of the tissues, and an increase in the oxygen-carrying capacity of the blood. The respiratory alkalosis resulting from the hyperventilation is compensated by renal excretion of bicarbonate. The cardiac output slowly returns to near normal values.

37.3 The effects of high environmental pressure

Exposure to high atmospheric pressures occurs during diving. The atmospheric pressure at sea level is 1 atmosphere, but for every 10 m (33 ft) of descent into the sea, the pressure rises by 1 atmosphere. Thus at a depth of 30 m (99 ft) the total pressure is $1 + 3 = 4$ atmospheres. Increased environmental pressures are also experienced by engineering workers when they are employed in tunnelling, as the air must be maintained under pressure to prevent water seeping into the workings. In both cases, the increased air pressure increases the amount of gas dissolved in the blood and tissues in accordance with Henry's law (see Section 32.2). Since the alveolar PCO_2 remains almost constant under these circumstances, the increases in the partial pressures and amounts of dissolved oxygen and nitrogen are of prime concern.

Oxygen toxicity

While it is essential for life, oxygen is a very reactive gas. It is therefore perhaps not surprising that certain hazards are associated with breathing pure oxygen. Breathing oxygen up to a partial pressure of 60 kPa (60 per cent in inspired air at normal pressure) is perfectly safe for an adult even for long periods. However, premature and newborn babies are particularly sensitive to the toxic effects of increased partial pressures of oxygen. For this reason, the oxygen pressure to which newborn babies are exposed must not exceed 40 kPa (c. 300 mm Hg or 40 per cent of normal atmospheric pressure). There is a risk of these infants becoming permanently blind if this PO_2 value is exceeded. This risk arises because high oxygen partial pressures cause an intense constriction of the immature retinal vessels, resulting in retinal ischaemia. This eventually leads to secondary pathological changes in the retina.

Even in adults, breathing oxygen at normal atmospheric pressure (101 kPa, 760 mm Hg) for more than 8 hours results in pharyngitis, tracheitis, and cough. Continued breathing of such a high PO_2 causes pulmonary congestion and sluggish mental activity. In moderate hyperoxia (increased alveolar PO_2), only the alveoli, pulmonary vessels, and systemic arteries experience significantly elevated PO_2 values (100–150 kPa), as the PO_2 rapidly falls once the blood reaches the tissues. It is therefore not surprising that the lungs are very vulnerable to the effects of prolonged hyperoxia.

Breathing oxygen at pressures above about 1.7 to 2 atmospheres (170–200 kPa) leads to overt signs of oxygen toxicity. These signs include nausea, dizziness, feelings of intoxication, tremor, and even convulsions or syncope. After a few hours breathing pure oxygen at a pressure of three atmospheres, people experience a reduction in their visual field and impairment of vision. It is therefore clear that the safe use of 100 per cent oxygen in diving is extremely limited (see Figure 37.8). For this reason, the partial pressure of oxygen must be carefully controlled in deep diving.

Problems associated with breathing compressed air during diving

Inhalation of compressed air at high pressure can also be dangerous. Although nitrogen is less soluble in water than either oxygen or carbon dioxide, it is about five times more soluble in fat than it is in blood. Prolonged exposure to compressed air (i.e. a high $P_I N_2$) will therefore lead to an accumulation of nitrogen in the tissues. Unless suitable precautions are taken when a diver returns to a normal ambient pressure, any dissolved gas will come out of solution and form bubbles in the tissues, causing traumatic damage and intense pain. This disorder, **decompression illness**, was once common in divers and caisson workers and is known as the 'bends' or 'chokes'. It may be avoided by slow ascent according to specific diving schedules. If decompression illness becomes evident on completing a dive, it should be treated by immediate recompression in a pressure chamber followed by slower decompression according to specific decompression tables.

It is often more convenient to house those regularly diving to great depths in a dry hyperbaric chamber on board a ship or oil rig. This is known as **saturation diving**, as the tissues become fully

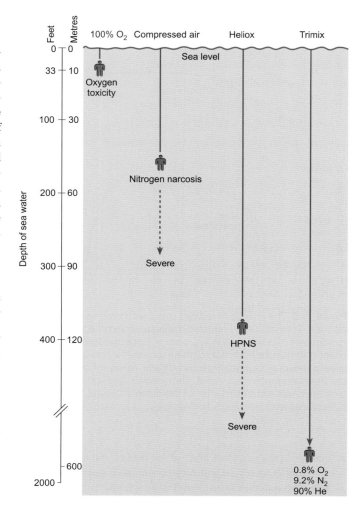

Figure 37.8 The use of various gases in diving. Oxygen toxicity limits the breathing of 100 per cent oxygen to dives of only 33 ft (10 m) of seawater. If air (c. 20 per cent O_2, 80 per cent N_2) is supplied under pressure, depths of about 150 ft (c. 50 m) of seawater are possible before there is a risk of nitrogen narcosis. (Note that with compressed air this sets in before oxygen toxicity becomes important.) Greater depths are possible with helium-oxygen mixtures (heliox) and helium-oxygen-nitrogen mixtures (trimix). See text for further information. (Courtesy of Professor M. de Burgh Daly.)

saturated with the required inert gas mixture (usually mixtures of helium and oxygen, discussed later in this section). The divers are kept at a pressure similar to that of the working environment. To reach the work site, they are quickly transferred to a smaller chamber that can be lowered to the required depth before they exit the chamber to commence work. They remain connected to the chamber, which provides the appropriate gas for breathing. Once they have completed their task, they return to the surface in the small chamber before being transferred to the holding chamber, where they rest before their next descent. The time in the pressure chamber normally lasts about three weeks. When they complete their tour of duty, divers spend several days in a decompression chamber before returning to life at normal atmospheric pressure.

Nitrogen narcosis

Breathing a gas mixture containing nitrogen at high pressure has a second danger—that of narcosis. Breathing air at a pressure greater than about 5 atmospheres (which would correspond to a depth of about 50 m) brings about the early signs of anaesthesia. There is a sense of euphoria ('rapture of the deep'), mental confusion, and a lack of proper motor coordination. These symptoms become more severe with increasing depth. For these reasons, the breathing of air during diving is limited to depths of less than 50 m (see Figure 37.8).

The use of helium-oxygen mixtures in diving

Nitrogen narcosis can be overcome by substituting helium for the nitrogen of the inhaled gas. Helium avoids the problem of narcosis and is less soluble in the tissues, so that decompression from a given depth can be somewhat faster. For these reasons, mixtures of oxygen and helium (known as **heliox**) are widely used in commercial diving (see Figure 37.8). Helium has the disadvantage that it is a very efficient conductor of heat, so precautions must be taken against hypothermia. It has the additional disadvantage that its low density leads to a rise in the pitch of the voice. As a result, speech becomes less intelligible and this can make communication between the diver and the surface difficult.

High pressure nervous syndrome

Exposure to very high pressures even with helium–oxygen mixtures is a hazard. To avoid oxygen toxicity, the fraction of oxygen in the inspired air is kept around 50 kPa (0.5 atmospheres) as explained above. In addition, there are direct effects of pressure on performance. Pressures above about 1200 kPa (corresponding to a depth of about 130 m or 400 ft) cause characteristic changes known as the **high-pressure nervous syndrome** or HPNS. There is tremor of the hands and arms, dizziness, and nausea. Intellectual performance is much less affected. The severity of the symptoms appears to relate to the rate of compression.

The excitatory effects of HPNS can be countered by a slow descent and by adding a small quantity of nitrogen (5–10 per cent) to the helium–oxygen mixture. This gas mixture (known as 'trimix') was developed to exploit the known depressant effect of nitrogen on the nervous system to offset the excitatory effects of high pressure. With this mixture, successful dives in excess of 650 m (c. 2000 ft) have been made (Figure 37.8).

Barotrauma

This is the commonest occupational disease of divers. It is caused by the contraction or expansion of gas spaces in the body that are for one reason or another unable to equilibrate with the ambient air. Air trapped in the sinuses and even the cavities of the teeth may cause pain as it is compressed and decompressed. If the Eustachian tube is blocked, the resulting pressure changes in the middle ear cause pain and may lead to rupture of the tympanic membrane. When a diver ascends, he must exhale continuously to ensure that the pulmonary gases do not over-distend the lungs. In addition, gases trapped in the intestinal tract may cause abdominal discomfort during decompression.

SCUBA and snorkel diving

Unlike conventional divers who have a continuous air supply pumped into their helmets, SCUBA divers (a term derived from the initials Self-Contained Underwater Breathing Apparatus) carry their own gas supply. They breathe via a demand valve that matches the pressure from the compressed gas bottles to the ambient pressure before delivering it to the diver. The system may either rely on releasing the exhaled air into the surrounding water as bubbles or it may employ a re-breathing circuit with a soda lime absorber. SCUBA divers generally operate at depths of 30 m (~100 ft) or less, although it is possible for them to operate at greater depths provided precautions are taken against decompression illness during their ascent. The great advantage SCUBA diving offers over fixed-line diving is mobility and the ability to manoeuvre in restricted places (e.g. inside wrecks).

Snorkel diving uses a small, J-shaped breathing tube with a mouthpiece on one end. There is sometimes a valve on the other end to prevent ingress of water. A snorkel allows a diver to breathe air while swimming face-down. It is sometimes used as an adjunct to SCUBA equipment to conserve air while locating the best position for a dive, but it is also widely employed with a face mask to examine marine life inhabiting shallow water. The tube is usually around 0.3 m in length (12–15 inches) with a diameter of 1.5–2.5 cm (0.6–1.0 inches). Since it is in series with the airways, a snorkel tube will increase dead space from around 150 ml to more than 250 ml. With a normal tidal volume of 600 ml, such an increase in dead space has the obvious disadvantage of limiting alveolar gas exchange. A longer tube with a narrower bore could hold the same volume of air, but this would increase total airways resistance and the work of breathing. Moreover, it would not allow breathing when snorkelling at depths much greater than about 0.5 m (~18 inches), since inspiration would be limited by the maximum inspiratory pressure that can be developed by the respiratory muscles when working against the hydrostatic pressure exerted on the chest by the surrounding water.

Summary

Breathing pure oxygen at pressures greater than 1 atmosphere leads to signs of oxygen toxicity. This limits its use in diving. In most instances, compressed air is used for diving at depths up to 50 m (c. 150 ft). However, at elevated pressures, the quantity of nitrogen dissolved in the tissues increases and rapid decompression can lead to bubble formation in the tissues, which causes pain and tissue damage. This can be avoided by progressive decompression. A second problem of breathing air at elevated pressure is nitrogen narcosis. This problem is largely overcome by breathing mixtures of helium and oxygen, but the partial pressure of oxygen is kept below 1 atmosphere. Deep diving (below 130 m) elicits the high-pressure nervous syndrome (HPNS), which is countered by adding a small quantity of nitrogen to the helium–oxygen mixture.

37.4 Breath-hold diving

In breath-hold diving, the body is subjected to a number of environmental changes. Respiration must be suspended, the ambient pressure is increased, the effects of gravity are offset by the buoyancy of the body, the contact with water increases heat loss, and normal somatosensory inputs are impaired. It is remarkable that, despite these handicaps, human beings can successfully carry out many activities under water.

Breath-hold diving is carried out by a large number of professional divers who operate off the Pacific coasts collecting pearls and sponges. They are able to dive to depths of 20 m (66 ft) and to stay submerged for up to a minute, diving as many as 20 times an hour. When they dive, they exhibit the **diving response**, which consists of a cessation of breathing, profound bradycardia, and selective peripheral vasoconstriction. The vasoconstriction occurs in those organs that can survive for short periods by utilizing anaerobic metabolism such as the skin, muscle, kidneys, and the gastrointestinal tract. The brain, however, relies entirely on oxidative metabolism and requires a constant supply of oxygen. During the diving response, the oxygen supply is maintained by the redistribution of the cardiac output to the cerebral circulation. A similar response is seen in aquatic mammals (e.g. seals and whales) where the circulatory changes are often very dramatic. A record of the cardiovascular changes produced by a simulated breath-hold dive is shown in Figure 37.9.

The initial stimulus for the diving response seems to be immersion of the skin of the face in water (especially cold water). The afferent fibres responsible for this response run in the trigeminal nerve. Immediately following face immersion, the heart rate falls by as much as a half. This bradycardia is the result of increased vagal activity. Despite the fall in heart rate, there is usually an increase in arterial blood pressure resulting from a profound peripheral vasoconstriction that is due to an increase in sympathetic nerve activity. The apnoea is induced partly by voluntary suppression of breathing and partly by a reflex inhibition of respiration elicited by stimulation of the trigeminal receptors.

During prolonged breath-holding, both hypoxia and hypercapnia occur. Levels of PaO_2 as low as 4–4.6 kPa (30–35 mm Hg) have been recorded in synchronized swimmers at the end of their performance. The fall in PaO_2 and rise in $PaCO_2$ is known as **asphyxia**. These changes in PaO_2 and $PaCO_2$ stimulate the carotid body chemoreceptors. This chemoreceptor activity does not, at least initially, lead to stimulation of breathing but it contributes to the bradycardia and peripheral vasoconstriction of the diving response. Under these conditions, the chemoreceptor stimulus to breathing is powerfully inhibited by the activation of the trigeminal receptors. Eventually, however, the hypoxia and hypercapnia become so intense that the trigeminal inhibition is overcome and the desire to breathe becomes compelling.

During the descent stage of a breath-hold dive, the lung volume will decrease and the alveolar gases will be compressed. This raises their partial pressure and facilitates their transport into the blood. Note that even for a dive of only 10 m (33 ft), the total gas pressure will approximately double. This will increase the partial pressure of both oxygen and carbon dioxide. The increased pressure will not, however, increase the oxygen content of the blood very significantly. The increased arterial PCO_2 that occurs during the dive acts to stimulate the desire to breathe (the 'break-point') and serves to limit the duration of the dive.

Before diving, many people hyperventilate. This has comparatively little effect on the alveolar PO_2 but greatly reduces the alveolar PCO_2. As a result, the respiratory drive from the central chemoreceptors is much reduced and this increases the time to the 'break-point'

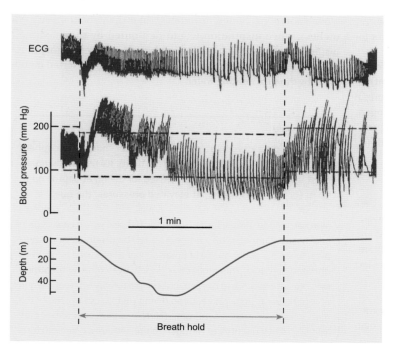

Figure 37.9 Recording of the ECG lead II and arterial blood pressure during a simulated dive to a depth of 50 m. The experiment took place in a hyperbaric chamber with a pool sufficiently deep to cover the subject to their waist. The water temperature was maintained at 25°C. Blood pressure was measured directly using a catheter in the radial artery. The subject in this case was a 30-year-old woman who regularly undertook deep breath-hold dives in the sea. The protocol was as follows: after an initial period of 5 minutes of hyperventilation the subject submerged and the pressure in the air above the pool was progressively increased (lower record) until the desired simulated depth had been reached. It was maintained at this pressure for 15 seconds before the subject began the simulated ascent, during which the ECG and blood pressure were monitored followed by the recovery phase. (Based on Fig 5 in M. Ferrigno et al. (1997) *Journal of Applied Physiology* **83**, 1282–90. Reprinted with permission of The American Physiological Society.)

where the desire to breathe becomes overwhelming. This practice is potentially dangerous. If the arterial PCO_2 does not rise sufficiently to stimulate breathing before cerebral hypoxia occurs, consciousness may be lost, possibly with fatal consequences. Remember that during the diving response the respiratory drive from the peripheral chemoreceptors (which sense PaO_2) is powerfully inhibited.

Summary

During breath-hold diving, there is a reflex slowing of the heart, peripheral vasoconstriction, and apnoea. The duration of a breath-hold dive is limited by the break-point, which is determined largely by the arterial PCO_2.

Checklist of key terms and concepts

Hypoxia and high altitude

- Acute hypoxia refers to short-term (hours to a few days) exposure to a low partial pressure of oxygen.
- Chronic hypoxia refers to longer-term exposure (days to weeks) to low partial pressure.
- Ascent to high altitude results in a low inspired partial pressure of oxygen.
- Above 3000 m, the fall in inspired PO_2 leads to an increase in ventilation and an increase in cardiac output.
- The increased ventilation results in a decrease in alveolar PCO_2 and a respiratory alkalosis.
- These physiological changes may be accompanied by the symptoms of mountain sickness.
- At high altitude the ability to exercise is compromised and any exertion is accompanied by dyspnoea.

Acclimatization

- Acclimatization to high altitude refers to the physiological changes that occur following a prolonged stay at high altitude. Except at very high altitude, these changes result in increased tolerance of the low inspired partial pressure of oxygen.
- Following acclimatization to chronic hypoxia, there is a progressive increase in red cell number and blood haemoglobin. These changes result in an increased oxygen-carrying capacity of the blood.
- At high altitude there is a maintained increase in ventilation that initially causes a respiratory alkalosis.
- This respiratory alkalosis is progressively compensated by renal excretion of bicarbonate.

- Cardiac output is increased on first arriving at high altitude but later it slowly returns to near-normal values.
- Prolonged sojourn at high altitude results in an increased vascularization of the tissues.

Physiological consequences of exposure to high ambient pressure

- At elevated ambient pressures, the quantity of dissolved gas is directly related to the pressure so that when divers or tunnelling engineers work in an atmosphere of compressed air, the quantity of nitrogen dissolved in the tissues increases.
- Rapid decompression can lead to bubble formation in the tissues, which causes tissue damage. This can be avoided by progressive decompression according to specified tables.
- A second problem of breathing air at elevated pressure is nitrogen narcosis.
- Breathing pure oxygen at pressures greater than 1 atmosphere leads to signs of oxygen toxicity so that it cannot be used as a safe alternative in diving.
- The problems associated with breathing air at high pressure are largely overcome by breathing mixtures of helium and oxygen. Even so, the partial pressure of oxygen must be kept below 1 atmosphere to avoid the problems of oxygen toxicity. (It is usually kept at 0.5 atmospheres or 50 kPa.)

Breath-hold diving

- During breath-hold diving, there is a reflex slowing of the heart, peripheral vasoconstriction, and apnoea.
- The duration of a breath-hold dive is limited by the break-point, which is determined largely by the arterial PCO_2.

Recommended reading

Hlastala, M.P., and Berger, A.J. (2001) *Physiology of respiration* (2nd edn), Chapters 12 and 13. Oxford University Press, New York.
  A useful introductory text with broad coverage.

Levitzky, M.G. (2013) *Pulmonary physiology* (8th edn), Chapter 11. McGraw-Hill, New York.
  An introductory account of the respiratory system under stress.

Lumb, A.B. (2016) *Nunn's applied respiratory physiology* (8th edn), Chapters 15–18. Butterworth-Heinemann, Oxford.
  An excellent introduction to applied respiratory physiology seen from the point of view of an anaesthetist.

Review articles

Grocott, M., Montgomery, H., and Vercueil, A. (2007) High-altitude physiology and pathophysiology: Implications and relevance for intensive care medicine. *Critical Care* **11**, 203, doi:10.1186/cc5142.

Grocott, M., et al. (2009) Arterial blood gases and oxygen content in climbers on Mount Everest. *New England Journal of Medicine* **360**, 140–49.

Leaf, D.E., and Goldfarb, D.S. (2007) Mechanisms of action of acetazolamide in the prophylaxis and treatment of acute mountain sickness. *Journal of Applied Physiology* **102**, 1313–22.

Lindholm, P., and Lundgren, C.E. (2009) The physiology and pathophysiology of human breath-hold diving. *Journal of Applied Physiology* **106**, 284–92.

Panneton, W.M. (2013) The mammalian diving response: An enigmatic response to preserve life? *Physiology* **28**, 284–97.

Westerterp, K.R. (2001) Energy and water balance at high altitude. *News in Physiological Sciences* **16**, 134–7.

 To check that you have mastered the key concepts presented in this chapter, complete the accompanying online self-assessment questions. Go to www.oup.com/uk/pocock5e/

CHAPTER 38

The physiology of exercise

This chapter should help you to understand:

- The classification of the intensity of physical work
- The sources of energy during exercise
- The adjustments to the circulation that accompany exercise
- The relationship between muscle work and ventilation
- The effects of training on performance

38.1 Introduction

The previous chapters have described the physiology of the endocrine, circulatory, and respiratory systems separately. This chapter is concerned with their interactions in the performance of everyday physical activity as the whole body responds to the stress of exercise. It will be concerned with four main issues.

1. How much energy is expended for different intensities of exercise?

2. What is the source of that energy?

3. How are the cardiovascular and respiratory systems adjusted to meet the demands of exercise?

4. How far can performance be improved by training?

38.2 Work and exercise

The intensity of exercise obviously varies from the very mild, a gentle stroll for example, to very severe, such as that encountered in sprinting and other athletic pursuits. It is a matter of everyday experience that the ability to sustain physical activity depends on its intensity. At rest, the skeletal muscles have relatively low metabolic needs. In adults, although the skeletal muscles account for about 40 per cent of total body weight, they utilize only about 20–30 per cent of the oxygen taken up by the body. During exercise, the muscles perform work and their metabolic requirements increase. The oxygen consumption of the skeletal muscle mass may rise from about 75 ml min^{-1} to as much as 3000 ml min^{-1} in severe exercise—a 40-fold increase. In addition, glucose and fats are mobilized from body stores for oxidation to yield the ATP required for muscle contraction. To meet these metabolic needs, there are major adjustments of the cardiovascular, respiratory, and endocrine systems.

The principal unit of measurement in exercise physiology is the **joule**, the SI unit of energy or work. One **joule** is the work done by a force of one newton acting to move an object through a distance of one metre. In older literature, energy is often expressed in calories or kilocalories (which are generally used to indicate the energy value of foods). These units are now obsolete and the energy used in specific tasks is given in joules, where one calorie is equal to 4.186 joules. **Power** is the ability to perform work, which is expressed as work per unit time (joules s^{-1} or **watts**—see Chapter 8 Box 8.1).

Since the energy for muscular work is supplied by the oxidation of foodstuffs, the amount of work carried out in the performance of different tasks can be assessed by measuring the increase in oxygen uptake over that for the resting body. The greater the amount of work performed, the greater the oxygen uptake. The metabolic rate for any period of time can be obtained by multiplying the oxygen consumption (in litres min^{-1}) by the **energy equivalent of oxygen** for the food being metabolized. This ranges from around 18.9 kJ per litre of oxygen for protein to 20.93 kJ per litre of oxygen for pure carbohydrate—see Chapter 47 p. 756.

Categories of work and exercise

To be able to compare the relative intensity of exercise, in people of different ages, sex, and size, the concept of the **metabolic** equivalent of task (MET) has been introduced. One metabolic equivalent (MET) is defined as the volume of oxygen consumed while sitting at rest and, by convention, was originally set at 3.5 ml O_2 per kg body weight per minute (equivalent to 245 ml O_2 for a 70-kg adult). To estimate the energy cost of any activity it is then possible to determine the resting metabolic rate and multiply that value by the appropriate MET value listed in tables compiled for the purpose. For example, a person walking at 4 km hr^{-1} is expending nearly three times as much energy as they would at rest (the MET value is actually 2.9). If their resting metabolic rate were to be 80 watts, they would be expending approximately 230 watts (~1.4 kcal min^{-1}).

Light and moderate work is that which requires an average oxygen uptake of less than four times the resting oxygen uptake (i.e. up to 4 MET). This corresponds to an oxygen consumption between about 300 ml min^{-1} for the lightest of tasks to 1 l min^{-1} during moderate work. This corresponds to an energy expenditure of 6–20 kJ min^{-1} (~100–300 W) and includes most everyday tasks such as dressing, washing, walking etc. Clearly, this type of work can be carried out for many hours without fatigue.

Heavy work requires an oxygen consumption of 1–2 l min^{-1} which is 4 to 8 times the MET and is equivalent to an energy consumption of 20–40 kJ min^{-1} (300–600 W). This category would include most of the labour-intensive jobs in heavy industry such as building and mining. While these levels of energy expenditure can be sustained by very fit individuals for an average eight-hour day, those who are not physically fit cannot sustain such levels of activity without periods of rest.

Severe work is defined as work rates requiring an oxygen consumption in excess of 2 l min^{-1} (approximately 8 MET). Such levels of work can only be sustained by very fit individuals. Very high levels of oxygen consumption can be attained for short periods (as in competitive athletics). For example, an oxygen consumption in excess of 5 l min^{-1} can be maintained for many minutes by elite cyclists. This is equivalent to a total energy consumption greater than 1000 W (16–20 MET).

In the course of daily life, exercise takes two forms, dynamic and static. In **dynamic exercise** such as walking, there is a rhythmical movement of the limbs with flexing of the joints and alternating periods of contraction and relaxation of the skeletal muscles. In **static exercise** such as lifting, specific muscles are maintained in an isometric contraction for a period of time and the muscles perform no external work. Nevertheless, the body responds to both kinds of exercise by adjustments to the cardiovascular and respiratory systems.

Summary

The SI unit of energy or work is the joule. The watt is a measure of work performed per unit time (joules s^{-1}). The work carried out in different tasks can be assessed by measuring the oxygen uptake. The greater the amount of work performed, the greater the oxygen uptake. Work can be classified as light to moderate (oxygen consumption of 0.3–1 l min^{-1}), heavy (oxygen consumption of 1–2 l min^{-1}), or severe (> 2 l O_2 min^{-1}). The exercise involved in any work may be either dynamic or static in nature.

38.3 Metabolism in exercise

Sources of energy in exercise

The energy for muscle contraction is provided by ATP, which is rapidly replenished by the transfer of a phosphate group from creatine phosphate (see Chapter 8). ATP can be produced by oxidative metabolism via the **Krebs cycle** and the electron transport chain, or by the anaerobic breakdown of glucose via glycolysis (see Chapter 4). Anaerobic metabolism, however, is much less efficient at generating ATP than oxidative metabolism and results in the production of large quantities of lactic acid.

The proportion of aerobic and anaerobic metabolism varies with the severity and duration of the exercise. The relative contributions of aerobic and anaerobic metabolism during maximal exercise lasting up to an hour are shown in Figure 38.1. Note the increasing proportion of energy contributed by aerobic metabolism with longer periods of exercise. In sustainable exercise, almost all the energy required is derived from aerobic metabolism, but in short periods of intense exercise such as sprinting, anaerobic metabolism (including the breakdown of ATP and creatine phosphate) may account for more than half the total energy expended. The ATP and creatine phosphate utilized during the anaerobic phase are subsequently replenished by aerobic metabolism.

Carbohydrates provide most of the energy for muscular exercise

The energy requirements of muscular exercise are met mainly by the oxidation of carbohydrates, with a smaller contribution from the oxidation of fats. In people who are well nourished, proteins are not used as a significant source of fuel in exercise.

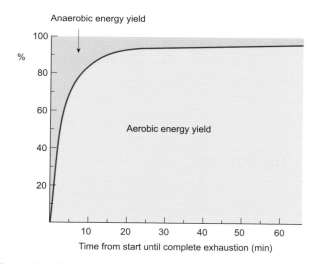

Figure 38.1 The relative contribution of anaerobic and aerobic metabolism to energy consumption during maximal efforts of various durations. Note that during short bursts of high-intensity exercise, the anaerobic energy yield is very high compared to its contribution during prolonged periods of exercise. (Redrawn from Fig 7.10 in P.-O. Astrand and K. Rohdal (1986) *Textbook of work physiology. Physiological basis of exercise*, 3rd edition. McGraw-Hill, New York.)

During exercise, plasma glucose falls very little unless the exercise is both severe and prolonged. Even after three hours of continuous exercise at half the maximal rate, the plasma glucose level falls by less than 10 per cent. The glucose utilized during exercise is derived from the glycogen stored in the skeletal muscles and the liver. The breakdown of glycogen into glucose (glycogenolysis) is stimulated by a rise in circulating adrenaline in both the liver and muscle (see Figure 6.6). In prolonged exercise, the glycogen reserves become depleted and glucose is generated in the liver from non-carbohydrate precursors (gluconeogenesis—see Chapter 4). This process is stimulated by the increased circulating levels of cortisol, adrenaline, and growth hormone that occur during exercise. Additionally, these hormones stimulate the breakdown of triglycerides to mobilize free fatty acids for oxidation (lipolysis).

The proportion of carbohydrates and fats used can be assessed by measuring the respiratory quotient (RQ) (see Chapters 32 and 47). Its value depends on the type of exercise (whether it is continuous or intermittent) and on the diet, the physical fitness, and the state of health of the individual. Nevertheless, with increasing severity of exercise, the oxidation of carbohydrates provides an increasing share of the energy needs. In endurance athletics, training can increase the utilization of fats during prolonged exercise.

Oxygen consumption rises in proportion to the work done

As soon as exercise begins, the muscles begin to expend energy in proportion to the work done. Oxygen consumption, however, does not rise immediately to match the energy requirements. Instead, it rises progressively over several minutes until it matches the needs of the exercising muscles. As the work continues, the oxygen uptake remains at a level appropriate to the severity of the exercise. Thus, at the commencement of exercise the body builds up an oxygen deficit ('oxygen debt'—see Figure 38.2a).

In the steady state, the oxygen consumption is proportional to the work done until the work rate approaches the maximum

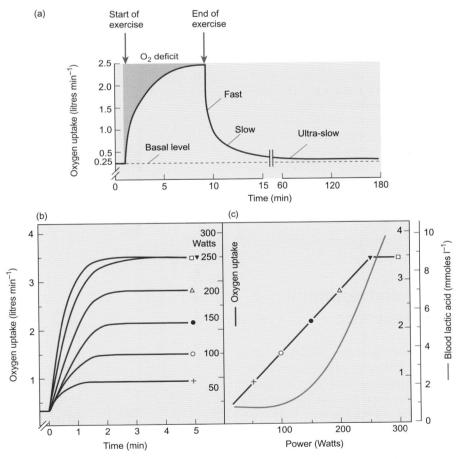

Figure 38.2 Oxygen consumption during and after a period of exercise. (a) The increase in oxygen uptake that follows the start of a period of exercise and the subsequent stages of recovery. (b) and (c) The relationship between oxygen uptake and energy expenditure. The highest levels of exercise are associated with large increases in blood lactate concentrations. (Redrawn from Fig 7.3 in P.-O. Astrand and K. Rohdal (1986) *Textbook of work physiology. Physiological basis of exercise*, 3rd edition. McGraw-Hill, New York.)

capacity. In the case of the example shown in Figure 38.2c, the work done was measured using a cycle ergometer and therefore represents the external work done, not the total energy expenditure. The maximal oxygen uptake ($\dot{V}O_2$ max) was 3.5 l min^{-1}, which provided the **maximal aerobic power** for this individual. The energy required for the further increase in work done (from 250 to 300 watts) was not accompanied by an additional oxygen uptake but was derived from anaerobic metabolism, as indicated by the steep rise in blood lactate. Blood lactate levels began to rise above 1–2 mmol l^{-1} (the normal plasma level) when the oxygen consumption rose above about 2 l min^{-1}. This is known as the **anaerobic threshold**.

At the end of a period of exercise, the oxygen consumption declines rapidly but does not reach normal resting levels for up to 60 minutes. The first phase of the decline in oxygen consumption is very fast, with a half-time of about 30 seconds. During this period, ATP and creatine phosphate are resynthesized via oxidative pathways. This is followed by a slower decline in oxygen consumption that has a half-time of about 15 minutes. Excess lactate is resynthesized into glucose and glycogen during this period. After severe and sustained exercise, oxygen consumption remains elevated for several hours, perhaps due to stimulation of metabolism as a result of the heat generated during the period of exercise.

Summary

During short periods of intense exercise, the ATP levels in the exercising muscles are maintained by the transfer of a phosphate group from creatine phosphate. In addition, glycogen is broken down to glucose, which can be metabolized either anaerobically or aerobically to generate ATP. Aerobic metabolism is much more efficient in generating ATP than anaerobic metabolism. At the onset of a period of aerobic exercise, oxygen consumption rises exponentially to its steady-state value, which is in direct proportion to the work rate. At the end of a bout of exercise, oxygen consumption falls quickly but may not reach resting values for an hour or more after the bout of exercise ended.

38.4 Cardiovascular and respiratory adjustments during exercise

When exercise is undertaken by a healthy person, the circulatory and respiratory systems are adjusted to meet the increased metabolic demands. Cardiac output rises as a result of an increase in both heart rate and stroke volume. There is a redistribution of the cardiac output, with a higher proportion going to the active muscles (see Figure 38.3), and oxygen uptake increases in proportion to the amount of work done. The high levels of oxygen utilization in heavy and severe exercise are made possible both by an increase in ventilation and by a greater extraction of oxygen from the circulating blood. In this section, the detailed mechanisms by which these changes are accomplished are considered.

Cardiovascular changes in exercise

At rest, cardiac output is about 5 l min^{-1}. Of this, only 15–20 per cent is distributed to the skeletal muscles (i.e. 0.75–1 l min^{-1}). In heavy exercise, cardiac output may rise to 25 l min^{-1} or more, of which approximately 80 per cent is distributed to the exercising muscles (more than 20 l min^{-1}). In contrast, the blood flow to the brain remains essentially constant, while that to the splanchnic and renal circulations declines. The blood flow in the splanchnic bed falls from 1–1.2 l min^{-1} at rest to about 0.75 l min^{-1} in exercise, while renal blood flow declines from about 1 l min^{-1} to less than half this amount (see Figure 38.3).

Effects on heart rate and stroke volume

At rest, the heart rate is kept low by the activity in the vagus nerves, and most blood vessels are partially constricted by activity in the sympathetic nerves (see Chapter 30). As exercise begins, vagal activity declines and sympathetic activity increases. This results in an increase in heart rate and mobilization of blood from the large veins. The increased sympathetic activity has a positive inotropic effect that leads to an increase in stroke volume. In addition to the inotropic response, stroke volume increases because the augmented venous return increases left ventricular end-diastolic pressure and stroke volume, by Starling's Law. The positive inotropic response allows an increase in stroke work at the same filling

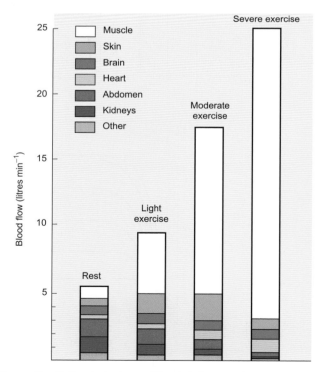

Figure 38.3 Estimated values of the distribution of cardiac output to various tissues at rest and during light, moderate, and severe exercise in humans.

pressure (see Figure 28.17). As a result, the end-diastolic volume of the heart does not increase with exercise. The increased stroke volume is achieved by a more complete emptying of the ventricles so that end-systolic volume falls. The increases in heart rate and stroke volume combine to increase cardiac output. At moderate levels of exercise (up to about 40 per cent of maximal oxygen uptake), both heart rate and stroke volume increase in proportion to the work done. Above this level, the increase in stroke volume begins to level off and further increases in cardiac output are mainly due to increases in heart rate (see panels (c) and (d) of Figure 38.4).

Effects on regional blood flow

The increase in sympathetic activity that both precedes and accompanies a period of exercise results in vasoconstriction in most vascular beds, partly as a result of the direct effects of the sympathetic nerves and partly as a result of the rise in circulating catecholamines secreted by the adrenal medulla.

At rest, the calibre of the arterioles of skeletal muscle is mainly regulated by the activity of their sympathetic nerves and by myogenic activity. Consequently, there is a high level of resting smooth muscle tone so that the arterioles are held in a state of partial constriction. Consequently, blood flow is low. During exercise, there is a marked vasodilatation of the arterioles of the active muscle. This is mainly due to the increased production of metabolites and represents an example of functional hyperaemia (see Chapter 30, Section 30.3). In addition, there is vasodilatation in response to circulating adrenaline secreted by the adrenal medulla. Overall, the arteriolar dilatation results in greatly increased perfusion of the capillary beds within the active muscle.

Exercising muscle has a problem similar to that of the myocardium with respect to its perfusion: it is not possible to maintain a steady blood flow. A strongly contracting calf muscle, for example, squeezes its blood vessels with each contraction, thereby reducing the amount of blood that can flow to it. In continuous rhythmical exercise such as walking, there will therefore be regular surges in blood flow. During the phase of muscular contraction, the blood within the muscles is pumped towards the heart to augment the venous return. This forward propulsion of blood is aided by the valves of the limb veins. The periodic changes in blood flow in an exercising muscle during such exercise are shown schematically in Figure 38.5.

Blood pressure during exercise

During exercise, the increased force of ventricular contraction causes an increase in systolic pressure, which becomes more

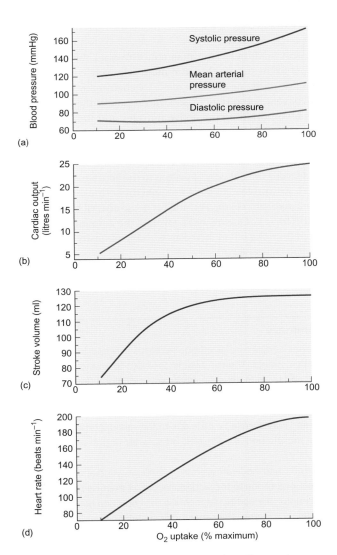

(a) (b) (c) (d)

Figure 38.4 Typical changes in the principal cardiovascular variables with increasing oxygen uptake (taken as a measure of the intensity of exercise).

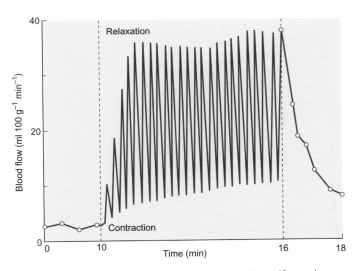

Figure 38.5 The changes in blood flow through the calf muscle during strong rhythmic contractions. Note the large increase in blood flow (functional hyperaemia) during the period of exercise and its rapid decline during the recovery phase. During the periods of contraction, the blood within the muscles is pumped towards the heart and inflow of blood to the muscles is greatly reduced. (Based on H. Barcroft and H.J.C. Swann (1953) *Sympathetic control of human blood vessels.* Edward Arnold, London.)

marked as exercise intensifies. In dynamic exercise, the diastolic pressure remains relatively stable and may even decline as the peripheral resistance falls due to the dilatation of the arterioles in the skeletal muscles. Consequently, under these conditions, the rise in mean arterial pressure is modest and there may even be a slight decline. In static exercise, however, the contraction of the muscles compresses the blood vessels and reduces blood flow. There is a marked pressor response, with an abrupt increase in heart rate. Peripheral resistance, diastolic pressure, and mean arterial pressure all rise. When large groups of muscles are engaged, as in weight-lifting, the systolic pressure can rise briefly to 40 kPa (300 mm Hg) and the diastolic pressure may reach 20 kPa (150 mm Hg).

During exercise, large molecules (notably glycogen and creatine phosphate) break down into smaller ones so that the osmotic pressure within the exercising muscle increases. In prolonged exercise, this results in a movement of fluid from the plasma into the exercising muscle cells and interstitial space. Consequently, during heavy exercise the haematocrit and the oxygen-carrying capacity of the blood are increased. The viscosity of the blood is increased at the same time. This phenomenon is known as **haemoconcentration**. Although this increases the oxygen capacity of the blood, the limited time for gas exchange in the lungs restricts the rise in the oxygen content of the arterial blood to a few per cent.

Ventilation increases in proportion to the work done

At rest, pulmonary ventilation is about 6–8 l min^{-1}, but in heavy exercise, it may increase to 100 l min^{-1} or more. When someone starts to exercise, their pulmonary ventilation increases as soon as the

exercise begins, and it increases progressively until a new steady state is reached that is appropriate to the work being done. At the end of exercise, ventilation rapidly falls, although it may not reach normal resting values for up to an hour if the period of exercise has been intense.

At moderate work rates, the steady-state ventilation is directly proportional to the work done as measured by the oxygen consumption. During very severe exercise, however, the increase in ventilation is disproportionately large in relation to the oxygen uptake. The maximum oxygen uptake ($\dot{V}O_2$ max) may be one limiting factor in the capacity for exercise (see Figure 38.6).

Blood gases in exercise

At rest, the oxygen content of the arterial blood is 19.8 ml dl^{-1} and the haemoglobin is about 97 per cent saturated. The mixed venous blood is approximately 75 per cent saturated and has an oxygen content of 15.2 ml dl^{-1}, showing that approximately 4.6 ml of oxygen is extracted from each dl of blood as it passes through the tissues.

At work loads below the anaerobic threshold, the PaO_2, $PaCO_2$, and pH of the arterial blood remain relatively constant, which raises the intriguing question of the mechanism by which the pulmonary ventilation is increased. Nevertheless, the PO_2 of the venous blood draining the active muscles and that of the mixed venous blood declines progressively as the intensity of the exercise increases. At the same time, the mixed venous PCO_2 rises from its normal value of 46 mm Hg. The rise in PCO_2 and the associated fall in pH favour the delivery of oxygen to the active tissues (the Bohr effect). At workloads above the anaerobic threshold, there is a gradual reduction in the PO_2 and pH of the arterial blood (see panels (a), (b), (c) of Figure 38.7). Overall, the amount of oxygen extracted from the blood increases with the intensity of exercise, as shown in Figure 38.7(c).

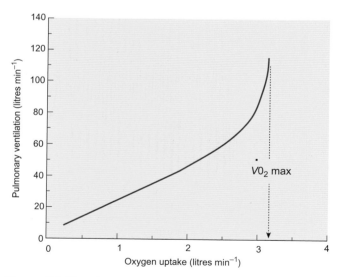

Figure 38.6 The relationship between pulmonary ventilation and oxygen uptake during exercise. Note the steep rise in ventilation as oxygen uptake approaches its maximum. (Drawn from data in Fig 5.9 in P.-O. Astrand and K. Rohdal (1986) *Textbook of work physiology. Physiological basis of exercise*, 3rd edition. McGraw-Hill, New York.)

Summary

In exercise, the cardiac output increases in proportion to the metabolic demand. The increase is due both to an increase in heart rate and to an increase in stroke volume. Blood flow is redistributed from the splanchnic circulation to the exercising muscles. Although systolic blood pressure rises, diastolic pressure is stable and may even fall a little due to the vasodilatation of the skeletal muscle bed. As a result, mean arterial pressure rises only slightly.

In mild and moderate exercise, pulmonary ventilation increases in direct proportion to the work done, but as exercise becomes more strenuous, the additional increase in ventilation becomes disproportionately large. The capacity for sustained exercise is probably limited by the maximum oxygen uptake ($\dot{V}O_2$ max). The oxygen requirement of the exercising muscles is met by an increase in cardiac output, increased blood flow, and increased extraction of oxygen from the circulating blood.

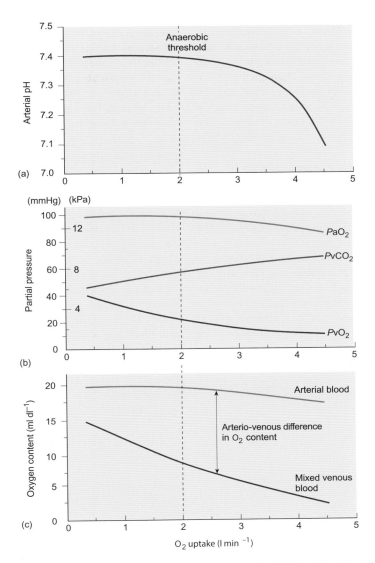

Figure 38.7 The changes in blood gases and arterial pH during exercise. (a) Arterial pH falls as the intensity of exercise increases. (b) As the oxygen consumption rises, the partial pressure of oxygen in the mixed venous (PvO_2) blood falls while the $PvCO_2$ rises. At high levels of exercise the partial pressure of oxygen in the arterial blood declines slightly. (c) The oxygen content of arterial blood falls slightly, but the arterio-venous difference in oxygen content increases markedly.

38.5 Matching cardiac output and ventilation to the demands of exercise

During exercise, both cardiac output and pulmonary ventilation are adjusted precisely to meet the metabolic demands until fatigue sets in. How is this remarkable feat achieved? This is the central problem of exercise physiology, which is still not completely resolved. Two principal factors seem to be involved: commands from the brain—the **central command**—and reflexes elicited in response to the exercise itself. In considering how these different processes interact it is convenient to divide a period of exercise into three phases.

- **Phase 1**, in which exercise starts, ventilation increases, and the partial pressures of oxygen and carbon dioxide in the mixed venous blood begin to change.

- **Phase 2**, during which ventilation, cardiac output and the partial pressures of the respiratory gases in the mixed venous blood approach their steady-state values.

- **Phase 3**, during which the steady-state levels of ventilation, cardiac output, PaO_2, $PaCO_2$, and arterial pH are maintained. This phase is not reached until the exercise approaches its maximal sustainable capacity. In severe exercise, pH continues to fall as lactate accumulates.

Regulation of the circulation

During phase 1, or even before exercise begins, the heart rate increases and there is an increased force of contraction. These changes are due to an inhibition of parasympathetic activity and an increase in sympathetic activity (including the secretion of adrenaline from the adrenal glands). The increased sympathetic activity causes vasoconstriction in most vascular beds. In some animals, there is a vasodilatation in the skeletal muscles due to activity in sympathetic cholinergic vasodilator fibres, although this probably does not occur in humans. These changes are believed to be due to signals from the higher levels of the brain to those regions of the brainstem concerned with regulation of the cardiovascular system. This central command also acts to reduce the sensitivity of the baroreceptor reflex.

During phase 2 and phase 3, the central command is reinforced by reflexes, which are triggered by increased activity in the afferent nerves in the exercising limbs. In animal experiments, passive movements of the hind limbs elicit an increase in heart rate and blood pressure (see Figures 35.4 and 38.8). This increase can be blocked by cutting the nerves from the joints, so it appears that afferent activity arising in the joints contributes to the maintenance of the cardiovascular response to exercise. In addition, **metaboreceptors** (a type of chemoreceptor) in the exercising muscles respond to the fall in extracellular pH and rise in extracellular potassium and reinforce the cardiovascular response. These reflexes are probably responsible for the matching of cardiac output to the metabolic requirements of the exercise.

The metabolites released by the active muscles and the increased circulating levels of adrenaline cause a vasodilatation in the arterioles that augments local blood flow. As exercise continues, body temperature begins to rise and this is sensed by the hypothalamic

thermoreceptors. The change in the activity of these receptors elicits a reflex vasodilatation in the skin vessels to aid in the dissipation of the heat generated by the active muscles (see Figure 38.3). In very severe exercise, the blood flow to the skin declines, as shown in the right-hand column of Figure 38.3. The resulting increase in body temperature is presumably a factor limiting the duration of the period of severe exercise.

Regulation of pulmonary ventilation during exercise

The increase in ventilation during exercise is believed to be due to both neural and humoral factors. The neural mechanisms activate the respiratory muscles, but the fine tuning of ventilation to match the oxygen utilization appears to be accomplished by various chemical agents.

Ventilation increases as soon as exercise begins. This can only be explained by a central command that is associated with the initiation of motor activity from the premotor area of the cerebral cortex. During phase 2, pulmonary ventilation rises in proportion to the intensity of the exercise. The factors responsible for this increase in ventilation have been difficult to establish. As the arterial blood gases and blood pH change very little during this phase, it is clear that changes in the partial pressures of the respiratory gases cannot easily explain the close matching between oxygen uptake and ventilation. However, the values of the arterial PO_2 and PCO_2

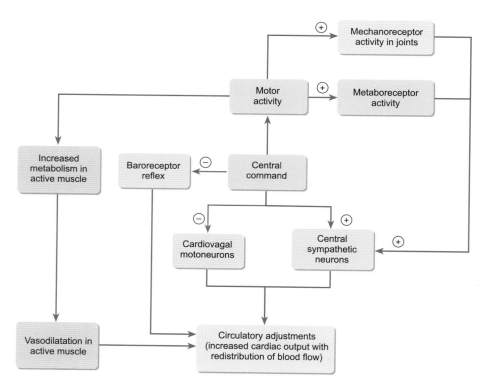

Figure 38.8 The factors that regulate the cardiovascular response to exercise: + indicates an increase in activity or an activation; − indicates a decrease in sensitivity or an inhibition.

are an average over a short period of time and take no account of the fluctuations in P_aO_2 that occur from breath to breath. Similar rapid fluctuations in arterial pH have also been observed. The discharge pattern of the carotid sinus nerve shows similar fluctuations during the respiratory cycle, so it is possible that some of the increase in ventilation during phase 2 is driven by signals from the peripheral chemoreceptors. In support of this conjecture, it is known that people with denervated carotid bodies have a slower ventilatory response to exercise than normal healthy subjects. In addition, there is an increased neural input arising from afferent activity in the joints, which is known to stimulate ventilation (see Figure 35.4) and may help to explain the delayed ventilatory response.

What other chemical signals might be involved? As metabolites build up in the exercising muscle, they will be transported to the peripheral chemoreceptors via the arterial blood. Moreover, it is known that during heavy exercise the potassium concentration of the blood is elevated. Indeed, in humans, the increase in ventilation during exercise is correlated with the rise in plasma potassium concentration. In anaesthetized animals, similar concentrations of potassium elicit a powerful ventilatory response from the carotid bodies. It is therefore possible that the plasma potassium provides an additional stimulus to the peripheral chemoreceptors.

In phase 3 of exercise, ventilation is in a steady state and matched to the metabolic requirements. As indicated in Figure 38.9, both chemical and neural stimuli are likely to be involved in maintaining the respiratory effort. The rise in body temperature that accompanies a period of exercise may also contribute to the respiratory drive. The $PaCO_2$ appears to be more closely regulated

than the PaO_2, and CO_2 elimination is closely matched to CO_2 production by the exercising muscles. How this is achieved is not known, but there is evidence that the relationship between ventilation and $PaCO_2$ is reset so that there is a higher ventilation for a given $PaCO_2$. The involvement of neural mechanisms in phase 2 and phase 3 of exercise is indicated by the matching of the respiratory rhythm to that of the exercise. The afferent barrage from the muscle spindles and the mechanoreceptors in the muscles and joints also contribute to the activation of the respiratory motoneurons.

Summary

During exercise, both cardiac output and pulmonary ventilation are adjusted precisely to meet the metabolic demands. The cardiovascular response to exercise is initiated by signals from the higher levels of the brain, which inhibit parasympathetic activity and increase sympathetic activity. As a result, heart rate and stroke volume increase and blood flow is preferentially distributed to the exercising muscles.

Afferent signals arising from the active joints and muscle metaboreceptors activate cardiovascular reflexes that act to maintain the cardiovascular and ventilatory responses at levels appropriate to the intensity of the exercise. The associated rise in body temperature initiates a reflex vasodilatation of the skin vessels that promotes heat loss, which is overridden in bouts of very severe exercise.

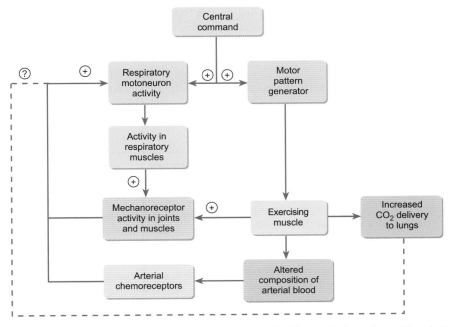

Figure 38.9 A flow chart summarizing the neural and chemical factors involved in the regulation of breathing during exercise: + indicates an increase in activity or an activation; ? indicates a possible, but unconfirmed, interaction.

38.6 Effects of training

The performance of exercise can be improved by training. This requires the regular undertaking of physical exercise that is of an appropriate intensity, duration, and frequency. To achieve optimal results, the intensity of training must increase as performance improves. Nevertheless, the load must be related to the fitness and strength of the individual. Frequent, regular exercise of an appropriate kind is important if the improvement in performance is to be maintained.

Regular training with strenuous exercise will lower the resting heart rate and increase the size of the heart and the thickness of the ventricular wall. As the end-diastolic volume and stroke volume increase, the resting cardiac output is maintained, despite the fall in resting heart rate. Maximal cardiac output can increase from $20–25 \, l \, min^{-1}$ in untrained young adults to values that may exceed $35 \, l \, min^{-1}$ in well-trained athletes. The cardiovascular changes increase the maximal oxygen uptake, and the capacity for physical work is thereby increased. In addition to its effects on the cardiovascular system, training improves the efficiency of gas exchange and athletes have a much greater ventilation for a given respiratory rate. For example, during heavy exercise a normal adult might achieve a minute volume of 50–55 litres at 40 breaths per minute while a trained athlete would have a minute volume of around 90 litres at the same respiratory rate.

Training also affects the bone, the connective tissues, and the muscle mass of the limbs involved in the type of exercise being undertaken. In response to the stress of exercise, the bone becomes remodelled so that the stressed areas have a greater degree of mineralization and therefore a higher strength. The importance of exercise in the maintenance of bone structure is shown by the effects of prolonged immobility, where the bone becomes progressively demineralized and more brittle. During training, the cartilage of the joints becomes thicker so that it is more compliant. Its contact area increases and, since pressure is force per unit area, the pressures generated within the joints are smaller for a given task.

In the early stages of training, the force generated by the muscles improves without hypertrophy of the muscle fibres themselves. This is believed to reflect an improvement in the recruitment of motor units during the performance of a specific type of exercise (i.e. the CNS learns to perform the task more efficiently). As training proceeds, both connective tissue and muscle mass hypertrophy. The growth in muscle mass does not reflect an increase in the number of muscle fibres, which remains the same. Rather, there is an increase in their diameter. The increased mass of the tendons and connective tissue of the muscles improves their ability to transmit the force generated by the muscle fibres to the skeleton. Further training leads to greater hypertrophy of the muscle fibres, and in dynamic exercise (e.g. running), there is a progressive increase in the capillary density of the muscle with the duration of training. This improves the transfer of oxygen from the blood to the tissues.

Factors that limit the performance of exercise

Different individuals have different capacities for physical work: for any person, their capacity for work depends on their size, age, sex, genetic makeup, and general state of health. In addition, the performance of strenuous exercise is greatly influenced by the level of motivation.

The ability to undertake muscular exercise ultimately depends on the ability of the muscles to generate sufficient ATP to sustain their contractile activity. In part, the metabolic needs of exercising muscles are met by their glycogen reserves and, for slow twitch muscle, by the small reserve of oxygen stored in myoglobin. Ultimately, however, the energy requirements of the muscles are dependent on the delivery of glucose and oxygen by the circulation. At low work rates, the energy requirements are met chiefly by aerobic metabolism, and exercise can be maintained for considerable periods of time. At high work rates, anaerobic metabolism contributes an increasing fraction of the energy requirements and fatigue sets in relatively quickly (see Figure 38.1). The fatigue associated with intense exercise is, at least in part, due to the accumulation of lactate.

Exercise reduces cardiovascular disease

Studies of a number of different groups of people have shown that a state of maintained physical fitness reduces the likelihood of coronary heart disease. In an extensive study of office workers, it was found that those who undertook some regular and vigorous exercise had an incidence of coronary heart disease that was less than half that of those who did not. The fit group also had a lower death rate from coronary heart disease. The benefits of exercise were found to extend to all groups, including the obese and smokers.

A recent study of sedentary individuals undergoing a regular modest exercise regime lasting three months found that their aerobic capacity increased and their systolic and diastolic blood pressures decreased. The regular exercise resulted in a lower resting heart rate and a noticeable exercise-induced increase in positive mood. These beneficial effects were achieved with only a modest reduction in body weight.

Summary

Physical performance is improved by regular training, which lowers the resting heart rate and increases the size and wall-thickness of the heart. Despite the fall in resting heart rate, the resting cardiac output is maintained; both end-diastolic volume and stroke volume are increased. Training also affects the bone, the connective tissues, and the muscle mass. Training for dynamic exercise leads to a progressive increase in the capillary density of the muscles. Maintained physical fitness reduces the likelihood of coronary heart disease.

(✱) Checklist of key terms and concepts

Work and exercise

- The unit of physical work is the joule, while power is the work done per unit time in joules s^{-1} or watts.
- Work rates are classified as light, moderate, heavy, and severe according to their metabolic demands.
- Exercise may be dynamic or static.
- In dynamic exercise there is a rhythmical movement of the limbs with flexing of the joints and alternating periods of contraction and relaxation of the skeletal muscles.
- In static exercise, specific muscles are maintained in an isometric contraction for a period of time and the muscles perform no external work.
- The body responds to both dynamic and static exercise by adjustments to the cardiovascular and respiratory systems.

Metabolism in exercise

- The energy for muscle contraction is provided by ATP derived from glycolysis and oxidative metabolism.
- Aerobic metabolism is much more efficient in generating ATP than anaerobic metabolism.
- The proportion of aerobic and anaerobic metabolism varies with the severity and duration of the exercise.
- During short periods of intense exercise such as sprinting, the ATP levels in the exercising muscles are maintained by creatine phosphate and by glycolysis.
- At the onset of a period of aerobic exercise, oxygen consumption rises exponentially to its steady-state value.
- During maintained exercise, oxygen consumption is in direct proportion to the work rate.
- At the end of a bout of exercise, oxygen consumption falls quickly but may not reach resting values for over an hour.
- The main sources of energy for sustained muscular exercise are the glycogen stores of the muscles and liver, with a small contribution from the oxidation of fats.

Cardiovascular and respiratory adjustments during exercise

- In exercise the cardiac output increases in proportion to the metabolic demand.
- The increase is due both to an increase in heart rate and to an increase in stroke volume.
- Blood flow is redistributed from the splanchnic circulation to the exercising muscles.
- Systolic blood pressure rises but diastolic pressure is stable so that mean arterial pressure rises only slightly.
- In mild and moderate exercise, pulmonary ventilation increases in direct proportion to the work done.

- As exercise becomes more strenuous, the additional increase in ventilatory effort becomes disproportionately large and may be a factor in limiting the duration of the most severe exercise.
- At workloads below the anaerobic threshold, the PO_2 and PCO_2 of the arterial blood do not change significantly.
- The oxygen requirement of the exercising muscles is met by an increase in cardiac output, increased blood flow, and increased extraction of oxygen from the circulating blood.

Regulation of cardiac output and ventilation during exercise

- During exercise, both cardiac output and pulmonary ventilation are adjusted precisely to meet the metabolic demands.
- The cardiovascular response to exercise is initiated by signals from the higher levels of the brain, which take effect prior to the commencement of exercise.
- Parasympathetic activity is inhibited, while sympathetic activity increases.
- As a result, heart rate and stroke volume increase and blood flow is preferentially distributed to the exercising muscles.
- Afferent signals, arising from the active joints and the muscle metaboreceptors, activate cardiovascular reflexes that act to maintain the cardiovascular response at a level appropriate to the intensity of the exercise.
- The rise in body temperature that occurs in exercise initiates a reflex vasodilatation of the skin vessels that promotes heat loss.
- The ventilatory response to exercise is initiated by signals from the higher levels of the brain.
- During exercise, the neural activation is supported by signals arising in the muscle spindles and the mechanoreceptors in the muscles and joints.
- The arterial partial pressures of the respiratory gases and arterial pH hardly change except in severe exercise.

Effects of training

- Physical performance is improved by regular training.
- Regular training lowers the resting heart rate and increases the size of the heart.
- Despite the fall in resting heart rate, the resting cardiac output is maintained; both end-diastolic volume and stroke volume are increased.
- Training also affects the bone, the connective tissues, and the muscle mass.
- Training for dynamic exercise leads to a progressive increase in the capillary density of the muscles.
- Maintained physical fitness reduces the likelihood of coronary heart disease.

Recommended reading

Astrand, P.-O., Rodahl, K., Dahl, H.A., and Stromme, S. (2003) *Textbook of work physiology: Physiological bases of exercise* (4th edn). Human Kinetics, Champaign, IL.

An authoritative account of the physiology of exercise; an excellent introduction to the classic literature of the field.

Levick, J.R. (2010) *An introduction to cardiovascular physiology.* (5th edn), Chapter 17. Hodder Arnold, London.

This text is eminently readable and remains the best introduction for advanced studies of the cardiovascular system.

Review articles

Joyner, M.J., and Coyle, E.F. (2008) Endurance exercise performance: The physiology of champions. *Journal of Physiology* **586**, 35–44.

Joyner, M.J., and Casey, D.P. (2015) Regulation of increased blood flow (hyperemia) to muscles during exercise: A hierarchy of competing physiological needs. *Physiological Reviews* **95**, 549–601.

Paterson, D.J. (2014) Defining the neurocircuitry of exercise hyperpnoea. *Journal of Physiology* **592**, 433–44.

Powers, S.K., and Jackson, M.J. (2008) Exercise-induced oxidative stress: Cellular mechanisms and impact on muscle force production. *Physiological Reviews* **88**, 1243–76.

To check that you have mastered the key concepts presented in this chapter, complete the accompanying online self-assessment questions. Go to www.oup.com/uk/pocock5e/

PART NINE

The regulation of the internal environment

CHAPTER 39

The renal system

Chapter contents

This chapter should help you to understand:

- The anatomy of the kidneys and the renal circulation
- The structure of the nephron and the organization of its blood supply
- The concept of autoregulation and the regulation of renal blood flow
- The formation of the glomerular filtrate
- The concept of renal clearance
- The transport processes in the kidney: reabsorption and secretion by the renal tubules
- The role of the distal tubule in the regulation of the ionic balance of the body
- The establishment of the osmotic gradient in the renal medulla and its role in the regulation of plasma osmolality
- Bladder function

39.1 Introduction

Human beings, like all animals, feed on other organisms both to provide energy and to provide themselves with the resources required for growth and reproduction. This lifestyle inevitably leads to the intake of variable quantities of essential body constituents such as sodium, potassium, and water and to the production of metabolic waste products. Throughout these processes, the body needs to maintain a close control over the composition of the body fluids, and this is the principal role of the kidneys. They achieve this by regulating the composition of the blood, and, since the plasma equilibrates with the interstitial fluid of the tissues, the kidneys effectively regulate the composition of the extracellular fluid. In doing so, the kidneys excrete the excess water, salts, and metabolic waste products. The production of urine of variable composition is, therefore, a necessary part of the homeostatic role of the kidney.

About 1–1.5 litres of urine containing 50–70 g of solids, chiefly urea and sodium chloride, are normally produced by an adult each day. Urine volume and osmolality vary both with fluid intake and with fluid loss through sweating and via the faeces. Moreover, the chemical composition of the urine is very variable and changes with the diet. Although the urine contains traces of most of the plasma constituents, some substances such as protein, glucose, and amino acids are not normally detected. Other substances are much more concentrated in the urine than they are in plasma (e.g. creatinine, phosphate, and urea—see Table 39.1).

The urine is normally somewhat acid compared to the plasma or extracellular fluid. Its pH may range between 4.8 and 8.0, but for people eating a normal mixed diet, it usually lies between 5 and 6 (cf. the range for plasma, which is 7.35–7.4). Normal, fresh urine has a slight odour that can readily be masked by aromatic compounds from certain foodstuffs (e.g. asparagus and garlic), but stale urine has an unpleasant fetid odour due to the bacterial production of ammonia and other amines. The typical yellow colour of urine is principally due to the presence of pigments known as urochromes. In various disease states the colour and composition of the urine changes in characteristic ways, giving important diagnostic clues (see Box 39.1). For example, it is almost colourless in diabetes insipidus (see Section 21.8) but is strongly coloured during infections and has a brownish-red colour when haemoglobin is present.

Table 39.1 Comparison between the composition of the plasma and that of urine

	Plasma	Urine	Units
Na^+	140–150	50–130	$mmol\,l^{-1}$
K^+	3.5–5	20–70	$mmol\,l^{-1}$
Ca^{++}	1.35–1.50	10–24	$mmol\,l^{-1}$
HCO_3^-	22–28	0	$mmol\,l^{-1}$
Phosphate	0.8–1.25	25–60	$mmol\,l^{-1}$
Cl^-	100–110	50–130	$mmol\,l^{-1}$
Creatinine	0.06–0.12	6–20	$mmol\,l^{-1}$
Urea	4–7	200–400	$mmol\,l^{-1}$
NH_4^+	0.005–0.02	30–50	$mmol\,l^{-1}$
Protein	65–80	0	$g\,l^{-1}$
Uric acid	0.1–0.4	0.7–8.7	$mmol\,l^{-1}$
Glucose	3.9–5.2	0	$mmol\,l^{-1}$
pH	7.35–7.4	4.8–7.5	$(-\log_{10}[H^+])$
Osmolality	281–297	50–1300	$mOsmol\,kg^{-1}$

Note: The data given in the table are the values that would encompass 95 per cent of a population of normal healthy adults. Note that while the concentration of the principal constituents of the plasma is maintained relatively constant, the composition of the urine is subject to considerable variation. Moreover, some important constituents of plasma such as protein, glucose, and bicarbonate are virtually absent from normal urine. Others, such as creatinine, NH_4^+, phosphate, and urea, are present in far higher concentrations in the urine.

In addition to their primary regulatory and excretory roles, the kidneys produce a hormone, **erythropoietin**, that regulates the production of red blood cells (see Section 25.4), and an enzyme, **renin**, which is important in the regulation of sodium balance via aldosterone secretion (see Section 40.3). They synthesize **1,25 dihydroxycholecalciferol** (also called calcitriol) from vitamin D, which stimulates the absorption of calcium from the gut and the calcification of bone (see Section 22.3). They also, together with the liver, are able to synthesize glucose from amino acids during fasting (gluconeogenesis—see Section 4.6).

39.2 The anatomical organization of the kidneys and urinary tract

The kidneys lie high in the abdomen on its posterior wall, either side of the vertebral column (T11–L3—see Figure 39.1). In adults, each kidney is about 11 cm long and 6 cm wide and weighs about 140 g. A simple diagram of the gross structure of the kidney is shown in Figure 39.2. Facing the midline of the body is an indentation called the **hilus**, through which the renal artery enters and the renal vein and ureter leave. Each kidney is covered by a tough, fibrous, inelastic capsule that serves to restrict changes in volume in response to any increase in blood pressure.

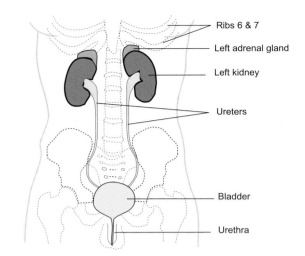

Figure 39.1 The principal structures of the male renal system, showing their position relative to the chest and pelvis.

If a kidney is cut open, two regions are easily recognized—a dark brown **cortex** and a pale inner region, which is divided into the **medulla** and the **renal pelvis**. The renal pelvis contains the major renal blood vessels and is the region where the ureter originates. The medulla of each human kidney is divided into a series of large conical masses known as the **renal pyramids** that have their origin at the border between the cortex and the medulla. The apex or **papilla** of each pyramid lies in the central space of the renal pelvis, which collects the urine prior to its passage to the bladder. The central space can be divided into two or three large areas known as the **major calyxes** (or calyces) that divide, in turn, into the minor calyxes that collect the urine from the renal papillae.

As befits their role as regulators of the internal environment, the kidneys receive a copious blood supply from the abdominal aorta via the renal arteries. Venous drainage is via the renal veins into the inferior vena cava. The renal circulation is regulated by nerves from both the parasympathetic and sympathetic divisions of the autonomic nervous system (discussed later in this section). The kidneys form the upper part of the urinary tract, and the urine they produce is delivered to the bladder by a pair of ureters. The bladder continuously accumulates urine and periodically empties its contents via the urethra under the control of an external sphincter—a process known as **micturition** (see Section 39.9). Like the kidneys, the lower urinary tract receives innervation from both divisions of the autonomic nervous system.

The nephron is the functional unit of the kidney

Each human kidney contains about 1.25 million **nephrons**, which are the functional units of the kidney. Each nephron consists of a **renal** or **Malpighian corpuscle** attached to a long, thin, convoluted tube and its associated blood supply (see Figures 39.3 and 39.7). The renal corpuscle is about 200 µm in diameter, while the tubules are about 55 µm in diameter and 50–70 mm in length. This arrangement provides a very large surface area for the exchange of

Box 39.1 The analysis of urine and its role in diagnosis

The analysis of urine (**urinalysis**) is a diagnostic tool used to detect a variety of metabolic and renal disorders. Routine analysis of urine by a simple 'dipstick' method is performed on admission to hospital, at antenatal appointments, and in GP surgeries and outpatient clinics. To minimize contamination of the urine by urethral debris, the tests are usually performed on a mid-stream urine sample (MSU). The colour, clarity, and odour of the sample provide initial information about the health of the individual concerned—for example, a fishy odour or a cloudy appearance strongly suggest the presence of a urinary tract infection.

Specific gravity is a measure of the number of dissolved particles in the urine. Strongly coloured urine with a high specific gravity indicates dehydration, while dilute urine with a low specific gravity indicates that the individual is well-hydrated. Low specific gravity coupled with symptoms of dehydration may indicate diabetes insipidus, in which there is a failure of ADH secretion by the posterior pituitary gland (see Section 21.8).

Normal urine has a pH that generally lies between 5 and 7, depending on the diet. However, certain disease states are associated with urine of extremely high or low pH. A very alkaline pH, for example, may indicate a failure of the renal tubules to secrete acid normally (renal tubular acidosis). Prolonged vomiting will also result in urine of high pH, as the kidneys compensate for the loss of gastric acid by secreting fewer hydrogen ions (see Chapter 44).

Glucose is not detectable in the urine of a healthy person, and the presence of glucose in the urine (**glycosuria**) is strongly suggestive of the condition diabetes mellitus, although blood tests and a glucose tolerance test are required to confirm the diagnosis (see Chapter 24). Protein is not normally present in the urine in more than trace amounts. Blood in the urine (**haematuria**) may have a number of origins: there may be contamination due to menstruation, catheterization, or any minor lesion of the urethra or surrounding tissue. For reasons that are not understood, heavy exercise may also lead to the transient appearance of blood in the urine (particularly in young males). More serious causes of haematuria include infections, tumours of the kidney or genito-urinary tract, or the presence of renal calculi (kidney stones). Nitrites are absent from the urine of a healthy person, and their presence suggests a urinary tract infection.

The presence of bilirubin in the urine may indicate a problem associated with liver function or with the normal delivery of bile to the small intestine (such as the presence of gallstones in the bile ducts). Gilbert's syndrome (also known as Gilbert–Meulengracht syndrome) is a common hereditary cause of increased plasma bilirubin which normally has no serious consequences. It is found in 5–10 per cent of the population. Urobilinogen appears in normal urine at concentrations below 17 μmol l^{-1}. It is derived from the actions of gut flora on bilirubin glucuronide and most of it is eliminated from the body via the faeces. An increased urinary urobilinogen concentration may indicate restricted liver function or excessive breakdown of red blood cells.

Ketones are derived from the metabolism of fats, and their presence in the urine will lower urinary pH, as these metabolites are acidic. Ketones appear in the urine (**ketonuria**) in measurable quantities only when fats are being used by the body as the principal source of energy, rather than carbohydrates. This may occur in a healthy fasting individual or in someone who is following a low-carbohydrate diet (such as the Atkins diet). It may also occur in the later stages of pregnancy, when large amounts of glucose are being used up by the fetus and the maternal energy supply has to come from fats. While these conditions represent normal physiological adaptations, prolonged or excessive ketonuria is more likely to be associated with diabetes mellitus, in which glucose cannot be utilized normally. In such cases, the level of ketones in the blood may become high enough to give rise to the serious and potentially life-threatening condition of **ketoacidosis**.

Although urinalysis is undoubtedly a very useful clinical tool, it is important to remember that, on its own, it is unable to confirm any diagnosis. Further detailed investigation will be required.

---

solutes between the tubular fluid and the cells of the tubular epithelium. The renal corpuscle consists of an invaginated sphere (**Bowman's capsule**) that envelops a tuft of capillaries known as a **glomerulus**. The glomerular capillaries originate from an afferent arteriole and recombine to form an efferent arteriole (see Figure 39.4). Between the glomerular capillaries are clusters of phagocytes called **mesangeal cells**.

The **proximal tubule** arises directly from Bowman's capsule. It is about 30–60 μm in diameter and 15 mm in length. The epithelial cells of the proximal tubule are cuboidal in appearance and rich in mitochondria. They are closely fused with one another via tight junctions near their apical surface, which is densely covered by microvilli giving rise to a prominent brush border (see Figure 39.5). As each square micron of the apical surface has approximately 100 microvilli, each of which is about 3 μm in height, the brush border increases the area available for transport by a factor of about 200. The lateral surfaces of the basolateral membranes also have an adaptation that greatly increases their surface area. In this case, it is an extensive series of deep infoldings of the plasma membrane. The space between the basolateral regions of the cells is called the **lateral intercellular space**.

The proximal tubule connects with the **intermediate tubule**, also known as the descending **loop of Henle**. Here the epithelial cells are thin and flattened (the wall thickness is only 1–2 μm). Compared to the cells of the proximal tubule, these cells have few mitochondria. The thin descending limb turns and ascends towards the cortex, finally merging with the thick segment, which is about 12 mm in length. The cells of the thick segment are cuboidal with extensive invaginations of the basolateral membrane. Like the cells of the proximal tubule, they are rich in mitochondria,

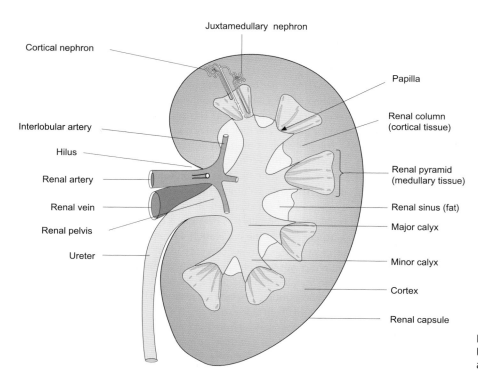

Figure 39.2 A cross-sectional view of the left kidney, showing the principal anatomical features.

indicative of a major role in active transport. The nephrons of the outer renal cortex (the **cortical** or **superficial nephrons**) have short loops of Henle (the thin segment is as little as 2 mm in length), some of which lie entirely within the cortex. In contrast, the nephrons nearest the medulla (the **juxtamedullary nephrons**) have long loops that penetrate deep into the medulla (Figure 39.3). Only about 15 per cent of the nephrons in humans have long loops, and in these nephrons, the thin segments (which may be as much as 14 mm in length) pass deep into the renal papillae.

The terminal part of the thick ascending limb of the loop of Henle and the initial segment of the distal tubule contact the afferent arteriole close to the glomerulus from which the tubule originated. The tubular epithelium is modified to form the **macula densa**, and the wall of the afferent arteriole is thicker due to the presence of **juxtaglomerular cells** (also called granular cells). These are modified smooth muscle cells containing secretory granules. The juxtaglomerular cells, macula densa, and associated mesangeal cells form the **juxtaglomerular apparatus** (Figure 39.4). The juxtaglomerular cells of the arteriole secrete **renin**, which has an important role in the regulation of aldosterone secretion from the adrenal cortex. In this way the juxtaglomerular apparatus plays an important role in the regulation of sodium balance (see Section 39.7 and Chapters 23 and 40).

The **distal tubule** arises from the ascending loop of Henle and is about 5 mm in length. Here the tubular wall is composed of cuboidal cells similar in appearance to those of the ascending thick limb. The distal tubules of a number of nephrons merge via **connecting tubules** to form **collecting ducts**, which are up to 20 mm in length and which pass through the cortex and medulla to the renal pelvis. The epithelium of the collecting ducts consists of two cell types: principal cells (P cells) that play an important role in the regulation of sodium balance, and intercalated cells (I cells) that are important in regulating acid–base balance. Electron micrographs of the proximal and distal tubules are shown in Figure 39.5.

The renal circulation is arranged in a highly ordered manner

The renal artery enters the hilus and branches to form the **interlobular arteries**. These subsequently give rise to the **arcuate arteries**, which course around the outer medulla. The arcuate arteries lead to cortical radial arteries (sometimes called the interlobular arteries) that ascend towards the renal capsule, branching en route to form the afferent arterioles of the Bowman's capsule (see Figure 39.6). The afferent arterioles give rise to tufts of capillaries within the Bowman's capsule that recombine to form the efferent arterioles.

The efferent arterioles of the outer cortex give rise to a rich supply of capillaries that covers the renal tubules (the **peritubular capillaries**—see Figure 39.7). Blood from the peritubular capillaries first drains into stellate veins and thence into the cortical radial veins and arcuate veins. In contrast, the efferent arterioles close to the medulla (the juxtamedullary efferent arterioles) give rise to a series of straight vessels known as the descending **vasa recta** (from the Latin for 'straight vessel') that provide the blood supply of the outer and inner medullary regions. Blood from the ascending vasa recta drains into the arcuate veins.

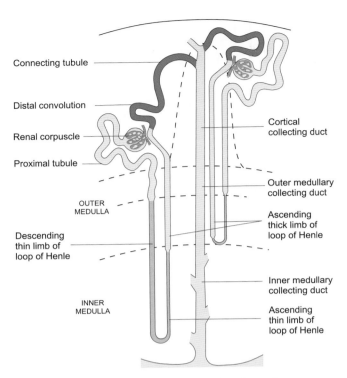

Figure 39.3 A short-looped (cortical) and long-looped (juxtamedullary) nephron, showing their basic organization. Note that the early distal tubule of each type of nephron is in contact with the afferent arteriole of its own glomerulus. (Based on Fig 1 in W. Kritz and L. Bankir (1988) *American Journal of Physiology* **254**, F1–F8.)

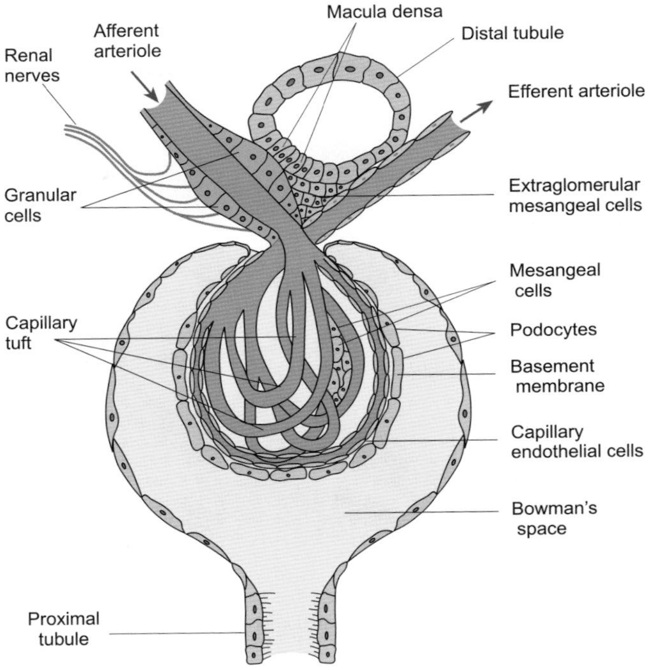

Figure 39.4 The principal features of a renal glomerulus and the juxtaglomerular apparatus. The wall of the afferent arteriole is thickened close to the point of contact with the distal tubule where the juxtaglomerular (or granular) cells are located. These cells secrete the enzyme renin in response to low sodium in the distal tubule.

The kidneys are innervated by sympathetic and parasympathetic nerve fibres

The kidneys have a rich nerve supply and are innervated both by sympathetic postganglionic fibres from the sympathetic paravertebral chain (T12--L2) and by efferent fibres from the vagus nerve. The postganglionic sympathetic nerve fibres travel alongside the major arteries, supplying the renal cortex as far as the afferent arterioles. The vagal parasympathetic fibres synapse in a ganglion in the hilus and appear to innervate the efferent arterioles. The sympathetic supply is adrenergic while the parasympathetic fibres are cholinergic. This innervation provides extrinsic control for the renal circulation that can override the intrinsic autoregulation of blood flow.

39.3 Renal blood flow is kept constant by autoregulation

In normal adults, the total renal blood flow (i.e. the blood flow for both kidneys) has been measured by a number of methods and is about 1.25 l min^{-1} or about 25 per cent of the resting cardiac output. The renal cortex has the highest blood flow—about five times that of the outer medulla and 20 times that of the inner medulla. If systemic arterial pressure is altered over the range 10–26 kPa (80–200 mm Hg), the renal blood flow remains remarkably constant (Figure 39.8). This stability of renal blood flow persists even after the renal nerves have been cut and can be observed in isolated perfused kidneys. It is therefore due to mechanisms intrinsic to the kidneys and is called **autoregulation**.

What are the mechanisms that underlie autoregulation? Two hypotheses have been advanced, the **myogenic hypothesis** and the **metabolic hypothesis**. The myogenic hypothesis proposes that autoregulation is due to the response of the renal arterioles to stretch. An increase in pressure will distend the arteriolar wall and stretch the smooth muscle fibres, which then contract after a short delay. The resulting vasoconstriction will increase vascular resistance and decrease blood flow (see Section 30.2). The metabolic hypothesis proposes that metabolites from the renal tissue maintain a degree of vasodilatation. An increase in perfusion pressure will lead to an increased blood flow, which, in turn, will leach out more metabolites and so decrease the vasodilatation. In addition to tissue metabolites, humoral factors such as prostaglandins and nitric oxide may also act as vasodilators. Additionally, the macula densa of the juxtaglomerular apparatus has been postulated to play a role in maintaining the vasomotor tone of the afferent and efferent arterioles of the glomerulus (see Section 39.5). In short, it is probable that both myogenic and humoral factors are responsible for the regulation of blood flow in the kidney.

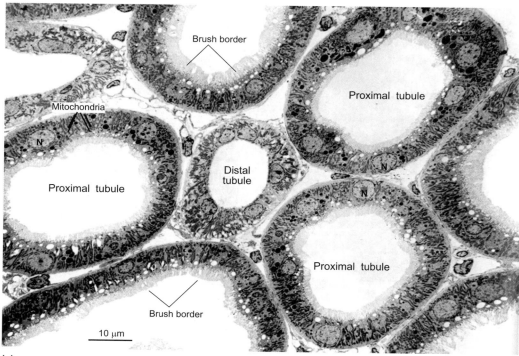

Brush border

Mitochondria

N

Proximal tubule

Proximal tubule

Distal tubule

N

N

N

Proximal tubule

Brush border

10 µm

(a)

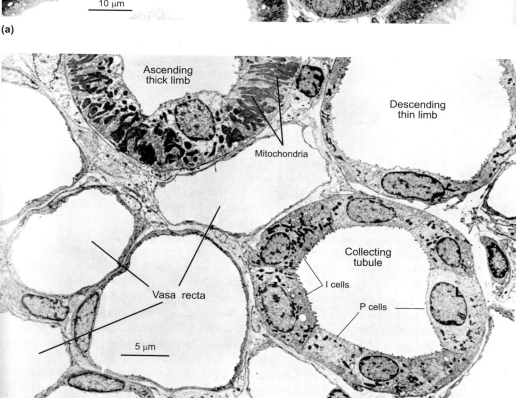

Ascending thick limb

Descending thin limb

Mitochondria

Collecting tubule

Vasa recta

I cells

P cells

5 µm

(b)

Figure 39.5 Electron micrographs showing the ultrastructure of the cells that constitute the nephron. (a) A section of the renal cortex. Note the prominent brush border of the proximal tubules, which is absent in the distal tubule. The proximal tubule cells have large numbers of mitochondria. (b) A section of the renal medulla in which the characteristic features of the thin descending and thick ascending limbs are clearly visible. Note the difference in staining between the I cells of the collecting tubule (strong with densely staining nuclei) and the P (principal) cells (pale with relatively lightly stained nuclei). Scale bars are 10 µm in (a) and 5 µm in (b). (From Fig 1, p. 236 and Fig 1, p. 237 in R.G. Kessel and R.H. Kardon (1979) *Tissues and organs—an atlas of scanning electron microscopy*. WH Freeman, New York.)

Despite powerful autoregulation, renal blood flow is also subject to modulation by extrinsic factors. This additional regulation is achieved both by the activity in the renal nerves and by humoral factors circulating in the blood. Sympathetic stimulation causes vasoconstriction of the afferent arterioles and thus reduces renal blood flow. Circulating adrenaline (epinephrine) and noradrenaline (norepinephrine) also cause vasoconstriction in the renal circulation, but norepinephrine acts mainly on the renal cortex. Intense sympathetic activation is responsible for the fall in renal blood flow observed during exercise. Angiotensin II

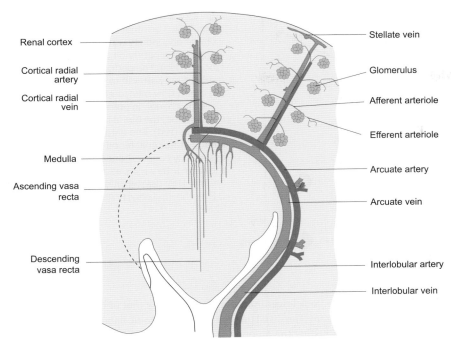

Figure 39.6 The principal features of the renal circulation, showing the circulation after the renal artery has branched to form the interlobular artery. Red: arterial blood; blue: venous blood. Note that the efferent arterioles of the superficial nephrons give rise to the peritubular capillaries, while those of the juxtamedullary nephrons give rise to straight vessels that pass deep into the renal medulla (the vasa recta). (Based on Fig 2 in W. Kritz and L. Bankir (1988) *American Journal of Physiology* **254**, F1–F8.)

and antidiuretic hormone (vasopressin) are other powerful vaso-constrictors, which play a significant role in regulating blood flow through the kidneys, particularly following severe haemorrhage. The role of the cholinergic parasympathetic nerve fibres is less clear, but they may act as vasodilators.

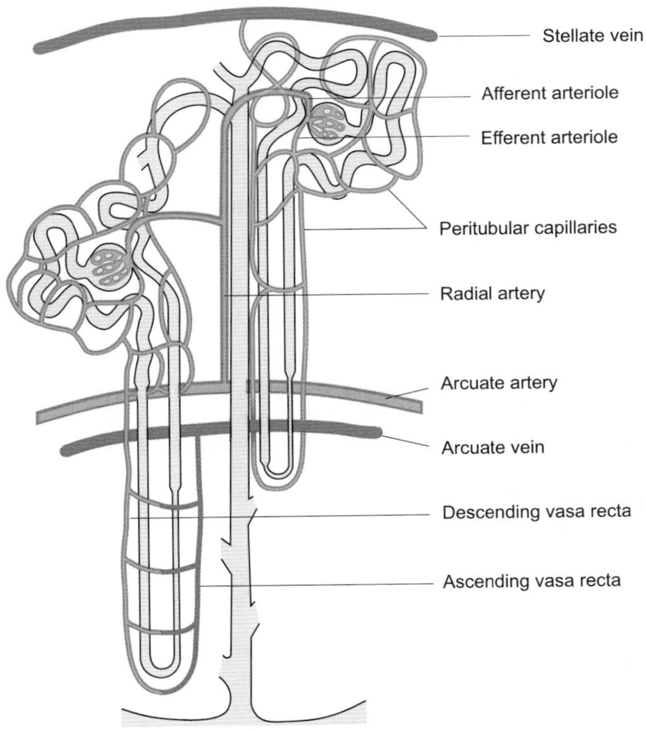

Figure 39.7 The blood supply of the cortical and juxtamedullary nephrons.

The mean pressures at key points in the renal circulation are illustrated in Figure 39.9. Note that while vasoconstriction in the afferent arterioles will reduce the pressure in the glomerular capillaries, vasoconstriction in the efferent arterioles will increase it.

39.4 The nephron regulates the internal environment by ultrafiltration followed by selective modification of the filtrate

Regulation of the plasma composition occurs via three key processes:

1. Filtration
2. Reabsorption
3. Secretion.

First, some of the plasma flowing through the glomerular capillaries is forced through the capillary wall and into Bowman's space by the hydrostatic pressure of the blood (filtration). Then, as this fluid—the **glomerular filtrate**—passes along the renal tubules, its composition is modified both by the reabsorption of some substances and by the secretion of others. **Reabsorption** is defined as movement of a substance from the tubular fluid to the blood, and tubular **secretion** is defined as movement of a substance from the blood into the tubular fluid. Both of these processes can occur via the tubular cells, the **transcellular route**, or between the cells, the **paracellular route** (see Figure 39.10). The reabsorption and secretion that occur via the transcellular route are largely the result of secondary active transport of solutes by the tubular cells. Paracellular reabsorption occurs as a result of concentration or electrical gradients that favour movement of particular solutes out

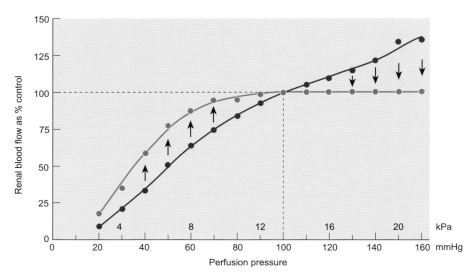

Figure 39.8 The autoregulation of renal blood flow. The renal blood flow for an isolated dog kidney was first allowed to stabilize at a perfusion pressure of 13.3 kPa (100 mm Hg). The perfusion pressure was then abruptly altered to a new value and the blood flow measured immediately after the change in pressure (blue). After a short period, the blood flow stabilized at a new level (red). The data show that the steady-state renal blood flow remains essentially constant once the pressure in the renal artery rises above about 10 kPa (75 mm Hg). (Based on data in Fig 6 in C.F. Rothe et al. (1971) *Am J Physiol* **220**, 621–6.)

of the tubular fluid. Paracellular secretion results when such forces favour movement into the tubular fluid.

The main stages of urine formation are illustrated in Figure 39.11 and may be summarized as follows. First, about a fifth of the plasma is filtered into Bowman's space from where it passes along the proximal tubule. Here many substances are reabsorbed while others are secreted. The proximal tubule reabsorbs all the filtered glucose and amino acids as well as most of the sodium, chloride, and bicarbonate. The reabsorption of these substances is accompanied by an osmotic equivalent of water so that, by the end of the proximal tubule, about two-thirds of the filtered fluid has been reabsorbed. As this phase of reabsorption is not closely linked to the ionic balance of the body, it is sometimes called the **obligatory phase** of reabsorption. The next limb of the nephron, the loop of Henle, is concerned with establishing and maintaining an osmotic gradient in the renal medulla. It does this by transporting sodium chloride from the tubular fluid to the tissue surrounding the tubules (the **interstitium**) without permitting the osmotic uptake of water. As a result, the osmolality of the fluid leaving the loop of

Henle is lower than the plasma, while that of the interstitium is higher. The distal tubule regulates the ionic balance of the body by adjusting the amount of sodium and other ions it reabsorbs according to the requirements of the body. It also secretes hydrogen ions, which acidify the urine. The fluid leaving the distal tubule is relatively dilute. As it passes through the collecting ducts, water is absorbed under the influence of antidiuretic hormone (ADH). If the osmolality of the body fluids is relatively high (> 290 mOsmol kg⁻¹), a concentrated urine is excreted. If the osmolality of the body fluids is relatively low (< 285 mOsmol kg⁻¹), little ADH is secreted and dilute urine is excreted.

In 1921, J.T. Wearn and A.N. Richards proved that the nephron works in this way by taking samples of the fluid in Bowman's capsule by micropuncture (see Box 39.2). Analysis of this fluid showed that it had the same ionic composition as plasma but contained very little protein.

The simplest explanation of this key observation is that the capsular fluid is formed by filtering out the plasma proteins while allowing the free passage of ions and small molecules such as

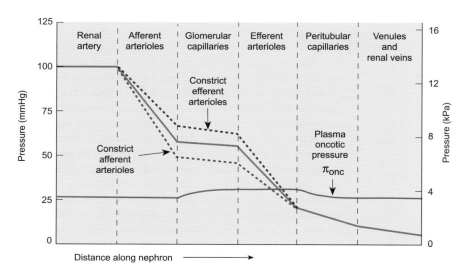

Figure 39.9 The fall in pressure across the renal circulation. Note the relatively high pressure in the glomerular capillaries and that the pressure in the peritubular capillaries (2 kPa or 15 mm Hg) is lower than the plasma oncotic pressure. Consequently, fluid reabsorption occurs along their length unlike the capillaries of other vascular beds. Note also that when the afferent arterioles constrict, the pressure in the glomerular capillaries falls, and when the efferent arterioles constrict, the pressure in the glomerular capillaries rises. The oncotic pressure of the plasma rises as fluid is filtered.

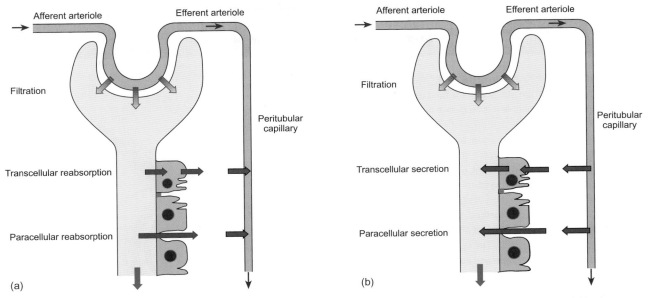

Figure 39.10 The key processes that occur as fluid flows through a nephron. The process begins with filtration of part of the plasma flowing through the glomerular capillaries. Subsequently, some substances are reabsorbed, as shown in panel (a), while others are secreted, as shown in panel (b). Note that both reabsorption and secretion can take place via the cells (the transcellular route) or via the junctions between the cells (the paracellular route). Both secretion and reabsorption may occur by either passive or active transport.

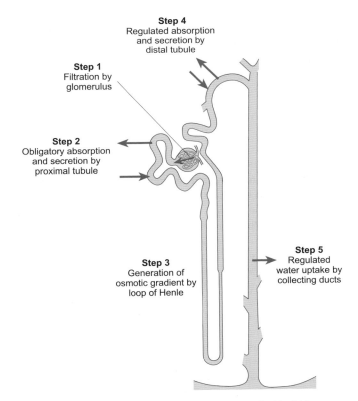

Figure 39.11 The main stages of urine formation by the kidney. Fluid is first filtered by the glomerulus. As it passes along the proximal tubule, it is modified by selective reabsorption and secretion. The reabsorption of ions by the loop of Henle without an osmotic equivalent of water leads to the generation of an osmotic gradient in the medulla, which is exploited to regulate the uptake of water by the collecting ducts.

sodium and glucose. The role of the subsequent segments of the nephron has been studied by obtaining small samples of tubular fluid by micropuncture, which can then be subjected to chemical analysis. In a variant of this approach, isolated segments of the nephron are perfused with a small volume of liquid, which can then be analysed to determine which substances have been reabsorbed or secreted.

Formation of the glomerular filtrate

As already described, the pressure in the glomerular capillaries forces a small proportion of the plasma into Bowman's space. During this process, small molecules and ions pass across the capillary wall, leaving the plasma proteins behind. This process is known as **ultrafiltration** and the fluid formed in this way is called the **glomerular filtrate**. The rate at which the two kidneys form the ultrafiltrate is known as the **glomerular filtration rate (or GFR)** and has units of millilitres per minute.

The barrier that restricts the passage of fluid from the glomerular capillaries into Bowman's capsule consists of three components (Figure 39.12). First, there is the capillary wall itself that consists of endothelial cells pierced by small gaps known as fenestrae that are around 60–80×10^{-9} m in diameter (i.e. much smaller than that of the platelets and blood cells). This arrangement makes the glomerular capillaries very much more permeable to water and to solutes than capillaries in other vascular beds. Second, the endothelial cells abut a basement membrane, which consists of fibrils of negatively charged glycoproteins. Finally, the epithelial cells or **podocytes** of the capsular membrane do not form a continuous layer but extend thin processes known as **pedicels** over the basement membrane, leaving gaps that provide the filtration slits.

Box 39.2 Direct measurements of the composition of tubular fluid can be achieved by micropuncture methods

Important insights into the function of the renal tubules can be gained by knowing the composition of the tubular fluid at different parts of the nephron. This can be achieved by penetrating the tubule wall with a very small micropipette and sucking up a sample of the fluid for chemical analysis. This is known as micropuncture. If the micropipettes contain pressure transducers, it is possible to measure the pressures in the afferent and efferent arterioles as well as that in Bowman's space. These micropuncture methods have been used extensively to study tubular function *in situ* in anaesthetized animals.

In the first application of this method, a droplet of oil was injected at the top of the proximal tubule and a micropipette inserted into Bowman's capsule (see Figure 1). The fluid in Bowman's space was aspirated and analysed for protein and electrolytes. Later variants of the technique have employed oil droplets to isolate sections of tubule; a fluid of known composition is then injected into the isolated segment and aspirated after a set period of time. Finally, this fluid is analysed for changes in its composition.

The methods outlined above have provided a great deal of information regarding the function of superficial nephrons but, by their nature, they cannot provide information about the deeper segments of the nephron or about the juxtamedullary nephrons. To overcome this difficulty, methods have been developed that permit the isolation of particular segments of individual nephrons. The transport processes occurring in these isolated segments can then be studied by microperfusion techniques similar to those employed in intact nephrons.

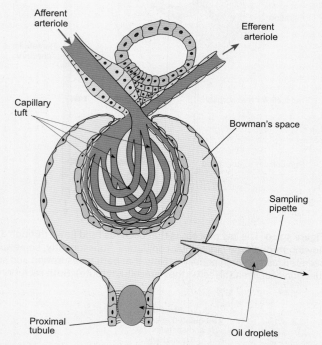

Figure 1 A simple schematic diagram to illustrate the principle of micropuncture.

Measurements of glomerular filtration have shown that substances with a low molecular weight are freely filtered, while the passage of large molecules is severely restricted (see Table 39.2). Myoglobin (M_r 17 kDa) passes through the filter relatively easily—it has about three-quarters the permeability of a small molecule like glucose (M_r 180 Da). Haemoglobin (M_r 68 kDa) passes through the filter only with great difficulty, and albumin, the smallest of the plasma proteins (M_r 69 kDa), has about a tenth of the permeability of haemoglobin, i.e. less than a hundredth of the permeability of a small molecule like glucose.

Even the largest of the plasma proteins are much smaller than the filtration slits that can be seen with the aid of the electron microscope. The barrier to the passage of proteins appears to be the meshwork of protein fibrils that form the basement membrane. Experiments with charged and neutral dextrans (carbohydrates with high molecular weights) show that the barrier depends both on the size of a molecule and on its electrical charge. The barrier is more permeable to neutral or positively charged molecules than it is to negatively charged ones. Thus albumin, which has a strong negative charge at physiological pH, is retained in the plasma both by its size and by mutual repulsion between its negative charge and that of the glycoproteins of the basement membrane. In contrast, haemoglobin, which has a similar molecular size but is not strongly charged at physiological pH, passes through the glomerular filter over five times more easily than albumin. Fortunately, unlike albumin, haemoglobin is contained within the red cells and is not normally present in the plasma.

The amount of fluid passing into Bowman's capsule of a single nephron is known as the single nephron glomerular filtration rate (snGFR). It is governed by the balance of the forces acting on the glomerular capillaries (see Figure 39.13 and Box 39.3). Hydrostatic pressure acts to force the plasma out of the capillaries, but, as already discussed, the plasma proteins cannot pass into the glomerulus and are retained within the blood. The osmotic pressure they exert (the **oncotic pressure**) opposes the hydrostatic force due to the pressure within the capillaries. In addition, there is a small pressure within the capsule itself that also opposes the hydrostatic pressure with the capillaries. The sum of these opposing pressures is called the **net filtration pressure**.

The pressure within Bowman's capsule (the capsular pressure) is about 2.6 kPa (20 mm Hg) and provides the force required to propel the filtrate through the nephron. It arises from the hydrostatic pressure forcing the ultrafiltrate from the glomerular capillaries and the restricted movement of the fluid through the tubules. (This is directly analogous to the pressure–flow relationship for blood flow—see Chapter 30).

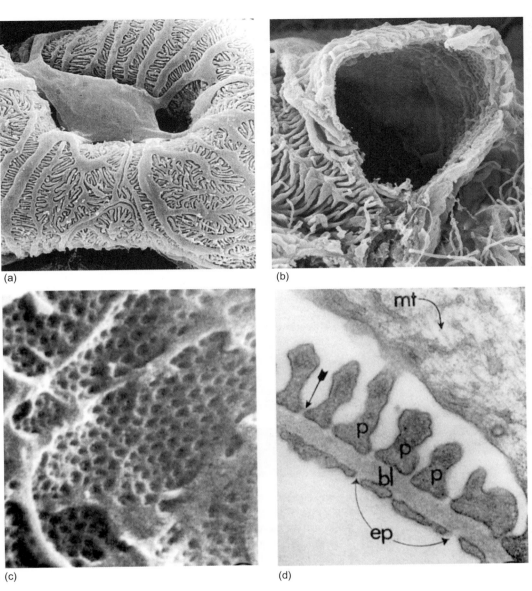

Figure 39.12 The ultrastructure of the glomerular capillaries. (a) A podocyte and its inter-digitating processes covering a glomerular capillary. (b) A ruptured glomerular capillary with its covering of pedicels (podocyte end feet). The fenestrae of the capillary endothelium can just be discerned but are more evident in (c). (d) The structure of the filtration barrier: ep = endothelial pores; p = pedicels; bl = basal lamina; mt = microtubule. The arrow points to a filtration slit. (Compiled from Fig 3 in P. Andrews (1988) *Journal Electron Microscopy Techniques* **9**, 115–44 (a); Creative commons, public domain (b); Plates 11 and 12 in P. Andrews and K.R. Porter (1975) *American Journal of Anatomy* **140**, 81–116 (c), (d).)

Table 39.2 The relationship between molecular radius and glomerular permeability

Substance	Molecular mass (Da)	Effective radius (nm)	Filtrate/ plasma
water	18	0.1	1.0
urea	60	0.16	1.0
glucose	180	0.36	1.0
sucrose	342	0.44	1.0
inulin	5200	1.48	0.98
myoglobin	17,000	1.95	0.75
haemoglobin	68,000	3.25	0.03
serum albumin	69,000	3.55	~0.005

Note: The ratio of the concentration of a substance in the filtrate to that found in the plasma is a direct measure of the ease with which it passes through the glomerular filter. A ratio of 1 indicates free passage, and a ratio of 0 would indicate a complete inability to pass the filter.

Inulin clearance can be used to measure the glomerular filtration rate

Measurement of the GFR is important to an understanding of renal physiology, but its value cannot be obtained directly except in isolated nephrons. It can, however, be estimated by measuring the rate of excretion of substances that are freely filtered but are then neither absorbed nor secreted by the renal tubules. In addition, such substances should have no influence on any physiological parameter that may alter renal function, such as blood pressure or blood flow. These criteria are met by the plant polysaccharide **inulin** (M_r 5.2 kDa), which is excreted by the kidneys in direct proportion to its plasma concentration over a very wide range (see Figure 39.14a).

The following explanation shows why the excretion of such substances can be used to measure the GFR. The rate at which a substance is excreted is simply its concentration in the urine (U_c) multiplied by the amount of urine produced per minute ($\dot{V}$) so:

$$\text{rate of excretion} = U_c \times \dot{V} \ \text{mg min}^{-1}$$

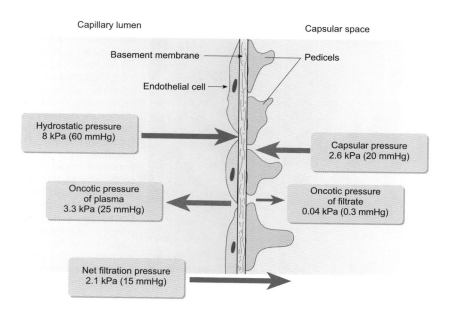

Figure 39.13 The filtration barrier in the glomerulus and the hydrodynamic forces that determine the rate of ultrafiltration. (See Box 39.3 for calculation of the Starling forces.) Fluid from the plasma is forced out by the pressure in the glomerular capillaries, but this is opposed by the pressure in the Bowman's capsule and by the osmotic pressure exerted by the plasma proteins (the oncotic pressure).

For a substance that is neither reabsorbed nor secreted by the renal tubules, the rate of excretion must be the same as the plasma concentration (P_c) multiplied by the rate at which it was filtered:

$$P_c \times GFR = U_c \times \dot{V}$$

$$\text{rate of filtration} = P_c \times GFR \text{ mg ml}^{-1}$$

This is known as the **filtered load**. Combining the two equations and rearranging:

$$P_c \times GFR = U_c \times \dot{V}$$

$$GFR = \frac{U_c \times \dot{V}}{P_c} \text{ ml min}^{-1}$$

Box 39.3 Calculation of the Starling forces governing the formation of the glomerular filtrate

The relationship between the glomerular filtration rate (GFR) and the hydrodynamic forces responsible for the formation of the filtrate is given by the relationship:

$$GFR = K_f \cdot ((P + \Pi_{BC}) - (P_{BC} + \Pi_{onc}))$$

where K_f is the *filtration coefficient* and represents the amount of fluid filtered for each unit of pressure in a minute. P is the hydrostatic pressure in the capillaries and Π_{BC} the osmotic pressure exerted by the protein in the capsular fluid. P_{BC} is the pressure within the Bowman's capsule and Π_{onc} the oncotic pressure of the plasma. The net filtration pressure (P_f) is the result of the forces tending to force fluid from the glomerulus minus those opposing filtration:

$$P_f = (P + \Pi_{BC}) - (P_{BC} + \Pi_{onc})$$

and

$$GFR = K_f \cdot P_f$$

The pressure in the glomerular capillaries is about 8 kPa (60 mm Hg)—significantly higher than that of the capillaries of other internal organs, where average capillary pressures range from 1.3 to 4 kPa (approximately 10–30 mm Hg). The pressure within Bowman's capsule is about 2.6 kPa (about 20 mm Hg) and the oncotic pressure in normal plasma is about 3.3 kPa (25 mm Hg) as the blood enters the afferent arteriole and about 4.7 kPa (35 mm Hg) as it leaves the glomerulus via

the efferent arteriole. The osmotic pressure attributable to the proteins in Bowman's capsule is about 0.04 kPa (0.3 mm Hg) in normal subjects. Thus the net filtration pressure at the afferent end of the glomerular capillaries (i.e. the pressure forcing fluid from the glomerular capillaries) is $(8 + 0.04) - (2.6 + 3.3) = 2.14$ kPa (c. 15 mm Hg). As the blood traverses the glomerular capillaries, the net filtration pressure declines as the oncotic pressure rises. By the time the blood leaves the glomerulus, the net filtration pressure is much less as the forces tending to drive fluid from the blood into the capsular space are opposed by the rise in oncotic pressure. In addition, the pressure in the glomerular capillaries falls progressively along their length and is about 7.75 kPa (58 mm Hg) by the time they merge with the efferent arteriole. Thus the net filtration pressure at the end of the glomerular capillaries is $(7.75 + 0.04) - (4.7 + 2.6) = 0.49$ kPa (c. 4 mm Hg). From these calculations it follows that if the capillary pressure falls below about 6 kPa (45 mm Hg), the forces tending to force fluid from the capillaries will be balanced by those opposing filtration. This can occur during constriction of the afferent arterioles (see Figure 39.9). Where the fall in capillary pressure is large, during haemorrhage for example, little or no plasma will be filtered and urine production will decline.

As the GFR in humans is about 125 ml min^{-1}, K_f has a value of about 100 ml min^{-1} kPa^{-1} (or 12.5 ml min^{-1} mm Hg^{-1}). This is more than a hundred times greater than the permeability of those vascular beds that do not have fenestrated capillaries.

The GFR measured by the rate of inulin excretion is called the **inulin clearance**. It is generally about 120–130 ml min^{-1} for adult men and about 10 per cent less than this for women of similar size. Despite its advantages, the use of inulin is not very convenient for clinical purposes, as a steady concentration needs to be maintained in the plasma for accurate measurement. For this reason it is desirable to use a substance that is normally present in the plasma and which is filtered, but neither secreted nor reabsorbed by the renal tubules. In addition, the plasma concentration of such a substance should not fluctuate rapidly. These criteria are largely met by creatinine (a metabolite of creatine), and the **creatinine clearance** is generally used to measure GFR in clinical practice. Nevertheless, as a small component of the excreted creatinine is secreted by the tubules, creatinine clearance overestimates the GFR when renal blood flow is low. In routine clinical examination, plasma creatinine is employed as a surrogate measure of GFR, though this is not very accurate as blood levels vary with age, diet, race, and body size. Values in excess of 0.12 mmol l^{-1} (1.4 mg dl^{-1}) in men and 0.10 mmol l^{-1} (1.2 mg dl^{-1}) in women are taken as early indicators of impaired renal function.

The concept of clearance can be extended to other substances that are secreted or reabsorbed by the renal tubules. As already discussed, the rate at which a substance is excreted is simply its concentration in the urine (U_c) multiplied by the amount of urine produced per minute ($\dot{V}$). The ratio of the rate of excretion to the plasma concentration (P_c) represents the *minimum* volume of plasma from which the kidneys could have obtained the excreted amount. This volume is called the clearance (C) and is expressed in ml per minute. Thus:

$$C = \frac{U_C \times \dot{V}}{P_C} \; \text{ml min}^{-1}$$

Renal clearance can therefore be defined as the volume of plasma completely cleared of a given substance in one minute. If a substance has a clearance smaller than the GFR, either the kidneys must reabsorb it (e.g. glucose) or it is not freely filtered (e.g. plasma

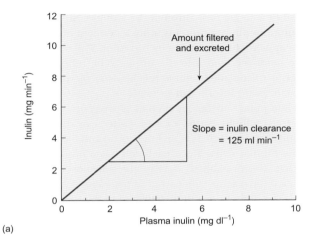

(a)

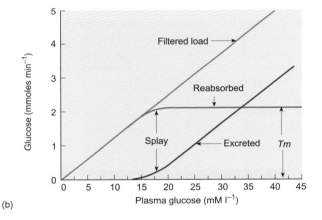

(b)

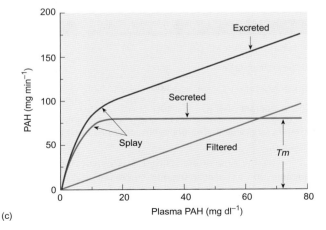

(c)

Figure 39.14 The relationship between the plasma concentration and the amount excreted in the urine for substances that are subject only to filtration (a), filtration followed by reabsorption (b), and filtration plus secretion (c). (a) The amount of inulin excreted plotted as a function of the plasma inulin concentration. The amount excreted is the urine concentration times the urine flow rate (U x $\dot{V}$) and the slope of the line is the clearance (C), which is 125 ml min^{-1} in this case. Note that the amount of inulin excreted is directly proportional to the plasma concentration. (b) The quantities of glucose reabsorbed and excreted plotted as a function of the plasma concentration. The blue line shows that glucose appears in the urine when the plasma concentration exceeds about 12 mmol l^{-1}. This is known as the renal threshold for glucose. Above about 18 mmol l^{-1}, the extra amount of glucose excreted is directly proportional to the plasma concentration. The red line shows the filtered load in mmol per minute, and this is calculated from the GFR (here taken as 120 ml min^{-1}) and the plasma concentration. The green line indicates the amount of glucose reabsorbed by the tubules, which is the difference between the filtered load and the amount excreted. Reabsorption reaches a maximum when the glucose load exceeds 2 mmol (360 mg) a minute. This is the transport maximum for glucose or Tm$_g$. (c) The amount of PAH excreted plotted as a function of the plasma concentration. The blue line indicates the amount excreted (in mg per minute) while the red line indicates the filtered load. The amount secreted by the tubules (indicated by the green line) is given by the difference between the filtered load and the amount excreted. Tubular secretion of PAH reaches a maximum when the plasma concentration is about 18 mg dl^{-1} and the Tm$_{PAH}$ is about 80 mg min^{-1}.

proteins). If a substance has a clearance larger than the GFR, it must be secreted by the renal tubules.

Summary

The nephron regulates the internal environment by first filtering the plasma and then reabsorbing substances from, or secreting substances into, the tubular fluid. The barrier that restricts the passage of fluid from the glomerular capillaries into Bowman's capsule consists of the capillary endothelial cells, a basement membrane, and the podocytes of the epithelial cells of the capsular membrane. These components prevent the passage of substances with a large molecular weight while allowing the free filtration of substances with a low molecular weight.

The amount of fluid passing into Bowman's capsule is governed by the net filtration pressure, which is determined by the balance of hydrodynamic forces acting on the glomerular capillaries. The sum of the opposing pressures is called the net filtration pressure. The rate at which the kidneys form the ultrafiltrate is known as the glomerular filtration rate or GFR and has units of ml per minute.

Renal clearance is defined as the volume of plasma completely cleared of a given substance in one minute. The clearance of inulin or creatinine is commonly used to estimate the GFR.

39.5 Tubular absorption and secretion

When a substance is simply filtered and excreted unchanged, the amount excreted is directly proportional to the plasma concentration (see Figure 39.14a) and the amount excreted represents the filtered load. If a substance is reabsorbed, the amount excreted will be less than the filtered load (see Figure 39.14b), and if filtration is followed by tubular secretion, the amount appearing in the urine will be greater than the filtered load (see Figure 39.14c). As discussed in Box 39.4, the difference between the filtered load and the amount excreted by the kidney is the amount that has either been reabsorbed or secreted. For example, glucose and amino acids are freely filtered but healthy people have virtually no glucose or free amino acids in their urine, so there must be tubular mechanisms for the reabsorption of these substances (remember: the filtered load is equal to the plasma concentration multiplied by the GFR). Conversely, as much as 70 per cent of the dye phenolsulphonphthalein is removed from the blood in a single pass through the kidneys. Since 75 per cent is bound to plasma proteins, only 5 per cent of the dye could have appeared in the urine by filtration. The remainder must have been secreted into the tubule. The following sections describe the detailed cellular mechanisms by which individual substances are absorbed or secreted.

When polar substances such as glucose are reabsorbed via the transcellular route, carrier molecules are required to permit their movement across the apical and basal membranes of the tubular cells. Since each cell has a limited number of carriers, it is possible to saturate the transport capacity of the tubule if the plasma concentration rises above its physiological level. The amount of solute delivered to the tubule per minute that just saturates its transport process is called the **transfer** or **transport maximum (Tm)**.

The concentration-dependence of glucose clearance provides a clear example of the Tm concept as applied to tubular transport. If glucose is infused into the blood, the plasma concentration rises above its normal level of about 4.5 mmol l^{-1}. Initially no glucose appears in the urine. However, when the plasma concentration exceeds 10–12 mmol l^{-1} it begins to do so. This value is known as the **renal threshold**. Above the renal threshold the amount of glucose appearing in the urine increases slowly at first. Then, as the transport process responsible for glucose reabsorption becomes fully saturated (above about 17 mmol l^{-1}),

---

(1₂3) Box 39.4 Calculating the amount of a substance transported by the renal tubules

The difference between the filtered load and the amount excreted by the kidneys is the amount of a substance that has been reabsorbed or secreted by the tubules, and it may be simply calculated using the following relationship:

$$T_s = (U_s \times \dot{V}) - (GFR \times P_s) \text{ (mg min}^{-1})$$

where T_s is the amount of substance (e.g. glucose) transported, GFR is the glomerular filtration rate, P_s and U_s are the plasma and urine concentrations (mg ml^{-1}), and $\dot{V}$ is the urine flow rate. When T_s is negative. the substance has been reabsorbed; when T_s is positive, the substance has been secreted into the tubular fluid.

Consider the data relating to the excretion of phosphate and PAH summarized in Table 1. The amount of phosphate transported is: $(30 \times 1) - (0.9 \times 100) = -60$ mg min^{-1}. So the amount of phosphate excreted is 60 mg min^{-1} less than the filtered load. This

Table 1 Summary of data relating to the excretion of phosphate and PAH

	Phosphate	PAH	Units
Plasma concentration (P)	0.9	0.05	mg ml^{-1}
Urine concentration (U)	30	25	mg ml^{-1}
Urine flow rate ($\dot{V}$)	1.0	1.0	ml min^{-1}
Inulin clearance (GFR)	100	100	ml min^{-1}

difference represents the phosphate that has been reabsorbed by the tubules.

For PAH, the amount transported is: $(25 \times 1) - (0.05 \times 100) = 20$ mg min^{-1}. In this case, 20 mg min^{-1} more PAH appears in the urine than can be accounted for by simple filtration. Thus it must have been secreted by the tubules.

the increase in urinary glucose becomes directly proportional to the increase in the plasma concentration, as shown in Fig 39.14b.

The transport maximum for glucose (Tm_g) can be determined by a simple calculation. Suppose that the GFR is 120 ml min^{-1} and that plasma glucose is 20 mmol l^{-1}. Further suppose that the urine flow rate is 1.6 ml min^{-1} and that the concentration in the urine is 200 mmol l^{-1}. Under these conditions, the glucose carriers are fully saturated and the amount reabsorbed is equal to the transport maximum. This is equal to the filtered load less the amount excreted in the urine:

$$\text{filtered load} = 20 \times 120 \times 10^{-3} = 2.4 \text{ mmol min}^{-1}$$

and

$$\text{amount excreted} = 200 \times 1.6 \times 10^{-3} = 0.32 \text{ mmol min}^{-1}$$

Therefore:

$$Tm_g = \text{amount reabsorbed} = 2.4 - 0.32 = 2.08 \text{ mmol min}^{-1}$$
$$\left(\text{or } 374 \text{ mg min}^{-1}\right)$$

Plasma glucose levels capable of saturating the transport capacity of the kidneys often occur in patients with inadequately controlled diabetes mellitus (see Section 24.6). This leads to the appearance of glucose in the urine (**glycosuria**). As the excreted glucose is accompanied by an osmotic equivalent of water, glycosuria is associated with an increase in urine production known as an osmotic diuresis. The line relating the amount of glucose absorbed to the plasma concentration is curved at its upper end (see Figure 39.14b). This curvature is known as **splay**. While differences in the transport capacity of different nephrons may partly account for the splay, it is an inherent feature of carrier-mediated transport.

The clearance of p-aminohippurate can be used to estimate renal plasma flow

As discussed earlier, tubular secretion (movement of a substance from the blood to the tubular lumen) may be either active or passive. Active tubular secretion occurs by carrier-mediated processes analogous to those discussed above for glucose reabsorption, which operate in the opposite direction. It is known that many organic substances are secreted by the tubules, including para-aminohippurate (PAH), penicillin, and other organic anions and cations. The relationships between filtration, secretion, and excretion of PAH are illustrated in Figure 39.14c. Moreover, analysis of arterial blood and of blood taken from the renal vein shows that PAH is almost totally removed from the plasma as it passes through the renal circulation. This property is exploited to estimate the renal plasma flow (RPF) in a relatively non-invasive way using the Fick principle (see Box 31.1).

$$RPF = \frac{\text{amount appearing in urine per minute}}{\text{arterio-venous difference for PAH}} \text{ ml min}^{-1}$$

Since both the amount of urine produced per minute and the concentration of PAH in the urine can be readily measured, the amount of PAH excreted per unit time can be calculated. The PAH content in a sample of venous blood taken from a peripheral vein can be measured and will be the same as that of arterial plasma. If the small amount of PAH in the venous blood leaving the kidney is neglected, the total RPF (i.e. the total for both kidneys) can be calculated. It should be noted, however, that this way of estimating renal plasma flow is reliable only while the tubular transport mechanism is not saturated—a situation that applies only when the plasma concentration is low (the Tm_{PAH} is about 80 mg min^{-1}; see Box 39.5). From estimates of GFR and renal plasma flow, we can

---

$\left(1_2 3\right)$ Box 39.5 The use of inulin and PAH clearance to estimate GFR and renal plasma flow

Two experiments were performed on the same subject to determine GFR and renal plasma flow. The data is summarized in Table 1, and Table 2 shows the calculation of the clearances for inulin and PAH.

In experiment 1, normal values are obtained for GFR (inulin clearance) and renal plasma flow (PAH clearance; see text), but in experiment 2 the GFR is normal but apparent renal plasma flow is 60 per

Table 1 Summary of data from two experiments performed on the same subject to determine GFR and renal plasma flow

		Inulin	PAH
Experiment 1	Plasma (mg ml^{-1})	0.25	0.08
	Urine (mg ml^{-1})	12.5	20.0
	Urine flow rate (ml min^{-1})	2.5	2.5
Experiment 2	Plasma (mg ml^{-1})	0.24	0.36
	Urine (mg ml^{-1})	6.0	25.0
	Urine flow rate (ml min^{-1})	5.0	5.0

Table 2 Calculation of the clearances for inulin and PAH in experiments 1 and 2

Experiment 1	Inulin clearance = (12.5 × 2.5)/0.25 = 125 ml min^{-1}
	PAH clearance = (20.0 × 2.5)/0.08 = 625 ml min^{-1}
Experiment 2	Inulin clearance = (6.0 × 5.0)/0.24 = 125 ml min^{-1}
	PAH clearance = (25.0 × 5.0)/0.36 = 347 ml min^{-1}

cent of normal. Obviously, the assumption that PAH clearance is a measure of renal plasma flow must be questioned. Inspection of the data shows plasma PAH to be nearly five times higher in experiment 2 than in experiment 1. Here is the clue to the problem. The amount of PAH delivered to the kidneys is 50 mg min^{-1} in experiment 1 and 225 mg min^{-1} in experiment 2. In experiment 1 all of this is excreted, but in experiment 2 only 125 mg min^{-1} is excreted. Of this, 45 mg min^{-1} are filtered. The difference is a measure of the total transport capacity—in this case 80 mg min^{-1}.

determine the proportion of plasma filtered. This is called the **filtration fraction**. If the GFR is 125 ml min^{-1} and renal plasma flow is 625 ml min^{-1}, the filtration fraction is $125 \div 625 = 0.20$.

The GFR is regulated by glomerulo-tubular feedback; this regulation may be overridden by sympathetic nerve activity

For efficient operation of the kidney, the GFR must be well matched to the transport capacity of the tubules. Too small a GFR and the kidneys may be unable to regulate the internal environment; too large a GFR and the transport capacity for amino acids, glucose, and ions will be exceeded, resulting in the loss of vital nutrients from the body.

Changes in the GFR will markedly alter the filtered load of sodium in particular. Under normal conditions, the filtered load of sodium is about 26 moles a day (180 l of plasma are filtered daily and the plasma sodium concentration is approximately 145 mmol l^{-1}; multiplying these together: $180 \times 145 = 26,100$ mmoles). Less than 1 per cent of this is usually excreted, and sodium balance is maintained (i.e. sodium intake via the diet equals sodium loss in the sweat, faeces, and urine). If GFR increases, the filtered load of sodium increases. For example, suppose the GFR rose from 125 to 130 ml min^{-1}, the total filtered load of sodium would increase to 27 moles a day. Unless the tubules can transport the extra mole of sodium that is filtered, it will be lost in the urine. Evidently, unless the filtered load is regulated or the transport capacity adjusted, sodium balance will be greatly disturbed.

The matching of the transport capacity to the filtered load is called **glomerulotubular balance**. Amongst the factors that contribute to glomerulotubular balance are the Starling forces in the peritubular capillaries. As the filtrate is formed, there is an increase in the oncotic pressure of the plasma (see Figure 39.9). The extent of this increase will depend on the filtration fraction. A larger filtration fraction will result in a greater oncotic pressure in the peritubular capillaries, and the Starling forces in the peritubular capillaries will favour greater movement of fluid from the tubular lumen into the capillaries via the lateral intercellular spaces.

Experiments with perfused single nephrons have shown that when an increase in the GFR for an individual nephron (the snGFR) is mimicked by an increased rate of perfusion of the loop of Henle, the snGFR for that nephron decreases. If the rate of perfusion is reduced, the snGFR increases. This relationship between the rate of fluid flow and glomerular filtration is known as **tubuloglomerular feedback**. This phenomenon, which probably contributes to glomerulotubular balance, is regulated by the cells of the macula densa, which form part of the juxtaglomerular apparatus (Section 39.2 and Figure 39.4). Whether it is the osmolality, the concentration of sodium, or some other variable that is sensed by these cells remains unclear—as is the mechanism that brings about the change in snGFR.

In response to an elevated systemic arterial blood pressure, the renal blood flow and GFR remain remarkably stable. As already explained, an increase in the diameter of the afferent arterioles following a rise in blood pressure would quickly lead to arteriolar constriction (autoregulation). This would offset the rise in pressure and maintain the net filtration pressure and GFR nearly constant. Conversely, following a modest fall in blood pressure, the renal blood flow would fall and this would be countered by a vasodilatation of the afferent arterioles, resulting in a compensatory increase in capillary pressure. The net effect would be to maintain the filtration pressure and GFR at near normal values as before. The autoregulation of renal blood flow is, therefore, essential to the maintenance of the GFR.

There are circumstances, however, in which renal blood flow and GFR are reduced. For example, during exercise, increased activity in the sympathetic nerves together with an elevation in the concentration of circulating catecholamines (adrenaline and noradrenaline) leads to vasoconstriction in the afferent arterioles. The net filtration pressure falls, and with it the GFR. Haemorrhage also leads to pronounced vasoconstriction in the afferent arterioles and a decreased GFR. In severe haemorrhage, the release of antidiuretic hormone (ADH) from the posterior pituitary gland greatly increases and adds to the vasoconstrictor effect of sympathetic activation (for this reason ADH is generally known as vasopressin). The resulting intense vasoconstriction can lead to a complete failure of urine production, which is called **anuria**. The actions of ADH during haemorrhage (increased water reabsorption and vasoconstriction) contribute to the maintenance of an adequate circulating plasma volume and so help to offset the fall in blood pressure that would otherwise ensue.

The proximal tubule absorbs solutes by active transport and by facilitated diffusion

The proximal tubule reabsorbs about two-thirds of the filtered water, sodium, potassium, chloride, bicarbonate, and other solutes. Under normal circumstances, it removes virtually all the filtered glucose, lactate, and amino acids. The sodium pump (Na$^+$, K$^+$-ATPase) of the basolateral surface of the epithelial cells of the proximal tubule provides the driving force for the reabsorption of all of these substances. Specific carrier proteins that are located in the brush border membrane take up glucose and other organic substances. Some of these carriers have now been cloned and their primary structure determined. They behave in many respects like enzymes: they can be saturated by large amounts of substrate and can be inhibited by appropriate agents. For example, glucose transport can be inhibited by other sugars such as galactose (an example of competitive inhibition). In addition to the glucose carriers, there are five different carriers for the amino acids, a lactate transporter, and at least one carrier for inorganic anions such as phosphate and sulphate. The existence of these carriers and their similarity to enzymes has provided a simple, logical explanation for the existence of the transport maximum for the reabsorption and secretion of many substances.

How do these carriers work? To take glucose as an example, it is clear that its uptake is not favoured by a transepithelial gradient, as the filtrate leaving the glomerulus has the same glucose concentration as the plasma. Moreover, as glucose is a neutral molecule it cannot be absorbed along an electrical gradient by itself. Experiments have shown that the uptake of glucose is sodium-dependent. Since the concentration of sodium in the cells of the

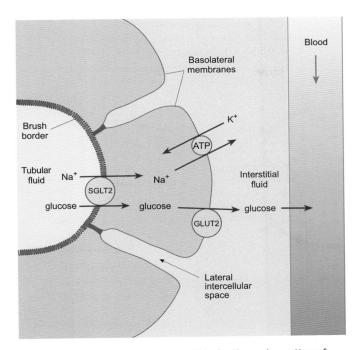

Figure 39.15 The processes responsible for the reabsorption of glucose in the proximal tubule. Glucose and sodium cross the apical membrane via a carrier protein. The sodium gradient that provides the energy for the uptake process is maintained by the activity of the sodium pump (Na$^+$, K$^+$-ATPase) of the basolateral membranes. In the first part of the nephron (illustrated here as a partial cross-section) glucose is transported from the lumen via a relatively low-affinity carrier called SGLT2 while a higher-affinity glucose transporter, SGLT1, is expressed by the cells of the final segment. The glucose leaves the tubule cells via different carrier proteins (GLUT2 in the first two-thirds of the proximal tubule and GLUT1 in the final segment), which are not linked to the transport of sodium. The renal absorption of glucose is therefore an example of secondary active transport.

transporters in the basolateral membrane: in the first part of the tubule, the cells express a low-affinity glucose transporter, GLUT2, while a higher-affinity glucose transporter, GLUT1, is expressed by the cells of the final segment. In short, SGLT2 works in conjunction with GLUT2 while SGLT1 works in conjunction with GLUT1. This sequential arrangement of low- and high-affinity carriers is an adaptation that permits almost complete recovery of glucose from the tubular fluid.

The reabsorption of amino acids also occurs via sodium-linked carrier molecules, as shown in Figure 39.16. At least five different amino acid carriers are expressed on the apical membranes: one for anionic (excitatory) amino acids, one for cysteine and cationic amino acids, one for glycine, alanine, and proline, one for neutral amino acids, and a specific transporter for proline. All are members of the solute carrier superfamily and require the cotransport of sodium. There are similar number of amino acid transporters in basolateral membrane that are not sodium-dependent (cf. GLUT1 and GLUT2 above). Small peptides are also reabsorbed by the proximal tubule, but in this case they are absorbed in association with hydrogen ions.

The active transport of glucose and the amino acids is so effective that virtually all of these solutes are normally removed from the tubular fluid during its passage along the proximal tubule. Essentially the same processes permit the efficient absorption of glucose and the amino acids by the small intestine (see Chapter 43).

proximal tubule is about 10–20 mmol l^{-1}, the movement of sodium into these cells from the tubular lumen (where the sodium concentration is c. 145 mmol l^{-1}) can occur along a favourable electrochemical gradient. It is this gradient that is exploited by the tubular cells to permit glucose uptake, even though there is no concentration gradient in its favour. Both glucose and sodium first bind to a carrier on the apical surface before being transported into the tubular cells. The inward movement of glucose is thus coupled to the movement of the sodium down its electrochemical gradient. The glucose leaves the cell via a carrier protein that is not sodium-dependent. The Na$^+$, K$^+$-ATPase of the basolateral membrane pumps the absorbed sodium into the lateral and basal extracellular spaces. The arrangement of these carriers on the cell therefore permits the secondary active transport of glucose (see Figure 39.15).

In the first part of the proximal tubule, both glucose and sodium bind to a relatively low affinity carrier on the apical surface known as SGLT2. In the final segment of the proximal tubule, the cells express a different, high-affinity carrier (SGLT1) which is able to transport the remaining 10 per cent of glucose from the tubular fluid. There is a similar sequential expression of glucose

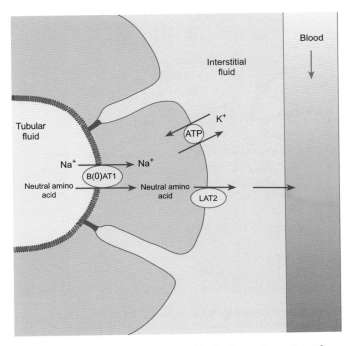

Figure 39.16 The processes responsible for the reabsorption of neutral amino acids in the proximal tubule. The neutral amino acids and sodium cross the apical membrane via a carrier protein called B(0)AT1 which is encoded by a gene named SLC6A19 (for **S**o**L**ute **C**arrier family 6, subfamily A, member 19). The sodium gradient provides the energy for the uptake process. The amino acids leave the cell via another carrier protein which is not linked to the transport of sodium, LAT2 (encoded by the SLC7A8 gene).

The reabsorption of bicarbonate ions is also linked to that of sodium. A Na^+/H^+ antiporter called NHE3 is present in the brush border membrane. This exchanges sodium ions in the tubular fluid for intracellular hydrogen ions. The resulting secretion of hydrogen ions acidifies the lumen and favours a shift of the carbonic acid–bicarbonate equilibrium towards carbonic acid. This is then rapidly converted into carbon dioxide and water by the enzyme carbonic anhydrase, which is located on the brush border of the proximal tubule cells. This locally raises the partial pressure of carbon dioxide in the tubular fluid and favours its diffusion into the tubular cells, where it is reformed into carbonic acid by intracellular carbonic anhydrase. The bicarbonate formed by this reaction leaves the cell via the basolateral membrane in exchange for chloride and passes into the circulation. The processes involved in bicarbonate absorption are summarized in Figure 39.17.

The uptake of sodium in the first half of the proximal tubule is coupled chiefly with the uptake of organic solutes and anions other than chloride (e.g. bicarbonate). As a result, the concentration of chloride in the tubular fluid passing through the second half of the proximal tubule rises to about 140 mmol l^{-1}, compared to about 105 mmol l^{-1} for the plasma. Some chloride is therefore able to diffuse together with sodium, through the tight junctions, down its concentration gradient and into the lateral intercellular spaces. In addition, sodium

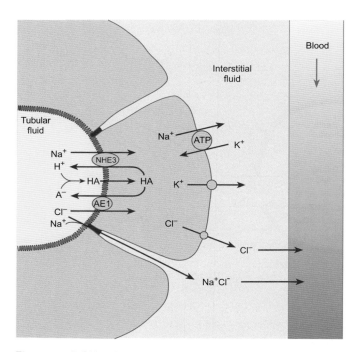

Figure 39.18 Chloride reabsorption in the proximal tubule. Sodium and chloride enter the cell across the apical membrane via the parallel activity of an Na^+–H^+ exchanger (NHE3) and an anion exchanger that couples chloride movement to the efflux of an organic anion (here represented as A^-), which is recycled. Chloride leaves the cell via chloride channels in the basolateral surface. Some sodium and chloride is reabsorbed via the paracellular pathway as a consequence of water movement (solvent drag).

and chloride are transported into the tubular cells via the parallel action of Na^+/H^+ and Cl^-/anion exchangers, as shown in Figure 39.18.

Calcium and phosphate absorption

Although the calcium permeability of the proximal tubule is low, around 70 per cent of the filtered calcium is reabsorbed in this segment of the nephron. This occurs mainly via the paracellular pathway, while the transcellular movement of calcium accounts for about a third of its uptake. In contrast, in the distal tubule all of the calcium uptake is via the transcellular route (see Section 39.7).

For a normal person with a plasma phosphate concentration of around 1 mmol l^{-1}, about 200 mmol of phosphate is filtered each day. Of this about 80 per cent is reabsorbed by the proximal tubule by a carrier-mediated process. Like the uptake of glucose and amino acids, phosphate reabsorption by the tubular cells is linked to sodium. Two kinds of carrier molecule have been implicated in the absorption of phosphate from the proximal tubule. Both belong to the solute carrier family SLC34 and are located in the brush border. The absorbed phosphate leaves the cells via the basolateral membrane by a poorly characterized anion exchange process. No phosphate is absorbed by the loop of Henle or collecting ducts and only half of the remaining phosphate is reabsorbed by the distal tubule, so that under normal circumstances about 10–15 per cent of the filtered load is excreted.

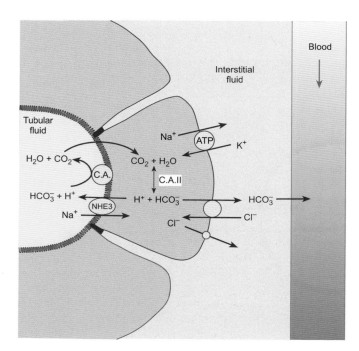

Figure 39.17 Bicarbonate reabsorption in the proximal tubule. Hydrogen ions secreted into the lumen lower the pH of the tubule fluid and this favours the conversion of HCO_3^- to carbonic acid, which is converted to CO_2 and water by the carbonic anhydrase (CA) of the brush border membrane. The CO_2 diffuses down its concentration gradient into the tubule cells where some is reconverted to carbonic acid by intracellular carbonic anhydrase and ionizes to form HCO_3^-, which leaves the basolateral surface of the cell in exchange for chloride. The H^+ that is generated is secreted into the lumen via Na^+–H^+ exchange to promote further HCO_3^- reabsorption.

Regulation of plasma phosphate

Phosphate ions are involved in many aspects of cellular function. A person eating a typical Western diet absorbs between 800 and 1500 mg (c. 25–50 mmol) of phosphate from the gut each day. To maintain phosphate balance, the kidneys excrete a similar amount. The excreted phosphate also serves to buffer hydrogen ions secreted by the distal tubule. Moreover, excess phosphate can bind to plasma calcium, which has a number of undesirable consequences: it decreases the ionized calcium in the plasma; it also leads to a decreased production of calcitriol that results in a decreased calcium uptake by the gut. The fall in the plasma calcium concentration leads to an increased secretion of parathyroid hormone and demineralization of bone (see Chapter 22). For these reasons, plasma phosphate must be closely regulated.

The mechanism by which the kidney regulates plasma phosphate is unusual: the renal threshold for phosphate ($\sim$1.0 mmol l^{-1}) is just below the normal plasma levels and the transport maximum (Tm_{phos}) is $\sim$0.1 mmol min^{-1}. When plasma phosphate is 1.1 mmol l^{-1} (i.e. about 10 per cent above the normal plasma level), the filtered load is $0.125 \times 1.1 = 0.1375 \approx 0.14$ mmol of phosphate min^{-1}. Since the transport maximum is 0.1 mmol min^{-1}, $(0.14 - 0.10) = 0.04$ mmol of phosphate will be excreted per minute. If plasma phosphate rises, all the excess phosphate is excreted. If it falls, less phosphate will be excreted.

The Tm_{phos} varies according to the physiological circumstances. Parathyroid hormone (PTH) regulates phosphate transport by the proximal tubule as follows: PTH binds to its receptors on the basolateral membrane of the tubular cells and these G protein linked receptors activate Gs which stimulates production of cAMP. In turn, protein kinase A is activated and this initiates an endocytotic process that removes transporters from the apical membrane. They are then transported to lysosomes for degradation. So, when the plasma level of PTH is high, phosphate reabsorption is inhibited and phosphate secretion is increased. In addition, various other factors known as **phosphatonins** are implicated in certain bone-wasting disorders and may be involved in the physiological regulation of phosphate balance.

Urate transport by the proximal tubule

Uric acid is the end product of purine metabolism. It circulates in blood as the anion urate, very little of which is bound to plasma proteins (< 5 per cent). There are considerable differences in the way different species excrete urate, and the following account is based on recent studies of renal urate transport in humans.

Urate is freely filtered by the glomerulus and is largely reabsorbed in the proximal tubule, only about 5–10 per cent of the filtered load being excreted. It is taken up across the brush border by an anion exchanger that can utilize other organic anions such as lactate and β-hydroxybutyrate as counter-ions. In addition, proximal tubule cells possess a voltage-dependent transporter for urate which may provide the mechanism for its transport across the basolateral membrane.

The clearance of urate in normal subjects lies in the range 6–9 ml min^{-1}. On an unrestricted diet, urate excretion can exceed 1 g day^{-1}, but it is normally around half this value. If plasma urate concentrations become persistently elevated above about 0.4 mmol l^{-1}, uric acid crystals may become deposited in the joints, giving rise to the painful condition known as **gout**. In addition, as uric acid is not very soluble at low pH, deposits may form in the renal pelvis as kidney stones (renal calculi). Indeed, around 5 per cent of all kidney stones are deposits of uric acid or its salts.

Water absorption in the proximal tubule is directly linked to solute uptake

The uptake of sodium, chloride, glucose, and other solutes by the tubular cells results in a transfer of osmotically active particles from the tubular lumen to the extracellular space. This transport is not accompanied by a full osmotic equivalent of water. As a result, there is a slight increase in the osmolality of the fluid surrounding the renal tubules. To maintain osmotic equilibrium, water moves from the tubular lumen to the extracellular space. While most of this water passes via aquaporin channels in the apical and basolateral membranes (the transcellular pathway), some water apparently reaches the lateral extracellular space via the tight junctions (the paracellular pathway). It is important to recognize that the water absorbed in the proximal tubule is not regulated independently of the reabsorption of solutes (unlike water reabsorption in the distal nephron, which is described in Section 39.8); consequently this is sometimes known as the obligatory phase of water reabsorption. There is some evidence that solutes such as potassium, magnesium, and calcium are partially reabsorbed via the paracellular pathway, along with the osmotically driven uptake of water.

Protein lost in the glomerular filtrate is reabsorbed by pinocytosis

Although ultrafiltration in the glomerulus effectively prevents proteins passing into the filtrate, the glomerular filtrate nevertheless contains a small amount of protein (about 40 mg l^{-1} compared to 65–80 g l^{-1} in plasma). However, as the kidneys form around 180 l of filtrate a day, about 7 g of protein reaches the proximal tubules each day. Such a loss has the potential to adversely affect the body's nitrogen balance. To conserve the filtered proteins, the cells of the proximal tubule engulf the proteins in the filtrate by pinching off a small volume of filtrate containing protein and absorbing it into the cell by **receptor-mediated endocytosis**. (This process is also called **pinocytosis**.) The filtered proteins bind to receptors on the apical membrane known as megalin and cubulin. These receptors have the ability to bind a wide variety of proteins and large peptides, and once they have bound a protein they undergo endocytosis to form vesicles within the tubular cells (endocytotic vesicles). The vesicles are then delivered to the lysosomes with which they fuse. The bound proteins are digested by the lysosomal enzymes, and their constituent amino acids are transported across the basolateral membrane and absorbed back into the bloodstream (Figure 39.19). The megalin and cubulin molecules are then recycled to the apical membrane for further rounds of endocytosis.

Under normal circumstances, only trace amounts of protein are found in the urine. If, however, the glomeruli become diseased,

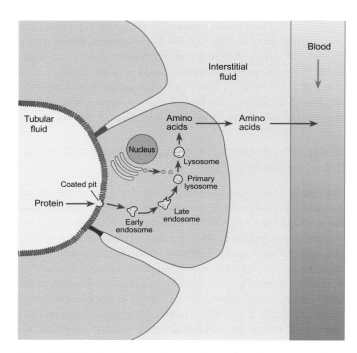

Figure 39.19 Endocytosis and protein breakdown in a proximal tubule cell. Plasma proteins escaping into the filtrate become bound to receptor proteins (megalin and cubulin) that are present at the apical membrane in coated pits. The bound proteins are absorbed into the tubule cells as endosomes prior to their breakdown by lysosomal enzymes.

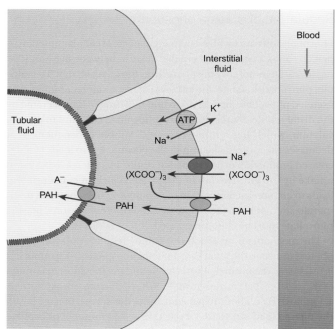

Figure 39.20 The process by which organic anions (e.g. PAH) are secreted into the lumen of the proximal tubule. PAH crosses the basolateral membrane in exchange for di- and tricarboxylate anions, here represented as $(XCOO^-)_3$, and is secreted into the lumen down its chemical gradient in exchange for another organic anion such as urate. The di- and tricarboxylates are reabsorbed into the cell via a sodium-dependent transporter.

significant amounts of protein may pass into the filtrate. Since the proximal tubule has a very limited capacity for reabsorbing protein, some will inevitably be lost to the urine, which will have a frothy appearance as the protein lowers the surface tension. This is known as **proteinuria** which may indicate the presence of a urinary tract infection (UTI), damage to the glomerulus, hypertension, or, in pregnancy, pre-eclampsia. As haemoglobin is small enough to be filtered by the glomeruli, proteinuria also occurs following haemolysis of the red cells, giving the urine a reddish-brown appearance.

Secretion by the cells of the proximal tubule

In addition to their reabsorptive activities, the cells of the proximal tubule actively secrete a variety of substances into the tubule lumen. Many metabolites are eliminated from the blood in this way, including bile salts, creatinine, hippurates (e.g. PAH), prostaglandins, and urate. In addition, the kidney also eliminates many foreign substances by secretion, including drugs such as penicillin, quinine, and salicylates (aspirin). These substances are ionized at physiological pH and two transport systems are involved, one for anions such as PAH and one for cations such as creatinine. Like those responsible for the transport of amino acids and glucose from the lumen into the tubular cells, these carriers are proteins and their transport capacity can be saturated. Since infusion of penicillin can depress the secretion of PAH and other organic anions (and vice versa), these molecules appear to be secreted by the

same transport system. As there is little structural similarity between the various organic anions, the transport is said to be of low specificity. Organic cations are secreted by a separate low-specificity carrier system. Active secretion provides the kidney with a very efficient means of elimination of protein-bound substances that could otherwise only be eliminated very slowly by filtration.

The secretion of organic anions such as PAH occurs by a two-stage process. The anion is taken up into the tubular cell across the basolateral membrane in exchange for α-ketoglutarate and other di- or tri-carboxylate anions that diffuse down their concentration gradients. As the concentration of PAH (or other organic anion) in the cell rises, it passes into the tubule lumen via an anion exchange protein located in the apical membrane. The di- and tri-carboxylates re-enter the cells of the proximal tubule via a sodium-dependent symporter that is located in the basolateral membrane (see Figure 39.20).

Summary

As the filtrate passes along the proximal tubule, all the protein, amino acids, and glucose contained in the fluid are reabsorbed. The absorption of amino acids and glucose is linked to the sodium gradient across the apical membrane, and the driving force for their uptake is the sodium pump of the basolateral membrane.

Almost all of the essential organic constituents in the tubular fluid are absorbed in the first half of the tubule. Additionally, about

80 per cent of the filtered bicarbonate is reabsorbed in the first half of the proximal tubule. As a result, sodium absorption in the second half of the proximal tubule is mainly coupled to that of chloride.

The proximal tubule actively secretes some organic anions and cations into the tubular fluid. Like the uptake of glucose and amino acids, these are carrier-mediated processes, each with its own transport maximum.

The movement of an osmotic equivalent of water accompanies the absorption of solutes so that the fluid leaving the proximal tubule is isotonic with the plasma. By volume, about two-thirds of the filtrate is absorbed in the proximal tubule. Water movement occurs both through the epithelial cells (the transcellular pathway) and via the tight junctions (the paracellular pathway).

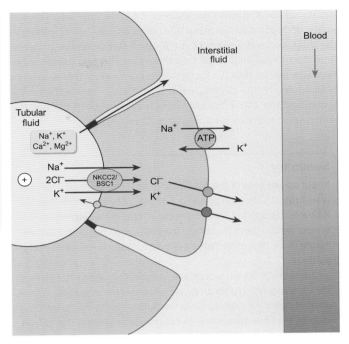

Figure 39.21 The transport processes responsible for the uptake of sodium and chloride in the thick ascending limb of the loop of Henle. Sodium, potassium, and chloride are transported into the tubule cells by an electroneutral cotransporter called NKCC2. (It is also called BSC1, which stands for bumetanide-sensitive sodium-(potassium)-chloride cotransporter 1.) The diffusion of potassium from the tubule cells into the lumen via potassium channels leads to the development of a lumen positive potential, which provides the driving force for the paracellular absorption of cations. Note that in this section of the nephron, the ion movements occur without the osmotically driven uptake of water.

39.6 Tubular transport in the loop of Henle

On entering the descending thin limb of the loop of Henle, the tubular fluid is still more or less isotonic with the plasma. As it passes down the descending limb, the tubular fluid becomes increasingly hypertonic. The cells of the descending thin loop are thin and flattened and do not actively transport significant amounts of salts. From this it follows that the change in the osmolality of the tubular fluid is the result of passive movement of water out of the tubule into the medullary interstitium and of sodium, chloride, and urea from the interstitium into the tubule. The water passes to the interstitium via aquaporin 1 (AQP1) channels, and specific urea transporters (UT-A2) allow urea to pass to the tubular fluid. The osmolality of the tubular fluid reaches its peak at the hairpin bend. For the longest loops, the osmolality may reach 1200 mOsmol kg^{-1} at the tips of the renal papillae. For loops that do not penetrate so deeply into the medulla, the peak osmolality at the hairpin bend will be lower.

The thin ascending limb does not actively pump sodium chloride across its wall. Moreover, its wall is impermeable to water so the fluid it contains does not equilibrate with the interstitium. Therefore, the fluid entering the thick portion of the loop of Henle is strongly hypertonic with respect to the plasma. The epithelial cells of the thick part of the ascending limb possess an electroneutral symporter (known as NKCC2 or BSC-1) that transports sodium, potassium, and chloride ions into the cell across the apical membrane (see Figure 39.21). As for the thin ascending limb, the wall of the thick ascending limb is impermeable to water. Consequently, the osmolality of the tubular fluid falls as the cells of the thick segment transport sodium, potassium, and chloride into the interstitium. By the time the tubular fluid has reached the beginning of the distal tubule, it has an osmolality of about 150 mOsmol kg^{-1} (i.e. about half that of the plasma). The way in which this transport is exploited to generate the osmotic gradient in the renal medulla will be discussed later in the chapter (Section 39.8) as part of the process of osmoregulation.

As with the other epithelial cells of the kidney, the driving force for the uptake of sodium, potassium, and chloride by the cells of the thick limb of the loop of Henle is provided by the sodium gradient established by the sodium pump of the basolateral membrane. Some potassium ions leak back into the tubular fluid through potassium channels and cause the tubular lumen to be positively charged with respect to the interstitial space. This positive electrical gradient provides the driving force for the reabsorption of sodium, potassium, calcium, and magnesium via the paracellular pathway.

Most filtered magnesium is reabsorbed in the thick ascending limb

Only about 10 per cent of plasma magnesium is bound to proteins, the remainder being present as ionized magnesium (Mg^{2+}) or chelated by anions such as citrate and phosphate. The ionized and chelated magnesium is freely filtered in the glomerulus, and most is reabsorbed as the filtrate passes along the nephron. Only about 2–6 per cent is excreted in the urine. About 20–25 per cent of filtered Mg^{2+} is absorbed in the proximal tubule, a further 5–10 per cent is reabsorbed by the distal convoluted tubule, but no significant reabsorption takes place in the collecting duct. The majority of Mg^{2+} reabsorption (c. 65 per cent) takes place in the thick ascending limb of the loop of Henle, where it passes from the tubular fluid to the interstitium via the tight junctions between the tubular cells (i.e. via the paracellular route). This movement is facilitated by the electrochemical gradient between the tubule lumen and the interstitium, the tubule being about 8 mV positive with respect to the surrounding interstitium.

Summary

About 20 per cent of the filtered sodium, chloride, and water are reabsorbed by the loop of Henle. Sodium, potassium, and chloride ions are transported from the tubular fluid by an electroneutral symporter located in the cells of the thick ascending limb. Calcium, magnesium, and other cations are absorbed via the paracellular pathway.

Unlike the proximal tubule, the transport of ions by the cells of the thick ascending limb is not accompanied by an osmotic equivalent of water. As a result, while the fluid entering the descending loop of Henle is isotonic with plasma, that leaving the loop is hypotonic.

39.7 The distal tubules regulate the ionic balance of the body

The reabsorption of sodium, potassium, and water by the proximal tubule and ascending loop of Henle largely takes place regardless of the ionic balance of the body. In the distal tubule and collecting ducts, however, the uptake and secretion of these ions is closely regulated. In addition, the distal tubule and collecting ducts play an important role in both acid–base balance and water balance.

Sodium ion uptake by the distal tubule is regulated by the renin–angiotensin system

The early part of the distal tubule reabsorbs sodium and chloride ions via a symporter similar to that described for the ascending limb of the loop of Henle. The apical membrane is impermeable to water so that the tubular fluid becomes progressively more dilute. In the later part of the distal tubule and in the collecting ducts, sodium reabsorption is linked to potassium secretion by principal cells (P cells). Sodium enters the P cells across the apical membrane via ion channels. It is then pumped out of the cell into the lateral intercellular space by the sodium pump of the basolateral membrane. The potassium that is taken up by the activity of the sodium pump leaves the cell via potassium channels in the apical and basolateral membranes.

About 12 per cent of the filtered load of sodium is reabsorbed in the distal tubule and collecting ducts, and the capacity of these segments to reabsorb sodium is regulated by the activity of the juxtaglomerular apparatus. When the sodium of the fluid in the distal tubule is low, the cells of the macula densa cause the granular cells of the afferent

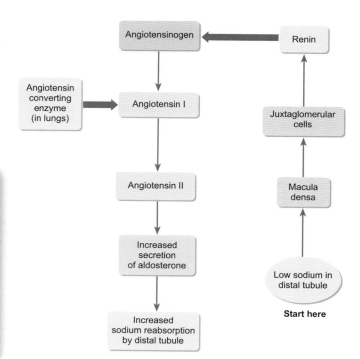

Figure 39.22 The regulation of aldosterone secretion by the sodium load in the distal tubule. The cells of the macula densa sense the sodium load delivered to the distal tubule and regulate the level of circulating renin according to the body's sodium requirements. If the sodium concentration in the tubular fluid is low, the juxtaglomerular cells increase their secretion of renin. This ultimately results in an increased sodium reabsorption.

arteriole to secrete the proteolytic enzyme renin into the blood. The exact process by which the macula densa cells stimulate the granular cells to secrete renin is not yet known. Renin converts a plasma peptide called angiotensinogen into angiotensin I. This, in turn, is converted to angiotensin II by **angiotensin converting enzyme**, which is found on the capillary endothelium of the lungs and some other vascular beds. Angiotensin II acts on the zona glomerulosa cells of the adrenal cortex to stimulate the secretion of the hormone aldosterone (Figure 39.22).

Aldosterone stimulates the production of sodium channels which become inserted in the apical membranes of the P cells. It also stimulates the synthesis of Na^+, K^+-ATPase molecules, which are inserted in the basolateral membrane. By increasing the number of available channels for sodium uptake, aldosterone promotes sodium reabsorption. Sodium moves into the cells down its concentration gradient and is transported to the interstitial fluid by the sodium pump of the basolateral membrane. Increasing the activity of the sodium pump also raises intracellular potassium, which can pass into the tubular fluid down its concentration gradient. These adjustments act to increase the ability of the nephron to reabsorb sodium and to secrete potassium. As this action of aldosterone requires the synthesis of new proteins, its effect is not immediate but is delayed by an hour or so and reaches its maximum after about a day (Figure 39.23).

When the plasma volume is expanded due to an increase in total body sodium, the secretion of renin is inhibited by atrial

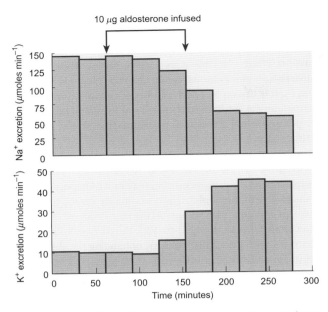

Figure 39.23 The effect of infusing 10 μg aldosterone on sodium and potassium excretion in the adrenalectomized dog. Note the time delay before significant changes in sodium and potassium excretion are seen and that the effect of aldosterone considerably outlasts the period of infusion.

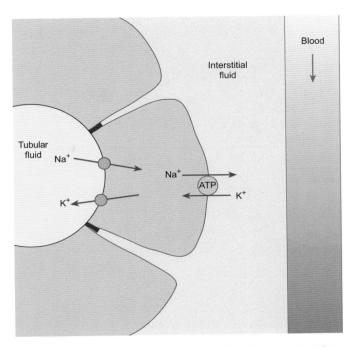

Figure 39.24 The mechanism of secretion of potassium ions into the fluid of the distal tubule and collecting ducts.

natriuretic peptide (ANP). This results in an increase in sodium excretion. The interplay between the renin–angiotensin system and ANP is important for the regulation of body fluid volume (see Chapter 40).

Body potassium balance is regulated by the distal tubules and collecting ducts

A balanced diet is rich in potassium, about 4 grams (100 mmol) normally being ingested each day. As the distribution of potassium ions across the plasma membrane of cells is the principal determinant of their membrane potential, there is a need for the extracellular potassium concentration to be closely regulated. This is achieved by the distal tubule and collecting duct, which actively regulate potassium excretion. Potassium reabsorption is increased in deficiency (**hypokalaemic**) and it is secreted in normal and potassium-retaining (**hyperkalaemic**) states.

By the time the filtrate reaches the distal tubule, nearly 90 per cent of the filtered potassium has been reabsorbed. About two-thirds is normally absorbed by the proximal tubule via the paracellular route, and some 20 per cent is absorbed by the ascending thick limb of the loop of Henle by cotransport with sodium and chloride ions. In the remainder of the nephron, both potassium absorption and potassium secretion can occur and the balance between them determines how much potassium is lost in the urine.

Potassium secretion into the tubular fluid occurs via a transcellular pathway (Figure 39.24). It is taken up into the P cells by the activity of the Na+, K+-ATPase located in the basolateral membrane and is secreted into the tubular fluid via potassium channels located in the apical membrane.

Under normal circumstances the amount of potassium secreted is determined by its concentration in the plasma. If plasma potassium is elevated, this will directly increase potassium uptake into the cells via the Na+, K+-ATPase. Additionally, if plasma potassium is elevated by as little as 0.2 mmol l^{-1}, the secretion of aldosterone by the zona glomerulosa cells of the adrenal cortex is increased. The aldosterone, in turn, stimulates the P cells to synthesize more sodium channels and Na+, K+-ATPase molecules, which become inserted in the apical membranes and basolateral membranes respectively. These changes augment the uptake of sodium from the tubular fluid and increase the secretion of potassium by the P cells. The effect is to restore plasma potassium to its normal level.

The intercalated cells regulate acid–base balance by secreting hydrogen ions

Although most of the filtered bicarbonate is reabsorbed in the proximal tubule and the loop of Henle, there is 1–2 mmol l^{-1} in the fluid entering the distal tubule. Under normal circumstances, all of this bicarbonate is reabsorbed and acidic urine is excreted. The reabsorption of bicarbonate by the intercalated cells differs from that of the proximal tubule. The type A intercalated cells of the collecting ducts actively secrete hydrogen ions into the lumen via an ATP-dependent pump, as shown in Figure 39.25. As in the proximal tubule, the decrease in the pH of the tubular fluid favours the conversion of bicarbonate ions to carbon dioxide and water. The liberated carbon dioxide diffuses down its concentration gradient into the intercalated cells, where soluble carbonic anhydrase (carbonic anhydrase II) catalyses the reformation of carbonic acid which then dissociates into hydrogen ions and bicarbonate ions. The hydrogen ions are secreted into

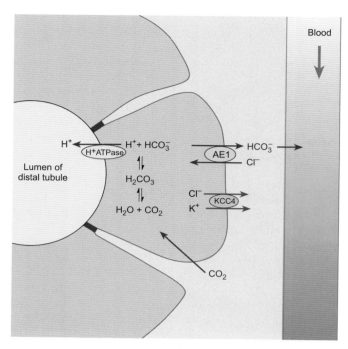

Figure 39.25 The mechanism by which the intercalated cells of the late distal tubule and cortical collecting ducts secrete hydrogen ions. Carbon dioxide and water are converted to carbonic acid by intracellular carbonic anhydrase (not shown) and the hydrogen ions formed by dissociation of carbonic acid are pumped into the tubule lumen by a H^+-ATPase. The excess bicarbonate is exchanged for chloride via an anion exchanger called AE1 which is present in the basolateral membranes. The excess chloride is exported from the cell together with potassium via a cotransporter called KCC4.

the lumen, and the bicarbonate ions exit the cells via a bicarbonate–chloride exchanger located in the basolateral membrane.

Active secretion of hydrogen ions into the lumen of the distal tubule and collecting ducts results in a fall in the pH of the tubular fluid, which can reach values as low as 4–4.5, much lower than elsewhere in the nephron. The apical membranes of the cells of the distal tubule and collecting ducts have a very low passive permeability to protons, which are thereby prevented from diffusing back into the tubular cells. As a pH of 4 corresponds to a free hydrogen ion concentration of only 0.1 mmol l^{-1}, only about 0.15 mmoles of free hydrogen ion can be excreted each day for a normal daily urine output of 1.5 l. Nevertheless, the body produces about 50 mmol of non-volatile acid a day that it needs to excrete. The excretion of this quantity of acid is achieved by the physico-chemical buffering of hydrogen ions with phosphate and by the secretion of ammonium ions (see Chapter 41).

Calcium ions are absorbed from the distal tubule by an active process that can be stimulated by parathyroid hormone

About 70 per cent of the filtered load of calcium is reabsorbed in the proximal tubule, 20 per cent is absorbed by the ascending loop of Henle, and most of the remainder is absorbed by the distal tubule and cortical collecting duct. Only about 1 per cent is normally excreted.

Calcium reabsorption occurs by both transcellular and paracellular routes in the proximal tubule and the ascending limb of the loop of Henle, but in the distal tubule it occurs via specific calcium transport channels known as TRPV5 and TRPV6 (short for **T**ransient **R**eceptor **P**otential cation channel subfamily **V** members 5 and 6). It is driven by passive influx of calcium down its steep electrochemical gradient into the cells of the distal tubule, which then pump calcium across their basolateral membrane into the interstitial space.

Parathyroid hormone plays a major role in calcium homeostasis (see Chapter 22). In the kidneys it activates the PTH receptor, which increases the expression of the calcium transport proteins in the distal part of the nephron and so directly stimulates calcium uptake. In addition, PTH stimulates the conversion of 25-hydroxy-cholecalciferol (calciferdiol) to 1,25-dihydroxycholecalciferol (calcitriol—see Figure 22.10), thereby enhancing the absorption of calcium from the small intestine.

Summary

The first part of the distal tubule continues the dilution of the tubular fluid that began in the ascending thick limb of the loop of Henle. In the second part of the distal tubule and in the collecting duct, the epithelium consists of two cell types, the P cells and the I cells. The P cells absorb sodium and water. They also secrete potassium into the tubular fluid. The I cells secrete hydrogen ions and reabsorb bicarbonate. The efficacy of sodium uptake and potassium secretion is regulated largely by the hormone aldosterone which is, in turn, regulated by the secretion of the enzyme renin from the juxtaglomerular cells of the afferent arterioles.

Although the transcellular movement of calcium accounts for about a third of the uptake of calcium in the proximal tubule, in the distal tubule the entire calcium uptake is via the transcellular route. The transcellular uptake of calcium is driven by passive influx down its electrochemical gradient into the cells of the distal tubule coupled to its active extrusion across the basolateral membrane. Calcium uptake is stimulated by parathyroid hormone (PTH), which also decreases phosphate reabsorption.

39.8 The kidneys regulate the osmolality of the plasma by adjusting the amount of water reabsorbed by the collecting ducts

In a normal individual, water intake varies widely according to circumstances. As a result the osmolality of urine can range from as little as 50 mOsmol kg^{-1} following a large water load, to around 1200 mOsmol kg^{-1} in severe dehydration. How do the kidneys produce urine with such a wide range of osmolality?

Three key facts are fundamental to understanding how the kidney regulates water balance.

- Between the outer border of the renal medulla and the papilla of the renal pyramids, the osmolality of the interstitium progressively increases from about 300 mOsmol kg^{-1} to 1200 mOsmol kg^{-1}.

- The flow of fluid in different parts of the nephron runs in opposite directions (i.e. there is a counter-current arrangement).
- The collecting ducts are impermeable to water unless antidiuretic hormone (ADH) is present.

In essence, the kidney generates an osmotic gradient within the medulla by active transport of sodium, potassium, and chloride from the lumen of the ascending loop of Henle into the interstitial space. This gradient is then used to reabsorb water from the urine as it passes through the medulla. The amount of water reabsorbed is regulated by the level of ADH (vasopressin) circulating in the blood. When the plasma osmolality is low, little ADH is produced and copious dilute urine is produced. Conversely, when plasma osmolality is high ADH secretion is increased. Under such circumstances, water reabsorption in the distal nephron is increased, with the result that a small volume of concentrated urine is produced. The next sections discuss the mechanisms responsible for establishing the medullary osmotic gradient before considering how water reabsorption in the distal nephron is regulated by ADH.

Salt transport by the ascending limb of the loop of Henle leads to the generation of a large osmotic gradient between the renal cortex and the inner medulla

While the fluid in the space between the tubules (the interstitium) in the renal cortex is approximately iso-osmotic with the plasma, the osmolality of the interstitial fluid of the medulla increases progressively from the border with the cortex to the renal papilla. In the outer medulla the osmolality is about 290 mOsmol kg^{-1}, chiefly attributable to sodium and chloride ions. In the inner medulla, the osmolality of the interstitium can reach 1200 mOsmol kg^{-1}, about half of which is attributable to sodium and chloride ions and half to urea. This remarkable osmotic gradient is formed primarily by the active transport of sodium, potassium, and chloride by the thick ascending limb of the loop of Henle without the reabsorption of an osmotic equivalent of water.

The crucial factors for the generation of the osmotic gradient are as follows.

- The fluid in the loop of Henle follows a counter-current arrangement so that fluid returning from the deeper reaches of the medulla flows in the opposite direction to that entering the medulla (see Figures 39.26 and 39.27).
- Sodium transport by the thick ascending limb of the loop of Henle can occur against an osmotic gradient of about 200 mOsmol kg^{-1}.
- The walls of the proximal tubule and descending thin limb are freely permeable to water and have a high passive permeability to sodium, chloride, and urea.
- In contrast, the ascending thick limb of the loop of Henle, the distal tubule and the collecting ducts have little passive permeability to ions or to urea.
- Although the ascending thin and thick limbs of the loop of Henle and the first third of the distal tubule are impermeable to water, they actively transport sodium, potassium, and chloride. As a result, water is separated from solute in this part of the nephron.

The mechanisms by which the osmotic gradient of the medulla is established are outlined in Figure 39.26. Before the gradient is established, the osmolality is the same throughout the nephron (Figure 39.26a). The active transport of sodium and chloride across the tubular epithelium of the ascending thick limb of the loop of Henle and the first part of the distal tubule occurs without concomitant movement of water. The result of this transport is a *decrease* in the osmolality of the tubular fluid in the thick ascending limb of the loop of Henle and an *increase* in the osmolality of the fluid surrounding the tubule (i.e. the fluid of the medullary interstitium) (Figure 39.26b).

As the tubular fluid passes down the descending limb of the loop of Henle, it loses water to the medullary interstitium and gains sodium and chloride ions, so that its osmolality progressively rises as it flows towards the hairpin bend (Figure 39.26c). Moreover, as it equilibrates with the interstitium, the flow of fluid carries sodium and chloride deeper into the medulla, raising its osmolality. Since the ascending limb is impermeable to water, the hypertonic fluid (400 mOsmol kg^{-1}) in the thin ascending limb now enters the thick limb, which transports sodium and chloride into the interstitium until there is a transepithelial osmotic gradient of 200 mOsmol kg^{-1}. This causes the osmolality of the interstitium to increase from 400 mOsmol kg^{-1} to 500 mOsmol kg^{-1} (Figure 39.26d). As more fluid enters the descending loop, it equilibrates with the interstitium, which is now more hypertonic than before. This hypertonic solution enters the ascending thick limb, which transports sodium and chloride until there is a transepithelial osmotic gradient of 200 mOsmol kg^{-1} once more and the osmolality of the interstitium rises from 500 mOsmol kg^{-1} to 600 mOsmol kg^{-1} (Figure 39.26e). This process continues until the osmolality of the tubular fluid at the hairpin bend reaches about 1200 mOsmol kg^{-1}. This high osmolality is chiefly attributable to sodium and chloride ions (1000 mOsmol kg^{-1}), but urea contributes about 200 mOsmol kg^{-1}.

Although the fluid entering the thick limb of the loop of Henle is hypertonic, sodium, potassium, and chloride are progressively removed from it as it flows towards the cortex. Consequently, the fluid entering the distal tubule is hypotonic with respect to the plasma. Indeed, by the time the fluid has reached the middle of the distal tubule, so much sodium chloride has been lost that its osmolality is less than 100 mOsmol kg^{-1} (i.e. less than a third of that of plasma). Overall, the progressive transport of sodium and chloride from the tubular fluid into the interstitium results in the establishment of a longitudinal osmotic gradient in the medulla. *The counter-current arrangement of the loop of Henle thus multiplies a relatively small transepithelial osmotic gradient into a large longitudinal gradient.*

Urea is concentrated in the medullary interstitium by a passive process

Chemical analysis shows that the osmotic pressure of the interstitial fluid of the inner medulla (which may be 1200–1400 mOsmol kg^{-1}) is almost equally attributable to sodium chloride and to urea. (The osmolality of the tubular fluid is, however, mainly due to sodium and chloride.) Like other small solutes, urea is freely filtered at the glomerulus, and a significant fraction of the filtered load is

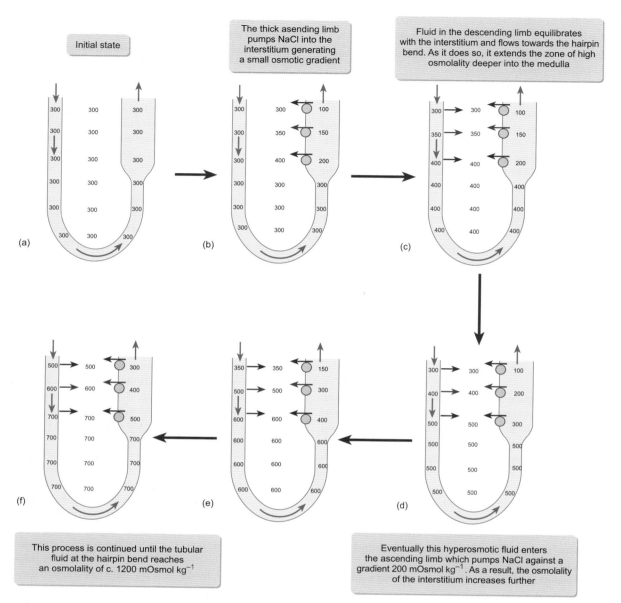

Figure 39.26 A schematic representation of the operation of the counter-current multiplier in the establishment of the osmotic gradient of the renal medulla (see text for further details).

reabsorbed passively together with its osmotic equivalent of water in the proximal tubule. As the tubular fluid reaches the loop of Henle, the urea concentration is still essentially the same as that of plasma, but by the time the fluid reaches the hairpin bend, urea contributes about 200 mOsmol kg⁻¹. This increase in urea concentration is due to passive secretion of urea from the medullary interstitium.

In the thick ascending limb, sodium and chloride transport occurs without the movement of water so that the fluid in the distal tubule is hypotonic with respect to plasma. When the urine is concentrated as a result of the action of ADH on the cortical collecting ducts, the osmolality in this part of the nephron can reach that of the plasma (290 mOsmol kg⁻¹). However, unlike the fluid entering the nephron, sodium and chloride ions account for much less of the osmolality and urea contributes much more because the powerful transport mechanisms of the thick ascending limb and distal tubule have removed most of the sodium and chloride ions.

As the fluid flows into the inner medulla, more of the water in the collecting ducts is reabsorbed under the influence of ADH and the urea concentration in the urine rises until, in the inner medullary collecting ducts, it exceeds that of the interstitium. Thus, urea is concentrated by the abstraction of water. When the urea concentration in the urine is high, its movement from the urine into the medullary interstitium is favoured, as shown in Figure 39.27. Moreover, in the inner medulla, ADH not only increases the permeability of the collecting ducts to water, it also increases their

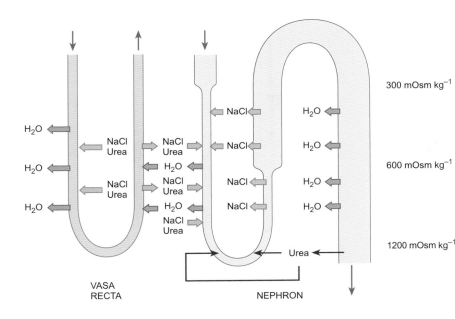

300 mOsm kg^{-1}

600 mOsm kg^{-1}

1200 mOsm kg^{-1}

VASA
RECTA

NEPHRON

Figure 39.27 The mechanism of urea concentration in the renal medulla and the role of the vasa recta in preserving the osmotic gradient. As urine flows towards the renal pelvis, water is reabsorbed by the collecting ducts under the influence of ADH. As a result, the concentration of urea in the urine rises until it exceeds that of the interstitium. In the inner regions of the medulla, ADH increases the permeability of the collecting ducts to urea, which can then be passively reabsorbed down its concentration gradient.

permeability to urea. This adaptation optimizes the conservation of osmotically active urea during dehydration and minimizes its loss from the interstitium during diuresis. There is a further advantage of this adaptation: although urea is the primary end product of nitrogen metabolism, it can be excreted in large amounts without large volumes of water. This situation arises because urea is in osmotic equilibrium across the wall of the collecting ducts when the urine is concentrated.

The vasa recta provide blood flow to the medulla without depleting the osmotic gradient

As described in Section 39.2, the renal medulla receives its blood supply by way of the vasa recta. The blood flow is much less than that of the renal cortex but is sufficient to provide the medullary tissue with nutrients and oxygen. The vasa recta play a further important role: they act to maintain the osmotic gradient of the medulla while removing the ions and water that have been reabsorbed. As the vasa recta are derived from the efferent arterioles of the juxtamedullary glomeruli, the blood entering them is isotonic with normal plasma (280–290 mOsmol kg^{-1}). Like the descending limb of the loop of Henle, the walls of the vasa recta are permeable to salts and water, so that the blood they contain progressively increases in osmolality as it passes into the inner medulla by gaining salt and by losing water (see Figure 39.27). By the time it reaches the deepest parts of the medulla, the blood has an osmolality equal to that of the surrounding interstitium. As the blood returns towards the cortex, the reverse sequence occurs and the blood leaving the vasa recta is only slightly hyperosmotic to normal plasma. The counter-current arrangement of the vasa recta together with their relatively low blood flow thus helps to maintain the osmolality of the renal medulla. During the course of its passage through the medulla, the blood has removed the excess salt and water that have been added by the transport processes occurring in the deeper regions of the medulla.

Antidiuretic hormone (ADH) regulates the absorption of water from the collecting ducts

The absorption of ions by the ascending limb of the loop of Henle results in the tubular fluid becoming hypotonic as it approaches the distal tubule. During its passage along the first third of the distal tubule, sodium, potassium, and chloride continue to be transported from the lumen to the interstitium. Moreover, the tubular epithelium is still relatively impermeable to water so the tubular fluid becomes progressively more dilute. By the time the fluid reaches the last third of the distal tubule, there is a substantial osmotic gradient in favour of water reabsorption. Both this part of the distal tubule and the collecting ducts are impermeable to water unless ADH (vasopressin) is present. This hormone is secreted by the posterior pituitary gland in response to increased plasma osmolality. Its secretion is regulated by sensors known as **osmoreceptors** which are located in the hypothalamus, close to the supraoptic and paraventricular nuclei which produce the hormone and transport it to the posterior pituitary (see Chapter 21). When plasma osmolality is below 290 mOsmol kg^{-1}, ADH secretion by the posterior pituitary is very low and plasma levels are less than 1 pg ml^{-1}. An increase in plasma osmolality of as little as 3 mOsmol kg^{-1} is sufficient to stimulate ADH secretion, and the degree of stimulation depends on the increase in osmolality above the threshold of around 290 mOsmol kg^{-1} (see Figure 39.28). This osmotic regulation of ADH secretion is central to the control of plasma osmolality. The secretion of ADH is inhibited by atrial natriuretic hormone and some drugs such as ethanol.

ADH increases the permeability of the last third of the distal tubule and that of the whole of the collecting duct to water. The result is a movement of water down its osmotic gradient into the tubular cells and thence into the interstitial fluid and the plasma. *This water movement is independent of solute uptake* and therefore results in an increase in urine osmolality and a fall in plasma osmolality that is directly related to the amount of water reabsorbed. Consequently,

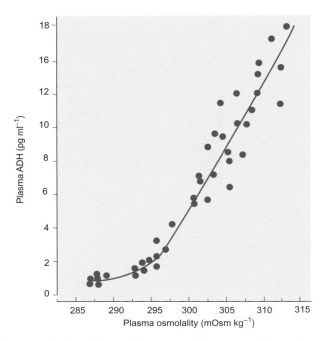

Figure 39.28 The relationship between plasma osmolality and plasma ADH. (Based on data in R.W. Schrier et al. (1979) *Am J Physiol* **236**, F32l–F332.)

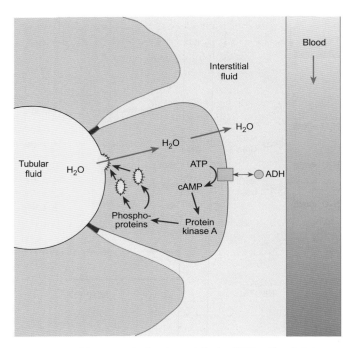

Figure 39.29 The control of water uptake by ADH in the distal tubule and collecting duct. When ADH activates V_2 receptors on the basolateral surface of the tubule cells, it activates an adenylyl cyclase and this leads to an increase in intracellular cAMP. The cAMP activates protein kinase A, which results in the fusion of vesicles containing aquaporins with the apical membrane. An increase in the capacity for water uptake is the result.

when the body has excess water and the plasma osmolality is less than 285 mOsmol kg^{-1}, ADH secretion will be suppressed. Under these circumstances, water will not be reabsorbed during its passage through the collecting ducts so that a large volume of dilute urine will be produced (giving rise to a **diuresis**). In contrast, ADH secretion is stimulated during dehydration as the plasma osmolality is greater than 285 mOsmol kg^{-1}. The secreted ADH acts on the collecting ducts to increase their permeability to water and a smaller volume of more concentrated urine is produced.

Only some 10–15 per cent of the total filtered load of water is subject to regulation by ADH, as the remainder has been reabsorbed along the nephron together with salts and other solutes. If the posterior pituitary is unable to secrete ADH or if the collecting ducts are unable to respond to it, a large volume of dilute urine is produced (**polyuria**). In excess of 15 l of urine can be excreted per day, which must be made up by increased water intake if life-threatening dehydration is to be avoided. This condition is known as diabetes insipidus (see Section 21.8).

Note that the osmolality of the extracellular fluid is controlled independently of its volume, which is determined by the total body sodium. The inter-relationship between the maintenance of osmolality and sodium balance will be discussed in Chapter 40.

ADH controls the insertion of water channels into the apical membrane of the P cells of the collecting ducts.

How does ADH regulate the permeability of the collecting ducts to water? It now appears that ADH circulating in the blood binds to receptors on the basolateral surface of the P cells of the collecting ducts. These receptors are coupled to adenylyl cyclase, and the resulting increase in cyclic AMP activates protein kinase A. The protein kinase A then initiates the fusion of vesicles containing water channels (**aquaporins**) with the apical membrane. When ADH levels fall, part of the apical membrane is endocytosed and the water channels are taken back into the cells to await recycling. The permeability of the apical membrane of the P cells to water is therefore regulated by the insertion and removal of specific water channels, as shown in Figure 39.29. The water that enters the cell across the apical membrane passes freely into the lateral intercellular space from where it enters the plasma. Similar mechanisms are thought to regulate the permeability of the membrane to urea, which crosses cell membranes via specific urea transporters.

The clearance of 'free water'

According to the prevailing water balance of the body, the kidneys excrete urine that may be hypertonic or hypotonic with respect to the plasma. The ability of the kidneys to excrete a concentrated or a dilute urine is sometimes expressed by the unsatisfactory concept of the 'free water' clearance. This is the amount of pure water that must be added to, or subtracted from, the urine to make it isotonic with the plasma. It is, therefore, not calculated in the same way as the clearance of other substances but is calculated as follows:

$$C_{H_2O} = \dot{V} - C_{osm}$$

where C_{H_2O} is the free water clearance, $\dot{V}$ the urine flow rate, and C_{osm} the osmolar clearance which, from the standard definition of clearance, is given by the following equation:

$$C_{osm} = U_{osm} \times \dot{V} / P$$

Suppose that the kidneys are excreting urine with an osmolality of 120 mOsmol kg^{-1} at a rate of 5 ml min^{-1} and plasma osmolality is 285 mOsmol kg^{-1}.

$$C_{H_2O} = 5 - (120 \times 5/285) = 5 - 2.1 = 2.9 \text{ ml min}^{-1}$$

So the free water clearance is 2.9 ml min^{-1} (i.e. the kidneys are excreting 2.9 ml min^{-1} more water than required to keep the urine iso-osmotic with the plasma).

If the urine osmolality was 450 mOsmol kg^{-1} and the urine flow rate was 1 ml min^{-1}:

$$C_{H_2O} = 1 - (450 \times 1/285) = -0.6 \text{ ml min}^{-1}$$

In this case, 0.6 ml min^{-1} less water is excreted than is required to maintain the urine iso-osmotic with the plasma. From these calculations, it is evident that if hypotonic urine is excreted, the kidneys have a positive free water clearance. If the urine is hypertonic, they have a negative free water clearance. When the free water clearance is zero, the urine is iso-osmotic with the plasma.

In the kidneys, 'free water' is generated by the absorption of salts in the ascending thick limb of the loop of Henle and distal tubule. It therefore represents the extent to which the fluid in the distal tubules is diluted with respect to the plasma. How much of the 'free water' is excreted will, of course, depend on the circulating level of ADH as discussed above.

Summary

The osmolality of the renal medulla is largely due to the high concentrations of sodium ions, chloride ions, and urea found in the fluid of the medullary interstitium. The osmotic gradient arises from the counter-current arrangement of the loop of Henle and the active transport of sodium and chloride from the lumen of the ascending limb into the medullary interstitium. This is augmented by passive urea movement from the tubular fluid to the medullary interstitium under the control of ADH.

Active transport of sodium and chloride ions by the ascending thin and thick limbs of the loop of Henle and by the initial part of the distal tubule leads to dilution of the tubular fluid. The circulating level of ADH regulates the subsequent reabsorption of water by the collecting ducts. When there is a significant water load, little ADH is secreted and a copious dilute urine is produced. Under these conditions, the 'free water' clearance is positive. During dehydration, ADH secretion is increased and the collecting ducts reabsorb water so that a small volume of concentrated urine is produced. The 'free water' clearance will be negative.

39.9 The collection and voiding of urine

The renal calyxes, ureters, urinary bladder, and urethra comprise the urinary tract which is concerned with collecting the urine formed by the kidneys and storing it until a convenient time occurs for the

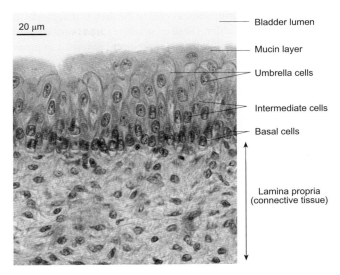

Figure 39.30 The appearance of the bladder epithelium (urothelium) in its unstretched state. This type of epithelium is also known as transitional epithelium. The apical cell membrane of the dome-shaped umbrella cells stains more densely and appears thicker than that of the other cells. This reflects the presence of uroplakin proteins, which are assembled into extensive plaques that cover most of the surface exposed to the urine. The overlying mucin layer is thought to prevent bacterial infection of the urothelium and underlying tissues.

bladder to be emptied. The urinary tract is lined by a specialized transitional epithelium—the **urothelium**—which consists of a layer 3–5 cells thick resting on a basement membrane (Figure 39.30). The basal cells appear to be the precursor cells for the overlying cell layers. The apical layer consists of large specialized cells known as umbrella cells, which are interconnected by tight junctions. Their apical membranes contain uroplakin proteins, which form extensive impermeable plaques that cover 80–90 per cent of the surface. Overlying the surface is a polysaccharide layer of mucin that acts as a defence against infection. The low permeability of the apical surface of the umbrella cells coupled with the tight junctions between the cells renders the urothelium relatively impermeable to water and solutes, protecting the underlying cell layers from the potentially damaging effects of the urine (which may be hypotonic or hypertonic depending on the state of water balance). The integrity of the urothelium is preserved even when it is fully stretched to accommodate a full bladder. Although mammals, including man, can transport sodium across the apical surface of the urothelium, the composition of the urine is not significantly modified during its passage through the lower urinary tract.

Urine passes from the kidney to the bladder by the peristaltic action of the muscle in the wall of the ureters

The ureters are tubes about 30 cm long that consist of an epithelial layer surrounded by circular and longitudinal bundles of smooth muscle. In addition, some muscle fibres are disposed in a spiral arrangement around the ureter. When the renal calyxes and upper regions of the ureters become distended due to the accumulation of urine, peristaltic contractions occur in the ureters that propel the urine towards the bladder. These contractions are almost certainly myogenic in origin and they are sufficiently powerful to propel urine towards the bladder against pressures of 6–13 kPa (50–100 mm Hg). They normally occur at intervals of 10 s to a minute, but their frequency can be modified by the activity of the pelvic nerves. The ureters pass for 2–3 cm obliquely through the bladder wall in a region known as the **trigone** before they finally empty into the bladder just above the neck. This arrangement closes the ends of the ureters when the pressure within the bladder rises above that in the ureter, so preventing the reflux of urine.

The bladder itself consists of two principal parts: the **body** or **fundus**, which serves to collect the urine, and the bladder neck or posterior urethra, which is 2–3 cm in length. The bladder is lined by a mucosal layer which becomes greatly folded into rugae when the bladder is empty. The bladder wall consists of an inner layer of urothelium overlying the lamina propria and three layers of smooth muscle which are surrounded by a layer of connective tissue, the adventitia (Figure 39.31). The muscle is of the single unit type and is known as the **detrusor muscle**. The wall of the bladder neck consists of a higher proportion of elastic tissue interlaced with detrusor muscle. The tension in the wall of the bladder neck keeps this part of the bladder empty of urine during normal filling, and so the posterior urethra behaves as an internal sphincter. The urethra passes through the urogenital diaphragm, which contains a layer of striated muscle called the external sphincter, which is under voluntary control via the **pudendal nerves**.

The micturition reflex is responsible for the voiding of urine

As the bladder fills it becomes distended, and the detrusor muscle becomes stretched and contracts. This basal tone results in a pressure within the bladder (the **intravesical pressure**) of about 200 Pa (2.0 cm H_2O; see Figure 39.32). Further filling results in little change in pressure until the bladder volume reaches 150–250 ml, when the first desire to void is experienced. Further increases in volume lead to increased intravesical pressure until at about 300–350 ml the pressure begins to rise steeply as more urine accumulates in the bladder, further stretching its wall. At these volumes the bladder undergoes periodic reflex contractions and there is an urgent need to urinate. The process by which the bladder normally empties is known as **micturition**. It is a reflex that is controlled by the sacral segments of the spinal cord.

The innervation of the lower urinary tract is shown in a schematic form in Figure 39.33. The ureters, bladder, and internal sphincters have stretch receptors in their walls. The afferent information from these receptors is carried in the visceral afferent fibres of the pelvic nerves to the spinal cord and brainstem. The motor fibres that control bladder function are derived from both divisions of the autonomic nervous system. Parasympathetic fibres from the sacral outflow (S2 and S3) innervate the bladder and internal sphincter via the pelvic nerves. Parasympathetic fibres also control the external sphincter (via the pudendal nerves), while sympathetic postganglionic fibres derived chiefly from spinal segment L2 run in the hypogastric nerve to innervate the bladder and posterior urethra. The sympathetic fibres act to inhibit micturition by decreasing the excitability of the detrusor muscle while exciting the muscle of the internal sphincter. The parasympathetic fibres act to initiate micturition by exciting the detrusor muscle and inhibiting the activity of the smooth muscle of the internal sphincter.

During the storage of urine, the progressive distension of the bladder stimulates the afferent nerve fibres of the pelvic nerves. At low pressures, this activity activates the sympathetic fibres of the hypogastric nerve which cause the smooth muscle of the internal sphincter (the posterior urethra) to constrict. The external sphincter is held closed by activity in the pudendal nerves. These responses are 'guarding reflexes' which act to promote continence. As the intravesical pressure rises, the increased activity in the afferent fibres from the bladder wall activates neurons in the brainstem that send impulses to the spinal cord to inhibit the guarding reflexes and permit voiding.

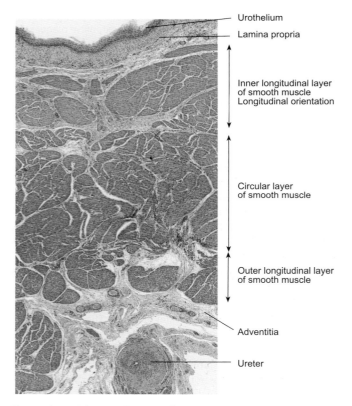

Urothelium
Lamina propria

Inner longitudinal layer
of smooth muscle
Longitudinal orientation

Circular layer
of smooth muscle

Outer longitudinal layer
of smooth muscle

Adventitia

Ureter

Figure 39.31 A low-power view of the bladder wall near to the bladder neck. The three layers of smooth muscle can be seen, although their boundaries are not clearly defined. The urothelium (top) lies on a layer of dense connective tissue, the lamina propria, which lies immediately above the innermost layer of smooth muscle. A thick layer of circular smooth muscle lies between the inner and outer layers of longitudinal smooth muscle. In this section, the outer adventitia abuts a ureter as it approaches the bladder wall.

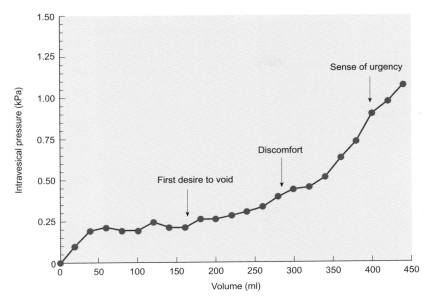

Figure 39.32 The pressure–volume relationship for a normal human bladder. Note that after the initial phase of filling, the volume increases three- to fourfold with little increase in intravesical pressure. As the volume increases further, the pressure rises progressively more and more steeply. Note that bladder capacity is related to body size and can be much greater than shown in this figure. (Based on data in Fig 1 in F.A. Simeone and R.S. Lampson (1937) *Ann Surg* **106**, 413–22.)

As the intravesical pressure rises above about 0.3 kPa, stretch receptors in the bladder wall initiate periodic reflex contractions of the detrusor muscle known as **micturition contractions**. These contractions involve the whole of the detrusor muscle and increase the intravesical pressure by positive feedback and so trigger the desire to urinate. Nevertheless, voiding will not occur unless the muscles surrounding the urethra relax. If voiding does not occur, the micturition reflex becomes suppressed and the intravesical pressure falls. The cycle will repeat itself after an interval of some minutes. If the micturition reflex succeeds in overcoming the tension in the wall of the posterior urethra, then a further stretch reflex is activated that inhibits the external sphincter to permit urination. During voiding, activity in the parasympathetic fibres stimulates the detrusor muscle and relaxes the internal and external sphincters.

Normally, micturition is a voluntary act that is controlled by corticospinal impulses sent to the lumbrosacral region of the spinal cord, and the basic spinal reflexes are inhibited by impulses from the brainstem. During urination, however, these reflexes become facilitated. Urination is further aided by contraction of the abdominal muscles, which raises intra-abdominal pressure, and by relaxation of the muscles of the urogenital diaphragm, which permits the dilation of the urethra. Once urination occurs the bladder is normally almost fully emptied, only a few millilitres of urine remaining.

Urinary incontinence is a common problem seen in the elderly. There are a variety of possible reasons for this, but the most common is that the muscles that control the release of urine from the bladder are weakened. In men, frequent urination with poor flow may result from an enlarged prostate, a condition seen in three-quarters of men over 55.

Section of the spinal cord causes loss of bladder control

In the period that immediately follows section of the spinal cord above the sacral segments (e.g. through a crush injury to the spine in the thoracic region), there is a complete suppression of the micturition reflex. This inhibition is due to spinal shock and the loss of descending control. Clinically, the first priority is to prevent damage to the bladder wall through over-distension. This is achieved by draining the bladder via a catheter. Over time, the bladder reflexes gradually return. Despite this recovery, such patients have no control over when they urinate and micturition is triggered when the pressure within the bladder reaches a threshold level. This is known as **automatic bladder**. In this situation, the urine does not dribble constantly from the urethra but is periodically voided as in a normal subject. A similar situation exists in infants before they have learnt voluntary control.

If the spinal cord is partially crushed, the fibres that normally inhibit the micturition reflex may be damaged while those that facilitate it may remain intact. In this situation, the micturition reflex is facilitated and small volumes of urine are frequently voided in an uncontrolled manner. This is known as **neurogenic uninhibited bladder**.

Some crush injuries and compression of the dorsal roots of the sacral region of the spinal cord can lead to loss of afferent nerve fibres from the bladder. If this happens, information from the stretch receptors of the bladder wall will not reach the spinal cord and the normal tone of the detrusor muscle is lost. In addition, the micturition reflex will be abolished even though the efferent fibres may be intact. This gives rise to a condition known as **atonic bladder**, in which the bladder cannot contract by either voluntary or reflex mechanisms. Instead, it fills to capacity and urine is lost in a continuous dribble through the urethra.

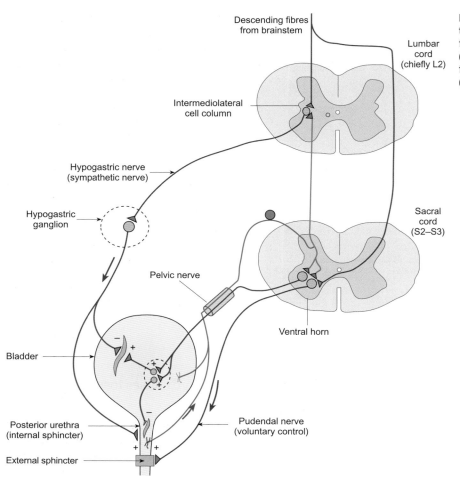

Figure 39.33 The principal nervous pathways that control micturition. Micturition is inhibited by activity in the hypogastric (sympathetic) nerves and pudendal nerves. It is facilitated by activity in the pelvic (parasympathetic) nerves.

Summary

Urine is transported from the kidneys to the bladder by the peristaltic action of the muscle in the wall of the ureters. The intravesical pressure (the pressure within the bladder) increases very slowly at first until 200–300 ml of urine has been accumulated. Thereafter the intravesical pressure rises progressively more steeply. This rise in pressure is associated with a sensation of fullness, and when intravesical pressure reaches about 250 Pa the stretch receptors in the wall of the bladder trigger contractions of the detrusor muscle of the bladder wall known as micturition contractions. If the micturition contractions succeed in overcoming the resistance of the posterior urethra, the micturition reflex is facilitated and voiding occurs. During normal voiding there is a voluntary relaxation of the external sphincter.

39.10 Changes in renal function with age and renal failure

The renal system shows significant changes with age and these may have important consequences for the ability of the kidneys to excrete the metabolites of drugs. When the vessels supplying the nephrons become atherosclerotic and narrow, parts of the kidney may experience a fall in blood flow or even complete ischaemia. This results in a corresponding fall in glomerular filtration rate (GFR) which, by the age of 80, may be as little as 50 per cent of its value at the age of 30. Even at 45 years of age, GFR is less than 90 per cent of its value in young adulthood (Figure 39.34). The chief consequence of these changes is a reduced ability to respond to homeostatic challenges—for example a sodium load or sodium depletion. It may also compromise the ability to respond to an acid or alkaline load so that acid–base balance may be disturbed. Drug dose regimes and timings may need to be altered to compensate for the reduction in the rate of renal clearance of drugs or their metabolites.

Renal failure occurs when the function of the kidneys is depressed to such an extent that they are unable to maintain the composition of the plasma within normal limits. Almost any factor that seriously impairs the function of the kidneys can cause renal failure. Broadly speaking, renal failure may develop rapidly (**acute renal failure**) or it may develop over a considerable period because of a significant loss of functional tissue (**chronic renal failure**). Acute renal failure may be caused by glomerulonephritis (inflammation of the glomerulus resulting from immune reactions to infection), renal ischaemia, or nephrotoxic poisons such as mercuric ions (Hg^{2+}). The first sign of acute renal

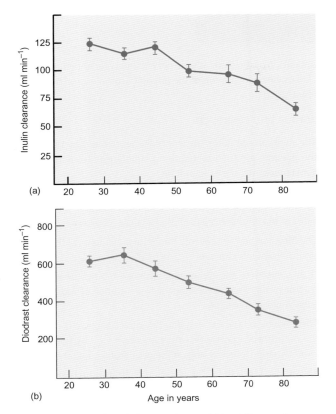

Figure 39.34 Age-related changes in renal blood flow and glomerular filtration rate. Renal blood flow was estimated from diodrast clearance. (Based on data in Fig 1 in D.F. Davies and N.W. Shock (1950) *J Clin Invest* **29**, 496–507.)

failure is a marked reduction in urine output. Chronic renal failure occurs when the proportion of damaged nephrons is so high that the kidneys are unable to perform their normal functions. This generally occurs when the glomerular filtration rate (GFR) falls below about 25 ml min^{-1}. (In healthy young adults it is normally about 120 ml min^{-1}.)

In **glomerulonephritis**, the inflammation of the glomeruli leads to their occlusion and a decrease in the GFR. As a result, the normal transport mechanisms of the proximal convoluted tubule are able to absorb a higher proportion of the filtered Na$^+$ and this results in a low Na$^+$ concentration in the fluid reaching the macula densa. Consequently, the secretion of renin increases and plasma angiotensin II levels become elevated. In turn, this leads to increased secretion of aldosterone from the adrenal cortex and a greater Na$^+$ absorption by the distal tubule. Since the absorbed Na$^+$ will be accompanied by its isotonic equivalent of water, there is fluid retention and oedema develops. This may be followed by hypertension, which persists until the condition subsides.

Aside from glomerulonephritis, the most common cause of acute renal failure is a sudden reduction in cardiac output resulting from heart failure or haemorrhage. This elicits the normal circulatory adjustments discussed in Chapter 40. There is a pronounced vasoconstriction of the afferent arterioles resulting from increased sympathetic activity in the renal nerves and increased plasma levels of adrenaline and vasopressin (ADH). This vasoconstriction overrides the normal autoregulatory response of the renal circulation, and the GFR falls sharply. Since the afferent arterioles are constricted, the pressure in the peritubular capillaries is reduced and there is increased absorption of the tubular fluid (which is mainly isotonic NaCl). This leads to a reduction in the concentration of Na$^+$ in the fluid reaching the macula densa of the distal tubule. In response to this there is an increased secretion of renin, leading to increased plasma levels of angiotensin II and further vasoconstriction. The increased plasma levels of angiotensin II leads to increased secretion of aldosterone from the adrenal cortex and further Na$^+$ absorption in the distal tubule. The decline in the concentration of Na$^+$ in the fluid of the distal tubule tends to decrease the secretion of H$^+$ and K$^+$ into the tubular fluid which, if not corrected, will lead to acidosis and hyperkalemia.

The increased plasma levels of ADH promote further water absorption in the collecting tubule. As a result, the kidneys produce a small volume of concentrated urine that is low in Na$^+$. This renal response will tend to correct the fall in the effective circulating volume (see Chapter 40). Prompt restoration of the effective circulating volume at this stage will restore the normal function of the kidneys. If the circulatory insufficiency is not corrected, the severe vasoconstriction will lead to tissue ischaemia, cell death, and tubular necrosis.

Chronic renal failure results when more than three-quarters of the functional renal tissue is lost. As a result, the GFR falls substantially and the concentration of urea in the blood rises (uraemia). The impaired tubular function leads to failure of normal ionic regulation, acidosis, and the accumulation of metabolites. Unless corrective measures are applied, the accumulation of metabolites (particularly nitrogenous metabolites from protein catabolism) and the disturbance of normal ionic balance lead to CNS depression, coma, and eventually death. In relatively mild chronic renal failure, dietary control of protein intake plus supplementation of the diet by administration of vitamins and essential amino acids may be adequate to maintain the patient in a reasonable state of health. If the GFR falls below about 5 ml min^{-1}, dialysis becomes necessary. A more or less normal lifestyle is possible following renal transplantation.

Summary

Renal failure occurs when the kidneys are unable to regulate the composition of the plasma. It may be acute in onset or it may develop over time (chronic renal failure). In either case, there is a marked reduction in the production of urine. In acute renal failure caused by glomerulonephritis there is activation of the renin–angiotensin system, which leads to fluid retention. Chronic renal failure occurs when more than three-quarters of the functional renal tissue is lost. This leads to a progressive failure of normal ionic regulation and an increased accumulation of urea and other metabolites in the plasma.

✳ Checklist of key terms and concepts

The anatomical organization of the kidneys and urinary tract

- The kidneys lie either side of the vertebral column on the posterior wall of the abdomen.

- Each kidney is covered by a tough, fibrous, inelastic capsule.

- The kidneys consist of an outer, dark brown, cortex and a pale inner region: the medulla and the renal pelvis.

- The functional units of the kidneys are the nephrons.

- Each nephron consists of a renal or Malpighian corpuscle attached to a long, thin, convoluted tubule.

- The renal corpuscle consists of an invaginated sphere (Bowman's capsule) that envelops a tuft of capillaries known as a glomerulus.

- The proximal tubule arises directly from Bowman's capsule. The proximal tubule connects with the descending limb of the loop of Henle.

- The descending loop of Henle passes from the inner border of the renal cortex into the medulla, where it turns and ascends towards the cortex.

- The distal tubule arises from the thick segment of the ascending loop of Henle.

- The distal tubules of a number of nephrons merge via connecting tubules to form collecting ducts, which pass through the cortex and medulla to empty into the renal pelvis.

- The kidneys are innervated by both sympathetic and parasympathetic nerve fibres.

The renal circulation

- The kidneys receive a copious blood supply from the abdominal aorta via the renal arteries.

- The renal artery enters the hilus of the kidney and branches to form the interlobular arteries. These subsequently give rise to the arcuate arteries, which course around the outer medulla.

- The arcuate arteries give rise to cortical radial arteries which supply the renal glomeruli.

- Blood leaving the renal glomeruli passes into the capillaries that surround the renal tubules (the peritubular capillaries).

- Blood from the peritubular capillaries of the renal cortex first drains into stellate veins and thence into the cortical radial veins and arcuate veins.

- At rest, the total renal blood flow (i.e. the blood flow for both kidneys) is about 25 per cent of the cardiac output.

- The renal blood flow remains remarkably constant during changes in arterial blood pressure. This stability of renal blood flow is due to mechanisms intrinsic to the kidneys and is called autoregulation.

- Activation of the sympathetic nerves and catecholamine secretion causes vasoconstriction of the afferent arterioles and thus reduces renal blood flow.

Filtration and clearance

- The nephron regulates the internal environment by first filtering the plasma and then reabsorbing substances from, or secreting substances into, the tubular fluid.

- The filtration barrier of the glomerular capillaries consists of the capillary endothelial cells, a basement membrane, and the podocytes of the epithelial cells of the capsular membrane.

- The filtration barrier prevents the cellular elements of the blood entering Bowman's capsule.

- It also restricts the movement of large molecules such as the plasma proteins while allowing the free filtration of substances with a low molecular weight.

- The amount of fluid passing into Bowman's capsule is governed by the net filtration pressure, which is determined by the balance of hydrodynamic forces acting on the glomerular capillaries.

- The rate at which the kidneys form the ultrafiltrate is known as the glomerular filtration rate or GFR and has units of ml per minute.

- Renal clearance is defined as the volume of plasma completely cleared of a given substance in one minute.

- The clearance of inulin or creatinine is commonly used to estimate the GFR, which is normally in the region of 120 ml min^{-1}.

- If the clearance of a substance is less than the GFR, the substance is neither filtered nor absorbed by the renal tubules.

- If the clearance is greater than the GFR, the substance is secreted by the tubules.

Tubular absorption and secretion

- As the filtrate passes along the proximal tubule virtually all the protein, amino acids, and glucose contained in the fluid are reabsorbed.

- The absorption of amino acids and glucose is linked to the sodium gradient across the apical membrane, and the driving force for their uptake is the sodium pump of the basolateral membrane.

- Almost all of the essential organic constituents in the tubular fluid are absorbed in the first half of the tubule.

- About 80 per cent of the filtered bicarbonate is also reabsorbed in the first half of the proximal tubule. As a result, sodium absorption in the second half of the proximal tubule is mainly coupled to that of chloride.

- In addition to reabsorbing solutes, the proximal tubule actively secretes some organic anions and cations into the tubular fluid. Like the uptake of glucose and amino acids, these are carrier-mediated processes.

- The movement of an osmotic equivalent of water accompanies the absorption of solutes so that the fluid leaving the proximal tubule is isotonic with the plasma.
- By volume, about two-thirds of the filtrate is absorbed in the proximal tubule.

Tubular transport in the loop of Henle

- About 20 per cent of the filtered sodium, chloride, and water are reabsorbed by the loop of Henle.
- Sodium, potassium, and chloride ions are transported from the tubular fluid by a symporter located in the cells of the thick ascending limb.
- Calcium, magnesium, and other cations are absorbed via the paracellular pathway.
- Unlike the proximal tubule, the transport of ions by the cells of the thick ascending limb is not accompanied by an osmotic equivalent of water. As a result, while the fluid entering the descending loop of Henle is isotonic with plasma, that leaving the loop is hypotonic.

The distal tubules regulate the ionic balance of the body

- The first part of the distal tubule continues the dilution of the tubular fluid that began in the ascending thick limb of the loop of Henle.
- In the second part of the distal tubule and in the collecting duct, the epithelium consists of two cell types: the P (principal) cells and the I (intercalated) cells. The P cells absorb sodium and water. They also secrete potassium into the tubular fluid. The I cells secrete hydrogen ions and reabsorb bicarbonate.
- The efficacy of sodium uptake and potassium secretion is regulated largely by the hormone aldosterone which is, in turn, regulated by the secretion of the enzyme renin from the juxtaglomerular cells of the afferent arterioles.
- Although the transcellular movement of calcium accounts for about a third of the uptake of calcium in the proximal tubule, in the distal tubule the entire calcium uptake is via the transcellular route.
- The transcellular uptake of calcium is driven by passive influx down its electrochemical gradient into the tubular cells, coupled to active extrusion of calcium across the basolateral membrane.
- Calcium absorption is stimulated by parathyroid hormone (PTH), which also decreases phosphate reabsorption.

The kidneys regulate the osmolality of the plasma

- The osmolality of the renal medulla is largely due to the high concentrations of sodium ions, chloride ions, and urea found in the fluid of the medullary interstitium.
- The osmotic gradient arises from the counter-current arrangement of the loop of Henle.

- The active transport of sodium and chloride from the lumen of the ascending limb into the medullary interstitium is augmented by passive urea movement from the tubular fluid to the medullary interstitium under the control of ADH.
- Active transport of sodium and chloride ions by the ascending thin and thick limbs of the loop of Henle and by the initial part of the distal tubule reduces the osmolality of the tubular fluid.
- ADH is secreted when plasma osmolality rises above about 290 $mOsmol\ kg^{-1}$ and regulates the subsequent reabsorption of water by the collecting ducts.
- When there is a significant water load, little ADH is secreted and a copious dilute urine is produced. Under these conditions, the 'free water' clearance is positive.
- During dehydration, ADH secretion is increased and the collecting ducts reabsorb water so that a small volume of concentrated urine is produced. The 'free water' clearance will be negative.

The collection and voiding of urine

- Urine is transported from the kidneys to the bladder by the peristaltic action of the muscle in the wall of the ureters.
- The intravesical pressure (the pressure within the bladder) increases very slowly at first until the bladder is about two-thirds full. Thereafter the intravesical pressure rises progressively more steeply.
- This rise in pressure is associated with a sensation of fullness, and when intravesical pressure reaches about 250 Pa the stretch receptors in the wall of the bladder trigger contractions of the detrusor muscle of the bladder wall.
- If these contractions succeed in overcoming the resistance of the posterior urethra, the micturition reflex is facilitated and voiding (micturition) occurs.
- During normal voiding there is a voluntary relaxation of the external sphincter and the bladder is fully emptied.

Renal failure

- Renal failure occurs when the kidneys are unable to regulate the composition of the plasma.
- Renal failure may be acute in onset or it may develop over time (chronic renal failure).
- In both acute and chronic renal failure there is a marked reduction in the production of urine.
- Chronic renal failure occurs when more than three-quarters of the functional renal tissue is lost.
- In chronic renal failure there is a progressive loss of normal ionic regulation and an increased accumulation of urea and other metabolites in the plasma.

Recommended reading

Anatomy of the urinary tract

MacKinnon, P.C.B., and Morris, J.F. (2005) *Oxford textbook of functional anatomy* (2nd edn), Volume 2: *Thorax and abdomen*, pp. 169–78. Oxford University Press, Oxford.

A clear, well-illustrated account of the anatomy and development of the urinary system. There is a short section on imaging of the urinary tract and some useful self-assessment questions with brief answers.

Histology of the urinary tract

Mescher, A. (2015) *Junquieira's basic histology* (14th edn), Chapter 19. McGraw-Hill, New York.

A well-illustrated account of the histology of the kidney with a brief account of the histology of the urinary tract.

Pharmacology and the kidney

Grahame-Smith, D.G., and Aronson, J.K. (2002) *Oxford textbook of clinical pharmacology and drug therapy* (3rd edn), Chapter 26. Oxford University Press, Oxford.

A useful account of the drugs available for treatment of disorders of the kidney and urinary tract. This chapter also includes discussions of drug-induced renal damage and dialysis.

Rang, H.P., Ritter, J.M., Flower, R.J., and Henderson, G. (2015) *Pharmacology* (8th edn), Chapter 29. Churchill-Livingstone, Edinburgh.

A useful account of the action of diuretics.

Renal physiology

Alpern, R.J., Caplan, M., and Moe, O. (2013) *Seldin & Giebisch's 'The kidney—physiology and pathophysiology'* (5th edn). Academic Press, London.

A detailed account of all aspects of renal physiology in health and disease.

Koeppen, B.M., and Stanton, B.A. (2012) *Renal physiology* (5th edn). Mosby, St Louis.

A well-established specialist introductory text.

Lote, C.J. (2012) *Principles of renal physiology* (5th edn). Kulwer Academic, Dordrecht.

An alternative specialist introductory text.

Valtin, H., and Schafer, J.A. (1995) *Renal function* (3rd edn). Little, Brown & Co, Boston.

Now rather old, but a beautifully written introduction to the principles of renal physiology with some useful student exercises.

Review articles

Birder, L., de Groat, W., Mills, I., Morrison, J., Thor, K., and Drake, M. (2010) Neural control of the lower urinary tract: Peripheral and spinal mechanisms. *Neurourology and Urodynamics* **29**, 128–39, doi:10.1002/nau.20837.

Bröer, S. (2008) Amino acid transport across mammalian intestinal and renal epithelia. *Physiological Reviews* **88**, 249–86, doi:10.1152/physrev.00018.2006.

Castrop, H., Höcherl, K., Kurtz, A., Schweda, F., Todorov, V., and Wagner, C. (2010) Physiology of kidney renin. *Physiological Reviews* **90**, 607–73, doi:10.1152/physrev.00011.2009.

Gamba, G., and Friedman, P.A. (2009) Thick ascending limb: The Na$^+$:K$^+$:2Cl$^-$ co-transporter, NKCC2, and the calcium-sensing receptor, CaSR. *Pflugers Archiv—European Journal of Physiology* **458**, 61–76, doi:10.1007/s00424-008-0607-1.

Hoenderop, J.G.J., Nilius, B., and Bindels, R.J.M. (2005) Calcium absorption across epithelia. *Physiological Reviews* **85**, 373–422, doi:10.1152/physrev.00003.2004.

Layton, A.T., Layton, H.E., Dantzler, W.H., and Pannabecker, T.L. (2009) The mammalian urine concentrating mechanism: Hypotheses and uncertainties. *Physiology* **24**, 250–6, doi: 10.1152/physiol.00013.2009.

Mandal, A.K., and Mount, D.B. (2015) The molecular physiology of uric acid homeostasis. *Annual Review of Physiology* **77**, 323–45.

Plans, V., Rickheit, G., and Jentsch,T.J. (2009) Physiological roles of CLC Cl$^-$/H$^+$ exchangers in renal proximal tubules. *Pflugers Archiv—European Journal of Physiology* **458**, 23–37, doi:10.1007/s00424-008-0597-z.

Vallon, V. (2009) Micropuncturing the nephron. *Pflugers Archiv—European Journal of Physiology* **458**, 189–201, doi:10.1007/s00424-008-0581-7.

Verrey, F., Singer, D., Ramadan, T., Vuille-dit-Bille, R.N., Mariotta, L., and Camargo, S.M.R. (2009) Kidney amino acid transport. *Pflugers Archiv—European Journal of Physiology* **458**, 53–60, doi:10.1007/s00424-009-0638-2.

To check that you have mastered the key concepts presented in this chapter, complete the accompanying online self-assessment questions. Go to www.oup.com/uk/pocock5e/

CHAPTER 40

Fluid and electrolyte balance

Chapter contents

This chapter should help you to understand:

- How body water is distributed between the various body fluid compartments

- How the fluid volumes can be measured

- The mechanisms involved in maintaining the fluid balance between body fluid compartments

- How total body fluid is sensed and regulated

- The importance of sodium in the determination of body fluid volume

- How water intake is regulated by thirst

- Some common disorders of fluid balance—dehydration, haemorrhage, and oedema—and the physiological principles that are the basis of their treatment

- Disorders of electrolyte balance

40.1 Introduction

Water is the principal constituent of the human body and is essential for life. In healthy individuals, the volume and osmolality of the tissue fluids are maintained within closely defined limits. This chapter is mainly concerned with the mechanisms that regulate the quantity of water that is present in the body and maintain its distribution between the different body compartments. It ends with a brief discussion of some common disorders of electrolyte balance. The detailed mechanisms by which the osmolality of the body fluids is regulated are discussed in Chapter 39.

40.2 The distribution of body water between compartments

Broadly speaking, body water is distributed between the intracellular fluid or ICF and the extracellular fluid or ECF—see Chapter 3. The extracellular fluid can be further subdivided into the interstitial fluid, the plasma, and the transcellular fluid (see Figure 3.2). The small contribution from the lymph is included in the interstitial fluid.

The proportion of total body weight contributed by water varies with the age and sex of an individual. In both men and women, the water content of the lean body mass (i.e. the non-adipose tissues) is about 73 per cent. Adipose tissue, however, only contains about 10 per cent water. For this reason the proportion of body weight contributed by water varies both between the sexes and with age. Since newborn infants possess little body fat, water accounts for nearly 75 per cent of their body weight. As adipose tissue forms and other tissues develop, this proportion declines so that, by the end of the first year, body water accounts for around 65 per cent of body weight. By the third decade of life, water makes up about 60 per cent of the total body weight of normal healthy adult males. However, as women of the same age have more adipose tissue than men, their body water accounts for a smaller proportion of body weight—about 51 per cent. By the seventh decade of life, body water accounts for about 50 per cent of body weight in males and about 45 per cent in females. Table 3.1 gives the average distribution of water between the major body compartments for males, females, and neonates.

The space between the cells (the interstitium) consists of collagen, hyaluronate, and proteoglycan filaments together with an ultrafiltrate of plasma. The water of the interstitial fluid hydrates the proteoglycan filaments to form a gel (much like a thin jelly) and in normal tissues there is very little free liquid. This important adaptation prevents the extracellular fluid flowing to the lower regions of the body under the influence of gravity. Nevertheless, while exchange of water and solutes between the cells and the tissue fluid occurs mainly through diffusion, there is a bulk flow of isotonic fluid between the capillaries and the interstitium, which is returned to the blood via the lymphatics (see Chapter 31 for further details). When the lymphatic drainage is obstructed, the tissues swell and free fluid is found within the interstitial space (this is known as oedema, see Section 40.6).

The amount of water in the main fluid compartments can be determined by the dilution of specific markers. For a marker to permit the accurate measurement of the volume of a particular compartment, it must be evenly distributed throughout that compartment and it should be physiologically inert (i.e. it should not be metabolized or alter any physiological variable). In practice, it is necessary to correct for the loss of the markers in the urine. Fortunately, the appropriate corrections are not difficult to make.

The plasma volume can be estimated from the dilution of the dye Evans Blue, which does not readily pass across the capillary endothelium into the interstitial space. Radio-labelled albumin has also been used to measure plasma volume. Since the amount of marker injected is known, it is a simple matter to calculate the volume in which it has been diluted (the principle is explained in Box 3.1).

To determine the total body water, a known amount of tritiated water (^3H_2O) or deuterium oxide (^2H_2O) is injected and sufficient time allowed for the label to distribute throughout the body. A sample of blood is then taken and the concentration of label measured. Measurement of the extracellular fluid volume requires a substance that passes freely between the circulation and the interstitial fluid but does not enter the cells. These requirements are met by mannitol (a metabolically inert saccharide) and by inulin, although other markers have been used. The volume of the intracellular fluid is simply the difference between the total body water and the volume of the extracellular fluid.

Osmosis and hydrostatic pressure determine the distribution of water between the fluid compartments of the body

With the exception of the apical membranes of the cells of the distal nephron (see Chapter 39), the plasma membrane of cells is very permeable to water. This permits the free movement of water from one body compartment to another. Two forces govern this movement: osmosis and hydrostatic pressure. The hydrostatic pressure is derived both from the pumping action of the heart and from the influence of gravity.

The movement of fluid between the plasma and the interstitial fluid is determined by the net filtration pressure and the capillary permeability. The net filtration pressure is determined chiefly by the difference between the capillary pressure and the plasma oncotic pressure (i.e. by Starling forces—see Box 31.3). Thus, when the hydrostatic pressure of the capillaries exceeds the plasma oncotic pressure, the hydrodynamic forces favour fluid movement from the capillaries to the interstitial space. When the oncotic pressure of the plasma exceeds the hydrostatic pressure, the hydrodynamic forces favour the absorption of fluid from the

interstitium into the capillaries. Any excess fluid in the tissues is drained by the lymphatic system and re-enters the blood via the subclavian veins (see Chapter 31).

The exchange of water between the cells and the interstitial fluid is governed by osmotic forces. To illustrate how osmotic pressure regulates the movement of water between the intracellular and extracellular compartments, consider what happens when a 70 kg man drinks a litre of water. As the water is absorbed, the osmolality of the extracellular fluid (ECF) falls. If there was no exchange of water between the intracellular fluid (ICF) and the ECF, the volume of the ECF (~20 per cent of 70 kg) would increase from 14 litres to 15 litres and its osmolality would fall from 285 mOsm kg^{-1} to 270 mOsm kg^{-1}—assuming that no solutes entered or left the ECF. This would result in a 15 mOsm kg^{-1} gradient in favour of water movement into the cells or of solute movement out of the cells. However, as the plasma membrane of the cells is much more permeable to water than it is to ions and other small hydrophilic solutes, water moves from the ECF into the ICF and total body water increases from 42 litres to 43, litres limiting the fall in osmolality to 2 mOsm kg^{-1} (rather than 15 mOsm kg^{-1}). Similar considerations govern the movement of water from the ICF to the ECF in response to water loss, e.g. during sweating. It should now be clear that the intracellular and extracellular compartments are normally in osmotic equilibrium and any departure from this situation will be transitory.

Changes in osmolality and the regulation of cell volume

Mammalian cells are highly permeable to water and swell or shrink when placed in solutions that are not isotonic. In hypotonic solutions the cells will swell while in hypertonic solutions they will shrink (see Chapter 3). When cells swell in response to a hypotonic challenge they regulate their volume by activating channels that permit the efflux of potassium and chloride ions as well as the efflux of taurine (2-aminoethanesulphonic acid—a derivative of the sulphur-containing amino acid cysteine). Conversely, cells respond to a hypertonic challenge by activating Na$^+$/H$^+$ exchange and the NKCC1 cotransporter which imports sodium, potassium, and chloride into the cytosol, so increasing its osmolality. By changing their internal concentration of small osmotically active particles, cells are able to adjust the osmolality of their cytosol to match the surrounding fluid.

When cells are exposed to hypertonic media for a considerable period of time (for example in the renal medulla) they take up small organic solutes (e.g. myo-inositol and N,N,N-trimethyl glycine—better known as betaine) and convert glucose to sorbitol. This enables them to achieve osmotic balance without disturbing their normal ionic equilibria. Within the body, changes in osmolality are usually gradual and cells are able to adjust the osmolality of their cytosol to maintain cell volume more or less constant (isovolumetric regulation).

Various pathological conditions have the potential to cause significant changes in cell volume. These include electrolyte disturbances (deficit or excess of sodium: hypo- and hypernatraemia), ischaemia, severe hypoxia, and dehydration. Exercise, which raises extracellular potassium, also causes changes in cell volume.

Summary

Body water is distributed between the intracellular fluid and the extracellular fluid. The extracellular fluid is further subdivided into the plasma and the interstitial fluid. Total body water can be estimated using specific markers. The volume of the intracellular fluid is given by the difference between total body water and the volume of the extracellular fluid.

The movement of fluid between the plasma and the interstitial fluid is governed by Starling forces, while that between the interstitial fluid and the intracellular fluid is governed by osmotic forces. Changes in osmolality affect cell volume. Cells respond by adjusting the osmolality of their cytosol to maintain their volume more or less constant.

40.3 Body fluid osmolality and volume are regulated independently

If a person drinks a litre of water, their urine output will increase rapidly (a diuresis). Urine output peaks within an hour of drinking and returns to normal about an hour later, by which time the excess water has been eliminated (Figure 40.1). If the same person were to drink a litre of isotonic saline, there would be no diuresis and only a very small rise in urine output. In this case, it takes many hours for the body to eliminate the excess fluid. In both situations, there is an initial increase in total body water but when there is a pure water load, the osmolality of the body fluids falls. To restore the normal osmolality of the plasma, only the excess water needs to be excreted. In contrast, when a litre of isotonic saline is drunk, the osmolality of the tissue fluids is unchanged but the total volume of the ECF is increased. To restore the normal ECF volume, the body must eliminate both the excess salt and the excess water. This simple

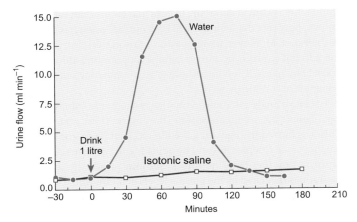

Figure 40.1 The effect of drinking 1 litre of water or 1 litre of isotonic saline on the urine flow rate of a normal subject in water balance. Drinking occurred at the arrow. The urine flow rate increased from about 1 ml min⁻¹ to nearly 15 ml min⁻¹ following the intake of a litre of water. Moreover, the excess water was excreted in about 2 hours, by which time the urine flow rate had returned to normal levels. In contrast, drinking 1 litre of isotonic saline had little effect on urine flow rate. (Adapted from H.W. Smith (1956) *Principles of renal physiology*. Oxford University Press, New York.)

experiment illustrates an important principle: the osmolality and volume of the body fluids are regulated by separate mechanisms. Excessive loss of water from the body is known as **dehydration** and results when water intake is not sufficient to balance water loss.

Body water balance is maintained by the activity of osmoreceptors in the hypothalamus

Water requirements depend on body size, specifically on body surface area, as this determines the extent of water loss via the skin and lungs. This loss, together with the water loss via the urine and faeces, is replaced by water in the diet and by that generated during metabolism. For a normal adult male with a body weight of 70 kg, a typical balance sheet could be:

Water gains (ml 24 h⁻¹)		Water losses (ml 24 h⁻¹)	
Fluid intake (drinking)	1600	Urine production	1500
Water content of food	500	Loss via skin	500
Water generated by metabolism	500	Loss via lungs	500
		Loss in faeces	100
Total	2600	Total	2600

The water loss from the lungs and skin obviously depends on prevailing conditions. In temperate climates, water is lost from the lungs and skin without sweating. This is called **insensible water loss** and cannot be reduced, so any restriction of fluid intake must be balanced by a decline in urine output. As the urine osmolality cannot be greater than about 1250 mOsm kg⁻¹ and as the quantity of solids excreted in the urine each day is between 50 and 70 g (chiefly as sodium chloride and urea), the minimum volume of urine required for excretion is about 700 ml per day. To balance this and the other losses, a minimum fluid intake of about 1.75 l each day is required just to maintain water balance. As noted above, this figure is for a 70 kg adult in a temperate environment. In hot environments there is additional water loss via sweating, which may amount to several litres a day and which must be matched by an appropriate intake of water.

The full extent of water exchange within the body is not revealed by these considerations. In addition to the obvious water exchanges that occur between the body and the environment, the gastrointestinal tract secretes and reabsorbs some 7–8 l of fluid each day and the kidneys form about 180 l of filtrate a day, of which about 178.5 l are normally reabsorbed. Any reduction in fluid reabsorption arising from a disturbance of gastrointestinal or renal function will therefore have dramatic consequences for water balance. This is discussed further in the following sections.

Thirst is the physiological mechanism for replacement of lost water

To maintain the osmolality of the body fluids, it is essential that the water lost from the lungs together with that lost in the urine, sweat, and faeces is replaced. Two sources provide water: oxidative metabolism of fats and carbohydrates, and water intake via the diet. In humans, the generation of water during metabolism is not sufficient to meet the needs of the body and drinking is essential for the maintenance of water balance. The stimulus for water intake is thirst, which may be defined as the appetite for water.

What factors stimulate drinking?

When the osmolality of the body fluids rises by about 4 mOsm kg^{-1}, the desire for water is stimulated. The state of water balance is monitored by the osmoreceptors of the anterior hypothalamus. Drinking behaviour can also be triggered by electrically stimulating the preoptic area of the hypothalamus, or by injecting hypertonic solutions into the same brain region. This suggests that the osmoreceptors that stimulate ADH secretion also play an important part in the regulation of water intake. Fluid loss through diarrhoea or haemorrhage both cause a loss of isotonic fluid which results in a reduction in the circulating volume and this also stimulates thirst—probably as a result of increased plasma levels of angiotensin II. Thus, an increase in the osmolality of the ECF or a decrease in the circulating volume will lead to an increased thirst and to an increase in water intake by drinking.

Drinking provides relief from thirst long before the gastrointestinal tract has been able to absorb the ingested water. Nevertheless, if water intake was not sufficient to satisfy the body's needs, drinking is resumed after a period of 15–20 min. As inflation of a balloon in the stomach can inhibit drinking behaviour, this regulation of water intake is probably mediated by stretch receptors in the stomach wall. The ability of the body to determine water intake by the degree of distension of the stomach avoids excessive intake of water and the consequent dilution of the body fluids.

During dehydration, water is initially lost from the extracellular compartment, but since there is an osmotic equilibrium between the extracellular and intracellular compartments, the loss of water will ultimately result in cellular dehydration. The increase in plasma osmolality during dehydration is detected by the osmoreceptors of the hypothalamus which, in turn, stimulate ADH secretion from the posterior pituitary. ADH then acts on the distal nephron to increase water reabsorption (see Chapter 39). There is a reduction in urine flow rate and an increase in urine osmolality. As a result, body water is conserved. Restoration of water balance requires an increase in water intake and the high plasma osmolality stimulates thirst. The increased water intake restores plasma osmolality to normal. Conversely, a water load causes a fall in plasma osmolality. This is detected by the osmoreceptors, which inhibit the secretion of ADH from the posterior pituitary (see Chapter 21). As a result, water reabsorption in the distal nephron declines and the urine flow rate increases while urine osmolality falls. The net effect is an increase in solute-free water excretion and the restoration of plasma osmolality to its normal value. These processes are summarized in Figure 40.2.

Alterations to the effective circulating volume regulate sodium balance via changes in the activity of the renal sympathetic nerves

As it is the most abundant ion in the extracellular fluid, sodium is the principal determinant of plasma osmolality, and since the osmolality of the plasma is closely regulated, total body sodium is

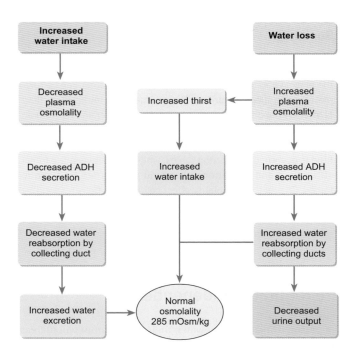

Figure 40.2 The principal mechanisms by which the osmolality of the body fluids is restored following a water load (left side) or pure water loss (right side).

also the principal determinant of body fluid volume. Moreover, as the equilibrium between the extracellular fluid and plasma volume is determined by Starling forces, any change in the total body sodium will affect both the volume of the extracellular fluid and that of the plasma.

In healthy people, the effective circulating volume or ECV (the degree of 'fullness' of the circulatory system) is essentially constant and the sodium chloride and water losses are balanced by dietary intake. While the regulation of osmolality is relatively well understood, little is known about the factors that control salt intake. It is known, however, that animals will seek out salt when it is deficient in their diet. Moreover, patients with Addison's disease (in which the adrenal cortex is no longer capable of secreting aldosterone) crave salty foods. Therefore, there is an appetite for salt which is regulated according to need.

The loss of body sodium is mainly governed by the kidneys, which can regulate the amount of sodium they excrete over a wide range. This is achieved by regulating the filtered load (glomerulotubular balance—see Chapter 39) and by adjusting the amount of sodium absorbed by the distal nephron. Sodium uptake by the distal nephron is regulated by the plasma levels of renin, which influence the circulating levels of aldosterone via the formation of angiotensin II, as discussed in the previous subsection. When the ECV is changed, various mechanisms operate to adjust whole body sodium content so that the normal situation is restored.

Where are the ECV receptors and how do they act to promote alterations to sodium balance?

Cardiac output, vascular tone, and the effective circulating volume determine both the systemic arterial blood pressure and the

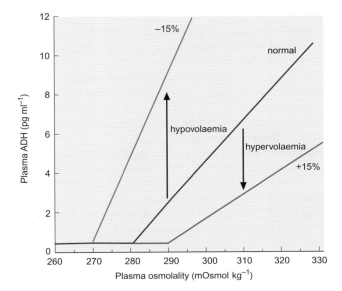

Figure 40.3 The influence of changes in the effective circulating volume on the secretion of ADH. Note that the slope of the relationship between ADH secretion and plasma osmolality is altered during deviations from normovolaemia. If the body fluid falls by 15 per cent (hypovolaemia), both the secretion of ADH and the sensitivity to changes in plasma osmolality are increased (the slope of the relationship is steeper). If the body fluid is increased by 15 per cent (hypervolaemia), the reverse applies, and ADH secretion is decreased and is less sensitive to changes in plasma osmolality. (Adapted from Fig 6 in G.L. Robertson et al. (1976) *Kidney International* **10**, 25–37.)

end-diastolic pressure. Systemic blood pressure is regulated by the baroreceptor reflex, whereby a reduction in pressure leads to an increase in cardiac output and total peripheral resistance as discussed earlier (Section 30.4). The end-diastolic pressure is sensed by low-pressure receptors in the great veins, the pulmonary circulation, and the cardiac atria (these are also known as the central volume receptors). When the ECV is diminished, central venous pressure falls and this results in a decline in the afferent activity of the central volume receptors. This decline elicits a sympathetic reflex that results in a peripheral vasoconstriction.

Atrial myocytes also play a role in regulating the ECV. In response to increased end-diastolic pressure, these cells secrete a hormone called atrial natriuretic peptide (ANP) which acts on the renal tubules to promote sodium excretion.

In addition to its regulation by plasma osmolality, the secretion of ADH by the posterior pituitary gland is also modulated by the arterial baroreceptors and the central volume receptors. As shown in Figure 40.3, a decrease in extracellular volume (**hypovolaemia**) leads to an increase in the secretion of ADH while an increase in volume (**hypervolaemia**) leads to a fall in the secretion of ADH.

When the ECV falls

When the ECV falls there is an increase in the activity of the renal sympathetic nerves which acts to promote sodium reabsorption and thus restore ECV. The principal mechanisms involved are illustrated in Figure 40.4.

Firstly, in response to the reduced filling pressures of the right side of the heart, there is vasoconstriction of the arterioles of the

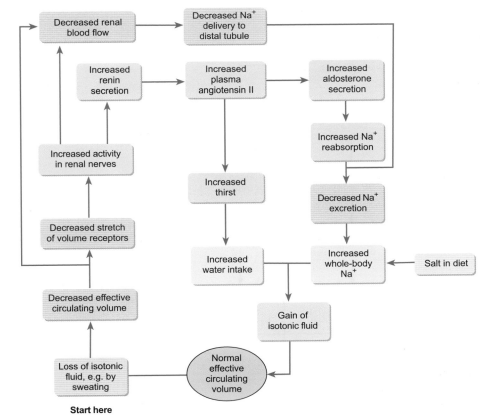

Figure 40.4 The principal mechanisms responsible for restoring the effective circulating volume (ECV) following loss of isotonic fluid (e.g. as a result of heavy sweating or diarrhoea).

muscle and splanchnic vascular beds. This is a reflex response that is mediated via the sympathetic nerves. This reflex is triggered by the unloading of the stretch receptors of the atria and great veins (the central volume receptors) and occurs before there is any significant change in arterial blood pressure. The afferent and efferent arterioles of the nephrons become constricted, reducing the glomerular filtration rate. Consequently, the filtered load of sodium declines and a higher proportion of the filtered sodium is reabsorbed (see Chapter 39).

Secondly, the increased sympathetic activity leads to an increase in renin secretion which, in its turn, leads to an increase in the amount of aldosterone in the circulation. As described in Chapter 39, aldosterone increases the ability of the distal nephron to reabsorb sodium. In addition, the increased activity of the renal sympathetic nerves directly stimulates the uptake of sodium by the cells of the proximal tubule. The secretion of ADH is increased, promoting the reabsorption of water in the distal nephron.

Finally, as these changes serve to increase sodium reabsorption by the renal tubules, there will be an increase in plasma osmolality which, together with the increased plasma levels of angiotensin II, will stimulate thirst. The increased water intake together with the increased retention of dietary sodium leads to an increase in ECV and the restoration of body fluid volume.

When the ECV rises above normal

When the ECV rises above normal, full correction requires elimination of both the excess sodium and the excess water. This is achieved by the following processes.

First, an increase in ECV activates both the arterial baroreceptors and the central volume receptors. This triggers the baroreceptor reflex and the activity of the renal sympathetic nerves is decreased. The afferent and efferent arterioles dilate, and both renal blood flow and GFR increase. This increases the filtered load of sodium and the delivery of sodium to the distal tubule.

Second, renin secretion is inhibited and the concentration of aldosterone in the plasma falls. As a result, less of the filtered sodium is reabsorbed. In addition, the secretion of ADH is decreased (Figure 40.3). These changes promote the excretion of both salt and water by the kidneys.

Third, the increased stretch of the atrial myocytes triggers the secretion of ANP which acts to increase sodium excretion. ANP causes dilatation of the renal afferent and efferent arterioles and thus augments both renal blood flow and GFR.

Fourth, ANP has actions that antagonize those of the renin–angiotensin system. It inhibits aldosterone formation and directly inhibits sodium chloride uptake by the distal tubule and collecting duct. Moreover, ANP also inhibits both ADH secretion and its action on the distal nephron, so promoting the loss of water. Overall, these effects lead to a loss of sodium chloride and water and re-establishment of the normal ECV and body fluid volumes. These events are summarized in Figure 40.5.

Summary

The osmolality of the body fluids is regulated by the activity of the osmoreceptors in the hypothalamus, which control the secretion of ADH from the posterior pituitary. Extracellular fluid volume is sensed via the effective circulating volume and it is regulated by adjustment of sodium intake and excretion. Thus, osmolality and fluid volume are regulated independently of each other.

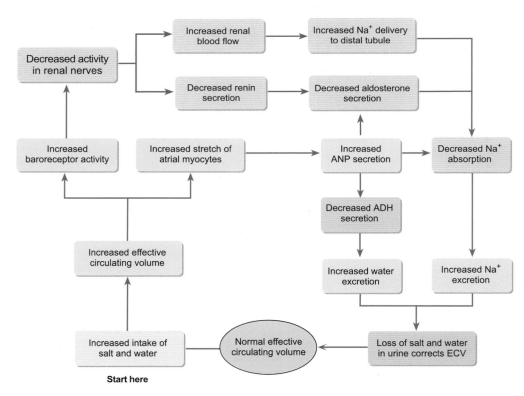

Figure 40.5 The principal mechanisms responsible for restoring the ECV after its expansion by the intake of salt and water. ANP is atrial natriuretic peptide.

During a water load, plasma ADH levels fall and there is less water reabsorption by the distal nephron. As a result, the excess solute-free water is excreted. When the body is dehydrated, the osmolality of the tissue fluids rises and ADH secretion is increased. This conserves body water by increasing water reabsorption from the distal tubule and collecting ducts. In addition, the increase in plasma osmolality stimulates thirst.

In response to a fall in the ECV, the kidneys increase sodium reabsorption via activation of the renin–angiotensin system. As angiotensin II stimulates thirst, water intake is increased and the ECV restored. When ECV is increased, there is a decrease in tubular absorption of sodium. An osmotic equivalent of water is also excreted and this results in the restoration of the normal ECV.

40.4 Dehydration and disorders of water balance

Dehydration has a number of causes:

- Excessive water loss from the lungs and sweat glands
- Excessive urine production
- Excessive loss of fluids from the gastrointestinal tract, either by persistent vomiting or as a result of chronic diarrhoea
- Inadequate intake of fluid and electrolytes.

In most of these situations both water and electrolytes are lost. For example, after heavy exercise significant quantities of water and salt are lost in the sweat and both need to be replaced. (Sweat contains 30–90 mmol sodium chloride per litre.) If the lost water is replaced by drinking but the lost salt is not replaced, there is a deficiency of sodium chloride and the osmolality of the tissue fluids falls. This fall in tissue fluid osmolality results in muscle cramps (known as heat cramp), which may be relieved by drinking a weak saline solution (0.5 per cent sodium chloride) or by drinking water and taking salt tablets. Many athletes use specially formulated sports drinks to aid rehydration and replace lost electrolytes.

Pure water loss can occur when drinking water is unavailable or in situations where the kidneys are unable to reabsorb water from the distal tubule and collecting ducts. The classical example of this is **diabetes insipidus**, where ADH secretion by the posterior pituitary is insufficient (central diabetes insipidus—see Chapter 21), although it can also arise if the kidneys are unable to respond to ADH (nephrogenic diabetes insipidus). Despite their deficiency, patients with diabetes insipidus remain in good health provided that enough water is available for them to drink. If they are rendered unconscious, however, their situation rapidly becomes perilous. In the normal course of events, pure water loss causes a rise in plasma osmolality and an increased thirst. Central diabetes insipidus can be controlled by adminstration of desmopressin (a synthetic analogue of ADH).

When water intake is persistently less than water loss, there is a progressive dehydration of the tissues. When more than 6–10 per cent of body water has been lost, the plasma volume falls and circulatory failure commences. The poor circulation causes a failure of urine production and a metabolic acidosis develops. In addition, during severe dehydration, the lack of water leads to a reduction in evaporative heat loss and fever may occur. The fever may be associated with drowsiness and delirium, and unless urgent corrective measures are taken and the lost fluid is replaced, these signs will be followed by coma and eventually death.

Over-hydration of the tissues (water intoxication) is much less common. Nevertheless, when it occurs, a new osmotic equilibrium between the plasma and the tissues is established and the cells swell. When the brain cells swell in cerebral oedema, the intracranial pressure increases and brain function becomes impaired. The clinical signs include nausea, headache, fits, and coma. The commonest cause is acute renal failure, where the ingested water cannot be excreted (**oliguria**, where < 400 ml of urine is produced per 24 hours). Inappropriate ADH secretion either from the posterior pituitary (perhaps as a result of a head injury) or from tumours can also lead to water retention. Failure of the anterior pituitary gland to secrete ACTH also leads to water intoxication, as the excretion of water depends on normal circulating levels of glucocorticoids. The exact mechanism of this effect is not known, but it may reflect a role of glucocorticoids in regulating the secretion of ADH.

Oral rehydration therapy

In vomiting and diarrhoea there is a large loss of water and electrolytes, as the fluids of the gastrointestinal tract are isotonic with the blood. This results in both dehydration and changes in acid–base balance (see Chapter 41). As the average person secretes about 7 litres of fluid into the gastrointestinal tract each day, persistent diarrhoea can lead to severe dehydration even in mild gastroenteritis. In cholera, the fluid loss is greater than with other causes of diarrhoea as the causative organism *Vibrio cholerae* stimulates intestinal secretions through the action of its toxin on cyclic AMP production. Consequently, fluid loss in the stools during an attack of cholera may be 10 litres or more each day. Clearly, unless the effects of the diarrhoea are rapidly countered, death will inevitably occur. Indeed, in many poor countries dehydration caused by fluid loss in the stools is a common cause of death, particularly in children.

Drinking water will not be sufficient by itself to combat the loss of fluid, as the lost fluid will have been isotonic with the plasma and both sodium chloride and water will be required to rehydrate the tissues. While intravenous infusions of Hartmann's solution (an isotonic physiological solution that contains 0.6 per cent sodium chloride, 0.32 per cent sodium lactate, and smaller quantities of other electrolytes) could be used to restore body fluid volume, this approach needs suitable resources. Nevertheless, it may be the only route of administration in cases of persistent vomiting and diarrhoea. The use of simple oral rehydration fluids has been found to provide a very effective alternative therapy. In diarrhoea the intestinal absorption of glucose is unimpaired, despite the high fluid loss in the stools. Oral rehydration is achieved by giving the patients a solution of salt and sugar to drink. The sugar (as glucose) is absorbed across the intestinal wall by co-transport with sodium, and water follows osmotically. The solution must not be significantly hypertonic, otherwise water loss will be enhanced. The use of sucrose or starch has the advantage that these sugars are readily available. Moreover, as they are broken down to glucose in the intestine before being absorbed, the amount of glucose available for absorption (together with sodium and water) can be increased without making the rehydration fluid hypertonic.

Summary

In dehydration there is a loss of water which is usually accompanied by a loss of electrolytes. Pure water loss can occur in diabetes insipidus or when drinking water is not available. If more than 6–10 per cent of body water is lost, plasma volume falls, resulting in circulatory failure. Water intoxication (over-hydration) is much less common than dehydration but can occur in renal failure. In vomiting and diarrhoea there is a large loss of both water and electrolytes. Restoration of body fluid balance therefore requires replacement of both salts and water.

40.5 Haemorrhage

If there is an acute loss of blood, the cardiovascular system rapidly adjusts to maintain blood pressure and to preserve blood flow to vital organs (see Figure 40.6). Nevertheless, the response of the body to blood loss depends on the amount lost. In conscious people, significant cardiovascular adjustments begin to occur as blood loss exceeds about 5 per cent (c. 250 ml). Initially the loss of blood leads to a diminished venous return and a reduced stimulation of the low-pressure receptors. This results in a vasoconstriction of the cutaneous, muscle, and splanchnic vessels, which occurs before

there is any significant change in either mean blood pressure or pulse pressure.

As blood loss increases, venous return falls further and there is a fall in both cardiac output and blood pressure. The fall in blood pressure activates the arterial baroreceptor reflex so that there is an increase in heart rate and in arteriolar tone. Together, these changes restore the blood pressure. The increase in sympathetic drive also increases venous tone. This results in a mobilization of the blood from the capacitance vessels to the distribution vessels and so helps to prevent the cardiac output falling further than it otherwise would. These adjustments occur rapidly following blood loss.

If blood loss continues, the cardiac output falls further and the activity of the sympathetic nerves becomes more intense. The heart rate continues to rise and the peripheral vasoconstriction increasingly diverts blood away from the skin, muscles, and viscera towards the brain and heart. As plasma levels of angiotensin II rise during the early stages of haemorrhage, this hormone may contribute to the early vasoconstriction. Despite the fall in cardiac output, these changes help to maintain blood pressure and they are the principal mechanisms by which the cardiovascular system adjusts to mild haemorrhage (loss of less than 10 per cent of the blood volume).

Further loss of blood intensifies the activation of the sympathetic nerves, but venous return is no longer sufficient to maintain blood pressure, which begins to fall. During this phase, catecholamine secretion by the adrenal medulla increases, as does the secretion of ADH by the posterior pituitary. These hormones further intensify the vasoconstriction. As a consequence of these adjustments, the pulse is weak and rapid, the skin is cold and clammy, and the mouth becomes dry as salivary secretion stops.

When blood loss reaches about 20 per cent the patient faints (known as syncope), there is a fall in heart rate, and blood pressure

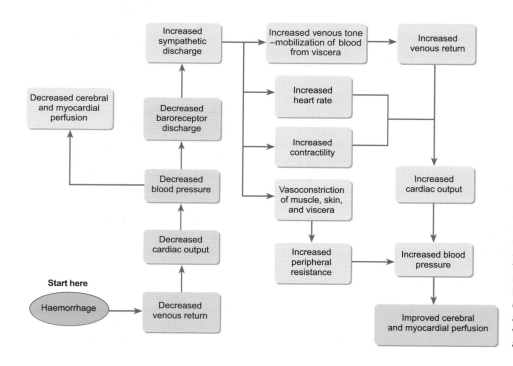

Figure 40.6 The haemodynamic changes that occur during moderate haemorrhage. Although cardiac output falls following blood loss, the compensatory mechanisms indicated act to restore the blood pressure so that perfusion of the brain and the heart muscle is not compromised. Note, however, that these cardiovascular adjustments result in a decrease in blood flow to other vascular beds, particularly the renal and splanchnic circulations.

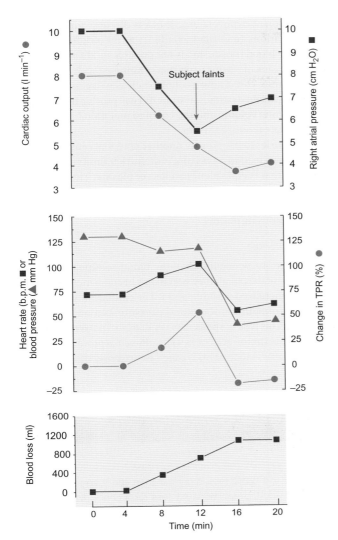

Figure 40.7 The cardiovascular changes that occur during acute blood loss. The bottom panel shows the amount of blood lost by a human subject. The centre panel shows the changes in heart rate, systolic blood pressure, and total peripheral resistance, and the top panel shows the changes in cardiac output and right atrial pressure. Note that cardiac output and right atrial pressure (equivalent to central venous pressure) progressively fall during the haemorrhage. The heart rate, systolic blood pressure, and total peripheral resistance show a biphasic response, rising during the early phase of blood loss and abruptly falling after 1080 ml of blood had been lost. At this point, the subject fainted. (Adapted from H. Barcroft et al. (1944) *Lancet* **243**, 489–91.)

falls precipitously. Note that in this stage of severe haemorrhage the patient has a bradycardia despite the extensive blood loss (see Figure 40.7). The bradycardia is the result of a vagal reflex, but the decline in peripheral resistance appears to result from inhibition of the central sympathetic drive. Catecholamine secretion from the adrenal medulla, ADH secretion, and angiotensin II all continue to increase. Curiously, cardiac output does not necessarily fall further and may even rise a little. Further blood loss results in intense sympathetic activation with pronounced vasoconstriction and tachycardia.

This description of the cardiovascular responses to haemorrhage assumes that there are no other factors influencing the autonomic responses. In practice, haemorrhage is often accompanied by traumatic tissue injury and the sympathetic response to the trauma tends to mask the bradycardia. The reflex bradycardia is also reduced by administration of general anaesthetics and may be complicated by drug therapy.

Following extensive blood loss, there is an urgent need to restore blood volume. Clinically, this can be achieved by blood transfusion or by infusion of fluids that expand the plasma volume, e.g. Hartmann's solution or a plasma substitute.

The physiological processes responsible for restoring the plasma volume are as follows.

1. Peripheral vasoconstriction lowers capillary pressure so that the oncotic pressure of the plasma draws fluid from the extracellular space to the circulation. In addition, there is reabsorption of fluid from the gastrointestinal tract. The increase in ADH secretion and diminished renal blood flow cause a marked fall in urine output (oliguria). These are the first steps towards restoration of blood volume.

2. The fall in renal blood flow increases renin secretion and thus increases the level of circulating angiotensin II, which increases the secretion of aldosterone by the adrenal cortex, which in turn increases sodium reabsorption by the renal tubules and colon. These events lead to the retention of sodium.

3. Loss of ECV also leads to intense thirst (probably as a result of the increased formation of angiotensin II) which increases water intake.

Taken together, these processes bring about the relatively rapid restoration of the ECV and permit adequate perfusion of the tissues provided that the haemorrhage has been stemmed. The secretion of both angiotensin and ADH aids spontaneous recovery from haemorrhage by minimizing water loss and increasing sodium retention by the kidneys. If antagonists of either angiotensin or ADH are administered, recovery is delayed.

Over the ensuing several days, the lost plasma proteins are replaced by the liver, but the absorption of fluid from the interstitial space leads to a further reduction in haemoglobin concentration (known as haemodilution). In the first two weeks after blood loss, the reticulocyte count is elevated (revealing a rapid increase in red cell production). Thereafter, the red cell count increases slowly, taking some weeks to return to normal. The increase in red cell production is stimulated by the hormone erythropoietin, which is secreted by the kidney in response to tissue hypoxia (see Chapter 25).

Circulatory shock

Circulatory shock exists when the circulation fails to provide adequate perfusion of the tissues. Once it has reached a critical stage, **circulatory shock** becomes irreversible. The failure of the circulation leads to the accumulation of metabolites and toxins within the vascular beds. These then cause more peripheral vasodilatation and a further fall in perfusion pressure until complete circulatory failure occurs and the patient dies.

There are four main types of circulatory shock.

1. Haemorrhagic or hypovolaemic shock, in which the circulating volume is inadequate to sustain the blood pressure.

2. Distributive shock due to excessive vasodilatation (e.g. in septicaemia), in which the capacity of the circulation exceeds the ability of the heart to pump sufficient blood to maintain an adequate blood pressure.

3. Obstructive shock, which is caused by restriction of the venous return or of ventricular filling.

4. Cardiogenic shock, in which the ability of the heart to pump blood is impaired. This is one aspect of heart failure (see Chapter 28).

Hypovolaemic shock

The normal circulatory adjustments that are made in response to haemorrhage have already been described. The clinical management should be to eliminate the cause of blood loss and replace the lost blood by transfusion where this is possible. A basic classification of hypovolaemic shock is discussed in Box 40.1. In some situations (e.g. road accidents and battlefield injuries), immediate blood transfusion is not possible. In these cases, intravenous infusion of plasma or plasma substitutes (such as 0.9 per cent saline solution containing a colloid such as polygelatin or a high-molecular-weight dextran, 'Dextran 70') will provide fluid to expand the circulating volume and thus help to maintain blood pressure and adequate tissue perfusion.

Burn injuries to the skin are another potential cause of hypovolaemia. These lead to a loss of a protein-rich fluid from the damaged area that is equivalent to a loss of plasma. The haematocrit rises as the ECV falls, and in cases of severe burns the fall in ECV may be sufficient to produce hypovolaemic shock, which can be treated by infusing plasma.

Distributive shock

Distributive shock occurs when the blood volume is normal but the volume of the circulatory system has become increased as a result of a generalized vasodilatation. The cardiac output is then insufficient to maintain the blood pressure and adequate tissue perfusion. This type of shock may result from a powerful emotional experience (extreme grief or fear) resulting in a powerful inhibition of sympathetic activity. This is called **neurogenic shock**.

Box 40.1 Classification of hypovolaemic shock

The clinical response to haemorrhage depends on the extent of blood loss. The assessment of blood loss is, however, not straightforward. Measurements of blood haemoglobin or the haematocrit are unreliable indicators of the extent of blood loss in acute haemorrhage. They do not change until there has been a significant redistribution of fluid from the interstitial space, which can take between 24 and 36 hours. No single measure gives a reliable indication of the extent of blood loss, but a combination of signs provides a means of classifying the degree of hypovolaemia into four categories.

Class I Blood loss is less than 15 per cent of blood volume (i.e. less than 750 ml for an adult). There are no signs of shock. The blood pressure and pulse pressure are normal but there may be a slight tachycardia. The capillary refill time is normal (< 2 sec). (This is assessed by pressing the subject's skin for 5 seconds with sufficient pressure to blanch the skin, releasing the pressure, and measuring the time taken for the colour of the skin to return to normal.) Respiration is normal and the subject is mentally alert. Normal healthy adults can sustain this level of blood loss without ill effect. In blood donation, 450 ml of blood (a blood unit) are routinely taken.

Class II. When blood loss is between 15 and 30 per cent of blood volume, the clinical signs of shock become evident. The pulse is rapid (the heart rate is greater than 100 b.p.m.) and because of the tachycardia the diastolic pressure is increased, although systolic blood pressure is normal. Consequently, the pulse pressure is reduced. Capillary refill time is slower than normal (i.e. > 2 seconds). The subject is pale, thirsty, anxious, and breathing rapidly (> 20 breaths min^{-1}). This extent of blood loss is rarely life-threatening. Nevertheless, if blood loss is greater than about 20 per cent (one litre for an adult, proportionately less for a child), transfusion of blood or a plasma replacement should be considered.

Class III. When blood loss is 30–40 per cent of blood volume, the pulse is weak and usually rapid. Be aware, however, that there may be a bradycardia (see text). Systolic blood pressure is very low and capillary refill time is slow. Breathing is fast (> 20 breaths min^{-1}) and shallow. The subject is confused, lethargic, pale, and has clammy skin. In white subjects, the skin may be cyanosed (blue). Urine output is low (10–20 ml h^{-1}).

Class IV If blood loss is greater than 40 per cent of blood volume, systolic blood pressure is very low, there is tachycardia, and the skin is pale and cold. Breathing is fast and shallow. Urine output is very low—less than 10 ml h^{-1}—and may fail altogether. The patient is drowsy and confused and may become unconscious. Following this degree of blood loss, the shock may quickly become irreversible and urgent measures are required to restore the circulating volume in order to prevent circulatory collapse.

If severe (i.e. class IV) hypovolaemic shock is suspected, the condition of the patient can be assessed by measuring the central venous pressure and its response to a rapid infusion of 100–200 ml of saline. If the patient is hypovolaemic, there will be little change in central venous pressure but there will be an improvement in cardiovascular performance shown by a fall in heart rate and a rise in blood pressure.

Distributive shock may also occur when a person is exposed to an antigen to which they have become previously sensitized. The resulting reaction releases histamine in large quantities and causes a profound peripheral vasodilatation. There is an increase in capillary permeability and fluid loss to the interstitial space. This is called **anaphylactic shock**.

Severe hypotension analogous to distributive shock can also arise following administration of local anaesthetics during epidural anaesthesia and spinal anaesthesia. The activity in the sympathetic nerves is blocked and the blood vessels dilate. The hypotension can be countered by administration of a sympathomimetic drug such as **ephedrine**, which acts to promote the release of noradenaline from sympathetic nerve endings.

In severe bacterial infection, the release of toxins into the blood causes both vasodilatation and an increase in capillary permeability. This type of distributive shock is known as **septic shock**. As with hypovolaemic shock, the treatment of distributive shock should be aimed at eliminating the source of the vasodilatation and restoring the circulating volume.

Obstructive shock

Obstructive shock arises when the venous return is inadequate or the heart is unable to fill adequately. Restricted venous return may occur as a direct consequence of a large pulmonary embolism, limiting the flow of blood in the pulmonary circulation. If the pericardium becomes inflamed (pericarditis), fluid may collect in the pericardial sac (a pericardial effusion) and the pericardium becomes stretched. As the fluid accumulates, the pressure outside the heart rises, preventing the proper filling of the ventricles during diastole. The result is a fall in cardiac output. This condition is known as **cardiac tamponade** and is associated with a raised jugular venous pressure (see Chapter 30).

Summary

The response of the body to haemorrhage depends on the extent of blood loss. In mild and moderate haemorrhage, the blood pressure is maintained by increased levels of sympathetic activity. Heart rate, peripheral resistance, and venous tone all increase and maintain a normal cardiac output and blood pressure. When blood loss increases beyond 20–30 per cent there is a fall in heart rate and blood pressure, and the individual faints. Any further fall in blood flow to the brain and heart may be fatal. After haemorrhage, the blood volume is progressively restored by a number of adjustments. Starling forces act in favour of fluid uptake by the plasma from the interstitium and the increase in angiotensin II and aldosterone promote both retention of sodium and an increased thirst. These changes act to restore the ECV. Over the ensuing days, the lost plasma proteins are replaced by the liver. Restoration of the red cell count may take several weeks.

The principal types of circulatory shock are hypovolaemic shock, distributive shock, obstructive shock, and cardiogenic shock. Unless circulatory shock is treated promptly, a vicious downward spiral occurs which will lead to death.

40.6 Oedema

Oedema is the abnormal accumulation of fluid in the interstitial space. It arises when alterations to the Starling forces occur as a result of various pathologies. As described in Chapter 31, fluid moves from the plasma to the interstitium when the capillary hydrostatic pressure exceeds the sum of the plasma oncotic pressure and the hydrostatic pressure within the tissues. Fluid moves from the tissues to the plasma when the sum of the plasma oncotic pressure and the tissue hydrostatic pressure exceeds the hydrostatic pressure in the capillaries. Under normal conditions, about 8 l of fluid pass from the circulation into the interstitium each day. Of this about half is reabsorbed by the circulation, either in the tissues or in the lymph nodes, and the remainder is returned to the circulation as lymph mainly via the thoracic duct which drains into the subclavian vein on the left side.

The hydrostatic pressure in the capillaries is normally closely regulated by the tone of the afferent arterioles. However, the average capillary pressure also depends on the venous pressure and the ratio of pre- and postcapillary resistance. A small increase in venous pressure has a disproportionately large effect on the capillary pressure and on the absorption of tissue fluid (see Box 40.2). Thus, when venous pressure is elevated as a result of a venous thrombosis, or of chronic right-sided heart failure, the average capillary pressure is increased and more fluid passes from the plasma to the tissues. The resulting fall in plasma volume is reflected in a fall in the ECV, and this in turn leads to the retention of sodium and water by the mechanisms discussed earlier in the chapter. A situation thus exists in which fluid can progressively accumulate in the tissues. As the fluid accumulates, the hydrostatic pressure in the tissues increases and opposes further accumulation of fluid in the interstitial space. A new equilibrium is established between the plasma and the volume of the interstitial fluid.

Since about half of the fluid passing from the capillaries to the interstitial space is returned to the circulation via the lymphatic drainage, any obstruction of the flow of lymph will lead to fluid accumulation in the affected region. In the industrialized countries of Western Europe and North America, lymphatic insufficiency is relatively rare but is seen when the lymph nodes have been damaged during radical surgery (Figure 40.8) or where cancerous growths have invaded the lymph glands (lymphomas). In developing countries, oedema resulting from obstruction of the lymphatic circulation is commonly the result of the invasion of the lymph nodes by parasitic nematode worms (filariasis). This results in obstruction of lymph flow from the limbs and scrotum that is manifest by a gross oedema (lymphoedema) known as elephantiasis.

Oedema also occurs when the plasma oncotic pressure is low. In this situation, the net filtration pressure rises and fluid accumulates in the tissues. This can occur during nephritis, when the glomerular capillaries become abnormally permeable to albumin and other plasma proteins, so that significant quantities of protein are lost in the urine. It may also arise when the liver is unable to synthesize adequate quantities of the plasma proteins. A similar situation arises during severe malnutrition, when the diet may be rich in carbohydrate but contains little or no protein. This gives

1_23 Box 40.2 Starling forces and oedema

The direction of fluid movement between a capillary and the surrounding interstitial fluid depends on four factors (Starling forces—see Box 31.3):

1. capillary pressure (P_c)
2. interstitial pressure (P_i)
3. oncotic pressure exerted by the plasma proteins (π_p)
4. oncotic pressure of the proteins present within the interstitial fluid (π_i).

The algebraic sum of the various pressures P_f is called the net filtration pressure. It is given by the following equation:

$$P_f = (P_c - P_i) - (\pi_p - \pi_i) \qquad [1]$$

The capillary pressure depends on the pressure in the arteries, the venous pressure, and the ratio of pre- and postcapillary resistance (R_a/R_v). This exact relationship can be determined as follows: the blood flow from an artery to the mid-point of a capillary depends on the pressure difference and the precapillary resistance (Darcy's law; see Box 30.1):

$$\text{blood flow} = \frac{(P_a - P_c)}{R_a}$$

Similarly, the blood flow from the mid-point of a capillary into the vein is:

$$\text{blood flow} = \frac{(P_c - P_v)}{R_v}$$

Since virtually all the blood entering the capillary leaves via the veins (only a very small proportion is returned via the lymphatic drainage):

$$\frac{(P_a - P_c)}{R_a} \cong \frac{(P_c - P_v)}{R_v}$$

which can be rearranged to give the capillary pressure:

$$P_c = \frac{\left(P_a + P_v\left(\dfrac{R_a}{R_v}\right)\right)}{\left(1 + \dfrac{R_a}{R_v}\right)} \qquad [2]$$

Normally the precapillary resistance (R_a) is about four times that of the post-capillary resistance (R_v) and the capillary pressure is determined mainly by the arteriolar resistance. An important consequence of this relationship is that a small increase in venous pressure has a similar effect on the filtration pressure (see worked example below). Thus, when venous pressure is raised, average capillary pressure (P_c) rises and filtration is favoured over absorption. Fluid accumulates in the tissues, giving rise to oedema. However, the swelling of the tissues raises the interstitial pressure and this reduces the net filtration pressure. A new balance becomes established and further accumulation of fluid in the tissues is prevented.

A fall in the oncotic pressure in the plasma raises the net filtration pressure, and fluid may accumulate in the tissues. In practice, oedema does not develop until the protein content of the plasma falls below about 30 g l^{-1}. This may arise because of liver failure, proteinuria, **malabsorption**, or malnutrition. If the colloid osmotic pressure of the proteins in the interstitium is increased (e.g. due to an increase in capillary permeability following an inflammatory reaction), this also has the effect of increasing the net filtration pressure, as the presence of additional protein in the interstitial fluid diminishes the osmotic force drawing fluid from the interstitium to the plasma.

A worked example

Assume that the arterial pressure is 13.3 kPa (100 mm Hg), that the venous pressure is 0.67 kPa (5 mm Hg), and that the precapillary resistance is four times that of the postcapillary resistance. Inserting these values into equation (2) gives the capillary pressure:

$$P_c = (13.3 + (0.67 \times 4)) \div (1 + 4) = 3.2 \text{ kPa (24 mm Hg)}$$

If venous pressure increases to 1.33 kPa (10 mm Hg), the capillary pressure increases:

$$P_c = (13.3 + (1.33 \times 4)) \div (1 + 4) = 3.73 \text{ kPa (28 mm Hg)}$$

Hence, the doubling of venous pressure results in a rise of capillary pressure of 0.53 kPa (c. 4 mm Hg). This entire rise contributes to the net filtration pressure, as the following calculation shows. Take the values of oncotic pressures as 3.5 kPa (c. 26 mm Hg) for plasma and 2 kPa (15 mm Hg) for interstitial fluid, and take the hydrostatic pressure of interstitial fluid (P_i) as about 0.2 kPa (1.5 mm Hg) below that of the atmosphere (i.e. −0.2 kPa or −1.5 mm Hg). The net filtration pressure is 1.9 kPa (14 mmHg) when capillary pressure is 3.2 kPa (24 mm Hg) and 2.4 kPa (18 mm Hg) when capillary pressure is 3.73 kPa (28 mmHg). The whole of the increase in venous pressure has resulted in a similar increase in capillary pressure, which directly increases the rate of fluid filtration. If the lymphatics become overloaded, the affected region will develop oedema.

rise to a disease known as **kwashiorkor**, which is common in children in the poorer parts of the world. A typical example is shown in Figure 40.9.

The release of certain local mediators such as histamine causes the capillaries to become permeable to albumin and other plasma proteins. This results in an increase in the net filtration pressure of the capillaries, and fluid accumulates in the affected area. The local swelling that accompanies insect bites arises in this way.

Systemic oedema first appears in the lower parts of the body (the dependent regions of the body), particularly in the ankles as the venous pressure in the legs is elevated during prolonged periods of standing. Oedema in the ankles may be distinguished from tissue

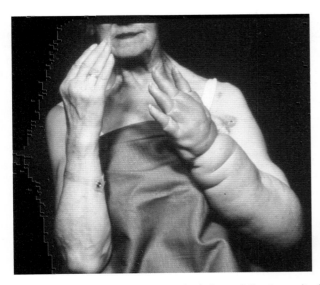

Figure 40.8 Severe lymphoedema in the left arm following radical surgery to treat breast cancer. (Courtesy of Professor R. Levick.)

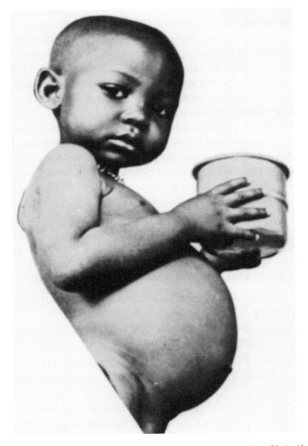

Figure 40.9 The appearance of a child with kwashiorkor. Note the widespread oedema, particularly the swollen abdomen. (Courtesy of WHO.)

fat by applying firm pressure to the affected area with a finger or thumb for a short period. If oedema is present, the pressure will have forced fluid from the area and a depression in the skin will remain for some time after the pressure has been removed (pitting oedema—see Figure 40.10). If the swelling is due simply to tissue fat, the skin springs back as soon as the pressure is removed.

In healthy people, hydrostatic oedema can arise when the leg muscles are relatively inactive and the muscle pumps contribute little to the venous return from the lower body. There is venous pooling and swelling of the ankles. This situation is exacerbated in persons with varicose veins, where the walls of the veins have become stretched rendering their valves incompetent. The accumulation of excess fluid is readily reversed by a short period of rest in a horizontal position or by mild exercise.

The fluids of the serosal spaces are separated from the extracellular fluid by an epithelial layer. These include the fluids of the pericardial, pleural, and peritoneal spaces. These fluids are essentially ultrafiltrates of plasma and their formation is governed by Starling forces. Normally the amount of fluid in these spaces is relatively small as the plasma oncotic pressure exceeds the hydrostatic pressure in the capillaries, but in certain disease states abnormal accumulations of fluid occur. For example, the volume of fluid between the visceral and parietal pleural membranes is normally only about 10 ml, but when fluid formation exceeds reabsorption, fluid accumulates between the pleural membranes in a process known as **pleural effusion**. The alveoli of the lungs are normally lined with a thin film of fluid, but in pneumonia and some other disease states fluid can accumulate in the alveoli (**pulmonary oedema**) and this impairs gas exchange.

An accumulation of excessive amounts of fluid in the peritoneal cavity known as **ascites** can arise when there is a rise in pressure within the hepatic venous circulation, when there is obstruction of hepatic lymph flow, or when plasma albumin is abnormally low (as in kwashiorkor). It also occurs during right-sided heart failure when the pressure within the systemic veins rises.

Treatment of oedema with diuretics

From the previous section, it is clear that oedema can arise as a result of various pathologies. Effective treatment requires identification and elimination of the underlying cause. It may be desirable to eliminate the oedema, and in many cases this can be achieved by treating the patient with drugs that promote the loss of both sodium and water in the urine. Since this results in an increase in urine output known as a diuresis, these drugs are called diuretics. They are classified according to their modes of action. Diuretics may act indirectly by exerting an osmotic pressure that is sufficient to inhibit the reabsorption of water and sodium chloride from the renal tubules, or they may act directly by inhibiting active transport in various parts of the nephron, as shown in Figure 40.11. Note that lymphoedema cannot be treated in this way.

Osmotic diuretics, such as the sugar mannitol, are filtered at the glomerulus but they are not transported by the cells of the proximal tubule. In consequence, as other substances are transported and the proportion of the original filtered volume falls, these substances accumulate and exert sufficient osmotic pressure to inhibit tubular

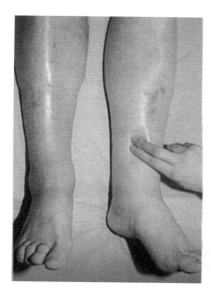
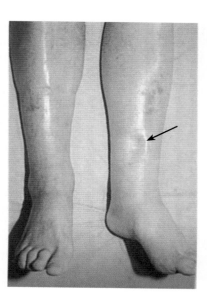

Figure 40.10 An example of pitting oedema.
(From plates 9 and 10 in R.A. Hope, J.M. Longmore, S.K. McManus, and C.A. Wood-Allum (1998) *Oxford handbook of clinical medicine*, 4th edition. Oxford University Press, Oxford.)

reabsorption of water. Since absorption by the proximal tubule is iso-osmotic, a decrease in fluid reabsorption allows more sodium to reach the distal nephron so that there is an increase in sodium excretion. There is, therefore, a loss of both sodium and water. Nevertheless, as the capacity of the distal nephron to transport sodium is increased when the sodium load is increased (glomerulo-tubular feedback, see Chapter 39), the increase in water excretion is not accompanied by an equivalent increase in sodium excretion. Consequently, the osmotic diuretics are more effective in increasing water excretion than they are in increasing sodium excretion.

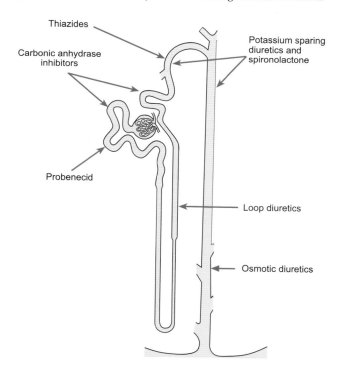

Figure 40.11 The sites of action of the different classes of diuretic drugs.

Diuretics that act by inhibiting active transport are exemplified by loop diuretics such as furosemide (also called frusemide). These compounds inhibit the co-transport of sodium, potassium, and chloride by the ascending thick limb of the loop of Henle (see Chapter 39). They appear to act only from the luminal side of the tubule. Inhibition of this transport decreases the ability of the nephron to concentrate urine. The effects of the loop diuretics are thus twofold: an increase in sodium excretion by inhibiting sodium chloride transport and an increase in water loss through impairment of the counter-current mechanism. They are the most potent diuretics in current clinical use and produce a pronounced increase in sodium excretion (known as a natriuresis).

One of the consequences of inhibiting the co-transport of sodium, potassium, and chloride in the ascending thick limb is an increase in potassium excretion. Unless this is carefully monitored, potassium balance will be disturbed and cardiac arrhythmias may result. To avoid this, a group of potassium-sparing diuretics has been developed. These drugs, of which amiloride is an example, act on the distal tubule and collecting ducts to inhibit both sodium absorption and potassium excretion. The diuretic spironolactone exerts its effect on the distal tubule by antagonizing the sodium-retaining action of aldosterone.

Summary

Oedema occurs when there is an abnormal accumulation of fluid in the tissues. Oedema may be caused by a disturbance of the Starling forces such as an increase in capillary pressure or a reduction in plasma oncotic pressure. It can also arise from lymphatic obstruction.

Several forms of oedema can be treated by administration of drugs known as diuretics that promote the excretion of both water and sodium. An increase in water excretion alone would not be sufficient to eliminate the accumulated fluid. Lymphoedema cannot be treated in this way.

40.7 Disorders of electrolyte balance

Disorders of sodium balance

Normal plasma sodium concentration lies between 135 and 145 mmol l^{-1}. A plasma sodium less than 130 mmol l^{-1} is known as **hyponatraemia**, which may be the result either of a lack of sodium (salt deficiency) or of an excess of water. Mild hyponatraemia (plasma sodium 115–130 mmol l^{-1}) is relatively common in hospital patients (around one in twenty patients will have some form of hyponatraemia). Severe hyponatraemia (plasma sodium < 115 mmol l^{-1}) is much more serious but, fortunately, it is rare. There are many different causes of hyponatraemia, a number of which are discussed here. Figure 40.12 shows the factors considered in determining the primary cause of the disorder. Note that in all cases there is a deficit of sodium relative to total body water.

Hypovolaemic hyponatraemia occurs in aldosterone deficiency (Addison's disease) and in acute renal tubular necrosis, where there is an enhanced loss of sodium from the kidneys with a reduction in total body water. Diarrhoea, vomiting, and excessive sweating all result in loss of sodium and water and, if prolonged, will also result in hypovolaemic hyponatraemia. If total body water is normal but plasma sodium is low, the condition is called **normovolaemic hyponatraemia**, which is usually caused by inappropriate secretion of ADH (the syndrome of inappropriate antidiuresis or SIAD), although it may be caused by glucocorticoid deficiency. In severe heart failure and renal failure there is fluid retention due to the reduced ability of the kidneys to excrete excess water, and this may result in **hypervolaemic hyponatraemia**.

A concentration of sodium in excess of 150 mmol l^{-1} is hypernatraemia, which is caused by an excess intake of sodium or an insufficient intake of water. **Hypervolaemic hypernatraemia** occurs when there is an excessive intake of sodium salts—often a result of inappropriate medical intervention. It is associated with severe thirst and irritability. **Hypovolaemic hypernatraemia** results when an individual has failed to drink enough to replace lost water (i.e. they are suffering from dehydration)—for example when water is not available or drinking is difficult or impossible because of coma or a physical handicap (e.g. following a stroke). Hypovolaemic hypernatraemia can occur in diabetes insipidus (both a failure of ADH secretion or a failure of the kidneys to respond to ADH will result in excessive loss of water in the urine). It may also be the result of an osmotic diuresis. However, the most devastating cause is the impairment of thirst following loss of the hypothalamic osmoreceptors. Remarkably, in hypovolaemic hypernatraemia there is little appetite for water (cf. hypervolaemic hypernatraemia).

Disorders of potassium balance

Most of total body potassium is within the cells, where intracellular potassium is around 100 mmol l^{-1} (the exact value depends on cell type). Plasma potassium is normally 3.5–4.5 mmol l^{-1}. The ratio of intracellular to extracellular potassium is of crucial importance in determining the membrane potential and excitability of nerve and muscle cells (see Chapters 7 and 8).

In **hypokalaemia**, plasma potassium is less than 3.5 mmol l^{-1}. If it is lower than 2.5 mmol l^{-1} (severe hypokalaemia) urgent treatment of the underlying disorder is necessary. Mild hypokalaemia is usually symptomless, but in severe hypokalaemia there is an

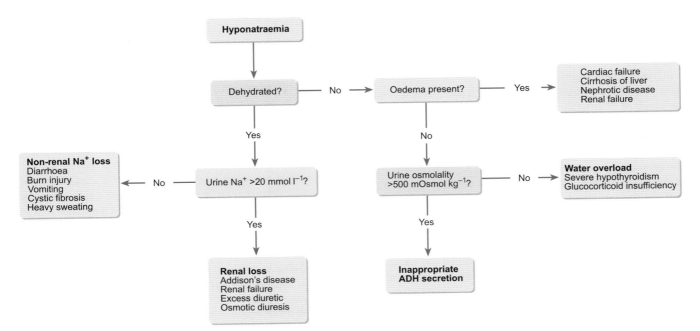

Figure 40.12 A flow chart for determining the cause of hyponatraemia. Note that both urine and plasma sodium concentrations need to be measured to arrive at the correct diagnosis. Other causes are also possible.

increased frequency of cardiac arrhythmias and the T wave of the ECG appears flattened (see Figure 29.16). Hypokalaemia commonly occurs as a consequence of the chronic administration of diuretics (especially thiazides) which results in excessive loss of potassium in the urine. Hypokalaemia can also indicate an excessive secretion of aldosterone (Conn's syndrome)—an excess of aldosterone stimulates potassium secretion and enhances sodium uptake by the cells of the distal tubule and collecting duct, causing loss of potassium in the urine. Excessive potassium loss may also result from prolonged vomiting or diarrhoea (the potassium content of the gastric juice is ~ 10 mmol l^{-1}, while that of the stools is 50–100 mmol l^{-1}).

A plasma potassium greater than 5.0 mmol l^{-1} prompts a diagnosis of **hyperkalaemia**. It is less common than hypokalaemia but it is a dangerous condition that causes muscle weakness and may result in sudden cardiac arrest. The ECG shows characteristic changes with an unusually large T wave (see Figure 29.16). A plasma potassium in excess of 6.5 mmol l^{-1} is a medical emergency. Hyperkalaemia may occur after a bout of very intense exercise, but in such cases it is short-lived and self-limiting. In other situations it may be the result of leakage of potassium from the intracellular compartment as a consequence of major tissue damage or ketoacidosis. Excessive potassium intake can overwhelm the capacity of the kidneys to excrete it, especially in acute renal failure.

Periodic paralyses affect plasma potassium

These are rare inherited diseases in which plasma potassium becomes unusually low (hypokalaemic periodic paralysis) or high (hyperkalaemic periodic paralysis) for a variable period of time. Both kinds are associated with weakness or even paralysis of certain muscle groups that may last some hours or days. The trigger for an episode is often rest after heavy exercise, although other factors have been implicated. These diseases are caused by ion channel mutations affecting the neuromuscular junction and skeletal muscle. (Such disorders are sometimes called **channelopathies**.)

In **hypokalaemic periodic paralysis** there are intermittent attacks of muscle weakness or paralysis which have a highly variable frequency of occurrence. Muscle strength is normal between attacks. The muscles of the shoulders and hips are most often affected, but the weakness may also affect the limb muscles and, in severe cases, the muscles involved in breathing and swallowing. The episodes of weakness last from a few hours to a day or two, during which plasma potassium is low (< 3.0 mmol l^{-1}), reflecting a shift of potassium into the intracellular compartment rather than a potassium deficiency (i.e. the low plasma potassium is a consequence of the disease, not its cause). Plasma potassium is normal between attacks. The defects linked to this condition are in two ion channels: a voltage-gated sodium channel called Na$_V$ 1.4 which is found at the neuromuscular junction, and a voltage-gated calcium channel called Ca$_V$ 1.1 which is found in the T-tubule system of the skeletal muscles.

The pattern of muscle weakness in **hyperkalaemic periodic paralysis** is similar to that of hypokalaemic periodic paralysis except that the respiratory and cranial muscles are not affected. Plasma potassium is either normal or slightly elevated during an attack. Attacks usually last 1–2 hours, although they may last as long as a day. The primary channel defects in this disorder are mutations of the Na$_V$ 1.4 voltage-gated sodium channel of the neuromuscular junction.

Disorders of magnesium balance

Normal plasma magnesium is 0.7–1.1 mmol l^{-1} and the daily intake is around 15 mmol per day. Around one-third of the intake is absorbed in the small intestine, and this is balanced by the excretion of a similar amount by the kidneys. Disorders of magnesium balance rarely occur in the absence of other electrolyte disturbances. **Hypomagnesaemia** is indicated when the plasma level of magnesium falls below 0.7 mmol l^{-1} and either reflects deficient uptake from the gut (poor absorption or low magnesium intake) or excessive loss in the urine. **Calcium deficiency** usually occurs at the same time (see Chapter 22 for a discussion of calcium regulation). Hypomagnesaemia is accompanied by neurological symptoms such as confusion and hyperreflexia (exaggerated reflexes), while the ECG shows a prolonged QT interval—all of which indicate a disturbance of nerve and muscle excitability.

Hypermagnesaemia most frequently occurs in renal failure, although it may occur in patients suffering from adrenal insufficiency. Symptoms usually become evident when plasma magnesium exceeds 1.2 mmol l^{-1} and are the reverse of those associated with hypomagnesaemia—cardiac and neurological depression with defects in cardiac conduction, hyporeflexia, and muscle weakness.

Summary

Plasma electrolyte balance is disturbed in diarrhoea, vomiting, and renal failure as well as in disorders of the adrenal cortex and pituitary gland. A plasma sodium less than 130 mmol l^{-1} is known as hyponatraemia, which may be caused by an enhanced loss of sodium in the urine and a reduction in total body water, inappropriate secretion of ADH, or a failure of the kidneys to excrete excess water. A plasma sodium greater than 150 mmol l^{-1} is known as hypernatraemia, which occurs when there is an excessive intake of sodium salts or dehydration.

A plasma potassium less than 3.5 mmol l^{-1} is known as hypokalaemia, which may be caused by an excess of aldosterone or by chronic administration of thiazide diuretics. It is associated with an increased frequency of cardiac arrhythmias. A plasma potassium greater than 5.0 mmol l^{-1} is hyperkalaemia. It is less common than hypokalaemia, but a plasma potassium in excess of 6.5 mmol l^{-1} is a medical emergency. Addison's disease is associated with a modest hyperkalaemia (5.0–6.0 mmol^{-1}), hyponatraemia, and a rise in blood urea.

✱ Checklist of key terms and concepts

The distribution of body water

- Body water is distributed between the intracellular fluid and the extracellular fluid.
- The extracellular fluid is further subdivided into the plasma and the interstitial fluid.
- The movement of fluid between the plasma and the interstitial fluid is governed by Starling forces.
- The exchange of water between the interstitial fluid and the intracellular fluid is governed by osmotic forces.
- Cells adjust the osmolality of their cytosol to maintain an optimal cell volume.

The regulation of body fluid osmolality and volume

- Osmolality and fluid volume are regulated independently of each other.
- Total body water is determined mainly by total body sodium, which is regulated by the renin–angiotensin–aldosterone system.
- Body water balance is maintained by the activity of osmoreceptors in the hypothalamus, which control the secretion of ADH from the posterior pituitary.
- Changes in extracellular volume are sensed by the arterial baroreceptors and the central volume receptors.
- Extracellular fluid volume is regulated by adjustment of sodium intake and excretion.

Dehydration and disorders of water balance

- In dehydration there is a loss of water that is usually accompanied by a loss of electrolytes, although pure water loss can occur when drinking water is not available.
- Water intoxication is less common than dehydration but it can occur in renal failure.
- In vomiting and diarrhoea there is a large loss of water and electrolytes, and replacement of both salts and water is necessary to restore body fluid balance.
- If more than 6–10 per cent of body water is lost, plasma volume falls, resulting in circulatory failure.

Haemorrhage

- The response of the body to haemorrhage depends on the extent of blood loss.
- After haemorrhage, the blood volume is progressively restored by a number of adjustments:
  - vasoconstriction

- an increased secretion of aldosterone, which promotes retention of sodium and an increased thirst
- replacement of plasma proteins by the liver
- increased activity in the bone marrow to restore the red cell count.
- If the circulating volume is insufficient, circulatory shock develops, which must be treated promptly to avoid a vicious downward spiral.
- The principal types of circulatory shock are:
  - hypovolaemic shock
  - distributive shock
  - obstructive shock
  - cardiogenic shock.

Oedema

- Oedema occurs when there is an abnormal accumulation of fluid in the tissues.
- It may be caused by:
  - an increase in capillary or central venous pressure
  - a reduction in plasma oncotic pressure
  - lymphatic obstruction.
- Several forms of oedema can be treated by administration of diuretics to promote the excretion of both sodium and water. Lymphoedema cannot be treated in this way.

Disorders of electrolyte balance

- Plasma electrolyte balance is disturbed in disorders of the adrenal cortex and pituitary gland and following diarrhoea or vomiting.
- A plasma sodium less than 130 mmol l^{-1} is hyponatraemia. There are different kinds:
  - hypovolaemic hyponatraemia
  - normovolaemic hyponatraemia
  - hypervolaemic hyponatraemia.
- A plasma sodium greater than 150 mmol l^{-1} is known as hypernatraemia. Two different forms may be distinguished:
  - hypervolaemic hypernatraemia
  - hypovolaemic hypernatraemia (dehydration).
- In hypokalaemia, plasma potassium is less than 3.5 mmol l^{-1}. It may be caused by an excess of aldosterone (Conn's syndrome).
- In hyperkalaemia, plasma potassium exceeds 5.0 mmol l^{-1}.
- Addison's disease is associated with a modest hyperkalaemia (5.0–6.0 mmol^{-1}), hyponatraemia, and a rise in blood urea.

Recommended reading

Physiology and pathophysiology

Koppen, B.M., and Stanton, B.A. (2006) *Renal physiology* (4th edn), Chapters 1, 5, 6, and 7. Mosby Year Book Inc., St Louis.
   A more detailed account of the topics covered in this chapter.

Levick, J.R. (2009) *An introduction to cardiovascular physiology* (5th edn), pp 351–4. Hodder Arnold, London.
   A characteristically lucid discussion of haemorrhage and shock.

Rolls, E.T. (2005) *Emotion explained*, Chapter 6: Thirst. Oxford University Press, Oxford.
   A more detailed discussion of the physiological and psychological aspects of thirst.

Clinical pharmacology

Grahame-Smith, D.G., and Aronson, J.K. (2002) *Oxford textbook of clinical pharmacology and drug therapy* (3rd edn), pp. 296–302. Oxford University Press, Oxford.
   A discussion of the clinical use of diuretics, their adverse effects, and the treatment of potassium depletion and hyperkalaemia.

Clinical physiology and medicine

Kumar, P., and Clark, M. (eds) (2009) *Clinical medicine* (7th edn), Chapter 12. Saunders, London.
   A useful introduction to clinical aspects of electrolyte balance.

Ledingham, J.G.G., and Warrell, D.A. (eds) (2000) *Concise Oxford textbook of medicine*, Chapters 12.2–12.4. Oxford University Press, Oxford.
   A comprehensive discussion of the disorders of water balance, sodium balance, and potassium balance.

Review articles

Castrop, H., Höcherl, K., Kurtz, A., Schweda, F., Todorov, V., and Wagner, C. (2010) Physiology of kidney renin. *Physiological Reviews* **90**, 607–73; doi:10.1152/physrev.00011.2009.

Greenlee, M., Wingo, C.S., McDonough, A.A., Youn, J.-H., and Kone, B.C. (2009) Narrative review: Evolving concepts in potassium homeostasis and hypokalemia. *Annals of Internal Medicine* **150**, 619–25.

Gutierrez, G., Reines, H.D., and Wulf-Gutierrez, M.E. (2004) Clinical review: Hemorrhagic shock. *Critical Care* **8,** 373–81, doi:10.1186/cc2851.

Hoffmann, E.K., Lambert, I.H., and Pedersen, S.F. (2009) Physiology of cell volume regulation in vertebrates. *Physiological Reviews* **89**, 193–277, doi:10.1152/physrev.00037.2007.

Wehner, F., Olsen, H., Tinel, H., Kinne-Saffran, E., and Kinne, R.K.H. (2003) Cell volume regulation: Osmolytes, osmolyte transport, and signal transduction. *Reviews of Physiology, Biochemistry and Pharmacology* **148**, 1–80, doi:10.1007/s10254-003-0009-x.

To check that you have mastered the key concepts presented in this chapter, complete the accompanying online self-assessment questions. Go to www.oup.com/uk/pocock5e/

CHAPTER 41

Acid–base balance

Chapter contents

This chapter should help you to understand:

- The physical chemistry of acids, bases, and hydrogen ions

- The pH scale of hydrogen ion concentration

- The physico-chemical factors that determine the hydrogen ion concentration of physiological solutions

- The role of the physiological buffers in maintaining the pH of body fluids

- The principal physiological mechanisms that regulate the pH of body fluids

- The common disorders of acid–base metabolism

- The compensatory mechanisms that minimize the effects of acid–base disorders

41.1 Introduction

Although the body continually produces CO_2 and non-volatile acids as a result of metabolic activity, the blood hydrogen ion concentration [H$^+$] is normally maintained within the relatively narrow range of 40–45 nmol (40–45×10^{-9} moles) of free hydrogen ions per litre. This corresponds to a blood pH between 7.35 and 7.4, and the extreme limits that are generally held to be compatible with life range from pH 6.8 to pH 7.7. This regulation is achieved in two ways: hydrogen ions are absorbed by other molecules in a process known as *buffering*, and acid products are subsequently eliminated from the body via the lungs and kidneys. *The concept of **acid–base balance** refers to the processes that maintain the hydrogen ion concentration of the body fluids within its normal limits.* An understanding of these processes provides the basis for a rational approach to the clinical treatment of acid–base disorders.

The importance of the hydrogen ion concentration in the regulation of body function may not be as readily appreciated as that of other ions such as sodium or potassium, which are present in far higher concentrations and play an important role in cellular physiology. Nevertheless, changes in the hydrogen ion concentration of body fluids can have profound consequences for cellular physiology as hydrogen ions can bind to charged groups on proteins (e.g. carboxyl, phosphate, and imidazole groups). Moreover, when hydrogen ions have been bound or lost from a protein its net ionic charge will change, and this in turn can lead to altered function. To take a dramatic example, the activity of the enzyme phosphofructokinase (a key regulatory enzyme in the glycolytic pathway) increases nearly 20-fold as the hydrogen ion concentration [H$^+$] decreases from about 80×10^{-9} to 60×10^{-9} mol l^{-1} (corresponding to a rise of about 0.1 pH unit from about 7.1 to pH 7.2). The functions of many other proteins are also affected by the prevailing hydrogen ion concentration although most are not as exquisitely sensitive as phosphofructokinase.

This chapter is organized so that the basic physical chemistry of acids and bases is discussed first, followed by the principles of hydrogen ion buffering. The physiological mechanisms of pH regulation are then discussed, and the chapter ends with a detailed discussion of acid–base disturbances and their clinical assessment.

41.2 The physical chemistry of acid–base balance

Hydrogen ions in solution and the pH scale

Historically, the concentration of hydrogen ions in the body fluids has been given in terms of pH units. As discussed in Chapter 2, the pH of a solution is the logarithm to the base 10 of the reciprocal of the hydrogen ion concentration:

$$pH = \log_{10}\left[\frac{1}{[H^+]}\right] \qquad [41.1]$$

Since, however,

$$\log_{10}\left[\frac{1}{[H^+]}\right] = -\log_{10}[H^+]$$

$$pH = -\log_{10}[H^+] \qquad [41.1a]$$

a common alternative definition is, therefore, that the pH of a solution is the negative logarithm of the hydrogen ion concentration. A change of one pH unit corresponds to a tenfold change in hydrogen ion concentration (because $\log_{10}10 = 1$). While the pH notation is convenient for expressing a wide concentration range, it is somewhat confusing as *a decrease* in pH reflects *an increase* in hydrogen ion concentration and vice versa. Acid solutions have a pH value less than the value for neutrality, i.e. less than 7.0 at 25°C and less than 6.8 at 37°C, while alkaline solutions have a pH value greater than neutrality.

The procedure for calculating the pH of a solution of known hydrogen ion concentration $[H^+]$ is detailed in Box 2.2, as is the calculation of free hydrogen ion concentration of a solution from its pH. For example, the pH of blood is normally close to 7.4 so its $[H^+]$ concentration is $10^{-7.4}$ moles l^{-1} or 39.8×10^{-9} mol l^{-1}. Similarly, if a urine sample has a pH of 5 the $[H^+]$ concentration is 10^{-5} mol l^{-1}. (Note that the 2.4 unit difference in pH between these samples of blood and urine corresponds to a 250-fold difference in hydrogen ion concentration.)

A neutral solution is one in which the concentrations of hydrogen and hydroxyl ions are equal. When pure water dissociates, both ions are produced in equal quantities:

$$H_2O \rightleftharpoons [H^+] + [OH^-]$$

Applying the law of mass action:

$$K_w = \frac{[H^+][OH^-]}{H_2O} \qquad [41.2]$$

where K_w is the dissociation constant for water. However, since undissociated water is present in much higher concentration (about 55.5 mol l^{-1}) than either $[H^+]$ or $[OH^-]$ (which are present at ~100×10^{-9} mol l^{-1}) the ionization of water has almost no effect on the concentration of non-ionized water, so equation 2 can be simplified to:

$$K'_w = [H^+][OH^-] \qquad [41.3]$$

where K'_w has a value of 10^{-14} (mol $l^{-1})^2$ at 25°C.
Since $[H^+] = [OH^-]$

$$[H^+] = \sqrt{K'_w}$$

so that for pure water

$$[H^+] = 10^{-7}\,\text{mol}\,l^{-1}$$

and the pH is

$$pH = -\log_{10}[10^{-7}]$$
$$= 7.0$$

Thus at 25°C a neutral solution has a pH of 7.0. The value of K'_w is dependent on temperature, and at body temperature it is $10^{-13.6}$ (mol $l^{-1})^2$ so that at body temperature neutral pH is 6.8 rather than 7.0.

Weak acids and weak bases are only partially ionized in aqueous solution

An acid is, in physiological terms, a substance that generates hydrogen ions in solution and a base is a substance that absorbs hydrogen ions. Acids and bases are further classified as weak or strong according to how completely they dissociate in solution. Strong acids such as hydrochloric acid (HCl) and bases such as sodium hydroxide (NaOH) are completely dissociated in aqueous solutions, while weak acids such as dihydrogen phosphate ($H_2PO_4^-$) and weak bases such as ammonium hydroxide (NH_4OH) are only partly dissociated. Thus, when sodium hydroxide is added to water only sodium and hydroxyl ions are present. There are no neutral sodium hydroxide molecules in the solution. Similarly, when hydrochloric acid is added to water only hydrogen ions and chloride ions are present. In contrast, the dissociation of a weak acid can be represented by the following equation:

$$HA \rightleftharpoons H^+ + A^-$$

Note that when a weak acid dissociates it generates an anion $[A^-]$, which is known as the *conjugate base* of that acid.

Similarly the dissociation of a weak base may be represented as follows:

$$BH^+ \rightleftharpoons B + H^+$$

The acid dissociation constant (K_a) is defined as

$$K_a = \frac{[H^+][A^-]}{[HA]} \qquad [41.4]$$

and, by analogy with pH, the pK_a value for an acid or base is defined as

$$pK_a = \log_{10}\frac{1}{K_a} = -\log_{10}K_a \qquad [41.5]$$

Acids have pK_a values less than neutrality (i.e. less than 7 at 25°C) and bases have pK_a values greater than neutrality (see Table 41.1).

Table 41.1 pK_a values of some common weak acids and bases

	Acid	Conjugate base	pK_a
Acetoacetic acid	CH_3COCH_2COOH	$CH_3COCH_2COO^-$	3.6
Lactic acid	$CH_3CH(OH)COOH$	$CH_3CH(OH)COO^-$	3.86
Acetic acid	CH_3COOH	CH_3COO^-	4.75
Carbonic acid	H_2CO_3	HCO_3^-	6.1
Dihydrogen phosphate	$H_2PO_4^-$	HPO_4^{2-}	6.8
Imidazole			6.95
Ammonium	NH_4^+	NH_3	9.25

Note: The table is arranged in order of acid strength so that the strongest acids, which have the smallest pK_a values, are at the top of the table. Note that while ammonium is a very weak acid (it has a large pK_a value) its conjugate base (ammonia) is a moderately strong base.

Equation (41.4) can be rewritten as:

$$\frac{1}{[H^+]} = \frac{1}{K_a}\frac{[A^-]}{[HA]} \qquad [41.6]$$

By taking logs, the following equation can be derived:

$$pH = pK_a + \log_{10}\frac{[A^-]}{[HA]} \qquad [41.7]$$

which may be written as

$$pH = pK_a + \log_{10}\frac{[Base]}{[Acid]} \qquad [41.8]$$

This important relationship is known as the **Henderson–Hasselbalch equation**. It shows that for any weak acid (or weak base) the ratio of base [A$^-$] to undissociated acid [HA] is determined by the pH of the solution. When the concentrations of acid and conjugate base are equal, the pH is equal to the pK_a (as [base]/[acid] = 1 and $\log_{10}1 = 0$). Moreover, where the pK_a is known, knowledge of two variables will define the third.

Summary

The acidity of a solution is determined by its hydrogen ion concentration. The degree of acidity is often expressed using the pH scale. Pure water is neutral in acid–base terms and has a pH of 7 at 25°C and, at this temperature, acid solutions have pH values below 7 and alkaline solutions have pH values above 7. Strong acids and strong bases in aqueous solutions dissociate completely into their constituent ions. Weak acids and bases are only partially dissociated and the degree of dissociation depends on the hydrogen ion concentration. The ratio of dissociated to undissociated weak acid can be calculated from the Henderson–Hasselbalch equation.

41.3 What factors determine the pH of an aqueous solution?

Before proceeding to discuss the detailed mechanisms that regulate the hydrogen ion concentration in plasma, it is important to establish the factors that determine the pH of aqueous solutions. At constant temperature, the hydrogen ion concentration of any physiological solution is determined by three factors.

1. The difference between the total concentration of fully dissociated cations ('strong' cations, e.g. Na$^+$) and that of the fully dissociated anions ('strong' anions, e.g. Cl$^-$). This difference indicates whether there is an excess of strong base or strong acid.

2. The quantity and pK_a values of the weak acids that are present (e.g. phosphate ions and the ionizable groups on proteins).

3. The partial pressure of carbon dioxide.

In all cases of acid–base disturbance, it is a change in one or more of these factors that is the underlying cause (see Box 41.1).

When other factors such as the PCO$_2$ are kept constant, changes in the difference between the sum of all the fully dissociated cations and the sum of all the fully dissociated anions will alter the hydrogen ion concentration of an aqueous solution. For example, in diarrhoea large quantities of sodium and bicarbonate are lost in the stools. As a result, the concentration of fully dissociated anions in the plasma (mainly chloride) is increased relative to that of the fully dissociated cations (chiefly sodium) and the plasma becomes more acid—giving rise to a condition known as a metabolic acidosis.

If the difference in the concentrations of the fully dissociated ions is kept constant and the concentration of weak acids is increased, as in uncontrolled diabetes mellitus, the plasma again becomes more acid.

Finally, if the PCO$_2$ is increased because of reduced alveolar ventilation, the plasma hydrogen ion concentration will increase, leading to a respiratory acidosis. Conversely, if, for a given level of activity, the alveolar ventilation is increased, the PaCO$_2$ will fall and so will the plasma hydrogen ion concentration (respiratory alkalosis). These conditions are covered later in this chapter.

The hydrogen ion buffers of the body fluids

From the previous section, it should be evident that, when an acid or a base is added to an aqueous solution, the magnitude of the change in hydrogen ion concentration (or the pH change) will depend on the nature and quantities of other ionizable substances present. When the change in hydrogen ion concentration is less than the quantity of acid or base added, the solution is said to be buffered and the substances responsible for this effect are called buffers (see Box 41.2 for a fuller explanation of how buffers work).

The buffers that have the greatest quantitative importance in whole blood are the bicarbonate of the plasma and the haemoglobin of the red cells (Table 41.2). The buffering of the extracellular fluid is principally due to bicarbonate and phosphate, and

(1₂3) Box 41.1   What determines the pH of an aqueous solution?

In a physiological solution, the free H^+ concentration is determined by a set of chemical equilibria that must be simultaneously satisfied. These are:

1) The dissociation of water:

$$H_2O \rightleftharpoons [H^+] + [OH^-]$$

The amount of free H^+ and OH^- is governed by the equilibrium constant so that:

$$K'_w = [H^+][OH^-]$$

2) The dissociation of the weak acids present. These reactions can be represented by the following equation:

$$HA \rightleftharpoons H^+ + A^-$$

The total amount of weak acid present $[T_a]$ is given by:

$$[T_a] = [HA] + [A^-]$$

and the degree of dissociation is determined by the equilibrium constant K_a:

$$K_a = \frac{[H^+][A^-]}{[HA]}$$

In physiological solutions the most important weak acid is carbonic acid derived from dissolved CO_2 and the amount of carbonic acid formed is directly related to the PCO_2.

3) Finally, as electroneutrality must be maintained, the concentration of all cations must be equal to that of all anions. As fully dissociated cations and anions do not liberate or bind hydrogen ions, it is convenient to define the difference between the total concentration of all the fully dissociated cations ('strong cations') and the total concentration of all the fully dissociated anions ('strong anions') as S, the strong ion difference with units of mol l^{-1}:

S = (sum of all 'strong' cations) − (sum of all 'strong' anions)

and $S + [H^+] = [OH^-] + [A^-]$

This relationship can be expressed as an equation with only $[H^+]$ as a variable. Since

$$[OH^-] = \frac{K'_w}{H^+}$$

and since

$$[A^-] = \frac{K_a[T_a]}{([H^+] + K_a)}$$

the free H^+ concentration is given by the solution of the following equation:

$$[H^+] - \frac{K'_w}{H^+} - \frac{K_a T_a}{([H^+] + K_a)} + S = 0$$

While in this simplified analysis only a single weak acid has been considered, the basic principle can be extended to include any number of weak acids or bases. In all cases, the basic relationship holds: *at constant temperature the hydrogen ion concentration of any aqueous solution is determined by the difference in concentration of the fully dissociated anions and cations and by the amounts and acid dissociation constants of the weak acids and weak bases that are present.*

While it is clear that adding a weak acid or weak base to a solution will alter the hydrogen ion concentration of that solution, the role of the strong ions in determining the hydrogen ion concentration may need a little further explanation. Consider the effect of adding NaOH to pure water: the OH^- ions react with H^+ ions to form water until the H^+ and the added Na^+ ions exactly balance the free OH^- ions. In effect, the Na^+ ions have replaced the H^+ ions as the principal carriers of positive charge. Note that the solution is alkaline, as the concentration of hydroxyl ions exceeds that of hydrogen ions. If HCl is subsequently added, the H^+ from the HCl reacts with OH^- to form water and the negative charge on the Cl^- ions replaces that of the OH^- ions. While $[Na^+]$ exceeds $[Cl^-]$ the solution will be alkaline, when $[Na^+] = [Cl^-]$ the solution will be neutral (as $[H^+]$ must also equal $[OH^-]$ to maintain electroneutrality), and when $[Na^+]$ is less than $[Cl^-]$ the solution will be acid.

Table 41.2 The amount of hydrogen ion absorbed by the principal buffers of the blood as blood pH falls from 7.4 to 7.0 following the addition of acid

Buffer	Amount per litre of blood	mmol H^+ absorbed per litre
Bicarbonate	25 mmol	18[†]
Phosphate	1.25 mmol	0.3
Plasma protein	40 g	1.7
Haemoglobin	150 g	8
Total		28

[†] This assumes that the bicarbonate buffer operates in an open system so that the CO_2 generated can be eliminated (see Box 41.2).

intracellular buffering is provided by proteins, organic and inorganic phosphates, and bicarbonate.

The effectiveness of any buffer depends on its concentration and on its pK_a value. In general, the closer the pK_a of a buffer is to the pH of the plasma, the more effective the buffer.

For phosphate the buffer reactions can be summarized as follows:

$$H_2PO_4^- \rightleftharpoons H^+ + HPO_4^{2-}$$

In this reaction the total amount of phosphate ($[HPO_4^{2-}] + [H_2PO_4^-]$) is unchanged. Only the degree of ionization alters as the hydrogen ion concentration varies. Thus, the effect of adding or removing hydrogen ions is to alter the $[H_2PO_4^-]/[HPO_4^{2-}]$ ratio.

$\left(1_2{}^3\right)$ Box 41.2 How buffers work: the buffer action of inorganic phosphate and bicarbonate

If 0.1 mmoles of a strong acid such as HCl were to be added to a litre of pure water, the total hydrogen ion concentration $[H^+]$ would be 0.1001 mmol l^{-1} (the amount added plus the amount due to the dissociation of water) and the pH of the solution would be:

$$pH = -\log_{10}[0.0001001] \approx 4$$

If, however the same quantity of HCl were to be added to a mixture of 0.5 mmol l^{-1} NaH_2PO_4 and 0.8 mmol l^{-1} Na_2HPO_4 the pH change would be far smaller. This can be shown by the following calculation.

From the Henderson–Hasselbalch equation the pH of the phosphate solution is:

$$pH = 6.8 + \log\frac{[HPO_4^{2-}]}{[H_2PO_4^-]}$$

$$= 6.8 + \log\frac{[0.8]}{[0.5]}$$

$$= 6.8 + 0.2 = 7.0$$

(where 6.8 is the pK_a for phosphate).

This pH corresponds to a free hydrogen ion concentration of 100 nmol l^{-1}—the same as that of pure water at 25°C. When 0.1 mmol of HCl is added to the mixture, the hydrogen ion reacts with the HPO_4^{2-} ion to form $H_2PO_4^-$.

$$H^+ + HPO_4^{2-} \rightleftharpoons H_2PO_4^-$$

If we assume that the reaction is complete, $[HPO_4^{2-}]$ has been reduced from 0.8 to 0.7 mmol l^{-1} and $[H_2PO_4^-]$ has been increased from 0.5 to 0.6 mmol l^{-1}. Entering these values into the Henderson–Hasselbalch equation:

$$pH = 6.8 + \log\frac{[0.7]}{[0.6]}$$

$$= 6.8 + 0.067 = 6.87$$

The pH of the solution is now 6.87 so that the presence of the phosphate buffer has reduced the pH change from 7.0 to 4.0 (i.e. 3 units) to 7.0 to 6.87—just 0.13 units! As all the reactants remain in the solution, this is called a closed buffer system.

For comparison, consider how the CO_2/HCO_3^- buffer system would react to the addition of 0.1 mmol of acid to a litre of solution. For direct comparison with the previous calculation for phosphate, let us assume that bicarbonate was 1.2 mmol l^{-1} and CO_2 0.15 mmol l^{-1}.

From the Henderson–Hasselbalch equation, the pH of this solution can be calculated to be 7.00. Addition of 0.1 mmol of acid would reduce bicarbonate to 1.1 mmol^{-1} and increase the CO_2 to 0.25, and the pH after the addition of acid would be 6.74—a pH fall of (7.0 – 6.74 = 0.26 pH units). Thus, mole for mole, bicarbonate is a less effective buffer than phosphate at neutral pH.

In normal plasma the PCO_2 of arterial blood is 5.3 kPa (40 mm Hg) which is equivalent to 1.2 mmol l^{-1} of carbonic acid and plasma bicarbonate is about 24 mmol l^{-1} so that the pH of arterial blood calculated from the Henderson–Hasselbalch equation is:

$$pH = 6.1 + \log_{10}\frac{24}{1.2} = 7.4$$

If 0.1 mmol of a strong acid were to be added to a litre of plasma, how large would the pH change be? Assuming all the $[H^+]$ reacts with HCO_3^-, the concentration of HCO_3^- after the addition of acid will be 23.9 mmol l^{-1} and the CO_2 would rise to 1.3 mmol l^{-1}. The net effect would be to change the pH from 7.4 to 7.36, a fall of only 0.04 pH units. The pH fall is far less than in the previous example, as the amount of HCO_3^- present in the solution is greater (24 mmol l^{-1} compared to 1.2 mmol l^{-1}).

If excess CO_2 is removed, the efficiency of bicarbonate as a buffer is dramatically enhanced. In the first example the CO_2 would be kept at 0.15 mmol l^{-1} and the bicarbonate would fall from 1.2 to 1.1 mmol l^{-1}. The pH of the solution would fall by only 0.035 of a pH unit to 6.965. In normal plasma, the bicarbonate would fall to 23.9 mmol l^{-1} and the fall in pH would be only 0.001 of a pH unit! (Use the Henderson–Hasselbalch equation to show this for yourself.) In these two situations, the bicarbonate is acting as an open buffer system.

To summarize, the effectiveness of any buffer depends on its concentration and on its pK_a value. For a closed buffer system where all the reactants remain in solution (e.g. phosphate), a buffer is at its most effective when the pH is within one pH unit of the pK_a for the buffer system. When CO_2/HCO_3^- buffering occurs in an open system and the PCO_2 is maintained constant, the buffering power of the CO_2/HCO_3^- system is greatly increased. Consequently, despite operating far from its pK_a the CO_2/HCO_3^- system is a very effective buffer because of its high concentration in plasma and because blood PCO_2 is maintained relatively constant by adjustment of the alveolar ventilation.

Similarly, the buffer reaction for haemoglobin (Hb) can be represented by the following reaction:

$$HHb^+ \rightleftharpoons H^+ + Hb$$

For the bicarbonate buffer system, there are two successive reactions:

First the dissolution of CO_2 in water to form carbonic acid:

$$H_2O + CO_2 \rightleftharpoons H_2CO_3$$

followed by the ionization of carbonic acid:

$$H_2CO_3 \rightleftharpoons H^+ + HCO_3^-$$

Since the concentration of H_2CO_3 depends on the partial pressure of CO_2:

$$[H_2CO_3] = \alpha \times PCO_2$$

where α is the solubility coefficient for CO_2 in plasma (Henry's Law—see Section 32.2). The Henderson–Hasselbalch equation for the bicarbonate buffer system can therefore be written as

$$pH = 6.1 + \log_{10} \frac{[HCO_3^-]}{\alpha \times P_{CO_2}}$$

where 6.1 is the pK_a for the dissociation of carbonic acid in plasma and α has the value of 0.225 mmol l^{-1} kPa^{-1} (or 0.03 mmol l^{-1} mm Hg^{-1}). When P_{CO_2} is 5.3 kPa (40 mm Hg), there are $5.3 \times 0.225 = 1.2$ mmoles CO_2 per litre of plasma. The amount of carbon dioxide in the plasma is directly related to the alveolar ventilation (panel (a) of Figure 41.1). Consequently, ventilation has a major influence on plasma pH. This significantly increases the buffering power of the CO_2/HCO_3^- system (see Box 41.2).

Buffers in combination: the isohydric principle

When a buffer reacts with hydrogen ions, the change in hydrogen ion concentration affects all the other buffer reactions in the same body compartment. This is known as the **isohydric principle**. Thus, knowledge of the status of one physiological

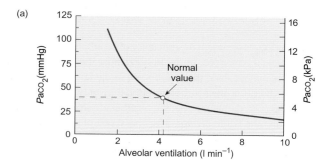

(a)

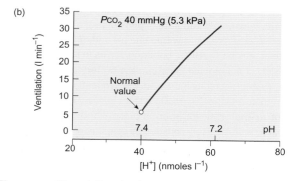

(b)

Figure 41.1 The relationship between P_{CO_2}, pH, and alveolar ventilation.(a) The effect of changes in alveolar ventilation on the P_{CO_2} of arterial blood. Note that as ventilation is increased P_{CO_2} falls, and vice versa. This relationship assumes a constant rate of CO_2 production. (b) The effect of changes in plasma hydrogen ion concentration [H⁺] on the rate of ventilation. In this experiment, the plasma P_{CO_2} was kept constant by allowing the subject to inhale gas mixtures containing varying amounts of carbon dioxide. An increase in plasma [H⁺] (fall in pH) stimulates ventilation independently of any change in P_{CO_2} and this increase in ventilation will lead to a fall in P_{CO_2} (a), which will tend to offset the rise in [H⁺].

buffer will serve to define the changes in [H⁺] concentration that affect all the other buffer systems in that compartment. In clinical practice, the state of acid–base balance (sometimes called the acid–base status) of patients is usually assessed by measurement of arterial pH, P_{CO_2}, and [HCO_3^-], and this approach will be adopted here.

Hydrogen ion buffering provides short-term stabilization of plasma pH

In blood, the principal buffers are inorganic phosphate (HPO_4^{2-} and $H_2PO_4^-$), bicarbonate, plasma protein, and the haemoglobin of the red cells. As Table 41.2 shows, the bicarbonate and haemoglobin are quantitatively the most important.

The large contribution made by bicarbonate towards the total buffer capacity of whole blood depends on the amount present (24–26 mmol l^{-1}) and on the fact that, when it absorbs hydrogen ions, bicarbonate becomes converted to carbonic acid which then dissociates to form carbon dioxide and water. Since the excess carbon dioxide will stimulate the central chemoreceptors, alveolar ventilation will increase and the excess carbon dioxide generated will be excreted via the lungs (see Chapter 32). Consequently, the partial pressure of carbon dioxide in the blood is maintained relatively constant. The amount of acid buffered in this way is equal to the amount of bicarbonate that is lost.

If acid is slowly infused into a vein, the fall in blood pH is remarkably small, far smaller than can be accounted for by the blood buffers alone. This difference is accounted for by the buffers of the interstitial fluid, those within the cells and the mineral component of bone. The total buffering capacity of the body is thus made up of four components:

- blood buffers—chiefly haemoglobin and bicarbonate
- buffers in the interstitial fluid—chiefly bicarbonate
- buffers in the cells—chiefly organic phosphates (e.g. ATP) and proteins
- the bone matrix of the skeleton.

Experimental evidence suggests that following infusion of acid, about half of the added hydrogen ions are buffered by the phosphates and protein within the cells. While this whole body buffering is capable of absorbing large quantities of hydrogen ions relative to the total amount of free hydrogen ion present in the body, it offers only a temporary defence against metabolic acid production. *Maintenance of blood pH within the normal range ultimately depends on the elimination of excess acid from the body.*

Summary

At constant temperature, the concentration of hydrogen ions [H⁺] in the plasma is determined by three factors: the difference between the total concentration of fully dissociated cations and that of the fully dissociated anions, the quantity and the pK_a values of the weak acids that are present, and the partial pressure of carbon dioxide. A change in any one of these will result in a change in plasma [H⁺].

The weak acids and bases of the plasma provide the first line of defence against changes in plasma [H$^+$] by buffering the hydrogen ions formed during metabolism. Of the buffers present in the plasma and interstitial fluid, the CO_2/HCO_3^- buffer system is quantitatively the most important.

41.4 How the body regulates plasma pH

In the course of a day, an average healthy person eating a typical Western diet produces between 12 and 15 moles of carbon dioxide and excretes about 50 mmoles of acid in the urine. From the previous section, it will be clear that the dissolved carbon dioxide of the body fluids leads to the formation of hydrogen ions in solution. Since it can be excreted via the lungs as a gas, carbon dioxide is often referred to as **volatile acid**. The acid that is excreted in the urine is chiefly sulphate derived from the metabolism of sulphur-containing amino acids (cysteine and methionine). This is known as **non-volatile acid** which must be excreted in the urine.

Under normal conditions no organic acids appear in the urine, but in severe uncontrolled diabetic ketosis, large quantities of β-hydroxybutyric acid and acetoacetic acid are produced each day and a significant portion of these acids may be eliminated in the urine as 'ketone bodies'.

Respiratory regulation of carbon dioxide

The concentration of carbon dioxide in the alveolar gas is governed both by the rate of carbon dioxide production by the body and by the rate of alveolar ventilation. Over a period of time, the alveolar PCO_2 will be directly related to the amount of carbon dioxide produced by the body ($\dot{V}CO_2$) and is inversely related to the alveolar ventilation ($\dot{V}_A$). This relationship can be written as follows:

$$P_A CO_2 = k . \frac{[\dot{V}_{CO_2}]}{[\dot{V}_A]}$$

Since the PCO_2 of alveolar air and arterial blood of healthy people is the same, a maintained doubling of the alveolar ventilation will halve the PCO_2 in the blood—provided that the rate of carbon dioxide production is unchanged. The relationship between $P_A CO_2$ and alveolar ventilation is shown in Figure 41.1(a) for a normal adult. If the arterial PCO_2 rises because of an increase in carbon dioxide production or if alveolar ventilation is insufficient, the chemoreceptors of the medulla and carotid bodies will detect the change and increase alveolar ventilation. The increase in ventilation then tends to bring the $P_A CO_2$ back to normal.

A fall in plasma pH stimulates the rate of respiration

Altering plasma [H$^+$] by injection of a weak solution of acid or base changes alveolar ventilation. Rapid injection of an acid solution into a vein causes an increase in the frequency and depth of respiration. Conversely, a rapid injection of an alkaline solution

depresses ventilation (hypoventilation) or briefly stops it (apnoea). By themselves, these observations do not prove that the plasma hydrogen ion concentration directly affects respiration, because an increase in the hydrogen ion concentration (lowering of pH) by injection of acid will result in an increase in PCO_2 and a fall in plasma bicarbonate. Conversely, a fall in the plasma hydrogen ion concentration (an increase in plasma pH) caused by injection of base will result in a fall in PCO_2 and a rise in plasma bicarbonate. These changes are a consequence of the chemical reactions that govern the formation of carbonic acid (see Section 41.3).

The role of hydrogen ions as an independent stimulus to ventilation can be examined by the continuous infusion of an acid solution. Initially, ventilation will increase and the PCO_2 of the plasma will tend to fall for the reasons discussed above. This fall in PCO_2 can be offset by allowing the subject to breathe a gas mixture containing carbon dioxide. Under these conditions, alveolar ventilation is found to increase progressively as the hydrogen ion concentration rises (i.e. as the pH falls—see panel (b) of Figure 41.1). As this increase in ventilation occurs without any change in PCO_2, the hydrogen ion concentration of the plasma must be capable of stimulating respiration independently of changes in PCO_2. Therefore, although ventilation increases in response both to a rise in plasma hydrogen ion concentration (a fall in plasma pH) and to a rise in arterial PCO_2, the two stimuli act independently.

When a subject is breathing air, the change in ventilation that follows a change in plasma [H$^+$] will itself alter the PCO_2 of the plasma over a period of time. The hyperventilation that follows an increase in the plasma hydrogen ion concentration will lead to a fall in plasma PCO_2 whereas the apnoea or hypoventilation that follows a fall in plasma hydrogen ion concentration will lead to retention of CO_2 and a rise in plasma PCO_2. These changes to PCO_2 tend to bring the plasma hydrogen ion concentration within the normal range. Final correction occurs when the excess acid or base has been removed from the body by the action of the kidneys.

The kidney excretes about 50 mmoles of non-volatile acid each day in the urine as ammonium or in combination with phosphate

In addition to its elimination of carbon dioxide, the body needs to excrete about 50 mmoles of non-volatile acid a day to maintain plasma pH within the normal range. The mechanisms responsible for the secretion of hydrogen ions into the proximal and distal tubules have already been discussed in Chapter 39. As the average volume of urine produced each day is 1–1.5 litres and as the pH of the urine is usually between 5 and 6, less than 0.05 per cent of this acid is excreted as free hydrogen ions. Far larger quantities of hydrogen ions are eliminated in combination with the urinary buffers, of which the most important is phosphate.

A normal person eating a typical Western diet excretes about 30 mmoles of phosphate a day. If the urine pH is 5, almost all of this will be in the form of $H_2PO_4^-$ (compared to only 20 per cent in the plasma). Thus, about 80 per cent of the excreted phosphate (24 mmoles a day) is available to buffer hydrogen ions in the urine.

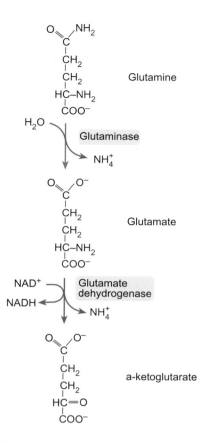

Figure 41.2 The steps involved in the deamination of glutamine to generate ammonium ions.

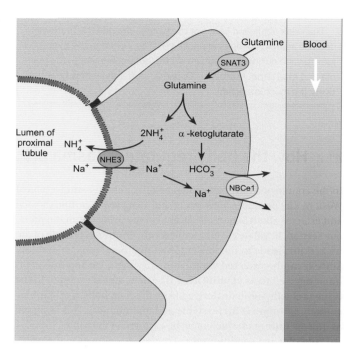

Figure 41.3 The secretion of ammonium by the cells of the proximal tubule: NH_4^+ is generated in the tubular cells by deamination of glutamine and is secreted into the tubular lumen in exchange for Na^+ via the sodium/hydrogen ion exchanger NHE3. The sodium entering the cell is subsequently removed by the Na^+, K^+-ATPase located on the basolateral surface. The complete metabolism of α-ketoglutarate generates two bicarbonate ions, which pass into the interstitial fluid via the electrogenic sodium/bicarbonate cotransporter NBCe1 located on the basolateral membrane.

This accounts for about a half of the hydrogen ion derived from non-volatile acids. The remainder is excreted as ammonium salts (NH_4^+). The ammonium is derived from glutamine in a process known as **ammoniagenesis**—see Figure 41.2.

The total amount of non-volatile acid that is excreted is the sum of that buffered by the urinary buffers (principally phosphate) and the amount of ammonium in the urine. It can be measured by titrating the urine with base until the urine pH is the same as that of the plasma (the **titratable acid**) and by separately measuring the total amount of ammonium excreted. The total acid excretion is the sum of the two figures.

The formation of ammonium ions and acid excretion

The generation of ammonium ions, (NH_4^+), from glutamine occurs primarily in the first part of the proximal tubules. Glutamine is the most abundant amino acid in plasma and is taken up across the basolateral surface via a specific transporter, SNAT3 (**S**odium **N**eutral **A**mino **A**cid **T**ransporter 3). The enzyme glutaminase catalyses the removal of the terminal amino group (deamination) by to form NH_4^+ and glutamate. The glutamate that is formed by this reaction is subsequently deaminated by glutamate dehydrogenase, resulting in the production of ammonium ions (NH_4^+) and

α-ketoglutarate. The sequence of reactions is shown in Figure 41.2. Chemically, the formation and excretion of a millimole of ammonium ions is equivalent to the titration of a similar quantity of hydrogen ions derived from non-volatile acids by the weak base ammonia. The non-volatile acids that are formed by the metabolism of amino acids are therefore excreted as their ammonium salts.

The ammonium ions produced by these reactions are secreted into the tubular lumen in exchange for sodium ions via the sodium/proton exchanger NHE3 (this isoform will utilize either H^+ or NH_4^+)—see Figure 41.3. The subsequent metabolism of the α-ketoglutarate via the TCA cycle and of NADH via the electron transport chain consumes two protons. In this way, the secretion of ammonium ions leads to a loss of hydrogen ions from the body. It is important to note that the key to the success of this process is the spatial separation of the NH_4^+ from the bicarbonate ions that are formed during the metabolism of the α-ketoglutarate. Some of the ammonium in the tubular fluid is reabsorbed in the later segments of the nephron, particularly the thick ascending limb of the loop of Henle. This results in the recycling of ammonium and an accumulation of ammonium in the renal medulla. The bulk of ammonium that eventually appears in the urine (~80 per cent) is secreted by the collecting ducts by the processes outlined in Figure 41.4.

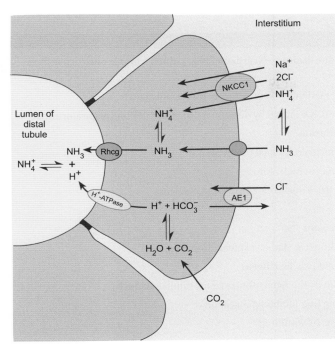

Figure 41.4 A simplified diagram of the process of ammonium secretion by the cells of the late distal tubule and collecting ducts. Ammonium ions are taken up by the NKCC1 cotransporter of the basolateral membrane and dissociate to form ammonia, which is secreted into the lumen via the Rhcg protein of the apical membrane. In the lumen it combines with hydrogen ions secreted by a H^+-ATPase, which is also located in the apical membrane. The resulting ammonium ions are then excreted in the urine. (Rhcg = Rhesus (Rh) family c glycoprotein.)

Summary

Each day a normal person produces about 15 moles of carbon dioxide and 50 mmoles of non-volatile acid (chiefly sulphate) which must be eliminated from the body. Carbon dioxide is excreted via the lungs and non-volatile acid is excreted via the kidneys.

The frequency and depth of respiration is stimulated by an increase in both plasma PCO_2 or its hydrogen ion concentration. An increase in either factor leads to an increase in alveolar ventilation. A rise in plasma hydrogen ion concentration leads to an increased loss of carbon dioxide, and this mechanism provides a rapid means of adjusting the plasma hydrogen ion concentration.

The excretion of non-volatile acid in the urine depends on the ability of phosphate to buffer free hydrogen ions and on the ability of the kidney to generate NH_4^+. Under normal circumstances about half of the non-volatile acid is excreted as NH_4^+ salts.

41.5 Primary disturbances of acid–base balance

When the pH of the arterial blood is less than 7.35 (i.e. when the plasma hydrogen ion concentration is greater than 45×10^{-9} mol l^{-1}) it is regarded as being acid with respect to normal. Patients with such a blood pH are said to have **acidaemia** (literally acid in the blood). The processes responsible for this increase in plasma hydrogen ion concentration can be attributed to increased amounts of non-volatile acid (**metabolic acidosis**) or to a failure to remove carbon dioxide from the blood (**respiratory acidosis**). Conversely, when the pH of the arterial blood is greater than 7.45 (i.e. when the plasma hydrogen ion concentration is less than 35×10^{-9} mol l^{-1}) it is regarded as being alkaline with respect to normal. Patients with such a blood pH are said to have **alkalaemia**. The underlying causes define the kind of **alkalosis** responsible for the condition (**respiratory** or **metabolic alkalosis**).

Since the pH of the body fluids is determined by the difference in the sum of the concentrations of strong cations and strong anions and by the quantity and pK_a values of the weak acids that are present (including carbon dioxide), any derangement in acid–base balance must arise from a change in one or more of these factors. Some of the common causes of acid–base imbalance are given in Table 41.3.

Respiratory acidosis is the result of an increase in the PCO_2 of the plasma due to inadequate ventilation or to the presence of significant amounts of carbon dioxide in the inspired air. The rise in plasma PCO_2 results in an increase in the formation of carbonic acid, which dissociates giving rise to H^+ and HCO_3^-. The increase in hydrogen ion concentration is directly related to the PCO_2 (see Section 41.3) and the plasma bicarbonate increases in proportion to the fall in plasma pH. These relationships are indicated in Figure 41.5 by the line linking point A to the normal value (P_aCO_2 40 mm Hg, pH 7.4, and HCO_3^- 24 mmol l^{-1}).

Respiratory alkalosis is the result of a fall in plasma PCO_2 due to an increase in alveolar ventilation. The fall in PCO_2 shifts the $[CO_2]$–$[HCO_3^-]$ equilibrium and leads to a decreased concentration of carbonic acid and so to a rise in plasma pH and a fall in plasma bicarbonate. The magnitude of the pH change is directly related to the increase in ventilation (see Section 41.4 and Figure 41.1). This condition commonly occurs when travelling from sea level to a high altitude, where the fall in atmospheric PO_2, and hence in arterial PO_2, stimulates respiration (see Chapters 35 and 37). The changes in plasma pH and bicarbonate as the P_aCO_2 falls are shown in Figure 41.5 by the line linking the normal value to point B.

In **metabolic acidosis** the fall in plasma pH is accompanied by a fall in plasma bicarbonate. There are many causes (see Table 41.3) including an increase in metabolically derived acids, a loss of base ($NaHCO_3$) from the gut during diarrhoea, and a failure of the renal tubules to excrete hydrogen ions. The pH and HCO_3^- changes in metabolic acidosis are indicated in Figure 41.5 by the line linking the normal value to point C.

When diabetes mellitus is inadequately controlled, energy metabolism shifts from carbohydrates to fats and the amounts of β-hydroxybutyric acid and acetoacetic acid in the plasma increase. As a result, there is a fall in plasma pH. These changes are known as **ketoacidosis**. The hydrogen ions react with plasma bicarbonate to form carbonic acid and the CO_2 liberated is excreted via the lungs.

The kidney excretes acid in three ways:

- Na^+–H^+ exchange across the apical membrane of the cells of the proximal tubule

Table 41.3 Some causes of acid–base disturbance

Disturbance	Cause
Respiratory acidosis (alveolar hypoventilation)	Impaired ventilation due to airway obstruction (e.g. asthma, chronic obstructive lung disease)
	Impaired alveolar gas exchange—ventilation-perfusion mismatch
	Decreased respiratory drive caused by disease or drugs that depress respiratory drive
	Inhalation of carbon dioxide
	Neuromuscular and musculo-skeletal problems
Respiratory alkalosis (alveolar hyperventilation)	Hypoxia (e.g. while living at high altitude)
	Increased respiratory drive due to cerebro-vascular disease
	Hepatic failure
	Effects of drugs and poisons
Metabolic acidosis	Endogenous acid loading (e.g. diabetic ketoacidosis)
	Loss of base from the gut (e.g. diarrhoea)
	Impaired acid secretion by the renal tubules (renal tubular acidosis)
	Exogenous acid loading (e.g. methanol ingestion)
Metabolic alkalosis	Loss of gastric juice (e.g. by vomiting)
	Excessive base ingestion
	Aldosterone excess
	Alkaline diuresis therapy of drug poisoning

- the secretion of NH_4^+ into the proximal tubule
- direct proton secretion via the H^+-ATPase of the intercalated cells of the distal tubule and collecting ducts.

As a consequence of this acid secretion, an equimolar amount of bicarbonate is generated and this 'new' bicarbonate replaces that lost by the reaction of plasma bicarbonate with metabolic acid during buffering. Clearly, if a constant plasma pH is to be maintained, the amount of acid secretion must be sufficient to match the generation of non-volatile acids. If this is not the case, both plasma pH and plasma HCO_3^- will fall and a type of metabolic acidosis known as **renal tubular acidosis** will result.

The bony skeleton contains a vast number of mineral crystallites bound together by cells, collagen, and ground substance rich in mucopolysaccharides. The mineral crystallites consist of calcium phosphate and calcium carbonate and their surface is negatively charged. Normally these charges are largely neutralized by sodium and potassium ions, but when plasma pH falls, these ions are displaced by protons so that the mineral phase of the skeleton provides an additional source of extracellular buffering. This additional buffering is bought at a price—the ions displaced (sodium, potassium, and calcium) are excreted in the urine. For this reason, there can be a significant loss of calcium during chronic metabolic acidosis, leading to a slow dissolution of the bone.

Metabolic alkalosis is caused by an excess of non-volatile base in the plasma, which may arise from a number of factors (see Table 41.3). Commonly, metabolic alkalosis arises as a result of vomiting the gastric contents and can be attributed to the loss of HCl, which is a strong acid. Since the PCO_2 is unchanged, the fall in hydrogen ion concentration that results from this loss is accompanied by an increase in plasma bicarbonate. In metabolic alkalosis, the rise in pH is associated with a rise in plasma bicarbonate. The pH and bicarbonate changes in metabolic alkalosis are summarized by the line linking the normal value to point D in Figure 41.5.

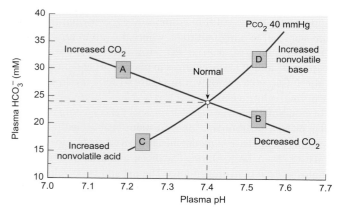

Figure 41.5 The pH-[HCO_3^-] diagram for blood plasma (sometimes called a Davenport diagram). The line A–B shows the plasma pH as PCO_2 is varied. If ventilation is less than normal, pH falls and [HCO_3^-] rises because CO_2 accumulates (normal to point A). This is respiratory acidosis. In hyperventilation plasma PCO_2 and [HCO_3^-] fall while pH rises (normal to point B). This is respiratory alkalosis. The line C–D shows the change in pH that occurs at a constant $PaCO_2$ of 40 mm Hg as non-volatile acid is gained in metabolic acidosis (normal to point C) or lost in metabolic alkalosis (from normal to point D).

Summary

Normal arterial blood pH is generally taken as 7.35–7.4. A blood pH less than 7.35 is acidaemia with an underlying acidosis; a blood pH is greater than 7.45 is alkalaemia with an underlying alkalosis. Respiratory acidosis and respiratory alkalosis are caused by changes in alveolar ventilation. All other disorders are classified as metabolic irrespective of their underlying cause.

41.6 Disorders of acid–base balance are compensated by respiratory and renal mechanisms

When acid–base balance has become disturbed, various mechanisms operate to bring plasma pH closer to the normal range in a process called **compensation**. The mechanisms that act to restore plasma pH can be grouped under two headings:

- Respiratory compensation, which is fast but not very sensitive (pH adjustment occurs in minutes but the changes in PCO_2 offset the original stimulus so the correction is incomplete).

- Renal compensation, which is sensitive but slow (pH adjustment takes hours to days).

While compensatory mechanisms operate to minimize the change in plasma pH, complete restoration of acid–base balance (i.e. correction) requires treatment or elimination of the underlying cause.

Compensation of chronic respiratory acidosis and alkalosis can occur only by renal means, as the primary deficit is due to a change in alveolar ventilation

While respiratory acidosis and alkalosis can be produced voluntarily by breath-holding or hyperventilation, this discussion is concerned with persistent alterations to ventilation arising from disease or from adaptation to a particular environment. The effects of short-term voluntary changes to ventilation are readily reversed by resumption of normal patterns of breathing.

Chronic respiratory acidosis

In chronic (i.e. long-term) respiratory acidosis, the plasma PCO_2 is elevated as the alveolar ventilation is insufficient to eliminate all of the carbon dioxide generated during metabolism. This leads to a fall in plasma pH and an increased hydrogen ion secretion into the proximal tubule and collecting ducts. Two mechanisms are responsible:

1. In the proximal tubule the increased hydrogen ion secretion occurs via Na^+–H^+ exchange (see Chapter 39). This process leads to increased reabsorption of the filtered bicarbonate. Normally the kidneys filter about 3 mmoles of bicarbonate each minute and the rate of hydrogen ion secretion by the proximal

tubule is sufficiently high to permit the absorption of about 85 per cent of the filtered load—the remainder is normally reabsorbed in the thick ascending loop and collecting duct. In respiratory acidosis the rate of hydrogen ion secretion by the proximal tubule is increased and up to 98 per cent of the filtered load can be reabsorbed in this part of the nephron. To preserve the electroneutrality of the plasma, the increase in bicarbonate reabsorption is associated with an increase in chloride excretion.

2. Excretion of excess hydrogen ions is performed by the intercalated cells of the distal tubule and collecting ducts where acid secretion occurs via a H^+-ATPase. This secretion of acid differs from that occurring in the proximal tubule, as it can take place against a steep pH gradient. Moreover, as the protons are derived from carbonic acid the secretion of acid leads to the generation of bicarbonate. This bicarbonate is reabsorbed across the basolateral membrane (see Figure 39.25) and represents new bicarbonate. As the line joining points A and B in Figure 41.6 shows, the plasma bicarbonate concentration increases progressively as the pH returns to normal.

Overall, the higher the PCO_2 the greater the secretion of hydrogen ions and the larger the quantity of bicarbonate generated and reabsorbed. Thus, in chronic respiratory acidosis increased renal hydrogen ion secretion leads to an increase in plasma bicarbonate. This, in turn, helps to limit the fall in plasma pH as the $[HCO_3^-]$/PCO_2 ratio increases towards normal.

Chronic respiratory alkalosis

In chronic respiratory alkalosis the situation is reversed. As the line joining points C and D in Figure 41.6 shows, if the kidneys are to restore plasma pH to normal, they must excrete bicarbonate. This is accomplished by a reduction in hydrogen ion secretion in the proximal tubule so that less of the filtered load of bicarbonate is reabsorbed. The proximal tubule cells reduce their secretion of

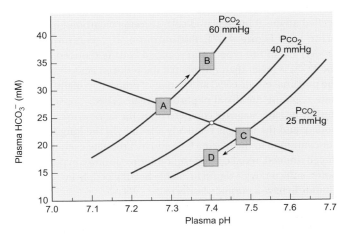

Figure 41.6 The changes in plasma $[HCO_3^-]$ that occur during renal compensation in a patient with respiratory acidosis (points A to B) and in one with respiratory alkalosis (points C to D). The white circle is the reference value (pH 7.4, 26 mmoles l^{-1} HCO_3^-).

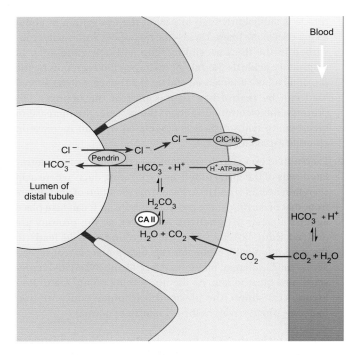

Figure 41.7 The process of bicarbonate secretion employed by type B intercalated cells of the late distal tubule and collecting ducts. These cells use the chloride/bicarbonate exchanger pendrin to excrete bicarbonate ions into the urine. The bicarbonate is generated from carbonic acid within the cells (catalysed by soluble carbonic anhydrase, CAII) and the hydrogen ions are pumped into the interstitial space by a H^+-ATPase. The chloride entering the cells by this process leave across the basolateral membrane via a CIC-Kb chloride channel.

hydrogen ions as plasma PCO_2 falls: the reduction in PCO_2 results in a decrease in carbonic acid formation within the tubular cells, fewer hydrogen ions are secreted, and less of the filtered bicarbonate is reabsorbed. In the late distal tubule and initial third of the inner medullary collecting ducts, a subtype of intercalated cell (called a type-B intercalated cell) secretes bicarbonate into the urine (see Figure 41.7). Furthermore, H^+ secretion by the H^+-ATPase of the type A intercalated cells is reduced because the PCO_2 is lower than normal (cf. the reduced acid secretion by the proximal tubule cells) and a smaller proportion of the bicarbonate reaching the distal nephron is reabsorbed. Consequently, during the early stages of compensation, the kidneys excrete bicarbonate and the urine is relatively alkaline. In the long term, the concentration in the plasma falls and bicarbonate excretion is reduced. Indeed, renal compensation for respiratory hyperventilation is sufficiently powerful to enable healthy people living at high altitude to have a normal plasma pH.

Metabolic acidosis

In metabolic acidosis and metabolic alkalosis, the changes in plasma pH are first minimized by respiratory compensation. Fine adjustment occurs over a longer period by altering the amount of

H^+ or HCO_3^- excreted by the kidneys. Metabolic acidosis is usually due to one of three factors:

1. increase in the production of metabolic acid
2. loss of base from the lower gut
3. reduced ability to excrete acid (in renal tubular acidosis).

The compensatory mechanisms employed will therefore vary according to the underlying cause. In the short term, the decrease in blood pH stimulates respiration and this increases the loss of carbon dioxide from the lungs. The resulting fall in PCO_2 causes the pH of the plasma to rise towards normal. This restoration of pH is relatively rapid (it occurs within minutes) but as the pH approaches 7.3 the stimulus to the central and peripheral chemoreceptors becomes less and the hyperventilation declines. Consequently, respiratory compensation will bring plasma pH within about 0.1 of a pH unit of its normal range but this is achieved at the cost of a fall in the plasma bicarbonate concentration, as shown by the change from point A to point B in Figure 41.8.

Secondly, the fall in plasma pH results in the secretion of a more acid urine. The effectiveness of this mechanism in eliminating the excess hydrogen ions is limited by the availability of the urinary buffers. A low plasma pH, however, also stimulates ammoniagenesis so that the kidneys are able to excrete more NH_4^+. This process helps to eliminate excess acid by the mechanism described in Figure 41.3. This stage of compensation can bring pH back to normal, but as it takes time to filter plasma and excrete excess acid, it takes hours or days for further compensation to occur. If metabolic

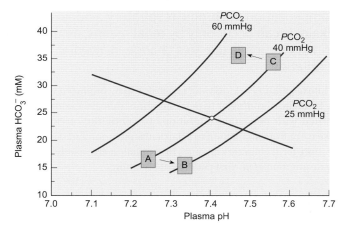

Figure 41.8 The changes in plasma [HCO_3^-] during metabolic acidosis and metabolic alkalosis and the effects of respiratory compensation. In metabolic acidosis (point A) the PCO_2 is initially normal (40 mm Hg) but the low plasma pH stimulates respiration, resulting in a fall in alveolar PCO_2 and thereby shifting the plasma pH closer to normal (point B). This adjustment requires a further fall in plasma bicarbonate. In metabolic alkalosis, the loss of non-volatile acid initially shifts the pH in the alkaline direction even though PCO_2 remains normal (point C). The increase in pH tends to depress respiration and leads to CO_2 retention and a fall in plasma pH (point D). In this situation the adjustment of plasma pH is accompanied by an increase in plasma bicarbonate. The white circle is the reference value.

acid production exceeds the ability of the kidneys to excrete NH_4^+, full compensation will not occur until the underlying cause is treated.

In acidosis due to loss of base from the gut, there is a fall in the filtered load of Na^+ which stimulates aldosterone production. This leads to an increase in Na^+ reabsorption by the mechanism discussed in Chapter 39 (p. 602).

Metabolic alkalosis

Metabolic alkalosis is commonly caused by vomiting gastric juice or by the intake of excessive quantities of alkali (e.g. self-administered antacid) when it cannot be excreted rapidly due to poor renal function. Initially, respiratory compensation occurs as the high plasma pH depresses respiration. In consequence, the P_aCO_2 rises and the plasma pH tends to fall towards normal, as shown by the shift from point C to point D in Figure 41.8. Nevertheless, as plasma pH approaches the normal range, the fall in pH offsets the depression of ventilation so that respiratory compensation is only partial. Final correction of metabolic alkalosis due to loss of gastric juice requires excretion of HCO_3^- and retention of Cl^- by the proximal tubule. Alkalosis due to ingestion of base is corrected by renal excretion of the excess base.

Summary

Following a disturbance of acid–base balance, compensatory mechanisms come into play to bring plasma pH within the normal range. Full correction requires treatment or elimination of the underlying cause and restoration of the plasma [HCO₃⁻] to normal levels.

Respiratory disorders are compensated by renal adjustments of plasma [HCO₃⁻], which may take several days to complete. Metabolic disorders are initially compensated by alterations to the rate of alveolar ventilation (respiratory compensation) but this is always insufficient to restore plasma pH to the normal range. Full compensation and correction occurs via renal mechanisms.

41.7 Clinical evaluation of the acid–base status of a patient using the pH-[HCO₃⁻] diagram

The state of acid–base balance in any patient can be deduced from the pH–HCO_3^- diagrams such as those already discussed. By measuring the plasma pH, bicarbonate, and P_aCO_2 an unambiguous interpretation can be made (see Table 41.4). There are a number of different ways of representing the relationship between plasma pH and plasma bicarbonate. In this chapter the pH-[HCO₃⁻] diagram (sometimes called a Davenport diagram) has provided the basis for this discussion, but an alternative representation is shown in Box 41.3. The pH-[HCO₃⁻] diagram can be divided into six zones, as shown in Figure 41.9. The numbers plotted on the graph show the pH and HCO_3^- values for various types of disorder. The primary

Table 41.4 The direction of the changes in plasma pH, PCO_2, and bicarbonate that characterize the primary metabolic and respiratory disturbances of acid–base balance

Condition	Plasma pH	Plasma PCO_2	Plasma HCO_3^-
respiratory acidosis	decreased	increased	increased
metabolic acidosis	decreased	normal	decreased
respiratory alkalosis	increased	decreased	decreased
metabolic alkalosis	increased	normal	increased

uncompensated changes are shown by the solid lines (see also Figure 41.5) and the vertical dotted line shows complete compensation (i.e. restoration of plasma pH to the normal range). As already discussed, complete correction requires that both the plasma pH and [HCO₃⁻] be returned to normal.

The difference between the total cation content and the total amount of chloride plus bicarbonate is known as the '**anion gap**' and is an indirect measure of the amounts of phosphate, sulphate, and organic acids present in plasma. Determination of the anion gap provides a quick way of assessing metabolic disorders of acid–base balance. Since blood pH is normally about 7.4, the fully dissociated cations (principally sodium) are present in

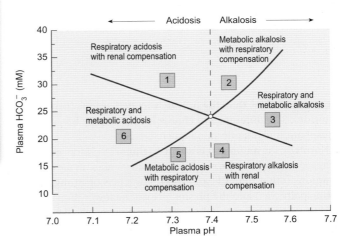

Figure 41.9 The pH-[HCO₃⁻] diagram for blood plasma can be used to determine the acid–base status of a patient. The white circle represents normal values. The diagram is labelled to show the underlying deficits for any given pair of pH-[HCO₃⁻] values. The numbers denote values that occur in six different conditions.

1. HCO_3^- 31 mmol l⁻¹, pH 7.30; partially compensated respiratory acidosis.
2. HCO_3^- 30 mmol l⁻¹, pH 7.45; partially compensated metabolic alkalosis.
3. HCO_3^- 24 mmol l⁻¹, pH 7.55; metabolic and respiratory alkalosis.
4. HCO_3^- 17 mmol l⁻¹, pH 7.42; compensated respiratory alkalosis.
5. HCO_3^- 16 mmol l⁻¹, pH 7.31; partially compensated metabolic acidosis.
6. HCO_3^- 20 mmol l⁻¹, pH 7.18; mixed metabolic and respiratory acidosis.

(1₂3) Box 41.3 The bicarbonate–PCO₂ diagram and its use in assessing acid–base status

The Henderson–Hasselbalch equation can be represented in a number of different ways, each of which gives a slightly different emphasis to the basic relationship. In the diagram shown in Figure 1, plasma bicarbonate is plotted on a logarithmic abscissa and PCO_2 is plotted as the ordinate on a linear scale. Plasma pH is represented by a diagonal axis. The advantage of this representation is that respiratory changes are shown in the vertical direction while metabolic changes are shown in the horizontal direction. The range of normal values is shown by the shaded ellipse in the centre of Figure 1.

In respiratory acidosis, PCO_2 rises and pH falls, as shown by point **a** for example. Restoration of normal pH (compensation) requires the excretion of acid, as indicated by the horizontal line **a–a'**. Conversely, in respiratory alkalosis, PCO_2 falls and pH rises, as shown by point **b**. Restoration of normal pH requires the excretion of excess bicarbonate, as shown by the line **b–b'**, and pH falls.

In metabolic acidosis, the fall in pH results in a decrease in plasma bicarbonate, as shown by point **c**. Short-term compensation occurs as hyperventilation lowers the alveolar PCO_2, as shown by line **c–c'**. Conversely, in metabolic alkalosis there is a rise in pH, as shown by point **d**. Short-term compensation occurs as respiration is inhibited and alveolar PCO_2 rises, as shown by the line **d–d'**.

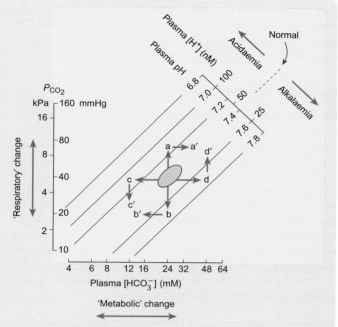

Figure 1 The bicarbonate–PCO_2 diagram. (Adapted from Fig 5.2 in E.J.M. Campell (1984). In E.J.M. Campbell, C.J. Dickinson, J.D.H. Slater, C.R.W. Edwards, and K. Sikora (eds) *Clinical physiology* 5th edition. Blackwell, Oxford.)

greater quantities than the fully dissociated anions (principally chloride). The difference in charge is made up by bicarbonate, phosphate, and organic anions such as lactate.

Normally the anion gap is about 15 mmol l⁻¹. If the anion gap increases, this will be the result either of an increase in the quantity of anions (other than Cl⁻ and HCO_3^-) or a loss of strong cations (principally Na⁺) and the plasma pH will be lower than normal. This situation typifies metabolic acidosis. Conversely, if the anion gap narrows this will be the result of ingestion of base or the loss of strong anion (e.g. loss of Cl⁻ from the stomach as a result of vomiting) and the plasma pH will be higher than normal. This situation typifies metabolic alkalosis.

Another common way of determining the extent of metabolic disturbance of acid–base balance is to measure the **base excess** or **base deficit**. The base excess is measured by titrating the blood or plasma to a pH of 7.4 with a strong acid or base while the PCO_2 is kept constant at 40 mm Hg. If strong acid (e.g. HCl) needs to be added to bring the pH to 7.4, there is a base excess, and if strong alkali (e.g. NaOH) is required, there is a base deficit. By definition, in normal people the base excess is zero but deviations of ±2.5 mmol l⁻¹ are considered to lie within the normal range.

The effect of changes to the base deficit or base excess can be understood by reference to Figure 41.10. The line A–B in Figure 41.10

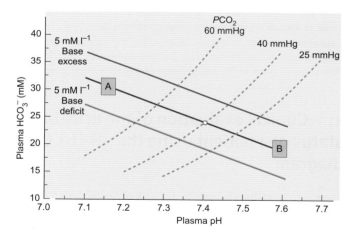

Figure 41.10 The effect of a base excess or a base deficit on the pH-[HCO_3^-] diagram for blood plasma. The line A–B is the normal buffer line and reflects the change in plasma pH as PCO_2 is varied. When there is a base deficit (as in metabolic acidosis), the buffer line is displaced downwards (shown here in green) and results in a lower plasma pH for a given PCO_2. When there is a base excess (as in metabolic alkalosis), the buffer line is displaced upwards (mauve line). The plasma pH will be higher than normal for a given PCO_2.

is called the blood buffer line. It represents the titration of the blood buffers with CO_2 (see also Figure 41.3). If there is a base excess or base deficit there is a parallel displacement of the buffer line. When there is a base excess, the line is displaced upwards as shown in Figure 41.10 by the mauve line and the blood pH becomes more alkaline for any given PCO_2. Conversely, if there is a base deficit the buffer line is displaced downwards as shown by the green line and the blood pH will be relatively more acid for any given PCO_2.

✳ Checklist of key terms and concepts

Physical chemistry of acids and bases

- The hydrogen ion concentration determines the acidity of a solution.

- The degree of acidity is often expressed using the pH scale. Pure water has a pH of 7 at 25°C and, at this temperature, acid solutions have pH values below 7 and alkaline solutions have pH values above 7.

- An acid generates hydrogen ions in solution and a base absorbs hydrogen ions.

- Strong acids and strong bases in aqueous solutions dissociate completely into their constituent ions.

- Weak acids and bases are only partially dissociated and the degree of dissociation depends on the hydrogen ion concentration.

- The hydrogen ion concentration of a solution is determined by three factors: the strong ion difference, the quantity and the pK_a values of the weak acids that are present, and the partial pressure of carbon dioxide.

- Buffering of hydrogen ions is provided by the weak acids and bases present in a solution.

- The CO_2/HCO_3^- buffer system of the plasma is quantitatively the most important in blood. This is partly because it is the most plentiful buffer and partly because the PCO_2 can be regulated by the respiratory system.

Acid–base regulation in the body

- To maintain the plasma hydrogen ion concentration within normal limits, the acid products of metabolism must be eliminated from the body.

- Carbon dioxide is excreted via the lungs, and non-volatile acid (chiefly sulphate) is excreted via the kidney.

- Respiratory control provides a rapid means of adjusting the plasma hydrogen ion concentration.

- The frequency and depth of respiration is stimulated by an increase in plasma PCO_2.

- A rise in plasma hydrogen ion concentration also stimulates ventilation and leads to an increased loss of carbon dioxide.

- The excretion of non-volatile acid in the urine depends on the ability of phosphate to buffer free hydrogen ions and on the ability of the kidney to generate NH_4^+.

- Under normal circumstances about half of the non-volatile acid is excreted as NH_4^+ salts, but if non-volatile acid production increases, the amount of NH_4^+ in the urine increases proportionately.

Disturbances of acid–base balance

- Normal arterial blood pH is 7.35–7.4.

- A blood pH of less than 7.35 is known as acidaemia.

- The process that leads to acidaemia is known as an acidosis.

- A metabolic acidosis is caused by an increased amount of non-volatile acid in the blood.

- A respiratory acidosis results from a failure to remove an appropriate amount of carbon dioxide from the blood.

- A blood pH greater than 7.45 is known as an alkalaemia.

- The alkalaemia that is the result of hyperventilation is called a respiratory alkalosis.

- Metabolic alkalosis is caused by a loss of acid from the stomach or by ingestion of non-volatile base.

- A metabolic acidosis can also develop if the production of non-volatile acids exceeds their rate of excretion via the kidneys, or if there is a loss of non-volatile base from the gut in diarrhoea.

- The difference in the plasma concentration of cations and the total amount of chloride plus bicarbonate is known as the 'anion gap'.

- It is an indirect measure of the amounts of phosphate, sulphate, and organic acids present in plasma.

- Determination of the anion gap provides a quick way of assessing metabolic disorders of acid–base balance.

- The anion gap is increased (> 15 mmol l^{-1}) in metabolic acidosis and decreased in metabolic alkalosis (< 15 mmol l^{-1}).

Compensation of acid–base disturbances

- Respiratory acidosis and respiratory alkalosis are compensated by renal adjustments of plasma [HCO$_3^-$].

- Such adjustments may take days to complete.

- Metabolic acidosis and metabolic alkalosis are initially compensated by alterations to the rate of alveolar ventilation.

- This respiratory compensation is always insufficient to restore plasma pH to the normal range.

- Full compensation and correction occurs via renal mechanisms over several days.

- Full correction of any acid–base disturbance requires treatment to eliminate the underlying cause and restore plasma [HCO$_3^-$] to normal levels.

Recommended reading

Cohen, R.D., and Woods, H.F. (2000) Disturbances of acid–base homeostasis, Chapter 6.17, in: *Concise Oxford textbook of medicine*, ed. J.G.G. Ledingham and D.A. Warrell. Oxford University Press, Oxford.
A very brief account of the clinical aspects of acid–base disturbances.

Holmes, O. (1993) *Human acid–base physiology*. Chapman & Hall, London.

Lowenstein, J. (1993) *Acid and basics: A guide to understanding acid–base disorders*. Oxford University Press, Oxford.
A somewhat unconventional but illuminating account of the regulation of acid–base balance.

Review articles

Koeppen, B.M. (2009) The kidney and acid–base regulation. *Advances in Physiology Education* **33**, 275–81, doi:10.1152/advan.00054.2009.

Roy, A., Al-bataineh, M.M., and Pastor-Soler, N.M. (2015) Collecting duct intercalated cell function and regulation. *Clinical Jounal of the American Society of Nephrology* **10**, 305–24, doi: 10.2215/CJN.08880914.

Wagner, C.A., et al. (2009) Regulated acid–base transport in the collecting duct. *Pflugers Archiv—European Journal of Physiology* **458**, 137–56, doi:10.1007/s00424-009-0657-z.

Weiner, I.D., and Hamm, L.L. (2007) Molecular mechanisms of renal ammonium transport. *Annual Review of Physiology* **69**, 317–40.

To check that you have mastered the key concepts presented in this chapter, complete the accompanying online self-assessment questions. Go to www.oup.com/uk/pocock5e/

CHAPTER 42

The skin and thermoregulation

Chapter contents

This chapter should help you to understand:

- Histological features of skin
- The principal functions of the skin
- The importance of the skin in thermoregulation
- The concept of core and shell temperatures
- The ways in which heat may be exchanged between the body surface and the environment
- The importance of the cutaneous circulation as a means of regulating heat exchange between the core and the surface of the body
- The responses of the body to cold, including behavioural changes, shivering, vasoconstriction, and non-shivering thermogenesis
- The responses of the body to heat, including behavioural changes, vasodilatation, and sweating
- The consequences of hypothermia and hyperthermia
- The initiation and consequences of fever

42.1 Introduction

The skin (or cutis) is the largest organ in the body, covering its entire outer surface. Together with the hair, nails, and glands, the skin forms the **integumentary system**. In an adult, the skin has a surface area of around 1.5–2.0 m^2 and accounts for around 15 per cent of body weight. It varies in thickness from 0.5 mm to 5 mm depending upon the region of the body and the age of the person. The thickest skin is found in areas that are subject to abrasion such as the soles of the feet, while the thinnest is found around the eyes.

Because the skin is on the outside of the body, and may therefore be seen and touched, a great deal of useful information regarding a person's age, nutritional status, hydration status, circulation, and emotional state can be obtained by careful inspection. Furthermore, the characteristics and colour of a person's skin can sometimes yield information regarding certain disease states (particularly in those whose skin is pale). Examples include the yellow hue of jaundiced skin or the blue skin colour (**cyanosis**) characteristic of a reduction in the amount of oxygen being carried by the blood. Furthermore, skin rashes and other changes may represent external manifestations of systemic disease or injury.

As the interface between the internal structures of the body and the external environment, the skin serves a number of important functions including protection against environmental hazards such as ultraviolet radiation and disease-producing agents (pathogens). It is also responsible for thermoregulation, sensation, and the synthesis of vitamin D—see Chapter 22 p. 327. Integrity of the skin is also vital for normal fluid balance—indeed, the severe fluid loss which occurs in patients who have suffered extensive burns constitutes a medical emergency. The structure and functions of the skin and its appendages will be discussed in this chapter, followed by a discussion of its role in the regulation of body temperature. The role of the skin in protecting the body against pathogens is described in Chapter 26, and a detailed discussion of its role in sensation is given in Chapter 13.

42.2 The main structural features of the skin

The skin is a complex structure made up of two principal layers of tissue that have different histological and functional characteristics and are of different embryological origin. These are illustrated in Figure 42.1. The thinner outer layer is the **epidermis** which overlies a thicker layer called the **dermis**. Beneath the dermis lies a third layer consisting largely of adipose tissue (fat). Although this is not part of the skin itself, this layer serves to bind the skin to the superficial fascia of the underlying tissues and supplies it with nerves and blood vessels. This layer is called the **hypodermis**, subcutis, subdermis, or subcutaneous layer.

Hairy skin is found over most areas of the body, but in some locations, such as the soles of the feet, the palms of the hands and parts of the external genitalia, the skin is hairless. Smooth skin of this kind is known as **glabrous skin**. Hairy skin varies in different parts of the body with regard to both the number of hairs present and their physical characteristics—compare, for example, the hairy skin of the scalp with that of the lower arm.

The epidermis

The epidermis forms the surface layer of the skin and is derived from ectodermal tissue in the embryo. It is a **keratinized stratified squamous epithelium** (see Chapter 4 for a description of the different kinds of epithelium) which varies in thickness from 0.1 mm

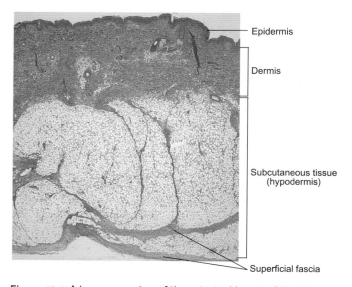

Figure 42.1 A low-power view of the principal layers of the skin.

Epidermis

Dermis

Subcutaneous tissue (hypodermis)

Superficial fascia

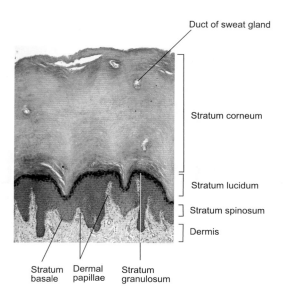

Duct of sweat gland

Stratum corneum

Stratum lucidum

Stratum spinosum

Dermis

Stratum basale   Dermal papillae   Stratum granulosum

Figure 42.2 Section through the outer layers of the skin showing the layers of the epidermis. This section is from skin taken from the sole of the foot.

around the eyes to around 1.4 mm on the soles of the feet. The epidermis contains no blood vessels and no nerve endings. Structurally, it is well-adapted to its key functions of preventing fluid loss and providing an effective physical barrier against infection.

As Figure 42.2 shows, the epidermis consists of five distinct layers of cells. Starting from the deepest layer and working towards the surface, these are:

- stratum basale (basal layer)
- stratum spinosum
- stratum granulosum
- stratum lucidum
- stratum corneum.

The two innermost strata, the stratum basale and stratum spinosum, are the sites of vitamin D synthesis from cholesterol (see Chapter 22). The **stratum basale** is formed by a single layer of cells that lie on a basement membrane (the basal lamina) that separates the epidermis and dermis. These cells may be cuboidal or columnar in character and they are the **germinal cells** of the epidermis, dividing rapidly to provide a continuous supply of new cells known as **keratinocytes**. As the cells migrate upwards towards the skin surface, they alter in shape and biochemical characteristics. In the **stratum spinosum**, the cells become irregular in shape and become separated by narrow gaps or clefts spanned by thin cytoplasmic processes or spines that give this layer its name. The cells of this layer are sometimes called **prickle cells**. Membrane-bound granules containing lipids (**lamellar granules**) begin to appear in their cytoplasm, together with accumulations (fine grains) of keratohyalin, a protein that will eventually form the keratin of the outer epidermis. As the cells enter the **stratum granulosum**, the contents of the lamellar granules begin to be released from the cells and to fill the interstitial space with lipid, which is important for the barrier function of the epidermis.

As cells reach the outer margins of the stratum granulosum their nuclei begin to degenerate and the cells begin to die as they move farther from their blood supply. The **stratum lucidum** consists of a few layers of flattened, dead cells with some filaments of keratin present inside them. The cells are difficult to visualize microscopically and appear unclear and poorly defined. Cells in the **stratum corneum** are known as 'horny cells'. They are completely filled with keratin filaments and the spaces between them are filled with lipids (released from the lamellar granules) that bind the cells together to form a continuous membrane. Towards the outer surface of the stratum corneum, the intercellular connections begin to loosen and cells slough off continuously from the skin surface (a process called **desquamation**), to be replaced by new cells migrating upwards from deeper layers. The process whereby cells gradually become filled with keratin is called **keratinization** and it normally takes around 30 days for cells to complete their journey from the stratum basale to the outer surface of the stratum corneum from which they are shed.

Other cells of the epidermis

Although the majority of cells within the epidermis are keratinocytes, three further cell types are also found. They are **melanocytes** (pigment cells), **Merkel cells**, and **Langerhans cells** (which are part of the immune system).

Melanocytes are derived from embryonic neural ectoderm. Typically, between 1000 and 2000 melanocytes are found per square mm of skin, predominantly in the stratum basale. They synthesize a brown pigment called **melanin** which is the primary determinant of skin colour. However, the depth of skin colour is determined principally by the concentration of melanin within the melanocytes rather than by the number of melanocytes present within the skin. Melanin levels in the skin vary between individuals and ethnic groups. Melanin is also found in the hair, iris, parts of the adrenal gland, and in certain regions of the brain such as the substantia nigra. As Figure 42.3 shows, melanocytes have fine processes which divide and make contact with surrounding keratinocytes.

The melanocytes synthesize melanin from the amino acid **tyrosine** within cytoplasmic organelles called melanosomes. Melanin is then transferred to the keratinocytes of the basal layer. The exact mechanism by which melanin is transferred to the epidermal cells is not clear, but it is thought that the fine processes of a melanocyte may actually invade a keratinocyte and then bud off to 'deposit' the melanin inside the cell.

The regulation of melanin synthesis

The synthesis of melanin by melanocytes is controlled by certain hormones and by light. **Melanocyte stimulating hormone (MSH)** is secreted by the anterior lobe of the pituitary gland and, as its name suggests, it increases the production of melanin by melanocytes. It is believed to work by stimulating the activity of tyrosinase. MSH is structurally related to adrenocorticotrophic hormone (ACTH—see Section 21.6), which also has a weak stimulatory

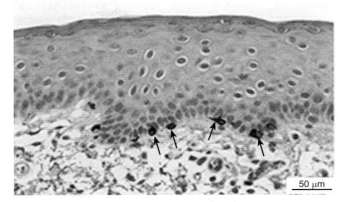

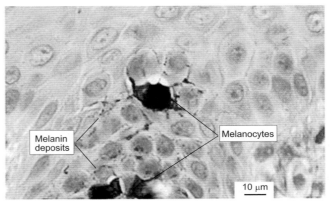

Figure 42.3 The upper image shows a section of skin with prominent staining of melanocytes in the basal layer of the epidermis (arrows). The lower image is a high-power view of melanocytes that shows the delicate processes that branch from the cell body to leave deposits of melanin amongst the neighbouring cells. The melanocytes were visualized using MART-1 immunostaining. (From Fig 3-15 in plasticsurgerykey.com/histology-of-the-skin/, by permission of Plastic Surgery Key.)

action on melanocytes. In conditions such as Addison's disease, in which circulating levels of ACTH are raised (see Chapter 23), there is often increased pigmentation of the skin and mucous membranes.

Melanin absorbs ultraviolet (UV) light, regulating the amount that enters the deeper layers of the skin and protecting the underlying structures from its harmful effects. Melanin synthesis is stimulated when the skin is exposed to UV light, which increases the activity of tyrosinase. This explains why people living in parts of the world where there is a lot of sunshine tend to have darker skins than those who live at higher latitudes. It also accounts for the 'tanning' of the skin that occurs after sunbathing. It has recently been shown that α-MSH binds to a receptor (MC-1) on the human melanocyte membrane and that this binding activates tyrosinase. Melanocytes taken from individuals who tan poorly (such as those with red hair) appear to have mutations in their MC-1 receptor.

Although over-exposure of the skin to UV light can be harmful, limited regular exposure to sunlight is necessary for the skin to synthesize vitamin D. Ultraviolet light penetrates the epidermis and acts on 7-dehydrocholesterol to convert it to vitamin D_3. By absorbing UV light, melanin not only protects the deeper layers of the skin, it also reduces the ability of the skin to synthesize vitamin D. For this reason, dark-skinned people, particularly those living in relatively sunless climates, may be unable to produce their daily requirement of vitamin D. For such individuals, dietary intake of vitamin D is essential.

Disorders of melanin synthesis may occur. For example, in the inherited condition known as **albinism** the ability to synthesize the enzyme tyrosinase is lacking, while the acquired condition, **vitiligo**, is characterized by progressive destruction of melanocytes. Here, large areas of skin lose their pigment and turn white. This may also happen to the hair growing on these areas. It is a relatively common condition and affects around 1 per cent of the population.

Excessive exposure to UV light is known to increase the risk of developing cancers of the skin in those areas of the body most frequently exposed to the sun such as the face, neck, hands, and, in some women, the lower legs. The three most common skin cancers are **basal cell carcinomas**, **squamous cell carcinomas**, and **malignant melanomas**. Melanomas are aggressive tumours derived from melanocytes and are the most lethal of the skin cancers because, if left untreated, they spread rapidly to other tissues (**metastasize**). Initially, the tumour spreads to regional (nearby) lymph nodes, but later tends to disseminate more widely via the blood.

Merkel cells are found in the deeper layers of the epidermis. They have small dense granules in their cytoplasm and are associated with free nerve endings (see p. 204). They are accessory cells that play a role in fine touch. They can give rise to a rare but highly aggressive form of skin cancer variously known as Merkel cell carcinoma, apudoma of the skin, primary small cell carcinoma, or small cell neuroepithelial tumour of the skin.

Langerhans cells form part of the immune system and behave rather like macrophages, responding to invading organisms in the skin. Langerhans cells are dendritic cells (see Figure 42.4) that bind to antigens in the epidermis before migrating to a regional lymph node where they then 'present' the antigen to the lymphocytes of the immune system so that an appropriate immune response can be initiated (see Chapter 26).

The dermis

The dermis lies directly beneath the epidermis. It ranges in thickness from 0.6–3 mm, being thickest on the soles of the feet and thinnest on the eyelids and prepuce. It contains blood vessels, lymphatic vessels, sensory receptors and their afferent nerve fibres, hair follicles, sebaceous glands, and sweat glands. It is made up of dense connective tissue which provides the skin with mechanical strength, and supports the structures within it. At its deep border, the dermis merges with the hypodermis without a sharply defined boundary. The dermis itself is composed of two layers, the **papillary layer** and the **reticular layer**. The papillary layer lies adjacent to the stratum basale of the epidermis and forms a series of ridges and hollows (the dermal papillae), as shown in Figures 42.1 and 42.2. Although under most normal circumstances this arrangement helps to prevent the dermis and the epidermis from separating, excessive or repeated shearing forces (e.g. from the rubbing of a tight

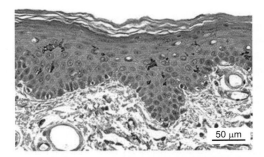
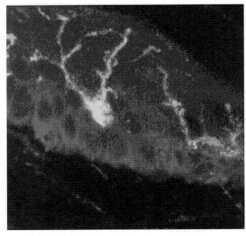

Figure 42.4 Langerhans cells in the skin. Top: Langerhans cells stained with an immunological marker are shown scattered throughout the epidermis. Bottom: a confocal image of Langerhans cells visualized using CD1a immunostaining (yellow-green) showing their characteristic dendritic structure. (From panels A and B of Fig 3-20 in plasticsurgerykey.com/histology-of-the-skin/, by permission of Plastic Surgery Key.)

shoe) can cause the epidermal layer to be torn off, creating a blister. The reticular layer is thicker and lies beneath the papillary layer, although the boundary between the two is indistinct.

Like all structural connective tissues, the dermis consists of a non-cellular matrix (or ground substance) within which lie the cells and fibres. The matrix is a gel-like substance that consists of hyaluronic acid, mucopolysaccharides, and chondroitin sulphate. Cells of the dermis include fibroblasts, which synthesize collagen and elastin fibres, macrophages, mast cells, and other cells of the immune system.

Collagen is a fibrous protein which has an extremely high tensile strength. It is arranged in bundles whose orientation lies predominantly parallel to the skin surface. Wherever possible, surgical incisions are made along these orientation lines (the line of cleavage), because the resulting wound is finer, heals more quickly and scars less noticeably than an incision made across the line of cleavage. Elastin fibres are branching protein fibres found in both the papillary and reticular dermal layers. They give the skin its flexibility (elasticity) but they can rupture if subjected to excessive stretch as in rapid weight gain, fluid retention, or pregnancy. Ruptured elastic fibres create silvery scars on the skin surface. These scars are known as **striae** or, colloquially, as stretch marks.

The hypodermis

Beneath the dermis, and attaching the skin to the underlying muscles, is a layer of tissue known as the hypodermis, subcutis, or subcutaneous layer. It consists largely of adipose tissue (fat) and provides the body with both mechanical and thermal insulation. The stored fat also forms a nutritional reserve which may be called upon to provide energy during periods of fasting.

Summary

The skin is the interface between the internal structures of the body and the outside world. It is a barrier to infection, a major sense organ, and a waterproof outer layer. It is also important in the regulation of body temperature. It consists of two main layers: the epidermis and the dermis. The dermis is bound to the underlying tissues by another layer, the hypodermis.

The epidermis consists of five main cell layers. The outermost layer (stratum corneum) consists of dead cells filled with keratin and a fatty matrix. This provides a waterproof, abrasion-resistant barrier. In addition to the keratinocytes that make up most of the epidermis, the epidermis contains melanocytes that are responsible for skin pigmentation, and Langerhans cells, which play an important role in combating infection. The epidermis has no nerve endings or blood vessels.

The dermis is made up of dense connective tissue containing collagen and elastic fibres. It also contains the skin blood vessels, sensory receptors, hair follicles, and sebaceous glands. The hypodermis binds the skin to the underlying tissues and consists largely of adipose tissue, which provides the body with both mechanical and thermal insulation.

42.3 The accessory structures of the skin—hairs, nails, and glands

Hairs (pili) are present over the entire body surface except for the plantar surfaces of the feet and toes, the palmar surfaces of the hands and fingers, and certain areas of the external genitalia. In some areas (such as the scalp) the hair grows thickly, while in others it is so sparse as to be virtually invisible. Hairs grow from **follicles** located at the lower border of the dermis. Typically, around 5 million follicles are distributed over the skin surface, of which about 1 million are in the scalp, and all are present within the skin at the time of birth. Figure 42.5 shows the anatomical features of a hair follicle and the hair contained within it.

The follicle itself is made up of an outer connective tissue sheath lined by an epithelial membrane that is continuous with the stratum basale (the germinal layer) of the epidermis. The lower portion of each hair follicle is formed by the **hair bulb**, at the base of which is an indentation called the **papilla**. This contains connective tissue and blood vessels that nourish the follicle. It also contains a population of cells (matrix cells) that arises from the stratum basale and is responsible for the growth of existing hairs as well as the development of new hairs.

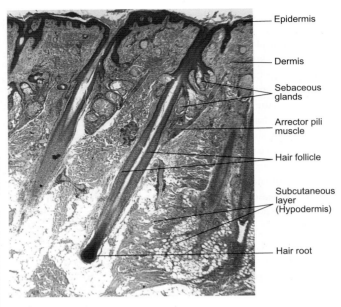

Epidermis

Dermis

Sebaceous glands

Arrector pili muscle

Hair follicle

Subcutaneous layer (Hypodermis)

Hair root

Figure 42.5 A section through the skin showing the main structures associated with the hair follicles. The arrector pili muscle of the right-hand follicle is clearly visible, as are the sebaceous glands of both follicles. Note that the hair root extends deep into the subcutaneous layer but is surrounded by a sheath of dermal tissue.

A network of nerve fibres surrounds each hair follicle (the hair root plexus). These fibres are stimulated by movements of the hair shaft and contribute to the touch sensitivity of hairy skin. A bundle of smooth muscle fibres (the **arrector pili muscle**) is also associated with the hair follicle and extends from the follicle to the upper border of the dermis (see Figure 42.4). This smooth muscle contracts in response to activation of its nerve supply, which is part of the sympathetic division of the autonomic nervous system. Contraction causes the hair to become erect, for example in response to cold or fright.

The hair itself is made up of a **root** and a **shaft**. The root forms the base of the hair and penetrates deep into the dermis, while the shaft is the more superficial part of the hair which projects above the surface of the skin. Both regions of a hair consist of three layers, the inner medulla, the middle cortex, and the outer cuticle. The cells of the medulla and cortex contain melanin granules that determine the hair colour, while the outer cuticle is formed by a single layer of flattened, highly keratinized cells arranged rather like the scales of a fish.

Hair growth

Hair tends to grow in cycles and, at any one time, about 85 per cent of the scalp hairs are likely to be growing actively. The rest will be in a state of rest or regression (prior to being lost). Between 25 and 100 hairs are shed from the scalp each day, pushed out by a new hair developing within its follicle. Hair growth and loss is affected by a number of factors including severe illness, nutritional status, childbirth, and exposure to radiation. Certain hormones, including testosterone and thyroxine, also influence the growth and texture of hair.

The nails

The nails are a form of modified hair and consist of tough translucent plates of keratin that protect the ends of the fingers and toes. Careful inspection of the nails may reveal information concerning nutritional deficiencies or certain disease states: for example, the 'clubbing' of the fingernails that occurs in some cardiovascular and respiratory disorders.

Figure 42.6 shows the different regions of the dorsal surface of a nail. The visible part of the nail is called the nail body, the free edge of which may extend beyond the end of the finger. The nail root is the part of the nail that lies under the fold of skin (the cuticle) situated at the base of the nail. Beneath the nail root is a layer of epithelial cells (the nail matrix) that divide as the nail grows. In humans and other primates, the fingernails play an important role in the manipulation of small objects.

The glands of the skin

Three types of glands are found within the skin. These are

- sebaceous (oil) glands
- sudoriferous (sweat) glands
- ceruminous glands.

Sebaceous (holocrine) glands are found over all parts of the skin except for the palms of the hands and soles of the feet. They are mostly found in association with hair follicles, as may be seen in Figures 42.5 and 42.7. These glands secrete an oily substance called **sebum**, either onto the hair shaft with which they are associated, or directly onto the skin surface. Sebum helps to keep the skin soft and flexible and reduces evaporative water loss. It also prevents the hair from becoming brittle and contributes to the slightly acidic pH

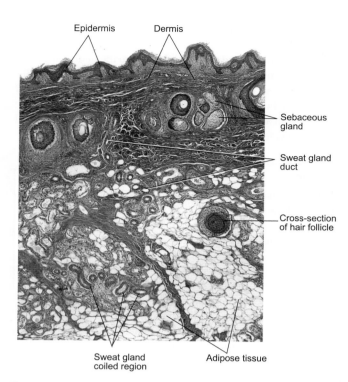

Figure 42.7 A cross-section through the skin showing the secretory zone and the duct of an eccrine sweat gland. The section also shows sebaceous glands and the adipose tissue of the subcutaneous layer. A number of hair follicles are shown in cross-section.

of the skin surface (pH 5.5–7.0), which helps to inhibit the growth of certain microorganisms.

Sudoriferous (merocrine) glands are further subdivided into two groups, according to their structure, their location, and the nature of their secretions: the eccrine glands and the apocrine glands. **Eccrine glands** are found in the dermis of the skin over the entire body. There are around 2.5 million of these, about half of which are situated in the skin of the back and chest. As Figure 42.7 shows, eccrine sweat glands are simple tubular structures consisting of a coiled portion deep in the dermis and an unbranched duct that opens on to the body surface via a sweat pore. In humans, **apocrine glands** are confined to the skin at specific locations, notably the axillae (armpits) and the genital area. Although their general structure is similar to the eccrine glands, the apocrine glands are larger (their secretory portion is 3–5 mm in diameter compared to c. 0.4 mm diameter for the eccrine glands) and their coiled region lies within the subcutaneous tissue, rather than in the dermis. Their ducts open into hair follicles, where their secretion mixes with that from the sebaceous glands as it flows to the skin surface. Apocrine gland secretions differ from those of the eccrine glands in that they contain proteins and lipids. The characteristic odour of sweat is produced by the action of skin bacteria on the fatty components of apocrine secretions. Eccrine sweat, by contrast, is normally odourless.

The function of the apocrine secretions in humans is uncertain but in other mammals the characteristic odour of the apocrine

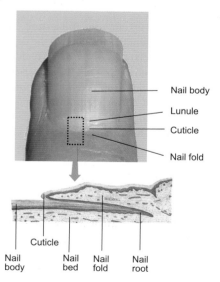

Figure 42.6 Dorsal view of a fingernail showing the main features. The lower part of the figure shows a diagrammatic representation of a sectional view of the area outlined in the upper panel.

secretions serves a chemical signal to identify individuals and to claim territory.

Ceruminous glands are specialized sweat glands found in the skin lining the ear canal. They secrete a thick, waxy material known as **cerumen** (or ear wax) which helps to prevent drying of the tissue and to maintain cleanliness of the ear canal by trapping dust, debris, and dead cells. Occasionally, over-production and impaction of cerumen can occur, leading to impaired hearing (see Chapter 15).

Summary

The accessory structures of the skin include the hairs, nails, and three types of gland: sebaceous (oil) glands, sudoriferous (sweat) glands, and the ceruminous glands of the ear canal. The sweat glands are subdivided into the eccrine glands, which play an important role in regulating body temperature, and the apocrine glands, which secrete a thick fluid that gives rise to body odour.

42.4 The cutaneous circulation

Figure 42.8 illustrates the principal features of the blood supply to the skin. The epidermis contains no blood vessels (it is avascular), but its basal layers receive nourishment from the rich blood supply of the dermis. Blood enters the dermis via arterioles which branch to form capillary loops that carry blood into the dermal papillae. The venous limbs of these capillary loops then drain into venules that form a venous plexus (network) within the dermis. Subsequently, blood is carried away from the dermis by veins as shown in Figure 42.8. In some areas of the skin, particularly the hands, feet, ears, nose, and lips, additional vessels are found. These are called arterio-venous (AV) anastomoses and they provide a route for blood to pass directly between the arterioles and the venules, thus bypassing the capillary loops. AV anastomoses are shown in Figure 42.8. They are richly innervated by sympathetic nerve fibres which can bring about powerful vasoconstriction.

The most important regulator of blood flow to the skin is not the supply of oxygen and nutrients, as it is in most tissues. Indeed, the nutritional requirements of the skin are small. Instead, blood flow to the skin is determined mainly by the level of sympathetic nervous activity which is continuously adjusted so that body temperature is kept largely constant, as discussed later in this chapter. If body temperature starts to fall, sympathetic activity increases, the arterioles constrict, and the AV anastomoses close. This diverts blood away from the body surface and acts to conserve body heat. When body temperature starts to rise, sympathetic activity is reduced and the arterioles and AV anastomoses dilate, blood flow to the body surface is increased, and this facilitates heat loss. In circumstances that require a greater degree of heat loss, sweating is stimulated. Sweat contains an enzyme whose action produces the powerful vasodilator **bradykinin**, which acts locally to bring about vasodilatation of the arterioles. The arterioles also dilate in response to increased local levels of metabolites, a situation likely to occur when body temperature is elevated and metabolic rate is increased (reactive hyperaemia—see also Chapter 30).

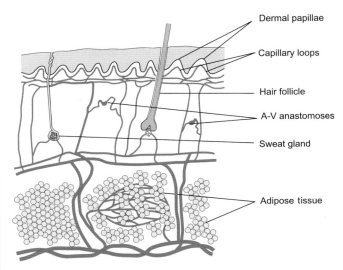

Figure 42.8 The arrangement of the vessels of the cutaneous circulation. There are three networks of blood vessels. One is very deep in the subcutaneous tissue, another lies below the dermis, and the third lies just below the epidermis. The outermost capillaries do not pass into the epidermis but remain within the dermis. The blood flow through the dermis determines the rate of heat loss from the skin. The arterio-venous anastomoses are found only in certain areas of the skin such as the tip of the nose and the fingertips.

Summary

The cutaneous circulation consists of three main networks of vessels: one in the subcutaneous tissue, one deep in the dermis, and a superficial network just below the epidermis.

The epidermis has no blood supply of its own and the oxygen and nutrients it requires are supplied by diffusion from the outermost dermal capillaries. Cutaneous blood flow is mainly controlled by the sympathetic nervous system and can be varied by more than a hundredfold. This ability is vital for the regulation of body temperature.

42.5 Heat exchange between the skin and surroundings

Almost all the cellular processes of the body will ultimately result in the production of heat. The more metabolically active a tissue is, the more heat it produces. The organs that produce the most heat are the brain, skeletal muscle (particularly during exercise), and the visceral organs such as liver and kidneys. To maintain normal body temperature, heat loss to the environment must be balanced by heat generated through metabolism. This is the situation when body temperature is stable. In a cold environment heat is lost continually from the body surface, and this must be minimized by appropriate physiological adjustments. Moreover, the lost heat must be balanced by some form of heat production. In a hot environment it is possible for the body to gain heat, and the mechanisms governing heat loss need to be activated to prevent a rise in body temperature.

Different regions of the body have different temperatures at rest. The brain and organs within the thoracic and abdominal cavities have the highest temperature. This is known as the **core temperature**. Under most conditions the body surface, the skin, has the lowest temperature, the **shell temperature**. The blood acts as the vehicle of heat exchange between the core of the body and its shell. It is the core temperature that is precisely regulated by the body's homeostatic mechanisms. By contrast, the shell temperature may vary substantially depending on the temperature of the surroundings. Despite wide variations both in their metabolic activity and in the temperature of their environment, people maintain a normal body temperature of about 37°C. (see Figure 42.9).

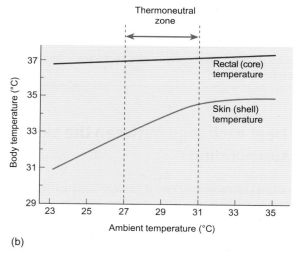

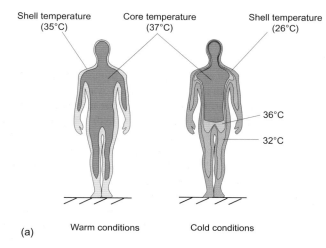

Figure 42.9 Body temperature variation under different environmental conditions. (a) The distribution of internal body temperatures under warm and cold conditions. (b) The effects of different ambient temperatures on rectal temperature (core body temperature) and skin temperature (shell temperature) in unclothed healthy adult male subjects. ((a) Adapted from Fig 6 in J. Aschoff and R. Weaver (1958) *Naturwissenschaften* **45**, 477; (b) Based on data in Fig 5 in J.D. Hardy et al. (1941) *J Nutrition* **21**, 383.)

For clinical purposes, core temperature is usually recorded from the ear canal (the tympanic temperature) or the oral cavity (in one of the sublingual pockets), although it can also be measured in the axilla (armpit) and rectum. When temperature is measured in the oral cavity, 95 per cent of people will have a temperature within the range 36.3–37.1°C. However, a small number of healthy people have core temperatures that lie outside these values so that a range of 35.5 to 37.7°C may provide a truer reflection of the population as a whole. In clinical practice, core temperature is measured routinely for a number of reasons including the following:

- to establish a baseline temperature on admission to a hospital or clinic
- to monitor the progress of those recovering from hypothermia
- to monitor fluctuations in temperature as an indicator of developing infection or the presence of a deep vein thrombosis (especially during the post-operative period)
- to monitor the temperature of people undergoing treatment for infections
- to detect signs of incompatibility during a blood transfusion.

Natural variations in body temperature in health

A number of factors influence core temperature in healthy individuals. These include genetic variability, exercise, circadian variation, the menstrual cycle, and pregnancy. Heavy exercise, for example, may elevate core temperature to 39° or 40°C for short periods. A healthy individual's body temperature fluctuates approximately 1°C in 24 hours, with the lowest values occurring in the early morning (around 2–3 a.m.) and the highest values in the late afternoon or early evening. The fall in core temperature during the early hours shows a seasonal variation and is smaller in summer than in winter. This 24-hour (circadian) rhythm is illustrated in Figure 42.10.

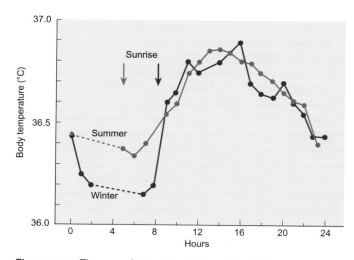

Figure 42.10 The normal circadian rhythm of core body temperature. The lowest temperature occurs in the early morning around sunrise in this individual and the highest in mid- to late afternoon. Note the seasonal difference. (Based on data in T. Sasaki (1964) *Exp Biol Med* **115**, 1129.)

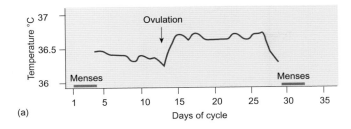

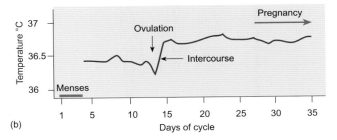

Figure 42.11 Natural variables affecting core temperature of healthy women of reproductive age. (a) The body temperature of a woman recorded each day at 7 a.m. throughout a complete menstrual cycle, illustrating the elevation brought about by ovulation. (b) A similar record for the same woman showing the persistence of the temperature rise once she became pregnant.

In women of reproductive age, core temperature fluctuates during each menstrual cycle. As Figure 42.11(a) shows, there is an increase in body temperature of about 0.5°C following ovulation, which persists until steroid levels fall just prior to the onset of menstruation (see Chapter 48). Couples practising natural contraceptive methods, or trying to maximize their chances of conceiving a baby, may use this rise in the core temperature as an indicator of ovulation. If a pregnancy occurs, the temperature remains elevated until delivery, as shown in Figure 42.11(b).

Why is body temperature regulated so precisely?

With each rise of 1°C, the rate of chemical reactions increases by around 10 per cent so that as temperature rises, the rate of enzymatic activity within cells is accelerated. However, different enzymes have different temperature sensitivities so changes in temperature alter the balance of reaction sequences. This balance is particularly important in energy metabolism. Moreover, most enzymes have an optimum temperature above which their activity declines as they begin to denature (degrade). The denaturation of enzymes and other cellular proteins accelerates when the temperature exceeds 40°C, with the result that cell damage will occur if body temperature rises much above this level. Prolonged exposure to temperatures in excess of this will result in cell death. Indeed, if core temperature exceeds 41°C, most adults will go into convulsions, while a temperature of 43°C appears to represent the absolute limit for human life.

In contrast, most cells can withstand marked reductions in temperature, although they function more slowly as their temperature

falls. In humans, physiological processes begin to be disrupted when body temperature falls below about 35°C, with a decline in muscle strength and a slowing of motor responses. The duration and refractory period of action potentials are markedly lengthened as the temperature falls, which has been attributed to delayed repolarization. In addition there is a slowing of the heart rate (bradycardia) with a widening of the QRS-complex in the ECG and an increased likelihood of the development of cardiac arrhythmias such as atrial or ventricular fibrillation. When core temperature falls below about 30–33°C, temperature regulation is impaired, consciousness is lost, and the heart may stop.

There is, therefore, a relatively narrow range of temperatures within which the body can function normally. Control of body temperature is therefore essential for survival. This process is known as **thermoregulation** and is one of the important homeostatic mechanisms of the body.

The role of the cutaneous circulation in thermoregulation

The range of environmental temperatures over which it is easy for the body to maintain its core temperature is known as the **thermoneutral zone**. It is between 27 and 31°C for a naked individual but can extend well below 27°C when appropriate clothing is worn. For ambient temperatures within the thermoneutral zone, skin temperature is maintained at about 33°C. Under these circumstances, thermoregulation is achieved solely by alterations in the blood flow to the skin, which is the surface from which the rate of heat loss from the body core can be regulated.

The skin is well endowed with blood vessels which provide it with an efficient means of regulating heat exchange between the body and its surroundings. As Figure 42.8 shows, there are three networks of blood vessels: one is very deep in the subcutaneous tissue, another lies below the dermis, and the third lies just below the epidermis. Arterioles supplying the dermis break up into dense capillary networks beneath the epidermis. These networks drain into a superficial venous plexus, which is able to accommodate a large volume of blood. In addition, in the most exposed areas of the body, such as the fingers, toes, ears, and nose, arterio-venous anastomoses are present which can open and close according to thermoregulatory requirements. When they are open, the blood flow through the superficial venous plexuses is augmented and the presence of a large volume of warm blood close to the skin surface increases the loss of heat from the body surface. Conversely, when the arterio-venous anastomoses are closed, blood flow through the dermis is reduced and heat is conserved.

The deep veins and arteries remain in close proximity as they course along the arms and legs. This allows heat to be transferred from the arteries to the veins. In cold conditions the blood in the arteries will be at, or near, core temperature while the blood returning from the extremities will be at a substantially lower temperature. So, as blood flows from the heart along a limb towards one of the extremities, it will be gradually cooled but, at the same time, the blood returning to the heart will be progressively warmed. This counter-current arrangement therefore helps to maintain the core temperature.

In a thermoneutral environment the resting skin blood flow is approximately 100 ml min^{-1} kg^{-1} or about 250 ml min^{-1}. When exposed to extreme cold, skin blood flow falls to around 10 ml min^{-1} kg^{-1}. In certain areas of the body where there are arteriovenous anastomoses, such as the finger tips, the ear lobes, and the tip of the nose, blood flow to the skin may virtually cease altogether for brief periods and these parts of the body will feel numb. As it has relatively low metabolic requirements, the skin is able to tolerate a severely reduced blood supply for fairly long periods but eventually, if circulation is not restored, frostbite will ensue in which tissue freezes and dies. In contrast, when body temperature is high (e.g. as a result of strenuous exercise in a warm environment), blood flow may reach 2.0 l min^{-1} kg^{-1}—a 20-fold increase over the resting level.

The large variations in skin blood flow are brought about by the activity of the sympathetic nerves that supply the arterioles (see Chapter 30). In warm conditions, sympathetic activity is low and the arterioles are dilated so that more blood is diverted to the superficial layers of the skin to increase heat loss. In cold conditions, the opposite happens. The arterioles supplying the superficial capillaries constrict, and blood flow to the skin surface decreases substantially. Although this adaptation conserves heat, it has the disadvantage that the extremities become rather cold, particularly the fingers and toes. Note, however, that the blood vessels of the head are much less subject to sympathetic vasoconstriction.

Mechanisms of heat exchange

Heat may be lost from the body by

- radiation
- conduction
- convection
- the evaporation of sweat.

These routes of heat loss are summarized in Figure 42.12.

Radiation

This is the loss of heat in the form of infrared rays (thermal energy). If a human sits naked at normal room temperature, **radiation** will account for around 60 per cent of their total heat loss. For a young adult male with a resting metabolic rate of 290 kJ h^{-1} this amounts to approximately 50 W. The body can also gain heat by radiation, as it does during sunbathing for example. Heat loss by radiation can be substantially reduced by appropriate clothing.

Conduction and convection

Conduction is the transfer of heat between objects that are in direct contact with one another. A very small amount of heat is lost from the body by conduction to objects (the warming of a chair by sitting in it for example). Heat may also be gained by the body through conduction, for example while lying in a warm bath. Much more significant heat loss occurs by **convection** as the heat is first conducted to the air immediately outside the skin, and then the warmed air is removed by convection. As the warmed air is replaced by new, cool air, further heat loss can occur. This form of

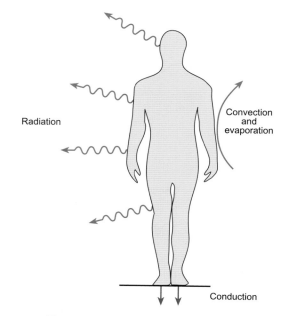

Figure 42.12 The routes by which heat is lost from the body. Heat can also be gained by convection, conduction, and radiation if the surroundings are at a higher temperature than the body.

heat loss becomes especially noticeable when there is a cool breeze blowing, as the air close to the body surface is being replaced by cold air more rapidly (the **wind chill** effect). Together, conduction and convection to the air account for 15–20 per cent of heat loss to the environment. Light clothing reduces loss by this route to about half that from a naked body, while arctic clothing can reduce it much further.

Because the specific heat of water is much greater than air, water in close contact with skin is able to absorb much more heat. Furthermore, the conductivity of heat through water is greater than that through air. Consequently, the body loses heat very rapidly indeed when immersed in water, even at moderate temperatures. A naked person will become hypothermic after only 1.5–2 hours in water at 15°C, while in arctic waters (c. 5°C), hypothermia occurs within 20 minutes.

Evaporation

Water evaporates when its molecules absorb heat from their environment and become energetic enough to escape as gas (water vapour). Approximately 2.4 MJ (570 kcal) of heat is lost for each litre of water that evaporates from the body surface. Thus as water evaporates continuously from the lungs, the mucosa of the mouth, and through the skin, there is a basal level of heat loss. Evaporative heat loss becomes obvious when body temperature rises and sweating provides increased amounts of water for vaporization. The sweating mechanism is initiated at an ambient temperature of 30–32°C in a resting individual. Once skin temperature rises above about 35°C, the rate of sweating increases substantially even in a resting individual (Figure 42.13).

The insensible water loss is about 600 ml a day, amounting to the loss of about 1.4 MJ (340 kcal) of heat per day. In cold conditions

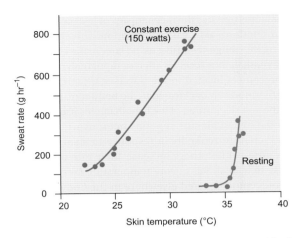

Figure 42.13 The steady-state values of sweat rate at rest (red symbols) and during maintained exercise (150 watts). The experiments with the resting subjects were carried out with environmental temperatures from 25–44°C. Their sweat rates begin to increase when the skin temperature rises above 35°C. The experiments on exercising subjects were carried out at environmental temperatures from 5–30°C. Their sweat rate progressively increases as skin temperature exceeds 24°C. (From data in B. Nielsen (1969) *Acta Physiol Scand* (Suppl 323), 1–74.)

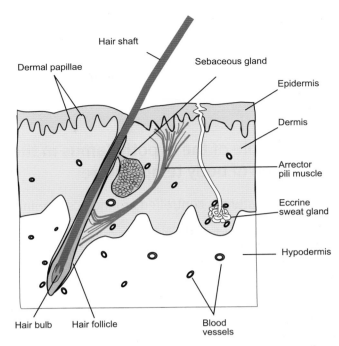

Figure 42.14 The principal structures of the skin, illustrating the location and appearance of an eccrine sweat gland. Note that the secretory portion lies deep within the dermis and that the initial secretion is modified as it flows along the duct leading to the surface of the skin.

around 500 ml of sweat are secreted each day. During vigorous muscular activity such as cross-country running, sweat may be produced at a rate of 1–2 litres h^{-1} thus removing about 2.4–4.8 MJ (570–1140 kcal) of heat from the body per hour. In some industrial and athletic activities, sweat production may reach 6 l h^{-1}—it can be more for short periods. As Figure 42.13 shows, there is a strong relationship between skin temperature and the sweat rate.

The formation of sweat

Sweat is secreted by the **eccrine glands**. Humans possess around 2.5 million of these, about half of which are situated in the dermis of the back and chest. As Figure 42.14 shows, the eccrine sweat glands consist of a coiled portion that lies deep within the dermis and a duct that opens on to the surface of the skin. The sweat glands are innervated by cholinergic sympathetic fibres and stimulation of these fibres leads to an increase in the rate of sweat production. This response can be blocked by atropine and is mediated by M_3 muscarinic receptors (see Chapter 11). Circulating catecholamines from the adrenal medulla are also thought to increase the rate of sweating via β-adrenoceptors. The flow of sweat to the skin surface is controlled by the contraction of the myoepithelial cells that surround the coiled region of the gland.

Sweating occurs when the coiled portion of the gland actively secretes a fluid called the precursor fluid, which is similar in composition to plasma. It is not, however, an ultrafiltrate as its secretion can be inhibited by ouabain and amiloride. As the fluid passes along the duct, sodium and chloride ions are reabsorbed so that, at low sweat rates, the fluid reaching the skin is significantly hypotonic with respect to plasma. However, the final composition of the sweat depends upon the rate at which it is being produced. At low

sweat rates, much of the sodium and chloride in the precursor fluid is reabsorbed by the duct cells and the sweat is very dilute. At higher sweat rates, however, there is less time for reabsorption as the sweat flows along the duct, and consequently more sodium and chloride is lost from the body.

Although sweating provides an effective means of heat loss from the body, it also represents a potentially dangerous loss of water and sodium chloride. The secretion of both ADH and aldosterone (see Chapter 39) is increased during heavy sweating and acts to conserve sodium and water. Nevertheless, it is important that the lost fluid and salts are replaced quickly if heat exhaustion is to be avoided.

Summary

The temperature of the central regions of the body is known as the core temperature, and that of the skin and superficial tissues is called the shell temperature. At rest the normal core temperature is in the range of 35.5–37.7°C but fluctuates by about 1°C over a 24-hour period.

Women of reproductive age have an increase in core temperature of about 0.5°C following ovulation, which is sustained in the event of a pregnancy.

The circulation of blood through the skin plays an important role in the regulation of body temperature and can be adjusted by more than 100-fold. The skin circulation is regulated according to the thermoregulatory requirements by activity in the sympathetic nerves. Heat loss is promoted by dilatation of the skin blood vessels, while constriction of the skin vessels reduces heat loss.

Heat is lost from the body surface to the environment by radiation, conduction, convection, and evaporation of sweat. The rate of heat loss is substantially increased by the evaporation of sweat.

42.6 The role of the hypothalamus in the regulation of body temperature

Body temperature is controlled around a 'set point'

The critical level at which the thermoregulatory mechanisms of the body try to maintain core temperature is known as the 'set point' of the system (see Chapter 1). While it is not clear how the set point is determined physiologically, it is under the control of the hypothalamus, which behaves as a kind of thermostat that maintains a balance between heat loss and heat production. The set point is normally close to 37°C but it can be altered in response to fever-producing agents (pyrogens, discussed later in this chapter) and by inputs from other parts of the nervous system (e.g. during deep sleep).

The neural basis of thermoregulation

Body temperature is chiefly regulated by neurons that lie within the hypothalamus, which is a part of the diencephalon. Of particular importance are the preoptic anterior hypothalamic area, the dorsomedial nucleus of the hypothalamus, the periaqueductal grey matter of the mid brain and the raphe pallidus nucleus of the medulla (see Figure 42.15). Together these brain areas control body temperature almost entirely via the sympathetic nervous system. When the core and shell temperatures are elevated (as in heavy exercise) the hypothalamus initiates the mechanisms of heat loss—skin vasodilatation and sweating via sympathetic vasodilator fibres. Conversely, when the core and shell temperature fall, the hypothalamus initiates the mechanisms of heat conservation and heat production—skin vasoconstriction, shivering, and **non-shivering thermogenesis** (discussed later in the chapter).

The hypothalamus receives afferent input both from peripheral thermoreceptors located in the skin (see Chapter 13) and from central thermoreceptors located in the body core, including some within the anterior portion of the hypothalamus itself. The preoptic area of the hypothalamus contains neurons whose rate of discharge increases markedly in response to a small rise in core temperature, while others increase their rate of discharge in response to a small fall in core temperature (see Figure 42.16). Thermoreceptors located elsewhere in the body (especially in the skin) provide additional information to the hypothalamic thermoregulatory neurons concerning the shell temperature. Figure 42.17 shows the steady-state patterns of discharge of afferents from cold and warm cutaneous thermoreceptors as a function of skin temperature. In a thermoneutral environment, both cold and warm cutaneous thermoreceptors show a moderate level of activity. The detailed mechanisms of the temperature sensitivity of the skin are discussed in Chapter 13.

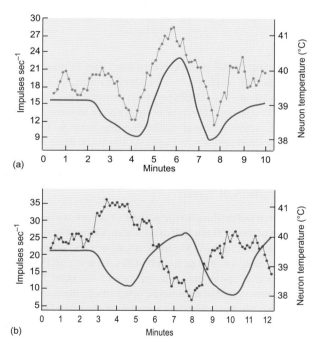

(a)

(b)

Figure 42.16 The firing rates of temperature-sensitive hypothalamic neurons as a function of core temperature. (a) The firing rate of a neuron that responds to an increase in temperature with an increase in firing rate. This neuron was cooled by 1°C and then warmed by just over 2°C before being returned to the initial temperature. The firing rate closely followed the changes in temperature—it almost doubled for a 2.2°C change in core temperature. (b) The firing rate of a neuron that responds to a fall in temperature with an increase in firing rate. In this example, during the initial period of recording, the neuron increased its firing rate from about 25 action potentials per second at 39.6°C to 34 action potentials when cooled by about 1°C. Most neurons in the hypothalamus are not sensitive to such small changes in core temperature. (From data in R.F. Hellon (1967) *J Physiol* **193**, 381–395.)

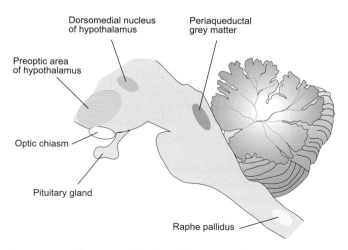

Figure 42.15 The areas of the hypothalamus and brainstem that are important in thermoregulation. All lie close to the midline.

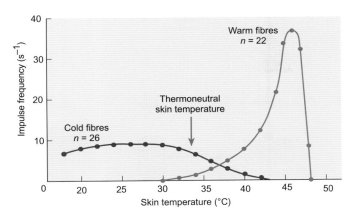

Figure 42.17 The average response of cold-sensing and warm-sensing cutaneous thermoreceptors. Note that at a skin temperature of 33°C, both sets of receptors are active. This is the skin temperature in a thermoneutral environment. (From Fig 12 in H. Hensel and D.R. Kenshalo (1969) *J Physiol* **204**, 99–112.)

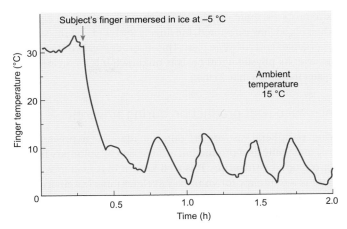

Figure 42.18 The 'hunting reaction' in a finger immersed in ice. Periodic vasodilatation raises the temperature of the finger and is thought to delay the onset of tissue damage. (Based on Fig 39.9 in J. Werner (1977) *Pflugers Archiv* **367**, 291–4.)

Summary

Under normal conditions the core temperature is close to 37°C, the set point of the thermoregulatory system. The hypothalamus receives information concerning body temperature from central and peripheral thermoreceptors and regulates the mechanisms of heat loss and heat generation via the sympathetic nervous system.

42.7 Thermoregulatory responses to cold

Under cold conditions, core temperature is relatively stable and rarely changes by much more than 0.5°C. In extreme hot or cold environments wider variations can occur, but even these changes are relatively modest. For example, in one naked volunteer exposed to a cold environment for several days, the shell temperature measured at one of the toes after 24 hours had declined to ambient temperature (8°C) while the core temperature had fallen by only about 1°C.

When the body's core temperature starts to fall, two kinds of homeostatic responses are initiated: those that increase the rate at which the body generates heat, and those that serve to reduce the rate at which heat is lost from the body surface. The *physiological* responses to low temperature include cutaneous vasoconstriction, shivering, and non-shivering thermogenesis. Furthermore, people will initiate appropriate behavioural responses to cold including seeking a warmer environment, putting on additional clothing, eating, and drinking warm fluids.

Cutaneous vasoconstriction

Skin has an extremely wide range of blood flow. When the ambient temperature falls within the thermoneutral zone, the blood flow to the skin is around 0.15 l min⁻¹ kg⁻¹. At very low temperatures it may fall as low as 0.01 l min⁻¹ kg⁻¹. The skin vessels, particularly the arterio-venous anastomoses, are richly supplied by sympathetic

noradrenergic fibres. Vasoconstriction in response to cold results from an increase in sympathetic activity and seems to be mediated chiefly by the action of noradrenaline at α-adrenoceptors. This vasoconstriction enhances the insulating properties of the skin and reduces the blood flow to the superficial venous plexuses, with the result that less heat is lost from the skin surface. The increase in sympathetic outflow to the cutaneous vessels is believed to be initiated by neurons in the posterior hypothalamus.

Paradoxically, during long periods of cold exposure, the cutaneous circulation of the extremities will often show intermittent periods of vasodilatation. This is known as the 'hunting reaction' (see Figure 42.18) which appears to be a safety mechanism that reduces the risk of ischaemic tissue damage (frostbite) in extreme cold. The underlying physiological mechanism is unclear but may result from a temporary loss of sensitivity to noradrenaline.

In up to 5 per cent of the population, the peripheral arterioles constrict excessively in response to cold. The fingers in particular may appear dead white (see Figure 42.19) and feel numb and, in

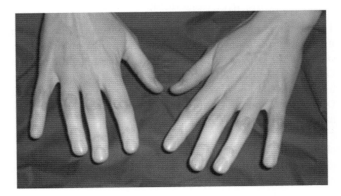

Figure 42.19 The hands of a person suffering from Raynaud's disease during a period of severe vasoconstriction in the cold. For sufferers, a slight change in temperature can cause the extremities of the body to become ischaemic. Restoration of blood flow is accompanied by severe pain. (Courtesy NHS Choices website.)

extreme cases, there may be local ischaemia and tissue damage. When blood flow is restored, the fingers flush red and the sufferer experiences considerable pain. This disorder is called **Raynaud's disease**. Its physiological basis is unknown but its incidence is greater in women than in men.

Increased heat production during shivering

When peripheral vasoconstriction is inadequate to prevent heat loss, metabolic heat production is increased by voluntary muscle contraction or shivering. It is well known that in cold weather there is a tendency for voluntary skeletal muscle activity to be increased. Foot-stamping, hand-rubbing, and a faster walking speed are all examples of such behaviour. When muscle cells contract, ATP is hydrolysed and heat is produced. This additional heat production will contribute to the restoration of a normal core temperature.

Shivering is a specialized form of muscular activity in which the muscles themselves perform no external work and virtually all of the energy of contraction is converted directly to heat. It is predominantly an involuntary activity consisting of the contraction and relaxation of small antagonistic muscle groups. Prior to the onset of overt shivering, there is an overall increase in the degree of muscle tone. Shivering begins in the extensor muscles and the proximal muscles of the trunk and upper limbs. The muscles contract in response to signals from the somatic motor neurons. These signals, which are believed to arise in the hypothalamus, are sustained by the response to afferent input from proprioceptors and stretch receptors in the joints and muscles.

Although shivering produces a considerable amount of additional heat, it is insufficient on its own to maintain body temperature for long if ambient temperature is very low. It also places a substantial burden on energy reserves and is exhausting for the individual. Furthermore, certain individuals have a reduced ability to shiver effectively. Examples are the elderly, the very young, those who are paraplegic, and patients undergoing surgery under general anaesthesia.

Non-shivering thermogenesis

This is the generation of heat through processes other than muscle contraction and includes the actions of a variety of calorigenic hormones and the stimulation of brown fat metabolism. Cold stress results in an increased secretion of thyroid hormones and of adrenaline and noradrenaline by the adrenal medulla. Other calorigenic hormones include glucocorticoids, insulin, and glucagon.

Tri-iodothyronine (T_3) increases the synthesis of the mitochondrial uncoupling protein UCP1 expressed in the inner mitochondrial membrane of brown adipose tissue. UCP1 uncouples electron transport from ATP synthesis so that heat is generated instead of the energy being stored as ATP. This uncoupling of oxidative phosphorylation stimulates oxygen consumption by the mitochondria, resulting in the dissipation of metabolic energy as heat that is reflected in an increase in BMR. The increase in the secretion of catecholamines during cold stress activates a hormone

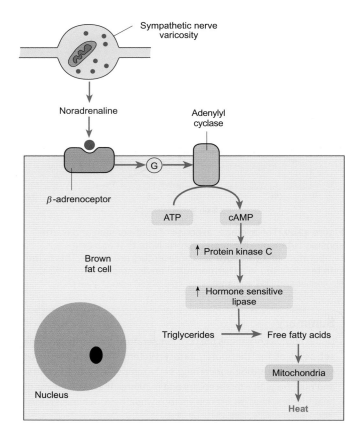

Figure 42.20 The heat-generating processes of a brown fat cell. Noradrenaline acts on β-adrenoceptors on the cell surface to initiate a signal cascade that results in an increased formation of free fatty acids. These are then metabolized in the mitochondria to generate heat rather than form ATP (see main text for further details).

sensitive lipase in the brown fat cells. The lipase releases glycerol and free fatty acids from the cellular stores of triglyceride, and the fatty acids are then used for heat production (see Figure 42.20). It is now well established that brown fat thermogenesis is of importance in neonates, and recent evidence supports a major role in determining whole-body metabolic rate and heat production in adults. Indeed, it is now considered probable that the action of thyroid hormones on whole-body metabolic rate is mediated via hypothalamic control of the sympathetic nerves that innervate the brown adipose tissue.

Summary

Physiological responses which act to conserve heat during exposure to cold include cutaneous vasoconstriction, shivering, and non-shivering thermogenesis. Under cold conditions, the sympathetic nerves cause vasoconstriction in the superficial vessels. Heat is generated by shivering—the asynchronous activity of motor units—and by non-shivering thermogenesis, particularly in neonates.

42.8 Thermoregulatory responses to heat

When the core temperature of the body starts to rise, physiological responses are initiated that increase the rate at which heat is lost from the body surface. A number of behavioural modifications are also seen, including a reduction in activity, the shedding of clothing, drinking cold fluids, and seeking a cooler environment. The major **physiological responses** to a rising core temperature in humans are cutaneous vasodilatation and sweating. The loss of heat by evaporation increases substantially as core temperature rises (see Figure 42.21).

Cutaneous vasodilatation

When the core temperature rises above its normal value, the blood vessels of the skin dilate and cutaneous blood flow may reach as much as $2 \, l \, min^{-1} \, kg^{-1}$. This dilatation is mediated by the autonomic nervous system, mainly through a reduction in vasomotor tone. As the vasculature swells with warm blood, heat is lost more readily from the body surface by radiation, conduction, and convection.

Enhanced sweating

When heat production exceeds heat loss and body temperature starts to rise, sweating is initiated. In extreme conditions, or in heavy exercise, large quantities of sweat may be produced. As long as the relative humidity of the air is low, the evaporation of sweat is an efficient means of losing heat from the body. When the relative humidity is high, however, evaporation occurs much more slowly and sweating becomes a less effective mechanism of cooling.

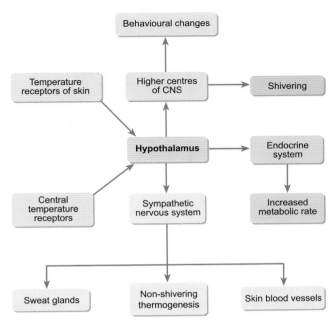

Figure 42.22 The factors that determine heat loss and heat production by the body. Note the central role of the hypothalamus.

Humans adapt (acclimatize) to hot climates relatively quickly (within a few weeks). The most important adaptation is an increase in the rate of sweating, which may double, and a lowering of the threshold temperature at which sweating is initiated. Furthermore, the sodium chloride content of sweat also appears to fall after prolonged exposure to heat, possibly as a result of increased aldosterone secretion. The factors that determine heat loss and heat production by the body are summarized in the flow chart shown in Figure 42.22.

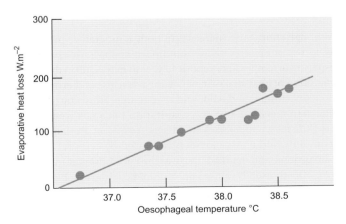

Figure 42.21 The relationship between oesophageal temperature and heat loss via sweating with a skin temperature of 30°C. These experiments were conducted on four young male subjects who were exercising on a bicycle ergometer at 90 per cent of their maximum capacity. (Based on data in B. Saltin et al. (1972) *J Applied Physiol* **35**, 635–43.)

Summary

Physiological responses to a rising core temperature include cutaneous vasodilatation as a consequence of a reduction in sympathetic vasomotor tone and sweating. Sweating brings about heat loss through evaporation. It occurs via the eccrine sweat glands, which are innervated by sympathetic cholinergic fibres. In heat stress, sweating can increase more than tenfold, but it is only effective as a heat loss mechanism if the sweat can evaporate from the skin.

42.9 Disorders of thermoregulation

Hypothermia

From the above discussion it is evident that the body employs a variety of strategies to prevent a fall in core temperature during exposure to a cold environment. Nevertheless, in extreme conditions, such compensatory mechanisms may prove inadequate, and under these circumstances hypothermia occurs (see Figure 42.23).

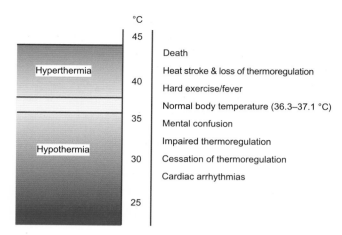

Figure 42.23 The effects of extreme core temperatures.

This is defined as a core temperature below 35°C. Below this temperature the muscles become weaker and both voluntary movement and shivering will be reduced. With the loss of these heat-generating mechanisms, the core temperature may start to fall quite rapidly.

As core temperature falls to 34°C, mental confusion is seen, with a loss of consciousness soon following. When the temperature falls below 28°C serious cardiovascular changes may occur, including a fall in heart rate and arrhythmias leading to ventricular fibrillation, which is fatal. However, complete recovery from even extreme hypothermia is possible, provided that the subject is warmed slowly, preferably 'from the inside out'. This is because the reduced temperature of the body tissues, especially the brain, considerably reduces their metabolic requirements, and enables them to be sustained by a severely restricted blood supply. In such cases it is important that efforts to resuscitate a patient should not be discontinued prematurely.

It is inadvisable to warm the surface of a hypothermic individual too rapidly, i.e. by hot blankets or vigorous rubbing, since the increased blood flow to the periphery may compromise blood flow to the body's vital organs and lead to further problems. Simple measures that may be used to warm a hypothermic person include covering them with layers of dry blankets and, if they are alert, giving warm, non-alcoholic drinks. In a clinical setting, extracorporeal warming of the patient's blood by dialysis, or peritoneal lavage with warmed fluids, are the preferred means of restoring body temperature. Airway re-warming by the administration of humidified oxygen via a mask or nasal tube may also be helpful.

Hypothermia in the elderly

The ability to regulate core body temperature declines with advancing age. There is a reduction in the awareness of temperature changes and impairment of thermoregulatory responses. While the young can detect a skin temperature change of 0.8°C, many elderly people are unable to discriminate a difference in temperature of 2.5°C—indeed some cannot discriminate a difference of 5.0°C. In these people, shivering in response to moderate cooling is reduced and the change in metabolic rate in response to cold is also small in comparison with that shown by young adults. In addition to these age-related changes, elderly people are often relatively immobile, suffering in many cases from orthopaedic and rheumatic problems, Parkinson's disease, or cardiovascular disease—all of which may make it difficult for them to increase their voluntary muscle activity. Therefore, old people should ideally live in surroundings where the temperature is maintained at a minimum of 20°C.

The clinical use of hypothermia

Intentional reduction of the body temperature is used in certain clinical situations. In surgery on the aortic arch, the heart must be stopped and the body temperature of patients undergoing such procedures is lowered in order to minimize the metabolic requirements of the tissues. This allows more time for the surgical procedures to be completed without the risk of hypoxic tissue damage.

Scalp cooling through use of a 'cold cap' may be effective in preventing hair loss (alopecia) in patients treated with some chemotherapy drugs. Cooling the scalp to a temperature of around 17°C constricts the blood supply to hair follicles, thus minimizing the dose of the drug delivered to the follicles and preventing or reducing hair loss.

The neurological damage that is caused by perinatal hypoxia sustained during or shortly after birth may be avoided by lowering the temperature of affected neonates to around 33–34°C for the first 72 hours or so of their lives. This injury (known as hypoxic ischaemic encephalopathy) appears to be caused, at least in part, by the return of blood flow to the affected brain areas when hypoxia is reversed ('re-perfusion injury'). The aim of lowering body temperature is to reduce the metabolic requirements of the brain and delay re-perfusion injury for long enough to allow cardiovascular stabilization to be achieved. Slow re-warming is then carried out over a period of about 6–8 hours.

Hyperthermia

Although short-term increases in body temperature to as much as 43°C can be tolerated without permanent harm, prolonged exposure to temperatures above 40°C or so may result in the serious condition of **heat stroke** in which there is a loss of thermoregulation. Sweating is reduced, and core temperature starts to soar. The skin feels hot and dry, respiration becomes weak, blood pressure falls, and reflexes are sluggish or absent. Initially there is irritability and a loss of energy, followed by cerebral oedema, convulsions, and neural damage as the core temperature exceeds 42°C. Death usually follows unless rapid cooling is achieved (see Figure 42.23).

Heat exhaustion is another potential consequence of hyperthermia and may be the result of either dehydration (a loss of body fluid) or salt deficiency (excessive loss of salt through sweating, that is not replaced in the diet). Dehydration is characterized by fatigue and dizziness in the early stages, progressing to intracellular dehydration and cellular damage. Salt deficiency produces a fall in tissue osmolality, which causes muscle cramps.

As with hypothermia, the elderly are also vulnerable to hyperthermia. Older people are less sensitive to thirst and have a decreased ability to sweat. Moreover, drugs such as anticholinergics (e.g. atropine) inhibit sweating. Each year, many people die during prolonged spells of hot weather and the majority of these are 50 years of age or older.

Hyperpyrexia

When body temperature is higher than 41.5°C the sufferer is said to have hyperpyrexia rather than hyperthermia. Hyperpyrexia is a medical emergency and death will rapidly ensue if it is not treated promptly.

Summary

Hypothermia occurs when core temperature falls below 35°C. As temperature drops, heat-conserving mechanisms start to fail; there is mental confusion, a fall in heart rate, and cardiac arrhythmias, followed by a loss of consciousness. The elderly are particularly at risk of hypothermia. Hyperthermia (a core temperature in excess of 40°C) may have grave consequences. As the body's heat loss mechanisms fail, cerebral oedema develops and later there is irreversible neuronal damage.

42.10 Special thermoregulatory problems of the newborn

Newborn babies are at greater risk of both hypo- and hyperthermia than older children or adults. Hypothermia is the more commonly encountered problem because infants lose heat readily, they have no control over their clothing and bedding, and their heat-conservation mechanisms are poorly developed. The mechanisms of heat loss from the body are discussed in Section 42.5. When the surface area to volume ratio is high, as in small babies, heat loss occurs more readily. This is exacerbated by the relatively thin layer of insulating fat forming the shell of the baby's body, which renders vasoconstriction rather ineffective at minimizing heat loss. Furthermore, the shivering mechanism of newborn infants is poorly developed so that shivering only occurs in response to extreme cold. Neonates are also unable to increase their voluntary muscular activity significantly in response to cold and therefore require an ambient temperature of 32–34°C to maintain their core temperature without increasing their metabolic rate.

However, the capacity of the newborn infant to generate heat by non-shivering (metabolic) pathways is 4–5 times greater (per unit body weight) than that of an adult. In the neonate the proportion of brown adipose tissue is very high relative to body weight (see Figure 50.8). Brown fat metabolism generates large amounts of heat in response to catecholamines from the adrenal medulla and the sympathetic nervous system.

Premature babies are even more susceptible to heat loss. They have an even larger surface area to volume ratio, an even thinner insulating layer, and incompletely developed brown fat reserves. Such infants need to be kept in incubators until they are sufficiently mature to regulate their own core temperature.

When a baby is exposed to a hot environment, its body temperature may rise above 37.5°C and hyperthermia may ensue. Although this is less common than hypothermia, it is equally dangerous as it increases the metabolic rate and water loss by evaporation. The loss of water can cause serious dehydration, which must be treated by getting the baby to drink or, if necessary, by providing fluid via a nasogastric tube or intravenous infusion. The commonest causes of hyperthermia in babies are wrapping the baby in too many layers of clothes for the climate, leaving the baby in direct strong sunlight, for example in a parked car, and putting the baby too near to a source of heat. Severe hyperthermia may cause circulatory shock, convulsions, and coma. Unless promptly treated, it may result in neurological damage or death.

42.11 Fever

This is the most common disturbance of thermoregulation and represents an elevation of normal body temperature that is not related to work or to exposure to a high ambient temperature. For most purposes a core temperature in excess of 38°C (100.4°F) is regarded as abnormal. Fever (or **pyrexia**) is most often associated with infectious disease, although it can also arise as a consequence of certain neurological conditions, cancer, or as a result of dehydration. It can also occur following a pulmonary embolism or the transfusion of incompatible blood.

During a fever there are significant changes to blood chemistry. Fibrinogen levels are increased and acute phase proteins are secreted by the liver (see Chapter 26 p 395). In addition, the heat stress of fever leads to an increase in the formation of heat shock proteins which protect host cell proteins from thermal damage. The acute phase and heat shock proteins enhance the response of the natural immune system to an infection before the adaptive immune system is fully activated (see Chapter 26). Other changes include a sharp fall in serum iron and serum zinc. Both are cofactors for a variety of enzymes that could be utilized by invading bacteria to enhance their proliferation. This decrease in plasma iron is therefore clearly an appropriate adaptation, as it inhibits the reproduction of pathogens. Indeed, if serum transferrin is infused the virulence of certain pathogenic bacteria is substantially increased.

The agents that cause fever are known as **pyrogens**. Infection by bacteria triggers an immune response via the natural immune system. As part of this response, the phagocytes secrete a variety of cytokines such as interleukin 1 (IL-1), interleukin 6 (IL-6), and tumour necrosis factor alpha (TNF-α). These cytokines reach the hypothalamus via the circumventricular organs that lie adjacent to the third ventricle (see Chapter 30), where the blood–brain barrier is relatively permeable to large molecules. Here they increase the production of prostaglandin E_2 via the arachidonic acid pathway (see Chapter 6). PGE_2 then acts on neurons in the preoptic area to

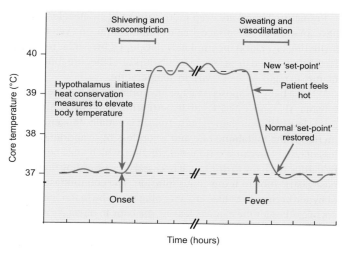

Figure 42.24 The time-course of a typical febrile episode.

initiate the febrile response. As a result, heat is conserved through the usual thermoregulatory mechanisms of the body. The maintained increase in core temperature seen during fever is associated with a greater chance of survival and elimination of an infection.

Figure 42.24 shows the time-course of a typical febrile episode during which an individual may feel cold, with peripheral vasoconstriction and shivering despite a raised core temperature. When PGE_2 synthesis returns to normal levels, the 'set point' returns to normal and the individual then feels hot. Peripheral vasodilatation and sweating occur (the 'flush' or 'crisis') as body temperature returns to normal.

Although the occurrence of a fever is correctly regarded as part of the body's natural response to infection, antipyretic drugs are often used to reduce the temperature so that a patient feels more comfortable. Nevertheless, there is some evidence from animal studies that suppressing a fever may reduce survival in severe infections. How far this applies to human populations remains unclear. Drugs used to lower temperature during an infection are called **antipyretics**. These include aspirin, paracetamol (acetaminophen), and other non-steroidal anti-inflammatory drugs (NSAIDs), which also have analgesic actions. They act by inhibiting the activity of cyclo-oxygenase which is required for the synthesis of prostaglandins (see Chapter 6).

In children under the age of five, fever can cause convulsions (presumably due to the relative immaturity of their nervous systems). While febrile convulsions rarely have any long-term consequences, they are distressing for both child and parents, and the appropriate course of action is to cool any small child with a high fever quickly and effectively.

Malignant hyperpyrexia is a relatively rare condition that can be triggered by volatile anaesthetics and by the neuromuscular blocker succinylcholine. In susceptible individuals there is a dramatic rise in body temperature, increased heart rate, rapid breathing, sweating, and muscle rigidity. The cause is an uncontrolled increase in oxidative metabolism by the skeletal muscles. Administration of the drug dantrolene is usually effective in terminating such episodes.

Susceptibility to malignant hyperpyrexia is commonly the result of a genetic defect of the ryanodine receptor (RYR1—see Chapter 8 pp. 133–5). Certain volatile anaesthetics increase the opening of the ryanodine channel, so increasing the leakage of calcium ions from the sarcoplasmic reticulum. The rapid pumping of this extra calcium by the Ca^{2+}-ATPase of the sarcoplasmic reticulum utilizes a large quantity of ATP that has to be replenished by oxidative metabolism. This results in the generation of an excessive amount of heat and the sharp rise in body temperature that characterizes the condition. Those people who have a family history of malignant hyperthermia can be tested for their susceptibility to the condition by administering a volatile anaesthetic such as halothane to a small sample of their skeletal muscle obtained by muscle biopsy, which entails a minor surgical procedure under local anaesthesia. If they are found to be susceptible it is then possible to select an alternative anaesthetic.

Summary

Fever is an elevation of body temperature usually associated with the presence of infectious agents. The hypothalamus initiates the mechanisms of heat generation and conservation in response to pyrogens. Although core temperature is higher than normal, fever is part of the body's defence against infection and is accompanied by an increase in plasma fibrinogen, a marked decrease in plasma iron and zinc, and enhanced immune responses. Provided body temperature is kept below about 41°C, fever is not harmful but beneficial, although febrile convulsions may occur in young children.

✳ Checklist of key terms and concepts

The structure of the skin

- The skin is the interface between the body and its environment.

- It consists of two main layers: the epidermis and the dermis.

- The dermis is linked to the underlying structures by the subcutaneous tissue of the hypodermis.

- There are two main types of skin: hairy skin and smooth (glabrous) skin.

- The thickness of the skin varies according to its location. It is thickest on the soles of the feet and thinnest around the eyes.

The epidermis

- The epidermis has no nerve endings or blood vessels.
- The outermost layer of the epidermis, the stratum corneum, consists of dead cells filled with keratin and a fatty matrix to provide a water-proof, abrasion-resistant barrier.
- Vitamin D is produced by the cells of the two innermost layers of the epidermis, the stratum basale and stratum spinosum.
- The stratum basale also contains pigmented cells (melanocytes) whose pigment (melanin) provides protection against ultraviolet radiation.
- The synthesis of melanin is controlled by melanocyte stimulating hormone (MSH) secreted by the anterior pituitary gland.
- The epidermis acts as a barrier to disease-producing agents (pathogens).
- Its immune cells (Langerhans cells) play an important role in com-bating infection.
- The epidermis forms a waterproof outer layer that is vital for the maintenance of normal fluid balance.

The dermis

- The dermis is made up of dense connective tissue containing collagen and elastic fibres.
- It also contains the skin blood vessels, sensory receptors, hair follicles, sebaceous glands, and the coiled regions of the eccrine sweat glands.
- The skin is a major sense organ and plays a crucial role in the regulation of body temperature.
- Sebaceous glands secrete an oily substance called sebum which keeps the skin flexible and reduces evaporative water loss.
- Sebum also inhibits the growth of microorganisms on the epider-mal surface.
- The hairs can be elevated to trap a thin layer of air which, in combi-nation with the adipose tissue of the subcutaneous tissue, reduce heat loss from the body.
- Sweat secreted from the eccrine glands evaporates from the skin surface to promote heat loss.

The importance of thermoregulation and the role of the cutaneous circulation

- To maintain optimal conditions for cellular activity, humans main-tain core body temperature between 36 and 38°C.
- Heat is generated within the body by metabolic reactions.
- The body may also gain a small amount of heat from the environ-ment, by conduction and radiation.

- Heat is lost from the body surface to the environment by radiation, conduction, convection, and evaporation.
- For effective thermoregulation, heat loss must be balanced by heat gain.
- The blood vessels of the skin play an important role in thermoregulation.
- Vasodilatation of the skin vessels promotes heat loss.
- Vasoconstriction of the skin vessels reduces heat loss as warm blood is diverted to the body's core.

The role of the hypothalamus in thermoregulation

- Core temperature is regulated through the operation of a negative feedback loop.
- The hypothalamus acts as the body's 'thermostat' to keep core temperature at the set point value of around 37°C.
- The hypothalamus receives input from thermoreceptors in the skin and the body core.
- The hypothalamus initiates appropriate mechanisms to conserve or lose heat.
- Physiological responses which act to maintain body temperature during exposure to cold include cutaneous vasoconstriction, shiv-ering, and non-shivering thermogenesis.
- Physiological responses to overheating include cutaneous vasodil-atation and sweating, which brings about heat loss through evaporation.

Disorders of thermoregulation

- Hypothermia occurs when core temperature falls below 35°C.
- As temperature drops, heat-conserving mechanisms start to fail.
- Consequences of hypothermia include mental confusion and car-diovascular complications, followed by a loss of consciousness.
- Newborn infants and the elderly are particularly at risk of hypothermia.
- Hyperthermia is defined as a core temperature in excess of 40°C.
- As the body's heat loss mechanisms fail, cerebral oedema develops and later there is irreversible neuronal damage.

Fever

- Fever (pyrexia) is an elevation of body temperature that is usually associated with the presence of infectious agents.
- Fever is associated with changes to blood chemistry that act to inhibit the proliferation of pathogens and to stimulate the immune system.
- In response to pyrogens, the body conserves heat inappropriately.
- If body temperature rises rapidly in young children, they may have febrile convulsions.

Further reading

Astrand, P-O., Rodahl, K., Dahl, H.A., and Stromme, S. (2003) *Textbook of work physiology, Physiological bases of exercise* (4th edn), Chapter 13. Human Kinetics, Champaign, IL.

An interesting slant on temperature regulation seen from the perspective of exercise physiology.

Cooper, K.E. (2008) *Fever and antipyresis. The role of the nervous system.* Cambridge University Press, Cambridge.

An overview of thermoregulation followed by a detailed consideration of fever and its treatment.

Levick, J.R. (2010) *An introduction to cardiovascular physiology* (5th edn), pp. 288–94. Hodder Arnold, London.

A good clear account of the basic organization and properties of the cutaneous circulation.

Histology

Mescher, A.L. (2013) *Junquieira's basic histology* (13th edn), Chapter 18. McGraw-Hill, New York.

Dermatology

Kumar, P., and Clarke, M. (2005) *Clinical medicine* (6th edn), pp. 1315–65. Saunders, Edinburgh.

A brief introduction to the physiology of the skin with a detailed account of common skin conditions and their treatment.

Review articles

Cooper, K.E. (2002) Molecular biology of thermoregulation: Some historical perspectives on thermoregulation. *Journal of Applied Physiology* **92**, 1717–24.

Evans, S.S., Repasky, E.A., and Fisher, D.T. (2015) Fever and thermal regulation of immunity: The immune system feels the heat. *Nature Reviews: Immunology* **15**, 335–46.doi: 10.1038/nri3843

Johnson, J.M., and Kellogg, D.L., Jr. (2010) Local thermal control of the human cutaneous circulation. *Journal of Applied Physiology* **109**, 1229–38, doi:10.1152/japplphysiol.00407.2010.

Mekjavic, I. B., and Eiken, O. (2006) Contribution of thermal and non-thermal factors to the regulation of body temperature in humans. *Journal of Applied Physiology* **100**, 2065–72, doi:10.1152/japplphysiol.01118.2005.

Nakamura, K. (2011) Central circuitries for body temperature regulation and fever. *American Journal of Physiology. Regulatory, Integrative and Comparative Physiology* **301**, R1207–28, doi:10.1152/ajpregu.00109.2011.

Nilsberth, C., Elander, L., Hamzic, N., Norell, M., Lönn, J., Engström, L., and Blomqvist, A. (2009) The role of interleukin-6 in lipopolysaccharide-induced fever by mechanisms independent of prostaglandin E_2. *Endocrinology* **150**, 1850–60, doi:10.1210/en.2008-0806.

Romanovsky, A.A. (2014) Skin temperature: Its role in thermoregulation. *Acta Physiologica* **210**, 498–507, doi: 10.1111/apha.12231.

Silva, J.E. (2006) Thermogenic mechanisms and their hormonal regulation. *Physiological Reviews* **86**, 435–64, doi:10.1152/physrev.00009.2005.

To check that you have mastered the key concepts presented in this chapter, complete the accompanying online self-assessment questions. Go to www.oup.com/uk/pocock5e/

PART TEN

The gastrointestinal system

CHAPTER 43

Introduction to the gastrointestinal system

Chapter contents

This chapter should help you to understand:

- The functions of the gastrointestinal system (the gut)
- The general anatomical organization of the gastrointestinal system (alimentary canal and accessory organs)
  - Anatomical components
  - Histological features
- The nervous and hormonal regulatory mechanisms operating within the gut that control
  - Motility
  - Secretion
- The organization and regulation of the splanchnic circulation

the body for cellular metabolism, it must first be broken down into simple molecules which can then be absorbed into the bloodstream for distribution to the tissues. This task is performed by the digestive or gastrointestinal (GI) system, whose major functions are:

1. the ingestion of food
2. the temporary storage of food prior to its transport through the gastrointestinal tract
3. the regulation of gastrointestinal motility to ensure optimal digestion of food and absorption of nutrients
4. the secretion of fluid, salts, and digestive enzymes
5. digestion by means of mechanical disruption and the activity of enzymes
6. absorption of the products of digestion
7. removal of indigestible remains from the body (defecation).

This chapter will describe the gross anatomical organization of the gastrointestinal system, the internal structure of the gut wall, and its blood supply. It will also provide an overview of the nervous and hormonal mechanisms that regulate the motility and secretory activity of the gut. The following chapter presents a more detailed examination of the structure and functions of the mouth and oesophagus, the stomach, and the small and large intestine, together with the digestive (exocrine) functions of the pancreas. The anatomy and physiology of the liver is discussed in Chapter 45, while the final chapter in this section outlines key aspects of nutrition and energy balance.

43.1 Introduction

Food is required by the body both for the production of energy and for the growth and repair of tissues. Each day an average adult consumes around 1 kg of solid food and 1–2 litres of fluid. As the majority of this material is in a form that cannot be used immediately by

43.2 The general organization of the gastrointestinal system

The major anatomical components of the gastrointestinal (GI) tract and its accessory organs are illustrated in Figure 43.1. The GI tract comprises the oral cavity, pharynx, oesophagus, stomach,

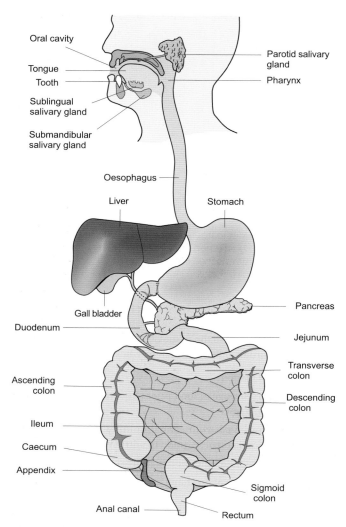

Figure 43.1 The gastrointestinal tract and its principal accessory organs.

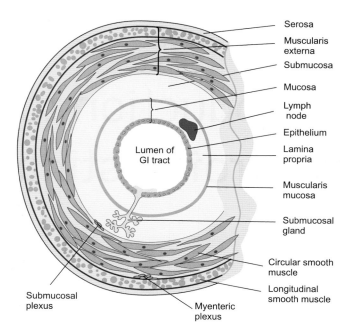

Figure 43.2 The gut wall seen in cross-section. Although this figure illustrates the basic four-layered organization, there are significant individual variations in structure in different regions of the gut.

small intestine (the duodenum, jejunum, and ileum), large intestine (the caecum and colon), rectum, and anal canal. Each section of the GI tract together with its accessory organs (the teeth, tongue, salivary glands, pancreas, appendix, liver, and gall bladder) is adapted to carry out a specific set of functions. Although the GI tract is located within the body, it is, in reality, a hollow convoluted tube that opens to the external environment at both ends (mouth and anus). The lumen of the tube is therefore an extension of the external environment. In life, the gastrointestinal tract is about 5.5 m long, although after death, its apparent length increases as smooth muscle tone is lost.

Histological features of the wall of the gastrointestinal tract

Although the detailed structure of the GI tract varies throughout its length, there are certain features of the gut wall that are the same in all regions. These are shown in Figure 43.2. From the outside

inwards the layers of the gut wall are: the serosa, a layer of longitudinal smooth muscle, a layer of circular smooth muscle, the submucosa, and the mucosa. A thin layer of smooth muscle fibres, the muscularis mucosa, lies between the mucosa and submucosa.

The outer covering of the GI tract is known as the **serosa** or **adventitia** according to its position. Most of the structures of the GI tract are covered by the serosa or visceral **peritoneum**, which is continuous with the parietal peritoneum that lines the abdominal cavity. It is a binding and protective layer that consists mainly of loose connective tissue covered by a layer of simple squamous mesothelium. The outer covering of those parts of the GI tract that do not lie within the peritoneal cavity is called the adventitia. These include the thoracic part of the oesophagus, the middle part of the rectum, the distal part of the duodenum, and the ascending and descending portions of the colon. The visceral peritoneum (i.e. the serosa) covers the transverse colon and sigmoid colon.

The space between the visceral and parietal layers of the peritoneum (the **peritoneal cavity**) normally contains very little free fluid. In certain diseases, for example cirrhosis of the liver or chronic heart failure, this space may become expanded by the abnormal accumulation of fluid. This fluid is known as **ascites** and its volume is often considerably more than 1.0 litre. The peritoneum possesses many large folds that pass between the abdominal organs and bind them together. The largest of these folds is called the **greater omentum**, which hangs down in front of the abdominal organs rather like an apron and attaches to the greater curvature of the stomach, the transverse colon, and the posterior body wall. It contains fatty tissue and collections of lymph nodes. The **lesser omentum** arises from the serosa of the stomach and duodenum and extends to the liver. Two additional thin folds of the peritoneum are the **mesentery**, which links the small intestine to the posterior abdominal wall, and the **mesocolon**, which links the large intestine to the posterior abdominal wall. Figure 43.3 illustrates the arrangement of the peritoneal folds. Blood vessels, lymphatics, and nerves reach the GI tract by way of these folds.

As Figure 43.2 shows, the smooth muscle of the gut wall is arranged in two main layers known as the **muscularis externa**. The smooth muscle cells of the outer layer run longitudinally while those of the inner layer have a circular arrangement. Contraction of these muscle layers mixes the food with digestive enzymes and propels it along the GI tract. The circular layer is 3–5 times thicker than the longitudinal layer. At intervals throughout the gastrointestinal tract, the circular smooth muscle layer is thickened and modified to form a ring of tissue called a **sphincter**. Sphincters control the rate of movement of the gastrointestinal contents from one part of the gut to another. They are present at the junctions between the pharynx and the oesophagus (the upper oesophageal sphincter), the oesophagus and the stomach (the lower oesophageal or cardiac sphincter), the stomach and the small intestine (the pyloric sphincter), and the ileum and the caecum (the ileo-caecal sphincter or ileo-caecal valve). The internal and external anal sphincters control the elimination of faeces.

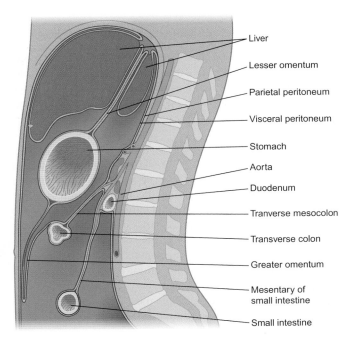

Figure 43.3 A midline section showing the arrangement of the peritoneal membranes that cover the various parts of the gastrointestinal tract: the greater and lesser omenta, the mesentery of the small intestine, and the mesentery of the colon—the sigmoid and transverse mesocolon. The shaded area indicates the extent of the peritoneal cavity. (Based on Fig 12.75 in *Gray's anatomy*, 38th edition (1995) eds L.H. Bannister, M.M. Berry et al. Churchill-Livingstone, Edinburgh. Republished with permission of Elsevier.)

The **submucosa** is made up of loose connective tissue with collagen and elastin fibrils, blood vessels, lymphatics, and, in some regions, submucosal glands. The innermost layer of the gut wall, the **mucosa**, is further subdivided into three regions: a layer of epithelial cells with their basement membrane, the lamina propria, and the **muscularis mucosa** which consists of two thin layers of smooth muscle—an inner circular layer and an outer longitudinal layer. The characteristics of the epithelium vary greatly from one region of the GI tract to another. For example, it is smooth in the oesophagus but is thrown into finger-like projections called villi in the small intestine.

The motility of the gastrointestinal tract

Motor activity is responsible for the mixing of food with digestive juices and the propulsion of food along the GI tract at a rate that allows the optimal digestion of food and absorption of the digestion products. With the exception of the mouth, tongue, and upper oesophagus, the motor functions of the GI tract are performed almost entirely by the action of smooth muscle. The structure and electrophysiological characteristics of smooth muscle are discussed fully in Chapter 8. Only the specific characteristics of the smooth muscle of the GI tract will be considered here.

The smooth muscle in the gut is of the **single unit** or **visceral** type. It operates as a functional syncytium, whereby electrical signals originating in one fibre are propagated to neighbouring fibres so that sections of smooth muscle contract synchronously. The smooth muscle fibres maintain a level of tone that determines the length and diameter of the GI tract. Various types of contractile responses are superimposed upon this basal tone, the most obvious of which are segmentation and peristaltic contraction.

Segmentation, which occurs principally in the small intestine, facilitates the mixing of food with digestive enzymes and exposes the digestion products to the absorptive surfaces of the GI tract. It is characterized by closely spaced contractions of the circular smooth muscle layer separated by short regions of relaxation, as shown in Figure 43.4(a). Segmentation occurs at a rate of around 12 contractions per minute in the duodenum and increases in both frequency and strength when chyme enters the small intestine. Segmental contractions are less frequent in the jejunum and ileum, where they seem to occur in bursts, in which there may be around eight contractions per minute interspersed with periods of inactivity.

Peristaltic contractions are concerned mainly with the propulsion of food along the tract and consist of successive waves of contraction and relaxation of the smooth muscle, as shown in Figure 43.4(b). The longitudinal smooth muscle contracts first and, halfway through its contraction, the circular muscle begins to contract. The longitudinal muscle relaxes during the latter half of the circular muscle contraction. This pattern of contraction is repeated, resulting in the slow but progressive movement of material along the GI tract. The normal trigger for a wave of peristalsis is local distension by food bulk.

The contractile properties of smooth muscle are determined by the underlying electrical activity of its cells. The resting membrane potential of the cells shows spontaneous rhythmical fluctuations

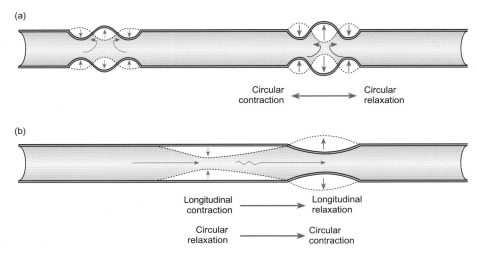

Figure 43.4 The principal patterns of contractile activity in the smooth muscle of the GI tract. (a) In segmentation movements, small regions of circular muscle first contract and then relax, thus promoting the mixing of the intestinal contents. (b) In peristalsis, the longitudinal muscle and circular muscle exhibit alternating cycles of contraction and relaxation. The net effect is to propel the intestinal contents along the GI tract. (Based on Fig 21.13 in D.F. Moffett, S.B. Moffett, and C.L. Schauf (1993) *Human physiology*. McGraw-Hill, New York.)

(the **basic electrical rhythm** or **slow wave rhythm**). The frequency of these fluctuations varies along the length of the GI tract, generally decreasing with distance from the mouth. For example, in the duodenum there are 11 or 12 slow waves per minute while in the colon there are only three or four. These differences in rate result in different degrees of contraction along the GI tract. In consequence, a gradient of pressure is created which contributes to the steady movement of the gut contents towards the ileo-caecal sphincter.

Although the smooth muscle layers of the gut can contract in the absence of action potentials, the depolarizing phases of the slow waves are sometimes accompanied by bursts of action potentials. Such bursts are associated with vigorous propulsive movements such as those seen in the antral region of the stomach.

Summary

The GI tract comprises the oral cavity, pharynx, oesophagus, stomach, small intestine (the duodenum, jejunum, and ileum), large intestine (the caecum and colon), rectum, and anal canal. It is served by a number of accessory organs: the teeth, tongue, salivary glands, pancreas, liver, and gall bladder. The wall of the GI tract has five basic layers: the serosa, a layer of longitudinal smooth muscle, a layer of circular smooth muscle, the submucosa, and the mucosa.

Ingested food is broken down by motor activity of the smooth muscle of the gut wall and by the action of digestive enzymes. The smooth muscle in the gut is of the **single unit** or **visceral** type which operates as a functional syncytium. The main motor activities of the GI tract are peristalsis and segmentation.

43.3 Nervous and hormonal control of the gastrointestinal tract

Innervation of the gastrointestinal tract

The complex afferent and efferent innervation of the GI tract allows for fine control of secretory and motor activity via intrinsic (enteric) and extrinsic (sympathetic and parasympathetic) pathways, which show a considerable degree of interaction. Figure 43.5 illustrates the extrinsic innervation of the gastrointestinal tract.

Afferent innervation of the GI tract

Although much of the activity of the GI tract is controlled through the intrinsic nerves of the **enteric nervous system**, the nerve plexuses are themselves linked to the central nervous system via afferent fibres. Chemoreceptor and mechanoreceptor endings are present in the mucosa and the muscularis externa. Chemoreceptors may be stimulated by substances present within foods or by products of digestion, while mechanoreceptors respond to distension or to irritation of the gut wall. Some of these receptors send afferent axons back to the central nervous system, to mediate reflexes via the CNS ('long-loop' reflexes), while the axons of others synapse with cells within the plexuses to mediate local reflexes. Extrinsic afferents relay information about the contents of the gut as well as sensations of discomfort and pain.

Many of the extrinsic sensory nerves are vagal afferents and have their cell bodies in the nodose ganglion of the brainstem (also called the inferior ganglion of the vagus). Some afferent fibres, particularly those forming part of the reflex arcs that control motility, travel to the spinal cord via the sympathetic nerves ('visceral afferents'). The cell bodies of these fibres lie within the dorsal root ganglia. It is important to realize that at least 80 per cent of fibres in

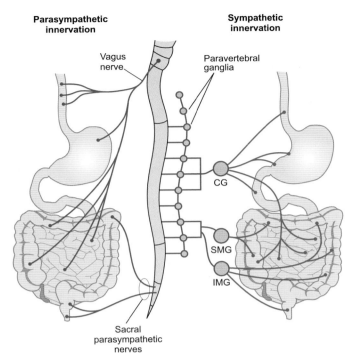

Parasympathetic innervation

Sympathetic innervation

Vagus nerve

Paravertebral ganglia

CG

SMG

IMG

Sacral parasympathetic nerves

Figure 43.5 The autonomic innervation of the gut. The left side of the figure shows the parasympathetic innervation originating from the cervical and sacral outflows. The right side shows the sympathetic innervation which reaches the stomach and small intestine via the postganglionic fibres of the coeliac ganglion (CG). The ascending colon and first part of the transverse colon receives sympathetic innervation via the superior mesenteric ganglion (SMG), while sympathetic postganglionic fibres of the inferior mesenteric ganglion (IMG) innervate the second part of the transverse and descending colon, and the rectum. Preganglionic fibres are shown in blue and postganglionic fibres in green.

the vagus and up to 70 per cent of splanchnic nerve fibres are afferent. Indeed, in vago-vagal reflexes, both the afferent and efferent fibres travel in the vagus nerve. Vago-vagal reflexes play a significant role in the control of motility and secretion in the GI tract (see Chapter 44).

Extrinsic control via sympathetic and parasympathetic nerves

Sympathetic innervation

The preganglionic fibres of the gastrointestinal sympathetic innervation arise from segments T8 to L2 of the spinal cord. The cell bodies of the postganglionic fibres lie within the coeliac, superior, and inferior mesenteric and hypogastric plexuses (see Chapter 11). Some of the sympathetic fibres innervate the smooth muscle of arterioles within the GI tract. When activated, these fibres cause vasoconstriction and redirection of blood away from the splanchnic bed. Other sympathetic fibres enter glandular tissue and innervate secretory cells. These responses are mediated by α-adrenoceptors. In addition, a large number of sympathetic fibres terminate within

the submucosal and myenteric plexuses where they appear to inhibit synaptic transmission, probably by presynaptic inhibition. There is also some innervation of the circular smooth muscle layers of the small and large intestines by sympathetic fibres. In general, increased sympathetic discharge reduces gastrointestinal activity. Indeed, powerful stimulation of the sympathetic innervation of the gut can produce almost complete inhibition of motility. By contrast, the sphincters of the GI tract are supplied with adrenergic fibres whose actions are usually excitatory and serve to constrict the circular smooth muscle of these regions.

Parasympathetic innervation

Parasympathetic input to the gut stimulates both its motility and secretory activity and thus opposes the actions of sympathetic stimulation. The vagus nerve relays the parasympathetic innervation to the oesophagus, stomach, small intestine, liver, pancreas, caecum, appendix, ascending colon, and transverse colon. The remainder of the colon receives parasympathetic innervation from pelvic nerves via the hypogastric plexus. All the parasympathetic fibres terminate within the myenteric plexus and are predominantly cholinergic. Further details of the autonomic innervation of the gut are given in Chapter 11.

Intrinsic innervation—the enteric nervous system

Two well-defined networks of nerve fibres and ganglion cell bodies are found within the wall of the GI tract from the oesophagus to the anus: the **myenteric plexus** and the **submucosal plexus**. These networks are called intramural plexuses and, together with other less well-defined neural networks, such as the much smaller mucosal plexus, they constitute the **enteric nervous system (ENS)** which is sometimes considered to be a separate division of the autonomic nervous system. The ENS is innervated by parasympathetic preganglionic axons and postganglionic sympathetic fibres from the coeliac, superior, and inferior mesenteric ganglia. This arrangement permits extrinsic neural regulation of the motility and secretory activity of the GI tract. In effect, the ENS controls the activity of the gut by integrating the central commands it receives via parasympathetic and sympathetic fibres with local information about the degree of stretch and the presence of nutrients within the gut lumen. Nevertheless, disrupting the parasympathetic and sympathetic innervation to the GI tract does not prevent the ENS from continuing its role as the main regulator of gut activity. For this reason the ENS is sometimes referred to as the 'mini brain' of the gut.

The myenteric plexus (also known as Auerbach's plexus) lies between the circular and longitudinal smooth muscle layers of the muscularis externa, while the less extensive submucosal plexus (or Meissner's plexus) lies within the submucosa (see Figures 43.6 and 43.7). The myenteric plexus is largely motor in function and regulates the motility of the gastrointestinal tract. In **Hirschsprung's disease**, an inherited developmental disorder affecting the rectum and a variable length of the bowel above it, ganglion cells are missing from part of the myenteric plexus. The condition is characterized by severe constipation caused by the resulting reduction in gut motility. In newborn babies, this condition may be life-threatening if not diagnosed and treated promptly.

The submucosal plexus receives sensory information both from the intestinal epithelium and from stretch receptors in the gut wall. Although it is chiefly concerned with controlling secretory activity and blood flow to the gut, it also regulates the activity of the muscularis mucosa. The axons of the enteric neurons form extensive networks (see Figure 43.6) as they synapse with the cells of other ganglia. They also innervate the blood vessels, glands, and smooth muscle of the GI tract. Figure 43.6(a) shows a whole mount preparation stained to reveal nerve fibres and ganglia of the myenteric plexus, and Figure 43.6(b) shows one of the ganglia at a higher magnification.

The enteric nervous system is responsible for coordinating much of the secretory activity and motility of the GI tract through intrinsic pathways often called 'short-loop' reflexes. These reflex pathways utilize many neurotransmitters including cholecystokinin (CCK), substance P, VIP (vasoactive intestinal polypeptide), serotonin (5-HT), somatostatin, and the enkephalins. Chemical signals play a role in the regulation of the gastrointestinal secretions, while stretching the gut wall elicits peristaltic contractions.

An outline of the organization of the nerve networks of the ENS is shown in Figure 43.7. Like the spinal cord, the enteric ganglia contain sensory neurons, interneurons, and motoneurones. The intrinsic sensory neurons respond to chemical signals from the lumen of the gut and to stretching of the gut wall. Some enteric motor neurons are secretomotor in function. Intrinsic primary afferent neurons are activated by local chemical mediators such as 5-HT and synapse with both secretomotor and vasomotor neurons in the submucosal plexus. These local circuits regulate both secretory activity and blood flow.

The motoneurons controlling muscle contraction may be excitatory or inhibitory. The excitatory neurons utilize acetylcholine, substance P, and neurokinin A as transmitters, while inhibitory synaptic transmission is mediated by nitric oxide, ATP, GABA, and neuropeptide Y. The inhibitory motoneurons are tonically active. It is their activity that determines both when contraction of the gut wall can occur and the directionality of gut movement. Inhibitory activity is normally less active in the direction facing away from the mouth (the aboral direction) so that contractile activity propagates towards the anus. During vomiting, however, the directionality of contraction is reversed (pp. 697–698).

Hormonal regulation of the gastrointestinal tract

In addition to its extensive innervation, the gastrointestinal tract is regulated by a number of peptide hormones that act through endocrine and/or paracrine pathways (see Chapter 6). The hormone-secreting or entero-endocrine cells are scattered throughout the mucosa. (Entero-endocrine cells are also known by the acronym APUD cells for **A**mine **P**recursor **U**ptake and **D**ecarboxylation.) The gastrointestinal tract utilizes at least 20 different regulatory peptides. Eight polypeptides that are known to act as circulating (endocrine) hormones are listed below, together with the region of the gut from which they are secreted:

- gastrin (antrum of the stomach)
- secretin (duodenum)
- cholecystokinin (CCK) (duodenum)
- pancreatic polypeptide (pancreas)
- gastric inhibitory polypeptide (GIP) (jejunum and duodenum)
- motilin (jejunum and duodenum)
- glucagon-like peptides GLP-1 and GLP-2 (ileum and colon), formerly known as enteroglucagons
- neurotensin (the lower small intestine).

CCK and neurotensin also exert a neurocrine action, producing their effects close to their site of secretion by nerve fibres. Paracrine agents acting on the gut include somatostatin and histamine. In general, the effects of these agents on motility and secretory activity supplement those of the gastrointestinal innervation, although the relative importance of nervous and hormonal influences differs throughout the tract. In the salivary glands, for example, nervous control is the dominant influence on secretion. In other areas such as the stomach, nervous and endocrine influences are of equal importance, while hormones are the principal regulators of secretion from the exocrine pancreas.

A further two peptide hormones, known as **ghrelin** and **obestatin**, have been identified more recently. These are derived from a

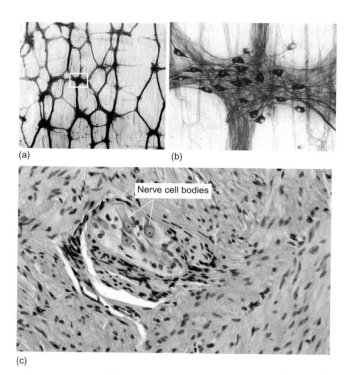

(a)

(b)

Nerve cell bodies

(c)

Figure 43.6 The intrinsic innervation of the GI tract. (a) A whole mount preparation of the myenteric plexus stained to reveal the nerve cell bodies and interconnecting fibres. (b) An enlarged view of the area bounded by the white square in (a). (c) A cross-section through a myenteric ganglion which is, in this case, surrounded by the smooth muscle of the stomach wall.

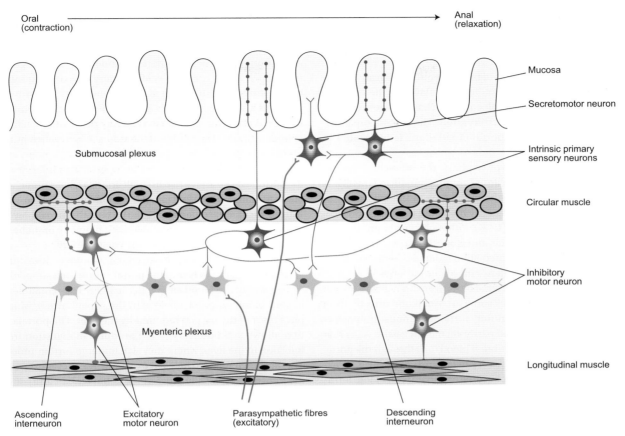

Oral
(contraction)

Anal
(relaxation)

Mucosa

Secretomotor neuron

Submucosal plexus

Intrinsic primary
sensory neurons

Circular muscle

Inhibitory
motor neuron

Myenteric plexus

Longitudinal muscle

Ascending
interneuron

Excitatory
motor neuron

Parasympathetic fibres
(excitatory)

Descending
interneuron

Figure 43.7 The nerve connections of the enteric nervous system. The two principal plexuses are the submucosal plexus, which plays a major role in the control of the secretory activity of the gut, and the myenteric plexus, which is largely motor in function. The parasympathetic preganglionic fibres synapse with neurons in both the submucosal and myenteric plexus and have an excitatory action. Sympathetic postganglionic fibres (not shown) exert an inhibitory effect on secretomotor and motor activity, and cause vasoconstriction. (Adapted from E.E. Benarroch (2007) *Neurology* **69**, 1953–17).

common precursor (pre-pro-ghrelin), and immuno-histochemical techniques using polyclonal antibodies have revealed their presence in cells of the gastrointestinal tract from the stomach to the ileum. They are also found in the pancreatic islets. Ghrelin is believed to exert a number of effects both within the gut and elsewhere, including stimulation of growth hormone secretion by the anterior pituitary, and stimulation of gastric emptying. Ghrelin also appears to stimulate appetite, probably through an effect on the hypothalamus. Experiments in humans have shown that ghrelin secretion is enhanced during fasting and is at its lowest immediately after a meal. Individuals with the genetic disorder Prader–Willi syndrome, which is characterized by a voracious appetite and obesity, show extremely elevated levels of ghrelin in their blood. The actions of obestatin were originally thought to oppose those of ghrelin, as experiments using laboratory rats indicated that obestatin suppressed food intake and slowed gastric emptying, possibly inducing a feeling of satiety. However, more recent work has cast doubt on this hypothesis and specific obestatin receptors have not been identified. Additional studies will be needed to confirm a role for obestatin in the mechanism of satiety.

Summary

The GI tract is regulated by both endocrine and neural factors. The gastrointestinal tract utilizes at least 20 different regulatory peptides including gastrin, secretin, and cholecystokinin. The relative importance of nervous and hormonal influences differs throughout the tract. The nerve pathways include both intrinsic networks via the ENS and extrinsic control via the sympathetic and parasympathetic divisions of the ANS. The ENS receives fibres from the parasympathetic preganglionic axons, via the vagus nerves, and postganglionic sympathetic fibres. The ENS coordinates secretory and motor activity via 'short-loop' reflexes. Visceral afferents relay information about the contents of the gut as well as sensations of discomfort and pain. They play a role in regulating the activity of the GI tract via 'long-loop' reflexes.

43.4 General characteristics of the blood flow to the GI tract

The various digestive functions of the gut require a rich and highly organized blood supply. The combined circulation to the stomach, liver, pancreas, intestine, and spleen (which has no digestive function) is called the **splanchnic circulation**. The general organization of the splanchnic and portal circulations is illustrated in Figure 43.8.

Throughout the post-absorptive phase, the splanchnic vessels receive 20–25 per cent of the cardiac output, but during the digestion and absorption of food, the blood flow to the gut increases considerably. This increase is partly mediated by the secretion of gastrin and cholecystokinin (CCK), although certain products of digestion, principally fatty acids and glucose, also act as powerful vasodilators. The splanchnic circulation is derived from the coeliac artery and the superior and inferior mesenteric arteries.

The **oesophagus** receives arterial blood from branches of the left gastric artery but also receives some arterial blood directly from small branches of the descending aorta. Its venous drainage is partly via the azygous vein and partly via the left gastric veins that join the hepatic portal circulation. If blood flow through the portal vein is restricted (as may be the case in cirrhosis of the liver for example), the oesophageal veins become engorged with blood, forming **oesophageal varices**, as blood backs up via the gastric veins.

As the veins engorge they gradually thin and may rupture, causing torrential bleeding.

The **stomach** receives blood from the coeliac artery, which arises from the aorta as it enters the abdomen. The left gastric artery, which arises from the coeliac artery, anastomoses with the right gastric artery (a branch of the hepatic artery) to form an arch that runs along the lesser curvature of the stomach. A similar arch is formed along the greater curvature by the right and left gastroepiploic arteries. The fundus of the stomach is supplied by a series of short gastric arteries.

The **intestine** is supplied by the superior and inferior mesenteric arteries. These give rise to small arteries whose branches penetrate the longitudinal and circular smooth muscle of the intestinal wall to form a vascular network in the submucosal layer. The **spleen** and **pancreas** are supplied with blood by branches of the splenic artery, which arises from the coeliac artery.

Venous blood leaving the stomach, pancreas, intestine, and spleen flows ultimately into the portal vein, which supplies about 70 per cent of the blood supply to the **liver**. The remainder of the hepatic blood supply is provided by the hepatic artery, which supplies most of the oxygen required by the liver. The portal circulation allows rapid delivery of the products of digestion from the intestine to the liver, where they will undergo further processing. This arrangement also helps to explain why cancers originating in the intestine (especially those of the colon and rectum) often metastasize to the liver. The liver is drained by the hepatic vein, which

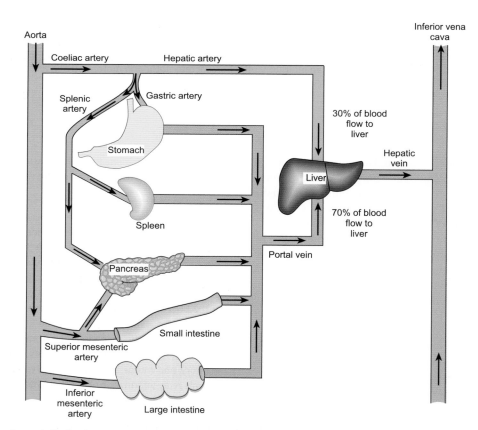

Figure 43.8 The blood supply to the liver and gastrointestinal tract. Note that the blood returning from the GI tract passes to the liver before reaching the inferior vena cava via the hepatic vein.

enters the inferior vena cava. **Acute ischaemia of the GI tract is a medical emergency.** It may arise as a result of circulatory shock, as a consequence of embolism or thrombosis, or following localized venous occlusion by a twisted bowel or strangulated hernia. The gastrointestinal mucosa is extremely sensitive to hypoxia, and tissue function is quickly disrupted as ischaemia progresses. The tips of the intestinal villi become necrotic first, and underlying tissue is exposed to the actions of digestive enzymes in the intestinal lumen. Bacteria and bacterial toxins from the gut lumen begin to enter the portal circulation, and septic shock may develop rapidly. There is also loss of fluid, electrolytes, and blood from the gut.

Management requires measures to maintain cardiac output and tissue oxygenation, replacement of lost fluid and electrolytes, and possibly surgery to remove the necrotic intestinal tissue.

Summary

The blood supply to the stomach, liver, pancreas, intestine, and spleen is supplied by the splanchnic circulation, which accounts for 20–25 per cent of the cardiac output. The activity of the GI tract is regulated by extrinsic autonomic nerves, by the enteric nervous system, and by hormones.

Checklist of key terms and concepts

The organisation and structure of the GI tract

- The GI tract consists of the oral cavity, pharynx, oesophagus, stomach, small intestine, large intestine, and anal canal.
- Accessory organs include the teeth, tongue, salivary glands, exocrine pancreas, liver, and gall bladder.
- The basic layers of the gut are (from the outside) the serosa, a layer of longitudinal smooth muscle, a layer of circular smooth muscle, the submucosa, and the mucosa.

Functions of the GI tract

- The major functions of the GI tract include the ingestion of food and its transport along the tract; the secretion of fluids, salts, and digestive enzymes; the absorption of digestion products; and the elimination of indigestible remains.
- The GI tract also protects itself from auto-digestion and contains a large quantity of lymphoid tissue, which forms part of the immune system.

- Contractile activity of the smooth muscle promotes mixing and propulsion of food and digestion products.

Regulation of the GI tract and blood flow to the GI tract

- The GI tract is regulated by extrinsic nerves and by the enteric nervous system.
- The enteric nervous system is a system of intramural plexuses which mediate a number of intrinsic reflexes that control secretory and contractile activity.
- Afferent and efferent extrinsic nerves and both endocrine and paracrine hormones play an important role in regulating the activities of the GI tract.
- The gut receives a rich blood supply. The combined circulation to the stomach, liver, pancreas, intestine, and spleen is called the splanchnic circulation.
- The splanchnic circulation receives 20–25 per cent of the cardiac output at rest.

Recommended reading

Anatomy

Mackinnon, P.M., and Morris, J. (2005) *Oxford textbook of functional anatomy*, Volume 2: *Thorax and abdomen*, pp. 131–66. Oxford University Press, Oxford.
    A clear description of the main structures of the gastrointestinal tract. Nice uncluttered anatomical illustrations.

Histology

Merscher, A. (2015). *Junquiera's basic histology* (14th edn), Chapter 15. McGraw-Hill, New York.
    Clear photomicrographs and text, well illustrated.

Review articles

Costa, M., Brookes, S.J.H., and Hennig, G.W. (2000) Anatomy and physiology of the enteric nervous system. *Gut* (Suppl IV) **47**, 15–19.

Gribble, F.M., and Reimann, F. (2016) Enteroendocrine cells: Chemosensors in the intestinal epithelium. *Annual Review of Physiology* **78**, 277–99.

Kunze, W.A.A., and Furness, J.B. (1999) The enteric nervous system and regulation of intestinal motility. *Annual Review of Physiology* **61**, 117–42.

Murthy, K.S. (2006) Signaling for contraction and relaxation in smooth muscle of the gut. *Annual Review of Physiology* **68**, 345–74, doi:10.1146/annurev.physiol.68.040504.094707.

To check that you have mastered the key concepts presented in this chapter, complete the accompanying online self-assessment questions. Go to www.oup.com/uk/pocock5e/

CHAPTER 44

The gastrointestinal tract

Chapter contents

This chapter should help you to understand:

- The role of the teeth and tongue in mastication (chewing)
- The salivary glands and saliva
  - Anatomy of the salivary glands and the composition of saliva
  - The regulation of salivary secretion
  - The role of saliva in chewing, swallowing, and digestion
- The Swallowing reflex
- The structure and function of the oesophagus and the oesophageal sphincters
- The anatomy and physiology of the stomach
  - Mechanical digestion by the stomach—the formation of chyme
  - Enzymes of the gastric juice
  - The secretion of hydrochloric acid by the oxyntic cells of the gastric glands
- Neural and hormonal control of gastric motility and secretion
- Gastric emptying
- Why the stomach does not digest itself
- The anatomy and physiology of the small intestine
  - Secretion of fluid and enzymes
  - Motility of the small intestine
  - The absorption of digestion products by the small intestine
  - Malabsorptive states
- The exocrine functions of the pancreas
  - The composition of pancreatic juice
  - The regulation of pancreatic secretion
- The large intestine
  - Functions of the colon
  - The rectum, the anus, and defecation

44.1 Introduction

As outlined in the previous chapter, the principal function of the gastrointestinal (GI) system is to convert the foods we eat into simple molecules that can be absorbed into the bloodstream and utilized by the cells of the body for energy and to build and repair body tissues. The structures that carry out these tasks may be divided into those that form the alimentary canal itself (the tube that extends from mouth to anus) and the accessory organs that contribute to mechanical digestion (teeth and tongue) or to chemical digestion through the secretion of various fluids and enzymes (salivary glands, liver, and exocrine pancreas). This chapter will provide a detailed explanation of the roles played by the different parts of the GI system in the processes of ingestion, digestion (both mechanical and chemical), absorption, and elimination of indigestible remains (**defecation**). It will also include discussion of the complex nervous and hormonal mechanisms that regulate motility and secretory activity.

44.2 Intake of food, chewing, and salivary secretion

Food is ingested via the mouth (also called the oral or buccal cavity), the only part of the GI tract that has a bony skeleton. The mouth is formed by the cheeks (which end anteriorly as the lips), the hard and soft palates, and the tongue. Here, the food is broken into smaller pieces by the process of chewing (mastication) and is mixed with saliva, which softens and lubricates the food mass. As food moves around the mouth, the taste buds of the tongue and palate are stimulated and odours are released from the food. The presence of food in the mouth and the sensory stimuli of taste and smell play a part in the stimulation of gastric secretion (see Section 44.5).

The mouth is divided into two regions, the **vestibule** and the **oral cavity**. The vestibule is the region between the teeth, lips, and cheeks. The oral cavity is the space that extends from the teeth and gums to the opening between the mouth and the oropharynx (the fauces). The stratified epithelium here is typically 15–20 layers of cells thick and is structurally adapted to withstand the frictional forces generated during chewing.

The roof of the mouth is formed by the palate, which separates the oral cavity from the nasal cavity. The hard palate is formed by the maxillae (upper jaw) and palatine bones. The soft palate (also called the velum) is made up of muscle and connective tissue; it can be raised or lowered to facilitate swallowing, sucking, and speech. A fold of tissue called the uvula hangs down from the soft palate in front of the oropharynx. It plays a role in producing certain speech sounds and also moves to close off the nasopharynx during swallowing.

Several small masses of lymphoid tissue are found in the oral cavity. The most prominent of these are the palatine tonsils located at the back of the mouth. Smaller tonsils are located at the base of the tongue (the lingual tonsils) and in the roof of the nasopharynx, posterior to the nasal cavity (the pharyngeal tonsils or adenoids). The tonsils form part of the gut-associated lymphoid tissue (GALT—see Chapter 26) and play a role in the immune response to pathogens in the oral cavity.

The teeth

A child has 20 deciduous (milk or primary) teeth, which normally erupt between the ages of 6–30 months. The incisors usually erupt first, followed by the first molars, the canines, and finally the second molars. This set has no premolars. Between the ages of about 6 and 15 the deciduous teeth are gradually shed and replaced by 32 permanent (adult or secondary) teeth—eight incisors, four canines, eight premolars, and twelve molars, as shown in Figure 44.1(a). The teeth are embedded in the alveoli or sockets of the alveolar ridges of the mandible (lower jaw) and maxilla (upper jaw). In adults, the upper and lower jaws each possess four incisors, two canines, four premolars and six molars. The incisors and canines are the cutting teeth, while the premolars and molars have broad flat surfaces for grinding or chewing food.

Although the shapes of the different teeth vary, they share a similar basic structure (see Figure 44.1b). The crown is the part of the tooth that protrudes from the gum (or **gingiva**), while the root is embedded in the bone of the mandible or maxilla. In the centre of the tooth is the pulp cavity, which contains blood vessels, lymphatics, and nerves, and this is surrounded by hard dentine. Outside the

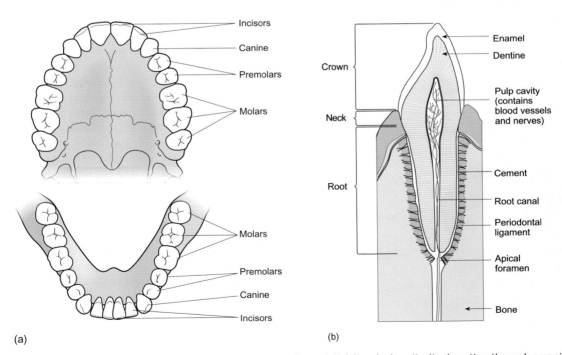

(a)   (b)

Figure 44.1 The arrangement of the teeth in the upper and lower jaws of an adult (a) and a longitudinal section through a canine tooth to show its internal structure and location in the jaw. (Adapted from Figs 4.4.9 and 6.4.10 in P.C.B.McKinnon and J.F. Morris (2005) *Oxford textbook of functional anatomy*, Vol. 3. Oxford University Press, Oxford. By permission of Oxford University Press.)

dentine of the crown is a layer of even harder, translucent enamel. The root of the tooth is surrounded by softer cement, which fixes the tooth into its socket. The periodontal ligament holds the tooth in place while still allowing a little movement during chewing. Blood vessels and nerves pass to the tooth through a small opening at the tip of each root. Branches of the maxillary nerves supply the upper teeth, while the lower teeth are innervated by branches of the mandibular nerves, both of which arise from the trigeminal nerve (cranial nerve V).

The tongue

The tongue is formed from skeletal (voluntary) muscle and occupies the floor of the mouth. Its upper surface is covered by keratinized stratified squamous epithelium, while its underside is covered with a thin, non-keratinized mucous membrane. Taste buds are found on the tip and dorsal surface of the tongue (see Chapter 17). Lingual glands (also called glands of Ebner), located in the region of the vallate papillae, secrete a serous fluid that plays an important part in taste sensation by moistening the taste buds. This fluid also contains a fat-digesting enzyme, lingual lipase.

The upper surface (or dorsum) of the tongue is divided into two main regions by a V-shaped depression called the terminal sulcus: an anterior (palatine) region and a posterior (pharyngeal) region. The tongue is inserted into the hyoid bone and attached to the anterior floor of the mouth, behind the lower incisor teeth, by a fold of its mucous membrane covering called the **frenulum**, which can easily be seen by raising the tongue. Occasionally the frenulum is very short and interferes with the ability of the tongue to facilitate speech ('tongue-tie'). In such cases the frenulum may be cut to free the tongue. In addition to its role in speech, the tongue is necessary for chewing, swallowing, and the perception of taste. Figure 44.2 illustrates the underside of the tongue. The location of the taste buds on the dorsum of the tongue is shown in Figure 17.5.

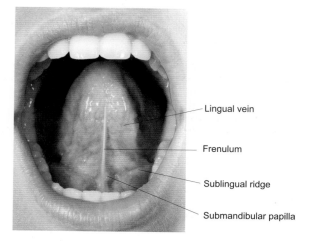

Figure 44.2 The underside of the tongue showing the sublingual ridge and submandibular papilla where the secretory ducts from the sublingual and submaxillary glands open into the mouth.

Mastication (chewing)

A crushing force of up to 50–80 kg can be generated on the molars during chewing, a value which far exceeds the forces generally required for a normal diet. In addition to the teeth, both the tongue and the cheeks also have important roles to play in the process of chewing—their movements help to keep the food in the correct position, while the sensory receptors on the tongue provide information regarding the readiness of the food for swallowing.

The secretion of saliva

Approximately 1500 ml of saliva is produced each day by the salivary glands. Saliva performs several important functions. It lubricates the food to facilitate swallowing, it aids in speech, and it contains the enzymes, salivary α-**amylase** and salivary lipase, that begin the process of digestion. The saliva dissolves certain substances in foods, making them available to the taste cells. It also contains IgA, haptocorrin (p. 714), and an enzyme called lysozyme, which acts on the cell walls of certain bacteria to cause their lysis and death. The bacteriostatic actions of saliva contribute significantly to oral comfort and reduce the risk of infection developing in the mouth, particularly after oral or dental surgery. Individuals who lack functional salivary glands, or in whom salivary secretion has been inhibited by irradiation or medications such as tricyclic anti-depressants, show a predisposition to dental caries (decay) and infections of the buccal mucosa and often experience a reduction in taste sensation. This condition is known as **xerostomia**.

There are three main pairs of salivary glands; the parotid, submandibular, and sublingual glands. Their locations are illustrated in Figure 43.1. Other smaller glands exist over the surface of the palate and tongue and inside the lips, though these do not appear to be under nervous control. Each gland consists of a number of lobules surrounded by a fibrous capsule. Each lobule or acinus is made up of balls of cells (acinar cells) which are drained by ductules that join to form larger ducts leading into the mouth. Each acinus consists of serous cells and mucous cells whose relative numbers vary from gland to gland. In the parotid and submandibular glands, serous cells greatly outnumber the mucous cells, while in the sublingual gland the proportion of mucous cells is much higher. The fine processes of myoepithelial cells extend over the outer surface of the acini. The cell bodies themselves lie on the basement membrane beneath the secretory cells and basal lamina. (See Figure 4.12 for a description of the main features of epithelia.) Myoepithelial cells are also found in the epithelia that line the intercalated ducts.

All the major salivary glands receive both sympathetic and parasympathetic innervation. Noradrenergic sympathetic fibres from the superior cervical ganglion are distributed to both blood vessels and acinar cells. Preganglionic parasympathetic fibres arrive by way of the facial and glossopharyngeal nerves and synapse with postganglionic neurons close to the salivary glands themselves. Both the secretory cells and the duct cells receive parasympathetic postganglionic fibres.

The **parotid glands** are situated at the angle of the jaw, lying posterior to the mandible and inferior to the ear. The main ducts from the parotid glands open into the mouth opposite the second molars on each side. They are the largest of the salivary glands and produce an entirely serous secretion, a watery fluid lacking mucus. Saliva from the parotid glands accounts for around 25 per cent of the total output of the salivary glands. It contains alpha-amylase and a small amount of immunoglobulin A.

The **submandibular glands** lie below the mandible of the lower jaw and discharge their secretions via the submandibular papilla, which is clearly shown in Figure 44.2. They secrete mucoprotein and serous fluid. Their secretion is therefore more viscid than that of the parotid glands. Overall, the submandibular glands secrete about 70 per cent of the daily output of saliva.

The **sublingual glands** lie in the floor of the mouth below the tongue and discharge their secretion via a row of small ducts along the sublingual ridge. They contribute the remaining 5 per cent or so of the total salivary output, producing a secretion that is rich in mucoprotein and gives the saliva its somewhat sticky character.

The mechanism of salivary secretion

Figure 44.3 illustrates the ion movements that occur during the two-stage formation of saliva. An isotonic fluid (primary secretion) is formed by the acinar cells as a result of the active transport of electrolytes followed by the passive movement of water. As in the thick ascending limb of the loop of Henle, a sodium pump in the basolateral membrane establishes a sodium gradient that is subsequently exploited to allow the secondary active transport of sodium, potassium, and chloride ions into the secretory cells via co-transporters known as NKCC1 (these are one of two kinds of Na-K-Cl co-transporter; the other, NKCC2, is found in the ascending limb of the loop of Henle—see Section 39.6). Chloride ions are secreted into the central lumen of the acinus through chloride channels. To maintain osmolality, water follows via aquaporin 5 channels. Sodium and potassium ions enter the central lumen by a paracellular pathway accompanied by more water molecules. The mechanism of bicarbonate transport is still unclear. The concentrations of sodium, chloride, and bicarbonate ions in the primary secretion resemble those of plasma. Secondary modification of this fluid then occurs by means of ion transport processes occurring in the epithelial cells lining the ducts, which absorb sodium and chloride and secrete potassium and bicarbonate as the saliva flows past them. As these later ion movements occur without the concomitant movement of water, the final secretion is hypotonic with respect to the plasma.

The final electrolyte content of saliva depends upon the rate at which it is secreted and flows along the salivary ducts. At low rates of secretion, there is ample time for ductal modification of the primary secretion. As the flow rate increases, there is less time for the reabsorption of electrolytes. Consequently, the salivary content of sodium, bicarbonate, and chloride increases, while there is a small drop in the potassium concentration (see Figure 44.4). The increase in the concentration of bicarbonate at higher rates of secretion is associated with a rise in the pH of the saliva from pH 6.2 to pH 7.4.

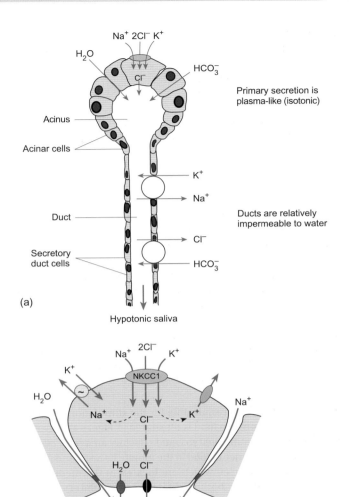

(a)

(b)

Figure 44.3 (a) Water and electrolyte transport leading to the formation of saliva by acinar and ductal cells of a salivary gland. The primary secretion of the acinar cells is subsequently modified as fluid passes along the ducts with the exchange of potassium and bicarbonate ions for sodium and chloride. (b) A simplified scheme showing the main electrolyte movements during the elaboration of the primary secretion. Chloride is transported into the acinar cells by the NKCC1 cotransporter along with sodium and potassium ions. Chloride ions exit the cells via chloride channels in the apical surface, while sodium is pumped across the basolateral surface by the Na⁺, K⁺-ATPase (the sodium pump) and potassium exits via basolateral potassium channels. The result is a negative potential in the duct lumen which draws in sodium ions via the paracellular pathway. Water follows via aquaporin 5 channels in the apical membrane and by paracellular movement.

The regulation of salivary secretion

The rate of salivary secretion is controlled primarily by reflexes mediated by the autonomic nervous system. The resting rate of salivary secretion is about 0.5 ml min⁻¹, but during maximal stimulation by sapid substances, the rate of secretion may increase to 7 ml min⁻¹.

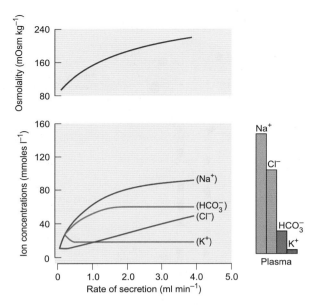

Figure 44.4 The variation of the ionic composition and osmolality of the saliva with flow rate. The composition of plasma is shown in the bar diagram on the right for comparison. Note that the saliva is hypotonic with respect to the plasma. (Adapted from J.H. Thaysen et al. (1954) *Am J Physiol* **178**, 155–9.)

The reflex pathway responsible for the stimulation of salivary secretion in response to food is illustrated in Figure 44.5. Sensory receptors in the mouth, pharynx, and olfactory area relay information about the presence of food in the mouth, its taste, and its smell to the salivatory nuclei, which are located in the medulla. The medulla also receives both facilitatory and inhibitory impulses from the hypothalamic appetite area and regions of the cerebral cortex concerned with the perception of taste and smell. Most salivatory responses are mediated by parasympathetic efferent fibres originating in the salivatory nuclei.

Parasympathetic stimulation promotes an abundant secretion of watery saliva that is rich in amylase and mucins. The transporting characteristics of the ductal epithelium are also altered so that bicarbonate secretion is stimulated while the reabsorption of sodium and the secretion of potassium are inhibited. These changes are mediated by muscarinic receptors and are inhibited by atropine (see Chapter 11).

The response to parasympathetic stimulation also includes a significant increase in blood flow to the salivary glands. This effect is not atropine-sensitive. Several mechanisms are thought to contribute to the increase in blood flow. These include the release of vasoactive intestinal polypeptide (VIP) from parasympathetic nerve terminals and the release of the proteolytic enzyme kallikrein into the interstitial fluid from the acinar cells themselves. Kallikrein, in turn, promotes the production of the powerful vasodilator, **bradykinin**, from plasma α-2 globulin.

The response of the salivary glands to sympathetic stimulation is variable. Sympathetic fibres stimulate the secretory cells and enhance the output of amylase. At the same time, however, blood flow to the glands is usually reduced through vasoconstriction and the net result is a fall in the rate of salivary secretion. Indeed, a dry mouth is a characteristic feature of the sympathetic response to fear and stress. A summary of the factors that control the secretion of saliva is shown in Figure 44.6.

Digestive actions of saliva

The digestive enzyme salivary amylase is stored within **zymogen granules** in the serous acinar cells. Once secreted, it is able to degrade complex polysaccharides such as starch and glycogen to maltose, maltotriose, and dextrins, working optimally at pH 6.9. Although food remains in the mouth for only a short time and the contents of the stomach are highly acidic, salivary amylase is believed to continue working within the food bolus for some time after the entry of food into the stomach. It is probably inactivated only after complete mixing of the bolus contents with the gastric juice.

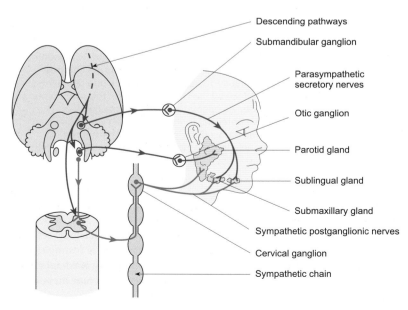

Figure 44.5 The secretomotor pathways subserving the salivary reflex. Salivation is initiated by a variety of stimuli including the sight and thought of food as well as by the presence of food in the mouth. Taste receptors in the mouth (predominantly on the tongue) signal the presence of food via the chorda tympani and glossopharyngeal nerves to the nucleus of the tractus soltarius in the medulla. The neurons of the NTS activate the secretomotor neurons in efferent nuclei of the salivary glands, which trigger the secretory responses of the salivary glands via the parasympathetic fibres of the facial (CN VII) and glossopharyngeal (CN IX) nerves (shown in blue). The submandibular and sublingual glands are innervated by the facial nerve, and the parotid glands by the glossopharyngeal nerve. Sympathetic fibres (green) reach the salivary glands via the blood vessels. (Adapted from Fig 7.7 in P.C.B. McKinnon and J.F. Morris (2005) *Oxford textbook of functional anatomy*, Vol. 3. Oxford University Press, Oxford.)

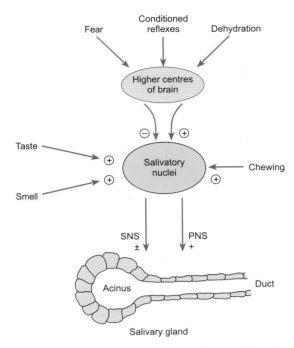

Figure 44.6 The major factors that influence the secretion of saliva. Stimulation of the parasympathetic nerves (PNS) results in a copious production of fluid saliva, while stimulation of the sympathetic nerves (SNS) results in the production of thick viscid saliva.

The saliva also contains a lipase that plays a significant role in fat digestion in the stomach and will be discussed later in the chapter (p. 692).

Swallowing (deglutition)

Once mastication is complete and the lubricated food bolus has been formed, it is swallowed. From the mouth, food passes into the oropharynx and then into the laryngopharynx, both of which are common passageways for food, fluids, and air.

The first (oral) phase of swallowing is voluntary but the subsequent pharyngeal and oesophageal stages of the process are under reflex autonomic (involuntary) control. Neurons in the medulla and lower pons mediate the involuntary phase of swallowing. During the voluntary oral phase, the tip of the tongue is placed against the hard palate and the tongue is then contracted to force the food bolus into the oropharynx (the part of the pharynx lying immediately behind the mouth). The pharynx is richly endowed with mechanoreceptors, and when food stimulates these receptors, a complex sequence of events is initiated which results in completion of the swallowing process. The major events of the swallowing reflex are illustrated in Figure 44.7.

Information from the mechanoreceptors passes via afferents in the glossopharyngeal nerve (CN IX) to sensory nuclei in the brainstem. It also spreads via interneurons to the motor nucleus ambiguus in the medulla. Motor impulses travel from the motor nuclei in the brainstem (see Figure 10.12) to the muscles of the pharynx, palate, and upper oesophagus via the glossopharyngeal and vagal nerves. Lesions of the area of the brain that controls swallowing, or

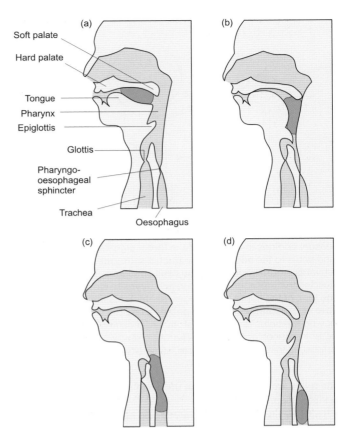

Figure 44.7 The events occurring during swallowing as a food bolus (shown in green) moves from the mouth to the oesophagus. As the bolus passes to the back of the throat (b), the epiglottis is moved downward to occlude the airways (c). (Redrawn from L.K. Knobel (1976) Fig 25B-1. In *Physiology* 4th edition, ed E.E. Selkurt. Little, Brown and Co., Boston)

of the glossopharyngeal and vagal nerves carrying the efferent impulses, result in difficulty in swallowing (**dysphagia**).

The efferent response begins with contraction of the superior constrictor muscle, which raises the soft palate towards the posterior pharyngeal wall to prevent food entering the nasopharynx. It also initiates a wave of peristaltic contraction that propels the bolus through the relaxed upper oesophageal sphincter into the oesophagus. The larynx is raised so that the epiglottis covers the opening of the nasopharynx into the trachea, while the opening of the oesophagus is stretched. At the same time, respiration is inhibited (**deglutition apnoea**). In this way, food is prevented from entering the airways. In the final phase of swallowing, the oesophageal phase, the wave of peristaltic contraction that was initiated in the pharynx continues along the length of the oesophagus. This wave of contraction lasts for 7–10 seconds and is usually sufficient to propel the bolus to the stomach. If it fails to do so, the resulting distension of the oesophagus initiates a vago-vagal reflex which triggers a secondary peristaltic wave.

The oesophageal submucosa contains glands that secrete mucus in response to pressure from a food bolus. This mucus helps to lubricate the oesophagus and facilitates the transport of food.

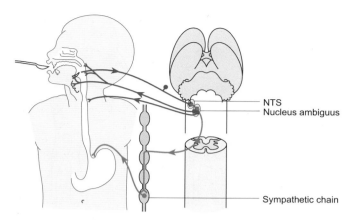

Figure 44.8 The pathways subserving the swallowing reflex. Sensory endings in the back of the throat signal the presence of a food bolus to the nucleus of the tractus solitarius NTS, which activates motoneurons in the nucleus ambiguus. These control muscles that raise the soft palate and larynx to close the epiglottis. The tone of the upper oesophageal sphincter is inhibited and a wave of peristaltic contraction passes along the oesophagus, so moving the bolus towards the stomach. (Adapted from Fig 7.8 in P.C.B. McKinnon and J.F. Morris (2005) *Oxford textbook of functional anatomy*, Vol. 3. Oxford University Press, Oxford.)

The reflex arc that brings about the act of swallowing is illustrated in Figure 44.8.

The time taken for food to travel down the oesophagus depends on the consistency of the food and the position of the body. Liquids take only 1–2 seconds to reach the stomach, but solid foods can take much longer. The process is normally aided by gravity. At the junction between the oesophagus and the stomach, there is a slight thickening of the circular smooth muscle of the GI tract called the lower oesophageal sphincter (also called the cardiac or gastro-oesophageal sphincter). This acts as a valve and remains closed when food is not being swallowed, thus preventing regurgitation of food and gastric juices into the oesophagus. Just before the peristaltic wave (and the food) reaches the end of the oesophagus, the lower oesophageal sphincter relaxes to permit the entry of the bolus into the stomach. Once the bolus has entered the stomach, the sphincter closes once more.

Gastro-oesophageal reflux of acidic stomach contents can occur under certain conditions. If the lower oesophageal sphincter remains relaxed for long periods or if the abdominal pressure exceeds that in the thorax, acidic stomach contents may enter the oesophagus (reflux). Prolonged or excessive reflux may cause inflammation of the oesophagus resulting in burning retrosternal pain in the region between the epigastrium and the throat (heartburn). Symptoms are often exacerbated by pregnancy, stooping, or lying down, or following a heavy meal, all situations in which intra-abdominal pressure is raised. Many older people with gastro-oesophageal reflux disorder (GORD) are found to have a **hiatus hernia**, a condition in which the proximal region of the stomach herniates (i.e. protrudes) through the opening in the diaphragm through which the oesophagus passes into the thoracic cavity (the oesophageal hiatus). Figure 44.9 illustrates the two main forms of this condition, the sliding hernia and the rolling hiatus hernia (or

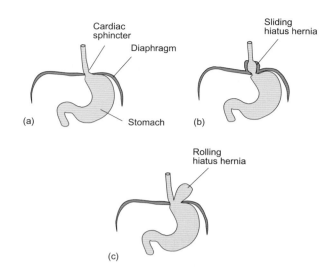

Figure 44.9 Diagrammatic representations of a sliding hiatus hernia (b) and a rolling hiatus hernia (c) compared to the normal arrangement of the diaphragm and cardiac sphincter (a).

para-oesophageal hernia). In the sliding hernia, the most common type, the gastro-oesophageal junction slides through the oesophageal hiatus (aperture) in the diaphragm. In the rolling hiatus hernia, the sphincter remains below the diaphragm but part of the stomach rolls up through the hiatus to lie alongside the oesophagus. Treatments for GORD include lifestyle advice; antacids such as sodium bicarbonate, calcium carbonate, magnesium, and aluminium salts (which neutralize the acidic chyme); and drugs such as ranitidine and omeprazole that inhibit gastric acid secretion.

Summary

In the mouth, food is mixed with saliva secreted by the parotid, submandibular, and sublingual salivary glands. The saliva contains mucus, which helps to lubricate the food, and an α-amylase that initiates the breakdown of carbohydrate. The saliva secreted by the salivary acinar cells undergoes modification as it flows along the salivary ducts, so its final composition depends upon its flow rate.

Salivary secretion is controlled by autonomic reflexes. Parasympathetic stimulation promotes an abundant secretion of watery saliva rich in amylase and mucus, while sympathetic stimulation promotes the output of amylase but reduces blood flow to the glands. The overall effect of sympathetic stimulation is generally a reduction in the rate of salivary secretion.

Swallowing occurs in three phases. The first oral phase is voluntary, but subsequent phases are under reflex autonomic control. As swallowing occurs, a wave of peristalsis is initiated which propels the food bolus through the upper oesophageal sphincter into the oesophagus. This wave of contraction continues for several seconds and moves the bolus down to the lower oesophageal sphincter, which relaxes to permit the entry of food into the stomach. The lower oesophageal sphincter normally remains closed to prevent reflux of acidic gastric contents into the oesophagus.

44.3 The stomach

Below the oesophagus, the GI tract expands to form the stomach, which lies in the left side of the upper abdominal cavity. The functions of the stomach include

- the temporary storage of food
- mechanical breakdown of food into small particles
- chemical digestion of proteins to polypeptides by pepsins
- regulated passage of processed food (chyme) into the small intestine
- the secretion of intrinsic factor, which is essential for the absorption of vitamin B_{12}
- the secretion of a mucus barrier that protects against auto-digestion.

The stomach is continuous with the oesophagus at the lower oesophageal (cardiac) sphincter and with the duodenum at the pyloric sphincter. Its position in relation to other abdominal organs is shown in Figure 43.1. It is around 25 cm long, and roughly J-shaped, although its exact size and shape varies between individuals and with its degree of fullness. When the stomach is empty, it has a volume of about 50 ml and its mucosa and submucosa are thrown into large longitudinal folds called **rugae** which flatten out as the stomach fills with food. This organization is similar to that seen in the urinary bladder (see Chapter 39). When fully distended, the adult stomach typically has a volume of around 4 litres.

The major regions of the stomach are illustrated in Figure 44.10. The cardiac region (cardia) surrounds the cardiac orifice, through which food enters the stomach. The part of the stomach that extends above the cardiac orifice is called the **fundus**, while the

mid-portion is called the **body**. This is continuous with the funnel-shaped pyloric region. The upper, widest portion of the pyloric region is the **antrum**, which narrows to form the pyloric canal terminating in the **pylorus** itself. The convex portion of the stomach is called the greater curvature, while the concave region is the lesser curvature.

The fundus and body of the stomach act as a temporary reservoir for food. As indicated above, they can accommodate large increases in volume without appreciable changes in intragastric pressure because the smooth muscle in these areas relaxes in response to the presence of food and the rugae flatten out. This reflex is mediated, at least in part, by afferent and efferent fibres running in the vagus nerve which cause an inhibition of muscle tone. Contractile activity in the fundus and body is relatively weak, so that food remains here largely undisturbed for fairly long periods (up to 4 or 5 hours). By contrast, vigorous contractions take place in the antral region, where food is broken down into smaller particles and mixed with gastric juice to form **chyme**, the semi-liquid form in which it is passed to the duodenum.

Blood supply and innervation of the stomach

Arterial blood is supplied to the stomach by the gastric arteries. These arise from the left coeliac artery and from the common hepatic and splenic arteries. They form a plexus of vessels within the submucosa, from where they branch extensively to provide the mucosal layer with a rich vascular network. Venous drainage is via the gastric veins. These empty into the portal vein, which carries blood to the liver. Blood flow to the gastric mucosa increases significantly when the stomach secretes gastric juice in response to a meal. The lymphatics of the stomach lie along the arteries and drain into lymph nodes around the coeliac arteries.

The stomach is richly innervated by both intrinsic and extrinsic nerves (see Figures 43.5–43.7). The intrinsic neurons of the enteric nervous system supply the gastric smooth muscle and secretory cells. They also receive numerous synaptic connections from the extrinsic nerves. The extrinsic innervation includes sympathetic fibres from the coeliac plexus and parasympathetic vagal neurons. Afferent fibres sensitive to a variety of stimuli including distension and pain reach the brainstem and spinal cord by way of the vagus and sympathetic nerves. Intrinsic primary afferent neurons form the afferent arms of reflex arcs mediated by the enteric nervous system (Section 43.3).

Structure of the gastric mucosa

The basic organization of the stomach wall resembles that shown in Figure 43.2. However, in addition to the longitudinal and circular muscle layers, there is an additional layer of muscle between the mucous membrane and the circular layer of the muscularis externa, on the anterior and posterior sides of the stomach. The smooth muscle cells in this layer are obliquely orientated and play a role in the grinding and churning movements displayed by the stomach. Figure 44.11 illustrates the three layers of the muscularis externa found in the stomach wall (note that the muscularis mucosa is a separate layer).

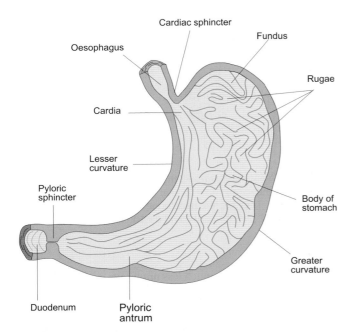

Figure 44.10 A longitudinal section of the stomach showing the major anatomical regions.

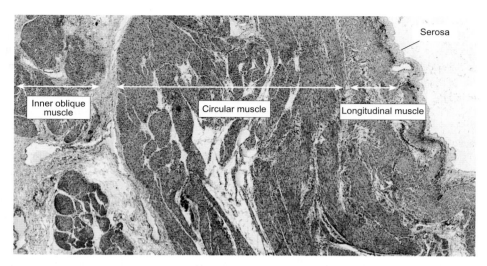

Figure 44.11 A section through part of the stomach wall showing the circular and longitudinal muscle layers of the muscularis externa together with the inner oblique layer.

The circular smooth muscle layer varies in thickness in different parts of the stomach. It is relatively thin in the fundus and body but thicker in the antral region where the most vigorous contractile activity occurs. Further thickening is seen in the pylorus, where the circular muscle forms the pyloric sphincter that regulates gastric emptying.

The surface epithelium of the gastric mucosa is of the simple columnar type, composed almost entirely of mucous cells (also called foveolar cells), which produce a protective alkaline fluid containing mucus. This epithelial layer is dotted with millions of deep **gastric pits** into which the secretions of gastric glands empty. There are around 100 gastric pits per square mm of the mucosa, occupying about 50 per cent of its total surface. The **gastric glands** themselves are simple tubular structures whose cellular composition differs according to the region of the stomach. A diagrammatic representation of a gastric gland, showing the locations of the various secretory cells, is shown in Figure 44.12.

The cells of the gastric glands include the following types:

1) **Mucous neck cells**, which are situated at the opening of the gastric glands. These cells secrete a mucus distinct from that secreted by the surface epithelial cells. Its special significance remains unclear.

2) **Chief cells**, which are located in the basal regions of the gastric glands. These cells secrete **pepsinogens**, the inactive forms of proteolytic enzymes collectively known as pepsin.

3) **Parietal cells** (also called **oxyntic cells**), which are relatively large cells scattered amongst the chief cells. They secrete hydrochloric acid and **intrinsic factor**.

4) **Entero-endocrine cells** (G-cells), which secrete the hormone gastrin. This enters the bloodstream and affects motility and secretory processes within the GI tract. The gastric mucosa secretes two further hormones, ghrelin and leptin, whose role in the control of food intake is discussed in Chapter 46.

In the glands of the fundus and body of the stomach, both chief and parietal (oxyntic) cells are numerous. In the antral and pyloric

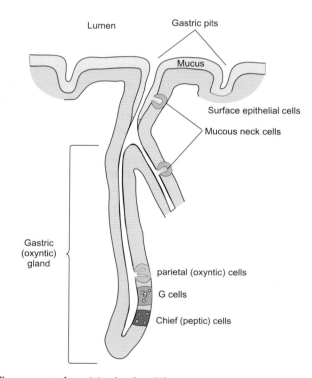

Figure 44.12 A gastric gland and the various secretory cell types. The gastric glands are simple tubular glands that discharge into the gastric pits. They secrete pepsinogen from the chief cells, acid from the parietal (oxyntic) cells, and gastrin from the entero-endocrine G cells.

regions, parietal cells are much less numerous, with mucus, pepsinogens, and gastrin the predominant secretions. In the cardiac region of the stomach, the gastric glands consist almost entirely of mucus-secreting cells. Because of these regional variations in structure, the exact effects of partial gastrectomy will depend upon the area that is removed.

Summary

The stomach stores food and mixes it with the gastric secretions (acid, enzymes, mucus, and intrinsic factor) to produce chyme. The surface epithelium of the gastric mucosa is composed almost entirely of cells that secrete an alkaline fluid containing mucus that protects it from digestion. The gastric glands contain mucous cells, chief cells, which secrete pepsinogens, parietal cells, which produce gastric acid, and intrinsic factor, as well as a variety of endocrine cells including G-cells which secrete gastrin.

44.4 The composition of gastric juice

Between 2 and 3 litres of gastric juice are secreted each day in adults. It contains salts, water, hydrochloric acid (HCl), pepsinogens, and intrinsic factor. The exact composition and flow rate of gastric juice is determined by the relative activities of the different types of cell within the gastric glands and will vary according to the time that has elapsed since the ingestion of food.

Clinical tests of gastric secretory function, involving the collection of gastric juice by means of a swallowed tube, have shown that, during fasting, there is little or no secretion of gastric juice and that the stomach contains only about 30 ml of fluid. Following the ingestion of a test meal (such as thin porridge), the pH of the stomach contents first rises, as the acid present in the stomach lumen is neutralized, and then falls progressively over the ensuing 90 minutes or so, as hydrochloric acid is secreted by the parietal (oxyntic) cells of the gastric glands. The output of all the other constituents of gastric juice also increases following a meal.

The electrolyte composition of gastric juice depends upon its rate of secretion. As the rate of secretion rises, the sodium concentration falls while that of hydrogen ions is increased. The level of potassium ions in gastric juice is always higher than that of the plasma so that prolonged vomiting may lead to hypokalaemia (low plasma potassium).

The formation of stomach acid

The pH of gastric juice is very low (pH 1–3). Although **hydrochloric acid** is not essential for the overall digestive process, the highly acidic environment it creates is important for several reasons:

1. It helps in the breakdown of the connective tissue and muscle fibres of ingested meat.
2. It activates inactive pepsinogens and provides optimal conditions for their activity.
3. By combining with calcium and iron to form soluble salts, the low pH of the gastric juice aids in the absorption of these minerals.
4. It acts as an important defence mechanism for the stomach, killing many of the microorganisms that may cause infection (e.g. typhoid, salmonella, cholera, and dysentery).

Gastric acid is secreted by the parietal cells of the gastric glands, predominantly in the fundus and body of the stomach. The majority of these cells secrete HCl only after they have been stimulated following the arrival of food. The cytoplasm of unstimulated cells is filled with an elaborate branching system of tubular structures derived from the endoplasmic reticulum. These are lined by microvilli, which possess the apparatus required for hydrogen ion secretion. When food enters the stomach and stimulates the parietal cells to secrete, the tubular structures fuse to form deep invaginations of the apical membrane. These invaginations are known as **secretory canaliculi**. The formation of canaliculi results in a large (more than tenfold) increase in the surface area of the parietal cell membrane and brings large numbers of hydrogen ion pumping sites into close proximity with the luminal fluid.

The metabolic steps involved in the production of acid by a parietal cell are shown in Figure 44.13. The secretion of both hydrogen and chloride ions occurs by active transport. Chloride ions are moved from plasma to lumen against an electrochemical gradient. Although hydrogen ions move down their electrical gradient, they are transported against a massive concentration gradient (as much as a million to one). The process of acid production is therefore highly energy-dependent and parietal cells contain numerous mitochondria.

A unique membrane transport system is now known to be located on the canalicular membrane of parietal cells. The system is driven by H^+, K^+-ATPase that uses energy derived from the hydrolysis of ATP to pump hydrogen ions out of the cell in exchange for potassium ions (Figure 44.13). Chloride ions may leave the cell by two routes. There is a chloride channel in the secretory canaliculi through which chloride ions diffuse. In addition, the canaliculi have a potassium–chloride symporter that transports chloride across the membrane. The potassium ions, therefore, shuttle in and out of the cells, leaving with chloride ions and re-entering in exchange for hydrogen ions. The hydrogen ions themselves are derived from the dissociation of water within the cell. Hydroxyl ions are also produced as a result of this reaction. These combine with carbonic acid to generate bicarbonate ions, which leave the cell in exchange for chloride ions at the basolateral membrane. This process provides the chloride ions, which will leave the cell at the canalicular membrane via the chloride channel and the potassium chloride symporter. As bicarbonate is added to the plasma during the secretion of acid by the stomach, the venous blood draining the stomach is more alkaline than the arterial blood. This is sometimes called the 'alkaline tide'.

The secretion of enzymes by the gastric glands

Gastric acid secretion is accompanied by the release of a number of proteolytic enzymes from the chief cells of the gastric glands. These are collectively known as pepsin. They are secreted in the form of inactive precursors called pepsinogens, which are contained in membrane-bound zymogen granules that are secreted when the gastric glands are stimulated. In the acid environment of the stomach, pepsinogens are converted to active pepsins, which show their greatest proteolytic activity at pH values below 3. The gastric pepsins are endopeptidases, i.e. they hydrolyse peptide bonds within the protein molecule to liberate polypeptides and a few free amino acids.

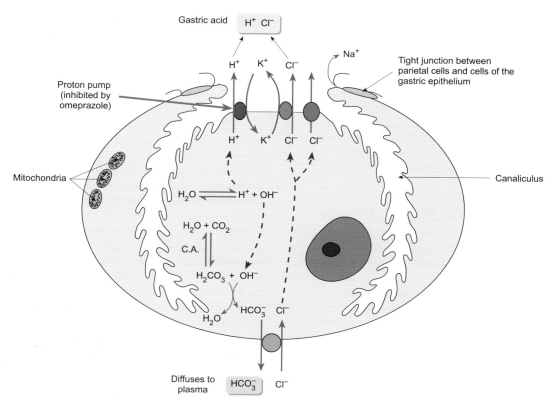

Figure 44.13 The steps involved in the secretion of gastric acid by a parietal cell. Hydrogen ions are generated from carbon dioxide via the action of carbonic anhydrase (C.A.) and the subsequent ionization of carbonic acid. The hydrogen ions are pumped into the lumen of the stomach in exchange for potassium ions by the H^+, K^+-ATPase (also called the proton pump). Chloride ions enter the cell via the basolateral membrane in exchange for bicarbonate ions and enter the stomach lumen via chloride channels or the potassium–chloride symporter. The widely used drug omeprazole reduces acid secretion by direct inhibition of the proton pump, which exchanges intracellular hydrogen ions for potassium ions.

Most fat digestion occurs in the small intestine, but the gastric glands do secrete a lipase, and lingual lipase accumulates in the stomach between meals. Both are active at low pH and this activity is responsible for about a third of all fat digestion. Unlike pancreatic lipase, whose action is discussed later in the chapter, gastric lipase and lingual lipase do not require colipase or bile for their action. However, they do not completely digest triglycerides but release one fatty acid residue from each molecule. The fatty acids are absorbed but the diglycerides require further digestion in the small intestine. In neonates, the action of the acid resistant lipases is more important than in adults and accounts for up to 60 per cent of fat digestion.

The secretion of intrinsic factor by the gastric glands

In addition to hydrochloric acid, the parietal cells of the stomach also secrete a glycoprotein, known as intrinsic factor, in response to the same stimuli that promote acid secretion. Intrinsic factor binds to vitamin B_{12} (**cobalamin**) in the upper small intestine and protects it from the enzymatic actions of the gut. The complex of B_{12} and intrinsic factor is absorbed by the mucosal epithelial cells of the lower ileum. Vitamin B_{12} is needed for the production of mature red blood cells, and its absence gives rise to **pernicious anaemia** (see Chapter 25). Lifelong treatment by intramuscular injections of vitamin B_{12} reverses this anaemia and can enable patients to survive even after total gastrectomy, when intrinsic factor is lost.

Why doesn't the stomach digest itself?

The gastric mucosa is exposed to extremely harsh chemical conditions. Gastric juice is corrosively acidic and contains protein-digesting enzymes. To protect itself, the stomach creates a **mucosal barrier**. Three factors contribute to this barrier. Firstly, the tight junctions between the cells of the mucosal epithelium prevent the gastric juice from leaking into the underlying layers of tissue. Secondly, mucus secreted by the surface epithelial cells and the neck cells of the gastric glands adheres to the gastric mucosa and forms a protective layer 5–200 μm in thickness. This mucus is alkaline because the surface epithelial cells secrete a watery fluid that is rich in bicarbonate and potassium ions. When food is eaten, the rates of secretion of both the mucus and the alkaline fluid from the surface epithelial cells are enhanced. Consequently, the surfaces of the gastric epithelial cells themselves remain bathed in their

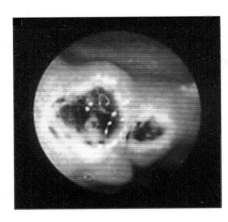

Figure 44.14 A gastric ulcer viewed through an endoscope. Note the unhealthy white areas and the dark regions, which indicate that the ulcer has been bleeding.

own protective fluid and are shielded from direct contact with the potentially damaging gastric contents. Finally, prostaglandins, particularly those of the E series, appear to play an important role in the protection of the gastric mucosa: they increase the thickness of the mucus gel layer, stimulate the production of bicarbonate, and cause vasodilatation of the microvasculature of the mucosa. This improves the supply of nutrients to any damaged areas of mucosa while the increased bicarbonate content of the fluid neutralizes the gastric acid, thus optimizing conditions for tissue repair.

The epithelial cells of the gastric mucosa are in a dynamic state of growth, migration, and desquamation (shedding), providing further protection against damage from the harsh environment. Damaged epithelial cells are shed and replaced by new cells derived from relatively undifferentiated stem cells, which migrate up from the necks of the gastric glands. Anything that breaches the mucosal barrier produces inflammation of the underlying tissue, a condition known as **gastritis**. Persistent erosion of the stomach wall can lead to the formation of gastric ulcers such as the one shown in Figure 44.14. Common predisposing factors include hypersecretion of acid and/or reduced mucus secretion.

Many drugs promote ulcer formation by altering the rates of acid and mucus production. These include caffeine, nicotine, and non-steroidal anti-inflammatory drugs such as ibuprofen and aspirin. The latter act by interfering with the production of prostaglandins. Occasionally, bile acids are regurgitated from the small intestine via the pyloric sphincter. Their detergent action may break down the mucosal barrier, rendering it susceptible to erosion by the gastric acid.

Stress may also contribute to the development of gastric ulcers in some individuals (p. 703). However, it is now known that many ulcers (up to 80 per cent) are caused by the flagellated spiral bacterium *Helicobacter pylori*. These bacteria survive the acidic environment of the stomach by using their flagella to swim quickly through to the epithelial cells of the gastric mucosa (where the pH is higher due to the alkaline secretions of the epithelial cells) and by using an enzyme (urease) to convert urea in the stomach to ammonia, which is alkaline. They then adhere to the gastric epithelium and destroy the mucosal barrier through the secretion of proteases such as mucinase, exposing large unprotected areas of mucosa. *H. pylori* may also stimulate gastrin secretion, which promotes acid secretion.

Summary

Gastric juice is secreted by the gastric glands in response to food. It contains salts, water, hydrochloric acid, pepsinogens, and intrinsic factor. The parietal cells secrete hydrochloric acid and a glycoprotein called intrinsic factor. The latter is essential for the absorption of vitamin B12 in the terminal ileum. A number of proteolytic enzymes are secreted by the chief cells of the gastric glands. They are released as inactive pepsinogens and activated to pepsin by the acidic environment in the gastric lumen. Pepsin hydrolyses peptide bonds within protein molecules to liberate polypeptides. The stomach also secretes an acid-resistant lipase which, together with lingual lipase, initiates the process of fat digestion.

The gastric mucosa creates a 'barrier' to protect itself from erosion by the gastric juice. Alkaline, mucus-rich fluid, secreted by the gastric epithelial cells, provides a protective coating for the mucosa. Prostaglandins of the E series increase the thickness of the mucus layer and stimulate the secretion of bicarbonate ions, thus contributing to the protection of the stomach lining. Factors that inhibit the production of the mucus barrier increase the vulnerability of the gastric mucosa to ulceration.

44.5 The regulation of gastric secretion

The secretion of hydrochloric acid and of pepsinogens by the gastric glands are regulated largely by the same factors. Both nervous and endocrine mechanisms are involved, and these interact at many levels. Gastric secretion is normally considered to occur in three phases, the timing of which overlaps considerably. These are the cephalic, gastric, and intestinal phases, and they are regulated by the various factors summarized in Figure 44.15.

The cephalic phase of gastric secretion

This takes place even before food enters the stomach and results from the anticipation of food, its sight, smell, and taste. The relative contribution of the cephalic phase to overall gastric secretion in response to a meal is variable, and dependent upon mood and appetite, but may amount to as much as 30 per cent. Neurogenic signals originating in the cerebral cortex, or in the appetite centres of the amygdala and hypothalamus, are relayed to the stomach via efferent fibres whose cell bodies lie within the dorsal motor nuclei of the vagus nerve.

Parasympathetic vagal activity influences gastric secretion both directly and indirectly, as indicated in Figure 44.15. Postganglionic parasympathetic fibres in the myenteric plexus release acetylcholine and stimulate the output of the gastric glands. Vagal stimulation also causes the release of gastrin from the G-cells of the

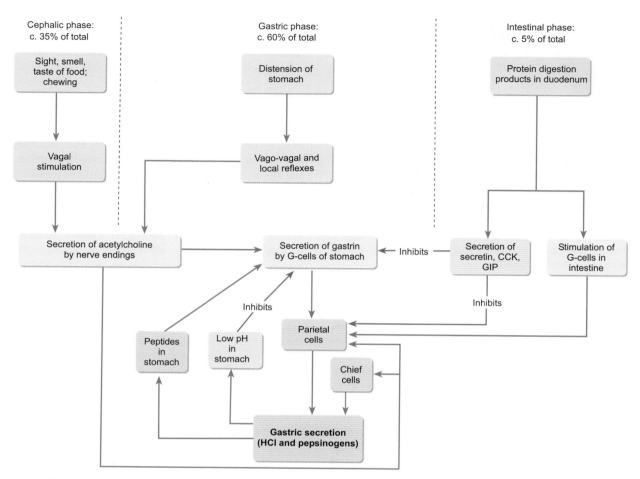

Figure 44.15 The major factors involved in the regulation of gastric secretion. Secretin, CCK, and GIP are secreted by enteroendocrine cells in the epithelium of the upper small intestine and have an inhibitory action on gastrin secretion, as does a low pH in the lumen of the stomach. The stimulatory action of gastrin on mucus and enzyme secretion is omitted for clarity.

antral glands. Gastrin reaches the gastric glands by way of the bloodstream and stimulates them to secrete acid and pepsinogens. Furthermore, both vagal activity and gastrin stimulate the release of histamine from mast cells and enterochromaffin-like (ECL) cells. Histamine binds to H_2 receptors on parietal cells to potentiate the actions of gastrin and acetylcholine on acid secretion.

The gastric phase of gastric secretion

The arrival of food in the stomach initiates the gastric phase of acid, pepsinogen, and mucus production, which accounts for more than 60 per cent of total gastric secretion. The two principal triggers are distension of the stomach wall and the chemical content of the food.

Distension of the stomach activates mechanoreceptors and initiates both local (short-loop) myenteric reflexes and long-loop vago-vagal reflexes. Both types of reflex lead to the secretion of acetylcholine, which stimulates the output of gastric juice by the secretory cells of the stomach. The importance of the vagally mediated reflexes is revealed by the 80 per cent reduction in acid production in

response to distension which is seen following vagotomy (cutting the vagus nerves). Emotional stress, fear, anxiety, or any other state that triggers a sympathetic response will inhibit gastric secretion because the parasympathetic controls over the GI tract are temporarily overridden.

In addition to its direct cholinergic action, the vagus nerve stimulates the output of gastrin from G cells in response to distension of the body of the stomach. Gastrin is a powerful stimulus for acid secretion from the parietal cells. It also enhances the release of enzymes and mucus from the gastric glands. Although intact proteins are without effect on the rate of gastric secretion, both peptides and free amino acids stimulate the output of gastric juice through a direct action on the G-cells. The amino acids tryptophan and phenylalanine are particularly potent **secretogogues**, as are bile acids and short chain fatty acids.

Gastrin secretion is inhibited when the pH of the gastric contents falls to between 2 and 3. Thus, gastrin secretion is maximal soon after the entry of food into the stomach, when the pH is relatively high, but declines as acid secretion and protein digestion get underway and the pH of the gastric contents falls. The inhibition of gastrin secretion is mediated by an increase in the secretion of somatostatin from cells (D-cells) of the gastric mucosa. In this way, gastric acid secretion is self-limiting and the gastric phase of gastric secretion normally only lasts for about an hour.

The intestinal phase of gastric secretion

A small proportion (5 per cent or so) of the total gastric secretion in response to food takes place as partially digested food (chyme) starts to enter the duodenum. This is believed to be due to the secretion of gastrin from G-cells in the duodenal mucosa, which encourages the gastric glands to continue secreting. However, this effect is short-lived. As acid chyme distends the duodenum, an enterogastric reflex is initiated whereby gastric secretory activity is suppressed. Several hormones contribute to this reflex:

* secretin
* cholecystokinin
* gastric inhibitory peptide.

Secretin is secreted by the S-cells of the duodenal mucosa in response to acid and reaches the stomach via the bloodstream. In addition to its role in pancreatic secretion (discussed later in the chapter), secretin also exerts a direct inhibitory action on the parietal cells to reduce their sensitivity to gastrin. Cholecystokinin (CCK) and gastric inhibitory peptide (GIP—also called glucose-dependent insulinotropic peptide) are released in response to the presence of products of fat digestion in the duodenum and proximal jejunum. Both inhibit the release of gastrin and gastric acid, though their relative importance is not clear.

Disorders of gastric acid secretion

Reduced gastric secretion is a relatively rare condition generally restricted to elderly patients with atrophy of the gastric mucosa. **Achlorhydria** (a decrease in hydrochloric acid secretion) may occur because of a loss of parietal cells. Although digestive processes are normally unaffected, achlorhydria impairs the absorption of substances requiring an acid environment. Patients with achlorhydria normally have a high circulating level of gastrin because the stomach contents are never acidic enough to inhibit secretion.

A number of disorders, including stress in some individuals, are associated with abnormally high rates of gastric acid secretion, and a variety of drugs and constituents of foods are known to stimulate the production of acid (e.g. caffeine and possibly alcohol). A rare condition, the Zollinger–Ellison syndrome, is caused by a gastrin-secreting tumour of the non-β cells of the pancreatic islets. Here, gastric acid secretion reaches such high values that erosion of the gastric mucosal barrier occurs, leading to ulceration of the stomach wall.

Several strategies have been developed to treat excessive acid production and promote healing of the gastric mucosa. Simple antacids, alkaline substances such as magnesium hydroxide and sodium bicarbonate, combine with HCl and neutralize it to raise the pH of the stomach above a value of 4. If these are unsuccessful, other drug treatments are available. Specific H_2 receptor antagonists such as cimetidine and ranitidine, which block parietal cell histamine receptors, may be used to inhibit acid secretion. Other agents such as benzimidazoles, which are weak bases, are known to inhibit the activity of the proton pumps on the apical surface of the parietal cells. Drugs based on agents of this kind, such as omeprazole and lansoprazole, are widely used to treat patients with ulcers caused by the hypersecretion of gastric acid. Individuals whose gastric ulcers are caused by bacterial infection with *H. pylori* are treated using a triple therapy in which omeprazole or another proton pump inhibitor is administered for one week, together with the antibiotic clarithromycin, and *either* amoxicillin *or* metronidazole. Such a regime normally eradicates the bacteria and cures 80–90 per cent of cases, but antibiotic resistance is emerging.

Summary

Nervous and endocrine mechanisms combine to regulate gastric secretion. Secretion occurs in three phases: cephalic, gastric, and intestinal. The cephalic phase of secretion occurs in response to the anticipation of food. The arrival of food in the stomach initiates the gastric phase of secretion, in which distension and the presence of amino acids and peptides stimulates the output of HCl and pepsinogen. Gastrin is an important mediator of this phase. Secretion is inhibited when the pH of the gastric contents falls to around 2 or 3. As partially digested food enters the duodenum, some gastrin is secreted from G-cells in the duodenal mucosa and this initiates the intestinal phase of gastric secretion.

44.6 The storage, mixing, and propulsion of gastric contents

In addition to its important secretory activity, the motor functions of the stomach play a significant role in the overall process of digestion. The stomach stores food until it can be processed by the lower regions of the GI tract. It also mixes the food with gastric secretions and, through its grinding action, breaks food particles down into smaller pieces to form semi-liquid chyme. The stomach contents are then delivered to the duodenum at a rate compatible with their complete digestion and absorption. Gastric motility and emptying are controlled by complex interactions between the enteric nervous system, the autonomic nervous system, and a number of hormones.

For the purposes of its motor function, the stomach may be divided into two parts: the proximal motor unit consisting of the fundus and body of the stomach, and the distal motor unit consisting of the antral and pyloric regions. The proximal motor unit carries out the reservoir functions of the stomach while the distal motor unit is responsible for the mixing of food and its propulsion into the duodenum.

Storage function of the stomach

The empty stomach has a volume of around 50 ml and an intragastric pressure of 0.6 kPa (5 mm Hg) or less. Although it can stretch to accommodate large volumes of food, little increase in intragastric pressure is seen until the stomach volume exceeds 1 litre. There are several reasons for this. The stomach exhibits receptive relaxation and its smooth muscle is able to increase its length significantly without altering its tone, a property known as **plasticity** (see Chapter 8). Moreover, as it is stretched, a vagal reflex is triggered which inhibits muscle activity in the body of the stomach and the rugae flatten as food accumulates. This reflex is coordinated by the regions of the brainstem responsible for swallowing. Finally, the shape of the stomach itself contributes to its effectiveness as a reservoir. As the diameter of the stomach increases during filling, the radius of curvature of its walls also increases. At a given pressure, the stretching force on the walls decreases in proportion to this radius of curvature (Laplace's law; see also Section 33.4). As a result, intragastric pressure rises only slightly despite significant distension.

Contractile activity of the full stomach—mixing and propulsion

Gastric motility results from the coordinated contraction of the three smooth muscle layers that lie within the stomach wall. The different orientations of the longitudinal, circular, and oblique layers allow the stomach to perform a wide variety of different movements including grinding, churning, and kneading as well as propulsion.

During fasting, the stomach shows only weak contractile activity (though in extreme hunger there may be short periods of intense contractile activity experienced as hunger 'pangs'). After a meal, peristaltic contractions begin in the body of the stomach. These are very weak, rippling movements but as the contractions approach the pylorus, where the musculature is thicker, they become much more powerful, reaching a maximum close to the gastro-duodenal junction. Thus, the contents of the fundus remain relatively undisturbed while those of the pyloric regions receive a vigorous pummelling and mixing. Although the intensity of these peristaltic contractions may be modified by many factors, their rate remains constant at around three contractions a minute. The rate of propagation of the peristaltic wave accelerates as it nears the distal regions so that the smooth muscle of the antrum and pylorus contracts virtually simultaneously, pushing the gastric contents ahead of the peristaltic wave. As pressure in the antrum rises, the pyloric sphincter opens and a few millilitres of chyme are squirted through it into the first part of the duodenum, the duodenal bulb. The sphincter closes almost immediately thus preventing further emptying, and as a result of the continued high pressure in the antrum, some of the gastric contents are forced back into more proximal regions. This is called retropulsion. It increases the effectiveness of mixing and breakdown of food within the stomach as food particles rub against each other.

Gastric motility is enhanced by many of the same nervous and hormonal factors that stimulate gastric secretion. Although intragastric pressure rises only a little, distension of the stomach by food activates stretch receptors and, as a result, both the force of peristaltic contractions and the overall level of smooth muscle tone are enhanced. As already mentioned, gastrin-secreting cells are stimulated by the presence of food in the stomach, and this hormone increases gastric motility as well as acid secretion. Consequently, there is an increase in movements that promote both mixing and emptying. The nervous control of gastric motility is not completely understood. Both parasympathetic (vagal) and sympathetic fibres supply the smooth muscle and, in general, it appears that parasympathetic activity increases motility while sympathetic activity decreases it.

The rate of gastric emptying is carefully controlled

If digestion and absorption are to proceed with optimum efficiency, it is essential that chyme is delivered to the duodenum at a rate which enables the small intestine to process it fully. Furthermore, it is important that the duodenal contents are prevented from being regurgitated into the stomach, as the gastric and duodenal secretions are very different. The gastric mucosa is resistant to acid but may be eroded by bile, while the duodenum is resistant to the effects of bile but is unable to tolerate low pH. Consequently, gastric emptying that is too rapid may result in the formation of duodenal ulcers, while regurgitation of duodenal contents may result in gastric ulceration.

Many factors contribute to the regulation of gastric emptying

Emptying of the stomach depends upon the factors that influence motor activity throughout the GI tract—the inherent excitability of smooth muscle, intrinsic and extrinsic nervous pathways, and hormones. In general, the stomach empties at a rate that is proportional to gastric volume, i.e. the fuller the stomach the more rapidly it empties. In addition, the physical and chemical nature of the gastric contents affects the rate of emptying. Fats and proteins in the ingested food, a very acidic juice, and a hypertonic mixture of juice and food will all delay emptying. In general, the closer the contents are to isotonic saline, the more rapidly they will leave the stomach. The half-time for liquids remaining in the stomach is about 20 minutes, as compared with about 2 hours for solids.

Receptors of various kinds are present within the duodenum and contribute to the regulation of gastric emptying via the so-called **enterogastric reflex**, a collective term used to describe all the hormonal and neural mechanisms that mediate the intestinal control of gastric emptying.

The presence of fatty acids or monoglycerides in the duodenum causes an increase in the contractility of the pyloric sphincter. This reduces the rate at which the gastric contents are propelled through the sphincter into the small intestine and ensures that fats are not delivered to the duodenum more quickly than the bile salts can emulsify them (see Section 44.9). How fats exert this effect is unclear, but both CCK and GIP are released from the small intestine in response to fats and their digestion products, and both of these hormones have been shown to delay gastric emptying.

The products of protein digestion are also believed to exert their inhibitory effect on gastric emptying through endocrine pathways. Gastrin is secreted from G-cells in both the antrum and duodenum in response to peptides and amino acids. The action of gastrin on motility is two-fold. Although it stimulates the motor activity of the antrum, it also increases the degree of constriction of the pyloric sphincter so that the net effect of gastrin secretion is normally to delay emptying of the stomach.

The delay in gastric emptying seen when acid enters the duodenum is probably mediated by both nervous and hormonal factors. Cutting the vagus (vagotomy) reduces the response. This suggests a role for the vagus nerve in the regulation of gastric emptying by acid. Secretin also appears to be involved. This hormone (which also stimulates the secretion of alkaline pancreatic juice: see Section 44.9) is released in response to acid in the duodenum and delays gastric emptying by inhibiting contraction of the antrum.

The importance of controlling the rate of gastric emptying is highlighted by the extremely unpleasant condition known as dumping syndrome in which gastric contents move very rapidly into the duodenum during and just after eating. The syndrome is characterized by abdominal bloating and pain sometimes accompanied by nausea, vomiting, diarrhoea, weakness, and dizziness. It is seen most often in patients who have undergone partial gastrectomy or bariatric surgery in whom the normal neuronal and endocrine control mechanisms have been disrupted. Careful regulation of diet and meal size is necessary to alleviate the symptoms.

The electrical activity underlying gastric contractions

The regular peristaltic contractions shown by the stomach are the mechanical consequence of the basic electrical rhythm (BER) of the smooth muscle cells. This basic rhythm is set by the spontaneous activity of pacemaker cells in the longitudinal smooth muscle layer of the stomach wall in the region of the greater curvature near to the middle of the body of the stomach (the pacemaker zone). The cells here show spontaneous depolarization and repolarization every 20 seconds or so to establish the BER or 'slow wave' rhythm of the stomach, as shown in Figure 44.16a. The pacemaker cells are electrically coupled to the rest of the stomach muscle sheet by means of gap junctions and their rhythm is therefore transmitted to the entire muscularis.

The change in membrane potential is triphasic and similar to a cardiac muscle action potential in appearance, although it is about ten times as long. The inward current responsible for initial depolarization is carried by calcium ions moving through voltage-gated channels, while the plateau is maintained by the inward movement of both sodium and calcium ions. Repolarization is associated with a delayed outward potassium current.

In the body and fundus of the stomach, action potentials are not normally associated with the gastric pacemaker potentials. Nevertheless, contraction of the smooth muscle occurs when the depolarization phase of the potential reaches the mechanical threshold. The relationship between the gastric slow wave and the contractile force generated by the smooth muscle following stimulation with acetylcholine is shown in Figure 44.16b. Tension development is slow and progressive. The force of contraction is related to the degree of depolarization and the time for which the potential exceeds threshold. In the distal antrum and pyloric regions of the

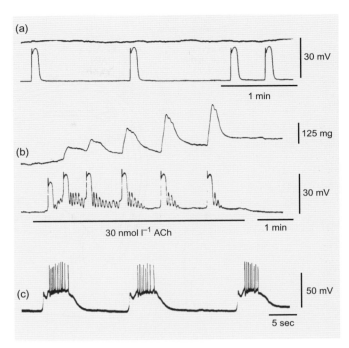

Figure 44.16 The relationship between the electrical activity and contraction of the smooth muscle of the stomach. In (a) and (b) the upper traces show the force generated by the strip of muscle and the lower traces show the basic electrical rhythm in the body of the stomach. Following stimulation with 30 nmol l⁻¹ acetylcholine (ACh) there is an increase in the frequency of the slow plateau potentials with the progressive development of tension. In the antrum of the stomach, fast action potentials are superimposed on the slow waves (c). (Adapted from Figs 1 and 8 in J.H. Szurszewiski (1975) *J Physiol* **252**, 335–61 (a and b) and from Fig 9 in T.Y. El-Sharkawy, K.G. Morgan and J.H. Szijrszewski (1978) *J Physiol* **279**, 291–307, (c). (a and b) J.H. Szurszewski (1975) Mechanism of action of pentagastrin and acetylcholine on the longitudinal muscle of the canine antrum. *J Physiol* **252** doi: 10.1113/jphysiol.1975. sp011147. (c) T.Y. el-Sharkawy K.G. Morgan and J.H. Szurszewski (1978) Intracellular electrical activity of canine and human gastric smooth muscle. *J Physiol* **279** doi: 10.1113/jphysiol.1978.sp012345.)

stomach, muscular activity is more vigorous and trains of fast action potentials are superimposed on the BER. These are associated with vigorous propulsive movements, which may result in gastric emptying (see Figure 44.16c).

Vomiting

Vomiting (or **emesis**) is the sudden and forceful oral expulsion of the contents of the stomach and sometimes the duodenum. It is frequently preceded by **anorexia** (loss of appetite) and **nausea** (a feeling of sickness). Immediately before vomiting, it is common to experience characteristic autonomic responses such as copious watery salivation (waterbrash), vasoconstriction with pallor, sweating, dizziness, and tachycardia. Retching often precedes vomiting and, during the process of vomiting itself, respiration is inhibited. The larynx is closed and the soft palate rises to close off the nasopharynx and prevent the inhalation of vomited material

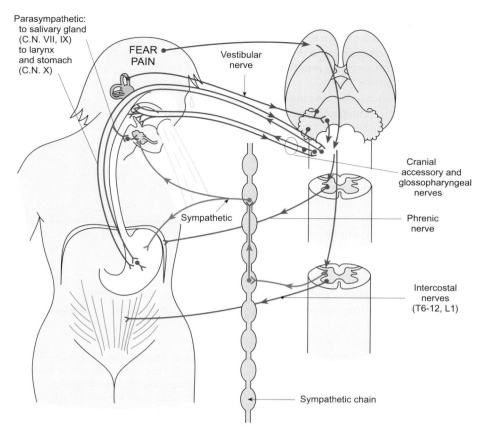

Figure 44.17 The neural pathways of the vomiting reflex. Vomiting is triggered by many different stimuli including gastric disorders, disagreeable odours, and motion sickness. These stimuli converge on the medulla of the brainstem resulting in a sensation of nausea. The motor responses are complex—there is activation of both parasympathetic and sympathetic nerves resulting in increased salivary secretion ('waterbrash'), pupillary dilatation, increased heart rate, sweating, and vasoconstriction of the skin vessels. Vomiting is usually preceded by retching, during which the abdominal muscles undergo strong contractions followed by a deep inspiration which lowers the diaphragm. The glottis is closed and powerful contractions of the muscles of the abdominal wall raise intra-abdominal pressure, causing the ejection of the gastric contents. (Adapted from Fig 8.4 in P.C.B. McKinnon and J.F. Morris (2005) *Oxford textbook of functional anatomy*, Vol. 2. Oxford University Press, Oxford.)

(**vomitus**). The stomach and pyloric sphincter relax, and contraction of the duodenum reverses the normal pressure gradient so that intestinal contents are allowed to enter the stomach (a process sometimes referred to as reverse peristalsis). The diaphragm and abdominal wall then contract powerfully, the gastro-oesophageal sphincter relaxes, and the pylorus closes. The resulting rise in intragastric pressure causes the expulsion of the gastric contents.

The vomiting reflex is coordinated by the dorsal portion of the reticular formation of the medulla, which lies close to the respiratory and cardiovascular control areas of the brainstem. The nervous pathways involved are shown diagrammatically in Figure 44.17. Afferent impulses arrive at the brainstem from many parts of the body including the pharynx and other areas of the GI tract; viscera such as the liver, gall bladder, urinary bladder, uterus, and kidneys; the cerebral cortex; and the semicircular canals of the inner ear (which are powerfully activated in motion sickness). Furthermore, a variety of chemical agents including general anaesthetics, opiates, and toxins produced by infectious agents can trigger vomiting by stimulating neurons in the area postrema, which is located on the floor of the fourth ventricle.

(This area is also known as the chemoreceptor trigger zone or CTZ.) A variety of antiemetic drugs (including cyclizine, ondansetron, and metoclopramide) act by blocking neurotransmitter receptors within the CTZ. The motor impulses responsible for the action of vomiting are transmitted from the medulla via the trigeminal, facial, glossopharyngeal, vagus, and hypoglossal nerves (cranial nerves V, VII, IX, X, and XII).

Although vomiting is, generally, a protective mechanism whereby potentially toxic substances are removed from the body, prolonged vomiting can lead to a state of metabolic alkalosis due to the continued loss of acid from the stomach (see Chapter 41). It may also result in hypokalaemia because gastric juice contains a higher concentration of potassium than plasma.

Absorption by the stomach

Very little absorption occurs in the stomach. Ethyl alcohol is the only water-soluble substance normally absorbed in significant amounts, and even this can only be absorbed because its lipid-solubility enables it to diffuse readily through the plasma membranes of the gastric mucosal cells. Certain organic substances, which are not ionized at the acidic pH of the stomach, may be absorbed here. An example is aspirin, which has a pK_a of 3.5 so that it remains largely un-ionized in the stomach. Molecules of aspirin diffuse through the mucosal barrier into the intracellular compartment in which the pH is closer to neutral. The aspirin molecules then become ionized and are therefore unable to diffuse back into the gastric lumen. Instead, they pass out of the cells and are absorbed into the circulation.

Summary

The stomach stores food, mixes it with gastric juice, and breaks it into smaller pieces to form a semi-liquid chyme. The stomach then delivers chyme to the duodenum in a controlled fashion.

The stomach is able to store large amounts of food since intra-gastric pressure rises very little despite significant distension of the stomach wall. The fasting stomach shows only weak contractile activity. After a meal peristaltic contractions begin, increasing in power as they approach the antrum, where mixing is most vigorous.

The stomach normally empties at a rate compatible with full digestion and absorption by the small intestine. Distension of the stomach increases the rate of emptying, while the presence of fats, proteins, high acidity, and hypertonicity in the chyme all delay the rate of emptying.

Vomiting is a protective mechanism whereby noxious or potentially toxic substances are expelled from the GI tract. The vomiting reflex is coordinated in the medulla of the brain.

Prolonged vomiting can cause metabolic alkalosis through the loss of gastric acid.

44.7 The small intestine

The small intestine is the major site of both digestion and absorption in the GI tract. In life, it is a tube about 4 metres long with a diameter of about 2.5 cm and is divided into three segments: the duodenum, jejunum, and ileum. The **duodenum** is about 25 cm long ('duo-denum' literally means 12 finger-widths) and has no mesentery. It takes the form of an arc within which lies the head of the pancreas, as shown in Figure 44.18. Chyme, produced by the chemical and mechanical actions of the stomach, is emptied into the duodenum where it is mixed with secretions from the liver (bile) and exocrine pancreas (pancreatic juice) as well as from the duodenum itself. The bile and pancreatic ducts unite close to the duodenum at the hepato-pancreatic ampulla, which opens into the duodenum at the major duodenal papilla. A ring of smooth muscle, the **sphincter of Oddi**, controls the entry of bile and pancreatic juice into the small intestine. The junction between the duodenum and the jejunum is formed by a sharp bend, the duodeno-jejunal flexure (Figure 44.18).

The **jejunum** is about 1.5 metres long and extends from the duodenum to the **ileum**, which is a coiled tube about 2.5 metres in length. There is no clear anatomical distinction between the jejunum and the ileum, but throughout the length of the small intestine, there is a gradual reduction in the thickness of the mucosal wall and subtle differences in its histological characteristics (see Figure 44.19). The jejunum and ileum are supported by a mesentery that contains branches of the superior mesenteric artery and the venous and lymphatic drainage vessels of the small intestine. The arterial blood supply to the small intestine and upper colon is illustrated in Figure 44.20.

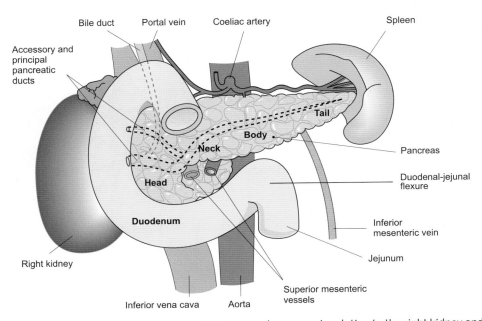

Figure 44.18 The anatomical arrangement of the duodenum, jejunum, and pancreas in relation to the right kidney and spleen. The left kidney, which lies beneath the tail of the pancreas, is not shown. (Adapted from Fig 6.4.14 in P.C.B. McKinnon and J.F. Morris (2005) *Oxford textbook of functional anatomy*, Vol. 2. Oxford University Press, Oxford.)

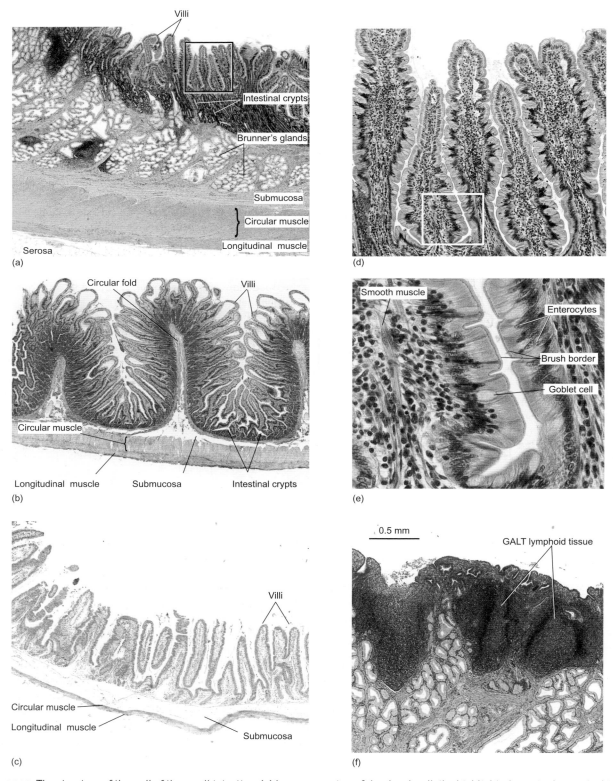

Figure 44.19 The structure of the wall of the small intestine. (a) Low-power view of duodenal wall; the highlighted area is shown at a higher magnification in (d). Note the prominent secretory tissue in the submucosa (Brunner's glands). (b) Low-power view of the wall of the jejunum. The circular folds are a prominent feature of the jejunal wall. Note the distended lacteals of the villi, which are especially prominent at the top of the circular folds. (c) Low-power view of the wall of the ileum. (d) A higher-power view of the highlighted area in panel (a) to show details of the duodenal villi. (e) The highlighted area of panel (d) at a high magnification to show the brush border and goblet cells as they appear in the light microscope. Strips of smooth muscle can be seen running longitudinally in the centre of the villus on the left. (f) A low-power view of an area of duodenal wall occupied by lymphoid tissue. Note the absence of villi. GALT = gut associated lymphoid tissue.

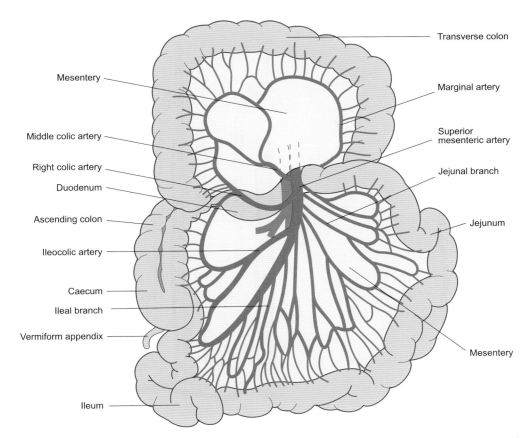

Figure 44.20 The arterial blood supply to the small intestine and upper colon. The superior mesenteric artery gives rise to a number of smaller arteries which branch to supply different segments of the small intestine and colon. The venous drainage follows a similar pattern. (Redrawn from Fig 13.70 in G.J. Romanes (ed.), (1981) *Cunningham's textbook of anatomy*, 12th edition. Oxford University Press, Oxford.)

Special histological characteristics of the small intestine

The small intestine is exquisitely adapted for nutrient absorption, particularly in the proximal portion. Its key structural characteristics are illustrated in Figure 44.21. The small intestine presents a huge surface area (estimated at around 200 square metres) both because of its length and because of the structural modifications of its wall. The mucosa and submucosa are thrown into deep folds called **circular folds**. These are a particular feature of the jejunum (Figure 44.19b). Because of their shape, these folds force the chyme into a spiral motion as it passes through the lumen. This spiralling slows down the rate of passage of chyme and facilitates mixing with intestinal juices, thereby optimizing conditions for digestion and absorption.

The folded surface of the small intestine is covered with finger-like projections or **villi** (see Figure 44.19 and panel (b) of Figure 44.21) which are between 0.5 and 1.5 mm long, depending upon their location. The surface of each villus is formed mainly by columnar absorptive epithelial cells (enterocytes) bound by tight junctions at their apical surfaces. The mucosal surface of these cells consists of many tiny processes or **microvilli** (roughly 1.0 μm long and 0.1μm in diameter) which constitute the **brush border**. This adaptation

increases the surface area of the small intestine still further. Other cells present in the intestinal epithelium are endocrine or paracrine in character. Although the functions of some are unknown, others have been shown to produce somatostatin (D-cells), secretin (S-cells), neurotensin (N-cells), CCK (I-cells), and 5-HT (enterochromaffin cells). CCK and secretin are secreted by cells in the wall of the upper small intestine in response to the presence of fat digestion products and low pH respectively.

The villi themselves differ in appearance throughout the small intestine. They are broad in the duodenum, slender and leaf-like in the jejunum, and shorter, more finger-like in the ileum. These differences are seen in Figure 44.19(a)–(c). Within each villus is a modified lymph vessel (lacteal) opening into the local lymphatic circulation, blood vessels, some smooth muscle (which enables the villus to alter its length), and connective tissue. The arterioles supplying the villi branch extensively to form a capillary network that collects into veins at the base.

Between the villi are simple tubular glands, 0.3–0.5 mm deep, called **crypts of Lieberkuhn** which are found throughout the small intestine but are most numerous in the mucosa of the duodenum and jejunum. A number of different cell types have been identified within the crypts, including Paneth cells, which secrete lysozyme, and undifferentiated stem cells that proliferate to replace lost

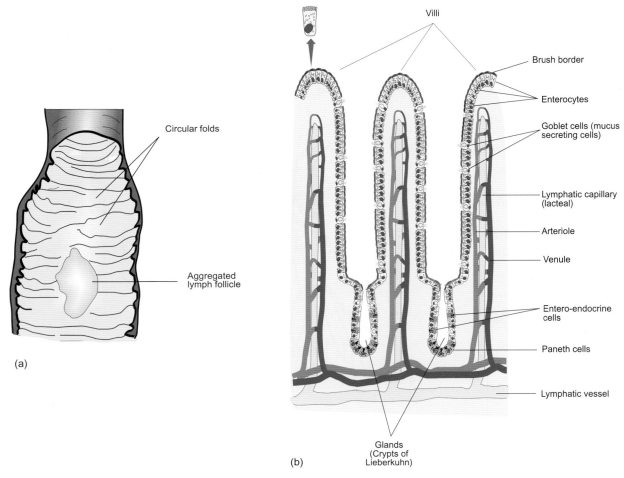

Figure 44.21 The structural characteristics of the small intestine. (a) A section of ileum to reveal circular folds (which are prominent in the jejunum) and the position of a lymph nodule (Peyer's patch). (b) A longitudinal section of intestinal villi: an enlarged view of one of the cells lining the small intestine is shown on the left. Note the microvilli on the apical surface which form the brush border. Mucus-secreting goblet cells are shown in yellow. Entero-endocrine cells are located in the intestinal glands (crypts of Leiberkuhn). These secrete a variety of hormones that regulate both the motility and secretory activity of the small intestine.

enterocytes. In addition to the crypts of Lieberkuhn, the duodenum (but not the jejunum or ileum) also contains submucosal **Brunner's glands**, which secrete a mucus-rich alkaline fluid. **Peyer's patches** are large isolated clusters of lymph nodules present in the wall of the ileum. They are similar to the tonsils and form part of the collection of small lymphoid tissues of the GI; they are referred to as the gut-associated lymphatic tissue (GALT—see Chapter 31). An example is shown in Figure 44.19(f).

The epithelium of the small intestine is self-renewing

The small intestine has a very rapid rate of cell turnover. In humans, the entire epithelium is renewed every six days or so. This rapid turnover is important because the epithelial cells are sensitive to hypoxia and to other irritants. Epithelial cells are formed by the mitotic proliferation of a population of undifferentiated stem cells within the crypts. These new cells then migrate upwards

towards the tip of the villus, from where they are shed into the lumen of the intestine. As they migrate and leave the crypts, they become fully mature and the brush border enzymes and transporter proteins develop. The rate at which cells proliferate can be altered by a number of factors. Irradiation, starvation, and prolonged intravenous feeding, for example, cause atrophy of the cells and a reduction in proliferation. Certain medications such as methotrexate and other drugs used in the treatment of cancer may also slow the rate of production of enterocytes.

Secretion of fluid and enzymes by the small intestine

Cells of the crypts are responsible for the secretion of about 1.5 litres of isotonic fluid each day. Secretion occurs because of transcellular chloride movement from the interstitial fluid to the lumen followed by paracellular movement of sodium and water. The principal stimulant of fluid secretion is distension of the intestine by acidic or hypertonic chyme. In the duodenum, the Brunner's glands secrete a bicarbonate-rich alkaline fluid containing mucus which, together with the secretions of the crypts, protects the duodenal mucosa from mechanical damage and erosion by the acid and pepsin contained within chyme arriving from the stomach. Although the glands secrete spontaneously when acid chyme enters the duodenum, their secretion may be further stimulated by vagal activity, endogenous prostaglandins, and the hormones gastrin, secretin, and CCK. Sympathetic stimulation, however, causes a marked decrease in the rate of mucus production, leaving the duodenum more susceptible to erosion. Indeed, three-quarters of peptic ulcers occur in this region of the gut, and many are related to stress, which is characterized by a generalized increase in sympathetic activity.

The fluid secreted by the small intestine is known as intestinal juice or **succus entericus** and it was once believed to contain most of the enzymes required for the complete digestion of food. It has now been established, however, that the only enzymes derived from the small intestine itself (rather than the pancreas) are the **brush border enzymes** which include disaccharidases (maltase, sucrase, etc.), peptidases, and phosphatases. One of the duodenal brush border peptidases, enteropeptidase (commonly, though wrongly, called enterokinase), converts pancreatic trypsinogen to trypsin, which is then able to activate other pancreatic enzymes.

Children under 4 years of age also express the enzyme lactase, which promotes the digestion of lactose (milk sugar). The enzyme is less active in older individuals. Lactose intolerance is caused by a lack of this enzyme and is more common in people with Asian, African, Native American, or Mediterranean ancestry than in those who have a northern European ancestry. The major hormones, electrolytes, and proteins secreted by the small intestine are listed in Table 44.1.

Table 44.1 The secretions of the small intestine

a) Hormones	
Hormone	**Cells of origin**
cholecystokinin (CCK)	I-cells of villus
motilin	M-cells of the crypts
neurotensin	N-cells of villus
secretin	S-cells of villus
serotonin (5-HT)	enterochromaffin cells
somatostatin	D-cells of villus

b) Other secretory products	
Secretory product	**Source**
lysozyme	Paneth cells of crypts
mucus	goblet cells of villus
isotonic fluid (1.5 L day^{-1})	crypts
alkaline mucus	Brunner's glands (duodenum only)

Summary

The small intestine provides a huge surface area for nutrient absorption. The folded mucosal surface is covered with projections called villi. The brush border membranes of the mucosal epithelial cells house enzymes. Between the villi lie simple tubular glands, the crypts of Lieberkuhn. These contain many different cell types including mucus-secreting goblet cells and immune cells. The epithelium of the small intestine renews itself completely every 6 days or so. Loss of cells at the tips of the villi releases enzymes from the brush border of the enterocytes into the intestinal lumen. One of these, enteropeptidase, activates pancreatic trypsin which then activates other proteolytic enzymes.

Crypt cells secrete around 1.5 litres of isotonic fluid each day. Chloride is transported out of the cells, and sodium and water follow passively by the paracellular route. In the duodenum, Brunner's glands secrete an alkaline mucus, which helps to protect the epithelium from the corrosive effects of the acidic chyme arriving from the stomach. Secretion is stimulated by the vagus nerves and by CCK, secretin, gastrin, and endogenous prostaglandins.

44.8 Motility of the small intestine

Typically, chyme traverses the length of the small intestine in 3–5 hours (although under certain conditions it may take as long as 10 hours). The rate of movement is normally such that the last part of one meal is leaving the ileum as the next meal enters the stomach. The most important types of movement in the small intestine are segmentation and peristalsis, which are described in Section 43.2. Segmentation is of great importance in mixing the chyme with the digestive enzymes present in the small intestine and in facilitating the absorption of the products of digestion. The villi and microvilli of the intestinal mucosa also exhibit mixing movements. Waves of peristalsis are initiated by distension of the small intestine and rarely travel more than about 10 cm. These are short-range peristaltic contractions which should be distinguished from the so-called 'housekeeper contractions' described below.

Segmentation and peristalsis are inherent properties of the intestinal smooth muscle

The basic electrical rhythm of the small intestine is independent of extrinsic innervation, and both segmentation and peristaltic contractions of the intestinal smooth muscle are inherent properties of the intramural plexuses of the enteric nervous system. However, the excitability of the smooth muscle and the strength of its contraction can be modified by extrinsic nerves as well as by the variety of hormones utilized as neurotransmitters by the intramural nerve plexuses. Parasympathetic stimulation increases the excitability of the smooth muscle while sympathetic stimulation depresses it. These autonomic effects are exerted principally via the enteric nerve plexuses (see Section 43.3).

Extrinsic nerves play a role in certain long-range intestinal reflexes. These include the so-called ileo-gastric and gastro-ileal reflexes, which describe the reflex interactions that operate between the stomach and the terminal ileum. The **ileo-gastric reflex** refers to the reduction in gastric motility that occurs in response to distension of the ileum. The **gastro-ileal reflex** describes the increase in motility of the terminal ileum (particularly of segmentation) that occurs whenever there is an increase in secretory and/or motor activity of the stomach. Both of these will have the overall effect of matching emptying of the small intestine with the arrival of chyme in the duodenum.

Movements of the mucosal villi contribute to absorption and mixing

The intestinal villi show piston-like contraction and relaxation movements, which are thought to facilitate the removal of the digestion products of fats from the lacteals (the lymphatic vessels which course through the villi). One possible sequence of events is that, when the villus is relaxed, absorption takes place via intercellular channels. As the villus contracts, these intercellular channels are cut off and the absorbed material is forced into the lymphatic collecting vessels. This is sometimes referred to as 'milking' the lacteals. Strands of smooth muscle within the lamina propria are thought to give rise to these pumping movements.

The villi also show pendular (swaying) movements that may contribute to the mixing of chyme within the intestinal lumen. These movements are enhanced by the presence of amino acids and fatty acids in the lumen.

Patterns of motility in the small intestine during fasting

The patterns of contractility described above relate to the behaviour of the small intestine following a meal. During periods of fasting, or once a meal has been processed, the smooth muscle of the small intestine shows a different characteristic pattern in which segmentation movements wane and waves of peristalsis, initiated at the duodenal end, sweep slowly along the length of the small intestine. Individual waves travel up to 70 cm before dying out, and the entire wave of contraction takes 1–2 hours to travel the length of the small intestine. The electrical activity that underlies this contractile behaviour is known as the **migrating motility complex** (MMC) and is repeated every 70–90 minutes. The purpose of these waves of peristalsis appears to be to sweep out the last remains of the digested meal together with bacteria and other debris into the large intestine. For this reason, the contractions are sometimes called 'housekeeper contractions'. The mechanisms that initiate and control the MMC are not understood. Both vagal and hormonal mechanisms (particularly another gut hormone, **motilin**) have been implicated. Motilin is secreted by the entero-endocrine M-cells of the duodenum and jejunum.

The ileocaecal sphincter controls emptying of the small intestinal

The first part of the large intestine is called the **caecum**, and the junction between the terminal ileum and the caecum is the **ileocaecal sphincter** (or valve). This normally regulates the rate of entry of chyme into the large intestine to ensure that water and electrolytes are fully absorbed in the colon. Its activity is governed by the neurons of the intramural plexuses. The sphincter is normally closed, but short-range peristaltic contractions of the terminal ileum cause the sphincter to relax and allow a small amount of chyme to pass through. The long-range reflexes ensure that the rate of emptying is matched to the ability of the colon to deal with the volume of chyme delivered. After a meal, for example, ileal emptying is enhanced through the operation of the gastro-ileal reflex.

Summary

The rate at which chyme moves through the small intestine is carefully controlled to ensure adequate time for the completion of digestion and absorption. Segmentation is responsible for the mixing of chyme with enzymes and for exposing it to the absorptive mucosal surface, while peristaltic contractions propel chyme along towards the ileocaecal sphincter. The motility of the intestinal smooth muscle is influenced by the activity of both intrinsic and extrinsic neurons. Parasympathetic activity enhances intestinal motility.

The intestinal villi exhibit both piston-like contractions and swaying pendular movements. The latter may contribute to the mixing of chyme, while the former serve to facilitate the removal of fatty digestion products from the lacteals of the villi. In the fasting intestine, periodic bursts of peristaltic activity are seen in which the contents of the gut are swept long distances along the tract. These are called 'housekeeper' contractions.

44.9 The exocrine functions of the pancreas

The pancreas performs two distinct functions in the body. It acts both as an endocrine gland, secreting the hormones insulin and glucagon into the bloodstream, and as an accessory digestive (exocrine) organ secreting enzyme-rich fluid into the duodenum. The endocrine role of the pancreas is discussed fully in Chapter 24. Only its exocrine function will be described here.

The anatomy and histology of the pancreas

The pancreas lies deep to the stomach and extends across the abdomen for about 20 cm. The tail of the pancreas lies close to the spleen, while its head is encircled by the duodenum. Its anatomical situation is illustrated in Figures 43.1 and 44.18.

A schematic representation of the structure of the exocrine pancreas is shown in Figure 44.22. Its organization is similar to that of the salivary glands, being made up of lobules of acinar cells that secrete enzymes and fluid into a system of microscopic ducts (intercalated ducts) which are lined with epithelial cells that secrete a bicarbonate-rich fluid. These drain into larger intralobular ducts which in turn empty into interlobular ducts. Finally, the interlobular ducts empty into the main pancreatic duct which extends along

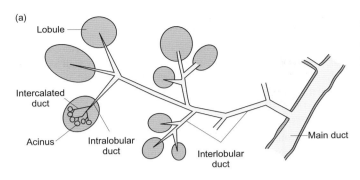

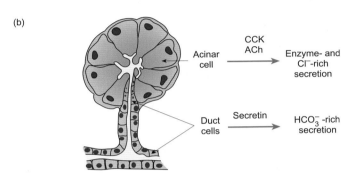

Figure 44.22 (a) The structure of the exocrine pancreas. (b) The sites of action and effects of secretin, cholecystokinin (CCK), and acetylcholine (ACh) on the acinar and duct cells of the exocrine pancreas. Secretin promotes the copious secretion of bicarbonate-rich serous fluid from the duct cells, while acetylcholine and CCK promote the secretion of enzymes by the acinar cells.

the axis of the pancreas. In most people, the main pancreatic duct fuses with the bile duct before emptying into the duodenum. There is also a smaller accessory pancreatic duct (duct of Santorini) that drains directly into the duodenum. Acinar cells occupy more than 80 per cent of the total pancreatic volume and duct cells about 4 per cent, while the endocrine islet cells only occupy about 2–3 per cent. The remainder of the pancreas consists of connective tissue, blood vessels, lymphatics, and nerves.

The pancreas is supplied by branches of the coeliac and superior mesenteric arteries and its venous drainage is via the portal vein. It is innervated by preganglionic parasympathetic vagal fibres, which synapse with cholinergic postganglionic fibres within the pancreas. Pancreatic blood vessels receive sympathetic innervation from the coeliac and superior mesenteric plexuses.

The composition of pancreatic juice

The exocrine pancreas secretes about 1500 ml of fluid each day, the aqueous component of which is rich in bicarbonate and has a pH of about 8. Together with the intestinal secretions, it helps to neutralize the acidic chyme as it arrives in the duodenum. The major enzymes needed to digest fats, proteins, and carbohydrates are also contained within the pancreatic juice (the enzyme component).

The aqueous component of pancreatic juice

This is formed almost entirely by the columnar epithelial cells that line the ducts. Resting secretion is chiefly from the intercalated and intralobular ducts but, during stimulation, the interlobular ducts also secrete pancreatic fluid. The fluid secreted by the duct cells is slightly hypertonic. It is rich in bicarbonate ions (80–140 mmol l^{-1}) and has sodium and potassium concentrations similar to those of plasma.

Precise details of the ionic mechanisms underlying the secretion of the alkaline fluid are still being clarified. A possible sequence of events is illustrated in Figure 44.23. Bicarbonate ions are formed from carbon dioxide and exchanged for chloride ions in the lumen of the duct. The hydrogen ions generated during the formation of bicarbonate are transported out of the cell into the interstitial fluid in exchange for sodium ions. Chloride ions cross the apical membrane via chloride channels and thus can shuttle between the duct lumen and the duct cells. Sodium ions diffuse from the interstitial fluid to the duct lumen via a paracellular pathway to maintain electroneutrality. Water follows osmotically, moving transcellularly or paracellularly into the duct lumen. Part of this fluid secretion appears to be regulated by cyclic AMP, which increases the probability of opening of a particular type of Cl^- channel known as the cystic fibrosis transmembrane conductance regulator or CFTR. It

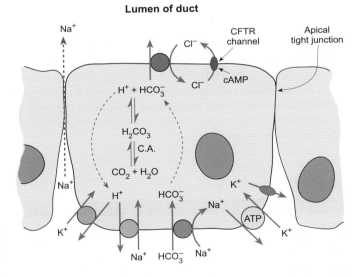

Figure 44.23 The ionic movements underlying the secretion of an alkaline fluid by duct cells of the exocrine pancreas. Bicarbonate is taken up from the extracellular fluid by a sodium-dependent symporter and is formed within the cells from carbon dioxide and water under the influence of carbonic anhydrase (C.A.). It is secreted into the duct in exchange for chloride ions, which shuttle across the apical membrane via two chloride channels, the CFTR channel which is regulated by cAMP and a calcium-dependent chloride channel (not shown). Excess hydrogen ions are pumped out of the cells via the K^+, H^+-ATPase and the Na^+, H^+ antiport NHE1.

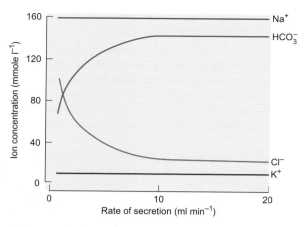

Figure 44.24 The electrolyte composition of rabbit pancreatic juice as a function of the rate of secretion. The greater the rate of secretion, the higher the bicarbonate concentration. (Based on data in F. Bro-Rasmussen et al. (1956) *Acta Physiol Scand* **37**, 97.)

may also stimulate the activity of the proton pumps in the basolateral membrane. Certain mutations of the CFTR channel result in a reduced flow of pancreatic juice as the pancreatic ducts become obstructed by a thick, sticky mucus that is characteristic of cystic fibrosis (see also Section 36.3).

The ionic composition of the pancreatic fluid depends upon its rate of secretion, as shown in Figure 44.24. As it flows along the ducts, the primary secretion of the ductal epithelial cells undergoes modification. Bicarbonate ions are reabsorbed from the fluid in exchange for chloride ions. The result of this is that at low flow rates, the bicarbonate content of the pancreatic fluid is significantly lower than it is at high flow rates, when the fluid spends little time in the ducts and is therefore scarcely modified. At maximal flow rates, the bicarbonate concentration of human pancreatic juice is around 140 mmol l^{-1} while during basal secretion it is generally 80–100 mmol l^{-1}. Figure 44.24 shows that the chloride levels rise as the bicarbonate concentration falls.

The enzyme components of pancreatic juice

Pancreatic juice contains a wide array of digestive enzymes including proteolytic, amylolytic, and lipolytic agents as well as many others such as ribonuclease, deoxyribonuclease, and elastases. A list of the principal enzymes and their actions is given in Table 44.2.

Proteolytic enzymes of the pancreas

Proteolytic enzymes including **trypsin**, a number of **chymotrypsins**, and **carboxypeptidases** are stored within the acinar cells as zymogen granules. They are secreted in this inactive form (as trypsinogen, chymotrypsinogens, and procarboxypeptidases) and then activated inside the lumen of the small intestine. In this way, the pancreas, like the stomach, avoids self-digestion. Activation of trypsinogen may occur spontaneously, in response to the alkaline environment of the small intestine, or in response to enteropeptidase, one of the brush border enzymes.

Chymotrypsinogens are activated by trypsin. Trypsin and chymotrypsins are endopolypeptidases, which hydrolyse peptide bonds within protein molecules to release some free amino acids and polypeptides of varying size. Carboxypeptidases, elastase, and aminopeptidases are then able to digest these further to release small peptides and amino acids. Figure 44.25 illustrates the actions of the major proteolytic enzymes, while Table 44.3 summarizes the sites at which the various proteolytic enzymes act to hydrolyse peptide chains.

It is important that trypsinogen is not activated within the acinar cells themselves or as it passes along the ducts. Activation is normally prevented by the maintenance of an acid environment within the zymogen granules (probably through the action of a proton pump) and by the presence of **trypsin inhibitor** in the pancreatic juice. The latter binds to any active trypsin that may be present to form an inactive complex. Acute necrotizing **pancreatitis** is a life-threatening disorder often caused by the reflux of bile into the pancreas, or because of alcoholism. It is characterized by autodigestion of the pancreatic tissue, with inflammation and tissue damage caused by the escape of activated enzymes from the pancreas.

Table 44.2 The pancreatic enzymes

Enzyme	Zymogen	Activator	Principal action
trypsin	trypsinogen	enteropeptidase	cleaves internal peptide bonds
chymotrypsin	chymotrypsinogen	trypsin	cleaves internal peptide bonds
elastase	proelastase	trypsin	cleaves internal peptide bonds
carboxypeptidase	procarboxy-peptidase	trypsin	attacks peptides at C-terminal end
amylase			digests starch to maltose and oligosaccharides
lipase			cleaves glycerides liberating fatty acids and glycerol
colipase	procolipase	trypsin	binds to micelles to relieve lipase from inhibition by bile salts
phospholipase A$_2$	phospholipase	trypsin	cleaves fatty acids from phospholipids
cholesterol esterase		trypsin	releases esterified cholesterol
RNAase			cleaves RNA into short fragments
DNAase			cleaves DNA into short fragments

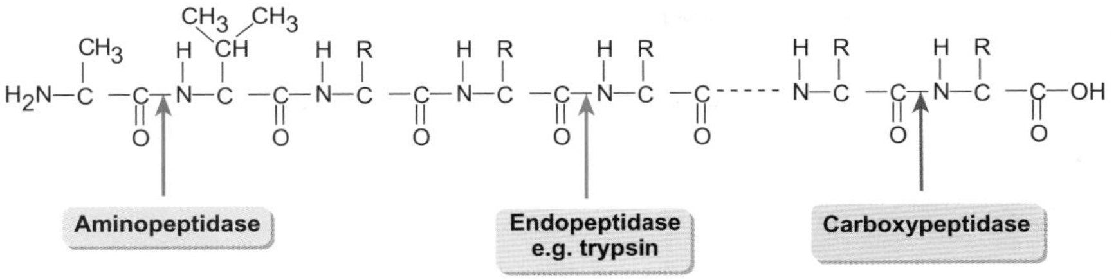

Figure 44.25 The sites at which various proteolytic enzymes digest proteins. Trypsin and chymotrypsin split peptide bonds within a protein molecule. Such enzymes are called endopeptidases. Aminopeptidase and carboxypeptidases attack peptides at their amino and carboxy terminals respectively. The specificity of the different enzymes is given in Table 44.3.

Pancreatic amylase and carbohydrate digestion

Much of the carbohydrate of plants is in the form of cellulose and hemicellulose, which the human gut is unable to digest. This material forms the dietary fibre. However, plants store carbohydrates in their seeds and tubers as starch and this is a source of energy that humans are able to exploit (see Chapter 4). Although salivary amylase may initiate the digestion of starch in the mouth and possibly the stomach, pancreatic α-amylase is responsible for the majority of starch digestion, which occurs in the duodenum. This enzyme is secreted in its active form and is stable between pH 4 and 11, although its optimal pH is 6.9. Like salivary amylase, it splits the α-1,4-glycosidic bond but not the 1-6 linkages between the chains of glucose residues (see Figure 44.26). Unlike the salivary enzyme, pancreatic amylase is able to digest both uncooked and cooked starch. Within 10 minutes or so of entering the small intestine, starch is entirely converted to various oligosaccharides, chiefly maltose, maltotriose, and limit dextrins. The intestinal brush border enzymes amylo-1,6-glucosidase (dextrinase), maltase, and sucrase then hydrolyse these saccharides into glucose.

In certain pathological conditions including acute pancreatitis, blood concentrations of pancreatic amylase start to increase and measurement of plasma amylase levels provides useful diagnostic information about the extent and progression of injury to pancreatic tissue.

Lipolytic enzymes of the pancreas

Because of their insolubility in water, fats pose a special problem for the GI tract in terms of both their digestion and absorption. As fatty chyme enters the duodenum, the gall bladder contracts in response to cholecystokinin (CCK) secreted by the I-cells of the duodenal villi. This contraction forces bile, via the bile duct, into the duodenum, where it mixes with the chyme arriving from the stomach. The fat globules within the chyme become coated with bile salts, which are amphipathic: that is, they have both hydrophobic and hydrophilic regions. The non-polar regions of the bile salts interact with the hydrophobic fatty acid side chains of the di- and triglycerides, while their hydrophilic polar regions allow them to interact with water. As the bile salts are large planar molecules, they disrupt the surface layer, causing fatty droplets to detach from the fat globules (see Figure 44.27) and form a stable emulsion. (An emulsion is an aqueous suspension of fatty droplets each about 1µm in diameter.) The **emulsification** of the dietary fats greatly increases the number of triglycerides exposed to the pancreatic lipases, facilitating their digestion. The manufacture and secretion of bile by the liver and the chemical properties of bile salts are discussed in more detail in Chapter 45.

Pancreatic juice contains a number of lipases, including pancreatic lipase (triacylglycerol hydrolase), which is probably secreted in its active form, and several others which are secreted in the form of inactive zymogens and are subsequently activated by trypsin in the duodenal lumen. This group includes colipase, cholesterol esterase, and phopholipase A_2. These enzymes can only access their substrate and digest the ingested triglycerides after the dietary fats have undergone the process of emulsification. Although pancreatic lipase is inhibited by physiological concentrations of bile salts, as well as by the phospholipids and

Table 44.3 The main proteolytic enzymes of the small intestine

Enzyme	Class	Source	Site of hydrolysis
trypsin	endopeptidase	pancreas	cleaves peptide bonds involving basic amino acid residues to yield oligopeptides that have a basic amino acid at the C-terminal
chymotrypsin	endopeptidase	pancreas	cleaves peptide bonds involving aromatic amino acid residues, glutamine, leucine, and methionine to yield oligopeptides with one of these at the C-terminal
elastase	endopeptidase	pancreas	cleaves peptide bonds involving neutral amino acid residues to yield oligopeptides with a neutral amino acid residue at the C-terminal
aminopeptidase A	exopeptidase	brush border	cleaves peptide bonds with acidic amino acids at their N-terminal end
aminopeptidase N	exopeptidase	brush border	cleaves peptide bonds with neutral and basic amino acids at their N-terminal end
carboxypeptidase A	exopeptidase	pancreas	cleaves peptide bonds with aromatic or neutral amino acids at C-terminal end
carboxypeptidase B	exopeptidase	pancreas	cleaves peptide bonds with basic amino acids at C-terminal end

Other brush border peptidases hydrolyse the peptide bonds of dipeptides and oligopeptides.

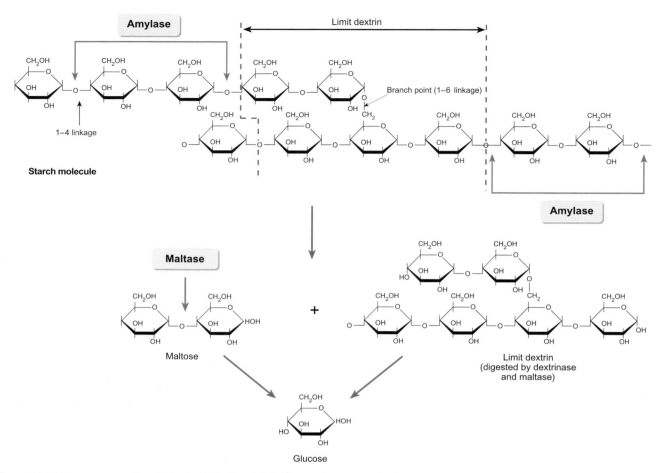

Figure 44.26 The processes by which starch is digested in the small intestine. Molecules of starch consist of branched polysaccharide chains; each chain consists of glucose molecules linked together via 1-4 glycosidic bonds, and these chains are linked via 1-6 glycosidic bonds. Pancreatic amylase splits the 1-4 bonds to release maltose and maltotriose, which are subsequently converted to glucose by enzymes located on the brush border of the enterocytes. Pancreatic amylase cannot hydrolyse the 1-6 bonds; instead, the limit dextrins are hydrolysed by dextrinase.

proteins that are associated with the lipid micelles, its activity is restored when it binds to colipase, with which it is secreted in a 1:1 molar ratio. The lipase–colipase complex is then able to digest the dietary fats effectively.

The steps by which water-insoluble triglycerides are digested by pancreatic lipase are illustrated in Figure 44.28. Note that this enzyme hydrolyses the ester linkages of the 1 and 3 positions but not that of the 2 position, so that for each triglyceride molecule two molecules of free fatty acid and one monoglyceride molecule are formed. Cholesterol esterase hydrolyses cholesterol esters as well as the esters of the fat-soluble vitamins (vitamins A, D, and E). It is also able to hydrolyse all the ester linkages of trigycerides (i.e. it is a non-specific lipase). Phospholipase A_2 requires bile salts to digest phospholipids, thus releasing free fatty acids and lysolecithin (a phospholipid lacking one of the fatty acid groups).

The regulation of pancreatic secretion

Like gastric secretion, pancreatic secretion is regulated both by the activity of the vagus nerves and by hormones. However, the

endocrine control of pancreatic secretion is the more important. A list of the chief chemical factors that affect the exocrine secretions of the pancreas is given in Table 44.4. As for the stomach, the process of secretion may be considered in three phases, cephalic, gastric, and intestinal. The cephalic phase is under nervous control, while the gastric and intestinal phases are controlled chiefly by hormones. The principal factors that regulate pancreatic secretion are summarized in Figure 44.29.

The cephalic phase of pancreatic secretion

The acinar cells and the smooth muscle cells of the pancreatic ducts and blood vessels are innervated by parasympathetic vagal efferent fibres. Stimulation of these fibres brings about the release of zymogen granules from the acinar cells into the ducts and an increase in local blood flow. The blood vessels also receive some sympathetic vasoconstrictor fibres whose activity causes a reduction in blood flow.

Parasympathetic vagal activity is enhanced by the sight, smell, and taste of food. The neurotransmitters acetylcholine and VIP are released and act synergistically to increase blood flow and promote

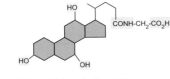

(a)   Bile salt (glycocholic acid)

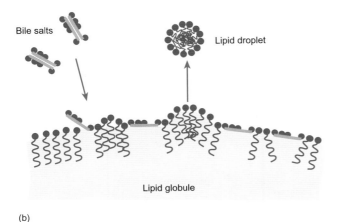

Figure 44.27 The emulsifying action of bile salts on dietary fats. (a) The structure of one of the bile salts (glycocholic acid) with the hydrophobic region coloured orange and the hydrophilic groups highlighted in light blue. (b) The action of bile salts on lipid globules. Bile salts penetrate the surface layer of the lipid and disrupt its structure, forcing aggregates of lipids to bulge out and eventually bud off as droplets. The polar head groups of the fats, together with molecules of bile salts, stabilize the dispersed lipid droplets and form a stable emulsion.

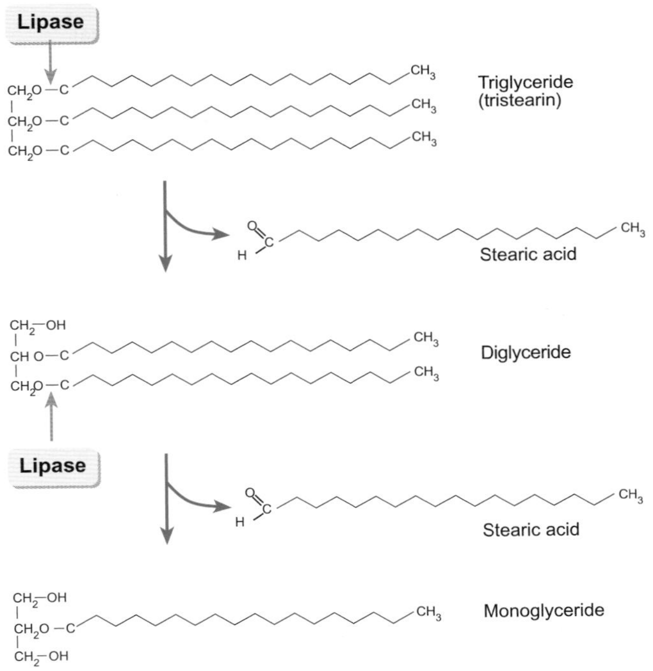

Figure 44.28 The steps by which fats are digested by pancreatic lipase. After dietary fats have been emulsified, pancreatic lipases hydrolyse the triglycerides and split off the two outer fatty acid residues as shown. The monoglycerides and the fatty acids are then absorbed.

the secretion of pancreatic juice. In addition to the direct action of vagal efferent nerves, a small part of the cephalic phase of pancreatic secretion is mediated by gastrin released from the antral cells in response to vagal stimulation.

The gastric phase of pancreatic secretion

Gastrin is chiefly responsible for this, relatively small, component of pancreatic secretion. It is secreted in response to distension of the stomach, as well as in response to the presence of amino acids and peptides in the antrum. Distension of the stomach also elicits pancreatic secretion via a vago-vagal reflex.

The intestinal phase of pancreatic secretion

This phase accounts for more than 70 per cent of total secretion by the exocrine pancreas and occurs in response to hormones secreted by the duodenal mucosa. Secretin is released in response to low pH and stimulates the secretion of bicarbonate-rich fluid from the ductal cells. CCK is secreted when the mucosal surface is bathed in monoglycerides, fatty acids, peptides, and amino acids (especially tryptophan and phenylalanine). It stimulates the production of an enzyme-rich fluid from the acinar cells. Furthermore, CCK appears to potentiate the secretory effects of secretin.

Summary

The exocrine portion of the pancreas consists of acinar cells, which secrete enzymes and fluid into a system of ducts lined with epithelial cells, which secrete alkaline fluid and modify the primary acinar secretion. The ionic composition of the pancreatic juice depends upon its rate of secretion. At high rates of secretion, the bicarbonate content of the juice is higher than at lower rates.

All the major enzymes required to complete the digestion of fats, carbohydrates, and proteins are contained within the pancreatic juice. Most are stored in the acinar cells as inactive precursors (zymogen granules), to avoid self-digestion. Activation of these enzymes takes place in the duodenum. Only pancreatic α-amylase is secreted in its active form. It is responsible for starch digestion. Several lipases are present in pancreatic juice that hydrolyse water-insoluble triglycerides to release free fatty acids and monoglycerides.

Control of exocrine pancreatic secretion is chiefly hormonal, although the initial cephalic phase of secretion is under the control of parasympathetic nerves. Gastrin contributes to the gastric phase of secretion, but about 70 per cent of secretion occurs during the intestinal phase in response to secretin and CCK. These hormones are released by the duodenal mucosa in response to H^+ ions and the products of fat and protein digestion.

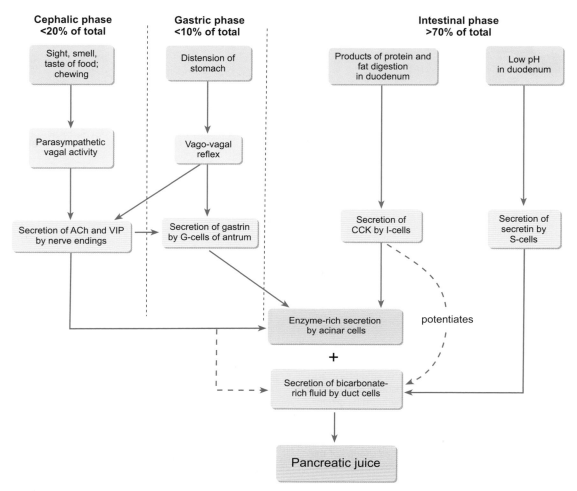

Figure 44.29 The major factors involved in the regulation of pancreatic secretion during the different phases of digestion. The intestinal phase of pancreatic secretion is far more significant and the gastric phase far less significant compared to the role of these phases in the regulation of gastric secretion.

Table 44.4 Chemical regulators of exocrine pancreatic secretion

Agents	Action on pancreas
cholecystokinin (CCK) gastrin acetylcholine substance P	increased secretion of pancreatic enzymes and chloride-rich fluid by acinar cells
secretin vasoactive intestinal polypeptide (VIP) histidine isoleucine	increased secretion of bicarbonate-rich fluid from duct cells
insulin insulin-like growth factors (IGFs)	increased enzyme synthesis and secretion trophic effects
somatostatin	inhibits secretion from acinar and duct cells

The stimuli are given in order of importance.

44.10 The absorption of digestion products in the small intestine

Absorption is the process by which the products of digestion are transported into the epithelial cells that line the GI tract and from there into the blood or lymph draining the tract. Each day about 8–10 litres of water and up to 1 kg of nutrients pass across the gut wall. Absorption through the gastrointestinal mucosa occurs by the processes of active transport and facilitated diffusion. Because the epithelial cells of the intestinal mucosa are joined at their apical (luminal) surfaces by tight junctions, products of digestion cannot move between the cells. Instead, they must move through the cells and into the interstitial fluid abutting their basal membranes if they are to enter the capillary blood. This process is called **transcellular transport**. The physical principles of active and passive transport are described together with **epithelial transport** in Section 5.3. The specific mechanisms by which the digestion products of the principal nutrients (carbohydrates, fats, and proteins) are absorbed will be considered first.

This will be followed by a discussion of the absorption of fluid and electrolytes, and the section concludes with a brief account of the absorption of vitamins. The mechanisms involved in the intestinal absorption of iron are described in detail in Section 25.5.

The absorption of monosaccharides, the digestion products of dietary carbohydrate

As described earlier, starch and other large polysaccharides are broken down by the pancreatic and brush border enzymes to form glucose. Galactose and fructose are produced by the breakdown of lactose and sucrose respectively. All three monosaccharides are largely absorbed in the upper small intestine, entering the blood of the hepatic portal vein. None remain in the chyme reaching the terminal ileum.

The process of glucose absorption depends on its concentration in the gut lumen. When its concentration is low, glucose is absorbed into the enterocytes across their apical surface against its concentration gradient by the sodium-dependent cotransporter SGLT1 described in detail in Chapter 5. The glucose is then transported across the basolateral surface via the glucose carrier GLUT2, as illustrated in Figure 44.30(a). The sodium gradient that drives this form of glucose transport is maintained by the Na^+, K^+-ATPase and the activity of SGLT1 enables the gut to scavenge all the available glucose.

After a meal, the glucose concentration in the gut lumen rises substantially and may exceed 100 mmol l^{-1}, a concentration that saturates the SGLT1 carriers. However, the increased activity of SGLT1 activates an intracellular signalling cascade that results in the insertion of preformed GLUT2 carriers from intracellular vesicles into the apical membrane (cf. the insertion of water channels into the apical membrane of the P-cells of the collecting ducts—see Section 39.8). The GLUT2 carriers of the apical membrane permit glucose to diffuse into the cells down its concentration gradient, as shown in Figure 44.30(b). As before, glucose leaves the enterocytes via the GLUT2 carriers of the basolateral surface. It then diffuses from the extracellular space into the blood and is transported to the liver via the portal vein. Galactose is absorbed by the same processes as glucose.

The mechanisms that result in the transient apical insertion of GLUT2 carriers in response to increased glucose levels in the gut lumen have not been clarified fully, but there is some evidence of a link between apical GLUT2 carriers and a newly recognized glucose-dependent calcium channel (Cav1.3). It has been suggested that when the intraluminal concentration of glucose is high, calcium entry through this channel is enhanced, and when intraluminal calcium rises, GLUT2 carriers are inserted into the apical membranes of the enterocytes. Such integration of glucose and Ca^{2+} absorption could represent a complex nutrient-sensing system allowing the absorption of both calcium and glucose to be regulated rapidly and precisely to match dietary intake.

In addition to the proposed link between the Cav1.3 calcium channel and apical GLUT2 insertion, recent studies have raised the possibility that stimulation of receptors for 'sweet taste'

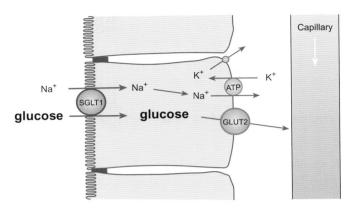

(a) Low glucose

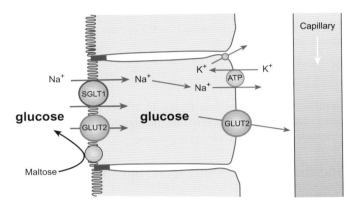

(b) After a meal

Figure 44.30 Glucose absorption by the small intestine. (a) When the glucose concentration in the lumen of the small intestine is low, glucose is absorbed by secondary active transport using the SGLT1 co-transporter. This permits effective absorption of all the available glucose. The excess sodium is pumped out of the cell by the sodium pump (Na^+, K^+-ATPase). (b) After a meal, the glucose concentration in the lumen is high (it may exceed 100 mmol l^{-1}). The glucose transporter GLUT2 becomes inserted in the apical membrane, allowing glucose to diffuse into the enterocytes down its concentration gradient (see text for further detail). In both situations, glucose is transported to the extracellular space across the basolateral membrane by the GLUT2 transporter. From there it diffuses into the blood and is transported to the liver via the portal vein. (Based on an original drawing of Dr E.S. Debnam.)

(similar to those found in the taste cells of the tongue—see Section 17.3) present on the membranes of enterocytes within the small intestine may also drive the insertion of GLUT 2 transporters into the apical membranes and thus facilitate the absorption of glucose by the GI tract.

Galactose is absorbed across the small intestinal epithelium by the same processes as glucose, but fructose is absorbed from the intestinal lumen by sodium-independent facilitated diffusion. It is taken up into the enterocytes mainly via a specific carrier called GLUT5 and leaves the cells via the GLUT2 carriers of the

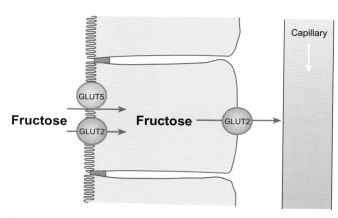

Figure 44.31 Fructose absorption by the small intestine occurs by facilitated diffusion and uses the carrier GLUT5, which is selective for fructose. Fructose can also enter the enterocytes via GLUT2 when this carrier has become inserted in the apical membrane. Fructose exits the enterocytes via GLUT2 carriers present in their basolateral membranes.

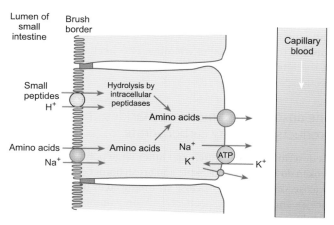

Figure 44.32 The mechanisms by which amino acids and small peptides are absorbed by the small intestine. The major classes of amino acids utilize separate carrier systems. Most amino acids are transported into the enterocytes by sodium-linked carriers but utilize uniports or sodium-linked antiports to exit via the basolateral membranes. The sodium pump (Na⁺, K⁺-ATPase) transports the sodium ions out of the cell. Small peptides are absorbed from the lumen in association with hydrogen ions. Once inside the cell, they are broken down into their constituent amino acids before leaving the cell by the amino acid carriers of the basolateral membrane.

basolateral membrane (see Figure 44.31). As the plasma concentration of fructose is very low, the gradient for the passive absorption of fructose is always favourable.

The absorption of peptides and amino acids

Dietary protein is broken down by gastric and pancreatic protease enzymes to small peptides and amino acids. Each day approximately 200 g of amino acids and small peptides are absorbed from the small intestine of an adult eating a normal mixed diet. In order to maintain a positive nitrogen balance, and to meet the needs of an adult body for tissue growth and repair, at least 50 g of amino acids must be absorbed each day. Large peptides and whole proteins are not normally absorbed, although some may enter the bloodstream. The immunoglobulins present in colostrum, for example (see Section 49.10) appear to be absorbed intact across the intestinal epithelium of the neonatal gut.

Amino acids are absorbed at the brush border of the intestinal epithelial cells by sodium-dependent cotransport mechanisms similar to those utilized in the renal tubules (see Figure 44.32 and also Figure 39.16). More than ten separate transporter systems have now been characterized. There are carrier systems for the neutral amino acids and methionine, for the cationic amino acids (arginine, histidine, and lysine), for the acidic amino acids (glutamate and aspartate), for proline, and for glycine. (These transporters are also present in the proximal tubule of the nephron, where they perform a similar function.) Once the amino acids have entered the enterocyte, they are exported into the extracellular space across the basolateral surface by a separate set of carriers. From the extracellular space, the amino acids diffuse into the capillaries of the villus and are transported to the liver via the portal vein. Most of the amino acids are absorbed in the first part of the small intestine. A few may enter the colon where they are metabolized by the colonic bacteria.

Small peptides (mainly dipeptides) are transported into the enterocytes by another carrier that is not linked to sodium but is linked to the influx of hydrogen ions. This carrier is also responsible for the rapid uptake of certain drugs by the gut, such as the anti-hypertensive drug captopril. Once the peptides enter the intracellular compartment, they are broken down to their constituent amino acids. These leave the enterocytes via the amino acid carrier systems of the basolateral surface (see Figure 43.32). Up to half of the ingested protein is now thought to be absorbed in this way.

The absorption of the digestion products of fats

Each day about 80 g of fat are absorbed from the small intestine, largely in the jejunum. The monoglycerides and free fatty acids liberated by the activity of the pancreatic lipases become associated with bile salts and phospholipids to form **micelles** (small colloidal particles in suspension). The non-polar core of the micelle also contains cholesterol and fat-soluble vitamins. The hydrophilic outer region of the micelle enables it to enter the unstirred aqueous layer covering the microvilli that form the brush border of the enterocytes. Monoglycerides, free fatty acids, cholesterol, phosphatidylcholine, and fat-soluble vitamins then diffuse passively into the duodenal cells while the bile salt portion of the micelle remains within the lumen of the gut until the terminal ileum, where it is reabsorbed. The majority (c. 90 per cent) of the bile salts entering the small intestine are recycled by the enterohepatic circulation (see Chapter 45).

While small amounts of short chain fatty acids are absorbed directly from the intestinal epithelial cells into the capillary blood

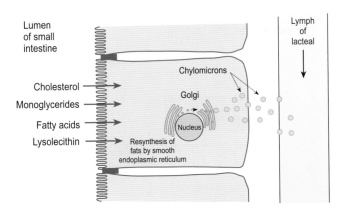

Figure 44.33 The key steps involved in the absorption of lipids by the small intestine. Fatty acids, monoglycerides, and other lipids pass across the brush border and are re-esterified by enzymes of the smooth endoplasmic reticulum. The reconstituted lipids are then secreted across the basolateral membrane as small fat droplets (chylomicrons) and enter the lymphatic system via the lacteals of the intestinal villi.

by passive diffusion, the majority of the products of fat digestion undergo further chemical processing inside the enterocytes. In the smooth endoplasmic reticulum, triglycerides are reformed by the re-esterification of monoglycerides, phospholipids are re-synthesized, and much of the cholesterol undergoes re-esterification. The lipids accumulate in the vesicles of the smooth ER to form **chylomicrons**, which are released from the cells by exocytosis at the basolateral membrane (see Figure 44.33). From here, they enter the lacteals of the villi and leave the intestine in the lymph, from where they are released into the venous circulation via the thoracic duct. Lipids thus avoid the hepatic portal vein and bypass the liver in the short term.

The faeces normally contain about 5 per cent fat, most of which is derived from bacteria. Greater amounts of fat are found in the faeces if bile production is diminished or if bile is pre-vented from entering the duodenum by a biliary obstruction such as a gallstone.

The absorption of fluid and electrolytes

Approximately 2 litres of fluid are ingested each day (although this may vary considerably depending upon thirst and social factors). The secretion of digestive juices and intestinal fluids adds a further 7–8 litres of fluid to the gastrointestinal lumen. Almost all of this fluid is absorbed from the small and large intestines; only 50–100 ml leaves the body with the faeces. About 5–6 litres a day are absorbed in the jejunum, 2–3 litres a day in the ileum, and between 400 ml and 1 litre a day in the colon. Figure 44.34 summarizes the overall fluid balance in the GI tract.

Absorbed electrolytes originate both from ingested foods and from gastrointestinal secretions. Most are actively absorbed along the length of the small intestine, though the absorption of calcium and iron is restricted mainly to the duodenum. As described earlier, the absorption of sodium ions is coupled with

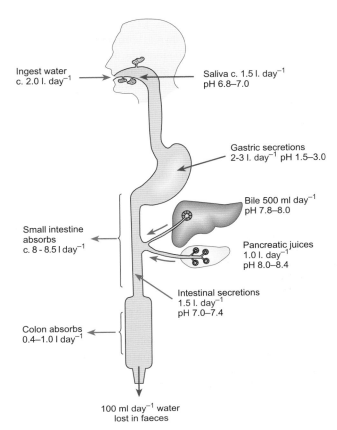

Figure 44.34 The overall balance between fluid intake, secretion, absorption, and loss by the GI tract. Fluid secretion is shown by green arrows and fluid absorption by red arrows. A small amount of water is lost in the stools. (Adapted from Fig 16.4 in A.J. Vander, J.H. Sherman, and D.S. Luciano (1990) *Human physiology*, 5th edition. McGraw-Hill, New York.)

the transport of both sugars and amino acids. There are numer-ous active sodium–potassium pumps in the basal membranes of the intestinal epithelial cells that pump sodium out of the cells, thereby creating a gradient that draws sodium passively into the cell from the intestinal lumen. Some potassium is actively se-creted into the gut, particularly in mucus. Potassium is absorbed passively along a concentration gradient set up by the absorption of water. For the most part, anions passively follow the electrical potential generated by the active transport of sodium. Chloride ions are also actively transported and, in the lower ileum, bicar-bonate ions are actively secreted into the intestinal lumen in exchange for chloride.

The intestinal absorption of calcium is similar to that described for the distal tubule of the kidney. Calcium ions enter the entero-cytes down a steep electrochemical gradient via epithelial calcium channels (TRPV5 and TRPV6). Once in the enterocytes, calcium binds to calbindin prior to diffusing to the basolateral membrane where it leaves the cell via the calcium pump (Ca^{2+}-ATPase). The active form of vitamin D, 1,25-dihydroxycholecalciferol (calci-triol), stimulates intestinal calcium absorption by increasing the expression levels of all of the proteins involved (i.e. the calcium channels, calbindin, and Ca^{2+}-ATPase). Calcium absorption is regulated over the long-term by parathyroid hormone (see Section 22.3).

The absorption of water by the intestine occurs by osmosis into blood vessels in response to the gradients established by the absorption of nutrients and electrolytes. Water can also be transported by osmosis from the blood to the intestinal lumen when the chyme entering the duodenum becomes hypertonic as a result of the digestion of nutrients. In this way, isotonicity of the chyme is rapidly established and is then maintained along the length of the intestine. As nutrients and electrolytes are progressively absorbed, water follows almost instantaneously. Fluid absorption along the gut plays a key role in overall body fluid balance—serious, or even fatal, dehydration can develop very quickly if insufficient fluid is absorbed. This is often the case in diseases such as dysentery and cholera, in which there is excessive diarrhoea.

The absorption of vitamins

The fat-soluble vitamins are absorbed in the same way as the products of fat digestion, partitioning into micelles and passing into the lymph as described above. Specific transport molecules have been identified for most of the water-soluble vitamins, and they may enter the intestinal epithelial cells by passive diffusion, facilitated diffusion, or active transport. Vitamin C, for example, is absorbed in the jejunum by sodium-dependent active transport via a mechanism similar to that described earlier for amino acids and monosaccharides.

The absorption of vitamin B_{12} (cyanocobalamin) is a complex process which is summarized in Figure 44.35. Vitamin B_{12} is released from food by the low pH of the gastric juice and becomes bound to **haptocorrin**, a glycoprotein secreted by the salivary glands. The haptocorrin protects vitamin B_{12} from the acid environment of the stomach but is digested in the duodenum, where the free vitamin B_{12} then binds to **intrinsic factor**, a glycoprotein secreted by the parietal cells of the gastric mucosa. The resulting complex is recognized by a receptor (cubulin) in the brush border membrane of cells in the lower ileum and the resulting complex is endocytosed by the enterocytes. Once in the cells, vitamin B_{12} is released from the cubulin–intrinsic factor complex and is transported across their basolateral membranes. It enters the capillary endothelial cells, where it binds to transcobalamin II prior to its transport in the blood.

Disorders of absorption (malabsorptive states)

A number of disease states may lead to a failure to absorb nutrients from the small intestine. In some cases there is direct impairment of the absorption of nutrients. In others, impaired absorption arises as a consequence of problems with digestion. A range of symptoms are common to many kinds of malabsorption including weight loss, flatulence with abdominal distension and discomfort, and glossitis (painful loss of the normal epithelium covering the tongue). Symptoms that are more specific are indicative of particular malabsorption states. Poor digestion and absorption of fats, for example, will produce greasy, soft, malodorous stools (steatorrhoea). Anaemia is often seen if the absorption of iron or folic acid

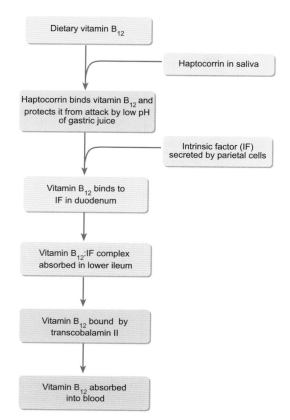

Figure 44.35 The main stages by which vitamin B_{12} is absorbed from the GI tract. The vitamin is bound by salivary haptocorrin to protect it from attack by the low pH of the gastric juice. The parietal cells of the gastric glands secrete a glycoprotein known as intrinsic factor, which binds to free vitamin B_{12} in the duodenum. The complex of vitamin B_{12} and intrinsic factor is then absorbed in the lower ileum by endocytosis.

is impaired, while calcium deficiency (manifest as demineralization of the skeleton, and possibly tetany), may be the result of vitamin D deficiency. Failure to absorb vitamin K may lead to an increased tendency to bleed. A detailed account of the many specific malabsorptive states is beyond the scope of this book, but a few important examples are described below.

Carbohydrate intolerance

An inability to digest carbohydrate may arise because of an inherited or acquired deficiency in one or more of the necessary intestinal enzymes. Several types of deficiency may occur. One of the most common is lactase deficiency, in which the enzyme responsible for digesting lactose (the disaccharide milk sugar) is lacking. Undigested disaccharide remains in the lumen of the bowel, causing fluid to be retained there by its osmotic effect. This causes distension and diarrhoea. The colonic bacteria ferment the lactose, and gaseous, acidic stools are produced. Children with lactase deficiency will fail to thrive and may suffer diarrhoea each time they ingest milk. Adults will suffer nausea, flatulence, and abdominal discomfort. The condition is readily controlled by the avoidance of foods containing lactose. A very rare congenital form of carbohydrate intolerance arises from an inability to absorb glucose and galactose. This is due to the absence of the normal intestinal sodium-dependent carrier protein for these sugars.

Coeliac disease (non-tropical sprue)

This is a hereditary chronic intestinal disorder of absorption caused by intolerance to gluten, a cereal protein found in wheat, rye, oats, and barley. Part of the gluten molecule forms an immune complex within the intestinal mucosa promoting aggregation of killer T-lymphocytes (see Chapter 26) which release toxins that promote lysis of the enterocytes. There is progressive atrophy of the villi (in particular of the duodenum and proximal jejunum) because enterocytes cannot be replaced quickly enough by the stem cells of the intestinal crypts. Furthermore, many of the cells that are present are relatively immature and cannot absorb nutrients effectively.

The symptoms of coeliac disease vary in severity and in presentation but steatorrhoea (fatty stools), bloating, and discomfort are common. Children fail to thrive and pass soft, pale, malodorous stools after eating foods containing gluten. Nutritional supplementation of a gluten-free diet may be necessary in extreme cases.

Crohn's disease

This condition may affect any region of the intestine but most commonly affects the terminal ileum and ascending colon. It is characterized by chronic inflammatory lesions with accumulation of macrophages (granulomatous deposits) and thickening of the bowel. There is enlargement of the lymph nodes and inflammation across the entire thickness of the gut wall. The causes of the disease are unclear but there may be a genetic component. The disease shows periods of remission and renewed activity, and symptoms are extremely variable both in type and in severity. There may be pain in the right iliac fossa, diarrhoea, steatorrhoea, and rectal passage of blood and mucus. Complications such as gallstones may arise because of malabsorption of bile acids in the terminal ileum. There may also be deficiencies of the fat-soluble vitamins A, D, E, and K. Treatment with anti-inflammatory drugs including steroids may prolong the periods of remission, but a high proportion of sufferers eventually require surgery to remove affected parts of the intestine.

Summary

Absorption is the process by which the products of digestion are transported into the epithelial cells of the GI tract and thence into the blood or lymph draining the gut. Almost all of the absorption of water, electrolytes, and nutrients occurs in the small intestine. Monosaccharides are absorbed in the duodenum and upper jejunum by sodium-dependent cotransport. Amino acids utilize similar mechanisms. The products of fat digestion are incorporated into micelles along with bile salts, lecithin, cholesterol, and fat-soluble vitamins. In this way, they are brought close to the enterocyte membrane and the fatty components of the micelle diffuse into the cells.

Bile salts are recycled and the fats form chylomicrons, which enter the lacteals of the villi.

Most water-soluble vitamins are absorbed by facilitated transport. A specific uptake process involving gastric intrinsic factor is responsible for the absorption of vitamin B_{12}. Fat-soluble vitamins are absorbed along with the products of fat digestion.

The GI tract absorbs 8–10 litres of fluid and electrolytes each day. The active transport of sodium and nutrients is followed by anion movement and the absorption of water by osmosis. A number of conditions result from malabsorption of nutrients. Specific syndromes include carbohydrate intolerance, coeliac disease, and Crohn's disease.

44.11 The large intestine

Around 500 ml of chyme passes via the ileocaecal valve from the ileum into the caecum every day. Material then passes in sequence through the ascending colon, transverse colon, descending colon, sigmoid colon, rectum, and anal canal (see Figure 44.36). Semisolid waste material (faeces) is eliminated from the body through the anus. In adults, the large intestine is approximately 1.3 metres long but, as its name implies, its diameter is greater than that of the small intestine. It has a variety of functions, which include the storage of food residues prior to their elimination, the secretion of mucus, which lubricates the faeces to ease their passage, and the absorption of most of the water and electrolytes remaining in the residue. In addition, bacteria, which live in the colon, synthesize vitamin K and some B vitamins.

Special histological features and innervation of the large intestine

Structurally, the wall of the large intestine follows the basic plan of the GI tract. However, the longitudinal smooth muscle layer of the muscularis externa is thickened to form longitudinal bands

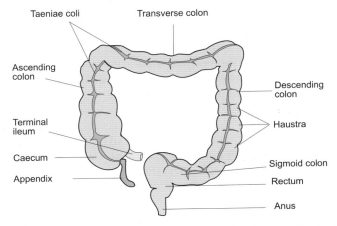

Figure 44.36 The large intestine, showing the bands of longitudinal muscle (the teniae coli) and haustra of the colon.

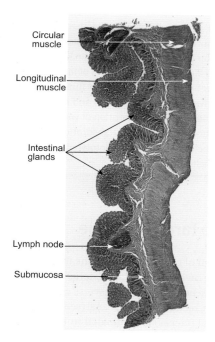

Figure 44.37 labels: Circular muscle, Longitudinal muscle, Intestinal glands, Lymph node, Submucosa

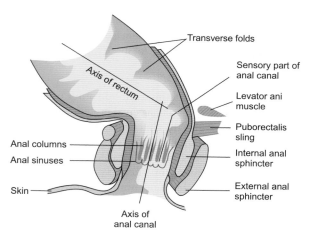

Figure 44.38 labels: Transverse folds, Sensory part of anal canal, Levator ani muscle, Puborectalis sling, Internal anal sphincter, External anal sphincter, Axis of rectum, Anal columns, Anal sinuses, Skin, Axis of anal canal

Figure 44.37 A section of the wall of the colon showing the main layers. Note the thin layer of longitudinal muscle and the darkly staining mucosa which contains the intestinal glands. These glands absorb fluid from the food residues and secrete mucus to aid the movement of faeces through the remainder of the GI tract.

Figure 44.38 A longitudinal section through the rectum and anal canal that shows the position of the internal and external anal sphincters. Note the pronounced angle between the axis of the rectum and that of the anal canal. (Based on Fig 6.2 in P.A. Sanford (1992) *Digestive system physiology*, 2nd edition. Edward Arnold, London.)

called **taeniae coli**. Three of these are present in the caecum and the colon. In between the taeniae the longitudinal muscle layer is relatively thin. The tone of the smooth muscle in the taeniae causes the wall of the large intestine to pucker into pocket-like sacs called **haustra**. The final part of the colon is called the sigmoid colon, which merges with the rectum, a muscular tube about 15 cm in length. In the rectum, there are two broad bands of longitudinal muscle but haustra are absent. In addition to the longitudinal smooth muscle, the large intestine also possesses a thick circular layer of smooth muscle. Figure 44.36 is a diagrammatic representation of the large intestine showing its main divisions and the taeniae coli.

Figure 44.37 is a section of the wall of the colon showing the main tissue layers. The mucosal surface of the caecum, colon, and upper rectum is smooth and has no villi. However, large numbers of crypts are present. The mucous membrane consists of columnar absorptive cells, as well as many mucus-secreting goblet cells.

The anal canal forms the final part of the GI tract. It is about 3 cm long and lies entirely outside the abdominal cavity. It possesses internal and external sphincters (Figure 44.38), which remain closed (acting rather like purse strings) except during defecation. The mucosa of the anal canal reflects the greater degree of abrasion that this area receives. It hangs in long folds called anal columns and is lined by stratified squamous epithelium. Between the columns are the anal sinuses. Two superficial venous plexuses are

associated with the anal canal. If these become inflamed, itchy varicosities called **haemorrhoids** form.

The large intestine receives both parasympathetic and sympathetic innervation. Vagal fibres supply the caecum and the colon as far as the distal third of the transverse region. Parasympathetic fibres supplying the rest of the colon, rectum, and anal canal are from the pelvic nerves of the sacral spinal cord (**nervi erigentes**). The parasympathetic fibres end chiefly on neurons of the intramural plexuses. The sympathetic input is from the coeliac and superior mesenteric ganglia (caecum, ascending and transverse colon), and from the inferior mesenteric ganglia (descending and sigmoid colon, rectum, and anal canal) (see Figure 43.5). The external anal sphincter also receives branches of somatic nerves arising from the sacral region of the spinal cord.

The caecum and appendix

The caecum is a blind-ended tube about 7 cm long leading from the ileocaecal valve to the colon. Although it is important in herbivores for the digestion of cellulose, it has no significant digestive role in humans. Attached to the posteromedial surface of the caecum is the vermiform appendix, a small blind pouch about the size of a finger, containing lymphoid tissue. Although it is part of the mucosa-associated lymphoid tissue (MALT), it has long been considered to have no essential function in humans. More recently, however, evidence has emerged to suggest that the appendix may be important in the overall maintenance of intestinal health, acting as a reservoir for commensal bacteria. Some studies also implicate the appendix in the development of ulcerative colitis. Inflammation of the appendix is known as **appendicitis**. To prevent its rupture it is necessary to remove the appendix surgically. If the appendix does rupture, faecal material containing bacteria will enter the abdominal cavity causing the more serious condition, **peritonitis**.

The colon

The colon acts as a reservoir, storing unabsorbed and unusable food residues. Although most of this residue is excreted within 72 hours of ingestion, up to 30 per cent of it may remain in the colon for a week or more. The colon is divided into four sections: the ascending, transverse, descending, and sigmoid colon. The ascending and descending colon are closely attached to the posterior abdominal wall behind the parietal peritoneum (such an attachment is said to be **retroperitoneal**). The transverse colon is attached to the posterior abdominal wall by a short mesentery. Suspended from the lower border of the transverse colon is the greater omentum, a large apron-like sheet of mesentery that contains deposits of fat. The ascending colon and the first part of the transverse colon receive blood via the superior mesenteric artery, while the rest of the colon is supplied by the inferior mesenteric artery as shown in Figure 44.39. Venous blood enters the superior and inferior mesenteric veins before draining into the hepatic portal vein. Blood vessels supplying the colon penetrate the circular smooth muscle layer, creating areas of potential mechanical weakness. Occasionally herniation of the mucosa in these areas leads to the formation of pouches or **diverticula**. Diverticulosis (the condition of having intestinal diverticula) is harmless but occasionally, diverticula can rupture leading to significant loss of blood via the rectum (diverticular bleeding). The diverticula may become infected and inflamed, causing considerable pain to the sufferer. This condition is known as diverticulitis.

Electrolyte and water absorption in the colon

Although large amounts of fluid are absorbed from chyme as it passes along the small intestine, the chyme entering the colon still contains appreciable quantities of water and electrolytes. Indeed,

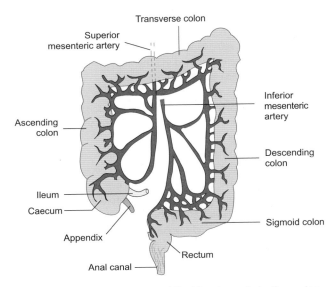

Figure 44.39 The arrangement of the blood supply to the major structures of the large intestine. (Based on Fig 6.1 in P.A. Sanford (1992) *Digestive system physiology*, 2nd edition. Edward Arnold, London.)

the colon absorbs 400–1000 ml of fluid each day and failure to do so results in severe diarrhoea. Sodium ions are transported actively from the lumen to the blood. This absorption is sensitive to aldosterone. As in the lower ileum, the absorption of chloride ions is linked with the secretion of bicarbonate. This bicarbonate may help to neutralize the acidic end products of bacterial action. Water is absorbed by osmosis.

The role of intestinal bacterial flora

A variety of bacteria colonize the large intestine, living symbiotically with their human host (a symbiotic relationship is one in which organisms of different species co-exist, often to the mutual benefit of each). Some of these, such as *Clostridium perfringens* and *Bacteroides fragilis*, are anaerobic species, while others like *Enterobacter aerogenes* are aerobic. The intestinal flora perform a number of functions within the large intestine. One of these is the fermentation of indigestible carbohydrates (notably cellulose) and lipids that enter the colon. As a result of these fermentation reactions, short chain fatty acids are produced, along with a number of gases (for example hydrogen, nitrogen, carbon dioxide, methane, and hydrogen sulphide), which form about 500 ml of flatus each day (more if the diet is rich in indigestible carbohydrates such as cellulose). Short chain fatty acids, including acetate, propionate, and butyrate, are absorbed readily by the colon, stimulating water and sodium uptake at the same time. The colonocytes appear to utilize the short chain fatty acids for energy. Another action of the intestinal flora is the conversion of bilirubin to the non-pigmented metabolites, the urobilinogens. They are also able to degrade cholesterol and some drugs. Finally, intestinal bacteria are able to synthesize certain vitamins, e.g. vitamin K, vitamin B_{12}, thiamine, and riboflavin. However, vitamin B_{12} can only be absorbed in the terminal ileum, so any vitamin B_{12} that is synthesized in the colon is usually excreted and is of no value to the body. The importance of the intestinal flora to overall gastrointestinal health is highlighted by the side effects (nausea, diarrhoea, constipation, abdominal discomfort, etc.) caused by many antibiotic medicines as they disrupt the balance between the different commensal bacterial strains.

Movements of the colon

The movements of the colon, like those in the small intestine, may be defined as either mixing or propulsive movements. Since the role of the colon is to store food residues and to absorb water and electrolytes, its propulsive movements are relatively sluggish. Characteristically, material travels along the colon at 5–10 cm hour^{-1} and typically remains within the colon for 16–20 hours.

Mixing movements (haustrations)

Contraction of the circular smooth muscle layer of the colon serves to constrict the lumen in much the same way as described for the small intestine above. This kind of segmental movement is called **haustration** in the colon because the segments correspond to the smooth muscle thickenings called haustra. It is the predominant type of movement seen in the caecum and the proximal colon. Haustration squeezes and rolls the faecal material around so that every portion of it is exposed to the absorptive surfaces of the colonic mucosa, thus aiding the absorption of water and electrolytes.

Propulsive movements—peristalsis and mass movements

In addition to haustral contractions, short-range peristaltic waves are seen in the more distal parts of the colon (transverse and descending regions). These serve to propel the intestinal contents, now in the form of semisolid faecal material, towards the anus. Several times a day, usually after meals, a more vigorous propulsive movement of the colon occurs in which a portion of the colon remains contracted for rather longer than during a peristaltic wave. This is called a mass movement and results in the emptying of a large portion of the proximal colon. Mass movements are also seen in the transverse and descending colon. When they force a mass of faecal material into the rectum, the desire for defecation is experienced.

Mass movements are initiated, at least in part, by intrinsic reflex pathways resulting from distension of the stomach and duodenum. These are termed the gastrocolic and duodeno-colic reflexes. These intrinsic motor patterns are modified by autonomic nerves and by hormones. Vagal stimulation, for example, enhances colonic motility while both gastrin and CCK increase the excitability of the colon and facilitate emptying of the ileum by causing the ileocaecal sphincter to relax. Opiate drugs such as morphine, codeine, and pethidine decrease the frequency of colonic mass movements. Other drugs, including aluminium-based antacids, have the same effect. People taking these drugs may therefore become constipated.

The role of dietary fibre in the large intestine

The time taken for food residues to be expelled from the body after eating varies considerably but appears to be directly related to the amount of dietary fibre ingested. Dietary fibre (or 'roughage') consists largely of cellulose. Humans are unable to digest this so it remains in the intestine, adding bulk to the food residues. The fibre tends to exert a hygroscopic effect, absorbing water; stools with a high fibre content tend to be bulkier and softer, making them easier to expel. A shorter mouth-to-anus transit time is also believed to reduce the risk of developing carcinoma of the large intestine and rectum. This may be due partly to a reduction in the time for which bacterial toxins and potentially harmful metabolites are in contact with the gut wall.

The rectum and defecation

The rectum is a muscular tube about 12–15 cm long. It is normally empty but when a mass movement forces faeces into the rectum, the urge to defecate is initiated. The rectum opens to the exterior via the anal canal, which has both internal and external sphincters. The internal sphincter is not under voluntary control. It is supplied by both sympathetic and parasympathetic nerves. Contraction of the smooth muscle of the internal sphincter is initiated by sympathetic stimulation and relaxation by parasympathetic stimulation. The external anal sphincter is composed of skeletal muscle. It is supplied by the pudendal nerve and is under learnt voluntary control from the age of about 18 months. Both the anal sphincters are maintained in a tonic state of contraction.

About 100–150 g of faeces are normally eliminated each day, consisting of 30–50 g of solids and 70–100 g of water. The solid portion consists largely of cellulose, epithelial cells shed from the lining of the GI tract, bacteria, some salts, and the brown pigment stercobilin. The characteristic odour of faeces is due to the presence of sulphides, and indole and skatole which are derived from tryptophan.

Defecation itself is a complex process involving both reflex and voluntary actions. It is possible to inhibit the reflex consciously if the circumstances are not convenient and, under such conditions, the urge to defecate will often subside until reinitiated by the arrival of further faecal material in the rectum. Eventually, however, the urge to defecate will become overwhelming and the reflex will proceed. Under the influence of the parasympathetic nervous system, the walls of the sigmoid colon and the rectum contract to move faeces towards the anus. The anal sphincters relax to allow faeces to move through the anal canal. Expulsion of the faecal material is aided by voluntary contractions of the diaphragm and the muscles of the abdominal wall as well as closure of the glottis, as in the Valsalva manoeuvre. As a result, intra-abdominal pressure rises and helps to force faeces through the relaxed sphincters. The muscles of the pelvic floor relax to allow the rectum to straighten, thus helping to prevent rectal and anal prolapse. The components of the defecation reflex are depicted diagrammatically in Figure 44.40.

Summary

The large intestine consists of the caecum, colon, rectum, and anal canal. Its main functions are to store food residues, secrete mucus, and absorb remaining water and electrolytes from the food residue. Faeces are eliminated via the anus. The colon absorbs 400–1000 ml of fluid each day.

Intestinal bacteria synthesize certain vitamins such as vitamin K and short chain fatty acids from indigestible food residues. The short chain fatty acids are absorbed by the colonocytes, stimulating salt and water uptake.

The colon exhibits mixing movements (haustrations) and sluggish propulsive movements. Mass movements occur several times daily, serving to move intestinal contents over longer distances. These contractions move faecal material into the rectum and the resulting stretch of its wall elicits the urge to defecate. Defecation involves both voluntary and involuntary contractions of the anal sphincters and muscles of the abdominal wall and diaphragm.

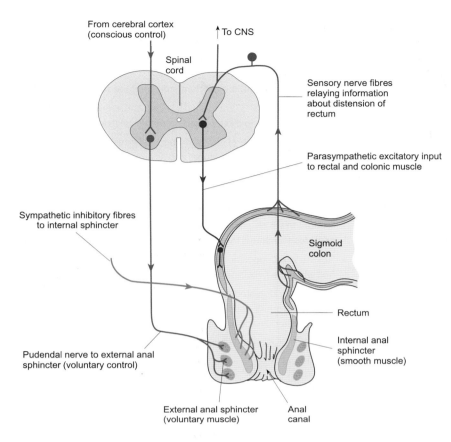

Figure 44.40 Neural pathways involved in the defecation reflex. Sensory afferents relay information about the distension of the rectum by the faecal mass to the sacral region of the spinal cord via the pelvic plexus. This information is then relayed to the brain, indicating the need to defecate. When the time is appropriate, the cerebral cortex initiates the process of defecation. Parasympathetic fibres excite contraction of the smooth muscle of the colon and rectum, while sympathetic fibres inhibit the tone of the internal sphincter, causing it to relax. The voluntary muscle of the external sphincter then relaxes, allowing the bowel to be evacuated.

✱ Checklist of key terms and concepts

The organization, structure, and functions of the GI tract

- The GI tract consists of the oral cavity, pharynx, oesophagus, stomach, small intestine, large intestine, and anal canal.

- Accessory organs include the teeth, tongue, salivary glands, exocrine pancreas, liver, and gall bladder.

- The major functions of the GI tract include the ingestion of food and its transport along the tract, the secretion of fluids, salts, and digestive enzymes, the absorption of digestion products, and the elimination of indigestible remains.

- The basic layers of the gut wall are (from the outside) the serosa, a layer of longitudinal smooth muscle, a layer of circular smooth muscle, the submucosa, and the mucosa.

- Contractile activity of the smooth muscle promotes mixing and propulsion of food.

- The GI tract is regulated by extrinsic nerves and by the enteric nervous system.

- The enteric nervous system is a system of intramural plexuses which mediate a number of intrinsic reflexes that control secretory and contractile activity.

- Afferent and efferent extrinsic nerves, and endocrine and paracrine hormones, also play an important role in regulating the activities of the GI tract.

- The gut receives a rich blood supply. The combined circulation to the stomach, liver, pancreas, intestine, and spleen is called the splanchnic circulation.

- The splanchnic circulation receives 20–25 per cent of the cardiac output at rest.

The oral cavity, saliva, and swallowing

- In the mouth, food is mixed with saliva as it is chewed.

- Three pairs of salivary glands (parotid, submandibular, and sublingual) secrete about 1500 ml of saliva each day.

- The saliva contains mucus that helps to lubricate the food and an α-amylase that initiates the breakdown of carbohydrate.

- Isotonic fluid is formed by the acinar cells of the salivary glands through the secretion of electrolytes and water.

- This undergoes modification as it flows along the salivary ducts so that the final composition of saliva depends upon its flow rate.

- Salivary secretion is controlled by reflexes mediated by the autonomic nervous system.

- Parasympathetic stimulation promotes an abundant secretion of watery saliva rich in amylase and mucus. Blood flow to the salivary glands is also enhanced.

- Sympathetic stimulation promotes the output of amylase but reduces blood flow to the glands.

- The overall effect of sympathetic stimulation is generally a reduction in the rate of salivary secretion.

- Swallowing occurs in three phases. The first oral phase is voluntary but subsequent phases are under reflex autonomic control.

- As swallowing occurs, a wave of peristalsis is initiated which propels the food bolus through the upper oesophageal (pharyngo-oesophageal) sphincter into the oesophagus.

- This wave of contraction continues for several seconds and moves the bolus down to the lower oesophageal (cardiac) sphincter, which relaxes to permit the entry of food into the stomach.

- The lower oesophageal sphincter normally remains closed except to permit entry of a food bolus into the stomach. This helps to prevent reflux of acidic gastric contents into the oesophagus.

The stomach

- The functions of the stomach include storage of food; mixing, churning and kneading the food to produce chyme; and the secretion of acid, enzymes, mucus, and intrinsic factor.

- In addition to the circular and longitudinal smooth muscle layers, the stomach wall possesses a third, obliquely arranged muscle layer that promotes churning movements.

- The surface epithelium of the gastric mucosa is composed almost entirely of cells that secrete an alkaline fluid containing mucus.

- Gastric glands empty into gastric pits in the epithelium. The glands contain mucous cells, chief cells, which secrete pepsinogens, and parietal cells, which produce gastric acid and intrinsic factor.

- A variety of endocrine cells are also present, for example the G-cells which secrete gastrin.

- Gastric acid is secreted by the parietal cells of the gastric glands in response to food.

- Hydrogen ions move out of the cells against a massive concentration gradient. Chloride ions move from blood to lumen against an electrochemical gradient.

- The parietal cells also secrete a glycoprotein called intrinsic factor, which is essential for the absorption of vitamin B_{12} in the terminal ileum.

- A number of proteolytic enzymes are secreted by the chief cells of the gastric glands. They are released as inactive pepsinogens and activated by the acidic environment in the gastric lumen.

- Pepsins hydrolyse peptide bonds within protein molecules to liberate polypeptides.

- The gastric mucosa creates an alkaline mucus 'barrier' to protect itself from erosion by the gastric juice.

- Prostaglandins of the E series increase the thickness of the mucus layer and stimulate the secretion of bicarbonate ions.

The regulation of gastric secretion

- Nervous and endocrine mechanisms combine to regulate gastric secretion.

- Secretion occurs in three phases: cephalic, gastric, and intestinal.

- The cephalic phase of secretion occurs in response to the sight, smell, and anticipation of food, and as a result of chewing.

- Parasympathetic vagal fibres stimulate secretion via the release of acetylcholine, both directly and by stimulating gastrin secretion.

- The arrival of food in the stomach initiates the gastric phase of secretion, in which distension and the presence of amino acids and peptides stimulates the output of HCl and pepsinogen.

- Gastrin is an important mediator of this phase. Secretion is inhibited when the pH of the gastric contents falls to around 2 or 3.

- As partially digested food enters the duodenum, some gastrin is secreted from G-cells in the duodenal mucosa. This stimulates some further gastric secretion.

- Secretin, CCK, and GIP all contribute to the inhibition of gastric secretion.

Storage and mixing of gastric contents and gastric emptying

- The stomach stores food, mixes it with gastric juice, and breaks it into smaller pieces to form a semi-liquid chyme.

- The stomach then delivers chyme to the duodenum in a controlled fashion.

- The stomach is able to store large amounts of food since intragastric pressure rises very little despite significant distension of the stomach wall.

- The fasting stomach shows only weak contractile activity. After a meal, peristaltic contractions begin, increasing in power as they approach the antrum where mixing is most vigorous.

- The stomach normally empties at a rate compatible with full digestion and absorption by the small intestine. Many factors contribute to this regulation.

- Distension of the stomach increases the rate of emptying. The presence in the chyme of fats, proteins, high acidity, and hypertonicity all delay the rate of emptying.

- Vomiting is a protective mechanism whereby noxious or potentially toxic substances are expelled from the GI tract.

- The vomiting reflex is coordinated in the medulla of the brain.

- Prolonged vomiting can cause metabolic alkalosis through the loss of gastric acid. It may also cause hypokalaemia.

The small intestine

- The small intestine is the major site of both digestion and absorption in the GI tract. Here chyme is mixed with bile, pancreatic juice, and the intestinal secretions.

- The small intestine provides a huge surface area for nutrient absorption. The folded mucosal surface is covered with projections called villi.
- The brush border membranes of the mucosal epithelial cells house enzymes.
- Between the villi lie simple tubular glands, the crypts of Lieberkuhn. These contain many different cell types including mucus-secreting goblet cells and phagocytic cells.
- The small intestinal epithelium is self-renewing, replacing itself completely every 6 days or so.
- Loss of cells at the tips of the villi releases enzymes from the brush border of the enterocytes into the intestinal lumen. One of these, enteropeptidase, activates pancreatic trypsin which then activates other proteolytic enzymes.
- Crypt cells secrete 1.5 litres of isotonic fluid each day. Chloride is transported out of the cells, and sodium and water follow passively by the paracellular route.
- In the duodenum, Brunner's glands secrete an alkaline mucus, which helps to protect the epithelium from the corrosive effects of acidic chyme arriving from the stomach.
- Secretion by Brunner's glands is stimulated by vagal neurons and by CCK, secretin, gastrin, and endogenous prostaglandins.

Motility of the small intestine

- The rate at which chyme moves through the small intestine is carefully controlled to ensure adequate time for the completion of digestion and absorption.
- Segmentation is responsible for the mixing of chyme with enzymes and for exposing it to the absorptive mucosal surface.
- Peristaltic contractions propel chyme towards the ileocaecal valve.
- The motility of the intestinal smooth muscle is influenced by both intrinsic and extrinsic neurons and the neurotransmitters of the intramural plexuses.
- Parasympathetic activity enhances intestinal motility.
- The intestinal villi exhibit both piston-like contractions and swaying pendular movements. The latter may contribute to the mixing of chyme, while the former serve to facilitate the removal of fatty digestion products from the lacteals of the villi.
- In the fasting intestine, segmentation wanes and periodic bursts of peristaltic activity are seen in which the contents of the gut are swept long distances along the tract. These are called 'housekeeper' contractions.

The exocrine pancreas

- The exocrine portion of the pancreas consists of acinar cells, which secrete enzymes and fluid into a system of tiny ducts lined with epithelial cells.
- The epithelial cells secrete alkaline fluid.
- The ionic composition of the pancreatic juice depends upon its rate of secretion. At high rates of secretion, the bicarbonate content of the juice is higher than at lower rates.

- All the major enzymes required to complete the digestion of fats, carbohydrates, and proteins are contained within the pancreatic juice.
- Most of the proteolytic enzymes (trypsins) are stored in the acinar cells as inactive precursors (zymogen granules), to avoid self-digestion. Activation of these enzymes takes place in the duodenum.
- Pancreatic α-amylase is responsible for starch digestion to oligosaccharides in the duodenum. It is secreted in its active form.
- Several lipases are present in pancreatic juice. They hydrolyse water-insoluble triglycerides to release free fatty acids and monoglycerides.
- Control of exocrine pancreatic secretion is chiefly hormonal, although the initial cephalic phase of secretion is under the control of parasympathetic nerves.
- Gastrin contributes to the gastric phase of secretion, but about 70 per cent of secretion occurs during the intestinal phase in response to secretin and CCK.

Absorption of the products of digestion

- Absorption is the process by which the products of digestion are transported into the epithelial cells of the GI tract and thence into the blood or lymph draining the gut.
- Almost all of the absorption of water, electrolytes, and nutrients occurs in the small intestine.
- Monosaccharides are absorbed in the duodenum and upper jejunum by sodium-dependent cotransport driven by the sodium-potassium pump.
- Amino acids utilize similar mechanisms, though at least 10 separate transporters exist.
- The products of fat digestion are incorporated into micelles along with bile salts, lecithin, cholesterol, and fat-soluble vitamins. The fatty components of the micelle diffuse into the cells.
- Bile salts are recycled and the fats are reprocessed by the smooth ER to form chylomicrons. These are exocytosed across the basolateral cell membrane and enter the lacteals of the villi.
- The GI tract absorbs 8–10 litres of fluid and electrolytes each day. The active transport of sodium and nutrients is followed by anion movement and the absorption of water by osmosis.
- Failure to absorb fluid results in potentially life-threatening diarrhoea.
- Fat-soluble vitamins are absorbed along with the products of fat digestion. Most water-soluble vitamins are absorbed by facilitated transport.
- A specific uptake process involving gastric intrinsic factor is responsible for the absorption of vitamin B12.
- A number of conditions result from malabsorption of nutrients. Specific syndromes include carbohydrate intolerance, coeliac disease, and Crohn's disease.

The large intestine and defecation

- The large intestine consists of the caecum (which plays no significant role in humans), colon, rectum, and anal canal.

- Its main functions are to store food residues, secrete mucus, and absorb remaining water and electrolytes from the food residue.

- Faeces are eliminated via the anus.

- The colon absorbs 400–1000 ml of fluid each day. Sodium is actively transported from the lumen to the blood. Chloride moves in exchange for bicarbonate, and water follows by osmosis.

- Intestinal flora perform fermentation reactions that produce short chain fatty acids and flatus. The short chain fatty acids are absorbed by the colonocytes, stimulating salt and water uptake.

- Intestinal bacteria also synthesize certain vitamins such as vitamin K.

- The colon exhibits mixing movements (haustrations) and sluggish propulsive movements.

- 'Housekeeper' contractions occur several times daily, serving to move intestinal contents over longer distances.

- These contractions move fecal material into the rectum, and the resulting stretch of its wall elicits the urge to defecate.

- Between 100 and 150 g of faeces are eliminated each day.

- Defecation involves both voluntary and involuntary contractions of the anal sphincters and muscles of the abdominal wall and diaphragm.

 ## Recommended reading

Anatomy

Mackinnon, P.M., and Morris, J. (2005) *Oxford Textbook of Functional Anatomy*. Volume 2 *Thorax and Abdomen*. pp 131–66. Oxford University Press. Oxford.

> A clear description of the main structures of the gastrointestinal tract. Nice uncluttered anatomical illustrations.

Biochemistry

Elliott, W.H., and Elliott, D.C. (2009) *Biochemistry and molecular biology* (4th edn), Chapters 9 and 10. Oxford University Press, Oxford.

Histology

Merscher, A. (2009) *Junquiera's basic histology* (12th edn), Chapters 15 and 16. McGraw-Hill, New York.

> Clear photomicrographs and text, well illustrated.

Physiology

Johnson, L.R. (2006) *Gastrointestinal physiology* (7th edn). Mosby Year Book, St Louis.

> A short introductory monograph covering the essentials; not too detailed.

Smith, M.E., and Morton, D.G. (2010) *The digestive system: Basic science and clinical conditions* (2nd edn). Churchill-Livingstone, London.

> An all-in-one account of the digestive system.

Medicine

Ledingham, J.G.G., and Warrell, D.A. (eds) (2000) *Concise Oxford textbook of medicine*, Chapters 5.1–5.42. Oxford University Press, Oxford.

> A very detailed account of the medical problems associated with the GI tract. Includes some basic physiology.

Pharmacology

Rang, H.P., Ritter, J.M., Flower, R.J., and Henderson, G. (2016) *Pharmacology* (8th edn), Chapter 30. Churchill-Livingstone, Edinburgh.

Review articles

Bröer, S. (2008) Amino acid transport across mammalian intestinal and renal epithelia. *Physiological Reviews* **88**, 249–86, doi:10.1152/physrev.00018.2006.

Carpenter, G.H. (2013) The secretion, components and properties of saliva. *Annual Review of Food Science and Technology* **4**, 267–76.

Cura, A.J., and Carruthers, A. (2013) The role of monosaccharide transport proteins in carbohydrate assimilation, distribution, metabolism and homeostasis. *Comprehensive Physiology* **2**, 863–914, doi:10.1002/cphy.c110024.

Depoortere, I. (2014) Taste receptors of the gut: emerging roles in health and disease. *Gut* **63**, 179–90, doi:10.1136/gutjnl-2013-305112.

Horowitz, B., Ward, S.M., and Sanders, K.M. (1999) Cellular and molecular basis for electrical rhythmicity in gastrointestinal muscles. *Annual Review of Physiology* **61**, 19–43.

Mansbach, C.M., and Siddiqi, S.A. (2010) The biogenesis of chylomicrons. *Annual Review of Physiology* **72**, 315–33.

Moestrup, S.K. (2006) New insights into carrier binding and epithelial uptake of the erythropoietic nutrients cobalamin and folate. *Current Opinion in Hematology* **13**, 119–23.

Proctor, G.B., and Carpenter, G.H. (2007) Regulation of salivary gland function by autonomic nerves. *Autonomic Neuroscience: Basic and Clinical* **133**, 3–18.

Roussa, E. (2011) Channels and transporters in salivary glands *Cell and Tissue Research* **343**, 263–87;,doi:10.1007/s00441-010-1089-y.

 To check that you have mastered the key concepts presented in this chapter, complete the accompanying online self-assessment questions. Go to www.oup.com/uk/pocock5e/

CHAPTER 45

The liver and gall bladder

Chapter contents

This chapter should help you to understand:

- The anatomy of the liver and gall bladder
- The hepatic circulation
- The formation and secretion of bile
- The formation and excretion of bile pigments
- Different types of jaundice
- The role of the liver in whole-body energy metabolism
- The synthesis of plasma proteins by the liver
- The endocrine functions of the liver
- Detoxification by the liver
- Liver failure

45.1 Introduction

The liver plays a crucial role in energy metabolism: it stores glucose as glycogen (glycogenesis), converts amino acids to glucose (gluconeogenesis), and metabolizes fatty acids. It plays an important part in regulating plasma cholesterol levels and is important in the biosynthesis of almost all the plasma proteins except for the immunoglobulins. The liver secretes a number of peptide hormones,

including the insulin-like growth factors IGF-1 and IGF-2, hepcidin, and thrombopoietin. It also acts as a store of a number of vitamins including vitamin A, vitamin D, and vitamin B_{12}. Indeed, the stores of vitamin B_{12} are often sufficient to last the body for several years. The liver breaks down haemoglobin from senescent red blood cells to form bilirubin, which is then conjugated with glucuronic acid prior to its elimination via the bile. It also detoxifies many exogenous chemicals (including therapeutic drugs) and hormones, so that they can be excreted via the bile or the urine. Hepatic macrophages (Kupffer cells) remove bacteria and other debris from the blood. Finally, the liver converts ammonia derived from deamination of amino acids to urea. These functions of the liver are summarized in Table 45.1.

At an average weight of 1.3 kg, the liver is the largest internal organ. In adults it accounts for around 2 per cent of body weight, but in infants it may account for as much as 5 per cent. Together with the gall bladder, the liver occupies a large proportion of the right upper quadrant of the abdomen. It receives and processes the nutrient-rich venous blood from the GI tract and performs many vital metabolic and homeostatic functions. It secretes bile, which is crucial for the emulsification of fats prior to their digestion and subsequent absorption by the intestinal villi. Between meals, the gall bladder stores and concentrates bile in readiness for its delivery into the small intestine.

The liver has a significant functional reserve and is able to regenerate efficiently following damage or injury. Indeed, as much as 75 per cent of the liver can be removed without obvious loss of function. However, the consequences of more significant injury and ultimately of liver failure are far-reaching and reflect the wide range of functions performed by this organ.

This chapter is concerned with the major functions of the liver and gall bladder outlined above. It also outlines the common causes of different types of jaundice and will conclude with a brief consideration of the consequences of liver failure.

45.2 The structure of the liver

Figure 45.1 illustrates the main structural features of the liver, which consists of four lobes surrounded by a tough fibro-elastic capsule called Glisson's capsule. The major right and left lobes are

Table 45.1 The functions of the liver

Function	Comments
Synthesis and secretion of bile	Bile is important for fat absorption and for the excretion of bile pigments, which are derived from the breakdown of haemoglobin.
Carbohydrate metabolism	The liver contains the body's reserve of carbohydrate in the form of glycogen. It is also able to form glucose by gluconeogenesis.
Fat metabolism	Absorption of fats and fat-soluble vitamins is facilitated by bile salts. The liver stores fat-soluble vitamins and synthesizes lipoproteins.
Protein metabolism	The liver is the major source of plasma proteins including albumin and the circulating clotting factors.
Endocrine	The liver synthesizes and secretes IGF-1 and IGF-2 in response to the secretion of growth hormone by the anterior pituitary. It also secretes hepcidin, which regulates iron absorption from the small intestine, and thrombopoietin, which regulates the number of platelets in the blood.
Detoxification	The liver inactivates hormones, conjugates various drugs and toxins for excretion, and converts ammonia to urea.
Iron storage	Essential for erythropoiesis.
Storage of vitamin B_{12}	Required for normal erythropoiesis.
General	As the liver is situated between the gut and the general circulation, it is able to protect the body by inactivating toxic materials absorbed from the gut.

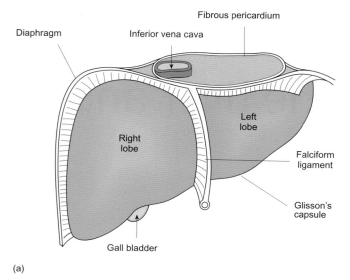

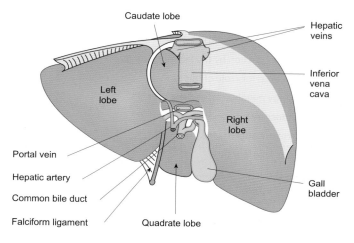

Figure 45.1 The gross anatomy of the liver. Panel (a) shows the view from the front and panel (b) the view from the back. Note the major division by the falciform ligament and the position of the gall bladder, here shown in green. (Based on Fig 6.6.2 in Vol. 2 of P.C.B. McKinnon and J.F. Morris (2005) *Oxford textbook of functional anatomy*, 2nd edition. Oxford University Press, Oxford.)

separated by the falciform ligament, which is attached to the diaphragm and anterior abdominal wall. The smaller visceral and quadrate lobes are on the posterior surface of the liver. A dorsal mesentery, the lesser omentum, attaches the liver to the lesser curvature of the stomach (see Figure 43.3). Its accessory gland, the gall bladder, is a pear-shaped sac 7–10 cm in length that rests in a recess on the inferior surface of the right lobe of the liver. The liver secretes bile which leaves the liver through the left and right **bile ducts**. These subsequently fuse to form the large common hepatic duct, as shown in Figure 45.2. As this duct passes towards the duodenum it fuses with the **cystic duct**, draining the gall bladder to form the **bile duct** which is separated from the duodenum by a ring of smooth muscle, the **sphincter of Oddi**, which regulates the entry of both bile and pancreatic juice into the duodenum.

The liver consists of between 50,000 and 100,000 structural and functional units called **lobules**. These are irregularly shaped polyhedral structures each bounded by a thin layer of connective tissue that, in humans, is not well-defined. When cut across, the lobules

are roughly hexagonal in shape and are 1–2 mm in diameter. As Figure 45.3a shows, a central vein lies at the centre of each lobule. Structures known as the **portal triads** are found at the boundaries between lobules. As Figure 45.3b shows, the portal triads comprise a branch of the hepatic artery, a branch of the portal vein, and a bile duct. The basic structure of part of a hepatic lobule is shown in Figure 45.4.

The cells that make up the bulk of the liver are the **hepatocytes**. These are large polygonal cells 20–30 μm in diameter which are arranged in sheets one or two cells thick that radiate from the central vein. The sheets of cells anastomose freely to form a sponge-like structure that is filled with blood from the portal vein and hepatic

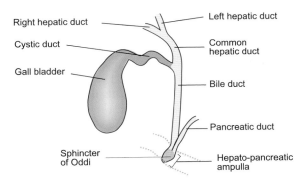

Figure 45.2 The anatomical relationships between the hepatic ducts, the cystic duct, and the bile duct. The pancreatic duct joins with the bile duct as it reaches the hepatopancreatic ampulla. The smooth muscle of the ampulla forms the sphincter of Oddi, which controls the secretion of bile and pancreatic juice into the duodenum. The grey dotted lines indicate the position of the duodenal wall.

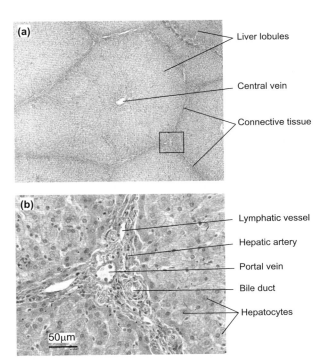

Figure 45.3 The histological structure of the liver. Panel (a) shows several liver lobules, each of which is bounded by connective tissue septa. The central vein of the principal lobule shown in the field of view is also clearly visible. Panel (b) shows the anatomical relationships of the hepatic artery, portal vein, and bile duct that constitute a portal triad.

artery. The irregularly shaped spaces between the sheets of cells are known as **sinusoids**. They are lined with a discontinuous fenestrated endothelium that is separated from the surface of the hepatocytes by a narrow cleft called the space of Disse (Figure 45.4). Stellate pericytes (also called Ito cells or fat-storing cells) surround the vascular epithelium. These are antigen-presenting cells that also store vitamin A.

Since the endothelium is discontinuous, there is almost no barrier between the sinusoids and the hepatocytes, thus allowing the free exchange of materials with a molecular mass below 250,000 Da. The hepatocytes themselves are richly endowed with rough and smooth endoplasmic reticulum, mitochondria, and lysosomes. Microvilli are present on that part of the plasma membrane that faces the blood sinusoids, an adaptation that provides a large surface area for the exchange of substances with the blood.

Adjacent hepatocytes are joined together by both gap junctions and tight junctions. The gap junctions provide a means of

signalling between cells, while the tight junctions seal off a narrow space to form a **bile canaliculus** into which the hepatocytes secrete bile. The bile flows towards the edge of the lobule, where it enters the bile ductules (also called the canals of Hering) which are lined with a cuboidal epithelium. From the bile ductule, the bile enters a bile duct at a portal triad. Note that within the lobule,

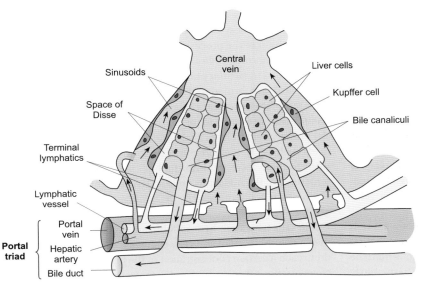

Figure 45.4 Part of a liver lobule and its vascular supply. The small arrows indicate the direction of flow for the blood, bile, and lymph. The central vein drains into the main hepatic vein which empties into the inferior vena cava. Note that the direction of blood flow is in the opposite direction to that of the bile (arrows). (Adapted from A.C. Guyton et al (1975) *Circulatory physiology*, Vol 2: *Dynamics and control of the body fluids*. Saunders, Philadelphia.)

the flow of bile is in the opposite sense to the flow of blood (see Figure 45.4). As the bile flows towards the duodenum it enters a sequence of progressively larger ducts until it reaches the common bile duct.

Summary

The liver consists of four lobes surrounded by a tough fibro-elastic capsule (Glisson's capsule). The major right and left lobes are separated by the falciform ligament which attaches the liver to the diaphragm and anterior abdominal wall. The basic structural unit of the liver is the hepatic lobule, which consists of sheets of liver cells (hepatocytes).

The liver performs a large variety of metabolic functions including the synthesis of most plasma proteins, glucose storage (as glycogen), gluconeogenesis, and detoxification. The liver also secretes several hormones, including the IGFs and hepcidin, and forms the bile, which is essential for the digestion of fat.

45.3 The hepatic circulation

At rest the liver normally receives about 25 per cent of the cardiac output and is unique among the abdominal organs in having a dual blood supply from the portal vein and hepatic artery. The gall bladder is supplied by the cystic artery, which arises from the right hepatic artery, and is drained by cystic veins.

The hepatic artery carries about 400 ml min^{-1} of oxygenated blood to the liver, while the portal vein supplies about 1000 ml min^{-1} of nutrient-rich, but deoxygenated, blood. The portal blood also contains breakdown products of haemoglobin from the spleen. Small portal venules lying in the septa between lobules receive blood from the portal veins. From the venules, blood flows into the hepatic sinusoids (see Figures 45.4 and 45.6) which empty into the central vein. Fluid in the space of Disse is drained by terminal lymphatic vessels which empty into the lymphatic vessels of the portal triads. The interlobular septa also contain arterioles derived from branches of the hepatic artery. Many of these arterioles drain directly into the sinusoids and supply them with fully oxygenated blood.

The sinusoids essentially form a leaky capillary network from which blood flows from the portal vein and hepatic artery to the central veins of the lobules. Ultimately, the deoxygenated blood from the central veins empties into the hepatic veins, which join the inferior vena cava just below the level of the diaphragm. Since the pressure in the portal vein is about 1.3 kPa (10 mm Hg) while that in the hepatic vein is only slightly lower (about 0.6 kPa or 5 mm Hg), the veins of the liver contain 200–400 ml of blood, providing a reservoir that can be shunted back into the systemic circulation during periods of hypovolaemia or shock. The hepatic veins therefore act as capacitance vessels (see Chapter 30).

The sinusoids are lined with two types of cells: endothelial cells typical of blood vessels, and Kupffer cells (macrophages) that remove particulate matter including debris and pathogens from the circulation by phagocytosis. This arrangement is particularly important for the removal of any intestinal bacteria arriving at the liver by way of the portal vein.

The blood flow to the liver is regulated both by sympathetic nerves and by circulating hormones. Cholecystokinin (CCK) and secretin enhance blood flow to the liver during digestion. Conversely, the activation of the sympathetic nerves to the liver that occurs in exercise causes an increase in the vascular tone of the arterioles and veins. This both decreases blood flow to the liver and mobilizes blood in the liver for use by the exercising muscle. A similar redistribution of blood occurs following haemorrhage.

Summary

At rest the liver normally receives about 25 per cent of the cardiac output and receives blood from the hepatic artery and the portal vein. The portal vein delivers nutrient-rich blood from the gut, and the hepatic artery supplies oxygenated blood. The blood flows from the portal vein and hepatic artery to the central veins of the lobules via venous sinusoids. Blood returns to the systemic circulation via the hepatic vein. Hepatic blood flow is regulated by sympathetic nerves and by hormones.

45.4 Bile production by the hepatocytes

The hepatocytes secrete a fluid, known as **hepatic bile**, into the bile canaliculi. Hepatic bile has a pH of between 7 and 8, is isotonic, and resembles plasma in its ionic composition. It also contains bile salts, bile pigments, cholesterol, lecithin, and mucus. As it passes along the bile ducts, the ductal epithelial cells modify this primary secretion by secreting a watery, bicarbonate-rich fluid that adds considerably to the volume of the bile. Overall, the liver produces 600–1000 ml of bile each day which may be discharged continuously into the duodenum or stored in the gall bladder, during which time its composition changes.

The chemical nature of bile acids and bile salts

Each day the liver synthesizes around 500 mg of bile salts, which are important in the processing of fats within the small intestine. The precursors of the bile salts, the bile acids, are synthesized from cholesterol. The **primary bile acids**, cholic acid and chenodeoxycholic acid, are formed in the hepatocytes themselves, but small amounts of the **secondary bile acids**, deoxycholic acid and lithocholic acid, are formed in the intestine by the action of bacteria on the primary acids. The primary bile acids are conjugated to amino acids such as glycine and taurine to form water-soluble **bile salts** prior to their secretion into the bile. Figure 45.5 illustrates the structures of the bile acids and the structure of glycocholic acid, one of the bile salts formed by the conjugation of cholic acid with glycine.

Bile salts have both hydrophobic and hydrophilic regions (i.e. they are amphipathic) and they aggregate to form micelles when they reach a certain concentration in the bile (the critical micellar concentration). The micelles are organized so that the hydrophilic

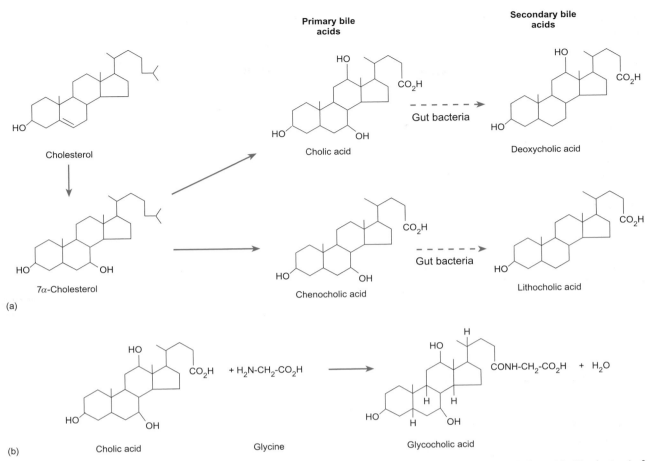

Figure 45.5 The structure of the bile acids and their formation from cholesterol (a) and the conjugation of cholic acid with glycine to form the bile salt glycocholic acid (b). The dotted arrows in (a) indicate transformations by the gut flora.

groups of the bile salts face the aqueous medium while the hydrophobic groups face each other to form a core. This amphipathic character of the bile salts is of key importance to their role in the emulsification of fats (see Chapter 44).

Bile acid-dependent and bile acid-independent components of bile secretion

The processes involved in the formation of hepatic bile are summarized in Figure 45.6. Two distinct secretory mechanisms are involved in the elaboration of bile by the liver, giving rise to the so-called bile acid-dependent and bile acid-independent components of bile.

- The rate at which bile salts are actively secreted into the canaliculi depends upon the rate at which bile acids are returned from the small intestine to the hepatocytes via the **enterohepatic circulation**. This component of bile secretion is therefore referred to as the bile acid-dependent fraction. The bile salts are actively taken up from the blood by sodium-dependent cotransporters (including a specific sodium-dependent taurocholate transport protein) and pass into the bile canaliculi via a bile salt export pump.

- The bile acid-independent fraction of bile secretion refers to the secretion of water and electrolytes by the hepatocytes and the ductal epithelial cells. Sodium is transported actively into the bile canaliculi and is followed by the passive movement of chloride ions and water. Bicarbonate ions are actively secreted into the bile by the ductal cells and are followed by the passive movement of sodium and water.

Bile salts are recycled via the enterohepatic circulation

About 94 per cent of the bile salts that enter the intestine in the bile are reabsorbed into the portal circulation by active transport from the distal ileum. Many of the bile salts return to the liver unaltered and are recycled. Some are deconjugated in the gut lumen and returned to the liver for reconjugation and recycling. A small fraction of the deconjugated bile acids may undergo modification by intestinal bacteria to secondary bile acids (see Figure 45.5). Some of these are relatively insoluble and are excreted in the faeces—particularly lithocholic acid. It is estimated that bile acids may be recycled up to 20 times before they are finally excreted. A schematic illustration of the enterohepatic circulation of the primary bile acids is shown in Figure 45.7.

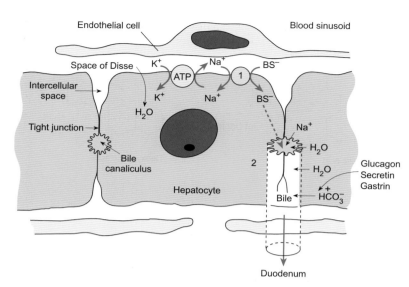

Figure 45.6 The processes involved in the formation of bile by the hepatocytes (BS = bile salt). The bile salts in the blood are transported into the hepatocytes by sodium-dependent organic anion transporters. They are actively transported into the bile canaliculi by an active bile salt export pump. The epithelia of the bile canaliculi transport sodium and bicarbonate ions into the lumen and water follows passively. This process is stimulated by secretin, glucagon, and gastrin.

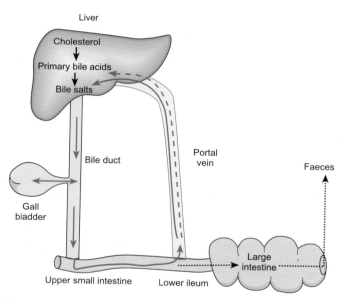

Figure 45.7 The enterohepatic circulation. Only the recirculation of the primary bile acids is illustrated. The transport of conjugated bile acids is shown by the solid green line and that of the deconjugated bile acids by the broken green line. A small fraction of the bile salts are excreted in the faeces.

The regulation of bile secretion

The major regulator of the rate of production of hepatic bile is the return of bile salts to the hepatocytes via the enterohepatic circulation, as it provides the driving force for fluid transport into the biliary system. Although the production of hepatic bile is not under hormonal control, the secretion of bicarbonate-rich watery fluid by the ductal epithelial cells is enhanced by secretin and, to a lesser extent, by glucagon and gastrin. These substances are called **choleretics**. A further stimulus to the secretion of hepatic bile is thought to be the increase in blood flow to the liver that follows a

meal. However, as a meal will increase the rate of reabsorption of bile salts via the enterohepatic circulation it will also stimulate the bile acid-dependent fraction of bile secretion.

The role of the gall bladder

The gall bladder stores bile that is not required immediately for digestion. Between meals, most of the bile produced by the liver is diverted into the gall bladder because of the relatively high level of tone in the sphincter of Oddi. The mucosa of the gall bladder, like that of the stomach, is thrown into folds (rugae) when the organ is empty. These can expand, allowing the gall bladder to accommodate 50–60 ml of bile during the period between meals.

The gall bladder concentrates the bile (to form so-called gall bladder bile) by absorbing sodium, chloride, bicarbonate, and water by the processes outlined in Figure 45.8. The extent of this

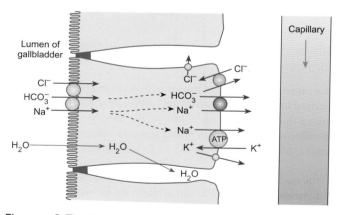

Figure 45.8 The absorption of salt and water in the gall bladder, illustrating the standing gradient hypothesis. The concentration of bile begins with the cotransport of sodium, chloride, and bicarbonate across the apical membranes followed by the efflux of sodium and bicarbonate across the basolateral membranes and into the intercellular space. Water follows passively.

Table 45.2 Typical values for the relative concentrations of some constituents of hepatic bile and gall bladder bile

Solute	Solute concentration ratio (gall bladder bile/hepatic bile)
Na^+	1.7
K^+	3.0
Ca^{++}	5.0
HCO_3^-	0.63
Cl^-	0.3
bile acids	15
bilirubin	4.5
cholesterol	6.0
lecithin	8.0
fatty acids	8.9
protein	2.5
bile osmolality (there is little difference between hepatic and gall bladder bile)	290–300 mosmol kg^{-1}
secreted volume of hepatic bile	250–1100 ml day^{-1}

concentration is shown by the relative concentration ratios of solutes in gall bladder and hepatic bile presented in Table 45.2. Note that the concentration of bile salts in the bile of the gall bladder may be increased 15- to 20-fold.

The mucosa of the gall bladder consists of a single layer of tall, columnar epithelial cells bound at their apical regions by tight junctions so that long lateral channels form between the cells. As salts are transported into these channels, local regions of high osmotic pressure are created, with tonicity highest at the apical regions of the channel. This sets up a standing osmotic gradient that permits the continuous absorption of water from the gall bladder to the interstitial fluid.

The primary mechanism responsible for the concentration of bile is the cotransport of sodium, chloride, and bicarbonate across the apical membranes and the subsequent efflux of sodium by the basolateral sodium pump. Water follows passively. Bicarbonate crosses the basolateral membrane by Na^+-HCO_3^- cotransport and Cl^-/HCO_3^- exchange. Chloride can equilibrate with the surrounding fluids via chloride channels. As water is absorbed, potassium concentrations rise but subsequently fall as a favourable electrochemical gradient is established for its uptake into the epithelial cells.

Contraction of the gall bladder forces bile into the duodenum

Within a few minutes of starting a meal, particularly one that is rich in fats, the smooth muscle of the gall bladder contracts, forcing bile towards the duodenum. This initial response is mediated via the vagal nerves, but the major stimulus for contraction is CCK, a hormone secreted in response to the presence of fatty and acidic chyme in the intestine (see Chapter 44). CCK also stimulates pancreatic enzyme secretion and relaxes the sphincter of Oddi, thereby allowing bile and pancreatic juice to enter the duodenum. Substances that increase the flow of bile by stimulating the gall bladder to contract are known as **cholagogues**. The gall bladder normally empties completely about an hour after a fat-rich meal, and its contents maintain the level of bile salts in the duodenum above the critical micellar concentration.

Summary

The liver secretes 600–1000 ml of bile each day. This bile is stored and concentrated in the gall bladder and delivered to the duodenum following a meal. Bile salts have both a hydrophobic and hydrophilic region so that at high concentrations, they aggregate together to form micelles. Bile salts are vital for fat digestion in the small intestine.

About 94 per cent of the bile salts that enter the small intestine return to the liver via the enterohepatic circulation. This stimulates the formation of bile. CCK, secreted in response to the presence of chyme in the duodenum, stimulates contraction of the gall bladder.

45.5 The excretory role of bile

The elimination of bile pigments

Bile pigments are the excretory products of the haem portion of haemoglobin and account for around 0.2 per cent of the total bile composition. The bile pigments are responsible for the characteristic colour of the bile. The major bile pigment is bilirubin which is formed when old red blood cells are broken down in the spleen. Bilirubin is relatively insoluble and is carried to the liver mainly in combination with plasma albumin. In the hepatocytes about 80 per cent of the bilirubin becomes conjugated with glucuronic acid to form bilirubin diglucuronide, which is water-soluble. The remaining bilirubin is conjugated with sulphate or with a variety of other hydrophilic agents.

In the large intestine, bilirubin diglucuronide is hydrolysed by bacteria to form three products: urobilinogen (which is highly water-soluble and colourless), stercobilin, and urobilin (which give the faeces their characteristic brown colour). Some of the urobilinogen is absorbed from the intestine back into the bloodstream. From there it is either re-secreted back into the bile by the liver or excreted by the kidneys into the urine. The processes of bilirubin formation, circulation, and elimination are shown diagrammatically in Figure 45.9.

In addition to the bile pigments, many other substances are excreted via the bile. They then enter the small intestine where they may be recycled via the enterohepatic circulation or excreted in the faeces. Examples are morphine and the anti-tuberculosis drug rifampicin.

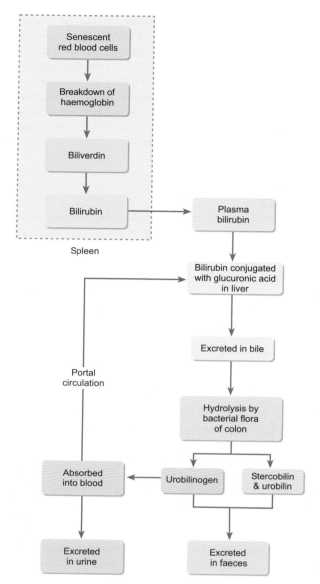

Figure 45.9 An outline of the formation and excretion of the bile pigments. Bilirubin is formed by the breakdown of the haem portion of haemoglobin derived from senescent red blood cells. It is excreted in the bile and ultimately in the faeces. A small quantity is metabolized to urobilinogen by the gut flora and excreted in the urine.

Accumulation of bilirubin in the blood causes jaundice

If the concentration of bilirubin in the blood exceeds a value of around 34 µmol l^{-1} (hyperbilirubinaemia), a condition called **jaundice**, or icterus, develops. It is characterized by yellow discoloration of the skin, the sclera of the eyes, and the deep tissues. Figure 45.10 shows the characteristic yellow discoloration of the eyes and skin of a patient suffering from severe jaundice.

There are many causes of jaundice, the most important of which are excessive haemolysis of red cells, impaired uptake of bilirubin by hepatocytes, and obstruction of bile flow either through the bile canaliculi or the bile ducts. Excessive haemolysis may occur following a poorly matched blood transfusion or in certain hereditary disorders. Jaundice is also seen in newborns whose fetal red cells are haemolysing more quickly than the immature liver can process the bilirubin. The jaundice that develops in these conditions is called **haemolytic jaundice**.

Jaundice resulting from a failure of the liver to take up or conjugate bilirubin is known as **hepatic jaundice**. Hepatitis and cirrhosis are the most common causes of this disorder. Due to a genetic defect, some people have a blood bilirubin that is higher than normal. This is relatively common (about 4 per cent of the population) and is known as **Gilbert's syndrome** or unconjugated hyperbilirubinemia. This condition is usually benign and requires no treatment, but those affected may suffer bouts of jaundice.

Obstructive jaundice occurs if bile is prevented from flowing from the liver to the intestine. Common causes are gallstones, strictures or tumours of the bile duct, and pancreatic tumours. Pruritus (itching) often accompanies this type of jaundice and is caused by the accumulation of bile acids in the blood. In a patient suffering from obstructive jaundice, the faeces are pale in colour due to the absence of bilirubin in the bile and often contain fatty streaks due to the lowered absorption of dietary fat. The urine, however, is darker than normal due to the increased excretion of bilirubin via the kidneys.

The regulation of plasma cholesterol

Cholesterol is an essential component of the plasma membrane (see Chapter 4) and is required for the synthesis of the bile acids, steroid hormones, and vitamin D. It can be obtained from the diet, particularly butter, eggs, and some offal, but most of the cholesterol used by the body is synthesized in the liver and intestine. Excess cholesterol is excreted via the bile along with phospholipids, particularly lecithin. Both are secreted as lipid vesicles, which are emulsified by the bile salts to form micelles in which the cholesterol partitions into the hydrophobic core while the lecithin, which is amphipathic, lies partly in the core and partly near the outer surface of the micelle.

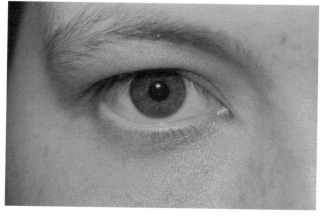

Figure 45.10 The characteristic yellow discoloration of the eyes and skin in a patient suffering from obstructive jaundice. (From http://www.nhs.uk/conditions/Jaundice/Pages/Introduction.aspx.)

Cholesterol circulates in the blood in combination with lipoproteins of various kinds, named according to the relative amounts of protein and lipid that they contain. Low-density lipoprotein (LDL), which contains a relatively high proportion of fat, is primarily responsible for carrying cholesterol from the liver to the tissues that use it as a biosynthetic precursor. High density lipoprotein (HDL), which contains relatively more protein, carries cholesterol away from tissues and to the liver where it may be utilized by the hepatocytes, incorporated into bile salts, or excreted via the bile.

Although cholesterol is vital to normal health, high plasma levels of cholesterol pose a risk. Excess circulating cholesterol may be deposited within arteries, where it leads to the formation of atheromatous plaques that can narrow or even block a vessel. Atherosclerosis of the coronary arteries may precipitate angina pectoris or a myocardial infarction, while blockage of arteries within the brain may lead to a stroke. It is therefore evident that the synthesis, metabolism, and excretion of cholesterol must be closely regulated in order to ensure that the plasma concentration remains at a desirable level (below 200 mg dl^{-1} or 5 mmol l^{-1}). Furthermore, it is important that plasma levels of LDL are kept relatively low (100 mg dl^{-1} or less) and that HDL is kept relatively high (60 mg dl^{-1} or more), as the former carries cholesterol to tissues such as the heart and skeletal muscle where it may cause vascular damage, while the latter carries cholesterol away from these tissues. A high LDL level has been linked to an increased risk of coronary artery disease even in patients whose total cholesterol levels are within the desirable range. Indeed, the ratio of LDL to HDL or of total cholesterol to HDL may be better indicators of health risk than total cholesterol alone. Currently, a LDL:HDL ratio below 3.5:1 and a total cholesterol:HDL ratio below 5:1 are considered desirable. (Optimal values are thought to be LDL:HDL ~2.5:1 and total cholesterol:HDL ~3.5:1.)

Since the liver is primarily responsible both for the synthesis and excretion of cholesterol, it plays a vital role in the overall maintenance of normal plasma cholesterol levels. This regulation is achieved mainly through alterations in the hepatic synthesis of cholesterol. The rate of cholesterol synthesis is determined principally by the activity of the enzyme 3-hydroxy-3-methylglutaryl CoA reductase (HMG CoA reductase), which catalyses an important step in the biosynthetic pathway. The activity of this enzyme appears to be significantly altered by dietary intake of cholesterol. When the dietary intake is high, the activity of HMG CoA reductase is inhibited, and vice versa. In this way, in health, a desirable plasma cholesterol level is maintained. Unfortunately, a number of environmental and physiological factors may contribute to a raised circulating cholesterol level. These include the ingestion of large amounts of saturated fats, smoking cigarettes, genetic factors, and oestrogen levels (oestrogens tend to lower plasma cholesterol). An inherited condition called familial hyperlipidaemia (or hypercholesterolaemia), caused by a single mutant gene on the short arm of chromosome 19, results in an extremely elevated plasma cholesterol level (especially in the form of LDL) and a markedly increased risk of coronary heart disease.

A group of drugs collectively known as **statins** are widely prescribed to individuals known to be at an increased risk of cardiovascular disease or who have a high plasma cholesterol level. These drugs act by inhibiting the action of the enzyme HMG CoA reductase and thus reducing the amounts of cholesterol synthesized by the liver. Examples of statins include simvastatin, atorvastatin, and pravastatin.

Gallstones

Any excess cholesterol that cannot be dispersed into micelles may form crystals in the bile and contribute to the formation of gallstones in the hepatic ducts or the gall bladder by acting as nuclei for the deposition of calcium and phosphate salts. If the common bile duct becomes blocked by a gallstone, bile cannot enter the duodenum and accumulates in the gall bladder and liver, resulting in jaundice. Gallstones can be imaged by ultrasound, as in the example shown in Figure 45.11. Although surgical removal of the gall bladder is often carried out in patients whose gallstones are obstructing the flow of bile and causing pain, other treatments may be useful in some cases. These include drug treatments such as ursodeoxycholic acid, which slowly dissolves gallstones made of cholesterol, and lithotripsy, in which ultrasonic waves are directed at the gallstones to break them into smaller fragments that can pass along the bile duct and into the gut without causing pain. The fragments then leave the body via the faeces.

Summary

Bilirubin is the principal breakdown product of the haem derived from the haemoglobin of senescent red cells. It is excreted in the bile as bilirubin diglucuronide, which is water-soluble. Failure to excrete bile pigments leads to their accumulation in the blood and the development of jaundice.

The liver synthesizes most of the cholesterol needed by the body. Cholesterol is transported in the plasma in combination with lipoproteins. Low density lipoproteins (LDLs) carry cholesterol to the tissues that use it for synthetic processes. High density lipoproteins (HDLs) transport cholesterol to the liver for excretion or incorporation into bile salts. Cholesterol is excreted in the bile and may contribute to the formation of gallstones in the hepatic ducts or the gall bladder.

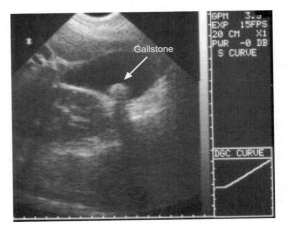

Figure 45.11 An ultrasound image of a gallstone in the gall bladder. Note the shadow cast by the gall stone. The outline of the gall bladder is clearly visible.

45.6 Energy metabolism and the liver

The control of whole-body energy metabolism is driven by the need to ensure an adequate supply of glucose for the brain. It is dependent both on an adequate food intake and on the ability of the body to store nutrients between meals. The following account provides a brief summary of the ways in which the liver contributes to the control of energy metabolism.

Glycogen storage and breakdown

Following a meal, blood glucose is high and this stimulates the secretion of insulin by the β-cells of the pancreas (see Chapter 24). Insulin promotes the uptake of glucose by the liver and muscle, which store it as glycogen. The glycogen within the liver represents an important store of carbohydrate for the whole body and amounts to an energy reserve of 600 kcals (2500 kJ) or more. This store can readily be mobilized when plasma glucose levels begin to fall, for example between meals. A fall in plasma glucose triggers the secretion of glucagon, while that of insulin declines. Glucagon stimulates the breakdown of glycogen to glucose (glycogenolysis), which is released into the blood for use by the brain and other tissues. During exercise, or other forms of acute stress, the secretion of adrenaline (epinephrine) from the adrenal medulla is increased. This hormone also increases the rate of glycogenolysis. The relationships between food intake, glycogen stores, and glucose utilization are summarized in Figure 45.12, together with the roles of the key regulatory hormones.

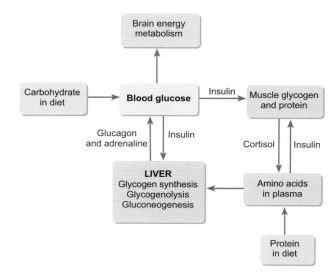

Figure 45.12 An outline of the central role of the liver in the regulation of plasma glucose. During a meal, plasma glucose rises and stimulates the secretion of insulin, which promotes the uptake of glucose by the liver (as well as other organs). Within the liver, glucose is stored as glycogen. As the plasma glucose begins to fall, the secretion of insulin also falls while that of glucagon rises. The glucagon promotes the breakdown of stored glycogen (glycogenolysis) and the release of glucose into the blood. The interplay between insulin and glucagon helps to maintain plasma glucose relatively constant. When the glycogen stores of the body are exhausted, muscle protein is broken down under the influence of cortisol and the amino acids are transported to the liver for conversion to glucose by gluconeogenesis.

Gluconeogenesis

The synthesis of glucose from non-carbohydrate precursors is known as **gluconeogenesis**, which is stimulated during fasting by a number of hormones including glucagon, growth hormone, and cortisol. The principal site of gluconeogenesis is the liver, although some also takes place in the kidneys. Before amino acids can be used for gluconeogenesis, they must first be deaminated (see Chapter 4). This process takes place in the liver and results in the production of ammonium ions (NH_4^+) which are subsequently metabolized to urea. Once deaminated, the carbon skeletons of most amino acids can be used for gluconeogenesis. The exceptions are the ketogenic amino acids leucine and lysine which are metabolized via the β-oxidation pathway (see Chapter 4).

The main stages of gluconeogenesis are outlined in Figure 45.13. The process begins in the mitochondria with the formation of oxaloacetate through the carboxylation of pyruvate. Subsequent steps take place in the cytoplasm. Here oxaloacetate is simultaneously decarboxylated and phosphorylated by the enzyme phosphoenolpyruvate carboxykinase to produce phosphoenolpyruvate, which is the starting point for the formation of glucose 6-phosphate. The final step involves the hydrolysis of glucose 6-phosphate to glucose. This takes place in the lumen of the endoplasmic reticulum, in which the membrane-bound enzyme glucose

6-phosphatase is located. Gluconeogenesis is of vital importance during periods of fasting, as the body's glucose reserves are only sufficient to meet the metabolic requirements of the brain and other glucose-dependent tissues for about a day.

The role of the liver in lipid metabolism

In addition to its regulation of glucose metabolism, the liver has a key role in the regulation of lipid metabolism. Under conditions of plenty, fatty acids produced by the liver together with those obtained from the diet are esterified and secreted into the blood as very low density lipoproteins (VLDLs). In adipose and muscle tissue, the VLDLs are hydrolysed by lipoprotein lipase to release free fatty acids and glycerol. The fatty acids are taken up by these tissues while the glycerol is transported to the liver or kidneys, where it enters the glycolytic pathway. In starvation, the liver converts fatty acids to acetoacetate and β-hydroxybutyrate (ketone bodies) which are released into the blood for utilization by the CNS and muscle tissue. (The liver itself cannot utilize the ketone bodies for energy metabolism.) The switch between the secretion of fatty acids and the production of ketone bodies is regulated by the metabolic intermediate malonyl CoA, which is present in high concentrations following a meal, but declines markedly during starvation.

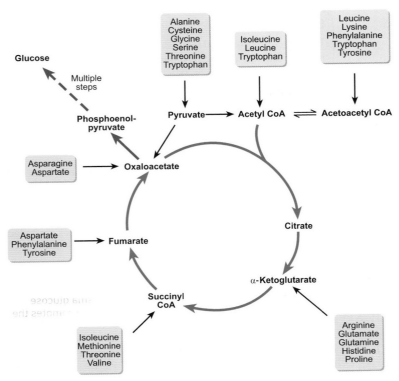

Figure 45.13 A diagram showing the routes by which amino acids can be converted into glucose (gluconeogenesis). After deamination, the glucogenic amino acids enter the biosynthetic pathway via oxaloacetate, fumarate, α-ketoglutarate, or succinyl CoA depending on their structure. Oxaloacetate is decarboxylated and phosphorylated to give rise to phosphoenol pyruvate, a key intermediate in the pathway for glucose synthesis. Those amino acids that directly give rise to acetyl CoA or acetoacetyl CoA are ketogenic rather than glucogenic, although several can be metabolized to oxaloacetate. Only leucine and lysine cannot be converted to glucose.

The synthesis of plasma proteins

The liver secretes more protein than any other organ in the body and produces most of the plasma proteins including albumins, globulins, and the protein clotting factors. The albumins are the smallest and the most abundant of the plasma proteins, accounting for about 60 per cent of the total. They are transport proteins for lipids and steroid hormones and provide most of the colloid osmotic pressure (the oncotic pressure—see Chapter 2) that determines the flow of water and solutes across the walls of the capillaries (see Box 31.3). Consequently, they also play an important role in body fluid balance.

Globulins account for about 40 per cent of total plasma protein (see also Chapter 25). The α- and β-globulins are made in the liver and transport lipids and fat-soluble vitamins in the blood, while the γ-globulins are antibodies produced by the lymphocytes (white blood cells) in response to exposure to antigens (see Chapter 26). Most of the clotting factors (e.g. fibrinogen and prothrombin) are synthesized by the liver, the main exceptions being Factor III (tissue factor, also called thromboplastin) and calcium ions.

Summary

The liver is able to store glucose as glycogen, and release it as required. Glycogen storage is promoted by insulin while its breakdown is promoted by glucagon and adrenaline. The liver is also able to form glucose from non-carbohydrate precursors including amino acids. The liver synthesizes and secretes most of the plasma proteins, including albumin, globulins, and most of the clotting factors.

45.7 Endocrine functions of the liver

The liver produces hepcidin, which regulates iron metabolism and the insulin-like growth factors. It synthesizes 25-hydroxycholecalciferol from vitamin D, which is an important step in the synthesis of calcitriol. It also synthesizes and secretes angiotensinogen, which is the precursor of angiotensins I and II, and thrombopoietin, which stimulates the differentiation of stem cells into megakaryocytes to increase the number of platelets circulating in the blood (see Chapter 25).

Insulin-like growth factors

Although growth hormone is secreted by the anterior pituitary gland, it exerts many of its actions indirectly via peptide hormone intermediaries known as insulin-like growth factors (IGF-1 and IGF-2), which are synthesized and secreted by the liver and some other tissues. IGF-1 and IGF-2 are structurally similar to proinsulin (hence their names). They stimulate the clonal expansion of chondrocytes, the formation and maturation of osteoblasts within the growth plates, and the formation of osteoid, the organic matrix of bone (see Chapter 51). The principal actions of IGF-1 are mediated via a specific receptor known as IGF-1R, although it is also known to bind to the insulin receptor with a much lower affinity. The blood level of IGF-1 is low in infancy, rises gradually until puberty, then increases rapidly to reach a peak that coincides with the peak height increase of the growth spurt (see Figure 51.2), after which it falls to its adult (and pre-pubertal) value. IGF-2 also binds to the IGF-1 receptor as well as to its own specific receptor, which acts as a binding

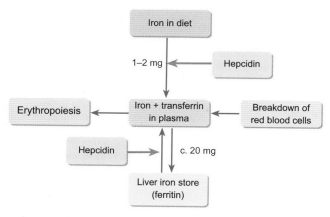

Figure 45.14 The role played by the liver in the metabolism of iron. Hepcidin is a hormone secreted by the liver that inhibits both the uptake of iron from the small intestine and the mobilization of iron from stored ferritin.

site to regulate the quantity of IGF-2 available in the blood. The vast majority of IGF-1 (> 97 per cent) is carried in the blood bound to specific binding proteins, which are also synthesized by the liver. The actions of the IGFs are discussed in more detail in Chapter 21.

Hepcidin and iron metabolism

Iron is an essential component of the body and an adult has a total of about 4.5 grams of iron. About two-thirds of this is within the haemoglobin of red blood cells. A further 5 per cent or so is contained within myoglobin and cellular enzymes such as cytochromes, while the remainder is stored in the form of ferritin, largely by the liver.

When red blood cells become senescent, they are removed from the blood by phagocytes in the liver and spleen. Much of the iron derived from haemoglobin is recycled by the body, as illustrated in Figure 45.14. The iron from the digested haemoglobin is either returned to the plasma where it binds to transferrin (an iron-carrying protein), or is stored in the liver as ferritin.

In iron deficiency, or following haemorrhage, the capacity of the small intestine to absorb iron is increased. After severe blood loss, there is a time lag of 3 or 4 days before absorption is enhanced. This is the time needed for the enterocytes to migrate from their sites of origin in the mucosal glands to the tips of the villi, where they are best able to participate in iron absorption. The enterocytes of iron-deficient animals absorb iron from the intestinal lumen more rapidly than normal, a process controlled by a hormone called hepcidin, a 25-amino acid peptide which is synthesized by the liver. Hepcidin reduces the capacity of the enterocytes to transport iron. When demand for iron is low, hepcidin levels are high and iron absorption decreases. However, when demand for iron is high, for example following haemorrhage, circulating levels of hepcidin fall and iron uptake from the small intestine is increased. Genetic haemochromatosis, an autosomal recessive disorder characterized by excessive iron absorption from the gut and organ dysfunction caused by iron deposition, has been linked in 80–90 per cent of cases to mutations in the gene responsible for hepcidin production (the HFE gene is located on the short arm of chromosome 6).

Summary

The liver synthesizes and secretes a number of hormones including hepcidin, IGF-1, IGF-2, and thrombopoietin. The IGFs mediate many of the actions of growth hormone, while hepcidin is a key regulator of iron metabolism.

45.8 Detoxification by the liver

The liver performs a number of important chemical reactions that modify or inactivate endogenous physiological substances such as hormones, as well as foreign substances such as drugs and other toxins prior to their removal from the body. Furthermore, the liver converts ammonia to urea. Collectively these processes are called **detoxification reactions**.

Urea synthesis

The deamination of amino acids that occurs during gluconeogenesis and some other metabolic pathways results in the formation of ammonia, which is a highly toxic strong base. Any rise in the concentration of ammonia in the plasma causes a disturbance of acid–base balance which may result in coma. To defend against this, the liver converts ammonia to urea, which is non-toxic and can be excreted by the kidneys. The liver is the only organ able to perform this function. The ammonia is converted to urea by the **urea cycle** shown in Figure 45.15. Ammonium ions and bicarbonate ions combine under the influence of an enzyme, carbamyl phosphate synthetase, to form carbamyl phosphate, which then combines with ornithine to form citrulline. In turn, citrulline is converted to argininosuccinate and arginine, which give rise to urea and ornithine. The ornithine is then able to re-enter the cycle to participate in the synthesis of more urea.

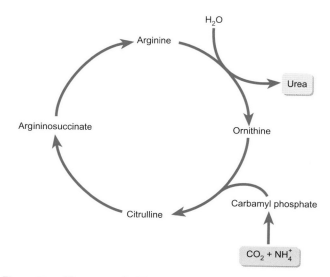

Figure 45.15 The urea cycle. The initial step is the formation of carbamyl phosphate from carbon dioxide and ammonium ions derived from the deamination of amino acids. The carbamyl phosphate enters the cycle by combining with ornithine to form citrulline. The splitting of urea from arginine results in the regeneration of ornithine, which is then able to participate in another turn of the cycle.

Biotransformation of drugs and poisons

Those substances that are not normally present in the body are known as **xenobiotics**. In addition to drugs administered for therapeutic purposes, xenobiotics may enter the body as constituents of plants that are not part of the normal diet (many plants are highly toxic to humans), or as a result of contamination of food by industrial chemicals such as dioxin and plasticizers. The liver is the main organ responsible for neutralizing xenobiotics. It achieves this by modifying their chemical constitution to render them inactive and more easily excreted. These changes are known as **biotransformation** or detoxification.

The biotransformation of most foreign substances occurs in two main stages known as phase I and phase II. In phase I, the foreign molecule is modified by oxidation, reduction, or hydrolysis. In phase II, the modified molecule is joined to another molecule that inactivates it and makes it easier to excrete. This is known as **conjugation**. Commonly, the molecules formed in phase I are conjugated with glucuronic acid, sulphate, or acetate, which increases their water solubility and thus facilitates their excretion via the kidneys. The well-known anti-inflammatory drug aspirin provides a simple example of biotransformation. The aspirin is first hydrolysed to salicylic acid, which is then conjugated with glucuronic acid to form salicylglucuronide. These steps are illustrated in Figure 45.16. Other substances may be oxidized by one of the family of cytochrome P450 enzymes prior to their conjugation.

In most cases the phase I reactions result in a reduction or loss of biological activity, but in a few unfortunate cases, the phase I reactions sharply increase the toxicity of the xenobiotic material. A familiar example is the toxicity of paracetamol, which is due to the transformation of the parent compound by cytochrome P450 to the toxic metabolite N-acetyl-p-benzoquinine. Liver failure occurs a few days after an overdose of paracetamol. This disastrous outcome highlights the importance of following the correct guidelines for drug administration.

In addition to the detoxification and removal of foreign molecules from the body, the liver plays an important part in the inactivation and excretion of hormones such as the steroids, which are oxidized by cytochrome P450 enzymes to make them more water-soluble and inactive. They are subsequently conjugated with glucuronic acid before being excreted by the kidneys.

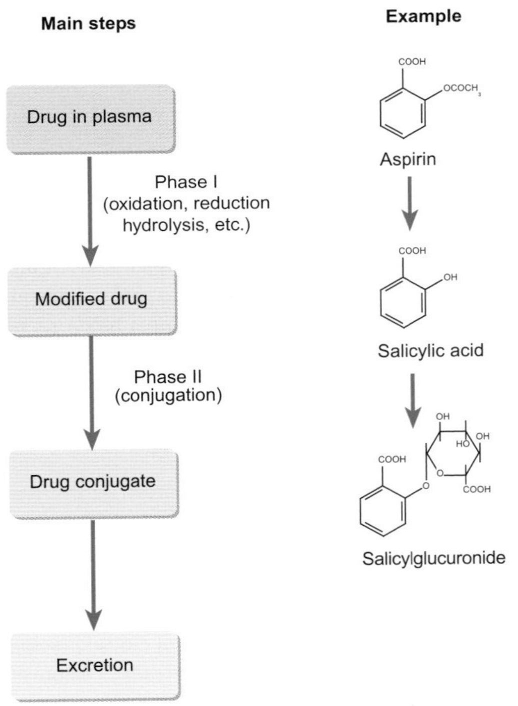

Summary

The liver is the only organ in the body able to convert the ammonia liberated during amino acid metabolism to urea, which is non-toxic. The liver is responsible for the metabolism of many hormones, and detoxifies many exogenous substances.

Figure 45.16 The main stages of drug metabolism. Although phase I and phase II are shown occurring sequentially, this is not always so. In the case of aspirin shown here, phase I occurs by hydrolysis of the bond between the acetate residue and a hydroxyl group on the benzene ring.

45.9 Liver failure

The liver is one of the most versatile organs in the body. Despite its great importance in metabolism, however, hepatic disease only rarely leads to serious illness. This fortunate circumstance arises because the liver has a great reserve capacity and a remarkable ability to regenerate. However, there are circumstances in which damage to the liver is so extensive that its function is impaired. This is known as **liver failure** which is diagnosed when there is evidence of jaundice, fluid accumulation in the peritoneal cavity (ascites), failure of blood clotting, and marked psychological changes (hepatic encephalopathy). The many functions of the liver summarized in Table 45.1 indicate that the consequences of hepatic failure or insufficiency will be widespread and extremely grave.

The liver is subject to most of the disease processes that can affect other body structures including inflammation, vascular disorders, metabolic disease, toxic injury, and neoplasms. Two of the most common liver diseases are hepatitis, which is characterized by inflammation of the liver, and cirrhosis, which is characterized by fibrosis and a loss of normal structure and function. Although several conditions can lead to cirrhosis of the liver, alcoholic cirrhosis is the most common form of the disease. The manifestations of liver failure reflect its various functions and, for the purposes of discussion, may be divided into digestive effects, effects on the blood, and effects concerning the excretion of toxic substances.

In liver failure, severe hypoglycaemia develops rapidly because of the central role played by the liver in the maintenance of plasma glucose. Galactose levels, however, tend to rise because galactose is no longer converted to glucose by the liver. Bile production declines and this may impair the digestion and absorption of fats. Since the ammonia formed by deamination of amino acids can only be converted to urea in the liver, the concentration of ammonia in the plasma will rise while that of urea will fall. These changes are likely to cause disturbances of the acid–base regulation of the body.

The protein profile of the blood will change radically as a result of liver failure. The production of plasma proteins by the liver will be impaired, leading to a loss of albumin and of certain globulins. If the total plasma protein level falls significantly there will be a reduction in the plasma oncotic pressure, which may lead to generalized oedema, or to fluid accumulation in the peritoneal cavity (ascites). Figure 45.17 shows a man suffering from severe abdominal ascites secondary to cirrhosis of the liver.

As liver damage progresses, many of the enzymes contained within the hepatocytes will leak out of the damaged cells and enter the blood. These enzymes include the aminotransferases (or transaminases), alkaline phosphatase, and lactate dehydrogenase, which catalyses the conversion of pyruvic acid to lactic acid. Liver function tests (LFTs) are often carried out to assess the presence and extent of liver damage. Tests include measurements of the plasma levels of liver enzymes, bilirubin, and albumin as well as blood clotting time.

The liver plays a vital role in the excretion of bilirubin derived from the breakdown of red blood cells. In liver failure this pathway is blocked, and so unbound, unconjugated bilirubin accumulates in the plasma. Gradually the patient will develop jaundice, a characteristic yellowing of the skin resulting from the presence of high levels of bilirubin (Figure 45.10). In neonates, bilirubin can accumulate in the basal ganglia of the brain resulting in a condition known as **kernicterus**, which is characterized by motor deficits.

As the liver starts to fail, the production of clotting factors will fall, leading to an impairment of blood clotting which may result in spontaneous bleeding from the skin and mucous membranes. The prothrombin clotting time will also increase from its normal range of 10–13 seconds.

In chronic liver disease, normal liver tissue is gradually replaced by fat and fibrous (scar) tissue. This causes an increase in the vascular resistance of the portal circulation resulting in **portal hypertension**. As the portal pressure rises above about 12 mm Hg, the venous anastomoses linking the portal and systemic circulations dilate and form 'varices' (swollen veins) that may project into the lumen of the oesophagus and stomach. These may rupture and cause gastro-intestinal bleeding. Varices may also develop in the umbilical region of the abdomen. These are a clearly visible sign of portal hypertension and are known as caput medusa from their supposed resemblance to the head of the mythological creature of Greek legend. Anorectal varices may also develop. Neither the umbilical nor the anorectal varices are likely to bleed.

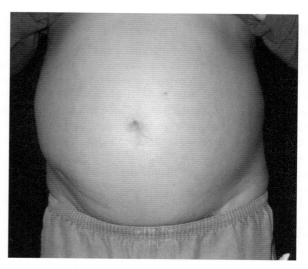

Figure 45.17 The abdomen of a patient suffering from severe ascites (accumulation of fluid in the abdominal cavity) as a result of cirrhosis of the liver. (Courtesy of Sciencephotolibrary.)

Finally, a failing liver is less able to carry out detoxification of toxic chemicals. For this reason, drug actions are likely to be prolonged in patients in liver failure. Disturbances in gonadal function and, in males, gynaecomastia (swelling of the breast tissue) may also arise as the metabolism of sex steroid hormones is impaired. There may also be signs of Cushing's syndrome, with sodium retention and subsequent oedema, and hypokalaemia arising from a reduction in the rate of metabolism of gluco- and mineralocorticoids.

The treatment of liver failure

The treatment of liver failure is, of necessity, complex and must address the many physiological processes throughout the body that are disturbed. If the initial problem was alcoholic cirrhosis, the first step in treatment will be to eliminate alcohol intake. Sufficient carbohydrate and calories will be required to prevent protein breakdown, fluid and electrolyte imbalances must be corrected, while protein intake may be limited to inhibit the production of ammonia. For patients in the final stages of liver failure, liver transplantation is now a realistic form of treatment.

Summary

During liver failure, many physiological processes are disrupted. There is hypoglycaemia, disordered lipid metabolism and decreased protein synthesis which leads to the formation of ascites and peripheral oedema. The loss of clotting factors leads to disorders of blood clotting. Liver failure leads to failure of detoxification mechanisms, which prolongs the action of the steroid hormones and gives rise to symptoms of endocrine disease. An increased level of bilirubin in the plasma results in jaundice.

✱ Checklist of key terms and concepts

Structure, functions, and blood supply of the liver

- The liver is the largest internal organ. It is formed from units called lobules, which are made up of sheets of liver cells (hepatocytes).

- The liver secretes bile which is essential for the digestion of fat and the excretion of bile pigments, cholesterol, and phospholipids.

- The liver performs a large variety of metabolic functions including protein synthesis, glucose storage (as glycogen), and gluconeogenesis.

- The liver carries out detoxification reactions that metabolize endogenous and exogenous substances and prepare them for excretion.

- The liver secretes several hormones including the IGFs, thrombopoietin, and hepcidin.

- The liver has a dual blood supply. It receives blood from the gut via the portal vein and from the hepatic artery. Blood returns to the systemic circulation via the hepatic vein.

- Hepatic blood flow is regulated by sympathetic nerves and by hormones.

- Macrophages (Kupffer cells) are present on the walls of the blood sinusoids. They phagocytose cell debris and bacteria.

The secretion and digestive functions of bile

- The liver secretes 600–1000 ml of bile each day. It is stored and concentrated in the gall bladder between meals.

- Bile is vital for the processing of fats by the small intestine.

- Bile salts are amphipathic, i.e. they have both hydrophobic and hydrophilic regions.

- The gall bladder contracts in response to the hormone cholecystokinin (CCK) to deliver bile to the duodenum following a meal.

- CCK is secreted in response to the presence of chyme in the duodenum.

- The enterohepatic circulation returns about 94 per cent of the bile salts that enter the small intestine to the liver.

- The formation of bile is stimulated by bile salts, secretin, glucagon, and gastrin circulating in the blood.

The excretion of bile pigments

- Bile pigments (the excretory products of haem) and other waste products are excreted in the bile.

- Bilirubin is the principal pigment. In the hepatocytes it is conjugated with glucuronic acid to form the water-soluble bilirubin diglucuronide which enters the bile.

- Failure to excrete bile pigments leads to their accumulation in the blood, and the development of jaundice.

- Jaundice may arise as a result of excessive breakdown of red blood cells (haemolytic jaundice), liver disease (hepatic jaundice), or blockage of the bile duct (obstructive jaundice).

- Extremely high circulating levels of bilirubin may lead to the development of hepatic encephalopathy or brain jaundice.

Excretion of cholesterol and phospholipids

- Cholesterol is a vital component of cell membranes and is the precursor for steroid hormone synthesis.

- The liver plays an important role in the regulation of plasma cholesterol levels.

- The liver synthesizes most of the cholesterol needed by the body.

- Cholesterol is transported in the plasma in combination with lipoproteins.

- Low density lipoproteins (LDLs) carry cholesterol to tissues that use it for synthetic processes. High density lipoproteins (HDLs) transport cholesterol to the liver for excretion or incorporation into bile salts.

- Bile is the major route for the excretion of cholesterol from the body. Phospholipids, especially lecithin, are also secreted into the bile as lipid vesicles, which then form micelles following emulsification with bile salts.

- Any excess cholesterol that cannot be dispersed into micelles may form crystals in the bile. These may contribute to the formation of gallstones in the hepatic ducts or the gall bladder.

- Excess plasma cholesterol may be deposited in the walls of blood vessels, leading to the formation of atheromatous plaques.

The role of the liver in whole-body metabolism

- The liver is able to store glucose as glycogen, and release it as required.

- The synthesis of glycogen is promoted by insulin, while the breakdown of glycogen and the release of glucose is promoted by glucagon and adrenaline.

- The liver is also able to form glucose from non-carbohydrate precursors including amino acids—a process called gluconeogenesis.

- The liver synthesizes and secretes many of the plasma proteins, including albumin, globulins, and most of the clotting factors.

- The liver synthesizes and secretes a number of hormones including angiotensinogen, insulin-like growth factors (IGF-1 and IGF-2), hepcidin, and thrombopoietin.

- The IGFs are vital to normal growth and mediate many of the actions of growth hormone.

- Hepcidin is a key regulator of iron metabolism.

- Thrombopoietin stimulates the stem cells of the bone marrow to differentiate into megakaryocytes. It regulates the number of circulating platelets (thrombocytes).

Detoxification reactions of the liver

- The liver converts the ammonia liberated during amino acid catabolism to urea, which is non-toxic. It is the only organ able to do this.

- The liver protects the body from toxins by modifying their chemical composition to render them inactive, or less active, and more easily excreted.

- Drug metabolism normally occurs in two steps, phase 1 (oxidation, reduction, or hydrolysis) and phase 2 (conjugation) reactions.

- The liver is responsible for the metabolic inactivation of many hormones.

Liver failure

- During liver failure, many physiological processes are disrupted. In acute hepatic failure there is hypoglycaemia, disordered lipid metabolism, and decreased protein synthesis.

- The diminished synthesis of albumin leads to the accumulation of fluid in the abdominal cavity (ascites) and peripheral oedema.

- The loss of those clotting factors that are synthesized in the liver leads to disorders of blood clotting and an increased clotting time.

- As the liver plays a central role in the recycling of iron from senescent red blood cells, in liver failure there is an increased level of bilirubin in the plasma which results in jaundice.

- Liver failure leads to failure of detoxification mechanisms, which prolongs the action of the steroid hormones and gives rise to symptoms of endocrine disease.

Recommended reading

Histology

Mescher, A.L. (2013) *Junquieira's basic histology* (13th edn), Chapter 16. McGraw-Hill, New York.

Physiology of the liver

Johnson, L.R. (2013) *Gastrointestinal physiology* (8th edn), Chapter 10, pp 97–106. Mosby, Philadelphia.

   A clear explanation of bile production and the control of its secretion.

Medicine

Burroughs, A.K. and Westerby, D. (2012) The liver, biliary tract and pancreatic disease. Chapter 7 in: *Clinical medicine* (8th edn), ed. P. Kumar and M. Clarke. Saunders Elsevier, Edinburgh.

   A brief introduction to the functions of the liver and biliary tract with a detailed account of liver diseases and appropriate clinical investigations.

Ramrakha, P., and Moore, K. (2010) Gastroenterological emergencies. Chapter 3 in: *Oxford handbook of acute medicine* (3rd edn). Oxford University Press, Oxford.

   A helpful guide to the diagnosis and management of liver and gall bladder diseases.

Review articles

Boyer, J.L. (2013) Bile formation and secretion. *Comprehensive Physiology* **3,** 1035–78, doi:10.1002/cphy.c120027.

Dowell, P., Hu, Z., and Lane, M.D. (2005) Monitoring energy balance: Metabolites of fatty acid synthesis as hypothalamic sensors. *Annual Review of Biochemistry* **74**, 515–34.

Gantz, T., and Nemeth E. (2012) Hepcidin and iron homeostasis. *Biochimica et Biophysica Acta* **1823,** 1434–43, doi:10.1016/j.bbamcr.2012.01.014.

Lefebvre, P., Cariou, B., Lien, F., Kuipers, F., and Staels, B. (2008) Role of bile acids and bile acid receptors in metabolic regulation. *Physiological Reviews* **89**, 147–91, doi:10.1152/physrev.00010.2008.

Ponziani, F.R., Pecere, S., Gasbarrini, A., and Ojetti, V. (2015) Physiology and pathophysiology of liver lipid metabolism. *Expert Review of Gastroenterology & Hepatology* **9**, 1055–67, doi:10.1586/17474124.2015.1056156.

Puche, J.E., and Castilla-Cortázar, I. (2012) Human conditions of insulin-like growth factor-I (IGF-I) deficiency. *Journal of Translational Medicine* **10**, 224–53, http://www.translational-medicine.com/content/10/1/224

Siddle, K. (2011) Signalling by insulin and IGF receptors: Supporting acts and new players. *Journal of Molecular Endocrinology* **47**, R1–10.

To check that you have mastered the key concepts presented in this chapter, complete the accompanying online self-assessment questions. Go to www.oup.com/uk/pocock5e/

CHAPTER 46

Nutrition and the regulation of food intake

Chapter contents

This chapter should help you to understand:

- The concept of nutrition and the importance of a mixed diet
- The role of the principal nutrients: carbohydrates, fats, proteins
- The importance of vitamins, minerals, and trace elements
- Causes and consequences of malnutrition and obesity
- The factors which regulate hunger, appetite, and satiety
- Eating disorders: anorexia nervosa and bulimia nervosa
- Physiological adaptations to starvation
- The measurements used to determine nutritional status
- Enteral and parenteral nutrition

46.1 Introduction

The selection of foods eaten by an individual is called the **diet**. A nutrient is any substance that is absorbed into the bloodstream from the diet and utilized to promote the various functions of the body. Nutrients include carbohydrates, proteins, fats, vitamins, mineral salts, and water. All are essential to health and a balanced diet contains appropriate amounts of each nutrient. In certain diseases, and during recovery from trauma or illness, nutritional requirements may be altered.

A key function of the diet is to provide the source of energy for cell metabolism. The energy content of the main foodstuffs is explored in Chapter 47, which also discusses the energy requirements of individuals in different circumstances. The chemical characteristics of carbohydrates, fats, and proteins are considered in Chapter 3, while the basic biochemical principles of energy metabolism are treated in Chapter 4. This chapter is concerned with the nutritional requirements of the body as a whole.

46.2 The principal requirements for a balanced diet

The bulk of the human diet is formed by carbohydrates, fats, and proteins—the **macronutrients**. In addition, a number of micronutrients are also needed by the body for cellular metabolism. Many of these cannot be made by the body and must be included in the diet. Examples are the vitamins, essential amino acids, and essential fatty acids. These are known as **essential micronutrients**. By convention, essential nutrient requirements are assessed by determining the amount per day needed to prevent clinical deficiency. An additional 30–100 per cent is added to this value to give a figure for the 'recommended daily amount' (RDA). Since for many nutrients, deficiencies develop progressively, it is often difficult to define an RDA with precision. For this reason, published figures often show a wide range and may differ significantly between different countries. Furthermore, a number of factors influence an individual's nutritional requirements. These include age, gender, size, activity level, pregnancy, lactation, and state of health.

The adequacy of a diet can be ascertained by following the weight gain of healthy children. With a well-balanced diet this follows a consistent pattern. In their early years (from birth to the age of 4 or 5) children show a very rapid weight gain which gradually becomes slower until the age of 8–10, when it accelerates. Around the age of 15 in young females, weight gain slows, but in young

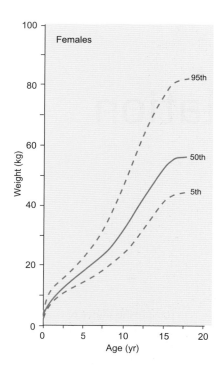

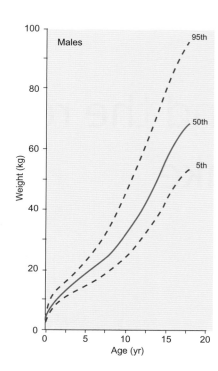

males, it continues beyond the age of 20 as they continue to add muscle mass (see Figure 46.1). In older adults, a well-balanced diet will be sufficient to maintain a steady weight.

Carbohydrates

These are found in a wide variety of foods such as cereals, bread, pasta, fruit, and vegetables such as potatoes. Animal carbohydrate is present in the glycogen of meat and liver and in milk in the form of lactose. Polysaccharides (which are complex molecules consisting of large numbers of monosaccharide residues) include starches, cellulose, glycogen, and dextrins (low-molecular-weight carbohydrates produced by the hydrolysis of starch or glycogen—see Chapter 3). However, some of these are not digested by the human gut and pass through virtually unchanged. Examples include celluloses and lignins, which are components of plant cell walls. Such carbohydrates are collectively known as **dietary fibre** (or roughage) and are important in facilitating the movement of material through the gut. Dietary fibre normally accounts for around a tenth of the total intake of carbohydrate. The remainder is digested in the gastrointestinal tract. After digestion, carbohydrates are absorbed in the form of monosaccharides which are utilized by the cells as a source of energy. Indeed, most people consume around half of their total energy requirement in the form of carbohydrate. The body also stores carbohydrate as glycogen, most of which is found in the liver. Once the glycogen stores are full, excess dietary carbohydrate is laid down in the form of fat.

Proteins

The body's requirement for protein is determined by its need for amino acids as the building blocks for structural protein (used for growth, maintenance, and repair of tissues) and for functional proteins such as enzymes and hormones. Most mixed diets provide about 15 per cent of the body's total energy requirements in the form of proteins, which are broken down into small peptides and amino acids before being absorbed by the small intestine (see Chapter 44).

Of the 20 α-amino acids that make up the proteins of the body, 12 can be synthesized by the body itself and need not, therefore, be included in the diet. These are alanine, arginine, asparagine, aspartic acid, cysteine, glutamic acid, glutamine, glycine, histidine, proline, serine, and tyrosine. They are called **non-essential amino acids**. The other eight amino acids cannot be synthesized by the body and must be included in the diet. These **essential amino acids** are isoleucine, leucine, lysine, methionine, phenylalanine, threonine, tryptophan, and valine.

The nutritional value of a protein depends upon the amino acids it contains. Protein foods that contain all the essential amino acids in the proportions required to maintain health are termed **complete proteins**. These include meat, fish, soya beans, milk, and eggs. **Incomplete proteins** do not contain all the essential amino acids in the correct proportions. They are mainly of vegetable origin and include cereals and pulses such as peas, beans, and lentils. However, by eating a wide variety of incomplete proteins, it is possible to avoid amino acid deficiencies. This is an especially important consideration for those following a strictly vegetarian or vegan diet.

During illness, there is an increase in total protein turnover as both protein synthesis and degradation are increased. In most cases, there is an overall loss of body protein, with muscle wasting, which is restored during recovery from the illness.

A diet deficient in both energy and protein may result in a range of clinical syndromes collectively known as protein-energy malnutrition (PEM). In western societies it usually only occurs in patients

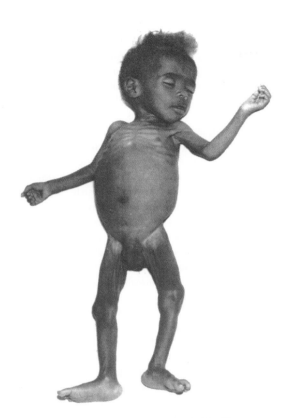

Figure 46.2 The typical appearance of a child with marasmus. Note the characteristic wrinkling of the skin and thinning of head hair. (From Fig 17.1 in J. A. Mann and S. Truswell (1998) *Essentials of human nutrition.* Oxford Medical Publications, Oxford.)

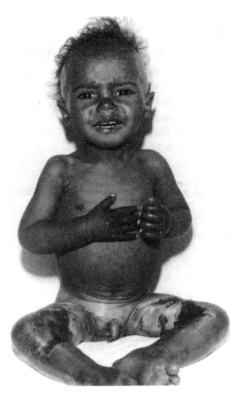

Figure 46.3 The typical appearance of a child with kwashiorkor in which oedema masks the loss of lean tissue. Note also the cracked, peeling, discoloured skin. (From Fig 17.2 in J. A. Mann and S. Truswell (1998) *Essentials of human nutrition.* Oxford Medical Publications, Oxford.)

who have or have had severe illness such as GI disease or surgery. In developing countries, it is more often caused by a poor diet and may be exacerbated by infections that cause diarrhoea. Chronic deprivation may result in stunting of height as well as inadequate weight gain. **Marasmus** and **kwashiorkor** are clinical syndromes resulting from a diet lacking in protein. In the case of marasmus, the energy content of the diet is also inadequate. This condition typically presents in infants under a year old. The child is less than 60 per cent of median body weight for its age (see Figure 46.1). Both lean tissue and subcutaneous fat are lost and the skin becomes characteristically wrinkled (see Figure 46.2). Because of the typical appearance of the child, marasmus may be confused with dehydration. Head hair is often thin but there may be long, fine, silky hair (lanugo) on the arms. The child is generally alert and keen to eat. Plasma albumin is usually normal but plasma sodium and potassium levels are low. Although internal organs are relatively unaffected in the early stages of the disease, atrophy of the heart and reduced brain weight may occur in advanced cases of marasmus.

Kwashiorkor most often affects children at the time of weaning (i.e. in the second year of life) when milk is replaced by solid foods, such as cassava, that contain very little protein. The condition is characterized by oedema that may mask the loss of lean tissue and create the false impression that the child is well nourished (see Figure 46.3). As protein intake is inadequate, there is a loss of

plasma proteins that results in a fall in plasma oncotic pressure. Fluid thus leaks out of capillaries and into the interstitium (see also Chapter 31). Oedema is first seen in the legs but later becomes generalized. The hair may become thin and discoloured and there are characteristic skin changes, with areas of pigmentation which later crack and peel. Also typical of kwashiorkor is an enlargement of the liver caused by deposition of fat. This occurs because the hepatocytes fail to synthesize sufficient very low-density lipoproteins (VLDL) normally required for the transport of fats out of the liver. Many children with kwashiorkor also have a co-existing infection. Some pathogens produce toxins that generate free radicals, which in turn may damage capillaries and increase the leakage of fluid into surrounding tissue, thus exacerbating the oedema.

Although protein-energy malnutrition in children is relatively common in developing countries, it is also seen in adults in the Western world. For example, protein-energy malnutrition is present in many patients on admission to hospital. Severe weight loss may be due to cancers of the gastrointestinal tract, malabsorption diarrhoea, pyrexia (which increases the demand for nutrients), or chronic conditions such as AIDS, Crohn's disease, and chronic obstructive pulmonary disease (COPD). However, in many cases weight loss is the result of inadequate food intake. This may occur for a variety of reasons including depression, pain, social deprivation, or anorexia nervosa.

Table 46.1 Recommended daily intake of dietary protein (in grams) for different age groups

Age (years)	Males	Females
1–3	15	15
4–6	20	20
7–10	28	28
11–14	42	41
15–18	55	45
19–50	56	45
over 50	53	47

If protein is eaten in excess of the body's requirements, the nitrogenous part is detached in the liver (deamination) and excreted by the kidneys as urea. The rest of the molecule may be converted to fat for storage in adipose tissue or used for the synthesis of glucose (gluconeogenesis)—see Chapter 45. Table 46.1 lists the recommended daily protein requirements of males and females at different ages.

Fats

Fats are divided into two groups, saturated and unsaturated (see also Chapter 3). Saturated fat is the principal type found in milk, cheese, butter, eggs, and meat. Unsaturated fats are found predominantly in vegetable oils but are also found in oily fish such as mackerel, sardines, and salmon. Linoleic, linolenic, and arachidonic acid are polyunsaturated fats that cannot be synthesized by the body. They are therefore known as **essential fatty acids** and must be included in the diet. Plant oils such as sunflower, corn, walnut, and linseed are good sources of linoleic and linolenic acid, while animal tissues contain small amounts of arachidonic acid. Cholesterol is synthesized in the body. It is also present in fatty meat, egg yolk, and full-fat dairy products.

Fats serve a number of important functions in the body. As well as forming the adipose tissue that supports and protects organs such as the kidneys and eyes, fat is an important constituent of nerve sheaths (myelin) and cell membranes. Cholesterol is required for the synthesis of steroid hormones. Certain membrane phospholipids (e.g. phosphatidylcholine) play an important role in cell signalling through arachidonic acid which is the precursor for the prostaglandins and leukotrienes (Chapter 6). Fat depots store the fat-soluble vitamins A, D, E, and K and are an important store of metabolic energy.

In a typical Western adult diet, fats contribute around 30 per cent to total energy intake, of which around one-third is in the form of saturated fats with the rest as mono- and polyunsaturated fats. However, even in populations whose diets are relatively low in fat (with fats contributing less than 15 per cent to total energy) subcutaneous stores remain adequate, indicating that many people's diets contain amounts of essential fatty acids which are considerably in excess of actual requirements. Consequently, fatty acid deficiencies are relatively rare.

Vitamins

Although required only in very small quantities by the body, vitamins are essential for normal metabolism and health. They are found in a wide range of foods and are subdivided into two categories, fat-soluble vitamins (A, D, E, and K) and water-soluble vitamins (C and B complex). As Figure 46.4 shows, the molecular structures of vitamins are very diverse.

Table 46.2 lists the different vitamins, their recommended daily requirements, major sources, and principal functions. Deficiencies of particular vitamins result in the progressive development of specific deficiency disorders. Some of these are also listed in Table 46.2. Furthermore, some vitamins, such as vitamin C and folic acid, are involved in cellular processes in a more generalized way and deficiencies will have more wide-reaching consequences. Some of the more important deficiency syndromes are discussed in the following paragraphs and, in cases where excess intake is known to be damaging to health, toxic effects are also considered.

The fat-soluble vitamins (A, D, E, and K)

Vitamin A (retinol) is required for the formation of visual pigments (see Chapter 14). Normally, 80–90 per cent of the body's vitamin A is stored in the liver but vitamin A deficiency can result from inadequate intake, impaired absorption of fats, or liver disorders. The condition **xerophthalmia** (Greek for 'dry eyes'), which is caused by vitamin A deficiency, is a major cause of blindness in developing countries. The first visual symptoms are normally a loss of sensitivity to green light, followed by a loss of acuity in dim light and then total loss of dark adaptation (night blindness). The conjunctival membranes of the eyes become dry and develop foamy oval or triangular spots (Bitot's spots) consisting of epithelial debris and secretions. The corneas become cloudy and soft, and in severe cases, keratinization of the cornea, with erosion and ulceration, may lead to total blindness. Other epithelia such as those of the respiratory tract, GI tract, and genito-urinary tract are also affected by vitamin A deficiency, and the resulting changes may cause respiratory disease and diarrhoea. Vitamin A is thought to play a part in the efficient operation of the immune system, and a mild deficiency seems to increase susceptibility to infectious disease. Vitamin A is needed for normal limb development so that deficiency during pregnancy may result in developmental abnormalities of the fetus (though see the next paragraph on toxicity).

Vitamin A toxicity is potentially serious. It was first seen in explorers who had eaten polar bear liver, which is very rich in vitamin A. Its acute effects include vomiting, vertigo, headache, blurred vision, loss of muscle coordination, and raised CSF (cerebrospinal fluid) pressure. The chronic effects are variable and include hyperlipidaemia, bone and muscle pain, and skin disorders. These symptoms normally disappear within a few weeks of reducing vitamin A intake. However, excess vitamin A is now known to cause fetal malformations (i.e. it is teratogenic) in the first trimester of pregnancy. It may lead to spontaneous abortion or to abnormalities of the cranium, face (harelip), heart, kidneys, thymus, and CNS (including deafness and learning difficulties). These changes cannot, of course, be reversed.

Vitamin D (calcitriol) is synthesized from dietary cholecalciferol or by the action of sunlight on 7-dehydrocholesterol in the skin.

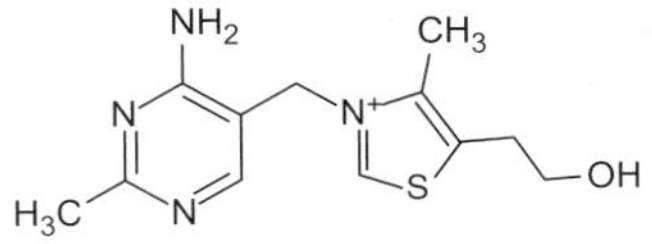
Thiamine (vitamin B_1)

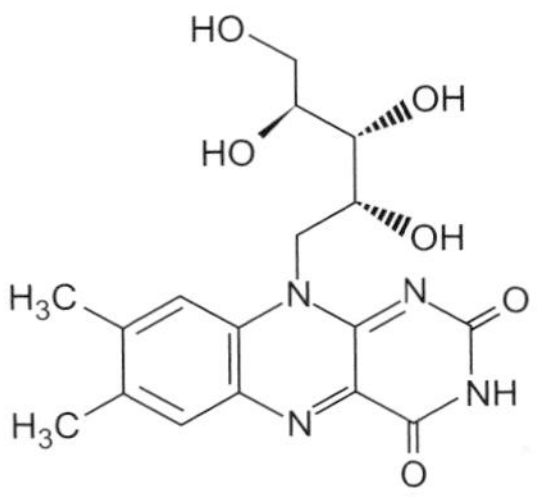
Riboflavin (vitamin B_2)

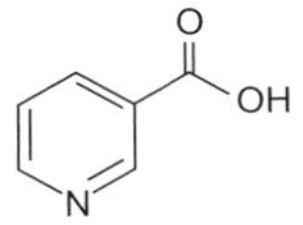
Niacin (vitamin B_3)

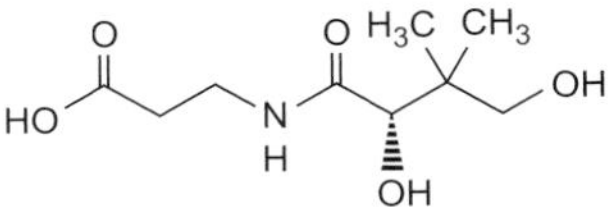
Pantothenic acid (vitamin B_5)

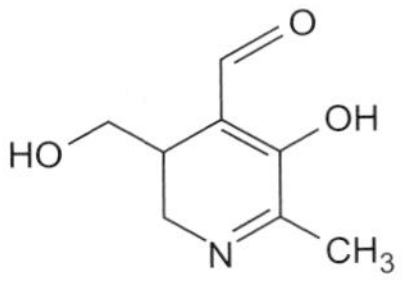
Pyridoxal (vitamin B_6)

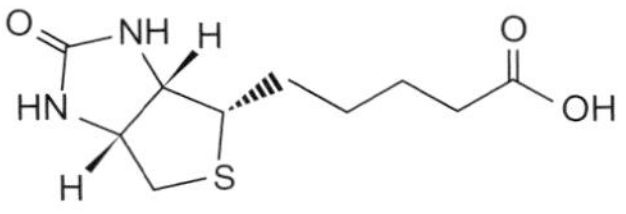
Biotin (vitamin B_7)

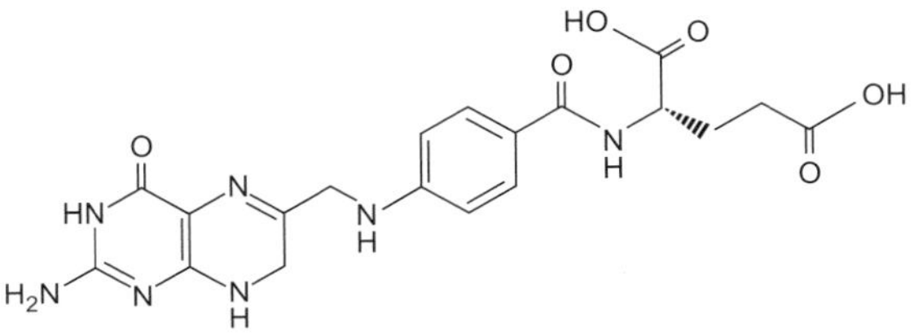
Dihydrofolate (vitamin B_9)

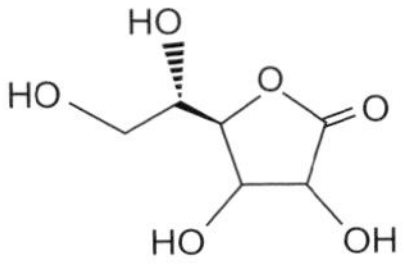
Ascorbic acid (vitamin C)

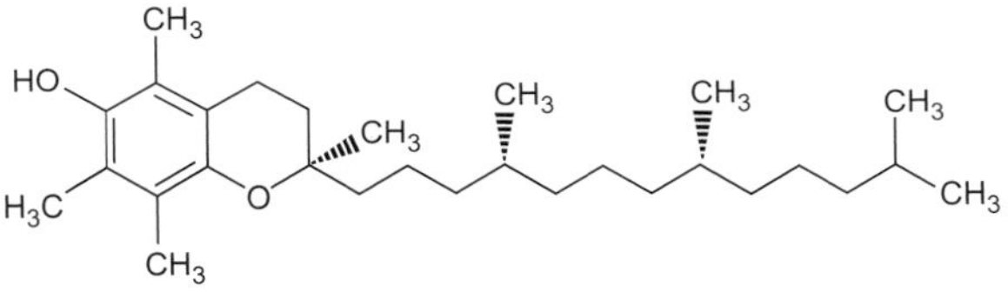
α-Tocopherol (vitamin E)

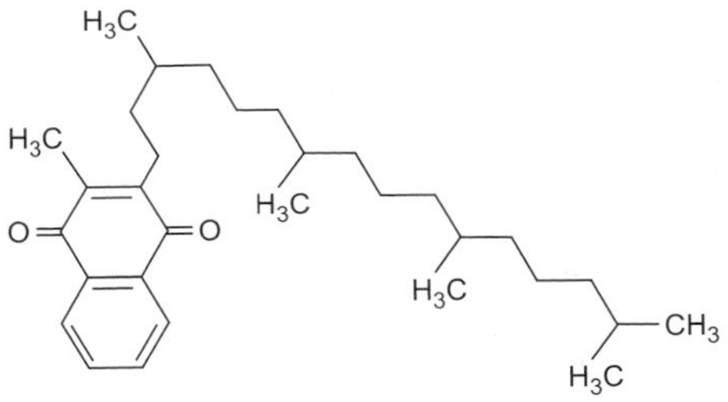
Phylloquinone (vitamin K_1)

Figure 46.4 The structure of some of the vitamins found in plants. The structures of vitamins A and D_3 can be found in Figures 14.16 and 22.10. Note that while the vitamins are all needed in various amounts to sustain health, they are chemically very diverse.

Its importance in the maintenance and health of the skeleton are described in Chapter 22. Vitamin D deficiency has different consequences at different stages of life. In toddlers, deficiency causes a condition known as **rickets** in which bones are under-mineralized because of poor calcium absorption. When the child starts to walk and bear weight on its legs, the long bones of the legs become deformed and bowed (see Figure 51.20). There may also be pelvic deformities and collapse of the ribcage. **Osteomalacia** is the adult equivalent of rickets and results from the demineralization of bone rather than failure to mineralize in the first place. The elderly, and those women who have had several pregnancies but little exposure to sunlight, are at most risk of experiencing skeletal problems caused by vitamin D deficiency.

In addition to its importance in the maintenance of the skeleton, vitamin D is also thought to possess anti-cancer properties and to improve muscle strength. Furthermore, recent studies have revealed that more than half of adults and children worldwide are deficient in vitamin D and may need to modify their diet or take supplements to avoid deficiency disorders.

Vitamin D is toxic if ingested in amounts far in excess of the recommended daily amount (RDA). Symptoms of toxicity are related to the elevated plasma calcium concentration resulting from enhanced intestinal calcium absorption (see Chapter 22). Hypercalcaemia may lead to contraction of blood vessels and dangerously high blood pressure, and to **calcinosis**—the calcification of soft tissues including the kidney, heart, lungs, and blood vessel walls.

Table 46.2 Actions and daily requirements of vitamins

Vitamin	Daily requirement (adults)	Major dietary sources	Functions	Deficiency disorders
A (retinol)	700 μg	Dairy products, oily fish, eggs, liver	Formation of visual pigments, development of bone cells, normal fetal development	Night blindness, epithelial atrophy, susceptibility to infection; fetal developmental abnormalities
D (cholecalciferol)	15 μg (600 I.U.)	Fish oils, dairy products; synthesized in skin	Normal bone development, absorption of calcium from gut	Rickets (children) Osteomalacia (adults)
E (α-tocopherol)	8 μg	Nuts, egg yolk, wheatgerm, milk, cabbage	Antioxidant	Haemolytic anaemias
K (coagulation vitamin)	100 μg	Green vegetables, cereals, pig liver; synthesized by intestinal flora	Formation of several clotting factors and some liver proteins	Bruising, bleeding
B$_1$ (thiamine)	1.6 mg	Lean meat, fish, eggs, legumes, green vegetables	Carbohydrate metabolism	Beriberi with many neurological symptoms, disturbances of metabolism
B$_2$ (riboflavin)	1.8 mg	Milk, liver, kidneys, heart, meat, green vegetables	Constituent of flavine coenzymes	Dermatitis, hypersensitivity to light
B$_3$ (niacin or nicotinamide)	15 mg	Most foods; can be synthesized from tryptophan	Constituent of nicotinamide coenzymes	Pellagra, listlessness, nausea, dermatitis, neurological disorders
B$_6$ (pyridoxine)	2 mg	Meat, fish	Amino acid metabolism, synthesis of haemoglobin, antibodies and a number of neurotransmitters	Irritability, convulsions, anaemia, vomiting, skin lesions
Pantothenic acid	5–10 mg	Most foods	Component of coenzyme A	Neuropathy, abdominal pain
Biotin (vitamin H)	100 μg	Liver, egg yolk, nuts, legumes	Fatty acid synthesis	Muscle pain, scaly skin, elevated blood cholesterol
B$_{12}$ (cyanocobalamin)	2.4 μg	Liver, meat, fish, dairy products (NOT plants)	Erythrocyte production and amino acid metabolism	Pernicious anaemia
Folic acid (B$_9$)	250–600 μg for women considering pregnancy and 700 μg for the first 12 weeks of pregnancy	Liver, dark green vegetables; synthesized by intestinal bacteria	Haematopoiesis, nucleic acid synthesis, development of neural tube	Anaemias, gastrointestinal disturbances, diarrhoea; neural tube defects—spina bifida and anencephaly
C (ascorbic acid)	50 mg	Fresh fruits (especially citrus fruits), vegetables	Protein metabolism, collagen synthesis	Scurvy, liability to infection, poor wound healing, anaemia

Note: I.U.= international units

Vitamin E (α-tocopherol) is a powerful anti-oxidant that protects cell membranes and plasma lipoproteins from free radical damage. Deficiency can lead to damage to nerve and muscle membranes. Children may develop ataxia, loss of reflexes, and changes in gait, with loss of proprioception. Vitamin E is found in a wide variety of foods including nuts, egg yolk, milk, and some green vegetables. As it is fat-soluble, vitamin E relies on normal fat digestion and absorption for its own absorption across the gut. Disorders such as cystic fibrosis, coeliac disease, and biliary obstruction all reduce the intestinal absorption of vitamin E. Premature infants are also at risk of deficiency as they are often born with inadequate vitamin E reserves. Their red blood cell membranes are very fragile due to attacks by free radicals, which could lead to haemolytic anaemia unless vitamin E supplements are given. Deficiency in experimental animals causes infertility, but there is no evidence of a similar role for vitamin E in humans.

Vitamin K is a fat-soluble vitamin that is found in green leaves, cereals, and pig liver. It is also made by bacteria found in the large bowel and absorbed in small amounts in the caecum. It promotes the synthesis in the liver of a special amino acid (γ-carboxyglutamic acid) which is an essential component of four of the coagulation factors of the clotting cascade (prothrombin and factors VII, IX, and X). It is thus needed for the normal coagulation of blood (see Chapter 25), and vitamin K deficiency causes bleeding disorders characterized by hypoprothrombinaemia. The widely used anti-coagulant drug warfarin is a vitamin K antagonist.

Vitamin K is not transported readily across the placenta from mother to fetus, and the neonatal gut is sterile. For this reason, vitamin K levels in the newborn may be very low. Haemorrhagic disease of the newborn is a risk, particularly for premature infants, and such babies are often given an intramuscular injection of vitamin K immediately after delivery as a precaution. Deficiency may also occur in patients with malabsorption of fats and also following prolonged treatment with antibiotics, which destroy the colonic bacteria. Elderly patients admitted to hospital with fractures of the wrist or femur are often found to be low in vitamin K, which suggests that deficiency may predispose to osteoporosis. This vitamin is known to be required for the synthesis of osteocalcin, a protein that enhances osteoblastic activity and is an important calcium binding protein of the bone matrix (see Chapter 51).

The water-soluble vitamins (B group and vitamin C)

Vitamin B$_1$ (thiamine) is present in a variety of foods of both plant and animal origin and plays an essential role as a coenzyme in the metabolism of carbohydrates. In its absence, pyruvic acid and lactic acid (products of carbohydrate digestion) accumulate in the tissues, leading to acidosis and triggering a variety of clinical responses collectively known as **beriberi**. Beriberi occurs in different forms classified as wet or dry. In wet beriberi, which is usually an acute response, symptoms are largely cardiovascular. Vasodilatation occurs which leads to right-sided heart failure and general oedema. In dry beriberi (usually a chronic deficiency state) a variety of neuropathies are seen including nystagmus, paralysis, or weakness of the extraocular muscles (ophthalmoplegia), ataxia, abnormal pupillary reactions, and altered consciousness. There may be headaches, vomiting, and confusion. If untreated, this may progress to an irreversible condition known as Korsakoff's psychosis in which there is an inability to form new memories. This condition occurs most often in alcoholics because ethanol metabolism requires thiamine. It may also be seen, though rarely, in patients suffering from hyperemesis gravidarum (extreme sickness of pregnancy).

Vitamin B$_2$ (riboflavin) is a constituent of flavine co-enzymes such as FAD (flavine adenine dinucleotide) which participate in a variety of oxidation and reduction reactions. Its main dietary sources are milk products, meat, and eggs. Deficiency is comparatively rare and non-fatal, though symptoms include cracking at the edges of the lips (**cheilosis**) and corners of the mouth (**angular stomatitis**), **glossitis** (loss of the epithelium covering the tongue), and skin lesions. The typical features of vitamin B$_2$ deficiency are illustrated in Figure 46.5.

Vitamin B$_3$ (niacin) is a constituent of nicotinamide co-enzymes, and deficiency gives rise to a clinical syndrome known as **pellagra** (literally sour skin), which is fatal if left untreated. Diarrhoea, dementia, and dermatitis are typical symptoms of pellagra. The skin becomes inflamed when exposed to sunlight, giving the appearance of severe sunburn (see Figure 46.6). Lesions become pigmented and later crack and peel. These changes most often involve the skin of the neck (**Casal's collar**). The tongue may

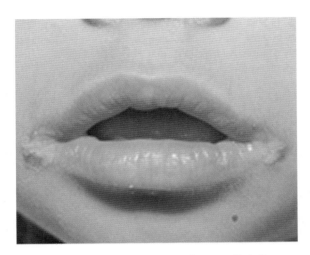

Figure 46.5 A patient with vitamin B$_2$ deficiency. Note the cracking at the edges of the lips (cheilosis) and corners of the mouth (angular stomatitis). (From www.cheilosis.net.)

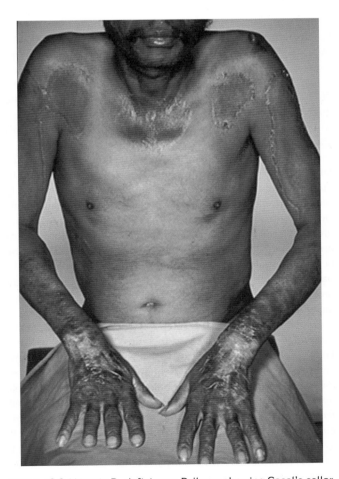

Figure 46.6 Vitamin B$_3$ deficiency. Pellagra showing Casal's collar and severe dermatitis on the back of each hand. (From Creative Commons, courtesy of Drs Herbert L. Fred and Hendrik A. van Dijk.)

also be inflamed. Neurological presentations of severe niacin deficiency include neuropathies, tremor, fits, depression, ataxia, and rigidity. Depressive psychoses may be due to a lack of serotonin as a result of tryptophan deficiency.

Niacin is present in many foods (see Table 46.2) and deficiency is normally restricted to populations whose chief food is maize, a cereal that contains very little tryptophan from which niacin can be synthesized. In developed countries with a rich protein diet and a variety of cereal crops, pellagra is very rare.

Vitamin B$_6$ (pyridoxine) exists in three forms that are interconvertible in the body. These are pyridoxine, pyridoxal, and pyridoxamine. The main form in the body is pyridoxal 5′-phosphate (PLP), which is involved in amino acid metabolism, in the release of glucose from glycogen stores, particularly those of muscle, and in gluconeogenesis. It may also be involved in the modulation of steroid hormone receptors by interacting with hormone–steroid receptor complexes and thereby reducing the amount of bound hormone. In vitamin B$_6$ deficiency, there is increased sensitivity of target tissues to the actions of low concentrations of hormones such as oestrogens, androgens, cortisol, and vitamin D. A possible link between vitamin B$_6$ deficiency and the development and progress of hormone-sensitive cancers such as breast and prostate cancer has been suggested. Vitamin B$_6$ is also needed for the synthesis of a number of neurotransmitters including serotonin, adrenaline, noradrenaline, dopamine, and GABA, an important inhibitory neurotransmitter of the CNS (see Chapter 7). Deficiency of this vitamin is often associated with convulsions, sleeplessness, and other neurological changes.

Extreme use of vitamin supplements may lead to a toxic overdose of vitamin B$_6$. Moderate overdose may cause tingling in the fingers and toes, which is reversible on cessation of supplements, while very high intake may result in damage to peripheral nerves and partial paralysis.

Vitamin B$_{12}$ is the collective name for a group of cobalt-containing substances known as cobalamins. They are present in meat, fish, and dairy products and deficiency is normally only seen in those following a strict vegan diet or in individuals deficient in intrinsic factor (due to gastrectomy or malabsorption disorders). Pernicious anaemia (see Chapter 25) and neuropathies are characteristic symptoms of vitamin B$_{12}$ deficiency. There may be loss of sensation in the lower limbs, some spinal nerve demyelination, and a loss of motor power. Infants breast-fed by vegan mothers may be at risk of impaired neurological development and anaemia.

Folate (Vitamin B$_9$) is the generic name for compounds chemically related to folic acid (from the Latin 'folia' meaning leaf). Folates are found in green leafy vegetables, liver, oranges, beans, and yeast. Folic acid, the primary vitamin, is essential for the normal synthesis of nucleic acids, for haematopoiesis, and for development of the neural tube. Deficiency leads to anaemias typified by large, fragile red cells, and during pregnancy, deficiency is associated with neural tube defects.

Pantothenic acid is required for the synthesis of coenzyme A. It cannot be synthesized by the body and must therefore be supplied by the diet. It is present in almost all foods, and dietary deficiency is not normally seen except in cases of starvation. The recommended intake is 6 mg per day. Pantothenic acid is sometimes called vitamin B$_5$.

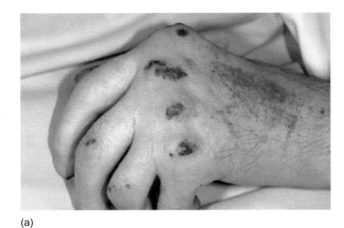

(a)

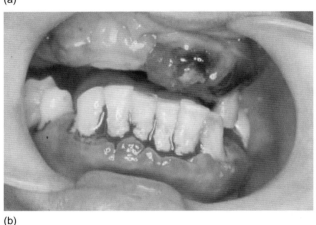

(b)

Figure 46.7 The typical symptoms of scurvy. Panel (a) illustrates the characteristic skin lesions while panel (b) shows the typical bleeding and swollen gums and the loss of teeth from the upper jaw. (Courtesy of the Wellcome Library, London.)

Vitamin C is needed for the synthesis of collagen, the principal connective tissue of tendons, arteries, bone, skin, and muscle. It is also required for the synthesis of noradrenaline from dopamine. The disorder caused by vitamin C deficiency is called **scurvy**. In this disease, new connective tissue cannot be made in sufficient amounts to replace ageing or injured tissue. Skin lesions are therefore commonly seen and even minor injuries may cause significant bleeding. There may be spontaneous bleeding from the gums, nose, and hair follicles or into joints, the bladder, or the gut. Other symptoms include listlessness, anorexia, weight loss, halitosis, gingivitis (gum disease), and loose teeth. Figure 46.7 illustrates the characteristic skin lesions, bleeding gums, and loss of teeth which are characteristic of scurvy.

Essential minerals

A wide variety of inorganic ions is required for normal cellular activities. These include calcium, phosphorus, sodium, potassium, magnesium, iron, iodine, and zinc. Furthermore, a number of other elements such as cobalt, copper, manganese, selenium, and vanadium are also required in tiny amounts. These are the so-called trace elements, many of which act as co-factors for enzyme-catalysed cellular processes. The recommended daily amounts of

Table 46.3 Reference values for daily requirements of minerals

Mineral	Sources	Daily requirement (adults)
Calcium	Bread, dairy products, eggs, fish, fruit, meat, nuts, pulses, vegetables	700 mg
Chromium	Eggs, meat, bread, vegetables	35 µg
Copper	Seafood, meat, nuts, whole grains	1.2 mg
Iodine	Iodized salt, seafood, bread, milk and dairy products	140 µg
Iron	Meat, fish, eggs, green vegetables	8.7 mg
Magnesium	Fish, fruit, meat, nuts, green vegetables	400 mg
Manganese	Cereal products, tea, vegetables	4.1 mg
Phosphorus	Dairy products, eggs, fish, fruit, meat, nuts, pulses	700 mg
Potassium	All fresh foods including fruits, vegetables, meat, milk	3.5 g
Selenium	Seafood, meat, whole grains, vegetables	75 µg
Sodium	Processed foods, bread, cheese, eggs, milk, meat, seafood; often added during cooking	1–2 g
Zinc	Fish, meat, poultry, pulses, whole grains	9.5 mg

Note: The figures given in the table have been compiled from various sources and refer to adult men; the amounts required for non-pregnant women are a little lower except for iron, where it is higher in women of child-bearing age.

many of these minerals are given in Table 46.3 together with the dietary sources for each.

Calcium is found in milk, eggs, green vegetables, and some fish. Sources of **phosphorus** include cheese, oatmeal, liver, and kidney. Calcium and phosphorus are needed for the normal mineralization of bone. Calcium is involved in secretion, muscle contraction, and blood clotting while phosphorus is an important component of cell membranes and of ATP.

Magnesium is widely involved in enzymic reactions, particularly those involving ATP. Intracellular magnesium concentrations are in the range 5–20 mmol l^{-1}. Only a small fraction is free Mg^{2+}; the remainder is bound to ATP, proteins, and other negatively charged molecules. Around half of the total body magnesium is bound to hydroxyapatite in bone and this acts to buffer changes in dietary magnesium. Most of the remainder is found in muscle and other soft tissues. Hypomagnesaemia is common in hospitalized patients, and particularly high incidence is observed in intensive care units. The signs of deficiency include anorexia, nausea, vomiting, lethargy, weakness, tremor, muscle fasciculation, and changes to the pattern of the ECG (see Chapter 29). Sources of magnesium include fish, green vegetables, meat, nuts, and whole grain bread.

Sodium is found in most foods, especially meat, fish, eggs, milk, bread, and as table and cooking salt. The adult daily requirement for sodium is about 1.6 g, though most people ingest much more. It is the major extracellular cation and plays a crucial part in volume regulation, muscle contraction, and nervous conduction. **Potassium** is widely distributed in all foods, especially fruit and vegetables. It is the major intracellular cation and is involved in many cellular processes.

Iron, as a soluble compound, is found in liver, kidney, beef, egg yolk, green vegetables, and wholemeal bread. About 1–2 mg of iron is required each day to replace that lost from the body. A higher intake is needed by women, particularly during pregnancy. Iron is essential for the formation of haemoglobin (see Chapter 25) and, as an important component of cytochromes, is necessary for cellular respiration.

Iodine is found in salt-water fish and in vegetables grown in iodine-rich soil. The daily requirement is 140 µg. In areas of the world in which naturally occurring iodine is deficient, small quantities may be added to table salt. Iodine is required for the synthesis of thyroid hormones. Individuals whose diets lack sufficient iodine develop an enlarged thyroid gland (goitre) in an attempt to trap any available iodine from the plasma (see Figures 22.7 and 46.8).

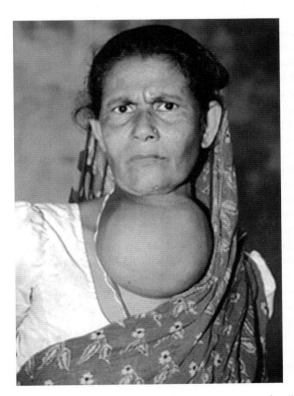

Figure 46.8 A Bangladeshi woman with a very pronounced goitre caused by iodine deficiency. (Courtesy of WHO).

Zinc is present in a wide variety of foods including red meat, poultry, beans, nuts, dairy products, oysters, and some other types of seafood. The RDA for zinc in adults is around 11 mg for men and 8 mg for women (rising to 12 or 13 mg during pregnancy and lactation). Zinc functions as a key structural or catalytic component of a great many enzymes and is believed to participate in many aspects of cellular signal transduction, regulating communication between cells, cell proliferation, differentiation, and survival. In recent years, the possible role of zinc in immune responses and inflammation has attracted much interest.

Summary

A balanced diet is essential for health. The main foodstuffs are the carbohydrates, fats, and proteins. These are known as the macronutrients. Carbohydrates and fats are major sources of energy. Certain fatty acids are classed as essential since they cannot be manufactured by the body. Dietary proteins are broken down into their constituent amino acids, before being absorbed and used to make structural protein, enzymes, hormones, etc. Eight essential amino acids must be included in the diet. In addition to the macronutrients, a wide variety of **micronutrients** are required for health. These are the vitamins and certain minerals.

46.3 Regulation of dietary intake

Although, to a large extent, the content and size of our meals is dictated by social factors and the daily pattern of activity, both hunger and appetite are important regulators of dietary intake. **Hunger** refers to a physiological sensation of emptiness, usually accompanied by contraction of the stomach. **Appetite** refers to the feelings associated with the anticipation of the forthcoming food. It may be affected by an individual's emotional state—nervousness or fear, for example, often suppresses the appetite. Hunger and appetite, therefore, although related, are different sensations.

The hypothalamus plays an important role in the regulation of food intake and energy expenditure. Its precise role, however, is not understood. Initial experiments suggested that two regions within the hypothalamus contribute to the regulation of appetite and satiety. Lesions in the lateral hypothalamus of rats produce **aphagia** (lack of feeding) leading to starvation and death. Consequently, this region is known as the 'feeding centre'. Lesions of the ventromedial hypothalamus, however, lead to increased food intake (**hyperphagia**) and is the 'satiety centre'. More recent studies, however, suggest that food intake is regulated by complex neuronal pathways involving many hypothalamic areas and a number of specific neurotransmitters, some of which are 'orexigenic' (they stimulate appetite and feeding) and others 'anorexigenic' (they inhibit feeding).

Being **overweight** or **obese** are common conditions which are increasing in prevalence, especially in affluent societies. Globally, the World Health Organization (WHO) estimates that over 1 billion people worldwide are currently overweight and that over 300 million people are obese. The growing numbers of obese children and adolescents is of particular concern. Overweight or obese people have excessive fat stores in relation to their height, gender, and race. Obesity has many consequences, both physical and physiological. These include metabolic changes such as diabetes, fatty liver, gallstones, infertility in women, changes in plasma lipid levels, cardiovascular problems such as hypertension, coronary heart disease, varicose veins, peripheral oedema, osteoarthritis, spinal problems, and obstructive sleep apnoea. Furthermore, obese individuals often have low self-esteem. The distribution of the excess fat is thought to be significant in relation to certain health risks. Abdominal fat deposition, with waist circumference measurements of over 100 cm (39 inches) in men and 95 cm (37 inches) in women, is particularly associated with an increased risk of cardiovascular and metabolic disorders such as heart attack and diabetes. Obese people are more likely to die while undergoing surgery and to suffer post-operative complications.

There is considerable interest in the likely causes of obesity and the possibility of a 'cure' for overweight and obese individuals. Obesity reflects an imbalance between energy expenditure and intake, meaning that the amount and type of food eaten and the amount of work done (both to maintain body functions and exercise) are key issues. High-fat diets predispose to obesity and some studies have shown that weight and body mass index can go on increasing if the proportion of fat in the diet is increased, even when the total energy consumed is reduced.

It is a commonly held belief that excessive weight can arise because of glandular disturbances. Endocrine disorders are not, however, generally associated with obesity, although the following conditions may contribute to weight gain (see also Chapters 21–24):

- hypothyroidism
- acromegaly
- Cushing's syndrome
- insulin resistance.

There is some evidence to suggest that the tendency to be overweight is at least partly determined by genetic factors, probably in a multi-gene, multi-factorial fashion. There is currently much interest in a gene (the *ob* gene discovered in 1994) that is responsible for the production of a satiety factor. Its protein product, **leptin** (from the Greek word 'leptos' meaning 'thin'), has been shown to reduce the body weight of genetically obese mice (*ob/ob* mice in whom there is a mutation of the leptin gene) and to reduce food intake in all species studied so far, including humans. Humans with a (rare) mutation of the leptin gene appear unable to regulate their food intake and are morbidly obese. In addition to reducing food intake, in rats, leptin also increases the rate at which energy is expended, possibly by stimulating brown fat metabolism (see Chapter 42). With the discovery of leptin and its effects on food intake came the inevitable suggestion that administration of leptin to obese humans could be used to modulate homeostatic mechanisms of **energy balance** and lead to weight loss. However, leptin is produced in adipocytes and its levels are related to the amount of fat tissue in the body. Women have higher levels of leptin than men, reflecting their higher proportion of body fat, and leptin is also elevated in obese individuals. These findings would indicate that control of appetite and therefore weight is not just a simple matter of altering leptin concentrations.

In recent years it has become clear that fat is an active endocrine tissue and that, in addition to leptin, it produces and secretes a number of other proteins that act as hormones and may play a role in the regulation of energy intake and expenditure. One of the most widely studied of these is **adiponectin**, which appears to increase the sensitivity of cells to insulin. Its levels are low in obese individuals and increase after weight loss. This might help to explain the insulin resistance that is often associated with obesity. Several other newly discovered compounds secreted by fat cells and other tissues, including **resistin**, **ghrelin** (see Chapter 43), and **obestatin**, have also been implicated in the regulation of appetite and satiety but their exact contributions remain unclear. Certainly more work is required in this important area of nutrition.

Anorexia nervosa and bulimia nervosa

Together, anorexia nervosa and bulimia nervosa represent a significant source of morbidity. Although seen occasionally in elderly people, **anorexia nervosa** is mainly confined to women aged between 10 and 30 years, with males accounting for less than 10 per cent of all cases. The incidence of the disorder in girls aged 11–18 years is thought to be between 0.2 and 1.1 per cent. Anorexia nervosa effectively represents a state of prolonged fasting or starvation and is associated with a number of characteristic symptoms. There is commonly a very small food intake, which is usually between 2.5 and 3.37 MJ a day (600–900 kcal day^{-1}). BMI (see Section 46.4) is generally below 17.5 and the patient will actively strive to maintain an unduly low weight, often exacerbating their condition by pursuing a vigorous exercise regime. There seems to be a fear of fatness and a distorted body image, often with a feeling of worthlessness and depression.

Amenorrhoea (cessation of menstrual periods) is also typical of postmenarcheal anorexic patients who are not taking oral contraceptives, presumably because body fat stores are no longer sufficient to sustain normal sex steroid production. Anorexia is associated with many other physiological abnormalities related to reduced energy intake and low body weight. These include an increased susceptibility to cold and gastrointestinal problems including constipation and abdominal pains. Blood pressure and heart rate are often low in anorexic patients as a result of cardiomyopathy (as cardiac muscle is degraded for gluconeogenesis). Indeed, heart failure is a common cause of death in these patients. Furthermore, there may also be skeletal changes such as osteoporosis. A severely anorexic woman is shown in Figure 46.9.

Bulimia nervosa has many similarities with anorexia in that the patient has extreme concerns about weight, a distorted body image, and often depression and anxiety. In bulimics, however, weight and BMI is usually in the normal range. Typically, bulimic patients starve themselves and then binge on enormous amounts of 'forbidden' foods, often consuming the energy equivalent of 8–12 MJ (c. 2000–3000 kcal) at a time. After this, they feel very guilty and may vomit or take large doses of laxatives in an attempt to purge the calories. Although there are few physical complaints associated directly with bulimia, persistent vomiting can cause

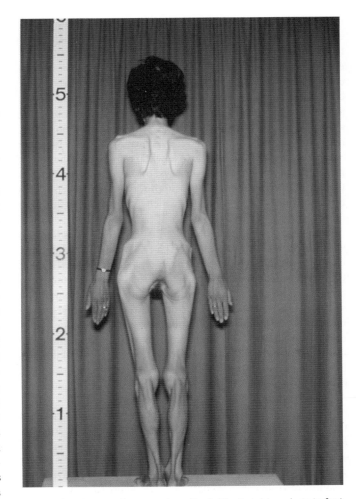

Figure 46.9 A severely anorexic patient. The height scale is in feet. (Courtesy of the Wellcome Library, London.)

erosion of the enamel on the lingual surface of the upper front teeth. Frequent vomiting or laxative abuse may also lead to electrolyte imbalances.

The metabolic response to starvation

The hormonal responses that occur during the normal fasting state, i.e. between meals, and in response to eating a meal, are described in Chapter 24. The following is a brief description of the metabolic responses made by the body during a period of more prolonged fasting (starvation).

The body of a healthy man weighing 70 kg will typically possess an energy store equivalent to around 670,000 kJ (160,000 kcals), an amount which would be sufficient to meet his calorific needs for between 4 and 12 weeks, depending on his activity level. Although some of this energy is in the form of stored carbohydrate (glycogen), the vast majority is in the form of fat. Indeed, carbohydrate reserves will be exhausted within 1–2 days. Since certain tissues, including the brain, are critically dependent on glucose for their metabolism, it is essential that the body maintains a plasma glucose concentration in excess of around 2–3 mmol litre^{-1} at all times. As fats cannot be converted to glucose, in periods of starvation glucose can only be obtained by the process of gluconeogenesis—the conversion of amino acids to glucose. These are not stored by the body but are obtained by the breakdown of body proteins. Loss of protein inevitably results in a weakening of the muscle mass and, eventually, loss of function and death. To minimize the possibility of this catastrophe, in prolonged starvation the body must maintain adequate levels of plasma glucose while minimizing the loss of protein. This is achieved in two ways: by ensuring that all tissues that are able to metabolize fatty acids do so (to spare glucose) and by the synthesis of ketone bodies which may be used by the brain for energy.

In the early stages of starvation, insulin secretion is low and glucagon levels are relatively high (see Chapter 24). In addition, GH levels rise and act to antagonize the systemic actions of insulin. GH also promotes lipolysis, making fatty acids available for oxidation (see Chapter 21). As a result, glucose uptake into tissues such as muscle is reduced. At the same time, fatty acids and glycerol are mobilized from adipose tissue and oxidized in preference to glucose by muscle and liver cells to produce energy. Furthermore, the liver is able to synthesize new glucose from the glycerol that has been released from the stored fat. These metabolic adjustments ensure the availability of glucose for those tissues that require it for ATP production, such as the CNS and germinal tissue.

As starvation proceeds, proteins will start to be catabolized and their carbon skeletons will be used for gluconeogenesis. Initially, proteins such as those of the intestinal epithelium and pancreatic secretions are utilized, but later, the proteins of skeletal muscle will start to be broken down. The liver begins to synthesize large amounts of acetoacetate and β-hydroxybutyrate (the ketone bodies) from acetyl CoA (see Chapter 45) and releases them into the blood. This helps to minimize the loss of muscle mass. The brain is able to utilize ketone bodies for ATP production and, by about the third day of starvation, it is obtaining around 30 per cent of its energy from this source. If starvation continues, ketone bodies eventually become the major fuel source for the brain. However, this situation can only be sustained while sufficient quantities of fat are available, so the time for which a person can survive starvation will depend on the extent of their fat reserves.

Summary

Hunger and appetite are important regulators of food intake. Both are controlled by neuronal 'centres' in the hypothalamus. A number of hormones synthesized by adipose tissue may contribute to the regulation of appetite and satiety. Inappropriate food intake leads to the health problems of obesity, anorexia nervosa, and bulimia nervosa.

In prolonged starvation, the energy metabolism of muscle and liver cells shifts away from glucose towards fatty acids, conserving glucose for use by the brain. Ketone bodies, produced by the liver, become the principal source of fuel for the brain if starvation is prolonged.

46.4 Measurements used to monitor nutritional status

In individuals who are being treated for malnourishment of any kind, it is important to have ways of monitoring their progress in terms of either weight loss or gain. In addition to physical examination and detailed medical history, a variety of measurements of nutritional status are employed.

Body mass index (BMI)

Although a single determination of weight is of relatively little value, regular sequential weighing over a period of time is more helpful. It is also important to view weight in the context of a patient's height. This is the BMI or body mass index, which is calculated as follows:

$$BMI = \frac{weight\ (kg)}{(height\ in\ m)^2}$$

Values between 18.5 and 25 kg m^{-2} are considered normal, while values below 18.5 are classified as underweight for height and may be associated with health problems in some individuals. Values in excess of 25 are considered increasingly overweight. When BMI exceeds 30, the individual is classified as obese. As BMI increases beyond this value, the obesity is further classified as moderate, severe, clinical, etc. Figure 46.10 illustrates the classification of BMI into underweight, normal weight, overweight, and increasing degrees of obesity.

Although the BMI can contribute to nutritional assessment in most adults, it may be misleading in those people who have a large muscle mass. This is because muscle weighs more than the same volume of fat. Highly trained athletes, for example, may well have a calculated BMI in excess of 25 because of their lean body mass but could not be considered overweight or obese. It is also important to realize that BMI is of much less value in children. Many healthy children have a calculated BMI well below 20 because of the changes in build which occur at various times during development (for example the growth spurt at puberty). Longer-term monitoring of growth (both height and weight) as well as assessment of cognitive function is of greater value. BMI is also of little value in patients suffering from excessive oedema, or in pregnancy, when weight increases significantly as the uterus and its contents increase in size and weight and body fluid volume expands. In clinical settings, it is also important to take into account any changes in weight caused by plaster casts or amputated limbs.

Skinfold thickness measurements

Additional information regarding nutritional status may be obtained from measurements of skinfold thickness at various sites such as the triceps, biceps, and subscapular and suprailiac regions. About one-third of body fat is subcutaneous, so skinfold thickness reflects the size of the subcutaneous fat depot, which reflects total body fat. Such measurements are made using calipers, and tables of normal ranges are published. Table 46.4 shows some typical values for men and women.

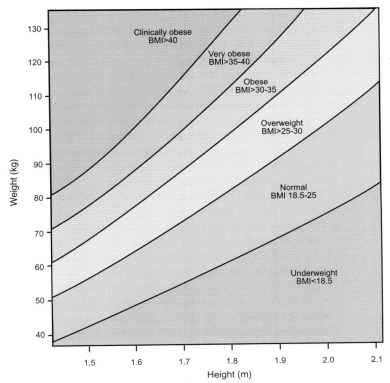

Figure 46.10 A body mass index (BMI) calculator showing the various categories of BMI from underweight to morbidly obese. The BMI units are kg m^{-2}. This figure is used to determine within which category an individual person lies. For example, someone who is 1.7 m tall weighing 60 kg would have a BMI of 20.8 which is normal (as shown by the green zone).

Table 46.4 Percentage of body fat estimated from skinfold thickness

Skinfold thickness (mm)	Percentage body fat		
	Age 17–19 years	Age 30–39 years	Age 50+ years
Male subjects			
20	8.3	12	12.5
30	13	16	18.5
40	16.5	19	23
50	19	21.5	26
60	21	23.5	29
70	23	25	32
Female subjects			
20	14	17	21
30	19	22	26.5
40	23	25.5	30
50	26	28	33
60	28	31	35.5
70	30.5	32.5	37.5

Notes: Skinfold thickness is measured at four locations: the biceps, the triceps, below the shoulder blade (the subscapular skinfold), and at the waist (the suprailiac skinfold). The values are added together to determine the final skinfold thickness. For males the desirable percentage of body fat lies between 15 and 21 per cent depending on age. For females, the desirable range is 17–25 per cent.

Other measurements

Measurements of mid-upper-arm circumference and mid-arm circumference can be helpful in monitoring the progress of severely undernourished individuals. These measurements are particularly useful indicators of the body's protein reserves. Coupled with such measurements, analysis of body composition may be helpful in determining nutritional status. Devices are available which determine the relative proportions of fat, lean tissue, and water in the body. It should be remembered, however, that all these measurements are subject to considerable errors in patients who have oedema.

Certain biochemical measurements may also contribute to an overall assessment of nutritional status. The appearance of ketones in the urine, for example, indicates that fats are being broken down for use in metabolism. This will reflect weight loss and may be desirable in obese people who are trying to lose weight—though not of course in malnourished individuals. Nitrogen balance is a further indicator of the extent to which protein synthesis and breakdown are taking place and can form part of a clinical evaluation of malnutrition.

The Malnutrition Universal Screening Tool

The Malnutrition Universal Screening Tool (MUST) has been developed to provide a standardized means of detecting and monitoring both over- and under-nutrition in hospital in-patients and out-patients and in other care settings. It is also used in GP surgeries and in community medicine. The assessment consists of five steps, which are as follows. Steps 1–3 are scored as indicated.

1. Calculation of BMI:

   BMI > 20 = 0 (BMI > 30 is obese)

   BMI 18.5–20 = 1

   BMI < 18.5 = 2

2. Estimation of unplanned weight loss in past 3–6 months:

   < 5 per cent = 0

   5–10 per cent = 1

   > 10 per cent = 2

3. Consider the effect of acute disease: if the patient is acutely ill and there has been or there is likely to be no nutritional intake for more than five days, score 2

4. Add scores from 1, 2, and 3 to give an overall risk of malnutrition

   Total score = 0—low risk

   Total score = 1—medium risk

   Total score > or = 2—high risk

5. Initiate appropriate nutritional management and reassess regularly.

46.5 Enteral and parenteral nutritional support

There are a number of circumstances in which people are unable to eat normally. These may include an inability to swallow (dysphagia), an inability to absorb food across the intestine, a reduced level of consciousness, pain, or the effects of therapy. Furthermore, the emotional and psychological factors associated with being in hospital may modify the appetite. Nutritional support may be required in these cases. Enteral nutrition (via the gastrointestinal tract) is possible if the gut is healthy, while parenteral nutrition (bypassing the gastrointestinal tract) may be necessary if it is not.

Enteral feeding can take the form of high-energy liquid feeds taken orally or formula feeds administered via a nasogastric tube. Such feeds will contain appropriate amounts of protein, carbohydrates (including glucose), fats, vitamins, and minerals. Different formulas are available to meet the requirements of different patients. Several different enteral feeding routes are used, including nasogastric (NG) feeding, nasojejunal (NJ) feeding, and percutaneous endoscopic gastrostomy (PEG) feeding.

If the nasogastric tract is obstructed or if the gut is unable to process food adequately, **parenteral feeding** may be needed. Here the gut is bypassed and nutrients are infused directly into a large central vein, usually the subclavian or jugular. Nutrients must be in a form in which they can be used without digestion, e.g. pure amino acids, glucose, and an emulsion of triglyceride. If parenteral feeding is prolonged, vitamins, minerals, and trace elements will also need to be included in the feed. Feeds of this kind are hypertonic and must be given along with appropriate amounts of fluid. Furthermore, regular monitoring of the patient's weight and hydration status is essential.

Summary

A variety of tests are used to assess nutritional status including measurements of skinfold thickness, arm circumference, body composition, and nitrogen balance. The Malnutrition Universal Screening Tool (MUST) provides a standardized means of detecting and monitoring over- and under-nutrition in a wide variety of settings.

(✳) Checklist of key terms and concepts

What constitutes a balanced diet?

- A balanced diet is essential for health.
- A balanced (mixed) diet contains adequate amounts of all the essential nutrients.
- The main foodstuffs are known as macronutrients, as they are required by the body in large amounts.
- The macronutrients are the carbohydrates, fats, and proteins.
- Carbohydrates are an important energy source.
- Fats are also a source of energy and are important constituents of cell membranes.
- Certain fatty acids are classed as essential since they cannot be manufactured by the body.
- Dietary proteins are broken down in the gut into their constituent amino acids which are absorbed and used to make structural protein, enzymes, hormones, etc.
- Eight essential amino acids must be included in the diet.

- The vitamins and minerals of the diet are known as micronutrients. They are vital to health.
- Vitamins A, D, E, and K are fat-soluble, while vitamin C and the B group vitamins are water-soluble.
- Vitamin deficiencies are associated with characteristic symptoms, e.g. rickets, pellagra, beriberi, and scurvy.
- Calcium, phosphorus, sodium, potassium, iron, iodine, and trace elements are required for health.

The regulation of food intake, inappropriate food intake, and starvation

- Hunger and appetite are important regulators of food intake. Feeding and satiety 'centres' are located in the hypothalamus.
- A number of hormones, including leptin and adiponectin, are synthesized by adipose tissue and may contribute to the regulation of appetite, satiety, and energy balance.

- Obesity, anorexia nervosa, and bulimia nervosa are examples of inappropriate food intake and carry a range of associated health problems.

- The metabolic responses to prolonged starvation include a shift in the metabolism of muscle and liver cells away from glucose, towards fatty acids. In this way glucose is conserved for use by glucose-dependent tissues such as the brain.

- In prolonged starvation ketone bodies, produced by the liver, become a major source of fuel for the brain.

Nutritional assessment

- A variety of anthropometrical and physiological tests may be carried out to assess nutritional status. These include measurements of BMI, skinfold thickness, arm circumference, body composition, and nitrogen balance.

- The Malnutrition Universal Screening Tool (MUST) provides a standardized means of detecting and monitoring over- and under-nutrition in a wide variety of settings

- Under some circumstances, it is necessary to provide nutritional support.

- Enteral feeding (via the gut) may be provided via a nasogastric tube, a nasojejunal tube, or percutaneous endoscopic gastrostomy (PEG).

- Parenteral nutrition is needed when the gut is unable to carry out its normal digestive functions. Here nutrients are infused directly into the bloodstream via a cannula inserted into a large vein.

Recommended reading

Bender, D.A. (2014) *Introduction to nutrition and metabolism* (5th edn). CRC Press (Taylor and Francis Group), London.
  A helpful introduction which provides clear explanations of all aspects of nutrition.

Elia, M., Ljungqvist, O., Stratton, R.J., and Lanham-New, S.A. (eds) (2013) *Clinical nutrition* (2nd edn). Wiley-Blackwell, Oxford.
  Examines nutrition in a clinical setting.

Mann, J.A., and Truswell, S. (eds) (2012) *Essentials of human nutrition* (4th edn). Oxford Medical Publications, Oxford.
  An up-to-date, authoritative text.

Webster-Gandy, J., Madden, A., and Holdsworth, M. (eds) (2011) *Oxford handbook of nutrition and diatetics* (2nd edn). Oxford University Press, Oxford.
  A useful pocket-sized guide to the practical aspects of nutrition and nutritional assessment.

Review articles

Carreiro, A.L., Dhillon, J., and Gordon, S. (2016) The macronutrients, appetite, and energy intake. *Annual Review of Nutrition* **36**, 73–103.

Mason, B.L., Wang, Q., and Zigman, J.M. (2014) The central nervous system sites mediating the orexigenic actions of ghrelin. *Annual Review of Physiology* **76**, 519–33.

Rolls, E.T. (2016) Reward systems in the brain and nutrition. *Annual Review of Nutrition* **36**, 435–70.

Sutton, A.K., Myers, M.G., Jr., and Olson, D.P. (2016) The role of PVH circuits in leptin action and energy balance. *Annual Review of Physiology* **78**, 207–12.

To check that you have mastered the key concepts presented in this chapter, complete the accompanying online self-assessment questions. Go to www.oup.com/uk/pocock5e/

CHAPTER 47

Energy balance and the control of metabolic rate

Chapter contents

This chapter should help you to understand:

- The general concepts of metabolism, metabolic rate, and basal metabolic rate

- The production of heat by cellular reactions

- The concept of the energy equivalent of oxygen for carbohydrates, fats, and proteins

- The relationship between oxygen consumption and carbon dioxide production—the respiratory quotient

- The energy requirements of different tasks

- The actions of circulating hormones on the metabolic rate

47.1 Introduction

The previous chapters have been concerned with the ingestion of food, its digestion to simple nutrients, and their absorption into the bloodstream prior to their participation in the chemical reactions of the body. This chapter is concerned with the broad concepts of energy balance and metabolic rate and the various factors that influence them.

47.2 The chemical processes of the body produce heat

The scale of the chemical reactions of the body can be gauged by considering the fact that an average human being uses about 360 litres of oxygen each day to 'burn' several hundred grams of carbohydrates and fats and generate about 7500 kJ of heat. This is roughly equivalent to a steady heat output of 90 watts (1 watt = 1 joule s^{-1}). The chemical processes of the body that are responsible for generating this heat constitute the **metabolism** of the body, which may be divided into two categories, anabolic and catabolic. **Anabolic metabolism** involves the synthesis of complex molecules from simpler ones, such as the synthesis of proteins from amino acids. **Catabolic metabolism** involves the breaking down of large, complex molecules to smaller, simpler ones—often to liberate energy for cellular processes. The oxidative metabolism of glucose to carbon dioxide and water is a typical example.

In cells, anabolic and catabolic reactions take place side by side. The structural components of the cell are continually being broken down and replaced. Some of the energy released by the catabolic processes is harnessed to drive the energy-requiring anabolic processes of the cell and important cellular activities such as muscle contraction, membrane ion pumps, and secretory processes. The rest is 'lost' as heat. This heat is, however, of vital importance in the maintenance and regulation of core body temperature (see Chapter 42).

ATP plays a crucial role in all these processes. It is manufactured in large amounts in the mitochondria during the oxidative metabolism of glucose, fats, and proteins as described in Chapter 4. Its function is to transfer, in the form of phosphate bonds, the energy

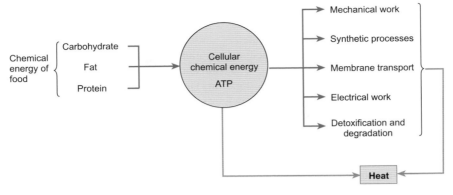

Figure 47.1 An overview of the biological transformation of various forms of energy.

liberated by catabolism to the anabolic reactions of the cells or to provide energy for the active processes carried out by the cells, as shown in Figure 47.1. Examples of the crucial role of ATP in active transport and in the contraction of skeletal muscle are given in Chapters 5 and 8 respectively.

Virtually all the work performed by the body eventually ends up as heat. This is strikingly evident during exercise, where heat production rises with the amount of work done. Only the work done outside the body, such as that involved in lifting a heavy object, is not directly converted to heat. To take another example, the heart does mechanical work in pumping blood against the pressures in the aorta and pulmonary artery. However, the kinetic energy imparted to the blood is converted to heat in overcoming friction as it passes through the circulatory system. Even in anabolic reactions, such as protein synthesis, the chemical energy stored within a protein is lost as heat when that protein is eventually broken down.

Summary

The chemical processes of the body constitute its metabolism. Metabolic processes in which large molecules are synthesized from smaller ones constitute anabolic metabolism. Those in which large molecules are broken down to simpler ones constitute catabolic metabolism. Catabolic metabolism provides energy in the form of ATP and is accompanied by the generation of heat.

47.3 Energy balance

The unit of energy used in the study of energy metabolism is the kilojoule (kJ). The joule is defined as the amount of work done when a force of one newton moves through a distance of one metre in the direction of the force. The rate at which work is done (power) is measured in watts, where 1 watt is equivalent to one joule of work per second. In the earlier literature, the unit of energy is given as kilocalories (or Calories). In terms of heat, one kilocalorie is the amount of heat needed to raise the temperature of 1 kg of water by 1°C. One kilocalorie is equivalent to 1000 calories or 4187 joules (4.187 kJ).

According to the First Law of Thermodynamics, energy can neither be created nor destroyed. Applied to the human body, this means that the total amount of energy taken in by the body must be accounted for by the energy put out by the body. Thus:

$$\text{ENERGY INPUT} = \text{ENERGY OUTPUT}$$

This may be expressed as:

$$\text{Chemical energy of food} = \text{Heat energy} + \text{Work energy} + \text{Stored chemical energy} \quad [1]$$

When the amount of energy ingested as food is sufficient to balance the amount of energy put out in the forms of heat and work, the chemical energy of the body remains constant. In reality, this is rarely the case. More often there is a small imbalance such that energy is either stored within the body or depleted by the catabolism of carbohydrates, fats, and, in more prolonged fasting, proteins. Even during the course of one day, there are small fluctuations in the chemical energy stores of a normal individual. At night, the glycogen stores of the liver are slowly consumed, and they are restored following the first meal of the next day. During growth, there is substantial gain in body mass due to the synthesis of new proteins. In later life, as activity declines, there is a tendency for excess food intake to result in the deposition of fat.

To study the energy balance of a person under the conditions normally encountered in life, three of the variables in equation [1] above would have to be measured so that the fourth may be calculated. This can be done in a whole-body calorimeter, which measures the total heat output of a subject. This technique is impractical for the clinical assessment of patients, but the measurement of energy balance can be simplified by eliminating some of the variables in equation [1]. By carrying out measurements on an individual in the post-absorptive state of metabolism (8–12 hours after a meal—see Chapter 24), the input of chemical energy from food may be taken as zero. If the person is at complete rest, energy output in the form of external work can also be disregarded. Under these conditions, all the chemical energy derived from the metabolic pool is used to maintain vital bodily functions and is ultimately completely converted to heat. Equation [1] now simplifies to:

$$\text{Loss of stored chemical energy} = \text{Heat energy} \quad [2]$$

The rate at which chemical energy is expended by the body is known as the **metabolic rate**. As equation [2] shows, the rate at which heat is produced by the body is equal to the rate at which chemical energy is expended. Consequently, the heat production is a direct measure of metabolic rate.

Although the metabolic rate will vary with activity, under the resting conditions described above, the rate of heat production is said to represent the **basal metabolic rate** (BMR) which is defined as the energy requirement of the fasting body *during complete rest* (generally sitting) in a thermoneutral environment (c. 20°C). It is usually measured early in the morning, after an overnight fast. The BMR is an index of metabolism under standardized conditions. It does not represent the minimum metabolic rate, which is generally achieved during deep sleep. As BMR varies with body size, it is usual to relate it to body surface area or to lean body mass. Abnormal values of BMR often result from disturbances of hormonal function, some of which will be discussed in Section 47.5.

Summary

Energy is measured in kilojoules (kJ) or kilocalories (Calories) (where 1 kilocalorie is 1000 calories or 4187 joules). The rate at which chemical energy is expended by the body is called the metabolic rate. The basal metabolic rate (BMR) is measured in a thermoneutral environment, following an overnight fast under conditions of complete rest.

47.4 How much heat is liberated by metabolism?

The answer to this question depends upon the nature and proportions of the foods being metabolized. If known amounts of pure fat, carbohydrate, or protein are burned in a calorimeter together with 1 litre of oxygen, then different amounts of heat energy are liberated from each foodstuff. Pure carbohydrate, for example, will liberate 20.93 kJ (5 kcal) of energy per litre of oxygen, whereas pure fat will liberate 19.68 kJ (4.7 kcal) per litre of oxygen and protein will liberate about 18.84 kJ (4.5 kcal) per litre of oxygen. These values represent the **energy equivalent of oxygen** for the complete oxidation of carbohydrate, fat, and protein.

Another way of looking at the energy value of different foods is to consider how much energy is liberated by the complete oxidation of a standard amount. Thus, the complete oxidation of 1 g of carbohydrate or protein will liberate 17.16 kJ (4.1 kcal), while 1 g of fat will liberate 38.94 kJ (9.3 kcal). The body's metabolism may be considered in much the same way, because carbohydrates, fats, and proteins liberate energy as they react with oxygen. The energy supplied by some common foodstuffs is listed in Table 47.1.

The measurement of oxygen consumption

In order to measure the metabolic rate of an individual it is necessary to know, or to calculate, two values:

1. the rate at which the individual is using up oxygen—his or her *oxygen consumption* (normally expressed in litres minute^{-1});

2. the energy equivalent of oxygen for the food being metabolized (expressed in kJ per litre of oxygen).

By multiplying these two values by 60 (the number of minutes in an hour), a figure for that individual's metabolic rate is obtained in kJ h^{-1}. Alternatively, if the values are divided by 60, the number of seconds in a minute, the person's energy consumption can be expressed in kW (kJ s^{-1}). A relatively simple way of measuring the rate of oxygen consumption of an individual is by means of a spirometer (see also Chapter 33). The spirometer bell is filled with oxygen, from which the subject inspires. Expired air passes back to the bell via a canister of soda lime, which removes all the carbon dioxide from it. As the subject consumes oxygen, individual respiratory cycles are recorded and the volume of the oxygen in the bell (represented by the height of the bell) gradually falls (see Figure 47.2). The spirometer is calibrated in terms of volume per unit distance fall of the bell. The difference between the height of the bell at the start and finish of the experimental period is therefore a measure of the volume of oxygen consumed during that time (Figure 47.3). Modern methods rely on measuring the airflow and the difference in the oxygen and carbon dioxide content between the inspired and expired air. As this can be measured for each breath, it is possible to obtain continuous measurements of a subject's work rate as they perform a variety of tasks.

The energy equivalent of oxygen

As explained above, the energy equivalent of oxygen will vary according to the relative amounts of protein, fat, and carbohydrate being utilized at the time of metabolic rate determination. These amounts cannot be measured directly. The energy equivalent of oxygen, therefore, has to be calculated indirectly from the respiratory quotient (RQ) or respiratory exchange ratio. The RQ is the ratio of the volume of carbon dioxide evolved from the lungs to the volume of oxygen absorbed from the lungs in one minute:

$$RQ = \frac{\text{volume of } CO_2 \text{ produced per minute}}{\text{volume of oxygen consumed per minute}} \quad [3]$$

The respiratory quotient depends upon the nature of the foodstuffs undergoing metabolism. If pure carbohydrate is oxidized, an RQ of 1.0 is obtained, while the specific oxidation of fat gives an RQ of 0.7 and that of protein an RQ of about 0.8. In reality, the body uses a variable mixture of all three metabolic fuels (although protein is usually a minor energy source). The RQ will then be a mean value weighted towards the principal fuel. The RQ of a subject on an ordinary mixed diet will be around 0.85, while that of a fasting individual is about 0.82, as relatively more fat (or even protein) is metabolized.

Once the RQ value for the subject under investigation has been obtained, it is possible to relate this to the energy equivalent of oxygen using standard values (see Figure 47.4). The metabolic rate may then be calculated by multiplying oxygen consumption by the energy equivalent of oxygen obtained from the tables of standard values. This simple indirect method is subject to a number of

Table 47.1 The calorific values of some common foodstuffs (per 100 g of fresh produce unless otherwise indicated)

Foodstuff	kcal	kJ	Foodstuff	kcal	kJ
Fruit			*Meat*		
Apple	58	242	Bacon (streaky)	353	1473
Banana	85	356	Beef (lean)	164	686
Dates (dried)	274	1147	Chicken	197	824
Figs (fresh)	65	272	Duck	326	1365
Grapes	67	280	Ham (smoked)	345	1444
Oranges	49	205	Lamb (leg)	239	1000
Strawberries	37	155	Pork (lean)	168	703
Vegetables			Turkey	218	912
Beans (green)	32	133	Venison	124	519
Broccoli	32	133	*Fish & shellfish*		
Cabbage	25	105	Cod	78	326
Carrots	40	137	Clams	70	292
Cauliflower	27	113	Crab	100	418
Celery	17	71	Halibut	126	527
Cucumber	13	54	Herring	243	1017
Lettuce	14	59	Mackerel	192	800
Mushrooms	22	92	Salmon	208	871
Peas	84	352	Scallops	79	331
Potatoes	76	318	Shrimps	97	406
Tomatoes	22	92	Trout	100	418
Cereals etc.			Tuna	290	1214
Barley	347	1148	*Dairy etc.*		
Bread (white)	270	1130	Butter	716	2997
Bread (wholemeal)	241	1009	Cheese (soft)	287	1200
Noodles (dry)	376	1574	Cheese (hard)	398	1666
Rice (whole)	360	1507	Cream (double)	288	1205
Vegetable oils			Milk (full fat)	64	268
Olive oil	883	3696	Eggs	162	678
Sunflower oil	883	3696	Margarine	720	3013
			Yogurt	71	297

Note: The values given are approximate—the calorific value of fruits will depend on their ripeness (i.e. on the amount of sugar they contain) and that of meat products will depend on their fat content.

assumptions and inaccuracies, but it can give information about the total amount of heat being generated by the body. Such information can be very useful as an aid to diagnosing conditions such as thyroid disorders (see Chapter 22).

Although the RQ is a useful index of the type of fuel being oxidized by the body at a particular time, it is not reliable during changes in the acid–base status of an individual. In heavy exercise in particular, the rise in blood lactate (and consequent fall in arterial pH) favours the formation of carbonic acid from plasma bicarbonate, with the result that more carbon dioxide is lost from the body than is produced during metabolism. Under these conditions the RQ value may exceed unity. During recovery following a period of exercise, much of the lactate that was produced is resynthesized into glucose and glycogen and blood pH rises. During this phase, the RQ may be less than normal as bicarbonate is reformed.

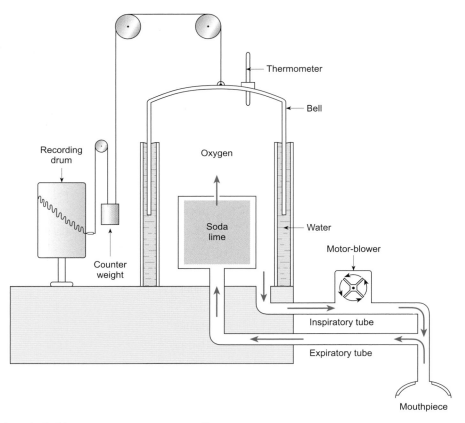

Figure 47.2 A spirometer adapted to measure oxygen consumption.

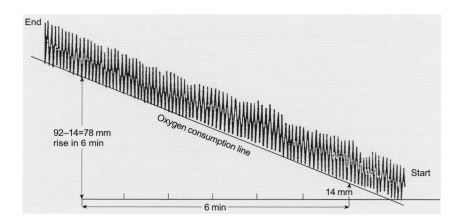

Figure 47.3 The measurement of oxygen consumption. The record is read from right to left and can be obtained from apparatus of the kind shown in Figure 47.2. Each millimetre fall of the spirometer bell (and corresponding rise in the experimental record) represents a known volume of oxygen used. The slope of the trace can then be used to calculate the oxygen consumption. (Adapted from Fig III.81 in C.A. Keele, E. Neil, and N. Joels (1992) *Samson Wright's applied physiology*, 13th edition. Oxford University Press, Oxford.)

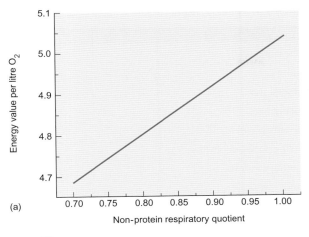

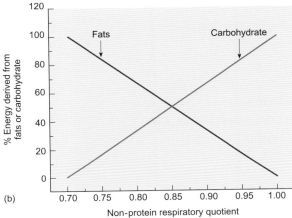

Figure 47.4 The relationship between the non-protein respiratory quotient and the energy equivalent of oxygen (a) and the proportion of energy derived from carbohydrates and fats (b). (From data in L. Lusk (1924) *J Biol Chem* **59**, 41–2.)

Summary

The body metabolizes variable quantities of fats, carbohydrates, and sometimes proteins, to produce energy. Different amounts of heat energy are liberated from each foodstuff. The amount of energy liberated per litre of oxygen consumed is known as the energy equivalent of oxygen. The volume of carbon dioxide produced each minute divided by the volume of oxygen consumed in the same period of time is called the respiratory quotient (RQ), or respiratory exchange ratio, and varies with the kind of food being metabolized. The metabolic rate may be calculated from the oxygen consumption and the RQ.

47.5 Basal metabolic rate (BMR) and the factors that affect it

In adults, BMR amounts to an average daily expenditure of 84 to 105 kJ (20 to 25 kcal) per kg body weight. This requires the consumption of some 200–250 ml of oxygen each minute. About 20 per cent of the BMR is accounted for by the central nervous system, 25 per cent by the liver, 20 per cent to 30 per cent by the skeletal muscle mass, and about 16 per cent by the heart and kidneys.

The BMR depends on many different factors. Based on evidence obtained from identical twins and families, it is believed that the BMR is determined at least partly by genetic factors. It is also affected by a number of physiological variables including body weight, body surface area, lean body mass, age, and the sex of the individual. Although there is no simple, direct relationship between BMR and body size, it is customary to relate metabolism to the surface area of the body (kJ or kcal per square metre per hour) since heat is produced in proportion to surface area. Since muscle has a much higher metabolic rate than adipose tissue, BMR is also related to the lean body mass. The surface area of an individual can be obtained from a standard nomogram or it can be calculated using the following empirical formula:

$$\log_{10}[\text{surface area in cm}^2] =$$
$$0.425 \times \log_{10}[\text{weight in kg}]$$
$$+ 0.725 \times \log_{10}[\text{height in cm}] + 1.8564$$

For a 70 kg man who is 1.8 m tall the logarithm of their surface area in cm will be:

$$(0.425 \times 1.84) + (0.725 \times 3.255) + 1.856 = 4.2755$$

Taking the inverse logarithm of 4.2755, the surface area is 18,856 cm^2 or 1.88 m^2.

Any comparison of the BMR of different individuals must take into account the factors that might alter it. As a result of numerous measurements of BMR in a wide range of subjects, tables are now available which give the expected normal range of BMR values for people of different body size, age, and sex. This information is valuable in clinical situations where it is necessary to determine whether the BMR of a patient lies outside the normal range, and by how much. BMR is increased during fever and is affected by changes in the circulating levels of thyroid hormones.

Figure 47.5 shows the effects of age on the normal metabolic rates of both males and females. The BMR of women is 6 to 10 per cent lower than that of men of comparable size and age. However, during pregnancy, the additional metabolic activity of the fetus results in a significant increase in the BMR. For adult males between 20 and 60 years of age, the normal BMR lies in the range 142–168 kJ m^{-2} h^{-1} (34–40 kcal m^{-2} h^{-1}). For non-pregnant women the range is 134–147 kJ m^{-2} h^{-1} (32–35 kcal m^{-2} h^{-1}).

The metabolic rate of a young child, in relation to its size, is almost twice that of an elderly person. The high BMR of childhood results from high rates of cellular reactions, rapid synthesis of cellular materials, and growth, all of which require moderate quantities of energy. Part of the decline in BMR seen in old age is due to the fall in lean body mass with age. The percentage of total body weight represented by fat more than doubles for men between the ages of 20 and 55.

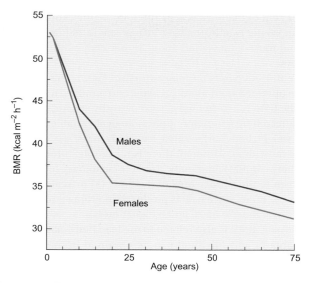

Figure 47.5 The average age-related changes in BMR for males and females corrected for body size. (From data in W.M. Boothby, J. Berkson and H.L. Dunn (1936) *Am J Physiol* **116**, 468–84.)

Table **47.2** The energy cost of various occupational activities

Activity	Energy cost (kJ min^{-1})
Office work	5.7
Ironing	6
Dishwashing	6.3
Cooking	6.7
Laboratory work	7.4
Working on assembly line	12.3
Cleaning windows	12.4
Laundry work	12.8
Driving heavy goods vehicle	14.2
Dusting	14.6
Cleaning and drying	19.8
Climbing stairs (72 steps/min)	20.1
Bed making	20.4
Labouring	21.0
Gardening	22.5
Forestry (hewing and trimming timber)	25.0
Coal mining (excluding rest periods)	29.7
Stone cutting	30.9

47.6 Physiological factors that affect metabolic rate

The **metabolic rate** is the rate at which chemical energy is expended by the body. It varies from minute to minute depending on the nature of the activities being undertaken by an individual. It is important to distinguish between the metabolic rate and the BMR, which refers to the metabolic activity of a subject during standard conditions of fasting and complete rest. The BMR provides a consistent way of comparing the metabolic rates of different subjects, which is of particular value in clinical situations. There are, however, very few times during everyday life when body metabolism is occurring this slowly—any activity that alters the chemical activity of cells will alter the metabolic rate of the body. Some of the physiological factors that alter the metabolic rate are discussed below.

The energy requirements of different tasks and different occupations have been investigated very thoroughly, and some examples are presented in Table 47.2. People involved in sedentary work have a modest energy requirement—about 8.25 MJ (c. 2000 kcal) per day—while manual workers in heavy industries such as mining may have an energy requirement three times greater than this. A typical calculation of daily energy expenditure is presented in Box 47.1.

Exercise

Any degree of exercise, from simply carrying out the normal daily activities of life such as walking, eating, and dressing, to the severe exercise involved in manual labour, produces an increase in metabolic rate. Walking slowly (4.2 kph) uses about 3 times as many kilojoules per hour as lying in bed asleep, while maximal muscular exercise can, in short bursts, increase the metabolic rate by around

20-fold. Table 47.3 shows some examples of the energy requirements of particular forms of physical activity.

Ingestion of food

After eating a protein-rich meal, there is an increase in the rate of heat production by the body, i.e. an increase in metabolic rate. Although part of this rise may be due to the cellular processes involved in the digestion and storage of foods, it is believed to result mainly from specific effects exerted by certain amino acids derived from the protein in the meal. This is known as **the specific dynamic action of protein**. Heat production also increases following the ingestion of a meal rich in carbohydrates or fats, but to a lesser extent. It is thought that the liver is the major site of the extra heat production because of its importance in the intermediary metabolism of absorbed foods (see Chapters 24 and 45).

Fever

Whatever its cause, fever elicits an increase in metabolic rate. For every half degree Celsius rise in body temperature, the BMR increases by about 7 per cent. Thus the metabolic rate of a person with a fever of 39°C will be nearly a third higher than normal. This can be explained largely by the fact that all chemical reactions, including those taking place within cells, proceed more rapidly as temperature increases. However, fever is also associated with changes to the composition of the blood and with enhanced production of white cells (see Chapter 42).

(1₂3) Box 47.1 Daily energy balance calculation

A man spends about 8 hours a day working, with an average energy consumption of 700 kJ h^{-1}. He spends another 8 hours sleeping or lying down (276 kJ h^{-1}). He walks to and from work at a brisk pace (1250 kJ h^{-1}), which takes 30 minutes a day. In his remaining time, he spends 7 hours sitting and attending to personal needs (dressing, eating etc.) at an average energy consumption of 376 kJ h^{-1} and exercises by jogging for 30 minutes each evening at an average energy consumption of 3000 kJ h^{-1}. His total energy requirement each day can be calculated as follows.

Activity	Time spent (hours)	Rate of energy consumption (kJ h^{-1})	Total energy requirement (kJ)
sleeping and lying	8	276	2208
sitting etc.	7	376	2632
walking	0.5	1250	625
jogging	0.5	3000	1500
working	8	700	5600
total	24	—	12,565

His total energy consumption is 12.56 MJ (3000 kcal) a day, which must be met by his diet. For optimal nutrition, about 60 per cent of the energy value of food should be in the form of carbohydrate, 30 per cent as fats, and the remainder as protein. Since each gram of carbohydrate or protein yields approximately 17 kJ of energy and each gram of fat yields approximately 39 kJ, his recommended dietary intake should be: as follows.

Carbohydrates $(12,560 \times 0.6) \div 17 = 443$ grams

Fats $(12,560 \times 0.3) \div 9.3 = 97$ grams

Protein $(12,560 \times 0.1) \div 4.1 = 74$ grams

Table 47.3 Approximate energy consumption of an adult male engaged in various physical activities

Activity	Energy consumed watts	Energy consumed kJ min^{-1}
Sleeping	77	4.6
Sitting	105	6.3
Standing	174	10.5
Slow walking (3.2 kph)	195	11.7
Washing, dressing etc.	230	13.4
Household chores (cleaning, ironing etc.)	250	15.1
Brisk walking (6.4 kph)	363	21.8
Gardening	400	23.9
Cycling (16 kph)	433	26.0
Mining/industrial loading	500	30.1
Running (10 kph)	712	42.7
Running (14 kph)	963	57.8
Swimming (fast crawl)	872	52.3

1 watt = 1 joule s^{-1}

Fasting and malnutrition

The basal energy requirement of a man with a surface area of 1.8 m^2 is about 7–7.5 MJ a day (1700 to 1800 kilocalories a day). As already mentioned, different physical activities demand the expenditure of additional energy (Table 47.2). The daily energy requirement is normally met by the appropriate intake of food. However, when little or no food is being eaten, the component tissues of the body are used for the production of energy. The hepatic glycogen stores are consumed first, followed by the mobilization of stored fats and the conversion of amino acids to glucose (see Chapter 46). Later, tissue protein is broken down to provide amino acids for glucose production, and creatine for skeletal muscle activity. A fasting person is said to be in a state of catabolism, in which the body's stores of fat, carbohydrate, and protein will all be diminishing. Prolonged starvation decreases the basal metabolic rate often by as much as 20–30 per cent, as the availability of necessary food reserves declines. This decrease in BMR serves to limit the drain on available resources.

Sleep

Although small variations in metabolic rate are seen during the different stages of sleep, overall there is a fall in metabolic rate of between 10 and 15 per cent during sleep as a result of a reduction in the level of muscular tone and a fall in the activity of the sympathetic nervous system.

Summary

BMR is partly determined by genetic factors and varies with the age, sex, and size of each individual. Metabolic rate is increased by any form of exercise, the ingestion of food, and fever. It is reduced in malnutrition and during sleep.

47.7 The actions of hormones on energy metabolism

A variety of hormones are able to alter the metabolic rate by changing the way in which fats, carbohydrates, and proteins are utilized by the cells and tissues of the body. The principal hormones of this category are the catecholamines (both from the adrenal medulla and the sympathetic nervous system), the thyroid hormones, and growth hormone. The male and female sex steroids also exert a small effect on metabolic rate. Full details of the actions of all these hormones are given in the appropriate chapters of this book (particularly Chapters 21–24 and 48), so only a brief description is given here.

Adrenaline and noradrenaline stimulate the metabolic rates of virtually all the tissues of the body. Their major effect is to increase the rate of glycogenolysis within cells. Maximal sympathetic stimulation is thought to increase metabolic rate by between 25 and 100 per cent.

The thyroid hormones exert a powerful effect on the metabolic rate. Loss of thyroidal activity gives rise to a condition known as myxoedema (see Chapter 22), in which metabolic rate may fall to half its normal value. In hyperthyroid individuals, metabolic rate is correspondingly enhanced. Because of these effects, hypothyroidism produces a reduced tolerance to cold conditions whereas increased thyroid hormone secretion often causes the person to feel hot. Thyroid hormones raise metabolic rate because they stimulate many of the chemical reactions taking place within cells.

Growth hormone and the male and female sex steroids influence metabolic rate to a minor degree, all having a mild stimulatory effect. Male sex steroids are more potent than female hormones in this respect and probably contribute to the higher BMR seen in males.

Summary

Catecholamines and thyroid hormones are potent stimulators of metabolism, while growth hormone and the sex steroids exert a mild stimulatory effect.

✳ Checklist of key terms and concepts

Energy metabolism and the measurement of metabolic rate

- The chemical processes of the body constitute its metabolism.
- Metabolism gives rise to heat and provides energy in the form of ATP.
- Energy is measured in kilojoules (kJ) or kilocalories (Calories) (where 1 kilocalorie is 1000 calories or 4187 joules).
- The rate at which chemical energy is expended by the body is called the metabolic rate.
- The body metabolizes variable quantities of fats, carbohydrates, and sometimes proteins, to produce energy.
- Different amounts of heat energy are liberated from each foodstuff.
- The amount of energy liberated per litre of oxygen consumed is known as the energy equivalent of oxygen.
- The respiratory quotient (RQ) is equal to the volume of carbon dioxide produced each minute divided by the oxygen consumed each minute.
- The respiratory quotient varies with the kind of food being metabolized.

- Metabolic rate may be calculated from oxygen consumption and the respiratory quotient.

Basal metabolic rate (BMR)

- The basal metabolic rate is measured after an overnight fast under conditions of complete rest.
- BMR is partly determined by genetic factors but also varies with the age, sex, and size of the individual.

Factors influencing metabolic rate

- Metabolic rate is increased by any form of exercise, by the ingestion of food, and by fever.
- Metabolic rate is reduced in malnutrition and during sleep.
- A variety of hormones can modify the metabolic rate.
- Catecholamines and thyroid hormones are potent stimulators of metabolism, while growth hormone and the sex steroids exert a mild stimulatory effect.

📄 Recommended reading

Astrand, P-O., Rodahl, K., Dahl, H.A. and Stromme, S. (2003) *Textbook of work physiology, Physiological bases of exercise* (4th edn). Human Kinetics, Champaign, IL.
An authoritative account of the physiology of exercise; an excellent introduction to the classic literature of the field.

Review articles

Carreiro, A. L., Dhillon, J., Gordon, S. (2016) The macronutrients, appetite, and energy intake. *Annual Review of Nutrition* **36**, 73–103.

Dowell, P., Hu, Z., and Lane, M.D. (2005) Monitoring energy balance: Metabolites of fatty acid synthesis as hypothalamic sensors. *Annual Review of Biochemistry* **74**, 515–34.

Lefebvre, P., Cariou, B., Lien, F., Kuipers, F., Staels, B. (2008) Role of bile acids and bile acid receptors in metabolic regulation. *Physiological Reviews* **89**, 147–91; doi:10.1152/physrev.00010.2008

Sutton, A. K., Myers, M. G. Jr., and Olson, D. P. (2016) The role of PVH circuits in leptin action and energy balance *Annual Review of Physiology* **78**, 207–2.

 To check that you have mastered the key concepts presented in this chapter, complete the accompanying online self-assessment questions. Go to www.oup.com/uk/pocock5e/

PART ELEVEN

Reproduction and growth

CHAPTER 48

The physiology of the male and female reproductive systems

Chapter contents

This chapter should help you to understand:

- The significance of sexual reproduction
- The principal structures of the male reproductive system and their functions
- The formation of mature sperm by spermatogenesis and spermiogenesis
- The hormonal regulation of testicular function
- The principal structures of the female reproductive system and their function
- The menstrual cycle and its hormonal regulation
- The role of hormones in the regulation of the female reproductive system
- Puberty and the menopause
- The peripheral actions of the testicular and ovarian steroids in the adult

48.1 Introduction

Reproduction—the ability to produce a new generation of individuals of the same species—is one of the fundamental characteristics of living organisms. Genetic material is transmitted from parents to their offspring to ensure that the characteristics of the species are perpetuated. The essential feature of sexual reproduction is the recombination of chromosomes from two separate individuals to produce offspring that differ genetically from their parents. At the heart of the process lies the creation and fusion of the male and female **gametes**: the **spermatozoa** (**sperm**) and ova (eggs).

Gametes are specialized sex cells produced by the gonads that provide a link between one generation and the next. Spermatozoa, the male gametes, are produced by the **testes**, while ova, the female gametes, are produced by the ovaries. The nuclei of these cells are **haploid**—i.e. they contain only a single set of 23 unpaired chromosomes. Haploid cells are created when a diploid cell divides by **meiosis**, a process in which the genes are parcelled out afresh in single chromosome sets (discussed later in the chapter). During meiotic division, old combinations of genes are broken and new combinations are formed by chromosomal exchange so that the genetic composition of each chromosome is modified, while the correct number of genes is maintained. At **fertilization**, the gametes fuse to form a new **diploid** cell possessing a full set of chromosomes (23 pairs), half of which originate from the sperm (the paternal chromosomes) and half from the **ovum** (the maternal chromosomes). This new diploid cell is called a **zygote**. The reshuffling of genes during sexual reproduction helps to create a genetically diverse population that is able to show greater resilience in the face of environmental challenges.

This chapter discusses the processes that lead to the production of the male and female gametes together with the neural and hormonal mechanisms that regulate reproductive activity. It concludes with a brief discussion of puberty and the menopause.

48.2 Reproductive physiology of the male

The anatomy of the male reproductive system

Figure 48.1 is a simple diagram of the adult male reproductive tract showing its major organs, while Figure 48.2 illustrates the internal structure of the testes, which are the male gonads. These have two main functions: the production of spermatozoa and the secretion of the male sex hormones (the **androgens**). The testes lie outside the abdominal cavity, within a thick-skinned sac, the scrotum. This is divided internally into two compartments by a fibrous septum. One testis lies within each compartment. In an adult man, each testis is about 4.5 cm long, 3.5 cm wide, and 40 g in weight. The germinal tissue is enclosed in a fibrous coat, the tunica albuginea, which extends into the body of the testis, dividing it into lobules (see Figure 48.2). Within these lobules lie the coiled **seminiferous tubules** which contain two types of cells, the **spermatogonia** (immature germ cells) from which sperm develop, and the large **Sertoli cells**. The Sertoli cells surround the developing gametes,

protecting and nourishing them. Each testis of a healthy young adult normally contains more than 800 seminiferous tubules. Between these tubules lies supportive connective tissue which contains clusters of **interstitial** or **Leydig cells** which are responsible for the synthesis and secretion of the testicular androgens, particularly testosterone.

The sperm produced by the testis are carried to the penis by a system of ducts. At the apex of each testicular lobule, the seminiferous tubules that lie within it merge and pass into the first section of the efferent ducts, the **tubuli recti**. These are short, straight tubes which enter the dense connective tissue of the central posterior portion of the testis (the mediastinum testis) to form a system of irregular, epithelium-lined spaces, the **rete testis**. At its upper end, the rete testis drains into a series of 15–20 tubules called the efferent ducts (**vasa efferentes**) which merge to form a tightly coiled tube, the **epididymis**. The epididymis is able to store and nourish sperm for up to five weeks. Its walls contain circular smooth muscle whose contractions help to propel sperm towards the **vas deferens** (also called the ductus deferens or sperm duct), a tubular structure 30–35 cm in length. The vas deferens terminates in the **ejaculatory duct**, close to the prostate gland which is a round structure about the size of a walnut and weighing 20–30 g. As Figure 48.1 shows, the prostate gland is situated just below the urinary bladder, where it surrounds the first part of the urethra. If the prostate gland enlarges (as a result of benign prostatic hyperplasia or prostate cancer), it will exert pressure on the base of the bladder and constriction at the

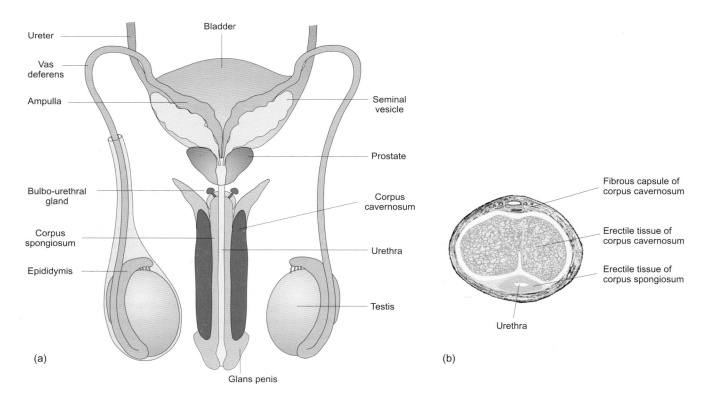

Figure 48.1 (a) A diagrammatic representation of a posterior view of the adult human male reproductive system to show the principal structures. (b) A cross-section of the middle part of the penis to show the anatomical relationships between the corpus cavernosum, the corpus spongiosum, and the urethra. (Adapted from Fig 26.1 in L. Weiss and R.O. Greep (1977) *Histology*, 4th edition. McGraw-Hill, New York.)

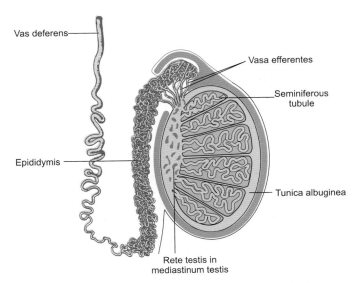

Figure 48.2 A diagrammatic representation of the adult testis, epididymis, and vas deferens. (Adapted from Fig 26.2 in L. Weiss and R.O. Greep (1977) *Histology*, 4th edition. McGraw-Hill, New York.)

top of the urethra. These effects lead to a range of urinary symptoms including frequency and urgency of micturition and poor urinary flow. Glandular tissue within the prostate secretes a fluid containing mucus, salts, and sugars. An enzyme, prostate specific antigen or PSA, is also produced by the prostate. This helps to liquefy clotted sperm in the vagina and is an important screening tool for prostate cancer. Normal blood levels of PSA lie between 0.1 and 4 ng ml^{-1} but significant increases are often seen in prostate malignancy. Located on either side of the prostate are the **seminal vesicles**, which secrete an alkaline fluid containing fructose that helps to nourish the sperm and to neutralize the acidity of the vaginal fluid. Fluid from the seminal vesicles is emptied into the ejaculatory duct together with the sperm and prostatic secretions to form **semen**. From the ejaculatory duct, the semen enters the urethra, which transports semen during ejaculation. The urethra is a membranous tube that runs the length of the penis and connects the bladder to the outside to allow the passage of urine during micturition.

As Figure 48.1 shows, the penis contains three cylindrical regions of erectile tissue. These are the two corpora cavernosa, which lie on either side of the medial plane, and the corpus spongiosum, which lies on the ventral side of the flaccid penis surrounding the urethra. The corpora cavernosa and corpus spongiosum contain vascular cavities called venous sinusoids. Engorgement of this tissue with blood during sexual arousal is responsible for erection of the penis (see Chapter 49). The corpus spongiosum extends beyond the corpora cavernosa to form the glans penis, a sensitive area possessing many nerve endings. Two small round glands, the **bulbo-urethral glands** (also called Cowper's glands), lie directly below the prostate gland and open via short (2.5 cm) ducts into the penile part of the urethra. During sexual arousal these glands secrete an alkaline fluid (sometimes called pre-ejaculate) that neutralizes any residual acidic urine in the urethra, rendering it more hospitable for sperm.

The development of the testis takes place within the abdominal cavity of the fetus (see Chapter 50). To reach their adult position

within the scrotal sac, the testes must migrate posteriorly through the abdominal cavity and over the pelvic brim via the inguinal canal. Migration begins during the last few weeks of gestation and descent of the testes into the scrotum is usually complete just before or very soon after birth. The inguinal canal normally closes after the testes have passed through but if closure is incomplete, or if the area is strained or torn, an inguinal hernia may result.

The location of the testes outside the abdominal cavity allows the sperm to form at a temperature which is some 2 or 3 degrees lower than the body's core temperature. Failure of the testes to migrate into the scrotum results in a condition known as **cryptorchidism** which, should it persist until puberty, leads to an arrest of spermatogenesis, and thus infertility, since it appears that the testes cannot function normally at body temperature. Undescended testes are also at increased risk of developing testicular cancer. Surgical correction is normally carried out during childhood to avoid this risk.

Summary

Spermatozoa, the male gametes, are produced by the testes. Septa from the outer fibrous coat of the testes, the tunica albuginea, divide the testes into lobules within which lie the seminiferous tubules which are the site of sperm formation. The sperm develop from spermatogonia, which are surrounded by large Sertoli cells which protect and nourish them. The connective tissue supporting the seminiferous tubules contains clusters of Leydig cells which secrete the male hormone, testosterone. The seminiferous tubules empty into a series of ducts that eventually merge into the epididymis, which can store sperm for some weeks. From the epididymis, the sperm pass to the vas deferens which terminates in the ejaculatory duct close to the prostate gland. The seminal vesicles are located on either side of the prostate.

The adult testis produces gametes and synthesizes androgens

The testes of a mature male perform two fundamental roles, both of which are vital to his fertility and sexual competency. These are:

- The production of sperm which will carry his genes and fertilize an ovum.
- The secretion of testicular androgens, particularly testosterone, which are required both for normal sperm production and for full masculine development.

The two major products of the male gonads, the spermatozoa and the androgenic steroid hormones, are synthesized in separate compartments within the testis. The sperm are produced in the seminiferous tubules themselves, while the androgens are synthesized and secreted by the Leydig cells that lie between the tubules (see Figure 48.3). Indeed, these two compartments of the testis appear to be separated functionally as well as anatomically since there is a barrier that prevents the free exchange of water-soluble materials between them. This is known as the blood–testis barrier and arises as a result of the extremely tight junctional complexes

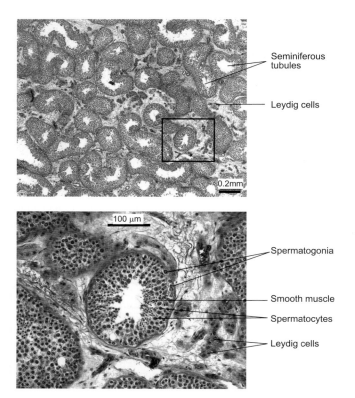

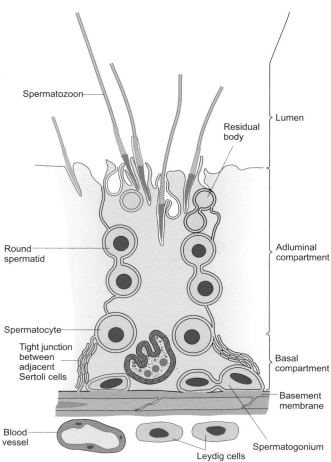

Figure 48.3 A section of the testis showing seminiferous tubules cut across at various angles. The dark clumps of cells scattered between the tubules are Leydig cells. The top panel shows a low-power view and the lower panel shows the area bounded by the square at a higher magnification. The section was stained with iron haemotoxylin and eosin.

Figure 48.4 The relationship between the Sertoli cells and the developing spermatozoa. Note the tight junctions between the basal regions of the Sertoli cells separating the basal compartment from the adluminal compartment. (Redrawn after Fig 26.30 in L. Weiss and R.O. Greep (1977) *Histology*, 4th edition. McGraw-Hill, New York.)

that exist between the basal regions of adjacent Sertoli cells (see Figure 48.4). This barrier protects the developing sperm from any damaging blood-borne agents and so maintains a suitable environment for their maturation. It also prevents antigenic materials (e.g. proteins) that arise in the course of spermatogenesis from passing into the circulation and triggering an autoimmune response to the sperm. If this should happen—for example as a consequence of traumatic injury to the testis—it can lead to male infertility. The importance of this sanctuary effect becomes clear when one considers that spermatozoa are produced at puberty and will therefore be classed as foreign by the immune system, which developed self-recognition during the first year of life (see Chapter 26). Although the manufacture of sperm and androgens occurs in separate compartments, their production is functionally related because the production of mature sperm is only possible if androgen secretion is normal.

Testosterone is the major testicular androgen

The Leydig cells of the testis synthesize and secrete testosterone (the principal testicular androgen) from acetate and cholesterol, as outlined in Box 48.1. Adult males secrete between 4 and 10 mg of testosterone each day, most of which passes into the blood. As

testosterone is a steroid hormone it is lipid-soluble, and a small amount is able to cross the blood–testis barrier by passive diffusion. Once in the seminiferous tubules, testosterone binds to androgen binding protein secreted by the Sertoli cells, where it plays a crucial role in the development of the spermatozoa.

The peripheral actions of testosterone

Testosterone circulates in the plasma bound either to sex-hormone binding globulin (SHBG) or to other plasma proteins, especially albumin. Only around 2 per cent circulates in unbound form. As it is lipid-soluble, testosterone freely enters cells, where it may be converted to **dihydrotestosterone, androstenedione,** or **oestradiol.** Testosterone and dihydrotestosterone bind to a specific intracellular receptor, the androgen receptor, to form a complex that interacts with chromosomal DNA to regulate gene expression (see Chapter 6). The androgen receptor can also be recruited by other transcription factors to modulate gene expression. In addition, testosterone can

Box 48.1 Biosynthesis of sex steroid hormones

A simple diagram to show the principal pathways involved in the biosynthesis of the major sex steroid hormones from cholesterol. As the diagram shows, extensive interconversions are possible although the exact biosynthetic routes for a particular tissue will depend on the enzymes present in that tissue.

Acetate → Cholesterol → Pregnenolone → Progesterone

Dehydroepiandrosterone   17α Hydroxyprogesterone

Androstenediol   Androstenedione — Aromatase → Oestrone

Dihydrotestosterone ← 5a reductase — Testosterone — Aromatase → Oestradiol-17β

exert relatively rapid effects by binding to receptors on the plasma membrane to influence ion transport and other cell functions.

Androgen receptors are most numerous in the tissues which are specific targets for testosterone and dihydrotestosterone, i.e. those which depend upon androgens for their growth, maturation, and/or function. Such tissues include the accessory organs of the male reproductive tract—the prostate gland, seminal vesicles, and epididymis—as well as non-reproductive tissues such as the liver, heart, and skeletal muscle.

Dihydrotestosterone is important in the fetus for the differentiation of the external genitalia and, at puberty, for the growth of the scrotum, prostate, and sexual hair. In addition to its role in the production of mature sperm, testosterone stimulates the fetal development of the epididymis, vas deferens, and seminal vesicles. At puberty, it is responsible for enlargement of the penis, seminal vesicles, and larynx as well as for the changes in the skeleton and musculature that are characteristic of the male (the secondary sexual characteristics). It also increases libido (sexual desire). In short, testosterone is the hormone chiefly responsible for the development of the male primary sexual characteristics and the development of mature sperm, while dihydrotestosterone is the hormone chiefly responsible for the development of the secondary male characteristics. However, interactions between androgens and oestrogen receptors (through the conversion of testosterone to oestradiol) may mediate some aspects of male sexual development.

Spermatogenesis—the production of sperm by the testes

A sexually mature man produces around 200–300 million spermatozoa each day, which is equivalent to around 300 sperm per gram of testis per second. **Spermatogenesis** is a complex process that involves the generation of huge numbers of cells by mitosis and the halving of their chromosomal complement by meiosis (see Figures 48.5 and 48.6). It culminates in the formation of a

highly specialized cell, the mature spermatozoon, which is adapted to carry the genetic material a considerable distance within the female reproductive tract to maximize the chances of fertilization.

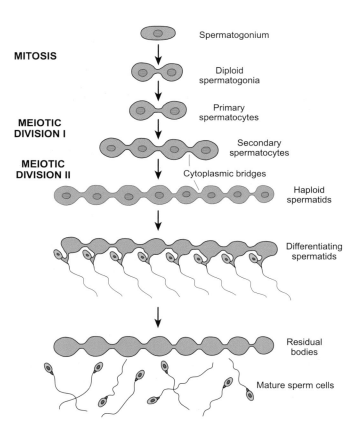

Figure 48.5 A simplified scheme to show the principal stages of spermatogenesis. The primordial germ cells divide to form spermatogonia, which undergo two more divisions to form primary spermatocytes. The primary spermatocytes undergo meiotic divisions to form secondary spermatocytes and spermatids. During meiosis the chromosome number is halved. Note the cytoplasmic bridges between the differentiating spermatids.

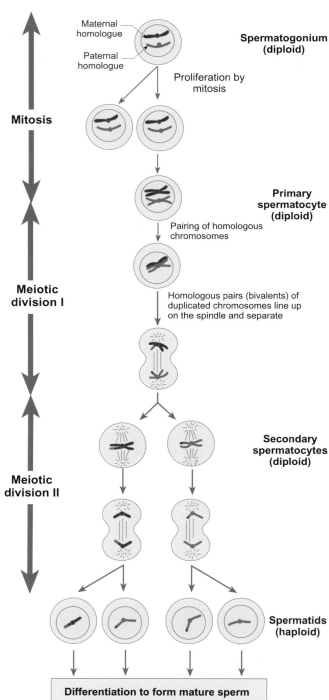

Figure 48.6 The chromosomal changes that occur in spermatogenesis. Initially the spermatogonia proliferate by mitosis to form primary spermatocytes. The primary spermatocytes then undergo two meiotic divisions to form spermatids, which are haploid cells. The spermatids then undergo spermiogenesis to form the mature sperm.

The major steps in the processes of mitosis and meiosis are described in Chapter 4 (Figures 4.9 and 4.10), while the specific chromosomal changes that occur during spermatogenesis are illustrated in Figure 48.6.

At puberty, the male germ cells commence mitotic division, an event which marks the beginning of spermatogenesis and which results in the formation of a population of spermatogonia that lies within the basal compartment of the seminiferous tubules. The first two mitotic divisions of each germ cell give rise to four cells which remain connected to one another by a thin bridge of cytoplasm (see Figure 48.5). Three of these undergo further division to form spermatogonia, while the fourth arrests at this stage and will later serve as a stem cell for a subsequent generation of sperm. The three active cells divide twice more to give rise to a population of so-called primary **spermatocytes**. Up until now, the cell divisions have taken place within the basal compartment of the tubule (see Figure 48.4), but at this point the primary spermatocytes enter the adluminal tubular compartment. They apparently achieve this by transiently disrupting the tight junctions between neighbouring Sertoli cells. Following a period of growth, each of the primary spermatocytes then undergoes two meiotic divisions (see Figure 48.6). The first of these gives rise to haploid secondary spermatocytes, which immediately divide again to form **spermatids**, each of which possesses 22 **autosomes** (the chromosomes not associated with the determination of sex) and either an X or a Y sex chromosome. The progeny of a single spermatogonium still remain connected by cytoplasmic bridges. The genetic events of spermatogenesis are now complete. The final stages of the process involve the conversion of the round spermatids to mature motile spermatozoa, a process known as **spermiogenesis**. These stages of sperm production are entirely dependent upon testosterone.

Spermiogenesis involves major cytoplasmic remodelling of the spermatid

Figure 48.7 shows the essential structures of a mature, motile human spermatozoon and from this it is clear that there are considerable differences between this and the round spermatid. The process of spermiogenesis involves reorganization of both the nucleus and the cytoplasm of the cell as well as the acquisition of a **flagellum** (tail).

The whole process of sperm manufacture takes place in close association with the Sertoli cells, which nourish and support the germ cells as they develop. The exact role of the Sertoli cells is unclear, but it is known that they possess high-affinity receptors for follicle stimulating hormone (FSH). The binding of FSH to these receptors induces the production of androgen binding protein, which is then secreted into the tubular luminal fluid. Binding of testosterone to this protein ensures that levels of androgen within the seminiferous tubules remain high enough to support spermatogenesis. Sertoli cells may also play a role in maintaining the vasculature, and thus the nutritional support, of the testis.

Newly-formed sperm consist of two regions that are morphologically and functionally distinct, the head region and the tail. The head contains a haploid nucleus (i.e. it possesses half the full

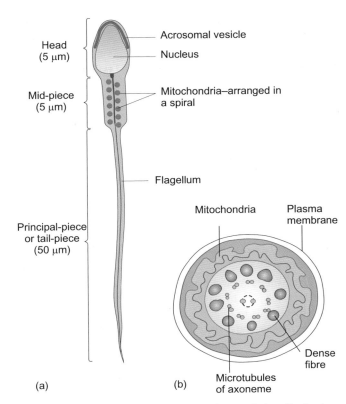

Figure 48.7 Diagram of a mature spermatozoon in longitudinal section (a) and transverse section through the midpiece (b). Note the spiral mitochondria and the arrangement of the nine pairs of microtubules around the central axoneme.

chromosomal complement) and a specialized secretory vesicle called the **acrosomal vesicle**, which contains hydrolytic enzymes that will help the sperm to penetrate the **oocyte** prior to fertilization.

The tail region of the sperm is motile. It is a long flagellum that has essentially the same internal structure as that seen in all cilia or flagellae from green algae to humans. It consists of a central **axoneme** originating from a **basal body** situated just behind the nucleus. The axoneme consists of two central microtubules surrounded by nine evenly spaced pairs of microtubules (see Figure 4.7). Active bending of the tail is caused by the sliding of adjacent pairs of microtubules past one another and movement is powered by the hydrolysis of ATP generated by mitochondria present in the first part of the tail, the **mid-piece**.

In humans the process of differentiation of a spermatocyte to a sperm takes approximately 70 days. After this, the newly formed sperm are released from the adluminal compartment of the Sertoli cells into the lumen of the seminiferous tubule and from there into the epididymis. As Figure 48.5 shows, connected residual bodies of cytoplasm are left behind by the released sperm. During their passage through the epididymis, sperm continue their maturation and acquire motility. This is the ability of the sperm to use its tail to propel itself forward and is vital if a sperm is to successfully make the journey through the female reproductive tract to fertilize an egg.

The formation of seminal fluid

The epididymis can serve as a reservoir for sperm, as their passage through this coiled tube takes anything from 1 to 35 days. From the epididymis, the sperm and other testicular secretions are transported along the vas deferens and into the ejaculatory ducts. The seminal fluid is markedly increased in volume by contributions from the seminal vesicles (about 60 per cent of total volume) and the prostate (about 20 per cent). The fluids secreted by these glands provide nutrients for the sperm. The seminal fluid is alkaline and helps to counteract the acidic fluid of the vagina as well as to enhance the motility and fertility of the sperm, both of which are optimal at a pH of around 6.5.

Summary

Spermatogenesis involves the generation of huge numbers of cells by mitosis followed by the halving of the chromosomal complement by meiosis to form haploid spermatids. Spermiogenesis involves remodelling of the cytoplasm of the immature spermatid and culminates in the formation of the mature, motile sperm. As they move through the male reproductive tract, the sperm are mixed with alkaline secretions from the seminal vesicles and prostate, which nourish the sperm and act to neutralize the acidity of the vagina.

The hormonal control of spermatogenesis— the hypothalamic–pituitary–testis axis

In Chapters 20 and 21 some of the general mechanisms involved in the regulation of hormone secretion within the body are discussed, including the importance of the hypothalamic releasing hormones and the concept of negative feedback control. These key regulatory processes have been shown to operate in the endocrine control of male reproductive function and are summarized in Figure 48.8.

Gonadotrophin releasing hormone (GnRH) is synthesized by neurons in the hypothalamus. It is secreted into the vessels of the hypophyseal portal tract and transported to the anterior pituitary gland. Here it stimulates the pituitary gonadotrophs to secrete the gonadotrophins follicle stimulating hormone (FSH) and luteinizing hormone (LH) into the systemic circulation. LH acts primarily on the Leydig cells to bring about the secretion of testosterone, while FSH acts mainly on the Sertoli cells to cause the secretion of androgen binding protein and a hormone called **inhibin**. It also promotes the synthesis of the aromatase complex that is responsible for the conversion of testosterone to oestradiol. In turn, testosterone inhibits the secretion of LH by exerting a negative feedback action at the level of both the anterior pituitary itself and the hypothalamus. It is thought that at the same time, inhibin and oestradiol also depress the further secretion of FSH by similar feedback mechanisms. These negative feedback loops provide an important internal control system for maintaining fairly constant levels of

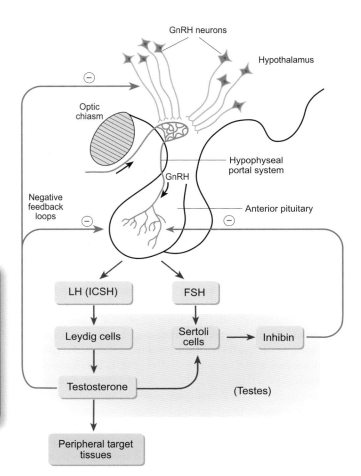

Figure 48.8 A schematic diagram to illustrate the relationship between the hormonal secretions of the hypothalamus, pituitary gland, and testes. Note the negative feedback control of GnRH secretion by testosterone and inhibin.

both gonadotrophic and androgenic hormones in the systemic circulation.

As discussed earlier, testosterone exerts a variety of important effects throughout the body of the male including the development of secondary sex characteristics at puberty. It is also essential for normal sperm production. As it is lipid-soluble, some testosterone secreted by the Leydig cells enters the intratubular compartment where it is bound by androgen binding protein secreted by the Sertoli cells. In some way, which is as yet poorly understood, this bound testosterone helps to maintain the production of spermatozoa. While testosterone is needed for the maintenance of spermatogenesis, pituitary FSH is required both to initiate the process and to mediate the differentiation of spermatids into spermatozoa.

Summary

Spermatogenesis is regulated by the hypothalamic gonadotrophins FSH and LH, and by testosterone. Hormone levels are regulated by negative feedback loops.

48.3 Reproductive physiology of the female

Like the testis in the male, the ovary produces haploid gametes and a variety of hormones. The production of gametes by the ovary is coordinated with its endocrine activity. Unlike the testes, however, which release an enormous number of gametes in a continuous stream, in humans the ovaries produce relatively few oocytes and their release normally occurs only once every four weeks or so at ovulation. This regular release of oocytes from the ovary is controlled by physical, neural, and, above all, endocrine mechanisms involving a complex interplay between the hypothalamic, anterior pituitary, and ovarian hormones. These will be discussed in some detail later, but briefly, the ovarian steroids (oestrogens and progesterone) are secreted in a cyclical fashion. A period of oestrogen dominance characterizes the first half of each cycle, during which one **ovarian follicle** reaches full maturity and the body is prepared for gamete transport and fertilization. This period culminates in ovulation roughly halfway through the cycle and is followed by a period of progesterone dominance during which the uterus is maintained in a state favourable for the implantation and development of a blastocyst (early embryo) and the body is prepared for possible pregnancy.

The anatomy of the female reproductive tract

Figure 48.9 shows a simplified diagram of the adult female reproductive organs seen in sagittal section, while Figure 48.10 shows the relationship between the major structures of the female reproductive system in coronal (frontal) section.

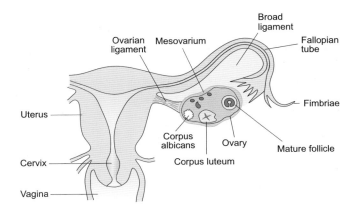

Figure 48.10 A simple diagram to show the relationship between the ovary, the Fallopian tube, and the uterus in coronal section. The section through the ovary shows a mature Graafian follicle, a corpus luteum, and a corpus albicans. In reality these would not all be present at the same time.

The **ovaries** are about 3–4 cm long and weigh 15 g each. They lie in the ovarian fossa of the pelvis and are attached to the uterus and Fallopian tubes by an extension of the broad ligament known as the mesovarium (or ovarian mesentery). (The broad ligament is a double layer of peritoneal membrane.) The adult organ is composed of stromal tissue, which contains the primary oocytes housed within primordial follicles, and glandular interstitial cells. The physiology of the ovaries will be considered in more detail in subsequent sections. The remainder of the reproductive tract is concerned not with gamete production but with the processes of fertilization and nurture of an embryo. To achieve this, both the male and female gametes must be transported to the site of fertilization in one of the Fallopian tubes and a favourable environment must be created, both for implantation and for the subsequent development of the embryo. Each ovarian cycle reflects these two roles.

The **Fallopian tubes** or oviducts are thin tubes about 12 cm in length that serve to transport the ovum released at ovulation from the ovary to the uterus. They also receive any sperm that have swum through the uterus. The opening of each Fallopian tube is expanded and split into fringes or fimbriae, which move nearer to the ovary at ovulation. There are numerous cilia on the fimbriae which create currents in the peritoneal cavity to direct an egg released from the ovary towards the mouth or **ostium** of the tube. As Figure 48.11 shows, the human Fallopian tube has three main layers: an outer covering of peritoneum (the serosa), a layer of circular smooth muscle covered by a layer of poorly defined longitudinal fibres, and an internal lining of mucosal membrane. The mucosal membrane is elaborately folded and covered by high columnar ciliated epithelial cells interspersed with glandular secretory cells. Movements of the cilia, coupled with contractile activity of the smooth muscle of the wall, facilitate gamete transport by the Fallopian tube.

The non-pregnant **uterus** is about 7.5 cm long and 5 cm wide (around the size and shape of a pear) but during pregnancy it increases enormously in size as it expands to accommodate the growing fetus throughout the 38-week gestation period. The uterus

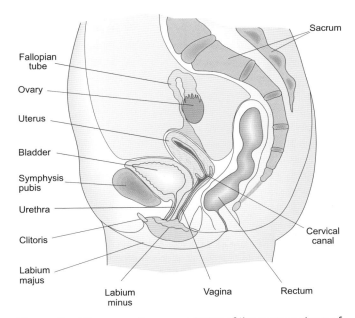

Figure 48.9 Diagrammatic representation of the gross anatomy of the female reproductive organs and their spatial relationship to other structures. (Adapted from Fig 6.9.3 in vol. 2 of P.C.B. McKinnon and J.F. Morris (2005) *Oxford textbook of functional anatomy*. Oxford University Press, Oxford.)

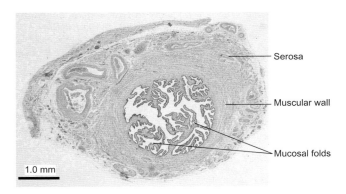

Figure 48.11 A cross-section through a human Fallopian tube showing its three main layers: the outer serosa, the muscular wall, and the highly folded mucosa. The section was stained with haematoxylin and OG erythrosin.

is therefore adapted to receive the early embryo (implantation), to allow placental formation, and to house the fetus as it develops. Moreover, it must remain quiescent throughout gestation to allow full fetal development but must contract powerfully to expel the fetus at birth.

Figure 48.12 shows a low-power view of a section of the wall of the human uterus, which consists of an outer covering or serous coat (the **perimetrium**), a middle layer of smooth muscle, the **myometrium**, that forms the bulk of the thick wall, and the inner endometrial layer or **endometrium**. This is composed essentially of epithelial cells, simple tubular glands, and the spiral arterioles that supply the cells. The characteristics of the endometrium are altered considerably during each ovarian cycle. The neck of the uterus is formed by the **cervix**, a ring of smooth muscle containing many mucus-secreting cells. The cervix forms the start of the birth canal and is traversed by sperm deposited in the vagina during sexual intercourse. During each menstrual cycle, the mucus-secreting cells undergo important changes in activity that optimize conditions for fertilization (discussed later in the chapter).

The final internal structure of the female reproductive tract is the **vagina**, a muscular tube that receives the penis at intercourse and allows delivery of a baby. It is therefore liable to be subjected to considerable frictional force and is lined by a stratified squamous epithelium, the cells of which show characteristic changes during each menstrual cycle. The vaginal epithelium is covered by a layer of fluid composed of secretions from the cervix, uterine endometrium, and Fallopian tubes. Although the pH of the vaginal fluid shows cyclical variations, it is normally acidic (pH 4–4.5) due to its high content of organic acids such as lactic acid. This low pH is thought to protect the vagina against invading pathogens. Observation of cyclical changes in the vaginal fluid and the vaginal epithelial cells can be useful in the treatment of infertility, as it can help to determine which stage of the cycle has been reached.

The remaining female sexual organs are known collectively as the external genitalia. They are illustrated in Figure 48.13. The vaginal orifice, urethral orifice, and clitoris are protected by folds of tissue called the **vulva** composed of the labia majora and minora. The labia majora consist largely of adipose tissue covered by a layer of skin. Their primary function is to protect and enclose the other external genitalia. The labia minora are thinner and usually hidden, at least partially, by the labia majora. They are rich in nerve endings and may swell and moisten during sexual arousal. The labia minora meet at the top of the vulva to form the clitoral hood beneath which lies the **clitoris** itself. This is a small erectile structure that is homologous with the male penis. Within the walls of the vulva lie the vestibular glands, which secrete mucus during sexual arousal and help to lubricate the movement of the penis within the vagina during sexual intercourse.

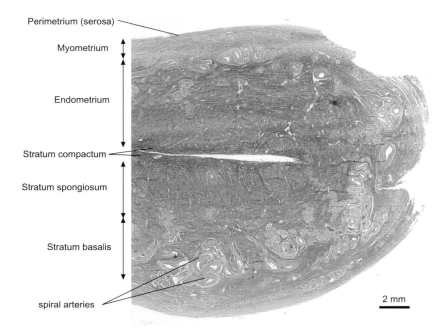

Figure 48.12 A low-power view of a section of the wall of a human uterus during the proliferative phase. The top half is labelled to show the three main layers: the outer serosa, the myometrium, and the endometrium. The bottom half is labelled to show the layers of the endometrium: the inner stratum compactum, the stratum spongiosum, and the outer stratum basalis, in which the developing spiral arteries are clearly visible. The section was stained with haematoxylin and eosin.

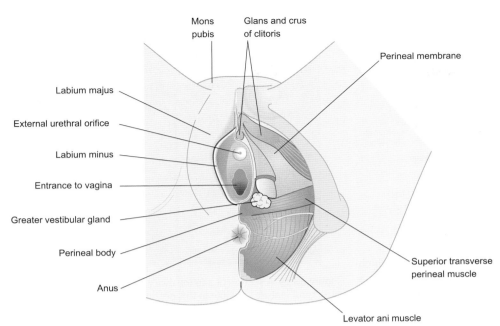

Mons pubis

Glans and crus of clitoris

Perineal membrane

Labium majus

External urethral orifice

Labium minus

Entrance to vagina

Greater vestibular gland

Perineal body

Anus

Superior transverse perineal muscle

Levator ani muscle

Figure 48.13 An illustration of the perineal region of the female to show the principal external genitalia. (Adapted from Fig 6.10.9 in vol. 2 of P.C.B. McKinnon and J.F. Morris (2005) *Oxford textbook of functional anatomy.* Oxford University Press, Oxford.)

Summary

The ova or eggs, the female gametes, are produced by the ovaries (the female gonads) which lie in the ovarian fossa within the pelvic cavity. They are attached to the uterus and Fallopian tubes by the ovarian mesentery of the broad ligament. Fertilization takes place in the Fallopian tubes.

The uterus houses the growing fetus throughout gestation and is adapted to permit implantation and the subsequent formation of a placenta. The vagina is a muscular tube that receives the penis at intercourse and forms the birth canal during delivery of a baby. The female external genitalia comprise the vaginal orifice, urethral orifice, and clitoris. They are protected by folds of tissue called the vulva (the labia majora and minora).

The physiology of the ovaries

Like the testes in the male, the ovaries play a fundamental role in the reproductive physiology of the female. They release mature, fertilizable ova at regular intervals throughout the reproductive years of a woman's life and secrete a number of steroid hormones including oestrogens and progesterone, which regulate both the ovaries and the rest of the reproductive tract. During the following account of ovarian function, the term 'oestrogen' will be used to refer to the oestrogenic hormones secreted by the follicular cells. It will also be used as a collective term for the variety of oestrogenic hormones of physiological importance secreted during pregnancy. In the non-pregnant woman, the predominant oestrogenic hormone is oestradiol-17β while oestrone and oestriol are also produced in large amounts during pregnancy (particularly by the placenta). In the following discussion of ovarian function, it will be important to bear in mind two questions:

1. What are the mechanisms that ensure the regular release of ova?

2. How does the endocrine activity of the ovary prepare the rest of the reproductive tract for successful fertilization and the ensuing pregnancy?

It is well known that during the fertile years of a woman's life the activity of her ovaries occurs in a cyclic fashion. The orderly sequence of events that underlies this cyclical behaviour is called the **ovarian cycle**. At the same time, under the influence of the ovarian hormones, the uterine endometrium undergoes a pattern of change known as the **uterine cycle**. Furthermore, many other body tissues are also influenced by the cyclical variations in ovarian hormone secretion. Together, the ovarian and uterine cycles form the familiar menstrual cycle during which there is remarkable coordination between the physical changes in various organs and the secretion of hormones. The onset of bleeding at menstruation is the visible sign that the ovaries and uterus have completed one cycle and that pregnancy has not occurred.

The interplay between the morphological and endocrine events that underpins the ovarian cycle is rather complicated. To simplify matters this account will be divided broadly into two sections: first, a description of the physical changes leading up to and following the release of an ovum at mid-cycle, and second, the changes that follow ovulation. The mechanisms that regulate each half of the cycle will be considered following the descriptions of the physical changes.

At birth the ovary already contains a large pool of immature follicles

The fundamental functional unit of the ovary is the follicle—indeed the bulk of the ovary is made up of follicles at various stages of development. This may be seen diagrammatically in Figure 48.14 and in the histological section of a feline (cat) ovary shown in Figure 48.15. During fetal life, the primordial germ cells of the ovary (known as oogonia) are laid down and continue their mitotic proliferation throughout the first three months of gestation. This

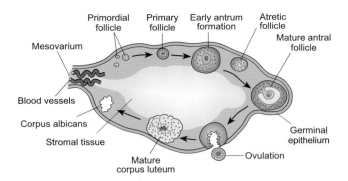

Figure 48.14 The internal structure of the ovary showing the stages of follicular development, ovulation, the formation of the corpus luteum, and its subsequent regression.

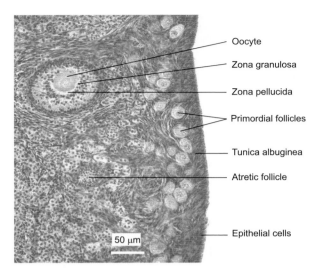

Figure 48.15 A section of a feline ovary showing the main structural features. Visible in this section are an antral follicle with a maturing oocyte, an atretic follicle, and a number of primordial follicles. The outer connective tissue layer (tunica albuginea) and its covering epithelium are clearly shown.

process is similar to that described earlier for the male, in which the germ cells (spermatogonia) divide by mitosis to generate a population of primary spermatocytes. In the male, new germ cells are produced throughout adult life. In females, however, the situation is somewhat different. Although recent evidence suggests that small numbers of follicles may continue to be laid down within the ovary after birth, it is believed that mitosis is largely complete by the end of the first trimester of the gestation period of a female fetus. Between this time and birth, around 70 per cent of these follicles will undergo atresia (degeneration). By the time she is born, a girl will typically possess between 150,000 and 500,000 follicles.

Once mitosis is complete, the oogonia enter their first meiotic division and become known as oocytes. At the same time, they become surrounded by mesenchymal cells on a basement membrane (the basal lamina) to form the primordial follicles. The oocytes arrest in the diplotene of the first meiotic prophase (see Chapter 4) and then remain in this arrested state until signalled to resume further development. This might occur at any time between menarche and the menopause (climacteric)—the reproductive years of a woman's life.

The pool of primordial follicles that is predominantly established during fetal life is gradually depleted throughout the reproductive years as follicles are recruited in a steady trickle to undergo further development. The classical view of the menstrual cycle is that it is a roughly 28-day process during which one of these follicles completes its maturation, with ovulation occurring at around the midpoint of the cycle (day 14). In reality this represents only the final four weeks or so of a maturational process that began at least two months earlier, when the primordial follicle was first recruited into the pool of developing follicles. Very little is known about the factors that control the recruitment of primordial follicles, but it appears to take place independently of hormonal control as removal of the anterior pituitary gland has no effect on the process.

Development of a follicle from the primordial to the preantral stage

During the 2 months or so following recruitment and prior to the start of the classically recognized 28-day menstrual cycle, the follicle progresses through several maturational stages as it develops from a primordial, to a primary, and then a preantral follicle.

During this growth phase, an enormous amount of synthetic activity takes place within the oocyte. This loads up its cytoplasm with the nutrient materials that the oocyte will require during its subsequent maturation. The developing follicle increases in diameter, from about 20 μm to 200–400 μm, and the primary oocyte within the follicle increases in size to around 120 μm.

The stromal cells surrounding the oocyte divide to form several layers of **granulosa cells** and secrete a glycoprotein that forms a cell-free region around the oocyte known as the **zona pellucida**. In addition, the cells adjacent to the basal lamina multiply and differentiate to form concentric layers around the primary follicle called the **theca**. The outermost layers of thecal cells are flattened and fibromuscular in nature (the theca externa) while the inner layers are more cuboidal (the theca interna). At this stage the follicle is known as a preantral follicle and is ready to begin the final stages of maturation that culminate in ovulation about 14 days later. Figure 48.16 shows the development and appearance of a preantral follicle. Although many primordial follicles begin the process of maturation to the preantral stage, it is likely that a great many of these fail to complete this maturational stage and undergo atresia.

From preantral follicle to luteolysis: the menstrual cycle

The following discussion will describe the events taking place within the ovary as a follicle undergoes the final 4 weeks of its developmental journey. In keeping with convention, this will now be referred to as the menstrual cycle.

In most women, the menstrual cycle is between 25 and 35 days in length, although wider variations occasionally occur. This represents the time it takes for a follicle to complete the final four weeks of its development and may, for the purposes of explanation, conveniently be split up into several distinct stages. These stages have names that reflect either the changing structure or the function of the follicle. Day 1 of the cycle is taken as the first day of menstrual

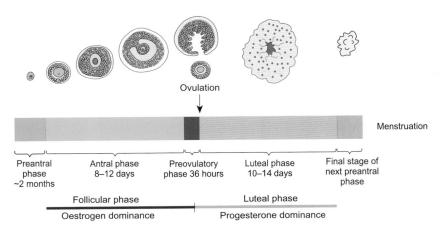

Figure 48.16 The principal phases of the ovarian cycle. The top part of the figure shows the sequence of follicular development, ovulation, and the formation of the corpus luteum. The lower part of the figure shows the relationship between the follicular and luteal phases of the cycle in relation to follicular development.

bleeding. The first 12 days or so of the four-week cycle consists of the progression of the follicle from the preantral through to the antral stage, a period of development that is concerned predominantly with follicular growth and development and is known as the **follicular phase**. The 24–48 hours immediately preceding ovulation is referred to as the **pre-ovulatory phase**. At mid-cycle, **ovulation** occurs, after which the collapsed follicle is converted to a **corpus luteum** (from the Latin meaning 'yellow body')—a process known as **luteinization.** The **luteal phase** of the menstrual cycle lasts for 10–14 days, after which, in the absence of pregnancy, the corpus luteum involutes and regresses to form a **corpus albicans** (from the Latin meaning 'white body'), a process called **luteolysis.** The onset of menstrual bleeding marks the first day of the next four-week cycle. These events are summarized in Figure 48.16.

The antral follicle

The menstrual cycle begins with a follicle that has completed the first stages of its maturation and become a preantral follicle. Towards the end of the preantral stage, an event occurs which is crucial for further follicular development. The follicular cells acquire receptors for certain hormones: the granulosa cells develop receptors for oestrogens and for pituitary FSH, while the thecal cells develop receptors for pituitary LH. This acquisition of hormone sensitivity is a prerequisite for continued follicular development since each subsequent stage is absolutely dependent on hormonal control. The continuous trickle of follicles through the hormone-independent preantral stage ensures that, at any one time, there are always several follicles that have completed their preantral growth and possess the appropriate receptors for anterior pituitary gonadotrophins and oestrogens. Further development depends upon the endocrine status of the body at the time. Provided that there are adequate levels of FSH and LH in the circulation, any follicles with the appropriate receptors will enter the next, **antral**, stage of development. Preantral follicles that do not possess hormone receptors degenerate and die—a process called **atresia**. An atretic follicle can be seen in the histological section of an ovary shown in Figure 48.15.

The anterior pituitary gonadotrophins FSH and LH convert preantral to antral follicles. The antral stage of development normally lasts for around ten days. During this time, the granulosa and thecal cell layers continue to increase in thickness. The granulosa cells also start to secrete follicular fluid all around the oocyte. This fluid forms the antrum, which gives the stage its name. Figure 48.17(c) is a diagrammatic representation of the general appearance of an antral follicle towards the end of this stage. The entire follicle is now much bigger (around 5 mm in diameter) although the oocyte itself remains much the same size (120 μm). The oocyte, surrounded by a few granulosa cells, is virtually suspended in follicular fluid and remains attached to the main rim of granulosa cells by a thin stalk. A fully developed antral follicle is also known as a **Graafian follicle**, an example of which can be seen in Figure 48.18(a).

Under the influence of gonadotrophins, the cells of the antral follicle start to secrete large quantities of hormones. Both the granulosa and thecal cells take on the characteristics of steroid-secreting tissue, with an extensive smooth endoplasmic reticulum, many lipid droplets, and microtubules. Pituitary LH stimulates the cells of the theca interna to synthesize and secrete the androgens testosterone and androstenedione. They also produce small amounts of oestrogens. The granulosa cells, which possess receptors for FSH, respond by converting androgens to oestrogens (particularly oestradiol-17β) as outlined in Box 48.1. The overall result of this secretory activity is a substantial increase in the circulating levels of both androgens and oestrogens, though especially the latter, during the antral phase of the menstrual cycle.

The oestrogens exert a significant effect within the follicle itself

As well as converting androgens to oestrogens, the granulosa cells of the antral follicle themselves possess receptors for oestrogens. The oestrogens produced by the follicular cells bind to these receptors and stimulate proliferation of further oestrogen-sensitive granulosa cells. There are thus more granulosa cells available for converting androgens to oestrogens, and this internal potentiation mechanism results in a substantial increase in circulating levels of oestrogens throughout the antral phase. Indeed, during the final two or three days of this stage (around days 10–12 of the menstrual cycle), oestrogen levels rise rapidly (the oestrogen surge). This peak is illustrated in the profile of oestrogen secretion during the menstrual cycle shown in Figure 48.19b. The oestrogens secreted during this time have many important actions throughout the reproductive tract, which will be discussed later in the chapter.

Early preantral follicle

Late preantral follicle

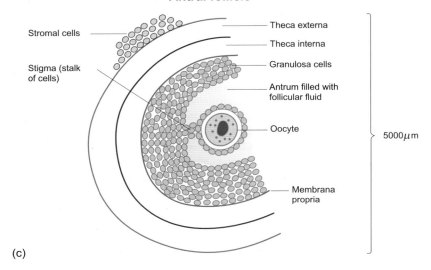

(a)

(b)

Antral follicle

5000 μm

(c)

Figure 48.17 A diagrammatic representation of the stages in the development of the ovum. Panel (a) shows an early preantral follicle; panel (b) shows a late preantral follicle and panel (c) shows a late antral (or Graafian) follicle. Note the proliferation of stromal and granulosa cells and the development of the fluid-filled antrum.

The pre-ovulatory follicle

As the follicle approaches the end of its antral phase of development, around the time of the oestrogen surge, two important events must coincide if the follicle is to progress further and enter the brief but dramatic pre-ovulatory stage. These are:

1. acquisition of receptors for pituitary LH by the granulosa cells
2. a rise in circulating levels of LH and FSH.

LH receptors are synthesized in response to pituitary FSH (for which the granulosa cells already possess receptors), and

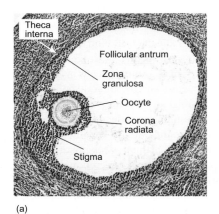

(a)

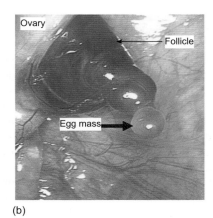

(b)

Figure 48.18 (a) A section through a mature antral follicle. Note the large fluid-filled space (the follicular antrum), the layer of granulosa cells, and the theca interna. The oocyte itself is covered with a layer of granulosa cells (the corona radiata), which link it to the zona granulosa by a stalk or stigma. (b) A remarkable photograph of ovulation occurring in a 45-year old woman observed while she was undergoing surgery. (Courtesy of Dr Jacques Donnez, Department of Gynaecology, Catholic University of Louvain, Belgium.)

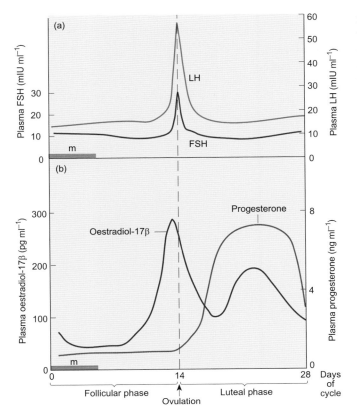

Figure 48.19 The changes in hormone levels during the menstrual cycle. Panel (a) shows the pattern of secretion shown by the gonadotrophins (FSH and LH), while panel (b) shows the changes in the plasma levels of oestradiol-17β and progesterone. The solid bars marked 'm' represents the period of menstruation. (Based on data in Thorneycroft et al. (1971) *Am J Obstset Gynecol* **111**, 950 and Ross et al. (1970) *Recent Prog. Hormone Res.* **26**, 19.)

oestrogen. An oestrogen surge also seems to be required for the rise in LH secretion.

If an antral follicle is to proceed to the pre-ovulatory stage, with subsequent ovulation at mid-cycle, its acquisition of appropriate receptors must coincide with high circulating gonadotrophin levels. Any follicles that do not have LH receptors at this time will become atretic. Therefore, although several primordial follicles begin to develop every day during fertile life, usually only one follicle (the so-called dominant follicle) proceeds to ovulation. Indeed, there is considerable wastage of follicles during each cycle as some follicles undergo atresia at each stage of development.

The pre-ovulatory stage lasts for only about 36 hours, but during that time the follicle shows marked changes that culminate in the rupture of the follicle and release of the oocyte, as shown in Figure 48.18(b). This is the process of **ovulation**, which occurs approximately halfway through the menstrual cycle. All the changes that characterize the pre-ovulatory stage are critically dependent upon pituitary gonadotrophins, particularly LH.

Soon after the rise in LH output at the start of the pre-ovulatory stage, the oocyte completes its first meiotic division. This culminates

in a rather peculiar division in which half the chromosomes but virtually all the cytoplasm is contained within one cell, the secondary oocyte, while the remaining chromosomes are discarded in the form of the first **polar body**. This is shown in Figure 48.20, which illustrates the meiotic events leading to mature egg formation. Meiosis then arrests again, in metaphase II, and only proceeds further after fertilization when the secondary oocyte is ovulated. (A more detailed account of the stages of meiosis is given in Chapter 4.) The mechanism by which LH initiates the recommencement of meiosis is not understood—perhaps it antagonizes the activity of a meiotic inhibitory factor.

During the antral stage, the granulosa cells of the follicle were mainly concerned with converting androgens to oestrogens under the influence of pituitary FSH. In the pre-ovulatory stage, LH stimulates these cells to start synthesizing progesterone instead. As a result, oestrogen levels begin to fall slightly while progesterone output rises. At the same time, the granulosa cells lose their receptors for FSH and for oestrogen.

At ovulation the follicle ruptures and the secondary oocyte enters the Fallopian tube

By the end of the pre-ovulatory stage of development, the volume of follicular fluid has increased substantially and the oocyte remains attached to the outer rim of granulosa cells only by a thin stalk (see Figure 48.17(c) and 48.18(a)). At the time of ovulation, under the influence of LH and probably also FSH, the cells of the stalk dissociate and the follicle ruptures. The detailed biochemistry of this process is not understood, but it is widely suspected that follicular rupture is in some way dependent upon the switch away from oestrogen production towards progesterone production that occurs in the granulosa cells just prior to ovulation. Furthermore, LH is believed to stimulate the synthesis of prostaglandins, particularly PGE_2, which appear to be necessary for follicular rupture.

At ovulation, the follicular fluid flows out onto the surface of the ovary carrying with it the secondary oocyte with a few surrounding cells. The egg mass is swept into the Fallopian tube by currents set up by the movements of cilia on the fimbriae of the ostium (see Figure 48.10). The first half of the ovarian cycle is now complete. Figure 48.18(b) shows ovulation occurring in a 45-year-old woman, in whom the process took about 15 minutes to complete.

After ovulation the follicle forms a corpus luteum which is regulated by LH from the anterior pituitary

After the departure of the oocyte and follicular fluid, the remainder of the follicle collapses into the space left behind, and a blood clot forms within the cavity. The post-ovulatory follicle, therefore, consists of a fibrin core surrounded by collapsed layers of granulosa cells enclosed within the fibrous thecal capsule. This collapsed follicle then undergoes transformation to become a corpus luteum which, in the event of fertilization, will secrete the appropriate balance of steroid hormones to ensure implantation and maintenance of the embryo during the early weeks of pregnancy. The second half of the ovarian cycle is often referred to as the **luteal phase**.

Formation of the corpus luteum is entirely dependent upon the surge of pituitary LH that occurs during the pre-ovulatory stage to

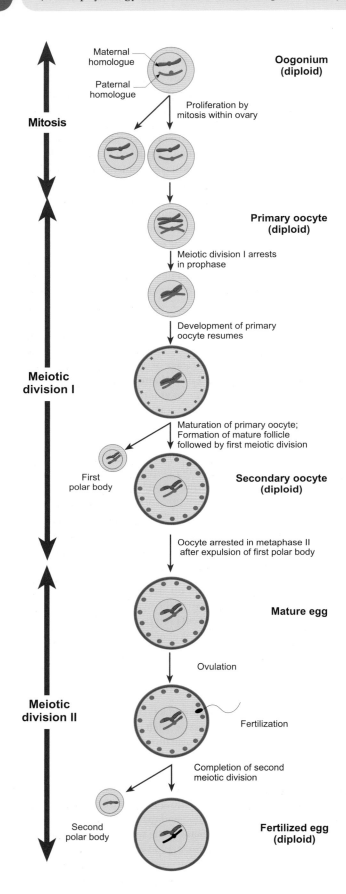

Mitosis

Meiotic division I

Meiotic division II

Maternal homologue

Paternal homologue

Oogonium (diploid)

Proliferation by mitosis within ovary

Primary oocyte (diploid)

Meiotic division I arrests in prophase

Development of primary oocyte resumes

Maturation of primary oocyte; Formation of mature follicle followed by first meiotic division

First polar body

Secondary oocyte (diploid)

Oocyte arrested in metaphase II after expulsion of first polar body

Mature egg

Ovulation

Fertilization

Completion of second meiotic division

Second polar body

Fertilized egg (diploid)

Figure 48.20 The meiotic events leading to the formation of a mature egg. Initially the diploid oogonia proliferate by mitosis to form primary oocytes. The primary oocytes enter the first phase of meiosis but are arrested in prophase I. At some time during the fertile years an oocyte resumes development and enlarges considerably in size. During this stage of development, the oocyte acquires an outer coat (shown in red) and cortical granules develop (represented by the small red circles). It then undergoes the first of two meiotic divisions in which the first polar body is extruded and the secondary oocyte is formed. The oocyte then arrests in metaphase II as a mature egg, which is ovulated at mid-cycle. If the egg is fertilized, it undergoes a cortical reaction to prevent a second sperm entering and completes its second meiotic division before the head of the sperm penetrates the nucleus to restore the diploid chromosome number. The second polar body is extruded at this time.

bring about ovulation itself. The factors that maintain the corpus luteum following the steep decline in gonadotrophin levels after ovulation are not clear. In some animals, a luteotrophic complex of LH, prolactin, and possibly other hormones seems to be important, but the situation in the human female is unclear and normal basal levels of LH may be sufficient for luteal function.

In the first hours following expulsion of the egg from the ovary, the remaining follicular cells undergo the process of **luteinization**. They enlarge and develop lipid inclusions that give the corpus luteum the yellowish colour from which it takes its name. The corpus luteum grows to 15–30 mm in size by about eight days after ovulation. At this time, it shows peak secretory capacity. The cells of the corpus luteum contain increased amounts of Golgi apparatus, endoplasmic reticulum, and mitochondrial protein, and they secrete large amounts of progesterone. Plasma progesterone levels, which showed a small rise just before ovulation, now increase dramatically from about 1 ng ml^{-1} to around 6 or 8 ng ml^{-1} (see Figure 48.19). A significant amount of oestrogen is also secreted by the corpus luteum, and a second oestrogen peak is seen around the middle of the luteal phase. Nevertheless, it is evident from the relative plasma concentrations of oestradiol and progesterone shown in Figure 48.21 that the dominant steroid during the luteal phase of the cycle is progesterone.

In the absence of fertilization, the corpus luteum has a finite lifespan

If the oocyte released at ovulation remains unfertilized, the corpus luteum degenerates after 10 to 14 days. This process is known as luteolysis. It involves collapse of the luteinized cells, ischaemia, and cell death, with a resulting fall in the output of oestrogens and progesterone. The time course of the decline in steroid output at the end of the cycle may be seen in Figures 48.19 and 48.21. The degenerated corpus luteum leaves a whitish scar within the ovarian stroma, the corpus albicans, which persists for several months. (The complete cycle of follicular development is illustrated in Figure 48.14.)

What brings about degeneration of the corpus luteum in the absence of fertilization?

The processes that account for the regression of the luteal cells in the absence of pregnancy remain unclear, but there is some evidence that oestrogens have a role to play in the human ovary. Firstly, the start of degeneration roughly coincides with the oestrogen peak seen six to eight days after ovulation (see Figure 48.21), and secondly, injections of oestrogens given prior to the naturally occurring peak hasten luteal decline. An alternative explanation, however, is that regression simply occurs naturally because the luteal cells gradually lose their sensitivity to LH as the cycle advances.

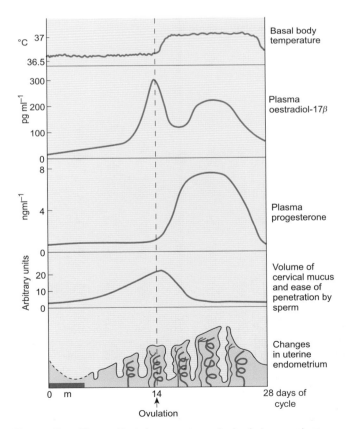

Figure 48.21 The cyclical changes shown by body temperature, cervical secretions, and the uterine endometrium in relation to the circulating levels of oestradiol-17β and progesterone. The red bar in the bottom panel labelled 'm' indicates the period of menstruation.

Summary

The classically recognized 28-day menstrual cycle represents the final four weeks of a maturational process that began some weeks or months before, when a primordial follicle began its development into a preantral follicle. During the first half of the menstrual cycle (the follicular phase) a preantral follicle undergoes the growth and development that culminates in ovulation. The physical changes occurring as the follicle develops are regulated by FSH and LH and by oestrogens produced by the follicle itself.

In the luteal phase, the post-ovulatory follicle is transformed into a corpus luteum under the influence of anterior pituitary LH and begins to secrete large amounts of progesterone. In the absence of fertilization, the corpus luteum degenerates after 10–14 days and steroid output falls to very low levels. This is the process of luteolysis, which marks the end of one cycle.

Hormonal regulation of the female reproductive tract

During the follicular phase, oestrogens prepare the reproductive tract for fertilization

The follicular phase of the ovarian cycle is characterized by the secretion of increasing amounts of oestrogens. This pattern of oestrogen output may be seen in Figures 48.19 and 48.21. Plasma levels of oestradiol-17β rise gradually during the antral stage of follicular development, reaching concentrations of up to 300 pg ml^{-1} just prior to ovulation (Figure 48.19). The oestrogens secreted during the first half of the cycle perform the crucial task of preparing the reproductive tract to receive and transport gametes, and to create a favourable environment for fertilization and implantation.

Oestrogens increase ciliary and contractile activity in the Fallopian tubes

The influence of ovarian hormones on the Fallopian tubes appears to be quite significant. The high levels of oestrogens seen during the follicular phase enhance tubal ciliary and contractile activity in preparation for recovering the ovum from the peritoneal cavity after ovulation. These actions also help to transport the ovum towards the uterus and to transport sperm towards the egg. Removal of the ovaries results in a loss of tubal cilia and a reduction in both secretory and contractile activity of the tubal cells. These effects can be reversed by the subsequent administration of oestradiol-17β, which is further evidence that oestrogens are important for ciliary and muscular activity in the Fallopian tubes. This is consistent with the reproductive role that the tubes perform.

Oestrogens stimulate endometrial proliferation and increase myometrial excitability

Both the myometrium and the endometrium of the uterus are extremely sensitive to the ovarian steroids, and the changes in appearance and function that occur in response to these hormones reflect the different roles that the uterus must fulfil during each cycle. The uterus is first prepared to receive sperm from the cervix and transport them to the site of fertilization in the Fallopian tubes.

In the event of fertilization, it must then be prepared to receive and nourish the embryo.

Steroids secreted from the follicular cells act on the uterus to enable it to fulfil these tasks. The oestrogens secreted during the follicular phase of the cycle exert a uterotrophic (stimulatory) effect on the endometrium (see Figure 48.22). As a result, the endometrial stroma proliferates, the surface epithelium increases in surface area, and the oestrogen-primed epithelial cells secrete an alkaline watery fluid. At the same time, the spiral arteries that permeate the stroma start to enlarge. By the time of ovulation the endometrial thickness has increased to around 10 mm (from about 2 or 3 mm just after menstruation). This phase of the endometrial cycle, corresponding to the oestrogen-dominated follicular phase, is known as the **proliferative phase**. During this phase the uterus is being prepared to receive a blastocyst (early embryo). Oestrogens also stimulate the development of progesterone receptors on the endometrial cells so that, by the end of the follicular phase, the endometrium is primed to respond to progesterone. Figure 48.12 shows a low-power view of a section of the wall of the human uterus during the proliferative phase. It illustrates the thickened endometrial layer and the developing spiral arteries.

The uterine myometrium is also under the influence of the ovarian hormones. Oestrogens appear to increase the excitability of the myometrial smooth muscle and therefore its spontaneous contractility. This may assist in the transport of sperm through the uterus, particularly as ovulation approaches.

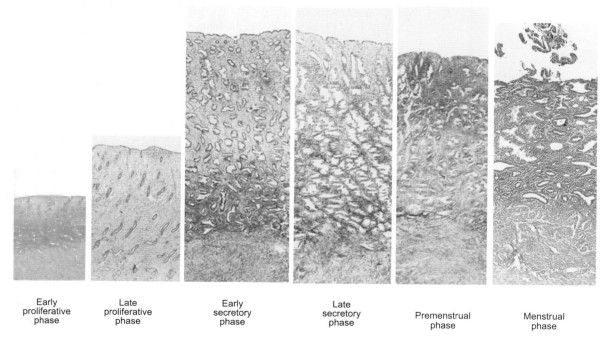

Early proliferative phase Late proliferative phase Early secretory phase Late secretory phase Premenstrual phase Menstrual phase

Figure 48.22 The changes seen in the endometrium of the uterus during the menstrual cycle. From left to right the changes in endometrium are shown for a complete cycle. Note the increased depth of tissue formed by the early secretory phase and the increased development of the endometrial glands (these are the irregular white spaces). At the onset of menstruation, the outer layers are shed into the central cavity of the uterus.

Oestrogens also affect non-reproductive tissues

In addition to those specific actions within the reproductive tract discussed above, the oestrogens have widespread and generalized effects throughout the body. In particular they exert effects on metabolism and the cardiovascular system which can be summarized as follows.

- Oestrogens are mildly anabolic and tend to depress the appetite.
- They reduce plasma levels of cholesterol, which may explain why pre-menopausal women have a lower risk of heart attacks than both post-menopausal women and men of a comparable age.
- They reduce capillary fragility.
- Oestrogens also appear to have profound effects on mood and behaviour, but the underlying mechanisms are not clear.
- Oestrogens cause proliferation of the ductal system of the mammary tissue (see Chapter 49).
- They have important effects on the maintenance of the skeleton (see Chapter 51).

Progesterone secreted by the corpus luteum optimizes conditions for implantation within the uterus

The uterus houses the embryo for the entire period of its development (gestation). There are two elements to this task:

1. the endometrial layer must permit implantation of the embryo and subsequently must participate in the formation of the placenta (**placentation**)
2. the myometrium must remain quiescent during gestation to guard against premature expulsion of the fetus.

Progesterone plays a key role in each of these elements. Indeed, adequate levels of progesterone are essential throughout the entire period of gestation to ensure a successful outcome.

During the follicular phase of the cycle, oestrogens, secreted by the antral follicle, bring about proliferation of the uterine endometrium along with an increase in the number of glandular structures. Oestrogens also stimulate the acquisition of progesterone receptors by the cells of the endometrium. As progesterone levels rise during the luteal phase, stromal proliferation continues and the spiral arteries develop fully. In the event of pregnancy, the spiral arteries will form the blood supply to the maternal side of the placenta. The endometrial glands start to secrete a thick fluid, rich in sugars, amino acids, and glycoprotein. For this reason, the second half of the uterine cycle is often referred to as the **secretory phase**. It coincides with the luteal phase of the ovarian cycle. All these progesterone-mediated changes help to create a favourable environment for an embryo and to optimize conditions for implantation and placental formation.

In the absence of fertilization, the corpus luteum regresses after 10–14 days and steroid output falls precipitously (see Figures 48.19 and 48.21). Once the endometrium is deprived of its steroidal support, its elaborate secretory epithelium collapses. The endometrial layers are shed, together with blood from the ruptured spiral arteries, which contract to reduce bleeding. This process is known as **menstruation**. The onset of menstrual bleeding marks the start (day 1) of a new ovarian cycle. Contraction of the spiral arteries can lead to the pain experienced by some women at the start of menstruation (**dysmenorrhoea**). Bleeding continues for three to seven days, during which the total blood loss is between 30 and 200 ml. After this time the endometrial epithelium has been repaired completely. Figure 48.22 shows the dramatic changes seen in the endometrium of the uterus throughout one complete menstrual cycle.

Progesterone relaxes the uterine myometrium

As described earlier, the oestrogen-dominated myometrium shows a fair degree of excitability and spontaneous contractility. Clearly, although this may be helpful in assisting gamete transport, it is highly undesirable once an embryo has entered the uterus. Too much excitability could result in spontaneous abortion ('miscarriage') of the fetus. Progesterone tends to relax the smooth muscle of the myometrium, probably by decreasing the expression of certain proteins involved in the formation of gap junctions (e.g. connexin 43). This reduces the excitability of the myometrium and reduces the likelihood of spontaneous contractions.

Some non-reproductive tissues are influenced by progesterone

Like oestrogens, progesterone exerts widespread effects throughout the whole body, most of which are poorly understood. For example, it is a mildly catabolic steroid that stimulates the appetite. Increased levels of progesterone during the luteal phase cause a rise in basal body temperature of 0.2 to 0.5°C (see the top panel of Figure 48.21). This rise is a useful indicator that ovulation has occurred for women who are trying to conceive—and for those who are trying not to! Progesterone promotes development of the lobules and alveoli of the breast (see Chapter 49) and causes the breasts to swell because of fluid retention by the mammary tissue. This is probably the reason for the breast discomfort experienced by many women during the pre-menstrual period.

The cervical secretions and vaginal epithelium show hormone-dependent cyclical changes

The endo-cervical glands secrete mucus whose characteristics vary considerably during the ovarian cycle. These changes are regulated by the ovarian hormones and have important consequences for fertility. Under the influence of the high circulating levels of oestrogens seen during the late follicular phase, the cervical epithelium increases its secretory activity and mucus is produced in large amounts, up to 30 times the quantity secreted in the absence of oestrogen (see Figure 48.21). The mucus is thin, watery, and clear and exhibits a characteristic 'ferning' pattern if dried on a glass slide and examined under a microscope (see Figure 48.23). Cervical mucus also shows increasing elasticity during the late follicular phase so that a drop of mucus may be stretched to a length of 10–12 cm. The peak volume and elasticity coincides with the oestrogen surge just prior to ovulation. Mucus with these characteristics is most readily penetrated by sperm and this action of oestrogen on the cervical glands is a good example of the way in which the endocrine activity of the ovary optimizes the conditions for successful reproduction: when an oocyte is likely to be present, the passage of sperm through the female tract is facilitated.

Figure 48.23 A photomicrograph of an air-dried specimen of cervical mucus obtained at mid-cycle, showing the characteristic crystalline ferning pattern. (From Fig 1 in M. Menarguez et al. (2003) *Human Reproduction* **18**, 1782–9.)

During the luteal phase, with its high progesterone levels, cervical mucus is produced in far smaller volumes and becomes much thicker, stickier, and relatively hostile to sperm. It is thus less likely that sperm will reach the uterus and Fallopian tubes during the luteal phase. This action of progesterone forms part of the mechanism of action of the progesterone-only contraceptive pill (see also Box 48.1).

The stratified squamous epithelium lining the vagina also changes in appearance in response to the ovarian hormones. Indeed, the histological appearance of the vaginal epithelial cells may be used as an indicator of the phase of the menstrual cycle. In the follicular phase, increased secretion of oestrogens stimulates proliferation of the epithelial layers. As the superficial layers move farther away from the blood supply, they keratinize and many slough off. At mid-cycle, a vaginal smear will show a preponderance of such cells.

Summary

The follicular phase of the menstrual cycle is dominated by oestrogens secreted by the developing follicle that prepare the reproductive tract for fertilization and implantation. During the luteal phase, the progesterone secreted by the corpus luteum maintains the endometrium in a favourable condition for implantation and placentation. Progesterone also renders the myometrium less excitable.

In the absence of an embryo, the corpus luteum degenerates after 10–14 days and steroid hormone output falls steeply. As progesterone levels fall, the endometrium is shed, together with blood from the spiral arteries. This process is called menstruation and its onset marks the beginning of a new menstrual cycle.

Why do the plasma concentrations of gonadotrophins and ovarian steroids vary during the ovarian cycle?

The previous sections have focused particularly on the structural and functional changes which occur throughout the 28 days or so that span one menstrual cycle. The cyclical alterations in the plasma levels of FSH and LH (shown in Figure 48.19) are crucial in controlling the cellular and endocrine activity of the ovary; i.e. the growth of follicles, the formation of the corpus luteum, and its hormonal activity. How do these fluctuations come about and how do they regulate the follicular cells? In answer to these questions it is important to realize that ovarian function is regulated both by gonadotrophins and by the ovarian steroid hormones themselves which, in turn, influence gonadotrophin secretion. This feedback interaction between the anterior pituitary gland, the hypothalamus, and the ovary is illustrated in Figure 48.24.

The ovarian steroids can exert both negative and positive feedback control on the output of FSH and LH, depending upon the concentration of hormone in the blood and the time for which it has been present. Low or moderate levels of oestrogens, particularly oestradiol-17β, tend to inhibit the secretion of FSH and LH (negative feedback). If, however, oestrogens are present in high concentrations for several days, the effect switches to one of

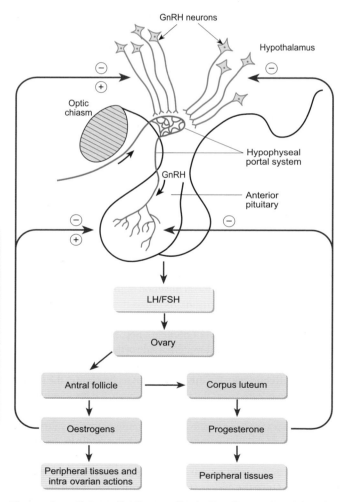

Figure 48.24 Schematic diagram illustrating the positive (+) and negative (−) feedback control of the hormonal secretions of the hypothalamus, pituitary gland, and ovaries.

positive feedback in which the output of FSH and LH is stimulated. The feedback actions of progesterone are roughly opposite to those of the oestrogens. High concentrations of progesterone inhibit gonadotrophin release, while low levels appear to enhance the positive feedback effects of oestrogens.

The feedback effects of the ovarian steroids are mediated primarily at the level of the anterior pituitary itself, probably by alterations in the sensitivity of the gonadotrophin-secreting cells to hypothalamic gonadotrophin releasing hormone (GnRH). There may also be a direct effect of the steroids on GnRH output by the hypothalamic neurons, although this is hard to establish conclusively because of the difficulty in measuring the tiny quantities of GnRH present in the hypophyseal portal blood.

The ovarian cycle begins on day 1 of menstruation. Just prior to this time, levels of both oestrogens and progesterone have fallen, as the previous corpus luteum declined. Released from the negative feedback inhibition of the ovarian steroids, FSH levels start to rise slowly, followed shortly afterwards by LH (see Figure 48.19). These events coincide with the initiation of the antral phase of follicular development. Towards the end of the hormone-independent preantral phase, the thecal cells gain receptors for LH while the granulosa cells become responsive to both FSH and oestrogens. The coincidence of receptor acquisition with steadily rising levels of FSH and LH allows the follicle to enter the hormone-dependent antral phase.

In the early antral phase of the cycle, ovarian steroid output does not change very much (see Figure 48.19). Over the next six to eight days or so, however, the maturing follicle starts to produce large quantities of oestrogens under the influence of FSH and LH. During this period, gonadotrophin levels themselves remain low because of the negative feedback effect of low and moderate levels of oestrogens. This steady rise in oestrogen secretion, however, culminates in an 'oestrogen surge' during the latter days of the antral phase, when plasma concentrations of oestradiol-17β reach values between 200 and 400 pg ml^{-1}. This oestrogen surge initiates important changes in the output of gonadotrophins. After about 36 hours, the negative feedback effects of oestrogens are replaced by positive feedback that results in a sharp increase in the output of both the gonadotrophins but especially of LH. This constitutes the so-called LH surge, which is responsible for the events occurring during the pre-ovulatory phase and for ovulation itself.

Once ovulation has taken place, the level of oestrogens falls sharply as the luteal cells switch to progesterone production. Consequently, the gonadotrophins are released from the positive feedback effects of high oestrogen levels and the output of FSH and LH drops, as negative feedback reasserts control. Although oestrogen levels may rise to values similar to those seen during the pre-ovulatory surge, this second, luteal, peak fails to elicit a further LH surge because the high circulating levels of progesterone seem to block the positive feedback effects of the oestrogens. Instead, negative feedback continues to predominate and gonadotrophin secretion remains low throughout the luteal phase.

In the absence of fertilization, the corpus luteum regresses after 10–14 days and steroid output declines quickly. Deprived of steroid support, the specialized uterine endometrial layers are shed during menstruation (as illustrated in Figure 48.22). Soon afterwards, FSH and LH levels begin to rise slowly as the anterior pituitary is released from the negative feedback inhibition of oestrogen, and a new cycle gets underway as preantral follicles enter the gonadotrophin-sensitive antral phase.

The menstrual cycle may be influenced by neural factors

Although the interactions between the ovarian steroids and the pituitary gonadotrophins are well documented, and appear to offer a reasonably complete explanation for the cyclical activity of the ovaries, it is known that a variety of other factors both neural and hormonal also affect the control of fertility.

The reproductive behaviour of many farm, domestic, and laboratory animals illustrates how environmental stimuli mediated by the CNS, such as olfactory, tactile, coital, and visual cues play a role in regulating gonadal function. For example, the fertile periods of certain animals such as sheep appear to be controlled by the relative lengths of day and night. Others, such as cats and rabbits, are reflex ovulators: i.e. they ovulate in response to coitus. Such regulatory mechanisms must be mediated via the CNS, the afferent information being integrated in the hypothalamus in order to control the output of GnRH and thus of the gonadotrophins.

While human gonadal function is clearly not subject to external control mechanisms in such a rigid fashion, the possibility of neurally mediated influences on gonadotrophin release cannot be ruled out. Indeed, the wealth of neural inputs to the GnRH-secreting neurons of the hypothalamus would argue strongly in favour of such mechanisms. It is certainly well established that factors such as anxiety and emotional stress can upset cyclical ovarian activity and fertility.

Physiologically, the role of prolactin in the normal menstrual cycle is unclear, but clinically it is well known that over-secretion of prolactin (hyperprolactinaemia) is a common cause of female infertility. The condition may be physiological, as in lactating women (see Chapter 49), or pathological, arising from a pituitary tumour. It is often associated with anovulatory cycles (cycles in which ovulation fails to occur) or a complete loss of cyclical ovarian activity. High levels of prolactin seem to impair the response of the anterior pituitary to GnRH so that LH surges are lost and ovulation fails to take place.

Summary

Cyclical variations in the levels of steroid and gonadotrophic hormones act together to ensure the regular release of mature ova and to prepare the body for fertilization and pregnancy. While the gonadotrophins control ovarian function, the oestrogens and progesterone secreted by the ovaries regulate the secretion of pituitary FSH and LH by both negative and positive feedback. Very high levels of oestrogens stimulate the anterior pituitary to initiate an LH surge, which is crucial to the events of the preovulatory stage and to ovulation itself. After ovulation, negative feedback prevails for the remainder of the cycle and gonadotrophin output is relatively low.

48.4 Activation and regression of the gonads—puberty and the menopause

Menarche and the menopause

In the female, the fertile years are defined by two events:

- the onset of menstruation at puberty (**menarche**)
- the cessation of cyclical ovarian activity (the **menopause** or **climacteric**), which occurs at around the age of 50.

Puberty is a broad term which includes the variety of changes taking place within the body of an adolescent girl as her ovaries mature. Menarche, the onset of menstruation, is the outward sign that these changes have taken place and that cyclical secretion of ovarian steroids has begun. The other changes that take place during the two or three years preceding menarche include the adolescent growth spurt, the development of secondary sexual characteristics (pubic hair and mammary development), and changes in body composition—adult females have about twice as much body fat as males and a smaller mass of skeletal muscle.

Circulating levels of the pituitary gonadotrophins, FSH and LH, rise gradually up to the age of around 10 years, probably as a result of increasing activity of the hypothalamic GnRH-secreting neurons. After this, with the approach of menarche, pulsatile release is established, with spurts of gonadotrophin secretion occurring during sleep. With this increase in gonadotrophin secretion there is also a rise in the output of ovarian oestrogens. Under the influence of these steroids, budding of the breasts occurs—usually the first outward physical sign of puberty. From two to three years prior to the onset of menstruation, androgen secretion from the adrenal cortex is increased (adrenarche) and these hormones are important in the stimulation of pubic hair growth. It has also been suggested that androgen secretion plays a part in the control of menarche, though no clear link has been established. What is clear, however, is that the increasing synthesis and secretion of FSH and LH eventually results in the onset of menstruation, although the first few cycles are usually anovulatory and progesterone is not produced in large amounts. Bleeding therefore tends to be light and to occur irregularly at first. The exact trigger for menstruation is still not fully understood, but various suggestions have been made:

- the ovaries may become more sensitive to gonadotrophins
- the anterior pituitary may become more sensitive to the positive feedback effects of oestrogens.

The average age for menarche in the United Kingdom is around 12 years, with the first ovulatory cycles occurring six to nine months later. The range of normality, however, extends from 10 to 16 years of age (see Table 48.1). There has been a trend towards earlier menarche in the last 150 years, possibly because of improved health care and nutrition. The latter may be particularly important since it is thought that menarche requires the attainment of either a critical body mass (around 47 kg) or possibly a critical ratio of fat to lean mass. It is certainly true that regular menstrual cycles are disrupted in girls who lose large amounts of weight either through anorexia, excessive exercise, or starvation. The continuing trend towards earlier puberty even in the last 20 years may also be linked with the increasing incidence of childhood obesity in Western societies.

The cessation of menstrual cycles (the menopause or climacteric)

The menopause or climacteric marks the end of a woman's fertile years and usually occurs between the ages of 45 and 55. However, for several years leading up to this point there is progressive failure of the reproductive system. The number of oocytes in the ovaries has been depleted by atresia (the ovary of a 50-year-old woman contains on average only about 1000 follicles) and the ovarian responsiveness to gonadotrophins declines. Cycles often become anovulatory and irregular before ceasing altogether. Pituitary FSH and LH levels are high in post-menopausal women because of the loss of negative feedback inhibition by oestrogens, though LH surges are no longer seen.

The years leading up to and just after the menopause itself are often referred to as the 'perimenopause' and are characterized by a variety of somatic and emotional changes. Among the more obvious changes are a loss of breast tissue, vaginal dryness, increased fibrosis of the uterine muscle, night sweats, hot flushes, and depression. An increased susceptibility to myocardial infarction accompanies the menopause and there is often also an increase in bone weakness due to increased bone resorption and loss of bone mass. As a result, fractures, particularly of the wrist and the neck of femur, are more frequently seen in post-menopausal women. Most of the changes that accompany the menopause are attributable to the loss of ovarian oestrogens and can be treated successfully by hormone replacement therapy (HRT) should they be sufficiently serious to warrant medical intervention.

Puberty in the male

Testosterone is the key to reproductive function in the male. It plays a crucial role in sexual differentiation during embryonic life (see Chapter 50) and its concentration in the plasma rises at the onset of puberty to reach adult levels by about 17 years of age. Between early infancy and the start of puberty, testosterone secretion by the testes is low. Pituitary gonadotrophin levels are also low. During the years between 10 and 16 (on average), boys develop their full reproductive capability. At the same time, they acquire secondary sexual characteristics and adult musculature and undergo a linear growth spurt, which is halted by closure of the epiphyses when adult height is reached (as discussed earlier in the chapter; see also Table 48.1).

The first endocrine event of puberty is an increase in the secretion of hypothalamic GnRH leading to a rise in output of pituitary LH. As a result, there is maturation of the Leydig cells and the initiation of spermatogenesis. Testosterone production is also enhanced and this hormone is responsible for the anatomical changes that are characteristic of male puberty. These include enlargement of the testes, growth of pubic hair starting at the base of the penis, reddening and wrinkling of the scrotal sac, and, later on, an increase in size of the penis. Facial hair begins to appear, there is

Table 48.1 Key points of the principal changes of puberty

Characteristic	Age range of first appearance	Principal hormones responsible for development
Girls		
Breast bud	8–13 years	oestrogens, progesterone, GH
Pubic hair	8–14 years	adrenal androgens
Menarche	10–16 years	oestrogens, progesterone
Growth spurt	10–14 years	oestrogens, GH
Boys		
Growth of testis	10–14 years	testosterone, FSH, GH
Growth of penis	11–15 years	testosterone
Pubic hair	10–15 years	testosterone
Facial and axillary hair	12–17 years	testosterone
Enlargement of larynx	11–16 years	testosterone
Growth spurt	12–16 years	testosterone, GH
Male pattern of skeleto-muscular development	12–16 years	testosterone, GH

deepening of the voice due to thickening of the vocal cords and enlargement of the larynx, and the scalp hair takes on the masculine pattern with the hairline receding at the temples. These maturational changes take place over a period of several years, and Table 48.1 shows the average timing of the major events of male puberty.

Is there a male menopause?

While there is no obvious event marking the end of reproductive capacity in the male comparable to the female menopause, sperm production does decline between the ages of 50 and 80. There is also a reduction in plasma testosterone levels in men over 70 and a parallel increase in plasma levels of FSH and LH, though this is much less marked than in women. This decline in sexual capacity, which is sometimes called the 'andropause', may be associated with emotional problems and a loss of libido in some men. Nevertheless, for most men these changes are relatively insignificant and many elderly men maintain active sex lives and are able to father children.

Summary

In the female, the fertile years are defined by menarche and the menopause, the commencement and cessation of cyclical ovarian activity. FSH and LH are secreted in increasing amounts prior to menarche, but the exact trigger for the onset of cyclical ovarian activity is unclear. At puberty, the secretion of oestrogens initiates the growth spurt and the development of secondary sexual characteristics. The menopause is due to depletion of the oocyte pool and marks the progressive failure of the reproductive system. Many somatic and emotional changes accompany the loss of ovarian steroids.

Between 10 and 16 years of age, boys show a growth spurt and develop their full reproductive capacity. They also start to produce sperm. Leydig cells mature under the influence of pituitary LH and testosterone output rises. Testosterone and dihydrotestosterone are responsible for the development of the male secondary sexual characteristics.

(✳) Checklist of key terms and concepts

Anatomy of the male and female reproductive tracts

- Successful reproduction requires the fusion of male and female gametes to form a zygote.

- Spermatozoa, the male gametes, are produced by the testes, the male gonads that lie outside the abdominal cavity within the scrotum.

- The testes are divided into lobules within which lie the seminiferous tubules, which are the site of sperm formation.

- The seminiferous tubules empty into a series of ducts that eventually merge into the epididymis, which can store sperm for some weeks.

- From the epididymis, the sperm pass to the vas deferens which terminates in the ejaculatory duct close to the prostate gland.

- Fluid from the seminal vesicles is emptied into the ejaculatory duct together with the sperm and prostatic secretions to form semen, which is discharged via the penis during ejaculation.

- The ova or eggs, the female gametes, are produced by the ovaries (the female gonads).
- The ovaries lie in the pelvic cavity within the ovarian fossa.
- The ovaries are attached to the uterus and Fallopian tubes by the ovarian mesentery of the broad ligament.
- The Fallopian tubes are the normal site of fertilization.
- The uterus houses the growing fetus throughout gestation and is adapted to permit implantation and the subsequent formation of a placenta.
- The vagina is a muscular tube that receives the penis at intercourse and forms the birth canal during delivery of a baby.

The formation of sperm

- Spermatogenesis involves the generation of huge numbers of cells by mitosis, followed by the halving of the chromosomal complement by meiosis to form haploid spermatids.
- Spermatogenesis and spermiogenesis both take place in close association with the Sertoli cells.
- Spermiogenesis involves remodelling of the cytoplasm of the immature spermatid and culminates in the formation of a highly specialized cell: the mature, motile sperm.
- As they move through the male reproductive tract, the sperm are mixed with alkaline secretions from the seminal vesicles and prostate. These secretions nourish the sperm and act to neutralize the acidity of the vagina.
- Spermatogenesis is regulated by a variety of hormones including the hypothalamic gonadotrophins FSH and LH, and by testosterone secreted by the testicular Leydig cells.
- Hormone levels are regulated by negative feedback loops operating within the hypothalamic–pituitary–testicular axis.

The menstrual cycle

- During the first half of the menstrual cycle (the follicular phase) a follicle undergoes the growth and development that culminates in ovulation.
- Follicular development is closely regulated by pituitary FSH and LH, and by oestrogens produced by the follicle itself.
- The second half of the cycle is known as the luteal phase, during which the post-ovulatory follicle is transformed into a corpus luteum under the influence of anterior pituitary LH.
- The corpus luteum secretes large amounts of progesterone.
- In the absence of fertilization, the corpus luteum degenerates after 10–14 days and steroid output falls to very low levels. This is the process of luteolysis, which marks the end of one ovarian cycle.

Hormonal regulation of the female reproductive tract

- The follicular phase of the ovarian or menstrual cycle is dominated by oestrogens secreted by the developing follicle.
- Oestrogens prepare the reproductive tract for gamete transport, fertilization, early embryonic development, and implantation.
- During the luteal phase (the second half of the ovarian cycle), the dominant hormone is progesterone which is secreted by the corpus luteum.
- Progesterone prepares the uterus to receive and nourish an early embryo in the event of fertilization.
- In the absence of an embryo, the corpus luteum degenerates after 10–14 days and steroid output falls steeply.
- As progesterone levels fall, the elaborate endometrium that was built up during the cycle is sloughed off and shed together with blood from the spiral arteries. This process is called menstruation and its onset marks the beginning of a new ovarian cycle.
- Cyclical variations in the levels of steroid and gonadotrophic hormones reflect interactions between the ovarian steroid hormones and the secretion of FSH and LH by the pituitary. This regulation employs both negative and positive feedback.
- Very high levels of oestrogens stimulate the anterior pituitary to initiate an LH surge, which is crucial to the events of the preovulatory stage and to ovulation itself.
- After ovulation, negative feedback prevails and gonadotrophin output is relatively low.
- The menstrual cycle is influenced by neural activity, pituitary prolactin, and nutritional status.

Puberty and the menopause

- In the female, the fertile years are defined by menarche and the menopause, the commencement and cessation of cyclical ovarian activity at around the ages of 12 and 50 years respectively.
- Many changes take place within the body of an adolescent girl in addition to the onset of menstrual cycles. These include a growth spurt and the development of secondary sexual characteristics.
- The menopause marks the progressive failure of the reproductive system and is due to depletion of the oocyte pool and a reduced ovarian responsiveness to pituitary gonadotrophins.
- Between 10 and 16 years of age, boys show a growth spurt and start to produce sperm.
- Leydig cells mature under the influence of pituitary LH and testosterone output rises.
- The increase in testosterone secretion is responsible for the development of the male secondary sexual characteristics.

Recommended reading

Anatomy

MacKinnon, P., and Morris, J. (2005) *Oxford textbook of functional anatomy* (2nd edn), Volume 2, pp 181–200. Oxford University Press, Oxford.

A clear, well-illustrated account of the anatomy of the male and female reproductive systems and their development.

Cell biology of germ cells

Alberts, B., Johnson, A., Lewis, J., Morgan, D., Raff, M., Roberts, K., and Walter, P. (2015) *Molecular biology of the cell* (6th edn). Garland, New York.

Histology

Junquieira, L.C., and Carneiro, J. (2005) *Basic histology* (11th edn). McGraw-Hill, New York.

One of the best accounts of the histology of the reproductive system.

Physiology

Ferin, M., Jewckwicz, R., and Warres, M. (1993) *The menstrual cycle*. Oxford University Press, Oxford.

Now rather old but a good reference source for detailed information.

Johnson, M.H. (2013) *Essential reproduction* (7th edn). Wiley-Blackwell, Oxford.

A useful account of the physiology of reproduction.

Kovacs, W.J., and Ojeda, S.R. (2011) *Textbook of endocrine physiology* (6th edn). Oxford University Press, New York.

A good source of detailed information relating to the endocrinology of reproduction.

Review articles

Drummond, A.E., and Fuller, P.J. (2010) The importance of ERb signalling in the ovary. *Journal of Endocrinology* **205**, 15–23.

Farage, M.A., Neill, S., and MacLean, A.B. (2009) Physiological changes associated with the menstrual cycle: A review. *Obstetrical and Gynecological Survey* **64**, 58–72.

Messinis, I.E. (2006) From menarche to regular menstruation: Endocrinological background. *Annals of the New York Academy of Science* **1092,** 49–56.

Ruwanpura, S.M., McLachlan, R.I., and Meachem, S.J. (2010) Hormonal regulation of male germ cell development. *Journal of Endocrinology* **205**, 117–31.

Sofikitis, N., Giotitsas, N., Tsounapi, P., Baltogiannis, D., Giannakis, D., and Pardalidis, N. (2008) Hormonal regulation of spermatogenesis and spermiogenesis. *Journal of Steroid Biochemistry & Molecular Biology* **109**, 323–30.

Tsutsumi, R., and Webster, N.J.G. (2009) GnRH pulsatility, the pituitary response and reproductive dysfunction. *Endocrine Journal* **56** (6), 729–37.

To check that you have mastered the key concepts presented in this chapter, complete the accompanying online self-assessment questions. Go to www.oup.com/uk/pocock5e/

CHAPTER 49

Fertilization, pregnancy, and lactation

This chapter should help you to understand:

- The sexual reflexes of males and females
- The processes of fertilization, implantation, and the maternal recognition of pregnancy
- The formation and function of the placenta
- The role of the placental hormones in the maintenance of pregnancy and the preparation for delivery and lactation
- Parturition
- The major changes in maternal physiology during gestation
- The nutritional demands of pregnancy

- The structure of the non-pregnant mammary gland
- The hormonal control of mammary development during puberty and pregnancy
- The hormonal control of milk synthesis and milk ejection
- The composition of milk and how it changes in the first weeks post-partum
- Involution of the mammary gland after weaning

49.1 Introduction

In the previous chapter, the discussion of female reproductive physiology assumed a situation in which cyclical ovarian activity continues uninterrupted by pregnancy, that is, with regression of the corpus luteum 10–14 days after ovulation and the onset of menstrual bleeding. If, however, an oocyte is fertilized, an entirely different set of events are initiated: the loss of the uterine endometrium is prevented, and the uterus is maintained in a quiescent state for the implantation of the embryo, the formation of the placenta, and gestation. This chapter is chiefly concerned with the events surrounding fertilization, pregnancy, and parturition (birth). It also discusses the physiological changes that take place in the body of a pregnant woman during gestation and concludes with a discussion of the physiology of lactation.

49.2 The sexual reflexes

Fertilization of an ovum requires that sperm are deposited high in the vagina of a woman close to the time of ovulation. Except for the case of fertilization by artificial insemination, this is achieved through the act of sexual intercourse. For successful intercourse to take place, the penis of the male must become erect and ejaculation of seminal fluid must occur within the vagina.

The sexual response in both males and females can be divided into four main phases: excitement, plateau, orgasm, and resolution. Each phase is under the control of autonomic and somatic nerves originating in the lumbar and sacral regions of the spinal cord. The innervation of the sexual organs is derived from both the parasympathetic (S2–S4) and the sympathetic (T11–L2) divisions of the autonomic nervous system, together with somatic fibres running in the pudendal nerves (see Figures 49.1 and 49.2).

The main stages of the male sexual act are:

1. erection of the penis
2. the secretion of alkaline fluid and mucus by the bulbourethral glands
3. the emission of fluid from the seminal vesicles, vas deferens, and prostate gland
4. the expulsion of seminal fluid from the penis—ejaculation.

Penile erection is the result of a parasympathetic reflex

Penile erection results from either descending nerve activity originating in the higher centres of the brain (**psychogenic erection**) or from stimulation of the skin in the genital region (**reflexogenic erection**). The afferent nerve fibres from the genital region run in the pudendal nerves to the sacral region of the spinal cord. The efferent fibres are parasympathetic in origin and are derived from spinal segments S2–S4. After leaving the spinal cord, the efferent fibres run in the pelvic nerve and synapse in the pelvic plexus. The neurons of the pelvic plexus also receive sympathetic fibres from

the hypogastric nerves. The cavernous nerve provides the final limb of the efferent pathway (see Fig 49.1).

The erectile tissue of the penis consists of the two **corpora cavernosa** and the **corpus spongiosum** which surrounds the urethra (see Figure 48.1). The erectile tissues are essentially large venous sinusoids surrounded by a coat of strong fibrous tissue. Erection is a simple hydraulic process which is mainly controlled by the parasympathetic nerves. Impulses in the cavernous nerve cause the internal pudendal artery and its main branches to dilate. Consequently, blood flow to the penis is increased but the venous outflow remains unchanged so that blood becomes pooled in the erectile tissue, causing the penis to enlarge and extend. As the volume of blood within the erectile tissue increases, the pressure rises, partially occluding the emissary veins (which provide the venous outflow), so that the penis becomes rigid and erect. Erection is also known as **tumescence**.

Dilatation of the internal pudendal artery and the associated arterioles is mediated chiefly by nitric oxide derived from fibres running in the parasympathetic nerves. Somatic afferent pudendal nerves provide sensory feedback to maintain erection during intercourse. Most of this sensory input is derived from touch receptors in the skin of the most distal part of the penis (the glans penis).

Emission and ejaculation are the final stages of the male sexual act. When sexual stimulation becomes very intense, rhythmic contractions in the vas deferens and ampulla begin to drive sperm towards the ejaculatory duct. This is followed by secretion of prostatic fluid and contraction of the seminal vesicles. This stage is called **emission**; the expulsion of semen from the penis is called **ejaculation**. During ejaculation, the internal urethra becomes tightly closed to prevent the seminal fluid entering the bladder, the

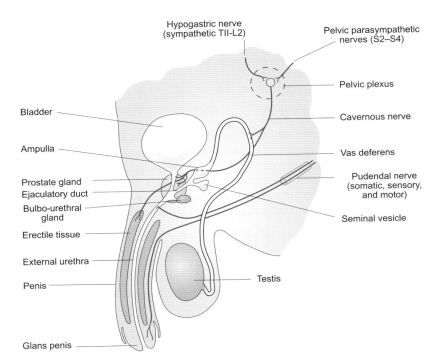

Figure 49.1 A schematic drawing to show the nerve supply to the penis and accessory male sex organs.

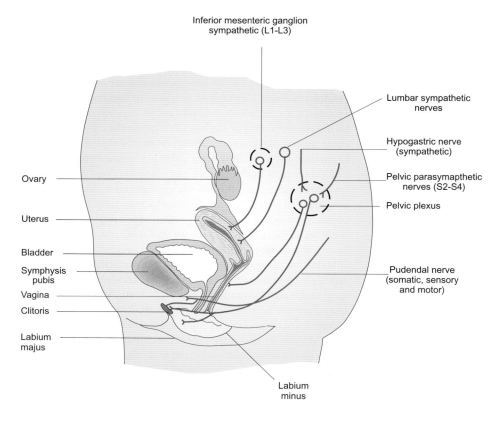

Inferior mesenteric ganglion sympathetic (L1-L3)

Lumbar sympathetic nerves

Hypogastric nerve (sympathetic)

Pelvic parasympathetic nerves (S2-S4)

Pelvic plexus

Ovary

Uterus

Bladder

Symphysis pubis

Vagina

Clitoris

Labium majus

Pudendal nerve (somatic, sensory and motor)

Labium minus

Figure 49.2 A schematic drawing to show the nerve supply to the female sex organs.

prostate, or the seminal vesicles. Contractions of the bulbocavernous and ischiocavernous muscles drive the seminal fluid along the urethra. Together, these adaptations result in forward movement of the semen and its expulsion from the penis. Ejaculation is accompanied by an intense sensation known as **orgasm**. Detumescence (reversal of erection) is probably mediated by sympathetic nerves arising from the sacral ganglia.

In the female, the afferent and efferent neural pathways involved in controlling the sexual responses are the same as for males. The innervation of the female genitalia is illustrated in Figure 49.2. Afferent impulses are relayed from the genitalia to the sacral region of the spinal cord, where sexual reflexes are integrated. The spinothalamic tract carries sensory information to the brain, from which descending pathways transmit impulses back to the sacral segments of the spinal cord. In this way, sexual reflexes are modified by cerebral influences. During sexual excitement, erectile tissue within the clitoris and around the vaginal opening becomes engorged with blood. As in the male, this engorgement is caused by parasympathetic stimulation of the blood vessels supplying these tissues. The nipples of the breasts may also become erect. Mucus secretions are provided both by **Bartholin's glands**, which are adjacent to the labia minora, and by the vaginal epithelium. These secretions facilitate entry of the penis into the vagina and ensure that intercourse is associated with a pleasurable massaging sensation rather than dry frictional irritation, which can inhibit both ejaculation and the female climax. The tactile stimulation of the female genitalia associated with sexual intercourse, coupled with

psychological stimuli, will normally trigger an orgasm. This is associated with rhythmic contraction of the perineal muscles, dilatation of the cervical canal, and increased motility of the uterus. It is probable that there is also an increase in motility of the Fallopian tubes. Although the female orgasm is not a requirement for fertilization, contractions may help to transport sperm towards the site of fertilization in one of the Fallopian tubes. In both males and females, orgasm is followed by a feeling of warm relaxation known as **resolution**.

Contraception

The prevention of unwanted pregnancies (contraception) is now of major importance in most societies as human fertility leads to inexorable population growth. The principal means of preventing conception are summarized in Box 49.1. Most rely either on preventing contact between sperm and ovum by a physical barrier (e.g. a condom or diaphragm) or on preventing ovulation.

Summary

Sexual intercourse involves penile erection, penetration of the vagina by the penis, and ejaculation of seminal fluid containing sperm. Both male and female sexual responses are mediated by sacral reflexes involving the autonomic nervous system.

Box 49.1 An outline of contraceptive methods

Method	Effectiveness (estimated as % of couples remaining childless after 1 year)	Comments
Oral contraceptives (the birth control pill)	99.5%	Contraceptive preparations are of two types: a) combination—consisting of oestrogens and progesterone in sufficiently high concentrations to exert powerful negative feedback on gonadotrophin output. They therefore mimic the luteal phase of the menstrual cycle and prevent ovulation. b) the mini-pill—this contains only progesterone and acts by modifying the secretions and internal environment of the reproductive tract. Side effects include minor symptoms of early pregnancy—nausea, breast tenderness, fluid retention, hypertension, and, in rare cases, thromboses (mainly in smokers over 35 years of age). There is still debate over whether or not the risks of endometrial, ovarian, and breast cancer are altered by these drugs.
Transdermal contraceptive patches	99.5% (lower in obese women)	Delivers oestrogens and progesterone into the bloodstream for 3 weeks followed by a week without the patch.
Norplant (2 silicone rods implanted just under the skin)	Better than 99.5%	Release progesterone over a five-year period
Depo-Provera (an injectible form of progesterone)	Better than 99.5%	Lasts for three months
Post-coital contraception (the 'morning after pill')	Around 75–95% (depending upon how soon after intercourse the pills are taken)	Taken within three days of unprotected sexual intercourse, these contraceptive pills contain high concentrations of hormones that prevent fertilization and/or implantation
Intrauterine device (IUD)	98.5%	Probably works by creating an environment within the uterus that is hostile to fertilization or implantation. Newer forms of IUD deliver a sustained low dose of progesterone to the endometrium. May cause uterine bleeding. Increased risk of inflammatory disease in the pelvic region.
Condom (sheath)	96%	A barrier method that is more effective if used with a spermicidal gel, sponge, or foam. Its chief advantage is that it protects against AIDS and other sexually transmitted diseases. It may also protect against cervical cancer. Its disadvantage is that it may split in use.
Diaphragm	98% when used with spermicide	An alternative barrier method. Needs to be inserted before intercourse. It fits over the cervix and blocks the entrance to the uterus. Its use carries a small risk of infection.
Rhythm method	Highly variable, dependent on regularity of cycle and accuracy with which mid-cycle is calculated. Lifespan of sperm within the female reproductive tract is variable but may be up to 5 days.	Couple must refrain from intercourse during the fertile period of the cycle (i.e. the two or three days on either side of ovulation). The time of ovulation must be calculated from the previous cycle assuming that ovulation occurred on day 14 before menstruation. Other indicators such as the change in body temperature at mid-cycle and the constitution of cervical mucus may also be used.
Sterilization (vasectomy in men and ligation of the Fallopian tubes in women)	c. 100%	Requires surgery and is not always reversible. Risks are similar to those of other minor surgical procedures.

49.3 Fertilization

Physiologically, pregnancy begins at the moment of fertilization or conception. Typically the pregnancy then continues for 38 weeks from fertilization. This is known as the **gestation period**. In the clinical setting, pregnancy is normally deemed to start from the first day of the woman's last menstrual period rather than from the day on which fertilization occurred. Using this method of reckoning, a pregnancy or gestation period will last for 40 weeks, and the woman will be 2 weeks 'pregnant' on the day of fertilization (assuming that this took place around mid-cycle). Throughout this chapter, the gestation period will be defined as the 38-week period of time between fertilization and delivery of the baby.

Around 3 ml of seminal fluid are released in each ejaculation and this will normally contain about 200 million sperm. Sperm deposited in the vagina during intercourse swim through the cervical mucus and through the uterus to the site of fertilization in a Fallopian tube. Progression of sperm through the female reproductive tract is assisted by contraction of the smooth muscle of the uterus and Fallopian tubes. Both the male and female gametes have a limited period of viability within the female reproductive tract. Although sperm are thought to retain their fertility for around 5 days, ova remain viable for only 24–48 hours after ovulation. There is therefore a relatively short time during each menstrual cycle in which intercourse must take place if a pregnancy is to be achieved.

To reach the ovum, sperm must travel from the upper vagina through the uterus and along the appropriate Fallopian tube. The cumulus cells of the egg mass signal their position by secreting progesterone. This progesterone activates a specific calcium channel known as CatSper which is present only in the plasma membrane of the principal piece (tail) of the sperm. The resulting calcium influx enhances the beating of the flagellum (known as hyperactivation) and helps to initiate a series of calcium-dependent changes within the sperm head known collectively as **capacitation**. Unless they undergo capacitation, newly ejaculated sperm are unable to fertilize an egg. Sperm are prevented from premature capacitation by the presence of decapacitation factors in the seminal fluid that coat and stabilize the plasma membrane of the sperm head. Capacitation occurs naturally while the sperm are travelling through the female reproductive tract. The biochemical processes involved are complex and not fully understood but depend on the high concentration of bicarbonate ions in the fluid of the uterus and Fallopian tube. Indeed, sperm that are intended for *in vitro* fertilization are unable to fertilize an egg unless they are first washed free of seminal fluid and placed in a bicarbonate-rich medium.

The process of capacitation takes around five hours in humans, during which time the acrosomal vesicle (see Figure 48.7) fuses with the adjacent plasma membrane—a process known as the **acrosome reaction**. This calcium-dependent membrane fusion process is believed to involve proteins from the SNARE (Soluble NSF [N-ethylmaleimide-sensitive factor] Attachment protein Receptor) family. The resulting exocytosis releases hydrolytic enzymes that, together with the force generated by hyperactivation of the sperm tail, aid the passage of the sperm through the zona pellucida of the ovum.

A number of signalling pathways appear to contribute to the preparation of sperm for fertilization. There is good evidence to suggest that a key initiating event is a bicarbonate influx into the sperm head. The subsequent rise in intracellular pH activates a soluble adenylyl cyclase that is not dependent on G protein activation. In turn, the rise in intracellular cyclic AMP activates protein kinase A which initiates a series of protein phosphorylation reactions. There is a loss of cholesterol from the plasma membrane of the sperm which is crucial for capacitation, the plasma membrane lipids are rearranged, and the plasma membrane becomes more fluid (see Figure 49.3). This fluidization promotes the acrosomal reaction which occurs via a calcium-dependent process as described above.

Should an activated sperm meet a viable egg in the Fallopian tube, the sperm first swims through the cumulus cells before binding to the zona pellucida and entering the egg itself. The fertilized

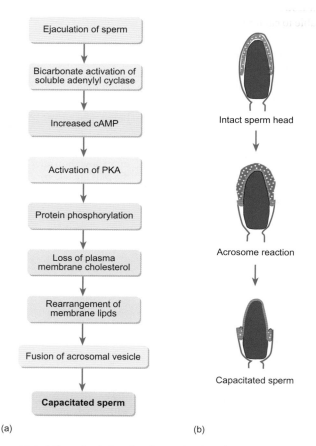

Figure 49.3 A flow chart showing the main steps in sperm capacitation (a) and the principal structural changes in the sperm head (b). The relatively high concentration of bicarbonate in the female reproductive tract compared to that in the epididymis activates a soluble adenylyl cyclase. The subsequent increase in cAMP in the sperm head activates protein kinase A (PKA) and initiates a series of events that lead to fusion of the acrosome membrane with the plasma membrane of the sperm head. The acrosomal enzymes then facilitate the penetration of the zona pellucida by the capacitated sperm.

egg must now complete three important tasks in order to ensure its continued successful development.

- It must complete its second meiotic division to avoid triploidy now that it has fused with a sperm (remember that the secondary oocyte arrested in the second meiotic metaphase: see Chapter 48, p. 780). This division is normally completed within 2 or 3 hours of fertilization and the second polar body is extruded.

- It must avoid fusing with any further sperm (polyspermy). To avoid polyspermy, a special reaction takes place within the newly fertilized egg, called the **cortical reaction**. Much of our information concerning this reaction has come from experiments using sea urchin or mouse eggs, and the extent to which it is applicable to humans is unclear. The key events are shown in Figure 49.4.

- Finally, the fertilized ovum must initiate changes that prevent regression of the corpus luteum and shedding of the endometrium.

After fertilization, the sperm head enters the oocyte. In doing so it introduces a form of phospholipase C (known as PLC zeta) that is able to catalyse the formation of IP$_3$ from membrane lipids. As the IP$_3$ concentration rises, it triggers the release of calcium from internal stores within the oocyte. The crucial role played by calcium in this process has been studied using calcium-sensitive fluorescent probes, which permit the visualization of the calcium waves that occur following fertilization (see Figure 49.5). These calcium waves trigger the cortical reaction during which cortical granules fuse with the oocyte membrane, releasing their contents into the perivitelline space (the space immediately surrounding the oocyte) by exocytosis (see Chapter 4). This is known as the 'fertilization membrane'. The enzymes released from the cortical granules promote the cross-linking of the glycoproteins ZP2 and ZP3 in the zona pellucida, rendering the region impenetrable to any further sperm.

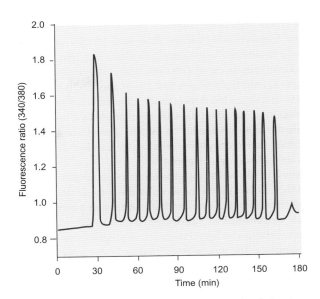

Figure 49.5 Calcium transients seen in a mouse egg following fertilization (which occurred at time zero). Changes in intracellular calcium were monitored with a fluorescent dye (FURA-2) that had been injected into the egg. An increase in the ratio of the light emitted at 340 nm and that emitted at 380 nm indicates a rise in intracellular calcium. (Fig 3A in H. Igarashi J.G. Knott et al (2007) *Developmental Biology* **312**, 321–30. Reproduced by permission of Elsevier.)

The maintenance of the structure and function of the corpus luteum beyond the end of the cycle is dependent on the secretion of a glycoprotein hormone known as **human chorionic gonadotrophin (hCG)**. This hormone is a potent luteotrophic agent that is secreted

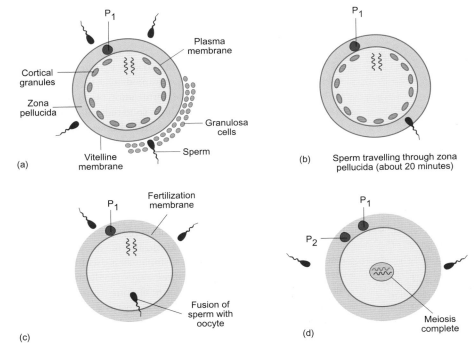

Figure 49.4 Diagrammatic representation of the events that accompany fertilization. For a sperm to fuse with an egg it must first penetrate the layer of granulosa cells and the zona pellucida, as shown in panels (a) and (b). Fusion with the egg leads to the formation of the fertilization membrane (c) in which the enzymes of the cortical granules are secreted into the space surrounding the oocyte. This is followed by the second meiotic division and the extrusion of the second polar body (d). P$_1$ and P$_2$ are the first and second polar bodies.

by the zygote. By its luteotrophic action, hCG is believed to maintain the corpus luteum beyond its normal lifespan so that it will continue to secrete the progesterone that is required for the continuation of pregnancy. hCG is structurally very similar to anterior pituitary LH but has a longer half-life. It appears in the maternal circulation within a few days of fertilization and may be detected in the urine by about 10–14 days after ovulation. Levels of hCG then continue to increase steadily up until 8–10 weeks of gestation before falling rather sharply over the next few weeks. This profile of secretion is illustrated in Figure 49.6.

Figure 49.7 shows a histological section through an ovary in early pregnancy, in which it can be seen that the corpus luteum occupies a large fraction of the ovarian tissue. After 6 or 8 weeks, the placenta is well established (as discussed later in the chapter) and is able to synthesize and secrete sufficient progesterone to maintain the pregnancy. The pregnancy is then said to be autonomous, and the fall in hCG secretion seen from about eight weeks after fertilization probably reflects this diminishing requirement for the hormones of the corpus luteum.

Clinically, hCG is a very important hormone, chiefly because of its early appearance in the maternal body fluids following fertilization. hCG can be detected in the maternal plasma as early as six days after ovulation. Its presence in the urine two weeks or so after ovulation is used as a reliable and simple test for pregnancy, indeed so simple that it can be carried out, using a kit, by a woman herself at home. More sophisticated assay techniques can also be used by clinicians to gain information about the pregnancy. Levels of hormone above the normal range, for example, suggest the presence of twins, a situation that may be confirmed later by ultrasound scans. In women who have suffered habitual miscarriages because of insensitivity of the corpus luteum to hCG, it may be useful to be able to detect the presence of an embryo within a few days of fertilization so that progesterone may be administered exogenously to prevent loss of the pregnancy when the corpus luteum regresses.

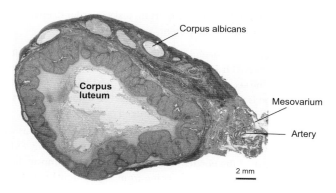

Figure 49.7 A histological section through a human ovary during pregnancy. This section shows how much space is occupied by the corpus luteum. The section was stained with haematoxylin and eosin.

Summary

A sperm is capable of fertilizing an egg only if it first undergoes capacitation followed by the acrosome reaction. Fertilization occurs in the Fallopian tube when an activated sperm fuses with the oocyte. The newly fertilized egg then completes its second meiotic division and undergoes the cortical reaction, which prevents any further sperm from fusing with it.

The early embryo secretes hCG, which prolongs the secretory life of the corpus luteum. This ensures that progesterone continues to be secreted and the specialized endometrial layers of the uterus are maintained until the pregnancy can be supported by progesterone of placental origin.

49.4 Implantation and formation of the placenta

At fertilization, the nuclei of a sperm and an oocyte fuse to produce a diploid cell called a **zygote**. Almost immediately, this cell starts to divide by mitosis and by day 4 after fertilization the zygote has become a cluster of 16–32 cells forming a solid ball called a **morula**. During the next 24 hours or so, the morula hollows out so that the embryonic cells surround a fluid-filled space. At this stage the organism is called a **blastocyst**. The cells of the blastocyst form a layer called the **trophoblast** that encloses a fluid-filled space. One end of the blastocyst possesses a thicker accumulation of cells called the **embryonic disc** or inner cell mass (panel (a) of Figure 49.8). The cells of this region will form the embryo. The outer cell layer will form the placenta and other extra-embryonic structures.

Six or seven days after fertilization, the blastocyst will have reached the uterus and will begin to implant itself within the endometrium. **Implantation** occurs in three stages. Firstly, the blastocyst makes contact with the receptive uterine endometrium at its embryonic disc region and becomes pressed up against it (**apposition** or adplantation). Microvilli on the surface of the blastocyst cells then interact with the endometrial cells via cell surface glycoproteins to

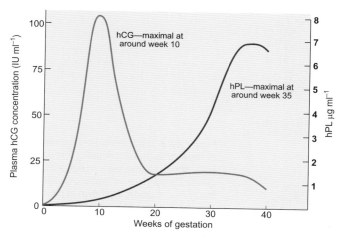

Figure 49.6 The changes in the plasma concentrations of human chorionic gonadotrophin (hCG) and human placental lactogen (hPL) that occur during gestation. (Based on data of Fig 10.6 in B.R. Carr and V.E. Beshay (2012) In W.J. Kovacs and S.R. Ojeda (eds) *Textbook of endocrinology.* Oxford University Press, New York.)

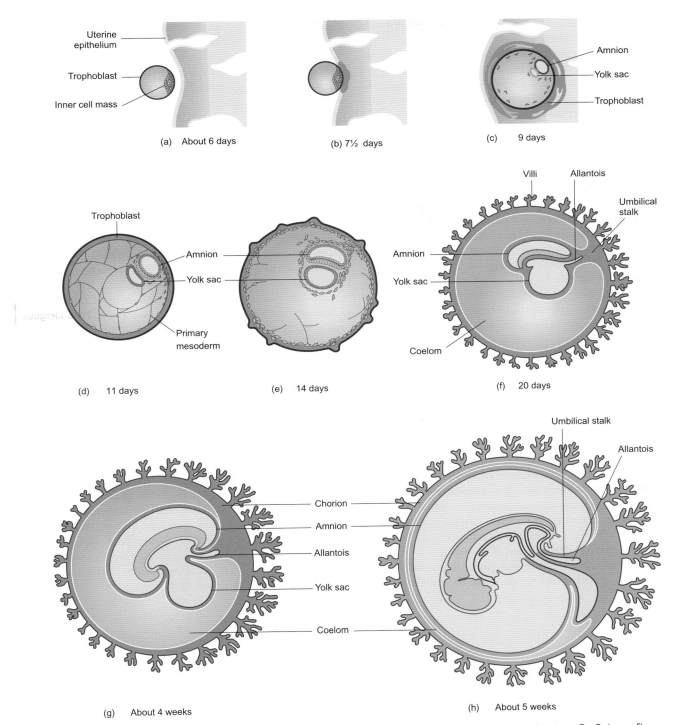

Figure 49.8 A series of diagrams to illustrate the events that take place in the first few weeks after fertilization. By 6 days after fertilization (a) the embryo (blastocyst) consists of a hollowed-out ball of cells with a thickening at one end (the inner cell mass). Over the next few days (b) and (c) the blastocyst implants into the uterine endometrium. By day 11 after fertilization the blastocyst is completely surrounded by maternal tissue (d). Subsequently (e) and (f) the fetal membranes (chorion and amnion) develop, and villi of embryonic tissue invade the uterine tissue to begin the formation of the placenta. Further invasion of the uterus by embryonic tissue and the development of the umbilical stalk are shown in panels (f)–(h). In this way the crucial interface between the maternal and fetal circulations is established. (Adapted from Fig 53.3 in J.Z. Young and M.J. Hobbs (1975) *Life of mammals.* Oxford University Press, Oxford.)

bring about **adhesion** of the blastocyst to the uterine lining. Invasion of the uterine endometrium by the trophoblast then begins.

Shortly before the blastocyst makes contact with the endometrium, the trophoblast differentiates into two regions of cells: the inner **cytotrophoblast**, a layer of cells undergoing rapid mitotic activity, and the outer **syncytiotrophoblast**, a multi-nucleated layer without cell boundaries, formed by the fusion of trophoblast cells. The cells in this region secrete enzymes that break down the endometrial tissue and the walls of capillaries. This invasive behaviour allows the embryo to penetrate the endometrium and become implanted. The syncytiotrophoblast develops further as implantation proceeds so that it surrounds the embryo completely by nine to ten days after fertilization, as shown in Figure 49.8(c). The uterine mucosa reacts to implantation by undergoing the so-called **decidual reaction**: the stromal cells enlarge and become filled with glycogen that provides nourishment for the embryo until the placenta becomes vascularized. Implantation is complete by around day 12 after fertilization, and over the next few weeks, the placental structures are established. The placenta is fully formed (the so-called 'definitive placenta') by the end of the third month of gestation (the first **trimester**).

All organisms throughout their embryonic development need a large and continuous supply of nutrients. They must also be able to respire and to dispose of the waste products of their metabolism. Most mammalian species, including humans, accomplish this by the process of placentation in which a specialized organ, the **placenta**, is developed. This provides an association between the uterine endometrium (at this stage called the **decidua**) and the embryonic membranes that are derived from the trophoblast (see Figure 49.8). The placenta allows the blood supply of the fetus to be brought into close proximity with that of its mother, thereby permitting the exchange of substances between the two circulations. The placenta performs those functions carried out by the lungs, kidneys, and gastrointestinal organs in the adult—indeed it is the only source of nourishment, gas exchange, and waste disposal available to the fetus. It is also an important endocrine organ. For this reason, normal development of the placenta is crucial to the success of a pregnancy. Furthermore, in the first months of gestation, placental growth must keep pace with the requirements of the growing fetus. Without an adequate surface area for transplacental exchange, the growth of the fetus will be impaired and its life may be threatened. The association between the maternal and fetal circulations that is established by the placenta allows for prolonged development within the uterus and, as a result, the delivery of a highly developed baby.

While the anatomical details of placental formation are beyond the scope of this book, the following brief account will explain how the placenta is adapted structurally for carrying out its vital role as an organ of exchange. Figure 49.9 shows an enlarged view of a section of human placenta soon after the start of embryonic implantation. This stage of development is known as the **stem-villus** stage because the fetal tissue grows up into the maternal endometrial tissue in the form of finger-like projections or **villi** (see also panels (f)–(h) of Figure 49.8). These villi are formed from a membrane called the **chorion** which is derived from trophoblastic tissue and the **extra-embryonic mesoderm**, i.e. the mesoderm that lies outside the developing embryo. Blood vessels form within this outgrowth to give rise to the fetal component of the placental circulation, which will later become the umbilical vessels in the umbilical cord. As the trophoblastic tissue invades the endometrium, it secretes digestive enzymes that break down the spiral arteries. As a result of this erosion, blood spills out of the maternal vessels to create blood-filled spaces between the chorionic villi. These are called the intervillous blood spaces. At the same time, the villi themselves become cored with mesoderm (see Figure 49.9), which subsequently becomes vascularized by fetal vessels. These vessels carry fetal blood into close proximity with the maternal blood spaces, and the essential interface between the maternal and fetal circulations is thus established.

This invasive behaviour of the trophoblast during implantation and early placentation is highly aggressive and reminiscent of that of a malignant neoplasm. It has indeed been shown that, occasionally, fragments of tissue break away from the trophoblast and become lodged in distant maternal tissues, particularly the lung, rather in the same way as malignancies tend to metastasize. Fortunately, regression of the ectopic trophoblastic tissue generally occurs quite rapidly following delivery.

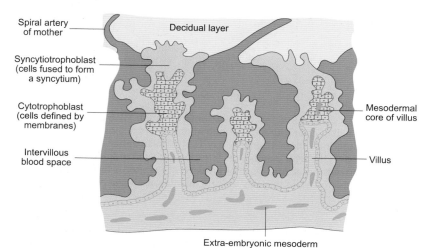

Figure 49.9 A simplified drawing to show the stem-villus stage of placental development at around 3 weeks gestation.

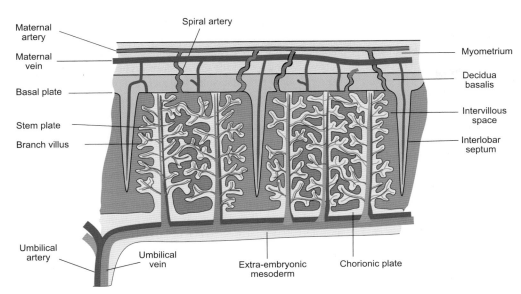

Figure 49.10 A diagrammatic representation of the definitive placenta at the end of the first trimester.

Invasion of the endometrium and the formation of villi take place during the first month after conception (see Figure 49.8). This establishes proximity between maternal and fetal blood. In the succeeding two months, the highly vascularized villi become much more highly branched, thereby increasing the surface area of the fetal capillaries available for transplacental exchange of nutrients and waste products. By the end of the first trimester of gestation the placenta is known as the **definitive placenta** because although it will increase in size, it will show few further changes to its basic structure for the remainder of the pregnancy.

A diagrammatic and highly simplified representation of the placenta at three months (the so-called definitive placenta) is shown in Figure 49.10. Trophoblastic cells at the outer margins of the villi fuse together to form a syncytium (the syncytiotrophoblast), while those further from the outer margins retain their individual membranes (the cytotrophoblast). There is also elaborate branching of the primary villi. The erosion of the maternal spiral arteries is now so extensive that their blood is simply discharged into the intervillous spaces. The fetal blood supply runs in the capillaries of the branched villi and the arrangement of the placental blood flow is such that the fetal capillaries essentially dip into the maternal blood spaces so that they are virtually surrounded by maternal blood. This arrangement of the two circulations within the placenta is termed a *dialysis pattern* and it allows movement of solutes in either direction according to the concentration gradient, over the entire surface of the fetal placental capillaries.

Summary

The placenta forms the interface between the maternal and fetal circulations. At implantation, the trophoblastic tissue of the embryo invades the endometrial tissue of the uterus. As a result, the spiral arteries of the uterus are eroded and spill their blood into the spaces between adjacent chorionic villi so that a dialysis pattern of blood flow is established.

49.5 The placenta as an organ of exchange between mother and fetus

The rate and extent of diffusion of a substance across any cellular barrier depends upon a variety of factors (see Fick's Law, Box 31.1). In addition to the chemical characteristics of the substance itself, these include:

- the nature and thickness of the barrier to diffusion
- the surface area available for exchange
- the concentration gradient of the substance.

Each of these factors will now be examined with reference to the human placenta as an organ of exchange.

For a substance to diffuse from the maternal blood space to the fetal capillary blood (or vice versa), it must cross the syncytiotrophoblast and the fetal capillary endothelial layer. The latter consists merely of a single layer of cells on a basement membrane and is therefore very thin, but the syncytiotrophoblast contains many layers of cells, with no paracellular pathways to act as shortcuts for diffusion. It therefore makes the placental barrier rather impermeable. However, this low solute permeability is largely offset by the enormous surface area available for placental exchange created by the extensive branching of the fetal capillaries within the villi, and the dialysis arrangement of the two circulations.

Placental blood flow

Within the placenta, the concentration gradient of any substance between the maternal and fetal blood will be influenced by the blood supply to the maternal and fetal circulations, particularly the relative blood flow rates on either side. The rate of entry of blood to the maternal intervillous spaces is rather difficult to measure since the uterine arteries supply both the uterus and the placenta. This makes it hard to differentiate between blood destined for the intervillous spaces and that supplying the uterus. Total blood flow in the uterine artery at full term is 600–1000 ml min^{-1}, measured just prior to delivery by Caesarean section. About half of this (or 10 per cent of the total maternal cardiac output at full term) is thought to perfuse the maternal side of the placenta. The maternal blood space has a total volume of about 250 ml, so the blood on the maternal side of the placenta is exchanged roughly twice each minute. The blood entering the intervillous spaces is at a relatively high pressure—about 13 kPa (100 mm Hg), as it is discharged from the eroded spiral arteries, thus ensuring a fair degree of turbulence and good mixing.

In a full-term fetus, perfusion of the fetal capillaries in the placental villi is estimated at around 360 ml min^{-1}, roughly half the fetal cardiac output. The total volume of blood in the capillaries is around 45 ml, so the blood in the fetal placental compartment is exchanged about eight times a minute. To summarize, the maternal blood spaces present a large volume of well-mixed blood with a moderate turnover, while the fetal capillaries have a much smaller volume but a greater turnover.

This pattern of circulation emphasizes the dialysis nature of the placental blood flow which optimizes the conditions for passive exchange of solutes by maximizing their concentration gradients. Consider the diffusion of a solute from maternal to fetal blood: as the fetal blood flow is very high, solute diffusing into the fetal blood from the maternal side will be removed from the placenta rapidly, keeping its concentration in the fetal capillaries low. The maternal blood, however, has a much larger volume and so, despite its lower flow, will not become depleted of solute readily.

The dialysis arrangement of the placental blood flow also helps to optimize conditions for the removal of waste products from the fetal circulation by maintaining a steep concentration gradient for the waste substance across the placental barrier. The rapid turnover of blood on the fetal side ensures a constant delivery of waste to the placenta, while the relatively large volume of blood in the intervillous spaces keeps the concentration of waste product low on the maternal side.

Towards the end of pregnancy, the exchange capacity of the placenta tends to diminish. This is chiefly due to changes in the perfusion of the organ. Maternal blood flow may be somewhat reduced as the spiral arteries become progressively occluded. At the same time, the fetal capillaries tend to become blocked with small clots and other debris. This leads to a progressive decline in placental perfusion towards term. Those parts of the chorionic villi that become poorly perfused can no longer participate efficiently in exchange, reducing the effective surface area for diffusion. As a result of this declining efficiency, the placenta is less and less able to meet the demands of the fetus and is said to become 'senescent' near to term. This may be one of the many factors involved in the triggering of parturition (see Section 49.7).

Gas exchange across the placenta occurs by diffusion

In addition to supplying oxygen to the fetus, the placenta must, in its capacity as the fetal organ of gas exchange, remove the carbon dioxide produced by fetal metabolism. Oxygen diffuses passively from the maternal to the fetal side of the placenta, while carbon dioxide diffuses in the opposite direction. How efficient is the placenta as an organ of gas exchange? To answer this question, it is necessary to consider the gas tensions on both the maternal and the fetal sides of the placenta. Figure 49.11 shows the values for oxygen and carbon dioxide tensions in each compartment. The normal arterio-venous differences in partial pressures seem to prevail on the maternal side, with a $Pa\mathrm{O_2}$ of around 12.6 kPa (95 mm Hg) in the uterine artery falling to around 5.6 kPa (42 mm Hg) in the uterine venous blood. The maternal $P\mathrm{CO_2}$ rises from 5–5.6 kPa (38–42 mm Hg) in the arterial blood to 6.1 kPa (46 mm Hg) on the venous side. Since the maternal blood space is large and the blood well mixed, equilibrium values of gas partial pressures are quickly reached and it is therefore reasonable to assume that the blood of the intervillous spaces has a $P\mathrm{O_2}$ of 5.6 kPa (42 mm Hg) and a $P\mathrm{CO_2}$ of 6.1 kPa (46 mm Hg).

On the fetal side, the blood in the umbilical artery, which is travelling to the placenta from the fetus, is both highly deoxygenated and hypercapnic, having a $P\mathrm{O_2}$ of around 2.1 kPa (16 mm Hg) and a $P\mathrm{CO_2}$ of about 7.3 kPa (55 mm Hg). Blood returning from the placenta to the fetus in the umbilical vein has a $P\mathrm{CO_2}$ of 5.4 kPa (41 mm Hg) and a $P\mathrm{O_2}$ of about 3.9 kPa (29 mm Hg). The fetal umbilical venous blood is, therefore, not in equilibrium with the maternal blood. In this respect, the placenta differs from the lung, in which complete equilibration is normally achieved between the pulmonary blood and the alveolar air. There are two important reasons for this failure to reach equilibrium. Firstly, not all the maternal blood is in direct contact with the villi (which is the area of gas exchange) so that 'shunts' exist which are analogous with a ventilation/perfusion mismatch in the lungs (see Chapter 34). Secondly, the placental tissue itself, which is highly metabolically active, uses around 20 per cent of the oxygen in the maternal blood before it

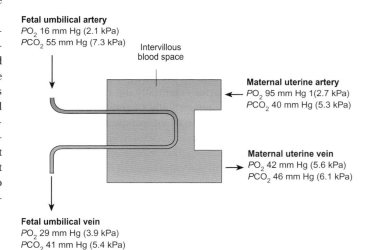

Fetal umbilical artery
$P\mathrm{O_2}$ 16 mm Hg (2.1 kPa)
$P\mathrm{CO_2}$ 55 mm Hg (7.3 kPa)

Intervillous blood space

Maternal uterine artery
$P\mathrm{O_2}$ 95 mm Hg 1(2.7 kPa)
$P\mathrm{CO_2}$ 40 mm Hg (5.3 kPa)

Maternal uterine vein
$P\mathrm{O_2}$ 42 mm Hg (5.6 kPa)
$P\mathrm{CO_2}$ 46 mm Hg (6.1 kPa)

Fetal umbilical vein
$P\mathrm{O_2}$ 29 mm Hg (3.9 kPa)
$P\mathrm{CO_2}$ 41 mm Hg (5.4 kPa)

Figure 49.11 Schematic drawing to show the dialysis organization of the placenta with typical blood gas values. Note that the maternal and fetal circulations are separate and that the blood leaving the placenta via the umbilical vein is not fully equilibrated with the maternal blood.

has a chance to reach the fetal capillaries. The placenta is therefore a less efficient organ of gas exchange than the lung. However, it is able to satisfy the oxygen demand of the fetus because of a variety of specific adaptations that ensure that the transfer of oxygen to the fetal tissues is maximized. These are discussed in Chapter 50. As for oxygen, the passive movement of carbon dioxide depends on blood flow and diffusion gradients, but the placental barrier is more permeable to carbon dioxide than it is to oxygen and exchange is more or less complete.

Placental exchange of glucose and amino acids is carrier-mediated

The placenta provides all the nutrients that are essential for the growth of the fetus. The most important source of metabolic energy is glucose. Because glucose is a polar molecule, and therefore rather lipid-insoluble, it cannot rely solely upon passive diffusion across the lipid-rich placental barrier. Instead it is transported from the maternal to the fetal side by facilitated diffusion (see Chapter 5). The process is mediated by a carrier (GLUT1) that is located in the membrane of the cells of the syncytiotrophoblast. Unless the carrier becomes saturated, fetal levels of glucose will be in equilibrium with those of the mother. While this would normally be desirable, there are circumstances in which it is not. Consider the case of poorly controlled maternal diabetes. Here, maternal glucose levels may be abnormally high, and, because of facilitated diffusion across the placenta, fetal levels will also be high. This can lead to over-nourishment and obesity of the baby—indeed, the babies of diabetic mothers are often bigger than normal for their gestational age.

Amino acids are also vital to the fetus during its development in the uterus. They are needed to support the high rate of protein synthesis that occurs during gestation. The concentrations of most amino acids are higher in the fetal plasma than in the maternal plasma. There is evidence that amino acids are actively transported across the placenta, and recent studies using the techniques of molecular biology have shown that the human placenta has specific transporters for all the essential amino acids.

Lipids are also of great importance in fetal development. While phospholipids do not pass readily across the placental barrier, free fatty acids are able to do so and it is in this form that the fetus receives most of its lipid. Phospholipid in the maternal blood is hydrolysed by enzymes on the placental surface to form free fatty acids, which diffuse passively down their concentration gradients to the fetal blood. They then pass to the fetal liver, where they undergo reconjugation to form new phospholipid.

The excretion of fetal waste products by placental exchange

Like carbon dioxide, which diffuses from the fetal to the maternal blood across the placenta, other fetal waste products are removed in a similar fashion and are then excreted along with those of the mother herself. One of the most important of these metabolic waste products is urea, the nitrogenous waste product of protein metabolism. Although much of the metabolism of the fetus is concerned with the synthesis of new structural protein, there is also a fair amount of tissue destruction throughout gestation, as fetal tissues are remodelled. Indeed, of the total nitrogen that enters the fetus in the form of amino acids transported from the maternal blood, about 40 per cent ends up in the form of urea, which must be disposed of. Excretion occurs by the passive diffusion of urea down its concentration gradient between the fetal and maternal blood.

Another fetal waste product of considerable clinical significance is bilirubin, which is produced by the breakdown of haemoglobin. In adults, bilirubin is conjugated by hepatic enzymes to form bilirubin glucuronide. This is water-soluble and can therefore be excreted without difficulty. The fetal liver, however, is relatively immature and does not possess sufficient amounts of the necessary conjugation enzymes. During fetal life there is significant destruction of red blood cells and the bilirubin produced crosses the placenta. It is then conjugated by maternal liver enzymes before being excreted in the bile. If it is not disposed of but builds up in the fetal blood, bilirubin can cross the blood–brain barrier and cause severe brain damage. The basal ganglia are the most commonly affected brain regions, giving rise to the condition known as **kernicterus,** in which there may be permanent impairment of motor function.

Summary

During fetal life, the placenta carries out the functions normally performed in the adult by the lungs, kidneys, and gastrointestinal tract. Although the placental barrier itself is relatively impermeable to polar molecules, the surface area available for exchange is immense, due to the considerable branching of the chorionic villi. Oxygen diffuses passively from maternal to fetal blood, although full equilibration is not achieved. Carbon dioxide diffuses in the opposite direction, normally to complete equilibration. Glucose and amino acids move across the placenta from the maternal to the fetal plasma by carrier-mediated transport, while free fatty acids diffuse passively across the lipid-rich placental barrier. Fetal waste products such as urea and bilirubin diffuse from fetal to maternal plasma down their concentration gradients.

49.6 The placenta as an endocrine organ

In addition to its crucial transport role described above, the placenta is also an extremely important endocrine organ. At full term it normally weighs around 650 g and is the largest endocrine organ in the body of a pregnant woman. During pregnancy it secretes a wide variety of peptide and steroid hormones that are important both for the maintenance of pregnancy and for the preparation of the mother's body for parturition and subsequent lactation. The major peptide hormones secreted by the placenta are:

- human chorionic gonadotrophin (hCG)
- **human placental lactogen** (hPL)

and the major placental steroids are:

- oestrogens
- progesterone.

The physiological role of each of these hormones will be considered in turn.

Human chorionic gonadotrophin

This hormone was discussed in some detail in Section 49.3. It is secreted by the trophoblastic tissue of the embryo from a very early stage of pregnancy and probably provides the signal that enables the mother's body to recognize the existence of a fertilized egg (the maternal recognition of pregnancy). It is a powerfully luteotrophic hormone which prolongs the life of the corpus luteum beyond the normal 12–14 days. As a result, progesterone secretion continues so that shedding of the uterine endometrium is prevented and spontaneous contractile activity of the myometrium is inhibited. Luteal progesterone is required for about the first six to eight weeks of pregnancy, and loss of the ovaries during this period will result in miscarriage. After this time, the placenta takes over as the main source of progesterone and the pregnancy is said to have become autonomous (the so-called luteo-placental shift). Indeed, levels of hCG output decline sharply after about 10 weeks, although the hormone continues to be produced by the placenta for the remainder of the pregnancy (see Figure 49.6).

In addition to its luteotrophic role, a number of other actions have been attributed to hCG. It has been suggested that it may exert a direct effect on the maternal hypothalamus to inhibit the synthesis of FSH and LH. If so, this might contribute to the suppression of ovulation during pregnancy. There is also some evidence that hCG has a direct effect on the myometrial smooth muscle, promoting uterine quiescence, possibly by inhibiting the actions of oxytocin. hCG is also thought to play a role in preventing rejection of the fetus by the mother. It also exerts a stimulatory effect on the Leydig cells of the testes in male fetuses and is believed to play a part in the differentiation of the male reproductive tract. Plasma levels of hCG are high during the period of Wolffian duct development and the start of differentiation of the external genitalia (see Chapter 50).

Recent studies suggest that a hyperglycosylated variant of hCG (known as hCG-H) is produced by the trophoblast in addition to regular hCG. It is secreted by undifferentiated cells of the cytotrophoblast and appears to exert an autocrine effect on the cells that produce it, promoting trophoblast invasion during implantation. Unlike hCG, which is secreted predominantly by the syncytiotrophoblast, hCG-H seems to have a minimal luteotrophic effect.

Human placental lactogen

The pattern of secretion of human placental lactogen (hPL), sometimes known as human chorionic somatomammotrophin (hCS), is shown in Figure 49.6. This hormone appears in the maternal circulation at the time that hCG levels are beginning to fall, at around eight weeks of gestation. The concentration of hPL in the maternal plasma then continues to rise during the pregnancy, reaching a peak at around week 35. It is secreted by the syncytiotrophoblastic tissue of the placenta, and unlike hCG, which appears equally in both the fetal and maternal circulations, hPL is secreted preferentially into the maternal blood.

Like hCG, hPL is structurally and functionally related to other peptide hormones. It has a high degree of homology with both growth hormone and prolactin, both of which are secreted by the anterior pituitary. Like them, hPL can stimulate both somatic growth and milk secretion, though only weakly. Its principal action is to encourage the proliferation of breast tissue during pregnancy in preparation for lactation following delivery.

In addition to its mammotrophic action, hPL exerts some important metabolic effects. These are mainly concerned with adjusting the maternal plasma levels of certain metabolites in order to favour fetal uptake via the placenta, without undue depletion of the maternal blood. hPL exerts a diabetogenic effect by reducing the sensitivity of maternal tissues to insulin. Maternal plasma glucose levels therefore tend to rise under the influence of hPL as glucose uptake into cells is inhibited. Gluconeogenesis (the synthesis of glucose from amino acids) also appears to be suppressed by hPL, leading to an increase in amino acid levels in the maternal plasma, while increased lipolysis causes an increase in plasma free fatty acids. By these actions it is thought that hPL counteracts the fall in metabolites that might otherwise occur as a result of fetal uptake and ensures the maintenance of favourable gradients for the placental transport of these important nutrients.

hPL can be measured accurately and simply by a variety of techniques including radioimmunoassay. The monitoring of hPL levels during pregnancy has clinical significance, as its concentration in the maternal plasma provides a valuable indication of placental sufficiency. Although it is not unknown for pregnancies to proceed successfully in the absence of hPL (in the case of specific genetic deficiencies for example), falling levels of this hormone are, in general, indicative of placental insufficiency, which may put the fetus at risk.

Other placental polypeptide hormones

In addition to the two major placental peptide hormones described above, a large number of other proteins are produced by the placenta. In most cases no specific actions have been ascribed to these hormones, but chorionic FSH and chorionic thyrotrophin (which has a similar action to TSH) have been identified. Clinically, measurement of these hormones may be of value in assessing possible risks to the fetus from placental insufficiency.

Relaxin is a large polypeptide hormone secreted by the corpus luteum of both pregnant and non-pregnant women. It is also secreted by the cytotrophoblast of the placenta, particularly during the first three months of gestation and then in increasing amounts again as delivery approaches. Relaxin softens the cervix and relaxes the pelvic ligaments in preparation for parturition.

The placenta secretes huge amounts of progesterone and oestrogens

For a pregnancy to proceed successfully to term, adequate amounts of progesterone are crucial. Indeed, there is no recorded case of a pregnancy continuing normally in the face of insufficient progesterone secretion. During the early weeks after conception, this progesterone is supplied by the corpus luteum, rescued from its declining phase by hCG as described in the previous sections. By about eight weeks after fertilization, however, the placenta has become well established and starts to produce large amounts of progesterone. For the remainder of gestation, placental progesterone output is extremely high. The pattern of progesterone secretion by the placenta is illustrated in Figure 49.12. During late gestation, progesterone is produced at a rate of 250–350 mg a day. This compares with about 20 mg a day during the luteal phase of the menstrual cycle. The placental tissue is capable of synthesizing progesterone without the need for precursors from elsewhere, so levels of this hormone during pregnancy are determined solely by the synthetic and secretory capacity of the placenta itself. Consequently, plasma progesterone provides another valuable clinical index of placental performance.

Why is progesterone so important during pregnancy? A number of different functions have been suggested, but the most important of these appear to be the maintenance of a quiescent myometrium and the prevention of endometrial shedding prior to placentation. Progesterone also plays a role in the stimulation of breast development and the suppression of ovulation, and in preventing immunorejection of the embryo.

In addition to progesterone, the placenta secretes large amounts of a number of oestrogenic hormones including oestrone, oestriol, and oestradiol-17β. The patterns of secretion of these hormones are also illustrated in Figure 49.12. By comparing the values of plasma oestrogen levels during gestation with those seen in non-pregnant women, it is possible to get some idea of the scale of oestrogen secretion during pregnancy. In the non-pregnant state, the principal oestrogen (secreted by the developing follicle) is oestradiol-17β and in pregnancy the levels of this hormone rise to about 100 times the non-pregnant values. Circulating levels of the other oestrogens of pregnancy are even higher.

What is the role of oestrogens in pregnancy? This question remains largely unanswered. Although it is usual for high levels of oestrogens to be present throughout gestation, pregnancies can continue successfully even when oestrogen levels are rather low. However, they do seem to have a role in preparing the body for giving birth and for lactation. They seem to contribute to relaxation of the symphysis pubis (along with relaxin) and to act alongside hPL to stimulate proliferation of the mammary tissue. They may also play a part in the initiation of parturition, discussed in the next section.

Summary

The placenta secretes a wide variety of peptide and steroid hormones including human chorionic gonadotrophin (hCG) and human placental lactogen (hPL). hCG prevents regression of the corpus luteum and ensures the continued secretion of progesterone during early pregnancy. hPL is secreted from around week 10 of gestation and contributes to the development of mammary tissue in preparation for lactation. It also increases the maternal plasma levels of glucose, amino acids, and free fatty acids, and thereby optimizes their transport from mother to fetus.

Oestrogens and progesterone are also produced in huge amounts by the placenta. The placenta takes over from the corpus luteum as the major source of progesterone at around week 10 of gestation. Progesterone maintains the endometrium, reduces myometrial excitability, and stimulates mammary development.

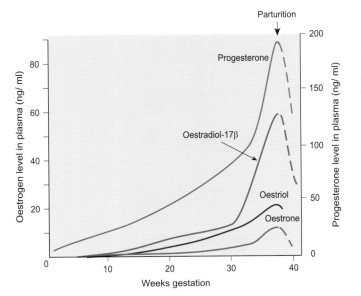

Figure 49.12 The plasma levels of various steroid hormones during pregnancy. Note that progesterone secretion dominates the period of gestation, falling only after parturition. Oestradiol-17β levels begin to rise rapidly towards the end of gestation. (Based on data in P. Coats et al. (1977) *Acta Obst Gynecol Scand* **56**, 453–7.)

49.7 The infant is delivered around 38 weeks after conception—what triggers parturition?

In humans, as in all mammals, the length of the gestation period is remarkably constant under normal conditions—40 weeks after the start of the last menstrual period or 38 weeks after conception. This constancy suggests that there is a well-coordinated trigger for the onset of **parturition** (the process of expulsion or delivery of the fetus). What are the critical signals that bring pregnancy to an end? What brings about the conversion from the stable maintenance of pregnancy with minimal myometrial activity to one in which the cervix dilates and the myometrium contracts powerfully to expel the fetus?

Over the last few years, a considerable research effort has gone into trying to shed more light upon this question. Most of this work has been carried out either on small laboratory animals such as the rat or on larger domestic animals, particularly sheep. It is now apparent that in these experimental animals there is no single trigger for the initiation of labour but rather a combination of different factors, both physical and endocrine. It is not clear, however, to what extent any information gathered from animal work may be applied to human parturition.

The physical factors that seem most likely to contribute to the onset of parturition are stretching of the myometrium and placental insufficiency. The uterine musculature (myometrium) is progressively stretched as the fetus grows during gestation. As it stretches it becomes thinner and its excitability increases. Once a certain level of excitability is reached, spontaneous contractions occur that tend to squeeze the contents of the uterus down towards the cervix. Small areas of myometrium act as pacemaker cells to initiate action potentials that are then conducted throughout the myometrium, which behaves as a syncytium.

The second physical factor that may play a part in the process of parturition is the increasing inability of the placenta to meet the ever-growing nutritional demands of the fetus. Growth of the fetus far outstrips that of the placenta after the first trimester of pregnancy, and the fetal capillaries tend to become clogged with clots and other debris as the pregnancy nears term. As a result, the placenta becomes less efficient as an organ of exchange and, although it is difficult to verify experimentally, this decline may contribute to the onset of labour.

The fetus may control the timing of its own birth

Whether or not the physical changes described above form a truly significant part of the trigger for labour is open to question. What is clear, however, is that a number of hormonal factors also have a role to play, at least in non-human species. A number of observations have led to the belief that the endocrine system of the fetus itself plays a key role in triggering its own delivery. For example:

- anencephalic fetuses (i.e. fetuses in whom the brain is absent or severely damaged) are frequently born postmature (i.e. after the normal gestation time)
- in sheep, an infusion of cortisol or ACTH to the fetus brings about premature delivery.

Such findings have led to the idea that maturation of the fetal adrenal cortex is in some way responsible for triggering the onset of parturition. Using a sensitive assay for cortisol, it has been shown in sheep that fetal plasma cortisol rises 15 to 20 days before full term. This rise correlates well with an increase in fetal adrenal enzyme activity and in the abundance of ACTH receptors on the adrenal cortical cells. As might be expected, fetal ACTH levels appear to rise as full term approaches and the paraventricular nucleus of the hypothalamus shows an increased content of both CRH and arginine vasopressin, the hormones which act as releasing factors for ACTH (see Chapter 21). The mechanisms that underpin this series of changes in the fetus at the end of gestation are unclear at present.

At the end of gestation the uterus is released from the 'progesterone block' which has dominated pregnancy

The experimental findings discussed above suggest very strongly that fetal cortisol plays a significant role in the onset of labour, at least in some species. It has been recognized for some time that a variety of other hormones also show marked changes in their pattern of secretion as pregnancy proceeds. Throughout most of the gestation period, placental progesterone secretion outstrips that of oestrogens (although as Figure 49.12 shows, these are also secreted in very large amounts). This is, of course, vital for the success of the pregnancy since progesterone acts as a myometrial relaxant that prevents premature expulsion of the fetus from the uterus. It also reduces the sensitivity of the uterine smooth muscle to other agents that increase uterine excitability. Oestrogens, on the other hand, increase myometrial excitability and enhance its sensitivity to other substances that may cause increased contractile activity. An oestrogen-primed uterus is highly sensitive to agents such as histamine, oxytocin, acetylcholine, and prostaglandins, reflecting alterations in resting membrane potential of the smooth muscle cells.

Fetal cortisol may initiate the switch from progesterone- to oestrogen-dominance

It is known that cortisol stimulates the conversion of progesterone to oestrogens in the placenta. Figure 49.13 shows a diagrammatic representation of events that could lead up to the onset of parturition, which can be summarized as follows.

- Fetal cortisol stimulates the conversion of progesterone to oestrogens in the placenta and so redirects placental steroidogenesis in favour of hormones that increase myometrial excitability.
- Oestrogens in turn stimulate the production of $PGF_{2\alpha}$ by the placenta, which may help to enhance rhythmical contractions of the uterus.
- These contractions, in their turn, stimulate the secretion of oxytocin from the posterior pituitary (discussed later in the chapter). Oxytocin increases the excitability of the musculature during labour itself.

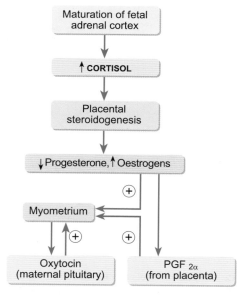

Figure 49.13 A schematic diagram illustrating some of the endocrine and mechanical factors involved in the initiation of parturition. ⊕ indicates a stimulatory action on the smooth muscle of the uterine wall.

Box 49.2 The stages of labour

Labour (parturition) is the process by which a pregnant woman expels the fetus at the end of gestation. It normally begins between 250 and 285 days after presumed ovulation, although during the last month of gestation the woman often experiences irregular uterine contractions. These are known as Braxton-Hicks contractions and are believed to occur in response to a gradual increase in sensitivity of the uterine smooth muscle to spasmogenic agents such as oxytocin. The process of labour takes place in 3 stages.

• Stage 1. Uterine contractions start to occur regularly, initially at intervals of 20–30 minutes and then more frequently until they are occurring every 2–3 minutes. The purpose of these contractions is to dilate the cervix to around 10 cm to allow the baby to move into the vagina (birth canal). Although the average time for this stage is about 15 hours for a first baby, full cervical dilation may be achieved much more quickly in subsequent deliveries. The amniotic sac that surrounds the baby in the uterus may rupture at any time during labour but this often occurs at the onset of labour, a process sometimes described as the 'waters breaking'.

The amniotic membranes may also be ruptured artificially to induce labour. Occasionally a 'show', consisting of mucus and a small amount of blood from the cervix (the 'mucus plug'), is passed from the vagina at the onset of labour.

• Stage 2. In this stage, the baby is delivered. Using the Valsalva manoeuvre to raise the intra-abdominal pressure during uterine contractions, the woman pushes to expel the baby through the cervix and vagina. Delivery is normally head first, though other presentations occasionally occur (e.g. breech delivery, in which the baby is delivered feet first). This stage may take anything from a few minutes to several hours. In some cases forceps or suction are required to facilitate expulsion of the infant.

• Stage 3. Within about 30 minutes after the birth of the baby, the placenta, fetal membranes, and any remaining amniotic fluid (the 'afterbirth') are expelled from the uterus in response to further uterine contractions. These also help to seal the blood vessels ruptured by separation of the placenta from the wall of the uterus.

It must be emphasized that this scheme is purely speculative—the nature of the interactions between the neural and hormonal changes that occur as pregnancy nears full term is far from clear even in well-studied species like the sheep. Human parturition is even less well understood, as obvious ethical difficulties surround research using human subjects. The stages of labour are described in Box 49.2.

Evidence from non-human primates suggests that there is a rise in fetal ACTH secretion just prior to delivery and that maternal oestrogen levels (especially oestriol) rise close to full term. Furthermore, removal of the fetus (but not the placenta) some days before the expected date of delivery delays expulsion of the placenta for up to 50 days after full term. This is highly suggestive of a role for the fetus itself in controlling the onset of labour. Premature birth remains a major cause of infant mortality and disability, but it is hoped that understanding gained from further research into the factors that control the timing of parturition will lead eventually to the development of drugs that can selectively inhibit myometrial contractility and prevent premature delivery.

Summary

Parturition is a multifactorial process involving both the maternal and the fetal nervous and endocrine systems. The trigger for parturition is still poorly understood, but evidence suggests that fetal cortisol initiates a switch in the placenta away from progesterone synthesis to oestrogenic hormones in the last days of pregnancy. Oestrogens, $PGF_{2\alpha}$ and oxytocin then increase the excitability of the myometrium to bring about delivery of the infant.

49.8 Changes in maternal physiology during gestation

Throughout gestation, the growing fetus makes considerable metabolic demands upon its mother, effectively plundering maternal resources to ensure its survival and successful development. Consequently, during the months of pregnancy, the anatomy, physiology, and metabolism of a pregnant woman's body undergo a number of significant alterations that create favourable conditions in which the fetus can grow. These changes also prepare the mother's reproductive tract and mammary glands for the delivery and subsequent nourishment of the baby.

There is considerable individual variation in the exact nature and extent of the physiological and anatomical changes shown by the mother's body during pregnancy. Typically, however, in the first trimester (three months) of pregnancy, the changes that occur are designed to prepare the body for the additional metabolic demands of the later stages of gestation. The latter part of pregnancy represents a state of 'accelerated starvation' for the mother's body in which nutrients and amino acids are conserved for use by the fetus. Parallel changes take place throughout pregnancy in all the major systems of the body, and most revert to normal after delivery of the infant. Many of these, particularly towards the end of pregnancy, occur because of the marked changes in spatial relationships of the maternal abdominal organs caused by the increasing amount of space occupied by the fetus. Figure 49.14 illustrates the growth of the fetus and the space occupied by the uterus within the abdomen during gestation. This shows how the organs of the mother's cardiovascular, respiratory, renal, and gastrointestinal systems are progressively displaced or compressed by the enlarging uterus.

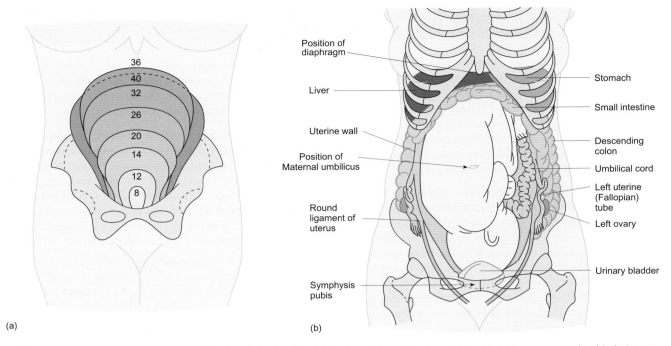

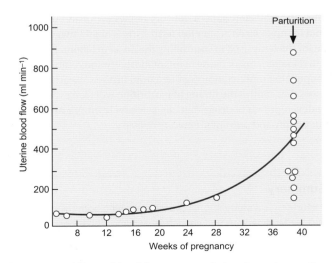

Figure 49.14 (a) A diagram illustrating the height of the fundus of the uterus throughout gestation. Heights vary considerably between women, but a general rule is that at 20 weeks the fundus is usually around the height of the umbilicus. (b) A diagram showing the position and space occupied by the fetus at full term. Note the displacement of the gastrointestinal structures, and pressure on the bladder and diaphragm.

Furthermore, the mechanical demands placed upon the structures of the musculo-skeletal system are significantly altered during pregnancy.

Cardiovascular changes during pregnancy

Cardiac output adapts to meet the requirements of the placenta. During pregnancy, blood flow through the uterine artery rises from around 50–70 ml min^{-1} to between 600 and 1000 ml min^{-1} at full term (see Figure 49.15). There is also a general increase in the maternal metabolic rate. As a consequence of these changes, maternal cardiac output increases by 30–50 per cent between weeks 6 and 28 of gestation, typically from around 4.5 to 6.0 l min^{-1}. This increase in cardiac output is the result of changes in both heart rate (from 70 b.p.m. to 80 or 90 b.p.m.) and stroke volume (which increases by around 10 per cent). Despite the high uterine blood flow (Figure 49.15), there may be a slight fall in resting cardiac output at the very end of pregnancy because the partial obstruction of the inferior vena cava by the uterus reduces venous return.

Cardiac output increases further during labour itself but then drops rapidly post-partum, so that pre-pregnancy values are reached by about six weeks after delivery. The increased cardiac output during gestation supplies the increasing demands of the uterus and the placenta for nutrients. It also ensures the removal of waste products. Functional murmurs are heard more frequently in pregnancy because the circulation is more dynamic (see Chapter 28).

Changes in volume and composition of the maternal blood

The changes in plasma volume and red blood cell mass occurring during pregnancy are illustrated in Figure 49.16. Plasma volume increases by about 50 per cent so that the total circulating blood

Figure 49.15 Uterine blood flow measured at various stages of pregnancy. (Based on data in F. Hytten and G. Chamberlain (1991) in *Clinical physiology in obstetrics*, Blackwell Scientific Publications, Boston.)

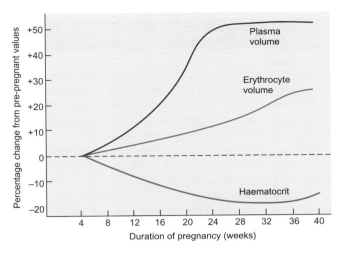

Figure 49.16 The changes that occur in plasma volume, erythrocyte volume, and haematocrit during pregnancy. (Based on data in R.M. Pitkin (1976) *Clinical obstetrics and gynecology* **19**, 489–513.)

volume is raised. However, red blood cell mass normally rises by only about 30 per cent and, as a result, there is often a fall in the haemoglobin content of the maternal blood. Published figures vary, but a typical drop in haemoglobin content might be from 13.3 to 12.1 g dl⁻¹ blood. Although some degree of dilutional anaemia is normal during pregnancy, if haemoglobin levels fall significantly (to less than ~11 g dl⁻¹), iron supplements may be required (see also Section 49.9).

Other changes in the composition of the maternal blood include a small increase in the white blood cell count and a fall in the platelet count (although leukocyte and platelet numbers generally remain within normal limits). Despite the fall in platelet count, pregnancy is sometimes regarded as a 'hypercoagulable state' because levels of fibrinogen and clotting factors VII to X are often increased. Consequently, there is an increased risk of deep vein thrombosis (see Box 25.1).

Changes in maternal arterial blood pressure are related to circulating volume and peripheral resistance

Maternal systemic arterial blood pressure shows some interesting changes throughout gestation as a consequence of changes in both blood volume and peripheral vascular resistance. Despite the increased cardiac output described above, arterial blood pressure often falls from the pre-pregnant value during the second trimester because the placental circulation is expanding and peripheral vascular resistance falls. A typical fall in mid-pregnancy would be about 0.66–1.32 kPa (5–10 mm Hg) for systolic pressure and 1.32–1.98 kPa (10–15 mm Hg) for diastolic pressure. As a result, there will be a small increase in the pulse pressure (the difference between the systolic and diastolic values). Arterial pressure may rise a little towards term and any significant increase, particularly if seen in conjunction with oedema or proteinuria (urinary excretion of proteins), may indicate a risk of pre-eclampsia, a potentially life-threatening

condition for both mother and fetus (see Box 49.3). For this reason, it is important that maternal blood pressure is monitored very carefully during the third trimester of pregnancy.

The maternal respiratory system is influenced by increasing fetal size

There are normally relatively few significant changes to respiratory function during pregnancy, but those changes that do occur may be attributed either to positional changes associated with increasing uterine size, or to the effects of placental progesterone. However, a number of respiratory parameters remain largely unchanged in a pregnant woman. These include vital capacity, respiratory rate, and inspiratory reserve volume. Furthermore, the partial pressure of oxygen in the arterial blood remains unaffected by pregnancy. Nevertheless, small reductions in the following respiratory volumes are commonly observed close to the end of pregnancy, probably because of the obstruction caused by the uterus as it pushes up under the diaphragm. Functional residual capacity, expiratory reserve volume, and residual volume all decrease by around 20 per cent, while the decrease in total lung capacity is much smaller at around 5 per cent.

By contrast, resting tidal volume may increase by as much as 40 per cent by the end of pregnancy and there may be a slight increase in thoracic circumference. Mild dyspnoea is commonly seen during exertion, particularly in late pregnancy, and this may be associated with a very slight respiratory alkalosis (plasma pH rising to about 7.44). Progesterone causes hyperaemia and oedema of the tissues of the respiratory tract which, in some women, leads to nasal stuffiness, obstruction of the Eustachian tube, and changes in the quality of the voice.

Maternal renal function and body fluid balance reflect cardiovascular changes

Many of the major changes in renal function and body fluid homeostasis occurring during pregnancy may be explained in terms of the cardiovascular changes discussed earlier. As cardiac output increases, there is a rise in renal blood flow and, as a result of this, an increase in glomerular filtration rate (GFR) of between 30 and 50 per cent. Peak GFR values are normally recorded at around weeks 16-24 of gestation. As renal function increases, the concentrations of both urea and creatinine in the maternal plasma fall from their pre-pregnancy values of 4–7 mmol l⁻¹ and 100 μmol l⁻¹ to around 3.6 mmol l⁻¹ and 60 μmol l⁻¹ respectively. Near to term, the expanding uterus starts to exert pressure on the inferior vena cava and renal blood flow may fall slightly. This effect is known as positional stasis and may lead to a slight drop in GFR in late gestation. As Figure 49.14 shows, the abdominal organs are displaced and compressed by the fetus as it enlarges and the increased pressure on the ureters may cause urine to back up rather than pass freely to the bladder. This effect, coupled with the direct relaxing effects of progesterone on smooth muscle, may cause dilatation of the ureters.

Many pregnant women experience an increase in the frequency of micturition. There are several reasons for this. In the first and

Box 49.3 Pre-eclampsia, eclampsia, and HELLP syndrome—serious disorders of pregnancy

Hypertensive disease in pregnancy is a major cause of perinatal morbidity and mortality. It displays a wide spectrum of severity from mild pre-eclampsia, with subtle, non-specific symptoms, through to full-blown eclampsia with life-threatening complications. A third, related condition of pregnancy is the so-called HELLP syndrome, in which there are damaging alterations in liver function and in certain properties of blood cells.

Pre-eclampsia (PE) occurs in around 3 per cent of pregnancies. It is characterized by high blood pressure, the appearance of protein in the urine (proteinuria), and often oedema. In the majority of cases it begins after week 20 of gestation and it is most common in the final trimester. Mild pre-eclampsia is defined as an arterial BP of 18.6/12 kPa (140/90 mmHg) or an increase of 4 kPa (30 mmHg) in systolic BP and of 2 kPa (15 mmHg) in diastolic BP compared with the pre-pregnancy values. The pre-eclampsia is classified as severe if an arterial BP of 21.3/14.6 kPa (160/110 mmHg) is maintained even during bed rest, and if proteinuria reaches a value of 5 g/24 hours.

Although pre-eclampsia has been the subject of extensive research, the underlying pathophysiology of the disease remains poorly understood and its precise causes unknown. Although both maternal cardiac output and peripheral resistance are increased in pre-eclampsia (both changes that could contribute to hypertension), the placenta is believed to have the primary role since delivery of this organ results in reversal of the signs and symptoms of pre-eclampsia. The disease also occurs more frequently with twin pregnancies and hydatidiform disease, in which there is an increased amount of trophoblastic tissue.

Microscopic examination of the trophoblastic tissue in patients with pre-eclampsia reveals some interesting changes. In normal pregnancies, the uterine spiral arteries are invaded by trophoblast at 14–16 weeks gestation. As a result of this invasion, the arteries are converted into low-resistance, high-volume vessels that do not respond to vasoconstrictors. These changes maximize perfusion of the maternal side of the placenta. In pre-eclampsia, trophoblastic invasion of the spiral arteries appears to be incomplete and the vessels remain responsive to vasoconstrictors. Some studies have also shown evidence of endothelial dysfunction in women with pre-eclampsia. There is a reduction in the secretion of endothelial vaso-

dilator substances but an increase in the concentration of the potent vasoconstrictor endothelin. These changes may further exacerbate the hypertension and hasten the progression of damage to the capillary beds of the placenta and elsewhere in the maternal circulation. The net effect of all these changes is likely to be a reduction in placental blood flow. Indeed, pre-eclampsia is frequently associated with intrauterine growth retardation of the fetus, presumably as a result of the reduction in nutrient supply.

Although pre-eclampsia can affect any pregnancy, there is an increased risk in women at the extremes of age for pregnancy, in those who are pregnant for the first time, and in women who have had normal pregnancies but who are now pregnant with the child of a new partner. These findings suggest that genetic or immunological factors may have a role in the development of the disease.

In a small proportion of cases, pre-eclampsia progresses to the potentially fatal condition known as eclampsia in which extreme hypertension results in changes in intracranial pressure, seizures, and coma. There is a significant risk to the mother of cerebral haemorrhage, renal failure, and abruptio placentae (in which the placenta starts to detach from the wall of the uterus, causing massive haemorrhage). Eclampsia occurs in roughly 0.05 per cent of all deliveries and in such cases the fetal mortality rate is between 13 and 30 per cent, while the maternal mortality rate is from 8 to 36 per cent depending upon the speed with which a diagnosis is made and the facilities available for treatment and immediate delivery of the infant.

Occasionally, pre-eclampsia may progress to another illness known as HELLP syndrome in which there is haemolysis (H), elevated liver function tests (EL) and low platelets (less than 100×10^9/litre) (LP). This condition, which is characterized by nausea, malaise, and abdominal pain, affects up to 10 per cent of women with severe pre-eclampsia and 30–50 per cent of those who have progressed to eclampsia. In a few patients, however, the HELLP syndrome presents without any prior hypertension or proteinuria. Although the symptoms can be vague and relatively mild, the underlying hepato-cellular necrosis and bleeding problems associated with thrombocytopenia are potentially life-threatening even in those women who do not have severe hypertension.

second trimesters of pregnancy, the increased urinary flow rate probably reflects increased renal function. In later weeks, as the fetus increases in both size and activity, there is likely to be increased pressure on the bladder, causing urgency and discomfort.

Many of the hormones that act on the renal tubules alter their output during pregnancy

In response to increased oestrogenic stimulation in pregnancy, there is an increase in the secretion of renin from granular cells of the afferent arteriole which, in turn, elicits a rise in the plasma level

of angiotensin II. This might be expected to produce an increase in arterial blood pressure. Fortunately, there is also a reduction in the sensitivity of blood vessels to angiotensin, which seems to cancel out any hypertensive effect. However, the increased concentration of angiotensin will enhance the secretion of aldosterone from the adrenal cortex. By the end of pregnancy, aldosterone secretion may be 6–8 times higher than in the non-pregnant state. Aldosterone stimulates the reabsorption of salt and water from the distal tubular fluid and this more than balances the increase in filtered water and salt resulting from the increased GFR.

Progesterone, which is secreted by the placenta in very large amounts, is both natriuretic and potassium sparing in its effect on the renal tubules. This is in direct contrast to the effects of aldosterone, which favour potassium excretion and sodium retention. Consequently, the two steroid hormones largely offset one another's actions during pregnancy. If anything, there is a small degree of both potassium and sodium retention which may contribute to the expansion of total body fluid volume that occurs during gestation.

Total body water increases by 6–8 litres during pregnancy

The increased volume of water in the body of a pregnant woman contributes significantly to the overall weight gain occurring during a normal pregnancy (see Section 49.9). Extracellular fluid volume is increased by about 3 litres at term, roughly half of which is in the plasma and the rest in the interstitial fluid. In some women, this expansion of the extracellular fluid compartment causes significant oedema. Provided it is not associated with hypertension or proteinuria, a small degree of oedema is considered a normal consequence of pregnancy.

In addition to the expansion of the extracellular compartment, plasma osmolality may fall during gestation by as much as 10 mOsm kg^{-1}. This fall may be accounted for, at least in part, by the drop in urea and creatinine levels described earlier. In the non-pregnant state, such a fall would normally be countered by a reduction in the secretion of ADH so that increased amounts of dilute urine would be excreted. During pregnancy this does not appear to occur, possibly because the hypothalamic osmoreceptors are reset at a lower value throughout gestation. Many pregnant women experience an increased thirst. This may be the result of increased circulating levels of angiotensin II, a hormone known to stimulate thirst (see Chapter 40), and an increased fluid intake will contribute to the increased extracellular volume of pregnancy.

Minor changes to gastrointestinal function are common features of pregnancy

Nausea and vomiting are experienced by the majority of pregnant women, particularly in the first 12–14 weeks of pregnancy. After this time, symptoms usually become less severe and often disappear completely, but in rare cases, problems may persist throughout gestation. This severe condition which is known as **hyperemesis gravidarum** affects around 1 in 200 pregnant women. In the most serious cases it can lead to dehydration, acid-base imbalances, and weight loss. The precise cause of nausea and vomiting in early pregnancy is not clear, but sickness seems to parallel the rising level of hCG secretion by the syncytiotrophoblast in the first 10–12 weeks after fertilization (see Section 49.4). Rising levels of placental steroids may also contribute to the problem by causing the stomach to empty more slowly. Pregnant women also have a heightened sense of smell, so various odours, such as strong-smelling foods, perfumes, and cigarette smoke, may cause waves of nausea.

Constipation is another common feature of pregnancy, and while it may occur at any stage, it is most frequently experienced during late pregnancy as the enlarging uterus presses against the rectum and lower colon. Certain other factors may also contribute to constipation during gestation. These include relaxation of the smooth muscle of the colon under the influence of placental progesterone, and a reduction in the water content of the stool brought about by increased water absorption from the colon in response to increased concentrations of angiotensin II and aldosterone.

Many aspects of gastrointestinal function are reduced or slowed during pregnancy, principally as a result of relaxation of the smooth muscle of the gut wall in response to progesterone. Gastric emptying, for example, occurs more slowly and there is an overall decrease in gastrointestinal motility. Relaxation of the lower oesophageal sphincter, particularly at night in late pregnancy, may result in acidic gastric contents being regurgitated into the oesophagus, causing belching and heartburn. Furthermore, the pressure of the fetus on the diaphragm and intra-abdominal organs, particularly when the mother is lying down, adds to the discomfort and the likelihood of reflux.

The secretion of gastric acid decreases and peptic ulcers may show an improvement during pregnancy. Gallstones are, however, more common because of the increasingly sluggish flow of bile caused by relaxation of the smooth muscle of the gall bladder. Increased levels of plasma progesterone may lead to swelling and oedema of the gums, which tend to become spongy and bleed more easily.

The functions of most endocrine glands are altered during pregnancy

The specific actions of placental progesterone, oestrogens, hCG, and hPL are described in Section 49.6. However, pregnancy also alters the functions of most of the other endocrine tissues throughout the body. These changes occur partly through the actions of the placental hormones themselves, and partly because many hormones (particularly steroids) circulate in the blood in combination with plasma proteins, whose levels are often altered in pregnancy.

The placenta secretes a variety of hormonal factors whose actions are similar to those of the trophic hormones of the anterior pituitary. One of these, which is similar to anterior pituitary TSH (see Chapter 21), increases thyroid function and stimulates the secretion of thyroxine. Occasionally this may lead to symptoms reminiscent of hyperthyroidism—tachycardia, palpitations, excessive sweating, and anxiety—although in most women, levels of thyroxine binding globulins in the plasma also increase so that free plasma thyroxine levels remain unchanged.

The placenta also secretes a substance similar to ACTH (adrenocorticotrophic hormone), which stimulates the output of adrenal cortical hormones, and a melanocyte stimulating hormone (MSH), which causes increased pigmentation of the skin. The latter probably accounts for the relatively common so-called mask of pregnancy (melasma), in which a blotchy, brownish pigment appears over the forehead and cheeks. Pigmentation of the **areolae** of the nipples also increases and there is sometimes a line of darkness down the midline of the lower abdomen. Furthermore, MRI scans show that the maternal pituitary gland itself increases in size during pregnancy. The enlarged gland secretes additional amounts of both ACTH and MSH, whose effects are added to those of the placental factors. In rare instances, enlargement of the pituitary during pregnancy may be sufficient to cause some visual disturbances as a result of increased pressure on the optic chiasm, which lies very close by.

Secretion of growth hormone (GH), follicle stimulating hormone (FSH), and luteinizing hormone (LH) by the anterior pituitary is reduced during pregnancy. This reduction is mediated by the negative feedback effects of hPL (a powerfully somatotropic hormone) on GH and of the placental sex steroids on FSH and LH. Prolactin secretion, by contrast, is enhanced, showing an eightfold increase by the end of gestation. This hormone is partly responsible for the preparatory changes which take place within the mammary glands in readiness for milk synthesis and secretion. Its presence also seems to be required for oestrogens and progesterone to exert their effects on the mammary glands (discussed later in the chapter).

Parathyroid hormone secretion increases during pregnancy

The fetus represents a significant drain on maternal stores of calcium, largely because of the demands of its developing skeleton. As a result of this, maternal plasma calcium is at risk of being reduced. To prevent any such fall, there is normally an increase in the rate of secretion of parathyroid hormone during gestation. This mild degree of hyperparathyroidism augments plasma levels of 1,25 dihydroxycholecalciferol (calcitriol—see Chapter 22), which in turn stimulates the intestinal absorption of calcium. These changes are especially important during the final weeks of pregnancy.

Gestational diabetes is a relatively common condition of late pregnancy

In early pregnancy, tissues typically show an increased sensitivity to insulin. Consequently, plasma glucose may fall slightly. Later in gestation, insulin sensitivity falls and plasma glucose may rise. These changes reflect the initial increase in the requirement for glucose of the maternal tissues and the later needs of the developing fetal tissues. Glycosuria is often seen in pregnancy and may be explained by the combination of raised plasma glucose, increased GFR, and a reduction in the tubular reabsorption of glucose (see Chapter 39).

Gestational diabetes occurs in between 1 and 3 per cent of pregnancies. In this condition, carbohydrate intolerance of varying severity develops and often (though not invariably) resolves once the baby has been delivered. Some of the women who develop gestational diabetes are obese, hyperinsulinaemic, and insulin-resistant while others are relatively thin but insulin-deficient. In either case, they are unable to respond effectively to the metabolic stress of pregnancy. Furthermore, the increase in plasma glucose caused by hPL may increase the risk of developing gestational diabetes in susceptible women.

The fetus of a diabetic mother with hyperglycaemia will also be hyperglycaemic, since glucose is in equilibrium between the fetal and maternal sides of the placental circulation. Not surprisingly,

therefore, the fetus of a diabetic mother often has a higher than normal weight-for-dates (which can pose difficulties during labour and delivery). Also associated with maternal gestational diabetes is an increased risk of fetal respiratory distress due to retarded lung maturation and surfactant production. The incidence of fetal abnormalities is also increased. Early recognition of gestational diabetes and good maternal plasma glucose control are therefore vital if such problems are to be minimized. Glucose-tolerance tests are routine in many antenatal clinics.

Summary

Throughout pregnancy, changes take place in all the major systems of the mother's body to create favourable conditions for fetal development and prepare for the delivery and subsequent nourishment of the baby. Other changes occur as the inevitable consequence of the displacement of the maternal abdominal organs by the growing fetus.

During gestation there is a rise in cardiac output that meets the increasing demands of the uterus and placenta for nutrients and waste disposal, while arterial blood pressure often falls in midpregnancy. Changes in maternal respiratory function result from the increasing size of the uterus, which compresses the diaphragm. Changes in cardiovascular function affect renal function, and glomerular filtration is increased along with the output of renin and aldosterone. Total body water increases and plasma osmolality falls. Minor changes in gastrointestinal function such as nausea, vomiting, and constipation are often seen in pregnancy. Changes in the insulin sensitivity of maternal tissues take place during gestation, with an increase in sensitivity in the early weeks, and a fall later on. Gestational diabetes occurs in 1–3 per cent of pregnancies.

49.9 Nutritional requirements of pregnancy

Adequate nutrition both prior to conception and during pregnancy itself is essential in order to ensure that optimal intrauterine conditions are maintained for fetal development and that the health of the mother is not compromised by the additional metabolic and nutritional demands of the growing fetus. Poor maternal nutrition is associated with an increased risk of low birth weight, fetal abnormalities, and neonatal mortality.

Weight gain during pregnancy is normally between 7 kg and 14 kg, averaging around 0.4 kg each week during the second and third trimesters. Twin pregnancies may result in a total weight gain of between 16 and 20 kg. Although women with a high body mass index (BMI in excess of 25 kg m^{-2}) are generally advised to gain less weight during pregnancy than those who are underweight (BMI less than 20 kg m^{-2}), even obese women will need to gain a minimum amount of weight (about 6 kg) in order to optimize the chances of delivering an infant of healthy birth weight. Excessive maternal weight gain is, however, undesirable. It is associated with

Table 49.1 The distribution of maternal weight gain at 40 weeks gestation

	Weight (kg)
Fetus	3.3–3.5
Placenta	0.65
Additional blood volume	1.3
Amniotic fluid	0.8
Weight gain of uterus and breasts	1.3
Additional fluid retention and fat deposits	4.2–6.0
Total	11.5–13.5

prolonged gestation (late delivery), problems during labour, and an increased risk that delivery will need to be performed by Caesarean section. Table 49.1 shows figures for the distribution of weight gained during a typical pregnancy, in which maternal weight increases by around 12 kg. Of the total weight gain, the fetus contributes around 3.5 kg, the placenta and amniotic fluid around 1.5 kg, and the breasts 0.5–1 kg. Increases in maternal fat and extracellular fluid account for the remainder.

During pregnancy there is an increase in the daily calorific requirement

In order to supply the demands of the fetus and to maintain adequate maternal nutrition, an increased daily energy intake of about 1050–1250 kJ (250–300 kcal) is required. This represents an increase of about 15 per cent in the total energy requirement. In very late pregnancy this need may be somewhat reduced because of a reduction in maternal energy expenditure.

Many women experience an increase in appetite during pregnancy. Contributory factors could include the presence of high circulating concentrations of progesterone and the fall in plasma glucose concentration, which is characteristic of early pregnancy (see Section 49.8). There may also be odd cravings, particularly for highly flavoured foods, perhaps related to dulled taste sensation, as well as aversions to certain foods and flavours. True **pica** (craving for inappropriate foods such as coal) is, however, rare in pregnancy.

During early pregnancy the mother's body prepares for the metabolic demands to come

During early pregnancy the metabolism of the mother is modified to ensure that adequate reserves are available to meet the requirements of both mother and fetus during later gestation. During the first half of a typical pregnancy about 3 kg of maternal fat is laid down, and this will provide a store of energy for the final trimester during which fetal growth is particularly rapid. In addition, the maternal tissues become more sensitive to insulin so that carbohydrate loads are readily assimilated. As discussed earlier, a small decrease in plasma glucose is often seen after about week 6 of gestation. At the same time, protein synthesis increases, the net effect of which is growth of the uterus, breasts, and essential musculature of the mother.

During later pregnancy, the metabolism of the mother adapts to accommodate the growing fetal demands

As the fetal requirements for oxygen and nutrients rise and the sheer physical size of the fetus increases during the second half of gestation, the metabolism of the mother enters a state similar to that occurring during starvation. A full-term fetus metabolizes up to 25 g of glucose each day and it is essential that the mother's plasma glucose concentration remains adequate to meet these needs as well as those of her own nervous system. In late pregnancy, maternal tissues show a fall in insulin sensitivity. This ensures that the utilization of glucose by insulin-dependent tissues is reduced so that glucose is spared for use by the fetus and the maternal CNS (remember that glucose uptake by neurons is not insulin-dependent). At the same time, mobilization of lipids is facilitated, possibly by hPL, which provides an alternative source of energy for the mother. Increased plasma levels of triglycerides are seen, as the hepatic synthesis of very low-density lipoprotein increases under the influence of placental oestrogens. Some of these triglycerides are stored by the mammary tissue in preparation for milk synthesis during lactation. During late gestation, the mother requires a small quantity of additional protein (no more than 6–10 g day^{-1}) mainly to provide the necessary amino acids for the synthesis of fetal protein. By the time of its birth, the fetus will have accumulated 400–500 g of protein. In most cases, the additional protein requirements of pregnancy are provided by a normal balanced diet.

The maternal requirement for some micronutrients increases during pregnancy

The need for certain vitamins and minerals also increases by a small amount during pregnancy, both to satisfy the demands of the fetus and to prepare the mother's body for lactation. A normal mixed diet will usually provide adequate concentrations of these micronutrients but in some women (for example those following a strict vegetarian or vegan diet) nutrient supplements may be necessary. An example of such a nutrient is vitamin B$_{12}$, which is found only in animal products. Table 49.2 shows the recommended daily nutrient intakes during pregnancy, although these recommendations often vary widely between different countries.

Folic acid is needed for normal cell proliferation and is now recognized as a very important dietary requirement of pregnancy—particularly during the first weeks after conception. Folate deficiency has been linked to an increased risk of abnormalities in the closure of the neural tube, which normally occurs between weeks 3 and 6 of gestation. Neural tube defects include **anencephaly** (in which the cerebral hemispheres fail to develop normally) and **spina bifida**, in which there is defective fusion of the vertebral arches, most often in the lumbar region. Women are encouraged to supplement their folic acid intake both in the weeks prior to conception and throughout pregnancy, by eating leafy green vegetables, wheat grains, and legumes.

A vitamin A intake of around 700 µg day^{-1} is recommended for adult females (see Table 49.2) and, in contrast to most vitamins, the

Table 49.2 Recommended daily protein and micronutrient intake in pregnancy

Nutrient	USA	Canada	UK	Australia
Protein (g)	60	75	51	51
Vitamin A (µg)	800	800	700	750
Vitamin D (µg)	10	5	10	Not set
Vitamin E (mg)	10	8	Not set	7
Vitamin C (mg)	70	40	50	60
Vitamin B$_1$ (thiamin) (mg)	1.5	0.9	0.9	1.0
Vitamin B$_2$ (riboflavin) (mg)	1.6	1.3	1.4	1.5
Vitamin B$_3$ (niacin or nicotinic acid) (mg)	17	16	13	15
Folate (µg)	400	385	300	400
Vitamin B$_{12}$ (µg)	2.2	1.2	1.5	3.0
Calcium (g)	1.2	1.2	0.7	1.1
Magnesium (g)	0.32	0.25	0.27	0.30
Iron (mg)	30	13	14.8	22–36
Zinc (mg)	15	15	7	16
Iodine (µg)	175	185	140	150

requirement for vitamin A remains unchanged in pregnancy. Indeed, there is some concern that a high daily intake of this vitamin may be teratogenic, leading to abnormalities of the fetal nervous system and heart. This effect may be linked with the role of vitamin A in cell differentiation. A normal intake of vitamin A during pregnancy, however, has been shown to be important in reducing the likelihood of transmission of the human immunodeficiency virus (HIV) from mother to fetus.

There is an increased need for calcium, iron, zinc, and magnesium during pregnancy

During the third trimester the fetus needs ~0.3 g of calcium a day to allow calcification of the skeleton. Infants are normally born with about 28 g of calcium, which must come from the maternal reserves. During pregnancy, maternal calcium absorption by the small intestine increases under the influence of raised levels of calcitriol, while urinary calcium loss is reduced by the increased secretion of parathyroid hormone. Nevertheless, to ensure that the increased calcium requirement is met, a pregnant woman needs to increase her daily intake of calcium by about 70 per cent (i.e. to about 1200 mg day^{-1}). If dietary calcium is not increased during pregnancy, the fetus will draw on calcium stored in the maternal skeleton. This may increase the mother's susceptibility to osteoporosis in later life, particularly if repeated in several pregnancies.

There is much debate surrounding the requirement for iron during pregnancy. Some studies have indicated that iron deficiency may be associated with low birth weight and premature delivery. Others indicate that the fetal demand for iron can be met by the maternal stores provided that the diet contains sufficient amounts of meat and vitamin C (which enhances the absorption of iron in

the gut) and that the mother was not iron-deficient prior to her pregnancy. The fetus and placenta use about 300 mg of iron during a typical pregnancy, while the increased red blood cell mass of the mother requires a further 500 mg. The average Western diet contains 10–15 mg of iron day^{-1}, of which about 10 per cent is normally absorbed. The efficiency of absorption appears to rise during pregnancy, reaching around 66 per cent by week 36 of gestation. This increase contributes to the ability of the maternal iron stores to supply the needs of the fetus. If, however, absorption of dietary iron falls short of the demands of pregnancy, maternal iron stores will become depleted and iron-deficiency anaemia may occur. This can be avoided by regular monitoring of the maternal haemoglobin content so that iron supplements can be given if necessary.

The normal requirement for dietary zinc in a non-pregnant woman is around 7 mg day^{-1}. Zinc is a constituent of a large number of enzymes including carbonic anhydrase, reverse transcriptase, and DNA/RNA polymerase. It therefore plays an important part in a number of metabolic processes including the synthesis of proteins and nucleic acids. Zinc is also essential for the synthesis and activity of insulin. Not surprisingly, therefore, there is a small increase in the requirement for this trace element during pregnancy. In most women, a normal mixed diet containing meat and fish will provide sufficient zinc to support the fetoplacental unit, but strict vegans may be unable to absorb enough zinc from their food. Women taking iron supplements may also become deficient in zinc because iron:zinc ratios greater than 3:1 have been shown to interfere with the intestinal absorption of zinc.

Adequate maternal intake of magnesium is needed for normal embryonic and fetal development. Maternal magnesium deficiency has been associated with premature labour and has also been implicated in the pathogenesis of Sudden Infant Death

Syndrome (SIDS). High doses of magnesium sulphate have been shown to be effective in preventing seizures associated with eclampsia and pre-eclampsia.

Summary

Adequate maternal nutrition during pregnancy is required to ensure the continued health of the mother, and to minimize the risks of fetal abnormalities and low birth weight. In most cases a normal mixed diet will be sufficient to meet the demand for extra micronutrients, although occasionally supplements, particularly of vitamin B_{12} and iron, are needed. Weight gain during pregnancy is typically 7–14 kg.

In early pregnancy lipogenesis is favoured and the insulin sensitivity of tissues is increased. Protein synthesis also increases as breast and uterine tissues develop. In late gestation, insulin sensitivity declines to maintain an adequate concentration of plasma glucose for placental transfer. Lipids are mobilized to provide an alternative energy source. The requirement for certain vitamins is enhanced, while there is an increased utilization of calcium, iron, and zinc.

49.10 Lactation—the synthesis and secretion of milk after delivery

Most placental mammals are born at a relatively advanced stage of development. This is the result of the extended gestation period, which is made possible by the direct link between the maternal and fetal circulations established in the first weeks of pregnancy. Consequently, the infant makes considerable nutritional demands upon its mother. While the fetus develops within its mother's uterus, it receives all the nutrients it requires via the placenta. Once it has been delivered, however, the baby needs a regular and plentiful supply of milk. The formation and synthesis of milk is also called **galactopoiesis**. Although in certain human communities, particularly in Western society, bottle-feeding with powdered 'formula milk' offers a suitable alternative, in many parts of the world the mother's breast is the only source of nourishment for the newborn infant. To ensure that sufficient milk of adequate calorific value is produced from the very start of lactation, preparatory changes must occur within the mammary glands during pregnancy. These changes are regulated by hormones from the placenta, pituitary, and adrenal glands.

The non-pregnant mammary gland is incapable of lactation

Until puberty, the immature breast consists almost entirely of ducts known as **lactiferous ducts**. Around puberty, gonadotrophins begin to be secreted in larger amounts from the anterior pituitary and, under their influence, the ovaries start to increase their production of oestrogenic hormones. These steroids initiate further breast development—in particular, the ducts begin to sprout and become more highly branched (see Figure 49.17). Once menstruation has commenced, the progesterone secreted during the luteal phase of each cycle stimulates the formation of small, spherical masses of granular cells at the end of each duct. These are known as immature **alveoli** and are the cells which will develop into the milk-secreting alveoli of the lactating gland in the event of a successful pregnancy. Once the mammary glands have been exposed successively to oestrogens and progesterone during the follicular and luteal phases of many cycles, the non-pregnant breast is said to be fully developed, as shown in Figure 49.17.

Throughout this adolescent period, there is deposition of fat and connective tissue, which, together with the ductal growth, brings about a considerable increase in breast size and results in a gland that is highly developed even though no pregnancy has yet occurred. This is in marked contrast to most other mammals, including non-human primates, in which very little mammary growth is seen at all until mid- or late pregnancy. Figure 49.18 shows the basic organization of the human mammary gland and illustrates the way in which the ducts separate the gland into lobes. In each gland there are between 15 and 20 lobes separated by fat. Each lobe consists of clusters of granular cells at the ends of the lactiferous ducts. The ducts dilate near to the areola (the area of brownish pigment surrounding the nipple) to form lactiferous sinuses, each of which runs up into the nipple and opens onto its surface. Dotted around the areola are small sebaceous glands called Montgomery glands.

Although the mammary gland is fully developed by the end of puberty, small changes do take place during each menstrual cycle, as first oestrogens and then progesterone influence the mammary tissue. Occasionally there is a small amount of secretory activity during the luteal phase of the cycle, and there is often an increase in both size and weight of the gland throughout the pre-menstrual period due to the retention of fluid. Despite its comparatively advanced stage of development, however, the breast is not at this stage capable of large-scale milk production (**lactogenesis**). Before that can happen, further development, particularly of the alveoli, must take place. This occurs during pregnancy, under the influence of a variety of hormones.

The development of the mammary gland during pregnancy

Most of the growth and structural changes that are essential for successful lactation take place during the first four months or so of pregnancy. By mid-term, the mammary gland is fully developed for milk secretion. The lobular ductal-alveolar system that was laid down during adolescence undergoes hypertrophy, the ducts proliferate further, and the alveoli mature. As the alveoli mature the balls of granular cells become hollowed out so that they surround a central lumen, which is drained by a branch of one of the lactiferous ducts—see Figure 49.17. The hormones thought to be responsible for these changes are the placental steroids oestradiol and progesterone and the placental peptide hPL. Progesterone in particular seems to be required for the alveolar changes characteristic of early pregnancy. In addition to the placental hormones, pituitary growth hormone and prolactin may also play a role in breast development (i.e. have a mammotrophic effect), although their

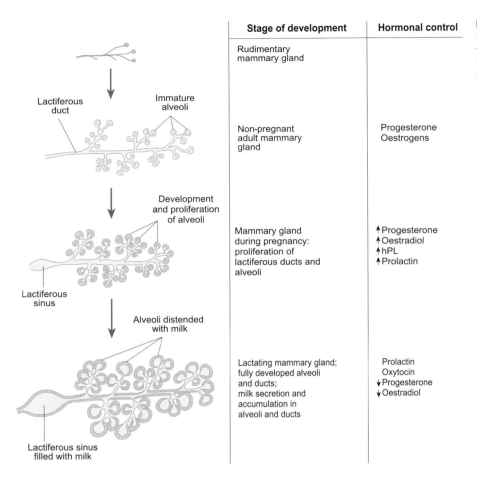

Stage of development	Hormonal control
Rudimentary mammary gland	
Non-pregnant adult mammary gland	Progesterone Oestrogens
Mammary gland during pregnancy: proliferation of lactiferous ducts and alveoli	↑Progesterone ↑Oestradiol ↑hPL ↑Prolactin
Lactating mammary gland; fully developed alveoli and ducts; milk secretion and accumulation in alveoli and ducts	Prolactin Oxytocin ↓Progesterone ↓Oestradiol

Figure 49.17 The development of the mammary gland from its immature state to lactation. The pituitary and steroid hormones that are involved at the various stages are shown on the right.

contribution at this stage is not clear. Additional adipose tissue is deposited between the lobules of the gland during early pregnancy, adding further to the size and weight of the breast. Figure 49.19 shows the appearance of the breast tissue as it develops during pregnancy and during lactation.

The alveoli are the primary sites of milk production

The mature alveoli develop under the influence of placental progesterone, prolactin, and hPL. Figure 49.20 is a highly simplified diagram showing the basic organization of a mature alveolus. The alveolar wall is formed by a single layer of epithelial cells, whose shape can vary from low cuboidal to tall columnar depending on the volume of secretory material filling the central lumen. During pregnancy, but before the onset of full-scale lactation, the epithelial cells are columnar in appearance. After delivery, once milk production is underway, the cells are usually squashed flat by material within the lumen. These epithelial cells are the cells that synthesize and secrete the constituents of milk, and they show all the classical characteristics of secretory cells: they possess microvilli on their luminal surfaces and their cytoplasm is rich in mitochondria, Golgi membranes, rough endoplasmic reticulum, secretory granules, and lipid droplets. Adjacent alveolar cells are connected by junctional complexes near to the luminal surfaces, and specialized

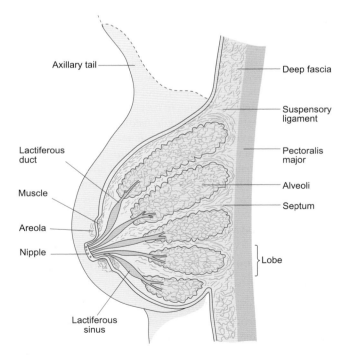

Figure 49.18 Sectional view of the mammary gland during pregnancy. Note the development of the alveoli.

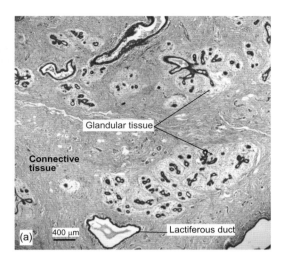

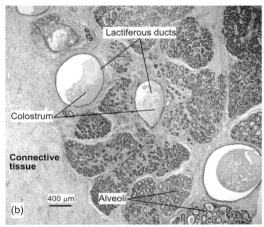

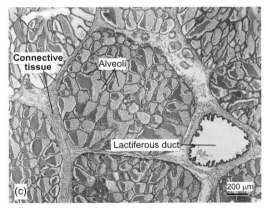

Figure 49.19 Histological sections illustrating the differences between resting breast tissue (a) and its appearance during pregnancy (b) and lactation (c). In panel (a) the alveoli are immature and there is a large amount of connective tissue between the lobules and around the glandular cells. In panel (b) note how the alveoli have proliferated and developed during pregnancy. Mature alveoli can be seen filled with colostrum in panel (b). During lactation, the alveoli have developed further to occupy most of the breast tissue and are now filled with milk, which appears as dark staining patches within the alveolar lumen (c). Bars, 400 μm (a), (b), 200 μm (c).

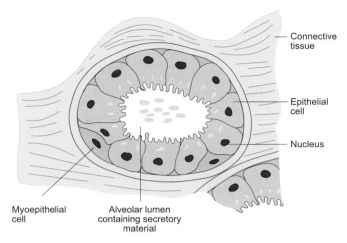

Figure 49.20 A diagrammatic representation of a cross-section of a mature (lactiferous) alveolus.

myoepithelial cells are found between the basement membrane and the secretory alveolar cells. These myoepithelial cells are, as their name suggests, contractile, and they are important for moving milk into the lactiferous ducts prior to ejection from the nipple when the baby suckles.

Summary

Lactation is the synthesis and secretion of milk by the mammary glands. The ovarian steroid hormones secreted from puberty are responsible for subsequent development of the mammary glands. Progesterone stimulates development of the spherical masses of cells that will produce milk—the alveoli. Under the influence of placental hormones (oestrogens, progesterone, and hPL), the alveoli mature and the breast acquires the potential for milk secretion.

49.11 Lactation is triggered by the fall in steroid secretion that follows delivery

Although the breast is fully developed for milk production by the middle of pregnancy, no significant lactogenesis takes place until the infant has been delivered. The endocrine changes that follow delivery are necessary to activate the prepared gland and trigger the synthesis and secretion of milk.

It is well established that the primary lactogenic hormone is prolactin, secreted by the anterior pituitary, and that high levels of prolactin must be maintained for sustained lactation. It is also known that this hormone is secreted in large amounts throughout gestation. Why then does lactation not occur during this time? The most widely accepted explanation is that the high circulating levels of the placental steroids, oestrogens, and progesterone exert a direct inhibitory effect on the secretory activity of mammary tissue. The gland therefore remains unresponsive to high levels of prolactin until the baby has been born. Figure 49.21 shows that while placental oestrogen and progesterone levels drop dramatically following delivery, prolactin levels remain high and are then able to initiate milk production by the fully prepared breast.

The composition of human milk changes gradually over the first weeks after delivery

The composition of the milk produced by the mammary gland varies with the time that has elapsed since parturition. So-called 'mature milk' is not secreted until about two or three weeks post-partum. Prior to this time, fluids of varying composition are produced. For the first week or so after delivery, a fluid called **colostrum** is secreted, at a rate of around 40 ml a day. Colostrum is a sticky, yellowish fluid, which, while relatively low in fats, lactose, and some B vitamins, is rich in protein, minerals, and the fat-soluble vitamins A, D, E, and K. It also contains significant quantities of immunoglobulins (IgAs) that provide the newborn infant with some resistance to infection. During the second and third weeks after birth, the composition of the fluid secreted gradually changes (see Table 49.3). Although the proportion of immunoglobulins and other proteins decreases, the milk becomes much richer in fats and sugars. Its calorific value increases as a result. At this time the fluid is known as 'transitional milk', but by the time the baby is three weeks

Table 49.3 The composition of human breast milk

	Colostrum	Transitional milk	Mature milk
Total fats (g l⁻¹)	30	35	45
Total protein (g l⁻¹)	23	16	11
Lactose (g l⁻¹)	57	64	71
Total solids (g l⁻¹)	128	133	130
Calorific value (MJ l⁻¹)	2.81	3.08	3.13

Note: These values are approximate as the composition changes both during a single feed and during the course of the day. In general, the fat content rises from the beginning to the end of a feed.

old, the milk has attained its 'mature' composition—it is high in fats, sugars, and essential amino acids, is iso-osmotic with plasma, and has a calorific value of about 3.1 MJ l⁻¹ (75 kcal per 100 ml).

The synthesis of breast milk is controlled by prolactin and insulin

Most of the fat in human milk consists of triglycerides. These are synthesized from glucose and fatty acids within the alveoli of the mammary tissue under the control of prolactin and insulin. The epithelial cell membranes of the alveoli are especially rich in prolactin receptors. Prolactin is also believed to stimulate the secretion of lipid from the cells into the central lumen of the alveolus. The mechanism by which this secretion takes place is rather interesting. The lipid is manufactured in the smooth endoplasmic reticulum of the alveolar cell, and lipid droplets migrate towards the luminal surface of the cell. As the microdroplets fuse within the cytosol, they increase in size. Once a droplet reaches the luminal surface, it pushes out against the cell surface membrane, causing a bulge. The area behind the lipid droplet gradually thins, and eventually the membrane 'pinches off' so that the membrane-bound droplet is released into the lumen.

The major milk proteins are casein, α-lactalbumin, and lactoglobulin

The three major milk proteins have both nutritional and immunological significance, while α-lactalbumin has an additional and specific role in the synthesis of milk sugar. The basic processes of milk protein secretion and synthesis are similar to those occurring in other protein-secreting tissues such as the pancreas and liver. Amino acids reach the mammary tissue via the circulation and pass from the blood into the alveolar cells via specific carrier systems. The milk proteins are synthesized in the usual way by the endoplasmic reticulum and Golgi membranes and are then packaged into vesicles, which bud off from the Golgi apparatus into the cytoplasm of the alveolar cell. These vesicles or granules move to the luminal surface, possibly by the action of microtubules, and release their contents into the alveolar lumen by the process of exocytosis. The release of protein vesicles, like the budding off of the lipid droplets, is controlled by prolactin. While the release of lipid droplets results in a loss of cell membrane as they pinch off from

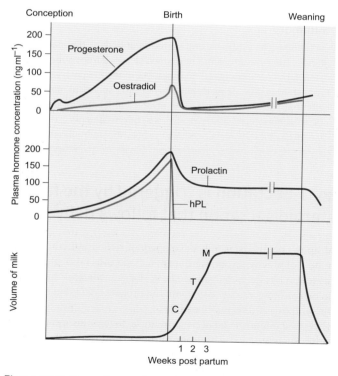

Figure 49.21 The patterns of secretion of hormones before and after birth in relation to their role in the control of lactation. When milk secretion is first initiated, colostrum (C) is secreted which gradually undergoes a transition (T) until mature milk is produced (M). Milk is not secreted until the level of steroid hormones falls following delivery, despite the high level of prolactin that prevails in the latter stages of pregnancy.

the alveolar cells, the exocytotic release of vesicles containing protein adds to the cell membrane by fusion of the vesicle membrane with the plasma membrane.

Lactose is the most abundant milk sugar

Lactose is synthesized within the Golgi apparatus of the alveolar epithelial cells. Its synthesis is dependent on the prior production of α-lactalbumin, which is made in the endoplasmic reticulum and passed to the Golgi. Once there, it combines with galactosyltransferase, an enzyme present within the Golgi membranes. This enzyme system metabolizes blood glucose to form lactose, which is packaged, together with the proteins, into vesicles that bud off from the Golgi and undergo exocytotic release into the alveolar lumen, as described above. The enzyme system comprising α-lactalbumin and galactosyltransferase is stimulated by prolactin but inhibited by the high levels of progesterone circulating throughout gestation.

Human breast milk contains more than 50 different oligosaccharides, most of which are synthesized from lactose. In addition to providing the source of many of the other milk sugars, lactose also promotes the growth of intestinal flora which is very important to the newborn infant. Furthermore, galactose, one of the digestion products of lactose, is an essential component of the myelin which surrounds many nerve axons (see Chapter 7).

Nutritional requirements of lactating women

During lactation, a woman needs to ingest sufficient nutrients to provide for her own bodily requirements together with those needed by the infant for its growth and development. Obviously, the extra requirement will depend on the amount of milk she is producing. As a guide, an average baby weighing 5–6 kg will probably drink about 750 ml of breast milk each day. This volume of mature milk has an energy equivalent of about 2.6 MJ (630 kcal). Although a proportion of the additional energy requirement will come from mobilized maternal fat stores, it is usually recommended that a lactating mother's energy intake should increase by around 2 MJ (400–500 kcal) a day. Table 49.4 illustrates some other important nutritional requirements of lactation. Of particular significance is the need for an adequate intake of calcium and phosphate. The normal requirement for a woman of childbearing age is about 800 mg a day of both calcium and phosphate. An extra 400 mg a day of each mineral is normally sufficient to match the quantities secreted in the milk.

Summary

During gestation, progesterone and oestrogens inhibit the lactogenic action of prolactin, but after delivery, this inhibitory influence is lost and lactation commences. The composition of milk changes during the first weeks after parturition. Colostrum is secreted in the first few days after delivery. This is rich in proteins, minerals, and immunoglobulins but low in fats and sugars. Gradually, the composition of the milk changes and, by three weeks post-partum, mature milk is produced. This is rich in fats, proteins, and sugars, particularly lactose.

Table 49.4 Reference values for the daily intake of protein and micronutrients during lactation

Nutrient	USA	UK
Protein (g)	65	56
Vitamin A (mg)	1.3	0.95
Vitamin D (µg)	10	10
Vitamin E (µg)	12	10
Vitamin C (mg)	95	70
Vitamin B$_1$ (thiamin) (mg)	1.6	1.0
Vitamin B$_2$ (riboflavin) (mg)	1.8	1.6
Vitamin B$_3$ (niacin or nicotinamide) (mg)	20	15
Vitamin B$_{12}$	2.6	2.0
Folate (µg)	280	260
Calcium (g)	1.2	1.25
Magnesium (mg)	355	320
Iron (mg)	15	15
Zinc (mg)	19	13
Iodine (µg)	200	140
Phosphate (mg)	1200	1350

After delivery, milk production is maintained by regular suckling

Lactation is initiated by the precipitous drop in steroid levels that occurs following expulsion of the placenta at the time of delivery. But why does the breast continue to secrete milk for as long as the baby requires it after this time? What is the hormonal basis underlying the maintenance of lactation? It is known that lactation will continue normally in women who have undergone removal of their ovaries, but not in those who have damaged or absent pituitaries. The critical hormone for continued milk secretion appears to be prolactin. Levels of this hormone must remain high for efficient lactogenesis, and the only way this can be ensured is through regular suckling by the infant. Indeed, the suckling stimulus is the single most important factor in the maintenance of established lactation—in the absence of suckling, milk production ceases after two or three weeks. Suckling, or more correctly nipple stimulation, induces the release of prolactin from the anterior pituitary gland via a neuroendocrine reflex in which the afferent limb is neural and the efferent limb is endocrine. Nerve impulses, set up by the mechanical stimulation of the baby suckling at the breast, pass via the spinal cord and brainstem to the hypothalamus. This results in a fall in the output of prolactin inhibitory hormone (PIH—which is now known to be dopamine) from the hypothalamic neurons and a subsequent increase in the secretion of prolactin—remember that prolactin release is usually suppressed by dopamine (see Table 21.2). The prolactin then stimulates the synthesis and secretion of milk as described earlier.

Prolactin output is a direct consequence of nipple stimulation

Denervation of the nipple abolishes the release of prolactin in response to suckling. It has also been shown that the amount of prolactin released depends directly on the strength and duration of the suckling stimulus. If both breasts are suckled together, for example during the feeding of twins, more prolactin is released than when a single infant is suckled, resulting in the production of larger quantities of milk. To summarize: once lactation is initiated by the removal of inhibitory steroidal influences, milk production is ensured by the release of bursts of prolactin occurring each time the infant suckles—the baby makes sure of its next meal while enjoying its current one.

Milk ejection (let-down) is a direct response to the suckling stimulus

The constituents of breast milk are produced by the alveolar epithelial cells and secreted into the alveolar lumen under the influence of prolactin. For this to be of any value to the baby, however, the milk must be moved from the lumen to the nipple. This is the process of milk let-down and is another example of a neuroendocrine reflex occurring as a direct response to the suckling stimulus. The hormone responsible for the let-down of milk is **oxytocin**, a peptide hormone synthesized within the hypothalamus but stored and secreted by the posterior pituitary gland (see page 313). When the baby suckles, afferent impulses are initiated in the nipple and areola and travel to the paraventricular and supraoptic nuclei of the hypothalamus. In response to this stimulation, both the synthesis and the secretion of oxytocin are enhanced. Once oxytocin is released into the general circulation, it reaches the breast, where it stimulates contraction of the myoepithelial cells that lie within the alveolar basement membrane (see Figure 49.20). When these cells contract, the contents of the alveolar lumen are squeezed out into the lactiferous ducts. As the ducts and sinuses fill with milk, the intra-mammary pressure rises. When it reaches a high enough level, milk is ejected from the nipple to the suckling baby.

Figure 49.22 illustrates, in a highly simplified diagram, the reflex control of both prolactin and oxytocin release during suckling. The output of both hormones rises in synchrony with the episodes of suckling. While suckling seems to be the only effective stimulus for prolactin release under normal circumstances, this is not the case for oxytocin, whose secretion may be enhanced by a number of other stimuli. Uterine contractions, for example, and mechanical stimulation of the cervix and vagina can initiate oxytocin release, e.g. during parturition. Furthermore, the milk let-down reflex is readily conditioned. In cows, the rattling of the milking equipment may be sufficient to start the release of milk, while in humans the cry of a hungry baby may induce the secretion of oxytocin. By the same token, the reflex seems particularly susceptible to inhibition by stress (both physical and psychological), a response that may be mediated by catecholamines.

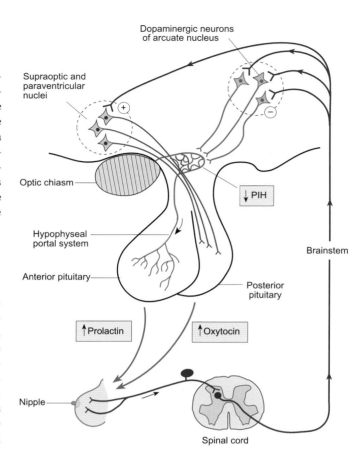

Figure 49.22 The principal neuroendocrine pathways responsible for the reflex release of prolactin and oxytocin during suckling.

After weaning, cessation of suckling suppresses milk production

Lactation normally ceases within two or three weeks of weaning the baby onto a bottle or solid foods. This is entirely due to the loss of the suckling stimulus. In the absence of mechanical stimulation of the nipple, prolactin secretion declines and lactogenesis gradually slows down. Although milk production itself stops relatively quickly, complete **involution** of the mammary gland takes about three months. At first, milk accumulates in the alveoli and small lactiferous ducts causing distension and mechanical atrophy of the epithelial structures. The alveolar cells are ruptured and hollow spaces form within the mammary tissue. The distension also causes compression of the capillary network supplying the alveoli and, as a result of the reduced perfusion, the alveolar cells become hypoxic and lack nutrients. This in turn depresses milk production. Desquamated alveolar cells and glandular debris are phagocytosed and the alveoli disappear almost completely. Consequently, the ductal system starts to dominate and the involuted alveolar epithelial cells revert to the granular, non-secretory type characteristic of the non-pregnant state. All these changes occur quite naturally as a direct result of removing the suckling stimulus at the time of weaning.

It is occasionally necessary to suppress lactation artificially and rather more quickly than would occur naturally. The human mammary gland is fully prepared for lactation by the fourth month of pregnancy. This means that in the event of a miscarriage or abortion after this time, milk production will commence because of the decline in steroid secretion following loss of the placenta. Under these circumstances, it is clearly desirable to inhibit lactation as rapidly as possible. Years ago this would have been assisted by the application of ice packs and tight bandages to the breasts. Since the discovery that PIH is the neurotransmitter dopamine, pharmacological suppression of lactation has been possible through the administration of dopamine agonists such as bromocriptine.

Summary

After delivery, milk production is maintained by regular suckling. The hormone responsible is prolactin, which is secreted in direct response to nipple stimulation. The processes by which milk is moved from the alveoli of the mammary gland to the nipple (milk 'let-down') occur in response to oxytocin secreted during suckling. Once the baby is weaned, and the suckling stimulus is lost, prolactin output falls and lactation ceases two or three weeks later.

(✱) Checklist of key terms and concepts

Sexual intercourse and fertilization

- At sexual intercourse about 3 ml of seminal fluid containing 200 million sperm is ejaculated into the female vagina.

- Male and female sexual responses are mediated by autonomic reflexes. Penile erection is mediated by parasympathetic nerves, while ejaculation is a sympathetic reflex.

- Before a sperm can fuse with an oocyte to bring about fertilization, it must first undergo capacitation followed by the acrosome reaction.

- Fertilization normally takes place in a Fallopian tube.

- The newly fertilized egg (zygote) completes its second meiotic division and undergoes the cortical reaction to create a fertilization membrane. This prevents further sperm from penetrating the oocyte.

- Human chorionic gonadotrophin (hCG) secreted by the zygote prevents regression of the corpus luteum, thus ensuring the continued secretion of progesterone and maintenance of the uterine endometrium.

Implantation and placental formation

- The placenta forms an interface between the maternal and fetal circulations.

- At implantation, the trophoblastic tissue of the fertilized egg invades the endometrial tissue of the uterus by means of chorionic villi containing the fetal capillaries.

- A dialysis pattern of blood flow is set up within the placenta such that fetal capillaries essentially dip into maternal blood spaces.

Transport functions of the placenta

- The placenta is the sole organ of exchange for the developing fetus. It carries out the functions normally performed in the adult by the lungs, kidneys, and gastrointestinal tract.

- Substances including oxygen, carbon dioxide, and essential nutrients cross the placenta by means of either passive diffusion or carrier-mediated transport.

- Oxygen diffuses passively from maternal to fetal blood, and carbon dioxide diffuses in the opposite direction.

- Glucose and amino acids move across the placenta from the maternal to the fetal plasma by carrier-mediated transport, while free fatty acids diffuse passively across the placenta.

- Fetal urea and bilirubin diffuse from fetal to maternal plasma down their concentration gradients.

Endocrine functions of the placenta

- The placenta secretes many peptide and steroid hormones.

- Human chorionic gonadotrophin (hCG) and human placental lactogen (hPL) are the major peptide hormones.

- hCG is a potent luteotrophic agent that prevents regression of the corpus luteum and ensures the continued secretion of progesterone during the early weeks of pregnancy.

- hPL contributes to the proliferative changes seen in the mammary tissue in preparation for lactation.

- hPL stimulates an increase in the maternal plasma levels of glucose, amino acids, and free fatty acids, ensuring optimal placental transport of essential nutrients from mother to fetus.

- Progesterone is essential for successful pregnancy, and the placenta takes over from the corpus luteum as the major source of this steroid at around week 10.

- A variety of oestrogenic hormones are secreted by the placenta. They appear to prepare the body for labour and lactation.

Parturition

- The nature of the trigger for parturition is still poorly understood, but it appears to involve a combination of maternal and fetal mechanisms.

- Fetal cortisol appears to initiate a switch in the placenta away from progesterone synthesis in favour of the synthesis of oestrogens.

- Oestrogens dominate the hormonal profile in the last days of gestation, and these hormones, together with PGF2α and oxytocin, increase the contractility of the myometrium to bring about delivery of the infant.

Changes in maternal physiology

- Weight gain during pregnancy is typically 7–14 kg and is associated with an increase in daily calorific requirement of 1050–1260 kJ (250–300kcals)/day.

- Throughout pregnancy, changes take place in all the major systems of the mother's body.
    - In early pregnancy these changes create favourable conditions for fetal development.
    - In the later stages of pregnancy, the mother's body is prepared for the delivery and breast-feeding of the infant.
    - Other changes are the result of spatial alterations brought about by the increasing size of the fetus.

- Maternal cardiac output increases to satisfy the demands of the uterus and placenta for nutrients and waste disposal.

- Arterial blood pressure often falls in mid-pregnancy as peripheral resistance falls, but may increase slightly towards term.

- Glomerular filtration is increased by up to 50 per cent, and the output of both renin and aldosterone is stimulated during gestation.

- Total body water increases by 6–8 litres during pregnancy, and plasma osmolality falls by about 10 mOsm/kg.

- Changes in maternal respiratory function are largely connected with the positional changes associated with increasing uterine size and compression of the diaphragm.

- During gestation the insulin sensitivity of maternal tissues increases in the early weeks, but then declines towards full term.

- Gestational diabetes occurs in 1–3 per cent of pregnancies.

Lactation

- Lactation is the synthesis and secretion of milk by the mammary glands.

- The pre-pubertal mammary gland is composed largely of lactiferous ducts.

- From puberty, progesterone stimulates the development of alveoli at the ends of the ducts.

- In pregnancy, under the influence of placental hormones (oestrogens, progesterone, and hPL), the alveoli mature and the breast acquires the potential for milk secretion.

- Colostrum is secreted in the first few days after delivery. This is rich in proteins, minerals, and immunoglobulins, but low in fats and sugars.

- By three weeks post-partum, mature milk is being produced. This is rich in fats, proteins, and sugars.

- After delivery, milk production is maintained by regular suckling. The hormone responsible for galactopoiesis is prolactin, secreted in direct response to nipple stimulation.

- Milk 'let-down' occurs in response to oxytocin secreted by the posterior pituitary during suckling.

Recommended reading

Alberts, B., Johnson, A., Lewis, J., Morgan, D., Raff, M., Roberts, K., and Walter, P.A. (2015) *Molecular biology of the cell* (6th edn), Chapter 17. Garland, New York.

A well-illustrated description of the main stages of cell division.

Ferin, M., Jewckwicz, R., and Warres, M. (1993) *The menstrual cycle*. Oxford University Press, Oxford.

Despite its age, this text provides helpful extra detail on the menstrual cycle for those students seeking a more in-depth understanding.

Johnson, M.H. (2013) *Essential reproduction* (7th edn). Wiley-Blackwell, Oxford.

A useful account of the physiology of reproduction.

Kovacs, W.J., and Ojeda, S.R. (eds) (2011) *Textbook of endocrine physiology* (6th edn). Oxford University Press, New York.

A very useful and authoritative introduction to the endocrinology of reproduction.

Review articles

Bachelot, A., and Binart, N. (2007) Reproductive role of prolactin. *Reproduction* **133**, 361–69.

Forbes, K., and Westwood, M. (2010) Maternal growth factor regulation of human placental development and fetal growth. *Journal of Endocrinology* **207**, 1–16.

Kovacs, C.S. (2016) Maternal mineral and bone metabolism during pregnancy, lactation, and post-weaning recovery. *Physiological Reviews* **96**, 449–47, doi:10.1152/physrev.00027.2015.

Neville, M.C., and Picciano, M.F. (1997). Regulation of milk lipid secretion and composition. *Annual Review of Nutrition* **17**, 159–84.

To check that you have mastered the key concepts presented in this chapter, complete the accompanying online self-assessment questions. Go to www.oup.com/uk/pocock5e/

CHAPTER 50
Fetal and neonatal physiology

Chapter contents

This chapter should help you to understand:

- The differences in the organization of the fetal and adult cardiovascular systems
- The carriage of oxygen in the fetal blood
- The first breath and the role of surfactant in lung inflation
- The cardiovascular changes that follow the first breaths
- The factors responsible for closure of the fetal shunts (the foramen ovale, ductus arteriosus, and ductus venosus)
- Respiration in the neonate
- The differences between the fetal, neonatal, and adult gut, kidneys, and adrenal glands
- Temperature regulation in the neonate
- The differentiation of the male and female reproductive tracts

50.1 Introduction

A fetus is totally dependent upon the placenta for gas exchange, nutrition, and waste disposal while it remains within the uterus of its mother. In a number of important ways it is adapted to life within a fluid-filled bag. The physiological changes that take place at, or soon after, birth enable a baby to make a successful transition from its intrauterine existence to a semi-independent air-breathing life. While the changes in the cardiovascular and pulmonary systems that occur around the time of birth are of paramount importance to its survival, a number of other organs must also adapt at birth to the different requirements of extra-uterine life. Of these, the changes in the adrenal glands, kidneys, thermoregulatory tissues, and gastrointestinal tract are significant. This chapter begins with some of the more important aspects of the physiology of the fetus and newborn and ends with a simple account of the differentiation of the male and female reproductive systems.

50.2 The fetal circulation is arranged to make the best use of a poor oxygen supply

The fetal heartbeat is detectable at four to five weeks of gestation, and by the eleventh week the cardiovascular system is fully developed in miniature. The fetal circulation differs from that of the adult in a number of important ways. The pattern of circulation is adapted for placental rather than pulmonary gas exchange, and organs which are virtually non-functional such as the lungs, gut, and liver are largely bypassed.

Figure 50.1 shows a simplified plan of the organization of the fully developed fetal cardiovascular system. It illustrates the three important shunts that differentiate the fetal from the adult circulations.

- The **foramen ovale** is a gap formed by incomplete fusion of the septum that separates the left and right atria. As shown in Figures 50.1 and 50.2, the foramen ovale provides a direct path for blood to pass between the inferior vena cava and the left atrium.
- The **ductus arteriosus** forms a direct link between the pulmonary artery and the aorta.
- The **ductus venosus** links the umbilical vein with the inferior vena cava.

These shunts, which normally close at birth, enable the two sides of the fetal heart to work in parallel, with mixing of the right

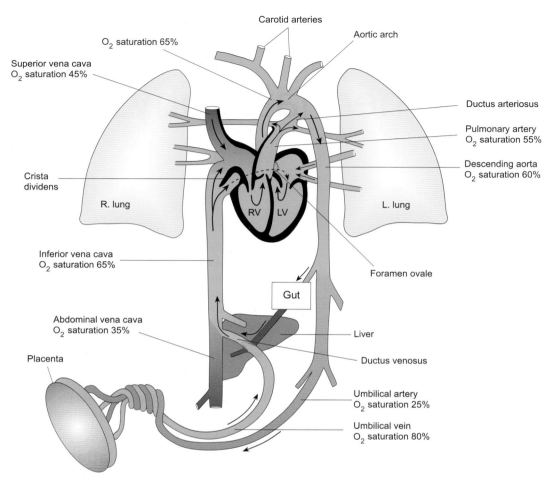

Carotid arteries

Aortic arch

O_2 saturation 65%

Superior vena cava
O_2 saturation 45%

Ductus arteriosus

Pulmonary artery
O_2 saturation 55%

Descending aorta
O_2 saturation 60%

Crista
dividens

R. lung

L. lung

RV LV

Inferior vena cava
O_2 saturation 65%

Foramen ovale

Gut

Abdominal vena cava
O_2 saturation 35%

Liver

Placenta

Ductus venosus

Umbilical artery
O_2 saturation 25%

Umbilical vein
O_2 saturation 80%

Figure 50.1 Diagrammatic representation of the organization of the fetal cardiovascular system. Note that the structure of the heart has been distorted to show the operation of the crista dividens. The dashed arrow indicates the direct flow of blood from the inferior vena cava through the foramen ovale to the left atrium.

ventricular and left ventricular outputs. This arrangement differs from that of the adult, in which the pulmonary and systemic circulations are perfused entirely separately (see also Figure 50.4 below).

What follows is a description of the route taken by the blood as it flows around the fetal circulation, and it will be helpful to refer to Figure 50.1 while reading this section. The fetus receives oxygenated blood from the placenta via the umbilical veins. A significant fraction of this blood (estimates vary between 20 and 50 per cent at mid-gestation) passes directly into the inferior vena cava via the ductus venosus, thus bypassing the immature fetal liver. The remainder, which is sufficient to support hepatic growth and development, enters the liver via the portal vein. This arrangement means that the oxygenated blood arriving from the placenta is mixed almost immediately with deoxygenated blood returning from the lower parts of the fetal body in the inferior vena cava. In the adult, blood in the inferior vena cava mixes with blood arriving from the upper parts of the body in the superior vena cava as it enters the right atrium (see Figure 27.1). If this were to happen in the fetus, the oxygenated blood from the umbilical veins, already mixed with deoxygenated blood in the inferior vena cava, would

also mix with deoxygenated blood from the superior vena cava. The oxygen saturation of the blood from the placenta would thus be reduced still further. In the fetus, complete mixing of blood from the inferior and superior vena cavae is avoided as a result of the anatomical position and mode of operation of two anatomical features: the **crista dividens** and the foramen ovale. The crista dividens is the upper part of the incomplete septum that divides the two sides of the heart. It essentially splits the bloodstream entering the right atrium, allowing most of the blood in the inferior vena cava to pass directly into the left atrium via the foramen ovale (which lies between the inferior vena cava and the left atrium—see Figure 50.2). Only a small amount of blood from the inferior vena cava enters the right atrium. By contrast, all of the blood in the superior vena cava enters the right atrium.

Blood from the left ventricle is pumped into the ascending aorta from which branch the carotid arteries that supply the brain (Figure 50.1). Blood from the right ventricle enters the pulmonary arterial trunk. However, soon after the start of the pulmonary trunk, a wide muscular blood vessel—the ductus arteriosus— branches off to provide a direct link between the pulmonary artery

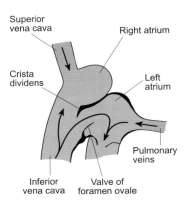

Figure 50.2 Detailed drawing to show the pattern of blood flow in the right atrium. Note how the crista dividens diverts the flow of oxygenated blood in the inferior vena cava through the foramen ovale to the left atrium.

and the descending aorta (see Figure 50.1). When blood reaches the point at which the ductus arteriosus branches off from the pulmonary artery, it may take one of two routes. It may travel to the fetal lungs in the pulmonary arteries, or it may bypass the lungs and travel directly to the descending aorta via the ductus arteriosus. Because the fetal lungs are collapsed and fluid-filled, there is a high resistance to blood flow in the pulmonary capillaries. By contrast, resistance in the ductus arteriosus and aorta is much lower, so around 75–85 per cent of the blood leaving the right ventricle takes this route; only 15–25 per cent perfuses the lungs. As a consequence of this arrangement, the blood in the ascending aorta has a somewhat higher oxygen content than that in the descending aorta: blood in the ascending aorta is around 65 per cent saturated with oxygen compared with 60 per cent in the descending aorta. As Figure 50.1 also shows, the carotid arteries branch off from the aorta before the point of entry of the ductus arteriosus. This ensures that the blood reaching the fetal brain has a higher oxygen content than that which perfuses other fetal tissues.

The control of the fetal circulation

By the 11th week of gestation, when the fetal cardiovascular system has been laid down, the fetal heart is beating at around 160 beats per minute (b.p.m.), a very high rate when compared with the average resting heart rate of the adult, which is about 70 b.p.m. Later, during the final three months (the third trimester) of gestation, the autonomic nervous system becomes functional and the parasympathetic innervation of the heart is established. At this stage the fetal heart rate slows to around 140 b.p.m.

This gradual development of the autonomic control of the cardiovascular system is also evident in the changes in blood pressure that occur during fetal life. Systemic arterial pressure (usually referred to as aortic pressure) is very low in the early weeks of gestation, when there is very little peripheral vascular tone and therefore a low total peripheral resistance. Values of around 20/5 mm Hg (2.5/1 kPa) have been recorded at 16 weeks gestation during sampling of fetal blood from the umbilical vein (umbilical cordocentesis). Blood pressure gradually rises as autonomic activity becomes

established and vascular tone is increased. It rises to 40/15 mm Hg (around 5/2 kPa) by 20–25 weeks gestation, the time at which the aortic and carotid baroreceptors begin to function. This increase in blood pressure continues throughout the remainder of gestation. At full term, blood pressure is around 70/40 mm Hg (9/6 kPa), with a mean systemic arterial pressure of around 50 mm Hg (7 kPa). Fetal pulmonary arterial pressure is a little higher than aortic pressure, with mean pressure values around 55 mm Hg (7.3 kPa) at full term. Systemic arterial blood pressure continues to rise after birth, reaching adult values by the age of about seven years.

The fetus depends upon the placenta for gas exchange as its lungs are collapsed and the alveoli filled with fluid

The role of the placenta in the transport of oxygen and carbon dioxide between the maternal and fetal blood was described in some detail in Chapter 49. As far as gas exchange is concerned, the fetal lungs are non-functional and the alveoli are almost collapsed and filled with fluid. This fluid is secreted by type I alveolar cells (epithelial cells which overlie the pulmonary capillaries) and its composition differs from that of the amniotic fluid. It first appears around mid-gestation, and by full term the lungs contain a total of about 40 ml of this fluid. Because the alveoli are collapsed, their capillaries are tortuous and offer a high resistance to blood flow. Consequently, the lungs are relatively poorly perfused.

Breathing movements develop before birth

Although the fetal lungs do not participate in gas exchange, ventilatory movements are seen during gestation. Ultrasound scans have revealed that fetal breathing movements begin at around the tenth week of gestation. They remain shallow and irregular up until around week 34 of gestation, after which they start to display a more rhythmical pattern with periods of activity, interspersed with periods when movements are absent. Occasionally, gasping movements are seen, especially if the fetus experiences hypercapnia— for example as a result of placental insufficiency or compression of the umbilical cord. This response suggests that chemoreceptors (see Chapter 35) are functional during the latter part of gestation. It is now believed that fetal breathing movements are important in the preparation of the respiratory system for its postnatal function of gas exchange. Fetal hiccups are also common, particularly during the later stages of gestation: these are felt by many pregnant women as distinct episodic movements of their unborn babies, which may go on for several minutes at a time. Non-invasive measurements of fetal diaphragmatic activity have demonstrated that these are due to spasmodic contractions of the diaphragm similar to those that cause hiccups in children and adults.

Fetal blood has a higher affinity for oxygen than adult blood

The partial pressure of oxygen in the blood travelling from the placenta to the fetus in the umbilical veins is around 30 mm Hg (~4 kPa), much lower than that of normal adult arterial blood, which is 95–100 mm Hg (~13 kPa). Moreover, PaO_2 in the fetal aorta is only about 20 mm Hg (2.6 kPa—see Figure 50.3), as the blood in the umbilical vein is mixed with blood returning from the lower part of the body before reaching the left side of the heart. Such a low value for PaO_2 would be expected to result in a very low value for blood oxygen saturation and content in the fetus. In fact, even at this low PaO_2, the arterial blood in the fetus is about 60–65 per cent saturated compared with 98 per cent or so in the adult arteries. Moreover, the oxygen content of the arterial blood of the fetus is about 16 ml dl^{-1} compared to 19.5 ml dl^{-1} in an adult. Two important factors are responsible for this state of affairs.

- Fetal blood has a higher haemoglobin concentration (around 14–18 g dl^{-1}) than adult blood (around 11.5–15.5 g dl^{-1} in females of reproductive age). The oxygen-carrying capacity of fetal blood is therefore generally higher than that of an adult.

- **Fetal haemoglobin (HbF)** has a higher affinity for oxygen than adult haemoglobin (the P_{50} for HbF is around 18–20 mm Hg compared to ~26 mm Hg for adult haemoglobin). This difference reflects the differences in structure between adult Hb and HbF. The α chains are identical but HbF has two γ-polypeptide chains instead of β chains as in adult Hb. Although the differences in oxygen affinity created by these structural differences are small, they are highly significant as the oxygen dissociation curve for fetal blood is shifted to the left in relation to that of the adult (see Figure 50.3).

Consequently, fetal blood has a relatively high oxygen saturation and total oxygen content, despite the low PO_2 values experienced by the fetus. Furthermore, the portion of the curve over which the HbF normally operates is very steep. Thus there is a large difference in oxygen saturation and oxygen is off-loaded readily to the tissues, even though the difference between the PO_2 of arterial and systemic venous blood is relatively small.

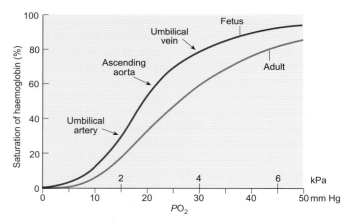

Figure 50.3 The oxygen dissociation curves for adult and fetal haemoglobin. Note that the dissociation curve for the fetus is displaced to the left, indicating a higher degree of saturation of fetal haemoglobin for a given partial pressure of oxygen. The approximate values for the partial pressure and percentage saturation in the ascending aorta of the fetus and in the umbilical artery and vein are indicated.

Summary

The fetus is dependent on the placenta for gas exchange, waste disposal, and nutrition. The fetal circulation is arranged so that the two sides of the heart work in parallel and those organs with little or no function are bypassed by the three fetal shunts: the ductus arteriosus, the ductus venosus, and the foramen ovale. The fetal lungs are fluid-filled and virtually collapsed, which ensures a high pulmonary vascular resistance. Consequently, pulmonary blood flow is only 20 per cent of the right ventricular output; the other 80 per cent passes through the ductus arteriosus. The fetal haemoglobin concentration is high and, as fetal haemoglobin has a high affinity for oxygen, fetal blood carries about 16 ml O_2 dl^{-1}.

50.3 Respiratory and cardiovascular changes at birth

After birth, the establishment of air breathing is accompanied by a series of cardiovascular changes that allow the circulation to adapt to pulmonary gas exchange. The changes that occur at, or very soon after, birth can be summarized as follows.

1. Initiation of the first breath.

2. Expansion of the lungs followed by a reduction in the pulmonary vascular resistance that results in a striking rise in pulmonary blood flow.

3. Gradual closure of the ductus arteriosus so that all the right ventricular output eventually passes through the pulmonary circulation.

4. Closure of the foramen ovale in response to the increase in pulmonary blood flow into the left atrium.

5. Closure of the ductus venosus so that all the blood in the portal vein passes through the liver.

These changes in the pattern of the circulation following birth are summarized in Figure 50.4.

The first breath is probably triggered by cooling and hypercapnia

Once the infant has been delivered, it is cut off from the placenta that has acted as its site of gas exchange for the previous nine months. Even if the cord is not clamped surgically in the delivery room, the umbilical vessels quickly shut down of their own accord. So, despite the fact that both the fetus and neonate are able to tolerate degrees of hypercapnia and hypoxia that would be likely to kill an adult, the infant has to start breathing independently if it is to survive for more than about 10 minutes.

The mechanisms responsible for triggering a baby's first breath are not known with certainty, but a number of possible factors may contribute. These include:

- The drop in ambient temperature experienced by the baby following delivery.

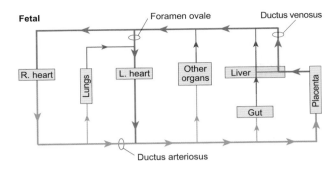

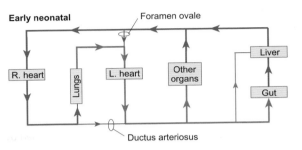

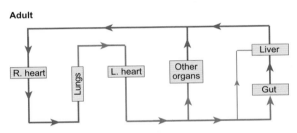

Figure 50.4 Schematic diagram to show how the pattern of the circulation changes following birth.

- Hypercapnia experienced during delivery. During labour and delivery, particularly if this has been long and difficult, the placental blood supply may be partially or wholly occluded for short periods of time. Certainly, once the baby has been born the placental blood

supply will be lost. Consequently, the baby will experience a considerable degree of hypercapnia. Indeed, measurements of the PCO_2 of scalp blood sampled during and just after delivery have revealed a marked increase, which may well provide the neonate with an important stimulus to gasp for air. This reflex is known to be functional in the fetus, whose breathing movements become more pronounced during hypercapnia.

- Increased sensory input. It is known that tactile and painful stimuli can stimulate breathing movements even in the fetus. After birth, the neonate is subjected to a barrage of sensory information, including tactile, auditory, and visual stimuli, from which it has been insulated while in the womb.

It is not known which, if any, of these factors is responsible for the initiation of ventilation following delivery—perhaps a combination of both physical and chemical stimuli is required.

Lung inflation is facilitated by surfactant

The lungs have remained collapsed and fluid-filled throughout embryonic life. In order to inflate its lungs for the first time, the newborn baby must overcome the enormous surface tension forces at the gas–liquid interface of the alveoli. Around week 20 of gestation, type II epithelial cells start to appear in the developing alveoli, and eight to ten weeks later, under the control of fetal cortisol, these cells start to secrete phospholipid surfactants, the most important of which are lecithin and sphingomyelin. Surfactant molecules are responsible for reducing the surface tension forces that oppose lung inflation (see Chapter 33). Although this reduction makes it possible for the newly born infant to inflate its lungs, a considerable ventilatory effort is nevertheless required in order to expand the lungs for the first time, as Figure 50.5 shows.

A very large negative pressure must be generated to make the initial inspiration possible. Figure 50.5 illustrates the pressure/volume relationships for the neonatal lung operating during the first and subsequent breaths. It also shows that an equally large positive pressure must be generated to bring about expiration because lung

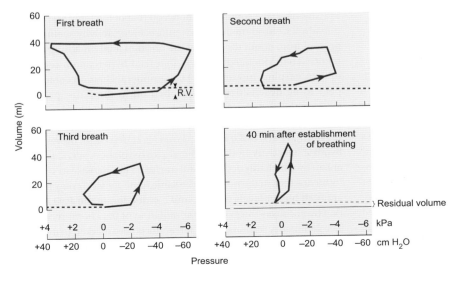

Figure 50.5 The lung pressure–volume relationships during the first, second, and third breaths together with that for breathing about 40 minutes after the first breath. Note that the residual volume (RV) is established with the first breath and that the compliance increases with subsequent breaths (i.e. the pressure change required for a given change in volume falls after the first breath). (Adapted from Fig 13.1 in D.J. Begley, J.A. Firth, and J.R.S. Hoult (1980) *Human reproduction and developmental biology*. Macmillan Press, London.)

compliance is still rather low. A great mechanical effort is therefore demanded from the baby. During the initial breaths the diaphragm contracts strongly and the ribs and sternum, which at this stage are very flexible, become slightly concave. After the first breath, the volume of the lungs does not return to zero after expiration but a little air remains in the lungs to form the beginnings of the residual volume (RV), which persists throughout life. Subsequent breaths are achieved with much smaller pressure changes and consequently require much less mechanical effort, indicating that lung compliance has increased (see Chapter 33 for more detail of the mechanics of breathing).

Once the fetal shunts have closed and the lungs have expanded, the pulmonary vascular resistance is much lower than before birth and the postnatal pulmonary blood flow is greatly increased. As a result, the fluid that filled the alveoli during fetal life is reabsorbed quickly into the pulmonary capillary blood, which has a higher oncotic pressure than the alveolar fluid.

What happens if surfactant is inadequate?

The presence of sufficient quantities of surfactant is crucial to the initiation of ventilation following delivery. Even then, the first breath requires a considerable mechanical effort on the part of the baby (see Figure 50.5). Imagine, therefore, the problems faced by babies born prematurely, i.e. before adequate surfactant secretion by type 2 cells has been established. If a baby is born before about week 28 of gestation, it will almost certainly have difficulty overcoming the surface tension forces opposing ventilation, and is very likely to suffer respiratory distress. If such infants are to have a chance of survival, they must be ventilated artificially until their lungs are sufficiently well developed to permit independent respiration.

Babies born to diabetic mothers often show delayed type 2 alveolar cell development, and even those born after 30 weeks gestation may suffer from respiratory distress. In some cases, administration of cortisol to the mother can accelerate the maturation of the fetal type 2 cells and prevent the occurrence of respiratory distress. Aerosol surfactants can also be administered to some newborn babies at risk of respiratory distress.

How does neonatal respiration differ from that of the adult?

Once the first few—rather difficult—breaths have been accomplished, respiration settles down into the 'neonatal' pattern, a somewhat erratic rhythm with certain characteristics that differ significantly from those of a more mature child or adult. The ventilatory rate of a newborn infant (or neonate—a baby less than about one month old) is rather high and extremely variable. Neonatal breathing often resembles the fetal pattern of respiratory movements, with episodes of shallow breathing (or even apnoea) interspersed with periods of normal respiration. There are, understandably, considerable difficulties associated with the study of lung function in very small babies, but some information has been obtained, much of which is contained in Table 50.1.

Table 50.1 Comparison of respiratory variables between the neonate and the adult

Variable	Neonate	Adult
Body weight (kg)	3.3	70
Ventilation rate (breaths min^{-1})	20–50	12–15
Minute volume (ml)	c. 500	c. 6500
Tidal volume (ml)	18	500
Vital capacity (ml)	120	4500
Surface area for gas exchange in lungs (m^2)	3	60
Compliance		
(l kPa^{-1})	0.05	1.7
(ml cm H_2O^{-1})	5	165
Bronchiole diameter (mm)	0.1	0.2
Oxygen diffusion capacity		
(ml s^{-1} kPa^{-1})	0.6	6
(ml min^{-1} mm Hg^{-1})	2.5	25
Energy expended in breathing as % of total O_2 consumption	6	2

Because of its high rate of ventilation, the infant's minute volume is relatively high in relation to its body weight. As Table 50.1 shows, the neonatal respiratory system has a rather low compliance, which indicates high airway resistance when compared with that of an adult. A number of factors contribute to this resistance. Firstly, during the early weeks of its life a baby tends to breathe mostly through its nose. Secondly, the bronchioles are very narrow, and thirdly, lung compliance is itself still rather low. These characteristics mean that the energy expenditure of breathing in the neonate is high.

Regulation of ventilation in the neonate

Regulation of ventilation during the first weeks of life represents a transitional stage somewhere between that of the fetus and that of the adult. Central medullary chemoreceptor activity seems to be present from about the time of the onset of fetal breathing movements. The evidence for this is that fetal hypercapnia seems to initiate 'gasping' movements. The peripheral chemoreceptors appear to be desensitized or switched off in the fetus, probably because the partial pressure of oxygen in the fetal blood is always low. They show slight tonic activity in the full-term fetus at the normal PaO_2 of 3.0 kPa (c. 23 mm Hg). After birth, however, the hypoxic sensitivity gradually changes to that of the adult. The mechanism by which this occurs is still unclear. This increase in sensitivity means that, in the neonate, the same tonic activity is now present at much higher levels of PaO_2 and a reduction in the PaO_2 below that level causes the expected stimulation of the chemoreceptors. The ventilatory response to hypercapnia is very marked in the newborn baby—the addition of only 2 per cent CO_2 to the inspired air produces an increase in the minute volume of around 80 per cent.

Summary

After delivery, the baby must begin to use its lungs to breathe air. The likely trigger for the first breath is the hypercapnia that occurs during delivery, but other physical factors may also stimulate breathing. When the lungs are inflated for the first time, pulmonary surfactant plays a vital role by reducing the surface tension forces at the gas–liquid interface in the alveoli. Nevertheless, a massive mechanical effort is still required to open the lungs, and babies born before adequate levels of surfactant have been produced (around week 28) are at risk of showing respiratory distress.

In neonates the ventilatory rate is higher and more erratic than in adults. Airway resistance is also higher, so the work of breathing is greater. As in the fetus, the ventilatory response to hypercapnia is well developed. The peripheral chemoreceptors start to respond to reductions in oxygen tension following delivery.

50.4 Following delivery, the fetal circulation adapts to pulmonary gas exchange

As Figure 50.4 shows, the anatomical arrangement of the fetal circulation differs from that of an adult in a number of ways. In essence, the pulmonary and systemic circulations work in parallel, with the three fetal shunts permitting blood to bypass those organs with little or no function. Such an arrangement is well adapted to gas exchange via the placenta, but would be quite inappropriate once a baby has begun to breathe for itself. Following delivery, the placental blood supply is lost and the lungs become the sole source of oxygen. As the infant takes its first breaths of air, the fetal circulation must start to adapt to the adult pattern so that blood no longer bypasses the pulmonary circulation. To achieve this it is essential that the three fetal shunts close.

Probably the most important step in the initiation of shunt closure is the increase in pulmonary perfusion that accompanies the establishment of ventilation. During fetal life no more than 20 per cent of the cardiac output enters the pulmonary circulation because resistance to blood flow through the pulmonary vessels is high in the collapsed lungs. After the first breath, pulmonary blood flow is dramatically increased. Two key factors play a part in bringing about this increase:

- Inflation of the alveoli reduces the tortuosity of the pulmonary capillaries, thereby reducing their resistance to blood flow.

- As a result of breathing air, there is a considerable increase in the partial pressure of oxygen in the blood perfusing the lungs. In response to this rise, there is vasodilatation of the pulmonary vessels (probably mediated by endogenous nitric oxide) and a corresponding fall in resistance. Both of these changes stimulate a large increase in the volume of blood circulating in the pulmonary vessels.

At the same time, the fetus is separated from its placental blood supply. Following delivery, the umbilical cord is normally clamped, but even if this is not done, the umbilical vessels appear to shut down spontaneously as a result of vasoconstriction in response to the raised systemic PO_2. (Note that this is the opposite reaction to that of the pulmonary arterioles, which dilate in response to a raised PO_2—see also Chapter 34.)

How do the changes in the pattern of blood flow following delivery bring about closure of the fetal shunts?

Consider first the foramen ovale—the shunt between the left and right atria. During fetal life, right atrial pressure is similar to, or just exceeds, left atrial pressure because of the relatively high pulmonary resistance and the relatively low systemic resistance. After birth and the onset of independent breathing, increased pulmonary perfusion results in an increase in venous return to the left atrium. At the same time, loss of the umbilical blood supply reduces the venous return in the inferior vena cava to the right atrium. Consequently, left atrial pressure rises above right atrial pressure. The foramen ovale consists of two unfused septa. When right atrial pressure exceeds left atrial pressure, the septa part and the shunt is open. Once the pressures are reversed, the septa will be forced against each other and the shunt will be closed. At first, closure is purely physiological, but within a few days the septa fuse permanently and closure is anatomically complete. Figure 50.6 shows how the pressure changes bring about the closure of the foramen ovale.

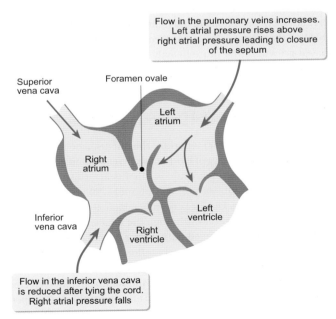

Figure 50.6 Diagrammatic representation of the changes in the pressures of the right and left atria that lead to closure of the foramen ovale. While the lungs are non-functional with respect to gas exchange, the pressure in the right atrium is greater than that in the left and blood passes through the foramen ovale. Following the first breath, the pressure in the left atrium becomes greater than that in the right and this leads to the closure of the foramen ovale.

Consider next the closure of the ductus arteriosus—the fetal shunt between the aorta and the pulmonary artery. This is an extremely wide channel, almost as large in diameter as the aorta itself (see Figure 50.1), and the mechanisms by which closure is achieved have not been established beyond doubt. The ductus arteriosus receives little or no innervation, so the most likely cause of closure following delivery and the onset of breathing is constriction in response to blood-borne factors. The smooth muscle of the ductus arteriosus, like that of the umbilical vessels, is thought to constrict in response to the substantial rise in PO_2 that occurs once the first breaths have been achieved. Permanent closure of the shunt normally occurs within 10 days or so as a result of fibrosis within the lumen of the vessel. The fibrous tissue remains as the ductus ligamentosum.

The third fetal shunt is the ductus venosus—the channel that bypasses the liver and carries a significant fraction of the blood in the umbilical veins directly to the inferior vena cava during fetal life. Once again, the exact mechanisms of closure are poorly understood, but it is believed to occur as a result of constriction of the umbilical vessels following delivery.

In essence, the parallel arrangement of the two sides of the heart, which is characteristic of the fetus, is converted to a serial arrangement soon after birth and the commencement of pulmonary gas exchange (see Figure 50.4). At the same time, the relative workloads of the two sides of the heart are altered. During fetal life the right ventricle works harder than the left and the thickness of the right ventricular wall is slightly greater than that of the left. After birth, however, as the resistance to flow in the vascular bed of the lung falls to only about 12 per cent of that of the systemic circulation, the workload of the right side of the heart is considerably less than that of the left. As a consequence there is an accelerated growth of the more heavily loaded left ventricle, which eventually develops a mass of muscle about three times that of the right.

Occasionally, the fetal shunts fail to close

While there have been no reports of the ductus venosus failing to close during the first days of life, persistent fetal connections in the form of a patent foramen ovale or ductus arteriosus are seen. Indeed, each accounts for probably about 15–20 per cent of congenital heart defects. During very early neonatal life, it is not unusual to see intermittent flow through the fetal shunts, but if they remain open for a prolonged period after birth, circulatory function is impaired and surgical intervention will be required to correct the defect. For example, if the foramen ovale remains patent, leaving an opening between the atria, the volume of blood ejected by the right ventricle is often increased and there is likely to be persistent admixture of oxygenated and deoxygenated blood (the blue baby syndrome). A patent foramen ovale may also lead to the formation of emboli which may reach the brain and cause an ischaemic stroke. Where the ductus arteriosus remains open, 50 per cent or more of the left ventricular stroke volume can be diverted into the pulmonary circulation, resulting in pulmonary hypertension and heart failure. This condition can be treated surgically by tying the ductus arteriosus.

Summary

After delivery, the circulation adapts to pulmonary gas exchange by the closing of the three fetal shunts, converting the parallel arrangement seen during fetal life to a serial arrangement. Closure of the shunts depends upon ventilation itself. As the lungs expand, pulmonary perfusion increases, while at the same time the umbilical vessels close. This causes left atrial pressure to rise above right atrial pressure, favouring closure of the septa that form the foramen ovale. The ductus venosus and the ductus arteriosus are thought to close as a result of vasoconstriction in response to a rise in arterial PO_2.

50.5 The fetal adrenal glands and kidneys

The fetal adrenal gland secretes large quantities of cortisol during development

The adrenal glands are vital endocrine organs in the adult. They consist of two distinct regions, the cortex, which synthesizes and secretes a variety of steroids (see Chapter 23), and the medulla, which produces the catecholamines adrenaline and noradrenaline (also known as epinephrine and norepinephrine). During fetal life, the adrenal glands appear to be, if anything, more important as they play a key role in the development of many of the fetal organ systems and are important in the initiation of parturition.

In relation to the overall body size of the fetus, the fetal adrenal gland is much bigger than that of the adult. Furthermore, it is organized in a different way. Unlike the adult gland which consists of a medulla and a zoned cortex, the fetal adrenal is divided into three areas with differing characteristics: a small region of medullary tissue derived from embryonic neural crest cells; a small region of cortical tissue—the so-called definitive cortex; and a third, very large region, the fetal zone. The relative sizes of these areas are shown in the lower series of panels of Figure 50.7.

The functions of the different regions of the fetal adrenal gland are not fully established, but the medullary tissue is capable of secreting catecholamines. This occurs particularly in response to hypoxic stress, and, immediately after delivery, to cold stress. The fetal zone seems to be very important in producing the precursors required for the placental synthesis of oestrogens (see Chapter 49), and its large size probably reflects the enormous output of these steroids during gestation. The 'definitive cortex' seems to carry out little in the way of steroid synthesis itself during fetal life, but it does perform one very important task—it converts progesterone to cortisol, especially during the last three months of pregnancy.

Fetal cortisol has a number of crucial functions.

- It is linked with the production of surfactant by type 2 alveolar cells.
- It accelerates functional differentiation of the liver and induces the enzymes that are involved in glycogen synthesis.
- It plays an important role in the triggering of parturition.

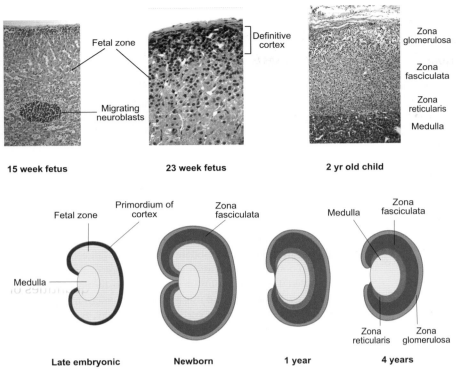

Figure 50.7 The fetal adrenal gland. The top panels show the appearance of the fetal adrenal gland at 15 and 23 weeks compared to that seen in a two-year-old child (right). The lower four diagrams illustrate the main changes in the organization of the adrenal gland during development. Note the complete regression of the fetal zone following birth.

After delivery, the fetal zone of the adrenal gland regresses rapidly while the definitive cortex grows quickly to establish the organizational pattern of the adult. Although the zona glomerulosa and fasciculata are present at birth, the zona reticularis is not fully established until the age of about three years.

Renal function and fluid balance in the fetus and neonate

Although the placenta is the major organ of homeostasis and excretion of metabolic waste products during gestation, the fetal kidneys do play a role in the regulation of fluid balance and the control of fetal arterial blood pressure. In addition, amniotic fluid volume is regulated chiefly by the formation of fetal urine. The human fetus begins to produce urine at about the eighth week of gestation. Its volume increases progressively throughout gestation and is roughly equivalent to the volume of amniotic fluid swallowed by the fetus (around 28 ml h^{-1} in late gestation). Fetal urine is usually hypotonic with respect to the plasma. Indeed, the ability of the kidney to concentrate the urine is not fully developed until after birth when the organ matures, the loops of Henle increase in length, and sensitivity of the tubules to ADH increases.

In the adult, virtually all the filtered sodium is reabsorbed by the renal tubules. In the fetus, sodium reabsorption is comparatively low (85–95 per cent of the filtered load). The exact reasons for this difference are not clear, but it may be explained partly by a low tubular sensitivity to aldosterone. Fetal glucose reabsorption is thought to occur by sodium-dependent transport, as in the adult. Furthermore, the tubular maximum for reabsorption (when corrected for the lower GFR) is higher than that of the adult, as is the renal plasma threshold for glucose.

The fetal kidney plays a part in the regulation of acid–base balance during gestation. Between 80 and 100 per cent of the filtered bicarbonate is reabsorbed by the tubules. The fetal response to metabolic acidosis is relatively poor, but in severe acidosis there is an increase in hydrogen ion excretion.

Renal changes occurring at or soon after birth

Although the changes in kidney function that accompany birth are less dramatic than those taking place in the respiratory and cardiovascular systems, they are just as important. Once the placenta is lost, the kidneys of the newborn infant become solely responsible for maintaining fluid balance and disposing of waste products. Babies will usually pass urine within the first 24 hours after birth.

GFR and urine output increase gradually over the first weeks of life—although adult levels (relative to body surface area) are not reached for 2–3 years. Tubular function is difficult to assess in neonates, but it is thought that while glucose and phosphate are reabsorbed efficiently, bicarbonate and amino acids are reabsorbed less well. Babies cannot concentrate their urine to the degree seen in adults. Possible reasons for this include immaturity of the tubules, shorter loops of Henle, lower sensitivity to ADH, and a low plasma concentration of urea as nearly all the amino acids derived from the protein in a baby's diet are used in the formation of new tissue—very little is metabolized in the liver to form urea.

Newborn babies are at risk from dehydration

The inability of young babies to concentrate their urine efficiently means that they can quickly become dehydrated, particularly during episodes of diarrhoea or vomiting. For this reason, it is essential that fluids are replenished by mouth and, if this is not possible, intravenous fluid replacement may be needed.

Summary

The fetal adrenal gland consists of a medullary region, a small 'definitive' cortex, and a larger fetal zone. The medulla secretes catecholamines; the fetal zone provides the precursors for the synthesis of oestrogens by the placenta, while the cells of the definitive zone convert progesterone to cortisol. Fetal cortisol has several key functions. It stimulates surfactant production, accelerates maturation of the liver, and has a role in the initiation of parturition.

The fetal kidneys play a part in fluid and acid–base balance. Although urine is produced from about week 8 of gestation, the fetal kidneys cannot concentrate it effectively. Therefore the urine they produce is generally hypotonic. At birth, the kidneys assume sole responsibility for fluid balance and waste disposal. GFR and urine output increase gradually, as does the ability to concentrate urine.

50.6 The gastrointestinal tract of the fetus and neonate

The role of the placenta in delivering essential nutrients to the fetus has been described in Chapter 49. The fetus obtains glucose, amino acids, and fatty acids from its mother. Towards the end of gestation, glycogen is stored in the muscles and liver of the fetus, while deposits of both brown and white fat are laid down. These stores are crucial to the survival of the infant immediately after its birth.

The gut of the fetus is relatively immature, with limited movements and secretion of digestive enzymes. Some salivary and pancreatic secretion commences during the second half of gestation. Gastric glands appear at around the same time, although they do not appear to be actively secretory as the gastric contents are neutral at birth. Most of the major gastrointestinal hormones are secreted during fetal life, although at a low level. Motilin secretion is especially low, possibly accounting for the low level of gut motility when compared with that of the adult.

The fetus passes little, if any, faecal material while it remains in the uterus. The contents of the large intestine accumulate as **meconium**, a sticky greenish-black substance. Meconium does not normally enter the amniotic fluid, although if the fetus becomes distressed, for example during a prolonged or difficult labour, motilin levels rise, gut motility increases, and meconium is passed. Meconium-stained amniotic fluid is recognized as a sign of fetal distress and can cause damage to the lungs if it is aspirated (meconium aspiration syndrome, MAS).

At birth, placental nutrients are lost but oral feeding is not yet established

With clamping of the umbilical cord soon after delivery, placental nutrition of the baby ceases. However, it may be several days before oral feeding is fully established. During this time the neonate must rely on the stores of fat and carbohydrate laid down during late gestation. This accounts for the typical weight loss seen in the first few days of life. A baby will normally regain its birth weight within 7–10 days of birth. Most importantly, glycogen is broken down to glucose under the influence of catecholamines secreted by the adrenal medulla. As milk feeds are established, the chief metabolic substrate switches from glucose to fat. Premature or low birth-weight babies may experience problems because of inadequate stores and may require intravenous nutrition.

As the first milk feeds are ingested by a baby, its gut increases rapidly in size to accommodate the relatively large volume of fluid it must now handle. At the same time, secretion of digestive juices is stimulated and gut motility increases. Mature human milk is rich in fat (see Chapter 49), and this becomes the major metabolic substrate. Lactose, the chief carbohydrate of milk, is hydrolysed by lactase, an enzyme located in the brush border of the small intestine. A specific lack of this enzyme, or a more generalized reduction in pancreatic enzymes, as occurs for example in cystic fibrosis, will result in a substantial reduction in digestion and absorption. The sticky meconium that was present in the fetal large intestine is usually passed during the first few days of life. After this, the semi-liquid stools change to green and then yellowish-brown in colour. Bowel movements are generally frequent in young babies, although there is also great variability—there may be as many as 12 stools a day or as few as one every three or four days.

Summary

The fetal gut is relatively immature and there is a limited degree of motility and secretory activity. The fetus is nourished solely by the placenta and its chief metabolic substrate is glucose. The contents of the fetal large intestine accumulate as meconium, which may be passed into the amniotic fluid during fetal distress.

After birth, but before the full establishment of oral feeding, the baby relies largely on stores of fat and carbohydrate laid down during late gestation. With the establishment of milk feeds, the major metabolic substrate switches to fats.

50.7 Temperature regulation in the newborn infant

The fetus has no problems with temperature regulation, as it is surrounded by amniotic fluid which is at body temperature. The mother is responsible for generating and dissipating heat. At delivery, however, the newborn infant has to make a rapid adjustment from the warm, moist, constant environment of its mother's uterus to an outside world in which the temperature is much lower and heat is readily lost by radiant, convective, and evaporative routes. A baby's core temperature normally drops to around 35°C during the first hours of its life. A number of factors combine to bring about this fall:

- a high surface area to volume ratio, which means that heat is readily lost from the skin's surface
- a high cardiac output in relation to surface area
- a comparatively thin layer of insulating fat
- a comparatively large head from which significant heat loss occurs (especially in the relative absence of hair).

Babies generate large quantities of heat through the metabolism of brown adipose tissue

Normally, when a critical temperature difference of 1.5°C is reached between the skin and the environment, thermogenesis begins and oxygen consumption increases, in order to restore the body temperature to normal. While thermoregulatory mechanisms are only partly functional at birth, newborns are capable of maintaining their body temperature above ambient temperature. They respond to a lowered ambient temperature by increased muscular movement (although this is limited) and by shivering, but only in a very minor way. These responses cannot account for all the heat generated in response to cold. The extra heat is generated by non-shivering thermogenesis via the metabolism of **brown adipose tissue** or brown fat, which is abundant in the infant. It is situated between the scapulae, at the nape of the neck, in the axillae, between the trachea and the oesophagus, and in large amounts around the kidneys and adrenal glands. In all, the neonate possesses about 200 g of brown fat which represents a relatively high proportion of the total body mass. Figure 50.8 illustrates the principal locations of brown adipose tissue within the neonate's body.

The brown fat is well vascularized and exhibits unique metabolic properties which are triggered either by increased plasma levels of circulating catecholamines or by noradrenaline released by sympathetic nerve endings. Cold stress results in an increase in sympathetic nervous activity and an increase in secretion of adrenaline and noradrenaline (epinephrine and norepinephrine) by the adrenal medulla. These hormones stimulate the metabolism of brown fat cells by interacting with β-adrenoceptors on the cell surface to activate a lipase (hormone sensitive lipase or HSL), which then releases glycerol and free fatty acids from cellular stores of triglyceride (see Figure 50.9). Most of these free fatty acids are resynthesized directly into triglyceride by the incorporation of α-glycerophosphate so the brown fat stores are not unduly depleted. The inner mitochondrial membrane of brown fat cells contains a protein that uncouples oxidation from ATP generation, and heat is generated instead of the energy being stored as ATP for subsequent use during cellular metabolism. Furthermore, the free fatty acids and glycerides which are not immediately resynthesized become available for oxidation by the usual biochemical pathway (see Chapter 4) to provide still more heat energy. Since the tissue is well supplied with blood, the heat that is generated by this pathway is quickly carried to the rest of the body and in this way the brown fat acts as a very effective source of heat for the newborn baby.

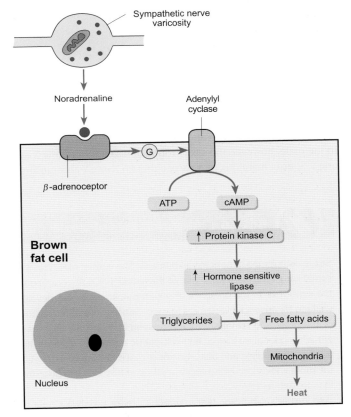

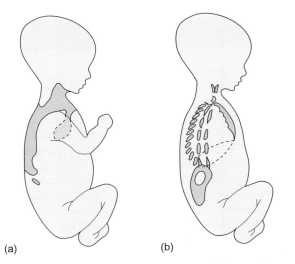

(a)   (b)

Figure 50.8 A diagram to show the distribution of brown adipose tissue in the body of a neonate. Panel (a) shows the interscapular pad and panel (b) the thoracic and abdominal depots.

Figure 50.9 Diagrammatic representation of the metabolism of a brown fat cell. Activation of β-adrenoceptors on the cell surface leads to a signal cascade that results in an increased activation of hormone-sensitive lipase and increased breakdown of triglycerides, which are metabolized in the mitochondria to generate heat (see text for further details).

Premature infants have special thermoregulatory problems

Premature babies have even greater difficulty in maintaining their body temperature than normal infants born at full term. Their surface area to volume ratio is even bigger, allowing more rapid heat loss, their insulating fat layer even thinner, and their brown fat stores less well developed. For this reason it is almost always necessary to keep premature babies in a thermally controlled environment—an incubator—until their thermoregulatory mechanisms develop sufficiently to permit independent temperature regulation.

Summary

A newborn infant can lose heat very rapidly and its high surface area to volume ratio, relatively high cardiac output, and lack of insulating fat combine to cause a drop in core temperature after birth. Babies can generate large quantities of heat through the metabolism of brown fat. The metabolism of brown fat is stimulated by catecholamines released in response to cold stress. Premature infants have even greater difficulty in maintaining their body temperature and frequently need to be kept in a thermally controlled incubator.

50.8 Development of the male and female reproductive systems

Humans have 23 pairs of chromosomes: 22 pairs of autosomes and one pair of **sex chromosomes**. In the female, both sex chromosomes are X chromosomes and all ova carry a single X chromosome. In the male, however, the sex chromosomes consist of one X and one Y. Sperm, therefore, may carry either an X or a Y chromosome.

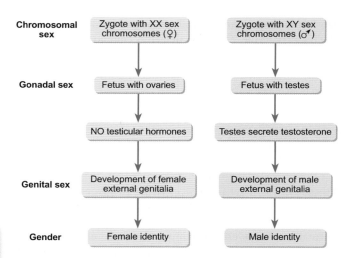

Figure 50.10 An outline of the sequence of prenatal development of gender including differentiation of the appropriate gonads and genitalia.

(Remember that the ova and sperm are haploid and carry half the full complement of chromosomes—see Chapter 4.)

In humans, the female is said to be the homogametic sex (XX), while the male is the heterogametic sex (XY). If an ovum is fertilized by a sperm carrying an X chromosome, the resulting baby will be a girl, while fertilization by a sperm carrying a Y chromosome will produce a boy (Figure 50.10). Studies of patients with a range of chromosomal abnormalities have revealed that the presence of a Y chromosome is the critical determinant of 'maleness', at least as far as gonadal development in the embryo is concerned (Box 50.1). If a Y chromosome is present, male gonads (testes) will develop, but in the absence of a Y chromosome, female gonads (ovaries) will form. Recently it has been shown that only a small part of the Y chromosome is actually required for the determination of 'maleness'. This is the so-called SRY gene (**S**ex determining **R**egion,

 Box 50.1 Abnormalities of sexual differentiation

Sex is determined genetically in humans. The normal male chromosomal karyotype is XY while that of the female is XX. As described in the main text, in the presence of a Y chromosome, testes develop, while in its absence ovaries are formed. Subsequent differentiation of the male and female genitalia depends upon the existence of either functional ovaries or testes. Genetic errors can result in anatomical aberrations and distortion of sexual differentiation. A few of the more widely occurring abnormalities of this kind are described in more detail here.

In Turner's syndrome (karyotype XO), there is only a single sex chromosome and, in the absence of either a second X chromosome to stimulate normal ovarian development or a Y chromosome to stimulate testicular formation, the gonad remains as a primitive streak. In the absence of functional testes, the external genitalia develop as the female type.

In Klinefelter's syndrome (karyotype XXY), the internal and external genitalia develop as male, because a Y chromosome is present, but the ability of the testes to carry out spermatogenesis is severely impaired by the presence of an additional X chromosome. Females who carry additional X chromosomes (e.g. karyotype XXX or XXXX) may also have a shortened or impaired reproductive life because of damage to germ cell function, although the mechanism for this is not understood.

Certain individuals with a normal XY (male) karyotype lack the capacity to respond to androgens due to a receptor deficiency. Such individuals will develop testes but show no growth or development of the Wolffian ducts nor masculinization of the external genitalia.

Certain enzymatic deficiencies in otherwise normal XX individuals can result in over-production of androgens during fetal life. In such cases there may be mild or severe masculinization of the external genitalia, despite the presence of normal ovaries.

Y-chromosome), in whose presence testes develop. Indeed, studies using mice have shown that the SRY gene can induce maleness in XX individuals otherwise lacking in all other genes normally carried by the Y chromosome.

In the early embryo, the gonads of males and females are indistinguishable

For the first five or six weeks of fetal life, the gonads of both males and females develop identically. They are made up of two different types of tissue: somatic mesenchymal tissue, which forms the matrix of the organ, and the primordial germ cells, which form the gametes. Ridges of mesenchymal tissue (the primitive sex cords) develop on either side of the dorsal aorta between three and four weeks of gestation. The primordial germ cells originate outside these ridges but migrate via the developing hindgut, the gut mesentery, and the region of the kidneys to lie between and within the sex cords by the sixth week of fetal life. At the same time the population of germ cells expands by mitosis.

The development of testes depends on the presence of a Y-chromosome

Up until around six weeks of gestation, the gonads of males and females are indistinguishable and are said to be 'indifferent'. The presence of a Y-chromosome within the mesodermal cells of the genital ridge initiates the conversion of an indifferent gonad into a testis. After completion of the migration of the germ cells to the primitive sex cords, divergence of the gonads resulting from Y-chromosome determination of 'maleness' starts to become apparent. The primitive sex cords of the male embryo undergo considerable proliferation to make contact with in-growing mesonephric tissue and form a structured organ surrounded by a fibrous layer, the tunica albuginea. The cells of the sex cords, incorporating primordial germ cells, secrete a basement membrane. They are then known as the seminiferous cords and will give rise to the seminiferous tubules of the fully developed testis. Within these cords, the primordial germ cells will eventually give rise to spermatozoa, while the mesenchymal cord cells will form the Sertoli cells. The specific endocrine Leydig cells form as clusters within the stromal mesenchymal tissue that lies between the cords.

In the absence of a Y-chromosome, the changes in gonadal organization do not occur—the developing female gonad appears to remain indifferent. The primordial germ cells continue to proliferate mitotically and the primitive sex cords disappear. A second set of cords arises in the cortical region of the gonad, and breaks up into clusters of cells that surround the germ cells. In this way, the primitive follicles that characterize the ovary are laid down—the germ cells form the oocytes while the cord cells form the granulosa cells of the follicles. Between the follicles, groups of interstitial cells are laid down.

To summarize, during the early development of the fetal gonads, activity of a small part of the Y-chromosome appears to play an essential role in triggering the divergence of the primitive sex organs. If the SRY gene is present, the indifferent gonad is converted to a testis with seminiferous cords containing primordial germ cells, which will form sperm, and mesenchymal tissue that will give rise to the Sertoli and Leydig cells. In the absence of a Y-chromosome the indifferent organ forms an ovary containing a population of primordial follicles.

Subsequent development of the male and female genitalia depends on the hormones secreted by the gonads

Once the fetal gonads are established, the role of the sex chromosomes in the determination of sex is largely complete. Subsequent steps in the development of the male and female genital organs seem to be determined by the nature of the gonads themselves. This is particularly so in the case of the male, in whom the fetal testes secrete two hormones which appear to play a key role in differentiation of the male genitalia. These are testosterone from the Leydig tissue and a protein known as Müllerian inhibiting hormone (MIH) from the Sertoli cells. In the absence of these hormones (i.e. when ovaries are present), female genitalia are formed as outlined in Figures 50.10 and 50.11.

The fetus possesses two primordial internal genital tissues, the Wolffian duct which forms male organs and the Müllerian duct which gives rise to female parts. In a female fetus in whom ovaries have developed, the Wolffian duct disappears (possibly as a consequence of the lack of testosterone), and the Müllerian ducts go on to develop into the Fallopian tubes, uterus, cervix, and upper vagina. In a male fetus, however, testosterone seems to stimulate development of the Wolffian ducts to give rise to the epididymis, seminal vesicles, and vas deferens. At the same time, the female Müllerian ducts regress under the influence of MIH secreted by the Sertoli cells.

Fetal androgens also play a part in the development of the male external genitalia. They bring about fusion of the urethral folds to enclose the urethral tube and fusion of the genital swellings to form the scrotum. There is also enlargement of the genital tubercle to form the penis. In the female, the urethral folds and genital swellings remain separate to form the labia while the genital tubercle forms the small clitoris. These stages of development are represented diagrammatically in Figures 50.11 and 50.12. As with the divergence of the fetal sex organs, the male pattern of differentiation must actively be induced. In the absence of intervention, the female pattern develops inherently.

Summary

Humans have 23 pairs of chromosomes, one pair of which are the sex chromosomes. The female (homogametic sex) has two X chromosomes while the male (heterogametic sex) has one X and one Y chromosome. In the presence of a Y chromosome the indifferent gonads of the fetus develop as testes, but in its absence, ovaries develop. Subsequent steps in the development of the male and female genital organs seem to depend on the gonads themselves. Androgens from the fetal testes play a particularly important role in stimulating the development of the internal male genitalia from the Wolffian ducts. In the presence of ovaries, the Müllerian ducts develop into the internal female genitalia.

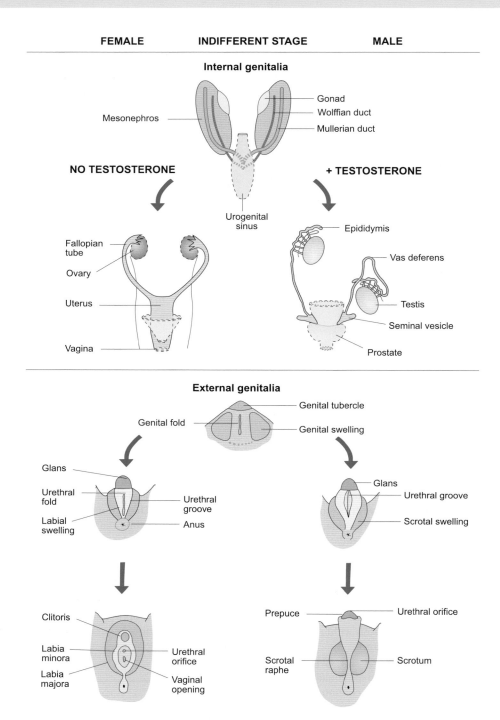

Figure 50.11 An outline of the role of testosterone in the development of the internal and external genitalia. In the upper panel one of the testes is shown in the process of descent. Note that the male pattern of differentiation is actively induced by testosterone. In its absence, the female pattern develops inherently. (Based on Fig 7.3 in N.E. Griffin and S.R. Ojeda (1995) *Textbook of endocrine physiology*, 2nd edition. Oxford University Press, Oxford.)

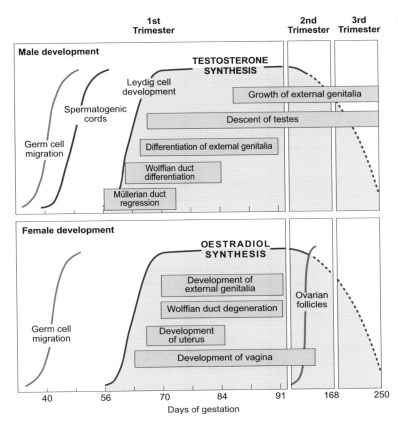

Figure 50.12 A diagrammatic representation of the timing of the differentiation of the internal and external genitalia of the human fetus. (Based on Fig 7.2 in N.E. Griffin and S.R. Ojeda (1995) *Textbook of endocrine physiology*, 2nd edition. Oxford University Press, Oxford.)

Fetal circulation

- The fetus depends on the placenta for gas exchange and waste disposal. Oxygenated blood is delivered to the fetus from the placenta via the umbilical vein.

- The fetal circulation is arranged so that the two sides of the heart work in parallel.

- Three fetal shunts permit the bulk of the circulating blood to bypass those organs with little or no function, while ensuring that the brain receives blood with the maximum possible oxygen content.

- The ductus venosus allows a significant fraction of the oxygenated blood arriving in the umbilical vein to bypass the liver and enter the inferior vena cava directly.

- The foramen ovale lies between the right and left atria. It provides a route for oxygenated blood from the inferior vena cava to pass directly into the left atrium.

- The ductus arteriosus connects the pulmonary arterial trunk with the descending aorta. It allows blood to pass directly into the aorta, bypassing the non-functional lungs.

- The fetal heart rate is high and the blood pressure low.

Fetal haemoglobin

- Despite its low PaO_2, fetal blood carries about 16 ml O_2 dl^{-1}.

- Fetal blood has a higher haemoglobin concentration than adult blood.

- Fetal haemoglobin (HbF) has a higher affinity for oxygen than adult haemoglobin (HbA).

The first breath and neonatal respiration

- After delivery and separation from the placenta, a baby must start to breathe air.

- Hypercapnia is a likely trigger for breathing, but physical factors such as temperature and tactile stimulation may also contribute.

- To inflate its lungs for the first time, a baby must overcome the surface tension forces at the gas–liquid interface in the alveoli.

- Surfactant plays a vital role in reducing these forces, but a massive mechanical effort is still required to generate the intrathoracic pressures needed to open the lungs.

- Babies born before adequate levels of surfactant have been produced may suffer respiratory distress.

- Ventilatory rate is high but erratic in neonates.
- Airway resistance is higher and the work of breathing is greater in the neonate than in adults.
- Central medullary chemoreceptors are active from about mid-gestation, and the ventilatory response to hypercapnia is well developed in both the fetus and neonate.
- Peripheral chemoreceptors have a very low level of activity in the fetus but start to respond to reductions in oxygen tension following delivery.

Circulatory adaptations following delivery

- After delivery, the circulation of the infant adapts to pulmonary gas exchange.
- The three fetal shunts close and the adult circulatory pattern of pulmonary and systemic circulations in series is established.
- Closure of the shunts depends upon ventilation itself.
- As the lungs expand, pulmonary perfusion increases and left atrial pressure rises above right atrial pressure, leading to closure of the septa that form the foramen ovale.
- The ductus venosus and the ductus arteriosus are thought to close as a result of vasoconstriction in response to a rise in PaO_2.

Fetal adrenal glands

- The fetal adrenal glands are relatively large, consisting of three regions, the medulla, a small 'definitive' cortex, and a large fetal zone.
- The medulla secretes catecholamines. Secretion is increased by stress.
- The fetal zone provides the precursors for the synthesis of oestrogens by the placenta.
- The 'definitive' cortex converts progesterone to cortisol.
- Fetal cortisol stimulates surfactant production, accelerates maturation of the liver, and has a role in the initiation of parturition.

Fetal kidneys and fluid balance

- The fetal kidneys play a part in fluid and acid–base balance.
- Hypotonic urine is produced from around eight weeks gestation.
- Glucose reabsorption is comparable to that of adults relative to glomerular filtration rate (GFR), but sodium reabsorption is comparatively low.
- From birth, GFR and urine output increase gradually, as does the ability to concentrate urine.

The fetal gastrointestinal tract

- The fetus is nourished solely by the placenta, which supplies it with glucose, and the fetal gut is relatively immature.
- There is a limited degree of gastrointestinal motility and secretory activity.
- The contents of the fetal large intestine accumulate as meconium.
- Meconium may be passed into the amniotic fluid during fetal distress. Aspiration of meconium-stained amniotic fluid can damage the lungs.
- With the establishment of milk feeds, the major metabolic substrate switches from glucose to fats.
- Digestive juice secretion and gut motility increase.

Regulation of core temperature in the neonate

- A newborn infant can lose heat very rapidly because of its high surface area to volume ratio, relatively high cardiac output, and lack of insulating fat.
- Babies can generate large quantities of heat through the metabolism of brown fat.
- Brown fat is located around the kidneys, at the nape of the neck, between the scapulae, and in the axillae.
- The metabolism of brown fat is stimulated by catecholamines released in response to cold stress.

Sexual differentiation

- Humans have 23 pairs of chromosomes, one pair of which are the sex chromosomes.
- The female has two X chromosomes, while the male has one X and one Y chromosome.
- The reproductive organs of the fetus remain undifferentiated (indifferent) until around five to six weeks gestation.
- The presence of a Y chromosome is required for the development of testes. In its absence, ovaries develop.
- Subsequent steps in the development of the male and female genital organs seem to depend on the presence or absence of testes.
- Androgens secreted by the fetal testes stimulate the development of the internal male genitalia from the Wolffian ducts.
- In the presence of ovaries (and the absence of androgens), the Müllerian ducts develop into Fallopian tubes, uterus, cervix, and upper vagina.

Recommended reading

Blackburn, S. (2007) *Maternal, fetal, & neonatal physiology: A clinical perspective* (3rd edn). Elsevier Health Sciences, London.

Harding, R., and Bocking, A.D. (2001) *Fetal growth and development.* Cambridge University Press, Cambridge.

Johnson, M.H. (2007) *Essential reproduction* (6th edn), Chapters 12 and 13. Blackwell Scientific, Oxford.

Kovacs, W.J., and Ojeda, S.R. (eds) (2012) *Textbook of endocrine physiology* (6th edn), Chapter 7. Oxford University Press, Oxford.

Chapter 7 provides a more detailed account of the principal steps in sexual differentiation, as well as some disorders of gonadal sex.

Review articles

Forbes, K. and Westwood, M. (2010) Maternal growth factor regulation of human placental development and fetal growth. *Journal of Endocrinology* **207**, 1–16, doi:10.1677/joe-10-0174.

Griswold, M.D. (2016) Spermatogenesis: The commitment to meiosis. *Physiological Reviews* **96**, 1–17, doi:10.1152/physrev.00013.2015.

Hillman, N., Kallapur, S.G., and Jobe, A. (2012) Physiology of transition from intrauterine to extrauterine life. *Clinical Perinatology* **39**, 769–83.

Ruwanpura, S.M., McLachlan, R.I., and Meachem, S.J. (2010) Hormonal regulation of male germ cell development. *Journal of Endocrinology* **205**, 117–31, doi: 10.1677/joe-10-0025.

Wilhelm, D., Palmer, S., and Koopman, P. (2007) Sex determination and gonadal development in mammals. *Physiological Reviews* **87**, 1–28, doi:10.1152/physrev.00009.2006.

To check that you have mastered the key concepts presented in this chapter, complete the accompanying online self-assessment questions. Go to www.oup.com/uk/pocock5e/

CHAPTER 51

The physiology of bone and the control of growth

Chapter contents

This chapter should help you to understand:

- Patterns of growth before and after birth

- The development and growth of bones during fetal life and childhood

- The physiology of bone

- The role of growth hormone in growth and development of the skeleton

- The consequences of over- and under-production of growth hormone in childhood and adulthood

- The role of other hormones in growth and the physiology of the skeleton

- The adolescent growth spurt and the influence of the sex steroids

- The factors which govern the overall size of tissues and organs

- The processes involved in carcinogenesis—the transformation of normal into malignant cells

51.1 Introduction

All biological tissues are made up of cells. Life begins as a single cell, the fertilized egg, from which all the diverse cell types of the body arise within a few weeks. Very early in development, cells begin to specialize and develop into particular types—liver cells, nerve cells, epithelial cells, muscle cells, and so on. Each cell type has its appropriate place within the organism. This development of specific and distinctive features is known as **differentiation**. Differentiated cells maintain their specialized character and pass it on to their progeny through the process of mitosis (see Chapter 4).

Overall growth of the body involves an increase in size and weight of the body tissues with the deposition of additional protein, and is thus a measurable quantitative change. In contrast, development occurs through a series of coordinated *qualitative* changes that affect the complexity and function of body tissues. Developmental change is most rapid while an individual is young.

Growth and development are complex processes that are influenced by a number of different factors, both genetic and environmental. Genetic and epigenetic factors set both the basic guidelines for the overall height that may be achieved (as indicated by the correlation of adult height between parents and children) and the pattern and timing of growth spurts. The major influence superimposed upon the genetic makeup of an individual is probably nutritional, although illness, trauma, and other socio-economic factors such as smoking can also modify the processes involved in growth. A child who has a diet that is inadequate either with regard to its quality or quantity will be unlikely to achieve his or her full genetic potential in terms of adult height. Indeed, improved nutrition is cited as one of the most important factors in the increase in average height that has been noted in Western societies over the last century.

Growth occurs at the level of individual cells, in populations of cells (i.e. in the tissues and organs), and at the level of the whole body. The underlying processes are regulated by a number of different hormones including growth hormone, thyroid hormones, and the sex steroids. The general properties of these hormones have been discussed in earlier chapters, and only those aspects of endocrine activity that relate specifically to growth will be considered here. Overall body growth will be considered first, and be given the greatest emphasis. While a detailed discussion of the many factors responsible for the development and maintenance of appropriately sized populations of differentiated cells is beyond the scope of this book, the importance of the factors that control the overall size of cell populations cannot be ignored and a brief overview, including abnormalities of cell and tissue growth, will be given at the end of the chapter.

Patterns of growth during fetal life

The period of prenatal growth is of great importance to an individual's future well-being. The development of sensitive ultrasound techniques makes it possible to monitor fetal size very accurately throughout pregnancy. Measurements of abdominal circumference, femur length, and biparietal diameter (the distance across the head measured from one ear to the other) are commonly taken to assess the increasing size of a fetus. Visualization and measurement of the internal organs can also help to detect abnormalities of growth and development during fetal life.

Figure 51.1 shows that the rate of increase in body length is at its greatest at 16–20 weeks gestation. Prior to this, particularly during the embryonic period (the first eight weeks after fertilization), growth velocity is lower, but during this phase the differentiation of the various body parts takes place. The head, arms, and legs begin to develop accompanied by the specialization of cells into tissues such as muscle and nerve. Each region of the body is moulded into a definite shape as a result of cell migration and differential growth rates (this is known as **morphogenesis**). Until weeks 26–28 of gestation, the increase in fetal weight is due largely to the accumulation of protein, as the cells of the body multiply and enlarge. During the last ten weeks or so, the fetus starts to accumulate a considerable amount of fat (up to 400 g) that is distributed both subcutaneously and deep within the body. Peak velocity of fetal weight gain occurs at around 34 weeks gestation. The rate of increase then declines towards the time of delivery. The exact mechanism for this slowing is not clear, but it seems likely that the placental blood supply is less and less able to meet the ever-increasing nutritional demands of the fetus (see Chapter 48).

A large number of factors may influence the rate of fetal growth, but their relative importance remains unclear. Genetic, endocrine, and environmental factors are likely to be as important in fetal life as they are in postnatal development, with the genetic constitution setting the upper limits of fetal size and the level of nourishment provided by the placenta determining the extent to which the genetic potential is achieved. Nevertheless, placental efficiency will be affected by numerous maternal influences such as smoking, medication, alcohol consumption, and nutritional status.

Patterns of growth and development during childhood and adolescence

The rapid rate of growth seen in fetal life continues into the postnatal period but declines significantly through early childhood. There is further deceleration prior to the growth spurt of puberty. This

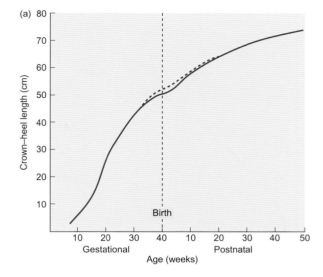

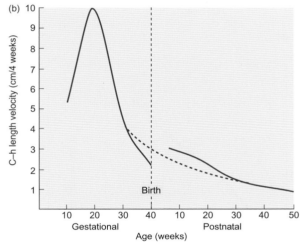

Figure 51.1 Fetal and early postnatal growth in boys. (a) The increase in length between the crown of the head and heel. (b) The rate of growth or growth velocity. In each case, the dotted lines represent the theoretical values expected in the absence of any uterine restrictions. In reality, growth commonly slows towards the end of gestation as the placenta becomes less able to meet the ever-growing demands of the fetus. 'Catch-up' growth is seen following delivery provided that the child is adequately nourished. (From Fig 16 in J.M. Tanner (1989) *Foetus into man*, 2nd edition. Castlemead, London.)

pattern is illustrated in Figure 51.2a, which shows the oldest known longitudinal record of growth, dating from the years 1759 to 1777.

The age at which the adolescent growth spurt takes place varies considerably between individuals. It occurs on average between 10.5 and 13 years in girls and between 12.5 and 15 years in boys. In general, the earlier the growth spurt occurs, the shorter will be the final stature. During this period, there is considerable variation in both stature and development between individuals of the same chronological age. The endocrine changes that accompany and contribute to the adolescent growth spurt will be considered further in Section 51.5.

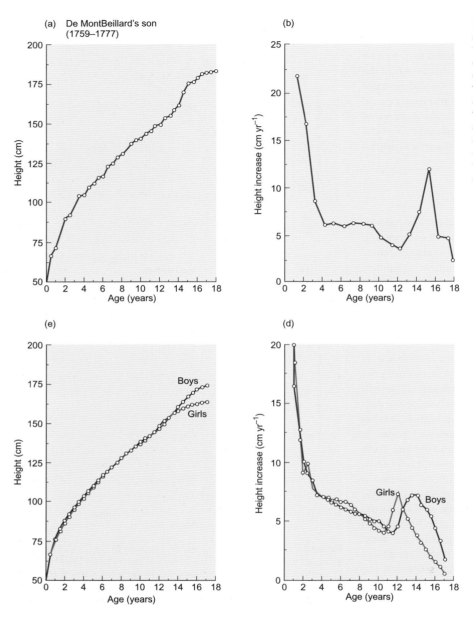

Figure 51.2 (a) and (b) The growth record of an individual child as recorded by his father in the eighteenth century. (c) and (d) The growth of boys and girls from early childhood to adulthood. Note the timing of the adolescent growth spurts, which begins around the age of ten years in girls and twelve years in boys. This difference in the age at which accelerated growth occurs accounts for the difference in final heights achieved. (Panels (a) and (b) are from Fig 1 in J.M. Tanner (1989) *Foetus into man*, 2nd edition. Castlemead, London.)

Most body measurements approximately follow the growth curves described for height. The skeleton and muscles grow in this manner, as do many internal organs such as the liver, spleen, and kidneys. Certain tissues, however, do not conform to this pattern and vary in their rate and timing of growth (see Figure 51.3). Examples include the reproductive organs (which show a significant growth spurt during puberty), the brain and skull, and the lymphoid tissue. The brain, together with the skull, eyes, and ears, develops earlier than any other part of the body and thus its development has a characteristic postnatal curve. The lymphoid tissue also shows a characteristic pattern of growth. It reaches its maximum mass before adolescence and then, probably under the influence of the sex hormones, declines to its adult value. In particular, the thymus gland, which plays a major role in the early development of the immune system, is a well-developed structure in children but involutes after puberty. In adults it is no more than a residual nodule of tissue.

Growth, even of the skeleton, does not cease entirely at the end of the adolescent period. Although there is no further increase in the length of the limb bones, the vertebral column continues to grow until the age of about 30 by the addition of bone to the upper and lower surfaces of the vertebrae. This gives rise to an additional height increase of 3–5 mm in the post-adolescent period. However, for practical purposes it may be considered that the average boy stops growing at around 17.5 years of age and the average girl at around 15.5 years of age, with a two-year variability range on either side.

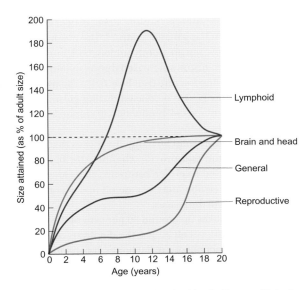

Figure 51.3 Growth curves for individual body tissues. Note the rapid growth of the head and brain. Peak growth of lymphoid tissues occurs around the age of ten years. The reproductive tissues develop much later.

Summary

Growth occurs at the level of the cells, the tissues, and the whole body. It is influenced both by genetic and by environmental factors. Hormones involved in the regulation of growth and development include growth hormone, thyroid hormones, and the sex steroids. Fetal growth is at its greatest between weeks 16 and 20 of gestation as new protein is laid down through the process of cell division. Before this time the emphasis is on morphogenesis—the differentiation and specialization of cells into tissues and organs. Fat is deposited predominantly during the last ten weeks of gestation. Growth velocity progressively declines until puberty, when there is a growth spurt. Adult height is achieved by the end of puberty, when growth in stature ceases. Individual tissues such as the brain and lymphoid tissue have characteristic growth patterns.

51.2 The physiology of bone

Bone is a specialized form of connective tissue that is made durable by the deposition of mineral within its infrastructure. In an adult, skeletal bone forms one of the largest masses of tissue, weighing 10–12 kg. Far from being the inert supporting structure its outward appearance might suggest, bone is a dynamic tissue with a high rate of metabolic activity, which is continuously undergoing complex structural alterations under the influence of mechanical stresses and a variety of hormones. Four main functions are ascribed to bone. These are:

1. to provide protection and structural support for the body

2. to provide an attachment for muscles, tendons, and ligaments and thus allow movement by means of articulations (joints)

3. the homeostasis of minerals (calcium and phosphate)

4. the formation of blood cells in the haematopoietic tissue of the red bone marrow (the red marrow is found mainly in the short, flat, and irregular bones—see Chapter 25).

Bone is formed from four elements. About 60 per cent of bone tissue is the mineral hydroxyapatite ($Ca_5(PO_4)_3$, which also forms the hard component of tooth enamel) in the form of a mineral matrix laid down by the action of mineralizing **bone cells**, and 25 per cent of the remaining tissue is fibrillar collagen. There is a significant water component (around 15 per cent by weight), and the remainder is composed of bone cells and extracellular proteins. The term 'osteoid' is used to describe the unmineralized component of bone tissue. Osteoid volume varies, but in normal bone of young children it is never more than 7–8 per cent and gradually decreases until it accounts for only 1–2 per cent of the total bone volume in adults. Osteoid fraction increases in some bone diseases, for example in Paget's disease (discussed later in the chapter), where it can be 10 per cent or more in spongy bone. The principal cells of bone are chondrocytes (found only in growth plates), osteoblasts (bone-forming cells), osteoclasts (bone-resorbing cells), osteocytes (mature bone cells), and fibroblasts. The adult skeleton contains between 1 and 2 kg of calcium (about 99 per cent of the body total) and between 0.5 and 0.75 kg of phosphorus (about 88 per cent of the body total).

The structure of bone

The anatomical features of a typical long bone are illustrated in Figure 51.4. The central shaft is called the **diaphysis**, while the regions at either end of the bone are the **epiphyses**. Between the diaphysis and epiphysis is a region of bone known as the **epiphyseal plate** or growth plate. Adjacent to this is the growing end of the diaphysis, known as the **metaphysis**. During growth, this region is made of cartilage, but once growth is completed at the end of puberty, the epiphyseal plate becomes fully calcified and remains as the **epiphyseal line**. Growth in length occurs by deposition of new cartilage at the metaphysis and its subsequent mineralization. Although the process by which bone becomes mineralized is not fully understood, it appears that calcium phosphate crystals become oriented along the collagen molecules of the organic matrix. Surface ions of the crystals are hydrated, forming a layer through which exchange of substances with the extracellular medium can occur.

The surfaces of the bones are covered by **periosteum**, which consists of an outer layer of tough fibrous connective tissue and an inner layer of osteogenic ('bone forming') tissue. A space runs through the centre of bones. This is the **marrow** (or medullary) space, which is lined with osteogenic tissue (the **endosteum**). The marrow spaces of the long bones contain mainly fatty yellow marrow that is not involved in haematopoiesis under normal circumstances. Red marrow containing haematopoietic tissue is found within the small, flat, and irregular bones of the skeleton such as the sternum, ilium, and vertebrae. It is here that blood cell production is carried out (see Chapter 25).

Long bones are supplied with blood by the nutrient artery, the periosteal arteries, and the metaphyseal and epiphyseal arteries.

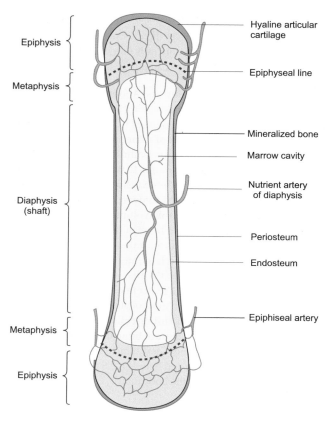

Epiphysis

Metaphysis

Diaphysis (shaft)

Metaphysis

Epiphysis

Hyaline articular cartilage

Epiphyseal line

Mineralized bone

Marrow cavity

Nutrient artery of diaphysis

Periosteum

Endosteum

Epiphiseal artery

Figure 51.4 The structural components of a typical long bone. The venous drainage is omitted for clarity but follows a similar pattern to that of the nutrient arteries.

The nutrient artery branches from a systemic artery and pierces the diaphysis before giving rise to ascending and descending medullary arteries within the marrow cavity. In turn, these give rise to arteries supplying the endosteum and diaphysis. The periosteal blood supply takes the form of a capillary network, while the metaphyseal and epiphyseal vessels branch off from the nutrient artery.

At rest, the skeleton accounts for around 12 per cent of the total cardiac output (2–3 ml 100 mg tissue^{-1} min^{-1}). The mechanisms that control skeletal circulation are poorly understood, but it is known that blood flow is significantly increased during inflammation and infection, and following fracture (see Box 51.1). The blood flow to the red bone marrow is also increased during chronic hypoxia, which enhances red blood cell production in response to erythropoietin secreted by the kidneys.

Bone is not uniformly solid but contains spaces that reduce the weight of the skeleton and provide channels for blood vessels. Bone may be classified as either **compact** (also called **dense** or **cortical**) or **spongy** (**trabecular** or **cancellous**) according to the size and distribution of the spaces.

Compact bone makes up around 80 per cent of the skeleton. It forms the outer regions of all bones, the diaphysis of long bones, and the outer and inner regions of flat bones. It contains few spaces and provides protection and support, especially for the long bones in which it helps to reduce the stress of weight bearing. The functional units of compact bone are the **Haversian systems** or **osteons**. As Figure 51.5 shows, osteons consist of a central canal, containing blood vessels, lymphatics, and nerves, surrounded by concentric rings of hard intercellular substance (lamellae) between which are spaces (lacunae) containing mature bone cells called osteocytes. Radiating from the lacunae are tiny canals (canaliculi) that connect with adjacent lacunae to form a branching network by which nutrients and waste products may be transported to and from the osteocytes. By contrast, **spongy bone**, which makes up the remaining 20 per cent or so of the skeleton, contains no true osteons. Instead, it consists of an irregular lattice of thin plates or spicules of bone called trabeculae between which are large spaces filled with bone marrow. Lacunae containing osteocytes lie within the trabeculae.

The osteocytes are nourished directly by blood circulating through the marrow cavities from blood vessels penetrating to the spongy bone from the periosteum. Figure 51.6 illustrates the different organization of dense and spongy bone. Spongy bone is found within the epiphyses of long bones (see Figure 51.6c) and at the

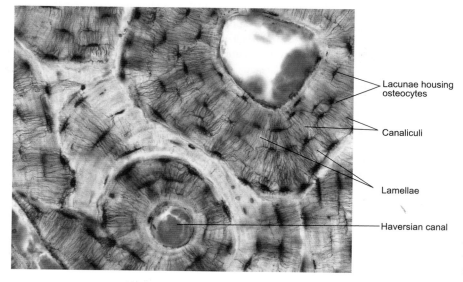

Lacunae housing osteocytes

Canaliculi

Lamellae

Haversian canal

Figure 51.5 The structure of cortical bone. Note the lamellar organization of the Haversian systems. The lacunae, which imprison the osteocytes, are connected via thin canaliculi which, in life, contain thin osteocytic processes.

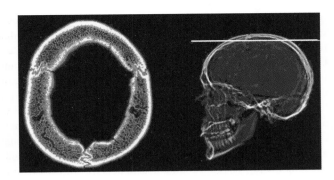

(a)

growth plates. It also makes up most of the mass of short, flat, and irregular bones such as those of the skull, which consist of thin plates of compact bone separated by a layer of spongy bone called the diploë (see Figure 51.6b).

The bone cells

Three major cell types are recognized in histological sections of bone. These are **osteoblasts**, **osteocytes**, and **osteoclasts**. Their general appearance is shown in Figure 51.7. Osteoblasts and osteocytes originate from progenitor cells within the osteogenic tissue of the bone. Osteoclasts differentiate separately from mononuclear phagocytic cells. Quiescent bone, which is not undergoing remodelling, is covered by lining cells which are derived from osteoblasts following bone formation.

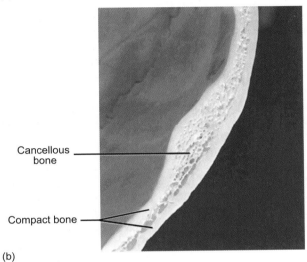

(b)

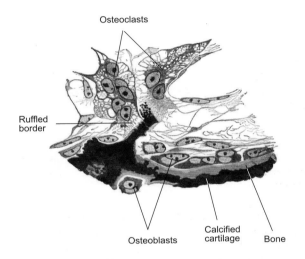

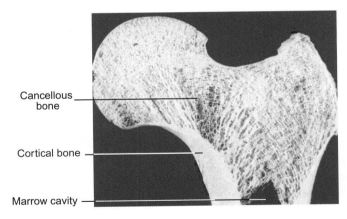

(c)

Figure 51.6 (a) The head of the femur showing the regions of dense and spongy bone. (b) A computed tomography x-ray image (CT scan) of the skull at the level shown by the white line of the inset picture. The bones of the skull consist of two outer layers of compact bone separated by a layer of trabecular bone (the diploë). (Panel (b) is from Fig 1 in J. Skrzat (2006) *Folia Morphologica* 65, 132–5.)

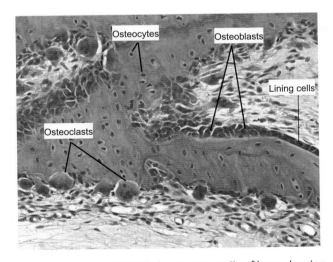

Figure 51.7 A section through the marrow cavity of bone showing calcified cartilage (here stained dark blue), osteoblasts, and two large multinucleate osteoclasts. Note the ruffled border next to the calcified cartilage. The lower panel shows a section of jaw in which all three bone cell types are clearly visible.

Osteoblasts are present on the surfaces of all bones and line the internal marrow cavities. They contain numerous mitochondria and an extensive Golgi apparatus associated with rapid protein synthesis. They secrete the constituents of osteoid, the organic matrix of bone, including collagen, proteoglycans, and glycoproteins. They are also important in the process of mineralization (calcification) of this matrix. Osteoblasts possess specific receptors for parathyroid hormone and calcitriol.

Osteocytes are mature bone cells derived from osteoblasts that have become trapped in lacunae (small spaces) within the matrix that they have secreted. Figure 51.8(a) shows the arrangement of osteocytes in the parietal bone of the skull and the inner and outer covering of osteoblasts. Note that adjacent osteocytes are linked by fine cytoplasmic processes that pass through tiny canals (canaliculi) between lacunae. This arrangement, clearly visible in Figure 51.5(a), permits the exchange of calcium from the interior to the exterior of bones, from where it can pass into the plasma. This transfer is known as osteocytic osteolysis and

may be used to remove calcium from the most recently formed mineral crystals when plasma calcium levels fall (for more details regarding whole-body calcium balance see Chapter 22). Osteocytes are mechanosensors and are responsible for initiating the process of bone remodelling in response to damage or changes in load.

Osteoclasts are giant multinucleated cells that are believed to arise from the fusion of several precursor cells and contain numerous mitochondria and lysosomes. They are highly mobile cells that are responsible for the resorption of bone during growth and skeletal remodelling. They are abundant at or near the surfaces of bone undergoing erosion. At their site of contact with the bone is a highly folded 'ruffled border' of microvilli that infiltrates the disintegrating bone surface (see panel (b) of Figure 51.8). Bone dissolution is brought about by the actions of collagenase, lysosomal enzymes, and acid phosphatase. Calcium, phosphate, and the constituents of the bone matrix are released into the extracellular fluid as bone mass is reduced. The activity of the osteoclasts is regulated by osteoblasts via various factors. In addition, hormones including parathyroid hormone, calcitonin, thyroxine, oestrogens, and the metabolites of vitamin D regulate the activity of osteoclasts via their action on osteoblasts.

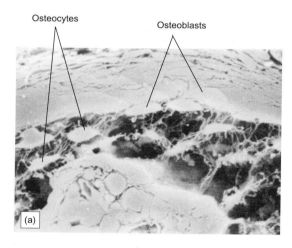

(a)

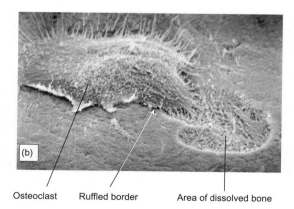

(b)

Osteoclast   Ruffled border   Area of dissolved bone

Figure 51.8 Panel (a) shows a scanning electron micrograph of parietal bone osteocytes which are sandwiched between the cells that line the outer and inner regions of bone. The bone itself has been dissolved to reveal the fine canaliculi. Panel (b) is a scanning electron micrograph of an osteoclast. ((a) is from Fig 1a in K. Abe et al. (1992) *Journal of Electron Microscopy* **41**, 113–15; (b) Courtesy of Professor Tim Arnett.)

Summary

Growth of the long bones is responsible for the increase in height seen during childhood and adolescence. It occurs through the proliferation and hypertrophy of cartilage cells at the growth plates of the long bones followed by their subsequent mineralization.

Bone consists of three types of bone cells (osteoblasts, osteocytes, and osteoclasts) and osteoid, an organic matrix strengthened by the deposition of complex crystals of calcium and phosphate (hydroxyapatite). The osteoblasts secrete the organic matrix, osteocytes are mature bone cells, and the osteoclasts are responsible for the resorption of old bone during growth and remodelling of the skeleton.

51.3 Bone development and growth (osteogenesis)

At six weeks' gestation, the fetal skeleton is constructed entirely of fibrous membranes and **hyaline cartilage**. From this time, bone tissue begins to develop and eventually replaces most of the existing structures. Although this process of osteogenesis (also called ossification) begins early in fetal life, it is not complete until the third decade of adult life, and bone remodelling and repair continues even after this time. The flat bones of the cranium, lower jaw, scapula, pelvis, and clavicles develop from fibrous membranes by a process called **intramembranous ossification**. For this reason, these bones are collectively known as membrane bones. The bones of the rest of the skeleton are called endochondral bones and develop from hyaline cartilage templates that are gradually replaced by bone (a process known as **endochondral ossification**).

The development of flat bones by intramembranous ossification

At around week 6–8 of fetal development, centrally located mesenchymal cells within the embryonic fibrous connective tissue membrane begin to cluster together (condense). Some of these cells develop into capillaries, while others differentiate into osteoblasts. These condensations of developing osteoblasts are known as **ossification centres**. The osteoblasts then start to secrete **osteoid**, the collagen–proteoglycan matrix for mineralization (or calcification), which takes place within a few days. Following mineralization those osteoblasts that are trapped within the matrix become osteocytes—bone cells. Bony spicules or trabeculae start to radiate outwards from the ossification centre in a fairly random fashion between developing blood vessels to form 'woven' bone, which is not normally found in adults as it is usually converted to lamellar bone. However, woven bone is present in adults in areas near the sutures of the flat bones of the skull, the tooth sockets, and the insertion sites of some tendons.

The bony spicules subsequently become surrounded by mesenchymal cells, which form the periosteum. Osteoblasts lying along the inside of the periosteum continue to secrete osteoid, which subsequently becomes calcified. In this way, layers of bone are built up beneath the periosteum. As these layers become thicker they eventually form a plate of compact bone. Deeper within the bone, the trabecular structure persists, giving rise to spongy bone. A simple schematic of the principal steps occurring during intramembranous ossification is shown in Figure 51.9.

The development of short and long bones by endochondral ossification

The short and long bones of the skeleton are formed by endochondral ossification, in which a cartilage template is gradually replaced by bone. This is a rather complex process with many steps, so the following will of necessity be a simplified account, using the growth of a long bone such as the femur (thigh bone) or radius of the forearm as an example. The series of diagrams shown in Figure 51.10 summarizes the principal steps involved.

- The bone is laid down first as a hyaline cartilage model formed by chondroblasts and covered by a membrane called the perichondrium. Once the chondroblasts have become surrounded by cartilage matrix, they are called chondrocytes. Chondrocytes continue to divide and to secrete cartilage matrix so that the model grows.

- During the second month of gestation, the bone will begin to develop from a region within the diaphysis called the **primary ossification centre**. The perichondrial membrane becomes infiltrated with blood vessels and, with their increased supply of nutrients, the underlying cells differentiate into osteoblasts, which secrete a layer of osteoid against the hyaline cartilage. This layer of newly secreted bone forms a cylindrical column around the middle of the cartilage model—the **periosteal bone collar**.

- As the cartilage model grows, cartilage cells in the centre of the model begin to undergo a degenerative process of programmed

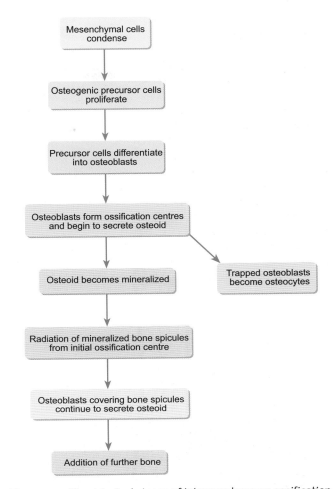

Figure 51.9 The principal stages of intramembranous ossification.

cell death (apoptosis) during which they first hypertrophy and then die. At the same time, the cartilage model begins to calcify. In this way a calcified cartilage matrix is formed within the diaphysis of the cartilage model beneath the bone collar. Death of chondrocytes creates spaces (lacunae) within the model, which later merge.

- During the third month of fetal development, periosteal capillaries begin to grow into the region of disintegrating cartilage. Associated with these capillaries are nerve fibres, osteoblasts, osteoclasts, and red bone marrow cells which form a **periosteal bud**. The osteoclasts in the bud begin to erode the calcified cartilage matrix, while the osteoblasts secrete osteoid around the remaining cartilage fragments. The resulting arrangement is of cartilage trabeculae covered by bone—in essence, fetal spongy bone.

- As the primary ossification centre enlarges and grows towards the ends of the bone, some of the spongy bone is destroyed by osteoclasts to create a space (the marrow cavity) in the centre of the diaphysis.

- Towards the end of fetal life, secondary centres of ossification develop, predominantly at the ends (epiphyses) of the bone. Bone

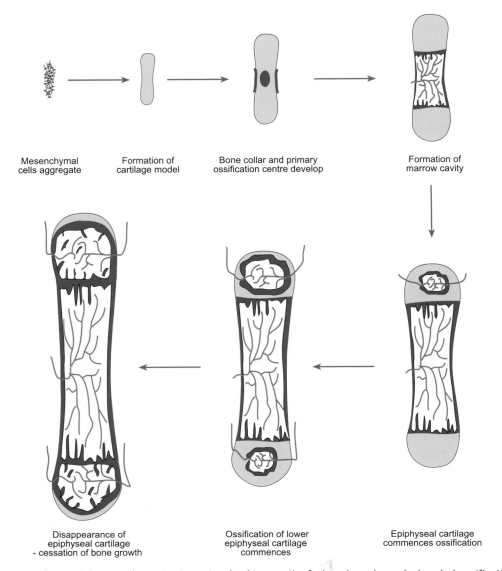

Mesenchymal
cells aggregate

Formation of
cartilage model

Bone collar and primary
ossification centre develop

Formation of
marrow cavity

Disappearance of
epiphyseal cartilage
- cessation of bone growth

Ossification of lower
epiphyseal cartilage
commences

Epiphyseal cartilage
commences ossification

Figure 51.10 A schematic diagram showing the main stages involved in growth of a long bone by endochondral ossification.

formation continues at these secondary sites, but here spongy bone remains within the epiphyses. Smaller bones such as the carpals and tarsals of the hands and feet develop from a single ossification centre.

- During childhood and adolescence, cartilage remains between the epiphyses and diaphysis of the bone. These areas of cartilage are known as the epiphyseal growth plates.

Growth of bone length

Elongation of the long bones shares many of the characteristics of endochondral ossification previously described. In the part of the growth plate immediately under the epiphysis is a layer of stem cells or chondroblasts. These give rise to clones of cells (chondrocytes) arranged in columns extending inwards from the epiphysis towards the diaphysis.

Several zones may be distinguished within the columns of chondrocytes. The outer zone is one of proliferation, in which the cells are dividing rapidly. Beneath this are layers in which the cells mature, enlarge, and eventually degenerate, as shown in Figure 51.11. The innermost layer of cells is the region of calcification. Here, the osteogenic cells differentiate into osteoblasts and lay down bone. In the radiographs of growing hands illustrated in Figure 51.12, the regions at which rapid calcification are occurring are clearly visible as areas of high density (which appears white in the radiographs).

Thus at one end of the epiphyseal plate cartilage is produced, while at the other end it is degenerating. Growth in length is therefore dependent upon the proliferation of new cartilage cells. In humans, it takes around 20 days for a cartilage cell to complete the journey from the start of proliferation to degeneration. Clearly, the bone marrow cavity must also increase in size as the bone grows and, to ensure this, osteoclasts erode bone within the diaphysis.

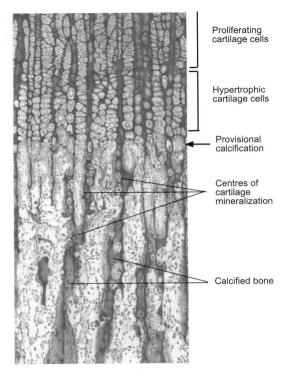

Proliferating cartilage cells

Hypertrophic cartilage cells

Provisional calcification

Centres of cartilage mineralization

Calcified bone

Figure 51.11 Decalcified bone showing the process of bone growth and ossification in a typical long bone. New cells formed in the proliferative region move down to the hypertrophic region to add to bone accumulating on the top of the diaphysis.

At the end of the growth period, the growth plate thins as it is gradually replaced by bone until it is eliminated altogether and the epiphysis and diaphysis are unified, a process known as **synostosis**. Following this 'fusion' of the epiphyseal plate, no further increase in bone length is possible at this site. Although growth in length of most bones is complete by the age of 20, the clavicles do not ossify completely until the third decade of life. The dates of ossification are fairly constant between individuals but different between bones. This fact is exploited in forensic science to determine the age of a body according to which bones have ossified and which have not.

Growth of bone diameter

The growth in width of long bones is achieved by appositional bone growth, in which osteoblasts beneath the periosteum of the bone form new osteons on the external surface of the bone. The bone thus becomes thicker and stronger. Rapid ossification of this new tissue takes place to keep pace with the growth in length of the bone. This process is similar to the mechanism of intramembranous ossification by which the flat bones grow.

Remodelling of bone

Even after growth has ended, the skeleton is in a continuous state of remodelling as it is renewed and revitalized at the tissue level. Large volumes of bone are removed and replaced and bone architecture continually changes, as 5–20 per cent of bone mass is recycled each year. Furthermore, following a break to a bone, self-repair takes place remarkably quickly (see Box 51.1). Remodelling allows

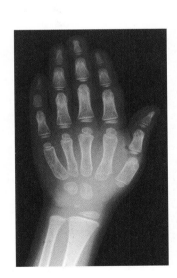

(a) Radiograph of hand of a 2 yr old child

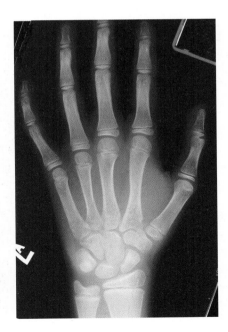

(b) Radiograph of hand of an 11 yr old child

Figure 51.12 Radiographs of the hands of two children aged (a) 2 years and (b) 11 years. Note the increase in the number of ossified carpal bones in the wrist of the 11-year-old. (Courtesy of Ruth Denton.)

Box 51.1 Bone healing following fracture

When bone is fractured, its original structure and strength is restored quite rapidly through the formation of new bone tissue. Provided that the edges of the fractured bone are repositioned and that the bone is immobilized by splinting, repair will normally occur with no deformity of the skeleton. There are three stages in the repair of a fractured bone. The first stage occurs during the first 4 or 5 days after injury and involves the removal of debris resulting from the tissue damage. This includes bone and other tissue fragments as well as blood clots formed by bleeding between the bone ends and into surrounding muscle when the periosteum is damaged. Phagocytic cells such as macrophages clear the area and granulation tissue forms. This is a loosely gelled protein-rich exudate that forms at any site of tissue damage and which later becomes fibrosed and organized into scar tissue. As it revascularizes from undamaged capillaries in adjacent tissue, it takes on a pink, granular appearance. Osteoblasts within the endosteum and periosteum migrate to the site of damage to initiate the second stage of healing, as in Figure 1 (a). During this stage, which normally lasts for the next three weeks or so, osteoid is secreted by the osteoblasts into the granulation tissue to form a mass between the fractured pieces of bone to bridge the gap. This tissue mass is also known as **soft callus**, shown in Figure 1(b). The soft callus gradually becomes ossified to form a region of woven bone (similar to cancellous bone) also called **hard callus**: see Figure 1(c). At this stage of healing there is normally some degree of local swelling at the site of the fracture caused by the hard callus deposit. During the final stage in the process of healing, the mass of hard callus is restructured to restore the original architecture of the bone. This stage may take place over many months and involves the actions of both osteoblasts and osteoclasts. During this time, the periosteum also reforms and the bone is able to tolerate normal loads and stresses. The radiographs shown in Figure 2 illustrate the main stages of healing following a fracture.

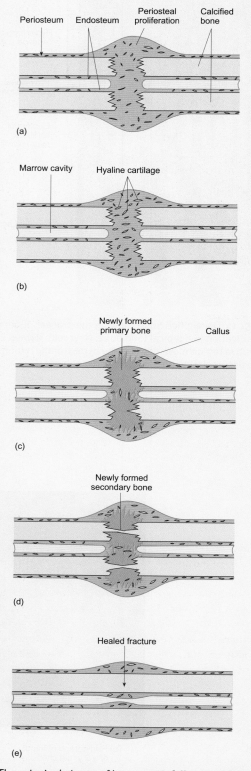

Figure 1 The principal stages of bone repair following a fracture.

Box 51.1 (*Continued*)

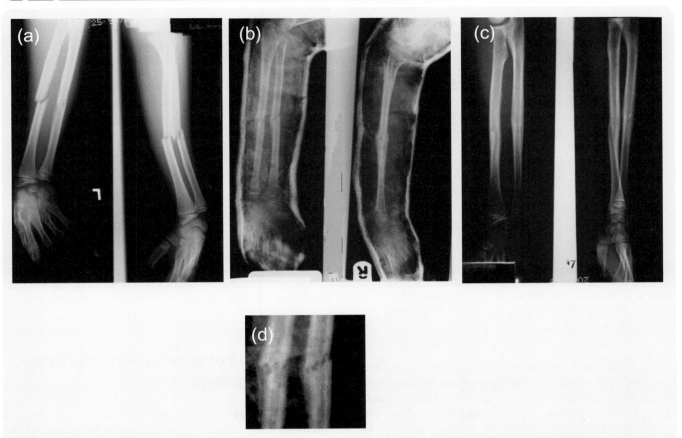

Figure 2 Radiographs showing the stages of repair following a fracture. (a) Simple fracture of humerus and radius of forearm. (b) Development of soft callus. (c) Final healing of fracture. (d) The soft callus in more detail. Note that the radiograph shown in (b) was taken through the immobilizing plaster cast. (Courtesy of Ruth Denton.)

bone to adapt to external stresses, adjusting its formation to increase strength when necessary and repairing any microscopic fractures caused by everyday stresses. The rate of bone remodelling is controlled by osteocytes and is lower in cortical bone (c. 5 per cent) and higher in trabecular bone (10–20 per cent). It is increased by disuse, notably during prolonged bed rest. Conversely, high levels of physical activity, such as that performed by elite athletes, will lead to an increase in both bone density and muscle mass. Normal exercise helps to maintain bone density, but does not significantly affect bone structure. Remodelling occurs in cycles of activity in which resorption precedes formation. First, bone is eroded by the osteoclasts. This erosion is followed by a period of intense osteoblastic activity in which new bone is laid down to replace that which has been resorbed.

Bone remodelling occurs in distinct cellular units called bone remodelling units. The signal that initiates a cycle of remodelling is not yet established with complete certainty, but mechanical stresses have been implicated. These appear to stimulate the osteocytes in the affected area to secrete IGF-1 (see Section 51.4) and other growth factors. The first stage of remodelling requires the recruitment of osteoclast precursors to the site of resorption. This is followed by retraction of the lining cells to expose the bone surface, and stimulation of osteoclasts by the osteoblasts. The osteoclasts bind to extracellular proteins via their integrin receptors and form a resorption compartment in which the ruffled border secretes hydrogen ions to dissolve the mineral component and proteolytic enzymes to digest the collagen (see Figure 51.13). A specific form of carbonic anhydrase found in bone (carbonic anhydrase II) catalyses the formation of carbonic acid which acts as the source of hydrogen ions. The hydrogen ions are secreted into the resorption compartment by a proton pump, where they act to dissolve the mineral component of bone. The osteoclasts eventually undergo apoptosis (programmed cell death—see Section 51.7) and disappear to be replaced by osteoblasts which then secrete fresh osteoid. In this way they reform new mineralized bone within the resorption cavity. Some osteoblasts remain trapped within the mineralized bone as osteocytes. Overall, the quantity of bone deposited matches that removed.

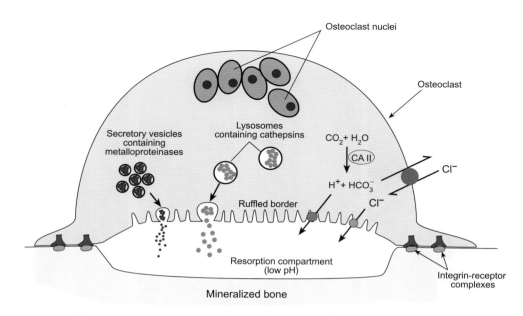

Figure 51.13 A diagram illustrating the mechanism of bone dissolution by an osteoclast. The cell is attached to the bone matrix via integrins to seal a small compartment beneath it. The membrane facing the bone is intensively folded to form the ruffled border. The cell secretes enzymes and hydrogen ions into the resorption compartment and takes up partially digested bone matrix by endocytosis (not shown in the figure). CAII is cytosolic carbonic anhydrase II.

In general terms, bone is deposited in proportion to the load it must bear. It follows therefore that in an immobilized person bone mass is rapidly (though reversibly) lost, a process known as **disuse osteoporosis**. Astronauts experiencing prolonged periods of weightlessness in space have also been shown to lose up to 20 per cent of their bone mass in the absence of properly planned exercise programmes. Similarly, appropriate exercise during childhood and adolescence is thought to enhance the development of bone and to result in a stronger, healthier skeleton in adult life, a factor that may be particularly important in women.

Summary

During fetal life, the flat and irregular bones of the skeleton are formed by intramembranous ossification, while the long and short bones of the skeleton are first laid down as cartilage with subsequent endochondral ossification. Ossification is not complete until the third decade of life.

The areas of cartilage between the diaphysis and epiphyses of long bones are called epiphyseal growth plates. Bones increase in length by the addition of new material at these growth plates, when stem cells (chondroblasts) give rise to proliferating clones of cells (chondrocytes) that extend into the diaphysis of the bone. As they progress, they mature and later degenerate. Calcification occurs within the innermost layer of cells. After puberty, the growth plate thins as it is replaced by bone and the diaphysis and epiphyses fuse. No further increase in length is possible after fusion. Around 5–20 per cent of total bone mass is recycled each year during adulthood as the skeleton responds to the changing demands placed upon it.

51.4 The role of growth hormone in the control of growth

Growth is the result of the multiple interactions of circulating hormones, tissue responsiveness, and the supply of nutrients and energy for growing tissues. Many hormones are known to be involved in the regulation of growth, at the various stages of life. Nevertheless, growth hormone is the hormone that undoubtedly exerts a dominant effect on normally coordinated growth.

The nature and secretion of growth hormone (GH) has been discussed fully in Chapter 21, but a brief summary may be helpful here. GH is a large polypeptide (M_r 22,000 Da) derived from the pituitary somatotrophs. Its secretion is controlled by hypothalamic hormones: GHRH (growth hormone releasing hormone) stimulates the output of GH, while somatostatin inhibits it. GH shows a marked irregular pulsatile pattern of release, which is influenced by a number of physiological stimuli. Stress and exercise, for example, both stimulate GH secretion and there is a significant increase in the rate of secretion during slow wave (deep) sleep, particularly in children. This pattern of secretion is illustrated in Figure 51.14, which also shows that both the pulsatile character and the sleep-induced patterns of release are lost in patients suffering from hypo- or hypersecretion of GH.

Other hormones and products of metabolism also influence the rate of GH secretion. Oestrogens, for example, increase the sensitivity of the pituitary to GHRH, an effect that contributes to the earlier growth spurt seen in adolescent girls compared with boys. GH secretion is decreased by the adrenal glucocorticoid hormones (e.g. cortisol) and stimulated by insulin. Oral glucose depresses GH release, while secretion is promoted by low levels of plasma glucose.

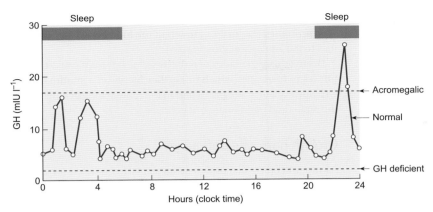

Figure 51.14 The pattern of secretion of GH in normal individuals and in patients suffering from acromegaly (GH-excess in adulthood) and GH-deficiency. Note that in each of the abnormal states, pulsatile secretion is lost. In acromegaly, there is a sustained high circulating level of GH. In normal healthy people, GH secretion shows significant spurts during phases of deep (non-REM) sleep. (Adapted from Fig 3.5 in C. Brook and N. Marshall (1996) *Essential endocrinology*. Blackwell Science, Oxford.)

In common with most endocrine systems, the secretion of GH is under negative feedback control. This is probably mediated both by GH itself (chiefly at the level of the hypothalamus) and by insulin-like growth factors (IGFs) that are thought to act both on the cells of the pituitary and on those of the hypothalamus. GH interacts with its target cells at the plasma membrane, where it binds to surface receptors. Synthesis of these receptors requires the presence of GH itself, while an excess of GH causes down-regulation of the receptors. The mechanisms of signal transduction have now been clarified. GH activates membrane-bound tyrosine kinases, which phosphorylate a group of proteins that activate gene transcription. The actions of GH may be divided into **metabolic** and **growth promoting** effects.

Metabolic actions

The metabolic actions of GH tend to oppose those of insulin and are largely direct in nature (see Chapter 21). GH exerts its direct actions on a variety of target tissues, principally liver, muscle, and adipose tissue. It depresses the rate of glucose uptake by muscle and adipose tissue but stimulates glycogenolysis by the liver, which raises plasma glucose and makes it available for use by the CNS—particularly in times of fasting or starvation. Furthermore, GH stimulates lipolysis, which increases the availability of fatty acids for oxidation. It also facilitates the uptake of amino acids into cells for protein synthesis.

Growth-promoting actions

The growth-promoting actions of GH embrace both direct and indirect effects. GH seems to exert a direct stimulatory effect on chondrocytes, increasing the rate of differentiation of these cells and, therefore, of cartilage formation. Many of the direct metabolic actions of GH, such as the increase in uptake of amino acids and the rate of protein synthesis, will also contribute to the overall processes of growth and repair.

The indirect actions of growth hormone are mediated by a family of peptide hormone intermediaries called **insulin-like growth factors (IGFs)**, formerly known as somatomedins. There are two forms, known as IGF-1 and IGF-2. They have a relative molecular mass of around 7000 and are structurally related to proinsulin, the precursor of insulin. The IGFs are synthesized in direct response to GH, chiefly by the liver but also by other tissues including cartilage and adipose tissue. Individuals who lack GH have low levels of IGF-1, but in healthy people, plasma IGF-1 is increased by the administration of GH, with a time lag of 12–18 hours. IGFs have plasma half-lives in excess of that of GH because they are carried in the blood bound to several proteins. The blood level of IGF-1 is low in infancy, rises gradually until puberty, then increases more swiftly to reach a peak that coincides with the peak height increase (see Figure 51.2) after which it falls to its adult (and pre-pubertal) value. IGF-2 plays an important role in fetal growth, when it promotes the growth and proliferation of cells in many developing tissues. Although IGF-2 is secreted in substantial amounts during fetal development, its secretion is much lower in adults. Inappropriate secretion of IGF-2 has been linked to certain cancers.

The actions of the IGFs, as their name suggests, tend to be insulin-like in character and account principally for the growth-promoting effects of GH. They act on cartilage, muscle, fat cells, fibroblasts, and tumour cells. More specifically related to bone growth is the action of IGFs in stimulating the clonal expansion of chondrocytes and the formation and maturation of osteoblasts in the growth plates of the long bones. All aspects of the functions of the chondrocytes are stimulated, including the incorporation of the amino acid proline into collagen and its subsequent conversion to hydroxyproline. Furthermore, GH (via IGFs) stimulates the incorporation of sulphate into chondroitin. Chondroitin sulphate and collagen, together, form the tough inorganic matrix of cartilage. Growth of soft tissue and the viscera is also attributed to the indirect actions of GH via the IGFs. A summary of the direct and indirect actions of GH, and of the factors regulating GH output, is shown in Figure 51.15.

The importance of GH in growth at different stages of life

Figure 51.16 illustrates the pattern of GH secretion throughout life. During the fetal period, GH itself is of little importance in the control of growth, and GH receptors do not appear until the final two months of gestation. The growth factors IGF-1 and IGF-2 appear to play a dominant role in the regulation of fetal growth.

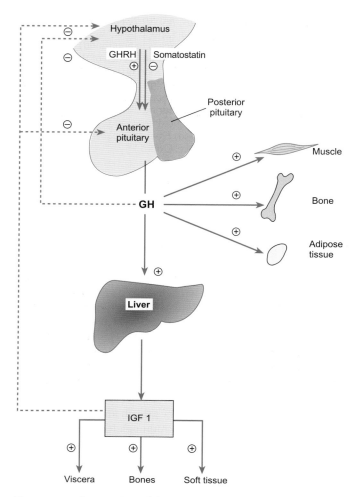

Figure 51.15 An overview of the regulation and effects of growth hormone on its target tissues. + indicates a stimulatory action, − an inhibitory one.

Following delivery, and in the early part of childhood, GH secretion increases considerably, and, during this phase, overall growth and increase in stature seems to depend almost entirely on the actions of GH itself and of IGF-1. At puberty, there is a further significant rise in GH secretion (probably associated with an increase in

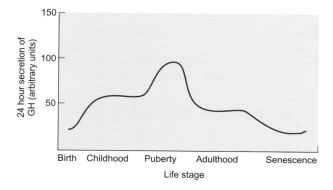

Figure 51.16 A diagram illustrating the lifetime pattern of GH secretion.

the output of sex steroids) with a parallel increase in IGF-1 output. This promotes the further growth of the long bones and contributes to the adolescent growth spurt.

During the final phases of puberty the sex steroids cause the epiphyses to fuse, after which no further increase in stature occurs. GH, however, may still play a part in the remodelling of bone and in the repair and maintenance of cartilage.

Growth hormone deficiency

As the preceding discussion suggests, GH is needed for normal growth between birth and adulthood. Individuals who lack GH (so-called pituitary dwarfs) will typically grow to a height of around 120–130 cm (about the height of a normal 6–8 year old) while remaining of normal proportions. This is in contrast to the disproportionate growth seen in achondroplasia, the congenital type of dwarfism in which growth of the bones is impaired due to defects in other local growth factors. A further type of growth impairment caused by defective GH receptors rather than a lack of the hormone itself is known as **Laron dwarfism**. People with this condition have the same physical appearance as those who lack growth hormone.

GH-deficient children may be treated by injections of human GH. After treatment, they usually achieve significant catch-up growth to within, or close to, the normal range. Figure 51.17 illustrates the 'catch-up' growth brought about by treatment of a GH-deficient girl with human GH. Children with Laron dwarfism will not benefit from GH-replacement therapy, as their cells are unresponsive to the hormone.

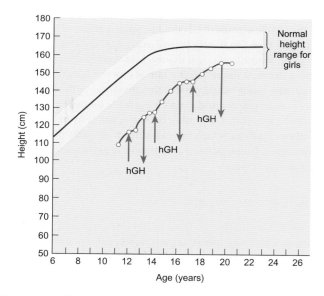

Figure 51.17 The pattern of growth in a girl with isolated pituitary GH-deficiency, treated with three periods of exogenous human growth hormone (hGH) administration. Note the 'catch-up' growth seen during the treatment periods. The downward-pointing arrows represent the end of each period of treatment. The range of normal heights for girls is indicated by the shaded area. (Based on data in Fig 33 in J.M. Tanner (1978) *Foetus into man*. Open Books, London.)

Unlike other hormones such as insulin and ACTH, growth hormone is species-specific: i.e. animal GH is without effect in humans. From 1958 until 1985, the GH administered to patients was extracted from the pituitary glands of human cadavers at post-mortem. Unfortunately a few of the children treated in this way have since become ill or died from the degenerative brain disease Creutzfeld–Jacob Disease (CJD). In recent years recombinant DNA technology has developed, and now human GH can be manufactured and used to treat GH-deficiency without risk of CJD.

Finally, short stature may be caused by a failure to produce the IGFs in response to GH rather than a simple lack of GH. In conditions of this kind, GH treatment will be of no value but such children can be treated with recombinant IGF-1.

Growth hormone excess

Although hypersecretion of GH may occur at any stage of life, the incidence of pituitary gigantism resulting from an excess of GH in childhood is extremely rare. Tumours of the pituitary gland or overgrowth of the GH-producing cells can, occasionally, cause vastly excessive (though proportionate) growth. A further condition characterized by extreme tallness is cerebral gigantism (Sotos' syndrome), which seems to be caused by an over-reaction to GH by its target tissues rather than an excess of GH itself. This is extremely rare. Figure 51.18 illustrates the extremes of height that may be caused by hyper- and hyposecretion of GH.

In adulthood, GH-secreting pituitary tumours are much more common and result in a condition called **acromegaly**. After epiphyseal fusion, no further increase in bone length can occur.

Instead, in response to the raised levels of plasma GH, the bones start to thicken. This gives the patient a characteristic appearance, with large hands and feet. There is also coarsening of the facial features, as illustrated in Figure 51.19. Furthermore, there is overgrowth of soft tissues and the anti-insulin-like metabolic effects of GH may cause diabetes. Patients with acromegaly are also at increased risk from cardiovascular disease and neoplasias, due to the action of GH on the heart and colon.

Summary

Growth hormone (GH) has a dominant effect on postnatal growth. GH secretion rises during early infancy and shows a further peak during puberty. It is increased during exercise, stress, fasting, and non-REM sleep. GH exerts both metabolic and growth-promoting effects: it directly stimulates the formation of cartilage by chondrocytes and enhances the rate of uptake and incorporation of amino acids into protein. GH exerts a number of indirect effects through IGFs (insulin-like growth factors), which stimulate the clonal expansion of chondrocytes and the formation of the organic bone matrix.

GH deficiency during childhood results in pituitary dwarfism, which can be treated by injections of genetically engineered human GH. Pituitary gigantism results from hypersecretion of GH before the age of puberty. After puberty hypersecretion of GH results in acromegaly.

51.5 The role of other hormones in the process of growth

Although growth hormone undoubtedly plays a pivotal role in the process of physical growth, many other hormones are also important. Indeed, the number of hormones involved in the normal growth and development of an individual is indicated by the range of abnormalities of hormone secretion that can result in disturbed growth and abnormal development. Hormones of particular significance include thyroxine and the sex steroids, but a number of other hormones may indirectly influence growth and development through their general metabolic actions or their actions on the physiology of bone. These include insulin, the metabolites of vitamin D, parathyroid hormone, calcitonin, and cortisol. Their growth-promoting actions are discussed in the relevant sections of Chapters 21–24, so only a brief summary will be given here.

Thyroid hormone

Thyroxine is necessary for normal growth from early fetal life onwards and for normal physiological function in both children and adults. Its secretion begins at around weeks 15–20 of gestation and it seems to be essential for protein synthesis in the brain of the fetus and very young children. It is also required for the normal development of nerve cells. As the brain matures, this action assumes less importance. Children born with thyroid hormone deficiency will

Figure 51.18 A pituitary giant and a pituitary dwarf standing next to the British television personality David Frost.

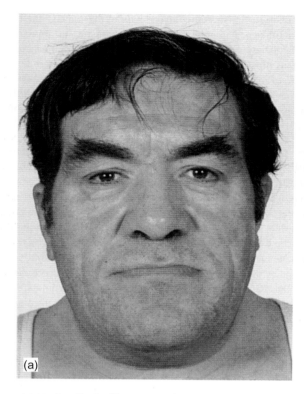

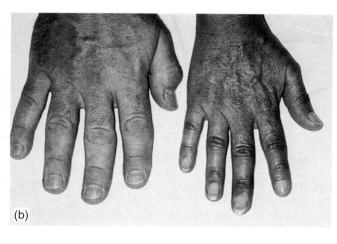

(a)

(b)

Figure 51.19 A patient with acromegaly caused by the over-production of GH during adulthood. Note the characteristic coarsened features and bone thickening of the face in panel (a) and the enlarged hand shown on the left in panel (b). (From Plates 4.1 and 4.2 in J. Laycock and P. Wise (1996) *Essential endocrinology*, 3rd edition. Oxford University Press Oxford.)

suffer severe learning difficulties unless treated quickly, a condition known as cretinism (see Chapter 22).

Children who develop thyroid hormone deficiency at a later stage have increasingly slowed bodily growth and delayed skeletal and dental maturity, but do not suffer obvious brain damage. Catch-up growth is achieved rapidly following treatment with exogenous thyroxine. Thyroid hormones appear to play a permissive rather than a direct role in growth, allowing cells (including the somatotrophs of the anterior pituitary) to function normally.

Corticosteroids

If present in excess of normal concentrations, hormones of the adrenal cortex, principally cortisol, appear to have an inhibitory action on growth. Such a situation may develop pathologically, for example in Cushing's syndrome, or following therapeutic administration of steroids to treat asthma, rheumatoid arthritis, kidney disease, or severe eczema. In such cases, the rate at which the skeleton matures is increased so that the potential for further growth is reduced.

Insulin

Insulin is produced by the islets of Langerhans in the pancreas (Chapter 24). It has no particular significance as far as growth is concerned except that it must be secreted in normal concentrations for normal growth to take place. The plasma level of insulin,

both in the fasting state and following a meal, rises during puberty, and falls back again at the end of puberty. Even small imbalances of plasma insulin and glucose levels can result in stunting and retardation of growth. However, diabetic children whose disease is well controlled by injected insulin and a suitable diet will grow normally.

Vitamin D metabolites and parathyroid hormone

The hormones that regulate plasma mineral levels have indirect effects on growth through their actions on the development and maintenance of the skeleton. Of particular importance are the metabolites of vitamin D (see Chapter 22). Calcitriol (1,25-dihydroxy-cholecalciferol) stimulates the intestinal uptake of calcium, thereby helping to maintain normal plasma levels of calcium. Calcitriol may also have a direct effect on bone to stimulate mineralization.

Vitamin D-deficiency causes the disorder of skeletal development known as **rickets** in children and **osteomalacia** in adults. Both conditions are characterized by failure of the matrix of bone (osteoid) to calcify. In children, whose bones are still growing, there is a reduction in the rate of remodelling which results in swelling of the growth regions of the bones, lack of ossification, and a thickened growth plate of cartilage which is soft and weak. The weight-bearing bones bend, leading to knock-knees or bowlegs as shown in Figure 51.20. In osteomalacia, layers of osteoid are produced which eventually cover practically the entire surface of the

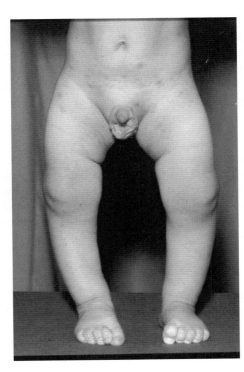

Figure 51.20 Typical curvature of the legs in a child suffering from rickets. (Courtesy of the Welcome Library, London.)

Virtually every aspect of muscular and skeletal growth is altered during puberty. Sex differences (e.g. in shoulder-growth) result in accentuation of sexual dimorphism (the differences between men and women) in adulthood. The hormonal mechanisms that underlie the growth spurt of puberty involve the cooperative actions of pituitary growth hormone and the gonadal steroids. At puberty, oestradiol-17β from the ovaries and testosterone from the testicular Leydig cells are secreted in increasing amounts under the influence of pituitary gonadotrophins. These steroids stimulate the secretion of GH, which in turn stimulates growth of the long bones resulting in an increase in height. Oestradiol-17β is also responsible for the development of the breasts, uterus, and vagina and for the growth of parts of the pelvis. Testosterone stimulates the development of male secondary sexual characteristics and has a direct action on the bones and muscles, which accounts for the differences in lean body mass and skeletal morphology seen between men and women. The increased secretion of sex steroids at puberty is important in triggering the process of epiphyseal fusion, limiting long-bone growth at the end of puberty.

Summary

Many hormones in addition to GH are involved in the regulation of growth. Thyroxine is required for growth from the early fetal period onwards and plays an important part in maturation of the CNS. Excessive secretion (or therapeutic administration) of corticosteroids can inhibit normal growth and maturation of the skeleton in children. Small imbalances in insulin secretion and plasma glucose also seem to interfere with normal development. Calcitriol (an active metabolite of vitamin D) stimulates the intestinal uptake of calcium, and parathyroid hormone stimulates the activity of the osteoblasts. Both are essential for normal growth of the skeleton.

The cooperative actions of the gonadal steroids and pituitary GH underlie the growth spurt of puberty. Both oestradiol-17β and testosterone stimulate the secretion of GH which in turn increases the rate of long bone growth. The male and female sex steroids also exert specific effects that give rise to the **secondary sex characteristics** and the differences in musculo-skeletal morphology between men and women.

skeleton. The main feature of the condition is pain, and bones may show partial fractures.

Parathyroid hormone (PTH) is important in whole-body calcium and phosphate homeostasis. Normal secretion of this hormone is needed for healthy bone formation. PTH is believed to bind to osteoblasts (possibly under the permissive influence of calcitriol) and to stimulate their activity. Calcitonin, secreted by parafollicular cells of the thyroid gland, is hypocalcaemic in its action, encouraging the binding of calcium to bone. Although its importance in adults is questioned, it is possible that calcitonin contributes to the growth or preservation of the skeleton during childhood, and possibly also throughout pregnancy, through an inhibition of osteoclast activity.

Sex steroids and the adolescent growth spurt

The growth velocity curves shown in Figure 51.2 illustrate the timing of the growth spurt that is evident in both girls and boys at puberty. The growth spurt may be divided into three stages. These are the age at 'take-off' (i.e. the age at which growth velocity begins to increase), the period of peak height velocity, and the time during which growth velocity declines and finally ceases at epiphyseal fusion. In general, boys begin their growth spurt 2 years later than girls. Boys are, therefore, taller at the time of 'take-off' and reach their peak height velocity two years later. During the growth spurt, boys increase their height by an average of 28 cm and girls by 25 cm. The average 10 cm difference in height between boys and girls is due more to the height difference at 'take-off' than the height gained during the spurt.

51.6 Disorders of the skeleton

Aside from congenital skeletal and developmental abnormalities, fractures caused by trauma, and endocrine abnormalities and nutritional deficiencies (see Section 51.5), a number of other disorders can affect the skeleton. These include osteoporosis, osteomyelitis, Paget's disease of bone, and bone tumours.

Osteoporosis

The skeleton changes throughout life as the body ages. Peak bone density is usually achieved around the age of 20–25 years. It then remains stable for around ten years as bone resorption is roughly matched by bone accretion. After this time, the rate of bone resorption begins to outstrip new bone production and bone density gradually declines throughout middle and old age. This progressive reduction in bone density is a normal consequence of ageing. In some people, however, bone resorption greatly exceeds production so that excessive amounts of bone mineral are lost and the architecture of the bone deteriorates significantly. Osteoporosis is a common metabolic bone disease characterized by structural fragility and low mineral density, which becomes more common as we age. The scanning electron micrographs in Figure 51.21 illustrate the difference in internal structure of normal and osteoporotic bone.

Although anyone, male or female, may suffer from osteoporosis, the condition is most often seen in post-menopausal women. Indeed, around a third of women between the ages of 45 and 80 suffer from some degree of osteoporosis. Oestrogen acts on osteoblasts to down-regulate osteoclast activity; consequently the reduction in oestrogen that occurs in postmenopausal women leads to accelerated bone resorption. As bone density falls, the fragility of the skeleton increases and there is an increased susceptibility to fractures. The most common fracture sites in osteoporosis are in the vertebrae, due to compression fractures of the vertebral column (particularly those of the weight-bearing regions below T6).

Oestrogen-replacement therapy (HRT) can be an effective treatment for many women, but carries with it significant risks

associated with cardiovascular dysfunction and cancer. Bisphosphonates, which inhibit the bone-absorptive function of osteoclasts, are now the most common treatment for osteoporosis and other metabolic bone diseases, while other agents that interfere with the signalling between osteoblasts and osteoclasts have recently been developed.

Osteomyelitis

This is a painful bacterial, or more rarely fungal, infection that causes significant destruction of bone and bone marrow. It most frequently develops as a result of a skin wound or a bone fracture. It causes fever and pain in the affected bone. Rapid treatment (with immobilization and antibacterial drugs) is important to halt the progress of the disease and to prevent further necrosis (death) of bone tissue.

Paget's disease of bone

After osteoporosis, Paget's disease of bone is the most common metabolic bone disorder. Indeed, its prevalence in the UK is the highest in the world. It is a chronic condition characterized by localized regions of accelerated bone turnover in which the activity of both osteoblasts and osteoclasts is increased. It may involve any part of the skeleton. In the affected areas, the normal bone matrix is replaced by patches of softened and enlarged bone. The disease often causes pain and gradual alterations of bone structure that lead to deformity. Drugs that suppress the activity of osteoclasts such as bisphosphanates, along with analgesic agents, are helpful in the treatment of this disorder.

Bone tumours

Most bone tumours arise as metastases from other sites (i.e. they are secondary tumours), but benign and malignant primary bone tumours do occur. These are usually described according to their behaviour and whether they form bone or cartilage. For example, an osteoid osteoma is a benign bone-forming tumour (occurring most frequently in children or young adults), while osteosarcomas and chondrosarcomas are malignant tumours that form bone and cartilage respectively.

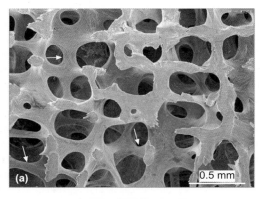
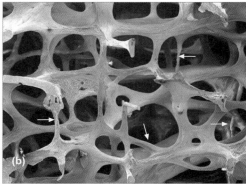

Figure 51.21 Scanning electron micrographs showing the difference in appearance between normal spongy bone (a) and bone from a person suffering from osteoporosis (b). Note the pronounced difference in the thickness of the trabeculae (some of which are indicated by the arrows). (Courtesy of Professor Tim Arnett.)

Summary

A number of disorders affect the skeleton, including genetic abnormalities, nutritional and endocrine disturbances, metabolic bone diseases, and tumours. The commonest bone disease is osteoporosis, which is characterized by a progressive loss of bone density with an increased susceptibility to fractures.

51.7 Growth of cells, tissues, and organs

The body consists of cells that are organized into populations that form the tissues and organs. As discussed earlier in this chapter, growth occurs through increases in both cell size and cell number, processes which are independently regulated. All cells continually renew their constituents by metabolism (see Chapter 4). Tissues, however, may be divided into three categories in terms of their growth characteristics. These are: non-regenerating tissues, regenerating tissues, and tissues that have significant powers of regeneration even though their cells do not normally divide continually.

In non-regenerating tissues (such as nerve and muscle), relatively few new cells are born once the period of growth is over. Once formed, many cells in these tissues last for most or all of the individual's life and growth occurs in three phases. First, the tissue increases its size through cell division and an increase in cell numbers. During the second phase, the rate of cell division falls but the cells themselves increase in size as proteins continue to be synthesized and enter the cytoplasm. In the third phase, cell division stops almost completely and the tissue expands only by increasing cell size. The age at which the cells stop dividing depends upon the individual tissue or organ; the neurons of the CNS are the first cells to stop dividing. In the case of the cerebral cortex, for example, neurogenesis is mostly completed by week 28 of gestation.

In regenerating tissues such as skin, blood, and the gastrointestinal epithelium, cells are continually dying and being replaced by new cells. Tissues such as these have a special germinative zone (the haematopoietic tissue in red bone marrow, for example) wherein new cells are born.

In those tissues whose cells do not normally divide continually but which have significant powers of regeneration, the cells are relatively long-lived and stable, but new cells can be generated if the tissue is damaged or when increased activity is required of it. Examples of these tissues are parts of the liver, the kidneys, and most glands.

An organ may enlarge in three ways:

1. the number of its constituent cells increases (**hyperplasia**)

2. the size of its constituent cells increases (**hypertrophy**)

3. the amount of substance between the cells increases.

Programmed cell death (apoptosis)

Each time a cell divides, two daughter cells are produced. If a cell and its daughter cells divide a further ten times, the initial parent cell will have given rise to over a thousand descendants. While this marked increase in numbers is desirable during growth and development, it would become a problem in an adult, where cell numbers must be regulated to maintain a more or less constant body size. Regulation of cell division is also important during wound repair and for the replacement of short-lived cells (e.g. the enterocytes of the small intestine). Even during embryonic development, more cells are produced than are required to form specific tissues. For example, a hand begins as a spade-like structure and the fingers develop by the selective removal of tissue.

The maintenance of normal cell numbers in adulthood and the removal of unwanted tissue during embryogenesis requires a regulated programme of cell death which is called **apoptosis** (from the Greek for 'falling off'—as in the leaves of trees in autumn). Programmed cell death also ensures that cells with damaged DNA are not permitted to progress through the cell cycle. By means of

this process, damaged or unwanted cells can be eliminated without local inflammation from leakage of cell contents.

Apoptosis is triggered by specific local signals, which activate caspases

If a cell's DNA is damaged (for example by radiation or chemicals) a protein known as p53 starts to accumulate inside the cell and then activates genes that lead to destruction of the cell. The first event is the stimulation by p53 of proteins that damage the membranes of mitochondria, allowing cytochrome c to leak out into the cytoplasm. Cytochrome c triggers a set of reactions that culminate in the activation of a group of proteins called caspases, which initiate a cascade of proteolytic enzymes that destroy the cell. The resulting cell fragments are taken up by neighbouring cells or macrophages which utilize them for energy production and the synthesis of new cell components. This sequence of apoptosis (mitochondrial damage leading to cell death) is intrinsic to each cell and is responsible for much of the cell death that occurs in neurodegenerative disorders such as Parkinson's Disease.

Apoptosis may also be mediated by receptors on the cell surface known as **death receptors**, an example of which is the protein Fas. If a cell becomes infected by a virus, for example, it will display viral antigens on its plasma membrane. These will be recognized by a killer T-cell (see Chapter 26) which will then bind to the Fas protein. This binding will in turn activate caspases, which will destroy the cell as described above.

Summary

Tissues are either non-regenerating (such as nerve) or regenerating. In non-regenerating tissues, cell division stops when the tissue has reached an appropriate size. Regenerating tissues, including the skin, blood cells, and intestinal cells, are in a continual state of renewal. Other tissues, such as the liver, can regenerate in response to tissue loss or damage.

Organs increase in size by cell division, by an increase in cell size, and by an increase in volume of intercellular material. In healthy adults, the balance between cell replication by mitosis and programmed cell death (apoptosis) maintains tissue homeostasis.

51.8 Alterations in cell differentiation: carcinogenesis

As discussed in Section 51.7, cells reproduce by cell division and many are programmed to die. The balance between cell proliferation and cell death within a tissue determines its overall size. Under normal circumstances, it seems that differentiated cells can continually sense their environment and adjust their rate of proliferation to suit the prevailing requirements. Although some cell populations, such as those of the haematopoietic tissue and germinal epithelium of the testes, show continual rapid division, most cells stop dividing once the tissue has reached an appropriate size. However, many cells can increase their rate of cell division if required. Liver cells, for example, are able to increase their rate of proliferation in response to loss of liver tissue caused by alcohol abuse. Growth factors and cytokines appear to regulate the rate of cell division in different tissues so that it remains appropriate to the needs of the body.

When cells fail to obey the normal rules governing their proliferation and multiply excessively, an abnormal mass of rapidly dividing cells is formed. This is called a **neoplasm** (new formation) and the process is called **neoplasia**. Neoplasms are composed of two types of tissue: parenchymal tissue, which represents the functional component of the organ from which it is derived, and stroma, or supporting tissue, consisting of blood vessels, connective tissue, and lymphatic structures.

Neoplasms are classified as benign or malignant according to their growth characteristics. Benign neoplasms are well-defined, local structures that usually grow slowly and do not **metastasize** (spread to distant sites to seed secondary tumours). Malignant neoplasms (cancers), however, are poorly differentiated, grow rapidly, and metastasize readily. Cancer cells consume large amounts of nutrients, thus depriving other cells of necessary metabolic fuels. This leads to the characteristic weight loss and tissue wasting (**cachexia**) which often contributes to the death of cancer patients. Cancers can arise from almost any cell type except neurons, but the most common cancers originate in the skin, lung, colon, breast, prostate gland, and urinary bladder. Each year in the UK over 250,000 people are diagnosed with cancer and about 20 per cent of all inhabitants of the prosperous countries of the world die from the disease.

Cancer cells differ from normal cells in many respects

Cancer develops when the mechanisms that normally regulate cell division are lost and there is uncontrolled multiplication of cells. Furthermore, the cells themselves possess very different characteristics to those of normal cells. Principal differences include the following.

- Unlike normal cells, which divide a finite number of times before becoming senescent, cancer cells appear to have limitless reproductive potential. In tissue culture they can divide almost indefinitely—they are 'immortal'. Normal cells possess segments of specialized DNA called **telomeres** at each end of their chromosomes that appear to control cell replication. Each time the cell completes one cell cycle, the telomere is shortened. Once the telomeres reach a certain length the cell enters senescence and stops dividing. Cancer cells, by contrast, maintain the length of their telomeres, in most cases by producing the enzyme telomerase, which replaces the region of telomeric DNA lost at each cell division.

- Normal cells divide in response to growth signals and stop dividing when instructed to do so by anti-proliferative signals acting at checkpoints within the cell cycle. Cancer cells appear to divide independently of growth signals and fail to respond to anti-proliferative signals so that they progress unchecked from G_1 to S of the cell cycle (see Chapter 4).

- Cancer cells evade apoptosis. Cells with damaged DNA are normally prevented from replicating and undergo apoptosis instead. This is controlled by p53 (see Section 51.7). Many cancer cells have loss or mutations of the p53 gene.

- Cells within a developing tumour require oxygen and nutrients. Cancer cells are able to promote sustained angiogenesis (growth of blood vessels) by releasing growth stimulators such as VEGF (vascular endothelial growth factor) and bFGF (basic fibroblast growth factor), which stimulate the growth of nearby capillaries into the tumour mass.

- Cancer cells look very different to normal cells when viewed under a microscope. As Figure 51.22 illustrates, cancer cells have variable shapes (pleomorphism), possess large nuclei, and often show prominent mitotic figures (Figure 51.23).

- Normal cells maintained in tissue culture tend to grow in flat sheets. A process known as contact inhibition prevents further division once cells have made contact with their neighbours. Cancer cells fail to exhibit contact inhibition and grow as piles of cells.

- Cancer cells are frequently transported from the primary tumour and travel via the blood or lymph, or within body cavities, to distant sites where they establish new (secondary) cancers. This is the process of metastasis. Of people with solid tumours, over half will have metastatic disease at diagnosis.

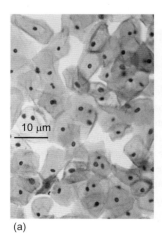
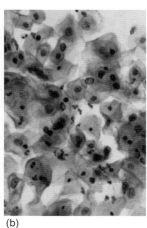
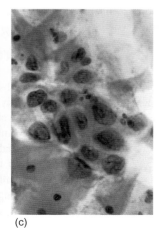

(a)　　　(b)　　　(c)

Figure 51.22 Typical appearance of cancer cells. This figure compares the appearance of cervical cells (obtained during pap smears) from: normal tissue (a); dysplasia (a pre-cancerous condition in which there is an increase in the number of immature cells) (b); and invasive carcinoma (c). (Courtesy Dr E. Walker, Science Photo Library.)

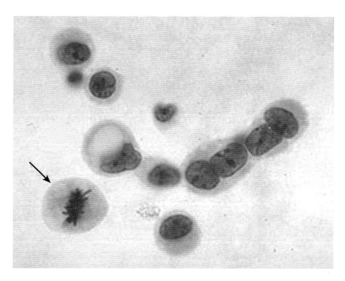

Figure 51.23 The appearance of malignant lymphocytes in CSF. Note the prominent mitotic figure (arrow) and the enlarged nuclei.

What are the factors that cause transformation of a normal cell into a cancer cell?

It is well known that certain physical and chemical factors, including irradiation, tobacco tars, and saccharine, can act as **carcinogens**. They do so by causing mutations—changes in the DNA that alter the expression of certain genes that are involved in cell growth, cell differentiation, or apoptosis. Mutations in two main classes of genes appear to be involved in the development of cancers. These are proto-oncogenes and tumour-suppressor genes.

Proto-oncogenes are found in normal cells. They code for the proteins that are essential for cell division, growth, and cellular adhesion and play a role in normal development and tissue maintenance. Many of these proto-oncogenes are active during embryonic development but are then switched off in differentiated adult cells. However, should proto-oncogenes undergo mutation (through the damaging effect of carcinogens) and become switched on once more, they become **oncogenes** (cancer-causing genes) whose activity stimulates cell division but inhibits differentiation and apoptosis. Cells are thus allowed to become invasive and to metastasize. More than 100 oncogenes have now been identified and may be detected in certain rapidly spreading tumours.

Examples of oncogenes include v-sis, the mutated form of the proto-oncogene c-sis that codes for platelet-derived growth factor, and v-eRbB, the mutated form of the proto-oncogene c-eRbB that codes for the epidermal growth factor receptor.

Tumour-suppressor genes (also called anti-oncogenes) effectively 'police' the processes that regulate cell numbers and ensure that new cells receive DNA that has been precisely replicated. Many tumour-suppressor gene products act as 'stop signs' to uncontrolled growth and therefore may arrest the cell cycle, promote differentiation, or trigger apoptosis. Others may protect cells against cancer by influencing processes that inactivate carcinogens, aid in the repair of DNA, or enhance the ability of the immune system to destroy cancer cells. Two examples of important tumour-suppressor genes are the retinoblastoma gene (Rb) and the p53 gene. The protein product of the retinoblastoma gene (pRb) regulates the cell cycle by inhibiting the progression of the cell from G_1 to S-phase (see Chapter 4). It also encourages cells to differentiate. If it is lost, the result is uncontrolled division of largely undifferentiated cells. The p53 gene is located on chromosome 17 and is the subject of intense investigation. It is activated by DNA damage, abnormal growth signals, or cell stress (such as hypoxia). It arrests the cell cycle, encourages DNA repair, inhibits angiogenesis, and stimulates apoptosis. Its activity stops the formation of tumours and has led to the use of the term 'guardian of the genome'. Mutated p53 is believed to play a key role in the pathological development of cancer, since it cannot bind effectively to DNA and thus its ability to act as a stop signal for cell division is lost. Mutations in p53 are found in most tumour types. A rare condition known as Li–Fraumeni syndrome is associated with the inheritance of only one functional copy of the p53 gene. People with this condition are predisposed to cancer and often develop multiple tumours in early adulthood.

Summary

Under normal conditions, differentiated cells adjust their rate of proliferation to the body's requirements in response to a variety of signals. Failure to do so results in the formation of a benign or malignant neoplasm. Malignant cells are poorly differentiated, have limitless reproductive capability, fail to exhibit contact inhibition, and can metastasize. A cell may be transformed into a cancer cell when its DNA undergoes mutation and the expression of certain genes is altered.

(✳) Checklist of key terms and concepts

The physiology of bone and bone growth

- Fetal growth is at its greatest between weeks 16 and 20 of gestation.
- Before 16 weeks the differentiation and specialization of cells into tissues and organs takes place.

- Growth velocity progressively declines until puberty, when there is a growth spurt. The adult height is achieved by the end of puberty.
- During fetal life, the flat and irregular bones of the skeleton are formed by intramembranous ossification.

- The long and short bones of the skeleton are laid down as cartilage 'models' which subsequently become ossified by the process of endochondral ossification.
- Growth of the long bones during childhood and adolescence occurs through the proliferation and hypertrophy of cartilage cells at the growth plates of the long bones.
- After puberty, the growth plate thins as it is replaced by bone and the diaphysis and epiphyses fuse, preventing further growth in stature.
- Bone consists of osteoid, an organic matrix, strengthened by the deposition of complex crystals of calcium and phosphate (hydroxyapatite).
- There are three types of bone cells: osteoblasts, which secrete the organic matrix; osteocytes (mature bone cells); and osteoclasts, which are responsible for the resorption of old bone during growth and remodelling of the skeleton.
- Around 5–7 per cent of total bone mass is recycled each week during adulthood as the skeleton responds to the changing demands placed upon it.

The role of hormones in growth

- The anterior pituitary peptide growth hormone (GH) is vital for normal growth in stature from the age of around three years to puberty.
- GH is under hypothalamic control and shows pulsatile release. Moreover, its secretion is stimulated by exercise, stress, and fasting. It is also increased during deep sleep.
- GH directly stimulates the formation of cartilage by chondrocytes and enhances the rate of uptake and incorporation of amino acids into protein.
- GH exerts a number of indirect effects through IGFs (insulin-like growth factors).
- IGFs stimulate the clonal expansion of chondrocytes and the formation of the organic bone matrix.
- GH deficiency during childhood results in pituitary dwarfism. Pituitary gigantism results from hypersecretion of GH before the age of puberty.
- Hypersecretion of GH in adulthood results in acromegaly, in which there are metabolic disturbances, thickening of bones, and overgrowth of soft tissues.
- Thyroxine is required for growth from the early fetal period onwards and plays an important part in maturation of the CNS.

- Calcitriol (an active metabolite of vitamin D) stimulates the intestinal uptake of calcium and seems to be essential for normal growth, calcification, and remodelling of the skeleton.
- Parathyroid hormone stimulates the activity of the osteoblasts and is important for the normal growth of bone.
- The cooperative action of the gonadal steroids and pituitary GH underlie the growth spurt of puberty.

Disorders of the skeleton

- These include genetic abnormalities, nutritional and endocrine disturbances, metabolic bone diseases, and tumours.
- Osteoporosis is the most common disorder of bone metabolism. It affects mainly post-menopausal women and is characterized by a progressive loss of bone density.

Cell and tissue growth

- Growth occurs at the level of individual cells, the tissues, and the whole body.
- Tissues are either non-regenerating (such as nerve), or regenerating. Regenerating tissues include those that are in a continual state of renewal (skin, blood cells, etc.) and those, such as liver, which can regenerate in response to tissue loss or damage.
- In non-regenerating tissues, cell division stops when the tissue has reached an appropriate size.
- In regenerating tissues, the balance between cell replication by mitosis and programmed cell death (apoptosis) maintains tissue homeostasis.

The pathological development of cancer

- If differentiated cells fail to respond to the normal signals that control division, a neoplasm will develop.
- Neoplasms may be benign or malignant.
- Malignant cells are poorly differentiated, have limitless reproductive capability, and metastasize readily.
- A cell may be transformed into a cancer cell when its DNA undergoes mutation and the expression of certain genes (proto-oncogenes and tumour-suppressor genes) is altered.
- Mutated forms of proto-oncogenes (oncogenes) trigger uncontrolled cell division and the loss of differentiation.
- Tumour-suppressor gene products act as 'stop signs' to uncontrolled growth.
- Two important tumour-suppressor genes are the retinoblastoma gene (Rb) and the p53 gene. Mutation of these genes results in uncontrolled division of largely undifferentiated cells.

Recommended reading

Alberts, B., Johnson, A., Lewis, J., Morgan, D., Raff, M., Roberts, K., and Walters, P. (2015) *Molecular biology of the cell* (6th edn), Chapters 17, 18, and 20. Garland, New York.

Keni, J., and Pawlikowska-Haddal, A. (2012) Growth regulation, Chapter 11 in: *Textbook of endocrine physiology* (6th edn), ed. J.E. Griffin and S.R. Ojeda. Oxford University Press, Oxford.

Pecorino, L. (2016) *Molecular biology of cancer: Mechanisms, targets and therapeutics* (4th edn). Oxford University Press, Oxford.

Tanner, J.M. (1989) *Foetus into man* (2nd edn). Castlemead, Hertford.

Review articles

Almeida, M., Laurent, M.R., et al. (2017) Estrogens and androgens in skeletal physiology and pathophysiology. *Physiological Reviews* **97**, 135–87, doi:10.1152/physrev.00033.2015.

Eriksen, E.F. (2010) Cellular mechanisms of bone remodeling. *Review of Endocrine and Metabolic Disorders* **11**, 219–27, doi:10.1007/s11154-010-9153-1.

Hill, P.A. (1998) Bone remodelling. *British Journal of Orthodontics* **25**, 101–7.

Kogianni, G., and Noble, B.S. (2007) The biology of osteocytes. *Current Osteoporosis Reports* **5**, 81–6.

To check that you have mastered the key concepts presented in this chapter, complete the accompanying online self-assessment questions. Go to www.oup.com/uk/pocock5e/

Appendix 1 Glossary of terms and common abbreviations

A

A-fibres A group of myelinated peripheral nerve fibres characterized by their high speed of conduction. They serve both motor and sensory functions and are sub-divided into Aα, Aβ, Aγ, and Aδ groups according to their function and conduction velocity.

absorption The uptake of water or solutes from one compartment to another, e.g. the uptake of water and solutes from the glomerular filtrate by the tubular cells of the kidney. The uptake of the products of digestion by the gastrointestinal tract.

accessory muscles The muscles of the neck and abdomen that assist in the process of forced breathing and in the vigorous breathing that accompanies severe exercise. During inspiration, the main accessory muscles are the scalene and sternocleidomastoid muscles of the neck. During expiration, the main accessory muscles are the rectus abdominus and the external and internal oblique muscles of the abdomen.

acclimatization, acclimation The process of adapting to a new environment, for example the increase in the number of red cells after moving from sea level to reside at a high altitude.

accommodation The physiological process that permits the eye to focus on objects at different distances.

acetyl coenzyme A A water-soluble molecule that participates in the transfer of acetyl groups (CH_3CO-) from one molecule to another. Acetyl CoA is produced during the glycolytic breakdown of glucose and the oxidation of fatty acids before they are metabolized via the citric acid cycle. It also participates in various biosynthetic pathways.

acetylcholine (ACh) A small molecule that acts as a neurotransmitter at synapses of the autonomic and central nervous systems and at the neuromuscular junction. Such synapses are known as **cholinergic synapses**.

acetylcholine receptor (AChR) A protein present in the postsynaptic membrane that binds acetylcholine. Two main types are found: ion channels that open when they bind acetylcholine—**nicotinic receptors**—and G protein linked receptors that activate second messenger cascades—**muscarinic receptors**.

achromatopsia Inability to discriminate colours caused by damage to visual association areas of the brain (i.e. areas concerned with interpreting information from the eyes).

acid A substance that liberates protons when it is dissolved in water. Examples are acetic acid and hydrochloric acid.

acidaemia Literally 'acid in the blood'. A condition in which there is a greater quantity of hydrogen ions per unit volume of blood than normal (i.e. greater than 45×10^{-9} moles per litre) or blood with a pH less than 7.35. See also **pH**.

acid–base balance Collective term for the processes that maintain the hydrogen ion concentration of the blood within its normal limits ($40–45 \times 10^{-9}$ moles per litre; pH 7.35–7.45).

acidosis Any process that causes the hydrogen ion concentration of the blood to be greater than normal. It may be caused by accumulation of carbon dioxide in the body (respiratory acidosis), or by increased amounts of non-volatile acid (metabolic acidosis).

acinus A cluster of cells found at the end of the ducts of an exocrine gland. It is where the primary secretion of the gland is produced in the salivary glands and pancreas, for example.

acromegaly An endocrine disorder in which there is excessive thickening of the bones of the hands, feet and jaw. It is caused by excessive secretion of growth hormone.

acrosome, acrosomal vesicle A vesicle found at the head of a mature sperm containing hydrolytic enzymes that enable the sperm to digest the protective coating around an oocyte (egg cell) as a prelude to fertilization.

ACTH adrenocorticotrophic hormone (corticotrophin).

actin A widely distributed protein that polymerizes to form part of the cytoskeleton of most cells. It also plays an important part in cell motility and muscle contraction. The monomeric form is known as G (globular) actin and the polymeric form as F (filamentous) actin.

action potential A transient fall in membrane potential (a depolarization) that is self-propagated along an axon or muscle fibre. Used by the nervous system for long-distance signalling. It is sometimes called the nerve impulse. The action potentials of muscle cells initiate the process of contraction.

active hyperaemia Alternative term for **functional hyperaemia**.

active transport The movement of an ion or molecule across a cell membrane that is against the prevailing electrochemical gradient.

adaptation The change in sensitivity of a sensory receptor in response to a steady stimulus. In general, receptors respond more strongly to changes in the environment than to constant stimulation. Subdivided into rapidly adapting receptors such as the carotid sinus baroreceptors and slowly adapting receptors such as muscle spindles.

adaptive control The ability of a regulatory system to change in response to prevailing circumstances. An example is the increased secretion of thyroid hormone in response to cold stress.

adaptive immune system That part of the immune system that responds to a specific antigen and generates an immunological memory, to permit a rapid response to any subsequent infection.

adenohypophysis Alternative term for the **anterior pituitary gland**.

adequate stimulus The physical or chemical stimulus to which a particular sensory receptor is most sensitive.

ADH **antidiuretic hormone**.

adherens junction (zonula adherens) Junction between cells in which the intracellular face of the plasma membrane is attached to actin and intermediate filaments. Important in the adhesion belt of epithelial cells.

adipose tissue Tissue largely made up of cells containing fat deposits (adipocytes).

ADP Adenosine diphosphate.

adrenal cortex The outermost part of the adrenal gland which is divided into three regions; from the outer to innermost they are zona glomerulosa, zona fasciculata, and zona reticularis. All three regions are concerned with the synthesis and secretion of steroid hromones.

adrenaline A catecholamine hormone secreted by the adrenal medulla. In North America it is more usually called epinephrine.

adrenal medulla The innermost region of the adrenal gland. It is concerned with the synthesis and secretion of the hormones adrenaline and noradrenaline.

adrenarche The increase in secretion of male hormones (androgens) by the adrenal cortex of females that occurs prior to the onset of menstruation (menarche).

adrenoceptor An adrenergic receptor; a member of a class of G protein coupled receptors that are targets of the catecholamines, especially noradrenaline (norepinephrine) and adrenaline (epinephrine). Two principal subgroups exist: alpha (or α)-adrenoceptors and beta (β)-adrenoceptors, both of which have various subtypes (e.g. $\alpha 1$, $\alpha 2$, etc.).

adrenocorticotrophic hormone (ACTH) A hormone secreted by the anterior pituitary gland to regulate the secretion of steroid hormones by the adrenal cortex.

aerobic Metabolic processes that require the presence of oxygen.

afferent Passing towards a structure, as in the afferent arterioles of the renal glomeruli or the afferent peripheral nerves (cf. **efferent**).

afferent nerves Those nerves that transmit sensory information from the periphery to the central nervous system. Frequently referred to as afferents.

afterload The tension in the wall of the left ventricle that occurs during the ejection phase of the cardiac cycle: the load against which the heart must work to eject blood into the aorta.

agglutination The clumping of red cells following transfusion of incompatible blood.

agnosia Inability to recognize objects, caused by damage to the parietal lobes of the cerebral cortex.

agonist In physiology and pharmacology it is a substance that initiates a physiological response when it binds to the appropriate receptor (cf. **ligand**). In anatomy it is a muscle whose contraction acts to move a part of the body.

agranulocytes White blood cells (leukocytes) that do not have prominent granules in their cytoplasm. The group includes the lymphocytes and monocytes, which have round or indented nuclei (cf. **granulocytes**).

AIDS Acquired immunodeficiency syndrome.

airways resistance The resistance offered by the airways to the flow of air. It is largely determined by the upper airways (the nose, pharynx, and larynx) and the diameter of the segmental bronchi.

aldosterone A steroid hormone secreted by the adrenal cortex to regulate the ionic composition of the blood. It is the principal mineralocorticoid.

alexia Word-blindness caused by interruption to the pathways linking the visual association areas to the angular gyrus of the temporal lobe.

algogen A pain-producing substance such as bradykinin.

alimentary canal Alternative term for the **gastrointestinal tract**.

alkalaemia Literally 'alkali in the blood'. A blood pH that is higher than normal (i.e. the hydrogen ion concentration is lower than normal). A blood pH greater than 7.45 is alkalaemia (equivalent to a plasma hydrogen ion concentration less that 35×10^{-9} moles per litre).

alkalosis Any process that causes the hydrogen ion concentration of the blood to be lower than normal. It may be caused by the excessive elimination of carbon dioxide from the body (respiratory alkalosis), or by a deficit of non-volatile acid (metabolic alkalosis).

allele One of two or more variants of a gene that have the same relative position on homologous chromosomes in different people. Alleles are responsible for alternative characteristics in an individual, such as eye colour. When the alleles on each of a pair of homologous chromosomes are identical they are said to be homozygous; when they are different they are said to be heterozygous.

allergen Any molecule that is capable of eliciting an immune reaction.

allergic reaction A hypersensitive reaction to an allergen. Such a reaction generally follows an initial exposure to the allergen, which sensitizes the immune system. Examples are hay fever and asthma.

altitude sickness Alternative term for **mountain sickness**.

alveolar pressure (intrapulmonary pressure) The pressure within the lungs.

alveoli (sing. alveolus) The small thin-walled air-containing compartments that are the main site of gas exchange in the lung. The term is also used to refer to cell clusters at the ends of the ducts of a compound gland, e.g. the alveoli of the mammary gland.

amenorrhoea Failure to menstruate.

amino acid A small organic molecule that possesses both an amino and a carboxyl group. Proteins are assembled from 20 alpha amino acids—those in which the amino and carboxyl group are linked to the same carbon atom.

ammoniagenesis The formation of ammonia by removal of the amino groups of glutamine. This occurs in the proximal tubule and permits excretion of excess hydrogen ions.

amnesia Loss of memory. Anterograde amnesia is loss of memory for events after a traumatic head injury; retrograde amnesia is loss of memory of events occurring prior to such an injury.

AMP Adenosine monophosphate.

amphipathic, amphiphilic Description of a molecule that has one part that is polar (soluble in water—hydrophilic) and one part that is non-polar (insoluble in water—hydrophobic).

amylase An enzyme that hydrolyses glycogen and starches to form simple sugars such as glucose.

anabolism, anabolic metabolism Metabolic processes in which large molecules are synthesized from smaller ones: for example, the synthesis of proteins from amino acids (cf. **catabolic metabolism**).

anaemia Abnormally low blood haemoglobin, often accompanied by a reduced number or size of red blood cells. It has many causes.

anaemic hypoxia Insufficient oxygen supply to the tissues, caused by an abnormally low haemoglobin content.

anaerobic Metabolism that occurs in the absence of oxygen.

anaerobic threshold The stage of exercise at which blood lactate begins to rise above $1–2$ mmol l^{-1}.

analgesic A substance that reduces the sensitivity to pain.

anaphase The stage of mitosis during which the two sets of chromosomes separate.

anaphylaxis A generalized inflammatory reaction to an antigen. If severe, it causes a circulatory collapse known as anaphylactic shock.

anastomosis An anastomosis is a connection between two tubular structures such as two blood vessels.

anatomical dead space see **dead space**.

anatomical position A term used to describe the upright body facing forwards with the feet slightly apart, the arms by the sides, the palms facing forwards, and the fingers extended.

anatomy The systematic description of the structures of the body. The fine microanatomy of the tissues is called **histology**.

androgen Any of the steroid hormones related to testosterone that promote the development and maintenance of the male sex characteristics.

anencephaly Absence of fully formed cerebral hemispheres resulting from a developmental failure.

angina pectoris Severe pain in the region of the chest and left shoulder, often radiating down the inside of the left arm. It is caused by a severe reduction in blood flow through the coronary circulation (cardiac ischaemia).

angiotensinogen A plasma protein that can be cleaved to form angiotensin I, the precursor of the hormone angiotensin II.

anion A molecule or ion that carries a negative charge (cf. **cation**).

anorexia Loss of normal appetite.

anorexia nervosa A severe psychiatric disorder in which the normal appetite is lost.

anovular cycle An ovarian cycle in which no egg cell is liberated for fertilization.

anoxia The absence of oxygen (cf. **hypoxia**).

ANP Atrial natriuretic peptide. A hormone that promotes the excretion of sodium. It is secreted by the atrial myocytes in response to expansion of the circulating volume.

antagonist A substance that inhibits the physiological action of another. Also a muscle that acts in opposition to another.

anterior pituitary gland The anterior portion of the pituitary gland, which is situated along the midline at the base of the brain. It is also called the anterior lobe, adenohypophysis, or pars anterior.

antibodies Proteins known as immunoglobulins that are secreted by the B-cells of the immune system in response to an invading microorganism or foreign molecule.

anticoagulants Substances that prevent coagulation (clotting) of the blood.

anticodon A sequence of three nucleotides in a molecule of transfer RNA that is complementary to a three-nucleotide codon in a molecule of messenger RNA.

antidiuretic hormone (ADH) A peptide hormone secreted by the posterior pituitary gland to regulate the osmotic pressure of the plasma.

antigen A molecule that is able to elicit an immune response.

antiport A carrier protein that exchanges an ion or molecule on one side of the plasma membrane for an ion or molecule on the other side.

antipyretic A substance that acts to reduce the increase in temperature resulting from a fever.

antral follicle Stage of follicle development characterized by the formation of a fluid-filled cavity around the oocyte which remains attached to the surrounding cells by the stigma. A fully developed antral follicle is known as a Graafian follicle.

aphasia Impaired ability to speak normally. Broca's aphasia is characterized by speech produced only with great effort. This is commonly caused by lesions in the left frontal lobe. Wernicke's aphasia indicates impaired ability to speak intelligibly, usually caused by lesions in the left temporal lobe.

apical surface The surface of an epithelium that faces towards the exposed free surface; the **basolateral** surface is the region facing inward toward the underlying tissue.

apnoea Interruption of normal breathing.

apocrine sweat glands Sweat glands that secrete an oily fluid into hair follicles. In humans, apocrine sweat glands are found only in certain region of the body such as the axillae (armpits), areola and nipples of the breast, and around the anus (cf. eccrine sweat gland).

apoprotein A protein that must combine with another molecule or prosthetic group to perform its function.

apoptosis Programmed cell death in which the cell constituents are broken down without causing any damage to neighbouring cells. Apoptosis plays an essential role in development and in the elimination of damage or infected cells.

appetite An instinctive desire to consume food; a craving (cf. **hunger**).

apraxia The loss of the ability to perform specific purposeful movements even though there is no paralysis or loss of sensation. Generally caused by damage in the parietal lobes of the cerebral cortex.

aquaporins Integral membrane proteins that permit the movement of water from one side of a cell membrane to the other. Also known as water channels.

aqueous humour The fluid that fills the anterior chamber of the eye.

arachidonic acid An unsaturated fatty acid present in cell membranes. It has four double bonds in a 20 carbon unbranched chain. The precursor of prostaglandins.

arachnoid membrane A delicate membrane lying beneath the dura mater that covers the outer surface of the brain and spinal cord; the middle layer of the three meninges (cf. **dura mater, pia mater**).

ARDS Adult respiratory distress syndrome.

areflexia Loss of reflex activity—normally the result of damage to the spinal cord.

areola The pigmented region surrounding the nipple of the breast.

arrector pili A small fan-shaped muscle attached to a hair follicle that causes the skin hairs to become erect.

arrhythmia Any irregular beating of the heart.

arteriole Small muscular blood vessels connecting the arteries to the various capillary networks. The arterioles regulate the blood flow through particular vascular beds and are the primary resistance vessels.

artery A muscular vessel that distributes the blood from the heart to the tissues. The arteries are the primary distribution vessels and may be subdivided into elastic arteries, which are large vessels with distensible walls, and muscular arteries, which are smaller in diameter and whose walls contain a high proportion of smooth muscle.

arterio-venous anastomosis A direct connection that permits the passage of blood from an artery to a vein without passing through the capillaries. Found in certain regions of the skin.

ascites An accumulation of fluid in the peritoneal cavity resulting in abdominal swelling.

asphyxia A lack of oxygen in the blood due to restricted respiration. If prolonged, it results in loss of consciousness and, eventually, leads to death.

association areas (cerebral cortex) These are the areas of the cerebral cortex that lie outside the areas concerned with direct motor control and the primary sensory areas. They are implicated in detailed interpretation of sensory impressions and in the initiation of motor activity.

astereognosia Inability to recognize an object by touch.

astrocytes (astroglia) Non-neuronal cells found in the brain and spinal cord. They have long, thin processes radiating from their cell body that form firm attachments to small blood vessels, especially the capillaries. These attachments are thought to form part of the **blood–brain barrier**.

asynergia A lack of muscle coordination caused by damage to the motor pathways.

athetosis Slow, snake-like writhing movements of the extremities especially the fingers, hands and arms, caused by damage to the brain, particularly the basal ganglia.

atoms The elemental particles that constitute all matter. Each atom comprises a dense nucleus of protons and neutrons surrounded by a cloud of electrons. The atomic nucleus of hydrogen consists of just a single proton.

atomic mass The mass of an atom is determined by the total number of protons and neutrons in its nucleus; the mass of the electrons is negligible in comparison. The atomic mass is sometimes called the atomic weight.

atomic mass unit One-twelfth of the mass of an atom of carbon 12.

atomic number The number of protons in the nucleus of the atom. This determines the chemical characteristics of the element concerned.

ATP Adenosine triphosphate, a nucleotide with three phosphate groups. The terminal phosphate group is used to transfer chemical energy in cells.

ATPase An enzyme that utilizes ATP hydrolysis to catalyse a cellular process. For example, the Na^+, K^+-ATPase (sodium pump) provides the energy for moving sodium and potassium ions against their concentration gradients.

ATPS Ambient temperature and pressure saturated with water vapour (with reference to respiratory gas).

atresia Failure of a tubular structure to develop normally, resulting in the absence of a normal opening—as in billiary atresia, where the bile duct is occluded or absent. **Ovarian follicle atresia** is the process in which immature ovarian follicles degenerate during the follicular phase of the menstrual cycle.

atrial flutter, atrial fibrillation Atrial flutter refers to a condition in which the atria depolarize more than about 220 times a minute. The ECG then has a characteristic sawtooth baseline. In atrial fibrillation, the atria depolarize more than about 350 times a minute and they fail to contract in a coordinated manner. The ECG baseline is then very irregular and P waves are absent.

atrioventricular block Impairment of the conduction of the cardiac action potential from the atria to the ventricles.

atrioventricular node The junction between the atrial tissue and the conducting system of the heart, the bundle of His.

atropine An alkaloid that blocks a class of acetylcholine receptor known as muscarinic receptors.

audiogram A graphical representation of the threshold of hearing for standardized frequencies as measured in a particular individual. It is obtained using a clinical audiometer and represents the minimum sound intensity that the individual is able to hear at each frequency compared to the average sensitivity of normal subjects.

Auerbach's plexus Alternative term for **myenteric plexus**.

auscultation Literally 'listening to'; investigation of the body by listening to the sounds emanating from its internal structures, generally by means of a stethoscope. Particularly useful for investigating the heart and lungs, also used to determine arterial blood pressure—see **Korotkoff sounds**.

autocrine Chemical signalling in which the chemical signal acts on the cell that secreted it.

autoimmunity A pathological state in which the immune system fails to recognize a host tissue and mounts an immune response against it. Diabetes mellitus type I is an example of autoimmunity.

autonomic nervous system That part of the nervous system that regulates the activity of the internal organs. Also called the involuntary nervous system.

autonomic reflexes Reflexes mediated by the autonomic nervous system that regulate the functions of the viscera, e.g. reflex control of the heart rate, blood pressure, and sweating.

autoregulation The ability of an organ to regulate its own blood supply.

autorhythmicity The ability of a cell to initiate its own rhythm, e.g. the pacemaker cells of the SA node of the heart.

autosomes Chromosomes that are not involved in the determination of the sex of an individual. The sex of an individual is determined by the X and Y (sex) chromosomes.

AV Atrioventricular, a-v arteriovenous.

AV node **atrioventricular node**.

Avogadro's number The number of atoms or molecules in one mole of a particular substance: 6.022×10^{23}.

axon The thin process that connects one nerve cell to its target site. Axons conduct action potentials, often over long distances, to their terminals which form the **presynaptic** side of a **synapse**.

axon reflex A reflex in which an afferent impulse travels along a sensory axon until it reaches a branch point, where it is propagated to an end organ before it reaches the cell body. As it does not involve a complete reflex arc, it is not considered to be a true reflex. The flare reaction of the skin in response to injury is the best-known example.

axonal transport The directional movement of organelles and large molecules along an axon. Anterograde transport is transport from the cell body to the periphery, while transport from the periphery to the cell body is retrograde transport.

axoneme The bundle of microtubules that form the core of a cilium or a flagellum.

B

B-cells Lymphocytes that secrete immunoglobulins in response to a specific infection.

B-fibres Thinly myelinated autonomic preganglionic fibres ~3 μm in diameter with conduction velocities of 3–15 m s^{-1}.

b.p.m. Beats per minute—the heart rate; also breaths per minute—the respiratory rate.

Babinski sign A clinical test used to determine the site of damage in cases of motor impairment. In a positive Babinski sign, the response to stimulation of the sole of the foot causes the big toe to extend upwards while the other toes fan out. In a normal person the toes flex downwards. A positive Babinski test is normal in a baby but in an adult it indicates damage to the motor cortex or the corticospinal tract (an upper motoneuron lesion).

bactericidins Inhibitory factors secreted by the resident bacterial population of the skin and gut to combat infection by invading pathogens.

baroreceptors, baroceptors Receptors that sense pressure: the dynamic high-pressure receptors of the carotid body and aortic arch.

barotrauma Tissue damage caused by changes in ambient pressure.

basal ganglia The deep nuclei of the cerebral hemispheres comprising the caudate nucleus, putamen, globus pallidus, and claustrum.

basal lamina The thin matrix of connective tissue fibres on which epithelial cells rest.

basal metabolic rate (BMR) The energy used by the body in the fasting state during complete rest in a thermoneutral environment.

base A substance that accepts protons in aqueous solutions. Also a purine or pyrimidine found in certain coenzymes, DNA and RNA.

basement membrane The thin region of connective tissue that separates an epithelial layer from the lamina propria of the underlying tissue. It is made up of two distinct regions: the **basal lamina** and the underlying reticular lamina. It provides physical support to an epithelium.

basic electrical rhythm (BER) A slow rhythmical fluctuation in the membrane potential of the smooth muscle cells of the gastrointestinal (GI) tract. The frequency of these fluctuations varies along the length of the tract, decreasing with distance from the mouth. The BER coordinates the activity and emptying of the stomach and the peristaltic movements of the small and large intestine.

basilar membrane The stiff membranous structure that supports the organ of Corti in the inner ear. It separates the scala media (which contains endolymph) from the scala tympani (which contains perilymph). The physical properties of the basilar membrane vary along its length: it is stiffer and narrower near the base of the cochlea than at the apex, so that the hair cells at different points along its length are excited by different frequencies of sound. The physical characteristics of the basilar membrane thus contribute to the perception of pitch.

basolateral surface The surface of an epithelial cell facing towards the tissues of the body. Functionally, the side opposite the **apical** membrane.

basophil A type of white blood cell that stains with basic dyes such as methylene blue. Basophils are the least common of the white blood cells. When stimulated they secrete histamine, prostaglandins, and cytokines.

BER **basic electrical rhythm**.

beriberi A disease caused by a deficiency of vitamin B$_1$ (thiamine). Marked by widespread metabolic disturbances and neurological disorders.

bicuspid valve Alternative term for the **mitral valve**.

bioassay The detection of a substance (such as a hormone) and its relative concentration by its effect on a biological system in comparison to that of a standardized preparation.

biochemistry That branch of science concerned with the substances and chemical processes occurring in living organisms. Also known as biological chemistry.

biotransformation The chemical alteration of a substance, such as a drug, by enzymic activity within the body.

bivalent 1. A complex of four chromosomes with one pair of homologous chromosomes becoming attached to an identical pair. This combination is seen near the beginning of meiosis (a stage known as zygotene). 2. Having the ability to form two covalent bonds.

blastocyst The stage of early embryonic development where the cellular mass of the morula becomes hollowed out so that the embryonic cells surround a fluid-filled space.

blood gases The oxygen and carbon dioxide found in the blood, either stated as their partial pressures (PCO_2, PO_2) in mm Hg or kPa or as their blood content (C_aO_2, C_aCO_2, etc.) in ml per decilitre.

blood pressure The arterial pressure measured in the large arteries during the cardiac cycle. In healthy young adults the arterial pressure is about 120 mm Hg (c. 14.7 kPa) at the peak of ventricular contraction (systole). During ventricular relaxation (diastole) the pressure falls to about 80 mm Hg (c. 10.7kPa). Significantly higher values are often seen in older people (see **hypertension**).

blood–brain barrier The barrier separating the circulating blood from the brain extracellular fluid. It is formed by the endothelial cells of the brain capillaries, which are connected by tight junctions. In addition, astrocytes extend processes that contact endothelial cells so helping to seal the interstitial space of the brain from the fluctuating levels of ions, hormones, and tissue metabolites of the general circulation.

BMI **body mass index**.

BMR **basal metabolic rate**.

body mass index Provides a simple index of the weight of an individual in relation to their body size measured by their height. It is widely used to determine whether an adult is underweight, overweight, or obese. It is defined as the body weight of an individual in kilograms divided by the square of their height in metres. The units are kg m^{-2}.

Bohr effect (Bohr shift) The rightward shift in the oxyhaemoglobin dissociation curve caused by a rise in the partial pressure of carbon dioxide or fall in pH of the blood.

Bohr equation Equation used to quantify the ratio of physiological **dead space** in the airways to the total tidal volume.

bone cells A collective term for the osteocytes, osteoblasts, and osteoclasts of the bone.

Bowman's capsule The double-walled capsule that encompasses the tuft of capillaries that constitute a renal glomerulus. Also known as the glomerular capsule.

Boyle's law A basic physico-chemical principle which states that the pressure exerted by a given quantity of gas is inversely proportional to its volume.

BP Blood pressure (arterial blood pressure).

2,3-BPG 2,3-bisphosphoglycerate. Formerly known as 2,3-DPG (2,3-diphosphoglycerate). It is produced during glycolysis in red cells from 1,3-BPG by an enzyme called BPG mutase which is specific to red cells. When 2,3-BPG is bound to haemoglobin,

the oxyhaemoglobin dissociation curve is shifted to the right (i.e. it decreases the affinity of haemoglobin for oxygen), aiding the offloading of oxygen in the tissues.

bradycardia An abnormally slow heart rate; a heart rate slower than 60 beats a minute.

bradykinesia Extreme slowness of movements and reflexes: a characteristic of certain neurological diseases such as Parkinson's disease.

bronchi, bronchioles The main conducting airways. From the trachea there are 11 generations (successive branches) of bronchi. From the twelfth generation to the alveolar ducts the airways are known as bronchioles. The walls of the bronchi are supported by rings or plates of cartilage. The bronchioles have no cartilage in their walls.

bronchodilatation The expansion of the bronchial air passages resulting in a fall in airways resistance.

bronchospasm A sudden contraction of the smooth muscle of the walls of the bronchioles, resulting in difficulty in breathing.

Brodmann's areas Different areas of the cerebral cortex classified according to the arrangement and sizes of their neurons. These areas, first distinguished by the neuroanatomist K. Brodmann in 1909, broadly reflect different brain functions.

brown fat Also known as brown adipose tissue; its cells are smaller in size than those of white fat and the tissue owes its colour both to the large number of mitochondria the cells contain and to the relatively large number of blood vessels present compared to white adipose tissue. Its chief function is the generation of heat for the maintenance of body temperature. Infants, especially newborns, rely on brown fat for heat generation and have relatively large brown fat deposits compared to adults.

BTPS Body temperature and pressure saturated (with water vapour).

buffer A substance that limits the change in hydrogen ion concentration following the addition of an acid or base to an aqueous solution.

bulimia nervosa An emotional disorder characterized by a distorted body image and an obsessive desire to lose weight, in which bouts of extreme overeating are followed by fasting or self-induced vomiting or purging (cf. **anorexia nervosa**).

bundle block Impairment of the conduction of the cardiac action potential through the bundle of His.

bundle of His Specialized cardiac muscle fibres that conduct the action potentials originating in the atria to the ventricles. The bundle of His begins at the AV node that lies at the junction of the atria and right ventricle. It passes through the right atrioventricular fibrous ring and travels along the membranous part of the interventricular septum, before dividing into the right and left bundle branches that spread the cardiac action potential across the walls of the ventricles via Purkinje fibres.

C

C The gas content of blood (e.g. the carbon dioxide content of venous blood—C_vCO_2).

C-fibre The smallest-diameter nerve fibres of peripheral nerves. They are 0.5–2.0 μm in diameter and are unmyelinated. They conduct action potentials at velocities of 0.5–2.0 m s^{-1}. In the peripheral nervous system they are afferent fibres, conveying information about touch, skin temperature, and pain. Autonomic postganglionic fibres are also C-fibres.

cadherins A group of proteins important in cell–cell adhesion.

calcinosis A disorder in which calcium deposits form in the soft tissues such as the skin.

calcitriol 1,25-dihydroxycholecalciferol: a hormone derived from Vitamin D$_3$ that acts to stimulate calcium absorption from the intestine.

calcium pump A transport protein, found in the plasma membrane and endoplasmic reticulum of cells, that acts to remove Ca^{2+} from the cytoplasm either across the plasma membrane or into intracellular stores. The calcium pump is a membrane-bound Ca^{2+}-ATPase.

calorie A unit of heat. The amount of energy needed to raise the temperature of 1 gram of water by 1°C. When spelled with a capital C it is the energy needed to raise the temperature of 1 kilogram of water by 1°C, more correctly called a kilocalorie or kcal. Kilocalories are generally used to indicate the energy value of foods. The calorie and kilocalorie are now obsolete units and quantities of energy are normally given in joules (J). One calorie is approximately equivalent to 4.186 joules.

calorigenesis, calorigenic hormones Calorigenesis is heat production, and calorigenic hormones are those that increase body heat production.

cancellous bone Alternative term for **spongy bone**.

cancer See **neoplasm**.

candela The luminous intensity of a light source producing single-frequency light at a frequency of 540 terahertz (THz) with a power of 1/683 watt per steradian, or 18.3988 milliwatts, over a complete sphere centred at the light source. This frequency corresponds to a wavelength of approximately 555 nanometres (nm), which is green light. The normal human eye is most sensitive to light of this wavelength.

capacitation A process normally occurring within the Fallopian tube during which the glycoprotein coat and the proteins of the seminal fluid are removed from the head of the sperm. This is the penultimate step in the maturation of mammalian spermatozoa and permits the **acrosome** reaction to occur. Both events are required to render sperm competent to fertilize an oocyte.

capillary The smallest of the blood vessels, linking the arterioles and venules. The walls of the capillaries allow the exchange of various substances between the blood and tissue fluid.

carbohydrates Chemical compounds that contain only carbon, hydrogen, and oxygen. They are part of a larger group of chemical compounds known as saccharides. Carbohydrates are a major source of energy in the normal diet.

carboxylic acids Alternative term for **fatty acids**.

carcinogen A substance or agent that is able to cause a cancer.

carcinogenesis The process by which cancerous cells develop from healthy cells.

cardiac cycle The series of events from the initiation of one beat of the heart to the initiation of the next, comprising both **systole** and **diastole**.

cardiac failure A situation in which the heart is unable to pump sufficient blood at normal filling pressures to meet the metabolic demands of the body. Also called heart failure.

cardiac index The cardiac output divided by the body surface area, usually expressed in l m^2 min^{-1}. The normal range is 2.6–4.2 l m^2 min^{-1}.

cardiac output The volume of blood pumped by the heart in a minute. It varies with both body size and physical activity. At rest it is around 4–7 l min^{-1} in healthy adults.

cardiac tamponade An accumulation of fluid in the pericardial sac that impairs the normal expansion of the heart during diastole.

cardiodynamics The study of the regulation of the heart by intrinsic and extrinsic factors.

cardiovascular system The heart and vessels involved in the circulation of the blood.

carotid bodies Small highly vascular organs lying between the internal and external carotid arteries just superior to the carotid bifurcation. They sense the partial pressures of oxygen and carbon dioxide in the arterial blood.

carotid sinus A dilation of the common carotid artery as it divides to form the internal and external carotids. It contains baroreceptors that regulate short-term changes in blood pressure.

carrier protein A transmembrane protein that has specific binding sites for a particular ion or molecule to transport that substance across the plasma membrane. See also **uniport**, **symport**, **antiport**.

cartilage Flexible connective tissue that lines the articulating surfaces of the joints, and forms the elastic framework of the outer ear, nose, larynx, and trachea. In young children it forms the bulk of the skeleton; it is progressively converted into bone as they grow into adulthood.

catabolic metabolism Biochemical processes in which large molecules are broken down into smaller ones, either to provide energy or to modify them prior to excretion.

catabolism **catabolic metabolism**.

cation An ion or molecule possessing a positive charge, such as sodium (Na$^+$) and ammonium (NH$_4^+$).

caudate nucleus One of the principal structures that make up the basal ganglia of the brain.

caveolae Sub-microscopic plasma membrane pits that are found in many types of mammalian cell. They play an important role in a variety of cell functions including endocytosis.

CCK **cholecystokinins**.

cell The basic functional unit of the body. All tissues and organs consist of cells and their products. Each cell is surrounded by a plasma membrane enclosing a space containing a nucleus, cytoplasm, and small structures called organelles.

cell cycle The sequence in which a cell grows and then divides. The cell cycle is conventionally divided into five phases: G$_0$ (the gap); G$_1$ (the first gap); S (the synthesis phase, during which the DNA is synthesized and replicated); G$_2$ (the second gap); and M (mitosis) followed by cytokinesis, the process of division proper.

cell division The processes by which a cell forms two daughter cells at the end of the cell cycle. It results in the distribution of identical genetic material between the two daughter cells. It may involve either **mitosis**, in which the number of chromosomes is preserved, or **meiosis**, in which the daughter cells formed after cell division II have half the normal complement of chromosomes.

cell-mediated immunity   The immune response that is mediated by T-lymphocytes (T-cells). Helper T-cells secrete cytokines to stimulate antibody secretion by B-cells, while cytotoxic T-cells secrete cytotoxins to kill target cells.

central nervous system (CNS)   The collective term for the brain and spinal cord.

central venous pressure   The hydrostatic pressure at the junction of the inferior and superior venae cavae as they enter the right atrium. It is measured to indicate right ventricular preload and is normally 0–6 mm Hg in healthy individuals.

centriole   A cylindrical array of microtubules found at the centre of a centrosome of a cell.

centromere   The centromere is the region at which the two sister chromatids that form a chromosome are joined. It is the region of attachment of a chromosome to the spindle fibre during cell division.

centrosome   An organelle found near the nucleus of a cell. It comprises two centrioles perpendicular to each other surrounded by a mass of proteins that are responsible for anchoring microtubules and initiating their formation. During cell division the centrosome is duplicated and spindle fibres develop from the centrosomes.

cerebellum   A small highly convoluted structure that lies above the brainstem and below the occipital lobes of the cerebral hemispheres. It plays an important role in motor control.

cerebral cortex   The outermost part of the cerebral hemispheres of the brain. It consists of grey matter containing neurons and glial cells.

cerebral hemispheres   The most prominent structures of the human brain. Originating from the embryonic telencephalon (forebrain), they enlarge greatly during development and envelop most of the major structures of the mid-brain (the diencephalon). The right and left hemispheres are separated by the longitudinal cerebral fissure. Each hemisphere is subdivided into four lobes: the frontal, parietal, occipital, and temporal lobes.

cerebral ventricles   The fluid-filled spaces in the centre of the brain. There are two lateral ventricles linked to a smaller ventricle, the third ventricle, which is located along the midline and is connected to the fourth ventricle by the aqueduct. The fourth ventricle lies between the brainstem and the cerebellum.

cerebrospinal fluid   The clear fluid found in the cerebral ventricles and the space surrounding the brain (the subdural space). It is not an ultrafiltrate of plasma but is formed by the choroid plexus of the lateral and fourth ventricles.

cerebrovascular accident   Alternative term for **stroke**.

channel protein   A protein that forms a pore that allows charged ions to pass across the plasma membrane.

channelopathies   Diseases affecting the function of one or more ion channels.

Charles's law   This states that the volume occupied by a given quantity of gas is directly related to the absolute temperature.

chemical compound   A combination of two or more chemical elements held together by chemical bonds.

chemical element   A substance consisting of a single type of atom characterized by its atomic number. There are 92 naturally occurring elements.

chemical senses   The senses of taste (gustation) and smell (olfaction).

chemical symbol   A one- or two-letter code representing a specific chemical element, e.g. carbon—C; calcium—Ca; chlorine—Cl; sodium—Na.

chemical synapse   A specialized junction between an axon terminal and a target cell that operates through the secretion of a small quantity of a chemical (a neurotransmitter) by the axon terminal to influence the activity of the target. The target cell may be another neuron, a muscle fibre, or a gland cell.

chemokine   Alternative term for **cytokine**.

chemoreceptor   Sensory receptor that responds to a local chemical change, such as the chemoreceptors of the carotid bodies.

chemotaxis   Movement of a cell towards or away from the source of a diffusible chemical.

Cheyne–Stokes breathing   A cyclical pattern of breathing in which respiratory movements gradually come to a complete stop (apnoea) before resuming after 10–20 sec. Each cycle repeats every 1–3 min. In adults, Cheyne–Stokes breathing occurs in various pathological conditions, and at high altitudes. Also called periodic breathing.

chloride shift   The exchange of chloride and bicarbonate ions between the plasma and the red blood cells as carbon dioxide is carried from the tissues to the lungs, where the process is reversed. The chloride shift helps to buffer changes in blood pH. Also called the Hamburger effect.

cholagogue   A substance that causes the gall bladder to contract, such as cholecystokinin.

cholecalciferol   Vitamin D_3.

cholecystokinins   Peptide hormones secreted and synthesized by the I-cells of the intestinal mucosa to promote the contraction of the gall bladder and the secretion of enzyme-rich pancreatic juice.

choleretics   Any substance that increases the flow of hepatic bile.

cholesterol   A steroid that is synthesized by the body and is an essential component of the plasma membrane of cells. It also acts as the precursor of the steroid hormones, vitamin D_3, and bile salts.

chondrocytes   Those cells that secrete and maintain cartilage.

chorea   A dyskinesia (neurological disorder) characterized by involuntary jerky movements affecting especially the shoulders, hips, and face.

chorion   An embryonic membrane derived from trophoblastic tissue and extra-embryonic mesoderm. As the chorion develops, it becomes invaded by blood vessels to form the fetal side of the placenta.

chromaffin cells   Cells that stain with chromic salts; the term is generally taken to mean the catecholamine-containing cells of the adrenal medulla, which secrete the hormones adrenaline and noradrenaline.

chromatid   One of the two identical chromosomes linked by a **centromere**.

chromosome Structure formed by the condensation of DNA and associated proteins in the early stage of cell division.

chronotropy, chronotropic effect Chronotropy is the action of affecting the heart rate. Factors influencing the heart rate have a chronotropic effect.

chylomicrons Lipoprotein particles found in the blood and lymph after a meal containing fats. They are formed by the cells of the intestinal mucosa to transport lipids from the small intestine to the tissues of the body.

chyme The semi-liquid material produced by gastric secretion and mechanical activity in the initial stages of the breakdown of food.

cilia (sing. cilium) Short hair-like extensions of certain epithelial cells that are found in airways from the trachea to the terminal bronchioles. They are also present in the epithelial cells of Fallopian tubes and on the free surface of the ependymal cells of the cerebral ventricles. The cilia have a characteristic core of microtubules known as the axoneme which is linked to the cytoskeleton via a structure called the basal body. Cilia beat with rhythmical movements that produce a wave-like motion across the ciliated surface.

circadian rhythm A biological rhythm that has a period of approximately one day.

circulatory shock A pathological situation that arises when the blood supply of the tissues is inadequate. If not treated promptly, circulatory shock leads to a vicious cycle that results in death.

citric acid cycle Alternative term for **tricarboxylic acid cycle**.

CJD Creutzfeld–Jakob disease: a neurodegenerative disease caused by prions (infectious protein particles).

clathrin-dependent endocytosis Alternative term for **receptor-mediated endocytosis**.

claudins Membrane proteins found at the tight junctions of epithelia.

claustrum The most lateral nucleus of the basal ganglia. It is a thin sheet of cells that lies in a sagittal plane between the insular cortex and the putamen. Its function is largely unknown.

clearance The volume of plasma completely cleared of a given substance in 1 minute. Generally used with reference to renal clearance, but also applicable to other systems such as the lungs. The units are volume per unit time.

climacteric Alternative term for **menopause**.

CLIP Corticotrophin-like intermediate peptide: an endogenous peptide derived from ACTH.

clone A population of cells arising from a single ancestor cell as a result of repeated divisions.

closing capacity The volume at which the airways begin to collapse during a forced expiration. In healthy young adults it is about 10 per cent of the **vital capacity**.

closing volume In the lungs and airways, the closing volume is the **closing capacity** less the **residual volume**.

clotting factors Proteins present in normal blood that can be converted from their inactive form to initiate the formation of a blood clot. The final step in the process is the conversion of fibrinogen to fibrin mediated by thrombin.

CN **cranial nerve**.

CNS **central nervous system**.

CO **cardiac output**.

CoA **coenzyme A**.

CRH **corticotrophin releasing hormone**.

coagulation factors Alternative term for **clotting factors**.

cobalamin The collective name for the group of cobalt-containing substances designated as vitamin B_{12}.

cochlea Part of the inner ear located in the temporal bone of the skull. It consists of three fluid-filled tubes arranged as a helix of 2½ turns. The cochlea houses the organ of Corti, the structure that converts sound waves into nerve impulses.

codon A sequence of three nucleotides of DNA or RNA that codes for a specific amino acid in a peptide sequence.

coenzyme A An important cofactor in the transfer of acetyl groups (CH_3CO-) in enzymatic reactions.

colostrum A yellowish fluid rich in protein and antibodies, secreted by the breast during the first few days of lactation, before the secretion of true milk begins.

colour blindness Something of a misnomer as affected individuals have no blindness but have a specific defect in their colour perception. Most people (> 90 per cent) are trichromats and can match a given colour by mixing the appropriate proportions of red, green, and blue light. Those who can match any colour with only two colours of light are known as **dichromats** and account for around 2.5 per cent of the population. However, around 6 per cent of the population match colours with abnormal proportions of the primary colours; they are known as anomalous trichromats.

commissure A tract of nerve fibres passing from one side of the spinal cord or brain to the other.

compact bone The dense outer portion of bone that provides most of the strength of the skeleton, also called cortical or dense bone. It consists largely of a series of osteons running parallel to the long axis of the bone—see **Haversian system**.

complement system A group of about 30 plasma proteins that are part of the natural immune system. They are activated by microbial cell surfaces and act together with the adaptive immune system to kill invading organisms.

complete and incomplete proteins Complete proteins are mainly of animal origin and contain all ten of the essential amino acids; incomplete proteins lack one or more. Examples of foodstuffs with complete proteins are eggs, cheese, and meat. Incomplete proteins are mainly of plant origin and are found in cereals and in pulses such as beans and lentils.

compliance (respiration) A measure of the ease with which the chest can be distended. It is measured when there is no movement of air into or out of the lungs (static compliance). Units are litres per unit pressure change with typical values of 1.0 l kPa^{-1}. The static compliance is determined by the elasticity of the chest wall and that of the lungs.

compound action potential An action potential recorded from a nerve trunk following electrical stimulation. It is the summed activity of many nerve fibres firing in synchrony.

conditioned reflex One type of learning known as associative learning, in which a neutral stimulus is associated with a more important matter such as feeding. First systematically explored by I.V. Pavlov in dogs which he trained to salivate by ringing a bell and then feeding them. After a number of such pairings the dogs began to salivate in response to the bell alone in anticipation of being fed. The normal reflex salivary response to food had become conditioned to the sound of the bell, hence the term conditioned reflex.

conducting airways The trachea, bronchi, and bronchioles that conduct the air to the respiratory surface but do not participate in gas exchange.

conduction (heat) The process by which heat is transferred from one medium to another by direct physical contact (cf. convection).

conduction velocity The rate at which an action potential is propagated along an axon or muscle fibre.

cones One of two classes of light-sensitive cells in the retina of the eye, the other being the **rods**. In most people there are three kinds of cones that sense different wavelengths of light, which permits full colour vision. The cones are active in relatively bright light, while the rod cells work better in dim light. Also called cone cells or retinal cones.

conformation The three-dimensional shape of a molecule, especially a protein molecule; one of the configurations of a molecule that can exist in equilibrium with molecules of different configuration.

conformational change A change in the configuration of a molecule, often of functional importance, such as the change in the state of an ion channel from open to closed or vice versa.

conjugation A chemical term for joining two molecules together, frequently seen in detoxification reactions, which inactivate poisons and prepare them for excretion.

connexon Part of a gap junction; connexons from two adjacent cells are linked to form a continuous aqueous pore allowing water-soluble materials to diffuse between the adjacent cells.

contralateral An anatomical term to denote structures on opposite sides of the midline of the body.

convection The main means of heat transfer in liquids and gases. It is the transfer of heat from a heat source to cooler regions by the movement of the liquid or gas driven by differences in density.

convergence 1. The merging of information derived from sensory receptors (cf. **divergence**). 2. The coordinated movement of the two eyes to maintain a single clear image of an object as it moves towards the head.

core temperature The temperature of the central part of the body, normally around 37°C. Unlike the skin temperature, core temperature is relatively constant and only varies by 1°C even though that of the surroundings undergoes much larger changes (cf. **shell temperature**).

coronal section Anatomical term for a vertical plane that divides the body or an organ into ventral (front) and dorsal (back) parts. Also called a frontal section.

corpus cavernosa, corpus spongiosum The spongy erectile tissue columns of the penis. The two corpora cavernosa form the dorsum and sides of the penis, while the corpus spongiosum is situated in the middle of the ventral surface near the urethra. As the three erectile columns fill with blood, the penis enlarges and becomes erect.

corpus callosum An extensive band of white matter linking the two cerebral hemispheres.

corpus striatum A major part of the basal ganglia of the brain. It comprises the caudate nucleus which is separated from the globus pallidus and putamen (the lentiform nuclei) by the internal capsule. Strands of grey matter run between the putamen and caudate running across the internal capsule, so giving the appearance of a striped mass of white and grey matter.

cortical reaction An event triggered by the entry of a mature sperm into a mature oocyte (egg cell). The moment of fertilization is followed by the fusion of the oocyte's cortical granules with its plasma membrane to form the fertilization membrane. This prevents further sperm from entering the egg.

corticospinal tract Tract that passes through the internal capsule before coursing over the ventral surface of the medulla. Also called the pyramidal tract. The fibres of the corticospinal tract originate mainly from neurons in the primary motor area of the frontal lobe, but a significant proportion arise from neurons in the parietal lobes. Around 80 per cent of the axons cross to the contralateral side at the level of the medullary pyramids (the pyramidal decussation) before passing down the spinal cord as the lateral corticospinal tract. Uncrossed fibres pass down the spinal cord as the ventral corticospinal tract.

corticotrophin A synonym for adrenocorticotrophic hormone (ACTH).

corticotrophin releasing hormone A hormone secreted by the hypothalamus to stimulate the secretion of ACTH by the anterior pituitary gland.

cortisol A steroid hormone secreted by the adrenal cortex. It is essential to life and plays an important role in the regulation of carbohydrate metabolism (i.e. it is a glucocorticoid).

cotransmission The secretion of two different neurotransmitters by the same nerve terminal.

cotransport, coupled transport The movement of a molecule or ion across the plasma membrane that requires the simultaneous or sequential movement of another molecule either in the same direction (cotransport) or in the opposite direction (counter transport).

cranial nerves The nerves arising directly from the brain. There are twelve pairs in all, serving both motor and sensory functions (cf. **spinal nerves**).

creatine phosphate A small organic molecule found in muscle that acts as a readily mobilized reservoir of metabolic energy, as its phosphate group is readily transferred to ADP to form ATP.

cretinism Abnormal brain development caused by a lack of thyroid hormone in early childhood.

crista A fold of the inner membrane of mitochondria (plural cristae).

crista ampullaris The sensory portion of the semicircular canals of the vestibular apparatus.

crista dividens A structure in the fetal heart that divides the right atrium in such a way that oxygenated blood flowing from the placenta is mainly diverted through the foramen ovale into the left atrium; the flap of the septum that forms the upper margin of the foramen ovale.

cryptorchidism The failure of one or both testicles to descend from the abdominal cavity to the scrotum.

crystalline lens See **lens**.

CSF **cerebrospinal fluid**.

cutaneous circulation The blood supply to the skin.

CVP **central venous pressure**.

cyanosis A blue discoloration of the skin and mucous membranes caused by poor circulation or inadequate oxygenation of the blood. It is seen when the blood concentration of deoxyhaemoglobin exceeds 5 g dl^{-1}.

cyclic AMP Nucleotide used as an intracellular signalling molecule. It is formed from ATP in response to activation of a G protein coupled receptor. It activates protein kinase A.

cyclic GMP A nucleotide formed from GTP which acts as an intracellular signalling molecule much like cyclic AMP. Cyclic GMP is produced in response to the formation of nitric oxide by the endothelial cells of the blood vessels. It also plays a role in the photoreceptor response to light.

cytokine A polypeptide that is secreted by cells of the immune system to act as a local signal in triggering an immune response. Examples are the interleukins and interferons. Sometimes called a chemokine.

cytokinesis The final stage of cell division when the cytoplasm is shared between the two daughter cells.

cytoplasm All the contents of a cell enclosed by the plasma membrane, except the nucleus.

cytoskeleton The protein filaments that give a cell its distinctive shape.

cytosol The contents of the **cytoplasm** except the membrane-bound organelles.

cytotrophoblast Cells that, together with the **syncytiotrophoblast**, cover the entire surface of the placental villi and are in direct contact with the maternal blood of the intervillous space.

D

DAG Diacylglycerol.

Dalton An alternative name for an atomic mass unit (abbreviated as D or Da). Used mainly in biochemistry in relation to large molecules such as proteins, where kilodaltons (kDa) are often employed.

Dalton's law of partial pressures This states that the total pressure exerted by a mixture of gases is equal to the sum of the partial pressures of all the gases of the mixture.

Darcy's law This states that the rate at which a fluid flows through a permeable medium is directly proportional to the pressure gradient between the points of entry and exit. When applied to the circulation, it implies that the blood flow through a capillary bed will depend on the difference between the pressure in the arterioles and that of the venules.

dark adaptation The process by which the eye increases in sensitivity when passing from a well-lit environment to one that is dark. It is primarily a process that occurs in the photoreceptors.

dB **decibel**.

dead space The anatomical dead space is the total volume of all those airways that do not participate in gas exchange and includes the upper airways, trachea, bronchi, and larger bronchioles. The physiological dead space is the volume of air that is taken in during a breath that does not participate in gas exchange. It includes the anatomical dead space and the volume of those alveoli that do not participate in gas exchange.

deamination The removal of an amino group. Enzymes that perform this function are called deaminases.

decibel (dB) One tenth of a unit known as the Bel, a relative measure of the intensity of sound pressure. The reference sound intensity is 20 µPa, which is close to the threshold of normal human hearing. The decibel scale is logarithmic, so that if the pressure of a sound is 10 × that of the reference pressure this corresponds to 20 decibels. A sound pressure difference of 1000 × corresponds to 60 dB, and so on.

decidua The endometrium of the pregnant uterus. The **decidual reaction** is the term used to describe the cellular and vascular changes that occur in the endometrium at the time of implantation.

deep An anatomical term meaning away from the body surface and towards its core.

defecation The act of eliminating solid waste from the anus.

deglutition The act of swallowing.

Dejerine–Roussy syndrome Alternative term for **thalamic pain**.

denaturation The modification of the three-dimensional structure of a native protein by chemical or physical means (e.g. exposure to a strong salt solution or an elevated temperature). It involves the breaking of the weak linkages such as hydrogen bonds that are necessary to maintain its highly ordered structure. Denaturation disrupts the normal function of a protein.

dendrites The fine, highly branched processes that originate from the cell body of a neuron and receive synaptic contacts from other neurons (cf. **axon**).

dendritic cell An antigen presenting cell of the immune system that has branching processes and is derived from the same line as the blood monocytes. They are found in the skin, mucosae, and lymphoid tissues and initiate a primary immune response by activating helper T-cells.

dense bone Alternative term for **compact bone**.

denticulate ligaments Thin bands of connective tissue that restrict the movement of the spinal cord within the spinal canal.

deoxyhaemoglobin Haemoglobin without any bound oxygen.

depolarization A decrease in the membrane potential of a cell.

dermatome An area of skin that receives sensory information via a single spinal nerve root. It is also the term for the lateral part of an embryonic body segment.

desmosomes A type of attachment plaque between epithelial cells and comprising specialized membrane proteins that link the cytoskeletons of neighbouring cells, so binding them together.

desquamation The natural loss of dead epithelial cells as in the shedding of the outermost layer of an epithelium, such as the epidermis of the skin.

detoxification Biochemical modification of a molecule to render it less toxic or more readily excreted.

deuteranope An individual who has defective colour vision ('colour blindness') in which green and red shades are confused. Such people have a relative insensitivity to green.

diabetes A disorder characterized by the excretion of large volumes of urine. The two commonest forms are **diabetes insipidus**, which is caused by lack of, or to insensitivity to, antidiuretic hormone (ADH), and **diabetes mellitus**, which is caused by an insufficiency of insulin (type 1 diabetes) or a lack of sensitivity to it (type 2 diabetes).

diakinesis The final stage prophase I of meiosis. It is characterized by dissolution of the nucleolus and nuclear membrane, the formation of the spindle, and repulsion of the chromosome pairs.

diapedesis The passage of white blood cells (leukocytes) through the intact capillary wall into the tissues. It is part of normal immunological surveillance and of inflammatory reactions.

diaphragm The main muscle of respiration; the diaphragm and associated fibrous tissue separate the thorax from the abdomen.

diaphysis The central shaft of a long bone such as the femur.

diastole The stage of relaxation of the heart muscle during which the chambers fill with blood (cf. **systole**).

diastolic pressure The lowest arterial pressure recorded during the cardiac cycle.

dichromat An individual lacking one type of colour sensitive visual pigment, usually either the red or the green pigment. A **colour blind** person, generally, but not exclusively, a male.

dietary fibre The indigestible cellulose and lignin derived from plants. Considered to play an important role in the passage of material through the gut.

dietary requirements The complete range of foodstuffs necessary to support a healthy life including carbohydrates, fats, and proteins as well as certain minerals and vitamins.

differentiation The process by which embryonic cells change as they form the individual tissues and organs of the body.

diffusing capacity The volume of a gas transferred from the alveoli to the capillary blood per second for a pressure difference of 1 kPa. Also called the transfer factor.

diffusion The passive movement of a substance from a region of relatively high concentration to an area of lower concentration. The amount of a substance moving from one region to another is expressed by **Fick's law of diffusion**, which states that the amount of substance diffusing depends on the area over which it diffuses, the concentration gradient, and a constant known as its **diffusion coefficient**. Diffusion also depends on the temperature: the higher the temperature, the faster the diffusion.

diffusion coefficient The mass of solute or gas diffusing in unit time between opposite faces of a unit cube of a system when there is a one-unit concentration difference between them. Hence the SI units of a diffusion coefficient are $m^2 s^{-1}$. In addition to temperature and pressure, the diffusion coefficient of a dissolved substance depends both on the properties of the solvent and those of the solute. For a gas it is dependent on molecule size.

digestive system The digestive tract and associated glands that form an integrated system able to ingest food, digest it, and absorb nutrients. The digestive tract proper begins at the mouth and extends through the oesophagus, stomach, small intestine, and large intestine, ending with the rectum and anus. The main associated structures are the salivary glands, pancreas, liver, and gall bladder.

dioptre A unit of measurement of the optical power of a lens or curved mirror, which is equal to the reciprocal of the focal length measured in metres. The units are therefore m^{-1}.

dioptric power Also called focusing power; the degree to which a lens is able to focus light.

diploid Having two copies of each chromosome, with one member of each chromosome pair derived from the ovum and one from the sperm (cf. **haploid**).

diplotene The fourth stage of the first cell division of meiosis: the stage at which the developing ova arrest their meiotic division and begin to accumulate material. For this reason it is also called the synthesis stage.

disjunction The normal separation of homologous chromosomes during anaphase.

distal Situated away from the centre of the body or some point of reference: e.g. the distal tubule of the kidney is that part of the renal tubule furthest from the renal glomerulus (cf. **proximal**).

DIT Di-iodotyrosine, a precursor of thyroid hormone.

diuresis An increased production of urine.

diuretic A substance that promotes an increase in urine production.

divergence The spreading of the influence of activity of a single nerve cell through multiple synaptic connections (cf. **convergence**).

DMT1 Divalent metal ion transporter 1.

DNA Deoxyribonucleic acid; a polynucleotide in which two deoxyribonucleotide chains are linked in a helical arrangement by hydrogen bonds in which adenine bases are paired with thymine bases while guanine bases are linked with cytosine bases. DNA is the nucleic acid constituent of the chromosomes and acts as the store of genetic information.

dominant hemisphere The cerebral hemisphere, usually the left, that controls the hand that is used preferentially in skilled movements and which controls the comprehension and production of speech. This is a somewhat simplistic concept, as it is now understood that each hemisphere is specialized to perform different functions.

Donnan equilibrium An equilibrium that exists between two ionic solutions separated by a semi-permeable membrane when one or more of the ionic species (usually a protein) cannot pass from one solution to the other. In such a situation the distribution of the diffusible ions is unequal, giving rise to a potential difference across the membrane. There is also a difference of osmotic pressure between the two solutions.

dopamine The usual name for 3,4-dihydroxyphenethylamine, which is a monoamine neurotransmitter derived from tyrosine. It plays an important role in motor control and in reward-motivated behaviour. Structurally, it is closely related to noradrenaline.

dorsal An anatomical term relating to the back of a person (cf. **ventral**); towards the back, as in the dorsal horn of the spinal cord. Also refers to the uppermost side of a four-legged animal.

dorsal root ganglion A non-synaptic ganglion found on the dorsal root of the spinal cord that contains the cell bodies of the primary afferent fibres.

Down syndrome A congenital condition caused by the presence of an extra copy of chromosome 21 (trisomy 21) that is manifest by intellectual and physical impairments.

dromotropic effect An effect influencing the speed of conduction in the heart muscle. A positive dromotropic effect is an increase in conduction speed, while a negative dromotropic effect is a decrease.

dual innervation Having a nerve supply from both divisions of the autonomic nervous system.

ductus arteriosus A direct connection in the fetal heart between the pulmonary artery and aorta, permitting much of the blood coming from the right ventricle to bypass the lungs.

ductus venosus A fetal shunt connecting the umbilical vein with the inferior vena cava, permitting a significant fraction of oxygenated blood from the placenta to bypass the immature liver.

dura mater The tough outer membranous covering of the brain and spinal cord. The outermost layer of the three meninges.

dyskinesia The impairment of voluntary movements, resulting in fragmented or jerky motions.

dysmenorrhoea Pain that is associated with menstruation.

dysphagia Difficulty in swallowing.

dyspnoea Difficulty in breathing, or a subjective feeling of breathlessness.

dysarthria Difficult or unclear articulation of speech.

dyspraxia Difficulty of coordinated movement.

dysrhythmia An abnormal physiological rhythm, usually of the heart.

E

eccrine sweat glands The sweat glands distributed widely over the skin that secrete an aqueous solution important for regulating body temperature in hot environments and during exercise through the process of evaporation (cf. apocrine sweat gland).

ECF **extracellular fluid.**

ECG **electrocardiogram.**

echocardiograph An image of the chambers of the heart obtained by an ultrasound scan. Used as a diagnostic test.

eclampsia A life-threatening condition that occasionally develops in late pregnancy and is characterized by extremely high blood pressure, oedema, proteinuria, and convulsions (cf. **pre-eclampsia**).

ectoderm The outer layer of embryonic tissue that develops into the epidermis and the nervous system.

ectopic beat A contraction of the heart originating from a site other than the sinoatrial node.

ECV Effective circulating volume.

EDRF Endothelium-derived relaxing factor, now known to be nitric oxide (NO).

EDTA Ethylenediaminetetraacetic acid: an organic compound that binds divalent ions, especially calcium. Used to prevent the coagulation of blood samples.

EDV End-diastolic volume: the volume of blood in the ventricles at the end of diastole, just before the onset of systole.

EEG **electroencephalogram.**

effector A gland or muscle that becomes active in response to nerve stimulation. Also a molecule that selectively binds to a protein and regulates its biological activity.

efferent A term used to indicate a nerve or nerve pathway that carries nerve impulses from the central nervous system to an effector—a gland or muscle. Also indicates a blood vessel leaving an organ, as in efferent vein, efferent arteriole.

eicosanoids Signalling molecules derived from 20-carbon unsaturated fatty acids, especially arachidonic acid. Physiologically important eicosanoid families include the prostaglandins, thromboxanes, and leukotrienes.

Einthoven's triangle A theoretical inverted equilateral triangle formed by the left and right shoulders and the pubic region, with the heart positioned at the centre. In practice, Einthoven's triangle is approximated by the triangle formed by the axes of the limb leads of the standard ECG, and the sum of the voltages in the three leads is zero at any instant. This concept is used to determine the axis of the heart and to provide a reference zero for the monopolar ECG leads.

electrical synapse A synapse that is formed between two neurons by way of a gap junction. Most synapses in the central nervous system are chemical synapses, but electrical synapses are now known to be present in the neocortex, hippocampus, inferior olivary nucleus, and retina, as well as the spinal cord. This is by no means an exhaustive list.

electrocardiogram (ECG) A visual record of the variations in electrical potential measured over the chest that reflect the electrical activity of the heart. Widely used as a diagnostic tool.

electroencephalogram (EEG) A graphical record of the electrical activity in the brain obtained by a set of electrodes attached to the scalp. The EEG is helpful in the diagnosis of epilepsy and other neurological disorders.

electrochemical gradient The tendency of an ion to move passively from one side of the cell membrane to the other, taking into account its electrical charge, the concentration gradient, and the potential difference between the two sides.

electrolyte Any substance that can dissociate into ions in the body fluids. A substance that conducts electricity in an aqueous solution; those substances that completely dissociate into their constituent ions such as $NaCl$ (sodium chloride) are called strong electrolytes, while those that only partially dissociate such as CH_3COOH (acetic acid) are weak electrolytes.

electron transport chain The collective term for a set of carrier molecules located in the inner membrane of mitochondria that transfer electrons from hydrogen to oxygen. The electrochemical energy generated via this electron transport is subsequently used to form ATP from ADP.

electron A subatomic particle that is present in all atoms and carries negative electrical charge. Electrons are the main carriers of electrical current in solids.

embolism An obstruction in a blood vessel caused by a blood clot, air bubble, or particulate matter that has travelled from a distant site.

emulsification The breakdown of fat globules into fine dispersed particles (an emulsion). In the small intestine this process is facilitated by bile salts.

enantiomers Molecules of the same composition with structures that are mirror images of one another so that they cannot be superimposed on one another. They can be distinguished by their ability to rotate plane-polarized light in opposite directions, but they do so by the same extent. An **optical isomer**.

endo- A prefix used to denote something that is inside or within the body, cell, or organ, as in the endoderm of an embryo, or the endoscope, an instrument for examining the internal features of a body cavity (cf. **extra-**).

endocardium The epithelial layer that lines the chambers of the heart.

endochondral ossification The formation of bone from a cartilage template that is gradually replaced by calcified bone. The process involved in the formation of the long bones (cf. intramembranous ossification).

endocrine gland A gland that secretes one or more hormones directly into the bloodstream (cf. **exocrine gland**); a ductless gland. Examples are the adrenal and pituitary glands.

endocytosis The uptake of material by a cell (cf. **exocytosis**).

endoderm The innermost of the three germ layers of an animal embryo. During development the endoderm forms the lining of the digestive and respiratory tracts.

endogenous A process that has its origin within an organism, such as an endogenous rhythm (i.e. a rhythm generated internally).

endolymph The potassium-rich fluid of the scala media and membranous labyrinth of the inner ear (cf. **perilymph**).

endometrium The lining of the uterus, which thickens under the influence of oestrogens and progesterone during each menstrual cycle in readiness for implantation, and is shed at menstruation if there is no pregnancy.

endomysium The delicate connective tissue covering the cell membrane (the sarcolemma) of individual muscle fibres.

endoneurium The innermost layer of connective tissue within a peripheral nerve; the delicate membrane that surrounds individual nerve fibres.

endoplasmic reticulum (ER) A cellular organelle that consists of a network of membranes that encloses a space distinct from the cytoplasm. The rough ER is the site of synthesis of secretory and membrane-bound proteins, while the smooth ER is the site of lipid synthesis. The smooth ER also provides a store of calcium within a cell. In muscle cells the ER is called the sarcoplasmic reticulum.

endosomes Large vesicles that are located either just beneath the plasma membrane (peripheral endosomes) or near the nucleus (perinuclear endosomes). They play a role in recycling materials within cells.

endosteum The vascular membrane that lines the marrow cavity of the long bones.

endothelium A continuous layer of squamous epithelial cells that lines the blood vessels, heart, and lymphatic vessels.

endplate potential The depolarization of a skeletal muscle fibre that results from the activity of its motor nerve. The depolarization is caused by acetylcholine, released from the nerve terminal, binding to nicotinic receptors on the postsynaptic (muscle) membrane and activating ligand-gated ion channels.

energy The capacity of a physical system to perform work. Energy itself exists in several forms such as electrical energy, heat, light, or mechanical energy.

energy balance The difference between the amount of energy taken up as food and the performance of work in the form of heat generation, muscular activity, secretion, or tissue growth. Positive energy balance occurs when the body lays down carbohydrate or fat reserves and there is tissue growth. Negative balance occurs when food intake is insufficient to meet the demands of the body's metabolism. This depletes the body's reserves and tissue mass falls.

energy equivalent of oxygen The amount of energy liberated when a foodstuff is burned in one litre of oxygen. Pure carbohydrate will liberate 20.93 kJ or 5 kcal per litre of oxygen, pure fat 19.98 kJ or 4.7 kcal per litre of oxygen, and protein about 18.84 kJ or 4.5 kcal per litre of oxygen.

energy metabolism The cellular processes involved in generating ATP (the currency of energy in cells) from the breakdown of carbohydrates, fats, or other organic compounds. It may occur both in the presence of oxygen (aerobic metabolism) and in its absence (anaerobic metabolism).

ENS **enteric nervous system.**

enteral nutrition The supply of nutrients to the digestive system via a nasogastric tube. The technique is used when an individual can digest and absorb nutrients but cannot eat, chew, or swallow their food (cf. **parenteral nutrition**).

enteric nervous system (ENS) The network of neurons and nerve fibres that regulate the activity of the digestive system. It is one of the four main divisions of the nervous system. The ENS has its own independent reflex activity but is linked to the autonomic nervous system.

epi- Prefix used to denote a structure that is above, on, or over another structure, as in epidermis, epineurium, and epithelium. It also indicates something that is near another structure (as in epiglottis) or attached to it (as in epiphysis). It is also used to indicate a process that is in addition to a main topic, as in epigenetics.

epicardium The membrane that surrounds the heart; the innermost layer of the pericardium.

epidermis The epithelial layer that covers the outer surface of the body; the outermost layer of the skin.

epidural space That part of the spinal canal lying between the vertebral column and the dura mater surrounding the spinal cord.

ependymal cells The ciliated cells that line the central fluid-filled spaces of the brain (the cerebral ventricles and aqueduct) and the central canal of the spinal cord.

epilepsy A neurological disorder that is characterized by recurrent episodes of abnormal electrical activity in the brain. Symptoms include sensory disturbances, loss of consciousness, and, in certain cases, overt convulsions. Often caused by traumatic brain injury. The site at which the abnormal electrical activity begins is known as an epileptic focus.

epimysium The layer of dense connective tissue that surrounds a muscle, protecting it from rubbing against neighbouring muscles and bones. Arising from the epimysium, thin sheets of connective tissue form the **perimysium**, which covers the bundles of fibres that make up the muscle. At the ends of a muscle the perimysium merges with the tendons attaching the muscle to the bone.

epinephrine The term used in North America for **adrenaline**.

epineurium The outermost layer of dense irregular connective tissue surrounding a peripheral nerve.

epiphyseal plate A layer of hyaline cartilage separating the shaft (**diaphysis**) of a long bone from the two ends (the **epiphyses**). An epiphyseal plate, also known as the growth plate, is found only in growing children, adolescents, and young adults. When a young person has stopped growing, the cartilage becomes ossified and its position is marked by a dense line of bone called the epiphyseal line.

epiphysis (pl. epiphyses) One of the ends of a long bone; the end of a bone that forms one side of a joint.

epithelium (pl. epithelia) A continuous layer of cells that separates one internal space from another. Epithelia line the hollow organs such as the gut, urinary tract, and lungs (cf. **endothelium** and **mesothelium**).

epithelial transport The process that moves a substance from one body compartment to another. It may be an active process requiring a source of energy, or it may be passive as in the case of a substance diffusing down its concentration gradient.

epitope That part of an antigen that is recognized by the immune system. Also known as an antigenic determinant.

epp End-plate potential.

epsp **excitatory postsynaptic potential**.

equilibrium potential The membrane potential at which there is no net electrochemical force driving the movement of an ion into, or out of, a cell. The potential at which the electrical and chemical forces acting on an ion are balanced. It may be calculated using the Nernst equation.

ER **endoplasmic reticulum**.

erythrocytes Alternative term for **red blood cells**.

erythropoietin A glycoprotein hormone secreted by the endothelial cells of the renal capillaries that regulates the production of red blood cells by the bone marrow.

escape rhythm Beating of the heart initiated by a part of the heart other than the sinoatrial node.

essential amino acids Those amino acids, ten in number, that cannot be synthesized by the body. They are arginine, histidine, isoleucine, leucine, lysine, methionine, phenylalanine, threonine, tryptophan, and valine.

essential fatty acids Polyunsaturated fatty acids that are necessary for health but cannot be synthesized by the body. They are linoleic acid and linolenic acid.

estrogen North American spelling of **oestrogen**.

ESV End-systolic volume: the volume of blood remaining in the heart at the end of the ejection phase of the cardiac cycle (systole).

euchromatin The state of chromosomal DNA which stains lightly with basic dyes. It is generally agreed to be the form of DNA that is partially or fully uncoiled and actively engaged in gene transcription (cf. **heterochromatin**).

eupnoea Normal, unstressed breathing (cf. **dyspnoea**).

evaporation The conversion of liquid water into water vapour, a process that requires a significant amount of heat and is exploited to regulate body temperature.

excitation–contraction coupling The process by which the electrical activity of a muscle initiates a contraction.

excitatory postsynaptic potential (epsp) A depolarization of a nerve cell resulting from synaptic activity. All epsps tend to increase the excitability of neurons and the likelihood that they will initiate an action potential (cf. **inhibitory postsynaptic potential**).

exercise Exertion of the skeletal muscles in order to perform a physical task such as walking, lifting etc.

exocrine gland A gland that secretes a product, such as a digestive enzyme, into a duct that empties onto an epithelial surface (cf. **endocrine gland**). Examples are the salivary glands, sweat glands, and pancreas.

exocytosis The process by which cells export material to the extracellular environment.

exogenous A term used to describe something that originates from outside the body, e.g. exogenous stimuli.

exon A segment of a gene separated from other parts by a region of DNA known as an **intron**.

exophthalmos Bulging of the eyes, characteristic of hyperthyroidism.

extensor A muscle that acts on a joint to straighten a limb (cf. **flexor**).

exteroceptor A sensory receptor adapted to detect changes in the external environment of the body (cf. **interoceptor**). The ear, eye, and touch receptors are all exteroceptors.

extra- A prefix meaning outside or beyond, as in extracellular.

extracellular Pertaining to the environment outside the cells, as in extracellular fluid, extracellular matrix (cf. **intracellular**).

extracellular fluid The fluid that surrounds the cells, characterized by a high sodium and chloride concentration and a low potassium concentration (cf. **intracellular fluid**). It includes the plasma, transcellular fluid, and interstitial fluid.

extrafusal fibre A muscle fibre that is not part of a stretch receptor; one of the fibres that make up the bulk of a muscle (cf. **intrafusal fibre**).

extraocular muscles The muscles of the orbit that control the position of the eye and therefore the direction of the gaze.

extrasystole A contraction of the ventricles that is not initiated by the sinoatrial node.

F

facilitated diffusion Carrier-mediated diffusion across cell membranes down a concentration gradient. Facilitated diffusion does not require the expenditure of metabolic energy.

FAD Flavine adenine dinucleotide.

$FADH_2$ Reduced flavine adenine dinucleotide.

fatty acids Also called carboxylic acids, the fatty acids have a general molecular formula of $CH_3(CH_2)_nCOOH$. Their glycerol esters are the principal constituents of fats and oils.

feedforward control A method of regulation in which a change in a physiological variable leads to an adjustment of the system to suit the new situation before the system responds to the change. In this it is unlike a feedback process, which detects a change in the responding system before making the necessary adjustment. See also **negative feedback** and **positive feedback**.

fertilization The process by which the male and female gametes fuse to form a zygote.

fetal haemoglobin (HbF) The form of haemoglobin found in the fetus—it has a higher affinity for oxygen than adult haemoglobin.

fetus An unborn human eight weeks or more after conception. At less than eight weeks it is called an embryo.

FEV$_1$ Forced expiratory volume at 1 second after a maximal inspiration. See also **FVC**.

fever A body temperature in excess of 38°C that is not caused by exercise or the environment. Fever usually occurs in response to infection by an invading organism. It is accompanied by shivering, headache, and in severe instances, delirium. Also known as pyrexia.

fibroblasts The most common cells of connective tissue. Fibroblasts secrete the materials that form the extracellular matrix including the collagen, elastin, and glycoproteins that constitute the connective tissue fibres.

Fick principle An important principle in physiology which states that blood flow to an organ or tissue is proportional to the difference in concentration of a substance in the blood as it enters and leaves that organ or tissue. It is also used to determine the cardiac output, which is calculated from the difference in oxygen content of the mixed venous blood and the systemic arterial blood and the rate of oxygen consumption by the whole body.

Fick's law of diffusion This is, strictly speaking, Fick's first law of diffusion, which is concerned with the rate at which a solute diffuses through a uniform sheet of tissue under steady-state conditions. The law states that the quantity transferred is proportional to the tissue area and the difference in concentration between the two sides, and is inversely proportional to its thickness. This means that the greater the concentration difference between the two sides, the higher the rate of diffusion, while the thicker the tissue barrier, the slower it becomes. Fick's second law of diffusion is concerned with non-steady-state conditions. See also **diffusion**.

filtration The separation of small and large particles by passing a fluid through a porous membrane. In physiology the pressure needed to do so is provided by blood pressure, and the pores of the membrane are often sufficiently small to permit the filtration of ions and small solutes (e.g. glucose) from proteins.

flaccid Describes a muscle that lacks normal muscle tone. Seen in the limb muscles after a lower motor neuron lesion and accompanied by weak or absent tendon reflexes.

flagellum (pl. flagellum) The long whiplike extension emerging from the head of a sperm that provides its means of locomotion. It shares the same internal structure of microtubules (axonemes) as the **cilia**.

flexor A muscle that acts on a joint to allow a limb to bend (cf. **extensor**).

fMRI Functional magnetic resonance imaging: an imaging technique that directly measures local blood flow in the brain to provide information on regional brain activity.

foetus Alternative spelling of **fetus**.

follicle A spherical mass of cells containing a fluid-filled cavity, as in the case of an ovarian follicle. Also a small cavity in the skin from which a hair emerges (a hair follicle) and an alternative name for a small lymph node.

follicle stimulating hormone (FSH) A glycoprotein hormone secreted by gonadotrophs of the anterior pituitary gland that acts synergistically with luteinizing hormone to regulate the reproductive processes of the body. In males, FSH induces the Sertoli cells of the testes to secrete androgen binding protein. In females, FSH stimulates the growth and development of ovarian follicles.

foramen ovale The opening in the septum that connects the two atria of the fetal heart. It normally closes after birth as the lungs inflate. It is also the term for the oval opening in the greater wing of the sphenoid bone that allows the passage of several nerves, arteries, and veins.

force A dynamic influence that changes the state of motion of a physical body. The magnitude (F) of such a force is equal to the mass of the body (M) multiplied by its acceleration (A), i.e. F = M × A. The SI unit of force is the newton, which has base units of joules per metre.

Frank–Starling mechanism See **Starling's law**.

FRC Functional residual capacity: the volume of air in the lungs at the end of a quiet spontaneous expiration. Also called the end-expiratory lung volume.

frontal section Alternative term for a **coronal section**.

FSH **follicle stimulating hormone**.

functional hyperaemia The increase in blood flow that accompanies the activity of a muscle or tissue. Also known as active hyperaemia.

Functional magnetic resonance imaging See **fMRI**.

fusimotor fibres The muscle fibres that lie within the capsule of the muscle spindles and regulate their sensitivity to stretching. Also called gamma-efferents.

FVC Forced vital capacity: the maximum volume of air that can be rapidly exhaled after a full inspiration. Also called the forced expired vital capacity (FEVC).

G

G protein A GTP binding protein. A member of a large family of proteins that regulate the biochemical activity of cells. When a G protein is bound to GTP it is active; when it is bound to GDP, it is inactive. There are two classes of G proteins: the monomeric small GTPases and the heterotrimeric G protein complexes. The heterotrimeric G proteins are involved in the regulation of activity within a cell via a signalling cascade. They consist of three separate subunits called alpha (α), beta (β), and gamma (γ), which dissociate when the complex is activated by the binding of a surface receptor to its ligand. Monomeric G proteins regulate the activity of many signalling cascades and can be thought of as enzyme second messengers.

GABA γ-aminobutyric acid.

galactopoiesis The formation and secretion of milk by the mammary gland cells.

GALT Gut-associated lymphoid tissue.

gamete An unfertilized mature egg cell (oocyte) or a mature sperm.

gamma-efferent Alternative term for **fusimotor fibres**.

gamma motoneuron A motoneuron that controls the activity of the muscle fibres within a muscle spindle. Often written as γ-motoneuron.

ganglion A collection of nerve cell bodies. In the case of autonomic ganglia, a site of synaptic activity.

gap junction A junction between cells that allows ions and small molecules such as cyclic AMP to diffuse from the cytoplasm of one cell to that of another.

gastrointestinal tract (GI tract) The body system concerned with the intake and processing of food. The tract comprises the oral cavity, oesophagus, stomach, small intestine, large intestine, and rectum. Also known as the alimentary canal.

GDP Guanosine diphosphate.

gene A region of DNA that determines a distinct heritable characteristic, usually corresponding to a single species of protein or RNA.

gene expression The process by which information from a gene is used in the synthesis of a functional gene product such as a protein or a non-protein molecule such as transfer RNA (tRNA).

gene transcription The process of transcribing the sequence of DNA bases that form a single gene into the sequence of bases of mRNA that codes for the amino acid sequence of a protein.

genotype The genetic constitution of an individual person, as distinct from their physical appearance (which is called their **phenotype**).

gestation The period of time from fertilization to birth. The time during which a fetus develops before birth.

GFR **glomerular filtration rate**.

GH **growth hormone**.

GHIH Growth hormone inhibiting hormone: see **somatostatin**.

GHRH Growth hormone releasing hormone.

GI Gastrointestinal, as in gastrointestinal tract.

GIP Gastric inhibitory peptide.

gland A collection of cells often forming a specific organ that synthesizes materials for secretion either into the bloodstream (an endocrine gland) or onto an epithelial surface such as the lining of the small intestine (an exocrine gland).

glial cells The principal non-neuronal cells of the nervous system. Four distinct morphological types are recognized: **astrocytes**, **oligodendrocytes**, **microglia**, and **ependymal cells**.

globus pallidus A part of the basal ganglia lying adjacent to the putamen and characterized by its pale staining with basic dyes. Together with the putamen it forms the lentiform nucleus. Also called the pallidum.

glomerular capsule **Bowman's capsule**.

glomerular filtration rate The volume of plasma filtered in a minute by the glomeruli of both kidneys. It is estimated from measuring the clearance of inulin or creatinine and has a value of around 120 ml min^{-1} in healthy young men and 110 ml min^{-1} in healthy young women.

glucagon A peptide hormone secreted by the alpha cells of the pancreatic islets in response to low blood glucose levels; it acts to mobilize glucose from glycogen to restore normal blood glucose levels.

glucocorticoids Steroid hormones secreted by the zona fasciculata of the adrenal cortex to regulate carbohydrate, protein, and fat metabolism. The glucocorticoids also perform anti-inflammatory activity. The best known glucocorticoid is cortisol (also called hydrocortisone).

glucogenic A term used to describe those amino acids that can be converted to glucose in the process of **gluconeogenesis**. All the amino acids found in proteins except leucine and lysine are glucogenic.

gluconeogenesis The metabolic process by which certain amino acids can be converted to glucose.

glucose A monosaccharide having the molecular formula $C_6H_{12}O_6$ that is the principal source of metabolic energy for most cells, especially those of the brain and testes.

GLUT2 Glucose transporter family member 2.

GLUT5 Glucose transporter family member 5.

glycation The non-enzymic addition of a glucose residue to a protein (see also **glycosylation**).

glycocalyx The outer filamentous covering of glycoproteins that protects the apical surface of epithelial cells.

glycogen A polysaccharide that forms the carbohydrate stores of the liver, muscle, and other organs. It consists of long branched chains of glucose residues which can easily be hydrolysed to release free glucose into the blood.

glycogenesis The formation of glycogen from glucose.

glycogenolysis The process of hydrolysing glycogen to release glucose into the blood for utilization by other tissues.

glycolipid An ester formed by the combination of a fatty acid with sphingosine to form a ceramide. There are two classes of glycolipid: the cerebrosides, in which the ceramide is linked to a monosaccharide, and the gangliosides, in which the ceramide is linked to an oligosaccharide.

glycolysis The pathway by which glucose is broken down prior to its oxidative metabolism via the tricarboxylic acid cycle. Unlike the TCA cycle, glycolysis can continue to generate ATP in the absence of oxygen (anaerobic metabolism).

glycoprotein A protein containing one or more carbohydrate chains. Glycoproteins are important constituents of bone and connective tissues.

glycosuria The presence of glucose in the urine.

glycosylation The process of adding a sugar residue to a protein to form a glycoprotein.

GMP Guanosine monophosphate.

GnRH Gonadotrophin releasing hormone.

Goldman equation An expression that relates the membrane potential to the concentrations of the principal ions (sodium, potassium, and chloride) on either side of the membrane and the permeability of the membrane to those ions.

Golgi apparatus A membranous complex of vesicles, vacuoles, and flattened sacs near the cell nucleus. Found in most cells, particularly those involved in secretion. The Golgi apparatus is involved in the post-translational modification of proteins prior to their transport to other parts of the cell where they perform their specific functions.

Golgi tendon organ A stretch receptor that signals the tension in the tendons of skeletal muscles.

gonadotrophins The anterior pituitary hormones that stimulate the growth and activity of the ovaries and testes: **follicle stimulating hormone** (FSH) and **luteinizing hormone** (LH).

gonads The organs that produce the female and male gametes: the ovaries and testes.

Graham's law of diffusion This states that the rate of diffusion of a gas in a mixture is inversely proportional to its molecular mass. Thus, the higher the molecular mass, the slower the rate of diffusion. In the case of the respiratory gases there is little difference between their rates of diffusion as their molecular masses are very similar.

granulocytes The white blood cells that contain large granules—the basophils, eosinophils, and neutrophils. The neutrophils are phagocytic cells. The granulocytes are also called polymorphonuclear leukocytes because their nuclei are divided into lobes or segments.

grey matter That part of the central nervous system that contains the nerve cell bodies.

growth hormone (GH) A large polypeptide hormone synthesized in and secreted by the somatotrophs of the anterior pituitary gland. It acts to promote growth and increases the synthesis of the proteins essential for growth. It increases blood glucose levels and promotes the breakdown of fats to release fatty acids into the blood. Both processes are important in fasting. Also called somatotrophin.

GTP Guanosine triphosphate.

GTP binding protein Alternative term for **G protein**.

gustation The sense of taste.

gut A collective term for the structures of the gastrointestinal tract.

gyrus (pl. gyri) A sinuous fold on the surface of the brain bounded by shallow grooves called sulci.

H

habituation The waning of an innate response to a stimulus; habituation occurs when the stimulus is repeated but has no particular significance.

haematocrit ratio The fraction of blood volume occupied by the red blood cells. Taken to be around 45 per cent in males and 42 per cent in females.

haematopoiesis The process by which new blood cells are formed in the red bone marrow. Also called haemopoiesis.

haematuria The presence of blood in the urine.

haemoconcentration An increase in haematocrit caused by a reduction in plasma volume.

haemodynamics The study of the forces that determine the circulation of the blood.

haemoglobin The oxygen binding protein of the red blood cells. It conveys oxygen from the alveoli to the tissues and assists in the transport of carbon dioxide from the tissues to the lungs.

haemolysis The rupture of red blood cells and the release of haemoglobin into the plasma.

haemophilia A disorder of blood clotting that results from a defective blood clotting factor, usually factor VIII. The condition is often inherited but may arise from a spontaneous mutation. It affects males more often than females.

haemopoiesis Alternative term for **haematopoiesis**.

haemorrhage The acute loss of blood; the circulatory response depends on the extent of blood loss.

haemostasis The processes that act to prevent the loss of blood, involving vasoconstriction of damaged arteries and blood clotting.

Haldane effect The increase in the quantity of carbon dioxide carried by the blood resulting from its deoxygenation.

Hamburger effect Alternative term for **chloride shift**.

haploid Having half the normal number of chromosomes; the male and female gametes are haploid; humans have 23 unpaired chromosomes in their gametes (cf. **diploid**).

haptic touch The use of the senses of touch and proprioception to ascertain the characteristics of an object.

haustra The small pocket-like sacs of the colon that give the colon its characteristic segmented appearance.

haustration Contraction of the circular muscle of the colon that mixes its contents to aid the absorption of water and electrolytes from the faecal material.

Haversian canal The central channel of an osteon of bone through which blood vessels and nerves pass.

Haversian system The cylindrical, columnar structures of the long bones consisting of a series of concentrically arranged lamellae surrounding a central channel known as a **Haversian canal**. The basic structural unit of compact bone; also called an osteon.

Hb **haemoglobin**.

HbF **fetal haemoglobin**.

HbS Sickle cell haemoglobin.

hCG **human chorionic gonadotrophin**.

heart block A defect in the conduction of the electrical activity of the heart between the sinoatrial node and the ventricles. Classified according to the severity of the block into first-, second-, or third-degree block.

heart failure Alternative term for **cardiac failure**.

heliox A mixture of helium (He) and oxygen (O_2) used in commercial diving to avoid the complications of nitrogen narcosis and decompression sickness. It is also used for certain patients who have breathing difficulties as the mixture requires less effort to breathe. It is mainly used where there is upper airway obstruction or vocal cord dysfunction.

hemianopia A term usually applied to denote defective vision or blindness in half of the visual field.

hemiballismus A rare movement disorder with sudden spontaneous flinging motions of the limbs on one side of the body. It is usually caused by a lesion in the basal ganglia of the opposite side of the brain.

hemidesmosome A protein complex similar to half a **desmosome** that serves to connect the basal surface of an epithelial cell to the basement membrane.

Henderson–Hasselbalch equation The relationship between the pH (hydrogen ion concentration) and the ratio of undissociated acid to its conjugate base. When the ratio is 1, the pH is the same as the dissociation constant of that acid (its pK_a).

Henry's law This states that the amount of gas dissolved in a given volume of liquid is proportional to the partial pressure in the gas phase. The proportionality constant is called the solubility coefficient, with units of ml $l^{-1}kPa^{-1}$.

hepatic insufficiency Alternative term for **liver failure**.

hepatocyte A polygonal epithelial cell of the liver which secretes bile. Hepatocytes form the bulk of the liver's parenchyma. Also called a hepatic cell or a liver cell.

Hering–Breuer reflex A respiratory reflex that acts to limit the depth of breathing. The reflex is initiated by slowly adapting stretch receptors in the bronchi and bronchioles, which send afferent impulses to the respiratory areas of the brain via the vagus nerve. The result is a decreased drive to the respiratory muscles.

heterochromatin The state of chromosomal DNA which stains intensely with basic dyes. It represents the tightly coiled form of DNA most abundant in cells that are less actively engaged in gene transcription (cf. **euchromatin**).

hGH **human growth hormone.**

high-pressure nervous syndrome (HPNS) A physiological disorder that affects divers at depths below 130 m when they are breathing helium and oxygen mixtures. The effects experienced, and the severity of those effects, depend on the rate of descent, the depth, and the percentage of helium. The symptoms include tremors, dizziness, and nausea. Mental performance is much less affected.

hippocampus A region of the cerebral cortex that forms the floor of the lateral ventricle. So named because its cross-section has the shape of a sea horse, it is part of the limbic system and is involved in the processing of emotions and memory.

histology The study of the microscopic structure of the tissues.

histotoxic hypoxia Hypoxia in the tissues resulting from an inability of the tissues to utilize the available oxygen—as in poisoning of the oxidative enzymes.

HIV Human immunodeficiency virus.

HLA Human leukocyte antigen.

holocrine A type of secretion that consists of the disintegrated secretory cells mixed with their secretory product. The sebaceous glands of the skin are holocrine glands, as the oily secretion sebum is mixed with the remnants of dead cells.

homeostasis The tendency for an organism to maintain a stable internal environment by suitable adjustment of its physiological processes.

hormone A blood-borne chemical signal secreted by endocrine cells to regulate the activity of another group of cells.

hPL **human placental lactogen.**

HPNS **high-pressure nervous syndrome.**

HRT Hormone replacement therapy.

HSL Hormone-sensitive lipase.

5-HT (5-hydroxytryptamine) Chemical name for **serotonin**.

human chorionic gonadotrophin (hCG) A glycoprotein hormone synthesized in the placenta that maintains the corpus luteum until the placenta is able to produce sufficient progesterone to maintain gestation.

human growth hormone (hGH) A hormone synthesized and secreted by the somatotrophic cells of the anterior pituitary gland. Although hGH is similar in structure to growth hormone isolated from other species, these cannot act as replacements as they do not have significant effects on the human growth hormone receptor. See also **growth hormone**.

human placental lactogen (hPL) A peptide hormone secreted by the placenta. It is structurally related to human growth hormone and acts to increase blood glucose levels in maternal blood and mobilize fatty acids for energy production to facilitate the energy supply of the fetus. Little hPL enters the fetal circulation. Also called human chorionic somatomammotrophin.

hunger The physiological sensation of emptiness resulting from lack of food. It is usually accompanied by stomach contractions (cf. **appetite**).

hyaline cartilage Cartilage that is found on joint surfaces, in the walls of the larger airways, and in the immature skeleton. It lacks blood vessels and nerves and consists of chondrocytes that synthesize a matrix of hyaluronic acid, collagen, and other proteins.

hydrogen bond A weak chemical bond in which a hydrogen atom of one molecule is attracted to an electronegative atom, especially an oxygen or nitrogen atom. Commonly seen in proteins, where the structure is stabilized by hydrogen bonds. It is also important as the basis of base pairing in DNA.

hydrogen ion A hydrogen atom that has lost its electron; a proton. Hydrogen ions (H^+) are found in aqueous solutions of acids, where they are solvated by one or more water molecules to form hydroxonium ions (H_3O^+).

hydrolysis The breakdown of a large molecule into smaller ones by the addition of water, usually catalysed by an enzyme. The breakdown of glycogen to release glucose is one example; the digestion of proteins to their constituent amino acids is another.

hydrophilic readily soluble in water. Examples of hydrophilic materials are the simple sugars such as glucose and fructose and inorganic salts such as sodium chloride.

hydrophobic Insoluble or only sparingly soluble in water. In biological systems, hydrophobic materials are characterized by the presence of long hydrocarbon chains. Examples are the long chain fatty acids such as palmitic acid and arachidonic acid.

hyper- A prefix indicating something that is excessive, as in **hypertension**—excessively high blood pressure.

hyperaemia A large increase in blood flow through a tissue or organ, as in functional hyperaemia—the increase in blood flow that accompanies a period of intense exercise.

hyperalgesia A marked increase in sensitivity to pain.

hypercapnia An abnormally high level of carbon dioxide in the arterial blood ($PCO_2 > 45$ mm Hg or 6 kPa).

hyperglycaemia An abnormally high blood glucose level (> 7 mmol l^{-1} or 126 mg dl^{-1}). Often seen in uncontrolled diabetes mellitus, where much higher values may be encountered.

hypermetropia (in UK) or hyperopia (in North America) A defect of vision that results in difficulty in focusing near objects. It is caused by an imperfection in the eye in which the eyeball is too short or the lens is too inelastic to permit near accommodation.

Also known as long-sightedness in the UK and far-sightedness in North America.

hyperphagia An obsessive craving for food. Sufferers often continue to gorge themselves long after they are full—as in Prada–Willi syndrome, for example.

hyperplasia Increase in the size of an anatomical structure caused by an increased proliferation of its cells.

hyperpolarization An increase in membrane potential, i.e. a state in which the membrane potential becomes more negative.

hyperreflexia Exaggerated tendon reflexes associated with the spasticity caused by lesions to the primary motor cortex. See also **spasticity**.

hypertension Excessively high blood pressure: systemic arterial blood pressure in excess of 140/90 mm Hg (18.7/12 kPa).

hyperthermia An increase in body core temperature above the normal range of 35.5–37.7°C.

hypertonic solution A solution with a sufficiently high solute content that it causes cells suspended within it to shrink (cf. **hypotonic solution, isotonic solution**).

hypertrophy Increased size of a tissue or organ resulting from an increase in the size of its constituent cells rather than by their proliferation.

hypo- A prefix indicating 1. a structure lying beneath another, as in hypothalamus, or 2. a physiological variable below the normal range, as in hypoglycaemia.

hypocapnia A low partial pressure of carbon dioxide (PCO_2) in the arterial blood—less than 35 mm Hg (PCO_2 < 5.33 kPa) (cf. **hypercapnia**).

hypoglycaemia A blood glucose concentration below 3.5 mmol l^{-1} (63 mg dl^{-1}).

hypotension Abnormally low blood pressure, generally taken to be less than 90 mm Hg (12 kPa) systolic pressure.

hypothalamic releasing hormones The hormones secreted by the hypothalamus to regulate the activity of the various pituitary endocrine cells. At present five different releasing hormones are recognized, as well as two inhibitory hormones.

hypothalamic–hypophyseal portal system A specialized system of blood vessels linking the hypothalamus with the anterior pituitary gland. Certain hypothalamic nuclei secrete small quantities of hormones that enter a capillary network in the hypothalamus and are carried to the anterior pituitary gland via the hypothalamic–hypophyseal portal vessels, where they divide to form a secondary capillary plexus.

hypothalamus The part of the forebrain that lies below the thalamus. Its role is to regulate the homeostatic processes of the body such as body temperature, thirst, and appetite. It also regulates the activity of the pituitary gland and is involved in regulating the sleep cycle.

hypothermia A state in which the core body temperature falls below 35°C.

hypotonia A state of low muscle tone often accompanied by reduced muscle strength. Hypotonia occurs in many different motor diseases.

hypotonic solution A solution with a sufficiently low solute content that it causes cells suspended within it to swell (cf. **hypertonic solution, isotonic solution**).

hypoxaemia An abnormally low oxygen partial pressure in the blood.

hypoxia Insufficient oxygenation of the tissues. Four types of hypoxia are distinguished: **hypoxic hypoxia**, **anaemic hypoxia**, **stagnant hypoxia**, and **histotoxic hypoxia**.

hypoxic hypoxia A low arterial partial pressure of oxygen (PO_2) in the arterial blood. The condition is normal following ascent to a high altitude; at low altitude it may reflect some pathology such as respiratory failure.

hysteresis A state that occurs when the magnitude of a physical property lags behind the change producing it as this change varies. Hysteresis is shown in the pressure–volume relations of the respiratory system, where a higher pressure is required to inhale a given volume of air than is required to exhale it.

I

ICF Intracellular fluid: the fluid inside the cells, characterized by a high concentration of proteins and potassium ions, and a low concentration of sodium and chloride ions.

ICSH Interstitial cell stimulating hormone, a male hormone identical to **luteinizing hormone**.

ideal gas law A law that summarizes the relationships between the pressure P, volume V, and temperature T of a given amount of an ideal gas. The mathematical equation summarizing the law is $PV = nRT$, where n is the number of moles of gas and R is the gas constant. Also called the universal gas law. See also **Boyle's law, Charles's law**.

IgA, IgD, IgE, IgG, and IgM Abbreviations for the different classes of **immunoglobulin**: A, D, E, G, and M.

IGF-1, IGF-2 **insulin-like growth factors**.

ileum That portion of the small intestine lying between the jejunum and the colon.

ilium The uppermost of the three fused bones that form each side of the pelvis. The other two bones are the ischium and the pubis.

immune system The body's defence system of cells and proteins that are able to identify invading organisms and then inactivate and eliminate them. Consisting of two main divisions: the innate or natural immune system, and the adaptive immune system.

immunoglobulins Antibodies: proteins of the adaptive immune system that bind to foreign particles such as bacteria prior to their elimination. They are globulins with the same basic structure of two identical light chains linked with two identical heavy chains to give a Y-shaped molecule. There are five main types of heavy chain which determine the particular class of an antibody: IgA, IgD, IgE, IgG, and IgM.

impedance matching The process by which the ossicles of the middle ear transfer the airborne vibrations of sound to the fluid of the inner ear. The matching of the low acoustic impedance of air to the high impedance of the scala media is achieved largely because the pressure collected by the large area of the tympanic membrane is transferred to the small area of the footplate of the stapes, thus gaining a 17-fold net pressure increase.

implantation The stage of pregnancy when the early embryo adheres to the wall of the uterus prior to the formation of the placenta. Implantation allows the embryo to receive the oxygen and nutrients that enable it to develop and grow. In humans, implantation of a fertilized ovum generally occurs 6–12 days after ovulation.

imprinting The rapid learning that occurs during the brief receptive period soon after birth. Imprinting establishes a long-lasting attachment to a parent, offspring, or place.

incisura The downward notch in the aortic pressure curve which occurs between the systolic maximum and diastolic minimum. It reflects the backflow of blood that occurs just before the aortic valve closes. Also a term for a notch, cleft, or fissure in an organ or part of the body.

incretins Hormones secreted by the entero-endocrine cells of the gut in response to ingestion of food that act to modulate the secretion of insulin to enhance the response to glucose. Two incretins are known: gastric inhibitory polypeptide (or glucose-dependent insulinotropic peptide, GIP) and glucagon-like peptide-1 (GLP-1).

infarction An interruption to the blood supply of an organ or part thereof, typically caused by a blood clot (thrombus) and resulting in damage to the affected tissue, as in a cardiac infarct: loss of blood supply to part of the heart, commonly known as a 'heart attack'.

inferior An anatomical term indicating that a structure lies below another, as in inferior colliculus and inferior vena cava.

inhibin One of two protein hormones (known as inhibins -A and -B) secreted by the Sertoli cells in the male and the follicular granulosa cells in the female. Inhibins, as their name implies, inhibit the production of follicle stimulating hormone (FSH) by the pituitary gland. They do not inhibit the secretion of GnRH from the hypothalamus. Inhibin is implicated in the regulation of spermatogenesis.

inhibitory postsynaptic potential (ipsp) Synaptic activity that results in a decrease in the likelihood that a nerve cell will initiate an action potential. It results from a hyperpolarization of the membrane potential, shifting it further from the threshold for action potential generation (cf. **excitatory postsynaptic potential**).

inotropic state (inotropy) The state of contractility of the heart, i.e. how forcefully it can beat. Inotropy is increased by adrenaline and similar agents (positive inotropic agents) and decreased by various drugs (negative inotropic agents).

INR International Normalized Ratio. A measure of the effectiveness of the extrinsic pathway of blood coagulation which is used to determine the clotting tendency of blood, primarily for monitoring warfarin dosage. The INR is based on the time it takes plasma to clot after addition of tissue factor, but as this varies with different batches of tissue factor, the results are standardized by comparison with an international reference standard.

insulin A peptide hormone of molecular mass of 5805Da that is secreted by the islet cells of the pancreas to regulate blood glucose levels. It acts to promote the uptake of glucose by the tissues.

insulin-like growth factors (IGF-1 and IGF-2) Proteins structurally related to insulin. Insulin-like growth factor 1 (IGF-1) is secreted by the liver and other target issues in response to growth hormone (GH) and is required for achieving normal growth. Insulin-like growth factor 2 (IGF-2) appears to be required for early development. IGF-1 is also called somatomedin C.

integrins A large family of transmembrane proteins that link cells to each other or to the extracellular matrix.

integumentary system The skin and its associated structures such as hair, nails, and the sweat glands.

intercalated discs Densely staining regions of attachment between cardiac muscle cells consisting of gap junctions, fascia adherens, which anchor the actin filaments of the terminal sarcomeres, and maculae adherens (desmosomes), which act to prevent the cardiac cells separating from each other during contraction.

interferons (IFNs) Cytokines that act to prepare the immune system to eliminate an infection, especially by a virus. They are named for their ability to minimize viral replication. IFNs also activate natural killer cells and macrophages and up-regulate antigen presentation.

intermediate filaments Cell filaments that range in diameter from 8–12 nm and play an important role in cell–cell and cell–substrate attachments (desmosomes and hemidesmosomes). Together with thin filaments of actin and microtubules, they constitute the cytoskeleton. All three types of filament act to stabilize cell shape and provide structural integrity.

interneuron A neuron that receives synaptic contacts from afferent nerve fibres and synapses with a motor neuron as part of a reflex arc. Occasionally called an internuncial neuron.

interoceptor A sensory receptor that responds to changes in the internal environment (cf. **exteroceptor**). Examples are the carotid chemoreceptors and the stretch receptors of the aortic arch (which respond to the arterial blood pressure).

interphase The stage between successive cell divisions, during which the nucleus is not dividing. It includes the phases known as G_1, S, and G_2.

interstitial cell stimulating hormone (ICSH) A male hormone identical to **luteinizing hormone**.

interstitium The space surrounding the cells of a tissue or organ.

interstitial fluid The aqueous fluid in the spaces between the cells (the interstitium) that bathes the cells of the tissues. As it is the medium through which materials are delivered to, and waste products removed from, the cells, its precise composition depends upon the exchanges between the cells of a tissue and the blood. Nevertheless its is similar in composition to an ultrafiltrate of plasma. Also called tissue fluid.

intra- A prefix meaning within or inside, as in intravenous.

intracellular Pertaining to the contents of cells, as in intracellular fluid (**ICF**)—the fluid within the cell itself (cf. **extracellular**).

intrafusal fibre A skeletal muscle fibre that is within the capsule of a stretch receptor (cf. **extrafusal fibre**).

intramembranous ossification The development of flat bones from embryonic fibrous connective tissue. Cells within the embryonic tissue membrane begin to cluster together (condense) and differentiate into osteoblasts to form ossification centres. The osteoblasts then start to secrete osteoid, the collagen-proteoglycan matrix of bone which becomes mineralized. Bony spicules start to radiate outwards from the ossification centre in a fairly random fashion between developing blood vessels to form 'woven' bone. The bony spicules subsequently become surrounded by mesenchymal cells which form the periosteum. Osteoblasts lying along the inside of the periosteum continue to secrete osteoid, which subsequently becomes calcified. In this way, layers of bone are built up beneath the periosteum.

intraocular pressure The pressure exerted by the fluid within the eye on its wall. The intraocular pressure serves to maintain the shape of the eye and thus plays an important role in image formation. Excess production of aqueous humour can, however, lead to the condition of glaucoma, which needs treatment to avoid excessive pressure on the retina with the loss of its blood supply. Normal intraocular pressure is around 2 kPa (15 mm Hg).

intrapleural pressure The pressure in the space between the visceral and parietal pleural membranes. It is normally below that of the atmosphere and considered to be equal to the mid-oesophageal pressure. At the end of quiet expiration it is roughly 0.5 kPa below that of the atmosphere. Sometimes called the intrathoracic pressure or pleural pressure.

intrapulmonary pressure Alternative term for **alveolar pressure**.

intrathoracic pressure Alternative term for **intrapleural pressure**.

intravesical pressure The pressure within the bladder. It acts as a signal to initiate urination and provides the driving pressure.

intrinsic factor A glycoprotein secreted by the parietal cells of the stomach that binds vitamin B_{12} (cobalamin) in the upper small intestine, so protecting it from the digestive enzymes of the gut. The complex formed by vitamin B_{12} and intrinsic factor is absorbed in the lower intestine.

intron A non-coding region of a gene that is removed during the production of messenger RNA (mRNA).

inulin clearance The rate at which inulin, a plant polysaccharide, is eliminated from the body. Inulin is not metabolized by the body but it is freely filtered by the renal glomerulus. As it is neither absorbed nor secreted by the renal tubules, it is used to determine the volume of plasma filtered each minute—the **glomerular filtration rate**.

involution 1. The infolding of a blastula during gastrulation. 2. The progressive decline of a normal physiological function that accompanies the aging process. 3. The restoration of the size of the uterus following childbirth.

ion An atom or molecule that has lost or gained one or more electrons and therefore carries an electrical charge.

ion channel A transmembrane protein that allows one or more species of ion to pass across the plasma membrane. Some ion channels are selective for a particular ion, for example sodium channels, potassium channels, and chloride channels; others allow similarly charged ions to pass, such as the nicotinic channels of the neuromuscular junction, which are non-selective cation channels.

ionization The process by which a neutral molecule or atom gains an electrical charge. In physiological systems it occurs when a weak acid partially dissociates to yield a hydrogen ion (H^+) and its negatively charged conjugate base.

IP_3 Inositol trisphosphate.

ipsilateral An anatomical term referring to a structure or nerve pathway that is on the same side of the body as another.

ipsp **inhibitory postsynaptic potential**.

iso- Prefix meaning equal or identical, as in isomer, isotope.

isoenzymes Two or more enzymes that have a similar function but differ slightly in their amino acid sequences (cf. **isoforms**).

isoforms Two or more functionally similar proteins that have similar properties and similar, but not identical, amino acid sequences.

isohydric principle Principle stating that in an aqueous system containing more than one buffer, any change in conditions that affects one buffer will also change the balance of all the others. The situation arises because all the acid/base pairs in a solution will be in equilibrium with one another and with the prevailing hydrogen ion concentration.

isomer Two chemical compounds with the same molecular composition but a different structure. (See also **optical isomer**).

isometric contraction Contraction of a muscle in which its tension is developed while its length is held constant (cf. **isotonic contraction**).

iso-osmotic Two solutions having the same osmolarity. Iso-osmotic solutions are not necessarily isotonic with cells.

isotonic contraction A contraction of a muscle under a constant load so that the muscle is able to shorten as it contracts (cf. **isometric contraction**).

isotonic solution A solution that is in osmotic equilibrium with the cells of the body (cf. **iso-osmotic**).

isotopes Atoms with the same atomic number but different atomic mass. Such atoms have differing numbers of neutrons in their nuclei. Some isotopes are unstable and therefore radioactive.

J

jaundice The yellow pigmentation of the whites of the eyes and of the skin due to high levels of bile pigments in the blood. It may be caused by an increased rate of destruction of the red cells, obstruction of the bile duct, or liver disease.

jejunum The section of the small intestine lying between the duodenum and the ileum. It is about 1.5 metres in length and can be distinguished from the ileum as it has a thicker wall and has more circular folds.

joule (J) This is the SI unit of work or energy, defined as the work done by a force of one newton moving an object through a distance of one metre.

juxtaglomerular apparatus A functional unit that is situated next to a renal glomerulus that regulates renin secretion. It consists of juxtaglomerular cells in the wall of the afferent arteriole, the macula densa cells of the distal tubule, and associated mesangial cells.

JVP Jugular venous pressure: the pressure within the jugular vein, normally in the range 0.59–0.79 kPa (6–8 cmH_2O).

K

kernicterus The brain damage caused by bilirubin to a newborn with severe jaundice.

ketoacidosis A low blood pH caused by an excessively high concentration of ketone bodies: a dangerous complication of type 1 diabetes mellitus that requires urgent treatment.

ketone bodies Acetoacetic acid and β-hydroxybutyrate synthesized by the liver hepatocytes from fatty acids during fasting or carbohydrate restriction.

ketonuria The presence of ketones in the urine. Although low levels of ketones are present in the blood of healthy individuals they are not normally found in the urine.

kilo- A prefix denoting one thousand of a given unit, e.g. kilocalorie, kilojoule, kilopascal.

kinaesthesia The sense of awareness of the position and movement of the limbs.

kinase An enzyme that facilitates the transfer of a phosphate group from ATP to a particular molecule, as when the transfer of a phosphate group from ATP to glucose is catalysed by hexokinase.

kinocilium A large cilium that emerges from the basal body of hair cells. Kinocilia are found in the mature hair cells of the vestibular system and in immature hair cells of the cochlea.

Korotkoff sounds Sounds heard during the indirect measurement of blood pressure with a sphygmomanometer. A stethoscope is applied to the brachial artery at the inside of the elbow below an inflatable cuff. The cuff is first inflated to occlude the artery and then the pressure is slowly released while listening to the sounds heard through the stethoscope. As air is released from the cuff, blood is heard pulsing through the occluded artery—this is the first Korotkoff sound and indicates the systolic arterial pressure. The pressure at which the sound disappears is taken as the diastolic pressure.

kph Kilometres per hour.

Krebs cycle Alternative term for the **tricarboxylic acid cycle**.

Kupffer cells Phagocytic cells found in the liver.

kwashiorkor A condition arising from severe malnutrition in young children caused by a low-protein diet. It is characterized by ascites, oedema, loss of skin pigment, and hair loss.

L

labyrinth The various chambers of the inner ear located in the temporal bone of the skull. The labyrinth consists of bony cavities (the bony labyrinth) within which is a series of membranous tubes (the membranous labyrinth).

lactation The secretion of milk by the mammary gland.

lactogenesis The process of synthesizing milk.

lamellipodium A flattened extension of a cell that enables it to adhere to, or move across, a surface.

lamina propria The layer of connective tissue lying immediately below the epithelium of a mucous membrane.

laminar flow A pattern of flow in which a fluid moves smoothly through a pipe such as a blood vessel without forming eddies (cf. **turbulent flow**).

Langerhans cells Dendritic cells in the skin (i.e. they are antigen-presenting immune cells).

language A means of communication between people consisting of a specific vocabulary and rules of expression (grammar).

Laplace's law A formal statement of the equilibrium relationship between the pressure difference (ΔP) across the wall of an alveolus or blood vessel, the tension in its wall (T), and its radius of curvature (R). For a spherical alveolus $\Delta P = 2T/R$; for a cylindrical vessel $\Delta P = T/R$. In each case, the tension is the force exerted per unit area of the wall that is tending to collapse the sphere or cylinder.

Laron dwarfism Also known as Laron syndrome: a genetic disorder in which the individual has an insensitivity to growth hormone (GH).

lateral Anatomical term referring to the side of the body or to an anatomical feature that is further from the midline than another.

lateral inhibition The inhibitory influence of an excited neuron on the activity of weakly activated neurons to which it is connected, so limiting the lateral spread of excitation.

LDL Low-density lipoprotein.

learning The acquisition of knowledge or skills through study or experience. Also a long-lasting change in behaviour born of experience.

lectins Proteins that tightly bind particular sugar residues.

lens Also called the crystalline lens: the biconvex body behind the iris of the eye that is able to change shape to focus light on the retina. More generally, a piece of transparent material used to converge or diverge light to form an optical image. See also **accommodation**.

leptotene The beginning of the first cell division of meiosis (prophase I) when the duplicated homologous chromosomes condense and can be seen down a microscope.

leucocyte Alternative spelling for **leukocyte**.

leukopenia A deficiency of white blood cells.

leukocyte A **white blood cell**.

leukotrienes Inflammatory mediators generated by the oxidation of arachidonic acid that have three conjugated double bonds.

Leydig cells Interstitial cells of the testes that secrete testosterone and other androgens.

LH **luteinizing hormone**.

LHRH Luteinizing hormone releasing hormone (cf. GnRH).

ligand Any molecule that binds to a specific site on a protein or other large molecule.

ligand-gated ion channel An ion channel that is opened following the binding of a ligand.

light adaptation The process by which the sensitivity of the retina is adjusted to bright light.

lipase An enzyme that breaks down fats to release fatty acids and glycerol.

lipid A member of a chemically diverse group of substances that are insoluble in water but are soluble in organic solvents. Examples are cholesterol and triglycerides.

lipolysis The breakdown of triglycerides to fatty acids and glycerol for ATP production.

lipoprotein A combination of a fatty acid with a specific carrier protein.

liver failure The inability of the liver to perform its normal synthetic and metabolic functions. Also known as hepatic insufficiency.

LTB_4 Leukotriene B_4.

LTP Long-term potentiation.

lusitropic effect The relaxation of heart muscle is called lusitropy. In positive lusitropy the relaxation of the heart is accelerated, while in negative lusitropy the rate of relaxation is slowed. Catecholamines such as adrenaline have a positive lusitropic effect.

luteinization The process by which a corpus luteum is formed after ovulation.

luteinizing hormone (LH) A peptide hormone produced by gonadotrophs of the anterior pituitary gland. In women of childbearing age, a surge in the secretion of LH triggers ovulation followed by the development of the corpus luteum. In males, LH stimulates the production of testosterone by the Leydig cells. See also **follicle stimulating hormone**.

luteolysis The regression of a corpus luteum at the end of the luteal phase of the menstrual cycle.

lymph The clear fluid of the lymphatic vessels. It has a composition similar to plasma but has little protein.

lymph nodes Small bean-shaped bodies that house large numbers of lymphocytes and are found along the lymphatic vessels, particularly in the neck, armpits, and groin. Also called lymph glands.

lymphatic system A series of thin-walled vessels and associated lymph nodes that drain excess fluid from the tissues and return it to the blood. The associated cells, the lymphocytes, play a major role in combating infection.

lymphocytes Small white blood cells 6–12 μm in diameter, each with a prominent nucleus but little cytoplasm (unless it has been transformed into a plasma cell). They play a central role in combating infections. They either produce antibodies (B-lymphocytes—B-cells) or are involved in cell-mediated immunity (T-lymphocytes—T-cells). They are amongst the most numerous of white cells, accounting for 20–45 per cent of all white blood cells (leukocytes).

lysosomes Membrane-bound organelles that contain proteolytic enzymes; these allow cells to digest materials that they have taken up during endocytosis or phagocytosis. Lysosomes have a low internal pH.

lysozymes Enzymes that are part of the innate immune system and can be found in tear fluid, saliva, and other body fluids. They are hydrolytic enzymes (glycoside hydrolases) that attack the complex carbohydrates of bacterial cell walls. Also called muramidase or N-acetylmuramide glycanhydrolase.

M

macro- A prefix meaning large or excessive, as in macromolecule, macrocyte etc.

macrocyte An abnormally large red blood cell—commonly seen in **pernicious anaemia**, which is also known as macrocytic anaemia.

macronutrients Carbohydrates, fats, and proteins—the major components of the diet.

macrophages Large phagocytic cells that develop from the blood monocytes which migrate from the blood to the spleen, lung, and lymph nodes where they mature. They are non-specific immune cells which will phagocytose both cellular debris and invading organisms.

macula densa A densely packed group of modified epithelial cells in the distal tubule of a nephron, adjacent to the juxtaglomerular cells.

macula lutea A small yellowish oval-shaped region near the centre of the retina that has a high density of cones.

major histocompatability complex (MHC) Cell surface molecules that bind to peptide fragments derived from pathogens and display them on the cell surface for recognition by the appropriate T-cells. The MHC complex is also known as the HLA (human leukocyte antigen) complex.

malabsorption A disorder in which there is difficulty in absorbing key constituents of the diet. There are many causes.

malignant hyperpyrexia A rapid rise in body temperature (fever) that is caused by a genetic disorder in which certain general anaesthetics cause severe muscle contractions. It is not a response to all anaesthetics but to specific agents.

MALT Mucosa-associated lymphoid tissue.

MAP Mean arterial pressure; calculated as diastolic pressure plus one-third of the pulse pressure.

marasmus A wasting of body tissues caused by starvation. It is seen in young children who have insufficient intake of carbohydrates and protein.

mast cells Cells formed in the bone marrow before migrating to become associated with nerves, blood vessels, and the epithelia of the airways and gut. When activated they degranulate to initiate an immune response by releasing histamine, leukotrienes, and other mediators.

mechanoreceptors Sensory receptors that respond to mechanical stimuli such as pressure, stretch, or vibration.

medial Anatomical term indicating a structure situated nearer the midline of the body than another (cf. **lateral**).

mediastinum The central part of the thoracic cavity that contains all the thoracic viscera except the lungs.

medulla oblongata The part of the brainstem that lies beneath the cerebellum between the pons and the spinal cord and plays a major role in the control of autonomic functions.

megakaryocyte A large multinucleated cell of the bone marrow that gives rise to the platelets of the blood.

meiosis Cell division that produces the eggs and sperm. In meiosis there are two successive cell divisions giving rise to four haploid daughter cells.

Meissner's plexus Alternative term for the **submucosal plexus**.

membrane potential The difference in potential that exists across the plasma membrane of a cell that results from the unequal distribution of ions between the cytoplasm and the extracellular fluid.

memory The store of information within the brain that is used to recall past events and guide future behaviour.

memory cells Lymphocytes that have been primed by prior exposure to an antigen. They reside in the lymph nodes and are able to rapidly respond to that antigen in any subsequent exposure.

menarche The onset of menstruation in puberty.

meninges The membranes covering the brain and spinal cord, consisting of the dura mater, the arachnoid membrane, and the pia mater.

menopause The time at which menstruation ceases permanently. It usually occurs between the ages of 45 and 55 years and is formally defined as the absence of menstruation for a year. Also called the climacteric.

menstruation, menstrual cycle The cyclical physiological changes that occur in the ovaries and uterus of a woman as an oocyte develops, is released, and degenerates. The period between the start of one menstruation and the start of the next is around 28 days. (See also **ovarian cycle**.)

mepps Miniature end-plate potentials: small depolarizations recorded at the neuromuscular junction that result from the spontaneous secretion of individual synaptic vesicles.

Merkel cells Sensory cells found in the basal layer of the epidermis. At the base of each Merkel cell is a disc-shaped nerve ending, a Merkel disc. The Merkel cell complex is thought to be a slowly adapting touch receptor.

merocrine A type of secretion that occurs without damage to the secretory cells (cf. **holocrine**).

mesangial cells Cells of the central stalk of the renal glomerulus.

mesaxon The longitudinal cleft of the plasma membrane of a Schwann cell that is wrapped around a nerve axon.

mesoderm The middle of the three primary germ layers in the very early embryo, lying between the ectoderm (outside layer) and endoderm (inside layer).

mesothelium The lining of abdominal and thoracic cavities which also surrounds the male internal reproductive organs. The mesothelium that covers the internal organs is called visceral mesothelium, while the layer that covers the body walls is called the parietal mesothelium.

metabolic acidosis See **acidosis**.

metabolic alkalosis See **alkalosis**.

metabolic pump A membrane transport protein that requires the direct expenditure of metabolic energy, such as ATP hydrolysis.

metabolism The biochemical processes that occur within a living organism. See also **anabolism** and **catabolic metabolism**.

metabolic rate The energy used by the body at any instant. See also **basal metabolic rate**.

metaboreceptors Sensory nerve endings that respond to metabolites that are produced by active muscles, e.g. hydrogen ions and lactate.

metaphase The stage of mitosis during which the chromosomes become aligned along the equatorial region of the mitotic spindle.

metaphysis The growing end of the diaphysis of a long bone adjacent to the growth plate.

MHC **major histocompatibility complex.**

micelles An aggregate of large molecules dispersed within a colloidal solution. The large molecules found in micelles have a polar headgroup that faces the aqueous phase and a hydrophobic chain oriented to the interior of the micelle. The lipid micelles formed by the action of bile salts on fatty acids are an important example.

microcirculation The circulation of the blood through the arterioles, capillaries, and venules.

microglia Small, migratory interstitial cells of the central nervous system that act as phagocytes to dispose of cellular debris.

micro- (μ) Prefix indicating one-millionth (10^{-6}) of a given unit, as in micromole (μmole), a millionth of a mole. Also indicates anything very small—examples are microscopic and microtubules.

micronutrients The minerals and vitamins that form an essential part of the diet but are required only in very small amounts.

microtubules Long tubular polymers of tubulin that are an integral part of the cytoskeleton. They transport secretory vesicles and organelles around the cell and are major constituents of mitotic spindles.

micturition The act of voiding urine.

MIH Müllerian inhibiting hormone.

milli- Prefix indicating one-thousandth (10^{-3}) of a given unit, as in a millimole—a thousandth of a mole, or a millimetre—a thousandth of a metre.

mineralocorticoid Any of the steroid hormones secreted by the adrenal cortex that are primarily involved in the regulation of electrolyte and water balance. The principal mineralocorticoid is aldosterone. The mineralocorticoids are essential to life.

minute volume The volume of air entering and leaving the lungs in one minute.

MIT Mono-iodotyrosine.

mitochondria Membrane-bound organelles that produce most of the ATP of a cell via oxidative phosphorylation. They also buffer intracellular calcium.

mitosis The division of a diploid cell into two daughter cells.

mitral valve The atrioventricular valve of the left side of the heart, separating the left atrium from the left ventricle. Also called the bicuspid valve.

mixed nerves Nerve trunks serving both motor and sensory functions.

MMC Migrating motility complex.

MODS Multiple organ system dysfunction syndrome.

molality The molality of a solution is the concentration of a substance expressed as the concentration per kilogram of solvent.

molarity The molarity of a solution is the concentration of a substance expressed in moles per litre of that solution.

mole One mole of a chemical compound is equal to its relative molecular mass expressed in grams. A mole of any substance contains 6.023×10^{23} molecules.

molecular biology The branch of the biological sciences concerned with the molecular basis of biological phenomena, especially the molecular basis of gene function.

molecular formula An expression indicating the number of atoms of various kinds that constitute a given molecule but that does not indicate their spatial arrangement. For example, the formula of glucose $C_6H_{12}O_6$ gives no indication of its ring structure.

molecule A combination of atoms joined by chemical bonds.

monocyte A white blood cell that originates in the bone marrow but migrates to the tissues to become a macrophage.

monosaccharide Any carbohydrate that cannot be hydrolysed to yield simpler carbohydrates. Monosaccharides have the general molecular formula $(CH_2O)_n$, where n = 3 to 8.

morphogenesis The development of the organs of the body as it grows from an embryo into an adult.

morula The solid ball of cells that arises during the successive divisions of a fertilized ovum.

motoneuron A neuron that directly synapses with a muscle fibre or gland cell.

motor cortex That region of the frontal lobe of the cerebral cortex that directs voluntary motor activity: the precentral gyrus.

motor end plate See **neuromuscular junction**.

motor nerve A nerve carrying impulses from the brain or spinal cord to a muscle or secretory gland.

motor system The parts of the nervous system that control the activity of the skeletal muscles, chiefly comprising the pyramidal and extra-pyramidal pathways.

motor unit A single motor axon and the muscle fibres to which it is connected.

mountain sickness Also called altitude sickness, mountain sickness is a debilitating effect of high altitude (greater than 2400 metres) caused by acute exposure to low partial pressures of oxygen. It usually occurs following a rapid ascent, and the likelihood of its occurrence can be reduced by a slow ascent.

M_r Relative molecular mass.

mRNA Messenger RNA. See **RNA**.

MSH Melanocyte stimulating hormone, also known as melanotrophin.

mucosa Alternative term for a mucous membrane.

murmur An abnormal sound heard on auscultation of the chest caused by regurgitation of blood through a heart valve. Often the result of deformities of the heart valves or narrowing of the aorta or pulmonary arteries.

muscarinic receptor A type of acetylcholine receptor that activates G protein linked second messenger systems in its target cells. It may be activated by the alkaloid muscarine.

muscle A tissue composed of bundles of elongated cells (muscle cells) that are capable of contraction to produce movement. A term used to describe a structure composed of muscle tissue.

muscle paralysis The inability of a muscle to contract. It is caused by damage to a nerve pathway or by neuromuscular block brought about by drugs such as curare.

muscle spindle A sensory organ sensitive to stretching of a muscle. Also called a stretch receptor.

muscularis mucosa The continuous thin layer of muscle located between the lamina propria and submucosa of the gastrointestinal tract.

MVV Maximum ventilatory volume or maximum voluntary ventilation. The maximum volume of gas that a subject can inhale deeply and exhale in a given time while breathing as quickly as possible. Also called the maximal breathing capacity (MBC).

myelin The white fatty material that forms an insulating sheath around certain nerve axons (called myelinated axons or myelinated fibres).

myenteric plexus (Auerbach's plexus) The nerve network lying between the layer of circular smooth muscle and the layer of longitudinal smooth muscle of the gastrointestinal tract.

myocardium The muscle of the heart.

myocyte A muscle cell.

myoepithelium An epithelium that contains contractile cells (myoepithelial cells) arranged around secretory glands and the alveoli of the mammary gland.

myofibril A long protein filament made from actin or myosin. The relative motion of actin and myosin filaments is the basis of all muscle contraction.

myogenic activity Muscle activity that has its origin within the muscle rather than resulting from stimulation by a nerve. The rhythmic activity of the heart is myogenic in origin.

myoglobin An oxygen binding protein related to haemoglobin that is found in muscle tissue.

myopathy Any disorder characterized by a primary structural or functional impairment of skeletal muscle.

myopia Also called near- or short-sightedness: an optical defect of the eye in which light is focused in front of the retina rather than on it. It is caused either by the eyeball being too long or by a weak lens (cf. **hypermetropia**).

myosin One of the family of ATP-dependent motor proteins that play a key role in muscle contraction. The main constituent of the thick filaments of muscle cells.

N

NAD Nicotinamide adenine dinucleotide. A coenzyme that is able to accept hydrogen from another molecule to form NADH (reduced nicotinamide adenine dinucleotide). NADP (nicotinamide adenine dinucleotide phosphate) is similar in structure to NAD but has an extra phosphate group. NADH plays an important role in the generation of ATP by the mitochondria, while NADP plays an important role in biosynthetic reactions.

nano- A prefix to indicate one-billionth (10^{-9}) of a unit, as in a nanomole (nmole—10^{-9} mole) or a nanometre (nm—10^{-9} metre).

natural immune system The natural or innate immune system is that part of the immune system that is effective at birth, does not require prior exposure to an invading organism, and is able to respond immediately to a foreign antigen. Although the natural immune system is only able to respond to a limited number of antigens, the same ones are found on many different invaders. Unlike acquired immunity, innate immunity has no memory of any previous exposure.

natural killer cells (NK cells) Natural killer cells are large granular lymphocytes that do not require prior sensitization to become activated (unlike cytotoxic T-cells). Their granules contain proteases as well as membrane-penetrating proteins called perforins that lyse their target cells. They play an important role in combating viral infections and eliminating tumour cells.

near point The nearest point at which the eye is able to form a clear image of an object.

negative feedback A process in which a change in a physiological parameter acts to return the system to its original state; a stabilizing influence. The secretion of many hormones is regulated by negative feedback (cf. **positive feedback**).

neonate, newborn These terms are interchangeable, and refer to a baby in the first month (28 days) following its birth.

neoplasm, neoplasia A neoplasm is an abnormal growth of tissue (a neoplasia) that often forms a mass or tumour.

nephron The functional unit of the kidneys consisting of a glomerulus, glomerular or Bowman's capsule, proximal tubule, loop of Henle, distal tubule, and collecting duct.

nerve A long, thin structure comprising bundles of axons which either carry impulses to the CNS (sensory nerves) or from the CNS to the effector organs (motor nerves). Most nerves have both motor and sensory fibres and are called mixed nerves.

nerve cell See **neuron**.

nerve fibre Alternative name for an **axon**.

nervous system A system that comprises the central nervous system, the brain and spinal cord, the autonomic ganglia, and the peripheral nerves. It uses electrical signals to transmit information rapidly to specific cells or groups of cells. The nervous system acts together with the endocrine system to control and coordinate the activities of the body.

neuralgia An intense pain caused by irritation of or damage to a nerve.

neurogenic Term used to describe a process having its origin in nerve activity (cf. myogenic); e.g. the respiratory rhythm is neurogenic in origin.

neuroglia See **glial cells**.

neurohypophysis Alternative term for the **posterior pituitary gland**.

neuroma A mass of nerve tissue resulting from the abnormal regrowth of the end of a cut nerve and consisting of a tangle of nerve fibres.

neuromuscular junction The region of contact between a motor nerve and a skeletal muscle fibre. As a motor axon approaches its terminal, it loses its myelin sheath and runs along the surface of the muscle membrane to form a complex nerve terminal called the motor end plate.

neuromuscular transmission The process of excitation of a skeletal muscle by an action potential in its motor nerve.

neuron The preferred name for a nerve cell, which are cells specialized for transmitting electrical signals over long distances within the body via processes called **axons**.

neurophysins Soluble proteins that bind vasopressin (ADH) and oxytocin in the hypothalamus and transport them to the posterior pituitary gland. They have no independent endocrine function.

neuropil The dense network of glial cell and neuronal processes found in the grey matter of the nervous system.

neurosecretion The secretion of a substance, such as a hormone, by the nerve terminals of neurons.

neurotransmitter A chemical signal that is released at a synaptic nerve terminal in response to an action potential. Neurotransmitters diffuse across the synaptic cleft to influence the activity of a specific target cell such as another neuron, a muscle fibre, or a secretory cell.

neutron One of the two particles that make up an atomic nucleus. Neutrons have mass but no electrical charge.

neutrophils The most numerous of the white blood cells whose cytoplasm does not stain with eosin or basic dyes. They are part of the natural immune system and actively phagocytose invading bacteria.

newton (N) The SI unit of force. A force of one newton will accelerate a mass of one kilogram at the rate of one metre per second per second. In more easily understood units, the newton is equal to about 0.102 kilograms of force.

nicotinic receptor A ligand-gated, non-selective cation channel that is activated by acetylcholine to initiate a rapid response, for example the depolarization and contraction of a skeletal muscle fibre. Nicotinic receptors can be activated by the alkaloid nicotine.

nitrogen balance The difference between nitrogen intake and nitrogen excretion. A positive balance indicates that more nitrogen is being taken in the diet than is being excreted, while a negative balance exists when more nitrogen is being excreted than consumed. A positive nitrogen balance is associated with tissue growth. See also **catabolic metabolism** and **anabolism**.

nociceptor A receptor that responds to injury to the body tissues to give rise to the sensation of pain.

node of Ranvier The gap between segments of the myelin sheath of a nerve axon. The presence of the nodes permits the rapid propagation of an action potential by **saltatory conduction**.

nondisjunction The failure of homologous chromosomes to separate normally in anaphase. This results in one of the daughter cells having fewer chromosomes than normal while the other has a greater number than normal. Nondisjunction causes a number of genetic diseases, of which **Down syndrome** is perhaps the best-known.

non-shivering thermogenesis The generation of body heat by increased metabolism, particularly in brown fat.

non-volatile acids Acids derived from the metabolism of sulphur-containing amino acids that cannot be excreted by the lungs but must be excreted in the urine. Non-volatile acids are also produced in severe uncontrolled diabetes due to the incomplete metabolism of carbohydrates and fats.

noradrenaline A catecholamine hormone secreted by the adrenal medulla. In North America it is more usually called norepinephrine.

NTS Nucleus of the tractus solitarius.

nuclear envelope The membrane surrounding the nucleus of a cell, also called the nuclear membrane.

nuclear pore A protein-lined opening in the nuclear envelope that allows the passage of molecules between the nucleus and the cytoplasm.

nucleic acid The principal nucleic acids are **DNA and RNA**, which consist of long chains of nucleotides. DNA consists of two complementary chains arranged in a helical pattern, while the various forms of RNA have a single chain.

nucleolus A densely staining feature of a cell nucleus that contains protein and RNA. The nucleolus is the site at which ribosomal RNA is synthesized and ribosome subunits are formed. For this reason it is prominent in rapidly dividing cells.

nucleosides Organic compounds consisting of a purine or pyrimidine base linked to a pentose sugar. Examples are adenosine and cytidine.

nucleosome The fundamental subunits of nuclear chromatin, composed of DNA wrapped around a core of proteins called histones.

nucleotides Organic compounds consisting of a purine or pyrimidine base linked to a pentose sugar and one or more phosphate groups. Examples are adenosine triphosphate (ATP) and cyclic GMP.

nucleus 1. Atomic nucleus; the dense core of an atom composed of neutrons and protons.
2. Cell nucleus; the largest and most prominent organelle of a cell. The nucleus contains the genetic material (DNA).
3. A group of nerve cell bodies in the CNS.

nystagmus A repeated involuntary movement of the eyes consisting of a smooth drift in one direction followed by a flick back. It occurs under various physiological situations, immediately after the body has been rotated (post-rotatory nystagamus) for example. It may also occur in disorders of the cerebellum.

O

oedema The abnormal accumulation of fluid in the tissues or serous cavities of the body.

oestrogens (estrogens) A generic term for steroid hormones that promote the development and maintenance of the female characteristics of the body.

olfaction The sense of smell, one of the chemical senses.

oligodendrocyte A glial cell with relatively few processes that forms the myelin surrounding axons in the central nervous system.

oliguria The production of low volumes of urine.

oncogenes Genes that normally regulate cell growth which may mutate and permit uncontrolled cell growth (i.e. cancer). The unmutated gene is also known as a proto-oncogene.

oncotic pressure The osmotic pressure exerted by the plasma proteins.

oocyte An immature ovum, or egg cell. Oocytes develop in the ovary from **oogonia**.

oogonia (sing. oogonium) Diploid cells that develop from primordial germ cells. They proliferate by mitosis. Once mitosis is complete they undergo meiotic division to become primary oocytes.

operant conditioning The linking of a specific behaviour to a reward or punishment. After a number of such pairings, an individual comes to associate the behaviour with the pleasure or distress caused by the reinforcing stimulus.

opsins The protein part of photoreceptor molecules which change their shape when a photon of light is absorbed. This is the fundamental process of phototransduction, the physiological response to light (cf. **rhodopsin**).

opsonization The process by which a foreign particle is labelled with an antibody prior to its elimination by a phagocyte. Molecules that activate the complement system are also considered opsonins.

optic disc The area of the retina through which the optic nerve leaves the eye. It is devoid of photoreceptors and is the blind spot.

optical isomers Compounds in which a carbon atom is bound to four different chemical groups. The resulting molecule can exist in one of two forms that are mirror images of each other (enantiomers) that cannot be superimposed. Their chemical and physical properties are virtually identical, but the two forms differ in their effect on the rotation of polarized light. One form, designated as the (+) or D form, rotates a beam of polarized light to the right; the other, the (-) or L form, rotates it to the left. However, as optical isomers have a different three-dimensional shape they frequently have different biological effects.

organelles Discrete structures having specific functions within a cell. Examples are the mitochondria, nucleus, and ribosomes.

osmoreceptors Sensory receptors in the hypothalamus that detect changes in the osmotic pressure of the blood. They regulate fluid balance via the secretion of vasopressin (ADH).

osmosis The movement of water from a less concentrated to a more concentrated solution via a semi-permeable membrane until both solutions are of the same concentration. The **osmotic pressure** is that pressure when applied to the less concentrated solution that is just sufficient to prevent such movement.

osmolality Osmolality is defined as the number of osmoles of a substance in 1 kg of water (i.e. osmol/kg). This definition is important because some compounds can dissociate in solution into their constituent ions and the number of osmotically active particles is proportionately greater than the molal concentration would suggest. The term osmolality is preferred to **osmolarity** in physiology for technical reasons.

osmolarity The osmolarity of a solution is the measure of solute concentration, defined as the number of osmoles of solute per litre of solution (osmol/L or Osm/L). Osmolarity measures the number of osmoles of solute particles per unit volume of solution.

ossicles The small bones of the middle ear—the malleus, incus, and stapes—that transmit airborne vibrations to the oval membrane of the inner ear.

ossification The natural process of bone formation from cartilage. Also the process leading to the calcification of soft tissues.

osteoblasts The bone cells that secrete the organic matrix of the bone (osteoid). They also play an important role in bone mineralization.

osteoclasts Large multinucleated cells that reabsorb bone during modification of the skeleton.

osteocytes Mature bone cells. They are osteoblasts that have been trapped in the small spaces left within the bone matrix.

osteoid The organic matrix of bone which consists chiefly of collagen, hyaluronic acid, chondrotin sulphate, and osteocalcin.

osteomalacia Bone disease of adults caused by a deficiency of vitamin D.

osteomyelitis A bacterial infection of bone.

osteon Alternative term for **Haversian system**.

osteoporosis A loss of bone mass resulting from the depletion of calcium and bone protein. A disease that is often seen in post-menopausal women, probably caused by a fall in oestrogen secretion.

otoliths Crystals of calcium carbonate found on the sensory membranes of the saccule and utricle of the inner ear. Movement of the otolith membrane stimulates the hair cells and provides information about the orientation of the head.

ovum (pl. ova) An unfertilized female gamete; an unfertilized egg cell.

ovarian cycle The cyclical changes in which an oocyte develops to form a mature follicle which then ruptures to release the ovum, followed by the formation of a corpus luteum.

ovarian follicle A maturing ovum, its associated granulosa cells, and its surrounding capsule (stroma).

ovary A female reproductive organ that produces the female gametes (ova) and secretes the hormones that regulate the female cycle.

ovulation The release of a mature egg (ovum) by the ovary, which takes place at the mid-point of the ovarian cycle.

oxyntic cells Alternative term for **parietal cells**.

P

P Pressure.

P_{50} Pressure for half saturation, e.g. the P_{50} for the oxygen binding of haemoglobin of normal red cells is 26 mm Hg at a PCO_2 of 40 mm Hg (or 3.4 kPa at a PCO_2 of 5.3 kPa).

pacemaker cells Cells that determine the electrical rhythm of the heart and some smooth muscles.

pacemaker potential The rhythmic change in membrane potential that sets the rate at which a tissue is activated. Classically the slow depolarization of the cells of the sinoatrial node that sets the heart rate.

pachytene The third stage of the first cell division of meiosis (prophase I) during which the homologous maternal and paternal chromosomes become aligned along their length and chromosomal segments recombine. Pachytene ends with the separation of the homologous chromosomes into four distinct **chromatids**.

Pacinian corpuscles Large encapsulated endings located in the subcutaneous tissue. They respond to mechanical vibrations. They are a type of rapidly adapting mechanoreceptor.

PAH p-aminohippurate, used to estimate renal blood flow.

pain producing substances Substances that cause the sensation of pain when applied to sensory nerve endings. Also known as algogens. Examples are bradykinin and serotonin (5-HT).

papilla The area of the eye through which the optic nerve exits.

paracellular transport Transport of solutes across an epithelium that takes place between cells rather than through them (cf. **transcellular transport**).

paracrine secretion The secretion of a hormone or other extracellular signalling molecule into the surrounding tissue to influence the activity of cells in the immediate vicinity.

parafollicular cells The secretory cells of the thyroid gland that lie outside the follicles. They secrete the hormone calcitonin.

paraplegia Paralysis of the legs and lower body resulting from a spinal injury or disease.

parasympathetic nervous system One of the two main divisions of the autonomic nervous system; the other is the **sympathetic nervous system**.

parathyroid hormone (PTH) The main hormone secreted by the parathyroid glands. It plays an important role in the regulation of plasma calcium.

parenchyma The tissue that constitutes the specialized part of an organ such as the lung or liver that is distinct from the associated blood vessels, connective tissue, lymphatic vessels, and nerves.

parenteral feeding (parenteral nutrition) Feeding an individual by infusing nutrients directly into a large vein, so bypassing the gut (cf. **enteral nutrition**).

paraesthesia A confused sensation generally arising from stimulation of a nerve by pressure, by traumatic damage, or arising from disease. Well known as 'pins and needles'.

parietal cells Those epithelial cells of the stomach that secrete hydrochloric acid (HCl) and intrinsic factor. Also called oxyntic cells.

pars anterior Alternative term for the **anterior pituitary gland**.

pars nervosa Alternative term for the **posterior pituitary gland**.

parturition The process of giving birth.

pascal The SI unit of pressure. A pressure of one pascal is equal to one newton per square metre.

PDGF Platelet-derived growth factor.

peptide bond The chemical bond formed between the carboxyl group (COOH) of one amino acid and the amino group (NH_2) of another.

perfusion pressure The difference in pressure between the arterial and venous sides of a vascular bed.

pericardium The membrane surrounding the heart.

pericytes Contractile cells embedded within the basement membrane that wrap around the endothelial cells of capillaries and small venules (pericytic venules). They appear to play a role in the regulation of blood flow. Also called Rouget cells.

perilymph The clear fluid of the scala vestibuli and scala tympani of the inner ear. Similar in composition to an ultrafiltrate of plasma (cf. **endolymph**).

perimetry Determination of the extent of the visual field.

perimysium The outermost connective tissue layer covering a muscle; see **epimysium**.

perineurium The outermost layer of connective tissue around a peripheral nerve.

periodic breathing Alternative term for **Cheyne–Stokes breathing**.

periosteum The layer of vascular connective tissue covering the surface of the bones other than the articulating surfaces of the joints.

peripheral nerve One of the motor and sensory nerves that connect the spinal cord to the rest of the body.

peripheral nervous system (PNS) System consisting of the nerves and ganglia that lie outside of the brain and spinal cord.

peristalsis The contractions of the oesophagus, stomach, and intestine that move the food and digestion products along the digestive tract.

peritoneum The serous membrane that lines the wall of the abdominal cavity and covers the abdominal organs.

pernicious anaemia A disorder characterized by the presence of red cells that are larger than normal but which are present in significantly reduced numbers. Commonly caused by lack of vitamin B_{12} or folate, both of which are required for normal red cell maturation. Also known as macrocytic anaemia.

peroxisomes Membrane-bound organelles found in the cytoplasm of most mammalian cells. Peroxisomes contain enzymes that oxidize fatty acids and amino acids with the production of hydrogen peroxide, giving rise to their name. However, they also contain enzymes such as catalase that convert hydrogen peroxide to water and oxygen. Peroxisomes contribute to the biosynthesis of membrane lipids known as plasmalogens.

Peyer's patches Aggregates of lymphatic cells that occur in the mucous membrane lining of the small intestine—mainly in the ileum. Part of the gut-associated lymphatic tissue (GALT).

PGE_2 Prostaglandin E_2.

$PGF_{2\alpha}$ Prostaglandin $F_{2\alpha}$.

PGI_2 Prostacyclin.

pH scale The pH scale ranges from 0 to 14. It is defined either as the logarithm of the reciprocal of the hydrogen ion concentration or as the negative logarithm of the hydrogen ion concentration, so that higher numbers indicate lower hydrogen ion concentrations. At room temperature, solutions with a pH less than 7 are acid, while those with a pH greater than 7 are alkaline.

phagocytosis A specialized form of endocytosis in which a cell, such as a white blood cell, takes up cell debris or invading microorganisms.

phagosome A vacuole formed by a phagocyte during the final stage of phagocytosis. Once formed, a phagosome fuses with a lysosome to give rise to a phagolysosome, which is then able to digest the ingested material.

phenotype The physical characteristics of an organism rather than its genetic makeup (its **genotype**).

phonocardiogram A record of the sounds made by the heart during the cardiac cycle.

phospholipids A group of phosphorus-containing compounds composed of fatty acids, glycerol, and another small molecule such as serine or ethanolamine. Also known as phosphatides. They are major constituents of cell membranes.

photon A discrete particle having zero mass, no electric charge, and an indefinitely long lifetime. It is the smallest particle that conveys electromechanical radiation (i.e. a quantum). Each photon carries a quantum of energy equal to the product of the frequency of the radiation and the Planck constant.

photopic vision Vision adapted to daylight. In photopic conditions, visual acuity is high and there is full colour vision (cf. **scotopic vision**).

physiological dead space See **dead space**.

physiology That branch of biology concerned with the functions of the body rather than its structure, which is the domain of its sister discipline, **anatomy**. Physiology attempts to account for the activity of the organs and tissues in terms of their chemical and physical properties.

pia mater The innermost of the three meninges, the pia is a delicate vascular membrane of connective tissue that completely covers the surface and infoldings of the brain and spinal cord. The pia mater also forms the **denticulate ligaments** of the spinal cord.

pica The persistent craving for and compulsive eating of nonfood substances (true pica).

PIH Prolactin inhibitory hormone (dopamine).

pilomotor Relating to the erection of hairs in the skin, as in pilomotor nerve, pilomotor reflex (see also **arrector pili**).

pinocytosis Also called fluid phase endocytosis, it involves the invagination of a small area of membrane which is then sealed off to enclose a small volume of the extracellular fluid and any small particles dissolved or suspended in it. The vesicles arising from this process subsequently fuse with lysosomes, which digest the material that has been engulfed.

pitch The perception of pitch depends on the frequency of a sound entering the ear. High-pitched sounds have higher frequencies than low-pitched sounds. The human ear can perceive sounds with frequencies extending from 20 Hz to 20 kHz.

placenta A highly vascular structure formed between the uterine endometrium and the embryonic membranes that has a dialysis arrangement to allow exchange of substances between the fetal and maternal circulations. It acts to provide for both the nourishment of the fetus and the elimination of its waste products.

placentation The process of forming the placenta.

plasma The fluid part of the blood, which amounts to 55–58 per cent of blood volume. Unlike **serum**, the plasma contains all the protein clotting factors.

plasma cell A B-lymphocyte that has been stimulated by an antigen to secrete large amounts of a single type of antibody.

plasma membrane The lipid membrane that covers the entire surface of a cell. Alternative names are the cell membrane and the plasmalemma.

plasticity The ability of a process to adapt to change. When applied to a synapse (synaptic plasticity) it refers to a change in the strength of the synaptic contact induced by prior activity.

platelets Small membrane-bound cell fragments found in whole blood that are derived from large multinucleated cells (megakaryocytes) found in the bone marrow. Platelets play a crucial role in blood clotting and in the maintenance of the blood vessels. Also called thrombocytes.

pleura The serous membranes that line the thoracic cavity (the parietal pleura) and cover the lungs (the visceral pleura).

pleural pressure Alternative term for **intrapleural pressure**.

plexus An intricate network or arrangement, as in a complex network of nerves or blood vessels.

pneumotachograph A device for recording the rate of airflow during breathing.

pneumothorax A condition in which air has entered the pleural cavity, causing one or both of the lungs to collapse.

Poiseuille's law The physical law governing the flow of a fluid through a tube such as a blood vessel. It states that the flow is proportional to the pressure applied and the fourth power of the radius of the tube, but is inversely proportional to the length of the tube and the coefficient of viscosity of the fluid.

polar In chemistry, this is the tendency of electrons in a chemical bond to be associated with one of the two atoms involved, e.g. in the bonds between carbon and oxygen the bonding electrons tend to associate with the oxygen atom. Molecules with polar bonds tend to be soluble in water.

polar body A small cell arising during the meiotic divisions that occur during the development of an oocyte. The polar bodies contain a nucleus and a small amount of cytoplasm but do not develop further and undergo apoptosis.

polycythaemia Excess production of red blood cells.

polymerase chain reaction (PCR) A technique used to amplify a few copies of a piece of DNA by many orders of magnitude to generate a large number of copies of a particular DNA sequence.

polymorphonuclear leukocytes Alternative term for **granulocytes**.

polysaccharide A large molecule consisting of a number of sugar molecules bonded together, e.g. glycogen and starch.

polyuria The production of an unusually large volume of urine.

pores of Kohn Small holes in the alveolar septum that allow communication between adjacent alveoli of the lungs.

positive feedback A process in which the effects of a small change in a system cause an increase in the magnitude of that change (cf. **negative feedback**)—as when a small depolarization of a nerve cell results in an action potential, for example.

postcentral gyrus The surface of the parietal lobe of the cerebral cortex located immediately posterior to the central sulcus. It is the primary somatosensory cortex.

posterior pituitary gland The part of the pituitary gland that arises directly from the nervous system and remains connected to it via the hypothalamo-hypophyseal tract. It secretes two peptide hormones: oxytocin and vasopressin (ADH). Also called the neurohypophysis or pars nervosa.

postganglionic fibres The nerve fibres that connect autonomic ganglia to their target organs.

postsynaptic A term used to describe structures located on or events occurring in the receiving side of a synapse; the distal side of a synapse. E.g. postsynaptic neurone, postsynaptic potential.

post-translational modification The enzymic modification of proteins that follows their synthesis by the ribosomes. Such modification alters the properties of a protein and plays an important role in cell signalling.

power The rate of performing physical work: the amount of energy consumed per unit time. In the SI system, the unit of power is the **watt**.

precentral gyrus The surface of the frontal lobe immediately anterior to the central sulcus; the primary motor area.

pre-eclampsia A disorder of pregnancy characterized by hypertension and proteinuria (cf. **eclampsia**).

preganglionic fibres The nerve fibres that connect the autonomic ganglia with the central nervous system.

pregnancy The period from conception to birth which in humans is around 38 weeks, or 40 weeks from the first day of the last menstrual period.

preload The tension in the wall of the heart immediately prior to systole. It is related to the end-diastolic pressure, the wall thickness, and the diameter of the chamber.

premotor area The part of the frontal lobe of the cerebral cortex anterior to the primary motor cortex that is concerned with motor activity.

presbyopia The gradual age-related increase in the closest distance at which the eyes can form a clear image; the receding of the near point is caused by a progressive loss of elasticity of the lens.

presynaptic Structures located on or events occurring in the proximal side of a synapse. E.g. presynaptic neurone, presynaptic nerve, presynaptic action potential.

presynaptic inhibition A type of inhibition mediated by a synaptic contact between an axon and an axon terminal. Such inhibition results in the selective blockade of a specific synaptic contact without altering the excitability of the postsynaptic cell.

PRL Prolactin; an anterior pituitary polypeptide hormone that stimulates and maintains milk production by the mammary glands after parturition.

progenitor cells Cells in the bone marrow that are able to divide to generate new blood cells but whose daughter cells have a predetermined developmental path and give rise to a single cell line (e.g. red cells).

prometaphase The stage of mitosis in which the chromosomes become attached to the mitotic spindle and the nuclear membrane disappears. It is followed by **metaphase**.

prone Lying face down. Pronation is the position of the forearm with the hand facing down.

prophase The first stage of mitosis during which the nuclear chromatin condenses to form the chromosomes followed by formation of the mitotic spindle.

proprioceptors Sensory receptors in the muscles, tendons, and joints that relay information about their length and tension to the central nervous system.

prostaglandin One of a group of naturally occurring unsaturated fatty acids derived from arachidonic acid. The prostaglandins are important local hormones.

protanope An individual with a relative insensitivity to red and who therefore tends to confuse green and red shades.

protein A large biological molecule consisting of one or more chains of α-amino acids linked by peptide bonds.

proteinuria The presence of significant quantities of protein in the urine.

proto-oncogene See **oncogene**.

proton A positively charged particle found in atomic nuclei; the nucleus of a hydrogen atom.

proximal Situated nearer to the trunk of the body or some point of reference: e.g. the proximal tubule of the kidney is that part of the renal tubule nearest the renal glomerulus (cf. **distal**).

pruritus Alternative term for itch.

PTH **parathyroid hormone**.

puberty The stage of development during which adolescents reach sexual maturity and become capable of reproduction.

pudendal nerves The main nerves of the perineum. They conduct sensory, autonomic, and motor signals to and from the genitals, anal area, and urethra.

pulmonary surfactant A complex of phospholipids and proteins secreted by the type 2 cells of the alveoli. It acts to lower the surface tension of the liquid lining the alveoli and reduces the pressure necessary to inflate the lungs. In doing so, it stabilizes the alveolar structure and reduces the work of breathing.

pure tone A sound wave consisting of a single frequency; a pure sine wave.

Purkinje cells Large neurons in the middle layer of the cerebellar cortex with an extensively branched dendritic tree. Their axons synapse with the deep cerebellar nuclei.

Purkinje fibres specialized cardiac muscle cells that are adapted to permit the rapid spread of the cardiac action potential across the left and right ventricles.

pursuit movement Slow movement of both eyes that enables the smooth tracking of an object travelling across the visual field.

putamen One of the main structures of the basal ganglia.

pyramidal tract Alternative term for **corticospinal tract**.

pyrexia Alternative term for **fever**.

pyrogens Substances that cause a fever.

Q

quadriplegia Alternative term for **tetraplegia**.

R

radiation A process in which thermal energy (heat) is emitted by one body, passes through an intervening space, and is absorbed by another body.

radioimmunoassay A sensitive immunological assay that makes use of radioactive tags on antibodies to detect and quantify the amount of a biologically important substance, such as a hormone.

ramus communicans (pl. rami communicantes) One of the bundles of nerve fibres that connect the sympathetic ganglia with the spinal nerves. Rami that connect preganglionic fibres are called white rami, while those that connect postganglionic fibres to spinal nerves are called grey rami.

raphe nuclei Unpaired serotonin-containing neuronal nuclei that lie along the median plane of the brainstem.

rapid eye movement See **REM**.

RBF Renal blood flow.

RDA Recommended daily amount (of foodstuffs).

receptive field A general term for the sensory range over which a receptor can respond. For the somatosensory system it is the area of the body surface over which a sensory receptor is capable of sensing stimuli. In the retina it is the area of visual space (the visual field) that influences the activity of a photoreceptor or neuron. In the auditory system it refers to the range of sound frequencies to which a hair cell or auditory neuron can respond.

receptor A specialized cell or nerve ending that responds to a specific sensory stimulus. Also a molecule in a cell membrane that responds to a specific ligand (such as a neurotransmitter, hormone, or antigen).

receptor-mediated endocytosis A process by which cells absorb molecules from the extracellular space by the inward budding of plasma membrane pits. Proteins bind to receptors before being absorbed, after which the membrane invaginates to form clathrin-coated vesicles. For this reason it is also called clathrin-dependent endocytosis.

receptor potential A change in the membrane potential of a sensory receptor that reflects the intensity of a stimulus. It is the first stage in sensory transduction, which is generally a depolarization that may elicit action potentials in the afferent nerve fibre to which the receptor is connected.

reciprocal inhibition An antagonistic arrangement of synaptic connections between motoneurons controlling pairs of antagonistic muscles. It operates so that the contraction of a flexor inhibits the simultaneous excitation of the extensor muscle operating at the same joint (and vice versa). This inhibition involves inhibitory interneurons in the spinal cord.

red blood cells The oxygen-carrying cells of the blood. They are biconcave discs about 7–8 μm in diameter and lack both a nucleus and mitochondria. They contain the oxygen binding protein haemoglobin. Also called erythrocytes or red cells.

red nucleus Part of the extra-pyramidal pathway of motor control which is located in the rostral midbrain. It is so named for its red tint, attributed to the presence of organic iron compounds.

referred pain Pain that is perceived as coming from a part of the body other than its actual source.

reflex The simplest motor act controlled by the nervous system: a rapidly executed automatic and stereotyped response to a particular stimulus. The nervous pathway mediating such a response constitutes a reflex arc which has a minimum of two neurons and one synapse.

refractory period The brief period after the passage of an action potential when the nerve or muscle membrane is less excitable than normal. It has two phases: first a brief period during which no action potential can be elicited, the absolute refractory period, which is followed by a longer period during which the cell can be excited only by stimuli stronger than normal, the relative refractory period.

relative molecular mass (M_r) The mass of a molecule in atomic mass units.

relaxin A protein hormone of about 6000 Da belonging to the insulin superfamily. It is produced by reproductive tissues in both males and females and is responsible for motility of sperm, for relaxation of the pelvic ligaments in pregnancy, and for the softening of the pubic symphysis during delivery.

releasing hormones Hormones secreted by various nuclei in the hypothalamus into the hypophyseal portal vessels to regulate the secretion of hormones from the anterior pituitary gland.

REM Rapid eye movement; REM sleep is one of the stages of deep sleep.

renal failure Acute or chronic malfunction of the kidneys resulting in major changes to blood chemistry, notably metabolic acidosis and uraemia. Also called renal insufficiency, acute kidney injury, or chronic kidney disease.

renal glomerulus The tuft of capillaries that is situated within a Bowman's capsule, the site of ultrafiltration within a nephron.

renal system The system that acts to regulate the composition of the extracellular fluid. It comprises the kidneys, ureters, bladder, and urethra.

renin An enzyme secreted by the juxtaglomerular cells of the kidney to cleave angiotensin I from plasma angiotensinogen. Angiotensin I is then converted to angiotensin II, which stimulates aldosterone secretion from the adrenal cortex to promote sodium retention by the distal tubule.

Renshaw cells Inhibitory interneurons found in the grey matter of the anterior (ventral) horn of the spinal cord that exert a negative feedback effect on the discharge of those alpha motoneurons to which they are connected.

repolarization Restoration of the resting membrane potential of a muscle fibre or nerve cell following an action potential.

residual volume (RV) The volume of air remaining in the lungs and airways after a maximal exhalation.

respiratory acidosis See **acidosis**.

respiratory alkalosis See **alkalosis**.

respiratory exchange ratio The ratio of the volume of carbon dioxide excreted via the lungs to the volume of oxygen taken up per unit time.

respiratory failure A pathological state in which the respiratory system fails to maintain normal arterial blood levels of oxygen and carbon dioxide.

respiratory quotient (RQ) The ratio of carbon dioxide production to oxygen consumption by metabolizing cells. It is numerically the same as the **respiratory exchange ratio** under steady-state conditions. For that reason it is often used where the term 'respiratory exchange ratio' would be more appropriate.

respiratory system The specialized organs of respiration comprising the nasal passages and oral cavity, trachea, bronchi, bronchioles, lungs, chest wall, and diaphragm.

reticular formation A complex network of neurons and axons distributed throughout the brainstem that has a major role in the central control of autonomic and endocrine functions. It also plays an important role in posture and the regulation of the sleep–wakefulness cycle.

reticulocyte An immature red blood cell in the circulation. Reticulocytes lack a nucleus but have a granular or reticulated appearance when suitably stained. They normally account for around one per cent of the red cell population.

reticulo-endothelial system A diffuse aggregate of phagocytic cells lining the sinusoids of the spleen, lymph nodes, and bone marrow. It is therefore part of the cardiovascular and immune systems rather than an anatomically discrete system.

retina The innermost layer of the posterior chamber of the eye consisting of light-sensitive cells, the rods and cones, which are connected to the optic nerve via bipolar cells and retinal ganglion cells.

retinal Vitamin A aldehyde; an orange pigment that is the photosensitive component of the visual pigment rhodopsin. Also called retinaldehyde or retinine.

Reynolds number A dimensionless number that indicates whether the flow of blood in a vessel is smooth (i.e. laminar) or turbulent.

rhesus factor (Rh factor) An antigen found on the surface of red blood cells. Red blood cells with the antigen are said to be Rh positive (Rh+)—about 85 per cent of the population have blood of this type. Those without the surface antigen are said to be Rh negative (Rh−). Rh factor is also known as D antigen.

rhodopsin A red pigment present in the rod cells of the retina which dissociates into **opsin** and **retinal** when exposed to light

in the first stage of phototransduction (the conversion of light into nerve signals). Also called visual purple.

ribosomes Small organelles found in the cytoplasm composed of RNA and protein. Ribosomes may be free in the cytoplasm or attached to the outer surface of the endoplasmic reticulum (rough endoplasmic reticulum). They are the site of protein synthesis.

rickets Disease of children caused by lack of vitamin D. It is characterized by imperfect calcification of the long bones, which become distorted. This typically results in bowed legs.

RNA Ribonucleic acid—a linear polymer composed of four types of nucleotide: adenine, cytosine, guanine, and uracil. There are three distinct forms: messenger RNA (mRNA), ribosomal RNA (rRNA), and transfer RNA (tRNA). Individual strands of mRNA form a copy of a segment of **DNA** (a gene) and migrate from the nucleus to the ribosomes, where proteins are synthesized from amino acids bound to specific transfer RNA molecules.

rods, rod cells One of two classes of light-sensitive cells in the retina of the eye, the other being the **cones**.

rouget cells Alternative term for **pericytes**.

rough endoplasmic reticulum The part of the membrane network of a cell that is studded with ribosomes. This part of the **endoplasmic reticulum** is concerned with the synthesis of proteins.

RPF Renal plasma flow.

RQ **respiratory quotient**.

rRNA Ribosomal RNA. See **RNA**.

rubrospinal tract This tract originates in the red nucleus of the midbrain. The efferent fibres cross over near their point of origin and then descend the spinal cord next to the lateral corticospinal tract. Neurons in the posterior portion of the red nucleus give rise to axons influencing motor neurons of the neck and upper limbs. Those of the anterior portion influence the motor activity of the lower limb muscles.

rugae A fold in the lining of the stomach. More generally, any fold, crease, or wrinkle in an organ.

RV **residual volume**.

S

SA node **sinoatrial node**.

saccades Rapid movements of the eyes between different points of fixation.

saccharides See **carbohydrates**.

saccule Part of the membranous labyrinth of the inner ear concerned with the sense of balance. Together with the utricle it responds to linear acceleration such as gravity.

sacral outflow Spinal nerves that supply the parasympathetic ganglia of the large intestine, bladder, and reproductive organs. They originate from the sacral segments of the spinal cord.

sagittal An anatomical term signifying a vertical plane that divides the body into right and left halves. Planes parallel to the sagittal are called parasagittal planes.

saltatory conduction The mode of propagation of action potentials along myelinated axons from one node of Ranvier to the next, rather than as a continuous wave of depolarization.

sarcolemma The plasma membrane of a muscle fibre.

sarcomere The repeating contractile unit of striated muscle consisting of the thick and thin filaments between two successive Z lines.

sarcoplasm The contents enclosed by the cell membrane (sarcolemma) of a muscle cell excluding the nucleus.

sarcoplasmic reticulum The endoplasmic reticulum of muscle fibres, especially that of skeletal muscle. A major store of calcium within a muscle fibre.

scala media, scala tympani, and scala vestibuli The fluid-filled canals of the inner ear.

Schwann cells Satellite cells that surround peripheral nerve axons. In the case of myelinated fibres, they form the myelin wrapping.

scotoma A blind spot in the visual field, other than the optic disc.

scotopic vision Vision adapted to low levels of light (cf. **photopic**).

scurvy A disease caused by lack of vitamin C in the diet, characterized by swollen and bleeding gums, loss of teeth, bruising, and poor wound healing.

sebaceous glands Exocrine glands in the dermis which produce and secrete an oily material (sebum) into hair follicles. Also called sebaceous follicles.

second messenger A chemical signal formed within a cell in response to a signal from a hormone or neurotransmitter after it has bound to a cell surface receptor. Second messengers generally initiate a series of changes within the cell known as a signalling cascade.

secondary sex characteristics The changes seen at puberty that are specific for each sex, such as the distribution of fat tissue and the development of breasts in women; the growth of a beard, the change in the pitch of the voice, and muscular development in men.

secretin A peptide hormone secreted by the wall of the duodenum to stimulate secretion of bicarbonate-rich pancreatic juice.

secretion The process by which substances are released by a cell or gland for a particular function. Also a material that is secreted, such as saliva, gastric juice, a hormone, or sweat.

secretogogues Agents that promote a secretion.

segmentation 1. In physiology, segmentation refers to contractions of the circular muscles that occur mostly in the small intestine, although similar contractions are seen in the large intestine. Segmentation contractions promote the mixing of chyme with the digestive juices. 2. In biology, the term is used to describe the subdivision of an organism into similar parts (segments).

selectins A family of membrane glycoproteins that promote cell–cell adhesion via calcium-dependent binding. Also known as cell adhesion molecules (or CAMs).

semen The thick whitish secretion of the male reproductive organs. It contains spermatozoa together with secretions from the seminal vesicles, prostate, and bulbourethral glands.

seminiferous tubules The coiled tubules of the testis in which the spermatozoa are produced.

sensory receptor A structure supplied by an afferent nerve that responds to a chemical or physical stimulus in the internal or external environment of the body. Some sensory receptors are bare nerve endings, while others are encapsulated structures such as Pacinian corpuscles. (See also **receptor**.)

sensory transduction The process initiated in a sensory receptor in response to an **adequate stimulus** that results in a graded change in membrane potential prior to initiating an action potential in the afferent nerve.

sepsis The pathological condition caused by the presence of bacterial pathogens or their toxins.

septum A sheet of tissue dividing a cavity, such as the nasal cavity, or two structures: the sheet of tissue dividing the left side of the heart from the right.

serotonin A monoamine synthesized from tryptophan that occurs in the brain, intestines, and blood platelets. It acts as a neurotransmitter, a vasoconstrictor, and a paracrine mediator. Also called 5-hydroxytryptamine (5-HT).

serous membrane Thin layer of tissue that lines certain internal cavities of the body, particularly the peritoneum, pericardium, and pleura. Serous membranes are transparent, two-layered sheets of tissue that are lubricated by a fluid derived from the blood plasma.

Sertoli cells Large cells that line the seminiferous tubules and form the blood–testis barrier. They secrete androgen binding protein as well as various hormones. Sertoli cells are essential for the development and maturation of spermatozoa.

serum The clear liquid that is exuded from clotted blood. Serum differs from blood **plasma** in that it lacks the clotting factors.

sex chromosomes The chromosomes that determine the sex of an organism. They are called X and Y chromosomes, a Y chromosome being necessary for the development of male characteristics. In normal human development, females have two X chromosomes while males have one X and one Y chromosome.

SGLT1 Sodium linked glucose transporter 1.

shell temperature The temperature of the outer parts of the body, especially the skin and subcutaneous tissues, whose temperature declines during exposure to a cold environment (cf. **core temperature**).

signal molecule A chemical that is secreted by one cell to influence the activity of other cells. See also **hormone, cytokine, neurotransmitter**.

signal transduction The response of a cell when a molecule such as a hormone or neurotransmitter attaches to a receptor on the cell membrane. In many cases the initial response of a cell is the generation of an internal **signal molecule** (a second messenger e.g. cyclic AMP and IP_3) that activates a cascade of biochemical reactions inside the cell, ultimately leading to a specific response such as muscle contraction, secretion, or cell division.

sinoatrial node (SA node) A cluster of specialized muscle cells located beneath the epicardium in the wall of the right atrium lateral to its junction with the superior vena cava. The SA node cells set the heart rate by spontaneously generating action potentials at a rate determined by the autonomic nervous system and some circulating hormones, especially adrenaline. Also called the sinus node.

sinus rhythm The normal rhythm of the heart, distinct from various irregularities of the heart beat (cf. **arrhythmia**).

small intestine That part of the alimentary canal that lies between the stomach and the large intestine, consisting of the duodenum, jejunum, and ileum. The small intestine is the part of the gut that completes the digestion of the food and absorbs the products of digestion.

SNAREs A large group of proteins that mediate the fusion of intracellular vesicles such as lysosomes or synaptic vesicles with the plasma membrane. The name is an acronym of **S**oluble **N**on-specific **A**ttachment protein **RE**ceptor.

sodium pump A carrier protein that utilizes ATP to transfer sodium ions across a cell membrane in exchange for potassium ions. The process consumes energy as ATP and is known as active transport. Its activity results in a high concentration of potassium ions and a low concentration of sodium ions within a cell. Also known as the Na$^+$, K$^+$-ATPase.

solute A substance that is dissolved in a liquid (a solvent) to form a solution. One of the lesser components of a solution.

solution A homogeneous mixture of two or more substances in which the individual molecules, ions, or atoms of those substances are completely dispersed. The constituents can be solids, liquids, or gases.

somatomedin See insulin-like growth factor.

somatostatin Growth hormone inhibiting hormone (GHIH): a peptide hormone that acts to inhibit the secretion of growth hormone from the anterior pituitary gland and as a paracrine hormone in the gastrointestinal tract. Somatostatin also inhibits the secretion of insulin and glucagon by the islet cells of the pancreas. It is also found in a scattered population of neurons within the brain.

somatotopic A term describing the orderly mapping of particular body regions in the motor and sensory areas of the brain. For example, the motor area controlling the tongue is adjacent to that controlling the jaw, which in turn lies next to the area controlling the muscles concerned with the lips, and so on.

somatotrophin Alternative term for **growth hormone**.

sound The pressure waves in the air or other conducting medium that are perceived by the ear. In young healthy adults the frequency of these vibrations ranges from 20 Hz to 20 kHz.

spasmogen A substance that causes the contraction of smooth muscle.

spasticity A state of increased muscle tone caused by damage to the primary motor cortex. Spasticity is characterized by a rigid paralysis without muscle wasting. The increased muscle tone is associated with exaggerated tendon reflexes (hyperreflexia).

spatial summation The summation of synaptic potentials from different afferent fibres over the dendrites of a neuron. If the synaptic inputs are excitatory, the summation may be sufficient to generate an action potential in the cell receiving the inputs. If the synaptic inputs are inhibitory, the summation may prevent the neuron generating an action potential.

specific dynamic action An increase in metabolic rate that follows the intake of food, especially food containing a relatively large amount of protein. It reflects the heat produced during metabolic conversions.

speech Vocal communication in which perceptions, ideas, and thoughts are expressed in language; the articulation of words to convey meaning.

sperm, spermatozoa The mature male gametes.

spermatids Haploid male cells arising from the meiotic division of secondary spermatocytes, which are diploid cells. Like their precursor cells, spermatids remain connected by cytoplasmic bridges and must undergo further maturation to develop into spermatozoa (the process called spermiogenesis).

spermatocyte A precursor cell for the formation of sperm that develops from immature germ cells called spermatogonia. Spermatocytes are found in the seminiferous tubules of the testes.

spermatogenesis The process leading to the formation of mature sperm.

spermatogonia Undifferentiated male germ cells.

spermiogenesis The process by which spermatids develop into mature sperm.

sphincter A ring of muscle that normally maintains constriction of a body passage or orifice but relaxes when required for normal physiological function. Examples are the pyloric sphincter between the stomach and the duodenum, and the anal sphincters which relax to permit elimination of the faeces.

sphygmomanometer An instrument used for measuring arterial blood pressure, consisting of a pressure gauge and an inflatable rubber cuff that wraps around the upper arm to constrict the brachial artery.

spinal cord That part of the central nervous system that extends along the spinal canal and gives rise to the peripheral nerves. It generally extends as far as the upper border of the second lumbar vertebra.

spinal nerves The 31 paired nerves that arise from the spinal cord. In humans, the spinal cord has 8 cervical pairs, 12 thoracic pairs, 5 lumbar pairs, 5 sacral pairs, and 1 coccygeal pair of spinal nerves.

spinal shock Loss of reflexes following traumatic injury to the spinal cord.

spirometer An instrument used to measure the volume of air entering and leaving the lungs. The measurement of such volumes is known as spirometery.

SPL Sound pressure level, the pressure of a sound relative to a reference value of 20 µPa. It is measured in decibels (dB SPL).

splanchnic circulation The circulation of the gastrointestinal tract consisting of the parallel circulations supplying the stomach, small and large intestines, pancreas, liver, and spleen. The blood supply to the splanchnic organs is via the celiac, superior mesenteric, and inferior mesenteric arteries.

spongy bone The type of bone found at the ends of the long bones and in the centre of the vertebrae. It is characterized by a latticework of mineralized bone spicules with the intervening spaces filled with bone marrow. Also called cancellous bone or trabecular bone (cf. **compact bone**).

stagnant hypoxia Insufficient oxygenation of the tissues resulting from a low blood flow.

Starling forces The forces that govern the movement of fluid between the circulation and the tissues. They are the hydrostatic pressure in the capillaries, the tissue pressure, and the difference in **oncotic pressure** between the plasma and tissue fluid.

stenosis A stricture: an abnormal narrowing of a hollow structure such as a duct or blood vessel, as in aortic stenosis—narrowing of the aorta; mitral stenosis—narrowing of the mitral valve; etc.

Starling's law This states that over the normal working range, the heart will eject the same volume of blood as it receives from the venous return. If the venous return increases, this distends the ventricles further (the end-diastolic volume increases) and the ventricles respond with a more forceful contraction. Also called the Frank–Starling mechanism.

stem cells Undifferentiated cells that give rise to specific cell lineages such as those of the bone marrow that generate the various types of blood cell.

stenosis A stricture: an abnormal narrowing of a hollow structure such as a duct or blood vessel, as in aortic stenosis—narrowing of the aorta; mitral stenosis—narrowing of the mitral valve; etc.

stereocilia Specialized microvilli found on the apical surface of hair cells. They are non-motile but respond to the movement of fluid within the inner ear and play a crucial role in hearing and in the sense of balance.

stereoselectivity The ability of an enzyme, a receptor, or a transmembrane carrier protein to differentiate between the **optical isomers** of an organic compound.

steroids Organic compounds with a rigid molecular structure consisting of three six-membered rings of carbon atoms linked to a five-membered ring of carbon atoms. Cholesterol and many hormones including the oestrogens and testosterone are steroids.

STP Standard temperature and pressure.

STPD Standard temperature and pressure dry (of a respiratory gas).

stroke a local failure of the cerebral circulation that causes the death of many neurons and results in a failure of some aspect of brain function. Stroke may reflect a lack of blood flow caused by blockage of an artery (ischaemic stroke) or bleeding caused by rupture of a cerebral blood vessel (haemorrhagic stroke). Also known as a cerebrovascular accident (CVA).

stroke volume (SV) The volume of blood pumped into the aorta and pulmonary arteries for each beat of the heart.

stroma The connective tissue that forms the framework of an organ or gland, as distinct from the tissues performing the special function of that organ.

structural formula A chemical formula that shows the spatial disposition of the atoms of a molecule.

submucosal plexus (Meissner's plexus) The nerve network lying in the submucosal layer beneath the muscularis mucosae in the gastrointestinal tract.

sugar See **carbohydrate**.

sulcus, pl. sulci A shallow groove on the surface of the brain that separates adjacent folds (**gyri**) of the cerebral cortex.

superficial An anatomical term indicating something at or oriented towards the body surface.

superior An anatomical term indicating a structure lying above another or towards the head or top of a structure.

supine An anatomical term indicating a person lying face up (cf. **prone**).

surface tension The force exerted on the surface molecules of a liquid that tends to draw them into the bulk of the liquid.

Surface tension causes a drop of a liquid to assume a shape with the smallest possible surface area. It has the dimensions of force per unit area.

surfactant See **pulmonary surfactant**.

SV **stroke volume**.

SWS Slow-wave sleep.

sympathetic nervous system One of the two divisions of the autonomic nervous system. It originates in the thoracic and lumbar segments of the spinal cord and, when activated, speeds up the heart, constricts the blood vessels, and dilates the airways. It tends to oppose the physiological effects of the **parasympathetic nervous system**.

sympathetic tone The continuous level of activity in the sympathetic nerves. It is varied according to the demands of the body: in times of stress sympathetic tone rises, and in repose it falls.

symport A protein that transports two different molecules or ions in the same direction through the plasma membrane. The uptake of glucose and sodium by the cells of the small intestine is an example. Also called a symporter.

synapse The junction between a nerve ending and its target cell such as another neuron, a gland cell, or a muscle. See also **postsynaptic, presynaptic**.

synaptic cleft The minute gap between a nerve ending and its target cell; also called the synaptic gap.

synaptic transmission The process by which nerve cells pass information to their target cells. Two kinds of synaptic transmission exist: 1. Chemical transmission, in which a small quantity of a chemical signal diffuses across the synaptic cleft to influence the activity of its target cell. 2. Electrical transmission, in which the presynaptic and postsynaptic cells are electrically connected via gap junctions.

synaptogenesis The formation of new synaptic connections.

syncytiotrophoblast The epithelial covering of the placental villi formed by the complete fusion of the cells of the outer layer of the trophoblast. It acts as a giant cell with no gaps, so that immune cells cannot invade the developing embryo (cf. **cytotrophoblast**).

syncytium A multinucleated cell formed by the complete fusion of many separate cells with the elimination of the cell membranes that separate them. The term also refers to cells interconnected by gap junctions to form a single-function unit, as in the heart and certain smooth muscles.

syndrome A combination of signs and symptoms characteristic of a particular disease or disorder.

synostosis The process by which the growth plates at the ends of the long bones are replaced by bone until the epiphyses become unified with the main shaft of the bone (the diaphysis). The term is also used to describe the fusion of two separate bones by the ossification of connecting tissues, as in the fusion of the separate cranial bones to form the skull.

synthase Any enzyme that catalyses the synthesis of a molecule from a precursor (whether or not it uses ATP or GTP).

systole The time during the cardiac cycle when the ventricles contract to force the blood into the aorta and pulmonary arteries.

systolic pressure The maximum arterial pressure attained during contraction of the left ventricle of the heart. Normally taken to be around 120 mm Hg (16 kPa) in a healthy young adult at rest (cf. **diastole**).

T

T Absolute temperature, which is measured in kelvin (K); formerly measured in degrees Kelvin (°K). One kelvin represents the same temperature difference as one degree Celsius.

T-cells A class of lymphocytes concerned with cell-mediated immunity. The classification includes helper T-cells and cytotoxic T-cells.

T_3 Tri-iodothyronine; a thyroid hormone that has a shorter half-life than thyroxine. Normal blood levels are 80–180 ng dl^{-1}.

T_4 Thyroxine; the principal thyroid hormone circulating in the blood. Normal blood levels are 4.6–12 µg dl^{-1}.

tachycardia A heart rate at rest that is above 100 beats per minute.

tachypnoea An abnormally high frequency of breathing at rest; rapid breathing. The normal range is 12–18 breaths a minute.

taeniae coli The three separate longitudinal ribbons of smooth muscle found on the outside of the ascending, transverse, and descending segments of the colon.

tectum Any rooflike structure of the body, especially the dorsal part of the midbrain (as in the optic tectum).

telomere The end of a chromosome that has a characteristic DNA sequence.

telophase The final stage of mitosis, in which the two sets of chromosomes become enclosed by their respective nuclear membranes.

temperature A fundamental physical property related to the average kinetic energy of the atoms or molecules that constitutes the heat energy of a substance. It is a measure of the ability of that substance to transfer that heat energy to another substance. Temperature is generally measured using a scale with fixed reference points, such as the Celsius scale where 0°C is the melting point of ice and 100°C is the boiling point of water.

temporal summation The incremental increase in a postsynaptic potential that occurs when the interval between the presynaptic action potentials is shorter than the duration of the individual postsynaptic potentials. As for **spatial summation**, if the synaptic inputs are excitatory, the summation may be sufficient to generate an action potential in the postsynaptic cell; but if they are inhibitory, the summation may prevent the neuron generating an action potential.

TENS Transcutaneous electrical nerve stimulation. Used in pain control.

testis The reproductive organ that produces the male gametes (sperm) and the male sex hormones (androgens).

testosterone The chief steroid hormone secreted by the Leydig cells of the testes. It is crucial for the fetal development of the male reproductive system. In adult life it is required for normal sperm development and the maintenance of the male secondary sex characteristics. It is also produced in smaller amounts by the ovaries and by the adrenal cortex of both males and females.

tetanus A prolonged contraction of a skeletal muscle caused by rapidly repeated stimuli (tetanic stimulation). The term also refers to a disease characterized by muscular rigidity and spasms, caused by infection with the bacterium *Clostridium tetani*.

tetany Intermittent muscular spasms as a consequence of hypocalcaemia (a low plasma calcium). It may be caused by a lack of calcium, an excess of phosphate, or a malfunction of the parathyroid glands.

tetraplegia Paralysis caused by injury to the cervical region of the spinal cord resulting in the partial or total loss of use of all the limbs. The injury usually affects both sensory and motor functions. Also known as quadriplegia.

TGF-α, TGF-β Transforming growth factors -α and -β.

TH Thyroid hormone (thyroxine or T_4).

thalamic pain Severe unremitting pain resulting from damage to the thalamus, usually caused by a small stroke (a thalamic infarct). Also known as the Dejerine–Roussy syndrome.

thalamus (pl. thalami) The largest subdivision of the diencephalon—the posterior part of the forebrain that connects the midbrain with the cerebral hemispheres. The two thalami are paired structures, each consisting of a large ovoid mass of grey matter that form part of the lateral walls of the third ventricle. The thalamus chiefly serves to relay sensory information to and from the cerebral cortex.

thermogenesis The generation of heat by metabolic activity.

thermoreceptor A sensory receptor that responds to changes in temperature over a specific range. Some respond to cooling of the skin (cold receptors) while others respond to warming (warm receptors). In addition there are thermoreceptors within the central nervous system—central thermoreceptors.

thermoregulation The processes involved in regulating body temperature.

threshold The value of a physiological variable that must be exceeded to trigger a response: for example, the membrane potential that must be reached to trigger an action potential in a nerve or muscle is the threshold for action potential generation.

thrombocyte Alternative term for **platelet**.

thrombocytopathia Rare disorders of platelet function in which the platelet count is normal or even increased but bleeding time is prolonged; von Willebrand's disease is an example.

thrombocytopenia A disorder in which the platelet count is severely reduced from its normal range of 150–400 × 10^9 platelets per litre. This reduction is accompanied by spontaneous bleeding into the tissues and bruising.

thyrotrophin A glycoprotein hormone also called thyroid stimulating hormone (TSH) that is synthesized and secreted by the thyrotrophs of the anterior pituitary gland. It is a member of the glycoprotein hormone family that includes follicle stimulating hormone (FSH) and luteinizing hormone (LH). TSH exerts an important influence on the thyroid, including stimulation of iodine uptake and the release of iodothyronines from the gland. It also promotes thyroid growth.

tidal volume The volume of air passing in and out of the airways with each breath.

tight junction A continuous region of contact between adjacent cells in an epithelium that eliminates the space between them, so sealing off the space above the apical surface from that surrounding the basolateral membrane. Also called a zonula occludens.

timbre The specific characteristics of a sound that distinguish it from other sounds of the same pitch and volume.

tinnitus An unremitting sound that has its origin in the auditory system itself. Sometimes described as 'ringing in the ear', although its character varies from subject to subject.

tissue fluid Alternative term for **interstitial fluid**.

tissues Groupings of similar cells that act to perform a specific function. The principal types of tissue are blood and lymph, connective tissue, nervous tissue, muscle tissue, and epithelial tissue.

Tm transport maximum.

tone 1. A sound with a particular pitch and quality. 2. The continuous level of activity in a nerve fibre. 3. The normal state of partial contraction seen in smooth muscle and in skeletal muscles at rest (as in abdominal tone, for example).

tonic activity The continuous activity of nerves, especially those of the autonomic nervous system and the muscles they supply. The state of partial contraction of many blood vessels is maintained by the tonic activity of their sympathetic nerves.

tonic reflexes Reflexes that maintain a steady level of muscle activity, especially those concerned with the maintenance of posture.

tonicity of solutions The osmotic pressure of aqueous solutions relative to that of the cells.

total peripheral resistance (TPR) The instantaneous resistance to the flow of blood offered by all the peripheral blood vessels. Together with the cardiac output, the total peripheral resistance determines the systemic arterial blood pressure.

TPA Tissue plasminogen activator.

transcellular fluids The fluids of the serosal spaces including the ocular fluids, cerebrospinal fluid, synovial fluid, and fluid of the abdominal cavity and intrapleural space. They are separated from the plasma and interstitial fluid by cellular barriers and account for around 5 per cent of total extracellular water.

transcellular transport The epithelial transport of solutes that occurs through cells rather than between them (cf. **paracellular transport**).

transcription The copying of the genetic code of one strand of DNA to a complementary strand of messenger RNA. Also called DNA transcription.

transcytosis The uptake of material at one face of an epithelial or endothelial cell by endocytosis, followed by its transfer to the other face and its secretion into the extracellular space.

transduction A process by which a biological cell converts one kind of signal or stimulus into another: for example, mechanotransduction—the conversion of a mechanical stimulus into a sequence of nerve impulses; auditory transduction—the conversion of sound into nerve impulses. See also **signal transduction**.

transfer factor Alternative term for **diffusing capacity**.

translation (RNA translation) The process by which the genetic code embedded in a strand of messenger RNA is converted into a sequence of amino acids to make a protein. This occurs in the ribosomes of a cell (cf. **transcription**).

transport maximum (Tm) The amount in mmole min^{-1} of a solute (e.g. glucose) filtered by the renal glomerulus that just saturates the tubular transport processes responsible for its reabsorption or secretion.

transverse An anatomical term indicating a section that divides the body or internal structure into an upper (superior) and a lower (inferior) portion. Also called a cross-section or horizontal section.

treadmilling The maintenance of the length of polymeric protein filaments (e.g. actin) by addition of subunits at one end and the removal of subunits from the other.

TRH Thyrotrophin releasing hormone: a hypothalamic hormone secreted into the hypothalamo-hypophyseal portal system to regulate the secretion of thyroid stimulating hormone.

triads Structures seen in electron micrographs of skeletal muscle fibres showing the region of contact between the transverse tubules (**T-tubules**) and the terminal cisternae of the sarcoplasmic reticulum on each side. The triads enable the muscle action potential to trigger the rapid release of calcium ions from the sarcoplasmic reticulum to initiate a contraction. In mammals, triads are located at the A-I junction.

tricarboxylic acid cycle (TCA cycle) This is a cycle of chemical reactions used to generate ATP by the complete oxidation of acetate (as acetyl CoA) to carbon dioxide. Also called the citric acid cycle or the Krebs cycle.

tricuspid valve The heart valve between the right atrium and right ventricle. As the name implies it has three membranous flaps, unlike the **mitral** (bicuspid) **valve**, which has two.

triglyceride A compound consisting of three fatty acids joined by ester bonds to glycerol. Triglycerides are the main store of energy in the body and usually contain fatty acids that have many carbon atoms, such as oleic acid, palmitic acid, and stearic acid.

trigone A roughly triangular region of the inner surface of the urinary bladder between the openings of the urethra and the two ureters.

trimester A period of three months, especially one of the three periods of a pregnancy (the first, second, and third trimester), in which different phases of fetal development take place.

triple response The local reddening, vasodilatation of the surrounding area, and weal formation at the site of an injury to the skin such as a burn. Also called the triple response of Lewis.

tritanopes Individuals with a form of colour blindness in which sensitivity to blue is reduced so that they have a tendency to confuse blue and green.

tRNA Transfer RNA. See **RNA**.

trophic (or tropic) hormones Those hormones secreted by the anterior lobe of the pituitary, such as thyroid stimulating hormone (TSH) and adrenocorticotrophic hormone (ACTH), that influence the growth, maturation, or function of other endocrine glands.

trophoblast The ectodermal cell layer of a blastocyst, which erodes the uterine mucosa to form the placenta. The cells do not contribute to the formation of the embryo.

tropomyosin and troponin Regulatory proteins of skeletal and cardiac muscle that complex with actin to prevent the binding of myosin during periods of inactivity. When troponin binds calcium ions it moves tropomyosin away from the myosin binding sites on actin, so permitting actin and myosin to interact, resulting in a muscle contraction.

TSH Thyroid stimulating hormone—see **thyrotrophin**.

T-tubules Deep tubular invaginations of the plasma membrane (sarcolemma) of skeletal muscle and cardiac muscle cells that allow the depolarization of an action potential to spread to the interior of the cell and initiate a contraction—see also **triads**. Also called transverse tubules.

tumour-suppressor genes Genes that restrain inappropriate cell growth, division, and apoptosis (programmed cell death). Certain of these genes are involved in DNA repair processes and so help to prevent the accumulation of mutations in cancer-related genes. When tumour-suppressor genes are damaged, the control of cell division may be impaired and result in abnormal proliferation of tumour cells.

tunnel vision A restricted field of vision in which the central region remains but there is a loss of peripheral vision. It has many causes, including glaucoma and compression of the optic chiasm due to the presence of a pituitary tumour.

turbulence The chaotic, haphazard secondary motion in a flowing gas or liquid caused by eddies.

turbulent flow A pattern of flow of a gas or liquid where its direction and speed at any point shows continuous variation (cf. **laminar flow**).

TXA_2 Thromboxane A_2.

U

ultrafiltration The physical separation of large molecules such as proteins from small ones by passing a solution through a filtration membrane. Of particular importance in the microcirculation, especially during the formation of the glomerular filtrate in the kidneys.

uniport (uniporter) A carrier protein that transports a single solute from one side of the cell membrane to the other (cf. **antiport, symport**).

unmyelinated A term used to describe those axons that lack a fatty sheath of myelin. These are generally classed as C-fibres, which conduct nerve impulses relatively slowly.

urinalysis The examination of a urine sample using chemical assays and microscopic examination of any sediment obtained by centrifugation.

uterine cycle The changes in the endometrium of the uterus seen during the ovarian cycle.

utricle Part of the organ of balance, the membranous labyrinth of the inner ear. Together with the saccule it responds to linear acceleration such as gravity and is therefore able to signal changes in the position of the head.

V

$\dot{V}$ Flow rate.

$\dot{V}/\dot{Q}$ Ventilation/perfusion ratio (in lungs).

V Volume (usually of gas; see Box 31.1 for explanation of subscripts).

vagal afferents Sensory nerve fibres of the vagus nerve; see **visceral afferents**.

vagus nerve The 10th cranial nerve (CN X). It is the longest of all the cranial nerves and has the most extensive distribution. It is a mixed nerve, with both motor (parasympathetic) and sensory functions. It supplies the larynx, pharynx, heart, bronchi, and various organs of the GI tract. Activation of the vagus causes slowing of the heart, constriction of the bronchi, and increased motility of the GI tract.

vagal tone The level of activity in the motor fibres of the vagus nerve.

Valsalva manoeuvre Forced expiration against a closed glottis.

vascular resistance The resistance offered by the circulation to the flow of blood. It is mainly determined by the arteriolar tone (i.e. the degree of constriction of the arteries).

vasoconstriction A decrease in the diameter of blood vessels caused by contraction of their smooth muscle.

vasodilatation (vasodilation) An increase in the diameter of blood vessels caused by relaxation of their smooth muscle.

vasomotion The spontaneous changes in the tone of the arterioles that are independent of the influence of hormones or nerve activity.

VC **vital capacity**.

veins The veins are thin-walled blood vessels, 0.5–3 cm in diameter, that carry blood towards the heart.

venous admixture The mixing of venous blood with oxygenated blood.

venous return The total amount of blood returning to the heart.

ventilation The change in the volume of air in the lungs relative to their resting volume during a single respiratory cycle. More broadly, the exchange of air between the lungs and the atmosphere. Also called external respiration.

ventral Relating to the anterior aspect (front) of the body (cf. **dorsal**); towards the front, as in the ventral horn of a spinal nerve. Also refers to the underside of a four-legged animal.

ventricular fibrillation An uncoordinated contraction of the ventricular muscle of the heart: a potentially life-threatening state.

venules Small thin-walled blood vessels, 50–200 μm in diameter, that link the capillaries to the smallest true veins.

vergence movements Movements that require the simultaneous convergent or divergent movement of the visual axis of each eye to obtain or maintain clear focus on an object near the head.

vesicle A membrane-bound sac in the cytoplasm of a cell that stores or transports the products of metabolism. In certain cases vesicles are the sites of digesting or recycling cellular constituents (see **lysosomes**).

vestibular system The structures of the inner ear and brain that act to provide the sense of balance and orientation in space.

villus (pl. villi) The finger-like projections of the mucous membrane that lines the small intestine. The term is also used to describe similar structures such as those of the placenta. Similar subcellular structures are called microvilli.

VIP Vasoactive intestinal polypeptide.

visceral afferents Afferent nerve fibres that provide information from sensory receptors in the internal organs, the membranes that cover them or their attachments to the body wall. They play an important role in the physiological regulation of the visceral organs.

viscosity A measure of the internal resistance of a fluid to flow. It is an intrinsic property that depends on the temperature of the fluid.

vision The sense of sight.

visual acuity The ability to see the fine detail of an object. It is measured as the angle subtended by two points that can just be distinguished as separate entities.

visual field The area of space that can be seen by one or both eyes at any moment of time.

visual illusions Images in which the perceived object differs in its characteristics from the actual object.

vital capacity The volume of air exhaled during a maximal expiration following a maximal inspiration. Unlike measurements of forced vital capacity (FEV_1, etc.), the exhalation is not time-limited.

vitamin A small organic molecule essential to a healthy life that cannot be synthesized by the body and must therefore be provided by the diet.

vocalization The production of a voiced sound that has no specific meaning, as in an alarm call by an animal.

volatile acid The term used in discussions of acid–base balance for carbon dioxide which is excreted by the lungs and not by the kidneys.

voltage-gated ion channel An ion channel that is opened or closed in response to a change in the membrane potential.

volume receptors Baroreceptors that respond to low pressures. They are located in the large systemic veins, in the pulmonary vessels, and in the walls of the right atrium and right ventricle. They play a major role in the regulation of blood volume.

voluntary apnoea Breath-holding.

voluntary hyperpnoea A deliberate increase in the rate and depth of breathing. Also known as volutary hyperventilation.

von Willebrand's disease A disease of blood clotting caused by a deficit of von Willebrand factor (a protein required for platelet adhesion) or by a mutation affecting its efficacy. Although the disorder is generally inherited, a deficiency of von Willebrand factor may result from an autoimmune response or from other disease processes.

vulva The external genitalia of the human female including the mons pubis, the external and internal labia, the clitoris, and the opening of the vagina.

W

warfarin An anticoagulant normally used to prevent the formation of blood clots in blood vessels. It acts by inhibiting the vitamin K-dependent synthesis of several of the clotting factors.

water channels Alternative term for **aquaporins**.

watt The SI unit of power (the rate at which work is done or expended). One watt is equal to one **joule** of work per second.

weak acid Acids are proton donors in aqueous solution, and a weak acid is one which does not ionize fully when it is dissolved in water. An example is acetic acid, which reacts with water to produce hydrogen ions and acetate ions but with some undissociated acetic acid molecules also remaining present.

weak base Bases are substances that accept protons, and a weak base is one that is not fully protonated in an aqueous solution.

Wernicke's area An area of the temporal lobe concerned with the processing of language. Damage to Wernicke's area causes a type of aphasia called receptive aphasia, in which there is poor comprehension of written and spoken language. Moreover, the speech of those affected is fluent but often meaningless.

white blood cells The mobile cells of the body's immune system. They are transported rapidly to specific areas of infection and inflammation to defend against invading organisms and protect the body against disease. Generally larger than the red blood cells, they possess a nucleus and are present in smaller numbers. They are also called leukocytes.

Windkessel effect As the left ventricle of the heart empties during systole, it stretches the elastic tissue of the aorta and large arteries so that the pulse wave rises smoothly (during stretching) and then falls smoothly during ventricular diastole as the energy stored in the elastic tissue is progressively released. The elastic tissues essentially act as a store of energy. This is the Windkessel effect.

work Work is done when a force moves a body through a distance. The SI unit of work is the joule (J), which is the work done by a force of one newton acting to move an object through a distance of one metre.

X

xenobiotic A substance present in the body that is foreign to it, such as a drug or carcinogen.

xerophthalmia A dry condition of the cornea commonly arising from vitamin A deficiency.

Z

zona pellucida The cell-free zone of glycoprotein surrounding an oocyte.

zonula occludens Alternative term for a **tight junction**.

zonule of Zinn Alternative term for the suspensory ligament.

zygote A diploid cell resulting from the fusion of two haploid gametes (sperm and oocyte): a fertilized ovum.

zygotene The stage of prophase I of meiosis during which homologous chromosomes become aligned along their length.

zymogen granules Densely staining vesicles containing inactive enzyme precursors (proenzymes). They are present in large numbers in secretory cells such as those of the salivary glands and pancreas.

Appendix 2 SI units

A system of units based on the metre, kilogram, and second has been adopted internationally. This system is known as the 'Système International des Unités' or SI system of units. There are seven base units, as listed below.

Physical quantity	Unit	Standard symbol
mass	kilogram	kg
length	metre	m
time	second	s
electric current	ampere	A
temperature	kelvin	K
light intensity	candela	cd
amount of substance	mole	mol

All other units are derived from these base units. The principal derived SI units are as follows.

Physical quantity	Unit	Standard symbol	Definition
electrical potential	volt	V	$J\,A^{-1}\,s^{-1}$
energy	joule	J	$kg\,m^2\,s^{-2}$
force	newton	N	$J\,m^{-1}$
frequency	hertz	Hz	s^{-1}
power	watt	W	$J\,s^{-1}$
pressure	pascal	Pa	$N\,m^{-2}$
volume	litre	$l\,(or\,dm^3)$	$10^{-3}\,m^3$

Each unit can be expressed as a multiple of ten or as a decimal fraction. The most important of these in physiology are as follows.

Multiple		Name	Symbol
1000	10^3	kilo	k
0.1	10^{-1}	deci	d
0.001	10^{-3}	milli	m
0.000001	10^{-6}	micro	μ
0.000000001	10^{-9}	nano	n
0.000000000001	10^{-12}	pico	p
0.000000000000001	10^{-15}	femto	f

Under the SI system the standard volume for expressing concentrations is the litre. Thus, a plasma protein concentration of 7 g per 100 ml should be expressed as $70\,g\,l^{-1}$, although $7\,g\,dl^{-1}$ (7 grams per decilitre) is equally correct. Where the molecular weight of a constituent of one of the body fluids is known, its concentration should be expressed as its molar concentration (moles per litre). Thus, the plasma sodium concentration should be expressed as $0.14\,mol\,l^{-1}$ or $140\,mmol\,l^{-1}$. The same rule applies for expressing cell counts in blood, so that a red cell count of 5×10^6 cells per microlitre on the old system is now expressed as 5×10^{12} cells per litre.

The unit of pressure in the SI system is the pascal, which is one newton per square metre ($N\,m^{-2}$), as pressure is force per unit area. The conventional unit of pressure is millimetres of mercury (mm Hg), which is still widely used. To convert from mm Hg to pascals, multiply by 133.325. For kilopascals, multiply by 0.133325. A pressure of 7.5 mm Hg is equivalent to $7.5 \times 0.133325\,kPa = 0.9999375\,kPa$.

Thus to a good approximation:

7.5 mm Hg	= 1 kPa
15 mm Hg	= 2 kPa
40 mm Hg	= 5.3 kPa
60 mm Hg	= 8 kPa
75 mm Hg	= 10 kPa
100 mm Hg	= 13.3 kPa
150 mm Hg	= 20 kPa
760 mm Hg	= 101 kPa

The unit of temperature is K (kelvin), but °C (degrees Celsius) are still commonly used. To convert from degrees Celsius to kelvin, add 273.15:

$$thus\ 37°C = 37 + 273.15$$
$$= 310.15\,K$$

The calorie is not an SI unit, as the joule is used as the unit for energy. Heat is merely one form of energy. To convert from calories to joules multiply by 4.185. For example: the energy equivalent of 100 g of bread is 240 calories; (240×4.185) joules = 1004 joules.

Index

Tables, figures, and boxes are indicated by an italic t, f, and b following the page number